英汉岩石力学与工程大辞典

ENGLISH-CHINESE DICTIONARY FOR ROCK MECHANICS AND ENGINEERING

《英汉岩石力学与工程大辞典》编委会

主编单位　中国岩石力学与工程学会

中国建筑工业出版社

图书在版编目（CIP）数据

英汉岩石力学与工程大辞典/中国岩石力学与工程学会主编．—北京：中国建筑工业出版社，2016.12
ISBN 978-7-112-19991-4

Ⅰ.①英… Ⅱ.①中… Ⅲ.①岩石力学-对照词典-英、汉②岩土工程-对照词典-英、汉 Ⅳ.①TU4-61

中国版本图书馆CIP数据核字（2016）第246434号

本辞典为岩石力学与工程相关行业专业词汇英译汉工具书，包括约143000个词条，涵盖：水利、水电、核电、天然气、石油、煤层气、页岩气、采矿、冶金、铁路、交通、国防、地球科学、城乡建设、环境保护、古遗址保护、CO_2地质封存、岩爆冲击地压、人工智能、工程管理等多个领域。除收录了大量已广泛应用的专业术语外，本辞典升级并更新了一系列近几年涌现出来的新词汇和新的表述，旨在解决专业技术人员在研究学习工作中遇到的英文专业词汇翻译问题。

本辞典可供岩石力学与岩石工程相关领域的学生、教师、科研人员、工程师、管理者及企业界人士参考使用。

责任编辑：魏　枫　国旭文　张莉英　李　雪　刘　燕
责任校对：关　健

英汉岩石力学与工程大辞典
ENGLISH-CHINESE DICTIONARY FOR ROCK MECHANICS AND ENGINEERING
《英汉岩石力学与工程大辞典》编委会
主编单位　中国岩石力学与工程学会
*
中国建筑工业出版社出版、发行（北京西郊百万庄）
各地新华书店、建筑书店经销
唐山龙达图文制作有限公司制版
北京圣夫亚美印刷有限公司印刷
*
开本：787×1092毫米　1/16　印张：79¼　字数：3000千字
2016年12月第一版　2016年12月第一次印刷
定价：**180.00**元
ISBN 978-7-112-19991-4
（29479）

版权所有　翻印必究
如有印装质量问题，可寄本社退换
（邮政编码100037）

编审委员会

名誉主编：钱七虎　J. A. Hudson

主　　编：冯夏庭　傅冰骏

副 主 编：蔡美峰　何满潮　刘大安

编辑委员会（按拼音首字母排序，排名不分先后）：

白　冰	贲宇星	卞海英	蔡美峰	常金源	陈　勉	陈志龙
程良奎	崔振东	底青云	方志明	冯夏庭	傅冰骏	郭青林
何满潮	赫建明	黄理兴	姜鹏飞	康红普	李贵红	李海波
李丽慧	李　宁	李　琦	李术才	李小春	李　晓	李仲奎
李最雄	廖红建	刘大安	刘汉龙	刘月妙	牛晶蕊	潘鹏志
尚彦军	盛　谦	石　磊	唐春安	王　驹	王明洋	王旭东
王学良	王　媛	韦昌富	魏　宁	邬爱清	吴锋波	伍法权
谢富仁	薛　雷	晏　飞	杨春和	杨国香	杨　强	杨晓东
杨晓杰	姚海波	张　娜	张强勇	张　群	张晓平	张勇慧
赵海军	郑　宏	周创兵	朱维申			

审查委员会（按拼音首字母排序，排名不分先后）：

廖红建	董陇军	冯吉利	冯夏庭	傅冰俊	黄理兴	姜鹏飞
康红普	李建春	李　琦	李仲奎	刘红元	尚彦军	王金安
王可钧	王旭东	杨　强	余　雄	张金才	朱万成	

秘 书 组：（负责协调并执行辞典编写的具体工作）

刘大安	傅冰骏	方祖烈	徐文立	张　维	张绍宗	白　冰
崔振东	冯　婷	牛晶蕊	周　妍			

序 言

我们生活在一个信息时代。人们越来越意识到信息技术的重要性，包括互联网＋、大数据、云计算、人工智能等。在岩石力学与岩石工程领域，我们更应该融入到国际交流的潮流中去，而语言是最重要的信息交流工具。

多年以来，中国的基础设施建设高速发展。除三峡工程和青藏铁路外，还有从长江向黄河引水的南水北调大型水利工程、一系列在建及已完成的深部采矿工程、深埋长隧道工程等。所有上述大型工程建设为中国岩石力学与岩石工程领域提供了巨大的机遇和挑战。

由于处在一个高速建设时期，这使得中国比其他国家拥有更多的岩石力学与岩石工程的实践者。此外，通过工程实践及相关科学问题的研究，中国的科学家和工程师在创造性的岩石工程建设及相关技术研发方面均取得了许多重要进展。

鉴于中国如此密集的岩石力学与岩石工程活动，中国岩石力学与工程学会及国际岩石力学学会中国国家小组，在国际岩石力学学会大家庭里越来越活跃，并且正扮演着越来越重要的角色，使得中国岩石力学与工程专家在国际岩石力学学会拥有了更多的朋友、会员资格、职位及话语权。在过去，中国国家小组与国际岩石力学学会其他国家小组之间的交流已经取得了丰硕的成果，相信将来的中国国家小组的国际交流会更加成功。

为了对国际岩石力学共同体做出更大贡献，我们非常荣幸地编写了这本《英汉岩石力学与工程大辞典》。这本辞典广泛收录了岩石力学与岩石工程领域相关专业词汇约 143000 个，受到国际岩石力学界的高度认可，并从中精选 1000 个重要核心词条，作为国际岩石力学学会多国语言技术辞典的蓝本。

与之前已经出版的辞典及类似出版物不同的是，我们在以往专业词汇的基础上升级并更新了大量词条，涵盖了岩石力学、岩石工程及地球科学等领域近年来的新词条或新的表述。

这本辞典对岩石力学与岩石工程相关领域的学生、教师、科研人员、工程师、管理者及企业界人士都大有裨益。作为一种为我国科技工作者提供的服务，本辞典的电子版本将可申请在学会官网进行下载：http://www.csrme.com。

在此，编者们要向中国科学技术协会、中国科学院武汉岩土力学研究所、中国科学院地质与地球物理研究所等单位致以诚挚的谢意，感谢上述单位在本辞典编写过程中做出的重要贡献。同时，向国际岩石力学学会前任主席 J. A. Hudson、国际岩石力学学会秘书长 L·Lamas 博士等一并表示感谢，感谢他们对本辞典编写工作的大力支持和关心。最后，感谢全体编审及工作人员对大辞典编纂工作付出的辛勤劳动。

欢迎各位阅读和使用本辞典的专业人士及社会公众批评指正并提出宝贵建议。

冯夏庭
《英汉岩石力学与工程大辞典》编委会主编
中国岩石力学与工程学会主席
国际岩石力学学会主席
2016 年 7 月

Preface

As is known to all, we are living in an age of information. The world is becoming more progressively aware of the information technology, including Internet +, big data, cloud computing, artificial intelligence and so on. Meanwhile, there is no doubt that language is a most important vehicle for communicating information.

Undoubtedly, in the field of rock mechanics and rock engineering, we should also adapt ourselves to the world trends, especially in China.

Over the years, the infrastructure construction in China is in full swing. In addition to the Three Gorges Project and Qinghai-Tibet Railway, there are also other projects such as the mega Hydro-Project transferring water from the Yangtze River to Yellow River, a series of deep mines and long tunnels, both undertaking and contemplated, especially in the west region. All the events mentioned above underline the enormous challenges and opportunities in rock mechanics in China.

In recent years, China has entered an accelerated construction period with the result that China now has more practitioners of rock mechanics and engineering than other countries. Moreover, through engineering practice and associated research, Chinese scientists and engineers have achieved a lot of great successes both in creative rock engineering and related technical development in rock mechanics.

As a result of the intense rock mechanics activity, the Chinese Society for Rock Mechanics and Engineering (CSRME) and the Chinese National Group (NG China) of the International Society for Rock Mechanics (ISRM), is becoming more and more active and playing an increasingly important role in the ISRM family with membership, with occupations, with voices and with many friends.

Accordingly, the communication and interaction between NG China and other National Groups ISRM have been extremely successful and fruitful in the past, and we are confident that it will continue to be so in the future.

Under these circumstances, for making a greater contribution to the international rock mechanics community, we are greatly honored to compile this English-Chinese Dictionary of Rock Mechanics and Rock Engineering.

This dictionary is highly recognized by ISRM, and 1000 important core terms were selected in the multilingual technical dictionary.

This dictionary contains a comprehensive collection of about 143,000 entries. Differing from similar literatures previously published, we upgraded and updated this dictionary with a large amount of words or expressions appeared in the past few years in the field of rock mechanics, rock engineering and related earth sciences.

It will be of major benefit to students, teachers, researchers, engineers, executives and businessmen working in our circle.

In keeping with our policy of releasing information which may be of general interest to the rock mechanics profession and the public, the content of this dictionary can be downloaded from the website of CSRME: http://www.csrme.com with application.

Although the editors assume responsibility for content as well as for form, they should express their sincere appreciation to the China Association for Science and Technology, Institute of Rock and Soil Mechanics as well as Institute of Geology and Geophysics, both affiliated to the Chinese Academy of Sciences. And also, to Prof. J. A. Hudson, former President ISRM and Dr. L. Lamas, Secretary General ISRM, for their generous support and consideration.

Any comments or suggestions from the reading public will be highly appreciated.

Xiating Feng
President, CSRME
President, ISRM
July, 2015

A

A blasting powder　A级炸药
A level reserve　A级储量
A truss　A型桁架
A type wooden ladder　A字木梯
A/D and D/A combined converter　A/D；D/A混合转换器
aa-field　块熔岩分布区
aa-lava　渣块熔岩，块岩熔
Aalenian　阿林阶
aarite　砷锑镍矿
Aarkansas stone　阿肯色岩
Aasby-diabase　阿斯柏辉绿岩
abac　坐标网，列线图
abaca　马尼拉麻
abaciscus　嵌工用的石瓦小顶板
abacus　顶板；柱冠；列线图；曲线图；Abaqus软件
abacus-bead stone　算珠石
abaft the beam　在正横后的方向
abamurus　扶壁，扶垛挡土墙；支墩
abandon　抛弃；放弃；废弃
abandoned　废弃的；残留的
abandoned blasting　扬弃爆破
abandoned channel　废河道；牛轭湖
abandoned cliff　崩塌崖
abandoned coal　遗留煤
abandoned coal mine　废弃煤矿
abandoned coal pillar　废弃煤柱
abandoned footage　报废进尺
abandoned hazardous waste site　废弃的危险废物场
abandoned heading　废弃平巷
abandoned hole　报废孔
abandoned land　弃耕地
abandoned meander　废牛轭湖
abandoned mine　废矿，废弃的矿山
abandoned mine land　废弃矿山土地
abandoned mine shaft　废弃的矿井
abandoned mine water　老窑水；老空水
abandoned mines' plan　废矿图
abandoned oil well　废弃油井
abandoned pillar　废弃煤柱；废柱；残柱
abandoned pillars method　残柱式开采法，遗留矿柱采矿法
abandoned pit　废井；废矿井
abandoned place　废弃区；采空区
abandoned project　废弃工程；废弃项目
abandoned range　报废井段
abandoned road　废路
abandoned shaft　废井
abandoned solid waste　弃置固体废物
abandoned stope　废采区
abandoned support　废弃的支撑物，损失的支架；废弃的支护
abandoned track　废线
abandoned underground coal mine　废弃地下煤矿
abandoned well　废孔；废弃井；报废井
abandoned workings　废巷道；采空区；老塘

abandoning party　弃权方
abandonment　废弃；报废
abandonment blasting　扬弃爆破
abandonment cap　弃井封盖
abandonment charge　废弃费用；报废支出
abandonment cost　废弃成本
abandonment of claim　放弃索赔
abandonment of contract　放弃合同
abandonment of water rights　水权放弃
abandonment oil production rate　废弃产油量
abandonment plugging　废弃封堵
abandonment pressure　废弃压力，油层枯竭压力；枯竭压力
abandonment water cut　废弃含水量
abandonment water oil ratio　废弃水油比
abapical　离开顶的
ABAQUS　一个大型有限元商业软件，一种有限元数值模拟器
abatement　减少；废除；冲销
abatis　障碍物，通风隔墙，风挡；三角形木架透水坝
abat-vent　同定百叶窗；通气帽；障风装置
abatvoix　吸声板；消声板；止响板
abaxial　离轴的，轴外的，远轴的
Abbe comparator principle　阿贝比长原理
Abbe double-diffraction principle　阿贝两次衍射原理
Abbe jar　阿贝瓷瓶
Abbe theorem　阿贝定理
abbertite　黑沥青
Abbevillian　阿布维尔期（旧石器时代初期）
Abbot compaction test　阿沃特压实试验
abbreviated analysis　简易分析，简项分析
abbreviated drawing　简图
abbreviated pumping test　简易抽水试验
abbreviated tender system　简易招标程序
ABC inventory control　ABC库存管理法
ABC method　ABC校正法
A-B-C process　污水三级净化处理过程
abeam　正横着；正横的
Abel flash point　阿贝耳闪点
Abel Fourier method　阿贝耳傅里叶法
Abel heat test　阿贝尔耐热试验
Abel test　阿贝尔试验
abelite　阿贝立特炸药
abelsonite　卟啉镍石；紫四环镍矿
aber　河口；汇合点；汇流点
Aberdeen sandstone　阿伯丁砂岩
abernathyite　水砷钾铀矿
aberrant　异常的
aberration　像差；光行差；偏差；畸变
aberration constant　像差常数
aberration free　无象差的
aberration of needle　磁针偏差
aberrational correction　像差校正
abherent　防黏材料
abhesion　脱黏

A

abies oil 松香油
ability of flood control 防洪能力
ability of self-compactability 自压实性
ability to monitor contract 合同监管能力
abime 无底洞
abiocoen 无机生境；生境非生物成分
abiogenesis 自然发生论；无生源论
abiogenetic gas 无机成因气
abiogenetic hydrocarbons 非生物成因烃类
abiogenic gas 无机成因气
abiogenic reaction 非生物反应
abiogeny 自然发生；无生源说
abioglyph 非生物层面印痕
abiologic process 非生物过程
abioseston 无机浮聚物质
abiotic degradation 非生物降解
abiotic environment 非生物环境
abiotic factor 非生物因素
abiotic origin 非生物成因
abiotic oxidation 非生物氧化
abiotic phase 无生命阶段
abiotic surround 无生命环境
abkhazite 透闪石棉
ablation 烧蚀
ablation area 消融区；消融带
ablation breccia 溶蚀角砾岩；消融角砾岩，剥蚀角砾岩
ablation cave 冰川消融穴
ablation cone 冰锥，消融锥
ablation debris 消融碎屑
ablation drift 融蚀冰碛；风化冰碛
ablation factor 消融因素；消融速率
ablation form 融蚀形态，消融地形
ablation funnel 溶蚀洼地；溶蚀漏斗
ablation gradient 消融梯度，溶蚀梯度
ablation layer 消融层
ablation material 烧蚀材料
ablation moraine 消融冰碛
ablation of meteorites 陨星烧蚀
ablation period 消融期
ablation season 消融期；消融季节
ablation skin 熔蚀皮
ablation spherule 消融球粒
ablation swamp 消融沼泽
ablation till 消融冰碛；冰碛
ablation valley 消融谷
ablation zone 消融带
ablative plastics 烧蚀塑料
ablykite 阿布石
Abney clinometer 阿布尼测斜仪
Abney level 阿布尼水准仪
abnormal 异常的，不寻常的；非正常的
abnormal anticlinorium 逆复背斜；上敛复背斜；倒扇状复背斜
abnormal beach cycle 海滩异常旋回
abnormal compaction 异常压实
abnormal contact 异常接触
abnormal current 反常流
abnormal curve 非正态曲线，异常曲线，非正规曲线
abnormal dip 异常倾斜
abnormal distribution 非正态分布
abnormal drainage 异常水系；反常水系
abnormal dump 异常转储
abnormal end 异常终止
abnormal erosion 反常侵蚀；反常冲蚀；异常侵蚀
abnormal events 非反射波，异常波，非反射波至
abnormal exposure 异常照射
abnormal fan-shaped fold 逆扇状褶皱
abnormal fault 逆断层；异常断层
abnormal fluid pressure 异常流体压力
abnormal formation pressure 异常地层压力
abnormal indication 反常指示；反常征候
abnormal intensity 烈度异常
abnormal load 不规则荷载
abnormal metamorphism 异常变质作用
abnormal pore fluid pressure 异常孔隙流体压力
abnormal pressure gradient 异常压力梯度
abnormal pressurized reservoir 异常压力储集层
abnormal profile 异常剖面
abnormal risk 异常风险
abnormal scour 异常冲刷
abnormal seismic intensity 异常地震烈度
abnormal shock 异常震动
abnormal soil 异常土
abnormal spoilage 非常损坏；异常损耗
abnormal stress 异常应力
abnormal synclinorium 逆复向斜
abnormal tide 稀遇潮；异常潮
abnormal water level 异常水位
abnormality of soil environment 土壤环境异常
Abo'er Formation 阿波尔组
aboideau 挡潮闸，水闸，坝，堰
aboral organ 反口器
aborane 乔木烷
aboriginal land 原生土地
abortion limit 破碎极限
above critical 临界以上的；超临界的
above curb 路缘石以上的
above freezing 零度以上的；零上；冰点以上的
above ground level 地平面以上；底面标高以上
above ground pipeline 地上管道；地面管道
above ground piping insulation 地面管道保温
above high water mark 高水位以上的
above sea level 海拔
above tide 海拔；高出海面
aboveground 地上，在地面上的
aboveground supported 地面支承的
abra 岩洞；地穴
abradability 磨蚀性；磨损性；耐磨度
abradant 研磨剂；研磨材料；金刚砂
abrade 研磨；磨蚀；磨耗；磨损
abraded bedrock surface 磨蚀基岩面；浪蚀基岩面
abraded depth 磨蚀深度
abraded perforation 受磨蚀射孔
abraded platform 浪成台地；海蚀台地
abrader 研磨机；磨损试验机
Abram's law 阿布莱姆定律
abrased glass 磨光玻璃
abrasion 磨蚀；海蚀；冲蚀；剥蚀
abrasion action 磨蚀作用
abrasion cycle 磨耗周期

abrasion degree 磨耗度
abrasion drill 回旋钻
abrasion geomorphy 海蚀地貌
abrasion graded coast 海蚀夷平海岸
abrasion hardness 耐磨度，磨损硬度
abrasion index 磨损指数；磨蚀指数
abrasion loss 磨耗
abrasion machine 磨损试验机
abrasion mark 磨痕；擦痕
abrasion of blown sand 风沙磨蚀
abrasion of sea and lake 海湖磨蚀作用
abrasion plane 浪蚀平面；磨蚀面
abrasion platform 浪蚀台地；海蚀台地
abrasion proof 耐磨的
abrasion quality 耐磨性
abrasion ratio 耐磨率
abrasion resistance 耐磨性，抗磨性
abrasion resistance index 抗磨指数；耐磨指数
abrasion resistance refractory 抗磨性耐火材料
abrasion resistance test 抗磨损试验
abrasion resistant pump 耐磨损泵
abrasion resisting rubber 抗磨橡胶
abrasion shoreline 磨蚀海岸
abrasion strength 研磨强度；耐磨强度
abrasion surface 海蚀面；浪蚀面
abrasion tableland 剥蚀台地
abrasion terrace 浪蚀阶地；海蚀阶地
abrasion test 磨损试验；耐磨试验
abrasion test machine 磨耗试验机
abrasion value 磨损值；磨损量
abrasion wave action 波蚀作用，浪蚀作用
abrasion wear protection 防腐蚀；防侵蚀
abrasion wheel 磨轮
abrasion wind erosion 磨损风蚀
abrasion-accumulative graded coast 复式夷平岸
abrasion-proof 耐磨的
abrasion-resistance 抗磨蚀能力
abrasion-resisting alloy 耐磨合金；抗磨合金
abrasite 刚铝石（人造金刚砂）
abrasive 磨蚀的；海蚀的；浪蚀的；研磨料
abrasive action 磨蚀作用
abrasive band finishing 砂带抛光
abrasive blast 喷磨处理
abrasive blast equipment 喷砂磨光设备
abrasive brick 研磨砖
abrasive cloth 砂布
abrasive compound 研磨材料
abrasive cut off machine 砂轮切割机
abrasive disk 砂轮
abrasive dust 磨屑
abrasive erosion 磨损性剥蚀；腐蚀性侵蚀
abrasive finishing machine 抛光机
abrasive floor 防滑地板
abrasive formation 研磨性地层
abrasive grain 磨粒；磨料粒度
abrasive grinding machine 研磨机
abrasive hardness 研磨硬度
abrasive jet bit 磨料喷射钻头
abrasive jet cleaning 喷砂清理
abrasive jet drilling 磨蚀喷射钻井

abrasive jet wear testing 喷砂磨损试验
abrasive jets 磨料喷嘴；喷射磨料钻头；磨料喷流
abrasive machine 砂轮机；研磨机
abrasive material 磨料
abrasive mineral 磨料矿物
abrasive paper 砂纸
abrasive particle 磨料颗粒；磨粒
abrasive perforating 喷砂射孔
abrasive property 研磨性
abrasive ratio 磨耗比
abrasive resistance 磨蚀抗力；耐磨性
abrasive rock 磨蚀性岩石；磨石
abrasive sand 磨料砂
abrasive solid 磨蚀性固体
abrasive stick 油石
abrasive substance 研磨材料
abrasive surface 磨损面；磨蚀面
abrasive suspension 悬浮磨料
abrasive tool 研磨工具
abrasive walkway surface 磨蚀面
abrasive water 磨蚀水
abrasive wear 磨损，磨耗量
abrasive wheel 砂轮
abrasiveness 磨损性；磨蚀性；磨耗；磨蚀
abrasiveness of coal 煤的研磨性；煤的耐磨性
abrasiveness of rock 岩石的磨蚀性
abrasives 研磨料；研磨物
abrasivity 冲蚀度；磨蚀度
abrasivity of ground 土壤冲蚀度；地面冲蚀度
abration index 磨蚀指数
abrator 抛丸清理机；喷砂清理
abraum salt 杂盐；层积盐
abrazite 水钙沸石
abreuvage 机械黏砂
abreuvoir 石块间隙缝；圬工缝
abri 山洞；庇护所；防空洞
abriachanite 镁铁青石棉
abridged drawing 略图；简图
abridged edition 缩编本
abridged general view 示意图
abridgment 概略
abrogation 废除；取消
abrolhos 蘑菇形堡礁
abrupt 急剧的；突然的；陡峭的
abrupt bend 急弯；突弯
abrupt change 突变
abrupt change in shear force 剪力突变
abrupt change of cross-section 截面突变
abrupt component 管路上流动截面有突变的部件
abrupt contraction 突缩
abrupt curve 急弯曲线
abrupt discharge 突然排放
abrupt expansion 突然膨胀
abrupt front 突变锋；锋面
abrupt geological hazard 突发地质灾害
abrupt interface 突变界面
abrupt slope 陡坡；削壁
abrupt tooth engagement 齿轮急剧啮合
abrupt translatory wave 行进陡波
abrupt wall 陡壁

abrupt wave 陡波；陡浪
abruption 拉断；隔段；断层；断裂，破裂，断裂
abruptness 急缓度，陡度
absarokite 橄榄粗玄岩
abscess 气孔；砂眼；脓肿
abscissa 横坐标；横线；脉横线
abscissa axis 横轴
abscission 截去；切除；脱离；切断
absence of restriction 无约束
absent phase 缺失相
absite 钍钛铀矿
absolute absorption 绝对吸收
absolute abundance 绝对丰度
absolute acceleration feedback 绝对加速度反馈
absolute accuracy 绝对精度
absolute age 绝对年龄；绝对年代
absolute age dating 绝对年龄测定
absolute age determination 绝对时代鉴定
absolute air pressure 绝对大气压力
absolute alcohol 无水酒精
absolute altimeter 绝对高度计
absolute altitude 绝对高度；绝对高程；海拔
absolute ambient pressure 绝对周围压力；绝对环境压力
absolute annual range of temperature 绝对年温差
absolute assembler 绝对地址汇编程序
absolute assembly language 机器汇编语言
absolute atmosphere 绝对大气压
absolute centigrade scale 绝对摄氏温标
absolute chronology 绝对年代；绝对年代学
absolute closing error 绝对闭合差
absolute construction 独立结构
absolute continuity 绝对连续性
absolute contour 绝对等高线
absolute control point 绝对控制点
absolute convergence 绝对收敛
absolute cost 绝对成本
absolute cross section 绝对截面
absolute damping 绝对衰减，绝对阻尼
absolute datum 绝对基准
absolute deflection 绝对变化；绝对垂度；绝对挠度；绝对偏移
absolute density 绝对密度
absolute deviation 绝对偏差
absolute dispersion 绝对离差
absolute displacement 绝对位移
absolute dry specific gravity 全干比重
absolute elevation 海拔；绝对高程
absolute elongation 绝对伸长
absolute energy scale 绝对能量标度
absolute error 绝对误差
absolute expansion 绝对膨胀
absolute extreme 绝对极值
absolute flow rate 绝对流量
absolute galvanometer 绝对检流计
absolute gas emission rate 绝对瓦斯涌出量
absolute gauge 绝对压力表
absolute geochronology 绝对地质年代学
absolute geopotential 绝对重力势
absolute gradient 绝对梯度
absolute gravimeter 绝对重力仪

absolute gravity 绝对重力
absolute gravity measurement 绝对重力测量
absolute gravity value 绝对重力值
absolute hardness 绝对硬度
absolute humidity 绝对湿度
absolute inclinometer 绝对倾斜仪
absolute index 绝对指标；绝对率
absolute instability 绝对不稳定
absolute instruction 绝对指令；完全指令
absolute intensity 绝对烈度；绝对强度
absolute location 绝对位置
absolute magnetometer 绝对磁强计；绝对磁力仪
absolute magnitude 绝对值，绝对量；绝对星等
absolute manometer 绝对压力计
absolute mass unit 绝对质量单位
absolute maximum stage 绝对最高水位
absolute minimum stage 绝对最低水位
absolute misclosure 绝对闭合差
absolute modulus 绝对模量
absolute moisture capacity 绝对容水量
absolute orientation 绝对定向；绝对定位
absolute out flow of gas 绝对瓦斯涌出量
absolute permeability 绝对渗透率，绝对渗透性
absolute perturbation 绝对摄动
absolute phase shift keying 绝对相移键控
absolute plotter control 全值绘图机控制
absolute porosity 绝对孔隙度；绝对孔隙率
absolute positioning 绝对定位
absolute pressure 绝对压力；绝对压强
absolute price 绝对价格
absolute rock stress 岩石绝对应力
absolute roof 绝对顶板；本煤层的全部上覆岩层
absolute roughness 绝对粗糙度
absolute settlement 绝对沉降，绝对沉陷
absolute sorption 绝对吸附
absolute specific gravity 绝对比重；真比重
absolute stoppage design 绝对阻挡设计
absolute strength 绝对强度
absolute strength value 绝对强度值
absolute stress 绝对应力
absolute structure 绝对构造
absolute surface 绝对表面
absolute symmetrical balance 绝对对称平衡
absolute system 绝对制；绝对系
absolute temperature 绝对温度；热力学温度
absolute thermometric scale 绝对温标；开氏温标
absolute threshold 绝对阈
absolute time 绝对年代；绝对时间
absolute time scale 绝对年代表
absolute topography 绝对地形
absolute unit system 绝对单位制
absolute vacuum 绝对真空
absolute valency 绝对价；最高价
absolute value error 绝对值误差
absolute variability 绝对变率；绝对变异度
absolute velocity 绝对速度
absolute viscosity 绝对黏滞度；绝对黏滞性；动力黏滞性
absolute water absorption of rock 岩石的绝对吸水量
absolute water content 绝对含水量
absolute weight 绝对重量

absolutely dry sample	绝对干燥样	abutment body	台身
absolutely horizontal	绝对水平的；绝对横向的	abutment coping	台帽
absorb	吸收；吸取；减震；缓冲	abutment crane	台座起重机
absorb coefficient	吸收系数	abutment deformation	坝座变形
absorb shocks	减震	abutment for jack	千斤顶后座
absorbed gas	吸附状态的气体	abutment fracture	支承压力带裂隙
absorbed layer	吸着层；吸附层	abutment joint	对接接头；平接缝；对接缝
absorbed striking energy	冲击吸收能量	abutment load	支撑负载
absorbed tension	吸附张力	abutment of arch	拱座
absorbed water	吸着水吸附水	abutment of corbel	翅托支座；悬臂支座
absorbed-in-fracture energy	冲击韧性；弹能；冲击功；冲击弹度；断裂功	abutment of dam	坝肩
absorbent formation	吸水地层	abutment piece	基石；垫底横木；桥台；支座；隔墙底槛
absorbent ground	吸水地层；吸水场地	abutment pier	岸墩；墩式桥台；对接栈桥
absorbent surface	吸收面	abutment pillar	承载煤柱
absorber	吸收器；吸收体；减震器；缓冲器；阻尼器	abutment point	支承点
absorbing well	吸水井；泄水井；渗水井	abutment pressure	支承压力；拱脚压力
absorption coefficient	吸收系数；吸收率	abutment pump	外啮合凸轮转子泵
absorption cross-section	吸收截面	abutment ring	连接环；定位环
absorption force	吸附力	abutment sleeve	定位套筒
absorption hologram	吸收全息图像；吸收全息照片	abutment slope	坝肩边坡
absorption method	吸收法	abutment span	桥台跨度
absorption of energy	能量吸收	abutment stone	桥台石；拱座石
absorption of shock	缓冲；减震	abutment stress	支承应力；支壁应力；集中应力
absorption of vibration	嗳振；减震；振动阻尼	abutment tank	坝肩槽
absorption profile	吸收剖面	abutment test wall	受反力试验墙；反力壁
absorption ratio	吸收率；吸收系数	abutment wall	拱座墙；翼墙；桥座；墩台墙；桥台墙
absorption reaction	吸收反应	abutment wingwall	桥台翼墙
absorption test	吸收试验	abutment with cantilevered retaining wall	耳墙式桥台
absorption thickness	吸收厚度	abutment zone	支承压力带；拱脚压力带
absorption water	吸附水；附着水	abutment	桥台墩，坝肩，坝座，拱座，岸墩
absorption well	吸水井；抽水井；泄水井	abuttal	桥台；支柱；邻接；接界
absorptive force field	吸附力场	abutted surface	邻接面
absorptive index	吸收指数，吸收率	abutting	邻接
absorptive lining	吸水衬里；吸声板	abutting beam	托梁
abstracted river	袭夺河	abutting building	邻接建筑物；毗连房屋
abstraction	袭夺；分离；提取；萃取	abutting end	对接端；相邻端
abstraction borehole	抽水井；吸水孔	abutting joint	端接接头；对接接头
abstraction of pillar	回收矿柱，回收煤柱	abutting structure	毗邻构筑物
abstraction rate	抽水速率	abysmal	深海的；深渊的；深成的
abstraction rights	抽取权	abysmal area	深海区
abstraction well	抽水井；吸水井	abysmal clay	深海黏土
abstraction well water level	抽水井水位	abysmal deposit	深海沉积
abtragung	剥蚀作用	abysmal facies	深海相
Abukuma metamorphic belt	阿武隈变质带	abysmal fault	深断层
abukumalite	阿武隈石；钇硅磷灰石	abysmal region	深海区
abundance measurement	丰度测定；分布量测定	abysmal rock	深海岩石；深成岩
abundance of elements	元素的丰度	abysmal sea	深海
abundance of isotopes	同位素丰度	abyssal	深海的；深成的
abundance of methane	沼气涌出量	abyssal activity	深源火山活动
abundance of minerals	矿物丰度；矿藏丰富	abyssal assimilation	深同化作用
abundance pattern	丰度模式	abyssal benthic zone	深渊底栖带
abundance ratio	丰度比	abyssal circulation	深渊环流；深海环流
abundance ratio of isotopes	同位素相对丰度	abyssal cone	深海扇；深海锥
abundance sensitivity	丰度灵敏度	abyssal current	深海流
abundance value	丰度值	abyssal deep	海渊；深海深处
abundance zone	富集带	abyssal depth	深渊
abundant mineral	丰富的矿物；广布的矿物	abyssal dune	深海砂丘
abutment area	岩压应力集中区；支壁带	abyssal environment	深海环境
abutment block	人造坝座；人造边墩	abyssal facies	深海相
		abyssal fan	深海扇

A

abyssal fault 深断层；深海断层
abyssal fauna 深海动物群
abyssal floor 深海底
abyssal fractionation 深海分异作用
abyssal gap 深海通道；海脊山口；深海裂隙
abyssal geology 深海地质学
abyssal hearth 深层生烃热源
abyssal heat recharge 深源热补给
abyssal hill 深海丘陵
abyssal hills province 深海丘陵区
abyssal injection 深成贯入
abyssal intrusion 深成侵入
abyssal oceanic basin 深海大洋盆地
abyssal oceanography 深海海洋学
abyssal ooze 深海软泥
abyssal pelagic zone 远洋深海带
abyssal peridotite 深海橄榄岩
abyssal plain 深海平原
abyssal pressure 深海压力
abyssal province 深海区
abyssal red earth 深海红土
abyssal region 深海区
abyssal rock 深成岩
abyssal sea 深海；深渊
abyssal sediment 深海沉积物
abyssal structure 深海构造
abyssal theory 深成理论
abyssal tholeiite 深海拉斑玄武岩
abyssal zone 深海带，深海区；深渊带，深渊区
abyssalpelagic zone 深渊带
Abyssinian pump 埃塞俄比亚式水泵
Abyssinian tube well 埃塞俄比亚管井
Abyssinian well 埃塞俄比亚井，管桩井
abyssoconite 深海钙质软泥
abyssolith 岩基；深层岩
abyssolithic 深成的；深成岩体的
abyssopelite 深海软泥
A-C powered mine 用交流电的矿山
AC soil classification system AC 土分类法；机场土分类法
academic stratigraphy 理论地层学
acadialite 红菱沸石
Acadian series 阿卡迪亚统
Acadian geosyncline 阿卡迪亚地槽
acanthconite 绿石
acanthite 螺状硫银矿
Acanthocordylodus 刺棘肿牙形石属
Acanthodus 荆棘牙形石属
Acanthograptus 刺笔石属
acanthoica 刺球石
acanthophyllia 漏斗针珊瑚属
acaustobiolith 非燃性有机岩；非燃性生物岩
accelerated agent 促进剂，促凝剂
accelerated alluvial deposit 加速冲积物；加速冲积泥沙
accelerated boundary element method technique 加速边界元技术
accelerated cement 速凝水泥；快凝水泥
accelerated completion 提前完工
accelerated concrete test 混凝土加速试验
accelerated consolidation 加速固结
accelerated corrosion test 加速腐蚀试验
accelerated creep of rock 岩石加速蠕变
accelerated curing of concrete 混凝土快速养护
accelerated delamination test 快速层离试验
accelerated erosion 加速侵蚀
accelerated flow 加速流
accelerated life test 加速寿命试验
accelerated load test 加速载荷试验
accelerated mission test 加速等效试验
accelerated motion 加速运动
accelerated speed 加速速度
accelerated stall 过载时失速；动态失速
accelerated weathering 加速风化；人工老化；风雨侵蚀试验
accelerating action 加速作用
accelerating admixture 速凝剂
accelerating and water reducing agent 促凝型减水剂
accelerating chemicals 速凝剂
accelerating convergence 加速收敛
accelerating type seismometer 加速型地震仪
accelerating well 补偿油井
acceleration 加速；加速度；促化作用
acceleration amplification coefficient 加速度放大系数
acceleration amplitude 加速度幅值
acceleration and velocity recordings 加速度和速度记录
acceleration attenuation 加速度衰减
acceleration cancelling hydrophone 消加速度海洋检波器
acceleration creep stage 加速蠕变阶段
acceleration detector 加速探测器；加速度（地震）检波器
acceleration meter 加速度计
acceleration of earthquake 地震加速度
acceleration of gravity 重力加速度
acceleration of subsidence 沉降加速度
acceleration of translation 平移加速
acceleration pedal 加速器踏板
acceleration pickup 加速度拾振器
acceleration potential 加速势
acceleration resonant frequency 加速度共振频率
acceleration response 加速度反应；过载反应
acceleration restrictor 加速度限制器；过载限制器
acceleration seismograph 加速度地震仪
acceleration sensitive transducer 加速度换能器
acceleration sensitive trigger 加速度式触发器
acceleration spectrum 加速度振动谱
acceleration tensor 加速度张量
acceleration time 加速时间；起动时间；存取时间
acceleration time history 加速度时程
acceleration torque 加速转矩
acceleration transducer 加速度传感器
acceleration type vibration pick-up 加速度型拾振器
acceleration vector 加速度矢量
acceleration versus time curve 加速度-时间曲线，加速度时程图
acceleration zone graph 加速区图；加速度区划图
acceleration-intensity correlation 加速度-烈度相关性
acceleration-sensitive 对加速度敏感的
acceleration-time data 加速度-时间数据
accelerator 加速器；促凝剂；催化剂；速滤剂
accelerator conveyor 高速胶带运输机
accelerator jar 加速式震击器
accelerator mass spectrometer 加速器质谱仪
accelerator mass spectrometry 加速器质谱分析法，加速

器质谱测定法
accelerator matrix 加速矩阵
accelerator pedal 加速踏板
accelerator rod 加速杆
accelerator spectrometer 加速器谱仪
accelerator theory 加速理论
accelerator-retarder 加减速顶
accelerators in shotcrete 喷射混凝土速凝剂
accelerogram 加速度图；加速度记录；加速度时程
accelerogram of real earthquake 真实地震加速度图
accelerogram of simulated earthquake 模拟地震加速度图
accelerograph 加速器；自记加速计；加速度记录仪
accelerograph array 加速度仪台阵
accelerograph station 加速度仪台站
accelerometer 加速度计；加速度记录器
accelerometer torque motor 加速度计转矩电动机
accelerometer type seismometer 加速度地震检波器；加速度地震计；加速计型地震计
accelerometer 加速器仪
accelofilter 加速过滤器
accented contour 指示等高线
accentuation contrast 反差增强；对比拉伸
accentuator 加强电路；加重器；音频放大器
acceptability of risk 风险的可接受性
acceptable 可接受的；容许的；验收的
acceptable accommodation 适中居所
acceptable age 采用年龄
acceptable concentration 容许浓度
acceptable daily intake 容许日摄入量；容许日输入量
acceptable dose 容许剂量
acceptable emergency dose 容许紧急剂量
acceptable live load 容许活载
acceptable material 合格材料
acceptable moisture 容许水分
acceptable noise level 容许噪声标准；容许噪声级
acceptable point 可接受点
acceptable reliability level 容许可靠性程度
acceptable risk 可接受的风险
acceptable sampling inspection 验收取样检查
acceptable seismic risk 可接受的地震风险
acceptable sensitivity 可接受的灵敏度
acceptable settlement 容许沉陷
acceptable variation 可接受的变更
acceptable water supply 符合要求的供水
acceptable wave condition 容许波浪条件
acceptance bill 承兑票据
acceptance certificate 合格证书，验收合格证
acceptance Certificate on Engineering Completion 工程竣工验收证明
acceptance check 验收
acceptance cone 接纳锥
acceptance contract 承兑合同
acceptance credit 承兑信用证；承兑汇票
acceptance criterion 接受准则；验收标准
acceptance domain 接受域
acceptance following construction 随工验收
acceptance for honor 接受承兑
acceptance gauge 验收规
acceptance inspection 验收检查
acceptance letter of credit 承兑信用证

acceptance liability 承兑责任
acceptance market 票据承兑市场
acceptance of a project 工程验收
acceptance of freight 货物承运
acceptance of grout 吸浆量，吃浆率
acceptance of locomotive 机车验收
acceptance of materials 材料验收
acceptance of offer 接受报价
acceptance of project 工程验收
acceptance of risk 承担风险
acceptance of tender 得标；中标
acceptance of the bid 中标；得标
acceptance of the tender 中标；得标
acceptance of work 工程验收
acceptance of works subelements 分项工程验收
acceptance on examination 工程验收
acceptance operation 预除废石后的选矿
acceptance probability 合格概率
acceptance rate 吸浆率
acceptance requirement 验收要求
acceptance sampler 验收抽样
acceptance specification 验收规范
acceptance summary report 验收总结报告
acceptance surveying of excavated volume 采剥量验收测量
acceptance test 验收试验；收料试验
acceptance test procedure 验收测试程序
acceptance threshold 认购率下限
acceptance tolerance 容许公差
acceptance value 合格判断值
accepted bid 得标；中标
accepted depth 采用深度；允许深度
accepted error 允许误差
accepted installation tolerance 允许安装公差
accepted load 承受荷载；承载能力
accepted risk 容许危险
accepted tolerance 容许公差；规定公差
accepting station 接收站
access arm 存取臂
access bend 检修弯管
access bit 存取位；取数位；访问位
access board 便桥；入口铺板；跳板
access bridge 进入栈桥；引桥
access capability 接入能力；接入容量
access cover 出入孔盖；检修盖
access door 检修门；通道门
access environment 存取环境
access eye 检查孔
access for residents 居民通道
access gallery 交通廊道
access grant 允许接入
access gully 交通沟；检查窨井
access hole 检查孔；检修孔；出入孔
access ladder 便梯；通道竖梯
access lane 进出路径
access line at depot 车辆段出入线
access line for double track 双线出入线
access line for single track 单线出入线
access manway 人行道
access method 取数方法
access mode 读出方式

access plug and cap 检修孔塞和孔盖
access port 输入孔；接入端口
access privilege 存取特权
access procedure 存取程序
access railroad 专用铁路
access ramp 进口坡道；进入匝道
access right 存取权限
access road 便道，对外交通道
access server 接入服务器
access shaft 交通竖井；进口竖井；进入井
access sleeve 放样套
access speed 存取速度
access spiral loop 螺旋式回旋通道
access staircase 通道楼梯
access step 上落踏板；出入踏板
access to site 进入现场
access to stope 采场联络道
access to the gob 采空区入口
access to the site 地盘出入口；工地出入口
access to work 地盘施工处通路；工地施工处通路
access tunnel 交通隧洞；进出隧洞
access type 存取类型
access unit 访问单元；存取单元
access valve 入口阀
access violation 存取违例；访问破坏
access way 通路；道路
access well 交通竖井；升降机井道
access width 存取位数
accessary 附件
accessary maceral 次要显微组分
accessibility 可达性；可接近性
accessible 易接近的
accessible depth 可及深度；可采深度
accessible duct 通行地沟
accessible pore 可接近孔
accessible resources 可及资源；可采资源
accessible roof 可到达的屋顶；有通道的屋顶
accessible state 可达状态
accession differentiation 岩浆上升分异作用
accessional ventilation 上行通风
accessory apparatus 附属仪器；附属装置
accessory building 附属建筑；辅助房屋
accessory constituent 副组分；次要组分
accessory drawing 附图
accessory ejecta 附生喷出物，副喷出物；早成同源抛出物
accessory element 痕量元素；副元素；伴生元素
accessory equipment 附属设备；辅助设备
accessory equipment market 附属设备市场
accessory equipment of fan 通风机附属装置
accessory gypsum 工业废石膏
accessory maceral 次要显微组分
accessory material 辅助材料
accessory mineral 副矿物；附生矿物
accessory parking area 附属停车场
accessory plate 补助片，补偿片；附加片
accessory power plant 辅助动力装置
accessory pyroclast 同源火山碎屑
accessory risk 附加风险
accessory rock 附生岩；次要岩石
accessory sample 副样

accessory shock 副震；附加振动
accessory structure 附属建筑物
accessory substance 副产品
accessory system 辅助系统
accessory tephra 同源火山碎屑
accident alarm 事故警报
accident analysis 事故分析
accident and sick benefit 事故和疾病津贴
accident averting 事故预防
accident data recording 事故数据记录
accident error 偶然误差
accident forecast 事故预测
accident free 无事故
accident frequency 事故率
accident hazard 事故危险
accident in shunting operation 调车事故
accident indemnity 事故赔偿
accident information management 事故信息管理
accident insurance 事故保险
accident investigation 意外调查，事故调查
accident liability 事故易发性；事故责任
accident lighting transfer board 事故照明切换盘
accident loading condition 意外荷载条件
accident mean square error of elevation difference 高差偶然中误差
accident mutation 突变；事变
accident of the ground 褶皱；地形不平
accident on duty 工伤事故
accident prevention 事故预防；故障预防；安全措施
accident prevention instruction 技术安全规程
accident prone 易出事故的
accident rate 事故率
accident record 事故记录
accident relief 事故救援
accident repeater 屡出事故者
accident report 事故报告
accident rescue 事故救援
accident severity 事故严重程度
accidental collapse 意外坍塌
accidental combination for action effects 作用效应偶然组合
accidental connection 偶然性关联
accidental cost 意外费用
accidental count 偶然计数
accidental damage 偶然破坏
accidental discharge 事故排放
accidental dispersal 偶然迁移
accidental eccentricity 偶然偏心
accidental ejecta 异源抛出物；偶然喷出物
accidental error 偶然误差；不可预测的误差随机误差
accidental explosion 意外爆炸
accidental form 意外地形；突变地形
accidental inclusion 捕房体；捕房岩；外源包体
accidental lake 偶成湖泊
accidental lithic 岩屑
accidental load 偶然荷载；突发荷载
accidental maintenance 事故维修
accidental operation 误操作
accidental pearl 天然珍珠
accidental relief 崎岖地形
accidental resonance 随机共振

accidental shutdown　事故停机
accidental situation　偶然状况
accidental species　偶见种
accidental spill　事故性溢漏
accidental tephra　异源火山碎屑
accidental torsion　偶然扭转
accidental variation　偶然变异
accidental xenolith　外源捕房岩；外源异晶体
accident-cause code　防止矿山事故规程
accidented　凹凸不平的
accidented relief　崎岖地形；起伏地形
accident-prone　易发生事故的
acclimatization　服水土；适应新环境
acclinal valley　顺斜谷
acclivitous　倾斜的；向上斜的
acclivity　向上的斜坡；上行坡；逆向坡
accommodate　调节；适应；接纳；供应
accommodation　调节；适应；供应
accommodation bill　空头票据
accommodation bridge　专用桥梁
accommodation coefficient　调节系数；适应系数
accommodation coefficient modulation factor　调节系数
accommodation highway　专用公路
accommodation increasing rate　可容空间增长速率
accommodation lane　专用车道
accommodation rig　半潜式居住和供应海洋平台
accommodation road　专用公路
accommodation space　起居间
accommodation train　普通旅客车；逢站皆停的区间铁路客车；慢车
accompanied layer　瓦斯渗透伴随层；瓦斯伴随煤层
accompaniment　伴随物；跟踪
accompany mineral　伴生矿物
accompanying bed　伴生层；伴生煤层
accompanying diagram　附图
accompanying element　伴生元素
accompanying figure　附图
accompanying gas　伴生气
accompanying mineral　伴生矿物
accomplished　竣工的
accordance　调和；协调；一致
accordance of summit levels　山峰高度一致
accordant　平齐的；与岩层倾斜一致的
accordant connection　整合连接
accordant drainage　协调水系；调和水系
accordant fold　同向褶皱；协和褶皱
accordant junction　交合汇流；平齐汇流
accordant river　顺向河
accordant unconformity　平行不整合
accordant valley　平齐谷
according to usual practice　按惯例
accordion　手风琴；可折叠的；折式地图
accordion fold　尖顶褶皱
accordion partition　折式隔屏
accordion plate　折叠式板
account　账，账目，账单；计算，估计；报告，报表；原因，理由；账户，会计科目
account of business　损益表
account of payments　支出表
account of receipt and payment　收支账目
account year　会计年度
accredited private laboratory　认可的私人实验室
accredited safety auditor　检定安全稽核员
accreted terrane　增生地体
accreting bank　淤积河岸
accreting plate　增长板块；增生板块
accretion　冲积层；冲积物；淤积物；表土
accretion beach　增长海滩；堆积海滩
accretion beach face　加积海滩面
accretion bed　加积层
accretion coast　加积海岸；增积海岸
accretion gley　结核潜育层
accretion ice sheet　堆积冰盾
accretion moraine　前进终碛
accretion of bed level　河床淤高
accretion of bottom　河底淤高
accretion of continent　大陆增生
accretion of levels　淤高
accretion of river bottom　河底淤高
accretion of terranes　地体增生；地体增置；地体增积
accretion plate boundary　增生型板块边界
accretion ridge　古海滩砂脊
accretion ripple mark　加积波痕
accretion terrane　吸积岩层
accretion theory　吸积理论
accretion topography　堆积地形；加积地形
accretion type orogeny　增生型造山运动；增积型造山运动
accretion unit　加积层；加积单位
accretion vein　增生矿脉
accretion wedge　增生楔；加积楔
accretion zone　加积层带
accretional prism　加积层系
accretionary bar　加积沙坝
accretionary basin　增生盆地
accretionary belt　加积带
accretionary coast　加积海岸；增积海岸
accretionary crystallization　增长结晶作用
accretionary lapilli　增生火山砾
accretionary limestone　堆积灰岩
accretionary meander point　凸岸边滩；曲流内侧沉积
accretionary phase　堆积期
accretionary prism　增生楔形体；增生楔；增积楔
accretionary ridge　外展滩脊；加积砂岭
accretionary subduction complex　加积俯冲复合体
accretionary terrane　增生地体
accretionary wedge　增生楔形体；增生楔；增积楔
accumulated angle　总角；合力角
accumulated damage　累积破坏；累积损伤
accumulated deficit　累计亏空
accumulated deformation　累积变形
accumulated density　堆积密度
accumulated depletion　累计折耗
accumulated error　累积误差
accumulated growth　层积生长
accumulated hydraulic power　累积水力；水压
accumulated inflow　累积进水量；累计入流量
accumulated island　堆积岛
accumulated loss　累计损失
accumulated mountain　堆积山
accumulated plastic strain　累积塑性应变

accumulated precipitation	累积降水量
accumulated profit	累积利润
accumulated snowmelt	累积融雪量
accumulated storage	累积蓄水量
accumulated strain	累积应变
accumulated stress	累积应力
accumulated surplus	累计盈余
accumulated temperature	积温
accumulation	堆积；存储；堆积；聚集
accumulation area	堆积区；积冰区
accumulation area ratio method	积累面积比率法
accumulation body	堆积体
accumulation cost	累计成本
accumulation curve	累积曲线
accumulation diagram of water demand	需水量累计图
accumulation factor	累加因子
accumulation form	堆积形式
accumulation horizon	堆积层；积聚层；聚积层
accumulation hydraulic system	带储能器的液压系统
accumulation landform	堆积地形
accumulation landslide	堆积层滑坡
accumulation moraine	堆积冰碛
accumulation of error	误差累积
accumulation of floating ice	浮冰集积
accumulation of heat	蓄热
accumulation of lime	石灰累积
accumulation of mud	淤积污泥
accumulation of pressure	压力积累
accumulation of rounding errors	累积舍入误差
accumulation of sea and lake	海湖堆积作用
accumulation of strain	应变累积
accumulation of stress	应力累积；应力集聚
accumulation of water	积水
accumulation platform	堆积台地
accumulation point	聚点
accumulation rate	堆积速率
accumulation register	累加寄存器
accumulation relief	堆积地形
accumulation season	累积季节
accumulation strain	累积应变
accumulation system	储能系统
accumulation terrace	堆积阶地
accumulation theory of volcano	火山堆积说
accumulation till	堆积冰碛
accumulation variogram	累积方差图
accumulation zone	聚集带；堆积带
accumulational landform	堆积地形
accumulational platform	堆积台地
accumulational relief	堆积地形
accumulational terrace	堆积阶地
accumulative crystallization	聚集结晶
accumulative damage	累积损伤
accumulative deposit	堆积沉积
accumulative effect	集聚效应
accumulative electric charge	积累电荷
accumulative error	累计误差
accumulative formation	堆积建造；堆积层
accumulative phase	堆积阶段；堆积相
accumulative plastic deformation	累积塑性变形
accumulative process	累积过程
accumulative rock	堆积岩
accumulative sampler	累积取样
accumulator	蓄能器
accumulator system	储能系统
accumulator tank	储罐；储蓄槽；蓄电池箱
accuracy	精度；准确性
accuracy adjustment	精度调整
accuracy assessment	准确度评价；精度评价
accuracy check	精度检验
accuracy control	精度控制
accuracy control character	精度控制符号
accuracy control system	精度控制系统
accuracy degree	精度；准确度
accuracy gradation of groundwater resources evaluation	地下水资源评价精度分级
accuracy horizontal control	平面控制精度
accuracy limit	精度限制
accuracy measurement	精度测量
accuracy of calibration	标定精度
accuracy of determination	测量精度
accuracy of geological reconnaissance	地质调查精度
accuracy of measurement	测量精度
accuracy of observation	观测精度
accuracy of reading	读数精度
accuracy rate	精确度
accuracy rating	额定准确度
accuracy requirement	精度要求
accuracy standard	精度标准
accuracy table	修正值表
accuracy test	精度检验
accuracy-degree	精确度
accurate	准确的；精确的
accurate adjustment	精密调整
accurate calibration	精确校准
accurate grinding	精磨
accurate measurement	精密测量
accurate pointing	精确定向
accurate position finder	精确测位装置
accurate position indicator	精确位置指示器
accurate seismic analysis	精细地震分析
accurate solution	精确解
accurate surveying	精确观测
accurate traverse survey	精密导线测量
AC-DC drive	交-直流传动
acendrada	白泥灰岩
acervate	成堆的
acetacetate	乙酰乙酸盐
acetal polymer	乙缩醛聚合物
acetate	醋酸盐；醋酸根；醋酸酯
acetate butyrate	醋酸丁酸盐
acetate fiber	醋酸纤维
acetate tracer	醋酸盐示踪剂
acetenyl	乙炔基
acetic acid	醋酸；乙酸
acetic aldehyde	乙醛
acetic anhydride	醋酸酐；无水醋酸
acetification	醋化作用
acetimeter	酸度计；醋酸计
acetince	醋精
acetoacetanilide	乙酰基乙酰替苯胺

acetol 丙酮醇；乙酰甲醇
acetolysis 醋解
acetometer 醋酸计
acetone 丙酮
acetone cyanohydrin 丙酮氰醇
acetone resin 丙酮树脂
acetonitrile 乙腈；氰化甲烷
acetophenone 苯乙酮；甲基苯基酮
acetpophenone 乙酰苯
acetyl 乙酰基；醋酸基
acetyl cellulose 醋酸纤维素
acetyl chloride 乙酰氯
acetyl peroxide 过氧化乙酰
acetyl ricinoleate 乙酰蓖麻酸酯
acetyl tetrabromide 四溴乙炔
acetyl value 乙酰值
acetylation 乙酰化
acetylene 电石气；乙炔
acetylene bucket lamp 大电石灯；大乙炔灯
acetylene burner 乙炔焊枪
acetylene cutter 乙炔切割器
acetylene cylinder 乙炔气瓶
acetylene gas 乙炔气
acetylene generating station 乙炔站
acetylene generator 乙炔发生器
acetylene hose 乙炔胶管
acetylene hydration 乙炔水合
acetylene lamp 电石灯
acetylene polymer 乙炔聚合物
acetylene residue 电石渣
acetylene series 炔烃
acetylene silver 乙炔银
acetylene tetrabromide 四溴化乙炔
acetylene welding equipment 氧气乙炔焊
acetylenic alcohol 炔醇
acetylenic hydrocarbon 炔属烃
acetylsalicylic acid 乙酰水杨酸
acetyltributyl citrate 柠檬酸乙酰三丁基酯
achavalite 硒铁矿
Acheson furnace 艾奇逊电炉
achiardite 坏晶石
achieved reliability 工作可靠性
Achimedean spiral 阿基米德螺线
achirite 透视石
achlusite 钠滑石
achnakaite 黑云钙长辉长岩
achoania 无孔鱼属
achondrite 无球粒陨石
achroite 无色电气石
achromaite 浅闪石
achromat 消色差；消色差透镜
achromatic body 消色物体
achromatic eyepiece 消色差目镜
achromatic film 色盲片
achromatic fringe 消色差条纹
achromatic image 消色差图像
achromatic lens 消色差透镜
achromatic light 全色光；白光
achromatic magnetic mass spectrometer 消色差磁性质谱仪
achromatic map 单色图
achromatic object 消色差物镜
achromatic sheet 单色图
achromaticity 消色差；非彩色
achromatism 消色差
achromatization 色差的消除；消色
acicular aggregate 针状集合体
acicular blastic texture 针状变晶结构
acicular crystal 针状晶体；针状结晶
acicular diabase 针状辉绿岩
acicular fabric 针状组构
acicular fracture 针状断口
acicular iron ore 针铁矿
acicular martensite 针状马氏体；针状炭甲铁
acicular shape 针状
acicular texture 针状结构
acid alpine meadow soil 配性高山草甸土
acid and alkaline gas treatment system 酸碱气体处理系统
acid attack 酸侵蚀
acid bog 酸性沼泽
acid brick 酸性砖
acid brittleness 氢化脆性；酸脆
acid brown forest soil 酸性棕色森林土
acid clay 酸性白土；酸性黏土
acid coke 酸性焦炭；固体沥青
acid corrosion 酸腐蚀
acid deasphalting 硫酸脱沥青法
acid degradable chalk mud 可用酸降解的白垩泥浆
acid degradable polymer mud 可用酸降解的聚合物泥浆
acid deposition 酸沉降；酸性沉积物
acid deposition or acid rain 酸雨
acid dissoluble cement 酸溶性水泥
acid earth 酸性白土
acid effluence 酸性废水；酸性废液酸性排水
acid egg 压气喷酸罐
acid electrode 酸性焊条
acid etch method 酸蚀法；酸刻法
acid forest soil 酸性森林土
acid geochemical barrier 酸性地球化学障
acid ground glass 毛玻璃
acid heat test 酸热试验
acid humification 酸性腐殖化
acid hydrolysis 酸解；加酸水解
acid igneous rock 酸性火成岩
acid inhibitor 酸缓蚀剂
acid leaching 酸法浸出
acid lining 酸性衬里；耐酸衬砌
acid mine drainage 酸性煤矿水；酸性废水
acid mine drainage waste 酸性矿坑废水
acid mine spoil 酸性矿山剥离物，酸性矿山废渣，酸性矿山废石堆
acid mine water 酸性矿水；酸性矿坑水
acid mineral reaction rates 酸性矿物反应速率
acid mist cleaner 酸雾净化机
acid moor 酸性沼泽
acid number 酸值
acid oil 高含硫原油
acid open-hearth steel 酸性平炉钢
acid ore 酸性矿石
acid out 酸析

acid oxide	酸性氧化物
acid pack	稠化酸砾石填充
acid peat	酸性泥炭
acid peat soil	酸性泥炭土
acid penetration	酸穿透
acid pickling	酸洗
acid plagioclase	酸性斜长石
acid plant	酸洗设备；酸洗机
acid potential	酸性潜能
acid preflush	酸预冲洗液
acid process	酸性法
acid proof lining	耐酸衬里
acid proof material	耐酸材料；防酸材料
acid proof refractory	防酸耐火材料
acid pump	酸泵；耐酸泵
acid rain	酸雨
acid reaction	酸性反应
acid recovery	酸回收
acid refractory	酸性耐火材料
acid regeneration	酸再生
acid resistant paint	耐酸漆，耐酸涂料
acid resistant paving	防酸砖地面
acid resisting concrete	耐酸混凝土
acid rock	酸性岩
acid rock drainage	酸性岩排水
acid rocks	酸性岩类
acid run off	酸性泾流
acid salt	酸式盐
acid slag	酸性炉渣
acid sludge	酸性淤渣；酸渣
acid soil	酸性土
acid solubility	酸溶解度
acid soluble magnesia cement	酸溶性氧化镁水泥
acid soluble screen protection	酸溶性筛管保护层
acid solution	酸溶液
acid spring	酸性泉
acid stain	酸性污点；酸性染料
acid steel	酸性钢
acid stimulation	酸化增产
acid store room	储酸室
acid sulfate soil	酸性硫酸盐土
acid sulfate spring	酸性硫酸盐泉水
acid sulphate	酸式硫酸盐
acid sulphate soil	酸性硫酸盐土；红树林沼泽土；热带沼泽林土
acid test	氟氢酸测定；酸性试验
acid thermal activity	酸性水热活动
acid treatment	酸性处理；酸化
acid tube	氢氟酸瓶
acid value	酸值
acid wash	酸洗
acid waste	酸性废水；酸性废物
acid waste liquid	酸性废液
acid waste sludge	酸性废污泥
acid waste water	酸性废水
acid wood	酸性木材
acid-base indicator	酸碱指示剂
acid-fast process	酸乳浊液爆破法
acid-forming substance	成酸物质
acid-gas injection	酸性气体回注
acidic corrosion	酸性侵蚀
acidic corrosion and filtration of rockmass	岩体酸性溶蚀
acidic geothermal water	酸性地热水域
acidic granulite	酸性麻粒岩
acidic group	酸性基
acidic igneous rock	酸性火成岩
acidic lava	酸性熔岩
acidic magmatic clasts	酸性岩浆碎屑
acidic oxide	酸性氧化物
acidic precipitation	酸沉降
acidic rock	酸性岩
acidic salt	酸式盐
acidic silica gel	酸性硅胶
acidic slag	酸性渣
acidic soil	酸性土
acidic volcanic	酸性火山岩；酸性火山岩的，酸性火山的
acidic waste water	酸性污水；酸性废水
acidic water	酸性水
acidite	酸性岩；硅质岩
acidity coefficient	酸性系数
acidity of groundwater	地下水的酸度
acidity quotient	酸度系数
acidity soil	酸性土
acidity test	酸度试验
acidize the hole	用酸处理钻孔
acidizing	酸化；酸处理
acidizing economics	酸化处理经济评价
acidizing of well	水井酸化处理；酸洗井
acidizing treatment	酸化作业
acid-mine drainage	酸性矿山排水
acid-proof brick	耐酸砖
acid-proof brick lining	耐酸砖衬
acid-proof hose	耐酸软管
acid-proof lining	耐酸衬
acid-proof material	耐酸材料
acid-proof paint	耐酸涂料
acid-proof pipe	耐酸管
acid-recovery plant	废酸回收厂
acid-resistant	耐酸的，抗酸的
acid-resistant alloy	耐酸合金
acid-resistant floor	耐酸地面
acid-resisting cement	耐酸水泥
acid-resisting concrete	耐酸混凝土
acid-resisting pump	耐酸泵
acid-resisting vat	耐酸桶
acid-rock interactions	酸岩相互作用
acid-treated residue	酸处理残渣
acidulae	碳酸矿质水
acidulant	酸化剂
acidulous spring	酸泉；酸性矿泉
acid-waste product	酸废料；酸渣
acidylation	酰化
acierage	表面钢化
acinose ore	粒状矿石
acinose texture	细粒结构
acisculis	石工小锤
aclimatic soil formation	非气候性成土
aclinal	非倾斜的；无倾角的；无倾角的；水平的
acline	磁赤道；无倾线；水平地层

aclinic line	零倾线；无倾线；磁赤道
aclinic structure	无倾斜构造
acme biozone	顶峰带
Acme thread	爱克米螺纹，梯形螺纹
acme thread tap	梯形丝锥
acme zone	顶峰带；富集带；盛期
acmite	锥辉石
acmite augite	霓辉石
acmite-granite pegmatite	绿辉花岗岩伟晶岩
acmite-trachyte	绿辉粗面岩
acnode	孤点
acolpate	无沟型；无槽的
acomb switch	艾柯姆型开关
acontiodus	矢牙形石属
acorite	锆石
acorn	橡树子；整流罩；橡实形管
acorn nut	螺母；盖帽式螺帽
acoustic absorbant	吸声材料
acoustic absorption	声吸收
acoustic absorptivity	声吸收系数
acoustic admittance	声导纳
acoustic alarm	声报警
acoustic amplitude log	声幅测井
acoustic attenuation	声衰减
acoustic attenuation constant	声衰减常数
acoustic barrier	隔音墙
acoustic basement	声波基底
acoustic beacon	水声信标
acoustic beacon on bottom	海底声标
acoustic blanking	声能的消隐；消声作用
acoustic blur	声斑
acoustic board	隔音板；吸声板
acoustic booth	隔声间
acoustic bop	声控防喷器
acoustic borehole television viewer	有声电视钻孔检查器
acoustic box	消声罩
acoustic buoy	声响浮标；音响浮标
acoustic caliper	声波井径曲线
acoustic capacitance	声容
acoustic cavitation	声波的空穴
acoustic ceiling	吸声吊顶
acoustic celotex board	隔音纤维板
acoustic cement bond log	声波水泥胶结测井
acoustic character log	声波特征测井
acoustic cloud	声波反射云
acoustic coagulation	声波凝聚
acoustic conductivity	传声性
acoustic construction	隔声构造
acoustic contrast	声阻抗比差
acoustic control	声控
acoustic couplant	声耦合剂
acoustic coupler	声音耦合器；音效耦合器
acoustic depth sounding	回声测深
acoustic design	声学设计
acoustic detection	声学探测
acoustic determination of pressure	声学法测压
acoustic diagraph	声波绘图仪
acoustic discontinuity	声学不连续
acoustic dispersion	声频弥散
acoustic Doppler current profiler	声学多普勒海流剖面仪
acoustic drying	声频干燥；超声干燥
acoustic emission	声发射技术，声发射
acoustic emission monitoring	声发射监测
acoustic emission of rock	岩石声发射
acoustic emission signal	声发射信号
acoustic emission source location	声发射源定位
acoustic emission spectrum	声发射频谱
acoustic emission system	声发射系统
acoustic emission testing	声发射检测
acoustic emission well	声测井
acoustic emission (AE)	声发射
acoustic energy	声能
acoustic energy density	声能密度
acoustic energy flux	声能通量
acoustic energy ratio	声能比
acoustic exploration	声波勘探
acoustic formation factor	声波地层因数
acoustic foundation	声波基底
acoustic frequency	声频
acoustic frequency geo-electric field method	声频大地电场法
acoustic frequency sounding	声波频率测深
acoustic frequency spectrum	声波频谱
acoustic generator	发声器
acoustic hologram	声波全息照相
acoustic holography	声学全息；声波全息照相
acoustic holography system	水声全息系统
acoustic imaging	声成像
acoustic impedance curve	声阻抗曲线；声阻曲线
acoustic impedance layering	声阻抗分层
acoustic impedance log	声阻抗测井
acoustic impedance section	声阻抗剖面
acoustic impedance step	声阻抗阶跃
acoustic inspection	声学检验
acoustic insulation	隔音；隔音设备
acoustic lining	隔音板
acoustic liquid level instrument	声波液面测定器
acoustic load	声波负载
acoustic location	声波定位
acoustic log	声波测井
acoustic marine speedometer	海洋声波测速仪
acoustic marker	水声信标
acoustic material	隔音材料
acoustic measurement	声学测量
acoustic memory	声存储器
acoustic methanometer	发音沼气检定仪
acoustic method	声学法
acoustic mobility	声迁移率
acoustic piezometer	声发射测压计
acoustic pinger	声脉冲发射器
acoustic plaster	吸声灰膏；吸声粉刷
acoustic porosity	声波孔隙度
acoustic position indicator	水声定位仪
acoustic positioning	声波定位；水声定位
acoustic positioning system	水声定位系统
acoustic property of rock	岩石声学性质
acoustic prospecting	声波探测；声波勘探
acoustic sand probe	声波探砂器
acoustic scanner tool	声波扫描测井仪
acoustic shock	声震
acoustic signature log	声波波列测井

acoustic simulation	声学模拟
acoustic sounder	回声测深仪；声波探测仪
acoustic sounding	回声测深；声波测深，声学探测
acoustic sounding apparatus	回声测深仪
acoustic strain gauge	钢弦应变计
acoustic stratigraphy	声波地层学
acoustic tank	声学试验水槽
acoustic telemetry	声波遥测技术
acoustic televiewer log	声波电视测井
acoustic tile	吸声面砖
acoustic tomography	声波层析成像法
acoustic transducer	声换能器
acoustic velocity log	声速测井；地震速度测井
acoustic velocity logging	声速测井
acoustic water level	声学水位计
acoustic wave acquisition	声波采集
acoustic well-logging	声波测井
acoustical	声学的，音响的
acoustical amplitude log	声幅测井
acoustical borehole television	声波井下电视
acoustical detector	声波探测器
acoustical feedback	声反馈
acoustical full-wave log	声波全波测井
acoustical log	声波测井
acoustical plaster	吸声灰泥
acoustical velocity log	声速测井
acoustical velocity logging	声速测井
acoustical well sounder	钻井液面声学测深器
acoustostratigraphy	声波地层学
acreage evaluation	探区评价
acreage rent	采矿地租
acree	岩屑锥
acrisol	强淋溶土
acro orogenic movement	顶部造山运动
acrobatholithic	露岩基的
across fault	横断层
across pitch	沿走向
across strike	横交走向
across the bedding seepage	横切层面渗流
acrox	高风化氧化土；强酸性氧化土
acrozone	极顶带
acrylamide	丙烯酰胺
acrylamide grout	丙烯酰胺浆
acrylic binders	丙烯黏结剂
actic	潮区；潮区的
actic region	（大陆架边缘）陡坡带，大地构造变化带
acting circles of blasting	爆破作用圈
acting flank	作用侧面
acting head	有效水头；工作水头
acting load	加载
actinolite	阳起石
actinolite asbestos	阳起石石棉
actinolite rind	阳起石壳
actinolite schist	阳起片岩
actinolitic greenschist facies	阳起绿色片岩相
actinolitite	阳起石岩
actinote	阳起石
action and reaction	作用力与反作用力
action of coal forming	成煤作用
action of gravity	重力作用
action of pile group	群桩效应群桩作用
action of sheet erosion	片状侵蚀作用
action of surge chamber	调压井功能
action radius	影响半径；作用半径
activated adsorption	活化吸附
activated alumina	活性矾土；活性铝土；活性氧化铝
activated bentonite	活性膨润土
activated calcium carbonate	活性碳酸钙
activated carbon	活性炭
activated carbon absorber	活性炭吸附塔
activated carbon fiber	活化碳纤维
activated carbon filtration	活性炭过滤
activated carbon process	活性炭处理法
activated carbon volatility	活性炭法挥发度
activated charcoal	活性炭
activated clay	活性土
activated complex	活化配合物；活化复体
activated earth	活性土
activated faults	活化断裂
activated material	激活材料
activated molecule	活化分子
activated montmorillonite clay	活性蒙脱土
activated plough	动力刨煤机，振动式刨煤机
activated ramp plate	梭动装煤犁
activated sewage	活性污水；活性泥
activated sludge culture	活性污泥培养
activated sludge floc	活性污泥块
activated sludge loading	活性污泥负荷
Activated Sludge Model NO. 2	活性污泥2号模型
Activated Sludge Model NO. 3	活性污泥3号模型
Activated Sludge Model NO. 1	活性污泥1号模型
activated sludge plant	活化污泥设施
activated sludge process	活性污泥法
activated sludge pump	活性污泥泵
activated sludge settling tank	活性污泥沉淀池
activated sludge system	活性污泥系统
activated state	活化态
activated surface	激活面
activated water	活化水
activated-sludge effluence	经活性污泥法处理后的出水
activating agent	活化剂
activating element	活化元素
activating radiation	活化辐射
activation	活化；激活；曝气处理法
activation analysis	活化分析；激活分析；辐射分析
activation cross section	活化截面
activation energy	活化能
activation enthalpy	活化焓
activation entropy	活化熵
activation experiment	活化试验
activation function	激活函数
activation gamma ray	活化伽马射线
activation grade	活化度
activation logging	活化测井
activation method	活化法；激发法
activation number	活化值
activation of platform	地台活化
activation of slide	启滑
activation overpotential	活化过电位
activation polarization	活化极化

activation rule 激活准则
activator 活化剂;活化器
active acidity 活性酸度
active agent 活性剂;活化剂
active and passive blocks 主动岩块与被动岩块
active and passive margins 主动和被动大陆边缘
active antenna 有源天线;激励天线
active aquifer 活跃含水区
active arch 余摆线形
active area 活化区;显示区;活动区
active aurora 活动极光
active barrier 主动式岩粉棚
active basin 活动盆地
active belt 活动带
active block 有源组件;有效网络;有效块;活动断块
active breaking load 主动断裂载荷
active capital 流动资本
active carbon 活性炭;放射性碳
active carbon adsorption method 活性炭吸附法
active carbon gas purification 活性炭法煤气净化
active card 现用卡片
active cave 活洞穴
active center 有效中心;活性中心
active channel 流水河槽
active charcoal 活性炭
active chimney 活动烟筒
active circuit 有源电路
active cirque 活动冰斗
active clay 活性黏土
active combustion 强烈燃烧
active component 主动部件;有效部分;作用分量;活性组分
active condition 活性状态
active constituent 活性组分
active constraint 起作用约束,有效约束
active containment alterative 主动抑压法
active continental margin 活动大陆边缘
active control 主动控制;有源控制
active corrosion 活性腐蚀
active cross section 有效断面;水流断面
active current 有功电流
active degrade voltage apparatus 有源降压装置
active deposit 活性沉积;放射性沉积
active depth of swelling 膨胀有效深度
active diapirism 主动底辟
active drainage area 有效排水面积
active drift 正阻力
active dust 活性尘末
active duty 现役
active earth pressure 主动土压力
active earth pressure coefficient 主动土压力系数
active earth thrust 主动土推力
active earthquake 主动地震
active electron 激活电子;活性电子;活跃电子
active element 放射性元素
active emitting material 放射活性材料
active emplacement 强力就位;主动就位
active energy dissipator 有源耗能器
active entry 生产煤巷
active erosion 强烈侵蚀

active face 生产工作面;切缘
active fault 活断层;活性断层
active fault block 活动断块
active fault system 活动断层系统
active fault trace 活动断层迹线
active fault zones 活动断裂带
active field shield 主动屏障
active filler 活性填料
active fire 烈火;仍在继续的火灾
active fissure 活动裂缝
active flexure 活动挠褶
active fold 活动褶皱
active force 主动力;作用力;有效力
active fracture 活断裂
active fracture plane 活动破裂面
active gas 活动气体
active geophysics method 主动源地球物理方法
active geothermal field 活跃地热田
active geyser 活间歇泉;现代间歇泉
active glacier 活动冰川
active growing period 生长活跃时期
active guidance 主动制导
active heading 施工中掘进头
active heave compensator 主动式升沉补偿器
active hydrothermal area 活动水热区,现代水热区
active infrared meter 有源红外仪;主动式红外仪
active infrared system 主动式红外系统
active ingredient 活性拌料
active injection-pressure 主动贯入压力;有效灌浆压力
active injector 现注井
active insulation 积极防振;主动绝缘
active intensity forecasting 活动烈度预测
active island arc 活动岛弧
active isolation 主动隔振
active landslide 活滑坡
active layer 活性层;冻融活性层
active length 有效长度
active line 作用线
active link 主用链路,激活的链路
active load 有效荷载;有效负荷
active margin 活动边缘
active mass 有效质量
active mass damper 有源质量阻力器
active mass driver 有源质量驱动器
active material 活性物质;放射性材料
active measured value 实测值
active mechanism 作用机制
active melt period 强烈融化期
active microwave 主动微波
active microwave system 主动微波系统
active mitigation strategies 主动缓解策略
active moraine 活冰碛
active mud pit 工作泥浆池
active mud system 在用泥浆系统
active navigation and positioning 自主式导航定位
active network 有源网络
active noise control 有源噪声控制
active notch filter 有源陷波滤波器
active nucleus 放射性核
active oil 可采石油

A

active oil calculation	石油可采储量计算
active optical network	有源光网络
active oxidation	活性氧化
active oxygen	活性氧；活化氧
active page	活动页
active part of time slot	时隙的活动部分
active period	活跃期；活动期
active permafrost	活动永冻层
active phase	活跃期
active pile	主动桩
active plate	活动板块
active plate margin	活动板块边缘
active porosity	有效孔隙率
active power	有效功率；有功功率
active pressure	主动压力
active producer	现产井
active profile	有效齿廓
active program	活动程序
active Rankine pressure	朗肯主动土压力
active Rankine zone	主动朗肯区
active reaction	活性反应
active recording device	有源的记录设备
active recreation area	动态康乐用地
active recreation facility	动态康乐设施
active region	活动区
active remote sensing	主动遥感；有源遥感
active reserve	可动用储量
active response	活性反应
active ridge	活动海岭
active ridge system	活动洋脊系
active rift	活动断裂；活动裂谷
active rig	工作钻机
active rudder	主动舵
active sand dune	流动沙丘
active satellite	有源卫星
active scanner	主动式扫描仪
active section	工作段
active sediment	活性沉积物
active seepage	主动渗流
active segment	活动段
active seismic area	地震活动区
active seismic measurement	有源地震测量
active seismic pressure	主动地震压力
active seismic zone	活动地震带
active seismic zones around the project area	工程区外围地震危险区
active seismological area	活动地震区
active seismometer	有效检波器
active sensor	主动遥感器；有源遥感器
Active Server Pages	动态服务器页面；激活服务器页面
active servo	有源伺服
active shear stress	主动剪应力
active side	主动盘
active silica	活性硅
active slide	活滑坡
active slide area	主动滑动区
active sliding surface	主动滑动面
active slope	活动斜坡
active sludge	活性污泥
active sludge bed	活性煤泥层
active soil	活性土
active soil formers	活性成土因素
active soil pressure	主动土压力
active solar house	主动式太阳房
active solid	活性固体
active sonar	主动声呐
active sound reduction	有源减噪
active source	放射源
active source bed	有效生油层
active source method	主动源法
active source rock	有效生油层
active spreading belt	活动扩张带
active springs	活动泉区；现代泉区
active stabilizer	活性加固剂；活性稳定剂
active state	活动状态；活动期
active state of plastic equilibrium	主动塑性平衡状态
active stope	生产回采工作面
active stope area	采区的有效面积；矿房有效面积
active storage of reservoir	水库有效库容
active stream sediment	水系活性沉积物
active stress	主动应力
active structural basin	活动性构造盆地
active structural control	主动结构控制
active structure	活动构造
active structure zone	活动构造带
active subsidence	活动性下沉
active substance	活性物质，活性材料
active sulphur	活性硫
active support	主动支护
active surface	有效面
active surface of sliding	主动滑动面
active system	有源系统；有外源设备系统
active tank	在用罐
active tectonic belt	活动构造带
active tectonic element	活动构造单元
active tectonic pattern	活动构造型式
active tectonic region	活动构造区
active tectonic system	活动性构造体系
active tectonic zone	活动构造带
active tectonics	活动构造
active thrust of earth	主动土推力主动土压力
active transducer	有源换能器
active tuned mass damper	有源调谐质量阻尼器
active type continental margin	主动型大陆边缘
active vapor control	有效蒸汽控制
active variable stiffness control	有源变刚度控制；主动变刚度控制
active variable stiffness system	有源变刚度系统
active vibration isolation	主动隔振
active volcano	活动火山；活火山
active volt ampere	有效伏安
active voltage	有功电压
active volume	有效容积
active water	活性水
active water circulation	活跃水循环
active water drive	活跃水驱
active water exchange	活跃水交替
active wedge	主动楔形体
active well	工作井
active working face	生产工作面

active zone 活动层
active zone of foundation 地基持力层；地基主动变形区
active-breaking load 主动断裂载荷
actively yielding coal 塑性流变煤
activity gradient 活动梯度
activity level 活度级
activity of soil 土的活性
activity quotient 活化系数
activity stability 活性安定性
activity test 活性试验
activity unit 放射性强度单位
activity-based costing 作业成本法
activity-on-the-arrow system 箭杆法；双代号法
activity-on-the-node system 节点法；单代号法
activization of platform 地台活化
activization platform block 活化台块
Actonian 阿克顿阶
actual additional cost 实际附加成本
actual address 实际地址
actual age 绝对年代学
actual amount 实际数量
actual behaviour 实际性状
actual bit on bottom time 纯钻进时间
actual bit wedge 实际刃角
actual braking time 实际制动时间
actual breach 实际违约
actual breaking strength 实际断裂强度
actual breaking stress 真实断裂应力
actual budget 决算
actual building cost 实际造价
actual burst pressure 实际破裂压力
actual capacitance 实际性状；实际电容
actual capacity 实际生产能力；实际容量
actual compression 实际压缩
actual computation 实际计算
actual construction sequence 实际施工程序
actual cost 实际成本
actual cost rent 实际成本租金
actual count 实际计数
actual damping 有效阻尼
actual delivery 实际排量
actual demand 实际需求
actual deviation 实际偏差
actual dimension 实际尺寸
actual dishonour 实际拒付
actual displacement 实际位移
actual draft 实际牵伸
actual earth fill 实际填土坡线
actual eccentricity 实际偏心
actual efficiency 实际效率
actual elevation 实际高度
actual error 真差
actual evaporation 实际蒸发；有效蒸发
actual evaporation rate 实际蒸发率
actual evapotranspiration 实际蒸散发；有效蒸散发
actual failure stress 实际破坏应力
actual fault pattern 实际断裂模式
actual field intensity 实际场强
actual fluid 实际流体；真实流体
actual flux density 有效磁感应；实际磁通密度
actual force 实际作用力
actual frame duration 实际框期间
actual gas 实际气体
actual gradient 实际梯度
actual grain size 实际粒径
actual horizon 实际层面
actual horsepower 实际马力；有效马力
actual hour worked 实际工时
actual investment 实际投资
actual landform 实际地形
actual length of turnout 道岔实际长度
actual liabilities 实际负债
actual life 实际寿命
actual load 实际载荷；有效负载
actual loading test 实际负荷试验
actual loss of reserves 实际损失储量
actual loss ratio 实际损失率
actual market volume 实际销售量
actual mean pressure 实际平均压力
actual measured value 实测值
actual measurement 实测；实物测量
actual mechanism 有效机制
actual mining 回采工作
actual mining roadway 回采巷道
actual model 事实模式
actual number 实数
actual numerical value 实际数值
actual output 实际产量，实际输出
actual overload factor 实际过载系数
actual parameter 实际参数
actual parking volume 实际停放量
actual path 实际路线
actual peak load 实际峰荷
actual performance 实际性能；实际工作
actual performance curve 实际性能曲线
actual plot ratio 实际地积比率
actual point of frog 辙叉心轨尖端
actual point of switch rail 尖轨尖端
actual porosity 有效孔隙度
actual power 有效功率；实际功率
actual pressure 实际压力
actual price 实际价格
actual production 实际产量，实际能力
actual profit 实际利润
actual purchase order price 实际购货订单价格
actual quantity 实际工程量
actual recovery 实际回收率
actual reduction ratio 真实破碎比
actual relative movement 滑距，实际相对运动
actual reserves 可靠储量，实际储量
actual resolution 实际分辨率
actual sea level 实际海平面；实际海水位
actual selsyn 实际自整角机；实际自动同步机
actual setting frequency 实际导频率；实际设定频率
actual size 实际尺寸
actual specific gravity 实比重
actual specific gravity of separation 实际分选比重
actual standard 现行标准
actual state 实际状况
actual stopping distance 实际制动距离

actual store 实际库存
actual strength 实际强度
actual stress 有效应力；作用应力；实际应力
actual stress at fracture 破坏的实际应力
actual stripping ratio 实际剥采比
actual structure 真实结构
actual stuff 现货
actual test 运转试验
actual throughput 实际输量
actual time 实际时间
actual time of observation 实际观测时间；实际观测时刻
actual tons of handling 装卸自然吨
actual total evaporation 实际总蒸发
actual value 实际价值
actual velocity 实际流速；真实速度
actual velocity of groundwater flow 地下水实际流速
actual volume 实际容积
actual wage 实际工资
actual yield 实际产量，实际回收率
actual yield point 实际屈服点
actual yield stress 实际屈服应力
actual zero-point 基点；图零点；绝对零点；水尺零点
actualistic paleontology 实证古生物学
actualization 现实化；实现
actualization function 现实化函数
actuals 现货
actuarial return 精算收益率
actuary 固定资产统计师；保险统计师
actuate 驱动；激励；使动
actuated roller switch 驱动滚柱开关
actuating 开动；收张；激励
actuating apparatus 促动器
actuating arm 力臂
actuating cam 推动凸轮，主动凸轮
actuating device 驱动装置
actuating lever 起动杆
actuating logic 执行逻辑
actuating mechanism 驱动机构，传动机构，引动机构
actuating medium 工作介质
actuating motor 驱动电动机；伺服马达；伺服电动机
actuating pressure 启动压力
actuating rod 作用杆
actuating screw 致动螺丝
actuating signal ratio 作用信号比；信号比；激励信号比
actuating sole 操纵电磁铁；控制电磁铁
actuating strut 作用支柱
actuating system 传动系统
actuating transfer function 动作传递函数；激励传递函数
actuating unit 动力机构；动力传动装置
actuation test 启闭试验
actuation time 促动时间
actuator 激振器；加载器；作动器；促动器
actuator capacity 激振器出力
actuator mechanism 执行机构；传动机构
actuator valve 致动器阀
actuopalaeontology 现实古生物学
acuity 敏锐度；分辨能力；尖锐，锐利
Acus 针刺藻属
acutance 锐度；锐度曲线
acute 锐的；尖锐的；急剧的

acute angle 锐角
acute angle intersection 锐角相交
acute arch 锐拱
acute bisectrix 锐角等分线
acute folding 强烈褶皱作用
acute frog 锐角辙叉
acute hazard event 急性危害事件
acute intersection 锐角交会
acute moving landslide 剧动式滑坡
acute peak abutment 矿柱的锐角应力峰值区
acute poisoning 急性中毒
acute projection 尖锐突起
acute silicosis 急性矽肺
acute toxicity 急性毒
acute triangle 锐角三角形
acute-angle interarea 锐角交互面
acutely folded 强烈褶皱的
acuteness 锐度
acutifoliate 尖叶的
acuturris 尖石
adakite 埃达克岩
Adaline adaptive linear element 阿达林自适应线性元件
adamant 硬石；金刚石
adamant plaster 纤维板
adamant steel 铬钼特殊耐磨钢
adamantine 坚硬的；金刚石似的；金刚合金
adamantine boring 冷铸钢粒钻进；钻粒钻进；钻粒钻眼
adamantine crown 金刚石钻头；金刚硼钻头；冷铸钢粒钻头
adamantine drill 金刚石钻头，金刚石钻进，金刚石钻机，金刚石钻具
adamantine lustre 金刚光泽
adamantine shot 钻井钢砂
adamantine spar 刚玉
adamas 金刚石
adambulacral 侧步带板
adam-diorite 石英二长闪长岩
adam-gabbro 石英二长辉长岩
adamic clay 红黏土
adamic earth 红黏土
adamite 水砷锌矿；人造刚玉；镍铬耐磨铸铁
adamite roll 高碳铬镍耐磨铸铁轧辊
adamsite 暗绿云母
adam-tonalite 英云二长闪长岩
adaptability 适应性；可用性
adaptability of soil 土宜；适土性
adaptability test 适应性测验
adaptable data base system 自适应数据库系统
adaptation 适应；匹配
adaptation coefficient 自适应系数
adaptation level 适应性水平
adaptation modification 适应变态
adaptation physiology 适应生理学
adaptation reaction 适应性反应
adaptative histogram adjustment 适应性直方图调整
adaptative radiation 适应变异；适应性辐射
adaptative regression 适应退化
adapter 转接器；接合器；拾波器；变换器
adapter bonnet 变径管；法兰短节
adapter converter 适配变换器；附加变频器

adapter coupling 转接；异径接头；异径连接管
adapter flange 变径法兰
adapter joint 异径接头
adapter junction bowl 分线盒
adapter kit 整套附件
adapter substitute 转换接头
adapter-booster 传爆药管
adaptibility 适应性
adapting pipe 管节；接合管
adaption 适应；配合；匹配；改进方案
adaptive 适应的
adaptive algorithm 自适应算法
adaptive analysis 自适应分析
adaptive and closed loop design 自适应闭环设计
adaptive array 自适应阵
adaptive behavior 适应行为
adaptive capacity 适应能力
adaptive character reader 自适应字符阅读器
adaptive control 自适应控制
adaptive control system 自适应控制系统
adaptive convergence 适应性收敛
adaptive cross approximation 自适应交叉估计；一种加速算法
adaptive data base manager 自适应数据库管理系统
adaptive deconvolution 自适应逆卷积
adaptive deconvolution algorithm 自适应反褶积算法
adaptive differential pulse code modulation 自适应差分脉冲编码调制
adaptive divergence 适应趋异
adaptive equalization 自适应均衡
adaptive estimation 自适应估计
adaptive evolution 适应性演化
adaptive faculty 适应能力
adaptive feedback control 自适应反馈控制
adaptive filtering 自适应滤波
adaptive form 适应类型
adaptive fuzzy control 自适应模糊控制
adaptive genetic algorithm 自适应遗传算法
adaptive grid 适应性网格
adaptive implicit method 自适应隐式方法
adaptive mesh refinement 自适应网格精化
adaptive migratory 适应迁移
adaptive mill control 轧机自适应控制
adaptive model 自适应模型
adaptive neural network 自适应神经网络
adaptive optimization 自适应最优化
adaptive peak 适应峰
adaptive prediction 自适应预测
adaptive processing 适应性处理
adaptive radiation 适应性辐射；适应性趋异
adaptive regression 适应性退化
adaptive servo 自适应伺服机构
adaptive servo system 自适应伺服系统
adaptive signal discrimination 自适应信号辨识
adaptive signal processing 自适应信号处理
adaptive stack 自适应叠加
adaptive step-size control 适应性的步长控制
adaptive system 自适应系统
adaptive telemetry 自适应遥测
adaptive testing 自适应测试

adaptive time stepping 自适应时步
adaptive value 适应值
adaptive zone 适应带
adaptivity 自适应性
adaptometer 适应性测量计
adaptor 接合器，接头，连接装置；应接器，应接管，适配器；拾音器，拾波器；座，架，附件联轴器齿套
adaptor flange 连接法兰
adaptor sleeve 定心套
adaptor spool 接头短管
adaptor trough 接合槽
adaptors and fittings 接头和连接件；接头和配件
adarce 石灰华，钙华
adatom 吸附原子
adaxial 近轴的
adcumulate 累积岩；补堆积岩
adcumulate texture 补堆积岩结构
add and drop multiplexer 分插复用器
add button 相加按钮
add gate 加法门
add instruction 加法指令
add operation 加法运算
add output 加法输出
add pulse 加法脉冲
add time 加法时间
addaverter 加法转换器
added carrier 注入的载流子
added diamonds 补加的金刚石克拉数
added drawdown 附加降深
added hours 加点，增加时数
added mass 附加质量
added twist 追加捻
added value 增加值
added-value tax 增值税
addend 加数；被加数
addend register 加数寄存器
addendum 附录，补遗；附加物
addendum circle 齿顶圆
adder 求和器；混频器
adder accumulator 加法累加器
adder subtracter 加减器
addigital 附指羽
adding device 加法器
adding machine 加算机；算术计算机
adding operator 加法运算符
adding water place 上水地点
adding water time 上水时间
addition 掺加料；拌合料；增设；加建
addition agent 掺合剂；外加剂；添加剂
addition condensation 加成缩合
addition constant 加常数
addition coverage 附加覆盖
addition effect 加法效应
addition formula 加法公式
addition hole 累加孔；补充孔
addition of clay 掺黏土
addition polyamide 加成聚酰胺
addition polycondensation 加成缩聚
addition polymer 加聚物
addition polymerization 加聚反应

addition reaction 加成反应
addition record 附加记录；附加资料；补充资料
addition rule 加法定律；魏斯晶带定律
addition theorem 加法定理
addition time 加算时间
additional 附加的；辅助的
additional acceleration 附加加速度
additional alarm 辅助报警
additional analysis 补充分析
additional anion 附加阴离子
additional base point 辅助基点
additional bit 附加位
additional borehole 补充孔
additional budget 追加预算
additional building 附加建筑
additional building works 增补建筑工程
additional buoyancy tank 辅助浮力罐
additional carrier 过剩载流子
additional character 专用符号；附加符号；特殊符号
additional charge 额外费用；附加费；补助炸药；附加炸药；补充充电
additional clause 附加条款；补充条款
additional coefficient 额外扣除
additional condition 附加条件
additional constraint 附加约束
additional contact pressure 基底附加应力
additional control point 辅助控制点
additional damping 附加阻尼
additional draft 意外牵伸
additional drag 附加阻力
additional equipment 辅助设备；外加设备
additional expenditure 追加支出；额外开支
additional expense 附加费用
additional expenses 追加费用
additional exploration 深入勘探
additional fixture 固定附物
additional force 附加力
additional free face 附加自由面
additional freight 附加运费
additional gravel 砾石附加量
additional hole 补充炮眼
additional horizontal force 额外横向力
additional imagery 辅助成像
additional information 可加信息；辅助信息
additional investment 追加投资
additional item 增添项
additional load 附加负载；增加负载
additional margin 追加押金；追加保证金
additional mass 附加质量
additional monitoring 附加监控
additional motion 附加运动
additional passenger train 临时旅客列车
additional phase 加成相
additional plan 增补图则
additional porosity 附加孔隙度
additional potential 附加位
additional premium 附加保险费；追加保费
additional pressure 附加压力
additional producer 补充生产井
additional project 补充方案

additional provision 附加条款
additional recording 补录
additional regulation 补充规定
additional rehousing criterion 附加安置资格
additional remark 补充说明
additional resistance 附加阻力
additional rules 附加法规
additional safety factor 额外安全因子
additional sampling 追加抽样；补充采样
additional sensor 辅助传感器
additional service 附加职责
additional stake 加桩
additional stop button 附加停止按钮；附加停车按钮
additional stress 附加应力
additional survey 补充勘测
additional tax 附加税
additional tax for education 教育费附加
additional thermal resistance 附加热阻
additional transformation 附加变换
additional vent 加设通风口
additional water 补给水
additional well 补充井
additional winding 附加绕组；辅助绕组
additions policy 加户政策
additive 添加剂；掺合料；外加剂
additive agent 添加剂
additive anomaly 累加异常
additive cement 有掺合料水泥
additive color image enhancement 加色影像增强
additive color process 加色处理
additive color projection 加色投影
additive color viewer 加色观察器
additive combination 添加剂组合
additive constant 积分常数；附加常数
additive effect 附加效应；相加效应
additive error 相加误差
additive false color 加假色
additive filter 加色滤光片
additive for clean burning of briquette 型煤清洁燃烧剂
additive for sulphur capture 固硫添加剂
additive group 加法群
additive metamorphism 加物变质作用
additive method 相加法；叠加法
additive noise 相加噪声
additive operation 加法运算
additive primary colors 加色法三原色
additive process 加色法
additive response 添加剂感受性
additive term 附加项
additive three-colour exposure 加色法三色曝光
additive viewing technique 加色观察技术
additive wire 附加导线
additives 处理剂；添加剂
additivity 加和性；相加性
additivity law 相加定律
add-on 附加
add-on interest 追加利息
add-ons 增加额
address 地址；存储器号码；致辞
address arrangement 地址阵

address assignment　地址分配
address blank　空地址
address bus　地址总线
address check　地址校验
address code　地址码
address comparator　地址比较器
address computation　地址计算
address constant　地址常数
address conversion　地址转换
address counter　地址计数器
address decoder　地址译码器
address field　地址段
address file　地址文件
address format　地址格式
address line　地址线
address mapping　地址变换
address Mask　地址掩码
address mask request　地址屏蔽码请求
address offset　地址偏移
address pattern　地址形式
address pipe　地址流水线
address portion　地址部分
address priority　地址优先级
address register　地址寄存器
address search　地址检索
address selection　地址选择
address selection circuit　地址选择电路
address space　地址空间
address stop　地址符合停机
address substitution　地址更换
address table　地址表
address trace　地址跟踪
address translation　地址转换
address translator　地址转换器；地址转换程序
address unmodifiable　地址不可变
address wire　地址线
addressable　可访问的；可寻址的；可编址的
addressable memory　可访问存储器
addressable register　可寻址寄存器
addressable storage　可访问存储器；可寻址存储器
addressed direct access　编址直接存取
addressed memory　编址存储器
addressing　访问；寻址；编址
addressing capability　寻址能力
addressing fault　寻址故障
addressing mode　寻址方式
addressing operation　寻址操作
addressing space　访问空间
addressing system　定址系统
addressing technique　定址技术；寻址技术
addressmarker　信息块首标记
adduct　加合物
adduction relay　感应继电器
adductor muscle scar　收肌筋痕
Adelaidean　阿德雷德阶
adelite　砷钙镁石
adelogenic　显衡隐晶质的
adelogenic texture　非显晶结构
adelpholite　铌铁锰矿
adenticulate　无锯齿的

adequate　适当的；足够的；胜任的；敷用的
adequate housing　适当居所
adequate relief　均匀地形
adequate sample　充足样本
adequately housed　居住面积足够的；有适当居所的
adequation　拉平
adergneiss　脉状片麻岩
A-derrick　A型转臂起重机
adetognathus　自由颚牙形石属
adfreeze　冰与物体黏附力
adfreezing　冻附
adfreezing force　冻附力
adfreezing strength　冻附强度
adglutinate　胶结物；烧结
adheive force　附着力
adhere　黏固；黏接；附着
adherence　吸附；黏着；严守
adherence limit　黏附范围
adherence surface　黏附面
adherent　胶黏体的；黏附的
adherography　胶印法
adherometer　附着力测定仪
adheroscope　黏附计
adhesion　黏附；黏附力；黏着力
adhesion action　胶结作用
adhesion agent　胶黏剂
adhesion bond　黏接
adhesion braking　黏着制动
adhesion character　黏附性
adhesion coefficient　黏附系数
adhesion degree　黏附程度；黏着度
adhesion factor of pile　桩的黏着系数
adhesion failure　黏合破坏
adhesion force　黏着力
adhesion heat　黏合热
adhesion kinetics　黏附动力学
adhesion loco　爬坡机车
adhesion promoter　助黏剂
adhesion ripple　黏附波痕
adhesion settling　吸持沉降
adhesion strength　黏附力，黏附强度
adhesion tapetest　胶黏带试验
adhesion tension　黏附张力
adhesion test　附着力试验
adhesion theory of friction　黏附摩擦理论
adhesion wart　黏附瘤痕
adhesion weight　黏着重量；附着重量
adhesion wheel　摩擦轮，摩擦盘
adhesional wetting　黏润作用
adhesive　接合剂
adhesive ability　黏着力
adhesive anchor bolt　胶结型锚杆
adhesive attraction　附着力
adhesive bond　黏着力
adhesive bonding　黏结
adhesive capacity　黏附能力；黏着性；胶黏度
adhesive coating　黏着层
adhesive disk　吸盘
adhesive effect　黏附作用
adhesive film　吸附膜；黏膜

adhesive force	附着力；黏附力
adhesive Glue	万能胶
adhesive joint	黏接合；胶黏接合；附着接头
adhesive layer	黏性层
adhesive material	黏结材料
adhesive moisture	黏附水分；附着水分
adhesive mortar	黏结砂浆
adhesive paper strip	胶带
adhesive plaster	橡皮膏
adhesive power	黏附力；黏着力
adhesive reflective warning tape	反光警告贴纸
adhesive retention	黏附滞留
adhesive seal	胶泥密封
adhesive separation	黏附分离
adhesive slate	黏板岩
adhesive station	井下测站
adhesive strength	黏着强度；黏合强度
adhesive strip	黏合带
adhesive tape	胶带；胶布
adhesive tension	黏附张力；界面吸引力
adhesive tractive effort	黏着牵引力
adhesive tyre	黏着式轮胎
adhesive water	黏附水；薄膜水
adhesive wear	胶黏磨损
adhesive weight	附着重力
adhesive weight utility factor	黏着重量利用系数
adhesivemeter	黏着力计
adhesives	黏合剂
adhint	黏合接头
adiabatic acceleration	绝热加速
adiabatic adjustment	绝热调节
adiabatic approximation	绝热近似
adiabatic bubble period	绝热气泡周期
adiabatic calorimeter	绝热量热计
adiabatic curve	绝热线
adiabatic demagnetization	绝热去磁
adiabatic efficiency	绝热效率
adiabatic envelope	绝热包络线
adiabatic equation	绝热方程
adiabatic equilibrium	绝热平衡
adiabatic exponent	绝热指数
adiabatic extrusion	绝热挤出
adiabatic heating	绝热增温
adiabatic horsepower	绝热马力
adiabatic modulus of elasticity	绝热弹性模量
adiabatic process	绝热过程
adiabatic reactor	绝热反应器
adiabatic temperature gradient	绝热温度梯度
adiagnostic texture	隐微晶质结构
adiathermal	绝热的
adiathermancy	不透红外线性
adigeite	镁蛇纹石
adinole	钠长英板岩
adion	吸附离子
adipocerite	伟晶蜡石
adit	平洞；支洞；平导；导洞
adit collar	平硐口
adit digging survey	巷道掘进测量
adit end	平硐工作面
adit entrance	平硐口
adit for draining	排水坑道；排水平硐
adit level	平硐水平；平硐水平上的平巷
adit mine	平硐矿
adit opening	坑道口；坑道入口
adit planimetric	坑道平面图
adit portal	平硐口
adit prospecting engineering survey	坑探工程测量
adit survey	坑道测量
adit test	试坑调查
adit window	从平硐通地面的天井
adit-cut mining	平硐开采
adjacency analysis	近邻性分析
adjacency matrix	邻接矩阵
adjacent	毗连的；相邻的
adjacent beach erosion	相邻海滩侵蚀
adjacent bed effect	围岩影响；邻层效应
adjacent blasting area	相邻爆区
adjacent boreholes	相临井眼
adjacent building	相邻建筑物
adjacent chock	邻架
adjacent construction	相邻建造物
adjacent control system	邻架控制系统
adjacent control valve	邻架控制阀
adjacent corner point solution	相邻角点解
adjacent formation	相邻地层
adjacent formation stress	临近地层应力
adjacent ground	相邻土地
adjacent hole	相邻孔
adjacent hole blasting	相邻孔爆破
adjacent level	相邻水平
adjacent map	邻接图
adjacent panel	相邻槽段
adjacent position	邻位；相邻位置
adjacent pounding	互撞
adjacent rock	围岩；相邻岩石
adjacent sea	毗邻海域
adjacent segment	邻接块
adjacent sheet	相邻图幅
adjacent site	相邻基地；相邻地盘
adjacent span	邻跨
adjacent stratum	邻接地层
adjacent street	相邻街道
adjacent unconformity	毗连不整合
adjacent unit	相连单位
adjacent water	相邻水域
adjacent wavelet	相邻子波
adjacent workings	邻接工作区
adjacent-bed effect	围岩影响
adjacent-channel Interference	相邻信道干扰；邻道干扰
adjacent-channel power	邻道功率
adjacent-channel selectivity	邻信号选择性
adjective	附属的
adjoin	邻接；接，连接；加，附加
adjoining area	毗邻地区
adjoining building	毗邻建筑物
adjoining course	连接层，邻接层
adjoining girder	邻接梁
adjoining hole	毗邻孔
adjoining land	毗邻土地
adjoining position	邻接位置

adjoining rock	围岩
adjoining sheet	相邻图幅
adjoining structure	毗邻构筑物
adjoining well	邻井
adjoint	伴随；共轭
adjoint boundary value problem	伴随边值问题
adjoint equation	伴随方程
adjoint functions and operators	伴随函数和运算符
adjoint matrix	伴随矩阵
adjoint operator	伴随算子
adjoint variable	伴随变量
adjudication fee	法庭费用
adjudication of bankruptcy	宣告破产；裁定破产
adjunct	附属物；附件；助手；副手
adjust method of intersection error	交点差调整法
adjust value	调整值
adjustability	可调整性；可调节性
adjustable	可调的；可校准的
adjustable air damper	风窗口调节门
adjustable anchor	可调锚
adjustable anchorage bar	可调锚杆
adjustable angle cam plate	可调倾角凸轮盘
adjustable angle turn	可调转向装置
adjustable attachments	建筑结构可调支撑；活支撑
adjustable bar screen	可调节式棒条筛
adjustable base anchor	活动底锚
adjustable bearing	可调整轴承
adjustable blade	可调式叶片
adjustable block level	可调气泡水准仪
adjustable bolt	可调螺栓
adjustable boom	移动式悬臂
adjustable bracket	可调整的托架
adjustable bush	可调轴衬
adjustable cam	可调式凸轮
adjustable cam-plate	变量斜盘
adjustable cistern barometer	调槽式气压表
adjustable clearance	可调间隙
adjustable crest	可调堰顶
adjustable discharge gear pump	可调卸载齿轮泵
adjustable discharge pump	变量泵
adjustable drive	可调传动装置
adjustable end stop	升降挡板
adjustable expanding reamer	可调式扩眼器；可调扩张式铰刀
adjustable filter	可调滤波器
adjustable flow bean	可调节流器
adjustable gauge	调整式卡规
adjustable grille	可调栅格
adjustable inlet guide vane	可调进口的导叶
adjustable jack	可调节千斤顶；可伸缩立柱
adjustable needle roller bearing	可调间歇滚针轴承
adjustable overload friction clutch	可调式过载摩擦离合器
adjustable pitch	可调式螺距
adjustable plumb bob	可调垂球，定中垂球
adjustable prop	可调式支柱，伸缩式支柱
adjustable resistance	可调阻力
adjustable undercarriage	可调底座
adjustable-aperture grizzly	可调孔式格筛
adjusted angle	平差角
adjusted datum	校正基准点
adjusted drainage	适应水系
adjusted earned income	调整后收入
adjusted elevation	平差高程
adjusted mean	调整平均数
adjusted net fill	调整后的净填方
adjusted replacement cost	经调整重置成本
adjusted river	平行走向河
adjusted stream	平行走向河；顺向河；调节河流
adjusted trial balance	调整后试算表
adjusted value	调整值；平差值
adjuster	修理工；装配工；调整工；调节器
adjuster bar	调节杆
adjuster bolt	调节螺栓
adjuster grips	抽油杆夹具
adjusting bolt	调整螺栓
adjusting character	调整特性
adjusting device	调整装置
adjusting distance of track lining	拔距
adjusting gasket	调节垫片
adjusting gear	调整齿轮
adjusting handle	调整手柄
adjusting joint gaps up to standard	整正规缝
adjusting key	调整键
adjusting knob	调节钮
adjusting lever	调整杆
adjusting links	调整悬挂装置长度的构件
adjusting mechanism	调整机构
adjusting needle	调准针
adjusting nut	调整螺母；调节螺母
adjusting of cross level	整正水平
adjusting of rail gaps	调整轨缝；均匀轨缝
adjusting pan	调节盘；调节槽
adjusting piece	调整片；调节片
adjusting pin	固定销；定位销；锁紧销；校正计
adjusting plate	调整板
adjusting procedure	调整程序
adjusting ring	调整环
adjusting rod	调整杆
adjusting screw	调节螺丝；调准螺丝
adjusting shop	修配车间
adjusting spring	调整弹簧
adjusting to zero	置零
adjusting washer	调整垫圈
adjusting wedge	调整楔
adjustinggear	调整装置
adjustingkey	调整键
adjustment	调整，调节；校正；调准；适应安排
adjustment by coordinates	坐标调准
adjustment by correlates	相关平差
adjustment by method of junction point	结点平差
adjustment by method of polygon	多边形平差法
adjustment calculation	平差计算
adjustment cone	调节锥
adjustment controls	调谐机构
adjustment curve	校准曲线；缓和曲线；过渡曲线
adjustment diagram	调整图
adjustment factors	修正系数
adjustment for collimation	视准改正
adjustment for definition	调整清晰度
adjustment in group	分组平差法

adjustment item	调整项目
adjustment jack	调节千斤顶
adjustment letter	海损理算书；海损精算书
adjustment method	平差法；调试方法
adjustment model	平差模型
adjustment network	调谐电路
adjustment of altitude	高度校正
adjustment of astro-geodetic network	天文大地网平差
adjustment of base station net	基点网平差
adjustment of car flow	车流调整
adjustment of car loading	装车调整
adjustment of condition equation	条件平差
adjustment of control network	控制网平差
adjustment of coordinates	坐标平差
adjustment of correlated observation	相关平差
adjustment of empty cars	空车调整
adjustment of errors	调整误差，修正错误
adjustment of figure	图形平差
adjustment of free network	自由网平差
adjustment of image	影像的调整；图像的调节
adjustment of leveling network	水准网平差
adjustment of measurement	测量平差
adjustment of model	模型修正
adjustment of navigation data	定位数据平差
adjustment of observation	观测平差；测量平差
adjustment of planimetry	平面坐标平差
adjustment of position	位置平差
adjustment of traverse network	导线网平差
adjustment of triangulation	三角网平差
adjustment of typical figures	典型图形平差
adjustment process	调整过程
adjustment table	调准表
adjustment to structure	适应构造
adjustment well	调整井
adjustor	调节器
adjutage	喷射管；喷水管；排水筒；延伸臂；接长臂
adligans	固着的
admantine spar	刚玉；氧化铝
admeasure	分配；度量；测量
admeasurement type of contract	计量型合同
administered price	管制价格
administration	管理局；管理；行政
administration building	行政办公楼
administration committee	管理委员会
administration expenses	管理费
administration fee	管理费
administration law	行政法
administration measure	行政措施
administration of assignment	委派管理
administrative accounting	行政会计
administrative analysis	行政分析
administrative area	办公区
administrative authority	管理部门
administrative budget	管理费预算；行政预算
administrative building	行政办公楼
administrative center	管理中心
administrative code	管理规章
administrative control	行政管理控制
administrative cost	行政管理费
administrative decision making	制订管理决策
administrative department	行政管理部门，行政管理处
administrative distance	管辖距离，管理距离
administrative expense	行政管理费
administrative Lawsuit	行政诉讼
administrative levels	管理层次
administrative management	行政管理
administrative map	行政区划图
administrative measure	行政措施
administrative office	管理处；办事处；行政办公室
administrative officer	总务管理员；高级官员
administrative organization	管理机构
administrative policy	管理政策
administrative procedure	行政程序
administrative regulations of environment	环境行政法
administrative Remedy	行政救济
administrative staff	管理人员
administrative standard	管理标准
administrative strategy	管理战略
administrative tribunal	行政法庭
administrator	管理人员
administrator of estate	遗产管理人
admirabilis	奇异叠层石
admiralty	海军部；制海权
admiralty brass	海军黄铜；船用黄铜
admiralty bronze	船用青铜
admiralty Chart	海军图表，航海图
admiralty coal	船用煤
admiralty gun metal	海军炮铜
admiralty mile	英制海里
admiralty tide table	潮位预测表
admissibility	可容许性
admissible	可采纳的
admissible average content	允许平均含量
admissible concentration	容许浓度
admissible condition	容许条件
admissible control	容许控制
admissible deviation domain	容许偏移域
admissible deviation region	容许偏移范围
admissible error	容许误差
admissible evidence	可接受证据
admissible failure mechanism	容许破坏机理；容许失效机理
admissible flow velocity	允许水流速度
admissible function	容许函数
admissible load	容许荷载
admissible mark	可允许符号；可允许数位
admissible parameter	容许参数
admissible set	容许集
admissible strain	容许应变
admissible stress	容许应力；许用应力
admissible structure	容许结构
admissible transmittance	容许透射率
admissible value	容许值
admission	接纳；承认；进气；进入
admission cam	进气凸轮
admission gear	进气装置
admission line	进气管道
admission of air	补气
admission opening	进口
admission passage	进路

admission piece	进气嘴
admission port	进气口
admission pressure	进口压力
admission section	进口断面
admission space	装填体积
admission stroke	进气冲程；进汽冲程
admission valve	进口阀
admission velocity	进气速度
admission zero	进给起点
admissive	容许有的
admit	进气；供给；容许，允许
admittance	导纳；进入权，进入许可；入场
admixer	混合器
admixture	掺合剂；外加剂；混合料
admixture defect	杂质缺陷
admixture liquor	混合溶液
admixture stabilization	拌合加固法；掺拌加固法
admixture volume	掺量
admixtures	掺合物
admontite	水硼镁石
adnate algae	并生藻类
adobe	砖坯；土坯；灰质黏土
adobe blasting	裸露装药爆破；糊炮爆破
adobe block	土坯块
adobe brick	风干砖坯
adobe building	土坯房
adobe clay	灰质黏土；制砖黏土
adobe clay soil	灰质黏土
adobe construction	土坯建筑
adobe house	土坯房
adobe masonry	土砖圬工
adobe shooting	糊炮爆破；外部装药爆破；覆土装药爆破
adobe soil	冲积黏性土壤；灰质黏土
adobe structure	土结构；生土结构
adobe wall	土坯墙
adolescence	少壮期；青年期
adolescent river	少壮河；青年河
adopt	采取
adopted latitude	纬度采用值
adopter	接受管
adpressed fold	紧压褶皱
adret	山的向阳面
adretto	阳坡
Adriatic Carbonate Platform	亚得里亚海碳酸盐岩台地
Adriatic foreland	亚得里亚海前陆
Adriatic plate	亚得利亚板块
adrift	漂流
adscititious	追加的
adsorb	吸附
adsorbability	吸附性；吸附能力
adsorbable	可吸附的
adsorbance	吸附量
adsorbate	被吸附物；吸附质
adsorbed air	吸附气体
adsorbed film of water	吸附水膜
adsorbed gas	吸附气
adsorbed hydrocarbons	吸附烃
adsorbed ion	吸附离子
adsorbed layer	吸附层
adsorbed methane	吸附的甲烷
adsorbed oil	吸附油
adsorbed phase	吸附相
adsorbed stage	吸附状态
adsorbed water	吸附水
adsorbent formation	吸附性地层
adsorbent ground	吸附性地表
adsorbent pressure	吸附压力
adsorbent regeneration	吸附再生；吸附剂再生
adsorbents	吸附剂
adsorber	吸附器
adsorbing bed	吸附层
adsorbing clay	吸附性黏土
adsorbing material	吸附剂
adsorbing medium	吸附介质
adsorbing substance	吸附剂
adsorption	吸附
adsorption action	吸附作用
adsorption analysis	吸附分析
adsorption band	吸附带
adsorption capacity	吸附容量；吸附能力
adsorption centre	吸附中心
adsorption chromatography	吸附色层分离法；吸附色谱法
adsorption cleaning	吸附式净化
adsorption column	吸附柱
adsorption effect	吸附效应
adsorption equilibrium	吸附平衡
adsorption equipment	吸附装置
adsorption film	吸附膜
adsorption heat	吸附热；吸着热
adsorption hysteresis	吸附滞后现象
adsorption indicator	吸附指示剂
adsorption isobar	吸附等压线
adsorption isotherm	吸附等温线
adsorption isotherm curve	吸附等温曲线
adsorption kinetics	吸附动力学
adsorption layer	吸附层
adsorption method	吸附法
adsorption modeling	吸附模型
adsorption potential	吸附势
adsorption power	吸附力
adsorption process	吸附法处理；吸附过程
adsorption properties	吸附性质
adsorption rate	吸附率
adsorption ratio	下伏水域
adsorption refining	吸附精制
adsorption site	吸附点
adsorption surface area	吸附表面积
adsorption time	吸附时间
adsorption water	附着水；吸附水；薄膜水；结合水
adsorption water of clay minerals	黏土矿物吸附水
adsorptive capacity	吸附能力
adsorptive force field	吸附力场
adsorptive layer	吸附层
adsorptive power	吸附能力
adsorptive value	吸附值
adsorptive water	吸着水
adsorptive weakening	吸附弱化
adtidal	潮下的
adular	冰长石
adularization	冰长石化

A

adult tomb 成人墓
adulterant 掺杂物
adulterated sample 混有异物的样品
adulteration 掺杂；伪造；劣货
adumbrate 前兆；预兆；画轮廓；预示
adumbration 轮廓；草图；素描
adustion 可燃性
advance 前进，超前；推进预付款
advance air wave 冲击波
advance along 沿着…掘进；沿着…推进
advance anchor bolt 超前锚杆
advance angle 前进角
advance angle of fuel supply 供油提前角
advance angle of influence 超前影响角
advance average value 进尺平均值
advance borehole 超前钻孔
advance by hand 人工掘进
advance by machine 机械掘进
advance by overdraft 透支
advance chart 进度表
advance coefficient 进速系数
advance constant 进速常数
advance copy 样本
advance deep-hole infusion 深孔预注水
advance developing 基建开拓；超前掘进
advance development 基建开拓；超前掘进
advance directional sign 前置指路标志；方向预告标志
advance drainage 超前抽放
advance drainage diversion 前期排水渠改道
advance earthwork 前期土方工程；预造土方工程
advance gate 超前平巷
advance grouting 超前灌浆，预注浆
advance heading 掘进方向；推进方向；超前推进平巷
advance heading method of roadway formation 巷道结构的超前推进平巷法
advance hole 超前钻孔
advance instruction station 超前指令站
advance line 推进线
advance mining 从井底向井田边界开采；前进式开采
advance of flood wave 洪水波推进
advance of glacier 冰川前进
advance of ice cover 冰层推进；冰层运动
advance of sea 海侵，海进
advance of the face 工作面推进
advance of working cycle 循环进度；循环进尺
advance on wage 预付工资
advance outwash 前展冰水平原
advance pack 预先充填带，超前充填带
advance payment 垫付款；预付款
advance payment for contract 合同预付款
advance payment guarantee 预付款保函；预付款保证金
advance per day 日进尺；日进度
advance per month 月进尺；月进度
advance per round 循环进尺
advance per shift 班进尺；班进度
advance per week 周进尺；周进度
advance rate 掘进速度，推进速度
advance reclamation works 早期填海工程
advance section 超前部分
advance sign 前置交通标志
advance slope grouting 斜向压力灌浆；预置集料横向压力灌浆
advance slope method 混凝土斜面浇灌法
advance speed 掘进速率
advance starting signal 总出站信号机；总出发信号机
advance stripping 超前剥离；初始剥离
advance support 超前支护
advance supporting 超前支护
advance timbering 前探支架；超前支护
advance warning sign 前置警告标志
advance water-detecting hole 超前探水孔
advance wave 超前波
advance works 前期工程；早期工程
advanced acknowledge 超前响应
advanced allocation scheme 提前配屋计划
advanced bore hole 超前钻孔
advanced boring 超前钻探
advanced casing 跟进套管
advanced charge 预付费用
advanced coal mining technique 先进采煤技术
advanced contract 前期合约
advanced control 超前控制；先行控制
advanced detection 先进的检测
advanced drilling 超前凿岩
advanced dune 前移沙丘
Advanced Dynamics of Structures 高等结构动力学
Advanced Engineering Mechanics 高等工程力学
advanced experience 先进经验
advanced exploration 初勘；早期勘探
advanced face 推进的工作面，前进式工作面
advanced forecast 超前预报
advanced gallery 超前小平巷
advanced grouting 超前灌浆
Advanced Handover Algorithms 多种切换算法，高级切换算法
advanced hole 超前钻孔
advanced ignition 提前点火
advanced money 预先付款；预付款项
advanced overburden 超额剥离量
advanced payment 预付款；预付的定金
advanced prediction 超前预报
advanced query 高级查询
Advanced Reinforced Concrete Structure 高等钢筋混凝土结构
advanced research 远景研究；探索性研究
advanced rheology 高等流变学
Advanced Rock Mechanics 高等岩石力学
advanced science 尖端科学
Advanced Soil Mechanics 高等土力学
advanced starting valve 预开起动阀
advanced technique 先进技术
advanced technology 先进工艺；先进技术
advanced technology mining 先进技术采煤法；先进装备采煤法
advanced unit 推进装置
advanced velocity package 先进速度程序包
advanced warning 提前告警
advancement 推进，掘进，进度；进步；预先支付，垫款
advancer 超前补偿器
advancer unit 推进装置

advancing angle　前进角
advancing boring　超前钻孔；预先钻孔
advancing casing　跟进套管
advancing coast　前进海岸；推进海岸
advancing consequent bedding rockslide　前进式顺层滑坡
advancing contact angle　前进接触角
advancing drive　前进驱动
advancing edge　上升沿
advancing face　推进中的工作面；前进式工作面
advancing failure　推进式破坏
advancing load stress　动载应力
advancing longwall　前进式长壁采煤法
advancing longwall face　前进式长壁工作面
advancing longwall mining method　前进式长壁采煤法；前进式长壁开采法
advancing mining　前进式开采
advancing movement　前进；掘进
advancing ram　推移千斤顶
advancing rock face　推进中岩巷
advancing shoreline　前进滨线
advancing side　紧张侧
advancing speed　推进速度
advantage analysis　优势分析
advent time　到达时间
adventitious　外来的；偶然的；不定的
adventitious deposit　附着沉积
adventitious plants　外来植物
adventitious species　外来种
adventitious stream　偶发性河流
adventive cone　寄生火山锥；侧火山锥
adventive crater　侧裂火山口；附属火山口
adventive volcano　寄生火山
adverse balance of international trade　国际贸易逆差
adverse chemical change　逆化学变化
adverse circumstances　恶劣环境
adverse condition　不利条件
adverse current　逆流
adverse environment effect　对环境不利影响
adverse factor　有害因素
adverse geologic action　不良地质作用
adverse geologic condition　不良地质条件
adverse geologic phenomena　不良地质现象
adverse geological action　不良地质作用
adverse geological condition　不良地质情况
adverse geological process　不良地质作用
adverse grade　反向坡度；反倾斜
adverse grade for safety　反坡安全线；安保逆坡
adverse physical condition　不利自然条件
adverse pressure gradient　反压力梯度；逆压梯度
adverse reaction　有害反应逆反应
adverse slope　反向坡
adverse title to land　相逆土地业权
adverse weather　不利天气；坏天气
adverse wind　逆风
adversity　逆境，灾难
advertisement for bids　公开招标；招标公告
advisory architect　顾问建筑师
advisory body　顾问团
advisory committee　咨询委员会；顾问委员会
Advisory Committee on Appearances of Bridges and Associated Structures　桥梁及有关建筑物外观咨询委员会
Advisory Committee on Marine Pollution（ACMP）　海洋污染咨询委员会
Advisory Council on the Environment　环境问题咨询委员会
advisory engineer　咨询工程师；顾问工程师
advisory Group on Greenhouse Gases　致温室效应气体问题咨询小组
advisory letter　劝喻信
Advisory Note on Buried Services　检查地下设施须知
advisory service　咨询服务
advisory system　咨询系统
adynamic soil　非动力土壤
Aegean Sea plate　爱琴海板块
aegirine　霓石
aegirine aplite　霓细晶岩
aegirine augite　霓辉石
aegirine augitization　霓辉石化
aegirine rhyolite　霓流纹岩
aegirine-granite　霓花岗岩
aegirinite　霓石岩
aegirinolite　霓辉石岩
aegirite　霓石
aegisodite　方钠霓石岩
aegista　滑口螺属
aegyrite-augite　霓辉石
aenigmatite　三斜闪石
aeolation　风蚀；风成作用
aeolian　风积的；风蚀的
aeolian accumulation　风成堆积
aeolian anomaly　风积物异常
aeolian basin　风成盆地
aeolian bedding　风成层理
aeolian clastics　风成碎屑岩
aeolian cross bedding　风成交错层理
aeolian deposit　风积土风成沉积
aeolian dune field　风成沙丘塬
aeolian energy　风能
aeolian erosion　风蚀
aeolian facies　风积相；风成相
aeolian flat　风坪
aeolian landform　风蚀地形
aeolian material　风成物；风积物
aeolian plain　风成平原
aeolian process　风积作用
aeolian ripple mark　风成波痕
aeolian rock　风成岩
aeolian sand　风积砂
aeolian sand ripple　风成沙纹
aeolian sand sea　风沙海
aeoLian sand transport　风运砂；飞砂
aeolian sediment　风积物；风成沉积
aeolian soil　风积土；风成土
aeolianite　风成沉积岩
aeolic erosion　风蚀
aeolic soil　风积土
aeon　十亿年
aerated concrete　加气混凝土；气孔混凝土
aerated conduit　曝气管道
aerated emulsion　充气乳状液

aerated filler	疏松填料
aerated flow	混气液流
aerated fluid	混气液
aerated formation water	含气地层水
aerated layer	松散层；风化层
aerated mud	充气泥浆
aerated plastics	充气塑料
aerated porosity	充气孔隙；含气孔隙
aerated solid	充气固体；气溶胶
aerated spring	含气泉
aerated trickling filter	曝气滴滤池
aerated water	饱气水；碳酸水
aerated zone	包气带
aerating agent	充气剂
aerating chamber	空气混合室
aerating mixture	充气混合物
aerating mud	混气泥浆
aerating system	通风系统
aeration basin	曝气池
aeration brush	曝气刷
aeration cell	充气电池
aeration chamber	充气室
aeration coefficient	曝气系数
aeration device	曝气设备
aeration distribution	充气分布
aeration drilling	充气泥浆钻井
aeration flotator	充气式浮选机；气升式浮选机
aeration homogeneity coefficient	充气均匀系数
aeration intensity	充气强度
aeration jet	充气射流
aeration of moulding sand	松砂
aeration period	曝气周期
aeration pipe	通气管
aeration porosity	充气孔隙；通气孔隙度
aeration rate	充气率
aeration tank	曝气池
aeration test	充气试验
aeration tissue	通气组织
aeration treatment	曝气处理
aeration zone	曝气区；饱气带
aeration zone water	包气带水
aeration-drying	通风干燥
aerator classifier	空气分级机
aerator pipe	曝气管
aerial arch	蚀顶背斜
aerial archaeology	航空考古学
aerial bracket	天线托架
aerial cable	天线电线
aerial cableway	架空索道
aerial cableway survey	架空索道测量
aerial chart	航空图
aerial conveyor	吊式输送机；架空运道
aerial crossing	管道架空穿越
aerial data	航空资料
aerial dump	空中卸料
aerial erosion	大气侵蚀
aerial fold	剥蚀褶皱
aerial geologic survey	航空地质调查
aerial geology	航空地质学
aerial geophysical exploration	航空地球物理勘探
aerial gravity	航空重力学
aerial gravity measurement	航空重力测量
aerial mapping	航空测绘
aerial method of geology	航空地质方法，航空地质调查方法
aerial mosaic	航空镶嵌图
aerial navigation map	领航图
aerial oxidation	空气氧化
aerial photograph	航空照片；航摄照片
aerial platform	高空工作升降台
aerial pollution	空气污染
aerial radioactivity measurement	航空放射性测量
aerial radioactivity survey	航空放射性调查
aerial railway	架空铁路；高架铁路
aerial rapid transit system	高架快速运输系统
aerial reconnaissance	航空勘测；空中探测
aerial reconnaissance photography	航测摄影
aerial remote reconnaissance	航空遥测
aerial remote sensing	航空遥感
aerial ropeway	高架索道
aerial ropeway haulage	架空索道运输
aerial sediment	大气沉积物
aerial sewer	露置下水管道
aerial spectrograph	航空摄谱仪
aerial spectrophotography	航空光谱摄影术
aerial stereogram	航空立体图
aerial strip survey	航线测量
aerial supervision	空中巡线
aerial surveillance	空中监测
aerial survey	航空测量；空中测量
aerial survey method	航空测量方法
aerial surveying alignment	航测选线
aerial system	天线阵
aerial thermography	航空测温法
aerial topographic map	航测地形图
aerial tram	架空索道缆车
aerial tramline	架空索道；吊车索道
aerial tramway	架空索道；架空电车道
aerial transport	航空运输；空运
aerial triangulation	航空三角测量
aerial view	鸟瞰图；航测图
aerialgeology	航空地质学
aerialmap	航测图
aerial-tramline	架空索道
aerobasilus	需氧杆菌
aerobe	需氧微生物；需氧菌
aerobic bacteria	需氧菌；好氧细菌
aerobic biodegradation	需氧生物降解作用
aerobic biological process	需氧生物处理法
aerobic biological treatment	好氧性生物处理
aerobic environment	含氧环境
aerobic lagoon	好氧湖
aerocartograph	航空测图仪；航空测量图
aeroconcrete	加气混凝土；泡沫混凝土
aerogel	气凝胶
aerogenic bacteria	产氧细菌
aerogenous rock	风成岩
aerogeochemical prospecting	航空地球化学勘探
aerogeography	航空地理学
aerogeological mapping	航空地质制图

aerogeological reconnaissance map 航空地质草图
aerogeology 航空地质学；航天地质学
aerogeophysical prospecting 航空地球物理勘探
aerogeophysical survey 航空地球物理勘探
aerogravimetry 航空重力测量；空中重力测量
aeroleveling 空中水准测量
aerolite 陨石；石陨星；硝酸钾
aerological measurement 高空测量
aeromagnetic map 航空磁力勘探图
aeromagnetic method 航空磁测法
aeromagnetic profile 航磁剖面
aeromagnetic prospecting 航空磁力勘探
aeromagnetic survey 航空磁测
aeromagnetic survey method 航空磁测量法
aeromagnetic triaxial gradiometer system 航空全轴磁力梯度系统
aeromagnetic vertical gradient survey 航空垂直磁力梯度测量
aeromagnetic vertical gradiometer system 航空垂直磁力梯度系统
aerophotogeodesy 航空摄影测量
aerophotogeological map 航空摄影地质图
aerophotogeology 航空摄影地质
aerophotogonometry 空中导线测量
aerophotogrammetrical survey 航空测量；航空摄影测量
aerophotography of geology 地质专业航空摄影
aerophototopography 航空摄影地形测量学
aerophysical survey 航空物理测量
aeroplane mapping 航空制图
aeropolygonometry 空中导线测量
aeroprojector 航测投影器；航测制图仪
aeroprospecting 航空勘探
aerosiderolite 铁陨石
aerosimplex 简单投影测图仪
aerosite 硫锑银矿；深红银矿
aerosurveying 航空勘测
aerotriangulation 空中三角测量
A'erpishimaibulake Formation 阿尔皮什麦布拉克组
A'ertemeishibulake Group 阿尔特梅什布拉克群
A'ertenghusu Group 阿尔腾呼苏群
A'ertengkashi Formation 阿尔腾卡什组
A'ertushileike Formation 阿尔图什雷克组
aerugite 块砷镍矿
aerugo 铜绿氧化锏
aesar 蛇丘
aeschynite 易解石
aethiopites 埃塞石
aethoballism 撞击变质作用
aetite 矿石结核；瘤状矿石；鹰石
aetna-basalt 阿伊特纳玄武岩
aetostreon 鹰蛎属
affected area 波及区
affected head 影响水头
affected overburden 受影响的覆盖层；扰动的覆盖层
affecting area of collapse tunnel top 洞顶塌陷影响范围
affecting factor 影响因素
affluent stream 支流
African Rift Valley 非洲大裂谷
africandite 钛黄云橄岩
afrodite 泡石；镁皂石
after condensation 后缩合

after contraction 残余收缩
after earthquake 余震
after effect 后效
after expansion 后膨胀
after flow 残余塑流
after hardening 后期硬化
after ignition 延迟着火
after immersion ratio 浸水抗压强度恢复率
after load 后载荷
after pack settling 充填后砾石的沉降
after pump gravel injection system 泵后加砂注入系统
after purification 后净化
after settling effect 后沉效应
after shock 余震
after shrink 后收缩
after shrinkage 后期收缩
after skirt 后缘
after stretching 后拉伸
after survey calibration summary 测后刻度记录
after treatment profile log 油井处理后生产剖面测井
after twist 复捻
after vibration 余振动
after working 后效
after-activity 余震活动
afterbay dam 尾水池坝；二道坝
afterbreak 后破坏
afterburst 后期崩塌
afterdamp 爆后气体
aftergases 爆后气体
aftergate 尾水闸门
afterhardening 后硬化
afterprecipitation 二次沉淀
aftershock activity 余震活动
aftershock sequence 余震序列
aftershock wave 余震波
afterslip 余滑
afterworking creep recovery 弹性后效
aftonian interglacial stage 阿夫唐间冰期
afwillite 柱硅钙石
against silicosis protection 防止矽尘危害
against the wind 逆风
agalite 纤滑石
agallospira 圆环螺属
agalmatoaster 首饰星石
agalmatolite 寿山石；滑石；块云母；叶蜡石
agaric mineral 岩乳；山乳
agate 玛瑙
agate mortar 玛瑙研钵
agate opal 玛瑙蛋白石
agathacopalite 化石树脂
agatized wood 硅化木
age allowance limit 楼龄减额上限
age dating 年龄测定
age determination 时代鉴定
age distribution function of groundwater 地下水年龄分配函数
age factor 年代因数
age group 年代界
age hardening 时效硬化
age of concrete 混凝土龄期

A

age of landform 地貌年龄
age of landslide 滑坡年代
Age of Man 人类纪
age of marine invertebrates 海洋无脊椎动物时代
age of mine 矿山寿命，矿山可采期
age of remanence 剩磁年龄
age of reptiles 爬行动物时代
age of the Earth 地球年龄
aged clay 老黏土
aged core 陈置岩心
aged product 老化油品
aged sample 陈置样品
aged soil 老年土
aged topography 成年地形；古地形
aged water sample 陈置水样；陈化水样
aged well 超龄期井孔；老年井孔
age-dating 测年，测定年代
agehardenability 可时效硬化性
ageing 老化；时效；养护
ageing effect 时效处理；老化效应
ageing index 老化指数
ageing of clay 黏土老化
ageing resistance 时效阻力
agent of erosion 侵蚀力
agents of metamorphism 变质营力
ageostrophic wind 非地转风
age-resistance 抗老化
agerite 弹性地蜡；地蜡
age-size law 油井采期与产量关系律；油井寿命与产量关系
age-strength relation 龄期-强度比
agglaciation 冰川加强作用
agglomerability 可黏结性；可附聚性
agglomerant 黏结剂；凝聚剂
agglomerate 集块岩烧结矿
agglomerate bomb 集块状火山弹
agglomerate foam concrete 烧结矿渣泡沫混凝土
agglomerate lava 集块熔岩
agglomerate texture 集块结构；集块组织
agglomerate tuff 集块凝灰岩
agglomerated cake 烧结块
agglomerated overburden 结块剥离物
agglomeratefoam concrete 烧结矿渣泡沫混凝土
agglomeratic 集块岩状的
agglomeratic ice 集块冰；附聚冰
agglomerating agent 烧结因素；凝结剂；胶凝剂
agglomerating coal 烧结煤；黏结煤
agglomerating flotation 团聚浮选
agglomerating index 集块指数；结焦指数
agglomerating value 凝结值；团结值；结焦值
agglomeration 凝聚作用
agglomeration capacity 烧结性
agglomeration process 附聚过程
agglomeration test 焦结性试验
agglomerative 凝聚的
agglomerator 团结剂
agglutinate 黏结集块岩；熔结集块岩；月壤集块岩
agglutinating power 烧结能力；烧炼能力
agglutinating property 黏结性；附着性
agglutinating test 烧结试验
agglutinating value 烧结值；黏结性
aggradated plain 加积平原；填积平原；山前冲积平原
aggradation 加积；加积作用；填积
aggradation coast 加积海岸
aggradation island 淤积岛
aggradation plain 加积平原；填积平原
aggradation reef 加积礁
aggradation stream 加积河
aggradation terrace 加积阶地；堆积阶地
aggradational 层序加积的
aggradational deposit 加积沉积
aggradational plain 加积平原
aggradational process 加积过程
aggrade 加积，淤积；填积；淤高
aggrading action 加积作用
aggrading braided stream 加积辫状河
aggrading continental sea 填积大陆海
aggrading neomorphism 递进新形作用
aggrading river 加积河流，填积河流
aggrading stream 加积河
aggradingstream 加积河
aggragate structure 团聚结构
aggreagative fluidization 聚式流态化
aggregate 骨料；集合土粒，集料
aggregate abrasion value 石料的耐磨值
aggregate abrasion value test 骨料抗磨试验
aggregate analyses 团聚体分析
aggregate area 总面积
aggregate averaging 集料平均粒径的测定
aggregate base course 粒料基层；碎石基层
aggregate bin 集料仓，集料存储斗
aggregate breaking force 综合破断拉力
aggregate capacity 合计容量；机组总功率
aggregate chips 石屑
aggregate coherence test 团粒黏聚性试验
aggregate combination 集合结构
aggregate composition 骨料成分；集合体组成
aggregate demand 总需求量
aggregate discharge 总流量；汇流
aggregate drain 集料盲沟
aggregate erection 组合安装
aggregate error 集合误差；累积误差
aggregate feeding 集料给料
aggregate for concrete 混凝土骨料
aggregate formation 团聚体形成
aggregate gradation 骨料级配
aggregate grading 砂石级配骨料级配
aggregate impact value 石料的冲击强度值
aggregate lime content 砂石的石灰含量骨料的石灰含量
aggregate of loess 黄土集合体
aggregate of soil 土壤团块
aggregate production system 砂石料生产系统
aggregate properties 骨料性质
aggregate ratio 集料水泥比
aggregate reclaiming plant 砂石料厂
aggregate size 骨料粒度
aggregate structure 砂石结构；骨料结构
aggregate superficial area 表面总面积
aggregate test 集料试验
aggregate thickness 总厚度；合计厚度

aggregate voidage	集料空隙度
aggregate washing plant	集料洗选厂
aggregated lump	聚集团粒
aggregated structure	集聚结构
aggregate-making property	集料形成特性
aggregates processing	骨料的制备
aggregate-to-cement ratio	砂石与水泥比例；砂灰比
aggregation	集合体；聚集作用
aggregation analysis	聚集分析
aggregation degree	团聚度
aggregation force	聚合力
aggregation of soil particles	土粒聚集土粒团
aggregation operator	聚类算子
aggregation plain	集积平原
aggregational rock	集合岩
aggregative index	综合指数
aggregative model	综合模型
aggregative planning	总体规划
aggregative state	聚集态
aggremeter	集料称量器
aggression	侵蚀
aggressive	侵蚀性的；腐蚀性的；积极的
aggressive action	侵蚀作用
aggressive agent	侵蚀剂
aggressive assembly	进取性钻具组合
aggressive carbon dioxide	侵蚀性二氧化碳
aggressive carbonic acid	侵蚀性碳酸
aggressive condition	腐蚀性环境
aggressive groundwater	侵蚀性地下水
aggressive intrusion	强力侵入作用
aggressive magma	侵入岩浆
aggressive substance	使质量变坏的物质
aggressive water	侵蚀性水；侵进水
aggressivity	侵蚀性
Aghil Formation	阿基尔组
Agidise filter	阿吉迪斯克型真空过滤机
Agilisaurus	灵龙属
aging	龄期；年代；时效；老化成熟的过程
aging action	老化因素
aging coefficient	老化系数
aging crack	时效裂纹
aging crust	老化的地壳
aging effect	时效
aging inhibitor	防老剂
aging period	老化周期
aging property	老化性
aging resistance	抗老化性能
aging resister	防老剂
aging test	老化试验
Agip Spa	阿吉普石油公司
agitair flotation cell	搅拌式浮选机
agitair machine	空气搅拌式浮选机
agitating lorry	汽车式混凝土搅拌车
agitating truck	混凝土搅拌汽车
agitating vane	搅动叶轮；搅拌浆
agitation	搅动，搅拌
agitation cone	搅拌漏斗
agitation dredging	搅动挖泥
agitation dryer	搅动干燥器
agitation leaching	搅拌浸出
agitation ratio	搅动比
agitation water	搅动水
agitation-froth machine	搅拌式泡沫浮选机
agitation-froth process	搅拌泡沫法
agitator cell	充气式浮选机
agitator shaft	搅拌器轴
agitator tank	搅拌桶，搅拌槽
agitator treating	搅动洗涤
agitator truck	搅拌运输车
agmatite	角砾状混合岩
Agnotozoic era	元古代
Agnotozoic erathem	元古界
agpaite	钠质火成岩类；钠质霞石正长岩类
agpaitic coefficient	碱性系数
agraphitic	非结晶的
agraphitic carbon	无定形碳
agravic	无重力的；无重力状态
Agreement and Conditions of Sale	卖地协议及条件
Agreement Board	鉴定委员会
agreement of frontier railway	国际铁路协定
Agrichnia	耕作迹
agricolite	闪铋矿
agricultural amelioration	农业土壤改良
agricultural drain	农用排水沟；农田排水管
agricultural drainage	农田排水
agricultural dwelling house	农地住屋；农舍
agricultural geology	农业地质学
agricultural geomorphology	农业地貌学
agricultural hydrology	农业水文学
agricultural land	农田，耕地
agricultural lime	农用石灰
agricultural lot	农地地段
agricultural sewage	农业污水
agricultural soil	农业土壤
agricultural tenancy	农地租赁；农地租约
agroecological geology	农业生态地质学
agroecology geological division	农业生态地质区划
agroforestrial geology	农林地质学
agrogeochemistry	农业地球化学
agrogeological hazard	农业地质灾害
agrogeological map	农业地质图；土壤地质图
agrogeology	农业地质学
agrogeomorphology	农业地貌
agrohydrologg	农业水文学
agrology	农业土壤学
agrometeorology	农业气象学
agronomy	作用点
agropedogenesis	农业土壤发生
agropedology	农业土壤学
agropolitics	农业政策
aground sweeping	拖底扫海
agstone	石灰石粉
aguada	集水洼地
aguilarite	辉硒银矿
Aguja Canyon	阿古哈峡谷
agustite	磷灰石
Ahati Group	阿哈提群
ahead fluid	前置液
ahead running	顺车
aheap	堆积

ahermatypic	异性型；非造礁型
ahermatypic coral	非造礁珊瑚
ahmuellerella	阿缪石
ahnfeltia	伊谷草属
A-horizon	A层，淋溶层；层位；矿质土表层
Ahuoquemake Limestone	阿霍却马克石灰岩
Aichuan Formation	爱川组
aid decision making	辅助决策
aided decision way	辅助决策方式
aided design	辅助设计
aided device	半自动装置
aided tachometer	半自动转数表
aided tracking	半自动跟踪；辅助跟踪
aided tracking mechanism	半自动跟踪装置
aider	辅助者；支援者
aids to navigation	航标
aidyrlite	杂硅铋铝镍矿
Ai'erjigan Group	爱尔基干群
Aierjisi Formation	爱尔基斯组
Aigeliumu Formation	艾格留姆组
Aijiashan Formation	艾家山组
Aiketike Group	艾克提克群
aikinite	针硫铋铅矿
Ailaoshan group	哀牢山群
Ailegan Formation	爱勒干组
aileron	副翼
Ailiaojiao Formation	隘寮脚组
Ailikehu Formation	艾里克湖组
aillikite	方解霞黄煌岩
ailsyte	钠闪微岗岩
aiming circle	测角器；测角罗盘
aiming head	照准头
aiming line	瞄准线
aimless drainage	紊乱水系；无目标排水
Ainishan Formation	爱尼山组
aiounite	辉云斜岩
Aipoli Formation	爱坡里组
air base	航空基地；空中基线；摄影基线
air basin	大气限域
air binding	气堵
air blast effect	空气冲击波效应；空中爆炸效应
air blast quenching	吹风淬火
air blasting wave in mine	矿山爆破空气冲击波
air bleed valve	排气阀；放气阀
air bleeding	放气
air bleeding valve	排气阀
air blocking area	风阻面积
air blown asphalt	氧化沥青
air blown filter press	吹风压滤机
air blown mortar	喷射水泥砂浆
air blowpipe	炮眼吹洗管
air bond	型芯砂结合剂；气障
air booster	压气助力器；风动助力器
air borne exploration	航空勘探
air borne geophysical prospecting	航空地球物理勘探
air borne geophysics	航空地球物理学
air borne gradiometer	航空梯度仪
air borne gravimeter	航空重力仪
air borne gravimetry	航空重力测量
air borne infrared survey	航空红外测量
air borne magnetic gradiometer	航空磁力梯度仪
air borne magnetic survey	航空磁测
air borne magnetometer	航空磁力仪
air borne magnetometer profile	航磁剖面
air borne radioactivity	大气放射性
air bottle	氧气瓶；空气瓶；压缩空气瓶
air bound	空气阻塞
air box	木风筒；风箱
air brake	气压制动器
air brake equipment	空气制动装置
air branch pipe	空气支管
air brattice	风障
air break contactor	空气断开接触器
air breaker	空气爆破筒；空气断路器
air breaker shell	空气爆破筒
air breaking installation	压气落煤装置
air breakwater	气动防波堤
air brick	空心砖；干砖坯
air bubble	气泡；砂眼
air bubble level	水准仪
air bubble mineralization	气泡的矿化作用
air bubble viscometer	气泡黏度计
air buffer	空气缓冲器软靠枕空气隔层
air camera	航空摄影机；空中摄影机
air canal	通风道
air cement gun	风动水泥枪；风动水泥喷射器
air cement separator	气灰分离器
air centrifuge	空气离心机
air circulation	空气循环洗井；大气环流
air coefficient of permeability	空气渗透系数；透气系数
air compressor	空气压缩机；空压机
air cooling installation	冷却装置；降温设备
air cooling plant	空气冷却器；空气冷却设备
air cooling system	空气冷却系统；风冷系统
air course	风巷；风道
air crossing	风桥
air curtain method for sinking caisson	空气幕沉井法
air cushion drilling	气垫钻进
air cushion drilling platform	气垫钻井平台
air cushion effect	气垫效应
air cushion mechanism	气垫层机制
air cushion seismic vibrator	空气缓冲垫地震振动器
air cushion shock absorber	气垫减震器
air cushion surge chamber	气垫式调压室
air cushion vehicle	气垫船；气垫运输工具
air cushioning	空气减震
air delivery duct	送风道
air dried basis analysis	风干基分析
air dried coating	风干层；风干硬皮
air dried moisture content	风干含水量
air dried sample	风干样
air drift	通风平巷
air drifter	风动凿岩机；风钻
air drill	风钻，风动凿岩机
air drilling	风动钻眼；风动凿岩
air driven hammer	风洞锤
air driven hammer drill	气动冲击钻机
air driven mine car loader	风动矿车装载机
air driven pump	风动泵
air driven rockerloader	压气式铲斗后卸装载机

air driven turbine	气动涡轮
air driving drilling machine	气腿式风钻
air drop hammer	气锤
air drum	储气罐
air dryer	空气干燥器
air drying loss	风干失重；风干损失
air drying period	空气干燥时期；空气干燥阶段
air drying varnish	风干清漆
air duct	风槽；通风管道；气槽
air duct area	风渠面积
air duct line	通风管道；通风路线
air dustiness	空气含尘性；空气含尘量
air dynamics	空气动力学
air eddy	空气涡流
air ejector	空气喷射器
air electromagnetic	航空电磁法
air endway	副风巷
air entrained cement	加气水泥
air entraining agent	加气剂
air entry	风巷
air entry permeameter	进气渗透仪
air entry pressure	进气压力
air entry value	进气值
air flow in mine	矿井风流
air flush	空气洗孔
air flush drilling	空气洗孔钻井；空气钻井
air flushing	压气吹洗
air foam stabilizer	空气泡沫稳定剂
air force aeronautic chart	空军航图
air force atmospheric model	空军大气模式
air forging hammer	空气锤
air free concrete	密实混凝土
air funnel	通风井；竖通风道
air gap	风口；空气隙
air gap flux density	气隙磁通密度
air gap sensitiveness	殉爆感度
air gate	风巷；通风调节门；气孔
air gauge	气压计；气压表
air gun	空气枪
air gun firing chamber	空气枪起爆室
air hammer	气锤
air hammer drill	风钻；风动冲击凿岩机
air hammer drilling	气动震击钻井
air hardening	空气硬化；气硬
air hardening lime	气硬石灰
air heading	通风平巷；风巷
air hoist	气动卷扬机
air hole	通气孔
air in soil	土中气
air induction opening	进气口
air infrared survey	航空红外调查
air inlet	进气；进气孔
air inlet duct	进风道
air jack	气压千斤顶
air jacketed thermometer	空气套温度计
air leakage in mine	矿井漏风
air leg	风动锤
air lift mining system	气力提升采矿法
air lift pump	气压泵
air manometer	空气压力表
air map	空中摄影地图
air mortar	加气灰浆；气硬砂浆
air motor drill	风钻
air opening	通风巷道；通风口
air pass	风巷；风道
air passage	风道；气路；空气道
air percussion drilling	风动冲击钻进
air permeability	透气性
air permeability test	透气性试验
air pick	风镐
air pile hammer	气压桩锤
air piston	纹泥
air pit	风井；显微小气泡
air placed concrete	气压浇注混凝土
air pollution in mining area	矿区大气污染
air porosity	孔隙率
air prospecting	航空勘探
air pulsated coal washer	空气脉动洗煤机
air pumping rate	抽吸速率
air puncher	风动凿孔机
air purge	气洗
air purification	空气净化
air raid shelter	防空洞
air raise	风动天井；通风上山
air ram	风动锤
air rammer	气动夯锤
air range	空气管道；飞行航程
air reamer	气压扩孔机
air receiver	储风筒；贮气箱
air reconnaissance	航空勘测
air refrigerating machine	空气冷冻机
air rig	风钻
air ring	通气环
air riveter	气压铆钉枪
air roadway	通风平巷
air room	联络风巷；空气室
air sac	气囊
air saddle	空鞍；气鞍；空背斜；剥蚀褶皱
air sample	空气取样
air sampler	空气采样器
air sand blower	气压喷砂机
air saturation ratio	空气饱和率
air scour	空气冲洗
air screw fan	轴流式扇风机
air scrubber	空气洗涤器；涤气器；除尘器
air sea interaction	大气海洋相互作用
air sea interface	大气海洋界面
air seal	气封
air seasoning	风干
air seepage	漏风
air separating tank	空气分离罐
air separation	风选；风力选
air separation plant	空气分离装置
air separator	气力分离器
air set pipe	空气冷凝管
air setting	气干
air setting refractory mortar	气硬性耐火砂浆
air shaft	风井；通风井
air sheet	风障；风幕；风布
air shock wave	冲击波

air shock wave intensity	冲击波强度
air shooter	压气爆破工
air shooting	空中爆破
air shot	空心爆破；气垫爆破
air shower	大气簇射
air shrinkage	干缩
air shut-off valve	关气阀
air shuttle valve	阻气阀
air siltometer	气拢泥沙分析仪
air sinker	手持风动凿岩机
air sizing	风力分级；风力分粒
air slaked	风化了的；消化的；潮化的
air slaked lime	气化石灰
air slaking	风化；潮解
air slide	气动滑阀；空气活塞
air slip forming	气滑成型
air slips	气动卡瓦
air slit	联络通风小巷
air slug	气塞
air sluice	空气槽；压气沉箱工作室
air sollar	通风隔墙
air source	气源
air source heat pump	空气源热泵
air space	空隙；气隙；空间；空气室
air space cable	空气绝缘电缆
air space ratio	含气量体积比；排水气隙比
air spacing	空气柱
air speed	风速
air sphere	大气圈
air spider	气动卡盘
air spinning	气流纺
air spinning cathead	气动旋螺纹猫头
air split	风流分支
air sprayer	喷涂器
air spraying	压力喷涂
air spring	气垫
air stack	通风竖筒
air starter	空气启动器
air station	航空站
air sterilization	空气消毒
air stone	石陨石
air stop glass	阻流玻璃片
air stopping	风障；风墙
air storage	储存库；露天堆放；空气中养护
air stower	风力充填机
air strainer	空气滤清器；空气粗滤器
air stream	气流
air stripper	空气吹提器
air suction	先张预应力构件
air suction inlet	吸气孔；抽气口
air suction pipe	吸气管
air suction valve	吸气阀
air supply	送风；供气
air supply duct	供气道
air supply outlet	供气出口
air supply pipe	供气管
air supply rate	充气速度
air supply system	送风系统
air supply valve	供气阀
air survey	航空测量；航空勘探；通风测量
air survey aircraft	航测飞机
air survey of damage	震害航测
air suspension	气悬浮体
air sweeping	空气抽吸清除
air sweetening	空气氧化脱硫醇
air switch	空气开关
air table reject	风力淘汰盘尾矿
air tack cement	封气黏胶水泥
air tamper	风动搗棒
air tank	压气箱；空气箱
air tap	气嘴
air temperature	空气温度；室温
air test	气密试验
air textured yarn	空气变形丝
air thermometer	气温表
air throttle	节流阀；风门
air tight	气密；不漏气的
air tight cabin	气密舱，气密室
air tight partition	气密隔墙
air tight test	气密试验
air tightness	气密性
air track drill	履带式风动钻机
air trammer	风动机车；压气机车
air transmission	配气；配风；输气
air transport	风力搬动作用；空运
air transport in pipes	风力管道运输
air trap	空气阱；防气阀
air triangulation	空中三角测量
air trunk	大通风管；主风管
air trunking	通风槽
air tube	通风管风筒
air tube clutch	气胎离合器
air tugger	气动卷扬机；气动绞车
air turbine	空气涡轮机；空气透平
air turbo lamp	气动透平发电机式灯
air turbulence	空气扰动
air valve	气阀；风阀；跳汰机风阀
air valve cage	气阀盖座
air valve control lever	气阀控制杆
air valve pit	进出气阀井；放气阀井
air valve seat	气阀座
air vane	风标翼；扇风机叶轮
air vapour mixture	空气油品蒸汽混合气
air velocity	气流速度；空气速度
air velocity in airway	井巷风速
air vent	排气口；通风眼；通风口
air vent cock	通风管旋塞
air vent manifold	通风多叉管
air vent valve	空气活门
air ventilator	空气通风器
air vessel	空气容器；气包；缓冲小气包
air vibration	空气振动
air vibrator	气动振动器
air void	空气空隙；气穴
air void ratio	孔隙比；含气率
air void soil	气隙泥土
air voids	含气率；空气率；气孔；气隙
air voids content	孔隙含气量
air voids line	孔隙曲线
air voids of a soil	土的气隙，土的空气孔隙

air volcano 气火山；喷气泥火山
air volume change 空气体积变化
air volume flow rate 风量流率
air volume in voids 孔隙中气体容积
air volume modulus 空气体积模量
air water table 空气-水界面
air wave 空气波
air way 风道；空气层
air way characteristic curve 风阻特性曲线
air weight 钻柱悬于空气中之重量
air well 通风井；风蚀穴
air winch 风动绞车；压气绞车
air-blast 空气冲击波
air-blast goaf-stowing machine 采空区风力充填机
air-blown concrete 喷射混凝土
airbond 气障；型芯砂结合剂
airborne 风成的；空气传播的
airborne apparent resistivity mapping 航空视电阻率填图
airborne concrete 风力浇注的混凝土
airborne contaminants 空气中的污染物
airborne contamination 空气污染物
airborne dust 气浮尘末；气载尘末；浮尘
airborne dust survey 气载尘末测定；浮尘测定
airborne electromagnetic method 航空电法；航空电磁法
airborne electromagnetic prospecting 航空电磁勘探
airborne electromagnetic survey 航空电磁法勘察
airborne electromagnetic system 航空电磁系统
airborne electromagnetic tow-bird system 航空电磁吊舱系统
airborne electromagnetic tow-boon system 航空电磁拖架系统
airborne electromagnetics 航空电磁法
airborne gamma survey 航空伽马测量
airborne gamma-ray spectrometer 航空伽马能谱仪
airborne gasoloid survey 航空气溶胶测量
airborne geochemical prospecting 航空化探；航空地球化学探矿
airborne geophysical device 机载地球物理装置
airborne geophysical exploration 航空地球物理勘探
airborne gravimeter 航空重力仪
airborne gravimetric gradiometer system 航空重力梯度系统
airborne gravitational acceleration correction 航空重力加速度校正
airborne gravity measurement 航空重力测量
airborne gravity survey 航空重力勘探
airborne ground mapping radar 机载测地雷达
airborne image sensor 机载成像传感器
airborne imagery 机载成像
airborne infrared equipment 机载红外装置
airborne infrared survey 空中红外探测
airborne instrumentation 航空测量仪工作；航空勘探仪工作
airborne laser profiler 航空激光剖面测绘仪
airborne laser radar 机载激光雷达
airborne laser sounding 机载激光测深
airborne laser terrain profiler 机载激光地形剖面记录仪
airborne location 航空定位
airborne magnetic and radiometric survey 航空磁测与辐射测量
airborne magnetic prospecting 航空磁法勘探
airborne magnetic survey 航空磁测
airborne magnetometer 航空磁力仪

airborne magnetometric gradiometer 航空磁力梯度仪
airborne magnetometry 航空磁力测量
airborne material 风成物质；风运物
airborne mercury vapor survey 航空汞蒸气测量
airborne mineralogical search 航空找矿；航空探矿
airborne particle 飘尘；大气尘
airborne photogrammetric camera 机载摄影机
airborne platform 海上航空勘测工作平台
airborne pollutant 风载污染物；气悬污染物
airborne profile recorder 航空剖面记录仪
airborne radar 机载雷达
airborne radiation detector 航空放射性检测仪
airborne radio phase method 航空无线电相位法
airborne radioactivity survey 航空放射性测量
airborne range 航空测距
airborne ranger 航空测距仪
airborne rapid scan spectrometer 机载快速扫描光谱仪
airborne remote sensing 航空遥感
airborne remote-sensing technique 航空遥感技术
airborne sand 风成砂
airborne scanner 机载扫描仪
airborne sensitivity 空中灵敏度
airborne sensor 机载遥感器；机载传感器
airborne survey 航空测量；航空勘探
airborne target location system 机载目标定位系统
airborne thermal survey 航空热辐射测量
airborne trace 航迹
airborne VLF electro-magnetics 航空甚低频电磁法
air-brick 风干砖；砖坯；空心砖
air-bridge 风桥
air-bubble viscometer 气泡黏度计
air-bubbler system 气泡防冻系统
air-conditioning and refrigeration installations 空调及制冷装置
air-containing zone 包气带
air-cooled riveting hammer 气冷式铆钉锤
air-cooling system 空气冷却系统
air-core coil 空心线圈
air-core sole 空心螺线管；空心圆筒形线圈
air-coupled Rayleigh wave 空气耦合瑞利波
air-crack 干裂
aircraft based 机载的
aircraft compass 航空罗盘
aircraft coverage 航天器感测范围
aircraft detector 飞机探测器
aircraft engine emission 飞机发动机排放物
aircraft engine fuel 航空发动机燃料
aircraft oil 飞机润滑油
aircraft photography 航空摄影学
aircraft station 机载电台
air-crossing 风桥
air-current 气流
air-curtain method 气幕法
air-curtain technique in blasting 减震气幕水下爆破法
air-density pressure 空气密度压力
airdent 喷砂磨齿机
air-detraining admixture 去气剂
airdoor 风门
air-door opener 风门开闭装置
airdraulic 气动-液压的

air-dried moisture content	风干含水率
air-dried sample	空气干燥煤样，分析煤样
air-dried soil	风干土
air-dried state	风干状态
air-dried strength	风干强度
air-dried wood	风干木材
air-drilled hole	空气洗井钻孔
air-drilling	空气洗井钻进；空气钻井；风动凿岩
air-driven	风动的
air-driven grout pump	气动灌浆泵
air-driven hammer	风动锤
air-driven hammer drill	风动冲击凿岩机
air-driven liquid pump	气动液体泵
air-driven mine car loader	风动装车机；风动矿车装载机
air-driven motor	风动马达
air-driven post drill	气动柱钻
air-driven pulsating jig	脉动跳汰机；无活塞跳汰机
air-driven pump	气泵；风动泵
air-driven rock breaker	气腿式凿岩机
air-driven roller	压缩空气传动滚轮
air-driven wrench	风动扳手
air-dry	风干
air-dry weight	风干重
air-duct	风管
air-earth interface	空气-地表分界面
aired up	气堵
air-end way	回风道
air-entrained cement	加气水泥
air-entrained concrete	加气混凝土
air-entraining admixture	加气剂
air-entraining chemical compound	掺气剂
air-entraining concrete	加气混凝土
air-entry permeameter	掺气渗透仪
air-escape pipe	泄气管
airfall deposit	降落式火山沉积物
airfall deposition	灰雨沉积；降尘沉积
airfall tuff	空降凝灰岩
air-fan	通风机
airfast	不透气的
air-feed drifter	风动推进的架式钻机
air-feed drill	压气推进式钻机
air-feed leg drill	气腿钻机；气腿凿岩机
air-feed stopper	风力推进伸缩式凿岩机
airfield classification System	飞机场分类法
airfield cone penetrometer	机场的锥式贯入仪
airfield index	机场指标
airfield landing mat	护顶网
airfield runway survey	机场跑道测量
airfield soil classification	机场土分类法
air-filled porosity	充气孔隙度
airflow	空气流
air-flow liquid level gauge	气流液面压力计
airflow measurement	气流测量，风流测量
airflow measurement by traversing method	往返移动测风法
air-flow meter	气流计
air-flow resistance	气流阻力
air-flush drilling	风吹钻眼；风吹凿岩
air-foam drilling	空气泡沫钻进
air-foam system	气雾除尘系统；空气泡沫灭火系统
air-free concrete	密实混凝土
air-fuel ratio	空气燃料比
airgun	气枪
airgun array source	空气枪组合震源
airgun bubble pulse	气枪气泡脉冲
airgun signature	空气枪特征
airgun subarray	气枪子阵列
airgun towing system	气枪拖曳系统
air-hardened quality	气硬性
air-hardening lime	气硬石灰
air-hole	通气孔
air-hose connection	贮木槽
airing	气爆；晾；烘干；通风
air-inlet valve	进气阀
air-jackleg drill	气腿凿岩机
air-jet dispersion	用空气喷射法抛散
air-jet screen	气喷式筛分机
air-jet sieve	气喷式筛
airleg	气腿
air-leg drill mounting	气腿式钻架
air-leg drilling	气腿钻眼，气腿凿岩
air-leg rock drill	气腿凿岩机
airless end	不通风的长壁工作面端；独头工作面
air-linking	用高压空气渗透贯通
air-liquid tension	气-液张力
airload	气动载荷；空运装载
air-motor drilling	风马达钻眼；风马达凿岩
air-motor driven jumbo	风动自行式钻车
air-operated	风动的；压气操作的
air-operated blasting device	压气爆破装置；压气爆破筒
air-operated cleaning table	风力分选摇床
air-operated crawler-type loader	风动履带式装载机
air-operated damper	气动风闸
air-operated drill	气动凿岩机；气动钻机
air-operated drill motor	风动钻机
air-operated gate	风动闸门
air-operated grab	风动抓斗
air-operated overcast loader	风动扬斗装载机
air-operated sander	气力操纵撒砂器
air-placed concrete	喷注混凝土
airport pavement	机场路面
air-powered cartridge loader	风力装药器
air-powered servo	气压伺服装置；气压伺服马达
air-powered winch	风动卷扬机
air-pressure	气压；空气压力
air-pressure drop	风压降
air-pressure drop formula	通风压降公式
air-pump	抽气泵
air-return shaft	回风井
air-rider jack hammer	气腿凿岩机
air-sand process	风砂分选法
air-sealed	气封的；空气封闭的；不透空气的
air-seasoned	风干的
airseasoning	通风干燥
air-set	气硬
air-setting mortar	气硬性砂浆
airshaft	通风竖井
air-slake	潮解；空气熟化；消化
air-slaked lime	气化石灰；潮解石灰；空气消化石灰
airslide	气力输送
air-slide feeder	气滑式给料器

airslusher	风动刮斗绞车；风动扒矿绞车
air-space ratio	孔隙比；含气率
airspeed	气流速度
air-sprayed hydrocyclone	气喷水力旋流器
air-starting valve	气动阀
air-storage tank	储气罐
air-stowing machine	压气充填机；风力充填机
airstrip	飞机跑道；简易机场
air-supported structures	充气结构物
air-swept ball mill	气吹式球磨机
air-swept mill	气吹式滚磨机
air-tanker	空中加油机
air-terminal	候机楼
airtight	不透气的；气密的；密封的
air-tight construction	气密结构
airtight cover	气密盖
airtight dust-exhausting system	密闭除尘系统
airtight joint	气密接合；密封接头
air-tight joint	密封接合；不透气接合
air-tight machine	密封式机器
air-tight packing	不透气填密
airtight partition	不透气隔墙
airtight seal	不透气密封；气密
air-tight stopping	气密风墙
air-track drill	履带式重型风动凿岩机组
air-void ratio	气隙比
air-water blast spray	放炮后气水式喷水
air-water jet	气水混合射流
air-water mixture	气水混合体
air-water ratio	气水比
air-water spray	气水式喷水
air-water surface	气水结合面
airway	风巷；风道
airway horizon	通风水平
airway optimization	风巷优化
airway repairer	回风道修理工；风巷修理工
airway stopping	风道风墙
Airy function of stress	应力艾里函数
Airy Heiskanen system	艾里海斯堪宁地壳均衡论
Airy isostasy	艾里地壳均衡说
Airy model	艾里模型
Airy phase	艾里震相
airy route for power transmission survey	架空送电线路测量
Airys stress function	Airys 应力函数
Airy's theory of isostasy	埃里地壳均衡理论
aisle	耳房走道；通道
ait	河心岛，湖中岛，江心洲
Aixi Formation	艾西组
Aizhuke Formation	艾主克组
Aizishan Shale	矮子山页岩
akalche	喀斯特沼泽
akaustobiolite	非可燃性有机岩
Akebulake Formation	阿克不拉克组
akenobeite	花岗闪长细晶岩
Akeqiayi Formation	阿克恰衣组
akerite	尖晶石；英辉正长岩
akle	沙丘网络；网状沙丘
akmolith	岩刃
akoafimite	含角闪石苏长岩
akremite	艾克雷麦特炸药
akrit	钴铬钨刀具合金
akrochordite	球砷锰矿
aksaite	阿氏硼镁石
Aksan	阿克萨阶
aktian deposit	陆坡沉积
aktological	近岸浅水的
aktology	近海学
akyrosome	杂岩体的附属部分
ala twin	轴双晶
Alabama design method	美国亚拉巴马州设计法
alabandine	硫锰矿
Alabasitao Formation	阿拉巴斯套组
alabaster	雪花石膏
alabastrite	雪花石膏
alabradorite	含碱性长石岩
alader	硅铝合金
aladra	含卤岩地蜡
alaiophyllum	阿莱珊瑚属
alalite	绿透辉石
alamosite	铅辉石
Alangiopollis	八角枫粉属
alar	有翼的；翼状的
alarm signal system	事故信号系统；警报信号系统
alary	翼状的
Alaska Earthquake	阿拉斯加地震
alaskaite	铅泡铋矿
Alaskides	阿拉斯加褶皱带；阿拉斯加带
alaskite	白岗岩
alaskite porphyry	白岗斑岩
Alatage Formation	阿拉塔格组
alate	有翼的
alaunstein	茂石
albanite	地沥青；暗白榴岩
albarium	白石粉
albasalt	碱性玄武岩
albedo	反照率；反射率，反照率
albedo of the Earth	地球反照率
albert coal	黑沥青
Albert shale	阿尔伯特页岩
albertite	沥青煤；阿柏特柏油
Albian stage	阿尔布阶；阿尔必期
albic horizon	白土层；假灰化层；漂白层
albite	钠长石
albite Carlsbad twin law	钠长石-卡尔斯巴德双晶律，卡钠双晶律
albite dolomite	钠长石；白云岩
albite jade	钠长石玉
albite law	钠长双晶律
albite twin	钠长石双晶
albite twin law	钠长石双晶律
albite-Carlsbad twin law	钠长石-卡斯巴德双晶律
albite-epidote-hornfels facies	钠长石-绿帘石-角岩相
albitite	钠长岩
albitization	钠长石化
albitolite	火山沉积钠长岩
albitophyre	钠长斑岩
albocarbon	萘
alboll	漂白软土
alboranite	拉苏安玄岩
alcomax	无碳铝镍钴磁铁

alcove 凹室
alcove lands 泉蚀地形
alcres 铁铝铬耐蚀耐热合金
alcrosil 铝铬硅耐酸钢
alcu washer 铜铝过渡垫片
alcu wire casing 铜铝过渡衬套
Aldanian 阿尔丹阶
Aldanian period 阿尔丹纪
Aldermac method 艾尔德麦克采矿法
Aldingan 艾丁加阶
Aldrey 阿尔德雷导线用铝合金；铝镁硅合金
Aldrin 艾氏剂
alectrosaurus 阿莱龙属
alembic 蒸馏器
alemite fitting 黄油嘴
alemite gun 黄油枪
alentegite 云英闪长岩
alert criteria 警戒准则
alert field 报警字段
alert level 预警水平
alert standard 警戒标准
alerting signal 报警信号
Aletonggou Formation 阿勒通沟组
aleurite 粉砂
aleuritic texture 粉砂结构
aleuritopelitic 粉砂泥质的
aleurolite 粉砂岩
aleuropelitic 粉砂泥质的
aleuropelitic texture 粉砂泥质结构
Aleutian arc 阿留申弧
Aleutian trench 阿留申海沟
Aleutians 阿留申群岛
aleutite 闪长辉长斑岩；易辉安山岩
aleuvite 粉砂岩
Alexandrian 亚历山大统
alexandrite 亚历山大石；金绿宝石
alexandrite effect 金绿宝石效应
alexeyevite 阿列沥青
alexoite 磁黄橄榄岩
Alfa Ridge 阿尔法海岭
alfalfa gate 螺旋式闸门
Alfenide 阿尔芬纳德镍黄铜；铜镍合金
alfenol 高导磁铝铁合金
alfisol 淋溶土
Alfven layer 阿尔文层
Alfven Mach number 阿尔文马赫数
Alfven perturbation theory 阿尔文微扰理论
Alfven velocity 阿尔芬速度
Alfven wave 阿尔芬波
algae 藻类
algal accretionary grain 藻类加积颗粒
algal anchor stone 锚状藻类岩
algal bench 藻阶地
algal bioherm 藻类生物岩礁
algal biscuit 海藻饼
algal body 藻类体
algal bound sand flat 藻结砂坪
algal bryozoan boundstone 藻苔藓动物黏结灰岩
algal cannel 藻烛煤
algal coal 藻煤

algal coating 藻类包壳
algal crust 藻壳
algal limestone 藻灰岩
algal mat biolithite 藻丛生物岩
algal paste 含藻泥晶灰岩
algal pellet 藻粒；藻泥
algal reef 藻礁
algal ridge 藻礁岩脊；藻脊
algal stromatolite 藻叠层石
algal structure 藻丛构造
algal-reef sediment 藻礁沉积
Algar dedusting unit 奥尔加型除尘机组
algarite 藻沥青；高氮沥青
algarvite 云霞霓岩
alginite 藻类体；藻质体
algo-clarite 藻质微亮煤
algodonite 微晶砷铜矿
algoflora 藻类区系；藻类志
algoid 藻类的
algology 藻类学
Algoman orogeny 阿尔戈马造山运动
Algoma-type iron formation 阿尔戈马型铁建造
algon 分解的有机屑
Algonkian period 阿尔冈纪
algovite 辉斜岩类
Ali Shan Formation 阿里山组
alidade 指方规；照准仪；照准架
alidade method 图解交会法
alidade rule 照准尺
alienation period 让与期；转让期
alienation restriction period 让与权的限制期；转让限制期
aliform 八字墙
alighting gear 起落架
alighting passenger volume 下车客流量
align 对准；调直；定线
align reamer 长铰刀
aligned current structure 定向构造；线状流动构造
aligned sample 均衡采样
aligner 准直仪
aligner ram 复位千斤顶；对直千斤顶
aligning ram 调架千斤顶
alignment 定线；线向；路线
alignment design 平面设计；线形设计
alignment diagram 列线图
alignment element 线形要素
alignment of road 公路定线
alignment plan 路线平面图
alignment stake 定位桩；路线桩
alignment survey 定线测量
alignment thread 对中丝扣
Alimak raise climber system 阿利马克反井开凿系统
alimentation area 补给区
alimentation facies 补给相
alimentation of glaciers 冰川补给
alimentation of river 河流补给；水源
alimentative 补给的
aliquot part 代表性部分
alisphaera 海石
alisporites 阿里粉属
alistillate gas 凝析气田气

alite	阿利特；A 矿，硅酸三钙
aliting	渗铅，铝化
alitizing	表面渗铝；镀铝
alkali accumulation	碱性累积
alkali aggregate reaction	碱集料反应
alkali aggregate reaction test results	砂、砾石骨料碱活性试验成果表
alkali alkyl	烷基碱金属
alkali amide	氨基碱金属
alkali amphibole	碱性角闪石
alkali andesite	碱性安山岩
alkali basalt	碱性玄武岩
alkali basalt magma	碱性玄武岩浆
alkali degreasing	碱法除油
alkali diorite	碱性闪长岩
alkali dolerite	碱性粒玄岩
alkali earth	碱土
alkali earth metal	碱土金属
alkali feldspar	碱性长石
alkali feldspar charnockite	碱长紫苏花岗岩
alkali feldspar foidite	碱长副长岩
alkali feldspar granite	碱长花岗岩
alkali feldspar rhyolite	碱长流纹岩
alkali feldspar syenite	碱长正长岩
alkali feldspar trachyte	碱长粗面岩
alkali flat	盐碱滩
alkali gabbro	碱性辉长岩
alkali granite	碱性花岗岩
alkali lake	碱湖
alkali land	碱地
alkali lime index	碱钙指数
alkali lime series	碱钙岩系
alkali olivine basalt	碱性橄榄石玄武岩
alkali pyroxene	碱性辉石
alkali rhyolite	碱性流纹岩
alkali rock series	碱性岩系
alkali salt	碱盐
alkali sodium chloride water	碱性氯化钠型水
alkali soil	碱土；碱性土
alkali syenite	碱性正长岩
alkali trachyte	碱性粗面岩
alkali wash	碱洗
alkali-affected soil	碱化土壤
alkalic basalt	碱性玄武岩
alkalic basalt series	碱性玄武岩系列
alkalic calcic rock series	碱钙性岩系
alkalic lavas	碱性熔岩
alkalic olivine basalt series	碱性橄榄玄武岩系列
alkalic rock	碱性岩
alkalic ultrabasic rock	碱超基性岩
alkali-calcic series	碱钙系列
alkali-earth metal	碱土金属
alkali-feldspar gneiss	碱性长石片麻岩
alkali-lime index	碱钙指数
alkaline aggregate reaction	碱性集料反应
alkaline amendment of acid soils	酸性土壤的碱化弥补
alkaline amphibolization	碱性角闪石化
alkaline barrier	碱性障
alkaline basalt	碱性玄武岩
alkaline cell	碱性蓄电池
alkaline cleaning	碱洗
alkaline damage	碱害
alkaline dissociation	碱式离解
alkaline earth	碱性土
alkaline earth element	碱土元素
alkaline earth metal	碱土金属
alkaline element	碱元素
alkaline fusion	加碱熔化
alkaline geochemical barrier	碱性地球化学障
alkaline hydrolysis	碱解
alkaline kaolin	碱性高岭土
alkaline land	盐碱地
alkaline leaching	碱法浸出
alkaline lignite	煤碱剂
alkaline magma	碱性岩浆
alkaline marsh	碱性沼泽
alkaline meadow soil	碱性草甸土
alkaline metasomatism	碱性交代酌；碱质交代作用
alkaline mine drainage	碱性矿井排水
alkaline minerals	碱性矿物质
alkaline mud	碱性沉渣；碱性滤泥
alkaline patch	碱斑
alkaline permanganate oxidation	碱性高锰酸盐氧化法
alkaline pump	碱液泵
alkaline pyroxenization	碱性辉石化
alkaline reaction	碱性反应
alkaline reagent	碱性试剂
alkaline reducer	碱性还原剂
alkaline reduction	碱性还原
alkaline resistant	耐碱的
alkaline rock	碱性岩
alkaline saline soil	盐碱土
alkaline saponification	加碱皂化
alkaline scale	碱性垢
alkaline sierozem	碳化灰钙土
alkaline silica gel	碱性硅胶
alkaline silicon kaoline	碱性硅质高岭土
alkaline soil	碱性土；碱化土；盐渍土
alkaline solution	碱性溶液
alkaline spring	碱性矿泉；碱性泉
alkaline substance	碱性物质
alkaline takyr	碱化龟裂土
alkaline water	碱性水
alkalinealkalic	碱性的
alkaline-calcareous soil	碱性钙质土
alkaline-earth metal	碱土金属
alkaline-saline soil	盐碱土
alkalinity coefficient	碱性系数
alkalinity of groundwater	地下水的碱度
alkalinity ratio	碱度率
alkalinity-acidity	酸碱度
alkalipicrite	碱性苦橄岩
alkaliplete	碱性暗色岩
alkali-proof	耐碱的
alkaliptoche	贫碱火成岩
alkali-reactive aggregate	碱活性骨料
alkali-reactive rock	碱反应岩石
alkali-resistant	抗碱的
alkali-saline soil	盐碱土；盐化碱土
alkali-sensitive	碱敏的

alkali-silica reaction 碱-硅反应
alkali-soluble pan 碱溶性盐磐
alkalitrophic lake 碱液营养湖
alkali-type 碱型
alkalized soil 碱性土；碱化土
alkalized solonchak 碱化盐土
alkall spot 碱斑
alkall tolerance 耐碱度；耐碱性
alkall-saline soil 盐碱土
alkorthosite 钠正长岩
alkyl xanthate 烷基黄基盐，烷基黄药
alkylammonium 烷基铵
alkylaryl 烷基芳基
alkyl-aryl sulphonate 烷基芳基磺酸盐
all hydraulic mine 全水力化矿井
all hydraulic operated drill 全液压钻机
all in aggregate 天然混合集料
all in one paver 联合铺路机
all inertial guidance system 全惯性制导系统
all levels sampler 多层取样器
all over work 长壁开采
all parallel AD converter 全并行模数变换器
all purpose bit 万能钻头
all purpose machine 万能工具机
all purpose room 多用途房
all radiant furnace 全辐射炉
all radiant type heater 纯辐射型炉
all rock breakwater 堆石防波堤
all round adsorbent 万能吸附剂
all round pressure 周围压力
all rowlock wall 通斗墙
all weather grade liquified petroleum gas 全天候级液化石油气
allactite 砷水锰矿
allagite 绿蔷薇辉石
allalinite 蚀变辉长岩
allanite 褐帘石
allantoid structure 似香肠构造
all-belt mine 全面采用胶带输送机的矿山
all-belt-drift mine 运输巷道全设胶带输送机的矿山
all-brick building 砖建筑物
all-concrete frame 全混凝土框架
all-conveyor 全输送机化的
Alleghenian 阿勒格尼阶
Allegheny orogeny 阿勒格尼造山运动
Allegheny series 阿勒格尼统
all-electric colliery 全电气化煤矿
all-electric shearer 电驱动采煤机
allemontite 砷锑矿
Allen equation 艾伦方程
Allen salt-velocity method 艾伦盐液测速法
Allerod 阿勒罗德间冰期
allevardite 钠板石
alley 巷
all-face-centred lattice 全面心格子
all-flotation process 全浮选法
allgovite 辉绿玢岩
all-haydite concrete 陶粒混凝土
allied rock 类似岩石；有关岩石
alligator 鳄鱼；鳄式碎石机；水陆军底军用车

alligator cracking 路面龟裂
alligator jaw 碎石机额板
alligatoring 鳄纹；裂痕
all-in aggregate 粒状筑路石料
allite 富铝土；铝质土
allitic soil 富铝土
allitic weathering 富铝风化
allitization 铝铁土化
allivalite 橄长岩
all-levels sample 平均样品
all-mains receiver 交直流两用接收机
allocation of frequency 频率分配
allocation of risks 风险分担
allocation of water 水的分配；水的布局
allocation of well rates 油井配产
allocation plan 拨地图则；批地图则
allocation plan of physical assets 物资分配计划
allocation range 编配幅度；编配范围
allochem 异化粒
allochemical constituent 异化组分
allochemical dolomite 异化粒白云岩；粒屑白云岩
allochemical element 异化粒；异化组分
allochemical limestone 异化粒灰岩
allochemical metamorphism 异化学变质作用；增减变质作用
allochemical rock 异化岩；异常化学岩
allochetite 霞辉二长斑岩
allochroite 粒榴石
allochthon 异地岩体；外来体；异地岩体
allochthonic groundwater 外来地下水；外源地下水；派生地下水
allochthonous 异地的；移地的
allochthonous anomaly 他源异常；异地异常
allochthonous block 异地岩块
allochthonous coal seam 异地生成煤；移置煤层
allochthonous deposit 异地沉积；移积
allochthonous fold 异地褶皱
allochthonous groundwater 外源地下水
allochthonous limestone 异地灰岩
allochthonous peat 移置泥炭
allochthonous river 外源河；外来河
allochthonous rock 移置岩；异地岩
allochthonous sediment 移置沉积物；外来沉积物
allochthonous soil 运生土；移积土
allochthonous stream 外源河流
allochthonous terrane 移置地体
allochthonous theorem 异地岩体学说；外来岩体学说；移置岩体学说
allochthonous thrust sheet 异地逆掩鳞岩席
allochthony 异地生成煤
alloclastic 火山碎屑的
alloclastic breccia 火山碎屑角砾岩
allocyclicity 他生旋回性
allodapic 异地砂屑
allodapic limestone 浊积灰岩
allogene deposit 异地沉积
allogene fold 异地褶皱
allogeneic cycle 异旋回
allogenic 他生的；外源的
allogenic material 外源物质；他生物质

allogenic mineral 他生矿物，外源矿物
allogenic river 异源河；外源河
allogenic succession 他生演替
allogenic water 外源水
allogenous stream 异源河
allograptus 奇笔石属
alloisomerism 立体异构
alloite 脆凝灰岩
allomer 异分同晶质
allomeric 异质同晶的
allomerism 异质同象现象
allometamorphism 他变酌；他变质
allometry 异速生长
allomigmatite 他混合岩
allomorph 同质多象变体；假象；副象
allomorphic 同质异晶的
allomorphism 同质异晶现象，同质异晶
allomorphite 贝状重晶石
allomorphous 同质异晶的；同质异象的；同质多象的
allopalladium 硒钯矿
allopatric polymorphism 同质多形
allophane 水铝英石
allophanite 水铝英石
allophanoid 水铝英石类
allophase metamorphism 他相变质
allophone soil 水合硅酸铝土
alloskarn 外成矽卡岩
allostropia 奇转螺属
allostylops 异柱兽属
allothi steromorphic 他生碎屑的
allothigene 他生物外源物
allothigenetic 他生的；外来的；外源的
allothimorph 原形晶
allothimorphic 变生的
allothogenic 他生的；外来的；外源的
allothrausmatic 异核球颗的
allotment 矿建用地，采矿用地；分配；份额；拨款
allotment area 采矿用地面积
allotriomorphic crystal 他形晶
allotriomorphic granular 他形粒状
allotriomorphic granular texture 他形晶粒状结构
allotriomorphic structure 他形构造
allotriomorphic texture 他形结构
allotriomorphism 他形性
allotrope 同素异晶体
all-over work 长壁开采
allowable amplitude 容许振幅
allowable bearing capacity 容许承载力
allowable bearing capacity of dam foundation 坝基容许承载力
allowable bearing capacity of foundation 地基容许承载力
allowable bearing capacity of single pile 单桩容许荷载
allowable bearing load 容许荷载
allowable bearing pressure 容许承载力；容许压力
allowable bearing value 容许承载力
allowable bending-resistance tensile stress 容许抗弯曲拉应力
allowable bond stress 容许黏着应力
allowable carrying capacity of a pile 单桩容许承载力
allowable combined stress due to external load 许用外载综合应力

allowable contact stress 允许接触应力
allowable creep strain 容许蠕变
allowable deflection 容许偏斜；容许挠度
allowable deformation 容许变形
allowable deformation of subsoil 地基允许变形值
allowable design 容许设计
allowable deviation 容许偏差
allowable dewatering depth of aquifer 含水层允许疏干深度
allowable dewatering rate of aquifer 含水层允许疏干率
allowable dewatering time limit period of aquifer 含水层允许疏干年限
allowable differential settlement 容许差异沉降
allowable displacement 容许位移
allowable drawdown 允许水位降深
allowable drift 容许变位
allowable emission rate 容许排放速率
allowable error 容许误差；公差
allowable exploitable modulus of aquifer 含水层允许开采模数
allowable factor 容许系数
allowable factor of safety 容许安全系数
allowable failure probability 容许破坏概率
allowable firedamp 容许瓦斯气量
allowable flexural stress 容许弯曲应力
allowable for deformation 预留变形量
allowable grade of slope 边坡坡度容许值
allowable gross bearing pressure 总容许荷载
allowable grouting pressure 容许灌浆压力
allowable heeling 容许倾斜度
allowable impact load 容许冲击负载
allowable leakage 容许渗漏；容许漏失量
allowable limit 容许极限
allowable load 允许荷载；容许载重
allowable loss 设计损失，设计损失
allowable maximum velocity 最大允许速度
allowable movement 容许位移
allowable net bearing pressure 容许净承载力
allowable period 允许开采期
allowable pile bearing capacity 桩的容许承载力
allowable pile bearing load 单桩容许荷载
allowable pile load 容许桩基荷载
allowable pressure drop 允许压力降
allowable production 允许产量
allowable range 容许界限
allowable rebound deflection 容许弯沉
allowable relative deformation 容许相对变形
allowable resultant stress 许用合成应力
allowable scour 容许冲刷
allowable seepage gradient 允许渗透坡降
allowable settlement 容许沉降，容许沉降量
allowable shear 容许剪力
allowable slenderness ratio 容许长细比
allowable slenderness ratio of steel member 钢构件容许长细比
allowable slenderness ratio of timber compression member 受压木构件容许长细比
allowable soil pressure 容许土压力
allowable storey deformation 容许层间变形
allowable strength 容许强度

allowable stress	容许应力；预计可承受的压力
allowable stress design	容许应力计算
allowable stress design method	容许应力设计法
allowable stress method	容许应力法
allowable stress range of fatigue	疲劳容许应力幅
allowable stresses method	容许应力法
allowable stripping ratio	允许剥采比
allowable subsoil deformation	地基变形允许值
allowable tensile stress	容许张应力
allowable tension	容许拉力
allowable thickness of residual frost layer	容许残留冻土层厚度
allowable tightness	容许紧密度
allowable tolerance	容限；公差；容许偏差
allowable total settlement	容许总沉降
allowable twisting stress	容许扭转应力
allowable twisting unit stress	容许单位扭转应力
allowable ultimate tensile strain of reinforcement	钢筋拉应变限值
allowable unbalance of pressure loss	压力损失允许差额
allowable value of crack width	裂缝宽度容许值
allowable value of deflection of structural member	构件挠度容许值
allowable value of deflection of timber bending member	受弯木构件挠度容许值
allowable value of deformation	允许变形值
allowable value of deformation of steel member	钢构件变形容许值
allowable value of deformation of structural member	构件变形容许值
allowable value of width-height of foundation steps	基础台阶宽高比的容许值
allowable variation	允许偏差
allowable vibration acceleration	容许振动加速度
allowable water leakage	容许渗水量
allowable withdrawal of groundwater	地下水可开采量；地下水允许开采量
allowable working load	容许使用荷载；容许工作压力
allowable working resistance	允许工作阻力
allowable yield of groundwater	地下水允许开采量
allowance above nominal size	尺寸上偏差
allowance below nominal size	尺寸下偏差
allowance error	允许误差
allowance for bad debts	呆账准备金
allowance for contraction	收缩留量
allowance for corrosion loss	预留锈蚀量
allowance for depreciation	备抵折旧；折旧提成
allowance for finish	精加工余量
allowance for machining	机械加工余量
allowance for shrinkage	收缩留量
allowance for under-pit work	矿山井下津贴
allowance in kind	实物津贴
allowance test	公差配合试验；容差试验
allowed direction	振动方向
allowed exposure	容许照射
allowed ground bearing capacity	地基容许承载力
allowed spectrum	容许谱
alloy bit	合金钻头；合金截齿
alloy drill bit	合金钻头
alloy for die casting	压铸合金
alloy material	合金材料
alloy steel bar	合金钢筋条
alloy steel rail	合金轨
alloy steel rod	合金钢抽油杆
alloy steels rail	合金钢轨
alloy throw-away bit	一次性合金钻头；一次性合金截齿
all-purpose adhesive	万能胶
all-purpose bit	万能活钻头
all-purpose dredger	通用挖泥船
all-purpose loader	修整钻头
all-purpose tester	综合测试仪；万能测试仪
all-rail transportation system	全铁路运输系统
all-round miner	全能矿工
all-round pressure	全周压力
all-round transporter	万能输送机
all-sky camera	全天空照相机
all-sky photometer	全天空光度计
all-sliming	搅动氰化法；矿石细磨
all-sliming process	泥化法
all-steel rotary hose	全钢旋转泥浆软管
all-timber	全木质的
allumen	锌铝合金
all-ups	原煤；混合煤
alluvia	冲积物；冲积层组
alluvia flat	河漫滩；泛滥平原
alluvial apron	冰水冲积平原；冲积扇；冲积裙
alluvial aquifer	冲积含水层
alluvial bed	冲积河床
alluvial bench	山麓冲积平原；砾质山麓平原
alluvial breccia	冲积角砾岩
alluvial brown soil	棕色冲积土
alluvial channel	冲积水道；冲积河床；冲积河道
alluvial clay	冲积黏土
alluvial cone	冲积锥；扇状地
alluvial conglomerate	河成砾岩
alluvial dam	冲积堤
alluvial delta	冲积三角洲；冲积锥
alluvial deltaic cycle	冲积三角洲旋回
alluvial deposit	冲积物；冲积层；冲积矿床
alluvial district	冲积区
alluvial epoch	冲积期；冲积世
alluvial facies	冲积相
alluvial fan	淤积扇；冲积扇
alluvial fan deposit	冲积扇沉积
alluvial fan sediments	冲积扇沉积物
alluvial fill	冲积填充
alluvial flat	河漫滩冲积滩
alluvial formation	冲积地层
alluvial gold	砂金
alluvial ground	冲积地
alluvial ground water	冲积层地下水
alluvial grouting	冲积层灌浆
alluvial horizon	冲积层
alluvial island	冲积岛
alluvial land	冲积地
alluvial laterite	冲积红土
alluvial layer	冲积层
alluvial meadow soil	冲积性草甸土
alluvial meander	冲积曲流
alluvial miner	砂矿矿工

alluvial mining	砂矿开采
alluvial ore deposit	冲积矿床
alluvial piedmont plain	山麓冲积平原
alluvial placer	冲积砂矿床
alluvial plain	冲积平原
alluvial proluvial fan	冲洪积扇
alluvial proluvial plain	冲洪积平原
alluvial prospecting	砂矿勘探
alluvial ridge	冲积脊
alluvial river	冲积河；滞流河
alluvial salinization	冲积平原盐渍化
alluvial sand	冲积砂层
alluvial sand bodies	冲积砂体
alluvial sand ripples	河成砂纹
alluvial sand wave	河成沙波
alluvial sandy soil	冲积砂土
alluvial sediment	冲积的沉积物
alluvial series	冲积系
alluvial silt	冲积粉土
alluvial slope spring	冲积坡泉；边界泉
alluvial soil	冲积土；冲积土壤
alluvial sorting	冲积分选
alluvial splay	河渠扩展淤积；河滩堆积
alluvial stream	冲积河
alluvial talus	冲积山麓堆积
alluvial terrace	冲积阶地；冲积台地
alluvial Tin	砂锡矿
alluvial tract	冲积河段；下游河段
alluvial valley	冲积河谷
alluvial veneer	冲积表层
alluvial water	冲积层水
alluvial well	冲积层井
alluvial working	砂矿开采
alluvial-drifted deposit	冲积-冰碛层
alluvial-pluvial deposit	冲积-洪积层
alluvial-proluvial plain	冲洪积平原；山前平原
alluviation	冲积
alluviation delta	三角洲淤积
alluviation of delta	三角洲增长；三角洲淤积
alluvideserta	冲积荒漠群落
alluvion	淤积层，冲积层；波浪的冲刷；泛滥，洪水；固结火山灰
alluvium	冲积层；淤积层
alluvium grouting	冲积层灌浆
Alluvium period	第四纪
alluvium splay	河渠扩展淤积；河滩堆积；河滩淤积
all-wave rectifier	全波整流
all-weather operation	全天候作业
all-weather port	不冻港
all-weather road	全天候道路
all-weather Service	全天候服务
all-wheel-drive scraper	全驱动轮铲运机
allwork	全面回采；连续回采
ally arm	架；托架；悬臂
almandine	铁铝榴石
almandine-amphibolite facies	铁铝榴石角闪岩相
almandite	铁铝榴石
almond structure	杏仁构造
almond-shaped structure	杏仁状构造
almost atoll	准环礁
almost periodic	准周期的
almost plain	准平原
almost-linear	准线性的
almost-plain	准平原
almucantar	地平纬圈；等高圈
alneon	铝锌铜合金
alnico	铝镍钴合金
alnoite	黄长煌斑岩
along slope	沿坡面等深线的
along slope coordinate	纵向坐标
along slope current	沿坡面等深线流
along the strike	平行走向；沿走向
along track resolution	纵向分辨力
alongface position transducer	沿工作面定位转换器
along-grade	顺坡
alongshore	顺岸的
along-shore current	沿岸流；顺岸流
alongshore drift	沿岸漂砂
alongshore feature	顺岸要素
alongshore interactions	近岸的相互作用
alongshore wind	海岸风
along-track coordinate	纵向坐标
along-track direction	沿航迹方
along-track resolution	纵向分辨力
aloxite	铝砂；人造刚玉
alp	高山；山峰
alpenglow	高山辉
alpestrine	高山的；林木线以下冷温坡地
alpha deconvolution method	α反褶积法
alpha energy loss	α能量损失
alpha granite	花岗岩
alpha mesohaline water	α中盐度水
alpha method of pile design	单桩摩阻力设计α法
alpha pile	α桩
alpha quartz	石英
alpha-quartz type	α石英组
alphitite	岩粉土
alphitolite	硬岩粉土
Alpides	阿尔卑斯造山带
alpine	高山的；阿尔卑斯式的
Alpine age orogeny	阿尔卑斯期造山运动
alpine animals	高山动物
alpine belt	高山带
alpine border lake	山前湖；冰蚀谷湖
alpine carbonate raw soil	高山碳酸盐粗腐殖土
alpine climate	高山气候
Alpine crustal type	阿尔卑斯地壳型
alpine diamond	黄铁矿
Alpine facies	阿尔卑斯相
Alpine fold	阿尔卑斯褶皱
alpine garden	高山植物园
Alpine geosyncline	阿尔卑斯地槽
alpine glacier	高山冰川
Alpine Himalayan belt	阿尔卑斯；喜马拉雅造山带
alpine humus soil	高山腐殖质土
Alpine hydrothermal system	阿尔卑斯山水热系统
alpine karst	高山喀斯特
alpine lake	高山湖
alpine marsh	高山沼泽
Alpine movement	阿尔卑斯运动

A

Alpine nappe 阿尔卑斯推覆体
Alpine orogeny 阿尔卑斯造山运动
Alpine peridotite serpentinite 阿尔卑斯型橄榄蛇纹岩
alpine planation terrace 山上阶地；高山夷平阶地
Alpine precision navigating equipment 阿尔卑精确航海设备
alpine road 山岭道路
alpine scenic spot 高山风景区
alpine silicated raw soil 高山硅酸盐粗骨土
alpine soil 高山土壤
alpine steppe soil 高山草原土
alpine tundra landscape spot 高山草甸风景区
Alpine type peridotite 阿尔卑斯式橄榄岩
Alpine type ultramafic associations 阿尔卑斯型超铁镁组合
Alpine type vein 阿尔卑斯型矿脉
alpine zone 高山带
Alpine-Himalayan belt 阿尔卑斯-喜马拉雅造山带
Alpine-Himalaya-type orogenic belt 阿尔卑斯-喜马拉雅型造山带
Alpine-type fold 阿尔卑斯式褶皱
Alpine-type glacier 阿尔卑斯型冰川
Alpine-type tectonics 阿尔卑斯型构造
Alpine-type ultramafic associations 阿尔卑斯型超镁铁组合
Alpine-type vein 阿尔卑斯型矿脉
alpinotype 阿尔卑斯型
alpinotype orogenesis 阿尔卑斯型造山作用
Alpinotype orogeny 阿尔卑斯型造山酉
Alpinotype tectonics 阿尔卑斯型构造
Alportian 阿尔波特阶
Alps 阿尔卑斯山脉
Alps glaciation 阿尔卑斯冰川作用
Alps mountain glacier 阿尔卑斯式山岳冰川
alramenting 表面磷化
alsbachite 榴云花岗闪长斑岩
Altaides 阿尔泰造山带
altait 碲铅矿
altaite 碲铅矿
altalf 黑淋溶土
Alta-mud 阿尔太泥
Altay Formation 阿尔泰组
Altay geosyncline 阿尔泰地槽
altazimuth 地平经纬仪；高度方位仪
altazimuth instrument 地平经纬仪
alteration 变异；蚀变
alteration and addition 改建及加建
alteration cost 更改费用
alteration envelop 蚀变晕；蚀变封套
alteration halo 蚀变晕
alteration index 蚀变指数；变异指数
alteration mineral 蚀变矿物
alteration of cross-section 截面改变
alteration of generations 世代变化
alteration of strata 互层
alteration petrology 蚀变岩石学
alteration product 变质产物
alteration rock 蚀变岩石
alteration texture 蚀变结构
alteration to a building 改建
alteration wall rock 蚀变围岩
alteration zone 蚀变带
alteration zoning 蚀变分带
altered aureole 蚀变晕
altered basaltic oceanic crust 改变玄武洋壳
altered granite 变质花岗岩
altered mineral 蚀变矿物
altered rock 蚀变岩石
altered volcanic rocks 蚀变火山岩
alternate angle 相反位置角；交错角
alternate bay 隔仓
alternate bending test 反复弯曲试验
alternate current 交流电
alternate drilling and mucking 钻眼装岩顺序作业
alternate fold 交互褶皱
alternate freezing and thawing 冰融交替
alternate fuel 代用燃料
alternate heading 半煤半岩巷槽；半煤半岩工作面
alternate Host address 修改主机地址
alternate joint 错列接缝；错列接头
alternate layer 交互层；夹层
alternate load 交替荷载；反复荷载；交变负荷；代用荷重
alternate method of charging 更迭装药法；分段装药法
alternate stopes and pillars 交错房柱开采法
alternate strain 交变应变
alternate stress 反复应力；交变应力
alternate sweep 交替扫描
alternate tension 交变拉力
alternate terrace 曲流阶地
alternate the connections 起钻错扣
alternate track 替换磁道
alternately inflow-outflow lake 吞吐湖
alternately pinnate 互生羽状的
alternate-shovel method 间隔铲取采样法
alternating acceleration peak 交变加速度峰值
alternating bending 交替弯曲；双向弯曲
alternating bending test 反复弯曲试验
alternating curing 交替养护
alternating deposit 交替沉积
alternating direction implicit 交替方向隐式
alternating direction implicit scheme 交替方向隐格式
alternating direction iterative 交替方向迭代法
alternating direction method 交替方向法
alternating field cleaning 交变场清洗
alternating fluid hydraulics 交流液压
alternating force 交变力
alternating freezing and thawing 交替冻融
alternating gradient 交变梯度
alternating hydraulic flow 交流液压传动
alternating induced polarization 交流激发极化法
alternating injection 交替注入
alternating lake 交替湖
alternating lateral load 交变侧向荷载
alternating layer 交互层
alternating load 交变载荷；交替载荷
alternating magnetization 交变磁化
alternating matrix 交错矩阵
alternating method 交错法
alternating motion 往复运动；交替运动
alternating of strata 地层的交互
alternating repeated loading 交变重复荷载
alternating shock load 交变冲击载荷

alternating strain　交变应变
alternating strain amplitude　更迭应变幅度
alternating stress　交变应力
alternating stress amplitude　交变应力振幅
alternating-current measurement　交流测量
alternation　交替，替换，替代物；交互层
alternation of bed　交互层
alternation of freezing and thawing　冻融循环；冻融交替
alternation of generations　世代交替
alternation of rock　岩石变异；岩石蚀变
alternation of stopes and pillars　矿房和矿柱的交替排列
alternation of strata　互层；交互层
alternation of stress　应力交替；应力交变
alternation wave　交替波；交换波
alternative　交替的；备选的；另外的
alternative arrangement　交错布置
alternative automated approach　交互自动法
alternative Channels　多种渠道
alternative cost　替换成本
alternative denial gate　与非门
alternative design　择一设计；比较设计
alternative energy　替代能源
alternative equipment　替代设备
alternative frequency　交变频率
alternative fuel　代用燃料；石油燃料代用品
alternative hypothesis　备择假设
alternative interpretation　择一解释
alternative joint　错缝
alternative land use　可采用的土地使用方案
alternative load　交变荷载
alternative maintenance　轮修
alternative material　替代材料
alternative mechanism　交变机制
alternative method　替换法
alternative path　可供选择的通道
alternative power station　替代电站
alternative project　比较方案；比较设计
alternative proposal　比较方案投标；代替方案投标
alternative provision　替代条款
alternative recharge well　备用回灌井
alternative requirement　任选要求
alternative route　迂回进路；比较线路
alternative scheme　替代方案；备用方案
alternative solution　替代解决方案
alternative stress　交变应力
alternative technology　替换技术
alternative test　交错检验；替换试验
alternative to explosive　炸药代用品
alternative trunking　迂回中继
altigraph　测高计；高度记录器
altimeter　测高仪；高度计
altimetric frequency curve　高程频率曲线
altimetric measurement　测高；高程测量
altimetric point　高程点
altimetry　高度测量术；高角测量法；测高法
altiplanation　高地夷平作用；高山剥夷作用；冰冻夷平作用
altiplanation terrace　山上阶地；高山夷平阶地
altiplano　上升高原
altiscope　潜望镜

altithermal　高热期，冰后期的高温期
altithermal period　高温期
altitude　海拔
altitude above sea level　海拔高程；海拔高度
altitude angle　仰角；高度角
altitude chamber　高空试验模拟室
altitude circle　竖直度盘；地平经圈
altitude control survey　高程测量
altitude correction　高度改正
altitude correction of zenith difference　异顶差
altitude difference　标高差；高程差
altitude effect　高程效应
altitude error　高度误差
altitude figures　高程数据
altitude gauge　测高仪；高度计
altitude grade gasoline　高海拔级汽油
altitude hole effect　高度死区效应
altitude indicator　高度指示器
altitude intercept　高度差距
altitude interpolation　高程内插
altitude level　测高程水准仪
altitude modulation　幅度调制；调幅
altitude of rocks　岩层产状
altitude of source　河源高度
altitude of terrain　地势高程
altitude of triangle　三角形高线
altitude scale　高程划分尺
altitude survey　高度测量；高程测量
altitude tint legend　高程色调符号；高程表
altitude valve　水位控制阀
altitude-tint legend　高程表
altitudinal belt　高度带；垂直带
altitudinal zonality　垂直分布带
altoll　黑软土
altometer　经纬仪
altonian　阿尔顿阶
aludel　梨坛
aludip　镀铝钢板
aludur　阿鲁杜尔合金；铝镁合金
aluflex　锰铝合金
alum　明矾；矾
alum clay　明矾黏土；铝土
alum coal　煤矸；白色氧化煤
alum earth　明矾土
alum schist　明矾片岩
alum Shale　明矾页岩
alum shear bushing　铝制剪切套
alum slate　明矾板岩
alum stone　明矾石
alumel　阿卢梅尔镍铝合金；镍基热电偶合金
alumian　无水矾石
aluminate　铝酸盐
aluminaut　潜水艇
aluminiferous　含铝的；含矾的
aluminiferous rock　铝质岩
aluminite　矾石
aluminium bronze　铝青铜
aluminium cable　铝芯电缆
aluminium cap　铝壳雷管
aluminium Capacitor　铝电容器

aluminium cell ceiling	铝格天花板
aluminium chloride	氯化铝
aluminium citrate	柠檬酸铝
aluminium coating	覆盖铝
aluminium detonator	铝壳雷管
aluminium dispersal	预防矽肺病的铝粉扩散
aluminium drill pipe	铝合金钻杆
aluminium dust	铝粉
aluminium effluent	铝排水含铝废液
aluminium enamel paint	铝粉磁漆
aluminium Flat Bar	铝扁条
aluminium hydroxide	氢氧化铝
aluminium isopropylate	异丙酸铝
aluminium killed steel	铝镇静钢
aluminium lead-azide detonator	铝壳叠氮化铅雷管
aluminium mennige	铝红
aluminium minerals	铝矿类
aluminium nitrate	硝酸铝
aluminium ore	铝矿石
aluminium oxide	氧化铝
aluminium paint	铝粉漆
aluminium panel	铝板
aluminium phosphate	磷酸铝
aluminium plated steel	包铝钢
aluminium pontoon type internal cover	铝浮筒式内浮顶
aluminium potassium sulfate	硫酸铝钾
aluminium primer	铝粉底漆
aluminium rich primer	富铝底漆
aluminium sacrificial anode	牺牲铝阳极
aluminium sepiolite	铝海泡石
aluminium shell	铝壳
aluminium sleeve	铝管；铝壳；铝管体
aluminium spray	喷铝
aluminium stearate	硬脂酸铝
aluminium sulphate	硫酸铝
aluminium tape	铝带
aluminium triethyl	三乙基铝
aluminium tube	铝管；铝壳
aluminium-soap grease	铝皂脂，铝基脂
aluminium-steel composite conductor line	钢铝复合接触轨
aluminized coat	镀铝上衣；铝镀覆层
aluminized explosive	含铝炸药
aluminized slurry explosive	用铝粉敏华的浆状炸药
aluminizing	镀铝；渗铝
aluminizing by hot dipping	热浸渍渗铝
aluminosilicate	铝硅酸盐
aluminothermic	铝热的
aluminothermic process	铝热法
aluminothermics	铝冶术
aluminothermy	铝热法
aluminous	铝土的；矾的
aluminous cement	含铝水泥；高铝水泥
aluminous fire brick	矾土耐火砖
aluminous laterite	铝质红土
aluminum alkyls	烷基铝
aluminum alloy	铝合金
aluminum armor rod	铝包带
aluminum bridge	铝桥
aluminum chloride	氯化铝
aluminum coating	铝涂层
aluminum fiber	铝纤维
aluminum foil sampler	铝箔取样器
aluminum link capsule gun	连杆式铝质弹壳射孔器
aluminum octahedron	铝氧八面体
aluminum powder	铝粉
aluminum steady arm	铝定位器
aluminum steady arm curved	软定位器
aluminum trihydrate	三水合铝
aluminum-silicate minerals	铝硅酸盐矿物
aluminum-steel cable	钢芯铝电缆
alumite	明矾石
alumite process	氧化铝膜处理法
alumochromite	铝铬铁矿
alumyte	变埃洛石
alundum	人造刚玉；铝氧粉；刚铝石
alundum cement	氧化铝水泥
alunite	明矾石；明矾石族
alunitization	明矾石化
alunogen	毛矾石
alurgite	淡云母；锰云母
alushtite	蓝高岭石
alusil	铝硅活塞合金
alustite	蓝高岭土
alveolar	蜂窝状的；气泡状的
alveolar erosion	蜂窝状侵蚀
alveolar texture	洞室结构；蜂窝状结构
alveolar weathering	蜂窝状风化
alveolus	小蜂窝
alvikite	细粒方解碳酸岩
alvite	铪锆石
Alxa Group	阿拉善群
Alxasaurus	阿拉善龙属
Alyna Sub	阿利纳接头
alyphite	轻沥青
amagmatic	非岩浆活动的
amagmatic formation	非岩浆岩层
amagmatic succession	非岩浆序列
amalgam	汞合金；汞齐
amalgam process	混汞法
amalgam reduction	混汞还原；汞膏还原
amalgam retort	混汞提金甑；混汞蒸馏甑
amalgam table	混汞台；混汞板
amalgam test	汞齐试验
amalgamable	可汞齐化的；可混汞的
amalgamated metal	混汞金属
amalgamated society of engineers	工程师联合会
amalgamating barrel	提金桶
amalgamation	汞齐化；混汞
amalgamation of terranes	地体拼合
amalgamation pan	混汞盘
amalgamation process	汞齐法；汞膏法
amalgamator	混汞器；提金器；混汞者
amalgam-barrel	汞齐化桶；混汞桶
amalgam-pan	汞齐化盘；混汞盘
amanthium	沙丘植物群落
amarantium	红铁矾
Amarassian	阿玛拉斯阶
amargosite	膨润土；斑脱岩
amateur earthquake prediction	业余地震预报
amateur service	业余业务

amatol	阿马托炸药
amazon stone	天河石；绿长石
Amazonia	亚马孙古陆
amazonite	天河石；微斜长石
amazonitization	天河石化
ambatoarinite	碳酸锶铈矿
amber	琥珀；琥珀色
amber flashing light	黄色闪光灯
amber insects	琥珀昆虫
amber placer	琥珀砂矿
amberite	阿比理特炸药；灰黄琥珀
amberlite	合成阴离子交换树脂
amberoid	人造琥珀
ambient air quality	环境空气质量
ambient air quality standard	环境空气质量标准
ambient conditions	外部条件；大气状态
ambient influence	环境影响
ambient light	环境光照度
ambient load	环境荷载
ambient measurement	周围大气检定
ambient microseismic noise	环境脉动噪音
ambient pressure	周围压力
ambient pressure effect	围压效应
ambient probability	环境概率
ambient rock	洞室围岩
ambient state of stress	原岩应力状态；初始应力状态
ambient stress	周围应力
ambient stress condition	环境应力条件环境应力场
ambient temperature	周围温度；环境温度
ambient vibration	环境振动
ambient vibration test	环境振动试验
ambient water quality	环境水质
ambiguity	模糊；不明确
ambiguity function	模糊函数
ambiguity resolution	模糊度解算
ambiguous classification	模糊分类
ambiguous error	模糊误差；多义性误差
ambiguous interpretation	模糊解释
ambitus	赤道部
ambivalence	两性化合能力
amblygon	钝角三角形
amblygonite	磷铝锂石
ambonite	堇青安山岩
ambroin	合成琥珀；绝缘塑料
ambulacral groove	步带沟
Ambulacral system	步带系
ambularcral zone	步带
Ambulate resin	安勃力特树脂
ambulator	测距仪
Ambursen dam	安布森式坝
ameboid fold	无定形褶皱
Amelia type structure	阿米里亚型构造
amenable ore	可调和的矿石，顺从药剂的矿石
amenity banks	环境保护堤
amenity forest	风景林
amenity planting strip	美化市容种植地带
amenity railing	美观栏杆
amenity strip	美化市容地带
Amerada pressure gauge	阿米雷达井下压力计
American Association for Artificial Intelligence	美国人工智能协会
American Association for Geodetic Surveying (AAGS)	美国大地测量协会
American Association of Engineers (AAE)	美国工程师协会
American Association of Petroleum Geologists	美国石油地质家协会
American Association of Quaternary Research (AAQR)	美国第四纪研究协会
American Association of Scientific Wo-rkers (AASW)	美国科学工作者协会
American Association of State Highway and Transportation Officials (AASHTO) Standard	美国公路及交通标准协会
American boring system	美国冲击式钻孔系统
American Cartographic Association (ACA)	美国制图学协会
American College of Radiology	美国放射线学会
American Concrete Institute	美国混凝土学会
American Congress of Surveying and Mapping (ACSM)	美国测绘联合会
American Documentation Institute (ADI)	美国文献学会
American Engineering Standards Committee (AESC)	美国工程标准委员会
American excavation method	美国式开挖法
American Federation for Information Processing (AFIP)	美国信息处理联合会
American Gas Association	美国天然气协会
American Geographical Society (AGS)	美国地理学会
American Geological Institute (AGI)	美国地质学会；美国地质研究所
American Geophysical Union (AGU)	美国地球物理联合会
American Institute of Architects	美国建筑师学会
American lining cycle	井壁混凝土浇灌的美国工作循环制
American Mining Congress	美国矿业会议
American National Standard Code for Information Interchange (ASCII)	美国信息交换标准代码
American National Standard Institute (ANSI)	美国国家标准协会
American oscillating table	美国风力摇床；美国风力淘汰盘
American Petroleum Institute (API)	美国石油学会
American Plate	美洲板块
American pump	砂泵；泥泵
American Quaternary Association (AQA)	美国第四纪协会
American Society for Testing and Materials (ASTM)	美国材料与试验学会
American Society of Civil Engineers and Architects (ASCEA)	美国土木工程师与建筑师学会
American Society of Civil Engineers (ASCE)	美国土木工程师学会
American Society of Foundation Engineers (ASFE)	美国基础工程师协会
American Society of Limnology and Oceanography (ASLO)	美国湖沼与海洋学会
American Society of Lubricating Engineers	美国润滑工程师学会

A

American Society of Photogrammetry and Remote Sensing (ASPRS) 美国摄影测量学遥感学会
American Society of Photogrammetry (ASP) 美国摄影测量学会
American Standard Building Code (ASBC) 美国标准建筑法规
American Standards Association (ASA) 美国标准化协会
American system of drilling 美国钻井方式；绳式顿钻
American Underground Construction Association (AUCA) 美国地下建筑协会
American Water Resources Association (AWRA) 美国水资源协会
American-Eurasian plate boundary 美洲-欧亚板块边界
amesdial 测微计；千分表
amesite 镁绿泥石
amethyst 紫晶；水碧；紫水晶
ametrometer 屈光不正测量器
Amex process 艾麦克斯法
Amgaian 阿姆加阶
amherstite 反纹正长闪长岩
amianthine 石棉
amianthus 白丝状石棉；石绒
amiatite 硅华
amicrobic 非微生物的
amicron 次微粒
amictic lake 永冻湖；常年封冻湖；永久封冻湖
amide imide polymer 酰胺酰亚胺聚合物
amide powder 氨化火药
amine gas purification 胺法煤气净化
amine hardener 胺固化剂
amino acid geochemistry method 氨基酸地球化学法
amino acid geothermometer 氨基酸地质温度计
amino oxyethylene sulfate 氨基氧乙烯硫酸盐
amino resin adhesive 氨基树脂胶黏剂
amino-acid racemization age method 氨基酸消旋法测年；氨基酸旋光法定年
amino-acid racemization dating 氨基酸外消旋断代
ammersooite 阿米水云母
ammite 鲕状岩
ammocolous 砂生生物；栖砂的
ammodyne 艾木达因炸药
ammomum nitrate 硝铵炸药
ammomum oxalate 草酸铵
ammomum perchlorate 高氯酸铵
ammomum powder 硝酸铵火药
ammon explosive 硝铵炸药
ammon gelatin 硝酸铵吉利那特炸药
Ammonal 阿莫纳尔
ammon-dynamite 硝铵炸药
ammon-gelatin-dynamite 铵胶炸药
ammon-gelignite 阿芒-杰利奈特炸药
Ammonia alum 铵明矾；硫酸铝铵
ammonia dynamite 硝铵炸药
ammonia gelatin 硝酸铵胶质炸药
ammoniated clay 氨化黏土
ammoniated coal 氨化煤
ammoniated peat 氨化泥炭
ammonioborite 水铵硼石
ammonite 菊石；阿芒奈特炸药
Ammonite Age 菊石时代

ammonium carbonate 碳酸铵
ammonium chloride 氯化铵，卤砂
ammonium gelatin 硝酸铵胶质炸药
ammonium lauryl sulphate 月桂硫酸铵
ammonium nitrate 硝酸铵
ammonium nitrate explosive 硝铵炸药
ammonium nitrate fuel oil mixture 铵油炸药
ammonium nitrate pill 硝铵与燃料油混合物炸药丸；粒状硝酸铵
ammonium nitrate powder 硝铵炸药
ammonium nitrate trinitrotoluene 硝酸三硝基甲苯
ammonium nitrate-fuel oil 干爆炸剂；铵油炸药
ammonium nitrate-fuel oil agent 铵油炸药
ammonium nitrate-fuel oil mixture 铵油炸药
ammonium powder 硝酸铵火药
ammonium soil 铵质土
ammonium thiocyanate 硫氰酸铵
ammonium-nitrate hose 硝铵炸药装药软管
ammonizator 加氨器
ammuntion 弹药；装弹药
amneite 角闪霞石岩
amnuralithus 暗弯角石
Amodyn 阿莫丁炸药
amoeba glacier 不定形冰川
Amogel 阿莫杰尔炸药
amor 人工防冲铺盖；护面
amorce 起爆剂
Amorican Orogeny 阿莫利坎造山运动
amorphic soil 幼年土；初成土；初型土
amorphism 非晶性；无定形
amorphogen 无定形岩
Amorphognathus 变形颚牙形石属
amorphous 非晶质的；无定形的
amorphous binder 无定形黏合剂
amorphous bitumen 无结构沥青；无定形沥青；沥青质体
amorphous black peat 无结构黑色泥炭
amorphous carbon 无定形碳
amorphous geomembrane 非晶土工膜
amorphous graphite 无定型石墨
amorphous kerogen 无定形干酪根
amorphous liptinite 无定形类脂体
amorphous material 无定形体
amorphous matrix 无定形基体
amorphous mineral 非晶质矿物
amorphous opaque matter 无定形不透明物质
amorphous organic matter 无定形有机质
amorphous peat 无结构泥炭
amorphous rock 非晶质岩
amorphous silica 非晶质硅石
amorphous solid 非晶固体
amorphous state 非晶态
amorphous structure 非晶质结构
amorphous substance 无定形物，非晶质
amorphous vitrinite 无定形镜质体
amortisseur 减震器；缓冲器；阻尼器；消声器
amortisseur winding 阻尼绕组
amortization cost 折旧费用；偿还资金
amortization factor 折旧率
amortization of equipment 设备折旧
amortization period 折旧期，折旧年限

amortization rate 分期偿还率；折旧率
amosite 铁石棉
amount of compensated thermal expansion 热胀补偿量
amount of compression 压缩量
amount of cone penetration 穿透孔径
amount of contraction 收缩量
amount of creep 蠕变量；蠕动量滑动量；滑动迹
amount of deflection 挠度；垂度
amount of deviation 偏差数值
amount of drilling 钻进工作量钻井数量
amount of inclination 弯曲度；倾斜度
amount of leakage 渗漏量
amount of loosening 松碎程度
amount of memory 存储量
amount of peeling rock 废石剥离量
amount of precipitation 降水量；降雨量；沉淀量
amount of pressurization 承压量；承压程度
amount of rainfall 雨量
amount of rise 潮升量
amount of scale buildup 结垢量
amount of sediment deposition 泥沙淤积量
amount of water pumped 总抽水量；抽出水总量
amount of water required 需水量
ampangabeit 铌链铁铀矿
ampasimenite 辉霞斑岩
ampelite 黄铁碳质页岩；硫铁黑土
ampelitic limestone 碳质灰岩
ampferer subduction 大陆俯冲作用
amphibious bulldozer 水陆两用推土机
amphibious operation 水陆联合作业
amphibious seismic operation 两栖地震勘探作业
amphibious tractor 水陆两用拖拉机
amphibious vehicle 水陆两用车
amphibole 闪石；角闪石
amphibole asbestos 角闪石石棉
amphibole eclogite 角闪石榴辉岩
amphibole schist 角闪石片岩
amphibole-gneiss 角闪石片麻岩
amphiboles 闪石类
amphibolic rock 角闪质岩石
amphibolite 闪岩；角闪岩
amphibolite facies 角闪岩相
amphibolite granulite facies 角闪麻粒岩相
amphibolite-schist 角闪片岩
amphibolization 闪石化；闪岩化
amphibololite 火成角闪石岩
amphicar 水陆两用车
amphidromic point 无潮点
amphidromic region 无潮区
amphidromic system 旋转潮波系统
amphidromic wave 无潮波
amphidromous migration 两侧洄游
amphigene 白榴石
amphigenite 白榴熔岩
amphitheater terrace 圆弧形阶地
amphizygus 双轭颗石
amphodelite 钙长石
amphogneiss 混合片麻岩
ampholytic surface active agent 两性表面活性剂
amphoteric 两性的

amphoterite 无粒古橄陨石
ample coal 大煤柱
amplification coefficient 放大系数
amplification coefficient of acceleration 加速度放大系数
amplification control 增益调整；放大控制
amplification factor 放大因数
amplification frequency 放大频率
amplified coefficient of eccentricity 偏心距增大系数
amplified curve 放大曲线
amplitude absorption 振幅吸收
amplitude adder 幅度加法器
amplitude adjusted indices 幅度调节指数法
amplitude anomaly 振幅异常
amplitude attenuation 振幅衰减
amplitude log 声幅测井；振幅测井
amplitude of fold 褶皱波幅
amplitude of ground motion 地震动振幅
amplitude of stress 应力幅度
amplitude to period ratio 振幅周期比率
amplitude uniformity 振幅一致性
amplitude value 幅值
amplitude variation with offset 振幅随偏移距变化
amplitudes of groundwater level fluctuation 地下水水位变幅
amplitudes probability density 振幅概率密度
amplitude-versus-offset 振幅随偏移量
amputation plane 截断面
Amsler planimeter 阿姆斯勒求积仪
Amstelian 阿姆斯特尔阶
Amushan Formation 阿木山组
amygdale 杏仁孔
amygdaloid 杏仁岩；杏仁状
amygdaloidal 杏仁状的
amygdaloidal basalt 杏仁状玄武岩
amygdaloidal nodule 杏仁状结核
amygdaloidal structure 杏仁状构造
amygdaloidal texture 杏仁状结构
amygdalophyllum 杏仁珊瑚属
amygdalophyre 杏仁状玢岩
amygdule 杏仁孔
AN slurry 硝铵浆状炸药
Anabar block 阿纳巴尔地块
Anabaria 阿纳巴尔叠层石属
anabasis salsa 盐生假木贼
anabatic differentiation 上升分异作用
anabohitsite 铁橄苏辉岩
anabranch 汊河；交织支流
anaclinal river 逆斜河；逆向河
anaclinal stream 逆向河
anaclinal valley 逆向谷
anacline 正倾斜
anadiagenesis 深埋成岩作用
anadiagenetic stage 深埋成岩期
anaerobic environment 缺氧环境
anaerobic filtration 厌氧滤池
anaerobic halophiles 厌氧嗜盐微生物
anaerobic methane oxidation 甲烷厌氧氧化
anaerobic methanotrophy 厌氧甲烷氧化菌
anaerobic micro organism 厌氧微生物
anagenite 杂砾岩；石英质砾岩
anaglacial 始冰期

anaglyphic map　互补色地图
anaglyphic method　互补色立体观测法
anaglyphic projection　补色立体投影
anaglyphic spectacles　补色立体眼镜
anaglyphic viewing system　互补色立体观察系统
anaglyphical stereoscopic viewing　互补色立体观察
Anahuce Stage　阿纳海斯阶
analbite　钠歪长石；三斜钠长石
analcime　方沸岩；方沸石
analcime basalt　方沸石玄武岩
analcime basanite　方沸石碧玄岩
analcime diorite　方沸石闪长岩
analcime gabbro　方沸石辉长岩
analcime monzodiorite　方沸石二长闪长岩
analcime monzogabbro　方沸石二长辉长岩
analcime monzosyenite　方沸石二长正长岩
analcime phonolite　方沸石响岩
analcime plagisyenite　方沸石斜长正长岩
analcime syenite　方沸石正长岩
analcime tephrite　方沸岩碱玄岩
analcimite　方沸岩
analcimization　方沸石化
analcimolith　方沸石岩
analcite　方沸石
analcite orthoclase lamprophyre　方正煌斑岩
analcite-basalt　方沸玄武岩
analcitite　方沸岩
analcitization　方沸石化
analemma　赤纬时差图
analog machine　模拟机
analog method　类比法
analog method for rock mechanics　岩石力学模拟方法
analogical inference　类比推理
analog-to-digital conversion　模拟-数字转换
analysis and calculation method　分析计算法
analysis and evaluation methodology　分析和评估方法
analysis bar charting　分析条形图法
analysis by absorption of gas　吸气分析
analysis by plasticity　塑性分析法
analysis by sedimentation　沉降分析
analysis geomorphology　动力地貌学；分析地貌学
analysis in situ　现场分析；原位分析
analysis map　分析图
analysis model　分析模式
analysis of a factorial experiment　析因实验分析
analysis of a network　网络分析
analysis of frequency distribution　频率分布分析
analysis of geomagnetic field　地磁场分析
analysis of grain composition　颗粒分析
analysis of ground vibration　地面振动分析
analysis of hydrologic data　水文资料分析
analysis of limit equilibrium　极限平衡分析
analysis of ocean graphical data　海洋资料分析
analysis of plane stress　平面应力分析
analysis of Quaternary sediments　第四纪沉积物分析
analysis of stability　稳定性分析
analysis of structural response to seismic excitation　地震响应分析
analysis of structure reliability　结构可靠度分析
analysis of symmetrical and asymmetrical geometries triangular roof prism　对称和非对称集合形状分析
analysis on interaction of superstructure and foundation　上部结构和基础的共同作用分析
analysis photogrammetry　解析摄影测量
analysis sample　分析试样
analysis stereo plotter　解析测图仪；解析型立体测图仪
analysis stratigraphy　分析地层学
analysis tectonics　解析构造学
analysis variance　分析方差
analytic curve　分析曲线；解析曲线
analytic data　分析资料
analytic demonstration　分析证法，解析证法；分析论证
analytic density　解析密度
analytic dynamics　分析动力学
analytic expression　分析式
analytic extrapolation　分析外推法
analytic function　分析函数
analytic geometry　解析几何
analytic geomorphology　分析地貌学；动力地貌学
analytic hierarchy process　层次分析法
analytic mapping control point　解析图根点
analytic method　分析方法；解析方法
analytic method of groundwater resources evaluation　地下水资源评价解析法
analytic method of selection　分析选择法
analytic model　分析模型，解析模型
analytic modeling　解析模型
analytic solution　解析解
analytic standard　分析标准
analytic stratigraphy　分析地层学
analytic structural geology　分析构造地质学
analytical aerial triangulation　解析空中三角测量
analytical function　解析函数
analytical geochemistry　分析地球化学
analytical geomorphology　分析地貌学
analytical hierarch process　层次分析法
analytical investigation　分析研究；分析试验
analytical liquid chromatograph　分析液相色谱仪
analytical map　分析地图；解析地图
analytical method of photogrammetric mapping　解析法测图
analytical model　解析模型
analytical modeling　分析模型
analytical orientation　解析定向
analytical photogrammetry　解析摄影测量
analytical photography　解析摄影学
analytical plate number　分析塔板数
analytical plotter　解析测图仪
analytical procedure　分析步骤；分析方法
analytical proof　分析证明
analytical rectification　解析纠正
analytical scale　分析度盘；分析标尺
analytical separation　分析分离
analytical solution　解析解；分析解
analytical solution of motion equation　运动方程分析解
analytical standard　分析标准
analytical stereoplotter　解析立体绘图仪
analytical stratigraphy　分析地层学
analytical structural geology　解析构造地质学
analytical trend　解析趋势

analytical ultracentrifugation 分析型超速离心
analyzing prism 分析镜
analyzing spot 扫描点
anamesite 细玄岩；中粒玄武岩
anamigmatism 深熔混合岩化
anamigmatization 深熔作用
anamorphic 深带合成变质的
anamorphic zone 深成变质带
anamorphism 合成变质；深带合成变质；深带复合变质
anapaite 三斜磷钙铁矿
anaseism 离源震
anaseismic onset 离派初动
anastomosed stream 网状河
anastomosing 河道交织作用
anastomosing branch 支汊
anastomosing cleavage 交织劈理；网状劈理
anastomosing deltoidal branch 交织三角洲汊河
anastomosing drainage 交织水系
anastomosing drainage pattern 交织状水系
anastomosing river 交织河；分汊河
anastomosing stockwork 网状脉
anastomosing stream 交织河；分汊河
anastomosing stream pattern 交织河网
anastomosis 网状的；交织
anastrophe 倒装法；倒置法
anatase 锐钛矿
anatectic 深源的；深熔的
anatectic batholith 深源岩基
anatectic earthquake 深源地震
anatectic magma 重熔岩浆；深熔岩浆
anatectic melting 深熔作用
anatectite 深熔混合岩
anatexis 深溶作用
anatexite 深熔岩
anathermal 冰后升温期；冰后升温的
anauxite 富硅高岭石；蠕陶土
anbauhobel 快速刨煤机
anbei Sand 安北砂
ancestral hydrosphere 原始水圈
ancestral petroleum 原石油
ancestral river 古水系
ancestral structural belt 古老构造带
ancheilotheca 无唇螺属
anchi metamorphism 近地表变质作用
anchi monomineralic rock 近单矿物岩
anchieutectic magma 近共结岩浆
anchieutectic rock 近共结岩
anchieutectic system 近结系
anchimetamorphism 近地表变质作用
anchi-monomineral rock 近单矿物岩
anchimonomineralic 近单矿物的
anchor 锚；锚固，固定；锚杆；牵绳加固；衔铁，簧片
anchor agitator 锚式搅拌器
anchor arm 锚臂
anchor arrangement 锚固装置
anchor bar 锚筋；锚杆
anchor barge 锚泊驳船
anchor beam 锚固梁；防滑梁
anchor bearing 锚承；锚座
anchor behavior 锚固状态

anchor block 锚块；锚墩；地锚；木楔
anchor body 锚定体
anchor bolster 锚架
anchor bolt 锚杆；地脚螺栓
anchor bolt support 锚杆支护
anchor bolt with grout 注浆锚杆
anchor bolt-net-spray support 锚定锚杆净喷雾支护
anchor bolt-spray support 锚喷支护
anchor boom 锚臂
anchor bracket 锚固托架
anchor bundle 锚筋束
anchor buoy 锚浮标；系船浮筒
anchor cable 锚索铰线；锚缆
anchor casing packer 锚定式套管封隔器
anchor catcher 锚捕捉器
anchor cell 锚箱
anchor chain 锚链
anchor chock ram 垛式支架锚固千斤顶
anchor clamp 锚夹
anchor column 锚柱；竖旋桥支柱
anchor cone 锚锥
anchor deformation 锚头变形
anchor device 锚固装置；固定装置
anchor ditcher 锚式挖沟机
anchor drill hole 锚孔；锚固孔
anchor drilling 锚孔钻进
anchor ear 抱箍
anchor element 锚定件
anchor eye 锚孔
anchor failure 锚杆破坏
anchor fitting 锚固装置
anchor fixed length 锚固段
anchor fluke 锚爪
anchor foundation 下锚基础
anchor free length 自由段
anchor gate 锚支闸门
anchor gear 起锚设备
anchor grip 锚杆夹具
anchor grout 锚固浆液；锚定灌浆
anchor grouting 锚固灌浆；封闭灌浆
anchor head 锚头
anchor holding capacity 锚抓力
anchor hole 锚定孔
anchor ice 底冰；锚冰；冰礁地下冰
anchor in rock gallery 山洞式锚碇
anchor jack 撑柱；锚柱
anchor lantern 锚灯
anchor latch 锚锁销
anchor length of bolt 锚杆锚固长度
anchor lift 抓钩提升绳器
anchor light 停泊灯
anchor line 拉线；板线；锚链
anchor line supporting 锚线配套
anchor load monitoring 地锚桩基负荷检查
anchor log 锚固圆木；锚定桩
anchor loop 锚箍；锚圈
anchor lowering 锚索下设
anchor maintenance clause 地锚桩基保养条款
anchor male cone 锚塞
anchor mast 下锚支柱

A

anchor mast stay wire	锚柱拉线
anchor mechanism	锚定机构
anchor member	拉紧构件；固定构件；锚固构件
anchor nut	锚定螺帽
anchor packer	卡瓦封隔器
anchor palm	锚爪
anchor parts	锚件
anchor pattern	底材表面粗糙类型
anchor peg	套栓
anchor picket	锚桩
anchor pier	锚墩
anchor pile	锚桩；抗拔桩
anchor pin	锚销；带动销
anchor pipe	锚管
anchor pitch	锚杆间距
anchor plate	锚定板
anchor point	锚固定；锚准点
anchor pole	撑杆；锚杆
anchor post	锚柱；路线标桩
anchor pressure	锚固压力
anchor projectile	投射锚
anchor prop	锚杆；支承杆
anchor rack	锚架
anchor ring	锚环；圆环面
anchor rod	锚杆
anchor rod retaining wall	锚杆式挡土墙
anchor rope	锚索
anchor screw	地脚螺丝；固定螺丝
anchor seal assembly	锁紧式密封总成
anchor shackle	锚钩环
anchor shank	锚柄；锚身
anchor shoe	锚床
anchor slab	锚定板；锚定板桩墙
anchor slab abutment	锚定板式桥台
anchor slab structure	锚定板结构
anchor slab wall	锚定板挡土墙
anchor slip	固定卡瓦
anchor spacing	锚杆间距
anchor span	锚跨；锚孔
anchor spike	锚尖端
anchor stake	锚橛；锚桩
anchor station	锚地；锚测站；锚地
anchor stay	锚定拉条
anchor stock	锚杆；锚座
anchor stone	底砾
anchor strap	锚定带
anchor stressing trial	锚索张拉试验
anchor strip	锚条
anchor strut	拉杆
anchor support	锚固支架
anchor swivel	锚转环
anchor system	锚固系统
anchor tendon	钢筋束锚杆
anchor test	锚杆试验
anchor testing	锚固试验
anchor tie	锚拉杆；锚条
anchor tower	锚塔
anchor trial	起锚试验
anchor unit	锚固装置
anchor wall	锚定墙；锚固墙
anchor washpipe spear	固定冲管式打捞矛
anchor winch	起锚绞车
anchor windlass	起锚机
anchor wire	锚线
anchor wire pile	锚索桩
anchor works	锚杆工程
anchor yoke	锚索环
anchorage	地锚，锚定作用，锚杆支护
anchorage after grouting	先注浆后插филь
anchorage agent	锚固剂
anchorage and shotcrete support	喷锚支护
anchorage and tensioning point	锚固和拉紧点
anchorage area	锚定区
anchorage bar	锚固杆；定位杆
anchorage beam	锚梁
anchorage bearing	锚座；锚杆
anchorage bend	锚定弯曲
anchorage berth	锚位
anchorage block	锚定块
anchorage bolt	锚杆；地脚螺栓
anchorage bond	锚固黏结
anchorage bond strength	锚固黏结强度
anchorage bond stress	锚固黏结应力；握裹应力
anchorage bulkhead	锚定岸壁；锚定挡墙
anchorage bulkhead retaining wall	锚定板式挡土墙
anchorage by bond	握裹锚定
anchorage cable	锚索
anchorage capacity	锚固能力
anchorage capacity test	锚固力试验
anchorage chamber	锚固洞
anchorage deformation	锚具变形；锚固件滑脱
anchorage device	锚定装置；锚具
anchorage distance	锚固距离
anchorage efficiency test	锚固效率试验
anchorage element	锚具
anchorage end	锚杆端
anchorage failure	锚固失效
anchorage fixing	锚定
anchorage force	锚固力
anchorage interface	锚固界面
anchorage length	锚固长度；锚固段
anchorage length of steel bar	钢筋锚固长度
anchorage loss	锚件应力损失锚固损失
anchorage of dam toe	坝趾加固
anchorage of reinforcement	钢筋锚固
anchorage picket	锚桩；固定水准点
anchorage pier	锚墩
anchorage point	锚固点
anchorage section	锚固段
anchorage shear test	锚固剪切试验
anchorage shoe	锚座，锚靴
anchorage slip	锚具变形
anchorage space	锚地；停泊区
anchorage spud	固船柱
anchorage stone	底砾
anchorage strength	锚固强度
anchorage stress	锚固应力
anchorage system	锚定系统；锚定体系
anchorage zone	锚固区；锚定区
anchorage-prohibited area	禁锚区

anchorages of rock bolts	岩石锚杆锚固
anchorage-shotcrete support	锚喷支护
anchorage-testing apparatus	锚固力试验器
anchor-cables supporting	锚配套电缆
anchored	被锚固的
anchored and jacked pile	锚杆静压桩
anchored bar	锚杆
anchored bar in rock	岩石锚杆
anchored bar in soil	土层锚杆
anchored beam	锚固梁
anchored bearing	锚固支承座
anchored bolt	锚固式锚杆
anchored bolt retaining wall	锚杆挡墙
anchored bond stress	锚固黏结应力
anchored bulkhead	锚定岸壁；锚定挡墙
anchored bulkhead abutment	锚碇板式桥台
anchored bulkhead retaining wall	锚碇板式挡土墙
anchored buoy	锚泊浮标
anchored cable	锚索
anchored concrete grid	加锚杆混凝土格构
anchored concrete pillar buttress	加锚杆混凝土扶壁柱
anchored diaphragm wall	锚定地下连续墙
anchored dune	固定沙丘
anchored earth	锚定土
anchored force	锚固力
anchored foundation	锚拉基础
anchored guy line	锚定绷绳
anchored hydrographic station	定泊水文站；锚定水文站
anchored in rock piles	嵌岩柱
anchored joint	锚定接头
anchored peg	锚桩
anchored pier	锚固墩
anchored plate	锚定板
anchored plate retaining wall	锚定板挡土墙
anchored retaining structure	锚杆支护结构
anchored retaining wall	锚定式挡土墙
anchored retaining wall by tie rods	锚杆式挡土墙
anchored section	锚同段
anchored sheet pile wall	锚定板桩墙
anchored sheet piling	锚定板桩墙
anchored steel trestle	锚固式钢栈桥
anchored structure	锚泊结构
anchored suspension bridge	锚固式悬索桥
anchored tension cable	锚定拉张钢缆
anchored tied-back shoring system	后拉锚定系统
anchored wall	锚定墙；锚固墙
anchored zone's invalidation form	锚定区的失效形式
anchoring	锚定；锚固
anchoring accessories	锚件附件；锚件加固筋
anchoring agent	结合剂；增黏剂
anchoring berth	锚泊泊位；系泊浮筒处
anchoring bond	锚固黏结
anchoring bond force	锚固力
anchoring bracket	锚固联板
anchoring disk	固根盘
anchoring fascine	锚定柴捆
anchoring force	锚固力
anchoring mechanism	锚固机理
anchoring organ	固着器官
anchoring picket	锚柱；路线标桩
anchoring plate	锚板；锚定板
anchoring plug	锚塞
anchoring screen plate	止动螺钉
anchoring strength	锚接强度；锚定强度
anchoring system	铺管驳船抛起锚系统
anchoring technique	锚固技术
anchorite	脉状闪长岩
anchor-mooring	锚系定位
anchor-plate retaining	锚喷支护
anchor-section lining	下锚段衬砌；接触网锚段衬砌
ancient	古代的；古老的；古代
ancient analogs	古类似物
ancient architecture	古建筑
ancient beaches	远古海滩
ancient borderland terrene	古陆缘地表
ancient building	古建筑
ancient channel	古河道
ancient Chinese garden	中国古代园林
ancient continental margin	古大陆边缘
ancient contourite	古等深积岩
ancient cultural relic	文化遗址
ancient earthen architecture	土遗址
ancient elephant	古大象
ancient erosion	古代侵蚀
ancient geothermal system	古地热系统
ancient land	古陆
ancient landform	古地形；旧地形
ancient landslide	古滑坡
ancient magnetic field	古磁场
ancient map	古地图
ancient planation surface	古夷平面
Ancient Pland Path	古栈道
ancient plate	古板块
ancient platform	古地台
ancient river course	古河道
ancient rock slide	古岩石滑坡
ancient sediment	古代沉积
ancient shoreline	古海岸线
ancient soil	古土壤
ancient strand line	古滨线
ancient stream channel	古河道
ancient triple junction	古三联点
ancient turbidite	古浊流层
ancient underground channel	地下古河道
ancient volcano	古火山
ancillary building	附属建筑物
ancillary data	辅助数据
ancillary depocenter	辅助沉积中心
ancillary equipment	外围设备；外部设备；补充设备
ancillary mineral	伴生矿物
ancillary shoring	辅助支撑
ancillary structure	附属构筑物
ancillary work	辅助工作
ancon	悬臂托架；肘托
ancudite	高岭石
ancylite	菱锶铬矿
Ancylus lake time	曲螺湖期
ancyrodella	小锚牙形石属
ancyrognathus	锚颚牙形石属
ancyroides	拟锚牙形石属

A

ancyrolepis 锚鳞牙石属
ancyropenta 五角锚牙形石属
andaluzite 红柱石
Andaman islands 安达曼群岛
Andaman ophiolite 安达曼蛇绿岩
Andaman-Nicobar island arc 安达曼-尼科巴岛弧
Andean igneous rocks 安第斯火成岩
Andean orogenesis 安第斯山型造山作用
Andean tectonism 安第斯构造运动
Andean type margin 安第斯型边缘
andelatite 二长安山岩
andendiorite 英辉闪长岩
andengranite 云闪花岗岩
andennorite 云闪苏长岩
andept 火山灰始成土
Anderson-Boyes double-ended ranging drum shearer-loader 安得逊-波依斯型可调高双滚筒采煤机
andersonite 水碳钠钙铀矿
Anderson's method 安德逊法
Anderson's model 安德逊模型
Anderton shearer 安德尔腾型滚筒采煤机
Anderton shearer loader 安德腾型滚筒采煤机
Andes 安第斯山脉
andes basalt 玄武安山岩
andesilabradorite 安山拉长岩
andesine 中长石
andesine anorthosite 中长斜长岩
andesine basalt 中长石玄武岩
andesinfels 中长角斑岩
andesinite 中长岩
andesite 安山岩；中长石
andesite basalt 安山玄武岩
andesite lava 安山熔岩
andesite line 安山岩线；马绍尔线
Andesite porphyry 安山斑岩
Andesite tuff 安山凝灰岩
andesite-basalt 安山玄武岩
andesite-porphyry 安山斑岩
andesite-tephrite 安山碱玄岩
andesite-tuff 安山凝灰岩
andesitic 安山岩的
andesitic arc volcanism 安山岛弧火山
andesitic texture 安山结构
andesitic tuff 安山凝灰岩
andesitoid 似安山岩
ando soil 暗包土
andorite 硫锑银铅矿
andosoil 火山灰土壤
andosol 暗色土；火山灰土
Andou Formation 安陡组
andradite 钙铁榴石；钙铁榴石宝石
Andresen pipette 安德瑞逊型吸管
anduoite 安多矿
Anduomaima Formation 安多买马组
anelastic 滞弹性的；弹性后效
anelastic absorption 滞弹性吸收
anelastic attenuation 滞弹性衰减
anelastic behaviour 滞弹性性能；滞弹行为
anelastic deformation 滞弹性变形
anelastic property 滞弹性

anelastic strain recovery 滞弹性应变恢复
anelasticity 滞弹性
anemoarenyte 风成砂岩
anemoclast 风成碎屑
anemoclastic 风成碎屑的
anemoclastic rock 风成碎屑岩
anemoclastics 风成碎屑岩
anemoclinometer 风斜表
anemogenic sediment 风成沉积物
anemolite 风成洞穴堆积；不规则石钟乳
anemosilicarenite 风成硅质砂屑岩
anemousite 三斜霞石；低硅钠长石
anergy 无效能
Angara shield 安加拉古陆
Angaraland 安加拉古陆
Angarides 安加拉古陆
Angaris 安加拉地盾
Angarophycus 安加拉藻属
angle and distance measurement 角度和距离测量
angle averaging 角平均法
angle azimuth indicator 倾斜度及方位指示器
angle bar 角铁；角钢；角材
angle bead 角条
angle beam 角铁梁
angle beam method 斜射法
angle belt 二角皮带；三角皮带传动
angle bender 钢筋弯曲机
angle blade 角刀片，L形刀片；角刮板，L形刮板
angle bolt truss 倾斜锚杆桁架
angle bond 啮合隅角
angle brace 水平斜撑；角撑
angle bracket 角形托架；角钢撑架
angle branch 弯管；肘管
angle brick 角砖
angle build 增斜
angle build up 造斜
angle building 造斜
angle chimney 拐角烟囱
angle cleat 连接角钢；短角钢
angle closing error of traverse 导线角度闭合差
angle cock 折角塞门
angle column 角柱
angle crane 斜座起重机
angle cross-cut 与矿层走向斜交的横巷
angle cut 斜孔掏槽；楔形掏槽；斜角切割
angle cutter 角铁切割机；角铣刀
angle dozer 斜板推土机
angle drift 锥铰刀；锥形心轴
angle drive 角驱动
angle drop 降斜
angle dropping 降斜；减斜
angle elevation 上升角
angle face 对角工作面；斜工作面
angle fitting 弯头
angle frame 角钢框
angle handle 开关拐柄
angle head 角机头
angle holding 稳斜
angle holding assembly 稳斜钻具组合
angle hole 定向钻孔

angle hole drilling 倾斜钻孔，钻斜孔
angle in 下钻时的井斜角
angle increment 角增量
angle iron 角铁；角钢
angle iron bracket 角铁支架；角铁托架
angle iron frame 角铁框架
angle iron stiffening 角铁支撑；角铁加固
angle iron track 角铁滑轨
angle jaw tongs 弯嘴钳
angle joint 角接；角接接头
angle measurement 测角
angle measurement grid 量角格网
angle measurement network 测角网
angle meter 量角仪
angle modulation 调角
angle momentum 角动量
angle mortar 斜射臼炮；带槽臼炮；角臼炮
angle needle valve 弯角针形阀
angle noise 角度噪声
angle number 角度号码
angle of advance 提前角
angle of altitude 仰角；高度角
angle of apparent internal friction 视内摩擦角
angle of application 作用角；加力角
angle of approach 接近角
angle of arrival 到达角落角
angle of aspect 视线角
angle of attachment 连接角
angle of attack 入射角；迎角；攻角；冲角
angle of azimuth 方位角
angle of backing off 刃后角
angle of band 倾斜角；侧倾角
angle of base friction 基底摩擦角
angle of bedding 层面倾角；地层倾角
angle of bench slope 台阶坡面角
angle of bend 弯度；弯角弯头角度
angle of bevel 坡口角度
angle of bite 轧入角
angle of branch 支河角；分汊角
angle of break 崩落角；裂缝角；塌陷角
angle of breakage 爆破漏斗张开角
angle of brush displacement 电刷位移角
angle of caving 切顶角
angle of circumference 圆周角
angle of contact 接触角；湿润角
angle of convergence 收敛角
angle of coverage 视场角；覆盖角
angle of critical deformation 移动角
angle of crossing 交会角
angle of current 流向角
angle of curvature 曲度角；曲率角
angle of cut off 截止角
angle of cutting 切削角
angle of cutting edge 刀前角
angle of declination 偏角
angle of deflection 偏斜角
angle of deformation 变形角
angle of departure 分离角出角
angle of depression 俯角低降角
angle of deviation 偏向角

angle of diffraction 衍射角
angle of dig 掺掘角；切割角
angle of dilatancy 剪胀角，扩容角
angle of dip 倾斜角；入射角磁偏角
angle of displacement 位移角
angle of divergence 发射角
angle of downwash 冲淘角；淘刷角
angle of draw 陷落角，塌陷角；边界角；落角
angle of eccentricity 偏心角
angle of elevation 仰角；高度角
angle of emergence 出射角
angle of entry 入射角
angle of external friction 外摩擦角
angle of extinction 消光角
angle of field 视场角
angle of field of vision 视场角
angle of final pit slope 最终边坡角
angle of firing 射角
angle of flare 承口角
angle of fracture 断裂角；破裂角
angle of friction 摩擦角；静摩擦角；垂直倾斜角
angle of friction for fractured rock 破裂岩石的摩擦角
angle of friction for internal friction 内摩擦角
angle of friction for mine fill 矿山充填料的摩擦角
angle of full subsidence 充分采动角
angle of gradient 坡度角；倾斜角；仰角
angle of hade 断层余角；伸角
angle of heel 横倾角
angle of helix 螺旋角；螺旋线升角
angle of highwall 顶帮坡角
angle of incidence 入射角
angle of inclination 倾角
angle of incurvature 凹角
angle of internal friction 内摩擦角
angle of intersection 交角
angle of intromission 全透射角
angle of lag 滞后角移后角
angle of lead 超前角前置角
angle of lean 倾角
angle of maximum subsidence 最大下沉角
angle of maximum subsidence velocity 最大下沉速度角
angle of minimum deviation 最小偏差角；最小偏向角
angle of minimum resolution 最小分辨角
angle of missetting 误置角
angle of nadir 天底角
angle of natural repose 自然休止角
angle of neutral plane 中性角
angle of nip 咬入角；错入角
angle of nonslip point 临界角
angle of obliquity 倾角
angle of orientation 方位角
angle of parallax 视差角
angle of pit 边帮角
angle of pit wall 边帮角
angle of pitch 俯仰角；坡度角
angle of polarization 偏振角；极化角
angle of pressure 压力角
angle of pull 滑动偏移角
angle of rake 后倾角
angle of reach 触及坡角

angle of refraction	折射角
angle of repose	休止角;安息角
angle of repose test	休止角试验
angle of resistance	抗滑角
angle of rest	静止角;休止角
angle of retard	减速角;推迟角
angle of roll	滚动角
angle of rolling friction	滚动摩擦角
angle of roof	屋面斜角
angle of rotation	转动角;旋转角
angle of run	去流角
angle of rupture	破裂角;土坡破裂角
angle of scattering	散射角
angle of shear	剪变角
angle of shear dilation	剪胀角
angle of shear resistance	抗剪角
angle of shearing resistance	抗剪角
angle of shearing strength	剪切强度角
angle of skew	削角斜角
angle of skew back	侧倾角
angle of skin friction	表面摩擦角
angle of slewing	转角
angle of slip	滑移角;斜坡角
angle of slope	倾角;坡度角;边坡角
angle of slope of the terrain	地面坡度
angle of solar irradiance	太阳入射角
angle of splitting	劈裂角
angle of spoil	排土场边坡角
angle of spread	展开角
angle of stress dispersion	应力扩散角
angle of strike	走向角;走向方位角
angle of subsidence	边界角陷落角
angle of supercritical mining	超临界开采角
angle of surcharge	物料安息角
angle of throw	投掷角
angle of tilt	倾角;倾斜角
angle of torsion	扭转角
angle of total reflection	全反射角
angle of trim	平衡角
angle of trimming	纵倾角
angle of true internal friction	真内摩擦角;天然内摩擦角
angle of turning flow	水流转角
angle of twist	扭角
angle of ultimate stability	最后稳定角
angle of unconformity	不整合角
angle of underlay	倾斜余角
angle of view	视场角;观测角
angle of visibility	视界角
angle of wall friction	墙摩擦角;壁面摩擦角
angle of wetting	湿润角
angle of winning-block	采块角
angle of working slope	工作帮坡角
angle of wrap	围包角;包容角;接触角
angle of yaw	偏航角
angle opening	斜向开切
angle out	起钻时的井斜角
angle pass	角钢孔型
angle pedestal bearing	斜架轴承
angle phase digital converter	轴角相移数字转换器
angle pier	角墩
angle pile	斜桩
angle pipe	角管;弯管
angle plate	角板
angle plug	弯插头
angle point	井斜角测量点
angle press	角压塑机
angle probe	斜探头
angle protractor	量角规
angle pulley	导轮;导辊;转向滑轮
angle purlin	角檩;角钢檩条
angle rafter	角椽;角铁椽
angle resolution	角分辨率
angle room	斜开煤房
angle rule	角尺
angle seam	角缝
angle seat	角钢支座
angle section	角形截面;角钢
angle sheave	导轮;转向滑轮
angle shot mortar test	斜射白炮试验;角白炮试验
angle shut-off valve	直角截止阀
angle station	输送机水平转弯处
angle steel	角钢
angle stop	角形挡车器
angle strain	剪应变
angle strut	角钢支柱
angle support	弯角件
angle table	角钢托座;三角桌;角撑架
angle test tube	氟氢酸测斜管
angle thermometer	直角温度计
angle tie	斜撑木;角铁柱杆
angle to digit converter	角数转换器
angle tolerance	角度公差
angle transducer	角传感器
angle trough	弯形转向槽
angle true net	等角网
angle unconformity	角度不整合
angle valve	角阀
angle wheel	斜齿轮
angle-bending machine	钢筋弯折机
angle-cut blasting	中心掏槽爆破
angled	倾斜的
angled aluminum steady arm	角型铝定位器
angled cut	巷道中心斜炮眼掏槽
angled drilling	倾斜钻炮眼;定向钻进
angled loading chute	V形截面的装载溜槽
angled pile	斜桩
angled type adaptor	弯接头;角度接头
angledozer	侧推式推土机
angledozing	用侧推式推土机推土
anglemeter	测角器;倾斜仪
angle-mortar test	定向白炮爆炸试验
angle-reciprocating compressor	缸体V形排列往复式压缩机
angles back to back	背对背角钢
angle-shot method	斜射白炮试验法;角白炮试验法
anglesite	铅矾;硫酸铅矿
angle-table	牛腿;托座;角撑架
angle-true net	吴尔夫网;吴氏网
Anglian glacial stage	安格利亚冰期
angling blade	可测斜式刮板

angling downwards 向下倾斜
angling dozer 斜板推土机
angling hole 倾斜炮眼
angling snubbing hole 倾斜掏槽炮眼
Anglo-Chinese style garden 中英混合式园林
Angola abyssal plain 安哥拉深海平原
angrite 钛辉无球粒陨石
anguclast 角碎屑
angular 棱角状的；有棱角的
angular acceleration 角加速度
angular accuracy 角精度
angular adjustment 角度调整
angular advance 提前角
angular altitude 仰角；高度角
angular aperture 孔径角
angular ball bearing 径向球轴承
angular belting 斜角皮带传动装置
angular bevel gear 斜交伞齿轮
angular bisector 角平分线
angular blocky structure 角块状结构
angular breadth 角宽度
angular brightness 角亮度
angular capital 角形柱头；斜角式柱头
angular change 角变化
angular coarse aggregate 有棱角粗骨料
angular cobble 碎石
angular conglomerate 棱角状砾岩
angular contact 斜接
angular contact ball bearing 向心止推滚珠轴承
angular contact bearing 向心推力轴承
angular coordinate 角坐标
angular cross bedding 倾斜交错层
angular deflection 角度偏移
angular deformation 角变形
angular dependence 角关系
angular derivative 角微商
angular deviation 角偏向
angular discordance 角度不整合
angular displacement 角位移；角旋转
angular distance 角距离
angular distortion 角挠曲；角度变形
angular divergence 角偏向，角误差
angular divergence method 角发散法；角度不整合法
angular domain 角域
angular drainage pattern 角状水系
angular epicentral distance 震中角距
angular error 角度误差
angular field 视场角
angular field of view 像场角
angular fine gravels 角砾
angular fishplate 角形鱼尾板
angular fold 尖棱褶皱
angular force 角动力
angular fragment 角碎块；角碎片
angular frequency 角频，角频率
angular gap 角形裂隙
angular gear 斜齿齿轮，人字齿轮
angular grain 多角形颗粒；角粒
angular grain chipping 角砾碎屑
angular grained 尖角颗粒状的

angular gravel 碎石；角砾
angular gravel soil 角砾土
angular harmonic motion 角谐运动
angular hole 斜钻孔；斜炮眼
angular instrument 测角仪
angular interstice 棱角状空隙
angular jointing 斜节理
angular magnification 角度放大率
angular measurement 角测
angular moment 角矩
angular momentum 角动量；动量矩
angular motion 角动
angular observation 角度观测
angular orientation 角定向；角取向
angular oscillation 角摆动
angular parallax 角视差
angular particle 多角形颗粒；棱角颗粒
angular pebbles 角砾
angular perspective 成角透视
angular phasing 角相位
angular pitch 倾斜角
angular point 角点
angular pole 角杆
angular Position 角度位置
angular projection 角投影
angular rate 角速度
angular resolution 角分辨率
angular rotation 角旋转
angular sand 角粒砂；棱角状砂
angular setting 角度调整
angular shear 角度剪切
angular shear strain 角剪应变
angular slip 角位移
angular spacing 角度间隔
angular speed 角速度
angular spreading 角扩散
angular step 步进角度
angular strain 角应变
angular surface 棱角表面
angular test 弯曲试验；挠曲试验
angular tolerance 角容限
angular transformation 角变换
angular travel 角行程
angular unconformity 角度不整合
angular unit 角度单位
angular variance 角方差
angular variation 角变化
angular velocity 角速度
angular velocity between plates 板块间的角速度
angular vertex 角的顶点
angularity 棱角度；有角；角状
angularity chart 棱角度对比图
angularity correction 角度校正
angularity factor 棱角度因素
angulate 有角的
angulate drainage 角状水系
angulate drainage pattern 角状水系
angulated 有角的；含角形颗粒的
angulation measurement 角度测量
angulator 变角器

angulisporites	菱环孢属
angulodus	角牙形石属
angulolithina	角颗石
angulometer	量角器
Angus Smith composition	安格斯-史密斯防腐蚀涂料
Angzanggou Formation	肮脏沟组
anh	无水的
anharmonic coupling	非谐耦合
anharmonic curve	交比曲线；非调和曲线
anharmonic force	非谐力
anharmonic ratio	非谐比例
anharmonic resonance	非谐共振
anharmonic vibration	非谐振动
anharmonicity	非谐性
Anhe Formation	安河组
anhedral crystal	他形晶
anhedral grain	他形晶粒
anhedral granular	他形粒状
anhedral texture	他形结构
anhedritite	硬石膏
anhedron	他形晶
anhistous	无结构的
anhydration	干化；脱水
anhydric	无水的
anhydride	酐；酸酐；脱水物
anhydrite	硬石膏；无水石膏
anhydrite cement	无水石膏水泥
anhydrite formation	硬石膏层
anhydrite precipitation	硬石膏沉淀
anhydrite rock	硬石膏岩
anhydrite sheath	硬石膏鞘
anhydrock	硬石膏岩
anhydroferrite	赤铁矿
anhydrokaolin	无水高岭土
anhydrone	无水高氯酸镁
anhydrous	无水的
anhydrous alcohol	无水酒精
anhydrous ammonia	无水氨
anhydrous calcium sulfate	硬石膏；无水硫酸钙
anhydrous facies	无水相
anhydrous gypsum	无水石膏
anhydrous lime	无水石灰
anhydrous method	无水法
anhydrous period	无水时代
anhydrous phase	无水相
anhydrous plaster	无水粉饰；无水石膏粉饰
anhydrous sodium carbonate	无水碳酸钠
anhydrous solvent	无水溶剂
anhydrous sulphate of calcium	无水硫酸钙
anhydrous Tin	脱水锡
anhyetism	缺水，无水；缺雨区缺雨性
anhysteretic	非磁滞；非滞后的
anhysteretic magnetization	无滞磁化
anhysteretic remanent magnetization	非滞后剩余磁化
anidex	阿尼迪克斯纤维
anidrite	硬石膏
anil	靛蓝
animal black	兽炭黑；骨炭
animal charcoal	动物煤
animal debris	动物残余
animal-drawn traffic	圆周拉应力
animated mapping	动画制图
animated perspective	动态鸟瞰
animated steering	动画引导
animated walk-through	动态漫游
animation sequence	动画序列
animikite	铅银砷镍矿
Anisian	安尼西阶
Anisic stage	安尼西阶
Anisodesmic structure	蛤稳变异构造；异结合结构
anisoelastic	非弹性的
Anisograptidae	反称笔石科
Anisograptus	反称笔石属
anisomeric rock	复杂结晶岩
anisomerous texture	不等粒结构
anisometric	不等轴的；不等粒的
anisometric breccia	不等粒角砾岩
anisometric crustal	非等轴晶体
anisometric deposit	不等粒沉积
anisometric growth	不等量生长；非等比生长；非同形生长
anisotropic	各向异性的
anisotropic absorption	非均质吸收
anisotropic apparent rotation angle	各向异性视旋转角
anisotropic aquifer	各向异性含水层
anisotropic body	非均质体；异向性体
anisotropic clay	各向异性黏土
anisotropic coefficient	各向异性系数
anisotropic coke	非均质焦炭；异向性焦炭
anisotropic conductivity	各向异性传导率；各向异性电导率
anisotropic consolidation	各向不等压固结；各向异性固结
anisotropic consolidation stress ratio	非均等固结应力比
anisotropic crystal	各向异性晶体
anisotropic damage	各向异性损伤
anisotropic elastic ground	各向异性弹性地基
anisotropic fabric	非均质组构；各向异性组构
anisotropic formation	各向异性地层
anisotropic friction	各向异性摩擦力
anisotropic hardening	各向异性硬化
anisotropic hardening model	各向异性硬化模型
anisotropic index	各向异性指数
anisotropic layer	非均质层；各向异性层
anisotropic mass	各向异性物质；各向异性体
anisotropic material	各向异性材料
anisotropic media	各向异性介质
anisotropic medium	各向异性电介质
anisotropic permeability	各向异性渗透性；各向异性渗透率
anisotropic porous foundation	各向异性多孔基础
anisotropic rock	非均质性岩石
anisotropic seepage	各向异性渗流
anisotropic softening model	各向异性软化模型
anisotropic soil	各向异性土；非均土
anisotropic strength criterion	各向异性强度准则
anisotropic stress	各向不等压应力
anisotropic stress loading	各向异性应力加载
anisotropic stress state	各向异性应力状态
anisotropic substance	各向异性物
anisotropy	各向异性；非均质性
anisotropy coefficient	各向异性系数

anisotropy dipmeter 各向异性地层倾角仪
anisotropy energy 各向异性能
anisotropy factor 各向异性因数
anisotropy index 各向异性指标
anisotropy of rock 岩石各向异性
anisotropy paradox 各向异性佯谬；各向异性反常
anisotropy parameters 各向异性参数
anisotropy ratio 各向异性率
anisotropy swell 各向异性膨胀
Ankangian 安康阶
ankangite 安康矿
ankaramite 橄榄辉玄岩
ankaranandite 紫苏碱长正长岩
ankaratrite 黄橄霞云岩
ankerite 铁白云石
ankeritization 铁白云石化
Ankou Formation 安口组
Anlong Limestone 安龙石灰岩
Anmen Beds 庵门层
Anmin Fomation 安民组
anmoor 矿质还原层；青灰色潜有层；矿质湿土
Annanba Group 安南坝群
annexed leveling line 附合水准路线
annexed traverse 附合导线
Anning Formation 安宁组
annite 铁云母；羟铁云母
annivite 铋铜矿
annual accumulation of earthquakes 年地震积累数
annual accumulation of sediment 年河泥淤积量；年淤泥量
annual advance 年进度；年推进
annual amount of sedimentation 泥沙年淤积量
annual amplitude 年变幅；年振幅；年较差
annual erosion 年侵蚀量
annual evaporation 年蒸发量
annual flood 年最大流量；年洪水量；年最大洪峰
annual flood peak series 年洪峰系列
annual flow 年径流；年流量
annual fluctuation 年波动；年变化；年变幅
annual frost zone 年冻结带；年冻融带
annual groundwater recharge 地下水年补给量
annual growth layer 年生长层
annual growth ring 年轮
annual heat budget 年热收支量
annual lamination 年纹理
annual layer 年层
annual maximum flood 年最大洪水量
annual maximum magnitude of earthquake 年最大震级
annual mean discharge 年平均流量
annual mean sea level 年平均海面
annual mean tidal cycle 年平均潮汐周期
annual natural replenishment 年度天然补给量
annual normal flow 年正常径流
annual output of mine 矿山年产量
annual peak load 年峰荷
annual production capacity 年生产能力
annual progress 年进度；年进尺
annual railway plan 铁路年度计划
annual temperature variation 年温度变化
annually flooded plain 季节性泛滥平原

annually frozen layer 年冻结层
annular anomaly 环状异常
annular auger 环孔钻
annular ball bearing 径向轴承
annular bit 环状钻头
annular borer 环状钎头
annular boring 环孔钻
annular cut mining method 环形工作线采矿法
annular drainage 环状水系；环形排水系统
annular drainage pattern 环状水系
annular eclipse 环蚀
annular flow 环空液流；环状流
annular fluid 环形流体
annular kiln 环形窑；轮窑
annular mud velocity 环空泥浆速度
annular nozzle 环状喷嘴
annular space 环状回填注浆间隙
annular support 环形支架；拱形支架
annulated bit 环形钻头
annulithus 环颗石
annulus pressure 环空压力
annulus profile 环形剖面
annulus test 环空试压
annulus top up cementation 环空满眼注水泥
anogene 火成的；深成变质的
anomalous aftershock decay 异常余震衰减
anomalous drainage pattern 不规则水系
anomalous earthquake 异常地震
anomalous stream sediment 异常河流沉积物
anomalous structure 反常结构
anomalous surface 异常面
anomalous thermality 热异常
anomalous underflow 异常地下流
anomalous upheaval 异常隆起
anomalous upper mantle 异常上地幔；异常上部地函
anomalous watershed 异常分水岭
anomalous zone 异常带
anomaly area 异常区
anomaly axial zoning 异常轴向分带
anomaly gradient zone 异常梯度带；异常梯阶带
anomaly high 异常高区
anomaly horizontal gradient 异常水平梯度
anomaly horizontal zoning 异常水平分带
anomaly in geomagnetism 地磁异常
anomaly intensity 异常强度
anomaly interpretation 异常解释
anomaly longitudinal zoning 异常纵向分带
anomaly map 等异常线图；异常等值线图；距平线图
anomaly of geological landscape in photo 相片地质景观异常
anomaly of gravity gradient 重力梯度异常
anomaly of seismic intensity 地震烈度异常
anomaly threshold 异常阈值；异常下限
anomaly transversal zoning 异常横向分带
anomaly trend 异常趋势；异常走向
anomite 褐云母
anorganic 非有机的
anorganogene 无机成因的
anorganolite 无机岩
anorganolith 无机岩

anorge 不有机性
anormal anticlinorium 逆复背斜
anorogenic 非造山运动的
anorogenic granite 非造山花岗岩
anorogenic period 非造山期
anorogenic pluton 非造山深成岩体
anorogenic time 非造山期
anorthic system 三斜晶系
anorthite 钙长石
anorthitissite 钙长角闪石岩
anorthitite 钙长岩
anorthoclase 歪长石；钠微斜长石
anorthoclase basalt 歪长玄武岩
anorthoclasite 歪长岩
anorthophyre 歪长斑岩
anorthose 斜长石
anorthosite 斜长岩
anorthosite norite troctolite 月岩基性岩套；月球斜长岩苏长岩橄长岩系
anorthositic gabbro 钙长辉长岩
anorthositization 斜长岩化
anorthosyenite 歪长正长岩
Anothermic 大洋平流层
Anqing Conglomerate 安庆砾岩
Anren Formation 安仁组
Ansha Sandstone 安砂砂岩
Anshan Group 鞍山群
ANSYS software 有限元软件
antaciron 硅铁合金
antarctic 南极的；南极区
antarctic air mass 南极气团
antarctic anticyclone 南极反气旋
antarctic bottom water 南极底层水
Antarctic Circle 南极圈
Antarctic circumpolar current 南极环流；西风环流；南极绕极流
antarctic climate 南极气候
antarctic convergence 南极辐合带
antarctic front 南极锋
Antarctic Geological Drilling 南极地质钻探
antarctic glacial isostasy 南极冰川地壳均衡
antarctic glacier 南极冰川
antarctic ice cores 南极冰心
antarctic ice sheet 南极冰盾
antarctic intermediate water 南极中层水
antarctic ionized layer 南极电离层
Antarctic Ocean 南冰洋
Antarctic Oscillation 南极涛动
antarctic ozone distribution 南极臭氧分布
antarctic plate 南极洲板块
Antarctic Pole 南极
antarctic realm 南极区
antarctic region 南极区
Antarctic Research Programmes 南极研究计划
antarctic sea floe 南极海冰
antarctic telluric magnetic force 南极地磁
Antarctica 南极洲
Antarctica meteorite 南极洲陨石
Antarctica plate 南极板块
antarcticite 南极石

antecedent consequent river 先成顺向河
antecedent consequent stream 先成顺向河
antecedent discharge 前期流量
antecedent drainage 先成水系
antecedent gorge 先成峡
antecedent moisture 前期水分，前期土壤湿度
antecedent precipitation 前期降水量
antecedent precipitation index 前期降水指数
antecedent river 先成河
antecedent sedimentary basin 先成沉积盆地
antecedent soil moisture 雨前土湿度；前期土湿度
antecedent stream 先成河
antecedent valley 先成谷
antecedent wetness 前期含水量
antechamber 沉沙室
antechamber pillar 矿房间矿柱；煤房间煤柱
anteclise 台背斜；陆背斜
anteconsequent river 先成顺向河
anteconsequent stream 顺向先成河
Antediluvial 前洪积世
anteklise 台背斜
ante-stream 先成河
anthophyllite 直闪石
anthracides 可燃矿产
anthraciferous coal 无烟煤
anthracite 无烟煤；白煤
anthracite basin 无烟煤煤田
anthracite coal 无烟煤
anthracite culm 无烟煤粉；无烟煤中的废渣
anthracite fines 无烟煤粉煤
anthracite mine 无烟煤矿
anthracite preparation 无烟煤洗选
anthracitic coal 无烟煤
anthracitic column mining 煤柱开采
anthracitization 无烟煤化作用
anthracography 煤相研究学
anthracolite 无烟煤
Anthracolithic 大石炭系
anthracolithic period 大石炭纪
anthracolithic system 大石炭系
anthracolithization 煤化；炭化
anthracology 煤岩学
anthraconite 沥青灰岩；黑方解石
anthraco-silicosis 煤矽肺病
anthracosis 煤肺病；矿工黑肺病
anthrafilt 过滤用无烟煤
anthrafilt filter 无烟煤介质过滤器
anthrafine 无烟煤细末
anthragenesis 煤化作用；碳化作用
anthraphyre 安斯拉斑岩
anthraxolite 碳沥青
anthraxylon 镜煤；凝胶物质；结构镜质体
anthraxylon attrital coal 镜煤质细屑煤；镜煤质暗煤
anthraxylon coal 镜煤；凝胶化煤
Anthrinoid group 炭质组；高镜组；高变镜质组
anthropic epipedon 耕作表层；人为表层
anthropic soil 人为土壤；熟土
Anthropogene 人生代；人类纪
anthropogenic discharge 人为排放
Anthropogenic Era 灵生代

anthropogenic erosion 人为侵蚀
anthropogenic factor 人为因素
anthropogenic karst 隐伏岩溶；隐伏喀斯特；耕土伏岩溶
anthropogenic process 耕作土壤发生过程；人为过程
anthropogenic soil 人为土壤；耕作土壤
anthropogenic waste 人为废弃物
anthropogenic-alluvial soil 灌淤土；人为冲积土壤；耕作冲积土壤
anthropogeography 人类地理学
Anthropognene 人类纪
anthropolithic age 石器时代
anthropostratigraphy 人类地层学
anthrosphere 人类生命圈
anti abrasive 耐磨损的
anti blocking agent 防粘连剂
anti clogging 防堵塞的
anti condensation lining 防凝水内衬
anti corrosion insulation 防腐蚀绝缘层
anti corrosive agent 防腐剂
anti corrosive oil 防锈油
anti corrosive paint 防锈漆
anti-abrasive 抗磨损的
anti-acid 抗酸的
anti-caking 抗结块的
anti-carburizer 渗碳防止剂
anticenter of earthquake 震中对点；反震中
anticentre 震中对跖点
anticentre of earthquake 震中对点
anti-chatter spring 抗震弹簧
anticipated earthquake 预期地震
anticipated load 预期负荷；预期负载
anticlinal 背斜的
anticlinal arch 背斜拱起
anticlinal axis 背斜轴
anticlinal belt 背斜带
anticlinal bend 背斜弯曲
anticlinal bowing 背斜鼻，背斜突出部分
anticlinal bulge 背斜脊的厚度和高程
anticlinal closure 背斜闭合度
anticlinal core 背斜核部
anticlinal crack 背斜张力裂纹
anticlinal crest 背斜脊
anticlinal dome 背斜穹隆
anticlinal fault 背斜断层
anticlinal flank 背斜翼
anticlinal flexure 背斜挠曲
anticlinal fold 背斜褶皱
anticlinal fold zones 背斜褶皱带
anticlinal folding 背斜褶皱
anticlinal high 背斜高点
anticlinal hinge region 背斜脊线区
anticlinal leg 背斜腿；背斜翼
anticlinal limb 背斜翼
anticlinal meander 背斜河曲
anticlinal mountain 背斜脊
anticlinal nose 背斜鼻；背斜突出部分；构造鼻
anticlinal nucleus 背斜核部
anticlinal pool 背斜油气藏
anticlinal recumbent 伏卧背斜层

anticlinal ridge 背斜脊
anticlinal rise 背斜隆起
anticlinal river 背斜河
anticlinal spring 背斜泉
anticlinal strata 背斜层
anticlinal stream 背斜河
anticlinal structure 背斜构造
anticlinal tension crack 背斜张力裂缝
anticlinal theory 背斜理论
anticlinal trap 背斜圈闭
anticlinal trend 背斜走向
anticlinal turn 背斜顶部，背斜鞍部
anticlinal uplift 背斜隆起
anticlinal upwarp 背斜翘起
anticlinal valley 背斜谷
anticlinal zone 背斜带
anticlinaloid 假背斜层
anticline 背斜；背斜层
anticline axis 背斜轴
anticline overthrust 背斜逆掩断层
anticline structure 背斜构造
anticlines 背斜
anticlinoria 复背斜
anticlise 台背斜
anti-coagulant 抗凝剂
anticoagulation 阻凝作用
anti-coal-breakage device 煤炭防碎装置
anticonsequent stream 逆向河
anti-corrosion agent 抗腐蚀剂
anti-corrosion by seawater 防海水腐蚀
anti-corrosive precaution 抗侵蚀性措施
anti-creep 防爬；防蠕动
anti-creep equipment 防滑装置；防止蠕动装置
anti-creep heap 挡车堆
anti-creep strut 防爬支撑
anticreeper 防爬器；反滑行装置
anti-detonation advance steel timbering 抗崩倒超前钢支架
antidetonation timbering 抗爆震支架
anti-detonator 防爆剂
antidip 反倾斜
anti-dip layered accumulation body 反倾层状堆积体
antidip layered slope 反倾层状结构边坡
antidip stream 反倾斜河；逆向河
antidune 逆行沙丘；反向沙丘
anti-dune 逆波痕
antidune cross bedding 逆行沙波斜层理
antidune movement 逆向沙丘运动
antidunes 逆行沙丘；反向沙丘；背转沙丘
anti-earthquake design 抗震设计
anti-epicenter 震中对跖点
antifenite pegmatite 反纹长石伟晶岩
antiform 背斜型构造；背形构造
antiformal anticline 背形背斜
antiformal axial trace 背形轴迹
antiformal first fold 第一期背形褶皱
antiformal flexure 背形挠曲
antiformal fold 背形褶皱
antiformal limb 背形翼
antiformal structure 背形构造
antiformal syncline 背斜型向斜

antigorite	叶蛇纹石
antigravity filtration	抗重过滤
antigravity screen	抗重筛
anti-heave measures	防止冻胀措施
antihomocline	非均斜层
antimonite	辉锑矿；亚锑酸盐
antimony mineral	锑矿类
antiperthite	反条纹长石
antiperthitic texture	反纹结构
antipertite	反纹长石
antiphase	反相位；逆相
anti-plane shear crack	反平面剪切裂纹
anti-seep diaphragm	防渗膜；截渗墙
anti-seepage mark number	抗渗标号
anti-seepage well	防渗井
antiseismic	抗震的
anti-seismic calculation	抗震计算
anti-seismic construction	抗震建筑；抗震构造
anti-seismic design	抗震设计
anti-seismic design of main power buildings	主厂房抗震设计
antiseismic engineering	抗震工程
anti-seismic joint	抗震缝
antiseismic structure	抗震建筑；抗震结构
anti-shear chamber	抗剪洞
anti-shearing	抗剪断
anti-skid-type tunnel without cover	抗滑明洞
anti-slew	防止旋转；防止回转
anti-slide pile	抗滑桩
anti-slide retaining	抗滑支挡
anti-slide section	抗滑段
anti-sliding key	抗滑键
anti-sliding pile	抗滑桩
antisliding stability	抗滑稳定性
antisliding stability of dam foundation	坝基坑滑稳定性
antislip	防滑
antisohite	云斑闪长岩；黑斑云闪长岩
antistress	反应力
antistress mineral	反应力矿物
anti-swelling	防膨胀
antithetic block fault	对偶块状断层
antithetic block rotation	对偶断块旋转
antithetic conjugate Riedel shears	里德尔对偶共轭剪切
antithetic drag	逆牵引
antithetic fault	反向断层
antithetic shear	对偶剪切
antithetic slip cleavage	相向滑动劈理
antithetic step fault	对偶阶状断层
anti-vibrating stability	抗震稳定性
antivibration	抗震；防震；阻尼
anti-vibration device	防震器
Antler orogeny	安特勒造山运动
antofagastite	水氯铜矿
antozonite	呕吐石
antracen	树脂质沥青
Anxiliao Beds	安溪寮层
Anyangou Formation	暗延沟组
Anyao Formation	鞍腰组
Anyuan Formation	安源组
Anyuan Group	安源群
Anziping Beds	庵子坪层
Aocheng Formation	敖城组
Aodi Formation	澳底组
Aogaosituquan Formation	敖高四土泉组
Aogen Gneiss	奥根片麻岩
Aojiatan Group	奥家滩群
Aoton Formation	坳头组
Aoxi Formation	敖溪组
apachite	闪辉响岩
Apada'erkang Formation	阿帕达尔康组
apaneite	磷灰霞石岩
A-parameter	孔隙压力系数 A
apatite	磷灰石
apatite iron ore	磷灰铁矿石
apatite-magnetite formation	磷灰石-磁铁矿建造
apatitolite	磷灰石岩
apatognathus	犁颚牙形石属
apatotrophic lake	有生物微咸湖
apenninic flysch	亚平宁复理层
aperture	隙缝；壁孔；裂缝开度；孔径，开口
aperture diameter	孔径
aperture law	采矿用地法
aperture of culvert	涵洞孔径
aperture of discontinuities	不连续面的张开度
aperture of discontinuity	不连续开度
aperture of rock fissure	岩石裂隙张开度
aperture of screen	筛孔
aperture of sight	观测孔
aperture ratio	孔径比；口径比
aperture size	缝隙尺寸
aperture time	时窗时间
aperture width	孔径宽度
aperture-detector technique	孔径探测器技术
aperturing	孔径作用
apex angle	顶角；全面钻进金刚石钻头唇部锥度内角
apex load	峰值荷载；顶点荷载
apex of arch	拱顶
apex of cave	穹形冒顶区的顶；陷落平衡拱的顶
apex of fold	褶皱顶；褶皱顶点
apex of shell	壳顶
apex opening	底流口
apex point	顶点
apex stone	顶石
apex-lobate fold	舌状褶皱脊线
apex-shaped rim	凸状轮壳；尖状轮壳
aphanerite	隐晶岩类
aphanesite	光线矿；砷铜矿
aphanic	显衡隐晶质的
aphanic texture	非显晶结构
aphaniphyric	显微隐晶斑状；显微霏细斑状
aphanite	隐晶岩；非显晶岩
aphanitic	隐晶的；非显晶的；细粒的
aphanitic basalt	隐晶玄武岩
aphanitic texture	隐晶结构
aphanitic variolitic texture	隐晶球颗结构
aphanocrystalline texture	隐晶结构
aphanophyre	隐晶斑岩
aphanophyric	隐晶斑状
aphototropism	背光性
aphrolite	泡沫岩

aphrolith	渣状熔岩
aphrosiderite	铁华绿泥石
aphthalose	钾芒硝
aphthitalite	钾芒硝
aphthonite	银铜矿；银黝铜矿
aphyric	无斑隐晶质；无斑非显晶质
aphyric texture	无斑隐晶结构
API method	API法
apical	顶点的；顶尖的
apical angle	顶角
apical axis	纵轴
apical disk	顶系
apical plate	顶板
apical pore plate	顶孔板
apical region	顶端
apical system	顶系
apices	坡顶；顶点，顶尖；背斜顶；露头的顶部；底流排出口
apiculatasporites	圆形细刺孢属
apiculate	细尖的
aplite	细晶岩；长英岩
aplite granite	细晶花岗岩
aplitic	细晶状的
aplitic dike	细晶岩脉
aplitic facies	细晶岩相
aplitic texture	细晶岩结构
aplodiorite	淡色闪长岩
aplogranite	淡色花岗岩
aploid	刮长细晶岩
aplosyenite	浅色正长岩
apo epigenesis	远后成岩作用
apo orogenic intrusion	造山期后侵入作用
Apo Reef	阿波礁
apoandesite	脱玻安山岩
apoapsis	远重心点；远点
apobasalt	脱玻玄武岩
apodolerite	脱玻粒玄岩
apo-epigenesis	远后成岩作用
apogee altitude	远地点高度
apogranite	变花岗岩
apogrite	杂砂岩
apokatagenesis	晚后生作用
apolar adsorption	非极性吸附
Apollo data bank	阿波罗数据库
Apollo lunar sounder	阿波罗月球探测器
Apollo spacecraft	阿波罗飞船
apomagmatic	外岩浆的
apomagmatic deposit	外岩浆矿床
apomecometer	测远器；测距仪；视距仪
apophyllite	鱼眼石
apophyse	岩枝；岩支
apoporphyry	变斑岩
aporate	无孔的
aposandstone	石英岩
aposedimentary	沉积后生的
apotaphral	向外扩张构造的
apotectonic	造山后的
apotectonic phase	构造后幕
apotroctolite	变橄长岩
Appalachia	阿帕拉契亚边缘古陆
Appalachian	阿巴拉契亚山脉
Appalachian Basin	阿巴拉契亚盆地
Appalachian orogenic belt	阿巴拉契亚造山带
Appalachian orogeny	阿帕拉契亚造山运动
Appalachian relief	阿巴拉契亚型地形
Appalachian revolution	阿巴拉契亚造山运动
apparent altitude	视高度
apparent angle	视角
apparent angle of friction	视摩擦角
apparent angle of internal friction	表观内摩擦角
apparent anisotropy	视各向异性
apparent azimuth	视方位角
apparent bed thickness	视地层厚度
apparent brightness	视亮度
apparent bulk density	视体积密度；表观堆密度
apparent bulk modulus	视体积模量；表观体积模量
apparent coefficient of friction	表观摩擦系数，似摩擦系数
apparent coefficient of viscosity	视黏滞系数；表现黏滞系数
apparent coherence	视黏结力；视内聚力
apparent compressional wave	视压缩波
apparent contact angle	视接触角
apparent correlation	视相关
apparent density	表观密度；视密度；松装密度
apparent dip	外观倾斜；视倾斜
apparent dip angle	视倾角
apparent dip azimuth	视倾斜方位
apparent dip slide	视向滑动
apparent directional angle	视方位角
apparent displacement	视位移；视断距
apparent distance	视距离
apparent downthrow	视断层下降盘
apparent dry specific gravity	视干比重；表观干比重
apparent earth conductivity	视在大地导电率
apparent earth pressure	表观土压力
apparent efficiency	表观效率
apparent fault displacement	视断层位移
apparent field	有效视场
apparent field of view	视场
apparent fluidity	视在流度；表现流度
apparent formation factor	视地层因数
apparent formation resistivity factor	地层视电阻率系数
apparent frequency	视频率
apparent frequency dispersivity	视频散
apparent friction angle	视摩擦角
apparent grain density	视颗粒密度
apparent gravity	视比重；视重力
apparent heave	视平错
apparent heave slip	视断距
apparent height	有效高度；视高度
apparent horizon	视地平线；视水平；表观水平
apparent horizontal overlap	视平超距
apparent impedance	视阻抗
apparent isochron age	视等时年龄
apparent length	视波长
apparent life	视年龄
apparent lifetime	表观寿命；近似寿命
apparent lineation	视线理
apparent mass	外表质量；表观质量
apparent migration velocity of seismic events	地震事件表观迁移速度

apparent mining damage 开采地面沉陷；露天采矿破坏
apparent modulus of elasticity 表观弹性系数表观弹性模量
apparent normal fault 视正断层
apparent particle density 表观颗粒密度
apparent penetration rate 表观掘进速率
apparent permeability 视渗透率；表观导磁率
apparent plunge 视倾伏角
apparent polar wander 视极移
apparent porosity 显气孔率；开口气孔率
apparent powder density 表观粉密度
apparent preconsolidation pressure 表观先期固结压力
apparent reduction ratio 表观破碎比
apparent relative density 视相对密度
apparent rock strength 岩石视强度
apparent shearing strength 表观抗剪强度
apparent slip 视滑距
apparent sonic porosity 视声波孔隙度
apparent specific gravity 表观比重；体比重
apparent storativity 视储水系数
apparent stratigraphic gap 视地层间断
apparent stratigraphic interval 视地层间隔视地层厚度
apparent stratigraphic overlap 视地层重叠
apparent stratigraphic separation 视地层分隔
apparent stratigraphical gap 视地层间断
apparent stratigraphical overlap 视地层重叠
apparent stratigraphical separation 视地层离距
apparent stress 表观应力；视应力
apparent structure 视构造
apparent thickness 表观厚度；视厚度
apparent throw 视纵断距
apparent trend 视构造方向
apparent unconformity 视不整合
apparent value 视值
apparent velocity theorem 视速度定理
apparent viscosity 视黏度；表观黏度
apparent volume 表观体积；视体积
apparent volumetrical efficiency 容积效率
apparent water injectivity index 视吸水指数
apparent water table 视水位
apparent wave length 视波长
apparent wave period 表观波周期
apparent wave velocity 视波速
apparent wellbore radius 视井眼半径；有效井径
apparent wellbore storage coefficient 视井筒储存系数
apparent width 视厚度；表观厚度
appearance fracture test 断口外观试验
appearance of crystal 结晶外貌
appinite 暗拼岩
applanation 加积夷平
apple coal 软煤，沥青煤
application efficiency 有效灌溉率
application extent 适用范围
application factor 应用系数；应用因素
application for milling license 申请采矿执照
application of force 加力，施力
application of load 施加荷载
application point 作用点；施力点
application program 应用程序
application program bag 应用程序包
Application Program Interface 应用编程接口

application program library 应用程序库
application programming 应用程序设计
application software 应用软件
application water efficiency 用水效率；灌水效率
applications of backfill to stope 采矿充填料应用
applied cartography 实用地图学
applied chemistry 应用化学
applied force 作用力；外加力
applied frequency 应用频率
applied geochemistry 应用地球化学
applied geology 应用地质
applied geomorphology 应用地貌学
applied geophysics 应用地球物理
applied geothermics 应用地热学
applied hydrodynamics 应用水动力学
applied hydrogeologic investigation 专门性水文地质勘查
applied hydrogeology 专门水文地质学
applied hydrology 工程水文学；应用水文学
applied load 施加荷载；作用荷载；附加荷载
applied mechanics 应用力学
applied moisture 应用水分；可用水分
applied moment 外加力矩；作用力矩
applied pressure 施加压力；作用压力；附加压力
applied seismology 应用地震学
applied shock 外施震动
applied strain 外加应变
applied stratigraphy 应用地层学
applied stress 作用应力；外加应力
applied thrust 作用推力；外施推力
apposite fault 归并断层
apposition fabric 原生组构
apposition tectonite 同沉积构造岩
appressed fold 两翼封闭的褶皱
approach angle 碎土角；挺进角
approach bridge 引桥
approach channel 进港航道；引渠
approach cutting 引桥挖方；桥头挖方；引道挖方；隧道两端挖方
approach embankment 引道路堤；引桥路堤
approach fill 填实引道
approach grade 引道坡度
approach jetty 原生碎屑结构
approach length 引路长度；入口距离
approach locking 接近锁闭
approach pit 导坑
approach ramp 公路引道；建筑物的坡道；引桥坡道
approach road 引道；进路
approach section 接近区段
approach section of crossing 道口接近区段
approach segment 趋近线段
approach slab 引道板
approach span 引桥；岸跨；引跨
approach table 输入辊道
approach taper 楔形引道路段
approach velocity 行近流速
approach viaduct 引道高架桥
approved apparatus 许可的井下安全型设备
approved cable 防爆电缆
approved concrete mix 认可混凝土拌合料；认可混凝土配合比

approved gas detector	安全瓦斯检定器
approved lamp	安全灯
approved land use zoning plan	核准土地用途分区图则
approved layout	核准图则
approved shotfiring apparatus	安全放炮器；防爆型放炮器
approximate contour	近似等高线
approximate evaluation	近似估算
approximate original contour	近似于原地貌
approximate railway location	铁路选线；铁道选线
approximate reasoning	近似推理
approximate relative error for bisection	二分法的近似相对误差
approximate reverberation deconvolution	近似混响反褶积；近似水振荡解回旋
approximate solution	近似解
approximate theorem	近似理论
approximate treatment	近似计算
approximating function	逼近函数
approximation error	近似误差
approximation of function	函数逼近
appurtenant structure	附属结构
apriori probability	先验概率
apron	护坦，冰川前砂砾层
apron area	屋前地台
apron plain	冰川沉积平原；冰前平原
apron plate	大型混汞板
apron raft	板筏
apron reef	石中住裙礁
apron stone	护脚石
apron wall	前护墙
Aptian	阿普特阶
apyre	红柱石
apyrous clay	耐火黏土
Aqikebulake Formation	阿其克不拉克组
Aqjiagou Formation	安家沟组
Aqjing Formation	安靖组
aqua ammonia	氨水
aqua distillate	蒸馏水
aqua filter	蓝色宝石滤色镜
aqua fortis	浓硝酸
aqua privy	乡村式厕所
aqua regia	王水
aqua storage tank	储水槽
aquadag	炭末润滑剂；胶体石墨；导电敷层
aquadrill	水井钻机
aquaeductus	导水管
aquafacts	水浸石
aquafer	含水层
aquagel	水凝胶
Aqua-gel	艾夸杰尔土
aquagraph	导电敷层
aqualf	潮淋溶土
aquamarine	海蓝宝石
aquamarine blue	蓝绿色
aquamarine filter	海蓝宝石滤色镜
aquamarsh	水沼地；沼泽
aquaseis	水震系统；水中爆炸索
aquatic soil	水生土
aquatillite	冰海碛沉积
aquation	水化作用；水合作用
aqueduct bridge	渠桥渡槽
aquent	潮新成土
aqueoglacial	冰水的
aqueoglacial deposit	冰水沉积
aqueo-residual sand	水蚀残沙
aqueous	水的；含水的；水成的
aqueous clay-based drilling fluid	水质黏土基洗井液
aqueous colloidal solution	水胶溶液
aqueous deposit	水成沉积
aqueous environment	水相环境；液相环境
aqueous fluid	水流体；水状流体
aqueous fracturing fluid	水基压裂液
aqueous gas	水气
aqueous gel	水凝胶
aqueous halo	水成晕
aqueous ion	水溶离子
aqueous lava	泥流岩
aqueous layer	水层
aqueous liquor	富铀溶液
aqueous oscillation ripple mark	浪成波痕
aqueous phase	水相；液相；含水相
aqueous pulling	水矿浆；泥水浆
aqueous ripple mark	水成波痕
aqueous rock	水成岩
aqueous sediment	水成沉积
aqueous slurry	翻浆
aqueous soil	饱水土；含水土
aqueous solifluction	水下泥流，水下土溜
aqueous solution	水溶液
aqueous stimulation fluid	水基增产液
aqueous stratum	含水层
aqueous surfactant solution	表面活性剂水溶液
aqueous suspension	水悬浮颗粒；水悬浮物
aqueous system	水系统；含水体系
aqueous transport fluid	水基携带液
aqueous turbidity	水下浊流
aqueous vapor	水蒸气；水汽
aqueous wetting film	水润湿膜
aquept	潮始成土
aquert	潮变性土
aquiclude	含水层；隔水层；弱透水层
aquifer	含水层；蓄水层
aquifer bed	含水层
aquifer boundary	含水层边界
aquifer cell	含水层单元
aquifer chain	链状水层
aquifer CO_2 sequestration	含水层二氧化碳储存
aquifer constant	含水层常数
aquifer contamination hazard map	含水层污染危害图
aquifer contamination potential map	含水层污染潜势图
aquifer degradation	含水层恶化；含水层降解
aquifer depletion	含水层枯竭
aquifer depressurization	含水层降压
aquifer diffusivity	含水层扩散性
aquifer encroachment	水侵
aquifer floor	含水层底板；含水层底
aquifer gas storage field	含水岩层储气库
aquifer group	含水岩组
aquifer hydraulic characteristics	含水层水力特性
aquifer influx	水侵量

aquifer injection 含水带注水
aquifer isopach map 含水层等厚线图
aquifer model 含水层模型；含水层模拟
aquifer parameter 含水层参数
aquifer potential 含水层潜力
aquifer pressure 水层压力
aquifer productivity 含水层产水能力；含水层富水性
aquifer protection 含水层防护
aquifer reservoir 含水层；热水储；水储
aquifer response 含水层效应；含水层反应
aquifer roof 含水层顶板；含水层顶部
aquifer sequence 含水层顺序；含水层系
aquifer simulation 含水层模拟
aquifer storage 含水层储量
aquifer stream interaction 含水层河流交互作用
aquifer system 含水岩系；含水层系；含水层组
aquifer test 含水层试验，抽水试验
aquifer test analysis 含水层试验分析；抽水试验分析
aquifer thermal extraction 含水层取热
aquifer transmissivity 含水层导水系数；含水层导水性
aquifer unit 含水层单元
aquifer water influx 水体进水量
aquifer yield 含水层储水量；含水层产水量
aquiferous 含水的；水成成因的
aquiferous karst 含水喀斯特岩溶；含水岩溶
aquiferous storage 含水层储水量
aquiferous stratum 含水层
aquiferous system 含水体系
aquiferous yield 含水层出水量
aquifuge 隔水层；不透水层
Aquitanian 阿启坦阶
Aquitanian stage 阿启坦阶
aquitard 隔水层；半封闭层；弱透水层；弱透水岩体
aqult 潮老成土
aquod 潮灰土
aquoll 潮软土
aquosity 潮湿；含水性
aquox 潮氧化土
arable 可耕的；耕地
arable area 可耕地面积
arable land 可耕地
arable layer 耕植层
arable takyr 耕种龟裂土
araeoxene 钒铅锌矿
aragon spar 霰石
aragonite 霰石；文石
aragonite compensation depth 文石补偿深度
aragotite 黄色结晶的天然沥青
arakawaite 磷锌铜矿
araldite 特制混凝土黏性砂浆
aramayoite 硫铋锑银矿
aramid fiber 聚芳基酰胺纤维
arandisite 硅锡矿
arapahite 磁铁玄武岩
arbitrary circulation distribution 任意环流分布
arbitrary constant 任意常数
arbitrary loading 任意荷载
arbitrary orientation 任意取向
arborescent drainage pattern 树枝状排水系
Arbuckle Ellenburger group 阿布克尔艾伦伯格群
Arbuckle group 阿布克尔群
Arbuckle movement 阿布克尔运动
Arbuckle orogeny 阿巴克尔造山运动
arc cutter 弧形钻孔钻头
arc dozer 弧形板推土机
arc drilling 电弧钻井法
arc magmatism 弧岩浆
arc of compression 压缩弧；褶皱弧
arc of contact 接触弧
arc of folding 褶皱弧
arc of parallel 纬圈弧
arc rear basin 弧后盆地
arc resistance 抗电弧性
arc reversal 岛弧回返
arc rifting 弧的张裂作用
arc shearer 弧形掏槽截煤机
arc shield 弧形掩护支架
arc suppression coil 灭弧线圈
arc transfixion crack 弧形贯通裂缝
arc transform fault 岛弧转换断层
arc trench couple 弧沟对
arc trench gap 弧沟间隙
arc trench system 弧沟体系
arc trench tectonic 弧沟构造
arc triangulation 大圆弧三角测量
arc trough 弧形地堑
arc uplift 弧形隆起
arcade quay wall 连拱式驳岸墙
arc-arc collision 岛弧-岛弧碰撞
arc-continent collision 岛弧-大陆碰撞
arc-continent suture 弧陆缝合
arch 群岛
arch abdomen dam 腹拱坝
arch abutment 拱脚；拱座；拱基；弧形矿柱的应力集中区
arch action 成拱作用
arch bend 背斜弯曲
arch block 拱块
arch bond 拱砌合
arch brace 拱形撑架；拱形支撑
arch brick 拱砖；楔形砖
arch bridge 拱桥
arch buttress 拱扶垛
arch buttress dam 拱式支墩坝
arch camber 拱势；起拱；拱矢
arch cantilever bridge 拱式悬臂桥
arch cave 拱洞
arch centre 碳胎，拱架
arch concrete 拱部混凝土
arch core 背斜中心，背斜轴心；拱心，拱架；拱核
arch crown 拱冠；拱顶
arch culvert 拱形涵洞
arch dam 拱面水闸墙；拱坝
arch feet 拱脚
arch first lining method 先拱后墙法
arch fold 拱形褶皱
arch form 拱形
arch gallery 拱形明洞
arch girder 拱形梁；铆接拱形支架构件
arch girder stabilizer 拱梁稳定撑

arch gravity dam	重力拱坝
arch hinged at ends	双铰拱
arch key	拱石
arch keystone	拱顶石
arch limb	顶翼；背斜翼；拱翼
arch limb roof	倒转褶皱上翼
arch lining	拱形衬砌；拱形支架；拱形砌碹
arch of a roof beam in stratified rock	层状岩石中顶板梁拱
arch of baked soil	烘焙土拱；烧结土拱
arch of vault	穹拱
arch open cut tunnel	拱形明洞
arch pillar	拱形护顶矿柱；水平矿柱
arch pitch	拱高
arch pressure	拱压
arch profile pick	弧形截齿
arch rib	拱肋；钢拱支撑
arch ring	拱圈
arch riprap	拱矢；拱高
arch rise	拱高
arch roof	拱顶
arch sand control	砂拱防砂
arch section	拱截面
arch setter	拱门定型机；拱门安装者
arch sides	拱帮
arch skewback	拱座
arch span	拱跨
arch springing	起拱点
arch springing hinge	拱脚铰
arch stability	砂拱稳定性
arch stilt	拱门支柱；拱门支撑物
arch stone	拱石
arch stress	拱应力
arch striking	拆除拱架
arch structure	背斜构造；拱形结构
arch support	拱形支架
arch theorem	成拱论
arch theory	成拱论
arch thrust	拱推力
arch timbering	拱形木支架；多边形木支架
arch truss	拱式桁架
arch tube	拱砖管
arch tunnel without cover	拱形明洞
arch viaduct	拱形栈桥
arch wall	拱墙
arch walling	砌拱墙
arch with three articulations	三铰拱
arch without articulation	无接头拱；无铰接拱；无关节拱
archaea	太古代
Archaean	太古的；太古的
Archaean era	太古代
Archaean group	太古界
Archaean nuclei	太古代陆核
Archaean rock	太古岩
Archaeo volcanism	古火山作用
archaeogeology	考古地质学
archaeogeophysics	考古地球物理勘探；考古物探
archaeo-hydrology	考古水文学
archaeolambda	古脊齿兽属
archaeological evidence	考古学证据
archaeological excavation	考古发掘
archaeological material	考古材料
archaeological photogrammetry	考古摄影测量
archaeological survey	考古探查
archaeologist	考古学家
archaeology	考古学
archaeomagnetic	古地磁的
archaeomagnetic dating	古地磁断代
archaeomagnetism	考古地磁学
archaeomorphic rock	原型侵入岩体
Archaeozoic	太古宙；太古宇
Archaeozoic era	太古代
Archaian	太古代的
Archaic ape-man	早期猿人
arch-beam bridge	拱形梁式桥
archbend	褶皱头部
arch-core	拱心，拱架中心；背斜中心，背斜轴心
archean basement	弧形基底
Archean craton	太古代克拉通
Archean Eon	太古宙
Archean Eonothem	太古宇
Archean Era	太古代
Archean erathem	太古界
Archean greenstone belt	太古代绿岩带
Archean group	太古界
arched	背斜组成的；拱形的
arched abutment	拱式桥台
arched airway	拱形风巷；砌拱风巷
arched area	隆起地带；背斜顶部
arched beam	拱副梁；拱形梁
arched bridge	拱桥
arched cantilever bridge	悬臂式拱桥
arched cap	拱形顶梁
arched cap piece	拱形横梁
arched concrete dam	混凝土拱坝
arched drivage	拱形巷道掘进
arched falsework	拱架；拱里的模架
arched floor	拱形楼板
arched girder	拱主梁
arched rail	钢轨拱
arched rail set	拱形钢轨棚子
arched reflector	拱形反射面
arched retaining wall	拱形挡土墙
arched road	砌拱巷道
arched roof	拱形屋顶
arched section	拱形断面
arched structure	隆起构造；穹状构造
arched tunnel	拱形隧道
arched type piling bar	拱形板桩
arched up fold	背斜褶皱
arched window	拱窗
arched-girder	拱形梁
arched-up fold	穹起褶皱；背斜褶皱
archeological drilling	考古钻探
archeological investigation	考古调查
archeology	考古学
archeomagnetism	古地磁
Archeophytic era	太古植物代
Archeozoic era	太古代
archerite	磷钾石

archetype 溢洪道前缘
arch-flat 平拱
arch-gird red roadway 拱梁支架巷道
arch-gravity dam 重力拱坝
archime dean drill 螺旋钻
Archimedean screw loading 阿基米德螺旋装载法
archimedean screw pump 螺旋抽水机
archimedean screw pump volute pump 螺旋泵
Archimedian Series 阿基米德统
arching 成拱作用
arching action 成拱作用
arching characters 成拱特性
arching effect 拱效应；拱作用；弯拱作用
arching factor 拱作用系数
arching force 成拱力
arching phenomenon 砂拱现象
arching strength 砂拱强度
arching theorem 成拱论
archipelagic apron 群岛沿边漫坡海底
archipelagic ocean 多岛洋
architectonic 大地构造的；地质构造的
architectonic division 大地构造区划
architectonic geology 构造地质学
architectonics 大地构造学
architectural concrete 装饰混凝土
architectural engineer 建筑工程师
architectural engineering 建筑工程
architectural mechanics 建筑力学
architectural working drawing 建筑施工图
archless bucket 无拱吊斗，无拱铲斗
arch-limb 拱翼
arch-type support 拱形支撑
archway 拱道；牌楼；拱门
archwise 成弓形
arcogenesis 拱曲运动；拱曲作用
arcogeny 拱曲运动；拱曲作用
arcose 长石砂岩
arcosic 长石的
arc-path sampler 弧路式取样机
arctic Ocean 北冰洋
arctic oil 北极地区用油；靠近北极区开采的石油
arctic pack 北极冰
Arctic Plate 北极板块
Arctic Pole 北极
arctic region 北极区；极地；极地地区
arctic rehabilitation 寒冷土地复垦
Arctic Sea 北极海
arctic soil 极地土壤
arctic subregion 北极亚区
arctic suites 北极岩套
arctic tundra 北极苔原
Arctica 北极蛤属
arctoalpine 北极高山的
arc-to-chord correction in Gauss projection 高斯投影方向改正
arc-trench gap 弧-沟间隙
arc-type illuminator 弧光灯
arcual 拱形的
arcual distance 弧距
arcuate 拱形的；弓形的

arcuate architecture 拱式建筑
arcuate delta 扇形三角洲；弧形三角洲
arcuate fault 弧形断层
arcuate island 弧形列岛
arcuate mountain 弧形山脉
arcuate ridge 弧形海岭
arcuate structure 弧形构造
arcuate trench 弧形海沟
arcuate welt 弧形缘地
arcuate zone 弧形带
arcuated architecture 拱式建筑
arcwall 弧形壁
arcwall face 弧形工作面
arcwall header 截弧形槽煤巷掘进机
arcwaller 水平弧形截槽截煤机
ardealite 磷石膏
Ardeer technique 阿尔狄尔试验法
Ardennian orogeny 阿登造山运动
ardennite 锰硅铝矿
ardmorite 膨润土
area blasting 多排孔爆破；面爆破；地区通风
area boundary 地区界线
area burst 大面积岩石突出
area colliery 地区煤矿
area division 区域范围
area drain 地面排水；地面排水沟
area excavation 区域发掘
Area Executive Committee 区域执行委员会
area exposed 揭露的区域；暴露的范围
area extraction ratio 面积开采比
area factor 面积系数
area filter 面积过滤器
area for ground true collection 地面实况搜集区
area geology 区域地质学
area histogram 面积直方图；面积频率分布图
area increment of face advance 作业面推进面积增量
area law 面积定律
area layout plan 地区详细蓝图
area load 面荷载
area map 区域图
area mean pressure 面积平均压力
area measurement 面积测量
area mining 倒推采矿法
area moment 面积矩
area navigation 区域导航
area normalization method 面积归一化法
area of ablation 消融区
area of artesian flow 自流水区
area of bearing 支承面积
area of catchment 流域面积；集流面积；集水面积
area of contact correction in shear tests 剪切试验中变化的接触面积
area of contour 投影面积
area of coverage 覆盖区；影响区；观测区
area of cross section 横截面积
area of discharge 排水区；排泄出
area of dissipation 消融区；融化区
area of draw 放矿区
area of explosion 爆破面积；爆炸区
area of extraction 采空区范围；采空区

area of face 工作面区
area of faulting 断层区
area of gas-lift flow 气鼓自流水区
area of ground water discharge 地下水出流区；地下水出流带
area of infection 受影响区；受影响面
area of influence method 影响域法
area of influence of extraction 采空区临界面积
area of intense shaking 强震区
area of mesh 筛孔面积
area of nature reserves 自然保护区
area of net pay 纯油区面积
area of occurrence 显示区；活动区
area of orebody 矿体面积
area of perceptibility 有感地震区
area of pressure 受压面积
area of pumping depression 抽水下降面积，地下水位下降区
area of recharge 补给区
area of reinforcement 配筋面积
area of section 截面面积
area of shear plane 剪面面积
area of slack water 平潮区
area of subsidence 陷落区；塌陷区
area of the stress-strain diagram 应力应变图面积
area of tolerance 公差范围；公差带
area of transformed section 换算截面面积
area of visibility 能见范围
area of wastage 消耗区
area of weakness 软弱区
area per unit volume 比表面积
area percent 面积百分比
area prone to outburst 突出危险区域
area ratio 面积比
area ratio of soil sampler 取土器面积比
area reconnaissance 区域踏勘；地区草测
area reduction 断面减缩率；面积缩小
area reduction coefficient 面积折减系数
Area Safety Committee 地区安全委员会
area sampling technique 面积取样法
area scaling 面积测定
area search 分区检索
area separator 区域分隔带
area slope 区域斜坡
area source 面源
area stripping method 台阶开采法；采场剥离法
area study 区域调查
area substation 井下区域变电所
area symbol 面状符号
area time prism 时空地质体
area triangulation 三角网测量
area type mining 分条区开采法
area union 地区工会
area velocity method 面积速度测流法
area volume ratio 面容比
area weighting 面积加权法
area-altitude curve 面积高度曲线
area-capacity sure 面积容积曲线
area-depth curve 面积深度曲线
area-elevation distribution curve 面积高程分配曲线

area-extraction ratio 开采面积比；矿床总面积和已经采空总面积之比
areal analysis 面积分析
areal area 分布区
areal array 面积组合
areal average pressure 面积平均压力；区域平均压力
areal block dimension 平面网格步长
areal data collection system 面积数据采集系统
areal degradation 区域剥蚀
areal density 表面密度
areal disjunction 间断分布区
areal distribution 区域分布
areal effect 面积效应
areal entity data 面状实体数据
areal eruption 区域喷溢
areal exploration 区域勘探
areal extent 面积延伸
areal geochemical anomaly 区域地球化学异常
areal geologic map 区域地质图
areal geologic structure 区域地质构造
areal geological map 区域地质图
areal geology 区域地质学
areal heterogeneity 平面不均质性
areal hydrogeochemical anomaly 区域性水文地球化学异常
areal hydrogeology 区域水文地质学
areal map 地区图；一览图；矿体界限图
areal mapping 区域填图
areal mean rainfall 地区平均雨量
areal model 平面模型
areal pattern 面积组合；面积井网
areal permeability 横向渗透率
areal pluton 区域深成岩体
areal precipitation 面降水量
areal productivity 面生产率
areal prospecting 区域勘探
areal rainfall 面雨量
areal reflection seismic 区域反射地震
areal resolution 面积分辨率
areal rockburst prediction 分布岩爆预测
areal sampling 面积采样；分区取样
areal seismic survey 区域地震观测
areal source 面源
areal strain 区域应变
areal stratotype 区域层型
areal structure 区域构造
areal study 区域性研究
areal suction effectiveness 表面质量吸收效应
areal surface water 区域地表水；地表水体
areal survey 面积调查
areal sweep 面积注水驱油；面积驱油；面积扫油
areal sweep efficiency 面积波及系数
areal sweep out behavior 面积驱扫动态
areal sweep out performance 面积驱扫动态
areal symbol 面状符号
areal temperature gradient 平面温度梯度
areal value estimation 面值估计法；区值估计法
areal variation 地区差异；地域差异
areal velocity 面积速度
areal well pattern 面积井网

areal yield method	面积产量法
areas developing	分区域开拓
areas development	分区域开拓
areas planning	区域规划
area-utilization rate	井田利用率
area-velocity method	面积-流速测流法
area-volume curve	面积容积曲线
area-volume ratio	面积体积比
area-wide	全区的；区域广布的
arecipites	槟榔粉属
areic	内流的；闭流的；无流的
arena stage	中心舞台
arenaceous	砂质的；松散的
arenaceous cement	砂质胶结物
arenaceous clay	砂质黏土
arenaceous conglomeratic facies	砂砾质相
arenaceous facies	砂质相
arenaceous hornfels	砂质角岩
arenaceous limestone	沙质石灰岩
arenaceous mud stone	砂质泥岩
arenaceous pelitic facies	砂泥质相
arenaceous quartz	石英砂
arenaceous rock	砂质岩
arenaceous sediment	砂岩；砂质沉积物
arenaceous sedimentary rocks	砂质沉积岩
arenaceous sequence	砂质层序
arenaceous shale	砂质页岩
arenaceous texture	砂质结构；松散结构
arenaceous-pelitic facies	砂泥质相
arenarious	砂质的
arenated	砂化的
arendalite	紫苏花岗岩类
arene	风化粗砂
arenicola	沙蠋
arenicolite	似海蚯蚓迹
Arenigian stage	阿伦尼克阶
arenilitic	砂岩状的
arenite	砂屑岩；砂质岩
arenopelitic	砂泥质的
arenose	粗砂质；多砂的
arenosol	红砂土
arenous	砂质的；含砂的
arenovidalina	砂质维达虫属
arenstritic pelitic turbidite	砂屑泥质浊积岩
arent	红砂质新成土
arenyte	砂屑岩；净砂岩；纯砂岩
areolation	形成网眼状空隙；网眼状结构
areopycnometer	稠液比重计
arete	尖薄山脊；刃脊
arfvedsonite	亚铁钠闪石；钠铁闪石
argeinite	橄闪石
argentiferous galena	含银方铅矿
argentiferous ore	银矿
argentite	辉银矿
argentobismutite	硫银铋矿
argentojarosite	辉银黄钾铁矾
argentometer	测银比重计
argentometry	银量滴定法
argentopyrite	含银黄铁矿
argic water	中间包气带水
argid	黏化旱成土
argilla	泥土；陶土；高岭土；铝氧土
argillaceous	泥质的；含黏土质的
argillaceous bottom	泥质底板；黏土质底土
argillaceous cement	泥质胶结物
argillaceous desert	黏土荒漠
argillaceous dolomite	泥质白云岩
argillaceous earth	黏土
argillaceous facies	泥质岩相
argillaceous fraction	泥质部分
argillaceous hematite	土状赤铁矿
argillaceous ingredient	泥质混杂物；黏土质混杂物
argillaceous limestone	黏土质石灰岩；泥质石灰岩
argillaceous marl	泥质泥灰岩
argillaceous matter	泥质
argillaceous red bed	红色黏层；红色黏土层
argillaceous rock	泥质岩；黏土岩
argillaceous sand	泥质砂层
argillaceous sand ground	泥质砂地
argillaceous sandstone	泥质砂岩
argillaceous schist	泥质片岩
argillaceous sediment	泥质沉积
argillaceous shale	泥质页岩
argillaceous siltstone	泥质粉砂岩
argillaceous slate	泥质板岩
argillaceous texture	泥质结构；黏土质结构
argillaceous top	泥质顶板
argillan	泥质胶膜；黏土胶膜
argillation	泥化；黏土化
argillic alteration	泥质蚀变
argillic horizon	黏化层
argillic zone	泥化带
argillicalcilutite	泥质泥屑灰岩
argilliferous	含黏土的；含泥的
argillite	泥质板岩；泥质岩；厚层泥岩
argillitization	泥化
argillization	泥化；黏土化
argillized seam	泥化夹层
argillized zone	泥化带
argillo arenaceous	泥砂质
argillo calcareous	泥灰质
argillo ferruginous	泥铁质的
argillo-arenaceous ground	泥砂地
argillo-calcite	泥灰方解石
argillolite	泥质层凝灰岩；硅质细粒凝灰岩
argillophyre	泥基斑岩
argillous	泥质的；含泥的
argillyte	厚层泥岩；泥板岩
argiloid	泥质岩类
argodromile	缓流河流
argol	粗酒石
argulite	沥青砂岩
argument of perigee	近地点角距
argyrodite	硫银锗矿
arid area	干旱区
arid basin	干燥盆地
arid climate	干燥气候
arid desert	干旱漠境
arid erosion	干旱侵蚀
arid fan	干旱地区冲积扇

arid intermontane basin 干旱山间盆地
arid land 旱地
arid landform 干燥地形
arid peneplain 干旱准平原
arid period 干季
arid plain 干旱平原
arid region 干旱地区；旱境
arid region karst 干旱区喀斯特
arid regions 干旱地区
arid season 干旱季节
arid shoreline 干旱滨线
arid soil 干旱区土壤；旱境土壤
arid tract 干旱地带
arid zone 干旱地带
arid zone hydrology 干旱区水文学
aridic 干燥的
aridisol 干旱土；旱成土
aridity 干燥度
aridity coefficient 干燥系数
aridity index 干燥指数
ariegite 尖榴辉岩
arien deposit 霜沉积
Arikareean 阿里卡利阶
aristarainite 硼钠镁石
arithmetic average grade 算术平均品位
arithmetic average thickness 算术平均厚度
arithmetic capability 运算能力
arithmetic device 运算装置；运算器
arithmetic logic unit 算术及逻辑运算部件
arithmetic logical unit 算术逻辑部件
arithmetic manipulations of computers 计算机的算法操作
arithmetic mean 算术平均
arithmetic mean permeability 算术平均渗透率
arithmetic mean strength 强度平均数
arithmetic register 运算寄存器
arizonite 正长脉岩；红钛铁矿
Arkatag Formation 阿尔喀塔格组
arkhangelskiella 阿氏颗石
arkite 黑白榴霞斑岩
arkose 长白砂岩
arkose conglomerate 长石砂岩质砾岩
arkose quartzite 长石石英岩
arkose-sandstone 长石质砂岩
arkosic 长石的
arkosic arenite 长石砂屑岩
arkosic bentonite 长石砂岩质班脱岩
arkosic conglomerate 长石砂岩质砾岩
arkosic graywacke 长石砂岩质杂砂岩
arkosic limestone 含长石灰岩
arkosic sandstone 长石砂岩；花岗质砂岩
arkosite 准脆沥青；长石石英岩
arkositite 嵌胶长石砂岩
arm mixer 叶片式混合机
armangite 砷锰矿
armchair geology 推想地质学
armor rock 护面岩石
armor stone 护面块石
armored cone 熔壳火山锥
armored flexible conveyor 铠装可弯曲输送机
Armorican movement 阿莫里克造山运动
Armorican orogeny 阿莫利卡造山运动
armour 护面；护面块体
armour block 护面块体
armour course 保护面层
armour excavation method 插管支护开挖法
armour layer 护面层
armour rock 护面块石
armoured 装甲的；铠装的
armoured belt 铠装胶带输送机
armoured bitumen 包皮沥青
armoured bubble 矿化气泡；带矿粒的气泡
armoured cable 铠装电缆；装甲缆
armoured chain conveyor 铠装链板输送机；铠装刮板输送机
armoured concrete 钢筋混凝土；钢筋水泥
armoured conveyor 铠装输送机
armoured face conveyor 铠装工作面输送机
armoured hose 钢丝包皮软管；铠装软管
armoured mud ball 硬皮泥球
armoured pliable cable 铠装软电缆
Armstrong process 阿姆斯特朗爆破法
Armstrong shell 阿姆斯特朗型空气爆破筒
Arnu-Audibert dilative ability test 阿奥膨胀度试验
arousing blast layer 起爆层
arrangement diagram 布置图
arrangement in parallel 并列
arrangement of bars 配筋图
arrangement of holes 钻孔布置
arrangement of piles 桩的布置
arrangement of reinforcement 钢筋排列
arrangement of sewerage system 排水管道布置
arrangement of soil particles 土粒排列
arrangement of station tracks 车站股道布置
arrangement of switch 道岔布置
array seismology 台阵地震学
array sonic tool 阵列声波测井仪
array sorting algorithm 阵列分类算法
array spectral 台阵谱
array station 组合站
arrest toughness 止裂韧度
arrested anticline 平缓背斜；不成熟背斜
arrested crushing 有限性破碎
arrested decay 阻止分解
arrested dune 稳定沙丘；固定沙丘
arrested evolution 滞留演化
arrested stoping 间歇回采
arrester bed 沙池
arris 边棱；尖脊
arrival shaft 到达竖井
arrival time 到达时间；波至时间
arrival time curve 波至时间曲线
arrival time difference 到时差
arrival time error 到时误差
arrival time of P wave P波到时
arrival time of S wave S波到时
arrival wave 到达波
arrow diagram 矢量图
arrow head 箭头
arrow point anchor 箭头式砂浆填充锚杆

A

arrowhead 箭头
arrowhead method 运动线法
arroyo 干涸沟壑；干涸河道；溪流；旱谷
arroyo running 暂时山洪
arsenate 砒酸盐；砷酸盐
arsenic minerals 砷矿类
arsenic reagent 砷试剂
arsenical antimony 砷锑矿
arsenical cobalt 辉钴矿
arsenical nickel 红砷镍矿
arsenical pyrite 毒砂
arsenical water 含砷矿水
arseniopleite 红砷铁矿
arseniosiderite 钙砷铁矿
arsenite 亚砷酸盐
arsenoclasite 水砷锰石
arsenoferrite 偶砷基铁酸盐
arsenolite 砷华；砒石
arsenomiargyrite 砷辉锑银矿
arsenopyrite 含砷黄铁矿；毒砂
arsenous acid 亚砷酸
arsensilver blende 淡红银矿
arsenuranospathite 铝砷铀云母
arsine 砷化二氢
arsine gas 砷化氢气体
arsoite 辉橄粗面岩
arterial drain 排水干渠；干线排水管
arterial drainage 排水干管；干渠
arterial road 主干道；主干路
arterial traffic 干线交通
arteries of communication 交通干线交通网
arterite 脉状混合岩
artesian 自流井的；喷水井的
artesian aquifer 承压含水层；承压蓄水层
artesian area 自流水区；承压水区
artesian basin 自流水盆地
artesian capacity 自流泉出水量；自流井出水量
artesian capacity of well 井的自流量；井的自流能力
artesian channel 自流水通道
artesian condition 承压状态；自流状态
artesian discharge 自流水排泄；自流量
artesian flow 承压水流；自流水流
artesian flow area 自流区；承压区
artesian flow well 自流井
artesian flowing well 自流井
artesian formation 自流含水层；有压含水层；承压含水层
artesian fountain 自流喷泉
artesian ground water 自流地下水承压地下水
artesian head 自流水头；承压水头
artesian hydrostatic pressure 承压水静压力；自流水静压力
artesian karst water 自流岩溶水；承压岩溶水
artesian leakage 自流渗漏
artesian level 承压水位；自流水位
artesian monoclinal strata 自流水单斜地层
artesian monoclinal stratum 自流单流层
artesian pressure 自流水压
artesian pressure head 承压水头；自流水压力
artesian pressure surface 自流压力面
artesian region 自流水区
artesian slope 自流斜地
artesian slope spring 承压水斜地泉
artesian spring 自流泉
artesian system 自流水系统
artesian water 承压水；自流水
artesian water circulation 细流
artesian water head 自流水头
artesian water leakage 自流水渗漏；有压水渗漏；承压水渗漏
artesian water region 自流水区
artesian water-yielding formation 自流含水层；自流出水层；承压含水层；承压出水层
artesian well 自流井；喷水井；深水井
artesian well capacity 自流井水量承压井水量
artesian-pressure decline 自流压力降低
artic front 北极峰
articulated chute 活接卸槽浇混凝土用的
articulated concrete 活节混凝土块
articulated concrete matting 活节混凝土块罩面
articulite 可弯砂岩
artifact 石器；人工制品
artificial aggregate 人工骨料
artificial asphalt 人造石油沥青
artificial beach 人工沙滩
artificial bed 人工床层
artificial bitumen 人造沥青
artificial block dike 人造方块堤
artificial borehole wall 人工井壁
artificial canal 人工渠道
artificial catchment 人工集水区
artificial cavern 人工洞室
artificial caving 人工崩落；人工放顶
artificial cavity 人工洞穴
artificial cement 人工胶结
artificial cementing 人工压浆处理
artificial channel 人工河道
artificial channel control 人工河槽控制；人工河道控制
artificial classification 人为分类
artificial clay 人造黏土
artificial coast 人工海岸；人工堤岩
artificial cognition 人工识别
artificial compression 人工压缩
artificial condition 仿真条件；人为条件
artificial consolidation of ground 地层的人为固结
artificial control 人工控制
artificial core 人造岩心
artificial corundum 人造刚玉
artificial crystal 人造晶体
artificial defence 人工防护
artificial deposit 人工堆积
artificial diamond 人造金刚石
artificial discharge 人工排泄
artificial disturbance 人工扰动
artificial drainage 人工排泄；人工排水
artificial drainage system 人工排水系统
artificial drying 人工干燥
artificial earth 人为接地；人工接地
artificial earth satellite 人造地球卫星
artificial earthquake 人工地震
artificial earthquake test 人工地震试验

artificial erosion 人为侵蚀	artificial reef 人工礁
artificial error 人为误差	artificial regulation of groundwater resources 地下水资源人工调蓄
artificial exposure 人工露头；人为露头	artificial release 人工放流
artificial field method 人工场方法	artificial replenishment 人工补给；人工回灌；人工填充
artificial fill 人工填土	artificial reserve 人工储量；人工储存
artificial fishing reef 人工渔礁	artificial reservoir 人工水库；人造水库
artificial flood wave 人为洪水波	artificial rich ore 人造富矿
artificial flooding 漫灌	artificial river-bed 人工河床
artificial floor 人工假顶	artificial roof 人工顶板；顶部支护
artificial flushing 人工冲洗	artificial sample 人工采样
artificial foundation 人工地基	artificial sand 人工砂
artificial fracturing 人工压裂	artificial seismic noise 人工地震噪声
artificial freezing of ground 地基人工冻结	artificial seismic source 人工震源
artificial freezing of soil 土壤人工冻结法	artificial side slope 人工边坡
artificial frozen pit 人为冻坑	artificial silting 人工淤填
artificial gas cap 人工气顶	artificial slope 人造斜坡
artificial gem 人造宝石；人工宝石	artificial soil 人工填土
artificial geothermal reservoirs 人工热储	artificial soil freezing method 冻结施工法
artificial geyser 人造间歇泉；人造喷泉	artificial stabilization technique 人工稳定技术
artificial graded aggregate 人工级配骨料	artificial stone 人造石；铸石
artificial ground 人工接地；模拟井下条件	artificial storage 人工蓄水
artificial ground motion 人工地面运动；人造地震动	artificial stream 人工河道；人工水道；运河
artificial ground water 人工地下水	artificial sub-irrigation 人工地下灌溉
artificial ground water recharge 人工下水补给	artificial support 人工支护；人为支架
artificial heavy mineral 人工重砂	artificial watercourse 人工水道
artificial horizon 人工水平仪；假地平；人为水平	artificial watering 人工灌溉
artificial hypocenter 人工震源	artificial waterway 人工河道；人工水道；运河
artificial impulse method 人工脉冲法；人工地震脉冲法	artificial-freezing process 人工冻结法
artificial intelligence 人工智能；智能模拟	artificially bored spring 人工泉；人工间歇泉
artificial island 人工岛	artificially excavated port 人工开挖的港口
artificial island at sea 海上人工岛	artificially fractured well 人工压裂井
artificial island method 人工筑岛法	artificially graded aggregate 人工级配骨料
artificial isotope 人工同位素；人工合成同位素	artificially improved soil 人工加固土
artificial lake 人工湖	artificially induced earthquake 人工诱发地震
artificial marble 人造大理石；仿云石	artificially lifted well 人工举升井
artificial mineral 人造矿物	artificially recharged groundwater 人工补给地下水
artificial model 人造模型	artificially supported mining methods 人工支护采矿法
artificial nourished beach 人工淤滩	artificially supported openings 需要支架的巷道
artificial nourishment 人工补给	artificially-recharged groundwater resources 地下水人工补给资源；地下水人工补给量
artificial oil 人造石油	artinite 纤维菱镁矿
artificial outcrop 人工露头	artothermal extrusion 自热挤出
artificial permeability 人工渗透率	asama-type flow 浅间型火山碎屑流
artificial pillar 薄矿脉充填柱	asbest 石棉
artificial pond 人工池	asbestic 石棉的
artificial porosity 人工孔隙度	asbestic half round tile 半圆形石棉瓦
artificial pozzolana 人造火山灰质材料	asbestic tile 石棉瓦
artificial precipitation 人工降雨	asbestiform 石棉状
artificial precipitation stimulation 人工催雨	asbestine 纤滑石；油漆填充料
artificial precompression 人工预压	asbestinite 微石棉
artificial product 人工制品	asbesto 石棉；防火布
artificial puzzolana cement 人造火山灰水泥	asbestonite 石棉制绝热材料
artificial radioactive tracer 人工放射性示踪剂	asbestophalt 石棉地沥青
artificial recharge 人工回灌	asbestor-rubber sheet gaskets for pipe flanges 管法兰用石棉橡胶板垫片
artificial recharge of aquifer 含水层人工回灌	asbestos 石棉；槐木
artificial recharge of ground water 人工回灌地下水	asbestos abatement 清除石棉
artificial recharge of groundwater 人工地下水灌注；人工回灌	asbestos abatement works 石棉拆除工程
artificial recharge regime 人工补给动态	asbestos ballast 石棉道床
artificial recharging technique 人工补给方法；人工补给技术	

asbestos bitumen	石棉沥青
asbestos blanket	石棉毡
asbestos board	石棉板
asbestos cement	石棉水泥
asbestos cement insulation board	石棉水泥保温板
asbestos cloth	石棉布
asbestos cord	石棉绳
asbestos diatomite	石棉硅藻土
asbestos fibre	石棉纤维，石棉纸板
asbestos flexboard	石棉板
asbestos float	石棉绒
asbestos gasket	石棉垫片
asbestos insulation	石棉绝缘
asbestos lagging	石棉外套
asbestos mica	石棉云母
asbestos minerals	石棉类
asbestos packing	石棉垫料；石棉密封
asbestos pipe	石棉管
asbestos plaster	石棉粉饰；石棉灰浆
asbestos protection	石棉护层
asbestos ring	石棉环
asbestos shingle	石棉瓦
asbestos textolite	石棉夹心胶合板
asbestos-cement interface	石棉水泥接口
asbestos-containing material	含有石棉的建材
asbestosis	石棉肺
asbolane	钴土矿
asbolite	锰钴土
ascanite	阿斯坎纳土
ascending grade	上升坡度
ascending ground water	上升地下水
ascending horizontal slicing	水平分层上行开采法
ascending method	上行开采法
ascending motion	上升运动
ascending order	上行次序；上行开采法
ascending pass	升轨
ascending pipe	上行管；注入管；压入管
ascending plate boundary	上升板块边界
ascending spring	上升泉
ascending stage grouting method	自下而上分段灌浆法
ascending workings	仰斜巷道
ascension	赤径；上升
ascensional ventilation	上行通风
ascent	登高
ascent angle	上升角
ascent curve	上升曲线
ascent of elevation	拔起高度；克服高度
ascent rate	上升速率
aschaffite	英云斜煌岩
ascharite	硼镁石
aschistic	岩浆同质的
aschistic dyke	未分异岩脉
aschistite	未分异岩
aseismatic	抗震的；耐震的
aseismatic design	抗震设计
aseismatic region	无震区
aseismatic structure	抗震结构
aseismic base isolation	抗震用的基底隔震
aseismic bearing	抗震支座
aseismic belt	无震带
aseismic brace	抗震支撑
aseismic building	抗震建筑
aseismic capability	抗震能力
aseismic code	抗震规范
aseismic construction	抗震构造
aseismic creep	无震蠕动
aseismic deformation	无震形变
aseismic deformation rate	无震形变率
aseismic design	抗震设计
aseismic design code	抗震设计规范
aseismic design code of oil and gas buried steel pipeline	输油埋地钢质管道抗震设计规范
aseismic detailing	抗震细部
aseismic fault	无震断层
aseismic fault displacement	无震断层位移
aseismic fracture zone	无震破碎带
aseismic ground deformation	无震地面形变
aseismic joint	抗震缝；抗震连接
aseismic lake	抗震湖
aseismic measure	抗震措施
aseismic period	地震平静期
aseismic plate	无震板块
aseismic region	无震区
aseismic ridge	无震海岭；无震洋脊
aseismic rise	无震隆起
aseismic safety	抗震安全性
aseismic slip	无震滑动
aseismic strength	抗震强度
aseismic strengthening	抗震加固
aseismic structure	抗震结构
aseismic test	抗震试验
aseismic wall	抗震墙；防震堵
aseismic zone	无震区
aseismicity	抗震性
ash acidity	灰的酸度
ash basicity	灰的碱度
ash bed	凝灰岩层
ash bonding strength	灰结合强度
ash coal	高灰分煤
ash coal ratio	灰煤比
ash concrete	灰渣混凝土
ash cone	凝灰火山锥
ash content	灰分；含灰量
ash content of coal	煤灰分
ash dam	灰坝
ash error	灰分误差
ash fall	降灰
ash flow	火山灰流
ash forming impurities	成灰杂质
ash free coal	不含灰分的煤
ash fusibility	灰熔度；灰熔融性
ash grain	火山灰；火山灰粒
ash layer	灰层；烧土层
ash removal	脱灰；除灰；排灰
ash rock	凝灰岩
ash shower	降落灰
ash sintering strength	灰渣烧结强度
ash slate	灰板岩
ash sluice	水流输渣道
ash structure	火山灰构造

ash test	灰分分析，灰分试验
ash tuff	火山凝灰岩
Ashan Formation	阿山组
ash-and-water free	无水无灰的
Ashantou Formation	阿山头组
Ashby	阿什拜阶
ashen-grey soil	灰色土
ash-forming impurity	成灰杂质
ash-forming minerals	成灰矿物
ash-free	无灰分的；不计算灰分的
Ashgillian	阿什及尔阶
Ashgillian stage	阿石极阶
ashlar	琢石；方石；石墙
ashlar wall	琢石墙
ashlaring	砌琢石
ashless	无灰分的
ashless detergent	无灰清净剂
ashless detergent dispersant	无灰清净分散剂
ashless dispersant	无灰分散剂
ashless paper filter	无灰滤纸
Ashley type groyne	艾什莱型丁坝
Ashman psychomotor	艾斯曼湿度计
ashore	在岸上
ashpan	灰箱
ash-specific gravity curve	基元曲线；灰分-比重曲线
ashstone	凝灰岩
ashtuff	火山灰凝灰岩
ashy grit	凝灰砂岩
Asia continent	亚洲大陆
Asian Regional Remote Sensing Center (ARRSC)	亚洲地区遥感中心
Asiatic land mass	亚洲陆块
Asiatic plate	亚洲板块
asif	干河谷
Asizha Formation	阿斯扎组
askanglina	碱土蒙脱石
askanite	蒙脱石
askew arch	斜拱
askew bridge	斜桥
aslope	斜，倾斜；倾斜地
asmanite	陨鳞石英
Asmari limestone	阿斯马里灰岩
asparagolite	黄绿磷灰石
asparagus stone	黄绿磷灰石
aspect angle	视线角；扫描角
aspect of slope	坡向
aspect ratio	宽长比；高宽比
asperite	粗泡沫熔岩
asperity source model	凹凸体震源模式
asphalt	沥青；柏油
asphalt base carbon fiber	沥青基碳纤维
asphalt base crude oil	沥青基原油
asphalt base petroleum	沥青基原油
asphalt based oil	沥青基原油
asphalt bearing shale	含沥青页岩；油页岩
asphalt binder	沥青结合料
asphalt blanket	沥青敷面
asphalt block	沥青块材
asphalt block pavement	沥青块路面
asphalt built up roof	沥青组合屋面
asphalt canal lining	沥青渠道衬砌
asphalt cement	地沥青胶结料
asphalt cemented ballast bed	沥青道床
asphalt clinker	沥青熔渣
asphalt coal	沥青煤；脉沥青
asphalt coated aggregates	涂有沥青的骨料
asphalt coating	沥青护层
asphalt concrete	沥青混凝土
asphalt concrete base	沥青混凝土底层；沥青混凝土基层
asphalt concrete pavement	沥青混凝土路面
asphalt concrete surface	沥青混凝土地面
asphalt covering	沥青铺面
asphalt damp course	沥青防水层
asphalt damp proof course	沥青防潮层
asphalt dip	沥青浸渍
asphalt ductility testing machine	沥青延性试验机
asphalt emulsion	沥青乳剂；沥青乳化液
asphalt felt	油毛毡；沥青毛毡
asphalt filler	沥青填料
asphalt grout	沥青浆
asphalt grouting	沥青灌浆
asphalt grouting method	沥青灌浆法
asphalt impregnated sandstone	沥青浸染砂岩
asphalt joint	沥青填缝
asphalt joint filler	沥青填缝料
asphalt jointed pitching	沥青砌石护坡
asphalt jointing	沥青填缝
asphalt lining	沥青衬里
asphalt macadam	沥青碎石路
asphalt mat	沥青垫层
asphalt mattress	沥青垫层
asphalt membrane	沥青防水膜
asphalt mix design	沥青配合比设计；沥青混凝土级配
asphalt mixing plant	沥青拌和厂
asphalt mixture	沥青拌合物
asphalt mortar	沥青砂浆
asphalt mound	沥青丘
asphalt nonskid treatment	沥青防滑处理
asphalt oil	沥青油
asphalt overlay	沥青涂层
asphalt paint	沥青漆
asphalt pavement	沥青铺面
asphalt paver	沥青混合料摊铺机
asphalt paving block	沥青铺块
asphalt paving machine	沥青铺面机
asphalt penetration index	沥青针入度指数
asphalt plant	沥青拌合设备
asphalt powder	沥青粉
asphalt prime coat	铺路面的头道沥青；沥青底漆
asphalt pump	沥青泵
asphalt remixer	复拌沥青混合料摊铺机
asphalt Residual Treating process	沥青渣油预处理过程
asphalt road	沥青道路
asphalt road burner	燖沥青路面机
asphalt roadbed	沥青道床
asphalt rock	石油沥青岩
asphalt roofing	沥青屋面；屋顶铺盖沥青工程
asphalt sand	沥青砂
asphalt sand mastic	沥青砂胶
asphalt seal	沥青封闭层；沥青露头

asphalt seal coat	沥青封层
asphalt sealing trap	沥青塞圈闭
asphalt slab	沥青板
asphalt slab mattress	沥青块沉排
asphalt softening point	沥青软化点
asphalt soil stabilization	土的沥青固化
asphalt spreader	沥青摊铺机
asphalt stabilization	沥青稳定；沥青稳定土；沥青稳定法
asphalt stone	沥青石
asphalt surface treatment	沥青表面处理
asphalt tanking	沥青防水层
asphalt tile	沥青砖
asphalt topping	沥青浇面
asphalt varnish	沥青清漆
asphalt-aggregate mixture	沥青集料混合料
asphalt-aggregate ratio	油石化
asphalt-base oil	沥青基油
asphalt-base petroleum	沥青基石油
asphalt-bearing petroleum	含沥青石油
asphalt-coated pipe	沥青刷面管子
asphalted paper	沥青纸
asphaltene	沥青烯
asphaltenic acid	沥青质酸
asphalter	沥青工
asphaltic	沥青质的；含沥青的
asphaltic acid	沥青酸
asphaltic adhesive	沥青系胶黏剂
asphaltic base	沥青基
asphaltic base crude	沥青基原油
asphaltic base crude oil	沥青基原油
asphaltic base oil	沥青基原油；沥青基油料；沥青冷底子油
asphaltic binder	沥青结合料
asphaltic coal	沥青煤
asphaltic coat	沥青涂层
asphaltic coating	沥青涂层
asphaltic concrete	沥青混凝土
asphaltic concrete batching plant	沥青混凝土配料装置
asphaltic concrete core earth-rock dam	沥青混凝土心墙土石坝
asphaltic concrete facing earth-rock dam	沥青混凝土面板土石坝
asphaltic concrete mixer	沥青混凝土拌合机
asphaltic concrete pavement	沥青混凝土路面
asphaltic crude	沥青质原油
asphaltic earth	沥青土
asphaltic grouting	沥青灌浆
asphaltic hydrocarbons	沥青烃类
asphaltic hyrobituminous shale	焦性地沥青页岩
asphaltic limestone	沥青灰岩
asphaltic lining	沥青衬砌
asphaltic membraneous materials	沥青的膜状材料；油毛毡
asphaltic mulch	沥青覆盖料
asphaltic pyrobitumen	焦性地沥青
asphaltic residual oil	沥青残渣油
asphaltic residues	沥青残余物
asphaltic resin	沥青树脂
asphaltic sand	沥青砂
asphaltic sands	沥青砂层
asphaltic sandstone	沥青砂岩
asphaltic saturated felt	沥青毡
asphaltic seal	沥青封闭；沥青塞
asphaltic sealing	沥青堵漏
asphaltic seepage	地沥青苗
asphaltic sludge	沥青质渣
asphaltic stone	沥青岩；沥青石
asphaltine	沥青质
asphalting	浇灌沥青
asphalting tool	浇沥青用工具
asphaltite	沥青岩
asphaltization	沥青化
asphalt-mattress revetment	沥青垫层护岸
asphaltogenic	成沥青的
asphaltos	地沥青
asphaltous acid	地沥青酸
asphalt-stabilized base course	沥青加固的底层
asphaltum	石油沥青
asphaltum oil	沥青油
aspherical lens	非球面透镜
aspirator pump	吸气泵；抽气泵
assay	试金；矿物分析；试样，样品；测定
assay grade	试金品级；分析品级
assay laboratory	试金实验室；分析实验室
assay limit	分析限度
assay map	试样图；采样图
assay plan	矿石试样采取地点图
assay split	规定分析平均值
assay walls	矿体可采界限
assay-content-grade	试金的含量及品级
assaying	试金；矿物分析
assaying of mineral	矿物鉴定
assay-ton	检定吨；化验吨
assay-wall stope	品位检定回采工作面
Asselian	阿瑟尔阶
assemblage	组装；装配；组件
assemblage of forces	力系
assemblage of switch and signal appliance	开关-信号联合装置
assemblage zone	群集带；组合带
assemble	装配；安装；集合；汇编；组装
assembled diamond	拼合钻石
assembled frog	钢轨组合辙叉
assembled length	组装长度
assembled lining	装配式衬砌
assembled milling cutter	装配式铣刀
assembled plant	装配式工厂
assembling bolt	装配螺栓
assembling Components	装配组件
assembling department	装配车间
assembling jig	装配夹具
assembling plant	装配车间
assembling sphere bulk method	球罐散装法
assembly and disassembly time	安装、拆卸时间
assembly average	集合平均值
assembly jig	装配架，装配夹具
assembly language	汇编语言
assembly line	装配线；装配线
assembly line production	流水线生产
assembly pier	拼装式桥墩
assembly stress	安装应力

assessment criteria for innovative technologies in mining industry 采矿业的创新技术的评估标准
assessment index 评价指标
assessment methods 评估方法
assessment of geothermal resource 地热资源评价
assessment well 估价井
assessment work 年度估定钻探工作
assignment of exploration rights 探矿权转让
assignment of mining rights 采矿权转让
Assipetra 阿西石
assise 系；层；地层
assist brake 辅助刹车
assistant driller 副司钻
assistant driver 副司机
assistant engineer 助理工程师
assistant general manager 副总经理
assistant manager 副矿长；副经理
assisting grade 加力牵引坡度
associate orebody 伴生矿体
associated bridge structure 跨桥结构
associated facies 共生相
associated fault 共轭断层
associated fold 共轭褶皱
associated gas 伴生气
associated gas oil ratio 伴生气油比
associated layer 伴生层
associated mineral 伴生矿物
associated minerals deposits of coal-bearing series 含煤岩系共生矿产
associated ore 伴生矿
Associated Research Centers for the Urban Underground Space (ACUUS) 城市地下空间联合研究中心
associated retaining wall 有关挡土墙
associated road and drainage works 有关道路及渠务工程
associated rock 伴生岩石
associated sandstone 伴生砂岩
associated sheet 伴生层
associated silicates 伴生硅酸盐类
associated structure 伴生构造
associated tensor 相伴张量
associated useful component 伴生有用组分；伴生有益组分
associated with cut and fill stoping 与分层充填采矿法相关的
associated works 相关工程；相关设施
Association for Geographic Information (AGI) 英国地理信息协会
Association International de Volcanology (AIV) 国际火山学协会
Association of African Geological Surveys (AAGS) 非洲地质调查协会
Association of American Geographers (AAG) 美国地理学家协会
Association of Consulting Engineers of Hong Kong (ACEHK) 香港顾问工程师协会
association of deposit 沉积物共生组合
association of elements 元素共生组合
Association of Engineering Geologists (AEG) 工程地质学家协会
Association of Environmental Engineering Professors 美国环境工程教授协会
Association of Geoscientists for International Development (AGID) 国际开发地球科学家协会
Association of Geotechnical and Geoenvironmental Specialists (Hong Kong) Ltd. 香港岩土及岩土环境工程专业协会
Association of Government Cartographic Staff 政府制图人员协会
Association of Government Land and Engineering Surveying Officers (AGLESO) 政府土地工程测量员协会
Association of Government Local Land Surveyors (AGLLS) 政府本地土地测量师协会
Association of Government Technical and Survey Officers 政府工程技术及测量主任协会
Association of International Engineering Geology 国际工程地质学会
association of magmatic rock 岩浆岩共生组合
associative storage 相联存储器
assorted 无分选的，未分选的
assorted sand and gravel 筛选的砂和砾石
assortment of particle 颗粒分配
assumed coordinate system 假定坐标系
assumed datum 假设起算值
assumed duration 假定的持续时间
assumed elevation 相对高程
assumed height 假定高程；相对高程
assumed latitude 选择纬度；假定纬度
assumed load 计算荷载；假定荷载
assumed longitude 选择经度；假定经度
assumed mean 假定平均值
assumed north 假定北
assumed origin 假定原点
assumed value 假定值
assumption load 计算荷重
assurance 确信；保证；保险；自信
assurance coefficient 保险系数；安全系数
assurance factor 安全系数
Assyntian orogeny 阿森特造山运动
assyntite 霓辉方钠正长岩
astatic gravimeter 助动重力仪；无定向重力仪
astatic instrument 无定向仪器
astatic lake 不定湖；内陆湖
astatic magnetometer 无定向磁强计
astatic regulation 无定位调节
astatic spring 助动弹簧
astatic type gravimeter 助动式重力仪
astel 平巷顶背板
asteria 星光宝石
asteriacites 似海星迹
asteriated 具星光效应的
asteriated stone 星光宝石
asterichnites 星形迹
asterisk 星号；星状物；加星号于
asterism 星群；星状图形；三星标；星芒
asteroconites 星锥箭石属
asteroid 星状的；小行星；星光宝石
asterolith 星状颗粒
asthenolith 软流圈岩浆体
asthenolith hypothesis 放射热成因岩浆假说
asthenosphere 上地幔；软流圈；软流层
asthenospheric bump 软流圈凸起
Astian 阿斯特阶

Astian stage	阿斯蒂阶
astillen	隔墙；脉壁；矿脉与围岩的间层
astite	红柱角页岩
ASTM Standards	美国材料与实验学会标准
astrakhanite	白钠镁矾
astral era	星云时代
astralite	星字炸药；阿斯特拉里特炸药
astridite	铬硬玉岩
astringency	黏性；收敛性；涩味
astringent substance	收敛性物质
astrobleme	古陨石坑；星疤构造
astrodynamics	天体动力学
astro-geodesy	天文大地测量学
astrogeodetic data	天文大地测量数据
astrogeodetic datum	天文大地基准
astrogeodetic datum orientation	天文大地基准定向
astrogeodetic deflection	天文大地测量偏差
astro-geodetic deflection of the vertical	天文大地垂线偏差
astrogeodetic leveling	天文大地水准测量
astro-geodetic network	天文大地网
astro-geodetic surveying	天文大地测量
astrogeodynamics	天文地球动力学
astrogeology	天体地质学；行星地质学
astrogeophysics	天体地球物理学
astrognathus	星颚牙形石属
astrograph	天体照相仪；天文定位器
astrogravimetric points	天文重力测点
astrolabe	等高仪
astrolithology	陨石学
astromagnetics	天体磁学
astrometric base line	天体测量基线
astrometry	天文测量学；天体测量
astronautics remote sensing	航天遥感
astronomic geology	天体地质学
astronomic latitude	天文纬度；黄纬
astronomic longitude	天文经度；黄经
astronomic meridian	天文子午线
astronomical azimuth	天文方位角
astronomical bearing	天文方位
astronomical fixation	天文定位
astronomical geodesy	天文大地测量学
astronomical horizon	天文地平
astronomical latitude	天文纬度
astronomical leveling	天文大地水准测量
astronomical orientation	天文定向
astronomical triangle	定位三角
astronomy longitude	天文经度
astrophyllite	星叶石
Asturian movement	阿斯图里运动
Asturian orogeny	阿斯图里造山运动
A-subduction	陆壳俯冲作用
asymbiotic nitrogen fixation	非共生固氮
asymmetric anticline	不对称背斜
asymmetric bedding	不对称层理
asymmetric building	非对称建筑物
asymmetric carbon atom	不对称碳原子
Asymmetric Digital Subscriber Line	非对称数字用户线
asymmetric drainage	不对称水系；不对称流域
asymmetric effect	东西效应；非对称效应
asymmetric faces	不对称面
asymmetric film mounting	不对称式底片安装
asymmetric fold	不对称褶皱
asymmetric loading	不对称荷载
asymmetric magnetosphere	不对称磁层
asymmetric multiple ray	非对称的多次波射线
asymmetric progress	非对称过程
asymmetric ripple mark	不对称波痕
asymmetric valley	不对称谷
asymmetrical anticline	不对称背斜
asymmetrical array	不对称组合
asymmetrical balance	不对称平衡
asymmetrical basin	不对称的盆地
asymmetrical bedding	不对称层理
asymmetrical characteristic	不对称特性
asymmetrical climbing ripple	不对称爬升波痕
asymmetrical crystal monochromator	不对称结晶单色仪
asymmetrical distribution	不对称分布；偏态分布
asymmetrical drainage	不对称水系
asymmetrical drainage pattern	不对称水系
asymmetrical effect	非对称效应
asymmetrical fold	不对称褶皱
asymmetrical rail head profile grinding	轨头非对称断面打磨
asymmetrical ridge	不对称山脊
asymmetrical ripple mark	不对称波痕
asymmetrical slope	非对称边坡
asymmetrical structure	非对称结构
asymmetrical syncline	不对称向斜
asymmetrical system	不对称系统
asymmetrical terminal voltage	终端不平衡电压
asymmetry coefficient	非对称性系数
asymmetry distribution	非对称性；分布；不对称分布
asymmetry parameter	非对称性；参数
asymmetry ratio	非对称比率
asymptotic analysis	渐近分析
asymptotic approximation	渐近逼近
asymptotic behavior	渐近特性
asymptotic convergence	渐近收敛
asymptotic convergence rate	渐进收敛速度
asymptotic curve	渐近曲线
asymptotic latitude	渐近维度
asymptotic line	渐近线
asymptotic loading	渐进加载
asymptotic longitude	渐近经度
asymptotic mean	渐近平均数
asymptotic method	渐近法
asymptotic point	渐近点
asymptotic ray theory	渐近射线理论
asymptotic relation	渐近关系
asymptotic series	渐近级数
asymptotic settlement	渐近沉降
asymptotic solution	渐近解
asymptotic stability	渐近稳定性
asymptotic surface	渐近面
asymptotic trend	渐近趋势
asymptotic value	渐近值
asymptotic velocity	渐近速度
asymptotically minimax test	渐近的极小极大检验
asymptotically stable solution	渐近稳定解
asymptotology	渐近学

asynchronism 不同时；异步
asynchronous algorithm 异步算法
asynchronous communication interface adaptor 异步通讯接口适配器
asynchronous computer 异步计算机
asynchronous data transfer 异步数据传输
asynchronous device 异步装置
asynchronous direct-coupled computer 异步直接耦合计算机
asynchronous flow 盐河灌溉计划
asynchronous oscillation 异步振动；异步波动
asynchronous parallelism 异步并行性
asynchronous protocol 异步规约
asynchronous reset 异步复位
asynchronous transmission 异步传输
asynchronous variable 异步变量
at grade intersection 平面交叉
at grass 在地面上；在地面上
at rest pressure 静止土压力
atacamite 氯铜矿
atactic 无规立构的
atactic polymer 无规聚合物
atatschite 线玻正斑岩
ataxic 不成层的
ataxic mineral deposit 不成层矿床
ataxite 角砾斑杂石；镍铁陨石；无结构铁陨石
ataxitic 角砾斑杂状的
Atdabanian 阿特达班阶
atectite 残留基性物质
atectonic 非构造的
atectonic dislocation 非构造变位
atectonic earthquake 非构造地震
atectonic fracture 非构造裂缝
atectonic granite 非构造花岗岩
atectonic mid-ocean ridge 大西洋中脊
atectonic Ocean 大西洋
atectonic pluton 非造山运动深成岩体
atelestite 砷酸铋矿
atelier 工作室；制作车间
at-grade intersection 平面交叉
Athens shale 阿森斯页岩
athermal growth 绝热生长
athwart ship force 横向水平分力
Atitlan-type caldera 阿蒂特兰型破火山口
Atlantic type coastline 大西洋型海岸线
Atlantic-type coast 大西洋型海岸
Atlantic-type continental margin 大西洋型大陆边缘
Atlantic-type geosyncline 大西洋型地槽
atlantite 暗霞碧玄岩
atlapulgite 活性白土
atlas 地图集
atlas information system 地图集信息系统
atlasspat 纤维石
atmoclast 大气碎屑岩
atmoclastic rock 残积碎屑岩；风化碎屑岩
atmoclastics 大气碎屑岩
atmodialeima 大气下形成的不整合
atmogenic deposit 大气沉积
atmogenic metamorphism 气生变质
atmogenic rock 大气成岩
atmogeochemistry 大气圈地球化学

atmolith 气成岩；风积岩
atmology 大气学；水汽学
atmophile element 亲气元素
atmoseal 气封；气封法；气密性
atmosilicarenite 风化崩解硅质砂
atmosphere absorption 大气吸收
atmosphere circulation 大气环流
atmosphere contamination 大气污染
atmosphere current 气流
atmosphere diffusion 大气扩散
atmosphere dust emissions 大气粉尘排放量
atmosphere environment 大气环境
atmosphere flow 气流
atmosphere lapse rate 大气温度递减率
atmosphere mass 大气圈质量；气团
atmosphere moisture 大气湿度
atmosphere monitoring 大气监测
atmosphere precipitation 大气降水
atmosphere pressure 大气压力
atmospheric disturbance 大气干扰
atmospheric diving 常压潜水
atmospheric drag perturbation 大气阻力摄动
atmospheric gas oil 常压瓦斯油
atmospheric gas oil ratio 地面气油比
atmospheric hammer 空气锤
atmospheric precipitation 大气降水；大气降水量
atmospheric pressure chamber 常压舱
atmospheric pressure method 真空预压法
atmospheric rock 气成岩
atmospheric sedimentation 大气沉积作用
atmospheric structure 大气结构
atmospheric surveillance 大气监测
atmospheric swamp 天然沼泽
Atokan 阿托卡统
atoll 环礁；环状珊瑚岛
atoll island 环礁岛
atoll lagoon 环礁泻湖
atoll lake 环礁湖
atoll moor 环形泥炭沼泽
atoll reef 环状珊瑚礁
atoll ring 环礁圈
atoll texture 环礁状构造；环状构造
atom cross section 原子截面
atomic mineral 放射性矿物
atomic ore 放射性矿石；铀矿石
atomizer 喷嘴；粉碎机；雾化器；喷雾器；乳化器
atomizer for aerating pulp 煤浆充气用乳化器
atomizing 雾化作用
atomizing air 雾化气；喷射气
atomizing steam 雾化蒸汽
atomizing wheel 雾化轮
atomology 原子论
atop 在…顶上
at-receiver mitigation measure 在受污染处采取的减轻污染措施
atreol 磺化油
at-rest condition 静止状态
atrio lake 火口原湖
atrium 天井；中庭
atrophy 萎缩；萎缩症

atrypa	无孔蜿，无穴蜿；无洞贝属
attached bubble	黏附气泡
attached building	附属建筑物
attached clause	附加条款
attached dune	附着沙丘
attached file	附件
attached green space	附属绿地
attached ground water	地下封存水
attached growth reactor	重视增长反应堆
attached island	陆连岛
attached mass	附加质量
attached moisture	黏附水分
attached type vibrator for concrete	附着式混凝土振捣器
attached water	结合水；附着水；吸附水；薄膜水；束缚水
attachment angle	连接角
attachment bolt	连接螺栓
attachment bracket	安装托架；附属托架
attachment coefficient	附着系数
attachment link	连接杆
attachment plough	连接式刨煤机
attachment ring	接合圈；联结环
attachment screen plate	紧固螺钉；装配螺钉
attachment screw	装合螺钉
attack	侵蚀；腐蚀；起化学反应
attack angle	钻进角
attack by ground water	地下水作用
attack crew	灭火作业班
attack drill	冲击式钻机
attal	矿石废料，矸石；废渣；充填物料
attapulgite	凹凸棒石
attapulgite clay	凹凸棒石黏土
attemper	回火；调节温度
attenuation band	衰减频带
attenuation coefficient	衰减系数
attenuation coefficient of spring flow	泉流量衰减系数
attenuation constant	衰减常数
attenuation curve	衰减曲线
attenuation distance	衰减距离
attenuation distortion	衰减畸变
attenuation equalizer	衰减补偿器
attenuation factor	衰减系数
attenuation frequency characteristic	衰减频率特性
attenuation law of ground motion	地震动衰减规律
attenuation length	衰减长度
attenuation measurement	衰减测量
attenuation of ground motion	地壳运动衰减
attenuation of hydromagnetic wave	磁流体波的衰减
attenuation of seismic wave	地震波衰减
atteration	表土；冲积土；冲积层；泥砂
Atterberg classification	阿特堡分类法
Atterberg consistency	阿特堡稠度
Atterberg consistency constant	阿特堡结持常数
Atterberg grade scale	阿特堡粒级标准
Atterberg limit	流限，阿特堡界限
Atterberg limit test	阿特堡极限试验
Atterberg liquid limit	阿特堡液限
Atterberg plastic limit	阿特堡塑限
Atterberg size classification	阿特堡粒级分类
attic	钻台地板及围板
attic base	座盘
attic floor	屋顶层
attic ladder	折叠梯
attic oil	阁楼油
attic oil recovery	阁楼油开采
Attic orogeny	阿蒂克造山运动
attic storey	屋顶层
attitor	超微磨碎机
attitude	姿态空间方位角产状
attitude of ore body	矿体产状
attitude of rock	岩石产状
attitude of strata	层态
attitude of stratum	地层产状
attle	矿石废料；矸石；废渣；充填物料
attractability	可吸引性
attracted continental sea	引缩大陆海
attracted mass	吸引质量
attracted water	吸着水
attraction lift force	吸浮力
attraction of gravitation	地心引力
attractive circle	吸引圈
attractive distance	吸引距离
attractive energy	引力能
attractive force	引力；吸引力
attractive resistance	牵引阻力
attrinite	细屑体；暗煤；细屑煤
attrital anthraxylon coal	暗质镜煤；细屑质镜煤
attrital coal	暗煤
attrition grinding	磨碎
attrition loss of catalyst	催化剂磨耗损失
attrition mill	对转圆盘式破碎机；擦碎机
attrition mixing	磨碎混合
attrition rate	磨损率
attrition resistance	抗磨损性
attrition test	磨损试验
attrition value	磨耗值
attrition vein	磨蚀脉
attrition wear	磨损
attritor	磨碎机
Atumushan Formation	阿图木山组
A-type granite	A 型花岗岩
A-type headframe	A 形井架
A-type mast	A 型塔
A-type section	A 型剖面
atypical	非典型的；不规则的；不正常的
atypical gravity sediment	非正常重力沉积物
aubrite	顽火辉石无球粒陨石
aub-satellite point	卫星星下点
Audibert tube	欧迪勃尔特管
Audibert-Arnu dilatation	奥亚膨胀度
Audibert-Arnu dilatometer test	奥亚氏膨胀度试验
audible alarm	发声报警信号；音响报警设备
audible and visual alarm	声光告警
audible noise	声频噪声
audible range	声达距离
audible signal	音响讯号
audio frequency magnetic field method	音频磁场法
audio logging	声频测井
audio-magneto-telluric method	音频大地电磁法
auerbachite	锆石

auerlite	磷钍矿
aufeis	冬季泛滥平原上的厚冰层
auganite	辉安岩
augelite	光彩石
augen chert	眼球状燧石
augen gneiss	眼球状片麻岩
augen migmatite	眼球状混合岩
augen schist	眼球片岩
augen structure	眼状结构；透镜状结构
augen-gneiss	眼状片麻岩
augensalz	石盐团块
auger	螺旋钻；钻孔机管道清除器
auger anchor	钻锚
auger and bucket drilling	屑斗螺钻钻井法
auger backfiller	螺旋回填机
auger bit	螺旋钻头
auger board	孔口板
auger boring	螺旋钻探
auger bulldozer	螺旋推土机
auger car	钻探车
auger coal mining	螺旋采煤
auger conveyor	螺旋输送机
auger drill	螺旋钻机；螺旋钻
auger drilling	螺旋钻孔
auger ground	软土层
auger head	螺旋钻头
auger hole	螺旋钻孔
auger hole method	钻孔注水法
auger injected pile	螺旋喷射桩
auger loading	螺旋装药
auger machine	钻孔采煤机
auger master	司钻的绰号
auger miner	螺旋钻采煤机
auger mining	钻采
auger pile	螺旋桩
auger rig	螺旋式钻井机
auger rock drill	螺旋钻机
auger rod	螺旋钻杆
auger sample	螺旋钻土样
auger sampler	螺钻采样器
auger sampling method	螺旋钻孔取样法
auger sinker bar guides	冲击钻杆导向器
auger spectroscopy	俄歇电子能谱学
auger steel	螺旋钻钢；麻花钎子钢
auger stem	钻杆；钻棒
auger string	螺旋钻杆组
auger tank	搅拌混合罐
auger twist bit	螺旋钻头；麻花柱孔钻头
auger vane	旋转板
auger with hydraulic feed	液压给进式螺钻
auger with valve	提泥螺钻
auger-bit	地螺钻头
augered pile	螺旋桩
augering	螺旋钻探
augering bench	螺旋钻采煤阶段
augering sampler	螺旋取样器
auger-type bit	螺旋式钻头
auget	雷管
augite	斜辉石；辉石
augite minette	辉石云煌岩
augite peridotite	辉橄岩
augite-diorite	辉石闪长岩
augite-porphyry	辉石斑岩
augitic	辉石的
augitite	玻辉岩
augitophyre	辉斑玄武岩
augmentation	增加；加强；扩张；增加率
augmented injection	增注
augmented Lagrangian form	增广拉格朗日形式
augmented matrix	增广矩阵
augmenting factor	增度因子
Augustine process	奥格斯丁炼银法
aulacogen	拗拉槽；拗拉谷；裂陷槽
aulacogen structure	拗拉谷构造
aulacogene	古断槽
aureole	接触变质带
aureole grouting	工作面灌浆；伞形晕轮状灌浆
aurichalcite	绿铜锌矿
auriferous conglomerate	含金砾岩
auriferous drift	含金沉积；含金冰碛
auriferous gravel	含金砾；金砂
auriferous ore	金矿
auriferous quartz vein	含金石英脉
auri-iron formation	金-铁建造
aurite	亚金酸盐
auri-uraniferous conglomerate deposit	金-铀砾岩矿床
aurostibite	方金锑矿
aurotellurite	针蹄金矿
ausannealing	奥氏体等温退火
auscultation	敲帮问顶；听诊
austausch coefficient	换量系数
austemper	奥氏体回火
austemper case hardening	等温淬火表面硬化
austemper stress	等温淬火表面应力
austempering	奥氏体回火；等温淬火
australasian tektite	澳洲玻璃陨石
australite	澳洲似黑曜岩
Austrian method	奥地利施工法
Austrian method for estimating rock jointing	奥地利岩石节理评估法
Austrian method for in situ tunnel tests	奥地利隧道现场试验法
Austrian method of timbering	奥地利式支撑法；上下导坑先墙后拱法
Austrian movement	奥地利运动
Austrian orogeny	奥地利造山运动
Austrian Society for Geophysics and Engineering Geology	奥地利地球物理与工程地质协会；爆破应力
authalic conical projection	等积圆锥投影
authalic map projection	等积地图投影
authentic sample	真实样品
authentication	鉴别；证实；认证
authiclast	自生碎屑
authigene	自生矿物
authigenesis	自生作用
authigenetic cement	自生胶结物
authigenetic feldspar overgrowth	自生长石增长
authigenic	自生的；自生
authigenic carbonate	自生碳酸盐
authigenic carbonate crust	自生碳酸盐壳

authigenic carbonate precipitation 自生碳酸盐沉淀
authigenic element 自生元素
authigenic kaolinite 自生高岭石
authigenic mineral 自生矿物
authigenic phyllosilicates 自生层状硅酸盐
authigenic sand size material 自生砂粒级物质
authigenic sediment 自生沉积
authigenic sediment in ocean 海洋自生沉积物
authigenous 自生的；本源的
authigenous constituent 自生成分
authineomorphic rock 再造岩
authorized architect 认可建筑师；核准建筑师
authorized capacity 核准负载
authorized land surveyor 认可土地测量师
authorized marine borrow area 认可挖砂区；已批准的挖砂区
authorized pressure 规定压力；容许压力
authorized total depth 规定钻井深度
authorized works 获授权进行的工程；批准进行的工程
auto compression 自动压缩
autobahn 高速公路；快车道
autobalance 自动平衡
autobarotropy 自动正压状态
autobreccia 原生角砾岩；自碎角砾岩；自生角砾岩
autobrecciated lava 自碎角砾熔岩
auto-brecciated lava 火山角砾岩
autobrecciation 自生角砾岩化
autobreccited lava 同生角砾岩熔岩
autochthon 原地岩体；原地残余沉积
autochthonous anatexite 原地重熔混合岩
autochthonous coal 原地生成煤
autochthonous coal seam 原地煤层
autochthonous cover sequence 原地覆盖层序
autochthonous deposit 原地淤积；原地沉积
autochthonous fold 原地褶皱
autochthonous granite 原地花岗岩
autochthonous groundwater 本区地下水
autochthonous limestone 原地生成石灰岩
autochthonous nappe 原地推覆体
autochthonous plankton 原地浮游生物
autochthonous river 原地河
autochthonous rock 原地岩
autochthonous sedimentation 原地沉积作用
autochthonous sediments 原地沉积物
autochthonous soil 原生土；原地土壤
autochthonous stream 本地河
autochthonous theorem 原地生成论；原地成煤说
autochthony （矿石、岩石等）原地（生成）的；本土性和独立性
autoclase 自生裂隙；自碎
autoclast 自碎屑；自生碎屑
autoclastic breccia 自生碎屑角砾岩
autoclastic melange 自生碎屑混杂岩
autoclastic rock 自生碎屑岩；碎裂岩
autoclastic texture 自碎结构
autoconvolution 自褶积
autocorrelation function 自相关函数
autofeed drifter 自动推进架式凿岩机
autofeeder 自动给料器；自动给矿器
autogenetic 自成水系；局域性自生地形
autogenetic cycle 自旋回
autogenetic drainage 自然排水系统
autogenetic river 内源河
autogenetic topography 地势自生
autogenic drainage 自成水系
autogenic river 内源河；自源河
autogenic stream 内源河；自生河
autogenous fragmentation 自生破碎作用
autogenous grinding 自生磨矿
autogenous healing 自动弥合
autogenous mill 自生磨矿机
autogenous shrinkage 自收缩量
autogenous volume change 自然容积变化
autogeosyncline 平原地槽；自地槽
autogonal map projection 等角地图投影
autogonal projection 等角投影
autograph 自动测图仪
autographic oedometer 自动绘图压密仪
autographic unconfined compression apparatus 自记无侧限压缩仪
autographometer 地形自动记录器
autohension 自黏性
autohydromorphic soil 自型水成土
autointrusion 自侵入
autolith 同源包体；自生包体
autoloader 汽车式装载机
autoloading 自动加载
automated drilling rig 自动化钻机
automated grout plant 自动制浆站
automated grouting equipment 自动灌浆设备
automated injection system 自动灌浆系统
automated interpretation 自动判读
automated mine 自动化矿山
automated mining 自动化开采
automated monitoring system 自动监测系统
automated refuse collection system 自动垃圾收集系统
automated slope maintenance hotline service 斜坡维修自动热线服务
automatic cartography 自动成图
automatic casing cutter 自动套管切割器
automatic centering 自动对中
automatic central refuse collection system 自动化中央垃圾收集系统
automatic chuck 自动夹盘
automatic coal feeding system 自动给煤系统
automatic coal tip 自动卸煤器
automatic compaction machine 自动击实机
automatic compactor 自动打夯机
automatic control engineering 自动控制技术；自动控制工程
automatic control of longhole loading 深孔装药自动控制
automatic control of shaft hoisting 矿井提升自动控制法
automatic control of underground haulage 井下运输自动控制
automatic control pump 自动控制泵
automatic core breaker 岩心自动卡断器
automatic correction 自动校正
automatic coupler 自动联结器；自动耦合器
automatic directional drilling 自动定向钻进
automatic discharge 自动排料

automatic drencher system　自动水幕系统
automatic drill steel handling　自动更换钻杆
automatic driller　自动送钻装置
automatic earthquake processor　地震数据自动处理机
automatic electronic biaxial tiltmeter　自动电子双轴倾斜度测量仪
automatic element subdivision　自动单元细分
automatic engineering design　自动工程设计
automatic engineer's level　自动找平工程水准仪
automatic feed　动进给；自动给料；自动推进
automatic feed drill　自动推进式钻机
automatic feed grate　自动加煤炉排
automatic firedamp alarm　自动沼气警报器
automatic flash board　重力墙式船闸
automatic floating station　自动浮站
automatic gain control　自动增益控制
automatic gas detector　自动泥浆气侵检测器
automatic grab　自动抓岩机；自动抓斗；自动夹具
automatic grouting parameter recorder　自动灌浆参数记录仪
automatic gun release sub　射孔器自动丢手接头
automatic header system　自动总管系统
automatic hydrographic survey system　水深测量自动化系统
automatic level control　自动水平控制；自动电平控制
automatic load limitation　自动减载装置
automatic loader　自动装卸机
automatic loading and dumping machine　自动装卸机
automatic loading device　自动装载设备
automatic lubrication　自动润滑
automatic lubricator　自动润滑器
automatic measurement　自动测量；自动量数
automatic migration　自动偏移
automatic milling machine　自动铣床
automatic mist control system　喷雾自控系统
automatic mode　自动操作方式
automatic modulation control　自动调制控制
automatic monitoring　自动监测
automatic picture transmission　图像自动传送
automatic pressure reducing valve　自动减压阀
automatic primer　悬臂托梁
automatic pump　消防贮水池
automatic pump control　水泵自动开关控制
automatic ram pile driver　自动冲锤打桩机
automatic ram sounding　自动落锤触探；自动落锤贯入
automatic range　自动测距仪
automatic range control　自动距离控制；自动遥控
automatic range finder　自动测距仪
automatic retraction device　自动提引器
automatic sampling device　自动采样装置；自动取样装置
automatic sanding　自动撒砂
automatic set　自动升柱；自动固定；自动凝固
automatic size control　尺寸自动检验；尺寸自动控制
automatic skip loader　自动箕斗装载装置
automatic slusher　自动扒矿机
automatic spider　自动夹紧十字叉
automatic triangulation　自动空中三角测量
automatic water level recorder　自记水位仪
automation of mine ventilation　矿井通风自动化
automation of pump station　水泵站自动化
automation of rotary drill　牙轮钻机自动化
autometamorphism　自变质作用

automolite　铁锌晶石
automorphous soil　自成土；自型土
automotive drilling rig　车装钻机
auto-oxidation　自氧化
autopatrol grader　维修平路机；巡视平路机
auto-percussive plough　自动冲击刨煤机
autopiracy　本流袭夺
autopneumatolysis　自气化变质作用
autopotamic　河流的
autoprecipitation　自沉淀；自动沉淀析出
autopsy report　尸检报告
autoset level　自动水准仪；自平水准仪；自动安平水准仪
autostoper　自持式风动凿岩机
autunite　钙铀云母
Auversian　奥维尔斯阶
auxiliary adit　辅助平硐；副平硐
auxiliary angle　补角；副偏角
auxiliary apparatus　辅助设备
auxiliary axis　副轴
auxiliary bar　辅助钢筋；辅助杆
auxiliary base　辅助基线
auxiliary bending shape　辅助弯曲形状
auxiliary bending shape function　辅助弯曲形函数
auxiliary breaking　准备破碎
auxiliary bridge　便桥
auxiliary contour　辅助等高线
auxiliary control block　辅助控制阀组
auxiliary cross-cut　辅助石门；辅助横巷
auxiliary crushing　辅助破碎；准备破碎
auxiliary dam　副坝
auxiliary fault　分支断层；支断层；次要断层
auxiliary feedwater pump　辅助给水泵
auxiliary intersection point　副交点
auxiliary joint　副节理
auxiliary measures to protect the roadway　保护巷道的辅助措施
auxiliary mineral　副矿物
auxiliary road　辅路；辅道
auxiliary shaft　副井
auxiliary shaft hoisting　副井提升
auxiliary station　辅助台站；副厂房
auxiliary storage　辅助存储器
auxiliary stress components　辅助应力分量
auxiliary structure　附属构筑物；附属结构；辅助建筑物
auxiliary support　辅助支承；辅助支护
auxiliary telemetry tool　辅助遥测下井仪
auxiliary telescope　经纬仪辅助望远镜；旁侧望远镜
auxiliary water discharge pumping station　辅助排水泵站
auxiliary water pump　辅助水泵
auxite　膨皂石
auxograph　体积变化自动记录器
availability factor　使用效率；运转因素
availability of coal　煤的利用
availability of land　可用土地
available defence　有效保护层
available flow　可用流量；有效流量
available gas　可用煤气
available groundwater　可用地下水；有效地下水
available head　可用压头；有效压头
available hydraulic head　有效水头

available land 可用土地
available moisture 有效水分；可用水分；有效壤中水分
available pore space 有效孔隙空间
available precipitation 可用降水；有效降水
available pressure head 有效压头；有效压力水头
available relief 有利地势
available reserve 可采储量；有效储量
available reserves of mine design 设计可采储量；设计生产能力；服务年限
available resources 可用资源
available runoff 可用径流
available soil moisture 土内有效水分
available soil water 土内有效水分
available water 可供水量，可用水分；可用水量；有效水分；有效含水量
available water resources 可用水利资源；可用水资源
available water supply 有效供水量；可用供水量；可供水量
available weight 有效钻压
avalanche 雪崩；山崩；塌方
avalanche baffle 坍方防御建筑物
avalanche blast 雪崩风
avalanche brake structure 山崩支挡结构
avalanche breakdown 雪崩击穿
avalanche breccia 岩崩角砾岩
avalanche chute 雪崩沟槽
avalanche cycle 山崩周期；岩崩周期
avalanche dam 崩坍形成坝；天然坝
avalanche damage 崩坍破坏
avalanche debris cone 倒石堆；岩崩锥
avalanche defence 防范崩落；崩落防护物
avalanche deposit 崩坍沉积
avalanche disaster 崩塌灾害
avalanche effect 泻落效应
avalanche face 塌落面
avalanche fan 岩崩扇
avalanche gallery 防塌硐
avalanche glacier 雪崩冰川
avalanche hazard 崩塌灾害
avalanche monitoring 崩塌监测
avalanche of rock 岩崩
avalanche prevention and cure 崩塌防治
avalanche prevention works 防坍工程
avalanche prone 雪崩倾向的
avalanche shed 坍方防御板
avalanche wind 山崩风；雪崩风
avalanches 雪崩
avalanching 磨球崩落
avalite 铬云母
Avalonian orogeny 阿瓦朗造山运动
avant port 前港
aven 落水洞；岩溶坑
aventurescence 砂金效应
aventurine 金星玻璃；砂金石；有金星的
aventurine feldspar 砂金长石
aventurine quartz 东陵石；砂金石英
average absolute error 平均绝对误差
average absolute gas emission capacity 平均绝对气体排放量
average acceleration method 平均加速度法
average acceleration response spectrum 平均加速度反应谱

average allowable error 平均容许误差
average amplitude 平均振幅
average angle 平均角
average annual flood 平均年洪水流量
average annual flow 年平均流量；年平均径流
average annual loss 年平均损失
average annual rate of earthquake occurrence 年均发震率
average annual runoff 年平均径流量
average available discharge 平均可用流量
average background value 背景平均值
average bond stress 平均黏结应力
average bucket load 平均铲斗装载量
average clause 平均条款
average closing error 平均闭合差
average closure 平均闭合差
average coefficient of contraction 平均线收缩系数
average composite sample 平均组合试样
average compressive strength 平均抗压强度
average compressive stress 平均压应力
average concentration method 平均含量法
average concrete 普通混凝土
average consistency 平均稠度
average contact pressure 平均接触压力
average content 平均含量
average daily pumpage 平均日抽水量
average daily rainfall 日平均雨量
average deviation 平均偏差
average diameter 平均直径
average diameter of particle 颗粒的平均直径
average dimension 平均尺度
average dipper load 平均铲斗装载量
average discharge 平均排水量；平均流量
average displacement 平均位移
average displacement response spectrum 平均位移反应谱
average distance between stations 平均站间距离
average divergence 平均散度
average drilling speed 平均钻速
average driving speed 平均行车速度
average duration 平均延停时间
average duration curve 平均历时曲线
average duration curve of water level 水位平均历时曲线
average dynamic equation 平均动态方程
average earthquake response spectrum 平均地震反应谱
average efficiency 平均效率；平均有效系数；平均生产率
average end area formula 平均末端截面积公式
average envelope detection 平均包络检波
average error 平均误差
average error area 平均误差区
average excavation round 平均开挖半径
average face length 工作面平均长度
average filling losses 平均灌注损耗
average firm output power 平均输出功率
average floor bearing pressure at yield 支架屈服时底板的平均承压力
average flow duration curve 平均流量历时曲线
average footage 平均进尺
average gradient 平均纵坡
average grading 平均级配
average grain diameter 平均粒径
average grain size 平均粒径

average head 平均水头
average modulus 平均模量
average modulus of deformation 平均变形模量
average normal stress 平均正应力
average overall efficiency 平均总效率
average overpressure 平均剩余压力
average particle size 平均粒度
average pore density 平均孔隙密度
average pore diameter 平均孔径
average potential water power 平均水力蕴藏量
average precipitation 平均降水量；平均沉淀
average pressure 平均压力
average rainfall 平均雨量
average rate of penetration 平均机械钻速
average rock condition 中等岩石条件
average roof bearing pressure at yield 支架屈服时顶板的平均支承力
average sample 平均煤样
average sample number 平均样本数
average settlement 平均沉降量
average slope 平均坡度
average speed 平均速度
average stage 平均水位
average stream flow 平均径流；平均流量
average strength 平均强度
average stress 平均应力
average stress error 平均应力误差
average stripping ratio 平均采剥比
average tensile strength 平均抗拉强度
average thickness 平均厚度
average thickness method 平均厚度法
average torque 平均扭矩
average trend 总观走向；平均趋向
average trip numbers 平均出行次数
average water flow 平均涌水量；平均流量
average weathering velocity 平均风化层速度
average weighted 加权平均的
average yearly rainfall 平均年降雨量
average Young's modulus 平均弹性模量
avezacite 钛铁辉闪岩
avignathus 鸟牙形石属
aviolite 堇云角页岩
avional 高拉应力铝合金
avogadrite 氟硼钾石
avoidable ore losses 可避免矿石损失
avulsion 冲裂；急流冲刷；决口
awaruite 铁镍矿
awash 浸湿；与水面齐平；浪刷；浪刷岩
awash rock 浪刷岩
Awatage Formation 阿瓦塔格组
axe stone 钺石；软玉
axes of principal stress 主应力轴
axial angle 光轴角
axial bearing capacity 轴向承载力
axial bush 岩枝
axial clearance 轴向间隙
axial compression 轴向压缩
axial compressive force 轴向压力
axial compressive load 轴向压力载荷
axial compressive stress 轴向压缩应力
axial compressor 轴流压缩机
axial corallite 轴向珊瑚石
axial culmination 轴部褶升区；轴线顶点
axial cutting waterflooding 轴向切割注水
axial deformation 轴向变形
axial depression 轴线下拗
axial extension test 轴向拉伸测试
axial fabric 轴组构
axial faulting 轴向断层作用
axial fields 轴向野外作业
axial foliation 轴向叶理
axial force 轴向力；轴向载荷
axial force effects 轴力效应
axial Force of Bolt 锚杆轴力
axial force of strut 支柱轴向力
axial fracture 顺轴裂缝
axial fracturing 轴向破坏
axial geanticlinal uplift 轴向地背斜隆起
axial glide plan 轴向滑动面
axial graben 轴地堑
axial hinge line 轴向脊线
axial hinge surface 轴向脊面
axial joint 轴节理
axial line 轴线
axial load 轴向载荷；轴向载荷
axial loading 轴向载荷；轴向加载
axial loading system 轴向加载系统
axial modulus 轴向模量
axial moment of inertia 轴惯性矩
axial plane cleavage 轴面劈理
axial plane folding 轴面褶皱
axial plane foliation 轴面叶理构造
axial plane fracture 轴面破裂
axial plane of fold 褶皱轴面
axial plane of symmetry 对称轴面
axial plane schistosity 轴面片理
axial plane-shears 轴面错动面
axial plunge 轴倾伏
axial section 轴向剖面
axial stiffness 轴向刚性
axial strain 轴向应变
axial stream 山谷主要河流
axial stress 轴向应力；轴应力
axial stress concentration 轴向集中应力
axial surface 轴面；顶面
axial surface cleavage 轴面劈理
axial surface of a fold 褶曲轴面
axial symmetric consolidation 轴对称固结
axial symmetric extension 轴伸长
axial symmetric shortening 轴缩短
axial symmetry fabric 轴对称岩组
axial tectonic belt 轴向构造带
axial temperature stress 轴向温度应力
axial tensile force 轴向拉伸力
axial tensile strength 轴向抗拉强度
axial tensile stress 轴向拉伸应力
axial tensile test 轴向抗拉测试
axial tension 轴心受拉；轴向拉力
axial thrust 轴向推力，轴压
axial thrust load 轴向推力负荷

axial torque	轴向扭矩
axial vibration	轴向振动
axial volcanic ridge	轴向火山山脊
axial vortex	轴向旋涡；轴向涡流
axially loaded column	轴心受压柱
axially loaded pile	轴向受荷桩
axially symmetric consolidation	轴对称固结
axially symmetrical body	轴对称体；回转体
axial-plane foliation	轴面叶理
axial-plane fracture	轴面断裂；轴面破裂
axial-plane slaty cleavage	轴面板劈理
axial-plane structure	轴面构造
axial-surface cleavage	轴面劈理
axial-tangential	轴向切向
axial-type multiple-position borehole deformer	钻孔孔长变形测定器
axifugal	离心的
axinite	斧石；钺石
axinitization	斧石化
axiolite	椭球粒
axis of anticline	背斜轴
axis of arch	拱轴；背斜轴
axis of fold	褶皱轴；褶皱脊线
axis of folding	褶皱作用的轴向
axis of principal stress	主应力轴
axis of stress	应力轴
axis of symmetry	对称轴
axis of syncline	向斜轴
axis of the fold	褶皱轴
axis of the principal stress	主应力轴
axis of tilt	倾斜轴
axis of tunnel	隧道纵轴方向
axis plane	轴平面
axis pole	轴极点
axisymmetric stress	轴向对称应力
axisymmetric stress condition	轴对称应力条件
axisymmetric thermal strain	轴对称热应变
axisymmetry plane strain	轴对称平面应变
axisymmetry plane stress	轴对称平面应力
axonometric drawing	轴测图
axonometric perspective	轴测透视法
axonometric projection	轴测投影；三向投影图
axonometry	轴测法；三向图
axoplasm	轴浆
axotomous	立轴解理
Ayakedakelake Formation	阿亚克达克拉克组
Ayigenkang Conglomerate	阿依根康砾岩
Ayilihe Formation	阿衣里河组
Ayusockanian	阿育索克康阶
azimuth of drilling hole	钻孔方位角
azimuth of epicenter	震中方位角
azimuth of fracture	断裂方位
azimuth of photograph	像片方位角
azimuth tool orientation	钻具定方位角
azimuthal chart	方位投影地图
azimuthal component	方位角分量
azimuthal correction	方位校正
azimuthal displacement	方位位移
azimuthal distribution	方位分布
azimuthal effect	方位效应
azimuthal equal area projection	等积方位投影
azimuthal error	方位误差
azimuthal measurement	方位测量
azimuthal orientation	方位定向；定方位
azimuthal projection	方位投影
azonal soil	不分层土；泛域土
azotification	固氮
azotine	艾若丁炸药
azotobacter	固氮菌
azran	方位距离
azure stone	青金石；石青
azurite	石青，蓝铜矿
Azygograptus	断笔石属

B

babel-quartz 塔状石英
babingtonite 铁灰石
Baca geothermal system Baca 地热系统
bacillite 杆锥晶束；棒状锥晶束
back abutment pressure 后方支承压力
back analysis 反分析；反算；逆分析
back anchor 副锚，串联锚，辅助锚
back and forth method 试算法；选择法
back angle 反方位角；后视角
back arch 拱背；墙内暗拱
back axle 后轴
back azimuth 反方位角
back basin 后盆地
back bay 后湾；封闭海湾
back beach 后滨；后滩；尾滩；里滩
back bolting 顶板杆柱支护；顶板锚杆支护
back bone 分水岭
back bracing 反斜撑
back calculation 反算法，反算；回算
back coil 后圈
back coming 后退回采
back coupling 反向耦合；反馈；回授
back cut 顶部掏槽；顶槽
back cutting 回挖；必要超挖部分
back deep 次生优地槽；后渊；岛弧后海渊
back deep basin 后渊盆地
back diffusion 反向扩散；反弥漫
back digger 反向铲；反铲挖掘机
back drain 背面排水；墙背排水设施
back elevation 背立面图；后视图
back elevation drawing 背视图
back entry 副平巷；无轨道的平巷
back erosion 逆蚀；向源侵蚀；溯源侵蚀
back facets 背面
back fall 山坡，斜坡；滑落
back feed 反馈，回授
back fill 回填，回填土；潜穴
back fill density 回填密度
back fill grouting 衬砌灌浆
back filled earth 回填土
back filled well 回灌井
back filler 回填机，回土机；回填物
back filling 回填
back filling tamper 回填压实机；填土夯实机
back flow 回流；逆流；反循环洗井
back flush 反冲洗
back flushed waste disposal 倒灌清淤
back flushing 反冲
back folding 背向褶皱
back furrow 蛇形丘
back geosyncline 后地槽
back grouting 回填注浆，壁后注浆；二次灌浆
back heading 副平巷；无轨道的平巷
back hoe 反铲；倒铲

back hub 后视桩
back land 腹地；后陆；后地；天然堤后低地
back lath 板桩；顶板背板，背板
back lead 沿岸砂矿
back leg 斜支柱；背面支柱；后支撑脚
back levee 背水堤
back levee march 堤后草本沼泽
back lining 背衬
back marsh 漫滩沼泽；低湿沼泽
back of dam 坝背；坝上游河段；坝的背水面
back of levee 堤防背水面；堤背
back of weir 堰的上游
back play 预制填缝料；余隙；齿隙
back post 支杆支柱
back pressure 反压力；背压
back pressure apparatus 反压装置
back pressure effect 反压效应
back pressure test 背压试验；回压试验；回压试井
back prop 支撑斜立柱；后撑；后支柱
back pull drawing 后张力拉伸
back pull stress 逆应力
back radiation 逆辐射；反向辐射
back rake angle 后倾角
back reaction 反向反应
back ripping 运输巷道挑顶；二次挑顶
back river 后河
back scatter densimeter 同位素反射密度仪
back scatter density 返回散射密度
back scattering angle 反向散射角
back scattering wave 反向散射波
back shattering 顶板碎裂
back shooting 反爆破
back sight 后视
back siphonage 倒虹吸作用；反虹吸；逆虹吸
back slip 向后倾斜的节理；下部劈理或节理
back slope 反坡，回坡，背坡，后坡；内坡
back stress effect 背应力效应
back strut 反撑
back substitution 回代
back substitution coefficient 回代系数
back swamp 漫滩沼泽；河漫滩沼泽，漫滩沼泽，河滩最低部分
back thrust 背冲逆断层；背冲，反冲
back thrusting 背向逆冲；背向冲断层作用
back tilt 反向倾斜
back view 后视图；背立面图
back wall 背墙；挡土墙；投料端墙
back wash 回流，反冲；向源侵蚀；回向冲刷
back water 壅水；回水；残水；筛下水
back wave 回流；反向波；反射波
backacter 反铲挖土机；反铲
back-acting shovel 反向铲；反铲挖土机
back-arc 后弧
back-arc basin 弧后盆地

back-arc basin basalt	弧后盆地玄武岩
back-arc spreading	弧后扩张
back-arc subduction	弧后俯冲
back-berm	后滩肩
back-berm trough	滩肩后凹槽
backboard	底板；后部挡板
backbone curve	骨干曲线；中轴线；主干线
backbone road	主干道；主要干路
backbreak	超挖；超爆
backdrop manhole	后级沙井；跌级沙井
backfeed loop	反馈回路
backfigured date	反算的数据
backfill compaction	回填夯实；回填土压实
backfill compactor	回填压实机
backfill consolidation	回填夯实；覆土压实；填土固结
backfill crown	回填土堆
backfill line	回填线
backfill machine	充填机
backfill material	回填物料；回填料
backfill mining	充填开采
backfill mining Technology	充填开采技术
backfill mining with solid	固体充填开采
backfill operations	充填工作
backfill pipe	回填土中排水管
backfill rammer	回填夯实机
backfill soil	回填土
backfill stability	回填材料的稳定性
backfill tamper	管沟回填夯实机；回填夯
backfill under pressure	加压回填；加压填空
backfill with mortar	灰浆回填；灰浆填空
backfilling body	回填体
backfilling hole	封孔
backfilling of boring	钻孔回填
backfilling soil	回填土
back-filling system	充填采矿系统
backfilling technology	回填技术
back-filling with rubble	回填乱石
backfin	裂缝；轧疤；夹层
backflow channel	回流水道
backflow valve	回流阀
backflow volume	回流液量
backflowing debris	流入井里的碎屑
backflowing water	抽水
backflushing chromatography	反冲层析；反冲反谱法
background activity	本底放射性；放射性背景值
background attenuation	背景衰减
background concentration	本底浓度；背景浓度
background contrast ratio	背影反差比
background earthquake	本底地震；背景地震
background field	背景场
background fluctuation	背景起伏；背景波动
background grid	背景网格
backgroundgroud level	背景水平；背景含量
background mineralization	背景矿化
background polarization	背景极化
background population	背景总体；背景全域
background radioactivity	本底放射性
background response	本底响应
background return	背景回波
background sample	背景样品；正常样品
background science	基础科学
background seismic wave	本底地震波
background seismicity	背景地震活动性
background value	背景值；本底值
background velocity	背景速度
back-grouting plant station	回填灌浆站
backhand drainage	倒转水系；逆向水系
backhaul cable	后曳索
backhoe	反铲；反铲挖土机；挖沟耙
backhoe digger	反铲挖土机
backhoe pull shovel	反向铲
backhoe shovel	反铲挖掘机
backing	回填土；背衬；后退；拱里壁
backing concrete	堤背混凝土
backing earth	填方土；填充土
backing pressure	托持压力；托持压强
backing sand	背砂；填充砂
backland deposit	天然堤后沉积
backlimb	缓翼；后翼
backlimb thrust	缓翼冲断层
backosmotic pressure	反渗压力
back-packing	回填
backpad	挤压垫
backpiece	后挡板
backplan	底视图；仰视图
back-pouring	补浇
backproject	幕后投影
backprojection	反向投射
backpropagation	反向传播
backpumping	反输
backreef	后礁；礁后区
backreef facies	后礁相
backreef lagoon	礁后泻湖
backreef sediment	后礁沉积
backreef zone	礁后带
back-refraction method	背反射法
back-ridge	洋脊背侧
backscatter	反向散射
back-scattered ground roll	反向散射地滚波
backscattering	反向散射；后散射
backscattering coefficient	反向散射系数
backscattering cross-section	后向散射截面
backscour	反向冲刷；后退冲刷；反冲；回卷
backset	逆流；涡流；障碍；收进
backset bed	逆流层；涡流层；逆流交错层
backset bedding	逆流交错层理；后积层层理
back-setting effect	逆流效应
backshore	岸后；滨后；后滨；后滩；尾滩
backshore beach	滨后滩
backshore beach ridge	滨后滩脊
backshore deposit	滨后沉积
backshore terrace	滨后阶地
back-sliding nappe	后滑推覆体
back-spreading	弧后扩张
backstay anchor	拉索锚定
backstay cable	拉索；后拉缆
backup	支撑；备份；支持；底板；阻塞
backup flooding	加压注水；保持压力注水
backup post	支柱；增力桩
back-up water level	回升水面；壅高水面；回升水位

壅高水位
Backus filter 巴克斯滤波器；去鸣震滤波器
backwall grouting 壁后注浆
backwall injection 井壁背后灌浆
backward 后边的；反向的；逆向的；向后的
backward analysis 反分析
backward deflation 向后降阶
backward difference approximation 向后差分近似值
backward difference finite difference method 反向差分有限差分法
backward difference four-step algorithm 向后差分四步算法
backward differences 向后差分
backward drainage 向源水系；逆向水系
backward erosion 向源侵蚀；溯源侵蚀；逆向侵蚀
backward extrusion 反挤压
backward flow 逆流；反流，回流；后退流
backward folding 背向褶皱作用
backward scattered field 反向散射波场
backward slip 后滑
backward station method 后向测点法
backward wave 回波；反波；反射波
backwash cycle 反冲洗循环
backwash efficiency 反洗效率
backwash limit 反冲界线
backwash mark 海滩回流痕
backwashing 回流冲刷；反冲；反洗；反溅
backwashing method 反洗法；反冲法
backwash-regeneration cycle 反冲洗-再生循环
backwasting 冰前消退作用
backwater area 回水面积；回水区
backwater curve 回水曲线，壅水曲线
backwater deposit 回水沉积
backwater effect 壅水效应，回水效应；回水作用，回水影响
backwater envelope curve 回水包络曲线
backwater evaluation 回水估算
backwater from ice jam 冰壅回水
backwater function 回水函数
backwater head 回水头
backwater height in front of bridge 桥前壅水高度
backwater length 回水长度
backwater level 回水高程，壅水位计
backwater limit 回水极限，回水界限
backwater profile 回水纵剖面；回水曲线，壅水曲线
backwater silting 回水区淤积
backwater slope 回水坡度
backwater storage 回水库容；总水头梯度；壅水库容；动库容
backwater trap 回水存水弯；反冲；反洗；回流
backwater valve 止回阀；逆止阀
backwater zone 回水区
backway 后退距离
backwearing 坡崖平行后退
backweathering 后退风化
backweld 封底焊缝
backwelding 反面焊接
backwoods 边远地区
bacon stone 冻石
bacteria bed 生化滤层

bacterial analysis 细菌分析
bacterial degradation 细菌降解作用
bacterial leaching 细菌浸出；细菌选矿
bacterial metallurgy 细菌冶金
bacterial oil recovery 微生物采油
bacterial prospecting 细菌勘探
bacterial reduction 细菌还原作用
bactericide 杀菌剂
bacteriogenic gas 细菌成因气
bacteriological analysis of water 水的细菌成分分析
bacterized peat 细菌化泥炭
baculite 杆菊石
bad ground 不稳定地层；复杂地层；复杂地基
bad land 瘠地；崎岖地
bad oil 含水原油
bad rock 破碎岩石
bad roof 不稳顶板
bad slip 强烈滑坡；严重滑坡
bad spot 受到损坏的部位
bad top 不稳固顶板
baddeleyite 斜锆石
badenite 镍铋砷钴矿
badly defined boundary 不清晰的边界
badly faulted 严重断裂的；被断裂严重破坏的
badly faulting 强烈断层活动
badly fractured ground 严重破裂地层
badly graded sand 级配不良砂
baectuite 白头岩
baeumlerite 盐氯钙石
bafertisite 钡铁钛石
baffle 隔板，缓冲板，挡板；导流板；遮护物；挑流物，分水墩
baffle pier 消力墩；分水墩；压电传感器
baffle plate 隔板；挡板
baffle wall 分水墙；遮挡墙；砥墙
baffle weir 砥堰
bafflestone 障积灰岩；生物滞积灰岩
bag dam 土袋状坝
bag of cement 袋装水泥
bag of ore 矿囊
bag pack 袋垒充填带
bagged cement 袋装水泥
bagged concrete 袋装混凝土
bagger 斗，铲；多斗电铲；挖泥机，挖泥船
bagrationite 铈黑帝石
bagwork 砂包；装袋作业；袋装干拌混凝土
bahada 山麓冲积扇；山麓冲积平原
bahamite 巴哈马灰岩
bahiaite 橄闪紫苏岩
bahr 深泉水
bailed sample 捞出的砂样
bailed sand 捞出砂
bailer 抽筒；泥浆泵
bailer boring 用钻泥提取器钻孔
bailer conductor 捞砂筒
bailer method 泥浆泵法
bailer sample 泥泵试样；捞砂筒中砂样
bailing 提水；抽汲；矿井吊桶排水；用泥浆泵清理钻机
bailing test 提水试验；舀水试验；抽水试验
bailing tube 抽筒；捞砂筒

bailing well 捞砂井；提捞井
bailing-out permeability test 抽水试验
Bairstow algorithm 贝尔斯托算法
Bairstow iteration method 贝尔斯托迭代法
bajada 山麓冲积扇；山麓冲积平原
bajada breccia 泥流角砾岩
bajada placer 荒漠砂积矿床
bajir 沙漠湖
baked contact test 烘烤接触检验
baked soil 烘焙土；烧结土
baked strength 干强度
baked test 烘烤试验
baken peat 烤干泥炭
bakerite 纤硼钙石
baking coal 黏结煤
balance 平衡，保持平衡；天平；使均等
balance arc 平衡弧
balance arch 平衡拱
balance area 均衡区
balance arm 平衡臂；秤杆
balance bar 平衡杆
balance cone 平衡锥
balance element 平衡要素；均衡要素
balance equation 平衡方程
balance force 平衡力
balance formula 平衡式
balance fractional crystallization 平衡分离结晶
balance level 水准仪；衡准仪
balance moment 均衡弯矩；平衡弯矩
balance pit 平衡重井筒
balance plane 自重滑行坡
balance test 平衡试验
balance weight 平衡锤；平衡重
balance weight retaining wall 衡重式挡土墙；衡重支承墙
balanced 平衡的
balanced action 平衡作用
balanced bridge 平衡电桥
balanced construction 均衡结构；平衡构造
balanced cross section 平衡剖面，平衡地质剖面；可复原剖面；平衡断面
balanced design 均衡设计
balanced earthwork 平衡土方工程
balanced eccentricity 界限偏心距
balanced filter 衡均滤波器
balanced grade 平衡坡度
balanced grading 填挖平衡的坡度；土方平衡
balanced growth 均衡增长
balanced load 对称负荷；平衡荷载；均衡荷载；平衡受力
balanced polyphase load 多相平衡载荷
balanced profile 均衡纵断面
balanced reaction 平衡反应
balanced rock 摇摆石
balanced section 平衡剖面
balanced type retaining wall 衡重式挡土墙
balanced weight 平衡重
balancing equation 平衡方程；配平方程
balancing force 平衡力
balancing load 平衡负载；对称负载
balancing speed 均衡速度

balancing tank 均衡池
balancing test 平衡试验
balancing weight 平衡载重；平衡重，平衡锤
balas 浅红晶石；玫红尖晶石
balas ruby 玫红尖晶石
bald mountain 秃山
bald peak 秃峰
bald-headed 蚀顶的
bald-headed anticline 秃顶背斜
baldite 辉沸煌岩
baliki 风化残积黏土；残积黏土
balipholite 纤钡锂石
balkstone 顶板岩石
ball bearing 滚珠轴承
ball burst test 圆球顶破试验
ball clay 球土
ball coal 球状煤
ball diorite 球状闪长岩
ball granite 球状花岗岩
ball inclinometer 球式测斜仪
ball indentation test 球压试验；布氏硬度试验
ball ironstone 球状铁矿石
ball joint 球状节理
ball level 圆水准器
ball mill 球磨机
ball quotient 球轴承座圈；滚珠轴承座圈
ball structure 球状构造
ball test 球压硬度试验；落球试验
ball testing 球压硬度试验
ball texture 球状结构
ball valve sample chamber 球阀取样器
ball viscometer 球式黏度计
ballas 工业用球面金刚石；玫红尖晶石；浅红晶石
ballast 道碴；石碴；堆载
ballast aggregate 道碴；石碴
ballast bed 石碴道床；道碴路基
ballast body 道碴床
ballast bowl 压载箱
ballast concrete 石碴混凝土；镇重用混凝土
ballast grading 道碴级配
ballast loader 清岩机
ballast mattress 沉渣垫层
ballast pit 采石场；采砂场
ballast road 石碴路
ballast rod 冲击式钻杆；冲击式钎子
ballast tamping 捣固道床
ballast tank 衡重水箱；压载罐
ballast trough 桥梁道碴槽
ballast weight 镇重；压重
ballasted anchor 压载锚具
ballasted condition 压载状态
ballasted pontoon 微小地震
ballasting mattress 柴排压沉
ballasting of mattress 柴排压沉
ballasting up 施加压重；铺完石碴
ballast-surfaced 麻面的
balled bit 泥包钻头
balled-up structure 球状构造
balling 糊钻；形成泥包；成球作用；球团作用
balling drum 球团滚筒；球磨机滚筒

balling formation	易泥包地层
balling up of sandstone	砂岩的成球作用
balloon densimeter	囊式密度计
balloon structure	轻型结构
balloon tectonics	气球膨胀构造
balloon volumeter	囊式体积仪
balloon-borne detector	球载探测器
balloon-borne sonde	球载探针
ballooning	鼓胀
ballooning effect	鼓胀效应
ballooning gash	鼓胀裂缝
ballstone	铁矿石；球石
ball-structure parting	球状节理
ball-up	阻塞，堵塞；钻孔堵塞；形成球状；形成黏土球
balm	凹形崖
balsam	树胶；香胶
bamboo basket	总蓄水量
bamboo bolt	竹锚杆
bamboo concrete	竹筋混凝土
bamboo framing	轴测投影三向图
bamboo reinforcement	淤沙地；竹筋
bamboo scaffold	轴对称射流；竹枝棚架
banados	浅沼泽
banana screening	等厚筛分
banana vibrating screen	等厚筛
banatite	辉英闪长岩；正辉英闪长岩
banco	牛轭湖；弓形湖
band	夹层，夹石层，夹矸；带；条；波段
band angle	变形带角
band clay	带状黏土；夹层黏土
band component	条带状显微组分
band drain	塑料排水板；带状排水板；排水板；排水肩带
band dryer	带式干化机
band filter	带式过滤机；带通滤波器
band foundation	带状基础
band iron	扁铁；扁钢；带铁
band pass	带通滤波器；带通
band pass filter	带通滤波器；带通滤光片
band sampler	间层取样
band sandstone	带状砂岩
band shaped prefabricated drain	带状预制排水板
band silicate	带状硅酸盐
band slate	带状板岩；夹层板岩
band stone	硬石夹层
band sulphur	硫夹层
band suppression filter	带阻滤波器
band viscometer	带式黏度计
band viscosimeter	带状黏度计
Banda arc	班达弧
bandaite	拉长英安岩
banded	有夹层的；带状的；有条纹的；连结的
banded agate	带状玛瑙
banded bedding	带状层理
banded clay	带状黏土；纹泥
banded coal	带状煤；带纹煤；条带煤
banded coal ingredient	条带状煤的煤岩成分
banded collar	加箍拉环；加箍圈
banded column	箍柱
banded constituent	条带组分；条带状成分
banded corrosion	层状腐蚀
banded deformation structure	条带状变形构造
banded differentiate	带状分异火成岩
banded echo	带状回波
banded encrustation structure	条带状皮壳构造
banded extinction	带状消光
banded gneiss	条带片麻岩；层状片麻岩
banded granite	带纹花岗岩；带状花岗岩
banded hematite quartzite	带状赤铁矿石英岩
banded iron formation	条带状含铁建造；条带状含铁岩层
banded iron ore deposit	条带状铁矿床
banded limestone	带状石灰岩
banded lithologic reservoir	带状油气藏
banded lode	带状矿脉
banded marble	带状大理岩
banded matrices	带状矩阵
banded matrix	带状矩阵
banded migmatite	条带状混合岩
banded ore	带状矿石
banded peat	带纹泥炭
banded porphyry	带状斑岩
banded precipitation	条纹沉淀
banded quartz hematite	带状石英赤铁矿；铁英岩
banded sandstone	层状砂岩
banded sediment	带状沉积；条带状沉积物
banded slate	带状板岩
banded structure	带状构造；条带构造；加箍结构
banded vein	带状脉
band-gap transition	带隙跃迁
banding	纹理，薄层理；带状构造；层状；条带状
band-like foundation	条形基础
band-limited functions	有限带宽函数
band-limited seismogram	限带地震记录
band-pass amplifier	带通放大器
bandpass analog filter	带通模拟滤波器
bandpass channel filter	带通道滤波器
bandy	带状的
bandy clay	带状黏土；纹泥
bandylite	氯硼铜矿
bank atoll	浅滩环礁
bank barrier	浅滩堡礁
bank breaching	堤岸冲毁
bank build-up	沙堤形成
bank caving	坍岸；堤岸坍陷；河岸淘空
bank coal	原煤
bank cutting	河岸切割；河岸侵蚀
bank deformation	堤岸变形；堤岸演变
bank deposit	河岸沉积物
bank digger	反向铲
bank drain	墙背排水设施
bank drill	杆钻；架式凿岩机
bank eddy	岸边涡流
bank edge	岸边；滩外缘
bank elevation	岸高
bank erosion	岸边侵蚀；河岸侵蚀；堤岸冲刷；沙滩侵蚀
bank failure	河岸淘空；坍岸
bank fire	矸石堆发火
bank gavel	河卵石
bank grading	河岸修坡；河岸修整
bank gravel	未选的砾石；采石坑砾石；岸滩砾石，河

卵石
bank high flow　高岸流量
bank infiltration　堤岸渗漏
bank inset reef　滩内礁
bank levee　河岸堤
bank line　河岸线；岸线
bank line profile　岸线纵断面；路基边线纵断面
bank material　土堤坝填筑材料；岸积物
bank measure　填方量；堤岸土方测量
bank of channel　堤岸；河畔；航道边坡
bank of gravel　砾石滩
bank of silt　泥滩；浅滩
bank of stream　河岸
bank pier　桥台；岸墩
bank protection　护岸，护堤；护坡；护岸
bank protection revetment　护坡
bank protection work　护岸工程
bank reef　岸礁；滩礁
bank regulation　平整河岸
bank reinstatement　复堤，复岸
bank reinstatement method　复堤法
bank revetment　堤岸护坡，护岸
bank run　岸边；岸滩砾石
bank run gravel　采石坑砾石
bank run sand　原砂；岸砂，河砂
bank sand　河岸沙；河岸砂
bank seat　碇桩
bank seismic facies　滩状地震相
bank seismic facies unit　滩地震相单元；滩震测相
bank settlement　堤的沉陷；河岸坍陷；河岸沉陷
bank shaft mouth　坚井口
bank shoal　岸滩
bank side　岸边；河岸斜坡
bank sill　岸坡底槛
bank sliding　河岸滑坡
bank slope　台阶边坡，阶段边坡；岸坡；路堤边坡
bank slough　河岸坍塌；坍岸；河岸沼地
bank sloughing　堤岸坍塌；岸坡表层脱落；坍岸
bank slump　塌岸
bank soil　岸土
bank stability　岸坡稳定性
bank stabilization　河岸加固，堤岸加固；岸坡稳定；岸坡稳定处理
bank stage　平岸水位
bank storage　河岸蓄水量，河岸蓄水；河岸调蓄量，河岸储量
bank storage discharge　河床调蓄流量
bank strengthening　堤岸加固，河岸加固；护坡
bank subsidence　岸堤沉陷；堤陷
bank suction　贴岸吸力
bank up　堆起，堆积；堵截
bank volume　实方体积
banka drill　冲积层勘探钻；砂矿钻机
banke　单斜崖；陡崖
banked　侧倾的；倾斜的
banked reef　含金砾岩层；含矿砾岩层
banked-up　堵截的，堆起的
banked-up water level　壅高水位
banker　堤防土工；满岸流；人工搅拌台；石灰池
banket　反压护道；弃土堆，填土；护坡道，护脚；含

金砾岩层
banket structure　板状节理；板状构造
bankette　弃土堆；填土；护坡道；运输段台
bankfull　平岸；齐岸；满槽
bankfull stage　漫滩水位
bankfull discharge　漫滩流量；齐岸流量；满槽流量
bankfull flow　满槽水流；平岸流
bankfull stage　齐岸水位；满槽水位；漫滩水位；平岸水位
bank-head　岸首
banking　填土，筑堤，堆积；填高，超高
banking construction　填筑堤的构造
banking indicator　磁倾计；测斜仪；倾斜计；重力陡度计
banking material　筑堤材料
banking process　堆积过程
banking structure　堤岸结构物；滑堆构造
bannock　顶部手工掏槽；褐灰色耐火黏土；耐火黏土
banquette　堤岸，堤；扩坡道，崖道；弃土堆；后戗
bantam　重矿物
banto faro　半沼泽
Baode Stage　保德期
baotite　包头矿；硅钡钛铌矿
bar　巴；沙坝；沙洲；加劲条，杆；气压计
bar beach　滨外滩
bar bender　钢筋弯曲机
bar bending machine　钢筋弯曲机；弯筋机
bar buoy　浅滩指示浮标
bar chart　直方图，条线图；横线工程图表；横线计划图表；柱形统计图
bar check　杆尺校正
bar chromatogram　直方色谱图
bar clamp　杆夹
bar cold-drawing machine　钢筋冷拉机
bar cropper　钢筋剪切机
bar diagram　直方图；柱状图解；图表；条线图
bar digging　河滩开采砂金
bar drill　杆钻；架式凿岩机；横梁架式钻机
bar extensometer　钢筋伸长计
bar formation　沙洲形成
bar frame　杆架；棒式车架
bar graph　条线图；柱形统计图
bar gravel　河流砾石浅滩；河滩砾石
bar grit　条筛
bar grizzly　格筛
bar lake　堰堵塞湖
bar lunate　新月形沙坝
bar magnetic compass　磁杆罗盘
bar mat　钢筋网片
bar mat reinforcement　网状钢筋
bar mining　河滩开采砂金
bar pile　扁钢桩
bar plain　沙坝平原
bar platform　沙坝台地
bar point　点状沙坝
bar scale　直线比例尺
bar screen　棒条筛，格筛
bar shear　钢筋切断机
bar spacing　钢筋间距
bar steel　粗钢筋；条钢；棒钢
bar tail　沙坝尾

bar tendon 钢筋
bar theory 沙洲说；沙坝说
baraboo 重现残丘
barachois 泻湖
baramite 菱镁蛇纹岩
barbed drainage 倒钩水系；倒钩水系统
barbed drainage pattern 倒钩状水系
barbed tributary 倒钩状支流；逆向支流
Barberton Greenstone Belt 巴伯顿绿岩带
barbierite 钠正长石
barbotage 起泡作用；起泡；鼓泡
barcan 新月形沙丘
barchan 新月形沙丘
barchan cluster 新月沙丘群
barchan dune 新月形沙丘
barchan swarm 新月沙丘群
bare 裸露，裸露的；手工刨煤，手工掏槽；沿断层采掘，沿边界采掘；贫，裸
bare area 裸露面积
bare bus bar 裸母线
bare cut slope 新开挖的边坡
bare exposed rock 岩石露头
bare field 裸露地面
bare foot 雨量分布系数
bare frame 空框架
bare ground 不毛之地；已采地区
bare hole 未下套管的钻孔；裸眼
bare ice 裸冰
bare karst 裸露喀斯特；裸露岩溶
bare land 白地，裸地；不毛之地；出露地区；基岩地区
bare log 钻孔柱状图
bare patch 秃块
bare rock 裸露岩石；明礁
bare rock soil slope surface 光秃的土坡面
bare weight 空重
bare well 裸井
barefoot completion 裸眼完井
barefoot well 裸井；未下花管生产井
bar-finger 指状沙坝
bar-finger sand 指状沙坝
bar-finger sandstone body 指状砂岩体
barge 卸泥船，趸船；驳船；平底船
barge canal 运河；船渠
barged-in fill 吹填土
bar-head 沙坝头
baria 重晶石
baric 气压的；含钡的
baricalcite 重解石
baring 剥露，剥离；剥离盖层；覆盖层；揭开
barite 重晶石；引线
barite cement 重晶石水泥
barite dollar 饼状重晶石结核
barite mud 重晶石泥浆
barite plug 重晶石塞
barite slurry 重晶石浆
barite-weighted mud 重晶石加重泥浆
baritic 重晶石的
baritic rock 重晶石岩
baritization 重晶石化
barium anorthitite 钡斜长岩

barium autunite 钡铀云母
barium carbonate 碳酸钡
barium chloride 氯化钡
barium cloud 钡云
barium dichromate 重铬酸钡
barium feldspar 钡长石
barium minerals 钡矿类
barium monoxide 氧化钡
barium nitrate 硝酸钡
barium nitrite 亚硝酸钡
barium oxide 氧化钡
barium plaster 钡灰浆；含钡灰泥；防射线抹灰
barium salt 层积盐；废盐
barium sulphate 硫酸钡
barium sulphate test 硫酸钡试验
barium sulphide 硫化钡
barium titanate 钛酸钡
barium vapour 钡蒸汽
barium-ferrite ceramic 铁酸钡陶瓷
bark coal 树皮煤
bark liptobiolith 树皮残殖煤
barkan 新月形沙丘
barkevikite 铁角闪石；钠铁角闪石；棕闪石
barkhan 新月形沙丘
barkhan chain 新月形沙丘链
Barkhausen effect 巴克霍森效应
bar-like lunate 新月形沙坝
barlike structure 刺状构造
Barlin stone 巴林石
Barlow point 巴洛点
barlte 天青重晶岩
Barnard's star 巴纳德星
barneite 棕闪霞石岩
Barnett shale 巴奈特页岩
baroceptor 气压传感器；气压敏感元件
barocline 压斜；斜压
baroclinic 斜压的
baroclinic condition 斜压条件
baroclinity 压斜状态
barodynamic experiment 重结构力学试验；重量力学试验
barodynamics 重结构力学；重力学
barograph 气压记录器；气压表
baroid 取心时加入泥浆之重晶石与水凝胶
barolite 天青重晶岩
barology 重力论
barometer 气压表；气压计
barometer altitude 气压高度
barometer level 气压表高度
barometer table 气压表
barometric 气压的
barometric altimeter 气压测高计
barometric change 气压变化
barometric coefficient 气压系数
barometric efficiency 气压效率
barometric elevation 气压高度
barometric fluctuation 气压扰动
barometric gradient 气压梯度
barometric height 气压高度
barometric hypsometry 气压计测高术

barometric maximum 最高气压
barometric minimum 最低气压
barometric pressure 气压；大气压
barometric pressure sensor 气压传感器
barometric tube 气压计管
barometric variation 气压变化
barometrical 气压计的
barometrical gradient 气压梯度
barometry 气压测定法
baropause 气压层顶
baroque cut 异型切工
baroque dolomite 变态白云岩
baroselenite 重晶石
barosphere 气压层
barostat 恒压器
barothermograph 气压温度计
barotolerancy 耐压性
barotropic 正压的；零压差的
barotropic atmosphere 正压大气
barotropic condition 正压情况
barotropic fluid 正压流体
barotropic instability 正压不稳定
barotropic state 正压状态
barotropic wave 正压波
barrage 拦河闸，堰，水闸，堤坝，坝坝，拦河坝；阻塞物；钟形桥墩
barrage power station 拦河坝发电厂
barrage type spillway 堰式溢洪道
barrage with stop plank 插板式拦河堰
barrage-type spillway 堰式溢洪道
barranca 峡谷；深谷
barrandite 磷铝铁石；铝红磷铁矿
barred basin 沙坝盆地；堰塞盆地；阻塞盆地，限制性盆地
barred river mouth 堵塞河口
barrel 桶；圆筒，筒形物；岩芯管；钻探取样器
barrel arch 筒形拱
barrel culvert 筒形涵洞
barrel drain 筒形排水渠
barrel of pipe 管筒
barrel pier 筒形桥墩；筒形支墩
barrel sampling 桶式取样方法
barrel vault 圆筒形穹窿；筒拱；边界矿柱；半圆形拱顶
barreler 高产油井
barrel-shaped structure 筒状构造；桶状构造
barren 不含矿物的，无矿的；不毛之地，脊地，瘠地；无矿物溶液；荒凉的
barren aureole of alteration 无矿蚀变带
barren bed 哑层；无矿层
barren coal-measures 无煤层
barren drill hole 无矿钻孔
barren field 无油气地区
barren gap 无油气地段
barren ground 非生产地层；无矿地层；空白地区
barren hole 空井；未下套管的井眼部分
barren intercalation 废石夹层；无矿夹层
barren interval 无矿区间；无矿段
barren interzone 哑间带；哑内带
barren land 不毛之地；荒地，裸地
barren layer 废石层；无矿层

barren lode 无矿岩脉
barren measures 无矿地层；无可采煤层的煤系
barren mine 无开采价值的贫矿
barren mineral 贫矿
barren of coal 不含煤的
barren rock 废石；无矿岩石；无油气的岩层
barren sand 无油砂层
barren shale 非可燃页岩
barren solution 水冶废液；贫液
barren spots 无油砂岩；废石
barren spotting 无矿地带；不生产地带
barren topography 荒芜地形
barren trap 无油气的圈闭
barren water 无矿水；贫瘠水域
barren waters 贫瘠海区；贫瘠水域
barren well 无水井；未出油井；无油气井
barren zone 无矿区间；无矿段；非生油带
Barrentheorie 巴伦理论
barrette 方形桩；壁桩；壁板桩；墙式桩
barricade 隔板，隔墙；屏蔽墙；防御墙；路障
barricaded 用墙隔开的；屏蔽的
barricading 围隔；防卫工事；障碍
barrier 阻隔物，障碍物；沙坝；沙洲；沙堤；护栏；屏障，障壁
barrier bar 潜坝型沙洲，沿岸沙坝；岩桥；岩石锚栓；岩石压缩
barrier bar-lagoon system 障壁沙坝-泻湖体系
barrier basin 堤堰水池；堤成盆地
barrier beach 沿滩沙埂；滨外沙坝，滨外滩；障壁滩
barrier bed 遮挡层
barrier berg 堡状冰山
barrier block 路障
barrier chain 沙岛群
barrier coast 沙坝海岸；堰洲海岸
barrier coat 阻挡层；防渗涂层
barrier complex 障壁岛复合体
barrier container 岩粉棚
barrier crossing 交叉屏障
barrier dam 堰塞坝；拦污坝
barrier effect 遮帘作用；势垒效应
barrier ice 冰堡；冰岸
barrier island 屏障岛
barrier island marsh 滨外岛沼泽
barrier lagoon 堡礁泻湖；障壁泻湖
barrier lake 堰塞湖
barrier layer 阻挡层
barrier method 柱式开采法
barrier pillar 安全煤柱，安全矿柱；隔离煤柱；防水煤柱；边界矿柱
barrier plain 深海平原
barrier platform 障壁台地
barrier potential 位垒；势垒
barrier power station 堤堰式电站
barrier reef 堡礁；堤礁
barrier screen 遮挡
barrier sediment 沙坝沉积；堰洲沉积
barrier sheet 隔板
barrier shoreline sequence 障壁滨线沉积序列
barrier source model 障碍体震源模式
barrier spit 障壁沙嘴；滨外沙嘴

barrier spring	堰塞泉；堤泉；堡坝泉
barrier wall	止水墙；防水墙；隔墙，围墙
bar-rigged drift	柱架式风钻
barringdown	撬落；挑顶；撬松仓内矿石
barroisite	冻蓝闪石
Barron's consolidation theory	巴隆固结理论
Barron's solution	巴隆解
Barrovian-type facies series	巴罗型相系
Barrovian-type metamorphic zone	巴罗型变质带
Barrovian-type metamorphism	巴罗型变质作用
barrow	手推车；弃石堆；山；矸石场
barrow area	取土坑
barrow truck	手推车
Barrow's zone	巴罗带
bar-shaped	骨棒状
barshawite	中长沸基辉闪斑岩
Barstovian	巴斯托夫阶
Barstovian Stage	巴斯托夫阶
barthite	砷锌铜矿
Barton failure criterion	巴顿破坏准则
bar-type grating	条式格栅
barycenter	重心；质心
barycentric coordinate	重心坐标
barye	微巴
barylite	硅钡铍矿
barysil	硅铅矿
barysilite	硅锰铅矿；硅铅矿
barysphere	地球核心；地心圈；溢水；重圈
baryta	氧化钡
baryta feldspar	钡长石
baryte	重晶石
barytocalcite	钡解石；补充
barytocelestine	钡天青石
barytolamprophyllite	钡闪叶石
basal	基层的；基础的；基底的；基本的
basal arkose	基底长石砂岩
basal bed	底部岩层
basal breccia	底部角砾岩
basal cement	基底胶结
basal cleavage	主劈理，主解理；底面解理；基解理；贮理
basal complex	基底杂岩；基岩
basal conglomerate	底砾岩
basal contact	底部接触
basal debris	底砾
basal disc	基盘
basal discontinuity	基层不连续面
basal disk	基盘
basal edge	底棱
basal erosion	底部侵蚀作用
basal excavation	基部凹陷，基底开挖
basal face	底面
basal granule	基粒
basal ground moraine	底碛
basal ground water	油层下部底水
basal heave	基坑底隆起
basal ice-flow	底冰流
basal instability	坑底不稳定性
basal joint	底面节理
basal lamina	海底纹理叠层
basal level	基准面；基线水准
basal moraine	底碛
basal orientation	基线定向
basal orifice	底孔
basal part	下盘，底帮，底板；底壁
basal pinacoid	底轴面
basal plane	基面；底轴面
basal plane slip	基底面滑动
basal planing	基底夷平作用
basal plate	基板
basal sandstone	基底砂岩，底部砂岩
basal sapping	挖掘；基部掏蚀；底部掏蚀
basal section	底切面；底面断面；基面剖面
basal shear surface	基底剪切面
basal sliding	基底滑动；底部滑动
basal slope	坡脚
basal structure	基底构造
basal surface	基岩顶面；基底顶面，底表面；风化面
basal tar mat	底部焦油垫层
basal thrust	基底冲断层
basal thrust plane	基底冲断层面
basal till	冰川；底碛
basal transgressive lithofacies	基底海侵岩相
basal tunnel	基底输水隧道
basal unconformity	底部不整合
basal water	底水；最低地下水层；主含水层
basal water table	底水面
basalatite	正长玄武岩；玄武安粗岩
basal-clastic phase	底碎屑相
basalt	玄武岩
basalt alteration	玄武岩蚀变
basalt carbonation	玄武岩碳化
basalt clay	玄武黏土；玄武土
basalt dyke	玄武岩岩墙
basalt eclogite transformation	玄武岩榴辉岩转化
basalt fitting	玄武岩衬垫
basalt flow	玄武岩流
basalt glass	玄武玻璃
basalt layer	玄武岩层
basalt structure	纵梁支撑
basalt wacke	玄武石
basalt-agglomerate tuff	玄块凝灰岩
basaltic	玄武岩的；玄武岩质的
basaltic achondrite	玄武质无球粒陨石
basaltic andesite	玄武安山岩
Basaltic aquifer	玄武岩含水层
basaltic augite	玄武辉石
basaltic clay	玄武质黏土
basaltic crust	玄武岩壳
basaltic crustal layer	玄武岩地壳层
Basaltic glassy substratum	玄武玻璃质底层
basaltic hornblende	玄武闪石；玄闪石
basaltic jointing	玄武岩节理；柱状节理；细脉
basaltic komatiite	玄武岩质科马提岩；玄武质科马提岩
basaltic lava	玄武质熔岩
basaltic layer	玄武岩层；硅镁层
basaltic magma	玄武岩浆；玄武质岩浆
basaltic meteorite	玄武质陨石
basaltic parting	玄武岩节理
basaltic rock	玄武质岩石

basaltic shell 玄武岩圈；玄武岩壳
basaltic structure 玄武岩构造
basaltic substratum 玄武岩基层
basaltic trachyandesite 玄武质粗安岩
basaltic tuff 玄武质凝灰岩
basaltic vitrophyre 玄武岩质玻基斑岩
basaltic wacke 玄武土
basaltiform 玄武岩状的
basaltine 玄武角闪石
basaltite 无橄玄武岩
basalt-like shaped 类似玄武岩柱状
basaltoid 像玄武岩的
basalt-porphyry 玄武斑岩
basalt-trachyte 玄武粗面岩
basanite 碧玄岩
basanitoid 玄武岩类，玻基碧玄岩
basanitoid foidite 碧玄质副长岩
basculating fault 翘断层；走向移动断层
basculating movement 翘起运动
base 基础，基底，基层，底座；墙基，柱脚，勒脚，踢脚板，打底层，底涂层；基质，基地，基线，碱；底边，底面
base adsorption 盐基吸附；基质吸附
base angle 基础边缘角铁；底角
base apparent viscosity 基液表现黏度
base band 基带
base bar 杆状基线尺
base block 基墩；基座垫块
base carriage 基线架
base centered lattice 底心晶格
base circle 坡底圆
base circle diameter 基圆直径
base concordance 底整合
base concrete 基础混凝土
base conditioning 基准条件
base contact 底部接触
base correction 基线改正；基线校正
base cost estimate 基本成本估计
base course 路面下层，底层，承重层，路基层，壁座；下垫层
base course drainage 基层排水；基底排水
base crude 原油
base data 原始数据；基本数据；原始资料；基本资料
base depth 基底深度
base desaturation 脱碱酌
base direction 基线方向
base discharge 基底流量
base discordance 底部不谐调
base exchange 盐基交换；碱交换；阳离子交换
base exchange capacity 碱性离子交换量；盐基交换容量
base failure 基底崩塌；底部破坏；基底破坏；基坍
base failure of slope 土坡基底破坏；坡脚坍塌；斜坡基底坍塌
base flow 基底流量；基本径流；流入排水系统的地下水
base flow hydrograph 基流过程线；地下径流过程线
base fluid 基液
base frame 底架，基架
base friction model 基础摩擦模型
base grid 基本坐标格网
base group length 组合基距

base heave 基底隆起
base heave failure 基底隆起破坏
base layer 底层；基层
base level 基准面；侵蚀基面
base level of abrasion 磨蚀基准面
base level of aggradation 填积基准面
base level of corrosion 溶蚀基准面；切蚀基准面
base level of denudation 剥蚀基准面
base level of deposition 沉积基准面
base level of erosion 侵蚀基准面
base level of stream 河流最低侵蚀面
base level peneplain 基准面准平原
base level plain 基面平原
base leveling epoch 均夷期
base levelling 基本水准测量
base line 基线；基准线；底线；扫描线；准线
base line adjustment technique 基线调整技术
base line correction 基线校正
base line crossing 基线穿越
base line dimensioning 基线尺寸
base line lane width 基线巷宽
base line measurement 基线测量
base line network 基线网
base line shift 基线偏移
base load 基础负荷；底面负荷；基极负载；基本负载
base load hydro-plant 基荷水电站
base loop map 基线网图
base map 基础图；基准图；工作草图；索引图
base map of topography 地形底图
base mark 基线标志
base material 垫层
base measurement 基线测量
base measurement cross over fault 跨断层基线测量
base measuring 基线测量
base measuring pressure 校准压力
base measuring tape 带状基线尺
base metal 母材；基底金属；碱金属
base mineral 基矿物；低值矿物
base moment 基底弯矩
base net 基线网
base network 基线网
base of a quay wall 岸壁基础
base of column 柱底；柱基座
base of corrosion 溶蚀基准
base of crude oil 原油的基类
base of dam 坝基；坝脚
base of excavation 开挖基线
base of foundation 基础底；地基
base of gas hydrate stability zone（BGHSZ） 天然气水合物稳定带底界
base of karstification 岩溶作用基底；溶蚀基准面
base of levee 堤基；堤底
base of permafrost 永久冻土基础
base of pillar 柱础
base of plate 板块底面
base of road 路基
base of slide 滑坡底面
base of sliding zone 滑动带底面
base of slope 边坡脚，坡底；斜面投影；大陆坡底
base of wall 墙基

base of weathering	风化层底面；低速带底面；风化基面
base oil	原油；基础油
base ore	贱矿石；低值矿石
base period	地下径流时期；基期
base pit	基坑
base plane	基面
base plate	基板；底盘；坡脚底塞
base plug	注水泥用下塞
base point	小数点；基点
base pressure	基底压力，基准压力；基压；底压
base price	标底
base projection	基线投影
base reflection	底面反射
base road	主巷道
base rock	基岩；底岩
base rock acceleration	基岩加速度
base rod	基线测杆
base rotation	基底转动
base run	基础测径
base runoff	基本径流；固定流量
base saturation	盐基饱和
base shear	基底剪切
base shear coefficient	基底剪力系数
base shearing resistance	基底抗剪强度
base sheet	原图；底图；基本图；片基
base shoe	井架大腿基脚
base slab	底板；平底板；基板
base slag	碱性矿渣
base slope	底坡
base soil	基底土；底土
base solution	基液；基本解
base stabilization	地基稳定
base station	基点，基线点；基准站；岸台；固定电台
base stock	基料；基本原料；基本组分；基本油料
base surface	基面；底面
base surge	基浪；底部涌浪
base surge deposits	底部涌浪沉积
base tankage	基本容量
base temperature	基础温度
base template	承框；基模台
base thickness	底厚；基本轴向厚度
base ties	基点连测
base tilt factor	基底倾斜因数
base value	基值
base volume	基准体积
base wall	底层墙
base width	底宽；基准宽度
base-flow storage	基流储量
base-height ratio	基-高比
baselap	超底；底覆；底部超覆
base-level lowland	基准面低地
baseline data	基本资料；基准数据；基线资料
baseline drift	基线漂移
base-line extension	基线延伸
baseline monitoring	本底监测；基线监测
base-line of section	断面基线
baseline of territorial sea	领海基线
baseline offset survey method	基线观测法
baseline survey	基线测量
baseline tape	基线尺
baseline vector solution	基线向量解算
baseline wander	基线漂移
basement	基础；基底；基岩；地下室
basement anisotropy	基底各向异性
basement block	基底断块
basement block faulting	基底块断作用
basement complex	基底杂岩
basement contour	基底等深线；基底等高线
basement controlled fracture zone	基底控制的断裂带
basement controlled trap	基底控制的圈闭
basement drain	地下室排水系统
basement drainage	地下室排水
basement extension	基础延伸部分
basement fault	基底断层
basement flexure	基底挠曲
basement floor	地下室层
basement fold	基底褶皱
basement folding	基底褶皱
basement fracture system	基底破裂系
basement high	基底隆起
basement horst	基底地垒；地垒式基底
basement lithology	基底岩性
basement mapping	基底绘图
basement migrate	基底漂移
basement nappe	基底推覆体
basement relief	基底起伏；基岩起伏
basement rift	基底断裂
basement rock	基底；基岩；基底岩石
basement slab	基础板
basement soil	基础土；地基；路基
basement structure	基底构造；地下室结构
basement style	基底型式
basement surge	基底岩涌
basement tanking	地库防水层；地库防水工程
basement tectonic grain	基底构造走向
basement tectonics	基底构造
basement terrane	基底地体
basement transcurrent fault	基底横推断层
basement uplift	基底隆起
basement wall	地库墙；地下室墙
basement-controlled structure	基底控制的构造
basement-cored nappe	以基底为核部的推覆体
basement-cover relationship	基底-盖层关系
base-point net	基点网
base-rock motion	基岩运动
base-vented hydrofoil	底面开孔水翼
bashing	充填采空区；充填废巷道
Bashkirian	巴什基尔阶
Bashkirian Age	巴什基利亚期
basic	基的；碱的；基本的
basic bearing capacity	基本承载力
basic benchmark	水准基点
basic border	镁铁边缘；基性岩边缘
basic bore	基孔
basic carbonate	碱性碳酸盐
basic cement	未加添加剂的水泥
basic control	基点控制；基本控制网
basic control survey	基本控制测量
basic data	原始资料；原始数据；基本数据
basic design	原图；原始草图

basic design criterion 基本设计准则
basic design data 基本设计资料
basic design load 基本设计荷载
basic deviation 基本偏差
basic earthquake intensity 地震基本烈度
basic element 基本元素
basic element of marine disasters 海洋灾害基本要素
basic exploratory line 基本勘探线
basic feasible solution 基本可行解
basic feature 基本要素
basic flow 基本气流
basic flow equation 基本流动方程
basic flowsheet 基本流程图；操作流程图；设备流程图；工艺原则流程图
basic frequency 主频；基频
basic friction angle 基本摩擦角
basic front 镁铁前缘
basic function 基函数
basic geochemical law 地球化学基本定律
basic geochemical map 基本地球化学图
basic geographic information system 基础地理信息系统
basic gneiss 基性片麻岩
basic granulite 基性麻粒岩
basic gravimetric point 基本重力点
basic hydrograph 基本水文过程线
basic hydrologic data 基本水文资料
basic hydrologic forecast 水文趋势预报
basic hydrologic information 基本水文情报
basic hydrological network 基本水文站网
basic hydrolysis 碱水解
basic hydrostatic time lag 基本静水压力滞后时间
basic igneous rock 基性岩浆岩；基性火成岩
basic intensity 基本烈度
basic intensity of earthquake 地震基本烈度
basic iron 碱性铁
basic lava 基性熔岩
basic line 底线；基线
basic load 基本荷载；主要荷载
basic motion equation 基本运动方程
basic net survey 控制导线测量
basic network 基本网络
basic ore 碱性矿石
basic parameter 基本参数
basic particle size 标准粒度
basic partitioned access method 基本分区存取法
basic period 基本周期
basic pig iron 碱性生铁
basic plagioclase 基性斜长石
basic plane 基面；基础平面；底图
basic point 基点
basic research 基础研究
basic reserve 基础储量
basic resistance 基本阻力
basic return rate 基准收益率
basic rock 基性岩，基性岩类；碱性岩
basic salt 碱性盐
basic sample 原始试样；基本试样
basic scale 基本比例尺
basic scale topographic map 基本比例尺地形图
basic sediment 碱性沉积物

basic sediment and water 底部沉积物和水
basic seismic parameter 地震基本参数
basic seismic wave 基本地震波
basic sequence 基本层序
basic shaft 基轴；主轴
basic soil 碱性土
basic spring 碱性泉
basic standard 基础标准
basic static correction 基本静校正
basic steel 碱性钢
basic stress 基本应力
basic structural index 基本结构指数
basic structure 基本构造；主要构造
basic sulphate 碱性硫酸盐
basic survey sheet 基本测量图
basic switching term 基本转换项
basic value 基本值
basic value of bearing capacity 承载力基本值
basic variables 基本变量
basic volcanics 基性火山岩
basic volume-mass relationship 基本体积-质量关系
basic water 碱性水；碱性泉水
basic Water Content 基本含水量
basic wavelet 基本子波
basic well pattern 基础井网
basicity 碱度；碱性
basicity constant 碱度常数
basification 碱化；基性岩化
basifier 碱化剂
basify 碱化
basin 煤田；池塘，洼地；盆地；流域
basin accounting 流域水量平衡；流域水账
basin analogy 流域相似
basin analysis 盆地分析
basin and Range 盆地和山脉
basin and range province 盆岭区
basin and range type structure 盆岭型构造
basin area 流域；流域面积
basin assessment 盆地评价
basin centered gas accumulations 盆地中心气藏
basin characteristics 盆地特征；流域特征
basin check method 畦灌法
basin configuration 流域外形；流域形状；流域轮廓
basin divide 流域分水岭；盆地分水岭；流域分水线
basin drainage area 流域
basin dry dock 池式干船坞
basin dynamics 盆地动力学
basin facies 盆地相
basin fill 盆地充填沉积；盆地沉积
basin flooding 畦田漫灌
basin fold 盆形褶皱；构造盆地
basin for artificial recharge 人工补给盆地
basin front 盆地边沿
basin interpretation 盆地解释
basin irrigation 淹灌
basin lag 流域滞延
basin length 流域长度
basin method 浅盆地渗入补给法
basin modelling 盆地模拟
basin models 盆地模型

basin mouth	流域出口；盆地出口
basin of combination type	复合型盆地
basin of stretch-sprain	张扭性盆地
basin outlet	流域出口
basin peat	盆地泥炭；潜育泥炭
basin perimeter	盆地周边；流域周长
basin plain	盆地平原；流域平原
basin plains deposit	深海盆地平原沉积
basin range	盆地山脉；盆岭；断块山岭
basin recharge	流域补给；流域再补给
basin shape	盆状；盆地形态；流域形状
basin shape factor	流域形状系数
basin storage	流域蓄水量
basin storage discharge	流域调蓄流量
basin structured	盆形构造的
basin swamp	湖缘沼泽
basin valley	盆状谷；宽浅谷地
basinal carbonate	盆地相碳酸盐
basinal facies	盆地相
basinal lake	内流湖；内壑；无出口湖；盆地湖
basinal reconstruction	盆地重塑
basinal subsidence	盆形沉陷
basin-analog	相似流域
basin-and-range structure	盆地山岭构造
basin-and-range terrain	盆岭区
basin-elongation ratio	流域伸长率
basin-floor fan	盆地底部之海下扇；盆底扇
basining	盆地形成作用
basin-like fold	盆地形褶皱
basin-margin facies	盆地边缘相
basin-microcontinent system	海盆-微大陆体系
basin-parallel normal fault	平行盆地的正断层
basin-range landform	盆岭地貌；盆峡岭地貌
basin-range structure	盆岭构造；盆地山脉构造；断块构造
basin-shaped strata	盆状层
basin-shelf fractionation	海盆陆架间分异作用
basin-wide	全流域的；流域规模的
basin-wide programme	流域整体规划
basin-wide structure map	整个盆地的构造图
basiophitic texture	基性辉绿结构
basipetal	由上向下的；向基式
basis	基底；基线；基础；基本；根据
basis function	基函数
basis matrix	基矩阵
basis meridian	首子午线
basis triangulation network	基本三角网
basis vector	基向量
basite	基性岩类
basite porphyry	基性斑岩
basket	钻头的岩芯爪；岩心管；打捞管；铲斗；出矿篮
basket barrel	带岩芯爪的岩芯管；笼式取心筒
basket bit	带取样筒的钻头；齿状钻头
basket centrifuge	过滤式离心机；筐式脱水离心机
basket core	捞抓岩芯；岩样
basket core lifter	爪簧式岩芯提断器
basket dam	篮形填石坝
basket filter	筐式过滤器；篮式过滤器
basket for centrifuge	离心机转筒；离心机吊篮；离心机筛篮
basket handle arch	三心拱
basket sampler	笼式取土器
basket scraper	筛篮刮刀
basket screen	篮式过滤器
basket strainer	笼式过滤器；笼式滤网
basket type sampler	笼式采样器；微波含水量测定仪
basket-type bed load sampler	笼式推移质采样器
bason	盆地
bass	劣煤，煤质黏土，炭质页岩；碳质黏土；黑矸石；秃椴木
bassanite	烧石膏
basse Oil-Measures	无油砂岩组
basset	露头；露出；地层露头部分
basset edge	煤层露头
basseting	露出；露头
bassetite	铁铀云母
bast	劣质煤；碳质黏土；碳质页岩；黑色碳质页岩
bast fibre	韧皮纤维
bastard	不纯的；异形的；劣质的；劣货；夹石，硬岩块
bastard coal	高灰分煤；薄煤夹层；硬煤；煤矸；劣质煤
bastard fireclay	不纯耐火黏土
bastard freestone	劣质毛石
bastard granite	片麻状花岗岩；片麻岩
bastard rock	不纯砂岩
bastard sand	灰砂层
bastard shale	烛煤页岩
bastite	绢石
bastitic	绢石状的
bastitie structure	绢石构造
bastnaesite	氟碳铈矿
bat	煤层中的碳质页岩夹层；铁石层中的页岩夹层；泥质页岩
BAT piezometer	BAT 型孔隙水压力计
batardeau	堤坝，围堰；堤；隔墙
batch	一次；一批；一次操作所需的原料量
batch capacity	一次拌合量
batch filter	间歇式过滤机；分批式过滤机
batch filter press	间歇式压滤机
batch leaching	间歇浸出；分批沥滤
batch load	批量加载
batch meter	分批定量器
batch mixer	分批式搅拌机
batch of concrete	分批混凝土
batch of mortar	分批砂浆
batch partial melting	批式部分熔融
batch plant	拌合厂；配料厂
batch production	批量生产；间歇生产
batch settling tank	间断沉降槽
batch system	一次配料法
batch test	单元试验；分批试验
batching	批次配料；配料；分批；计量；选择混凝土配合比
batching and mixing plant	配料及搅拌设备
batching by volume	按体积配料
batching equipment	配料设备
batching mixer	拌合机
batching plant	搅拌厂；混凝土配料机
bate	假节理；巷道卧底
batea	淘金盘；淘砂盘
bateau bridge	浮桥

bateque 泉水沉积物；泉水沉积
bath 槽，池；浴；电解液
bath carburizing 液体渗碳
Bath stone 巴斯岩
bath tank 电镀槽；洗涤槽
bathe 冲刷
bathimetry 水深测量
bathmeter 深度计
bathoclase 水平节理
bathograd 深变质级
batholite 最终沉降；岩基；岩盘
batholith 基岩，岩磐；岩基；盐湖；中间带
batholith mineralization hypothesis 岩基成矿说
batholithic granite 岩基花岗岩
batholithic joint 岩基节理
batholithic mass 岩基体
bathometer 测深计；测深仪；测深器；采样器
bathometers string 水深仪测量绳
bathometric chart 海洋深度图
bathothermograph 周期流
bathroclase 水平节理
bathtub chart 澡盆状波前图；等反射时间波前图
bathyal 次深海的；半深海的；半深海
bathyal deposit 次深海沉积；半深海沉积；半深海沉积物
bathyal environment 深海环境；半深海环境
bathyal facies 半深海相；次深海相
bathyal lake zone 半深水湖带
bathyal marine zone 半深海带
bathyal milieu 半深海环境
bathyal region 半深海区
bathyal sediment 半深海沉积物
bathyal zone 半深海带；半深海区
bathybic 深海层生物；深海生活的；深海底的
bathyderm 深部硅铝层
bathydermal 硅铝层下部的
bathydermal class 硅铝层下部运动级
bathydermal deformation 硅铝层深部地壳形变
bathydermal gliding 硅铝层下部滑动
bathygenesis 构造沉降；下降运动；负向运动
bathygenic movement 陆沉运动
bathygraphic chart 海深图
bathygraphic map 水深地形图
bathylimnion 湖底静水层；湖底滞水层；最深湖水层
bathylite 岩基
bathylith 岩基
bathymeter 测深器，测深仪，海洋测深器
bathymetric 测深的；等深的；深海的
bathymetric biofacies 深海生物相
bathymetric chart 海底等深线图，等深图；海洋深度图；海底地形图；水深图
bathymetric contour 等深线
bathymetric curve 等深曲线
bathymetric data 测深数据
bathymetric fascia 深海带
bathymetric high 水深高度；海底高地
bathymetric investigation 测深调查
bathymetric line 等深线；等水深线
bathymetric low 海底低地
bathymetric map 水深图；海图
bathymetric measurement 深海水深测量

bathymetric models 测深模型
bathymetric navigation system 海洋测深导航系统
bathymetric profile 等深剖面图
bathymetric slope 等深斜坡
bathymetric survey 水下测量；水下高程测量；海深测量
bathymetric surveying 海底地形测量；水下地形测量
bathymetric system 深海测深系统
bathymetriccontour 海底等深线
bathymetry 海洋测深学；测深学；测深法
bathymetry using remote sensing 遥感海洋测深
bathyorographic map 水深地形图；海底地形图
bathyorographical map 水深地形图
bathypelagic 半深海的
bathypelagic Zone 深海区；海洋半深海带；远洋深层带；下均匀层
bathyphotometer 深水光度计
bathyrheal underflow 地壳下部补偿物质流
bathyscaphe 深潜水器
bathyseism 深源地震
bathysphere 探海球；海底观测球；潜水球；深海潜球
bathythermogram 深度温度记录
bathythermograph 温深仪，温深计；深度温度仪；自计温度-深度仪
batt 页岩夹层；薄煤夹层；黏土质页岩；沥青质页岩；硬黏土
batten plate 斜坡推进法；缀板
batten plate column 缀合柱；空腹柱
batten sheet piling 打条板桩
batten underdrain 木板暗沟
batter 坡度，坡面，斜坡；锤打，打碎，槌薄，冲击；倾斜度；混合；揉黏土
batter brace 斜撑
batter chamber wall 液压传感器
batter drain 斜坡排水沟
batter drainage 溢流；斜坡排水；斜水沟
batter leader pile driver 斜导架打桩机；斜缆道
batter level 测斜仪；测坡仪
batter of face 阶段坡面
batter peg 斜坡标桩；斜桥墩；支撑桩；角撑桩
batter pile cluster 斜桩群
batter piling 打斜桩；斜排桩
batter post 斜支柱；斜撑；屋角防护柱；门道防护柱
batter rule 斜度尺；定斜器
batter wall 斜面墙；挡土墙
battered 倾斜的；斜削的；有坡面的；磨损的
battered bank system 倾斜护岸工程
battered columnar joint 倾斜柱状节理
battered pile 斜桩
battered prop 斜柱
battering 倾斜；挤压；磨损；打碎；捶薄，混合；冲击
battering ram 冲击夯；打桩锤；斜撞锤
battery assay 捣碎试样分析
battery of wells 井群；井组
battery stull 成排横撑；排柱；密集支柱
battery traction 柱槽筋
battery-type locomotive 原始固结
battie 炭质页岩；黑矸石
bature 河滩地；淤高河床；心滩
batukite 暗白榴玄武岩；暗色白玄岩
batymetric chart 等深线图

baulite	包斜流纹岩
Baume	波美；波美计
Baume density	波美度
baume gravity	波美比重；波美比重计
Baume hydrometer	波美比重计
Baume scale	波美比重标；波美比重计；波美标度
Baumescale	波美比重标度
Baumgarte ray-stretching method	鲍姆加特射线延伸法
baumhauerite	硫砷铅矿
bauxite	铝土矿，矾土
bauxite cement	矾土水泥；高铝水泥
bauxite clay	铝质黏土
bauxite deposit	铝土矿床
bauxitic clay	铝土
bauxitic laterite	铝土质红土
bauxitic ore	铝土矿石
bauxitite	铝土岩
bauxitization	铝土矿化
bavenite	硬沸石
Baveno twin	巴维诺双晶；斜坡双晶
bay	跨度；海湾，沼泽；间距；台，框
bay bar	湾内沙坝；海湾沙洲
bay barrier	湾口沙洲；湾口沙坝；海湾口障壁坝；拦湾坝
bay cable	水下电缆；海底地震电缆
bay crossing	海湾穿越
bay delta	海湾三角洲
bay disturbance	湾扰
bay harbour	海湾港
bay head	湾头
bay head bar	湾头坝
bay head barrier	湾头障壁坝
bay head beach	湾头滩
bay head delta	淤积平原；湾头三角洲
bay lake	湾状湖
bay mouth	湾口
bay mouth bar	湾口沙洲
bay sediment	湾沉积物
bay side beach	湾侧滩
bay water	海湾水
bayerite	三羟铝石；人工制备的氢氧化铝晶体
Bayes criterion	贝叶斯准则
Bayes Rule	贝叶斯法则
Bayesian statistics	贝叶斯统计
Bayesian uncertainty	贝叶斯不确定性
bayldonite	乳砷铅铜矿
bayleyite	碳镁铀矿
baylissite	水碳镁钾石
baymouth barrier	湾口沙坝
baymouth beach	湾口滩
bayou lake	浅滩海湾；牛轭湖；长沼
bayshon	挡风墙
bay-sound facies	港湾相
bay-water solid	海湾水中的固体物质
Bazin's average velocity equation	拜齐恩平均速度方程
bc-joint	纵节理
beach	海滩；湖滩；海滨
beach accretion	海滩冲积，海滩淤积，海滩堆积；海滩增长；海滩加积作用
beach accretion bedding	海滩加积层理
beach advance	海滩推进
beach and dune sands	海滩和沙丘砂
beach barrier	海滩暗礁；海滩沙堤；海滩沙埂；堡坝
beach berm	滩肩；滩边阶地
beach bottom	滩底
beach breccia	海滩角砾岩
beach building	海滩形成
beach chute	海滩流槽
beach complex	海滩复合体
beach concentrates	海滩重砂富集体
beach concentration	海滩重砂
beach crest	高滩脊
beach cusp	滩角；滩尖；滩嘴
beach cycle	海滩旋回；海滩循环
beach deposit	湖滩沉积；海滨沉积；湖滩沉积
beach drainage	海滩排水
beach drift	沿滩漂移
beach drifting	海滨漂流，沿海漂移；海滩漂积物
beach dune	海滩沙丘
beach embayment	海滩湾畔
beach environment	滩环境
beach erosion	海滩冲刷；海滩侵蚀
beach erosion control	海滩侵蚀控制
beach etching	海滩蚀刻
beach face	海滩面；海滩
beach facies	海滩相；湖滩相
beach fill	海岸淤填土；海滩填积
beach firmness	海滩坚固性
beach formation	海滩建造，滨海建造；海滩形成
beach formation by waters	浪成海滨；浪成滩；海浪冲积滩地
beach gradient	海滩梯度
beach gravel	海滨砾石，海滩砾石；海岸砾石
beach lamination	海滩纹理
beach line	海岸线；滩线
beach material	海滩物质
beach metal placer	海滨金属砂矿
beach mineral resources	滨海矿产
beach mining	海滨采矿
beach morphodynamics	海滩地貌
beach nonmetal placer	海滨非金属砂矿
beach nourishment	沙滩补给；阻水性；海滩淤涨；海滨淤填
beach pad	滨岸三角沙坝
beach placer	海滨砂矿；海滨漂砂矿床；海滨砂矿
beach placer exploitation	海滨砂矿开发
beach placer exploration	海滨砂矿勘探
beach placer mining	海滨砂矿床开采
beach plain	海滩平原；湖滩平原
beach platform	滩头浪蚀台地，浪蚀台
beach price	海底原油在海上处理后之陆上价格
beach profile	海滩剖面；海滨剖面
beach profile of equilibrium	海滩均衡剖面
beach protection and accretion promotion	促淤保滩
beach race	滨海水道
beach rampart	滩脊
beach reclamation	滩涂开垦；滩涂围垦；陷落地块
beach reef	滩礁
beach replenishment	海滩填补
beach retreat	海滩后退

beach retrogression 海滩后退
beach ridge 滩脊；沿岸堤；海滩堤
beach ridge facies 海滩脊相
beach rock 滩岩；洋滨石灰簧砂岩
beach sand 海滩砂；海滨砂；海滩泥沙
beach sandstone 海滩砂岩
beach scarp 海滩崖；浪蚀海滩崖
beach sediments 沙滩沉积物
beach slope 滩坡；海滨比降；海滩坡度
beach stability 海滩稳定
beach strand 滩岸
beach surveying 海滨测量
beach terrace 海滩阶地
beach width 海滩宽度
beach zone 海滨区；海滩区；海滩带
beach-backshore 滨后滩
beach-barrier lagoon 海滩-障壁泻湖
beach-comber 拍岸浪，滚浪
beach-combing 海滨开采砂矿
beached bank 铺石护岸
beach-head 滩头
beaching 海岸堆积；海滩堆积；砌石护坡
beaching of bank 护岸工程
beach-sand placer deposit 海滨砂矿床
beachy 近岸的；岸边浅滩的
beacon 航标；浮标；灯塔；讯标
beacon delay 信标延迟
beacon navigation and positioning 信标台导航定位
bead pack 玻璃珠子人造岩心
bead reaction 珠球反应；熔球反应
bead size 珠子大小；珠滴大小
bead storage 垫底库容
bead welding 向海防波堤
beaded cementation 串珠状胶结
beaded drainage 连珠小河
beaded esker 串珠蛇形丘
beaded glass 玻璃珠
beaded linear ridge 线形连珠山脊
beaded stream 深切融沟河
beaded texture 串珠状结构
beaded vein 珠串式脉
beading drainage 串珠状水系
beadlike 串珠状的
beads structure 串球状构造
beads-shaped 串珠状；串珠状的
beads-shaped structure 串珠状构造
beak 喙，鸟嘴；钩形鼻；柱的尖头
beak head 海角
beak shaped 喙状的
beaker sampler 杯形取样器；杯选取样
bealock 垭口；分水岭山口
beam 梁；桁条；射线，射束；波束；梁，横梁
beam anchor 梁端锚栓
beam angle 波束角；射束孔径角
beam balance 韦斯特法耳比重秤
beam bender 液压升降机
beam bending stress 悬梁弯曲应力
beam bridge 梁桥
beam building 锚固成梁
beam building effect 组合梁效应

beam column joint 梁柱节点
beam compasses 长杆圆规；横杆
beam culverts 梁涵洞
beam curvature 梁的曲率
beam deflection 梁垂曲；梁弯曲；梁挠曲
beam deviation 光束偏移；横臂偏转
beam divergence 光束发散度
beam draft ratio 宽度吃水比
beam drill 摇臂钻床，悬臂钻床
beam efficiency 波束效率；射束效率
beam on elastic foundation 弹性基础梁；弹性地基梁
beam on elastic support 弹性支承梁
beam pitman 锚筋；拉杆
beam position 横臂位置；摆位
beam post 梁柱
beam principle 梁原理
beam riding 束导
beam rock 岩梁；方位
beam roof timber 顶梁
beam slab 梁式板
beam slab system 梁板体系
beam soffit 梁拱腹
beam span 梁跨
beam splitter 分光束镜；光束分离器
beam splitting mirror 光束分裂镜
beam stack 倾斜叠加
beam statics 梁静力学
beam steering 延时组合；时差叠加法；时间倾角扫描叠加
beam test 抗弯试验；梁试验
beam texture 梁结构；梁状结构
beam theorem 岩梁理论
beam under foundation 地基梁
beam vibration 梁振动
beam wave 束状波
beam wave-guide 波导
beam well 抽油井
beam width 波束宽度；波束角；射线宽度
beam with both ends built in 固端梁
beam with both ends fixed 两端固定的横梁
beam with central prop 三托梁；三支点梁
beam with double reinforcement 双筋梁 双面配筋架
beam with fixed ends 固端梁
beam with one overhanging end 悬臂梁
beam with overhanging ends 两端悬臂梁
beam with simply supported ends 简支梁
beam with single reinforcement 单筋梁；平面配筋架
beam with variable cross section 变截面梁
beam-and-column construction 梁柱结构
beam-column 梁柱结构
beam-foil system 梁板体系
beaming 辐射
beam-to-beam connection 梁与梁的连接
beam-to-beam moment connection 梁的刚性连接
beam-to-column connection 梁柱连接
beam-to-column moment connection 梁-柱抗弯接头
beam-to-column stiffness ratio 梁-柱刚度比
beam-to-depth ratio 幅深比
beam-type gravimeter 横臂式重力仪
beam-type roof 梁式顶板

bean ore	豆铁矿
bear	承受；支承；掏槽，打孔器；小型冲孔机
bearable load	可承负载；可承载重；最大允许载荷
bearer	承木；垫块；托架；井筒框架支梁
bearer cable	承载索；受力绳
bearing	支承；承载；方位；方位角，象限角
bearing ability	承载力
bearing accuracy	定位精度
bearing anchor	轴承座
bearing angle	方位角；象限角
bearing area	支承面积；承重面积
bearing axle	支承车轴；承压车轴
bearing backing	轴承衬
bearing ball	轴承钢珠
bearing bar	承杆；支承杆
bearing beam	承载梁
bearing block	矿柱；煤柱；支承块；承重块
bearing calibration	方位校准
bearing capacity	支承力；承载力
bearing capacity equation	承载力方程
bearing capacity factor	承载力系数；承载力因数
bearing capacity failure	承重能力破坏
bearing capacity formula	承载力公式
bearing capacity of a pile	单桩承载力
bearing capacity of a single pile	单桩承载能力
bearing capacity of a structure	结构承载力
bearing capacity of anchor	锚杆允许承载力
bearing capacity of foundation	地基承载力
bearing capacity of ground	地基承载力
bearing capacity of multi-layered ground	多层地基的承载力
bearing capacity of pile	桩的承载力
bearing capacity of pillars	矿柱承载力
bearing capacity of props	支柱的承载能力
bearing capacity of roof and floor rocks	顶板和底板岩体的承载力
bearing capacity of single pile	单桩承载力
bearing capacity of soil	土的承载力；地基承载力
bearing capacity of the ground	土地承载力
bearing capacity under eccentric load	偏心荷载承载力
bearing capacity under inclined load	倾斜荷载承载力
bearing course	承压垫层；承压层
bearing deformation	承压变形
bearing effect by base course	下垫层的承载效果
bearing face	承压面；支承面
bearing failure	承压损坏；承载破坏
bearing fatigue point	轴承疲劳极限
bearing force	承重能力；承载力，支承力，承压力
bearing force of soil	土的承压力
bearing graph	反应曲线；承载曲线图
bearing indicator	方位指示器
bearing inspection platform	预计沉陷
bearing layer	支承层；持力层
bearing line	象限角线；走向线；方位线
bearing liner	轴瓦；轴衬
bearing lining	轴承衬
bearing load	轴承负载
bearing mat	承压垫板
bearing mechanism of pile	桩的承载机理
bearing meter	方位指示器；方向显示器
bearing of base line	基线方位角
bearing of subgrade	地基承载力
bearing of trend	走向；延伸方向
bearing operator	测向员
bearing pad	支承垫片；承重垫片；承压垫板；支座垫
bearing partition	承重壁；承重隔墙
bearing pile	支承桩；端承桩；承重桩
bearing pile through structure	贯通桩
bearing pile wall	承重桩墙
bearing piling	支桩
bearing plate	承重板，垫板；承压板；载荷板
bearing plate technique	承压板法
bearing plate test	承压板试验；承压板法
bearing platform	承台
bearing point	方位点；支承点
bearing post	支承柱
bearing power	承载能力；支力；承重能力
bearing power of soil	土壤承载力
bearing pressure	承压力；轴承压力；承载力；轴承比压
bearing pressure on foundation	地基承载力
bearing ratio	承载比；承重比
bearing ratio test	承重系数试验；载重比试验
bearing reaction	支承反力
bearing resistance	承压强度；抗压力；轴承阻力；承压抗力
bearing ring	支承框架
bearing rod	支柱；承重杆
bearing strain	承压变形；承压应变
bearing stratum	承载层；含矿层；持力层，承重层
bearing strength	承压强度；荷重强度；承载强度；承压力
bearing stress	支承应力；承压应力；挤压应力
bearing structure	支承结构；承重结构
bearing surface	支承面，承压面；承载面
bearing surface of foundation	地基支承面；基础支承面
bearing test	载荷试验；承压试验；承载力试验；承重试验
bearing test of soil	土壤承载试验
bearing value	承载力值；支承值；支承量
bearing value of soil	土壤支承量；土壤承载力
bearing wall	承重墙；结构墙
bearing wall and frame structure	内框架结构；组合结构
bearing wall construction	承重墙构造
bearing wear constant	轴承磨损常数
bear's grease	严重浸软的泥炭
bear-trap sill	隐晶岩
beat cob work	捣土工作
beat count	锤击数
beater	冲击式破碎机；搅拌机；打浆机；锤
beater pick	夯土镐；捣固道碴镐
beater pulverizer	锤碎机
beating machine	打浆机
beaumontage	填土料
beaver type timber dam	堆木坝；木笼填石坝
beaverite	铜铅铁矾
beaver-type timber dam	堆木坝
bebedourite	钙钛云辉岩
bech scarp	滩坎；滩崖
beche	打捞母锥
bechilite	硼酸钙石
beck	山涧

Beck hydrometer	贝克比重计
beck index	贝克指数
Becke line	贝克线
Becke line test	贝克线法
beckelite	方钙饰镧矿
becken	盆地
beckite	玉髓燧石
Becquerel	贝可
becquerelite	深黄铀矿
bed	路床；河床；基层；地基，路基；层，垫层，基层
bed accretion	河床淤高
bed bottom	层底；河床
bed boundary	地层界面；岩层界线；岩层边界
bed characteristics	河床特性
bed coke	底焦；焦床
bed complex	基底杂岩
bed condition	河床状况
bed configuration	河床地形；河床形态，海底形态；河槽形状
bed conglomerate	底砾岩
bed contour	河底等高线；河床等高线
bed correction	层厚校正
bed course	矿层走向；底层；垫层；泻湖循环
bed crack	底面裂隙
bed cracking	床层裂化
bed current	底层流
bed deposit	层状沉积
bed deposite	层状沉积
bed depth sensor	床层厚度传感器
bed development	河床演变
bed dune	河底沙丘
bed elevation	河底沙丘
bed erosion	河床冲刷
bed extension	矿层延长
bed filtration	多层过滤
bed form	河床形状；河槽形状
bed form roughness	底床糙度
bed form superimposition	底形叠加作用
bed formation	河床形成；海底形成
bed forms	河床
bed frame	基座架
bed frame of pile driver	桩架座基
bed geometric factor	地层几何因子
bed geometry	地层几何形态
bed gradient	河床比降；河床斜度
bed groin	河底潜堤
bed groyne	河底潜堤
bed interface	层界面
bed irregularity	河床不平整度
bed irrigation	应力体系
bed joint	顺层节理；水平砌缝
bed layer	基层；河床底层
bed level	河床高程
bed load	床沙；床沙载荷；推移质；河床负载
bed load discharge	推移质输沙率；推移质输送量
bed load equation	推移质方程
bed load formula	推移质公式
bed load function	推移质函数
bed load movement	推移质运动
bed load rate	推移质输沙率
bed load sampler	推移质采样器；底沙取样器
bed load sediment	推移质沉淀物
bed load transport	推移质输送
bed material	河床质
bed material load	推移质；底部物质负荷
bed material sampler	河床质采样器
bed material size	河床质粒径
bed moisture	煤层水分；内在潮湿；内在水分
bed movement	河床移动
bed mud	底泥
bed of debris	岩屑和泥饼垫层
bed of interest	目的层
bed of passage	过渡层
bed of river	河床；河底
bed of vein	脉层；层状脉
bed outcropping	层露头
bed phase	河床相
bed plane	层理面；层面
bed plate	座板；底板；基板
bed plotter	平板绘图仪
bed profile	沙床断面；河床纵剖面；河底纵剖面
bed resistance	河床阻力
bed resolution	分层能力
bed response	地层响应
bed ripple	河底高低不平；河底波痕
bed rock	基岩，底岩；围岩；砂矿基底；岩床
bed rock accumulation	基岩油气藏
bed roughness	河床粗糙度；河床糙率
bed sample	煤层试样；矿层试样；地层标本
bed scour	河床冲刷
bed separation	离层；分层作用；层状剥落；煤层离距，离层；层状节理化
bed separation cavities	离层腔；层状空洞
bed series	层群；岩系；地层序列；层系
bed set	层组
bed shear	层面剪切；河床剪切
bed shear stress	底床剪切应力
bed silt	底部淤泥；底粉砂
bed slope	河底坡降；河床坡降；河床坡度；河床比降
bed stabilization	河床加固
bed stone	基石；座石；底石
bed stratum	岩层
bed succession	地层顺序；地层序列；层序
bed thickness	地层厚度
bed thinning	煤层变薄；矿层变薄
bed timber	垫木；枕木
bed top	矿层顶板；煤层顶板
bed vein	层状脉；板状脉
bed velocity	河底流速
bed waves	河底波纹状起伏；沙浪；沙波
bed width	河床宽度
bed-building stage	造床水位
bed-by-bed analysis	逐层分析
bedded	层状的，成层的；平整层状的
bedded chert	层状燧石；层状燧石
bedded cherts	层状硅质岩
bedded deposit	层状矿床，层状沉积
bedded fissure water	层状裂隙水
bedded formation	层状建造，层状构造；层状地层；层状岩组；层状岩系

bedded iron formation	层状含铁建造	bedding-in	磨合；研配
bedded iron ore	层状铁矿	bedding-plane markings	层理面标志；层面印痕
bedded medium	成层介质	bedding-plane movement	顺层位移；顺层变位；层面滑动
bedded ore deposit	层状矿床	bedding-plane parting	层面裂理
bedded plane	层面；层理面	bedding-plane power spectrum	层面功率谱
bedded quartz	层状石英	bedding-plane shear surface	层面剪切滑面
bedded reef	层状礁	bedding-plane structure	层面构造
bedded rock	层状岩；成层岩	bedding-plane thrust fault	顺层冲断层
bedded rock slope	层状岩石斜坡	bedding-step model	层阶式
bedded rockfill	分层堆石；分层抛石；成层填石	bed-forming discharge	造床流量
bedded structure	层状构造；层状结构	bed-load calibre	河床推移质输送能力
bedded surface	层面	bed-load channel fill	底荷河道沉积
bedded vein	层状脉；顺层状脉	bed-load sampling	底沙取样
bedded volcano	成层火山	bedload segregation	泥沙分离
bedding	层理；成层，衬垫，基底；跳汰机的人工床层	bed-normal stylolite	立层缝合线
bedding material	垫层材料	bed-parallel fracture	顺层裂缝
bedding angle	层面方向；层面角	bed-parallel stylolite	顺层缝合线
bedding architecture	层理结构	bed-plate foundation	板式基础
bedding cave	层面溶洞；层面洞穴	bedrock	基岩
bedding cleavage	层面劈理，顺层劈理	bedrock acceleration	基岩加速度
bedding concrete	垫层混凝土	bedrock attenuation	基岩衰减
bedding correction	层面改正	bedrock cut	底岩槽沟
bedding course	基岩层；下伏层；垫层，褥垫，铺底	bedrock geochemistry	基岩地球化学
bedding cushion	垫层	bedrock geology	基岩地质
bedding error	端垫误差	bedrock geology map	基岩地质图
bedding fault	层面断层，顺层断层	bedrock intensity	基岩烈度
bedding fissibility	层面裂开性；顺层裂开性；岩石层理裂开性	bedrock map	基岩地质图
bedding foliation	层面叶理，顺层叶理	bedrock monitoring well	基岩监测井
bedding glide	层面冲断层；掩冲断层；逆掩断层；顺层冲断层	bedrock motion	基岩运动
		bedrock outcrop	基岩露头；底岩露头
bedding in layers	分层铺垫	bedrock plan	基岩地质图
bedding joint	层面节理；顺层节理	bedrock relief	基岩起伏
bedding lamination	层理	bedrock seated terrace	基座阶地
bedding landslide	顺层滑坡	bedrock slack zone	基岩松弛带
bedding layer	垫层	bedrock soil	基岩土壤
bedding material	垫底材料	bedrock spur	基岩突出部，暗礁
bedding mortar	砂浆垫层；垫层砂浆	bedrock structure interaction	基岩-结构相互作用
bedding parting	层理；层面裂理；顺层面	bedrock surface	基岩顶面
bedding plane	层面；层理面	bedrock surveying pole	基岩标
bedding plane cleavage	顺面劈理	bedrock topography	基岩地形
bedding plane fault	顺层断层；层面断层	bedrock topography map	基岩地形图
bedding plane slip	层面滑动	bedrock weathering zone	基岩风化区
bedding plane weakness	沉积弱面	bedrock-soil layer model	基岩-土层模型
bedding recumbent folds	顺层掩卧褶皱	beds of passage	过渡层
bedding rock	基岩；矿层底岩	beds of precipitation	沉淀层；化学沉积层
bedding rockslide	层状滑坡	beds pattern	河床型式
bedding sag structure	层兜构造	bedsoil	支承土壤
bedding sand	地基砂层；垫层砂；基层砂	bedway	床身导轨，滑板
bedding schistosity	层面片理；顺层片理	beegerite	辉铋铅矿
bedding screed	地台砂浆底层	beehive	蜂房；蜂窝状的
bedding slip	层面滑动，顺层滑动	beekite	玉髓
bedding slip fault	顺层滑动断层	bee-line	最短距离
bedding stratification	层理	beerbachite	辉长细晶岩；辉长细岩；微晶辉长岩
bedding structure	层状构造	beetle head	送桩锤
bedding surface	层面；顺层面；层理面	beetling cliff	悬崖
bedding surface parting	层面裂理	beetling wall	绝壁
bedding thrust	层面冲断层，顺层冲断层	befanamite	锆钪钇石
bedding value	基床系数；地基系数	beforehand withdrawal of gas	预抽瓦斯
bedding void	层间空隙	beforsite	镁云碳酸岩；细白云岩
		behavior	性状；特性；行为；效应

behavior index 流体流态指数
behavior of surrounding rock 围岩特性
behavior pattern 流型
behaviors of strata 地层的行为
behaviour of a river 河流性态
behaviour of rock mass 岩体性状
behaviour of soil 土的性状
behaviour of structure 结构性能；构造性能
behaviour of well 井的动态；井的生产特性
behavioural model 性状模型
beheaded river 断源河；断头河；夺流河
beheaded stream 断头河；被夺流河；被夺河；被袭夺河
beheaded valley 断头谷
beheading 袭夺
beheading of river 夺流
behind the lining 衬砌背面
behindarc 弧后
beidellite 贝得石
bekinkinite 霞岩辉
belaying pin 系索栓；套索桩
belching well 间歇井
belemnite 箭石
belemnite carbonate 箭石碳酸盐
bell and spigot bend 承插式弯头
bell and spigot joint 承插接头；钻探坑；套筒接合
bell caisson 钟形沉箱
bell dolphin 新月形沙丘
bell out 使成漏斗形
bell pit 扩底探坑；钟形坑；小探井
bell rope hand pile driver 人工拉索式打桩机
bell section 钟形断面
bell shaped function 钟形函数
bell socket 钟形打捞工具；打捞筒；母锥；套接
bell tap 打捞母锥
bell valve 周边出水堰
belled 漏斗形的
belled caisson 扩底桩；扩底墩
belled chute 钟形口溜槽
belled excavation 原木导墙
belled out 扩大成漏斗形的
belled pier 扩底墩；扩底桩
belled pile 扩底桩
belled shaft 扩底墩
belled-out 扩大成漏斗形的
belled-out cylinderical pile 大头圆柱桩
belled-out pile 扩底桩；大头桩
belling 基底做大放脚；凸出的形状；扩大放矿漏斗
belling bucket 扩底挖斗
belling tool 扩底工具
bellite 铬砷铅矿；贝尔炸药
bellmouth inlet 喇叭形进水口；漏头形进入口
bell-mouth intake 漏头形进水口；喇叭形进水口
bell-mouthed 有喇叭口的；漏斗形的
bell-mouthed opening 套口孔；承口孔
bellout 扩底；扩大巷道口
bellout diameter 扩底直径
bell's bund 导堤
bell-shaped 钟形的
bell-shaped assembly anchorage 钟形组合锚头
bell-shaped distribution 钟状分布；正态分布

belly 煤层变厚；矿脉变厚
belly brace 曲柄钻
beloeilite 长霓方钠岩
belonite 针雏晶；棒雏晶
belonosphaerite 针雏晶球粒
belonospherite 针皱晶球粒
below datum 地表下基准面
below grade 不合格的；标线以下的
below ground 地面以下
below-grade digging 向下挖掘
below-ground 地下的
below-ground masonry 地下圬工；地下砌体
below-ground storage 地下储存
below-order leveling 等外水准测量
belt 区，带；带状物；胶带；传动带
belt conveyer 传送带；带式输送机
belt course 带状层；圈梁
belt highway 城市环路；外环公路
belt meander 带状河曲
belt of cementation 胶结带
belt of erosion 侵蚀带
belt of fault 断层带
belt of fluctuation 波动带；变动带；升降带
belt of fluctuation of water table 水位变动带
belt of folded strata 褶皱带
belt of marshes 湿地带
belt of no erosion 无侵蚀带
belt of phreatic fluctuation 地下水位波动带；潜水位波动带
belt of seismic activity 地震活动带
belt of soil water 土壤水带
belt of totality 全食带
belt of transition 过渡带
belt of wandering 河流摆动带；河道蜿蜒带
belt of water table fluctuation 地下水位波动带
belt of weathering 风化带
belt sampler 胶带采样机
belt sand body 带状砂体
belt screen 带状拦污栅；带状筛；带筛机
belt separator 带式磁选机；带式分选机
belt speed sensor 带速传感器
belt stacker 胶带排土机；皮带装料机
belt stowing 胶带充填
belted coastal plain 带状沿岸平原
belted metamorphics 变质岩条带
belted outcrop plain 条带状露头平原
belugite 闪长辉长岩
bementite 蜡硅锰矿；硅锰矿
bench 台阶，阶段，台地，台坎；阶地；岩滩；工作台；海蚀平台
bench angle 台阶坡面角；台阶坡面线
bench antivibrator 台钻
bench beacon 水准点标志
bench blasting 阶段爆破，台阶爆破
bench cut 台阶式开挖；阶梯开挖；台阶法；台阶状斜坡
bench cut method 台阶开挖法；分层切割法；垂直切片开采法
bench cutting 台阶开掘；台阶式采掘；台阶式挖土
bench drill 台钻
bench driliing 阶段钻眼，梯段钻眼，台阶钻眼

bench edge	台阶边缘，坡顶线；眉线
bench elevation	台阶标高，平盘标高
bench excavation	台阶式开挖
bench excavation method	分断面掘进法
bench face	台阶工作面
bench floor	阶地底；台阶底；梯底；阶段底
bench flume	低渡槽；台架式渡槽
bench gravel	阶地砾石；阶地砂矿
bench land	阶地
bench lava	阶状熔岩
bench length	台阶长度
bench magma	半固结岩浆
bench mark built in wall	墙上水准点
bench mark elevation	基准点高程；水准点高程；水准点标高
bench marks	基准桩；水准点
bench method	台阶式挖土法
bench mining	台阶式开采；阶梯式开采；煤和夹矸层分采
bench out	掏槽
bench placer	阶地砂矿；高地砂矿
bench section	横剖面
bench slope	台阶坡面；台阶破面角
bench slope angle	台阶坡面角
bench slope face	台阶式边坡面
bench step	台阶
bench stoping	台阶式开挖；梯段形开采
bench system of sampling	重叠堆锥缩分取样法
bench terrace	阶形地埂；梯田；阶地
bench test	基准试验；台架试验；试验台试验；实验室研究
bench toe	坡底线；台阶坡底
bench toe brow	坡底线
bench toe rim	坡底线
bench wall	承拱墙
benched	台阶式的，阶段式的
benched excavation	阶梯形开挖
benched foundation	台阶式基础；阶梯形基础
benched quarry	阶段采石场
benched subgrade	台口式路基
bench-face driving method	台阶工作面掘进法
benching	沟肩；台阶式开采；用楔破裂底煤
benching bank	阶段，台阶形边坡
benching cut	台阶式挖土
benching-up	采掘煤层上部
bench-like form	阶状地形
benchmark	水准点，测量标点，测点；基准点，基准；基准测试程序；基准标法
bench-mark basin	参证流域；参考流域
benchmark crude	标准原油
benchmark data unit	标准测试数据单元
bench-mark lake	参证湖泊
benchmark leveling	基平；水准点高程测量
bench-mark soil	标准土
bench-mark station	参证站；参证水文基站
benchmarking	基准；标杆管理；标杆
benchs crust of the earth	风化壳
bench-scale test	小型试验
benchscale testing	实验室试验
bend	弯曲；挠曲；弯头，弯管；硬黏土
bend angle	弯角
bend bar	元宝筋；弯起钢筋
bend drill pipe	弯钻杆
bend ductility	弯曲延性
bend flattening	河曲变缓
bend folding	隆曲褶皱
bend gliding	挠曲滑动
bend improvement	相对含水量
bend in river	河曲
bend loss	弯道损失
bend metre	相对刚度比；硬质面层
bend of strata	层弯曲；弯曲地层
bend over	折转
bend pipe	弯管
bend radius	弯曲半径
bend strength	抗弯强度
bend summit	河弯顶点
bend test	弯曲试验
bender	钻井岩屑；弯曲器；泵缸上提环；挠曲机；
bend-glide folding	曲滑褶皱
Bendigonian	本迪阶
bending	弯曲，挠曲；配曲调整；偏移，折曲；无线电波束曲折
bending and unbending test	曲折试验；反复弯曲试验
bending angle	弯曲角
bending axis	弯曲轴
bending brittle point	弯曲脆点
bending cap	盖梁
bending coefficient	弯曲系数
bending crack	弯曲裂缝
bending deflection	挠度；挠曲变位
bending deformation	弯曲形变；挠形变
bending deformation tensor	曲率变形张量
bending failure	受弯破坏；挠曲毁坏
bending fatigue	挠曲疲劳；弯曲疲劳
bending fatigue test	弯曲疲劳试验
bending flexure	挠曲，弯曲
bending fold	拱曲褶皱；横弯褶皱作用
bending force	弯力
bending formula	弯曲公式
bending glide	挠曲滑动
bending line	挠曲线；总渠
bending machine	弯筋机
bending modulus	弯曲模量
bending moment	挠矩，弯矩；弯曲力矩；抗弯性能
bending moment area	弯矩图面积
bending moment arm	弯曲力臂
bending moment diagram	弯矩图
bending moment envelope	弯矩包络线
bending moments	弯矩
bending of bed	地层挠曲
bending of strata	层弯曲
bending of vault	拱顶弯曲
bending point	弯曲点；挠曲点
bending press	压弯机
bending radius	挠曲半径；弯曲半径
bending ratio	弯曲比
bending resistance	弯曲刚度，抗弯强度；抗弯力
bending rigidity	抗弯刚度
bending roof	弛垂顶板；弯曲顶板

bending stiffness	弯曲刚度；抗弯劲度；抗弯刚度
bending stiffness factor	抗弯刚度系数
bending strain	弯曲应变；挠应变
bending strength	抗弯强度；弯曲强度；抗挠强度
bending stress	弯曲应力；屈曲应力；挠应力
bending subsidence	弯曲下沉
bending test	抗弯试验；弯曲测试
bending vibration	弯曲振动
bending wave	弯曲波
bending yield point	抗弯屈服点
bending-swelling out	弯折内鼓
bending-tensile fracturing of slope	斜坡弯曲拉裂
bend-proof	防弯的
bends	潜水病；沉箱病
bendway	两河湾间河段；深水槽
benefication	选矿
beneficial element	有益元素
beneficiate	选，对进行预处理；增效
beneficiated bentonite	增效膨润土
beneficiated ore	精矿
beneficiating method	精选法；选矿法；选煤法
beneficiating ore	精选矿石
beneficiation	选矿；精选
benefit conservancy	水土保持；保护；黄土
Benioff fault plane	贝尼奥夫面
Benioff plane	贝尼奥夫面
Benioff seismic zone	贝尼奥夫地震带
Benioff short-period seismograph	贝尼奥夫短周期地震仪
Benioff strain meter	贝尼奥夫应变计
Benioff type subduction	B型俯冲
Benioff zone	贝尼奥夫带；毕鸟夫带；俯冲带；板块消亡带
benitoite	蓝锥矿；硅酸钡钛矿
benjaminite	银硫铋铅铜矿；铜银铅铋矿
Benkelman beam	弯曲仪；挠曲仪；杠杆弯沉仪；彭柯曼梁
benmoreite	歪长粗面岩
Benoto boring machine	贝诺特钻孔机
Benoto cast-in-place pile	贝诺特灌注桩
Benoto method	贝诺施工法
Benoto method boring machine	全套管钻孔机
Benoto piling system	贝诺特桩工系统
bent bar	弯起钢筋，挠曲钢筋；弯筋
bent bar anchorage	弯筋锚固
bent clamp	弯压板
bent cleavage	弯曲劈理
bent crystal monochromator	弯曲结晶单色仪
bent entry	弯曲平巷
bent reinforcement bar	弯起钢筋
bent rod	弯曲抽油杆
bent shaft	曲轴；曲柄轴
bent steel	弯起钢筋
benthal deposit	河床堆积物
benthic	海底的；海底生物底栖生物；水底的，海底
benthic disturbance	海底扰动
benthic division	海底区
benthic ecology	底栖生态；海底生态学
benthic ecosystem	海底生态系统
benthic observatories	海底观测
benthic organism	底栖生物
benthic seabed grab survey	海底抓斗勘测
benthic zone	底栖带
benthic-pelagic coupling	海底-水层耦合
benthonic	海洋的；底栖的
bentho-potamous	河底的
bentonite	斑脱岩，膨润土；膨土岩；皂土；浆土
bentonite binder	膨润土黏结剂
bentonite buffer	膨润土缓冲区
bentonite cement	膨润土水泥；胶质水泥
bentonite cement grout	膨润土水泥浆液
bentonite cement pellets	水泥膨润土团
bentonite clay	膨润土；皂土；浆土
bentonite clay particle	膨润土颗粒
bentonite content	膨润土含量
bentonite debris flow	斑脱岩冻融泥石流
bentonite degree	膨润度
bentonite gel	膨润土凝胶
bentonite grout	膨润土浆；皂土浆
bentonite grouting	膨润土灌浆
bentonite liner	膨润土衬垫；膨润土垫层
bentonite mud	膨润土泥浆
bentonite mud	膨润土泥浆
bentonite pellets	膨润土团
bentonite powder	膨润土粉
bentonite shield	泥膜加压式盾构
bentonite slurry	膨润土泥浆
bentonite slurry trench	膨润土泥浆槽
bentonite suspension	膨润土悬浮液
bentonite treatment	膨润土处理
bentonite value	膨润值
bentonite-type shield	泥水盾构
bentonite-water solution	膨润土水溶液
bentonitic clay	膨润土；皂土；斑脱土
bentonitic mud	膨润土泥浆
bentonitic shale	斑脱岩质页岩；膨润土页岩
bentonitic slurry	膨润土泥浆
bent-up	弯起的；向上弯的；上升的
bent-up bar	弯起钢筋
benzol	苯；粗苯
beraunite	簇磷铁矿
Berea core	贝雷岩心
Berea sandstone	贝雷砂岩
Berek compensator	贝瑞克补色器
Berek method	贝瑞克法
Berek slit microphotometer	贝瑞克裂隙光度计
beresite	倍利岩；黄铁长英石；黄铁细晶岩；黄铁绢英岩
beresitization	黄铁绢英岩化；黄铁细晶岩化
beresowite	铬铅矿
berg	冰山；山；丘
berg crystal	水晶
bergalite	蓝方黄长煌斑岩
bergbalsam	石油
bergmannite	假钠沸石
Bergmann's rule	伯格曼定律
bergmehl	岩乳；硅藻土
bergschrund	大冰隙；背隙隆；冰川后大裂隙；冰后隙
berg-till	浮冰碛
bergy bit	浮冰群；小冰山
Bering land bridge	白令海陆桥

beringite	棕闪粗面岩
berkeyite	天蓝石
berm	滩肩；护堤，围护带；护道；反压平台；截水沟
berm crest	滩肩脊；岸堤顶；后滨阶地外缘
berm ditch	傍山排水沟；护道排水沟；边坡截水沟
berm edge	滩肩外缘
berm elevation	平盘标高
berm for back pressure	反压护道
berm in cutting	路堑平台
berm scarp	滨后陡坎；滩肩崖
berm spillway	崖道排水沟；阶段排水沟
berm width	平盘宽度
berm with superloading	反压护道
berme	戗台；护道
bermstone	护脚石
bermudite	黑云碱煌岩；黑云霞石岩
Bernoulli	伯努利
Bernoulli equation	伯努里方程
Bernoulli's distribution	伯努利分布
Bernoulli's equation	伯努利方程
Bernoulli's law	伯努利定律
Bernoulli's theorem	伯努利定理
berondrite	辉闪霞斜岩
Berriasian	贝利阿斯阶
berth	停泊处；泊位；码头
berth occupancy factor	泊位利用率
berthierite	辉铁锑矿
berthing condition	靠泊条件；靠泊；靠码头
berthing energy	靠船能量
berthing facility	靠泊设备，码头设施；补充
berthing load	靠泊荷载
berthing reaction	靠泊反力
bertrand lens	勃氏镜
bertrandite	羟硅铍石；硅铍石
beryl	绿柱石
beryllate	铍酸盐
beryllium minerals	铍矿类
beryllonite	磷酸钠铍石
berzelianite	硒铜矿
berzeliite	黄榴砷矿
beschtauite	石英二长斑岩
bessel function	贝塞尔函数
Bessel transform	贝塞尔变换
bess-metal ore	黄锡矿
best approximation	最佳逼近
best available technique（BAT）	最佳可用技术
best bound rule	最佳界法则
best curve	最佳曲线
best density	最佳密度
best efficiency point	最佳效率点
best fit	最佳拟合
best fit stresses	最适应力
best hydraulic cross section	最佳水力断面
best load	最大效率时的负载
best moisture content	最佳含水量
best performance curve	最佳曲线
best slope	最佳斜率；最佳坡度
best value	最佳值
best-estimate parameter	最佳估计参数
best-fit criteria	最佳拟合标准

best-fit line	最佳拟合曲线
best-fit regression line	最佳拟合回归线
best-fitting	最佳拟合
best-fitting plane	最佳拟合面
best-path algorithm	最佳路径算法
beta decay	β衰变
beta diagram	β图解
beta method	β法
beta particle	β粒子；β质点
beta particle amplitude	β粒子射程
beta quartz	β石英
beta ray	β射线
beta spectrum	β射线能谱
beta testing	β试验
betafite	贝塔石；铌钽铀矿
beta-quartz type	β石英组
betoken	表示；指示；预示
beton	混凝土
beton armee	钢筋混凝土
beton brut	混凝土粗表面；混凝土未加工表面
betrunked river	断尾河
betrunking	河流断尾
betrunking of stream	河流干涸
better-drained	排水较好的
betterment	改善；修缮和扩建；修缮经费
bettle	夯实；捣棒；木槌
between-batch bias	批次间偏倚
between-laboratory bias	实验室间偏倚
between-quadrangle bias	图幅间偏倚
between-streams variation	水系间变异
betwixt land	盾地；马蹄形盾地；中介地区
betwixt mountain	中间地块
beudantite	菱铅矾
bevel	斜面，削面；斜削；斜的；斜角规
bevel angle	斜角
bevel edge	倒角，斜边，倒角边
bevel face	斜面
bevel sheet pile	斜角板桩
bevel square	斜角规；分度规
bevel wall bit	斜面环状钻头
beveled hill top	斜切的山顶
beveling of strata	地层斜削
bevelled	斜切的，开坡口的；锥形的
bevelled hill top	平削丘顶
bevelled surface	斜削面
bevelment	斜剖面，斜断面；削平
bevelwall bit	斜壁钻头；带内锥度的环状钻头
bevelways	斜着；倾斜地；斜向地
bewel	挠曲；预留曲度
bezel facet	梨形多面型顶侧翻光面
b-glide	b-滑移
bhel	河床沙岛
B-horizon	残积层，淋余层，B层；底土；淀积层
bias	斜线；偏置，偏压；偏流偏差；斜纹，斜的
bias angle	斜角；偏度
bias distortion	偏畸变；偏移失真
bias distribution	偏差分布
bias error	偏差；系统误差
biasar	侧面蛇丘，冰砾堤
biased	偏的；偏压

biased amplifier 偏置放大器
biased error 偏置误差；有偏误差
biased estimator 有偏估计量
biased exponent 偏置指数
biased field 偏移磁场
biasing 加偏压
biat 支承木；井筒横梁
biaxial 二轴的；双轴的
biaxial apparatus 双轴仪
biaxial compaction 双向压缩
biaxial compression test 双轴压缩试验
biaxial compressive strength 双轴抗压强度
biaxial coordinates 双轴坐标
biaxial creep 双轴蠕变
biaxial crystal 双轴晶体；二轴晶
biaxial deformation 双轴向变形
biaxial eccentricity 双向偏心距
biaxial effect 双轴向影响
biaxial extension 双轴伸长
biaxial extensional flow 双轴拉伸流
biaxial interference figure 二轴晶干涉图
biaxial lateral load 双轴向侧力
biaxial load 双轴荷载
biaxial loading 双轴载荷
biaxial orientation 双轴取向
biaxial orienting 双轴取向
biaxial pressure 双轴压力；双向压力
biaxial shaking table 双向振动台
biaxial state of stress 双轴应力状态
biaxial strain 双轴应变；二轴应变
biaxial stress 双轴应力；双向应力
biaxial stress-strain relationship 双轴应力-应变关系
biaxial stretch 双轴伸长
biaxial stretching 双轴向拉伸
biaxial tensile test 双轴抗拉试验；双向拉伸试验
biaxial viscosity 双轴向黏度
biaxial winding 双轴缠绕法
biaxiality 二轴性
biaxiality factor 双向因子
bibbley-rock 砾石岩；砾岩
bibeveled 双斜面的
bibliolite 纸状页岩
bicapillary pycnometer 双毛细管比重计
bicarbonate 重碳酸盐；碳酸氢盐
bicarbonate water 重碳酸水；碳酸盐水
bichromate 重铬酸盐
bicirculation 双重循环
biconcave 双凹的；两面凹的
bicontinuous function 双连续函数
bicontinuous porous media 双连续多孔介质
bicontinuous structure 双连续结构
biconvex 双凸的；两面凸的
bi-coordinate system 双坐标系
bicrystal 双晶体
bidding document 投标文件；标书
bidimensional 二维的
bidimensional shaking motion 双向振动；二维振动
bidirectional 双向的
bi-directional cross-bedding 双向交错层理
bidirectional flux 双向通量

bidirectional reflectance 双向反射
bidirectional reflectance distribution function（BRDF） 二向性反射分布函数
bidirectionality 双向性
bieberite 赤矾
bielenite 顽剥橄榄岩
Bieler-Watson method 皮勒-华生电法勘探
bifacial 双面
bifacies 双重岩相；氧化还原岩相
bifid 二裂的
bifurcated 分叉的；分成双支的
bifurcation 河道分叉口；分叉；矿脉分支；分流
bifurcation of coal seam 煤层分叉
bifurcation phenomenon 分岔现象
bifurcation ratio 分枝比率
bifurcation structure 分叉结构；分水结构
bifurcation theory 分叉理论
bifurculapes 双叉迹
big aperture well 大口井
big area caving 大面积垮落；大面积垮落法
big diameter pile 大直径桩
big distance well 大位移井
big ditch 史前潜河
big explosion cosmology 宇宙大爆炸学说
big extended well 大位移井
big floe 大浮冰川块
big hole drilling 大直径钻进
big muck 大块崩落岩石
big opened void 粗大开启空隙
big opened void rate 粗大开启空隙率
big panel form construction 大块模板施工法
big Red 大红页岩层
big sand 大砂层
big trough 大地槽
big unit weight 大容重样品
big volume weight 大容重样品
big yield 大水量
big-bang hypothesis 大爆炸宇宙论；大爆炸假说
bight 大湾；开展海湾；小海湾；绳环
bigraph 偶图
big-scale work 大型工程
bigwoodite 钠长微斜正长岩
Biharian 比哈尔阶
biharite 黄叶蜡石
biharmonic 双调和的；双谐的
biharmonic equation 双调和方程
biharmonic function 双调和函数
biharmonic operator 双调和算子
bijective 双射的
bijective mapping 双射映射
biland 半岛
bilateral 两面的；双方的；双向的；双通的
bilateral bar 双向钢筋
bilateral boundary 两侧边界
bilateral clearing 双边清算；双边结算
bilateral constraint 双侧约束；双边约束
bilateral control 双向控制
bilateral extraction 矿田双面开采；双面采区
bilateral fault 双侧断裂；双向扩展断层
bilateral faulting 双侧断裂

bilateral geochemical barrier	两侧地球化学障碍
bilateral geosyncline	双侧地槽
bilateral leveling	对向水准测定
bi-lateral pack	双面充填带
bilateral rupture	双向破裂
bilateral symmetry	两侧对称；左右对称
bilateral tolerance	双向公差
bilateralism	左右对称
bilaterally symmetrical fossil	两侧对称化石
bilayer	双层
bilge ways	底边板；滑板
biliminal geosyncline	双侧对称地槽
bilinear	双直线的；双线性的
bilinear element	双线性单元
bilinear flow	双线性流动
bilinear form	双线性型
bilinear function	双线性函数
bilinear interpolation	双线性插值法
bilinear mapping	双线性映射
bilinear model	双线性模型
bilinear parametrization	双线性参数化
bilinear relation	双线性关系
bilinear shape function	双线性形函数
bilinear system	双线性系统
bilinear transform	双线性变换
bilinear transformation	双线性变换
bilinear trial function	双线性试验函数
bilinearity	双线性
billabong	死水河；死水洼地；旧河床；故河道；死河
billietite	黄钡铀矿
billow	小型流速仪；卸载期；狂涛；山崩
biloculate	二室的
bilogarithmic diagram	双对数图
bilogarithmic graph	双对数图
bimagmatic	两代岩浆的
bi-Maxwellian distribution	双麦克斯韦分布
bimetal	双金属
bimetallic	双金属的
bimetasomatism	双交代作用；双交代酌
bimodal	双峰的；双模态的；双峰式的；双向的
bimodal bipolar palaeocurrent pattern	双峰反向古水流模式
bimodal cleavage	双型劈理
bimodal cross-bedding	双向交错层理
bimodal current rose	双向水流玫瑰图
bimodal curve	双峰型粒径曲线
bimodal diffusion	双式扩散
bimodal distribution	双峰分布
bimodal igneous activity	双峰式火成活动
bimodal palaeocurrent direction	双向古水流方向
bimodal pattern	双峰模式
bimodal rift volcanism	双峰裂谷火山活动
bimodal sediment	双峰粒径沉积物
bimodal size distribution	双众数粒度分布；双峰型粒度分布
bimodal suite	双套；双模式
bimodal volcanic complex	双向火山杂岩
bimodal volcanic rock	双峰式火山岩
bimodal volcanism	双峰火山
bimodal-bipolar herringbone cross-bedding	双峰反向人字形交错层理
bimodal-bipolar rose diagram pattern	双峰反向玫瑰花图解模式
bimodality	双向；双峰
bimoment	双弯矩；双力矩
bin	矿仓；料仓
bina	坚硬黏土岩，硬黏土岩
binary	二叉；二元的；二进制的；二相的
binary acid	二元酸
binary diagram	二元相图
binary diffusivity	二元扩散系数
binary gain control	二进制增益控制
binary granite	二云母花岗岩
binary image	二值图像；两色图像
binary magma	二元岩浆
binary matrix	二元矩阵
binary medium	双重介质
binary mixture	二元混合物
binary number	二进制数
binary number system	二进制
binary ore	双合矿物
binary representation	二进制表示法
binary rock	含两种矿物的岩石；两相岩石
binary sediment	二相沉积物
binary system	二进制；二元系；双重星系
binary-state variable	二值变量
binate	二分取样；重取样
bind	泥质夹层；结合，黏合，胶结，束缚，捆绑，系杆；使…凝固；煤系中的页岩或泥岩
binder	胶凝材料，系杆，托梁；丁砖，拉结石；黏合剂，黏料，胶结剂；联结搁栅，连系筋；拉结料，夹子
binder briquetting	黏结剂成型
binder clay	胶结性黏土
binder course	结合层；拉结层；黏合料
binder metal	黏结金属
binder modification	黏结剂改性；黏结剂的活化
binder Mortar	黏合灰浆
binder soil	黏性土；黏合土；胶结土
binderless briquetting	无黏结剂煤砖制造；无黏结剂团矿法
bindeton	结合黏土
bindheimite	水锑铅矿
binding	黏合；黏结；结合；有约束力的；束缚
binding agent	黏合剂；黏结料；胶结料
binding beam	连系梁；联结梁
binding bolt	连接螺栓；接合螺栓
binding coal	黏结煤
binding course	结合层；联系层
binding element	黏结剂
binding energy	结合能；束缚能；黏结能
binding force	结合力；黏结力；束缚力
binding gravel	掺黏结料的砾石
binding layer	结合层
binding material	胶结材料；黏合料；黏结剂
binding medium	黏合剂
binding power	黏合力；内聚力；约束力
binding property	黏合性；黏结性
binding reinforcement	绑扎钢筋
binding strength	黏结强度
binding wire	绑扎钢丝；扎线

bindstone 黏结灰岩；生物黏结灰岩
bineite 黑云霞石岩
binemelite 黑云黄长霞石岩
bing 废石堆，矸石堆；堆，堆场；垛
Bingham body 宾汉体
Bingham equation 宾汉方程
Bingham fluid 宾汉流体
Bingham grout 宾汉体浆液
Bingham model 宾汉模型
Bingham plastic 宾汉塑性
Bingham plastic body 宾汉塑性体
Bingham plastic fluid 宾汉塑性流体
Bingham plastic model 宾汉塑性流模型
Bingham plastic oil 宾汉塑性油
Bingham plastics 宾汉塑性流体
Bingham substance 滨汉体
Bingham viscosity 宾汉黏度
Bingham yield value 宾汉屈服值
Bingham's model 宾汉姆模型
Binglingsi Grottoes 炳灵寺石窟
binnite 淡黝铜矿
binocular 双目的；双筒的；双目镜；双筒镜
binocular microscope 双目显微镜
binodal 双节的；双节点的
binodal curve 双结点曲线
binodal lateral wave 双节横波
binomen 双名
binomial coefficient 二项式系数
binomial coefficient series 二项式系数的级数
binomial distribution 二项式分布
binomial expansion 二项展开式
binomial expression 二项式
binomial formula 二项式公式
binomial nomenclature 二名法
binomial theorem 二项式定理
binominal nomenclature 双名法
binomiual distribution 二项分布
bin-type retaining wall 仓式挡土墙
bin-wall 隔仓式挡土墙
bio clogging 生物填充
bioaccumulated limestone 生物滩灰岩；有机生成石灰岩；生物灰岩
bioarenite 生物砂屑岩
bioassay 生物测定；生物鉴定
bioastronautics 生物航天学
bioaugmentation 生物强化
biocalcarenite 生物碎屑灰岩
biocalcarenite decay 生物砂屑石灰岩衰变
biocalcilutite 生物泥屑石灰岩；生物泥屑灰岩；生物砾屑灰岩
biocementation 生物胶结
biocementstone 生物胶结灰岩
biocenosis 生物群落
biochemical 生物化学的
biochemical coalification 生物化学煤化作用；生物化学煤化酌
biochemical conversion 生物化学转化
biochemical cycle 生物化学循环
bio-chemical decomposition 生化分解
biochemical degradation 生物化学降解作用；生物化学降解酌
biochemical deposit 生物化学沉积
biochemical engineering 生物化学工程
biochemical evolution 生物化学进化
biochemical factor 生物化学因素
biochemical gelification 生物化学凝胶作用
biochemical mineralization 生物化学矿化作用
biochemical origin 生物化学起源
biochemical oxygen demand 生化需氧量
biochemical rock 生物化学岩；生物化学岩
biochemical sedimentary deposit 生物化学沉积矿床
biochemical sedimentary ore formation 生物化学成矿作用
biochemical sedimentary rock 生物化学沉积岩
biochemical sedimentation mineral deposit 生物化学沉积矿床
biochemical treatment 生物化学处理；生化处理
biochemical weathering 生物化学风化
biochemigenic rock 生物化学岩
biochemistry 生物化学
biochemistry reaction 生化反应
biochron 生物年代
biochronologic 生物年代的
biochronologic unit 生物年代单位
biochronology 生物年代学
biochronostratigraphic 生物年代地层的
biochronostratigraphic unit 生物年代地层单位
biochronostratigraphy 生物年代地层学
biochronotype 生物年代型
bioclast 生物碎屑岩；生物碎屑
bioclastic 生物碎屑的
bioclastic calcarenite 生物碎屑砂屑灰岩
bioclastic grainstone 生物碎屑颗粒灰岩
bioclastic limestone 生物碎屑灰岩
bioclastic packstone 生物碎屑泥粒灰岩
bioclastic pile 生物碎屑堆积
bioclastic rock 生物碎屑岩
bioclastic wackestone 生物碎屑粒泥灰岩
bioclastics 生物碎屑岩
bioclayey texture 生物黏土结构
biocolloid 生物胶体
bioconstructed facies 生物建造相
bioconstructed limestone 生物灰岩
bioconstruction 生物建造；生物建造作用
bio-contact oxidation 生物接触氧化
bio-contact oxidation process 生物接触氧化法
biocorrelation 生物对比；生物关联
biocycle 生物循环；生物带
biodatum 生物基准面
biodegrability 生物降解能力
biodegradability 生物降解性；生物退化性
biodegradable 可生物降解的
biodegradation 生物降解；生物降解作用
biodegradation of solid wastes 固体废物的微生物降解
biodeposition 生物沉积作用
biodeposition structure 生物沉积构造
biodetritus 生物碎屑
bioecological 生物生态的
bioecology 生物生态学
bioecozone 生物生态区
bioencrusted grain 生物包壳颗粒

bioengineering 生物工程
bioerosion 生物侵蚀
bioerosional structure 生物侵蚀构造
biofacies 生物相
biofacies map 生物相图
biofacies-paleogeographic map 生物相古地理图
biofilter bed treatment 生物滤池法
biofiltration 生物过滤
biofiltration process 生物过滤法
biofragmental rock 生物碎屑岩
biogas 生物瓦斯；生物气；沼气
biogenesis 生物成因；生源说
biogenetic 生物起源的；生物成因的
biogenetic anomaly 生物成因异常
biogenetic deposit 生物沉积
biogenetic law 生物发生律
biogenetic origin 生物成因
biogenetic reworking 生物再造作用
biogenetic rock 生物岩
biogenetic sediment 生物沉积物
biogenetic texture 生物结构
biogenic 生物生成的；生物起源的；生物成因的
biogenic accumulation 生物堆积
biogenic chain 生物链
biogenic coal bed gas 生物成因煤层气
biogenic community 生物群落
biogenic deposit 生物沉积
biogenic element 生物元素
biogenic gas 生物气
biogenic imprint 生物痕迹
biogenic incrustation 生物结壳作用
biogenic limestone 生物灰岩
biogenic magnetite 生物磁铁矿
biogenic migration 生物迁移
biogenic mineralization 生物矿化作用
biogenic mud mound 生物成因泥丘
biogenic ooze 生物软泥
biogenic process 生物过程
biogenic rock 生物岩
biogenic sediment 生源沉积物；生物源沉积物
biogenic sedimentary structure 生物成因沉积构造
biogenic surface layer 生物形成的表面层
biogenic theory 油气生物成因学说
biogenochemical anomaly 生物地球化学异常
biogenous 生物起源的；生物成因的
biogenous rock 生物岩
biogeocenosis 生物地理群落
biogeochemical 生物地球化学的
biogeochemical anomaly 生物地球化学异常
biogeochemical classification 生物地球化学分类
biogeochemical cycle 生物地球化学旋回；生物地球化学循环
biogeochemical enrichment 生物地球化学富集
biogeochemical prospecting 生物地球化学探矿；生物地球化学勘探
biogeochemical province 生物地球化学省
biogeochemical survey 生物地球化学测量；生物化探
biogeochemistry 生物地球化学
biogeographic province 生物地理区
biogeographic realm 生物地理区

biogeographic region 生物地理区
biogeography 生物地理学
biogeologic time scale 生物地质年代表
biogeology 生物地质学
bioglyph 生物遗迹
bioherm 生物岩礁；生物丘
biohermal complex 生物岩丘复合体
biohermal facies 生物礁相
biohermal limestone 生物礁灰岩
biohermal reservoir 生物礁储集层
biohermite 礁屑灰岩
biohorizon 生物面；生物层
bio-hydrogeochemical method 生物-水文地球化学方法
biohydrography 生物水文学
biohydrology 生物水文学
bioleaching 生物浸矿；生物提炼法；生物淋滤
biolevel 生物基准面；生物水平带
biolime 生物软泥
biolimiting elements 生物限制元素
biolite 生物岩；生物成因矿物
biolith 生物岩
biolithite 生物骨架灰岩；生物岩；黏结灰岩
biolithogenesis 生物成岩作用
biolitllite limestone 生物灰岩；礁灰岩
biolog 生物测井
biologic 生物的
biologic absorption coefficient 生物吸收系数
biologic agent 生物营力
biologic association 生物群丛
biologic barrier 生物阻限
biologic chain 生物链
biologic community 生物群落
biologic cycle 生物循环
biologic facies 生物相
biologic group 生物类群
biologic monitoring 生物监测
biologic necessary nutrient element 生物必需营养元素
biologic nitrogen fixation 生物固定氮
biologic oceanography 生物海洋学
biologic ooze 生物软泥
biologic oxygen demand (BOD) 生物需氧量
biologic phosphorite 生物磷块岩；生物磷灰岩
biologic pollution 生物污染
biologic prevention and treatment 生物防治
biologic purification 生物净化
biologic relative absorption coefficient 生物相对吸收系数
biologic resource crisis 生物资源危机
biologic sedimentology 生物沉积学
biologic self-purification 生物自净
biologic time scale 生物年代表
biologic treatment 生物处理法
biologic weathering 生物风化
biological 生物学的；生物的
biological action 生物作用
biological activity 生物效能
biological aerobic treatment of wastewater 废水好氧生物处理
biological agent 生物制剂
biological analysis 生物分析
biological assay 生物鉴定

biological augmentation batch enhanced（BABE） 生物强化；间歇富集
biological benchmark 生物水准基点
biological beneficiation 生物选矿
biological chain 生物链
biological chemistry 生物化学
biological clarification 生物净化
biological cleaning 生物净化
biological clock 生物钟
biological colonization 生物侵染
biological concentration 生物富集
biological concentration factor 生物浓集因素
biological contaminant 生物污染
biological context 生物演化关系
biological control 生物防治
biological cycle 生物循环；生物周期
biological decay 生物衰变；生物分解
biological decomposition 生物分解
biological degradability 生物降解性；生物退化性
biological degradation 生物降解作用
biological detritus 生物碎屑
biological filter 生物过滤器；生物滤池
biological filtration 生物过滤
biological geology 生物地质学
biological group 生物类群
biological pollution 生物污染
biological process 生物处理法；生物过程
biological sedimentation 生物沉积作用
biological settlement 生物沉降
biological slime 生物黏泥；生物膜
biological sludge 生物污泥
biological soil crust 生物土壤结皮
biological solid mechanics 生物固体力学
biological time scale 生物地质年代表
biological transformation 生物转化
biological weathering 生物风化；生物风化作用
biology 生物学；生态学
biolysis of sewage 污水生物分解
biomagnetism 生物磁性
biomarker 生物标志化合物
biomass 生物量；生物质
biome 生物群落
biomechanics 生物力学
biomedical engineering 生物工程
biometrical method 生物统计学方法
biomicrite 生物微晶灰岩；生物泥晶灰岩
biomicrosparite 生物微亮晶灰岩
biomicrudite 生物泥晶砾屑灰岩；含砾生物泥晶石灰岩
biomineralization 生物成矿作用；生物矿化作用
biomineralogy 生物矿物学
bionomics 个体生态学
bionomy 生态学；生理学
biont 生物
biopelmicrite 生物球粒泥晶灰岩
biophile element 亲生物元素
biophysics 生物物理学
bioplasm 原生质
biopolymer 生物聚合物
bioremediation 生物治理
biosiliceous ooze 生物硅质软泥

bioslime 生物软泥
biosome 生物岩体；生物体
biospararenite 生物亮晶砂屑灰岩
biosparite 生物亮晶灰岩
biosparitic calcirudite 生物亮晶砾屑灰岩
biosparrudite 生物亮晶砾屑灰岩
biosparry 生物亮晶的
biosparudite 生物亮晶砾屑灰岩
biostatistics 生物统计学
biostrata 生物层
biostratic unit 生物地层单位
biostratification 生物层理
biostratification structure 生物成层构造；生物地层层理构造
biostratigraphic 生物地层的
biostratigraphic correlation 生物地层对比；生物地层学对比
biostratigraphic unit 生物地层单位
biostratigraphic zone 生物地层带
biostratigraphical 生物地层的
biostratigraphy 生物地层学
biostratinomy 生物埋藏学；生物造林沉积学；生物遗体沉积学；生物层积学
biostratum 生物地层
biostromal 生物层的
biostromal carbonate 生物层碳酸盐岩
biostromal limestone 生物层灰岩；壳灰岩
biostrome 生物层
biosurface 生物面
biot 黑云母
Biot coefficient 毕奥系数
Biot consolidation theory 比奥固结理论
Biot constant 毕奥常数
biotechnical slope protection 植物护坡
biotic 生物的；生命的
biotic balance 生物平衡
biotic barrier 生物阻障；生物阻限
biotic community 生物群落
biotic control 生物防治
biotic environment 生物环境
biotic factor 生物因素
biotic index 生物指数
biotic marker 生物标志物
biotine 钙长石
biotite 黑云母
biotite andesite 黑云安山岩
biotite gneiss 黑云母片麻岩
biotite granite 黑云母花岗岩
biotite isograd 黑云母等变度
biotite phyllite 黑云母千枚岩
biotite schist 黑云母片岩
biotite-muscovite granite 含黑白云母花岗岩
biotitite 黑云母岩
biotitization 黑云母化
Biot's equation 比奥公式；比奥方程
bioturbate 生物扰动的
bioturbate structure 生物扰动构造
bioturbation 生物扰动；生物扰动作用
bioturbation heaving 生物扰动隆起
bioturbation structure 生物扰动构造

bioturbite 生物扰动岩
biotype 生物型；纯系群
biounlimiting elements 非生物限制元素
biozeolite 生物沸石
biozonation 生物分带性
biozone 生物带
biparted hyperboloid 双叶双曲面
bipartite 两部分构成的；双向的
bipartite cubic 双枝三次曲线
bipartite texture 二分结构
biphase 双相，双相的
biphase equilibrium 二相平衡
bipolar 双极的；偶极的
bipolar coordinate 双极坐标
bipolar current rose 双极水流玫瑰图
bipolar distribution 极地分布；双极性分布；两极分布；两极同源
bipolar gravity anomaly 两极重力异常
bipolarity 双极性；两极分布；两极同源
bipole 双极
bipyramid 双锥；双锥体
biquartz 双石英；双石英片
biquartz plate 双石英试板
birbirite 蚀变橄榄岩
bird foot delta 鸟足状三角洲；桨叶状三角洲
bird-eye structure 鸟眼构造
birdfoot-lobate delta 鸟足-舌形三角洲
bird's eye 鸟眼状
bird's eye limestone 鸟眼状灰岩
bird's eye structure 鸟眼构造
birds' eye view 鸟瞰图
bird's eye view drawing 鸟瞰图
bird's eye view map 鸟瞰图
bird's-eye boundstone 鸟眼黏结灰岩
bird's-eye gravel 细砾石
bireflectance 双反射；双反射率
bireflection 双反射
birefracting 双折射的；重折射的
birefraction 双折射率
birefractive 双折射的；重折射的
birefringence 双折射率；重折射率；双折射
birefringence chart 双折射图；双折射率图
birefringence method 双折射法
birefringent 双折射；重折射的
birefringent effect 双折射效应
birkremite 石英正长岩；紫苏石英正长岩
birnessite 钠水锰矿；水钠锰矿
Birrimian orogeny 比理姆造山运动
bischofite 水氯镁石
biscuit cutter 短管心取心筒；笨拙管工
biscut cutter 岩心管钻头
bisect 平分；二平分；二等分
bisecter 平分线；等分线
bisectrix 等分线
biserial 双列的
Bishop and Morgenstern slope stability analysis method 毕肖普与摩根斯坦法
Bishop consolidometer 毕肖普固结仪
Bishop method of slope stability analysis 毕肖普斜坡稳定性分析法

Bishop piezometer tip 毕肖普孔隙水压力计测头；毕肖普测压头
Bishop sampler 毕肖普取土器
Bishop simplified method 简化毕肖普法
bishopite 水氯镁石
Bishop's equation 毕肖普方程
Bishop's method 毕肖普法
Bishop's method of slices 毕肖普条分法
Bishop's routine analysis method 毕肖普常规分析法
Bishop's simplified method of slice 毕肖普简化条分法
bisilicate 二硅酸盐；偏硅酸盐
bismalith 岩柱
bismuth glance 辉铋矿
bismuthic ocher 铋华
bismuthine 辉铋矿
bismuthinite 辉铋矿
bismutite 泡铋矿
bismutosmaltite 钴砷铋矿石
bismutospherite 球泡铋矿
bismutotantalite 铋钽铁矿
bistagite 纯透辉石；透辉岩
bisymmetric 两侧对称的
bisymmetry 两侧对称
bit adapter 钻头接头
bit and mud socket 钻头及泥浆吸取筒
bit brace 手摇钻；钻孔器
bit burnt 烧钻
bit clearance 孔壁与钻头间隙
bit cone 钻头牙轮
bit configuration 钻头构形
bit constant 钻头常数
bit consumption 钻头消耗量
bit cutting angle 钻头切削角；刃角；修磨角
bit cuttings 钻屑
bit deflection 钻头偏斜
bit density 位密度
bit diameter 钻头直径
bit dresser 磨钎机；钻头修整机
bit grinder 磨钎机；钻头磨锐机
bit loading 钻头荷载；钻头压力
bit of carbide insert 合金钻头
bit of diamond 金刚石钻头
bit penetration 钻头贯入度；钻头穿透；钻头吃入深度；钻头进尺速度
bit pressure 钻头压力；钻头钻进压力
bit pressure drop 钻头压力降
bit reaction force 钻头反作用力
bit reamer 扩孔钻；扩孔器；扩孔钻头
bit side cutting action 钻头侧向切削作用
bit side force 钻头侧向力
bit size 钻头尺寸
bit skew 钻头偏转量
bit speed 钻速
bit stock 钻柄；手摇钻；曲柄钻
bit thrust 钻头钻进压力；钻头推力
bit tilt 钻头倾斜
bit tilt mechanism 钻头倾斜机理
bit wear equation 钻头磨损方程
bit wear index 钻头磨损率
bit wear test 钻头磨损试验

bit weight	钻头负载；钻具组重量；钻压
bit wing	钻头刀
Bitaunian	比陶恩阶
bit-rod	钻杆
bittem	卤水；天然盐水；盐卤
bitter earth	苦土；氧化镁
bitter lake	苦水湖；盐湖
bitter spar	纯晶白云石；白云石
bitter spring	苦泉；卤泉
bittern	卤水；天然盐水
bitulith	沥青混凝土
bitulithic pavement	沥青混凝土路面
bitumastic	沥青的
bitumen	沥青；天然地沥青
bitumen binder	沥青黏合剂
bitumen caprock	沥青盖层
bitumen coated pile	沥青涂面桩
bitumen emulsion	沥青乳液；乳化沥青
bitumen grouting	沥青灌浆
bitumen jointing	沥青封接
bitumen limestone	沥青石灰岩
bitumen peat	沥青泥炭
bitumen reflectance	沥青反射率
bitumen rock	沥青岩
bitumen shale	沥青页岩
bitumen-aggregate ratio	油石比
bitumencarb	沥青质；沥青炭
bitumen-coated	沥青敷面的
bitumen-concrete mixture	沥青混凝土混合物
bitumenite	芽孢油页岩
bitumenization	沥青化作用
bituminite	油页岩；沥青质体；烟煤
bituminization	沥青止水法；沥青化；煤化
bituminous	沥青的；含沥青的；熘煤
bituminous clay	沥青黏土
bituminous concrete	沥青混凝土
bituminous concrete mixture	沥青混凝土混合料
bituminous concrete pavement	沥青混凝土路面
bituminous cover	沥青盖层
bituminous dike	沥青脉
bituminous emulsion	沥青乳液；乳化沥青
bituminous grouting	沥青灌浆
bituminous limestone	沥青灰岩
Bitaunious luster	沥青光泽
bituminous macadam	沥青碎石；沥青碎石路；沥青碎石层
bituminous macadam mixture	沥青碎石混合料
bituminous macadam pavement	沥青碎石路面
bituminous mastic	沥青砂胶；沥青铺料
bituminous mat	沥青层垫
bituminous mortar	沥青砂浆
bituminous oakum	沥青麻筋
bituminous paint	沥青涂料；沥青漆
bituminous pavement	沥青路面；沥青面层
bituminous peat	沥青质泥炭
bituminous penetration	沥青灌注
bituminous penetration pavement	沥青贯入式路面；沥青贯入碎石路面
bituminous plaster	沥青灰泥
bituminous rock	沥青岩
bituminous sand	沥青砂
bituminous Sand Slurry	沥青砂浆
bituminous sandstone	沥青砂；沥青砂岩
bituminous seal	沥青封层；沥青止水
bituminous shale	沥青页岩；油母页岩
bituminous shield	沥青覆盖层；沥青护面
bituminous stabilization	沥青加固
bituminous surface	沥青路面
bituminous surface treatment	表面处治
bituminous texture	沥青结构
bituminous water-proof coating	沥青防水层
bituminous waterproofing	沥青防水层；沥青防水薄膜
bitummen ratio	沥青比值
bitumogel	沥青凝胶
bitumogene	沥青成因
bitusol	天然沥青；胶溶沥青
bivariant system	双变物系；双变体系
bixbyite	方铁锰矿
bizardite	霞黄煌斑岩
bizeul	辉绿闪长岩
bjerezite	霞辉中长斑岩
black	黑色；软质黑色页岩
black adobe soil	黑色冲积黏土
black alkali soil	黑碱土
black amber	煤玉；黑色琥珀；致密黑褐煤
black ash mortar	黑砂浆；石灰煤渣灰浆
black band iron carbonate ore	黑泥碳酸铁矿
black base	沥青基层
black batt	黑色沥青页岩；薄煤夹层
black bog	黑色泥沼；泥炭沼泽；低沼
black carbon	黑碳
black cat	煤质页岩；黑页岩
black chalk	黑垩
black chert	黑燧石
black cobalt	钴土矿
black copper	黑铜矿；粗钢
black cotton soil	黑棉土；膨胀土
black diamond	黑金刚石；黑赤铁矿
black diamond slate	黑金刚石板岩
black dirt	黑色杂质
black dot	黑斑
black earth	黑土
black frost	黑霜；黑灾
black fundamental matrix	黑色基质；不透明基质
black gold	黑铋金矿
black iron ore	磁铁矿
black jack	黑色闪锌矿；粗黑焦油；煤层炭质软黏土；一种烛煤，瘦煤
black Jura	黑侏罗统
black Jura Series	黑侏罗统
black lead	墨铅
black lead ore	黑铅矿
black limestone soil	黑色石灰土
black loam	黑垆土
black manganese	黑锰矿
black mica	黑云母
black mud	黑色泥
black nephrite	墨玉
black oil reservoir	重质油储油层
black opal	黑欧泊

black ore deposit	黑矿型矿床
black power plant	单岸式水电站
black quartz	墨晶
black sand	黑砂
black sands	黑沙滩
black schist	黑色片岩
black shale	黑色页岩
black shale deposit	黑色页岩沉积
black shale formation	黑色页岩的形成
black silver	脆银矿
black slab stone	黑板石；青板石
black soil	黑土
black stone	高碳页岩；炭质页岩
black turf	黑泥炭
black turf soil	黑色草炭
blackband	黑矿层；黑菱铁矿；碳质铁矿
black-band ironstone	泥质碳酸铁矿；黑菱铁矿；碳质铁石
blacking	炭粉；黑色涂料
Blackriverian	布莱克里威尔亚阶
Blackriverian substage	布莱克河亚阶
blacks	软质黑色页岩
blacktop	沥青道路面层
blacolite	绢石蛇纹岩
blade	刮刀；翼片；刀片；桨叶
blade bowl	铲斗
blade machine	平路机；推土机
blade paddle mixer	桨叶式搅拌机
blade shield	叶片式盾构
bladed	刃状的；叶片状的
bladed structure	刃状构造；叶片状构造；刀片状结构
blader	平路机；推土机
blae	灰青碳质页岩
blaes	硬质砂岩；灰青碳质页岩；煤层底部黏土；煤系页岩
Blaine fineness	布莱恩细度
Blaine indicator	布莱恩型表面面积测量计
blairmorite	方沸斑岩
blaize	硬砂岩
Blake	布雷克期
Blamne fineness	勃氏细度；布莱恩细度
Blamne specific area	勃氏比表面积
Blamne value	勃氏细度值
Blan	布兰卡阶
blanch	铅矿石；漂白；使变白
blank	空白的
blank analysis	空白分析；空白试验
blank arch	假拱；圆木桩；装饰拱
blank casing	无孔套管；实套管；无眼套管
blank liner	无眼衬管；封隔衬管
blank map	底图；空白地图
blank test	空白试验
blanket	铺盖；覆盖物；表土；表层；矿层
blanket bog	覆被泥炭沼泽；覆盖沼泽
blanket core tube	表土取心管
blanket course	透水铺盖层；垫层
blanket deposit	平伏矿床；席状矿床；大面积的层状沉积；层状矿床
blanket drain	铺盖排水
blanket effect	覆盖效应
blanket formation	平伏层
blanket grouting	铺盖灌浆；护面灌浆
blanket moss	毡状泥炭
blanket of graded gravel	级配砾石铺盖
blanket peat	藻类泥炭
blanket peat mire	泥炭沼泽；覆被泥炭沼泽
blanket sand	冲积砂；平伏砂岩层；平伏砂层；冲积覆盖砂层
blanket sandstone body	毡状砂岩体
blanket sluice	织物底衬洗矿槽
blanket vein	平伏脉；板状脉
blanking plate	封板
blast	燃爆；爆破；喷砂机
blast bedding structure	变余层理构造
blast cement	高炉渣水泥
blast cleaning	喷砂清理；喷砂处理
blast damper	爆炸减震器
blast densification	爆破振密
blast depth	爆破进尺
blast draught	锥形钻头
blast finishing	喷丸处理；喷砂处理
blast furnace cement	高炉水泥；炉渣水泥
blast furnace coke	冶金焦炭，高炉焦炭
blast hole	炮眼；爆破眼；风洞
blast hole drill	凿岩机
blast hole drilling	爆破法钻进
blast nozzle	喷砂嘴
blast pressure	爆炸波压力；风压；鼓风压力；暴风
blast ratio	爆炸比
blast vibration	爆破振动
blast wave	爆破波；冲击波
blastability	可爆破性
blastability coefficient	可爆性系数
blastability of rock	岩石可爆性
blastation	吹蚀
blasted rock	爆破岩石；爆碎石料
blastesis	变晶
blast-furnace cinder	高炉矿渣
blast-furnace ore	高炉矿石
blast-furnace Portland cement	矿渣硅酸盐水泥
blast-furnace slag	高炉熔渣
blast-furnace slag cement	高炉矿渣水泥
blast-furnace slag sand	高炉渣砂
blasthole bit	炮眼钻头
blast-hole blasting	浅眼爆破
blasthole burden	钻孔最小抵抗线
blastic	变晶的
blastic deformation	爆破变形
blastic texture	变晶结构
blasting	爆破；爆炸；碎裂
blasting belled pile	爆扩桩
blasting borehole	爆破孔
blasting by abrasive	磨料喷砂
blasting cap	起爆雷管；引信；起爆筒
blasting compaction	爆破压实
blasting compaction method	爆炸挤密法
blasting concussion	爆炸冲击作用
blasting consolidation	爆破加固
blasting crater	爆破漏斗
blasting displacement method	爆破换土法

blasting drift	爆破平巷
blasting efficiency	爆破效率
blasting for loosening rock	松动爆破
blasting for throwing rock	抛掷爆破
blasting force	爆破力
blasting fragmentation	爆破
blasting fuse	起爆引线；导火线
blasting hole	爆破孔；炮眼
blasting machine	喷砂机；放炮器
blasting of profiles	光面爆破，周边爆破
blasting parameter	爆破参数
blasting pattern	爆破模式
blasting power	爆炸力；爆破力
blasting procedure	土石方爆破
blasting site	爆石工地；爆破工地
blasting stress wave	爆炸应力波
blasting vibration	爆破振动
blasting vibration observation	爆破地震观测
blasting vibration wave	爆破震动波
blasting works	爆石工程；爆破工程
blast-loosening	松碎爆破
blastoamygdaloidal structure	变余杏仁状构造
blastoaplitic texture	变余细晶结构
blastobreccia	变余角砾岩
blastocataclastic	变余碎裂结构
blastocataclastite	变余碎裂岩
blastocrystal	变晶
blastodiabasic texture	变余辉绿结构；变余花岗状结构
blastogranite	变余花岗岩
blastogranitic	变余花岗状
blastogranitic texture	变余粒状结构；变余花岗状结构
blasto-holoclastite	变晶全碎裂岩
blastohypidiomorphic texture	变余半自形结构
blastolaminous texture	变余页状结构
blastomylonite	变余糜棱岩；变晶糜棱岩；变晶磨岭岩
blastomylonitic	变余糜棱
blastomylonitic texture	变余糜棱结构
blastopelitic texture	变余泥质结构
blastophitic	变余辉绿岩的
blastophitic structure	变余辉绿状结构
blastophitic texture	变余辉绿结构
blastoporphyric	变余斑状的
blasto-porphyritic	变余斑状
blastoporphyritic leptite	变余斑状长英麻粒岩
blastoporphyritic structure	变余斑晶构造
blastoporphyritic texture	变余斑状构造
blastopsammitic	变余砂状
blastopsammitic structure	变余砂状结构
blastopsammitic texture	变余砂状构造
blastopsephitic	变余砾状的
blastopsephitic structure	变余砾状结构
blastopsephitic texture	变余砾状构造
blastopyroclastic texture	变余火山碎屑构造
blast-oriented	定向爆破
blasto-ultramylonite	变晶超糜棱岩
blastovesicular structure	变余气孔构造
blatt	裂缝；横推断层；平推断层
blatt flaws	平推断层
blatterstein	球颗玄武岩
blazed iron	高硅生铁
bleached bed	褪色层
bleached rock	漂白岩石
bleached sand	漂白砂
bleached zone	淋溶带；褪色带
bleaching	漂白；淋溶酌
bleaching clay	漂白土
bleb	气泡；空洞；矿物内的异矿包裹体；气泡，气孔
bleed	自流；渗出；涌出；自喷；泌水
bleed effect	流失效应
bleed hole	排水孔；活塞行程限制孔；放泄孔；放出孔
bleed of gas	瓦斯喷出
bleed off	放掉
bleed off pipe	溢流管
bleed path	泌水路径
bleed pressure	屈服压力；溢流压力
bleed test	吸水试验
bleed water	泌水；析水；废水；剩余水
bleeder	取样管
bleeder well	排水井；减压井；渗沥井
bleeding	逸出；渗出；析出；泌浆
bleeding capacity	泌水能力；灰浆泌水率
bleeding cement	水泥浆沫；浮浆
bleeding rate	析水率；泌浆率；泌水率
bleeding rock	渗水砂岩
bleeding surface	渗出面
blend	混合料；参合；拌合
blend composition	掺混料
blend gravity	混合比重
blend ratio	掺合比
blende	闪矿类矿物
blended	混合的；融合的
blended cement	合成水泥；掺合水泥；混合水泥
blended sand	混合砂
blended soil	混杂土
blended unconformity	混合不整合；模糊不整合
blender	搅拌器；掺合机；混合器；松砂机
blending	混合；掺合；配料
blending functions	弯曲函数
blending of wastes	废水掺合；废物掺合
blind	无露头的；不通的；暗井；堵塞；隐蔽
blind apex	掩盖矿脉露头；隐伏露头
blind catch basin	截留盲井；暗截流井
blind chimney	盲脉
blind creek	干涸河床；间歇河床；雨季河；吸收床
blind deposit	无露头矿床；掩蔽矿床；盲矿床；隐伏矿床
blind ditch	暗沟，暗渠；填碎石或砾石的排水沟；地下排水沟；渗水暗沟
blind drainage	盲沟排水；暗沟排泄；闭流水系；内流水系
blind drainage area	闭流区
blind drainage conduit	盲沟排水沟道
blind drift	采石平巷；独头平巷；不通主水平的平巷；独头巷道
blind fastener	隐蔽式扣件；膨胀螺栓
blind fault	隐伏断层；盲断层
blind front	隐伏前锋
blind hole	盲孔；不通孔；不返泥浆的漏井；瞎井
blind image	隐蔽像
blind in	封堵

blind inlet	排水暗沟进水口	block failure	整体破坏
blind island	隐蔽岛	block failure strength	整体强度
blind joint	盲节理；隐节理	block fault	块状断层
blind layer	地震记录无反射层	block fault pattern	块状断裂模式；块状断裂型
blind lead	无露头矿脉	block faulted area	块断区
blind level	排水平巷；冰间死水道；水平暗管	block faulted coast	断块海岸
blind lode	无露头矿脉	block faulting	块状断裂；块断作用；块断运动
blind ore	盲矿	block faulting tectonics	块状断裂构造
blind ore body	盲矿体	block field	砾海；石海；块砾场
blind outcrop	盲露头；隐伏露头	block flexure toppling	岩体弯折倾倒
blind pass	暗道	block folding	地块褶曲；块断褶皱作用
blind pit	暗井	block folds	块褶曲
blind shaft	暗井；盲井	block footing	块型底脚；块形大方脚
blind spot	静区；盲点	block foundation	块式基础；砌块基础；实体式基础
blind structure	隐伏构造	block glacier	块状冰川；石流；石河
blind subdrain	盲沟；填石阴沟	block glide	断块滑动；整体滑移
blind thrust	隐伏逆冲断层	block hammer	落锤
blind unconformity	隐蔽不整合；渐变不整合	block ice-movement	块状冰川运动
blind valley	盲谷	block lava	块状熔岩
blind vein	盲脉；隐脉	block level	气泡水准仪
blind zone	隐蔽层，盲区，盲带；屏蔽区	block like structure	块状构造
blinded unconformity	盲不整合	block line	钻井钢丝绳
blinded-pipe pumping	盲管抽水	block load	块状荷载
blinding	补路石砂；填封，基础垫层，闭塞，堵塞，淤塞；盖土；铺石屑	block map	框图
		block masonry	砌块砌体
blinding concrete	混凝土护层；盖面混凝土；基础垫层混凝土；地基垫层混凝土	block masonry structure	砌块砌体结构
		block mass	大块；岩块；地块；古陆；团块
blinding course	封层；垫层	block mica	厚片云母
blinding layer	基础垫层；地基垫层；"草鞋底"	block mining	块段开采；分块开采
blinding material	填塞材料；透水材料	block model	块体模式
blinding valley	盲谷	block moraine	块砾碛
blister	气泡，水泡，气孔；起皮；发泡；局部隆起，小丘；折叠，皱折，凸纹	block mosaic	块状镶嵌；断块镶嵌
		block mountain	断块山
blister cracking	折叠破裂	block movement	断块运动；块体移动
blister hypothesis	局部隆起假说	block of stone	磴石
blistering test	圆孔刺破试验	block off	封堵；隔开；砌防护隔墙；用套管等保护钻孔孔壁
bloat	膨胀；熏制		
bloated clay	膨胀黏土	block off water	堵水
bloater	膨胀器；膨胀剂	block oilfield	断块油田
bloating	膨胀；肿胀	block overthrusting	断块上冲；断块逆掩
bloating agent	膨胀剂	block packing	终碛堆；砾块堆
bloating test	膨胀试验	block panel system	分区切块开采法；分区短柱开采法
blob foundation	独立浅基础	block pavement	砌块路面
blob of slag	火山渣块	block pitching	块石铺砌
block	砌块；块料；岩块；块石	block placement	分块浇筑；分区填筑
block anchorage	块体锚固	block plan	规划图；区划图；分段图；楼宇平面图
block and tackle	滑轮组；吊车	block power plant	单岸式电站
block arch	砌块拱	block power station	单岸式电站
block basin	断块盆地	block protection	砌块石护岸
block beam	拼块梁；砌体梁	block ramp	断块对冲
block breakwater	石块防波堤；混凝土防波堤	block reef	不稳定矿脉；珠串状矿脉
block bridging	横撑	block ridge	断块海岭
block caving	大块坍落	block sag	块状下陷；断块凹陷
block clay	泥石混杂物	block sample	块体试样；块状试样
block curve	实线；连续曲线	block scheme	方块图；方框图
block diagram	立体地质剖面图；框图；部件图；框式图解	block shear test	断块剪切试验；块体剪切试验
		block size	岩块尺寸；岩块大小
block disintegration	块状崩解	block slide	块体滑动；块体滑坡
block disruption	岩块崩解	block slope	块石坡
block dome	断块穹隆	block stamp	冲锤

block stone	块石
block stone lining	块石衬砌
block stream	块状岩流；岩块流；石河；砾河
block structure	块状构造；断块构造
block subsidence	块体沉陷；块状断陷
block tectogene	断块构造带
block tectonics	断块构造
block theory	块体理论；阻滞学说
block thrust	块状冲断层
block tin	纯锡块
block tipping	岩块倒落
block topple	岩体倾覆
block up	阻塞；隔断；堵塞；封锁
block up water grouting	堵水注浆
block uplift	断块上升
block upwarping	地块隆起
block volume	岩石块体
blockage	堵塞；淤塞
blockage of drain	排水堵塞；渠道淤塞；渠道堵塞
block-block covariance	岩体对岩体协方差
block-bounding faults	断块分界断层
block-caved area	崩落法开采区
block-caving method	分段崩落采矿法
block-cut method	分区回填开采法
blocked channel	向心式水泵；淤塞河道
blocked stone	块石
blocked stream	堵塞河；阻断河
block-edge flexure	断块边棱挠曲
block-fault differential uplift	断块差异上升
block-fault structure	块状断裂构造
block-faulted mountain range	块断山脉
block-faulted oilfield	块断油田
block-faulted structure	块状断裂构造
blockgebirge	断块山
blockhole	炮眼
blockhole blasting	爆破地面大块岩石
block-in-course	嵌入楔状层；成层砌石块体
blocking	撑木；垫木；阻塞；模块化
blocking against water	堵水
blocking circulation	阻塞环流
blocking coefficient	堵塞系数
blocking drainage	阻塞水系
blocking efficiency	整流效率
blocking extension	阻碍臂
blocking factor	堵塞系数
blocking flow	阻塞流
blocking layer	阻挡层；闭锁层
blocking leakage	堵漏
blocking leaking	堵漏
blocking of drainage	排水阻碍
blocking of ice	冰塞；冰坝
blocking permeability	很差的渗透率
blocking temperature	间歇温度；封闭温度；中断温度
blocking up	堵塞；泥沙填塞；淤沙；淤塞
blocking-out of deposit	划分矿床；矿床分区
block-lifter	升高机；千斤顶
blockness	块度
blockout	打草图；切开采区；方形槽口
block-precondition	块预处理
blocks	砌块
blocks active force	活动块体作用力
blocks dislocation	块体错动
blocks of elements	单元块
blocks of nodes	节点块
blockslump	地滑
blockstone pavement	块石铺砌的路面
blockwork seawall	方块海堤
blockwork wall	方块体墙
blocky	块状的；短而粗的；结实的
blocky cement	块状胶结物岩
blocky coke	块状焦
blocky granular shape	块状颗粒形状
blocky joint	块状节理
blocky lava	块状熔岩
blocky lava surface	块状熔岩面
blocky pattern	块状图像
blocky rock	块状岩石
blocky soil structure	土壤块状结构
blocky stones	块石
blocky structure	块状构造
blockyard	预制混凝土构件场
blodite	钠镁矾
bloedite	白钠镁矾
Blondeau method	布朗多法
bloodstone	鸡血石；赤铁矿；血滴石
blooey-line	排粉管
blooie line	气体钻井的排岩屑管
bloom	风化煤层露头；起霜，煤华；石油荧光；大钢坯；盐华
bloom out	起霜
blossom	风化煤层，风化矿层露头；华，煤华，矿华；铁帽
blossom rock	岩华
blot	墨水渍；污渍；污点；涂污
blotchy	斑点的
blotter coat	吸收层
blow	爆破；冲击；顶板崩落；岩石隆起
blow count	锤击数
blow count of SPT	标准贯入击数
blow down	放水，送风；排气，放气；强制崩顶；爆破落煤
blow fill slope	吹填坡
blow gun	喷枪
blow hole	砂眼；气泡；铸孔；吹穴
blow land	风蚀地
blow of dynamic sounding	动力触探锤击
blow of penetration test	贯入试验锤击
blow off	沉箱放气；喷出
blow off valve	放水阀；排水阀
blow out preventer (BOP)	防喷器
blow over	吹积
blow pile	爆破桩
blow sand	喷砂
blow stress	冲应力
blow test	冲击试验
blow up	拱胀；炸毁；爆破
blow well	自喷井；自流井
blowaut	井喷
blowed fill	风力充填
blower	鼓风机；空气压缩机

blow-in	涌入；吹入	bluff	悬崖；陡峭的；变质围岩；陡岸
blowing	崩解；爆裂；溅出	bluff body	阻流体
blowing a well	放喷除水与砂	bluff erosion	陡壁冲刷
blowing cone	熔岩锥；火山渣锥	bluff failure	陡壁坍毁
blowing hole	风穴	bluff sand	陡砂层
blowing power	冲击力	bluff work	边坡整平作业
blowing pressure	吹灰压力	bluish	带蓝色的；浅蓝色的
blowing sand	高吹沙	blunge	揉黏土；用水搅拌
blowing well	自喷井；喷泉井	blunger	圆筒掺和机；揉土机
blown sand	风成砂；飞砂	blunt angle	钝角
blown sand damage	风沙破坏	blunt arch	垂拱；平圆拱
blown tip pile	爆扩桩	blunt delta	钝三角洲
blown-out concrete	加气混凝土；气孔混凝土	blunt drill	钝钻；钝钎
blown-out land	风蚀区	blunt end of pile	桩的粗端
blow-off chamber	排泥井；吹泄室	blunt jumper	钝钎子；钝冲击钻杆
blow-off water	排污水	blunt nosed pier	平头墩
blowout	突涌；井喷；突喷；溃喷；风蚀坑	blunt pile	钝桩
blowout basin	风蚀盆地	blunting of spur	山嘴削磨
blowout boil	井喷沸腾	bluntness	钝度
blowpipe test	吹管试验	blythite	锰石榴石
blows	标贯击数	board	板；薄板
blows foot	每英尺深锤击数	board and pillar	房柱式开法
blows per-min	每分锤击数	board and pillar method	房柱式开采法
blst	压载；压载物；压舱物	board and pillar work	房柱式开法
blub	浇筑灰浆；灰泡	board and wall method	房柱式开采法
blue asbestos	蓝石棉	board butt joint	喷射施工缝
blue aura	蓝辉	board chairman	董事会主席；董事长
blue bands	蓝色冰带	board coal	纤维褐煤；木质褐煤
blue base	蓝图	board gate	通采区的巷道；切割巷道
blue bind	煤系中的页岩或泥岩	board hammer	夹板捣碎机；重力式落锤
blue brittleness	蓝脆	boarding	背板；板条；安装木板；钉板
blue clay	青土；蓝黏土	boar's back	猪背岭
blue copper	铜蓝；蓝铜矿	boart	工业用圆粒金刚石；不纯的金刚石；低质钻石；金刚砂；粗凿石
blue copper ore	蓝铜矿		
blue copperas	胆矾	boaster	阔凿；榫槽刨
blue gas	水煤气；蓝煤气；蓝水煤气	boasting	粗堑石头
blue green algae	蓝绿藻	boasting chisel	破石凿；阔凿
blue ground	含金刚石橄榄岩；硬黏土层或页岩层	boat basin	锚地
		boat bridge	浮桥
blue key	蓝底图	boat hoist	吊艇
blue lead	蓝铅；金属铅	boat-shaped mass	船形块体
blue malachite	石青；蓝铜矿	bob	向上式凿岩机；浮子；摆锤；铅锥
blue marl	青泥灰岩	bocca	喷火口；熔岩口
blue metal	蓝铜锍；硬质页岩；硬黏土；一种灰蓝色长石砂岩	bodden	古水盆海湾
		Bode's law	波得定律
blue Monday sand	蓝色蒙特砂岩	bodily seismic wave	地震体波
blue mud	蓝泥；深海泥；青泥	bodily tide	固体潮
blue occurrence	蓝铁矿	bodily wave	体波
blue powder	蓝粉	body	体；物体，机体；底质，质地；支架
blue print	蓝图	body axis	物体轴线；固定轴；联系轴
blue quartz	蓝水晶	body centered lattice	体心晶格
blue spar	天蓝石	body fluid	流体
blue stone	胆矾	body force	体积力；质量力；体力
blue vitriol	胆矾；蓝矾；五水硫酸铜	body force components	体力组成
blue whistler	喷气井	body force potential	体力势
blueground	青地；青脉石	body fossil	实体化石
blue-john	纤维状或柱状荧石	body load	自重；本身重量；静重；覆盖地层的自重
blueprint	蓝图，设计图	body lock ring	封隔器主体锁紧圈
blueschist	蓝片岩	body measurement method	实体测定法
blueschist belt	蓝片岩带	body of a dam	坝体
bluestone	长石砂岩；硬黏土；页岩；胆矾		

body of masonry	砌体
body of river	主河床
body of the bit	钻头体
body of water	水体；水域；贮水池
body pile	桩身
body plan	横剖面型线图；船体型线图；船体正面图
body slaking	体积崩解
body stress	有效应力；体应力；体积应力
body water	水体
body wave	体波
body wave magnitude	体波震级
body wave radiation	体波辐射
body wave spectrum	体波谱
body wave velocities	体波速度
body-chamber	体腔
body-fixed coordinate system	地固坐标系
body-strain rotation components	体应变旋转分量
body-wave motion	体波运动
body-wave phase	体波相位
Boehm lamellae	勃姆纹；勃姆层纹
boehmite	软水铝石；勃姆石
bog	沼泽，泥炭地；泥塘
bog blasting	土方爆破开挖
bog body	沼泽矿体
bog burst	沼泽丘
bog bursting	沼泽涌出
bog butter	奶油沥青
bog coal	土状褐煤
bog drainage	沼泽排水
bog earth	沼泽土；沼泽地
bog flow	沼泽泥流
bog formation	沼泽层；沼泽建造
bog ground	沼泽地；沼泥地
bog iron	褐铁矿
bog iron ore	沼铁矿
bog land	沼泽地
bog lime	沼灰土；灰岩；湖白垩
bog mine ore	褐铁矿
bog muck	泥炭；沼泽腐殖土；沼泽腐泥；泥炭
bog ore	沼铁矿
bog peat	沼泽泥炭；沼泽泥炭土
bog soil	沼泽土
bog specificity	沼泽特征
bog surface area	沼泽面积
bog water	沼泽水；泥沼水
bogan	滞水湾
bogaz	深岩沟；溶蚀坑
bogen structure	弧状构造
bogen texture	弧形玻屑结构
bogginess	沼泽化；泥沼状态
bogging down	下陷
boggy	多沼泽的；沼泽的
boggy ground	沼泽地
boggy soil	沼泽土
boghead	藻煤
boghead cannel	藻烛煤
boghead cannel coal	藻烛煤
boghead cannel shale	沼油页岩
boghead coal	藻煤；深褐色烛煤
boghead shale	藻油页岩

bogheadite	泥沼；藻煤
boghedite	藻烛煤
bogie bolster	转向架支承梁
bogue	河口
bogusite	淡沸绿岩
bohemian garnet	深红色石榴子石
Bohemian gemstone	波希米亚宝石
bohemian ruby	玫瑰水晶
boil	冒水翻砂；冒泡；水浪翻花；沸腾
boiling	翻浆，翻浆冒泥；砂浴；沸腾；喷溅
boiling flask	烧瓶
boiling geyser	高温间歇喷泉；激喷间歇喷泉
boiling ground	冒汽地面
boiling inclusion	沸腾包裹体
boiling inclusion geobarometry	沸腾包裹体地质压力计
boiling lake	沸湖；沸热塘；高温湖
boiling mud	沸泥；沸泥塘
boiling mud pot	沸泥塘
boiling of sand	砂沸；涌砂
boiling pit	沸泉塘
boiling point	沸点
boiling pool	沸水塘；沸泉塘
boiling process	沸腾法
boiling sea-water spring	海水沸泉
boiling spouter	沸喷泉
boiling spring	沸泉；高温泉
boiling tub	化灰池
boiling up	底鼓
boiling up in the floor	地鼓；底板鼓裂
boiling up soil	冻胀土
boiling water	沸水
boil-off	蒸发；汽化；煮掉
bojite	角闪辉长岩
boke	小细脉
bold	陡的
bold cliff	绝壁
bold coast	陡峻海岸；高海岸
bold relief	粗糙地形
bold shore	峭岸；高岸；陡岸
bole	胶块土；红玄武土；黏土
boleite	银铜氯铅矿
bolide	火流星
Bolindian	波林德阶
bollard	系缆柱；系船柱；矮柱；护柱
bollard plinth	护柱柱基；护柱基座
bollard pull	系缆桩拉力
Bolling Stage	伯灵期
bologna spar	重晶石
bologna stone	重晶石
bolred rigid frog	钢轨组合辙叉
bolson	河流汇交盆地；荒芜盆地；沙漠盆地；干湖地；封闭洼地
bolson plain	山间平原
bolster	横撑；支承梁；垫木
bolt	锚杆；螺栓；楔形木块；短圆木
bolt anchor	锚杆
bolt anchorage	锚杆锚固
bolt and anchoring substance	锚杆和锚固物质
bolt and cable-bolt parameters	锚杆和锚杆参数
bolt bar	螺栓杆

bolt class	锚杆分类
bolt core	螺栓杆
bolt density	锚杆密度
bolt fastening	螺栓连接
bolt grouting machine	锚杆孔注浆器
bolt gun	锚枪
bolt head	螺栓头；锚杆头
bolt invert	底锚杆反拱
bolt load	锚杆承载力；锚杆荷载
bolt mesh support	锚网支护
bolt plate parameters	锚杆板参数
bolt pretension	螺栓预紧
bolt pulling force	锚杆拉力
bolt reinforcement	锚杆加固
bolt rows space	锚杆排距
bolt shaft force tester	螺栓轴力测试仪
bolt shank	锚杆杆体
bolt sleeve	螺栓套管
bolt spacing	锚杆间距
bolt supporting	锚杆支护
bolt wall	锚杆式墙
bolt yield load	锚杆屈服荷载
bolt-anchorage test	锚杆锚固强度试验
bolted joint	螺栓联接
bolted splice	螺栓接头
bolt-grouting support	锚杆注浆支护的
bolthole	联络小巷；锚杆孔
bolting	锚固；锚栓支护；锚杆支护；螺栓连接
bolting and meshing support	锚网支护
bolting and shotcrete	锚喷支护
bolting and shotcrete lining	锚喷支护
bolting and shotcreting	锚喷支护
bolting and shotcreting with wire mesh	锚喷网支护
bolting investigation	锚杆研究
bolting with mesh	锚网支护
bolting with wire mesh	锚网支护
bolt-mesh-anchor support	锚网，锚喷支护
boltting	锚固
boltwoodite	硅钾铀矿
Boltzman's theological model	波尔兹曼流变模型
bolus	红玄武土
bomb	火山弹；量热弹；还原钢弹；弹性容器
bomb sag	火山弹坑；层面坑
bombiccite	晶蜡石
bomite	斑铜矿
Bomolithus	簇石
bonanza	富矿脉；丰饶砂；大矿囊；富矿体
bonanza pool	富油藏
bond	黏结，黏结力，结合力；黏着；联结；联结力；砌合
bond anchorage	黏结锚固
bond beam	结合梁
bond cement	固结水泥
bond clay	黏合土；胶结的黏土；黏土
bond coat	黏合层；打底层
bond course	砌砖层；黏结层
bond creep	黏着蠕变
bond failure	黏结破坏；结合损坏
bond force	黏结力；结合力
bond index	胶结指数
bond length	锚固长度；黏结长度；锚固段；键长
bond resistance	黏着抗力
bond slip relation	黏滑关系
bond slippage	黏结滑动
bond strength	黏合强度；结合强度
bond stress	黏结力；握裹力；黏结应力；结合面的剪应力
bond stress of mortar	砂浆握裹力
bond value	黏结值；黏结力；黏着度
bond water	结合水；束缚水
bondability	握裹力；握裹性能
bonded concrete	胶结混凝土
bonded length	锚固段
bonded post tensioning	黏结后张
bonded prestressed concrete structure	有黏结预应力混凝土结构
bonded resistance strain gage	黏贴式应变计
bonded seal	黏合密封；接头密封；保险密封；焊接密封
bonded seam	黏结缝
bonded stone	拉结石
bonded strain gage	黏贴式应变计
bonded type	黏贴型
bonded type strain gauge	黏贴式应变计
bonder	连接器；砌墙石；连接石；结合物
bonding	加固；冰冻胶结；连结；黏结
bonding admixture	黏合剂
bonding character	黏结性；黏合性
bonding characteristic	黏结特性
bonding clay	胶结性黏土
bonding coat	黏合层
bonding course	结合层
bonding force	耦合力；黏结力；键力，结合力
bonding property	黏结性
bonding strength	结合强度；黏接强度
bonding water	结合水
bondless	无联系的；松散的；不结合的
bone bed	骨层；含骨化石地层
bone black	骨炭
bone breccia	骨屑角砾岩
bone coal	骨煤；高灰煤；碳质页岩
bone dust	骨粉
bone tool	骨器
bone-coal	骨炭；骸炭
boning	水准测量
boning board	测平板；标杆；砧板；测平杆
boninite	玻古安山岩
bonny	矿囊；矿巢
Bononian	博诺阶
bononian stone	重晶石
bonstay	暗井
bony	页岩质的，骨煤的，劣质的
bonze	末精选的铅矿石
boojee pump	气压灌浆泵；压浆泵
book clay	纹泥；带状黏土；书页黏土
book structure	页状构造
bookhouse structure	片堆构造
book-leaf clay	页状黏土
bookshelf fault	书架式断层
bookshelf faulting	书架断层

bookstone	书状页岩
boole's rule	布尔规则
boom	悬臂，吊杆；桁，构架；起重臂
boom and bucket delivery	吊杆和斗输送
boom angle	吊机臂转角；箍用角条；吊臂角
boom breaker	悬臂式破碎机
boom clutch	吊杆离合器
boom crane	臂式起重机
boom derrick	带吊臂的起重塔
boom derrick crane	支臂桅杆起重机
boom equipment	悬臂装置
boom excavator	臂式挖土机
boom foot	悬臂底座
boom head shield	钻臂式掘进盾构
boom header	部分断面隧道掘进机
boom hoist	悬臂起重机；扒杆绞车；臂式吊车
boom hoist cable	提升悬臂的钢丝绳
boom lift	悬臂升程
boom lifter	悬臂式起重机
boom line	悬臂拉绳
boom loader	悬臂装载机
boom miner	悬臂式联合开巷机
boom net	拦网
boom out	悬臂伸出
boom pendent	悬臂吊环
boom pitch	悬臂斜角
boom plus underframe valve	摇臂和底架阀
boom point	悬臂头
boom point sheave	悬臂首端滑轮
boom reach	悬臂长度；吊杆长度
boom ripper	悬臂式挑顶机；悬臂式平巷掘进机
boom ripping machine	悬臂截割头掘进机
boom road header	悬臂式平巷掘进机
boom sheave	天轮；悬臂天轮
boom sprinkler	长臂喷灌机
boom stacker	悬臂堆积机
boom support gutter	悬臂牵索
boom support guy	悬臂牵索
boom support type pile driver	支架式打桩机
boom suspension rope	悬臂绳
boom swing	悬臂回转
boom swinging gear	悬臂的转动机构
Boomer	布麦尔震源；布麦尔波；水中振源激发器
boomerang corer	自返式取样管
booming	放水冲刷法
booming dune	鸣砂沙丘
booming sand	鸣砂；响砂
boom-mounted drifter	钻车式钻机
boom-mounted wheel excavator	轮式吊杆挖土机
boom-type drill rods carrier	笼式钻杆箱
boom-type road header	悬臂式掘进机
boose	矿石内的脉石
boost	升举；升压
boost motor	助推器；加速器
boost pressure	升压力；吸入管压力
boost pressure control	泵入口增压
boost pump	增压泵；辅助泵
boost ratio	增压比
booster	助推器；局扇，局部通风机，催爆用少量炸药，爆管；升压器，增压器，助力器，放大器
booster disk	传爆管垫片
booster explosive	传爆药
booster fan	辅助鼓风机；辅助通风机
boosting	助推；提高，升高；升高电压
boosting cable	吸上线
boosting wire	吸上线
boot	进料斗；给料斗，靴；保护罩
boot and latch jack	带闩捞钩打捞工具
boot basket	靴式打捞篮
boot cap	油罐车底卸油阀帽
boot jack	打捞钩
boot pulley	底轮
boot truck	洒沥青卡车
boot-cleaning machine	刷靴装置
bootee pump	灌浆泵
booth	小室，小亭；工作台
boothite	七水胆矾
bootleg	炮窝；私自开采；瞎炮眼，靴筒；炮眼底
bootstrap generator	仿真线路振荡器
bootstrap memory	引导存储器
bootstrap routine	辅助程序；引导指令
bootstrap test	自展测试
bootstrapping	引导指令
booze	铅矿
bop	防喷器
boracic water	含硼水
boracite	方硼石
borasca	博拉斯科雷暴
borate	硼酸盐；硼酸酯
borated water storage tank	含硼水贮存箱
borates	硼酸
borax	丹石；硼砂
borax bead	硼砂珠
borax bead test	硼砂球试验
boraxand borates	焦硼酸钠
borazon	博拉任
bord	巷道；矿房
bord and pillar method	房柱式开采法
bord and pillar work	房柱式开采法
bord and wall method	房柱式开采法
bord cleat	巷道夹板
bord face	垂直主经理推进的工作面
Bordas mouthpiece	博达营
bord-down	顺倾斜巷道
border	边缘；界限；边界线；邻接
border arc	边缘弧
border belt	漂砾带
border check	畦；畦田；土埂
border Density	边界密度
border dike	边堤；无污染；围堤
border ditch	岩体
border effect	边界效应；边缘相；边行影响
border facies	边缘相
border fault	边界断层
border figure	图廓数据
border flooding	畦田漫灌
border furrow	外缘沟
border information	图廓标注说明；图廓注记
border irrigation	畦灌
border irrigation method	畦灌法

border lake	边湖
border line	边界线；图廓线
border link	边境连接道路
border matching	边缘匹配；接边
border moraine	侧碛；边碛
border pile	边桩；边缘桩；边界桩
border ring	反应边
border river	边界河
border stone	镶边石；桩墩；界石
border strip	畦条
border zone	边缘带
bordering	炮泥；转弯
bordering fault zone	边界断裂带
bordering mountain chains	边缘山脉
borderland	边区；边缘地带
borderland basin	陆缘盆地
borderland slope	大陆边缘坡
borderline detonation point	边界爆震点
border-line risk	难以确定的风险
borderline spring	边界泉
borderline tectonite	边界构造岩
borderline well	边界井；边缘井
bordersea	边缘海
bord-face	与主解理成直角前进的采煤工作面
bord-up	逆倾斜的巷道
bore	孔，镗；内径；打眼，钻进；石门，隧道
bore bit	钎头，钻头；岩柱
bore borings	钻粉
bore casing	套管
bore climb	高潮冲击
bore core	岩芯
bore diameter	孔径；内径；净径
bore diameter of anchor cable	锚索孔径
bore discharge	井孔流量；井孔排放
bore frame	井身结构；钻架
bore hammer	钻机；凿岩机
bore hole	钻孔；井眼
bore hole gravel packer completion	裸眼砾石填充完井
bore hole sealing	钻孔止水
bore hole seismic logging	地震测井
bore hole specimen	钻孔样品
bore journal	钻孔记录表
bore liner	井孔套管
bore log	钻孔柱状图，钻井记录；测井记录；测井曲线
bore machine	镗床
bore meal	钻粉；钻屑；细粒的；钻粉浆
bore mining	钻孔开采
bore mud	钻探泥浆；钻泥
bore of water meter	水表口径
bore out	镗孔
bore pile	钻孔灌注桩
bore pipe	井管
bore pit	探井；钻探坑；钻凿小井
bore plug	钻孔土样
bore pump	钻孔泵
bore rod	钻杆；镗杆；凿岩杆
bore sample	岩芯；钻探土样
bore size	孔径；内径
bore specimen	钻孔岩样；无因次系数；岩芯标本
bore well	钻井
boreability	可钻进性
bored cast-in-place pile	钻孔灌注桩；钻孔桩
bored concrete pile	混凝土灌注桩
bored hole diameter measurement	钻孔直径检测
bored hole verticality measurement	钻孔垂直度检测
bored pile	钻孔桩；钻孔灌注桩；钻孔墩；钻孔插入桩
bored pile wall	钻孔桩墙
bored piles	钻孔桩
bored precast pile	钻孔预制桩
bored spring	人工泉；自流井
bored tube well	管井
bored tubular well	往复移动式缆索道；细粒料止水
bored tunnel	钻挖的隧道
bored underreamed pile	扩底钻孔灌注桩
bored well	井孔；钻井；浅钻打的井；席状节理
borefield	井场；井田；井区
borehole accelerometer	井下加速度计
borehole audio tracer survey	噪声测井
borehole axial strain indicator	钻孔轴向应变记录指示器；钻孔轴向应变指示计；永久积雪作用
borehole axis	井眼轴线
borehole blasting method	炮眼爆破法；浅眼爆破法
borehole bottom	钻孔底；井底；孔底
borehole breakout	井筒爆裂
borehole breakout method	钻孔崩落法
borehole breakouts	钻孔塌陷
borehole bumps	钻孔凸点
borehole caliper	钻孔规；钻孔测径器
borehole camera	钻孔照相机；孔内摄像；压应力轨迹线；井内照相机
borehole camera	钻孔孔壁摄影机
borehole camera inspection	钻孔摄影检查
bore-hole cameras and television	钻孔机与电视
borehole casing	压力强度；钻孔套管
borehole cavity	井眼空穴
borehole characteristics	井眼特性
borehole charge	钻孔装药；炮眼装药
borehole check	钻孔核对
borehole cleaning	清孔
borehole clinometer	钻孔测斜仪
borehole collapse	井塌
borehole collar	钻孔口
borehole columnar section	钻孔柱状图
borehole compensated mode	井眼补偿方式
borehole compensated sonic log	井眼补偿声波测井
borehole compensated sonic tool	井眼补偿声波测井仪
borehole compensation	井眼补偿
borehole compensation system	井眼补偿装置
borehole condition	井眼条件
borehole core	钻孔岩心；消力孔洞
borehole core logging	岩芯编录
borehole core logging	钻孔岩心编录
borehole corrected flow	经过井眼校正的流量
borehole correction	井眼校正
borehole cost	钻眼费；打眼费
borehole curvature	井眼曲率
borehole deflectometer	钻孔挠度计
borehole deformation	钻孔变形
borehole deformation gauge	钻孔形变计；钻孔变形测量仪

borehole deformation method 钻孔孔径变形法
borehole deformation strain cell 钻孔变形式应变计
borehole deformer 钻孔变形测定器
borehole depth 钻孔深度
borehole depth correction 校正孔深
borehole deviation 钻孔偏斜；井斜
borehole deviation surveying 钻孔弯曲测量
borehole device 自由水面
borehole diameter 钻孔直径
borehole diametral strain indicator 钻孔径向应变指示计；钻孔径向应变计；咸水域
borehole dilation test 钻孔膨胀试验
borehole dilatometer 钻孔膨胀计
borehole direct shear apparatus 钻孔直剪仪
borehole direct shear device 钻孔直剪仪
borehole director 钻孔引导器
borehole drilling time 钻眼时间
borehole effect 井孔效应；井筒效应
bore-hole electric load cell 钻孔电阻应变测力计
borehole electromagnetic method 钻孔电磁法
borehole electromagnetic wave penetration meter 钻孔电磁波透视仪
borehole enlargement 井眼扩大；扩孔
borehole environment 井眼环境
borehole extensometer 钻孔引伸计；钻孔伸长计；钻孔位移计
borehole face 井孔内壁；井壁
borehole failure 井筒破坏
borehole flow modeling 钻孔流量建模
borehole flow-rate profile 井眼流量剖面
borehole flow-velocity measurement 钻孔流速测量
borehole fracture 井眼破裂
borehole gas emission law 钻孔瓦斯涌出规律
borehole gauge 钻孔仪
borehole geometry 井眼几何形状
borehole geometry log 井眼几何形状测井
borehole geometry tools 井眼几何形状测井仪
borehole geophone 井中检波器
borehole geophysical logging 地球物理测井记录
borehole geophysical prospecting 钻孔地球物理勘探
borehole geophysical software 井中地球物理软件
borehole geophysics 钻孔地球物理；地球物理测井学
borehole gravimeter 钻孔重力仪；井下重力仪
borehole gravimetry 井中重力测量
borehole gravity density logging 井下重力密度测井
borehole gravity gradiometer 井下重力梯度仪
borehole gravity meter 井下重力仪
borehole gravity sensor 井下重力传感器
borehole gravity survey 井下重力测量
borehole heater 井内加热器
borehole heating 井内加热
borehole hydraulics 钻孔水力学
borehole images 钻孔图像
borehole impression device 钻孔打印器；钻孔印像器
borehole impression packer 钻孔打印器
borehole inclination 井斜
borehole inclinometer 钻孔倾斜仪
borehole inclusion stressometer 钻孔包体式应力计
borehole induction log 筑路沥青
borehole inspection 钻孔检查

borehole instability 井壁失稳
borehole instability mechanism 井眼失稳机理
borehole jack 钻孔千斤顶
borehole jack method 钻孔塞孔法
borehole layout 布孔
bore-hole line 井中套管柱
borehole lining 钻眼加固
borehole loading 炮孔装药
borehole loading test 钻孔载荷试验
borehole location plan 钻孔位置平面图
borehole log 钻孔测井；测井曲线；录井；钻孔记录；钻孔柱状图
borehole logger 钻孔记录器
borehole logging 钻孔测井，钻孔编录
borehole lubricant 井眼解卡液
borehole magnetometer 井下磁力仪
borehole measurement 井中测量
bore-hole mining 钻孔开采
borehole mining method 钻孔法采矿
borehole multi-point extensometer 钻孔多点伸长仪
borehole navigation 井眼导向
borehole noise 井眼噪声
borehole nomenclature 炮眼名称
borehole numerical simulator 井筒数值模拟器
borehole observation of crustal strain 井下地应变观测
borehole observation of crustal tilt 井下地倾斜观测
borehole operation 井下作业
borehole orientation 钻孔方向
borehole orientation tool 井眼定向仪
borehole parameter 炮孔参数
borehole pattern 逐日流量
borehole penetrometer 钻孔针穿计；钻孔贯入器
borehole periscope 钻孔潜望镜
borehole permeameter method 钻孔渗透仪法
borehole photo 井中摄影
bore-hole position survey 钻孔位置测量
borehole pressure 炮眼压力；孔壁压力
borehole pressure recorder 钻孔压力测定器
borehole pressure recovery test 钻孔压力恢复试验
borehole problems 井下复杂问题
borehole profile 钻孔柱状图
borehole pump 钻井泵；深井泵
borehole radiowave method 井中无线电波法
borehole radio-wave penetration method 井中无线电波透视法
borehole recharge 钻孔回灌
borehole reciprocating pump 往复式井下泵
borehole record 钻孔记录
borehole resistivimeter 井内铃电阻计
borehole resistivity meter 井内电阻计
borehole response 井眼响应
borehole rugosity 井眼不规则度
borehole sample 完全掺气水流；涡流作用；钻孔土样
borehole scanner 钻孔扫描仪；钻孔扫描器；井下扫描器
borehole seal 镗孔密封垫；镗孔密封垫；钻孔封孔
borehole seal packer 钻孔封孔工
borehole seismic 井下地震
borehole seismic data 井中地震资料
borehole seismic logging 钻孔地震测井

borehole seismic profile	井中地震剖面
borehole seismograph	井下地震仪
borehole Seismology	钻孔地震
borehole shaft driven pump	井下长轴传动泵
borehole shear apparatus	钻孔剪切仪
borehole shear test	钻孔剪切试验
borehole shooting	钻井爆破
bore-hole sinking	钻孔
borehole size	井眼大小；井眼尺寸
borehole sonde	钻井探头
borehole spacing	钻孔间距
bore-hole springing	孔底扩大；钻孔底扩大
borehole stability	井眼稳定性；井筒稳定性；井壁稳定
borehole status	井眼状况
borehole strain cell	钻孔应变盒
borehole strain gauge devices	钻孔应变计
borehole strainmeter	钻孔应变计
borehole strata orientator	钻孔地层定位仪
borehole stress gauge	钻孔应力计
borehole stressmeter	钻孔应力计
borehole survey	钻孔精密测量；钻孔岩层测量；钻孔偏斜测量；钻孔测量
bore-hole surveying	钻孔勘探；钻孔测量；井下勘探；钻探；井下测量
bore-hole surveying instrument	钻孔检测仪
borehole televiewer	井下声波电视；钻孔电视；井中电视；井下声波电视
borehole television	井中电视；超声成像测井；钻孔电视
borehole television camera	蛙式夯；钻孔电视设备；钻孔电视摄影机
borehole television（BHTV）	井下电视
borehole test	钻孔测试
borehole thermometer	井温仪
borehole throw	钻孔偏距
borehole traverse	井斜
borehole trepanning gage	钻孔环向仪
borehole TV	钻孔电视
borehole volume	井径测井体积
borehole wall	孔壁；井壁
borehole wall contact device	贴井壁装置
borehole wall sloughing	井壁坍塌
borehole wall strain method	钻孔孔壁应变法
borehole washout	井眼冲蚀
borehole waves	钻孔波
borehole well	钻孔；钻井；自流井
borehole yield	钻孔出水量；钻孔流量
borehole-borehole system	跨井系统
borehole-borehole variant	井中-井中方式
boreholes- exploratory	钻孔-探索性
bore-holes for exploration	勘探钻孔
boreholes- logging	钻孔-日志
boreholes slotter	钻孔槽
borehole-surface system	井地系统
borehole-surface variant	井中-地面方式
borehole-to-borehole seismic measurement	井间地震测量
borer	钻探工；钻孔器；镗床；风钻
borescope	钻孔检查显示器；管道镜；井内探测镜
boresight	瞄准线；视轴
bore-springing experiment	原地连续爆破试验
boric acid	硼酸
boric acid injection	硼酸注入
boric anhydride	硼酸酐
borickite	磷钙铁矿
boride	硼化物
boring	打眼，钻机，钻探；镗孔；钻孔，炮眼；钻粉
boring and blasting method	钻爆法
boring and mortising machine	循环性振动
boring animal	钻孔动物
boring apparatus	钻探设备
boring arm	碎岩杆；碎煤杆
boring bar	钻杆；镗杆
boring barge	转臂式固定起重机；延爆剂
boring bit	钎头；镗孔钻头
boring by percussion	冲击式钻进
boring by rotation	旋转式钻进；旋转钻井
boring casing	钻孔套管；套管；钻探套管
boring clamp	提升钻杆的夹具
boring cock	钻头
boring core	钻孔岩心
boring crown	钻头
boring cutter	钻孔器
boring cycle	掘进循环
boring depth	有效流量
boring diameter	掘进直径
boring dust	钻屑
boring equipment	钻机设备
boring for drainage, dewatering boring, drain boring	排水钻孔
boring for oil	石油钻深
boring frame	涡流方程；钻架；钻塔
boring gauge	孔径规
boring head	钻头；钎头；钻具的切削头
boring inclination measurement	扬压力折减系数
boring journal	钻探报表；钻进日志；钻进记录
boring log	钻孔记录；钻孔柱状图；钻井日志；自由溢流堰；自应力水泥
boring log columnar section	钻孔柱状图
boring machine	管道穿越打洞机；镗床；钻机；镗缸机；钻机
boring master	钻机长
boring method	打钻法；钻井法
boring mud	钻泥
boring of shaft	钻井
boring organism	钻孔生物
boring pattern	钻孔排列法；炮眼组型式和放炮次序图
boring period	掘进循环周期，掘进期
boring pipe	钻管
boring pore	生物钻孔
boring porosity	生物钻孔孔隙度
boring principle	钻进原理
boring process	掘进过程
boring profile	屋脊瓦；钻孔断面图；井下剖面图；网络特性线
boring programme	钻探进度计划；钻探方法；钻探程序
boring pump	打钻用泵
boring record	钻孔记录；钻探报表
boring resistance	钻探阻力
boring rig	钻塔；钻架；钻机；钻探设备
boring rod	下尾闸门；钻杆
boring rod clamp	围岸浅滩

boring rod set	钻杆装置	bort	圆粒金刚石；不纯金刚石；金刚石屑
boring rope	钻机用钢丝绳	bort bit	金刚石钻头
boring sample	钻探试样；岩芯取样	bort-set bit	细粒金刚石钻头
boring sampling	钻探取样	bortz	钻用不纯金刚石粒；圆粒金刚石；细粒金刚石
boring section area	掘进断面面积	bortz powder	金刚石粉
boring site	钻探工地	borway bit	齿状钻头，旋转式环状钻头；齿状取心钻头
boring spacing	主要土类	boryslawite	硬脆地蜡
boring specimen	转向轴套	Bosch-Omori seismograph	玻什-大森地震仪
boring spindle	钻轴；镗杆	boschung	重力坡
boring stem	钻杆	bosh	炉腰；炉腹；浸冷水管
boring stick	土钻	bosh angle	炉腹角
boring stroke	掘进行程	bosh jacket	炉腰外套
boring test	钻孔调查；钻孔试验	boshing	浸水冷却
boring through	穿孔；钻穿	bosk	压缩空气储存器；矮林；丛林
boring time	掘进时间	bosom	矿藏；蕴藏
boring tool	钻具，钻探工具；圆酒精水准器	bosphorite	含水磷铁矿
boring tower	钻塔；钻架；自由流压力	boss	外壳；岩瘤，岩钟；轮壳，节销孔座，凸出部；捣矿机锤身；采空区，掏槽
boring tube	钻管	boss hammer	大锤；石工锤
boring verticality	钻孔垂直度	boss head	凸头
boring well	钻井	Bossak-Newmark algorithm	波萨克-纽马克算法
boring winch	钻进用绞车	bossed	凸形的
boring with line	索钻	bossing	厚层切底；大掏槽
borings	钻孔碎屑	bosslike	穹状
boring-type continuous miner	钻进型连续采煤机	bossy	凸起的
boring-typemining machine	钻进式联合采煤机	Boston blue clay	波士顿蓝黏土
borland type fish pass	闸式鱼道；鱼闸	bostonite	浅歪细晶岩
Born equation	玻恩方程方程	bosun chair	吊椅
born fuse	角式保险器	Boswell classification	包氏维尔分级法
bornemanite	磷硅铌钠钡石	bot	底，船底；泥塞；始点
bornhardt	残山；岛山	bot canal	浅水运河
bornite	斑铜矿；蚀变斑岩	both side drift method	双侧壁导坑法
borocalcite	硼钙石；硼酸方解石	Bothell's process	拜塞尔木材浸渍法
borolanite	霞榴正长岩	botryogen	赤铁矾
boromagnesite	硼镁石	botryoid aragonite	葡萄状霰石
boron carbide	碳化硼	botryoidal	葡萄状；葡萄状的
boron counter	硼计数器	botryoidal agate	葡萄玛瑙
boron fibre	硼纤维	botryoidal aggregate	葡萄状集合体
boron isotope	硼同位素	botryoidal structure	葡萄状构造
boron nitride	氮化硼	botryolite	葡萄硼石
boron trichloride	三氯化硼	botryose inflorescence	总状类花序
boron trifluoride catalyst	三氟化硼催化剂	botrys	总状类花序
boron trifluoride proportional counter	三氟化硼正比计数管	bott	泥塞
		bott stick	渣塞杆
boron water	含硼水	bottle agitation	瓶搅动
boronatrocalcite	钠硼钙石	bottle chock	轨道转弯处的导绳轮
boronizing	渗硼处理；硼化	bottle coal	气煤；瓦斯煤
borosilicate	硼硅酸盐	bottle float	瓶式浮标
Borros point	波罗杆；参考标杆	bottle jack	瓶式千斤顶；瓶式起重器
borrow	取土；采料；采料场	bottle method	比重瓶法
borrow area	采泥区；采料区；取土区；挖泥区	bottle neck	交通拥挤区；交通道路狭窄区；瓶颈谷；峡谷
borrow bank	借土边坡；挖土边坡	bottle neck control	瓶颈路段管制法
borrow cut	借土挖方	bottle Neck Road	瓶颈段；瓶颈路
borrow earth	借土	bottle neck slides	瓶颈状滑动
borrow excavation	取土挖掘；挖土方	bottle oiler	瓶式加油器
borrow exploration	料场勘探	bottle post	漂浮瓶
borrow fill	借土填方	bottle roller	转瓶器
borrow material	取土坑材料；填方土料；挖方取得的泥石料	bottle sampler	瓶法取样；取样瓶
borrow pit	取土坑，采石场	bottle silt-sampler	瓶式泥沙采样器；泥沙采样瓶
borrow site	取土地点；料场		

bottle spring 淡水泉
bottle test 瓶内混汞试验；瓶试验
bottle top mould 瓶口锭模
bottle trap 瓶型气隔
bottle type suspension load sampler 瓶式悬移质采样器
bottleneck 瓶颈；交通频繁的窄道路，阻碍场所；难关，难以解决的问题；妨碍，从中作梗；薄弱环节
bottle-neck 薄弱环节；瓶颈；峡谷
bottleneck area 堵塞区
bottleneck pipe 渐缩端管
bottle-neck pore 瓶颈状孔隙
bottleneck problem 瓶颈问题
bottleneck slide 瓶颈状滑坡
bottlenecking 管端缩径
bottler 灌注机
bottom 山脚；山麓；底部；井底；低地
bottom accretion 河底淤高
bottom anchorage 底部锚具
bottom aquifer 底水层
bottom aquifer drive 底水驱
bottom articulated joint 底部铰接缝
bottom ash 底灰
bottom assemblies 底部钻具组合
bottom bank 沙洲；浅滩
bottom banksman 井底把钩工
bottom bed 底层
bottom belt 回程段胶带；回程段皮带；底部皮带
bottom bench 下台阶，底部台阶
bottom blade 底刀
bottom blowing 底吹
bottom blown converter 底吹转炉
bottom board 井底桩板；底板
bottom boom 下弦杆
bottom bottle 底层浮标瓶；瓶式浮标
bottom bounce 海底多次反射
bottom boundary 底界面
bottom bracing 底撑
bottom bullnose 下部喷管
bottom bush 底衬；下轴瓦
bottom canch 卧底
bottom cargo 底货载
bottom case 底座
bottom casting 下铸法；底铸法
bottom chain track 底链导槽；底履带
bottom characteristic 底质
bottom characteristics exploration 底质调查
bottom characteristics sampling 底质采样
bottom charge 孔底装药
bottom chart 海洋底质图
bottom choke 井底油嘴
bottom chord 下弦；底弦
bottom clay 淤泥
bottom clearance 底部净距；补充
bottom coal 底层煤；煤层底煤
bottom collar 钻铤
bottom configuration 海底地形
bottom contact platform 海底钻井平台；海底采矿工作平台
bottom contour 水底地形；水下地形；海底地形；水底等高线；等深线

bottom contour chart 海底等深图
bottom contour map of coal bed 煤层底板等高线图
bottom contraction 底部收缩
bottom coring 海底取岩心
bottom course 底层；基层
bottom crawler 海底履带式钻进机
bottom current 底层流；海底强流；底流
bottom cut 下部淘空；底部淘刷；底槽；底部掏槽
bottom cutting 底部截槽；掏槽底
bottom dead centre 下死点
bottom dead point 下止点
bottom density current 密度底层流
bottom deposit 海底沉积
bottom diameter 底径
bottom discard 锭尾切头
bottom discharge 底部卸载，活底卸料
bottom discharge bit 底冲式钻头
bottom door 底门
bottom drag 河底拖力；河底曳引力；河底推移力
bottom drainage 底部排水
bottom drift 海底漂砂；海底漂移；底部掏槽；底部导洞
bottom drift method 下导坑超前先拱后墙法；底部导坑法
bottom drift method of tunnel mininging 隧洞开挖的下导洞推进法
bottom drift method, bottom heading method 下导坑超前先拱后墙法
bottom drift, bottom heading 下导坑
bottom driller 钻井井底深度
bottom drive 井底驱动
bottom dump 经舱底卸置；舱底式卸置
bottom dump bucket 底卸式铲斗；底卸料斗
bottom dump skip 底卸式箕斗
bottom dump truck 底部卸料车；底卸卡车
bottom dumping 底部卸载
bottom dumping car 底卸式车
bottom dumping door 底卸门
bottom end 底端
bottom end of the face 作业面底端
bottom equipment 孔内设备
bottom feed 底部加料；下送料；底部供液；底部进给或进刀
bottom fill 底部充填
bottom flange 下翼缘
bottom flash 底部飞边
bottom flow 底流；底层流；回卷浪；回头浪
bottom flowing pressure 井底流动压力
bottom flush bolt 底部埋入插销
bottom flushing of sediment 底部排沙
bottom formation 底层
bottom founded rig 坐底式钻井平台
bottom friction 底部摩擦；海底摩擦力；筛底粉
bottom frictional layer 底摩擦层；底埃克曼层
bottom gangway 运输道；底导坑
bottom gas 积聚在低注处的爆炸性气体
bottom gate 运输道；底闸门；底导坑；底注式浇注系统
bottom girder 底梁
bottom glade 河滩沼地；底谷
bottom gravimeter 水底重力仪
bottom gravity current 底部重力流
bottom gravity measurement 海底重力测量

bottom guide	下导板
bottom head	底盖
bottom heading	下导坑；底导坑；下部平巷；底部导洞
bottom heading method	下导坑超前先拱后墙法
bottom heave	坑底隆起；底鼓；底部隆起；基底隆起
bottom heaving	底板隆起
bottom hold down	井底防上滑器；底部压住
bottom hole	孔底；井底；底眼；底部炮眼
bottom hole assembly（BHA）	井底钻具组合
bottom hole back torque	井底反扭矩
bottom hole bit locator	井底钻头定位器
bottom hole choke	井底油嘴
bottom hole circulating pressure	井底循环压力
bottom hole circulating temperature	井底循环温度
bottom hole circulation pressure	井底循环压力
bottom hole coverage	对井底的覆盖程度
bottom hole dynamic condition	井底动态条件
bottom hole flow bean	井底喷嘴
bottom hole flowing pressure	井底流动压力
bottom hole heater	井底加热器
bottom hole hydraulic parameter	井底水力参数
bottom hole location	井底位置
bottom hole orientation sub	井下定向接头
bottom hole packer	井底封隔器
bottom hole pattern	井底形状
bottom hole plug packer	塞式井底封隔器
bottom hole pressure	钻孔底部压力；井底压力
bottom hole temperature	井底温度
bottom hole thermometer	井温仪
bottom hopper barge	底开式驳船；底卸式船
bottom hull	沉垫；下部船体
bottom ice	底冰；锚冰
bottom impermeable layer	底部不透水层；下封闭层
bottom installation	井底车场设备
bottom intake rack	底孔进水口拦污栅
bottom joint	水平节理；底节理
bottom kerf	底槽
bottom land	低洼地；谷底；河漫滩；滩地
bottom land forest	河谷林；组织水
bottom layer	底层
bottom layer fold	底层褶皱
bottom layer of subgrade	基床表层
bottom layer of subgrade bed	基床表层
bottom layout	井底布置
bottom level	下部水平；井底车场标高；底面标高；层底标高
bottom lift	扬程；最下水平
bottom lifting	开凿底板；卧底
bottom line	向斜褶皱轴；盆地轴
bottom liner	底衬
bottom load	底负载；推移质；地层负载；底沙
bottom loading belt	底带装载式胶带输送机
bottom material	底质
bottom moraine	底碛
bottom mounted breakwater	混合式防波堤
bottom mud	底层泥浆
bottom nose	尾端
bottom of beach	滩底
bottom of formation	层底
bottom of pass	溜眼底
bottom of the groove	凹沟底
bottom operation coupler	下作用车钩
bottom out	探明
bottom outlet	深孔；底孔；泄洪底孔
bottom outlet diversion	底孔导流
bottom outlet hole	底排水孔
bottom outlet orifice	底部放水孔口
bottom packer	井底封隔器
bottom pan	最底层的灰土
bottom perforation	底部穿孔；底部射孔
bottom photography	海底照相
bottom pillar	井底保安煤柱；井底保安矿柱
bottom plate	主夹板；底板
bottom plate screen plate	底板螺钉
bottom plug	底塞
bottom pressure	底压力；底部压力
bottom price	底盘
bottom priming	底部点火；孔底起爆；反向起爆
bottom probe	河底取样钻
bottom product	残留产品；塔底产物
bottom profiler	海底地层剖面仪
bottom pull	海底牵引
bottom pump	最低水平的泵；沉没泵
bottom quotient	底槽；底座圈；底水道
bottom rail	下冒；下冒头
bottom raising	卧底；底鼓
bottom reflection	海底声反射；海底反射
bottom reflection coefficient	海底反射系数
bottom relief	水底地形；水下地形；海底地形；底面起伏
bottom relief map	海底地形图
bottom reverberation	河底交混回响
bottom ring	沉井下部的刃脚圈；底环
bottom ripping	卧底
bottom ripple	河底波痕
bottom road	井下大巷
bottom roll	下罗拉；下轧辊
bottom roller	底辊
bottom sample	底部样品；储罐底原油样
bottom sampler	底泥采样器；孔底取样器；水底取土钻；水底采样器
bottom sampling	海底取样；海底采样；底泥采样
bottom sampling device	底层取样器
bottom sampling tube	底质柱状取样器
bottom scanning sonar	海底扫描声呐
bottom scattering	海底散射
bottom scavenging	井底清洗
bottom scour	底部冲刷
bottom seal	底水封
bottom section	底部截面；井底部分；下部半断面
bottom sector gate	底部扇形闸门
bottom sediment	底沉积物；底泥；底部泥沙
bottom sediment chart	底质分布图
bottom sediment transport	河底泥沙搬运
bottom sedimentation tube	沉淀管
bottom sediments	底沉积物
bottom set	底部沉积
bottom set bed	底积层
bottom set bed of delta	三角洲底积层
bottom settiings	底部沉淀

bottom shear stress	底切应力；底流剪切应力
bottom shot	井底爆炸作业
bottom side	下侧，底侧
bottom side of casing	套管的下半侧
bottom simulating reflector	海底仿拟反射器
bottom size	粒度下限；入料下限；下限
bottom skip	底卸式箕斗
bottom slant	底部斜巷；底斜
bottom slice	底分层
bottom slicing	上行分层开采法；分层陷落采矿法
bottom slicing method	底部分层法
bottom slope	底坡；河底坡降；河底比降
bottom sludge	残渣
bottom sluice	底部泄水闸；污水净化设备
bottom sluice gate	底部泄水闸门
bottom soil	下层土；底土
bottom sprag	底部支撑
bottom spring	湖底泉；河底泉
bottom squeeze	底板隆起；底鼓
bottom stone	基石；耐火黏土
bottom stope	下向梯段工作面；下向回采；正台阶工作面
bottom stoping	下向梯段回采
bottom sub	堵头
bottom subsidence	底板塌陷；底板沉陷
bottom surface	底面
bottom survey	海底测量
bottom suspension	底部悬挂装置
bottom taking	卧底
bottom tap ladle	塞棒式铸钢桶
bottom temperature	底部温度；底层水温；井底温度
bottom terrace	河底阶地
bottom topographic survey	海底地形测量
bottom topography	水底地形；水下地形；海底地形；底面
bottom tow	底牵法
bottom trace	底部记录线
bottom traction	底部牵引；底部牵引力
bottom tractive force	河底拖力；河底曳引力；河底推移力
bottom transverse	底部横材
bottom trash rack	底部拦污栅
bottom valve	底阀
bottom velocity	底部流速；河底流速
bottom view	底视图；底视
bottom wall	下盘，底帮
bottom wall and anchor packer	井壁及尾管封隔器
bottom water	底部水；底层水；底水
bottom water coning	底水锥进
bottom water drive	底水驱动
bottom water drive reservoir	底水驱油藏
bottom water drive type reservoir	底水驱动型储层
bottom water encroachment	底水水侵
bottom water injection	底水注水
bottom water sand	底部含水砂层
bottom water warming	底水变暖
bottom wave	底波；水底波；海底波
bottom width	底宽
bottom wing	堆石坝截水墙下翼
bottom-bole sample taker	孔底取样器
bottom-charting fathometer	海底地貌测绘仪
bottom-discharge bucket	活底铲斗，底卸式铲斗；底卸式吊桶
bottom-discharge skip	底卸式箕斗
bottom-discharge-type	底卸式的
bottom-discharging truck	底卸式载重汽车
bottom-draught bin gate	底部放料仓闸门
bottom-drift excavation method	漏斗棚架式隧道开挖法
bottom-dump barge	开底运泥船
bottom-dump body truck	底卸式载重汽车
bottom-dump car	底卸车；底卸式拖运机
bottom-dump platform	活底板吊框
bottom-dump tractor trailer	底卸式拖运车
bottom-dump trailer	底卸拖车
bottom-dump wagon	底开式料车
bottom-dumping gismo	吉斯莫底卸式万能采掘机
bottom-dumping semi-trailer	底卸式半拖车
bottom-dwelling	底栖生物
bottom-emptying	底卸式的
bottomer	井底把钩工
bottom-friction factor	底摩擦系数
bottom-grab	斜桩
bottom-hole cleaning	井底清洗
bottom-hole coverage rate	井底遮盖系数
bottom-hole flowmeter	井底流量计
bottom-hole fracturing pressure	井底破裂压力
bottom-hole inclinometer	井底测斜仪
bottom-hole pressure device	井底压力计
bottom-hole pressure drawdown	井底压降
bottom-hole pressure gauge	井底压力计
bottom-hole sample taker	井底取样器
bottom-hole sampler	井底取样器
bottom-hole sampling	井底取样
bottom-hole scavenging	井底清洗
bottom-hole treating pressure	井底施工压力
bottom-hole wellbore pressure	井底压力
bottoming	铺底；到达底部；输出下限；从下面切断信号；石块铺底
bottoming bit	可卸式钻头，活动钻头
bottomland meadow	泛滥地草甸
bottomless bridge	敞口卸矿台
bottomless scraper bucket	无底耙斗
bottom-loading pressure	底部压力
bottommost	最深的
bottommost layer	最深层
bottom-pull method	底拖法
bottom-pull unlocking mechanism	底拉开锁机构
bottoms	底部沉积物；塔底残留物；残渣；煤层底部
bottoms up	井底至井口的泥浆行程
bottomset deposit	底部沉积
bottom-side ionogram	底部频高图
bottom-side ionosphere	底部电离层
bottom-side sounder	底视探测仪
bottom-slicing system	底部分层崩落采矿法
bottom-supported	底部支承的
bottom-supported fixed platform	底部支撑的固定式平台
bottom-supported mobile rig	坐底式活动钻井装置
bottom-supported movable platform	底部可支撑的可移式平台
bottom-supported platform	坐底式平台

bottom-tension	海底张力
boudinage	石香肠构造；布丁构造；石香肠；香肠构造
boudinage structure	布丁构造；香肠构造
boudinage-mullion	香肠-窗棂构造
Bougainville Trench	布干维尔海沟；北所罗门海沟
Bouger effect	布格尔效应
Bouguer anomaly	布格异常
Bouguer anomaly	布格异常
Bouguer anomaly map	布格异常图
Bouguer correction	布格校正；布格改正；布格校正值
Bouguer effect	布格效应
Bouguer gravity anomaly	布格重力异常
Bouguer gravity map	布格重力图
Bouguer layer	布格层
Bouguer plate	布格板
Bouguer reduction	布格归算；布格改正；布格校正
Bouguer slab concept	布格平板概念
Bouguer-Beer's law	布尔-比尔定律
boulangerite	硫锑铅矿
boulder	漂石；漂砾，散석；滚石；砾石
boulder barricade	砾石堤；砾石埂
boulder barrier	漂砾堤
boulder base	蛮石地基
boulder beach	巨砾滩；砾滩；砾石滩
boulder belt	漂砾带；冰砾带
boulder blaster	大块爆破工
boulder blasting	岩石小爆破；放小炮；改炮；二次爆破；爆破巨砾
boulder buster	大块石破碎器
boulder channel	漂砾河槽；漂砾河
boulder clay	冰碛物；冰碛土；粗砾土；漂砾土；漂石黏土
boulder conglomerate	巨砾岩
boulder cracker	大块石破碎器
boulder crusher	粗碎机
boulder dam	石坝
boulder ditch	卵石沟；法式排水沟；盲沟
boulder extractor	漂石提取器
boulder facet	砾棱面
boulder fall	砾石下坠；孤石崩落；孤石下坠；巨砾泻落
boulder fence	防石栅；防石栏
boulder field	砾海
Boulder Field Inventory	富孤石区目录
boulder flat	平坦砾块区
boulder flint	燧石
boulder fragmentation	巨砾爆破
boulder gravel	含漂砾石；漂砾；巨砾层
boulder mud	冰砾泥；泥砂
boulder of disintegration	崩解巨砾
boulder of weathering	风化巨砾
boulder orientation	砾石排列方向；砾石定向
boulder pavement	漂砾地；漂石路面
boulder prospecting	漂砾勘查
boulder quarry	布满裂隙的采石场
boulder sand	漂砾砂岩
boulder setter	蛮石铺砌层
boulder stone	漂石；巨石
boulder stream	砾河
boulder stream canyon	蛮石峡谷
boulder strip	条石
boulder tracing	漂砾追踪
boulder train	漂石列；漂砾流
boulder well	底穴井；砂砾层井；漂砾井
boulder-controlled slope	砾块控制坡
bouldereet	小漂砾；中砾
boulder-fan	漂砾扇
boulder-field	巨砾原；漂石原
bouldering	铺筑巨砾；铺筑大圆石
boulder-like structure	漂砾状构造
boulders	巨石；卵石；漂石；井中软硬交替层
boulderstone	巨砾岩
bouldery	含巨砾的；漂砾的；巨砾的
bouldery colluvium	漂石坡积物；砾石坡积物
bouldery ore	砾状矿石
boulet	小块蛋形煤砖；小煤球
boulis	岩屑堆
boult	筛
Bouma cycle	鲍玛旋回
Bouma rhythmite	鲍玛韵律
Bouma sequence	鲍玛序列；鲍玛层序
bounce	岩石突出；跳动；反跳；界限；限制
bounce dive	速返潜水
bounce diving limit	速返潜水深度极限
bounce impact elasticity	冲击回弹性
bounce mark	弹跳痕迹
bounce on error	误差范围
bounce table	冲击台；振动台
bounce-back	回跳
bounce-back effect	回弹效应
bouncing	跳动；跳跃的；大的
bouncing pin	跳针；弹跳杆
bouncing vibration	浮沉振动
bound	界限；限制；连结的，跳动；结合的
bound energy	结合能
bound energy	束缚能
bound gravel	胶结砂砾体
bound greatest lower	最大下界
bound layer	固定层；边界层
bound least upper	最小上界
bound of aggregation	聚集边界
bound on error	误差范围
bound pile	带铁箍桩
bound radon	束缚氡
bound rupture	链断裂
bound stress	黏合应力
bound surface	邻接面
bound term	约束项
bound to failure probability	破坏概率限值
bound variable	约束变量
bound vector	有界向量
bound vortex	附体涡流
bound water	附着水；封存水；结合水；束缚水；薄膜水
bound water saturation	束缚水饱和度
boundary	边界，地界，限度，范围；范围，界线；分层面；交汇面；分界线
boundary adjustment	地界调整
boundary angle	边界角
boundary block weakening	矿块边界的削弱

boundary breakthrough 边界联络横巷
boundary caving drift 边界崩矿平巷
boundary cavitation 边界区空蚀
boundary collocation 边界配置
boundary compensation 边界补偿
boundary condition 极限条件；边界条件
boundary condition data 边界条件数据
boundary condition of sliding 滑移边界条件
boundary conduction layer 边界导热层
boundary constraint 边界约束
boundary current 边界流
boundary curvature influence on boundary stresses 边界曲率对边界应力的影响
boundary curve 边界曲线；界限曲线
boundary cutoff 沿边界切割矿块；沿崩落矿块边界的分割巷道
boundary deformation condition 边界变形条件
boundary demarcation of surface mine 露天矿境界圈定
boundary diffusion 界面扩散
boundary dimension 外形尺寸；轮廓尺寸
boundary displacement 界壁位移；边界位移
boundary effect 边界效应；边际效应
boundary element 边界元
boundary element method 边界元法
boundary element method for rock engineering 岩体工程边界元法
boundary energy 边界能
boundary enhance 边缘增强
boundary equation 边界方程
boundary fault 界断层；边界断层
boundary fence 边篱
boundary film 边界膜
boundary film lubrication 边界油膜润滑
boundary force vector 边界力向量
boundary fracture 边界破裂；界面断裂
boundary friction 边界摩擦
boundary geometry 边界几何形状
boundary identification plan 地界辨认图
boundary impedance functions 边界阻抗函数
boundary integral equation 边界积分方程
boundary integral equation method 边界积分方程法
boundary integral method 边界积分法
boundary integral terms 边界积分项
boundary interference 边界干扰
boundary layer 边界层；界面层；边界图
boundary layer control 边界层控制
boundary layer effect 边界层效应；界面层效应
boundary layer model 边界层模式
boundary layer of laminar flow 层流边界层
boundary layer separation 边界层分离现象
boundary layer theory 边界层理论
boundary layer thickness 边界层厚度
boundary layer transition 边界层过渡区；边界层过渡
boundary layers 边界层
boundary line 边界线，界限；界线
boundary line of general survey 普查范围线
boundary line of preliminarily investigated reserves 初查储量计算范围线
boundary line of road construction 道路建筑限界
boundary line problem 边界问题
boundary line survey 境界线调查
boundary load 边界载荷
boundary lubricating 边界润滑；境界润滑
boundary lubrication 边界润滑
boundary map 行政区划图；边界图
boundary mark 地界标志；界址标记；地界标志
boundary marker 界标
boundary methods 边界方法
boundary monument 界碑；界标
boundary mutation 边界变异
boundary network node (BNN) 边界网络结点
boundary node 边界结点
boundary node method 边界点法
boundary of aquifer 含水层边界
boundary of borrow area or quarry area 料场范围线
boundary of capillarity 毛管作用
boundary of cutting and filling 挖方、填方间界线
boundary of deviation 偏差极限
boundary of fixed flow 定流量边界
boundary of fixed water level 定水头边界
boundary of grout zone 灌浆区范围
boundary of known flow 已知流量边界；二类边界
boundary of known water level 已知水位边界
boundary of plate 板块边界
boundary of property 建矿地界，区段边界；矿界
boundary of refilling 回填边缘线
boundary of relative confining bed 相对隔水层界线
boundary of section 采区边界，区段边界
boundary of specified head 给定水头边界
boundary of subsidence trough 沉陷盆地边界
boundary of weathered zones 风化界线
boundary of works area 施工范围
boundary opening 边界巷道
boundary parameter 边界参数
boundary peg 地界桩钉
boundary perturbation problem 边界扰动问题
boundary piezometer 边界压力计
boundary pillar 分界矿柱
boundary plane 边界面；分界面
boundary point 边界点；界址点；地界点
boundary pore 边界孔隙
boundary porosity 边界孔隙度
boundary post 界柱；界标
boundary pressure 油水或气水界面之平均流体压力
boundary pressure cell 边界压力盒
boundary problem 边界问题
boundary process 边界面际法
boundary rectification 地界修正
boundary re-establishment survey 地界重整测量
boundary reflection of seismicity 地震的界面反射
boundary register 边界寄存器
boundary relationship 边界关系
boundary repulsion term 边界排斥项
boundary resistance 边界阻力
boundary response 边界反应
boundary saturation 饱和界面
boundary scattering 边界散射
boundary science 边缘科学
boundary segment format 周边线段格式
boundary settlement 划界

boundary shear 边界剪应力
boundary shrink 留矿边界切开口
boundary sign 界标
boundary similarity 边界相似
boundary slots 边界切割槽
boundary solution 边界解
boundary solution method 边界解法
boundary spring 边界泉；接触泉；冲积坡泉；溢出泉
boundary state of stress 边界应力态
boundary stone 界石；界碑
boundary stope 矿体边部采场
boundary stratotype 界线层型；地层界线；标准剖面；地层分界线
boundary stress concentration 边界应力集中
boundary stresses 边界应力
boundary stresses for a circular opening 圆形硐室边界应力
boundary strip 田界
boundary stripping ratio 境界剥采比
boundary surface 界面
boundary surface active agent 界面活性剂
boundary surface model 边界面模型
boundary survey 边界测量；勘界；桩界测量
boundary tension 边界张力
boundary traction 边界力；补充
boundary treatment 边界处理
boundary Trefftz-type elements 边界 Trefftz 型单元
Boundary Tribunal 地界审裁处
boundary value 边界值；临界品位值；临界品位
boundary value problem 边值问题
boundary velocity 边界速率；界面速度
boundary ventilation 边界式通风；对边通风
boundary wall 边界墙
boundary wave 界面波；边界波
boundary well 边界井
boundary work 边界功
boundary zone 边界带；边缘带
boundary zone of capillarity 毛细管界限区；毛细管作用带
boundary-drag coefficient 边界阻力系数
boundary-film-forming material 形成界限膜的物质
boundary-layer acceleration 边界层加速
boundary-layer flow 边界层水流；边界层流动
boundary-layer growth 边界层增厚
boundary-layer oscillation 边界层振动
boundary-layer skin friction 边界层表面摩擦
boundary-layer velocity profile 边界层速度剖面
bounded 有界的
bounded difference 有界差
bounded function 有界函数
bounded naturally fractured reservoir 封闭式天然裂缝性储层
bounded reservoir 封闭储层
bounded sum 有界和
bounded sum operator 有界和算子
bounded system 封闭系统
bounded variable 有界变量
boundedness 有界性；局限性；限度
bounder 矿地丈量员；矿山定界人；小土地占有者
bound-free transition 束缚-自由跃迁

bounding capacity 黏着力
bounding depth 极限深度
bounding erosion surface 边界侵蚀面
bounding fault 分界断层；边界断层
bounding force 黏合力
bounding medium 黏合介质
bounding surface active agent 界面活性剂
bounding surface of fold 褶皱界面
boundless 无限的
bounds 边界
bounds on displacements 位移界限
bounds on vibration frequency 振动频率界限
boundstone 生物黏结灰岩；黏结岩
bounge 煤的挤出；煤的突出
bouquet 钙华
bourdon gage 波登规
Bourdon gauge 波登管式压力计；波登应力计
Bourdon manometer 波登压力表
Bourdon pressure gauge 波登压力计
Bourdon pressure vacuum gage 波登压力真空计
Bourdon tube 波登管；波登型管式压力计
Bourdon tube gage 波登管压力计
Bourdon tube pressure gage 波登管压力计；弹簧管压力计
Bourdon-type pressure gauge 波登型压力计
bourn 细流；小河
bourne 边界；小河；小溪；间歇河流
Bournonia 波褶蛤属
bournonite 车轮矿
bourock 大块石头；石堆
bouse 矿石内的脉石
Boussinesq coefficient 布辛涅斯克系数
Boussinesq solution 布辛涅斯克解
boussinesq theory 布辛涅斯克理论
Boussinesq's equation 布辛涅斯克方程
boussingaultite 六水铵镁矾
boustay 暗井；盲井
bout 绕筒上的线圈
boutgate 通地面的人行道；井底通道；煤层间的通道
bovey coal 褐煤
bow 弧；拱梁；船头；弄弯
bow anchor 船首锚
bow area 褶皱带；褶皱区
bow beam 弓形梁
bow check 弓面检查
bow compasses 两脚规；小圆规；外卡钳
bow drill 弓形钻
bow gantry 前架；高架；拉力构架
bow girder 弓形大梁
bow member 拱架；弓形构件
bow pen 小圆规
bow saw 弓形锯
bow shaped hyaloclastic texture 弓形玻屑结构
bow tie 弓形交叉；蝴蝶结；领结状记录；回转现象
bow type centralizer 弓形扶正器
bow valley flat 弓形河漫滩
bow wave 冲激波；船头波
Bowditch's rule 包迪奇闭合导线误差校正定则
bowel 内部；内脏
bowels of the earth 地球内部

Bowen ratio	鲍恩比率
bowenite	鲍文玉；透蛇纹石
Bowen's reaction series	鲍文反应系列
bower	船头锚；涡线；主锚
bower anchor	主锚；船头锚
Bower-Banff process	鲍沃尔-巴尔夫钢面氧化法
bowk	地层破裂声；瓦斯泄放声；凿井用吊桶
bowl	盆状落水洞；凹穴；浮槽；圆锥壳；离心机的转筒
bowl and slips	卡瓦座及卡瓦
bowl centrifuge	无孔截头锥形筒离心机；沉降过滤式离心脱水机
bowl classifier	浮槽式分级机；分级槽
bowl cover packing	杯盖密垫
bowl grab	锅锥
bowl hole	瓯穴
bowl line	弓形曲线；船首缆
bowl mill	盆式辊磨机；球磨机
bowl mixer	碗形搅拌机
bowl overshot	圆锥壳式打捞筒
bowl scraper	铲运机；斗式耙装机
bowlder	巨砾；蛮石
bowl-gudgeon	木质绞车轴头
bowlingite	包林皂石
bowl-rake classifier	浮槽耙式分级机
bowl-shape basin	碗形盆地
bowl-shape settlement	碟形沉降
bowl-shaped depression	碗状坳陷
bowralite	透长伟晶岩
bowser	加油车
bow-shaped dune	弓形沙丘
bow-shaped structure	弓形构造
bowstring	弓形的；弓弦；绞索
bowstring arch	弓弦拱
bowstring truss	弓弦式桁架
bow-tie effect	蝴蝶结效应
box	盒，箱；轴箱，洗箱；井下煤车
box abutment	消融区；箱形桥台
box and needle	罗盘
box annealing	闭箱退火；封闭退火
box antenna	箱形天线
box auger	套筒式螺旋钻
box barrier	匣式岩粉棚
box beam	匣形梁；箱形梁
box breakwater	柱塞泵
box bridge	箱式桥；电阻箱电桥；吸湿性
box bubble	球形水准器
box caisson	沉箱；箱式沉箱；匣式沉箱
box caisson foundation	沉箱基础
box canyon	箱形峡谷
box car	棚车
box carburizing	装箱固体渗碳
box chute	箱形溜槽
box classification method	盒式分类法
box cofferdam	未干燥木材
box column	自动磅秤；箱形柱
box compass	罗盘
box conduit	淹没水舌
box cooler	箱式冷却器
box corer	箱式取样器；盒式取样器
box coupling	套筒联轴节；函形联轴节；接箍；箱形联轴节
box culvert	匣形涵洞；箱涵；盒形排水渠；阻尼系数
box cut	开段沟；箱形剥离法；箱切；开挖槽
box dam	重力；箱式围堰；箱式坝；整体围堰
box dock	阻尼作用
box down	箱体回落
box drain	木沟渠；匣形沟渠；方形沟；阻尼常数
box effect	框限作用
box end	母扣
box eye bolt	环头螺栓
box face	阴螺纹端面
box filter	箱式过滤器
box floor	溜口底
box flume	尾座
box fold	穹形褶皱；箱形褶皱；状态转移矩阵
box footing	箱型基础
box foundation	箱型基础；叶片状构造
box frame	小冰川；匣形构架
box frame construction	阻尼力
box gauge	轴流泵
box girder	箱形主梁；箱形大梁；下轴承
box groove	箱形轧槽
box gutter	箱形槽
box heading	小坑道
box holding	溜口架
box hole	漏口；溜口；箱穴
box inlet	压气打桩
box iron	槽钢；箱形钢
box junction	接线盒；分线盒
box keel	箱形龙骨
box key	套筒扳手；开箱钥匙
box launder	箱形流槽；矩形断面流槽
box link	夹板滑环
box loop	环形天线；箱形天线
box model	箱式模型
box out	预留箱形凹位
box packing	盒式密封
box pallet	箱式托盘
box pass	矩形孔型；箱形孔型
box pile	箱型桩；新火山岩；箱型钢桩
box plot	框图
box raise	漏斗天井
box ramp	耙斗卸料台
box regulator	矩形滑动风窗
box sampler	箱式取样器；盒式取样器
box section	方形管；方通；箱型断面；方框图
box segment	箱形管片；中空弧形模板
box sextant	盒式六分仪
box shape	箱形
box shear apparatus	盒式剪切仪；直剪仪
box shear test	盒式剪切试验；直剪试验
box spanner	套管扳手；套筒扳手
box spanner box wrench	套筒扳手
box spinning	罐式纺丝
box spreader	箱形撒料机
box staff	塔式标尺
box steel sheet piling	压土辊
box stop	耙斗阻挡器
box structure	箱结构；箱状构造

box switch	匣式开关；密封型开关
box system	箱形系统
box tap	母锥
box test	铁箱试验
box thread	内螺纹
box timber	矩形井框
box timbering	板框支护
box to pin	公母接头
box tong space	母接头上打大钳处
box truss	箱形桁架
box tunnel	箱形隧道
box type	匣型；箱型
box type cofferdam	微滴
box type scraper	箱形刮斗
box type substructure	箱式底座
box unit	箱形单件
box up	箱体翻起
box wrench	套筒扳手
box-bill	打捞工具
boxcar function	矩形波函数
box-car loader	棚车装载机
boxed	装箱的；隔离的
boxed-off	钉板的，钉隔板的，隔开的
box-hole method	溜眼放矿开采法
box-hole mucking	在溜井口用装载机装载
box-hole pulling	溜口放矿；溜口放石
boxing	装箱；制箱木料；箱状罩；绕焊
boxing-up	装箱；填道碴
boxlike	箱形的
box-like fold	箱状褶皱
Box's formula	包克斯公式
box-thread bit	内螺纹钻头
box-trough compass	方框罗盘
box-type dryer	箱式干燥器
box-type frame	箱形架
box-type sampler	桩式码头
box-type scraper	箱形耙斗；箱形矿耙
boxwork	窗孔构造；蜂窝状构造；蜂窝状沉积褐铁矿
Boyden radial outward flow turbine	博伊顿式辐向外流水轮机
Boylan automatic cone	博伊兰型自动调节圆锥分级机
Boyle's law	波义耳定律
Boyle's law porosimeter	波义耳定律孔隙仪
Bozhenpollis	泊镇粉属
bozo line	机械索具
B-parameter	孔隙压力系数 B
Braarudosphaera	布拉鲁德球石
braccianite	富白碱玄岩
brace	支撑，横撑，斜撑；拉条，井口，井颈，井口出车场；井架上的永久工作台；加固，拉紧；手摇曲柄钻
brace bit	曲柄钻孔器；手摇曲柄钻；弓钻；装卸设备
brace blocking	钢枕
brace bracket	斜撑座
brace head	钻杆组顶端转动装置
brace pile	支撑桩；斜桩；角撑桩
brace rod	联结杆
brace socket	拉筋连接用螺丝筒
brace strut	支撑；斜撑
brace summer	双重梁；支撑梁
brace tee	丁字撑杆
brace wrench	曲柄头扳手
braced	有主撑的；联结的；加固的；拉牢的
braced arch	桁架式拱
braced beam	桁架式梁
braced chain	桁链
braced door	联结门
braced excavation	支撑开挖；支撑式挖掘
braced frame	刚性构架；联结架；撑系框架；加固构架
braced frame construction	杆系框架结构
braced framing	联结构架
braced girder	桁架梁；联结大梁
braced pier	联结墩
braced quadrilateral	对角线四边形
braced reinforced concrete flume	桁架式钢筋混凝土渡槽
braced rib arch	桁架式肋拱
braced sheeting	支撑挡板
braced structure	受横向支撑的结构；支撑结构
braced strut	联结支撑；桁架支柱
braced wall	内撑式挡土墙
bracehead	钻杆组的转动把手
bracelet	箍套；镯
bracelet anode	手镯状阳极
braceman	井口把钩工
brachiopod	腕足动物
Brachiosaurus	腕龙属
brachistochrone	最小时程；反射波垂直时距表；最速落径
brachistochronic principle	最短时间原理
brachy	短的
brachy fold	短褶皱
brachy-anticlinal fold	短背斜褶皱
brachyanticline	短背斜
brachy-anticline fold	短轴背斜褶皱
brachy-anticlinorium	短复背斜
brachy-axis	短轴
brachydome	短轴穹隆
brachyfold	短轴褶皱
brachygeosyncline	短轴地槽
brachylinear	短轴线性特征
brachyprism	短轴柱
brachysynclinal fold	短轴向斜褶皱
brachysyncline	短向斜
brachy-synclinorium	短复向斜
bracing	支撑；加固；联结；拉条；拉紧
bracing arrangement	支撑装置
bracing balance	加强板
bracing beam	支撑梁
bracing bolt	连接螺栓
bracing boom	加劲杆
bracing cable	拉索；拉缆
bracing column	支撑柱
bracing frame	加劲框架；带斜撑的框架；加固构架
bracing in open cut	明挖支撑
bracing in tunnel support	顶撑
bracing member	支撑构件；拉紧连件拉筋
bracing piece	斜撑；斜梁；横梁；加劲杆
bracing pile	支护桩
bracing plane	支撑面
bracing ring	加固圈
bracing structure	支撑结构

bracing strut 支杆；斜梁；撑脚
bracing support 斜撑支护；加强支护
bracing system 支撑系统
bracing system support system 支撑体系
bracing the bit 钻头加固
bracing wire 拉线；拉索
bracing, collar bracing 系杆，斜撑
brack soil 微盐土
brackebuschite 锰铁钒铅矿
bracket 托架；底座；悬臂；亚稳状态
bracket block 托架垫块
bracket bracing 托座支撑
bracket clamp 托架夹
bracket connection 托架结合
bracket Support 支架
bracketed board gate 托架式板闸门
bracketing method 交叉法
bracketing spacing 阀配置的范围
brackish 盐渍的；碱化的；微咸的
brackish deposit 半、微咸水沉积
brackish environment 微咸环境
brackish flow 钻孔型式
brackish groundwater 微咸地下水
brackish lake 半咸水湖；微咸水湖
brackish swamp 微咸水沼泽
brackish water 半咸水；微咸水
brackish water lake 半咸水湖
brackish water sediment 微咸水沉积
brackish water shale 半咸水页岩
brackishness 微咸性
brackish-water lagoon facies 淡化泻湖相
brackish-water lake 微咸湖
bracteal 苞的
brad 角钉
bradawl 打眼钻
braden head 盘根接头；盘根式油管头
bradford breaker 选择性破碎机；布拉福德型碎选机；滚筒碎选机
Bradford gauge 布拉福德雨量斗
Bradfordian Age 勃拉特福德期
Bradley Hercules mill 勃兰德雷-赫克里斯型水平辊磨机
bradleyite 磷碳镁钠石
bradyseism 缓震
bradytelic evolution 慢速进化
brae 山坡；坡；斜井；冰川
Bragg's law 布拉格定律
bragite 褐钇铌矿
braid 编织层；编织物；编织
braid index 辫状指数；网状指数；游荡性指数
braid perthite 辫状条纹长石
braid river 游荡性河；辫状河
braided 网状的；辫状的
braided cable 编缆
braided channel 分汊河道；网状河道；游荡性河道
braided course 涌沙
braided drainage 辫状水系
braided drainage pattern 网状水系型；网状水系
braided packing 编织填料
braided reach 分汊河道；分汊河段；游荡性河段
braided river 分汊河流；网状河；游荡性河流
braided river course 分汊河道
braided stream 辫状河流；网状河流；交织河流；分汊河
braided-fan-delta deposit 网状冲积扇三角洲沉积
braiding 河道分支；河道分叉
braiding anastomosis 河道交织
brail 斜撑
brait 未加工的金刚石；粗金刚石
braize 煤尘；煤粉；焦粉
brake beam pull rod 制动梁拉杆
brake beam strut 制动梁支柱
brake beam tension rod 制动梁弓形杆
brake shoe gauge 闸瓦间隙规
brake shoe key 闸瓦插销
brake shoe lining 闸瓦衬层
brake shoe pressure 闸瓦压力
brake slipper 制动滑块
brake system 制动系统
brake test 制动试验
brakeage 制动作用；制动力
brakeing force 制动力
braking absorption 阻尼吸收
braking bracing 制动撑架
braking device 闸；制动装置
braking power 闸力；制动力；制动功率
brakpan jockey 连接抓叉
brale 金刚石锥头；金刚石压头
Bramletteius 布兰利特石
brammallite 钠伊利石
branch 校，支，分支，支路，无分支的通风线路；支脉，支管，支线，支部；转移；划分
branch and bound method 分枝限界法
branch angle 分叉角
branch canal 支渠
branch channel 支流河槽；分支河道
branch channel flow 分汊水流
branch conduit 支管；分支导管
branch connection 分支接线；分叉连接
branch crack 分支裂缝
branch cut 分支切割
branch drain 排水支管
branch entry 支巷
branch fault 分支断层；支断层；次要断层
branch flow 分流
branch gate 分支浇口
branch haulageway 分支运输道
branch heading 错车巷道；分支巷道
branch line 支线
branch nozzle 溢流管
branch order 分支指令
branch pipe 支管；三通；套管
branch point 分叉点；分支点
branch raise 分支天井
branch river 支流；支河
branch road 支巷；副巷
branch roadway 中间巷道；支巷道
branch track 支线
branch unconditionally 无条件转移
branch vein 支脉
branch water 支流水；溪水

branched 分支的，枝状，树枝状；分岔的
branched chute 支溜槽
branched lode 支矿脉
branched structure 分支结构；枝状结构
branching 分支；分叉；分路；分线
branching bay 河口湾；三角湾
branching fault 分支断层
branching of a roadway 水平巷道的岔道；矿山巷道的分支
branching operator 分支算子
branching stream 分汊河流
branching structure 树枝状构造
branchite 晶蜡石
branch-like loess land 枝状黄土地
branch-off 支管
branch-raise system 分支上山崩落采矿法
branchy 多枝的
Branden interglacial stage 勃兰登间冰期
brandisite 葱绿脆云母
brandtite 砷锰钙石
brannerite 钛铀矿；钛酸铀矿
brash 易碎的；脆的；崩解石块；碎块
brash ice 碎冰
brashy 易碎的；脆的；崩解石块
brasque 浆质衬料；浆质炉衬
brass balls 肾状黄铁矿
brassil 黄铁矿；含黄铁矿的煤
brassy 黄铜的；含黄铁矿的煤
brat 含黄铁矿薄煤层；含方解石薄煤层
brat coal 不纯煤
brattice 风障；风帘；风墙；隔板
brattice balance 风降；风幕；风布
brattice or brattice cloth 风幛布，风帘布
brattice prop 挂风布的支柱
brattice sheeting 风障；风布；风帘
brattice ventilation 风障通风
brattice wall 风墙；风幕
brattice way 采空区巷道；上部巷道
brattice-stoppings 临时风墙
bratticing 风障；隔风墙
Braun pulverizer 布劳恩型粉磨机
braunerde 棕壤；褐土
braunite 褐锰矿
braunkohle 褐煤
braunkohle stage 褐煤阶段
Bravais lattice 布拉维空间格子；布拉维点阵
Bravais orientation 布拉维定向
bravaisite 漂云母
Bravais-Miller symbol 布拉维-米勒符号
Bravias rule 布拉维法则
bravoite 镍黄铁矿；方硫铁镍矿
Bravos process 巴尔沃依斯重介选煤法
bray 捣碎；研碎
Bray diagram 布雷图
bray stone 多孔砂岩；碎石
Bray's solution for stresses around an elliptical excavation 椭圆形巷道周围应力的 Bray 解
braze 转速计
braze welding process 钎焊法
brazier 制铜技工；火盆

brazil 黄铁矿；含黄铁矿的煤
Brazil Basin 巴西海盆
Brazil test 巴西试验
Brazil twin 巴西双晶；石英双晶
Brazilian disc 巴西盘
Brazilian gem 巴西宝石
Brazilian tensile strength test 巴西抗拉强度试验
Brazilian tensile test 劈裂试验；巴西抗拉试验
Brazilian test 巴西试验；劈裂试验
Brazilian test of rock 岩石的巴西劈裂拉伸试验
brazilianite 磷铝钠石
brazing solder 黄铜焊料
brea 沥青砂；沥青土；矿物焦油
brea bed 沥青砂土层
breach 缺口；豁口；破裂；碎波；深沟
breached 有缺口的；有裂口的
breached anticline 削峰背斜
breached cone 裂口火山锥
breached crater 裂火山口
bread crust bomb 面包皮火山弹
breadboard 模拟板；实验板；实验性的
bread-crust bomb 层剥火山弹
bread-crust boulder 层剥岩块
bread-crust structure 层剥构造；皮壳状构造
breadth 宽度，厚度
breadth length ratio 宽长比
breadth of section 截面宽度
breadth ratio method 宽度比法
breadth-first search 广度优先搜索
break 断裂，断层；裂面，裂缝；冲断；突变，拐角；破坏
break angle 折角
break away 破裂；冒顶；崩落；脱离
break barrow 碎土机
break circulation 中途循环泥浆
break criterion 破坏准则
break detector 裂缝探测器
break down 落煤；崩矿；落岩；障碍；破损
break earth 破土动工
break fault 间断断层
break friction 裂缝摩擦
break gallery 断裂巷道
break gallery test 平硐爆破试验
break ground 破土，动工；创办
break head layer 断顶
break hiatus 间断
break in 接入；磨合；跑合
break in declivity 倾斜变化
break in grade 坡度变化；坡折点
break in grain size 粒径中断
break in slope 坡折
break in the new bit 新钻头磨合
break in the new tool joint 新钻具或接头磨扣
break in the profile 剖面中断
break in the succession 地层缺失；层序间断
break into 打通
break joint 断裂节理；格状裂隙；棋盘式裂隙；错缝接合；断缝
break line 放顶线；截断线；倾斜变换线；断线
break loose pull 提松力

break mechanism	破坏机理
break of an earth bank	路堤滑坡
break of slope	坡折点；坡折
break off	弄断，打断；崩落
break out	卸扣；卸开；破裂
break phenomenon	断裂现象；崩落现象
break point	转效点；转折点；拐点；断点
break setting method of intermediate pile	防止中间桩下沉作业
break slide	背斜破裂面断层；破裂滑动
break strain	断裂应变
break strength	破坏强度
break stress	破裂应力
break tank room	减压水缸房
break tenacity	断裂韧度
break test	断裂试验；破损性试验
break through	突破；冲破；穿透
break through survey	贯通测量
break thrust	背斜上冲断层；切穿逆断层
break to	破坏；破裂
break up	破碎
break water	防波堤；挡水板
breakability	破碎性；脆性
breakage	崩落；破坏；损毁；破碎
breakage angle	爆破漏斗角
breakage face	回采工作面
breakage heading	回采巷道
breakage line	断裂线
breakage of rock by explosives	爆破引起的岩石破坏
breakaway	破裂；脱离；摆脱；拆毁
breakdown	破坏；故障；分类；粉化；磨碎；击穿
breakdown mechanism	破坏机理
breakdown path	击穿路径
break-down point	破坏点；断点
breakdown pressure	破碎压力；断裂压力
breakdown strain	断裂应变
breakdown strength	断裂强度；破坏强度
breakdown stress	断裂应力
breakdown survey	崩塌调查；塌陷调查
breakdown test	破坏试验；耐压试验；击穿试验
breaker	破碎机；无烟煤选煤厂；顶板大裂缝；落煤工；破碎工
breaker bar	碎浪坝
breaker chock	切顶支垛
breaker depth	破浪深度；碎波深度；碎波深度
breaker height	破波高度；浪高
breaker line	破波线；碎波线
breaker plate	轧碎板；破碎机衬板；电流开关板
breaker prop	支撑；支柱；崩矿支柱
breaker row	排柱；切顶支柱
breaker wave	拍岸浪；破波
breaker wave height	破波波高
breaker zone	破浪带；碎波区
break-hiatus	间断
break-in	插入，嵌入；打断，轧碎；滚动，碾平；试车，试运转
break-in grade	纵坡折变；纵坡变更点
breaking	破裂，断裂，破碎，轧碎；分层；分解；澄清
breaking and screening of dredging	采砂船碎散分级
breaking by nuclear blasting and dredging under sea consolidate deposits	用核爆破和挖掘船挖掘海底固结矿床
breaking coefficient	碎裂系数
breaking disk	破碎盘
breaking down	断裂；分解；落矿；落煤
breaking down test	耐压试验
breaking edge	切顶线；崩落线；截断边
breaking elongation	致断延伸率
breaking extension	断裂伸长
breaking failure	断裂破坏
breaking force	顶破强度；断力力
breaking ground	爆破；破土
breaking in curve	曲线折点
breaking joint	断缝；错缝；沉降缝；断裂节理
breaking level	破碎水平
breaking limit	断裂极限；破裂极限
breaking limit circle	破坏极限圆；破坏应力圆
breaking line	破裂线；断裂线；崩落线
breaking link	脆性连杆
breaking load	破坏荷载；断裂荷载；极限荷载
breaking moment	断裂力矩
breaking of a dike	堤防溃决；决堤；破堤
breaking of contact	断接卡
breaking of emulsion	乳胶体分层；乳浊液的破坏
breaking of ground	落岩；破土；奠基
breaking off	剥落；剥蚀
breaking plane	破坏平面；破裂面
breaking point	破波点；碎波点；破浪点；断裂点
breaking property	断裂性；破裂性
breaking resistance	破碎阻力；抗断强度
breaking shot	炮眼
breaking strain	断裂应变
breaking strength	抗断强度；断裂强度；破坏强度
breaking strength test	抗断强度试验
breaking stress	破坏应力；断裂应力
breaking stress circle	破裂应力圆
breaking sub	压的亚阶段；分段开采
breaking test	破坏试验；断裂试验
breaking through	穿透
breaking through of water	突然涌水；透水
breaking up	破裂；破碎
breaking up ice floe prevent collision	防撞破凌
breaking up ice run	防撞破凌
breaking up of sand	拌砂
breaking up sand	松砂
breaking water	突水
breaking water level	跌落水面
breaking wave	破碎波
breaking wave depth	破波水深
breaking wave height	破波波高
breaking wave pressure	碎波压力
breaking weight	破坏荷载
breaking-down point	破坏点；断裂点
breaking-in hole	掏槽炮眼
breaking-in shot	掏槽炮眼；开槽炮眼；开口炮眼
breaking-off bar	切割梁
breaking-off props	切顶排柱
breaking-out	崩落；爆发；卸开
breaking-out in bulk	大块崩落；大量崩落；大崩矿
breaking-out of ore	崩矿；大量矿石崩落

breaking-up 破裂；破碎；崩落
breakoff 联络小巷；矿脉变蚀
break-off load at setting 支架初撑时折断载荷
break-off load at yield 支架屈服时折断载荷
breakout 涌出；打开；突发；崩落
breakout force 爆发力；突然涌出力；劈开力；切割力
break-out tong 外钳
breakover 圆脊
break-point 断点；停止点
break-point instruction 断点指令
breaks 巷道顶、底板和两帮破裂；地形突变；劣地；陡坎；岩层中之小断层
breaks of mine openings 顶底板和巷帮等破裂
breakstone 碎石；石碴
breakthrough 突破；联络横巷；通风横巷
breakthrough capacity 突破能力；突破能量；饱和容量
breakthrough of glide 滑移贯通
breakthrough of water 突水；突然涌水
breakthrough pattern 冒水井排；见水井排
breakthrough point 穿透点；泄露点；临界点；贯通点
breakthrough pressure 突破压力
breakthrough survey 贯通测量
breakthrough well 冒水井；见水井
break-thrust 背斜上冲断层
break-up 破裂；破碎；崩落；崩溃
breakup phase 破裂阶段；破裂期
breakup unconformity 裂开不整合；间断不整合
break-up, decompose 分解
breakwater 防波堤
breakwater capstone 防波堤压顶石
breakwater end 防波堤端
breakwater gap 防波堤缺口；堤头口门
breakwater of submerged reef type 潜礁式防坡堤
breakwater pier 防波突堤
breakwater quay 防波堤堤岸
breast 工作面前壁；工作面；煤房；回采工作面
breast abutment 岩浆熔蚀
breast board 挡土板；开挖面支护
breast hole 掏底槽后爆破的炮眼
breast pile 挡土桩；坡脚桩
breast wall 挡土墙；胸墙；防浪墙；蓄水系数
breasting 超前短煤房；宽巷道；从平巷工作面上采矿；局部临时支护
breasting dolphin 承冲护船桩
breasting method 矿房全高单一工作面开采法
breasting mining method 全面采矿法
breasting wheel machine 切削型掘进机
breast-shot water wheel 异常断层
breast-stoping operation 矿房梯段回采工作
breast-wall 防浪墙，挡土墙
breathing of gob 采空区漏气
breathing well 透气井；呼吸井
breathing zone 伸缩区；呼吸区
breccia 角砾岩；角砾岩；断层角砾岩
breccia dike 角砾岩脉；角砾岩墙
breccia fragment 角砾岩碎片
breccia marble 角砾大理岩
breccia nappe 角砾推复体
breccia pipe 角砾岩筒
breccia pore 灰岩内的角砾孔隙

breccia sandstone 角砾砂岩
breccia structure 角砾构造
breccia vein 角砾岩脉
breccia-conglomerate 角砾砾岩
breccial 角砾的
brecciaous 角砾状的
breccia-sandstone 角砾砂岩
brecciated 成角砾岩的；角砾岩状的
brecciated coal 角砾状煤
brecciated structure 角砾构造
brecciated texture 角砾结构
brecciated vein 角砾脉
brecciated zone 角砾岩带
brecciation 角砾岩化
brecciform 角砾状
breccio-conglomerate 角砾状砾岩
brecciola 角砾灰岩
breciated structure 角砾构造
bredigite 白硅钙石
breeches pipe 叉管
brick baffle 砖隔墙
brick base 总表面积；砖基础；压重填土
brick bate 主应力轨迹
bridge arch 桥拱
bridge bearing 桥梁支承
bridge foundation 桥梁基础
bridge of fault 断层脊
bridge span rise 桥跨结构拱度
bridge wall 坝墙
bridge-action time 顶板自行支撑的期限
bridge-approach fill 岩心取样
bridged flume 渡槽
Bridgerian 布里杰阶
Bridges and culverts 桥涵
bridges in cascade 多段桥
bridge-tunnel work 桥隧工程
bridge-type movable dam 桥式活动坝
bridge-type pontoon 浮桥
bridging 桥连；跨接；橱撑；拥挤颗粒流中拱的形成
bridging beam 渡梁；横梁；架接梁
bridging joist 渡梁
bridging line 桥接线；跨线
bridging of model 模型连接
bridging-action 桥撑作用
brief dip-wash 短暂浸洗
briefly frozen soil 短暂冻土
Brigantian 布里根阶
Briggs clinograph 布雷格斯型钻孔测斜仪
brilliance 亮度；光洁度；耀度
brilliant 光辉的；卓越的；钻石；金刚石
brimstone 硫黄石；硫
brine cavity 盐层空穴
brine deposit 卤水矿床；卤水矿
Brinell apparatus 布氏硬度试验器
Brinell hardness 布氏硬度；布氏硬度；压球硬度
Brinell hardness number 布氏硬度数
Brinell hardness test 布氏硬度试验
Brinell hardness tester 布氏硬度试验机
Brinell instrument 布氏硬度计
Brinell machine 布氏硬度测定器

Brinell number	布氏硬度值；布氏硬度值
Brinell tester	布氏硬度计
brining	盐浸作用；用盐水处理
brink	陡峭边缘；陡缘
brink depth	边缘水深
brisket	含褐煤夹层的黏土
britholite	硒磷灰石
brittle	易碎的；脆性的
brittle behavior	脆性特征
brittle clay	脆性黏土
brittle coating	脆性涂料
brittle crack	脆性裂缝
brittle damage	脆性损伤
brittle deformation	脆性变形
brittle extensional tectonic system	脆性伸展构造系统
brittle failure	脆性破坏，脆断；脆性失效
brittle fashion	脆性破坏方式
brittle fault	脆性断层
brittle faulting	脆性断裂作用
brittle fracture	脆性断裂；脆性破坏；脆度；脆断
brittle heterogeneous material	脆性非均质材料
brittle index	脆性指数
brittle material	脆性材料
brittle mica	脆云母
brittle model	脆性模型
brittle pan	脆磐
brittle plasticity	脆塑性
brittle point	脆化点
brittle rock	易碎岩石；脆性岩石
brittle roof	脆性顶板；易碎顶板
brittle rupture	脆性破坏；脆性断裂
brittle sandstone	脆性砂岩
brittle shale	脆性页岩
brittle shear failure	脆性剪切破坏
brittle shear zone	脆性剪切带；脆性变形带
brittle silver ore	脆银矿
brittle structure	脆性结构；脆性材料
brittle tectonic	脆性构造
brittle-ductile transition	脆性-延性转换
brittle-ductile transition pressure	脆性-韧性过渡压力
brittle-elastic	脆性的
brittleness	脆性；变脆；脆度
brittleness index	脆性指数
brittleness index of rock	岩石脆性指数
brittleness material	脆性材料
brittleness test	脆性试验
Brno excursion	布尔诺漂移
broad angle	钝角
broad band seismograph	宽频带地震仪
broad brush approach	广义划分方法；概括分组方法
broad ocean	轧石
broad top anticline	宽顶背斜
broad top basin	宽顶盆地
broad valley	宽谷
broad warp	平缓弯曲
broad-band digital seismograph	宽频带数字地震仪
broadly graded soil	宽级配土
broadness	广阔
broadside	非纵排列；旁线观测
broadside railway dry dock	软道干坞
broadside slipway	侧边滑道
broadside spread	非纵排列
brocatel	彩色大理石
brockram	坡积灰岩角砾岩
broggerite	钍铀矿
broil	矿脉指示碎石；露头；铁帽
Broinsonia	布罗因索石
broke stone	碎石
broken	破碎的；中断的；已部分开采的；矿脉错断
broken line	破线
broken ashlar masonry	不等形琢石圬工
broken belt	破碎带
broken block mass	破碎块度
broken concrete groin	碎块混凝土丁坝
broken country	沟谷交错地带；起伏地区；丘陵地
broken course	断层
broken curve	虚线
broken flexure	破裂挠曲
broken fold	破裂褶皱
broken foreland basin	破裂前陆盆地
broken formation	起伏不平的岩层；断续的地质建造；破碎地层
broken ground	地面破裂；破碎地层；断裂地带；破裂地层
broken height	断高
broken joint	错缝接合；顺砖式砌合；顺砌法；断缝
broken line	虚线，折线；破折线；断链
broken material	矸石场；破碎物料
broken pile	爆堆
broken plateau	断裂高原
broken range masonry	断层石砌
broken reserves	破碎待运矿
broken rock	破碎岩石
broken roof	破碎顶板
broken round	磨圆砾石
broken sand	含页岩砂岩
broken sandstone	残破的砂岩
broken slide	解体滑坡
broken sliding surface	不连续滑动面
broken stone	碎石
broken stream	隐现河；间歇性河流
broken terrain	沟谷交错地带；起伏地形
broken zone in surrounding rock	围岩松动区；围岩破坏区
broken-in face	有爆破槽的工作面
broken-rock pressure	松动压力
broken-rock subsurface	碎裂地下岩层
bromargyrite	溴银矿
bromellite	铍石
bromhead ring shear	环剪
bromite	溴银矿
bromlite	碳酸钙钡矿
bronsted acid	质子酸；B酸
bronze guide	青铜导板；青铜导承
bronze luster	古铜光泽
bronze mica	金云母
bronzite	古铜辉石
bronzite andesite	古铜安山岩
bronzite chondrite	古铜辉石球粒陨石
bronzitfels	古铜岩

bronzitite	古铜辉岩
brood chamber	育室
brook	小河；小溪；溪流；锥形桩
Brookfield viscometer	布氏黏度计
brookite	板钛矿
brooklet	小河；小溪
broom head	击坏的桩头；发毛的桩头
broom like veins	帚状矿脉
broom-like fold	帚状褶皱
brown	褐色；茶色；棕色
brown andean soil	棕色火山灰土
brown asbestos	褐石棉
brown calcareous soil	棕色石灰性土
brown clay	褐色黏土
brown clay iron ore	泥质褐铁矿
brown coal	褐煤
Brown correction	布朗改正；二次项改正
brown desert-steppe soil	棕漠草原土；褐土；棕壤
brown earthened soil	棕壤化土
brown earthy coal	土状褐煤；岩石学；棕色森林土
brown hematite	褐铁矿
brown iron ore	褐铁矿
brown jade	糖玉
brown jura	褐侏罗；褐侏罗统
Brown Jura Series	褐侏罗统
brown lateritic soil	棕色砖红壤性土
brown lignite	棕褐煤
brown lime	棕石灰
brown limestone soil	棕色石灰土
brown meadow steppe soil	棕色草甸草原土
brown mica	金云母
brown mud	褐泥
brown ocher	褐铁矿；褐铁矿粉
brown podzolic soil	棕色灰化土
brown print	棕色图
brown rendzina	棕色石灰土
brown sandstone	褐砂岩
brown semi-desert soil	棕色半漠土
brown soil	棕钙土
brown spar	铁菱镁矿；铁白云石
brown steppe soil	棕色草原土
brown stone	褐砂岩；氧化黄铁矿
Brown tongs	布朗钳
brown turf	褐色泥炭
brown water quality index	布朗水质指数
brown-glazed brick	褐釉砖
Brownian motion	布朗运动
Brownian motion process	布朗运动过程
Brownian movement	布朗运动
browning	褐色氧化处理；呈褐色的
brownish oxisiallitic soil	棕色氧化硅铝土
brownish yellow brick	土黄色面砖
brownmillerite	钙铁石；钙铁铝石
brownstone	深棕色砂岩；褐石
broyl	矿脉指示碎石；露头；铁帽
Bruchfaltung	断褶构造
brucite	水镁石；氢氧镁石；水滑石
brugnatellite	次碳酸镁铁矿
brunisolic soil	棕壤
Bruns formula	布隆斯公式
brunsvigite	纹绿泥石
brunton	矿用袖珍罗盘
Brunton compass	袖珍罗盘仪
brush	富赤铁矿；采空区巷道
brush joint	帚状节理
brush mattress protection	柴捆护坡；柴排护岸
brush revetment	柴排加固边坡；柴排护坡
brushite	钙磷石
brush-type structure	帚型构造
brushwood check dam	梢捆挡水坝
bryle	矿脉指示碎石；露头；铁帽；导脉
bubble	气泡；水准器气泡；起泡；发泡
bubble attachment	气泡黏附
bubble-column machine	气泡柱式浮选机
bubble-column process	气泡柱法
bubble-mineral attachment	气泡和矿物的黏附
bubbler tide gauge	气室潮汐计
bubble-type pneumatic gauge	水泡式气压计
bubbling	冒泡；起泡
bubbling ability	气泡能力
bubbling polymerization	气泡聚合
bubbling pressure	始沸压力；气泡压力
bubbling spring	泡沸泉；溢气泉；珍珠泉；碳酸泉
bubbly	发泡的；泡多的
Buchan-type facies series	巴肯型相系
buchite	玻化岩
buchnerite	二辉橄榄岩
bucholzite	硅线石
buck quartz	不含金的石英脉
buckboard	研样板
bucked ore	富矿石
bucker	碎矿锤；碎矿工；溜槽推煤工；破碎机
bucket height	挑坎高度
Bucket wheel stripping	轮斗铲倒堆剥离
bucking	人工磨矿；二次爆破；电压降；纵向弯曲
bucking board	磨矿板
bucking hammer	矿样研磨手锤
bucking load	压曲荷载
bucking plate	磨矿平铁板
bucking stress	弯折应力
Buckingham's theorem	白金汉定理
bucklandite	黑帘石
buckle	箍，扣；扣；弯曲；使变形
buckle fold	挠曲褶皱；液流；弯曲褶皱
buckle folding	弯曲褶皱；侧压褶皱；挤压褶皱；弯曲褶皱作用
buckle plate	凹凸板；圬工拦沙坝
buckled folding	弯曲褶皱；纵弯褶皱
buckled layer	弯曲层
buckling	纵弯褶皱作用，纵弯作用；压曲，弯曲；地壳弯曲；屈曲；紊动强度
buckling amplitude	压曲幅度
buckling analysis	屈曲分析
buckling bifurcation	屈曲分叉
buckling coefficient	压曲系数
buckling component	弯曲分力
buckling deformation	翘曲变形
buckling failure	形态测量学
buckling fold	弯曲褶皱
buckling force	压曲临界力

buckling load 压曲临界荷载
buckling mode 屈曲型式
buckling of a roof span 顶板层跨度的翘曲
buckling of rods 钻杆压弯
buckling resistance 抗弯强度；屈曲抗力
buckling stability 压屈稳定；压曲稳定性；抗弯稳定性
buckling strength 纵向弯曲强度；抗屈曲强度；抗弯强度；压屈强度
buckling stress 屈曲应力；纵弯曲应力
buckling test 屈曲试验
bucksaw 架锯
buckshot 粒状黏土；熔岩粒；粒状结核
buckshot aggregate 大团粒
buckshot sand 含褐铁结核砂
buckstay 拱边支柱
buckwall 安全挡墙
buckwheat slate 脆性板岩
bud 芽
buddling 淘选
buff 抛光轮；擦光轮；抛光，擦光；减振，缓冲
buffer 减振器；缓冲器；缓冲
buffer action 缓冲作用
buffer disk 缓冲盘；减震盘
buffer distance 缓冲距
buffer function 缓冲作用
buffer head 缓冲头
buffer layer 缓冲层
buffer shooting 缓冲爆破
buffer storage area 贮存区；贮料区
buffer strip 缓冲带
buffer substance 缓冲物质
buffer system 缓冲系
buffered reaction 缓冲变质反应
buffering 缓冲作用；减震；阻尼
buffering trench 缓冲沟
buffeting 冲击；振动
bug dust 粉尘
Buganda-Toro-Kibalian orogeny 布干达-托罗-基巴连造山运动
bugite 紫苏英闪岩
buhrstone 砂质多孔石灰岩；细砂质磨石；油石
build joint 构造缝
build up factor 积累因子
building grade 房屋地面标高；设计标高
building settlement 建筑物沉降
building slope engineering 建筑边坡工程
building-up 建起；装配；增加；上升
building-up period 成长时间；建立时间
build-up 生长，增长；积累，加厚；建起，建筑，建立；树立
build-up analysis 压力恢复分析
build-up area 增阻区
build-up in the hole 钻孔堵塞；钻孔内钻粉的积聚
build-up of dust 尘末积聚；积尘
build-up of fluid 液面恢复
build-up of load 载荷的增长
build-up of pressure 压力恢复
build-up of stress 应力的增大
build-up of stresses 应力增大；应力积累
build-up of water production 出水量增加

buildup test 增压试井
built 建成的；造成的
built pile 组合桩
built platform 堆积台地
built terrace 浪成阶地；冲积阶地
built-in 装入的；嵌入的；内装的
built-in arch 固端拱
built-in arch theorem 固定拱的理论
built-in beam 嵌固梁
built-in check 机内检验；自动检验
built-in component 嵌入部分
built-in edge 嵌固边缘
built-in end 嵌入端；固定端
built-in fitting 预埋件
built-in lighting 预防措施
built-in-place pile 就地浇筑桩；现场浇筑桩
built-up 组合；组装；建立；装配
built-up arch 最高蓄水位；组合拱
built-up frame 逐步逼近法
built-up pile 组合桩；预制桩；岩柱
built-up rigid frame 组合管
built-up section 组合截面
built-up steel section 悬浮颗粒
built-up terrace 堆积附地
built-up truss 组合桁架
built-up welding 堆焊
Bulalo geothermal field Bulalo地热田
bulb 球；球管；球形物；玻璃球
bulb bar 圆头型材；圆头钢钎
bulb of pressure 应变记录器；压力泡
bulb pile 扩端桩；球根桩
bulb stopper 球塞
bulb turbine 灯泡式水轮机
bulbous zone of stress 球面应力区
bulbs of pressure 膨胀压力
buldymite 蛭石
bulge 隆起，凸起，膨胀；隆起；熔岩膨胀
bulge effect 隆起效应
bulge of the loop 回线凸出程度
bulge size 凸出尺寸
bulge test 打压试验；扩管试验；打压试验
bulged blade 凸起叶片
bulging 膨胀突起；鼓起；爆裂
bulging failure 膨胀破坏
bulging force 膨胀力；隆起力
bulging projection 凸出
bulgram 煤层中的黑黏土夹层
Bulitian 布里特阶
bulk 整体；大块大量；体积；容量
bulk analysis 总分析；整体分析
bulk bag 袋装扰动土样
bulk caving 大冒顶；大崩落
bulk check 整体检查
bulk composition 总体成分；全岩成分
bulk compressibility 体积压缩性；总体压缩系数；总体积压缩率
bulk concentrate 整体精矿；整体精煤
bulk constitutive equation 体积本构模型
bulk density 容积密度；体积密度；松密度；整体密度；单位体积密度；堆密度；散容重

bulk density of soil	土的容量；土的天然容重；土的湿容重
bulk effect	体积效应
bulk elastic modulus	体积弹性模量
bulk elasticity	体积弹性
bulk elements	常量元素
bulk factor	体积因数；紧缩率；松散系数
bulk fill	松填土
bulk filtration	散装过滤；外部过滤
bulk flow	体流变
bulk grouting	大体积灌浆
bulk handling	散装运输；散装货物装卸；体积淬火
bulk increase	容积增大
bulk index	体积指数
bulk isotope composition	全同位素组成
bulk material	散料；大块材料
bulk method of mining	大量开采法
bulk minerals	矿物块体
bulk modulus	体积弹性模量；体积模数
bulk modulus of elasticity	体积弹性模量
bulk modulus	体积弹性模量
bulk motion	整体运动
bulk plane strain	总平面应变
bulk plant pump	油库泵
bulk polymerization	整体聚合
bulk porosity	体积孔隙性；大孔隙性
bulk relaxation	体积松弛
bulk rock	岩体；整体岩石
bulk rock analysis	全岩分析
bulk rockfill	抛石体；抛填堆石
bulk rougher flotation	全部粗浮选
bulk sample	大量采取的试样；体样
bulk sampling	大量取样
bulk shrinkage	体缩
bulk solid	松散固体
bulk sound speed	体积声速
bulk specific gravity	散比；容重；堆比重；毛体积比重
bulk specimen	大试样
bulk storage	大容量存储器；散装储存
bulk strain	体积应变；总应变；体积应变
bulk strain component	总应变分量
bulk strain regime	总应变状态
bulk strain	体积应变
bulk strength	体积强度
bulk tank	灰罐
bulk unit weight	体积单位重量；体容重；体重度
bulk volume	总体积；容积；松散体积；扩张体积
bulk water	自由水；重力水
bulk wave	体波
bulk weight	散重；散料容重；体重
bulk weight of ground	岩土容重
bulk-caving method	大崩落开采法
bulk-cleaning method	混合入选法
bulk-density factor	物料体积-密度系数
bulked volume	体积胀量；胀体
bulked yarn	膨体纱
bulkhead	堤岸，驳岸；防护堤；岸壁；堵水闸门；围堰
bulkhead building	挡土墙；堤岸；岸壁
bulkhead connector	穿过隔板的接头；穿墙接头
bulkhead gate	检修闸门
bulkhead line	堤岸线
bulkhead quay wall	岸壁码头
bulkhead slot	堵水闸门槽
bulkhead wall	岸壁，挡土墙
bulkhead wharf	堤岸式码头
bulkiness	庞大
bulking	土体积的增大；湿胀；体胀；隆起
bulking agent	填充料
bulking capacity	膨化能力
bulking curve	容积增大曲线
bulking effect	湿胀性；膨化效应
bulking factor	岩石松胀系数；岩石碎胀系数；湿胀率
bulk-packed	箱装的
bulk-rock sample	全岩样品
bulky	体积大的；庞大的；松散的
bulky grain	大颗粒
bulky refuse	大块矸石
bull	铁制捣泥棒；车轮制动棒；斜井防跑车装置
bull a hole	用黏土封闭孔壁
bull bit	凿形钎头；一字形钎头
bull mica	巨晶白云母
bull plug	临时管道堵头
bull quartz	白色粗结构不含矿的石英；白色致密石英；无矿石英
bull ring	研磨圈
bull riveter	大型铆机
bull rod	钻杆；盘条；炮棍
bull shaker	摇动溜槽
bull stretcher	露边侧砖
bull wheel	绕绳轮；凸轮轴轮；大齿轮
bull wheel derrick	翼坝
Bullard discontinuity	布拉德不连续面
Bullard Dunn process	波拉-杜恩电解镀锌法
Bullard probe	布拉德型探针；斯克里普斯型探针
Bullatosauria	膨颅龙类
bullet connector	插塞接头
bullet valve	球阀；球形阀
bullet wave	弹射波
bullet-and-flange joint	插塞凸缘连接
bullet-nosed vibrator	球头式振捣器
bullet-shot drill	弹丸钻粒钻井
bullfrog	向上推矿车用的小车；单钩提升上山用的平衡重；平衡锤
bullgrader	平路机；推土机
bullhead	平面孔型
bullheaded down	整体向下推进
bullheaded sand consolidation	不均匀地层砂胶结
bull-heading method	下推压井法
bulling	岩缝装药爆破；钻孔壁糊泥；用热法扩大炮眼
bull-kelp	巨藻
bullnose	有圆头的；圆端的
bullnose bit	凸出的无岩心钻头
bullnose clamp	绳卡
bull-nose plane	牛头刨
bull-nosed piezometer tip	牛鼻式测压头
bullport	引导孔
bullrail	码头护缘；安全缘
bull's eye	牛眼状；黄铁矿结核
bull's eye level	完整水平

bull's eye-type zoning	牛眼状分带；环状分带
bully	沿倾斜掘的钢丝绳提升斜井或下山；矿用锤；矿工锤
bult	固定沙丘
bulten	土丘
bulwark	防浪堤
bump	煤突出，岩石突出，煤岩突出；冲击地压，冲击；碰撞，凸出，凸度；凸起处；破裂，爆裂
bump disaster	岩爆灾害
bump down	光杆下行撞击
bump joint	扩口接合；异径接头
bump plug	碰压
bump seat	岩石突出源地
bump stress	撞击应力
bumped head	凸形底
bumper bar	缓冲杆；保险杆
bumper beams	保险梁；缓冲梁；缓冲杆；保险杆
bumper gusset	保险杆角撑板；缓冲挡角撑板
bumper post	缓冲柱；挡车柱
bumper stop	缓冲阻挡器
bumpiness	颠簸
bumping	冲撞，碰撞；岩石突出，造成凹凸，凸出；剧沸，溅沸
bumping block	撞击块
bumping face	碰撞面；缓冲面
bumping trough	撞击式溜槽
bump-prone area	易发生突出区域
bumps on dislocations	碰撞位错
bumpy	不平的；坎坷不平的；容易脱落的；容易突出的
bumpy coal	自落煤；自崩煤
bunch	束，簇，捆；小矿巢；管状矿脉膨大部分
bunch blasting	成束导火线爆破
bunch holes	束状炮眼
bunched fuses	成束导火线
buncher	调制腔
bunching	束；聚束；成组
bunching machine	合股机；束线机
bunching of broken rock	采落岩石的成堆排列
bunchy	分散小矿窝组成的矿体
bund	河岸；海岸；岸墙；堤
bund wall	墙
bunded area	堤保护区
bunding	岸堤
bundle iron	绑扎的铁丝
bundle of folds	褶皱群
bundle of parallel capillary tube	平行毛细管束
bundle pillar	集柱
bundled tubes	管束
bundline	海岸线；河岸线
bung	塞头；塞住
bung bord	企口板桩
bunion pocket	梁窝；罐梁窝
bunker	煤仓；矸石仓；仓库
Bunker Hill screen	邦克希尔型筛
bunker sampler	贮仓自动取样机
bunkerage	贮煤设施；贮存设施
bunkering capacity	煤仓容量
Bunsen flame	本生灯火焰
Bunsen photometer	本生光度计
Bunsen reactions	本生反应
bunsenite	绿镍矿
bun-shaped	面包形的
bunter	杂色砂岩
Bunter sandstone	本特砂岩
Bunter Stage	本特阶
bunton	横梁；罐道梁；横撑
bunton hole	井筒梁窝
Buntsandstein	斑砂岩统
buoy	浮标；浮起；浮升；浮筒
buoy anchor	浮锚
buoy floats	浮标浮体
buoy light	灯浮标
buoy mooring	浮筒系泊；浮标系泊
buoy mooring lines	浮标锚链
buoy mooring mark	浮筒系泊标志
buoyage	浮标装置；浮标系统
buoyancy	漂浮性；浮力；扬压力；浮箱，浮托压力
buoyancy chamber	浮舱
buoyancy compartment	浮力舱
buoyancy compensation	浮力补偿
buoyancy correction	浮力改正
buoyancy correction factor	浮力校正系数
buoyancy curve	浮力曲线
buoyancy effect	浮力效应；上浮效应
buoyancy float	浮筒
buoyancy flow	浮泛流；上流水流；浮力流
buoyancy force	浮力
buoyancy force on fluid	流体浮托力
buoyancy foundation	浮筏基础；浮筏地基
buoyancy method	浮力法
buoyancy module	浮舱
buoyancy moment	浮力矩
buoyancy of water	水的浮力
buoyancy pile	拉力桩；抗浮桩
buoyancy pump	浮力泵
buoyancy raft	浮筏基础；浮筏地基
buoyancy tank	浮力罐
buoyancy-driven instability	浮力不稳定性
buoyancy-supported	浮力支承的
buoyant	能浮的；有浮力的
buoyant apparatus	救生器；救生器材
buoyant ascent	浮升
buoyant density	浸没密度；浮密度
buoyant diverter	悬浮型转向剂
buoyant force	浮力
buoyant foundation	筏基础
buoyant jet	预制混凝土构件
buoyant mass	漂浮块体
buoyant medium	浮力介质
buoyant pile	悬浮桩；摩擦桩
buoyant proppant	悬浮支撑剂
buoyant raft	浮筏
buoyant unit weight	浮容重；浮重度
buoyant uplift	漂浮上升
buoyant weight	浮重
buoy-type seismic station	流动地震台
bur	块燧石；磨石；毛刺；飞边
bur cut	平行空炮眼掏槽法
buratite	绿铜锌矿

burble 起泡
burbling 气流分离
burbling cavitation 扰流空化；局部空泡
burbling point 气流分离点
Burckhardt sampler 包契尔特型取样机
burden 负载，表土，覆盖岩层；上部岩层，最小抵抗线，崩矿的层厚；配料，矿石与熔剂之比
burden calculation 配料计算
burden depth 爆破纵深
burden descent 炉料下降
burden distance 抵抗线；爆破距离
burden distribution 炉料分布
burden gauge 炮孔导向架
burden line 负载线；抵抗线
burden line on the toe of the hole 炮眼底端负载线
burden material 炉料
burden of hole 最小抵抗线长度
burden of river 河流携带物
burden of river drift 河流夹沙；河流冲积层
burden on the toe of a hole 炮眼底部的负载
burden per hole 每炮眼的负载
burden pressure 上覆压力
burden removing 剥离
burdened stream 含悬浮土粒水流
burdening 负担；配料；炉料计算
burden-to-spacing ratio 炮眼最小抵抗线与炮眼间距之比
buret 滴定管；量管
burette holder 滴定管架
burette viscometer 滴定管黏度计
Burg deconvolution 伯格反褶积
burga 布加风
Burgers body 伯格斯体
Burgers circuit 伯格斯回路
Burgers model 伯格模型
Burgers vector 伯格斯矢量
Burgess Shale 布尔吉斯页岩
Burg's deconvolution 伯格反褶积
Burg's method 伯格法
burgy 细粉；煤屑；薄煤
burial 埋葬；埋藏
burial depth 埋藏深度
burial diagenesis 埋藏成岩作用
burial ground 墓地
burial hill 潜山；埋伏丘
burial jade 葬玉
burial layer 掩埋层
burial metamorphism 埋藏变质；埋深变质作用
burial water 封存水
buried 潜在的
buried abutment 埋入式桥台；埋入式岸墩
buried and waterlogged environments 地下和饱水环境
buried anomaly 埋藏异常
buried arc 潜弧；埋弧
buried body 掩蔽体
buried cable 地下电缆
buried cap pile 低承台桩
buried channel 地下通道；地下沟渠；埋藏河道
buried culvert 暗涵
buried depth 埋深；埋置深度
buried depth of groundwater table 地下水埋藏深度
buried depth of route 线路埋深
buried depth of tunnel 隧道埋置深度；埋深
buried drain 暗沟排水；地下排水沟
buried dump 掩埋的废土；废物坑
buried dune 埋藏沙丘
buried erosion surface 埋藏侵蚀面；地下冲蚀面
buried explosion 地下爆炸
buried fault 隐伏断层
buried focus 地下焦点
buried focus effect 地下聚焦效应
buried fossil peneplain 潜伏古准平原
buried glacier 埋藏冰川
buried highs 埋藏高地
buried hill 潜丘；埋藏山；潜山
buried hill accumulation 潜山油气藏
buried hill type reservoir 潜山型油藏
buried ice 埋藏冰
buried karst 埋藏喀斯特；潜伏岩溶
buried landform 埋藏地形
buried landscape 隐伏景观
buried landslide 埋藏滑坡
buried layer 埋藏地层
buried line 埋藏线；地下线
buried mountain 潜山
buried ore body 埋藏矿体
buried outcrop 埋没露头；隐藏露头
buried paleosol 埋藏古土壤
buried peat 埋藏泥炭
buried penstock 埋藏式压力水管
buried pile cap 低桩承台
buried pipe 地下管；暗管
buried pipe and culvert 地下埋设物
buried pipeline 填埋管
buried pipeline detector 地下管线探测器
buried placer 埋藏砂矿；深砂矿床
buried placer deposit 埋藏砂矿床
buried pluton 隐伏深成岩体
buried relief 潜起伏
buried ridge 埋藏山脊
buried river 地下河；暗河
buried river channel 埋藏河槽
buried shelter 地下防空洞
buried soil 古土壤；埋藏土壤
buried soil horizon 埋藏土层
buried source 埋藏源
buried storage 地下埋存
buried stream 埋藏河
buried structure 埋入式结构物，隐蔽式结构物；潜伏构造
buried suture 埋藏裂隙；隐伏裂隙
buried tank 地下储罐
buried terrace 埋藏阶地
buried terrain 掩蔽地形
buried thrust fronts 埋藏的冲断层前缘
buried topography 埋藏地形；隐伏地形
buried valley 埋藏谷；掩埋谷
buried water 埋藏水
buried weathered crust 埋藏风化壳
buried well 潜埋井
buried wiring 暗线

buried-hill highs 潜山高地异常
buried-ridge oilfield 潜山油田
burk 矿脉内的硬块
burkeite 碳钠石矾
burleigh 重型凿岩机
Burlington limestone 布林顿灰岩
burn 垂直空炮眼掏槽；火伤；烧伤；燃烧
burn cut 直孔掏槽；直眼掏槽；平行空眼掏槽
burn cut borehole 燃烧切钻孔
burn hole 空炮眼
burn hole drilling machine 中孔凿岩机
burn in 烙上；腐蚀；老化
burn in image 残留影像
burn out 曝光过度
burn out area 烧尽区
burn pit 燃烧坑
burn shoe 燃烧鞋
burn through 烧穿；焊穿
burnable 可燃的
burnable bone 可燃性骨煤
burn-back 炉衬烧损
burnback resistance 抗烧能力
burn-back resistance test 抗烧结试验
burn-cut 空眼掏槽
burn-cut blasting 空眼掏槽爆破
burned clay 烧结土
burned lime 烧石灰
burned out zone 已燃带
burned residue 灼烧残渣
burner 燃烧器；喷灯；燃烧嘴；燃烧炉
burner block 燃烧炉体
burner capacity 燃烧器容量
burner cup 喷头；喷嘴头
burner nozzle 燃烧器喷嘴
burner port 燃烧器喷口
burner-reamer-blowpipe assembly 燃烧器、扩大器和吹管
burnetizing 氯化锌防腐
burn-in 老化；预烧；烧焊
burning 燃烧；焙烧；燃烧的；空炮眼掏槽
burning area 火区；燃烧区
burning characteristic 燃烧特性
burning coal-mining dumps 燃烧煤炭开采转储
burning efficiency 燃烧效率
burning face 燃烧工作面
burning flame front 燃烧焰锋
burning front 燃烧前缘；燃烧带
burning gas 燃烧气
burning halo 焰晕
burning on 烧涂法
burning performance 燃烧性能
burning point 燃烧点；发火点
burning process of heterogeneous diffusion 多相扩散燃烧过程
burning rate 燃烧率；燃烧速度；燃速
burning shoe 打捞用旋转鞋
burning speed 燃烧速度
burning tip 火花熄灭接点
burning torch 气割炬
burning velocity 燃烧速度；燃速
burning volcano 燃烧火山；活火山

burning well 燃烧井
burning-on method 熔接法
burning-out 烧光；烧尽
burning-quality index 燃烧质量指数
burnish 摩擦抛光；磨光；擦亮；光泽
burnisher 磨光器；抛光器
burnishing 磨光；抛光
burnishing oil 抛光油
burnishing powder 抛光粉；磨光粉
burn-off 熔化；烧穿；焊穿
burn-off rate 熔化速度
burn-out 烧断；熔断
burnout altitude 燃尽高度；熄火高度
burn-round jumbo 平行炮眼钻车
Burnside apparatus 伯恩赛德型井下探水钻
Burnside boring machine 伯塞德探水钻机
burnstone 硫黄石；天然硫
burnt bit 烧损的金刚石钻头
burnt coal 烧变煤；天然焦
burnt deposit 燃烧沉积；过烧镀层
burnt gas 燃烧废气
burnt gault 煅烧白垩；烧黏土
burnt gypsum 烧石膏
burnt ingot 过热钢锭
burnt lime 煅石灰；氧化钙
burnt metal 金属过烧现象
burnt metamorphism 燃烧变质作用
burnt pyrite 焙烧黄铁矿
burnt rock 烧变岩；火烧岩
burnt sand 焦砂；烧成砂
burnt shale 烧页岩
burnt steel 过烧钢
burnt stone 火法变色宝石
burnt-cut holes 中心掏槽不装药的炮眼组；间隔空眼的炮眼组
burntlime 煅石灰；生石灰
burnt-out zone 烧尽区
burn-up 烧毁；烧尽
burozem 棕壤
burping well 间歇自喷井
burr 磨石；焊瘤；毛口；过火砖
burr free perforation 无毛刺射孔孔眼
burr height 毛刺高
burr mill 磨碎机
burr scraper 毛刺刮除器
burrel gas detector 少量沼气检定仪
Burrell apparatus 伯罗尔型气体分析器
Burrell indicator 伯勒尔型瓦斯检定器
burr-free 无毛刺
burrhstone 硅质磨石
burring 毛口磨光；去毛刺；去毛口
burring of taper prop 让压支柱尖端
burring prop 尖端支柱
burrless perforating 无毛口射孔
burrock 堤堰
burrow 洞穴；潜穴；掘洞；脉石；废石
burrow lining 潜穴衬里
burrow pore 生物钻孔
burrow porosity 虫掘孔隙
burrow structure 洞穴构造

burrow system 潜穴系；潜穴系统
burrowed pelmicrite 挖穴球粒微晶灰岩
burrowed structure 穿穴构造
burrowing type 潜穴类型
burrows 潜穴
burrstone 粗刻边；硅质磨石
burst 爆炸，爆裂，破裂；突出，猝发；脱落块；暴雨期
burst accident 突出事故
burst forecasting 突出预报
burst hub 轮毂破裂
burst internal yield 屈服内压力
burst into pieces 爆裂成碎块
burst load 爆破的负荷
burst mechanism 突出机理
burst mode 猝发方式
burst noise 脉冲噪声
burst of dam 溃坝
burst of mortar 溃浆
burst of rail base 轨底崩裂
burst of rail bottom 轨底崩裂
burst out 猝发；冒出；迸发
burst pressure 崩裂压力
burst pressure test 破坏压力试验
burst pulse 窄脉冲
burst rate 猝发传输率；猝发速度
burst risk 突出危险
burst source 爆发源
burst stage 突出阶段
burst strength 爆裂强度；耐破强度；胀破强度
burst tearing test 胀破试验；爆裂测试
burst test 胀裂试验；顶破试验
burst transmission 突发传输
burst-chamber 爆破药室
burst-collapse pressure 破坏-坍塌压力
burster 爆破筒；无截槽爆破
burster barrel 爆破筒
bursting 爆破；破裂；突出；猝发；爆开
bursting cone 破煤锥
bursting crack 崩裂
bursting diaphragm 防爆板
bursting energy 爆破能量
bursting force 炸力
bursting liability 突出倾向性
bursting Liability of Rock 岩石冲击倾向性
bursting of a dam 溃坝；垮坝
bursting potential 爆破潜力
bursting pressure 岩爆压力；爆破压力；胀破压力
bursting space 爆裂空间
bursting strength 脆裂强度；抗破裂强度；抗爆强度
bursting stress 爆裂应力；崩坏应力
bursting tool 破碎工具
bursting water from apical bed 顶板突水
bursting water gushing 突发式突水
bursting-charge 炸药
burstone 磨石
Burt revolving filter 伯特型转筒过滤机
Burt solar compass 伯特真子午线罗盘
burton 滑车组
burtoning system 双杆联吊
bury 埋；隐藏；软黏土；黏土页岩

bury barge 埋管驳船
bury method 填埋法
bury threads 暗螺纹
burystone 硅质磨石
bus bar blanking plate 母线盖板
bus system 汇流排系统；母线系统
bus tunnel 母线洞
bus turnover rate 客车周转率
busbar 汇流排；汇流条；母线
busbar adapter 母线接头；接母线盒
bus-bar wire 汇流条；母线
busduct 密封式汇流排
bush 轴衬；轴瓦；衬套；灌木
bush hammer 齿纹锤；汽动凿毛孔
bush hammering 錾凿加工；表面琢磨；粗面石工
bush log 灌树原木
bush neck 主轴衬；轴承座
bush plate 导向
bush roller chain 套筒滚子链
bushing 轴衬，衬套；加金属衬套里；套管
bushing elevation 补心高度
bushing extractor 衬套拔出器
bushing handling tool 方补心取出装置
bushing pin 衬套销
bushing retainer 套筒保持器
bushing tool 衬套装卸工具；套筒装卸工具
bushing transformer 套管式互感器
bus-organization 总线式结构
Buss table 巴斯型摇床
bustamente furnace 蒸馏汞的圆竖炉
bustamite 钙蔷薇辉石
buster 爆破筒；落煤机；落煤楔；风镐；双壁开沟犁
buster hole 空炮眼
bustite 顽火辉石无球粒陨石
butler finish 软抛光加工
butlerite 基铁矾
butment 桥台；拱座
butt 平接；对接；铰链；根端
butt and collar joint 套筒接合
butt and strap hinge 丁字铰链
butt and strapped joint 管子对口再用套筒铆住的连接
butt block 对接缝衬垫
butt cable 连接电缆
butt cap 前探支架的顶梁；悬臂梁
butt cleat 短而不清楚的解理面；尾割理
butt contact 对接触点；对接
butt cover plate 并合盖板
butt diameter 粗端直径
butt end 对接端；平头端
butt entry 与次解理成直角的平巷；与主平巷垂直的煤巷
butt heading 与次解理垂直的平巷；回采平巷
butt hinge 对接铰链
butt joint 端接；对接；次解理面；平缝
butt joint fitting 对接管接头
butt joint, butt connection 对接
butt junction 对接
butt muff 对接套筒
butt of pile 桩的大端
butt pin 铰链销钉
butt plate splice 钢板对头拼接

butt pocket 平行次解理面的通路
butt ramming 对头捣实；对头击
butt resistance welder 电阻对焊机
butt resistance welding 对接电阻焊
butt rivet joint 对头铆接
butt riveting 对接铆
butt seam welder 对接滚焊机
butt seam welding 对接缝焊
butt splice 对头拼接
butt strap 对接板
butt strapped joint 搭板对接
butte 孤山；孤峰；地垒
butte temoin 外露层；老围层
butte temoin ground 外露层
butted 平接的；对接的
butt-end packing 端面密封
butt-entry 与次解理垂直的平巷；与主平巷垂直的煤巷
butterfly bolt 双叶螺栓
butterfly cock 蝶形旋阀
butterfly dam ding 蝶闸坝
butterfly damper 蝶阀
butterfly diagram 蝶形图
butterfly effect 蝴蝶效应
butterfly filter 蝶形滤波器
butterfly gate 蝶形闸门
butterfly governor 蝶形节流阀；蝶形节气阀
butterfly hinge 蝶形铰链
butterfly nut 蝶形螺母
butterfly operation 蝶形操作
butterfly screen plate 元宝螺钉
butterfly throttle 蝶形风门
butterfly twin 蝶形双晶
butterfly valve 蝶形阀
butterfly valve of lock 船闸蝶形阀
butterflying trowel 涂灰镘
buttering 预堆边焊；涂灰浆；隔离层
buttering trowel 涂灰镘
Butters filter 巴特斯型多叶真空过滤机
Butterworth bandpass 蝶值带通；巴特沃恩氏通频带
Butterworth filter 巴特沃斯滤波器
buttgenbachite 氯铜硝石
butting 顶撞
butting face 平接面；对接面；紧接面；抵触面；
buttock 煤壁拐角；回采工作面缺口、壁龛；工作面侧壁；煤帮
buttock face 台阶式回采工作面；和长壁工作面成直角的工作面；切口的端面；壁龛的端面
buttock getter 壁龛开采机
buttock line 纵剖线
buttock shearer 爬底板滚筒采煤机；截深内滚筒采煤机
button bit 嵌珠钻头
button bottom 圆头桩尖；桩端
button control 按钮控制
button cutter 镶齿滚刀；镶齿钻头
button die 板牙
button drainage 串珠水系；串珠河
button head 圆头；圆钮头；圆钻头
button head rivet 自记蒸发计；自生收缩
button head screen plate 圆头螺钉
button up 扣紧

button welding 点焊
button-electrode 纽扣电极
button-head anchorage 圆头锚定
butt-prop 立柱的撑杆
buttree 拦石墙
buttress 撑墙；支墩；扶壁；扶垛；
buttress brace 支撑刚性梁；加劲梁；横撑
buttress bracing 垛间支撑
buttress centres 坝垛中心距
buttress coupling 加强接箍
buttress dam 支墩坝
buttress drain 扶壁排水孔
buttress head 垛头
buttress out 尖灭
buttress pack 预先充填带；超前充填带
buttress powered support 扶壁式液压支架
buttress retaining wall 支墩式挡墙
buttress sand 支撑砂层
buttress spacing 坝垛间距
buttress thread 锯齿螺纹；直三角螺纹
buttress thread screen plate 斜方纹螺钉
buttress unconformity 陡壁不整合
buttress wall 扶垛墙
buttress weir 垛堰
buttressed abutment 撑式桥台墙
buttressed retaining wall 扶壁式挡土墙
buttressed wall 外支墙；扶垛墙
buttresseddam 扶壁式坝
buttressing slope 扶壁支撑斜坡
Buxton test 安全炸药检定试验
Buys Ballot's law 白贝罗定律
BUZ 蜂鸣器
buzz 蜂鸣；噪声
buzzard 不可采煤层
buzzer 轻型凿岩机；蜂鸣器；蜂音器；磨轮
buzzerphone 蜂鸣器
buzzy 伸缩式风钻
by and large 一般说来；基本上；从各方面来看
by components used 按所用材料区分
by gravity 自溜；自坠；自落
by hand level 辅助平巷
by mass of cement 占水泥重量；占洋灰重量
by pass seepage 绕渗；绕流渗透
by passage 旁路
by pit 辅助竖井
by wash channel 排水沟；排水槽
by weight 按重量；按重量计算
byat 井筒横梁；撑木
by-channel 旁侧溢洪道；旁通水道；支渠
byerite 黏结沥青煤
Byerly discontinuity 拜尔利不连续面
by-hand level 辅助平巷
byland 半岛
by-level 辅助平巷
by-pass 绕道；支路；旁通管；泄流
bypass area 旁流面积
bypass belt 旁路胶带
bypass canal 分水渠；侧流渠
bypass channel 分洪道；减河；侧路渠；溢流渠
by-pass chute 支溜槽

by-pass conductor	迂回导线
by-pass culvert	旁通涵洞
bypass damper	跨越闸门；旁通管
by-pass filter	旁路滤波器
bypass flow	迂回水流；旁通水流
by-pass flowmeter	旁路式流量计
by-pass gate	旁通闸门
by-pass governing	旁通调节
by-pass intermediate depots	旁接的中间油库
by-pass loop line	环形绕道
by-pass operation	旁通运行
bypass pilot	旁通导阀
bypass pipe	旁通管
by-pass plug	旁泄塞
by-pass port	旁通孔
by-pass relief valve	旁通安全阀；旁通减压阀
bypass seepage	绕渗
by-pass shunt meter	旁路分流流量计
bypass stack	旁通排气管
by-pass switch	旁路开关
bypass tool	旁通工具
by-pass tunnel	旁通隧道
by-pass valve	旁通阀
by-pass vent	分支排气孔
by-pass weir	旁通堰
by-passed stone	拣出的废石
by-passing	旁泄；绕过；旁通
bypassing effect	旁通效应
by-passing of river	河道支流
bypassing region	未波及区
by-path	小道；侧道；绕道；支路
bypath mechanism	回踪机构
by-pit	副井；通风井
byproduct	副产品；意外结果；副作用
by-product coke	副产焦炭
by-product coke oven	干馏炼焦炉；副产炼焦炉
by-product coke process	副产炼焦法；干馏炼焦法
by-product coking	副产炼焦
by-product gas	副产气体；副产煤气
by-product recovery	副产品回收
by-road	小路；侧道
bysmalith	岩柱；垂直岩体；岩柱
byssolite	绿石棉
byssus	足丝；菌丝
bystream	支流
bystream deposit	近河沉积
by-terrace	漫滩阶地；近河阶地
bytownite	倍长石
by-volume	按体积；按容积
by-wash	排水管；河岸排水道；排水沟；溢洪道
by-water	牛轭湖；旧河道；旧河床
byway	偏路；小路；绕道

C

cable 缆索;锚索
cable anchor 锚索;钢索锚固
cable anchorage area 钢索锚固区
cable and tool rig 钢绳冲击式钻机
cable band 索夹
cable belt 缆摔带运输机;钢绳带
cable belt conveyor 钢绳牵引带式输送机
cable bent 钢丝绳垂度;缆索垂度
cable bent tower 索塔;支绳铁塔
cable bit 钢绳冲击钻头
cable bolt 钢丝绳锚杆
cable bolt support of shaft-drift 井巷锚索支护
cable bolting 锚索支护
cable bolting rig 锚索支护钻机
cable boring 冲击钻进;钢绳钻进
cable brace 钢索拉筋;锚索支撑
cable canal 电缆沟
cable car 吊车;缆车
cable carriage 缆车
cable carrier 电缆托架
cable chain 锚链
cable channel 电缆槽;电缆沟
cable churn drilling 钢绳冲击式钻进
cable clamp 索夹
cable clamshell 索式抓斗
cable conductor 缆芯
cable conduit 电缆管道
cable controlled scraper 索铲
cable core tube assembly 绳索取心装置
cable crane 缆索起重机
cable cutter 电缆剪;钢丝绳割刀
cable dowel 钢索锚栓
cable drag scraper 拉送斗机;索铲
cable dragging device 钢丝绳牵引装置
cable draw pit 电缆井
cable dredger 绳索式挖掘船
cable drift 电缆漂移
cable drill 钢绳冲击式钻机
cable drilling 钢绳冲击钻进
cable drilling system 钢丝绳钻眼法
cable drilling tool 钢绳冲击钻具
cable drive 绳索传动
cable duct 电缆导管;电缆槽
cable excavator 索缆式挖掘器
cable fencing 钢索栏栅
cable fishing tool 钢丝绳冲击钻进打捞工具
cable grab 钢丝绳抓斗
cable grip 电缆夹;电缆钳
cable guide 缆索导向器
cable hanger 钢丝绳吊架
cable head overshot 绳帽打捞筒
cable hoist 钢丝绳绞车;缆索吊
cable input plug 电缆输入接头
cable installation 电缆线路敷设

cable interface module 电缆接口模块
cable jacking test 锚索千斤顶试验
cable jerk 电缆牵动
cable kinking 电缆扭结
cable lay 缆索捻向;线绳螺旋缠绕;电缆敷设
cable laying ship 海底电缆敷设船
cable layout 电缆布置图
cable line of underground mine 地下矿电缆线路
cable lubricant 钢丝绳润滑油
cable percussion boring 钢绳冲击钻探
cable pulley 绳轮
cable railway 缆索道
cable reel 电缆滚筒
cable rig 钢丝绳冲击式钻机
cable road 缆道;索道
cable roadway 钢丝绳运输巷道
cable rope 电缆;缆索
cable run 电缆线路;索道线路;电缆敷设路径
cable scraper 缆索挖土机
cable scraper-planer 绳牵引耙式刨煤机
cable shaft 电缆竖井
cable sheathing alloy 电缆包皮合金
cable sheave 缆索滑轮;电缆绞轮
cable sheave bracket 钢丝绳轮支架
cable ship 海底电缆敷设船
cable slide 缆索滑车
cable socket 电缆插座;绳帽
cable speed panel 电缆速度面板
cable splice 电缆接头
cable spreading room 电缆分布室
cable stayed bridge 斜拉桥
cable structure 悬索结构
cable strum 拖缆振动
cable supported conveyor 钢绳承载式胶带机
cable suspended bridge 钢索桥
cable suspension bridge 吊桥;悬索桥
cable take-up 钢丝绳弹性收缩
cable tank 电缆舱
cable telemetry system 电缆遥测系统
cable tension 钢丝绳张力
cable tool 冲击钻具;钢绳冲击钻具;顿钻
cable tool barrel 顿钻捞筒
cable tool coring 顿钻取芯
cable tool drill 钢绳冲击钻
cable tool hole 顿钻的井眼
cable tool jack 顿钻千斤顶
cable tool joint 顿钻接头
cable tool rig 顿钻钻机
cable tool underreamer 顿钻管下扩眼器
cable tool wrench 顿钻大扳手
cable tower 缆索塔
cable tramway 缆索电车道
cable trench 电缆沟
cable trough 电缆暗渠

cable truss	悬索桁架
cable truss system	缆索桁架系统
cable tunnel	电缆隧道
cable wheel	锚链轮
cable winch	钢丝绳绞车
cable wire	钢索
cable work	钢丝绳绞车
cable wrapping machine	绕缆索机
cable-actuated excavator	索动挖土机
cable-dump truck	索式自卸卡车
cable-guide method	钢丝绳导向法
cable-hung clamshell bucket	绳索悬挂瓣式抓斗
cable-laid rope	多股钢丝绳
cable-supported belt	悬索胶带
cable-supported structure	悬索结构
cable-tool bit	钢绳冲击钻头
cable-tool boring rig	钢绳冲击式钻机
cable-tool core barrel	钢绳冲击钻进岩心管
cable-tool drilling outfit	钢绳冲击钻进设备和工具
cable-tool drilling system	钢绳冲击钻进方法
cable-tool drilling unit	钢绳冲击钻机装置
cable-tool well	冲击钻井
cableway	电缆管道；索道
cableway bucket	架空索道运料桶；索道吊罐
cableway erecting equipment	缆索吊装设备
cableway excavator	缆索挖掘机
cableway measurement	缆索测量
cableway power scraper	缆道动力刮土机
cableway tower	索道塔
cableway transporter	缆道起重运送机
cabling	电缆敷设
cacciatore	水银地震计
cache	高速缓冲存储器；地窖；窖藏
cache buffer memory	超高速缓冲存储器
cache coherence	高速缓存相干；高速缓存一致性
cacheutaite	硒铜铅矿
cacoxenite	黄磷铁矿
cactus grab	多爪抓斗
cadaster	地籍册；河流志
cadastral lists	地籍册
cadastral map	地籍图
cadmia	菱锌矿，锌渣，碳酸锌
cadmium	镉
cadmium electrolyte	镉电解液
cadmium filter	镉过滤器
cadmium plated steel	镀镉钢
cadmium pollution	镉污染
cadmium shield	镉屏蔽
cadmium sulfoselenide	硫硒化镉
cadre	骨干；基础结构
cadwaladerite	氯羟铝石
caen stone	浅黄色海相灰岩
caenozoic	新生代的
caesium	铯
caesium iodide	碘化铯
cafemic	钙铁镁质
cage	吊篮；升降车；施工骨架罐笼
cage adapter	深井泵罩的接头；外壳连接器
cage bar	罐笼的锁门杆；罐庞中的稳车杆
cage bumping	蹾罐事故
cage clearance	罐笼间隙
cage compartment	罐笼隔间
cage construction	骨架结构；罐笼结构
cage cover	罐笼顶盖
cage dog	罐笼防坠器；断绳保险器
cage door	罐笼门
cage equipment	罐笼设备
cage gate	罐笼门
cage guide holder	固定罐道的托架
cage hoist	罐笼提升机
cage hoisting mine	罐笼提升井
cage hoop	罐笼环箍
cage of reinforcement	钢筋笼
cage positioned ball valve	笼式定位球调节阀
cage rack	笼形拦污栅
cage safety catch	断绳保险器；罐笼稳车器
cage safety catch gear	罐笼断绳保险器；罐笼断绳防坠器
cage screen	笼形拦污栅；笼筛
cage set	罐笼座
cage shaft	罐笼提升井
cage shooting	笼中爆炸法
cage structure	笼状构筑物
cage tender	井底司罐工
cage type	笼式
cage type mill	笼式破碎机
cage winder	罐笼提升机
cage winding machine	罐笼提升机
caged man	笼屋居民
cager	井底司罐工
cageway	罐笼间
caging machine	装罐笼推车机
caging unit	装罐机
Cagniard apparent resistivity	卡格尼亚德视电阻率
Cagniard impedance	卡格尼亚德阻抗
cagniard resistivity	卡尼亚电阻率
cahemolite	腐殖煤
cahnite	水砷硼钙石
cailloutis	砾石；卵石；碎石
cainosite	钙钇铒矿
cainozoic	新生代
cairngorm	烟晶
caisson	沉井井壁；沉箱；挖孔桩
caisson adhesion	沉井外壁黏着力
caisson bell	沉井扩大部分
caisson box	沉箱
caisson breakwater	沉箱防波堤
caisson cap	沉箱盖
caisson chamber	沉箱室
caisson disease	沉箱病；潜水员病；减压过急病症
caisson drill	沉井桩孔钻机
caisson foundation	沉箱地基；沉箱基础
caisson gate	沉箱式坞闸
caisson launching	沉箱下水
caisson method	沉井法；沉箱法
caisson monolith construction	整体沉箱结构
caisson operation	沉箱作业
caisson pier	沉箱墩
caisson pile	沉管桩；沉箱桩；管柱桩
caisson pile seabed interaction	沉箱-桩-海床相互作用
caisson platform	沉箱型平台

caisson quay wall	沉箱岸壁
caisson retaining wall	沉井挡墙
caisson ring segment	沉箱环形块段
caisson shaft construction	沉井施工
caisson sinking method in submerged water	淹水沉井法
caisson sinking method under forced vibration	震动沉井法
caisson system	沉箱系统
caisson toe	沉井刃脚
caisson type pile	沉井式桩
caisson wall	沉箱挡墙
caisson work	沉箱作业
caisson yard	沉箱制作场地
caisson-set	沉箱结构
caisson-sunk shaft	沉井法开凿的立井
cajon	峡谷，窄峡谷
cajuelite	金红石
cake	结块煤；滤饼；泥饼；泥浆钻孔井壁结块；烧结块
cake breaker	滤饼破碎机
cake consistency	泥饼致密度
cake copper	铜锭
cake indicator	滤饼指示器
cake of salt	盐饼
cake permeability	泥饼渗透性
cake porosity	泥饼孔隙度
cake thickness	泥饼厚度
caked mass	结块体
caking	结块；烧结
caking ability	黏结度
caking capacity	结焦性；黏结性
caking coal	炼焦煤；黏结性煤
caking index	结焦性指数；黏结指数
caking number	黏结指数
caking property	结焦性；黏结性
caking test	结焦性试验；黏结性试验
calacata	有灰色纹理的大理石
calafatite	水钾铝矾，明矾石
calaite	绿松石
calamine	菱锌矿；炉甘石；异极矿；绿透闪石；透闪石
calamity	灾害；灾难
calander	研光机
calandria	排管式堆容器；蛇管冷却器
calaveras	含煤变质沉积岩
calaverite	磅金矿；碲金矿
calc alkali granite	钙碱花岗岩
calc alkalic rock	钙碱性岩石
calc aphanite	钙质隐晶岩
calc gneiss	钙质片麻岩
calc silicate gneiss	钙硅酸盐片麻岩
calc silicate rock	钙硅酸盐岩
calc sinter	钙华；钙华；石灰华
calc spessartine	钙锰铝榴石
calc-alkali	钙碱性
calc-alkali basalt	钙碱性玄武岩
calc-alkali gabbro	钙碱辉长岩
calc-alkali magma	钙碱性岩浆
calc-alkalic index	钙碱指数
calc-alkaline volcanics	钙碱性火山岩
calc-aluminosilicate rock	钙铝硅酸盐岩
calcar	煅烧炉；熔炉
calcarenaceous orthoquartzite	钙质沉积石英岩
calcarenaceous sandstone	钙屑砂岩
calcarenite	钙质岩；砂屑石灰岩
calcarenitic facies	钙屑沉积相
calcarenitic limestone	含砂屑石灰岩
calcarenyte	砂屑石灰岩
calcareobarite	钙重晶石
calcareous	钙质的；含钙的
calcareous aggregate	钙质集料
calcareous alga	钙藻
calcareous alluvial soil	石灰性冲积土
calcareous aquifer	钙质含水层
calcareous binding	钙质胶结
calcareous brown soil	石灰性棕色土
calcareous cement	钙质胶结物；含钙水泥
calcareous cementing material	石灰类黏结料
calcareous chert	钙质燧石
calcareous clay	钙质黏土；石灰质黏土
calcareous concretion	钙质结核
calcareous dolomite	钙质白云岩
calcareous dolostone	钙质白云岩
calcareous earth	石灰质土
calcareous facies	钙质相
calcareous green algae	钙质绿藻
calcareous grit	钙质粗砂岩
calcareous incrustation	钙质结壳
calcareous layer	钙质层
calcareous lithosol	石灰性石质土
calcareous marl	钙质泥灰岩
calcareous microfacies	钙质微相
calcareous microfossil	钙质微体化石
calcareous nannofossils	钙质超微化石
calcareous nannoplankton	钙质超微浮游生物
calcareous ooid	钙质鲕粒
calcareous ooze	钙质软泥
calcareous ore	灰质矿石
calcareous paddy soil	石灰性水稻土
calcareous peat	钙质泥炭
calcareous phyllite	钙质千枚岩
calcareous plate	钙板
calcareous platform	钙质平台
calcareous rock	钙质岩；灰岩
calcareous sand	钙质砂；石灰质砂
calcareous sandstone	钙质砂岩；灰质砂岩
calcareous schist	钙质片岩
calcareous sediments	钙质沉积物
calcareous shale	钙质页岩；灰质页岩
calcareous shell	钙质介壳
calcareous silex	钙质石英；石灰质石英
calcareous silicate rock	钙质硅酸盐岩
calcareous siliceous sinter	钙硅华
calcareous silt	钙质粉砂
calcareous sinter	钙华；石灰华
calcareous skeleton	钙质骨骼
calcareous slate	钙质板岩
calcareous soil	钙质土；石灰质土
calcareous spar	方解石
calcareous spring	碳酸钙泉
calcareous stone	石灰石
calcareous tufa	石灰华，钙华
calcareous tuff	灰质凝灰岩

calcareous water	石灰水；硬水
calcarinate	钙质胶结
calcarneyte	钙质碎屑岩
calccrust	钙质层
calc-dolomite marble	钙质白云石大理岩
calcedony	玉髓
calci amphiboles	钙角闪石类
calci-alkalic	钙碱性的
calcian dolomite	钙质白云岩
calciborite	硼钙石
calcibreccia	钙质角砾岩
calcic	钙质的
calcic brown soil	棕钙土
calcic concretion	钙质结核
calcic horizon	钙积层
calcic mull	钙质腐殖质
calcic rock	钙性岩
calcic skarn	钙质矽卡岩
calcic travertine	钙华
calcicalathina	钙兰石
calciclase	钙长石
calciclasite	钙长岩
calciclasitite	钙长岩
calciclastic	碎屑碳酸盐岩的
calcicole	钙生植物
calcicrete	钙结砾岩，钙质结砾岩，钙质壳
calcidiscus macintyrei	麦氏钙盘石
calciferous	钙质的；含钙的
calcific	钙化的
calcification	钙化作用
calcification process	钙化过程
calcified	钙化的
calcify	钙化
calcigranite	钙质花岗岩
calcilith	钙质岩
calcilization	方解石化
calcilutite	灰泥岩；泥屑石灰岩；灰泥岩；灰质碎屑岩
calcimeter	石灰测定器；钙定量器
calcimetry	碳酸盐含量测定法
calcimicrite	钙质微晶灰岩
calcimine	石灰浆；刷墙粉
calcimonzonite	钙质二长岩
calcimorphic soil	钙成土
calcinate	焙烧；煅烧；煅烧产物
calcinated sediment	煅烧沉积物
calcination	焙烧；煅烧
calcination loss	灼烧损失
calcinator	焙烧炉；煅烧炉
calcine	焙烧；煅烧；煅烧产物
calcined bauxite	煅烧矾土
calcined coke	焙烧焦；焙烧焦炭
calcined dolomite	煅烧白云岩
calcined flint chip	煅烧燧石片
calcined gypsum	烧石膏；熟石膏
calcined lime	烧石灰；生石灰
calcined magnesite	煅烧的菱镁矿
calcined natural pozzolan	煅烧天然火山灰
calcined ore	焙烧矿石
calcined plaster	煅石膏；熟石膏
calciner	煅烧工；煅烧炉

calcining	煅烧；焙烧
calcining furnace	焙烧炉；煅烧炉
calcining kiln	焙烧窑；煅烧窑
calcio ancylite	钙菱锶铬矿
calciobiotite	钙黑云母
calcioferrite	钙磷铁矿
calcion	钙离子
calciosamarskite	钙铌钇矿，钙铌钇铀矿
calciothermics	钙热法
calciothorite	钙钍石
calcipelite	泥屑石灰岩
calciphyre	斑花大理石
calcisilicarenite	钙质石英砂岩
calcisiltite	粉砂屑灰岩
calcisol	钙质土
calcisphere	钙球
calcite	方解石
calcite carbonatite	方解石碳酸岩
calcite cement	方解石胶结物
calcite cleavage	菱形解理
calcite compensation depth	方解石补偿深度
calcite crystallisation	方解石结晶
calcite deposition	方解石沉积
calcite dissolution index	方解石溶解指数
calcite dolomite	方解白云石
calcite flour	石灰石粉
calcite limestone	方解石灰岩
calcite marble	方解大理石
calcite mudstone	泥屑灰岩
calcite saturation level	碳酸钙饱和度
calcite streak	方解石夹层；方解石条纹
calcite syenite	方解石正长岩
calcite twinning	方解石双晶
calcite-cemented formation	钙质胶结地层
calcitic	方解石的
calcitic cementation	钙质胶结
calcitic dolomite	方解白云岩；钙质白云岩
calcitic dolostone	方解石白云岩
calcitite	方解石岩；方解碳酸岩
calcitization	方解石化
calcitrant	耐火的；耐酸的；难熔的
calciturbidite	钙质浊积岩
calcium	钙
calcium acetate	醋酸钙
calcium age	钙法年龄
calcium alkalinity	钙碱度
calcium alloys	钙合金
calcium aluminate cement	矾土水泥；铝酸钙水泥
calcium aluminum garnet	钙铝榴石
calcium base mud	钙基泥浆
calcium bicarbonate	碳酸氢钙
calcium bicarbonate water	重碳酸钙型水
calcium bichromate	重铬酸钙
calcium bioxalate	草酸氢钙
calcium bleach	漂白粉
calcium bromide	溴化钙
calcium carbide	电石；碳化钙
calcium carbide process	电石法
calcium carbonate	白垩；碳酸钙
calcium carbonate compensation depth	碳酸钙补偿深度

calcium carbonate polymorphs 碳酸钙晶型
calcium carbonate scale 碳酸钙垢
calcium carbonate stones 石灰石
calcium chloride 氯化钙
calcium chromium garnet 钙铬榴石
calcium citrate 柠檬酸钙
calcium clay 钙质黏土
calcium compounds 钙化物
calcium contamination 钙侵
calcium cyanamide 氰氮化钙
calcium feldspar 钙长石
calcium fluoride 氟化钙；萤石
calcium hardness 钙质硬度
calcium humate 腐殖酸钙
calcium hydrate 熟石灰
calcium hydroxide 熟石灰
calcium hydroxide nanoparticles 纳米级氢氧化钙粒子
calcium hydroxide nanosuspension 纳米氢氧化钙悬浮液
calcium hypochlorite 次氯酸钙
calcium lime 钙石灰；生石灰
calcium metatitanate 偏钛酸钙
calcium molybdate 钼酸钙
calcium nitrite 亚硝酸钙
calcium oxide 生石灰；氧化钙
calcium oxychloride 氯氧化钙；漂白粉
calcium phosphate 磷酸钙
calcium plaster 石膏浆
calcium plumbate primer 铅酸钙底漆
calcium quicklime 生石灰
calcium remover 除钙剂
calcium requirement 石灰需要量
calcium salt 钙盐
calcium silicate 硅酸钙
calcium silicate concrete 硅酸盐混凝土
calcium soddyite 钙水硅铀矿
calcium soil 钙质土
calcium subgroup 含钙水亚组
calcium sulfate 硫酸钙；无水石膏
calcium sulfate cement 硫酸钙水泥；石膏胶凝材料
calcium sulfate hemihydrate 半水石膏
calcium sulfate spring 硫酸钙泉
calcium sulfates 硫酸钙
calcium sulfide 硫化钙
calcium sulfite 亚硫酸钙
calcium sulphate 硫酸钙
calcium sulphate hemihydrate 硫酸钙半水化物；烧石膏；熟石膏
calcium sulphate plaster 石膏灰浆；石膏抹灰面
calcium sulphoaluminate 硫铝酸钙
calcium sulphur ratio 钙硫比
calcium titanate 钛酯钙
calcium tungstate 钨酸钙
calcium-acrylate treatment 丙烯酸钙处理
calcium-alkalic granite 钙碱花岗岩
calcium-chloride spraying system 氯化钙喷雾系统
calcium-magnesium bicarbonate water 重碳酸钙镁水
calcium-potassium dating 钙-钾法测定年龄
calcium-rich coal 富钙煤
calcium-rich peat 富钙泥炭；富钙泥炭
calcium-rich soil 多钙土

calcium-treated mud 钙处理泥浆
calclithite 碎屑灰岩
calcolistolith 滑塌灰岩块体
calcomalachite 钙孔雀石
calcrete 钙结层；钙质结砾岩
calcrete uranium 钙积层铀
calcrudite 砾屑石灰岩
calcrudyte 砾屑石灰岩
calc-sapropel 钙藻腐泥
calc-schist 钙质片岩
calc-series 钙碱性系列
calc-silicate hornfels 钙质硅酸盐角岩
calcspar 方解石
calcsparite 亮晶
calc-tonalite 钙质英云闪长岩
calculate point of hump height 峰高计算点
calculated feed 计算入料
calculated live load 计算活荷载
calculated raw coal 计算原煤
calculated reserves 计算储量
calculated self-weight collapse 计算自重湿陷量
calculated settlement 计算沉降
calculated span 计算跨径
calculated strength 计算强度
calculating area of compression member 受压构件计算面积
calculating scale 比例尺；计算尺度
calculation map of reserves 储量计算图
calculation of load-carrying capacity of member 构件承载能力计算
calculation of pillar size 矿柱尺寸计算
calculous 坚硬的；结石的；微积分
calculous soil 砾质土
calculus 微积分学；牙结石
calcurmolite 钼钙铀矿
caldasite 杂斜锆石
caldera 火山口；火山洼地
caldera collapse 破火山口塌陷
caldera complex 火口组合体
caldera fault 破火山口断层
caldera lake 巨火山口湖；破火山口湖
caldera of explosion 爆裂火山口
caldera ring fault 火口环形断裂
caldera subsidence 火口陷落
caldo 硝卤水
caldron 锅状盆地；火口沉陷
caledonite 铅绿矾
calender 研光机
calender line 国际日期变更线
calendered film 压延薄膜
calenderstack 研光机
calescence 渐增温
caleta 小溺谷湾
calf line 游动大绳
calf reel 大绳滚筒
calf wheel 大绳滚筒
calf wheel box 大绳滚筒轴承
calfdozer 小型推土机
calgon 六偏磷酸钠
caliber 测径器；管径
caliber gauge 测径规；厚薄规

caliber square	测径尺
calibrate	校准
calibrated	校准的；有刻度的
calibrated disc	刻度盘
calibrated feeder	校准计量给料机
calibrated flow metre	校准流量计
calibrated flux	校准流量；校准通量
calibrated gate	校准量水闸门
calibrated load gauge	经率定的测力计
calibrated pressure gauge	经率定的压力计
calibrated resistance	校准电阻
calibrated sand	标定砂
calibrater	校准器
calibrating device	校准器；校准装置
calibrating instrument	校准仪器或仪表
calibrating resistor	校准电阻
calibrating stem	校准杆
calibration	刻度；校准
calibration accuracy	校准精度
calibration block	校准试块
calibration chart	校准表
calibration check	标定检验；格值检定
calibration coefficient	标定系数
calibration constant	检定常数；校准常数
calibration curve	标定曲线；校准曲线
calibration curve of anemometer	风表校正曲线
calibration curve of current metre	流速仪率定曲线
calibration equation	换算公式
calibration error	刻度误差
calibration factor	刻度因子；率定系数；校准系数
calibration function	校准作用
calibration function of magnitude	震级起算函数
calibration gauge	刻度计
calibration graph	校正图
calibration instrument	校准仪
calibration interval	检定周期
calibration line	校准线
calibration loop	刻度环
calibration marker	标准指示器
calibration measurement	校准量测
calibration method	率定法
calibration of current meter	流速仪率定
calibration of seismograph	地震仪校准
calibration pit	刻度井
calibration plate	校准板
calibration point	标定点；校准点
calibration problem	校正问题
calibration procedure	标定方法
calibration pulse	校准脉冲
calibration resistor	校准电阻器
calibration ring	校准环
calibration sample	校准试样
calibration scale	校准表
calibration screw	校准螺旋
calibration station	率定站
calibration system	检定装置
calibration tail	刻度线
calibration tank	校准罐
calibration test	校准试验
calibration well	刻度井
calibration zero	刻度零线
calibrator	测厚仪；定径规；校准器
calibre	测径器；管径；卡规测径器
calibre circle	保径圆
calibre compasses	弯脚圆规
calibre gauge	测径规
calibre insert	保径镶齿
caliche	钙质层
caliche crust	钙结层；钙质壳
caliche nodule	钙质结核
calico rock	印花岩
caliduct	暖气管道
californium	锎
caliper	卡尺；卡钳；测径器；双脚规
caliper brake	卡钳式制动器
caliper data	井径数据
caliper dipmeter	井径倾角仪
caliper gauge	井径仪
caliper log	井径测井；井径纪录图；钻孔柱状剖面
caliper measure	测径
caliper method	卡规法
caliper reading	测径读数
caliper ring	井径仪刻度环
caliper rule	测径器；卡尺
caliper shield	圆弧式掩护支架
caliper square	游标规
caliper survey	井径测井；井径测量
caliper type brake shoe	卡钳式闸瓦
calipered hole size	测量的井径
calipered id	测量的内径
calipers leg	卡钳腿
calite	镍铝铬铁耐热合金
calk	生石灰；嵌缝；填隙
calker	紧缝凿
calking metal	填隙金属
call-back marker buoy	重返声控定位浮标
calliper pig	测径清管器
calliper rule	卡尺测径器
callophane	荧光分析器
Callow flotation cell	凯卢型浮选槽
Callow screen	卡路筛
calm buoy	悬链锚腿系泊浮筒
calmalloy	铜镍铁合金
Calmet	铬镍铝奥氏体耐热合金
calobiosis	同栖共生
calomel	氯化亚汞
caloric	热量；卡路里
caloric balance	热量平衡
caloric value	热值；卡值
calorie	卡路里
calorie metre	热量计
calorific	热量的；生热的
calorific capacity	热值；卡值
calorific equivalent	发热当量
calorific power	热量功率
calorimeter	测热计；热量计
calorimeter for gaseous and liquid fuels	气体、液体燃料量热器
calorimeter room	热量测定室
calorimetric	量热的

calorimetric assay	热量测定	cambered ceiling	拱形顶棚；拱形天花板
calorimetric bomb	量热弹；卡计弹	cambered girder	拱形梁；弧形梁
calorimetric box	量热箱	cambered inwards	内倾的
calorimetric test	量热试验	cambered outwards	向外凸的
calorimetry	量热学	cambered plane	弓形面
calorite	镍铬铁锰耐热合金	cambered plate	弓形板
calorization	铝化作用	cambered road	弓形顶巷道
calorize	铝化处理；铝处理表面	cambered truss	弓形桁架
calorstat	恒温箱	cambering	拱起；上拱度；向上弯曲
calotte	圆顶；圆拱	cambering gear	凹面加工装置
cal-silicate	钙硅酸盐	cambic horizon	过渡层；新生层土壤
Calstronbarite	钙锶重晶石	cambisol	始成土
calthrop	海绵骨针	cambium	形成层
caltonite	方沸碧玄岩	cambrian	寒武世
calutron	铀同位素分离器；卡留管	cambrian geodynamics	寒武纪地球动力学
calve	崩解；下层开采时崩落	cambrian rock	寒武纪岩石
calx	金属灰；矿灰；生石灰	cambric schist	含铜片岩；细纺片岩
calx chlorinata	含氯石灰	Cambridge drive-in tip	剑桥打入式头
calymmian	盖层纪	Cambridge model	剑桥模型
calyptolite	锆石	Cambridge pressure meter	剑桥旁压仪
calyx	粗钻屑收集筒；岩屑导筒	Cambridge sampler	剑桥取土器
calyx bit	钢粒钻头	cambro-ordovician	寒武-奥陶纪
calyx boring	钻粒钻进	cam-cutter	凸轮式割刀
calyx core drill	筲状取芯钻	came	有槽铅条
calyx drill	筲状取芯钻；岩心管钻粒钻机	camel	打捞浮筒；起重浮箱；骆驼；驼色的
calyx drill holes	筲状钻孔	camelback	翻转箕斗装置
calyx lobe	筲片	camel-back truss	驼背式桁架
calyxshot drill	岩心管钻粒钻机	camera caliberation	摄影机检校
calzone	运河地带	camera lucida	转绘仪
cam adaptor	可拆式凸轮联轴器	camerate	分室的；隔开的
cam case	凸轮箱	Cameron's hypothesis	卡米隆假说
cam curve	凸轮曲线	camkometer	剑桥旁压仪
cam disk	偏心盘；凸轮盘	camlock	凸轮锁紧器
cam drive	凸轮传动	ca-montmorillonite	钙蒙脱石
cam grinder	凸轮磨床	camouflet	地下空洞；地下爆炸
cam packer	凸轮式充填机	camp equipment	野外作业设备
cam plate type axial piston pump	斜盘式轴向柱塞泵	camp sheeting	板桩挡土岸壁
cam pump	凸轮型旋转泵	camp shop	野外修理站
cam pusher	装罐推车器	campanite	碱玄白榴岩
cam reduction gear	凸轮减速装置	camphor oil	樟脑油
cam shaft thrust bearing	凸轮轴推力轴承	campiler beds	坎坡层
cam sleeve	凸轮套筒；牙嵌离合器	campos basin	坎波斯盆地
cam stick	凸轮托棍	campshedding	护岸
cam-adjusting gear	凸轮调整装置	campshot	堤防
cam-and-lever mechanism	凸轮杠杆机构	camptonite	康煌岩；闪煌岩
camber	坝顶超填高度；起拱度；预拱度；预留沉落量；反挠度	camptospessartite	钛辉斜煌岩
		camptovogesite	斜闪正煌岩
camber arch	弯拱；弯拱	campylite	磷砷铅矿
camber beam	反挠梁；弓背梁；上拱梁	campylosphaera	弯曲球石
camber block	拱架垫块	camsellite	硼镁石
camber board	曲面板	camshaft case	凸轮轴箱
camber curve	拱度曲线	camshaft housing	凸轮轴箱
camber height	反挠高度	cam-type axial piston pump	斜盘式轴向柱塞泵
camber of bearing	支承反挠度	can buoy	罐形浮筒
camber of ceiling	预拱度	can filler	装罐机
camber piece	砌拱垫块	can hoisting system	井口操纵吊桶提升
camber ratio	拱度比	can rinser	油罐冲洗机
camber slip	砌拱垫块	canadite	钠霞正长岩
cambered	拱形的；弧形的；曲面的；外凸的	canadol	坎那油
cambered axle	弯轴	canal	沟槽；管道；运河

canal alignment	渠道定线
canal appurtenant structure	渠道附属建筑物
canal aqueduct	跨渠渡槽；输水渡槽
canal bank	运河河堤
canal basin	渠漕；水渠沟
canal bed	渠底
canal capacity	渠道过水能力；渠道流量
canal cross section	渠道断面；运河断面
canal crossings	渠道交叉口
canal ditch	泥浆槽
canal dryer	管道干燥器
canal freeboard	渠堤超高
canal grade line	渠道纵坡线
canal head	渠道起点；运河起点
canal headworks	渠道工程；渠首建筑物
canal in cut	挖方渠道
canal inlet	渠道进水口
canal intake	渠道进水口；渠首
canal irrigation	渠灌
canal leakage	渠道渗漏
canal lining	渠道衬砌
canal lock	渠闸
canal loss	渠道损失
canal offlet	渠道放水口
canal on embankment	填方渠道
canal permissible scouring velocity	渠道容许冲刷流速
canal pond	船闸间河段；渠首前池
canal rapids	陡槽
canal reach	运河渠段
canal revetment	渠道护面
canal rider	诱生裂隙理论
canal river	运河
canal rotation	渠道轮灌
canal seepage	渠道渗漏
canal seepage loss	渠道渗漏损失
canal side slopes	渠道岸坡
canal silting	渠道淤积
canal slope	渠道比降
canal structure	渠道构造
canal theory of tide	水道潮汐论；潮道理论
canal transition	渠道渐变段
canal trimmer	渠道修整机
canal tunnel	渠道隧洞
canal turnout	渠道分叉口
canal wall	渠道边墙
canal with earth section	无衬砌渠道
canal wrench	管沟锹或铲；管钳
canalization	管道系统；开掘运河；渠道网；运河化
canalization dam	渠化坝；通航用坝
canalized river	渠化河流
canalized river reach	渠化河段
canalized river section	梯级化河段；通航化河段
canalized river stretch	渠化河段；梯级化河段
canal-lay construction	开运河铺管法
canary	金丝鸟
cancalite	橄榄煌斑岩
cancarixite	霓英煌岩
cancellated structure	网状构造
cancelled structure	格构式结构；格组构架
canch	刷帮；挑顶；卧底；狭底水沟
canch hole	挑顶炮眼；卧底炮眼
canch work	挑顶卧底；凿石
cancrinite	钙霞石
cancrinite diorite	钙霞石闪长岩
cancrinite gabbro	钙霞石辉长岩
cancrinite monzodiorite	钙霞石二长闪长岩
cancrinite monzogabbro	钙霞石二长辉长岩
cancrinite monzosyenite	钙霞石二长正长岩
cancrinite plagisyenite	钙霞石斜长正长岩
cancrinite syenite	钙霞石正长岩
candidate frac fluid	待选压裂液
candidate reservoir	候选油藏
candidate sequence	候选层序
candidate well	候选井
candle coal	长焰煤；烛煤
candle turf	烛泥炭
candy bottoms	捞砂阀件
canebreak	植物碎屑
canfieldite	黑硫银锡矿
canga	铁角砾岩
Canif-mixer	卡尼夫砂浆拌合机
canine	大齿
canister	活性炭罐
canister respirator	罐型防毒面具
canite	钾盐镁矾
canker	腐蚀锈；煤矿矿水中的赭色沉淀
cankerous	腐蚀的
canned	罐贮的；密封的
canned oil	听装油
canned program	定型成套程序；现成程序
canned pump	密封轻便泵
canned sample	罐装样品
canned software	现成软件
cannel	烛煤
cannel and boghead oil shale	烛煤和藻煤质油母页岩
cannel boghead coal	烛藻煤
cannel coal	烛煤
cannel shale	粗油页岩；烛煤页岩
cannelite	微烛煤；烛煤
canneloid	烛煤的；烛煤质煤
canneloid coal	烛煤式煤
canneloid shale	烛煤质页岩
cannelure	环形沟槽；纵向槽
cannibalization	零件拆用；拼修；同型装配
canning	罐藏
cannizzarite	辉铅铋矿
cannon	臼炮；空心轴
cannon gallery test	试验平硐内臼炮试验
cannon plug	圆柱形插头
cannon test	臼炮试验
cannonball	巨型结核
cannon-fenske viscometer	坎农-芬斯克黏度计
cannon-type machine	炮式充填机
cannula	套管
canoe fold	舟状褶皱；紧闭向斜
canoe valley	向斜谷
canoe valley mountains	向斜谷山
canoe-shaped ridge	舟状向斜脊
canonical correlation coefficient	典型相关系数；正则相关系数

canonical correspondence analysis	典范对应分析
canonical profile	标准剖面
canonical transformation	典型变换
canopy	防护罩；顶盖
canopy extending jack	顶梁升降千斤顶
canopy guard	护顶板梁
canopy side flap	顶梁活动侧防护板
canopy window	上旋窗
cant	超高；翻转；角隅；斜面
cant bay	切口；缺口
cant beam	斜横梁
cant column	多角柱
cant course	倾斜走向
cant dog	端柄搬钩
cant frame	艉端斜肋骨；斜横梁
cant hook	钩杆；平头搬钩
cant of curve	曲线超高
cantalite	疗松脂岩；松脂流纹岩
canted wall	斜交墙
cantilever	悬臂梁
cantilever abutment	悬臂式桥台
cantilever arch bridge	悬臂拱桥
cantilever arch truss	悬臂桁架
cantilever arm	悬臂距
cantilever bar	悬臂梁
cantilever bar technique	前探梁设计方法
cantilever base	伸臂底座
cantilever beam	挑梁；悬臂梁
cantilever beam bridge	悬臂梁桥
cantilever bracket	悬臂托架；悬臂支架
cantilever bridge	悬臂桥
cantilever buttress dam	悬臂式支墩坝
cantilever cofferdam	悬臂式围堰
cantilever column	悬臂柱
cantilever conveyor	长臂式撒料机；悬臂式排土机
cantilever crane	悬臂起重机
cantilever crib	带托梁的垛式支架
cantilever deck dam	悬臂平板坝
cantilever derrick	悬臂式起重机；折叠式井架
cantilever discharge end	卸载探头
cantilever drilling platform	悬臂式钻井平台
cantilever footing	悬臂基脚；悬臂梁式基础
cantilever foundation	挑梁式基础；悬臂式基础
cantilever girder	悬臂梁
cantilever girder bridge	悬臂梁桥
cantilever grizzly	外伸式格筛
cantilever load	悬壁荷载
cantilever mast	悬臂式桅杆；折叠式轻便井架
cantilever moment	悬臂力矩
cantilever pile	悬臂桩
cantilever pivoted	旋转腕臂
cantilever platform	悬臂平台
cantilever portion of center sill	中梁悬臂部
cantilever pump	悬臂泵
cantilever retaining wall	悬臂式挡土墙
cantilever roof bar	前探梁；悬臂顶梁
cantilever shear wall	悬臂剪力墙
cantilever shed gallery	悬臂式棚洞
cantilever shed tunnel	悬臂式棚洞
cantilever sheet pile	悬臂式板桩
cantilever sheet pile quay wall	悬臂式板桩岸壁
cantilever sheet pile wall	悬臂式板桩墙
cantilever sheet piling	悬臂式板桩；悬臂式板桩墙
cantilever slab	悬臂板
cantilever span	悬臂跨度
cantilever stone on eave	挑檐石
cantilever structure	悬臂结构
cantilever support	前探式支架；悬臂支架；悬臂座
cantilever swivel bracket	腕臂底座
cantilever timbering	前探支架
cantilever troughed idler	悬臂式槽型托辊
cantilever truss	悬臂桁架
cantilever type inclinometer	悬臂式井斜仪
cantilever walkway	悬臂式行人道
cantilever wall	悬臂式墙
cantilevered	突出的；悬臂式的
cantilevered assembling construction	悬臂拼装法
cantilevered element	悬臂式构件
cantilevered stairway	悬挑楼梯
cantilevered well module	悬臂式井模
cantilevering	悬臂
canting	倾斜
canvas air conduit	帆布通风管道
canvas hose	帆布水龙带
canvas line brattice	纵向帆布风障
canvas stopping	帆布风障；帆布风布
canvas table	帆布衬垫洗矿槽
canvas ventilation pipe	帆布通风管
canvas ventilation tubing	帆布通风筒
canvas-apron unloader	输送带式卸载机
canyon	河谷；峡谷
canyon bench	峡谷阶地
canyon delta	谷口三角洲
canyon dune	峡谷沙丘
canyon fill	峡谷淤积
canyon profile	峡谷剖面；峡谷轮廓
canyon topography	峡谷地形学
canyon wall	峡谷壁
canyoned stream	深谷河流
canyon-fan system	海底谷-扇系
canyonside	峡谷陡壁
cap	爆破用雷管；顶板岩石；盖梁；罩盖
cap and fuse firing	雷管导火线爆破
cap bead	表面焊道
cap block	柱头垫木；桩帽；龙骨墩顶垫；大斗
cap bolt	盖螺栓
cap cable	短钢索；钢丝束
cap clamp	盖夹
cap copper	带状黄铜
cap crimper	雷管封口器；导火线封口器
cap effect	盖岩影响
cap face	顶面
cap formation	盖层
cap grouting	顶盖灌浆
cap height	焰晕高度
cap inserting	雷管插入
cap key	闭口扳头；套筒扳手
cap lamp	矿灯；帽灯
cap lamp glass	帽灯玻璃罩
cap lid	顶梁；柱帽

cap light	安全帽灯；矿灯；帽灯
cap lug	帽耳
cap model	帽盖模型
cap nut	螺帽
cap of end wall of portal	洞门墙顶顶帽
cap of pile	头箍；桩帽
cap pass	面焊道
cap piece of sheet-pile	板桩联系帽梁
cap pillar	顶柱
cap plug	盖塞
cap rock failure	盖层失效
cap rock integrity	盖层完整性
cap rock of reservoir	油气藏盖层
cap screw	有头螺钉；戴帽螺钉
cap shell	雷管壳
cap shot	石面爆破
cap sill	顶梁；帽梁
cap stone	顶盖石；盖顶石；拱冠石；压顶石
cap tool	夹压雷管钳子
cap wire	电雷管导线
capable fault	能动断层
capacitance	容量
capacitance meter	电容测量仪
capacitance potentiometer	电容电位计
capacitance probe	电容探针
capacitance strain gauge	电容式应变仪
capacitive plug	电容插头
capacitive-displacement transducer	电容-位移传感器
capacitor oil	绝缘油；电容器油
capacitor reactance	容抗
capacity	容量；搬运量
capacity bridge seismograph	电容电桥地震仪
capacity factor	容量系数
capacity of a stream	河流搬运能力；挟沙量
capacity of an artesian well	自流井出水量
capacity of artesian well	自流井出水量
capacity of body	载重量；车皮容量
capacity of driven pile	沉桩的承载力
capacity of drum	滚筒绕绳量
capacity of groundwater reservoir	地下库容
capacity of grout acceptance	吸浆能力
capacity of pump	泵的排量
capacity of reservoir	水库库容
capacity of saturation	饱和量
capacity of spillway	溢洪能力
capacity of surface mine	露天矿生产能力
capacity of trap	圈闭容量
capacity of well	井出水量；井出油量；井的生产能力
capacity rating	额定生产率，额定功率；额定容量
capacity seismometer	电容式地震检波器
capacity test	能率试验
capacity to stand	承载能力
capacity-head curve	排量-压头曲线
cape	吊台，斗篷；海角；岬角
cape blue asbestos	青石棉
cape diamond	淡黄色金刚石
cape foot	南非尺
cape front	岬角锋
cape ruby	红柘榴石
cape seal	石浆封层
capel	钢丝绳接头；鸡心环；嵌环；脉壁黏土带；石英质岩石
capillar migration	毛细管迁移
capillarimeter	毛管测液器；毛管检液器
capillarimetric method	毛细测量法
capillarity	毛细管现象；毛细管作用
capillarity absorption	毛细管吸收作用
capillarity constant	毛细常数
capillarity correction	毛细现象校正
capillarity phenomenon	毛管现象
capillarity-correction chart	毛细误差校正表
capillary	毛细管；毛细现象
capillary absorbed water	毛细吸附水
capillary absorption	毛细管吸水作用；毛细吸湿
capillary action	毛细作用
capillary analysis	毛细分析法
capillary ascension	毛细上升
capillary ascent	毛管水上升高度
capillary attraction	毛管吸力
capillary barrier	毛细屏障；毛细阻滞覆盖层
capillary bond	毛管联结
capillary bore	毛细管孔
capillary bound water	毛细结合水
capillary breakthrough pressure	突破时毛细管压力
capillary bulb	毛细球
capillary canal	毛细通道
capillary capacity	毛管容量；毛细容量
capillary capacity of soil moisture	土壤毛管水容量
capillary channel	毛细孔道
capillary cohesion	毛细内聚力
capillary column gas chromatograph	毛细柱气相色谱仪
capillary concentration	毛细浓缩作用
capillary condensation	毛管凝缩作用；毛细凝聚作用
capillary condensed water	毛细管凝结水
capillary conductivity	毛细管传导度
capillary constant	毛细常数
capillary crack	发状裂缝；毛细裂缝
capillary cut-off	毛细隔离层
capillary depression	毛细下降
capillary desaturation	毛细管驱替
capillary diaphragm	毛细隔膜
capillary diffusivity	毛管扩散系数
capillary discontinuity	毛细管不连续性
capillary driving force	毛细管驱动力
capillary effect	毛细管效应
capillary electrode	毛细管电极
capillary elevation	毛细上升高度
capillary end barrier	毛细管末端屏障
capillary end effect	毛细终点效应
capillary entry pressure	毛细管吸入压力
capillary exposed surface	毛细管暴露面
capillary film	吸附水膜
capillary fissure	毛细裂纹
capillary fittings	毛细接合配件；渗锡溶接配件
capillary flow	毛细运移
capillary flowmeter	毛细管流量计
capillary force	毛细力；毛细作用力
capillary force balance	毛细力平衡
capillary fracture	毛细裂缝
capillary fringe	毛管水边缘；毛细管上升带；毛细管酚带

capillary fringe zone 毛细管上升边缘区
capillary front 毛管湿锋
capillary head 毛细管水头
capillary height 毛细管水上升高度
capillary hydraulic-head 毛管测压水头
capillary hypothesis 毛细管假说
capillary imbibition 毛细管渗吸作用
capillary imbibition waterflood 毛细管自吸注水
capillary infiltration 毛细管浸润
capillary interaction 毛细互作用
capillary interstice 毛细间隙
capillary jet viscometer 毛细管射流黏度计
capillary leak 毛管渗漏
capillary lift test 毛细管水上升高度试验
capillary lubricant 毛细管作用润滑剂
capillary membrane 毛细管膜
capillary migration 毛细管运移
capillary model 毛细模型
capillary moisture 毛细管水分
capillary moisture capacity 毛管持水量
capillary moisture zone 毛细上升带
capillary movement 毛细管水运动
capillary neck 毛细管颈
capillary opening 毛细管孔
capillary packed column 毛细管填料柱
capillary penetration 毛细管渗透
capillary phenomena 毛细管现象
capillary piezometric head 毛细管测压水头
capillary pore 毛管孔隙
capillary porosity 毛细管多孔性;毛细管孔隙度
capillary porous body 毛细多孔体
capillary potential 毛细管水势
capillary potential gradient 毛细管梯度
capillary pressure 毛管压力
capillary pressure head 毛细管压力水头
capillary pressure hysteresis 毛细管压力滞后现象
capillary pressure permeability 毛细孔压力渗透性
capillary pressure withdrawal curve 退汞毛细管压力曲线
capillary pressure-saturation curve 毛细管压力-饱和度关系曲线
capillary proximal 毛细管始端
capillary pull 毛细管拉力
capillary pulling power 毛管张力;毛细拉力
capillary pyrite 白铁矿;针镍矿
capillary radius 毛细作用半径
capillary repulsion 毛细斥力
capillary response 毛细反应
capillary rheometer 毛细管流变仪
capillary ripple 毛细波痕
capillary rise 毛细管水上升高度
capillary saturating 毛管饱水
capillary saturation 毛细管水饱和度
capillary saturation zone 毛细水饱和带
capillary seal 毛细管封闭
capillary seepage 毛细渗流;毛细渗透
capillary siphoning 毛细虹吸
capillary soil moisture 土壤毛细水
capillary stage 毛细管水位
capillary suction 毛细吸引
capillary suction head 毛管吸升高度;毛管吸水头

capillary tension 毛细管水张力
capillary tension theory 毛细管张力理论
capillary threshold pressure 临界毛细管压力
capillary tip 毛细管尖端
capillary trapping 毛细管捕获
capillary trapping force 毛细管捕获力
capillary tube 毛细管
capillary vessel 毛细管
capillary viscometer 毛细管黏度计
capillary viscometry 毛细管测黏法
capillary viscosimeter 毛细管黏度计
capillary viscosity 毛细管黏度
capillary void 毛细孔隙
capillary water 毛细管水
capillary water capacity 毛细管水容度
capillary watering 毛管灌水
capillary wave 表面张力波;毛管波;毛细波
capillary wetting method 毛管湿润法
capillary yield 毛细出水量
capillary zone 毛细管水带
capillary-fringe belt 毛细管水边缘带
capillary-height method 毛细管高度法
capillary-pressure cell 毛细管压力计
capillary-retained oil 毛细管滞留油
capillary-size pore 毛细孔隙
capillometer 毛细管试验仪
capillose 针镍矿
caplastometer 毛细管黏度计
caple 脉壁黏土带;石英质岩石;套环嵌环鸡心环
capless stand 开口式机架
capnite 铁菱锌矿
caporcianite 红粒浊沸石
capote 带风帽的斗篷;活动车篷
capped concrete core 加帽混凝土芯
capped fuse 带有雷管的导火线
capped pile 带箍桩;帽箍桩
capped pipe 帽管
capped primer 起爆药包
capped rock 顶盖岩石;上覆岩石
capped steel 加盖钢;压盖钢
capped wellhead 扣装井口
capped yield model 盖帽屈服模型
cappel 钢丝绳接提升容器的接头箍;钢傈环头
cappelenite 硼硅钡钇矿
capper 封口机
cappiece 顶梁;横梁
capping 覆盖层;矿帽;封顶;桩帽;盖层岩;顶梁横梁
capping bead 盖面焊道
capping bed 盖层
capping curb 井壁各段的顶环
capping formation 盖层
capping hookup 防喷器组井控装置
capping layer 上封层
capping mass 表土;浮土;覆盖岩石
capping material 表土;覆盖物
capping metal 浇灌合金
capping operation 封井作业
capping piece 垫木;压檐木
capping plate 盖板
capping thickness 剥土厚度;覆盖层厚度

caprinid biosparite	双壳类生物亮晶灰岩
caprock	顶盖岩石；覆盖岩层；盖层；冠岩
caprock damage	盖层破坏
caprock effect	盖层效应
caprock fracture connectivity	盖层裂缝连通性
caprock geology	盖层地质
caprock or seal	盖层
caprock quality	盖层品质
caprock reactivity experiments	盖层反应性试验
caprock seal	盖层密封
caprock structure	冠岩结构
caprocks	盖层
caprock-type geothermal aquifer	盖层型地热含水层
caprock-type geothermal reservoir	盖层型地热储
cap-shoe-type hinged bar	套鞋式铰链横梁
capsizing moment	倾覆力矩
capstan	起锚机
capstan engine	绞盘机
capstan lathe	车床转塔刀架
capstan plough	挖沟机
capstan plow	挖沟机
capstan pulley	竖式绞绳轮
capstan winch	绞盘
capsulate	封装在里面的
capsulated	胶囊包裹的
capsulation	封装
capsule	密封仓
capsule aneroid	空盒气压计
capsule loaded forward bar	设压力传感器的前探梁
capsule metal	铅锡合金
capsule-type pressure gauge	囊式压力计
captance	容抗
captation	集水装置；筑坝壅水
captivated track	导向轨道
captive air bubble	捕获的空气泡
captive bolt	固定螺栓
captive bubble method	捕泡法
captive drop technique	俘滴法
captive droplet technique	俘滴法
captive float	栓定浮筏
captive mine	产品不出售；自销矿山
captive rail car	卡轨车
captive screen plate	固定螺钉
captive stand	固定架；专用架
captive wedge	栓楔
captivity	捕获；俘房；监禁
captor stream	袭夺河
capture	捕获；捕集；吸收；袭夺
capture and recapture experiment	标识放流
capture area	捕捉区
capture cross section	俘获截面
capture efficiency	采油效率；俘获效率
capture elbow	桩脚
capture hypothesis	捕获假说
capture river	断头河；袭夺河
captured river	袭夺河
captured source	袭夺源
captured stream	被夺河
capturing	袭夺
capturing river	截夺河；袭夺河
capwise	垂直走向
car arrester	阻车器
car block	阻车器
car cleaner	车辆清理工；矿车清除机
car coupling	矿车联接
car greaser	车辆润滑器；矿车润滑工
car gripe	减速装置
car hauler	调车绞车；推车器
car level	矿车高度的水平
car lift	升车机
car loader	装车机
car pincher	煤仓装车工
car puller	牵车绞盘
car puller hook	拖车钩
car putting	矿车运输；推矿车
car runaway prevention device	跑车防护装置
car safety chain	矿车安全链
car safety dog	阻车器
car sampler	采样车
car scale	车辆秤；地磅
car spotting	调配矿车
car spotting hoist	调车绞车
car spragger	矿车制动棒；减速装置
car stop	稳车器
car stopping device	挡车装置
car tipper	翻车机；翻斗车
car tripper	翻车器
car unloader	卸车机
car vibrator	装车振动台
car volume	车辆容量
car whacker	矿车修理工
caracole	螺旋形楼梯
caracolite	氯铅芒硝
carapace	背壳；硬盖
carat loss	金刚石损失
carat weight range	钻头上的克拉数范围
carballoy	碳化钨硬质合金
carbankerite	碳铁白云石；微碳酸盐质层
carbapatite	碳酸磷灰石
carbargilite	碳质泥岩
carbargillite	碳质页岩；微泥质层
carbene	炭青质
carbide	碳化钙；碳化物；碳化合金
carbide alloy	硬质合金；碳化物合金
carbide banding	碳化物带
carbide bit	碳化物刀头；硬质合金钻头
carbide black	碳化黑
carbide blade	硬质合金刀片
carbide cutter	硬质合金刀具
carbide die	硬质合金模具
carbide drill	硬质合金钻头
carbide extension	硬合金齿高度
carbide hob	硬质合金滚刀
carbide insert	碳化钨镶嵌物；镶硬合金齿
carbide insert bit	合金钻头；镶硬合金齿的牙轮钻头
carbide insert blade type bit	片状合金钻头
carbide insert drag bit	合金刮刀钻头
carbide insert pyramid type bit	棱锥形合金钻头
carbide lamp	电石灯
carbide mandrel	硬质合金心棒

carbide miner 硬质合金截煤机
carbide mud residue 电石渣
carbide network 网状碳化物
carbide of calcium 碳化钙
carbide percussion drilling 硬质合金冲击钻探；碳化物冲击钻探
carbide precipitation 碳化物析出
carbide segregation 碳化物偏析
carbide slag 电石渣；碳化物炉渣
carbide slug 碳化钨镶嵌
carbide tool 硬质合金工具
carbide type bit 镶硬合金齿的牙轮钻头
carbide-chlorination 碳化物氯化
carbidefeed generator 乙炔发生器
carbide-insert burial 硬合金齿吃入深度
carbide-tipped cutting tool 硬质合金刀具
carbide-tipped drill 硬质合金钻头
carbide-tipped milling cutter 硬质合金镶刃铣刀
carbide-tipped multiple-rotating cutter 硬质合金多头转割机
carbide-tipped slip 硬质合金牙尖的卡瓦
carbide-tipped steel 碳化物镶尖钻头
carbide-tipped tool 硬质合金刀具
carbo 碳；煤
carbo lapideus 烟煤
carboanimalis 动物碳
carboapatite 碳酸磷灰石
carbobitumen 焦油沥青；煤系沥青
carboborite 水碳硼石
carbocer 稀土沥青
carbochain polymer 碳链聚合物
carbocoal 半焦
carbocycle 碳环
carbocyclic compound 碳环化合物
carbofloc process 碳化絮凝法
carbofossil 化石炭
carbofrax 金刚硅耐火料
carbogel 煤胶体
carbohumin 棕腐质
carbohydrate 碳水化合物
carboid 焦沥青；油焦质
carbolic 碳的
carbolineum 蒽油；焦油木材防腐剂
carboloy 碳化钨硬质合金
carbometer 定碳仪；二氧化碳计
carbominerite 碳矿质
carbomorphism 碳化
carbon 碳精电极
carbon activity 碳素活度
carbon adsorbents 碳吸附剂
carbon adsorption 碳吸附
carbon and oxygen isotopes 碳氧同位素
carbon arc air gouging 碳弧气刨
carbon arc cutting 碳弧切割
carbon arc welding 碳弧焊
carbon arrester 炭精避雷器
carbon assimilation ratio 碳同化系数
carbon balance 碳平衡
carbon bisulfide 二硫化碳
carbon black 炭黑；黑烟末
carbon bond 碳键

carbon brush 碳刷；碳电刷
carbon budget 碳收支
carbon buildup 碳堆积
carbon burial 碳埋藏
carbon burning 烧碳
carbon capture 碳捕集
carbon capture and sequestration 碳捕获与储存
carbon capture and storage 碳捕获和储存；碳捕获与封存
carbon capture and utilization 碳捕集与利用
carbon capture project 碳捕集项目
carbon case hardening 渗碳硬化
carbon chain 碳链
carbon chain fibre 碳链纤维
carbon chloroform extract 碳氯仿抽提物
carbon circulation 碳循环
carbon collector strips 碳滑板
carbon combustion 碳燃烧
carbon composite iron ore briquette 碳纤维复合材料铁矿石型煤
carbon composite iron ore hot briquet 碳纤维复合材料铁矿石热蜂窝煤
carbon composite pellet 碳纤维复合材料颗粒
carbon compound 碳化合物
carbon containing mineral 含碳矿物
carbon content 含碳量
carbon content in coal 煤中碳
carbon conversion 碳转化
carbon crucible 石墨坩埚
carbon cycle 碳循环
carbon cycle-climate feedbacks 气候反馈碳循环
carbon dating 碳年龄测定
carbon depletion 脱碳
carbon dichloride 二氯化碳
carbon dioxide 二氧化碳
carbon dioxide cannon 二氧化碳大炮
carbon dioxide capture 二氧化碳捕获
carbon dioxide capture and storage 二氧化碳捕获和封存
carbon dioxide corrosion 二氧化碳腐蚀
carbon dioxide disposal 二氧化碳处置
carbon dioxide dissolution 二氧化碳溶解
carbon dioxide drive 二氧化碳驱油
carbon dioxide emissions 二氧化碳排放
carbon dioxide equivalent coefficients 二氧化碳当量系数
carbon dioxide exchange 二氧化碳交换
carbon dioxide fixation 固定二氧化碳
carbon dioxide gas 碳酸气
carbon dioxide gas reservoir 二氧化碳气藏
carbon dioxide ice 干冰
carbon dioxide injection 二氧化碳灌注；二氧化碳灌注
carbon dioxide leakage 二氧化碳泄漏
carbon dioxide process 二氧化碳处理法
carbon dioxide reduction 二氧化碳减排
carbon dioxide removal 二氧化碳清除
carbon dioxide separation 二氧化碳分离
carbon dioxide sequestration 二氧化碳封存；二氧化碳隔离
carbon dioxide spring 二氧化碳泉；碳酸泉
carbon dioxide storage 二氧化碳封存；二氧化碳封存
carbon dioxide storage capacity 二氧化碳封存量
carbon dioxide utilization 二氧化碳利用
carbon dioxide water 碳酸水

carbon dioxide well flushing　二氧化碳洗井
carbon disulfide　二硫化碳
carbon disulphide　二硫化碳
carbon drill steel　碳钢钻杆；碳钢钻钢
carbon emission　二氧化碳排放；碳排放
carbon emission inventory　碳排放清单
carbon emission source　碳排放源
carbon equivalent　碳当量
carbon fiber　碳纤维
carbon fiber reinforced concrete　碳纤维加筋混凝土
carbon fiber reinforced plastics　碳纤维增强塑料
carbon fiber reinforced polymer　碳纤维增强复合材料
carbon fiber sheet　碳纤维布
carbon fibre　碳纤维
carbon fibre plastic composite　碳纤维塑料复合材料
carbon filament　碳灯丝，碳素长丝
carbon fixation　碳固定作用
carbon flux　碳通量
carbon flux and sink　碳通量与碳汇
carbon footprint　碳足迹
carbon formation　积炭形成
carbon freezing　用二氧化碳冷冻
carbon group analysis　碳族分析
carbon hydrogen ratio　碳氢比
carbon ice　干冰
carbon iron composite　铁碳复合材料
carbon isotope　碳同位素
carbon isotope chemostratigraphy　碳同位素化学地层学
carbon isotope composition　碳同位素组成
carbon isotope ratio　碳同位素比
carbon isotope stratigraphy　碳同位素地层学
carbon isotopes　碳同位素
carbon isotopes mutation　碳同位素突变
carbon isotopes of individual hydrocarbon　单体烃碳同位素组成
carbon isotopic signature　碳同位素标识
carbon knock　积炭爆震
carbon limestone　碳质石灰岩
carbon moderator　碳慢化剂
carbon monoxide　一氧化碳
carbon monoxide emission　一氧化碳排放
carbon monoxide monitor　一氧化碳测定仪
carbon monoxide self-rescuer　一氧化碳自救器
carbon offset　碳补偿
carbon oil　碳油
carbon on regenerated catalyst　再生剂含炭量
carbon oxide　碳的氧化物
carbon pile　炭堆；碳柱
carbon plant　碳黑厂
carbon predominance index　碳优势指数
carbon preference index　碳优势指数
carbon production factor　产炭因素
carbon ratio　定碳比；碳同位素比
carbon refractory　碳质耐火材料
carbon rejection process　脱碳过程
carbon residue　炭渣
carbon resistance thermometer　碳电阻温度计
carbon restoration　复碳处理；复碳法
carbon rheostat　碳质变阻器
carbon rode　炭棒

carbon sequestration　碳封存；碳隔离；碳回收
carbon sequestration long term monitoring　碳储存长期监测
carbon shale　碳质页岩
carbon silicide　硅化碳
carbon sink　碳汇
carbon speciation　碳形态
carbon spot　金刚石中黑点
carbon stable isotopes　碳稳定同位素
carbon steel　碳钢；碳素钢
carbon steel coupons　碳钢测腐蚀挂片
carbon steel pipe　碳钢管
carbon steel pipe screen　碳钢筛管
carbon stock　碳库
carbon storage　碳存储；碳储量
carbon storage and sequestration　碳储存与隔离
carbon storage in soils　土壤中碳储存
carbon test　含碳量试验
carbon tetrachloride　四氯化碳
carbon tool steel　碳素工具钢
carbon tracer　碳同位素示踪剂
carbon transfer　碳转移
carbon tube furnace　碳管炉
carbon tungsten alloy　碳化钨合金
carbon turnover　碳周转
carbon unit factor　生炭的装置因素
carbon utilization efficiency　碳利用率
carbonaceous　含碳的；炭质的
carbonaceous adsorbent　碳质吸附剂
carbonaceous backfill　用碳质物回填
carbonaceous biochemical oxygen demand　碳生化需氧量
carbonaceous chondrite　碳质球粒陨石
carbonaceous coal　半无烟煤；贫煤
carbonaceous deposit　煤烟；碳渣；碳质沉积
carbonaceous foundry sand　碳质型砂
carbonaceous fragment　碳质碎屑
carbonaceous inclusion　碳质包裹体
carbonaceous material　含碳物料；碳质物料
carbonaceous matter　含碳物质
carbonaceous meteorites　含碳陨石
carbonaceous mudstone　钙质泥岩
carbonaceous oil shale　碳质油母页岩
carbonaceous particles　碳颗粒物
carbonaceous refuse　碳质废弃物；碳质矸石
carbonaceous rock　炭质岩；碳质岩
carbonaceous sandstone　碳质砂岩
carbonaceous sediment　碳质沉积
carbonaceous shale　石煤；碳质页岩；碳质页岩
carbonaceous slate　碳质板岩
carbonaceous substances　碳质物质
carbonado　黑金刚石
carbonado bit　黑金刚石钻头
carbonatation　碳酸盐化
carbonate　碳酸盐
carbonate accumulation　碳酸钙堆积
carbonate algal stromatolite　钙藻叠层石
carbonate analysis log　碳酸盐岩分析录井
carbonate apatite　碳酸磷灰石
carbonate aquifer　碳酸盐含水层
carbonate balance　碳酸平衡

carbonate bank	碳酸盐岩滩
carbonate breccia	碳酸盐岩角砾石
carbonate brine	碳酸盐型卤水
carbonate buildup	碳酸盐建隆；碳酸盐岩隆
carbonate buildup seismic reflection configuration	碳酸盐岩隆地震反射结构
carbonate caprock	碳酸盐岩类盖层
carbonate carbon dioxide	碳酸盐二氧化碳
carbonate cement	碳酸盐胶结物
carbonate chemistry	碳酸盐化学
carbonate compensation depth	碳酸盐补偿深度
carbonate compensation level	碳酸盐补偿面
carbonate concretion	碳酸盐结核
carbonate conglomerate	碳酸盐岩砾岩
carbonate content	碳酸盐含量
carbonate critical depth	碳酸盐临界深度
carbonate cycle	碳酸盐旋回；碳酸盐循环
carbonate deposition	碳酸盐沉积
carbonate deposits	碳酸盐质沉积
carbonate diagenesis	碳酸盐成岩作用
carbonate dissolution	碳酸盐溶解作用
carbonate dunes	碳酸盐沙丘
carbonate equilibrium	碳酸盐平衡
carbonate facies	碳酸盐岩相
carbonate facies iron formation	碳酸盐相铁建造
carbonate formation	碳酸盐岩地层
carbonate hydroxylapatite	碳羟磷灰石
carbonate iron	碳酸铁
carbonate islands	碳酸盐群岛
carbonate lake	碳酸盐湖
carbonate line	碳酸盐线
carbonate looping	碳酸盐循环
carbonate lump	碳酸盐块
carbonate lysocline	碳酸盐溶跃面
carbonate mass	碳酸盐岩体
carbonate mineral	碳酸盐矿物
carbonate mound	碳酸盐丘；碳酸盐土堆
carbonate mud	碳酸盐泥
carbonate mud mound	碳酸盐泥丘
carbonate nodules	碳酸盐结核
carbonate of lime	石灰岩；碳酸钙
carbonate of seamount	海山碳酸盐台地
carbonate ooze	碳酸盐软泥
carbonate petroleum accumulation	碳酸盐岩油气藏
carbonate petroleum deposit	碳酸盐岩油气藏
carbonate petroleum pool	碳酸盐岩油气藏
carbonate platform	碳酸盐岩台地
carbonate precipitation	碳酸盐沉淀
carbonate ramp	碳酸盐缓坡
carbonate ramp facies	碳酸盐缓坡相
carbonate reef	碳酸盐岩礁
carbonate reservoir	碳酸盐储油层；碳酸盐储层
carbonate reservoir rock	碳酸盐储油岩石
carbonate reservoir rock pore system	碳酸盐岩油层岩石孔隙系统
carbonate rise	碳酸盐岩隆
carbonate rock	碳酸盐岩
carbonate sealant	碳酸盐胶结物
carbonate sediment	碳酸盐沉积物
carbonate sedimentary facies	碳酸盐岩沉积相
carbonate shelf	碳酸盐陆架；碳酸盐岩陆棚
carbonate solution level	碳酸盐溶解层
carbonate sorption	碳酸盐吸附
carbonate stones	碳质岩
carbonate stratum	碳酸盐地层
carbonate tidal-flat sediment	碳酸盐潮坪沉积
carbonate water	碳酸水
carbonate-cemented sand	碳酸盐胶结砂岩
carbonate-compensation surface	碳酸盐补偿面
carbonated concrete	碳化混凝土
carbonated concrete thickness	碳化混凝土厚度
carbonated hardness of water	水的碳酸盐硬度
carbonated mortar	碳化砂浆
carbonated spring	碳酸泉
carbonated water	苏打水；碳酸水
carbonated water flooding	碳酸水驱油
carbonated water injection	碳酸水注入
carbonate-evaporite	碳酸盐蒸发岩
carbonate-fluorapatite	碳氟磷灰石
carbonate-free basis	无碳酸盐基础
carbonate-platform sediment	碳酸盐台地沉积
carbonates	碳酸盐
carbonate-type residuum	碳酸盐型风化壳
carbonation	碳化作用；碳酸饱和
carbonation depth	碳化深度
carbonation front of concrete	混凝土碳化表层
carbonation process	碳化过程
carbonatite	碳酸岩
carbonatite magma	火成碳酸盐岩浆
carbonatite weathering-crust-type deposit	碳酸岩风化壳型矿床
carbonatization	碳酸盐化作用
carbonator	碳酸化器
carbon-bearing	含碳的
carbon-burning capacity	烧焦能力
carbon-burning load	烧焦负荷
carbon-carbon bond	碳-碳键
carbon-carbon link	碳-碳键
carbondeposit	积碳；碳沉积；煤烟附着
carbon-dioxide flooding	二氧化碳驱油
carbon-dust resistor	碳末电阻
carbon-grain seismometer	传声地震计；碳粒地震计
carbonhydrate	碳水化合物
carbon-hydrogen link	碳-氢键
carbonic	石炭纪；石炭系；碳的
carbonic acid	碳酸
carbonic acid gas	含碳酸气
carbonic acid hardening	碳酸硬化法
carbonic limestone	石炭纪石灰岩；碳质石灰岩
carbonic oxide	一氧化碳
carbonic period	石炭纪
carbonic sandstone	碳质砂岩
carbonic snow	干冰
carbonide	碳化物
carbonide spring	碳酸泉
carboniferous	含碳的；石炭纪
carboniferous basin	石炭纪盆地
carboniferous glaciation	石炭纪冰期
carbonification	炭化作用；碳化酚
carbonification series	煤化系列

carbon-in-pulp 炭浆法
carbon-isotope geochemistry 碳同位素地球化学
carbonite 碳质炸药；天然焦炭
carbonitride 碳氮化物
carbonitriding 气体氰化；碳氮共渗
carbonium 阳碳；正电离子
carbonization 碳化酌；碳化作用
carbonization characteristic temperature 焦化特征温度
carbonization degree 碳化程度
carbonization induration 碳化固结
carbonization of coal 煤的碳化
carbonization process 碳化过程
carbonization theory 碳化说；碳化论
carbonize 碳化
carbonized briouette 焦化团煤；碳化团煤
carbonized carbonization 碳化作用
carbonized plant tissue 碳化植物组织
carbonizer 碳化剂；碳化器
carbonizing 渗碳；增碳过程
carbonnitriding 碳氮共渗
carbon-nitrogen cycle 碳-氮循环
carbon-nitrogen ratio 碳-氮比
carbonolite 碳质岩
carbonolith 碳质沉积岩
carbonolyte 碳质岩
carbonometer 碳酸计
carbonous 含碳的
carbon-oxygen log 碳氧比测井
carbon-oxygen ratio log 碳氧比测井
carbon-steel bit 碳素钢截齿；碳素钢钻头
carbon-to-carbon linkage 碳-碳链合
carbon-to-carbon rupture 碳-碳键断裂
carbophyre 碳质火成岩
carbopolyminerite 微复合质煤
carbopyrite 碳黄铁矿
carborne detector 车载探测器
carborne radioactivity survey 自动放射性测量
carborundum 碳化硅
carborundum paper 砂纸
carborundum paste 金刚砂研磨膏
carborundum stone 金刚石
carborundum wheel 金刚砂轮
carbosilicate 含碳硅酸盐；碳硅质
carbosilicite 微硅质煤
carbothermy 碳热法
carboy 酸瓶；玻璃瓶
carbuncle 红玉
carburan 铀铅沥青
carburane 碳铀矿
carburant 增碳剂
carburated hydrogen 矿井气；沼气
carburation 渗碳作用
carburator 渗碳器
carburator detergent 汽化器清净剂
carburator engine 汽化器式发动机
carburet 碳化物；与碳化合；增碳
carbureted gas 加烃煤气
carbureted water gas 加烃水煤气
carburetion 汽化；增碳
carburetor engine 汽化器式发动机

carburetted water gas 增碳水煤气
carburization 渗碳作用
carburization of coal 煤的高温干馏；煤炭焦化
carburized layer 渗碳层
carburizing 碳化
carburizing furnace 渗碳炉
carburizing sintering 碳化烧结
carburizing steel 渗碳钢
carcase work 预埋工程；骨架工程
carcass 框架；房屋骨架；外壳；龙骨
carcassing works 毛体工程；预埋工程；主体工程
carclazyte 白土；陶土；高岭土
car-cleaning device 矿车清扫装置；清车器
card board drain 排水板带；排水纸板
card board drain method 排水板法
cardan 万向接头
cardan axis 卡登轴；万向节轴
cardan axle 铰接轴；万向轴
cardan gear 万向接头传动装置
cardan joint 卡登接头；通用接头；万向接头
cardan shaft 万向轴
cardan spider 万向十字形接头
cardboard wick 排水纸板
cardhouse structure 盒式构造；片架构造
cardinal angle 主角
cardinal axis 主轴
cardinal line 主线
cardinal point 基点；方位基点
cardinal rule 基本原则
cardinal theorem 基本定理；取样定理
cardox blaster 二氧化碳爆破筒
cardump 翻车机
careen 倾倒
careless working 滥采
carex peat 苔草泥炭
cargo crane 船货起重机
cargo handling system 货物装卸系统
cargo lift 载货升降机
car-gravity road 缓坡巷道；矿车自重滑行道
ca-rich protodolomite 富钙原白云岩
carina 龙骨
carinate 脊状的
carinate anticline 脊状背斜
carinate fold 脊状褶皱
carinate syncline 脊状向斜
caring power 承载能力
Carinolithus 卡里农颗石
carload 车载量；满载量
car-loading control 装车控制
carlosite 柱星叶石
Carlsberg ridge 卡尔斯伯格洋脊
Carlson compass 卡尔逊罗盘
Carlson resistance gauge 卡尔逊电阻计
Carlson sensor 卡尔逊传感器
Carlson strain gauge 卡尔逊应变计
Carlson strain meter 卡尔逊应变计
Carlson-type piezometer 卡尔逊式渗压计
Carlton chock 卡尔顿型自移垛式支架
Carlton coupler 卡尔顿车钩
Carlton joint 卡尔顿管接头

Carmeloite	伊丁玄武岩
carmine	洋红
carminite	砷铅铁矿
carmor hollow drill steel	卡莫空心钻钢；碳化钨空心钻钢
carn	堆石标记
carnallite	光卤石
carnallitite	杂光卤石岩；杂盐
carnallitolite	光卤石岩
carnat	珍珠陶土
carnauba wax	巴西棕榈蜡；加诺巴蜡
carnegieite	三斜霞石
carnelian	光玉髓；红玉髓
carneule	角砾状白云岩
carniole	多孔白云岩；角砾状白云岩
Carnot cycle	卡诺循环
carnotite	钒钾铀矿；钾钒铀矿
carousel	固盘传送带
carousel buoy-floating barge storage	转盘式浮标驳船贮舱
carpathite	黄地蜡
carpet	地毯；路面；毯状粗糙织物
carpet form structure	毡状构造
carpholite	纤锰柱石
carphosiderite	草黄铁矾
carpolite	石果
carpolithus	化石果；种子化石
Carr bit	卡尔钻头；冲击式钻头
carriage	车辆；承重装置；楼梯搁栅
carriage arbor	铣刀架
carriage bolt	车架螺栓
carriage control	托架控制
carriage feed	钻床钻头进给器
carriage jack	车辆起重器
carriage mounting	钻机架
carriage slide track and slide frame	大车滑轨和滑架
carriage wrench	套筒扳手
carrier	载体；运送者；支座托架
carrier amplifier	载波放大器；载频放大器
carrier amplitude regulation	载波幅度调整
carrier bar	承载环
carrier bed	含油层；传导层；运矿层
carrier bucket	斗式运输机
carrier cable	承重索
carrier canal	供水干渠
carrier channel	载波信道
carrier gas	运载气体
carrier leak	载漏
carrier material	运载物质
carrier mobility	载体迁移率
carrier oil	携砂油
carrier oscillator	载波振荡器
carrier phase measurement	载波相位测量
carrier pipe	输送管
carrier ring	导环；垫圈
carrier rock	输导岩
carrier solvent	载体溶剂
carrier stop	导向架停止器
carrier supply system	载供系统
carrier synchronization	载波同步
carrier viscosity breakback	携砂液降黏
carrier voltage	载波电压
carrier wire	载波电缆
carrier-current transducer seismograph	载波电流换能器地震仪
carrier-to-noise ratio	载波噪声比
carrockite	卡洛克岩
carrollite	硫铜钴矿
carrot	杵体；套管或衬管的碎片
carrot-free	无杵体；无毛刺的
carrotless charge	不填塞的炸药包；无杵堵射孔弹
carrot-shaped hole	胡萝卜形井眼
carryall	轮式铲运机
carryall scraper	轮式铲运机
carry-in terminal	输入端
carrying agent	运载介质
carrying area	承压面积
carrying bar	顶梁
carrying belt	传送带；承载胶带
carrying block	承重滑车
carrying cable	承重索；运输索
carrying capacity	承载能力；负荷能力；载运能力
carrying capacity of pipe	管道泄水能力；管道过水能力
carrying fluid	携砂液
carrying frame	托架；支承架
carrying member	载送构件
carrying pan	泥浆槽
carrying plate	承压板
carrying power	承载能力；泄水能力
carrying roller	移动滚柱
carrying rope	承载索；吊重绳
carrying scraper	铲运机
carrying sheave	承载滑轮
carrying shoe	承载靴
carrying tongs	搬运钳
carrying wall	承压墙；承重墙
carry-out terminal	输出端
carry-over	带出；携带
carry-over efficiency	承转效率
carry-over factor	传递系数
carry-over flow	前期降雨径流
carryover storage	多年调节水库库容
carryover storage plant	多年调节电站
carryover storage reservoir	多年调节水库
carse	冲积平原
carst	喀斯特；岩溶
carstification	喀斯特作用；岩溶作用
carstone	含铁砾质砂岩
Cartesian axes	笛卡儿坐标轴；直角坐标轴
Cartesian coordinate	直角坐标
Cartesian paper	方格纸
Cartesian plot	直角坐标曲线图
Cartesian tensor	笛卡尔张量
Cartload	车载量
Cartogram	统计图
cartographic	地图的；制图的
cartographic expression	地图表示法
cartographic projection	地图投影；制图学投影
cartographic surveying	地形测量
cartographic symbolization	地图图例
cartography	地图学；制图法
cartology	地图学

cartridge brass	弹壳黄铜
cartridge clip	套夹
cartridge count	药包量
cartridge coupler	药卷连接装置
cartridge density	炸药卷密度
cartridge fuse	保险丝管；管形熔断器
cartridge strength	药卷强度
cartridge tube	装药筒
cartridge type respirator	滤罐型呼吸器；筒型防毒面具
carve out a lot	分割地段
carvoeira	巴西电英岩
car-washing plant	矿车洗涤装置
caryinite	砷锰铅矿
caryocerite	褐稀土矿
caryopilite	肾硅锰矿
Casagrande drive-in piezometer	卡萨格兰德打入式测压计
Casagrande liquid limit apparatus	卡萨格兰德液限仪；碟式液限仪
Casagrande method for liquid limit	卡萨格兰德液限测定法
Casagrande osmometer	卡萨格兰德渗压计
Casagrande piezometer tip	卡萨格兰德测压头
Casagrande plasticity chart	卡萨格兰德塑性图
Casagrande soil classifcation system	卡萨格兰德土分类法
Casagrande system	卡萨格兰德法
Casagrande's soil classification	卡萨格兰德土分类法
cascade	串联；级联；层叠；小瀑布
cascade aeration	梯级曝气法
cascade amplification	级联放大；梯级放大
cascade battery	级联电池组
cascade chute	分段式溜槽
cascade connection	串联
cascade control	串级控制
cascade dams	梯级坝
cascade development	梯级开发
cascade fold	叠褶；滑落褶皱
cascade hydroelectric station	梯级水电站
cascade hydropower station	梯级水电站
cascade machine	梯流式浮选机
cascade method	逐级测量法
cascade migration	级联偏移
cascade mill	梯落式辊磨机
cascade planning	梯级规划
cascade portion	急流段
cascade power stations	梯级电站
cascade protection	分级保护
cascade reservoirs	梯级水库
cascade ring packing	不锈钢阶梯环填料
cascade sampler	梯落式取样机
cascade shower	级联簇射
cascade zone	梯落带
cascaded carry	串联进位；逐位进位
cascade-type dryer	梯流式干燥机
cascadia land	喀斯喀迪古陆
cascadian revolution	喀斯喀迪运动
cascading	串级；串联
cascading glacier	瀑布冰川
cascading method	串级连接法
cascading of coal	地压破煤；自动落煤
cascading speed	梯落速度
cascading system	梯流系统
cascadite	橄辉云煌岩
cascajo	含金砂
cascode	共源共栅
case	外壳；箱盒；下套管；情况；事件
case adapter	套管异型接头
case bay	梁间距
case depth	渗碳层深度；表面深度；硬化层深度
case hardened glass	钢化玻璃
case hardening	表面渗碳硬化；岩石灌浆
case hardening box	渗碳箱
case in	下套管
case off	套管隔开；下管入井
case pile	套管桩
case-based design	基于实例的设计
cased	下套管的；加套的；装箱的；装在外壳内的
cased and perforated productivity	下套管射孔井的产能
cased beam	箱形梁
cased bore	安上套管的钻孔
cased bore pile	套管钻孔桩
cased bored pile	套管钻孔桩
cased borehole	下套管的井
cased column	箱形柱
cased concrete pile	套管混凝土桩
cased depth	下套管深度
cased frame	箱形框
cased hole	套管井；下套管井
cased hole completion	下套管完井
cased hole fishing	已下套管井的打捞
cased hole gravel packing	管内砾石充填
cased off	套管隔住的；用套管保护
cased pile	钢壳混凝土桩；套管桩；沉箱桩
cased seal	带壳密封；骨架密封
cased through well	全部下套管井
cased well	下套管井
cased-hole formation tester	套管井地层测试器
cased-hole log	套管井测井
cased-in pile	带套桩
cased-muff coupling	刚性联轴节；套筒联轴节
case-hardened	表面渗碳硬化的
case-hardened face	渗碳表面
case-hardened steel	表面渗碳硬化钢
case-hardening steel	渗碳钢
casein glue	酪素胶
caseless shaped charge	无壳聚能弹
casement	空型
caser	大套管的专业队
cash	煤层中的软夹层；软页岩
cash joint	平齐接缝
casing	井框支架；套管；外壳
casing accessories	套管附件
casing adapter	套管接头；套管异径接头
casing anchor	套管锚
casing anchor packer	套管锚定封隔器
casing and cementing	固井工程
casing annulus	套管环空
casing appliances	下套管工具
casing barrel	套管筒
casing bit	打套管钻孔用的钻头
casing block	套管用滑车组；下套管滑车
casing bowl	套管打捞筒

casing box	砂箱	casing head	螺旋管塞；套管头
casing bridge plug	套管桥塞	casing head gas	套管头气体
casing buckling	套管弯曲	casing head gasoline	天然汽油
casing calipering	套管内径检测	casing hold-down ring	套管压紧环
casing cap	堵头；管堵；螺旋管塞	casing hole-boring method	套管护壁钻孔法
casing catcher	套管防坠器	casing hook	套管大钩；下套管钩
casing cement	固井水泥；套管固结	casing imperfection	套管缺陷
casing cementing job	套管注水泥作业	casing injection	套管注浆
casing centralizer	套管扶正器	casing inspection	套管探伤
casing check valve	套管止回阀	casing inspection log	套管检查测井
casing clamp	套管夹子；套管卡子	casing installation	下套管；钻井套管下放
casing clean-out operation	管内冲砂作业	casing jack	套管千斤顶
casing clearance	套管柱外环隙	casing jar hammer	拔套管撞锤
casing collapse	套管挤坏；套管压裂	casing job	下套管作业
casing collar	套管接触；套管接箍	casing joint	套管接箍；套管接头
casing collar kick	磁力钻孔测深器	casing knife	切管刀；套管割刀
casing collar locator	套管接箍定位器	casing landing	套管联顶
casing collar log	套管接箍位置测井	casing leak	套管泄漏
casing collection system	套管气收集系统	casing lifter	套管提升器
casing column	套管柱	casing lifting equipment	套管提升设备
casing connection	套管连接	casing line	套管钢丝绳；下套管钢丝绳
casing corrosion monitoring	套管腐蚀监测	casing mandrel	套管整形器
casing coupling	套管接箍；套管接头	casing method of cementing	套管灌浆法
casing coupling locator	套管接箍定位器	casing milling tool	套管铣磨工具
casing coupling log	套管接箍测井	casing off	下套管封隔
casing cutter	套管割刀；套管内切刀；套管切割器	casing oil saver	套管顶部临时出油装置
casing cutter jar	套管割刀震击器	casing oscillator	扭动式套管驱动装置
casing cutter sinker	套管割刀加力器	casing pack	管内充填
casing cutter wedge	套管割刀楔	casing packer	套管封隔器
casing damage	套管损伤	casing patch tool	套管补孔器
casing deformation	套管变形	casing penetration	套管深度
casing depth	套管深度	casing perforation	套管冲孔；套管孔眼
casing disk	套管盘	casing perforator	套管冲孔器；套管射孔器
casing dog	套管打捞矛；捞管器	casing pipe	井壁管；套管
casing down drilling	套管钻进	casing pipe in borehole	套管；钻孔套管
casing drift swage	套管整形器	casing point	下套管深度
casing drilling in	下套管	casing pole	套管紧扣杆
casing drive adapter	套管传动联接器	casing pressure	套管压力
casing drive hammer	套管打入锤	casing procedure drilling	套管钻进
casing drive head	套管帽；套管撞击头	casing profile	套管截面
casing drive shoe	套管靴	casing program	安装套管程序；井身结构
casing driving device	套管驱动装置	casing protection of hole	套管护孔
casing elevator	套管吊卡；套管卡；套管式提升机	casing protector	套管保护圈；套管护箍
casing failure	套管断裂	casing puller	拔套管机；套管起拔器
casing fishing tap	套管打捞公锥	casing pump	套管泵
casing fitting	套管配件	casing rack	套管架
casing float	套管浮箍	casing repair	套管修复
casing float collar	套管浮箍	casing ripper	套管切割器
casing float plug	套管浮塞	casing roller	套管滚子；胀管器
casing float shoe	套管浮鞋	casing rotator	回转式套管驱动装置
casing float valve	套管浮阀	casing running	下套管
casing floating coupling	下套管漂浮接箍	casing salvaging	套管修补
casing fracturing	套管压裂	casing scraper	套管清刮器
casing gauge	套管丝扣规	casing screw head	套管母扣护丝
casing grab	套管打捞爪	casing section mill	套管铣鞋
casing grade	套管钢级	casing separator	套管隔板
casing grip	套管卡子	casing setting depth	套管下入深度
casing grouting	套管灌浆	casing severing	套管割断
casing guide shoe	套管引鞋	casing shoe	套管鞋；套管靴
casing gun	套管射孔器	casing side tracking	套管侧钻

casing size   套管尺寸
casing snubber   下套管导引工具
casing spear   套管打捞矛
casing spider   套管卡盘；辐射架
casing split   套管裂纹
casing splitter   剖管器；套管割刀
casing spool   大绳滚筒；套管防喷法兰短节
casing squib   套管爆炸装置
casing starter   最下一节套管；入井第一节套管
casing sticking   套管遇卡
casing string   套管柱
casing string design   套管柱设计
casing sub   套管异径接头
casing substitute   套管异径接头
casing suspender   套管悬挂器
casing swab   套管抽子
casing swage   套管整形器
casing swivel   套管旋转头
casing tap   打捞套管公锥
casing thickness detection   套管厚度检验
casing threads   套管螺纹
casing tongs   套管钳；套筒钳
casing tube   套管
casing underreaming system   套管扩孔系统
casing wall   套管壁
casing water swivel   套管进水转座联接器
casing wear   套管磨损
casing well   管井
casing-actuated intermitter   套管控制的间歇出油井
casing-bearing formation   下套管地层
casing-blow condensate   套管排放的凝析液
casinghead pressure   套管头压力
casinghead spool   套管头四通
casinghead squeeze   套管头挤水泥
casinghead top   套管头顶
casing-hole annulus   套管-井壁环空
casingless completion   无套管完井
casing-screw protector   套管公扣护丝
cask buoy   桶形浮标
caspian   干旱内陆盐水体
Cassagrande dot   卡萨格兰德孔隙压力测头
Cassagrande liquid limit apparatus   卡萨格兰德液限仪
Cassagrande's soil classification   卡萨格兰德土分类法
cassianite   卡申煤
Cassie's equivalent contact angle   卡西等效接触角
Cassini projection   卡西尼投影法
cassinite   钡正长石
cassiterite   锡石
Casson model   卡森模式
Casson yield stress   卡森屈服应力
Casson's viscosity   卡森黏度
cast   浇铸；井周声波扫描测井仪；抛掷；印模
cast aluminium   铸铝
cast basalt   铸玄武岩
cast basalt chute   铸石溜槽
cast bearing plate   铸铁轴承板
cast bit   孕镶式金刚石钻头；铸造钻头
cast blade   铸铁叶片
cast brass   生黄铜
cast coating   铸造包覆

cast concrete   浇灌混凝土；模铸混凝土
cast hard metal   硬质合金
cast iron   生铁；铸铁
cast iron ballast weight   铸铁压重
cast iron conductor   铸铁导管
cast iron electrode   铸铁焊支
cast iron fittings   铸铁接头配件
cast iron gear   铸铁齿轮
cast iron grill   铸铁笼
cast iron lining   铸铁井壁
cast iron pipe   铸铁管；生铁管
cast iron rainwater pipe   铸铁雨水管
cast iron segment   铸铁圆弧模板
cast iron waste pipe   铸铁废水管
cast matrix   钻头铸嵌基体
cast primer   散装起爆剂
cast product   铸造制品
cast set diamond bit   金刚石钻头
cast skin   铸件表层
cast spheroidal graphite iron   球墨铸铁
cast steel   铸钢
cast stone   铸石
cast temperature   浇筑温度
castability   可铸性
castable   可铸的
castable refractory   浇灌耐火材料；耐火浆料
castanite   褐铁矾
castaway slag   废弃渣
casted overburden   被抛掷的剥离物
castellated beam   堞型梁
castellated nut   槽形螺母；蝶形螺母
castellated shaft   花键轴
castellite   铝钛片晶石
cast-in anchorage   浇筑锚固
cast-in fixing bolt   预埋锚栓
cast-in-drilled-hole   钻孔灌注
cast-in-drilled-hole pile   钻孔灌注桩
casting   剥离倒堆；浇筑；抛掷；岩石搬运
casting basin   浇灌场；预制件工厂
casting bed   砂床
casting blast   抛掷爆破
casting bowl   砂箱；型箱
casting box   砂箱
casting cut   高卸式开采；上卸式采掘
casting cycle   浇筑期
casting die   金属铸模；压铸铸模
casting head   铸造冒口
casting mould   铸型
casting mould material   铸造材料
casting of rocks   矸石倒堆；岩石搬运
casting pattern   铸造木模；铸造模型
casting slip   浇筑用泥浆；铸型涂料
casting surface defect   铸件表面缺陷
casting yard   浇制场地；砂床；预制件工厂
casting-on   浇补
castings fecal pellets   粪化石
casting-up   铸型浇注
cast-in-place   就地灌注；现场浇筑；现场预制
cast-in-place anchor   现浇锚固
cast-in-place cantilever construction   悬臂灌注法

cast-in-place cantilever method 悬臂浇筑法
cast-in-place cased concrete pile 现浇混凝土管柱桩
cast-in-place concrete 现浇混凝土
cast-in-place concrete column 就地浇筑混凝土支柱
cast-in-place concrete foundation 灌注混凝土基础
cast-in-place concrete pile 就地灌注桩;混凝土灌注桩
cast-in-place concrete pile rig 灌注桩钻孔机
cast-in-place diaphragm wall 就地灌注地下连续墙
cast-in-place method 就地贯注法
cast-in-place mortar pile 灌注砂浆桩
cast-in-place pile 灌注桩;现浇桩
cast-in-place pile underpinning 灌注桩托换
cast-in-place reinforced concrete 现浇钢筋混凝土
cast-in-place shellless pile 就地灌注无壳桩
cast-insert bit 铸嵌式金刚石钻头
cast-in-site 现场浇捣
cast-in-site concrete 现浇混凝土
cast-in-situ 就地浇筑;现场浇筑
cast-in-situ bored pile 钻孔灌注桩
cast-in-situ concrete construction 现浇混凝土施工
cast-in-situ concrete lining 二次衬;内衬砌;永久衬
cast-in-situ concrete pile 就地灌注桩;现浇混凝土桩
cast-in-situ concrete structure 现浇混凝土结构
cast-in-situ concrete unit 现浇混凝土组件
cast-in-situ diaphragm wall 现浇防渗墙;现浇连续墙
cast-in-situ lining 就地浇注衬砌
cast-in-situ method 就地贯注法
cast-in-situ pile 灌注桩;现浇桩
cast-in-situ pile by excavation 挖孔桩
cast-iron flat-gate-type sluice 铸铁平板门式闸
cast-iron ribbed washer 铸铁肋垫圈
cast-iron valve 铸铁阀门
cast-iron washer 铸铁垫圈
castle nut 槽顶螺母;带顶槽的六角螺母
castolin 卡斯托林低熔铸铁合金
casual 不规则的;随机的
casual inspection 不定期检查;临时检查
casual pillar 临时矿柱;临时煤柱
casual sand 不规则砂岩层
cat 吊锚;驴蹄形绳结;起锚滑车;套管衰减厚度;硬耐火土
cat coal 含黄铁矿煤
cat dirt 含黄铁矿的煤;硬质耐火黏土
cat fall 吊锚索
cat gold 金色云母
cat ladder 便梯;爬梯;竖梯
catabiosis 老化现象
catabolism 分化酶;分解代谢
catabolite 降解产物
catacausis 自燃
cataclase 破碎
cataclasis 破裂作用;碎裂作用
cataclasite 碎裂岩;压碎岩
cataclasm 碎裂作用
cataclast 动力变质破裂碎屑
cataclastic 碎裂的;压碎的
cataclastic breccia 碎裂角砾岩
cataclastic cleavage 碎裂劈理
cataclastic conglomerate 碎裂砾石;压碎砾岩
cataclastic fabric 碎裂组构

cataclastic failure 碎裂破坏
cataclastic flow 碎裂流动
cataclastic fluidal texture 碎裂流动结构
cataclastic foliation 碎裂叶理
cataclastic lineation 碎裂线理
cataclastic matrix 碎基
cataclastic metamorphism 动力变质作用;碎裂变质作用
cataclastic rock 破碎岩石
cataclastic rudite 碎裂砾屑岩
cataclastic structure 碎裂构造;压碎构造
cataclastic texture 碎裂结构;压碎结构
cataclastic zone 碎裂带
cataclinal 下倾的
cataclinal river 顺向河
cataclinal stream 顺向河
cataclinal valley 顺斜谷
cataclysm 地震;洪水;灾变
cataclysm theory 灾变说
catadioptric apparatus 反折射器
catadromous migration 降海洄游;降河洄游
catafront 下滑峰
catagenesis 后生作用;后退演化;退化作用
catagenetic 后成作用的
catagenetic breccia 后生角砾岩
catagenetic gas 退化阶段成因气
catagenetic stage 深成热解阶段
catagenic 后成作用的;碎裂的
catagraphite 菌蚀孔道构造
catalan process 凯特兰熟铁锻炼法
catalastic breccia 破碎角砾岩
catalectic rock 碎裂岩
catalogue of abandoned mines 废矿图集
catalogue of slopes 斜坡记录册;斜坡目录
catalysant 被催化物
catalysate 催化产物
catalysed gasification 催化气化
catalysis 催化剂;催化酶
catalyst 触媒;催化剂
catalyst activity stability 催化剂活性安定性
catalyst aging 催化剂老化
catalyst carrier 催化剂载体
catalyst deactivation 催化剂失活
catalyst matrix 催化剂基体
catalyst microstructure 催化剂微观结构
catalyst oil ratio 催化剂油比
catalyst poisoning 催化剂中毒
catalyst reduction 催化剂还原
catalyst support 催化剂载体
catalytic action 催化酶;催化作用
catalytic agent 催化剂
catalytic combustion 催化燃烧
catalytic cracking 催化分解
catalytic decomposition 催化分解
catalytic dewaxing 催化脱蜡
catalytic distillation 催化蒸馏
catalytic exchange 催化交换
catalytic gas 催化气化
catalytic gas sensor 催化气体传感器
catalytic hydration 催化水合
catalytic hydropyrolysis 催化加氢热解

catalytic liquefaction　催化液化
catalytic methanometer　催化沼气测定仪
catalytic oxidation　催化氧化
catalytic polymerization　催化聚合
catalytic purifier　催化提纯器
catalytic reaction　催化反应
catalytic reflector　催化反射器
catalytic reforming　催化重整
catalytic scrubber　催化脱毒洗涤器
catalyzator　催化剂
catalyze　催化
catalyzed polymerization　催化聚合
catalyzer　催化剂
catamorphism　风化变质；破碎变质现象
cataphoresis　阳离子电泳
cataphorite　红钠闪石
catapleiite　钠锆石
catapleite　单斜钠锆石
cataracting speed　瀑泻速度
cataracting zone　瀑泻带
catastrophe　大灾难；灾变
catastrophe point　突变点
catastrophe set　突变集
catastrophe theory　突变理论
catastrophic　灾变的；灾难性的
catastrophic collapse　灾害性塌陷
catastrophic creep　致毁蠕变
catastrophic earthquake　灾害性地震
catastrophic error　灾变性错误
catastrophic event stratigraphy　灾变事件地层
catastrophic evolution　灾变演化
catastrophic failure　严重故障；灾难性破坏
catastrophic flood　特大洪水；灾害性洪水
catastrophic landslide　灾害性滑坡，灾害性山崩
catastrophic prediction　灾变预测
catastrophic theory　灾变论
catastrophism　灾变论
catastrophist　灾变论者
catathermal　降温期的
catawberite　滑石磁铁岩
catazone　高温变质带；深变质带
catch　捕捉；接收器
catch all　打捞工具；装杂物的器具
catch basin　汇水盆地；流域；截流井；沉泥井
catch bench　拦截台阶
catch bowl　集液器；收集箱
catch cam　凸轮夹持器；抓爪
catch collecting well　集水井
catch dam　节流堰
catch ditch　集水沟
catch drain　集水沟；截水沟
catch fan　防护栅；斜棚
catch feeder　灌溉沟
catch fence　拦截围墙
catch frame　挡泥板；格栅
catch gear　闭锁装置
catch gutter　排水渠
catch handle　键柄
catch hook　掣子爪；打捞钩
catch inlet　集水井

catch net　拦石网
catch pawl　挡爪
catch pin　带动销；挡杆
catch pit　沉砂井；汇水盆地；集水井；排水井；截流井；聚泥坑
catch plate　导夹盘
catch platform　坠台
catch plug　挡住螺栓
catch point　支撑点
catch pot　除凝析液和岩屑的罐；油气分离罐
catch prop　警戒木支柱
catch reversing gear　棘轮回动装置
catch scaffold　保护盘
catch sleeve　打捞筒
catch spring　挡簧
catch tray　集液盘；收集盘
catch wall　拦石挡墙
catch water basin　集水盆地；集水区
catch water channel　集水槽
catch water drain　汇水沟；集水沟；截水沟
catch-all steam separator　截液器
catcher　捕捉器；收集器；岩芯提断器；岩芯抓；制动装置
catcher sub　捕捉接头
catching　捕集；捕收；捕捉
catching device　捕捉器；阻车装置
catching piece　捕捉器；提引器
catchment　流域；储集范围；汇水盆地；集水区
catchment area　储油面积；含水层补给区；集水区；流域面积；汇水面积
catchment area survey　汇水面积测量
catchment basin　汇水盆地；汇水区；集水盆地；集水区；流域
catchment basin of debris flow　泥石流流域
catchment boundary　集水区边界；流域边界
catchment channel　截水渠；引水道
catchment development　流域扩展
catchment ditch　集水沟
catchment of dirty water　集中污水
catchment of sewage disposal works　污水处置设施引水区
catchment of underground water　地下水的汇集
catchment of water　汇水；集水；截水
catchment sewage system　流域排水系统
catchment tray　汇水槽；集水槽
catchment water channel　集水槽
catchment water drain　截水沟
catchment well work　集水井工程
catchment yield　流域出水量
catchwater　排水沟；截水沟
catchwork　集水工程
catchwork irrigation　漓漫灌溉
categories of reserve　储量等级
categorization of geotechnical engineering　岩土工程分级
categorization of geotechnical investigation　岩土工程勘察分级
categorization of geotechnical project　岩土工程分级
categorizer　分类器
category　范畴；分类
category index　分类索引
category of damage　震害类别

category of earthquake data	地震数据分类
category of earthquake-related data	地震数据分类
category of project security	工程安全等级
category of reserve	储量级别
category of site	场地土类别
category texture	晶簇状结构
catena soil	土链
catenary	垂曲线；吊线；接触网悬挂；悬链线
catenary action	悬链作用
catenary correction	垂曲改正
catenary flume	悬链形渡槽
catenary length	悬链线长
catenary suspension	链式悬吊；架空悬吊
catenary wire	悬链线；承力索
catenary wire chainette	悬链绳
catenary wire clip	单承力索吊弦线夹
catenary wire clip support	承力索线夹支持座
catenary wire electric connecting clamp	承力索电连接线夹
catenary wire support clamp	承力索双线支撑线夹；绝缘双线支撑线夹
catenation	链锁
catenulate	串珠状的
caterpillar	履带式牵引机
caterpillar bar	有履带装置的梁
caterpillar bulldozer	履带式推土机
caterpillar crane	履带式起重机
caterpillar drive	履带传动
caterpillar excavating machine	履带式挖掘机
caterpillar excavator	履带式挖掘机
caterpillar gate	履带式闸门
caterpillar guide	履带导承
caterpillar loader	履带式装载机
caterpillar loading	履带装载机装载
caterpillar machinery	履带式机械
caterpillar shovel	履带式机铲；履带式挖掘机
caterpillar track	导轨
caterpillar tractor	履带式拖拉机
caterpillar truck	履带式载重汽车
cathaysia	华夏古大陆
cathaysia massif	华夏地块
cathaysian structural system	华夏构造体系
cathaysian tectonic system	华夏构造体系
cathaysoid structure	华夏式构造
cathaysoid tectonic system	华夏构造体系
cathead	吊锚架；猫头
cathead line	猫头钢丝绳
cathead shaft	猫头轴
cathetometer	测高计；测高仪
cathode-ray tube refraction seismograph	阳极射线管折射地震仪
cathodic deposition	阴极淀积
cathodic protection	阴极防腐；阴极保护
cathysia	华夏古陆
cathysian	华夏式的
catilever erection	悬臂架设法
cation	正离子
cation exchange and adsorption	阳离子交替吸附作用
cation exchange capacity	阳离子交换量
cation exchanger	阳离子交换器
cation geothermometer	阳离子地温计
cationic	阳离子的
cationic emulsified bitumen	阳离子乳化沥青
cationic floatation	阳离子浮选
cationic polymer drilling fluid	阳离子聚合物钻井液
catline	猫头钢丝绳
catlinit	烟斗泥
cat-mounted	安装履带的
cat's silver	银云母
cat's-paw knot	钢丝绳拴钩结
catstep	滑坡阶地
cattermole process	凯特尔莫尔油浮选法
cattle-guard	铁丝网
catty	斤
catwalk	天桥
catwalks pea-gravel backfilling	回填豆砾石桥行通道；桥行通道
catwall	轻便栈道
catworks	辅助绞车
caucasite	歪长花岗岩
cauce	河床
Cauchy distribution	柯西分布
Cauchy elastic model	柯西弹性模型
Cauchy machine	柯西机
Cauchy stress	柯西应力
Cauchy-poisson relations	柯西-泊松关系
Cauchy's convergence criteria	柯西收敛准则
Cauchy's deformation tensor	柯西变形张量
Cauchy's mid-value theorem	柯西中值定理
Cauchy's residue theorem	留数定理
Cauchy-schwarz inequality	柯西-施瓦茨不等式
cauda	震尾
caudal fin	尾鳍
caul	填块
cauldron	锅形陷坑；锅状盆地；火山凹陷；陷落火山口
cauldron bottom	煤层顶板结核体
cauldron subsidence	锅形塌陷；火口沉陷
caulk	填缝；填密；填塞
caulk weld	封口焊缝
caulked joint	嵌缝
caulked rivet	嵌缝铆钉
caulked seam	嵌缝
caulker	敛缝锤；密缝凿
caulker's oakum	嵌缝麻絮
caulking	堵缝；堵塞；填隙
caulking bush	嵌缝衬套
caulking chisel	嵌缝凿
caulking compound	嵌缝填料；填隙料
caulking gun	堵缝枪；油灰枪；压胶枪
caulking hammer	填缝锤
caulking material	堵缝材料；敛缝材料；填隙料
caulking nut	嵌缝螺母
caulking piece	捻缝填隙片
caulking ring	垫圈；隔环；加固圈；密封圈
caulking set	嵌缝工具
caulking strip	嵌缝条
caulking tool	嵌缝工具；填缝凿
caulking-butt	填缝对接
caulorrhizous	茎状根的
caunche	岩层的采掘部分，挑顶；卧底；刷帮
caunter	交错矿脉

causal mechanism	发震机制
causality fault	发震断层
causative fault	发震断层
cause analysis	成因分析
cause of earthquake	地震成因
cause of formation	构造成因
cause of landslide formation	滑坡成因
cause-result analysis	因果分析
causes of rock bursts	岩爆起因
causeway	长堤；堤道；埋藏地垒
causse	灰岩高原，喀斯特地形
caustic	腐蚀性的；碱性的；焦散的
caustic alkalinity	苛性碱度
caustic ammonia	苛性氨
caustic ash	苛性苏打灰
caustic embrittlement	碱脆；碱蚀致脆
caustic filter washdown	碱性滤器冲洗废料
caustic flooding	碱水驱油
caustic lime	苛性石灰；生石灰
caustic lime mud	碱质淤泥；石灰泥浆
caustic metamorphism	腐蚀变质酌
caustic mud	石灰浆
caustic potash	氢氧化钾
caustic soda	烧碱
caustic solution	苛性碱溶液
caustic stain	碱性锈蚀
caustic washing	碱洗
caustic waterflooding	碱水驱
caustical	腐蚀性的
causticity	碱性
causticization	苛化作用
causticized lignite	苛性化褐煤
causticized starch	苛化淀粉
causticizer	苛化剂
caustics	苛性剂
caustobiolite	可燃性生物岩；可燃性有机岩
caustobiolith	可燃性生物岩；可燃性有机岩
caustolith	可燃岩
caustophytolith	可燃性植物岩
caustozoolith	可燃性动物岩
cauterization	烧灼
cautery	腐蚀
caution board	警告牌
caution sign	警告标志
cautionary zone	警戒区
cautious blasting	谨慎爆破
cavability	矿体可崩性
cavability index	落顶性指数
cavability number	落顶性指数
cavability of ore	矿石的塌落性
cavability of orebody	矿石可崩性
cavalorite	更长正长岩；钠钙斜长岩
cave	凹槽；洞穴；塌落；坍塌
cave biota	洞穴生物
cave biotic deposit	洞穴生物堆积
cave boundary	洞穴界面
cave breakdown	洞穴崩塌堆积
cave breccia	洞穴角砾岩
cave cavern	溶洞
cave collapse deposit	洞穴崩塌堆积
cave coral	洞穴珊瑚；石珊瑚
cave deposit	洞穴沉积
cave deposit chronology	洞穴沉积年代学
cave deposits	洞穴沉积物
cave dwelling	窑洞
cave dwelling site	洞穴遗址
cave erosion	洞穴侵蚀
cave exfoliation	洞穴鳞剥作用
cave filling	洞穴充填
cave flower	石膏花
cave formation	崩落地层；洞穴沉积层；洞穴矿床；有洞穴的岩层
cave ice	洞穴冰
cave in	冒顶；塌落；坍陷
cave in of roof	顶板陷落
cave in soil	土洞
cave inlet	地下河入口；溶洞入口
cave marble	洞穴华
cave microclimate	石窟微环境
cave onyx	洞穴石华
cave paintings	石窟壁画
cave passage	洞穴通道
cave roof	陷落顶
cave sediments	洞穴沉积
cave shield	石盾
cave soil	洞穴土
cave storage	地下储存
cave system	洞穴系统
cave temple	石窟；石窟寺
cave tomb	崖洞墓
cave-chamber tomb	土洞墓
cave-collapse breccia	洞穴塌陷角砾岩
caved	塌落的
caved arch	坍落拱
caved area	崩落区；落顶区
caved chamber	塌落的硐室
caved debris	崩落的废石；塌落的碎屑
caved goaf	垮落了的采空区；塌陷的采空区
caved ground	坍陷的地面；陷落地层；陷落地带
caved hole	有洞穴的井眼
caved material	崩落岩石；坍落体
caved stope	崩落回采
caved workings	冒顶工作区；坍塌巷道
caved zone	垮落带；冒落带
caveology	岩洞学，洞穴学
cavern	洞室；洞穴；岩洞；石窟
cavern breakdown	岩洞崩塌堆积
cavern breccia	洞穴角砾岩
cavern deposit	洞穴沉积
cavern design	洞室设计
cavern disposal	岩洞处置
cavern filling	洞穴充填；空洞回填
cavern floor	洞底
cavern flow	洞穴水流
cavern limestone	多孔石灰岩
cavern porosity	洞穴状孔隙；溶洞孔隙
cavern rock	多孔岩石；溶洞岩石
cavern spring	洞穴泉；溶洞泉
cavern storage	洞穴式贮存；洞穴储存
cavern storage tank	洞穴式贮存缸

cavern survey	洞穴调查；溶洞测量
cavern water	洞穴水；溶洞水
cavernicolous	穴栖的
caverning	洞穴性
cavernment	洞穴性
cavernment index	洞穴指数
cavernous	洞穴的；凹的
cavernous aquifer	洞穴状含水层
cavernous body	海绵体
cavernous dolomite	孔状白云岩
cavernous fissure zone	洞穴裂隙带
cavernous formation	洞穴性地层
cavernous ground	洞穴地基
cavernous limestone	孔状灰岩
cavernous marble	含溶洞大理石
cavernous opening	孔穴状缝隙
cavernous porosity	洞穴孔隙系统；孔洞孔隙度
cavernous rock	多孔岩石；孔状岩石
cavernous rock formation	多孔岩层；洞穴岩层
cavernous structure	洞穴状构造
cavernous terrain	孔洞地体
cavernous vein	孔穴矿脉
cavernous water	洞穴水；溶洞水
cavernous weathering	洞穴风化；孔状风化
cavi-jet	气蚀射流
caving	崩落开采法；落顶开采法；冒顶；塌落
caving and subsidence	垮落与沉陷
caving arch	冒落拱
caving bank	崩塌河岸；塌岸
caving block	崩落式采矿法
caving break-off line	陷落崩裂线
caving by raising	漏斗天井放矿崩落法
caving design of layouts	自然崩落采场结构设计
caving distance	放顶步距
caving formation	坍塌的岩层
caving ground	不稳定地层；塌陷地
caving ground condition	易塌孔地质条件
caving height	崩落高度
caving in	顶板冒落；冒顶；塌陷
caving interval	放顶距
caving line	放顶线；冒落线；坍落线
caving method	垮落法；崩落开采法
caving mining method	崩落采矿法
caving of overlapping	重叠放
caving operation	崩落作业
caving pressure	崩落压力；坍井压力
caving processes	放顶煤工艺
caving sequence	放煤顺序
caving shale	坍塌的页岩
caving shield	掩护梁；掩护支架
caving space	放顶步距
caving space interval	放顶距；落顶距离
caving stoping method	崩落采矿法
caving system	崩落开采法
caving the roof	放顶
caving thickness	放顶煤厚度
caving time interval	垮落时间间隔
caving unit	放顶区段；放顶组
caving without breaker props	无特种柱放顶
caving without specific props	无特种柱放顶
caving zone	垮落带；冒落带；塌落带
caving-height ratio	垮采比
cavings	坍塌的落石
cavitating	气穴作用
cavitation	成洞；空化作用；气蚀作用；气穴现象
cavitation aeration system	空蚀曝气系统
cavitation character	空穴特性；气蚀特性
cavitation coefficient	空蚀系数
cavitation corrosion	空蚀；气蚀
cavitation criterion	气蚀标准
cavitation damage	混凝土表面气穴损害；空化剥蚀；气蚀损坏
cavitation degree	空蚀度
cavitation erosion	空蚀；气蚀
cavitation jet	空穴射流
cavitation metre	气蚀仪
cavitation noise	气蚀噪声
cavitation range	气蚀区域
cavitation resistance	抗气蚀性
cavitation test	气蚀试验
cavitation-resistant alloy	抗气蚀合金
cavity	洞室；空洞；冒顶；溶洞
cavity area	冒落洞穴面积
cavity block	空心块体
cavity caused by the extraction	采空区
cavity charge	空穴装药；锥孔装药
cavity effect	空穴效应
cavity expansion theory	空穴膨胀理论；扩孔理论
cavity filling	洞隙充填；空穴填料
cavity filling deposit	洞穴填充矿床
cavity flow	空腔水流
cavity flushing	冲洗洞壁
cavity fracture	空隙断口
cavity grouting	洞穴灌浆；空腔灌浆
cavity in the stowing	充填死角；充填体中的空洞
cavity radius	孔道半径
cavity theory	空穴理论
cavity wall	空斗墙；空心墙；洞穴井；空穴井
cavity working	空洞开挖
cavity-filled ore deposit	洞穴填塞矿床
cavy	易坍塌的
cay	沙礁；沙洲；珊瑚礁
cay rock	礁岛岩
cay sandstone	礁岛砂岩
cayman trench	开曼海沟
ceag detector	西格型瓦斯探测器
cebollite	纤维石
cecilite	黄长白榴岩
cedarite	松脂石
cedar-tree laccolite	杉木型岩盖
cedricite	透辉白榴岩
cegamite	水锌矿
ceiling	顶板
ceiling board	舱顶板
ceiling channel	天沟
ceiling coal layer	护顶煤层
ceiling duct	屋顶通气管
ceiling girder	顶梁；排梁
ceiling height	室内净高
ceiling hole	顶板炮眼；上向炮眼

ceiling joist	吊顶龙骨；天花板搁栅	cellular construction	多孔构造；蜂窝式构造；格形构造
ceiling load	平顶载重	cellular core wall	格形岸壁；箱格形心墙
ceiling margin	上极限	cellular dam	格箱式坝
ceiling panel	平顶镶板；天花板	cellular dolomite	多孔状白云岩
ceiling pocket	天锅	cellular girder	格形梁；空心梁
ceiling river	天河；悬河	cellular glass	泡沫玻璃
ceiling slab	顶板；天花板	cellular gravity dam	格箱式重力坝
ceiling stick	单杠挂钩短梯	cellular plastic	泡沫塑料
ceiling suspension hook	天花吊钩	cellular porosity	蜂窝状孔隙
ceiling suspension mount	悬吊式天花板	cellular quartz	多孔石英
ceiling without trussing	无梁平顶	cellular quaywall	圆墩岸壁
ceilometer	测云仪；云高计	cellular raft	格状筏基
celadon green	灰绿色	cellular retaining wall	格孔式挡土墙
celadonite	绿鳞石	cellular sheetpile breakwater	格型板桩围堰防浪堤
celestial body	天体	cellular soil	蜂窝状土壤
celestial coordinate system	天球坐标系	cellular steel sheet pile	格状钢板桩
celestial equator	天球赤道	cellular structure	多孔状构造；蜂窝状结构；网格结构；细胞状结构
celestial horizon	真地平；真水平线		
celestine	天青石	cellular texture	网格组织；细胞状结构
cell	传感器；浮选机	cellular-bulkhead type quaywall	格排式岸壁码头
cell action factor	盒作用因数	cellular-type cofferdam	蜂窝式围堰
cell concrete	多孔混凝土	celsian	钡长石
cell pressure	室压；围压	celsius thermometer	摄氏温度计
cell quarry wall	格式采石场墙	celtium	铪
cell quartz	多孔石英	cement	水泥
cell quay wall	格式岸壁	cement accelerator	水泥速凝剂
cell sampling	格子采样	cement activity	水泥活性
cell state	元胞状态	cement additive	水泥浆外加剂
cell stiffness	细胞刚度	cement aggregate bond	水泥-集料黏结
cell structure	多孔状构造；蜂窝状构造；细胞状构造	cement aggregate ratio	水泥骨料比
		cement and sand cushion	水泥拌砂垫层
cell test	压力室试验	cement and sand screed	水泥砂浆底层
cell texture	多孔状结构；细胞状结构	cement asbestos board	水泥石棉板
cell thickness	堆层厚度	cement asbestos pipe	石棉水泥管
cellar	地窖；地下室	cement bacillus	水泥杆菌
cellar connection	井口装置	cement baffle collar	水泥塞挡圈
cellar control gate	井口控制阀门	cement bag opener	水泥开包机
cellar cover	圆井盖	cement basket	灌浆筒；水泥伞
cellar deck	下部甲板	cement bentonite grout	水泥膨润土浆液
cellar flooding	单井注采地窖油	cement bentonite milk	膨润土掺水泥乳浆
cellar floor	底层地板	cement bentonite pellet	水泥膨润土球
cellar hole	地下坑洞	cement binder	黏合剂
cellar hook up	钻井井口装置	cement block	混凝土砌块；水泥砌块
cellar oil	地窖油	cement blower	水泥浆喷枪
cellar stone	碎石	cement bond log	水泥胶结测井
cell-less bucket wheel	无格式斗轮	cement bonded concrete overlay	水泥黏结混凝土罩面层
cell-type bucket wheel	有格式斗轮	cement bound granular base	水泥胶结碎石基底
cellular	多孔状的；蜂窝状的；格子状的；细胞状的	cement brand	水泥标号；水泥牌号
		cement brick	水泥砖
cellular abutment	格形桥台；框格式边墩	cement brick pavement	水泥砖铺面
cellular automata	细胞自动机；元胞自动机	cement briquette	水泥标准试块
cellular automaton evolution	细胞自动机演化	cement bunker	水泥仓
cellular bauxite	多孔状铝土矿	cement burn	水泥烧伤
cellular beam	格形梁	cement cap	悬空水泥塞
cellular breakwater	格构式防波堤	cement carbide tool	硬质合金工具
cellular brick	多孔砖	cement carbon	渗碳
cellular bulkhead	格排式支护；格型岸壁；格型堤岸	cement casing head	水泥头
cellular buttress	格型支墩；格型支墩	cement casing shoe	注水泥用套管鞋
cellular caisson	格式沉箱；框格式沉箱	cement channelling	水泥窜槽
cellular cofferdam	格形围堰	cement clay grout	水泥黏土浆
cellular concrete	多孔混凝土；加气混凝土		
cellular concrete block	格排式混凝土块		

cement clinker	水泥熔渣；水泥熟料
cement coating	水泥盖层；水泥抹面
cement colours	水泥染料
cement column	水泥浆液柱
cement compounds	水泥复合物
cement concrete	水泥混凝土
cement concrete mark	水泥混凝土标号
cement concrete mixture	水泥混凝土混合料
cement concrete pavement	水泥混凝土路面
cement consistency	水泥浆稠度
cement contamination	水泥侵
cement content	水泥含量；洋灰含量
cement conveyer	水泥输送机
cement copper	沉积铜；渗碳铜
cement covering	水泥护面；水泥罩面
cement curing time	水泥候凝时间
cement cut	水泥侵
cement deep mixing	水泥土搅拌法
cement deep mixing composite foundation	深层搅拌桩复合地基
cement deep mixing method	水泥深层搅拌法
cement degradation	水泥降解
cement dehydration	水泥浆脱水
cement deliver truck	散装水泥运输车
cement deposit	胶结沉积物；水泥岩层
cement displacement	水泥替置
cement dressing	水泥粉刷；水泥敷面
cement dump	在井底倾倒水泥浆的筒；注浆泵
cement equipment	注水泥设备
cement evaluation log	水泥评价测井
cement evaluation tool	水泥评价测井仪
cement facing	水泥敷面；水泥护面
cement factor	水泥系数
cement family coating material	水泥系涂料
cement fibrolite plate	水泥纤维板
cement filtrate	水泥滤液
cement fineness	水泥细度
cement finish	水泥罩面
cement flag pavement	水泥板路面
cement float collar	注水泥浮箍
cement float shoe	注水泥浮鞋
cement floor	水泥地板；水泥地面
cement flour	细粉水泥
cement flow	水泥浆液流
cement fly-ash gravel pile	水泥粉煤灰碎石桩
cement flyash grout	水泥粉煤灰浆
cement fondu	矾土水泥；高铝水泥
cement gel	水泥凝胶
cement gold	沉淀金
cement grade	水泥标号
cement grain	水泥颗粒
cement gravel	黏接砾石
cement grit	粗粒水泥
cement grout	水泥薄浆；水泥浆；英泥油
cement grout filler	水泥灌浆填料
cement grouted bolt	水泥注浆锚杆
cement grouted zone	水泥灌浆区
cement grouter	水泥灌浆机
cement grouting	水泥灌浆
cement grouting process	水泥灌浆法
cement grouting technology	水泥灌浆技术
cement gun	水泥浆喷枪
cement hardener	水泥速凝剂；水泥早强剂
cement hydration	水泥水化作用
cement in bulk	散装水泥
cement injection	水泥灌浆；水泥注浆
cement injector	水泥灌浆机
cement jet	水泥喷枪
cement joggle	水泥接榫
cement joint	水泥接缝；水泥接头
cement kiln	水泥窑
cement limestone	水泥用石灰岩
cement lined ditch	水泥砂浆衬砌沟渠
cement lining	水泥衬砌
cement manufacture	水泥制品；水泥制造
cement mark	水泥标号
cement matrix	水泥浆
cement milk	水泥浆
cement mill	水泥厂
cement minerals	水泥矿料
cement mixer	水泥搅拌器
cement mixing pile	水泥搅拌桩
cement mixture	水泥混合物
cement modified soil	水泥改良土
cement mortar	水泥灰浆；水泥砂浆
cement mortar canal lining	水泥砂浆渠道衬砌
cement mortar cushion	水泥砂浆垫层
cement mortar mark	水泥砂浆标号
cement mortar mixing machine	水泥净浆搅拌机
cement mortar plastering	砂浆抹平
cement mortar rubble	水泥砂浆砌片石
cement mortar surface	水泥砂浆地面
cement needle	水泥硬度检验计；水泥凝固检验针
cement packer	水泥封隔器
cement paint	水泥涂料；雪花英泥；水泥涂料
cement paint finish	水泥涂面
cement particle	胶结颗粒
cement paste	水泥浆
cement paving	水泥砖地面
cement penetration	水泥贯透
cement pipe	水泥管
cement pit	水泥池
cement placement	水泥充填
cement plaster	水泥灰浆；水泥抹面
cement plaster finish	水泥灰粉饰面
cement plastometer	水泥稠度仪
cement plug	水泥塞
cement product	水泥制品
cement pump	水泥输送泵；水泥输浆泵
cement putty	水泥灰泥；水泥油灰
cement ratio	水泥骨料比
cement reef	胶结物生物礁
cement rendering	水泥抹面
cement rendering dado	水泥护壁
cement replacement material	水泥代用材料
cement retainer	灌浆不透水层；水泥挡墙
cement retarder	水泥缓凝剂
cement rock	胶结岩；水泥岩
cement sampler	水泥取样器
cement sand bed	水泥砂浆层

cement sand cushion	水泥砂浆垫层
cement sand grout	水泥砂浆
cement sand mix	水泥砂浆
cement sand moulding	水泥砂造型
cement sand plaster	水泥砂浆抹面
cement sand ratio	灰砂比
cement sand rendering	水泥砂浆抹面
cement sandfill	胶结充填砂
cement scale	水泥垢
cement scouring	水泥侵蚀
cement screed	水泥砂浆地面
cement screeding flooring	水泥批荡地面
cement screw feeder	螺旋式水泥进料器
cement seal	胶结物封闭；水泥封闭
cement set	水泥凝固
cement setting	水泥凝固
cement setting retarder	水泥缓凝剂
cement sheath	水泥护层；水泥环
cement sheath bonding quantity	水泥胶结质量
cement shed	水泥仓库
cement sheeting	水泥板块
cement shovel	水泥铲
cement silo	水泥密封仓；水泥筒仓
cement skin	水泥表皮
cement slab revetment	水泥板护岸
cement sludge	水泥沉淀物；水泥砂浆
cement slurry	英泥油，水泥浆
cement slurry hydrating period	水泥浆水化期间
cement slurry shrinkage	水泥浆收缩
cement slurry volume	水泥浆量
cement slurrypump	水泥浆泵
cement soil stabilization	水泥土加固
cement solidification	水泥固化
cement soundness	水泥安定性
cement space ratio	灰隙比；水泥量与骨料孔隙比
cement spacer	水泥隔离层
cement squeeze work	挤水泥作业
cement stabilization	水泥加固
cement sticky and vibrating platform	水泥胶砂振动台
cement sticky sand mixing machine	水泥胶砂搅拌机
cement stone	水泥灰岩；水泥石
cement storage	水泥仓库
cement storage platform	水泥存放平台
cement store	水泥仓
cement strength	水泥标号；水泥强度
cement substance	黏合质
cement suspension	水泥悬浮液
cement tester	水泥试验器
cement testing	水泥试验
cement texture	胶结结构
cement throwing jet	水泥喷浆机
cement tile	水泥瓦
cement treated base	水泥处治基层；水泥加固基础
cement treated soil	水泥处理过的土；水泥加固土
cement unloading equipment	水泥卸料设备
cement value	黏结力；黏结力值
cement veneer	水泥饰面
cement volume contraction	水泥体积收缩
cement wash	水泥浆；水泥浆刷面
cement water factor	水灰比
cement water glass grout	水泥水玻璃浆
cement water ratio	水灰比
cement water thin slurry	稀水泥浆
cement well pipe	水泥井管
cement works	水泥厂
cement-aggregate reaction	水泥骨料反应
cementation	水泥灌浆；水泥胶结；渗碳；硬化
cementation belt	胶结带
cementation bond	胶结连结
cementation bore	灌浆用钻孔
cementation borehole	水泥灌浆钻孔
cementation chargehand	水泥灌浆工
cementation effect	胶结效应
cementation exponent	胶结指数
cementation factor	胶结系数
cementation index	硬化指数；胶结指数
cementation launder	渗碳槽
cementation method	水泥灌浆法
cementation of fissures	灌缝
cementation operation	灌浆工作；注浆法
cementation pipe	灌浆管
cementation power	胶结能力
cementation process	灌浆法；渗碳法；水泥灌浆工艺；硬结法
cementation pumping equipment	灌浆泵设备
cementation reef	胶结作用生物礁
cementation sinking	注浆法凿井
cementation steel	渗碳钢
cementation strength	凝固强度
cementation structure	胶结结构
cementation tank	置换沉淀槽
cementation zone	胶结带；渗碳层
cementation-off	用水泥封隔住
cementatory	水泥的；胶结的
cement-base waterproof coating	水泥基防水层
cement-based grout	水泥基浆液
cement-bound granular material	水泥结粒料
cement-chipping hammer	混凝土打碎机
cement-clay grouting	黏土水泥灌浆
cement-clay mortar	水泥黏土砂浆
cemented	胶结的；渗碳的；灌水泥的
cemented belt joint	胶带胶接；皮带胶接
cemented carbide	烧结碳化物；硬质合金
cemented carbide bit	硬质合金钻头
cemented carbide matrix	烧结碳化物胎模
cemented carbide tip	硬质合金刀片
cemented carbide tool	硬质合金刀具
cemented casing	注水泥的套管
cemented fill	胶结充填
cemented formation	胶结地层
cemented fracture	胶结裂缝
cemented gravel	胶结砾石
cemented hardpan	胶结硬盘土
cemented joint	胶黏接缝；水泥接缝
cemented metal	渗碳金属
cemented multicarbide	多元碳化物硬质合金
cemented rock	胶结岩石
cemented roof bolt	钢筋混凝土锚杆
cemented sand	胶结砂
cemented sand fill	胶结充填砂

cemented soil	胶结土
cemented steel	表面硬化钢；渗碳钢
cemented together	水泥浇筑；用胶黏结
cemented up	管材内壁的水泥；注完水泥
cementer	灌浆装置；黏结剂
cement-filled inflatable casing packer	注入水泥的膨胀式套管封隔器
cement-filling system	胶结充填系统
cement-formation interface	固井水泥与地层的界面
cementing	胶结；水泥灌浆
cementing agent	胶结剂；黏合剂
cementing between two moving plugs	双塞法注水泥
cementing bond	胶结连接
cementing casing head	水泥头
cementing collar	注水泥接箍
cementing composition	注水泥组分
cementing engineer	固井工程师
cementing formulation	注水泥配方
cementing furnace	渗碳炉
cementing head	灌浆头；喷浆头
cementing job	固井作业；注水泥作业
cementing line	固井管线
cementing material	胶凝材料
cementing medium	胶结介质
cementing method	固井方法
cementing mineral	胶结矿物
cementing operation	固井施工
cementing outfit	注水泥设备
cementing phase	黏结相
cementing power	黏接能力
cementing practice	水泥灌浆法
cementing process	渗碳法
cementing property	黏合性；黏结性
cementing pump	注水泥泵
cementing section	固井段
cementing substance	胶结材料
cementing tank	水泥罐
cementing tool	灌浆工具
cementing value	黏结力；黏结值
cement-injection pipe	水泥浆注浆管
cementitious	胶结的；水泥质的
cementitious agent	胶结物；黏结剂
cementitious content	水泥成分
cementitious corrugated panel	水泥波纹板
cementitious ingredient	胶结成分
cementitious material	灌浆材料
cementitious matter	胶结料
cementitious sheet	石棉水泥板
cementitiousness	黏结性；胶结能力
cement-lime mortar	三抠灰；水泥石灰砂浆
cement-lime plaster	水泥石灰抹面
cement-lined pipe	水泥衬砌管
cement-mixture strength	水泥混合物强度
cement-modified aggregate	水泥处治集料；水泥稳定集料
cement-mortar lining	水泥砂浆衬里
cement-rubble masonry	水泥毛石砌体；水泥毛石圬工
cement-rubble retaining wall	水泥毛石挡土墙
cement-sand mixture	水泥砂混合物
cement-sand-water ratio	配合比
cement-soil base	水泥土基层
cement-soil pipe	水泥土管
cement-stabilized	浆砌的；水泥稳定的
cement-stabilized soil	水泥加固土
cement-supported reefs	胶结物支撑的生物礁
cement-testing sand	水泥试验砂
cement-treated grout	水泥灌浆
cement-treated soil slurry	水泥土稀浆
cement-treated-soil grout	水泥土浆
cementum	水泥质
cement-weighing machine	水泥称量机
cenogenesis	新生成因
cenology	地面地质学
cenomanian stage	赛诺曼期
Cenophytic era	新生植物代
cenosite	钙钇铒矿
cenosphere	煤胞；新生球体
cenote	锅穴；天然井；岩洞陷落井
cenotypal rock	新相火山岩；新相岩
cenotype	新相
Cenozoic	新生代
Cenozoic fold	新生代褶皱
Cenozoic group	新生界
Cenozoic sediment distribution	新生代沉积物分布
centenary wire dropper clip	承力索吊弦线夹
centenary wire support clamp	双线支承线夹
center bearing bracket	中心支架
center bit	中心钻头
center block	槽升子
center bore	中心孔
center bright-dim method	中心明暗法
center column	山柱
center counter	中心密度计
center coupling	中间接头
center cracked panel specimen	中心裂纹板试件
center cracked tension specimen	中心裂纹拉伸试件
center cross diagram method	交叉中隔壁法
center cutter	中心刀
center cutting	中部掏槽
center diagram method	中隔壁法
center diaphragm method	单侧壁导硐法
center drain	中心排水沟
center drift	中心掏槽；中央导坑
center drift excavation method	中洞法
center filler	心盘座
center heading	中央导坑
center hole	中心孔
center hole jack	穿心式千斤顶
center jar socket	震击打捞筒
center keelson	竖龙骨
center level	中心标高
center line	中线
center line of tunnel	隧道中线
center line of ventilation adit	风道中线
center line stake	中线桩
center line survey	线路中线测量；中线测量
center mixer	集中搅拌机
center of borehole	钻孔轴线
center of curvature	曲率中心
center of dispersal	分布中心
center of disturbance	扰动中心

center of embankment 路堤中心
center of gravity 重心
center of immersion 浸心
center of mass 质心
center of moment 弯矩中心
center of origin 震源
center of rotation 转动中心
center of slip circle 滑动圆心
center of springing 起拱中心
center of subsidence 沉降中心
center of symmetry 对称中心
center of the most dangerous slip circle 最危险滑动圆心
center peg 中心桩
center pier 中心墩
center pile 中间桩
center point load 中点载荷
center point of extraction 回采中心；开采中心
center position 正常位置
center rest 中心架
center section 中心截面；中心剖面
center shaft drive 中轴驱动
center sill 中梁
center spear 打捞钩；定心打捞矛
center split pipe 对开管
center split supporting ring 半圆支承环
center stake 中桩
center stake leveling 中平；中桩高程测量
center station 中心桩号
center support system 中心支撑系统
center to center distance between main girder 主梁中心距
center to edge 中心至边缘
center tool 中心刀具
center tube 中心管
center valve 中心阀
center wedge 中央楔块
center well 中央井
center-discharge mill 中心排料式磨矿机
center-hole bit 带中心水眼的钻头
center-hole type load cell 中心孔式荷载盒
centering 定中心；对中
centering adjustment 对准中心；中心调整
centering device 对心装置；扶正器
centering error 对心误差
centering grinding machine 定心磨床
centering guide 定心器
centering method 中线法
centering piece 对中件
centering punch 中心冲头；中心铳
centering rod 对中杆
centering under roof station 点下对中
center-jet drill bit 中心喷射式钻头
centerline method 四线法；中线法
centerline of arch 拱轴线
center-line stake leveling 中平
center-prick basin 中心锥盆地
center-to-center separation 中心距
center-to-center spacing 中心距
centesimal 百分的；百分之一的
centesimal system 百分制；百进制
centigrade 摄氏温度的

centigrade degree 摄氏温度
centigrade scale 百分温标；摄氏温标
centigrade thermal unit 百分度热单位
centigrade thermometer 百分温度计；摄氏温度计；摄氏温度计
centile 百分位数
centipoise 厘泊
centistoke 厘沲
central air supply 中央送风
central body 中央体
central chute 中央溜槽
central coal preparation plant 中心选煤厂
central concrete core 混凝土心墙
central concrete membrane 混凝土心墙
central concrete mixing plant 混凝土中心搅拌厂
central cone 中央锥
central core pillar 井筒中心柱
central core type 心墙型
central cylinder 中心柱
central cylindrical projection 圆柱中心投影
central dehydration complex 集中脱水装置
central difference finite difference method 中心差分有限差分法
central difference method 中心差分法
central distribution cone 中央分料漏斗
central dome 中央圆丘
central dry mixture 中心站制备的干混合物
central earthquake 中心震
central electrode configuration 中心梯度法电极布置
central engineering establishment sampler 中央工程处自动横向截流取样机
central firing 集中爆破；集中点火
central flux 中心通量
central force 中心力
central gathering station 集输泵站
central geanticline 中央地背斜
central generating station 中央发电厂
central geoanticline 中央地背斜
central granite 中心花岗岩
central impact 对心碰撞
central inclusion 中心包体
central indian basin 中印度洋海盆
central indian ridge 印度洋中脊
central induced sounding 中心感应测深
central induction sounding 中心感应测深
central load 轴心荷载
central loading 集中装车
central loop system 中心回线系统
central lubrication 集中润滑；中央润滑
central mix method 集中拌合法
central mixer 集中搅拌站
central mixing method 集中搅拌法
central mixing plant 中央搅拌厂
central mixing plant type stabilizer 集中拌合型加固装置
central pier 中墩
central plain 湖心平原
central plant mixing method 集中拌合法
central plug 泥石流中央堵塞
central producing well 中央生产井
central projection 中心投影

central pumping well	中心抽水井
central radiator	中心辐射器
central ray	中心射线
central rift	中心断裂谷；中央裂谷
central sample recovery drilling	中心取样钻探
central shaft	中心轴；中央井筒
central span	中跨距
central strip	中间带
central tendency	集中趋势；居中趋向
central track	中央航道
central transantarctic mountains	横断山脉中部
central type volcano	中心式火山
central uplift	中央隆起
central uplift belt	中央隆起带
central valley	中央谷
central value	代表值
central vent eruption	中心式喷发
central vent volcano	中心喷口火山
centralised filtering system	集中过滤系统
centralization	集中化
centralized concrete mixing	集中混凝土搅拌
centralized concrete mixing plant	混凝土中心搅拌厂
centralized control	集中控制
centralized control of switches	道岔集中控制
centralized ventilation system	中央通风系统
centralizer	对中器；定心夹具；套管扶正器；中心器
centralizer lug	扶正器叶片
centralizer sub	扶正器接头
centralizing gear	定中装置
centrallassite	白硅钙石
centrally mixed concrete	集中拌制混凝土
central-mixed concrete	集中拌制混凝土
central-typle volcanic edifice	中心式火山机构；中心式火山体
centrate	离心分离液
centre adjustment	定中调整；置中
centre anchorage	中心锚结
centre bearer bush	中轴承套
centre bit	中心钻头
centre chain conveyor	中间链输送机
centre compartment	中央隔间
centre distance	中心距
centre drift method	中心漂移的方法
centre drift tunnelling	中心漂移隧道
centre gate	中间控制级；中心浇口
centre hole	中心炮眼；中心钻孔
centre lifter	中心升降机
centre line	轴线
centre line of survey	测量中线
centre line stake	中线标桩
centre machine	定心机；定心钻
centre mixer	中心拌合站
centre moment	中心矩
centre of buoyancy	浮力中心
centre of crest circle	坝顶曲线中心点
centre of curvature	曲率中心
centre of equilibrium	平衡中心
centre of impact	撞击中心
centre of inertia	惯性中心
centre of mean distance	均距中心
centre of origin	震源
centre of parallel forces	平行力系中心
centre of percussion	撞击中心
centre of pressure	压力中心；压强中心
centre of sedimentation	沉积中心
centre of subsidence	沉降中心
centre peg	中心桩
centre point	中心点
centre post	中间立柱
centre refuse extraction chamber	中间排渣室
centre rotating grab	中心回转式抓岩机
centre section	中心截面
centre stake	中心桩
centre strand conveyor	中间链输送机
centre tap	中点引线；中心抽头
centre to centre	中至中
centre wire	中心铅垂线；中心线
centre-cut method	中心开挖法
centre-hole bit	中心冲孔式钻头
centre-line grade	中线坡度
centre-point control	中点控制
centre-to-centre method	心对心法
centric diatom	辐射硅藻；中心硅藻
centric load	轴向荷载
centric texture	向心状结构
centriclone	锥形除渣器
centrifugal	离心力
centrifugal acceleration	离心加速度
centrifugal action	离心作用
centrifugal air separation	离心空气分选
centrifugal autoclaved concrete pile	离心蒸压混凝土桩
centrifugal blender	离心式搅拌机
centrifugal blower	离心式送风机
centrifugal brake	离心闸式制动器
centrifugal caster	离心浇铸机
centrifugal casting process	离心式浇铸法
centrifugal clarifier	离心沉降器；离心澄清器
centrifugal classifier	离心分级机
centrifugal cleaning	离心选煤
centrifugal clutch	离心式离合器
centrifugal collector	离心收集器
centrifugal compacting	离心压坯
centrifugal compressor	离心式压缩机
centrifugal concrete	离心法混凝土
centrifugal concrete pile	离心混凝土桩
centrifugal deep-well pump	离心式深井泵
centrifugal dehydrator	离心脱水机
centrifugal dense medium separator	离心重介质选矿机
centrifugal dewatering screen	离心脱水筛
centrifugal drainage pattern	离心状水系
centrifugal drier	离心脱水机
centrifugal drying	离心法脱水
centrifugal effect	离心效应
centrifugal fan	离心式风机
centrifugal filter	离心过滤机
centrifugal force field	离心力场
centrifugal governor	离心式调速器
centrifugal gravitational method of coal preparation	重力离心选煤法
centrifugal grindability	离心式研磨机

centrifugal head	离心水头
centrifugal impeller mixer	离心叶轮式混合器
centrifugal jig	离心式跳汰机
centrifugal lubrication	离心润滑
centrifugal lubricator	离心润滑器
centrifugal machine	离心机
centrifugal mill	离心式磨机
centrifugal model of rock mass	岩体离心模型
centrifugal model test	离心模型试验
centrifugal moisture	离心含水量
centrifugal movement	离心运动
centrifugal mud machine	离心式泥浆分离机；离心式泥浆净化器
centrifugal oiler	离心加油器
centrifugal pendulum	离心摆
centrifugal pitot pump	离心皮托泵
centrifugal prestressed concrete pile	离心预应力混凝土桩
centrifugal pump	离心泵
centrifugal pump of horizontal axis	卧式离心泵
centrifugal pump of multistage type	多级离心泵
centrifugal pump of vertical axis	立式离心泵
centrifugal purification	离心净化
centrifugal reinforced concrete driven pile	离心式钢筋混凝土打入桩
centrifugal reinforced concrete pile	离心钢筋混凝土桩
centrifugal relay	离心式继电器
centrifugal replacement	离心交代
centrifugal rougher	离心粗选机
centrifugal screen	离心筛
centrifugal screw pump	离心螺旋泵
centrifugal scrubber	离心洗涤器
centrifugal sedimentation method	离心沉降法
centrifugal separation	离心分选法
centrifugal separator	离心分离器；离心机
centrifugal sewage pump	离心污水泵
centrifugal sizing	离心分级
centrifugal slimer	离心摇床
centrifugal stirrer	离心搅拌器
centrifugal stowing machine	离心式充填机
centrifugal stress	离心应力
centrifugal supercharger	离心式增压器
centrifugal table	离心摇床
centrifugal thickener	离心式浓缩机
centrifugal thrower	离心式抛掷充填机
centrifugal ventilator	离心扇风机
centrifugal-discharge bucket elevator	离心式排料提斗机
centrifugalization	离心分离作用
centrifugally compacted reinforced concrete pile	离心法制成的钢筋混凝土桩
centrifugally spun concrete pipe	离心浇制混凝土管
centrifugally-cast concrete driven pile	离心法浇筑的混凝土打入桩
centrifugation	离心分离作用
centrifuge	离心干燥机
centrifuge effluent	离心机排出液；离心液
centrifuge method	离心分离法
centrifuge model studies	离心机模型研究
centrifuge model test	离心模型试验
centrifuge modeling	离心机模拟
centrifuge moisture equivalent	离心持水当量；离心含水当量
centrifuge test	离心试验
centrifuged coal	离心脱水精煤
centrifuger	离心机
centrifuging of algae	藻类离心分离
centring	定中心；对中点
centring bracket	定心支架
centring device	定心装置
centring flange	固定法兰
centring sleeve	定心套
centring under point	点下对中
centripetal	向心状的
centripetal acceleration	向心加速度
centripetal blastic texture	向心变晶结构
centripetal drainage	辐合状水系
centripetal drainage pattern	向心水系
centripetal elements	向心元素
centripetal fault	向心断层
centripetal pump	向心泵
centripetal replacement	向心交代
centroclinal	向心倾斜的；向心倾斜的；四周向中心倾斜
centroclinal dip	向心倾斜；向心倾斜
centroclinal fold	向心褶皱；向心褶皱；同心褶皱
centroclinal structure	同心向斜构造
centrocline	同心向斜；向心倾斜
centroid	重心
centroid of a wedge	岩楔的矩心
centroid of drainage basin	流域形心；流域重心
centroidal axis	形心轴；质心轴
centroidal distance	形心距离
centrosphere	地核
centrosymmetric	中心对称的
centrosymmetry	中心对称
centrum	震源；中心；震源
centuple	百倍的
cenuglomerate	泥流角砾岩；泥流碎屑岩
ceofficient of impact	冲击系数
cepstrum	倒谱；对数逆谱；逆谱
ceramal	合金陶瓷；金属陶瓷
ceramic arrester	陶瓷避雷器
ceramic capsule charge	陶瓷射孔弹
ceramic clay	陶瓷黏土
ceramic coat	陶瓷涂层
ceramic disk	陶瓷板
ceramic fiber blanket	陶毡
ceramic filter	陶瓷过滤器
ceramic flux	陶瓷熔剂
ceramic insulator	陶瓷绝缘子
ceramic magnet	烧结磁铁；陶瓷磁石
ceramic material	陶瓷材料
ceramic metallurgy	陶瓷冶金法
ceramic microporous vacuum filter	陶瓷微孔真空过滤机
ceramic mosaic	陶瓷地砖；陶瓷马赛克
ceramic piezometer tip	陶瓷孔压测压头
ceramic pipe	陶瓷管
ceramic sensor	陶瓷传感器
ceramic structural member	陶瓷结构构件
ceramic tile	瓷砖
ceramic tip	金属陶瓷刀具

ceramic transducer	陶瓷换能器；陶瓷传感器
ceramic type insulation	瓷绝缘
ceramicite	陶土岩
ceramic-metal combinations	金属陶瓷法
ceramics	陶瓷；陶器
ceramic-to-metal seal	陶瓷金属封焊
ceramsite	陶粒
ceramsite concrete	陶粒混凝土
cerargyrite	角银矿
cerasite	樱石
ceresin	白地蜡；纯地蜡
ceresine	地蜡
cerinite	腊质体
cerite	硅铈石
cerium	铈
cerium carbide	碳化铈
cerium minerals	铈矿类
cermet	含陶合金；金属陶瓷
cerolite	蜡蛇纹石
certified reference coal	标准煤样
cerussite	白铅矿
cervanite	黄锑华
cervantite	白安矿；黄锑华
cesarolite	铅锰矿
cesium	铯
cesium iodide	碘化铯
cesium magnetometer	铯磁力仪
cesium standard	铯钟
cesium vapor magnetometer	铯磁力仪
cess	多孔排水管
cessation of deposition	沉积作用停止
ceylonite	镁铁尖晶石
ceyssatite	硅藻土
chabasite	菱沸石
chabazite	菱沸石
chad	粗砂；金砂；砾石；碎石
chafing	摩擦；磨损
chafing plate	防擦板
chaidamuite	柴达木石
chain	链条
chain anchorage	牵引链锚固装置
chain and attachment	链和附件
chain and bucket	链斗式
chain and bucket conveyor	链斗式输送机
chain and bucket type excavator	链斗式扩掘机
chain and flight conveyor	刮板输送机
chain and pan conveyor	链板式输送机
chain bit	链式钻头
chain block	起重机
chain bond	链式搭接
chain bridge	链索吊桥
chain bucket	提料斗
chain bucket dredger	链斗式挖泥船
chain bucket elevator	链斗式提升机
chain bucket excavator	链斗式扩掘机；链斗式挖土机
chain bucket loader	链斗式装载机
chain cable	链索
chain cager	链式装罐机
chain car hauler	链式升车机；链式推车机
chain car lift	链式升车机
chain carrier	链反应活性中心；链锁载体
chain claw	链爪
chain coal cutter	链式截煤机
chain coal saw	链式锯煤机
chain coupling	链式联轴节；链形连接器
chain crab	链起重绞车
chain creeper	链式爬车机
chain curtain	链幕；链式闸门
chain cutter	链式截煤机
chain cutter bar	链式截盘
chain cutting machine	链式截煤机
chain dredge	链式挖掘船
chain dredger	链式挖泥机
chain drive	链条驱动；驱动链
chain drive section	截链传动部分
chain elevator	链式升降机；链式提升机
chain feed	链式给料机
chain feeder	链式给料机；链式加料机
chain gage	链式水位计
chain gate	链式闸门
chain gauge	链式水位计
chain grab	链式抓具
chain grip jockey	矿车的无极绳抓链
chain grizzly	链筛
chain haulage	链牵引
chain hauled shearer loader	链牵引截装机
chain hoist	链式起重机
chain hook	链钩
chain initiation	连锁开始；链引发作用
chain iron	铁链环
chain jack	链式起重器
chain jib	链式截盘
chain joint	链接
chain lacing pattern	截链上截齿排列
chain like veins	链状矿脉；锁键状矿脉
chain link fence	铁丝网围栏
chain link mesh	铁丝网
chain loader	链式装载机
chain lubricant	链润滑油
chain mat	链幕
chain measure	测链测量
chain mesh of roof bolt	护顶锚杆金属网
chain of buckets	链斗式排水设备
chain of locks	梯级船闸
chain of mountain	山脉；山系
chain of power plants	梯级电站
chain of triangulation	三角锁
chain of volcano	火山链
chain of volcanoes analysis	火山链分析法
chain oiler	链加油器
chain oiling	链润滑
chain pillar	巷道矿柱；巷道煤柱
chain pin	测钎；链销
chain pipe wrench	链式管钳
chain plough	链式刨煤机
chain post	链柱
chain pre-tensioning	链预张力
chain pulley	链条滑轮
chain pump	链泵；链斗提水机
chain pusher	链式推车器

chain reaction 连锁反应；链式反应
chain reaction type source model 链式反应型震源模型
chain retarder 链式阻车器
chain ring 链环
chain riveting 并列铆接
chain saw 链锯
chain scraper conveyor 链板输送机
chain scraper planer 链牵引耙式刨煤机
chain silicate 链状硅酸盐
chain sling 链式吊索
chain sounding 链式水深测量
chain sounding line 测深链
chain structure 链状结构
chain suspension bridge 链索吊桥
chain table 链式拣矸输送机
chain tension 链条张力
chain tightener 紧链器
chain tractive effort 链牵引力
chain transmission 链条传动
chain winch 链绞车；链绞盘
chain windlass 链式卷扬机
chainage 链测长度；桩号
chain-and-tooth machine 链齿式挖掘机
chain-back bucket 链网底式铲斗
chain-driven powered cowl 链传动装煤板
chain-driving sprocket 传动链轮
chained sequential operation 链锁程序操作
chain-structure 链式结构
chain-grate stoker 链炉篦式加煤机
chaining arrow 测点插钎
chainless hauled shearer 无链牵引采煤机
chain-like structure 链状结构
chain-oiled 链润滑的
chain-oiled bearing 链润滑的轴承
chain-tensioning device 链条张拉设备
chain-track tractor 履带式拖拉机
chain-type stacking-raking machine 链式堆放耙装两用机
chainwall 巷侧带状煤柱
chainway 斗链导向装置
chair 罐托；罐座
chair-lift 索道升降椅
chalcanthite 胆矾
chalcedonic chert 玉髓状燧石
chalcedonite 玉髓
chalcedonization 玉髓化
chalcedony 石髓；玉髓
chalcoalumite 铜明矾；铜茂
chalcocite 辉铜矿
chalcocitization 辉铜矿化
chalcography 矿相学
chalcolite 铜铀云母
chalcomenite 蓝铜硒矿
chalcomorphite 硅铝钙石
chalcophanite 黑锌锰矿
chalcophile 亲铜的
chalcophile affinity 亲铜性
chalcophile element 亲铜元素
chalcophyllite 云母铜矿
chalcopyrite 黄铜矿
chalcosiderite 磷铜铁矿

chalcosine 辉铜矿
chalcostibite 硫铜锑矿
chalcotrichite 毛赤铜矿
chalicosis 石末肺
chalk 白垩岩
chalk black 黑色碳质黏土
chalk clay 白垩质黏土；泥质岩
chalk drilling fluid 白垩土钻探冲洗液
chalk emulsion 白垩乳状液
chalk marl 白垩泥灰岩
chalk ooze 白垩软泥
chalk pit 白垩矿
chalk rock 白垩岩
chalk slate 白垩板岩
chalk soil 白垩土
chalk sphere 白垩球体
chalk test 白垩试验；粉末渗透探伤法
chalk tuff 白垩凝灰岩
chalk-crust soil 石灰结壳土
chalking 白垩现象
chalkstone 白垩石；石灰岩
chalky 白垩的
chalky clay 白垩质黏土
chalky porosity 白垩状孔隙
chalky soil 白垩质土
chalmersite 方黄铜矿
chalybeate 含铁的；铁剂；铁泉
chalybeate spring 铁质泉
chalybeate water 铁质水
chalybite 球菱铁矿
chamber 硐室；矿房；沙井
chamber and pillar method 房柱开采法
chamber bench excavation 坑道台阶式挖掘
chamber blasting 硐室爆破；药室爆破
chamber blasting method 硐室爆破法
chamber charge 硐室装药
chamber detritus 碎屑室
chamber dock 箱式船坞
chamber excavation 硐室掘进
chamber gate 闸室
chamber lift 汇集室式气举
chamber mining 房式开采
chamber of ore 矿袋
chamber plant 铅室硫酸厂
chamber press 箱式压滤机
chamber pump 箱式泵
chamber support 硐室支护
chamber surveying 采场空硐测量
chamber test 模型槽实验
chambered 分成格室的
chambered deposit 囊状矿层
chambered residuum 囊状风化壳
chambered vein 囊状矿脉
chambering 扩孔；炮眼掏壶
chambersite 锰方硼石
Chambishi mine 谦比希矿
chameleon-type corundum 变色刚玉
chameleon-type diamond 变色钻石
chameolith 腐殖煤
chamfer 倒角；倒棱；切角；切面；斜面

chamfer bit	倒角钻头
chamfer cut	楔形切面；斜切面；辙尖斜刀
chamfered edge	削边；削角边
chamfering	倒角
chamockite	麻粒岩
chamois	油鞣革
chamoisite	鲕绿泥石
chamosite	鲕绿泥石
chamotte	耐火黏土
chamotte brick	耐火黏土砖
champaign	平原
champion lode	巨矿脉
champion vein	巨矿脉
chance cone	强斯型圆锥洗煤机
chance cone agitator	漏斗式搅拌器
chance distribution	随机分布
chance failure	随机破坏
chance process	强斯砂浮法；强斯砂浮法
chance sample	随机样品
chaneling machine	截煤机
changbaiite	长白矿
change bit	换钻头
change gear bowl	变速齿轮箱
change gear box	变速器
change house	地面打钻棚；地面钻机房；更衣室；交接班房；矿山浴室
change in gauge	直径差
change in safeyield	安全涌水量变化
change in type of geosyncline	地槽类型转化
change of a course	航运改向；河道改向
change of Coulomb stress	库伦应力变
change of facies	相变
change of ground water level	地下水位变化
change on one method	不归零法
change point	转变点
change rate	变化率
change side of double line	换边
change tape	记录修改带
change valve	三通阀
changeability	变易性
changement	转换设备；换向机构
changeover button	换向按钮
change-over cock	换向转心阀
change-over key	换向按钮
change-over lever	操纵杆；转换杆
changeover switch	转换开关
changing channel	变动河槽；变动河床
changing stage	变动水位
changjiang diluted water	长江冲淡水
channel	槽钢；沟渠；航道；河道
channel accretion	航道淤积；渠道淤积
channel alignment	渠道定线；水道定线
channel axis	河槽线；渠道轴线
channel back	回流
channel bank	渠岸
channel bar	槽钢；槽铁；河心沙坝；心滩
channel basin	河床盆地；河谷盆地
channel beam	槽钢；槽形梁
channel bed	河槽；河床
channel black	槽法炭黑
channel boundary	孔道边界
channel buoy	航道浮标
channel cast	槽痕迹
channel centre bunton	中间槽钢罐道梁
channel change	河道演变
channel characteristic	河床特性
channel column	槽形柱
channel contraction	河道缩狭段
channel control	测流控制断面；河槽控制
channel cover	泥浆槽盖；渠盖
channel cross-section	过水断面；河槽断面
channel cutting	河槽割切；渠道割切
channel deposit	河槽沉积物；河成矿床；河床淤积
channel detention	河槽滞蓄量
channel diameter	孔道直径
channel diversion	明渠导流
channel dredging	河道疏浚
channel effect	间隙效应；沟槽效应；管道效应；河槽效应
channel enlargement	河道拓宽
channel entrance	航道入口；水道入口
channel erosion	河床冲刷；河道侵蚀
channel excavation	沟渠开挖
channel expansion	渠道扩大
channel fill	河槽淤积
channel fill deposit	河槽沉积；河床沉积
channel floor lag	河底滞留沉积
channel flow	河槽径流；河床径流
channel flow accretion	渠道流量增长
channel frequency	河道密度
channel friction head loss	河槽摩擦水头损失
channel geometry	河槽形态；孔道几何形态
channel girder	槽形梁
channel glacier	河床冰川
channel gradient	河床坡度；河道坡降
channel gradient ratio	河床坡降比率
channel groove	沟槽
channel guide	槽钢罐道
channel improvement	航道整治；河槽整治
channel inflow	河槽入流量；河道来水量
channel intake	渠道进水口
channel interception	拦截渠道
channel iron	槽钢；槽铁
channel island	河间岛；江心洲
channel lag	河槽汇流滞后时间
channel lag deposit	河道滞后沉积；河底滞留沉积
channel lead	航线
channel leak	毛细漏孔；通道漏孔
channel line	河道主流线；轨道线
channel linearity	通路线性
channel lining	河床铺砌
channel loading machine	截煤装载机
channel loss	渠道渗流损失；渠道渗漏量
channel marking	航道标志
channel method	刻槽法
channel migration	河道迁移
channel morphology	河槽形态；河床地貌
channel mouth bar	河口沙洲
channel net	河道网
channel net loss	通路净衰竭

channel of approach	引槽；引渠
channel of flow	水流汇聚；水流渠道
channel of main streams	主河槽
channel of naval ports	军港航道
channel order	河流等级
channel pattern	河流类型；河网类型
channel permeability	渠道渗透性
channel pin	槽楔钉
channel plate	槽钢垫板；带肋板
channel porosity	通道孔隙度
channel precipitation	河床沉淀作用；河道沉淀
channel prop	槽钢支柱
channel properties	河槽特性；河床特性
channel protection	河道防护
channel rail	沟形轨条
channel reactivation	河道复活作用
channel realignment	河流改道
channel rectification	航道整治
channel regulation	航道整治
channel relocation	河流改道
channel restriction	河道束缩；通道狭窄处
channel roughness	沟床糙度
channel routing	河槽洪水演进
channel sample	槽样
channel sampling	刻槽取样
channel sand	河道砂
channel sand deposit	河床沙沉积
channel scour	河床冲刷作用
channel section	槽形截面；渠道断面
channel segment	河段；河流分支
channel set	河道组
channel shaped wall section	槽形墙截面
channel slide rail	槽式滑轨
channel slope	河床坡度；河床坡降
channel span	航道跨度；河槽跨度
channel splay	河流泛滥冲扇
channel spring	渠道泉；沿河泉
channel stability	河床稳定性；信道稳定性
channel steel	槽形钢
channel steel arch	槽形钢拱
channel step	河道梯级；渠道梯级
channel storage	河槽储水量
channel straightening	河道裁弯取直
channel strait	海峡
channel stringer	槽形钢纵梁
channel structure	沟渠构造
channel survey	航道测量
channel thread	河道主流
channel tile	槽瓦
channel time delay	河槽汇流时间
channel training wall	引水导墙
channel wave	槽波；沟道波；河槽波；通道波
channel wave seismic method	槽波地震法
channel way	河槽；河床；通道
channel width	海峡宽度；河道宽度
channel-bank slumping	河岸滑塌作用
channeled plate	槽纹板
channeled runoff	沿槽径流
channeled strap	波纹钢顶梁
channeled upland	沟蚀高地
channeler	挖沟机；凿沟机
channel-fill cross-bedding	水道充填交错层理
channel-fill swamp	河道填积沼泽
channel-forming process	河床演变；造床过程
channeling	成沟作用；沟道效应；掏槽，挖槽；凿沟
channeling effect	沟道效应
channeling machine	辊式弯曲机
channelization	河道取直加深；河道疏浚
channelization island	导流岛
channelization project	河道渠化工程
channelization ratio	沟壑比
channelized debris flow	沟壑控制的泥石流；沿沟泥石流
channelized delta plain	渠化三角洲平原
channelized flow	渠化水流
channelized river	渠化河流
channelled	有凹槽的
channeller	开渠机；挖槽机；凿沟机
channelling	敷设管道；敷设渠道；开渠；掏槽
channelling character of ore	矿石的成沟性
channelling karren	溶痕
channelling machine	截煤机；滚槽机
channelling of flow	水流汇聚；引集水流
channelling of seismic waves	地震波导作用
channeltest	成沟试验
channery	碎石块；岩石碎片
chaos	混乱；混杂角砾岩；巨屑混杂堆积
chaos breccia	混杂角砾岩
chaos motion	不规则运动
chaotic	不规则的；混乱的；混杂堆积的
chaotic boulder clay	混杂漂砾黏土
chaotic deposit	混杂沉积
chaotic facies	混合相；混杂相
chaotic geological body	混杂地质体
chaotic landform	混杂地貌
chaotic melange	混杂堆积
chaotic motion	不规则运动
chaotic sediment	混杂沉积物
chaotic seismic reflection configuration	杂乱地震反射结构
chaotic slump block	不规则滑动断块
chaotic structure	不规则构造；混乱构造
chaotic-fill	混杂填充
chap	龟裂；裂缝；裂纹；敲帮问顶；敲帮测距
chap knock	敲帮问顶
chapada	高地
chapapote	软沥青
chaparral	灌木群落
chape	夹子；卡钉；线头焊片
chapeau de fer	铁帽
chapeirao	孤立珊瑚礁；礁堆
chapelet	斗式提升机；链斗式提水机
chapiter	柱头
chaplet	泥芯撑；项圈
chapman layer	查普曼层
chapman production function	查普曼生成函数
char	河心沙坝；木炭
char gasification	煤焦气化
char yield	半焦产率
chara limestone	轮藻灰岩
chara marl	轮藻泥灰岩

character 特征
character log 声波测井
character of char residue 焦渣特性
character of discontinuities 不连续面的特征
character profile 示性剖面
characteristic ash curve 灰分特性曲线
characteristic based split 特征分割
characteristic concentration 特征浓度
characteristic concrete cube strength 混凝土立方块特征强度
characteristic constant 特征常数
characteristic curve 特征曲线
characteristic curve of a prop 立柱工作特性曲线
characteristic curve of support 支架工作特性曲线
characteristic depth 特性水深
characteristic dimension 基准尺寸；特性尺寸
characteristic discharge 特性流量；特征流量
characteristic distortion 特性畸变
characteristic earthquake 特征地震
characteristic feature 特征要素
characteristic focal depths 特征震源深度
characteristic fossil 标准化石
characteristic frequency 特征频率
characteristic friction angle 特征摩擦角
characteristic head 特性水头
characteristic hydrograph 特性水位图；特性水文曲线
characteristic impedance 特性阻抗；特征阻抗
characteristic length 特性长度；特征长度
characteristic line 特征谱线
characteristic load 特征荷载
characteristic matrix 特征矩阵
characteristic method 特征线法
characteristic mineral 特征矿物
characteristic number 特征值
characteristic of a channel 河槽特征
characteristic of earthquake 地震特征
characteristic of seismic phase 震相特征
characteristic oscillation 特征振动
characteristic parameter 特性参数
characteristic pattern 特征模式
characteristic period of response spectrum 反应谱特征周期
characteristic point 特征点
characteristic quadratic form 特征二次型
characteristic resistance 特性阻力
characteristic scalar 特征标量
characteristic scale 特征尺数
characteristic sensitivity constant 灵敏度特性常数
characteristic sheet 样图
characteristic site period 场地特征周期
characteristic size curve 粒度特性曲线
characteristic strength 抗压强度；特性强度
characteristic test 特性试验
characteristic value of a material property 材料性能标准值
characteristic value of actions 作用标准值
characteristic value of an action 作用标准值
characteristic value of earthquake action 地震作用标准值
characteristic value of permanent action 永久作用标准值
characteristic value of strength of steel 钢材强度标准值
characteristic value of strength of steel bar 钢筋强度标准值
characteristic value of variable action 可变作用标准值
characteristic value of windload 风荷载特征值

characteristic vector analysis technique 特征向量分析法
characteristic vibration 特征振动
characteristic water level 特征水位
characteristic wave 特征波
characteristic wave height 特征波高
characteristics of pump 水泵特性
characterization 界定方法
characterizing accessory 表征性次要矿物
characterizing accessory mineral 特征副矿物
character-unknown anomaly 性质不明异常
charco 水注
charcoal 木煤；木炭
charcoal adsorption process 活性炭吸附法
charcoal filter 木炭滤器
charcoal gray 炭灰色
charcoal saturation time 活性炭饱和时间
charcoal test 活性炭吸附试验
charge anchor 炸药包锚
charge coupled device 电荷耦合器件
charge density 装药密度
charge efficiency 给料效率
charge indicator 充电指示器；带电指示器
charge model 充注模式
charge of rupture 破坏荷载
charge of safety 安全荷载
charge of surety 安全荷载；容许荷载
charge off 充电指示器；带电指示器
charge pattern 药包排列，电荷图型，充电曲线
charge potential 充电电位
charge pressure 充气压力
charge pump 充电泵；供给泵
charge quantity 装药量
charge ratio 满载系数；配料比
charge relief valve 辅助泵的溢流阀；热交换溢流阀
charge stock 进料
charge transfer device 电荷转移装置
charge volume 充注体积
charge-amplitude relation 炸药量-振幅关系
charge-coupled device camera CCD摄影机
charged 装药的
charged shale 压缩页岩
chargehand 抓岩机工
charger 充电器
charger valve 充气阀
chargiability 可充电性；光电率
charging 充电；负载；加料
charging and release valve 注液-放泄阀
charging apparatus 装料设备
charging area 装料台
charging bay 充电台
charging by batches 分批装料
charging by conduction 传导充电；接触充电
charging by induction 感应充电
charging capacity 装载量
charging construction 装药结构
charging crane 转装机；装卸机；装料起重机
charging efficiency 灌注效率
charging equipment 充电设备
charging floor 加料台；装料台
charging hopper 装料仓；装料漏斗

charging machine	装料机
charging pump	进料泵
charging scoop	装料铲
charging skip	料车
charging spout	给料槽；料斗
charging stand	充电台
charging station	充电站；压气机；收费站
charging stick	炮棍
charging unit	充电车间；充电装置
charging valve	充电管；充气阀；给水阀
charging-up	充电；给料；装料；装药
chark	焦炭
charnockite	紫苏花岗岩
charnockite gneiss	紫苏花岗片麻岩
charoitite	查罗石岩
charpy impact test	单梁式冲击试验
charred	焦化的；炭化的
charred coal	焦炭
charred peat	炭化泥炭
charred pile	焦头桩
chart datum	海图深度基准面
chart folio	海图集
chart large correction	海图大改正
chart of marine gravity anomaly	海洋重力异常图
chart of ore drawing	放矿计划图表
chart of stability analysis	稳定性分析图表
chart paper	图表纸
chart projection	海图投影
chart scale	海图比例尺
chartered engineer	特许工程师
chartered surveyor	特许测量师
chartographic unit	制图单位
chartometer	测距仪
chaser	螺纹梳刀；碾压机；揉泥碾；石棉碾；送桩器
chaser mill	干式辊碾机；碾碎机
chasing	雕镂；螺旋板；切螺纹；沿走向探查矿脉；铸件最后抛光
chasing tool	螺纹刀具；梳刀
chasles theorem	透视旋转定律
chasm	断层；裂口；深渊；峡谷；陷坑
chasmic stage	裂陷阶段
chasmy	裂口多的
chassignite	纯橄无球粒陨石
chassis	底板；底架；底座
chassis connection	车底盘的连接
chassis lubricant	底盘润滑剂
chat roller	焙烧矿石辊碎机
chathamite	砷镍铁矿
chatoyant	闪光石
chatoyant stone	猫眼石
chats	钎石；矿山废石；燧石砾岩
chattel	动产
chatter	振动；振碎；震颤
chatter bump	不平整路面上的凸起处
chatter detection	震颤检测
chatter mark	颤动擦痕；振痕；擦痕
chatter-free finish	无颤痕光洁度
chattering	颤动；震颤
chean coal	洁净煤
cheap	交错层理
chebka	古干水道
check	单向阀；细裂纹；校核
check action	单流作用
check ball	止回球
check bar	校验棒
check baseline	校正基线
check bearing	检验方位
check borehole	检验钻孔
check by sampling	抽查检查
check by sight	肉眼检查
check cable	安全绳
check clamp	固定压板
check crack	龟裂
check curtain	风帘；风障
check dam	拦砂坝；拦水坝；风挡
check damper	止回阀
check door	阻风门
check experiment	对比试验；对照试验
check flooding irrigation	分畦淹灌
check for zero	校对零点
check gate	节制闸门；配水闸门
check gauging	验证计量
check hole	检查孔
check indicator	校验指示器
check irrigation method	畦灌法
check jump	阻抑水跃
check level	校核水准
check lock lever	锁紧杆
check measurement	校核测定
check meter	校验仪表
check method of irrigation	分畦灌溉法
check off	合格；检验
check pawl	止回棘爪
check piece	制速器
check plate	挡板；垫板；止动板
check point	检验点
check ring	锁环
check sampler	核对取样；校对取样
check screen	超粒控制筛
check shot	校验炮；校验地震测井
check sluice	逆止水闸
check specimen	对照标本；核对标本
check spring	调整弹簧；复原弹簧；止动弹簧
check station	监测台
check structure	拦沙建筑物；拦水建筑物
check surface	表面龟裂
check system of irrigation	畦灌系统
check test	对照试验；校核试验
check valve	单向阀；防逆阀
check valve ball	单向阀活门球；止回阀球
check valve lubricator fitting	止回阀加油嘴
check valve pump	单向阀配流的泵
check valve seat	止回阀座
check washer	止动垫圈；防松垫圈
check weight	校准锤，校核重量
checked flood lever	校核洪水位
checked mine capacity	矿山核定生产能力
checked surface	裂纹面
checked-up lake	堰塞湖
checker	校验员；棋盘式排列；复合材质

checker brick 格子砖
checker coal 长方粒无烟煤
checker plate 网纹钢板
checkerboard 方格盘；棋盘
checker-board method 棋盘式短壁开采法；棋盘式房柱开采法
checker-board system 房柱式开采法；棋盘式开采法
checkerboard topography 棋盘式地形
checkerboarding 棋盘式的；棋盘式短壁开采法
checkered 方格的；棋盘式的
checkered iron 格纹铁
checkered plate 网纹钢板
checkered shear wall 格纹抗剪墙
checkered steel plate 花纹钢板
checking 表层裂纹；细裂缝；龟裂；核对；检验
checking and accepting 验收
checking bollard 防松式系船柱；码头带缆桩
checking of basic seismic intensity 地震基本烈度复核
checking of building line 验线
checking of seismic ground motion parameter 地震动参数复核
checking pile 验桩
checking procedure 检验程序
checking routine 检查程序
checking shot point 爆破检查点
checknut 防松螺帽；锁紧螺帽
check-off list 查讫单，核对清单
checkout 尖灭；检测；调整校正
checkout console 检验台
checkshot survey 校验炮观测
checkweigher 检重秤；监秤人
cheddite 谢德炸药
cheek 中型箱
cheek board 边模板
cheek plate 颊板，破碎机
cheese head bolt 圆头螺栓
cheese head screen plate 圆头螺钉
cheese tester 深部取样器
cheese weight 导绳拉紧坠锤
chelant 螯合剂
chelate 螯合体；螯合物
chelate compound 螯形化合物
chelate fibre 螯合纤维
chelate initiator 螯合引发剂
chelate polymer 螯合聚合物
chelate ring complex 螯合环状络合物
chelate stabilizer 螯合稳定剂
chelating agent 螯合剂
chelating ligand 螯合配位体
chelating media 螯合介质
chelation 螯合酐；螯合作用
chelation group 螯合基团
chelatometric titration 螯合测定法
chelatometry 螯合测定法
chelator 螯合剂
chelogenic cycle 地盾形成旋回
chem-fix process 化学加固处理法
chemiadsorption 化学吸附
chemical absorption 化学吸收
chemical accelerating admixtures 速凝化学掺和剂

chemical action 化学酐；化学作用
chemical activation 化学活化
chemical additive 化学添加剂
chemical admixture 化学添加剂
chemical adsorption 化学吸附
chemical agent 化学试剂
chemical alteration 化学蚀变
chemical analysis liquid from compressive soil 土壤压出液化学分析
chemical analysis of water 水质分析
chemical anchor 化学锚固
chemical attack 化学侵蚀
chemical beneficiation 化学选矿
chemical blowing agent 化学起泡剂
chemical board drain 化学板排水；排水化学板
chemical bonding 化学胶合
chemical breaker 化学破坏剂
chemical burn 化学品灼伤
chemical cartridge respirator 化学防毒面具
chemical churning pile 旋喷桩；化学搅拌桩
chemical churning pile or chemical churning pattern 化学旋转喷射桩法，化学旋喷法
chemical churning process 旋喷法
chemical churning under high pressure 高压旋喷法
chemical cleaning process 化学选矿过程
chemical coagulation 化学凝聚
chemical combination 化合作用
chemical combined water 化学结合水
chemical combustion 化学燃烧
chemical comminution 化学破碎
chemical compatibility 化学相容性
chemical competence 化学搬运力
chemical composition 化学组分
chemical composition analysis of rock and soil 岩土化学成分分析
chemical composition of pyrite 黄铁矿的化学成分
chemical compound 化合物
chemical concentrate 化学浓缩；化学提浓物
chemical concentration 化学选矿；化学提浓
chemical conditioning of mud 泥浆的化学处理
chemical consolidant 化学加固剂
chemical consolidation 化学固结作用；化学加固作用
chemical constituents in groundwater 地下水化学成分
chemical contaminant 化学污染
chemical conversion film 化学处理膜
chemical corrosion 化学腐蚀；化学侵蚀
chemical creep 化学蠕变
chemical cure 化学处理；化学同化；化学养护
chemical damage 化学性损伤
chemical deaeration 化学除气
chemical decay 化学分解
chemical decomposition 化学分解
chemical degradation 化学剥蚀；化学降解
chemical demulsifier 化学破乳剂
chemical denudation 化学剥蚀
chemical deposit 化学沉积；化学堆积
chemical deposition 化学淀积
chemical deposition method 化学沉淀法
chemical derivative 化学衍生物
chemical derusting 化学除锈

chemical desalting 化学脱盐
chemical detection 化学检测
chemical diffusion 化学扩散
chemical dispersant 化学分散剂
chemical dispersion 化学分散
chemical dissolution 化学溶解
chemical dosing 化学剂量
chemical durability 化学耐久性
chemical effect 化学效应
chemical eluvial mineral deposit 化学残积矿床
chemical eluviation 化学淋滤
chemical equilibrium 化学平衡
chemical equivalent 化学当量
chemical erosion 化学侵蚀；化学溶蚀
chemical erosion drill 化学腐蚀钻井
chemical etching 化学腐蚀；化学侵蚀
chemical evolution 化学演化
chemical exfoliation 化学剥离作用
chemical exploration 化探
chemical explosives 化学炸药
chemical extraction 化学萃取
chemical feed pump 注化学剂用泵
chemical feeder 化学品加料机；化学药剂投加器
chemical films 化学薄膜
chemical flocculation 化学絮凝
chemical flood simulator 化学驱油模拟模型
chemical flooding 化学剂驱
chemical foam extinguisher 化学泡沫灭火器
chemical fossil 化学化石
chemical fragmentation 加化学溶液破碎岩石
chemical gaging 化学测流速法
chemical geodynamics 化学地球动力学
chemical geohydrology 化学水文地质学
chemical geology 化学地质学
chemical geothermometer 化学地球温度计
chemical ground solidification 化学剂加固地层
chemical grout 化学灌浆
chemical grouting 化学灌浆；化学加固；化学注浆
chemical grouting material 化学灌浆材料
chemical heterogeneity 化学不均一性
chemical hydrogen consumption 化学耗氢量
chemical hydrology 水文化学
chemical improvement 化学改良
chemical index of alteration 化学蚀变指数
chemical indicator 化学指示剂
chemical inertness 化学惰性
chemical injection 化学灌浆；化学注浆
chemical injection process 化学注浆法
chemical interaction 化学相互作用
chemical intermediate 化学中间体
chemical leaching 化学淋滤
chemical limestone 化学成因石灰岩；化学灰岩
chemical logging 化学录井
chemical milling 化学蚀刻
chemical mineral 化学矿物
chemical mining 化学采矿法
chemical modification 化学改性
chemical mud 化学泥浆
chemical oxygen demand 化学需氧量
chemical petrologic classification 岩石化学分类
chemical petrology 化学岩石学
chemical pipe cutter 化学割管器
chemical piping 化学潜蚀；化学管涌
chemical plugging 化学封堵
chemical polish 化学抛光剂
chemical potential energy 化学势能
chemical precipitation 化学沉淀作用；化学淀积作用
chemical precipitation ore deposit 化学沉淀矿床
chemical precipitation process 化学沉淀法
chemical preservation 化学防腐
chemical property 化学特性
chemical property of rock 岩石化学性
chemical pulp 化学浆
chemical pump 化学剂泵
chemical purification 化学净化法
chemical reaction power 化学反应力
chemical reactor 化学反应器
chemical refuse 化学垃圾
chemical regeneration tower 化学再生塔
chemical reinforcement 化学加固法
chemical remanent magnetization 化学剩余磁化
chemical removal of sand 化学除砂
chemical residue 化学风化残留物
chemical resistance 耐化学性
chemical rock 化学岩
chemical sampling 化学取样
chemical sand control 化学防砂
chemical seal 化学密封
chemical sediment 化学沉积物
chemical sedimentary deposit 化学沉积矿床
chemical sedimentary differentiation 化学沉积分异作用
chemical sedimentary rock 化学沉积岩
chemical sediments 化学沉积物
chemical segregation 化学分离作用
chemical sequestration 化学封存
chemical shift 化学位移
chemical softener 化学软化剂
chemical softening 化学软化
chemical soil consolidation 化学法固结土壤
chemical soil stabilization 土的化学加固
chemical solidification 化学固结；化学加固
chemical solution 化学溶液
chemical solvent 化学溶剂
chemical stability 化学稳定性
chemical stabilization 化学加固
chemical stabilizer 化学稳定剂
chemical test 化学测试
chemical tracer 化学示踪剂
chemical treatment 化学处理
chemical type of groundwater 地下水化学类型
chemical unconformity 化学不整合
chemical union 化合
chemical valence 化合价
chemical vapor deposition method 化学气相沉淀法
chemical vapour deposition 化学汽相淀积
chemical vapour deposition process 化学气相沉积
chemical variation 化学变化
chemical waste disposal 化学废料处理
chemical water 结合水
chemical water plugging 化学堵水

chemical water treatment	化学水处理
chemical waterflooding	化学注水
chemical weakening and hardening	化学软化和硬化作用
chemical-electric treater	电化学脱水器
chemical-flooding model	化学驱油模型
chemical-generation process	化学发气法
chemical-grade coke	化学级焦炭结焦性用
chemical-grouted soil	化学灌浆土
chemically active	化学活性的
chemically bound water	化合水；化学束缚水
chemically combined water	化合水
chemically formed rock	化学成因岩石
chemically neutral oil	化学中性油
chemically precipitated mineral	化学沉积矿物
chemically precipitated sludge	化学沉淀污泥
chemically stabilized earth lining	化学加固的土质衬砌
chemico-pile method	生石灰桩施工法
chemi-ionization	化学电离
chemiluminescence	化学发光；冷焰光
chemise	衬；衬墙；复面层；土堤护面墙
chemism	化学机理；化学历程
chemisorbed film	化学吸附膜
chemisorbed layer	化学吸附层
chemisorption	化学吸附作用；化学吸着
chemistry analysis in infiltration liquid from soil	土壤淋滤液分析
chemistry mechanics coupling	化学力学耦合
chemo osmotic effect	化学渗透效应
chemocline	化变层
chemofacies	化学相
chemogenic deposit	化学沉积
chemographic	化学图解的
chemolysis	化学分解
chemoorganotrophic	化学有机营养
chemopause	化学层顶
chemorheology	化学流变学
chemosmosis	化学渗透
chemosphere	臭氧层
chemostratigraphy	化学地层学
chemosynbiosis	化学共栖关系
chemosynthesis	化能合成；化学合成
chenevixite	绿砷铁铜矿
chengdeite	承德矿
chenier	沼泽沙丘
chenier plain	海沼沙脊平原
cheralite	富钍独居石；磷钙钍矿
cheremchite	乔仑油页岩
chernovite	砷钇石
chernozem	黑钙土；黑土
chernozem meadow solonetz	黑钙土状草甸碱土
chernozem meadow-steppe solonetz	黑钙土状草甸草原碱土
chernozem rendzinas	黑钙土状黑色石灰土
chernozem soil	黑钙土
chernozem steppe solonetz	黑钙土状草原碱土
chernozem-like alluvial soil	黑钙土状冲积土
chernozem-like meadow soil	黑钙土状草甸土
chernozem-like soil	黑钙土状土壤
chernykhite	钡钒云母
cherokite	致密褐砂
cherry coal	樱煤
cherry picker	车载升降台；车载式吊车
cherry picker hoist	调车绞车；车载式吊车起重机
cherry red heat	樱桃红热
chert	黑硅石；燧石；角岩
chert aggregate	燧石骨料
chert argillite contact	燧石泥板岩接触
chert nodule	燧石结核
chertification	燧石化作用
chert-shale	燧石页岩
cherty	燧石的
cherty ground	含燧石的地层
cherty limestone	硅质灰岩；含燧石的石灰岩
cherty loam	石英质壤土
cherty soil	石英质土
chessboard face	棋盘式工作面
chessboard manner	交错排列法；棋盘形排列法
chessboard structure	交错构造；棋盘构造
chessom	散粒的；疏松土壤
chessylite	蓝铜矿；石青
chest trouble	肺病
chestnut coal	栗级无烟煤
chestnut soil	栗钙土；栗土
cheuch	沟谷；峡谷
chevkinite	硅钛铈钇矿
chevron cross-bedding	人字形交错层理
chevron drain	人字形排水沟
chevron fold	尖顶褶皱；尖棱褶皱
chevron gear	人字齿轮
chevron mark	尖楞刻痕；锯齿痕
chevron pattern	人字形图形
chevron profile	尖顶形剖面
chevron ring	V形圈；人字密封圈
chevron structure	人字形构造
chevron-shaped fishplate	人字形夹板
chevron-style fold	尖顶式褶皱
chewed	粉碎的；磨碎的
chews	中粒煤
chian	柏油；沥青
chiasmolithus	叉心颗石
chiastolite	空晶石
chiastozygus	交叉轭石
chibinite	胶硅铈钛矿；粒霞正长岩
Chicago caisson	小型墩台沉井；芝加哥式沉箱
chicken blood stone	鸡血石
chicken grit	大理石碴；大理石屑
chicken ladder	防滑跳板；有刻痕的圆木梯或方木梯
chicken wire structure	鸡笼网状构造；网状构造
chicken wire texture	网格状结构
chicken-wire fabric	网状组构
chicken-wire pattern	鸡笼状井网；网状
chicksan	高压旋转接头
chicksan line	固井管线
chief axis	主轴
chief geostatistician	首席地质统计师
chief resident engineer	驻段工程师
childrenite	磷铝铁石
chile bar	粗铜棒
chile niter	智利硝石
chile rise	智利隆起
chile saltpeter	钠硝石；硝酸钠；智利硝石

chilean mill	智利式辊碾机
chileite	智利石
chill	冷冻
chill block	冷铁；三角试块
chill down area	冷却区
chill hardening	激冷硬化；冷硬化
chill point	冻结点；凝固点
chillagite	钼钨铅矿
chilled border	冷凝边
chilled casting	冷硬铸造
chilled contact	冷凝接触带
chilled distillate	冷冻馏出液
chilled iron	冷铸生铁
chilled iron shot	冷铸钢粒
chilled margin	冷却边缘；冷凝边
chilled shot bit	冷铸钢粒钻头
chilled shot drill	冷钢钻粒钻机
chilled shot drilling	冷钢粒钻进；钻粒钻进
chilled soil	冻土
chilled steel	淬火钢；冷铸钢
chilled-shot system	钻粒钻进法
chiller	冷冻装置；冷却器；致冷装置
chilling	淬火；冷凝
chilling effect	致冷效应
chilling layer	冷硬层
chilling machine	冷凝器
chilling method	冻结法
chilling pipe	顶接管
chilling press	冷压机
chilling tank	冷却槽
chill-pressing	低温压制
chiltonite	葡萄石
chiluite	赤路矿
chimney	冰川竖坑；冰川井；出风筒；管状矿体
chimney caving	筒状陷落
chimney coping	烟囱盖顶
chimney drain	竖向截水体；垂直排水系统
chimney forming	形成管洞
chimney ore body	柱状矿体
chimney radius	井下核爆冲峒半径
chimney rock	柱状石
chimney sand drain	沙质排水竖井
chimney ventilation	烟囱抽风
china clay	瓷土；高岭土
China continent	中国大陆
China intensity scale	中国烈度表
China plate	中国板块
China white marble	汉白玉
China wood oil	桐油
China-pacific-philippine triple junctions	中国太平洋菲律宾板块三合点
chinastone	瓷石；瓷土石；高岭土化花岗岩
chine	缝隙；隆起
Chinese earthquake information network	中国地震信息网
Chinese platform	中国地台
Chinese pump	中国式水车
Chinese seismic intensity	中国地震烈度
Chinese seismic intensity zoning	中国地震烈度区划图
Chinese wax	中国蜡
chink	缝隙；龟裂；裂缝
chinking	裂开；塞缝
chinkolobwite	硅镁铀矿
chinless process	无屑加工
chinley coal	高级烟煤；块煤
chinney-rubble zone	柱状破碎带
chip	金属屑；晶片；片石；切屑；石屑；切片
chip ballast	碎石碴
chip blasting	浅层爆破；浅孔爆破
chip catcher	岩粉收集器
chip clearance	切屑间隙
chip guard	防屑罩
chip hold down effect	压持效应
chip morphology	岩屑形态
chip of stone	碎石片
chip rejecter	排除碎屑装置
chip removal	清除碎屑；消除碎屑
chip sample	岩屑样品
chip sampler	碎片取样
chip sampling	岩屑取样
chip screen	木片筛
chip sealing	石屑封盖
chip slip rate	岩屑滑落速度
chip stone	碎石片
chipboard	刨花板
chipbreaker	断屑槽；断屑器
chip-channel method	刻槽取样法
chiphragmalithu	隐叠颗石
chipless machining	无屑加工
chipless working	无屑加工
chipped	粉碎的；磨碎的
chipped ashlar	条石
chipped stone tool	打制石器
chipped tool	打制石器
chipper	削片机；切片机
chipping	碎石片；碎屑
chipping action	凿削作用
chipping chisel	凿刀；平头凿
chipping process	切削过程
chipping type bit	切削型钻头
chipping type wear	自锐式磨损
chipping-crushing action	切削-破碎作用
chippings	碎片；碎屑
chippings spreader	碎石撒布机
chippy	碎片的；多屑片的
chippy cage	辅助罐笼
chippy hoist	辅助提升机
chippy shaft	辅助竖井
chips	石屑；碎石
chipway	钻粉排出槽道
chipway space	钻粉排出空隙
chirvinskite	含硫沥青；贫硫沥青
chisel	錾子；凿子；扁铲；錾子；凿子
chisel bit	扁凿钻头；冲击式钻头；凿形钻头
chisel bit tool	凿形钻头
chisel edge	凿锋；凿尖；凿切削
chisel edge angle	凿尖角
chisel insert	楔形镶齿
chisel point	凿刃
chisel shaped	凿尖的
chisel shaped bit	一字钎头；凿头钻头

chisel shoe	楔形管鞋
chisel steel	凿钢
chisel teeth	楔形齿
chisel tool	凿形工具
chisel type insert bit	楔形硬合金齿钻头
chiseling	凿边；凿平
chisel-like	凿形的
chiselling	用凿削切
chiselly	粗粒的；砾石的
chisley soil	石质土
chitin	几丁质；甲壳质
chitinous	几丁质的，壳质的
chitter	泥质薄铁矿夹层
chiviatite	硫铅铋矿
chladnite	顽无球陨石
chlamydospore	厚垣孢子
chloanthite	复砷镍矿
chlopinite	赫洛宾矿
chloraluminite	氯铝石
chlorapatite	氯磷灰石
chlorargyrite	角银矿；氯银矿
chlorate	氯酸盐
chlorating agent	氯化剂
chloration	加氯作用；氯化作用
chlorazotic acid	王水
chlorhydrate	盐酸盐
chlorhydric acid	盐酸
chloride	氯化物；漂白剂
chloride as tracer	氯化物示踪剂
chloride brine	氯化物型卤水
chloride contaminated concrete	受氯化物损害的混凝土
chloride content	氯化物含量
chloride diffusion	氯化物扩散
chloride extraction	除氯；氯化物萃取
chloride flux	氯离子浓度
chloride ion titration method	氯离子滴定法
chloride log method	氯化物录井法
chloride of lime	漂白粉
chloride stabilization	氯盐土稳定法
chloride storage battery	氯蓄电池
chloride stress cracking	氯化物应力破裂
chloride trends	氯化物含量趋势
chloride water	含氯化物水；氯化物型水
chloride-magnesium type	氯化镁型
chloride-sodium water	氯化钠水
chloridization	氯化作用
chloridizing roasting	氯化焙烧
chloridometer	氯量计
chlorimet	耐热镍基合金
chlorinate	氯化
chlorinated lime	氯化石灰；漂白粉
chlorinated paraffin	氯化石蜡
chlorinated solvent	氯化溶剂
chlorinating	氯化
chlorination	氯化酌；氯化作用
chlorination chamber	氯化池
chlorination roasting	氯化焙烧
chlorinator	氯化器
chlorinator installation	加氯设备
chlorine	氯；氯气
chlorine compounds	氯化合物
chlorine concentration	氯含量
chlorine dating method	氯测年法
chlorine dosage	加氯量
chlorine generator	氯气发生器
chlorine kill effectiveness	氯杀菌效果
chlorine log	氯测井
chlorine method	氯气法
chlorine water	氯水
chlorine-bromine ratio	氯溴比值
chlorine-sensitive logging	氯敏测井
chlorinity	含氯量；氯度
chlorinolog	含氯量录井
chlorion	氯离子
chlorite	绿泥石；亚氯酸盐
chlorite facies	绿泥石相
chlorite jade	绿泥石玉
chlorite law	亚氯酸盐法
chlorite phyllite	绿泥石千枚岩
chlorite schist	绿泥石片岩
chlorite shale	绿泥页岩
chlorite slate	绿泥板岩
chlorite zone	绿泥石带
chloritization	绿泥石化
chloritoid	硬绿泥石
chloritoid schist	硬绿泥石片岩
chlormanganokalite	钾锰盐
chloroantimonate	氯化锑酸盐
chloroethane process	氯乙烷分选法
chloroform bitumen	氯仿沥青
chloroform ratio	氯仿比率
chloromelanite	暗绿玉
chloromelanitite	浓绿玉岩
chlorometry	氯量滴定法
chloropal	绿蛋白石
chlorophaeite	蠕绿泥石
chlorophane	绿萤石
chlorophyll coal	叶绿素腐泥褐煤
chlorophyre	绿英玢岩
chlorosity	体积氯度
chlorospinel	绿尖晶石
chlorothionite	钾氯胆矾
choanoid	漏斗形
chock	垛式支架；木楔；垫木；制动楔子；制动垫块
chock advancing wave	楔超前波；楔推进波
chock block	捣矿机筛垫块；垛架垫楔
chock cycle time	支架动作循环时间
chock filter	支架过滤器
chock mat	楔垫
chock pile	填塞桩
chock point	堵塞点
chock prop	丛柱；加强柱；木楔支柱；棚柱脚撑杆
chock release	木楔拆除
chock shield	支撑掩护式支架
chock shield support	支撑掩护式支架
chock support	垛式支架
chock timber	木垛；楔形木材
chock timbering	垛式支架
chock wedge	木垛楔
chocked screen	塞孔的筛子

chocking up	塞紧
chocking-up degree	堵塞程度
chock-shield-type power support	支撑掩护式液压支架
chock-type support	楔形支架
chocolate	红褐色；深褐色
chocolate block structure	巧克力块构造
chocolate tablet structure	巧克力块构造
chocolate-table structure	巧克力方盘构造
choice	抉择
choke	节流器；阻气门
choke blasting	堵塞爆破
choke coil	扼流圈；抗流圈
choke feeding	堆挤装料；滞塞给料
choke flow line	节流管线
choke manifold	节流管汇
choke material	填塞材料
choke plug	塞头；阻流塞
choke stone	拱顶石；拱心石
choke valve	节流阀；阻流阀
choke-capacity coupling	扼流圈电容耦合
choke-condenser filter	电感电容滤波器
choke-crushing	破碎机的挤塞破碎
choked crushing	阻塞破碎
choked layer of sand	砂填层
choked opening	堵塞巷道
chokedamp	窒息瓦斯
choke-fed	滞塞给料的
choke-fed rolls	挤塞给料辊碎机
choke-flow connection	节流连管
chokepoint	阻塞点
choker rope	吊索环
choking action	阻塞作用
choking cavitation	阻塞性空化
choking factor	扼流系数
choking flow	淤泥覆盖的
choking gas	窒息性毒气
choking resistance	扼流电阻
choking section	临界截面
choking turns	扼流圈
choking up	堵塞；淤塞
choky	窒息性的
chomata	隔壁褶皱
chondrin	软骨胶
chondrite	球粒陨石
chondrite earth model	球粒陨石地球模型
chondrite-normalized	球粒陨石标准化的
chondritic meteorite	球粒陨石
chondritic unfractional reservoir	未分馏球粒陨源区
chondrocranium	软骨颅
chondrodite	粒硅镁石
chondrule	陨石球粒
chondrus	陨石球粒
chonolite	畸形岩盘
chonolith	岩铸体
chooser	选择器
chop	风浪突变；港口；切碎；裂缝；裂口
chop coring	劈取岩心
chop feeder	箱式计量给料器
chop hill	沙丘
chopped fiber	短纤维
chopped mode	斩波式
chopped pulse	削顶脉冲；缩短脉冲
chopper	断续装置；切碎机
chopper amplifier	斩波放大器
chopper modulation	断续器调制
chopping	破碎；切砍；斩波
chopping bit	冲击式钻头
chopping disk	断续器转盘
choppy	裂缝多的；有皱纹的
choppy cross-bedding	皱纹状交错层理
choppy grade	锯齿形纵断面
chord	弦
chord member of truss	桁架弦杆
chord modulus	弦线模量
chord offset method	弦线支距法
chord splice	桁弦接合板
chord stress	桁弦应力
chordal thickness	弦齿厚
chorismite	混合岩
chorismitization	混合岩化作用
chorisogram	等值图
chorisopleth	等值线图
c-horizon	母质层
chorogram	等值图
chorography	地方地理
choroisotherm	等温线
chorologic unit	分布单位
choropleth	等值线图
choroplethic map	等值区域图
chott	沙漠中浅盐水湖或洼地
christianite	钾沸石；碱花岗岩
christman method	克里斯曼法；骤然加剧的方法
christmas tree	采油树；井口带有阀门的管道；井口装置
christmas tree gage	采油井口压力表
christmas tree receiver plate	采油树接收板
christmas-tree assembly	采油树总成
christmas-tree-type reaming head	塔型扩孔钻头
christobalite	方石英
Christoffel equation	克里斯托弗尔方程
Christoffel matrix	克里斯托弗尔矩阵
christograhamite	脆沥青；硅质中铁陨石
christophite	黑闪锌矿；铁闪锌矿
chroma	色度
chromadiaing	铬酸盐钝化处理
chromang	克拉曼格不锈钢
chromate	铬酸处理；铬酸盐
chromate coating	铬酸盐涂层
chromatic	色彩的；着色的
chromatic aberration	色像差
chromatic difference	色差
chromatic distortion	彩色失真
chromatic resolution	色分辨率
chromatic sensitivity	感色敏感度
chromatogram	色层谱
chromatograph	层析仪；色谱
chromatographic	色层的
chromatographic contact print method	色谱接触印相法
chromatography	层析法；色层分离法；色层分析法
Chromax	克罗马克铁镍铬耐热合金
Chromax bronze	克罗马克青铜

chrome	铬
chrome alum	铬矾
chrome chert	铬燧石
chrome clinochlore	铬斜绿泥石
chrome diopside	铬透辉石
chrome garnet	铬榴石
chrome iron ore	铬铁矿
chrome iron stone	铬铁矿
chrome nickel steel	铬镍钢
chrome ocher	铬华
chrome plated	镀铬的
chrome plated mold	镀铬模具
chrome plating	镀铬
chrome sand	铬砂
chrome spinel	铬尖晶石；镁铬铁矿
chrome stainless steel	铬不锈钢
chrome tungsten steel	铬钨钢
chrome vanadium steel	铬钒钢
chromel	镍铬合金
chrome-manganese steel	铬锰钢
chrome-molybdenum potentiometer	铬-钼合金膜电位器
chrome-molybdenum steel	铬钼钢
chrome-nickel alloy	铬镍合金
chrome-nickel-molybdenum steel	铬镍钼钢
chromic acid	铬酸
chromic alum	铬矾
chromic anhydride	铬酐
chromic hydroxide	氢氧化铬
chromic oxide-ferrous oxide ratio	铬铁比
chromite	铬铁矿；亚铬酸盐
chromite deposit	铬铁矿床
chromite-dolomite	含铬铁矿白云岩
chromitite	铬铁岩
chromium	铬
chromium humate	腐殖酸铬
chromium minerals	铬矿类
chromium nitride precipitation	氮化铬析出
chromium plating	镀铬
chromium polish	铬抛光剂
chromium stainless steel	铬不锈钢
chromium steel	铬钢
chromium tungsten steel	铬钨钢
chromium vanadium steel	铬钒钢
chromium-iron lignosulfonate	铁铬盐
chromium-plated brass	镀铬黄铜
chromizing	铬化；渗铬
chromogenic reaction	显色反应
chromogenic reagent	显色试剂
chromosome	染色体
chromosorb	红硅藻土
chromotropic acid	铬变酸
chromotype	彩色石印图
chromous compound	亚铬化合物
chromow	铬钼钨钢
chromowulfenite	铬钼铅矿
chrompicotite	硬铬尖晶石
chron	年代
chronic	长期的
chronic poisoning	慢性中毒
chronic sander	经常出砂井
chronic tilting	缓慢偏斜
chronic water shortage	长期缺水；持续缺水
chronicity	长期
chronicle	年代
chronocline	年代差型；时变生物群
chronocorrelation	年代对比
chronogenesis	时序
chronogeochemistry	年代地球化学
chronograph	计时器
chronographic chart	年代地层图
chronography	时间记录法
chronohorizon	年代层位
chrono-interferometer	计时干涉仪
chronolith	年代地层单位；时间地层单位
chronolithologic	年代地层的
chronolithologic unit	年代地层单位
chronologic age	地层年代
chronologic scale	地质年代表
chronologic subspecies	年代亚种
chronologic table	编年表；年代表
chronologic time scale	地层年代表
chronological	按照年代的；编年的；年代学的
chronological order	年代次序
chronological scale	地质年代表
chronological table	地质年代表
chronological time scale	地层年代表；地层时序表
chronology	年表；年代学；时序；地质年代；年表；年代学
chronomere	时间单位
chronometer	精密测时计；精密计时计
chronometric data	精密计时数据
chronometric tachometer	计时式转速器
chronometry	年代测定学；年龄测定；时间测定法
chronopher	报时器
chronosequence	年代序列；年龄系列
chronostratic	年代地层的
chronostratic scale	年代地层表
chronostratic unit	地层年代单位
chronostratigraphic	年代地层的
chronostratigraphic boundary	年代地层界线
chronostratigraphic classification	年代地层划分
chronostratigraphic correlation chart	年代地层对比表
chronostratigraphic framework	年代地层格架
chronostratigraphic unit	年代地层单位；时代地层单位
chronostratigraphic zone	年代地层带
chronostratigraphy	年代地层学
chronotaxis	时间相似性
chronotopocline	时空渐变群
chronotron	延时器
chronozone	年代带；时带
chrysitis	致密硅页岩
chrysoberyl	金绿宝石
chrysocolla	硅孔雀石
chrysocollite	硅孔雀石
chrysodor	似大理石
chrysolite	贵橄榄石
chrysoprase	绿玉髓
chrysotile	温石棉；纤维蛇纹石
chrysotile asbestos	温石棉；纤维蛇纹石
chthonic	深海碎屑沉积的

chthonic sediment　深海碎屑沉积物
chthonisothermal line　等地下温度线
chuck　卡盘；夹盘；钻头夹头
chuck assembly　卡盘装置
chuck bare　凿岩机导向衬套
chuck block　捣矿机筛垫块
chuck bushing　卡头衬套；夹钎器衬套
chuck drilling　冲击钻进
chuck drilling tool　冲击钻具
chuck handle　卡盘扳手；钻夹头钥匙
chuck nut　夹盘螺母
chuck on　复轨器
chuck plate　夹盘
chuck ring　夹圈；锁紧环；压环
chuck wrench　夹盘扳手
chucker　卡盘车床
chucker on　复轨器
chucking machine　卡盘式转塔车床
chuco　芒硝层
chugging　嚓嘎声；功率振荡
Chukchi shelf　楚科奇海陆架
chum drill　冲击钻
chump　木块；木片
chunam　灰泥；朱南
chunam cover　朱南护面；灰泥护面
chunam plaster　灰泥抹面；灰泥批荡
chunam surface　灰泥表层
chunam surface slope　灰泥面斜坡
chunk　大块；大量；碎石
chunk method　土块试验法
chunk reduction　大块破碎
chunk sample　块状试样
chunk structure　碎块构造
chunky rock　大块岩石
churchite　水磷铈筒
churn　搅拌；搅筒
churn bit　旋冲钻头
churn drill　冲击钻；旋冲钻；顿钻
churn drill rig　钢绳冲击式钻机
churn drill sludge　钢丝绳冲击钻钻泥
churn drilling　冲击钻进；钢丝绳冲击式钻进；无游梁绳式顿钻
churn drilling method　冲击钻井方法
churn flow　涡流
churn of wave　猛冲海岸
churn type percussion drill　钢绳冲击钻机；无游梁绳式顿钻钻机
churn-drill bit　冲击钻头；钢绳冲击式钻头
churn-drill blasting　钢丝绳冲击钻眼爆破
churner　手持式长钎子；手摇式长钻
churning　搅拌
churning stone　石渣；碎石
churning time　搅拌时间
chute　斜槽；滑道；瀑布；急流
chute air-door　溜口风门
chute bar　急流沙坝
chute blaster　消除溜井堵塞的爆破工
chute blasting　溜井爆破
chute block　陡槽消力墩
chute blockage protection　溜眼堵塞防护

chute board　滑道板
chute bottom　溜口底
chute breast　溜槽煤房
chute caving　天井放矿采矿法
chute closure dike　串沟锁坝
chute compartment　贮存和下放矸石的隔间
chute cutoff　流槽截止
chute deposit　串沟沉积
chute discharge　倾斜槽卸载；斜槽出料
chute door　溜槽闸门
chute draw　溜槽放矿
chute drawer　溜槽放矿工
chute drop　溜槽浇筑
chute feeder　溜槽式给料器
chute gap　溜井；溜眼
chute gate　溜槽闸门
chute gate operator　溜放矿工
chute jaw　溜子口颚板
chute line　溜槽线
chute lining　溜槽衬里
chute lip　溜槽嘴
chute loading　溜槽装载
chute method of loading　溜槽装载法
chute mouth　溜槽口；溜井口
chute mouth control　溜槽口控制
chute opening　溜眼或溜道的放矿口
chute pitch　溜槽斜角
chute puller　溜子口装车工
chute pulling　操纵溜口闸门；溜槽口放矿
chute sampler　溜槽取样机
chute sill　溜口底
chute spacing　溜口间距
chute spillway　陡槽式溢洪道；河岸式溢洪道
chute structure　流槽构造
chute system　平硐溜井开拓
chute tapper　溜子口放矿工
chute tapping　溜口装车
chute template　漏口框的托底横梁
chute throat　溜槽口
chute work　溜口装车
chute-and-pool structure　槽泊构造
chute-breast method　仓储采煤法；贮煤式房柱开采法
chute-breast system　溜槽运煤的房柱式采煤法
chute-served stope　溜口放矿回采工作面
chuting　溜槽送料；溜槽运送
chuting arrangement　溜道布置
chuting concrete　溜槽输送的混凝土
chuting system　溜槽系统
chysiogenous　熔岩的
ciffusion coefficient　扩散系数
cigarette wrapping　直缝缠绕
cigarette-filter tow　卷烟过滤嘴丝束
Cilgel　西尔杰尔炸药
cimentfondu　铝土水泥
ciminite　橄辉粗面岩；正边粗面粗玄岩；漂砾黏土；水土
cinch anchor　加固锚栓
cinder　灰烬；矿渣；炉渣；煤渣；火山渣
cinder aggregate　煤渣骨料
cinder bed　煤渣床

cinder block	煤渣砖
cinder brick	炉渣砖；煤渣砖
cinder catcher	集尘器
cinder coal	劣质焦炭；天然焦
cinder concrete	炉渣混凝土；煤渣混凝土；煤渣混凝土
cinder concrete brick	煤渣混凝土砖
cinder cone	火山口；火山渣锥
cinder dump	废渣堆
cinder fall	渣坑
cinder fill	煤渣填土
cinder mill	煤渣粉碎机；碎渣机
cinder soil	褐色土
cinder tub	渣车
cinder wool	玻璃棉；矿渣棉
cinder yard	堆渣场
cindery	灰烬的；煤渣的
cindery coal	灰烬煤；炼焦煤
cindery rock	火山渣岩
cine-camera	电影摄影机
cineholomicroscopy	显微全息电影照相术
cinephotomicrography	显微摄影
cinerite	玻屑火山碴凝灰岩；火山灰；火山凝灰岩
cinetheodolite	高精度光学跟踪仪
cinglos	单斜崖面
cinley coal	块煤
cinnabar	辰砂；朱砂
cinnabarit	辰砂
cinnamon garnet	肉桂榴石
cinnamon soil	褐土
ciplyite	磷硅钙石
cipolin	云母大理石
cipolino	云母大理石
cipollin	云母大理岩
cippoletti weir	梯形堰
circalittoral zone	环岸带
circarc gear	圆弧齿轮
circle	度盘；循环
circle bend	圆曲管
circle brace	圆拉条
circle coefficient	圆常数
circle coordinates	圆坐标
circle cutter	环刀
circle diagram	圆图
circle distribution	环形配水
circle division	度盘分划
circle graduation	度盘分划
circle jack	圆轨千斤顶
circle location	圆弧定位法
circle of failure	破坏圆
circle of influence	影响范围
circle of Mohr	莫尔圆；摩尔圆
circle of principal stress	主应力圆
circle of rupture	破裂圆面
circle of stress	应力圆
circle setting	度盘定位
circle shear	圆盘剪
circle sliding	圆弧形滑坡
circle sprinkler	环动喷灌器
circle test	回转试验；循环试验
circle vector diagram	圆向量图
circle-hyperbolic system	圆双曲线系统
circlet	小环；小圈
circline	环形
circline lamp	环状荧光灯
circlip	簧环
circuit	环路
circuit analysis	电路分析
circuit branch	支路
circuit closure	环线闭合差
circuit diagram	电路图
circuit equivalent	等效电路
circuit error	环线闭合差
circuit feed	返砂量；回路给料；循环给料
circuit grouting	循环灌浆
circuit guard	电路保护；电路闭塞
circuit header	环路集管
circuit logic	电路逻辑
circuit railroad	环形铁道
circuit sensitivity	电路灵敏度
circuit test	爆破网路试验；爆破网路试验
circuit tester	电路测试器；万用表；电路测试器
circuitation	旋转
circuit-breaker carriage	断路器小车
circuit-breaker oil-storage tank	断路器油箱
circuition	绕轴转动
circuitousness	绕行
circuity	环绕
circulant	循环行列式
circular	通告；通知；循环的；圆形的；周节齿轮
circular airway	圆形风巷
circular aperture	圆孔
circular aquifer	环形含水层
circular arc	圆弧；圆弧拱
circular arc analysis	圆弧分析法
circular arc motion	圆弧运动
circular arch	钢制环形支架；弧拱；圆拱
circular are analysis	圆弧分析法
circular base slab	圆形支撑板
circular beam	圆梁
circular bit	旋转钻孔器
circular borehole	圆形井眼
circular boundary	圆形边界
circular bubble	气泡水准器；圆水准器
circular cage	圆形罐笼
circular casing pump	圆形套管泵
circular cell cofferdam	圆形格栅围堰
circular clamp	环形夹具
circular closed support	环形支架
circular cofferdam	圆形围堰
circular coiled spring	圆形螺旋弹簧
circular cone	圆锥
circular conical fold	圆锥形褶皱
circular conical surface	圆锥面
circular constant	圆周率
circular control weir	循环控制堰
circular convolution	循环褶积；圆周卷积
circular crack	环状裂隙
circular cross section	圆截面
circular culvert	圆形涵洞
circular cut	环形掏槽；圆形掏槽

circular cut joint	圆槽接头	circular rotating field	圆形旋转磁场
circular cutter	圆柱铣刀	circular saw	圆锯
circular cylinder	圆柱	circular saw blade	圆锯片
circular cylinder function	圆柱函数	circular scan	圆周扫描
circular cylindrical coordinate	圆柱坐标	circular screen	圆孔筛
circular drainage system	环行排水系统	circular seam	环缝
circular earthquake	环形地震	circular seam welder	环路滚焊机
circular excavation	圆形巷道	circular seam welding	环缝焊接
circular face	圆形工作面	circular section	圆形断面
circular fault	环形断层	circular section tunnel	圆形断面隧道
circular feature	环形要素；环状特征	circular sector	扇形
circular feed	回转进给	circular set	圆形支架
circular file	圆锉	circular sewer	圆形污水管
circular flood	环状注水	circular shaft	圆形井筒；圆轴
circular flow	环行流动	circular shaped shield	圆形盾构
circular footing	圆形基脚	circular shears	圆盘式剪断机
circular foundation	圆形基础	circular sliding method	圆弧滑动法
circular gate	环形内浇口	circular sliding plane method	圆形滑动面法
circular grading table	圆形转动拣矸台	circular sliding surface	圆弧滑动面
circular grindability	外圆可磨性	circular slip	圆弧滑动
circular groove	环形槽	circular slip surface	圆弧滑动面
circular ground-loop	环形接地回路	circular sluice	循环冲洗
circular helix	圆柱螺旋线	circular soil cutter	切土环刀
circular histogram	圆形直方图	circular spirit level	圆形气泡水准仪
circular hole	圆孔	circular stairs	盘旋楼梯
circular hoop	圆箍	circular stationary buddle	固定圆形淘汰盘
circular index	圆分度头	circular steel support	钢圈支架
circular jack	环形千斤顶	circular stone	磨盘；磨石
circular jet	圆截面射流；圆射流	circular support	圆形支架
circular kiln	圆筒式干燥炉	circular surface	圆弧面
circular level	圆水准器；万向水泡	circular track test	环道试验
circular lining	圆形衬砌	circular track-type grab loader	环行轨道式抓岩机
circular loaded area	圆形荷载面积	circular tunnel	圆形隧道
circular magnet	环形磁铁	circular type heater	圆筒炉
circular main	环形干管	circular uniform load	圆形均布荷载
circular measure	弧度法	circular valve	环形阀
circular metal flume	圆弧形金属水槽	circular ventilation shaft	圆形通风井
circular mil	圆密尔	circular waveguide	圆波导
circular motion	圆周运动	circular weir	圆形堰
circular movement	旋转运动	circular weld	环形焊缝
circular orbit	环形轨道	circular wood-stave flume	圆弧形木板条水槽
circular order	循环次序	circularity	环状
circular orifice	圆形孔板；圆隔板	circularity ratio	圆度比
circular pan-mixer	圆盘式搅拌机	circularly polarized shear	圆偏振切变波
circular path	圆周轨迹	circular-slide analysis	圆弧滑动分析
circular pattern	圆形组合	circular-throw screen	偏心振动筛
circular picking table	圆形转动拣矸台	circulate-and-weight method	循环加重压井方法
circular pier	圆形桥墩	circulated gas-oil ratio	回注气油比
circular pipe	圆管	circulating	环流
circular pitch	圆周齿距	circulating bottom current	河底环流
circular pitch error	周节误差	circulating control device	循环控制装置
circular plane	圆刨	circulating current	环流
circular polarization	圆偏振	circulating device	循环装置
circular rail	圆形导轨	circulating differential fill collar	循环压差式灌浆接箍
circular rammer	圆夯	circulating differential fill shoe	循环压差式灌浆套管鞋
circular ray	圆弧射线	circulating fluid	泥浆钻探循环液；循环流体
circular representation	圆形表示法	circulating gravel packing	循环砾石充填
circular ring	圆环；转动式闸门	circulating groundwater	地下循环水
circular roadway with uniform and local support 圆形巷道统一和本地支护		circulating head	循环冲洗管头
		circulating hot oil	循环热油

circulating jar	循环震击器	circulation washer	循环冲洗工具
circulating line	循环管路	circulative load	循环载荷
circulating liquid	循环液	circulator	循环器
circulating load	循环荷载	circulatory flow	环流；循环流动
circulating loop	环路	circum	毗邻；周边
circulating lubrication	循环润滑	circum pacific	环太平洋的
circulating medium	冲洗液；清孔介质；循环介质；通货	circum pacific geothermal belt	环太平洋地热带
circulating memory	循环存储器	circum pacific island arc	环太平洋岛弧
circulating mud	循环泥浆	circum pacific seismic zone	环太平洋地震带
circulating oil pump	循环油泵	circum pacific volcanic belt	环太平洋火山带
circulating oiling	循环润滑；循环注油	circumbasin material	环盆地物质
circulating opening	泥浆循环孔	circumcenter	外心；外接圆心
circulating out	排出	circumcircle	外接圆；外切圆
circulating out fill	循环带出井底沉淀物	circum-continental	环大陆的
circulating overshot	可循环泥浆的打捞筒	circum-continental geosyncline	环陆地槽
circulating path	循环通道	circum-continental mobile belt	环陆活动带
circulating pressure	循环压力	circum-continental ring	环大陆褶皱带
circulating pump	吸水井；循环泵	circumcrust	环壳
circulating rate	流量；循环率；循环速度	circumdenudation	环状剥蚀；环状侵蚀
circulating reflux	循环回流	circumerosion	环状侵蚀
circulating sleeve	循环套筒	circumference	圆周
circulating slip socket	可循环泥浆的卡瓦式打捞筒	circumference fraise	圆周刃铣刀；圆周爪棱钻
circulating storage	循环存储器	circumference gauge	圆周规
circulating suspension	循环悬浮液	circumference weld	环向焊道
circulating system	冲洗系统；循环系统	circumferential	环绕的；圆周的；周围的
circulating tank	贮浆罐	circumferential acoustilog	井眼环形声波测井
circulating tube	循环管	circumferential corrugation	圆周波纹
circulating valve	循环阀	circumferential crack	环状裂缝；周边裂隙
circulating washer	循环冲洗工具	circumferential crack in plate	辐板圆周裂纹
circulating water	活水；冷却水；循环水	circumferential crest tooth	弧形齿顶的镶齿
circulating water consumption rate	循环用水量	circumferential dispersion	环形散射
circulating water pump	循环水泵	circumferential displacement	周向位移；环向位移
circulating water system	循环水系统	circumferential fault	环形断层；周边断层
circulating whipstock	循环式造斜器	circumferential flow	环流
circulating-type shoe	循环型管鞋	circumferential force	环向力；切向力
circulation	井漏；浚通；流通量；环流量；循环	circumferential hoop stress	圆周应力；环向应力
circulation ad converter	循环模数转换器	circumferential line load	圆周切向线荷载
circulation boring method	正循环钻孔法	circumferential microscopic sonde	环行微声波测井探头
circulation convection	环流式对流	circumferential microscopic tool	环行微声波测井仪
circulation drill	正循环钻机	circumferential motion	圆周运动
circulation flow	环流	circumferential pitch	圆周齿距；周向节距
circulation flow reactor	环流反应器	circumferential position	周边位置
circulation fluid	循环液	circumferential pressure	环向压力；围压；圆周压力
circulation grouting	循环灌浆	circumferential prestressing	环向预加应力
circulation loss	循环漏失	circumferential reinforcement	环向钢筋
circulation mode	循环模式	circumferential road	环回道路；环形道路
circulation of air	井下受控风流	circumferential seam	环形焊缝
circulation of water	水循环	circumferential spread of channels	孔道沿外围分布
circulation pack	循环充填	circumferential strain	圆周应变；环向应变
circulation passage	循环通道	circumferential stress	环向应力；圆周应力；周向应力
circulation path	循环通路	circumferential structure	同心圆形结构
circulation pattern	环流模式	circumferential tensile stress	环向拉应力
circulation process	循环流程	circumferential velocity	切向速度；圆周速度
circulation return	循环回水	circumferential wave	表面波；环形波；周边波
circulation reversing shoe	反循环清洗管鞋	circumferential weld	环缝；环形焊缝
circulation shoe	循环冲洗管鞋	circumferentor	测角仪；矿山罗盘仪；地质罗盘
circulation squeeze	循环法挤水泥	circumflent air-current	循环风
circulation stinger	循环插入管	circumflexion	弯曲；弯曲度
circulation under well closed in condition	关井后循环	circumfluence	环流；回流；绕流
circulation valve	环流阀	circumfluent	周流的；环流的

circumflux	环流
circumfusion	周围灌注
circumgranular	颗粒周围的
circum-island arc	环太平洋岛弧
circumjacent	邻接的；周围的
circumlittoral	海滨的；沿海的
circum-mediterranean region	环地中海区
circum-metallogenic belt	环太平洋成矿带
circum-metallogenic domain	环太平洋域成矿域
circum-oceanic	环大洋的
circum-oceanic andesite	环大洋安山岩
circum-oceanic basalt	环大洋玄武岩
circum-oceanic geosyncline	环大洋地槽
circum-orogenic belt	环太平洋造山带
circum-orogenic zone	环太平洋造山带
circum-pacific belt	环太平洋带
circum-pacific calc-alkaline petrographic province	环太平洋钙碱性岩区
circum-pacific mobile belt	环太平洋活动带
circum-pacific orogenic zone	环太平洋造山带
circum-pacific plate boundary	环太平洋板块边界
circum-pacific province	环太平洋岩区
circum-pacific seismic belt	环太平洋地震带
circum-pacific seismic zone	环太平洋地震带
circum-pacific type orogeny	环太平洋型造山运动
circumpolar	极地附近的
circumpolar current	环极海流
circumpolar glaciation	环极冰川作用
circumpolar water	环极水域
circumradius	外接圆半径
circumscribe	划边界线；外接；外切；限定；限制
circumscribed circle	外接圆
circumscribed cone	外切圆锥
circumscription	界限
circum-seismic zone	环太平洋地震带
circumstance	环境；情况
circumvallation	蚀原造山作用
circum-volcanic belt	环太平洋火山带
circumvolution	卷缠
cire fossile	地蜡
cirque	凹地；冰斗；冰坑；圈椅状谷；山凹
cirque cutting	冰斗切刻；冰斗削蚀
cirque erosion	冰斗侵蚀
cirque floor	冰斗底
cirque glacier	冰斗冰川
cirque lake	冰斗湖
cirque moraine	冰斗冰碛
cirque platform	冰斗台地
cirque step	冰斗阶地
cirque terrace	冰斗阶地
cirque-cupped pond	冰斗湖
cirrhosis	硬化
cirrolite	黄磷钙铝石
cis-orientation	顺向定位
cistern	水槽；天然水库
cistern barometer	水银槽气压计
cistern car	罐车
cistern rock	岩盖
citrine	黄水晶
city-lot drilling	市区钻井
civic survey	城市勘察
civil air defense engineering	人防工程
civil briquette	民用型煤
civil engineering fabric	土工布；土工织物
civil engineering structure	土木工程结构
clabber	变酸，凝结
clack	瓣阀；单向阀
clack seat	瓣阀座
clack valve	瓣阀
clad	金属包层
clad lining	金属衬里
clad metal	包层金属；复合金属
clad plate	复合板
clad sheet steel	复合钢板
clad steel	包层钢；复合钢
cladding	包层；衬板；覆盖层；骨架外墙
cladding balance	骨架填充板材
cladding diameter	包层直径
cladding material	镀层；覆盖材料
cladding steel	包层钢
cladding wall	围护墙
cladism	分支理论
cladistic systematics	分支系统学
cladogram	分支图；亲缘图
claggett clay shale	克拉格特黏土页岩
claggy	紧黏顶板的；泥质煤
claim	采矿权申请；采矿用地；开采权
claim system	矿权申请制度
claiming oven	煅烧炉
clam	夹钳；夹绳器；抓斗挖土机；蛤壳式抓斗
clam bucket	夹子
clamber	攀登
clammer	夹绳工
clamp	管子止漏夹板；夹板；夹紧装置；夹具
clamp bucket	抓斗挖掘机；抓岩机
clamp connection	卡箍连接
clamp connector	紧固连接器
clamp coupling	壳形联轴节；纵向夹紧联轴节
clamp device	固定装置；夹紧装置
clamp for connecting earth wire to rail	接地线夹
clamp for double headspan cross messenger wire	双横承力索线夹
clamp for headspan cross messenger wire	横承力索线夹
clamp for synthetic wire	用于合成纤维绳线夹
clamp frame	夹钳
clamp holder	夹柄；钻杆夹持器
clamp hose connection	用夹子夹住的软管连接
clamp nut	花螺母；紧固螺母
clamp operator	抓岩机司机
clamp piston	推靠器
clamp prop	夹紧支柱；楔紧支柱
clamp ring	压紧环
clamp ring prop	急增阻恒阻摩擦支柱
clamp screen plate	夹筛板
clamp screw	紧固螺钉
clamp slice	夹片
clamp sub	夹头
clamped	固支撑；夹紧的
clamped amplifier	箝位放大器
clamped beam	固支梁；夹紧梁

clamped edge	夹紧边
clamped end	固定端；夹紧端
clamped moment	固端弯矩
clamped plate	固支板；夹紧板
clamped plate theorem	固定梁理论
clamped ring	夹紧环
clamped supported beam	固定简支梁
clamped terminal	接线夹式引出端
clamped-in style	夹紧式
clamped-insert joint	插入夹紧连接
clamper	接线板；钳位器
clamping	夹具；夹紧；钳位；嵌固
clamping apparatus	夹具
clamping arrangement	夹紧装置
clamping bolt	夹紧螺栓
clamping chuck	夹头
clamping device	夹具；制动装置；夹紧装置
clamping disk	卡盘
clamping fish plate with lugs	夹紧带耳状柄的导轨连接板；夹紧带耳状柄的鱼尾板
clamping force	夹紧力；握裹力
clamping handle	制动手柄
clamping head	夹头
clamping jack	夹紧千斤顶
clamping length	夹持长度；夹具的间隔
clamping load	楔紧初撑力
clamping lug	压耳
clamping mechanism	卡紧机械
clamping pad	支撑靴；夹紧板
clamping piece	夹件；夹片
clamping plate	夹板
clamping plate for fixing of I-beam	工字梁固定板
clamping ring	夹板环；夹圈；锁环
clamping ring stop	止动夹环
clamping stirrup	夹紧卡箍
clamping strap	夹板；压板
clamping surface	夹紧面
clamping system	支撑系统
clamping tongs	固定大钳
clamping yokes	夹具
clamp-on	夹紧
clamp-on design	卡箍式设计
clamp-on vibrator	附着式振动器
clamps	夹子；钳夹；皮带绊
clamp-type end fitting	卡壳式软管终端接头
clams	蛤蜊
clamshell	抓斗；蛤壳式抓斗；蛤壳状挖泥机；合瓣式抓斗；抓斗式挖土机
clamshell attachment	蛤壳抓斗装置的附件
clamshell bucket	抓斗；蛤壳式铲斗
clamshell bucket dredge	合瓣式挖泥船
clamshell crane	抓斗式起重机
clamshell dipper dredger	抓斗式挖泥机
clamshell dredge	合瓣勺斗式挖掘船；蛤斗式挖泥机；抓斗式挖泥船；合瓣式挖泥船
clamshell excavation	抓斗式挖掘
clamshell excavator	蛤斗式挖掘机；抓斗式挖掘机
clamshell grab	蛤壳式抓斗；合瓣式抓斗
clamshell grabbing crane	蛤壳式抓斗；抓斗式挖土机
clamshell loading	双瓣铲斗装载；抓斗式装载
clamshell marks	贝壳状纹理
clamshell mucking	合瓣抓岩机抓岩
clamshell scoop	蛤斗；合瓣勺；抓斗
clamshell shovel	挖掘机；抓斗挖土机
clamshell type dredge	合瓣勺斗式挖掘船
clamshell-equipped crane	抓斗吊车
clam-type loader	抓斗装载机
clanny lamp	克兰尼火焰安全灯
clap	拍打
clap valve	球阀
clapboard	隔板；护墙板；楔形板
clapper	拍板
clappet valve	瓣阀
claquage grouting	致裂灌浆法
clarain	亮煤
clarifiant	澄清剂
clarification of water	澄水
clarification zone	澄清区
clarificator	澄清器
clarifier	沉淀池；澄清器
clariflocculator	澄清絮凝剂
clarifying agent	澄清剂
clarifying basin	沉淀池；澄清池
clarifying centrifuge	澄清离心机；净化离心机
clarifying tank	沉淀槽；沉淀池
clarifying thickener	澄清浓缩机
clarinite	亮煤素质
Clarion fracture zone	克拉里翁断裂带
clarite	亮煤；硫砷铜矿
clark value	克拉克值
clarke	地壳元素百分比；克拉克值
clarke beam	组合式木梁
clarke ellipsoid	克拉克椭球
clarke number	克拉克值
clarke of concentration	浓度克拉克值
clarke projection	克拉克投影
clarke value	克拉克值
clarkeite	水钠铀矿
clark's process	克拉克软水法
clarocollain	亮煤质无结构镜煤
clarocollite	亮煤质无结构镜煤
clarodurain	亮暗煤
clarodurite	微亮暗煤
clarofusain	亮丝炭
clarofusite	微镜质丝炭；亮质丝炭
clarotelain	亮煤质结构镜煤
clarotelite	微亮煤质结构镜煤
clarovitrain	亮镜煤
clarovitrinite	富氢镜质体
clash gear	滑动齿轮
clasolite	碎屑岩
clasp	扣紧物；扣子
clasp clamp	抱合钩；弯脚钩
clasp handle	键柄
clasp hook	拖合钩；弯脚钩
clasp nut	对开螺母
class boundary	分组界限
class frequency	组频率
class interval	分组间隔；组区间
class limits	组限

class midpoint	组中值
class of buildings	建筑等级
class of fit	配合级别
class of hardness	硬度分类
class of soil liquefaction	砂土液化等级
class of station	车站等级
class of washability	可选性等级
class separability	类别可分离性
class symbol	组符号
classfication of floor	底板分类
classfication of roof	顶板分类
classfication of top-coal caving characteristics	顶煤冒放性分类
classic hydrodynamics	古典流体动力学
classic method	传统方法
classic order	古典柱式
classic set	经典集
classical dynamical system	经典动力系统
classical earth pressure theory	经典土压力理论
classical geology	古典地质学
classical Horner plot	经典霍勒图
classical mortar	典型砂浆
classical theory	经典理论
classification	分级
classification by rank	按碳化程度分类
classification criterion	分类标准
classification efficiency	分级效率
classification image	分类影像
classification of damaged area	受损区域分级
classification of earth materials	土料定名统计表
classification of earthquakes	地震分类
classification of filling material	填料分类
classification of foundations	基础分类
classification of gaseous mine	瓦斯矿井等级
classification of gassy mine	瓦斯矿分级
classification of ground	场地分类；地基分类
classification of hydraulic tunnel	水工隧洞类型
classification of insitu rock	原地岩石分级
classification of landslides	滑坡分类
classification of loess	黄土分类
classification of methane	矿井沼气等级
classification of minerals	矿物分类
classification of mining method	开采法分类
classification of organisms	生物分类
classification of reserves	储量级别
classification of rock	岩石分类
classification of rock masses	岩体分类
classification of rock particles	岩石的颗粒分类
classification of soil	土的分类
classification of soil and rock	土石分类
classfication of solonetz	碱土分类
classification of stations	车站分类
classification of subsidence damage	沉陷破坏分类
classification of surrounding rock	围岩分类
classification of tunnel surrounding rock	隧道围岩分级
classification of water quality	水质分级
classification of waterproof	防水等级
classification of weathering	风化分类
classification parameter	分类指标
classification results of adit rock mass	平洞岩体分类成果表
classification screen	分级筛
classification standard	分类标准
classification system	分类系统
classification test	分级试验；分类试验
classification with water	湿法分级；水力分级
classified	分类的
classified feed	分级给料；分级入选
classifier	分级机；分类器；分离器；分选机；上升水流洗煤机
classifier cyclone	分级旋流器
classifier overflow	分级机溢流
classifier section	分级车间
classifier separator	分级机式分选机；上升水流分选机
classifier tank	槽式分级机
classifier washer	分级机式洗选机；上升水流洗选机
classify	分级；分类
classifying pool	分级池
classifying screen	分级筛
clast	碎屑物；碎屑岩
clastation	碎裂作用
clastic	碎屑的；碎屑状的
clastic anomaly	碎屑异常
clastic block model	碎块体模型
clastic block theory	碎块体理论
clastic breakdown	弹性变形破坏
clastic breccia	碎屑角砾岩
clastic carbonate	碎屑碳酸盐
clastic constituent	碎屑成分
clastic debris	碎屑；岩屑
clastic deformation	碎屑变形
clastic deposit	碎屑沉积物；碎屑矿床
clastic dike	碎屑岩墙
clastic dolomite	碎屑白云岩
clastic dyke	碎屑岩脉；碎屑岩墙
clastic ejecta	碎屑喷出物
clastic ejectamenta	喷屑
clastic facies	碎屑沉积相
clastic flow mechanism	碎屑流机制
clastic fragment	碎屑
clastic impulse	弹性脉冲
clastic limestone	碎屑灰岩
clastic phosphorite	碎屑磷钙土；碎屑磷灰岩
clastic pipe	碎屑岩筒
clastic pressure	形变压力
clastic ratio	碎屑比
clastic reservoir	碎屑岩储集层
clastic rock	碎屑岩
clastic sediment	碎屑沉积
clastic sedimentary rock	碎屑沉积岩
clastic sequence	碎屑层序
clastic spike	弹性道钉
clastic structure	碎裂构造；碎裂结构
clastic texture	碎屑结构；碎屑组织
clastic tidal facies	潮汐碎屑相
clastic weathered crust	碎屑型风化壳
clastic wedge	碎屑楔状体；碎屑岩楔
clastichnic	具原碎屑结构的
clasticity	碎屑性
clasticity index	碎屑度指数

clastizoic 含生物碎屑的
clastizoichnic 有生物碎屑构造痕迹的
clastocrystalline 碎屑结晶质
clastogene 碎屑成因
clastogenic flow 碎屑成因流
clastogranitic 碎屑花岗质的
clastogranitic texture 碎屑花岗状结构
clastolepidoblastic texture 碎屑鳞片变晶结构
clastomorphic 碎屑侵蚀变形的
clastration 碎屑
clasts 碎屑
clast-supported conglomerate 碎屑支撑砾岩
clathrate 格子状的；笼形化合物；天然汽水化合物
clathrate complex 笼形配合物
clathrate separation 络合物分离
clathrate texture 格子状组构
clathration 笼合作用
clatholithus 格子颗石
clatter 巨砾堆；咔嗒声
claudetite 白砷华；白砷石；砒霜
claugh 峡谷
clausius unit 克劳瑟斯单位；熵单位
clausthalite 硒铅矿
clavalite 哑铃雏晶
clavicle 锁骨
clavis 拱顶石；楔石
clavodiscoaster 柄盘星石
claw 钳；爪
claw beam 钳杆
claw clutch 牙嵌离合器；爪形离合器
claw coupling 矿车连接锁；爪形连接器
claw crane 钳式起重机
claw finger 斗爪；抓斗指
claw hammer 拔钉锤；羊角锤
claw locker 爪形锁扣
claw setting 爪镶
claw stop 止爪
claw tool 钩爪铁铤
clay 黏粒；黏土
clay-spectroscopic analysis 黏土-光谱分析
clay acid treatment 土酸处理
clay alteration 黏土改性；黏土蚀变
clay and straw plaster 草泥抹面
clay auger 黏土螺旋钻；黏土麻花钻
clay ball 黏土球；黏土团
clay band 黏土层；黏土夹层
clay barrier 黏土隔层；黏土防渗层
clay base mud 黏土基泥浆
clay bed 黏土层；黏土地基
clay binder 胶黏土；黏土胶结物
clay blanket 黏土封层；黏土覆盖层；黏土护层；黏土铺盖
clay boulder 泥砾
clay brick 黏土砖
clay brown loam 黏质棕色壤土
clay burner 黏土炉
clay cement 黏土胶结物
clay cement grout 黏土水泥浆
clay cement grouting 黏土水泥灌浆
clay chunk 黏土块

clay classification 黏土分类
clay coast 黏土海岸
clay coating 泥糊；泥皮；黏土盖层；黏土胶膜
clay concentration 黏土含量
clay contacting 黏土接触
clay contamination 黏土浸
clay content 黏粒含量；黏土含量
clay control 黏土控制
clay control neutralizer 控制黏土的中和剂
clay core 黏土防渗层；黏土心墙
clay core earth-rock dam 黏土心墙土石坝
clay core wall 黏土防渗墙
clay course 黏土夹层；黏土脉壁
clay crucible 黏土坩埚
clay cutter 黏土薄壁取土器；黏土切削器
clay damp blanket 黏土隔水层
clay deposit 黏土矿床；黏土层
clay desert 黏土荒漠
clay diagenesis 黏土成岩作用
clay diaphragm 黏土隔墙
clay diapir 黏土刺穿构造
clay diapirism 泥底辟
clay digger 掘土铲
clay disintegrator 黏土破碎机
clay drape 黏土盖层
clay dyke 黏土岩脉
clay electrohardening 黏土电化固结
clay element 黏土单元
clay extender 黏土增效剂
clay facing tile 饰面瓷砖
clay filled cutoff 黏土填筑的截水墙
clay filler 黏土填料
clay filling 黏土填料
clay film 泥糊；黏土膜
clay floor 黏土地面
clay flow 黏土流；黏土塑变
clay flushing 泥浆冲洗；泥浆洗孔
clay foundation 黏土地基
clay fraction 黏粒粒组
clay gall 黏土片
clay gel 黏土胶凝体
clay gouge 薄黏土层；断层泥
clay gouged intercalation 泥化夹层
clay grain 黏粒；黏土颗粒
clay gravel 砾石土
clay grease 白土基润滑脂
clay ground 黏土层
clay grounting 黏土化
clay grout 泥浆；黏土浆
clay grouting 黏土灌浆
clay gun 泥炮；喷浆枪
clay hydration 黏土水合
clay inclusion 黏土夹杂物
clay indicator 黏土含量指示
clay infillings 黏土充填物
clay inhibitor 黏土抑制剂
clay inrush 黏土涌入
clay iron ore 褐铁矿；泥铁矿
clay iron stone nodule 泥铁矿石结核
clay ironstone 不纯菱铁矿；泥铁岩；泥质铁矿石

clay kneading machine	泥土搅拌机；捏泥机
clay lamination	黏土夹层
clay landslide	黏土滑坡
clay landslide with locked patch	含锁固段的黏土滑坡
clay layer	黏土层
clay lens	黏土透镜体
clay liner	黏土衬层
clay lining	黏土衬砌
clay loam	黏壤土；黏土质壤土
clay lump	土块；土团
clay lumps and friable particles in aggregate	黏土块及易碎颗粒含量
clay marl	黏质泥灰岩
clay material	黏土物质
clay matrix	黏土基质；黏土质母岩
clay medium	黏土介质
clay membrane	黏土膜
clay migration	黏土运移
clay mill	黏土磨碎机；黏土拌合机
clay mineral	黏土矿物
clay mineral assemblages	黏土矿物组合
clay mineral crystallinity	黏土矿物结晶度
clay mineralogy	黏土矿物学
clay minerals	黏土矿物
clay mixer	黏土搅拌器
clay mixing consolidation	黏土搅拌加固
clay mobilization	黏土运移
clay mortar	黏土灰浆；黏土砂浆
clay movement	黏土运移
clay mud	黏土泥；黏土泥浆
clay of high plasticity	高塑性黏土
clay outrush	黏土涌出
clay overburden	黏土覆盖层
clay pan	不透水黏土层；隔水黏土层
clay pan chernozem	黏盘黑钙土
clay pan soil	黏盘土
clay parings	黏土夹层
clay particle	黏土颗粒
clay parting	黏土夹层
clay pipe	陶土管；瓦管
clay pipe drainage	瓦管排水
clay pit	采黏土场；泥浆池；黏土坑
clay platelet	黏土片晶
clay plug	黏土塞
clay pocket	黏土囊
clay porosity	黏土孔隙度
clay primary	原生黏土
clay product	黏土质材料；黏土制品
clay puddle	捣实黏土；胶土；黏土膏；黏土坑
clay refining	白土精制
clay region	软土地区路基
clay rich rocks	富黏土岩
clay rock	黏土岩
clay rush	黏土突出
clay sampler	黏土取样器
clay sand	黏质砂土
clay sandstone	黏土砂岩
clay seal	黏土密封；黏土止水
clay seam	黏土夹缝；黏土层
clay sensitivity	黏土敏感性
clay sewer pipe	轴承阻力
clay shale	泥页岩；黏土页岩
clay silt	黏质粉土
clay silty loam	黏质粉砂壤土
clay size	黏土粒度
clay skeleton	黏土骨架
clay skin	泥糊；黏粒胶膜
clay slaking	黏土水解
clay slate	黏板岩；黏土板岩
clay slip	泥浆；黏土滑动；黏土滑坡
clay slurry	黏土泥浆
clay slurry jacket	泥浆润滑套
clay slurrying plant	黏土浆制备厂
clay smear	黏土涂抹
clay sod	黏土草皮
clay soil	黏土；黏性土
clay solid content	黏土固相含量
clay solution	黏土胶体溶液
clay spade	风动铲；土铲
clay spreading	黏土膨胀
clay stabilization	黏土稳定
clay stabilizing polymer	使黏土稳定聚合物
clay stable agent	黏土稳定剂
clay stemming	黏土炮泥
clay stratum	黏土层
clay streak	黏土薄夹层
clay strength anisotropy	黏土强度的各向异性
clay substance	黏土物质
clay suspension	泥浆；黏土悬浮液
clay swelling	黏土膨胀
clay tamping	黏土炮泥；黏土填塞
clay texture	黏土结构
clay tile	陶瓦；黏土瓦
clay transformation	黏土转换
clay treating chemicals	黏土处理剂
clay treatment	黏土处理
clay vein	黏土脉
clay volume	黏土体积
clay wall	土墙
clay wash	白土洗涤；黏土纯化
clay-ball method	泥球法
clay-band ironstone	不纯泥质菱铁矿
clay-bearing formation	含黏土地层
clay-bearing rock	黏土岩
clay-bearing soil	含黏土土壤；亚黏土
clay-bound macadam pavement	泥结碎石路面
clay-bound water	黏土束缚水
clay-cement mortar	黏土水泥砂浆
clay-chemical grout	加有化学药剂的泥浆
clay-control additive	黏土控制添加剂
clay-cutter dredge	切土式挖泥船
clay-cutting machine	黏土切割机
clayed	黏土的
clayed bottom	黏质底土
clayed podzol	黏质灰壤
clayed soil	黏性土壤
clayey	泥质的；黏土的；黏土质的
clayey aquitard	黏土隔水层
clayey breccia	泥质角砾岩；黏土质角砾岩
clayey desert	黏土沙漠

clayey fine sandstone	黏土质细粒砂岩	clean coal screen bowl	精煤离心脱水机
clayey formation	黏土质地层	clean coal storage bin	精煤贮仓
clayey gravel	泥砾；黏土质砾石	clean coal technology	洁净煤技术
clayey ground	黏土地	clean coal yield	精煤回收率；洗精煤量
clayey loam	亚砂土；黏质壤土	clean coal-based fuel	纯净煤基燃料
clayey loess	黏黄土	clean cut	冷切削；精确切割
clayey marl	黏土泥灰岩	clean cutting formation	无岩屑地层
clayey material	黏性材料；黏性土	clean development mechanism	清洁发展机制
clayey mud	黏土质淤泥	clean drilling	净化钻井；正常泥浆钻进
clayey rock	黏土岩	clean dumped rock fill	纯石块填石；抛石
clayey sand	黏土质砂；黏质砂土	clean energy	清洁能源
clayey sand ground	泥质砂地	clean energy resources	清洁能源
clayey sandstone	黏土质砂岩	clean energy via cryogenic technology	低温技术清洁能源计划
clayey sediment	黏土质沉积物		
clayey silt	黏质粉砂；黏质粉土	clean form of energy	洁净能源
clayey soil	黏性土；黏质土壤	clean formation	纯地层；含泥很少的地层
clayey stratum	黏土层	clean fuel	洁净燃料
clay-filled	黏土填塞的	clean fuel from coal	洁净燃料煤
clay-filled joint	黏土充填的节理	clean fuel gas	洁净燃料气
clay-filtered oil	白土精制油	clean gap graded aggregate	精选集料
clay-free drilling fluid	无固相钻井液	clean gas	净煤气
clay-hydration water	黏土水合水	clean mining	高回收率开采法；全采
claying	糊泥；抹泥	clean mud	净泥浆
claying bar	涂泥棒	clean oil	轻质油；透明油；未加裂化油的油料；无添加剂润滑油
claying of soil	土壤黏化		
clay-in-thick	厚层黏土	clean oil tank	纯油储罐
clayish	泥质的；黏土的；黏土质的	clean ore	纯净矿石；精矿；洗矿
clayite	高岭石	clean proof	清样
clayization	黏土化	clean river	清水河
claylike	黏土状的	clean sand	纯砂；净砂层
clay-like fines	黏土状细粉；塑性细粉	clean sand laminae	纯砂层夹层
clay-lined	薄层黏土的	clean sand line	纯砂层线
clay-organic complex	黏土-有机复合体	clean sand streak	纯小砂层
clay-pocketing method	黏土袋穴法	clean sandstone	纯砂岩
clay-size particle	泥级颗粒	clean side	净侧
clayslide	黏土滑动；黏土滑坡	clean uniform perforation	干净等径射孔孔眼
claystone	泥岩；黏土岩；黏土岩	clean water reservoir	给水池
claystone porphyry	黏土化斑岩	clean water sand	纯含水砂层
clay-water mud	黏土-水泥浆	clean water tank	清水箱
clay-water property	含水黏土性质	clean-cut separation	清晰分离；准确分离
clay-water system	清水泥浆	cleaned coal	精煤
clay-with-flints	含燧石黏土	cleaner	清洁剂；清洁器；选煤机
clay-with-race	含钙质结核黏土	cleaner cell	精选机
clayzation	黏土化	cleaner flotation	精浮选
cleading	护壁板；挡板；护壁板	cleaner for cuttings	钻屑和割屑清除器
clean	精选的；无杂质的	cleaner production	清洁生产
clean aggregate	清洁骨料	cleaner roller	清洗器滚轮
clean air	新鲜空气	cleaner tailings	精选尾矿
clean annealing	光亮退火	cleaness	清洁度
clean blast	完全爆破；有效爆破	clean-gap-graded	精密粒度分级的
clean channel	清水河	cleaning	清洗；选矿；精选
clean coal	精煤；无矸煤	cleaning agent	清洁剂
clean coal belt plow	精煤胶带机刮板	cleaning and priming machine	除锈-涂底漆联合作业机
clean coal bunker	精煤仓	cleaning berm	清扫平盘
clean coal liberation	精煤解离	cleaning bottom of hole	扫孔
clean coal loss	纯煤损失	cleaning cell	精选机
clean coal mining	清洁煤开采	cleaning circuit	精选流程
clean coal product	精煤产品	cleaning coal	清洗煤
clean coal ratio	精煤率	cleaning device	清扫装置
clean coal recovery	精煤回收率	cleaning eye	清理孔

cleaning flotation	清洁浮选
cleaning fluid	清洗液
cleaning gas	清洗气
cleaning machine	清除器
cleaning machine for scaffolding	脚手架清扫机
cleaning magnetic	磁清洗
cleaning of coal	选煤
cleaning of forms	铲去灰浆渣，清底
cleaning of hole	洗井
cleaning of medium	介质的净化；介质的再生
cleaning of ore	洗矿
cleaning of well	清洗钻孔
cleaning operation	洗煤操作；洗煤作业
cleaning plant	清选装置；洗选设备
cleaning plant reject	洗煤厂废渣
cleaning port	清除门；掏灰门
cleaning rejects	选煤厂废渣
cleaning rod	清理棒
cleaning scraper	清管刮刀
cleaning screw	放空螺丝
cleaning shift	岩石清理班；装运班
cleaning solution	洗涤液
cleaning solvent	洗涤溶剂
cleaning strainer	滤净器
cleaning table	选矿摇床
cleaning tool	清孔机具
cleaning unit	净化设备；清除单位；选矿设备
cleanliness	纯净；清洁
cleanly mining	清洁开采
cleanout	清理孔；清扫口
clean-out auger	抽汲筒；清孔螺纹钻头；清孔钻
clean-out bailer	捞砂筒
clean-out bit	清除井底钻头
cleanout blow	洗井；洗井性放喷
clean-out box	排污箱；油罐清积孔
clean-out drain	排出口
clean-out machine	反循环钻机
clean-out of hole	洗孔
clean-out of well	洗井
clean-out opening	清扫孔
clean-out operation	清洗作业
clean-out plate	入孔盖板
clean-out plug	放油塞
clean-out port	清洗孔
clean-out string	洗井用钻柱
clean-out system	清洗剂
clean-out tool	清孔机具
cleans	精煤
cleans ash	精煤灰分
cleanse	纯化；净化
cleanse pollutant	净化污染物质
cleanser	清洁剂；清洁工
cleansing	纯化
cleansing blower	喷气净化器；喷砂器；喷砂清理机
cleansing oil	清洗用油
cleansing solution	洗涤液
clean-surface coal	表面清洁的煤；净煤
cleanup	净化；清井；提纯
cleanup additive	助排剂
clean-up cyclone	精选旋流器
clean-up effect	净化作用
clean-up pit	排液池
clean-up pump	清扫用泵
clean-up radius	挖掘净半径
clean-up range	装载范围；装载面宽度
clean-up sump	洗涤水水池
cleanup system	净化系统
clean-up trip	为清除岩屑的起下钻
clean-up width	装载面宽度
cleap	交错层理
clear	归零
clear area	净面积；有效面积
clear clay	纯泥
clear cover	保护层；覆盖层
clear cutting	清坡
clear distance	净距
clear effective length	净有效长度
clear face	节理面
clear headroom	净高
clear height	净高度
clear ice	纯净冰
clear interval	净间隔；净间距
clear lacquer	透明漆
clear mineral	淡色矿物
clear octane number	不加铅辛烷值
clear off	清除
clear opening	净开口；有效面积
clear operation	清零操作
clear plastic	透明塑胶
clear point	澄清点
clear relief	显地形
clear section	净断面
clear sheet glass	透明平板玻璃
clear space	净空；畅通空间
clear space between bars	钢筋间净距
clear spacing	净间距
clear span	净跨距；净孔距
clear up sludge	清淤
clear water	净水
clear width	净宽度
clearage	空隙；清除
clearance	间距；净空；清除；限界
clearance above bridge floor	桥面净空
clearance adjustment	孔隙调整
clearance blasting	空隙爆破
clearance cavitation	空隙气蚀
clearance channel	缝隙槽
clearance detector	间隙探测器
clearance diagram	净空图
clearance distance	净距离
clearance fit	间隙配合
clearance gauge	测隙规；测厚规；千分垫
clearance height	净高
clearance leakage	缝隙漏泄
clearance limit survey	净空区测量
clearance plane	间隙面
clearance ratio	间隙比
clearance ring	外圈
clearance road	井下储煤仓巷道
clearance size	净空尺寸

clearance space 间隙空间
clearance system 通关系统；清除系统
clearance under bridge 桥下净空
clearance way 净空道
clearance width 缝隙宽度
clearcole 打底明胶
clearing 纯化；清除
clearing and grubbing 清理施工现场
clearing basin 沉积池
clearing device 清除装置；洗涤器
clearing locomotive 露天矿运矸机车
clearing mark 导航标
clearing shift 岩石清理班，装运班
clearing silt 清淤
clearing-away 清除
clearway 超高速公路
cleat 防滑木；管子止漏夹板；加劲木条；夹具；楔子；系缆枕
cleat compressibility 割理压缩性
cleat direction 夹板方向；解理方向
cleat face 节理面；劈理面
cleat of coal 煤裂隙
cleat spar 裂隙晶石
cleatering 漆膜凹坑
cleavability 抗劈性；可解理性；可裂性
cleavable 可劈裂的
cleavage 节理；解理；劈理
cleavage banding 劈理条带
cleavage brittleness 晶间脆裂
cleavage crack 劈理裂缝
cleavage direction 劈理方向
cleavage dislocation 解理位错
cleavage domain 劈理域
cleavage face 解理面；劈理面
cleavage fan 扇状劈理
cleavage fissure 解理缝
cleavage flake 解理片
cleavage fold 劈理褶皱
cleavage foliation 解理分层；劈理面理
cleavage form 劈理形
cleavage fracture 解理破裂；劈裂
cleavage in one direction 一组解理
cleavage in trace 劈理痕迹
cleavage in two directions 两个方向解理；两组解理
cleavage lamellae 解理薄片
cleavage maximum extension direction 劈理最大延伸方向
cleavage model 解理模型
cleavage modulus 劈裂模量
cleavage mullion 劈理窗棂构造
cleavage plane 解理面；劈理面
cleavage plate 解理片
cleavage refraction 劈理折射
cleavage reliefs 劈理浮雕
cleavage sandstone 可劈砂岩
cleavage slice 劈理薄片
cleavage slip 假劈理
cleavage strength 劈裂强度
cleavage structure 解理结构；劈理构造
cleavage surface 解理面；劈理面
cleavage test 劈裂试验

cleavage vergence 劈理降向
cleavage with the bedding 顺层劈理
cleave 裂开；劈开解理
cleavelandite 叶钠长石
cleaving 劈开；劈理
cleaving chisel 分叉钎头；分叉凿
cleaving stone 板岩；页岩
cleaving timber 劈开的木材
cleaving way 解理方向；顺层面方向
cledge 漂白土上层；黏土
cleek coal 原煤
cleet 煤的内生裂隙割理
cleft 裂缝；裂隙
cleft girdle 分裂环带
cleft timber 顺纹劈开的木材
cleft water 裂隙水
cleft water pressure 裂隙水压力
cleft water pressure on tunnels 隧道裂隙水压力
cleft welding 裂口焊
cleftness 裂隙
cleiophane 纯闪锌矿
Clemens vacuum flotation cell 克莱门斯型真空浮选机
clerici solution 克勒里西溶液；轻重矿分离液
cleuch 沟谷；峡谷
cleve 陡坡；悬崖
cleveite 结晶沥青铀矿；钇铀矿
Cleveland iron ore 克利夫兰鲕铁矿
clevis 连接叉；马蹄钩
clevis pin U形夹销
cliche frame 组合模板框
click 插销；掣子；定位销；棘爪
click pulley 棘轮
clicker die 冲模；带刀切割模具
clicker press 冲床；带刀切割压力机
client acceptance test 用户验收测试
cliff 陡岸；峭壁；悬崖
cliff base 悬崖脚
cliff breeder 岩壁滋生
cliff coast 陡岸；峭壁海岸
cliff debris 坡积物；山麓碎石；岩堆
cliff edge 崖边
cliff erosion 悬崖的侵蚀
cliff face 悬崖壁
cliff glacier 悬崖冰川
cliff landslide 崖滑
cliff line 悬崖线
cliff of differential degradation 差别侵蚀悬崖
cliff of displacement 断崖
cliff recession 陡崖后退
cliff section 陡剖面
cliff spring 峭壁泉；悬崖泉
cliffed 陡的；陡峭的
cliffed coast 陡岸；浪蚀海岸
cliffed headland 陡崖岬角
cliff-fall deposit 崖崩堆积物
cliffing 成崖作用
cliff-maker 造崖层
cliffy 峭壁的
clift 硬泥岩
cliftonite 方晶石墨

climactic orogeny	造山作用最强烈时期
climactichnites	栅形迹
climafrost	季节冻土；气候冻土
climate stratigraphic unit	气候地层单位
climatic condition in mine	矿井气候条件
climatic soil formation	气候性土壤形成
climatic stratification	气候地层法
climatic stratigraphic unit	气候地层单位
climatic terrace	气候阶地
climatochronology	气候测年学
climato-isophyte	气候等植物生长线
climatostratigraphy	气候地层学
climax	顶点
climax avalanche	大雪崩
climax community	生态顶极群落
climax soil	顶极土壤
climax vegetation	顶极植被
climb	上漂；爬高速度；攀登
climb cut grinding	同向磨削
climb cutting	顺铣
climb form technique	提升模板技术
climb hobbing	顺滚
climb of dislocation	位错攀移
climber	爬罐；爬山者；提升机
climbing	爬升；上漂
climbing ability	爬升能力
climbing capacity	爬升能力
climbing dune	上叠沙丘
climbing film evaporator	升膜蒸发器
climbing form	滑升模板
climbing lane	爬坡车道
climbing rate	上升率；上升速度
climbing ripple bedding	爬升沙纹层理
climbing ripple lamination	爬升波痕纹理
climbing shuttering	爬升模板
clinch	夹紧；钳住
clincher	紧钳
cline	渐变群
cline strata	倾斜地层
cling	依附；黏着
clingage	计量罐罐壁附着油量
clinging	黏着的
clinging power	黏着力
clinism	倾斜
clink	叮当声；劈楔；裂纹
clink bolt	弯头螺栓
clinker	缸砖；炉渣；接触变质煤；熟料
clinker bed	熔结块层
clinker brick	熔渣砖；烧结砖
clinker cement	矿渣水泥
clinker concrete	矿渣骨料混凝土；熔渣混凝土
clinker concrete block	熔渣混凝土块
clinker field	块熔岩地
clinker whiteness	熟料白度
clinker-bearing slag cement	矿渣硅酸盐水泥
clinker-free cement	无熟料水泥
clinkering	熔结；烧结
clinkering coal	熔结煤
clinkering property	结渣性
clinkering rate	结渣率
clinkery	渣状的
clinkstone	响石
clino	海底斜坡；水下岸坡
clino axis	斜轴
clino diagonal axis	斜轴
clino pyramid	斜轴锥
clino-amphibole	单斜闪石
clinochlore	斜绿泥石
clinoclase	光线矿；光线石；砷铜矿；斜解理；斜羟砷铜矿
clinoclasite	光线矿
clinocrocite	铁钾茂
clinodome	斜轴坡面
clinoenstatite	斜顽辉石
clinoferrosilite	斜铁辉石
clinoform	斜坡沉积；斜坡地形
clinoform surface	斜坡地形面
clinoform zone	斜坡沉积带
clinograph	孔斜计；钻孔测斜仪
clinographic	倾斜的
clinographic curve	坡度曲线
clinographic projection	斜射投影
clinohedrite	斜晶石
clinohedron	坡面
clinohumite	斜硅镁石；斜硅镁石
clinohypersthene	斜柴苏辉石；斜紫苏辉石
clinoklase	光线矿
clinometer	测角器；测斜仪；地质罗盘
clinometer rule	测角器
clinometric projection	斜投影
clinophone	测斜仪
clinopinacoid	斜轴面
clinoplain	倾斜平原
clinoplane	倾斜平面
clinoprism	斜轴柱
clinoptilolite	斜发沸石
clinopyroxene	斜辉石
clinopyroxene norite	单斜辉石苏长岩
clinopyroxenite	单斜辉石岩
clinorhombic	单斜的
clinorhombic system	单斜晶系
clinorhomboidal	三斜的
clinorhomboidal system	三斜晶系
clino-rule	倾斜规
clinoscope	水平钻孔测斜仪；旋斜视计
clinosol	坡积土
clinostat	稳斜器
clinothem	斜坡沉积
clinothem deposit	斜坡沉积
clinothem facies	斜坡沉积相
clinothen	斜坡岩层
clino-unconformity	斜交不整合；角度不整合
clinoungemachite	斜菱碱铁矾
clinozoisite	斜帘石；斜黝帘石
clint	石灰岩参差面；石芽；岩沟
clintheriform	板状
clinton ore	克灵顿矿石
clintonite	绿脆云母
clinunconformity	斜交不整合
clip	夹住；剪短；剪断；锚固件

clip gauge	夹住式变形计
clip hook	双抱钩
clip on	夹住矿车；夹住
clip plate	夹板
clip pulley	夹绳轮
clip ring	弹簧挡圈；开口环
clip-in	活动安装；嵌入
clip-on charging	夹卡式充电
clipped	限幅的
clipped signal	限幅信号
clipped trace	限幅道
clipped wave	限幅波
clipper	大剪刀；尖嘴钳；吊桶挂钩；削波器
clipper-limiter	削波限幅器
Clipperton fracture zone	克利伯顿断裂带
clipping circuit	限幅器
clipping factor	削波系数
clipping lever	限幅电平
clisis	倾斜；摄吸
clitter	巨砾堆
clival	斜坡的
cliver	吊钩；连接钩环；钻管吊环
clivis	山坡
clivus	斜坡
clivvy	安全吊钩；吊环；钩环
clivvy hook	弹簧钩
clnt	冷却剂
cloaca	下水道
cloak	外套
cloak-like superposition	套状被覆
clob	泥炭田
clock brass	钟表黄铜
clock gauge	千分表
clock wavelet	钟形子波
clockwise	顺时针方向的；右旋的
clockwise shearing	顺时针扭动
clod	岩块；土块；煤层的软泥土顶板；煤层顶底板页岩；黏土质岩石
clod coal	块煤；硬煤
clod crusher	碎土块机
clod smasher	土块破碎机
clod test	土块试验
cloddy	土块似的；团块状的
cloddy structure	块状构造
clog	堵塞
clog up	堵塞；阻塞
clogged	堵住的
clogged chute	溜眼堵塞
clogged point	堵塞点
clogged sand layer	堵塞砂层
clogging	堵塞；堵塞作用；淤塞
clogging drain	排水障碍
clogging of oil screen	滤油网堵塞
clogging of river sediment flow	浆河现象
clogging test	淤堵试验
cloggy	易堵塞的；易黏住的
clogproof	防堵塞的
cloisonne	景泰蓝
close annealing	密闭退火
close butt joint	紧密对接
close check	精密校核
close circuit	闭合线路
close circuit crushing	闭路碎矿
close clearance	狭窄间隙
close cluster reef	紧密丛状生物礁
close coefficient	紧密系数
close coupling	强耦合
close cut fraction	窄馏分
close cut gravel	精筛砾石
close difference of pressure	压力闭合差
close drilling	密集钻进；密集钻眼
close earthquake	近震
close fault	闭合断层
close fit	固定配合；紧配合
close fit method	精密拟合法
close fitting cover	紧合封盖
close fittings	紧合配件
close flume	封闭式运输水槽
close fold	闭合褶皱
close foliation	薄片理
close fractionation	精密分馏
close geological investigation	详细地质调查
close graded asphaltic concrete	密级配地沥青混凝土
close grain	密木纹
close grained texture	细粒结构
close hole spacing	密布炮眼
close impeller	闭合式叶轮
close investigation	密切调查；详细调查
close joints cleavage	闭合劈理；滑劈理
close level	低矮平巷；窄狭平巷
close limit	窄范围
close mapping	碎部测量；详测
close nipple	螺纹接口
close of anticline	背斜闭合度
close off	封堵
close packed plane	密集面
close packed structure	密集结构
close packing	紧密装填；密堆积
close pile	密排桩
close port	不开放港
close push-button	闭合按钮
close regulation	精确调节
close return bend	回弯头
close riveting	密铆
close running fit	紧动配合
close sampler	密集取样
close sampling	密集取样
close sand	密实砂；致密砂层
close scanning	细密扫描
close set timber	紧密支架
close sheeting	挡土排桩；锁口钢板
close sizing	精密筛分；细致筛分
close spacing	紧密间距；密井网
close spacing of flights	密排刮板
close spread	接近排列法
close standing supports	紧密排列；密集支柱
close streaming method	紧密流水法
close substance	致密物质
close texture	密实结构；致密质地
close timber	紧密支架；密集支架

close timbering 密集支撑；拼板木模
close work 平巷掘进工作；狭窄采煤区
close-boarded platform 密合封板平台
close-boiling cut 窄馏分
close-burning 结焦的；黏结性的
close-coiled spring 闭式盘簧
closed 闭合的
closed air 密闭气体
closed anticline 闭合背斜
closed aqueduct 封闭式输水管道
closed arch 闭合拱；圆拱
closed area 闭合面积；封闭区
closed basin 闭合流域；闭合盆地；封闭流域；内流流域
closed bay 闭合海湾；封闭海湾
closed boundary 封闭边界
closed cage 闭式阀罩
closed caisson method 密闭沉箱法
closed capillary fringe 完全毛细上升带
closed center valve 零位封闭的换向阀
closed chain 闭锁链系
closed chain compound 闭链化合物
closed chamber 密室
closed circuit 闭合线路；闭式回路；通路；循环路线
closed circuit crushing 闭路破碎
closed cirque 封闭冰斗
closed colliery 关闭的煤矿
closed conduit 暗管道
closed conduit flow 暗管流
closed construction 封闭结构；隐蔽构造
closed container 密闭容器
closed contour 闭合等值线
closed country 沟谷交错地带；起伏地形；丘陵地带
closed coupling 固定联轴节
closed crosshead 封闭十字头
closed crossover valve 零位封闭的换向阀
closed cutterhead with peripherical buckets 带有周边铲斗的封闭式刀盘
closed depression 封闭洼地
closed discontinuity 闭合裂隙
closed drain 排水暗管
closed drainage 暗管排水；封闭水系；内流水系
closed drainage basin 闭合流域；封闭流域
closed end pile 闭端桩；桩端封闭桩
closed face 密闭型
closed face shield 密闭型盾构
closed fault 闭合断层；封闭断层
closed filter 封闭式过滤器
closed flash point 封闭闪燃点
closed fluid-pellet system 钻液-钻粒闭路系统
closed fold 闭合褶皱
closed foliation 闭叶理
closed force polygon 力的闭合多边形
closed form 闭形
closed form solution for simple excavation shapes 简单形状硐室的封闭解
closed fracture 闭合裂缝；闭合裂隙
closed frame 密封棚子
closed gaslift string 闭式气举管柱
closed grained 密纹的；细粒的
closed groove 闭口式孔型
closed impeller 封闭式叶轮
closed interval 闭区间
closed jet 闭合射流
closed joint 闭式节理；接头瞎缝；密缝接头
closed lake 封闭湖；内陆湖
closed lamp colliery 安全灯煤矿
closed lamp mine 多尘瓦斯矿
closed length 闭合长度
closed level 封闭圈
closed level circuit 闭合水准环
closed leveling line 闭合水准路线
closed line 闭合线
closed lock 闭合船闸
closed loop 闭合环路；闭合回路
closed loop adjustment 闭环调节
closed loop feedback system 闭环反馈系统
closed loop field 闭合回线场
closed loop gain 闭路增益
closed marginal geosyncline 封闭陆缘地槽
closed on itself traverse 闭合导线
closed outer boundary 封闭外边界
closed pass 闭口式孔型
closed path 闭路
closed piezometer 闭口压力计
closed pipe string 末端封闭的管柱
closed pipeline system 闭式管道输送系统
closed polygon of force 闭合力多边形
closed pore 封闭孔；内部空隙
closed porosity 闭口气孔率；封闭气孔
closed power fluid system 闭式动力液系统
closed power oil type pump 闭式动力油泵
closed pressure 封闭压力；关井压力
closed region 封闭区域
closed reservoir 圈闭储层
closed riser 暗冒口
closed river system 闭合河系
closed shears 闭式剪切机
closed shell 封闭壳
closed shield 闭胸式盾构
closed slot 闭口槽
closed solution 闭合解
closed spiral state 闭旋状态
closed stirrup 封闭箍筋
closed structure 闭合构造；隐伏构造
closed support 密集支护；密排支护
closed syncline 闭合向斜
closed system 封闭系统
closed topped housing 闭口式机架
closed traverse 闭合导线
closed trickling filter 封闭式滴滤池
closed tube analysis 闭管分析
closed type 封闭式
closed type mixture 最小孔隙度混合物
closed valve gear 封闭式阀动装置
closed ventilated container 封闭式通风集装箱
closed ventilation system 密闭式通风系统
closed void 封闭空隙
closed volume 闭合容积
closed water circuit 洗水闭路循环
closed-air-circuit motor 有闭合风冷系统的电动机

closed-bell diving procedure	封闭钟潜水法
closed-cast contour	闭合等高线
closed-cast depression	闭合凹陷
closed-cast drainage	闭合水系；内陆水系
closed-cast fault	闭合断层
closed-cast fold	闭合褶皱
closed-cast form	闭形
closed-cast isotope system	封闭同位素体系
closed-cast movement	封闭形变运动
closed-cell foamed plastics	闭孔泡沫塑料
closed-chain hydrocarbon	闭链烃
closed-circuit comminution	闭路破碎循环
closed-circuit comminution operation	闭路破碎操作
closed-circuit grinding	闭路研磨
closed-circuit grouting	闭合压力灌浆；环流式灌浆
closed-circuit ventilation	闭路式通风
closed-circuit water supply	闭路供水
closed-flow	密闭流动
closed-form equation	闭合形式公式
closed-in bottom hole pressure	关井井底压力
closed-in pressure	关井压力
closed-in production	关井产量；关井停产
closed-in time	关井时间
closed-in tubing pressure	关井油管压力
closed-in well	关闭井；停产井
closed-in wellhead pressure	关井井口压力
closed-loop alarm disposal process	闭环报警处理过程
closed-loop control system	闭环控制系统
closed-loop hydraulic loading system	闭环液压加载系统
closed-loop path	闭合回路
closed-loop pipe viscometer	闭路管式黏度计
closed-loop reservoir	圈闭储油层
closed-loop rotary drilling system	闭环钻井系统
closed-loop servo-accelerometer	闭合回路式伺服加速度计
closed-loop structure	闭合结构
closed-loop system	封闭系统
closed-path vibrator	闭路轨迹运动的振动器
closed-pipe system	闭管法
closed-spiral auger	闭式螺旋钻具
closed-system separator	闭路分选机
closed-system test	闭式试验法
closed-system trap	封闭型圈闭
closed-tank process	封闭桶处理法
close-graded	密级配的
close-grained	纹理细密的；细粒的
close-grained structure	密晶结构
close-grained wood	密纹木材；细纹木
close-in bottom hole pressure	关井井底压力
close-in data	近场数据
close-jointed	闭合节理；节理密集的
close-knit surface	密实面层；密织表面
closely cemented	紧密胶结
closely coincide	完全相等；完全重合
closely drilled area	密集钻孔区
closely folded	紧闭褶皱
closely graded	粒度成分均一的
closely graded soil	粒径均匀土
closely packed	紧密堆积的
closely packed grain	填充紧密的颗粒
closely spaced faults	密集断层
closely spaced joint	密集节理
closely spaced perforation	加密射孔
closely timbered area	稠密支架区
closely-coupled system	紧密耦联系统
closely-spaced fine fracture	密集的细裂隙
closely-spaced fracturing	短距碎裂；密集压裂
close-meshed wire net	细孔铁丝网
closeness of fissures	裂隙密度
close-quarter work	狭窄场所的工作
closer	闭合器；闭塞器；封闭板桩；镶墙边的砖石
closer cap	密封帽
close-set	密布的；密集的
close-set rolls	小开口辊碎机
close-spaced wells	密布井；密网布置
closest packing	最密充填
close-style winning	闭式落煤顺序
close-textured	结构细密的
close-type winning sequence	闭式落煤顺序
close-up	闭合
close-up view	近视图；特写镜头
closing	断开
closing appliance	关闭装置
closing cam	闭合凸轮
closing cock	闭锁旋塞
closing date	截止日期
closing device	闭合装置；关闭装置
closing dike	合龙堤；截流堤
closing disk	圆盖板
closing dyke	合龙堤；截流堤
closing error	闭合差
closing error in coordinate increment	坐标增量闭合差
closing error in departure	横距闭合差
closing head	铆钉上头
closing jaw	活颚板
closing levee	堵口堤；截流堤
closing line	开闭绳；封闭线
closing net	闭锁网
closing of fracture	裂隙闭合
closing pile	封桩；闭口桩；末桩
closing plug	封闭塞
closing ratio	井喷压力与关闭心子操作液压比
closing reading	结束时读数
closing relay	合闸继电器
closing ring	挡圈；卡圈；止动环
closing rope	关闭绳
closing stage	封闭阶段
closing stock	期末存货
closing switch	闭合开关
closing the top of lining	拱圈封顶
closing the top of tunnel lining	衬砌封顶
closing the top of wall	拱圈封顶
closing trap	井盖门
closing valve	隔离阀；节制阀
closterite	库克油页岩；皮拉藻烛煤
clostridium	梭状芽孢杆菌
closure	截流；闭合差；闭合构造；背斜层的背合度；上盘下盘的闭合
closure against fault	断层圈闭
closure age	闭合年代
closure criteria	封闭标准

closure dam	合龙坝；截流坝
closure discrepancy	闭合差
closure error	闭合差
closure error of elevation	高程闭合差
closure error of traverse	导线闭合差
closure error of triangle	三角闭合差
closure gap	截流龙口
closure gate	封堵门
closure measurement	闭合测量
closure member	闭合件
closure method	截流方法
closure mines	关闭矿井
closure of anticline	背斜层的背合度；闭合度
closure of freezing wall	冻结壁交圈
closure of ice wall	冻结壁交圈
closure of landfill	填埋场封场
closure of mining	闭坑
closure of premises	封闭楼宇
closure of roadway	巷道收敛；巷道的闭合
closure of structure	构造圈闭
closure operation	闭合运算；合龙作业；截流作业
closure pressure	闭合压力
closure rail	导轨；合龙轨
closure rate	压缩率
closure section	合拢段；截流段
closure simulation technique	巷道收敛量模拟技术
closure stress	闭合应力
closure test	弹簧压缩试验
closure time	闭合时间
closure work	截流工程；合龙工程
clot	块凝物；凝块
cloth discharge belt for belt filter	带式过滤机排料带
cloth filter	滤布
cloth filtering area	滤布面积
cloth opening	筛布孔
cloth tape	布卷尺
clotted	凝结成块的
clotted cementation	凝块状胶结
clotted limestone	凝块灰岩
clotted texture	凝块结构
clotting	块凝；凝结
cloud point	浊点
cloud seeding	人工降雨
cloud test	浊点试验
cloudburst flood	暴雨洪水
cloudburst hardening	喷丸硬化处理
cloudburst process	喷丸处理
clouded	不透明的；混浊的；阴暗的
clouded glass	毛玻璃
cloudiness	混浊性
cloudy	混浊的
clough	沟谷；谷崖
cloup	灰岩坑；落水洞
cloustonite	发气沥青
clout	垫片；垫圈
clove hook	双抱钩
clove-hitch joint	卷结式导爆线连接
cloverleaf	苜蓿叶；四叶苜蓿形的立体交叉公路
cloverleaf buoy	三叶浮标
cloverleaf cam	三星凸轮
clover-leaf crossing	四叶式交叉
cloverleaf interchange	四叶式交汇处；蝶式交汇处
clover-leaf lining	三叶草叶形衬里
clover-leaf loop	三叶草叶形环道
club	棒；棍；俱乐部
club footed pile	扩底桩
clubbing	拖锚
clubfoot roller	羊脚碾
clubmoss	石松
clue	线索
clump	块；群
clump anchor	丛锚
clump of piles	群桩
clumper	严重冒顶
clumpy	笨重的；成块的；块状的
clumpy conglomerate	块状砾岩
clumpy structure	块状构造
clunch	煤层底黏土；耐火黏土；硬白垩；硬化黏土
cluse	横谷
cluster	聚合；聚集；团聚体；桩群
cluster analysis	聚类分析
cluster analysis dendrogram	聚类分析树形图
cluster bent	群桩排架
cluster cracking	密集裂缝
cluster criterion	聚类准则
cluster crystal	晶簇
cluster diagram	集群图
cluster dome	簇状穹丘
cluster drilling	丛式钻井
cluster floodlight	簇泛光灯
cluster gear	塔式齿轮；组合齿轮
cluster lamp	簇灯
cluster matching	聚类匹配
cluster model	聚类模型
cluster of cones	火山锥群
cluster of fender piles	护桩群
cluster of grains	颗粒团
cluster of particles	颗粒团
cluster of piles	群桩
cluster of pore	孔隙簇
cluster of springs	弹簧集群；泉群
cluster pine	海滨松
cluster point	聚点
cluster porosity	簇状气孔
cluster post	丛柱
cluster props	丛柱
cluster reefs	群礁
cluster rock	丛礁
cluster sampling	分组取样
cluster seeking	聚类搜索
cluster set	聚类集
cluster setting	群镶
cluster spray	簇式喷雾器
cluster structure	团粒结构
cluster switch	组开关
cluster value	聚值
cluster well	丛式井
cluster well drilling	丛式钻井
cluster wells	丛式井
clustered	群生的；集群的

clustered aggregate	集聚体；群集体
clustered column	集柱
clustered pier	集墩；群墩
clustered pile	群桩
clustering	集群；聚集
clustering algorithms	聚类算法
clustering criteria	聚类准则
clustering of earthquakes	震群
clustering of seismic events	震群
clustering procedure	聚类法
clustering sand phenomenon	聚砂现象
cluster-type residual oil	簇状残余油
clutch	扳手；钩爪；夹紧装置；联动器；凸轮
clutch bearing	离合器轴承
clutch blocks	摩擦块
clutch lining	离合器摩擦片衬片
clutch release bearing	离合器分离轴承
clutch roller bearing	离合器滚柱轴承
clutch shaft	离合器轴
clutch stop	止动凸爪
clutch throwout lever	离合器推杆
clutch thrust bearing	离合器推力轴承
clutch-drum winder	离合器滚筒式提升机
clutcher	司钻
clutter noise	杂波噪声
clutter suppression	杂波抑制
clyburn spanner	活扳手
CO boiler	一氧化碳锅炉
CO log	碳氧比测井
CO_2 accumulation	二氧化碳富集
CO_2 adsorption	二氧化碳吸附
CO_2 capture	二氧化碳捕集
CO_2 capture and sequestration	二氧化碳捕集与储存
CO_2 capture and storage	二氧化碳捕获与封存
CO_2 coal seam sequestration	二氧化碳煤层封存
CO_2 compressibility	二氧化碳压缩性
CO_2 compressor	二氧化碳压缩机
CO_2 consumption rate	二氧化碳消耗率
CO_2 containment	二氧化碳控制
CO_2 corrosion	二氧化碳腐蚀
CO_2 credits	二氧化碳学分
CO_2 deep geological storage	二氧化碳深层地质封存
CO_2 desorption	二氧化碳解吸
CO_2 diffusion	二氧化碳扩散
CO_2 disposal	二氧化碳处置
CO_2 dissolution	二氧化碳溶解
CO_2 effects on wells	二氧化碳对井筒的影响
CO_2 emission	二氧化碳排放
CO_2 emission and injection	二氧化碳排放量和注射
CO_2 emission density	二氧化碳排放密度
CO_2 enhanced coalbed methane	二氧化碳驱煤层气
CO_2 enhanced coalbed methane recovery	二氧化碳驱替煤层气；二氧化碳提高煤层气采收率
CO_2 exsolution	二氧化碳出溶结构
CO_2 fertilization	二氧化碳肥效作用
CO_2 field	二氧化碳气田
CO_2 fixation	二氧化碳固定
CO_2 flooding	二氧化碳驱油
CO_2 flux	二氧化碳通量
CO_2 gas leaks	二氧化碳气泄漏
CO_2 geological storage	二氧化碳地质封存
CO_2 geosequestration	二氧化碳地质隔离
CO_2 hydrate	二氧化碳水合物
CO_2 infrared gas analyzers	二氧化碳红外气体分析仪
CO_2 injection	二氧化碳注入
CO_2 injection and sequestration	二氧化碳注入与隔离
CO_2 injection and storage	二氧化碳注入与封存
CO_2 injection effectiveness	二氧化碳注入效果
CO_2 injection into coal seams	二氧化碳注入煤层
CO_2 injection test	二氧化碳注入试验
CO_2 injectivity and storage capacity assessment	二氧化碳注入能力及封存容量评估
CO_2 leakage	二氧化碳渗漏
CO_2 metasomatism	二氧化碳交代作用
CO_2 migration	二氧化碳运移
CO_2 mineral sequestration	二氧化碳矿物碳酸化固定
CO_2 mineralization	二氧化碳矿化
CO_2 mitigation	二氧化碳减排
CO_2 modelling	二氧化碳模型
CO_2 monitoring	二氧化碳监测
CO_2 natural analogues	二氧化碳自然类似物
CO_2 negative balance	二氧化碳负差额
CO_2 permeability	二氧化碳渗透率
CO_2 point sources	二氧化碳排放源
CO_2 properties	二氧化碳特性
CO_2 recovery	二氧化碳回收
CO_2 reduction	二氧化碳减排；减少二氧化碳的排放量
CO_2 removal	二氧化碳脱除
CO_2 reservoir	二氧化碳储层；二氧化碳气藏
CO_2 reservoir integrity	二氧化碳储层完好性
CO_2 saturation	二氧化碳饱和
CO_2 seismic response	二氧化碳地震反应；二氧化碳地震响应
CO_2 separation	二氧化碳分离
CO_2 sequestration	二氧化碳封存
CO_2 sequestration in coal	二氧化碳煤层封存
CO_2 shielded welding	二氧化碳气体保护焊
CO_2 solubility	二氧化碳溶解度
CO_2 sorption	二氧化碳吸附
CO_2 sources	二氧化碳源
CO_2 stability	二氧化碳稳定性
CO_2 storage capacity	二氧化碳封存容量
CO_2 storage in aquifers	含水层二氧化碳封存
CO_2 storage in geological strata	二氧化碳地质岩组封存
CO_2 storage potential	二氧化碳封存潜力
CO_2 storage security	二氧化碳封存安全性
CO_2 storage sites	二氧化碳封存场地
CO_2 sublimation	二氧化碳升华
CO_2 toxicity	二氧化碳毒性
CO_2 transport	二氧化碳运移
CO_2 trapping mechanisms	二氧化碳捕集机制
CO_2 well integrity	二氧化碳井筒完整性
CO_2-acidified seawater	二氧化碳致酸海水
CO_2-brine	二氧化碳-盐水
CO_2-enriched water	富含二氧化碳的水
CO_2-rich fluid	富含二氧化碳流体
CO_2-rock interaction	二氧化碳-岩相互反应
CO_2-rock-water reactions	二氧化碳-岩-水反应
CO_2-water-basalt interaction	二氧化碳-水-玄武岩反应
CO_2-water-rock interaction	二氧化碳-水-岩石相互作用

CO₂-water-sandstone interaction　二氧化碳-水-砂岩相互反应
coabsorption　共吸收
coacervate　凝聚层；凝聚的
coacervation　凝聚作用；团聚作用
coacervation resistance　抗凝聚力
coach bolt　方头螺栓
coach clip washer　方头夹垫圈
coach screw　方头螺钉
coaction　相互作用
coactivation　共活化作用；共激活作用
coactivator　共活化剂
coadjacent　互相连接的；邻接的
coadsorption　共吸附
coagel　凝聚胶
coagulability　凝固性；凝结力；凝结性
coagulant　凝结剂；凝聚剂
coagulant agent　促凝剂
coagulant aid　助凝剂
coagulate　凝固；凝结
coagulated sediment　凝结沉积物
coagulating agent　凝结剂
coagulating basin　凝结沉淀池
coagulating bath　凝固浴
coagulation　凝结；凝聚
coagulation accelerator　促凝剂
coagulation bond　凝聚联结
coagulation point　凝结点
coagulation sediment　凝聚沉淀
coagulation stability　聚结稳定性
coagulation structure　凝聚构造
coagulation tank　凝聚池
coagulation threshold　凝结极限；凝结阈值
coagulative power　凝结力
coagulator　凝结剂；凝结器
coagulum　凝结块；凝结物
coal　煤
coal-additive compounds　煤-添加剂化合物
coal-elastoplasticity　煤-弹塑性
coal-freezing　煤-冻结
coal accumulating area　聚煤区
coal accumulating basin　聚煤盆地
coal accumulating palaeostructure　聚煤古构造
coal accumulating process　聚煤酌；聚煤作用
coal accumulating theory　煤炭积累理论
coal accumulation　聚煤作用
coal accumulation process　聚煤作用
coal acid mine drainage　煤炭矿山酸性排水
coal adustion　煤的可燃性
coal age　煤时代
coal agglomerating value　煤的凝结值
coal analysis　煤质分析
coal and gas burst　煤与瓦斯突出
coal and gas bursting　煤与瓦斯突出
coal and gas outburst accident　煤与瓦斯突出事故
coal and gas outburst prevention　煤与瓦斯突出防治
coal and rock mass　煤体和岩体
coal apple　煤层中的石球；煤结核
coal ash analysis　煤灰成分分析
coal ash and slag　煤灰渣
coal ash fusibility　煤灰熔融性

coal ash fusion temperature　煤灰熔点
coal ash monitoring　煤灰分监测
coal ash viscosity　煤灰黏度
coal auger　采煤螺旋钻
coal ball　煤层石球；煤结核
coal band　薄煤夹层；煤条带；煤线
coal bank　劣质煤堆
coal barge jetty　煤驳码头
coal basin　含煤盆地；聚煤盆地；煤田
coal bearing formation　含煤地层；煤系
coal bearing property　含煤性
coal bearing ratio　含煤系数
coal bearing region　含煤区域；聚煤地区
coal bearing strata　含煤地层
coal bed　煤层
coal bed methane　煤层气；瓦斯
coal bed washout　煤层冲刷
coal beneficiation　选煤
coal beneficiation process　煤的精选过程
coal between two slips　两个滑带之间的煤
coal bin　煤箱；煤仓
coal biogasification　煤炭生物气化
coal bit　煤钻钻头
coal blasting　爆破落煤
coal blende　煤中闪锌矿
coal blending　混合煤；配煤
coal block　块煤；煤柱
coal borer　煤钻
coal bounce　煤的突出；煤柱被压碎突出
coal brass　煤中黄铁矿结核
coal breakage　落煤；煤的破碎
coal breakage parameter　落煤参数
coal breaker　破煤机；碎煤机
coal breaking　落煤；破煤
coal brick　煤砖
coal briquette　煤块；煤砖
coal bump　煤炭突出；煤柱被压碎突出
coal bunker　煤仓；煤库；煤炭燃烧器；燃煤炉
coal burning pollution　燃煤污染
coal burst　煤的挤出；煤的突出
coal burster　落煤爆破筒；水压爆煤筒
coal bursting　煤的突出
coal byproducts　煤炭副产品
coal cake　煤饼；煤砖
coal caking properties　煤黏结性
coal car　煤车
coal carbonization　焦化；煤的碳化
coal carbonization gas　焦化煤气
coal cement tube　煤-水泥管
coal char reactivity　煤焦反应
coal character　煤的特性
coal characteristic　煤的特性
coal charge　炼焦炉料
coal charger　煤炭装载装置
coal chemicals　煤化学制品
coal chemistry　煤化学
coal chute　放煤溜槽；输煤管
coal cinder　煤渣
coal classification　煤的分类
coal classification on hydroscreen　湿法分级

coal classification system 煤炭分类系统；煤炭分类法
coal clay 耐火黏土
coal cleaning 洗煤；选煤
coal cleaning plant 洗煤厂
coal clearance 煤炭贮运；清出存煤
coal clearing 装清落煤
coal cleat performance 煤储层割理性能
coal coke 焦炭；煤焦
coal coking properties 煤结焦性
coal collapsed column 煤柱倒塌
coal collector 集煤器
coal column 煤柱
coal combine 联合采煤机
coal combustibles 煤可燃成分
coal combustion 燃煤
coal combustion byproducts 煤炭燃烧的副产品
coal combustion byproducts solubility 煤燃烧副产品的溶解度
coal combustion products 煤炭燃烧产物
coal compactor 煤夯机
coal conglomerate 煤砾；煤砾岩
coal constitution 煤的成分；煤的组分
coal consumption 煤炭消费量；煤耗
coal containing gas 含瓦斯煤
coal content 煤含量
coal conversion 煤转化
coal conveyor 运煤输送机；运煤装置
coal core 煤芯
coal core recovery percentage 煤心采取率
coal core sample 煤心煤样
coal coring tool 煤心采取器；取煤管
coal corridor conveyor tunnel 运煤胶带廊
coal cracker 碎煤机
coal crane 装煤起重机
coal crusher 碎煤机
coal crushing 煤的破碎
coal crystallite 煤晶核；煤微晶
coal culm 煤粉
coal curtain 隔离煤柱
coal cutter 采煤机；割煤机
coal cutter loader 联合采煤机
coal cutting 落煤；破煤
coal cutting equipment 采煤设备；截煤设备
coal data base 煤炭数据库
coal deformation 煤变形
coal degradation 煤的粒度减小；煤的碎裂
coal deposit 煤层；煤矿床；煤炭矿床
coal depot 贮煤场
coal derived 煤基
coal derived particles 煤基颗粒
coal diagenesis 煤成岩酌；煤成岩作用
coal discard dumps 煤炭丢弃转储
coal district 采煤区；煤产地
coal draw control 煤炭抽奖控制
coal dredger 挖煤船
coal dressing 选煤
coal drift 煤巷
coal drill 煤钻
coal drilling 煤田钻探
coal drop 溜煤槽；卸煤机

coal dry separation using fluidized bed 空气重介选煤；流化床干法选煤
coal dryer 煤炭干燥机
coal dump 卸煤门；卸煤站；煤堆
coal dumper 煤车翻车机
coal dust 煤尘；煤粉
coal dust cementing agent 煤尘黏合剂
coal dust deposition 抑尘
coal dust explosion concentration 煤尘爆炸浓度
coal dust furnace 煤粉燃烧炉
coal dust index 煤尘指数
coal dust interaction 煤粉尘相互作用
coal dust test 煤尘试验
coal economy 煤炭经济
coal electrical process 煤田电法勘探
coal electrical prospecting 煤田电法勘探
coal epoch 成煤期；聚煤期
coal equivalent 标准煤；煤当量
coal excavation 煤炭挖掘
coal excavation face 煤炭开挖面
coal exploitation 煤炭开采
coal exploration 煤田地质勘探
coal exploration type 煤田勘探类型
coal exposure procedures 露煤程序
coal expulsion 喷煤
coal extraction 采煤
coal extracts 煤的萃取物；煤的馏出物
coal face 采煤工作面
coal face airstream 采煤工作面的风流
coal face distribution substation 工作面配电点
coal face haulage 采煤工作面运输
coal face layout 采煤面布置
coal face monitoring equipment 采煤工作面监控设备
coal face of blasting mining 采煤工作面爆破开采
coal face recovery coefficient 采煤工作面回采率
coal face supervision 采煤工作面监控
coal face support 采煤工作面支护
coal face survey 采煤工作面测量
coal facies 煤相
coal fall 煤的崩落
coal feed regulator 给煤调节器
coal field 煤田
coal field gas 煤田气
coal field prediction 煤田预测
coal filling 装煤
coal fine tailings 煤炭精尾矿
coal fines 煤粉
coal fines pelletization 粉煤团矿；粉煤造粒
coal fire gas minerals 煤火气矿产
coal fires 煤火
coal flap 煤仓入口盖板
coal flotation 浮选选煤
coal fluidization technology 煤炭液化技术
coal fly ash 粉煤灰
coal for coking 炼焦用煤
coal for gasification 气化用煤
coal formation 含煤岩系；煤层
coal forming process 成煤作用；煤的形成过程
coal fractures and cleats 煤层裂缝与割理
coal fuel ratio 煤燃料比

coal fueled furnaces 燃煤炉
coal gangue 煤矸石
coal gangue concrete 煤矸石混凝土
coal gas 煤气
coal gas outburst 煤层瓦斯突出
coal gas scrubber 煤气洗涤器
coal gasification 煤的气化
coal gasification by unclear heat 核能煤气化
coal generated gas 煤成气
coal geochemistry 煤地球化学
coal geological exploration 煤田地质勘探
coal geology 煤田地质学
coal geophysical log 煤田测井
coal geophysical prospecting 煤田地球物理勘探
coal getter 采煤工
coal getting 采煤
coal getting machine 采煤机
coal goaf 煤矿采空区
coal grab 抓煤机；装煤机
coal grade 煤的品位
coal grading 煤的分级
coal granule 煤颗粒
coal gravity prospecting 煤田重力勘探
coal grit 煤质砂岩
coal gush 煤喷出；煤涌出
coal hammer 采煤风镐
coal handing 煤炭搬运
coal handler 煤装卸工
coal handling 输煤设备
coal handling and storage 煤炭处理及存储
coal hardiness coefficient 煤的坚固性系数
coal haulage 煤的运输
coal haulage ramp 运煤坡道
coal hauler 运煤卡车
coal heading 煤巷
coal heaver 煤炭装卸工
coal heel 小煤柱
coal height 煤层厚度
coal hewer 采煤工
coal hoister 煤矿提升机；主井提升机
coal hopper 煤仓；煤漏斗；煤斗
coal hutch 煤车
coal hydraulicking 水力采煤
coal hydrogenation 煤的加氢作用
coal ignitability 煤的可燃性
coal in solid 煤层完整部分；煤层未采部分
coal injection 喷吹煤粉
coal inrush 煤的突然崩落
coal jigging 跳汰选煤
coal lash 煤灰分
coal laundering 流槽选煤
coal layer 煤层
coal lead 煤线；断层带煤线
coal levelling bar 平煤杆
coal lifting pump 煤水泵
coal line 煤线
coal liquefaction 煤液化
coal lithology 煤岩学
coal lithology log 煤田岩性测井
coal lithotype 煤岩类型

coal loading 煤装载
coal loading and unloading 煤炭装卸
coal local ventilator 煤炭局部通风机
coal log 煤田测井
coal longwall 煤工作面
coal losses 煤损失
coal maceral 煤岩显微组分
coal maceral composition 煤素质成分
coal maceral group 煤素质组
coal machinery 采煤机械
coal magnetic prospecting 煤田磁法勘探
coal mark 煤牌号
coal mass 煤体
coal mass strength 煤体强度
coal mass thickness 煤体厚度
coal matrix 煤基质
coal matrix shrinkage 煤基质收缩
coal measure facies 煤系相；煤系岩相
coal measure strata 煤系地层
coal measure unit 煤系单位
coal measures 含煤岩系；煤层组
coal metamorphic series 煤的变质系列
coal metamorphism 煤变质酌；煤变质作用
coal microbiology 煤微生物学
coal microstructure 煤的微观结构
coal mill 磨煤机
coal mine 煤矿
coal mine bumps 煤突出
coal mine drainage 煤矿排水
coal mine dust 煤矿粉尘
coal mine dust exposure 煤矿粉尘暴露
coal mine engineering exploration 煤矿工程勘探
coal mine environment 煤矿矿区环境
coal mine excavation 煤矿开挖
coal mine expansion 煤矿改扩建
coal mine fire 矿井火灾
coal mine gas monitoring 煤矿瓦斯监测
coal mine geological map 煤矿地质图
coal mine heaps 煤矿堆
coal mine impact 煤矿的影响
coal mine methane 煤层气；煤矿瓦斯
coal mine methane drainage 煤层甲烷抽采
coal mine monitoring 矿井监测
coal mine overburden spoil 煤矿覆土糟蹋
coal mine permitted detonator 矿用电雷管
coal mine powder 煤矿用炸药
coal mine reclamation 矿山复垦
coal mine roof rating 煤矿顶板评级
coal mine safety evaluation 煤矿安全评价
coal mine safety monitoring and control 煤矿安全监测监控
coal mine security 煤矿安全
coal mine shaft 煤矿矿井；煤矿井筒
coal mine solid waste 煤矿固体废物
coal mine spoils 煤矸石
coal mine sties 煤矿经纬网
coal mine streams 煤矿流
coal mine subsidence 煤矿塌陷
coal mine underground environment gases 煤矿井下环境气体
coal mine ventilation system 煤矿通风系统

coal mine wastewater 煤矿废水
coal mine water 矿井水
coal mine water management 矿井水管理
coal miner 煤矿工
coal mineralogy 煤炭矿物学
coal miner's lung 煤肺病；煤矿工尘肺病
coal mines 煤矿
coal mining 采煤；采煤学
coal mining above aquifer 承压含水层上采煤
coal mining activities 煤炭开采活动
coal mining ammonium nitrate explosives 硝铵炸药
coal mining and selection industry 煤炭采选业
coal mining area 煤炭采空区
coal mining by the longwall method 长壁采矿法采煤
coal mining collapse 采煤塌陷
coal mining combine 联合采煤机
coal mining dumping area 煤矿堆填区
coal mining effluent 煤炭开采出水
coal mining explosive 煤矿用炸药
coal mining face 采煤工作面
coal mining field 煤矿区；煤田
coal mining geology 煤矿地质学
coal mining goaf land 采煤采空区土地
coal mining machine 采煤机
coal mining method 采煤方法
coal mining mode 煤炭开采模式
coal mining monitoring system 煤炭开采监测系统
coal mining residues 煤炭开采残留
coal mining seeper subsidence district 煤炭开采积水塌陷区
coal mining serious accident 煤炭开采严重事故
coal mining shearer 煤矿采煤机
coal mining subsidence area 采煤塌陷区
coal mining subsidence wetland 采煤沉陷区湿地
coal mining technique 采煤工艺
coal mining under buildings 建筑物下采煤
coal mining under buildings, railway and water bodies 三下采煤
coal mining under railways 铁路下采煤
coal mining under water-bodies 水体下采煤
coal mining with gangue backfilling 煤炭开采；煤矸石回填
coal mining without pillars 煤炭开采无支柱
coal mining work-out area 采煤工作区
coal mining zone 煤炭开采区
coal mixing plant 配煤厂；配煤车间
coal modification 煤改质
coal mud 煤泥
coal nest 煤窝
coal oil 煤焦油；煤油
coal oil coprocessing 煤油共处理
coal oil mixture 煤焦油混合物
coal or rock dynamic disaster 煤岩动力灾害
coal or rock electromagnetic emission 煤与岩石电磁辐射
coal origin 煤的成因
coal outbreak 煤层出露；煤层露头
coal outburst 煤的挤出；煤突出
coal output 煤炭产量
coal outrush 煤喷出；煤涌出
coal oxidation 煤炭氧化；煤-氧化

coal paleogeography 聚煤古地理
coal particles 煤粒
coal paste 煤糊
coal pavement 煤台道；煤炭路面
coal pellet 煤球；型煤
coal permeability 煤层渗透率
coal permeability model 煤层渗透率模型
coal petrographic composition 煤岩成分
coal petrography 煤相学；煤岩学；煤岩
coal petrology 煤岩学
coal physical characteristics 煤物理特性
coal pick 采煤风镐；刨煤镐
coal pile 煤堆
coal pillar 煤柱
coal pillar factor of safety 煤柱的安全系数
coal pillar failure 煤柱失稳
coal pillar load 煤柱负荷
coal pillar thrust 煤柱压碎
coal pillar width 煤柱宽度
coal pipe 很薄的煤层
coal pipeline 输煤管
coal pipelining 管道运煤
coal pit 矿井；煤矿；煤窑
coal pitch 煤沥青
coal planer 刨煤机
coal planning 刨煤；刨煤机采煤
coal plant 成煤植物
coal plough 刨煤机
coal ploughing 刨煤机回采
coal pocket 煤仓
coal pollutants 煤炭污染物
coal pore 煤孔隙
coal pore structure 煤孔隙结构
coal porosity 煤孔隙度
coal port 运煤港口
coal powder conveyor 煤粉输送机
coal preparation 煤炭洗选；洗煤
coal preparation-screening 选煤-筛选
coal preparation-separation 选煤-分离
coal preparation-tailings disposal 选煤-尾矿处理
coal preparation-waste utilization 选煤-废物利用
coal preparation facilities flow balance 选煤设备流程图
coal preparation plant 煤浆制备厂；洗煤厂；选煤厂
coal preparation process 煤炭洗选法；煤炭洗选工序
coal preparation shift 采煤准备班；选煤转变
coal preparation waste water 选煤废水
coal preserving structure 贮煤构造
coal processing 煤炭加工
coal processing plant 选煤厂；煤炭加工厂
coal producing regions 产煤地区
coal production 煤产量
coal production and logistics system 煤炭生产和物流系统
coal production base 煤炭生产基地
coal projection 煤突出
coal property 煤质
coal prospecting 煤田地质勘探
coal province 含煤区；聚煤区
coal pulling 采煤
coal pulverizer 碎煤机
coal pump 煤水泵

coal puncher 采煤风镐；冲击式截煤机
coal pusher 推煤机
coal pyrolysis 煤热解
coal quality 煤质
coal quality control 煤质控制
coal quality log 煤质测井
coal quota 煤的限额
coal rank 煤的品级；煤级
coal reclamation 煤矿复垦
coal recovery 采煤回收率
coal recovery drill 螺旋采煤机；螺旋采煤机
coal re-extraction 煤炭再提取
coal reference 煤层基准
coal refuse 煤矸石
coal refuse fires 煤矸石火灾
coal reserves 煤炭储量
coal reserves variation 煤炭储量变化
coal reservoir 煤储层
coal reservoir pressure 煤储层压力
coal reservoir properties 煤储层物性
coal residues 煤渣
coal resource conformity 煤炭资源整合
coal resource type city 煤炭资源型城市
coal resources assets 煤炭资源资产
coal rich center 富煤中心
coal rich zone 富煤带
coal road 煤巷
coal roadway 煤巷
coal roadway deep well 深井煤巷
coal roadway excavation 煤矿巷道掘进
coal roadway surrounding rock 煤巷围岩
coal rock 煤岩
coal rock and gas outburst mine 煤与瓦斯突出矿井
coal rock creep deformation 煤岩蠕变变形
coal rock deformation 煤岩变形
coal rock dynamic disasters 煤岩动力灾害
coal rock electromagnetic radiation 煤岩电磁辐射
coal rock reinforcement 煤岩加固
coal rock strength 煤岩强度
coal roof interface detector 煤岩分界探测器
coal safety pillar 煤矿安全支柱
coal sample 煤样
coal sample division 煤样的缩分
coal sample for gas test 瓦斯煤样
coal sample reduction 煤样破碎
coal sampler 取煤样器
coal samples 煤样
coal sampling 煤炭取样
coal saw 锯煤机
coal scatter 分煤设备
coal science 煤炭科学
coal seam 煤层
coal seam bifurcation 煤层分岔
coal seam correlation 煤层对比
coal seam exposure 煤层露头
coal seam fire 煤层火
coal seam floor 煤层底板
coal seam floor contour map 煤层等高线图
coal seam gas 煤层气
coal seam geometry 煤层几何

coal seam group 煤组
coal seam hole 沿层钻孔
coal seam infusion 煤层注水
coal seam liable to dust explosion 有煤尘爆炸危险的煤层
coal seam mining 煤层开采
coal seam of rock burst-prone 煤层冲击地压倾向的
coal seam parameters 煤层参数
coal seam permeability 煤层透气性
coal seam productive capacity 煤层产出能力
coal seam prone to spontaneous combustion 自燃发火煤层
coal seam roof 煤层顶板
coal seam sample 煤层煤样
coal seam series 煤系
coal seam water infusion 煤层注水
coal seam-roof stability 煤层顶板稳定性
coal seat 底黏土；煤层底耐火黏土
coal section 煤薄片
coal seismic prospecting 煤田地震勘探
coal selection 煤的选采
coal self-ignition 煤炭自燃
coal separator 选煤机
coal series 煤化系列；煤种系列
coal series gas 煤系气
coal shaft 出煤井；提煤井
coal shearer 采煤机
coal shift 采煤班
coal shoot 放煤溜槽
coal shooting 煤层爆破
coal shovel 铲煤机；煤铲
coal shrinkage 煤收缩
coal shrinkage and swelling 煤收缩与膨胀
coal shrinking 煤收缩
coal silo 煤仓
coal silt 煤泥渣
coal size distribution 煤的粒度分布
coal size separation 煤-大小分离
coal skip 提煤箕斗
coal slack 煤矸石；末煤
coal slacking 风化煤
coal slag 煤渣
coal slaking 煤炭粉化
coal slate 煤质板岩
coal slice 煤炭片
coal slide 片帮
coal slime 煤泥
coal sludge 煤泥
coal slurry 煤浆；煤泥
coal slurry ability 煤炭成浆性
coal slurry electrolysis 煤浆电解
coal slurry preparation room 煤水洞室
coal slurry pump 煤水泵
coal slurry sedimentation 煤泥沉淀
coal slurry sump 煤水仓
coal slurry treatment 煤泥处理
coal slurry water 煤泥水
coal smut 煤华；煤层露头；泥质劣煤；煤矿煤尘
coal sorter 手选工
coal split 分裂煤层
coal spontaneous combustion 煤炭自燃
coal spontaneous combustion tendency 煤的自燃倾向性

coal stacker reclaimer 贮煤场装载输送机
coal station 贮煤站
coal step 煤阶
coal stock yard 储煤场
coal stone 无烟煤；硬煤
coal storage 存煤量；贮煤场
coal storage pond 储煤池
coal storage silo 圆筒煤仓
coal storage yard 煤场
coal store 储煤仓；贮煤场
coal streak 煤线
coal stream 煤流
coal structure 煤结构
coal substance 煤有机质
coal sudden fall 煤层倾出
coal surface treatment 煤面处理
coal swamp 聚煤沼泽；煤沼泽
coal swelling 煤膨胀；煤岩膨胀
coal tailings 煤炭尾矿
coal tar 煤焦油；焦油沥青
coal tar asphalt 煤焦油沥青
coal tar coating 煤焦油涂层
coal tar enamel 煤焦油磁漆
coal tar epoxide paint 环氧煤焦油漆
coal tar epoxy 煤焦油环氧树脂
coal tar felt coating 煤焦油毡涂层
coal tar mixture 煤焦油混合物
coal tar pitch 煤焦油沥青
coal terminal 煤码头
coal thickness 煤厚
coal to liquid technology 煤制油技术
coal topographic-geological map 煤田地形地质图
coal transport 煤运输
coal trimmer 平舱工人，平舱机
coal tub 煤车
coal type 煤型；煤岩类型
coal type gas 煤型气
coal valuation 煤质评价
coal vein 煤层
coal volatiles 煤挥发物
coal wall 采煤工作面
coal wall spalling 煤壁片帮
coal wash 选煤
coal washability curve 煤炭可洗选性曲线
coal washability test 煤可洗性试验
coal washer 洗煤机
coal washery 洗煤厂
coal washing 洗煤
coal waste 煤矸石
coal waste pile 煤矸石堆
coal waste utilization 煤炭废物利用
coal water fuel 水煤浆
coal water mixture 煤水浆
coal water mixture stabilizing agent 水煤浆稳定剂
coal water ratio 煤水比
coal wedge 落煤楔
coal wettability 煤润湿性
coal winning machine 采煤机
coal winning method 采煤方法
coal winning technique 采煤工艺

coal winning technology 采煤工艺
coal work 采煤工作
coal workers' pneumoconiosis 煤工尘肺
coal yard 储煤场；贮煤场
coal yield 出煤量；煤产量；煤的回收率
coal-alkali reagent 煤碱剂
coal-and-gas outburst mine 煤与瓦斯突出矿井
coal-and-methane outburst mine 煤与沼气突出矿井
coal-aqueous mixture 水煤浆
coal-bearing 含煤的
coal-bearing area 含煤区
coal-bearing basin 含煤盆地
coal-bearing coefficient 含煤系数
coal-bearing cycle 含煤岩系旋回结构
coal-bearing density 含煤密度
coal-bearing deposit 含煤沉积
coal-bearing sediment 含煤沉积
coal-bearing series 含煤岩系；煤系
coal-bearing structure 赋煤构造
coalbed gas 煤层气；煤层瓦斯
coalbed gas content 煤层气含量
coalbed gas drilling 煤层气钻探
coalbed gas geology 煤层气地质学；瓦斯地质
coalbed gas heating value 煤层气热值
coalbed gas occurrence 煤层气赋存状态
coalbed gas pressure 煤层瓦斯压力
coal-bed gas well 煤层气井
coalbed gases 煤层气
coalbed methane langmuir pressure 煤层甲烷兰格缪尔压力
coalbed methane langmuir volume 煤层甲烷兰格缪尔体积
coalbed methane production 煤层气生产
coalbed methane recovery 煤层气回采
coalbed methane reservoir simulation 煤层甲烷储层模拟
coalbed methane reservoirs 煤层气储层
coalbed methane simulation 煤层气模拟
coalbed rocks 煤层的岩石
coal-black 墨黑的；像炭一般的黑
coal-body structure 煤体结构
coal-burning generating plant 燃煤发电厂
coal-burning rotary kiln 燃煤回转窑；烧煤旋转干燥炉
coal-cleaning method 选煤方法
coal-cutting machine 截煤机
coal-derived fuel 煤基燃料
coal-derived gas 煤成气
coal-derived oils 煤源油类
coal-dust brick 煤灰砖
coal-dust explosibility 煤尘爆炸性
coal-dust explosion seam 煤尘爆炸危险煤层
coal-dust explosion wave 煤尘爆炸波
coaled gas pressure 煤层瓦斯压力
coaler 运煤车辆
coalesced copper 聚结铜
coalescence 合并；聚结；凝聚
coalescence force 聚结力
coalescence kinetics 聚结动力学
coalescence rate 聚集速率
coalescent 聚结的
coalescent debris cone 复合碎石锥
coalescent pack 聚结填料
coalescer 聚结剂

coalescing 凝聚
coalescing alluvial fan 复合冲积扇；接合冲积扇
coalescing fan 接合冲积扇
coalescing of fine particles 细粒凝聚
coalescing pediment 接合麓原；联合山麓侵蚀面
coalescing section 聚结段
coalescing separator 聚结分离器
coalescing submaring fan 接合海底扇
coalescive neomorphism 聚结新生变形
coaleum 煤烃
coal-face extraction loss 回采工作面采矿损失
coal-face mechanization 采煤工作面机械化
coal-face wining aggregate 采煤联动机
coal-feeding machine 给煤机
coalfield analysis 煤田分析
coal-field electric substation 矿区变电所
coalfield hydrogeology 煤田水文地质
coal-fired 烧煤的
coal-fired gas turbine 烧煤燃气涡轮机
coal-fired steam generator 燃煤的蒸汽机
coal-fired steam plant 燃煤蒸汽发电厂
coal-formation epoch 成煤地质时代
coal-formed gas 煤成气
coal-forming facies 造煤相
coal-forming material 成煤物质
coal-forming period 成煤期；聚煤期
coal-forming plant 成煤植物
coal-forming stage 成煤阶段
coal-gas flow 煤-气体流量
coal-gas interactions 煤-气相互作用
coal-gas pressure 煤制气压力
coal-handling plant 给煤机
coal-head 煤巷掘进头
coalification 煤化酌；煤化作用
coalification break 煤化间断；煤化转折
coalification equilibrium 煤化作用平衡
coalification gradient 煤化梯度
coalification jump 煤化间断；煤化转折
coalification pattern 煤化型式
coalification process 煤化过程
coalification ratio 碳化率
coalification series 煤化系列
coalification stage 煤化阶段
coalification track 煤化轨迹
coalified 煤化的
coalified wood 煤化木
coalignment 调准装置
coaling 加煤；装煤
coaling crane 装煤起重机
coaling gear 采煤和上煤装置
coalite 半焦炭；科莱特无烟燃料
coalitenessity 低温焦
coalition 结合
coalless column 无煤柱
coal-measure rock 煤系岩石
coal-mining geological condition 煤矿地质条件
coal-mining system 采煤方法；煤矿系统
coal-oil mixing liquefaction 煤油共炼液化
coalplex 煤的综合利用
coal-plough classification 刨煤机的分类

coal-plough haulage 刨煤机的牵引
coal-plough winch 刨煤机绞车
coal-producing area 产煤区
coal-pulverizing plant 磨煤车间
coal-raising system 提煤系统
coal-recovery auger 采煤螺钻；螺旋钻采煤机
coal-related gas 煤型气
coal-rock combinations 煤-岩组合
coal-rock drift 煤岩巷
coal-rock drivage 半煤岩巷道掘进
coal-rock interface transducer 煤岩界面传感器
coal-rock pillar against water 对水煤岩柱
coal-running pitch 煤块自溜的倾角
coal-sawing machine 锯煤机
coal-seam correlation section 煤层对比图
coal-seam uncovering 煤层剥离
coal-sensing probe 底板留煤的核探测器
coalshed 煤夹层；不可采的极薄煤层；很薄的煤层；煤夹层
coal-sizing analysis 煤的筛分分析
coal-storage reloader 贮煤场转载机
coal-tar product 煤焦油产品
co-altitude 天顶距
coal-to-liquids 煤制油
coal-type scraper loader 耙斗装煤机
coal-washing table 选煤淘汰盘；选煤摇床
coal-water cross-cut 溜煤石门
coal-water mixing chamber 煤水仓；煤水混合室
coal-water pump 煤水泵
coal-water pump hoisting 煤水泵提升
coal-water separation 煤的脱水
coal-water slurry 煤水浆；水煤浆
coaly 含煤的；煤状的
coaly debris 煤屑
coaly facies 煤相
coaly inclusion 煤包体
coaly kerogen 煤质干酪根；煤油母岩质
coaly organic matter 煤型有机质
coaly polymeric 煤质聚合物
coaly rashings 软炭质页岩；软页岩
coaly ratings 碳质小块软页岩
coaly shale 煤质板岩
coaming 舱口栏板；挡水围墙；天窗围坎
Coanda air curtain 降尘气幕；柯安达气幕
Coanda air curtain system 附壁效应式气幕系统
co-antiknock agent 抗爆助剂
coarctate 密集的
coarray 并合台阵
coarse 粗糙的；未加工的；粗粒的
coarse adjustment 粗调节装置；粗调整
coarse aggregate 粗大粒集合体；粗集料
coarse aggregate concrete 粗骨料混凝土
coarse analysis 粗筛分
coarse ash tuff 粗粒凝灰岩
coarse asphaltic concrete 粗骨料沥青混凝土
coarse bar screen 粗格栅
coarse bed 粗骨料层
coarse bedload transport 粗泥沙运输
coarse bentonite 粗粒膨润土
coarse breaking 粗碎

coarse break-up	粗粒分散体	coarse sand	粗砂
coarse cherty	粗燧石的	coarse sand beach	粗砂滨
coarse chip	粗石片	coarse sandy loam	粗砂壤土
coarse clastic	粗屑的	coarse sandy soil	粗砂质壤土
coarse clay	粗黏土	coarse scanning	疏扫描
coarse coal	大块煤	coarse screen	粗孔筛
coarse coal slime	粗煤泥	coarse screening	粗粒筛除；粗筛
coarse concentrate	粗粒精矿	coarse sediment	粗泥沙；粗粒沉积物
coarse control	粗调	coarse sieve	粗筛
coarse crusher	粗碎机	coarse silt	粗粉砂；粗粉土
coarse crushing	粗碎	coarse sizing	粗粒筛分
coarse crystallization	粗结晶	coarse slime	粗煤泥
coarse deposit	粗粒沉积	coarse sluice	粗粒溜槽
coarse fibre	粗纤维	coarse slurry	粗泥
coarse filament	粗丝	coarse soil	粗质土；大粒土壤
coarse fill	粗料充填	coarse spoil	粗矿渣
coarse filter	粗过滤器；粗滤池	coarse staple fibre	粗短纤维
coarse filtration	粗过滤	coarse stone	粗碎石
coarse fraction	粗粒部分；粗粒组	coarse strainer	粗滤器；粗滤网
coarse fraction content	粗粒组含量	coarse stuff	粗填料；粗涂料
coarse fragment	粗碎屑	coarse supporting screen	粗孔支承筛板
coarse gold	粗粒金	coarse suspension	粗粒悬浮液
coarse graded	粗粒的；粗纹理的	coarse synchronizing	粗同步
coarse grading	粗级配；粗粒径筛分	coarse texture	粗粒结构
coarse grain	粗晶粒；粗颗粒	coarse thread	粗螺纹；粗牙螺纹
coarse grain sand	粗粒砂	coarse topography	粗切地形
coarse grain soil	粗粒土	coarse trash rack	粗格拦污栅
coarse grained fracture	粗晶断口	coarse vibrating screen	粗筛振动格栅
coarse grained soil	粗粒土	coarse washed coal	粗粒精煤
coarse grained texture	粗粒结构	coarse waste	粗废石；粗岩屑
coarse grained wood	粗纹理木材	coarse work	粗糙工作；普查工作
coarse granitoid texture	粗粒花岗状结构	coarse zero	粗调零
coarse granular	粗团粒	coarse-aggregated	粗粒的；粗聚合的
coarse granular fracture	粗晶断口	coarse-banded	粗条带状
coarse gravel	粗砾石	coarse-crystalline	粗晶状的
coarse grid	粗网格	coarse-crystalline dolomite	粗晶白云石
coarse grind	粗磨	coarse-fibered	粗纤维的
coarse grinding	粗磨	coarse-fine tuning	粗微调
coarse ground cement	粗研水泥	coarse-fragmental ground	大块碎屑类土
coarse jigging	粗粒跳汰	coarse-grained	粗晶的；粗粒的
coarse launder washer	块煤洗煤槽	coarse-grained clastics	粗粒碎屑
coarse loam	粗壤土	coarse-grained paving	粗粒石铺
coarse material	粗粒物质	coarse-grained soil fill	粗粒土填料
coarse mesh	粗孔筛；大网筛孔	coarsely crystalline	粗晶质
coarse meshed sieve	粗目筛	coarsely disseminated	粗浸染的
coarse metal	粗冰铜；粗皮金属	coarsely graded	粗粒级的
coarse middling	块中矿；块中煤；中级块煤	coarsely granular	粗粒状的
coarse mixture	粗混合物；浓水泥浆	coarsely granular particle size	粗颗粒粒度
coarse ore	粗矿石	coarsely ringed timber	粗纹木材
coarse particle	粗颗粒	coarse-meshed	大筛孔的
coarse pebble	粗中砾	coarse-meshed filter	粗网滤器
coarse pitch	大螺距	coarseness	粗度；粒度
coarse pore network	粗孔隙网络	coarseness of grading	级配粒度；粒度组成
coarse porosity	粗孔隙	coarsening	变粗；粗化
coarse pulling	粗提取	coarsening-upward sequence	向上变粗层序
coarse rack	疏拦污栅	coarse-ore bin	粗矿仓
coarse reading	粗读数	coarse-pored	粗孔的
coarse reduction gyrator	粗碎圆锥破碎机	coarse-tail grading	下部粗粒向上变细现象
coarse regulation	粗调整	coarse-textured drainage	粗切水系
coarse rolls	粗碎辊碎机	coarse-textured pattern	粗稀结构模式

coarse-textured soil 粗质土壤
coarse-textured topography 粗切地形
coarse-threaded joint 粗纹接头
coarse-to-fine aggregate ratio 粗细骨料比
coarsness 粗粒度
coast area 海岸带
coast avalanche 海岸崩塌
coast chart 海岸图
coast current 沿岸流
coast defense engineering 海防工程
coast deposit 海岸沉积
coast deposit and spread 海岸淤进
coast deposition 海岸淤积
coast depot 海岸油库
coast dune 海岸沙丘
coast elevation 海岸上升
coast ice 沿岸冰
coast landslide 海岸滑坡
coast material 海岸泥沙
coast mountains 海岸山脉
coast mud 海岸泥
coast of dalmatian type 达尔马堤亚式海岸
coast of elevation 隆起海岸；上升海岸
coast of emergence 上升海岸
coast of mobile region 变动地域海岸
coast of stable region 稳定区域的海岸
coast of submergence 海侵海岸；下沉海岸
coast onlap 海岸上超；海岸上覆
coast pilot 航路图志
coast plain 浪蚀平原
coast plane 侵蚀基准面
coast prominences 海岸突起
coast protecting wall 防波墙
coast protection works 海岸防护工程
coast range orogeny 海岸山脉造山作用
coast ranges 海岸山脉
coast recession 海岸后退
coast reef 海岸礁
coast refraction 海岸折射
coast riff 海岸礁
coast salinization 滨海盐渍化
coast sand 海岸砂
coast shelf 近岸大陆架；下沉海岸平原
coast terrace 海岸阶地
coast topographic survey 海岸地形测量
coast trend 海岸走向
coast zone 海岸带
coastal 沿海的
coastal accretion 海岸加积
coastal aggradation 海岸加积
coastal alluvial plain 沿海冲积平原
coastal and offshore hazards 沿海及近海灾害
coastal aquifer 滨海含水层
coastal archaeology 沿海考古
coastal bar 沙洲；沿岸沙坝
coastal barrier 滨海障壁坝
coastal basin 海岸盆地
coastal bays 沿海海湾
coastal beach 滨海沙滩
coastal belt 海岸地带；沿岸带

coastal bitumens 海岸沥青
coastal canal 沿海运河
coastal cave 海岸洞穴
coastal changes 沿海的变化
coastal classification 海岸分类
coastal cliff 海崖
coastal collapse hazard 海岸坍塌灾害
coastal cordillera 沿海山脉
coastal crude oil 沿岸区原油
coastal defense 海岸防护工程
coastal deposition 海岸沉积
coastal deposits 海岸沉积；海岸堆积
coastal dike 防潮堤；海堤
coastal domain 沿海相范围
coastal domes 海滨盐丘群
coastal drifting 海岸迁移；沿岸漂移
coastal drifting map 海岸变迁图
coastal dune 海岸沙丘
coastal dune rock 海岸沙丘岩
coastal dynamic map 海岸动态图
coastal dynamics 海岸动力学
coastal embayment 滨岸海湾
coastal encroachment 海岸前侵
coastal engineering 海岸工程
coastal environment 沿海环境
coastal erosion 海岸冲蚀；海岸侵蚀
coastal erosion disaster 海岸侵蚀灾害
coastal erosion survey 海岸侵蚀调查
coastal evolution 沿海进化
coastal facility 海岸设施
coastal feature 海岸地形
coastal friction 海岸摩擦
coastal front 海岸锋
coastal geology 海岸地质学
coastal geomorphology 海岸地貌学
coastal groin 海岸丁坝；海岸防波堤
coastal ground water 海岸地下水
coastal groyne 海岸丁坝；海岸防波堤
coastal habitat 海岸栖息地
coastal harbour 海港
coastal hazards 滨海灾害
coastal karst 滨海岩溶
coastal lagoon 沿海泻湖
coastal lagoon facies 海岸泻湖相
coastal lake 滨海湖；沿岸湖
coastal landform 海岸地貌地形
coastal landslide 海岸滑坡
coastal levee 防潮堤；海岸堤防
coastal marine coal measures 近海相煤系
coastal marsh 海岸沼地；海滨沼泽
coastal morphology 沿海形态
coastal mudflat 海岸泥坪
coastal nonmarine deposits 非海相海岸沉积
coastal observation station 海岸观测站
coastal ocean dynamics 近岸海洋动力学
coastal oceanography 沿海海洋学
coastal onlap 海岸上超
coastal plain 滨海平原；滩地
coastal plain estuary 海岸平原河口
coastal plain swamp 海岸平原沼泽地

coastal platform	海岸台地	coated emerald	镀层祖母绿
coastal pollution	近岸污染	coated grain	包壳颗粒
coastal profile of dynamic equilibrium	海岸动态平衡剖面	coated grain dolomite	包粒白云岩
coastal progradation	海岸进积作用	coated grain limestone	包粒灰岩
coastal protection	海岸防护	coated gravel	含壳砾石
coastal protection area	沿岸保护区	coated jadeite	镀膜翡翠
coastal range	海岸山脉	coated material	有涂面的材料
coastal reclamation	围海造地	coated particle	涂敷粉粒
coastal reef	沿海礁	coated pile	涂面桩
coastal refraction	海岸折射	coated surface	涂镀表面
coastal resources	海岸资源	coating	镀膜；保护层；涂层；覆盖层；喷涂饰面
coastal restoration	海岸修复	coating agent	涂层剂
coastal rock cliffs	海岸岩石峭壁	coating compound	涂料
coastal rock platform	沿海的岩石平台	coating effect	覆盖效应
coastal sand	海岸沙	coating engineering	被覆工程；涂装工程设计
coastal sand balance	沿海砂平衡	coating flaw	涂层中的空隙
coastal sand dune	海岸沙丘	coating jade	璞玉；石包玉
coastal sediment	海岸沉积物	coating machine	涂抹机；喷镀机
coastal sediment wedge	沿海沉积楔	coating material	涂层材料；涂盖物质
coastal sedimentation	沿海沉淀	coating of cement	水泥抹面
coastal seismic belt	海岸地震带	coating ratio	阳极氧化膜生成比
coastal shelf	沿海大陆架	coating reaction	被膜反应
coastal shoaling	沿岸浅滩	coating resin	涂布树脂
coastal stream	沿岸流	coating techniques	涂层技术
coastal structure	海岸防护结构；海岸建筑	coating test	涂布试验
coastal studies institute for Hull	英国赫尔海岸研究所	coating tubing	涂层油管
coastal superphosphate	沿海过磷酸钙	coax	同轴电缆
coastal survey	岸线地形测量	coax signal cable	同轴信号电缆
coastal swallow hole	海岸落水洞	coaxality	共轴性
coastal swamp	海岸木本沼泽	coaxial	共轴的；同轴的
coastal terrace	海岸阶地	coaxial cable	同轴锚索
coastal tide	近岸潮汐	coaxial configuration	同轴结构
coastal toplap	海岸顶超	coaxial cylinder apparatus	同轴圆筒仪
coastal topography	海岸地形	coaxial cylinder viscometer	同轴圆筒式黏度计
coastal uplift	沿海隆起	coaxial deformation	共轴变形
coastal vegetation	海滨植被	coaxial drive	同轴驱动
coastal vulnerability	海岸易损性	coaxial flow	共轴流动
coastal weathering	海岸风化	coaxial folds	共轴褶皱
coastal wetlands	沿海湿地	coaxial fractures	共轴断裂；同轴裂隙
coastal zone development	海岸带发育；海岸带演化	coaxial line	同轴线
coastal zone mineral resources	海岸带矿产	coaxial loop system	同轴线圈系统
coastal zone resources	海岸带资源	coaxial magnetic field	同轴磁场
coaster	惯性运转装置；沿海航船	coaxial nozzles	同轴喷嘴
coasting	惯性滑行；海岸线；滑翔；溜放调车；沿海岸航行	coaxial progressive deformation	共轴递进变形
		coaxial rainfall runoff relation	降雨径流共轴关系
coasting grade	溜坡	coaxial rotational viscometer	同轴旋转黏度计
coasting resistance	惰行阻力	co-axial seal	同轴密封
coastland	沿海地区	coaxial socket	同轴电缆插座
coastline effect	海岸效应	coaxial spiral	套管的螺旋形扶正器
coastline shift	海岸线移位	coaxial strain	共轴应变
coastline survey	沿岸测量	co-axial stream	同轴水流
coastlining	海岸线测量	coaxial wall	同轴管壁
coastward	朝着海岸的；向海岸的	coaxiality	共轴性
coastwise	近海的；沿海	coaxswitch	同轴开关
coat	保护层；面层；抹灰层；涂层	coazervate	聚析液
coated abrasive	涂敷磨料；砂纸	coazervation	聚析
coated abrasive working	砂纸磨光	cob	草筋泥；干打垒；夯土建筑；糊墙土；脚煤
coated chipping	裹有沥青的石屑	cob brick	土砖
coated diamond	包壳钻石；镀膜钻石	cob wall	土墙
coated electrode	包剂焊条；涂料焊条	cob walling	板筑墙；夯土墙；土坯墙

cobalt 钴
cobalt benzoate 苯甲酸钴
cobalt bloom 钴华
cobalt compounds 钴化物
cobalt crust 钴地壳；钴华
cobalt glance 辉砷钴矿
cobalt hexacyanide 六氰化钴
cobalt minerals 钴矿类
cobalt molybdate catalyst 钼酸钴催化剂
cobalt naphthenate 环烷酸钴
cobalt pyrite 硫钴矿
cobalt radiation source 钴辐射源
cobalt-base alloy 钴合金
cobalt-bonded tungsten carbide 钴钨硬质合金
cobaltic compound 高钴化合物
cobaltine 辉钴矿；辉砷钴矿
cobalt-iron-bismuth-molybdenum catalyst 钴-铁-铋-钼催化剂
cobaltite 辉钴矿；辉砷钴矿
cobalt-nickel pyrite 钴镍黄铁矿
cobaltocyanide 氰钴酸盐
cobaltomenite 硒钴矿
cobaltous sulphate 硫酸钴
cobalt-rich crusts 富钴结壳
cobalt-rich manganese crust 富钴锰结壳
cobaltsmithsonite 钴菱锌矿
cobb system of sampling 采样科布系统
Cobbe pan 混汞磨盘；磨盘；科布盘
cobber 磁选机；选矿机
cobbing 粗粒分选；人工敲碎；手选矿石块
cobbing hammer 拣矸锤；手选用锤
cobble 砾石；卵石；铺路石
cobble beach 砾石海滩
cobble boulder 铺路用卵石
cobble canal 卵石渠道
cobble conglomerate 粗砾砾岩
cobble gravel 卵砾石
cobble pavement 卵石路面
cobble soil 粗砾质土
cobble stone 粗砾岩；砾石；卵石
cobbled canal 卵石衬护渠道
cobbled gutter 卵石边沟
cobbly soil 粗砾质石；卵石土；粗砾质土
cobcoal 成团煤炭
cobi pile 哥毕桩
Coble creep 柯勃尔蠕变
co-carbonization 共焦化；共炭化
cocatalyst 助催化剂
coccinite 硒汞矿
coccoconite 粒状深海灰泥
coccodes 粒状体
coccoid 球状物
coccolite 粒状辉石岩
coccolith 颗石；颗石藻
coccolith ooze 白垩泥；颗石藻软泥
coccolithes 颗石藻
coccolithophore 钙板藻；颗石藻
coccolithus ooze 超微体钙质软泥
coccosphere 颗壳；颗石球
cochleoid 螺旋曲线
cocinerite 杂硫银铜矿

cocite 橄辉白榴斑岩
cock 扳机；风标
cock orifice 旋塞口
cock spanner 旋塞扳手
cock valve 旋塞阀
cockade fabric 鸡冠组构
cockade ore 环状矿石
cockade structure 鸡冠构造
cocker 人字形支架
cockering 人字形支架
cockermeg 采煤工作面临时支柱；人字形斜撑
cockerpole 顶底板斜撑间的横梁
cocking 击发
cockpit 灰岩盆地；漏斗状渗水井；驾驶座舱
cockpit karst 麻窝状岩溶
cockscomb ridge 鸡冠状山脊
cockscomb structure 鸡冠状构造
cocnentration factor 浓缩比
coco mat 椰毛垫
co-content 同容积
coconut nets 椰纤维网
cocos plate 科科斯板块
cocrystallization 共结晶
co-cumulative spectra 共积谱
cocurrent flow 并流
cocurrent laminar flow 共层流
cocurrent system of forces 汇交力系
cocycle 闭上链
coda 地震尾波；结论；结尾
coda amplitude 尾波振幅
coda length 尾波长度
coda pattern 尾波图形
coda spectrum 尾波谱
coda wave 地震尾波
codan 载频控制的干扰抑制器
codazzite 菱钙铈矿
code converter 代码转换器
code minier 矿山法则
code revision 修订规范
codeclination 同轴磁偏角
coded pulse 编码脉冲
coded sequence 编码序列
codeposition 共存沉积物；共同沉积作用
coderdecoder 编码译码器
co-disposal landfill 共同处理堆填区
cod-piece 井框节盘连接板
coefficient affecting fundamental period 基本周期影响系数
coefficient between layers 层间系数；管涌比
coefficient index 压缩指数
coefficient of abrasiveness 磨砺性系数
coefficient of absolute viscosity 绝对黏滞系数
coefficient of absorption 吸收系数
coefficient of active earth pressure 主动土压力系数
coefficient of adhesion 黏附系数；黏着系数
coefficient of admission 充填系数
coefficient of adsorption 吸附系数
coefficient of alienation 不相关系数
coefficient of amplification 放大系数
coefficient of anisotropy 非均质系数；各向异性系数
coefficient of attenuation 衰减系数

coefficient of autocorrelation 自相关系数
coefficient of axial deformation 轴向伸缩系数
coefficient of basin shape 流域性状系数
coefficient of basin slope 流域坡度系数
coefficient of bearing capacity 承载力系数
coefficient of blasting 爆破系数
coefficient of brittleness 脆性系数
coefficient of bulk compressibility 体积压缩系数
coefficient of bulk increase 松散系数;碎胀系数;体积膨胀率
coefficient of capillary suction 毛管吸水系数
coefficient of coherence 相干系数
coefficient of cohesion 内聚系数;黏聚系数
coefficient of colloid 胶体率
coefficient of collapsibility 湿陷系数
coefficient of collapsibility due to overburden pressure 自重湿陷系数
coefficient of compaction 夯实系数;压实系数
coefficient of compressibility 压缩系数
coefficient of compressibility of soil 土的压缩系数
coefficient of compression 压缩系数
coefficient of concentration 富集系数;凝聚系数;集中系数
coefficient of condensation 凝结系数
coefficient of conductivity 传导系数
coefficient of consolidation 固结系数
coefficient of consolidation for radial flow 径向固结系数
coefficient of contraction 收缩系数
coefficient of convergence 收敛系数
coefficient of correction 校正系数;修正系数
coefficient of correlation 对比系数;相关系数;相干系数
coefficient of coupling 耦合系数
coefficient of creep 蠕变系数
coefficient of cubic elasticity 体积弹性系数
coefficient of cubic expansion 体积膨胀系数
coefficient of cubical compressibility 体积压缩系数
coefficient of cubical elasticity 体积弹性系数
coefficient of cubical expansion 体积膨胀系数
coefficient of curvature 曲率系数
coefficient of damping 阻尼系数
coefficient of decay 衰减系数
coefficient of deflection 挠度系数
coefficient of deformation 变形系数
coefficient of deformation due to leaching 溶滤变形系数
coefficient of diffusion 扩散系数
coefficient of dilatation 体膨胀系数
coefficient of discharge 流量系数;排料系数
coefficient of dispersion 弥散系数;弥散系数
coefficient of dissociation 分解系数
coefficient of ditch rebound 基坑回弹系数
coefficient of divergence 发散系数
coefficient of driving 掘进率
coefficient of drying sensitivity 干燥敏感系数
coefficient of ductility 延性系数
coefficient of dynamic subgrade reaction 地基反力动力系数;基床动反力系数
coefficient of dynamic viscosity 动力黏滞系数
coefficient of earth pressure 土压力系数
coefficient of earth pressure at rest 静止土压力系数

coefficient of eddy diffusion 涡流扩散系数
coefficient of elastic non-uniform compression 弹性非均匀压缩系数
coefficient of elastic recovery 弹性恢复系数;回弹系数
coefficient of elastic resistance 弹性抗力系数
coefficient of elastic shear 弹性剪切系数
coefficient of elastic uniform compression 弹性均匀压缩系数
coefficient of elasticity in shears 剪切弹性模量
coefficient of elongation 伸长系数
coefficient of energy 能量系数
coefficient of energy dissipation 能耗系数
coefficient of energy utilization 能量利用系数
coefficient of equivalence 当量系数
coefficient of erosion 冲刷系数;侵蚀系数
coefficient of excentralization 偏心系数
coefficient of expansion 膨胀系数
coefficient of extension 伸延系数
coefficient of filling setting 充填压实系数
coefficient of fineness 细度系数
coefficient of fissuration 裂隙系数
coefficient of fissure 裂隙系数
coefficient of flood recession 退水系数
coefficient of flow 流量系数
coefficient of foaming 泡沫倍数
coefficient of foundation ditch's rebound 基坑回弹系数
coefficient of free swelling 自由膨胀率
coefficient of freezing resistance 抗冻系数
coefficient of friction 摩擦系数
coefficient of frictional resistance 摩擦阻力系数
coefficient of fullness 充满系数
coefficient of geochemical mobility 地球化学移动性系数
coefficient of gradation 级配系数
coefficient of ground reaction 地基反力系数
coefficient of groundwater head loss during movement 运动时地下水水头损失
coefficient of hardness 硬度系数
coefficient of heat conductivity 导热系数
coefficient of heat emission 放热系数;散热系数
coefficient of heat insulation 绝热系数
coefficient of heat perception 受热系数
coefficient of heat supply 供热系数
coefficient of heat transfer 传热系数
coefficient of heat transmission 传热系数
coefficient of homogeneity 均匀系数
coefficient of horizontal consolidation 水平固结系数
coefficient of horizontal pile reaction 桩的水平向反力系数
coefficient of horizontal pressure 侧压系数
coefficient of horizontal soil reaction 水平土反力系数
coefficient of horizontal subgrade reaction 水平地基反力系数
coefficient of humidity 湿润系数
coefficient of hydraulic conductivity 导水系数
coefficient of hydraulic stability 水力稳定性系数
coefficient of hydrodynamic dispersion 水动力弥散系数
coefficient of ignorance 未知率
coefficient of impact 冲击系数;动力系数
coefficient of imperviousness 不透水系数
coefficient of indicated pressure 指示压力系数
coefficient of infiltration 渗透系数

coefficient of influence 影响系数
coefficient of interference 干扰系数
coefficient of internal friction 内摩擦系数
coefficient of irregularity 不规则系数
coefficient of joint roughness 裂隙面粗糙系数
coefficient of kinematic viscosity 动黏滞系数
coefficient of kinetic friction 动摩擦系数
coefficient of lateral earth pressure 侧向土压力系数
coefficient of lateral expansion 侧膨胀系数
coefficient of lateral pressure 侧压力系数
coefficient of leakage 渗漏系数；越流系数；漏失系数；漏损系数
coefficient of linear expansion 线膨胀系数；线性膨胀系数
coefficient of longitudinal dispersion 纵向弥散系数
coefficient of loss 损失系数；耗损率
coefficient of macrodispersion 宏观弥散系数
coefficient of magnetization 磁化系数
coefficient of magnification of mapping 测图放大系数
coefficient of mechanical dispersion 机械弥散系数；水力弥散系数
coefficient of mechanical efficiency 机械效率系数
coefficient of migration 转移系数
coefficient of mine air leakage 矿井漏风系数
coefficient of mineralization 矿化系数
coefficient of mining 开采率
coefficient of moisture absorption 吸湿系数
coefficient of moisture transition 变湿系数
coefficient of molecular diffusion 分子扩散系数
coefficient of mutual inductance 互感系数
coefficient of nonuniformity 不均匀系数；开挖不均匀系数
coefficient of nozzle 管嘴系数
coefficient of numerical dispersion 数值弥散系数
coefficient of orifice 孔板系数
coefficient of overburden 剥离系数
coefficient of partial correlation 偏相关系数
coefficient of passive earth pressure 被动土压力系数
coefficient of percolation 渗透系数
coefficient of performance 工作效率；性能系数；性能系数
coefficient of permeability 导水系数；渗透系数
coefficient of permeability variation 渗透率变异系数
coefficient of piezo conductivity 压电传导系数
coefficient of planar permeability 平面渗透系数
coefficient of plasticity 塑性系数
coefficient of pore pressure 孔隙压力系数
coefficient of precipitation intensity 沉淀强度系数
coefficient of pressure conductivity 压力传导系数
coefficient of proportional similarity 比例相似系数
coefficient of purification 净化系数
coefficient of radial consolidation 径向固结系数
coefficient of rainfall and penetration 雨量系数
coefficient of rainfall infiltration 降水入渗系数
coefficient of recovery 回采率
coefficient of reduction 折减系数
coefficient of refraction 折射系数
coefficient of regression 回归系数
coefficient of relative roughness 相对粗糙系数
coefficient of relative subsidence 相对下沉系数

coefficient of relaxation 松弛系数
coefficient of resilience 弹性系数；回弹系数
coefficient of resistance 电阻系数
coefficient of restitution 恢复系数；抗冲系数
coefficient of rigidity 刚度系数；刚性系数
coefficient of rock resistance 岩石抗力系数
coefficient of rock strength 岩石强度系数
coefficient of rock weathering 岩石风化程度系数
coefficient of rolling friction 滚动摩擦系数
coefficient of roof weighting 顶板来压强度系数；动载系数
coefficient of roughness 粗糙系数；粗糙度系数；粗糙系数
coefficient of runoff 径流系数
coefficient of safety 安全系数
coefficient of saturation 饱水系数
coefficient of scouring 冲刷系数
coefficient of secondary compression 次压缩系数
coefficient of secondary consolidation 次固结系数
coefficient of sectional form 断面形状系数
coefficient of seepage 渗漏系数
coefficient of seismic action 地震作用效应系数
coefficient of seismic effect 地震影响系数
coefficient of self-induction 自感系数
coefficient of self-weight collapsibility 自重湿陷系数
coefficient of sensitivity 灵敏度系数
coefficient of settlement 沉降系数；沉陷系数
coefficient of shaft utilization 井筒的利用率
coefficient of shear 剪力系数；切变系数
coefficient of shear resistance 抗剪系数
coefficient of shear viscosity 剪切黏度系数
coefficient of shock resistance 局部阻力系数；耐冲击性系数
coefficient of shrinkage 干缩系数；收缩系数
coefficient of shrinkage of soil 土的收缩系数
coefficient of silt transport 泥沙输送系数
coefficient of similarity 相似系数
coefficient of skid friction 滑移摩擦系数
coefficient of sliding friction 滑动摩擦系数
coefficient of sliding friction of cover plate 路面板滑动摩擦系数
coefficient of sliding resistance 抗滑系数
coefficient of softening 软化系数
coefficient of soil reaction 土反力系数；地基系数
coefficient of sorting 分选系数
coefficient of sound transmission 传声系数
coefficient of stability 稳定系数
coefficient of static earth pressure 静止土压力系数
coefficient of stiffness 刚度系数；刚性系数
coefficient of storage 贮水系数
coefficient of strain-hardening 应变硬化系数
coefficient of stress concentration 应力集中系数
coefficient of stripping 剥离系数
coefficient of structure stability 结构稳定性系数
coefficient of subgrade reaction 地基反力系数；地基刚度系数
coefficient of sudden contraction 骤缩系数
coefficient of surface runoff 地表径流系数
coefficient of swelling 膨胀系数
coefficient of tamping 填塞质量系数

coefficient of thaw 融化系数
coefficient of thaw compression 融化压缩系数
coefficient of thaw setting 融陷系数
coefficient of thaw subsidence 融沉系数
coefficient of thermal conductivity 导热系数；热传导系数
coefficient of thermal diffusion 热扩散系数
coefficient of thermometric conductivity 导热系数
coefficient of thermometric transmission 热传导系数
coefficient of tide 潮汐系数
coefficient of torsion 扭曲系数；扭转系数
coefficient of tortuosity 曲折度系数
coefficient of traction 牵引系数
coefficient of transfer in water 水中迁移系数
coefficient of transmissibility 传导系数；导水系数
coefficient of transmission 传递系数；透射系数
coefficient of transverse dispersion 横向弥散系数
coefficient of turbulence 湍流系数；紊流系数
coefficient of twist 捻系数
coefficient of uniaxial compaction 单轴压实系数
coefficient of uniformity 均匀系数
coefficient of utilization 利用率
coefficient of variation 变异系数；不均匀系数
coefficient of variation of grade 级配变异系数
coefficient of velocity 速度系数
coefficient of vertical consolidation 竖向固结系数
coefficient of vertical permeability 垂直渗透系数
coefficient of vertical pile reaction 桩的竖向反力系数
coefficient of viscosity 黏性系数；黏滞系数
coefficient of viscous traction 黏滞牵引系数
coefficient of volume 容积系数
coefficient of volume change 体积变化系数；体积压缩系数
coefficient of volume compressibility 体积压缩系数
coefficient of volume decrease 体减系数
coefficient of volumetric expansion 松胀系数；体积膨胀系数
coefficient of volumetric thermal expansion 热体积膨胀系数
coefficient of water balance 水平衡系数
coefficient of water content change 含水量变化系数
coefficient of water lines 侵水系数
coefficient of water saturation 饱和吸水率
coefficient of water-level conductivity 水力扩散系数；水位传导系数
coefficient of waterplane 水线面系数
coefficient of weathering 风化系数
coefficient of weight 加权系数
coefficient of weighting 权系数
coefficient of wet subsidence 湿陷系数
coefficient of wetted surface 湿面积系数
coefficient of winding resistance of mine 矿井提升阻力系数
coelom 体腔
coelosphere 坐标仪
coercibility 可凝性；可压缩性
coercible 可压凝的；可压制的
coercible gas 可压缩气体
coercimeter 矫顽磁力计
coercion 强迫
coercitive force 矫顽磁力；消磁强度
coercive field strength 矫顽磁场强度

coercive force 矫顽力
coercivity 抗磁性
coercivity magnetic 矫顽磁性
coeruleite 铜矿
coeruleolactite 微晶磷铝石
coes wrench 活动扳手
coesite 人造石英；柯石英
coessential 同素的
coetaneous 同年代的；同时期的
coeval 同年代的
coevolution 共同演化
coexist 共处；共存
coexistence approach 共存分析方法
coexisting element 共生元素
coexisting fluid 共存流体
coexisting minerals 共生矿产
coexisting phase 共存相；共生相
coextraction 共萃取
coextru-lamination 共挤复合
coextrusion 共挤塑
coffer 围堰
coffer dam bracing 围堰支撑
coffer work 砌片石墙；围堰工程
cofferdam 沉箱；围堰坝；围堰
cofferdam construction 围堰施工
cofferdam dewatering 围堰排水
cofferdam piling 围堰板桩
cofferdam removal 围堰拆除
cofferdamming 修筑围堰
coffered foundation 沉箱基础；围堰基础
coffered wall 围堰墙
coffering 防水围堰；沉箱法；防水井壁
coffer-wall 围墙
coffin 屏蔽罐；熔炼坩埚炉
coffin hoist 起管吊车
coffinite 水硅铀矿；铀石
cofinal 共尾
coflexip 伴热柔性管
cog 齿轮；岩脉
cog building 架设木垛
cog drive 嵌齿传动
cog of bags 砂袋垛
cog of round timber 圆木垛
cog wheel 嵌齿轮
co-gasification 共气化
cogbell 冰柱；垂冰
cogelled 共凝胶的
cogeneration 热电联产
cogeneration mode 联合生产方式
cogeneration plant 热电厂
cogeneration technology 热电联供工艺
cogenerator 热电联供装置
cogenetic 同成因的；同源的
cogenetic gas 同生气
cogenetic mineral 同源矿物
cogenetic pluton 同源深成岩体
cogenetic rock 同成因岩；同源岩
co-geoid 共大地水准面；通用大地水准面
cogged belt 三角皮带
cogged bit 冲击式凿岩器

cogged crown	齿状钻头
cogged joint	榫齿接合
cogged rail	齿轨
cogging	钝齿啮合；架设木垛
cogging joint	齿节
cognate	同性质的；同族的
cognate ejecta	同源喷出物
cognate fissures	同源裂隙
cognate inclusion	同源包体
cognate lithic	同源岩层的
cognate xenolith	同源包体；同源捕虏体
cognitive	认知的
cognitron	认知机
cogradient matrices	同步矩阵
cograft	共接枝
cog-wheel coupling	齿形联轴器
cogwheel gearing	齿轮传动装置
cohabitation	共栖
co-hade	补伸角；断层倾角
cohenite	陨碳铁；陨碳铁矿
cohere	附着；相干
coherence	相互叠合；内聚力；黏性
coherence criterion	相干准则
coherence effect	相干效应
coherence emphasis	相干加强
coherence enhancement	相干性加强
coherence filtering	相干滤波
coherence function	相干函数
coherence matrix	相干矩阵
coherence stack	相干叠加
coherence structure	凝聚结构
coherency	相干性
coherent	内聚的；相干的；连贯的
coherent alluvium	凝聚性冲积层
coherent analysis	相关分析
coherent arrival	相干波至
coherent beam	相干光束
coherent boundary	连贯边界
coherent carrier	相干载波
coherent collision	相干碰撞
coherent crustal plate	黏合地壳板块
coherent detection	相干检波
coherent detector	相干检测器
coherent distribution	相干分布
coherent disturbance	相干扰动
coherent element	相关元素
coherent filtering	相干滤波
coherent frequency	相干频率
coherent imaging	相干成像
coherent lamellae	连贯条纹
coherent light beam	相干光束
coherent material	黏附物材
coherent model	相关模型
coherent noise	相干噪声
coherent optical radar	相干光雷达
coherent precipitate	黏合性沉淀物；共格沉淀
coherent radiation	相干辐射
coherent ray	相干射线
coherent reflection	相干反射
coherent rift belt	连贯裂谷带
coherent rock	胶结岩石
coherent scattering	相干散射
coherent separation	黏附选矿
coherent signal	相干信号
coherent sliver	黏合丝条
coherent slump	黏结滑塌
coherent soil	黏性土
coherent source	相干光源
coherent system	凝聚系统；相干系统
coherent unit	相关单位
coherent wave	相干波
coherer	粉末检波器；金属屑检波器
cohesible	可黏合的
cohesiometer	黏度计；黏聚力仪
cohesion	凝聚力；黏结力
cohesion assessment	黏聚力评定
cohesion intercept	黏聚力截距
cohesion loss	凝聚力损失
cohesion number	黏聚数
cohesion of discontinuity	岩体弱面黏结力
cohesion of rock	岩石的黏结
cohesion of soil	土壤黏聚力
cohesional resistance	黏结阻力；黏抗力
cohesionless	非黏结性的；松散的
cohesionless grain flow	非黏结性颗粒流动
cohesionless material	松散材料；无黏性材料
cohesionless medium	松散介质
cohesionless sand	松散砂
cohesionless soil	非黏性土；松散土
cohesive	内聚的；黏着的；黏结的
cohesive action	黏结作用
cohesive affinity	凝聚力
cohesive bed	黏性河床
cohesive energy	内聚能
cohesive failure	内聚衰坏
cohesive force	内聚力；黏结力；黏合力
cohesive material	黏性土；黏性材料
cohesive property	黏结性
cohesive rock	黏结性岩石
cohesive seal	黏封
cohesive sediment	黏性泥沙；黏结性淤积物
cohesive soil	黏性土
cohesive soil foundation	黏结土地基
cohesive strength	黏结强度
cohesive zone	内聚区
cohesive zone model	黏聚力模型
cohesiveness	凝聚力；黏结性
cohesiveness degree	黏聚度
cohesiveness test	黏结性试验
coho	相干振荡器
cohobation	回流蒸馏
Cohron shear graph	柯朗剪切图
cohydrolysis	共水解作用
coign	外角；隅石
coignet pile	楔桩；隅桩
co-ignimbrite breccia	共熔结凝灰岩
coil	盘管
coil array	线圈系
coil brake	盘簧制动器
coil capacitance	线圈电容

coil car	有加热盘管的油槽汽车
coil coating	卷材涂料
coil condenser	旋管冷凝器
coil constant	线圈常数
coil detector	线圈式检波器
coil drag	钻孔用螺旋捞矛
coil ends	绕组端
coil heat exchanger	盘管换热器
coil loss	线圈损耗
coil magnetometer	线圈式磁力仪
coil power	线圈功率
coil pulser	线圈式脉冲发生器
coil spring drain	弹簧排水管
coil spring retainer	盘簧底圈
coil tubing reel	软管滚筒
coil winding	线圈绕组
coil-coated strip	卷缠涂料带
coiled	螺旋的
coiled column	盘旋柱
coiled expansion pipe	蛇形膨胀管
coiled pipe	盘管
coiled radiator	盘管散热器
coiled rod	连续抽油杆
coiled spring	螺旋弹簧
coiled tubing	挠性管
coiled tubing unit	挠性管作业机
coiled tubing workover system	挠性管修井系统
coiled-tubing logging system	挠性管测井系统
coiled-tubing-conveyed wireline system	挠性管传送测井系统
coiler	卷绕机；盘管
coiling	卷绕
coiling action	卷绕作用；扭转作用
coil-mounted conveyor	弹簧支座输送机
coincide	符合
coincidence	叠合
coincidence amplifier	重合放大器
coincidence analyzer	符合分析器
coincidence micrometer	重合测微器
coincidence rangefinder	叠像测距仪
coincidence rate	符合率
coincidence selectivity	重合选择性
coincident	符合的；一致的；重合的
coincident basin	重合盆地
coincident spectral density	共谱密度
coining	模压；压花；压印加工
coinstone	货币岩
coir rope	棕绳
cokability	焦化性；结焦性
coke	焦；焦炭
coke battery	炼焦炉
coke bed	底焦；焦炭床层
coke blast furnace	焦炭高炉
coke breeze	焦炭屑；碎焦炭
coke breeze concrete	焦煤灰混凝土；煤渣混凝土
coke briquette	焦炭块
coke build-up	碳堆积
coke button	坩埚焦炭；焦渣
coke car	熄焦车
coke charge	炼焦炉料
coke cleaning	清焦
coke coal	天然焦
coke cooling	熄焦
coke cutter	切焦器
coke degradation	焦炭块度减小
coke discharging machine	推焦车
coke drum	焦炭塔
coke expense ratio	耗焦比
coke fines	粉焦；焦屑
coke forming property	生焦倾向
coke gas	焦炉煤气
coke mosaic size index	焦块大小指数
coke oven	炼焦炉
coke oven gas	焦炉煤气
coke oven tar	焦炉焦油
coke packed bed	焦炭填充床
coke packed bed shaft furnace	焦炭盒装床竖炉
coke pig iron	焦炭生铁
coke plant	焦化厂
coke porosity	焦炭孔隙率
coke quenching	焦炭熄火
coke ratio	焦比
coke reactivity	焦炭反应性
coke residue	焦炭渣
coke stability	焦炭稳定性
coke stability factor	焦炭稳定因素；焦炭稳定系数
coke tar	焦油
coke water gas	炼焦水煤气
coke yield	焦炭产量；焦炭回收率
cokeability	成焦性
coke-burning rate	烧焦速率
coke-cement concrete	焦炭屑混凝土
coked pitch	焦沥青
cokeite	天然焦
coke-like coating	焦炭状涂层
coker	焦化设备
coker gasoline	焦化汽油
cokery	炼焦炉；炼焦厂
coking	焦化；炼焦；炼焦的；黏结的
coking ability	结焦能力
coking blend	炼焦配煤
coking capacity	结焦能力；结焦性
coking chamber	炼焦室；炭化室
coking coal	焦煤；炼焦煤
coking coal cleans	炼焦精煤
coking fat coal	焦肥煤
coking heat	结焦热
coking lean coal	焦贫煤
coking plant	炼焦厂
coking power	成焦率；焦结力
coking property	结焦性
coking quality	结焦性
coking residue	焦渣
coking resistance	抗生焦能力
coking still	焦化釜；炼焦炉
coking test	焦化试验
cokriging system	协同克里金系统
col	山坳
cola eddy sediments	冷涡沉积物
cola extraction analysis	冷提取分析

cola hardening	加工硬化；冷加工
cola ice sheet	冷冰盖；冷冰原
cola lake	极地湖；冷湖
cola loess	冰缘黄土；冷黄土
cola spring	冷泉
cola vent	冷喷口
cola volcano	凉火山
colander	滤器
colas	沥青乳胶体
colasmix	沥青砂石混合物
colation	粗滤；过滤
colature	粗滤产物；过滤液
colcothar	褐色氧化铁粉
colcrete	胶体混凝土；岩浆
cold and hot brittleness	冷脆热脆性
cold applied waterproofing	冷施工防水工程
cold belt	寒带
cold bend inspection of steelbar	钢筋冷弯试验
cold bend test	冷弯试验
cold calking	冷填缝
cold cap	寒带
cold cast	冷铸
cold caulking	冷填缝
cold charge	低温装料
cold chisel	冷錾
cold compacting	冷压
cold condensation	冷凝聚
cold cracking	低温开裂
cold crushing strength	低温压碎强度；冷挤压强度
cold curing paint	低温固化涂料
cold deformation	冷变形
cold drawn bar	冷拉钢筋
cold elbow	冷弯弯头
cold endurance	耐寒性
cold extrusion	冷挤压
cold finishing	冷加工精整
cold hammering	冷锻
cold hardening	冷加工硬化
cold hardiness	耐寒力
cold header	冷锻机
cold heavy oil production with sand	出砂冷采
cold joint	冷接缝；施工缝
cold jointed masonry	冷接砖石砌体
cold lahar	冷泥石流
cold lake	极地湖；冷湖
cold loess	冰缘黄土；冷黄土
cold mix recycling	冷拌再生法
cold mud-flow	冷泥流；冷泥石流
cold nosing	盲目钻探
cold plasma	冷等离子体
cold plume	冷焰
cold pressing	低温压榨；冷压合
cold pressure welding	冷压焊
cold quenching	冰冷处理
cold reducing	冷压缩
cold resistance	耐寒性
cold rolling	冷轧
cold setting adhesive	常温固化胶黏剂
cold settler	低温沉淀器
cold shallow water	浅层冷地下水
cold spring	冷泉
cold starting performance	低温起动性能
cold test	耐低温试验
cold waste	低放射性废物
cold water flooding	冷水驱油
cold weather concreting	冬季混凝土浇筑
cold weld	冷焊
cold-applied coating	常温涂敷涂层
cold-bending machine	冷弯弯管机
cold-drawing shop	冷拉车间
cold-extractable metals	冷提取金属
cold-molding	冷压
cold-proof	防寒
cold-rolled deformed bar	冷轧带肋钢筋
cold-rolled steel pipe	冷轧钢管
cold-test oil	低凝点石油
cold-wall slide valve	冷壁滑阀
cold-water drive	冷水驱动
cold-water geyser	冷间隙喷泉；冷水间歇泉
cold-weather pipelining	冬季管道敷设；冬季管道输油
cole screen	柯尔筛
cole-cole diagram	柯尔-柯尔图
colemanite	硬硼钙石；硬硼酸钙石
coleoptile	胚芽鞘
coleorhiza	根鞘
colerainite	透绿泥石
colgrout	预填骨料压灌砂浆
colidar	相关光雷达
colinear drainage pattern	共线状水系
colinear flow	共线流动
colinearity	同线性
colitic texture	鲕状结构
collaboration	合作
collaboration of steel and concrete	钢筋与混凝土共同受力
collagen	胶原
collagenous	胶原的
collagenous fibre	胶原纤维
collain	无结构镜煤
collaps lake	塌陷湖
collapsable	可分解的；可折叠的
collapsable air duct	可折式风管
collapse	湿陷；塌崩；塌陷；陷坑；崩落；坍陷
collapse acceleration	崩塌加速度
collapse arch	塌落拱
collapse area	崩落区
collapse basin	塌陷盆地
collapse breccia	崩塌角砾岩；塌陷角砾石
collapse breccia pipe	塌陷角砾岩筒
collapse brecciation	崩塌角砾岩化
collapse by wetting	湿陷
collapse caldera	塌陷破火山口
collapse column	陷落柱
collapse depth	坍坏深度
collapse doline	塌陷漏斗
collapse earthquake	陷落地震
collapse fault	塌陷断层
collapse feature	崩落特征；冒落特征
collapse fissure	塌陷裂隙
collapse job	取出井中破裂管子
collapse karst gorge	岩溶嶂谷

collapse lake	塌陷湖
collapse limit state	倒塌极限状态
collapse load	极限荷载；破坏荷载；损毁荷载
collapse loess	湿陷性黄土
collapse mechanism	塌陷机理；塌陷机制
collapse of casing	套管的破坏
collapse of mine	挤拢；巷道崩塌
collapse of pillars	矿柱破坏
collapse of rock	岩石坍落
collapse of roof	顶板崩落
collapse of sand and water	崩溃的沙子和水
collapse of slope	坡面坍塌
collapse of structure	结构物倒塌
collapse of tunnel roof or wall	隧道冒顶或片帮
collapse phase	塌陷期
collapse pit	塌陷坑
collapse post	陷落柱
collapse potential	坍陷潜能
collapse pressure	崩裂压力；极限承载力；破坏压力
collapse prevention	倒塌预防
collapse resistance	抗塌陷强度；抗破坏强度
collapse resistant	抗倒塌
collapse resistant coefficient	抗倒系数
collapse settlement	湿陷量；坍陷量
collapse sink	塌陷落水洞
collapse sinkhole	塌陷落水洞
collapse sloughing	坍塌
collapse state	湿陷状态
collapse strength	抗挤强度；破裂强度
collapse structure	塌陷构造
collapse test	破坏性试验；塌陷试验
collapse theory	塌陷学说
collapse threshold earthquake intensity	倒塌阈限的地震烈度
collapsed casing	挤扁的套管
collapsed coal basin	侵蚀煤盆地
collapsed diameter	收拢直径；塌陷直径
collapsed doline	塌陷漏斗
collapsed material	崩积物
collapsed monadnock	溶塌残丘
collapsed riser	折叠提升管
collapse-type mine earthquake	塌陷型矿震
collapsibility	沉陷性；湿陷性；易坍塌性
collapsibility coefficient	湿陷系数
collapsibility grading index	分级湿陷量
collapsibility of loess	黄土湿陷性
collapsibility test of loess	黄土湿陷试验
collapsible	可拆卸的；可收缩的；可折叠的
collapsible bit	伸缩式钻头；伸缩式钻头
collapsible blade	折叠式叶片
collapsible cage	可折叠罐笼
collapsible cantilever platform	可折悬臂平台
collapsible derrick	折叠式钻塔
collapsible drilling bit	可拆式钻头
collapsible drum	可折桶
collapsible elevator	折叠式提升机
collapsible filter	自落填缝反滤层
collapsible flash board	可拆卸闸板
collapsible form	活动模板；装配式混凝土模板
collapsible fuel tank	可折叠油箱
collapsible gate	可拆卸闸门
collapsible life boat	折叠式救生艇
collapsible loess	湿陷性黄土
collapsible mast	折叠式轻便井架
collapsible orifice	可收缩的孔板
collapsible package	管芯可抽的卷装；松软卷装；无纱芯卷装
collapsible prop	让压支柱；伸缩式支柱
collapsible roof	易塌顶板
collapsible rubber dam	可折叠的橡胶坝
collapsible settlement	湿陷量
collapsible shuttering	活动模板；铰接金属模板
collapsible soil	湿陷性土
collapsible stilt	让压性支柱
collapsible storage tank	折叠式油罐
collapsible stretcher	折叠式担架
collapsible subsoil	湿陷性地基
collapsible timbering	可缩性支架；让压支架
collapsible tool	组合式工具
collapsible top	易冒落的顶板
collapsible tube	可胀缩的筒管
collapsible whipstock	伸缩式造斜器
collapsing	压扁
collapsing and sliding response	崩滑效应
collapsing bucket	活底吊桶
collapsing cavity	溃灭空穴
collapsing gas saturation	油层溃缩气体饱和度
collapsing loess	湿陷性黄土
collapsing pressure	挤毁压力；破坏压力；坍塌压力
collapsing soil	崩塌土；湿陷性土
collapsing strength	抗挤强度；破坏强度
collapsing test	湿陷试验
collar a hole	开钻；钻孔口安装钻模
collar beam	圈梁；系梁
collar bearing	环形轴承
collar brace	撑木；连接杆
collar bracing	系杆；斜撑
collar bushing	补心短节
collar buster	切管器；套管切割工具
collar cave	锁口盘塌落；井颈塌落
collar clamp	卡箍
collar connection	接箍连接
collar control gate	钻井口调节闸阀
collar corker	管线接箍堵漏工
collar doors	封口盘门；井盖门
collar excavation	井颈开凿
collar finder	接箍探测器
collar flange	接箍法兰
collar grab	打捞卡套
collar house	井口建筑物
collar joint	竖缝
collar level	井口水平
collar locator device	接箍定位装置
collar log	套管接箍测井
collar mark	接箍标记
collar of a drill hole	护圈；束套；套环
collar of a thrust bearing	止推轴承环
collar of anchorage device	锚口
collar of shaft	竖井口
collar of the hole	炮眼的外端；炮眼口

collar oiler	轴环式加油器
collar oiling	轴环润滑
collar pipe	接头前后错开方式排放管子；钻井锁口管
collar piping	地面套管
collar plate	卡箍；轴环
collar priming	孔口起爆；正向起爆
collar screen plate	环筛板；领筛板
collar set	锁口盘
collar sleeve	钻孔口套筒；钻模
collar socket	接箍打捞工具
collar step bearing	环形阶式轴承
collar stop	接箍挡环
collar thrust bearing	环形止推轴承
collar tie beam	圈梁
collar wall	井颈壁；井筒锁口壁
collar-bound pipe	接箍被卡钻杆
collared hole	炮眼定位孔；装好地面导管的井
collared shaft	环轴
collared steel	有肩的钢钎
collarine	柱颈
collaring	标定炮眼位置；缠辊；打眼；开钻
collaring hole	炮眼的开孔
collar-initiated charge	正向装药
collarleak clamp	防漏卡箍
collar-oiled bearing	轴环润滑的轴承
collar-pull elevator	挂接箍吊卡
collar-type tubing elevator	接箍式油管吊卡
collate	核对；校对
collateral	并行的；附属的；次要的
collating sequence	整理顺序
collation	核对；整理
collation of data	数据整理
collectible size	可收集的粒度
collecting agent	捕收剂
collecting area	补给区；地面集水区；汇集区；汇水面积；集流面积；集油面积
collecting arm type loader	集爪式装载机；立爪式装载机
collecting band	矸石手选带；拣矸带
collecting bar	汇流排
collecting basin	补给区；集聚盆地；集水池；集水区
collecting belt	集料胶带
collecting bowl	集料箱；收集箱
collecting box	集料箱；收集箱
collecting brush	集流刷
collecting bucket	收集斗
collecting channel	汇水沟；集水沟
collecting conveyor	收集输送机
collecting ditch	集水沟
collecting drain	集水沟
collecting electrode	集电极
collecting funnel	汇集漏斗指圆形山谷
collecting gas method	集气法
collecting gutter	集水沟
collecting line	集油管线
collecting main	集流管
collecting passage	集水沟
collecting pipe	集水管；收集管
collecting reagent	捕收剂
collecting sand	集油砂层
collecting sewer	污水支管
collecting strength	捕收力；捕收强度
collecting sump	集料仓；集水仓
collecting system	集水系统；收集系统
collecting tank	集液槽
collecting vat	收集槽
collecting well	集水井
collection area	汇水区
collection design	采集设计
collection hopper	集料斗
collection melting	收集熔融
collection of property abroad	调汇
collection pipe	集水管
collective	集合的；聚合的
collective bargaining	集体谈判
collective concentrate	混合精矿
collective control grout plant	集中控制制浆站
collective drawings for gravity retaining wall	重力式挡土墙通用图集
collective flotation	混合浮选；全浮选
collective subsidence	干涉沉降
collective-selective flotation	混合优先浮选
collector belt conveyor	集中胶带输送机
collector ditch	集水沟
collector drain	集水沟
collector film	捕收剂薄膜
collector filter	集尘器
collector pipe	集水管
collector promoter	捕收剂
collector ring	滑环；汇流环
collector strip	滑板
collector well	集水井
collegial placer	塌积沉积砂矿
collemanite	硬硼钙石
collenia-type stromatolites	聚环藻式叠层石
collet	套筒夹头；套爪；筒夹
collet cam	筒夹控制凸轮
collet chuck	弹簧夹头；筒夹
collet connector	套筒连接器
collet ring	套爪环
collets	绝缘块；有缝夹套
collet-type locator	爪套式定位器
collet-type packer	夹套式封隔器
coll-exinite	无结构壳质体
collieite	红磷氯铅矿
collier	矿工；运煤船
colliery	煤矿
colliery arch	煤矿用金属拱形支架
colliery explosion	煤矿爆炸
colliery products	矿产品
colliery screened	在矿场筛选的
colliery self consumption	煤矿自用煤消耗量
colliery shale	煤矸石
colliery spoil	煤矿渣
colliery spoil heaps	煤矿煤矸石堆积
colliery warning	气压突降警报
colliery wire rope	矿用钢丝绳
collimate	照准；瞄准
collimated	平行的；照准的
collimated gamma ray source	定向伽马射线源
collimating adjustment	视准改正

collimating aperture	准直孔径
collimating apparatus	准直器
collimating axis	视准轴
collimating cone	准直锥体
collimating line	视准线
collimating point	视准点
collimating slit	准直缝
collimation	瞄准；平行校正
collimation axis	视准轴
collimation correction	视准改正
collimation error	视准误差
collimation line	视准线
collimation line method	视准线法
collimation method	视准法
collimation plane	视准面
collimator	准直仪；瞄准仪
collinear	共线的；同线的
collinear array antenna	直排天线阵
collinear forces	共线力
collinear noise	共线噪声
collineation	直射变换；共线
collinite	无结构腐殖体；无结构凝胶质
Collins miner	柯林斯型采煤机
collinsite	淡磷钙镁石；三斜镁铁磷灰石
colliquation	熔化
collision	冲击；碰撞
collision avoidance radar	防撞雷达
collision avoidance system	防撞装置
collision belt	碰撞带
collision boundary	碰撞边界
collision coast	碰撞海岸
collision fault zone	碰撞型断裂带
collision hypothesis	碰撞说；碰撞假说
collision mat	堵漏毡；防撞毡
collision mean-free path	碰撞平均自由程
collision of plates	板块碰撞
collision orogenesis	碰撞造山运动
collision orogens	碰撞造山带
collision post	防撞柱
collision prevention around pier	墩台防撞
collision process	碰撞过程
collision rift	碰撞裂谷
collision strut	防冲支撑
collision suture	碰撞接合
collision tectonics	碰撞构造
collision theory	碰撞学说
collision type orogeny	碰撞型造山运动
collision zone	板块碰撞带；碰撞带
collisional	碰撞的
collisional accretion	碰撞吸积
collisional suture	碰撞缝合带
collisional tectonics	碰撞构造
collision-deformed	碰撞变形的
collision-type	碰撞型
collite	微无结构镜煤
collobrierite	铁闪橄榴岩
colloclarain	无结构亮煤
colloclarite	无结构亮煤
colloclast	胶体碎屑岩；加积碎屑集合体
collodion	火胶
colloform	胶粒结构
colloform banding	胶体状条带
colloform fabric	胶体组构
colloform structure	胶状构造
colloid	胶粒；胶体；胶质
colloid adsorption	胶体吸附作用
colloid ageing	胶体陈化作用
colloid bearer	胶质载体
colloid content	胶质含量
colloid form structure	胶状结构
colloid fraction	胶粒粒组
colloid grain	胶粒
colloid mill	胶体研磨机
colloid mineral	胶体矿物
colloid particle	胶粒
colloid process	胶体作用
colloid size	胶粒尺寸
colloid transport	胶体运输
colloid valency	胶质价态
colloidal	胶体的；胶质的
colloidal activity of clay	黏土胶态活动性
colloidal agglutination mineral deposit	胶体化学沉积矿床
colloidal bond	胶质黏结剂
colloidal carbon	胶态碳
colloidal cement grout	胶体水泥薄浆
colloidal chemistry	胶体化学
colloidal clay	胶质黏土
colloidal complex	胶质复合体
colloidal concrete	胶质混凝土
colloidal deposite	胶体沉积
colloidal electrolyte	胶态电解质
colloidal fuel	胶质燃料
colloidal graphite	石墨乳；胶体石墨
colloidal grout	胶体浆液
colloidal grout concrete	胶态水泥砂浆混凝土
colloidal grout mixer	胶态水泥浆搅拌机
colloidal humus	胶状腐殖质
colloidal load of stream	河流胶质携量
colloidal lubricant	胶态润滑剂
colloidal material	胶体物质
colloidal medium	胶态介质
colloidal metal	胶体金属
colloidal mill	胶体磨碎机
colloidal mineral	胶体矿物
colloidal mixer	胶浆搅拌机
colloidal mud	触变泥浆，胶体泥浆；胶体钻泥
colloidal nucleus	胶核
colloidal quality	胶体质量
colloidal silica	硅胶体；胶体氧化硅
colloidal silica particles	硅胶粒子
colloidal solid	胶体颗粒
colloidal solution	胶体溶液
colloidal suspension	胶态悬浮体
colloidal water treatment	胶体处理硬水法
colloidality	胶体性
colloidization	胶化作用
collophane	胶磷矿
collophanite	胶磷矿
collose	木质胶
collosol	溶胶

colluvial	崩积的；崩积物
colluvial clay	崩积土
colluvial deposit	崩积物
colluvial fan	坡积扇
colluvial landform	崩积地形
colluvial lobe	坡积舌
colluvial mantle	塌积盖层
colluvial mass	崩积堆；崩积体
colluvial slope	崩积坡
colluvial soil	崩积土
colluvial wedge	崩积楔；断错崩积楔
colluvial-deposit landslide	堆积层滑坡
colluviation	崩积作用
colluvium	崩积层；崩积土；崩积物；坡积物
colluvium slide	崩积物滑动
colluvium soil	崩积土
collyrite	微光高岭土
colmatage	冲积层；堵塞；放淤；淤灌
colmatation zone	堵塞层
colmation	放淤
Colmol mining machine	科尔莫尔型联合采煤机
Coloma miner	科尔马型煤巷掘进机
colonial	集群的；群体的
colonial coral	群体珊瑚
colonization experiments	定植实验
colonnade	柱廊
colonnade foundation	管柱基础
colonnade foundation process	管柱基础施工法
colonnette	小柱
colony	集群
colony-forming bacteria	群集型细菌
colophene	松香
colophonic	松香的
colophonic acid	松香酸
colophonium	松香
colophony	松香
color addition	彩色叠加
color aerial photograph	彩色航空照片
color anomaly	彩色异常
color combination	彩色合成
color correction filter	校色滤光片
color difference threshold	色差阈
color distortion	彩色失真
color fidelity	彩色保真度
color filter	滤色镜
color hologram	彩色全息图；彩色全息相片
color infrared photography	彩色红外摄影
color mineral	有色矿物
color photomicrography	彩色显微照片；彩色显微照相术
color seismic display	彩色地震显示
color triangle	基色三角
colorability	着色性能；可着色性
colorado clay shale	科罗拉多黏土页岩
colorado impact screen	科罗拉多冲击筛
coloradoite	碲汞矿；石英粗安岩
colored cement	彩色水泥
colored concrete	彩色混凝土
colored equidensity image	彩色等密度图像
colored sweep	有色扫描
colorimetric estimation of silica	游离二氧化硅的比色测定
colorimetric titration	比色滴定法
color-painting stake	粉喷桩
color-transfer process	彩色传输过程
colosseum	角斗场
colour composite image	彩色合成影像
colour contrast banding	色彩分明的带状图案
colour infrared photograph	彩色红外相片
colour infrared photography	彩色红外摄影
colour mineral	有色矿物
colour palette	调色板
colour photo micrograph	彩色显微相片
colour reaction	显色反应
colour seismic display	彩色地震图
colour strainer	滤浆器
coloured aerial photograph	彩色航空相片
coloured fibre	有色纤维
coloured portland cement	彩色硅酸盐水泥
coloured sand experiment	染色砂试验
coloured test water	显色试验水
colouring	镜面加工；抛光；染色；着色
colouring stabilizer	固色剂
colourless mineral	无色矿物
colpate	沟
colse	接通
columbite	钶铁矿；铌铁矿
columbium	铌；钶
columbretite	白榴粗安岩
column	柱状图
column analogy method	柱比法
column and tie construction	穿斗式构架
column arm	柱式钻架的横托
column bar	柱筋
column base	柱础；柱基
column base block	钻架柱脚垫
column beam	柱型梁
column blasting	柱状装药爆破
column cap	柱帽；柱头
column cartridge	柱状药包
column charge	连续装药；柱状装药
column chromatography	色谱柱分离法；柱层析
column clamp	岩心夹子；柱夹
column crane	塔式起重机
column crystal	柱状晶体
column deflection	柱纵向弯曲
column detailing	柱子细部构造
column diagram	柱状图
column diagram of stratum	地层柱状图
column drill	柱架式钻机；架式风钻；取心钻；柱架式钻机
column drilling machine	柱式钻床
column extractor	柱式萃取器
column face	柱面
column filling	柱填充
column floatation of coal	浮选柱选煤
column flotation	浮选柱
column foot	塔脚
column footing	柱底脚；柱基脚
column frame	柱架
column head	柱帽；柱头

column hoist	塔式起重机	column-jib crane	塔式起重机；塔式旋臂起重机
column jack	柱式千斤顶	column-mounted drill	架柱式凿岩机；柱装钻机
column jacket	外柱；柱管	column-mounted stand drill	柱装立式钻机
column layout	柱网布置	column-plate retaining wall	柱板式挡土墙
column length	柱长	column-supported	柱支承的
column load	圆柱装药；柱荷载	column-type diaphragm wall	柱列式地下连续墙
column mixer	柱式混合器	column-type insulator	支柱绝缘子
column moment	柱上弯矩	column-type support	立柱
column mounted	装在柱架上的	colusite	锡砷硫钒铜矿
column mounting	柱架	coma scattering pattern	彗形象差散射图形
column of air	空气柱	comagmatic	同源岩浆的
column of angle	角钢柱	comagmatic evolution succession	同源岩浆演化序列
column of built channels laced	槽钢缀合柱	comagmatic region	同源岩浆区
column of clay	黏土柱	comagmatic rock	同源岩浆岩
column of coarse backfill	粗粒回填柱	comb	蜂窝；浪峰；梳状物
column packing	填料；柱填充物	comb filter	梳形滤波器
column pad	柱基	comb fold	梳状褶皱
column performance	柱性能	comb function	梳状函数
column pier	柱墩	comb generator	梳状波发生器
column pile	端承桩；柱桩	comb landform	梳状地形
column section	柱截面	comb pitot	梳状皮托管
column shaft	柱身	comb ridge	锯状山脊；梳状山脊
column sidesway mechanism	柱侧倾机理	comb sorter	梳片式纤维长度分析仪
column socle	柱基座	comb structure	蜂窝状结构；梳状构造
column still	冷塔式蒸馏器；蒸馏柱	combating fire	灭火
column strip	柱带	combating noise	反干扰
column structure	柱状构造	combe	冲沟；峡谷
column tray	塔盘	combed tops	黏梳毛条
column turbodrill	取心涡轮钻具	comber	拍岸浪；碎浪
column volume	柱容积	combe-rock	泥流混杂沉积
column with constant cross-section	等截面柱	combinability feature	组合测井性能
column with lateral reinforcement	配有箍筋的柱	combinate form	聚形
column with spiral hooping	螺旋钢筋柱	combination	化合；聚合
column with steel hooping	螺旋钢筋柱；配有箍筋的柱	combination advance and retreat method	前进后退式联合开采法
column with variable cross-section	变截面柱	combination arc	联合弧
column-and-panel wall	镶板式柱墙	combination beam	组合梁
columnar	柱状的	combination bearing	组合轴承
columnar aggregate	柱状集料	combination bit and mud socket	钻掏泥砂泵
columnar basalt	柱状玄武岩	combination buoy	组合浮标
columnar cleavage	柱状解理；柱状劈理	combination burn	混合锥楔式空炮眼掏槽
columnar coring tube	柱状取样管	combination casing string	组合式套管柱
columnar crystal	柱状晶体	combination collar	异径接箍
columnar fissure	柱状裂隙	combination column	组合柱
columnar fold	圆柱形褶皱	combination connector	万能连接器；通用连接器
columnar fracture	柱状裂缝	combination construction	混合结构
columnar grouting	柱状灌注	combination coupling	组合接头
columnar joining	柱状节理	combination cutting	混合掏槽
columnar joint	柱状节理	combination cutting-and-conveying unit	采煤机组
columnar order	柱型	combination dial pressure gauge	复式刻度压力计
columnar ore body	柱状矿体	combination directional valve	组合定向阀
columnar ore shoot	矿柱	combination displacement mechanism	混合驱替机理
columnar pile	端承桩	combination drill	冲击回转两用钻机；复合钻机；两用钻机
columnar plume	热柱；柱状羽	combination driller	两会司钻
columnar profile	钻孔柱状图	combination drilling	联合钻井；联合钻眼
columnar section	柱状剖面；钻孔柱状图	combination drilling concentric completion rig	钻井完井两用钻机
columnar strata section	地层柱状图	combination drilling equipment	两用钻井设备
columnar structure	柱状构造；柱状结构	combination drilling outfit	两用钻井装置
columnar zone	柱状结晶区		
columned pneumatic flotation machine	浮选柱		
columniation	列柱式		

combination drive 混合驱动
combination factor 综合系数
combination feeder 联合给料机
combination firing head 组合点火装置
combination flooding 复合驱
combination flow regulator 组合式流量调节阀
combination for action effects 作用效应组合
combination frame 组合框架
combination frequency 组合频率
combination gamma ray-neutron-laterolog 自然伽马-中子-侧向组合测井
combination gas 伴生气；富天然气；混合气；油井气
combination gas-lift 混合气举
combination gear 联合齿轮；组合齿轮
combination jumbo loader 钻架装载联合机
combination locomotive 联合电机车；蓄电池架线两用电机车
combination logging tool 组合测井仪
combination machinery 联合机械
combination multiclone precipitator unit 多管旋流联合集尘器组
combination of actions 作用的组合
combination of dynamic loads 动荷载组合
combination of exploration with mining 探采结合
combination of load 荷载组合
combination of particle size 粒度配成
combination of structural systems 构造体系的联合
combination of triangulation and trilateration 边角测量
combination of zones 地层连通
combination optimization 组合优化
combination packer 综合封隔器
combination picture 复合图像
combination pile 组合桩
combination pliers 钢丝钳；万能钳
combination pressure regulator 组合压力调节器
combination process 联合作业
combination production logging 组合生产测井
combination property 综合性能
combination resistivity-acoustilog method 电阻率-声波组合测井
combination rig 冲击旋转联合钻机；两用钻机；联合钻井设备
combination set 组合角尺；万能测角器
combination socket 联合打捞器；综合打捞筒
combination solution 复配溶液
combination striation 聚形条纹
combination string 复合钻柱
combination sub 组合接头
combination system 联合系统；综合钻进法
combination system of drilling 混合钻井法
combination thermal drive 综合热驱
combination tool 组合下井仪
combination trap 复合圈闭
combination travelling block 联合游动滑车
combination valve 复合阀；组合阀
combination wall and anchor packer 综合悬挂
combination wash tool 组合冲洗工具
combination water 结合水
combination well 组合井
combination whirler float and guide shoe 旋流浮阀引鞋

combination wrench 多用扳手
combinational cut 混合掏槽
combination-riffle sampler 格槽式分样器
combination-type bullet 组合型井壁取心弹
combination-type reservoir 混合驱型储层
combinatorial statistics 综合统计资料
combinatorial topology 组合拓扑学
combinatorics 组合学
combine 结合；联合；联合采煤机
combine stoping 联合式开采
combined action 联合作用
combined active/passive anchoring 主被动组合锚固
combined adjustment 联合平差
combined aggregate grading 混合骨料级配
combined anchoring technique of tiebacks and soil nails 锚索锚杆组合锚固技术
combined artesian-gravity flow 承压水-潜水混合流
combined artesian-gravity well 承压水-潜水混合井
combined blasting 混合爆破
combined blowing 顶底吹炼
combined blown converter 顶底吹转炉
combined bolting and shotcrete 喷锚支护
combined bridge 两用桥
combined cage and skip 罐笼箕斗；两用罐笼
combined carbon 化合碳；结合碳
combined casing column 混合套管柱
combined centrifugal axial-flow fan 离心轴流联合扇风机
combined characteristic 综合特性曲线；组合特性
combined cleans 混合精煤
combined column 复合柱
combined column tooting 联合式柱脚
combined cooling 混合冷却
combined covering stratum 复合盖层
combined crushing and screening plant 石料破碎筛分厂
combined cut 复合切削
combined cutter loader 截装机；冷采煤联合机
combined cycle 联合循环
combined cycle power generation 联合循环发电
combined cycle power plants 联合循环电厂
combined cycle process 联合循环工艺
combined cylinder block 组合缸体
combined dead load 组合恒载
combined developing 综合开拓
combined development 综合开拓
combined development of open pit 露天矿联合开拓
combined drain 综合排水
combined draw-off 联合抽水
combined drive 混合驱动
combined drive conveyor 多传动输送机
combined earthquake coefficient 组合地震系数
combined earthquake response spectrum 组合地震反应谱
combined echo ranging and echo sounding 回声测距与测深综合系统
combined efficiency 综合效率
combined efforts 紧密合作
combined end-bearing and friction pile 端承摩擦桩
combined feed 总进料
combined filament yarn 混纤丝
combined flexure 联合弯曲
combined flushing and ramming 水冲锤击混合法

combined footing 复合基柱；联合基础
combined force-feed and splash lubrication 压力-飞溅复合润滑
combined foundation 联合基础
combined fracturing fluid coefficient 压裂液综合系数
combined gas-oil ratio 综合气油比
combined girder 组合梁
combined heat and power 热电联产
combined heat and power plant 热能发电站
combined humic acid 结合腐殖酸
combined hydrogen 化合氢
combined influence 综合影响
combined injection-production history 综合注采曲线
combined load 联合荷载；组合荷载
combined maceral-microlithotype analysis 煤岩组分-显微煤岩类型综合分析法
combined method 联合开采法
combined mines 联合矿
combined mining 联合式开采
combined mining method 联合开采法
combined mining technology 综合开采工艺
combined model 联合模型
combined modulus calculation 组合模量计算
combined modulus nomography 组合模量图解法
combined moisture 化合水分；结晶水
combined mooring and warping bollard 码头带缆桩
combined opencut-and-underground mining 露天地下联合开采
combined opening 综合开拓方式
combined overhand and underhand stoping 上下向梯段联合回采法
combined plate margin 组合板块边缘
combined polymerization 共聚合
combined portal-to-shaft-to-portal ventilation 吹吸式通风方式
combined roof 复合顶板
combined sample 综合煤样
combined scheme 混合式布置
combined search of dangerous blocks 危险块体组合搜索
combined sewage 混合污水；合流下水道
combined sewer 合流下水道
combined sewerage system 合流排水系统
combined shrinkage-and-caving method 留矿崩落联合开采法
combined shrinkage-and-caving system 留矿与崩落联合采矿法
combined silica 结合硅石
combined static and dynamic 静-动组合
combined static and dynamic loading 动静组合加载
combined steel and concrete column 钢筋混凝土柱
combined steel formwork 组合钢模板
combined stoping 联合式回采法
combined strength 组合强度；复合强度
combined stress 复合应力；合成应力；复合应力
combined stress creep 复合应力蠕变
combined stress test 复合应力试验
combined string 组合管柱
combined sulfur 化合硫
combined supply air fan and exhaust fan system 吹吸式通风方式
combined support 混合支架；联合支架
combined support using shotcrete and bolting wire 锚喷金属网联合支护
combined surface and underground mining 露天地下联合开采
combined thermal-electromagnetic relay 热动电磁联合继电器
combined top slicing and ore caving 下行分层崩落联合采矿法；下行水平分层和矿体崩落联合采矿法
combined top slicing and shrinkage stoping 下行水平分层留矿联合采矿法
combined top slicing-and-shrinkage method 顶部分层下向留矿联合开采法
combined top slicing-and-shrinkage system 下向分层留矿联合采矿法
combined transportation of open pit 露天矿连续运输
combined type 复合型
combined underhand-overhand method 下向上向梯段联合开采法
combined underhand-overhand system 下向上向梯段联合采矿法
combined ventilation system 混合式通风系统
combined water 化合水；结合水
combined weighing and mixing machine 联合称量混合机
combined-bench of open-pit 露天矿组合台阶
combined-flow 复合流
combined-force anchorage 合力锚固装置
combiner unit 混合器
combing 梳刷
combing effect 梳理作用
combining affinity 化合亲和势
combining capacity 联结力；黏结力
combining chamber 混合室
combining zone 化合区
comb-notching filter 梳齿陷波滤波器
combosweep technique 混合扫描技术
comb-shaped fold 隔挡式褶皱；梳状褶皱
combust 燃料；燃烧
combustibility 可燃性
combustibility of coal 煤的燃烧性
combustible 可燃的；易燃的
combustible carbon 可燃碳
combustible component 可燃成分
combustible fibre 可燃纤维
combustible gas 可燃气体
combustible gas detector 可燃气体检测器
combustible gas monitor 可燃气监测器
combustible liquid 可燃液体
combustible mass 可燃物质
combustible material recovery 可燃体回收率
combustible matter 可燃物质
combustible mixture 可燃混合物
combustible natural gas 可燃天然气体
combustible organic rock 可燃性有机岩
combustible shale 可燃性页岩；油页岩
combustible sulfur 可燃硫
combusting delay period 滞燃期
combustion additive 燃烧添加剂
combustion adjuvant 燃烧助剂
combustion chamber 炉膛；燃烧室

combustion channel　燃烧通道
combustion drive　火烧驱油
combustion engine　内燃机
combustion furnace　燃烧炉
combustion gas　废气；燃气
combustion improver　燃烧促进剂
combustion knock　燃烧震动
combustion lag　燃烧滞后
combustion lagging period　滞燃期
combustion limit　燃烧极限
combustion method　燃烧法
combustion pressure curve　燃烧压力曲线
combustion rate　燃烧速率
combustion ratio　燃烧比
combustion residue　燃烧残余
combustion space　燃烧室
combustion tube　燃烧管
combustion value　热值
combustion-supporting gas　助燃气体
combustion-supporting layer　引火层
combustion-supporting material　引火料
combustion-tube ash burning furnace　燃烧管烧灰炉
combustion-tube furnace　燃烧管炉
combustion-tube lamp　燃烧管型安全灯
combustor　燃烧器
come down　崩落
come to grass　从井下来到地面
come up to the standard　达标；符合标准
comendite　碱性疗岩；碱性流纹岩；钠闪碱流岩
co-metabolic reaction　协同代谢反应
comfort　设备；舒适
coming-back of face　工作面的后退式回采
co-mining　合作开采
command of the sea　制海权
commensal　共生体
commensalism　共栖现象；偏利共生
commensurable　可公度的；可以同一单位度量的
commensurate　同量的；相称的；相应的
commercial accumulation　有开采价值的油藏
commercial bed　可采矿层；有商业价值的油气层
commercial blast cleaning　工业级喷砂清理
commercial blasting　工业爆破
commercial burner oil　工业用燃油
commercial cement retardant　商品水泥缓凝剂
commercial coal　商品煤
commercial cyclone　工业旋风分离器
commercial deposit　可采矿床；商业油气藏
commercial field　可采矿床；有工业价值的油气田
commercial grade　商品等级
commercial grade of coal　商品煤等级
commercial gravel　工业砾石
commercial iron　工业生铁
commercial natural gas　商品天然气
commercial oil　工业用油；商品原油
commercial oil deposit　工业油气藏
commercial oil pool　商业油藏
commercial oil reservoir　工业油层
commercial oil well　具开采价值的油井
commercial ore　工业矿石；可采矿体
commercial orebody　商品矿石

commercial petroleum accumulation　工业油气藏
commercial petroleum reservoir　工业油气藏
commercial pipeline　商业用管线
commercial production　商业性开采
commercial recovery　工业回收率
commercial recovery project　商业性开采方案
commercial reservoir　有商业价值的储层
commercial rock gas　石油气；天然气
commercial seam　可采煤层
commercial size　工业规模
commercial trap　有商业价值的储层圈闭
commercial value　商业价值
commercial well　有工业价值的油气井
commercially disposable coal　商品煤
commercial-scale operation　商业性开采
commercial-scale trial　工业规模试验
commercial-scale trials　商业规模的试验
commingled crude　混合原油
commingled multizone completion　多层混合完井
commingled producing well　合采井
commingled production　合采
commingled water　混合水
commingled water injection　合注
commingled well completion　混合完井
commingling　混合
commingling production　多层合采
comminute　碎磨
comminuted powder　粉碎粉末
comminuted rock　压碎的岩石
comminuter　粉碎机
comminuting machine　粉碎机
comminution　粉碎；磨碎
comminution of fuel　粉碎燃料
comminution of screenings　筛余物碾碎
comminution test　可磨性试验
comminutor　粉碎机
commissure　合缝处，接合处
commix　混合
commixture　混合物
commodity ore　商品矿石
common abundance value　共同丰度值
common aquifer　总水动力系统
common battery system　共电制
common beam　共用梁；普通梁
common block　共有段
common brass　普通黄铜
common bus-bar equipment　共用母线设备
common carrier line　公用载波线
common concrete　普通混凝土
common depth point　共深度点
common depth point gather　共深度点道集
common depth point grid　共深度点网格
common depth point shooting　共深度点爆炸
common depth point stack　共深变点叠加
common detector gather　共检波点道集
common drift　共用平巷
common ducts　共同管道
common element in groundwater　地下水常见元素
common emitter　共发射极；共射极
common excavation　普通开挖

common feldspar	正长石
common geophone stack	共检波点叠加
common grade	普通等级
common grid circuit	共栅极电路
common hornblende	普通角闪石
common ion	同离子
common ion effect	共离子效应；同离子效应
common lathe	普通车床
common lay brick	常见的铺砖
common lead correction	普通铅校正
common lead dating	普通铅测年
common leak	常见漏失
common mica	白云母
common mid point stacking	共中心点叠加；水平叠加
common midpoint	共中心点
common midpoint gateway	共中心点选排
common midpoint gather	共中心点道集
common midpoint stack	共中心点叠加
common mode	同模
common normal	公法线
common pile driver	普通打桩机
common range gateway	共射程集合
common reactance	互感
common receiver point gateway	共接收点集合
common reference point	共参考点
common reflection	共反射点
common reflection point	共反射点
common reflection point gateway	共反射点集合
common roadway	共用平巷
common salt	食盐
common shareholder	普通股股东
common shot direction	共炮点方向
common solvent	一般溶剂
common source	共震源
common source point gather	共炮点道集
common stack depth stack point stack	共深度点叠加；水平叠加
common stack offset stack	共炮检距叠加；同距叠加
common storage	共用库容
common straight carbon steel	普通碳钢
common strontium correction	普通锶校正
common-battery switchboard	共电式交换机
common-collector	共集电极
common-depth-point stacking	共深度点叠加
common-device resistivity log	普通电极系电阻率测井
common-geophone gather	共检波点道集
common-ground-point	共地面点
common-mode rejection	共模抑制
common-offset gateway	同距网关；同距通道
common-offset gather	共炮检距选排；共偏移距道集
common-offset migration	同炮检距偏移
common-offset stack	同距叠加
common-range gather	同偏移距道集
common-salt spring	氯化钠泉；普通含盐泉
common-shot-gather	共炮点道集
common-shot-migration	共炮点偏移
common-source point	共炮点
commotion	电震
communicating hole	串浆孔
communicating interstices	连通的空隙
communicating pipe	连通管
communicating pore	连通孔隙
communicating tube	连通管
communicating vessel	连通器
communication band	通信频带
communication between zones	层间窜流
communication blind district	通信盲区
communication chain	联络链
community	共同体
community noise	外界噪声
commutate	换向；整流
commutation	换向
commutation switch	换向开关
commutation value	转换阀
commutator insulating segment	整流子绝缘隔片
commutator motor	整流子电动机
commutator segment	整流子片
comol	铁钴钼永磁合金
compact	碾压；致密的；坯块
compact accelerator	小型加速器
compact battery	小型电池
compact chain	紧密型链条
compact design	紧凑设计
compact embankment	密实路堤
compact fill	密填
compact form	紧凑结构
compact grain structure	密纹组织
compact gravel	致密砾石
compact grouting	压实灌浆
compact gypsum	雪花石膏；致密石膏
compact heat exchanger	紧凑型换热器
compact land	镶嵌体
compact layer	紧密层
compact limestone	致密灰岩
compact machine	紧凑型机
compact material	密实材料
compact medium	致密介质
compact ore	致密矿石
compact rock	致密岩石
compact sand	密实砂
compact shearer	短机身采煤机
compact soil	坚实土
compact soil fabric	紧实土壤结构
compact structure	密实构造
compact texture	致密结构
compact wavelet	压缩子波
compact wheel	紧凑型轮斗
compact wheel excavator	小型轮斗挖掘机
compactability	紧实性
compacted	夯实的
compacted backfill	夯实回填土
compacted bed	紧密床层
compacted clay	夯实黏土
compacted clay liner	压实黏土衬垫
compacted column	挤密桩
compacted concrete	捣实混凝土；碾压混凝土
compacted density	夯实密度
compacted depth	夯实深度；压实深度
compacted earth pile	压实土桩
compacted expanded base concrete pile	夯实扩底混凝土桩

compacted fill	夯实填土	compaction degree	压实度
compacted fill density	压实填土密度	compaction delayed	压实滞后
compacted formation	压实地层	compaction density	紧密密度
compacted gravel placement	密实砾石充填	compaction device	击实仪
compacted impervious fill	压实填土防渗	compaction drive	压实驱动
compacted layer	坚实层	compaction effect	压实效应
compacted lift	夯实分层；夯实分层厚度	compaction effort	压实力
compacted measure	压实方	compaction energy	夯实能；击实能
compacted pyroclastic texture	压实火山碎屑结构	compaction equipment	碾压设备
compacted rock	致密岩石	compaction factor	压实系数
compacted rockfill	压实填石	compaction flexure	挤压挠曲
compacted sand matrix	致密砂岩体	compaction fluid	充填携砂液；压实流体
compacted sandstone	压实砂岩	compaction fold	压实褶皱；致密褶皱
compacted shale	压实页岩	compaction grouting	挤密灌浆；压密灌浆
compacted soil	夯实土	compaction grouting method	挤密喷浆法
compacted soil fabric	压实土的结构	compaction in layer	分层压实；分层夯实
compacted soil layer	压实土层	compaction index	压实指数
compacted standard	压实标准	compaction limit	压实极限
compacted thickness	夯实厚度	compaction log	压实曲线
compacted volume	压实体积	compaction machine	击实机；压实机械
compacted-earth lining	夯土衬砌	compaction material	充填材料
compactedness	紧密性	compaction measurement	夯实量测量
compact-grain	致密晶粒	compaction method	碾压方法
compact-grained	致密颗粒的	compaction methodology	充填法
compactibility	可夯实性；可压实性	compaction mold	击实模；击实筒
compactible interval	可压缩层段	compaction of concrete	混凝土的捣实
compacting	夯实；压实	compaction of deep bed	深层压实
compacting coefficient	压实系数	compaction of granular soil	粒状土壤的压实
compacting crack	夹层；压裂；压实	compaction of soil	土的压实
compacting criteria	压实标准	compaction parameter	压实参数
compacting effect	挤土效应	compaction percentage	充填率
compacting equipment	压实机具	compaction pile	挤密桩
compacting factor	夯实系数	compaction pile method	挤密桩法
compacting factor test	压实系数试验	compaction piling	打桩挤密
compacting force	成形压力；压实力	compaction plant	碾压设备
compacting machinery	碾压机械	compaction pressure	夯实压力
compacting method	压实方法	compaction rammer	击实锤
compacting pass	压实遍数	compaction rate	充填量；密实程度
compacting process	压实过程	compaction recorder	压实记录仪
compaction	夯实；击实；密实度；压固作用	compaction sand pile	挤密砂桩
compaction band	压实带	compaction settlement	压实沉降
compaction by compaction	置换挤密；压实加密	compaction test	击实试验；压密试验；压实测试
compaction by driving	夯实	compaction test apparatus	击实仪
compaction by explosion	爆破压实	compaction test equipment	击实仪
compaction by layer	分层碾压；分层填土夯实	compaction theory	压实原理
compaction by rolling	滚碾压实；碾压	compaction trend line	压实趋势线
compaction by tamping	夯实	compaction weight	压实重量
compaction by vibrating roller	振动碾压	compaction works	击实功
compaction by vibration	振动捣实；振动压实	compaction-mold permeameter	压密模渗透计
compaction by watering	注水压实	compactive effort	击实功；击实效果
compaction characteristic	压实特性	compactness	击实度；密实度；压实度；致密性
compaction coefficient	压实系数	compactness of fills	充填沉实率
compaction control	压密控制	compactness of soil	土的密实度
compaction control by degree of saturation	饱和度控制压实	compactness test	压实度试验
compaction control by density	密度控制压实	compactor	捣固锤；夯实机；碾压机
compaction control by moisture content	含水量控制压实	compactor pass	压夯遍数
compaction control by strength	强度控制压实	compact-state rock	致密状岩石
compaction criterion	压实准则	companion	同伴
compaction curve	击实曲线；压实曲线	companion blasting	共同爆破
		companion coupling	配对联轴节

companion drift	并行平巷
companion fault	伴生断层；副断层
companion flange	成对法兰
companion heading	平行副巷
companion method	配合法
companion mineral	伴生矿物
companion specimen	对比试样
companion structure	伴生构造
comparable	可比较的
comparable data	参照数据
comparable horizon of river	河流比降
comparative biochemistry	比较生化学
comparative cartography	比较地图学
comparative indicator	差动电流计
comparative interpretation	对比解释
comparative lithological method	岩性对比法
comparative observation	比较观测
comparative operator	比较算子
comparative petrology	比较岩石学
comparative sampling method	比较取样法
comparative scale	比较计
comparative sedimentology	比较沉积学
comparative structural geology	比较构造地质学
comparative tectonics	比较大地构造学
comparatively regular coal seam	较稳定煤层
comparator	比测仪；比较仪；比值器
comparer	比较仪
comparing rule	比例尺
comparison base	比较基线
comparison diagram	对比图
comparison potentiometer	标准电位计
comparison test	比较检验法
comparison-and analog method	比较类推法
compartment	梯子间；隔室；壳板；填埋库区
compartment ceiling	井口天花板
compartment dryer	分室干燥机；间格干燥器；箱式干燥器
compartment mill	多室式磨机；分室磨碎机
compartment wall	分隔墙
compartmentalization	分隔作用
compass	罗盘；指南针
compass adjuster	罗盘校准器
compass adjustment	罗经校正
compass adjustment beacon	罗经标
compass adjustor	罗盘校准器
compass bearing	罗盘方位
compass card	罗盘标度板；罗盘方位盘
compass circle	罗盘分度圈
compass clinometer	罗盘倾斜仪
compass declination	磁偏角
compass deflection	指南针偏转
compass error	罗盘误差
compass of proportion	比例规
compass roof	半圆形屋顶；圆弧形屋顶
compass rose	方位圈
compass saw	圆锯
compass sketch	罗盘草测图
compass survey	罗盘仪测量
compass theodolite	罗盘经纬仪
compass traverse	罗盘测量导线；罗盘仪导线
compass-variometer	罗盘磁变仪
compatibility	并存性；相容性
compatibility equation	相容方程
compatibility equation of strain	应变协调方程
compatibility of cement-aggregate	水泥和集料的相容性
compatibility test	相容性试验
compatible	兼容的；相容的；协调的
compatible computer	兼容计算机
compatible element	相容元素
compatible mineral	可共存矿物；相容矿物
compatible scale	兼容比例尺
compatibleness	相容性
compatilizer	相容剂
compensable	可补偿的
compensate	补偿
compensate ratio	补偿率
compensated acceleration	补偿加速度
compensated acidosis	代偿性酸中毒
compensated alkalosis	代偿性碱中毒
compensated densilog	补偿密度测井
compensated dewatering method	补给疏干法
compensated dual neutron log	双源距补偿中子测井
compensated footing	补偿式基础
compensated formation densilog curve	补偿地层密度测井曲线
compensated formation density log	补偿地层密度测井
compensated formation density logger	补偿地层密度测井仪
compensated formation density tool	补偿地层密度测井仪
compensated foundation	补偿式地基
compensated linear vector dipole	补偿线性向量偶极
compensated log	补偿测井
compensated neutron log	补偿中子测井
compensated neutron logging	补偿中子测井
compensated pendulum	补偿摆
compensated regulator	补偿稳压器
compensated relief valve	平衡式安全阀
compensated spectral density log	补偿能谱密度测井
compensated spectral natural gamma tool	补偿自然伽马能谱测井仪
compensated utilization of resources	资源有偿使用
compensating basin	补偿区
compensating beam	平衡梁
compensating circuit	补偿电路
compensating coil	补偿线圈
compensating current	补偿流
compensating density	补偿密度
compensating depth	补偿深度；均衡深度
compensating device	补偿装置
compensating factor	补偿系数
compensating field	补偿磁场
compensating filter	补偿滤波器
compensating flow	补偿水流
compensating gear	补偿装置；差动齿轮装置；调整装置
compensating master cylinder	带补偿贮油槽的主油缸
compensating measure	补偿测量
compensating mechanism	补偿机制
compensating method	补偿法
compensating pipe	平衡管；伸缩管
compensating piston	补偿活塞

compensating reservoir	补偿调节水库
compensating resistance	补偿电阻
compensating signal	补偿信号
compensating surface current	补偿表层流
compensation	补偿；补整；对消
compensation current	补偿流
compensation curve	补偿曲线
compensation depth	补偿深度
compensation fees of resource exhaustion	资源耗竭补偿费
compensation grade	折减坡度
compensation grade in tunnel	隧道坡度折减
compensation grouting	补偿灌浆
compensation isomorphism	补偿类质同象
compensation joint	补强接头
compensation level	补偿深度
compensation of gradient	坡度折减
compensation of undulation	波浪补偿
compensation pan	补偿溜槽
compensation panel	补偿面板
compensation point	补偿点
compensation reservoir	补偿水库
compensation valve	补偿阀；平衡阀
compensation water	补偿水
compensation-dewatering method	补给疏干法
compensator	补偿器
compensator balancer	补偿平衡器
compensator level	自动安平水准仪
compensator protector	钻头压力补偿器的保护杯
compensator-amplifier unit	补偿放大器
competence	搬运力；强固性
competence of rock	坚固度
competent	强硬的；胜任的
competent barrier	坚硬遮挡层
competent bed	坚硬岩层；强岩层
competent degree of rock	坚固度
competent fold	强性褶皱
competent folding	强褶皱作用
competent formation	坚硬地层
competent layer	能干岩层；强硬层
competent river	挟沙河流
competent rock	固结岩；坚固稳定岩石；坚硬岩石
competent sand	致密砂层
competent structure	强构造
competent velocity	起动流速
competitive	竞争的
competitive adsorption	竞争吸附
competitive cementation	优先胶结作用
competitive decay	竞争衰变
competitive desorption	竞争解吸
competitive experiment	对照试验
compiler	自动编码器
complanation	变成平面；平面化
complementary acceleration	补充加速度
complementary rock	余岩；互补岩；余岩
complementary scale	补充刻度
complementary shear plane	共轭剪切面
complementary shearing stress	余剪应力
complementary slackness	互补松弛性
complementary strain energy	应变余能
complementary structure	共轭构造；互补构造
complementary wave	副波
complete analysis	全分析
complete arch	整拱
complete assay	完全试金
complete backfilling	全部充填
complete Bouguer anomaly	完全布盖异常
complete circular lining	全部圆形衬砌
complete class	完全族
complete closure	完全封闭
complete combustion	完全燃烧
complete contraction	全收缩
complete coverage	完全覆盖
complete cycle	全循环；完全周期
complete detonation	完全爆发；完全起爆
complete differential	全微分
complete disorder	完全无序
complete dissociation	完全离解
complete drainage spring	全排泄型泉
complete elastic theory solution	全弹性理论解
complete elliptic integral	完全椭圆积分
complete equilibrium	不可逆平衡；完全平衡
complete equipment	成套设备
complete expansion	完全膨胀
complete explosion	完全爆发；完全爆炸
complete extraction	全部采出；全部回采
complete failure	完全失效
complete fill	全部充填
complete film lubrication	完整油膜润滑
complete fold	完全褶皱
complete fully mechanized coal face equipment	综合机械化采煤工作面成套设备
complete fully mechanized coal mining equipment	综合机械化采煤成套设备
complete function test	全面功能试验
complete gelation	完全胶凝
complete heterogeneous equilibrium	完全不均匀平衡
complete ionization	完全电离
complete isomorphous series	完全类质同象系列
complete lattice	完全晶格
complete linkage method	全联法
complete liquefaction	完全液化
complete loop	闭合导线
complete mechanization	全盘机械化
complete miscibility	完全混溶性
complete modulation	全调制
complete moment-resisting frame	完全抗弯框架
complete monopoly	完全垄断
complete overhaul	全面检修；全面大修
complete overstep	完全掩覆；全超覆
complete packing	全部充填
complete penetrating well	完整井
complete penetration	焊透；井的完全穿透
complete penetration butt weld	贯穿对焊
complete protection	充分保护；全防护
complete recovery	全部回收；完全复原
complete reflection	全反射
complete resonance	全谐振
complete robbing	全部回采
complete shear failure	纯剪切破坏
complete shut-off	全封闭

complete size analysis	全粒度分析
complete sliding	整体滑动
complete solid solution	完全固溶体
complete solubility	完全混溶性
complete stowing	全部充填；完整充填
complete stress relief technique	应力全解除技术
complete stress-strain curve	应力应变全过程曲线
complete structure	完整结构
complete subsidence	完全塌陷
complete synthesis	全合成
complete time of oscillation	振动周期
complete turbulence	完全紊流
complete water analysis	水样全分析
complete well	完整井
completed well	完成的井
completed works	已完成的工程
completely decomposed granite	全风化花岗岩
completely decomposed rock	全风化岩石
completely decomposed volcanic rock	全风化火山岩
completely elastic	完全弹性的
completely elastoplastic hysteresis loop	完全弹性塑性滞回线
completely grouted rock bolt	全胶结式锚杆
completely hydrated cement	完全水化水泥
completely miscible liquid	完全可溶混液体
completely penetrated well	完整井
completely penetrating artesian well	完整承压井
completely penetrating gravity well	完整潜水井
completely penetrating well	完整井
completely weathered granite	全风化花岗岩
completely weathered volcanic rock	全风化火山岩
completely weathered zone	全风化带
completeness	完整性
completeness of combustion	燃烧完全度
completion	结束；完成交割
completion acceptance	竣工验收
completion brine	盐水完井液
completion contract	完井合同
completion damage	完井损害
completion design	完井设计
completion design consideration	完井设计要素
completion design criteria	完井设计标准
completion efficiency	完井效率
completion evaluation	完井评价
completion factor	完井指数
completion failure	完井作业失败
completion fluid	完井液
completion fluid bridging material	完井液暂堵材料
completion fluid filtering	完井液渗滤
completion fluid formulation	完井液配方
completion interval	完井层段
completion job	完井作业
completion life	完井后的油井寿命
completion logging panel	完井测井面板
completion mechanism	完井机理
completion method	完井方法
completion of drilling	终孔
completion of well	钻井完成
completion optimization	完井方案优化
completion parameter	完井参数
completion perforating	完井射孔
completion period	完井工期
completion phase	完井阶段
completion platform	完井平台
completion practices	完井作业
completion program	完井方案
completion rig	完井钻机
completion riser	完井立管
completion sequence	完井工序
completion system	完井装置
completion technique	完井技术
completion test	竣工检验
completion tool	完井工具
completion tree	完井采油树
complex	杂岩
complex accelerator	复合早强剂
complex alloy steel	多合金钢
complex alluvial fan	复合冲积扇
complex alteration	复杂蚀变
complex anomaly	复合异常
complex aquifer system	复杂含水岩系
complex arc	复式岛弧
complex balance	综合平衡
complex barchan dune	复式新月形沙丘
complex beltconveyor	复合皮带运输机
complex builder	配位组分
complex building unit	复合构造单元
complex coal	多煤层；复煤层
complex coal seam	复杂煤层
complex coherent noise train	复杂相干噪声系列
complex compound	络合物
complex conduit	复合水道
complex correction	复合校正；混合改正
complex cover	复合覆盖
complex cross-bedding	复杂交错层理
complex curve	复杂曲线
complex cuspate foreland	复杂三角岬
complex delta	复合三角洲
complex deoxidizer	复合脱氧剂
complex deposit	多金属矿床；复合矿床
complex dome	复式穹丘
complex drainage pattern	复式水系
complex driving team	综合掘进队
complex dune	多风向沙丘；复合沙丘
complex earthquake	复合地震
complex element	混合元素
complex extinguishing	综合灭火
complex fault	复断层
complex film	复合膜
complex flow	复杂流动
complex fluid	复杂流体
complex fold	复式褶皱
complex formation	络合物形成
complex foundation	复杂地基
complex fracture zone	复杂破裂带
complex frequency	复频率
complex gear train	复合齿轮系
complex geologic condition	复杂地质条件
complex geology	复杂地质
complex heat transfer	综合传热

complex hydrograph	综合水文过程线
complex impedance	复阻抗
complex landslide	复合滑坡
complex lining	复合衬砌
complex liquid	复合液
complex lithology sequence	复杂岩性剖面
complex materials mechanics	复合材料力学
complex mineral	复合矿物
complex mixture	复杂混合物
complex mobile belt	复合活动带
complex modulus	复模量
complex mountain	复合山
complex multicycle basin	复式多旋回盆地
complex multifolded layer	复式褶皱的复层
complex multiphase thrust mass	复杂多期逆掩体
complex ore	多金属矿石；复合矿石
complex pipeline	复合管道
complex placer deposit	复式砂矿床
complex precaution against dust	综合防尘
complex process	复合作用
complex refraction index model	复折射系数法
complex refractive index	复折射率
complex refractive index method	复折射系数法
complex reservoir	复杂岩性油藏
complex reservoir analysis	复杂岩性储集层分析
complex resistance	复电阻
complex resistivity	复电阻率
complex resistivity method	复电阻率法
complex river	复循环河；复源河
complex rock	复合岩；非均质岩
complex salt	络合盐
complex seismic trace	复合地震记录道
complex site	复杂场地
complex slope movement	复杂斜坡运动
complex soap lubricating grease	复合皂基润滑脂
complex solution	多元溶液
complex spectral magnitude	复谱幅度
complex spectrum	复谱
complex steel	合金钢
complex stream	复循环河；复源河
complex stress	复合应力；综合应力
complex stress condition	复杂应力条件
complex structure	复杂构造
complex texture coal seam	复杂结构煤层
complex truss	复式桁架
complex twins	混合双晶
complex utilization	综合利用
complex valley	复式谷
complex volcano	复火山
complex wave	复合波
complexation reaction	络合反应
complex-crossed lamellar structure	复杂交错微层构造
complexing agent	配位剂
complexity	多元性
complexity of earth system	地球系统复杂性
complexity of site	场地复杂性
complex-pulse generator	复合脉冲发生器
compliance	符合；一致
compliance test	验证试验
complicated	复杂的
complicated fault	复杂断层
complicated geological condition	复杂地质条件
complicated structure	复式构造
complied column	综合柱状剖面图
compo	混合涂料；水泥砂浆
compo mortar	水泥砂浆
component	部件；分潮；分向量
component analysis	分量分析；组分分析
component assembly	零件装配
component compatibility	构件互换性
component concentration	组分含量
component current	分潮流；海流分量
component deformation	组成变形
component efficiency	局部效率
component element	部件；构件
component force	分力
component library	部件图库
component of acceleration	加速度分量
component of blends	配料组分
component of block motion	块状运动分量
component of displacement	位移分量
component of elongation	伸长变形分量
component of force	分力
component of movement	运动分量
component of strain	应变分量
component of stress	应力分量
component part	组成部分
component screening	组分筛选
component selection	部件选择
component solution	组分溶液
component solvent	混合溶剂
component stratotype	组分层型
component stress	分应力
component test	零件检验；组件试验
component tide	分潮
component wave	分波；组成波
componential differential movement	部分差异运动
componential movement	组分相对运动
components	成分；分量
components of hydraulic tunnel	水工隧洞构造
components of stress	应力分量
components of stress vector	应力矢量分量
components of traction	牵引力分量
composed fan structure	复杂扇形构造
composed peak	合成峰
composite	复合材料；合成材料；复合的；合成的
composite action	组合作用
composite amplitude	复合振幅
composite analogy	复合模拟
composite anomaly	组合异常
composite anticline	复背斜
composite apron reef	复合裙礁
composite arch	复合拱
composite basin	复合盆地
composite batholith	复合岩基
composite beam	混合梁；组合梁
composite beam bridge	组合梁桥
composite beams and girders	复合房梁
composite bed	复合矿层
composite board	合成板

composite boss　复成岩瘤
composite breakwater　混合式防波堤
composite bridge　组合结构桥梁
composite cable　合成电缆
composite calibration　综合标定
composite car　混合车厢
composite cement grout　组合水泥薄浆
composite charge　复合装药
composite chart　综合图
composite cleans　混合精煤
composite coast　复成海岸
composite coatings　复合涂层
composite column　组合柱
composite columnar section　综合柱状剖面图
composite compact　层状组合压坯；复合坯块
composite concentrate　混合精矿
composite cone　复合火山锥；混合锥
composite construction　混合结构
composite cross section　复式横断面
composite crustal plate　复合地壳板块
composite cut　分阶掏槽
composite cycle　复合循环
composite dam　土石坝
composite decay curve　复合衰减曲线
composite decline curve　综合递减曲线；组合耗竭曲线
composite deformation　复合变形
composite diagram　合成图；综合图
composite dike　复合岩脉；集岩脉；复合岩墙
composite drawdown curve　合成降深曲线
composite drawing　混合图
composite earth dam　混合式土坝
composite error　总误差
composite evaluation　综合评价
composite explosive　组合炸药
composite fan　复合扇状地
composite fault　复合断层
composite fault scarp　复合断层崖
composite fault-plane solution　综合断层面解
composite feed　不分级入选；混合入选
composite films　复合薄膜
composite filter　复合滤波器；复式反滤层
composite fissure vein　复合裂缝脉
composite fluid bank　复合流体汇集带
composite focal mechanism　复合震源机制
composite focal mechanism solution　复合震源机制解
composite fold　复合褶皱
composite foundation　复合地基；复合基础
composite foundation pile　复合基桩
composite foundation with settlement-reducing piles　减沉复合桩基础
composite frankie　法兰基复合桩；复合型法兰基灌注桩
composite frequency characteristics　综合频率特性
composite fuel　组合燃料
composite gang　混合工子
composite gear　组合齿轮
composite geobelt　复合土工带
composite geodrain　复合土工排水板
composite geological map　综合地质图
composite geomembrane　复合土工膜
composite geosynthetic　复合土工合成材料

composite geotextile　复合土工织物
composite girder　组合大梁
composite glacier　复合冰川
composite gneiss　复片麻岩
composite grain size distribution　复合粒径分布
composite grid　组合网格
composite ground　复合地基
composite ground with lime-fly ash columns　二灰土桩复合地基
composite grouting　复合灌浆
composite halo　组合晕
composite hard rock mass　复合硬岩体
composite hologram　复合全息图
composite hydrogeological map　综合水文地质图
composite image　合成图像
composite index method　综合指数方法
composite instrument　组合下井仪
composite intrusion　复侵入体
composite joint　混合连接；组合联结
composite laccolith　复合岩盖；复合岩盘
composite lava flow　复合熔岩流
composite liberation　混合分离
composite liner　复合衬垫
composite liner system　复合衬垫系统
composite lining　复合衬砌；组合衬砌
composite log　合成测井曲线
composite magma　复合岩浆
composite map　合成矿图
composite massif　复合岩体；复合地块
composite material　复合材料；合成材料
composite membrane　复合膜
composite metal　复合金属
composite meter factor　仪表综合系数
composite mixing　组合混渣
composite mobile belt　复合活动带
composite model　组合模型
composite mold　复合模
composite mortar　混合砂浆
composite neck　复合集岩颈
composite ore deposit　复合矿床
composite oxides　复合氧化物
composite parameter　综合参数
composite patching　综合修补
composite permeability　综合渗透率
composite photograph　合成照片
composite pile　复合桩；混合桩
composite pile foundation　复合桩基
composite pipe line　综合管道
composite plan　地层平面图；合成图；综合平面图
composite product　混合产品
composite production profile　综合生产剖面
composite profile　集成剖面图；综合剖面图
composite profiling　联合剖面法
composite profiling method　联合剖面法
composite prognostic chart　综合预测图
composite projection　合成投影
composite recording　合成记录
composite reflection　复合反射
composite resins　复合树脂
composite resistance　组合阻力

composite response spectrum	合成反应谱
composite river	复式河道
composite rock	复合岩；复合岩石
composite rockfill dam	混合土石坝
composite rod string	组合抽油杆
composite roof	复合顶板
composite roof gates	复合顶板关口
composite roughness	综合粗糙度
composite rupture	复合断裂
composite sample	混合试样
composite sampler	复合取土器
composite sampling	组合采样
composite sampling scheme	综合取样法
composite sandwich construction	复合夹层结构
composite schistosity	复合片理
composite seam	复合矿层；复煤层
composite seam thickness	复合层厚度
composite section	复合剖面；综合剖面
composite sediment	复合沉积
composite segment	组合管片
composite seismogram	合成地震记录
composite sequence	复合层序
composite shaft lining	复合井筒支护；复合井筒衬砌
composite sheet	编绘原图；复合岩席
composite sill	复合岩床
composite slide surface	复合滑面
composite sliding surface	复合滑动面
composite slip surface	复合滑动面
composite slope	复式边坡
composite slope movement	复合斜坡运动
composite soil	复合地基
composite specimens	复合样品
composite stalactite	复合钟乳石
composite standard reference section	复合标准参考剖面
composite steel plate	复合钢板
composite stock	复合岩干
composite stratigraphic column	综合地层柱状图
composite stratotype	复合层型
composite stream	合流河；原油和天然气混输
composite structural section	综合构造剖面
composite structure	复合结构；混合结构
composite subgrade	复合地基
composite support	复合支护
composite surface	合成曲面；复合面
composite surface of sliding	复式滑动面
composite syncline	复合向斜
composite system testing	复合地层试井
composite tectonic profile	综合构造剖面图
composite terrane	组合地体
composite thermally stable cutter	复合热稳定切削齿
composite timbering	混合支架
composite time-distance curve	合成时距曲线
composite topography	复旋回地形
composite trace	合成地震记录线
composite truss	组合桁架
composite type rockfill dam	混合式堆石坝
composite unconformity	复合不整合
composite variable	组合变量
composite vein	复合矿脉；组合矿脉；复合脉
composite vein of ore	复合脉；复合矿脉
composite volcano	复合火山
composite wall	复合墙；组合墙
composite waste	混合废水
composite wave	复合波；合成波
composite wave filter	复式滤波器
composite well	复合井
composite work	组合工作
compositing	叠加
composition	成分；组成；合成；混合物；构成
composition equation	组分方程
composition fiber	复合纤维
composition floor	组合地板
composition halo	组合晕
composition history curve	组分变化过程曲线
composition loading	组分载荷
composition of coal	煤的成分
composition of concurrent forces	汇交力系的合成
composition of earth	土的成分
composition of force	力的合成；力的组成
composition of forces in plane	平面力系的合成
composition of parallel forces	平行力系合成
composition of rock	岩石的成分
composition pile	组合桩
composition plane	复合面
composition profile	组分分布
compositional analysis	组分分析
compositional banding	成分条带；带状组成
compositional convection	成分对流
compositional diagram	组合图
compositional gradient	组分梯度
compositional layering	成分层理
compositional material balance	组分物质平衡
compositional maturity	成分成熟度
compositional profile	成分剖面
compositional reservoir simulator	油藏组分模拟软件
compositional simulation	组分模拟
compositional variable	合成变量
compositionally immature sediment	组分未成熟沉积物
compositor	合成器；混波装置
compound	复合的；合成的；合成物；化合物
compound adjustment	复合调整
compound air lift	复式串联空气升液器
compound alluvial fan	复合冲积扇
compound anticline	复背斜
compound arch	复合拱
compound beam	组合梁
compound bending	双向弯曲
compound body	合成体
compound bowl	进刀箱
compound breakwater	混合式防波堤
compound cam	组合凸轮
compound casting	复合铸件
compound chain	组合链系
compound cirque	复合冰斗
compound coal dry-cleaning machine	复合式干选机
compound coastline	复合海岸线
compound coil	复合线圈
compound column	组合柱
compound compressor	复式压气机
compound cone	复合火山锥

compound construction	混合结构
compound conveyor	复式长距离输送机
compound coral	复体珊瑚
compound cross stratification	复合交错层理
compound cross-bedding	复合交错层理
compound cross-section	复式断面
compound curve	复曲线
compound cuspate bar	复合三角沙坝
compound delta coast	复式三角洲海岸
compound distribution	复合分布
compound dredger	复式挖泥机
compound drift correction	混合漂移改正
compound drive	联合驱动
compound fabric	混合组构
compound fan	复合扇形地
compound fault	复合断层；组合断层
compound fold	复合褶皱
compound foreset bedding	复成前积层理
compound gear	复式齿轮
compound gear-driven rolls	复式齿轮传动辊碎机
compound gearing	复合传动装置
compound geophysical survey	综合地球物理勘探
compound girder	组合梁
compound glacier	复合冰川
compound gneiss	混合片麻岩
compound graphic log	组合钻探剖面图
compound hard cast pipe	复合式冷硬铸铁管
compound hydrograph	复合水文过程线
compound indicator	复合指示剂
compound intrusion	复合侵入体
compound lamp	复合式安全灯
compound lever	复杆
compound light	复合光
compound loading	复合加载
compound logging	综合录井
compound magnet	复合磁铁
compound manometer	复式压力计
compound microscope	复显微镜
compound modulation	多重调制
compound motion	复合运动
compound motor	复激电动机
compound nucleus reaction	复核反应
compound oblique-bedding	复合斜层理
compound oil	复合油
compound pellet	复合球粒
compound pendulum	复摆
compound pile	组合桩
compound pipe	复式管道
compound Poisson distribution	复合泊松分布
compound Poisson process	复合泊松过程
compound post	合成柱
compound pressure gauge	复式压力计
compound probability	合成概率
compound profile method	综合剖面法
compound pulley	复滑轮
compound pump	复合泵；双缸泵
compound rectification coast	复式夷平海岸
compound recurved spit	复折型砂嘴
compound reinforcement	复式配筋
compound rest	复式刀架
compound ripple	干涉波痕
compound river	合流河；汇流河
compound river channel	复式河槽
compound rock	多矿物岩石
compound rod string	复合抽油杆柱
compound roof	复合顶板
compound running	联合运转，复合运转
compound sample	混合样
compound seal	复合密封
compound section	复合断面；复合剖面；组合截面
compound shaft	复式井
compound shoreline	复合滨线
compound slide	复式滑坡
compound steam working	复合汽缸工作
compound stress	复合应力；合成应力
compound structure	复合构造
compound substance	复合材料
compound support	复方支护
compound surfaces	复合表面
compound tide	复合潮
compound trommel	多段滚筒筛
compound truss	复式桁架
compound turbodrill	复合涡轮钻具
compound vein	多矿物脉；复合矿脉
compound ventilation	混合式通风
compound vibration	复合振动
compound volcano	复合火山
compound vortex	复合旋涡
compound wall	组合墙
compound weir	复式断面堰；组合堰
compound well	复管片
compound-balanced pumping unit	复合平衡式抽油机
compounded	复合；混合的
compounded latex	复合胶乳
compounded luboil	复合润滑油
compounded lubricating oil	复合润滑油
compounded mineral oil	复合矿物油
compounded oil	复合油
compounding	配料
compounding in parallel	并联
compounding in series	串联
compounding of tectonic systems	构造体系的复合
comprehensive	广泛的；有理解力的
comprehensive analysis and judgement	综合分析判断
comprehensive anomaly map	异常综合图
comprehensive appraisal	综合评价
comprehensive atlas	综合地图集
comprehensive correlative degree	综合关联度
comprehensive design	综合设计
comprehensive earthquake prediction	地震综合预测
comprehensive evaluation index system	综合评价指标体系
comprehensive exploration	综合勘查
comprehensive feasibility study	综合可行性研究
comprehensive geological map of mine area	矿山综合地质图
comprehensive geophysical exploration	综合物探
comprehensive geophysical method	综合物探方法
comprehensive geophysical survey	综合地球物理勘探
comprehensive graphs of borehole	钻孔水文地质综合成果表

comprehensive grouting method	综合灌浆法
comprehensive index	综合指数
comprehensive integration	综合集成
comprehensive investigation	综合查勘
comprehensive mechanized caving coal mining	综合机械化放顶煤开采
comprehensive ocean atmosphere data set	海洋大气综合数据集
comprehensive planning	综合规划
comprehensive prediction	综合预测
comprehensive pressing water test	综合压水实验
comprehensive prevention	综合防治
comprehensive prevention and treatment of mine water	矿井水的综合预防和处理
comprehensive treatment	综合处理
comprehensive type of rock burst	综合型冲击地压
comprehensive utilization	综合利用
comprehensive well log analysis system	综合测井分析系统
compress	打包机；敷布；挤压
compressal-type foundation	夯实土地基
compress-crushed zone	挤压破碎带
compressed air accumulator	贮气罐
compressed air chuck	气动夹头；气动卡盘
compressed air delivery	压缩空气供量
compressed air drill	风钻；压缩气钻
compressed air fan	风动扇风机
compressed air hammer	风镐
compressed air jack	压气千斤顶
compressed air locomotive	压气机车
compressed air mixer	压缩空气式搅拌机
compressed air motor	气动发动机
compressed air pick	风镐
compressed air plant	压缩空气机车间；压缩空气设备
compressed air receiver	压缩空气储气罐
compressed air shield	全气压盾构
compressed air station	空压站；压缩空气站
compressed air system	压缩空气系统
compressed air tamper	风动掏锤；气压夯
compressed air tunnelling method	压缩空气开挖隧道法
compressed air-fuel oil burner	压缩空气燃油喷燃器
compressed air-operated	压缩空气驱动的
compressed bar	受压杆
compressed layer	压缩层
compressed mixture	密实混合料
compressed natural gas	压缩天然气
compressed pile	压挤桩；压注桩
compressed structural zone	挤压构造带
compressed zone	挤压带
compressed-air blasting	压气爆破
compressed-air caisson method	压气沉井法
compressed-air drive	风动
compressed-air feed pipe	压气推进管
compressed-air foundation	气压沉箱基础
compressed-air hoist	风动绞车
compressed-air injector	压缩空气喷射器
compressed-air lighting	井下风动发电机照明
compressed-air linkage	用压缩空气渗透贯通
compressed-air method	气压法
compressed-air quenching	压缩空气淬火
compressed-air rotary drill	风动旋转式钻机
compressed-air sampler	压气式取土器
compressed-air shield	全气压盾构
compressed-air shield method	气压盾构施工法
compressed-air winch	压缩空气绞车
compressed-air-driven generator	压气驱动发电机
compressed-air-driven lamp	压气驱动电灯
compressed-air-driven pump	风动泵
compressed-air-operated rod puller	风动拔钻杆机
compressibility	可压缩性
compressibility coefficient	压缩系数
compressibility degree	可压度
compressibility equation	压缩性方程
compressibility factor	压缩系数
compressibility foundation	压缩性地基
compressibility index	压缩指数
compressibility influence	压缩效应
compressibility modulus	压缩模量
compressibility of aquifer	含水层的可压缩性
compressibility of matrix	骨架压缩
compressibility of pore air	孔隙气的压缩性
compressibility of pore fluid	孔隙流体的压缩性
compressibility of pore space	孔隙压缩性
compressibility of soil skeleton	土骨架的压缩性
compressibility ratio	压缩比
compressibility wave	压缩波
compressible	可压缩的
compressible filler	压缩掺合料；压缩填料
compressible flow	可压缩流
compressible fluid	可压缩流体
compressible foundation	压缩性地基
compressible insert	压缩性嵌入物
compressible mudrock	可压缩泥岩
compressible prop	可缩性支柱
compressible shale	压缩页岩
compressible soil	压缩性土
compressible stratum	可压缩层
compressible support	可缩性支架
compression and back expansion	剪缩回胀性
compression and recovery test	压缩和复原试验
compression area	受压面积；压缩区
compression bar	受压杆
compression blasting	压缩爆破
compression capacity	受压承载力
compression chamber	加压舱；加压室
compression chord	受压弦杆
compression clamp	抗压夹具
compression coefficient	压缩系数
compression coefficient of mine-fills	矿山充填物压缩系数
compression coefficient of soil	土的压缩系数
compression connection	加压连接
compression consolidation	压缩固结
compression coupling type joint	压紧接箍型连接
compression crack	压缩裂隙
compression creep	压缩蠕变
compression curve	压缩曲线
compression curve of soil	土的压缩曲线
compression damage	压缩损伤
compression deep fracture	压性深断裂
compression deformation	压缩变形
compression diagonal	受压斜杆

compression dynamometer	压缩测力计	compression side	受压侧
compression efficiency	压缩效率	compression splice	压接
compression elasticity	抗压弹性	compression spring	压缩弹簧
compression face	受压面	compression steel	受压钢筋
compression factor	压缩系数	compression strain	压应变
compression failure	受压破坏	compression strength	抗压强度
compression fault	压性断层	compression stress	压应力
compression fissure	挤压裂缝	compression stress wave	压缩应力波
compression flange	受压翼缘	compression stroke	爆炸行程；压缩冲程
compression fracture	挤压断裂；受压破裂	compression strut	压杆
compression gap	压缩缝	compression subsidence	压缩沉陷
compression gasoline	气态汽油	compression support skirt	承压支承筒
compression gauge	压力计	compression test	抗压试验；压缩试验
compression grease cup	压缩加料润滑脂杯	compression test of loess	黄土压缩试验
compression grouting	压实注浆	compression testing machine	压力试验机
compression heat	压缩热	compression text	压缩试验
compression ignition engine	压燃式发动机；压缩点火引擎	compression valve	压缩阀
compression index	压缩指数	compression wave	压缩波
compression index of soil	土的压缩指数	compression zone	受压带；受压区
compression intensity	压缩率；压缩强度	compression zone depth	受压层深度
compression joint	承压缝；挤压节理；压缩接头；承压缝	compressional	挤压的；压缩的
compression law	压缩定律	compressional anticline	挤压背斜
compression layer	受压层	compressional axis	压缩轴
compression limit	压缩极限	compressional basin	挤压盆地
compression load	压缩荷载	compressional component	纵波分量
compression member	承压构件	compressional decollement	挤压滑脱
compression member with large eccentricity	大偏心受压构件	compressional deformation	挤压变形
compression member with small eccentricity	小偏心受压构件	compressional diapir	挤压底辟
compression metamorphism	挤压变质作用	compressional diffraction	压缩波绕射
compression meter	压力表	compressional dilatational wave	压缩膨胀波；疏密波
compression modulus	抗压模量；压缩模量	compressional displacement field	压缩位移场
compression modulus of rubber	橡胶压缩模量	compressional faulting	挤压断裂作用
compression molding	压缩模塑	compressional fissure	挤压裂缝
compression nut	压紧螺母	compressional fold	挤压褶皱
compression of soil	土的压缩	compressional force	压缩力
compression of the earth	地球扁率	compressional graben	挤压性地堑
compression packer	压缩式封隔器	compressional megasuture	挤压巨型接合带
compression pad	压缩垫	compressional member	受压构件
compression parallel to grain	顺纹压力	compressional mobile belt	挤压活动带
compression perpendicular to grain	横纹压缩	compressional motion	压缩运动
compression pile	抗压桩	compressional movement	挤压运动
compression point	压缩点	compressional orogenesis	压性造山作用
compression pressure	压缩压力	compressional orogenic arc	挤压造山弧
compression property	受压特性	compressional phase	挤压幕
compression pulse	压缩脉冲	compressional plate	挤压板块
compression pump	压缩机	compressional point source	压缩波点源
compression rate	压缩率	compressional potential	压缩波电势
compression ratio	压缩比	compressional structure	压性构造
compression reinforcement	受压钢筋	compressional tectonics	压缩构造
compression ring	压力环	compressional uniaxial strength	单轴压缩强度
compression rubber	压缩橡胶	compressional velocity	压缩波速度；纵波速度
compression seal	预铸压力填缝材；压力封接；压接	compressional vibration	纵振动
compression set	压力定形；压缩形变；压缩永久变形	compressional wave	压缩波
compression settlement	压缩沉降	compressional wave field	压缩波场
compression settling	压缩沉降	compressional wave front	压缩波波前
compression shock	压缩冲击；压缩激波	compressional wave velocity	压缩波速度
		compressional wave vibrator	纵波可控震源
		compressional window	纵波窗口
		compressional-to-shear conversion	纵-横波转换
		compression-fitting joint	压紧连接

compression-induced reverse faulting　挤压诱发逆断裂活动
compression-type load cell　压紧式承载元件
compression-type refrigerating machine　压缩式制冷机
compressive and torsion fracture　压扭性断裂
compressive apparatus　压缩仪
compressive belt　挤压带；挤压带
compressive breaking strength　抗压破坏强度
compressive capacity　受压承载力
compressive cleavage　压劈理
compressive creep　压缩蠕变
compressive deformation　压缩变形；压缩变形
compressive failure　压缩破坏
compressive failure of a roof beam　压缩顶板梁压缩破坏
compressive fault　挤压断层；挤压断层
compressive flange　受压翼缘
compressive fold　挤压褶皱
compressive force　压力
compressive fracture　挤压裂隙；压裂
compressive layer　受压层
compressive load　压缩荷载
compressive loading　压力荷载
compressive modulus　压缩模量
compressive modulus of elasticity　压缩弹性模量
compressive nappe　挤压推覆体
compressive nonlinearity　非线性压缩
compressive oscillation　压缩振荡
compressive plane　挤压面
compressive principal stress　主压应力
compressive property　抗压性能
compressive region　受压区
compressive reinforcement　受压钢筋
compressive resistance　抗压强度
compressive setting　受压沉降
compressive shear failure　压剪破坏
compressive stiffness factor　抗压刚度系数
compressive strain　压应变
compressive strain pulse　压应变脉冲
compressive strain wave　压缩应变波
compressive strength　抗压强度
compressive strength after immersion　浸水抗压强度
compressive strength for line contact　线抗压强度
compressive strength for point contact　点抗压强度
compressive strength for surface contact　面抗压强度
compressive strength of rock　岩石抗压强度
compressive strength rate　抗压强度比
compressive strength test　抗压强度试验
compressive stress　压应力；抗压应力
compressive stress constraint　压应力约束
compressive stress strain curve　压缩应力-应变曲线
compressive structural plane　压性构造面
compressive sub-stratum　地基压缩层
compressive tectonic regime　挤压构造状态；收缩构造体制
compressive tension failure　压致拉裂破坏
compressive traction　压缩牵引
compressive ultimate strength　极限抗压强度
compressive wave　压缩波
compressive yield point　抗压屈服极限；压缩屈服点
compresso-crushed zone　挤压破碎带
compressor　压缩机
compressor capacity　压缩机容量
compressor delivery　空压机气供量
compressor discharge rate　压缩机排量
compressor gun　润滑油增压机
compressor pressure　空压机压力
compressor truck　移动压缩机牵引车
compressor wire　预应力钢丝
compresso-shear structure plane　压扭性构造面
compresso-torsional fault　压扭性断层
compressure　压缩，抑制；压缩力
comprise　包含
compromise boundary　协和边界
compromise joint　异形接头
compromise joint bar　异形接头夹板
compromise potential　平衡电位
compu-log　计算机控制测井
compulsory land acquisition　土地征用
computation of ore reserve　矿石储量计算
computation of safe yield　安全涌水量计算
computation simulation　模拟计算
computational domain　计算区域
computational fluid dynamics　计算流体动力学
computational structural mechanics　计算结构力学
computed axial tomography　轴向计算层析成像技术
computed discharge　计算流量
computed log　计算机处理的测井曲线
computed log analysis　计算机测井分析
computed porosity　计算的孔隙度
computed tomography　计算机断层扫描
computed tractive effort　计算牵引力
computer controlled grouting　计算机控制灌浆
computer geology　计算机地质学
computer interpretation log　计算机测井解释图
computer servo-system　计算机伺服系统
computer supported mapping　数控绘图
computer tomography　计算机层析成像术
computer-aided　计算机辅助的
computer-aided leak detection　计算机辅助检漏
computer-aided log analysis　计算机辅助测井分析
computer-controlled well logger　计算机控制测井仪
computerized drilling control　计算机控制钻进
computerized log analysis　计算机测井分析
computerized logging unit　数控测井仪
computerized tomography　计算机层析成像
computerized well model　计算机油井模型
computerized well planning program　计算机钻井设计程序
computer-processed log interpretation　计算机处理测井解释
computing oscilloscope　计算示波仪
concatenated data sets　并置数据集；连续数据集
concave　凹的；凹面的；陷穴
concave bank　凹岸；冲刷河岸
concave base　凹底
concave bit　凹形凿；凹形钻头
concave camber　凹度
concave coastline　凹岸线
concave convex bank　凹凸岸
concave cross bedding　凹面交错层理
concave crown　凹形无岩心钻头；拱底
concave curvature　凹曲度

concave edge	凹刃
concave fillet weld	凹角焊
concave flow	凹岸水流
concave fracture	凹面断口
concave head piston	凹顶活塞
concave inward bit	中心凹陷形钻头
concave joint	凹缝
concave lens	凹透镜
concave levelling disk	凹面调平圆盘
concave mirror	凹面镜
concave pan	凹形溜槽；凹盘
concave refractor	上凹折射层
concave shore	凹岸
concave slope	凹形坡
concave tile	底瓦
concave trace tube	凹形伴热管
concave upward	向斜的
concave weld	凹面焊
concave-concave	双凹的
concave-convex	凹凸的
concave-convex lens	凹凸透镜
concave-sided indentation	凹边压痕
concealed anomaly	隐伏异常
concealed arch	暗拱
concealed bedrock	未露头基岩；隐蔽基岩
concealed bracing	暗支撑
concealed coalfield	无露头煤田；隐伏煤田
concealed conduit	暗线导管
concealed deposit	无露头矿床；隐伏矿床
concealed discharge	隐蔽式排放
concealed engineering	隐蔽工程
concealed fan	隐伏冲积扇
concealed fastening	隐蔽式锚固
concealed fault	隐伏断层
concealed fracture	隐伏断裂
concealed joint	暗缝；盖板接合；交叠接合
concealed mineralization	隐蔽矿化
concealed nailing	藏钉
concealed ore body	隐伏矿体
concealed outcrop	隐蔽露头
concealed pediment	隐蔽麓原
concealed piping	隐蔽喉管；暗管敷设
concealed structure	隐伏构造
concealed unconformity	隐蔽不整合
concealed water	潜藏水；隐伏水
concealed wire	暗线
concealed work	隐蔽工程
concealment	隐蔽所；隐蔽物
concentrate	富集；精矿；精煤；浓缩
concentrate bunker	精煤或精矿仓
concentrate grade	精矿等级；精矿品位
concentrate handing	精矿搬运；精矿处理
concentrate handling	精矿搬运；精矿处理
concentrate sampler	精矿取样
concentrate yield	精煤回收率
concentrated	集中的；浓的
concentrated acid	浓酸
concentrated alkali	浓碱
concentrated borrow bank	集中取土场
concentrated cartridge	集中药包
concentrated charge	密集装药
concentrated drainage	集中排水
concentrated earth movement	造山运动
concentrated field	富集场
concentrated flow	槽流；河床径流
concentrated force	集中力
concentrated grout-made station	集中制浆站
concentrated hydrochloric acid	高浓度盐酸
concentrated leak	集中渗漏
concentrated load	点荷载；集中荷载
concentrated load/point load	集中荷载
concentrated matte	浓缩冰铜
concentrated sludge	浓缩污泥
concentrated solution	浓溶液
concentrated spoil bank	集中弃碴场
concentrated spring	集中弹簧
concentrated sulphuric acid	浓硫酸
concentrated tension and full-bonded anchor cable	拉力集中全长黏结型锚索
concentrated wash	槽蚀；河道冲蚀
concentrated water	高矿化度水；浓缩水
concentrated weight	集中负载
concentrated zone of karst water flow	岩溶强迳流带
concentrating machine	精选机；选矿机
concentrating mill	选矿厂
concentrating operation	浓缩富集工作；原煤分选的加工工作
concentrating ore	待选矿石
concentrating pool	浓缩池
concentrating table	选矿摇床
concentration	富集；精选
concentration basin	浓缩池；选矿槽
concentration buffer	浓度缓冲剂
concentration by evaporation	蒸发浓缩作用
concentration clarke	浓度克拉克值
concentration coefficient	浓集系数
concentration coefficient of holes	炮眼密集系数
concentration constant	浓度常数；浓度常数
concentration covering stratum	浓度盖层
concentration criterion	分选判据；精选判据
concentration criterion of failure	浓度破坏准则
concentration curve	浓度曲线
concentration detector	浓度检测器
concentration difference	浓度差
concentration diffusion	浓差扩散
concentration effect	集中效应
concentration factor	集中因子；富集系数
concentration field	浓度场
concentration gradient	浓度梯度
concentration index	浓缩指数
concentration layout	集中布置
concentration of charge	集中装药
concentration of collector	捕收剂浓度
concentration of contamination	污染浓度
concentration of foaming agent	发泡剂浓度
concentration of loading	集中装载
concentration of mining	回采集中化；集中开采
concentration of operation	集中生产
concentration of output	集中开采；集中生产
concentration of rigidity	集中刚度

concentration parameter	浓度参数
concentration plant	选矿厂
concentration point	浓缩点
concentration profile	浓度剖面
concentration range	浓度范围
concentration ratio	浓缩比；选矿比
concentration recovery percentage	选矿回收率
concentration smelting	富集熔炼
concentration table	富集台；选矿摇床
concentration wave	浓度波
concentration-time curve	浓度-时间曲线
concentrator	浓缩器；选矿机；精选机
concentrator flow balance	选矿机流量平衡
concentrator table	精选台
concentric	同心的；同轴的
concentric action vibrating screen	同心动作振动筛
concentric adjustable bearing	同心可调轴承
concentric arch	同心拱
concentric banded structure	同心带状构造
concentric charge	集中装药
concentric cleavage	同心劈理
concentric column load	集中柱荷载
concentric costellae	同心壳线
concentric cylinder rheology measurement	同心圆筒流变性测量
concentric cylinder rotation viscometry	同心圆筒旋转式测黏法
concentric cylinder viscometer	同心圆筒式黏度计
concentric cylinder viscosimeter	同心圆筒黏度计
concentric dial	同心刻度盘
concentric domed mires	同心圆状沼泽
concentric drainage pattern	同心型水系
concentric exfoliation	同心剥离
concentric fold	同心褶曲；同心褶皱
concentric gas lift mandrel	同心气举阀工作筒
concentric gas lift valve	同心气举阀
concentric groove	同心纹
concentric insulated tubing	同心保温油管
concentric joint	同心节理
concentric load	集中荷载；轴心荷载
concentric reducer	同心异径管
concentric rig	钻井、完井两用钻机
concentric screen	双层筛管
concentric slotted liner	同心割缝衬管
concentric stria	同心纹
concentric string gravel pack	同心管柱砾石充填
concentric structure	同心环状构造
concentric tubing work string	同心油管工作管柱
concentric tubing workover	同心油管修井
concentric weathering	球状风化
concentric wire wrapped screen	同心绕丝筛管
concentric workover rig	同心管修井机
concentric zonal structure	同心带状构造
concentric zonation	同心带状
concentrical	集中的；同轴的
concentrically braced timber frame	同心支撑木框架
concentrically zoned	同心环状分带式
concentrically zoned structure	同心带状构造
concentricity	同心度；同轴度
concentricity tester	同心度检验仪
concentric-lamellar	同心片状
concentric-ring treater	同心环式脱水器
concept of conjugate prior	共轭先验概念
conceptual aseismic design	抗震概念设计
conceptual basis	基本原理
conceptual concession	矿山开采权；许可
conceptual cross section	理想剖面
conceptual design	概念设计；方案设计
conceptual design for earthquake resistance	抗震概念设计
conceptual framework	概念框架
conceptual geological model	概念地质模型
conceptual hydrogeological model	水文地质概念模型
conceptualization of hydrogeological condition	水文地质条件概化
concertina fold	棱角褶皱
concession	采矿用地；矿山开采权
concha	半圆穹顶
conchiferous	有贝壳的
conchilite	贝状褐铁矿
conchiolin	壳基质
conchite	多孔霞石
conchitic	含贝壳多的
conchitic sand	贝壳砂
conchoidal	蚌线的；贝壳状的
conchoidal briquette	贝壳状型煤
conchoidal fracture	贝壳状断口；贝壳状断裂
conchoidal shale	介壳页岩
conchoidal structure	贝壳状构造
conclinal	顺倾斜的
conclinal valley	顺向谷
concluding stage	终期；最后阶段
concoct	调制；混合
concoction	调合
concomitant	伴生的；相伴的
concord	一致
concordance	整合
concordance of summit level	峰顶面等高
concordancy	一致；整合
concordancy line	整合线
concordant	一致的；整合的
concordant batholith	整合岩基
concordant bedding	平行层理；整合层理
concordant body	整合岩体
concordant coastline	顺向海岸线
concordant curve	谐和曲线
concordant fault	同向断层
concordant fold	整合褶皱
concordant injection	顺层注入；整合贯入
concordant intrusion	整合侵入
concordant intrusive	整合侵入的
concordant massif	整合岩体
concordant morphology	平齐地貌
concordant strata	整合地层
concordant summit level	等高峰顶面
concordant thrust	同向冲断层；整一冲断层
concrete	混凝土；三合土；水泥
concrete accelerate curing box	混凝土加速养护箱
concrete accelerator	混凝土速凝剂
concrete additives	混凝土添加剂
concrete age	混凝土龄期

concrete aggregate	混凝土集料
concrete aggregates-fly ash	混凝土骨料-粉煤灰
concrete agitation	混凝土搅拌
concrete agitator	混凝土拌合机；混凝土搅拌车
concrete apron	混凝土护墙；混凝土护坦
concrete arch	混凝土拱
concrete arch dam	混凝土拱坝
concrete area	混凝土截面积
concrete backfill	混凝土回填
concrete baffle pier	混凝土消力墩
concrete bagwork	袋装混凝土护岸工程
concrete ballast	混凝土压舱块；混凝土压载
concrete base	混凝土底座；混凝土基础
concrete base slab	混凝土基础板
concrete batcher	混凝土配料器
concrete batching	配制混凝土；混凝土配料
concrete batching and mixing plant	混凝土配料拌合厂
concrete batching plant	混凝土配料厂；混凝土配料设备
concrete beam	混凝土梁
concrete beam extension	混凝土墙支撑法
concrete beams and girders	混凝土房梁；混凝土梁和主梁
concrete bedding course	混凝土垫层
concrete bent construction	混凝土构架结构；混凝土排架结构
concrete bleeding	混凝土泌水现象
concrete block	混凝土砌块；混凝土砖块
concrete block and rock-mounded breakwater	混凝土方块堆石防波堤
concrete block breakwater	混凝土块防波堤
concrete block column	混凝土砌块支柱
concrete block cutter	混凝土方块切割机
concrete block lining	混凝土砌块衬砌；混凝土块井壁
concrete block retaining wall	混凝土块挡墙
concrete block revetment	混凝土块护岸
concrete block seawall	混凝土海堤
concrete block walling	混凝土砌块井壁
concrete blocks	混凝土块
concrete blockyard	混凝土块制造场
concrete blower	混凝土风力压送机
concrete blowing	吹注混凝土
concrete box culvert	混凝土箱涵
concrete breaker	混凝土破碎机
concrete brick	混凝土砌块
concrete bricking	混凝土砖支护
concrete bridge	混凝土桥
concrete bucket	混凝土浇筑吊罐
concrete buffer	混凝土缓冲
concrete buildings	混凝土建筑物
concrete bulkhead	混凝土隔墙
concrete bunton	混凝土横梁
concrete buttress	混凝土扶壁；混凝土支壁
concrete caisson	混凝土沉箱
concrete caisson breakwater	混凝土沉箱防波堤
concrete caisson sinking	混凝土沉箱法
concrete canal lining	混凝土渠道衬砌
concrete casing	混凝土饰面；混凝土外壳
concrete check	混凝土配水闸
concrete check dam	混凝土谷坊；混凝土拦沙坝
concrete chimney	混凝土烟囱
concrete chute	混凝土溜槽；混凝土输送槽
concrete class	混凝土等级
concrete coating	混凝土覆盖层；混凝土抹面
concrete cofferdam	混凝土围堰
concrete collar	混凝土圈梁
concrete collar surface mat	井颈混凝土圈
concrete column	混凝土柱
concrete compactor	混凝土镇压器
concrete composition	混凝土成分
concrete compression strength	混凝土抗压强度
concrete compressive strength	混凝土抗压强度
concrete conduit	混凝土管道
concrete consistence	混凝土稠度
concrete consolidating	混凝土捣固
concrete construction	混凝土结构；混凝土施工
concrete contraction	混凝土的收缩
concrete core	混凝土心
concrete core wall	混凝土心墙
concrete cover	混凝土保护层；混凝土盖板
concrete cover to reinforcement	钢筋的混凝土保护层
concrete cracks	混凝土裂纹
concrete cradle	混凝土管座
concrete creep	混凝土徐变
concrete crib breakwater	混凝土木笼防波堤
concrete crib work	混凝土格排作业
concrete cribbing	混凝土筐笼；混凝土箱格
concrete crushing strength	混凝土抗压强度
concrete cube	混凝土立方块
concrete culvert	混凝土涵洞
concrete curb	混凝土支架；混凝土路缘
concrete curing	混凝土养护
concrete curing blanket	混凝土保温覆盖
concrete curing compound	混凝土养护剂
concrete curing mat	混凝土养护层
concrete cushion	混凝土垫层
concrete cutoff wall	混凝土防渗墙
concrete dam	混凝土坝
concrete deadman	混凝土锚桩
concrete defect	混凝土缺陷
concrete deformation	混凝土变形
concrete deliver truck	混凝土搅拌车
concrete delivery pipe	混凝土输送管
concrete delivery truck	混凝土载运车
concrete densifier	混凝土速凝剂
concrete deposit	混凝土浇筑物；浇筑混凝土
concrete design	混凝土配合比设计；混凝土设计
concrete desintegration	混凝土混合料离析
concrete diaphragm wall	地下连续墙；混凝土防渗墙
concrete disintegration	混凝土离析
concrete distributor	混凝土分配器
concrete drain tile	混凝土排水管
concrete durability	混凝土耐久性
concrete dwarf wall	混凝土矮墙
concrete encasement	混凝土包壳
concrete enveloped	混凝土包裹的
concrete equipment	混凝土设备
concrete excavation	混凝土开挖
concrete face	混凝土面板
concrete face rockfill dam	混凝土面板坝
concrete facing	混凝土护面；混凝土面板

concrete fatigue	混凝土疲劳
concrete feeder	混凝土喂料器
concrete filled caisson	混凝土充填沉箱
concrete filled steel pipe pile	钢管混凝土桩
concrete filler block	混凝土填块
concrete fillet	混凝土内补角
concrete fin	混凝土条
concrete finger	混凝土指状闸门
concrete finish	混凝土抹面
concrete finisher	水泥混凝土混合料整面机
concrete finishing	混凝土表面磨光
concrete finishing machine	混凝土整面机
concrete fireproofing	混凝土防火性
concrete floor	混凝土底板；混凝土楼板
concrete flowability	混凝土流动性
concrete flume	混凝土渡槽
concrete footing	混凝土基础；混凝土基脚
concrete for overbreakage	超挖衬砌
concrete forcing mixing machine	混凝土强制式搅拌机
concrete form	混凝土模板
concrete form oil	混凝土模板用油
concrete fortifications	混凝土防御工事
concrete foundation	混凝土基础
concrete foundation block	混凝土基础块体
concrete frame	混凝土构架；混凝土排架
concrete frame support	钢筋混凝土框式支架
concrete girder bridge	混凝土高架桥；混凝土梁桥
concrete grade	混凝土级别
concrete gravity dam	混凝土重力坝
concrete gravity oil platform	混凝土重力式石油平台
concrete gravity platform	混凝土重力式钻井平台
concrete gravity quaywall	混凝土重力式岸壁
concrete gravity wall	混凝土重力式挡墙
concrete grillage	混凝土格床
concrete grille	混凝土花格砖
concrete grip	混凝土握固力
concrete grout	混凝土浆
concrete grouting machine	混凝土灌注车
concrete grouting pad	混凝土止浆垫
concrete guard wall	混凝土挡墙；混凝土护墙
concrete guide wall	混凝土导墙
concrete gun	混凝土喷枪
concrete handling	混凝土吊运
concrete hardening	混凝土硬化
concrete hauling container	混凝土运送容器
concrete haunching	混凝土外壳
concrete headframe	钢筋混凝土井架
concrete headgear	混凝土井架
concrete hollow block	空心混凝土块
concrete hopper	混凝土料斗
concrete ingredient	混凝土成分
concrete injector	混凝土喷射泵
concrete in-situ	现浇混凝土
concrete inspection	混凝土检验
concrete jacket	混凝土壳板；混凝土套
concrete joint cleaner	水泥混凝土路面清缝机
concrete joint cutter	水泥混凝土路面切缝机
concrete joint sealer	水泥混凝土路面填缝机
concrete key trench	混凝土截水槽
concrete layer	混凝土层
concrete ledge	岩壁梁
concrete lift	混凝土浇筑层
concrete lifting bucket	混凝土吊斗
concrete liner	混凝土衬砌
concrete lining	混凝土衬砌；整体混凝土支护
concrete lining canal	混凝土衬砌水渠
concrete lintel	混凝土过梁
concrete lock	混凝土船闸
concrete lock floor	混凝土船闸底板
concrete lorry	混凝土搅拌车
concrete machine	混凝土喷射机
concrete mason	混凝土圬工
concrete masonry	混凝土砌体；混凝土圬工
concrete masonry unit	混凝土砌块
concrete mat	混凝土垫层
concrete material	混凝土配合料
concrete mattress	混凝土沉排
concrete mattress roll	混凝土排辊
concrete member	混凝土构件
concrete membrane	混凝土薄层
concrete mix	混凝土拌合料；混凝土混合物；混凝土配合比
concrete mix barge	混凝土拌合船
concrete mix design	混凝土配合设计
concrete mixer	混凝土搅拌机
concrete mixer heater attachment	混凝土搅拌机加热装置的附件
concrete mixer plant	混凝土搅拌站
concrete mixer truck	混凝土搅拌车
concrete mixing	混凝土搅拌
concrete mixing and transporting car	混凝土搅拌运输车
concrete mixing machine	混凝土搅拌机
concrete mixing plant	混凝土搅拌站；混凝土搅拌装置
concrete mixing station	混凝土搅拌站
concrete mixing system	混凝土拌合系统
concrete mixing water ratio	混凝土的水灰比
concrete mixture	混凝土拌合物
concrete mobility	混凝土流动性
concrete moist room	混凝土湿养护间
concrete monolith	混凝土大块体
concrete nail	混凝土钉
concrete nullah	混凝土防洪渠
concrete of low porosity	少孔隙混凝土；密实混凝土
concrete of stiff consistency	干硬性混凝土
concrete on architecture	混凝土结构；建筑用混凝土
concrete orifice turnout	孔口式混凝土斗门
concrete overflow dam	混凝土溢流坝
concrete pad	混凝土垫
concrete pancake column	圆形混凝土砌块柱
concrete patching	混凝土面修补
concrete pavement	混凝土护面；混凝土路面
concrete pavement restoration	混凝土路面修复
concrete paver	水泥混凝土混合料摊铺机；混凝土铺路机
concrete paving	混凝土护面
concrete paving block	混凝土铺路砖
concrete pedestal	混凝土基座
concrete ph	混凝土酸碱度
concrete pier	混凝土墩

concrete pile	混凝土桩	concrete running	浇注混凝土
concrete pile driving	打混凝土桩	concrete saddle	混凝土鞍座
concrete pile follower	混凝土桩帽	concrete sample	混凝土试件
concrete pile foundation	混凝土桩基础	concrete saw	路面锯缝机
concrete pile jetting	用水冲法下沉混凝土桩	concrete scaling	混凝土剥落
concrete piling	混凝土排桩；混凝土桩	concrete screw-pile	混凝土螺旋桩
concrete pillar	混凝土标石；混凝土墩；混凝土柱	concrete section	龙抬头连接段
concrete pipe	混凝土管；混凝土管	concrete segment	混凝土管片；混凝土组件
concrete pipe pile	混凝土管桩	concrete segregation	混凝土离析
concrete piping	混凝土管道输送	concrete set	混凝土框架；混凝土座盘
concrete placeability	混凝土可灌注性	concrete setting	混凝土凝固
concrete placement	混凝土浇筑	concrete shear key	混凝土抗剪键
concrete placement machine	混凝土浇筑机	concrete sheath coat	混凝土涂层
concrete placement machinery	混凝土浇筑机械	concrete sheet piling	混凝土板桩
concrete placer	混凝土浇筑机；混凝土摊铺机	concrete sheetpile breakwater	混凝土板桩防波堤
concrete placing	混凝土浇筑	concrete shelf	混凝土柜架
concrete placing gun	混凝土浇筑器；混凝土喷枪	concrete shell	混凝土薄壳
concrete placing installation	混凝土浇筑设备	concrete shell pile	混凝土薄壳桩
concrete placing plant	混凝土浇筑设备	concrete shell structure	混凝土壳体结构
concrete placing skip	混凝土浇筑斗；混凝土摊铺斗	concrete shrink mixing	混凝土搅拌混凝土收缩搅拌
concrete placing trestle	混凝土施工栈桥	concrete shrinkage	混凝土干缩；混凝土收缩
concrete plant	混凝土搅拌站	concrete skeleton	混凝土骨架
concrete plasticizer	混凝土增塑剂	concrete slab	混凝土板
concrete plinth	混凝土基脚；混凝土基座	concrete slab pavement	混凝土板护面
concrete plug	混凝土塞	concrete slab revetment	混凝土板护坡
concrete pole	水泥电杆	concrete sleeper	混凝土轨枕；混凝土枕
concrete pour works	混凝土浇灌工程	concrete slipforming	滑模法混凝土技术
concrete pouring	混凝土浇筑	concrete slope paving	混凝土铺砌坡
concrete pouring machine	混凝土浇筑设备	concrete slope protection	混凝土护坡
concrete precast element	混凝土预制件	concrete sluice	混凝土节制闸
concrete procedure	混凝土浇筑程序	concrete slump test	混凝土坍落度试验
concrete product	混凝土制品	concrete snow	固结雪
concrete profile barrier	混凝土纵向护栏	concrete spalling	混凝土剥落
concrete property	混凝土特性；混凝土特性	concrete specification	混凝土规范
concrete proportioning	混凝土配比；混凝土配合比	concrete specimen	混凝土试件
concrete protection of slope	混凝土护坡	concrete spiral casing	混凝土蜗壳
concrete pump	混凝土输送泵	concrete splitter	混凝土分离器
concrete pump placing	泵送混凝土浇筑	concrete spouting	混凝土浇筑
concrete pump truck	混凝土泵车	concrete spray	混凝土喷射
concrete pumping	泵送混凝土	concrete sprayer	混凝土喷射机
concrete pump-man	混凝土泵工	concrete spraying	混凝土喷射
concrete rammer	混凝土捣实器	concrete spraying machine	混凝土喷射机
concrete reactor	水泥电抗器	concrete spraying-mixing piling machine units	浆液喷射搅拌成桩机组
concrete re-alkalization	混凝土再碱性化		
concrete rebound	混凝土回弹	concrete spreader	混凝土平仓机；混凝土摊铺机
concrete rebound tester	混凝土弹模测定仪	concrete steel	混凝土钢筋；劲性钢筋
concrete rehandling silo	混凝土再处理罐	concrete stone	混凝土块
concrete reinforcement	混凝土配筋	concrete stopping	混凝土隔墙
concrete reinforcing bars	混凝土用钢筋	concrete strain meter	混凝土应变计
concrete remix kettle	混凝土再拌机	concrete strength	混凝土强度
concrete replacement	混凝土置换	concrete stress	混凝土应力
concrete retarder	混凝土缓凝剂	concrete stress after 28 days	混凝土 28 天强度
concrete retempering	混凝土重塑；混凝土二次搅拌	concrete strip foundation	混凝土条形基础
concrete revetment	混凝土护岸	concrete structure	混凝土结构
concrete ring	混凝土圈	concrete supply	混凝土下料
concrete road	混凝土道路	concrete support	混凝土支架
concrete road paver	混凝土路面铺设机	concrete support plate	混凝土支撑板
concrete roof	混凝土顶板	concrete support pole	混凝土支撑杆
concrete rotary smoother	混凝土旋转平机	concrete surface	混凝土地面
concrete rubble	混凝土碎石	concrete surround	混凝土围绕

concrete tank	混凝土贮水池
concrete technology	混凝土工艺
concrete tension pole	混凝土锚柱；混凝土拉杆
concrete terrazzo	混凝土水磨石；水磨石面混凝土
concrete test	混凝土测试；混凝土试验
concrete tie	混凝土枕
concrete timber	混凝土木材
concrete topping	混凝土覆盖层
concrete train	混凝土浇筑列车
concrete transfer pump	混凝土输送泵
concrete truss	混凝土桁架
concrete tube pile	混凝土管桩
concrete tubular pile	混凝土管桩
concrete unit masonry	混凝土块砌体
concrete vault	混凝土穹顶
concrete vibra column	振动混凝土柱
concrete vibrating machine	混凝土振捣机；混凝土夯实机
concrete vibrating screed	振动式混凝土模板
concrete vibrating stand	混凝土振动台
concrete vibrating tube	混凝土振动棒
concrete vibration	混凝土振捣
concrete vibrator	混凝土振捣器
concrete vibrator equipment	混凝土振捣器
concrete vibratory machine	混凝土夯实机；混凝土振捣器
concrete vibro column pile	振动混凝土柱桩
concrete wall	混凝土墙
concrete walling	混凝土井壁
concrete waterproofing	混凝土防水性
concrete waterproofing compound	混凝土防水剂
concrete water-proofing oil	混凝土防水油
concrete weight-coating	混凝土加重层
concrete work	混凝土工程；混凝土工事
concrete workability	混凝土和易性
concrete works	混凝土工程
concrete-aggressive action	对混凝土侵蚀作用
concrete-block experiment	混凝土试块试验
concrete-block gravity wall	混凝土块重力式墙
concrete-bound	混凝土固结的
concrete-conducting hose	混凝土输送软管
concrete-consistency meter	混凝土稠度计
concrete-filled tube column	混凝土充填管柱
concrete-handling equipment	混凝土处理设备
concrete-lined canal	混凝土衬砌渠道
concrete-lined channel	混凝土衬砌渠道
concrete-lined shaft	混凝土壁立井
concrete-lined tunnel	混凝土衬砌隧道
concrete-mattress revetment	混凝土沉排护岸
concreter	混凝土工
concreter piles	混凝土桩
concrete-rolled dam	碾压混凝土坝
concrete-spliced wood pile	用混凝土套接的木桩
concrete-spouting plant	混凝土灌注设备；混凝土喷射装置
concrete-stave silo	混凝土板制筒仓
concrete-steel building	钢架混凝土建筑
concrete-taper pile	混凝土锥形桩
concrete-timber pile	混凝土木桩
concreting	浇筑混凝土
concreting boom	浇筑混凝土用的桁架；浇筑混凝土起重机臂
concreting in cold weather	混凝土冬季施工
concreting in freezing weather	冻期浇筑混凝土；混凝土冬季施工
concreting in lifts	混凝土分层浇筑
concreting in water	水下浇筑混凝土
concreting method	混凝土浇筑方法
concreting outfit	混凝土施工设备
concreting process	混凝土浇筑过程
concreting program	混凝土浇筑程序；混凝土浇筑计划
concreting rate	混凝土浇筑速率
concreting tower	混凝土浇筑塔
concreting under drilling mud	钻孔泥浆下浇筑混凝土
concreting underwater	水下浇筑混凝土
concreting with tremie method	灌注水下混凝土
concretion	结核；固结作用；凝岩作用；凝结物
concretion nodule	结核
concretion principle	聚结原理
concretionary	结核的；凝结的
concretionary body	结核体
concretionary brown soil	结核棕色土
concretionary gravel	结核状砾石
concretionary limestone	结核状灰岩
concretionary sand	结核状砂
concretionary structure	结核状构造
concretionary texture	结核状结构
concurrence	并发；同时发生
concurrence control	并发控制
concurrency	并行性
concurrent	并发的；共存的；顺流的；一致的；重合的
concurrent computer	并行计算机
concurrent engineering	并行工程
concurrent flow	并行流；平行流
concurrent force	共力点；汇交力
concurrent gas-liquid flow	气液混合流
concurrent glaciation	共存冰川作用
concurrent line	共点线；同时线
concurrent operation	并行运算
concurrent processor	并行处理程序
concurrent production	同时开采
concurrent twist	同向捻
concurrent waterflooding	同时注水
concussion	冲击；震动
concussion blasting	松动爆破；震动性放炮
concussion crack	冲击裂缝
concussion fracture	冲击破裂
concussion of blasting	爆破冲击作用；爆破震动
concussion table	振动式倾斜摇床
concyclic	共圆
condeep platform	水下混凝土平台
condeep-type structures	康迪普型结构
condemnation deformation spectrum	倒毁变形谱
condemnation threshold earthquake	倒毁阈限地震
condensability	可压缩性
condensable	可冷凝的；可压缩的
condensable gas	可凝气体
condensable gasoline	可冷凝汽油
condensate	冷凝液
condensate bottle	冷凝瓶
condensate drain	冷凝液排出管

condensate field	凝析气田	condenser gauge	冷凝器计
condensate fractionator	凝析油分馏塔	condenser lens	聚光透镜
condensate gas reservoir	凝析气藏	condenser pipe	冷凝管
condensate knockout vessel	凝析油分离罐	condenser reactance	容抗
condensate oil	凝析油	condenser tube	凝器管
condensate oil reservoir	凝析油藏	condenser type blasting machine	电容式放炮器
condensate outlet	冷凝液出口	condensing agent	冷凝剂
condensate petroleum accumulation	凝析油气藏	condensing coil	冷凝盘管
condensate polisher	冷凝液纯化槽	condensing lens	聚光透镜
condensate pump	冷凝水泵	condensing rate	冷凝速度
condensate recovery	凝析油开采	condensing steam drive	凝结蒸汽驱
condensate reserve	凝析油储量	condensive	电容的
condensate return	冷凝水回流	condie	废料；废物；矸石
condensate well	凝析油气井	condiment	附加物
condensation	表面结露；冷凝作用；凝结作用；浓缩；缩合作用	condition equation of relative control	相对控制条件方程
condensation aerosol	凝聚型气溶胶	condition for closing the horizon	圆周角条件
condensation area	凝结面积	condition for slip on weak plane	软弱面滑动条件
condensation catalyst	缩合催化剂	condition for sum of angles	组合角条件
condensation cycles	冷凝循环	condition for validity	有效性条件
condensation gas drive	凝析气驱	condition monitoring	状态监测
condensation heat	凝结热	condition of convergence	收敛条件
condensation hypothesis	凝聚假说	condition of equilibrium	平衡条件
condensation in dynamics	动力凝聚	condition of groundwater occurrence	地下水埋藏条件
condensation in micropores	微孔内的冷凝	condition of groundwater recharge	地下水补给条件
condensation level	凝结高度；凝结面	condition of loading	荷载条件
condensation method	缩合法	condition of plasticity	塑性状态
condensation nucleus	凝结核	condition of rectification	矫正条件
condensation on a surface	表面上发生的冷凝	condition of rest	静止状态
condensation polymer	缩聚物	condition of shear strength	剪切强度条件
condensation process	冷凝过程	condition of similarity	相似条件
condensation product	冷凝物；缩合产品	condition of stability	稳定性条件
condensation rate	凝结速率	condition of steady flow	稳定流状态
condensation ratio	聚合比	condition of strength	强度条件
condensation recharge	凝结水补给	conditional	受限制的；有条件的
condensation return pump	冷凝回水泵	conditional instability	条件性不稳定性
condensation temperature	凝结温度	conditional reserves	暂定储量
condensation theory	凝结理论；冷凝说	conditional simulation	条件模拟
condensation trail	凝结痕迹	conditioned	有条件的，受制约的
condensation water	冷凝水	conditioner	搅拌筒；条件槽
condensation zone	冷凝带	conditioning agent	调节剂
condensational wave	密波；凝聚波	conditioning fluid	处理液
condense	聚光；冷凝；浓缩	conditioning hole	改善井孔状态；修整钻孔
condense gas	凝析气体	conditioning pulp	调和矿浆
condense oil	凝析油	conditions for stress and displacement distributions	应力和位移分布条件
condensed bed	凝结层	conduct pipe	导管；管道
condensed deposit	凝聚沉积物	conductance	传导性；电导性；热导性
condensed film	凝聚层；缩合膜	conductance measurement	电导率测定
condensed fluid	冷凝液	conductance ratio	传导比
condensed gas	冷凝气体	conductance ratio of well pattern	井网传导率
condensed liquid	凝析液	conductance titration	电导滴定
condensed monolayer	凝聚单层	conductimetric method	电导率测定法
condensed nucleus	稠核	conducting	传导的
condensed oils	稠合油	conducting bridge	电桥
condensed phase	凝聚相	conducting drain	主要排水管
condensed polymer	缩聚物	conducting layer	传导层
condensed section	浓缩段	conducting orebody	导电矿体
condensed sequence	缓慢堆积层序；压缩层序	conducting probe	导电探示器
condensed succession	缓慢堆积层序；凝聚层序	conducting stratum	传导层
condenser	电容器；聚光器；冷凝器	conducting surface	传导面

conducting-paper analogy　导电纸模拟
conduction　传导性
conduction resistance　传导电阻
conductive bed　低阻层；传导层
conductive coating　导电涂层
conductive coupling　电导耦合
conductive discharge　传导放电；电阻放电
conductive drilling mud　导电泥浆
conductive formation　低阻层
conductive heat flow　传导型热流
conductive heat loss　导热损失
conductive heating　传导加热
conductive host　导电围岩
conductive medium　传导介质
conductive mineral　导电矿物
conductive overburden　导电覆盖层
conductive shale　导电页岩
conductive solid　导电固体
conductive zone　低阻层
conductivity　导电性；传导性
conductivity anomaly　电导率异常
conductivity coefficient　传导系数
conductivity log　感应测井
conductivity logging　电导率测井
conductivity loss　电导率损耗
conductivity meter　电导仪
conductivity modulation　电导率调制
conductivity of crystals　结晶的传导性
conductivity of medium　介质传导性
conductivity water　导电水
conductometer　导电计；导热计
conductometric　测量导热率的
conductometry　电导测定法
conductor　避雷针；导体；导向管
conductor box　井口木制导管
conductor bracing　导管支撑
conductor cable　电缆
conductor casing　导引套管
conductor hole　导管井孔
conductor housing　导管头
conductor installed in conduit　导线穿管敷设
conductor line　电缆
conductor pipe　导向管；导向套管
conductor rail support　导轨支撑
conductor sag　导线垂度
conductor string　导管柱
conductor wire　缆芯线
conductor-piling　导管内打桩
conduit　风道；火山口；预应力丝孔道；导管；沟渠；导水管
conduit bowl　电缆管道分线盒
conduit flow　管流
conduit jacking　顶管法；顶管装置
conduit joint　管道接头
conduit pipe　导管；管道
conduit pit　管道坑
conduit run　电缆管道路线
conduit section　管道断面
conduit slope　管道坡度
conduit spring　管泉

conduit system　地下管道系统；地下管道系统
cone　锥状地形；海底锥形冲积扇；圆锥破碎机；圆锥选煤机
cone anchorage device　锥形锚具
cone backface　钻头牙轮底平面
cone balance method　漏斗均衡法
cone bearing capacity　圆锥承载力
cone bearing test　圆锥承重试验
cone bit　牙轮钻头；锥形钻头
cone bit design layout　牙轮钻头啮合图
cone calorimeter　锥形量热仪
cone classifier　圆锥分级机
cone contour　锥形牙轮外形
cone crusher　圆锥破碎机；圆锥式碎石机；圆锥破碎机
cone crushing　圆锥破碎机破碎
cone cut　锥形掏槽
cone cutterhead　锥面刀盘；锥面切削头
cone cutting　锥形掏槽
cone delta　锥状三角洲
cone dewaterer　锥形脱水机
cone dike　锥状岩墙；锥形岩脉
cone distance　锥顶距
cone drop test　落锥试验
cone feeder　锥形给料器
cone fisher　牙轮打捞器
cone foundation　锥形基础
cone fracture　圆锥形裂缝
cone friction coupling　锥形摩擦离合器；锥形摩擦联轴节
cone friction drum　锥形摩擦鼓轮
cone friction gear　摩擦减速器；锥形摩擦传动装置
cone gear　锥形齿轮
cone geyser　锥式间歇泉
cone grate　锥形格条
cone in cone　叠锥
cone in cone structure　叠锥构造
cone in cone volcano　叠锥火山
cone index gradient　锥尖指数梯度
cone intrusion　锥状侵入体
cone karst　漏斗形喀斯特；锥状喀斯特
cone nose　牙轮锥顶
cone of dejection　冲积锥；洪积锥
cone of depression　沉陷锥；下降漏斗
cone of detritus　冲积锥；洪积锥；岩屑锥
cone of eruption　喷发锥
cone of exhaustion　降落漏斗；下降漏斗
cone of ground water depression　地下水下降漏斗
cone of groundwater influence　水位降落漏斗
cone of influence　降落漏斗；影响漏斗
cone of influence of well　井的浸润漏斗
cone of pressure relief　等测压面下降漏斗；减压锥
cone of radiation　辐射锥
cone of recharge　补给漏斗
cone of salt water　盐水漏斗
cone of transition　过渡锥
cone of uncertainty　不定度圆锥
cone of water table depression　水位下降漏斗
cone offset　牙轮轴移
cone overhang　超顶
cone packer　锥形封隔器
cone penetration　锥体贯入度

cone penetration impact test	锥击试验
cone penetration method	圆锥贯入法
cone penetration resistance	圆锥贯入阻力
cone penetration test	圆锥静力触探试验
cone penetration test with pore pressure measurement	孔压静力触探试验
cone penetrometer	圆锥触探仪
cone penetrometer for liquid limit test	锥式液限仪
cone pin	锥形销
cone pinion	小伞齿轮
cone pulley	塔轮；锥形轮
cone quartering method	堆锥四分缩样法
cone resistance	圆锥触探头阻力
cone retaining ring	牙轮挡圈
cone rheology measurement	锥板流变性测量法
cone rock bit	锥形牙轮钻头
cone roll piercing mill	菌式轧辊穿孔机
cone roof	锥形顶
cone roof reservoir	锥顶油罐
cone roof tank	锥顶罐
cone rope socket	锥形电缆座
cone sampler	圆锥式试样缩分器
cone section	圆锥剖面
cone separator	圆锥分选机
cone settler	锥形沉淀箱
cone shaped wedges	锥形夹具
cone sheet	锥状岩席
cone shell	牙轮壳体
cone surface	锥面
cone test	锥体坍落度试验；锥探试验
cone thickener	锥形浓缩机
cone type roller bearing rock bit	滚柱轴承锥形牙轮钻头
cone valve	锥形阀
cone wedge	锥形楔
cone-and-plate sensor system	锥板传感系统
cone-and-plate viscometer	锥板黏度计
cone-and-plate-rheometer	锥板流变仪
cone-and-sleeve attachment	锥体套筒联接
cone-faced joint	锥面连接
conehead	锥头
cone-in-cone limestone	叠锥灰岩
cone-jet mixer	漏头射流式搅拌器
conelet	小锥
cone-like karst	锥状岩溶
cone-like of dejection	冲积锥；洪积锥
cone-like of depression	降落漏斗
cone-like penetrometer	锥状贯入器
cone-like rock bit	牙轮钻头
cone-like sheet	锥状岩席
cone-like volcanic edifice	锥状火山机构
cone-pulley lathe	塔轮皮带车床
cone-shaped anchorage	锥形锚
cone-shaped body	锥形体
cone-shaped shear zone	锥形剪切带
conette type viscometer	科内特型黏度计
conferva coal	丝状藻泥炭
conferva peat	丝状藻泥炭
confidence surface	置信面
configuration	地形；构造；结构；排列形式；原生层理
configuration line	轮廓线
configuration of ground	地形
configuration of the openings	巷道外形
configurational energy	配位能
confined	有侧限的；侧限的
confined aquifer	层间含水层；承压含水层；封闭含水层
confined area	封闭区；约束面积
confined bed	承压层；封闭层；隔水层
confined blasting	约束爆破
confined brine	承压卤水
confined compression strength	侧限抗压强度
confined compression test	侧限压缩试验
confined compressive strength	侧限抗压强度
confined concrete	侧限混凝土；约束混凝土
confined detonation velocity	密闭爆速
confined flood pattern	封闭井网
confined flow	承压水流；密闭流量
confined gap test	密闭殉爆试验
confined geothermal reservoir	圈闭式地热储
confined grain diameter	限制粒径
confined groundwater	承压地下水
confined hot-water system	承压型热水系统
confined injection	填塞注浆
confined jet	受限射流
confined karstic aquifer	承压岩溶含水层
confined length	约束长度
confined lid	盖层；圈闭层
confined mode	前端密闭模式
confined nonleaky aquifer	非越流承压含水层
confined pressure	侧限压力；封闭压力；围压
confined pressure effect	围压效应
confined pressure seepage flow	有压渗流
confined pumping test	承压抽水试验
confined reservoir	地下水库；封闭储水层
confined scar	封闭断崖
confined seepage	承压渗透
confined space	有限空间
confined stress	侧限应力
confined type geothermal reservoir	圈闭式地热储
confined water	层间水；承压水；自流水
confined water basin	承压水盆地
confined water head	承压水头
confined well	承压井
confinement	闭塞；封闭；约束
confinement pressure	侧限压力；围压
confinement zone	约束区
confining bed	不透水层；封闭层；隔水层
confining bed of aquifer	封闭含水层
confining boundary	隔水边界
confining fluid pressure	封闭液压
confining force	封闭力
confining layer	不透水层；封闭层；隔水层
confining load	围岩负荷
confining overlying bed	隔水顶层
confining pressure	侧限压力；围压
confining pressure effect	围压效应
confining reinforcement	约束钢筋
confining stratum	不透水层；封闭地层；隔水层
confining stress	侧限应力
confining underlying bed	隔水底层
confining unit	隔水层

confining wall	围墙；药室壁
confining water level	承压水头
confining zone	封闭带
confirmation well	证实井
confirmatory measurement	核实测量
confirmed test	确认性试验
confix	连接牢固
conflagration	爆发；大火；快速燃烧
conflow	合流；汇流
conflow gauge	康弗洛表
conflow meter	康弗洛流量计
conflow stop valve	康弗洛截止阀
confluence analysis	合流分析
confluence basin	汇流盆地
confluence ice	汇流冰
confluence monitoring	汇流监测
confluence of rivers	汇流点
confluence plain	汇流平原
confluent	汇合的；汇流的；支流
confluent streams	合流河
confocal ellipsoid	共焦椭球
conformability	整合性；适应性；一致性
conformable bedding	整合层理；整合地层
conformable contact	整合接触
conformable fault	整合断层
conformable fold	整合褶皱
conformable intrusion	整合侵入
conformable reservoir volume	可波及的储层体积
conformable strata	整合地层
conformable stratification	整合层理
conformal	保角的；正形的
conformal chart	正形投影地图
conformal conic projection	保形圆锥投影
conformal cylindrical projection	保形圆柱投影
conformal double projection	保形双重投影
conformal map	保角映像
conformal mapping	保角变换；保角映射；正形映像
conformal projection	保角映射；等角投影
conformal representation	保角映像；保形变换
conformal transformation	保角变换
conformality	正形性
conformance	一致性
conformance testing	一致性测试
conformation	构造
conformity	一致；整合；整合接触
conformity contact	整合接触
confriction	摩擦力
confucius reef	丘礁
confused sea	九级风浪；浪涛汹涌的海面
confused swell	暴涌；九级涌浪；乱涌
congeal	冻结；冷冻测温法；凝结
congealed ground	冻土
congealed lava crust	凝结熔岩壳
congealing point	冻结点
congealing process	冷凝过程；凝结酌
congelation	冻结物；凝结物
congelifraction	冻劈作用；寒冻风化；融冻崩解作用
congeliturbate	融冻堆积物
congeliturbation	融冻泥流；融冻泥流作用
congener	同种类的；同族元素
congeneric	同源的；同族的
congenetic granite	同源花岗岩
congenetic rock	同源岩
congeniality	同性质
congeries	堆积；聚集
congested	密集的
congestion	聚积；填充；阻塞
congestion of ore	矿石堵溜井；矿石聚积
congestion of reinforcement	钢筋密集
congl	砾岩
conglobate	球形的；圆的
conglobation	团聚
conglomerate	合成物；聚合物；砾岩
conglomerate fan	砾石扇
conglomerate quartz	石英砾岩
conglomerate reservoir	砾岩储集层
conglomerate rock	砾岩
conglomerate stratum	卵石层
conglomerate test	砾石检验；砾石试验
conglomerated pack	凝聚浮冰
conglomerate-gneiss	砾石片麻岩
conglomerates	砾岩
conglomeratic	含砾的；砾岩状的
conglomeratic limestone	砾状灰岩
conglomeratic sand	含砾砂层
conglomeratic sandstone	砾砂岩
conglomeration	堆集；聚集；凝聚；团块
conglomerite	变砾岩
conglutinate	愈合；黏住
conglutination	固结；胶结
congressite	淡粗霞岩
congruence	叠合；全等；适合；一致
congruent	叠合的；符合的；全等的；一致的
congruent dissociation	一致离解
congruent dissolution	一致溶解
congruent drag fold	同斜拖曳褶皱
congruent forms	左右相反形
congruent melting	一致熔融
congruent melting point	一致熔融点
congruent pattern	叠合图案
congruent solution	一致溶液
congruity	一致
congruous	褶皱同斜的；一致的
congruous drag fold	同斜拖褶皱
congruous fold	同斜褶皱
conic	圆锥曲线；圆锥的；圆锥
conic chart	圆锥投影地图
conic cylindrical viscometer	锥筒黏度计
conic fold	圆锥形褶皱
conic node	锥顶
conic quartering	四分取样法
conic reducer	锥齿轮减速器
conic section	圆锥截线；二次曲线
conical	锥体；锥形的
conical ball mill	锥形球磨机
conical beaker	锥形烧杯
conical bearing	锥形轴承
conical bottom tank	锥底罐
conical bushing	锥形轴衬
conical contact	锥形接触

conical cowl	锥型风帽	coniosis	尘肺
conical crater	圆锥形爆破漏斗	conjecture	假设；推测
conical cut	锥形掏槽	conjugate	共轭的；联合的
conical cutters	锥形牙轮	conjugate angle	共轭角
conical depression	降落漏斗；锥形洼地	conjugate beam method	共轭梁法
conical die	锥孔模	conjugate continental margin	共轭陆缘
conical dry blender	锥形干混器	conjugate curve	共轭线
conical dune	锥形沙丘	conjugate diameters	共轭直径
conical equidistant projection	等距圆锥投影	conjugate direction search method	共轭方向搜索法
conical fit	圆锥连接配合	conjugate displacement	共轭位移
conical fold	锥形褶皱	conjugate faults	共轭断层
conical fracture	圆锥状破裂	conjugate fissures	共轭裂缝
conical gear	锥齿轮	conjugate fold	共轭褶皱
conical helix	锥形螺旋线	conjugate fracture system	共轭破裂系
conical hill	锥形丘	conjugate fractures	共轭破裂
conical hillock	圆锥形小丘	conjugate gradient	共轭梯度
conical lens	圆锥透镜	conjugate gradient algorithm	共轭梯度算法
conical mill	锥形球磨机	conjugate gradient method	共轭梯度法
conical orthomorphic projection	正形圆锥投影	conjugate joints	共轭节理
conical pendulum	锥形摆	conjugate kink	共轭膝折
conical penetrometer	圆锥触探器	conjugate pressure	共轭压力
conical pick	圆锥形截齿；锥形镐	conjugate set of joints	共轭节理组
conical pile	锥形桩	conjugate shear fault	共轭剪切断层
conical pinion	小伞齿轮	conjugate shear fracture	共轭剪切裂缝
conical plug	锥形塞	conjugate shear pattern	共轭剪切模式
conical projection	锥形投影	conjugate shear planes	共轭剪切面
conical ring	锥形环	conjugate shear zone	共轭剪切带
conical roller bearing	锥形滚柱轴承；圆锥滚柱轴	conjugate sheer element	共轭剪切要素
conical screen	锥形棒条筛	conjugate slip	共轭滑移
conical scrubber	锥形滚筒擦洗机	conjugate stage	共轭水位
conical section	圆锥剖面	conjugate stress	共轭应力
conical slope	锥坡	conjugate structure	共轭构造
conical spring	圆锥螺旋形弹簧；锥形弹簧	conjugate system	共轭系
conical striker	圆锥形冲头	conjugate veins	共轭脉
conical tank	锥形顶储罐	conjugate zone	共轭带
conical trommel	锥形洗矿筒	conjugated compound	共轭化合物
conical tube	锥管	conjugated faults	共轭断层
conical volcano	锥形火山	conjugated fissure	共轭裂缝
conical wall niche	锥形凹壁	conjugated fracture	共轭断裂
conical wave	锥面波；锥形波	conjugated joint	共轭节理
conical wing valve	锥形节流阀	conjugated kink band	共轭扭折带；共轭膝折带
conical-bottom bin	锥形底仓	conjugated structure	共轭构造
conical-drum hoist	锥形滚筒提升机	conjugated vein	结合脉
conical-tower platform	锥塔形平台	conjugation fold	共轭褶皱
conichalcite	砷钙铜矿	conjunct	混合的
conicity	圆锥度；锥削度	conjunction	接头；连接；联合
conicograph	二次曲线规	conjunction gate	与门，接合闸
coniform	锥形	conjunction of tectonic systems	构造体系的联合
conimeter	空中悬浮矿尘测量器	conjunctive water use	地表地下水联合应用
coning	堆锥四分取样法；锥进	Conklin process	康克林选煤法
coning and quartering sample	锥堆四分法缩分的试样	connate	同生的；同源的；先天的；原生的
coning angle	圆锥角	connate deposit	原生沉积
coning behavior	锥进动态	connate fluid	原生流体
coning characteristic	锥进特征	connate formation water	原生地层水
coning control	锥进控制	connate salts	共生盐
coning curve	锥进曲线	connate water	封存水；古水体；原生水
coning effect	锥进效应	connate water saturation	原生水饱和度
coning model	锥进模型	connect in parallel	并联
coning quartering	四分取样法	connect in series	串联
coning simulation	锥进模拟	connect time	连接时间

connected drainage pattern	连通式水系
connected graph	连通图
connected pore channel	连通孔道
connected porosity	连通孔隙率
connected space	连通空间
connected to ground	接地的
connected yoke	连接横木；连接夹；轭架链接
connectedness	连通性
connecting	套管
connecting angle	连接角钢
connecting bar	道岔连接杆；连接杆
connecting beam	结合梁；连接梁
connecting bolt	连接螺栓
connecting bowl	电缆接头箱；接线盒
connecting bridge	联络桥；天桥
connecting busbar	连接母线
connecting commutator	连管；换向器链接
connecting cross-cut	连接两个井筒的石门
connecting device	连接装置
connecting drift	连通道
connecting experiment	连通试验
connecting flange	连接法兰
connecting frame	连接架
connecting hose	连接软管
connecting link	连接杆
connecting mole	连接防波堤
connecting nut	连接螺母
connecting pipe	连接管
connecting reducer	异径接头
connecting roadway	联络巷道
connecting rod	活塞杆；连杆
connecting rod pin	连杆销
connecting shaft	连轴
connecting slot	联络道
connecting strip	连接片；连接条
connecting survey allowable error	贯通测量容许误差
connecting survey through raise	天井联系测量
connecting terminal	夹具；接头；接线端子
connecting test	连通试验
connecting traverse	附合导线
connecting-tripping device	脱接器
connection	联络巷道
connection band	结合箍；连接带
connection between front shield and rear shield	前后护盾的连接
connection block	接线板；十八斗
connection cable	连接缆索
connection clip	结合扣
connection detail	连接细部
connection eccentricity	接头偏心度
connection flange	连接法兰
connection gang	管道安装队
connection gas	渗入的天然气
connection of freezing column	冻结壁合拢；冻结壁交联
connection plate	连接板
connection point for shaft orientation	定向连接点
connection rebars	连接钢筋
connection rod bearing	连杆轴承
connection sheet pile	连接板桩
connection survey for shaft orientation	定向连接测量
connection survey in mining panel	采区联系测量
connection weight	连接权值
connectivity	连接性；连通性
connectivity structure	连接结构
connector	接线夹；连接器
connector bend	管子弯头
connector lug	接线衔套；连接接头
connector pan	连接斜槽
connector plug	连接插头
connector sub	连接器
connector well	连接井
connellite	绿盐铜矿
conny	碎煤
conodont	牙形刺；牙形化石
conodont animal	牙形动物
conodont assemblage	牙形石集合
conoid	壁锥曲面；圆锥体；圆锥形
conoidal dam	锥形支墩坝
conoscope	锥光镜；锥光偏振仪
conoscopic	锥光镜
Conrad discontinuity	康拉德不连续面；康拉德界面
Conrad layer	康拉德层；下地壳层
Conrad machine	管式挖坑机
Conradson carbon residue	康拉特逊残碳值
conrotatory	顺旋
consanguineous	同源的
consanguinity	同源；岩浆同源
consanguinity association	同源组合
consecration ceremony	开光典礼
consecution	连贯
consecutive	连贯的；连续的；依次相继的
consecutive firing	顺序爆破
consecutive operation	连续运转
consecutive order	连续顺序
consecutive point	相邻点
consecutive reaction	连锁反应
consecutive shock	连续地震
consecutive terrace	连续阶地
consecutive-range zone	接续延限带
consedimental	同沉积的
consenco table	康森科型摇床
consequence analysis	后果分析
consequence of earthquake	地震后果
consequence of failure	崩塌后果；破坏后果
consequence-based design	后果设计；基于后果的设计
consequence-based seismic design	基于后果的抗震设计
consequent bedding rockslide	顺层岩质滑坡
consequent coast	顺向海岸
consequent divide	顺向分水岭
consequent drainage system	顺向水系
consequent fall	顺向瀑布
consequent fault	顺向断层
consequent lake	顺向湖泊
consequent landslide	顺层滑坡
consequent layered accumulation body	顺层层状堆积体
consequent relief	顺地形
consequent ridge	顺向山脊
consequent river	顺向河流
consequent shoreline	顺向滨线
consequent stream	顺向河；顺向河

consequent valley	顺向谷
consertal	缝合的；有缝的
consertal fabric	等粒岩组
conservancy	水土保持；资源保护区
conservation	守恒；保存
conservation and collection of water	水的保护与采集
conservation area	资源保护区
conservation district	保护区
conservation farming	保持水土耕种
conservation master plan	保护规划
conservation measure	保护措施
conservation of angular momentum	角动量守恒
conservation of cultural heritage	文化遗产保护
conservation of historical monuments	历史古迹保护
conservation of momentum	动量守恒
conservation of monuments	古迹保护
conservation of natural resources	自然资源保护
conservation of stone	石材保护
conservation practice	保护工作
conservation principle	守恒原理
conservation process	保护程序
conservation storage	蓄水库容；有效库容
conservation tillage	水土保护耕作
conservative	保守的
conservative design	保守设计
conservative element	保守元素
conservative factor	保险系数
conservative margin	保守边缘
conservative plate boundary	稳定板块边界
conservative plate margin	保守板块边缘
conservative slope	平缓坡
conservator	油枕
conserve	保存
conserving agent	除腐剂
conshelf	大陆架
consideration	困难地区补贴金
consistence limit	稠度界限
consistence state	稠度状态
consistency	稠度；连续性；密实度；相容性
consistency check	相容性检查
consistency coefficient	稠度系数
consistency condition	相容条件
consistency controller	稠度控制器
consistency criterion	一致性标准
consistency factor	稠度系数
consistency gauge	稠度计
consistency index	稠度指数
consistency indicator	稠度指示器
consistency limit	稠度界限
consistency meter	稠度计
consistency of bitumen	沥青稠度
consistency of soil	土的稠度
consistency test	稠度试验
consistent	一贯的；始终如一的
consistent condition	恒定条件
consistent convergence	一致收敛
consistent estimation	相容估计
consistent lubricant	固体润滑剂；黄油；润滑脂肪
consistent packing	坚实充填
consistometer	稠度计
console	控制台
console control desk	落地式控制台
console power	控制台电源
console section	操纵部分
console switch	操作台开关
console timbering	悬臂支架
consolette	小型控制台
consolidate	固结；加固；压密
consolidate by injection	喷射加固
consolidated	固结的；压实的
consolidated anisotropically drained test	各向不等压固结排水试验
consolidated anisotropically undrained test	各向不等压固结不排水试验
consolidated aquifer	固结含水层
consolidated balance sheet	合并资产负债表
consolidated deposit	固结的沉积；固结的矿床
consolidated depth	固结深度
consolidated drain condition	固结排水条件
consolidated drained direct shear test	固结排水直剪试验
consolidated drained shear test	固结排水剪切试验
consolidated drained test	固结排水试验
consolidated drained triaxial compression test	固结排水三轴压缩试验
consolidated drained triaxial test	固结排水三轴试验
consolidated fill	胶结充填
consolidated formation	胶结地层；压实地层
consolidated gravel	胶结砾石
consolidated gravel pack	胶结砾石充填
consolidated grouting	固结注浆
consolidated ice cover	固结冰盖层
consolidated isotropically undrained test	各向等压固结不排水试验
consolidated pack material	树脂砂浆充填材料
consolidated pack squeeze	树脂砂浆充填挤压
consolidated profit and loss	总损益
consolidated quick compression test	快速固结压缩试验
consolidated quick direct shear test	固结快剪试验；固结快剪直剪试验
consolidated quick shear strength	固结快剪强度
consolidated quick shear test	固结快剪试验
consolidated quick shear value	固结快剪值
consolidated quick shearing reslstance	固结快剪强度
consolidated quick sheer test	固结快剪试验
consolidated quick test	固结快剪试验
consolidated rig	强力钻机
consolidated rock	固结岩
consolidated rock aquifer	固结岩石含水层
consolidated sand	固结砂
consolidated sandstone	固结砂岩
consolidated sediment	固结沉积物
consolidated settlement	固结沉降
consolidated slow compression test	慢速固结压缩试验
consolidated slow direct shear test	固结慢剪直剪试验
consolidated slow shear test	固结慢剪试验
consolidated soil	固结土
consolidated subsoil	加固地基
consolidated thickness	固结厚度
consolidated triaxial test	固结三轴压缩试验
consolidated undrained compression test	固结不排水压缩

试验
consolidated undrained quick triaxial test 固结不排水三轴快压缩试验
consolidated undrained shear test 固结不排水剪切试验
consolidated undrained test 固结不排水试验
consolidated undrained triaxial compression test 固结不排水三轴压缩试验
consolidated undrained triaxial test 固结不排水三轴试验
consolidated volume 固结体积
consolidating fluid 胶结液
consolidating material 胶结材料
consolidating pile 强化桩；加固桩
consolidating stratum 固结地层
consolidation 沉积；固结；加固；胶结
consolidation action 固结作用
consolidation apparatus 固结仪；渗压仪
consolidation by dewatering 排水固结法
consolidation by electroosmosis 电渗加固
consolidation by increasing load 加荷加固
consolidation by injection 注浆固结
consolidation by preloading and drainage 预压排水固结
consolidation by the vacuum method 真空固结
consolidation by vertical drain 竖向排水加固
consolidation by vibration 振动压密
consolidation cell 固结盒
consolidation characteristic 固结特性
consolidation chemical 胶结剂
consolidation coefficient 固结系数
consolidation condition 固结条件
consolidation curve 固结曲线
consolidation deformation 固结变形
consolidation degree 固结度
consolidation device 固结设备
consolidation due to desiccation 干燥引起的固结
consolidation effect 加固效果
consolidation failure 固结破坏
consolidation grouting 固结灌浆
consolidation holes 固结灌浆孔
consolidation line 固结曲线
consolidation load 同结荷载
consolidation method 固结法
consolidation model 固结模型
consolidation of collapsing soil 湿陷性土地基加固
consolidation of earth dam 土坝固结；土坝压实
consolidation of expanding soil 膨胀土地基加固
consolidation of inhomogeneous ground 不均匀地基加固
consolidation of karst 岩溶地基加固
consolidation of liquefiable sandy soil 液化砂土地基加固
consolidation of mud 泥浆固结
consolidation of rock masses by cades 岩体锚索加固
consolidation of rock masses by grouting 岩体注浆加固
consolidation of rock masses under load 载荷作用下的岩体加固
consolidation of shaft wall 固井
consolidation of slope 斜坡加固
consolidation of soft foundation 软弱地基加固
consolidation of soft subsoil 软基加固
consolidation of soil 土的固结
consolidation parameter 固结参数
consolidation phenomenon 固结现象
consolidation pile 加固桩
consolidation pressure 固结压力
consolidation process 固结过程
consolidation radius 胶结半径
consolidation rate 固结速率
consolidation ratio 固结比
consolidation sedimentation 固结沉淀作用
consolidation settlement 固结沉降
consolidation strength 胶结强度
consolidation stress 固结应力
consolidation technique 固结技术
consolidation test under constant loading rate 等速加荷固结试验
consolidation test under constant rate of strain 等应变速率固结试验
consolidation test under controlled gradient 控制比降固结试验
consolidation theory 固结理论
consolidation time 固结时间
consolidation under constant stress ratio 等应力比固结
consolidation under k0 condition K0 固结
consolidation without axial strain 无轴向应变固结
consolidation yield stress 固结屈服应力
consolidometer 固结仪；渗压仪
consolute 共溶性的；会溶质的
constancy 恒定性
constancy of angle 面角守恒
constancy of composition 成分恒定
constant 常量；固定的；恒定的
constant acceleration 常加速度
constant acceleration method 常加速度法
constant amplitude sine-wave train 等幅正弦波列
constant background velocity 常背景速度
constant boiling mixture 恒沸混合物
constant concentration 固定含量
constant control valve 恒定控制阀
constant creep 恒速蠕变
constant cross-section 等截面
constant curvature 常曲率
constant deflection line method 常变位线法；常挠曲线法
constant dimension column 等形柱
constant discharge 恒定流量；恒定排量
constant discharge pumping test 定流量抽水试验
constant displacement pump 均匀送料量泵；定量泵
constant drawdown 定降深
constant drawdown pumping test 定降深抽水试验
constant duration 等时间间隔
constant entropy 等熵
constant fault 固定性故障
constant flow 稳定水流
constant flow control valve 恒流量控制阀
constant flow injection pump 恒流量注入泵
constant flow rate 恒定流量
constant force 恒力
constant geometric accuracy contours 等几何精度线
constant gradient test 等梯度固结试验
constant gradient velocity profile 等梯度速度剖面
constant head 常水头
constant head field test 常水头现场渗透试验
constant head permeability test 常水头渗透试验

constant head permeameter	常水头渗透仪
constant head test	常水头试验；恒定水头试验
constant humidity	恒湿
constant hydraulic gradient consolidation test	常水力梯度固结试验
constant hydrostatic head device	静水头调节装置
constant lead screw	等螺距螺杆
constant level device	恒液面装置
constant loading rate consolidation test	等速加荷固结试验
constant of earthquake epicentre	震中常数
constant of gravitation	引力常数
constant of gravity	重力常数
constant offset	恒定炮检距
constant parameter linear system	常系数线性系统
constant phase front	等相位面
constant phase shift	恒定相移
constant pitch	等螺距
constant pressure	恒压
constant pressure check valve	定压单流阀
constant pressure device	恒压装置
constant pressure line	等压线
constant pressure map	等压图
constant pressure shear	等压剪切
constant pressure surface	恒压面
constant pressure valve	定压阀
constant proportion	定比
constant quantity	常量
constant rate	不变速率；常产量；恒定流量
constant rate filtration	恒速过滤
constant rate loading	等速加荷
constant rate of flow	恒定流量
constant rate of loading test	等速加载试验
constant rate of penetration test	等速贯入试验
constant rate of uplift test	等速上拔试验
constant rate penetration test	等速贯入试验
constant ratio code	定比码
constant ratio of equivalent damping	等值衰减定比
constant return	固定收益率
constant scanning	等速扫描
constant section	等截面
constant shear strain	等剪应变
constant slope line	等坡度线
constant spectral density	恒定谱密度
constant speed	等速
constant speed drive	常速驱动
constant speed motor	恒速电动机
constant strain rate consolidation test	等应变速率固结试验
constant temperature	恒温
constant temperature zone	恒温层
constant tension winch	等张力绞车
constant throughput	恒定输送量
constant time slice	等时间切片
constant value	恒定值
constant velocity	等速；恒速
constant velocity migration	常速偏移
constant velocity stacks	常速叠加
constant vibration	恒定震动
constant voltage	恒定电压
constant volume combustion	定容燃烧
constant volume reservoir	等容储集层
constant volume sampler	定容取样器
constant volume shear test	等体积剪切试验
constant volume test	等体积试验
constant water content shear test	常含水量剪切试验
constant wave	等幅波
constant weight	不变权数；恒重
constant zone of subsurface temperature	恒温层
constant-amplitude carrier	恒振幅载波器
constant-amplitude stress cycle	等幅应力循环
constantan	康铜
constant-closed	常闭的
constant-flow lubrication	恒流润滑
constant-force-amplitude excitation	恒力幅激振
constant-frequency source	固定频率震源
constant-gain seismogram	固定增益地震记录
constant-level lubrication	恒定油位润滑
constant-porosity line	等孔隙度线
constant-potential generator	定压发电机
constant-pressure test	定压测试
constant-pressure type curve	定压型曲线
constant-rate drawdown test	保持产量恒定的压降试井
constant-rate evaporation	恒速蒸发
constant-rate pumping test	恒速抽水试验
constant-separation traversing	固定极距剖面法
constant-velocity gather	常速汇集
constant-velocity prestack migration	常速叠前偏移
constant-velocity scan	等速扫描
constituent	成分；要素
constituent element	组元
constitution diagram	组合图
constitution of earth	地球的组成
constitution water	化合水；结构水
constitutional diagram	相图
constitutional equation	本构方程
constitutive behaviour	本构特性；本构行为
constitutive equation	本构方程
constitutive law	本构定律
constitutive model	本构模型
constitutive relation	本构关系
constitutive surface	构造面
constrained	抑制的
constrained diameter	限制粒径
constrained grain size	限制粒径
constrained inversion	约束反演
constrained modulus	侧限模量；压缩模量
constrained motion	约束运动
constrained optimization	约束最优化
constrained oscillation	强迫振荡
constrained type	约束类型
constraint arc	限制弧
constraint condition	约束条件
constraint inversion	约束反演
constraint of motion	运动约束
constraint reacting force	约束反力
constraints	约束
constricted fold	收缩褶皱
constricted section	束窄段面
constriction	压缩；压缩物；阻塞物
constrictive strain	收缩应变；压缩应变

constrictor 压缩物
constringency 收缩
construct 构造；建造
constructability 可施工性；宜建性
constructed profile 示意剖面图
construction 施工；堆积
construction bend 现制弯头
construction bolt 安装螺栓
construction control survey 施工控制测量
construction cost 建筑成本
construction debris 建筑垃圾
construction design 施工组织设计
construction details 施工详图
construction document design 施工图设计
construction engineer 施工工程师
construction environmental impact assessment 建筑工程环境影响评估
construction fabrics 建筑用织物；土工织物
construction fibers 建筑用纤维
construction form 建造模板；结构形式
construction general layout 施工总平面图
construction hinge 建筑铰链
construction history 施工历史
construction in process 在建工程
construction investigation 施工勘察
construction joint 沉降缝；构造缝；伸缩缝；施工缝
construction layout 施工放样
construction load 施工荷载
construction management plan 施工组织设计
construction material 建筑材料
construction materials contract 包料
construction monitoring 施工监测
construction noise 建筑噪声
construction of piled foundation 地基打桩工程
construction period 施工工期
construction pit 施工基坑
construction quality control 施工质量控制
construction road 施工用道路
construction schedule 施工进度表
construction scheme 施工方案
construction sequence 施工程序
construction shaft 施工导洞
construction site 施工现场
construction survey 施工测量
construction terraces 堆积阶地；建筑阶地
construction timber 建筑木材
construction unit 构件；结构单元
construction vibration 施工振动
construction waste 建筑废料
construction waste recycling facilities 建筑废料再造设施
construction water level 施工水位
construction work 施工作业
construction wrench 井架安装用扳手
constructional 堆积的；建造的
constructional boundary 建设性板块边界；构造边界；施工边界
constructional column 构造柱
constructional delta 堆积三角洲
constructional drawing 构造图；施工图
constructional iron 建筑用钢铁；结构铁件
constructional landform 堆积地形
constructional limestone 结构灰岩
constructional measure for earthquake resistance 抗震构造措施
constructional plain 堆积平原
constructional plain coast 堆积平原海岸
constructional reinforcement 构造配筋
constructional steel 结构钢
constructional stream 堆积河流；构造河流
constructional surface 堆积面
constructional terrace 堆积阶地
constructive interference 相长干涉
constructive metamorphism 接力变质；增温变质
constructive phase 构造相
constructive plate boundary 建设型板块边界；扩张型板块边界
constructive plate margin 建设型板块边缘
constructive wave 堆积波；堆积浪
consultation of seismicity 地震趋势会商
consumable 可损耗的
consume 熔融消失；消耗
consume zone 消失带
consumed plate 消减板块
consumed plate boundary 消减板块边缘
consumer 消耗装置；消费者
consumering plate 消减板块
consumering zone 消失带
consuming plate 消减板块
consuming zone 消减带
consummation of errors 平差法
consumption 消亡作用
consumption indicator 消耗指示器
consumption per unit 单位消耗量
consumption zone 消减带
consumptive use of water normal consistence 标准稠度用水量
contact 接触；接触圈
contact action 接触酌；接触作用
contact adhesive 接触黏合剂
contact aeration 接触曝气
contact aerator 接触式曝气池
contact agent 接触剂
contact alteration 接触蚀变
contact analysis 接触分析
contact angle 接触角
contact angle aging effect 接触角老化效应
contact area 接触面
contact aureole 接触变质带
contact beam 接触剂
contact bed 界面层
contact biological filter 接触生物滤池
contact blasting 外部装药爆破
contact breccia 接触带角砾岩
contact cement 接触胶结物
contact cementation 接触胶结
contact column 接触塔
contact corrosion 接触腐蚀
contact deformation zone 接触变形带
contact deposit 接触矿床
contact dynamics 接触动力学

contact electromotive force	接触电动势
contact erosion	接触面冲蚀；接触侵蚀
contact field	接触区
contact filtration	接触过滤
contact force	接触力
contact goniometer	接触测角仪
contact grouting	环隙灌浆
contact halo	接触晕
contact line	地质分界线；接触线
contact load	接触荷载；触点荷载
contact lode	接触矿脉
contact log	接触式测井
contact material	触头材料
contact mechanics	接触力学
contact metamorphic	接触变质的
contact metamorphic aureole	接触变质带
contact metamorphic deposit	接触变质矿床
contact metamorphic rock	接触变质岩
contact metamorphic zone	接触变质带
contact metamorphism	接触变质作用
contact metasomatic	接触交代作用的
contact metasomatic deposit	接触交代矿床
contact metasomatic metamorphism	接触交代变质作用
contact metasomatic rock	接触交代岩
contact metasomatism	接触交代作用
contact mineral	接触变质矿物；接触矿物
contact moisture	接触带湿度
contact moulding	触压成型
contact noise	接触噪声
contact of the batholith	岩基接触
contact phenomena	接触现象
contact plane	接触面
contact potential	接触电位
contact pressure	基底接触应力；接触压力
contact printing	印痕法
contact probe	探头
contact ratio	啮合系数
contact relation	接触关系
contact scouring	接触面冲蚀
contact spring	接触泉
contact strain	接触应变
contact stress	接触应力
contact stress transducer	接触应力传感器
contact surface	接触面；界面
contact twin	接触双晶
contact type displacement gauge	接触式位移计
contact washing	接触冲刷
contact well logging	贴井壁测井
contact zone	接触带
contact-altered rock	触变岩
contacting gear	啮合齿轮
contacting sedimentation tank	接触沉淀罐
contact-less diagnosis	非接触式诊断
contactolite	接触变质岩
contagious distribution	不均匀分布
contained fragment	包块
contained nuclear explosion	密封核爆炸
container	容器
container horizon	含油层
container rock	储层
containing mark	容量刻度
containment	防漏；密封；容积
containment boom	海上柔性浮式拦油栅
containment liner	防渗漏层
containment method	封拦法
containment system	密封系统
contaminant	污染物；杂质
contaminant source	污染源
contaminant travel	污染物运移
contaminated	被污染的
contaminated aquifer	受污染含水层
contaminated catalyst	受污染催化剂
contaminated chemical	受污染化学剂
contaminated land	受污染土地
contaminated material	受污染物料
contaminated mud dredging	污泥疏浚
contaminated mud pit	污染泥弃置池
contaminated plug	污染段
contaminated rock	混杂岩
contaminated runoff	受污染径流
contaminated soil	受污染土
contaminated water	污染水
contaminated zone	污染带
contaminating fluid	污染液
contaminating metal	含杂质金属
contamination	混染；沾染；致污物
contamination facility	污染油品处理设施
contamination gas	污染气
contamination particle	污染颗粒
contamination precipitation	杂质沉淀
contamination sampling	污染采样
contaminative	污染的
contemporaneity	同期发生；同时代
contemporaneous	同时代的；同时发生的
contemporaneous deformation	同生形变
contemporaneous deformation structure	同生变形构造
contemporaneous delta-submarine fan couples	同期三角洲-海底扇对
contemporaneous deposit	同期沉积
contemporaneous difference facies	同期异相
contemporaneous erosion	同期侵蚀
contemporaneous fault	同生断层
contemporaneous fault accumulation	同生断层油气藏
contemporaneous fold	同生褶皱
contemporaneous glaciation	同期冰川作用
contemporaneous growth faulting	同生断层活动
contemporaneous heterotopic facies	同时异相
contemporaneous peat swamp	同生泥炭沼泽；原始泥炭沼泽
contemporaneous structure	同生构造
contemporaneous vein	同期脉
contemporaneousness	同时期
contemporary fill	同时充填
contemporary glaciation	现代冰川作用
content	成分；品位
content cement	水泥含量
content of gas	瓦斯含量
content of organic matters	有机质含量
content of readily soluble salts	易溶盐含量
content of water soluble salts	水溶盐含量

conterminous	邻接的
contiguity	接触
contiguous	接触的
contiguous bored pile	密排钻孔桩
contiguous zone	毗连区
continent	大陆
continent making movement	造陆运动
continental	大陆的
continental accretion	大陆增生
continental accretion theory	陆地增生说
continental alluvium	陆相冲积层
continental apron	陆隆
continental basin	陆相盆地；内陆盆地
continental block	大陆地块
continental borderland	大陆边缘地带
continental brake hypothesis	大陆车阀说
continental bridge	陆桥
continental coal measures	陆相含煤岩系
continental collision	大陆碰撞
continental collision zone	大陆碰撞带
continental connection	大陆连接；陆桥
continental convergence	大陆汇聚
continental convergence belt	大陆汇聚带
continental craton	大陆克拉通
continental crust	大陆地壳
continental crust stem	大陆壳套
continental delamination	岩石圈脱壳沉降
continental deposit	大陆沉积；陆相沉积
continental dispersion	大陆分离；大陆离散
continental displacement	大陆漂移；大陆迁移；大陆移位；大陆位移
continental drift hypothesis	大陆漂移假说
continental drift theory	大陆漂移理论
continental earthquake	大陆地震
continental edge	大陆边缘
continental embankment	大陆堤
continental emergence	陆隆
continental environment	大陆环境
continental epicenter	大陆震中
continental escarpment	陆崖
continental evaporite	陆相蒸发岩
continental facies	陆相
continental flexure	大陆挠褶带
continental flood basalt	大陆泛布玄武岩
continental flysch formation	陆相复理式建造
continental freeboard	大陆相对高度；大陆干舷
continental geosyncline	大陆地槽
continental glaciation	大陆冰川作用
continental glacier	大陆冰川
continental glacier isostatic process	大陆冰川均衡过程
continental growth	大陆生长
continental heat flow	大陆热流
continental ice sheet	大陆冰盖；大陆冰原
continental intraplate volcanism	大陆板内火山活动
continental island	大陆岛；陆边岛
continental isostatic hypothesis	大陆均衡假说
continental line	陆上管线
continental lithosphere	大陆岩石圈
continental lithospheric plate	大陆岩石圈板块
continental magnetic field	大陆磁场
continental mantle	大陆地幔
continental margin	大陆边缘
continental margin igneous rock	大陆边缘弧火成岩
continental margin subsidence	陆缘沉陷
continental marginal arc	大陆边缘弧
continental margin-sea	陆缘海
continental mass	大陆块体
continental migration	大陆漂移
continental nucleus	大陆核；地盾；陆核
continental origin of petroleum	陆相生油
continental plate	大陆板块
continental plate boundary	大陆板块边界
continental plateau	大陆高原；大陆台地
continental platform	大陆架；大陆台地；陆架
continental progressive overlap	大陆渐进超覆
continental raft	大陆筏
continental reconstruction	大陆复原
continental replacement	陆地移位
continental rift	大陆裂谷
continental rifting	大陆断裂
continental rise	大陆基；大陆隆起；大陆裙；大陆台
continental rise apron	陆隆裙
continental rise cone	陆隆锥
continental river	大陆河；内陆河
continental rotation	大陆旋转
continental rupture rift	大陆裂谷
continental salinization	大陆盐渍化
continental salinized groundwater	大陆盐化潜水
continental salting process	大陆盐碱化作用
continental sandstone	陆相砂岩
continental scientific drilling	大陆科学钻探
continental sea	陆表海；内海海
continental sediment	陆相沉积
continental sedimentary basin	陆相沉积盆地
continental sedimentation	陆相沉积
continental segment	陆块
continental shelf	大陆架；大陆棚
continental shelf break	大陆断层；大陆架断裂带；大陆架坡折；陆架坡折
continental shelf ecosystem	大陆架生态系统
continental shelf facies	陆棚相
continental shelf mineral resources	大陆架矿产资源
continental shelf sediment	陆架沉积物；浅海沉积物
continental shield	大陆盾；地盾
continental shore	大陆边缘
continental shore plain	陆架平原
continental shore terrace	陆边阶地
continental shoulder	陆肩
continental slab	大陆板块
continental slope	大陆斜坡
continental slope facies	大陆坡相
continental sphere	陆圈
continental spread hypothesis	大陆扩张说
continental stable massif	大陆稳定地块
continental stratum	陆相地层
continental structure	大陆构造
continental subduction zone	大陆俯冲带
continental super region	大陆大区
continental suture zone	大陆缝合带
continental tectonics	大陆构造

continental terrace	大陆阶地；大陆台地
continental transgression	陆相沉积扩展
continental trap structure	陆相圈闭构造
continental unconformity	大陆不整合
continental uplift	大陆上升
continental-bearing plate	大陆板块
continentalization	大陆形成
continental-shelf act	大陆架法
continent-bearing lithosphere	大陆岩石圈
continent-continent interference	大陆间相互作用
continentization	大陆化作用
continual loading test	连续加荷载荷试验；连续加荷固结试验
continuation of exploration right	探矿权延续
continuation of magnetic anomaly	磁异常延拓
continuation of mining rights	采矿权延续
continued sand production	连续出砂
continued waterflood recovery	连续注水开采
continued waterflooding	连续注水
continuing diastrophism	持续地壳变动
continuity	连续
continuity condition	连续条件
continuity equation	连续方程
continuity equation of seepage	渗流连续方程
continuity index	连续性指数
continuity of discontinuity	不连续面的延续性
continuity requirement	连续条件
continuous	连续的
continuous ageing	连续老化
continuous air injection	连续注空气
continuous analog	连续型模拟
continuous aquifer	连续水层
continuous area	连续区
continuous attenuation log	连续衰减测井
continuous back drain	连续排水孔道
continuous back-projection	连续背投影
continuous beam	连续梁
continuous beam bridge	连续梁桥
continuous beam on elastic support	弹性支承连续梁
continuous beam on many supports	多跨连续梁
continuous belt	传动皮带
continuous bored piling	连续钻孔桩
continuous bucket ditcher	多斗挖沟机
continuous bucket excavator	多斗挖掘机
continuous buckle detection	连续铺管屈曲检测
continuous butt weld mill	连续式炉焊管机组
continuous canalization	连续渠化；连续梯级化
continuous centrifuge	连续离心机
continuous chain bit	连续链式钻头
continuous chlorination	连续加氯处理
continuous cleavage	连续劈理
continuous concrete wall	地下连续墙；连续混凝土墙
continuous concreting	连续浇筑混凝土
continuous coring	连续取岩芯；连续取样
continuous coverage	连续测量
continuous creep	连续蠕变
continuous cycle	连续循环
continuous deformation	连续变形
continuous degassing	连续脱气
continuous delivery	连续输水；持续交付
continuous dipmeter	连续记录地层倾角仪
continuous distillation	连续蒸馏
continuous drafting	连续牵伸
continuous drive	连续驱动
continuous dryer	连续干燥机
continuous energy source	连续能源
continuous ester interchange	连续酯交换
continuous extraction	连续提取
continuous face	连续工作面
continuous face mining	连续工组开采
continuous field	连通油气田
continuous filament	连续长丝
continuous flight auger	连续螺纹钻
continuous flood irrigation	连续漫灌
continuous flow	连续流
continuous flow profile	连续流量剖面
continuous flowmeter	连续记录流量计
continuous folding	连续褶皱
continuous footing	连续基础
continuous foundation	连续基础
continuous Fourier transform	连续傅里叶变换
continuous fractional crystallization	连续分离结晶作用
continuous fractional melting	连续分离熔融
continuous frame	连续框架
continuous furnace	连续作业炉
continuous gas	钻屑天然气
continuous gas injection	连续注气
continuous gas-lift	连续气举
continuous girder bridge	连续梁桥
continuous gradient	连续梯度
continuous grading	连续级配
continuous gravimeter	连续记录重力仪
continuous head plug container	连续式水泥头
continuous heading machine	掘进康拜因；掘进联合机
continuous heat test	连续加热试验
continuous helical auger	连续螺纹钻
continuous hydrolytic polymerization	连续水解聚合
continuous immersion zone	全浸区
continuous impost	连续拱墩；连续拱基
continuous inclinometer	连续测斜仪
continuous injection	连续注入
continuous isomorphous series	连续类质同象系列
continuous layer	连续层
continuous line	连续作业线
continuous line bucket mining system	连续绳斗采矿系统
continuous line bucket system	连续戽斗链采矿
continuous liquid phase	连续液相
continuous load	持续负载；连续荷载
continuous measurement	连续计量
continuous media field	连续介质场
continuous medium	连续介质
continuous miner	连续采煤机
continuous mining	连续开采
continuous modification	连续修正
continuous monitoring	连续监测
continuous motion	连续运动
continuous operation	连续运转
continuous permafrost	永久冻土，永久冻地；连续永冻土带
continuous phase	连续相

continuous piezometric logging	连续孔压测井
continuous piled wall	连续排桩墙
continuous point source	连续点源
continuous polymerization	连续聚合
continuous porosity	连通孔隙度
continuous production	连续生产
continuous profiling	连续剖面法
continuous pumpage	持续抽水量
continuous pumping	连续抽水
continuous pumping method	连续送浆法
continuous pumping test	连续抽水试验
continuous raft	连续筏基础
continuous reaction	连续反应
continuous reaction series	连续反应系
continuous recording	连续记录
continuous ripening	连续熟化
continuous rotation	连续转动
continuous running	连续运转
continuous sample	连续取样
continuous seismic profiling	连续地震剖面法
continuous seismic reflection profiling	连续地震反射剖面法
continuous seismic source	连续震源
continuous seismic zone	连续地震带
continuous settlement gauge	连续沉降计
continuous shooting	连续放炮
continuous sidewall core cutter	连续式井壁岩心切割器
continuous slab deck	连续式挡水面板
continuous soil sample	连续土样
continuous solid solution	连续固溶体
continuous source	连续震源
continuous spectrum	连续谱
continuous stirred tank reactor	连续搅拌槽式反应器
continuous stratification profiler	连续地层剖面仪
continuous stream	常流河;连续流
continuous stress	长期应力;持久应力
continuous string coiled tubing unit	连续油管作业机
continuous strip pickler	连续式带材酸洗装置
continuous strip pickling plant	连续式带材酸洗装置
continuous terrace	连续阶地
continuous thickener	连续浓缩机
continuous time series	连续时间序列
continuous topology	连续拓扑学
continuous transformation	连续变形;连续转变
continuous treatment	连续处理
continuous trenching machine	连续挖沟机
continuous truss	连续桁架
continuous vacuum filter	连续真空过滤机
continuous velocity log	连续速度测井
continuous viscosimetry	连续黏度测量法
continuous wave	等幅波
continuous wetting phase	连续润湿相
continuous whipstock	连续造斜器
continuous wire mill	连续式线材轧机
continuous withdrawal	连续开采
continuous sweep	连续扫描
continuous-fixed base	连续固定底座
continuous-flexible base	连续柔性底座
continuously loaded oedometer	连续加荷固结仪
continuous-wave modulation	连续波调制
continuous-wave response	连续波响应
continuum approach	连续介质法
continuum mechanics	连续介质力学
contorted	扭歪的
contorted bed	扭曲层
contorted bedding	褶皱层理
contorted fold	扭曲褶皱
contorted mass	扭曲块体
contorted strata	扭曲层
contorted structure	扭曲构造
contortion fissure	扭曲裂缝
contortion folding	扭曲褶皱作用
contour	地势;地形;等高线;等值线
contour bank	等高堤
contour basin	等高水池
contour blasting	轮廓爆破
contour bowl centrifuge	异形沉淀式离心机
contour bund	等高堤
contour chart	等高线图
contour current	等深海流
contour design	轮廓设计
contour diagram	等值图
contour ditch	等高沟
contour extraction	轮廓抽取
contour flooding	边缘注水
contour gradient	等倾斜线
contour hole	周边眼;周边钻孔
contour horizon	标准层位
contour integral	围道积分
contour integration	围道积分
contour interval	等高线间隔;等值线间隔
contour line	等高线;等值线
contour map	等高线图
contour map of groundwater	地下水等水位线图
contour map of piezometric surface	等水压线图
contour map of water table	等水位线图
contour microclimate	地形性小气候
contour migration	等值线偏移
contour mining	等高线采矿法
contour of equal stress	等应力线
contour of oil pool	含油外边界
contour of oil sand	油砂层等高线图
contour of recharge	补给边界
contour of water table	等水位线
contour plan	等高线图
contour tracing	轮廓跟踪;等值线追踪
contoured bathymetric map	水深线图
contoured map	等高线图
contoured velocity	速度等值线
contour-following geostrophic current	等深地转流
contourgraph	轮廓仪
contouring	绘等值线;画等高线;轮廓圈定
contouring method	圈定法
contourite	等深线沉积物;平流沉积
contourite deposit	等深流沉积物
contourite facies	等深流沉积相
contours of groundwater	地下水等高线
contra rotating fan	反旋式通风机
contracid	镍铬铁耐酸合金
contract	收缩;合同;订约

contract boring	承包打钻
contract construction	承包施工
contract crack	收缩裂隙
contract demand	合同需求
contract depth	合同井深
contract design	定约设计
contract drilling	承包钻井
contract fault	收缩断层
contract footage rate payment	按进尺包工计价
contract for labour and material	包工包料合同
contract of earth	地球收缩
contract with dry rate	干缩率
contracted channel	收缩河道
contracted drawing	缩图
contracted flow	束窄水流
contracted jet	收缩射流
contracted section	收缩截面
contractibility	收缩性
contractile skin	结合膜；收缩膜
contracting duct	束狭渠道
contracting earth hypothesis	地球收缩说
contraction	缩短率；干缩
contraction allowance	收缩余量
contraction cavity	缩孔
contraction crack	冰冻裂隙；收缩裂缝
contraction curve	含沙量曲线；浓度曲线；集流段；涨洪段
contraction deformation	收缩变形
contraction fault	收缩断层
contraction fissure	收缩裂缝
contraction hypothesis	收缩假说
contraction joint	收缩缝；收缩节理
contraction joint grouting	收缩缝灌浆
contraction limit	收缩极限；缩限
contraction of area	断面收缩
contraction test	收缩试验
contraction theory	收缩说
contractive soil	收缩性土
contractive-dilative	收缩-膨胀
contractor's hole	草率完工的井
contra-dip drainage pattern	逆倾向水系
contradope	反添加
contraflexure	反弯；回折
contraflow	反向流动；逆流
contragradation	阻塞堆积
contraposed coast	叠置海岸
contraposed shoreline	叠置滨线
contraposition	对照
contrarotation	反向旋转
contrast control	对比度调整
contrast stretching	对比度增强；反差增强
contrast test	对照试验
contrast value	背景值；本底
contrasting fluid	对比液
contrasting region	对比范围
contrate gear pair	端面齿轮副
contrate wheel	端面轮
contratest	对比试验
contributary	支流；支渠
contributing pore volume	有效孔隙体积
contributing region	补给区；水源区；影响区
contributing zone	生产层
contributory area	补给区；水源区；影响区
contributory drainage volume	供油体积
contributory pore volume	供油孔隙体积
contributory zone	补给区；供水区
control	调整；控制
control access	控制存取；控制通路
control agent	调节介质
control amplifier	控制放大器
control area	控制范围；控制区
control base	检核基线；控制基点；控制基准面
control borehole	控制孔
control buoy	可控浮标
control bus	控制总线
control cabin	操纵室
control cabinet	操纵台
control cable	控制电缆
control casing head	带阀的套管头
control choke	节油嘴
control countermeasure	防治对策
control earthquake	控制地震
control earthquake value	控制地震值
control fluid	压井液
control hand wheel	调节手轮
control head	井口控制装置
control injection	控制注水
control instrument	控制仪表
control joint	控制缝
control level	控制液面
control lever	操作杆
control of flood	防洪
control of geological hazard	地质灾害防治
control of landslide	滑坡防治
control of underground water	地下水控制
control of vibration	振动控制
control on slope stability	斜坡稳定性控制；边坡稳定性控制
control panel	操纵台；控制面板
control point	控制点；水准基点
control point inside tunnel	隧道内控制点
control point outside tunnel	隧道外控制点
control rod	控制棒
control station	调度站
control well	控制井
controllability	可控性
controllable	可控的
controllable drilling variable	可控钻井参数
controllable pitch	可调螺距
controlled aggregation	适度絮凝
controlled amplitude processing	可控振幅处理
controlled angle drilling	定向钻进；定向钻井
controlled atmosphere	惰性气保护
controlled bit weight	控制钻压
controlled blast	控制爆破
controlled blasting	控制爆破；管制爆破
controlled blasting technique	控制爆破技术；收控爆破技术
controlled caving	控制放顶
controlled co-mingling	控制合采

controlled cooling line 控制冷作业线
controlled crest 设有调节水位装置的坝顶
controlled directional drilling 受控定向钻探
controlled directional well 定向井
controlled draft drawing frame 单程并条机
controlled drilling 定向钻进；定向钻井
controlled drilling rate 控制钻速
controlled fill 质量控制填土
controlled filling experiment 控制注水试验；控制灌注试验
controlled footage 控制进尺
controlled gradient 控制梯度
controlled interval 定距；控制井段
controlled pressure completion 控制压力完井法
controlled production 控制开采
controlled source seismology 可控源地震学
controlled tubing completion system 控制油管完井系统
controlled volume surge 适当反冲液体
controlled waterflooding 控制注水
controlled-electromagnetics 可控源电磁法
controlled-source audiomagnetotelluric method 可控源音频大地电磁法
controlled-strain test 应变控制试验
controlled-stress test 应力控制试验
controlling and monitoring system 控制和监测系统
controlling depth 控制水深
control-sleeve 油量调节套
convection 对流作用
convection cell 对流胞；对流元
convection current 对流；对流洋流
convection current hypothesis 对流假说
convection fractionation 对流分馏
convection of pollutant 污染物迁移
convection plume 对流羽
convection tectonics 对流构造
convection theory 对流说
convection transport 对流迁移
convection zone 对流层
convective circulation 对流循环
convective geothermal system 对流型地热系统
convective heat flow 对流型热流
convective velocity scale 对流速度尺度
convenience receptacle 插座
convenient synthesis 简便合成法
conventional 常规的
conventional abundance 惯用丰度
conventional acid breaker 常规酸破胶剂
conventional age 常规年龄
conventional analysis 常规分析
conventional approach 常规方法
conventional back pressure test 常规回压试井
conventional barrel 常规取心筒
conventional bit 常规钻头
conventional circulating technique 常规循环技术
conventional concrete 普通混凝土
conventional consolidation job 常规地层胶结作业
conventional coordinates 标准坐标系；通用坐标
conventional core 常规岩心
conventional core analysis 常规岩心分析
conventional core barrel 常规岩心筒

conventional creep limit 公称蠕变极限
conventional crude 常规原油
conventional derrick 常规井架
conventional diagram 示意图
conventional diving 常规潜水
conventional draft 普通牵伸
conventional drilling 常规钻井
conventional electric logging device 常规电测仪
conventional electroslag welding 普通电渣焊
conventional energy resources 常规能源
conventional fault 正常断层
conventional fault drag 正常断层拖曳
conventional gamma ray log 普通自然伽马测井
conventional gas 常规天然气
conventional geochemical data 传统地球化学数据
conventional gun 普通射孔器
conventional international origin 国际协议习用原点
conventional jet perforating 常规聚能喷流射孔
conventional log analysis 常规测井分析
conventional logging suite 常规测井系列
conventional logging technique 普通测井技术
conventional method 常规方法
conventional method of slices 常规条分法
conventional migration 普通偏移
conventional mud 常规泥浆
conventional neutron log 普通中子测井
conventional normal fault 标准正断层
conventional oil 常规石油
conventional open hole packing method 常规裸眼充填方法
conventional overbalanced perforating 常规正压射孔
conventional petroleum 常规油气
conventional piled steel platform 常规桩承钢质平台
conventional power plant 常规电站
conventional predictive deconvolution 常规预测反褶积
conventional processing 常规处理
conventional pumping unit 普通抽油机
conventional quenching 一般淬火
conventional scale 常规标度
conventional scanning diffraction 常规扫描绕射
conventional seismic method 常规地震勘探法
conventional seismic processing package 常规地震处理程序包
conventional sign 标准图例；图例
conventional simulator 常规模拟模型
conventional slurry 常规砂浆
conventional spread 常规排列
conventional stacking 常规叠加
conventional survey 常规勘探
conventional sweep 常规扫描
conventional test 常规试验
conventional tool 标准工具
conventional triaxial test 常规三轴试验
conventional triaxial test for rocks 岩石常规三轴试验
conventional triaxial testing machine for rocks 岩石常规三轴试验机岩芯钻探
conventional two zone completion 常规双层完井
conventional type gravel packing 常规砾石充填
conventional water injection 正注
conventional waterflood 常规注水
conventional well 常规井

conventional well-flushing 正洗
conventional workover rig 常规修井作业机
conventional x-ray tomography 常规 X 射线层析成像
conventionality 照例
conventional-type lubricant 习用润滑油
converged fold 收敛褶皱
converged terrace 辐合阶地；汇合阶地
convergence 汇合；矿体复合；交会
convergence correction 收敛校正
convergence current 辐聚流
convergence deformation 净空变形；净空收敛
convergence indicator 收敛仪
convergence master fault 会聚主断层
convergence measure 净空变形量测；收敛量测
convergence measuring device 收敛测量装置
convergence of meridian 子午线收敛角
convergence pressure 会聚压力
convergence rate 收敛速度
convergence ray 收敛射线
convergence zone 会聚区
convergent 会聚的；趋同的；收敛的
convergent angle 交向角；收敛角
convergent boundary 会聚边界
convergent continental margin 聚合大陆边缘
convergent drainage 会聚水系
convergent drainage pattern 辐聚水系
convergent fault 聚会断层；收敛断层
convergent geothermal belt 会聚型地热带
convergent homeomorphy 趋同同型
convergent margin 收敛边缘
convergent oscillation 减幅振荡
convergent plate 会聚板块
convergent plate boundary 板块汇合边界
convergent response 收敛性反应
convergent seismic reflection configuration 收敛型地震反射结构
convergent sequence 收敛序列
convergent series 收敛级数
convergent wrench faulting 会聚扭断层
convergent-divergent flow 汇-散型流动
convergent-divergent pattern 聚会-离散模式
converging 收缩水流；复合的；合流；交汇的；聚合的
converging flow 聚集水流
converging fold belt 辐合褶皱带
converse 逆的；相反的
converse metamorphism 逆变质作用
converse theorem 逆定理
converse twist 反向捻
converse-Labarre equation 拉贝尔转换公式
conversion 转化；转换
conversion accuracy 转换精度
conversion characteristic 转换特性
conversion chart 换算图表
conversion coating 转化型带锈底漆；转化膜；转化层
conversion efficiency 转换效率
conversion factor 换算系数
conversion moment 换算力矩
conversion of units systems 单位制换算
conversion paint for rusting surface 转化型带锈底漆
conversion parameter 换算参数

conversion per pass 单程转化率
conversion pig iron 炼钢生铁
conversion rate 转换速度
conversion resolution 转换分辨力
conversion standard 转换标准
conversion well 转注井
conversion works 改建工程
convert 转换
converted 改装的；修改的
converted production well 转注井
converted shear wave 转换横波
converted top 纤维条
converted wave 转换波
converter 转炉
convertibility of expansion and shrink age 胀缩可逆性
convertible crane 可更换装备的起重机
convertible hydrocarbons 可转化烃类
convertible shovel 两用铲
convertible stand 可换机座
converting 吹炼；转炉炼钢
convex 凸面；凸起的；凸形的
convex bank 凸岸
convex cross-bedding 凸交错层理
convex fillet weld 凸面填角焊缝
convex flood plain 凸形河漫滩
convex oil surface 凸油表面
convex plane 凸面型
convex profile 背弧型面
convex refractor 上凸折射层
convex slope 凸形斜坡
convex wall 凸形挡土墙
convex weld 凸面焊缝
convex-concave 凸凹形的
convexity 凸度；凸面
convexo-concave 凸凹形
convexo-convex 两面凸的
convey tubing perforation 油管传送射孔
conveyance 提升；运输工具
conveyance factor 输水率；输水系数
conveyance tunnel 输水隧道
conveyer bridge 排矸桥；输送机支架
conveyer bucket 输送料斗
conveyer chute 运输机槽
conveyer drying machine 带式干燥机
conveyer pan 运输机槽
conveyer pipe 输送管
conveying belt 传送带
conveying picker 拣选带
conveying picking table 拣选带
conveying plant 运输设备
conveying rope 牵引钢丝绳
conveying screw 螺旋输送机
conveying tunnel 输水隧洞
conveyor bridge 排矸桥
conveyor chute 运输机槽
conveyor drying machine 带式干燥机
conveyor extension 接长输送机
conveyor mine 输送机化煤矿
conveyor mining 输送机化的开采
conveyor roller 输送机滚筒

conveyor separator 输送机式分级机
convinced test 确认试验
convolute 回旋状的；旋卷的
convolute bedding 包卷层理；旋卷层理
convolute fold 翻卷褶皱；旋卷形褶皱
convolute lamination 旋卷纹理
convolution 回旋；卷积
convolution bedding 旋卷层理
convolution filtering 褶积滤波；卷积滤波
convolution structure 包卷构造
convolutional 回旋的；卷曲的
convolutional ball 旋卷球
convulsion 灾变；震动
convulsionism 灾变论
cookeite 锂绿泥石
cool 冷却
cool explosive 低爆热炸药
cool spring 冷泉
cool water source 冷水源
coolant 冷冻剂
cooling age 冷却年龄
cooling crack 冷缩裂缝
cooling fissure 冷缩裂缝
cooling joint 冷缩节理
cooling mantle 冷却套
cooling plant 冷却装置
cooling rate 冷却速率
cooling water 冷却水
coom 煤粉；煤烟；碎煤
coombe 峡谷；冰斗；冲沟
coombe rock 混杂泥砾；峡谷岩
cooperation of concrete and steel 混凝土和钢筋共同工作
cooperative 合作的
cooperative exploitation 合作开采
cooperative production 合作生产
cooperite 硫铂矿；天然硫砷化铂
Cooper-Jacob analysis 库泊-雅可比分析法
coordinator 同等物
coorongite 弹性藻沥青
co-oscillation 共振
co-oxidation 共氧化
copalite 黄脂石
cope 处理；应对
Copel 科普尔铜镍合金
Copenhagen water 国际哥本哈根标准海水
Copernican system 哥白尼体系
copf 渗透率公式的系数
cophasal 同相的
cophenetic correlation 同表象相关
copiapite 叶绿矾
coping 盖顶；压顶；墙帽
coping stone 墩台石；盖顶石
coplasticizer 辅塑剂
coplusory oil stock obligations 强制石油储备
copolar 共极的
copolymerization 共聚合反应
coppaelite 辉云黄长岩
copper 铜；紫铜
copper alloy 铜合金
copper arsenite 亚砷酸铜

copper bar 铜杆
copper bit 钎焊烙铁
copper blast furnace 炼铜鼓风炉
copper clad steel wire 包铜钢丝
copper concentrate 铜精矿
copper corrosion 铜片腐蚀；铜锈蚀
copper crucible 铜坩埚；铜结晶器
copper cylinder compression test 猛度试验；铜缸压缩试验
copper dish gum test 铜皿胶质试验
copper electrode 铜电极
copper electrolyte 铜电解液
copper glance 辉铜矿
copper loss 铜损
copper matte 冰铜
copper mine 铜矿山
copper nickel 红砷镍矿
copper ore 铜矿
copper ore deposit 铜矿床
copper oxide 氧化铜
copper pipe 铜管
copper plate 铜板
copper plated steel 包铜钢；镀铜钢板
copper plating 镀铜
copper porphyrin 铜卟啉
copper pyrite 黄铜矿
copper sandstone 铜砂岩
copper smeltery 铜冶炼厂
copper strip test 铜片试验
copper sulfate eletrode 硫酸铜电极
copper tube 铜管
copper vitriol 胆矾
copper weld steel wire 铜焊钢丝
copper wire 铜丝
copper-alloy arc welding electrode 铜合金焊条
copper-alloy steel 铜合金钢
copperas 硫酸亚铁；绿矾；水绿矾
copper-clad 包铜的
coppered wire 镀铜线；铜线
copperization 镀铜
copper-nickel alloy 铜-镍合金
copper-silicon alloy 硅-铜合金
coppersmithing 铜锻造
copperweld 包铜钢丝
coppite 铜矿；重烧绿石
coprecipitate 共沉淀
co-precipitated catalyst 共沉淀催化剂
coprecipitation 共同沉淀
coprecipitator 共沉淀剂
co-product 副产品
coprolite 粪化石；粪粒体
coprology 粪石学
copromicrite 粪粒微晶灰岩
coprosparite 粪粒亮晶灰岩
copula 连系物
copulation 配对
copying 晒图
coquimbite 针绿矾
coquina 贝壳灰岩
coquinite 硬壳灰岩

coquinoid limestone 贝壳灰岩	cordillera 山脉；雁列山脉
coracite 水钙铅铀矿	cording 绳索
coral 珊瑚虫	cordite 线状无烟火药
coral agate 珊瑚玛瑙	cordless coring tube 无缆取样管
coral algal facies 珊瑚藻相	cordylite 氟碳钡铈矿
coral atoll 珊瑚环礁	core 土芯；心墙；岩芯
coral beach 珊瑚滩	core analysis 岩芯检验；岩芯分析
coral bleaching 珊瑚白化	core analysis data 岩心分析数据
coral cap 珊瑚岩帽	core and mantle structure 核幔构造
coral cay 珊瑚小礁	core axis 岩心轴
coral crust 珊瑚薄盖层	core barrel 套管；岩芯管；岩芯钻管
coral growth line 珊瑚生长线	core barrel catcher 岩心筒抓取器
coral head 珊瑚丘	core barrel head 取心筒上盖
coral island 珊瑚岛	core barrel of double tube swivel type 双层取心筒
coral knoll 珊瑚丘	core barrel ring 取心筒垫圈
coral limestone 珊瑚灰岩	core barrel shoe 取心筒鞋
coral mud 珊瑚泥	core barrel support 取心筒支座
coral patch 珊瑚块礁	core barrel trajectory 取芯筒轨道
coral polyp 珊瑚虫	core barrel with rubber core container 橡胶套岩心筒
coral rag 珊瑚礁屑角砾岩	core basket 取心爪
coral reef 珊瑚礁	core binder 砂芯黏结剂
coral reef coast 珊瑚礁海岸	core binder pitch 型心黏结沥青
coral reef community 珊瑚礁生物群落	core bit 取芯钻头
coral reef ecosystem 珊瑚礁生态系统	core bit wing 刮刀取心钻头的刀翼
coral reef oil pool 珊瑚礁油藏	core block 岩心块
coral reef shoreline 珊瑚礁滨线	core blocked 岩心堵塞
coral reef soil 珊瑚礁土	core blower 芯型吹砂机
coral rock 珊瑚岩	core boring 取芯钻进
coral sand 珊瑚砂	core boring method 取心钻探法
coral shoal 珊瑚洲	core borings 岩粉；岩屑；钻粉
coral tableland 珊瑚丘	core boundary 地核边界
coral thicket 珊瑚丛	core breaker 岩芯切断器；岩芯提取器
coralgal micrite 珊瑚藻微晶灰岩	core breaking 割断岩心
coralgal ridge 珊瑚藻类边缘脊	core breaking by hydraulic pressure 投球液压割心
coralgal unit 珊瑚藻单元	core breaking by mechanical loading 机械加压割心法
coralgal zone 珊瑚藻沉积带	core catcher 采芯器；岩芯夹具；岩芯提取器；岩芯抓
corallaceous 珊瑚质的	core catcher spring 岩心卡簧
coralliferous 含珊瑚的	core chamber 岩心室
coralline facies 珊瑚相	core cone bit 牙轮取心钻头
coralline limestone 珊瑚灰岩	core container 岩心筒
corallite 珊瑚单体	core cross section 岩心断面
corallium 红珊瑚	core cutoff trench 芯墙截水槽
coralloid 珊瑚状的	core cutter 岩心切割机；岩心提取器
corallum 珊瑚体	core cutter method 环刀法
coral-stromatoporoid reefs 珊瑚-层孔虫礁	core dam 心墙
corange line 等潮差线	core data 岩心分析数据
corannulus 瞳仁环颗石	core data bank 岩心样本资料库
corbel 梁托；牛腿式；托架	core diameter 焊芯直径；土芯直径；岩芯直径
corbel-out 牛腿式；悬挑；支托	core displacement test 岩心驱替试验
corcovadite 花岗闪长玢岩	core distillation apparatus 岩心蒸馏仪
cord 绳索	core district 核心地区
cord fabrics 帘子布	core drawing 带芯冷拔
cord wire 绳索用钢丝	core drill 取心钻具；岩芯钻
cordage 绳索；索具	core drill rig 岩芯钻机
cordage oil 钢丝绳用油	core drilling 取心钻进；取芯钻探；钻取土芯；钻取岩芯
cordaitotelinite 科达木结构镜质体	core drilling bit 取心钻头
corded lava 绳状熔岩	core drilling machine 岩心钻机
cordierite 堇青石	core drilling method 岩心钻进法
cordierite norite 堇青苏长岩	core drilling rig 取芯钻机
cordierite-amphibolite facies 堇青石-角闪岩相	

core earth dam	心墙土坝
core entry	岩心入口
core equipment	取心设备
core erosion	岩心冲蚀
core examination	岩心检验
core extraction	岩芯采取
core extractor	推岩心杆；岩心提取器
core facies analysis	岩心相分析
core facility	核心设施
core factor	岩心系数
core fisher	岩心打捞工具
core grabber	岩心打捞工具
core grid	堆芯格架；泥心骨
core grinding	岩心磨损
core gripper	岩心爪
core gripper with slip	卡瓦式岩心爪
core gripper with slip-collar	卡箍式岩心爪
core gripper with slip-spring	卡簧式岩心爪
core grouting	岩心卡住；卡料
core head	取心钻头
core holder	岩心夹持器
core holding unit	岩心夹持器
core house	岩心房
core inlet face	岩心入口面
core interpreting device	鉴定岩心方向装置
core intersection	产油层；岩心断面；油层厚度
core iron	型心铁
core jig	型芯夹具
core library	岩心库
core lifter	岩心提取器
core log	岩心记录
core logging	测井
core material	心墙材料
core objective	核心目标
core of anticline	背斜中心
core of concrete	混凝土芯样
core of fold	褶皱核
core of rock	岩芯
core of syncline	向斜中心
core of the breakwater	防波堤堤心
core of the earth	地核
core oil saturation	岩心含油饱和度
core orientation	岩心定向
core orientator	岩心定向仪
core outlet face	岩心出口面
core penetration test	样孔贯入度测试
core permeability	岩心渗透率
core picker	落井岩心打捞工具
core pin	穿孔杆；销钉；芯子
core pin plate	穿孔杆板
core plug	岩心栓
core plunger	岩心推取杆
core porosity	岩心孔隙度
core preparation	岩心制备
core pusher	塞子
core recover percent	岩心采取率
core recovery	岩心采取率
core recovery horizon	岩心采取层
core recovery percent	岩心采取率
core repository	岩心库
core retainer plate	芯套
core run	取心进尺
core sample taker	井壁取心器
core sampler	岩芯取样器
core sampling	岩芯取样
core section	岩心截面
core shack	岩心贮藏室
core shanty	岩心贮藏室
core shell	岩心衬筒
core shoe	取心筒鞋
core slicer	切割式井壁取心器
core snatcher	岩心爪子
core specimen	岩芯标本
core splitter	岩心劈开机；岩心切断器
core spring	岩心卡簧
core stone	石芯；芯石；岩芯石
core storage	岩心房
core stub	残留岩芯；岩芯残段
core test	岩心测试；岩心试验
core testing	芯样测试
core testing pump	岩心试验泵
core texture	环礁结构
core tray	岩心盒
core trench	心墙截水槽
core trimming	岩心修整
core tube	岩芯管
core wafer	岩心薄片
core wall	心墙；隔水墙
core wall of dam	不透水的坝心墙
core wall type rockfill dam	心墙式堆石坝
core wash	岩心冲蚀
core wettability	岩心润湿性
core-barrel bit	硬砂岩钻头
core-bit tap	打捞金刚石取心钻头的公锥
core-catcher case	岩心爪外套
cored beam	空心梁
cored footage	岩心钻进进尺
cored hole	取芯孔
cored interval	取芯井段
cored screw	空心螺杆
cored steel	中空钢
cored verification hole	取芯检查孔
cored well	取芯井
core-description graph	岩心描述图
core-drill fittings	取心钻具零件
core-drill method	取芯钻探法
core-drilling exploration	取芯钻井勘探
core-drying oven	岩心烘箱
coreduction	同时还原
coreflooding experiment	岩心驱替试验
core-flow efficiency	岩心渗流效率
coregraph	岩心图
core-gripper case	岩心爪外套
coregun	井壁取心器
corehole	岩芯钻孔；钻孔
corehole extensometer	岩孔伸缩计
corelab	岩心分析实验室
core-leaving method	旁侧导坑法
core-lifter wedge	岩心提取器楔形块
core-mantle boundary	核幔界面

core-pusher plunger 岩心推取塞推杆
corequake 核震
corer 取心器
core-shell structure 核-壳结构
core-to-sludge ratio 岩心岩粉比
core-transitional zone 核幔过渡带
corf 矿车
corgun 井壁取心器
coring 取芯作业；钻取岩芯
coring bit 取芯钻头
coring bullet 取心弹
coring crown 取心钻头
coring depth 取心深度
coring device 取芯设备
coring equipment 取心设备
coring footage 取心进尺
coring formation 取心地层
coring gun 井壁取心器
coring in oil base mud 油基泥浆取心
coring machine 岩心钻机
coring operation 取心作业
coring programme 土芯钻取计划
coring system 取心钻具
coring tool 取心工具；岩心提取器
coring tube 底质柱状取样器
coring type turbodrill 取心涡轮钻具
coring vessel 取心桶
coring weight 取心钻压
coring-type bit 取芯钻头
Coriolis acceleration 科里奥利加速度
Coriolis effect 科里奥利效应
Coriolis force 科里奥利力
cork 软木
cork board 软木板
cork float 软木浮子
cork screwed wellbore profile 螺旋状井筒剖面
corkite 磷硫铅铁矿
corkscrew 螺丝起子；软木塞拔起器；扭弯的钻杆
corkscrew core 螺旋状的岩心
corn 颗粒
cornbrash 粗粒介壳灰岩
corncob bit 金刚石锥形钻头
cornean 隐晶岩
corned powder 粒状炸药
cornelian 光玉髓
corner bead 墙角护条
corner beam 角梁
corner break 角隅断裂
corner cube 角视立体图
corner effect 锐角效应；角隅作用
corner foundation 基墩
corner foundation bolt 基墩螺栓
corner joint 角接接头；弯管接头
corner post 井架大腿底柱
corner producer 角生产井
corner reflection 角反射
corner reflector 角形反射器
corner well 角井
corner-points method 角点法
cornetite 蓝磷铜矿

cornification 角质化
cornoid 牛角线
cornstone 结核状玉米灰岩
cornubianite 粒状角页岩
cornuite 土硅铜矿
cornwallite 翠绿砷铜矿
corollary equipment 配套设备
corollithion 花冠颗石
corona texture 冠状结构
coronadite 方锰铅矿
coronate faults 花冠状断层
coronite 反应边岩；镁电气石
coroutine 联立子程序；协同程序
corpocollinite 团块镜质体
corpohuminite 团块腐殖体
corposclerotinite 浑圆菌类体；浑圆硬核体
corpovitrinite 团块镜质体
corpuscle 微粒
corpuscular radiation 太阳微粒辐射
corrading river 侵蚀河流；下切河流
corrading stream 侵蚀河流；下切河流
corrasion 磨蚀；动力侵蚀；磨蚀；侵蚀
corrasion valley 刻蚀谷
corrasional gully 冲沟
corrasive 侵蚀性的
corrected 校正的
corrected area weighting 修正面积权重
corrected buildup pressure 校正后的恢复压力
corrected datum-level pressure 折算基准面压力
corrected density porosity 校正了的密度测井孔隙度
corrected depth 校正过的深度
corrected drawdown 校正降深值
corrected neutron porosity 校正了的中子测井孔隙度
corrected oil 合格油
corrected parameter 修正参数
correction 修正
correction calibrator 校正台；修正校准器
correction curve 修正曲线
correction curve for drawdown 降深修正曲线
correction data 修正数据
correction factor 修正系数
correction factor for inclination 倾角修正系数
correction for relief 地形起伏校正
correction for sampling disturbance 取样扰动校正
correction for the ionospheric refraction 电离层折射改正
correction for the tropospheric refraction 对流层折射改正
correction for topography 地形校正
correction of magnetic declination 磁偏角校正
correction of river 河流整治
correlation approach 相关方法
correlation ghost 相关伴随波
correlation interferometry 对比干涉测量
correlation logging 对比测井
correlation of coal seams 煤层对比
correlation of oil layers 油层对比
correlation of strata 地层对比
correlative sediments 相关沉积
correlative surface 对应面；关联面
correspond 符合
corresponding crest 相应洪峰水位

corrie	冰斗	corrosive action	腐蚀作用
corrie glacier	冰斗冰川	corrosive agent	腐蚀剂
corroded	侵蚀的	corrosive air	腐蚀性空气
corroded crystal	溶蚀晶	corrosive carbon dioxide	侵蚀性二氧化碳
corroded depression	溶蚀洼地	corrosive concentration	腐蚀介质浓度
corroded fissure	溶蚀裂隙	corrosive effects	腐蚀作用
corroded grain	被腐蚀颗粒	corrosive fluid	腐蚀性流体
corroded gully	溶沟	corrosive fume	腐蚀性烟气
corroded hollow	溶洞	corrosive ground	腐蚀土壤
corroded valley	溶谷	corrosive medium	腐蚀介质
corrodent	腐蚀剂；有腐蚀力的	corrosive salt	腐蚀性盐
corrodibility	可腐蚀性	corrosive sublimate	氯化汞；升汞
corroding electrode	腐蚀电极	corrosive substance	腐蚀性物质
corroding metal	腐蚀金属	corrosive water	侵蚀性水
corrodkote test	涂膏耐蚀试验	corrosive wear	腐蚀磨损
corronil	铜镍合金	corrosiveness	侵蚀作用
corrosimeter	腐蚀检定计	corrosiveness of groundwater	地下水腐蚀性
corrosion	腐蚀；侵蚀；溶蚀	corrosivity	腐蚀性
corrosion accelerator	腐蚀加速剂	corrosometer	腐蚀计；测蚀计
corrosion action	侵蚀作用	corrugate	起波纹状的；揉皱的；使起波纹
corrosion analysis	腐蚀分析	corrugated bar	竹节钢筋
corrosion behavior	腐蚀作用	corrugated bedding	波纹状层理
corrosion border	熔蚀边	corrugated coal	揉皱煤；波纹煤
corrosion cap	防腐盖	corrugated elbow	皱褶弯头
corrosion cell type	蜂窝型腐蚀	corrugated form	波形
corrosion control	侵蚀控制	corrugated friction socket	皱纹摩擦打捞筒
corrosion cracking	腐蚀断裂	corrugated furnace tube	波纹炉管
corrosion creep test	腐蚀蠕变试验	corrugated lamination	波纹状纹层
corrosion damage	腐蚀损害	corrugated mould	波浪边锭模
corrosion inhibitor	防蚀剂；阻锈剂	corrugated pipe	波纹管
corrosion mantle	熔蚀壳	corrugated ripple mark	波纹状波痕
corrosion margin	腐蚀外缘	corrugated sheet	波纹板
corrosion mitigation	缓蚀	corrugated steel	波纹钢板
corrosion of well	井腐蚀	corrugated tube	波纹管
corrosion prevention	防腐	corrugation	起皱；揉皱作用；灌水沟
corrosion probe	腐蚀探针	corrugation structure	煤的揉皱构造
corrosion products	腐蚀产物	corry	冰斗
corrosion proof	防腐的	corsilite	绿辉长岩
corrosion protection	防腐	corsite	球状辉长岩；球状闪长岩
corrosion rate	腐蚀速度	cortex	皮层
corrosion reaction	腐蚀作用	cortical	皮层的
corrosion remover	防蚀剂	cortlandtite	角闪橄榄岩
corrosion resistance	耐腐蚀性	corundellite	珍珠云母
corrosion resistant	防腐的	corundite	刚玉
corrosion resistant alloys	耐蚀合金	corundolite	刚玉岩
corrosion resistant steel	不锈钢；耐腐蚀钢	corundophilite	脆绿泥石
corrosion resisting steel	不锈钢	corundum	刚玉；金刚砂
corrosion resistivity	抗腐蚀性	corvusite	水复钒矿；氧钒多矾酸盐矿
corrosion rim	熔蚀边；熔蚀缘	cosalite	斜方辉铅铋矿
corrosion shortness	腐蚀脆性	cosedimentation	同时沉积作用；共沉积作用
corrosion sound meter	超声波腐蚀测定器	coseism	等烈度线；同震；同震线
corrosion spring swamp	溶蚀泉湿地	coseismal	同震线
corrosion treatment	防蚀；防腐蚀处理	coseismal line	同震线
corrosion water	腐蚀性水	coseismic	同震的
corrosion zone	溶蚀带	coseismic area	同震区
corrosional plain	侵蚀平原	coseismic circle	同震图
corrosion-resistant material	耐蚀材料	coseismic curve	同震曲线
corrosion-resisting valve	耐蚀灌浆阀	coseismic deformation	同震形变
corrosion-retarding	缓蚀的	coseismic dislocation	同震位错
corrosive	腐蚀性的；侵蚀的	coseismic displacement	同震位移

coseismic effect 同震效应
coseismic line 同震线
coseismic movement 同震运动
coseismic strain 同震应变
coseismic terrace 同震阶地
coseismic uplift 同震隆起
coseismic zone 同震带
coseparation 共分离
coset 复层组；交错层组
coslettizing 磷化处理；磷酸盐被膜防锈法
cosmic 宇宙的
cosmic dust 宇宙尘
cosmic geology 宇宙地质学
cosmic iron 陨铁
cosmic magnetic field 宇宙磁场
cosmic particle 宇宙粒
cosmic water 宇宙水
cosmical abundance 宇宙丰度
cosmical abundance of elements 宇宙元素丰度
cosmical background radiation 宇宙背景辐射
cosmical dust flux 宇宙尘通量
cosmical mineral 宇宙矿物
cosmical radiation 宇宙辐射
cosmochronology 宇宙年代学
cosmogenic hypothesis of mineralization 宇宙源成矿说
cosmogenic isotope 宇宙成因同位素
cosmogenic nuclide dating 宇宙成因核素测年法
cosmogenous sediment 宇宙沉积物
cosmogeology 宇宙地质学
cosmogeoscience 宇宙地学
cosmoline 防腐油
cosolvency 共溶性
cosolvent 共溶剂；助溶剂
cospectrum 同相谱
cost categories 储量成本分级
cost of drilling 钻井费用
cost of wasting 弃土费
cost of winning 开采成本
costean 井探；水冲勘探；探槽
costeaning pit 探坑
cosurfactant 助表面活性剂
cotectic 共结的；共溶的
coterminous 共边界的
cotter 栓；销
cotter bolt 带销螺栓
cotterite 珠光石英
cotton fabric multiply belt 棉织多层胶带
cotton rock 白垩状镁质灰岩
cotton sockfilter 棉套过滤器
cotton soil 棉土
cotunnite 氯铅矿
couch 河床
coulee 干河谷；熔岩流；深冲沟
coulee flow 舌状泥石流；黏熔岩流
coulisse 滑槽
couloir 峡谷
Coulomb analysis 库仑分析
Coulomb condition 库仑条件
Coulomb criterion 库仑准则
Coulomb damping 库仑阻尼
Coulomb failure criterion 库仑破裂准则
Coulomb force 库仑力
Coulomb friction 库仑摩擦
Coulomb friction damping 库仑摩擦阻尼
Coulomb line 库仑线
Coulombic interaction 库仑相互作用
Coulomb-Mohr shear failure theory 库仑-莫尔剪切破坏理论
Coulomb-Navier criterion 库仑-纳维尔强度准则
Coulomb-Navier strength theory 库伦-纳维强度理论
Coulomb's earth pressure 库仑土压力
Coulomb's earth pressure theory 库仑土压力理论
Coulomb's equation 库仑方程
Coulomb's equation for shear strength 库仑抗剪强度公式
Coulomb's failure criterion 库仑破坏准则
Coulomb's failure theory 库仑破坏理论
Coulomb's force 库仑力
Coulomb's law 库仑定律
Coulomb's law of friction 库仑摩擦定律
Coulomb's shear strength equation 库仑抗剪强度公式
Coulomb's soil-failure prism 库仑土破坏楔体
Coulomb's theory 库仑理论
Coulomb's theory of earth pressure 库仑土压力理论
Coulomb's wedge theory 库仑楔体理论
coulsonite 钒磁铁矿
counter 反的
counter balance 平衡块；配衡
counter balancing valve 反平衡阀；背压阀
counter bore 平底扩孔钻；钻平底孔
counter ceiling 平顶
counter collapse 抗倒塌
counter current 反流；逆流
counter curve 反曲线
counter dam 副坝；前坝
counter deflection 反挠度；反挠结构
counter diffusion 反向扩散
counter drain 堤脚排水沟
counter drive shaft 副传动轴
counter entry 平行平巷
counter field 反向场
counter flange 对接法兰
counter flooding system 对称浸水系统
counter flush 反洗孔
counter flush boring 反向冲洗钻进
counter gangway 中间运输平巷
counter head 平行平巷
counter level 中间水平
counter motion 反向运动
counter nut 埋头螺母
counter poise 衡重体
counter potential 反电位
counter pressure 反压力；平衡压力
counter radiation 反辐射
counter reservoir 反调节水库；平衡水库
counter rotating 反向旋转
counter rotation 反向旋转
counter septum 对隔壁
counter shaft 副轴
counter shear 反向剪切
counter sink drill 埋头钻；锥口钻

counter sinking bit 埋头钻；锥口钻
counter solvent 反萃溶剂
counter torque 反力矩；均衡扭矩
counter vein 交错矿脉；交错脉；交切脉
counter weight 平衡锤
counter weight fill works 反压填土工程
counteraction 反作用
counteractive 反对的
counteragent 中和力
counterbalance moment 抗衡力矩
counterbalance system 平衡力系
counterbalanced cage 带平衡锤的罐笼；平衡重笼
counterblast 逆流
counterblow hammer 锻锤；对击锤
counterbonification 反补偿
counterbuff 减震器
countercirculation-wash boring method 反循环洗井钻井法
counterclockwise 逆时针方向；左旋的
counter-clockwise twist 反手捻
counter-collapse skeleton 抗倒塌骨架
counter-current capacitor 逆电流电容器
countercurrent extraction 逆流萃取
countercurrent leaching 逆流浸出
counterflow 倒流；反向流；迎面流；逆流
counterflow underground burning 反向地下燃烧
counterflush drilling 反循环钻井
counterforce 反作用力
counterfort 扶壁；护墙
counterfort abutment 扶壁式桥台
counterfort dam 支墩坝
counterfort drain 排水扶垛
counterfort retaining wall 扶壁式挡土墙
counterfort wall 扶壁式挡土墙
counter-inclined fault 反倾断层
counterlode 交错矿脉
countermeasure 补救措施；防范措施
countermeasure against earthquake disaster 防震对策
countermeasure against landslide 滑坡防治措施
countermeasure against leakage of levee 防止堤坝渗漏措施
countermoment 均衡力矩
countermove 对抗措施
counterpart 副本
counterpoising 配重；压重
counter-regional fault 反向区域断层
counter-rotating mass 反旋转质量
counter-shots 反向放炮
countersinking 锥形扩孔
countersunk 埋头孔
countersunk bolt 埋头螺栓
countersunk nut 埋头螺母
countersunk rivet 埋头铆钉
countersunk screw 埋头螺钉
countertwist 反捻
counterweight fill 压重填土
counterweight fill work 压重填土作业
country rock 围岩；原岩
country rock alteration 围岩蚀变
couplaut 耦合剂
couple 连接
couple back 反馈

couple mode 耦合模式
couple of force 力偶
coupled 成对的
coupled frame shear wall structure 框架-剪力墙耦联结构
coupled impedance 耦合阻抗
coupled orocline 对偶变向造山带
coupled pendulum 耦合摆
coupled phenomena 耦合现象
coupled shear wall 并联剪力墙
coupled torsional and sway vibration 扭转和侧移的耦联振动
coupled translation and rotation 平移旋转耦合
coupled truck 拖挂式载重车
coupling 连接；耦合
coupling arrangement 连接装置
coupling beam 并联梁
coupling bend 弯管接头
coupling bolt 连接螺栓
coupling clutch 牙嵌离合器
coupling coefficient 黏着系数；耦合系数
coupling compound 耦合化合物
coupling degree 耦合度
coupling face 接箍端面
coupling failure 连接破坏
coupling flange 连接法兰
coupling hose 连接软管
coupling lever 离合器拉杆
coupling link 联结杆
coupling machine 管接头拧装机
coupling nut 连接螺帽
coupling of energy 能量耦合
coupling of modes 振型耦联
coupling orogeny 耦合造山作用
coupling shaft 连接轴
coupling sleeve 连接套管
coupling term 耦合项
coupling tubing connection 油管接箍连接
coupling with plain ends 无螺纹接箍
coupling zone of metamorphism 双变质带
coupon 试件
coupon corrosion test 挂片腐蚀试验
course 测线；走向；巷道；过程
course bearing 轨迹方向
course deviation 轨迹偏移
course length 轨迹长度
course of crystallization 结晶过程
course of seam 岩层走向
course vertical depth 井身垂直深度
coursed rubble wall 分层毛石墙
court dock 封闭谷
courtzilite 地沥青
couterclockwise rotation 反时针旋转
covalency 共价
covar 柯伐合金
cove 凹口；凹圆线；陡壁小谷；河湾；山凹
cove beach 湾头滩
covelline 铜蓝
covellite 铜蓝
covenant 契约
cover 保护层；表土；覆盖物；盖板

cover annealing	罩式炉退火	cow dung bomb	牛粪状火山弹
cover bolting	紧盖螺栓及螺帽	cow sucker	钢绳端悬重
cover caving	顶板崩落	coyote blast	洞室爆炸
cover coat	盖层	coyoting	不规则小窑开采；滥采
cover depth	埋深	CR figure	反旋干涉图
cover glass	防护玻璃罩	crab bucket	抓斗
cover hole	按井网钻的井；超前探水钻孔；勘探孔	crab derrick	抓斗式起重机
cover layer	防护层；覆盖层	crab winch	卷扬机；起重绞车
cover meter	面层测厚仪	crabbing	侧向飞行
cover nut	套帽	crack	开裂；裂缝
cover of unconformable strata	不整合盖层	crack and seat	破碎稳定工法
cover plate	阀盖	crack arrestor	止裂器
cover rock	盖层；上覆岩石	crack cutting	龟裂掏槽
cover thickness	剥离厚度；覆盖层厚度	crack deformation	裂缝变形
cover tile	盖瓦	crack depth	裂缝深度
cover type	覆盖类型	crack detection	裂纹检测；探伤
coverage	覆盖范围；室内面积	crack detector	探伤仪
coverage area	上盖面积	crack extension criterion	裂纹扩展判据
coverage coefficient	波及系数，覆盖系数	crack extension force	裂缝扩展力
coverage control	覆盖控制	crack formation	裂纹形成
coverage pattern	可达范围图形；覆盖区图形	crack gauge	测缝计
coverage scale	覆盖率；上盖面积比例	crack grouting	裂缝灌浆
coverage thickness	覆盖层厚度	crack growth	裂纹扩展
coveralls	工作服	crack growth rate	裂纹增长速率
cover-degree	覆盖度	crack initiation	裂纹萌生；起裂
covered	隐蔽的；有盖的	crack initiation energy	起裂能
covered area	室内面积；有盖地方；有盖面积；建筑面积	crack length	裂隙长度
covered canal	封闭式渠道	crack meter	裂纹探测仪
covered channel	暗渠；有盖排水渠	crack monitoring	裂缝监测
covered conduit	暗沟	crack morphology	裂纹形貌
covered digging method	盖挖法；暗挖法	crack opening	裂缝开度
covered drain	暗渠	crack pattern	裂隙模式
covered drainage	暗沟排水	crack penetration	裂纹穿透
covered electrode	涂料焊条	crack propagation	裂纹扩展
covered gutter	暗沟	crack propagation velocity	裂纹扩展速度
covered karst	覆盖型岩溶；隐伏岩溶	crack reflection	裂缝反射
covered nullah	大型暗渠	crack resistance	抗裂度
covered pedestrian walkway	有盖行人道	crack saw	裂缝切割机
covered plain	覆盖平原	crack seal vein	裂缝愈合脉
covered region	覆盖区	crack sealing mechanism	开裂愈合机制
covered reservoir	有盖配水库；地下水池	crack survey	裂缝调查
covered sewer	暗置污水渠	crack tip	裂纹尖端
covered structure	潜伏构造	crack treatment	裂缝处理
covered working area	有盖作业地方	crack water	裂隙水
covering	覆盖；消声；掩盖	crack-central type eruption	裂隙中心式喷发
covering body	封堵体；培土器	cracked gas	裂解气
covering graph	覆盖图	cracked tension zone	张裂区
covering layer	覆盖层	cracked valve	油嘴
covering power	覆盖能力	cracked zone	破碎带
covering slag	保护渣	cracker	裂化装置；破碎机
covering strata	盖层	cracker assembly	柔杆钻具总成
covering stratum	盖层	crack-free	无裂纹
covering up of works	工程的盖封	cracking	开裂；裂缝；破裂
covering work	护面作业	cracking coil	裂化炉管
covert	隐蔽处	cracking gas compressor	裂解气压缩机
coverture	覆盖物	cracking level	裂化深度
coveyor drive	输送机传动装置	cracking load	开裂荷载；破坏荷载
covibration	共振	cracking moment	开裂力矩
coving	凹蚀	cracking of concrete	混凝土开裂
covite	闪霞正长岩；暗霞正长岩	cracking of wall	墙开裂

cracking per pass	单程裂化率
cracking pressure	开裂压力
cracking resistance	抗裂性；抗裂强度
cracking resistant coefficient	抗裂系数
cracking severity	裂化苛刻度
cracking stock	裂解原料
cracking zone	裂化段
cracking-resistant material	防裂材料
crackle	裂纹；噼啪声
crackle breccia	裂纹角砾岩
crack-relief layer	开裂缓和层
cradle	洗矿槽；支座
cradle bedding	管子的基墩
cradling	支承
crag	壳质砂层；峭壁；砂质泥灰岩；岩石碎片
cragged	崎岖的
craig and tail	鼻尾丘
Craig water-flood prediction method	克雷格注水预测法
craigmontite	淡霞正长岩
craignurite	富玻英安岩
cramp	夹线板；扣钉；铁夹钳
crampon	吊钩
crandallite	纤磷钙铝石
crane	吊车；起重机；升降设备
crane arm	起重机吊杆
crane barge	起重机驳船；浮吊
crane beam	吊车梁
crane boom	起重机吊杆
crane driver	起重机司机
crane hook	起重机吊钩
crane rope	吊索
crane tower	起重机塔
crane truck	汽车吊
crane winch	起重绞车
craneman	起重机司机
crane-mounted	装于吊管机上的
crane-runway girder	起重机行车大梁
crank mechanism	曲柄机构
crank press	曲轴压力机
cranny	裂缝
craquele	网状裂纹
craquelure	龟裂缝
crash flood	快速进水，紧急注水
crash pad	防震垫
crassicutinite	厚壁角质体
crassisporinite	厚壁孢子体
crater	夯坑；火山口
crater basin	火口盆地
crater bottom	火山口底
crater crack	火口裂纹
crater effect	切口效应
crater fill	火山口充填物
crater floor	火口原；火山口底
crater formation	凹坑形成
crater formed by blasting	爆破漏斗
crater index	爆破漏斗指数；爆破作用指数
crater lake	火山口湖
crater magma	火山口岩浆
crater of elevation	高海拔火山口
crater of eruption	喷溢火山口
crater radius	爆破漏斗半径
crater rim	火山口边缘
crater ring	低平火山口沿；火山口环
crater terrace	火山口阶地
crater test	爆破漏斗试验
crater vent	火山口孔道
crater wall	火山口墙
crateriform	火山口状
cratering	形成坑穴
cratering action	造坑作用
cratering well	在表层套管外面漏气的井
crater-like depression	火山口式陷坑
cratogene	地盾
craton	古陆核；克拉通；稳定地块
craton basin	克拉通盆地
craton margin	克拉通陆缘
cratonic block	克拉通地块
cratonic cover	克拉通盖层
cratonic crust	克拉通地壳
cratonic embayment	克拉通支地槽
cratonic margin	克拉通边缘
cratonic shelf edge	克拉通陆棚外缘
cratonization	克拉通化
craven fault	小断层
crawl	爬行
crawl cutter	爬行切管机
crawler	履带式车辆；爬行物
crawler asphalt paver	履带式铺沥青机
crawler crane	履带式起重机
crawler drill	履带式钻机
crawler excavator	履带式挖掘机
crawler hydraulic bench drill	履带式液压凿岩台车
crawler loader	履带式装载机
crawler shovel	履带式铲土机
crawler single bucket excavator	履带式单斗挖掘机
crawler soil stabilizer	履带式稳定土的机械
crawler tractor	履带式拖拉机
crawler type pile driver	履带式打桩机
crawler-mounted	装在履带式拖车上的
crawler-mounted drilling rig	履带式钻机
crawler-mounted excavator	履带式挖土机
crawler-tractor-mounted bulldozer	履带式拖拉-推土机
crawler-type loader	履带式装载机
crawling	蠕变；涂漆不均
crawling burrow	爬行潜穴
crawling trace	爬行迹
crawling traction	履带牵引
cray	蠕变
craze	细裂纹
crazing	龟裂
creaming	形成乳状液
creams	钻探用特种金刚石
crease	折痕；皱痕
crease structure	皱纹构造
crease-flex test	折叠试验
creasing	折缝
created fracture	人工裂缝
credence	凭证；信任
credibility	可靠性
credible debris flowpath	可能的泥石流通道

crednerite	锰铜矿
creedite	铝氯石膏
creek brook	溪流
creel stand	筒子架
creeling	换筒子
creep	底板隆起；频率漂移；蠕变；探测海底；蠕动
creep behavior	蠕变性
creep behaviour test	蠕变性状试验
creep buckling	蠕变压曲
creep cell	蠕变传感器
creep coefficient	蠕变系数
creep collapse	蠕变破坏
creep contraction	蠕变收缩
creep curve	蠕变曲线
creep damage	蠕变损伤
creep deformation	蠕动变形
creep down	滑落
creep effect	蠕变效应
creep failure	蠕变破坏
creep flow	蠕动流
creep fracture	蠕动断裂
creep function	蠕变函数
creep instability	蠕变失稳
creep limit	蠕变极限
creep load	蠕变荷载
creep mechanism	蠕变机理
creep motion	蠕变运动
creep movement of fault	断层蠕动
creep observation	蠕变观测
creep of concrete	混凝土蠕变；混凝土徐变
creep of debris	碎屑物蠕动
creep of recovery	蠕变恢复
creep of rock	岩石蠕变
creep of soil	土层蠕动
creep point	蠕变屈服点
creep potential	蠕变势
creep pressure	蠕变压力
creep rate	蠕变率
creep ratio	蠕变比
creep resistance	蠕变强度；蠕变阻力
creep rupture	蠕变断裂
creep rupture strength	蠕变断裂强度
creep rupture test	蠕变破损试验
creep settlement	蠕变沉降
creep slippage	蠕变滑动
creep speed	蠕变速度
creep strain	蠕应变
creep strength	蠕变强度
creep stress	蠕变应力
creep structure	蠕动构造
creep test	蠕变试验
creep value	蠕变量
creep well	蠕变观察井
creepage	蠕动；徐变
creeper	定速运送器；链式爬车器；爬行物
creeper chain	爬车机带角链
creeper derrick	履带式起重机
creeping	底板隆起；蠕变；塑流
creeping flow	蠕动流
creeping motion	蠕动
creeping pressure	蠕变压力
creeping rock mass	蠕变岩体；蠕动岩体
creeping speed	蠕变速度
creeping waste	蠕动坡积物
creeping wave	蠕波
creepmeter	蠕变仪
creep-path length	爬径长度
creep-time curve	蠕变时间曲线
crenogenous	矿泉的
crenulate	细褶皱
crenulated bedding	细褶皱层理
crenulation	细褶皱；细皱纹；小圆齿
crenulation cleavage	细褶皱劈理；褶劈理
crenulation lineation	皱纹线理
creodont-like teeth	似食肉齿
creosoted pile	油浸桩
crepage	退捻回缩率
crescent	新月；新月形的
crescent and mushroom	新月蘑菇形
crescent beach	新月形海滩
crescent cast	新月形水流痕；新月形铸型
crescent dune	新月形沙丘
crescent lake	弓形湖；牛轭湖；月牙湖
crescent mark	新月形痕
crescent scour mark	新月形冲蚀痕
crescent type cross-bedding	槽状交错层理
crescent wrench	钩扳手
crescentic bar	新月形沙坝
crescentic barchan	新月形沙丘
crescentic crack	新月形裂缝
crescentic depression	新月形凹陷
crescentic dune	新月形沙丘
crescentic fracture	新月形裂缝
crescentic lake	新月形湖；月牙湖
crest	峰顶水位；山脊；斜坡顶部；褶皱顶部
crest access	坝顶公路
crest clearance	顶部间隙；端部间隙
crest contraction	堰顶收缩
crest depth	堰顶水深
crest discharge	洪峰流量
crest elevation	波峰高；顶高
crest flood	洪峰
crest forecast	洪峰预报
crest height	波峰高度
crest length	坝顶长度
crest level	堤顶高程；顶高程
crest line	峰线；脊线；褶皱脊线
crest of flood	洪峰
crest of fold	褶皱脊
crest of screw thread	螺纹牙顶
crest of slope	坡顶
crest of tide	潮峰
crest of tooth	齿顶
crest of weir	堰口
crest plane	脊面
crest plane of fold	褶皱脊面
crest segment	洪峰段
crest stage forecast	峰顶水位预报
crest surface	脊面
crest surface of fold	褶皱脊面

crest thicken fold 顶厚褶皱
crestal attenuation 脊部衰减
crestal collapse fracture 背斜顶部的塌陷裂缝
crestal culmination 脊顶
crestal development 顶部开发
crestal gas injection 顶部注气
crestal injection well 构造顶部注入井
crestal plane 脊顶面；脊面
crestal waterflood 顶部注水
crestal well 构造顶部井
crest-arc chord 拱坝坝顶弦长
crestmoreite 单硅钙石
crest-stage indicator 洪峰水位指示器；最高水位指示器
creta 白垩；漂白土
cretaceous 白垩纪
cretaceous sandstone 白垩纪砂岩
cretaceous transgression 白垩纪海进
cretadiscus 白垩盘石
creter crane 胎带机
creusot loire uddeholm process 蒸汽氧脱碳法
crevasse 决口；冰裂隙；冰隙；裂隙
crevasse channel 决口水道
crevasse crack 龟裂；裂缝
crevasse deposit 冰隙沉积；决口沉积
crevasse filling 裂隙填充
crevasse in glacier 冰川裂隙
crevasse of glacier 冰川裂隙
crevassed glacier 裂隙冰川
crevasse-splay deposit 决口扇沉积物
crevasse-splay forest 石林
crevice 缝隙；裂隙
crevice blow 爆炸气体从裂隙中喷出
crevice corrosion 缝隙腐蚀
crevice karst 石林
crevice oil 裂隙油
crevice water 裂隙水
creviced formation 裂隙地层
crib 格排；框形物
crib bed 基础垛盘
crib building 架设木垛
crib check dam 木笼拦沙坝
crib cofferdam 木笼围堰
crib dam 木笼坝；水闸
crib retaining wall 框格式挡土墙
crib ring 井框
crib timbering 垛式支架
crib wall 格状挡墙；网格式墙
cribbing 叠木；井框支架
cribbing and matting 叠架和排列；方木纵横叠架成的底座
cribble 筛
cribellate 具筛孔的
cribellum 筛板
cribwork 木笼井架混凝土桥墩；木笼围堰桥墩；石笼框垛
cribwork wall 木笼岸壁
cricondenbar 临界凝析压力
cricondentherm 临界凝析温度
crimp 褶皱
crimp contraction 皱缩率

crimp gauge 卷曲度测定仪
crimp rigidity 卷曲刚度
crimper 卷边机
crimple 皱痕
crinanite 橄沸粗玄岩
crinkle 皱
crinkle mark 细皱痕
crinkled 成波状的；皱的
crinkled bedding 旋卷层理
crinkling 微褶皱
crinoid bed 海百合层
crinoid ossicle 海百合骨板
crinoidal limestone 海百合灰岩
crioscopic method 冰点降低法
crippling 断裂；局部失稳
crippling load 临界荷载；弯曲荷载
crippling stress 断裂应力；压屈应力
criquina 海百合屑灰岩
criquinite 致密海百合屑灰岩
crisis export 转嫁危机
crispation 卷缩
criss-cross 十字形的
crisscross bedding 交错层理
criss-cross construction 十字交叉结构
criss-cross grouted vein 交错灌浆脉
criss-cross method 方格计数
criss-cross pattern 方格图像
criss-cross shaft 十字轴
crisscross structure 方格构造
cristate 鸡冠状的
cristobalite 方英石；方石英
cristograhamite 脆沥青
criteria for ore prospecting 找矿标志
criterion 标准；判据；准则
criterion for rupture 破裂判据
criterion function 准则函数
criterion of convergence 收敛判据
criterion of failure 破坏准则
criterion of intensity 烈度标准
criterion of perfect fidelity 高保真度准则
criterion of strengthening 加固准则
critical 临界的；临界质量
critical acceleration 临界加速度
critical angle 临界角
critical angle of incidence 临界入射角
critical angle of refraction 临界折射角
critical angle of total reflection 全反射临界角
critical area of formation 井眼周围的油层地带
critical band 临界频带
critical bar 险滩
critical bottom-hole pressure 井底破裂压力
critical breakdown potential 临界击穿电位
critical buckling load 临界弯曲荷载
critical building 情况严重楼宇
critical characteristic 临界特性
critical charge depth 临界装药深度
critical circle 临界圆
critical coefficient 临界系数
critical cohesion 临界黏聚力
critical composition 临界组成

critical compression pressure	临界压缩压力
critical compression ratio	临界压缩比
critical compressive stress	临界压应力
critical concentration	临界浓度
critical condition	临界条件；临界状态
critical configuration	临界构型
critical coning rate	临界锥进速度
critical constant	临界常数
critical constraint	临界约束
critical coupling	临界耦合
critical crack length	临界裂纹长度
critical curve	临界曲线
critical damping ratio	临界阻尼比
critical density	临界密度
critical depth	临界深度；临界水深
critical depth of excavation	临界开挖深度
critical depth of salinization	盐碱化临界深度
critical desorption pressure	临界解吸压力
critical deviator stress	临界偏应力
critical diameter	临界直径
critical differential pressure	临界压差
critical dip	临界倾角
critical discharge	临界流量
critical displacement	临界位移
critical domain	临界区
critical drawdown	临界降深
critical duration	临界持续时间
critical earth pressure	临界土压力
critical earthquake risk area	地震重点危险区
critical edge pressure	临塑荷载
critical energy	临界能量
critical envelope curve	临界包络曲线
critical erosion velocity	临界侵蚀速度
critical exit gradient	临界逸出梯度
critical factor	关键因素
critical failure surface	临界破坏面
critical floatation gradient	临界浮动梯度
critical flow	临界流；临界流动；临界流量
critical flow nozzle	临界截面喷管
critical flow prover	临界流检测计
critical flow rate	临界流量
critical flow velocity	临界流速
critical flowmeter	临界流量计
critical fluidization velocity	临界流化速度
critical frequency	临界频率
critical friction velocity	临界摩擦流速
critical gradient	临界梯度
critical grid current	临界栅极电流
critical ground motion	临界地面运动
critical head	临界水头
critical height of slope	斜坡临界高度
critical hole depth	临界井深
critical humidity	临界湿度
critical hydraulic gradient	临界水力梯度
critical inclination of slope	斜坡临界倾斜度；边坡临界倾角
critical length of pile	临界桩长
critical level	临界水平；临界高度
critical limit	临界限度
critical load	临界荷载
critical load of subsoil	地基临界荷载
critical locus	临界轨迹
critical mass	临界质量
critical micelle concentration	临界胶束浓度
critical mineral	临界矿物
critical mineral concentration	临界矿化度
critical moisture	临界湿度
critical moisture content	临界含水量
critical nucleus radius	临界晶核半径
critical parameter	临界参数
critical part	主要零件
critical particle size	临界粒度
critical path method schedule	关键路线法进度
critical penetration frequency	临界穿透频率
critical phase	临界相
critical phenomenon	临界现象
critical pitting potential	临界点蚀电位
critical point	临界点
critical point of explosion	临界爆点
critical pore ratio	临界孔隙比
critical porosity	临界孔隙度
critical pressure	临界压力
critical pressure ratio	临界压力比
critical pump rate	临界泵量
critical range	临界范围；临界区
critical rate of production	临界产量
critical ratio	临界比
critical reflection	临界反射
critical refraction	临界折射
critical region	临界域
critical relaxation time	临界松弛时间
critical release factor	临界排因子；临界排放系数
critical rotary speed	临界转速
critical saturation	临界饱和度
critical section	临界截面；临界区
critical shear stress	临界切应力
critical slide curve	临界滑弧
critical slide surface	临界滑动面
critical slip surface	临界滑动面
critical slope	临界边坡；临界坡度；临界坡降
critical slope angle	临界坡角；休止角
critical softening	临界软化
critical solution temperature	临界溶解温度
critical spacing	极限井距
critical speed	临界速度
critical stable state	临界稳态
critical state	临界状态
critical state design	临界状态设计
critical state energy theory	临界物态能量理论
critical state line	临界状态线
critical state model	临界状态模型
critical state soil mechanics	临界状态土力学
critical state strength	临界状态强度
critical strain	临界应变
critical stress	临界应力
critical surface	临界面
critical surface tension	临界表面张力
critical temperature	临界温度
critical tractive force	临界牵引力
critical tractive velocity	起动流速；临界牵引速度

critical twist	临界捻度
critical value	临界值
critical velocity	临界速度
critical void density	临界孔隙度
critical void ratio	临界孔隙比
critical volume	临界容积
critical water content	临界含水量
critical water level	临界水位
critical wave length	临界波长
critical weight on bit	临界钻压
critical well	易出问题的井
critical work	关键工作；临界功
critical-depth flume	测流槽；临界量水槽
criticality	临界性
criticality parameter	临界参数
critically damped seismograph	临界阻尼地震仪
critically limit	临界极限
critique	批评
crivaporbar	临界蒸气压力
crocidolite	蓝石棉
crockery	陶器
crocodile	鳄鱼；形成交叉裂缝；轧体前端的分层
crocodile clip	鳄鱼夹
crocodile shears	杠杆剪切机；鳄鱼剪
crocoite	铬铅矿
crocus	磨粉
crocus cloth	细砂布
crolite	陶瓷绝缘材料
cromaltite	黑榴霓辉岩；碱性辉岩
Cromerian interglacial stage	克罗默尔间冰期
croning process	壳型铸造法
cronstedtite	弹性绿泥石
crook	挠曲；弯曲
crooked	挠曲的
crooked drill pipe	弯曲钻杆
crooked hole	歪斜钻孔
crooked hole country	钻井容易钻弯的地区
crooked line	弯曲测线
crooked pipe	绕曲管
crooked stem	弯曲的钻杆
crooked well	钻弯的井
crooked zone	弯曲带
crooked-hole tendency	井眼弯曲趋势
crookedness	弯曲
crookesite	硒铊银铜矿
crook-veined	曲脉穿插的
crop	露头
crop bucket	切头箱
crop shears	剪头机
crop up	突然发生
cropping	出露；露头
cross	横断；横越；交叉的；十字的
cross adit	横通道
cross anticline	横背斜
cross arrangement	十字排列
cross array	十字排列
cross axis sensitivity	横向灵敏度
cross axle	横轴
cross bar	横撑木；十字杆件
cross beam	端梁；横梁
cross beam system	交叉梁系
cross bed	交错层
cross bedding	交错层理；交错层面
cross bit	十字扁铲；十字形钻头
cross bond	交叉砌合；交联键
cross brace	交叉斜撑
cross bracing	交叉联结
cross breaks	横折
cross bridging	交叉撑
cross bunker	横煤舱
cross calibration	交互标定
cross channel	横向水道
cross chipper hammer	爪形冲击锤
cross chisel	掏槽凿
cross coil	交叉线圈
cross component	侧向分量
cross connection	交叉连接
cross conveyor	横向输送机
cross correlation	互相关
cross correlation filter	互相关滤波器
cross correlation flowmeter	互相关流量计
cross coupling effect	交叉耦合效应
cross crack	横裂纹
cross current	横流；交叉水流；涡流
cross curve	十字线
cross cutter	十字形牙轮；横切切割器
cross cutting	穿脉；交叉平巷；石门掘进
cross dam	丁坝；横坝
cross dip	横向倾角
cross direction	横向，交叉方向
cross ditch	横沟
cross drain	横向排水沟
cross drainage	横向排水
cross drill bit	十字钻头
cross dyke	丁坝
cross eddy	横向涡流
cross equalization	互均化
cross fan	交叉扇形排列
cross fault	横断层
cross flexure	横向挠曲
cross flow	横向流动；漫流
cross fold	横褶皱
cross folding	横褶皱
cross force	横向力
cross fracture	横断裂
cross gangway	斜向平巷
cross garnet butt	十字铰链
cross gate	横流道；横巷；联络道；人行通道
cross girder	横梁
cross girdle	交叉环带
cross grain	横纹
cross hair	十字丝；十字准线
cross hatch	交叉影线
cross helical gear	螺旋齿轮
cross hole method	跨孔法
cross hole shooting	跨孔激振法
cross hole technique	交叉孔法
cross hole test	跨孔试验
cross iron	十字形铁
cross joint	横节理；交错节理；十字接头；四通

cross lamination	交错纹理
cross lay	交叉捻向；交叉设置
cross levee	横堤
cross level	横向水准仪；横巷；纵横水准仪
cross lineament	交错线性构造
cross link	交联
cross linking agent	交联剂
cross machine direction	横机向
cross measures	横向测量；交错层组
cross migration of fluid	流体横向运移
cross modulation	交叉调制
cross mouthed drill	十字形钻；十字形钻泥船
cross osar	横向蛇丘
cross over walking	过零晃动
cross peen sledge	地质手锤
cross piling	交叉堆垛
cross pin type joint	万向节头
cross pipe	十字管
cross pipeline	穿越管线
cross pitch	走向
cross pitch mining	沿走向开采
cross plot	交会图
cross poling	横撑木
cross power spectrum	互功率谱
cross profile	横断面
cross ripple mark	交错波痕
cross road duct	横越道路管道
cross roll straightener	斜辊矫直机
cross roller rock bit	十字形四牙轮钻头
cross sea	暴涛；横浪
cross section	横截面；横切面
cross section method	断面法
cross section of river	河玲断面
cross section profile	横断面图
cross sectional area	横截面面积
cross sectional view	横断面视图
cross set	交错层组
cross shaft	横轴
cross shaped	十字形的
cross slide	横向溜板
cross slip	交叉滑移
cross slope	横坡
cross spacer	横向间隔物
cross spectral power	互相关谱功率
cross spread	相交排列
cross stratification	交错层理
cross stratum	交错层
cross strike direction	垂直于走向的方向
cross superposition	横跨叠加
cross swell	横向海流；交错海浪；逆涌
cross tee	四通接头
cross threading	螺纹错扣
cross tide	逆潮；横向潮流
cross timber	横木
cross track scanning	垂直扫描
cross traverse winding	交叉卷绕
cross valley	横向谷
cross valve	十字阀；转换阀
cross vein	交错矿脉
cross ventilation	对流通风
cross wall	横墙；交叉墙
cross wave	三角浪
cross wind	侧风
cross wire	十字丝
cross-anisotropic soil	横观各向异性土
crossarm	横臂；横木
cross-arm settlement gauge	十字形沉降计
cross-axis fracture	横轴裂缝
cross-band twist	顺手捻
crossbar network	纵横开关网络
crossbedded rock	有交错层理的岩石
cross-bedded strata	交错层
cross-bedded structure	交错层构造
cross-beds	交错层
crossbinding	横向连接
cross-bladed chisel bit	十字刮刀楔形齿钻头
crossbolt	对扣螺栓
cross-borehole electromagnetic probing	井间电磁探测；孔间电磁探测
cross-borehole information	井间资料；孔间资料
cross-borehole seismic probing	井间地震探测；孔间地震探测
cross-borehole transmission	井间透射；孔间透射
crossbreaking	横断
cross-bridge	横桥
cross-check	相互检验
cross-chopping bit	十字冲击钻头
crosscut	横切；横巷；捷径；联络眼
crosscut method	横巷采矿法
crosscut shears	横向剪切机
cross-cutting fault	横切断层
cross-cutting relationship	横切关系；正交关系
cross-dating	交叉年代测定法
cross-drifting	小型迁移交错层理
crossed filter process	交叉滤色法
crossed nicols	正交偏光镜
crossed polarizers	正交偏光镜
crossed strip foundation	十字交叉条形基础
cross-edged bit	十字形钻头
crossed-girdle fabric	交叉环带组构
cross-energy spectrum	交互能谱
cross-equalization filter	相互均衡滤波器
cross-examination	盘问，交互讯问
cross-faulted anticline	横断背斜
crossfeed	互馈
crossflooding	改变方向注水；横贯注水
crossflow effect	窜流效应
cross-flow force	水流横向力
cross-flow separator	交叉流分离器
crossfoot	用不同计算方法核对总数交叉结算
cross-hatched area	断面线区域
crosshead beam	十字头联杆
crosshead die	直角机头模
crosshead extension rod	介杆
crosshead guide	十字头导板
crosshead inspection plate	十字头视孔盖
crosshead oil plug	十字头润滑油塞
crosshead oil tray	十字头油槽
crosshead pin	十字头销
crosshead pin bushing	十字头销圈

crosshead retainer 十字头护圈
crosshead shoe 十字头闸瓦
cross-heading 联络巷道
cross-hole acoustic logging 井间声波测井
crosshole electromagnetic tomography 井间电磁波成像
cross-hole region 井间区域
cross-hole resistivity tomography 井间电阻率层析成像
cross-hole section 井间剖面
cross-hole seismic investigation 井间地震测量
cross-hole seismic scanning 井间地震扫描
cross-hole seismogram 井间地震记录
cross-hole tomography 井间层析技术
cross-identification 交叉识别
crossing angle 交会角
crossing permit 穿跨越许可
crossing pit 河流穿越槽
crossing point 交点
crossing structure 穿越结构
crossing wave 横波
crossing zero 过零
cross-interaction 交叉作用
cross-interrogate 交叉询问
crossite 铝铁闪石；青铝闪石
crossline 穿越管线；横侧线；交叉线；十字丝
cross-line migration 横向测线偏移
crossline section 横向测线剖面
crosslink density 交联密度
crosslinkable plasticizer 可交联增塑剂
crosslinked action 交联作用
crosslinked aqueous gel 交联水基冻胶
crosslinked fluid 交联液
crosslinked foams 交联泡沫
crosslinked gel 交联冻胶
crosslinked guar 交联瓜尔豆胶
cross-linked high polymer 交联高聚物
crosslinked hydroxypropyl guar gel 交联羟丙基瓜尔胶
cross-linked polyadducts 交联加成聚合物
crosslinked polymer fluid 交联聚合物压裂液
cross-linked resin 交联树脂
crosslinked viscosity 交联黏度
crosslinker 交联剂
crosslinking 交联
crosslinking chemical 交联剂
cross-linking density 交联密度
crosslinking feature 交联特性
crosslinking reaction control 交联控制
crossmember 横构件；横梁
crossover 跨越
crossover assembly 转换装置
crossover bend 转换管道弯脖
crossover circulation 反循环
crossover circulation technique 反循环技术
crossover coupling 转换接头
crossover distance 超前距离
crossover flange 转换法兰
crossover frequency 分隔频率；交叉频率
crossover gravel packing method 转换砾石充填法
crossover joint 交叉接头
crossover kit 转换工具
crossover line 跨线；重叠的输送管

crossover placement method 转换充填法
crossover point 交点
crossover port 转向孔
crossover region 交叠区
crossover seal assembly 转换密封总成
cross-over seat 转换阀座钻屑天然气
cross-over shoe 交叉液流汇合装置；转向接头
crossover sub 过渡接头；转换接头
crossover tool 转换工具
crossover valve 转换阀
crosspiece 过梁；横挡；绞车横杆；四通管
cross-plane permeability 横面渗透性
crossplot point 交会点
crossplot porosity 交会孔隙度
crossplot technique 交会图技术
crossrail 横轨
crossrange 侧向
cross-reference 相互参照
cross-reference table 相互对照表
cross-resistivity method 跨井电阻率法
cross-ripple 交叉波痕
cross-section cone bit 四牙轮钻头
cross-section cutter 十字形排列的牙轮
cross-section display 剖面显示
cross-section paper 方格纸
cross-section survey 横断面测量
cross-sectional area 横截面积；断面面积
cross-sectional cone bit 十字形牙轮钻头
cross-sectional dimension 横断面尺寸
cross-sectional drawing 横断面图
cross-sectional flow area 流动截面积
cross-sectional method 断面法
cross-sectional model 横截面模型
cross-sectional profile 横断面图
cross-sectional shape 截面形状
cross-shaped 交叉形的；十字形的
cross-tie 横向拉杆
cross-track 联络测线
cross-verification 交互检验
cross-well data 井间数据
cross-well image 井间成像
crosswise 斜；交叉地
crosswise lamination 交错层压
crotch 叉状物；岔口
crotchet 小钩
crotovina 掘土动物穴
crow 铁挺
crowbar 撬棍
crowbar connection 撬杠连接
crowbar switch 短路开关
crowd loading 集束荷载
crowd shovel 正铲
crowder 沟渠扫污机
crowding 加密
crowding effect 集聚效应
crowdion 挤列
crow-fly distance 直线距离
crowfoot 打捞钻杆工具；吊索；防滑三脚架；鸟足构造
crowfoot bar 带拔钉子叉头的大撬杠
crowfoot cracks 皱裂

crow-foot drainage pattern	鸟足状水系	crude hydrocarbons	天然碳氢化合物
crowfoot guided valve	爪子扶正阀	crude lead dating	粗铅测年法
crown	齿冠；拱顶；井架顶部；隆起；路拱；钻冠	crude lifting procedure	提油程序
crown bit	环状钻头；活头钎子；金刚石钻头	crude material	原料
crown block	拱顶石；起重定滑轮；钻井架顶部滑轮组	crude metal	粗金属
crown buoy	锚顶浮标	crude oil	原油
crown chamfer	冠部倒角	crude oil allocation	原油分配
crown ditch	截水沟；天沟	crude oil demand	原油需求
crown drill	顶钻；顿钻；活头钎子	crude oil emulsion	原油乳化液
crown elevation	堤顶高程	crude oil forward contract	原油期货合同
crown for chilled shot	研钢砂钻头	crude oil forward market	原油期货市场
crown gate	进水闸门；引水闸门	crude oil heater	原油加热炉
crown hole	冒顶洞	crude oil lifting procedure	原油提取程序
crown line	锚顶浮标系绳	crude oil production	原油生产
crown of arch	拱顶	crude oil recovery process	原油回收方法
crown of landslide	断崖顶	crude oil sample analysis	油样分析
crown of levee	堤顶	crude oil shrinkage	原油收缩
crown pillar	阶段间矿柱	crude oil stabilizer	原油稳定装置
crown plate	顶板	crude oil term processing contract	原油加工合同
crown platform	天车平台	crude ore	粗矿；未选的矿石；原矿石
crown plug	顶部堵塞器	crude ore bin	原矿舍
crown profile	钻头冠的外形	crude output	原油产量
crown pulley	天车	crude petroleum	原油
crown safety platform	天车安全台	crude production	原油生产
crown section	拱冠断面	crude production rate	原油生产速度
crown settlement measurement	拱顶下沉测量	crude runs	原油加工量
crown sheave	天车滑轮	crude sampler	原油取样器
crownblock beam	天车梁	crude scale wax	粗汗蜡
crownblock protector	天车防碰装置	crude shale oil	粗页岩油
crowned bit	阶梯式钻头	crude specification	原油规格
crowngear	伞形齿轮	crude stabilization fundamental	原油稳定原理
crowning	中凸	crude still	原油蒸馏塔
crozzling coal	炼焦煤	crude stream	原油流
cruciate	十字形的	crude trunk line	原油输送干管
crucible	坩埚	crude wax	原石蜡
crucible assay	坩埚试金	crude-distillation stabilizer	原油蒸馏稳定塔
crucible swelling number	坩埚膨胀序数	crumb	碎屑
cruciform	十字形的	crumb rubber modified asphalt	改质橡胶沥青
cruciform bit	十字钻头	crumb structure	团粒结构
cruciform clay cutter	十字形黏土冲切器	crumbing crew	清沟班
cruciform vane	十字板；十字形叶片	crumble	崩碎；粉碎；破碎
crude	粗糙的；原生的；原油	crumble peat	松散泥炭
crude asphalt	粗沥青	crumbliness	破碎性
crude asphaltic petroleum	沥青基原油	crumbling	崩解；剥落；破碎；皱纹
crude assay	原油分析；原油鉴定分析	crumbling rock	崩解岩石；风化岩石；松散岩石
crude base	油基	crumbly	疏松的
crude bedding	原层理	crumbly rock	松软岩石；碎石
crude benzol	粗苯	crumbly soil	团块状土壤；屑粒土
crude bitumen	天然沥青	crumby soil	团粒状土
crude collecting line	原油集油管线	crumming	人工清理管沟
crude dam	天然坝	crumple	揉皱；碎裂；压碎
crude data	原始数据	crumpled	起皱的；揉皱的；压碎的，碎裂的
crude dehydrating plant	原油脱水装置	crumpled bedding	揉皱层理
crude diluent	原油稀释剂	crumpled mud-crack cast	揉皱泥裂铸型
crude distillation	原油蒸馏	crumpled structure	皱纹状构造
crude enroute	在途原油	crumpled texture	揉皱结构
crude flash-stabilization	原油闪蒸稳定	crumpling	揉皱作用；褶皱作用
crude gas	原煤气	crush	破碎；压碎
crude grading	粗选	crush belt	破碎带；压碎带
crude hauler	原油拖车	crush block	让压块

crush board	让压板
crush border	压碎边
crush breccia	压碎角砾岩
crush conglomerate	压碎砾岩
crush index	压碎指标
crush plane	压碎面
crush resistance	抗压
crush rock	压碎岩
crush structure	破裂构造
crush texture	压碎结构
crush zone	压碎带
crushability	可破碎性
crushability coefficient	可碎性系数
crushability factor	可碎性系数
crushed	破碎的
crushed aggregate	碎料
crushed ballast concrete	碎石混凝土
crushed boulder	碎漂石；碎砾石
crushed breccia	压碎角砾岩
crushed brick mortars	碎砖砂浆
crushed broken stone	粗碎石
crushed coke	破碎了的焦炭
crushed conglomerate	压碎砾岩
crushed gravel	碎砾石
crushed pellet	破碎球团
crushed rock	碎石
crushed stone	碎石
crushed stone base course	碎石底层
crushed stone concrete	碎石混凝土
crushed stone soil	碎石土
crushed stone soil foundation	碎石土地基
crushed timbering	压坏的支架
crushed zone	破碎带；压碎带
crushed-run aggregate	破碎料
crusher	碎石机
crusher chamber	破碎机房
crusher head	破碎机锥头
crusher jaw	破碎机颚；轧碎机颚板
crusher worm	破碎蜗杆
crusher-run aggregate	未筛碎石
crushing	破碎；压
crushing and screening flowsheet	破碎筛分流程
crushing and screening plant	破碎筛分厂
crushing and screening plants	轧筛机
crushing efficiency	破碎效率
crushing failure	压碎破坏
crushing in the wall rock	井壁岩石破碎
crushing load	断裂荷载；压碎荷载
crushing machine	破碎机
crushing machinery	破碎机械
crushing mill	捣矿机
crushing plant	破碎车间；破碎装置
crushing resistance	抗压瘪性
crushing rolls	辊碎机
crushing strain	挤压应变
crushing strength	挤压强度；抗碎强度；破碎强度
crushing strength across the grain	横纹抗压强度
crushing strength of grains	颗粒压碎强度
crushing strength parallel to the grain	顺纹抗压强度
crushing stress	挤压应力
crushing surface	破碎面
crushing test	破碎试验；压碎试验
crust	表层；地壳；硬表层
crust adjustment	地壳校正
crust breccia	断层角砾岩
crust cupola	地壳隆起
crust density	地壳密度
crust magnetic field	地壳磁场
crust movement	地壳运动
crust of the earth	地壳
crust of weathering	风化壳
crust source	壳源
crust tectonics	地壳构造
crust torsion	地壳扭动
crusta	外壳
crustal	地壳的；外壳的
crustal abundance	地壳丰度
crustal abundance of elements	地壳元素丰度
crustal accretion	地壳增生
crustal anatexis	地壳深熔作用
crustal arching	地壳隆起
crustal architecture	地壳隆起结构；地壳结构
crustal attenuation	地壳变薄作用
crustal bathyderm	地壳深层
crustal block	地块
crustal boundary	地壳边界
crustal circulation	壳内循环
crustal collision	地壳碰撞作用
crustal composition	地壳成分
crustal creep hypothesis	地壳蠕动说
crustal deformation	地壳形变
crustal deformation anomaly	地壳形变异常
crustal deformation field	地壳形变场
crustal deformation measurement	地壳形变测量
crustal deformation precursor anomaly	地壳形变前兆异常
crustal delamination	地壳分层
crustal derm	地壳表层
crustal disturbance	地壳变动；地壳扰动
crustal down-buckling	地壳下弯
crustal downwarping	地壳下挠
crustal dynamics	地壳动力学
crustal earthquake	壳内地震
crustal evolution	地壳演化
crustal fault	地壳断裂
crustal flexure	地壳挠曲
crustal fracture	地壳断裂
crustal garnetiferous ultrabasic rock	地壳型石榴子石超基性岩
crustal gravity bouguer anomaly	地壳布格重力异常
crustal growth	地壳增长
crustal inclination	地壳倾斜
crustal instability	地壳不稳定性
crustal interface	地壳分层界面
crustal layering	地壳分层
crustal magnetic anomaly	地壳磁异常
crustal magnetotelluric measurement	大地电磁测量
crustal motion	地壳运动
crustal movement	地壳形变；地壳运动
crustal plate	地壳板块
crustal residence age	地壳存留年龄

crustal reverberation	地壳混响
crustal rupture	地壳破裂
crustal seismic velocity	地壳地震波速度
crustal shock	地壳震动
crustal shortening	地壳收缩
crustal spreading	地壳扩张
crustal stability	地壳稳定性
crustal strain	地壳应变
crustal strain field	地壳应变场
crustal stress	地壳应力
crustal stress drive	地壳应力驱动
crustal stress field	地壳应力场
crustal stress state	地壳应力状态
crustal structure	地壳构造
crustal subsidence	地壳沉降
crustal tension	地壳张力
crustal thickening	地壳增厚
crustal thickness	地壳厚度
crustal tilt	地壳倾斜
crustal tilt observation	地倾斜观测
crustal type	地壳类型
crustal unrest	地壳不稳定
crustal uplift	地壳隆起
crustal warping	地壳翘曲
crustal wave movement	地壳波浪运动
crust-block	地壳断块
crust-dragging hypothesis	地壳牵引假说
crust-forming event	成壳事件
crustificated vein	带状脉
crustification	皮壳积结；矿壳层
crustification cement	结壳胶结物
crustified cementation	丛生胶结
crusting soil	结皮土
crustization	结壳
crust-mantle boundary	地壳-地幔界面
crust-mantle differentiation	壳-幔分异
crust-mantle system	地壳-地幔系统
crust-thermal structure	壳-幔热结构
crusty structure	皮壳状构造
crut	水平巷道
crutcher	拌合机
crybaby	需上油的接头
cryergic	冰缘
cryodrying	低温干燥
cryogenian	覆冰纪
cryogenic engineering	低温工程
cryogenic fluid	低温流体；冷却剂
cryogenic sampler	冻结取样器
cryogenic tiltmeter	超低温测斜仪
cryogenics	低温学
cryogeology	冻土地质学
cryolite	冰晶石
cryolite bath	冰晶石电解液
cryolithionite	锂冰晶石
cryolithology	冰冻岩石学
cryology	冰冻学
cryomorphology	冰冻地貌学；冻土地貌学
cryonival	冰雪作用的；冰缘的
cryopediment	冰冻麓原
cryopedology	冰土学；冻土学
cryoplanation	冰融夷平作用；融冻剥夷面
cryoscopic constant	冰点降低常数
cryoseism	冰冻地震；霜震
cryosixtor	冷阻管
cryosorption	深冷吸收
cryosphere	冰冻圈
cryostat	低温恒温器
cryostatic	冰静压力
cryostructure	冰冻构造
cryotectonic	冰川构造的；冰碛构造的
cryotexture	冰冻结构
cryoturbation	冰冻翻浆；融冻扰动
cryoturbation structure	冻裂搅动构造
crypt	地窖；地穴
cryptalgal	隐藻的
cryptalgalaminate	隐藻层
cryptalgalaminite	隐藻纹层岩
cryptand	穴状配体
cryptate	穴状化合物
cryptic	隐造礁生物
cryptic layering	隐蔽层状构造；隐层理
cryptic suture	隐蔽缝合
cryptic zoning	隐蔽分带；隐环带
cryptite	泥晶灰岩
cryptobedding	隐层理
cryptobiolith	隐生物岩
cryptobiotic soil crusts	隐生土结皮
cryptobioturbation	隐性生物扰动
cryptoblastic texture	隐变晶结构
cryptoclase	钠长石
cryptoclastic	隐屑的
cryptoclastic rock	隐屑岩
cryptoclastic texture	隐屑结构
cryptocorpocollinite	隐团块镜质体
cryptocrystal	隐晶
cryptocrystalline	无斑隐晶的；隐晶质；隐晶质的
cryptocrystalline allotriomorphic granular	隐晶他形的
cryptocrystalline texture	隐晶结构
cryptodacite	隐晶英安岩
crypto-depression	潜洼；潜隐陷落
cryptodiapiric fold	隐底辟褶皱
cryptodome	潜圆丘；潜伏火山丘
cryptoexplosion structure	隐爆发构造
cryptoflow state	潜流状态
cryptofluorescene	隐荧光
cryptogamic	隐花植物的
cryptogelocollinite	隐胶质镜质体
crypto-grained	隐晶粒状的；隐粒的
cryptograined texture	隐晶结构
cryptographic	隐晶文象的
cryptoland form	隐地形
cryptomaceral	隐显微组分
cryptomagmatic deposit	隐岩浆矿床
cryptomagmatic mineral deposit	隐岩浆矿床
cryptomere	隐晶岩
cryptoolitic	隐鲕状的
cryptoperthite	隐纹长石
cryptoplastic state	潜塑状态
cryptorheic	地下河排水的；潜流的
cryptotelinite	隐结构镜质体

crypto-texture	隐粒结构
cryptothermal deposit	隐温矿床
cryptovitrodetrinite	隐碎屑镜质体
cryptovolcanic caldera	潜火山型破火山口
cryptovolcanic earthquake	潜火山地震
cryptovolcanic rock	潜火山岩
cryptovolcanic structure	隐火山构造
crypto-volcanism	潜火山作用
cryst	结晶的
crystal	晶体；晶体检波器；水晶
crystal ash	晶质火山灰
crystal defect	晶体缺陷
crystal flotation	浮选法；晶体浮升作用
crystal fractionation	结晶分异
crystal fragment	晶屑
crystal geometry	晶体几何学
crystal gliding	晶体滑移
crystal grain limestone	晶粒灰岩
crystal habit	晶体习性
crystal imperfection	晶体不完整
crystal lapilli	结晶火山砾
crystal lattice defect	晶格缺陷
crystal microstructure	晶体微观结构
crystal morphology	晶体形态学
crystal nucleus	晶核
crystal oil	未脱蜡的油；中等黏度不易挥发的油
crystal reference gauge	晶体压力计
crystal settling	晶体沉淀
crystal silica sand	石英砂
crystal size	晶粒大小；晶体粒度
crystal structure analysis	晶体结构分析
crystal tuff	结晶凝灰岩
crystal water	结晶水
crystal-controlled oscillator	晶体控制振荡器
crystal-controlled transistor oscillator	晶体控制晶体管振荡器
crystalline bond	结晶连接
crystalline carbonate	结晶碳酸盐
crystalline clock	晶体钟
crystalline dolomite	结晶白云岩
crystalline limestone	结晶灰岩
crystalline material	结晶物质
crystalline reservoir	结晶岩储层
crystalline rocks	结晶岩
crystalline substance	结晶质
crystalline texture	结晶结构
crystallinoclastic rock	晶屑岩；晶质碎屑岩
crystallinohyaline	玻基斑状
crystallite	微晶；晶粒；雏晶
crystallite orientation	微晶定向
crystallization differentiation	结晶分异作用
crystallization differentiation mineralization	结晶分异成矿作用
crystallization fabric	结晶组构
crystallization fractionation	晶体分离作用
crystallization kinetics	结晶动力学
crystallization remanent magnetization	结晶剩余磁化
crystallization schistosity	结晶片理
crystallization test	结晶试验
crystallization texture	结晶结构
crystallize	使结晶
crystalloblastesis	变晶作用
crystalloblastic	变晶质的
crystalloblastic fabric	变晶组构
crystalloblastic series	变晶系列
crystalloblastic texture	变晶结构
crystallogeometry	晶体几何学
crystallothrausmatic texture	晶核球状结构
crystic texture	非纹层结构
ctenoid cast	栉状铸型
ctenoid scale	栉鳞
cubanite	方黄铜矿
cube compressive strength	立方体抗压强度
cube concrete test specimen	混凝土立方块试样
cube crushing strength	立方体试件破坏强度
cube spar	硬石膏
cube strength	立方强度
cube test	立方体强度试验
cubic boron nitride	立方氮化硼
cubic capacity	立方容积
cubic closest packing	立方最紧密充填
cubic content	容量
cubic deformation	体积变形
cubic dilatation	体积膨胀
cubic dilation	体积膨胀
cubic drafter	立体牵伸法并条机
cubic gravel packing	立方砾石堆积
cubic strain	体积应变
cubical cleavage	三向互相垂直劈理
cubical dilatation	体积膨胀
cubical drafting	立体牵伸
cubical expansion	体积膨胀
cubical pack	立方排列充填
cubical triaxial test	真三轴试验
cubichnion	停歇迹
cubicite	方沸石
cubit	库比特腕尺
cubital vein	肘脉
cuboid	长方体
cuesta	单面山；内向崖
cuesta face	单面山陡崖
cuesta scarp	单面山陡崖
cuff	袖口
cul-de-sac	独头巷道；死胡同
culm	煤尘；碳质页岩
culm bank	废渣堆
culm dump	废渣堆
Culmann construction	库尔曼图解法
Culmann line	库尔曼线
Culmann's method of slope stability	库尔曼斜坡稳定性分析法
Culmann's procedure	库尔曼方法
culmination	顶点；褶隆区；轴隆
culmophyre	聚斑岩
culsageeite	水金云母
cultivate reclamation rate	土地复垦率
cultivated land	耕地
cultural age	文化期
cultural layer	文化层
cultural objects	文物

cultural relic	文物	cumulative probability	累积概率
cultural remains	文化遗迹	cumulative production	累积产量
culture bottle	培养瓶	cumulative production-time relationship	累积产量-时间关系
culture dish	培养皿	cumulative profile	累积剖面
culture flask	培养瓶	cumulative property curve	累积特性曲线
culture medium	培养基	cumulative pump stroke	渐增泵冲程
culture noise	人为噪声	cumulative rainfall	累积降雨量
culture tube	培养管	cumulative reduction	总压下量
culture yeast	培养酵母	cumulative runoff	累积径流
culvert	排水渠	cumulative sand production	累积出砂量
culvert bridge	涵式桥	cumulative screen analysis	累积筛分分析
culvert inlet	涵洞进水口	cumulative sieve analysis	累积筛分分析
cum rights	附属机构	cumulative solid produced	累积出砂量
cumberlandite	钛铁橄榄岩	cumulative stock tank oil recovery	累积采油量
cumbraite	培斑安山岩	cumulative throughflow	累积流过量
cummingtonite	镁铁闪石	cumulative time shift	累积时间漂移
cumularsharolith	团粒	cumulative voidage replacement ratio	累积注采比
cumularspharolith	团粒	cumulative water encroachment	累积水侵
cumulate	堆积的；堆积岩	cumulative withdrawal	累积产量
cumulate complexes	堆积杂岩	cumulative yield curve	累计出水量曲线
cumulate texture	堆积结构	cumulite	玻质岩中包体
cumulates	堆积岩	cumulophyre	聚斑岩
cumulation	堆积	cumulophyric	联合斑状的
cumulative balances	累计余额	cumulose deposit	碳质堆积层；泥炭土
cumulative curve	累积曲线	cumulose soil	腐殖土
cumulative departure	累积偏差	cumulosphaerite	聚球粒
cumulative discharge	累计流量	cumulo-volcano	累积火山
cumulative distribution curve	累积分布曲线	cumulus	堆积晶体；堆积物
cumulative dose	累积剂量	cumulus rock	堆积岩
cumulative effect	累积影响	cuneatic arch	楔形拱
cumulative environmental effect	累积环境效应	cunico	铜镍钴永磁合金
cumulative fatigue damage	积累疲劳损伤	cunife	铜镍铁永磁合金
cumulative fluid production	累积产液量	cup anchor	碗式锚
cumulative frequency	累积频率	cup and ball jointing	关节状节理；球窝节理
cumulative gas evolved	累计脱出气量	cup and ball viscometer	杯球式黏度计
cumulative gas influx	累积气体流入量	cup and ball viscosimeter	杯球黏度计
cumulative gas injection volume	累积注气量	cup and cone arrangement	钟斗装料装置
cumulative gas oil ratio	累计气-油比	cup and vane anemometer	转杯风叶式风速风向仪
cumulative gas production	累积产气量	cup flow test	杯溢法流动试验
cumulative grading curve	累计粒度曲线	cup grease	润滑脂
cumulative grain-size distribution curve	累积粒度分布曲线	cup head bolt	半圆头螺栓
		cup head rivet	半圆头铆钉
cumulative gravity effect	累积重力效应	cup heater	帽式加热器
cumulative grout volume	累积灌浆量	cup leather	密封皮碗
cumulative hydrocarbon production	累积产烃量	cup leather packer	皮碗灌浆塞
cumulative index	累积索引；累积指数	cup leather packing	皮碗密封
cumulative injection	累积注入量	cup packer	皮碗式封隔器
cumulative injection in pore volumes	以孔隙体积计量的累计注入量	cup ring nut	皮碗环形螺母
		cup spring	盘形弹簧
cumulative liquid production	累积产液量	cup type retrievable packer	皮碗式可回收封隔器
cumulative log diagram	累积对数曲线图	cup valve	皮碗阀
cumulative measured reserves	累计探明储量	cup washing tool	皮碗式冲洗工具
cumulative oil recovery	累积采油量	cupel	灰皿
cumulative percent by weight	累积重量百分比	cupellation	灰吹法
cumulative percentage	累积百分比	cupferron	铜铁试剂
cumulative percentage plot	累积百分含量图	cup-forming packing element	皮碗式密封件
cumulative physical recovery	累积自然采油量	cupholder	杯式绝缘子螺脚
cumulative plot	总体曲线图	cuplock joint scaffold	碗扣式脚手架
cumulative porosity percentage	累积孔隙度百分数	cupola	冲天炉；钟状火山
cumulative potential	累积电位	cupola furnace	冲天炉

cupped washer	凹垫圈
cuprate	铜酸盐
cupreous	含铜的；铜色的
cupreous sandstone	含铜砂岩
cupric	二价铜的；铜的
cupric chloride	氯化铜
cupric compound	二价铜化合物
cupric oxide	氧化铜
cupric sulfate	硫酸铜
cupriferous	含铜的
cupriferous shale	含铜页岩
cuprite	赤铜矿
cupro	一价铜
cupro manganese	铜锰合金
cuproammonium solution	铜铵液
cuproautunite	铜钙铀云母
cuprobismutite	铜辉铋矿
cuprobond	硫酸铜处理
cuprocalcite	杂赤铜方解石
cuprodescloizite	铜钒铅锌矿
cuproiodargyrite	铜碘银矿
cupromagnesite	铜菱镁矿
cupron	科普隆铜镍合金
cupronickel	铜镍合金
cuproplatinum	铜铂矿
cuproplumbite	铜硫铅矿
cuproscheelite	铜白钨矿
cuprosklodowskite	硅铜铀矿
cupro-solvent	溶铜水
cuprotungstite	铜钨华
cuprous	亚铜的
cuprous chloride	氯化亚铜
cuprum	铜
cuprum nickeliferous	含铜镍的
cup-type current meter	旋杯式流速仪
cup-type grinding wheel	凹砂轮
cup-type packer	皮碗式封隔器
curb	井栏；约束
curb ramp	缘石坡道；路缘坡道
curb ring crane	转盘起重机
curbed modulation	约束调制
curbing	设置路缘石
curbstone	路边石，路牙
curd	凝乳
curdling	凝固
cure	养护；固化；硫化
cured pack	已胶结充填体
cured rubber	硫化橡胶
curf	金刚石取心钻头切削环；取心钻井的环状井底；取心钻头切削面
curienite	钒铅铀矿
curing	固化
curing accelerator	固化促进剂
curing agent	固化剂
curing blanket	养治被覆层
curing condition	养护条件
curing rate	固化速度
curing schedule	固化步骤
curing temperature	养护温度
curing time	养护时间
curite	板铅铀矿
curium	锔
curl	翘曲；旋度
curl field	旋度场
curled bedding	卷曲层理
curling	卷曲
curly bedding	旋卷层理
curly schist	卷曲片岩
current	流；倾向；势
current average rate	目前平均产量
current average reservoir pressure	目前平均产层压力
current bedding	水流层理
current bottom-hole pressure	目前井底压力
current channel data	现行通道数据
current crescent	新月形水流痕
current density	电流密度
current direction	流向
current disc	电流层
current drag	水流拖曳
current drain	漏电流
current drainage test	排流试验
current drift	海流；泥沙流
current efficiency	电流效率
current expenses	杂费
current field production	目前油田产量
current fluid saturation	目前流体饱和度
current focusing log	聚焦电流测井
current indicator	现行指示符；洋流标志；电流指示器
current influx rate	目前水侵量
current injection rate	目前注入速率
current international practice	当前国际惯例
current limiting reactor	限流电抗器
current lineation	水流线理
current loss instrument	井漏仪
current maintenance	日常维修
current mark	流痕
current measurement	水流量度
current money	通货
current oil production level	目前产油量水平
current oil saturation	目前含油饱和度
current oil-water contact	目前油水接触面
current producing gor	目前生产气油比
current repair	现场修理
current reservoir pressure	目前油层压力
current ripple mark	流水波痕
current rose	流向玫瑰图
current scour	水流冲刷
current sensitivity	电流灵敏度
current vector	水流矢量
current velocity	流速
current yield	日产量
current-carrying space	载流空间
current-carrying tube	载流管
current-emitting electrode	供电电极
current-laid	水流沉积的
cursor	游标
cursory appraisal	粗略评价
curtailment	缩短，长度不足
curtailment of drilling	钻井工作减少
curtain	隔板；石幔；帷幕

curtain boom	幕帘型构架	curved ray seismic tomography	曲射线地震层析成像法
curtain coater	帘幕式淋涂机	curved reach	弯曲河段
curtain dam	活动闸门坝	curved spit	弯曲沙嘴
curtain fold	幕式褶皱	curved spool	弯短管
curtain grout	帷幕灌浆	curved structure	曲型结构
curtain grouting	帷幕灌浆	curved surface	曲面
curtain hole	帷幕灌浆孔	curved three-dimentional wellbore	弯曲的三维井眼
curtain line	帷幕线	curved tooth	弧齿
curtain of piles	板桩排；帷幕桩	curved well	弯曲井
curtain wall	玻璃幕墙；护墙；幕墙；帷幕墙	curve-matching method	量板法；曲线对比法
curtain wall method	帷幕法	curves tracing scale	曲线绘制的标度
curtain wall support	幕墙承托物	curvic structure	弧形构造
curtain wall type breakwater	护墙式防波堤	curvilinear fault displacement	曲线断层位移
curtate	横向穿孔区	curvilinear fold	曲线形褶皱
curtisite	绿地蜡	curvilinear grid	曲线网格
curvature	曲度	curvilinear motion	曲线运动
curvature anomaly	曲率异常	curvilinear regression	曲线回归
curvature at first yield	初始屈服曲率	curvimeter	曲率计
curvature coefficient	曲率系数	curvity	曲率
curvature correction	曲率校正	curvity coefficient	曲率系数
curvature correction factor	曲率校正因子	cuselite	云辉玢岩
curvature of space	空间曲率	cushion	垫层；垫子；缓冲器；减震垫；桩垫
curvature parameter	曲率参数	cushion blasting	缓冲爆破
curvature pressure	曲面压力	cushion block	垫块
curvature radius	曲率半径	cushion coat	垫层
curvature relaxation	曲率松弛	cushion fluid	缓冲液
curvature vector	曲率向量	cushion gas	垫气
curve	弯曲	cushion head	桩头帽
curve drilling	曲线井段钻进	cushion material	垫层材料
curve family	曲线簇	cushion pile	垫桩；缓冲桩
curve fitting method	曲线拟合法	cushion socket	防震插座
curve generator	波形发生器	cushion stress factor	垫层应力系数
curve match	曲线拟合	cushion sub	缓冲器接头
curve of equal intensity	等烈度线	cushion support	可缩性支架
curve of equal settlement	等沉陷曲线	cushioning	缓冲器；软垫
curve of loads	荷载特性曲线	cushioning ability	减震能力
curve of output	生产能力曲线	cushioning effect	垫层作用；缓冲效应
curve of solubility	溶度曲线	cushioning material	垫承物料
curve resistance	非线性电阻	cusp	岬；尖顶；滩尖嘴
curve scale	曲线比例尺	cusp beach	海滩嘴
curve segment	弧段	cusp cast	浅新月形铸型
curve separation	曲线的幅度差	cusp catastrophe	尖点突变
curve stake	曲线桩	cuspate	尖的
curved	弯曲的	cuspate bar	三角沙坝；三角沙洲
curved bar	曲杆；弯曲沙坝；弯曲沙洲	cuspate boundary	尖状边界
curved beam	曲梁	cuspate delta	尖形三角洲
curved bedding	扭曲层理；旋卷层理	cuspate fold	尖顶褶皱
curved bridge	弯桥	cuspate foreland	尖头前陆；三角岬；三角沙洲
curved chord truss	折弦桁架	cuspate lobate fold	尖圆褶皱
curved conductor	弯导管	cuspate offshore bar	三角形滨外沙坝
curved cross-section arch bridge	双曲拱桥	cuspate reef	尖头岩礁
curved crystal	弯晶	cuspate sandkey	尖头小沙岛
curved drill guide	导向弯曲钻具	cuspate spit	尖头沙嘴
curved fault surface	弯曲断层面	cuspate-arcuate delta	尖弧三角洲
curved girder	曲梁	cuspidate	尖角的
curved hole	弯曲井	cuspidine	枪晶石
curved jack	弯曲千斤顶	cusping	形成水舌
curved path	曲线途径	custerite	灰枪晶石；锌黄锡矿
curved plate	曲板	custodian	保管人
curved portion	弯曲井段	custody	保管

custom clearance	清关
customary quenching	一般淬火
custom-built machine	专用机器
customer-furnished	由用户提供的
customer-originated change	由买主引起的更改
custom-made	专门定制
cut	宝石切工；馏分；路堑；切割；挖方；岩屑
cut and built platform	蚀积台地
cut and cover method	明挖法；明挖回填法
cut and fill	边冲边淤；充填开采；随挖随填
cut and fill excavation	随挖随填；挖方和填方；充填开采挖掘
cut and fill mining	充填开采
cut and fill stoping	充填采矿法
cut and paste	剪贴
cut and replacement method	换土法
cut and try method	渐近法，试凑法
cut bank	凹岸；淘蚀岸
cut blasting	掏槽爆破
cut both ways	模棱两可
cut diameter	分割粒径
cut fibres	短纤维；切断纤维
cut film	薄膜切片
cut off	截断；截距；截流；切断
cut off grade	边界品位；品位下限
cut out flaw	挖脏
cut plain	侵蚀平原
cut plane	剖面
cut platform	海蚀台地
cut point	截切点；两种油品间的分开点
cut ring	环刀
cut section	路堑断面；挖方断面
cut short	打断
cut slope	路堑边坡；切坡；削坡
cut spreader hole	扩槽孔
cut staple	短纤维
cut surface	开挖面
cut terrace	浪蚀阶地；岩质阶地
cut the slope	削坡
cut to pieces	切碎
cut up	切碎
cutability	可切性
cutan	胶膜
cut-and-cover	大开槽施工法；随挖随填；挖方填方
cut-and-cover tank	半埋半覆盖油罐
cut-and-fill design	挖填设计
cut-and-fill sedimentation	冲淤沉积作用
cut-and-fill slope	半填半挖式斜坡
cut-and-fill structure	冲蚀构造
cutaway	剖面的；剖视图
cutaway drawing	断面图；剖视图
cutaway view	剖视图
cutback	缩减
cutback asphalt	稀释沥青；油溶沥青
cutback in mud weight	泥浆比重因气侵下降
cutdown	削减
cut-fill contact	挖方与填方连接处
cut-fill height	挖填高度
cut-fill subgrade	半填半挖路基
cut-fill transition	土方调配；挖填方平衡
cutfit	备用工具
cuthole	掏槽孔；掏槽炮眼
cutic acid	角质酸
cuticle	表皮；角质层
cuticle oil	角质层油
cuticoclarite	角质微亮煤
cuticodurite	角质微暗煤
cuticular	表皮的
cuticular analysis	角质层分析
cuticular transpiration	角质层蒸腾
cuticular wax	角质蜡
cutin	角质；角质素；接通
cutinic anthracite	角质无烟煤
cutinite	角质煤素质；角质体
cutinite liptobiolith	角质残殖煤
cutinite-clarite	角质体微亮煤
cutinitic liptobiolith	角质残植煤
cutinization	角质化
cutlery	刀具
cutlery steel	刃具钢
cutoff angle	截止角
cut-off apron	底板截水墙；坝工截水护底墙
cut-off basin	闭合流域；封闭盆地
cut-off blanket	隔离层；截水层
cutoff characteristic	衰减特性曲线
cut-off collar	防渗环；截流环
cut-off depth	进入开挖面以下的深度
cut-off device	断流器
cutoff dike	截渗堤；截水堤
cut-off drain	截水沟；排水截槽
cut-off elevation	断桩高程
cutoff error	截断误差
cutoff frequency	截频；截止频率
cutoff lake	牛轭湖
cut-off level	截止高程
cut-off limiting	截止限幅
cutoff line	截层线；截断线
cutoff meander	割断曲流；弓形湖；牛轭湖
cut-off meander core	离堆；曲流弧山
cutoff meander spur	离堆山；曲流弧山
cutoff of anomaly	异常截止点
cut-off oxbow	割断曲流；截断牛轭湖
cutoff permeability	截止渗透率
cut-off piling	隔水板桩
cut-off plate	截断板；闸板
cutoff porosity	截止孔隙度
cutoff rate	截止产量
cut-off saturation	截止饱和度
cut-off spur	割断山嘴；离堆山
cutoff trench	齿槽；截水槽
cut-off trench of paddled clay	黏土填实的防渗隔离槽
cut-off valve	关闭阀
cutoff velocity	截止速度
cut-off voltage	截止电压
cut-off wall	齿墙；隔水墙；截水墙
cutoff wavelength	截止波长
cut-open view	剖视图
cut-out	切断
cut-plane test	切面试验
cuttability	可切割性；可挖性

cutter	刀具；切割器；倾斜节理
cutter arm	扩孔器牙轮伸展臂
cutter bar	刀杆；切割机
cutter bit	截煤机割齿
cutter body	刀体
cutter exposure	切削齿出刃度
cutter head	刀盘
cutter jib	截煤机截盘
cutter knife	割刀
cutter loader	联合采煤机；切割装载机
cutter mining machine	截煤机
cutter offset	牙轮移轴距
cutter pick	截煤机割齿
cutter ram	剪切闸板
cutter shell	钻头牙轮体
cutter suction dredge	绞吸式挖泥船
cutter teeth	牙轮齿
cutterhead dredge	铣轮式疏浚机；铣轮式挖土机
cutter-wear rate	切削齿磨损速度
cut-through	开凿
cutting	采掘；开挖；路堑；切割；掏槽；岩屑
cutting ability	切削能力
cutting action	切割作用
cutting angle	切削角
cutting blade	切削片
cutting blasting	切割爆破
cutting blowpipe	截割吹管；喷割器
cutting burr	切断毛刺
cutting curb	沉井筒脚；沉箱刃脚；挖路缘槽
cutting depth	切削深度；掏槽深度
cutting down	研磨修整
cutting edge	刃脚；刃口
cutting effort	切削力
cutting element	切削元件
cutting face	切削面；挖掘面
cutting fluid	切削液
cutting force	切削力
cutting head	取心钻头切削刃
cutting hole	掏槽眼
cutting jet	切削射流
cutting machine	截煤机；切割机
cutting nose	刃尖
cutting nozzle	割炬嘴
cutting of mud by gas	泥浆气侵
cutting oil	切削用油
cutting output	削减产量
cutting plane	切割面
cutting pliers	剪钳
cutting point	交点
cutting property	切削能力
cutting rate	切削速度
cutting resistance	切削阻力
cutting ring	环刀
cutting shoe	切刃；刃脚
cutting slope deformation	路堑边坡变形
cutting speed	切削速度
cutting square	方形切刀
cutting stroke	切削行程
cutting structure	齿及齿面结构
cutting structure dullness	切削结构钝度
cutting teeth	切削齿
cutting the head of contamination	切割混油头
cutting the tail of contamination	切割混油尾
cutting thickness	切削厚度
cutting tip	割炬嘴
cutting tool	切削工具；钻具
cutting torch	割炬
cutting under water	水下切割；水下挖土
cutting value	切削值
cutting vertical pipe	切割立管
cutting window	井中套管开孔
cutting-in	冲入
cutting-off piles	隔断桩群；隔水板桩
cutting-out piece	短木
cutting-plane method	割平面法
cuttings bed	岩屑垫层
cuttings density	岩屑密度
cuttings discharge	排屑
cuttings disposal system	岩屑处理系统
cuttings evaluation	岩屑评价
cuttings gas	岩屑气
cuttings hold-down effect	岩屑压持效应
cuttings logging	岩屑录井
cuttings of boring	岩粉；钻屑
cuttings removal	清除岩屑
cuttings samples	屑试样
cuttings shape	岩屑形状
cuttings size	岩屑大小
cuttings transport ratio	岩屑输送比
cut-type sealing element	密封皮碗
cutup line	带材横剪作业线
cutwater	分水角
cuyamite	蓝方沸绿岩
cyanaloc grout	氰基树脂灌浆
cyanobacteria	蓝藻细菌
cyanometer	蓝度计
cyanophycin	蓝藻颗粒体
cyanophyte	蓝藻植物
cyanotrichite	绒铜矿；线铜矿
cyberlook	计算机解释结果
cybernetics	控制论
cyberproduct	井场计算机处理成果
cybotactates	群聚体
cybotaxis	群聚
cycle	旋回
cycle breadth	周期
cycle clips	周跳
cycle compound	环化合物
cycle control	循环控制
cycle count	循环计数
cycle criterion	循环判据
cycle of carbon	碳循环
cycle of erosion	侵蚀旋回；侵蚀循环
cycle of fluctuation	波动周期
cycle of fluvial erosion	河蚀旋回
cycle of freezing and thawing	冻融循环
cycle of glacial age	冰期旋回
cycle of glaciation	冰期旋回
cycle of karst	岩溶旋回
cycle of load	荷载循环

cycle of marine erosion 海蚀旋回
cycle of nitrogen 氮循环
cycle of river erosion 河蚀轮回
cycle of sedimentation 沉积旋回
cycle of stress 应力循环
cycle of stress reversal 应力交变周期
cycle of vibration 振动周期
cycle of volcano 火山旋回
cycle of weathering 风化旋回
cycle ratio 周期率
cycle simulator 循环模拟器
cycle timer 测周计
cycler 周期计
cycle-skip 周波跳跃
cyclewelding 合成胶黏剂焊接法
cyclic 环的
cyclic action 环化作用
cyclic bedding 旋回层理
cyclic change 周期改变
cyclic compound 环状化合物
cyclic convolution 循环褶积；循环卷积
cyclic deformation 周期性变形
cyclic depletion 周期性耗竭
cyclic deposition 旋回沉积
cyclic evolution 完整演化周期；周期进化
cyclic gas injection 循环注气
cyclic heat test 周期加热试验
cyclic injection 周期性注入
cyclic injection-production 周期性注采
cyclic iterative 循环迭代法
cyclic load test 周期载荷试验
cyclic loading 循环荷载；周期性荷载
cyclic loading test 循环加载试验
cyclic microbial recovery 周期性微生物采油法
cyclic movements 周期运动
cyclic order 循环次序
cyclic oscillation 周期性振动
cyclic oxidation-reduction 循环氧化还原过程
cyclic planation 旋回均夷作用
cyclic polynomial 循环多项式
cyclic quasi-static loading 循环准静力荷载
cyclic recovery process 周期性开采法
cyclic regeneration 循环再生
cyclic sedimentation 沉积旋回；旋回沉积作用
cyclic sequence 旋回层序
cyclic shear resistance 周期抗剪强度
cyclic shear strength 循环抗剪强度
cyclic shift 循环移位
cyclic simple shear test 循环单剪试验
cyclic steam injection 周期注蒸汽
cyclic steam injection volume 周期注汽量
cyclic steam stimulation 周期注蒸汽
cyclic strain softening 周期应变软化
cyclic stress 周期应力
cyclic stress corrosion cracking 交变应力腐蚀开裂
cyclic structure 环状结构
cyclic terrace 旋回阶地
cyclic test 循环试验
cyclic torsional shear test 循环扭剪试验
cyclic transformation 循环变换
cyclic triaxial test 周期加荷三轴试验
cyclic variation 周期性变化
cyclic vector 循环向量
cyclic waterflooding 周期注水
cyclical 旋回的；周期的
cyclical construction 循环施工
cyclical degradation 周期性退化
cyclical deposition 旋回沉积作用
cyclical downturn 周期性下降
cyclical graded bedding 韵律性递变层理
cyclical hollow cylinder torsion test 空心柱体扭剪试验
cyclical integrating 循环积分
cyclical load 循环周期性荷载
cyclical loading test 循环载荷试验
cyclical motion 循环周期运动
cyclical movement 周期性运动
cyclical number 循环次数
cyclical operation 循环操作
cyclical pumping 周期性抽水
cyclical recharge 周期性补给
cyclical ring torsion test 圆环式循环扭力试验
cyclical sedimentation 旋回沉积作用
cyclical sequence 旋回层序
cyclical shear resistance 循环抗剪力
cyclical shear strain amplitude 循环周期剪应变幅
cyclical shear stress 循环周期剪应力
cyclical shear stress ratio 循环剪应力比
cyclical shear test 循环周期剪切试验
cyclical simple shear test 循环周期单剪试验
cyclical strain softening 循环周期应变软化
cyclical stratification 旋回层理
cyclical stress 循环周期应力
cyclical stress ratio 循环周期应力比
cyclical stress-strain behavior 循环周期荷载下应力应变特性
cyclical sulfide 环状硫化物
cyclical terrace 旋回阶地
cyclical test 循环试验
cyclical triaxial test 循环周期性加荷三轴试验
cyclical upturn 周期性上升
cyclical variation 周期性变化
cyclicity 旋回性；周期性
cyclicity of tectonic movement 构造运动的旋回性
cyclic-reciprocal sedimentation 正反旋回沉积作用
cyclics 环状化合物
cycling degassing 真空提吸脱气法
cycling plant 残气回注装置
cycling program 循环程序
cyclogenesis 气旋生成
cyclograph 圆弧规；转轮全景照相机
cyclogyro 旋翼机
cycloidal 摆线的
cycloidal gear 摆线齿轮
cyclolith 圆形杂岩体
cyclometer 记转器；记程计
cyclometry 测圆法
cyclone 旋风分离器
cyclone centrifugal separator 旋流离心分离器
cyclone classifier 旋流分级机
cyclone cleaner 旋流清洁器

cyclone collector	离心除尘器；旋风集尘器	cylindrical boundary	圆柱界面
cyclone desander	旋流除砂器	cylindrical buoy	圆筒形浮标
cyclone dust extractor	旋风除尘器	cylindrical charge	柱状装药
cyclone settler	旋风沉降器	cylindrical combustion chamber	筒形燃烧室
cyclonic storm	气旋性风暴	cylindrical coordinate	圆柱坐标；柱面坐标
cyclopean	大块石；蛮石	cylindrical coordinate system	柱面坐标系
cyclopean concrete	蛮石混凝土；毛石混凝土	cylindrical drum	圆筒形滚筒
cyclopean riprap	乱石堆	cylindrical equal-angle projection	等角圆柱投影
cyclopean texture	银嵌状结构	cylindrical equidistant projection	等距圆柱投影
cyclophon	族调管	cylindrical fault	圆柱状断层
cyclopite	钙长石	cylindrical fold	圆柱状褶皱
cyclo-polymerization	环化聚合作用	cylindrical furnace	圆筒炉
cyclorectifier	循环整流器	cylindrical gear	圆柱齿轮
cycloscope	转速计	cylindrical grain	圆形药柱
cyclosilicate	环状硅酸盐	cylindrical hydrophone	筒式压敏检波器
cyclosizer	连续水析仪	cylindrical jack	圆筒形千斤顶
cyclostratigraphy	旋回地层学	cylindrical jointing	柱状节理
cyclothem	旋回层；旋回沉积；韵律层	cylindrical lug	圆筒式吊耳
cyclothemic deposition	韵律沉积	cylindrical model	圆柱状模型
cyclothemic sedimentary sequence	韵律沉积层序	cylindrical opening	圆筒形开口
cyclothemic sedimentation	旋回沉积作用	cylindrical orthomorphic projection	正形圆柱投影
cyclotomy	分割圆法	cylindrical pier	圆柱墩
cyclotron	回旋加速器	cylindrical piezometer tip	圆柱形测压头
cylinder	圆柱	cylindrical pile	筒形桩
cylinder abutment	圆柱形桥台	cylindrical plot	圆柱图
cylinder and cone grinding machine	外径磨床	cylindrical plug	柱状岩心栓
cylinder block	气缸体	cylindrical polar coordinates	柱面极坐标
cylinder caisson	沉井；圆筒沉箱	cylindrical pore	圆柱形孔隙
cylinder compressive strength	圆柱体抗压强度	cylindrical powder pellet	圆形药柱
cylinder crushing strength	圆柱体抗压强度	cylindrical projection	圆柱投影
cylinder cut	圆柱形开挖；直眼掏槽	cylindrical reducer	正齿轮减速器
cylinder deposits	气缸沉积物	cylindrical screen	筛筒
cylinder drainage	汽缸排水	cylindrical slide	圆柱面滑坡
cylinder drill	筒形钻	cylindrical slide surface	圆柱形滑动面
cylinder face	圆柱面	cylindrical sonde	柱形探棒
cylinder gas	钢瓶压缩气体	cylindrical specimen	圆柱试件
cylinder helix method	圆柱螺线法	cylindrical spreading	柱面扩展
cylinder jacket	汽缸套	cylindrical structure	圆柱状构造
cylinder lid	缸盖	cylindrical tank	圆筒形储罐；柱形罐
cylinder liner	缸套	cylindrical thread	柱形螺纹
cylinder method	柱状法	cylindrical trommel	圆筒筛
cylinder oil	气缸油	cylindrical vessel	圆柱形容器
cylinder penetration test	圆筒贯入试验	cylindrical worm gearing	筒形蜗轮传动装置
cylinder pile	大直径桩；圆柱状	cylindricity	圆柱度
cylinder pile foundation	管桩基础	cylindrite	圆柱锡矿
cylinder piston	汽缸活塞	cylindroid	圆柱形面
cylinder shaft foundation	管柱基础	cylindroidal fold	圆柱形褶皱
cylinder specimen	圆柱形试件	cylindrometer	柱径计
cylinder strength	圆柱体强度	cymatogeny	地壳上隆
cylinder strength of concrete	混凝土圆柱体强度	cymograph	波频记录仪
cylinder stroke	汽缸冲程	cymometer	频率计
cylinder weir	圆筒堰	cymoscope	检波器
cylinder wrench	圆筒扳手	cymoscope wave detector	检波器
cylinder-head gasket	汽缸盖密封垫	cyprine	青符山石
cylindric jointing	圆柱状节理	cyprusite	垩状铁矾
cylindric structure	圆柱状构造	cyrtolite	曲晶石
cylindrical	圆柱状的	cyst pearl	囊珍珠
cylindrical arch dam	圆柱拱坝	cyst plate	泡沫板
cylindrical bearing	滚柱轴承	cystose	泡沫状的

D

dacite liparite 英安流纹岩
dacitoid 似英安岩
dacron 涤纶；聚酯纤维
dactylitic texture 指状构造；指状结构
dactylotype intergrowth 指状交生
dactylotype texture 指纹构造；指状结构
daily account 日计表
daily advance 日进尺；日进度；日掘进量
daily allowance 每日定额；日计工资；日计津贴
daily amount 日总量
daily amplitude 日变幅
daily and shift traffic plans 调度日班计划
daily average 日平均
daily average current 日平均电流
daily average injection rate 平均日注入量
daily average value 日平均值
daily average water injection 平均日注水量
daily balance 日计表
daily capacity 日出水量；日产量；日输送量
daily check 每日检查；日常检查
daily compensation 日补偿
daily construction report 施工日报表
daily consumption 日耗量
daily contracted hours 每日限定工时
daily cycle 每日工作循环，昼夜循环
daily demand 日需要量
daily detailed report 每日详报表
daily drilling progress 钻井日进尺
daily extreme 日极值
daily flood peak 日洪峰
daily flow 日流量
daily fluid production rate 日产液量
daily grouting report 灌浆日报
daily high tide 日高潮位
daily hydrograph 日流量过程线
daily inspection 日查；日常检查
daily letting return 每日出租报表
daily load 日负荷
daily load curve 日负荷曲线
daily load factor 日负荷系数
daily load fluctuation 日负荷变动
daily load prediction 日负荷预测
daily manning balance 工作情况日报表
daily maximum 日最大量
daily mean 日平均；昼夜平均
daily mean sea level 日平均海面
daily mean temperature 日平均温度
daily mine record 矿山日记录
daily minimum 日最小量
daily minimum temperature 日最低温度
daily morning report 每日晨报告
daily observation 日观察，日检查，日观测
daily output 昼夜生产量；日产量；日输出量
daily output per face 每个工作面的日产量

daily peak load 日峰荷
daily planning 日计划
daily pondage 日调节容量
daily precipitation 日降水量
daily prediction index 每日预测指标
daily price limit 每日价格限幅
daily production 日产量
daily production report 生产日报
daily progress 日进度
daily rainfall 日降雨量
daily range 日变化；日差程
daily rate of flow 日流量
daily record of construction 施工日志
daily regulating pond 日调节水池
daily regulation 日调节
daily report 日报
daily requirement 日需量
daily routine 日常工作；每天的例行公事
daily sample 逐日取的试样
daily span 每日间距
daily statement 日报表
daily station precipitation 测站日降水量
daily stint 日定额
daily storage 日蓄能
daily storage capacity 日调节库容；日蓄水量
daily storage plant 日调节电站
daily summary report 每日总报表
daily supply tank 日常供应罐
daily temperature fluctuation 日温变化；日温波动
daily temperature range 日温变幅
daily test 日检
daily throughput 日输送量
daily tonnage 日产吨数
daily total passenger volume 日客运总量
daily unbalance factor 日不均系数
daily variation 日变化；日变动；日变幅
daily variation coefficient 日变化系数
daily variation diagram 日变化图
daily variation factor 日变化系数
daily wage 每日工资；日薪
daily water consumption 日用水量
daily water-injection rate 日注水量
daily wave 日波
daily weather chart 每日天气图
daily yardage 日产量；日挖掘量
dale 宽谷；谷；峪；槽
dale-head 谷头；谷源
D'Alembert inertial force 达朗贝尔惯性力
D'Alembertian 达朗伯算符
D'Alembert-Lagrange's equation 达朗贝尔-拉格朗日方程
D'Alembert's principle 达朗贝尔原理
D'Alembert's ratio test 达朗伯比例试验法
dalles 峡谷急流；峡谷峭壁

Dalton-Raoult's law	道尔顿-拉乌尔定律
Dalton's law	道尔顿定律
dam	隔墙；密封墙；堤；坝；筑堤；筑坝
dam abutment	坝座；坝肩
dam abutments-stability analysis	坝座稳定性的分析
dam accessories	坝体附属结构物
dam ageing	坝的老化
dam appurtenances	坝体附属结构物
dam axis	坝轴线
dam base	坝基
dam body	坝体
dam breach	溃坝
dam built by dumping soil into water	水中倒土坝
dam bursting	溃坝；垮坝
dam cement	大坝水泥
dam construction	建坝；筑坝；坝体施工
dam construction survey	堤坝施工测量
dam core	坝心
dam crest	坝顶
dam crest level	坝顶高程
dam cut-off	坝基防渗帷幕；坝基截水墙
dam damage	大坝损坏
dam deflection	坝体挠度
dam deformation observation	大坝变形观测
dam deformation survey	大坝变形测量
dam deterioration	大坝损坏
dam embankment soil	坝堤土
dam face	堤坝迎水面
dam failure	坝失事；溃坝
dam foundation	坝基
dam foundation grouting	坝基灌浆
dam foundation investigation	坝基勘察
dam foundation leakage	坝基渗漏
dam foundation sliding	坝基滑移
dam gradient	坝坡度
dam heel	坝踵；坝的上游坡脚
dam height	坝高
dam heightening	堤坝加高
dam in	筑坝堵水；筑坝挡水
dam inspection	大坝检查
dam leakage	坝渗漏
dam length	坝长
dam location	坝位；坝址
dam out	筑坝排水
dam raising	坝体加高
dam roadway	坝上道路
dam root	坝根
dam seismic stability	坝的抗震稳定性
dam sheeting	坝面护板
dam shell	坝壳
dam site	坝址
dam site investigation	坝址勘查
dam site selection	坝址选择
dam site survey	坝址测量
dam slope	坝坡
dam stability	坝的稳定性
dam thickness	坝的厚度
dam toe	坝趾；坝的下游坡脚
dam type power plant	坝式发电站
dam up ravine	闸山沟
dam volume	坝体积
dam width	坝宽
dam with core wall	心墙坝
damage	损害；损坏；损伤；破损
damage accumulation	损伤累积
damage analysis	损坏分析
damage and fracture	损伤与断裂
damage and deterioration	损伤；残损；破坏；损坏
damage angle	破坏角
damage appraisal	损坏评价
damage beyond repair	无法修复的损坏
damage by tide	潮灾
damage categories	损坏种类
damage claim	损害索赔
damage class	破坏程度
damage contour map	损坏等值线图
damage cost factor	破坏代价因数
damage criterion	损伤准则
damage curve	损耗曲线
damage deformation spectrum	损坏变形谱
damage degree of building	房屋损害程度
damage detection	损坏探测；损伤检测
damage diagnosis	损坏诊断
damage distribution	损害分布
damage dosimeters	损坏测试仪
damage earthquake	破坏性地震
damage effect	损坏效应
damage evolution	损伤演化
damage evolution equation	损伤演化方程
damage evolution mechanism	损伤演化机制
damage factor	堵塞系数
damage feature	损坏特点
damage function	损伤函数
damage identification	破坏识别
damage increment	损害增量
damage index	损坏指数
damage initiation threshold	损伤起始阈值
damage intensity	损害率；震害烈度
damage level	损坏程度
damage mechanics	损伤力学
damage mitigation	损坏抑制
damage monitoring	损坏监测
damage of coastal water conservancy project	海岸水利工程损毁
damage of cultural heritage	文化遗产损坏
damage parameter	损伤参量
damage pattern	损坏形式
damage performance	损伤特性
damage phenomena	损坏现象
damage potential	损害势
damage potential index	损伤势指数
damage prediction	损伤预测；震害预测
damage prevention	损坏预防
damage probability distribution function	破坏概率分布函数
damage probability matrix	破坏概率矩阵
damage radius	损害半径
damage ratio	受灾率；损坏率
damage resistance	抗破坏性
damage rheology	损伤流变学

damage scenarios 损伤场景
damage situation of coal mining subsidence land 采煤塌陷土地的受害情况
damage skin factor 损害表皮因数；损害表皮系数
damage softening 损伤软化
damage state 损伤状态；破损状态
damage statistics 损害统计
damage strengthening 损伤强化
damage surface 破坏面
damage tensor 损伤张量
damage threshold earthquake 破坏阈限地震
damage threshold earthquake intensity 地震烈度损伤阈值
damage to persons and property 人身与财产损害
damage to works 对工程的损害
damage tolerance 损伤容限
damage tolerance design 破损容限设计；损伤容限设计
damage type 损伤类型，破坏类型
damage variable 损伤变量
damage vector 损伤矢量
damage zone 损伤区
damageability 易损性；易破坏性
damaged formation 受污染地层
damaged well productivity 受污染井的产能
damaged wellbore area 井眼受污染区
damage-driven numerical model 基于损伤的数值模型
damage-frequency curve 损害频率曲线
damages for default 违约赔偿金
damage-vulnerable 易损的；易破坏的
damaging deposition 严重淤积
damaging effect 损害效应
damaging element 损害因素；损害性成分
damaging flood 灾害性洪水
damaging flux 损伤通量
damaging impact 破坏性冲击；溢流水层
damaging phenomenon 破坏性现象；震害现象
damaging scour 严重冲刷
damaging sedimentation 严重淤积
damaging stress 破坏应力
damaging vibration 破坏性振动
damaging workover 损害性修井
damaging workover fluid 有损害性的修井液
dam-board 挡板
dam-break 溃坝
dammed entry 隔离巷道
dammed lake 堰塞湖
dammed water 壅水；回水
damming 筑坝；壅水
damming limit 壅水界限；回水界限；回水范围
damming lock 拦江船闸
damming off 截水；截流；断流
damming shield 挡水防护体
damming value of groundwater table 地下水位壅高值
damourite 水白云母
damp 湿度，潮湿；阻尼，缓冲；减幅；衰减；减震
damp atmosphere 湿空气；湿大气
damp course 防水层；防湿层
damp marsh 湿沼泽
damp proof course 防潮层
damp proof membrane 防潮表层
damp proofing 防潮

damp room 雾室
damp snow avalanche 湿雪崩坍
damp zone 含水层
damp-cure 潮湿养护
damped 阻尼的；衰减的；减震的；被窒息的
damped balance 阻尼天平
damped circular frequency 有阻尼圆频率
damped exponential 衰减指数
damped free vibration 有阻尼自由振动
damped geophone 阻尼地震检波器
damped harmonic motion 阻尼谐和运动
damped inductance 阻尼电感
damped lake 堰塞湖
damped least square method 阻尼最小二乘法
damped matrix 阻尼矩阵
damped natural circular frequency 有阻尼自振圆频率
damped natural frequency 有阻尼自振频率
damped oscillation 减幅振荡；衰减振荡；阻尼振动；阻尼振荡
damped vibration 有阻尼振动
damped wave 阻尼波；减幅波
damped wave equation 阻尼波动方程
damped-up 壅高的；拦蓄的
dampen 润湿；增湿；变湿；阻尼；阻抑；制动
damper 减振器；阻尼器；缓冲器；挡板
damper coefficient 阻尼系数
damper cylinder 减振筒
damper regulator 阻尼调节器
damper test stand 减振器实验台
damper upper seat 减振器上座
damper valve 阻尼阀；调节阀
damper weight 挡板配重；挡板平衡锤
damper winding 阻尼器绕组
damping 阻尼；减振；缓冲；湿润；潮湿
damping action 阻尼作用
damping angle 阻尼角度
damping apparatus 阻尼装置
damping area 减振面；阻尼面
damping arrangement 阻尼设备
damping by frication 摩擦减振
damping by frication of liquids 液体摩擦阻尼
damping capacity 阻尼量；减振能量
damping characteristic 阻尼特性
damping circuit 阻尼电路
damping coefficient 阻尼系数；衰减系数
damping component 阻尼元件
damping constant 阻尼常数
damping device 阻尼器；减振装置
damping disk 减振盘
damping distribution 阻尼分布
damping effect 阻尼效应；阻尼作用；减振作用；衰减作用
damping element 减振装置
damping energy dissipation 阻尼耗能
damping error 阻尼误差
damping extent 衰减度
damping factor 减幅因数；阻尼系数
damping fin 阻尼片
damping fluid 缓冲液
damping force 减振力；阻尼力

damping index 减震指数
damping influence coefficient 阻尼影响系数
damping intensity 阻尼强度
damping material 阻尼材料；减振材料；缓冲材料
damping matrix 阻尼矩阵
damping medium 阻尼介质
damping method 减振方法
damping model 阻尼模型
damping modulus 阻尼模量
damping moment 阻尼力矩；减震力矩
damping of waves 波浪衰减；波浪阻尼；波浪平息
damping orthogonality 阻尼正交性
damping parameter 减震参数；阻尼参数
damping piping 减震管路，减震管道系统
damping property 阻尼性质
damping rate 阻尼率
damping ratio 阻尼比
damping resistance 阻尼电阻；阻尼阻力；衰减阻力
damping resistor 熄灭电阻；阻尼电阻
damping ring 阻尼环
damping roller 减振滚柱
damping rope 罐笼间防撞钢丝绳
damping screen 缓冲瓶；稳水栅；整流网；整流栅
damping spring 减振弹簧；阻尼弹簧
damping time 阻尼时间
damping value 阻尼值
damping vibration attenuation 阻尼减振
damping washer 减震垫圈
dampness 潮湿；湿度；含水量
dampness penetration 湿度贯透；潮湿渗透
damp-proof insulation 防湿绝缘
damp-proofing admixture 防潮剂
damp-proofing coating 防潮层
damp-proofing foundation 防潮基础
damp-storage closet 高湿度养护室
dam-type power station 堤坝式水电站
danger angle 危险角
danger board 警告牌
danger buoy 障碍物浮标
danger class 危险等级
danger level 危险程度；危险水平
danger line 警戒线；警戒水位
danger rock 险礁
danger rupture surface 危险破坏面
danger sign 危险信号
danger warning 危险警报
danger zone 危险地带；危险区
dangered-off area 封闭的危险区
dangerous 危险的
dangerous area 危险区
dangerous atmosphere 危险大气环境
dangerous building 危楼；危险建筑物
dangerous coefficient 危险系数
dangerous deformation period 危险变形时期
dangerous goods 危险品
dangerous goods store 危险品仓库
dangerous hill 险坡道
dangerous influence 危险影响
dangerous movement boundary 危险移动边界
dangerous occurrence 危险现象；危险事故
dangerous rock 危岩
dangerous rockmass 危岩体
dangerous section 危险剖面；危险断面
dangerous slope 危险斜坡；危险边坡
dangerous stage 危险阶段，危险时期
dangerous structure 危险结构
dangerous terrain 危险地域；危险地势
dangerous zone 危险地带
dangers forecast 危险预测
danubite 闪苏安山岩
dapped joint 互嵌接合
dapped shoulder joint 互嵌肩接合
dapple 斑点；有斑纹的；有斑纹
Darcy flow 达西流
Darcy flow coefficient 达西流动系数
Darcy flow regime 达西流态
Darcy laminar-flow equation 达西层流方程
Darcy law 达西定律
Darcy linear flow equation 达西线性流动方程
Darcy unit 达西单位
Darcy velocity 达西速度；渗透速度
Darcy viscosity 达西黏度
Darcy's equation 达西方程
Darcy's formula 达西公式
Darcy's law 达西定律
Darcy's radial flow formula 达西径向流公式
dark layer 暗层；深色层
dark light 不可见光
dark mineral 暗色矿物
dark red 暗红；赭色的
dark red ferralsol 暗红色铁铝土
dark red latosol 暗红色砖红土
dark room 暗室
dark signature 深色特征
dark-colored mineral 暗色矿物
dark-colored soil 暗色土
darken 使变黑；弄模糊；变黑
dark-field illuminator 暗场照明灯
dark-field method 暗场法
dark-line spectrum 暗线光谱
darkness adaptation 暗适应
dark-red heat 暗红热
dart valve 突进阀；镖阀
dash 冲击；巷道通风；控制板；操纵板；挡泥板；遮水板
dash adjustment 缓冲调节
dash board 仪表板；仪器板；挡泥板
dash control 按钮控制；缓冲控制
dash line 虚线；短划线
dash plate 缓冲板
dash pot 缓冲器；减振器；阻尼器
dash-and-dot line 点划线
dashboard 挡泥板；仪表板
dashed area 阴影部分；虚线部分
dashed contour 虚线等高线
dashed contour line 虚线等高线
dashed curve 虚曲线
dashed line 虚线
dashpot 阻尼器；减振器
dashpot lever 缓冲杆

dashpot mechanism 油阻尼机构；油压延时机构
dash-pot plunger 缓冲活塞
dass 采掘带；采条；层；分层采掘
data 资料；数据；已知数
data access 数据存取
data access control 数据访问控制
data access line 数据存取线
data accessibility 资料可达性
data accumulator 数据累加器
data accuracy 数据精确性
data acquisition 数据采集；数据获得
data acquisition and interpretation system 数据采集和解释系统
data acquisition circuit 数据采集电路
data acquisition code 数据采集代码
data acquisition control system 数据采集和控制系统
data acquisition system 数据收集系统；数据采集系统
data acquisition terminal 数据采集终端
data acquisition unit 数据采集单元
data address 数据地址
data aggregate 数据集合体
data aggregation 数据聚合
data analysis 数据分析
data analysis centre 数据分析中心
data analysis system 数据分析系统
data analyst 数据分析人员
data analyzing and retrieval system 资料分析检索系统
data aquisition 数据采集
data array 数据数组
data attribute 数据属性
data bank 数据存储单元；数据组；数据库
data base 基础数据；基本资料；数据库
data base maintanance 数据库维护
data base management system 数据库管理系统
data base package 数据库程序包
data base procedure 数据库过程
data block 数据块
data buffer 数据缓冲区
data bus 数据总线
data capacity 信息容量；数据容量
data capture 数据收集
data carrier 数据载体
data cell 数据单元
data centre 数据中心
data channel 数据通道
data check 数据检查
data circuit 数据电路
data cluster 数据集群
data code 数据代码
data collecting 收集数据；回收数据
data collection system 数据收集系统
data collector 数据采集装置
data communication 数据通信
data communication network 数据通信网络
data communication system 数据传输系统
data compaction 数据紧缩
data compression 数据压缩；信息压缩
data compression factor 数据压缩引子
data compression routine 数据压缩程序
data concentrator 数据集中设备

data consistency 数据一致性
data control block 数据控制块
data conversion 数据变换；数据转换
data converter 数据转换器
data correlation 数据相关
data correlation method 数据关联方法
data counter 数据计数器
data coverage 数据范围
data curve 数据曲线
data date 数据日期
data declaration 数据说明
data decode 数据解码
data definition name 数据定义名
data delay 数据延迟
data description language 数据描述语言
data descriptor 数据描述符
data dictionary 数据字典
data dimensionality 数据维数
data direction register 数据方向寄存器
data directory 数据目录
data display 数据显示
data edit 数据编辑
data element 数据元素
data encryption 数据加密
data entry procedure 数据输入程序
data envelope analysis 数据包络分析
data envelopment analysis model 数据包络分析模型
data error vector 数据误差向量
data exchange 数据交换
data extraction 数据提取；资料提取
data fitting 数据拟合；数据选配
data flow 数据流；信号通过
data flow chart 数据流程图
data format 数据格式
data fusion 数据融合
data gathering 数据收集
data gridding 数据网格化
data handling 数据处理
data handling component 数据处理组件
data handling system 数据处理系统
data hierarchy 数据分层
data identification 数据识别
data in 数据输入
data input 数据输入
data input bus 数据输入总线
data input module 数据输入模块
data inserter 数据输入器
data integration 数据集成
data integrity 数据完整性
data interconnect transfer 数据互连传送
data interpretation 数据解释
data interpretation mode 资料解释模式
data item 数据项
data item type 数据项类型
data layout 数据格式
data library 数据库
data line 数据线
data link layer 数据链路层
data logger 巡回检测装置；数据记录器；数据输出器
data logging 数据记录

data logging system	数据自动记录系统
data management	数据管理；数据处理
data management system	数据管理系统
data manipulating language	数据操作语言
data manipulation	数据操作
data manipulation and display software	数据操作与显示软件
data mining	数据挖掘
data model	数据模型
data network	数据网络
data normalization	数据规格化
data on earthquake forecasting	地震预报数据
data optimization	数据优化
data ordering	数据排列
data organization	数据构造，数据编制
data output	数据输出
data output option	数据输出选择
data packet	数据组；数据包
data plot	数据图
data plotter	数据绘图仪
data point	数据点
data preprocessing	数据预处理
data presentation	数据显示
data processing	数据处理
data processing center	数据处理中心
data processing machine	数据处理机
data processing system	数据处理系统
data processing techniques	数据处理技术
data processor	数据处理机
data quality control	数据质量控制
data rate	数据速率
data recovery	数据恢复
data reducer	数据简化器
data reduction	数据简化；资料缩减
data register	数据寄存器
data reliability	数据可靠性
data resource	数据资源
data retrieval	数据检索；资料检索
data routing	数据路由
data safety	数据安全
data sample	数据样品
data sampling	数据采样；数据抽样
data sampling rate	数据采样率
data security	数据安全
data set	数据器；数据集
data sharing	数据共享
data sheet	数据表
data smoothing	数据信号平滑
data snooping	数据探测法
data source	数据源
data standard	数据标准
data station	数据站
data storage	数据存储；信息存储
data stream	数据流
data structure	数据结构
data surface	数据面
data switching equipment	数据交换设备
data system	数据系统
data terminal	数据终端
data transducer	数据传感器
data transfer	数据传输
data translation	数据转换；数据交换
data transmission block	数据传输块
data transmission device	数据传输装置
data transmission line	数据传输线路
data transmission radio	数传台
data transmission system	数据传输系统
data transmission terminal equipment	数据传输终端设备
data transmit pipeline	数据传输流水线
data trend	数据倾向
data type	数据类型
data unit	数据单位；数据机
data update	数据更新
data validity	数据有效性
data visualization	数据可视化
data volume	数据量
data warehouse	数据仓库
data-acquisition system	数据采集系统
data-adaptive filter	数据自适应滤波器
database	数据库
data-base administration	数据库管理
database atlas	数据库图集
database design	数据库设计
database of earth stress in China	中国地应力数据库
data-carrier store	数据存储器
data-carrying capacity	数据承载量
data-collection interval	采集数据的时间间隔
data-directed inference	数据制导推理
data-directed input-output	直接数据输入输出
data-driven modeling	数据驱动的建模
data-driven numerical model	基于数据的数值模型
datamation	数据处理自动化
data-processing equipment	数据处理设备；数据处理计算机
data-switching	数据转换
datatron	数据处理机
date back to	回溯到
date exchange	数据交换
date for inspection	检查日期
date of agreement	协议日期
date of assignment	转让日期
date of balance sheet	资产负债表日期；决算日
date of completion	竣工日期
date of delivery	交货日期
date of expiration	有效日期
date of payment	交付日期
date of possession	管有日期
date of settlement	决算日期
date of termination	终止日期
date of validity	有效日期
dating	年龄测定；断代
dating error	测年误差
dating of groundwater	地下水测龄；地下水计时
dating of rock	岩石年龄测定
dating pulse	同步脉冲；控制脉冲
dating technique	断代技术；断代法
datum	基准面；基准线；基准点；基标；读数基准；数据；资料；已知数
datum bed	基准层；标准层
datum correction	基准面改正

datum depth	基准面深度
datum drift	基准漂移；基准偏差
datum elevation	基准高程
datum ellipsoid	基准椭球
datum for heights	高程基准
datum for reduction of sounding	测深折算基点
datum horizon	标准层位；标志层；基准面；基准地平
datum level	基准；基准面；基准水位
datum levelling	基准水准测量
datum line	基准线；基准
datum line beam	基准梁
datum location	基准地点
datum mark	基准标志；基准点；水准点
datum of tidal level	潮位基准面
datum plane	基准面；基面
datum plane of gauge	测站基面
datum plate	基准板
datum point	基准点；基点
datum pressure	基准压力
datum quantity	基准量；参考量
datum static correction	基准面静校正
datum station	基准站
datum surface	基准面；基面
datum transformation	基准变换
datum velocity	基准速度
datum water level	基准水平面；水准面；水准零点
datum well	基准井
datum-centered ellipsoid	参数共轴椭球
datumized section	基准化剖面
datum-level pressure	基面压力
datum-level temperature	基准面温度
daub	涂；抹；胶泥
dauk	韧性的；硬的；紧密的；砂质黏土
daunialite	蒙脱石；硅质蒙脱岩
dawk	黑色碳质页岩
day coal	上部煤层；近地面煤层
day drift	通地面的平硐
day eye	通地面斜井
day fall	矿山地面塌陷
day hole	通地面的平硐
day of snow lying	积雪日
day rate payment	计日工资
day stone	自然露头；外露岩石
day to day construction activities	每日施工作业
day to day operation	日常业务
day water	地面水
day without frost	无霜日
day work	零工；计日工；散工
day work rate	计日工资；每日单价
daya	积水洼地
day-by-day operations	天天的作业，每天的作业或操作
daylight effect	昼光效应
daylight factor	采光系数；日光照明率
daylight illumination	日光照明
daylight lamp	日光灯
daylight mine	平硐矿山
daylight opening	采光口
daylight operation	亮室操作
daylight press	多层压机
daylight ratio	天然照明率
daylight saving time	日光节约时
daylight standard	日照标准
daylight time	白昼时间
daylight uniformity	采光均匀度
daylighting	采光
daylighting area	采光面积
day-to-day evaluation	逐日评价
day-to-day mine planning and design	矿的日常开采规划与设计
day-to-day traffic working plan	运输工作日常计划
day-to-day variation	日际变化
deacidification	中和酸性；去酸
deactivation	减活；去活；钝化
deactivator	减活化剂
dead abutment	隐蔽式桥台，隐蔽式岸墩
dead accurate	不准确
dead air	停滞的空气
dead angle	死角；辐射等盲区
dead annealing	完全退火
dead area	死角区；死水区；截面的不受力部分
dead band	死区；静区；不灵敏区
dead banking	长期封炉
dead beat pendulum	无周期摆
dead belt	盲区；静区；死区
dead cave	死洞穴；干洞穴
dead center	固定中心
dead channel	残遗河段；废河段
dead circuit	空路
dead cliff	古海蚀崖
dead end	封闭或堵塞端；死端；尽头
dead end roadway	独头巷道
dead ends ventilation	盲巷通风
dead face	独头工作面
dead fault	死断层
dead fold	残余褶皱；死褶
dead forms	古地形
dead glacier	停滞冰川；死冰川
dead ground	无矿地层；无煤带；盲区；死带；遮蔽地区
dead heading	放空；车船空回
dead heading back	绝汽后转向冲程
dead hole	过深炮眼；未穿透孔；死孔
dead lake	死湖
dead landslide	死滑坡
dead length	固定长度
dead level	绝对水平；绝对高程
dead lime	失效石灰
dead load	静荷载；恒荷载；恒载量
dead load of derrick	钻架自重
dead load test	静载荷试验
dead pit	关闭的矿；采尽的矿；废弃的矿
dead place	死区；尽头处；不通行的地方；独头工作面
dead point	死点；静点；哑点
dead position	静止位置
dead pressure	死压
dead resistance	吸收电阻；消耗电阻；整流电阻
dead rock	废石；尾矿；脉石；围岩
dead shore	垂直支撑；固定支撑
dead short	完全短路
dead size	净尺寸

dead slow	微速
dead small	小块料
dead soft steel	极软钢
dead space	静区；死区；无效空间；死角
dead steam	废气
dead storage	死库容；垫底库容
dead storage of sedimentation	淤积库容
dead stream branch	死河汊
dead take	死水湖
dead tide	最低潮
dead time	死时间；空耗时间
dead valley	干谷；死谷
dead water	死水；停滞水；船尾旋涡；死水区；死水域；静水区
dead water level	死水位
dead water region	死水区域
dead water space	死水区；静水区；死水域
dead water zone	死水区
dead weight	自重；静重
dead weight brake	重锤制动器
dead weight capacity	载重量；静载重量
dead weight gage tester	静重压力校正器；静重压力标表仪
dead weight pressure gauge tester	砝码式压力表检验仪
dead weight safety valve	重锤限压安全阀
dead weight tester	静重压力标定台
dead well	枯井；废井；渗水井；吸水井
dead work	非直接生产工作；正作
dead wraps	摩擦圈数
dead zone	空白带
dead-band regulator	静区调节器；非线性调节器
deadbeat	非周期性的；无振荡的
dead-beat	稳定的；不动的；不摆的
dead-beat discharge	非周期放电
dead-burn	透烧；烧坏
dead-burnt gypsum	硬石膏；无水石膏；烧石膏
dead-end	尽头巷道，独头巷道
dead-end anchor	终端锚定
dead-end anchorage	固定端锚具
dead-end capillary	不连通毛细管
dead-end clamp	耐张线夹
dead-end effect	空圈效应
dead-end fracture	不连通裂缝
dead-end insulator	耐张绝缘子
dead-end main	闷头主管
dead-end pore	不连通孔隙；封闭孔隙；盲端孔隙
dead-end pore volume	不连通孔隙体积
dead-end shaft station	尽头式井底车场
dead-end siding	尽头岔线；死岔子
dead-end stowing	从采石巷取石充填
dead-end terminal	尽头线；尽头终点站
dead-end tower	终点塔
dead-end trench	折返式堑沟
dead-end turnback track	尽头折返线
dead-ended	不通的
deadening	消去；衰减；下降
dead-fall	翻斗机；翻车机
dead-glacier	不动冰川
dead-hard steel	高强钢
deadhead area	尽头区

deadhead pressure	无输出流量时的压力
dead-hole	盲孔；不通孔
deadline	安全界线；行人止步线；不流通的管道；不通电的线路；限期；截止时间
deadline switch	死绳开关
dead-load moment	静重力矩；恒载力矩
dead-load power	静重动力；恒载动力
dead-load stress	静重应力；静载应力
dead-load test	恒载试验；静荷载试验
deadlock	停顿；僵局；关闭厂矿；死锁
deadman	绳锚；桩橛；井口挡车木；拉杆锚桩
deadman anchorage	锚定物；锚定桩；拉杆锚桩
deadman control	事故自动刹车控制；安全控制
deads	井下废石；围岩；尾矿
dead-weight brake	重锤制动器
dead-weight capacity	载重吨位；恒载容量
dead-weight ton	载重吨位
deadweight-to-payload ratio	静重-有效载重比
deadwork	岩石掘进工作；岩石巷道；非直接生产的工作；包价外的工作
deaerated water	脱气水
deaeration	去气；去除空气
deafening device	隔音装置
deair	脱气
deaired water	脱气水
deairing	去气；除气；脱气
deairing mixer	去气搅拌机；真空搅拌机
dealkalization	脱碱
deamplification	衰减；折减；缩小
deaquation	脱水
deashing	脱灰；去灰分
deasphalting	脱沥青作用
death rate	死亡率
debacle	解冻；崩溃；山崩
debalance of preparation and winning work	采掘失调
debase	质量变坏；降低纯度
debit	借方；借记；借项；记入借方
debit and credit	借贷
deboost	减速；制动
debooster	限动器；减压器
debouch	河口；流出
debouchment	支流汇合处；河口
debris	岩屑，碎石；尾矿，水力采矿流出的砂砾等；废石
debris apron	山麓冲积平原
debris aqueduct	泄流槽
debris avalanche	岩屑崩落
debris basin	贮砂库；沉沙池；拦沙池；漂流物沉淀池
debris chute	废石溜槽；矸石溜道
debris cleaning	清除碎屑；清除碎石
debris collection area	碎屑收集范围
debris cone	冲积锥；冲积扇；岩屑锥；冲出锥
debris dam	碎屑坝；冲积坝；拦砂坝，拦砂谷坊
debris disposal area	水力排土场；碎屑处置场
debris dump	废渣堆；尾矿堆
debris facies	碎积相
debris fall	碎屑崩落
debris flood	碎屑洪流；乱石洪流
debris flow	泥石流；岩屑流
debris flow deposit	泥石流坡积物

debris flow fan	泥石流堆积扇	decay distance	衰减距离
debris from heading	巷道掘进面矸石；巷道迎头的矸石	decay energy	衰变能量
debris guard lemniscate mechanism	防矸式三角区	decay factor	衰变因数；衰变系数
debris hazard	土石灾害；泥石流灾害	decay graph	衰变图
debris hazard mitigation	泥石流灾害缓解	decay index	衰变指数
debris kibble	岩石吊桶	decay law	衰变定律
debris laden stream	含大量砂砾的河流	decay mode	衰变方式
debris of diatom	硅藻碎屑	decay module	衰变模数；衰变模量
debris plain	岩屑平原	decay of pollutants	污染物衰减
debris run-out	泥石流伸展距离	decay of stones	石材衰变
debris scree	山麓岩屑	decay of wave	波浪衰减；波浪减幅
debris shield	掩护梁	decay ooze	软腐泥
debris slide	岩屑滑动；泥石滑移	decay parameter	衰减参数
debris storage capacity	淤积库容	decay pattern	衰减模式
debris stream	岩屑流，泥石流	decay pattern of anomaly	异常衰减模式
debris thickness	泥石流厚度	decay period	衰减周期
debris tipping	翻卸废石	decay process	衰变过程；衰化过程；腐败过程
debris torrent	泥石湍流；碎屑湍流；岩屑急流	decay product	分解产物；衰变产物
debris transportation	泥石运输，泥石搬运	decay rate	衰变率；风化速度
debris utilization	废品利用	decay resistance	抗腐强度
debris-free completion fluid	无固相完井液	decay rock	风化岩层
debris-laden stream	含大量岩屑的河流	decay scheme	衰变系统；衰变图
debt ceiling	债务限额	decay scheme diagram	衰变系统图
debt certificate	债务证明书	decay sequence	衰变序列
debt crisis	债务危机	decay series	衰变系；放射系
debt financing	负债融资，举债筹资，债券周转信贷	decay spectrum	衰变谱
debt instrument	债务证券；债务工具	decay stress	衰减应力
debt outstanding	未偿债务	decay time	衰减时间；衰变时间
debt ratio	负债率	decay zone	衰减区；衰退区；衰变区
debt relief	免除债务	decayed	衰变的；被破坏的；分解的；下降的
debt rescheduling	重订还债期限	decayed basalt	风化玄武岩
debt service	偿债	decayed gravel bed	衰变砾石层
debt service ratio	偿债比率	decayed rock	风化岩
decadent	衰落，减幅	decay-time	衰变时间；衰减时间
decadent wave	衰减波；减幅波	decay-time integral	衰变时间积分
decagram	十克	decay-to-resonance effect	衰变-共振效应
decalcification	脱钙作用	decelerated motion	减速运动
decalcified sample	脱钙样品	decelerating flow	减速流
decalcified soil	脱钙土壤	deceleration	减速
decaliter	十升	deceleration lane	减速车道
decameter	十米	deceleration radiation	减速辐射
decantation	迁移；调迁	deceleration rate	减速速率
decantation method	沉淀分离法	deceleration speed	减速速度
decantation test	倾析试验	deceleration stress	减速应力
decanting arm	泄水管	deceleration valve	减速阀
decanting centrifuge	沉降式离心机	decelerator	减速器
decapitation	削顶；解雇	decelerometer	减速计
decarbonation	脱碳酸作用	decementation	脱胶结；去胶结作用；分解作用
decarbonizer	脱碳剂	decentralization	不同心；偏心度；分数
decarburization	脱碳	decentralized air supply system	压缩空气分散供给法
decauville railway	轻便铁路	decentralized casing	偏心套管
decauville truck	轻轨料车	decentralized computer system	分散式计算机系统
decay	腐朽；风化；毁坏；剥落；衰减；衰变	decentralized control	分散控制方式
decay area	无风区；风浪衰减区；风浪平息区	decentralized drill collar	偏心钻铤
decay branching ratio	衰变分支比	decentralized gun	偏心射孔器
decay chain	衰变链；放射系	decentralized management	分权管理
decay characteristic	衰减特性	decentralized stochastic control	分散随机控制；非集中随机控制
decay coefficient	衰减系数	decentralized tool	偏心的仪器
decay constant	衰变常数	decentralized wellhead	偏心井口
decay curve	衰减曲线；减缩曲线；余辉曲线		

decentralizer	偏心器
decentralizing device	偏心装置
deceptive conformity	假整合
deceptive cordance	假整合
deceptive fold	假褶皱
decibel	分贝
decidability	可判定性
decigram	分克
deciliter	分升
decimal	十进位；小数；十进位；小数
decimal base	十进制
decimal carry	十进制进位
decimal fraction	小数；十进制小数
decimal multiple	十进倍数；十进的
decimal place	小数位
decimal point	小数点
decimal scale	十进制
decimal sequence	十进序列
decimal system	十进制
decimeter	分米
decipher	译码；解译
deciphering image	解译图像
deciphering photograph	解译像片
deciphering process	解译程序；解译方法
decision	决策；判定；决定
decision accounting	决策会计
decision analysis	决策分析
decision boundary	判定边界
decision criteria	决策准则；判定准则
decision design	决策设计；优选设计
decision function	决策函数；判定函数
decision making	决策
decision making capability	判别能力；决策能力
decision making network	决策网络计划
decision making package	综合决策
decision making through operations decision making under certainty	运筹决策
decision making under risk	风险性决策
decision making under uncertainty	不确定型决策
decision matrix	决策矩阵
decision model	决策模型
decision node	决策节点
decision package	决策组
decision point	决策点
decision policy	决策策略
decision procedure	判定过程；决策过程
decision process	决策过程
decision region	判定区域
decision rule	决策规则
decision science	决策学
decision sequence	决策序列
decision set	决策集
decision space	决策空间
decision support	决策支持
decision support system	决策支持系统
decision theory	决策论
decision tree	判别树；决策树
decision variable	决策变量
decision-making body	决策机构
decision-making management	决策管理
decision-making model	决策模型
decision-making process	决策过程
decision-making system	决策系统
decisive factor	主要因素；决定性因素
deck	盖层；岩床；桥面；层；台；甲板
deck arrangement	甲板设备
deck beam	上承梁
deck bridge	上承式桥
deck charge	分段装药
deck construction	桥面构造
deck covering	甲板覆材
deck crane	甲板吊车
deck curb	甲板边缘，桥面边缘
deck dam	平板坝
deck form	平台式钢模板
deck girder	上承梁
deck hand	甲板水手；甲板船工
deck head building	井口建筑物
deck house	舱面室
deck joint	面板接缝
deck ladder	甲板扶梯
deck light	甲板窗
deck load	面荷载
deck machinery	舱面机械
deck plate	筛面板；台面板；甲板；脚踏板；溜槽中隔板
deck reinforcement	面板钢筋
deck roof	台式屋顶
deck slab	上承板；桥面板
deck span	桥面跨度
deck spillway	盖板溢洪道
deck stringer	甲板边板
deck structure	上承结构
deck support	筛面支架；盘面支架
deck support structure	甲板支承结构
deck transverse	甲板横材
deck truss	上承桁架
deck-bridge	上承桥
decke	推覆体；盖
decked explosion	分层爆炸
decked nullah	铺面渠
decken structure	推覆构造；叠瓦构造
decker	分层装置；脱水装置
decker structure	叠瓦构造；推覆构造
deck-girder steel dam	板梁式钢坝
deckhead	出车台
decking	装罐；铺垫板；铺假顶；装模板；分段装药
decking equipment	罐笼装车设备；装罐设备
decking gear	装罐机；装罐笼推车机；倾卸台
decking level	装罐水平
decking platform	装罐台
decking ram	罐笼装车器
decking speed	停车速度；装卸罐笼速度
decking support	过滤介质支架；滤布支架
decking time	装罐笼时间
deckle	调型器
deckname	叠名称
deckplate	支撑架盖板；铠装可弯曲输送机溜槽
declaration	声明
declarative language	说明性语言

declarative statement	说明语句
declare	申报；声明
declared efficiency	申明保证的效率
declension	倾斜；偏差
declensional ventilation	下行通风
declinater	磁偏仪；测斜仪；方位计
declination	倾斜；偏差；偏角；磁偏角
declination angle	磁偏角；俯角
declination axis	偏差轴
declination chart	磁偏图
declination compass	倾角计；偏角仪；磁偏计
declination needle	磁针
declination of magnetic needle	磁偏角
declination of the boundary	边界偏差
declination parallel	赤纬圈
declination variation	倾角变化
declinational tide	赤纬潮
declinator	磁偏计；偏角仪
decline	下降；下倾；偏角；衰落
decline belt	倾斜向下胶带
decline characteristic	递减特性
decline curve	下降曲线；递减曲线
decline factor	递减系数
decline fraction	递减段
decline history	递减曲线
decline limit	递减界限
decline of ground water level	地下水位的降落
decline of piezometric surface	水压面下降
decline of production	产量递减
decline of underground water level	地下水位下降
decline of water table	地下水位下降；潜水位下降
decline period	递减期
decline period of water table	水位下降期
decline rate	下降速度；下降速率
decline stage	衰亡期
decline trend	递减趋势
decline well	产量递减的井
declined	下降的；下斜的；递减的
declining balance method	余额递减法
declining pit	接近采尽的矿；产量日减的矿
declining rate	递减速度
declinometer	磁偏仪；测斜仪；方位计
declivity	倾斜；倾斜度；斜坡；降坡
declivous	下向的；倾斜的；下坡的
decoiffement	蠕动
decoiler	开卷机；拆卷机
decollement fault	拆离断层；基底逆掩断层
decollement fold	滑脱褶皱；脱底褶皱
decollement horizon	滑脱逆掩层位
decollement nappe	脱顶推复体
decollement structure	滑脱构造
decollement thrust	脱底断层；挤离冲断层
decollement zone	拆离带
decolour	脱色
decommissioning	解除运作；停止运作；关闭
decompaction	振松
decomposable	可分解的
decomposable matrix	可分解矩阵
decompose	分解
decomposed coal	分解的煤；风化煤
decomposed degree	分解度
decomposed explosive	分解炸药，变质炸药
decomposed form	分解形式
decomposed geometry matrix	分解几何矩阵
decomposed granite	风化花岗岩
decomposed granite soil	风化花岗岩土壤
decomposed outcrop	分解露头，风化露头
decomposed rock	风化岩
decomposed schist	风化片岩；分解片岩
decomposed volcanic rock	风化火山岩
decomposition	分解作用
decomposition course	分解过程
decomposition decay	分解衰变
decomposition gas	分解气体
decomposition intensity	分解强度
decomposition of force	力分解
decomposition of rock	岩石风化
decomposition point	分解点
decomposition potential	电解电压；分解电势
decomposition product	分解产物
decomposition reaction	分解反应
decompound	分解；多回分裂
decompressed zone	减压区
decompression	减压；降压；解压；卸荷
decompression and recompression loop	卸压再压循环，卸压再压曲线
decompression chamber	减压室
decompression curve	卸荷曲线
decompression drilling tool	减压钻具
decompression modulus	卸压模量
decompression time	减压时间
decompression valve	泄压阀
decompression velocity	减压波速度
decompression zone	减压区
decompressional expansion	降压膨胀
decompressor	减压器；减压装置
decontamination	去杂质；去污；净化
decontamination chamber	净化室
decontamination facilities	清除污染装置；消毒装备
decontamination unit	除污室
decontrol	解除管制
deconvolution	反褶积；重叠合法
deconvolution filtering	反褶积滤波
deconvolution of near surface motion	近地表运动的反褶积
deconvolution phase error	反褶积相位误差
decoration	装修；装饰
decorative concrete	饰面混凝土
decorative stone	饰面石材
decorrelation	解相关；去相关
decouple	去耦；解耦
decoupling	退耦；解耦；拆离
decoupling device	去耦装置；脱钩装置
decoupling factor	解耦因数
decoupling filter	去耦滤波器
decoupling loading	非耦合装药
decrease by degree	递减
decrease in cost	费用的减少
decreasing axial pressure fracture test	轴向减压断裂试验
decreasing coefficient	降低系数

decreasing function	递减函数；下降函数
decreasing hole angle	井身递减角度，递减钻孔角度
decreasing pressure	递减压力
decreasing series	递减级数
decreasing temperature gradient	减温梯度
decrement	减量；衰减量；减缩
decrement curve	衰减曲线
decrement measurement	衰减量测量
decrement of velocity	减速
decrementation	分级卸荷
decremeter	减缩量计；衰减计；减幅器
decrepitation	烧爆；爆裂作用；毕剥作响；爆碎声
decrepitation method	爆裂法
decrepitation temperature	爆裂温度
decuple	十倍；使增加到十倍
decussate	交叉呈十字形
decussate structure	交错构造；十字形构造
decussate texture	交叉结构
decussation	十字交叉
dedicated channel	专用信道
dedicated chemical plant	专用化学剂厂；专用化工厂
dedicated line	专用线路
dedicated machining system	专用加工系统
dedicated memory	专用存储区
dedicated pipe prover	专用标准体积管
dedicated port	专用接口
dedicated processor	专用处理机
dedolomitization	去白云石化
deduce	推演；演绎
deduction	推论；推导；演绎
deduction rule	推理原则
deductions and exemptions of tax	税收减免
deductive inference	演绎推理
deductive method	演绎法
dedust	除尘；脱尘
dedusted coal	脱尘煤
deduster	除尘器
dedusting	除尘
dedusting agent	除尘剂
dedusting curve	除尘曲线；脱尘曲线
dedusting efficiency	除尘效率
deed of covenant	契约
deed of exchange	交换契
deed of grant	批地契据；批约
deed of indemnity	赔偿契约
deed of mortgage	抵押契约
deed of mutual covenant	公契；公共契约
deed of release	解除扣押契据；解除责任契约
deed poll	分割契据；单边契约
deed registration system	契据登记制度
deed restriction	使用限制权
deenergization	断开；释放；去能；去激励
deenergize	去能；切断电流；解除激励
deenergized period	释放期间；脉冲间隔
deep	深处；深槽；深坑；深水区；深渊；海渊；海沟
deep adit	低层平硐；深平硐
deep air cell	深型充气式浮选槽
deep and thick overburden	深厚覆盖层
deep anode bed	深阳极地床
deep aquifer	深层含水层
deep basin	深水盆地
deep basin gas	深盆气
deep basin gas accumulation	深盆地气藏
deep bay	深水湾
deep beam	深梁
deep blasting	深孔爆破；深层爆炸
deep bore well pump	深钻井泵
deep borehole	深孔
deep boring	深钻孔；深层钻探
deep bucking current attenuation	深侧向屏蔽电流的衰减
deep burial metamorphism	埋深变质作用
deep burialism	深埋
deep buried bimetal benchmark	深埋双金属标
deep buried clay	深埋黏土
deep calcareous clay	深钙质黏土
deep cast	沉入深水；深测
deep channel	深河槽；深水航道
deep circular tunnel	深埋圆形隧道
deep circulation	深层环流；深水环流
deep clay	深部黏土
deep coal mine	深部煤矿
deep coal roadway	深部煤巷
deep compaction	深层夯压；深层压实
deep compaction by vibration	深层振动压实
deep consolidation	深层加固
deep convection	深部对流
deep current	深层流；深海流
deep cut	深掏槽；掏槽眼；深切
deep densification	深层加密
deep digging	下向挖掘
deep dip angle	深倾角
deep displacement	深滑移；深部位移
deep displacement monitoring	深部位移监测
deep diving	深水潜水
deep diving system	深海潜水系统
deep dome	深部穹窿
deep drawing	深井提升
deep drawing sheet	深冲薄板
deep drilling	深钻；深部钻进
deep earthquake	深层地震
deep erosion	深向侵蚀；深向冲刷；强度侵蚀；强度冲刷
deep excavation	深挖方，深开挖
deep face	下挖工作面
deep fault	深断裂
deep fill	深填；深填方
deep flow	深层流
deep focus	深震源
deep focus earthquake	深源地震
deep fold	基底褶曲
deep footing	深基础
deep formations	深层
deep foundation	深层地基；深基础
deep foundation pit	深基坑
deep fracture	深断裂
deep fracture zone	深部破裂带
deep freezing	低温冻结
deep furrow open pit	深凹露天矿
deep gas	深层气
deep geological mapping	深部地质填图

deep geological repository	深地质处置库	deep penetrating	深穿透的；深探测的
deep geophysical exploration	深部地球物理探测	deep penetrating charge	深穿透射弹
deep geophysical prospecting	深部地球物理勘探；深部物探	deep penetrating fracture	深穿透压裂
deep geothermal development	深部地热能开发	deep penetration curve	深探曲线
deep geothermal penetration	深部地热探测	deep penetration electrode	深熔焊条
deep geothermal power plant	深部地热电厂	deep penetration test	深层触探试验
deep geothermal well	深地热井	deep percolation	深层渗透
deep girder	深梁	deep permeability damage	地层深部渗透性损害
deep gravel	深埋砂矿	deep phreatic water	深井水；深源水
deep gravitational deformations	深重力变形	deep placer	埋藏砂矿；深砂矿床
deep groove	深槽	deep pressure grouting	深压灌浆
deep groundwater	深部地下水；深层地下水	deep probing	深部探测；深部探查
deep heat	深部热能；深部热资源	deep production	深层开采
deep heat flow	深部热流	deep propagation tool	深电磁波传播测井仪
deep hoisting	深井提升	deep prospecting	深层钻探
deep hole blasting	深孔爆破	deep resistivity log	深电阻率测井
deep hole borer	深孔镗床	deep river-cut valley	深切河谷
deep hole drill	深孔钻机	deep rock mass engineering	深部岩体工程
deep hole grouting	深孔灌浆	deep rotational failure	深圆弧滑动破坏
deep hole infusion	深孔注水	deep rupture	深断裂
deep hole method	深眼爆破法	deep saline formation	深部咸含水层
deep hole pre-split blasting	深孔预裂爆破	deep sample	深部样品；深层样品
deep hole shooting	深井爆炸	deep scattering layer	深海散射层
deep hot reservoir	深部热储库	deep scraping	深撇泡沫
deep inclined thin coal seam	急倾斜薄煤层	deep sea	深海；深水的
deep induction	深感应	deep sea basin	深海盆地
deep induction log	深感应测井图	deep sea channel	深海槽
deep inflow	深部补给	deep sea deposit	深海沉积；深海堆积
deep injection well	深喷射井；深注水井	deep sea drilling	深海钻进
deep investigation	深探测	deep sea drilling project	深海钻探计划
deep investigation characteristic	深探特性	deep sea facies	深海相
deep investigation induction log	深感应测井	deep sea fauna	深海动物群
deep investigation laterolog	深侧向测井	deep sea floor	深海底
deep investigation resistivity log	深探电阻率测井	deep sea floor geology	深海底地质学
deep jet mixing pile machine	深层混合搅拌桩机器	deep sea graben	深海地堑
deep karst	深岩溶	deep sea measurement	深海测量
deep karst channel	深岩溶洞	deep sea mining	深海采矿
deep landslide	深层滑坡	deep sea nodule	深海结核
deep laneway	深部巷道	deep sea sediment	深海沉积物
deep laterolog	深探侧向测井	deep sea sedimentation	深海沉积
deep layer	深层；下均匀层	deep sea trench	深海沟
deep lead	深部砂矿	deep sea trench structure	海沟结构
deep level	深部层位，深处	deep sea trench-island arc system	深海沟-岛弧系统
deep level excavation	深开挖	deep sea trough	深海槽
deep level pressure	深部地压；深部压力	deep sea zone	深海区
deep lineament metallogeny	深部线性构造成矿作用	deep seabed mining	深海海底采矿
deep litter system	积聚法废物处理	deep seal	深层密封
deep load mark	重载水线标志	deep seated aquifer	深部含水层；深埋含水层
deep manhole	深井式进入孔	deep seated landslide	深层滑坡
deep marine	深海	deep seated weathering	深层风化
deep marine sediment	深海沉积物	deep seepage	深层渗漏；深层渗透
deep mine	深矿；深采矿山；地下矿	deep seismic	深地震
deep mine working	深井开采	deep seismic reflection profiling	深部地震反射剖面探测
deep mixed pile	深层搅拌桩；土水泥混合法	deep seismic sounding	深部地震测深
deep mixing	深层搅拌	deep settlement	深层沉降
deep mixing method	深层搅拌法	deep settlement gauge	深层沉降仪
deep ocean ooze	深海软泥	deep sewage collection tunnel	深浚污水收集隧道
deep open pit	深露天矿	deep shaft	深井
deep pay	深部产油层	deep slide	深层滑动；深层滑坡
		deep snow measurement	未碎波

deep socket wrench	深套筒扳手
deep soil	深层土
deep soil stabilization	深层土加固
deep sound channel	深海声道
deep sounding apparatus	测深仪；深层触探设备；深水探测器
deep sounding test	深度触探试验
deep source effect	深源效应
deep source map	深源图
deep space	深空
deep space probe	深空探测器
deep stabilization	深层加固
deep structure	深部构造
deep structure exploration	深构造勘探
deep submergence rescue vessel	深潜救生艇
deep submergence vehicle	深潜器
deep subsoil	深层土，深地基土
deep subsoil water	深层土中水
deep surface soil	深表层土壤
deep thermal water	深部热水
deep thread	深螺纹
deep through open pit	凹陷露天矿
deep toothed cutting structure	长齿切削结构
deep trench excavation	深沟挖掘
deep tube well	深管井
deep tunnel	深埋隧道
deep tunnel intercepting sewer system	地下深层污水截流系统
deep underground structure	深埋地下结构
deep underground water	深层地下水
deep unloading	深卸荷
deep valley	深谷
deep vibration compaction	深层震荡式压实
deep vibration technique	深层振动技术；深层振动法
deep water	深水；深源水；深层水
deep water corals	深水珊瑚
deep water dock	深水港坞
deep water drilling	深水钻井；深水钻探
deep water evaporitic environment	深水蒸发环境
deep water isotopic analyzer	深水同位素分析仪
deep water muck	深水腐泥土
deep water pier	深水码头
deep water platform	深水承台
deep water production	深水采油；深水开采
deep water seismic	深海地震勘探
deep water table	深地下水位
deep water transducer	深水换能器
deep water wharf	深水码头
deep water-bearing zone	深部含水带
deep water-table	深层地下水位
deep weathering	深风化；深层风化
deep web shearer	深截式采煤机
deep well	管井；深井
deep well array	深井台阵
deep well drainage	深井排水
deep well drilling	深井钻法
deep well elevator	深井吊梯
deep well fracturing	深井压裂
deep well functional group	深井作业组
deep well jet pump	深井喷射泵
deep well method	深井法
deep well plunger pump	深井柱塞泵
deep well point	深井点
deep well pump	深井泵
deep well turbine pump	深井透平泵
deep winding	深井提升
deep workings	深巷道；深工作区
deep zone	深层
deep-bed formation plugging	地层深部堵塞
deep-columnar solonetz	深位柱状碱土
deep-deposit	深部矿床
deep-depth tunnel	深埋隧道
deep-deviation type	深部造斜型
deep-digging dredge	深挖式挖掘船
deepdraft navigation	深水航运
deepdraft vessel	深水货轮
deep-drilling equipment	深孔钻进设备
deepen	延深；加深
deepening	加深；向下侵蚀；下切
deepening program	延伸程序；延伸计划；加深程序
deep-fault	深部断层
deep-fault structure	深断裂构造
deep-fault trough	深断裂谷
deep-fluid	深部流体；深位流体
deep-focus	深源
deepfocus earthquake	深发地震；深源地震
deepfreeze	冷藏箱；冷藏；冷处理
deep-gas	深部气
deep-hole	深孔
deep-hole infusion	深孔注水
deep-hole method	深孔爆破开采法
deep-hole mining	深炮眼开采
deep-hole presplitting blast	深孔预裂爆破
deep-hole prospecting	深孔钻探；深钻进勘探
deep-hole seismic detection	深孔地震探测
deep-level	深层的
deep-lime-mixing method	深层石灰搅拌法
deeply buried tunnel	深埋隧道
deeply eroded river bed	深蚀河床
deeply founded footing	深基础
deeply mining	深部开采
deeply weathered	深层风化；深度风化
deep-lying	深藏的；深埋的
deep-lying foundation	深埋基础
deep-lying seam	深煤层
deep-marine-environment	深海环境
deep-metamorphism	深成变质作用
deep-pit sewage pump	深井污水泵
deep-pocket classifier	深槽分级机
deep-prospecting	深部勘探
deep-reaching fault	深断裂；深断层
deep-reading device	深探测仪器
deep-rock	深成岩
deep-root zone	深陆根带
deep-rupture	深断裂
deep-sea bed	深海底
deep-sea canyon	深水峡谷；深海槽
deep-sea channel	深海水道
deep-sea clay	深海黏土
deep-sea deposit	深海沉积

deep-sea engineering	深海工程	deep-water current	深层水流
deep-sea fan	深海扇；海底扇	deep-water delta	深水三角洲
deep-sea fan sediment	深海扇沉积	deepwater development of oil and gas field	深水油气田开发
deep-sea mineral resources	深海矿产	deep-water evaporite	深水蒸发岩
deep-sea mining	深海采矿；大洋采矿	deep-water facies	深水相
deep-sea ooze	深海软泥	deepwater port	深水港口
deep-sea plain	深海平原	deepwater production	深水采油
deep-sea sand	深海砂	deepwater quay	深水岸壁
deep-sea sedimentation	深海沉积作用	deep-water wave	深水波
deep-sea sounding	深海测深	deep-water zone	深水区；深水域
deep-sea terrace	深海阶地	deep-well disposal	深井处理；深井处置；深井排放污水
deepsea thermometer	深海温度计	deep-well injection	深井灌注；深井注水
deepsea tide	深海潮汐	deep-well pump	钻井泵；深井泵
deep-sea velocimeter	深海流速仪；深水海流计	deep-well recharge	深井补给；深井回灌
deepsea wave	深海波	deep-well submersible pump	深井潜水泵
deepsea wave recorder	深海波浪计	deepwell turbine	深井涡轮
deep-seated	位于深部的；深部的	deepwell turbine pump	深井涡轮水泵
deep-seated anchor	深位锚固	deep-well working barrel	油井深泵筒；深井泵
deep-seated charge	深部装药	deep-zone orogeny	深带造山运动
deep-seated clay flowage	深层黏土流	defeated river	改道河流；弃置河
deep-seated crustal process	深部地壳活动作用	defect	缺陷；缺点；故障；损伤；裂缝；不合格品
deep-seated deformation	深部变形	defect liability	缺陷责任
deep-seated deposit	深成矿床；深海沉积	defect repair	缺陷修补
deep-seated dome	深部穹隆	defective building	不合格的建筑物
deep-seated earthquake	深源地震	defective design	设计不当
deep-seated fault	深部断层	defective insulation	绝缘不良
deep-seated fault structure	深断裂构造	defective material	有缺陷的材料
deep-seated faulting	深部断层作用	defective slope	有问题的斜坡
deep-seated flowage	深层流	defective tightness	不紧密
deep-seated fluid	深部流体	defective working condition	不良工作条件
deep-seated grouting	深层灌浆	defective works	有缺陷的工程；不合规格的工程
deep-seated heat	深部热能	defects correction period	缺陷改正期
deep-seated landslides	深部滑坡	defects detection	缺陷检测
deep-seated magma intrusion	深部岩浆侵入体	defects liability certificate	缺陷责任证书；养护合格证书
deep-seated metamorphism	深成变质作用	defects liability period	保养期；保修期；维修责任期
deep-seated origin	深部成因	defense	防护；防护物；保护层
deep-seated rock	深成岩	defer	推迟；延期
deep-seated salt dome	深成盐丘	deferment	押后；推迟；递延项目
deep-seated shear failure	深层剪切破坏	deferred charges	延期费；预付款；递延借项
deep-seated slide	深层滑动	deferred grouting	延缓灌浆；分段灌浆
deep-seated slip	深部滑动	deferred maintenance	逾期养护
deep-seated spring	深泉	deferred possession area	延迟移交区
deep-seated structure	深部构造；深成构造	deferred production	延迟生产
deep-seated thermal system	深部地热系统	deferred reaction	缓慢反应
deep-seated tunnel	深埋隧道	deficiency	缺乏；不足；缺失
deep-seated wrench-fault	深部扭断层	deficiency account	亏损报表
deep-sedimentation	深水沉积	deficiency curve	亏格曲线
deep-seismic reflection profiling	深地震反射剖面	deficiency gas	短缺气量
deep-seismic zone	深源地震带	deficiency of rain	缺雨
deep-set	深陷的	deficiency of saturation	饱和差
deep-shelf facies	深水陆架相	deficient draft	缺水量
deep-slot effect	深槽效应	deficient flow	不足流量；缺欠流量
deep-sounding apparatus	深水测深仪器	deficient place	亏空的工作区
deep-sourced gas	深源气	deficit	亏空，赤字；逆差
deep-water capability	深水工作能力	defile	隘道；峡谷
deep-water circulation	深层环流；深水环流	defined	规定的；明确的；定义的
deep-water circulation pattern	深水环流模式	defined area	划定地区
deep-water clapotis	深水驻波；深水激浪	definite	明确的；一定的；肯定的
deep-water cofferdam	深水围堰	definite conditions	定解条件
deep-water coral reefs	深水珊瑚礁		

definite extrapolation	有限外推	deflection magnet	偏转磁体
definite integral	定积分	deflection magnetometer	偏差地磁仪
definition domain	定义域	deflection matrix	挠度矩阵
definition of algorithm	算法定义	deflection measurement	偏斜测量
definition of failure	破坏定义	deflection method	位移法；变位法
definition of rock mechanics	岩石力学定义	deflection modulus	弯曲模量
definition of the image	影像清晰度	deflection moment	偏转矩；变曲力矩
definitive	确定的；最后的	deflection observation	挠度观测
definitive map	精确地图	deflection of plate	板的挠曲
definitive zoning	划定用途	deflection of bore	钻孔偏斜
deflagration point	爆燃点	deflection of borehole	钻孔偏斜
deflation	风蚀；吹蚀；吹；放气；抽气；降阶	deflection of pipe line	管线的挠度
deflation basin	风蚀盆地	deflection of the bit	钻头偏斜
deflation cave	风蚀穴	deflection of the vertical	垂线偏差
deflation flat	风蚀坪	deflection plate	折流板；折向板
deflation lake	风蚀湖；吹蚀湖	deflection point	交点；造斜点
deflation plane	风蚀面	deflection polarity	致偏极性
deflation valley	风蚀谷	deflection pulley	导向轮
deflect	偏斜；转向；偏转	deflection sensitivity	偏转灵敏度
deflected bore hole	偏斜钻孔	deflection separator	折流分离器
deflected drilling	偏斜孔钻进	deflection sheave	转向滑轮；导向天轮
deflected hole	偏斜井	deflection slip	偏转滑移
deflected pile	偏位桩	deflection survey	挠度测量
deflected river	偏转河	deflection temperature under load	载荷挠曲温度
deflected well	偏斜井	deflection test	挠曲试验
deflecting	偏斜	deflection test setup	挠曲试验装置
deflecting angle	造斜角	deflection tool	造斜工具
deflecting bar	转向杆	deflection-type storage tube	偏转式存储管
deflecting bucket	挑流鼻坎	deflective	偏斜的
deflecting field	致偏磁场	deflectivity	可偏性
deflecting force	偏转力	deflectometer	挠度计
deflecting gate	偏导闸门	deflector	偏导器；偏转板；致偏板
deflecting surface	造斜面	deflector baffle	折向挡板
deflecting tool	造斜工具	deflector balance	折板；转向板；转向风帘
deflecting wedge	转向楔	deflector bar	偏导杆；偏转条
deflecting well	斜井	deflector brattice	转向风障
deflection angle	偏转角；偏移角；转角；变位角	deflector bucket	鼻坎反弧段
deflection angle method	偏角法	deflector drum	转向滚筒
deflection angles and drifts	偏转角和漂移	deflector plate	挡板
deflection at break	裂断变位	deflector pulley	偏导轮；转向托轮
deflection at rupture	破裂变位	deflexion	偏转；偏斜；挠曲；偏转度；偏差
deflection basin	偏斜盆地	deflocculant	反絮凝剂；悬浮剂；胶体稳定剂
deflection behavior	挠曲性能	deflocculated colloid	不凝聚胶体
deflection bit	偏转钻头；纠偏钻头	deflocculated particle	反团聚作用
deflection coefficient	挠度系数	deflocculating agent	反絮凝剂；黏土悬浮剂；胶体稳定剂
deflection criterion	变位准则		
deflection curve	挠度曲线；变位曲线	defloculation	反絮凝；解絮凝；絮散
deflection curve or sag curve	挠度曲线	defluent	向下流的河段
deflection distance	偏距	deformability	可变形性；变形能力
deflection ductility factor	挠度延性系数	deformability measurement	变形量测
deflection error	方向误差	deformability modulus	变形模量
deflection experiment	偏转度测定	deformability of rock	岩石的变形
deflection factor	挠度系数	deformability of rock masses	岩体的变形
deflection field	偏转场	deformable	可变形
deflection fold	转向褶皱	deformable baffle	可变形的阻流隔板
deflection gauge	偏角仪	deformable body	变形体
deflection generator	扫描振荡器	deformable coordinate	可变形坐标
deflection inclinometer	测斜仪	deformable porous medium	变形多孔隙介质
deflection indicator	偏斜指示器；弯度计	deformable proppant	可变形支撑剂
deflection line method	变位曲线法	deformable rockmass	变形体

deformable structure 变形构造
deformat 反编排
deformation 变形
deformation allowance 预留变形量
deformation analysis 变形分析
deformation and failure 变形和破坏
deformation and failure of floor 底板变形和破坏
deformation area 变形区
deformation at failure 破坏时变形
deformation band 滑移带；形变带
deformation behavior 变形性状；变形行为
deformation blending 变形融合作用
deformation body 变形体
deformation boundary condition 边界变形条件
deformation calculation 变形计算
deformation characteristic 形变特征
deformation circle 变形图；形变圆
deformation coefficient 变形系数
deformation coefficient of rock mass 岩体变形系数
deformation condition 变形条件
deformation crack 变形裂缝
deformation curve 变形曲线
deformation cycle 变形旋回
deformation defect 变形缺陷
deformation detection 变形检测
deformation distributive function 变形分布函数
deformation domain 变形域
deformation drag 变形阻力
deformation due to shrinkage of soil 土的收缩变形量
deformation ellipsoid 变形椭球体
deformation energy 形变能；形变功
deformation fabric 变形组构
deformation field 变形场
deformation flowage 形变流动
deformation form 变形形态
deformation fracture 变形裂隙
deformation gauge 变形计
deformation gradient 变形梯度
deformation impact factor 变形的影响因素
deformation increase 变形增加
deformation instability of filling belt 变形失稳的填充带
deformation intensity 变形密度，变形强度
deformation inversion 变形反演
deformation joint 变形缝
deformation lamella 变形纹
deformation law 变形规律
deformation limit 变形极限
deformation locus 变形轨迹
deformation map 变形图
deformation measurement 变形测定
deformation mechanism 变形机制
deformation meter 变形测定计
deformation method 形变法
deformation modulus 变形模量
deformation modulus of rock mass 岩体变形模量
deformation monitoring 变形监测
deformation observation 变形观测
deformation of coal seam 煤层形变
deformation of expansion of soil 土的膨胀变形量
deformation of river bed 河床变迁；河床演变

deformation of roadbed 路基变形
deformation of rock 岩石变形
deformation of rock mass 岩体变形
deformation of soil 土壤变形
deformation of soil mass 土体变形
deformation of strata 岩层变形
deformation of surface 表面变形
deformation of surrounding rock 围岩变形
deformation of timbering 支架变形
deformation of wave 波的变形
deformation parameter test 变形参数试验
deformation partitioning 变形分解作用
deformation path 变形迹线；变形路径
deformation pattern 变形图像；形变图像；变形模式
deformation petrofabric 变形岩石组构
deformation plan 变形平面图
deformation plane 应变面，变形面
deformation pressure 形变压力
deformation process 变形；形变
deformation rate 形变率；变形速率
deformation relation with Rock Mass Rating 变形与岩体分级的关系
deformation resistance 变形阻力；变形抗力；变形回弹
deformation resistant structure 抗变形建筑物
deformation resolution 变形分辨率
deformation response factor 变形反应系数
deformation retract 变形收缩
deformation seismograph 形变地震仪
deformation sensitivity 变形灵敏度
deformation sequence 变形顺序
deformation set 残留变形，变形装置
deformation softening 应变软化
deformation spectrum 变形谱
deformation stability of dam foundation 坝基变形稳定性
deformation stage 变形阶段
deformation stage of rock 岩石变形阶段
deformation state variable 变形状态变量
deformation stress 变形应力
deformation structure 变形构造
deformation style 变形型式
deformation surface 变形面
deformation survey 变形测量
deformation temperature 变形温度
deformation tensor 形变张量
deformation test 变形试验
deformation test of rock mass 岩体变形实验
deformation texture 变形构造；变形结构
deformation thermometer 变形温度计
deformation time series 变形时间序列
deformation transducer 变形传感器
deformation value 变形值
deformation velocity 变形速率；变形速度
deformation vibration 变形振动
deformational energy of rock 岩石的变形能
deformational eustatism 地动性海面升降
deformational event 变形事件
deformational media 变形介质
deformational nonlinearity 形变非线性
deformational pressure of surrounding rock 围岩变形压力
deformational property 变形性质

deformational strain energy	变形应变能	degradation hazards	劣化危险
deformational stress	形变应力	degradation in size	粉碎；磨细；粒度减小
deformational structure	形变构造	degradation in water	泥化；水中降解，水中软化
deformational till	变形冰积物	degradation index	衰变指数；衰减指数
deformational trap	构造圈闭；变形圈闭	degradation magnitude	刷深程度
deformational zone	变形区	degradation of energy	能量退降
deformation-resisting assembly	抗变形装置	degradation of geosynthetics	土工合成材料的退化
deformation-time curve	变形-时间曲线	degradation of stream	河流切蚀
deformed accumulative formation	堆积层变形体	degradation of tensile properties	拉伸性能退化
deformed band	变形条带	degradation parameter	退化参数
deformed bar	变形钢筋	degradation particle	粉碎的颗粒
deformed belt	形变带	degradation processes	劣化过程；降解过程
deformed concrete caisson	异形混凝土沉箱	degradation product	降级产品
deformed configuration	变形构形	degradation rate	粉碎率
deformed conglomerate	变形砾石	degradation recrystallization	退变重结晶
deformed cross-bedding	变形交错层理	degradation regression	退化
deformed fabric	变形组构	degradation terrace	剥蚀阶地
deformed lamella	变形纹；变形纹层	degradation theory	降解说
deformed pebble	变形砾石	degradation threshold	降解阈值；劣化阈值
deformed tunnel	变形隧道	degradation vacuity	剥蚀缺失
deformeter	应变仪；变形测量仪	degradational process	递降分解过程，递降分解作用，降解过程
deforming force	变形力	degrade	递降；剥蚀；降解
deforming rate	变形速度	degraded lands	退化土地
deforming wedge	变形楔	degrading river	夷低河；下切河；退化河
deformity	变形；畸变	degrading stream	夷低河流；下切河流
defrostation	解冻；融化	degranitization	去花岗岩化作用
defrosting	溶解；溶化；解冻	degreasing fluid	脱脂剂；去油剂
defuzzification	逆模糊化	degreasing liquid	去脂液
degasification	脱气；除气；去气；脱气作用	degree	度；程度；等级；次；学位
degasification boreholes	抽放钻孔	degree centigrade	摄氏度数
degasser	除气机；脱气机	degree Fahrenheit	华氏温度
degassing equipment	除气设备；脱气设备	degree of abrasion	剥蚀度
degassing hole	排放瓦斯钻孔	degree of activity	活度
degassing screen plate	除气筛板	degree of adaptability	配合度；适应性；适应程度
degassing tower	脱气塔	degree of alkalisation	碱化度
degassing unit	除气装置；脱气器	degree of alteration	蚀变程度
degaussing	去磁；退磁	degree of approximation	近似度
degaussing effect	去磁效应	degree of arching	拱作用的程度
degeneracy	退化；蜕化；简并性	degree of association	关联度
degenerate distribution	退化分布	degree of balance	平衡度
degenerate flow	退落水流	degree of bed development	河床发育程度
degenerate isoparametric	退化等参	degree of branching	支化程度
degenerate matrix	退化矩阵	degree of bridging	桥架度
degenerate process	退化过程	degree of cementation	胶结度；胶结程度
degenerate solution	退化解	degree of coal metamorphism	煤变质程度
degenerated soil	退化土壤	degree of coal oxidation	煤的氧化程度
degeneration	退化；负反馈	degree of coalification	煤化程度；煤化度；碳化程度
degeneration coefficient	简并系数	degree of communication	连通度
degeneration control	负反馈调整	degree of compaction	压实度
degeneration factor	负反馈系数；退化因素	degree of compactness	致密度；紧密度
degenerative feedback	负反馈	degree of compression	压缩度
deglaciation	冰川消退	degree of confidence	置信度
degradable	可降解的；可退化的	degree of confinement	填封度
degradable fluid	可降解的液体	degree of consistency	稠浓程度
degradant additive	降解添加剂	degree of consolidation	固结度
degradation	降级	degree of convergence	收敛度
degradation below dam	坝下河底刷深	degree of conversion	转化度
degradation characteristic	退化特性	degree of correlation	关联度
degradation curve	降解曲线	degree of corrosion	溶蚀程度，侵蚀程度
degradation degree	冲刷程度；降解程度		

degree of coverage	覆盖度	degree of layering	分层程度
degree of creaming	膏化度	degree of liberation	解离程度
degree of crosslinking	交联度	degree of liquefaction	液化度
degree of crushing	破碎程度	degree of loading	负荷度
degree of curvature	曲度；曲率；弯曲度	degree of longitude	经度
degree of damage	破坏程度	degree of mapping	映射度
degree of damping	阻尼度	degree of medium contamination	介质污染程度
degree of dangers	危险程度	degree of membership	隶属度
degree of degeneration	退化度	degree of metamorphism	变质程度
degree of density	密实度	degree of mineral liberation	矿物解离度
degree of deviation	偏差度	degree of mineralization	矿化度
degree of dip	倾斜度；倾角	degree of mobilization	流动度
degree of disintegration	分解度	degree of modulation	调制度
degree of dispersion	弥散度；分散度	degree of non-membership grade	非隶属度
degree of distortion	失真度；畸变度	degree of non-uniformity	不均匀度
degree of drainage	疏干程度	degree of order	有序度
degree of drawing	拉伸度	degree of orderliness	规则程度
degree of dryness	干度	degree of organic metamorphism	有机质变质程度
degree of eccentricity	偏心度	degree of orientation	取向度
degree of elasticity	弹性度	degree of oxidation	氧化程度
degree of elevation	仰角	degree of packing	装药密度；填实度
degree of exactitude	精度；准确度	degree of pitch	倾斜度；倾角
degree of exhaustion	抽空度	degree of plasticity	塑性度；范性度；可塑度
degree of expansion	膨胀度	degree of pollution	污染程度
degree of exploration	勘探程度	degree of polynomial kernel	多项式核函数次数
degree of exposure	日照程度	degree of porosity	孔隙度等级
degree of extraction	回收率，回采率	degree of preciseness	准确度
degree of fill	充填度；充填密实度；充填密度	degree of precision	精密度
degree of fineness	细度	degree of primary consolidation	主固结度
degree of finish	光洁度	degree of proof	证实程度
degree of fissility	易剥裂程度；易裂度；可裂变性	degree of prospecting	勘探程度
degree of fissuration	裂隙频数；裂隙度；裂隙密度	degree of radial uniformity	径向均匀度
degree of fissure	裂隙频度；裂隙率	degree of reaction	反动度；反作用度，反力度
degree of fitting	拟合度	degree of redundancy	赘余度
degree of flatness	扁平程度	degree of regulation	控制程度
degree of fluctuation	起伏程度	degree of reliability	可靠性程度
degree of fractionation	分馏度	degree of remoulding	重塑度
degree of free swelling	自由膨胀率	degree of restraint	约束程度
degree of freedom	自由度	degree of ripeness	成熟度
degree of freedom count	自由度计算	degree of risk	风险度
degree of functionality	功能度	degree of rock weathering	岩石风化程度
degree of functioning	有效程度	degree of rounding	磨圆度
degree of geological complexity	地质复杂程度	degree of rounding of grains	颗粒圆度
degree of geological studiedness	地质研究程度	degree of roundness	圆度
degree of grains	粒度	degree of safety	安全程度
degree of grind	研磨程度	degree of saturation	饱和度
degree of grinding	研磨度	degree of sensitivity	灵敏度
degree of hardness	硬度	degree of settlement	沉降度
degree of heat	温度；热度	degree of sharing	共享度
degree of heterogeneity	不均质程度	degree of shrinkage	收缩度
degree of hydrolysis	水解度	degree of size breakage	破碎比
degree of impregnation	嵌布程度；浸染程度；灌注程度	degree of size reduction	破碎物料与给料的表面积比；表面积破碎比；粒度减小程度
degree of inclination	倾度；倾角	degree of slope	坡度
degree of indeterminacy	超静定次数	degree of sophistication	灵敏度；分辨率
degree of insensitiveness	不灵敏度	degree of sorting	分选度；分选程度
degree of integrity	完整性	degree of soundness	坚固程度
degree of ionization	电离度	degree of stability	稳定度
degree of irregularity	紊乱程度；不规则程度	degree of substitution	取代度
degree of karstification	岩溶率	degree of success	有效程度
degree of latitude	纬度		

degree of supersaturation	过饱和度；超饱和度
degree of swelling	膨胀程度
degree of syndiotacticity	间同规整度；交规度
degree of tightness	隔水度；隔气度，紧密度
degree of tilt	倾斜度
degree of trend surface	趋势面的阶
degree of turbulence	湍流度
degree of uniformity	均匀程度
degree of utilization	利用率
degree of vacuum	真空度
degree of variation	变异程度
degree of viscosity	黏滞度
degree of water wettability	亲水程度
degree of weathering	风化度
degree of wetness	湿度
degree of wettability	润湿度
degree of wetting	润湿度
degree scale	标度
degree variance of gravity anomaly	重力异常阶方差
degumming agent	脱胶剂
degypsification	去石膏化
dehalogenation	脱卤
dehardening	软化
dehumidifier	抽湿机；除湿器
dehumidifying heater	除潮加热器
dehumidizer	减湿剂
dehydrant	脱水剂
dehydrate	脱水；去水；脱水物
dehydrate consolidation settlement	脱水固结沉降
dehydrated air	脱湿空气
dehydrated analysis	脱水分析
dehydrated crude	脱水原油
dehydrated lime	生石灰；氧化钙
dehydrating agent	脱水剂
dehydrating slurry	脱水砂浆；脱水泥浆，脱水浆液
dehydration	脱水
dehydration curve	脱水曲线
dehydration melting	脱水熔融作用
dehydration plant	脱水装置
dehydration process	脱水工艺流程
dehydration product	脱水产物
dehydration rate	脱水率
dehydration reaction	脱水反应
dehydration test	脱水试验
dehydration tower	脱水塔
dehydrator	脱水器，除潮器
dehydrite	脱水剂
dehydrolysis	脱水
deicing device	破冰装置；碎冰器；防止结冰装置
deicing fluid	防冻液
deicing sluice	排冰闸
deionization	去电离；消电离
deionized water	无离子水
dejection cone	洪积锥；冲积锥
dejection rock	喷出碎屑岩
dejective loess	冲积黄土
del	倒三角形
delamination	分层；剥离；脱层
delamination of concrete cover	混凝土保护层层状剥落
delanovite	锰蒙脱石

delapsional deposit	塌陷沉积
delay	推迟；延缓；迟滞；延期
delay action	延期作用
delay action blasting	迟发爆破
delay action detonator	迟发雷管；延迟起爆剂
delay action fuse	延发导火线；延期引线
delay angle	延迟角
delay automatic volume control	延迟式自动容量调节
delay blast	迟发爆破
delay cable	延迟线
delay cap	迟发雷管
delay cell	延时元件
delay charge	延期药
delay circuit	延迟电路
delay compensation	滞时补偿
delay composition	延期混合炸药
delay connector	微差起爆器；毫秒继爆管；延期继爆管
delay distribution	延迟分布；延滞分布
delay duration	延滞持续时间
delay elastic deformation	滞弹性变形
delay element	延时元件
delay equalization	延迟均匀化
delay equalizer	延迟均衡器
delay filling	延期充填；采完后充填
delay in system	系统延滞
delay index	延迟指数
delay interval	迟发时间；延迟时间
delay interval of priming	起爆间隔
delay network	延迟网络
delay number	延时段
delay of voyage	航程延误
delay outburst	延迟突出
delay period	迟发期；延滞期
delay powder	延期药
delay rate	延滞率
delay relay	延时继电器
delay signal	滞后信号
delay study	延滞研究
delay switch	延时开关
delay time	延迟时间；滞后时间
delay yield	滞后出水量；滞后出水；延迟给水
delayed blast	迟误爆破；延迟爆破
delayed compaction	延迟压实
delayed completion	拖期竣工
delayed compression	次固结压缩
delayed consolidation	次固结
delayed crazing	后期龟裂
delayed deformation	延时变形
delayed detonation	迟发起爆
delayed development	延迟开发
delayed drainage	延迟排水
delayed elastic strain	滞后弹性应变
delayed elasticity	弹性后效
delayed explosion	延期爆炸；缓爆
delayed failure	滞后破坏
delayed fill	随后充填；延迟充填
delayed finish	延迟完成
delayed flow	延迟流，延迟水流
delayed formation	延时变形
delayed fracture	滞后破坏

delayed ignition 延迟点火
delayed index 延迟指数
delayed junction 缓延汇合
delayed mixing 延迟搅拌；延时搅拌
delayed recovery 迟滞恢复
delayed relay 缓动继电器
delayed runoff 延迟径流；迟滞径流；滞后径流
delayed setting 延迟凝固
delayed setting cement 缓凝水泥
delayed settlement 次固结沉降；延时沉降
delayed snow avalanche 有效含水量
delayed start 延迟起动
delayed storage 延迟给水
delayed storativity 延迟给水系数
delayed strain 鱼道辅助入口
delayed subsidence 再次沉陷；再次下沉；延期沉陷
delayed subsurface runoff 延迟地下径流
delayed sweep 延时扫描
delayed time system 延时系统
delayed trigger 延迟触发
delayed viscosity breakback 延迟降黏
delayed water exchange 缓慢水交替
delayed works 延误工程
delayed yield 延迟屈服；滞后屈服
delay-fuse 延迟导火索
delaying elasticity 滞弹性
delaying sweep 延迟扫描
delaying water gushing 滞后式突水
delay-line filtering 延迟线滤波
delegation of power 委托权；授权
deleterious 有毒的；有害的
deleterious chemical 有毒化学物
deleterious compositions abnormity 有害成分异常
deleterious dust 有害尘末
deleterious effect 有害作用；不良影响；有害效应
deleterious gas 有害气体
deleterious material 有害材料
deleterious mineral impurity 有害矿物杂质
deleterious reaction 有害反应
deleterious substance 有害杂质
deletion 删去；删除；消失；抹杀
develeling 基准面变化
delfation column 风蚀柱
deliberate reconnaissance 周密的勘查
deliberately 慎重地
delicacy 灵敏度；敏感；精密
delimit 定界；分界
delimitation 定界；分界
delimitation of ore deposit 矿体固定
delimitation of orebody 圈定矿体边界
delimitation of the frontier 划定边界
delineate 描绘轮廓；叙述；描写
delineation 固定；划界；轮廓；概图；清绘；略图；草图
delineation drilling 探边钻井；圈定探井，圈定钻探
delineation of ore bodies 矿体圈定
delineation of zones of failure 破坏区的确定
delineation of zones of rock failure 岩石破坏区的确定
delineation well 探边井；划界井，圈定井
delineator 道路标线；路线反光标记

deliquate 冲淡；稀释
deliquescent 潮解的；溶解的；吸湿的
deliver up possession 交回管有权
deliverability 供应量；交付能力；可采程度；生产能力；流量；产量
deliverability curve 产能曲线
deliverability equation 产能方程
deliverability of gas 天然气的供应量
delivery 交付；供应；补给；排出；流量；排水量；引水
delivery and measuring box 配水量水箱
delivery canal 输水渠；引水渠
delivery capacity 输送能力；生产额；排量；交货额
delivery cock 泄放旋塞；放水龙头
delivery conduit 排水沟；排水管；送风道
delivery drift 排水硐
delivery flap 放出瓣；放出阀
delivery head 供水水头
delivery lift 压升高度；输水高度；扬程
delivery line 导出管；递送管
delivery of energy 能量供给
delivery of materials 材料运送
delivery of pump 泵排量
delivery orifice 流通孔；放流孔
delivery outlet 输出口
delivery pipe 递送管；输出管
delivery pipe line 输送管线
delivery pressure 输送压力
delivery pressure head 供水压头
delivery pump 输送泵；供给泵
delivery rate 输送速度；输送速率
delivery rate of erosion 冲刷输沙率；冲刷输送泥沙率
delivery report 交货报告
delivery space 排出间隙；扩散器
delivery system 输送系统
delivery track 输送轨道
delivery value 输送能力
delivery valve seat 输送阀座
delivery valve stop 输送阀停止器；输油阀停止器
delivery volume 供给量；排量
dell 河源谷地；深坑
delta 三角洲；河口三角洲
delta arm 三角洲汊河
delta bar 三角洲坝
delta basin 三角洲盆地
delta bedding 三角洲层理
delta branch 三角洲支汊
delta building 三角洲形成
delta cap 三角洲上的冲积锥
delta channel 三角洲水道；三角洲汊流
delta coast 三角洲海岸
delta coastal plain 三角洲沿海平原
delta connection 三角形接法
delta deposit 三角洲沉积
delta estuary regulation 三角洲河口整治
delta facies 三角洲相
delta fan 三角洲扇形地
delta foreset stratification 三角洲前积层理
delta formation 三角洲形成
delta fringe deposit 三角洲边缘沉积

delta front	三角洲前缘
delta front facies	三角洲前缘相
delta front sheet sand	三角洲前缘席状砂
delta function	δ函数
delta growth	三角洲扩大；三角洲增长
delta lake	三角洲湖
delta measurement	三角洲测量
delta plain	三角洲平原
delta plain facies	三角洲平原相
delta project	三角洲整治计划
delta regime	三角洲体系
delta river mouth	三角洲河口
delta shoreline	三角洲岸线
delta structure	三角洲构造
deltageosyncline	三角洲地槽，准地槽
deltaic area	三角洲地区
deltaic bedding	三角洲层理
deltaic coast	三角洲海岸
deltaic coastal plain	三角洲海岸平原
deltaic cone	三角洲冲击锥
deltaic cross bedding	三角洲交错层
deltaic deposit	三角洲沉积
deltaic fan	三角洲沉积扇
deltaic levee lake	三角洲自然堤湖
deltaic plain	三角洲平原
deltaic progradation	三角洲前积
deltaic reclamation	三角洲的填海工程；三角洲整治工程，三角洲填筑工程
deltaic river	三角洲河流
deltaic river mouth	三角洲河口
deltaic seal	三角形密封圈
deltaic succession	三角洲层序
deltaic-plain complex	三角洲-平原复合体
deltalogy	三角洲学
delta-marginal plain	三角洲边缘平原
deltic cross bedding	三角洲交错层
deluge	大洪水；大雨，暴雨
deluge system	密集洒水系统；集水系统
deluge valve	集水阀
deluvial	洪积的
deluvial material	洪积物
deluvial placer	坡积砂矿
deluvial soils	冲积土
deluvium	坡水堆积物；坡积物
delve	凹地；穴；坑，掘；挖；刨
demagnetisation curve	去磁曲线
demagnetising effect	消磁效应；退磁效应
demagnetization	消磁作用
demagnetization force	去磁力；脱磁力
demagnetize	退磁；去磁
demagnetizing factor	退磁系数
demagnetizing field	去磁场；退磁场
demagnetizing switch	消磁开关
demagnification	缩小
demand analysis	需求分析
demand curve	需求曲线
demand deposit	活期存款；即期存款
demand draught	需要的通风
demand elasticity	需求弹性
demand factor	功率需求量系数；集中因数
demand forecast	需求预测
demand function	需求函数
demarcated boundary	标定边界
demarcated section	测流断面
demarcated section line	施测断面线
demarcated site	测流站址；测流现场
demarcation	划界；定界；标界；区分
demarcation district lot	丈量约份地段
demarcation district plan	丈量约份图则
demarcation district sheet	丈量约份地图
demarcation line	界线
demec gauge	缝隙测量计
demigration	反偏移；反迁移
demineralized water	软水
demixing	分层；反混合
demodulator	解调器；检波器
demolish	拆卸
demolition	拆卸；拆卸工程
demolition action	清拆行动
demolition work	拆除工作
demonstrated reserves	探明储量
demonstration	论证；证明；说明；表明；示范；示教
demonstration plant	示范装置
demonstration project	示范项目
demorphism	岩石分解；风化变质作用
demultiplication	倍减；缩减
demurrage	滞留期；滞期费
den	密度
denary	十的；十进制的
denaturation	变性作用
dendritic	树枝状的；多枝的
dendritic drainage	枝状水系；枝状河网；羽状水系；羽状河网；枝状排水系
dendritic drainage pattern	枝状水系；枝状河网；羽状关系；羽状河网
dendritic glacier	树枝状冰川
dendritic river system	树枝状水系；树枝状河系
dendritic structure	树枝状结构
dendritic type glacier	树枝状冰川
dendritic valley	树枝状河谷
dendrogram	树形图
dene	砂层；沙丘；幽谷
denitration	脱硝；脱硝酸盐
denitrification	脱氮；脱硝作用
denivellation	高差，水位差，不平度
denomination	命名；名称；名目；单位；种类
denotation	表示；符号；外延
dense	致密的；重的；具高折射率的；深色的
dense concrete	密实混凝土
dense fissuration	致密裂隙
dense fissure	致密裂隙
dense fluid	高黏液
dense fluidization	稠密流动度
dense formation	致密地层
dense gneiss	致密片麻岩
dense gradation	密级配
dense grout	浓浆；稠浆
dense jointing	密集节理
dense limestone	致密灰岩
dense liquid	重液

dense medium	重介质；重悬浮液
dense pack	密实充填
dense packing	致密堆积
dense packing arrangement	致密充填结构
dense phase	致密相
dense mortar	稠密矿浆；浓厚矿浆
dense rock	致密岩石
dense sand	密实砂
dense slurry	高密度砂浆；高密度泥浆，高密度浆液
dense soil	密实土
dense solution	重溶液
dense state	密实状态
dense structure	紧密结构
dense suspension	浓悬浮液
dense texture	致密结构
dense well pattern	密井网
dense well spacing	密集布井
dense-graded	密级配的
dense-graded aggregate	密级配集料；密级配骨料
densely populated area	稠密居住区
denseness	密度；稠度；稠密
densification	浓缩；稠化；稠集；密封；封严
densification by explosion	爆炸加密法
densification by sand pile	砂桩挤密
densification control point	加密控制点
densified cement	稠化水泥
densified control network	加密控制网
densified slurry	增稠水泥浆；增稠泥浆，增稠砂浆，增稠浆液
densified zone	加密区
densifier	浓缩机；稠化机；浓缩剂
densimeter	密度计；比重计；浓度计
densimetric curve	密度曲线；比重曲线
densimix	加重混合物
densinite	密屑体
densitometer	显像密度计；光密度计；深浅计
densitometric analysis	光密度分析
densitometry	密度测定法
density	密度；浓度；比重
density accuracy	密度精度
density analysis	密度分析
density anisotropy	密度异向性
density anomaly	密度异常
density bottle	密度瓶；比重瓶
density boundary	密度边界
density brine	高密度盐水
density composition	密度成分，密度组成
density contrast	密度差
density control	密度控制
density control fracture	密度控制压裂
density control method	密度控制法
density control valve	调浓阀；浓度控制阀
density controlling material	密度调节材料
density correction	密度校正值
density current	密度流；异重流；重流
density current bed	密度流沉积；密度流底层
density current in estuary	河口密度流
density difference	密度差
density differentiation	密度分异
density discontinuity	密度界面；密度间断面
density distribution	密度分布
density effect	密度效应；异重效应；密集效应
density flow	密度流；异重流
density fraction	密度分离
density fractionation	密度分离法
density function	密度函数
density gauge	密度测定器
density gradient	密度梯度
density in situ	现场密度
density index	密度指数
density interval	比重间隔；密度间隔
density level	密度级
density log	密度测井
density logger	井下密度测定仪
density logging	密度测井
density loss	浓度下降
density measurement	密度测量
density meter	密度计；比重计
density of collapse	塌陷密度
density of dry rock	干岩石密度
density of dry soil	干土密度
density of fluid	流体密度
density of liquid	液体密度
density of medium	介质密度
density of packing	填集密度
density of road network	道路密度
density of runoff	径流深度
density of saturated	饱和密度
density of saturated soil	饱和土密度
density of seams	煤层间隔密度；煤层群中煤层总厚与岩层总厚之比
density of separating medium	分选介质密度
density of soil	土壤密度；土密度
density of solid particles	固体颗粒密度
density of solids	固体密度
density of spacing	间隔密度
density of strain energy	应变能密度
density of submerged soil	浸水土密度；浸没土密度
density of volume	体积密度，容量密度
density of water	水的密度
density parameter	密度参数
density porosity	密度孔隙度
density pressure	密度压力
density probe	密度探测仪
density profile	密度剖面；密度纵断面
density ratio	密度比
density reconstruction	密度重建
density recorder	密度记录器
density regulation tank	比重调节槽
density regulator	密度调节剂；密度调节器
density repeatability	密度重复性
density response	密度响应
density separation	密度分离
density slicing	密度分割；密度分层
density stratification	密度分层；密度差异分层
density surface inversion	密度界面反演
density test	密实度试验
density tool	密度测井仪
density underflow	密度底流
density upper limit	密度值上限

density value	重度值
density zoning	发展密度分区
density zoning policy	发展密度分区政策
density-control system	比重控制系统
density-depth profile	密度-深度剖面
density-driven convection	重力驱动对流
densograph	黑度曲线
densometer	密度计；比重计
dental treatment	补缝填坑处理
dentition formula	齿式
dentition on rock slope	石坡补缝；石坡补隙
dentoid	齿状的
denudation	剥蚀作用；剥蚀
denudation chronology	剥蚀年代学
denudation landform	剥蚀地貌
denudation mountain	剥蚀山
denudation plain	剥蚀平原
denudation rate	剥蚀率
denudation slope	剥蚀坡
denudational fault	剥蚀断层
denuded	剥蚀的；变光的
denuded area	剥蚀地；剥蚀地区
denuded soil	剥蚀土壤
denumerable	可数的
deoscillator	阻尼器；减振器
deoxidizing agent	脱氧剂
deoxygenation	脱氧作用
department	部分；门类；部门；部；局；科；车间
departure	偏离；偏差；异常；离开；出发
departure curve	离差曲线
departure map	偏差图
departure port	始发港
departure rate	离开率
departure track	出发线
departure yard	发车场
depauperization	萎缩
depend on	决定于
dependability	可靠性
dependable discharge	可靠流量
dependable flow	保证流量
dependable hydroelectric capacity	可靠水电容量
dependable water source	可靠水源
dependable yield	可靠供水量；可靠产量
dependence	依赖；从属；相关性
dependency	依赖性
dependent	有关的；依赖的
dependent block	附属建筑
dependent criterion	相关性指标；相关准则
dependent data	相关数据
dependent function	相依函数
dependent linear equation	相关线性方程
dependent observation	相关观测
dependent population	被抚养人口
dependent program	相关程序
dependent variable	因变量；因变数；他变量；他变数
depergelation	冻土融化作用
dephlegmation	分馏；分凝
depict	叙述；描述
depillaring	采掘煤柱；采掘矿柱
depinker	抗爆剂
deplanation	夷平作用；夷低作用
depleted	耗尽；衰竭
depleted area	衰竭区
depleted field	枯竭油田
depleted formation	枯竭层
depleted reservoir	枯竭油层；枯竭水库
depleted soil	贫瘠土
depleting sphere	枯竭区域
depleting-layer	贫乏层
depletion	减少；亏损；贫乏；枯竭；耗尽；消耗
depletion allowance	消耗限额
depletion characteristics	衰竭特征
depletion coefficient	疏干系数
depletion curve	枯竭曲线；消耗曲线
depletion drive	溶解气驱
depletion drive index	衰竭驱动指数
depletion drive pool	溶解气驱油藏
depletion effect	耗尽效应
depletion factor	损耗因素；衰竭因素
depletion hydrograph	退水过程线
depletion layer	耗尽层
depletion mechanism	衰竭机理
depletion rate	衰减速度；枯竭速率
depletion region	耗尽区；采空区
depletion stage	采竭阶段
depletion type gas reservoir	内能消耗式气藏
depletion type of drive	衰竭式驱动
depletion yield	疏干开采量
depletion zone of slide	滑坡土移区
deployment	疏开；展开；调度；部署
deposit	沉积，矿床；沉淀，淀积；沉积物，沉淀物；存放
deposit apex	矿床顶部
deposit attack	沉积侵蚀
deposit concrete	浇注混凝土
deposit contact	沉积接触
deposit corrosion	沉积腐蚀
deposit discovery	矿床开拓
deposit division	矿床分割
deposit formation	沉积物的形成
deposit in situ	原地沉积
deposit inundation	矿床淹没
deposit modeling	矿床模式；矿床模型法
deposit position	矿床位置
deposit string	矿脉
deposit type	矿床类型
deposited	沉积的；沉淀的
deposited dust	沉积矿尘
deposited soil	沉积土
deposited-film	沉积膜
deposit-free	无沉积物
depositing	浇注；存储；存放
depositing concrete	浇注混凝土
depositing of concrete	混凝土浇筑
depositing substrate	堆积场基底
depositing tank	沉淀池；沉淀槽
deposition	沉积作用；沉淀；沉积物，沉淀物
deposition by flood	洪水淤积
deposition coast	堆积海岸
deposition cycle	沉积旋回

deposition fabric	沉积组构	depreciation period	折旧期限；折旧年限
deposition factor	沉积因子	depreciation rate	折旧率
deposition flux	沉降通量	depress	降低；压下；抑制
deposition interface	沉积界面	depressed area	沉陷区；沉降区；低地；洼地
deposition nucleus	凝华核	depressed block	陷落地块
deposition of suspended material	悬浮体沉积	depressed coast	沉降海岸
deposition period	沉积期	depressed water table	降低的地下水位
deposition potential	析出电位	depressed zone	下沉带
deposition rate	沉淀速度；熔敷速度	depression	沉降；低气压
deposition time	沉积时间	depression angle	俯角
deposition velocity	沉积速度	depression belt	低压带
depositional architecture	沉积架构	depression cone	水位降落漏斗
depositional area	沉积面积	depression cone method	降落漏斗法
depositional basin	沉积盆地	depression contour	洼地等高线
depositional break	沉积间断	depression curve	下降浸润曲线
depositional build-up	沉积加积	depression earthquake	陷落地震
depositional condition	沉积条件	depression meter	负压表；负压计
depositional control	沉积控制	depression of ground	地面沉降
depositional cycle	沉积循环；沉积轮回；沉积旋回	depression of level	水位下降；水面凹陷
depositional dip	原始倾斜；最初倾斜角	depression of order	降阶法
depositional energy	沉积能量	depression of water table	地下水位的降落
depositional environment	沉积环境	depression spring	洼地泉；下降泉
depositional fabric	沉积组构	depression storage	凹地蓄水
depositional fault	沉积断层，生长断层	depression-cone	降水漏斗
depositional feature	沉积形态	depressive coefficient of soil	土壤抗陷系数
depositional form	沉积地形	depressive earthquake	陷落地震
depositional fractionation	沉积分异	depressuring valve	减压阀
depositional framework	沉积格架	depressurization	降压；减压
depositional gradient	沉积坡度；天然坡度；自然坡度	depressurization belt	降压带
depositional history	沉积史	depth	深度；深处
depositional interface	沉积界面；沉积间面	depth adjuster	深度调节器
depositional landform	沉积地形	depth area curve	深面曲线，等深面曲线
depositional model	沉积模式	depth average	平均深度
depositional pattern	沉积样式	depth chart	水深图；测深图
depositional phase	沉积阶段	depth column	深度柱
depositional plane	沉积面	depth contour	等深线
depositional porosity	沉积期孔隙	depth contour interval	等深线间距
depositional remanence	沉积剩磁	depth contour map	深度等值线图
depositional remanent magnetization	沉积剩磁	depth controller	拖缆深度控制器
depositional sequence	沉积层序；沉积序列；沉积系列	depth conversion	深度转换
depositional setting	沉积背景	depth correction	深度校正
depositional site	沉积场所	depth counter	深度计数器
depositional stage	沉积期	depth creep	深层蠕动
depositional strike	沉积走向	depth cross-section	深度横剖面
depositional strike deposit	侧向连续沉积	depth curve	深度曲线
depositional surface	沉积面	depth datum	深度基准面；海图基准面
depositional system	沉积体系	depth dial gauge	深度千分表
depositional terrace	堆积阶地	depth digitizer	水深数字化器
depositional texture	沉积结构	depth displacement	高程差
depositional topography	沉积地形	depth distribution	深度分布
depositional trap	岩性圈闭	depth drill	深孔钻
depositional unconformity	沉积不整合	depth echo sounding	回波测深
depositional zone	沉积带	depth effect	深度影响
depositive	沉积的	depth facies	深度相；深部相；深成相
depot deduction	折旧费；折旧额	depth factor	采深系数
depot of ore	矿石贫化	depth filter	深度过滤器
depreciation	贬值，跌价；折旧	depth finder	测深仪
depreciation charges	折旧费	depth gauge	深度计；水位计；井底压力计；高度计；检潮标；水位尺
depreciation factor	折旧率		
depreciation fund	折旧基金；折旧费	depth indicator	深度指示器；深度探测仪

depth line	深度线
depth map	深度图
depth measurement	水深测量；测深
depth measurer	量深器
depth micrometer	深度测微计
depth migration	深度偏移
depth model	深度模型
depth molded	型深
depth of beam	梁的高度
depth of bedrock	基岩深度
depth of bore	钻孔深度
depth of bottom hole measurement	孔底测量深度
depth of breaking	破波水深；碎波水深
depth of burial	埋藏深度；埋深
depth of case	渗碳深度；表面硬化层深度
depth of chill	冷硬深度
depth of compensation	补偿深度
depth of compressed layer	受压层深度
depth of compression	压缩深度，受压深度
depth of compression zone	受压区高度
depth of cover	覆盖层厚度；盖层厚度
depth of cover to reinforcement	钢筋在混凝土的深度
depth of cracking	开裂深度
depth of draining	排水深度
depth of embedment	埋入深度；入土深度
depth of excavation	挖掘深度；采掘深度；开采深度
depth of exploration	勘探深度
depth of fill	填方高度；填土深度
depth of focus	震源深度；聚焦深度；焦深；焦点深度
depth of folding	褶皱深度
depth of foundation	基础深度
depth of freezing	冻结深度
depth of frost penetration	冻结深度
depth of groove	凹槽深度；孔型深度
depth of groundwater occurrence	地下水埋藏深度
depth of hardening	淬硬深度
depth of impression	刻痕深度
depth of influence	影响深度
depth of invasion	侵入深度
depth of investigation	探测深度
depth of laying	埋深
depth of lift	炮眼深度
depth of navigable channel	航道水深；通航水深
depth of origin	震源深度
depth of overburden	覆盖层深度，覆盖层厚度
depth of penetration	渗入深度；贯入深度
depth of pit	蚀孔深度
depth of prospecting	勘探深度
depth of pull	一次爆破深度
depth of rainfall	雨量；降雨深度
depth of rockburst failure	岩爆爆坑深度
depth of round	炮眼深度
depth of runoff	径流深度
depth of sampling	取样深度
depth of scour	冲刷深度
depth of seismic focus	震源深度
depth of setting	下入深度
depth of shock	震源深度
depth of shotcrete	喷混凝土厚度
depth of snow	雪深
depth of soil overlying bedrock	岩基上覆土层厚度
depth of stratum	层厚
depth of stress influence	应力影响深度
depth of the instrumental burial	仪器埋置深度
depth of tooth	齿深；齿高
depth of tunnel	隧道埋深
depth offset	深度偏差
depth perception	深度感
depth phase	深度相位
depth phase information	深相信息；深相资料
depth point	共中心点；深度点
depth pressure	单位深度岩层压力；覆盖层压力
depth probe	深度探测仪
depth profile	深度剖面
depth range	深度限度；深度范围
depth ratio	埋深比
depth record section	深度剖面
depth recorder	深度记录器
depth reference	深度基准
depth reference point	深度基准点
depth registration	深度记录
depth resistance curve of pile	桩的深度-阻力曲线
depth resolution	深度分辨率
depth round-off	深度化整
depth rule	深度规律
depth sampling interval	深度采样间隔
depth scale	测深标尺
depth section	深度剖面
depth selection	取样深度
depth shift	深度偏移
depth signal pole	水深信号杆
depth sounder	测深仪
depth sounding	测深
depth sounding sonar	测深声呐
depth survey	深度测量
depth telemetry	深度遥测
depth time conversion	深时转换
depth track	深度道
depth transducer	深度传感器
depth velocity curve	水深流速关系曲线
depth zone	深度带；深度分带
depth-area curve	雨量-面积关系曲线
depth-area formula	深度-面积公式
depth-area relationship	深度-面积关系
depth-area-duration analysis	雨量-面积-历时分析
depth-area-duration relation	深度-面积-历时关系
depth-control log	深度控制测井
depth-current recorder	深层自记海流器；深层自记海流计；深层自记流速仪
depth-dependent	与深度有关的
depth-dependent Riccati system	与深度有关的黎卡蒂系统
depth-discharge relation	水深-流量关系
depth-duration curve	深度历时曲线
depth-duration-area value	时面深值
depth-extent	延深
depth-first search	深度优先搜索
depthometer	深度计
depth-sampling	深度采样
depth-span ratio	高跨比
depth-temperature relation	深度-温度关系

depth-time curve	深-时曲线
depth-to-span ratio	高跨比
depth-to-width ratio	高宽比
depth-velocity curve	深度-速度曲线；垂直速度曲线
depth-width ratio	深宽比；高宽比
derailing switch	脱轨装置；脱轨转辙器
derailment	脱轨，掉道
deranged drainage	紊乱水系
derangement	扰乱；紊乱
derated	降级的；已降低额定值的；修改设计的；重新设计的；变形的
derelict	遗弃的；丢弃的；遗弃物
derelict area	荒废地区；荒废地
derelict land	废地；冲积地；荒地；海退遗地
derelict mine	荒废的矿山
dereption	水下剥蚀
derivation	起源；推导；衍生；偏差
derivational process	递降分解过程；递降分解作用；降解过程
derivative	导数；衍生物；变形，改型
derivative map	微商图；导数图
derivative of gravity	重力导数
derivative of higher order	高阶导数
derivative order	导数的阶
derivative polarography	导数极谱
derivative restrictor	差动开关；差动式限动器
derivative rock	转生岩；沉积岩
derivative structure	衍生构造；导生构造
derivatization	衍生
derivatograph	示差热分析仪；测偏仪
derive	推导，求导数；分路，分支
derived equation	导出方程
derived fossil	衍生化石；移积化石
derived function	导出函数
derived geological disaster	次生地质灾害
derived lipid	衍生脂类
derived map	资料地图
derived number	导数
derived plan	衍引地图
derived resistance	并联电阻
derived unit	导出单位
derived value	派生值
derrick	起重机，钻塔；临时井架
derrick stone	巨石块
derrick support	井架底座；钻塔底座
derrick telescoping	井架起升
derrick tower	起重吊塔
desalination	淡化；海水淡化
desalting of soil	土脱盐作用
desanding device	除砂装置
desanding equipment	除砂装置
desaturation	减饱和作用；稀释
desaxe	不同心度；轴心偏移
descend	下降；下行；下倾
descending bed	沉降床；下倾地层，下倾层沉降河床
descending current	下降流；下沉流
descending grade	降坡；下坡
descending inclined slicing	倾斜分层下行采煤法
descending lithosphere	沉降岩石圈
descending mining	下行式开采
descending plate	下降板块
descending slicing	下行分层开采法
descending spring	下降泉
descending water	下降水；渗流水
descension deposit	下降水矿床
descension theory	下降说；下降溶液成矿说
descensional deposit	下降矿床
descensional ventilation	下行通风；下向通风
descent	下降；下坡；降坡；斜坡；倾没；下潜
descent method	下降法
descent of water	水渗透
descent propulsion system	降落推进系统
describe	描述；画图；制图
description	叙述；图形；说明书
description card	描述卡
description of core	岩芯描述
description of equipment	设备说明
description of profile	剖面描述
description of project	工程说明书
description of property	房地产项目
description of station	测站说明书；测站图说；点之记
description of works	工程描述
descriptive	描述的；说明的
descriptive analysis	描述性分析
descriptive classification of soil	土壤描述性分类法
descriptive geometry	画法几何
descriptive hydrology	记述水文学
descriptive mineralogy	描述矿物学
descriptive model	描述模型
descriptive petrology	描述岩石学
descriptive statistics	描述统计
descriptive stratigraphy	描述地层学
descriptive text	说明书
desensitize	钝化
desensitized	敏感度减低的
desensitizer	钝化剂；减敏感剂
desert crust	沙漠盐壳；沙漠岩漆；沙漠砾石滩
desert deposit	沙漠沉积
desert drilling	沙漠钻井
desert dry valley	漠境干谷
desert dust soil	漠境尘暴土壤
desert ecosystem	荒漠生态系统
desert encroachment	沙漠扩侵
desert erosion feature	荒漠侵蚀地形
desert facies	沙漠相
desert lake	漠境湖
desert meteorite	沙漠陨石
desert pavement	沙漠砾石滩，沙漠砾石盖层
desert plateau	漠境高原
desert rig	沙漠钻机
desert rock plain	漠境石质平原
desert steppe	沙漠草原；干旷草原
desert storm	沙暴
desertification	沙漠化
desertification control	沙漠治理；沙漠化控制
desiccant	干燥剂；干燥的
desiccated crust	干缩表土层；干缩表层
desiccated soil	干燥土；失水土
desiccated wood	晒干的木料；烘干的木料
desiccating basin	疏干盆地

desiccation 干燥；晒干；烘干；脱水
desiccation breccia 干裂角砾岩
desiccation conglomerate 干裂砾岩
desiccation crack 干缩裂痕；干裂
desiccation fissure 干缩裂隙；干缩裂缝
desiccation joint 干裂节理
desiccation mark 干裂痕
desiccation polygon 干裂多边形土
desiccative 干燥的；干燥剂
design altitude 设计高度
design analysis 设计分析
design approval 设计批准
design assumption 设计假定
design audit 设计审查
design automation 设计自动化；计算机辅助设计
design basis earthquake 基本地震设计
design capacity 设计能力
design change 设计变更
design chart 设计图表
design code 设计准则；设计规范
design concept 设计方案；设计概念
design condition 设计条件
design configuration 设计图形
design constraints 设计制约
design criteria 设计标准；设计准则
design curve 设计曲线
design development 设计开发
design development phase 技术开发阶段
design diagram 设计图表
design discharge 设计流量
design documentation 设计文件
design draft 设计草案；设计草图
design drawing 设计图
design duty 设计产量；设计生产能力
design earth pressure 设计土压力
design earthquake 设计地震
design earthquake intensity 抗震设防烈度
design effort 设计工作，计划工作
design elevation 设计高程
design elevation of subgrade 路基设计高程
design error 设计误差
design event 设计后果
design experiment 设计试验
design factor 设计因数，设计要素
design far earthquake 设计远震
design fault 设计故障
design feasibility test 设计可行性测试
design filling 设计充满度
design fixed number of year 设计年限
design flood 设计洪水
design flood composition 设计洪水组成
design flood hydrograph 设计洪水过程线
design flood inflow 设计入库洪水流量
design flood level 设计洪水位
design flood occurrence 设计洪水出现率
design flow 设计流量
design for earthquake resistance 抗震设计
design for practical construction organization 实施性施工组织设计
design force spectrum 设计荷载谱

design formula 设计公式
design frequency 设计频率
design goal 设计目标
design handbook 设计手册
design head 设计水头
design height 设计高度
design high water level 设计高潮位
design highlight 设计要点
design hydrograph 设计水文过程线
design impedance 设计阻抗
design index 设计指标
design intensity 设计烈度
design item 设计项目
design level 设计水位
design load 设计载重；设计荷载
design load factor 设计负载因数
design mix 配料设计；设计成分
design near earthquake 设计近震
design norm 设计标准
design objective point 设计观点
design objects 设计目标
design of a reinforced concrete beam 钢筋混凝土梁设计
design of excavations in jointed rock 节理岩体中巷道设计
design of excavations in massive elastic rock 弹性岩体中的巷道设计
design of experiment 试验设计
design of mine ventilation 矿井通风设计
design of mines 矿山设计
design of remedial works 防治设计
design of repair mortars 修补砂浆设计
design of rock support and reinforcement 岩石支护和加固设计
design of station and yard 站场设计
design of steel structure 钢结构设计
design of stope and pillar layout 矿房矿柱布置的设计
design of underground excavations of in stratified rock 层状岩石地下巷道设计
design of vertical alignment 纵断面设计
design paper 图纸
design parameter 设计参数
design payload 设计有效载荷
design phase 设计阶段
design philosophy 设计原理；设计思想
design population 计划人口；预定人口；预计人口
design pore volume 设计孔隙体积
design power 设计功率
design precipitation 设计降水
design pressure 设计压力
design pressure diagram 设计压力示意图
design principle 设计原理
design procedure 设计程序；设计步骤
design procedure for plane strain for roof of an excavation 巷道顶板平面应变设计方法
design process 设计过程
design professional 专业设计师
design program 设计任务书；设计计划书；设计程序
design proposal 设计方案
design rainfall 设计雨量
design range 设计允许范围
design rate 设计速率

design recommendation 设计建议
design recurrence interval 设计重现期
design reference period 设计基准期
design requirement 设计上的规定
design reserve 设计储量
design response spectrum 设计反应谱
design return period 设计重现期
design review 设计审查
design safety factor 设计安全系数
design schedule 设计进度表；计算表；核算表
design scheme 设计方案
design seismic coefficient 设计地震系数
design seismic load 设计地震荷载
design seismicity 设计烈度
design sensitivity analysis 设计灵敏度分析
design simulation 设计模拟
design specification 设计规范；设计任务书；设计说明书
design spectrum 设计谱
design standard for rank and grade classification of hydraulic engineering complex 水利水电枢纽工程等级划分及设计标准
design standards 设计标准
design storm 设计暴雨；径流量
design strength 设计强度
design strength of concrete 混凝土设计标号
design stress 设计应力
design table 设计计算表
design team 设计组
design tide level 设计潮位
design vacuum degree 设计真空度
design value 设计参数；计算值
design value of a load 荷载设计值
design variable 设计变量
design water lever 设计水位
design wave 设计波
designate 指定；选派；任命
designated area 指定地区
designated size 选定粒度；规定粒度；合格粒度
designation 指定；指示；定名；名称；代表符号；图名
designation number 指定数；赋值数；标准指数
designation of rock 岩石定名
designed capacity 设计能力；设计容量；设计生产率
designed depth 设计深度
designed dilution 设计贫化
designed discharge 设计流量
designed drawdown 设计水位降深
designed earthquake 设计地震
designed elevation 设计高程
designed experiment 设计试验
designed flood frequency 设计洪水频率
designed flow 设计流量
designed flow per second 设计秒流量
designed flow velocity 设计流速
designed fuel consumption 设计上的用油量
designed groundwater withdrawal 地下水设计开采量
designed horsepower 设计马力
designed life 设计寿命
designed lifetime of water well 水井设计使用年限
designed load 设计荷载
designed mine capacity 矿井设计生产能力
designed mine reserves 矿井设计储量
designed mix concrete 配料设计混凝土
designed output 设计产量
designed period 设计期限
designed production 设计能力；设计产量
designed surface mine capacity 露天矿设计生产能力
designed surface mine reserves 露天矿设计储量
designed time of groundwater mining in well field 地下水源地设计开采时间
designed traffic volume 设计交通量
designed underground mine reserves 地下矿井设计储量
designed velocity 设计速度；设计流速
designed water line 设计水线
designing of blast 爆破设计
designing of concrete 混凝土配合设计
designing of section 剖面图设计，断面图设计
designing treatment 设计处理
desilication 脱硅；去硅；硅氧淋失
desilting dam 拦沙坝
desilting manhole 隔沙沙井
desilting material 疏浚物料
desilting sand pit 隔沙池；隔沙井
desilting sand trap 隔沙池；隔沙井
desilting strip 放淤畦条
desintegration 崩解；机械破坏；机械分解；蜕变；裂变
desired location 预定位置
desired output 期望输出值
desired pack pressure 预定充填压力
desired sequence 要求的程序
desired strength 预期强度，期望强度
desired tolerance 期望限度
desired trajectory 预定轨迹
desired value 期望值
desk study 资料研究；室内研究；资料搜集
desk training 就地培训
desolation 荒地
desorption 退吸作用；解吸附作用；解吸收作用
desorption curve 解吸曲线
desorption isotherm 解吸等温线
desorption pressure 退吸压力
desorption rate 解吸速率
desquamation 剥落；片落；层剥
destabilization 不安定
destabilization of subgrade 路基失稳
destabilizing effect 失稳效应
destabilizing moment 不稳定力矩
destaticization 脱静电作用
destination 目的地
destination button 终端按钮
destination node 终结点
destratification 去层理作用；水层混合
destrengthening 强度消失，软化
destress 放松应力
destress blasting 去应力爆破
destressed condition 卸压状态；去应力状态
destressed zone 应力释放区；卸压区；去应力带
destressed zone in roof strata 顶板垮落形成的减压带
destressing 去应力，卸压
de-stressing method 解除应力法；卸压法
de-stressing of peripheral rock 围岩的减压

destressing slot 卸压槽
de-stressing use in the control of rockbursts 减压在岩爆控制中的应用
destressing zone 卸压带
destruct 自毁
destructibility 破坏性
destruction 破坏；毁灭
destruction belt 破坏带
destruction test 破坏性试验
destructional 破坏的
destructional bar 侵蚀沙坝
destructional bench 侵蚀平台；侵蚀阶地
destructional delta 侵蚀三角洲
destructional delta deposit 破坏性三角洲沉积
destructional form 破坏形式
destructional plain 侵蚀平原
destructional terrace 侵蚀阶地
destructional valley 侵蚀谷
destructive 有破坏性的
destructive boundary 破坏性边界
destructive capability 破坏能力
destructive earthquake 破坏性地震
destructive effect 破坏效应
destructive hydrogenation 破坏加氢
destructive interference 相消干扰
destructive metamorphism 破坏性变质作用；等温变质作用
destructive motion 破坏性运动
destructive pitting 毁坏性点蚀
destructive plate boundary 消减型板块边界；会聚型板块边界
destructive plate margin 消减型板块边缘
destructive power 破坏力
destructive tension test 张力破坏试验
destructive test 破坏试验；断裂试验
destructive topographic form 侵蚀地形；破坏地形
destructive wave 侵蚀性波浪；破坏性波浪
destructiveness 破坏程度；破坏能力
destructor 破坏器，自炸装置
destructured soil 结构破坏土
desulfation 脱硫；脱硫作用；脱硫酸盐作用
detach 分离；采落
detach from ground 从整体采落
detachable 可分离的
detachable bit 活钻头；可卸式钻头
detachable core barrel 可拆式岩芯筒
detachable drill head 可拆卸钻头
detachable fixing 可拆卸装配
detachable head 可卸式盖；可卸式头
detachable point 活络触探头；可拆卸触探头
detachable top plate 可拆卸顶板
detached 独立的；分离的
detached anticline 挤离背斜
detached arch core 挤离背斜核部
detached block 外来岩块；飞来峰
detached breakwater 岛式防波堤；独立式防波堤
detached building 独立式房屋
detached coefficient 分离系数
detached column 独立柱；单立柱
detached mass 外来岩块；飞来峰；孤残岩块

detached pier 独立支墩
detached rock 悬石
detached structure 挤离构造
detachment 分离；拆卸；解开
detachment fault 拆离断层；推覆体底基逆掩断层；滑脱断层
detachment fold 滑脱褶皱；断滑褶皱
detachment gravity slide 滑脱重力滑动
detachment horizon 挤离层位
detachment structure 拆离构造；滑脱构造
detachment surface 拆离面；底基逆掩面；最下冲断面；滑褶面；滑脱面；浮褶面
detachment thrust 挤离冲断层
detachment zone 挤离带
detail 详图；大样图；零件图；细目；细节；细部；详细
detail design 细部设计；详细设计
detail drawing 细部图；详图；零件图
detail estimate 详细预算，详细估算
detail exploration 详细勘探
detail file 详细资料
detail map 详细图；详图
detail of construction 构造详图；施工详图；零件图
detail of design 设计详图
detail point 碎部点；细部点
detail prospecting 详细普查
detail requirements 详细技术要求，详细规格
detail shooting 地震详查；详细爆破
detail specifications 详细说明书
detail survey 细部测量；分图测量
detail survey for mine workings 矿内巷道细部测量
detail survey of workings 巷道细部测量
detail working drawing 施工详图
detailed cost 详细费用
detailed description 详细说明
detailed design 设计详图；详细设计
detailed design stage 详细设计阶段
detailed design study 详细设计研究
detailed drawing 详图；细节图；细明图
detailed exploration 详细勘探
detailed field data 详细现场资料
detailed field performance 详细现场使用情况，详细野外操作情况
detailed geological investigation 详细地质勘察
detailed geotechnical investigation 工程地质详细勘察，详细勘察
detailed geotechnical stability study 详细岩土稳定性研究
detailed gravity survey 重力详查
detailed hydrogeologic investigation 水文地质详细勘察
detailed investigation 详测；详查
detailed investigation stage 详细勘察阶段
detailed planning proposal 详细规划建议
detailed program 详细计划
detailed project report 工程详细报告
detailed prospecting 详细勘探
detailed report 详细报告
detailed rules and regulations 细则
detailed schedule 详细计划表
detailed specification 详细说明，详细规格
detailed survey 详细测量；详细勘测

detailing	细部设计；细部构造	determinate structure	静定结构
details	零件；部分；详图；细目	determination	确定；决定；测定；鉴定
details of apron	垂裙大样	determination coefficient	确定系数
detect	发现；检查；探测；侦察	determination method	确定法
detectability	探测能力；检波能力；可探测性	determination of azimuth	方位角测定
detecting device	探测器；探测设备	determination of control point	控制点测定
detecting head	探测头；探针；指示器	determination of density	密度测定
detecting instrument	探测仪器	determination of epicentre	震中测定
detecting scale	监测标尺	determination of lease	终止契约；解除契约
detection	探测；检查；检测；检出；检波	determination of quantities	确定工作量
detection coefficient	检波系数	determination of scale	比例尺的确定
detection curve	探测曲线	determination of strain	应变测量
detection device	检测装置	determination of tenancy	终止租赁；解除租约
detection efficiency	探测效率	determination of tilt	倾斜测定
detection equipment	探测装置	determining factor	决定性因素
detection instrument	探测仪	deterministic	定值的；定数的；确定的
detection limit	检测限；检出限	deterministic analysis	定值分析；确定性分析
detection logging	井下电测；勘探测井	deterministic automation	确定性自动机
detection of slip planes	滑面探测	deterministic calculation	确定性计算
detection probability	检测概率	deterministic characteristics	定数特性
detection range	探测距离；探测范围	deterministic data	确定性数据；定值数据
detection time	探测时间	deterministic deconvolution	确定性反褶积
detection width	探测宽度	deterministic demand	确定性需求
detectivity	探测能力；探测灵敏度	deterministic dynamic response analysis	定值动力反应分析
detector	检波器；检查器；探测器；测定器	deterministic evaluation	定值估计方法；确定性估计
detector arrangement	探测设备；检波设备	deterministic excitation	确定性激励
detent pin	定位销；止动销	deterministic fracture	确定性裂隙
detent plate cover	盖板	deterministic function	定值函数
detent plate stop	制动架；制动片；停臂	deterministic geological model	确定性地质模型
detention	滞留；阻滞；截留	deterministic intensity function	定值强度函数
detention dam	拦砂坝	deterministic method	确定性方法
detention effect	滞留作用；滞洪作用	deterministic model	确定性模型
detention of flow	水流滞留	deterministic modeling	确定性建模
detention period	延迟时期	deterministic network	肯定型网络
detention reservoir	滞洪水库；拦洪水库	deterministic policy	确定性方针
detention storage	拦洪库容；滞洪库容	deterministic prediction	确定性预报
detention tank	滞留池，调节池	deterministic schedule	确定性计划
detention time	停留时间	deterministic simulation	确定的模拟；定数模拟；测定模拟
detention volume	拦洪容量；滞洪容量	deterministic system	确定性系统；决定性系统
detergence rock	次生岩；后成岩	deterministic tuning curve	决定性调谐曲线
detergent factor	洗净系数	deterministic-cum-probabilistic procedure	定值法与概率法程序
deteriorate	弄坏；恶化；退化	deterrence	制止
deteriorating roof	易塌落顶板；易变坏顶板	deterrent	制止物；制止因素
deterioration	退化；降低；损伤；消耗	detided gravity	潮汐改正后重力值
deterioration failure	老化形成的故障	detonate	起爆；发爆
deterioration level	退化程度	detonating	起爆的；发爆的；起爆；发爆
deterioration map	劣化地图	detonating cap	雷管
deterioration mechanism	退化机理	detonating capacity	爆轰感度
deterioration of groundwater	地下水水质恶化	detonating charge	起爆炸药；起爆药包
deterioration of reinforcement	钢筋损蚀	detonating composition	起爆剂成分；起爆剂
determinability	可确定性	detonating compound	起爆剂成分
determinable error	可测误差	detonating cord	导爆索；导爆管；非电导爆管
determinacy	确定性	detonating explosive	起爆炸药
determinant	决定因素；行列式；决定性的	detonating fuse	导爆索
determinant equation	行列式方程	detonating fuse blasting	导爆索起爆
determinant factor	行列式因子	detonating gas	爆轰气；爆鸣气；爆炸瓦斯；爆炸气体
determinant of a matrix	矩阵的行列式	detonating mixture	爆炸混合物
determinant of coefficient	系数行列式	detonating powder	起爆炸药
determinant rank	行列式秩		
determinantal expansion	行列式展开式		

detonating pressure	爆炸压力	detritus	岩屑；碎屑；碎石
detonating relay	起爆迟发器；导爆替续器	detritus chamber	泥石沉淀室
detonating tube	导爆管	detritus cone	岩屑锥
detonating velocity	爆速	detritus equipment	碎石设备；制碴设备
detonation	引爆，爆燃	detritus reservoir	碎屑储油层
detonation cord	爆炸索	detritus rock	碎屑岩
detonation factor	爆炸因素	detritus rubbish	岩屑
detonation front	爆炸波前锋	detritus sand	碎屑砂
detonation inhibitor	防爆剂	detritus slide	碎屑滑动
detonation interval	炮眼起爆间隔	detrusion	外冲；剪错；滑动变形
detonation meter	爆震仪	detrusion ratio	剪切比
detonation power	爆力	deuteric action	岩浆后期活动
detonation pressure	爆燃压力	deuteric alteration	岩浆后期的蚀变作用
detonation product	爆炸产物	deuteric ore deposit	岩浆后期矿床
detonation property	起爆性	deuteric reaction	岩浆后期反应
detonation rate	起爆速率	deuterogene	后生岩；次生岩
detonation reaction	起爆反应；爆炸反应	deuterogene rock	后成岩；衍生岩
detonation seal	阻爆器	deuteromorphic	后生变形
detonation state	起爆态	deuteroxide	重水；氧化氘
detonation velocity	起爆速度	devastating earthquake	破坏性地震
detonation velocity test	爆速试验	devastating flood	毁灭性洪水
detonation wave	传爆波；爆震波	developable surface	可展曲面
detonation wave front	爆轰波阵面	developed area	经济发达区
detonation wave propagation	起爆波传播	developed field	已开发的油田
detonative	起爆的；发爆的	developed ore	开发矿量，开准矿量
detonator	雷管；发爆器	developed reserves	开发储量，开拓储量
detonator bowl	雷管箱；爆破器材贮存箱	developed view	展开图
detonator box	雷管箱；爆破器材贮存箱	developing butt	开拓巷道
detonator cap	雷管帽盖	developing by adit	平硐开拓
detonator lead	雷管导线	developing by combined methods	综合开拓方式
detonator signal	爆炸信号	developing by shaft	立井开拓
detour drift	迂回坑道	developing by slope	斜井开拓
detour matrix	迂回矩阵	developing chart	展示图
detour road	迂回路；绕道	developing chart of exploratory drift	探硐展示图，勘探巷道展示图
detouring section	绕行地段	developing entry	采准平巷；准备巷道
detraction	挖蚀作用；降低	developing face	掘进工作面
detrimental	有害的；不利的	developing heading	采准巷道
detrimental effect	损害效应	developing of land	土地开发
detrimental expansion	有害膨胀	developing of main level	主要水平开拓
detrimental settlement	有害沉降；不稳定沉降	developing per block	每盘区的开拓量
detrimental soil	不稳定土	developing section	开拓区
detrital	岩屑的；碎屑的	developing with easy cut	缓沟开拓
detrital clay	碎屑黏土	developing with steep cut	陡沟开拓
detrital clay matrix	碎屑黏土基质	developing within deposit	矿床内开拓；矿床内采准
detrital component	碎屑组分	development and mining scheme	采掘方案
detrital deposit	碎屑沉积；碎屑矿床	development and utilization	开发和利用
detrital fan	冲积扇；碎屑扇形地形	development area	拓展区；发展区
detrital laterite	残存砖红壤	development block	开发单元
detrital magnetic particle	碎屑磁颗粒	development cost	开发费用；研制费用
detrital mantle	碎屑风化层	development curve	地表沉陷曲线；过程曲线；演变曲线
detrital mineral	碎屑矿物	development drift	开拓平巷
detrital phase	碎屑相	development drilling	开发钻探
detrital precipitated mineral	碎屑沉积矿物	development effect	显影效应
detrital ratio	岩石碎屑比	development effort	研制计划
detrital reservoir	碎屑岩储集层	development end	开拓巷道掘进头
detrital rock	碎屑岩	development evaluation	开发评价
detrital sand	碎屑砂	development examination	加密探测
detrital sediment	碎屑沉积	development excavation	掘进巷道
detrital slope	岩屑坡；碎石坡	development face	采准工作面；开拓工作面
detrition	风化破碎；成屑；磨损；耗损		

development factor	显影因素	deviated hole	斜井
development gallery	准备平巷；开拓平巷	deviated well	斜井
development heading	开拓平巷；准备平巷	deviated well bore	斜井井筒
development machine	掘进机	deviating force	造斜力
development method	开拓方法	deviating hole	偏斜孔
development method of surface mine	露天矿开拓方法	deviating stress	偏斜应力
development method with adit	平硐开拓法	deviation	偏差；偏向；偏离；偏位
development method with inclined shaft	斜井开拓法	deviation absorption	偏移吸收
development method with vertical shaft	竖井开拓法	deviation angle	偏角
development model	开发模型	deviation bit	偏心钻头
development of a function	函数的展开	deviation compensator	偏差补偿器
development of pressure	压力发展；压力变化	deviation control	偏差控制
development of the field	油田开发	deviation correction	纠斜
development of underground mine	地下矿开拓	deviation curve	偏差曲线
development of water resources	水资源开发	deviation factor	偏差系数
development opening	开拓巷道	deviation from mean	均值离差
development operation	开发作业	deviation index	偏差指数；离差指数
development package	整套发展计划；发展综合计划	deviation log	井斜测井
development parameter	开发特性，开发参数	deviation map	偏差图
development period	基建期	deviation of inflection	拐点偏移距
development phase	开发阶段	deviation of inflection point	拐点偏移距
development plan	发展计划；发展计划图；发展蓝图	deviation of measuring instrument	量测仪器偏差
development plan of mine area	矿区开发规划	deviation of reading	读数偏差
development planning	发展规划	deviation sensitivity	偏位灵敏度
development procedure of mine field	矿区开发程序	deviation sensor	井斜传感器
development process	开发过程	deviation standard	偏差标准
development program	发展规划；开发方案	deviation survey	井斜测量；钻孔弯曲测量
development project	开发项目	deviational survey	偏斜测量
development raise	开拓上山；开拓天井	deviative absorption	偏移吸收
development ratio	开拓比；采掘比	deviator	偏差器；致偏装置
development research	开发研究	deviator of stress	应力偏量
development reserves	开拓矿量；开拓储量	deviatoric	偏量的
development risk	开发风险	deviatoric component	偏离分量
development road	开拓巷道	deviatoric component of stress	应力偏量，应力偏分量
development roadway	开拓巷道	deviatoric creep	偏斜蠕变
development sampler	沿开拓巷道取样	deviatoric pressure	偏应力
development scenarios	发展方案	deviatoric projection matrix	偏投影矩阵
development section	开拓区	deviatoric state of stress	偏应力状态
development seismology	开发地震学	deviatoric strain	偏应变
development shaft deepening	延深开拓矿井	deviatoric stress tensor	偏应力张量
development speed	发展速度	deviatoric tensor	偏张量
development stage	开发期	deviatoric tensor of strain	应变偏张量
development statement	发展纲领	deviatoric tensor of stress	应力偏张量
development status	开发状态	device	装置；设备；设计；方法；手段
development strategy	发展策略	device availability	设备完善率
development system	开发系统	device control	设备控制
development system of underground mine	地下矿开矿系统	devil water	废水
development technique	开发技术	devise	设计；计划；发明；创造；发生；产生
development tendency	发展趋向	devolution	崩坍；偏差；退化；衰落
development test	研制试验；发展试验	dew point	露点
development type	开发型式，开发类型	dew point control	露点控制
development way	开拓巷道	dew point corrosion	露点腐蚀
development well	开发井	dew point hygrometer	露点湿度计
development well success ratio	开发井成功率	dew point limit	露点极限
development work	开拓工程；上机准备工作	dew point method	露点法
development workings	开拓巷道	dew point pressure	露点压力
development zone	开拓区	dew point temperature	露点温度
developmental drilling program	开发钻井方案	dew pond	湿沼泽
deviate	偏离；偏差	dewater	脱水；排水
deviated flow	偏斜水流	dewaterer	脱水器

dewatering 降低地下水位；脱水
dewatering adit 疏干巷道
dewatering agent 脱水剂
dewatering apparatus 脱水设备
dewatering area 疏干区
dewatering boring 排水钻孔
dewatering bowl 脱水箱；脱水仓
dewatering coal 脱水煤
dewatering coefficient 脱水系数；疏干系数
dewatering conduit 排水管；泄水底孔
dewatering drift 迂回排水施工法
dewatering effect 排水效果；疏干效果
dewatering elevator 脱水提升机
dewatering equipment 排水设备；脱水设备
dewatering excavation 排水开挖；排水工程
dewatering extent 水位降低范围；疏干范围
dewatering facilities 去水设施；脱水设施
dewatering factor 疏干因数
dewatering gallery 排水廊道
dewatering hole 排水孔；排水井点
dewatering influence 排水影响
dewatering inspection 降水效果检验
dewatering method 降水法
dewatering of back fill 充填料脱水；回填料脱水
dewatering of excavation 开挖中排水
dewatering of ore-deposit 矿床疏干
dewatering project 降水工程
dewatering pump 排水泵
dewatering screen 脱水筛
dewatering system 脱水系统
dewatering test 脱水试验
dewatering through well 地表钻孔抽水
dewatering time 脱水时间
dewatering trough 脱水槽
dewaxing method 清蜡方法
dewing 结露
dewpoint temperature 露点温度
dextral 右旋的；右边的；右旋壳；右旋螺
dextral displacement 右旋位移
dextral fault 右行断层；右旋断层
dextral fold 右移褶皱
dextral horizontal displacement 右旋水平位移
dextral motion 右旋运动
dextral shell 右旋壳
dextral strike slip 右旋走向滑动
dextral strike-slip fault 右旋走向滑动断层；右旋平移断层
dextrorotatory form 右旋型
dextrorotatory substance 右旋物质
dextrorsal 右向的
diachronous 穿时的；时进的；跨时代的；不同时代的
diachronous reservoir rock 跨时储集岩
diachronous sedimentary rock 跨时沉积岩
diaclase 节理裂隙；构造裂隙；正方断裂线；压力裂缝；张开裂隙
diaclasite 黄绢石
diaclinal 横切褶皱的；垂直走向的
diagenesis 成岩作用，岩化作用
diagenesis facies 成岩相
diagenesis of fossils 化石成岩学

diagenetic 成岩的；沉积的
diagenetic bond 成岩连接；成岩键
diagenetic breccia 成岩角砾岩
diagenetic concretion 成岩结核
diagenetic degradation 成岩降解作用
diagenetic deposit 成岩沉积物
diagenetic differentiation 成岩分异作用
diagenetic environment 成岩环境
diagenetic fabric 成岩组构
diagenetic facies 成岩相
diagenetic fissure 成岩缝
diagenetic fracture 成岩裂缝
diagenetic gas 成岩阶段成因气
diagenetic gas window 成岩阶段生气窗
diagenetic grade 成岩程度
diagenetic metamorphism 成岩变质
diagenetic pattern 成岩型式
diagenetic plugging 成岩填塞
diagenetic process 成岩过程
diagenetic rock 成岩岩石
diagenetic state 成岩状态
diagenetic structure 成岩构造
diagenetic texture 成岩结构
diagenetic trap 成岩圈闭
diagenetic water 成岩水
diagenism 成岩作用；沉积变质作用
diaglyph 成岩痕迹
diagnosis 诊断；识别；判断；鉴定
diagnostic 诊断的；特征；症候；症状；诊断法；诊断学
diagnostic capability 诊断能力；识别能力
diagnostic check 诊断检验
diagnostic error processing 诊断错误处理
diagnostic feature 判别要素
diagnostic fossil 标准化石
diagnostic function 诊断功能
diagnostic gravity effect 鉴别性重力效应
diagnostic horizon 判别层
diagnostic message 诊断信息
diagnostic mineral 特征矿物；标志矿物
diagnostic mode 诊断方式
diagnostic package 诊断程序包
diagnostic program 诊断程序
diagnostic property 诊断症状
diagnostic test 诊断试验
diagonal 对角线；斜的；对角的
diagonal bar 对角钢筋
diagonal bedding 斜层理
diagonal brace 对角斜杆；斜撑杆；斜支撑；剪刀撑
diagonal cleavage 斜劈理
diagonal crack 斜裂缝；斜裂纹
diagonal displacement 对角线位移
diagonal dominance 对角占优
diagonal dominant matrix 对角占优矩阵
diagonal drift 对角平巷
diagonal element 对角元素
diagonal entry 伪倾斜巷道；对角元素
diagonal face 对角工作面；斜工作面
diagonal fault 斜断层；斜交断层
diagonal grid 对角网格

diagonal joint　斜节理
diagonal layout　对角布置；伪倾斜布置
diagonal layout system　伪倾斜开采法；对角布置法
diagonal line　对角线
diagonal lines section　对角测线剖面
diagonal longwall work　对角长壁工作面开采法
diagonal matrix　对角矩阵
diagonal member　对角构件；斜构件
diagonal network　角联网络
diagonal plane　对角面
diagonal scale　斜线尺
diagonal scaling　对角线缩放比例
diagonal slice　对角倾斜分层；斜向分层
diagonal slip fault　斜滑断层；斜移断层
diagonal stratification　假层理；交错层；流水层
diagonal strut　斜撑杆
diagonal tension　斜张力；斜拉应力
diagonal tensor　对角张量
diagonal tie　斜拉杆
diagonal transformation　对角变换
diagonal tube　斜撑管
diagonal ventilation　对角式通风
diagonal ventilation system　对角式通风系统
diagonalization　对角化
diagonalization of matrix　矩阵的对角化
diagonally dominant matrix　对角占优矩阵
diagonal-slotted screen　斜缝筛
diagram　图；简图；线图
diagram of particle distribution　粒度分布图
diagram of stresses　应力图
diagrammatic　图解的；概略的
diagrammatic drawing　草图；示意图
diagrammatic layout　原理图；原则性布置，概略型规划，概略型草图
diagrammatic map　简图
diagrammatic plan　简图；示意平面图；示意规划
diagrammatic profile　示意剖面
diagrammatic section　截面图；剖面图；图解剖面
diagrammatic sectional drawing　断面草图；断面图解
diagrammatic sketch　示意图，草图
diagrammatic view　图解表示图；示意图
diagraph　分度尺；绘画器；放大绘画器；仿型仪
dial gauge　量表；千分表
dial gauge micrometer　千分表测微计
dial gravity　度盘重力值
dial indicator　刻度盘指示器；千分表
dial plate　拨号盘；标度盘；指针板
dial pointer　度盘指针
dial reading　刻度盘读数
dial scale　度盘；标度
dial survey　罗盘测量
dialysis　渗析；透析
diamagnet　抗磁体
diamagnetic body　抗磁体
diamagnetic effect　抗磁效应
diamagnetic material　抗磁体
diamagnetic substance　抗磁质；反磁质
diamagnetism　抗磁性
diameter　直径
diameter clearance　径向间隙

diameter distribution　粒径分布
diameter gauge　量径规；直径规
diameter increment　直径增量
diameter of axle　轴径
diameter of bore　钻孔直径
diameter of finally formed hole　终孔直径
diameter of impression　印痕直径
diameter of invasion　侵入直径
diameter of particle　颗粒直径
diameter of sediment gain　泥沙粒径
diameter-to-height ratio　直径-高度比
diametral　直径的
diametral compression test　径向受压试验
diametral compression testing　径向压缩试验
diametral pitch　径节
diametral plane　切径平面；径向平面
diametric growth　径向增大
diametrical clearance　直径间隙
diametrical compression　对径压缩；正反向压缩
diametrical dimension　直径方向的尺寸
diamond core drill　金刚石岩芯钻头
diamond coring　金刚石钻机采取岩芯
diamond coring bit　金刚石取芯钻头
diamond crown　金刚石钻头
diamond crown bit　金刚石钻头
diamond drill　金刚石钻机
diamond drilling　金刚石钻头钻探
diamond drilling bit　金刚石钻头
diamond head　金刚石钻头
diamond impregnated bit　孕镶金刚石钻头；细粒金刚石钻头
diamond interchange　菱形立体交叉
diamond plug bit　金刚石取芯钻头
diamond rotary drilling　金刚石回转钻进
diamond spar　刚玉；金刚砂
diamond tipped casing shoe　金刚石刃套管靴
diamond tool　金刚石钻具
diamond-anvil pressure cell　钻石砧压力盒
diamond-point bit　菱形尖钻头；菱形尖钎头；金刚石菱形尖钻头
diaphaneity　透明度
diaphragm　隔膜；隔板；薄膜
diaphragm cell　连续墙格仓
diaphragm earth pressure cell　薄膜式土压力盒
diaphragm meter　膜片计量器
diaphragm method　隔膜法
diaphragm packing　隔膜充填
diaphragm plate　横隔板
diaphragm pressure gauge　膜片式压力计
diaphragm pressure span　膜片压力范围
diaphragm protector　隔膜保护器
diaphragm pump　隔膜泵
diaphragm valve　隔膜阀
diaphragm wall　截水墙；地下连续墙
diaphragm wall excavator　地下连续墙挖掘机
diaphragm wall method　地下连续墙法
diaphragm wall technique　帷幕技术
diaphragm-type compressor　薄膜式压缩机
diaphragm-type detector　膜片式检波器
diaphragm-type weight unit　膜片式钻压测量装置

diaphthoresis	退化变质
diapir	底辟；挤入构造；刺穿构造；盐丘
diapir core	底辟中心
diapir dome	底辟穹隆
diapir fold	挤入褶皱；冲顶褶皱
diapir penetration	底辟刺穿作用
diapir salt structure	底辟盐体构造
diapir structure	挤入构造；冲顶构造
diapire	底辟
diapiric	刺穿的；挤入的
diapiric anticline	底辟背斜
diapiric core	底辟体核部
diapiric fold	底辟褶皱；挤入褶皱；刺穿褶皱
diapiric intrusion	底辟侵入
diapiric material	底辟物质
diapiric orogenesis	挤入造山作用
diapiric rise	底辟上升
diapiric structure	刺穿构造
diapiric trapping mechanism	刺穿圈闭机理
diapiric uplift	底辟上升
diapiric uprise	底辟上升
diapirism	底辟作用；刺穿作用；挤入作用；挤入褶皱
diapirite	底辟岩；重熔岩
diara	心滩；沙洲
diastatite	角闪石
diastem	沉积暂停期；小间断
diastimeter	测距仪
diastrome	层面节理
diastrophic activity	地壳活动
diastrophic belt	地壳变动带
diastrophic block	地块；古地块，古陆
diastrophic cycle	地壳运动旋回
diastrophic eustatism	地壳变动；海面升降
diastrophic force	地壳变动作用力
diastrophic lake	上升湖泊，隆起湖泊
diastrophic movement	地壳运动
diastrophism	地壳变动；构造变动
diataphral tectonics	底辟重褶构造
diatectic structure	部分重熔构造
diatexite	高级重熔混合岩；高度深熔岩；高度熔合岩
diathermy	电热疗法；透热法
diatom earth	硅藻土
diatom ooze	硅藻软泥
diatom pelite	硅藻泥质岩
diatom shale	硅藻页岩
diatomaceous earth	硅藻土
diatomaceous earth block	硅藻土块
diatomaceous earth filter	硅藻土过滤器
diatomaceous ooze	硅藻软泥
diatomaceous silica	硅藻土；硅藻类硅石
diatomaceous soft rock	硅藻质软岩
diatomaceous soil	硅藻土
diatomaceous strata	硅藻层
diatomeae	硅藻类
diatomite	硅藻土；硅藻岩
diatomite insulating layer	硅藻土保温层；硅藻土绝缘层，硅藻土隔离层
diatreme	火山爆发口；火山道；火山角砾岩筒
dice	小方块；微型半导体材料；切成小方块；油页岩
dichroscope	二色镜
dictyodromous	网状脉
dictyonite	网状混合岩
dictyonitic structure	网纹构造
didactic structure	粒级构造，粒度递变构造
die away	兴灭；衰减
die casting	拉模铸造；压力铸造
die nipple	打捞公锥；捞管器
die out	灭绝
die plate	模板
die reduction angle	拉模孔圆锥角
die relief angle	模孔出口角
die rolled section	周期断面型材
die scratch	压模划痕
die semiangle	拉模半圆锥角
die set	成套模具；成套冲模
die slide	模座；模具滑移装置
die slotting machine	冲模插床
die spider	辐架；多脚架
die-away curve	衰减曲线
die-away test	渐息检验
die-away time	衰落时间
die-casting	压铸
die-formed	冲压的；模锻的
dielectric after effect	介电后效
dielectric analysis	介电分析
dielectric body	电介质体
dielectric breakdown	电介质击穿
dielectric capacitance	电容率
dielectric capacity	电容率；介电常数
dielectric coefficient	介电系数；电容率
dielectric conductance	电介质电导
dielectric constant	电介质常数；介电常数
dielectric dispersion	介电分散
dielectric displacement	介电位移
dielectric drier	高频烘干炉；电介质干燥机
dielectric fatigue	电介质疲劳
dielectric field	电介质场
dielectric field intensity	电介质场强
dielectric fluid	电解质液体；绝缘液体
dielectric flux density	电介质通量密度
dielectric induction	介电感应
dielectric isolation	介质隔离
dielectric log	介电测井
dielectric loss	介电耗损
dielectric loss angle	介电损耗角
dielectric loss coefficient	介电损耗系数
dielectric medium	电介质
dielectric permittivity	介电常数
dielectric polarization	电介质极化
dielectric properties	介电性质
dielectric resistance	介质电阻；绝缘电阻
dielectric separation	电介质分选；介电分离
dielectric strength	绝缘强度
dielectric surface	电介质表面
dielectric susceptibility	电介质极化率
dielectric test	介电性能试验；绝缘强度试验
dielectric tool	介电测井仪
diesel traction	内燃牵引
difference contour map	差分等值线图
difference determinant	差分行列式

difference differential equation	差分微分方程
difference equation	差分方程
difference expression	差分表达式
difference format	差分格式
difference frequency	差频
difference gate	异门
difference image	差值图像
difference in elevation	高程差
difference in gradients	坡度差；梯度差
difference in height	高差
difference in pressure	压力降；压力差
difference method	差分法
difference of elevation	高程差；标高差
difference of latitude	纬差；纵距
difference of longitude	经差；横距
difference of parallax	视差差数
difference of phase	相位差
difference of potential	电位差；电势差；位差；势差
difference of settlement	沉陷差异；沉降差
difference operator	差分算子
difference percentage	差值百分比
difference quotient	差商
difference scheme	差分格式
difference table	差分表
difference threshold	差异阈
differences method	差异法
different speed	不同速度等级
different speed train	不同速度等级列车
different time constants	差分时间常数
differentiability	可微性
differentiable	可微的
differentiable function	可微函数
differential	微分；微分的；差别的，差示的；差速的，差动的；差速器，差动器
differential ablation	差异消融
differential absorption	差异吸收
differential accumulation	差异聚集
differential accumulation theory	差异聚集原理
differential action	差动作用
differential adsorption	微分吸附
differential amplifier	差动放大器；差分放大器；推挽放大器
differential analysis	微分分析
differential and integral calculus	微积分
differential angular velocity	差动角速
differential attachment	差动机构附件
differential barometer	差动气压计
differential bit	差动式钻头
differential block	差动滑车
differential calculus	微分学
differential centrifugation	差速离心
differential compaction	差异压实
differential compensation	差动补偿
differential compound generator	差复励发电机；差动复激电动机
differential compound motor	差绕复激电动机；差动复激电动机，差动复激马达
differential condensation	差异冷凝
differential contraction stress	差异收缩应力
differential control	微分控制
differential correction	微分校正
differential cost	差别成本
differential cross-section	微分截面
differential curvature	曲率差
differential curve	微分曲线
differential deformation	不均匀变形
differential depletion	差异衰竭
differential device	差动机构；差动装置
differential dike	分异岩墙
differential disintegration	优先破碎；差别分裂
differential displacement	不均匀位移
differential divisor	微分因子
differential drive	差速器传动；差动装置
differential effect	差动效应；微分效应
differential effective medium theory	微分有效介质理论
differential element	微分元件
differential entrapment	差异圈闭作用
differential equation	微分方程
differential equation of constitution	本构微分方程
differential equation of continuity	连续性微分方程
differential equation of first order	一阶微分方程
differential equation of geodesic	大地线微分方程
differential equation of higher order	高阶微分方程
differential equation of motion	运动微分方程
differential equation simulation	微分方程模拟
differential equation solver	微分方程解算机
differential equation system	微分方程组
differential equations for particle motion	质点运动微分方程
differential equations of equilibrium	平衡微分方程
differential equations of equilibrium in two dimensions	二维平衡微分方程
differential equations of static equilibrium	静态平衡微分方程
differential equilibrium equation of Euler	欧拉平衡微分方程
differential erosion	差异侵蚀
differential excitation	差动激励
differential expansion	差异膨胀
differential expansion stress	差异膨胀应力
differential expression	微分式
differential extraction	差异萃取
differential fault	差动断层
differential feed gear	差动给进装置；差动推进装置
differential fibre	差别化纤维
differential flotation	优先浮选
differential flow	差异流动
differential flow-in and out	进、出口流量差
differential flowmeter	压差流量计
differential force	差异应力；定向压力
differential foundation settlement	基础不均匀沉降；差异沉降
differential frost heave	不均匀冻胀
differential fusion	差异熔融
differential galvanometer	差动电流计
differential gauge	差示流速规；差式流速计；差动压力计；差压计
differential geometry	微分几何
differential grinding	优先研磨；选择性研磨；选择性磨矿

differential ground potential　接地电位差
differential head　水头差
differential heat of adsorption　微分吸附热
differential heating　差温加热
differential heave　不均匀隆胀
differential injection profile　微差注入剖面
differential input　差动输入
differential interferometric synthetic aperture radar　雷达差分干涉测量
differential intrusion　差异侵入
differential layering　分异层理
differential leveling　水平测量；水准测量；高程差测量
differential liberation　差异分离
differential loading　差异载荷
differential log　微差测井曲线
differential manometer　差示压力表
differential material balance　微分物质平衡
differential measurement　微差测量
differential melting　差异熔融
differential method　微分法；差动法
differential method mapping　微分法测图
differential micromanometer　差动式微压计
differential migration　差异运移
differential mode voltage　差模电压
differential movement　不均匀运动；差异运动
differential of high order　高阶微分
differential of pressure analysis　压力导数分析
differential of pressure curve　压力导数曲线
differential of terrigenous deposit　陆源沉积分析
differential operator　微分算子
differential oxygen availability　差异氧化效应
differential parameter　微分参数
differential permeability　微分导磁率；差异渗透率
differential phase　微分相位
differential phase change　差示相更换
differential photo　微分法测图
differential photocell　差分光电池
differential polynomial　微分多项式
differential porosity　随地层走向而变化的孔隙度
differential positioning　差异定位
differential pressure　压力降；压差
differential pressure flowmeter　差压式流量计
differential pressure gauge　差动气压计；微分气压计
differential pressure recorder　差压记录仪
differential pressure type flowmeter　压差式流量计
differential pulley　差动滑车
differential pump　差动泵
differential push system　差压推移系统
differential quotient　微分系数；微商
differential radiometer　微分辐射计
differential rectification　微分纠正
differential refractometer　差示折光计
differential resistance　微分电阻；内阻
differential resistivity　电阻率差
differential rock bolt extensometer　差动式岩石锚杆引伸仪
differential sagging　差异沉陷作用
differential scanning calorimeter　差示扫描量热计
differential screen analysis　差示筛析
differential screen plate　差动过滤板，差动筛板
differential screw　差动螺旋
differential screw jack　差动螺旋起重器
differential sedimentation　差异沉积
differentiating sensitivity　差示灵敏度
differential separation　微分分离
differential settlement　不均匀沉降；差异沉降
differential shading　差异屏蔽
differential shaft　差动轴
differential shrinkage　不均匀收缩
differential shrinking　差示收缩；差异收缩
differential shut of water　分层止水
differential sill　分异岩床
differential similarity　微分相似性
differential slip　差异滑动
differential slope　差异倾斜
differential smoothing device　微分平滑装置
differential soil movement　不均匀土壤运动；差异性土体运动
differential statics　微分静校正量
differential strain curve　微差应变曲线
differential strain curve analysis　微差应变曲线分析
differential stress　应力差
differential stress field　应力差场
differential subsidence　差异性沉陷；差异性沉降
differential temperature log　微差井温测井
differential thermal analysis　差热分析
differential thermal curve　差热曲线
differential thermometer　差示温度计
differential time　差值时间
differential type　差动式
differential uplift　差异隆起；差异上升
differential voltage　差动电压
differential weathering　选择风化；差别风化；差异风化
differential weathering correction　微分风化层校正
differential-motion table　差动摇床
differential-pressure flow-meter　压差流量表；压差流量计
differentials　差价
differential-screw jack　差动千斤顶
differentiate　区分；分异；分异产物；分异岩；求微分
differentiated dike　分异岩脉
differentiated intrusion　分异侵入作用
differentiated pulse　微分脉冲
differentiated rock　分异岩
differentiating effect　辨别效应
differentiating operator　微分算子
differentiation　分异；沉积分异；分选；优先浮选；求导；微分
differentiation index　分异指数
difficult country　困难地区；起伏很大的地区
difficult foundation　难处理地基
difficult grain　难粒
difficult ground　问题地区；特殊土地区；施工困难地区
difficult ground condition　不良地质条件
difficult particle of screening　难筛粒
difficult waste　难于处理的废物
difficult water exchange zone　水交替困难带
difficult weather　恶劣天气
difficulty curve　分选密变±0.1曲线
diffluence step　分流坎

diffluent river　出渗河流；分流河
diffluent stream　出渗河流；分流河
diffracted intensity　衍射强度
diffracted multiple　绕射多次反射
diffracted P wave　绕射P波
diffracted ray　衍射线
diffracted reflection　绕射反射
diffracted wave　衍射波；绕射波
diffraction chart　绕射图
diffraction coefficient　绕射系数
diffraction contrast　衍射衬度
diffraction corona　衍射晕
diffraction current　回折流
diffraction curve　绕射曲线；最大凸率曲线
diffraction diagram　衍射图
diffraction ellipses　绕射椭圆
diffraction figure　绕射图
diffraction formula　衍射公式
diffraction fringe　衍射条纹
diffraction grating　衍射光栅
diffraction hyperbola　绕射双曲线
diffraction image　衍射图像
diffraction index　衍射指标
diffraction overlay　绕射重叠；绕射曲线透明图版
diffraction pattern　绕射图
diffraction peak　衍射峰
diffraction propagation　绕射传播
diffraction region　绕射区
diffraction ring　衍射环
diffraction scattering　衍射散射
diffraction spectroscopy　衍射光谱法
diffraction spectrum　衍射光谱
diffraction stack migration　绕射叠加偏移
diffraction stacking　绕射叠加
diffraction tail　绕射尾部，绕射末端
diffraction theory　绕射理论
diffraction tomography　绕射层析成像
diffraction traveltime curve　绕射旅行时曲线
diffraction wave　绕射波
diffraction zone　绕射区
diffractogram　衍射图
diffractometer　衍射仪
diffractometry　衍射测定法
diffusance　扩散性
diffusant　扩散剂
diffusate　渗出液
diffuse　渗出；扩散；漫射
diffuse anomaly　弥散异常
diffuse bedding　扩散层理
diffuse constant　扩散常数
diffuse diffraction　漫散衍射
diffuse discharge　弥散式排放
diffuse dispersion pattern　弥散分散模式
diffuse distance　扩散距离
diffuse flow　扩散流
diffuse layer　扩散层
diffuse light　漫射光
diffuse maximum　漫射峰
diffuse migmatite　扩散生成混合岩
diffuse peak　扩散峰值

diffuse penetration texture　扩散渗透结构
diffuse percolation flow　扩散渗流
diffuse plate boundaries　弥漫性板块边界
diffuse pollution　扩散污染
diffuse radiation　漫辐射
diffuse reflectance　漫反射率
diffuse reflectance spectra　漫反射谱
diffuse reflecting power　漫反射能力
diffuse reflection　扩散反射
diffuse reflector　漫反射体
diffuse relaxation time　扩散弛豫时间
diffuse scattering　漫散射
diffuse seepage　扩散渗流
diffuse series　漫射系
diffuse sound field　扩散声场
diffuse source　扩散源
diffuse spectrum　漫光谱
diffuse spot　漫斑
diffuse transmission　漫透射
diffuse transmission density　漫透射密度
diffuse-controlled growth　扩散控制生长
diffused air　扩散空气
diffused air aeration　扩散掺气
diffused area　传播地区
diffused illumination　漫射照明
diffused light　漫射光
diffused ray　漫射线
diffused resistor　扩散电阻；集成电阻
diffusedness　扩散性
diffusely reflecting surface　扩散反射面
diffuseness　扩散；漫射
diffuser area ratio　扩散器面积比
diffuser chamber　扩散室
diffuse-reflection factor　漫反射系数
diffusibility　扩散性；扩散率
diffusibleness　扩散度
diffusing effect　扩散效应
diffusing phenomena　扩散现象
diffusing well　渗井
diffusion　扩散；散射；漫射；渗滤
diffusion analogy　扩散模拟
diffusion anomaly　扩散异常
diffusion aureole　扩散环；扩散接触变质带
diffusion coefficient　扩散系数
diffusion constant　扩散常数
diffusion control　扩散控制
diffusion couple　扩散偶合
diffusion creep　扩散蠕变
diffusion dialysis　扩散渗析
diffusion differentiation　扩散分化
diffusion effect　扩散效应
diffusion equation　扩散方程
diffusion equilibrium　扩散平衡
diffusion film coefficient　扩散膜系数
diffusion flow flux　扩散通量
diffusion flux　扩散通量
diffusion gradient　扩散梯度
diffusion halo　扩散晕
diffusion heat　扩散热
diffusion impedance　扩散阻抗

diffusion in liquids	在液体中扩散	diffusivity coefficient	扩散系数
diffusion index	扩散指数	diffusivity equation	扩散方程；传导方程
diffusion law	扩散律	diffusivity of aquifer	含水层扩散率
diffusion layer	扩散层	diffusivity of porous media	多孔介质扩散性
diffusion length	扩散长度	diffusivity of unsaturated flow	非饱和水流扩散率，非饱和水流扩散性
diffusion light	扩散光		
diffusion loss	扩散损失	diffusivity potential	扩散势
diffusion matrix	扩散基质	dig	挖掘
diffusion mean-free path	扩散平均自由程	dig through	挖穿；挖通
diffusion mechanism	扩散机理	digenite	蓝辉铜矿，方辉铜矿
diffusion metasomatism	扩散交代作用	digest	文摘；摘要
diffusion method	扩散法	digestion period	老化周期
diffusion model	扩散模拟系统	digestion time	老化时间
diffusion of air	空气扩散	digger	挖掘机
diffusion of contaminant	污染物扩散	digger plough	犁式挖沟机
diffusion of interstitials	间隙扩散	digging ability	挖掘能力
diffusion of vacancy	空位扩散	digging angle	切入角；挖掘角
diffusion of water vapour	水汽的扩散	digging cycle	挖掘循环
diffusion path	扩散途径；扩散路径	digging depth	挖掘深度，采掘深度
diffusion porosity	扩散气孔	digging efficiency	挖掘效率
diffusion potential	扩散电位	digging force	挖掘力
diffusion pressure	扩散压力	digging height	铲挖高度；阶段高度
diffusion process	扩散法；扩散过程	digging out	掘出
diffusion pump	扩散泵	digging program	开采计划，采掘计划
diffusion range of recharge water	补给水扩散范围	digging reach	挖掘半径
		digging resistance	挖掘阻力
diffusion rate	扩散率	digging speed	挖掘速度
diffusion ratio	扩散比	digimigration	数字偏移
diffusion relaxation	扩散弛豫	digital acquisition	数字采集，数据采集
diffusion resistance	扩散阻力	digital analysis	数字分析
diffusion ring	扩散环	digital approximation	数值近似，数值逼近
diffusion theory	扩容说；扩散说	digital cartography	数字化制图，数字地图学
diffusion time	扩散时间	digital computation	数字计算
diffusion transfer	扩散转印；扩散传递	digital data	数字资料，数字数据，数据
diffusion transformation	扩散转换	digital display	数字显示器
diffusion treatment	扩散处理	digital earth	数字地球
diffusion velocity	扩散速度	digital echo sounder	数码回声测深仪
diffusion ventilation	扩散通风	digital electronic display	数字式电子显示
diffusion washing	扩散洗涤	digital elevation model	数字高程模型
diffusion water	浸提用水	digital field system	野外数字采集系统
diffusion well	扩散井；注入井	digital file	数字化文件
diffusion zone	扩散带	digital filling	数字填充
diffusional creep	扩散蠕变	digital form	数字形式
diffusional effect	扩散效应	digital geophone	数字检波器
diffusional resistance	扩散阻力	digital graphic processing	数字图形处理
diffusional separation	扩散分离	digital image	数字图像
diffusional stream	扩散流	digital image analysis	数字图像分析
diffusion-convection equations	扩散-对流方程组	digital image processing	数字图像处理
diffusion-ventilated	扩散通风的	digital information display	数字信息显示
diffusive deposition	扩散沉降	digital integration	数字积分
diffusive equilibrium	扩散平衡	digital log	数字测井
diffusive flux	扩散量	digital map	数字地图
diffusive force	扩散力	digital mapping	数码绘图
diffusive heterogeneous system	非均匀扩散系统	digital mine	数字化矿山
diffusive layer	扩散层	digital model	数字模型
diffusive mass transfer	扩散质量迁移	digital photogrammetry	数字摄影测量
diffusive metasomatism	扩散交代作用	digital photography	数字摄影
diffusive radiation	散射辐射	digital plot	数字绘图
diffusive sampling	扩散取样法	digital processing	数字处理
diffusive velocity	扩散速度	digital rectification	数字纠正
diffusivity	扩散能力；扩散率；扩散系数；扩散性		

digital representation	数字显示
digital sampling	数位采样
digital scale	数字比例尺，数字标尺
digital scanning	数字扫描
digital seismic amplifier	数字地震放大器
digital seismic recording	数字地震记录
digital seismic survey	数字地震勘探
digital seismics	数字地震
digital seismogram	数字地震记录
digital seismograph	数字地震仪
digital seismology	数字地震学
digital sensor	数字传感器
digital simulation	数字模拟，数字仿真
digital sonar	数字声呐
digital surface model	数字表面模型
digital surveying	数字测量
digital telemetry	数字遥测
digital terrain analysis	数字地形分析
digital terrain data	数字地形资料
digital terrain model	数字地面模型
digital topographic map	数字地形图
digitalization table	数字化平台
digitate delta	桨叶状三角洲；鸟足状三角洲
digitation	指状突起
digitization effect	数字化效应
digitized image processing	数字化图像处理
digitized mapping	数字化测图
digitizing map	数字化的图幅
digonal bar	斜列砂坝
dihydrite	翠绿磷铜矿
dikaka	固定沙丘
dike	岩脉，岩墙；堤
dike breach	决堤
dike failure	决堤
dike footing	堤防底脚
dike rock	脉岩，脉石
dike set	岩脉组
dike spreading	岩墙扩张
dike swarm	岩墙群；岩脉群
diking	筑堤；围堤
Dikou Formation	迪口组
diktyogenesis	小型垂直运动
diktyonite	网状混合岩
dilaceration	撕开，撕裂
dilapidated building	危房，破旧建筑物
dilapidation	崩塌
dilatability	膨胀性
dilatable soil	膨胀土
dilatancy	剪胀，扩容，膨胀
dilatancy and shear strength	剪胀与抗剪强度
dilatancy angle	剪胀角
dilatancy effect	剪胀效应
dilatancy equation	剪胀方程
dilatancy fissure	膨胀裂缝
dilatancy hardening	膨胀硬化
dilatancy of rock	岩石扩容
dilatancy test	扩容试验，膨胀试验
dilatancy-diffusion model	膨胀-扩散模式
dilatant	膨胀物质；膨胀的
dilatant fluid	胀流性流体，膨胀性流体
dilatant rheology	胀流型流变学
dilatation	膨胀，剪胀
dilatation fissure	膨胀裂缝
dilatation hardening	膨胀硬化
dilatation joint	膨胀缝；伸缩缝
dilatation modulus	膨胀模量
dilatational disturbance	膨胀扰动
dilatational strain	膨胀变形，膨胀应变
dilatational wave	膨胀波；疏密波
dilated bed	松散层
dilated interface	膨胀界面
dilation angle	剪胀角
dilation pressure	膨胀压力
dilation rate	膨胀率，剪胀率
dilative material	剪胀材料
dilative response	剪胀反应，膨胀反应
dilative soil	剪胀性土
dilatometer	膨胀仪，膨胀计
dilatometer method	膨胀计法
dilatometry	膨胀测定法，膨胀法
dilutability	稀释度
dilute concentration	稀释浓度
diluting effect	稀释作用；稀释效应
dilution	稀释，冲淡；稀释度
dilution effect	稀释效应，稀释作用
dilution factor	稀释系数，稀释因子
dilution limit	稀释极限
dilution metering	稀释测定
dilution of ore	矿石贫化
dilution profile	稀释剖面
dilution rate	稀释率，稀释比例
dilution shearing	剪切稀释
diluvial	洪积的；洪积层
diluvial clay	洪积黏土
diluvial deposit	洪积层，洪积物
diluvial epoch	洪积世
diluvial fan	洪积扇
diluvial formation	洪积层
diluvial hypothesis	洪积说
diluvial layer	洪积层
diluvial placer	洪积砂矿
diluvial soil	洪积土
diluvial theory	洪积论，洪积说
diluvion	洪积层
diluvium	洪积层
dilvar	镍铁合金
dim spot	暗点
dimension analysis	维度分析
dimension drawing	比例绘图
dimension survey	尺寸测量
dimensional analysis	量纲分析
dimensional constant	量纲常数
dimensional ratio	量纲比
dimensional split	维数分解
dimensional transformation	量纲变换
dimensionless factor	无量纲因数，无量纲因子
dimensionless number	无量纲数
dimensionless parameter	无量纲参数；无维参数；无因次参数
dimensionless pore volume	无因次孔隙体积

dimensionless setting velocity	无因次沉降速度	dip pole	磁极，磁倾极
dimensionless solution	无因次解	dip resolution	倾角分辨率
dimensionless thickness	无因次厚度	dip reversal	反拖曳；反牵引，逆牵引
diminishing pressure	递减压力	dip section	倾向剖面
dimorphite	硫砷矿	dip separation	倾向离距，倾向断距，倾向断离
dimpled current mark	交错波痕	dip separation fault	倾离断层，倾向隔距断层
dingy yellow horizon	暗黄色土层	dip shift	倾向移距
Dingyuan Formation	定远组	dip slip	倾向滑距；倾向滑动
dint	巷道卧底处；压痕	dip slip fault	倾向滑断层
dintheader	巷道掘进机	dip slope	反坡；倾向坡
dint-heading	巷道卧底；巷道掘进	dip test	倾角测定；钻孔偏斜测定
dinting depth	下切深度	dip valley	倾向谷
diopside	透辉石	dip vector	倾角矢量
diopsidite	透辉石岩	diploite	钙长石
diopsidization	透辉石化	dipmeter	倾角仪，测斜仪
dioptase	透视石；绿铜矿	dipmeter log	倾角测井
diorama	透视图	dip-pattern	地层倾角模式；地层倾角图形
dioride	角闪闪长岩	dipping anisotropy	倾斜各向异性
diorite	闪长岩	dipping aquifer	倾斜含水层
diorite porphyrite	闪长斑岩	dipping boundary	倾斜边界
diorite-laterite	闪长岩红土	dipping compass	矿用罗盘；倾角仪
dioritic	闪长岩状的；闪长岩的	dipping confined aquifer	倾斜承压含水层
dioritic aplite	闪长细晶岩	dipping fault	倾向断层
dioritic porphyrite	闪长玢岩	dipping fold	倾向褶皱
dioritite	闪长细晶岩	dipping formation	倾斜地层
dioritoid	似闪长岩；闪长岩类	dipping fracture	倾斜裂缝
dioritophyrite	闪长细晶斑岩	dipping layer	倾斜层
dip	倾向	dipping reservoir	倾斜储层
dip angle	倾角	dipping slip	倾向滑动；倾向滑距
dip angle of drilling hole	钻孔倾角	dipping strata	倾斜地层
dip at high angle	急倾斜，陡倾角	dipping working	沿倾斜下向开采
dip at low angle	缓倾斜，缓倾角	dip-slip displacement	倾向滑动位移
dip azimuth	倾向方位，倾斜方位	dip-slip drag	倾向滑移拖曳
dip circle	测斜仪；磁倾仪	dip-slip earthquake	倾向滑动地震
dip cleat	劈理；倾斜割理	dip-slip fault	倾向滑动断层
dip compass	测斜仪；倾角仪；矿山罗盘仪	dip-slip offset	倾向滑动
dip compass survey	矿山罗盘测量	dip-slip reverse fault	倾滑逆断层
dip computation	倾角计算	dipslope	倾向坡
dip deviation	倾斜偏差	dipyre	针柱石
dip direction	倾向；倾向方位角	dipyrization	针柱石化作用
dip direction structure	倾向构造	direct acting	直接施加的；直接作用的
dip drift	斜井	direct against	针对
dip entry	下山	direct analysis	直接计算
dip face	倾斜面；沿煤层倾斜向下推进的工作面	direct approach	直接近似；直接法
dip fault	倾向断层；倾斜断层	direct axis	纵轴
dip fold	倾向褶皱	direct breakage	直接断裂
dip heading	向…倾斜，朝…方向倾斜；下山	direct circulation	正循环冲洗
dip isogon	等倾线，倾角等斜线	direct compression	单纯受压
dip joint	倾向节理	direct conductivity	直接电导率
dip logging	岩层产状测井；地层倾角测井	direct cost	直接成本
dip meter	钻孔岩层倾斜仪；地下水位测量仪	direct current apparent resistivity	直流视电阻率
dip migration	倾角偏移	direct current measurement	直流测量
dip mining system	倾斜开采法	direct current resistance	直流电阻
dip moveout	倾角时差	direct current resistivity method	直接电阻法
dip needle	磁倾计，磁倾针	direct digital logging system	直接数字测井系统
dip needle work	磁法勘探	direct digital transfer	直接数字交换
dip offset	倾向滑动	direct digitizing seismograph	直接读数地震仪
dip orientation	倾向定位	direct engineering cost	直接工程费
dip out	挖出	direct flotation	直接浮选
dip plane	倾斜面	direct flushing	正循环洗孔，正循环冲洗

direct foundation 扩展基础；直接基础
direct imaging 直接成像
direct indication 直接显示
direct initiation 直接起爆
direct injection 直接喷射
direct integration 直接积分
direct interpolation 直接插值法
direct interpretation 直接解释
direct iteration 直接迭代
direct liquefaction 直接液化
direct measurement 直接测定，直接测量
direct method 直接法
direct modeling 直接模拟
direct numerical integration 直接数值积分
direct pressure 直接压力；定向压力
direct priming 孔口起爆
direct projection 直接投影
direct ratio 正比
direct recording seismograph 直接记录地震仪
direct runoff 地表径流；定向径流
direct seismic interpretation 直接地震解释
direct shear 直剪
direct shear apparatus 直剪仪
direct shear equipment 直剪装置，直剪设备
direct shear method 直剪法
direct shear strength test 直剪强度试验
direct shear test 直剪试验
direct simple shear test 直接单剪试验
direct stiffness method 直接刚度法
direct strain measurement 直接应变测量
direct stratification 原生层理
direct stress 直接应力；正应力，法向应力
direct surface runoff 直接地表径流
direct tensile method 直接拉伸法
direct tensile test 直接拉伸试验
direct vertical uplift 正垂直隆起
direct water filling aquifer 直接充水含水层
direct wave 直达波
direct-current measurement 直流测量
direct-dip-reading chart 井斜直读图
directed drilling 定向钻进
directed graph 有向图；定向图
directed pressure 定向压力
directed-drilling technique 定向钻井技术
direction adjustment 方向调节
direction angle 方位角；定向角
direction change 方位变化
direction coefficient 方向系数
direction cosine 方向余弦
direction dip 倾斜
direction drilling 定向钻孔
direction error 方向误差
direction finder 定向仪
direction finding 方位测定，测向，定向
direction instrument 方向仪
direction jet grouting 定喷注浆
direction of current 流向
direction of deflection 井斜方向
direction of digging 采掘方向
direction of dip 倾斜方向

direction of drill 钻眼方向
direction of earthquake 地震方向
direction of excavation 采掘方向
direction of extraction 回采方向，开采方向
direction of face advance 工作面推进方向
direction of fissures 裂缝方向
direction of flow 流向
direction of magnetization 磁化方向
direction of mining 开采方向
direction of motion 运动方向
direction of rock fissures 岩石裂隙方向
direction of sliding 滑动方向
direction of slope 坡向
direction of strata 地层走向
direction of strike 走向
direction of tilt 倾斜方向
direction of travel 运行方向
direction peg 导向桩
direction rose 方向玫瑰图
direction sensor 方向传感器
direction shear cell 定向剪切盒
direction theodolite 方向经纬仪
direction well 方向井；定向井
directional 定向的
directional angle 方位角
directional bit 定向钻头
directional blasting 定向爆破
directional circulation drilling 正循环钻进
directional clinograph 井斜测定仪
directional control 定向控制
directional data 方向数据
directional depth migration 定向深度偏移
directional diagram 方向图
directional drainage 定向排水
directional drill tool 定向钻井工具
directional drilling 定向钻进
directional drilling crossing method 定向钻穿越法
directional drilling machine 定向钻机
directional drilling technology 定向钻井技术
directional effect 方向效应
directional explosion 定向爆破
directional fabric 定向组构
directional filtering 定向滤波
directional fracturing 定向压裂
directional frequency diagram 方向频率图
directional gyro indicator 回转方位仪；定向陀螺仪
directional hardness 定向硬度
directional hole 定向孔
directional hydraulic conductivity 定向水力传导系数
directional hydraulic fracturing 定向水力压裂
directional jet grouting 定喷灌浆
directional method 定向法
directional observation 方向观测
directional orientation tool 定向工具
directional permeability 定向渗透率；方向渗透系数
directional pressure 定向压力
directional property 方向特性
directional scanning 定向扫描
directional shear cell 定向剪切盒
directional shooting 定向爆炸

directional source	定向震源	disassimilation	异化作用
directional structure	定向构造	disassociation	离解
directional tendency	方位趋势	disaster	灾害
directional tool	随钻测斜仪；定向钻具	disaster activity strength	灾害活动强度
directional well	定向井	disaster area	灾区
directionality	方向性；定向性	disaster evaluation	灾情评估
directionality effect	定向效应	disaster geology	灾害地质学
directionally drilled crossing	定向钻孔穿越	disaster mitigation	减灾
direction-finding	方位测定，探向，测向	disaster monitoring	灾害监测
direction-finding and ranging	定向和测距	disaster prediction	灾害预测
direction-finding chart	测向图	disaster prevention	防灾
directionless pressure	无定向压力	disaster prevention and control	灾害防治
directive action	定向作用	disaster prevention and mitigation system	灾害减少和防御系统
directive effect	方向效应		
directive erosion	定向侵蚀	disaster reduction	减灾
directive structure	定向构造	disaster science	灾害科学
directive texture	定向结构	disastrous	灾难性的
directivity characteristics	方向特性	disc boot	圆盘开沟器
directivity curve	方向曲线	disc cutter incline angle	盘形滚刀倾角
directivity factor	方向因数，方向性系数	disc cutter offset angle	盘形滚刀偏角
director circle	准圆	disc friction	圆盘摩擦力，圆盘摩阻力
director sphere	准球面	disc shaped fault	盘状断层
director surface	准曲面；准表面	disc test	圆盘，圆盘对径压缩试验
directory name	目录名称	discard	废弃，丢弃；矸石废渣
direct-recording seismograph	直接记录地震仪	discard disposal	矸石处理
directrix	准线	discard removal capacity	排卸能力
direct-viewing	直观的	discernable fracture	显裂隙
dirigibility	灵活性	discernible	可识别的，可辨别的
dirt	夹石；矸石	discernible foliation trend	可辨别叶理走向
dirt band	冰川碎石层	discernible image	可辨别的图像
dirt bed	有机质层	discharge	流量
dirt cone of glacier	冰川碎堆	discharge amount	排放量
dirt content	灰分；含矸量	discharge area	泄水区，排水区，排泄区
dirt disposal	废石处理，矸石处理	discharge area of groundwater	地下水排泄区
dirt extraction	拣矸	discharge bay	泄水池
dirt gas	粗天然气	discharge bell	卸载漏斗
dirt inclusion	夹石；石质结核；杂质包体	discharge canal	排沙渠；泄水渠
dirt pack	废石充填	discharge capacity	排水能力
dirt pile	矸石堆；废石堆	discharge channel	泄水渠
dirt room	矸石场，矸石硐室	discharge character	流量特性
dirt slip	黏土脉	discharge coefficient	排泄系数
dirty arkose	含杂质长石砂岩	discharge cross section	过水断面
dirty bench	脏煤带	discharge culvert	排水涵洞
dirty channel	泥沙河槽	discharge curve	流量曲线；泄水曲线
dirty coal	脏煤；高灰煤	discharge diagram	流量图
dirty formation	含泥质地层	discharge ditch	排水沟
dirty sand	淤积砂	discharge duration curve	流量历时曲线
dirty sandstone	泥质砂岩	discharge efficiency	排泄率
dirty seam	劣质煤层；劣质矿层	discharge end	卸载端，排料端，排出端
disabling injury	残障性伤害	discharge equipment	卸料装置，卸载装置，排出装置
disaccord	不协调		
disadjust	失调	discharge head	压头；扬程；供水水头
disadvantage factor	不利系数	discharge height	卸载高度
disagglomeration	解结；瓦解	discharge hydrograph	流量曲线；流量过程曲线
disaggregation	分散作用；崩解作用	discharge intensity	流量强度
disalignment	偏离中心线；轴线不重合	discharge into foundation pit	基坑涌水
disappearance of outcrop	露头缺失	discharge launder	卸载槽；排泄槽
disappearing stream	潜流，伏流	discharge level	排泄水位
disarrangement	无序	discharge liquid	废液；排出液
disassembled schematic	分解简图	discharge mass curve	流量累积曲线
		discharge observation	流量观测

discharge of drilling water	钻孔冲洗水排放
discharge of ground water	地下水排泄
discharge of water	排水量
discharge opening	泄水孔，排水孔
discharge orifice	排水孔，泄流孔
discharge outlet	卸料口；排气口；放水口
discharge path	排泄途径
discharge point	排放点
discharge port	排料口，排出口，放出口
discharge pressure	排水压力；排泄压力
discharge pressure head	排出压头
discharge rate	涌水量
discharge rating curve	流量率定曲线
discharge ratio	排泄率
discharge record	流量记录
discharge section	过水断面；过水面积
discharge section area	过水断面面积
discharge time	卸载时间
discharge velocity	泄流速度
discharge water	排出水
discharge zone	排泄带
discharging by wash	水冲卸载；水冲排料
discharging condition	排放条件
discharging well	排泄井
disciform	椭圆形的
discoaster	盘星石
discolith	盘状石
discolithus	盘颗石
discolouration	变色；褪色
discoloured clay	变色黏土
discone	盘锥形
disconformable	假整合的
disconformable intrusion	假整合侵入
disconformable plane	假整合面
disconformity	假整合
disconnect	断开
disconnect pore	不连通孔隙，孤立孔隙
disconnected	不连通的
disconnected region	不连通区域；非连通域
discontent	不平
discontinued stress	不连续应力
discontinuities	结构面
discontinuities aperture	不连续面张开度
discontinuities bedding planes	不连续层面
discontinuities dip direction	不连续面倾向
discontinuities infilling	不连续面的充填物
discontinuities rose diagram	不连续面的玫瑰图
discontinuities roughness	不连续面的粗糙度
discontinuities spacing	不连续面的间距
discontinuities stiffness	不连续面的刚度
discontinuities strike	不连续面的走向
discontinuities type	不连续面的类型
discontinuities unevenness	不连续面的不平整度
discontinuity	不连续面
discontinuity aperture	不连续面张开度；结构面张开度
discontinuity condition	不连续条件
discontinuity displacement	不连续位移
discontinuity filling material	不连续面充填物；结构面充填物
discontinuity focusing	不连续震源
discontinuity glide	间断滑移
discontinuity in rock mass	岩体不连续性
discontinuity lattice	不连续面晶格，不连续面点阵
discontinuity layer	不连续层
discontinuity of material	材料的不均匀性
discontinuity of rock	岩石非连续性
discontinuity orientation probability	不连续面的方位概率
discontinuity pattern	不连续面形式；结构面形式
discontinuity roughness	不连续面粗糙度；结构面粗糙度
discontinuity roughness classification	不连续面粗糙度分类；结构面粗糙度分类
discontinuity set	不连续面组；结构面组
discontinuity spacing	不连续面间距；结构面间距
discontinuity stress	间断性应力
discontinuity structural plane	不连续构造面
discontinuity surface	不连续面
discontinuity trace	结构面迹线
discontinuity type	不连续面类型；结构面类型
discontinuity waviness	结构面起伏度
discontinuous	不连续的；间断的
discontinuous cleavage	不连续劈理
discontinuous construction	不连续结构
discontinuous deformation	非连续变形
discontinuous deformation analysis	非连续变形分析
discontinuous dipmeter	不连续记录地层倾角仪
discontinuous displacement	不连续位移
discontinuous distribution	不连续分布
discontinuous excavating technique	非连续开采工艺，间断开采工艺；非连续开挖工艺
discontinuous factor	不连续因子
discontinuous folding	不连续褶曲
Discontinuous Galerkin Method	不连续伽辽金法；间断伽辽金法
discontinuous glide	不连续滑移；间断滑移
discontinuous gradation	不连续级配
discontinuous interstice	不连续间隙
discontinuous lineament	不连续线性构造；不连续区域构造线
discontinuous lithostratigraphic unit	不连续岩石地层单位
discontinuous load	间歇负荷，不连续载荷
discontinuous mining technology	非连续开采工艺，间断开采工艺
discontinuous motion	不连续运动
discontinuous oil phase	不连续油相
discontinuous permafrost	不连续永冻层，不连续多年冻土
discontinuous phase	非连续相
discontinuous plane	不连续面
discontinuous porosity	不连续孔隙
discontinuous pressure	不连续压力
discontinuous rock mass	不连续岩体；岩体非连续性
discontinuous running	间歇运转，不连续运转
discontinuous slip	不连续滑动
discontinuous slope	非连续边坡
discontinuous structure	不连续结构
discontinuous subsidence	不连续下沉
discontinuous surface	不连续面
discontinuous terrane	不连续地体；不连续岩带，不连续岩区
discontinuum mechanics	非连续介质力学；非连续体力学
discordance	不整合

discordance of stratification	地层不整合	disengage	松开，释放
discordant	不整合的	disengage nitrogen	脱氮
discordant basin	不整合盆地	disequilibrium	不平衡
discordant batholith	不整合岩基	disequilibrium permafrost	不平衡永冻层
discordant bedding	不整合层理；不平行层理	disfigurement	外形损伤
discordant contact	不整合接触	disfigurement of surface	地表起伏；路面损坏
discordant fold	不整合褶曲；不协调褶曲	dish channel	碟渠
discordant intrusion	不谐和侵入	dish drainage channel	碟形排水渠
discordant joint	不调和节理	dish structure	碟状构造；盘状构造
discordant morphology	不协调地貌	disharmonic faulting	不一致断裂作用
discordogenic fault	分界断层	disharmonic feature	不和谐地形
discovered reserves	探明储量	disharmonic fold	不和谐褶皱
discrasite	锑银矿	disharmonic folding	不谐和褶皱作用；不谐和褶皱作用
discrepance	偏差	dished sluice gate	圆盘状冲沙闸门，圆盘状泄水闸门
discrepancy	差异，偏差；矛盾；不符合	disintegrate	崩解，分裂
discrepancy in closing	闭合差	disintegrated	碎裂的
discrepancy in elevation	高差	disintegrated condition of rock	岩石崩解状态
discrete	松疏的，松动的；分立的，分离的，不连续的，离散的	disintegrated granite	崩解花岗岩
		disintegrated slag	碎渣
discrete analog	离散模拟	disintegration	崩解作用；剥蚀；裂变
discrete analysis	离散分析	disintegration constant	衰变常数
discrete approximation	离散逼近	disintegration of core	岩心破碎
discrete block	松散块体	disintegration time	崩解时间
discrete data	离散数据	disintegrative action	崩解作用
discrete distribution	离散分布，不连续分布	disinterment	掘出
discrete element method	离散元法	disjunctive cleavage	断离劈理；脱节劈理
discrete elements	离散单元	disjunctive fold	碎裂褶皱，断裂性褶皱；间断褶皱
discrete fracture model	分布裂纹模型	disk foundation pile	盘底桩；扩底桩
discrete fracture network	离散裂隙网络	disk pile	盘底桩，盘头桩
discrete loading	离散荷载	disking	成饼作用
discrete material	松散材料	disk-like rock core	饼状岩芯
discrete media	松散介质	disk-punching test	圆盘冲孔试验；圆盘冲剪试验
discrete model	离散模型	dislocated deposit	断错矿床
discrete parameter	离散参数	dislocated down-warping	坳断
discrete particle	离散颗粒	dislocated seam	错位煤层；错位矿层
discrete phase	分散相；不连续相	dislocated upwarping	隆断
discrete random process	离散随机过程	dislocated upwarping region	隆断区
discrete random variable	离散型随机变量	dislocation	位错
discrete sampling	不连续取样	dislocation basin	断层盆地
discrete series	不连续序列	dislocation boundary	错位边界
discrete state	离散状态	dislocation breccia	断层角砾岩
discrete system	离散系统	dislocation climb	位错上升
discrete type	离散类型	dislocation creep	位错蠕变
discrete variable	离散变量	dislocation creep strain	断错蠕动应变
discrete-continuous medium coupling model	离散连续介质耦合模型	dislocation cycle of active fault	活断层错动周期
		dislocation density	位错密度
discreteness	不连续性	dislocation discordance	断层不整合
discretization	离散化	dislocation earthquake	断层地震
discretization error	离散误差	dislocation fault	错位断层
discriminability	鉴别力；分辨力	dislocation glide	位错滑移，断错滑移
discriminant	判别式	dislocation lake	断层湖
discriminant analysis	判别分析	dislocation layer	位错层
discriminant critical value	判别临界值	dislocation line	错位线；断层线
discriminant function	判别函数	dislocation metamorphism	断错变质；碎裂变质
discriminant function criteria	判别函数准则	dislocation model	位错模型
discriminant relation	判别关系式	dislocation motion	位错运动
discriminant vector	判别向量	dislocation mountain	断层山
discriminating value	判别值	dislocation movement	位错运动
discrimination	判别，鉴别，辨别	dislocation node	位错节点
discriminatory	能鉴别的；能选择的	dislocation of river	河流错位

dislocation plateau	断层高原	dispersion slaking	分散崩解
dislocation reaction	位错反应	dispersion structure	分散构造
dislocation structure	位错构造，错位构造	dispersion system	分散系统，分散系
dislocation surface	位错面	dispersion test	分散试验
dislocation theory	位错理论	dispersion zone	扩散带，分散带
dislocation valley	断层谷，错断谷	dispersity	分散度；分散性
dislocation vein	断层脉	dispersive action	分散作用
dislocation wall	位错壁	dispersive capacity	分散能力
dislodged sludge	沉积泥渣	dispersive clay soil	分散性黏土
dislodged strata	已松动岩层；已松动地层	dispersive effect	频散效应，分散效应
dismembered geosyncline	解体地槽	dispersive medium	频散介质
dismembering	解体作用	dispersive region	弥散区
disomatic	包裹晶；捕虏晶	dispersive soil	分散性土
disordered layer	无序层	dispersivity	分散性；反絮凝
disordered state	无序状态	dispersoid	分散胶体，分散体
disordered structure	无序结构	displace mud	驱替泥浆
dispellet	扰动球粒	displaced beds	位移岩层
dispellet limestone	扰动球粒灰岩	displaced fold	异地褶皱
dispergated	分散状的	displaced mass	移位岩体；推覆体
dispersancy	分散力；分散性	displaced material	位移物质
dispersant	分散剂	displaced outcrop	移位露头
disperse soil	松散土	displacement	位移
disperse state	分散状态	displacement amplitude	位移振幅
disperse system	分散体系	displacement angle	位移角
dispersed	分散的	displacement back-deduction analysis	位移反分析
dispersed clay	分散的黏土	displacement bound	极限位移；位移范围
dispersed condition	分散状态	displacement boundary conditions	位移边界条件
dispersed creep	分散蠕变	displacement calibration	位移校正
dispersed medium	分散介质	displacement component	位移分量
dispersed phase	分散相	displacement conservation	位移守恒
dispersed shale	分散页岩	displacement constraint	位移约束
dispersed shale model	分散页岩模型	displacement continuity	位移连续性
dispersed structure	分散构造，分散结构	displacement curve	位移曲线
dispersed substance	分散质	displacement detector	位移检波器
dispersed texture	分散结构	displacement discontinuity	位移不连续
dispersed unconformity	分散不整合	displacement discontinuity element	位移非连续单元
dispersibility	分散能力；分散性	displacement discontinuity function	位移非连续函数
dispersing agent	分散剂	displacement discontinuity model	位移不连续模型
dispersing coefficient	分散系数	displacement dislocation	位移断错
dispersing medium	分散介质	displacement disturbance	位移扰动
dispersion	分散作用	displacement ductility	位移延性
dispersion coefficient	弥散系数；扩散系数	displacement effect	位移效应
dispersion constant	分散常数	displacement fault	平移断层；移位断层
dispersion curve	频散曲线	displacement field	位移场
dispersion degree	分散度，离散程度	displacement force	变位力
dispersion index	分散指数	displacement formulation	位移公式
dispersion law	离散定律	displacement function	位移函数
dispersion layer	扩散层	displacement gauge	位移计
dispersion loss	弥散损耗	displacement geophone	位移检波器
dispersion mechanism	分散机理	displacement gradient	位移梯度
dispersion medium	分散介质；分散剂	displacement gradient field	位移梯度场
dispersion method	分散法	displacement grouting	排水灌浆
dispersion modulus	分散模数	displacement increment	位移增量
dispersion of additional stress	附加应力扩散	displacement intensity scale	位移烈度表
dispersion of groundwater artery	地下水扩散	displacement limit	位移极限
dispersion of seismic wave	地震波频散	displacement line	位移线
dispersion pattern	分散模式	displacement method	位移法，替置法
dispersion phase	分散相	displacement mode	位移方式
dispersion phenomenon	分散现象	displacement monitoring	位移监测
dispersion radius	扩散半径	displacement observation	位移观测

displacement of oil by water	水驱油
displacement of pile top	桩顶位移
displacement of rock masses	岩石移动
displacement of surrounding rock	围岩位移
displacement pattern	位移模式
displacement pile	排土桩
displacement plane	位移面
displacement rate	位移速率
displacement rate of active fault	活断层错动率
displacement seismograph	位移地震仪
displacement seismometer	位移地震计
displacement sensitivity	位移灵敏度
displacement sensor	位移传感器
displacement shear	位移切变
displacement stress	位移应力
displacement theory	大陆漂移说
displacement transducer	位移传感器
displacement vector	位移矢量，位移向量
displacement velocity	运移速度
displacing force	移动力
displacive concretion	推移性结核
displacive precipitation	推移沉淀作用
display drawing	展览图
display mode	显示方式
display sensitivity	指示灵敏度，显示灵敏度
disposal	处理；清除
disposal ditches	导水沟
disposal dump	垃圾倾卸站
disposal optimization	处置优化
disposal outlet	倾倒场所
disposal pit	排水坑
disposal point	卸载区
disposition of water leakage	涌水处理
disproportion	不均衡；不相称；不成比例
disproportionate withdrawal	不均衡开采
disquisition	专题论文；学术讲演
disregistry	错合度
disrupted	不连续，中断的；分裂的，断裂的
disrupted anomaly	脱节异常
disrupted bed	断裂岩层；断裂地层
disrupted drainage system	不规则水系
disrupted fold	断裂褶皱
disrupted horizon	破坏层；变位层
disrupted peneplain	切割准平原
disrupted plateau	切割高原
disrupted porosity	破裂孔隙
disrupted seam	断裂煤层；断裂矿层，断裂夹层
disrupted slide	解体滑坡
disrupted terrane	分裂地体
disruption plane	破碎面
disruptive	分裂性的；破裂性的
disruptive strength	破坏强度，穿击强度
disruptive test	击穿试验；耐压试验
disrupture	分裂
dissected basin	切割盆地
dissected delta	切割三角洲
dissected fan	切割扇状地
dissected low plain	切割低平原
dissected map	拼幅地图
dissected peneplain	切割准平原
dissected plain	切割平原
dissected plateau	切割高原
dissected topography	切割地形
dissected valley	切沟
dissected volcano	切割火山
dissection of seepage field	渗流场剖分
disseminated	浸染的；分散的，散布的
disseminated clay sand	浸染的黏土质砂
disseminated deposit	浸染矿床
disseminated replacement	浸染交代作用
disseminated structure	浸染状构造
dissemination deposit	浸染矿床
dissimilation	异化
dissipated energy	耗散能
dissipation	消散
dissipation factor	损耗因数
dissipation function	分散函数
dissipation of energy	能量损耗；能量散逸
dissipation of pore-water pressure	孔隙水压力消散
dissipation process	消散过程
dissipation structure	耗散结构
dissipation system	耗散系统
dissipation test	消散试验
dissipative dynamic system	耗散动力系统
dissipative force	耗散力
dissipative structure	耗散结构
dissipative structure theory	耗散结构理论
dissociation chemisorption	离解化学吸附作用
dissociation pressure	离解压力
dissolubility	可溶性，溶解性
dissolution	溶解
dissolution basin	溶蚀盆地
dissolution cycle	溶解旋回
dissolution feature	岩溶现象；溶蚀现象
dissolution index	溶解指数
dissolution porosity	溶蚀孔隙度
dissolution rate	溶解速率
dissolution sequence	溶解序列
dissolvable	可溶的
dissolve out	析出，溶出
dissolved constituent	溶解组分
dissolved deposit	溶解矿床
dissolved fissure	溶蚀缝
dissolved natural gas	溶解天然气
dissolved oxygen	溶解氧
dissolved pore	溶解孔隙，溶孔
dissolvent	溶剂
dissolving capacity	溶解能力
dissonance	不协调
dissymmetric	不对称的
dissymmetrical structure	不对称结构
distal bar	远端沙坝
distal contourite	远源等深积岩
distal strata	远端地层
distal turbidite	远端浊积岩，远源浊积岩
distance finder	测距计
distance gauge	测距规，测距仪
distance measurement	距离测量
distance measuring error	测距误差
distance meter	测距仪

distance of epicenter 震中距
distance of settlement influence 沉降影响深度
distance of thrust 逆掩断距
distance theodolite 测距经纬仪
distance weighting method 距离加权法
distance-drawdown curve 距离-降深曲线
distancer 测距仪
distant action instrument 遥测仪表
distant earthquake 远震
distant lateral migration 长距侧向油气远移
distant measurement 遥测
distant reading 遥测读数；远距离读数
distant shock 远震
distensibility 膨胀性
disterrite 葱绿脆云母
disthene 蓝晶石
distilled water 蒸馏水
distinct boundary 清晰的边界，明显边界
distinct cleavage 显明劈理
distinct element code 离散元程序
distinctive mineral 特征矿物
distomodus 异牙形石属
distorted area 变形区
distorted bedding 扭曲层理
distorted grid 畸变格网
distorting stress 扭转应力,扭应力
distortion effect 畸变效应,失真效应
distortion factor 失真系数
distortion isogram 等变形线
distortion settlement 扭曲沉降；变形沉降
distortion shift 畸变位移
distortional 变形的
distortional deformation 扭曲变形
distortional failure 畸变破坏
distortional strain energy 畸变应变能
distortional stress 变形应力
distributary bar 分流沙洲
distributary channel deposit 分流河道沉积
distributary fan 分流冲积扇
distributed fault 断层带
distributed force 分布力
distributed load 分布荷重,荷载分布
distributed-parameter system 分布参数系统
distributing bar 分布钢筋,配力筋
distributing ditch 配水沟
distributing reinforcement 分布钢筋
distribution 分布
distribution angle 分布角
distribution characteristics 分布特征
distribution coefficient 分布系数；分配系数
distribution curve 分布曲线
distribution graph 分配曲线
distribution law 分配定律,分布律
distribution load 分布荷载
distribution map 分布图
distribution method 分配法
distribution modulus 分配系数；分配模量
distribution of contact pressure 接触压力分布
distribution of earthquake 地震分布
distribution of error 误差分布

distribution of grain size 颗粒级配；粒径分布
distribution of pressure 压力分布
distribution of secondary stress 次应力分布
distribution of sizes 粒度分布
distribution of strain 应变分布
distribution of stress 应力分布
distribution parameter 分布参数
distribution plan 分布图
distribution reinforcement 分布钢筋
distribution steel 分布钢筋
distributive fault 断层带
distributive step faults 断裂带
district cross-cut 采区石门
district dip 采区下山
district rise 采区上山
district sublevel 区段
district sublevel entry 区段平巷
disturb 干扰
disturbance 扰动
disturbance degree 扰动度
disturbance factor 扰动因素
disturbance index 扰动指数
disturbance ratio 扰动比
disturbed area 受扰区；受震区
disturbed bedding 扰动层理
disturbed belt 错动带,扰动带
disturbed clay 扰动黏土
disturbed deposit 扰动矿床
disturbed flow condition 紊流状态
disturbed force 扰动力
disturbed fractured zone 受扰动的破裂区
disturbed profile 扰动剖面
disturbed sample 扰动样品
disturbed sand 扰动沙
disturbed soil 扰动土
disturbed state concept 扰动状态理念
disturbed strata 扰动地层,扰动岩层
disturbed stress condition 扰动应力条件,扰动应力状态
disturbed zone 扰动带,扰动区
disturbed-terrain failure 非原状山体滑坡
disturbing acceleration 干扰加速度
disturbing force 扰动力
disused tunnel 废置隧道
ditch 截水沟，明沟
ditch cut 沟渠开凿
ditch diversion 导流沟
ditch edge 沟缘
ditch erosion 沟蚀
ditch excavation 挖沟
ditch for falling rock 落石沟
ditch grade 沟底坡度
ditch lining 沟渠衬砌
ditetrahedron 双四面体
ditroite 方钠二霞正长岩
divergence 散度
divergence analysis 散度分析
divergence plate boundary 离散板块边界
divergent boundary 离散边界；分离板块边缘
divergent fault 离散断层
divergent plate 离散板块；背离型板块

divergent structure	辐射构造；发散结构
divergent unconformity	角度不整合
diverging fault	树枝状断层
diversion	分流；改道；转向
diversion cut	泄洪口；泄洪道
diversion dam	分水坝
diversion dike	导流堤
diversion ditch	分水沟
diversion excavation	导流开挖法
diversion tunnel	导水隧硐
diversity factor	差异因数，差异系数
diverter	截夺河；袭夺河
diviation adjustment	偏差调整
divide line	分水线
divided difference	均差
dividing crest	分水岭
dividing dial	分度盘
dividing range	分水岭
diving bell	潜水钟
diving current	潜射流
diving drill	潜孔钻机
diving plate	气压式盾构前后仓分离板
diving slab	俯冲板块
divining device	探测设备
division dike	导流堤；分水堤；隔流堤
division surface	分界面
divisional plane	断面，节理面
Djungarica	准噶尔介属
doak	黏土膜
dobschauite	辉砷镍矿
document of engineering geological investigation	工程地质勘察报告
dodecahedron	十二面体
dofemic	多铁镁质
dogleg graben	折曲地堑
dogleg normal fault	折曲正断层
dog-legged strike-slip fault	折线形走向滑动断层
dohyaline	多玻质
dolarenaceous	砂屑白云岩状
dolarenite	砂屑白云岩
dolerite	粗粒玄武岩，辉绿岩
doleritic texture	粒玄结构
dolerophanite	褐铜矾
dolina	落水洞
doline	漏斗；渗穴
doll	黄土结核；钙质结核
doloclast	白云岩屑
dololithite	碎屑白云岩
dolomite	白云石
dolomite carbonatite	白云碳酸盐岩
dolomite dyke	白云岩脉
dolomite limestone	白云灰岩
dolomite marble	白云大理岩
dolomite reservoir	白云岩储集层
dolomite-sandstone	白云石质砂岩
dolomitic	含白云石的
dolomitic facies	白云岩相
dolomitic quartzite	白云石质石英岩
dolomitic siltstone	白云石质粉砂岩
dolomitisation	白云岩化作用
dolomitite	白云岩
dolomitization	白云石化
dolorudite	砾屑白云岩
dolosiltite	粉砂屑白云岩
domain decomposition method	区域分解法
domain fabric	团粒组构
domain structure	磁畴状结构
domal flank	穹隆翼部
domal structure	穹隆构造，丘状构造
domal uplift	穹状隆起
dome	穹地；丘
dome fold	穹状褶皱，穹隆
dome foundation	穹隆式基础
dome mountain	穹隆山；穹形山
dome of natural equilibrium	自然平衡拱
dome roof	拱顶
dome shaped	圆顶的；穹隆形的
dome shaped fold	穹状褶皱
dome shaped volcano	穹状火山
dome structure	穹窿构造
domed cutterhead	球面刀盘
domed plate	圆顶状锚杆托板
domelike	穹形的；圆顶形的
domeykite	砷铜矿
dominant fault	主断层
dominant mechanism	主要机理
dominant period	卓越周期
dominant wave	优势波；主波
doming	成拱；隆起
doming mechanism	穹窿形成机制
domite	奥长粗面岩
domitic	多铁矿质
Dongjiagou Formation	董家沟组
Dongjiagou Shale	董家沟页岩
Dongting Beds	洞庭层
Dongting Formation	洞庭组
Dongwen Formation	东汶组
donut	环形地震
dopebook	油井记录；测井曲线；测井记录
Doppler effect	多普勒效应
doquaric	多石英质
dorag dolomite	混合白云石
dorgalite	多橄玄武岩
dormant fault	休眠断层
dormant spring	休眠泉，假死泉
dormant volcano	休眠火山
dormant window	老虎窗
dorsal view	背视图
dosemic	多斑晶状
dosimeter	量筒
dot and dash line	点划线
dot density	点密度
dotted picture	点状图像
double arch dam	双曲拱坝
double barrel	双层岩芯管；复式岩芯管
double box collar	双母接头钻铤
double bridge type penetrometer	双桥式触探仪
double cap casing head	双顶套管头
double core barrel	双层岩芯管
double cribbing	双框架式支护

double curved surface	双曲面
double diaphragm pressure gauge	双膜式压力盒
double disc cutter	双刃滚刀
double drainage	双向排水
double drawing	双区拉伸
double drum hoist	双滚筒绞车
double esker	重叠蛇形丘
double folding	双重褶皱作用
double gallery	双平巷
double hull	双壳体
double hydrometer method	双比重计法
double integral	二重积分
double interpolation	双内插法
double iron	工字钢；工字铁
double lining	二次衬砌；永久衬砌，内衬砌
double logarithm	双对数
double logarithmic chart	双对数坐标图
double moving average method	二次移动平均法
double oedometer method	双固结仪法
double perthite	复纹长石
double pitch	双螺距的；双向坡
double plunging anticline	双倾伏背斜
double plunging fold	双倾伏褶皱
double porosity	双重孔隙
double porosity reservoir	双重孔隙介质储层
double raft foundation	双层筏式基础
double reflection pattern	双反射图形
double refracting	双折射
double ridge	双重山脊
double root	二重根
double row curtain	双排帷幕
double screw pump	双螺旋泵
double shaft extension	双轴伸长
double shear test	双面剪切试验
double shield	双护盾
double spike method	双同位素指示剂法
double stope	双翼回采工作面
double thread	双头螺纹
double tube	双层管
double tube core barrel	复式岩芯管
double tube core barrel sampler	双重岩芯管取样器
double tube sampler	双层取样器
double valley	双向谷
double vein	双脉
double wall breakwater	双墙防波堤
double wedge analysis	双楔形体分析法
double-arc	双弧
double-bridge probe	双桥探头
double-chisel bit	双錾头钻头
double-claw grab	双瓣抓岩机
double-column-mounted percussing drill	双柱架式冲击钻
double-cone bit	双牙轮钻头，双锥形钻头
double-contact	双触头的；双触点的
double-drained layer	两面排水层
double-duty	两用的
double-entry mining method	双平巷开采法
double-faulted anticline	双重断错背斜
double-jointing pipe	双根钻杆
double-layer	双层的
double-layer structure	双层结构
double-layer theory	双电层理论
double-line rock cutting	双线石质路堑
double-lining method	双层衬砌法
double-packer grouting	双塞灌浆
double-porosity model	双重孔隙介质模型
double-porosity system	双重孔隙介质
double-pyramid cut	双锥形掏槽
double-specimen oedometer test	双样固结试验
double-T rail	工字梁
double-taper bit	双斜刃钻头
double-thrust bearing	双向推力轴承，双向止推轴承
double-tracer technique	双指示剂法
double-track haulage roadway	双轨运输巷道
double-track road	双轨平巷；双轨巷道
double-track tunnel	双线隧道
double-tracked incline	双轨斜井
double-tube core barrel	双筒取芯筒
double-tube method	双管法
double-tube rigid-type core barrel	双动双层岩芯管
double-V cut	双 V 形掏槽
double-wall drill pipe	双壁钻杆
double-wedge cut	双楔形掏槽
doubling method	双重法
doubly plunging anticline	双倾伏背斜
doubly plunging fold	双倾伏褶皱
Doucun Group	窦村群
dough deformation	柔性变形
doughnut structure	环形构造
doughnut-shaped orebody	环状矿体
douglasite	绿钾铁盐
Douposi Formation	陡坡寺组
dowel bar	传力杆
down angle	俯角，下视角
down digging	下挖
down dip advance	沿倾向推进；横向推进
down dip sliding	沿倾向下滑
down direction	下行方向
down fault	正断层
down grade	下坡
down grade excavation	下坡开挖
down hole seismometer	井下地震仪
down slip fault	下落断层
down slope	下坡
down throw fault	下落断层
downbuckling theory	地壳弯曲说
downcast column	陷落柱，无碳柱
downcast fault	下落断层
downcast shaft	进风井
downcast side	下降盘，下降翼
downcurrent	顺流
downcutting	下切侵蚀，向下侵蚀
down-cutting stream	下切河流
downdip	向下倾，下倾
downdip block	下倾断块
downdip direction	下倾方向
downdip extent	下倾范围
down-dip limit	下部可采极限
down-dip lineation	倾向线理
down-dip mining method	下山开采法
downdip shooting	下倾放炮

downdip trap	下倾圈闭
downdrag	负摩擦力；下拉荷载
downdraw	水位下降
downdropped block	下降盘
downfall	坍方；下落
downfall earthquake	陷落地震
downfaulted basin	断陷盆地
downfaulted block	下降断块
downfaulted rift valley	断陷裂谷
downfaulted trough	断层凹槽
downflexure	向下挠曲
downflow well	注水井
downfold	向斜；低洼地
downglide motion	下滑运动
downgoing	俯冲；下行的
downgoing plate	下沉板块；俯冲板块
downgoing signal	下行信号
downgoing slab	俯冲板块
downhill creep	移动滑坡
downhole	向下打眼；向下钻进
Downhole Coaxial Heat Exchange System DCHE	潜孔钻井同轴热交换系统
downhole completion equipment	井下完井装置
downhole drill	潜孔钻机
downhole erosion	井下冲蚀
downhole geophysical test	孔内地球物理探测
downhole gradient	井下梯度
downhole gravity meter	井下重力仪
downhole measurement	井底测量；孔内测量
downhole orientation	井下定向
downhole pressure gauge	井底压力计
downhole seismometer	井下地震计
downhole sensor	井下传感器
downhole television	井下电视
downland	丘陵地
downlap	底部不整合；下超
downpunching	沉陷
downsagging	向下挠曲
downslide	下滑
downslide surface	下滑面
downslip fault	下落断层；正断层
down-slope displacement	下滑位移
down-slope movement	顺坡运动
downstream	下游
downstream face	下游面；背水面
downstream injection	顺流喷射
downstream method	下游法
downstream section	下游段，下流段
downstream slope	背水坡，下游坡
downstrike	沿走向
downstructure location	沿构造下倾部位
down-the-hole bit	潜孔钻头
down-the-hole drill	潜孔钻机
downthrow	塌陷，陷落
downthrow block	下落地块
downthrow fault	下落断层，正断层
downthrow side	下降盘，下落翼
downthrown block	下落断块
downthrown graben	下落地堑
downthrust	下冲断层
downthrust slab	下冲板块
downtrend	下降趋势
downward	下降的；向下的；下坡的
downward bulge	山根带
downward enrichment	次生富集
downward force	下向力
downward mining	下行式开采
downward movement	向下运动
downward seepage	向下渗流
downward traveling wave	下行波
downward working	下行开采
downwarped basin	下挠盆地，坳陷盆地
downwash	坡面冲刷，冲刷
drab clay	暗色黏粒
draft line	吃水深度线
draft zone	牵伸区
drafting accuracy	绘图精度
drafting by shield	平巷盾构法掘进
drafting scale	绘图比例尺
drag	负摩擦力；拖曳
drag anchor	拉锚
drag and slippage zone	牵引滑动带
drag angle	滞后角
drag calculation	阻力计算
drag coefficient	牵引系数；阻力系数
drag fault	拖曳断层
drag flexure	拖曳挠褶
drag fold	拖曳褶皱
drag force	拖曳力
drag hysteresis	拖曳滞后
drag landslide	牵引式滑坡
drag mark	拖痕
drag movement	拖拉运动
drag reducer	减阻剂
drag striae	拖曳条痕
drag strike	拖曳走向
drag structure	拖曳构造
drag thrust	推覆大断层，拖曳逆掩断层
drag zone	滞后带
drag-and-slippage zone	牵引滑动带
drag-folding	拖曳褶皱
drain cavern	泄水洞
drain discharge	排水量
drain ditch	排水沟
drain gutter	排水沟，泄水沟
drain hole	排水孔
drain outlet	泄水口；排水口
drain pile	排水砂桩；砂井
drain pit	排水坑
drain pressure	排泄压力
drain trench	排水沟
drain tunnel	泄水洞，排水隧道
drainability of soil	土壤排水能力
drainable area	泄水面积，排水面积
drainable porosity	排水孔隙率
drainage	排水法；排水；排水设备
drainage area	流域
drainage basin	流域盆地；流域
drainage basin morphology	流域形态学
drainage basin of groundwater	地下水流域

drainage blanket	排水垫层	drainage pipe line	排水管线
drainage boring	排水钻孔	drainage porosity	排水孔隙率
drainage by consolidation	固结排水	drainage pressure	排泄压力
drainage by desiccation	疏干	drainage radius	排水半径
drainage by electro-osmosis	电渗排水	drainage sand blanker	排水砂垫层
drainage by gravity	重力排水	drainage shaft	排水井
drainage by open channel	明渠排水	drainage sump	集水井；排水集水坑
drainage by surcharge	加载排水	drainage surface	排泄表面
drainage by well point	井点排水	drainage system	排水系统
drainage by wells	井群排水	drainage terrace	排水阶地
drainage canal	排水渠	drainage trench	排水沟
drainage capacity	排泄能力，排水能力	drainage triaxial test	排水三轴试验
drainage chamber	排水沙井	drainage tunnel	排水隧道
drainage channel	排水渠	drainage well	排水井
drainage characteristics	排渗特性	drainage works	排水工程
drainage coefficient	排水系数	drained condition	排水条件
drainage condition	排水条件	drained creep	排水蠕变
drainage consolidation	排水固结	drained repeated direct shear test	排水反复直剪试验
drainage crossings	交叉排水渠	drained shear	排水剪切
drainage culvert	排水涵洞	drained shear strength	排水剪切强度
drainage curtain	排水幕	drained shear test	排水剪切试验
drainage curvature	排泄曲率	drained slope	疏干边坡
drainage cut	排水截槽，排水槽	drained test	排水试验
drainage density	河网密度；排水密度	drained test condition	排水试验条件
drainage design	水系样式	drained triaxial test	排水式三轴试验
drainage device	排水装置	drained zone	排水区
drainage ditch	排水沟	draining trench	小排水沟
drainage divide	流域分界线；分水岭	drakonite	闪云粗面岩
drainage engineering	排水工程	drape fold	披盖褶皱
drainage excavation	排水沟开挖	drape structure	披盖构造
drainage facility	排水装置	drape trap	披盖圈闭
drainage factor	排水系数	drape unconformity	披盖不整合
drainage filter	排水滤层	draught marks	吃水标志
drainage funnel	排水漏斗	dravite	镁电气石
drainage gallery	排水平硐；排水廊道	draw angle	移动角
drainage hole	排水孔	draw area	放矿区域
drainage in embankment area	填方区排水	draw cone	锥形放矿溜眼；放矿漏斗
drainage in excavation area	挖方区排水	draw crack	拉延裂缝
drainage in open-pit	露天矿排水	draw hole	放矿口
drainage in underground mine	地下矿排水	draw level	放矿水平
drainage intensity	排水强度	draw shaft	溜井
drainage layer	排水层	drawbar test	牵引试验
drainage level	排水平巷	drawdown analysis	压降分析
drainage loading	排水加荷	drawdown curve	水位下降曲线
drainage manhole	排水沙井	drawdown deformation	收缩变形
drainage map	水系图；流域图	drawdown of groundwater	地下水位下降
drainage material	排水材料	drawdown pressure	生产压差
drainage measure	排水措施	drawdown profile	降深剖面
drainage method	排水法	drawdown range	降落范围
drainage modulus	排水模数	drawing back of pillars	后退式回采矿柱
drainage network	排水网	drawing breakage	拉裂
drainage of foundation	地基排水	drawing interval	放煤步距
drainage of landslide	滑坡排水	drawing out methane	抽放瓦斯，排放瓦斯
drainage opening	泻水孔	drawing pillar	回采矿柱；回采煤柱
drainage outlet	排水出口	drawing program	放矿计划
drainage path	排水路径	drawing sequence	放煤顺序
drainage pavement	排水铺面	drawing-out fault	拖引断层
drainage perimeter	流域周界	drawn ore grade	出矿品位
drainage period	排水期	drawn ore tonnage	出矿量
drainage pipe	排水管道	dredge level	疏浚标高

dredge spoil	疏浚弃土
dredged area	疏浚区
dredged trench	疏浚沟
dredger fill	吹填土；冲填土
dredging	疏浚
dredging engineering	疏浚工程
dredging of river course	河道疏浚
dredging operation	疏浚作业，挖泥作业
dredging works	疏浚工程，挖泥工程
dreelite	石膏重晶石
dreikanter	三棱石
driblet cone	熔岩滴锥，滴丘次生熔岩喷气锥
drift	冰碛堆积物
drift bed	冰碛层
drift bedding	迁移层理
drift block	漂流石块
drift boulder	漂砾
drift breccia	冰川角砾
drift clay	冰砾泥
drift deposit	冰川沉积
drift detritus	漂移岩屑
drift developing	平硐开拓
drift epoch	冰期
drift face	平巷工作面；水平巷道工作面
drift force	漂移力
drift indicator	井偏指示器；倾角测量仪
drift lake	冰碛湖；冰川湖
drift landform	冰碛地形
drift log	井斜测井
drift map	覆盖层地质图
drift material	冰积物；洪积物
drift mound	冰碛岗陵
drift of continents	大陆漂移
drift peat	冰碛泥炭
drift pillar	平巷矿柱
drift plain	冰碛平原
drift sand	流沙
drift sheet	冰碛层
drift stratigraphy	冰碛地层学
drift structure	冰碛构造
drift terrace	冰碛阶地；冲积阶地
drift theory	漂移说
drift topography	冰碛地形
drift way	水平巷道
driftance	漂移度
drift-barrier lake	冰碛阻塞湖
drift-border features	边碛特征
drift-chute	溜槽
drift-dam lake	冰碛湖
drifted bore	偏斜孔
drifted material	冰积物；冰碛
drifting beach	移动海滩，漂移海滩
drifting method	掘进法
drifting sand	流沙
driftless area	无碛带
drill ability scale of rock	岩石可钻性标度
drill bit	钻头
drill bit run	钻头行程
drill bortz	钻用金刚石
drill capability	钻孔能力
drill collar	钻铤；加重钻杆
drill column	钻孔柱状图
drill cuttings	钻粉，钻屑
drill diamond	钻用金刚石
drill hole	炮眼；钻孔
drill hole columnar section	钻孔柱状设计图
drill hole depth	钻孔深度
drill hole design	钻孔设计
drill hole deviation	钻孔弯曲
drill hole sampling	钻孔取样
drill hole structure	钻孔结构
drill hole survey	钻孔测量
drill hole trajectory	钻孔轨迹
drill hole wall	钻孔壁
drill log	钻井记录，钻井编录
drill pipe frozen	卡钻
drill pipe sticking	卡钻
drill pipe wobbling	钻杆摆动
drill pressure	钻头钻进压力
drill record	钻井记录
drill site	井场，井位
drill stem buckling	钻柱弯曲
drillability	可钻性
drillability classification number	可钻性分类指数
drillability index	可钻性指数
drillability of rock	岩石可钻性
drillable	可钻的
drill-blasting method	钻爆法
drilled caisson	钻孔沉井；管柱；钻孔桩
drilled drain hole	排水钻孔
drilled grout hole	灌浆钻孔
drilled pier	钻孔墩；钻孔桩
drilled pier foundation	钻孔墩基础
drilled pile	钻掘桩
drilled shaft	钻孔桩；钻孔竖井
drilled well	钻井
drill-footage	钻进进尺
drill-hole exploration	钻孔勘探
drill-hole wall	钻孔壁
drilling	钻探，岩芯钻探
drilling analytics	钻井分析学
drilling anchor	钻孔锚杆
drilling and blasting method	钻爆法
drilling and wireline coring	钻探绳索采样系统
drilling by crushing	压碎法钻进
drilling cake	钻浆泥皮
drilling chip	钻机岩屑
drilling coefficient	钻进系数
drilling damage zone	钻井污染带
drilling data	钻井数据
drilling data acquisition	钻井数据采集
drilling data analysis	钻井数据分析
drilling depth	钻孔深度
drilling diameter	成孔直径
drilling direction coring	钻探定向岩芯
drilling dust	钻粉
drilling engineering	钻井工程
drilling fluid	钻井液冲洗介质
drilling fluid circulating system	钻井液循环系统
drilling fluid cleaning system	钻井液净化系统

drilling fluid constant 钻井液常数
drilling fluid hydraulics 钻井液水力学
drilling fluid rheology 钻井液流变学
drilling footage 钻进进尺
drilling level 钻进水平
drilling log 钻探记录
drilling meal 钻粉
drilling mechanics 钻井力学
drilling mechanism 钻进机理
drilling medium 钻进介质
drilling mud 钻井泥浆
drilling operation 钻井作业
drilling parameter 钻井参数
drilling performance curve 钻井特性曲线
drilling period 钻井周期
drilling philosophy 钻井原理
drilling post 钻机支柱
drilling powders 钻进岩粉
drilling pressure 钻进压力
drilling report 钻探报告
drilling rig 钻机
drilling sequence 钻眼顺序
drilling site 孔位；钻孔工地
drilling sludge 钻屑
drilling slurry 钻进泥浆
drilling speed 钻进速度
drilling time log 钻时录井
drilling well 钻孔
drilling with counterflow 反循环冲洗钻进
drilling with sound vibration 声频振动钻进
drilling with water 清水钻进
drilling-in-balance 压力平衡钻进
drilling-induced cracks 钻生裂纹
drillings 钻屑
drillmobile 钻车
drill-rod size 钻杆尺寸
drip impression 滴痕
dripstone 滴水石，钟乳石，石笋
drivage 掘进
drivage efficiency 掘进效率
drivage height 巷道掘进高度
drivage ratio 掘进率
drive direction 掘进方向
drive head 承锤头
drive point penetrometers 打入式尖锥
drive sampler 击入式取土器
drive shoe 桩靴
driven cast-in-place pile 就地灌注桩
driven in-situ pile 打入灌注桩
driven pile 打入桩
driven pile foundation 打入式桩基础
driven test pile 打入式试验桩
driven underpinning piles 打入式托换桩
drive-pipe drilling 冲积层勘探钻
driving 掘进
driving by vibration 振动沉桩
driving cycle 掘进循环
driving energy 打桩能量
driving face 掘进工作面
driving force 下滑力

driving mechanism of plate motion 板块运动的驱动机制
driving record 打桩记录
driving resistance 贯入阻抗
driving speed 掘进速度；进尺速度
driving stress 打桩应力
driving test 打桩试验
drop 落距
drop caisson 开口沉箱
drop crusher 冲击式破碎机
drop crushing 落锤破碎
drop fault 下落断层
drop hammer 落锤
drop hammer pile driver 落锤打桩机
drop hammer test 落锤试验
drop height 落距；落锤高度
drop of pressure 压力降
drop penetration test 落锤贯入试验
drop sampler 沉落取样器
drop shaft 沉井
drop stowing 重力充填；自重充填
drop weight tear test 落锤扯裂试验
drop-impact penetrometer 落锤冲击贯入仪
dropped caisson 沉箱
dropping cut slice 水平切片
dropping head 水头降
drop-shaft method 沉井法
drossy coal 劣质煤
drought and waterlogging 干旱与内涝
drought period 枯水期
drowned coast 沉溺岸
drowned lime 熟石灰
drowned reef 沉没礁，溺礁，暗礁
drowned valley 溺谷
Drucker-Prager criterion 德鲁克-普拉格准则
drumlin 鼓丘；冰堆丘
drumlin field 鼓丘原
drumming 问顶，敲击检查顶板
druse 晶洞
drusite structure 晶腺构造
drusitic 晶洞状的
drusy 晶腺状的，晶簇状的
drusy cellular 晶腺蜂房状的
drusy structure 晶簇状构造
dry analysis 干法分析；干法试验
dry and hard loess cement 干硬性水泥黄土
dry avalanche 干雪崩
dry boring 干钻
dry box 干燥箱
dry bulk density 干重度
dry burning coal 非黏结性煤
dry chernozem 干黑钙土
dry coal preparation 干法选煤
dry compaction 偏干压实
dry concentration 干选；风选
dry concrete 干硬性混凝土
dry core sample 干岩样
dry density 干密度
dry deposition 干沉降
dry drill cuttings 干钻粉；干钻屑
dry drilling 干钻

dry friction	干摩擦
dry gap	废河道；风口
dry hard clay	干硬黏土
dry hot rocks	干热岩体
dry ice	干冰
dry karst	喀斯特干岩溶；干溶洞
dry mining	干法开采
dry mortar	干砂浆
dry mulch	干燥层
dry oven	烘箱
dry pellet	干团粒
dry permafrost	干永冻层
dry permeability	干透气性
dry river bed	干河床
dry rodding	干捣法
dry sand	无油砂层
dry screening	干筛
dry separation	干选
dry sieving	干筛
dry stone wall	干砌石墙
dry strength	干强度
dry tamping	干击法
dry unit weight	干重度
dry valley	干谷
dry wash	干河床
dry weathering	干风化作用
dry weight	干重
dry-drilled pile	干成孔灌注桩
drying agent	干燥剂
drying cabinet	干燥箱
drying crack	干裂
drying oven	烘箱
drying rate	干燥速率
drying shrinkage	干燥收缩
drying speed	干燥速度
drying split	干裂
drying test	干燥试验
dry-mix concrete	干拌混凝土
dry-out sample	无水试样
dry-placed fill	干式充填
dry-stone drain	干砌石排水沟；盲沟
dual completion	双层完井
dual gravel pack	双层砾石充填
dual induction focused log	双感应聚焦测井
dual induction log	双感应测井
dual laterolog	双侧向测井
dual lining	双层衬砌；双层支护
dual metamorphism	双重变质作用
dual plate tectonics	双板块构造
dual pore structure	双重孔隙结构
dual porosity	双孔隙度
dual-scale system	双比例尺法
dubiofossil	可疑化石
dubuissonite	红蒙脱土
duct	暗渠；管道
duct friction	管道摩擦力
duct grouting	导管灌浆
ductile	有延展性的
ductile behavior	延展性，韧性
ductile brittle transition	延性脆性转移
ductile concrete	延性混凝土
ductile crustal spreading	地壳延性扩张
ductile damage	延性损伤
ductile deformation	延性变形
ductile design	延性设计
ductile extensional tectonic system	韧性伸展构造系统
ductile failure	延性破坏
ductile fault	塑性断层，延性断层
ductile faulting	延性断层作用
ductile fracture	延性断裂，韧性断裂
ductile rock	韧性岩石，延性岩石
ductile rupture	韧性断裂
ductile shear belt	韧性剪切带
ductile shear zone	塑性剪切带
ductile support	柔性支护
ductile-brittle deformation	延性-脆性变形
ductility	可塑性；延展性
ductility capacity	延性能力
ductility coefficient	延性系数
ductility factor	延性系数
ductility limit	屈服极限；屈服点
ductility potential	延性势
dufrenite	绿磷铁矿
dufrenoysite	硫砷铅矿
duftite	砷铜铅矿
dug pile	挖孔桩
dull coal	暗煤
dull hard coal	暗硬煤
dull lustre	暗光泽，弱光泽
dumb well	枯井
dummy joint	假接合，假缝
dummy load	等效荷载
dummy shaft	暗井
dumortierite	蓝线石
dump	堆积
dump injection	自流注水
dump mud-rock flow	排土场泥石流
dump slide	排土场滑坡
dumped rock embankment	抛石堤，堆石堤，填石堤
dumped rockfill	抛填堆石
dumping	倾泻，倾倒
dumpy level	定镜水准仪
Duncan-Chang model	邓肯-张模型
dune	沙丘
dune bedding	沙丘层理
dune formation	沙丘建造
dune movement	沙丘状移动
dune plain	沙丘平原
dune range	沙丘链
dune ridge	沙脊；沙岭
dune sand	沙丘砂
dune sediments	沙丘沉积物
dune slack	沙丘间湿地
dune slope	沙丘坡
dune wandering	流动沙丘
dung	黏结物
dungannonite	刚玉中长岩
Dunhuang Formation	敦煌群
duning effect	沙丘效应
duning phenomenon	沙丘现象

dunite 纯橄榄岩
dunite ore 纯橄榄岩矿石
dunn bass 泥质页岩
dunnet shale 油页岩
Duojiatan Phyllite 多家滩千枚岩
duplex fault zone 双断层带
duplex structure 二相显微结构；复相组织
duplication of coal 煤层的重叠
duplication of marking 双重标志
durability 耐久性
durability factor 耐久系数
durability index 耐用指数；耐久率
durability of structure 结构耐久性
durability test 耐久性试验
durain 暗煤
duralumin 硬铝
duration 持续时间
duration characteristic 持续时间特征
duration curve 持续曲线；历时曲线
duration of settlement 沉降历时
duration of shaking 震动持续时间
duration performance 耐久性能
durbachite 暗云正长岩
durdenite 磅铁矿
durometer 硬度计
durothermic 耐温性的
durovitrain 暗镜煤
dust barrier 岩粉棚
dust explosion accident 煤尘爆炸事故
dust ore 粉状矿石
dust sampling 粉末取样
dust-removing by wet-drilling 湿式凿岩除尘
Dutch cone 荷兰锥
Dutch cone penetrometer 荷兰式圆锥触探仪
Dutch cone penetrometers 荷兰式尖锥
Dutch cone test 荷兰圆锥试验
Dutch triaxial cell 荷兰式三轴仪
Dyassic 二叠纪；二叠系
dying out 尖灭；消失
dyke boring 堤防钻探
dyke deposit 脉状沉积
dyke failure 堤防溃决
dyke reinforcement 堤防加固
dykes 岩墙
dynamic 动力的；动力学的
dynamic absorber 减震器
dynamic action 动力作用
dynamic analysis 动态分析；动力学分析
dynamic balance 动态平衡
dynamic bearing capacity 动承载力
dynamic behavior 动力特性，动力学性状
dynamic boundary condition 动力边界条件
dynamic breccia 动力角砾岩；构造角砾岩
dynamic buckling 动力屈曲，动态压曲，动态失稳
dynamic buffer 动态缓冲
dynamic bulk density 动重度
dynamic bulk elastic modulus 动态体弹性模数
dynamic calibration 动态校准
dynamic capability 动态性能
dynamic capillary pressure 动力毛细管压力

dynamic cataloging 动态分类
dynamic characteristic 动力特征，动态特性
dynamic check 动态检验
dynamic coefficient 动力系数
dynamic coefficient of friction 动摩擦系数
dynamic coefficient of viscosity 动力黏度系数
dynamic compaction 动力压实；动态夯实；强夯
dynamic condition 动态条件
dynamic cone penetration test 圆锥动力触探试验
dynamic cone penetrometer 动力圆锥触探仪
dynamic consolidation 动力固结；强夯
dynamic consolidation method 动力固结法
dynamic constraint 动力约束
dynamic contact angle 动力接触角
dynamic contact force 动态接触压力
dynamic control 动态控制
dynamic correction 动态校正
dynamic correlation 动态相关
dynamic cyclic loading 动力循环加载
dynamic damage 动态损伤
dynamic deformability 动力变形性
dynamic deformation 动变形
dynamic deformation modulus 动态变形模量
dynamic deviation 动态偏差
dynamic diagram 动态曲线
dynamic dilatation 动力扩容
dynamic displacement 动态位移
dynamic disturbance 动态扰动
dynamic ductility 冲击韧性
dynamic effect 动力效应，动态效应
dynamic effective stress method 动力有效应力法
dynamic elastic analysis 动力弹性分析
dynamic elastic constant 动态弹性常数
dynamic elastic limit 动力弹性极限
dynamic elastic modulus 动弹性模量
dynamic elastic property 动弹性性质
dynamic equalization 动态均衡
dynamic equilibrium 动力平衡
dynamic error 动态误差
dynamic evaluation 动态评价
dynamic experiment 动力学实验
dynamic factor 动力因数
dynamic failure stress 动破坏应力
dynamic fatigue failure 动疲劳破坏
dynamic flow condition 流体动力条件
dynamic force 动力
dynamic fracture strength 动力破裂强度
dynamic fracture toughness 动态断裂韧度
dynamic fragmentation 动力破碎
dynamic friction 动摩擦，动态摩擦
dynamic friction angle 动摩擦角
dynamic friction coefficient 动摩擦系数
dynamic geological analysis 动力地质分析
dynamic geology 动力地质学
dynamic geomorphology 动力地貌学
dynamic head 动力水头
dynamic instability 动力不稳定性
dynamic interaction 动力相互作用
dynamic lateral stiffness 动侧向刚度
dynamic level 动水位

dynamic level regime	动水位动态
dynamic load	动荷载
dynamic load coefficient	动载系数；顶板来压强度系数
dynamic load test of pile	桩动载荷试验
dynamic loading	动荷载
dynamic loading test	动载荷试验
dynamic magnetic induction	动磁感应
dynamic magnification	动力放大
dynamic magnification factor	动力放大因数
dynamic map	动态地图
dynamic measurements of pile	桩的动测法
dynamic metamorphic ore formation	动力变质成矿作用
dynamic metamorphism	动力变质
dynamic method	动力学方法
dynamic model	动力模型
dynamic modulus of elasticity	动力弹性模量
dynamic modulus of shear	动力剪切模量
dynamic monitoring	动态监测
dynamic oedometer	动力固结仪
dynamic optimization	动态优化
dynamic osmometer	动力渗透压力计
dynamic overload	动力过载
dynamic parameter	动态参数，动力学参数
dynamic pattern	动力模式
dynamic penetration	动力贯入度
dynamic penetration test	动力触探试验
dynamic performance	动态特性
dynamic photoelasticity	动力光弹性
dynamic pile driving formula	动力打桩公式
dynamic pile driving resistance	打桩动阻力
dynamic pile test	动力测桩
dynamic piled driving resistance	打桩动阻力
dynamic plastic buckling	塑性动力屈曲
dynamic point resistance	动力探头阻力
dynamic pore pressure	动孔隙压力
dynamic prediction forecast	动态预报
dynamic pressure	动压力
dynamic pressure behavior	动压力特性
dynamic pressure gradient	动压力梯度
dynamic probing test	动力触探试验
dynamic property of rock	岩石动力学性质
dynamic range	动态范围
dynamic relaxation	动力松弛
dynamic relaxed rock mass	动力松动岩体
dynamic resistance	动阻力
dynamic resistance curve	动阻力曲线
dynamic response	动态响应，动力特性
dynamic response analysis	动力响应分析
dynamic response approach	动力反应法
dynamic rock mechanics	岩石动力学
dynamic rock triaxial apparatus	岩石动力三轴仪
dynamic rupture	动态破裂
dynamic rupture process	动力破坏作用
dynamic sensitivity	动态灵敏度
dynamic shear elastic modulus	动切变弹性模数
dynamic shear modulus	动剪切模量
dynamic shear rheometer	动态剪切流变仪
dynamic shear strength	动抗剪强度
dynamic shock	动力冲击
dynamic similarity	动力相似
dynamic simple shear test	动单剪试验
dynamic simulation	动态模拟
dynamic slip approach	动滑移法
dynamic sounding	动力探测
dynamic stability	动力稳定性
dynamic stiffness matrix	动力刚度矩阵
dynamic stiffness ratio	动刚度比
dynamic stiffness test	动力刚度试验
dynamic strain	动应变
dynamic strain meter	动应变仪
dynamic stratigraphy	动力地层学
dynamic strength	动力强度
dynamic stress	动应力
dynamic stress concentration	动应力集中
dynamic stress wave	动应力波
dynamic stress-strain relationship	动应力-应变关系
dynamic structural analysis	结构动力分析
dynamic structural geology	动力构造地质学
dynamic subgrade reaction	动基床反力
dynamic surface pressure	动表面压力
dynamic surface tension	动表面张力
dynamic survey	动态观测
dynamic tensile strength	动抗拉强度
dynamic test	动力试验
dynamic theory	动力学理论
dynamic triaxial test	振动三轴试验
dynamic unloading	动态卸荷
dynamic variable	动态变量
dynamic viscosity	动力黏度，动态黏滞性
dynamic water level	动水位
dynamic water pressure	动水压力
dynamic well pressure	动态井压
dynamic wetting behavior	动态润湿特性
dynamic wetting effect	动力润湿效应
dynamic yield strength	动力屈服强度
dynamical action	动力作用
dynamical analysis	动力分析
dynamical breccia	构造角砾岩
dynamical rupture	动力破裂
dynamical stiffness method	动力刚度法
dynamical structure factor	动力结构因子
dynamical system	动力系统
dynamics of the earth	地球动力学
dynamo thermal metamorphism	动热变质
dynamochemical	动力化学的
dynamochemical activity	动力化学作用
dynamometamorphic deposit	动力变质矿床
dynamometamorphic rock	动力变质岩
dynamometamorphism	动力变质作用
dynamometer	测力计，压力盒，压力计
dynamometric study	测力分析
dynamo-regional metamorphism	动力区域变质作用
dynamo-relaxed rock mass	松动岩体
dynamo-thermal metamorphism	动热变质
dyscrasite	锑银矿
dysgeogenous	不易风化的
dyssyntribite	斑块云母；不纯白云母
dystric arenosols	不饱和红砂土；酸性红砂土

E

EAAI 人工智能的工程应用学报
eaglestone 鹰石；凝结小球
eagre 涛；涌潮；高潮；怒潮；潮流
eagre tidal bore 潮水上涨
eakleite 硬硅钙石
ear wall tunnel portal 耳墙式洞门
earlandite 水柠檬钙石；水碳氢钙石
earliness of forecast 预报期；预见期；预报时效
early age strength of concrete 混凝土早期强度
early mature valley 早成年期谷
early stage diversion 初期导流
early warning 早期报警；预警的；预警
early warning marker 预警指标
early warning radar 预警雷达
early warning system 预先警报系统；早起预警系统；预警系统
early-formed mineral 先成矿物
early-warning index 预警指数
earth 地球；土，土壤，土地，泥土；地面；大陆；接地
earth alkali metal 碱土金属
earth anchor 土层锚杆；地层锚杆；土锚
earth anchorage 地锚固定
Earth and Planetary Science Letters 地球与行星科学通讯
earth and rockfill dam 土石混合坝；土石坝
earth and stone works 土石方工程
earth attraction 地球引力
earth auger 螺旋钻；土钻；地螺钻；麻花钻
earth axis 地轴
earth backing 还土；回填土；复填土；填土
earth bank 土堤；路堤；土堤岸
earth base 土基
earth bearing strength 地基承载强度
earth bending test 土梁抗弯试验
earth blanket 土壤覆盖层；黏土防渗层
earth borer 地钻；螺旋土钻，土钻
earth boring machine 隧道钻进机；平巷钻进机；钻土机
earth boring outfit 钻土工具；平巷掘进装备
earth bund 土堤
earth canal 土渠
earth cave 土洞
earth circle 土环
earth column 土柱
earth compacted layer by layer 分层夯实；逐层夯实
earth compaction factor 土料的压实系数
earth concrete 掺土混凝土
earth conductivity 大地电导率
earth construction 土建筑
earth coordinate system 地球坐标系
earth coordinates 大地坐标系
earth core 地核，地心；黏土心墙
earth core rock-fill dam 土质心墙堆石坝
earth coupling 大地耦合
earth covering 覆盖土，覆土；覆盖法
earth creep 土体蠕动，土蠕变；土滑，地滑

earth crust 地壳
earth crust stress 地壳应力
earth crust structure 地壳构造
earth current 大地电流；地电流；接地电流
earth curvature 地球曲率；地面曲率
earth curvature correction 地球曲率校正
earth curve 地壳弯曲
earth cutting 挖土
earth dam ageing 土坝老化
earth deformation 地球形变；土体变形
earth density 土体密度
earth deposit 土堆；弃土堆
earth dike 土堤
earth din 地鸣；地震
earth dyke 土堤
Earth dynamic system 地球动力学系统
earth dynamics 地球动力学；地球动力状况
earth electrode system 地电极系统
earth ellipsoid 地球椭球体；地球体
earth embankment 土堤
earth excavation 开挖土方，挖方；土方工作，挖土工作；挖土
earth fall 土崩；地滑；土体坍塌
earth feature 地面特征
earth fill 填方；填土；填土方
earth fill cofferdam 土围堰
earth fill dam 土坝；堆土坝
earth fill dam paving 土坝铺面
earth fill road embankment 填土路堤
earth fill slope 填土斜坡；填土边坡
earth finger 小土柱
earth flax 石棉；石质纤维
earth flow 泥流；土流；土石流
earth force 地球营力
earth foundation 土质地基；土基
earth gas 土壤气体；天然气
earth globe 地球；地球仪
earth gravity field 地心引力场；地球重力场
earth gravity model 地球重力场模型；地球引力模型
earth heat 地热能；地热
earth heat well 地热井
earth heat well blowout 地热井喷
earth heaving 地面隆起
earth history 地质史，地质历史；地史，地史学
earth holography 大地全息摄影术；大地全息学
earth hummock 土丘；土核
earth inductor 地磁感应器；接地电感线圈
earth interior 地球内部
earth job 土方工程，土工
earth lab program 地球科学实验室计划
earth lateral pressure 土的侧压力；侧向土压力
earth light 地光
earth load 土压力；土荷载
earth magnetic effect 地磁效应；地磁场电效应

earth magnetic field 地磁场
earth magnetism 地磁
earth mantle 地幔
earth manual 土工手册
earth mass 土体；地球质量
earth material 土石，土料
earth medium 大地介质
earth membrane 黏土防渗层
earth model 地球模型
earth mortar 土灰浆
earth motion 地动；地震动；地球运动
earth mound 土丘；土墩；土堤
earth movement 地动；岩层移动；地层移动；地壳运动；地球运动
earth mover 运土机械；挖土机械；推土机；挖掘机；大型推土机
earth moving 土方搬运；运土；土方工程；翻动泥土
earth moving equipment 土石方机械；运土机械
earth nucleus 地核
earth observation 地球观测
earth observation satellite 地球观测卫星；对地观测卫星
earth oil 石油；原油
earth orbital imagery 地球轨道成像
earth orbital operation 绕地球轨道运行
earth orbital photography 地球轨道摄影
earth orientation parameter 地球定向参数；地球定位参数
earth pad 土基；土垫层
earth physical exploration technique 地球物理勘探技术
earth physics 大地物理学
earth pile 土挤密桩；土桩
earth pit 取土坑；地下温室；地窖；土坑；土圹墓
earth pitch 地沥青；软沥青
earth platform 土平台
earth pressure 地压，土压；土压力
earth pressure at rest 静止土压力；静土压力
earth pressure balance shield 土压平衡盾构
earth pressure balance shield machine 土压平衡盾构机
earth pressure cell 土压力盒；土压力计；土压计
earth pressure coefficient 土压力系数
earth pressure distribution 土压力分布
earth pressure during earthquake 地震土压
earth pressure gauge 土压力计，地压力计；土压计
earth pressure measurement 土压力测定
earth pressure wedge 土压力楔
earth protective device 接地保护装置
earth pyramid 土锥
earth quantity 土方量；土方工程量
earth rammer 夯土机；夯土器；夯土锤
earth reinforcement 土加筋；土加固；加筋土壤
earth remote technology satellite 地球遥感卫星
earth resistance test 地抗力探测法
earth resistance tester 地阻仪
earth resource technology satellite 地球资源技术卫星
Earth Resources Experiment Package 地球资源试验综合计划
earth resources information system 地球资源信息系统
earth resources observation satellite 地球资源观测卫星
earth resources satellite 地球资源卫星
earth resources survey program 地球资源调查计划

Earth Resources Technology Satellite image 地球资源卫星图像
earth retaining structure 护土结构；挡土结构；挡土墙
earth ridge 土垄
earth road 土路
earth rock cofferdam 土石围堰
earth rock dam 土石坝
earth rod 接地棒
earth rotation parameter 地球自转参数
earth run 土壤流失；土流舌
earth science 地学；地球科学
earth scoop 挖土铲斗；铲运机
earth scraper 刮土机；铲土机
earth screw 土钻；麻花钻
earth sensor 地球传感器
earth shape 地球形状
earth shift 地位移
earth shock 地震
earth sieve 土筛
earth silicon 硅石；二氧化硅
earth simulator 地球模拟器
earth slide 土层滑动，土崩，土滑，滑坡，塌方
earth slip 土滑；土层滑动；地滑，塌方
earth slope 土坡
earth slope stability analysis 土坡稳定性分析
earth slump 滑坡，错落，坍落；岩土滑塌
earth sound 地声
earth spheroid 地球扁球体；地球椭球体
earth stress 地应力
earth structure 土工建筑物；土工构筑物；土工结构物；地球构造；土结构
earth surface 地面；地表；地球表面
earth surface reflected wave 地面反射波
earth synchronous orbit 地球同步轨道
earth system 地球系统
earth system modelling 地球系统模拟
earth system science 地球系统科学
earth tamper 夯土机
earth tide 地潮；固体潮；陆潮；地球潮汐
earth tilt 地倾斜
earth tunnel 土质隧道
earth volume 土方量
earth vortex 地旋涡
earth water 硬水；含钙质水；地球水
earth wax 石蜡；地蜡
earth work 土方工程；土方工作；路基土石方
earth work operation 土石方工程作业
earth-based coordinate system 地球坐标系
earth-current 地电流，大地电流
earth-dam compaction 土坝压实
earth-dam paving 土坝护面
earthday 地球日
earthen heritage site 土遗址
earthen shaft tomb 竖穴土坑墓
earthen structure 土工构筑物；土工建筑物
earthen ware 陶器
earthenware clay 陶土
earthenware drain pipe 排水瓦管
earthenware pipe 瓦管；陶管
earth-filled 填土的

earth-filled dam 土坝
earth-fixed coordinate system 地球固定坐标系；站心坐标系
earth-flattening approximation 地球变平近似
earth-flattening transformation 地球变平换算
earthflow 土流；泥石流
earthflow tongue 泥石流舌部
earthing pole 接地极；接地安全棒
earthing system 接地系统
earthing terminal 接地终端
earth-like planet 类地行星
earthly heat 地热；地热能
earth-moon system 地月系
earth-moving machine 掘土机；动土机；推土机
earth-observation system 地球观测系统
earth potential 地电位
earth-pressure balanced machine 土压平衡盾构机
earth-pressure line 土压力线
earth-probing radar 地质探测雷达；探地雷达；地质雷达
earthquake 地震；天然地震
earthquake accelerogram 地震加速度图；地震加速度记录图
earthquake action 地震作用
earthquake analysis 地震分析
earthquake area 地震区
earthquake attenuation 地震衰减
earthquake axis 地震轴
earthquake belt 地震带
earthquake bulge 地震鼓包
earthquake catalogue 地震目录
earthquake catastrophe 地震灾害；震害
earthquake center 震源；震中
earthquake cluster 震群
earthquake coda 震尾
earthquake consequence 地震后果；地震影响
earthquake control 地震控制
earthquake countermeasure 地震对策；防震措施
earthquake cycle 地震周期
earthquake damage 震害；地震破坏；地震震害
earthquake damage index 震害指数
earthquake damage insurance 地震损失保险
earthquake damage investigation 震害调查
earthquake damage lesson 震害教训
earthquake damage prediction 震害预测
earthquake damage ratio 地震破坏率
earthquake damage susceptibility 震害敏感性
earthquake database 地震数据库
earthquake design 抗震设计
earthquake design spectrum 地震设计谱
earthquake destructiveness 地震破坏性；地震破坏程度
earthquake disaster 地震灾害；震害
earthquake disaster affection 地震灾情；地震灾害影响
earthquake disaster assessment 地震灾害评估
earthquake disaster potential 发生地震灾害可能性
earthquake disaster preparedness 地震灾害预防；地震防灾准备状态
earthquake dislocation 地震位错
earthquake displacement 地震位移
earthquake displacement field 地震位移场
earthquake distribution 地震分布
earthquake due to collapse 陷落地震；塌陷地震
earthquake dynamic earth pressure 地震动土压力
earthquake dynamic water pressure 地震动水压力
earthquake dynamics 地震动力学
earthquake earth pressure 地震土压力
earthquake effect 地震效应；地震影响
earthquake emergency management 地震应急管理
earthquake energy 地震能量；地震能
earthquake engineering 地震工程；地震工程学
Earthquake Engineering and Structural Dynamics 地震工程和结构动力学
earthquake environment 地震环境
earthquake epicentre 震中
earthquake fault 地震断层
earthquake fault plane 地震断层面
earthquake fault plane solution 地震断层面解
earthquake filtering 地震滤波
earthquake fire hazard 地震引起火灾危险；地震火灾危险
earthquake foci 震源
earthquake focus 震源；地震震源
earthquake force 地震力
earthquake foreshock 前震
earthquake fortification 抗震设防
earthquake fortification intensity 抗震设防烈度
earthquake fortification level 抗震设防标准
earthquake fortification zone 抗震设防区
earthquake frequency 地震频度；地震频率
earthquake generating stress 地震发生应力
earthquake geology 地震地质学
earthquake ground motion 地震地面运动
earthquake ground motion characteristic 地震地面运动特性；地震动特性
earthquake hazard 震害；震灾；地震危险性；地震灾害
earthquake hazard assessment 地震灾害评估
earthquake hazard evaluation 地震危险性分析；地震灾害分析
earthquake hazard map 地震危险区划图
earthquake hazard mitigation 地震灾害减轻措施
earthquake hydrodynamic force 地震动水作用力
earthquake hypocenter 震源
earthquake inactivity in intermittent period 地震平静期；地震活动间歇期
earthquake induced by deep-well water injection 深井注水诱发地震
earthquake induced geological hazard 地震诱发的地质灾害
earthquake input 地震输入
earthquake insurance 地震保险
earthquake intensity 地震烈度；地震震度；地震强度
earthquake intensity engineering standard 地震烈度工程标准；烈度工程标准
earthquake intensity resistance 抗震设防烈度
earthquake intensity scale 地震烈度表；烈度表
earthquake interaction 地震的相互作用
earthquake isoseismal 等震线
earthquake level models of crust 地壳地震分层模型
earthquake light 地光；地震光
earthquake line 地震线

earthquake liquefaction 地震液化
earthquake loading 地震荷载
earthquake magnetic field 地震磁场
earthquake magnitude 地震震级；地震规模；地震等级
earthquake mechanics 地震力学
earthquake mechanism 地震机制；地震机理
earthquake migration 地震迁移
earthquake model 地震模型
earthquake mole track 地震鼓包；地震隆起带
earthquake moment 地震矩
earthquake monitoring 地震监测
earthquake monitoring and prediction 地震监测预报
earthquake monitoring network 地震监测台网
earthquake motion 地震运动
earthquake nonresistive construction 非抗震建筑
earthquake nucleation 地震成核
earthquake observation 地震观测
earthquake occurrence model 发震模型
earthquake of distant origin 远震；远源地震
earthquake of intermediate focal depth 中等震源深度的地震
earthquake origin 震源
earthquake parameter 地震参数；震源参数
earthquake participation factor 地震成因系数；地震参与因子
earthquake period 地震周期
earthquake periodicity 地震周期性
earthquake phase 震相
earthquake phenomena 地震现象
earthquake precursor 地震前兆，地震预兆；地震先兆现象
earthquake precursor observation data 地震前兆观测数据
earthquake prediction 地震预报；地震预测
earthquake pregnant tectonic systems 孕震构造体系
earthquake premonitory phenomenon 地震前兆现象
earthquake preparation 地震孕育；孕震
earthquake prevention 地震预防
earthquake probability 地震概率
earthquake process 地震过程
earthquake prone area 易震区；多震区；地震活动区
earthquake proof 防震；抗震
earthquake proof construction 抗震建筑；抗震构造；防震建筑；抗震工程
earthquake proof foundation 抗震基础；抗震地基
earthquake proof joint 防震缝；抗震缝；防震接缝
earthquake proof structure 抗震结构
earthquake proofing technique 抗震技术
earthquake protection 抗震设防
earthquake protection of monuments 古迹的地震防护
earthquake protective measure 抗震措施
earthquake protective policy 抗震政策；抗震对策
earthquake province 震区；地震区
earthquake rattle 地震声
earthquake record 地震记录
earthquake recurrence map 地震重复性图
earthquake region 地震区域；震区
earthquake resistance code 抗震规范
earthquake resistance design 抗震设计
earthquake resistant building 抗震建筑
earthquake resistant capability 抗震能力

earthquake resistant code 抗震规范
earthquake resistant construction technique 抗震建筑技术
earthquake resistant constructional measure 抗震结构措施
earthquake resistant design 抗震设计
earthquake resistant design code 抗震设计规范
earthquake resistant feature 抗震特性
earthquake resistant level 抗震设防水平
earthquake resistant material 抗震材料
earthquake resistant regulation 抗震条例
earthquake resistant structure 抗震结构
earthquake resistant test 抗震试验
earthquake resisting component 抗震部件
earthquake resisting element 抗震构件
earthquake resisting structure 抗震结构
earthquake resistivity 抗震性
earthquake response 地震反应；地震响应
earthquake response function 地震反应函数；地震响应函数
earthquake response spectrum 地震反应谱；地震响应谱
earthquake risk 地震危险性；地震风险
earthquake risk region 地震危险区
earthquake risk zoning 地震危险区划；地震危险性分区
earthquake rupture mechanics 地震破裂力学
earthquake rupture surface 地震破裂面
earthquake safety assessment 地震安全性评价
earthquake safety facility 防震安全设施
earthquake scale 地震级；地震震级
earthquake scarp 地震陡坎；地震断坝
earthquake sea wave 地震海啸
earthquake segment 地震破裂段
earthquake seismology 地震学；天然地震学
earthquake separation 抗震缝；隔震
earthquake sequence 地震序列
earthquake sequence of main shock type 主震型地震序列
earthquake sequence of solitary shock type 孤立型地震序列
earthquake sequence of swarm shock type 震群型地震序列
earthquake series 地震序列；震群
earthquake shaking 地震摇动，地震动
earthquake shear distributing factor 地震剪力分配系数
earthquake shelter 地震避难场所
earthquake shock 地震冲击；地震活动
earthquake simulation 地震模拟
earthquake simulator 地震模拟器；地震模拟装置
earthquake situation 震情；抗震救灾情况
earthquake sound 地声；震声
earthquake source 震源
earthquake source dynamics 震源动力学
earthquake source mechanism 震源机制
earthquake source observation 震源观测
earthquake spectrum 地震谱
earthquake statistical prediction 地震统计预报
earthquake statistics 地震统计学
earthquake stimulation 地震激发
earthquake stress 地震应力
earthquake stress triggering 地震应力触发
earthquake structure 地震构造
earthquake subsidence 地震沉陷；震陷
earthquake swarm 群发地震；群震；震群；地震群
earthquake telemetering system 地震遥测系统
earthquake time series 地震时间序列

earthquake trace	地震记录迹线
earthquake tremor	地震颤动
earthquake triggering	地震触发
earthquake triggering tectonic systems	发震构造体系
earthquake under sea bottom	海底地震
earthquake uplift	地震隆起
earthquake vibration	地震颤动；地震振动
earthquake volume	震源体积
earthquake warning	地震警报
earthquake warning apparatus	地震警报装置
earthquake watch	地震监视
earthquake wave	地震波
earthquake zone	地震带
earthquake zoning	地震区划；地震分带
earthquake zoning map	地震区划图
earthquake-damage insurance	地震损失保险
earthquake-felt area	有感地震范围；震感区域
earthquake-generating mechanism	地震机制；发震机制
earthquake-generating stress	引震应力；发震应力
earthquake-induced landslide	地震诱发滑坡
earthquake-like aftershock sequences	似地震的余震序列
earthquake-like excitation	激发地震
earthquake-prone area	地震多发区；地震频发区
earthquake-proof	抗地震的；耐地震的；抗震的；防震的
earthquake-proof building	抗震建筑
earthquake-proof construction	抗震建筑
earthquake-proof foundation	防震基础
earthquake-proofness	抗震性
earthquake-resistance	抗震
earthquake-resistant construction	抗震构造；抗震工程
earthquake-resistant design	抗震设计
earthquake-resistant measure	抗震措施
earthquake-resistant structure	耐震结构
earthquake-response spectrum	地震反应谱；地震响应频谱
earthquake-source parameter	震源参数
earthquake-warning system	地震警报系统；地震预警系统
earth-reflected wave	地面反射波
earth-resistivity	地电阻率
earth-retaining structure	挡土构筑物；护土结构
earth-retaining support	挡土支护；挡土支撑
earth-retaining wall	挡土墙
earth-rock cofferdam	土石围堰
earth-rock dam	土石坝
earth-rock excavation	土石方开挖
earth-rock works	土石方工程
earth-rockfill dam	土石坝；堆石土坝
earth's average surface albedo	地球表面平均反射率
earth's axis	地球轴线；地球坐标系；地轴
earth's barodynamics	地球重型结构力学
earth's core	地心；地核
earth's crust	地壳
earth's ellipsoid	地球椭球体
earth's ellipticity	地球扁率；地球椭圆率
earth's field	地球重力场
earth's formation	地层；岩层
earth's gravity field	地球重力场
earth's magnetic dip angle	地磁倾角
earth's magnetic field	地球磁场；地磁场
earth's magnetism	地磁
earth's mantle	地幔
earth's orbit	地球轨道
earth's orbital plane incline	地球轨道面倾角
earth's rhythms	地球韵律
earth's rotation axis	地球自转轴
earth's shadow	地球阴影
earth's sphere	地球层圈；地球
earth's spheroid	地球椭球体
earth's spin vector	地球自转矢量
earth's surface	地球表面；地面
earth's temperature regime	地球温度状态；地球热状态
earth-scraper	铲土机
earth-shift	地层移动
earth-sound telemetry	地声遥测
earth-space path propagation	地-空路径传播
earth-synchronous orbit	地球同步轨道
earthwork	土石方工程
earthwork calculation	土方计算
earthwork engineering	土工工程
earthwork house	土筑房
earthwork operation	土方工程工序；土方工程作业
earthwork plant	土方设备
earthwork quantity sheet	土方表
earthwork specifications	土方工程规格；土方工程标准
earthwork volume	土方工程量
earthy	土状的；土质的；土的
earthy aggregate	土状集合体
earthy cobalt	钴土矿
earthy fracture	土状断口
earthy iron ore	泥状铁矿
earthy lignite	土状褐煤
earthy luster	土状光泽
earthy spring	泥水泉；泥泉
earthy tripolite	硅藻土
earthy water	泥水；硬水
easer hole	辅助钻孔；辅助炮孔；周边炮孔；辅助炮眼
easer reliever	辅助炮眼
easily hydrolyzable	易水解的
easily weathered	易风化的
east African graben	东非地堑
east African rift belt	东非裂谷带
east China sea	东海
east Pacific mid-ocean ridge	东太平洋洋中脊
east Pacific rise	东太平洋隆起；东太平洋海隆
east Pacific rise geothermal belt	东太平洋中脊地热带
east Shandong-South liaoning land	胶辽古陆
eastern fracture zone	东部断开带
eastonite	镁铝云母；富镁黑云母；水碱黑云母
easy coal	软媒，沥青煤；易选煤
easy magnetization axis	易磁化方向
easy-to-drill formation	易钻地层
ebb axis	落潮流轴
ebb duration	落潮历时
ebb gate	落潮闸门
ebb surge	落潮涌浪
ebb tide	退潮；落潮
ebb volume	落潮量
ebb-and-flow spring	潮汐泉；涨落泉
ebbing and flowing spring	潮汐泉
ebbing well	随潮汐波动的井
ebb-tide gate	落潮闸门

ebullient pool 鼓泡泉塘；沸泉塘
ebullition 沸腾；起泡
e-business 电子商务
eccentered gun 偏心射孔器
eccentric angle 偏心角
eccentric compression 偏心压力；偏心受压；偏心压缩
eccentric compression method 偏心受压法
eccentric dipole 偏心偶极子
eccentric factor 偏心系数
eccentric force 偏心力
eccentric jaw crusher 颚式偏心碎石机
eccentric load 偏心荷载；偏心负载
eccentric load with moment 有弯矩的偏心荷载
eccentric loaded foundation 偏心荷载基础
eccentric mass 偏心质量
eccentric observation 偏心观测
eccentric radius 偏心半径
eccentric ratio 偏心率
eccentric screen 偏心筛
eccentric tension 偏心受拉；偏心拉伸
eccentric well 偏心井；非中心井；偏心钻井
eccentrical loaded footing 偏心荷载的基础
eccentrical loading on abutment 桥台偏心受压
eccentrically compressed lining 偏压衬砌
eccentrically compressed member 偏心受压杆件
eccentrically loaded footing 偏心荷载基础
eccentrically loaded rigid footing 偏心荷载刚性基础
eccentrically-precompressed precast-concrete pile 偏心预压混凝土预制桩
eccentricity 偏心率；偏心距；偏心度；偏心性；偏离
eccentricity angle 偏心角
eccentricity coefficient 偏心系数
eccentricity correction 偏心率改正；偏心率修正
eccentricity of ellipsoid 椭球偏心率
eccentricity of foundation loading 基础荷载偏心距
eccentricity of loading 荷载偏心距
eccentricity ratio 偏心率；偏心比
eccentricity tester 径向跳动检查仪；偏心度测试仪；偏心距检查仪
eccentric-type vibrating screen 偏心振动筛
eccentric-type vibrator 偏心振动器
ecdemite 氯砷铅矿
echellite 钠沸石；白沸石
echelon faults 雁行断层；斜列断层；平行梯状断层；阶状断层
echelon folding 雁行褶皱；斜列褶皱
echelon folds 雁行褶皱；斜列褶皱；阶状褶皱
echelon fracture 雁行式裂缝
echelon like veins 斜列式矿脉
echelon pattern 雁行构造型式
echelon structure 雁行构造；阶梯构造
echelon tension joint 雁行式张节理
echelon veins 斜列式矿脉；雁行矿脉
echo depth sounder 回声测深仪
echo pulse 回波脉冲
echo sounder 回声测深仪；测声仪；声呐探测仪
echo sounding apparatus 回声测深仪
echo survey 回声探测
echo time 回声时间；回波时间
echo wave 回波；回声波

echoed signal 回波信号
echo-fathom 回声测深
echogram 音响测深图；回声测声曲线；回声波图
echo-image 回波图像
echoing 回波现象
echoing characteristics 回波特性
echometer 回声测距仪；回声测深仪；回声探测仪
echometry 测回声术
echo-pulse 回波脉冲
echo-ranging system 回声测距系统
echo-resonator 回波谐振器
echo-sounding 回声测深法，声呐；回声探测；回声测深
echo-sounding device 回声测深仪
echo-sounding instrument 回声测深仪
echo-sounding receiver 回声接收器；回声测深接收机
echo-strength indicator 回波强度指示器
eckermannite 镁铝钠闪石；氟镁钠闪石
eclipse factor 阴影率
eclipse of the moon 月食
eclipse of the sun 日食
ecliptic coordinate 黄道坐标
ecliptic coordinate system 黄道坐标系
ecliptic latitude 黄道纬度
ecliptic longitude 黄道经度
ecliptic obliquity 黄道斜度
eclogite facies 榴辉岩相
eclogite layer 榴辉岩圈；榴辉岩层
eclogite sphere 榴辉岩圈
ecoclimate forecasting 生态气候预测
ecoclimatology 生态气候学
ecocline 生态变异；生态倾差；生态差型
ecocycle 生态循环
ecological assessment study 生态评估研究
Ecological changes caused by mine development 矿井开发引起的生态变化
ecological character 生态性状
ecological deterioration of mining area 矿区生态破坏；矿区生态恶化
ecological distribution 生态分布
ecological facies 生态相；环境相
ecological factor 生态因素
ecological genetics 生态遗传学
ecological habitat 生态生长环境
ecological health risk assessment 生态健康风险评估
ecological impact 生态影响；生态冲击
ecological map 生态地理图
ecological pollution limit 生态污染极限
ecological restoration 生态恢复
ecological risk assessment 生态风险评估
ecological science 生态科学
ecological structure 生态结构
ecological system 生态系统
ecological type 生态型；生态类型
ecologically sensitive area 生态易受破坏地区
e-commerce 电子商务
econometric analysis model 计量经济分析模型
econometric model 经济计量模型；计量经济模型；计量经济学模式
econometrics 计量经济学；经济计量学
economic and efficiency analysis 经济效益分析

economic appraisal 经济评估；经济评价
economic benefit 经济效益
economic contract 经济合同
economic cost 经济成本
economic effect 经济效应
economic efficiency 经济效率
economic evaluation of project 工程经济评价
economic investigation 经济调查
economic section of airway 通风井巷经济断面
economic site 经济选址
economic studies 经济比较或经济设计
economic survey 经济调查
economic thermo resource 有经济价值地热资源
economic worth of project 工程经济价值
economic yield 经济合理抽水量；经济采收率；经济产量
economical loss 经济损失
economical ratio of reinforcement to concrete 混凝土的经济配筋率
economics of energy 能源经济学
economics of management 管理经济学
economics of scale 规模经济学
economill 轻便钻头；便携钻头
econtone 生态混合群落；生态过渡区；溶泌区
ecosphere 生物域；生物圈；生物大气层；生态圈
ecostratigraphic 生态地层学的
ecostratigraphic classification 生态地层分类
ecostratigraphic unit 生态地层单位
ecostratigraphy 生态地层学
ecosystem 生态系统
ecosystem component 生态系统要素
ecosystem dynamics 生态系统动力学
ecosystem function 生态系统功能
ecotelemetry 生态遥测术；生态遥测学
ecotype 生态类型；生态型
ecoulement 重力滑动；重力滑落
ecronic 港湾；河口湾；感潮河口
ectect 泌出变熔体
ectinite 非混合岩类
edaphic control 底土控制
edaphic factor 土壤因素
eddies frequency 涡旋频率
eddy 涡，旋涡；涡流；回流
eddy current 涡流；回流；涡电流
eddy current effect 涡流效应
eddy current examination 涡流检验
eddy flux 涡流通量；涡动通量
eddy loss 涡流损失
eddy marking 涡流痕迹
eddy motion 涡流运动；涡动
eddy resistance 涡流阻力
eddy scale 涡旋尺度
eddy shedding 涡流分离
eddy stress 涡流应力
eddy thermal conductivity 涡流热传导
eddy transfer 涡流传递
eddy velocity 涡流速度
eddy wind 旋风
eddy with horizontal axis 横轴旋涡
eddy with vertical axis 立轴旋涡
eddy zone 涡流区

eddy-current gauge 涡流测量计
eddy-current loss 涡流损耗
eddy-current type geophone 涡流式检波器
eddy-deposited silt 涡流沉积淤泥
eddying flow 涡流，旋流
edelite 葡萄石
edenite 浅闪石
edge action 边缘作用
edge angle 棱角；边缘角
edge aquifer 边部水体，边水
edge coal 边缘煤
edge coal group 边缘煤群
edge condition 边缘条件；边界条件
edge crack 横裂隙；边缘裂隙
edge damage 破边，边缘损伤
edge failure 边缘破坏，边际破坏
edge gradient 边缘梯度
edge limestone 竹叶状石灰岩；立砾石灰岩
edge of continental shelf 大陆架边缘
edge orientation 晶棱定向
edge radius 棱角半径
edge reflection 边界反射
edge resolution 边缘分辨率
edge ring 环刀
edge trace 边缘迹线
edge water 边水；边缘水；层边水
edge water drive 层边水驱动
edged ring 环刀
edgetone effect 尖劈效应
edgewise breccia 竹叶角砾岩
edgewise conglomerate 竹叶砾岩；竹叶状砾岩
edgewise limestone 竹叶状石灰岩
Ediacara fauna 埃迪卡拉动物群
edifice 大型建筑物；火山机构，火山筑积物
Edison effect 爱迪生效应；热电放射效应
edisonite 斑点绿松石；斜方金红石
edmonsonite 镍纹石；铁镍陨石
EDM-trigonometric leveling 电磁波测距三角高程测量
edolite 长云角页岩
eduction 出流，排水；出口；露头；推断；析出
eduction pipe 气举管；排泄管；排气管
eduction tube 气举管；排泄管
eduction valve 排气阀；泄流阀
eductor 气举管；喷射器；喷射器械
eductor well point 喷射井点
eductor well point system 喷射井点系统
Edward balance 爱德华天然气比重天平
edwardsite 独居石
Eemian interglacial stage 埃姆间冰期
EESD 地震工程和结构动力学
effect of anisotropy 各向异性效应
effect of cumulative energy 聚能效用
effect of delayed gravity drainage 重力疏干滞后效应；重力排水滞后效应
effect of depth 深度效应
effect of dragging 拖曳作用
effect of earth tide 地球潮汐影响；地体潮效应
effect of fault 断层影响；断层作用
effect of inertia 惯性作用
effect of investment 投资效果

effect of load	荷载效应
effect of matrix suction	基质吸力影响
effect of overconsolidation	超固结效应
effect of pore-water pressure	孔隙水压力效应
effect of sample disturbance	试样扰动影响
effect of seaboard	海岸效应
effect of self-purification	自净效应
effect of shock wave	冲击波效应;冲击波的影响
effect of temperature	温度效应
effect of vibration	振动效应
effect of void ratio	孔隙比效应
effect of wall friction	壁面摩擦影响
effect of wellbore storage	井筒储存效应
effect of yielding	屈服效应
effect on water table	对地下水位的影响;对潜水面的影响
effective abstraction	有效吸水量;有效降水差;有效退流
effective acceleration	有效加速度
effective acceleration of gravity	有效重力加速度
effective anchor diameter	有效锚固直径
effective anchored length	有效锚固长度
effective angle of friction	有效摩擦角
effective angle of inner friction	有效内摩擦角
effective angle of internal friction	有效内摩擦角
effective angle of shearing resistance	有效剪切角;有效抗剪角
effective angular field	有效视场;有效角视场
effective anomaly	有效异常
effective aperture	有效孔径;有效口径
effective area of concrete	混凝土有效面积
effective area of orifice	孔口有效面积
effective area of reinforcement bar	配筋的有效面积
effective bearing area	有效支承面积
effective bulk modulus	有效体积模量
effective burden	有效载重
effective bursting energy	有效爆破能量
effective capacity	有效容量;有效库容;有效功率
effective cohesion	有效黏聚力
effective cohesion intercept	有效黏聚力截距
effective compressibility	有效压缩性;有效压缩系数
effective compression ratio	有效压缩比
effective confining pressure	有效围压
effective consolidation pressure	有效固结压力
effective consolidation stress	有效固结应力
effective cross section	有效横截面;有效截面
effective cross-section	有效横截面;有效截面
effective damping	有效阻尼
effective damping ratio	有效阻尼比
effective deformation	有效沉降量,有效体变;有效变形
effective density of soil	土的有效密度
effective depth	有效高度;有效深度;有效厚度
effective diameter of grain	有效粒径
effective diameter of void	有效孔隙直径
effective discharge	有效排量
effective dose	有效剂量
effective drainage	有效排水系统
effective drainage area	有效排水面积
effective drainage porosity	有效排水孔隙率
effective dynamic modulus	有效动弹性模量
effective elastic modulus	有效弹性模量
effective emanation coefficient	有效放射系数
effective energy	有效能量
effective equilibrium coefficient	有效平衡系数
effective evapotranspiration	有效蒸散
effective field intensity	有效场强
effective filter length	滤水管有效长度
effective flooding permeability	有效驱动渗透率
effective flow	有效流量
effective footing length	有效基础长度
effective force	有效作用力;有效应力;有效力
effective foundation area	有效基础面积
effective friction angle	有效摩擦角
effective grain diameter	有效粒径;有效直径
effective grain size	有效粒径
effective grain-diameter	有效粒径
effective groundwater velocity	地下水实际流速;地下水有效流速
effective horizontal stress	有效水平应力
effective hydrodynamic dispersion coefficient	有效水动力弥散系数
effective influence distance	有效影响距离
effective in-situ overburden pressure	有效原位上覆压力
effective lateral pressure	有效侧压力
effective lateral stress	有效侧向应力
effective layer	有效产油层
effective length of a track	线路有效长度
effective length of platform	站台有效长度
effective life	有效寿命;有效使用期,使用期限
effective magnetization	有效磁化
effective matrix parameter	有效基质参数
effective minor principal stress	有效最小主应力
effective modulus of elasticity	有效弹性模数
effective molecular weight	有效分子量
effective multiplication factor	有效放大系数;有效增殖系数;有效倍增系数
effective noise temperature	等效噪声温度
effective normal stress	有效法向应力
effective opening	有效筛孔;有效孔隙
effective opening size	有效孔径
effective output	有效产量
effective overburden pressure	有效覆盖压力;有效上覆压力
effective particle density	有效颗粒密度
effective pay	有效产层
effective pay factor	生产层有效因素
effective pay thickness	产层有效厚度
effective perforation	有效炮眼
effective permeability	有效透水性;有效渗透性;有效渗透率;有效渗透系数
effective phreatic line	有效的浸润线;有效地下水位线
effective pillar length	有效支柱长度
effective placement	有效充填
effective pore diameter	有效孔隙直径
effective pore volume	有效孔隙体积;有效孔隙容积
effective porosity	有效孔隙率;有效孔隙度;给水度;岩石或岩层孔隙的含水量
effective precipitable water	有效可降水量
effective precipitation	有效降水量;净雨量
effective pressure	有效压力;有效力
effective prestress	有效预应力

effective principal stress 有效主应力
effective radiation 有效辐射
effective radius 有效半径
effective rainfall 有效降雨量
effective recharge boundary 有效补给边界
effective roughness angle 有效粗糙度角
effective saturation 有效饱和度
effective saturation line 有效浸润线
effective screen cut-point 实际的筛分粒度
effective screening area 筛面工作面积；筛面有效筛分面积
effective section 有效截面；有效段
effective shear strength parameter 有效抗剪强度参数
effective size 有效大小；有效尺寸
effective size of grain 有效粒径
effective size of sand 砂的有效粒径
effective slit 有效缝隙
effective source rock 有效生油岩；有效烃源岩
effective spacing 有效间距
effective specific fuel consumption 有效燃油消耗率
effective spreading radius 有效扩散半径
effective squeeze pressure 有效挤压力
effective stiffness 有效刚度
effective stiffness theory 有效刚度理论
effective storage 有效库容；有效蓄水量
effective storage of reservoir 水库有效库容
effective strain 有效应变
effective strength 有效强度
effective strength envelope 有效强度包线
effective strength parameter 有效强度参数
effective stress 有效应力
effective stress analysis 有效应力分析
effective stress circle 有效应力圆
effective stress concentration factor 有效应力集中系数
effective stress difference 有效应力差
effective stress equation 有效应力方程；有效应力公式
effective stress law 有效应力定律
effective stress method 有效应力法
effective stress parameter 有效应力参数
effective stress path 有效应力路径
effective stress strength index 有效应力强度指标
effective stress tensor 有效应力张量
effective stress theory 有效应力原理
effective superficial porosity 有效表面孔隙率
effective surcharge 有效超载，有效附加荷载
effective surface 有效表面积
effective terrestrial radiation 有效大地辐射；有效地面辐射
effective track length 线路有效长；有效轨道长度
effective unit weight 有效重度；有效容重；有效单位重
effective velocity 有效速度
effective vertical pressure 有效垂直压力
effective viscosity 有效黏度
effective volume 有效容积；有效容量
effective well radius 井的有效半径
effective width 有效宽度
effective width coefficient 有效宽度系数
effective-water-holding capacity 有效持水能力
effects of actions 作用效应
effervescence 起泡作用；发泡，沸腾现象
effervescent 起泡的
effervescent spring 鼓泡泉；珍珠泉；碳酸泉
efficacy 效率；效能；功效；效力；有效性
efficiency 效率；效能，性能；实效；功率；有效系数；供给量
efficiency coefficient method 功效系数法
efficiency coefficient of strengthening 加固效率系数
efficiency diagram 效率图
efficiency distribution coefficient 有效分布系数
efficiency estimate 有效估计量
efficiency extraction 有效提取
efficiency factor 效率因数；筛分效率系数；效率因子
efficiency factor of pile group 群桩效率系数
efficiency formula 效率公式
efficiency index of curing compound 养护药品的效率指标；保养药品的效率指标；养护剂的效率指标
efficiency of conversion 转换效率
efficiency of core penetration 岩芯穿透深度
efficiency of disaster defence 防灾效果
efficiency of displacement 驱替效率
efficiency of estimator 估计量的有效性
efficiency of heat utilization 热利用效率；热效率
efficiency of hydraulic jump 水跃的效率
efficiency of imbibition 自吸效率
efficiency of management 管理效率
efficiency of plant 电站效率
efficiency of pump 泵效率
efficiency of rectification 整流效率
efficiency of stream 河流搬运效率
efficiency of underground gasification 地下气化效率
efficiency rate of mine air quantity 矿井有效风量
efficiency ratio 效率比
efficiency reduction 效率降低；功效降低
efficiency statistic 有效统计量
efficiency test 有效试验；效率试验
efficiency value 效率值
efficient compaction 有效压实
efficient distribution coefficient 有效分布系数
efficient drilling 高效钻进
efflatum 火山喷出物
efflorescence 渗斑；风化，风化物；凝霜，霜化，盐霜；盐析，泛碱；盐华，粉化
efflorescent 粉化的；风化的
efflorescent crust 盐华壳
efflorescent ice 花状冰
effluent cave 出水溶洞；出水洞穴
effluent channel 流水沟；排水渠；排泄道
effluent concentration 流出物浓度；废水集中；排放浓度
effluent discharge 污水排放
effluent disposal system 污水处理系统
effluent flow 溢流；表面水流；潜水补给流
effluent injection 废水灌注；污水回灌
effluent lava flow 侧熔岩流
effluent outfall 污水出口管；污水排放管
effluent per unit length 单位长度出流
effluent river 地下水补给河；潜水补给河；吸声衬砌
effluent seepage 渗出；渗漏；渗流；坡面渗流；自内渗出
effluent sewage 废污水；流出污水

effluent treatment plant	污水处理车间；污水处理厂
effluent tunnel	污水隧道，排污隧道
effluent waste water	废水
effluent weir	出流堰
effluent-impounded body	地下水补给水体
efflux area	出流面积
efflux equation	出流方程
efflux of heat	热流；热流量
efflux pump	射流泵
efflux ventilation for tunnel	隧道射流式通风
effusion	渗出，泻出，渗漏物
effusion meter	流量计
effusive explosive eruption	爆裂溢出；爆裂喷出
effusive mass	喷发体
effusive period	喷发期
effusive rock	喷出岩；喷发岩；火山岩
effusive stage	喷发期
EFM	工程断裂力学
efremovite	铁磷钙石
egeran	符山石
egg coal	蛋级无烟煤；小块煤
egg coke	小块焦炭；蛋级焦炭
egg stone	鲕石
eggbeater PDC bit	打蛋器型 PDC 钻头
eggette	煤球
egg-like hypothesis	浑天说
eggonite	水磷铝石
egg-shaped cross section	卵形截面
eglestonite	褐氯汞矿；氯汞矿
egress of heat	热传导；传热；散热；热泉露头
egress pressure	出口压力
eidograph	缩放绘图仪
eidophor	大图像投射器
eigen space	特征空间
eigen value	特征值；本征值
eigen vector	特征矢量
eigencone	特征锥面
eigencurve	特征曲线
eigenfrequency	特征频率；本征频率
eigenfrequency spectrum	固有频谱
eigenfunction	特征函数；本征函数
eigen-function	特征函数；本征函数
eigenmatrix	特征矩阵；本征矩阵
eigenmode	正则型；本征型；固有模态
eigenperiod	固有周期；特征周期；本征周期
eigenstate	特征状态
eigentone	本征音；固有振动频率
eigenvalue	本征值；特征值
eigenvalue computation	特征值计算
eigenvalue extraction	特征值提取
eigenvalue problem	特征值问题
eigenvalue vector	特征值向量
eigenvector	特征向量；本征矢量；本征向量
eigenvibration	固有振动；本征振动；特征振动
eigenwavefront	特征波前
eigenwert	特征值
eikonometer	光像测定器；物像计
Einstein's equation	爱因斯坦方程
ejected rock	喷出岩
ejected scoria	喷出火山渣
ejecting action	喷射作用
ejection	喷出，喷射，排出，挤出；喷出物，排出物
ejection curve	退出曲线
ejection efficiency	退出效率
ejection test	喷射试验
ejection velocity	喷出速度
ejective fold	隔挡褶皱
ejector air pump	喷射空气泵；气流喷射泵；喷射抽气泵
ejector condenser	喷射冷凝器；喷射凝汽器
ejector priming	喷射泵启动；启动泵
ejector pump	喷射泵；射流泵；喷射式水泵
ejector return pin	复位杆
ejector rod	推顶柱
ejector sleeve	推顶套
ejector vacuum pump	喷射真空泵
ejector well point	喷射井点
ejector well point system	喷射井点系统
ejector-type through-tubing tool	喷射式过油管下井仪
ekanite	硅钙铁铀钍矿
eka-silicon	准硅
Eke Andesite	俄柯安山岩
ekerite	钠闪花岗岩
ekistics	城市与区域计划学；人类群居学
Ekman dredge	埃克曼采泥器
Ekman drift current	埃克曼漂流
ekmannite	锰叶泥石
ekzema	盐穹；盐垒
El Nino phenomenon	厄尔尼诺现象
elaeite	叶绿矾
elaeolite	脂光石
elaeometer	验油比重计；油脂比重计
elapsed time	经过的时间
elastic	弹性的，易伸缩的；橡皮线，橡皮带，松紧带
elastic absorption	弹性吸收
elastic abutment	弹性拱座
elastic acoustical reactance	弹性声阻抗
elastic aftereffect	弹性后效
elastic aftershock	弹性余震
elastic afterworking	弹性后效
elastic analogy	弹性模拟
elastic analysis	弹性分析
elastic anisotropy	弹性各向异性
elastic aquifer	弹性含水层
elastic artesian aquifer	弹性自流含水层
elastic beam	弹性梁
elastic bed	弹性地基
elastic behavior	弹性特性；弹性行为
elastic bending	弹性弯曲
elastic bitumen	弹性沥青
elastic body	弹性体
elastic boundary	弹性边界
elastic boundary condition	弹性边界条件
elastic breakdown	弹性破坏；弹性失效；弹性断裂
elastic buckling	弹性屈曲
elastic buffer	弹性缓冲器
elastic calculation	弹性计算
elastic capacity coefficient	弹性容量系数
elastic central impact	弹性中心碰撞；弹性对心冲击
elastic centre	弹性中心，弹心

elastic coefficient	弹性系数
elastic coefficient matrix	弹性系数矩阵
elastic collision	弹性碰撞；弹性撞击
elastic compaction	弹性挤压作用
elastic component	弹性构件
elastic compression	弹性压缩
elastic confined aquifer	弹性承压含水层
elastic constant	弹性常数
elastic constitutive equation	弹性本构方程
elastic continuum	弹性连续介质
elastic creep	弹性蠕动
elastic curve	弹性曲线
elastic cylinder	弹性柱体
elastic damping	弹性阻尼
elastic deflection	弹性挠度；弹性挠曲
elastic deformation	弹性变形；弹性形变
elastic deformation aquifer	弹性变形含水层
elastic design	弹性设计
elastic discontinuity	弹性不连续性；弹性不连续面
elastic dislocation	弹性位错
elastic displacement response	弹性位移响应
elastic distortion	弹性畸变；弹性扭曲
elastic distortion settlement	弹性畸变沉降量
elastic disturbance	弹性扰动
elastic domain	弹性域；弹性区；弹性范围
elastic drift	弹性后效；弹性残余变形
elastic drive	弹性驱动
elastic effect	弹性效应
elastic elongation	弹性拉伸
elastic embankment	弹性土堤
elastic energy	弹性能；弹性能量
elastic energy index	弹性能量指数
elastic energy release	弹性能释放
elastic equilibrium	弹性平衡
elastic expansion joint	弹性膨胀接头
elastic extension	弹性延伸；弹性伸长
elastic failure	弹性破坏；弹性失效
elastic fastening	弹性扣件
elastic fatigue	弹性疲劳
elastic fibre	弹性纤维
elastic field	弹性场
elastic flow	弹性流；弹性流动
elastic fluid	弹性流体
elastic force	弹力；弹性力
elastic formation	弹性地层
elastic formula	弹性公式
elastic foundation	弹性地基；弹性基础
elastic foundation beam method	弹性地基梁法
elastic gas drive	弹性气压驱动
elastic half space	弹性半空间
elastic half space theory	弹性半空间理论
elastic heave	弹性隆起
elastic heterogeneity	弹性非均匀性
elastic homogeneous material	弹性均质材料
elastic hysteresis	弹性滞后
elastic impedance	弹性阻抗
elastic imperfection	弹性缺陷
elastic instability	弹性不稳定性；弹性失稳
elastic isotropic material	弹性各向同性材料
elastic isotropy	弹性各向同性
elastic lag	弹性滞后；弹性惯性；弹性滞回
elastic layer	弹性层；弹性垫层
elastic limit	弹性极限；弹性限度
elastic line	弹性线；弹性轴线；弹性曲线
elastic line method	弹性线法
elastic liquid	弹性液体
elastic load	弹性荷载
elastic loss	弹性损失
elastic mass	弹性体
elastic material	弹性材料
elastic mechanics	弹性力学
elastic method	弹性法
elastic migration theory	弹性波偏移理论
elastic mineral pitch	弹性沥青；弹性柏油
elastic model	弹性模型
elastic moduli	弹性模数；弹性模量
elastic modulus	弹性模量；弹性模数；弹性系数
elastic multilayer theory	弹性层状体系理论
elastic normal stress component	弹性法向应力分量
elastic peak	弹性峰值
elastic permanent deformation	弹性永久变形；弹性残余变形
elastic phase	弹性阶段
elastic pile	弹性桩
elastic plan	弹性计划
elastic plate method	弹性板法
elastic potential energy	弹性势能
elastic potential theory	弹性势理论
elastic property	弹性性质；弹性特性
elastic range	弹性范围；弹性区；弹性限度
elastic ratio	弹性比
elastic reaction	弹性反应；弹性抗力
elastic rebound	弹性回跳；弹性回弹
elastic rebound hypothesis	弹性回跳假说
elastic rebound mechanism	弹性回跳机理
elastic rebound theory	弹性回跳理论
elastic rebound theory of earthquake	地震的弹性回跳理论
elastic recoil	弹性反冲；弹性回位
elastic region	弹性界限；弹性区，弹性区域
elastic removal	弹性迁移；弹性消除，弹性移除
elastic reserves	弹性储量
elastic resilience	回弹性；弹性回弹能；弹性回复能力
elastic resistance	弹性阻力；弹性抗力
elastic response	弹性反应；弹性响应
elastic response analysis	弹性反应分析；弹性响应分析
elastic restoring force	弹性回复力
elastic restraint	弹性约束
elastic rock	弹性岩石
elastic rock mass	弹性岩体
elastic scattering	弹性散射
elastic scattering cross-section	弹性散射横截面
elastic seal	弹性密封；弹性填料
elastic section modulus	弹性截面模量
elastic seismic coefficient	弹性地震系数
elastic seismic energy	弹性地震能量
elastic seismic response of structure	结构弹性地震响应
elastic semi-infinite body	弹性半无限体
elastic semi-infinite foundation	弹性半无限地基
elastic settlement	弹性沉降；弹性沉降量

elastic shell equation	弹性薄壳方程
elastic shock	弹性振动；弹性冲击
elastic side wall	弹性边墙
elastic similarity	弹性相似
elastic soil wedge	弹性土楔
elastic solid	弹性体；弹性固体
elastic squeeze-out	弹性挤开
elastic stability	弹性稳定；弹性稳定性
elastic stage	弹性阶段
elastic state	弹性状态
elastic state of equilibrium	弹性平衡状态
elastic stiffness	弹性刚度
elastic stiffness constant	弹性刚度常数
elastic storage	弹性储存量
elastic strain	弹性应变
elastic strain energy	弹性应变能；弹性能
elastic strain energy density	弹性应变能密度
elastic strain rebound	弹性应变回跳
elastic strain recovery	弹性应变恢复
elastic stress condition	弹性应力状态；弹性应力条件
elastic subgrade	弹性地基；弹性路基
elastic subgrade reaction	弹性地基反力；弹性路基反力
elastic support	弹性支座；弹性支持装置
elastic supported edge	弹性支承边
elastic system	弹性系统
elastic tangential strain component	弹性切向应变分量
elastic tangential tensile stress	弹性切向张应力
elastic theory	弹性理论
elastic visco-plastic model	弹黏塑性模型
elastic wall	弹性墙
elastic wave	弹性波
elastic wave exploration	弹性波探测法
elastic wave method	弹性波法
elastic yield	弹性屈服
elastic zone	弹性区；弹性带；弹性变形区
elastically supported beam	弹性支承梁
elastic-brittle material	弹脆性材料
elasticity	弹性，弹力；伸缩性；弹性力学
elasticity constant	弹性常数
elasticity equation	弹性方程
elasticity figure	弹性图形
elasticity form	弹性形式
elasticity matrix	弹性矩阵
elasticity mechanics	弹性力学
elasticity modulus	弹性，弹性模量
elasticity number	弹性数
elasticity of aquifer	含水层弹性
elasticity of compression	压缩弹性
elasticity of extension	拉伸弹性
elasticity of rock	岩石弹性
elasticity of shearing	剪切弹性
elasticity of torsion	扭转弹性
elasticity ratio	弹性比
elasticity test	弹性试验
elasticizer	增塑性；弹性增进剂；增韧剂
elastico-viscosity	弹性黏度
elastico-viscous	弹黏性的
elastico-viscous behavior	滞黏性；弹黏性状态；黏弹性行为
elastico-viscous body	弹黏体
elastico-viscous liquid	弹黏性液体
elastico-viscous property	弹黏性
elastico-viscous solid	弹黏性固体
elastic-perfectly plastic material	弹性-完全塑性材料
elastic-plastic analysis	弹塑性分析
elastic-plastic behavior	弹塑性性状；弹塑性行为
elastic-plastic boundary zone	弹塑性边界带；弹塑性边界区
elastic-plastic deformation	弹塑性变形
elastic-plastic domain	弹塑性区；弹塑性范围
elastic-plastic equilibrium	弹塑性平衡
elastic-plastic fracture mechanics	弹塑性断裂力学
elastic-plastic material	弹塑性材料
elastic-plastic matrix	弹塑性矩阵
elastic-plastic medium	弹塑性介质
elastic-plastic model	弹塑性模型
elastic-plastic region	弹塑性区
elastic-plastic solid	弹塑性体
elastic-plastic state	弹塑性状态
elastic-plasticity	弹塑性
elastic-rebound theory	弹性回跳说；弹性回跳理论
elastic-viscous flow	弹黏性流
elastic-viscous system	弹黏体系
elastic-wave theory	弹性波理论
elastodynamic	弹性动力学的
elastodynamic displacement	动力弹性位移
elastodynamic model	弹性动力模型
elastodynamics	弹性动力学
elastohydrodynamic lubrication	弹性流体动力润滑
elastohydrodynamics	弹性流体动力学
elastokinetics	弹性动力学
elastomeric bearing yielding support	弹性支承
elastomeric polymer	弹性聚合物
elastometer	弹性计；弹力计
elastoplastic	弹塑性的
elastoplastic analysis	弹塑性分析
elastoplastic beam	弹塑性梁
elastoplastic behavior	弹塑性特征
elastoplastic bending	弹塑性弯曲
elastoplastic cellular automation	弹塑性细胞自动机
elastoplastic continuum	弹塑性连续介质
elastoplastic coupling	弹塑性耦合
elastoplastic deformation	弹塑性变形
elastoplastic fracture	弹塑性断裂
elastoplastic fracture mechanics	弹塑性断裂力学
elastoplastic law	弹塑性法则
elastoplastic mass	弹塑性体
elastoplastic material	弹塑性材料
elastoplastic model	弹塑性模型
elastoplastic range	弹塑性范围；弹塑性区
elastoplastic response	弹塑性反应；弹塑性响应
elastoplastic response spectrum	弹塑性反应谱；弹塑性响应频谱
elastoplastic system	弹塑性体系
elastoplastic theory	弹塑性理论
elastoplastic wedge	弹塑性楔体
elastoplastic yielding	弹塑性屈服
elastoplasticity	弹塑性
elastoplasticity of soil	土的弹塑性
elastoplasticity theory	弹塑性理论

elastostatics	弹性静力学
elasto-viscoplastic	弹黏塑性的
elastoviscosity	黏弹性；弹黏性
elastoviscous behavior	黏弹性性状
elasto-viscous field	弹性黏性场
elastoviscous flow	弹黏性流
elastoviscous system	黏弹性体系
elaterite	弹性沥青；矿质橡胶
elaterometer	气体密度计
elbaite	锂电气石
electric altimeter	电测高度计
electric analog computer	电模拟计算机
electric blasting machine	电爆机
electric blasting unit	电力放炮机
electric cable	电缆；电线
electric cap	电雷管
electric capstan	电动绞盘；电动起锚机
electric cement bending resistance machine	水泥电动抗折机
electric circular saw	电动圆锯
electric conductivity	导电性；电导率
electric conductivity measurement	电导率测量
electric corundum	电熔刚玉
electric crane	电动起重机
electric curing of concrete	混凝土电热养护
electric current meter	电测流速仪
electric cutter	切割机
electric delay detonator	延发电雷管
electric detonator	电雷管
electric disintegration drilling	电破碎钻进
electric distribution box at workings	采掘工作面配电点
electric doublet	电偶极子
electric dynamometer	电测力计；电测功率计
electric elevator	电力升降机；电梯；电机驱动电梯
electric excavator	电铲
electric exploder	电爆器
electric eye device	电眼装置
electric fault	电气故障
electric field gradient	电场梯度
electric field intensity	电场强度
electric field meter	电场强度计
electric firing	电引爆；电发火；电放炮
electric fluviograph	电测水位计
electric heater drilling	电热钻进
electric high-pressure oil pump	高压电动油泵
electric hoist	电动吊车；电动葫芦；电力提升机；电绞车；电力起重机
electric ignition	电点火
electric illumination of mine	矿山电气照明
electric impact drill	强力冲击电钻
electric lift	电动升降机；电梯
electric linear transducer	线性电测传感器
electric locomotive for narrow gauge	窄轨电机车
electric locomotive standard gauge	准轨电机车
electric log	电测井；电测井曲线；电测记录；电测井图
electric logger	电测井记录仪
electric logging	电测井
electric motor operated conveyer	电动传送设备
electric motor train unit	电动车组
electric multiple unit	电动车组
electric oil pump	电动油泵
electric overhead line	架空电缆
electric piezometer	电测孔隙水压力计
electric pilot	电测仪
electric potential log	自然电位测井
electric power distribution cabinet	配电柜
electric power distribution of open-pit mine	露天采矿场配电
electric power distribution of underground mine	地下矿配电
electric power transmission	电力输送
electric powered load-haul-dump unit	电动铲运机
electric priming circuit	电爆网络；电动起爆回路
electric probing device	电测装置；电探针，电探头
electric pump	电动泵；电动抽水机
electric railway	电气化铁路；电力铁路
electric resistance-wire strain gauge	电阻丝弹性应变计
electric resistivity	电阻率
electric resistivity method	电阻法
electric resistivity thermometer	电阻测温仪
electric riveting machine	电动铆钉机
electric rock drill	电动凿岩机；岩石电钻
electric rock hammer	电动凿岩机
electric salinity meter	电测盐度计
electric semaphore signal	电动臂板信号机
electric shaker	电动摇筛器
electric shovel	电铲；电动挖土机
electric sounding	电测深，电测水深
electric sounding curve	电测深曲线
electric spark	电火花；电火花震源装置
electric stabilization	电加固
electric surveying	电法探测；电法勘探；电测井
electric tape gauge	电测水位计；电动带尺潮位计
electric traction	电力牵引
electric traction interference	电力牵引干扰
electric traction remote control system	电力牵引遥控系统
electric traction telemechanical system	电力牵引遥控力学系统
electric travelling crane	电动起重机；电力移动起重机
electric twinning	电双晶
electric two beam bridge type crane	电动双梁桥式起重机
electric type dip meter	电测磁倾角测量仪；电测倾斜仪
electric walking dragline	电动步行式拉铲挖土机
electric water pump	电动水泵
electric water sounder	电测水深器
electric wave recorder	电测波浪计
electric well log	电测钻井记录；电测钻井剖面
electrical analog method	电模拟法
electrical analogue for groundwater flow	地下水流的电模拟
electrical analogy	电模拟
electrical capacitance tide gauge	电容潮位计
electrical conductivity test	导电率测试
electrical displacement transducer	电位移传感器
electrical drainage	电动排水；电排流
electrical energy	电能；电力
electrical engineering	电气工程
electrical exploration	电法勘探
electrical haulage	电气牵引
electrical logging	电测井；电法测井

electrical mapping 电子绘图
electrical measurement 电测；电气测量
electrical measuring instrument 电测仪表
electrical method 电法；电法勘探
electrical piezometer 电压力计，电测压计
electrical prospecting 电法勘探
electrical prospecting apparatus 电法勘探仪
electrical resistance 电阻
electrical resistance inclinometer 电阻测斜仪；电阻式井斜仪
electrical resistance strain gauge 电阻应变计
electrical resistance tide gauge 电阻潮位计；电阻验潮仪
electrical resistivity 电阻率；电阻系数
electrical resistivity method 电阻率法；电阻系数法
electrical resistivity sounding 电阻测深
electrical resistivity surveying 电阻勘测
electrical soil moisture meter 电传土壤湿度计
electrical sounding 电测深
electrical submersible pump 电动潜水泵
electrical surveying 电法勘探；电法测量
electrical traverse 电测剖面
electrical two beam bridge type crane 电动双梁桥式起重机
electrical well logging 电测井
electrical works 电气工程
electrically-operated pump 电动泵
electrical-magnetic prospecting 电磁勘探
electric-capacity moisture meter 电容湿度计
electric-powered equipment 电动设备
electric-resistivity method 电阻法
electrified railway construction 电气化铁道工程；电气化铁路
electroacoustical transducer 电声换能器
electro-analysis 电解分析；电分析
electrocalorimeter 电量热计；电热计
electrochemical 电化学的
electrochemical activity 电化学活动性
electrochemical affinity 电化学亲和力；电化亲和势
electrochemical corrosion 电化学腐蚀
electrochemical effect 电化学效应
electrochemical gauging 电化学测流法
electrochemical grouting 电化学灌浆
electrochemical hardening 电化学加固
electrochemical injection 电化学注浆
electrochemical process 电化学过程
electrochemical property 电化学特性
electrochemical stabilization 电化学加固
electrochemical treatment 电化学处理；电化学加固
electrochromatography 电色谱；电色层分析法；电色层分离法
electroconductibility 电导率；导电性
electroconductivity 电导率
electrocoppering 电镀铜法
electrocorrosion 电腐蚀
electrodeionization 电去电离作用
electrodeplating 电解溶解
electrodeposit 电镀；电镀层
electrodeposition 电镀；电解沉积
electrodialysis 电渗析
electrodialysis plant 电渗析设备

electrodialysis reversal 反向电渗析
electrodispersion 电分解作用
electro-drainage 电渗排水
electrodrill 电动钻具
electrodynamic repulsion system 电动排斥式系统
electrodynamic suspension 电动悬浮
electrodynamics 电动力学
electrofacies 测井相
electrofacies analysis 测井相分析
electrofacies zonation 划分测井相
electrofiltration 电滤作用
electrofiltration potential 过滤电位
electrofiltration potential field 过滤位场
electrogasdynamics 气电动力学
electrogeochemical method 电地球化学方法
electrogoniometer 相位指示器；电测角计
electrographic 电谱法的
electrography 电记录术；电谱法
electrohydraulic cabinet actuator 电动液压整体型调速控制器
electro-hydraulic control 电液控制
electrohydraulic drilling 电动液压钻进
electro-hydraulic valve 电液阀
electrohydrometer 电湿度计
electro-induction 电感应
electro-injection 电化学灌浆
electro-kinetic injection 电动注浆液法
electrokinetic phenomena 电动现象
electrokinetograph 动电测流计；电动流速仪，电动测速仪
electro-level 电水准仪
electrolog 电测井
electrolysis 电解作用
electrolyte 电解质；电解液；电离质
electrolytic 电解的
electrolytic action 电解作用
electrolytic effect 电解效应；电解作用
electrolytic flotation 电解浮选
electrolytic fractionation 电解分离；电解分馏法
electrolytic ionization 电离
electrolytic logging 电解测井
electrolytic process 电解法；电解过程
electrolytic treatment of wastewater 废水电解处理法
electromachining 电加工
electromagnet 电磁铁；电磁体
electro-magnet elutriation 电磁淘洗法
electromagnetic 电磁的
electromagnetic absorption 电磁吸收
electromagnetic array profiling method 电磁阵列剖面法
electromagnetic attraction 电磁引力
electromagnetic attraction system 电磁吸引式系统
electromagnetic clutch 电磁离合器
electromagnetic coagulation 电磁凝结
electromagnetic compatibility 电磁兼容性
electromagnetic contactor 电磁接触器
electromagnetic coupling effect 电磁耦合效应
electromagnetic current meter 电磁流速仪
electromagnetic damper 电磁阻尼器
electromagnetic damping 电磁阻尼
electromagnetic dispersion 电磁扩散

electromagnetic distance measurement 电磁波测距
electromagnetic distance measuring instrument 电磁波测距仪
electromagnetic distance meter 电磁波测距仪
electromagnetic emission 电磁发射
electromagnetic emission index 电磁发射指标
electromagnetic energy 电磁能
electromagnetic environment 电磁环境
electromagnetic feeder 电磁给料机
electromagnetic field effect 磁场效应
electromagnetic field system 电磁场法勘探系统
electromagnetic fishing tool 电磁打捞工具
electromagnetic force 电磁力
electromagnetic gantry crane 电磁龙门吊；电磁龙门起重机
electromagnetic hypothesis 电磁说
electromagnetic inspection 电磁检验
electromagnetic interference 电磁干扰
electromagnetic logging 电磁波测井
electromagnetic noise 电磁噪声
electromagnetic oscillator 电磁式振荡器
electromagnetic prospecting 电磁法勘探
electromagnetic pulse 电磁脉冲
electromagnetic radiation anomaly 电磁辐射异常
electromagnetic radiation method 电磁辐射法
electromagnetic radiation monitoring technology 电磁辐射监测技术
electromagnetic seismograph 电磁式地震仪；电磁地震检波器
electromagnetic sensor 电磁传感器；电磁能传感器
electromagnetic subsurface probing 电磁地下测探
electromagnetic survey 电磁法测量；电磁勘探
electromagnetic surveying 电磁测量
electromagnetic suspension 电磁悬浮
electromagnetic suspension system 电磁悬浮系统
electromagnetic type sensor 电磁型传感器
electromagnetic velocity meter 电磁速度计
electromagnetic video pulse radar 电磁视频脉冲雷达
electromagnetic wave 电磁波
electromagnetic wave emission 电磁波发射
electromagnetic wave reflection 电磁波反射
electromagnetic well-logging 电磁法测井
electromagnetics 电磁学
electromagnetization 电磁化
electromechanical logging cable 机电测井电缆
electrometric 电测的
electrometric analysis 量电分析；电测量分析
electrometrical 电测的
electromotive 电动的
electromotive method 电测法；电动势法
electromotor 电动机
electron computer 电子计算机
electron density 电子密度
electron density log 电子密度测井
electron distance measuring instrument 电子测距仪
electron emission microscopy 电子发射显微镜检查法；电子发射显微镜学
electron energy loss spectroscopy 电子能量损失谱分析
electron energy spectrometry 电子能谱测定
electron magnetic resonance 电磁共振

electron micrograph 电子显微照片
electron microprobe 电子探针
electron microprobe analyser 电子显微探针分析仪
electron microscope 电子显微镜
electron microscopic analysis 电子显微镜分析
electron microscopy 电子显微镜学；电镜学
electron optics 电子光学
electron pair effect 电子对效应
electron paramagnetic resonance 电子顺磁共振；电子自旋共振
electron probe 电子探针
electron probe microanalysis 电子探针微区分析
electron probing analysis 电子探针分析
electron spectrometer 电子分光计；电子光谱仪
electron spectroscopy 电子能谱学；电子光谱学
electron spin resonance 顺磁共振
electron spin resonance analysis 电子自旋共振分析
electron spin resonance dating 电子自旋共振测年法
electron spin resonance spectroscopy 电子自旋共振波谱学
electron-beam drill 电子束钻井
electron-beam drilling 电子束钻进
electron-beam pumping 电子束泵
electronic 电子的
electronic casing caliper 套管内壁腐蚀电子检查仪；套管电子卡尺
electronic chart 电子图表；电子海图
electronic chart database 电子海图数据库
electronic clinometer 电子倾斜仪
electronic commerce 电子商务
electronic control system 电子控制系统
electronic data processing 电子数据处理
electronic data processing centre 电子数据处理中心
electronic detector 电子探测器
electronic digital display caliper 电子显示卡尺
electronic distance 电子测距仪
electronic distance measurement 电子测距法
electronic distance meter 电子测距仪
electronic distance-measuring technique 电子测距法
electronic dynamometer 电子测力计
electronic humidistat 电子恒湿器
electronic ionizaton 电子电离
electronic lens 电子透镜
electronic map 电子地图；电子海图
electronic measurement 电子仪器量测；电子测量
electronic oven 高频电炉；电烤箱，电子炉
electronic plane-table 电子平板仪
electronic planimeter 电子求积仪；电子面积仪，电子测面器
electronic printer 电子印像机；电子打印机
electronic scanning 电子扫描
electronic serial digital computer 串行电子数字计算机
electronic tilt sensor 电子测倾器
electronic tiltmeter 电子倾斜仪
electronic vibration recorder 电子振动记录器
electronograph 电子显像；电子显微照片
electronography 电子显像术
electron-probe microanalyser 电子探针微量分析仪
electro-optical distance measuring instrument 光电测距仪
electro-optical range finder 光电测距仪
electro-osmosis 电渗排水法；电渗土加固；电渗现象；

电渗作用
electro-osmosis drainage 电渗排水
electro-osmosis method 电渗法
electro-osmosis pressure 电渗透压力
electro-osmosis stabilization 电渗加固
electro-osmosis transmission coefficient 电渗系数
electro-osmotic coefficient of permeability 电渗透系数
electro-osmotic consolidation 电渗固结
electro-osmotic dewatering 电渗排水
electro-osmotic drainage 电渗排水
electro-osmotic permeability 电渗渗透性
electro-osmotic potential 电渗势
electro-osmotic process 电渗作用；电渗过程
electro-osmotic stabilization 电渗加固
electro-osmotic transmission coefficient 电渗传导系数；电渗系数；电渗透射系数
electrophotography 电子照相术
electrophysics 电物理学
electropneumatic conveyer 电控传送设备
electropositive ion 阳离子
electropsychrometer 电测湿度计；电湿度计
electropyrometer 电阻高温计
electroshock 电震
electrosilicification 电动硅化法
electrosmelting 电炉熔炼
electrostatic attraction 静电引力；静电吸引
electrostatic balance 静电平衡
electrostatic coalescing process 静电聚合法；静电聚合过程
electrostatic effect 静电效应
electrostatic filter 静电过滤器
electrostatic force 静电力；静电作用力
electrostatic induced current 静电感应电流
electrostatic precipitation 静电沉淀；静电沉积
electrostatic precipitator 电集尘器
electrostatic protection 静电防护
electrostatic repulsion 静电斥力
electrostatic seismometer 静电地震计；静电地震检波器
electrostatically-precipitated flyash 静电吸尘粉煤灰
electrum 银金矿；琥珀金；电金
element 元素；要素；元件，零件，部件，构件；单元，单元体
element abundance 元素丰度
element area 单元面积；元件区域
element array computation 单元数组计算
element boundary 单元边界
element characteristics 单元特性
element connection 单元连接
element damping matrix 单元阻尼矩阵
element design 构件设计
element discontinuity 单元非连续
element displacement 单元位移
element geochemistry 元素地球化学
element interface 单元交界面
element invariance 单元不变性
element mass matrix 单元质量矩阵
element matrices 单元矩阵
element migration 元素迁移
element modes 单元模式
element of a matrix 矩阵元素

element of absolute orientation 绝对定向元素
element of attitude 产状要素
element of circular curve 圆曲线要素
element of coal seam occurrence 煤层赋存要素；煤层产状要素
element of construction 结构单元；结构组件
element of crystal 晶体要素
element of exterior orientation 像片外方位角元素
element of fold 褶皱要素
element of groundwater regime 地下水动态要素
element of hydrologic balance 水均衡要素
element of integration 积分元素
element of interior orientation 像片内方位角元素
element of matrix 矩阵单元
element of mining 采矿学原理；采矿学基础；采矿概论
element of morphostructure 形态结构要素
element of morphotexture 形态构造要素
element of relative orientation 相对定向元素
element of water balance 水均衡要素
element pattern 井网单元
element property matrix 单元属性矩阵
element residual 单元残留
element shape function 单元形函数
element size parameter 网格尺寸参数
element stiffness 单元刚度；单元劲度
element stiffness matrix 单元刚度矩阵
element strain matrix 单元应变矩阵
element stress 单元应力
element stress matrix 单元应力矩阵
element subdivision 单元剖分；单元细分
element substitution 元素置换；元素取代
element transfer 元素运移
element vector 向量元
element yield 元素产额
element zoning 元素分带
elemental analysis 元素分析
elemental array 初等台阵；基本台阵
elemental composition 元素成分
elemental logging 元素测井
elementary analysis 初步分析；元素分析
elementary composition 基本组成；元素成分；元素组分
elementary cosmic abundance 元素宇宙丰度
elementary function 初等函数
elementary geodesy 普通测量学
elementary gravity wave 基本重力波
elementary instruction 基本指令
elementary layer 单元层
elementary matrix 初等矩阵
elementary nuclear synthesis 元素核合成
elementary particle 基本粒子
elementary proving 基本标定
elementary reflection analysis 基本反射分析
elementary rheology 初等流变学
elementary surveying 普通测量学
elementary volume 单元体积；体积单元
elementary wave 元波；基波
element-free method 无单元法
element-migration coefficient in water 元素迁移系数；水迁移系数
elements of symmetry 对称要素

eleolite	脂光石；霞石
eleolite syenite	脂光正长岩
elephant hunt	大油田勘探
eleutheromorph	自由形晶
elevated beach	上升海滩；上升阶地；岸边台地
elevated block	上升地块
elevated coast	上升海岸
elevated delta	上升三角洲
elevated ditch	填土高渠
elevated line	高架线路；高架道路
elevated peneplain	上升准平原
elevated pile foundation	高桩承台
elevated plain	上升平原；台地
elevated pressure	高压
elevated railroad	高架铁道
elevated railway	高架铁路；高架铁道
elevated reef	升礁；上升礁；上礁
elevated shore	上升海岸
elevated shore cliff	上升滨崖；上升岸崖
elevated shoreface terrace	上升滨岸阶地；升岸阶地；抬升海岸；海岸抬升阶地
elevated shoreline	上升滨线
elevated structure	高架构筑物；高架结构
elevated temperature	高温
elevated track	高架轨道
elevated trackway	高架轨道
elevated tramway	高架电车道
elevated wave-cut bench	上升浪蚀平台
elevated wave-cut terrace	上升浪蚀阶地
elevating dredge	提升式挖泥机；升降式挖泥船
elevating platform	升降式平台
elevation	上升，隆起；高地；高度，高程；标高；海拔；正视图，正面图，立视图
elevation above sea-level	海拔高度
elevation amplification effect	高程放大效应
elevation and subsidence	升降
elevation angle	仰角；高程角；竖角；高度角
elevation bearing	仰角方位
elevation change	高程变化
elevation control	高程控制
elevation correction factor	高程校正系数
elevation datum	高程水准面；高程基准
elevation datum plane	高程基准面
elevation difference	高差；标高差
elevation drawing	立面图；正视图；前视图
elevation frequency curve	高程频率曲线
elevation gradient	水头梯度；海拔梯度
elevation mark	高程标志；高标
elevation meter	高程计；测高计
elevation number	高程注记；海拔，标高
elevation of basin	流域中数高程
elevation of borehole	孔口标高
elevation of coast	海岸高程
elevation of curve	曲线超高
elevation of dam top	坝顶高程
elevation of hole	孔口标高
elevation of land	陆地上升
elevation of sight	视线高程
elevation of top of dam	坝顶高程
elevation of water	水面高度；水位；水位高度
elevation of well	井口标高
elevation point	高程点
elevation point with notes	高程注记点
elevation potential energy	高程位势能
elevation profile	垂直剖面；竖剖面；纵剖面
elevation scale	仰角刻度；高程比例尺
elevation shoreline	上升滨线
elevation tectonics	地块升降构造
elevation theory of volcano	火山隆起说
elevation transfer survey	导入高程测量
elevation treatment	立面处理
elevation view	立视图；正视图
elevation zero	高程零点
elevational point	高程点
elevation-capacity curve	高程库容关系曲线
elevation-capacity relation	高程容积关系
elevation-distance curve	高程距离曲线
elevationmeter	测高仪
elevationnet storage curve	高程净蓄量曲线
elevation-relief ratio	标高整体；高程地形比
elevator	吊卡；升降机；电梯；提升机
elevator bail	吊卡提环
elevator chamber	沉浮箱
elevator dredger	多斗挖泥机
elevator groove	吊卡槽
elevator hoistway	电梯间；电梯井；升降机井
elevator latcher	吊卡门；井架工
elevator plug	提升短节
elevator system	升运系统
elevator tectonics	起落构造
elevator type loader	提升式装料机
elevator-spider	吊卡-卡盘
elevon	升降副翼
eliminable	可消除的
elimination	除去；消去；排除；消元法；切断电路；删除
elimination by addition or subtraction	加减消元法
elimination by comparison	比较消元法
elimination by substitution	代入消元法
elimination method	消去法
elimination of burst noise	消除突发噪声
elimination of scaling	去除结垢；除垢
elimination of singularity	消除奇异性
elimination of water	脱水
eliquation	液析
elision	沉积间断作用；地层削蚀作用
elision of cycle stage	旋回阶段的缺失；沉积缺失
elitist strategy	最优继承策略
elkerite	埃尔克沥青
elkhornite	拉辉正长岩
elkonit	艾尔柯尼特钨铜合金
elkonium	埃尔科尼姆接点合金
ellestadite	硅磷灰石
ellipse	椭圆；椭圆体
ellipse area	椭圆面积
ellipse of error	误差椭圆
ellipse of vibration	振动椭圆
ellipse-hyperbolic system	椭圆-双曲线系统
ellipsis of stress	应力椭圆；应力椭球面
ellipsograph	椭圆规

ellipsoid 椭面；椭圆体；椭球；椭圆面
ellipsoid of reference 参考椭球
ellipsoid of revolution 回转椭面；旋转椭球
ellipsoid of rotation 旋转椭球面
ellipsoid of strain 应变椭球体
ellipsoid projection 椭球投影
ellipsoidal 椭球体的；椭球的
ellipsoidal chord distance 椭球弦距
ellipsoidal coordinate 椭球坐标
ellipsoidal geodesy 椭球面大地测量学
ellipsoidal harmonics 椭球调和函数
ellipsoidal parameter 椭球参数
ellipsoidal structure 椭球形构造
ellipsometer 偏振光椭圆率测量仪
elliptic arch 椭圆拱
elliptic arch dam 椭圆拱坝
elliptic coordinates 椭圆坐标
elliptic curve 椭圆曲线
elliptic difference equation 椭圆差分方程；椭圆型差分方程
elliptic equation 椭圆方程；椭圆型方程
elliptic fault surface 椭圆断层面
elliptic hyperboloid 椭圆双曲面
elliptic shearing surface 椭圆剪切面
elliptic structure 椭圆形构造
elliptical 椭圆形的
elliptical arch 椭圆拱
elliptical arch dam 椭圆形拱坝
elliptical coordinates 椭圆坐标
elliptical hill 鼓丘
elliptical integral 椭圆积分
elliptical opening 椭圆孔口
elliptical orbit 椭圆轨道
elliptical pointed arch 椭圆尖头拱
elliptical polarization 椭圆偏光；椭圆极化
elliptical projection 椭球投影
elliptical ring 椭圆环
elliptical section 椭圆截面
elliptically polarized wave 椭圆极化波；椭圆偏振波
elliptical-type weight 椭圆型重锤
ellipticity 椭圆率；扁率；椭圆度
ellipticity of ellipse 椭圆扁率
ellipticity of spheroid 椭球体扁率
ellipticity of the earth 地球椭率
Ellis fluid 艾利斯流体
ellitoral zone 浅海底；浅海带
ellonite 粉硅镁石
ellsworthite 钠烧绿石；钙铌水石
E-log 电测井；电测记录
elongate 伸长
elongate channelized scar 纵长沟豁状断崖
elongate delta 纵长三角洲
elongate mudflow 延伸状泥流
elongate scar 纵长断崖
elongated anticline 延伸背斜；狭长背斜
elongated charge 长条状爆炸震源；长条形炸药
elongated fold 伸长褶皱；狭长褶皱
elongated four-spot pattern 伸长四点井网
elongated lens 长条形透镜体
elongated structure 纵长构造

elongation 延伸率；拉长；伸长；延长
elongation at break 断裂伸长；断裂伸度
elongation at failure 破坏伸长；扯断伸长率
elongation at rupture 断裂伸长；断裂延伸；破坏伸长
elongation at yield 屈服伸长
elongation due to tension 拉伸伸长
elongation factor 伸长系数
elongation index 延伸指数
elongation modulus 伸长模量
elongation per unit length 伸长率
elongation rate 延伸率
elongation ratio 伸长比
elongation recorder 伸长记录仪
elongation sign 延长符号
elongation strain 拉伸应变；延伸应变
elongation test 拉长试验；拉伸试验；伸长试验
elongational flow 伸展型流动；拉伸流动
elongational viscosity 伸展黏度
elpasolite 钾冰晶石
elpidite 斜钠锆石；钠锆石
Elster glacial stage 埃尔斯特冰期
elusive reservoir 隐蔽油藏
elution chromatography 冲洗色谱法；洗脱色谱法
elutriation 淘洗，淘析，淘选；冲洗；清洗；淘净；洗净分析；粗化作用
elutriation method 淘洗法；水析法；冲洗法
elutriator 土粒级配仪；冲洗器；淘析器
elutriometer 含砂量测定瓶
eluvial deposit 淋溶堆积；残积物；淋溶沉积；残积矿床
eluvial facies 残积相
eluvial horizon 残积层；淋滤带；淋溶层；土壤溶提层
eluvial landscape 残积景观
eluvial laterite 残积红土
eluvial layer 淋溶层
eluvial placer 残积砂矿
eluvial sand 残积砂
eluvial soil 淋溶土；浮土；残积土
eluviate 残积作用；淋溶；淋滤
eluviation 淋溶作用；淋滤作用；残积作用
eluviation zone 淋溶带；残积层
eluvium 残积层；残积物；淋溶层；风化残积层；风积沙土
elvan 脉斑岩；白色英斑岩；淡英斑岩
elvanite 浅英斑岩
emagram 埃玛图；温压热力学图
emanation 射气；喷气；放射；辐射；发散；泄出
emanation prospecting 射气测量
emancipation 释放；脱离
emanium 射气
emanometer 射气仪；测氡仪
embanked area 围堤岸
embankment 堤；坝；围堰；路堤；堤防
embankment collapse 路堤塌陷
embankment crossing reservoir 水库路基
embankment dam 土石坝；填筑坝；堤坝
embankment failure 土堤破坏
embankment fill 路堤填筑；筑堤；填土
embankment fill material 路堤填料
embankment filler 路堤填料
embankment filling 路基填料

embankment foundation 堤基
embankment load 海堤荷载
embankment material 坝体土石料；堆筑材料；填筑材料
embankment of old sinter 老泉华堤
embankment on plain river beach 河滩路堤
embankment on river bank 滨河路堤
embankment piezometer 坝体渗压计
embankment pile 护堤桩
embankment protection 堤岸防护；护堤
embankment section 路堤截面；路堤地段
embankment slope 路堤边坡；堆筑体边坡；填土边坡
embankment strain meter 堤坝应变计
embankment top 路堤顶
embankment wall 堤墙
embayed coast 湾形海岸；多湾海岸；斜水沟
embayed coastal plain 湾形沿海平原
embayed mountain 海湾山地
embayed shore 海湾海岸；多湾海滨
embayment 港湾，河弯；形成港湾
embedability 埋置性；压入性，嵌入性，挤入性
embedded bolt 预埋螺杆；地脚螺杆
embedded construction 隐蔽工程；埋设工程
embedded depth 埋置深度
embedded footing 埋入式基脚；嵌入式基脚；嵌固基础
embedded hook 预埋吊钩
embedded instrument 埋设仪器；预埋仪器
embedded length 埋置长度；入土长度；埋入长度
embedded length of bar 钢筋埋入长度；埋杆长度
embedded matrix 嵌入矩阵
embedded penstock 埋藏式压力水管
embedded pile 埋入桩
embedded pipe cooling 埋管冷却
embedded stone pitching 埋藏式砌石护坡
embedded strap 预埋条状物
embedded surface area 埋藏表面积
embedded temperature detector 埋入式温度计
embedded wavelet 嵌入子波
embedding medium 嵌入介质
embedding crack 埋藏型裂纹
embedment 埋置；埋置深度；嵌入
embedment depth 埋置深度；埋深
embedment length 埋入长度；埋置长度
embedment method 埋入法
embedment of footing 基础埋深
embedment of foundation 基础埋深
embedment pressure 埋入压力
embedment strength 嵌入强度
embolus 插入物
embossed 隆起的
embouchure 河口
embracing rock 围岩
embranchment 支流；支脉
embrasure 炮眼
embrechite 残层混合岩
embrittlement cracking 脆裂
embryogeosyncline 萌地槽，雏地槽
embryonic crystal 雏晶
embryonic folding 雏形褶皱
embryonic geosyncline 萌地槽，雏地槽
embryonic ore formation 雏形成矿；雏形矿层

embryonic platform 萌地台
embryonic volcano 雏形火山；雏火山；胎火山
emendation 校勘；订正
emerald 祖母绿；绿宝石；绿柱石；翡翠
emerald green 鲜绿色
emerge 浮现
emerged bog 出水沼泽；浮现沼
emerged coast 上升海岸；出露海岸
emergence 出露；出现；上升
emergence coast 上升海岸
emergence shoreline 离水海岸线；上升岸线；上升滨线
emergency 紧急情况；突然事件；危急，紧急，意外事故
emergency access 紧急通道
emergency acoustic system 紧急声控系统
emergency alarm bell 报警铃
emergency alarm system 紧急警报系统
emergency brake valve 紧急制动阀
emergency braking 紧急制动
emergency braking tone 紧急制动信号音
emergency call 紧急呼叫；呼救信号
emergency capability 应急能力
emergency capacity 应急储备；安全储备
emergency channel 紧急频道
emergency closing device 事故关闭装置
Emergency command 应急指挥
emergency command system 应急指挥系统
emergency control centre 紧急事故控制中心
emergency counter measures 应急防范措施
emergency dam 防险坝
emergency delivery 紧急输送
emergency drain 紧急排水道
emergency escape 紧急逃生途径；应急出口
emergency evacuation 紧急疏散
emergency evacuation system 应急避难系统
emergency exercise 应急演习
emergency exit door 紧急出口门；安全出口门
emergency flood fighting 防汛抢险
emergency gate 事故闸门；应急闸门
emergency gate shaft 事故门井
emergency gate slot 事故闸门槽；事故门槽
emergency generator room 紧急发电机机房
emergency grouting 应急灌浆
emergency handwinding equipment 紧急人手绞动器
emergency management 应急管理
emergency manual 紧急事故手册
emergency material 应急备料
emergency monitoring and support 紧急事故监察及支援中心
emergency parking area 紧急停车带
emergency plan 应急计划
emergency planning zone 应急计划区
emergency portion 紧急部
emergency preparedness 应急准备
emergency project 应急方案
emergency push button 紧急按钮
emergency rail stored along the way 备用轨
emergency relief 事故放水口；紧急救援
emergency repair 急修；事故检修；临险抢护
emergency repair of road 抢修
emergency repair order 紧急维修通知

emergency repair team	紧急维修小组
emergency response system	应急系统；应急响应系统
emergency rush engineering	临险抢护
emergency sanding	紧急撒砂
emergency satellite communication	应急卫星通信
emergency service	救援；紧急服务
emergency set	应急装置
emergency shower	紧急喷淋
emergency shutdown	紧急停工；紧急停机；紧急停车
emergency shut-down	紧急关闭
emergency shutdown circuit	紧急关闭线路
emergency shutdown line	紧急关闭线路
emergency shut-in valve	紧急关闭阀
emergency shutoff	事故关闭
emergency signal	危急信号；紧急信号；紧急信令
emergency standby	应急待命
emergency station service transformer	事故厂用变压器
emergency stop mechanism	紧急停车装置
emergency stop switch	紧急停机掣
emergency switch	紧急开关；事故开关
emergency tank	事故储罐
emergency telephone	应急电话；紧急电话
emergency telephone service	紧急备用电话
emergency tender	急救车
emergency torque release	紧急倒扣释放
emergency treatment after failure	故障办理
emergency valve	紧急阀；事故阀；紧急通风
emergency vehicle	紧急服务车辆；救急车辆
emergency vehicular access	紧急车辆通道
emergency weir	临时应急堰
emergency works	紧急工程
emergent coast	上升海岸
emergent form	上升地形；浮出地形
emergent gas	逸出气体
emergent pressure	临界压力
emergent temperature	临界温度
emergent thrust front	上升的冲断层前缘
emergent wave	出射波
emerging coast	上升海岸
emerging ray	出射线
emerging temperature	泉口温度
emerging terrace	上升阶地
emerging water	自流水；涌出水；泉水
emerging wavefront	出射波波前
emery block	金钢砂块；磨石；油石
emery buff	金刚砂磨光机
emery cell	埃默里型浮选机
emery cloth	砂布；金刚砂布
emery cutter	砂轮
emery dust	金刚砂粉；磨料粉
emery grit	金刚砂
emery paper	金刚砂纸
emery powder	金刚砂粉
emery sand	刚玉砂层
emery sharpener	砂轮
emery stone	刚玉
emerylite	珍珠云母
eminent cleavage	极完全解理
emission angle	发射角
emission behavior	排放性能
emission cathode	发射阴极
emission criteria	排放标准
emission monitoring	排放监测
emission of gas	瓦斯涌出
emission point	排放点
emission source	排放源；发射源
emission standard	发射标准；排放标准
emissions trading	排放权交易
emissivity	发射率；比辐射率；辐射系数
emissivity factor	发射；发射系数
emit output	发射输出
emitted energy	发射能量
emitted wave	发射波
emitting antenna	发射天线
emitting area	发射面积
emitting electrode	发射电极
emitting light	发射光
emitting material	发射物质
emitting surface	发射表面
emmonite	钙菱锶矿
emmonsite	绿铁碲矿
empire cloth	绝缘油布；绝缘丝布
empire paper	绝缘纸
empiric	经验的；实验的
empiric formula	经验公式
empiric value	经验值
empirical assumption	经验假定
empirical coefficient	经验系数
empirical coefficient of settlement calculation	沉降计算经验系数
empirical constant	经验常数
empirical correction	经验校正法；经验校正值
empirical correlating constant	经验对比常数
empirical correlation	经验关系
empirical curve	经验曲线
empirical data	经验数据
empirical design	经验设计
empirical discriminate function	经验判别函数；经验判别公式
empirical discrimination method	经验判别法
empirical distribution function	经验分布函数
empirical earth pressure	经验土压力
empirical equation	经验方程；经验公式
empirical evaluation	经验评价
empirical exponential decline curve	经验指数递减曲线
empirical factor	经验系数
empirical fit constant	经验拟合常数
empirical formula	经验公式
empirical function	经验函数
empirical grid residual system	经验网格剩余系统
empirical law	经验定律
empirical method	经验方法
empirical orthogonal function	经验正交函数
empirical parameter	经验参数
empirical prediction method	经验预测法
empirical probability	经验概率
empirical production rate	经验产量
empirical rate-time equation	产量-时间经验公式
empirical relationship	经验关系式；经验关系
empirical risk minimization principle	经验风险最小准则

empirical river formula　经验河流公式
empirical shear-displacement curve　经验剪切位移曲线
empirical solution　经验解
empirical strength criteria　经验强度准则
empirical value　经验值
empirical waterflood prediction method　注水效果的经验预测法
emplacement age　侵位年龄
emplacement hole　放置孔穴；吊装竖井
emplectite　硫铜铋矿
employed reserves　动用储量
employment contract　雇用合同
employment dismissal　解雇
emptied stope　采空区；留矿放空的矿房
empty hole　空穴；充满空气或天然气的井；空井；未装药的炮眼
empty land　荒地；空地
empty return　空车回道
empty running　空转
empty set　空集
empty space　真空
empty speed　卸载速度
empty stope　放矿矿房；留矿放空的矿房
empty strand　回空段；回行段
empty traffic　空载行车
empty trip　空车运行；空车列车
empty tub　空矿车
empty way　空车道
empty-car haulage drift　空车运行平巷
emptying　卸载，卸空；排空，放空；卸载的，排空的；沉积；残留物
emptying culvert　放空涵洞
emptying device　卸载装置；翻车器
emptying of reservoir　水库放空
emptying point　卸载点；卸料点
emptying position　倾卸位置
emptying pump　排放泵
emptying time　放空时间；新生期
emptying tunnel　放空隧洞
emptying valve　放空阀；排油阀
emulation program　仿真程序
emulator　仿真器；仿真程序；仿真装置；模拟器
emulsification declining viscosity　乳化降黏
emulsified asphalt　乳化沥青
emulsified explosives　乳化炸药；乳化油炸药
emulsified oil　乳化油
emulsion colloid　乳胶体
emulsion grouting　乳化灌浆
emulsion structure　乳胶状结构；乳浊液
en block movement　大块运动，大块体移动
en echelon arrangement　雁行排列
en echelon fault　雁行断层
en echelon fault block　雁行断块；雁列断块
en echelon fissure　雁行裂隙
en echelon fold　雁形褶皱；雁列褶皱
en echelon fracture　雁行裂缝
en echelon joints　雁列节理；雁形节理；斜列节理
en echelon lenses　雁行透镜体
en echelon line　雁列线
en echelon offset　雁行断错

en echelon structure　雁行构造
enabling pulse　起动脉冲
enabling signal　启动信号
enalite　变铀钍石；铀钍石；水硅钍铀矿
enamel　瓷漆；珐琅质；搪瓷，珐琅
enamel covered wire　漆包线
enamel leather　漆皮
enamel paint　瓷漆
enamel paint for road　马路漆油
enameled and cotton-covered wire　纱包漆包线
enameled wire　漆包线
enameling　上涂料；上釉
enamelled brick　瓷砖
enamelled fire clay　釉瓷耐火黏土
enargite　硫砷铜矿
encapsulated pitting　囊状点蚀
encapsulated tree　隔水采油树
encapsulating compound　封铸用的混合料
encapsulating mud　囊护泥浆
encapsulation　密封；封补；包封；封装
encapsulation fitting　密封装置
encased conveying　集装箱运输
encased pile　有壳桩
encased structures　钢管混凝土结构
encasement medium　包封介质
enceladite　硼镁钛矿
encipherer　编码器
encirclement　包围；圈闭
encircling arcuate fracture　旋卷弧形断裂
encircling reef　堤礁
enclosed balcony　围封的露台
enclosed basin　封闭盆地
enclosed bay　封闭港湾
enclosed body of water　封闭水域
enclosed building　围封式建筑物
enclosed cutter　带罩切削器
enclosed electric machine　封闭式电机
enclosed environment　封闭环境
enclosed gearing　封闭式齿轮传动装置
enclosed lake　内湖
enclosed meander　深切曲流；封闭河区；环形河湾；环形曲流
enclosed sea　内海；封闭海
enclosed shield　封闭式防护罩
enclosed space of a unit　单位内部的面积
enclosed water　封闭水域
enclosed-type motor　封闭型电动机
enclosing external wall　围绕单位的外墙
enclosing rock　围岩
enclosing wall　围墙
encode　编码
encoded sweep technique　编码扫描技术
encoder matrix　编码矩阵
encoding scheme　编码方案
encrinite limestone　海百合灰岩
encroach　侵入；侵占；侵蚀；侵害
encroached water　侵入水
encroaching water advance　水侵前缘推进
encroachment　侵入；水浸；障积作用；矿区越界地段；侵蚀地

encroachment area	侵入面积
encroachment line	边水前侵线；侵蚀线；侵入线
encroachment of sea water	海水入侵
encrustation	结壳；皮壳；包壳；结垢；垢物；泉华
encrustment	结壳作用
encryption	编密码；加密
encryption description	编码术语
encyclopedia	百科全书
end anchor block	端部地锚；端部锚固墩
end anchorage	端部锚固
end anchored	底锚
end angle	终止角
end bearing	端承；端支座；端支承；端阻力；端面轴承
end bearing pile	端承桩；支承桩
end bent	端部弯曲
end bevel	坡口
end cap	管端盖板；端盖；桩帽；端帽
end cock	端部塞门
end door	端门
end fixing	固定端
end hole	周边炮孔；边缘孔；边眼；边井；开帮炮眼
end joint	层状节理；层状裂隙；次要节理；次解理
end lap	超覆；叠覆；重复
end load	端荷载；末端负载；终端载荷
end manhole	端部检查孔；端部检查井
end moment	端力矩
end moraine	终碛；尾碛；端碛
end moraine lake	终碛湖
end of consolidation	固结终点
end of sheet pile	板桩尾部；插板尾部
end of side span intercepted	截断边跨端
end of sun screen	遮光栅终点
end of turnout	道岔终端；岔跟
end off	尖灭；结束；停止
end partition wall	挡头墙
end peneplain	终期准平原
end plan	最终平面图；最终计划；端立面
end point	终点；端点；终沸点；终馏点
end resistance	桩尖阻力；桩端阻力
end resistance of pile	桩端阻力
end restraint	端部约束
end restraint effect	端部约束效应；末端约束效应
end sheathing	端墙包板
end sill	端梁
end slope of groin	丁坝头部坡度
end slope of groyne	丁坝头部坡度
end span	端跨；边跨
end stiffener	端部加劲条
end stiffener angle	端部加劲角钢
end view	端视图；侧视图
end wall	端墙；尾墙；端壁
end wall tunnel portal	端墙式洞门
end-anchorage type bolt	端头锚固型锚杆
end-anchored reinforcement	端部锚固钢筋
end-around carry	循环进位
end-around shift	循环移位
end-around-carry shift	循环进位
end-bearing capacity of pile	桩端承载能力
end-bearing member	端承构件
end-bearing pile	端承桩
end-bearing resistance	端承阻力
end-bearing steel H-pile	端部承载的工字钢桩
end-bearing stress	端承应力
end-coincidence method	端点符合法；端点重合法
end-effect zone	末端效应带
endeiolite	硅铌钠矿
endellite	水埃洛石
endemia	地方病
endenite	浅闪石
enderbite	紫苏花岗闪长岩；斜长紫苏花岗岩
endexoteric	内外因的
endfire magnetic dipole	端射磁偶极子；纵向磁偶极子
endiopside	顽透辉石
endless belt	环带；应变仪；环形带
endless chain dredger	轮链式挖泥机
endless chain trench excavator	环链式多斗挖沟机
endless rope haulage	无极绳运输；无端绳运输
endless screw	蜗杆；螺旋杆
endless tubing	无接头油管
endless-belt conveyor	环形皮带运输机
endless-bucket trencher	轮式挖沟机
endless-cable system	环索系统
endlichite	砷钒铅矿
endochronic plastic model	内时塑性模型
endochronic theory	内时理论；自连续理论；内时塑性理论
endoconch	内壳
endodyke	内成岩墙；内生岩脉
endodynamomorphic soil	内动力型土壤
endoenergic	吸能的
end-of-run temperature	运转末期温度
endogene effect	内生效应
endogene force	内力
endogenetic action	内成作用；内力作用
endogenetic deformation	内成形变
endogenetic deposits	内生矿床
endogenetic fissure	内成裂隙
endogenetic force	内营力；内力
endogenetic geology	内动力地质学
endogenetic rock	内生岩；内成岩
endogenetic sand	内生砂
endogenetic sediment	内成沉积
endogenetic sedimentary rock	内源沉积岩
endogenic	内成的；内生的；内源的
endogenic cleat	内生裂隙
endogenic deposit	内生矿床
endogenic force	内力；内营力
endogenic geological process	内动力地质作用
endogenic halo	内生晕
endogenic mineralization	内生成矿作用
endogenic process	内成作用
endogenic rock	内成岩；内生岩
endogenic texture	内生结构；内成结构
endogenous dome	内成穹窿
endogenous ejecta	内成喷出物
endogenous steam	内生蒸汽；地热蒸汽；天然蒸汽
endogenous vein	内成脉
endogenous water	内生水；内部水；内在水
endogeny	内成；内生
endoglyph	层内痕

endokinetic fissure 内生裂隙；自裂缝；内成裂缝
endokinetic joint 内成节理；内因节理；内生节理
endolithic breccia 内成角砾岩
endolithic brecciation 内成角砾作用
endomagmatic hydrothermal differentiation 内成岩浆热液分异作用
endomigmatization 内质混合岩化作用
endomomental 脉冲吸收的
endomorph 内容体；包裹晶；内容矿物
endomorphic 内变质的；接触变质的
endomorphic metamorphism 内变质作用
endomorphic zone 内变质带；接触变质带
endomorphism 内变质作用；接触变质作用
endomorphosed 内变质的
endoradiosonde 体内无线电探头
endorheic basin 内陆流域；内流盆地
endorheic drainage 内陆河流域；无泄水区；内陆水系
endorheic drainage basin 内陆河流域
endorheic system 内流水系
endorheism 内陆流域；内陆水系
endoskarn 内矽卡岩
endosmose 内渗；内渗透
endosmosis 内渗
endosmotic 内渗的
endostratic breccia 层内角砾岩
endosymbiosis 内共生
endotectonic 内力构造的
endothermal 吸热的
endothermal reaction 吸热反应
endothermic effect 悬臂起重机
endothermic process 吸热过程
endovolcanic structure 内火山构造
endpiece 出口端
end-point analysis 最后结果分析
end-point data 端点数据
end-point mobility ratio 端点流度比
end-point normality 末端当量浓度
end-point recovery 端点采收率
end-point relative permeability 端点相对渗透率
end-point saturation 端点饱和度
end-tipped fill slope 顶坡倾倒土料而成的填土坡
end-tipping lorry 尾卸式卡车
end-tipping of fill 车尾倾卸填土法；倾斗填土法
end-to-end arrangement 纵向排列
end-to-end discharge 双端卸载
end-to-end flood 从一端向另一端注水
end-to-end profile 连续剖面
end-upset anchorage system 镦头锚体系
endurance 耐久性；持久性；持续时间；强度；抗磨度；耐疲劳度
endurance expectation 估计使用期限
endurance failure 疲劳破坏
endurance life 使用年限；耐久寿命
endurance limit 疲劳极限；持久限度；耐久限度
endurance limit stress 耐久极限应力
endurance period 持续时间
endurance quality 耐久性能
endurance ratio 耐久比；疲劳比；疲劳系数
endurance test 耐久试验；疲劳试验
endurance trial 耐航试验；续航能力试验

enduring surface 抗风化面；稳定面
en-echelon 雁行式的；雁列式的；梯阵式的
en-echelon crack 雁行裂缝
en-echelon fault 雁列断层
en-echelon fold 雁行褶皱
en-echelon fracture 雁行裂隙
en-echelon joint 雁行节理
en-echelon offset 雁列断错；斜列式断错位移
en-echelon structure 雁行式构造；雁行构造
en-echelon zone 雁行带
energetic 高能的
energetic recoil atom 高能反冲原子
energetic reflection 高能反射
energetics 力能学；水能学；动能学；唯能论；热力学
energized acid 增能酸
energized liquid 增能液
energized packing element 增能密封元件
energizing gas 激发气体
energy absorbing 吸收能量
energy absorption coefficient 能量吸收系数
energy accumulation 能量积累
energy amplification 能量放大；功率放大
energy analysis 能量分析
energy attenuation 能量衰减
energy auditing 能源审计
energy balance equation 能量平衡方程
energy band 能带
energy budget 能量平衡
energy budget method 能量平衡法
energy capital 能源资本
energy coefficient 能量系数
energy conservation equation 能量守恒方程
energy conservation law 能量守恒定律
energy conservation principle 能量守恒定律；能量守恒原则
energy correction factor 能量校正系数
energy crisis 能量危机；能源危机
energy decay 能量衰减
energy degradation 能量降级
energy demand trend 能源需求趋势
energy density 能量密度
energy density function 能量密度函数
energy density spectrum 能量密度谱
energy deposition 能量沉积
energy development 能量开发；能源开发
energy diagram 能量图；热力图
energy difference 能差
energy dispersal 能量扩散
energy disperser 能量扩散器
energy dispersive X-ray detector 能量色散X射线探测仪
energy dissipater 消能装置
energy dissipating sill 消能槛
energy dissipation 能量消散；能量耗散；消能
energy dissipation baffle 能量耗散挡板
energy dissipation in hydraulic jump 水跃能量消耗
energy dissipation rate 能量耗散率
energy dissipator 消能器；消能装置
energy distribution 能量分布
energy distribution curve 能量分布曲线
energy economy 能源经济

energy flux 能通量，能量通量
energy flux rate 能量流率
energy future 能源期货
energy gap 能级距离
energy geotechnical engineering 能源岩土工程
energy grade line 水能线；压头线；能坡线；能量坡降线；能量均衡线
energy gradient 能量梯度
energy gradient line 能量梯度线
energy head 能量水头；动水头；能头
energy head loss 能头损失；动水头损失
energy index 能量指数
Energy Information Administration 能源信息管理局
energy intensive industry 能源密集工业
energy interchange 能量交换
energy interchange in the rock mass 岩体中能量交换
energy liberation 能量释放
energy line 能量线
energy loss 能量损失；能量损耗
energy management 能源管理
energy meter 电度表；电能表；能量计
energy method 能量法
energy mix 能源结构
energy of activation 活化能
energy of adsorption 吸附能
energy of attachment 附着能
energy of comminution 破碎能量
energy of deformation 变形能
energy of distortion 畸变能
energy of earthquake 地震能量
energy of flow 流动能量
energy of motion 动能
energy of position 位能
energy of strain 变形能；应变能
energy of vibration 振动能
energy option 能源交易选择权
energy output 能量输出；输出能量；电能生产量；发电量
energy path 能量途径
energy potential 能势
energy product curve 能积曲线
energy production 能源生产
energy production elasticity 能源生产弹性系数
energy ratio 能量比
energy recovery 能量回收；能源再生
energy recycling 能量循环利用
energy reflectivity 能量反射率
energy release 能量释放
energy release during rockburst 岩爆期间的能量的释放
energy releasing amount 能量释放量
Energy Research and Development Administration 能源研究和发展管理局
energy resolution 能量分辨率
energy resources 能源；能源资源
Energy Resources Conservation Board 节能局
energy response function 能量响应函数
energy saving product structure 节能型产品结构
energy security 能源安全
energy slope 能量梯度
energy source 能源

energy source boat 震源船
energy source controller 震源控制器
energy source synchronizer 震源同步装置；震源同步器
energy spectrum 能量谱；能谱
energy state 能态；能量状态
energy storage 能量储存
energy storage capacity 储能量
energy storage of aquifer 含水层储能
energy storage volume 储能容积
energy structure 能源结构
energy supply 能源供应
energy theory 能量理论
energy trace 能量曲线
energy transfer 能量转换；能量传递
energy transformation 能量转换
energy transmission in rock 岩石中能量的传播
energy transport 能量输送
energy utilization 能源利用；能源利用率
energy value 热值
energy window 能量窗
energy yield 发电量；能量产额；能量输出
energy-balance climatology 能量平衡气候学
energy-delivering system 能量传送系统
energy-dispersive analysis 能散分析
energy-dispersive X-ray analysis 能量色散X射线分析
energy-dissipating dent 消能凹槽
energy-dissipating sill 消能槛
energy-economic system 动能经济体系
energy-efficient equipment 节能设备
energygram 能量图
energy-intensive industry 能量密集型工业
energy-loss factor 能量损失系数
energy-saving 节能
energy-storing brake 蓄能制动
enerjet gun 半消毁式过油管射孔器
enforced meander 强迫蛇曲；强化曲流
enforced settlement 强迫下沉
engagement process 啮合过程
engaging and disengaging gear 离合装置
engaging angle 啮合角
engelburgite 橌杂花岗闪长岩
engine and pump assembly 泵机组
engine driven pump 机动抽水机；机动泵
engine exhaust 发动机排气
engine exhaust system 引擎排气系统
engine mud sill 钻机动力机下纵向底梁
engine-drive generator 机动发电机
engine-driven supercharging 机械增压
engineer in chief 总工程师
engineer inspection 工程师检查
engineered fill 控制回填土；质控回填土
engineered fracture network 人工裂隙网络
engineered geothermal system 工程地热系统；增强型地热系统
engineered slope 工程边坡
engineering 工程学；工程技术；设计；方案设计；策划
engineering acceptance 工程验收
engineering analysis 工程分析
engineering analysis report 工程分析报告

engineering and construction 设计与施工
Engineering Applications of Artificial Intelligence 人工智能的工程应用学报
engineering approximation 工程近似法
engineering assessment report 工程评估报告
engineering behavior 工程性质；工程特性
engineering behavior of clay 黏土工程性状；黏土技术特性
engineering calculation 工程计算
engineering change order 设计变动命令
engineering change request 工程更改申请
engineering change statement 工程更改说明书
engineering characteristics of rock 岩石工程特性
engineering classification of rock mass 岩体工程分类
engineering classification of soil 土的工程分类
engineering company 工程公司
engineering component parts list 工程部件清单
engineering concept 工程概念
engineering condition 工程条件
engineering construction 工程建设；工程施工
engineering construction standard 工程建设标准
engineering consultancy firm 工程咨询公司
engineering consultation unit 工程咨询单位
engineering control 工程控制
engineering control survey 工程控制测量
engineering cost 工程费用，工程造价；施工费用；技术性成本
engineering cybernetics 工程控制论
engineering data processing 工程数据处理
engineering decision 工程决策
engineering design plan 工程设计方案
engineering drawing 工程图；工程制图；工程图样
engineering dynamic geology 工程动力地质学
engineering electromagnetics 工程电磁学
engineering environment 工程环境
engineering environment of orebody 矿体工程环境
engineering estimation of rock mass strength 岩体强度工程计算法；工程岩体强度估算
engineering evaluation 工程评价
engineering example 工程实例
engineering facilities 工程设备；工程设施
engineering factor 工程因素；技术条件
engineering feasibility 工程可行性；技术可行性
engineering fluid mechanics 工程流体力学
Engineering Fracture Mechanics 工程断裂力学
engineering function 技术功能；工程职能
engineering geodynamics 工程动力地质学
engineering geologic analogy 工程地质类比法；工程地质比拟法
engineering geologic columnar profile 工程地质柱状图
engineering geologic condition 工程地质条件
engineering geologic drilling 工程地质钻探
engineering geologic exploration 工程地质勘探
engineering geologic investigation 工程地质勘察
engineering geologic investigation report 工程地质勘察报告
engineering geologic location of line 工程地质选线
engineering geologic map 工程地质图
engineering geologic mapping 工程地质测绘；工程地质填图
engineering geologic process 工程地质作用
engineering geologic profile 工程地质剖面图
engineering geologic unit 工程地质单元
engineering geologic zoning 工程地质分区
engineering geological analogy 工程地质类比法
engineering geological analytical map 工程地质分析图
engineering geological character 工程地质特征；工程地质特性
engineering geological classification 工程地质分类
engineering geological columnar profile 工程地质柱状图
engineering geological condition 工程地质条件
engineering geological condition map 工程地质条件图
engineering geological cross section 工程地质剖面图
engineering geological drilling 工程地质钻探
engineering geological evaluation 工程地质评价
engineering geological exploration 工程地质勘察
engineering geological exploratory drift 工程地质坑探；工程地质探槽
engineering geological geophysical exploration 工程地质物探
engineering geological horizontal profile 工程地质水平剖面图
engineering geological investigation 工程地质调查
engineering geological investigation report 工程地质勘察报告
engineering geological long-term observation 工程地质长期观测
engineering geological map 工程地质图
engineering geological map of ore field 矿区工程地质图
engineering geological mapping 工程地质测绘
engineering geological mechanics 工程地质力学
engineering geological modelling 工程地质模拟
engineering geological monitoring 工程地质监测
engineering geological plan of damsite area 坝址区工程地质图
engineering geological plan of power house 厂房工程地质图
engineering geological plan of project area 枢纽区工程地质图
engineering geological problem 工程地质问题
engineering geological problems of bridge 桥梁工程地质问题
engineering geological problems of reservoir 水库工程地质问题；储层工程地质问题
engineering geological process 工程地质作用；工程地质方法
engineering geological profile of dam axis 坝轴线工程地质剖面图
engineering geological profile upstream cofferdam 上游围堰工程地质剖面图
engineering geological prospecting 工程地质勘探
engineering geological section 工程地质剖面图
engineering geological survey 工程地质调查；工程地质测绘；工程地质填图
engineering geological terminology 工程地质术语
engineering geological testing 工程地质试验
engineering geological unit 工程地质单元
engineering geological work in mine 矿山工程地质工作
engineering geological zonation map 工程地质分区图
engineering geological zoning 工程地质区划；工程地质分区
engineering geologist 工程地质师；工程地质学家；工程地质工作者

engineering geology 工程地质学
Engineering Geology 工程地质学报
engineering geology for underground works 地下工程地质
engineering geology graduate 见习工程地质工程师
engineering geology in shafting and drifting 井巷工程地质
engineering geology of side slope 边坡地质工程
engineering geomechanics 工程地质力学
engineering geomechanics model 工程地质力学模型
engineering geomorphology 工程地貌学
engineering geophysical exploration 工程地球物理勘探
engineering geophysical prospecting 工程地球物理勘探；工程物探
engineering geophysical sounding 工程地球物理测深
engineering geophysics 工程地球物理学；工程物探
engineering geothermics 工程地热学
engineering graphics 工程图学
engineering graphics solution 工程图解法
engineering hydraulics 工程水力学
engineering hydrology 工程水文学
engineering index 工程索引
engineering infrastructure 基础建设工程
engineering intensity 工程烈度
engineering intensity scale 工程烈度表
engineering investigation study 工程勘察研究
engineering investigations 工程勘察
engineering judgment 工程判断；工程评价
engineering legislation 工程法制
engineering machinery 工程机械
engineering machinery plant 工程机械厂
engineering management 工程管理
engineering mathematics 工程数学
engineering measurement 工程测量
engineering mechanics 工程力学
engineering model 工程结构模型
engineering oceanography 工程海洋学
engineering of exploration 勘探工程
engineering operation 工程实施；工程程序；工程运作
engineering performance 工程绩效；工程特性曲线；技术性能
engineering petrology 工程岩石学
engineering philosophy 技术原则；设计原则；设计思想
engineering photogrammetry 工程摄影测量
engineering planning 工程计划；工程规划
engineering preliminaries 工程前期工作；工程准备事项
engineering project 工程项目
engineering property 工程特性，工程性质
engineering property of rock 岩石的工程性质
engineering quality 工程质量
engineering quality index of rock 岩石工程质量指标
engineering quality index of rock mass 岩体工程质量指标
engineering remote sensing 工程遥感
engineering report 工程技术报告
engineering rock and soil 工程岩土学
engineering seismics 工程地震
engineering seismometry 工程测震；工程测震学
engineering ship 工程船舶
engineering structure 工程结构；工程建筑
engineering supervision 技术监督；工程监理
engineering survey 工程测量
engineering survey of missile test site 导弹试验场工程测量
engineering surveying 工程测量学；工程勘测
engineering thermoplastic elastomer 工程热塑性弹性体
engineering transportation 工程运输
engineering treatment 工程处理
engineering triangulation 工程三角测量
engineering vessel 工程船舶
engineering well 工程井
engineering works 工程；工程建筑；工程设施
Engineering Structures 工程结构学报
engineering-triggered slide 工程引发的滑坡
engineer's level 工程水准仪
engineer's theodolite 工程经纬仪
engineer's transit 工程经纬仪
englacial 冰川内的；冰内的
englacial debris 冰内碎屑物
englacial drainage 冰川水系
englacial drift 冰内碛
englacial environment 冰内环境
englacial lake 冰内湖
englacial stream 冰内河
englacial zone 冰内带
English bond 英式砌合法；丁顺隔皮砌式
englishite 水磷钙钾石；水磷钾石
enhanced contrast 增强反差
enhanced gas recovery 强化天然气采收率
enhanced geothermal system 增强型地热系统；工程地热系统
enhanced image 增强图像
enhanced line 增强谱线
enhanced oil recovery 提高采收率法采油；强化采油
enhanced planned survey scheme 加强分区勘察计划
enhanced prop treatment 高浓度支撑剂压裂
enhanced recovery 强化开采；提高采收率；增强开采
enhanced recovery pump 强化采液泵
enhanced seismic profile 加强的地震剖面
enhanced spectral line 增强谱线
enhanced strain element 增强应变单元
enhanced weathering 增强风化
enhanced-gas injection 注富气
enhydrite 包水岩石；包水矿物
enkindle 点燃
enlarge hole opener 扩孔器
enlarged base 扩底
enlarged base pile 扩底桩
enlarged detail 大样；放大图
enlarged drawing 放大图
enlarged mosaic 放大镶嵌图
enlarged preliminary design 扩大初步设计
enlarged section 放大断面
enlarged side wall 扩大边墙
enlarged toe pile 扩底桩
enlarged view 放大图
enlargement expansion 扩大
enlargement factor 放大因数；放大倍数
enlargement loss 扩张能量损失；扩大损耗
enlargement scale 放大比例尺
enlarger 扩孔器
enlarging 放大，扩大

enlarging hole 扩孔
enlarging type TBM 扩孔式掘进机
enlargment 扩大
enneagon 九角形；九边形
enneahedron 九面体
enneri 干河谷；旱谷
ennoyage 倒转地形
enophite 绿蛇纹石
enquiry data 调查数据
enrich 富化；富集；浓缩
enriched gas 富化气
enriched mantle 富集地幔
enriched oil 富化油；增加了轻馏分的原油
enriched oxidation zone 富集氧化带
enriched shape function 增强形函数
enriched uranium 浓缩铀
enriched zone 富集带
enriched-gas drive 富化气驱
enriching 浓缩
enriching agent 富化剂
enrichment degrees of freedom 增强自由度
enrichment factor 富集因素；浓缩系数；富集率；富化系数
enrichment function 增强函数
enrichment ratio 富集比
enrichment zone 富集带
ensemble based optimization 基于集成的优化
ensemble Kalman filter 集合卡尔曼滤波
ensialic 硅铝层上的
ensialic basin 硅铝层盆地
ensialic belt 硅铝质带；硅铝层上的
ensialic geosyncline 硅铝层地槽；内硅铝地槽
ensimatic 硅镁层上的
ensimatic geosyncline 硅镁层优地槽；硅镁质地槽
ensonified area 水声仪器监听海区
enstatite 顽火辉石；顽辉石
enstatite basalt 顽火玄武岩
enstatite chondrite 顽火辉石球粒陨石
enstatolite 顽火辉石岩；顽火岩
enstenite 斜方辉石
ensuing earthquake 继发地震；续发地震
enterclose 通道
enterolithic structure 盘肠构造；肠状构造
enterprise administrative expense 企业管理费
enterprise standard 企业标准
enterprise type 企业性质
entexis 注入混熔作用；注入混合作用
entexite 注入混合岩；注入熔合岩；注入变熔岩
enthalpy balance 焓平衡
enthalpy diagram 焓图表
enthalpy potential method 焓差法
enthalpy-entropy diagram 焓-熵图
enthalpy-temperature diagram 焓-温图
entire depth 整个深度，总深度
entire function 整函数，整体函数
entire injection interval 总注入层段
entire optimization 整体优化
entire rational function 整有理函数
entire wavelet 完整子波
entirely oil-wet 完全亲油

entisol 新成土
entoolitic 内呈鲕粒状的；充填鲕状构造
entrail chroismite 肠状混合岩
entrained catalyst loss 催化剂夹带损失
entrained oil 夹带的油
entrained sand particle 夹带砂粒
entrained solid 夹带的固体
entraining 吸入；夹带
entrainment coefficient 携带系数；逸入系数；卷吸系数；夹卷系数
entrainment mechanism 携带机理
entrainment rate 夹带速率；夹卷率
entrainment separator 雾沫分离器
entrainment trap 雾沫分离器；捕沫器，雾沫阱
entrainment velocity 携带速度
entrance branch 入口分支；进口支管
entrance buoy 入口浮标，入口浮筒；进口浮标，进口浮筒
entrance capacity 进入容量；最大入渗量；入渗能；入口容水量
entrance capillary pressure 入口毛细管压力
entrance channel 入港航道；进港航道；进口航道；进水渠
entrance drop 水跃值
entrance effect 入口效应；进口段效应
entrance gate 进水闸
entrance head 进口水头
entrance head loss 进口水头损失
entrance jetty 进港导堤；入口导堤
entrance lighting 进口照明
entrance lock 船坞进口闸；入口船闸
entrance of tunnel 隧道洞口
entrance packing 密封环；密封圈
entrance point 输入点
entrance pressure 进口压力
entrance speed at retarder 缓凝剂入口速度；减速器入口速度
entrance test ratio 入口检验比
entrance to a river 河流入口；河道入口
entrance velocity 入口速度
entrance width 港口宽度；入口宽度
entrap 圈闭
entrapped basin 圈闭海盆，圈闭盆地
entrapped gas 滞留气
entrapped phase 圈闭相
entrapped pressure 内部压力
entrenched meander 嵌入曲流；下切曲流；嵌入曲流深切河湾
entrenched meander valley 嵌入曲流谷
entrenched river 嵌入河；深切河
entrenched stream 嵌入河；下切河
entrenched valley 嵌入谷；嵌入河谷
entrenchment 刻槽；下切；深切；挖壕
entrepreneurial decision 企业家决策；创业决策
entropy 熵；平均信息量
entropy function 熵函数
entropy minimization 熵最小化
entropy of dilution 稀释熵
entropy of evaporation 汽化熵，蒸发熵
entropy of information 信息熵

entropy of superheating	过热熵
entropy ratio map	熵比图
entropy spring	熵跃
entropy wave	熵波
entropy-temperature curve	熵-温曲线
entrustment letter	委托书
entry	进口，入口；水平巷道；河口；表列值；条目
entry condition	入口条件
entry driving	平巷掘进；平硐掘进；主水平巷道掘进
entry hole size	入口直径
entry mask	入口屏蔽
entry neck	入口孔颈
entry pore	入口孔隙
entry pressure	挤入压力
entry profile	注入剖面
entry radius	入口半径
entry ramp	入口斜路；入口坡道
entry terminal	输入终端
entry value	入口值
enumerable	可数的
enumerate	计算；列举，枚举
enumerative technique	枚举法
envelope	包络面，包迹；包封，外套；机壳
envelope amplitude	包络振幅
envelope curve	包络线，包线
envelope distribution	包络分布
envelope function	包线函数
envelope of curves	曲线的包络线
envelope of failure	破坏包线；失稳包络
envelope of Mohr's circles	莫尔圆包线
envelope of strain curves	应变曲线的包线
envelope of strength testing	强度试验包线
envelope surface	包络面
enveloping	包络
enveloping line	包络线
enveloping solid	包络体
enveloping space	包络空间
enveloping surface	包络面
environgeology	环境地质学
environics	环境学
environment adaptability	环境适应性
environment agent	环境作用因素
environment appraisement	环境评价
environment assessment	环境评估；环境评价
environment change	环境变化
environment condition	环境条件；周围条件
environment containment standard	环境保护标准
environment determinism	环境决定论
environment effect	环境影响
environment friendly geotechnical engineering	环境友好岩土工程
environment health	环境卫生
environment hydrogeology	环境水文地质
environment impact assessment	环境影响评估
environment influence assessment report	环境影响评估报告
environment modification	环境改造
environment monitoring	环境监测
environment mutagenesis	环境诱变
environment of deposition	沉积环境
environment of heat in mine	矿井热环境
environment of sedimentation	沉积环境
environment pollution	环境污染
environment possibilism	环境可能论
environment prediction	环境预报
environment recovery guarantee fund	环境恢复治理保证金
environment science	环境科学
environment self-purification	环境自然净化，环境自净
environment sequence stratigraphy	环境层序地层学
environment system	环境系统
environment temperature	周围温度；环境温度
environment water quality	环境水质
environmental adaptation	环境适应性
environmental administration	环境管理
environmental agent	环境动因，环境因素
environmental amenity	环境适宜
environmental analytical chemistry	环境分析化学
environmental appraisal	环境评价；环境鉴定；环境评核
environmental archaeology	环境考古学
environmental architecture	环保型建筑
environmental assimilating capacity	环境同化能力
environmental association	生态组合；生态关系
environmental capacity of groundwater	地下水环境容量
environmental characteristics	环境特征
environmental climate	环境气候
environmental complex	环境综体，环境总体
environmental conditioning	环境改善；环境调节
environmental consideration	环境研究
environmental contingency planning	环境应急计划
environmental correlation	环境相互关系；环境关联作用
environmental decay	环境衰退；环境破坏；环境恶化
environmental design assessment	环境设计评估
environmental determinism	环境决定论
environmental determinism environmentalism	环境决定论环保主义
environmental deviation	环境差异
environmental disaster control	环境灾害监测
environmental disfunction	环境功用丧失
environmental edaphology	环境土壤学
environmental electromagnetism	环境电磁学
environmental engineering geology	环境工程地质学
environmental entity	环境实体；环境本质
environmental evaluation	环境评价
environmental fate of contaminants	环境污染物的归宿
environmental gap	环境间隔
environmental geochemistry	环境地球化学
environmental geography	环境地理学
environmental geology work of mine	矿山环境地质工作
environmental geotechnical engineering	环境岩土工程
environmental geotechnics	环境岩土工程学
environmental hazard	环境公害；环境危害
environmental health engineering	环境卫生工程
environmental hydrobiology	环境水生物学
environmental hydrogeologic map	环境水文地质图
environmental hydrogeology	环境水文地质学
environmental impact	环境影响
environmental impact analysis	环境影响分析
environmental impact assessment	环境影响评估
environmental impact assessment report	环境影响评估

报告
environmental impact monitoring 环境影响监测
environmental improvement 环境改善
environmental improvement area 环境改善区
environmental influence 环境效应；环境影响
environmental intrusion 环境干扰
environmental lapse rate 环境温度递减率；气温垂直递减率
environmental limit 环境限度；环境极限；环境限制条件
environmental map 环境地图
environmental medium 环境介质
environmental message 环境信息
environmental meteorology 环境气象学
environmental microbiology 环境微生物学
environmental monitoring quality 环境监测质量
environmental oceanography 环境海洋学
environmental optics 环境光学
environmental organic geochemistry 环境有机地球化学
environmental parameter 环境参数
environmental pathology 环境病理学；环境病状
environmental pedology 环境土壤学
environmental perception 环境感知；环境感觉
environmental periodicity 环境周期性
environmental permit 环境许可
environmental physics 环境物理学
environmental pollution control measure 环境污染对策；环境污染控制措施
environmental pollution in coal mine 煤矿环境污染
environmental program 环境规划；环境方案
environmental project 环境规划
environmental protection 环境保护
environmental psychology 环境心理学
environmental quality assessment 环境质量评价
environmental quality assessment of groundwater 地下水环境质量评价
environmental quality coefficient 环境质量系数
environmental quality objective 环境质素指标
environmental quality parameter 环境质量参数
environmental quality standards 环境质量标准
environmental reform 环境改造
environmental resource 环境资源
environmental return 环境恢复
environmental risk 环境风险，环境危险
environmental risk assessment 环境风险评价
environmental sample 环境样品
environmental satellite 环境卫星
environmental science 环境学；环境科学
environmental security 环境安全
environmental selection 环境选择
environmental status 环境状况
environmental stratigraphy 环境地层学
environmental stress 周围应力；围压；环境应力
environmental test 环境试验；环境监测
environmental theory 环境理论
environmental trace analysis 环境痕量分析
environmental transition 环境变迁
environmental variable 环境变量
environmental variation 环境变异
environmental-ecological revolution 环境生态演化
environmentalism 环境保护论；环境保护主义

environmentalist 环境学家；环境工作者；环境保护工作者
environmentally friendly 环境友好的
environmentally sensitive area 环境易受破坏地区；环境敏感区
environmentally sound 对环境有益的；对环境无害的
environment-impairing activity 环境损害行为
environment-stratigraphy 环境地层学
envitrinite 非结构镜质体
enzymic structure analysis 酶结构分析
Eoarchaean 始太古界；始太古代
Eocambrian 始寒武系；始寒武纪
Eocambrian period 始寒武纪
Eocambrian system 始寒武系
Eocene 始新世；始新统
Eocene epoch 始新世
eocene era oil 始新世石油
eocrystal 早期斑晶
eodacite 变质英安岩
Eogene 古近系；古近纪；早第三纪；下第三系
Eogene system 古近系；下第三系
eogenetic 早期成岩的；成岩早期的；始成岩期的
eogenetic carbonate cement 成岩初期碳酸盐胶结物
eogenetic carbonatization 早期碳酸盐化作用
eogenetic glauconite 成岩早期海绿石
eogenetic porosity 成岩早期孔隙
eogenetic stage 始成岩阶段
eogenetic syntaxial quartz overgrowth 成岩早期共生石英增生物
eogenetic zone 早期成岩带
eohypse 古地面等高线
eoisohypse 古地面等高线
eolation 风蚀作用；风化作用
eolian cross-bedding 风成斜层理；风成交错层理
eolian deflation 风蚀作用
eolian deposit 风积土，风积物；风成沉积
eolian deposition 风力沉积；风成沉积；风积作用
eolian differentiation 风力分选作用
eolian erosion 风蚀作用
eolian hypothesis of loess 黄土风成说
eolian loess 风成黄土
eolian ore deposit 风成矿床
eolian placer 风成砂矿
eolian reservoir 风积储集层
eolian rock 风成岩
eolian sand 风成砂；风积砂；流砂
eolian sediment 风积物；风成沉积物
eolian sedimentation 风成沉积作用
eolian soil 风成土
eolian texture 风成结构
eolian transport 风力搬运
eolian transportation 风力搬运
eolian-environment 风成环境
eolianite 风成岩
eolic 风成的；风积的
eolite 风成岩
eolotropic 各向异性的
eolotropy 各向异性
eometamorphic 早期变质的
eometamorphism 早期变质作用；始变质作用

eoorogenic phase 潜伏造山幕
Eopaleozoic 始古生代；早古生代
eoplatform 始地台
Eopleistocene 早更新统；早更新世
eoposition 风蚀沉积作用
eorhyolite 变流纹岩；脱玻流纹岩
eosite 钒钼铅矿
eosphorite 磷铝锰石；磷铝锰矿
eovolcanic 原始火山的
Eozoic era 始生代
Eozoic erathem 始生界
Eozoic group 始生界
epanticlinal fault 背斜边缘断层；背斜上部断层
epeiric 大陆边缘的；陆表的
epeiric sea 陆表海；浅海；陆缘浅海
epeiric-pelagic facies 陆缘-远洋相
epeiroclase 地台裂缝
epeirocratic 陆地克拉通的
epeirocratic condition 造陆条件
epeirocratic craton 陆地克拉通
epeirocratic period 造陆期
epeirocraton 陆地克拉通；稳定陆块
epeirogenesis 造陆作用；造陆运动
epeirogenetic 造陆的；造陆运动的
epeirogenetic movement 造陆运动
epeirogenetic sedimentation 造陆沉积作用
epeirogenetic unconformity 假整合
epeirogenetic uplift 造陆上升；造陆隆起
epeirogenic 造陆的
epeirogenic facies 造陆相
epeirogenic movement 造陆运动
epeirogenic phase 造陆阶段；造陆相
epeirogenic sedimentation 造陆沉积作用
epeirogenic taphrogenesis 造陆地裂运动
epeirogenic unconformity 假整合
epeirogenic uplift 造陆上升；造陆隆起
epeirogeny 造陆作用；造陆运动
epeirophoresis 大陆漂移；大陆移动
epeirophoresis theory 大陆漂移说
EPFM 弹塑性断裂力学
ephebic 成年期；青壮年期
ephemeral channel 暂时河床；季节性河床；雨源河床
ephemeral data 瞬息数据；临时数据
ephemeral drainage 暂时性流水
ephemeral lake 季节性湖泊；雨季湖；间歇湖；季节湖
ephemeral stream 季节性河流；间歇河；短暂河流
ephemeral system 暂时性水系；暂时性河流；季节性水系
ephemeris 星历表；天体位置表
ephemeris calendar 星历
ephemeris error 星表误差
ephemeris time 历书时；天文历时
epi-Algonkian orogenic period 后阿尔岗造山期
epi-anticlinal fault 背斜边缘断层
epibelt 浅成带
epibenthic 浅水底的；湖沿岸的
epibenthic environment 浅海底环境
epibolite 顺层混合岩；间层状混合岩
epiboulangerite 过硫锑铅矿
epibugite 浅苏英闪长岩
epicenter area 震央区；震中区域

epicenter coordinates 震中坐标
epicenter deviation 震中偏差
epicenter distance 震央距离；震中距
epicenter distribution 震中分布
epicenter location 震中定位；震中位置
epicenter map 震中图
epicenter of earthquake 地震震中
epicenter shift 震中迁移
epicentral acceleration 震中加速度
epicentral area 震中区；震中面积
epicentral distance 震中距
epicentral intensity 震中烈度
epicentral location 震中位置
epicentral zone 震中带；震中区
epicentre 震中
epicentre bias 震中偏差
epicentre coordinates 震中坐标
epicentre distribution 震中分布
epicentre intensity 震中烈度
epicentre location 震中定位；震中位置
epicentre map 震中图
epicentre migration 震中迁移
epicentre shift 震中迁移
epicentrum 震中；震源地
epichlorite 次绿泥石
epiclast 表生碎屑；外来碎屑
epiclastic conglomerate 外力碎屑砾岩；表生碎屑砾岩
epiclastic debris 表生碎屑；外力碎屑
epiclastic rock 表生碎屑岩；异地碎屑岩；外力碎屑岩
epiclastic sand 表生砂
epiclastic sedimentation 外力沉积作用
epiclastic volcanic fragment 外力火山碎屑
epicontinental basin 陆缘盆地
epicontinental deposit 陆表海沉积
epicontinental facies 陆表相
epicontinental geosyncline 陆缘地槽
epicontinental marginal sea 陆缘海；陆缘浅海
epicontinental neritic facies 陆表浅海相；陆缘近岸相
epicontinental platform 陆缘地台
epicontinental sediment 陆缘沉积
epicontinental sedimentation 陆海沉积；浅海沉积
epicontinental warp 陆缘翘曲
epicratonic 克拉通期后
epicyclic 亚旋回的
epicycloid 外摆线；圆外旋轮线
epidemic-type aftershock sequence 流行型余震序列
epiderm 浅硅铝层；表皮
epidermal deformation 硅铝壳表层变形
epidermal fold 表皮褶皱
epidermal glide tectonics 表层滑动构造
epidermic 地球表层的
epidermic fold 表层褶曲；表层褶皱
epidesmine 红辉沸石；斜方束沸石
epidiabase 变辉绿岩
epidiagenesis 表生成岩作用；后生成岩作用
epidiagenetic 后生成岩期的
epididymite 斜板晶石，斜方板晶石
epidiorite 变闪长岩
epidolerite 变粒玄岩；变粗玄岩；变煌绿岩
epidosite 绿帘岩；绿帘石岩

epidote amphibolite facies	绿帘石闪石岩相
epidote vein	绿帘石脉
epidote-amphibolite facies	绿帘石角闪岩相；绿帘角闪岩相
epidote-mica schist	绿帘石云母片岩
epidote-tremolite schist	绿帘石-透闪石片岩
epidotite	绿帘岩
epidotization	绿帘石化
epieugeosyncline	次生地槽
epigene	外成的；表成的；后成的
epigene action	外力作用；外生作用；外成作用
epigene relief	外成地形
epigenesis	后生作用；表生作用；外生作用；渐成说
epigenetic	表成的；外生的；后生的；次生的
epigenetic anomaly	后生异常
epigenetic concentration	后成富集作用
epigenetic concretion	后成结核；次生结核
epigenetic dolomite	后生白云岩
epigenetic drainage	叠置水系；上遗水系
epigenetic geochemical anomaly	后生地球化学异常
epigenetic halo	后生晕；次生晕
epigenetic interstitial water	后生间隙水；后生填隙水
epigenetic mineral	次生矿物；后生矿物
epigenetic mineralization	后生成矿作用
epigenetic phase	后生相
epigenetic river	上层遗留河；上遗河
epigenetic rock	后生岩
epigenetic scour	后生冲刷
epigenetic structure	浅层构造；表皮构造；表面构造
epigenetic valley	上层遗留谷；后成谷
epigenetic volcanism	后期火山作用
epigenetic volcano	后期火山
epigenetic water	浅层水；后生水；再生水
epigenic dolomite	后生白云岩
epigenic sediment	外变沉积物
epigenite	砷硫铜铁矿
epigeosyncline	陆表地槽
epiglaubite	钙磷石
epiglyph	层顶痕
epigneiss	浅带片麻岩
epigranite	浅成花岗岩
epigranular	等粒的；粒度均匀的
epi-Hercynian	海西期后的
epi-Huronian orogenic period	后休伦造山期
epikarst	表层岩溶
epikarst zone	表层岩溶带
epikote	环氧类树脂
epilimnion	变温层；表水层；湖面温水层
epilittoral zone	潮上带
epimagma	后期岩浆残液；浅部岩浆；外岩浆
epimagmatic	浅部岩浆的；岩浆后期的；浅岩浆的
epimarble	浅变质大理岩；浅成大理岩
epimer	差向异构体
epimerization	差向异构化
epimetamorphism	浅变质作用
epimigmatization	浅混熔作用
epineritic	半浅海的；中浅海的
epiorogenic	造山期后的
epipedon	表层；表土层；表土特征
epipelagic zone	深海浅层水带；海洋水上层
epiplatform	地台浅部；边缘地台；外地台
Epipleistocene	上更新统；晚更新世
epipolar line	核线
epipolar plane	核面
epipolar ray	核线
epipole	核点
Epiproterozoic	上元古代；上元古界
epirelief	表浮痕；上浮雕
epirock	浅带变质岩
epirogenesis	造陆运动
epirogenetic	造陆的
epirogenic	造陆的
epirogenic movement	造陆运动
epirogeny	造陆作用；造陆运动
episode	幕，期，阶段
episodic	幕的
episodic erosion	侵蚀幕
episodic evolution	幕式演化
episodic miscible flow	幕式混相运移；幕式混相流动
episodic movement	短期地壳运动
episodic sedimentation	幕式沉积
episodic subsidence	阶段性沉降
epistratal	后成地层的
epistrophic movement	表层地壳运动
episutural basin	接合带盆地
epitaxy	面衍生；定向附生；取向附生
epi-tectonic	浅层构造
epithallium	表层
epithermal	低温热液的；浅成热液的
epithermal absorption	超热中子吸收
epithermal activation	超热中子活化
epithermal activity	超热中子放射性
epithermal capture	超热中子俘获
epithermal deposit	浅成低温热液矿床
epithermal neutron	超热中子
epithermal neutron activation analysis	超热中子活化分析
epithermal neutron log	超热中子测井
epixenolith	浅源捕虏体
epizonal	浅变质带的
epizonal magmatism	浅成岩浆作用
epizonal metamorphism	浅带变质作用
epizonal pluton	浅带深成岩
epizone	浅带；浅变带；浅成带；浅变质带
epoch	世，新纪元；时期；时代
epoch of mineralization	矿化期
epoch of neogenicum	新地时期
epoch of partial tide	分潮迟角
epoch-making	划时代的
epoxide resin	环氧树脂
epoxy adhesive	环氧树脂黏合剂
epoxy asphalt	环氧沥青
epoxy asphalt concrete	环氧沥青混凝土
epoxy asphalt paving	环氧树脂沥青铺路面
epoxy bitumen	环氧沥青
epoxy coat	环氧树脂涂层；环氧涂层
epoxy coated	环氧树脂涂层
epoxy compound	环氧化合物
epoxy glue	环氧树脂胶
epoxy mortar bed	环氧砂浆垫座
epoxy resin	环氧树脂

epoxy resin coat	环氧树脂搪层
epoxy resin grout	环氧树脂浆；环氧树脂薄浆
epoxy resin grouting	环氧树脂灌浆
epoxy resin mortar	环氧树脂砂浆
epoxy-tar coating	环氧焦油涂层
epsilon type	山字形构造
epsilon-shaped tectonic system	山字形构造体系
EPSL	地球与行星科学通讯
epsom salts	泻利盐
equal angle projection	等角度投影
equal area chart	等面积图
equal area map	等积投影地图
equal area projection	等面积投影；等积投影
equal balance	均衡；地壳均衡说
equal compacting	均匀压实
equal depth map	等深度图
equal energy depth	等能量水深
equal falling factor	等沉系数
equal hardness value	等硬度值
equal heterodyne	等幅外差法
equal loudness curve	等声响曲线
equal observation	等精度观测
equal risk curve	等风险图法
equal settlement	等量沉陷
equal settling ratio	等降比；等落比；等沉比
equal settling velocity	等沉速度
equal vertical strain	等竖向应变；等垂直应变
equal wear theory	等磨损理论
equal-alternation wave	等交变波
equal-altitude method	等高法；等高度角法
equal-altitude method of multi-star	多星等高法
equal-angle map	等角投影地图
equal-angle projection	等角度投影
equal-area conical projection	等积圆锥投影
equal-area map	等面积投影地图
equal-area map projection	等积地图投影
equal-area net	等积网；施密特网
equal-area projection	等面积投影
equal-area stereonet	等面积赤平投影网
equal-azimuth projection	等方位投影
equal-contrast gray scale	等比灰阶
equal-interval peak	等时距洪峰
equality	相等；等式
equality constraint	等式约束条件
equality gate	同门
equalization diagram	均值图
equalization neutralization	均衡中和
equalization of discharge	平衡出水
equalization of pressure	压力均衡
equalization pond	稳流池
equalized	平衡的
equalized delay line	补偿延迟线
equalized heat distribution	等热分配
equalized histogram	均衡化直方图
equalized pressure	平衡压力
equalizing	均衡的
equalizing culvert	平水涵洞；平压涵洞
equalizing device	均衡装置
equalizing effect of lake	湖泊调节作用
equalizing port	平衡孔；均衡口
equalizing prong	均衡支架
equalizing pulse	均衡脉冲
equalizing pulse interval	均衡脉冲间隔
equalizing reservoir	调压水库；调节水库；平衡水池；平衡水库
equalizing system	均衡系统
equalizing tank	调压水箱；调压水塔；平水塔
equal-length code	等长码
equally spaced	等距的；等间隔的；等齿距的
equally spaced reference	等间距基准点
equally tilted photography	等倾摄影
equal-precision measurement	等精度测量
equal-ripple filter	等波纹滤波器
equal-spaced	等距的
equal-surface projection	等面积投影
equal-value gray scale	等值灰度；等值灰阶
equal-value map	等值图
equant	等轴状的；等径的；等维的
equant element	等维要素
equant grain	等轴颗粒
equant micrite crust	等厚微晶外壳
equant-pore porosity	等径孔隙度
equant-shaped	等形的
equant-shaped basin	形状相同的盆地
equation	方程；等式；公式；等分；平衡，平均；时差
equation derivation	方程推导
equation group	方程组
equation of condition	条件方程
equation of conservation	守恒方程
equation of constraint	约束方程
equation of continuity	连续方程式；连续性方程
equation of dynamic	动力方程
equation of equilibrium	平衡方程
equation of equilibrium of stress	应力平衡方程
equation of error	误差方程
equation of finite difference	有限差分方程
equation of first variation	一次变分方程；第一变分方程；初级变分方程
equation of groundwater balance	地下水均衡方程
equation of higher degree	高次方程
equation of hydrologic equilibrium	水均衡方程
equation of line	直线方程式
equation of linear regression	线性回归方程
equation of mass conservation	质量守恒方程
equation of material balance	物质平衡方程
equation of motion	运动方程
equation of regression	回归方程
equation of satellite motion	卫星运动方程
equation of state	状态方程
equation of state of gases	气体状态方程
equation of state of liquids	液体状态方程
equation of state of solids	固体状态方程
equation of static	静力学方程
equation of straight line	直线方程
equation of stress transformation	应力变换方程；应力转换方程
equation of the influence line	影响线方程
equation of variation	变分方程
equation of volumetric balance	体积平衡方程
equation of water balance	水均衡方程

equation solver	方程解算器；解方程器
equator	赤道；昼夜平分线
equator undercurrent	赤道潜流
equatorial	赤道仪；赤道的
equatorial acceleration	赤道加速度
equatorial angle	赤平角
equatorial anomaly	赤道异常
equatorial atlantic mid-ocean canyon	赤道大西洋中央峡谷
equatorial climate	赤道气候
equatorial coordinate system	赤道坐标系
equatorial coordinates	赤道坐标
equatorial day	赤道日
equatorial depression	赤道低压
equatorial eccentricity	赤道偏心率
equatorial electrojet	赤道电集流；赤道电射流
equatorial ionosphere	赤道电离层
equatorial low	赤道低压
equatorial parallax	赤道视差
equatorial plane	赤道平面
equatorial projection	赤道投影；赤平投影
equatorial radius	赤道半径
equatorial space	赤道空间
equatorial system of coordinates	赤道坐标系统
equatorial tide	赤道潮
equatorial undercurrent	赤道潜流；赤道下层海流
equatorial vortex	赤道涡旋
equatorial westerlies	赤道西风带
equiaccuracy chart	等精度图
equi-amplitude	等幅
equiangular	等角的
equiangular hyperbola	等角双曲线
equiangular positioning grid	等角定位格网
equiangular projection	等角投影
equiangular spiral	等角航线；等角螺线
equiangular spiral antenna	等角螺形天线
equiangular triangle	等角三角形
equiareal mapping	等积映射
equiareal projection	等面积投影
equiasymptotical stability	等度渐近稳定性
equibalance	平衡
equibalent track kilometerage	换算线路长度
equidensity	等密度
equidensity film	等密度胶片
equidensity image	等密度影像
equidimension	等尺寸；同大小
equidimensional	等轴状的；等径的；等维的；等量纲的；等大的；大小相等的
equidimensional grid	等维网格；正方网格
equidimensional halo	等量度晕
equidisplacement chart	等位移量曲线图
equidistance	等距离
equidistance projection	等距离投影
equidistant	等距的
equidistant curve	等距曲线
equidistant projection	等距投影
equidistant surface	等距曲面
equidrawdown line	等降深线
equiform	相似；相似的
equiform transformation	相似变换
equigranular rock	等粒岩
equigranular texture	等粒结构
equilater	等边形；等面
equilateral	等侧的；等边的；轴对称的
equilateral arch	等边二心拱
equilateral hyperbola	等边双曲线，等轴双曲线；直角双曲线
equilateral polygon	等边多边形
equilateral triangle	等边三角形
equilateral turnout	单式对称道岔；双开道岔
equilibrant	平衡力
equilibrant force	平衡力
equilibrating stress	平衡应力
equilibration	平衡；均衡
equilibrator	平衡机；平衡装置
equilibrium	平衡式；平衡；均衡
equilibrium activity	平衡作用
equilibrium adsorption	平衡吸附
equilibrium angle	平衡角
equilibrium anomaly	均衡异常
equilibrium at rest	休止平衡；静平衡
equilibrium axis	平衡轴线
equilibrium bank	平衡砂堤
equilibrium boiling point	平衡沸点
equilibrium boundary layer	平衡边界层
equilibrium capillary pressure	平衡毛细管压力
equilibrium catalyst	平衡催化剂
equilibrium coefficient	平衡系数
equilibrium compaction	平衡压实
equilibrium composition	平衡组分
equilibrium compression	平衡压缩
equilibrium concentration	平衡浓度
equilibrium condensation	平衡冷凝
equilibrium condition	平衡条件
equilibrium constant	平衡常数
equilibrium copolymerization	平衡共聚作用
equilibrium crystallization	平衡结晶作用
equilibrium curve	平衡曲线
equilibrium dew point	平衡露点
equilibrium differential equation	平衡微分方程
equilibrium distribution	平衡分布
equilibrium drainage	均衡排泄
equilibrium equation	平衡方程；稳定流方程
equilibrium failure	平衡破坏
equilibrium filtration rate	平衡滤失速度
equilibrium gas	平衡气
equilibrium limit	平衡极限
equilibrium line	平衡线
equilibrium method	平衡法
equilibrium model of stress	应力平衡模型
equilibrium moisture content	平衡含水量；平衡湿量；平衡水汽含量
equilibrium of force	力平衡
equilibrium pressure	平衡压力
equilibrium process	平衡作用；平衡法；平衡过程
equilibrium profile	平衡剖面；平衡断面
equilibrium rebound	均衡反弹
equilibrium size	平衡粒径
equilibrium slope	平衡坡度；自然坡度
equilibrium state	平衡状态
equilibrium stress state	应力平衡状态

equilibrium swelling 平衡膨胀
equilibrium swelling ratio 平衡膨胀比
equilibrium system 平衡系统
equilibrium temperature 平衡温度
equilibrium tendency 平衡趋势
equilibrium theory of tide 潮汐均衡理论
equilibrium tide 平衡潮
equilibrium vapor phase 平衡蒸汽相
equilibrium vaporization 平衡蒸发
equilibrium viscosity 平衡黏度
equilibrium water 平衡含水量
equilibrium water surface 平衡水面
equilibrium water-advancing contact angle 水的平衡前进接触角
equilibrium well discharge 均衡井出水量
equilibrium wetting 平衡润湿
equilong 等距的
equilong circle arc grid 等距圆弧格网
equilong transformation 等距变换
equimolal diffusion 等分子扩散
equinoctial 昼夜平分线；天球赤道；昼夜平分的；二分点的
equinoctial gale 二分点风暴
equinoctial spring tide 分点大潮
equinoctial storm 二分点风暴
equinoctial tide 二分潮；分点潮
equinox 昼夜平分时；二分点
equipartition 均分；匀布；匀隔；匀配
equiphase 等相位
equiplanation 高纬度陆地夷平作用；冰冻夷平作用
equipluve 等雨量线；等雨量系数线；等降水量线
equipment 设备；装备；仪器；器械，器材；部件
equipment and materials 设备材料
equipment arrangement 设备布置
equipment calibration procedure 设备校准程序
equipment change request 设备变更申请
equipment compatibility 设备互换性；设备兼容性
equipment component list 设备元件明细表；全套设备部件零件表
equipment cost 设备成本；设备费
equipment error 设备误差
equipment factor 仪器因素
equipment failure 设备故障
equipment foundation 设备基础
equipment in tunnel 洞内设备
equipment investment 设备投资
equipment item 单项设备
equipment layout 设备布置
equipment list 设备清单
equipment malfunction 设备失灵
equipment out of use 设备停用
equipment package 整套设备；设备组合
equipment performance log 设备工作日志；设备运转状况记录
equipment requirement specification 设备要求规格
equipment set 整套设备
equipment specification 设备技术规范；设备说明书
equipment specification manual 设备说明书
equipoise 平衡力；平衡；相称；平衡物，平衡锤；使相称；使平衡
equipolarization 等配极变换
equipollence 相等
equipollent 均等物
equipollent load 等价荷载
equiponderance 等重；均衡，平衡
equipotent 等力的
equipotential 等势的；等电位的
equipotential boundary condition 等势边界条件
equipotential layer 等势层
equipotential line 等势线；等位线
equipotential plane 等电位面；等位面；等势面
equipotential point 等电位点
equipotential prospecting 等位势电法勘探
equipotential slope 等电位倾斜面
equipotential surface 等势面；水准面；水平面；液面；等位面
equipotential survey 等势测量
equipotential system 等电势系统
equipotential zone 等电位区域
equipresence 共存性
equipressure 等压
equipressure contour 等压图；等压线
equipressure line 等压线
equipressure surface 等压面
equiprobability 等概率；等机率
equiprobability curve 等概率曲线
equirectangular projection 正方形投影；等距柱状投影图；等量矩形投影
equiripple response 等涟波响应
equisaturation 等饱和度
equi-shaliness line 等泥质含量线
equitant 相重叠的
equitemperature 等温度
equitemperature metamorphism 破坏性变质作用；等温变质作用
equitime 等时
equivalence 相当；等效性；等值性；等价性；当量
equivalence law 等价定律
equivalence of coal 煤的技术当量
equivalence relation 等价关系
equivalence transformation 等价变换；等效变换
equivalency 相当；等效性；等值性；等价性；当量；等同性
equivalent 等值的，等价的；当量，对应的；相当的，等效的
equivalent acidity 当量酸度
equivalent alternative 等效方案
equivalent amount 当量数
equivalent angle of internal friction 等效内摩擦角
equivalent area 等效面积；换算面积；当量面积
equivalent basicity 当量碱度
equivalent beam method 等值梁法；等效梁法
equivalent bed 同位地层；同位层
equivalent bending moment 等效弯矩；等效弯曲力矩
equivalent bending rigidity 等效抗弯刚度
equivalent circulating density 当量循环密度
equivalent coefficient 等效系数
equivalent compressive force 等效压力
equivalent concentration 当量浓度
equivalent concentration unit 当量浓度单位

equivalent conductivity 等效电导率；等效导流能力；等效传导率
equivalent conical projection 等积圆锥投影
equivalent consolidation pressure 等效固结压力
equivalent contact angle 等效接触角；等效交会角
equivalent cross section 等效截面
equivalent curve 等值曲线；等效曲线
equivalent cylindrical projection 等积圆柱投影
equivalent damping ratio 等效阻尼比
equivalent depth 等效深度；当量深度
equivalent detector 等效检波器
equivalent diameter 当量孔径；等效直径；换算直径；等效粒径
equivalent distance of the oblique exposure 斜接近段的等效距离
equivalent domain integral 等效区域积分
equivalent elastic modulus 等效弹性模量
equivalent elastic parameter 等效弹性参数
equivalent electron density 等效电子密度
equivalent energy 能当量；等效能量
equivalent equation 等价方程；同解方程
equivalent faces 等效面
equivalent factor 换算系数；等效因素
equivalent flocculation 当量絮凝作用；等效絮凝产物
equivalent fluid 等效流体
equivalent focal length 等效焦距
equivalent focus 等效焦点
equivalent footing analogy 等效基础模拟
equivalent force system 等效力系
equivalent fracturing 等效压裂
equivalent fresh air supply 等量送风；等量新鲜空气输送
equivalent grade 等效粒级；当量粒度；换算坡度
equivalent grain size 等效粒径；等值粒径
equivalent height 等值高度；相当高度；等效高度
equivalent homogeneous system 等效均质系统
equivalent injection time 等效注水时间
equivalent isomorphism 等价类质同象
equivalent layer 同期地层；同位地层；等效层
equivalent linear method 等效线性化方法
equivalent loading 等效荷载
equivalent loudness level 等效响度级
equivalent lumped system 等效集总系统
equivalent map 等积投影地图，等积图
equivalent mass 等效质量；等价质量；变位质量；当量质量
equivalent material 相似材料；相等材料；等效材料；替代材料
equivalent material model 相似材料模型
equivalent matrix suction 等效基质吸力
equivalent mean diameter of aggregate 骨料等效平均粒径
equivalent method 等值法；等效法
equivalent moisture 相当湿度；等价含水量
equivalent moment factor 等效弯矩系数
equivalent mud density 当量泥浆密度
equivalent mud weight 当量泥浆比重
equivalent network 等效网络
equivalent nodal force 等价节点力；等效节点力
equivalent nodal load 等效结点荷载
equivalent noise 等效噪声

equivalent observation 等效观测；等值观测
equivalent of fuel 标准燃料；当量燃料
equivalent of heat 热当量
equivalent opening size 等效孔径；当量孔径；当量筛孔尺寸
equivalent parameter 等值参数；等价参数；等效参数
equivalent particle diameter 等效粒径；等值粒径
equivalent pendulum 等效摆
equivalent per million 百万分之几当量数；百万分之当量
equivalent periodic line 等周期线
equivalent permeability 等效渗透性；等效渗透率
equivalent photograph 等效相片
equivalent point-system 等效点系
equivalent pore pressure 当量孔隙压力；等效孔隙压力
equivalent porosity 当量孔隙率；等效孔隙度
equivalent porous media 多孔介质等效性
equivalent potential temperature 等效位温，等效位势温度
equivalent pressure 等效压力
equivalent production time 等效生产时间
equivalent profile 等效断面
equivalent projection 等值投影；等面积投影；等积投影
equivalent radius 等效半径；折算半径
equivalent rectangular stress distribution 等值矩形应力分布
equivalent relation 等价关系
equivalent relative density 当量相对密度；等效相对密度
equivalent replacement method 等量置换法
equivalent resistivity 等效电阻率
equivalent resistivity method 等效电阻率法
equivalent rigidity 等效刚度
equivalent roentgen 伦琴当量
equivalent roughness 等价粗糙度；当量粗糙度
equivalent roughness coefficient 当量粗糙度系数；等值粗糙系数
equivalent seismic load 等效地震荷载
equivalent sine wave 等效正弦波
equivalent size 当量尺寸
equivalent soil layer method 等值层法
equivalent soil reaction 等效地基反力；等效基底反力
equivalent sound level 等效声级
equivalent source method 等效源法
equivalent span length 当量跨距
equivalent specific heat 等效比热
equivalent specific weight 当量比重
equivalent spectrum curve 等效谱曲线
equivalent sphere diameter 等球粒直径
equivalent spherical diameter 当量球面直径
equivalent static criteria 等效静力准则
equivalent static force analysis 等效静力分析
equivalent static load 等效静载
equivalent static loading 等效静力加载
equivalent static method 等效静力法
equivalent stationary motion 等效稳态运动
equivalent stationary response 等效稳态反应
equivalent stiffness 当量刚度；等效刚度
equivalent strain 等效应变
equivalent stratified porous medium 等效层状多孔介质
equivalent stress 等效应力；折算应力

equivalent system 等效系统
equivalent temperature 当量温度
equivalent time 等效时间
equivalent time horizon 等时面
equivalent transfer function 等效传递函数
equivalent transverse shear force 等效横剪力
equivalent trend line 当量趋势线
equivalent uniform load 等效均布荷载
equivalent uniformly distributed load 等效均布荷载
equivalent value 等效值；换算值；当量值
equivalent vapor volume 当量蒸气体积
equivalent viscosity 等效黏度
equivalent viscous damping 等效黏滞阻尼
equivalent volume 等效体积
equivalent water level 相当水位；等值水位
equivalent wavelet 等效子波
equivalent weight 当量；等效重量
equivalent weight of ions 离子当量
equivalent weight replacement method 等重量代替法
equivalent wheel load 等效轮压；等效轮载
equivalent width 等效宽度；等值宽度；换算宽度
equivalent working height 等效开采高度
equivalent Young's modulus 等效杨氏模量
equivoluminal 等体积的
equivoluminal wave 等体积波；等容波
equiwavelength 等波长
equiwavelength pattern 等波长图
era 代；纪元；时代；阶段
eradiate 放射，辐射
erasability 可擦度；可擦除性
erasable 可擦除的
erasable memory 可擦存储器
erasable programmable read-only memory 可擦可编程序只读存储器
erasable storage 可擦存储器
eraser 橡皮擦；擦除者；消除用具；消磁器
erasing head 消磁头
erasure 消去，消除；消磁；删去，删掉
erasure signal 消除信号
erathem 界
erect anticline 直立背斜
erect fold 直立褶皱
erect image 正像，正立像
erect lay down derrick 起放井架
erectable whipstock 活动式造斜器
erected cost 安装费用
erecting 架设；组装；插入
erecting by floating 浮运架桥法
erecting by overhang 吊挂架
erecting crane 安装用的起重机；装配吊车
erecting frame 安装架
erecting platform 安装平台
erecting scaffold 搭脚手架
erecting stage 安装现场；安装平台
erecting tool 架设工具；安装工具
erecting yard 装配现场；装配场，安装场
erection bar 架立钢筋
erection bolt 安装螺栓
erection by bridge girder erecting equipment 架桥机架设法

erection by incremental launching 顶推式架设法
erection by longitudinal pulling method 纵向拖拉法
erection by protrusion 悬臂拼装法；悬臂架设法
erection by swing method 旋转法施工；转体施工
erection clearance 安装净空
erection diagram 装配图
erection drawing 装配图；装配图样
erection engineer 安装工程师
erection equipment 安装设备
erection jack 安装用千斤顶
erection loop 安装环
erection method 架设方法
erection of derrick 立井架
erection of water turbine 水轮机安装
erection schedule 安装进程；安装计划
erection team 安装工程队；安装队
erection tolerance 安装容许误差；安装允许偏差
erection with cableway 缆索吊装法
erection with scaffolding 赝架式架设法
eremology 沙漠学
ergodic 遍历性；各态历经的
ergodic hypothesis 各态历经性假定
ergodic process 遍历过程
ergodic state 遍历状态；各态历经状态
ergodic theorem 遍历定理
ergodicity 各态历经性
ergometric 测力的；测功的
ergonometrics 工效学；人类工程计量学
ergonomic 人类工程学的；人机控制的
erikite 硅磷铈石；多水磷铈石
erinite 翠绿砷铜矿
eriochalcite 水氯铜矿
eriometer 衍射测微计；微粒直径测量仪
erionite 毛沸石
eriophorum peat 苔藓泥炭；棉草泥炭
eriscope 光电摄像管
eritrosiderite 红砷铁矿
ermakite 褐蜡土
erode 冲刷；侵蚀；冲蚀；腐蚀
eroded canyon 侵蚀峡谷
eroded crater 侵蚀坑
eroded sediment 冲刷物；冲刷堆积物
eroded slope 受冲刷边坡；受冲刷坡
eroded soil 侵蚀土
eroded spot 冲蚀点
eroded valley 侵蚀谷
eroded volcano 侵蚀火山
erodibility 冲蚀性，侵蚀度；腐蚀度；易蚀性；可侵蚀能力
erodibility factor 冲刷系数
erodible 易的；易侵蚀的，可侵蚀的
erodible bank 冲刷性岸坡
erodible channel 可侵蚀河槽，可冲刷河槽，侵蚀河槽；不定河床，活动河床
erodible material 易冲刷物质；易受侵蚀的材料
erodible soil 易受侵蚀土壤
eroding bank 侵蚀岸
eroding channel 侵蚀河道；侵蚀海峡；侵蚀河槽
eroding force 侵蚀力
eroding shore 冲刷海岸

eroding velocity 侵蚀速度；冲刷速度
erosibility 可侵蚀性；可侵蚀程度；可侵蚀能力
erosion 侵蚀，冲刷侵蚀；侵蚀作用；磨蚀；熔蚀；风化
erosion action 侵蚀作用
erosion activity 侵蚀作用
erosion attack 冲蚀破坏
erosion bank 侵蚀河岸；冲刷河岸
erosion base 侵蚀基面，侵蚀基准面
erosion base level 侵蚀基面，侵蚀基准面
erosion basin 侵蚀盆地
erosion basis 侵蚀基准；侵蚀基面
erosion belt 侵蚀地带
erosion class 侵蚀等级
erosion column 侵蚀柱
erosion control 侵蚀防治；侵蚀控制
erosion control below dam 坝下冲刷防治
erosion control mattress 防冲沉排
erosion control measure 防冲措施
erosion control of land 陆地水土保持
erosion control works 防冲刷工程；防冲刷构筑物
erosion corrosion 冲蚀腐蚀；侵蚀性腐蚀
erosion crater 侵蚀坑
erosion cycle 侵蚀循环；侵蚀旋回
erosion depth 冲刷深度
erosion device 冲蚀探测装置
erosion drill 冲蚀式钻机；喷蚀式钻机
erosion drilling 冲蚀钻井
erosion equation 侵蚀方程
erosion factor 侵蚀因素
erosion fault scarp 侵蚀断崖；断层线崖
erosion form 侵蚀地形；侵蚀形态
erosion gap 侵蚀间断；侵蚀山口
erosion groove 侵蚀槽；侵蚀沟
erosion gully 侵蚀冲沟
erosion hollow 冲刷穴，冲刷坑
erosion index 冲刷指数；侵蚀度；侵蚀指数
erosion intensity 侵蚀强度；冲蚀强度；剥蚀强度
erosion lake 侵蚀湖
erosion landform 侵蚀地形
erosion level 侵蚀面；侵蚀剥削面
erosion loss 冲蚀流失；侵蚀流失
erosion mark 侵蚀痕
erosion mountain 侵蚀山
erosion of glacier 冰川侵蚀
erosion of levee slope 堤坡冲刷
erosion pavement 防冲刷铺砌层；护坡
erosion pilot 冲蚀监测器
erosion pit 侵蚀坑；剥蚀坑
erosion plain 侵蚀平原
erosion plateau 侵蚀高原
erosion platform 海蚀平台，海蚀台地；侵蚀台地
erosion point 冲蚀点
erosion prediction 侵蚀预报；侵蚀预测
erosion probe 冲蚀探测器
erosion procedure 侵蚀过程
erosion process 侵蚀过程；侵蚀作用
erosion protection 侵蚀防护；侵蚀保护；防冲
erosion protection works 防止侵蚀工程；侵蚀防治工程；冲刷防护桩
erosion rate 侵蚀速率；冲刷速率

erosion ratio 侵蚀率
erosion remnant 侵蚀残迹；冲蚀遗迹
erosion residual hill accumulation 剥蚀残丘堆积物
erosion resistance 耐冲蚀；抗侵蚀性
erosion resistant 耐冲蚀的
erosion sand plug 冲蚀砂柱
erosion sand probe 砂粒冲蚀探头
erosion spring 侵蚀泉
erosion surface 侵蚀面；风化面
erosion terrace 侵蚀阶地
erosion thrust 侵蚀冲断层；侵蚀逆断层
erosion trough 冲刷槽
erosion truncation 削蚀作用
erosion unconformity 侵蚀不整合
erosion velocity 侵蚀速度；冲刷速度
erosional accumulation basin 侵蚀堆积盆地
erosional agent 侵蚀营力
erosional basin 侵蚀盆地
erosional basis 侵蚀基面
erosional bedding contact 侵蚀层面接触
erosional channel reservoir 侵蚀河道油气藏；侵蚀河道型水库
erosional coal basin 侵蚀煤盆地
erosional competency 侵蚀能力
erosional depression 侵蚀洼地；侵蚀洼陷
erosional drilling 冲蚀钻井
erosional flood plain 侵蚀泛滥平原
erosional force 冲蚀力
erosional hiatus 侵蚀间断
erosional lake 侵蚀湖
erosional landform 侵蚀地形；侵蚀地貌
erosional level 侵蚀水平面；侵蚀面
erosional plain 侵蚀平原
erosional plateau 侵蚀高原
erosional process 侵蚀过程；侵蚀作用
erosional retreat 侵蚀后退
erosional scarp 侵蚀崖
erosional shore 侵蚀海岸
erosional spring 侵蚀泉
erosional stage 侵蚀期
erosional surface 侵蚀面
erosional terrace 侵蚀阶地；浪蚀阶地
erosional transgression 侵蚀海进
erosional truncation 削蚀作用
erosional unconformity 侵蚀不整合
erosional valley 侵蚀谷；侵蚀沟
erosional-accumulative relief 侵蚀堆积地形
erosion-control dam 防冲坝
erosion-ravaged area 侵蚀破坏地区
erosion-resistant channel 抗侵蚀河槽
erosion-resistant material 抗侵蚀材料
erosion-resisting characteristics 抗冲性能；抗侵蚀性能
erosive 冲蚀的；侵蚀的
erosive action of sand 砂蚀作用
erosive agent 侵蚀力
erosive capacity 侵蚀能力
erosive force 侵蚀力
erosive power 侵蚀能力
erosive velocity 侵蚀速度；冲刷流速
erosive velocity in canal 渠道冲刷流速

erratic 无规律的；不定的，反常的，不均一的；漂游的，漂移的；漂积的；漂砾，漂块，外来石块，煤层中树干化石
erratic behavior 不规律性
erratic boulder 漂砾；外来岩块
erratic curve 无规律曲线
erratic deposit 漂砾沉积
erratic deposition 不规则沉积
erratic drift 不规则漂移
erratic feeding 不均匀给进
erratic flow 涡流；紊流
erratic form 漂砾层；漂积层
erratic function 不规则函数
erratic high grade 特高品位，特高品级
erratic load 不规则荷载
erratic pebble 漂砾
erratic permeability 不规则渗透率
erratic raft 冰川巨砾
erratic soil 无规律土
erratic soil profile 不规则土层剖面
erratic soil structure 土的不规则结构
erratic stone 漂石；孤石
erratic subsoil 不均一土；不均质土；杂乱土层
erratic till 漂砾碛
errite 褐硅锰矿
erroneous analysis 错误的分析
erroneous correlation 不正确的对比
error analysis 误差分析；错误分析
error and accuracy 误差与精度
error at measurement 测定误差
error bar 误差条线
error bound 误差范围；误差界限
error check and correction 误差检查和校正
error checking 误差校验
error compensation 误差补偿
error condition 误差条件
error control 误差控制
error correcting code 纠错码
error correcting routine 错误校正程序
error correction 误差校正；修正误差
error correction by feed-back repetition 反馈重发纠错
error correction code 误差校正码
error covariance matrix 误差协方差矩阵
error criterion 误差准则
error curve 误差曲线；特劳姆普误差曲线
error detecting code 错误检测码
error detection 错误检测；差错检测；检错
error distribution 误差分配；误差分布
error ellipse 误差椭圆
error ellipsoid 误差椭球
error energy 误差能量
error envelope 误差范围
error equation 误差方程
error equation for interpolating 内插误差方程式
error estimate 误差估计
error evaluation 误差估计
error field 误差场
error free 没有误差的
error frequency limit 错误频率极限
error function 误差函数

error graph 误差图
error in address 地址误差
error in label 符号部分出错；标记误差
error in measurement 测量误差
error in reading 读数误差
error indication 误差指示
error interrupt 错误中断
error limit 误差极限；误差范围
error logger 误差记录程序
error magnitude 误差幅度
error message 误差信息；出错信息
error norm 误差范数
error of alignment 定线误差；对准误差
error of approximation 近似误差
error of calculation 计算误差
error of centering 对心误差
error of closure 闭合误差；闭合差；条件闭合差
error of closure in leveling 水准闭合差
error of collimation 视准误差
error of division 分度误差
error of estimation 估计误差；估算误差
error of mean square 均方误差
error of measurement 计量误差；量测误差
error of observation 观测误差
error of observation data 观测数据误差
error of sighting 照准误差
error of survey 测量误差
error of tilt 倾角误差；倾斜误差
error of traverse closure 导线闭合差
error on the safe side 安全误差
error propagation 误差传递；误差传播
error pulse 误差脉冲
error range 误差范围
error rate 出错率；差错率；误差率
error ratio 误差比
error recovery 错误校正；差错恢复
error routine 查错程序
error rule 误差法则
error sensitivity number 误差敏感度数
error sensor 误差敏感元件；误差传感器
error severity code 严重错误码
error signal 误差信号
error source 误差来源；误差源
error state 异常状态；疑误状态
error synthesis 误差综合法
error term 误差项
error test 误差检验
error theory 误差理论
error variance 误差方差；误差偏差；误差变量
error vector 误差向量
error-checking 错误校验
error-circular radius 误差圆半径
error-correcting code 错误校正码
error-detecting 错误检测；检测误差
error-detecting routine 错误检测程序
error-free operation 无错操作
error-free running period 无误差运转期
error-prone 易于出错的
ersbyite 钙柱石
erupting geyser 间歇喷泉

erupting spring 喷泉
erupting wave 突发波
eruption 喷发；爆发
eruption canal 喷发管道
eruption cloud 喷发灰云；火山云雾；火山灰气体云
eruption cycle 喷发周期；喷发旋回
eruption cylinder 喷发柱
eruption deposit facies 喷发沉积相
eruption fissure 喷发裂缝
eruption intensity 喷发强度
eruption laccolith 喷发岩盖；喷发岩盘
eruption magnitude 喷发规模
eruption proluvium 喷发洪积物
eruption rain 喷发雨
eruption symptom 喷发征兆；火山喷发前兆
eruption time 喷发年代；喷发时间
eruption type 火山喷发类型
eruptive 喷出的；爆发的；喷出岩，火成岩
eruptive activity of volcano 火山喷发活动
eruptive body 火山岩体；喷发岩体；火成岩体
eruptive breccia 火山角砾岩；喷发角砾岩
eruptive deposit 喷发沉积；喷发矿床，火成矿床
eruptive facies 喷发岩相
eruptive laccolith 喷发岩盖；喷发岩盘
eruptive material 喷发物
eruptive rock 喷出岩；火成岩；火山岩
eruptive stock 火成岩株
eruptive symptom 喷发征兆
eruptive tuff 喷发凝灰岩；凝灰岩
eruptive vein 火成岩脉
eruptivity 喷发活动
erythrite 赤藻醇；赤丁四醇；钴华
erythrosiderite 红钾铁盐
erythrozincite 锰纤锌矿
erzbergite 文石-方解石互层；霰钙华
Erzgebirgian orogeny 埃尔格堡造山运动
escalating rate 上升率
escape 逃逸，漏失，外泄，排泄；放出；逃生门，太平门
escape canal 排水沟；排水渠
escape character 换码符
escape hatch 应急的通道
escape hole 排水孔；泄水孔；排气口
escape ladder 脱险梯
escape orifice 排泄口；泄放孔
escape peak 逃逸峰
escape rate 逃逸速率
escape route 排出管路；排出通道；运移通道；流水口
escape stair 太平梯
escape structure 逃逸构造；逸出构造
escape trace 逃逸迹
escape works 泄水构筑物
escaped product 漏失的油品
escapeway of underground 矿井安全出口
escaping structure 逸逸构造；逸出构造
escarpment 悬崖，峭壁，陡坡；单斜山
escharotic 苛性剂；腐蚀性的
escherite 褐黄帝石
Eshka method 埃斯卡测硫法
eskar 蛇丘；蛇形丘；冰河沙堆

eskebornite 铜硒铁矿
esker 蛇丘；蛇形丘；冰河沙堆
esker delta 蛇行丘三角洲；滨河沙滩
esker fan 蛇丘扇形地
esker knobs 蛇丘丛
esker ridge 蛇丘脊
esker trough 蛇丘谷；蛇丘槽
eskerine 蛇丘的
esmeraldite 云英花岗岩；石英白云岩
espalier drainage 格状水系
espalier drainage pattern 格状水系；方格水系
espichellite 橄闪煌斑岩；橄闪斑岩
essential boundary 基准边界
essential boundary condition 基准边界条件
essential component 主要组分；必需组分；主要成分
essential parameter 基本参数
essential property 主要性质；基本性质
essential water 必要的水；组成水分
essentially positive matrix 本性正矩阵
essexite 碱性辉长岩
essonite 钙铝榴石
established dune 稳定沙丘
established reserves 确定的储量
establishment 建立，设立；确定；编制；估算；企业机构，机关，研究所
estavel 落水洞泉；涌泉；地下河
estavelle 水洞
esterellite 英闪玢岩；英微闪长岩
estimate of cost 估价；成本估算
estimate of seismic risk 地震危险性估计
estimate variance 估计方差
estimated amount 估计量；预算数量
estimated capacity 估计容量
estimated damage ratio 估算损害比
estimated design load 预估设计负荷
estimated discharge 估计流量；估计出水量
estimated maximum load 估计最大负荷
estimated performance 估测动态；估计性能
estimated settlement 预计沉降量
estimated value 估算值；估计值
estimated weight 估计重量
estimates and engineering budget works 估价及工程预算制作
estimation error 估计误差；估测误差；估算误差
estimation method 估计方法；估算方法
estimation of error 误差估计
estimation of population parameters 总体参数估计
estimation of reserves 储量估计
estimation of seismic intensity 地震烈度评定；地震烈度估计
estimation variance 估计方差
estuarial 河口的；港湾的
estuarial circulation 河口环流
estuarial model 河口模型
estuarial sediment 河口泥沙
estuarine 河口湾的；三角湾的；港湾的；河口，江口
estuarine and coastal dynamics 河口海岸动力学
estuarine delta 河口三角洲；三角湾状三角洲
estuarine deposit 港湾沉积；河口湾沉积
estuarine ecology 河口生态学

estuarine environment	河口环境；海湾环境	Euclidean algorithm	欧几里德算法
estuarine facies	港湾相；河口湾相	Euclidean space	欧氏空间
estuarine flat	港湾低洼地	eucrite	钙长辉长无球粒陨石；钙长辉长岩
estuarine harbor	港湾	eucryptite	锂霞石
estuarine lagoon	港湾泻湖	eucrystalline texture	全晶质结构
estuarine mud	河口软泥	eudialyte	异性石
estuarine pollution	河口污染	eudidymite	双晶石
estuarine sand	河口沙	eudiometer	量气管，测气管；空气纯度测定管
estuarine sediment	河口湾沉积物；港湾沉积物	eugeogenous	易风化的；风化岩屑的
estuarine sediment dynamics	河口沉积动力学	eugeosynclinal furrow	优地槽海谷
estuarine sediment movement	河口泥沙运动	eugeosynclinal ridge	优地槽海脊
estuarine soft soil	海积软土	eugeosynclinal terrane	优地槽地体
estuarine works	河口工程	eugeosyncline	优地槽
estuary	港湾；河口湾，河口；入海河口；三角湾	eugranitic	花岗岩状的
estuary coast	三角湾岸；海湾型海岸	euhaline water	高盐水
estuary dam	河口坝	euhedral crystal	自形结晶；自形晶
estuary delta	河口三角洲	euhedral granular	自形粒状
estuary deposit	河口湾沉积；河口沉积	euhedral granular texture	自形粒状结构
estuary ecosystem	河口生态系统	euhedron	自形晶
estuary facies	港湾相	eukairite	硒铜银矿
estuary harbor	河口港	euktolite	橄金云斑岩
estuary model	河口模型	Euler	欧拉
estuary mud bar	河口泥滩	Euler cripping stress	欧拉断裂应力
estuary pollution	河口污染	Euler deconvolution	欧拉反褶积
estuary port	河口港	Euler equation	欧拉方程
estuary region	河口区域	Euler explicit method	显式欧拉方法
estuary regulation	河口整治	Euler implicit schemes	隐式欧拉方法
estuary station	河口测站	Euler number	欧拉数
etch method	侵蚀法	Euler pole	欧拉极
etch period	刻蚀时间	Euler theorem	欧拉定理
etch pit	蚀坑；侵蚀麻点	Euler-Cauchy method	欧拉-柯西法
etch ring	刻蚀环痕	Eulerian angle	欧拉角
etch test	侵蚀试验；腐蚀试验	Eulerian equation	欧拉方程
etched channel	酸蚀孔道	Eulerian free period	欧拉自由周期
etched dimple	侵蚀陷斑	Eulerian scheme	欧拉法
etched ellipse	刻蚀椭圆痕	Euler-Lagrange equation	欧拉-拉格朗日方程
etched groove	蚀沟	Euler's approach	欧拉法
etched hill	蚀丘	Euler's equation	欧拉方程
etched like finish	侵蚀型光洁度	Euler's identity	欧拉恒等式；欧拉公式
etching method	刻蚀法	Euler's theorem	欧拉定理
etching pattern	侵蚀形态	eulysite	榴辉铁橄岩
etching power	刻蚀能力	eulytite	硅铋石；闪铋矿
etching structure	侵蚀构造	euosmite	水脂石
etchplain	刻蚀平原	eupelagic clay	远洋黏土
etch-proof resist	防蚀剂	eupelagic deposit	远洋沉积
eternal frost climate	永冻气候	eupelagic sediment	远洋沉积物
eternal frozen ground	永冻土	eupholite	滑石辉长岩
eternit	石棉水泥	euphotide	糟化辉长岩
etesian climate	地中海气候	euphyllite	钠钾云母
ether spectrum	电磁波谱；以太波谱	euporphyric texture	全斑结构
ether wave	电磁波；以太波	euprofundal	深湖底的
ethernet	以太网	Eurasia belt	欧亚带
ethiopsite	硫化汞	Eurasian plate	欧亚板块
ethmolith	漏斗状岩盘；岩漏斗	eurite	霏细岩
etindite	白榴霞石岩；白霞岩	European remote sensing satellite	欧洲遥感卫星
etnaite	碱橄玄岩	European standard	欧洲标准
ettringite	钙矾石；钙铝矾	eu-sapropel	成熟腐泥
eucairite	硒铜银矿	eustatic event	海面升降事件
euchroite	翠砷铜矿	eustatic fluctuation	海面变动
euclasite	蓝柱石	eustatic hypothesis	海面变动假说

eustatic movement 海面升降运动
eustatic regression 海退变化
eustatic river terrace 海面升降形成的河流阶地
eustatic terrace 海面变动形成的阶地
eustratite 透长辉煌岩
eutaxic 成层的；条纹斑状
eutaxic deposit 成层矿床
eutaxite 条纹斑杂岩
eutaxitic 条纹斑状
eutaxitic structure 条纹斑状结构
eutectic 共晶的；共溶的
eutectic curve 共结线；共晶线；低共熔线
eutectic mixture 低共熔混合物
eutectic point 低共熔点；共晶点；共结点
eutectic salt 低熔盐；共晶盐，易熔盐
eutectic structure 共晶结构
eutectic temperature 初熔温度；低共结温度；低共熔温度
eutectic texture 共结结构
eutecticum 共晶
eutectite 低温共结岩
eutectofelsite 共结霏细岩
eutectoid 类共熔体；似共物；似共晶；类共晶的；共析体
eutectoperthite 共结条纹长石
eutectophyre 共结斑岩
eutectophyric 共结斑状的
euthalite 密方沸石
eutropy 异序同晶现象
euvitrain 纯镜煤；无结构镜煤
euvitrinite 无结构镜质体
euxenite 黑稀金矿
euxinic 静海的；闭塞环境的；滞水环境的
euxinic basin 滞海盆地
euxinic deposit 静海沉积；闭流海沉积；滞海沉积
euxinic deposition 静海沉积；滞海沉积
evacuated pore 真空孔隙
evacuated space 真空
evacuation works 挖掘工程
evaluation 评价，评定，鉴定；估算，求值；核算
evaluation by thumb rule 粗估法
evaluation criteria 评估标准
evaluation criterion 评价标准
evaluation exploration well 评价井
evaluation groundwater quality 地下水水质评价
evaluation map 矿产评价图；预测图；数据整理图；估价图
evaluation of bid 评标
evaluation of deposit 矿床评价
evaluation of determinant 行列式求值
evaluation of engineering quality 工程质量评定
evaluation of groundwater quality 地下水水质评价
evaluation of opening stability 开挖稳定性评价；洞室稳定性评价
evaluation of ore deposit 矿床评价
evaluation of rock slope stability 岩质边坡稳定性评价
evaluation of water resources 水资源评价
evaluation procedure 评价步骤；评估程序
evaluation test 评价试验
evanescent wave 损耗波；消散波

evansite 核磷铝石
evapocrystic texture 蒸发晶质结构
evapolensic texture 蒸发纹层结构
evapoporphyrocrystic texture 蒸发斑状结构
evaporability 挥发性
evaporable 易蒸发的
evaporant 蒸发物
evaporate 蒸发；蒸发盐；蒸发岩；脱水
evaporate mineral 蒸发矿物
evaporated product 挥发的油品
evaporation 蒸发作用；脱水；消散；发射；蒸发量
evaporation area 蒸发面积
evaporation capacity 蒸发量；蒸发能力
evaporation coefficient 蒸发系数
evaporation discharge 蒸发排泄；蒸发消耗；蒸发量
evaporation from ice 冰面蒸发
evaporation from land 地面蒸发
evaporation from snow 雪融
evaporation from vegetation 植物蒸腾；植被蒸发
evaporation from water surface 水面蒸发
evaporation front 汽化前缘
evaporation gauge 蒸发量测量计
evaporation gum test 蒸发胶质试验
evaporation heat 蒸发热
evaporation latent heat 蒸发潜热
evaporation loss 蒸发损耗
evaporation loss test for asphalt 沥青挥发分试验
evaporation method 蒸发法
evaporation of groundwater 地下水蒸发
evaporation of water 水分蒸发
evaporation pan 蒸发皿
evaporation pit 蒸发坑
evaporation potential 蒸发势
evaporation rate 蒸发速度；蒸发率
evaporation residue 蒸发残留物
evaporation rock 蒸发岩
evaporation salt 蒸发盐
evaporation sediment 蒸发盐沉积
evaporation surface 蒸发面
evaporation test 蒸发试验
evaporative 蒸发的
evaporative cooling system 蒸发冷却系统
evaporite 蒸发岩；蒸发盐
evaporite deposit 蒸发沉积矿床
evaporite-solution breccia 蒸发岩溶解形成的角砾岩；盐溶角砾岩
evapotranspiration 蒸发蒸腾作用；总蒸发量；蒸散
even 平坦的；均匀的；偶数的
even carbon number predominance 偶碳数优势
even decompression 均匀减压
even front 均匀前缘
even function 偶函数
even grained 等粒状
even number 偶数
even parity 偶数奇偶校验；偶同位，偶数奇偶性
even parity check 偶数奇偶校验
even predominance 偶碳优势
even pulse 偶脉冲
even response 偶脉冲响应值
even settlement 均匀沉降

even wear 均匀磨损
even-crested upland 平顶高地
even-granular 等粒状
evening tide 夜潮，晚潮
even-integral number 偶整数
evenkite 鳞石蜡
evenly distributed load 均布荷载
even-odd 偶-奇的
even-odd check 奇偶校验
even-odd predom index 正构烷烃奇偶优势指数
even-order harmonic 偶次谐波
event counter 信号计数器
event deposit 事件沉积
event detection 事件检波
event flag 事件标记
event flag cluster 事件标记组
event of small probability 小概率事件
event stratification 事件层理
event time 事件时间
event-extinction 事件性绝灭
even-textured 均一构造的
eventual flood 特大洪水
ever frost 永冻土；多年冻土
Everett's projection 埃弗雷特投影
ever-frozen layer 永冻层；多年冻层
ever-frozen soil 永冻土
everglade 湿地；沼泽地
evergreenite 流英正长岩
everlasting 永久的
every second well 每隔一口井
everyday wave base 正常波基面
evidence of faulting 断裂证据；断裂标志
evisite 碱性花岗岩类；霓闪花岗正长岩
Evison wave 埃弗森波
evoked response 诱发响应
evolution cellular automation 演化细胞自动机
evolution characteristics 演化特征
evolution equation 演化方程
evolution of gas 放出气体
evolution of heat 放热
evolution of tectonic systems 构造体质演化
evolutional paleontology 进化古生物学
evolutionary algorithms 进化算法
evolutionary neural network 进化神经网络
evolutionary pattern 演化模式
evolutionary process 演变过程；演化过程；进化过程
evolutionary series 演化序列
evolutionary systematics 演化系统学；演化分类学
evolutionary trend 演化趋向
evolutionary zone 演化带
evolutionism 进化论
evolutoid 广渐屈线
evolvent 渐开线
evorsion 涡流侵蚀作用
exact analysis 精密分析
exact integration 精确积分
exact position 精确位置
exact scale 精确刻度；精确尺度
exact solution 精确解
exactitude 正确性；严格；精准；精密度

exactness 精度；精确度；精密度；正确；严格
exaggerated test 超常试验；强化试验
exaggerated vertical scale 放大垂直比例
exaggeration model 放大模型
examinant 检查人
examination and acception 验收
examination and approval 审批工作
examination and approval authority 审批单位；审批机构
examination in depot 库检
examination of discovery 矿点检查
examination of water 水质检查
examination of water quality 水质检验
examine 检查；查看；检验；考试
examine and approve 审批
excavate 挖掘；开挖；挖土
excavate with timbering 有支撑挖掘
excavate without timbering 无支撑挖掘
excavated cross-section 开挖断面；挖掘横截面
excavated depth 开挖深度
excavated pile 挖孔桩
excavated section 掘进断面
excavated surface 开挖工作面；挖掘表面
excavated tank 开挖的蓄水池
excavated volume 挖掘体量；土石方量；溢流段断面
excavated work face 开挖工作面
excavating 挖掘
excavating chute 开挖斜槽；开挖陡沟
excavating engineering 开挖工程
excavating loader 开挖装料机
excavating machine 挖掘机；挖土机
excavating shaft 开挖竖井
excavating technology 掘进工艺
excavation 挖掘；挖土，开挖；挖方工程；基坑；矿山巷道，铲掘，掘进
excavation base 开挖基础
excavation by blasting 爆破开挖
excavation by well-point method 井点排水开挖法
excavation cycle 掘进循环
excavation damaged zone 开挖损伤区
excavation depth 开挖深度
excavation depth for dam foundation 坝基开挖深度
excavation driving 掘进
excavation effect 挖掘效应
excavation engineering 掘进工程；基坑工程
excavation equipment 开挖设备
excavation face support 掘进工作面支护；开挖掌子面支护
excavation in the dry 干挖
excavation intensity 开挖强度
excavation laneway 开挖巷道
excavation limit 开挖边界；开挖范围
excavation line 开挖线
excavation line of tailwater channel 尾水渠开挖线
excavation material 开挖料
excavation method 开挖方法
excavation method using Messer sheet-pile 梅赛尔施工法
excavation method using Ranze sheet pile 长矛式钢插板施工法
excavation of slope surface 开挖坡面
excavation permit 挖掘准许证；掘路许可证

excavation pit 开挖基坑
excavation process simulation of rock slope 岩质边坡开挖过程的模拟
excavation prospecting 挖探
excavation protection 开挖防护
excavation roadway 掘进巷道
excavation section 开挖断面
excavation slope 挖方坡度；开挖边坡
excavation surface 开挖面
excavation treatment 开挖处理
excavation unloading 挖方卸载
excavation unloading destruction 开挖卸荷破坏
excavation with timbering 有支撑开挖
excavation without support 无支撑开挖
excavation without timbering 无支撑开挖
excavation works 开挖工程；挖方工程；挖掘工程
excavation-stability condition 开挖稳定条件；挖土稳定条件
excavator 挖掘机，开凿者，打洞机
excavator shield 挖掘式盾构
excavator with planning operation 刨铲挖掘机
excavator-mounted pneumatic breaker 挖土机上装配的气动破碎机
excellent bond 强键
excellent diamond 高质量金刚石
excellent picture 优质图像
excenter 外心
excentralization 偏心
excentralizer 偏心器
excentric bit 偏心钻头
excentric shaft 偏心轴；偏心竖井
exception condition 异常条件
exception dispatcher 异常调度程序
exception vector 异常向量
exceptional structure 特殊构筑物
exceptional well 特殊井
excess air factor 空气过剩系数
excess capacity 超额容量；过剩产能
excess clearance 预留沉落量
excess component 过剩组分；过剩成分
excess draught 过量吃水；过量气流
excess energy 超额自由能；多余能量
excess excavation 超挖；过度开挖
excess flow device 溢流装置
excess flow protection device 溢流保护装置
excess flow test 过流量试验；溢流量测试
excess head 过剩水头
excess heat 余热；废热
excess hydrostatic water pressure 超静水压力
excess load 超负荷；超荷载
excess moisture 过多水分；超湿度
excess pore pressure 超孔隙压力
excess pore water pressure 超静水压力；超孔隙水压力
excess pore-air pressure 超孔隙气压力
excess pressure 超压；过大压力；超额压力；多余压力
excess rainfall 超渗雨量
excess surface water 地面积水
excess water 余量水
excess weight 超重
excessive air pressure 过剩气压

excessive burden 过量荷载；超载
excessive deformation 过量变形；附加变形
excessive discharge 过量排水；过量排放
excessive grade 过大坡度
excessive gradient 过大坡度
excessive groundwater withdrawal 过度抽取地下水；地下水超采
excessive high salinity water 含盐过多的水；过高矿化水
excessive loss 过大损耗
excessive mining of groundwater 地下水过量开采
excessive moisture 多余水分；过度湿润
excessive pore pressure 超孔隙压力
excessive precipitation 非常降水量；过量降水
excessive rainfall 雨水过多
excessive settlement 过大沉降；过度沉陷；严重下陷
excessive snowfall 过量降雪
excessive temperature sensor 过热感应器
excessive wear 过度磨损
excess-oil return line 剩油回输管线
exchange acidity 交换性酸度
exchange address register 更换地址寄存器
exchange adsorption 交换吸附
exchange capacity 交换量；交换容量；交换能力
exchange energy 交换能
exchange of heat 热交换
exchange of ions 离子交换
exchange reaction 置换反应；交换反应
exchangeable ion 可交换离子
excitation coefficient 激振系数
excitation earthquake 激发地震
excitation energy 激发能
excitation force 激发力
excitation frequency 激励频率；激发频率
excitation function 激励函数
excitation level 激发强度；激励能级
excitation mechanism 激发机制；激励机理
excitation parameter 扰动参数
excitation source 激发源
excited atom 受激原子
excited condition 激发条件
excited molecule 受激分子
excited state 激发状态；受激态；激发态
exciting agent 激发剂
excluder 封隔器；排除器；渠首排沙设施
exclusion area 禁入区；非居住区；禁区；隔离区
exclusion chromatography 排阻色谱法；筛析色谱法
exclusion of water 堵水；隔水；止水
exclusion principle 不相容原理；相斥原理
exclusively-owned technology 专有技术
excurrent canal 出水沟
excurrent siphon 出水弯管；出水虹吸管
execution of contract 履约；执行合同；合同的签署
execution of grant 签立批约
execution of works 施工
execution surveying 执行测量
executive summary 简要报告；摘要
exemplar 典型；标本；试样；模范
exergy 熵
exergy analysis 熵析
exfiltration 渗漏；泄漏

exfoliation 叶状剥落；叶理；页状剥离；片落
exfoliation corrosion 层蚀；剥离腐蚀
exfoliation desquamation 剥落
exfoliation dome 页状剥落丘
exfoliation joint 页状剥落节理
exfoliation mountain 页片剥离山
exhalant canal 出水沟；出水通道
exhalant orifice 排水孔口
exhalant siphon 排水虹吸管；出水弯管
exhalation 喷流；喷气；岩浆射气；蒸发
exhalative deposit 喷流矿床；喷气矿床
exhalite 火山喷流岩；喷气岩
exhalogenic 喷气成因的
exhausr port 排气口
exhaust air 排气；废气
exhaust air fan 排气风机
exhaust back pressure 排气背压
exhaust baffle 排气隔板
exhaust blower 排气机；排气风箱；抽风机
exhaust box 消声器；排气箱
exhaust branch pipe 排气支管
exhaust brine 废卤水
exhaust bypass valve 排气旁通阀
exhaust cam 排气凸轮
exhaust chamber 排出室
exhaust elbow 排气肘管
exhaust fan 排气扇；抽气扇；排风机
exhaust fume collecting hood 集烟罩
exhaust lap 排汽余面
exhaust manifold reactor 排气支管反应器
exhaust muffler 排气消声器；排气管减声器
exhaust noise 排气噪声
exhaust nozzle 排气喷口；排气口
exhaust plant 抽气设备
exhaust pressure at turbine inlet 涡轮入口废气压力
exhaust purification 废气净化；排气净化
exhaust purifier 废气滤清器；尾气净化器
exhaust regulator 排气调节器
exhaust smoke density 排气烟度
exhaust stack 排气竖道
exhaust steam pipe 废气管
exhaust suck 排风塔
exhaust temperature 排气温度
exhaust temperature at cylinder head outlet 缸盖出口废气温度
exhaust temperature at turbine inlet 涡轮入口废气温度
exhaust turbine 废气涡轮
exhaust turbocharging 废气涡轮增压
exhaust velocity 排气速度
exhaust ventilation 抽出式通风；抽出通风
exhaust water pipe 排水管
exhaust-air duct 废气管道
exhausted area 采空区；采空地区
exhausted landslide 能量衰竭的滑坡；停止下滑的滑坡
exhausted liquid 废液
exhausted solution 废液
exhausted well 枯竭井；已开采过的井
exhausted workings 采空区
exhaust-heat 废热
exhaustible resource 可耗竭资源；有限资源；不可再生资源
exhausting and blowing combined system 吸压混合式通风
exhausting-type ventilation 吸出式通风
exhausting-type ventilation system 抽出式通风系统
exhaustion 用尽，耗竭，衰竭；抽空，抽出；排气；资源采尽
exhaustion degree 抽空度
exhaustive yield 枯竭性抽水量
exhumation mechanism 剥露机制
exhumed karst 剥露的岩溶
exhumed landscape 剥露景观
exhumed monadnock 裸露残丘
exhumed topography 剥露地形
exichnia 外痕迹；外生迹
exinite 壳质组；壳质素；壳质体；膜煤素；稳定组
exinoid 类壳质的
exinoid group 壳质组
exinonigritite 壳质煤化沥青；孢壳碳化沥青
existent gum 实际胶质
existing corrosion 现存腐蚀
existing data 现有数据
existing gallery 既有巷道
existing glacier 现代冰川；现有冰川
existing ground level 原地面高程
existing perforation 老炮眼
existing railway 既有铁路
existing slope 现有边坡；原有边坡
existing use 现有用途
existing well 现有井
exit angle 出射角；出口角
exit channel 出水渠
exit gradient 逸出梯度；逸出坡降
exit head loss 出口水头损失
exit loss 出口损失
exit point 出水点；逸出点；出口点
exit ramp 出口斜路；出口坡道
exit road 出口道路
exit route 出口路线；出路通道
exit section 出洞段
exit stream 泄流河；排水河
exobeam nucleon microprobe 外束核子显微探针
exodiagenesis 外成岩作用；表生成岩作用
exodynamic phenomena 外动力现象
exogenesis 表成作用；外生作用
exogenetic action 外力作用；外成作用
exogenetic deformation 外成变形
exogenetic force 外力；外营力
exogenetic inclusion 外源包体；外来包体；捕房体；外成包体
exogenetic process 外成作用；外力过程
exogenetic rock 外成岩
exogenetic sediment 外成沉积
exogenic 外成的；外生的
exogenic action 外力作用；外成作用
exogenic force 外力；外营力
exogenic geological hazard 表生地质灾害
exogenic mineral deposit 外生矿床
exogenic mineralization 外生成矿作用
exogenic ores 外生矿床
exogenic process 外成作用；外力过程

exogenic water 外生水
exogenite 外成岩；外生矿床
exogenous 外成的；外源的
exogenous clastic rock 外生碎屑岩
exogenous dome 外成穹隆；外成穹丘
exogenous ejecta 外成喷出物
exogenous enclosure 外源包体
exogenous inclusion 外生包裹体；外源包体；外来包体；捕房体
exogenous process 外力作用；外部营力；外动力过程
exogenous water 外生水
exogeology 外空地质学；天体地质学；星际地质学
exogeosynclinal 外地槽的
exogeosyncline 外地槽；外准地槽；横切盆地；外枝副地槽；外枝准地槽
exoglyph 层面痕
exograph X光照片
exokinematic 外运动的；外动力的
exokinetic 外动力的；外运动的；外成的
exokinetic fissure 外动力裂缝；外应力裂缝
exokinetic geological disaster 外动力地质灾害
exokinetic joint 外成节理；外因节理
exo-mantle 外地幔
exometamorphism 外接触变质作用
exomigmatization 外质混合岩化作用；外成混合岩化
exomomental 发射脉冲的
exomorphic 外接触变质的
exomorphic metamorphism 外接触变质作用
exomorphic zone 外接触变质带
exomorphism 外接触变质作用
exoolitic 外生鲕状的
exo-Pacific tectonic belts 外太平洋构造带
exorheic 外流水系的；外流的
exorheic basin 外流盆地；外流流域
exorheism 外流水系；外洋流域
exornate 有饰纹的
exoskarn 外矽卡岩
exosmose 外渗现象
exosmosis 外渗；外渗现象
exosphere 外大气圈；外大气层；外逸层；逸散层
exotectonic 外力构造的
exotherm 放热线，放热曲线
exothermal 放热的
exothermal gas 放热气体
exothermal reaction 放热反应
exothermic 放热的
exothermic chemical reaction 放热化学反应
exothermic reaction 放热反应
exotic block 外来岩块
exotic breccia 外来角砾岩
exotic river 外来河
exotic stream 外来河
exotic terrane 外来地体
exotic xenolith 异源包体；异源捕房体
exovolcanic structure 外火山构造
expand 膨胀；扩张；张开
expand shoe packer 支撑式封隔器
expand with wet and contract with dry 干缩湿胀
expandability 膨胀性
expandable 可膨胀的；膨胀的

expandable anchor 胀壳式锚杆
expandable blade plunger 膨胀叶片型柱塞
expandable material 膨胀性材料
expandable open hole packer 膨胀式裸眼封隔器
expandable tip pile 扩端桩
expanded cement 膨胀水泥
expanded clay 黏土陶粒；膨胀黏土
expanded drainage 扩大到排水面积；扩大的排水系统
expanded excavation 扩挖
expanded foot glacier 扩足冰川；尾部扩大冰川；宽尾冰川
expanded graphite packings of various section 柔性石墨变截面组合填料
expanded joint 伸缩缝；胀接
expanded line scale 放线比例尺
expanded packer 膨胀式封隔器
expanded pearlite 膨胀珍珠岩
expanded preliminary design 扩大初步设计
expanded range 延伸范围
expanded shale 膨胀页岩
expanded slag 膨胀矿渣；多孔矿渣
expanded slag concrete 膨胀矿渣混凝土
expanded slate 膨胀板岩；膨胀页岩
expanded slate aggregate 膨胀板岩骨料
expanded vermiculite mortar 膨胀蛭石灰浆
expanded-base pile 扩底桩
expander 扩孔器；扩管器；扩张器；膨胀剂
expander-booster compressor 膨胀机-增压压缩机
expanding 扩张
expanding admixture 膨胀剂
expanding agent 膨胀剂
expanding arbor 可胀心轴；扩管器
expanding auger 扩孔钻；扩孔螺旋钻
expanding bit 扩孔钻头；扩孔钻
expanding bubble 膨胀气泡
expanding cement 膨胀水泥
expanding cement slurry 膨胀水泥浆
expanding chuck 弹簧筒夹
expanding current 扩展水流
expanding earth 膨胀的地球学说
expanding earth theory 地球膨胀理论
expanding grout 膨胀性水泥浆
expanding mandrel method 膨胀心轴法
expanding mortar 膨胀性水泥砂浆
expanding reach 扩大段
expanding section 扩大段
expanding shell bolt 胀壳式锚杆
expanding speed 膨胀速度
expanding support 伸展支撑
expanding universe hypothesis 宇宙膨胀说
expanding universe theory 宇宙膨胀理论
expanding-shell bolt 胀壳式锚栓
expansibility 膨胀性；延伸率
expansible packer 膨胀灌浆塞；膨胀式封隔器
expansion 扩张，扩大；膨胀；扩口；胀管；地层厚度增加
expansion agent 膨胀剂
expansion allowance 膨胀容许量
expansion anchor 膨胀锚固；扩张锚杆；扩大式壁锚
expansion apparatus 膨胀仪

expansion bearing 活动支座；伸缩轴承
expansion crack 膨胀裂缝；膨胀裂纹
expansion curve 回弹曲线；膨胀曲线
expansion cutter 扩孔器
expansion degree 膨胀度
expansion dilatation 伸展
expansion factor 膨胀系数；扩展系数；膨胀率
expansion fissure 膨胀裂缝
expansion flow 扩张水流；扩大流动
expansion force 膨胀力
expansion index 膨胀指数；回弹指数
expansion joint 胀缝；伸缩缝；补偿器；伸缩接头
expansion joint sealant 伸缩缝止水层；伸缩缝填缝料
expansion pipe bend 伸缩管弯头
expansion plug 膨胀塞；防冻塞
expansion pressure 膨胀力；膨胀压力
expansion property 膨胀性；扩展性
expansion rail joint 钢轨伸缩调节器；温度调节器
expansion rock bolt 胀壳式岩石锚栓
expansion shell bolt 胀壳式锚杆；胀壳式顶板锚杆
expansion shell rock bolt 胀壳式张拉锚杆
expansion soil 膨胀土
expansion stage 膨胀阶段
expansion stroke 膨胀行程
expansion terminal pressure 膨胀终点压力
expansion terminal temperature 膨胀终点温度
expansion test 膨胀试验；延伸度试验
expansion theory 膨胀说；扩张说
expansion thrust 膨胀冲断层
expansion type anchor bolt 胀壳式锚杆
expansion type bolt 膨胀型锚杆
expansion type roof bolt 膨胀式顶板锚杆
expansion volume 膨胀量；膨胀体积
expansion work 膨胀功；体积功
expansion-pressure test 胀压试验
expansion-producing admixture 膨胀性掺合料；膨胀剂
expansion-sleeve bolt 胀壳式套筒锚杆
expansion-type bolt 膨胀式锚杆
expansive agent 膨胀剂；发泡剂
expansive clay 膨胀土；膨胀黏土
expansive concrete 膨胀性混凝土
expansive element 膨胀件；胀塞
expansive force 膨胀力
expansive grout 膨胀性浆液
expansive pressure 膨胀压力
expansive rock 膨胀岩
expansive soil 膨胀土
expansive-cement concrete 膨胀水泥混凝土
expansiveness 膨胀性；可膨胀性
expansivity 膨胀系数；膨胀性
expectancy life 概率寿命；预期寿命
expectancy map 预期图；预测图；预报图
expectant output value 期望产值
expectation criterion 期望值准则
expectation curve 期望曲线
expectation value 期望值
expectation variance 期望值方差
expected duration 预期持续时间
expected loss 预期损失
expected normal frequency 期望正态频数

expected reserves 预计储量
expected risk 预期风险；预计风险
expected service time 预期运营时间
expected value 期望值
expected yield 预期出水量；预期产量
expendable carrier 消耗式载体
expendable gun 消毁式射孔器
expendable plug 一次性桥塞
expendable through-tubing perforator 消毁式过油管射孔器
expendable-retrievable gun 可回收的消毁式射孔器
expending beach 吸收波能海岸
expenditure incurred 发生费用
expenditure item 开支项目
expenditure pattern 开支模式；费用构成
expenditure to acquisition 购置费
expenditures outside the budget 预算外开支
expenditures reduction 削减支出
expense for labor protection 劳动保护费
expense for survey and design 勘察设计费
expense of workshop 车间经费
expensive rock 膨胀岩
expensive soil 膨胀土
expensive soil subgrade 膨胀土路基
experience curve 经验曲线
experience factor 经验系数
experiment of tectonic simulation 构造模拟试验
experiment rock mechanics 实验岩石力学
experiment site 试验场
experiment soil mechanics 实验土力学
experiment station of soil and water conservation 水土保持实验站
experiment with clay model 泥巴试验；黏土模型试验
experimental 实验的；经验的
experimental analog 实验模拟
experimental balance plot 实验平衡图
experimental banking 试验性筑堤；试验性填土
experimental basin 实验池；实验槽
experimental blasting 实验爆破
experimental boring 试验性钻探
experimental catchment 试验流域；试验集水区
experimental condition 实验条件
experimental constant 实验常数
experimental data 实验资料；试验数据
experimental drilling 试验性钻探
experimental field 试验地
experimental flume 实验水槽
experimental geology 实验地质学
experimental geomorphology 实验地貌学
experimental geothermics 实验地热学
experimental grouting 实验灌浆
experimental mechanics 实验力学
experimental metallogeny 实验成矿学；实验矿床成因论
experimental mineralogy 实验矿物学
experimental model 试验模型；实验模型
experimental parameter 经验参数；实验参数；试验参数
experimental petrology 实验岩石学
experimental petrophysics 实验岩石物理学
experimental platform 试验平台

experimental plot 试验区
experimental power station 试验电站
experimental provision 实验装置
experimental report 实验报告
experimental research 实验研究
experimental rock mechanics 实验岩石力学
experimental sedimentology 实验沉积学
experimental seismology 实验地震学
experimental soil engineering 实验土工学；实验土壤工程学
experimental stage 实验阶段
experimental station 试验站
experimental structural geology 实验构造地质学
experimental study 试验研究；实验研究
experimental table 实验台
experimental tank 试验池；试验槽
experimental tunnel 试验平硐
experimental value 实验值；试验值
experimental watershed 实验性集水区；试验流域
expert GIS 专家地理信息系统
expert mode 专家模式
expert system in mining 采矿专家系统
expert system of tunnel 隧道专家系统
expert's statement 专家鉴定；专家阐述；专家意见
expert's survey 专家调查；专家信息反馈
expiration of the contract 合同期满
expired patent 过期专利
explanation 说明，解释；图例
explanatory guide 阐释指引
explanatory legend 图例；说明图例
explanatory pamphlet 说明图册；说明书
explanatory plan 说明图则
explanatory text 图注；图例
explicit address 显地址，显式地址
explicit difference formula 显式差分公式
explicit difference scheme 显式差分格式
explicit finite difference method 显式有限差分法
explicit finite difference scheme 显式有限差分格式
explicit finite element method 显式有限元法
explicit formulation 显式
explicit function 显函数；显式函数
explicit integral method 显式积分法
explicit method 显解法；显式法；显式方法
explicit program 显程序
explicit relaxation 显式松弛
explicit residual error estimator 显式残差估计量
explicit saturation calculation 显式饱和度计算
explicit solution 显解，显式解
exploded pile 爆扩桩
exploding bridge-wire initiator 桥式爆炸发火器
exploding gun 起爆枪
exploding reflector 爆炸反射面
exploding silt-drainage 爆破排淤法
exploding wire 爆炸导线；爆炸丝，爆丝
exploit 开发；利用
exploitability 可开发性，可利用性；可采出量
exploitable aquifer 可开采含水层；可开发含水层
exploitable boundary 矿体可采边界线
exploitable thickness 可采厚度
exploitable yield 可采水量

exploitation 使用，利用；采掘，勘探；开发，开采
exploitation dilution rate 开采贫化率
exploitation drilling 开采钻探
exploitation modulus 开采模数
exploitation ore block method 开采块段法
exploitation technical economic parameter 开采技术经济参数
exploited well 开采井；开发井
exploration 勘探，探测，探矿；勘察，调查；试采
exploration benefit 勘探效益
exploration bidding 勘探投标
exploration boring 勘探钻进；钻探
exploration cost 勘探成本
exploration ditch 探沟
exploration drilling 勘探钻进；钻探
exploration engineering 探矿工程；勘探工程
exploration geologist 地质勘探人员；地质勘探工作者；勘探地质学家
exploration geology 找矿地质学；勘探地质学
exploration geophysics 勘查地球物理学；物探
exploration guide 勘探指南；勘探标志
exploration hole 勘探孔；探测孔
exploration investigation 探索性研究；勘探调查
exploration map 勘探图；勘测图
exploration of ore deposit 矿床勘探
exploration phase 勘察阶段；找矿阶段；勘探阶段
exploration point 勘探点
exploration priority 勘探优选地段；勘探重点地段
exploration procedure 勘察程序；勘探程序
exploration program 勘探计划
exploration prospecting instrument 勘测器具；勘测仪器
exploration seismology 勘察地震学；应用地震学；勘探地震学
exploration stage 勘探阶段；勘察阶段
exploration target 勘察靶区；找矿靶区；勘探目标
exploration trench 探槽；探坑
exploration tunnel 勘探洞；勘探导坑
exploration well 探井；勘探井，预探井
exploration work 勘探工作
exploration-production well 探采结合孔
exploratory 勘探的；考察的；调查的
exploratory adit 勘探平洞，勘探导洞；探洞
exploratory borehole 勘探孔，勘探钻孔
exploratory degree of deposit 矿床勘探程度
exploratory drift 探硐；勘探巷道；勘探平巷
exploratory drill hole 勘探钻孔
exploratory drilling 勘探钻井；探井
exploratory geochemistry 勘探地球化学
exploratory grid 勘探网
exploratory heading 勘探平硐；勘探导洞
exploratory hole 勘探孔，探孔
exploratory investigation 探索性研究
exploratory line 勘探线
exploratory map 勘探图
exploratory method of deposit 矿床勘探方法
exploratory net survey 勘探网测量
exploratory oil well drilling 勘探石油钻井
exploratory phase 勘察阶段；勘探阶段
exploratory pit 勘探井；探坑；探井
exploratory point 勘探点

exploratory probe hole 超前钻孔
exploratory profile of mine area 矿区勘探线剖面图
exploratory program 勘探计划
exploratory sampling 探查抽样
exploratory search 考察；调查
exploratory shaft 探井；探坑
exploratory spot 勘探点
exploratory study 勘察研究；探索性研究
exploratory technique 勘察技术
exploratory test 探查试验；勘测试验；探索性试验
exploratory tunnel 探硐
exploratory well 探井；勘探钻井
exploratory work 勘探工作；勘探工程
explored 探明的；勘探过的
explored reserves 探明储量；勘定储量
explorer 勘察人员，勘测者；侦察机；探测线圈，测试线圈；探测器，探针；探索者，考察者
explorer's route 勘探路线
exploring drift 勘探坑道；调查导坑
exploring drill 勘探钻孔
exploring heading 勘探导坑，导坑；勘探巷道
exploring mining 勘探性开采；边探边采
exploring opening 坑探工程；坑探；勘探坑洞；勘探巷道
exploring shaft 探井
exploring team 勘探队
explosion 爆炸；爆破；爆发；爆裂
explosion accident 爆炸事故
explosion breccia 喷发角砾岩；爆破角砾岩
explosion caldera 爆发破火山口；爆裂巨火山口
explosion crater 爆破漏斗；爆裂火山口
explosion funnel 爆破漏斗
explosion lake 爆裂湖
explosion pit 爆破坑
explosion relief measure 防爆泄压设施
explosion seismic observation 爆破地震观测
explosion seismology 爆破地震学；人工地震学
explosion site 爆炸位置
explosion sound 爆炸声
explosion test ground 爆炸试验场
explosion wave 爆炸冲击波；爆炸波
explosion-calibrated seismic mode 爆破校准地震模型
explosion-proof construction 防爆结构
explosion-proof motor 防爆马达；防爆电机
explosive breccia 爆发角砾岩
explosive compacting 爆炸密实砂基；爆破加固砂基；爆炸成型
explosive compaction 爆炸挤密；爆炸压缩；爆炸压实
explosive compound 爆炸化合物
explosive drill 爆炸钻
explosive energy 爆炸功；爆炸能，爆炸能量
explosive eruption 爆发性喷发
explosive factor 单位炸药消耗量；单耗；炸药爆破系数
explosive index 爆破指数；爆炸指数
explosive loading equipment 装药机械
explosive loading truck 装药车
explosive signal 爆炸信号
explosive volcano 爆裂式火山
explosives 火药；炸药；爆破材料
exponent 指数；幂

exponential curve 指数曲线
exponential damping 指数衰减；指数阻尼；指数减幅
exponential decay 指数衰减
exponential decaying rate 指数衰减率
exponential distribution 指数分布；指数分配
exponential fitting 指数拟合
exponential Fourier transform 指数傅里叶变换
exponential function 指数函数；幂函数
exponential transform 指数变换
exposed conduit 明管；明渠
exposed downpipe 明雨水管
exposed outcrop 露头
exposed peat 裸露泥炭；表露泥炭
exposed penstock 露天式压力水管
exposed pipe 露天式管道；明管
exposed rock 暴露岩石；裸露岩石
exposed rock surface 岩石出露面
exposed structure 裸露构造；显式结构
exposed surface 自由表面；出露面；暴露面
exposed water surface 自由水面
exposure 暴露，暴露面；露头；坡向；朝向
exposure age 暴露年龄；表面年龄
exposure assessment 暴露评价
exposure level 剥蚀面；剥蚀平面
exposure measurement 曝光量测量；曝光测定
exposure of mineral occurrence 矿点揭露
exposure rate 照射量率；照射强度
exposure site 出露位置；露头位置
express highway 快速公路；高速公路
express passenger train 特快客运列车
express pile 大头桩
express train 旅客特别快车；特别快车
express way 快速道路；高速路
expression 表示，表达；表达式，公式
expressway 高速公路；快速道路；快速公路
expulsion amount of hydrocarbons 排烃量
expulsion threshold of hydrocarbons 排烃门限
exsiccator 干燥器；保干器；除湿剂
exsolution inclusion 出溶包裹体
exsolution texture 固溶体分解结构；出溶结构
exsolved 出溶的；固溶体分解的
extended Casagrande classification system 扩展卡萨格兰德土分类法
extended cell 扩展晶胞
extended completion date 延迟竣工日期；延迟完工日期
extended consequent stream 延展顺向河
extended core barrel 超前式取芯管
extended forecasting 中期预报
extended foundation 扩展式基础
extended Mohr diagram 广义莫尔图
extended Mohr-Coulomb failure envelope 广义莫尔-库仑破坏包线
extended river 延长河，延伸河
extended shear strength method 广义抗剪强度方法
extended stream 延长河，延伸河
extended structure 扩建物；扩展结构；拉伸构造，张性构造
extended succession 延续岩序；持续岩序；延续层序
extended Tresca failure criteria 广义特莱斯卡破坏准则
extended Tresca yield criterion 广义特莱斯卡屈服准则

extended valley 延长谷；伸长谷
extended Von Mises failure criteria 广义冯·米塞斯破坏准则
extended von Mises yield criterion 广义冯·米赛斯屈服准则
extensibility 伸展性；延长性；伸长率；可扩充性
extensible belt transporter 可伸缩胶带输送机
extensible boom 伸缩式臂杆；伸缩式悬臂
extension 延伸，延长，拉伸；附加物，套筒加长节
extension core barrel 加长岩芯管；可接长的取芯筒
extension fracture 张性破裂；拉伸破裂；伸长断裂
extension joint 张节理；张性节理
extension of field 域的扩张；油气田扩展；矿区扩展
extension of tender validity period 延长投标书有效期
extension of time for completion 延长竣工期；延长完工期；竣工时间的延长
extension of time limit 延展时限
extension sleeve 伸缩套筒
extension sliding tripod 伸缩三脚架
extension socket 伸缩套筒；延伸套筒
extension structure 张性构造；拉伸构造
extension theory 可拓学理论；扩张理论
extension unit 扩展单元；扩充单元
extension wing 延伸翼
extensional 伸展的；拉伸的；张性的
extensional fault 张性断层
extensional faulting 张性断裂；拉伸断裂作用
extensional geothermal system 扩展地热系统
extensional ridge 扩张脊
extensional stiffness 抗拉刚度；伸长刚度
extensional strain 张性应变；拉伸应变
extensional stress 张应力；拉伸应力
extensional structure 张性构造；拉伸构造
extensional tectonics 伸展构造
extensity 延伸率
extensive indicator 拉伸式位移计；伸缩式压力计
extensive quantity 广延量；外延量
extensive roof collapse 区域性冒顶，大面积冒顶；大面积顶板崩落
extensive sampling 扩大取样，扩大抽样；分散抽样法
extensometer 引伸计，伸长仪，位移计；应变计
extensometer measurement 引伸计测量
extent of mineralization 矿化度；矿化范围
exterior drainage 外部排水系统；外流水系
exterior dune 沿海沙丘
exterior orientation element 外部定向元素
exterior repair 外部修复
exterior sewer system 污水管道系统；外部排水系统
exterior waterproofing 外防水
external access 外部沟；外部存取
external agency 外营力；外力作用；外部因素；外部作用
external agent 外营力；外部因素
external basin 外流盆地
external boundary condition 外部边界条件
external bracing 外部支撑；外部横向支撑
external casing 外套管；表层套管
external collapse 表面塌陷；地表塌陷
external cooling 外部冷却
external diameter 外径；外直径
external dimension 外形尺寸
external dump 外部排土场
external environment 外部环境
external facility 户外设施
external factor 外因；外部因素
external force 外力；外动力
external friction 外摩擦
external hydraulic pressure 外周水压
external hydrostatic pressure 外静水压力
external instability 外部失稳
external load 外荷载；外负载
external moment 外力矩
external observation 外部观测
external perspective projection 外心投影；外部透视投影法；外部中心投影法
external pressure 外压力，外部压力
external process pipe line 室外工艺管线
external protection 外保护层
external secant 外矢距；外割线
external shell 外壳
external spraying 外喷雾
external stability 外部稳定性
external storage 外存储器；外存储；外部存贮器
external strain 外应变
external strengthening works 外部巩固工程
external stress 外应力
external structural bracing frame 外部支撑结构架；外部横向支撑结构架
external subsidence 表面沉陷
external support 外支护；外支撑
external surface 外层；外表面
external vibration 外振动；外部振动
external vibrator 附着式振捣器；外部振动器
external water attachment 冲洗水接头
external water pressure 外水压力
external wave 表面波；外波
externally applied force 外部作用力；外力，外加力
externally flush-coupled rods 外平连接钻杆
externally heated 外加热
externally heated pressure vessel 外热高压容器
extinct lake 干涸湖；死湖
extinct nuclide 灭绝核素；衰减核素
extinct volcano 死火山
extinction angle 消光角
extinction position 消光位
extinction recycle operation 全循环操作
extinction rule 消光法则
extinction value 消光值
extinguished volcano 熄火山，熄灭火山
extra fine grinding 超细研磨
extra heavy drive pipe 超重冲击管
extra hole 补充钻孔；补充炮孔
extra long-range forecast 超长期预报
extra orientation element 外定向元素
extra stress 附加应力
extra water pressure 附加水压力
extra-banking 附加填土；超高填土
extrabasinal river 盆外河流
extrabasinal rock 盆外岩
extract ratio 提取率，提取比例；回采率；回收率

extractant	提取剂；萃取剂；提炼物
extracted ore	采出矿石
extracted water	出水；出水量
extracted water temperature	出水温度
extracting block coal-water rise	采段溜煤上山
extracting block coal-water roadway	采段溜煤巷
extracting seam thickness	采煤层厚度
extraction	抽取，提取；抽提法，萃取法，回采率，回收率；采掘
extraction drift	回采平巷
extraction efficiency of geothermal energy	地热能汲取效率
extraction jack	上拔千斤顶；拔桩千斤顶
extraction method	提取法；回采方法
extraction of casing	起拔套管
extraction of information	信息提取；信息摘录
extraction of point feature	点特征提取
extraction of water-contained ore body	含水矿体开采
extraction structure	抽水结构物；引水结构物
extraction unit	回采单元
extraction ventilation system	抽气式通风系统
extraction well	抽水井；抽气井；回采井
extraction-development plan of mine	采掘技术计划
extraction-reinjection cycle	抽取回灌循环；抽灌周期
extractor	拔桩机；提取器，拔出器；拔桩机
extractor force	拔桩力
extrados	外表面，外弧；拱的外圈，拱背，拱背线；外褶皱面
extrados of the cast-in-situ concrete lining	现浇混凝土衬砌层的拱背
extrados radius	拱外弧半径
extrados springing	拱背起拱线点
extrados springing line	拱外弧起拱线
extra-fill	附加填土；超高填土
extraglacial deposit	冰川外沉积；外冰川沉积
extra-heavy drill rod	加重钻杆；超重钻杆
extra-load bearing capacity	额外承载能力
extra-long bolt	超长锚杆
extra-long mast	特长钻架
extramagmatic	外岩浆的
extraneous ash	外来灰分；外在灰分
extraneous force	外力
extraneous material	外来物
extraneous moisture	外在水分
extraneous risk	附加风险；外来风险
extraneous rock	外来岩石
extranormal soil	异常土；特殊土
extraordinary flood	特大洪水
extraordinary high tide	特大高潮
extraordinary scour	特大冲刷
extraordinary storm	特大暴雨；非常暴雨
extraplanetary	行星外的；地外的
extrapolated limit	外推极限
extrapolated value	外推值
extrapolation	外推法，外插法，归纳，推论
extrapolation distance	外推距离
extrapolation of discharge curve	流量曲线外延
extrapolation of rating curve	标定曲线外延；水位流量关系曲线外延
extra-rapid-hardening portland cement	特快凝硅酸盐水泥
extra-sensitive clay	超灵敏黏土；超塑性黏土
extraterrestrial causation	地外成因；天外成因
extraterrestrial geology	天体地质学；地球外地质学；星际地质学；宇宙地质学
extraterrestrial impact	地球外的冲击
extraterrestrial photograph	卫星照片；从外层空间拍的地面照片
extraterrestrial process	地外营力作用
extraterrestrial research	太空探索；外星探索
extravasation	外喷，喷出，溢出；熔浆外喷
extreme boundary lubrication	极限边界润滑
extreme downsurge	溢流量；水面下降极限
extreme earthquake motion	强烈地震运动
extreme environment	极端环境；极端恶劣的环境
extreme event	极端情况；极端事件
extreme fibre stress	极限纤维应力
extreme high water	最高水位
extreme hydrologic year	非常水文年
extreme limit	极限值
extreme loading condition	极端荷载条件
extreme low water	最低水位
extreme method	极值法
extreme pressure	极限压力；超高压
extreme rainstorm	罕见暴雨；特大暴雨；极大暴雨
extreme sea level	极限海水位
extreme temperature	极限温度；极端温度
extreme upsurge	最高上涌浪；极限上涌浪
extreme value index	极值指数
extremeline casing	极线套管
extremely high frequency	极高频
extremely irregular coal seam	极不稳定煤层
extremely low frequency	极低频
extremely low-density sampling	极低密度采样
extremely structure	极复杂构造
extreme-pressure additive	极压添加剂，极压剂
extremum	极值
extremum condition	极值条件
extremum principle	极值原理
extremum value	极值
extrinsic	非本征的；非固有的；外来的
extrinsic absorption	非固有吸收
extrinsic factor	外因；外界因素
extrude lining	挤压衬砌
extruded concrete lining	挤压混凝土衬砌
extruding concrete tunnel lining	挤压混凝土衬砌
extruding swirl	挤压涡流
extrusion	挤出；喷出；流出；排出
extrusion flow	喷出岩流；熔岩流；挤出流
extrusion pressure	挤压力
extrusion stress	挤压应力
extrusion tectonic model	挤出构造模型
extrusion test	挤出试验
extrusive	喷出的；喷出岩体
extrusive body	喷出体
extrusive rock	喷出岩；火成岩；火山岩
exudation basin	渗流盆地
exudation vein	分凝脉；分泌脉
exudation-pressure test	渗水压力试验
eye lens	目镜；透镜

eye measurement 目测
eye method 肉眼法
eye sketch 目测草图；草测图；略图；目测图
eye survey 目测
eye tunnel 双孔隧道
eyebase 目测基线

eyed structure 眼球状结构
eyepiece screw 目镜螺丝
eyepiece with micrometer 测微目镜
eyestone 眼石
eyot 河心岛；湖心岛；河洲；湖洲

F

fabaceous 豆状的
fabianite 法硼钙石
fabiform 豆形
fabric 土工织物，织物，织品；组构；构造，结构，组织，组成；建筑物钢筋网；纤维
fabric analysis 岩组分析，组构分析；结构分析
fabric anisotropy 组构各向异性
fabric axis 组构轴
fabric characteristics 组构特征
fabric component 组构成分
fabric coordinate 组构坐标
fabric data 组构数据
fabric diagram 岩组图；组构图
fabric element 组构单元；组构要素
fabric element unit 组构单元体；基本单位
fabric filter 织物过滤机；织物过滤层；织物过滤器；网状滤器；纤维过滤器
fabric form 织物模板；组构形式
fabric habit 组构惯态
fabric joint 软性万向节；组构结合线；织物结合缝
fabric lineation 组构线理
fabric of soil 土的组构
fabric orientation 组构定向；组构方位
fabric parameter 组构参数
fabric pattern 组构型式；组构形式
fabric peak strength 土工织物峰值强度
fabric reinforcement 编网钢筋；厂制钢筋；预扎钢筋；网状钢筋；钢丝网配筋；钢筋网
fabric retaining wall 土工布挡土墙；土工布挡墙
fabric sheet reinforced earth 铺网法
fabric structure 织物结构；组织结构
fabric symmetry 组构对称性
fabric symmetry of sediment 沉积物的组构对称
fabricable 可成形的；可塑造的
fabricant 制造人；制作者
fabricated 装配式的
fabricated bridge 装配式桥
fabricated building 装配式房屋
fabricated dam 装配式坝
fabricated gemstone 拼合宝石
fabricated parts 现成构件
fabricated soil 混合土
fabricated steel body 装配式钢体
fabricated steel bridge 装拆式钢桥
fabricated structure 装配式建筑物；焊接结构；装配结构，装配式结构
fabricated universal steel members 万能杆件
fabricating cost 制造成本
fabricating yard 加工场地
fabrication 制造，加工；装配；冷压加工；建造；浆液的配制；生产
fabrication cost 造价；生产成本
fabrication of steel pipe specials 特别钢管的制造
fabrication yard 预制件工场；施工现场

fabric-enclosed sand drain 袋装砂井
fabric-filled plastics 以织物为填料的塑料
fabric-reinforced seal 夹布密封
fabric-to-fabric bonding 织物与织物黏合
fabric-to-fabric lamination 织物与织物层压黏合
fabric-to-foam laminating 织物与泡沫塑料层压黏合
fabridam 合成橡胶坝；尼龙坝
fabulite 锶钛矿；钛锶矿；神石
facade 面墙；外墙；正面外墙；表面；外观；正面，立面
facade column 房屋正面立柱
facade design 外部设计；房屋正面设计
facade girder 房屋正面梁
facade panel 外墙板
facade treatment 外部装饰
facade wall 面墙
face 工作面，开采工作面，采煤工作面，工祖；面，前面，正面，饰面；主解理；晶面，层面，表面，界面，面向；开挖面，掌子面
face advance 掌子面推进；工作面推进，工祖推进；采掘进度，工作面进度，回采进度
face air velocity 工作面风流速度
face airing 工作面通风
face and slab work 台阶式开采
face angle 面角
face arch 前拱
face ashlar 表面琢石
face belt conveyor 工作面胶带输送机
face belt track 工作面胶带输送机道
face blower 工作面扇风机
face board 工作面前壁的护板
face boss 工作面班长
face breaks 工作面断裂
face brick 面砖，饰面砖
face bump 工作面突出；掌子面隆起
face capacity 工作面生产能力
face centered lattice 面心晶格
face centred cubic lattice 面心立方晶格
face cleat 主裂理；面割理，工作面割理；主解理面
face coal bin 回采工作面煤仓
face communication 工作面通讯
face concentration 回采工作面生产率
face conveyor 工作面运输机；工组运输机
face conveyor system 工作面输送机系统
face crew 工作面工作组；工作面采煤组；工祖工组
face cutter 正刀（掌子面切割机）
face cutting 端面切削；断面切削
face cycle 工作循环
face discharge bit 唇面排水孔钻头；底喷式钻头；底冲式钻头，钻头面冲洗式钻头；层面喷水钻头；低喷式钻头
face down 面向下；面朝下
face drain 表面排水；表面排水系统；坡面排水沟
face drill 工作面钻机

face dual operation 工作面工作；回采工作
face ejection 钻孔底冲洗
face ejection bit 底唇喷射式钻头；低层喷射式钻头
face end 工作面出口，掌子面，开挖端头，工作面端头；长壁工作面两端与巷道衔接的地段
face equipment 工作面设备；工祖设备
face excavation 工作面挖掘；工作面掘进
face fall 工作面塌落，工祖塌落；工作面冒顶
face firing 工作面放炮
face floor 工作面底板，工作面地板
face foreman 工作面工长
face gear 平面齿轮；盘状齿轮
face gradient 工作面坡度
face hammer 平锤；琢面锤
face hardening 表面硬化；渗碳
face haulage 回采工作面运输
face heading 与主解理垂直的平巷
face height 工作面采高
face jack 开挖面千斤顶，挡土千斤顶
face joint 表面接合；表面接缝；出面接缝
face labour 工作面工人；工作面工作组
face left 经纬仪左侧位置；盘左；正镜
face length 工作面长度；工祖长度
face life 工作面开采期限
face line 长壁工作面位置；工作面中线；面线
face loader 工作面装载机
face loading pan 工作面装载槽
face machinery cable 工作面机械电缆
face machinery power distribution 工作面机械供电
face mechanization 工祖机械化，工作面机械化
face milling 端面铣
face move 工作面搬迁；转场
face normal 面的法线
face of approach （坝）迎水坡面；接近工作面
face of bed 地层头部
face of cleavage 劈理面；解理面
face of coal 采煤工作面；煤层主解理面
face of fault 断层崖
face of fissure 裂隙面
face of heading 掘进工作面
face of hole 孔底
face of shaft 井底
face of stope 回采工作面前壁；回采梯段前壁
face of well 井底；油井底
face operation 工作面工作，回采工作
face piece 开挖面横撑；挖槽端面横撑
face pillar 工作面煤柱；工作面矿柱
face plate 花盘；划线平台；面板，锚杆托盘
face pocket 平行主解理面的通路
face preparation 工作面准备；工祖准备
face recovery coefficient 工作面生产率
face replacement 工作面接替
face right 经纬仪右侧位置；盘右；倒镜
face roof breaker 工作面顶板破碎机
face roof collapse with cavity 漏顶
face room 全矿工作面总长度；全矿工作面总生产能力
face run 采煤机移动时间；采煤机在工祖移动的时间；移动时间
face scraping 工作面耙运
face shield 面罩；手持焊接面罩
face shovel 工作面机铲；正铲挖掘机；正向铲，正铲
face shoveler 正铲司机
face side pump 工作面水泵
face signal device 工作面信号装置
face signal system 工作面信号系统
face signalling 工作面输送机开关信号
face slab 面板
face slip 向前倾斜的节理；上部劈理或节理；工祖滑动；工作面滑动；表面滑动
face sloughing 工作面煤壁片帮
face spalling 工作面煤壁片帮
face sprag 工作面斜支柱
face stability 开挖面稳定
face stone 面石；磨面石；底面金刚石
face support 工作面支护；工祖支架；开挖面支护
face timbering 工祖支架；工作面支撑；工作面支护
face toe 阶段下盘；坡面底部
face track 工作面轨道；工祖轨道
face trial installation 工作面试验装备
face veneer 表层装饰薄板；面板
face ventilation 工作面通风
face waling 开挖面横撑；挖槽端面横撑
face width 控顶距；工作面宽度，面宽；采煤工作面中煤壁至末排支柱的顶梁后端或至放顶柱之间的距离
face without support 无支护工作面
face work 工作面工作；抹面工作；工祖工作
face worker 工作面工人
face working 工作面作业
face zone 工作面区
faceadvance 工作面进度，回采进度
face-airing 工作面通风
face-and-slab plan 分段回采程序
face-area haulage 工作面运输；采区运输
face-centered cubic array 立方面心立方排列
face-centred crystal 面心晶体
faced surface 削光面
face-end machine 工作面端头采煤机
face-end support 工作面端头支架；端头支架
face-face contact 面-面接触；面接触
face-fall accident 工作面片帮事故
facelifting 改建；改装；翻新
face-line 面线
face-loader conveyor 工作面装载机的输送机
face-loading equipment 工作面装载设备
face-on 工作面平行于煤层主解理的掘进方向
face-on room 煤面与主解理面成直角的煤房
face-plate 面板；花盘；卡盘；荧光屏
face-pole 面极点
face-power system 工作面动力系统
faceside 靠工作面一侧
facet 小面 刻面；磨蚀面 磨光面；断层崖 溶蚀痕；磨光面；小平面；复眼；柱槽筋；凸线；中间成果
facet pebble 磨面砾石；棱石
faceted pebble 棱石；棱面
faceted spur 截切山嘴
faceted stone 刻面宝石
facetted peneplation 交切准平原
facetted spur 切削坡；切削嘴
facial sutures 面线
facieology 岩相学；相研究；相分析

facies 相；主拉应力；外观；外表；外形；岩相
facies analysis 相分析
facies angle 相角
facies architecture 相结构
facies association 相组合
facies belt 相带
facies change 相变；岩相变化
facies contour 等相线
facies development 相发展
facies diagram 相图解
facies differentiation principle 相分异原理
facies evolution 相演化
facies fossil 相化石
facies group 相组
facies map 相图
facies mapping 相图的绘制
facies mark 相标志
facies model 相模式，岩相模式；相模型
facies pattern 相型；相模式
facies ratio 层相比
facies sequence 相序列，相序
facies series 相系
facies strike 相走向
facies suite 沉积岩相套；岩相套；相组；相群
facies tract 相域；相带
facies unit 岩相单元；岩相单位
facilitate 促进，使容易；使便利
facilitate maintenance 便于维修；简化维修
facilities of tunnel portal 隧道洞口设施
facility 设施；设备；便利；器材；研究室；工厂；机关
facility for earthquake monitoring 地震监测设施
facing 面板，面层面向；饰面，铺面，衬面，砌面，贴面，护面；面对；端面车削，端面切削，刨削；平面加工；主要垂直裂隙，断层崖
facing brick 面砖；露面砖；明口砖；见光砖
facing concrete 面层混凝土；护面混凝土
facing decoration 面层装饰
facing direction 面向；变新的方向
facing joint 面层接缝
facing material 护面材料；表面装饰料
facing of fold 褶皱面向
facing point 对向辙尖；对向道岔
facing sand 模面砂
facing stone 护面岩石；护面石，护面石料；面层石，面石；饰面石材
facing thrusts 对冲式逆断层
facing tile 饰面瓷砖；面砖
facing wall 护壁；开挖面衬墙
faciostratotype 相地层典型剖面；相层型
facsimile crystallization 拟态结晶；后构造结晶
factor 系数，系数倍，乘数；因数，因子，要素；系数率；率，比，商
factor analysis 因子分析；因素分析
factor cluster analysis 因子聚类分析
factor comparison method 因素比较法
factor loading 因子载荷；因子负荷
factor matrix 因子矩阵
factor of adhesion 黏着系数
factor of assurance 安全因数；保证率
factor of bearing capacity 承载力因子
factor of drainage 疏干因数
factor of expansion 膨胀系数
factor of foundation bearing capacity 地基承载系数
factor of full extraction 采动系数
factor of global safety 总体安全系数；整体安全系数
factor of karst 岩溶率；喀斯特率
factor of merit 优良率；灵敏值；灵敏因数；质量因子
factor of partial safety 单元安全系数；局部安全系数
factor of porosity 孔隙率
factor of proportionality 比例因数；比例因子
factor of reduction 折减因子；折减系数
factor of rigidity 刚性系数
factor of runoff 径流系数
factor of safety 安全系数
factor of safety against chimney caving 防止筒状塌落的安全系数
factor of safety against cracking 抗裂安全系数
factor of safety against failure 抗破坏安全系数
factor of safety against floatation 抗浮动安全系数
factor of safety against overturning 抗倾覆安全系数
factor of safety against roof collapse 防止顶板破坏的安全系数
factor of safety against rupture 抗裂安全系数
factor of safety against shear failure 抗剪破坏安全系数
factor of safety against sliding 抗滑安全系数
factor of safety against slip 抗滑安全系数
factor of safety of pillars （矿）柱安全系数
factor of saturation 饱和率
factor of slope safety 斜坡安全系数
factor of stability 稳定性系数
factor of stress concentration 应力集中系数
factor of stress-concentration in surrounding rock 围岩应力集中系数
factor of utilization 利用系数
factor out 析出因数
factor pattern 因子模型
factor score 因子得分
factor set 因子组
factor structure 因子结构
factor-comparison job evaluation 因素比较工作评价法
factorial 阶乘阶乘积；因子的，因数的；阶乘的；工厂的；代理店的；析因
factorial design 析因设计；因子设计
factorial discriminant analysis 因子判别式分析
factorial experiment 析因实验
factorial sample design 因素抽样设计
factories and mines road 厂矿道路
factors influencing in-situ state of stress 影响现场应力状态的因素
factory external transportation line 对外道路
factory for beam 制梁场
factory layout 工厂布置
factory load 工厂负荷
factory made house 工厂预制房屋
factory planks 板材
factory-in road 厂内道路
factory-out road 厂外道路
factory-set bit 厂镶金刚石钻头
facts survey 事实调查

factual analysis 根据事实的分析；确凿的分析
factual data 确实数据；可靠事实资料
factual information 实际资料
factual material 实际材料；实际资料
factual observation data 实际观测资料；实际观测数据
fade-away 消失；衰减
fading bandwidth 衰减带宽
fading depth 衰减深度
fading envelope 衰减包络
fading of settlement 沉降衰减
fading rate 衰减率；衰退率
fading slate 褪色板岩
fag-end 废渣
faggot dam 柴捆坝
faggot dike 柴堤
fahlband 黝矿带
fahlore 铜矿；黝铜矿
fahlunite 褐堇青石；褐块云母
fah-off test 压降试井；衰减试井
fahrenheit 华氏；华氏温度
fahrenheit degree 华氏温度
fahrenheit scale 华氏温标
fahrenheit temperature 华氏温度
fahrenheit thermometer 华氏温度计；华氏寒暑表
fail 崩塌，塌毁，毁坏；故障，破坏，损坏；失败；破产；违约
fail in bending 受弯破坏；弯曲破坏
fail in bond 黏结破坏；黏着破坏
fail in compression 压力破坏；压缩破坏
fail in shear 剪力破坏；剪切破坏
fail in tension 拉力破坏；拉伸破坏
fail safe 可靠的；安全的；不间断的安全型的；事故保险装置
fail safety 系统可靠性
failed annulus extent 破裂圈；松动圈
failed forecast of earthquake 地震漏报
failed hole 拒爆炮眼；失效炮眼，已失效炮眼；拒爆残留孔眼；不爆炮眼；未爆炮孔
failed rift 夭折裂谷；衰亡裂谷
failed shot 拒爆；哑炮
failed slope 失稳斜坡
failed test sample 不合规格的试样
failed well 废井；枯井
failed zone 破坏带；破坏区
failing gradient 下坡，下向坡，下降坡度
failing load 破坏荷载
failing mass 滑落岩体
failing stress 破坏应力
fail-passive 工作可靠但性能下降
fail-safe 故障-安全；安全的；可靠的；不间断的；故障保险；故障自动保险
fail-safe system 安全系统；失效保险系统；故障保险系统
fail-safety 个别部件发生故障时的系统可靠性
fail-soft 故障弱化；有限可靠性
fail-to-safe 安全的；防故障的；可靠的
failure 破坏，毁坏，损坏，失事，破裂，破损，断裂；失败，破产；故障，失效，失灵，事故，缺点；无支付能力；倒闭，不合格，不履行；崩塌，塌毁
failure accumulation 故障积累

failure analysis 破坏分析；失效分析；故障分析
failure angle 破坏角
failure arc 滑动弧；破坏弧
failure at the free surface 自由面破坏
failure bearing strength 破坏支承强度
failure bearing stress 破坏支承应力
failure by deformation 变形破坏
failure by heaving 隆起破坏
failure by lateral spreading 侧向扩展破坏
failure by piping 管涌破坏
failure by plastic flow 塑流破坏
failure by rupture 破裂变形；断裂破坏；破裂破坏
failure by shear 剪切破坏
failure by sub-surface corrosion 地下侵蚀破坏
failure by subsurface erosion 地下侵蚀破坏；湾没
failure by tilting 倾斜破坏
failure characteristics 破坏特征；故障特点；失效特征
failure condition 破坏条件；破坏状况
failure crack 破坏性裂缝
failure creep 破坏型蠕变
failure criteria 失效判据；破坏准则；故障判据
failure criterion 破坏准则，断裂准则；破裂准则，失效准则；破裂判据；故障判据
failure deformation 破坏变形
failure density 失效密度
failure detection 故障探测
failure diagnosis 故障诊断
failure diagnostic 部件故障检测；故障诊断
failure discontinuity 破坏面的不连续性
failure distribution 破坏分布
failure envelope 破坏包线
failure field 破坏现场
failure forecast 破坏预测
failure form of rock and soil 岩土的破坏形式
failure histogram 故障或损坏的频率直方图
failure hypothesis 破坏假说
failure identification 故障标识；破坏识别
failure index 破坏指数
failure isolation 故障隔离
failure limit state 破坏极限状态
failure line 破坏线
failure load 破坏荷载；损毁荷载；膨胀土层；断裂负荷
failure locus 破坏轨迹
failure logging 故障记录；失效记录；破坏记录
failure mechanism 破坏机理，破坏机制；失效机制，失效机理；导致倒塌的机制；故障机理，破损机理
failure mode 破裂模式；失效模式；破坏方式
failure mode distribution 失效模式分布；破坏模式分布
failure modes 失效模式
failure moment 破坏力矩
failure monitoring 故障监控；故障监测
failure of anchorage system 锚固系统的破坏；锚定系统的破坏
failure of earth slope 土坡破坏；土坡滑塌，土坡滑坍；土坡坍毁
failure of foundation 地基破坏
failure of ground 地面破坏
failure of rock mass 岩体破坏
failure of shot 拒爆

failure of slope 斜坡破坏
failure of soil 土的破坏
failure of tunnels and caverns 隧道与峒室破坏
failure pattern 破坏形式，破坏样式；破坏图式
failure plane 破坏面；破坏平面；崩塌面
failure plane angle of inclination 破坏面的倾角
failure plane geometry 破裂面几何形状
failure plots 破坏点
failure point 破裂点；破裂点
failure prediction 故障预测；失效预测
failure probability 破坏概率；失效概率
failure process 破坏过程；破裂过程
failure progress 破坏过程
failure rank 破坏级别；破坏等级
failure rate 失效率；失败率；破坏率；故障率
failure ratio 破坏比
failure scar 崩塌痕迹；崩塌残壁；崩塌残痕；崩塌痕，破坏痕迹
failure state 破坏状态
failure strain 破坏应变；破裂变形；破坏变形
failure strength 破坏强度；破裂强度
failure stress 破坏应力；破裂应力
failure stress envelope 破裂应力包络面
failure stress level 破坏应力水平
failure surface 破坏面；破裂面
failure test 破坏试验；破损性试验
failure theory 破坏理论
failure tolerance 失效容许误差
failure trend 破坏趋势
failure type 破坏类型
failure types of foundation 地基破坏类型
failure types of landslides 滑坡类型
failure volume 坍塌体积
failure wedge 破坏棱体；破裂楔体
failure well 废井；事故井
failure zone 破坏区，破坏范围；滑落区
failure-critical height 破裂临界高度
failure-free 可靠的；无故障的
failure-free dual operation 无故障运行
fair cleavage 清晰的劈理；平滑劈理
fair current 顺流
fair curve 光滑曲线；平滑曲线；匀滑曲线
fair drawing 清绘；清绘原图
fair face concrete 光面混凝土
fair faced brick 饰面砖
faired strut 减阻支柱
fair-faced concrete 原身混凝土；清水面混凝土
fair-faced finish 原身终饰；无瑕面
fairfieldite 磷钙锰石
fairing 整流装置；整流的；流线型的；减阻装置，减阻物；流线型罩，流线型外壳；整形光滑
fairlead 导线管；导索板；导缆孔
fairness limit 容许量
fairway arch 通航拱
fairy stone 十字石；黄土中的异形结核；海胆化石
falaise 低海崖；古海崖
Falang Formation 法郎组
falcate 镰状的；弯月形的
falciform 镰状的；弯月形的
falkenhaynite 辉锑铜矿

falkmanite 针硫锑铅矿
fall 崩塌，塌落，倒塌，落下，坍落，崩落，降落，掉落，陷落；下斜，斜度；降雨量，降雪量；煤石碎渣下落；落差，落距，比降；瀑布
fall area 陷落区
fall belt 陷落带
fall breccia 崩塌角砾岩
fall depth of water level 水位降深
fall diameter 降落直径
fall flood 秋汛洪水，秋季洪水；秋汛；秋季；洪水
fall from height 从高处坠下
fall hammer test 落锤试验
fall head 落差；跌落水头，水头；压头
fall head of water 下降水头；落差
fall in compression 压力破坏
fall in level 水位下降；水位降落
fall in shear 剪力破坏
fall in tension 拉力破坏
fall in water level 水位降落
fall irrigation 秋季灌溉
fall line 瀑布线
fall measurement 倾斜测量；落差测量
fall meteorite 坠落陨石
fall of channel 渠道坡度
fall of girder 落梁
fall of ground 地层陷落；冒顶，岩体冒落；顶板岩石崩落；放顶
fall of ground accident 冒顶事故
fall of pipe 管道坡度
fall of potential method 电位下降法
fall of pressure 压降
fall of rock 落石；岩石冒落，冒顶；剥落
fall of roof 冒顶；顶板崩落；顶板陷落
fall of stage 水位下降；水位降落
fall of stream 河流落差
fall of tide 落潮；退潮
fall of water level 水位降落；水位下降
fall off 衰减，下降；脱落
fall pipe 落水管
fall rate 沉降速度；下沉速度
fall sensitivity 冒落灵敏度
fall survey 落差测量
fall table 井口防坠盖板
fall through machine 机内的落差
fall time 下降时间；衰变时间
fall trap 陷阱
fall velocity 沉降速度
fall way 吊物竖道；楼面井
fall wind 下坡风；下降风
fall zone 陷落区
fallback 回落
fallback breccia 崩塌角砾岩
fall-cone test 落锥试验
fallen 倒塌的，陷落的
fallen bit 掉钻
fallen earthquake 陷落地震
fallen-in 冒落了的
fallers 吊罐笼座；吊罐托；吊罐座
falling 倾斜；坠落，降落，沉陷，下落，流下，沉降；下降，落下；落差；弯头

falling after mining 采矿滞后垮落
falling apron 堆石防冲护坦
falling away 泥土倾泻
falling ball viscosimeter 落球黏度计
falling body 坠落体；落体；坠体
falling cone method 沉锥法
falling curve 退水曲线；退水段；落洪段；退水肢
falling cut method 降落时切割法；水平切片开采法
falling debris 垮落碎石
falling dune 下沉沙丘
falling dust 落尘；降尘
falling edge 下降沿
falling flood stage 落水期
falling funnel 降落漏斗
falling glacier 崩落冰川
falling gradient 坠落坡度；下坡；下向坡；下降坡度；屋顶式闸门
falling ground 陷落地层
falling groundwater level 降低地下水位
falling head 下降高度；变水头
falling head of water 降落水头
falling head permeability test 变水头渗透试验；降落水头渗透试验
falling head permeameter 变水头渗透仪；变水头渗透率仪
falling head test 降水头试验；变水头试验
falling limb 退水曲线；退水段；落洪段；退水肢
falling of cutting slope 边坡塌落，路堑边坡崩塌
falling of water-table 地下水位降落；水位降落
falling off of drill string 跑钻
falling pin seismometer 落针式地震仪
falling rate 沉降速率；下沉速率
falling ratio 沉缩率
falling rock 落石
falling rock formation 崩塌岩层
falling segment 退水段；落洪段；消落段；退水肢
falling sluice 跌落式泄水闸
falling soil 崩落土
falling sphere viscometer 落球式黏度计
falling stage 水位下降期；落洪期；退水期；消落期
falling star 流星
falling stone 陨石
falling test 落锤试验
falling tide 落潮；退潮
falling time 退水时间；退水历时
falling track 溜放线
falling velocity 坠落速度；降落速度；下沉速度；沉降速度
falling water level method 下降水位法
falling weather 雨季；雪季
falling weight 落锤；打桩锤；下落重量
falling weight test 落锤试验
falling with mining 及时垮落
falling-ball impact test 落球冲击试验
falling-body viscometer 落体黏度计
falling-in 塌方，滑塌；坍方；陷落
falling-sphere method 落球法
falling-weight test 落锤试验
fall-into step 同步
fall-off curve 降压曲线；下降曲线

fallout 散落物，散落；沉降灰；沉降物；放射坠尘；坠落碎屑；放射性降落物
fallout area 放射性沉降区
fallout breccia 散落角砾岩
fallout damage 落下灰损伤
fallout injury 落下灰损伤
fallout measure 错检率
fallout meter 散落物测定仪；辐射量测试器
fallout monitoring 散落物监测
fallout particles 落下灰粒子
fallow field 休闲田
fallow ground 休耕地；休闲地
falls 陨石；瀑布
false alarm of earthquake 地震误报
false anomaly 假异常
false beam 假梁；不承重梁
false Becke line 假贝克线
false bedding 伪层面
false bedrock 假底岩；假基岩
false bottom 假底层；装景的活动底，活底；假底板；假底岩
false cap 临时顶梁；板桩支架的辅助顶梁
false cleavage 假劈理；假解理
false color composite image 假彩色合成影像
false color density sliced 假彩色密度分割
false color density split 假彩色密度分割
false color image 假彩色图像；假彩色影像
false color imagery 假彩色成像
false color rendition 假彩色还原；假彩色再现
false colour composite 假彩色合成
false colour film 彩色红外片；假彩色片
false colour infrared 假彩色红外
false colour photography 假彩色摄影
false cropping 假露头
false diamond 假金刚石
false dip 假倾角；视倾角；假倾斜
false dynamite 低硝甘炸药
false equilibrium 假平衡
false folding 假褶皱
false form 假象；假晶
false galena 闪锌矿
false gradient 前后风压差
false granite 假花岗岩
false image 假象；虚像
false leader 索桅打桩架
false middling 夹矸煤
false oil seepage 假油苗
false proscenium 假台口
false quay 辅助码头
false reference axis 假参考轴
false roof 伪顶；假顶；位于煤层之上随采随落的极不稳定的岩层
false roof rock 伪顶岩石
false roof with wire mesh 金属网假顶
false ruby 假红宝石
false set 假终贯；临时支架，轻型临时支架；临时棚子；假凝现象，假凝
false specific gravity 假比重；视比重
false stratification 假层理；交错层理
false stull 临时横撑；临时支柱

false timber roof 木板假顶
false timbering 临时支护；临时支架
false topaz 假黄晶；黄萤石；假黄玉
false work 脚手架；工作架
false work plate 脚手架板
false-bedding 假层理
false-bottom 活底的
false-bottom bucket 活底料斗
false-color cloud picture 假彩色云图
false-inclined flexible shield mining system 伪倾斜柔性掩护支架采煤法
false-roof rock 伪顶岩石
falsework 临时支撑，临时支架；脚手架；拱架；膺架；工作架
faltung 褶合；褶合式；褶积
faluns 砂质泥灰岩
famatinite 块硫锑铜矿；脆硫锑铜矿
family 科；系；种；类；族；单种群落
family of characteristics 特性曲线族
family of crystal planes 晶面族
family of curves 曲线族；曲线组
family of fragmental rocks 碎屑岩族
famp 软韧的薄页岩层；分解的石灰石；细粒硅质层
fan 扇形地，扇状地；冲积扇；扇状海底；通风机，风机，扇风机；扇；风扇；扇风；排成扇形；叶片；风轮
fan advance 扇形推进
fan apex 冲积扇顶端
fan apron 山麓冲积扇
fan cleavage 扇状劈理；劈理扇
fan delta 扇形三角洲；舌状三角洲；扇状三角洲
fan delta facies 扇三角洲相
fan deposit 扇形沉积
fan diagram 扇形图
fan drift 风硐；扇形堆积物；扇风机风硐；扇风机引风道
fan dynamometer 风扇测力计
fan filter 扇形滤波器；扇形滤波
fan fold 扇形褶皱；扇形褶曲
fan folding 扇形褶皱作用
fan glacier 宽尾冰川；扇形冰川
fan jumbo 向上打扇形眼的钻车；扇形孔凿岩台车
fan of outwash sediment 冰水沉积扇
fan out 使水流分散；使流束扩散；扇形扩大；输出；输出端数
fan pattern 扇形排列方式；扇形布置
fan pattern holes 扇形炮眼组
fan pipe 风管；扇风机管
fan pulley gasket 风扇皮带轮衬垫
fan sewer system 扇形下水道系统
fan shaft 扇风机井；风井；通风井
fan shape 扇形
fan shaped 扇形的
fan shaped fold 扇形褶皱
fan shaped round 扇形炮眼组；扇形炮眼
fan shaped structure 扇状构造
fan shooting 扇形爆破；扇形排列；扇形地震勘探；扇形排列法地震勘探；扇形爆炸
fan structure 扇形构造
fan talus 扇形岩堆；扇形塌磊

fan terrace 冲积扇阶地；扇形阶地
fan turbidite 扇形浊积岩
fan vault 扇形拱
fancy coal 精煤；上等煤；精选煤
fancy lump coal 精选块煤；上等块煤
fan-filtering 扇形滤波
fang （巷道掘进时）用风墙或风帘隔开的进风和回风道；爪，牙，齿，铁柄
fang bolt 板座栓；地脚螺栓；地脚螺钉
Fangliao Formation 枋寮组
fanglomerate 扇砾岩；扇积砾
Fangniugou sandstone 放牛沟砂岩
Fangniuling limestone 放牛岭石灰岩
Fangshan stone 房山石
Fangshanzhen volcanics 芳山镇火山岩
fanholes 扇形布置炮眼；扇形炮孔
fanion 测量旗
Fanjiamen Formation 樊家门组
Fanjiatang Formation 范家塘组
Fanjiatun Formation 范家屯组
fanlight 楣窗；门头窗；气窗
fanned fold 扇状褶皱
fanned structure 扇状构造
fans in combination 混合式通风
fan-shaped alluvium 扇形冲积物；冲积扇
fan-shaped borehole 扇形钻孔
fan-shaped delta 扇形三角洲
fan-shaped dock 扇形码头
fan-shaped fold 扇状褶皱
fan-shaped round 扇形布置炮孔
fan-shaped routed 扇形掏槽炮眼组
fan-shaped shooting 扇形排列；扇形地震勘探
fan-shaped structure 扇状构造；扇形构造
fan-shaped terrace 冲积扇阶地
fan-shaped valley 冲积扇谷
Fanshengbao Formation 樊盛堡组
Fansteel process 范史提尔生产法
fantail 扇形尾；矿工帽；扇形帽
fan-topped pediment 冲积扇覆盖的山前侵蚀平原
fan-type fold 扇形褶皱
Fanxia Formation 翻下组
Fanzhao Formation 番召组
Fanzhuang Formation 范庄组
far and by 一般说来；大体上
far field area 远场区
far field boundary condition 远场边界条件
far field distribution 远场分布
far field pattern of laser 激光远场图
Faraday cage 法拉第笼式结构
Farallon plate 法拉隆板块
faratsihite 铁高岭石；黄高岭石
farewell-rock 磨石粗砂岩；粗砂岩；残山
far-field 远场
far-field approximation 远场近似
far-field body wave 远场体波
far-field boundary condition 远场边界条件
far-field displacement 远场位移
far-field earthquake motion 远场地震地面运动
far-field shear stress 远场切应力
far-field spectrum 远场谱

far-field stress 远场应力
far-field stress state 远场应力状态
far-field surface wave 远场面波
far-field temperature 远场温度
far-future stage 远期
fargite 钠沸石；红钠沸石
far-infrared 远红外
far-infrared ray 远红外光
far-infrared transmission filter 远红外透射滤光片
farm building 农业生产建筑；农场建筑物
farm dam 农用小土坝
farm ditch 毛渠；农沟
farm percolation loss 田间渗漏损失
farm survey 农田测量
farm water supply engineering 农田给水工程；农业供水工程
farmed cleavage 劈理扇；扇状劈理
faroelite 杆沸石
farrisite 透辉闪煌岩；闪辉黄煌岩
far-source zone 远源带
farsundite 闪苏花岗岩；淡花岗长岩
fascia 过梁横带
fascia beam 圈梁
fascia board 封檐板
fascicular 束状
fasciculite 角闪石
fascine 砂门子；梢料；柴笼；柴捆，梢捆；柴排
fascine contracting works 梢料束狭工程
fascine dam 梢捆坝
fascine dike 柴捆堤；柴捆护堤
fascine dyke 柴笼堤坝
fascine foundation 梢料基础；柴排基础
fascine groyne 梢捆丁坝
fascine layer 梢料填层
fascine mattress 柴排；沉排；柴排席；梢蓐
fascine revetment 柴排护岸；沉排护岸；梢蓐护岸
fascine roll 梢笼
fascine weir 梢料堰
fascine whip 梢鞭；梢笼
fascine work 木排填基工作
fascinebarge 梢料船
faserserpentine 纤蛇纹石
fashion 款式；形式；样式，式样；铸造，图，模型
fashion parts 异形配件
fashioned iron 型铁；型钢
fasibitikite 负异钠闪花岗岩；负异钠花岗岩
fasinite 橄云霞辉岩；辉霞岩
fasiostratotype 相层型
fassaite 深绿辉石
fast and hard 固定不动的；固定不变的
fast cap 快速雷管；地震勘探用电雷管
fast detector 快速探测器
fast digital filter 快速数字滤波器
fast drilling formation 快速钻进地层
fast extraction 快速回采
fast footage 快速推进；快速进尺
fast ground 原体岩层；整体岩层
fast hardening cement 快硬水泥
fast hardening concrete 快硬混凝土；速凝混凝土
fast ice 固定冰

fast junking 煤柱中纵向窄道
Fast Lagrangian Analysis of Continua 连续介质快速拉格朗日分析
fast needle survey 矿山罗盘测量
fast place 超前平巷；超前工作区
fast plough 快速刨煤机
fast powder 快速起爆炸药
fast roof 坚固顶板；稳固顶板
fast setting cement 快凝水泥
fast setting concrete 快硬混凝土；速凝混凝土
fast setting grout 速凝浆液
fast shot 空炮；瞎炮
fast side 工作面端露出的整体煤壁；工作面侧壁；煤房的煤柱面
fast step bearing 稳固立轴承
fast top 坚固顶板；稳固顶板
fast tracking 边设计边施工
fast-delay detonation 毫秒迟发延迟起爆
fasten with bolt 螺栓固定
fastener 紧固件
fastening bolt 紧固螺栓
fastening by pinning 销钉紧固
fastening pile 锁定桩
fastening technique 紧固技术
fastening wire 绑扎用钢丝
fastigium 尖顶；屋脊；山墙；高峰期
fastland 大陆；高地；高潮面以上陆地
fast-moving depression 快速移动的低气压
fastness 牢固性；稳定性；紧牢；坚牢度；抗拒性
fast-setting 速凝的；快凝
fast-setting cement 快凝水泥
fat 脂肪；富的；高质的；高挥发物的；肥的；含量高的
fat clay 肥黏土，富黏土；可塑性黏土；油性黏土；宫黏土
fat coal 肥煤；长焰煤；变质程度中等的烟煤
fat coking coal 肥焦煤
fat concrete 肥混凝土；富混凝土
fat gas 肥气；含油气
fat gas coal 肥气煤；富烟煤
fat lens 厚透镜体
fat lime 富石灰；高钙石灰
fat mortar 黏性砂浆；富灰浆
fat sand 黏砂
fatal damage 毁灭性破坏；致命破坏
fatal gas accident 致命的瓦斯事故
fathmeter 测深仪
fathogram 水深图；回声测深剖面图
fathom 英寻；测量深度，测深；方噚
fathom curve 等深线；等海深线；等英寻线
fathom line 等深线；等英寻线
fathomage 用英寻计的长度；用平方英寻计的面积；用立方英寻计的体积；按英寻支付
fathometer 回声测深仪，回声探测仪；水深仪；探测仪；测深仪，测深计
fathometer sounding 测深仪测深
fatigue 疲劳；疲乏；劳累；软弱化
fatigue at high temperature 高温疲劳
fatigue behaviour 疲劳行为
fatigue bending test 疲劳弯曲试验

fatigue bond strength	疲劳黏结强度
fatigue break	疲劳断裂；疲劳破坏
fatigue capacity	疲劳承载能力
fatigue crack	疲劳裂缝；疲劳裂纹；疲劳裂隙
fatigue crack initiation	疲劳裂纹的萌生
fatigue crescent	月牙状疲劳纹理
fatigue curve	疲劳曲线
fatigue damage	疲劳损伤
fatigue deformation	疲劳变形
fatigue effect	疲劳效应
fatigue endurance limit	疲劳持久极限
fatigue factor	疲劳因素
fatigue failure	疲劳破坏，疲劳断裂；疲劳失效；疲劳损毁；疲劳裂纹；疲劳破裂
fatigue fracture mechanics	疲劳断裂力学
fatigue fracture toughness	疲劳破坏韧性
fatigue hardening	疲劳硬化
fatigue index	疲劳指数
fatigue life	受疲劳荷载的使用期限；疲劳试验循环次数；疲劳寿命
fatigue life analysis	疲劳寿命分析
fatigue limit	疲劳极限；疲劳界限；疲劳限度
fatigue loading	疲劳荷载
fatigue mechanics	疲劳力学
fatigue mechanism	疲劳机理
fatigue meter	疲劳强度计
fatigue of material	材料疲劳
fatigue of rock	岩石的疲劳；岩石疲劳
fatigue point	疲劳点
fatigue prediction	疲劳预测；疲劳预估
fatigue property	疲劳特性
fatigue range	疲劳限度，疲劳界限；疲乏范围
fatigue ratio	疲劳强度比；疲劳比；疲劳系数
fatigue resistance	抗疲劳性；抗疲劳强度；疲劳强度；疲劳抵抗力；疲劳阻力
fatigue rupture	疲劳破坏
fatigue shear apparatus	疲劳剪切仪
fatigue strength	疲劳强度；抗疲劳强度；耐久试验
fatigue stress	疲劳应力
fatigue stress ratio	疲劳应力比值
fatigue striations	疲劳辉纹
fatigue test	疲劳测试
fatigue test rig	疲劳试验台
fatigue testing	疲劳测试；疲劳试验
fatigue threshold	疲劳阈值
fatigue yield limit	疲劳屈服极限
fatigue-cracking moment	疲劳-开裂力矩
fatigue-resistant structure	抗疲劳结构
fatigue-testing machine	疲劳试验机
fatigue-ultimate moment	疲劳极限力矩
fatty coal	肥煤
faujasite	八面沸石
fault	断层，断裂，破坏；高差；层错；故障；缺点，缺陷；失灵；错误
fault activity	断层活动性
fault amplitude	垂直断距；纵断距
fault analysis	故障分析
fault and fracture density	断层和破裂密度
fault angle	断层角
fault angle valley	倾斜盆地；断层角谷地；断斜谷；层谷
fault associated with earthquake	地震断层
fault band	断层带
fault basin	断层盆地；断陷盆地
fault bedding plane	错动层面
fault bench	断层阶地；断阶
fault bend fold	断弯褶皱
fault bend folding	断弯褶皱作用
fault block	断块；断裂地块；断层石块
fault block basin	断块盆地
fault block interdigitation	断块互冲
fault block mountains	断块山
fault block movement	断块运动
fault block overthrust	断块仰冲作用；断块逆掩断层
fault block reservoirs	断块油气藏
fault block tectonism	断块构造作用
fault block valley	断块谷
fault boundary	断层边界
fault branch	断层分叉；分支断层
fault breakout	断层崩落
fault breccia	断层角砾岩，断层面碎岩；断裂带；断层带
fault bridge	桥构造；断层桥
fault bundle	断层群；断层束
fault bus	故障母线
fault cause	故障原因
fault cave	断层洞穴
fault clay	断层黏土；断层泥
fault cliff	断层崖
fault coal	劣质煤
fault coal basin	断陷煤盆地
fault coast	断层岸；断层海岸
fault complex	断层群；断层组合
fault contact	断层接触
fault control stream	断层控制河流
fault creep	断层蠕动；断层稳滑
fault crevice	断层裂隙
fault crossing design	跨越断层设计
fault dam spring	断层堰塞泉；断层阻截泉
fault debris	断层岩屑
fault deformation	断层形变
fault depression	断层拗陷；断陷
fault depth	断层深度
fault detection	故障探测，探伤；断层检测
fault detector	故障探测器；探伤仪
fault diagnosis	故障诊断；故障检查
fault diagnostics	故障诊断
fault dimension	断层尺寸；断层规模
fault dip	断层倾斜；断层倾角
fault directivity effect	断层方向性影响
fault dislocation	断层位错
fault displacement	断层位移；断层移距；断距
fault drag	断层拖曳；断层扭曲
fault earthquake	断裂地震
fault effect	断层作用；断层效应
fault escarpment	断层崖
fault excavation	断层开挖
fault extension length	断层延伸长度
fault face	断层崖；断层面
fault facet	断层三角面；断层凸面

fault fall 断层落差
fault finder 故障探测器；探伤器；故障寻找器
fault finding 故障探测；找毛病
fault fissure 断层裂缝
fault flexure 断层挠曲
fault fluccan 断层泥
fault fold 断层褶皱
fault fold structure 断层褶皱构造
fault fold system 断褶系
fault folded mountain 断层褶皱山
fault folding 断层褶皱活动
fault formation 断层建造
fault forming stage 断层形成阶段
fault fracture 断层破裂；断裂
fault fractured zone 断层破碎带
fault friction slip 断层摩擦滑动
fault gap 断层隘口；断层峡谷
fault gas 断层气
fault geometrization 断层几何制图
fault gouge 断层泥
fault graben 地堑
fault graded 粒序断层
fault groove 断层擦痕，断层沟；断层刻槽；断层泥
fault group 断层群
fault heave 断层隆起；断层平错
fault horse 断片；马石
fault impedance 故障阻抗
fault influenced zone 断层影响带
fault information 断层信息
fault inlier 断层内围层；断层内露层
fault interaction 断层相互作用
fault ledge 断层崖
fault length 断层长度
fault line 断层线
fault line coast 断层线海岸
fault line gap 断层线峡谷；断层线垭口
fault line saddle 断层线鞍部
fault line scarp 断层线崖
fault line scarp shoreline 沿断层线陡崖滨线
fault line valley 断层线谷；断线谷
fault localization 断层局部；故障定位；使障碍局限化；故障点标定
fault location 断层位置
fault mechanism 断层机理
fault modeling 断层建模
fault motion 断层运动
fault motion transformation 断层运动转换
fault mountain 断层山
fault movement 断层运动；断裂运动；断层错动
fault movement and combination analysis 断层运动组合分析法
fault muck 断层岩屑
fault nose 断鼻构造
fault notch 断层鞍部
fault of parallel displacement 平行错动断层
fault of periodicity 周期性层错
fault of shear-compressive type 压剪型断层
fault of shear-tensile type 剪张型断层
fault orientation 断层方位
fault outcrop 断层露头

fault outlier 飞来峰
fault parameter 断层参数
fault pattern 断层型式；断层类型
fault permeability 断层渗透
fault piercing 过断层
fault pit 断层坑
fault plane 断层面；断层平面；断面
fault plane reflection 断面反射（波）
fault plane solution 断层面解
fault plane solution method 断层面解法
fault plateau 断层高原
fault polish 断层磨光面
fault pond 断陷塘
fault population 断层群
fault position 断层位置
fault propagation 断层传播；断层扩展
fault propagation fold 断展褶皱
fault proximity index 断层接近度指数
fault radius 断层半径
fault reactivation 断裂活化
fault region 断层区
fault ridge 断层岭；地垒
fault rift 断裂谷；裂谷
fault rock 断层岩石；断层岩
fault rubble 断层碎块
fault rupture 断层破裂；断裂；地层错动
fault rupture length 断层破裂长度；断裂长度
fault saddle 断层鞍状构造；断层鞍部
fault sag 断层洼陷
fault sag pond 断陷塘
fault scarp 断层崖；断崖
fault scarp coast 断层崖海岸
fault scarp shoreline 断崖滨线
fault scarplet 断层陡坎
fault scratch 断层擦痕
fault screened reservoir 断层遮挡油气藏
fault seal 断层封闭性
fault segment boundary 断层段落边界
fault segmentation 断层分段性
fault separation 断距
fault set 断层组
fault shoreline 断层岸线
fault side 断层翼
fault size 断层规模
fault slice ridge 断片脊
fault slip 断层滑移
fault slip cleavage 断层滑动劈理；破劈理
fault slip rate 断层滑移速率
fault slippage 断层滑动
fault sole 底基逆掩面；最下冲断面
fault space 断层间距；断层间隔，开口断层间隔
fault splinter 参差断层崖
fault spreading 断层扩展
fault spring 断层泉
fault stability 断层稳定性
fault stability models 断层稳定性模型
fault step 断层阶台，断层阶梯；断层坎
fault stepovers 断层跨距
fault stick slip 断层黏滑
fault strand 断层线

fault strength	断层强度
fault striae	断层擦痕
fault striation	滑动擦痕；断层擦痕；滑动镜面
fault strike	断裂走向
fault structure	断层构造
fault subsidence	断层沉陷；断陷
fault surface	断层面；断面
Fault surface slip structure	断面滑动构造
fault system	断层系
fault tectonic levels	断层构造层次
fault tectonics	断层构造
fault terrace	断层阶地
fault throw	断层落差；断层垂直位移；断距，断层断距
fault tip line	断端线
fault tolerance	容错
fault trace	断层迹
fault trap	断层圈闭；断层油捕
fault tree analysis	故障树分析
fault trench	断层沟
fault trend	裂层走向；断层方向
fault trough	断层槽；地堑；错断的
fault trough lake	地堑湖；断陷湖
fault type	断层类型
fault valley	断层谷；断裂谷
fault vehicle	断层脉
fault vein	断层脉
fault vent	断裂火山喷发口
fault wall	断层壁；断盘；断壁
fault warp	坳块
fault wedge	断层楔；楔形断块
fault zone	断层带；断裂带；断层区；断裂面
fault zone damage	断层区损伤
fault zone rheology	断层区流变
fault zone structure	断层区结构
fault-angle basin	断层角盆地
fault-block	断块
fault-block basin	断块盆地
fault-block mountain	断块山
fault-block movement	断块运动
fault-block valley	断块谷
fault-crossing crustal deformation measurement	跨断层地壳形变测量
fault-dam spring	断层壅水泉
faulted	断层的；断裂的；错断的；触头
faulted and folded corridor	断裂和折叠走廊
faulted bedding plane	错动层面；错层面；断折层面
faulted body	断错矿体；断层错失的矿体
faulted contact	断层接触
faulted deposit	断层矿床；断错矿床；断裂油层；断裂沉积
faulted episode	断层幕
faulted joint	断层缝
faulted mountain	断层山；断块山
faulted overfold memder	断裂倒转裙皱段
faulted overturned fold	断裂倒转褶皱
faulted pool	断层油气藏
faulted reservoir	断块油层；断裂油层
faulted rock	断裂岩石
faulted seam	断裂岩层；断裂煤层；断裂矿层
faulted structure	断裂构造；断块构造
faulted trough	断裂槽；断裂谷；断裂破坏的坳陷；地堑
faulted volcanic cone	断裂火山锥
faulted zone	断层带；断裂带
fault-fold	断层褶皱
fault-fold structure	断层-褶皱构造
fault-folding	断层褶皱
faulting	断层，断裂作用；断裂运动，断层活动；造成断层；断裂垂直间距
faulting activity	断层活动性
faulting mechanism	断层作用机理；错断机制
faulting of slab ends	板端错台
faulting plane	断层面
faulting process	断层作用
faulting recurrence	断裂复活
fault-line scarp	断层线崖；断层崖
fault-line scarp shoreline	断层崖滨线
fault-line scrap	断线崖
fault-line valley	断线谷；断层谷
fault-plane	断层面
fault-plane solution	断层面解；断层面溶解
fault-plane solution of earthquake focus	震源断层面解
faults and blocks	地质断块
faults group	断层群
faults in rock	岩石中的断层
fault-scarp shoreline	断层崖滨线；断层滨线
fault-screened hydrocarbon reservoir	断层遮挡油气藏
fault-slip cleavage	断层滑动劈理面
fault-structure lake	断层构造湖
fault-subsidence furrow	断层沉陷沟
fault-tilted structure	断挠构造
fault-trace rift	断裂谷
fault-trellis drainage pattern	断层格状水系
fault-trough coast	地堑海岸
fault-trough lake	地堑湖；断陷湖，断陷塘；裂谷湖
fault-trough polje	地堑坡立谷；断陷盆地
fault-wedge mountain	楔形断块山
faulty coal	劣质煤
faulty concrete	不合规格混凝土；劣质混凝土
faulty drilling	不合格钻孔；有缺陷钻孔
favas	蚕豆矿
faveolate	蜂窝状的
favorable geologic condition	良好的地质条件
favourable area	有望地区
favourable atmospheric condition	有利大气条件
favouring grade	下坡；顺坡
fayalite	铁橄榄石
faying surface	搭接面
feasibility	可行性；可行性设计；可能性；现实性
feasibility analysis	可行性分析
feasibility design	可行性设计；可行性方案设计
feasibility evaluation	可行性评估
feasibility investigation stage	可行性勘察阶段
feasibility stage	可行性论证阶段
feasibility study	可行性研究；技术经济论证；顺从性研究；远景评价
feasibility survey	可行性调研
feasibility test	可行性试验
feasible direction	可行方向

feasible extreme point	可行极点
feasible path	可行路径
feasible pumping rate	可抽水速度；可抽水量
feather	滑键；砭石楔；落煤楔；羽毛；羽饰斑
feather checking	微裂；细裂
feather cracking	羽状破裂；细裂纹；发丝裂缝
feather edge	楔形石块或煤块的薄边；薄缘；羽状薄边；尖灭；削边
feather edge boarding	薄边板
feather joint	羽状节理；压力隧洞；压头线
feather key	落煤楔；破石楔；滑键
feather structure	羽状构造
feather tension joints	羽状张节理
feather-checking	发裂
feathered structure	羽状构造
feather-edged	刀刃状的；薄边的；锐边的
feathering out	（地层）尖灭；（地层）变细；（地层）变薄
feathering vane	羽状叶
feather-like drainage pattern	羽毛状水系
feathery structure	羽状构造
feature	特点，特征，特性；特色，性能；要素；地势，地形，地貌；构造；零部件
feature rock bolt	随机张拉锚杆
feature rock dowel	随机砂浆锚杆
feature rock reinforcement	随机加固
feder joint	羽状节理
Fedorov goniometer	费多洛夫测角仪
fedorovite	霓透辉石
feeble shock	微弱地震
feebly cohesive soil	弱黏性土
feebly magnetic	弱磁性的
feed ditch	引水沟，进水沟，灌溉渠；供水沟
feed of drill	进尺；钻头进程
feed pitch	输送间隔
feed plate	给料板，给矿板
feed speed	给进速度；推进速度
feed stock coal	原料煤
feed water	给水；供水
feed water pipe	给水管
feed water pump	给水泵
feed well	给水井；供水井
feedback neural network	反馈神经网络
feeder	支流，支脉；补给河流；河源，泉源；运道，通道，岩浆侵入通道；馈给
feeder area	补给区
feeder beach	补给海滩
feeder canal	给水渠
feeder channel	供水渠道
feeder current	补流
feeder dike	补给岩墙；馈浆岩墙
feeder gas	流出的瓦斯；喷出的瓦斯；流出的气体
feeder highway	支线公路
feeder structure	补给构造
feeder trough	给矿槽
feeder zone	补给带
feedforward neural network	前馈神经网络
feeding	供应，给料，进给；给矿；馈电，供电；进给的，给料的
feeding area	补给区；给水区
feeding canal	给矿沟
feeding conduit	供水道
feeding ground	排泄盆地；泄水盆地；流域盆地；补给盆地
feeding of a river	河流补给
feeding pressure	推进压力
feeding reservoir	供水水库；蓄水池
feedpump	给水泵；上水泵
feedwater	给水；饮用水
feed-water treatment reservoir	给水处理池
feeler	测隙器，测隙规，探测器，探测杆，探测仪；厚薄规；塞尺；接触传感器，灵敏元件；触头，探头；探针
feeler gauge	塞尺；测隙规；厚薄规；千分垫；触杆规；量隙规
feeler hole	超前探孔；探查孔
feeler inspection	测隙规检验，探针检验；探测器检验；触探
feeler pin	探针
feeler plug	测孔规
feeling dilation	感触膨胀计
feeling dilatometer	触头膨胀仪
feet	英尺；脚
feet centre	英尺计中心间距
feet per bit	钻头进尺；单钻进尺
feigh	铅矿尾矿；废渣
Feilaifeng Limestone	飞来峰石灰岩
felder	镶嵌地块
feldsarenite	长石砂屑岩
feldspar	长石；长石类；钾长石
feldspar bed	长石床层；人工床层
feldspar dissolution	长石溶解
feldspar mineral	长石矿
feldspar phyllite	长石千枚岩
feldspar porphyry	长石斑岩
feldspar porphyry dyke	斑状长石岩墙
feldspar sandstone	长石砂岩
feldspar-basalt	长石质玄武岩；长石玄武岩
feldsparization	长石化
feldspar-knot gneiss	珍珠状片麻岩
feldsparphyric rock	长石斑岩
feldspathic	长石质的，长石质；含长石的；长石砂岩
feldspathic arenite	长石质砂岩；长石粗屑岩
feldspathic graywacke	长石质杂砂岩
feldspathic grit	长石粗砂岩
feldspathic quartz sandstone	长石石英砂岩；长石石英砂
feldspathic quartzite	长石质石英岩；长石石英岩
feldspathic rock	长石岩
feldspathic sandstone	长石质砂岩；长石砂岩
feldspathic shale	长石页岩；长石质页岩
feldspathic subgraywacke	长石质亚杂砂岩
feldspathic wacke	长石质瓦克岩
feldspathization	长石化
feldspathized	长石化的
feldspathoid	似长石类，似长石；副长石类
feldspathoidal minerals	似长石类矿物
feldspathoidite	副长石岩
feldspatholite	长石质岩
feldspatite	长石岩；长石质
feldspatlute	石英二长细晶岩

feldspattavite 长石质霓方钠岩
felitfe β 塑硅灰石
fell 筛下产品；一次放炮崩落的岩石；铅矿；荒山，丘陵；筛下产品；砍伐
Fellenius method 费伦纽斯法
Fellenius method of slices 费伦纽斯条分法
Fellenius solution 费伦纽斯解法
felloe band 钢带；载重带
felseid 浅色霏细岩
felseid porphyry 浅色霏细斑岩
felsenmeer 残积碎屑；石海
felsic 长石矿物的；含长石矿物的；长英矿物；长英质的，长英质
felsic gneiss 长英质片麻岩
felsic igneous rock 长英质火成岩
felsic magmatism 长英岩浆酌
felsic minerals 长英质矿物
felsic models 长英质模型
felsic rock 长英质岩石，长英质岩；长英岩
felsic volcanics 酸性火山岩
felside 无斑浅色霏细岩
felsilic 霏细状的；霏细岩的
felsiphyric 无斑隐晶的
felsite 霏细岩；霏细状长英岩；硅长石；致密长石；矽长岩；长英岩
felsite porphyry 霏细斑岩
felsitic 霏细状；硅长质的；霏细的
felsitic texture 霏细结构；霏细组织；霏细构造
felsoandesite 霏细安山岩
felsobanyite 斜方矾石
felsogranophyre 霏细花斑岩
felsophyre 霏细斑岩
felsophyric texture 霏细结构；霏细斑状结构
felsophyrite 霏细玢岩
felsovitrophyre 霏细玻基斑岩
felspar 长石
felsparite 富长花岗岩
felstone 致密长石；霏细岩
felt 毡；油毡
felt area 有感区
felt earthquake 有感地震
felt plug 毡塞
felt retainer 毡护圈；毛毡油封护圈
felt ring 毛毡垫圈；毡圈
felt roofing 油毛毡天台工程
felt seal 毡垫密封
felt washer 毡垫圈
felted groundmass 毡状基质
felting 垫毡；垫绒
feltless earthquake 无感地震
felty 毡状；针状的；毡状的
felty texture 毡状结构
femag 镁铁质
female joint 套筒接口
female screen plate 阴螺丝；阴螺旋；螺帽；螺母
female screw thread 内螺纹
female spanner 套筒扳手
female tap 打捞母锥
female tapered socket 内螺纹锥形接头
female union 管内接头

female-quick coupling 内丝快速接头
femic 铁镁质
femic mineral 铁镁质矿物
femic minerals 含铁和镁的矿物类
femic rock 铁镁质岩石
femmer 薄弱的
fen 沼，沼泽，泽地；沼泽地；沼泽低地；沼池
fen clay 沼泽软黏土；沼地黏土
fen land 沼泽地
fen peat 沼泽泥炭；低位泥炭；低沼泥炭
fen soil 低位沼泽土；沼泽土
fence diagram 立体投影图；透视断块图；立体透视图；立体剖面图；栅状图解，栅状图
fence off 用栅栏隔开；隔开
fence post 围柱
fence row method 凹形工作面全面开采法；排柱支护法
fence wall 围墙
fenced opening 围护缺口；围护入口
fencing 围栏，围墙；围网；铁网；棚栏；安设排柱；围上篱笆，砌上挡墙
fencing wall 护墙；圆环
fend off 挡开；架开；避开
fender 防护物，挡泥板，保护板，挡板；排障器；护木；护舷材，缓冲材，护舷设备；矿柱；碰垫
fender course 防撞层
fender log 护木
fender pier 防冲突堤；码头防冲桩；护墩
fender pile 护桩；防御桩；防撞桩；防护桩
fender post 围护桩；码头堤岸护舷桩；防撞柱；防护柱
fender restraint beam 防撞护梁；岩石墩
fender shield 挡泥板
fender system 防撞系统
fending groin 防冲堤
fenestral 格状
fenestral porosity 蚀窗孔深；格状孔隙
fenestrated 窗孔的；网孔的
Fenghuoshan Group 风火山群
fenite 长霓岩；霓长岩
fenitization 霓长岩化作用
fenitize 霓长岩化
fenland 低险地；沼泽；沼泽地，干沼泽地；干沼泽；排水沼泽
Fenner's formula 芬纳公式
fenny 沼泽的；生于沼泽地带的
fenster 构造窗；蚀窗；蚀穿掩冲体；网格状；格状的
ferberite 钨铁矿；钨铁石
Ferett triangle 三角坐标图
ferganite 水钒铀矿；铀钇钽矿；斜辉石岩
Fergenbaum universality constant 费根鲍姆普适常数
ferghanite 铀钇钽矿；铀锂钒矿；磷矾铀矿；水钒铀矿
fergusite 橄榄白榴岩；等色岩-碱性辉长岩系列；假白榴等色岩
fergusonite 褐钇铌矿；褐钇钽矿
Fermat's principle 费马原理；费马法则
fermented soil 熟土
fermentor 发酵场
fermentor with forced aeration 强制曝气发酵场
fermentor without aeration 非曝气发酵场
Fermi 费米
Fermi acceleration 费米加速

fermorite	锶磷灰石
fern leaf crystal	锯齿状晶体
fernandinite	纤钒钙石
fernico	铁镍钴合金；铁钴镍合金
ferralite	铁铝土；铁铝岩
ferrallitization	铁铝土化作用
Ferrel's law	费雷尔法则
ferreous	铁的；铁质的
Ferrero's formula	菲列罗公式
ferri halloysite	多水铁高岭土
ferrian	含高铁的；含铁的
ferric	铁的；三价铁的；高铁的；正铁的
ferric ammonium sulphate	硫酸铁铵
ferric chloride	氯化铁
ferric cyanide blueprint	铁盐晒图
ferric facies	铁质相
ferric induction	铁磁感应
ferric iron	三价铁
ferric oxide	氧化铁，氧化高铁；三氧化二铁
ferric sulfide	硫化铁
ferric sulphate	硫酸铁
ferric-cement	高铁水泥
ferricpozzolan cement	铁质火山灰水泥
ferricrete	铁质壳；铁质砾岩
ferricrust	铁质结壳；铁结核硬壳；硬铁质结土层
ferricyanide	铁氰化物
ferrierite	镁碱沸石
ferrifayalite	涞河矿；高铁铁橄榄石
ferriferous	正亚铁的；含铁的
ferrihydrite	水铁矿
ferrilite	铁岩
ferrimagnetic	亚铁磁质
ferrimagnetic minerals	亚铁磁性矿物
ferrimagnetism	铁氧体磁性；亚铁磁性
ferrimolybdite	铁钼华
ferrimontmorillonite	高铁蒙脱石；铁质蒙脱石
ferri-sericite	铁绢云母
ferrisymplesite	纤维砷铁矿
ferrite	铁索体；铁氧体；铁酸盐；纯粒铁；铁素质，砂岩中的铁质胶结物；自然铁
ferrite zone	自然铁带
ferrithorite	高铁方钍石；铁钍石
ferritic	铁索质的；铁氧体的
ferritic-steel	铁素体钢的
ferritization	褐铁矿化
ferritizing annealing	铁素体化的退火
ferritungstite	高铁钨华
ferro-	表示：铁；含铁的；含亚铁的
ferroalloy	铁铝合金，吕铁；铁合金
ferroaluminium	铁铝合金；铝铁
ferro-anthophyllite	铁直闪石
ferroaugite	铁辉石
ferrobasalt	铁质玄武岩
ferrocalcite	铁方解石
ferrocarbonatite	铁碳酸岩
ferrocerium	铈铁；发火合金
ferrochrome	铬铁合金
ferrocobaltine	铁辉钴矿
ferrocoke	炼铁的焦炭；冶金焦
ferrocolumbite	铁铌矿
ferroconcrete	钢筋混凝土
ferroconcrete plate	钢筋混凝土板
ferroconcrete prop	钢筋混凝土支柱
ferroconcrete shell	钢筋混凝土壳
ferrocrete	一种速凝波特兰水泥
ferrocyanide	氰亚铁酸盐；亚铁氰化物；亚铁氧化物
ferrodiorite	铁闪长岩
ferrodisulfide sulphur	硫化铁硫；黄铁矿硫
ferrodolomite	铁白云石
ferroelastic	铁弹性的
ferroelectric	铁电体的；铁电性的
ferroelectricity	铁电
ferroelectrics	铁电体；铁电材料
ferrofluid	铁磁流体
ferrogabbro	铵辉长岩
ferrography	铁粉记录术；图像的磁性记录
ferrohastingsite	低铁钠闪石；铁绿钙闪石
ferrohornblende	铁角闪石
ferrohydrite	褐铁矿
ferrohypersthene	铁紫苏辉石
ferrolite	铁矿岩；铁屑集料；铁岩
ferrolsol	铁铝土
ferromagnesian	含铁镁的；铁镁质
ferromagnesian component	铁镁组分
ferromagnesian mineral	铁镁矿物
ferro-magnesians	硅酸盐铁镁矿物类
ferromagnet	铁磁体
ferromagnetic	铁磁性的；铁磁的；铁磁质
ferromagnetic attraction force	铁磁吸力
ferromagnetic element	铁磁性元素
ferromagnetic material	铁磁性材料；铁磁性物质
ferromagnetic mineral	铁磁性矿物
ferromagnetic rock	铁质岩
ferromagnetic substance	铁磁质；铁磁物质；铁磁体
ferromagnetics	铁磁质体；铁磁质学
ferro-magnetism	铁磁性；铁磁学
ferromangandolomite	低铁锰白云石
ferromanganese	锰铁；锰铁合金
ferromanganese coating	铁锰覆盖层
ferromanganese crust	铁锰结壳；锰结壳
ferromanganese deposits	铁锰矿床
ferromanganese nodule	铁锰结核
ferromanganese pavement	铁锰敷积层
ferronatrite	针钠铁矾
ferro-nickel	镍铁合金
ferroplete	铁质火成岩
ferroplumbite	磁铅铁矿
ferroprobe	铁磁探测器
ferro-silicon	硅铁
ferrosilite	斜铁辉石；铁辉石
ferrospinel	铁尖晶石
ferrostatic pressure	铁水静压力
ferrotantalite	钽铁矿
ferrouranothorite	铁铀钍石
ferrous	铁的；含铁的，低铁的；亚铁的；二价铁的
ferrous alloy	铁基合金；铁合金
ferrous iron	低价铁；二价铁；亚铁；亚铁离子
ferrous metal	铁金属，黑色金属；铁类金属
ferrous pseudobrookite	低铁假板钛矿
ferrous sulfide	硫化亚铁

ferrous sulphate	硫酸亚铁
ferro-vanadium	钒铁合金
ferroxcube	立方结构铁氧体
ferroxdure	永久磁铁
ferruccite	氟硼钠石
ferruginization	铁化
ferruginous	含铁的；铁质的
ferruginous boulder	含铁的巨砾
ferruginous breccia	铁质角砾岩
ferruginous cement	铁质胶结物
ferruginous clay	含铁黏土
ferruginous deposit	含铁沉积物；铁矿床
ferruginous earth	含铁土
ferruginous lateritic soil	含铁红壤
ferruginous nodule	含铁结核
ferruginous nodule rock	铁质岩
ferruginous nodule sandstone	铁质砂岩
ferruginous quartz	铁石英
ferruginous quartzite	含铁石英岩
ferruginous rock	铁质岩
ferruginous sandstone	铁质砂岩
ferruginous shale	铁质页岩
ferruginous spring	含铁泉
ferruginous water	含铁矿水
ferry slip	轮渡引线；轮渡斜引道
ferry trestle bridge	轮渡栈桥
fersilicite	硅铁矿
fersmanite	硅钠铌石
fersmite	铌钙矿
fertile field	优质煤田
fertile mantle	富集地幔
fertile peat pot	含肥泥炭盆
fertility erosion	肥力侵蚀
fertilizer mineral	矿物肥料；肥料矿物
fertilizer yard	积肥场
fervanite	水钒铁矿
festoon	花彩；花彩弧
festoon cross-bedding	花彩交错层理
festooned pahoehoe	翻花绳状熔岩
fetid bituminous limestone	臭沥青石灰岩
fetid limestone	臭灰岩
fettling	修补；修整；铸件清理；涂炉材料
fettling room	修整工段
feu	下伏岩石；黏土层；永租地；永租权
fiamme	火焰石
fiant	制成，做成
fiard	低浅峡湾；低峡湾
fiasconite	钙白碧玄岩；钙长白榴碧玄岩
fiber	纤维
fiber board	纤维板；木丝板
fiber concrete	纤维混凝土
fiber reinforced concrete	纤维增强混凝土
fiber stress	纤维应力
fiberglass	玻璃纤维
fiber-glass bolt	玻璃纤维锚杆
fiberglass epoxy	玻璃纤维树脂板；玻璃钢板
fiberglass roof bolt	玻璃纤维锚杆
fiberglass-reinforced plastics	玻璃纤维增强塑料
fiber-optic bundle	光导纤维束
fiber-reinforced cement slurry	纤维水泥浆
fiber-reinforced concrete	纤维加强混凝土
fiber-reinforced material	纤维增强材料
fibre building board	纤维薄板
fibre concrete	纤维混凝土
fibre packing	纤维衬垫；纤维集束
fibre soil	纤维土
fibreboard	纤维板
fibred	纤维状的
fibreglass reinforced concrete	玻璃纤维混凝土
fibro-	表示：含纤维的；纤维状的
fibroblastic	纤维变晶状的；纤维状变晶的
fibroblastic texture	纤维变晶构造；纤状变晶结构
fibrocrystalline	纤维结晶状的；纤维结晶质
fibroferrite	纤铁钒
fibroid phthisis	纤维化结核；硅肺
fibrolite	夕线石；细硅线石；细矽线石；纤状矽线石
fibrous	纤维状的；纤维的；纤维状的；纤维质的
fibrous aggregate	纤维状集合体
fibrous coal	丝煤；纤维质煤
fibrous concrete	纤维性混凝土；纤维混凝土
fibrous fracture	纤维状断口；纤维裂面
fibrous gypsum	纤维石膏；纤石膏
fibrous insulation	纤维绝缘
fibrous loam	黏土质土；松软壤土
fibrous loam soil	成层黏土质土壤
fibrous mulch	纤维盖土
fibrous parallel	平行纤维状的；平行纤维状的
fibrous peat	纤维状泥炭，纤维性泥炭；木质泥炭，木质状泥炭；纤维泥煤
fibrous plaster	纤维灰泥
fibrous red iron ore	纤维赤铁矿
fibrous rock	纤维状岩石
fibrous shotcrete	纤维喷射混凝土
fibrous silicotic mineral	纤维状的硅质矿物
fibrous soil	纤维性土
fibrous structure	纤维构造
fibrous texture	纤维结构；羽状组织
fibrous turf	纤维状泥炭；纤维泥炭
fichtelite	白脂晶石；费希德尔石
ficinite	柴苏辉石
Fick's law	斐克定律
fiction braking	摩擦虚拟制动
fictitious displacement	虚位移
fictitious drag coefficient	假阻力系数
fictitious force	虚力
fictitious graticule	虚拟经纬网；虚拟经纬网格
fictitious isochron	假等时线
fictitious load	虚荷载；虚拟荷载
fictitious load distributions	虚拟载荷分布
fictitious power	虚功率
fictitious spectral density	虚拟谱密度
fid	销；销钉；螺柱；楔铁；镶条
fiducial	置信
fiducial axis	基准轴
fiducial interval	置信区间
fiducial limit	可靠界限；置信限
fiducial line	基准线
fiducial mark	基准标志；基准点；框标；柜标
fiducial point	零点；基准点；参考点；标准点；基准线
fiducial time method	标时法

fiedlerite 水氯铝石
field 现场,野外,田野;田间,田地,场地,煤田,油田,工地,矿区,矿田,土地;场,磁场,电场;字段,信息组,符号组;外业;领域,域,范围,区域
field adjusting device 野外校正装置
field analysis 野外分析;现场分析
field and laboratory test 现场和实验室测试
field array of seismic sensors 地震仪野外台阵
field assembly 工地装配;现场装配
field automatic topographical surveying device 野外地形自动测绘仪
field barrier 两矿间的分界煤柱;现场分隔线
field bearing test 现场载荷试验;现场承载测试;实地承载测试
field blasting test 现场爆炸试验
field board 野外作业板
field book 外业手簿,野外手簿;野外记录簿;工地记录本
field bus 现场总线
field calibration 野外检定;野外校准
field capacity 田间持水量,田间含水量;土壤毛细含水量;填埋场容量;自然土壤湿度
field capacity of soil 土壤田间持水量
field capillary moisture capacity 田间毛管含水量
field carrying capacity 土壤毛细含水量;田间含水能力
field characteristics of compression 现场压缩特性
field classification 分类;现场分类
field coefficient of permeability 场地渗透系数;田间渗透系数
field compaction behavior 现场压实性状
field compaction curve 现场压实曲线
field compaction test 现场碾压试验
field completion 野外校正完善;野外补点;野外补充工作
field compressibility 现场压缩性
field compression curve 现场压缩曲线
field compressometer (螺旋)压板载荷试验仪
field computation 现场计算
field concrete 现浇混凝土;现场拌制混凝土
field condition 现场条件;野外条件
field construction 现场施工
field control 地面控制;场控制;野外控制;外业控制;现场控制
field control point of aerophotogrammetry 航测外控点
field curing 工地养护
field damped 野外用阻尼摆
field data 野外数据;野外资料;现场数据
field data card 野外数据卡
field data collection 野外数据收集
field deciphering 现场解译;现场辨认
field deformation 场形变
field density 场强度;磁感应密度;磁通密度,磁通量密度
field density determination 现场密度测定
field density test 现场密度试验
field description 野外描述;现场描述
field determination 野外测定
field developing 现场开拓;现场发展
field development 油田开发;地热田开发
field difference 场差
field direct shear text 野外直剪试验;大型直剪试验

field distribution 场分布
field ditch 田间灌水沟;田沟;毛渠
field ditch efficiency 田间沟渠效率能
field ditch system 田间渠系
field document 外业资料;野外资料
field drainage system 油田排水系统;现场排水系统
field dual operations 野外工作;室外操作
field effect 场效应
field emission 场致发射
field engineer 野外工程师;安装工程师;现场工程师
field engineering 场地工程;安装工程
field epicenter 现场(震)中
field equation 场方程
field equipment 野外装备;野外设备;移动式设备
field erection 工地架设;现场安装
field evaluation 现场鉴定;现场评估;野外评价
field evidence 野外证据;现场证据;野外迹象
field examination 野外检查
field experience 现场经验;生产经验;野外经验
field experiment 野外试验;现场试验;田间试验
field exploration 野外勘察
field extension well (油田)探边井;(油田)扩边井
field extinction 灭磁
field fabricated 现场制造的;现场装配的
field formula 实用公式;装药量计算公式
field gas 矿场天然气
field geological logging 野外地质数据编录
field geological map 野外地质图
field geological work 野外地质工作
field geologist 野外地质工作者
field geology 野外地质;野外地质学
field gradient 场梯度
field groundwater velocity 地下水流速
field guide 野外指南;现场指南;现场指导
field ice 冰原;冰原冰
field identification 现场鉴定;实地鉴定
field identification of soil 土的现场鉴别
field identification procedure 野外鉴定法
field infiltration 田间下渗
field initial compression curve 现场初始压缩曲线
field inspection 野外检查,野外观测;现场视察,实地视察;工地视察;实地调查,踏勘;外业调绘
field intensity (磁)场强度;场强
field intensity sensibility 场强灵敏度
field investigation 现场勘察;野外调查,实地调查,实地勘测,实地勘察;现场试验,运转试验,野外试验;野外研究
field investigation group 野外调查队
field investigation of soils and rocks 岩土现场测试
field laboratory 工地实验室;现场试验室;野外实验室
field lateral 田间支渠
field leveling 野外水准测量
field life 矿床寿命;煤田寿命
field loading condition 现场载荷条件
field loading test 现场载荷试验;实地荷载测试
field log 野外记录;现场记录
field manual 野外手册;外业指南;现场作业手册
field map 实际材料图;野外原图;野外草图;外业原图;矿区图
field mapping 野外填图;外业测图;野外制图;野外

测绘
field measurement 野外测量，野外测定；野外观测；现场测量；实地量度
field measurement of permeability 渗透性的现场量测
field measurements 现场测量
field method 野外方法
field mode 地热田模型；现场模式
field moisture 天然湿度；天然含水量；田间含水量
field moisture capacity 天然含水量；现场含水量；田间保水量；田间持水量
field moisture equivalent 现场等价含水量；工地含水当量；现场含水当量，室外含水当量；现场水分当量值；田间含水当量
field moisture-density test 现场含水量密度试验
field monitoring 现场监测，现场监控；野外监测
field natural gas 气田天然气
field notes 现场记录；野外记录；外业记录
field observation 野外观察；野外观测；实地观察；现场观察，现场观测；外场观测
field observations of performance on pillars 矿柱性能现场观测
field of attraction 引力场；万有引力场
field of deformation 形变场
field of flow net 流网场
field of force 力场
field of gravity 重力场
field of load 受力范围
field of stress 应力场
field of view 视场；视界；视野；视域
field of water-head drawdown 水头降深场
field operation 现场作业；野外操作
field output 油田产量；地热田流量
field party 野外勘探队；野外工作组；外业队；勘测队
field pattern 场型式
field performance 野外性能；油田特征；油田状态
field performance testing 现场性能测试
field permeability coefficient 野外渗透系数
field permeability test 现场渗透试验
field permeability test methods 现场渗透试验法
field photogrammetric apparatus 地面摄影测量仪
field photogrammetry 地面摄影测量
field pilot 现场先导试验
field plane table map 野外平板图
field plate loading tests 现场载荷板试验
field plotter 绘图板
field plotting 野外标绘；野外测图
field potential 场强；油田潜在产能
field pumping test 现场抽水试验
field railway 轻便铁路；临时铁道；工地轻便铁道
field recognition 野外鉴定
field recompression curve 现场再压缩曲线
field reconnaissance 野外调查，野外普查；野外踏勘，踏勘，野外查勘，现场踏勘
field record 现场记录；外业记录
field recording card 野外记录卡
field rectifier 励磁整流器
field reduction 场衰减
field relation 野外关系
field reliability 野外可靠性
field report 野外报告

field reporting 工地报告
field resistivity measurement 现场电阻率测定
field review 现场复查
field ridge 畦垄；输水沟；田垄
field running test 实地运行试验
field sample 野外样品
field sampling kit 野外采样箱
field sampling technique 野外采样方法
field scanning 场扫描
field seminar 野外研讨会
field service 野外作业；现场作业；现场服务
field setup 野外观测装置，观测系统
field shear box 野外剪力仪；野外剪切盒
field sheet 现场记录表；外业记录表
field sketch 野外草图，外业草图；野外素描；压碎带
field sketch map 野外手图
field sketching 野外素描；草测；野外草图绘制，野外略图绘制
field sounding test 现场触探试验
field spectroscopy 野外光谱学
field station 野外工作站
field stone 散石；卵石
field strength 场强度；场强，场强
field strength coverage 场强覆盖区
field stress 天然应力，原地应力；场地应力；场应力；原岩应力
field study 实地研究；现场研究；实地考察；野外调查
field supervision of construction 现场施工监督
field survey 野外测量，勘探；野外调查；实地调查，实地测量，实地勘察；现场调查
field systems 野外（勘探）系统
field technique 野外方法
field test 现场试验，工地试验；现场测试；现场试验室；野外试验
field test data 野外试验数据
field test for volume stability 体积稳定性的现场试验
field test kit 野外分析箱
field test methods of subgrade and pavement for highway engineering 公路路基路面现场测试规程
field test site 野外试验场地
field testing 实地测试；工地测试；现场试验；野外试验
field theory 场论
field theory of earthquake 地震场论
field time break 现场起爆信号；现场定时起爆
field timing 野外计时信号
field tolerance 野外容差
field trial 野外试验；生产试验；现场试验；实地测验
field triangulation 野外三角测量
field usage 野外使用；现场使用
field use 现场使用；现场用途
field vane 现场十字板剪力仪
field vane test 现场十字板试验
field velocity 实地流速；实际速度
field virgin compression curve 现场初始压缩曲线
field water （油田）地下水
field water content 天然含水量；野外含水量
field water retaining capacity 野外持水率
field waterflood experiment （油田）现场注水试验
field waterflood performance （油田）现场注水性态
field well 生产井

field work	现场工作，野外工作；外业
field work order	工地工作通知
field-acquired geological profile	实测地质剖面图
field-aligned irregularity	场向不规则结构
fieldbook	野外工作记录本
field-confirmation well	油田证实井
field-cured cylinders	现场养护的圆柱形试件
field-intensity meter	场强计
field-reversal	场反向
field's producing life	油田开采期；油田开采寿命
fieldstone	粗石
field-strength meter	场强计
field-tested	经过现场试验的
field-to-field survey	分界测量
fieldwide pattern	全（油田）井网
fieldwide preferential flow direction	全（油田）主流向
fieldwide project	全（油田）方案
fieldwide rate	全（油田）产量
fieldwide simultaneous injection	全（油田）同时注入
fieldwide voidage	全（油田）亏空
fieldwide waterflood	全（油田）注水开发
fieldwork	野外工作；现场工作
fiery coal	气煤；瓦斯煤
fiery colliery	瓦斯煤矿
fiery gas	爆炸性气
fiery heap	自燃矸石堆
fiery mine	瓦斯矿，瓦斯煤矿；多煤尘矿；容易发生爆炸事故的矿山
fiery seam	瓦斯煤层；含气层
figuline	陶土；瓷土
figurative progress of construction work	施工形象进度
figure	数字，数值，数码；位数，计算；图形，形体，形象，图，插图；图表；金额，价格
figure adjustment	数值调整；图形平差
figure of merit	质量因数，质量指数；品质因素；性能系数；优值系数
file hardness testing	锉削硬度试验
filiform	丝状的
filiform corrosion	丝状腐蚀
filiform lapilli	火山毛；丝状火山砾
filiform texture	丝状结构
fill	填方，填土，填料，填塞，人工填土，垫板；填充填，充满，回填，充填材料，充填物；淤积；装车，装载；路堤，填筑
fill block	填块
fill compaction	填土压实
fill compaction test	填料压实试验
fill concrete	回填混凝土；充填混凝土；填筑混凝土
fill construction	填土施工；填土工程；填筑
fill dam	土坝，土石坝；填土；堆筑坝，填筑坝，填筑式坝；填土坝
fill dry density	填土干密度
fill excavation	采掘充填料；填方开挖；充填挖掘
fill factor	装满系数
fill height	填土高度
fill in well	插入井；补密井；注水井；注水孔
fill insulation	填塞绝缘
fill material	填料；填充材料；填土材料
fill moisture content	填土含水量
fill off	装煤；装走
fill pass	充填天井；充填溜眼
fill pit dual operation	填坑开采法
fill pit operation	填坑开采法
fill placement record	填筑记录
fill placement technique	填筑技术
fill platform	填土平台
fill raise	充填用天井；充填天井
fill resources	填料资源
fill rock	填石
fill section	填土部分；填土截面；充填区；路堤截面
fill seismic reflection configuration	充填地震反射结构；充填反射震测架构
fill setting	充填物沉实
fill settlement	填土沉降；填土沉陷；充填物沉实
fill slope	填土坡，填土边坡；填坡；路堤边坡
fill slope talus	路堤边坡
fill source	填料来源
fill stratum	填土层
fill structure	填土结构；填土构筑物
fill terrace	堆积阶地；填充阶地
fill tightening	填紧
fill toe	堆积底；填方坡脚，填土坡脚
fill type dam	填筑式坝
fill up ground	充填地
fill up to grade	按坡度充填
filled band	充填地区
filled bitum	填充沥青
filled bitumen	加填料沥青
filled cavity	充填穴
filled cell	填土格仓
filled column	填料柱
filled crib	填石木垛
filled deposit	填充矿床
filled die	满料压模
filled discontinuity	充填的不连续面；充填结构面
filled flat-back stope	水平向上梯段充填回采工作面
filled fracture	充填缝
filled frame reef	填充骨架生物礁
filled ground	填土；填土基础；填土地基；回填地基
filled land	填筑地
filled pile	灌注桩
filled pipe column	混凝土管柱；充填式管柱；填实管柱
filled plate	填板
filled rill stopes	充填沟采场（上向倒V形梯段充填回采工作面）
filled soil	填土
filled spandrel arch	实腹拱
filled spandrel arch bridge	实腹拱桥
filled stope	充填回采工组；充填回采工作面
filled stope method	充填回采工作法
filled stoping	充填回采法
filled valley	淤塞谷；淤积谷
filled wall	充填墙
filled-in ground	填筑地；填土地基
filled-up ground	填筑地；填土地基
filler	充填材料，填料，填隙料，掺合料，模板填孔料，填充剂，掺合料；装载设备；装煤厂，注入器；充填工，填料工，填充物；采煤学徒工；垫板，垫片，间隔铁，垫圈，衬片，雨水进口，淤积湖的河流
filler aggregate	填充骨料

filler clay	黏土填料
filler course	充填层
filler discharge orifice	水砂充填出砂口
filler for cement	水泥惰性掺合料；水泥填料
filler gate	充水闸门
filler line	充填管道；充灌管
filler metal	填充金属
filler piece	填隙片；垫片
filler ring	模板上段；垫圈
filler rod	焊条芯；焊条
filler sand	填充砂
filler-binder mortar	充填黏合灰浆；填料黏结砂浆
fillers	填料
fillet	填角；内圆角；平边条，轴肩；嵌缝条，角焊缝；倒圆角，倒角，畦；筋条，轮缘
fillet joint	角接合；角焊缝接头
fillet saw	割石手锯
fill-in fill terrace	填塞堆积阶地
fill-in well	补密井；注水井；注水孔
filling	充填物，填充物，填塞物；填充，充填，填塞，充水，充气，装满，灌注，装填；填料，填土，装料，装载，填方，装药；填筑；存档
filling and emptying system	充水和放水系统
filling behind abutment	台后填方
filling body	充填体
filling by flushing	水砂充填；湿法充填
filling compound	填料
filling crevasse	填塞裂缝
filling culvert	充水涵洞
filling cycle	装载循环；充填循环
filling degree	充填度
filling density	充填密度
filling deposit	填充矿床
filling device	充水装置
filling dual operation	充填工作
filling earth	回填土砂
filling element	填料
filling for groin	丁坝填筑
filling for subgrade	路堤填料
filling for subgrade bed	基床填料
filling grouting	回填灌浆
filling hole	注入孔
filling in fissure	裂隙充填物
filling interval	充填步距；充填间隔
filling line	满载线；装油管线
filling load	填筑加载；填土荷载
filling material	充填物，填充材料；填料；充填物质
filling material in fault	岩石断层中的充填材料
filling material of embankment	路堤填料
filling material of joints	节理充填物
filling material preparation	充填材料制备
filling method	充填采矿法；充填法
filling mineral	充填矿物
filling mining	充填采矿
filling of a depression	低压填塞
filling of discontinuity	不连续面充填
filling of reservoir	水库淤积；水库淤填；水库蓄水
filling operation	充填工作
filling orifice	充填注砂孔；注入孔
filling paste	填充糊，充填浆；活性物质
filling piece	填隙片
filling pile	填补桩；灌注桩
filling pipe	灌注桩
filling pressure	充填压力
filling raise	充填天井
filling rate	灌浆速率
filling ratio	充填率
filling shaft	充填井
filling shift	充填作业班次；充填班
filling stope	充填回采工祖；充填回采工作面
filling stoping	充填（采矿）法
filling strainer	加油过滤器
filling strength	填筑强度
filling surface	充填面
filling system	充填开采法；充填法；充填系统
filling test	填料试验
filling time	充水时间
filling times line	充填倍线
filling type	填料类别
filling valve	充液阀；注液阀；充气阀；充水阀
filling velocity	填土速率
filling water test	充水试验
filling with sands	填砂
filling work	充填工作
filling-extracting ratio	充采比
filling-in	充填；填入
filling-in block	填充砌块
filling-in brick	插入砖；混凝土插入砖
filling-in paste	充填糊，活性物质；填充糊状物
filling-setting coefficient	充填沉实系数
filling-up system	上向梯段充填采矿法
fillister	凹槽；开榫
fillister joint	凹槽接合
fillmaterial	充填料
fill-off	完成装煤工作
fillowite	锰磷矿
fill-type dam	填筑式坝
fill-up	充填；填充；装填；注水；灌满
fill-up ground	充填地带
film condensation	膜式冷凝
film flotation	表膜浮选；表层浮选
film flow	薄层流动；薄膜水流
film pressure	薄膜压力
film strength	油膜强度
film thickness	薄膜厚度
film water	薄膜水
film-capillary flow	薄膜毛细水
filsinitization	丝炭化作用
filter aid	助滤剂；过滤用硅藻土
filter basin	过滤池
filter bed	渗透层；过滤层；滤层；渗水层；反滤层
filter blanket	反滤铺盖；滤水垫层；反滤层；滤毡
filter bulkhead	过滤隔墙
filter cake thickness	（滤网上）泥皮厚度；滤饼厚度
filter core	滤波器用铁心
filter crib	滤水框
filter criteria	反滤层标准
filter dam	透水坝；滤水密封墙
filter design	过滤器设计

filter drain 滤水沟；盲沟；滤水暗管
filter fabric mat 土工布反滤层；反滤织物层
filter fabric soil retention test 土工布反滤层试验；滤层织物阻土试验
filter gallery 滤槽
filter gate 过滤网浇口
filter layer 反滤层；过滤层，滤层；疏水层
filter loading 滤池负荷
filter loss 滤失量
filter mat 滤水毡垫；滤水垫
filter material 过滤材料；过滤物料；反滤料；滤料
filter off 滤掉
filter opening 滤孔
filter out 滤出
filter pack 砾石填料层；反滤层
filter pressing 压滤作用
filter pressing differentiation 压滤分异作用
filter sandstone 滤水砂石岩；多孔砂岩；渗滤砂岩
filter strip 滤水畦条
filter subbase 地下滤水基层
filter tank 过滤池；过滤槽
filter toe 滤水坡脚
filter velocity 渗透速度；透水速度；过滤速度
filter wash water consumption 滤池冲洗耗水量
filter well 滤水井；反滤井；渗水井
filter zone 反滤带
filteration differentiation 过滤分异
filtering flow 渗流；滤流
filtering flow rate 渗滤流速
filtering layer 过滤层；滤层；反滤层
filtering velocity 滤速
filtering water 过滤水；渗滤水
filter-paper method 滤纸法
filter-paper-strain method 滤纸过滤法
filter-press action 压滤作用
filter-pressing 压滤脱水
filter-thickener 过滤-浓缩机
filter-well 渗水井
filtrate 滤液，渗流；过滤
filtrate loss 滤失量；失水量
filtration 渗滤，渗流，渗透，失水；过滤，过滤器；滤波；隔滤；反滤；渗漏
filtration bed 滤水层；渗水河床
filtration coefficient 渗透系数
filtration constant 过滤常数
filtration degree 过滤度
filtration differentiation 压滤分异
filtration effect 过滤效应
filtration erosion test 渗透变形试验；渗透侵蚀试验
filtration flow 渗透水流
filtration force 渗透力
filtration loss 渗漏损失；渗透损耗；过滤损失
filtration method 过滤法
filtration path 渗流线；渗流路径
filtration plant 滤水池设备（厂）
filtration process 过滤过程
filtration rate 过滤速度
filtration spring 滤清泉；渗流泉；渗出泉
filtration sump 滤液池
filtration system 过滤系统

filtration test 过滤试验
filtration theory 渗流理论
filtration theory of consolidation 固结渗流理论
filtration velocity 渗透速度；过滤速度
fimmenite 含孢子的泥炭
final acceptance 竣工验收；最终验收，最后验收
final acceptance test 最后验收测试
final assembly 最后装配；总装配图
final bending moment diagram 最终弯矩图
final blading 最后平整土方
final building cost 工程决算
final check 最终检验
final circulating pressure (FCP) 循环结束压力
final classification 最终分级
final concentrate 最后精矿；最后精煤
final construction survey 竣工测量
final contraction value 最终收缩度
final cross-section 净断面
final crushing 最终破碎
final crushing strength 极限抗压强度；极限压碎强度
final design 技术设计；最终设计；施工设计
final dimension 成品尺寸
final division sample 最后缩分试样
final draw 最终放矿
final dry density 最终干密度
final exploration 最终勘探；详勘；详查
final form 最终形态
final grading 最后平整
final grouting pressure 注浆终压力
final hole diameter 终孔直径
final injection pressure 注浆终压
final inspection 最终检查
final lining 二次衬砌；永久衬砌；内衬砌；二次支护
final load 最终荷载
final location 最后定位；最后定测
final log 成果记录
final map 成图；最后成图
final measurement 最终测量结果
final pass 终轧孔型
final penetration 最终贯入度
final pit slope 露天矿最终边坡；最终边帮
final pit walls 采场最终边坡
final plan 审定计划
final refuse bin 最终尾煤仓；最终废渣仓
final robbing 最后回采煤柱
final scheme 最终方案
final screening 最终筛分
final set 混凝土的终凝，终凝；最终贯入度；终锤
final setting 最后凝固
final settlement 最终沉降量；最终沉降；施工决算；竣工决算
final settlement of account 决算
final shut-in pressure 最终关井压力
final slope wall angle 最终边坡角
final soil moisture 终结土壤水分
final sounding survey 最后（海床）测深，最后（海底）测深
final state 最终状态；终态
final statement 最终报表；决算表
final static level 最终静止水位

final strength	最终强度；终凝强度
final stress	有效应力；最终应力值
final structural appraisal	最终结构勘测评估报告
final subsidence	最终沉陷；极限下沉量；最终沉降量
final support	永久支架
final survey	竣工测量；定测；定线测量
final survey stage	定测阶段；终测阶段
final tailings	最终尾矿
final void ratio	最终孔隙比
final water saturation	最终水饱和度；残余水饱和度
finalized run	出版原图；清绘原图
finally formed hole	终孔
finandranite	微斜正长岩
finder	探测器；选择器；选线机；瞄准装置，观察装置；检影器，取景器；中介人，经纪行，发现者
finding	发现，找出；试验结果，调查结果；探测；定位，测定；观察结果；发现数据
fine	细的，细粒的，细微的；精致的，好的；优质的，优良的；粉煤；细粒；粉砂
fine adjustment	精调装置；细调；细调准；精密校正；精密调整；微调
fine adjustment screw	微调螺丝；精调螺钉
fine aggregate	细集料，细粒集合体；细骨材，细骨料；细粒料；细砂石石粉
fine analysis	细颗粒分析
fine and microstructure of ocean	海洋细微结构
fine breaking	细碎
fine break-up	细粒分散
fine breeze	细焦末；细煤粉；煤尘；煤粉
fine clay	细黏土；幼黏土
fine coal	粉煤；末煤；细粉煤；煤粉；碎煤
fine coke	焦屑；碎焦
fine concentrate	细粒精煤
fine concrete	细骨料混凝土
fine controller	精密控制器；精密调整器
fine crack	细裂缝，细裂纹；发裂
fine crushing	细碎
fine crystalline	细晶粒
fine cuttings disposal	截粉钻屑处理
fine deposit	细粒沉积
fine drawing	清绘
fine dust	粉尘
fine earth	细粒土；细土
fine extraction	粉煤脱出量
fine fissure	细裂隙
fine fraction	细粒级；细粒部分
fine fuel	细燃料；细煤
fine gage screen	细孔筛
fine gauze filter	细网目过滤器
fine gold	纯金；粉金
fine grade	二级精度
fine grade filter	微孔过滤器
fine grading	细粒分级；细级配；细粒级分等
fine grain	细粒；细粒状的；小品粒
fine grain development	微粒显像
fine grain sediment	细粒沉积物
fine grained clastic	细粒碎屑
fine grained fracture	细纹裂面；细晶断口
fine grained sand	细粒砂
fine grained sandstone	细粒砂岩
fine grained soil	细粒土
fine grained texture	细粒结构
fine grain-size	细粒尺寸；细粒大小
fine granular	细粒状
fine granulated coal	细粒煤
fine gravel	细砾；细砾石；细砾细粒料；圆形控制堰；圆砾
fine grinding	细磨；精磨
fine mesh filter	小筛孔隔滤网
fine mesh sieve	细孔筛
fine ore	细矿石，细矿
fine regulation	细调整；细蝶
fine sand aquifer	细砂含水层
fine sand cement mortar	水泥细砂浆
fine silt bond	细粉粒黏结；细粉料黏结
fine soil	细粒土壤；细粒料
fine structure	精细结构
fine stuffing	细填料
fine veinlet	细切地形
fine-crystalline grain	细晶粒
fine-fibrous peat	细粒级分等泥煤；细纤维泥炭
fine-grained aggregate	预制板结构细骨料
fine-grained cement grout	细粒水泥浆
fine-grained coaly inclusion	细粒煤包体
fine-grained extrusive rock	细粒喷出岩；细粒火山岩
fine-grained sandstone	细砂岩
fine-grained siliciclastic rocks	细粒碎屑岩
fine-grained soil	细粒土
fine-grained soil fill	细粒土填料
fine-granular	细团粒的
fine-ground cement	细磨水泥
finely crystalline	细晶质
finely detritus groundmass	细碎屑基质
finely dispersed coal particles	细散煤粒
finely dissected topography	细切地形
finely divided	细碎的
finely granular	细粒状的
finely laminated clay	薄层状黏土
finely porous	细孔质的
finely stratified	薄层理的；薄层的；细分层
finely stratified clay	薄成层黏土
finely-divided silt	细泥砂
finely-granular	细粒的
finely-ground coke	细焦粉
finely-pulverized coal	细粉煤
fine-mesh	细眼网目；细网眼筛孔
fine-mesh sieve	细孔筛
fine-meshed	细筛孔的；细粒的
fineness	细度；纯度；成色，金的成色；光洁度；粒度；斜向扩散波
fineness factor	细度系数；纯度系数
fineness module	细度模数
fineness modulus	细度模量；细度模数
fineness of aggregates	骨料细度
fineness of cement	水泥细度
fineness of grain	颗粒细度
fineness of grinding	磨碎细度；磨细度
fineness of scanning	扫描密度
fineness ratio	细度比；粒度比
fineness test	细度试验

fine-ore bin	细矿仓
fine-pored	细孔的
fine-roughened	细微粗糙的；略微粗糙的
fines	细颗粒土，细粒土；细粒；细粉；粉矿，粉煤，细粉煤；细料；碎屑
fines content	细粒含量
fines desorption	粉煤解吸
fines migration	颗粒迁移
fines removal	脱除细粒
fine-slotted screen	细缝筛；条缝筛
finest grade	最细粒的
fine-textured soil	细粒结构土
finger bar	指状沙坝；稳车杆；捣矿锤的悬杆
finger bit	指形刮刀
finger board	指板
finger chute	悬轨条溜口
finger drains	指状排水系统
finger gully	指状冲沟
finger gullying	初期沟蚀
finger lake	指状湖系；指状湖
finger lifter	指状岩芯提取器
finger out	尖灭
finger pier	指形突码头；突堤码头
finger point	峰
finger raise	指状格天井
finger shape shield tail	指形盾尾
finger spacing	指状天井间距
finger stone	细石；小石块
finger veined	掌状脉的
finger-chute gate	指状溜口闸门；栅式溜槽闸门
fingered chute	指状天井
fingernailing	半月形焊接变形
fingerprint fossil	指纹化石
finger-type cutter head	指状切割头
fingery grain	指状颗粒
finish construction survey	竣工测量
finish grinding	最后研磨；最终磨矿
finished breaking	最终破碎
finished concrete	饰面混凝土
finished cross-section	净断面；完工后的净横截面；竣工断面
finished crushing	最终破碎
finished diameter	最终直径；净直径
finished floor level	楼地面标高
finished grade	修整过的坡度；竣工坡度
finished ground level	竣工室外地面标高；竣工地面高程
finished hole	钻孔完成；终孔
finished ore	精矿
finished section	净截面；有效断面
finished stope	采空区
finished surface	精加工表面
finished well	成井；完井
finishing bit	精加工的钻头
finishing coat	面层；外层；表面修饰层；终饰层；表层批荡；面漆
finishing level	阶段底盘标高
finishing steel	末尾钻钢超孔深的钻杆；结尾钻杆
finishing work	装修工程
finish-turn inspection	完工检查
finite amplitude wave	有限振幅波
finite analysis method（FAM）	有限分析法
finite aquifer	有限含水层
finite arc method	有限圆弧法
finite beam	有限梁；有限长梁
finite compressible layer	有限压缩层
finite compressive stratum	有限压缩层
finite deformation	有限形变；有限变形
finite difference	有限差分；有限差；有限差分法
finite difference approximation	有限差分近似式；有限差分逼近；有限差分近似
finite difference energy method	有限差分能量法
finite difference equation	有限差分方程
finite difference method（FDM）	有限差分法
finite difference migration	有限差分偏移
finite difference model	有限差分模型；差分模式
finite difference network	有限差分网格
finite difference nonlinear equation	非线性有限差分方程
finite difference schemes	有限差分格式
finite difference solution	有限差分解
finite differential	有限差分
finite discontinuity	有限不连续性
finite displacement	有限位移
finite displacement method	有限位移法
finite divided difference	有限差
finite domain	有限域
finite elastic layer	有限弹性层
finite element	有限单元；有限元；有限元素
finite element analogy	有限元模拟
finite element analysis program	有限元分析程序
finite element analysis（FEA）	有限元分析
finite element analysis-accelerations	有限单元分析：加速问题
finite element analysis-band width	有限单元分析：带宽问题
finite element analysis-convergence	有限单元分析：收敛性问题
finite element approximation	有限元近似；有限元逼近
finite element discretization	有限元离散化；有限元离散
finite element equation	有限元方程
finite element grid	有限元网络
finite element mesh	有限元网格
Finite Element Method（FEM）	有限元法
finite element model	有限元模型；有限单元模型
finite element modeling	有限元模拟
finite element net	有限元网络
finite element numerical modeling	有限元数值建模
finite element solution modules	有限元解模块
finite element technique	有限单元技术；有限元技术
finite extension	有限扩张
finite extraction	有限开采
finite extrapolation	有限外推
finite field	有限域
finite increment calculus	有限增量计算
finite interval	有限区间
finite iteration	有限迭代
finite length-fault	有限长断层
finite pile	有限桩
finite slice method	有限条分法；条分法
finite solution	有限解
finite strain	有限应变

finite strain axis　有限应变轴
finite strain theory　有限应变理论
finite strip method　有限条分法
finite volume method　有限体积法
finite volume（FV）　有限体积
finite-difference method　有限差分法
finite-difference network　有限差分网格
finite-difference nonlinear equation　非线性有限差分方程
finite-difference solution　有限差分解
finite-differences　有限差分
finite-element analogy　有限元模拟
finite-element analysis　有限元分析
finite-element method　有限元法
finite-element model　有限元模型
finite-element technique　有限单元技术
finite-strain tensor　有限应变张量
finite-strain theory　有限应变理论
finite-strain trajectory　有限应变轨迹
finlandite　芬兰辉长岩
finnemanite　砷氯铅矿
Finsterwalder prestressing method　以自重桥梁结构预应力法；芬斯特瓦尔德预应力法
fiord　海湾；峡湾
fire a hole　点火放炮
fire avalanche　火山崩流
fire barrier　防火墙；防火煤柱；防火隔离带
fire brake　防火墙
fire break　挡火墙；防火带；隔火道
fire brick　防火砖
fire bulkhead　防火墙；隔火墙；挡火墙
fire cement　耐火水泥
fire check　热裂纹
fire clay　耐火黏土
fire coal　取暖用煤；燃煤
fire curtain　防火幕；防火卷帘
fire dam　筑堤，筑坝；防火墙
fire damp　爆炸气体；爆炸性矿井瓦斯；沼气；瓦斯
fire district　防火分区；火区，火灾区
fire division wall　隔火墙
fire drift　起爆巷道
fire endurance　耐火极限
fire engineering contractor　防火工程承建商
fire explosion　瓦斯爆炸；沼气爆炸
fire face　煤层燃烧面
fire fountain　熔岩喷泉
fire gallery　起爆巷道
fire gas　可燃气体；易燃气体；火灾区瓦斯
fire grate　炉栅；炉床；炉箅
fire hearth　灰坑
fire in tunnel　洞内火灾；巷道火灾
fire load　燃烧负荷
fire opal　火欧泊；火蛋白石
fire outburst　瓦斯突出；沼气喷出
fire pillar　防火煤柱；防火矿柱
fire proof　耐火的
fire proof bulkhead　防火墙
fire proof cement　耐火水泥
fire proof concrete　耐火混凝土
fire quartz　火水晶
fire rating　抗火等级

fire resistance　耐火性；抗火性
fire resistance degree　隔火程度
fire resistance test　耐火试验
fire resistant　耐火的
fire resistant material　耐火材料
fire resisting construction　耐火建筑物
fire resisting curtain　防火帘
fire resisting shield　防火障板；防火挡板
fire retardant material　阻燃材料
fire rib　采区防火煤柱
fire setting　火烤法采掘
fire shot　爆破
fire shutter　防火闸
fire stone　耐火石
fire test　着火试验；防火试验
fire trier　沼气检查员
fire wall　隔火墙，防火墙
fire window　防火窗
fire zone sealing　火区密封
fireboss　瓦斯检查员；通风员；瓦斯检定
firebox　燃烧室；火箱
firebreak　防火墙；防火障；防火空隙地带
fireclay　耐火黏土
fireclay mineral　耐火黏土矿物
fired magnesite　烧成菱镁矿
fired shot　已爆破炮眼
fire-damaged concrete　火伤混凝土，火损混凝土
firedamp　沼气；矿井瓦斯；瓦斯，碳化氢；甲烷
firedamp accumulation　瓦斯积聚
firedamp content　瓦斯含量；沼气含量
firedamp detector　沼气检定器
firedamp drainage　沼气抽放；沼气排放；排放瓦斯
firedamp drainage by depression　抽放瓦斯
fire-damp drainage from virgin coal seam　未开采煤层的瓦斯抽放；未采煤层瓦斯抽放
firedamp emission　沼气泄出
firedamp evolution　沼气泄出
firedamp explosion　矿内瓦斯爆炸；沼气爆炸
firedamp fringe　瓦斯边界区；工作面风流和采空区污浊气体的接触地带
firedamp layer　瓦斯积聚层；沼气积聚层；瓦斯层；瓦斯爆炸界限
firedamp limit　沼气极限
firedamp migration　沼气迁移；沼气移动
firedamp outburst　沼气突出
firedamp probe　沼气检定器；瓦斯探针；沼气探示器
firedamp reforming process　瓦斯重整法
fire-damp tester　沼气检定器
firedamp testing　沼气测定；沼气检查
firedamp-proof　防爆炸的；隔爆的
fired-to　有自由面的爆破
fire-exploratory drift　探火巷
fireman　爆破工；瓦斯检验员；通风员；消防员；司炉
firepower intensity　火力强度
fireproof　防火的；耐火的；耐火；防火
fireproof building　防火建筑
fireproof bulkhead　防火墙
fire-proof concrete　耐火混凝土
fire-proof construction　耐火构造；防火构造；防火建筑
fireproof lining　耐火盖层；耐火衬层；耐火支架

fire-proof material	防火材料；耐火材料
fire-proof wall	防火墙
fireproofness	耐火度；耐火性
fire-resistance test	耐火试验
fire-resistant	耐火的；抗燃的
fire-resistant material	耐火材料
fire-resisting concrete	耐火混凝土
fire-resisting construction	耐火结构；耐热结构；耐火建筑
fire-resisting layer	防火层
fire-resisting material	耐火材料
firestone	火石；燧石；黄铁矿结核；耐火岩；耐火黏土；耐火岩石
firewall	隔火墙；防火墙；防火隔板
firing by power current	电流点火
firing by radio	无线电点火
firing cable	放炮母线
firing capability	发爆能力
firing of mine	矿井着火
firing order	放炮次序；发火次序；点火次序；爆破顺序
firing pattern	炮孔布置
firing sector	烧火间
firing sequence	爆破顺序；放炮次序
firing shrinkage	烧缩；烧成收缩
firing shrinkage coefficient	烧成收缩率
firm base	坚硬地基
firm bottom	坚固底板；坚固巷道底；坚硬地层；坚固地基；硬土；稳固底板
firm clay	硬黏土
firm demand	固定负荷
firm discharge	固定流量
firm fill	结实填土
firm ground	坚固地基；稳定地基；坚硬基岩；硬地面；坚固地层，坚实地层
firm load demand	固定负荷
firm rock	硬岩；稳固岩石
firm site	硬土场地
firm stratum	硬层；稳定地层；坚固地层
firm top	稳定顶板
firm wall	坚实围岩，稳固围岩；坚实岩墙
firmly bound water	强结合水
firmness coefficient	坚固性系数
firmness coefficient of rock	岩石坚固性系数
firmness of rock	岩石坚固性
firmness of soil	土壤压密性；土坚固性
firm-viscous material	稳黏性材料；刚黏性材料
firn crash	冰原崩溃
firn edge	粒雪界；永久积雪线；冰雪界线
firn field	永久积雪原
firn ice	粒雪冰
firn limit	粒雪界线
firn line	永久雪线，永久积雪线；雪线
firn wind	冰川风
firn zone	粒雪区
firn-ice avalanche	粒雪冰崩
firnification	粒雪化作用
first advance	超前工作面；超前工祖；超前掘进；超前推进
first appearance datum plane	首现基准面
first arrival	初至
first arrival refraction survey	初至折射法勘探
first arrival time	初至时间
first arrival wave	初至波
first azimuth method	第一方位角法
first bit	开眼用的钻头或钎子，开眼钎子；钻头的最前部
first blasting	初次爆破；原体爆破；首次爆破；原矿体爆破
first bottom	河漫滩；泛滥地；滩地；一级河床
first break	初至；初至时间
first category gassy mine	一级瓦斯煤矿
first caving	直接顶初次垮落；初次放顶
first caving distance	初次放顶距
first class standard meter	一级线纹米尺；日内瓦尺
first coal mining face	首采工作面
first crushing	初碎
first cut	起始掏槽；切槽；开切巷道；粗切削；初始开段槽
first departure section	第一离去区段
first failure	第一次破坏
first fill grouting	一次注浆
first filling	初次蓄水
first finite divided difference	一阶有限插分
first flooding surface	初始海泛面
first harmonic	一次谐波；基波
first ice	初冰
first meridian	本初子午线
first mining	第一次采；初采；采区准备；首次回采
first mode	第一振型
first mode force	第一振型力
first mode of vibration	第一振型
first mode resonance	第一振型共振
first motion	初动；直接运动；无变速的传动
first order leveling	一级水准测量
first order projection	一级突起
first order shear	一级剪切
first order station	一级台站；一等台站
first order triangulation	一等三角测量
first phase	初相
first pour	第一期浇筑
first preliminary tremor	第一初期微震
first principal plane	第一主平面
first principal stress	第一主应力
first roof caving	直接顶初次垮落
first roof fall	初次冒顶
first roof-contact point	顶梁第一接顶点
first several modes	最前几个振型
first shotcrete	初喷混凝土
first slice	第一分层
first stage cofferdam	一期围堰
first stage concrete	一期混凝土
first strength theory	第一强度理论；最大应力理论
first stress level	初应力大小
first terrace	一级阶地
first time slides	首次滑动
first water	最纯光泽；最高品质
first watering	初灌
first wave arrival	初至波
first weight	顶板初次来压；初始荷重，初始载荷；初始压力
first weighting	初次来压；基本顶初次来压
first working	采准工作；第一分层的采准工作；第一次

回采煤房
first yield 首次屈服
first-appearance datum 首次出现基准面
first-break intercept-time method 初至截距时间法
firsthand material of hydrogeological investigation 水文地质勘查原始资料
first-motion wave 初至波
first-order fold 一级褶皱
first-order hydraulic servo-mechanism 一阶液压伺服机构
first-order instrument 精密仪器
first-order phase-locked loop 一阶锁相环
first-stage cofferdam 第一期围堰
fir-tree bit 多刃式扩孔旋转钻头
fischerite 柱磷铝石；水磷铝石
Fischer's projection 费舍尔投影
fischer-schrader assay 费舍尔-施拉德验定法；铝甑干馏试验
fish 井下落物，打捞物；鱼尾板；接合夹板；钻孔异物；接板；拖轮
fish barrier 拦鱼栅
fish barrier dam 拦鱼坝
fish bellied beam 鱼腹梁
fish passage facility 过鱼设备
fish piece 夹接板；接合板
fish tail structure 鱼尾状构造
fish-belly sill 鱼腹梁
fished beam 接合梁
fisher 打捞器
fishing 打捞工作；夹板接合
fishing baseplate 鱼尾板；连接板；盖板
fishing basket 捞取管
fishing grab 打捞叉子
fishing haven 渔堰
fishing tap 打捞公锥；打捞丝锥；打捞母锥钻
fishtail bit 鱼尾式钻头；鱼尾钻头
fishtail bolt 鱼尾螺栓
fishtail structure 鱼尾构造
fissibility 可劈性；易裂性；可裂变性
fissible 可分裂的；裂变的，碎裂的；可劈的
fissile 页状的；片状的；易裂的；裂变的；易剥裂的；剥裂的
fissile element 可裂变元素
fissile material 裂变物质；可裂变物质
fissile shale 易裂页岩；劈理状页岩
fissility 可劈性，可裂性，可裂变性；裂理；易裂变性，易裂性，易劈裂性，易分离性；劈度；岩石洞室；分裂性
fission 裂变，核裂变；裂殖；分裂；剥离
fission action 分裂；裂变
fission parameter 裂变参数
fission product 裂变产物，核分裂产物
fission ratio 裂变强度比
fission reaction 裂变反应
fission spectrum 裂变谱
fission track 裂变径迹
fission width 裂变宽度
fissionable 可裂变的；可裂的；可分裂的
fissionable analysis 裂变径迹分析
fissionable counting 裂变径迹计数
fissionable dating 裂变径迹测年法

fissionable material 可裂变物质；可分裂物质；可裂变材料；裂变材料
fissioning 裂变；分裂
fission-product debris 裂变碎片
fission-track age 裂变径迹年龄
fission-track dating 裂变径迹法断代；裂变径迹测年
fissle 岩层微裂声
fissura 裂隙
fissurae filling 裂隙充填物
fissuration 碎裂；龟裂；裂隙形成，形成裂隙；裂缝，裂隙；裂开
fissuration porosity 裂隙孔隙率
fissure 裂缝；裂隙；龟裂，坼裂；裂纹
fissure angle 裂隙角
fissure aperture 裂隙张开度
fissure aquifer 裂隙含水层
fissure artesian groundwater 裂隙承压水
fissure cave 裂隙洞
fissure cementation 缝隙灌浆；裂隙胶结
fissure closing 裂隙闭合
fissure coefficient 裂隙系数
fissure collapse structure 裂隙塌陷构造
fissure conduit 裂缝通路
fissure direction 裂隙方向
fissure displacement 裂隙位移；缝隙移；缝错
fissure drainage 裂隙排水
fissure eruption 裂缝喷发；裂缝喷溢；裂隙式喷发
fissure fault 裂缝断层
fissure filling 裂缝充填物；填塞裂缝或隙缝；裂缝填充，裂隙填充；开裂
fissure filling deposit 裂隙填充矿床；脉矿
fissure flow 裂隙流，裂隙水流；裂隙式熔岩流；裂隙喷溢熔岩流
fissure flow volcano 裂隙式火山
fissure groundwater 裂隙潜水
fissure grouting 裂缝灌浆
fissure in ground 地裂缝；土裂缝
fissure karst water 裂隙岩溶水
fissure length 裂隙长度
fissure medium 裂隙介质
fissure observation 裂缝观测
fissure occupation 裂隙填充；裂缝填实
fissure of displacement 位移裂缝；错缝
fissure of retreat 退缩冰隙；收缩裂缝
fissure percentage 裂隙率
fissure permeability 裂隙渗透性
fissure plane 裂隙面
fissure polygon 裂缝多边形
fissure porosity 裂隙空隙度；裂隙率
fissure ratio 裂隙率
fissure reservoir 裂隙型油气藏
fissure separation 裂缝间距
fissure spacing to span ratio 隙跨比
fissure spacing-span ratio 隙跨比
fissure spring 裂隙泉；裂缝泉
fissure system 裂缝系统；裂隙系统
fissure theory 裂隙迁移油气说
fissure vein 裂隙脉；裂缝脉
fissure volcano 裂隙式火山
fissure water 裂隙水

fissure water filling deposit	裂隙充水矿床
fissure water reduction factor	裂隙水折减系数
fissure water zone	裂隙含水带
fissure zone	裂隙带
fissure-cave system	缝洞系统
fissured	裂开的
fissured aquifer	裂隙含水层
fissured clay	裂缝化黏土；裂隙土；裂隙黏土；龟裂黏土
fissured ground	裂缝地层
fissured rock	裂隙岩石，裂缝岩石；裂隙性岩石；有裂缝的岩石破碎岩
fissured stratum	裂缝地层
fissured water	裂隙水
fissured zone	裂隙带
fissure-filling deposit	裂缝填充矿床
fissure-karst aquifer	裂隙-岩溶含水层
fissures closeness	裂隙密集度
fissures model	裂隙模型
fissures system	裂隙系
fissures-defined by spacing	以间距表示的裂隙
fissures-microscopic study	裂隙的显微镜研究
fissures-porosity	裂隙-孔隙度
fissure-type volcanic edifice	裂缝型火山堆积
fissure-water	裂隙水
fissuring	形成裂隙；裂隙；节理
fissus	裂开的
fit goodness	拟合优度
fitchering	钻头卡住
fitting of water-head field	水头场的拟合
fitzroyite	白榴钾镁煌斑岩
five-compartment shaft	五隔间立井
five-leg tile	五脚砖
fix a bar	安装梁；安装杆
fix a plate	限位板
fix position	定位；测定船位
fix with plugs	采用栓塞固定
fixation method	固定方法
fixation of shifting Sand	流沙的固定
fixed action	固定作用
fixed anchor	固定锚杆
fixed anchored contact line	固定下锚接触线
fixed arch	重力锤；固定拱
fixed arcing contact	固定弧触头
fixed ash	内在灰分；固定灰分
fixed at one end	一端固定
fixed at one end beam	一端固定梁
fixed base structure	固定基底结构
fixed beam	固定梁
fixed bearing	固定承座；固定支座
fixed bed model	定床模型
fixed bed model test	定床模型试验
fixed boundary	固定界面；固定边界
fixed boundary condition	固定边界条件
fixed carbon	固定碳
fixed carbon matter	固定碳物质
fixed casement window	固定窗
fixed cavitation	固定空穴
fixed channel	固定河槽
fixed chute	固定溜槽
fixed component	固定组分；固定组件
fixed composition ratio	固定配比
fixed cone cradle	固定锥式钻架
fixed constraint	固定约束
fixed control area	固定控制区
fixed dam	固定坝；封闭型隔墙
fixed datum	固定基准；固定基准线；已知数；固定标高
fixed drilling platform	固定式钻探平台
fixed dune	固定沙丘
fixed end	固定端；定端；固端
fixed end anchorage	固定端的锚定；固定端锚具
fixed end moment	固端弯矩；固端力矩
fixed end pile	底端固定的桩
fixed end restraint	固端约束
fixed form paving	固定模板铺筑
fixed framed arch	构架式固端拱
fixed frequency	固定频率
fixed gas	固定气体；不凝气
fixed ground water	固定地下水；非重力式地下水；约束地下水；束缚水；结合水
fixed guide	刚性罐道；固定罐道
fixed joint	刚性连接
fixed layer	束缚层；固定层
fixed load	固定荷载，固定载荷；定负载
fixed measuring point	固定测点
fixed moisture	固定水分；固定湿度；束缚水；结合水
fixed overflow dam	固定式溢流坝
fixed pan	固定溜槽
fixed peg	固定标桩
fixed phase	固定相位
fixed phase drift	固定相移
fixed pile	嵌固桩
fixed pile head	嵌固桩顶；嵌固桩头
fixed piston sampler	固定活塞式取样器
fixed point	观测点；固定点；不变点；定点；不动点
fixed point leveling	定点水准测量
fixed point observation	定点观测
fixed position	支撑点；固定点
fixed pump	固定泵
fixed pump installation	固定水泵装置
fixed reference line	固定参考线
fixed reference point	固定参考点
fixed reference system	固定参考系
fixed retaining wall	锚固挡土墙
fixed screen	固定筛
fixed section	固定断面
fixed seismometer	固定地震仪
fixed source field	定源场
fixed source method	定源法
fixed span	固端跨；固定跨
fixed staff	固定标尺
fixed sulfur	固定硫
fixed support	固定支点；固定支架；固定支承
fixed switch	固定道岔
fixed symmetrical arch	固定对称拱
fixed termination	硬锚
fixed trace	固定迹线；基线
fixed water	固着水
fixed wavelength filter	固定波长滤波器
fixed weir	固定式堰；固定堰
fixed zone	固定区

fixed-blade turbine	定桨式轮机
fixed-crest spillway	固定堰顶式溢洪道
fixed-crest weir	固定堰顶式堰
fixed-cut ploughing method	固定截深刨煤法
fixed-earth support	固定端支承；地下固定支承
fixed-end arch	固端拱；无铰拱；定端拱
fixed-end beam	固定端梁
fixed-end moment	固定端弯矩，固端弯矩
fixed-end piles	底端固定的桩
fixed-in place inclinometer	固定式倾角计；固定式测斜仪
fixed-line observation	定线观测
fixed-method	定源法
fixedness	固定；不变；确定；凝固
fixed-pin	固定销
fixed-ring consolidometer	固定环式固结仪
fixed-source field	定源场
fixing bolt	固定锚杆；固定螺栓
fixing brake	止动闸；制动闸
fixing device	固定装置
fixing dimension	装配尺寸；规定尺寸
fixing mortar	固定砂浆
fixing pin	固定销
fixing plate for busbar	汇流排固定压板；母线压板
fixing wire	固定铁丝
fixity	固定；不变；确定；凝固；嵌固；固定度
fixity depth	嵌固深度
fixture	夹具；紧固；固定物，装置器，工作夹具；固定装置，装置，附属装置，固定附着物；设备；支架
fjeld	冰蚀高地；冰蚀高原；冰被山原
fjell	冰蚀沼地；冰蚀沼泽
F-joint	F节理
fjord	海湾；峡湾
fjord coast	峡湾海岸
fjord strait	海峡湾
fjord type coast	峡湾型海岸带
flabellate	扇形的
flabelliform	扇形的
fladen	扁平状火山弹；扁平陨击岩
flag	板层，薄层，易劈的平薄石层；石板，板石；信号旗，旗标，标旗，旗帜；标记，标志，特征；薄层砂岩
flag ore	板状赤铁矿
flag stone	板石；石板
flagged drill hole cover	有标示旗的钻孔盖
flaggy	板层状；薄层状；薄板状；薄砂岩的；可劈成板的
flagman	测旗工；信号旗手；司旗员
flagpole	花杆；旗杆
flagstone	石板，板石；薄层细粒砂岩，薄层砂岩；板状砂岩，板层砂岩；可劈成石板的岩石；铺路石板
flajolotite	锑铁矿
flake	薄片；碎片，石片，岩石碎片；鳞片，鳞；大片滑坡体；成片剥落，片落；将剥落，剥落
flake gold	薄片金；片状金
flake graphite	片状石墨
flake mica	鳞片云母
flake off	成片剥落；离层；呈鳞片状剥落
flake stone	石板；扁石
flake stone tool	石片石器
flake structure	片状构造；鳞片构造
flake sulphur	片状硫
flake texture	鳞片结构；片状结构
flake to	使成薄片；成片状剥落；片落
flakes	片状砂页岩
flake-shaped particle	片状颗粒
flakiness	成片性；片层分裂
flakiness ratio	宽厚比
flaking	剥落；剥成鳞片；成片剥落；冒顶；表面片落；破损
flaky	片状；叶片状；薄片状；鳞片状的；片状的
flaky constituent	片状成分；网状构造
flaky grain	片状颗粒
flaky graphite	鳞片状石墨
flaky material	片状材料
flaky structure	鳞片状构造
flaky texture	片状结构；鳞片状结构
flamboyant structure	辉耀构造；火焰构造
flame assay	火焰试验
flame black	炭黑；焰黑
flame coal	焰煤；长焰煤；火焰煤
flame coloration test	焰色试验
flame cutting	火焰切割；气割；氧炔切割
flame drilling	火焰钻进；热力钻进；火钻钻进
flame jet drilling	火焰喷射钻井
flame lighten	点火器
flame like	焰状的
flame mining	热力开采
flame monitor	火焰监测器
flame photometric method	火焰光谱法
flame photometry	火焰光谱法
flame proof bush	隔爆衬套
flame resistance	防爆性；耐火性；防火性
flame shape	火焰状
flame structure	火焰状构造
flame test apparatus	火焰测定仪
flame throwing drill	热力钻机
flamed granite tile	烧面花岗岩砖
flameproof	隔爆的；防爆的；耐火的；隔火的
flame-proof construction	防爆结构；防火结构
flameproof motor	隔爆式电动机
flameproof testing	防爆试验；防火试验
flame-throwing drill	火力钻机；热力钻机
flamper	煤层中的泥铁岩
flamy structure	火焰构造
flandrian transgression	佛兰德海侵
flange	凸缘，突缘，法兰；翼缘；矿脉变向处；矿脉变厚；法兰盘；溪谷
flange beam	工字钢
flange buckling	翼缘屈曲
flange bush	凸缘套筒；法兰套筒
flange connection	凸缘接合；法兰连接
flange coupling	法兰联结；凸缘喉套
flange cover plate	梁翼盖板
flange joint	凸缘接头，凸缘接合；凸缘接驳位；法兰接头；法兰盘连接
flange of beam	梁翼缘
flange pipe	凸缘管；法兰管
flange plate	凸缘板；翼缘板
flange test	弯边试验；翻边试验
flanged beam	工字梁；加翼梁

flanged expansion joint 法兰盘；伸缩接头
flanged packing 凸缘密垫；凸缘衬垫
flanged shear wall 带翼缘抗剪墙
flanged spigot 法兰盘孔堵
flanged wheel 凸缘轮
flanger 弯缘机
flanging test 卷边试验
flank 侧；侧面，侧翼，帮；翼；主壳区
flank bore 探水侧孔；探瓦斯侧孔
flank collapse 侧翼崩塌
flank eruption 山侧喷发；侧翼喷发
flank hole 侧面钻孔，侧向钻孔；探水侧孔；探瓦斯侧孔；侧向炮眼；帮眼；侧向探眼
flank landslides 侧翼山体滑坡
flank moraine 侧碛石；侧碛
flank rift zones 侧翼裂谷区
flank separation zone 侧分带
flank wall 边墙；侧面墙；山墙
flanking 侧向开切
flanking deposit 周边沉积物
flanking-hole method 侧孔法
flap 褶翼，座盖；片状悬垂物；重力崩塌构造；卷滑构造；挡水板，铰链板，防护板；瓣
flap bottom 活底；下放底
flap door 风门；车底吊门
flap top 悬顶
flapper topped air crossing 双门式风桥；顶门式井下交叉风道
flare angle 张角；扩散角；圆锥角
flare opening 喇叭口
flare surge 耀斑激浪
flare wing wall abutment 八字形桥台
flared-head gyratory 头部呈喇叭口形的回旋破碎机
flare-wall 翼墙
flaring head 外张破碎头
flaring wall 外张墙；八字墙
flaser 薄层；条带；压扁状，压扁状的；条带状；泥波体
flaser bedding 压扁层理；脉状层理
flaser gneiss 压扁片麻岩；薄层片麻岩
flaser structure 压扁构造；伪流动构造
flaser texture 压扁结构
flash 闪光；地面下沉；暴涨；泄水；堰；快泄闸，快调闸板
flash coat 喷浆盖层
flash evaporation 急骤蒸发
flash flood 山洪暴发，山洪；暴发洪水，暴洪；突发水灾；突然涌水
flash flow 暴发洪水流
flash plate 闪熔镀层；锚链垫板；闪光板
flash point 闪点；燃点
flash set 速凝；瞬时凝结，瞬凝；急凝
flash setting 瞬时凝固
flash tank 冷凝槽；汽化箱；扩容器；闪蒸膨胀箱
flash test 高压闪络试验；闪点试验
flashing point 闪点；燃点
flashing shaft 洗井
flashing system 闪蒸系统
flash-set cement 瞬凝水泥
flash-setting agent 速凝剂；促凝剂
flashway 注水通道；泄水道；冲沙道

flashy load 瞬间荷载
flashy regime of river 河流的暴涨性
flashy stream 山洪暴发
flat 扁平的，扁平结构，缓倾斜的；水平层，近水平层，平地，平坦，平坦区，平坦的，平面，扇形，坪，断坪，浅滩，平滩，井下车场，单位；平底火车，平底船；适应性；隔离采区
flat angle 平角
flat arch 平拱
flat back cut and fill method 水平分层充填开采法
flat back method 上向梯段开采法
flat bank revetment 缓坡护岸
flat bed 水平层，水平煤层，平层，水平矿藏；平台，平底，平坦河底；平板车，平板运输车，运梁车，运梁平板车；绘图平台
flat bed plotter 平板绘图仪
flat bedding 水平层理
flat bit 一字形钎头；平式钻头
flat bog 平地沼泽，平地沼；低位沼泽，低沼泽
flat bottom 平坦河底
flat bottom ditch 平底沟
flat channel 平渠；扁槽钢
flat chisel bit 凿形钻头；一字形钻头
flat coal 平扁煤粒；平层；水平煤层
flat coast 低平海岸；低平湖岸；低海岸；平坦海岸
flat country 平原；低地；平坦地区
flat cut 向下钻孔爆破
flat deposit 缓平矿床；缓斜矿床
flat dilatometer 扁式松胀仪；扁铲侧胀仪；扁式旁压仪
flat dip 微倾斜；平倾斜
flat dipping bed 缓倾斜层
flat dolly 扁头铆顶
flat dome 扁圆顶
flat drill 平钻；扁钻；平冲头
flat drill steel 扁平形钻钢
flat face 平面
flat factor of safety 静力安全系数
flat fading 按比例衰减；平衰落
flat fault 缓倾断层；平缓断层
flat filter 宽频滤波器
flat flood plain 平坦河漫滩
flat floor of square sets 方框支架的上部水平分层；方框架的平肋板
flat grade 平缓坡度；缓和坡度
flat grade mine 缓倾斜层矿山
flat gradient 缓倾斜；平缓斜坡，缓坡；平缓坡度；平缓比降
flat gradient for starting 起动缓坡
flat ground 平地
flat head rivet 扁头铆钉；平头铆钉
flat hole 水平炮眼；平炮眼，水平孔
flat incline 缓坡
flat jack 扁千斤顶；压力枕；平板千斤顶；液压枕
flat jack method 扁千斤顶法
flat jack technique 液压枕法；扁千斤顶法
flat jack test 扁千斤顶试验；狭缝法
flat joining 平缝，平节理
flat joint 平缓节理，平节理；平缝
flat lying 水平产状的；平卧的；平伏的
flat mass 平伏矿床

flat measures	平缓层系；平伏层
flat muck pile	平面废石堆
flat or gently rolling terrain	平缓起伏地形
flat pick	扁截齿；平头镐
flat pile	平桩
flat pitch	缓倾斜场地；低螺距
flat plate	平板；方形托盘
flat plate foundation	平板基础
flat plate structure	平板结构
flat raise	缓倾斜上山
flat rammer	平捣锤；平夯
flat region	平原区
flat rolling	扁平孔型轧制
flat roof	平顶；平屋顶；平台
flat roof area	平台范围
flat seam	近水平煤层，近水平矿层；平矿层；平合缝
flat sheet	平面图；平板，平盘
flat slab	无梁板；平板
flat slab buttress dam	平板坝；平板式支墩坝
flat slab construction	无梁板建筑
flat slab dam	平板坝
flat slab deck	平板式挡水面板
flat slab floor	无梁楼盖
flat slab structure	无梁板结构
flat slope	缓坡；平坡；平坦的坡度
flat spot	平点
flat stope	水平回采工作面
flat straight coast	平直海岸
flat structural plane	平直结构面
flat terrain	平坦地带
flat topography	平坦地形
flat topped culvert	平顶涵洞；平顶管路
flat topped fold	箱状褶皱
flat topped weir	平顶堰
flat type piling bar	平板桩
flat valley	平谷；浅谷
flat valving surface	平配流面
flat vein	平脉
flat wall	下盘；底帮；平直钢锭模壁；平直模壁
flat washer	平垫圈
flat wave	平缓波；平顶波
flat web piling	平腹板桩
flat working	缓倾斜开采
flat-back cut-and-fill method	水平分层充填开采法
flat-back cut-and-fill system	上向水平分层充填采矿系统
flat-back method	上向水平分层开采法；上向梯段开采法
flat-back shrinkage	水平留矿法
flat-back square-set method	水平分层方框支架开采法
flat-back square-set system	上向水平分层方框支架采矿法
flat-back stope	水平上向梯段工作面；长壁上向梯段回采工作面
flat-back system	上向水平分层采矿法
flat-bed plotter	平板绘图仪
flat-bed scanner	平板扫描仪
flat-bedding	水平层理；平坦层理
flat-bottom bin	平底仓
flat-bottom ditch	平底沟
flat-bottom land	平地；平原
flat-bottomed	平底的
flat-bottomed flume	平底测流槽
flat-bottomed valley	平底谷
flat-crested measuring weir	平顶量水堰
flat-crested spillway	平顶溢洪道
flat-deck buttress dam	平板支墩坝
flat-dipping bed	缓倾斜层；平缓倾斜岩层
flat-face cutter head	平面刀盘
flat-faced bit	平端钻头；平端面钻头
flat-grade mine	缓倾斜层矿山
flatiron	平顶脊；熨斗形山
flatjack	扁千斤顶；油压枕
flatjack measurements of stress	扁千斤顶应力测量
flatland	平原；平地
flatland river section	平原河段
flat-layer approximation	平层近似
flat-laying lake sand	平伏湖砂
flatly inclined	缓倾斜的；近水平的
flatly inclined seam	微倾斜煤层
flatly-dipping joint	缓倾角节理
flat-lying	平缓的；平躺的；平伏的；平卧的；平伏
flat-lying bed	平伏层；缓倾层
flat-lying joint	平缓节理
flat-lying seam	近水平煤层；近水平矿层
flatness	扁平度，平整度；平滑性；均匀性
flatness ratio	扁平度；扁平比
flat-nose bit	平端面钻头
flat-pitching seam	缓倾斜煤层
flat-ripple mark	平顶波痕
flats	水平矿层；缓斜矿层；平脉
flat-scale	顶板下沉标尺
flattened section	已拉平的剖面；展开的剖面
flattening	变平缓，陡度减小；扁平率，扁率，地球扁率；压扁，弄平；压扁酌，压扁作用
flattening deformation	压扁变形
flattening factor for the earth	地球扁率
flattening of ellipsoid	椭球扁率
flattening of the earth	地球扁平率
flattening plane	压扁面
flatter	平巷挂车工；平面锤
flat-to-flat survey	逐户勘测
flat-top junction	平顶接头
flat-topped	平顶的
flat-topped crest	平顶脊
flat-topped fold	平顶褶皱；平顶褶曲
flat-topped hill	平顶山
flat-topped pulse	方脉冲
flat-topped ridge	平顶山脊
flat-topped seamount	海底平顶山；桌状山；盖约特
flat-topping	削平
flaw	裂痕；裂纹，裂缝，断裂，破裂；瑕疵，缺陷，缺点，毛病；走滑断层，平移断层，捩断层；横推断层，平推断层
flaw detector	裂缝检查器；探伤器，探伤仪
flaw fault	平移断层
flaw size	裂缝尺寸
flawy	破裂的；裂隙的；有裂缝的；有缺点的
flaxseed coal	亚麻子级煤；细粒无烟煤
flaxseed ore	红色扁豆形晶体铁矿
fleckschiefer	斑点板岩
flection	弯曲
flection structure	弯曲构造

fleet angle 偏角
Flemish bond 荷兰式砌合法；梅花砌砖法
Flemish brick 铺面黄色硬砖
flerry 劈裂
flerry fracture 劈裂
fletz 层系中的异质夹层
flex 弯曲，褶曲；挠曲
flex cracking 挠裂
flex joint 挠性连接
flexboard 挠性板
flexed 弯曲的
flexed beam 弯曲梁
flexed foliation 弯曲叶理
flex-hone 软磨石
flexibility 可曲性，易弯性；挠性，弹性，韧性，可缩性，伸缩性；柔性，揉曲性，柔韧，柔顺性，灵活性，可变性，适应性；柔度；熔性，熔度
flexibility coefficient 挠度系数；柔度系数
flexibility factor 柔性因数；挠曲系数
flexibility first story construction 柔性底层结构
flexibility influence coefficient 柔性影响系数
flexibility matrix 挠度矩阵；柔度矩阵
flexibility method 柔度法
flexibility of fracture 断裂韧性
flexibility ratio 挠度比
flexibility tester 挠性试验器
flexibilizer 增塑剂；增韧剂；软化剂
flexible 可曲的，可挠曲的，易弯的，可弯曲的，易弯曲的；挠性的，柔性的，柔韧的，柔韧性的，能变形的；柔软的；可塑造的；灵活的；适应性强的
flexible arch bridge 柔性拱桥
flexible arches 柔性拱
flexible base 柔性基础；柔性基层
flexible base of hydraulic chock 液压垛式支架的可变形底座
flexible bearing 挠性轴承
flexible bend 柔性弯头；铰弯头
flexible block 柔性垫块
flexible bodies 柔性体
flexible conduit 柔性管道，柔性导管；软管；软导管
flexible connection 挠性连接，柔性连接；挠性联轴节，软连接，软联结；弹性接头
flexible connector 软接头
flexible constraint 柔性约束
flexible contact 柔性接触
flexible core wall 柔性心墙
flexible cross-span 软横跨
flexible dam 柔性坝
flexible drill 可伸缩钻机
flexible drill rod 柔性钻杆
flexible drill stem 柔性钻杆
flexible duct 柔性导管；软管
flexible ductility 软导管
flexible dummy bar 可挠引锭杆
flexible facing 柔性面层
flexible facing joint 柔性护面接缝
flexible fender 柔性冲垫；柔性挡板
flexible fender cushion 柔性防冲衬垫
flexible first storey 柔性底层
flexible floor 柔性楼盖

flexible footing 柔性基础；柔性基座
flexible foundation 柔性基础；柔性地基
flexible frame 弹性构架；柔性构架；柔性框架
flexible grilled dam 柔性格栅坝
flexible grout pipe 灌浆软管
flexible hose 胶皮管；挠性管；软管
flexible joint 弹性接缝，柔性接缝；挠性连接，柔性连接；柔性接头，挠性接头，软接头，柔性节点；万向接头，万向联轴节；柔性短节
flexible layer 柔性层
flexible lining 可缩性支架；柔性支护
flexible load 柔性荷载
flexible loading plane 柔性加荷面
flexible member 柔性构件
flexible membrane liner 柔性薄膜衬垫
flexible mortar 柔性砂浆
flexible mud hose 泥浆胶管
flexible pavement 柔性路面；沥青路面；柔性铺面
flexible pier 柔性桥墩；柔性墩
flexible pipe 软管；柔性管；蛇管；挠性管
flexible protection 柔性防护
flexible retaining structure 柔性支挡结构；柔性挡土结构
flexible retaining theory 柔性拦挡理论
flexible retaining wall 柔性挡土墙
flexible ring 柔性圆环
flexible rock 可塑性岩石；柔性岩石；挠性岩石
flexible rod 挠性钻杆
flexible roof 挠性顶板
flexible rule 软尺；卷尺
flexible ruler 卷尺；皮尺
flexible seal 柔性止水；活动封条
flexible shaft 挠性轴；软轴
flexible shafting 挠性轴系
flexible shield 柔性掩护支架；柔性防护罩
flexible shield support 柔性掩护支架
flexible shielding method 柔性掩护支架采煤法
flexible spout 挠性槽
flexible structural precautions 柔性结构预防措施
flexible structure 柔性结构
flexible structure measure 柔性结构措施
flexible support 柔性支架；柔性支承；可缩性支架
flexible tank 柔性罐
flexible trestle 柔性支架
flexible tube 柔性管；软管
flexible unit 通用装置；可挠部分
flexible ventilation ductility 通风软管
flexible ventilation line 柔性通风管线
flexible volume method 柔性体积法
flexible wall 柔性墙；柔性壁
flexible wire rope 柔性钢丝绳
flexibly supported end 柔性支承端
flexibly supported system 柔性支承系统
fleximeter 挠度计；曲率计
flexing 挠曲；弯曲
flexing action 挠曲作用；弯折作用
flexing endurance 耐挠曲，抗弯性
flexing test 挠曲试验
flexion 弯曲；屈曲；曲率；拐度
flexion of surface 曲面的拐度

flexion strain	弯曲应变
flexion test	挠曲试验；弯曲试验
flexion torsion	弯曲扭转
flexirope	弗莱克西钢丝绳锚杆
flexishaft	软轴；挠性轴
flexivity	挠度
flexometer	挠度计；挠度器；曲率计；挠曲计
flexotir	笼中爆炸，震源装置
flexural	弯曲的；受弯压曲；挠曲的
flexural beam	弯曲梁；挠曲梁
flexural buckling	弯曲失稳
flexural building	受弯建筑物
flexural capacity	抗弯承载能力
flexural centre	弯曲中心
flexural compliance	弯曲柔度
flexural constant	挠曲常数
flexural cracking	挠曲破裂
flexural damping	弯曲阻尼
flexural deformation	弯曲变形；挠曲变形
flexural failure	弯曲破坏
flexural fault	挠曲断层
flexural flow	挠曲流动
flexural fold	挠曲褶皱；弯曲褶皱
flexural glide	弯曲滑移
flexural loading	挠曲载荷；挠曲荷载
flexural member	受弯构件
flexural meter	挠度仪
flexural mode	弯曲模态
flexural mode of vibration	弯曲振型
flexural modulus	弯曲模量；弯曲系数
flexural rigidity	挠曲刚度；抗挠刚度；弯曲强度；弯曲刚度；抗弯刚度
flexural rigidity of a plate	岩板的抗弯刚度屈曲
flexural rigidity of section	截面弯曲刚度
flexural slip	挠曲滑动；弯曲滑动，弯滑；层面滑动
flexural slip folding	挠曲滑动褶皱作用
flexural stiffness	抗挠刚度；弯曲刚度
flexural stiffness of member	构件抗弯刚度
flexural strain	挠曲应变；弯曲应变
flexural strength	抗挠强度；抗弯强度，弯曲强度
flexural stress	挠曲应力；弯曲应力
flexural tensile strength	弯拉强度
flexural test	反复弯曲试验；挠曲试验
flexural toppling	挠曲式倾倒
flexural toppling failure	弯曲倾覆滑落
flexural wave	弯曲波
flexural yielding	弯曲屈服
flexural-slip fold	曲滑褶皱
flexural-torsional instability	弯曲-扭转失稳
flexure	挠曲，单斜挠曲，单斜绕褶；弯曲；屈曲，弯曲部分，打褶；受弯力；揉曲；褶皱
flexure block	挠曲断块
flexure centre	弯曲中心
flexure coast	单褶海岸；弯曲海岸；挠曲海岸
flexure fault	挠曲断层
flexure graben	弯曲地堑
flexure hinge	柔性铰
flexure line	挠曲线
flexure member	柔性构件
flexure metre	挠曲计
flexure moment	挠矩；变矩；弯矩
flexure plane	挠曲面；弯曲面
flexure reinforcement	受弯钢筋
flexure scarp	弯曲悬崖
flexure strain	挠曲应变
flexure strength	抗挠强度；抗弯强度
flexure stress	挠曲应力；弯曲应力
flexure test	挠曲试验
flexure-flow fold	弯流褶皱
flexure-shear interaction	弯剪相互作用
flexure-slip fold	层面滑动褶皱；曲滑褶皱
flexure-slip folding	曲滑；层面滑动
flexure-torsion vibration	弯-扭振动
flexure-wave	弯曲波
flight auger	螺旋钻；链动螺钻
flight auger drill	多级螺旋钻机；螺旋钻机
flight auger drilling	连续螺旋钻进
flight sewer	跌落式排水管道
fling	滑冲；突然的移动；突发的滑动；投掷
fling-step	滑冲幅度；滑冲留下的永久性位移
flinkite	褐水砷锰矿
flint	燧石；火石
flint aggregate	坚硬的骨料
flint brick	燧石砖；坚硬砖
flint clay	硬质黏土；燧石状黏土，燧石黏土，燧土
flint crush rock	超糜棱岩
flint nodule	燧石结核
flint stone	燧石；打火石
flinty	坚硬的，强硬的；燧石状；硅质的；燧石的
flinty crush rock	燧石状碎裂岩
flinty fracture	贝壳状断口
flinty ground	含燧石地带；砂砾土壤；燧石质土
flinty slate	燧石板岩
flip angle	翻转角度
flip bucket	挑流鼻坎；跳流鼻坎
flirting post	临时顶柱
flit	搬移；转场；拆迁和安装输送机；滚筒采煤机调头
flitch beam	合板梁
flitch dam	木板坝
flitch girder	组合板大梁
flitching	巷道加宽
flitch-plate girder	贴板梁；钢木组合梁
float and sink analysis	浮沉试验；浮沉分析
float and sink bath	浮沉试验筒
float and sink density	浮沉密度
float and sink drain rack	浮沉物泄水架；浮沉式排水栅
float and sink process	重悬浮液分离过程；重介法；浮沉法
float and sink sampling	浮沉试验取样
float and sink specific gravity	浮沉比重
float and sink test	浮沉试验
float barograph	浮子气压计
float bridge	浮桥
float coal	浮煤；精煤
float fill	水力充填
float fraction	浮物
float gauge	浮子水位计，浮标水位计；浮子水尺；浮筒式水位计；浮式水标尺；浮子验潮仪；浮标，浮规
float gauging	浮标测流法
float meter	浮子式流量计；浮尺

float ore　漂移矿石
float rod　浮标杆；浮杆
float run　浮标行程；浮标测距
float sample　浮煤样
float sand　浮砂；流砂
float stone　浮石；漂石；磨石
float tape　浮尺
float test　浮标试验
float tester　浮子测定器
float type flow meter　浮子式流量计
float type pressure gage　浮子式压力计
float well　浮尺井
floatability　可浮性；可浮选性；浮动性
floatable　可浮的
floatage　浮力；漂浮物；火车轮渡费
float-and-sink　浮沉
float-and-sink method　浮沉试验法
float-and-sink sample　浮沉试样；选煤试样
float-and-sink sampler　浮沉试验取样
float-and-sink test　浮沉试验
floatation　浮动；浮力；漂浮；浮选；浮游选
floatation centre　浮体水面形心
floatation depth　浮体吃水深度
floatation stability　浮体稳定性
floatative　浮游的；浮起的
floated concrete　流态混凝土；抹面混凝土
floater　脱散块；转石；浮体，流体，漂浮物浮标，浮顶油罐，浮球；浮式装置，浮式生产系统，浮箱，浮运沉箱，浮式设备；抹灰工具；临时工
floater continents　漂浮的大陆
floater oil-production-equipment barge　浮式石油钻井平台
floater sand　流砂
floating　提浆抹光；漂浮的；浮动的，浮的，浮游的，浮起的；流动的；不固定的；抹平
floating anchor　浮锚
floating auxiliary datum plane　浮动辅助基准面
floating baseline　浮动基线
floating bearing　浮动轴承；回转轴承
floating boundary condition　浮动边界条件
floating box foundation　浮箱基础
floating breakwater　浮式防波堤
floating bridge　浮桥
floating buffer　浮动缓冲区
floating caisson　浮式沉箱
floating caisson foundation　浮式沉井基础
floating charge　上浮炸药包；浮动押记；浮充电
floating control　无定向调节；无定位调节；无静差控制；浮动控制
floating cover　漂浮顶盖；浮顶；浮动盖板
floating crane　浮吊；水上起重机；浮式起重机
floating dam　浮坝；浮闸
floating datum　浮动基准面
floating drilling barge　钻探浮船
floating drilling platform　浮式钻井平台
floating drilling rig　浮动式钻探装置
floating drilling vessel　浮式钻井装置；浮式钻井船
floating driver　水上打桩机
floating driving plant　水上打桩设备
floating earth　浮土；流砂；流土；浮动接地
floating embankment　浮堤

floating foundation　浮筏基础；浮式基础，浮式地基；浮基，浮基础；补偿基础
floating gauge　浮标尺
floating ground　浮动接地
floating harbour　浮动港
floating head　浮头；浮盖
floating in casing　下部装回压阀的套管柱
floating jetty　浮式突码头
floating landing stage　浮动平台；浮码头；浮动栈桥
floating marsh　沿岸沼泽
floating offshore drilling rig　浮式海洋钻井装置
floating oil production　浮式采油
floating packing　活动密封垫
floating partition　活动隔墙
floating passenger landing stage　人行浮式码头；人行浮动栈桥
floating peat　漂浮植物泥炭；腐殖质泥炭
floating pier　浮墩；浮码头，浮动码头；浮式栈桥，栈桥
floating pile　摩擦桩；悬浮桩
floating pile driver　浮承桩打桩机；浮式打桩机；水上打桩机打桩船
floating pile foundation　摩擦桩基础；悬浮桩基；浮桩基础
floating piston　自动活塞；浮动活塞
floating platform　浮式平台；浮台
floating post　移动支柱
floating production　浮式采油；浮式生产
floating production facilities　浮式采油设施
floating production platform　浮式采油平台
floating production system　浮式生产系统
floating rig　浮式钻机
floating sand　流砂
floating sludge　浮泥
floating structure　浮式结构
floating the casing　浮法下套管
floating third axle　浮动第三桥
floating tip　移动式翻笼
floating-in of casing　下端有逆止阀的套管
floating-ring shaft　浮环套管
float-losing and load-increasing effect　失托加荷作用；浮力消失造成的加荷效果
float-ring consolidometer　浮环式固结仪
float-stone　多孔石英，浮石英；浮石，多孔蛋白石；磨砖石；悬粒灰岩
float-type flow metre　浮筒式流量计
floc　絮状物；絮凝物，絮凝体；使絮凝；溢流槽；网筛；蓬松物质
floc formation　絮凝物形成
floc point　絮凝点
floc test　絮凝试验
floc unit　絮凝体
flocculability　絮凝性
flocculant settling　絮凝沉淀
flocculate　絮凝；絮凝物
flocculated　絮凝的
flocculated clay　絮凝黏土
flocculated clay buttress bond　絮状支托黏结
flocculated sludge　絮状残渣
flocculated soil　蜂窝土

flocculated state 絮凝状态
flocculated structure 絮凝结构
flocculating 絮凝
flocculating colloid 絮凝胶体
flocculating constituent 絮凝成分；絮凝体；绒毛状组织
flocculating effect 絮凝效应；絮凝作用
flocculating settling 絮凝沉降
flocculation 絮凝；凝聚作用；絮化；絮凝物
flocculation effect 絮凝效应
flocculation energy 絮凝能；凝聚能
flocculation factor 絮凝系数
flocculation for tailings 尾煤的凝聚
flocculation test 絮凝试验
floccule 絮凝物；絮凝沉淀物；絮凝粒；绒毛状沉淀
flocculence 絮凝状
flocculent 絮凝的；羊毛状的
flocculent deposit 絮凝沉积
flocculent soil 絮凝土
flocculent structure 絮凝结构，絮凝状结构
flock 絮状物；絮状体；絮凝体；絮团；短纤维；绒屑；群
flockenerz 砷铅矿
floe 浮冰块；大片浮冰
floe belt 浮冰带
floe berg 浮冰堆
floe ice 漂冰；浮冰
floeberg 冰山
floetz 水平平行岩层
flogopite 金云母
floitite 帘片状岩
flokite 发光沸石；丝光沸石
floocan 脉壁黏土；脉节理黏土
flood 泛滥；洪水，水灾；涨潮，淹没；溢流，溢出，洪流；注水；富矿体
flood ability 可注水性
flood absorption capacity 蓄洪能力
flood amplitude 洪水波幅；洪水幅度
flood and ebb 潮汐
flood and waterlog disaster 洪涝灾害
flood area 浸水地
flood axis 洪水轴线；洪水流向
flood bank 防洪堤，挡洪堤；洪水河岸
flood barrier 拦洪坝
flood basalt 溢流玄武岩；高原玄武岩
flood basin 河漫盆地；洪泛盆地，洪泛区，洪泛面积，洪水面积，洪积盆地，泛洪区；集水盆地；汇水盆地；泛滥地，泛滥盆地
flood bed 河漫滩；洪水泛滥地；泛滥地
flood benefit 防洪收益
flood breadth of river 洪水河面宽度
flood bridge 洪水桥
flood by-pass 分洪道
flood calamity 水灾
flood carrying capacity 泄洪能力；排洪能力
flood catastrophe 水灾
flood channel 洪水河槽；行洪河道；涨潮水道；疏洪渠
flood coefficient 洪水系数
flood control barrage 防洪闸
flood control basin 调洪洼地
flood control channel 防洪渠；调洪河槽
flood control dam 防洪坝

flood control facility 防洪设施
flood control level (FCL) 防洪水位
flood control map 防洪地图
flood control project 防洪工程
flood control reservoir 调洪水库
flood control standard 防洪标准
flood control structure 防洪建筑物
flood control works 防洪建筑物；防洪工程
flood crest stage 洪峰水位
flood current 洪流；涨潮流
flood dam 防洪坝；蓄洪坝
flood damage 水毁；洪水灾害，水害，水灾；淹没损失
flood deposit 洪积物；泛滥沉积物，洪泛沉积物
flood deposition 洪水沉积
flood detention 滞洪；拦洪
flood detention dam 拦洪坝；滞洪坝
flood detention reservoir 滞洪水库
flood detention storage 滞洪库容
flood development 注水开发
flood dike 防洪堤
flood directionality 注水方向；洪水方向性
flood disaster 洪灾，洪水灾害；水灾
flood discharge 洪水量；泄洪；洪水流量；洪水泄流
flood discharge forecast 洪水流量预报
flood discharge level 洪水位
flood ditch 防洪沟
flood diversion 分洪
flood diversion area 分洪区
flood diversion channel 分洪渠
flood diversion gate 分洪闸门
flood diversion project 分洪工程
flood duration 洪水历时
flood effectiveness 注水效果
flood efficiency 注水效率
flood elevation 洪水位
flood embankment 防洪堤
flood erosion 洪水侵蚀；洪水冲刷
flood estimation 洪水估算
flood event 洪水事件
flood fall 退洪
flood flanking 堤防
flood flat marsh 河漫滩沼泽
flood flow 洪流；洪水流量
flood flow formula 洪流计算公式
flood forecast service 洪水预报机构
flood forecast system 洪水预报系统
flood forecasting 洪水预报
flood formula 洪水计算公式
flood frequency 洪水频率
flood front 洪水前缘
flood front advance 注水前缘推进
flood gate 水闸；防洪闸，防洪闸门，泄洪闸；泄水平巷；节水闸门；溢水门；溢洪道
flood geometry 注水井网
flood hazard 水灾；洪水险情
flood height 洪水位
flood intensity 洪水强度
flood interval 涨潮间隙；洪水间隙
flood inundation period 洪水淹没期
flood investigation 洪水调查

flood land　漫滩；河滩地；河漫滩，漫滩河槽；洪泛地，洪泛平原；泛滥区，洪水泛滥区；淹没地
flood level　洪水位；溢流水位
flood level mark　洪水痕迹
flood life　注水期限
flood line　洪水线；高潮线
flood mark　洪水痕迹；洪水标记；高潮标记；洪水线
flood mitigation　减洪
flood mobility ratio　注水流度比
flood monitoring　水浸监测
flood operation　流体注入作业
flood out　水淹
flood out flow　洪水下卸流量
flood out pattern　水淹形态
flood pattern　注采井网
flood pattern geometry　注采井网几何形态
flood pattern modification　注采井网调整
flood peak　洪峰
flood peak forecast　洪峰预报
flood peak rate　洪峰流量
flood peak reduction　洪峰消减
flood peak runoff　洪峰流量
flood peak stage　洪峰水位
flood performance　洪水动态
flood period　汛期；洪水期
flood periphery　洪水边缘
flood plain　漫滩，河漫滩，漫滩河槽，漫滩地，河滩地；泛滥平原，洪泛平原，洪积平原；滩地；洪泛区；河槽
flood plain bench　漫滩阶地
flood plain deposit　泛滥平原沉积；洪泛平原淤积土；河漫滩沉积
flood plain discharge　洪泛平原流量
flood plain lobe　舌瓣状河漫滩；洪泛蛇曲带
flood plain marsh　洪泛平原沼地
flood plain regulation　洪泛平原整治；河漫滩整治
flood plain scroll　迂回扇
flood plain sediment　河漫滩沉积物
flood plain splay　漫滩冲积扇
flood plain splays sediments　决口扇沉积物
flood plain terrace　泛滥平原阶地；漫滩阶地；洪泛平原阶地
flood plain type　漫滩式
flood plain zoning　洪泛区分区
flood plane　洪水水面
flood pool　防洪水库；蓄洪塘
flood potential　洪水势
flood prediction　洪水预报；洪水预测
flood prevention　防水，防洪；防汛工作
flood prevention measure　防洪措施
flood probability　洪水机率
flood profile　洪水历时过程曲线
flood proofing　防洪
flood protection　防水；防洪；防汛
flood protection during construction　施工期渡汛
flood protection embankment　防洪堤
flood protection scheme　防洪计划
flood protection wall　防洪墙
flood protection works　防洪工程
flood protective embankment　防洪堤

flood recession　洪水退落；退洪
flood record　洪水记录；洪水实测资料
flood regime　洪水情况；水情
flood region　洪水淹没范围
flood region reduction　蓄滞洪区
flood regulating level　调洪水位
flood regulation　洪水调节；调洪
flood regulation capacity　调洪能力
flood retarding project　拦河工程；防洪工程
flood retention　滞洪；阻洪；抑洪；拦洪
flood rise　涨洪；洪水上涨
flood risk regionalism　洪水危险区划
flood routing　洪水推测；洪水分布；洪水演算
flood routing process　洪水演算法
flood routing through reservoir　水库调洪演算
flood runoff　洪水径流；洪水流量
flood scroll　蛇曲带沉积
flood season　洪水季；洪水季节；洪水期；汛期
flood section　洪水断面
flood slack　高平潮；涨平潮
flood spillway　溢洪道
flood spreading　洪水散布
flood stage　洪水位；洪水期
flood stage forecast　洪水水位预报
flood storage　蓄洪；防洪库容，调洪库容
flood storage capacity　防洪排涝的能力；防洪库容；调节库容
flood storage pond　蓄洪池
flood storage work　蓄洪工程
flood stream　洪水河流；涨潮流
flood strength　最大涨潮流速
flood subsidence　洪水消退
flood subsidence rate　落洪率
flood suction　自流注水
flood surcharge　洪水超高
flood surge　洪水涌浪；涨水涌浪
flood survey　洪水调查
flood synchronization　洪水同步
flood synthesis　洪水组合
flood tidal delta　涨潮三角洲；潮汐三角洲
flood tide　涨潮
flood valve　溢流阀
flood velocity　洪水流速；洪流速度
flood volume　洪水量，洪量；洪水总量；洪水体积
flood wall　防洪堤；防洪墙；挡水墙
flood warning　洪水警报
flood warning service　洪水报警机关；防汛机构
flood water　注入水；洪水；汛水
flood water level　洪水位
flood water overflow dike　洪水漫堤
flood wave　洪水波；洪水浪
flood wave recession　洪水波消退
flood wave subsidence　洪水波消退
flood wave transformation　洪水波转换
flood wave velocity　洪水波速度
flood way　分洪道，分洪河道；泄洪道
flood years　洪涝年份；洪涝时期
flood zone　泛滥带；洪峰带
floodability　可注水性
floodable net sand volume　可水驱的砂层有效体积

floodable pay　可注水的产层
floodable pore　可注入的孔隙
floodable reservoir　可注水开发油藏
flood-bank　防洪堤
floodcontrol　防洪
flood-control capacity　防洪库容
flood-control engineering　防洪工程
flood-control project　防洪工程
flood-control reservoir　防洪水库；调洪水库
flood-control storage　防洪库容；防洪蓄水
flood-control works　防洪工程；防洪设施
flood-crest travel　洪峰行径；峰顶行进
flood-depth　洪水深度
flood-depth of river　河流的洪水深度
flood-discharge level　洪水位
flood-discharge tunnel　泄洪隧道
flooded area　洪泛面积，淹没面积，受淹面积；水淹区域，泛滥区，洪泛区
flooded connection　涌进连接
flooded ice　覆水冰
flooded land　水淹地区；水泛地
flooded mine　淹没的矿山；矿井淹没
flooded region　水淹区
flooded shaft　淹没矿井
flooded strata　水淹地层
floodgate　排洪闸门
flooding　洪水泛滥，泛滥；淹没，水浸，灌水，淹水，漫灌，注水，漫滩，溢流，涌水，发水；水驱，拥入，扩散；水灾
flooding alarm　水淹告警
flooding duration　洪泛期，溢流期
flooding efficiency　注水效率
flooding irrigation　洪水灌溉；浸灌；漫灌
flooding irrigation method　漫灌法
flooding land　河漫滩；河滩地
flooding level　洪水位
flooding limit　淹没界限
flooding line　洪水线
flooding method　淹没法；漫灌法；洪水法
flooding network　注采井网
flooding of mine　矿区水灾
flooding pattern　注水井网；水驱井网；洪水型式
flooding pipe　溢流管
flooding point　溢流点，液泛点
flooding rate　注水速度
flooding recurrence probability　洪水重现概率
flooding stage　注水驱替阶段
flooding unit　注采单元
flooding velocity　溢流速度；注水速度
flooding well network　注采井网
floodland　漫滩河槽；漫滩地
floodometer　水量记录计；潮洪水位测量仪；洪水计；涨潮水位计；水量计
floodout　水淹
flood-peak stage　洪峰水位
floodplain　泛洪区；泛洪平原；漫滩；泛滥平原；滩区
flood-plain bench　漫滩阶地
floodplain deposite　泛滥平原沉积，洪泛平原沉积；垂直加积沉积物；漫滩堆积；河漫滩沉积
flood-plain facies　洪泛平原相

flood-plain lobe　洪泛河曲
flood-plain regulation　河滩整治；泛滥平原整治
flood-plain scroll　蛇曲带沉积；洪泛漩涡带
floodplain sedimentation　河漫滩沉积
floodplain soft soil　漫滩软土
flood-plain splay　漫滩冲积扇
flood-plain terrace　漫滩阶地；洪泛阶地
flood-pot experiment　注水试验
flood-pot sample　注水试验样品
flood-pot-test　注水试验
flood-prone area　易受水浸影响的地区
floodproofing　防洪；防汛抢险
flood-protection measures　防洪措施
flood-protection works　防洪工程
flood-regulating structure　调洪建筑物
flood-response time　注水见效时间
flood-retention basin　滞洪区
flood-scroll　蛇曲带沉积
flood-source area　成洪地区
flood-stage forecast　洪水位预报
flood-storage project　蓄洪工程；预报径流
flood-storage works　蓄洪工事
flood-tide current　涨潮流
floodwall　防洪石堤；防洪岸壁
floodwater　洪水
floodwater pumping scheme　洪泛抽水系统；洪水抽排体系
floodway　分洪道，分洪河道；泄洪道，泄洪渠，排洪道；排水渠
floodway channel　分洪道；分洪河槽；泄洪渠
floor　底，海底，谷底，底板，底盘，巷道底；海床，河床；基岩，地基，岩盘，井下板台，楼面，层面，平面，地面，桥面，楼盖，楼层，地层，楼面层，阶段，水平，水平矿体；地板，地坪，垫板，铺板；最低额，底价
floor arch　底板拱
floor area　楼面面积；地板面积；占地面积
floor area of a room　房间面积
floor area of building　房屋面积；建筑楼层面积；建筑面积
floor bar　底梁；枕木；横梁
floor beam　楼面横梁，楼面梁；横梁
floor board　底板，地板；巷道底部铺板；假顶；楼面板
floor bolt　底板锚杆
floor bolting　底板锚杆支护
floor boundary line　底部境界线
floor breaks　底板破坏，底板裂隙，底板断裂；地板裂隙；底板破裂
floor buckling　底鼓
floor burst　底板瓦斯突出
floor casting　楼板浇筑
floor coal　煤层底部煤，底板煤
floor construction　楼面构造
floor crack　底板裂隙
floor cut　截底槽
floor digging　底板挖掘；起底
floor dinting　挖底；卧底；起底
floor drain　地面排水渠；地漏
floor drain outlet　地面排水口；地台去水
floor dust　底板尘末；地槽
floor elevation　底板高程

floor finish	面层
floor heave	底板隆起；底鼓
floor heave of roadway	巷道底鼓
floor hole	底部炮眼
floor hump	底板隆起；底板隆起
floor jack	底板支柱；托底千斤顶；卧式千斤顶
floor joist	钻台托梁；楼板搁栅
floor lay	底层
floor level	楼面，楼盖水平面；楼面标高，底板高程，楼面高度，地面高度；地台水平，基底水平，拉底水平，楼面水平
floor lift	底板隆起，底面隆起；巷道底部隆起；地板抬升
floor limb	下翼
floor live load	楼面活载
floor live loading	楼面活荷载
floor load	楼面荷载
floor load intensity	底板载荷集度；底板比压
floor loading capability	箱底承载能力
floor lowering	落底
floor margin	下极限；底板下缘
floor of coal seam	煤层底板
floor of laccolith	岩盖底盘；岩盘底板
floor of trench	沟底
floor opening	楼板洞
floor outburst	底板瓦斯突出
floor pack	充填底板；伪底；底板分块
floor penetration	底板穿透
floor pillar	底柱；阶段间的煤柱；阶段间的矿柱
floor plan	楼面平面图；楼宇平面图
floor plan layout	平面布置
floor plate	底垫；铺板
floor platform	楼面
floor plug spad	巷道底板测站钉
floor pocket	地面框
floor pressure	底板压力
floor pressure arch	底板压力拱
floor projection	水平投影
floor propping	支护底；支护底板
floor response spectrum	楼面反应谱；地面响应谱
floor rock	底板岩石
floor sampler	底板取样
floor sampling	底板取样
floor shrinkage stope	下盘留矿回采工作面
floor sill	地梁；底梁，钻台底梁
floor slab	地板；楼板；底板垫板；二层楼板标高
floor spring	门脚铰链；磨地铰；地弹簧
floor station	底板测点
floor strata	底板岩层
floor strength	底板承压力
floor strength against injection	底板抗压入强度
floor stringer	辅助梁
floor thrust	底板逆冲断层；底冲断层
floor tile	地砖；地台瓦；地面砖
floor undulation	底板起伏不平；地板起伏不平；底板起伏
floor wall system	楼板墙体系
floor water inrush	底板突水
floor water invasion	地板水侵
floor water irruption	底板突水
floor-bar	底梁；枕木；底座；横梁
floor-beam	横梁
floored intrusion	具底板的侵入体；岩盘
floor-engaging structure	接底框架
floor-framing system	楼板框架体系
floor-heave	底鼓；底板隆起
floor-lift	底鼓；底板隆起
floor-on-grade	地坪
floorplan	平面布置图
floor-plate	底板
floorstand	立架
floor-type	落地式的
floran-tin	嵌在岩石中的难见锡矿石；捣碎细锡矿
florencite	磷铝铈矿；磷铝铈石
Florida clay	漂白土
Florida earth	漂白土
floridin	漂白土
florinite	云沸煌岩
floss	河流；丝棉
flour gold	粉金；最细砂金矿
flour sand	粉砂
floury soil	粉土；细砂土；粉砂
flow	流动，水流，流水，流，径流，河流，塑变；潮流，流纹，涨潮；流量，水量；塑性变形；沼泽，低地
flow amount	流量
flow analogue	渗流的模拟
flow analysis log	流量分析测井
flow and plunge structure	流状倾伏构造
flow area	过水面积
flow balancing reservoir	流量均衡水库
flow banding	流状条带；流纹带
flow barrier	流动屏障
flow bean choke	节流嘴
flow behavior	流动特性
flow bog	浮沉泥炭沼泽；沼泽泥炭
flow breccia	流状角砾岩
flow by gravity	自流；重力流动
flow by heads	间歇自流；自喷
flow by port	旁流孔
flow capacity	排水能力；流量；流通能力
flow capacity of control valve	调节阀流通能力
flow capacity of water meter	水表的流通能力
flow cascade	梯级跌水；跌水
flow cast	流动铸型
flow cell	岩心夹持筒
flow channel	流槽；流道；水道；水渠，流经渠道；流程；工作顺序
flow character	流动特性；流性
flow characteristic of control valve	调节阀流量特性
flow characteristics	水流特征；流河特征；流动特性
flow characteristics coefficient	水流特性系数
flow chart	流程图；程序框图，程序方框；洪水图；洪量图
flow charting	流程图；程序框图
flow cleavage	流劈理；板状劈理
flow coating	流涂；浇涂；淋涂
flow coefficient	流动系数；量系数
flow column	油管柱
flow commutator	测流管
flow component	径流成分；水流成分
flow concentration	汇流；集流

flow concentration curve 汇流曲线
flow condition 流态；水流情况
flow conductivity 导流能力
flow cone 窦锥；流动锥
flow constant 流动常数
flow continuity 渗流中的连续性；水流连续性
flow control 流量控制；进料控制；流量调节；水流控制
flow convergence 汇流
flow cross-section 液流横断面；过水断面；流截面
flow cross-sectional area 过流面积
flow curve 流动曲线；流量曲线；塑流曲线，流变曲线
flow deficiency 流量不足
flow depth 水流深度；水深
flow diagram 流程图；操作程序图；流量图；流线图；作业图；流程框图
flow diffuser 流量扩散器
flow direction 流向；水流方向
flow direction measurement 流向测量
flow discontinuity 渗流的不连续性
flow distortion 水流变形；水流畸变
flow distribution 水流分布；流量分布；流量分配
flow domain 流域
flow duration 流量历时
flow duration curve 水流量历时曲线；流量历时曲线
flow duration diagram 流量历时图
flow dynamics 流体动力学
flow effect 流动效应
flow efficiency 流动系数；完善系数
flow equation 渗流方程；流动方程；流量方程
flow event 渗流过程
flow failure 流动破坏；流动滑坡
flow field 流场
flow filaments 流束；流线
flow flexure fold 流动挠曲褶皱
flow fluctuation 水流脉动；水流起伏
flow fold 流状褶皱；塑流褶皱
flow foliation 流叶理
flow form 流态
flow friction 水流摩阻；水流阻力；流动摩擦；流动摩阻
flow friction characteristics 流动摩阻特性
flow function 流动函数；流函数
flow gain 流量增益
flow gas 气流
flow gauge 流量计
flow gauging 流量测定
flow gauging weir 测流堰
flow gradient 流动梯度
flow gravel pack 循环法砾石充填
flow impairment 流动能力降低
flow improver 流动改进剂
flow in 流入
flow in open air 明流
flow in vortex 涡流
flow index 流动指数
flow initiation 流动的产生；水流起始
flow injection analysis 流动注射分析
flow injection system 流动注射系统
flow instability 流动不稳定性；水流不稳定性
flow intensity 水流强度
flow intensity parameter 流动强度参数

flow interactions 流动相互作用
flow into 流入；涌进
flow irregularity 水流不规则性；立流不均匀性；水流不均匀性
flow joint 流线节理；流层节理
flow lane 流槽
flow law 流动定律
flow layer 流动层；流层
flow layering 流动分层
flow limit 流限；液限
flow limitation 流动限制
flow line 流送管，出水管；流线，流程线，流纹，径流线，吕德斯线；蠕变线；蠕变图；输油管，出油管线，采气管线；流水作业线
flow line method 动线法
flow magnitude 水流强度
flow mark 流纹；流痕
flow mass curve 累积流量曲线；水文积分曲线；径流积分曲线；量累积曲线
flow matrix 地层中渗流孔道系统
flow measurement 测流；流量测定；流量测量
flow measurement during freezing 冰期测流
flow mechanics 流体动力学；流体力学
flow mechanism 流动机理；流动机制
flow meter 流量表；流量计；流速计
flow model 流动模型；水流模型
flow momentum 水流动量
flow monitoring 流量监测；水流监测
flow movement 水流运动
flow net 流网；水网；流网；渗流网
flow nonuniformity 水流不均匀性
flow of debris 碎屑流
flow of fluids 流体流动
flow of fluids-porous materials 流动的液体-多孔材料
flow of ground water 地下水流
flow of ore 矿石流
flow of rock mass 岩体流动
flow of rocks 岩石流动；地层移动
flow of solid matter 泥沙径流
flow of solids 固体流
flow of thickened medium 浓缩介质流
flow of water 涌水量；水流
flow of water in fissured rock 裂隙岩石中水的流动
flow off 径流；流出，流走；溢出
flow out 流出
flow out diagram 出流图
flow over 溢出
flow pack 循环充填
flow parameter 流动参数
flow partly full 半满流
flow passage 通流部分；液流通道
flow path 水流轨迹，流线；流动通道，流路；流动路程；流径；渗径
flow pattern 流动型，水流模式；流态，水流流态；流谱；流型
flow performance 流动特征
flow period 自喷期
flow phenomenon 流现象，水流现象；流动现象；涌流现象；水流现象；蠕变现象；塑流现象
flow pipe 水管；送水管；压力水管；输水管；排出管

flow plane	流面；水流面，水流平面
flow plastic strain	流动塑性应变
flow plug	流通塞；自喷油嘴；阻流器
flow point	流动点；屈服点；流点；倾点
flow potential	流动位能；流动势；最大流量
flow pressure	动水压力；水流压力
flow pressure response	流体压力反应
flow production	自喷开采；流水作业；连续生产；流水生产
flow production period	自喷开采期
flow profile	流量剖面；水流曲线；水流纵断面；回水曲线
flow property	流动性；流动性能；水流特性
flow proportional sample	按流量比例取样
flow quantity	流量
flow range	流量范围
flow rate	流量率；水流速度，水流速率；流量；流速；流率
flow rate coefficient	流速系数
flow rate log	流量测井
flow rate meter	流速计
flow rate profile	流量剖面
flow rating curve	流量率定曲线
flow ratio	流量比；流比
flow ratio method	流量比法
flow reactor	连续反应器；流动反应器
flow record	流量记录
flow recorder	流量计；流量记录仪；流量表；流量记录器
flow recovery ratio	流量恢复系数
flow regime	流动区域；流态，流动状态，水流动态；流动体制；水流情势，水流情况；水情
flow resistance	水流阻力；流动阻力，流阻
flow reversal	水流反向；逆流
flow rock	流岩；流砂；流石
flow roll	流卷层；流卷构造；流卷体
flow rule	流动法则
flow scheme	流程图
flow section	过水断面
flow shape factor	流形系数
flow sheet	流程表，流程图；流程
flow simulation	流动模拟
flow slide	泥流型滑坡，流滑；塑流型滑坡，流动滑块；流动滑坡泥流
flow sliding	流滑
flow speed	流动速度
flow spread	水流宽度；水面宽度
flow stability	流动稳定性；水流稳定性
flow state	水流状态；流动状态
flow stowing	水砂充填；水力充填；自溜充填
flow stream pressure	液流压力
flow stream velocity	液流速度
flow streamline	流线；流线
flow strength	流动强度
flow stress	流变应力；屈服应力；塑性变形应力；流动应力
flow string	自喷管柱；采油管；流水管柱
flow structure	流动构造；流状构造；流型结构；塑变组织；怜结构
flow superposition	渗流叠加
flow surface	过水面；流动面
flow synchronization	径流同步
flow system	流动系统；水流系统
flow tank	计量罐；沉降罐；油田储罐
flow temperature	液流动温度
flow test	流量测试；涌水试验；自流试验；流动试验；水流试验
flow test network	测流管网
flow tester	流量计；混凝土流动试验器；油层产量测定器
flow texture	流状结构
flow through	流过
flow through period	水流流动时间
flow through porous media	多孔介质流
flow till	运动冰碛；水流浮碛；表流碛；流碛
flow tube	流管
flow turbulence	水流紊动
flow uniformity	水流均匀性
flow uniqueness	水流唯一性
flow value	流值；流量值；流动值
flow vector	流向量；流矢量
flow velocity	流速；水流速度；渗流速度
flow velocity distribution	流速分布
flow velocity measurement	流速测定
flow velocity meter	流速仪
flow velocity valve	流速阀
flow with water table	潜水流；非承压流
flow zone	水流区；流动带
flowability	流动性；怜性
flowability coefficient	流动系数
flowability of concrete	混凝土的流动性
flow-after-flow test	流量调整试井
flowage	泛滥；流出，流变，流动，；柔流；积水，积涝，淹没，怜
flowage cast	流动底模；流动铸型
flowage differentiation	流动分异
flowage fold	流状褶皱；流动褶皱
flowage friction	流动摩擦
flowage land	洪泛地；泛滥地
flowage line	流线
flowage structure	流动构造；流状构造
flow-and-plunge structure	流状倾伏构造
flowback	返排
flowback water	返排水
flow-carrying capacity	流动携带能力，流动携砂能力
flow-chart	流程图；生产过程图解；程序方框图
flowchart symbol	流程图符号；程序图符号
flow-current	涨潮流
flow-drawing	流动拉伸
flowed sample	产出砂样
flower arm	华拱
flower bed	花台；花坛；花圃
flower hedge	花篱
flower of iron	铁华；霰石华；文石华
flower of sulfur	硫华；硫磺花
flower structure	花状构造
flower-like lava	翻花熔岩
flow-induced stress	流动次生应力
flowing	自流；自喷；流动；平滑；排液；喷涌
flowing artesian	自流承压水
flowing artesian water	自流承压水

flowing artesian well 自喷井；自流井
flowing avalanche 流动崩坝；流动崩坍
flowing bottom hole pressure 井底流动压力
flowing buging head pressure 自喷井井口压力
flowing casing pressure 开井套压
flowing coal 煤流
flowing deformation 流动性变形
flowing front 流动前缘
flowing gas 自喷天然气
flowing gas-oil ratio 自喷气-油比
flowing ground 流动性地基；流动地基；流动地层；湿流土
flowing landslide 流动滑坡
flowing life 自喷期
flowing line 放出管；引出管
flowing medium 流动介质
flowing of well 油井的自喷；井喷
flowing oil 自喷原油
flowing oil production 自喷采油
flowing phase 流动相；自喷阶段
flowing porosity 流动孔隙度；流动孔隙率
flowing pressure 流动压力，动水压力；流压；自喷压力
flowing pressure gradient 流压梯度；流动压力梯度
flowing producer 自喷井
flowing production 自喷采油；自喷开采；自喷产量
flowing property 流动性能
flowing proportion 流相比例
flowing rock formation 塑性岩层
flowing sand 流砂
flowing sheet water 地表径流
flowing slope 溜土坡
flowing spring 自流泉
flowing temperature factor 流动温度系数
flowing test 流动测试
flowing tide 涨潮
flowing tubing head pressure 自喷井井口压力
flowing tubing pressure 自喷油压
flowing water 活水；动水；流水
flowing water-oil ratio 自喷水油比
flowing well 自流井；自喷井
flowing yield 涌水量；溢流量
flow-line pressure 流送管内的压力
flowmeter 流量表；流量计；流速计；测流计
flowmeter logging 流量计测井；流量测井
flowmeter survey 流量计测量
flownet 流网
flow-net analysis 流网分析
flow-off 径流
flow-pressure character 流量-压力特性
flow-pressure diagram 流量-压力图
flowrate 流量；流率；流速
flow-resistance factor 水流阻力系数
flowstone 流石
flowstream sample 出油管流体样品
flow-table test 稠度台试验
flow-test line 测流管线
flowthrough 径流；径流量；流动；流过
flow-through centrifuge 无逆流离心机
flowtill 流冰碛物；表流碛
flow-type gravel packing 循环式砾石充填

flucan 脉壁黏土；脉节理黏土
fluctuate 波动；涨落
fluctuating 脉动
fluctuating flow 波动流；脉动流动；脉动水流；波动水流
fluctuating load 波动荷载，波动载荷；不稳定荷载；变化荷载；脉动荷载
fluctuating nappe 波动水舌
fluctuating pressure 脉动压力
fluctuating stress 波动应力；变化应力；交变应力；脉动应力
fluctuating water pressure 波动水压力
fluctuating water table 波动水位
fluctuation 波动，变动，脉动；升降；涨落；起伏；振幅
fluctuation amplitude 波动幅度
fluctuation belt 波动带
fluctuation belt of water table 地下水位变动带
fluctuation cycle of groundwater regime 地下水动态变化周期
fluctuation of climate 天气变动
fluctuation of glacier 冰川变动
fluctuation of medium 介质起伏
fluctuation of piezometric surface 测压水位变化
fluctuation of pressure 压力波动
fluctuation of sea level 海平面变动
fluctuation of static level 静水位变化
fluctuation range 波动范围
fluctuation ratio 波动率；波动比；变动率
fluctuation texture 波状结构
fluctuation velocity 脉动速度
fluctuation zone 变动区
fluctuation zone of groundwater level 地下水位变动区
fluctuations in discharge 流量变化
fluent 水流；流束
fluent lava 流动熔岩
fluent material 液态物质
fluffy mud 充气泥浆
fluid 流体，液体，流质；流体的；流动的，液体的；液态的；易变的；液
fluid analysis 流体分析
fluid at rest 静止流体
fluid bank 流体汇集带
fluid bearing 流体轴承
fluid bed 流化床层；流化床
fluid body 流动体；流体
fluid breakthrough 流体突破
fluid capacity 冲洗液量
fluid cement grout 水泥灌浆
fluid chamber 液腔
fluid circulation 冲洗液循环
fluid coefficient 流体系数
fluid column 液柱；流体柱
fluid column pressure 液柱压力
fluid compass 液体罗盘
fluid compressibility 流体压缩系数
fluid conductivity 水力传导系数；流体渗流强度
fluid conductivity of well 井底流过能力
fluid contact 流体接触面；流体接触
fluid course 流体流道
fluid crustal layer 流壳层；韧性流层
fluid cutting 流体侵蚀

fluid cutting damage	液流冲蚀破坏
fluid delivery	流体输出
fluid density	流体密度
fluid density difference	流体密度差
fluid density log	流体密度测井
fluid displacement	流体驱替
fluid drag force	流体阻力
fluid draw-texturing unit	流体拉伸变形装置
fluid drive	液压传动；流体传动；液力耦合器；液力传动
fluid driving	流体传动
fluid dune	流动沙丘
fluid dynamic theory	流体动力学
fluid dynamics	流体动力学；流体力学
fluid efficiency	压裂液效率
fluid element	流体单元；流体微分体
fluid erosion	流体冲蚀
fluid flow	水流；流体流动，流体运动；液流；流量；移动沙丘
fluid flow mechanics	流体流动力学
fluid flow simulation	流体流动模拟
fluid flow through fractures	裂隙介质流体流动
fluid fracturing	液力碎裂；流体压裂
fluid friction	流体摩擦；流体阻力
fluid front	流体前缘
fluid genic	流体成因的
fluid gradient tool	流体梯度测量仪
fluid head	流体压头
fluid head loss	液体压头损失
fluid impedance	流体阻抗
fluid inclusion	液包体；液体包体；流体夹杂；流体包裹体；流体进入
fluid inflow	液体流入
fluid injection	液体注射；流体注入；注液
fluid interface	流体界面
fluid interface log	流体界面测井
fluid kinematics	流体运动学
fluid kinetic	流体动力学
fluid layer	流体层
fluid level	液面
fluid load	流体负载；流体荷载
fluid loss	失水量；冲洗液漏失；流体损耗；失水
fluid loss coefficient	滤失系数
fluid mass	流体质量
fluid mechanics	流体力学；水力学
fluid medium	流体介质
fluid migration	流体运移
fluid mobility	流体流动性
fluid motion	流体运动
fluid mud	淤泥；浮泥；流泥
fluid over-pressure	流体超压
fluid over-pressuring	流体超压
fluid pack the wellbore	用液灌满井筒
fluid particle	流体质点
fluid penetration	液体渗透
fluid percolator	过滤器
fluid permeability	流体渗透率
fluid phase	液相
fluid pore pressure	流体孔隙压力
fluid potential	流体势；流体位能；水位差
fluid pound	液面撞击
fluid power	液力；液压
fluid power system	液压驱动系统
fluid pressure	流体压力；流体静压
fluid pressure-induced faulting	流体压力诱发的断层作用
fluid producing layer	产液层
fluid production force	供液能量
fluid productivity index	采液指数
fluid property	流体性质
fluid pumpability	流体可泵性
fluid redistribution	流体再分布
fluid resistance	流体阻力
fluid sampler	流体采样器
fluid saturation	流体饱和度
fluid seal	液封；油封
fluid sealing	液封
fluid state	流态
fluid static pressure	静水压力
fluid statics	流体静力学；静液压力
fluid strata	流体层
fluid stream	液流
fluid stress	流体应力
fluid structure interaction	流体-结构相互作用
fluid test	流体试验
fluid texture	流状结构
fluid trap	流体圈闭
fluid twisting	流体扭转
fluid underrunning	液体潜流
fluid velocity	流体速度；冲洗液流速；流动速度；液流速度
fluid viscosity	流体黏度
fluid volume	流体体积；孔内循环液量；冲洗液量；泵量；排量
fluid wave	流体波
fluid wedge	流楔
fluid withdrawal	流体抽取；流体流出；流体流量
fluidable particle size	可流化的颗粒尺寸
fluidal	流态的；流体的
fluidal geodynamic movement	流体地质动力运动
fluidal structure	流状构造；流纹构造；流动构造
fluidal texture	流动结构；流状结构；流纹构造
fluid-bearing aquifer	含水层
fluid-bearing inclusion	气液包裹体；流体包裹体
fluid-bearing infiltration	流体渗入
fluid-dynamics	流体动力学
fluid-elastic vibration	流体-弹性振动
fluid-escape structure	流体逸出构造
fluid-filled fracture	含流体的裂缝；含水断裂
fluid-filled pore space	含流体的孔隙空间
fluid-flow	流体流
fluid-fluid contact	流体-流体接触面
fluid-fluid interface	流体-流体界面
fluid-geochemistry	流体地球化学
fluidic	射流的；流体的
fluidics	射流技术；流控技术；应用流体学；流体学
fluidification	液化
fluidify	流化；流态化
fluidifying	流化；流化
fluidimeter	流度计；黏度计
fluid-index	流动性指数；流度指数；流体指数
fluid-induced vibration	流体诱发振动；流致振动

fluidisation	流态化；流化
fluidise	流化
fluidised bed	流砂地层；流砂层
fluidised tuff	流化凝灰岩
fluidity	流动性；流度；流性，液流性；流体状态；液性
fluidity index	流动性指数；流度指数
fluidity of grout	灌浆液流动性
fluidity point	流化点；流动点
fluidization	液体化作用，流体化；流态化；扩流作用；流质化；液化
fluidization quality	流化质量
fluidize	流化
fluidized bed	流砂地层；流化床
fluidized sediment flow	液化沉积物流；液状沉积流
fluidizing point	流化点
fluid-mechanics	流体力学
fluid-medium	流体介质
fluidmeter	黏度计
fluid-mineral equilibria	流体-矿物均衡
fluidometer	流度计
fluid-overpressure	流体超压
fluid-permeability	液体渗透率
fluid-potential	流体势；水位差
fluid-pressure	流体压力
fluid-resistivity log	流体电阻率测井
fluid-retaining structures	挡液体结构物；阻流结构
fluidrive	液压传动
fluid-rock boundary	流体-岩石界面
fluid-rock interaction	流体/岩石相互作用；流体岩石相互作用；水流-岩体相互作用
fluid-rock ratio	流体岩石比
fluid-saturation	流体饱和度
fluid-sensitive shales	液敏页岩
fluid-solid	流化固体
fluid-solid coupling	流-固耦合
fluid-solid coupling method	流固耦合方法
fluid-solid interaction	流体-固体相互作用
Fluid-solid interfaces	流-固界面
fluidstatic pressure	流体静压力
fluid-structure interaction	流体-结构相互作用
fluid-suppressing effect	流体缓冲作用；阻流作用
fluid-velocity survey	流速测量
fluiodeltaic tongue	冲积三角洲舌
fluke anchor	爪锚
fluke like drainage pattern	倒钩状排水系
flume	峡流；流水槽，水槽，引水水槽，运输，渡槽，测流槽，涧，斜流槽，水闸槽；尾矿；水滑道；沟，沟底川，滑动沟，流路；中性滨线
flume crossing	渡槽交叉口；跨河渡槽
flume experiment	水槽试验
flume with both side contractions	两侧束狭测流槽
flume with side contraction	侧向测流槽
flumping well	双层分采井
fluoborite	氟硼镁石
fluocerine	氟碳铈矿
fluocerite	氟铈矿
fluodensitometry	荧光密度测定法
fluoform	氟仿
fluolite	松脂岩
fluor	荧石；氟石；氟；荧光
fluor spar	荧石；氟石
fluorapatite	氟磷灰石；磷灰石
fluorate	氟化；氟酸盐
fluoration	氟化
fluoremetry	荧光测定；荧光测定法
fluorescence	荧光
fluorescence analysis	荧光分析
fluorescence chromatogram	荧光色谱
fluorescence microscope	荧光显微镜
fluorescence microwave double resonance	荧光微波双共振法
fluorescence spectroscopy	荧光光谱学
fluorescence spectrum	荧光光谱
fluorescence thin layer	荧光薄层
fluorescent	荧光的；发荧光的；有荧光性的
fluorescent bitumen	荧光沥青
fluorescent logging	荧光录井
fluorescent microscope	荧光显微镜
fluorescent mineral	荧光矿物
fluorescent scanning technique	荧光扫描技术
fluorite	荧石；氟石
fluorologging	荧光测井
fluosolid	流化固体
fluosolid bed	流态化固体物床层
fluosolids system	流化固体焙烧法
fluposition	河流沉积
flurosion	河流侵蚀；河流搬运侵蚀
flush	潮湿地，潮湿的，泛滥的，注满的；冲洗，激流，冲水，冲沙，泄水，冲刷，奔流，涌，水流，奔流；暴涨，泛滥，淹没；出渣，发红，变红；湿法充填，水砂充填；齐平面，齐平的；嵌平；填缝
flush away	冲去
flush boring	水冲钻
flush coupled casing	平接式套管
flush coupling	锥形接头；齐平连接
flush curb	齐平路边石；平埋路缘；平缘石
flush drill	水冲钻机
flush drilling	环陵井
flush flow	泄流
flush fluid	冲洗液；洗井液；替置液
flush gallery	冲砂廊道
flush gate	冲砂闸门；冲淤闸门；冲洗闸门
flush head	通水接头；平接头
flush joint	平头接合；平缝；平式接头
flush joint casing	平接式套管
flush off	排渣
flush out	析出；彻底冲洗；浸出；冲掉，冲走
flush perforation	失败的射孔
flush phase	自喷期
flush stage	自喷期；喷涌期
flush system	冲水系统
flush weir	泄水堰；溢水堰；泄沙堰；潜堰
flush weld	削平补强焊缝；齐平焊缝；平焊
flush with ground	齐地面；放于地面
flush-coupled casing	等径联接节套管；平齐联接节套管
flushed pool	被冲刷的油藏；受冲刷的油藏
flushed production period	水洗采油期
flushed sand volume	砂岩冲洗带体积
flushed zone	洗油带；渗入带；冲洗带
flushed zone water saturation	冲洗带含水饱和度
flushed-out trap	受冲刷圈闭

flush-filled joint 平填接合；平灰缝
flushing 冲洗，注水；湿洗充填，水砂充填；洗孔，冲淤，清洗；油井注水驱油；出渣；填缝
flushing arrangement 挡矸装置
flushing culm 水力充填用废渣
flushing effect 冲洗作用；冲洗效应
flushing efficiency 驱油效率
flushing fluid 冲洗液；洗井液
flushing gully 冲洗槽
flushing gutter 冲砂沟
flushing hole 冲洗孔
flushing medium 冲洗介质；冲洗液
flushing method 水力冲洗开采法；冲洗法；水力充填开采法
flushing out of core 取岩心
flushing pipe 水力充填管；冲洗管
flushing port 冲洗孔；进水孔
flushing shield 掩护梁
flushing sluice 排砂闸；冲砂闸
flushing system 水砂充填法；水力充填法；水力充填系统
flushing test 冲水试验；通水试验
flushing tunnel 冲砂隧洞
flushing water 冲洗水
flush-joint casing 平齐接头套管；内平接套管
flush-jointed 平接的；灰缝的
flute 槽，凹槽；舌形沟槽构造；冲沟，沟槽，槽沟
flute mark 凹槽痕迹；槽痕
flute or rib on sheathing 墙板压筋
fluted 有槽纹的；有凹槽的；槽形的
fluted column 凹槽柱
fluted core 外表有螺旋凹槽的岩心
fluted coupling 花键式接头
fluted drill 直槽扩孔钻
fluted drill collar 带槽钻铤
fluted grief stem 带槽方钻杆
fluted hill 冰槽丘
fluted moraine 槽脊冰碛
fluted moraine surface 槽脊冰碛面
fluted shaft 槽轴
fluted sheet 波纹钢板
fluted sub 带槽接头
fluted swedge 带槽的胀管器
fluted till 槽脊冰碛
fluted twist drill 麻花钻
fluting 成槽作用，刻槽作用；槽痕风化；沟槽，凹槽，槽，开槽，切槽，凿槽；柱槽
flutter 震动，振动；翼；颤振，颤动；脉冲干扰；图像跳动
flutter failure 颤振破坏
fluvial 冲积的；河成的，河的，河床的；河流的，流的；冲积物
fluvial abrasion 水流冲蚀；河流磨蚀；河流冲刷磨蚀
fluvial action 河流作用
fluvial aggradation 河流淤积
fluvial basin 河流流域
fluvial bog 滩地沼泽；河边低地
fluvial channel 河槽；河道
fluvial cycle of denudation 流水剥蚀旋回
fluvial cycle of erosion 河流侵蚀循环；河蚀旋回；河流侵蚀旋回

fluvial denudation 河流冲刷作用；河流剥蚀作用，河流剥蚀；流水剥蚀
fluvial deposit 河流沉积；冲积物；冲积型矿床
fluvial erosion 河流侵蚀，河蚀，河流冲蚀；河流冲刷；流水侵蚀
fluvial facies 河流相；河成相；河相
fluvial fan 冲积扇
fluvial flood-plain 河漫滩
fluvial geological process 河流地质作用
fluvial geomorphology 河流地貌；河霖貌学
fluvial hydraulics 河流水力学
fluvial incision 河流切口
fluvial karst 流水岩溶
fluvial lake 河成湖
fluvial landform 河成地貌；流水地貌
fluvial loam 冲积壤土
fluvial loess 冲积黄土
fluvial morphology 河川地貌学；河川形态学；河川地形学
fluvial outwash 河流冲刷；河流冲积
fluvial plain 冲积平原；河成平原
fluvial process 成河过程，河成过程；造床过程；河床演变，河道演变
fluvial sand 河砂
fluvial sediment 流水沉积物；河流沉积物；河流沉积
fluvial sedimentation 河流沉积
fluvial sedimentology 河流沉积
fluvial soil 河流积土
fluvial terrace 河成阶地；河流阶地；冲积阶地
fluvial transport 河流搬运
fluvial type 河流类型
fluvial-deltaic 河成三角洲的
fluvial-dominated delta 河控三角洲
fluvial-environment 河流环境
fluvial-tidal cycles 河流潮汐周期
fluvial-wave interaction 河流-波浪相互作用
fluviatile 河流的；河成的；冲积的；河川活动
fluviatile action 河流作用
fluviatile bed 河积层
fluviatile dam lake 河流堰塞湖
fluviatile delta 冰水三角洲
fluviatile deposit 冰水沉积；河流沉积
fluviatile drift 冰水漂积；冰水沉积
fluviatile facies 河相；河流相
fluviatile flood-plain 河漫滩
fluviatile gravel 河砾石
fluviatile lake 河成湖；活水湖；河中湖；冰水湖
fluviatile loam 冲积垆姆
fluviatile sediment 河流沉积
fluviatile terrace 冰水阶地
fluviatile transport 河流搬运
fluviation 河流作用；河成作用；河川作用
fluvio-aeolian interaction 河湖风沙相互作用
fluvioclastic rock 河成碎屑岩
fluviodeltaic 河成三角洲的
fluviodeltaic complex 河成三角洲复合体
fluvioeolian deposit 河风沉积；河风沉积物
fluvioglacial 冰水的；河冰的；水成因的；冰水形成的；冰水作用的
fluvioglacial accumulation 冰水堆积

fluvio-glacial deposit 冰水沉积
fluvioglacial drift 冰水漂积；冰水漂砾；冰水漂积
fluvioglacial drift stone 冰水漂砾
fluvioglacial gravel 冰水砾石
fluvioglacial stream 冰川水河流
fluvioglacial terrace 冰水阶地
fluviokarst 河成喀斯特
fluviolacustrine 湖成因的
fluviology 水道学；河川学；河流学
fluviomarine 河海沉积；河海的
fluviomarine deposit 河海沉积
fluviomarine sediment 河海沉积
fluviomorphology 河相学；河流形态学
fluvioterrestrial 陆地河流的，陆上河流的；陆水的
fluvio-terrestrial deposit 河陆堆积
fluvisol 冲积土
flux and reflux 涨落潮
flux boundary condition 流量边界条件
flux-gate magnetometer 饱和铁芯式地磁仪；磁通门磁变仪；饱和式磁力仪
fluxie 滑坡浊积岩；滑塌浊积岩
fluxion banding 流状条带
fluxion gneiss 流状片麻岩
fluxion structure 流动构造；流状构造；挠曲构造
fluxion texture 流纹结构
fluxoturbidite 滑坡浊积岩；滑塌浊积岩
fly drill 手拉钻
flyash 粉煤灰
flyash cement grout 粉煤灰水泥灌浆
flyash concrete 粉煤灰混凝土
flyash Portland cement 粉煤灰硅酸盐水泥
flyash stabilized soil 粉煤灰加固土
fly-in rig 可折卸空运的小钻机
flying ant 断钻杆打捞钩；钻杆打捞钩
flying arch method 先拱后墙法；比利时法
flying bridge 架空小桥
flying buttress 拱式支墩；拱拱垛
flying buttress arch 拱扶垛
flying leveling 初步测平；预先测平
flying reef 不连续矿脉；断续矿脉
flying scaffold 悬空脚手架；悬挂式脚手架
flying shore 无立柱支撑；横撑；飞撑
flying spot scanner 飞点扫描器
flyover 立体交叉；跨线桥；高架公路；天桥；行车天桥；飞机编队低空飞行
flyover crossing 立体交叉
fly-over crossing 高架桥
flyover junction 立体交叉
flyover roundabout 跨线桥
fly-past 跨度
flyrock 飞散石块；爆破飞石
flysch 复理石；复理层，复层理；浊积岩；厚砂页岩夹层
flysch deposit 复理石沉积
flysch facies 复理石相
flysch formation 复理式建造
flysch nappe 复理式推复体；复理石推覆体
flysch sandstone 复理石砂岩
flysch wedge 复理石岩楔
flyschoid 类复理石；类复理层；准复理石
flyschoid formation 类复理石建造；准复理石建造

flysh 复理层
foam block 泡沫砌块
foam concrete 泡沫混凝土
foam concrete block 泡沫混凝土块
foam dam 泡沫挡板
foam drilling 泡沫钻井
foam drilling fluid 泡沫钻井液
foam expansion ratio 泡沫膨胀比
foam feed 泡沫进给
foam fracturing 泡沫压裂
foam fracturing fluid 泡沫压裂液
foam gravel packing 泡沫砾石充填
foam injection 泡沫灌注
foam injection ratio 泡沫注入比
foam line 破波线；碎波线
foam liquid 泡沫液
foam nozzle 泡沫喷嘴
foam texture 泡沫结构
foamed fracturing fluid 泡沫压裂液
foamed mortar 泡沫灰浆；泡沫砂浆
foamed mud 泡沫泥浆
foamed pearlite 泡沫珍珠岩
foamed-slag concrete 泡沫熔渣混凝土；泡沫矿渣混凝土
foaming 起泡作用
foamy structure 泡沫构造；多泡构造
focal 震源的；焦点的
focal aperture 聚焦孔径
focal area 聚焦面；震源区
focal coordinates 震源坐标
focal depth 震源深度
focal dimension 震源尺度；震源大小
focal distance 震源距；焦距
focal force 震源力
focal hypocentral volume 震源体积
focal length 震源距；焦距
focal mechanism 震源机制；震源机理
focal point 焦点；震源点
focal point of subsidence 沉陷中心点；沉陷盆地中心；采动影响重心
focal point of working 开挖面中心
focal process 震源过程
focal radius 焦半径
focal region 聚焦区；震源区
focal sphere 震源范围；震源球
focal volume 震源体积
focal zone 震源区
focometer 焦距计
focus 震源；焦点；震焦；定焦点，调焦，聚焦；震中，中心，中心点，集中
focus depth 震源深度；聚焦深度
focus of earthquake 地震震源
focused current device 聚焦电流测井仪
focused dipmeter 聚焦地层倾角仪
focused image hologram 聚焦像全息图；聚焦像全息相片
focused induction log 聚焦感应测井
focused log 聚焦测井
focused log zero 聚焦测井零线
focused logging 聚焦测井
focused logging device 聚焦测井仪
focused microresistivity log 聚焦微电阻率测井

focused resistivity log	聚焦电阻率测井
fodichnia	觅食迹
fodinichnia	觅食构造
fodinichnion	觅食迹
fog	雾，雾点；影像模糊；徐变曲线
fog carbinet corrosion test	盐雾室腐蚀试验
fog cured	岩石出露面；喷雾养护的
foid	似长石；副长石
foid diorite	副长石闪长岩
foid dioritoid	副长石闪长岩类
foid gabbro	副长石辉长岩
foid gabbroid	副长石辉长岩类
foid monzodiorite	副长石二长闪长岩
foid monzogabbro	副长石二长辉长岩
foid monzosyenite	副长石二长正长岩
foid plagisyenite	副长石斜长正长岩
foid syenite	副长石正长岩
foid syenitoid	副长石正长岩类
foid-bearing	含副长石的
foid-bearing diorite	含副长石闪长岩
foid-bearing gabbro	含副长石辉长岩
foid-bearing latite	含副长石安粗岩
foid-bearing monzodiorite	含副长石二长闪长岩
foid-bearing monzogabbro	含副长石二长辉长岩
foid-bearing monzonite	含副长石二长岩
foid-bearing syenite	含副长石正长岩
foid-bearing trachyte	含副长石粗面岩
foidite	副长岩；似长石；副长石岩
foiditoid	副长石岩类
foidolite	副长深成岩；似长深成岩；副长石深成岩
foig	顶板裂缝
foil	箔；薄片
foil gauge	应变片
foil sampler	衬片取样器；薄环刀取土器；薄片取样器
foil-type gauge	箔式应变片
fold	褶曲；褶皱；折叠；起伏地面；合拢；倍
fold amplitude	褶皱幅度
fold and thrust belt	褶皱冲断带
fold angle	褶皱角；褶角
fold apex	褶皱顶
fold arc	褶皱弧
fold attitude	褶皱产状；褶皱状态
fold axial trace	褶皱轴迹
fold axial trend	褶皱轴向
fold axis	褶皱轴；褶皱脊线；褶曲轴
fold basin	褶皱盆地
fold belt	褶皱带；造山带
fold block mountain	褶皱块状山
fold breccia	褶皱角砾岩
fold building	褶皱构造
fold closure	褶皱闭合度
fold coast	褶皱海岸
fold curvature	褶皱曲率
fold earthquake	褶皱地震
fold episode	褶皱幕
fold facing	褶皱面向
fold fault	褶皱断层
fold fissures	褶皱裂隙
fold hinge	褶皱枢纽
fold intensity	褶皱强度
fold interference pattern	褶皱干扰型式
fold inwards	向内褶曲
fold limb	褶皱翼
fold morphology	褶皱形态
fold mountain	褶皱山；褶皱山脉
fold movement	褶皱运动
fold nappe	褶皱推覆体；褶皱推复；褶皱表面
fold of sedimentary cover	沉积盖层褶皱
fold phase	褶皱幕；褶皱期
fold plagisyenite	副长石斜长正长岩
fold plunge	褶皱倾伏角
fold profile	褶曲剖面
fold region	褶皱区
fold set	褶皱组
fold structure	褶皱构造
fold style	褶皱型式；褶皱样式
fold symmetry	褶皱对称性
fold system	褶皱系
fold tectonics	褶皱构造
fold test	褶皱检验
fold thrust zone	褶皱冲断层带
fold train	褶皱列
fold vergence	褶皱倒向
fold wave length	褶皱波长
fold zone	褶皱带
fold-and-thrust belt	褶皱-冲断层带
foldbelt	褶皱带；造山带
foldcarpet	褶皱盖层
fold-curve	褶皱弯曲
folded	褶皱的；折叠的；合拢的；褶曲的；折叠的
folded and faulted	经历褶皱和断层；褶皱及断裂的
folded belt	褶皱带
folded chain	褶皱山脉
folded doming-up region	隆褶区
folded downwarping region	拗褶区
folded fault	褶皱断层
folded fold	重褶褶皱
folded mountain	褶皱山
folded plate structure	折板结构
folded region	褶皱区
folded rock	褶皱岩石；褶皱岩层
folded section	褶皱剖面
folded structure	褶皱构造；褶皱地形
folded substructure	底层褶皱构造
folded tectogene	褶皱构造带
folded thrust	褶皱冲断层
folded trap	褶皱圈闭
folded zone	褶皱区；褶皱带
folded-plate	折板
folded-plate structure	折板结构
folder-type rock beam	折叠型岩梁
folding	褶皱作用；褶积；褶皱；褶叠；折叠，折叠的，可弯折的；合拢
folding belt	褶皱带
folding board	罐托；罐座；折叠板
folding earthquake	褶皱地震
folding fissure	褶皱裂缝；褶皱裂隙
folding horst	褶皱地垒
folding layers	褶叠层
folding nappe	褶皱推复体，褶皱推覆体

folding period 褶皱期
folding phase 褶皱幕
folding tectonics 褶皱构造
folding test 折弯试验
folding theorem 折叠定理
folding wedges 对垒式楔块
foldkern 褶皱核
fold-line sliding method 折线滑动法
fold-modified 褶皱变形的
foldover 折叠
fold-pluge 褶皱倾没
folds 褶皱
fold-thrust 褶皱冲断层
fold-wedge 褶皱楔体
folgerite 镍黄铁矿
folia 叶状层；纹层；薄层
foliaceous 层状的，分成薄层的；叶状的；叶片状的
foliaceous structure 叶片状构造
folial trace 叶迹
foliar 叶的；叶状的；似叶的
foliate 裂成薄片，薄片状，叶片状；打成薄片；层状的；叶状的，叶理状的；叶理岩，薄叶岩
foliated 叶状的，叶理的；叶片的；层状的；薄层状的
foliated coal 层状煤；层状褐煤；叶片状煤
foliated granite 叶状花岗岩，片状花岗岩；片麻岩
foliated rock 面理化的岩石；叶状岩石，叶片状岩石；层状岩石
foliated structure 层状构造；叶片状构造；薄片状构造；叶理构造
foliated talcum 片状滑石
foliated texture 叶片状结构
foliation 叶理，片理，板理，面理，层理；剥理；褶纹；条纹；成层，成片
foliation structure 面理构造；叶理构造
foliose 叶状的；叶理状的；面理化的
folium 薄层；纹层；叶理，面理；叶状层；叶片，薄片；叶形线
follow-down drilling 跟管钻进
follower 送桩；从板；接着的；下列的；其次的
follower chart 钻孔结构图
follower pile 送桩
follower plate 挤压垫；仿形圆盘
following 假顶；以下的，下列的；跟踪，观测；追随者；随动
following dirt 煤层的松散页岩假顶
following settlement 后续下沉
following stone 随煤落下的顶板；伪顶
following the dip 沿倾斜方向
following the strike 沿走向
following-up bank 回采煤带
follow-up action 跟进行动
follow-up hole 接着钻的井眼；跟进井眼
fondo 洋底
fondoform 基底型；洋底地形；洋底平原；洋底沉积
fondothem 洋底沉积；洋底岩层；滞流岩
fondothem facies 洋底岩相
fontology 温泉学
foot 墙脚，脚，基脚，地脚，山脚，足；底部，底座，基座，基础，底；英尺；下盘，下部；山麓
foot glacier 山麓冰川

foot hill 丘陵地带；山麓丘陵；山麓
foot mat 底垫层
foot of drift 导洞尽端
foot of hole 钻孔底部
foot of incline 斜井底
foot of landslide 滑坡足部；滑坡脚
foot of mountain 山麓；山脚
foot of pile 桩基
foot of slope 斜井底；边坡脚，坡脚
foot piece 底梁
foot rill 直达工区的巷道
foot slope 底帮；麓坡
foot stall 基脚；柱墩
foot stall of column 柱基墩脚，墩柱
foot stone 基石
foot wall 下盘，底帮；基础墙
footage 进尺，进度；总长度；尺码；英尺长
footage bid 按包钻进尺计费的报价
footage block 标志块；岩芯记录牌；岩芯隔板
footage drilled 钻进总长；进尺数；总钻探进尺
footage driving cycle 循环进尺
footage measurement of roadway 巷道验收测量
footage measurement of workings 巷道验收丈量；巷道验收测量
footage measuring day 收尺日
footage per bit 每个钻头进尺
footage-rate contract 包进尺钻井合同
footeite 绿盐铜矿
foothill 山麓小丘；丘陵地带，山麓丘陵带；山麓，山麓丘陵；山麓丘陵地带
foothill belt 山麓带
foothill freeway 山麓公路
foothill region 山麓区
foothold 支柱；支架；支点；立足点；据点
foothold of driller 钻工站台；钻工工作台
footing 浅层基础，基础；基脚；底脚；垫层；地基，墙基，基座，底座
footing analogy 等效基础模拟
footing area 基础面积
footing base additional stress 基底附加应力
footing base pressure 基底压力
footing beam 基础梁；底脚梁；基脚梁；地脚梁
footing contact pressure 基础接触压力
footing course 垫层；基底层；底层，基层
footing deck 基脚板；基脚台
footing depth 基础深度
footing design 基础设计
footing dressing 墙基处理；基础整形
footing elevation 基础标高
footing excavation 基础开挖；基脚开挖
footing form 基脚模板
footing foundation 基础；底脚地基；台阶式基础；底座基础
footing kern 基础截面核心；基础中心
footing load 基础荷载
footing moment 基础弯矩
footing movement 基础位移
footing of sunscreen wall 遮光墙基础
footing of wall 墙底脚；墙基
footing parameter 基础参数

footing reinforcement mesh	基础钢筋网
footing rotation	基础转动
footing shape factor	基础形状系数
footing socket	杯形基础
footing step	基础台阶
footing thickness	基础厚度
footing width	基础宽度
foot-of-hole	炮眼；炮眼进尺
footpath	人行道；行人径；行人路
foot-piece	支架底梁
footrail	露头矿
footrill	平硐；水平坑道；倾斜运煤道；斜坡道
footstep bearing	立轴承
footstep pivot	球面枢轴
footstock	顶座；尾架；定心座
footwall	底帮，底壁；断层下盘，下盘；底板
footwall alteration	底板蚀变
footwall blasting	底帮爆破；卧底
footwall block	下盘断块；断层下盘
footwall drift	底盘平巷
footwall drive	沿矿脉下盘掘进
footwall failure	底帮破坏
footwall haulage	下盘巷道运输
footwall place	下盘工作区；下盘
footwall remnant	底帮上未采的煤
footwall shaft	在下盘中的斜井
footwalled	卧底的
footwalling	卧底；挖底；拉底
footway	人行道；行人路；梯子间
Foraky freezing process	弗拉基冻结凿井法
foraminiferal limestone	有孔虫灰岩
foraminiferal ooze	有孔虫软泥
foraminiferal zonation	有孔虫成带现象
foraminiferous	含有孔虫的
foraminite	有孔虫岩
forbesite	纤砷钴镍矿
force	力，力量，效力，约束力；强制，迫使，强迫；强迫度，力度；射流作用力；职工，人力；影响
force application	施力
force arm	力臂
force balance	力平衡
force between slices	条间力
force boundary	力边界
force causing airflow	风流驱动力
force causing sliding	滑动力
force cell	力传感器
force centre	力心
force coefficient	力系数
force couple	力偶
force displacement curve	力-位移曲线
force due to viscosity	黏性力
force equilibrium	力平衡
force equilibrium factor of safety	力平衡安全系数
force field	力场
force gauge	测力计；测力器；功率表
force lift	压力泵；压力唧筒
force lift pump	增压泵
force line	力线
force main	压力干管；压力主管道
force moment	力矩
force of attraction	引力
force of bind	约束力
force of cohesion	附着力；黏合力；内聚力；凝聚力
force of compression	挤压力，压缩力；压力
force of crystallization	结晶力
force of explosives	炸药力
force of flocculation	絮凝力
force of friction	摩擦力
force of gravity	引力；重力；地心吸力
force of inertia	惯性力
force of penetration	穿透力
force of periphery	圆周力
force of support	支承压力；支承的反力
force origin	力的起点
force out	挤出；压出
force piece	斜撑木
force pipe	压送管；压力管；压力水管
force polygon	力多边形
force potential	势
force pulling	牵引力
force pump	压力泵；手摇泵
force resisting sliding	抗滑力
force resolution	力的分解；求分力
force resultant	合力
force system	力系
force transducer	测力传感器；力传感器
force transfer system	力传递系统
force triangle	力三角形
force vector	力矢量
force vibration	强制振动，受迫振动
force-apart structure	强力分离构造
force-balance model	力平衡模型
force-balanced	力平衡式的；伺服式的
force-balanced accelerometer	力平衡式加速度计
force-caving system	强制崩落法
forced	强迫的；施力的；强制的
forced block caving	强制分段崩落开采法；强制分段崩落开采
forced block caving method	阶段强制崩落采矿法
forced boundary condition	外力边界条件；强制边界条件
forced caving	强制放顶
forced caving the roof	顶板强制垮落
forced convection	强迫对流；强制对流；人工对流；受迫对流
forced drop shaft	强迫沉井
forced drop-shaft method	强迫沉井法
forced electrical drainage	强制电力排流
forced failure plane	强制破裂面
forced feed	压力推进；压力进给
forced folding	强制褶皱
forced inclined shear plane	强迫斜剪面
forced injection	强制注入
forced layer	受力层
forced non-linear oscillation	非线性强迫振动
forced oscillation	强迫振动；强迫振荡
forced production	强化开采
forced release	强迫释放
forced releasing	强拆
forced surface wave	强制表面波
forced tidal wave	强制潮汐波

forced vibration 强迫振动；受迫振动
forced vibration test 强迫振动试验
forced wave 强制波
forced-caving system 药室爆破采矿法；强制崩落采矿法
force-deflection curve 力-挠度曲线
force-deformation relationship 力-变形关系
force-displacement behavior 力-位移特性
force-displacement relationship 力-位移关系
forced-rotation method 强迫旋转法
forces composition 力的合成
forces parallelogram 力平行四边形
force-summing element 受力元件
Forchheimer's law 福熙海麦定律
forcing dissectible 挤切
forcing function 应力函数；扰动函数；强迫函数；力函数
forcing hammer foundation 锻锤基础
ford 渡口；浅滩；水滩；过水路面
fore and aft inclinometer 前后倾斜计
fore breast 巷道掘进工组，巷道掘进工作面；超前工作面；掘进工作面；煤房工作面；采煤工作面
fore canopy 前顶梁；前顶盖
fore deep 前渊；陆外渊
fore drift 超前平巷
fore earthquake 前震
fore effect angle 超前影响角
fore end 超前工作面，超前工组；前端
fore plate 前板；导板
fore poling 插板支撑
fore reef 前礁；礁前
fore reef escarpment 前礁悬崖
fore reef slope 前礁坡
fore reef terrace 前礁阶地
fore set 工作面棚子
fore shaft 井颈；锁口
fore shaft sinking 在凿井前预凿的一段井；井颈开凿；锁口下卧
fore shock 前震
fore support 前部支架；自移式支架前部；前探式支架
fore-and-aft clearance 前后间距
fore-and-aft tilt 航向倾斜
forearc 岛弧前，前弧；弧前
fore-arc basin 弧前盆地
forearc geology 弧地质
forearc slope 弧坡
forearc tectonics 弧构造
fore-barrier 前礁堤
forebasin 前盆地；盆地前
forebay 前湾，前池；水轮机进水道的缓冲渠；水闸前的缓冲渠
forebay channel 前池渠道
forebay reservoir 前池水库
forebeach 前滩
fore-breast 掘进工作面；煤房工作面
forebulge unconformity 前缘隆起不整合
forecast 预报；预测
forecast accuracy 预报准确率
forecast amendment 预报修正
forecast and prediction 预测预报
forecast area 预报区
forecast completion 预测完成量

forecast curve 预测曲线
forecast district 预报区
forecast for construction 施工预报
forecast model 预测模型
forecast of earthquake 地震预报
forecast of outburst danger 突出危险性的预测
forecast of runoff 径流预报
forecast of settlement 沉降预测
forecast value 预测值
forecast variable 预测变量
forecast verification 预报检验
forecasting 预报；预测
forecasting and early-warning 预警预报
forecasting efficiency 预测效率
forecasting just before sliding 临滑预报
forecasting of reservoir inflow 进库流量预报
forecasting of stream flow 河川流量预报
forecasting procedure 预报方法
forecasting risk 预测风险
forecasting technique 预测技术
forecasting water level 预测水位
forecourt 前庭
foredeep 前渊；陆外渊；外地槽；峡窄深海槽
foredeep basin 前渊盆地
foredeep inclusion 外来包体；捕房体
fore-drift 超前掘进巷道
foredune 岸前沙丘；水边低沙丘
foreeshore berm 前滨阶地
forefield 生产工作面；超前工作面；工作面
forefield end 巷道最远端；最远端的工作面
forefront pressure 前端压力
forehead 一个矿或一个水平的工作面
foreign inclusion 外来包体
foreland 前麓地；前沿地，山前低地；海角，海岸地，岬；前陆，前岸；滩地；山麓
foreland area 前陆区；前陆地带
foreland basin 前陆盆地
foreland dipping duplex 倾向前陆式双冲构造
foreland facies 前陆相；陆架相
foreland fold 前陆褶皱
foreland fold belt 前陆褶皱带
foreland grit 前陆粗砂岩
foreland sequence 前陆地层序列，前陆地层序
forelimb 前翼，陡翼；缓翼
forelimb thrust 前翼冲断层
forellenstein 橄长岩
foremountain basin 山前盆地
forensic engineering 法律工程学
forepart 前部；前段
forepole 护坡桩；超前支架；前探梁
forepoling 超前支架，超前支护；护坡桩支撑，插板支撑，插板支护；超前伸梁掘进法，前探梁掘进法；预支掘进；搭接插板挡土法
forepoling arm 超前支架的前伸梁
forepoling bar 前探梁；前插梁
fore-poling board 超前板桩
forepoling bracket 前探梁托架
forepoling method 前支护法
forepoling plate 密排插板
forepoling set 前伸背板支架

forepoling support 及时支撑
fore-reef complex 礁前复合体
fore-reef facies 礁前相
forerun 前兆；预报
forerunner 前兆；预兆；前震；先驱
forerunner earthquake 前震；预震
forerunner of strong earthquake 强震前兆
forerunning effect of earthquake 地震前兆
forerunning phenomena 火山喷发前兆
foreseeability 预知性；预见性
foreseeable 可预见的
foreseeingly 有预见地
foreset 前积层；前积斜坡；前积层的；临时超前支架
fore-set 临时超前支护；顶柱
foreset bed 前积层
foreset bed of delta 三角洲的前置层
foreset bedding 前积层理；交错层理
foreset cross-bedding 前积交错层理
foreset deposit 前组沉积
foreset laminae 前积纹层
foreshaft 井口；井颈
foreshock 前震
foreshock activity 前震活动
foreshock wave 前震波
foreshock-aftershock pattern 前震-余震型
foreshock-mainshock type 前震-主震型
foreshore 前滩；前滨
foreshore and seabed 前滨及海床
foreshore and seabed ordinance 前滨及海床条例
foreshore facies 前滨相
foreshore slope 前滨坡
foreslip 前滑
foreslope 前坡；前缘斜坡
foreslope facies 前缘斜坡相
forest bed 森林底层
forest bog 林沼；森林沼泽
forest highway 林区公路
forest humus soil 森林腐殖质土
forest land 林地；森林地区
forest moor 森林沼泽
forest peat 森林泥炭；木质泥炭
forest peat mire 森林泥炭沼泽
forest railway 森林铁路
forest road 林区道路
forest survey 林业测量；森林调查
forest swamp 森林沼泽
forest trail 林荫小径
forestage 前级的；前台
forested area 林区；绿荫面积
forest-moss peat 森林苔藓泥炭；森林沼泽泥炭
forestope 前探梁
forestry remote sensing 林业遥感；森林遥感
forestswamp facies 森林沼泽相
forest-to-coal process 森林成煤过程
fore-syncline 前缘向斜
forethrust 前逆冲断层，前冲断层
foretoken 预兆
fore-trough 前海槽
forewarning 警报；预先警告
forewinning 采房；第一阶段回采

forge crack 锻裂
forging bursting 锻裂
fork beam 半横梁；分叉梁
fork junction 叉形路口
fork pockets 叉槽
forking 河流分岔作用
forking bed 分叉岩层
form 形状，形式，形，方式，方法，晶形，形态；形成，构成，造成；模板；型，型式；格式；表格
form anchor 模板锚定
form bracing 模板支撑
form cage 模板骨架；模槽，框架拢
form clamp 模板夹
form coefficient 形状系数
form contour 地形等高线；模板轮廓
form design 框图设计；形状设计
form drag 形状阻力；形态阻力
form effect 形状效应
form energy 形面能；结晶力
form factor 形状因素，形态因子；形状系数，状态因子；波形因素；波形系数
form hanger 模板支撑
form joint 模板接缝；模板节缝
form laggings 模板板用木板；模板；拱模方木板
form lateral pressure 模板侧压力
form line 轮廓线；地形线
form line contour 形态等高线
form lining 模板衬垫，模板衬砌；模板涂层
form modulus 形状模数
form of bit 钻头形状
form of coal seam 煤层形态
form of structure 结构形式；构造形态
form on traveler 移动模架
form panel 模板
form placing 模板安装
form ratio 水流深宽比
form removal 拆除模板；拆除模壳；拆模
form removing 拆卸模板
form roughness 形态糙度
form set 形态层组
form stress factor 形状应力因数
form stripping 模板拆除；拆模
form surface map 构造地形图
form vibrator 模板振动器；附着式振动器
formability 可成形性，成型性，成形性
formable period of freezing wall 冻结壁形成期
formal fabric 形态组构
formal lithostratigraphic unit 正式岩石地层单位
formal road system 规整式道路系统
formanite 黄钇钽矿；钽钇铌矿
formant 共振峰
formant frequency 共振峰频率
formation 组；地层，层，岩层，层系，路基面，基床，建造；形成，生成，组成，构成，结构，平整，开拓，组织；系群，群系
formation pressure 地层压力
formation age 地层年代；生成年龄
formation analysis 地层分析
formation analysis log 地层分析测井；地层分析记录
formation and servicing 平整工程及辅助工程

formation anomaly simulation trace	地层层面与井壁相交的模拟迹线
formation base layer	基床表层
formation blanked off	地层堵塞
formation boundary	地层边界
formation breakdown	地层破裂；地层压裂
formation breakdown gradient	地层破裂梯度
formation breakdown pressure	地层破裂压力
formation brine concentration	地层盐水浓度
formation bulk density	地层体积密度
formation bulk volume	地层总体积
formation capacity	地层系数
formation capture cross section	地层俘获截面
formation catalog	地层表
formation caving	井壁坍塌
formation cementation	地层胶结
formation characteristics	地层特性
formation chunk	地层砂块
formation classification	地层分级
formation collapse sticking	地层坍塌卡钻
formation compaction	地层压实
formation complex	岩组；建造组合
formation compressibility	岩层压缩率；地层压缩性
formation compression	地层压缩
formation conductivity	地层电导率；地层系数
formation consolidation	地层固结
formation control	地层控制
formation core plug	地层岩心柱状样品
formation crookedness	地层弯斜度
formation crude	地层原油
formation cutting	岩屑
formation damage	储层损害；地层损害
formation damage control	地层损害控制
formation damage index	地层损害指数
formation damages	地层损害
formation deflecting force	地层造斜力
formation density log	地层密度测井
formation density logging	密度测井
formation density tool	地层密度测井仪
formation deviating force	地层造斜力
formation drillability	地层可钻性
formation drillability parameter	地层可钻性参数
formation electrical parameter	地层电性参数
formation entry	油砂侵入井中
formation evaluation	地层评价；地层估价；生产层评价
formation facctor-porosity relationship	地层因数-孔隙度关系
formation face	井底地层面
formation factor	地层因数，建造因素；产状要素；孔隙参数；结构因数
formation factor log	地层因数测井
formation fines	地层微粒
formation fluid	地层流体
formation fluid chamber	地层流体取样室
formation fluid pressure	地层流体压力
formation fluid sampler panel	地层流体取样器面板
formation fluid sampling	地层流体取样
formation fluid test	地层流体测试
formation fracture	地层裂缝
formation fracture gradient	地层破裂压力梯度；地层压裂压力梯度
formation fracture pressure	地层破裂压力
formation fracture toughness	地层断裂韧性
formation fracturing	地层压裂
formation gain size	地层颗粒大小
formation gas	地层天然气；地层瓦斯
formation gas-oil ratio	地层油气比
formation geology	建造地质学
formation grain density	地层颗粒密度
formation hardness	岩石硬度
formation heterogeneities	地层非均质性
formation integrity test	地层完整性试验
formation interval tester	地层间隔测试器
formation interval velocity	地层成层速度；地层间隔速度
formation leak off test	地层压漏试验
formation level	施工基面；岩层面；路基面；平整面标高，平整面水平；路肩高程
formation line	层面；地层线；施工线
formation lithology	地层岩性
formation map	地层图
formation match	地层匹配
formation mean direction	建造平均方向
formation mechanism	成因机制；形成机制
formation median grain size	地层砂粒度中值
formation microscanner	地层微扫描器
formation multi-tester	多次地层测试器
formation of coal	煤的成因；煤的生成
formation of disk-shaped rock cores	岩芯饼化现象
formation of fissure	裂纹的形成；裂隙构造
formation of image	地层成像
formation of interest	目的层
formation of land	土地平整；辟拓土地；开拓土地
formation of sand dune	沙丘形成
formation parameter	地层参数
formation particle	地层砂粒
formation parting factor	地层破裂系数
formation permeability	岩层渗透性
formation permeability channel	地层的可渗透通道
formation plugging	地层堵塞
formation pore pressure	地层孔隙压力
formation porosity	岩层孔隙率；地层孔隙度
formation pressure	油层压力；地层压力
formation pressure gradient	地层压力梯度
formation pressure prediction	地层压力预测
formation reaction	地层反作用力
formation resistance	地层阻力
formation resistivity	地层电阻率；地电阻率
formation resistivity factor	地层电阻率系数；地层电阻率因数
formation resistivity index	地层电阻率指数
formation resistivity logging	地层电阻率测井
formation restoration	地层复原
formation restressing	地层应力恢复
formation sample	岩样
formation sampling	地层取样
formation section	地层断面
formation sensitivity	地层敏感性
formation shut-in pressure	关井地层压力
formation shut-off	地层止水封闭；地层封堵
formation skeleton	地层骨架

formation solid	地层固体（颗粒）
formation solid control	地层防砂控制
formation solid matrix	地层固体骨架
formation stability	地层稳定性
formation storage	地层储存能力
formation strength	地层强度
formation strength index	地层强度指数
formation subgrade	路基
formation subsoil ripping	翻松底土；翻松地基面
formation swelling sticking	地层膨胀卡钻
formation temperature	地层温度；成矿温度
formation tendency	地层趋势
formation test	地层测试
formation tester log	地层测试录井记录
formation thermal conductivity	地层导热系数
formation thickness	地层厚度
formation time constant	地层时间常数
formation to break down	地层破裂
formation top layer	基床表层
formation transit time	地层传播时间
formation travel time	地层旅行时间
formation velocity	地层速度
formation volume factor	地层体积系数
formation water	地层水
formation water map	地层水图
formation water rate	地层水产量
formation water resistivity	地层水电阻率
formation width	路基面宽度
formation works	地盘平整工程；地盘开拓工程；平整工程
formation-abrasiveness parmeter	地层研磨性参数
formational control	地层控制
formational geology	地层学；建造地质学
formational sequence	地层层序
formation-bit force	地层-钻头力
formation-parting pressure	地层破裂压力
formation-to-gravel size ratio	地层砂与砾石粒度比
formation-wellbore pressure balanced	地层-井眼系统压力平衡
formed wire gun	成型钢索射孔器
former glacier	古冰川
former island arc	先存岛弧
former river valley	古河谷
former snow line	古雪线
former stream channel	古河道
former valley bottom	古谷底
forming of pillars	矿柱成形；开切煤柱
forming process	成形过程
formline	地形线
formline structure map	地形线构造图
formset	形态层组；交错层组
form-tie assembly	混凝土模板接合装置
formula	公式，式；方程式；算式，计算式；结构式；化学式；配方，药方
formula of pile driving	打桩公式
formulation of the cement blend	水泥混合料的组成；水泥骨料配比
formwork	模板；模板工程；制模工作；模板工作；模架
formwork uprighting survey	立模测量
fornix	穹隆
forsterite	镁橄榄石
fortunite	橄榄金云煌斑岩；橄榄钾镁煌斑岩
forward abutment	前支承带
forward abutment pressure	前支壁压力
forward angle	前方位角
forward azimuth	前向方位角；前方位角
forward bar	前探顶
forward beam	前梁；前探梁
forward bearing	前象限角；正象限角
forward bias	正向偏压
forward canopy	前探梁；前顶梁
forward cantilever	前探梁
forward flow	平直水流；向前流动
forward flow zone	射流区
forward hading breaks	向前方倾斜的断裂
forward initiation	正向起爆
forward intersection	前方交会
forward model	正演模型
forward modeling	正演模拟
forward motion	前向运动；前向移动
forward moving stack	前移迭加
forward osmosis	正向渗透
forward power-operated extensible roof bar	液压可伸缩前探梁
forward probe	超前探测
forward problem	正问题；正演问题
forward propagation	正向传播；前展式扩展
forward rate constant	正反应常数
forward reading	前视读数
forward reinforcement jet grouting	预加固喷射灌浆；超前注浆加固
forward roof	前探梁
forward runaway arrester	前向跑车防坠器
forward shield	前护盾
forward slope	前坡
forward thrust force	前推力；推压力
foshagite	变针硅钙石
foshallassite	水硅灰石
fossa	坳陷带；大海沟带；海渊；海沟
fosse	河沟，沟，海沟，壕，冰川边沟，海渊，渠道，坑，槽；冰川前缘狭长槽；齿；齿沟；型腔
fosse moat	壕沟
fossick	旧矿中采矿
fossicking	不规则采矿
fossil	化石；陈旧事物；化石的；古的
fossil arc-trench gap	古岛弧-海沟断谷
fossil basin	古盆地
fossil bed	化石层
fossil botany	古植物学
fossil boundary	古边界
fossil calcareous algae	钙藻化石
fossil cave	古洞穴；古溶洞
fossil channel	古河道
fossil charcoal	化石炭；含植物结构的软夹层；丝煤；焦炭
fossil cloudburst	泥石流堆积物
fossil consuming boundaries	古消减边界；古消亡线
fossil delta	古三角洲
fossil depression	古洼地
fossil diagenesis	化石成岩作用

fossil dopplerite 石化弹性沥青
fossil energy 化石能源；矿物能源
fossil epicontinental sea 古陆表海
fossil erosion surface 古侵蚀面；化石侵蚀面
fossil farina 硅藻土
fossil flood plain 古河漫滩；古洪泛平原
fossil flour 硅藻土
fossil fracture 古断裂
fossil fragment 化石碎片
fossil geochronometry 化石地质时代测定法，化石地质年代定法
fossil geyserite 古硅华；古泉华
fossil glacier 化石冰川；古冰川
fossil groundwater 古地下水
fossil ice 化石冰；古代冰，古冰
fossil in situ 原地化石
fossil interstitial water 古间隙水
fossil island-arc succession 古岛弧系列
fossil karst 古喀斯特；古岩溶
fossil lake 古湖
fossil landform 古地形
fossil landscape 古地形；埋藏侵蚀面
fossil landslide 古滑坡
fossil lebensspur 遗迹化石
fossil magnetism 古磁性；化石磁性
fossil magnetization 化石磁化
fossil marginal sea 古陆缘海
fossil microalgae 微藻化石
fossil mid oceanic ridge 化石洋脊
fossil mineralization 成矿石化作用
fossil molds and casts 铸模化石；模铸化石
fossil mud volcano 古泥火山
fossil oil reservoir 古油藏
fossil oil seepage 古油苗
fossil organism 生物化石
fossil peneplain 古准平原
fossil plain 古平原
fossil plate 古板块
fossil plate boundary 化石板块边界
fossil plate tectonic 古板块构造
fossil pressure 埋藏压力
fossil record 化石记录
fossil reef 古礁；化石礁
fossil relief 古地形
fossil residuum 古风化壳；古残积层
fossil ridge 古洋脊
fossil river bed 古河床
fossil river course 古河道；故河道
fossil river valley 古河谷
fossil salt 岩盐
fossil seepage 古油苗
fossil shelf sea 古陆架海
fossil soil 古土壤，古壤，古土；化石土
fossil soil method 古土壤法
fossil stream channel 古河道
fossil subduction zone 古消减带；古俯冲带；古沉没带
fossil time 化石地质年代；化石时间
fossil trace 遗迹化石
fossil valley 埋藏谷
fossil water 原生水；古水，古地下水；化石水；矿物水

fossil weathering 古风化作用
fossil weathering crust 古风化壳
fossil wood 木化石
fossilation 化石化作用
fossilbearing 含化石的
fossil-carbon 化石碳
fossildiagenesis 化石成岩学
fossiliferous 含化石
fossiliferous micrite 含化石泥晶灰岩；含化石微晶灰岩
fossiliferous porosity 含化石孔隙
fossiliferous strata 含化石地层
fossilification 化石化
fossilizaiton 化石化；化石作用
fossilize 变成化石
fossilized brine 古卤水
fossilized dune 古沙丘
fossilized water 化石水；封存水；古水
fossilogy 化石学
fossilology 化石学；古生物学
fossils volcano 埋藏火山；古火山
fossorial 掘土的
fossula 内沟
fother 量煤重量单位
foucherite 土磷铁钙矿
foul coal 劣质煤；不能销售的煤
foul drainage 污水渠道；脏水渠道
foul mine 瓦斯矿；多煤尘矿；易出爆炸事故的矿山
foul sewer 脏水渠；污水渠
foul sewer system 污水系统；脏水系统
foul water drain 污水渠；脏水渠
foul water sewer 污水渠
foul water system 污水排放系统；脏水排放系统
foul waters 危险海区
foulness 煤层杂质；煤层的不规则性；污染；堵塞；读数误差
foul-water sewer 污水管
foul-water system 污水系统
found 遇到煤层；铸造，建立，浇铸；建造基础；翻砂；熔制
foundation 基础，基底；地基，地脚，建造，座，基座，机座，奠基；根据；基础墙；基金，基金会
foundation analysis 基础分析
foundation anchoring 基础锚固
foundation area for rig 钻机占用地
foundation base 岩层面；路基面，建基面；基础底；基础底面，基底；地基底
foundation beam 地梁；基础梁；地基梁
foundation bearer 地基梁
foundation bearing capacity 地基承载力
foundation bed 基础垫层，地基垫层；基底层，基底土层；基底；基床
foundation block 基础块体
foundation bolt 地脚螺栓，地脚螺丝；基础螺栓
foundation brake rigging 基础制动装置
foundation by means of injecting cement 基础灌浆加固
foundation by pit sinking 沉井基础；挖坑沉基
foundation by timber casing with stone filling 木框石心基础
foundation capability 地基承载能力
foundation characteristics 基础特征
foundation coating 基础涂层

foundation coefficient	地基系数；基础系数
foundation compliance	地基可塑性；地基柔度
foundation concrete	基础用混凝土
foundation condition	地基条件
foundation construction	基础施工
foundation course	基础垫层，基层，勒脚层
foundation cushion course	基础垫层
foundation damp proofing course	基础防潮层；地基防潮层
foundation deformation	基础变形；地基变形
foundation deformation modulus	地基变形模量
foundation depth	基础深度
foundation design	地基设计；基础设计
foundation design code for building	建筑地基基础设计规范
foundation detailing	基础细部构造
foundation dewatering	基础降水；地基降水
foundation dimension	基础尺寸
foundation displacement	基础位移
foundation displacement indicator	基础位移指示器；基础位移标识
foundation ditch	基础沟；基坑；基槽
foundation drain	基础排水管
foundation drain hole	基础排水孔
foundation drainage	地基排水
foundation drainage system	基础排水系统
foundation drawing	基础图
foundation dredging	地基开挖
foundation elastic modulus	地基弹性模数
foundation embedment	基础埋置
foundation embedment depth	基础埋置深度
foundation engineering	基础工程；基础工程学；地基工程；地基工程学
foundation engineering contractor	基础工程承包商
foundation examination by drilling	（墩台）基础钻探
foundation excavation	地基开挖；基础开挖
foundation exploration	基础勘探；地基勘探；地基勘查；基础勘查
foundation failure	地基失效；基础破坏；地基破坏
foundation flexibility effect	基础柔性影响
foundation folding	基底褶皱
foundation frame	基架；机座底架
foundation frame-work	基础框架
foundation girder	基础梁
foundation grade beam	地基梁
foundation grill	基础格床；格排基础
foundation grouting	地基灌浆；基础灌浆
foundation grouting technique	基础灌浆技术；地基灌浆技术
foundation improvement	地基加固；基础加固；地基改良
foundation investigation	地基勘察；地基调查；地基查勘
foundation investigation program	地基勘察计划
foundation isolation	基础隔振；地基隔振
foundation layout plan	地基布置平面图，地基平面布置图；基础平面布置图，基础布置平面图
foundation leakage	地基渗漏
foundation level	基底标高，基础水平，基础标高，标高，地基水平，地基基面，地基等级
foundation line	基础线；开挖线
foundation load	基础荷载；地基荷载
foundation location	基础定位；基础位置
foundation made of materials	三合土基础
foundation map	基础图
foundation mat	片筏基础；基础板；基础底板；基褥垫，基褥；基垫层
foundation mattress	沉排基础
foundation medium	基础介质
foundation member	基础构件
foundation modeling	基础建模
foundation modulus	基床反力系数；地基模量，基础模量；地基反力系数；地基系数
foundation of dam	坝基
foundation of masonry	砌体基础；砖石砌体基础
foundation on raft	浮筏基础；筏基
foundation on wells	井筒基础
foundation pad	基础垫层
foundation perimeter	基础周边
foundation pier	桥墩；桥台；基墩；基座
foundation piezometer	基础渗压计
foundation pile	基础桩柱，基础桩；基桩；基础管桩；地基桩柱
foundation pile diagnosis system	基桩病害检测系统
foundation pile hole	基桩孔
foundation piping	地基管涌现象
foundation pit	基坑
foundation pit drainage	基坑排水
foundation pit well point drainage method	基坑井点排水法
foundation plan	基础平面图；地基平面图；基础图则
foundation planting	基础种植
foundation plate	基础板；底板
foundation platform	基础承台
foundation practice	基础施工技术
foundation pressure	基底压力；基础底面压力
foundation resilience test	地基回弹测试
foundation restraint	基础箍固；基础约束
foundation ring	底圈
foundation rock	基岩
foundation rocking	基础摇摆
foundation rotation	基础转动
foundation roughness	基底粗糙度；基础粗糙度
foundation sampling	基岩取样
foundation scouring	基础冲刷
foundation set	基础组
foundation settlement	地基沉降，地基下沉，地基沉陷，基础沉陷，基础沉降；基底沉降
foundation slab	基础板；基础底板，底板；护底；承台
foundation soil	地基土；基土；地基；持力层
foundation soil replacement by sand	地基的挖土换砂
foundation soil science	土质学；基础土质科学
foundation soil system	基础-土壤体系
foundation spring stiffness	基础弹簧刚度
foundation stability	地基稳定性
foundation stabilization	地基加固
foundation stay	基座
foundation stone	基石；屋基石
foundation strata	地基土层
foundation stratum	地基土层；基础层
foundation stress	基础应力
foundation stripping	清基
foundation structure	基础结构；基础

foundation structure interaction	基础-结构相互作用	four stage deck sinking stage	四层吊盘
foundation structure system	基础-结构体系	four wing rotary bit	十字钻头
foundation support	基座	four-abutment arch	十字形拱顶
foundation surface	建基面	four-arm caliper	四臂井径仪
foundation surface treatment	地基表面处理	four-bar mechanism	四连杆机构
foundation swaying	基础摇摆	four-bladed vane	十字板仪
foundation terrace	基座阶地	fourble	下部立杆；立根
foundation tie-beam	基础系梁	fourble board working platform	四节立根架工工作台
foundation treatment	地基处理；基础处理	four-centred arch	四心拱
foundation trench	基槽；基坑	fourchite	钛辉沸煌岩；无橄沸煌岩
foundation under rail	轨下基础	four-cutter bit	四齿轮钻头；四牙轮钻头
foundation under water	水下基础	four-decker cage	四层式罐笼
foundation underpinning	基础换托；基础托换	four-drill rig	四钻机钻车；四钻凿岩台车
foundation underwater	水下基础	four-edged bit	十字钻头
foundation uniformity	地基的均匀性	four-entry system	四巷式系统
foundation unit	基础单元	Fourier	傅里叶
foundation uplift	基础隆起	Fourier amplitude spectrum	傅里叶振幅谱
foundation vibration	基础振动	fourmarierite	红铀矿
foundation wall	基础墙；基墙；基脚	four-noded element	四节点单元
foundation washer	地脚垫板	four-noded interpolation	四节点插值
foundation weighting rectification	基础加荷纠偏法	four-post type headframe	四柱型井架
foundation well	基础井筒	four-roller bit	四牙轮钻头
foundation works	地基工程；基础工程	four-section support	四构件棚子，四构件支架；框形支架
foundation yielding	地基屈服；地基变形	four-sheet mineral	四片层矿物
foundational fault	基底断裂	four-spot flooding network	四点注采井网
foundational fault block	基底断块	four-spot pattern	四点法井网
foundation-bed	基床；基础底层	four-spot well network	四点法井网
foundations of solid mechanics	固体力学基础	four-surface prop	四面支柱
founder	首达矿脉的井筒；铸工；创始者，创立者，奠基者；缔造者；沉没；沉陷	fourth order sequence	第四级层序
		fourth-order leveling	四等水准测量
founder breccia	塌积角砾岩	four-way dip	十字倾角计算
foundering	地下水侵蚀作用；沉陷，陷落，沉没；岩浆蚀顶作用；倒塌	four-way drag bit	四翼刮刀钻头；四通切削钻头
		four-wing bit	四翼刮刀钻头
founding level	沉降基准面；基底高程	four-wing pattern bit	四翼钻头
founding stratum	持力层	four-wing rotary bit	十字形钻头；十字钻头
foundry pit	铸坑	foveolate corallite	孔穴珊瑚石
fountain	喷泉，泉；源泉；喷水，喷水池；喷水器；自流井，自喷井；中心注管	fowlerite	锌锰辉石
		fox bolt	端缝螺栓；开尾螺栓
fountain effect	喷泉效应	foyaite	流霞正长岩；良正长岩
fountain engineering	喷泉工程	Fraass breaking point	弗拉斯沥青破裂点
fountain equipment	喷水池装置	frac fluid	压裂液
fountain failure	冲毁；涌象；管涌破坏	frac instrument van	压裂作业监控车
fountain geyser	间歇喷泉	frac job	水力压裂含油层作业；水力压裂矿层作业；压裂作业
fountain head	喷泉水头；喷水头		
fountain spring	喷泉	frac pack	压裂充填
fountain type	多轴型；喷泉型	frac pressure	破裂压力
fountain well	自喷井；喷泉井	frac sand	压裂砂
fountain-type activity	喷泉式活动	frac shot	压裂用的支撑粒
fountain-type well	自喷井；自流井	frac width development	裂缝宽度扩展
fouqueite	绿帘石	fracking	压裂
four arm instrument	四臂下井仪	frac-monitor	压裂监视仪
four bladed vane	十字板仪	frac-sand jet	压裂喷砂器
four electrode dipmeter	四电极地层倾角测井仪	frac-string	压裂管柱
four groove drill	十字形钎子；十字钻头	fractal geometry	分形几何
four link mechanism	四连杆机构	fractal interpolation method	分形插值法
four point bit	十字形钎头，十字形钻头；十字钻头	fractal study	分形研究
four point test	四点法试井	fraction	组，粒组；粒级，碎片，破片，细粒，粒径组合，粒径级配；部分，小部分；分馏，馏分；分数；级别，比重级别
four pole derrick	四脚钻塔		
four roller bit	四牙轮钻头		
four sides supported plate	四边支承板	fraction of grain size	粒级

fraction of partial size	粒径组
fraction porosity	相对孔隙度
fraction toughness of rock	岩石断裂韧度
fraction void	疏松度
fractional analysis	粒级分析；筛析；分组分析
fractional coagulation	分级凝固
fractional crystallization	分离结晶；结晶分异；分步结晶
fractional flooding	部分注水
fractional flow	分相流动；分流量
fractional load	部分载荷，部分荷载；部分负载
fractional magma	部分岩浆
fractional melting	分离熔融
fractional oil wettability	斑点状示油润湿性
fractional pore volume	用小数表示的孔隙体积
fractional porosity	相对孔隙率；部分孔隙率
fractional precipitation	分级沉淀；分步沉淀；分段沉淀
fractional pressure	分压
fractional recovery	部分回收率；部分开采
fractional resorption	部分熔蚀；部分再吸收
fractional sampler	分级取样
fractional sampling	分部取样；分级取样
fractional saturation	部分饱和
fractional shoveling	分铲取样法；分铲缩分法
fractional wettability	部分湿润
fractional-flow formula	分相流动公式
fractional-shoveling sampler	用铲取样
fractionation of magma	岩浆分离
fractoconformity	破碎整合；断裂整合
fractography	断口分析；断口显微观察；金属断面显微镜观察
fracto-stratus	碎层云
fracturation	断裂；破裂
fracturation zone	裂隙带；破碎带；断裂带
fracture	裂隙，裂缝；断裂，破裂，破碎；断口，裂口；裂面，断面；裂缝端口；骨折
fracture acidizing	压裂酸化
fracture angle	破裂角；破坏角；裂隙角
fracture anisotropy	裂缝各向异性
fracture aperture	裂缝开度；裂隙宽带
fracture appearance transition temperature	裂纹扩展转变温度
fracture azimuth	裂缝方位
fracture behaviour	破裂性状
fracture belt	断裂带；破裂带
fracture capacity	裂缝导流能力
fracture character of rock	岩石破裂特性
fracture characteristic of rock	岩石破裂特性
fracture characteristics	裂缝特征
fracture cleavage	破劈理
fracture closure	裂缝闭合度
fracture closure pressure	裂缝闭合压力
fracture cloud	断裂云
fracture composite display	裂缝综合显示
fracture conductivity	裂缝导流能力
fracture confinement	裂缝限制
fracture control	断裂控制
fracture criterion	断裂准则；破裂准则
fracture criterion of rock	岩石断裂准则
fracture deformation	裂隙变形
fracture density	裂缝密度
fracture density index	裂隙密度指数
fracture detection	裂缝探测；裂缝检测
fracture development during cut-and-fill stopping	分层充填采矿期间的破裂发展
fracture dimension	破裂尺寸
fracture dip	裂缝倾角
fracture direction	裂缝方向
fracture distance	断裂距
fracture distribution	裂缝分布
fracture dome	松裂穹
fracture doming	破裂成穹
fracture dynamics of rock	岩石断裂动力学
fracture energy release rate	断裂能释放率
fracture envelope	裂隙包络
fracture evaluation	压裂评价
fracture extension	裂缝延伸
fracture extension pressure	裂缝延伸压力
fracture fabric	断裂组构
fracture face	裂缝面
fracture failure	裂隙破坏；断裂破坏
fracture filling	裂隙充填；断裂填充物
fracture finder	裂缝探测仪
fracture flow	裂隙流；裂缝流；流体在裂缝中的流动
fracture flow area	裂缝导流面积
fracture flow capacity	裂隙泄水能力；裂隙排水能力；裂缝导流能力
fracture fluid	压裂液
fracture frequency	裂隙频度；断裂频度
fracture front	破裂前锋
fracture geomechanics	断裂地质力学
fracture geometry	裂缝几何形状；裂缝形状
fracture geometry parameter	裂缝几何参数
fracture gradient	压裂梯度；破裂梯度
fracture gradient profile	裂隙梯度剖面
fracture groundwater	裂隙地下水；裂缝地下水
fracture grouting	压裂灌浆；劈裂灌浆
fracture growth	裂缝生长
fracture half length	裂缝半长
fracture height	裂缝高度
fracture hypothesis	破裂假说
fracture identification log	裂缝识别测井
fracture index	岩芯破裂指数；裂缝指数
fracture indicator	裂缝指示器
fracture induced perturbation	裂缝诱导微扰；裂缝导致的扰动
fracture initiation	裂隙产生；断裂发生；开始破裂
fracture initiation pressure	开始破裂压力；造缝压力
fracture intensity	断裂烈度；破裂强度
fracture intensity index	裂缝强度指数
fracture interpretation	断裂推断；断裂解绎
fracture interval	裂缝间距
fracture isotherm	破裂等温线
fracture joint spring	裂隙泉
fracture length	裂缝长度
fracture line	裂线；破裂线；断裂线
fracture lineament	区域断裂构造线；断裂轮廓
fracture load	破坏荷载；断裂荷载
fracture log	裂缝测井；破裂带钻井记录；裂缝测录
fracture mechanics	断裂力学
fracture mechanism	破裂机制；断裂机理

fracture medium 裂缝介质
fracture metamorphism 碎裂变质；断裂变质
fracture mode 破裂模式；断裂类型
fracture model experiments 破裂模拟实验
fracture modelling 裂隙模拟
fracture monitor 压裂监测仪
fracture morphology 裂缝形态（学）
fracture network 裂隙网络；裂缝网
fracture network volume 裂缝网络体积
fracture number 裂缝数
fracture of rock 岩石裂缝
fracture opening 裂缝张开度
fracture orientation 裂缝方向
fracture orientation method 定向压裂法
fracture orogeny 断裂造山
fracture pattern 破裂型式；断裂型式，断裂模式
fracture penetration 裂缝穿透深度
fracture performance 裂隙性能
fracture permeability 裂隙透水性；裂隙透水能力；裂缝渗透率
fracture phenomenon 破裂现象
fracture plane 破裂面；断裂面；裂缝面；裂面
fracture plane inclination 裂缝面倾斜度
fracture pole plot 裂隙极点图
fracture population characteristic 裂隙分布特征
fracture pore 裂缝孔隙
fracture porosity 裂隙率；裂缝孔隙度；破裂孔隙
fracture porosity reservoir 裂缝孔隙性储层
fracture position 裂缝形成部位
fracture pressure 裂压；破裂压力
fracture pressure gradient 破裂压力梯度
fracture pressure normalization 破裂压力标准化
fracture probability 断裂概率
fracture process 断裂过程
fracture produced wave source 断裂波源
fracture propagation 断裂扩展；裂缝延伸
fracture propagation model 裂纹扩展模型
fracture propagation pressure 裂缝延伸压力
fracture radius 碎裂半径；爆裂半径
fracture reservoir 裂隙型油气藏
fracture response 断裂响应
fracture sand jet 压裂喷砂器
fracture section 断裂剖面
fracture set 断裂组；裂隙组
fracture shape 裂缝形状
fracture slip 裂隙滑移
fracture spacing 裂隙分布；裂隙间距；裂缝间距
fracture spacing index 裂隙分布系数；裂隙间距系数；裂隙指数
fracture speed 裂纹扩展速度
fracture spring 裂隙泉；断层泉；裂缝泉
fracture stage 破裂阶段
fracture stiffness 断裂刚度
fracture strain 断裂应变
fracture strength 破裂强度；断裂强度；抗破强度
fracture stress 断裂应力；破裂应力；抗破强度
fracture strike 裂缝走向
fracture structural plane 破裂性结构面
fracture structure 断裂结构；断裂构造
fracture surface 破裂面

fracture surface roughness 裂缝面粗糙度
fracture swarm 裂缝群
fracture system 破裂系统；裂缝系统；裂隙系统
fracture tectonic element 破裂构造要素；破裂构造单元
fracture test 断裂试验，破裂试验；裂缝测试
fracture theory 断裂理论
fracture toughness 断裂韧性；断裂韧度；破坏韧性
fracture toughness of rock 岩石断裂韧性
fracture trace 裂缝迹线
fracture treatment 压裂处理
fracture treatment size 压裂处理规模
fracture type 断裂型
fracture water 裂隙水；破碎带水
fracture water of weathering crust 风化带裂隙水
fracture width 裂缝宽度
fracture zone 破裂带，裂缝带；破碎带，断裂破碎带；断裂带
fractured 裂开的；破裂的，断裂的；有裂缝的；裂缝性的
fractured and destressed strata zone 破碎与卸荷带；破碎卸压带；破碎应力释放区
fractured and jointed rock 破裂岩石；破碎与节理岩体
fractured annulus around a circular opening 圆形硐室周围的破坏环
fractured annulus around circular excavation 圆形巷道周围的环状破裂带
fractured aquifer 裂隙含水层；破碎含水层
fractured bedrock aquifer 基岩裂隙含水层
fractured carbonate 碳酸盐岩裂缝
fractured crystalline rock 裂隙性结晶岩
fractured drift 裂隙的错动
fractured fault zone 破裂断层区
fractured formation 断裂地层；裂隙地层；破碎地层；破裂地层；裂缝层
fractured gas field 裂缝性气田
fractured granite 破裂花岗岩
fractured ground 断裂地层；裂隙地层；破碎地层；破裂地层
fractured horizontal well 破裂的水平井；压裂水平井
fractured hydrothermal reservoir 破裂热储；裂隙形成的地热储层
fractured interval 压裂层段；断裂间距
fractured limestone 裂缝性灰岩
fractured media 裂隙介质
fractured model 裂缝模型
fractured porous media 裂缝性孔隙介质
fractured porous medium 破碎多孔介质；裂隙多孔介质
fractured pressure 破裂压力
fractured reservoir 裂缝性储层；断裂储层，破裂储层；裂隙型储集层
fractured rock 有裂隙的岩石；破碎岩石；裂隙岩体；破裂岩体；裂隙岩体
fractured rock aquifer 裂隙岩体含水层
fractured rock masses 裂隙岩体
fractured roof 破碎顶板
fractured shale bedrock 有裂隙的页岩基岩
fractured surface 断裂面；破裂面
fractured trend 裂缝走向
fractured weak zone 软弱破碎带
fractured well 压裂井

fractured zone	断裂破碎带，断裂带；断层破碎带，破碎带；破裂带；裂隙带
fracturefilling	裂隙充填
fracture-induced anisotropy	裂缝诱导各向异性；破裂引起的各向异性
fracture-matrix porosity	裂隙基岩孔隙度
fracture-orienting	裂缝方位测定
fractures identification	裂隙识别
fracture-type reservoir	裂缝性储层
fracture-veined	裂脉状；叶脉状裂隙
fracturing	破裂；碎裂；压裂；断裂，断裂作用；龟裂
fracturing agent	压裂剂
fracturing equipment	压裂设备
fracturing evaluation log	压裂评价测井
fracturing fluid	压裂液
fracturing fluid additive	压裂液添加剂
fracturing fluid loss property	压裂液滤失性
fracturing fluid recovery	压裂液返排量
fracturing fluid residue	压裂液残渣
fracturing grouting	劈裂灌浆
fracturing height	压裂高度
fracturing liquid	压裂液
fracturing location	破裂定位
fracturing mechanics	断裂力学
fracturing mechanism	断裂机制
fracturing models	裂缝模型
fracturing of reservoir formations	储层压裂
fracturing of separate layers	分层压裂
fracturing operation	压裂操作
fracturing parameter	压裂参数
fracturing pressure	破裂压力
fracturing process	破裂过程
fracturing process irreversibility	压裂不可逆性
fracturing scale	压裂规模
fracturing segment	断裂活动段
fracturing string	压裂管柱
fracturing structure	断裂构造
fracturing technology	压裂工艺
fracturing tool	压裂工具
fracturing treatment	压裂处理
fracturing unit	压裂设备
fragile	易碎的；脆的；易损坏的
fragile material	脆弱材料；脆性材料
fragile mineral	脆性矿物
fragileness	脆性
fragility	脆性；脆度；易碎性；易裂性；脆弱性
fragility test	脆性试验
fragipan soil	脆磐土
fragment	碎屑；碎块，分段；碎片；断片；破片
fragment cone	碎屑锥
fragment flow	碎屑流
fragment size	碎块尺寸
fragmental	碎屑的，零碎的；碎屑状的；碎屑状；不完全的；不连续的；碎块的
fragmental block structure	碎块状结构
fragmental debris	岩屑；碎屑
fragmental deposit	碎屑沉积
fragmental material	碎屑物
fragmental products	碎屑产物；碎屑物
fragmental reservoir	碎屑岩储集层
fragmental rock	碎屑岩
fragmental soil	粗骨土；砾质土；碎屑土
fragmental structure	碎屑构造；碎片状结构
fragmental texture	碎屑结构
fragmental volcanic rock	碎屑火山岩
fragmentals	碎屑岩
fragmentary	碎块的；碎屑的，碎屑状的；由碎屑组成的；不完整的；不连续的
fragmentary clay	零碎的黏土
fragmentary ejecta	喷屑
fragmentary hydrologic date	间断水文资料
fragmentation	破碎，碎裂，分裂；破碎块度，碎片；爆破；裂解，裂殖；存储区未用满部分；报文分割
fragmentation and draw control	块度和放矿控制
fragmentation degree	块度；破碎程度
fragmentation efficiency	破碎效率
fragmentation hypothesis	分离说
fragmentation model	碎裂模型
fragmentation of grains	细晶化
fragmentation of rock	岩石破碎
fragmentaulic	碎屑矿体
fragmented	支离破碎；成碎片的；打碎；碎裂的
fragmented bortz	不纯金刚石砂；碎粒金刚石
fragmented rock	碎破的岩石；碎屑岩；浮石，浮动块；碎裂岩；危石；松石
fragmenting	震碎；破碎
fragmenting of continents	大陆分裂
fragmention	分裂；破碎
fragments	碎屑；废物，垃圾
fraidronite	云母正长脉岩；云煌岩
frail	脆弱的；易碎的
frailty	弱点
framboid	微球丛；莓状体；莓球粒
framboidal aggregate	莓状集合体；球丛状集合体
framboidal pyrite	煤层中的细粒黄铁矿；草莓状黄铁矿
framboidal texture	草莓状结构
frame	框架，构架，相框，架，车架，支架，构梁，边框，机框，图框；基座，底座，骨架，结构，机座，体制，构成，机构，系统；制定，编制；固定式摇床，旋转式摇床；帧，画面，像幅
frame and brick veneer construction	框架嵌砖结构
frame and panel construction	框架墙板结构
frame bridge	框架桥，框架式桥
frame building	框架建筑；构架建筑
frame construction	框架结构；构架结构；构架建筑
frame crane	龙门吊
frame dam	木制水坝
frame diagram	框架图
frame floor interaction	框架-楼板相互作用
frame foundation	框架式基础
frame girder	构架梁
frame member	构架；车架构件；构件
frame of axes	坐标系统
frame of reference	参考格架，参考标架；参考系统，参考系；参考坐标
frame of the bit	钻头架
frame reef	骨架礁；骨架生物礁
frame set	框式棚子；框式支架；肋骨样板
frame shear wall	框架剪力墙
frame shear wall interaction	框架剪力墙相互作用

frame stiffness	框架刚度；框架劲度
frame strength	骨架强度
frame structure	框架结构；构架
frame support	框式支架；框架支护；节式支架；车架支撑
frame support advancing in successive arcs	连续弧形的车架支撑
frame timbering	框形支架；木支架；板支架
frame type bridge	框架式桥
frame type power support	框式液压支架
frame wall	构架墙
frame-covered structure	全框架结构
framed barrage	构架式拦河堰
framed bent	排架；框式桥台；框架桥台
framed dam	框架坝
framed floor	构架地板；构架桥面
framed partition	有边框隔墙
framed revetment	框架式挡土墙；框架式护岸
framed shed gallery	刚架式棚洞
framed shed tunnel	刚架式棚洞
framed structure	构架结构；框架结构
framed weir	框架堰
frame-panel interaction	框架-墙板相互作用
framer	预制支架工；预制支架机；成帧器；制造者
frame-shear wall structure	框架剪力墙结构
framestone	骨架灰岩；生物构架灰岩；生物骨架灰岩
frame-structure medium bridge	框架式中桥
frame-tube structure	框筒结构
frame-type powered support	节式液压支架
frame-type structural analysis	框架式结构分析
frame-wall interaction	框架-墙相互作用
frame-wall structure	框架-墙结构
framework	格架，骨架；框架，构架，结构，机构；体制，组织；测网；脚手架
framework geology	框架地质学
framework grains	骨架颗粒；构架颗粒
framework of control	控制网
framework of faults	断裂系统
framework porosity	构架孔隙度
framework silicate	架状硅酸盐；网硅酸盐
framework silicate structure	架状硅酸盐结构；架状硅酸盐水泥
framework skeleton	骨架
framework structure	架状结构
framing	构架，骨架，框架，构筑框架；受力结构；结构，组织，龙骨；编制，构思；计划；图像定位
framing coefficient	结构系数
framing factor	结构系数；结构因子
framing member	框架构件
framing scaffold	脚手架，框式脚手架
framing sheet pile	木板桩；框架板桩
framing steel	结构钢
francevillite	黄钒铀钡铅矿
franckeite	辉锑锡铅矿
francolite	细晶磷灰石；碳氟磷灰石
frange plate	翼缘板
frangibility	脆度；易折性；脆性，易脆性，易碎性；可碎性
frangite	碎变岩；崩解岩
Franki bored pile	法兰基钻孔桩
Franki displacement caisson	法兰基位移沉箱
Franki drilled pile	法兰基钻孔桩
Franki driven-in pile	法兰基冲孔灌注桩
Franki Miga pile	法兰基-米盖灌注桩；法兰基静压预制混凝土小桩
Franki pile	法兰基桩
Franki piling system	法兰基灌注桩成桩方法
franklandite	硼钠钙石
franklinite	锌铁尖晶石
fray out	尖灭；熄灭；磨损
frazil river	片冰河
freboldite	六方硒钴矿；软硒钴矿
freckled basalt	斑状玄武岩
free angle of breakage	无侧限的爆破漏斗张角
free aquifer	潜水含水层；非承压含水层
free area	有效截面；自由区域
free ash	游离灰分
free asphalt	游离沥青
free azimuth	自由方位
free beam	简支梁
free bearing	铰座；球形支座；自由支座；活动轴承
free bitumen	游离沥青
free body	自由块体；分离体；自由体；孤立体；隔离体
free body diagram	分离体图；自由体受力图
free boundary	自由边界；无荷载边界
free breakage	无挤压崩落；自然崩落
free bulkhead	活动挡土墙；活动岸壁
free burden	从自由面爆落的岩石层；临空面崩落岩层；悬空岩体
free burning coal	不黏结煤；长焰煤
free carbon	单体碳；游离碳
free carbon dioxide	游离二氧化碳
free casing pump	套管型活动式井下泵
free caving	易塌陷的
free channel	明渠
free circulation	自由循环
free clearance	间隙；余隙
free coal	易分离煤；软煤
free column	空柱
free convection	天然对流，自然对流；自由对流
free crushing	自由破碎
free damping	自由衰减；无阻尼
free deformation	自由形变
free diffusion	自由扩散
free discharge	自由排出；雨水冲蚀
free discharge valve	自由泄流阀
free dislocation	自由位错
free dissolved CO_2	游离二氧化碳
free distance	间隙；空隙；净距；余隙
free docks	自由港
free draining	自然排水；自由排水
free drop	无侧限圆柱体试样；自由落物
free eddy	自由涡流
free edge	自由边缘；无支承边
free edge condition	自由边界条件
free end	工作面的自由面；自由端；活动支座
free end bearing	松端支承；简单支承，简支
free end of beam	梁的自由端
free energy of activation	自由活化能

free energy of formation 生成的自由能
free expansion 自由膨胀
free expansion coefficient 自由膨胀率
free face 临空面；自由面；无支柱的长壁工作面
free face comprehensive geological logging 自由面综合地质编录
free fall 直坠；自由下坠；顿钻钻井法；顿钻钻具下击
free fall boring 自由降落冲击式钻进
free field 自由场；自由声波场；自由字段；自由信息组
free field design spectrum 自由场设计谱
free field stress 原岩应力；自由场应力；采前应力；自由应力
free floating pontoon 溢流式闸门；自由浮筒
free flooding irrigation 自由漫灌
free flow 对空排放，排放，放空；地表排放，自由流，明流；无压流，无压流动，自然流动；非控制流
free flow channel 明渠渠道
free flow chute 明流泄槽
free flow conduit 无压管道
free flow tunnel 明流洞
free flowing contour 自然起伏的地形
free flowing test 涌水试验；自流试验
free flowing well 不控制的自喷井，不加控制的自喷井；畅流油井；自流井
free fluid 自由流体
free fluid index 自由流体指数
free fluid log 自由流体测井
free fluid porosity 自由流体孔隙度
free form 游离态
free frequency 固有频率；自由频率
free front in face 工作面的自由面
free gas 游离气，游离状态气体；自由气体
free gas bank 游离气带
free gas cap 游离气顶
free gas drive 游离气驱动
free gas drive pool 游离气驱油藏
free geodetic network 自由大地控制网
free gliding 自由滑移
free gold 自然金；砂金
free groundwater 自由地下水，自由水；非承压地下水；自由潜水
free hand profile 信手剖面图，手绘地质剖面；素描剖面图
free hand sketch 徒手草图；手绘草图；手勾草图
free harmonic vibration 自由谐振
free head 自由水头；无压水头
free height 净高度；自由高度
free hydrocarbon 游离烃
free hydrocarbon gas 游离烃气
free interstitial water 自由隙间水
free iron 游离铁
free jet 自由喷流；自由射流
free length 自由长度；锚杆非黏结段长度；自由段；净长度
free length of bolt 锚杆未锚固段
free level 自由水位
free lime 生石灰
free machining 易切削的；易加工的
free meander 自由曲流
free moisture 游离水分；表面水分；自由水分；外在水分

free molecular flow 自由分子流
free molecule 自由分子
free motion 自由运动；自动
free nappe 弯曲段
free needle survey 罗盘测量
free number swelling 自由膨胀序数
free of void 无空隙的
free of void dense pack 无空隙的致密充填
free open-textured 松散的；疏散的
free oscillation 自由振荡；自由振动
free oscillation of the earth 地球自由振荡
free oscillations of nonconservative systems 非保守系统的自由振动
free period 自由周期；固有周期；自由振动周期
free point locator 测卡仪；自由探卡仪
free point log 卡点测井
free radical 自由基；游离基
free retaining wall 无超载挡墙；自由式挡土墙
free settling 自由沉降；自由下沉；自然沉降
free shrinkage 自由收缩
free silica 游离硅石；自由硅石；游离二氧化硅
free slope 自然边坡；自然斜坡
free space 自由空间；自由间隙
free state 游离状态；自由状态，自由态；游离态单体态
free subsidence 自由沉陷；自由下沉；自由沉降
free subsidence method 缓慢下沉法；自然下沉法
free supports 自由支座
free surface 临空面；自由面，自由表面；潜水面；自由液面，自由水面
free surface amplification 自由表面放大作用
free surface energy 自由表面能
free surface flow 自由表面流
free surge 自由涌浪；无阻尼涌浪
free suspension test 悬吊试验
free swell 自由膨胀；自由膨胀率
free swell odometer tests 自由膨胀固结试验
free swell test 自由膨胀试验；自由膨胀率试验
free swelling index 自由膨胀指数；坩埚膨胀序数；自由膨胀系数
free swelling rate 自由膨胀率
free swelling ratio 自由膨胀率
free swing period 自摆周期；自由摆动周期
free table 自由液面
free to deflect 自由变位
free torsional vibration test 扭转自由振动试验
free vertical strain 自由的竖向应变
free vibration 自由振动
free vibration characteristics 自由振动特性
free vibration column test 自振柱试验
free vibration of undamping systems 无阻尼系统的自由振动
free vibration test 自由振动试验
free water 自由水；游离水；重力水；自由流动水
free water body 自由水体
free water content 游离水含量；自由含水量；自由水含量
free water drilling 清水钻进
free water elevation 地下水位；自由水位；潜水面
free water exchange zone 自由水交替带
free water level 自由水位；自由水面
free water resistivity 自由水电阻率

free water saturation	自由水饱和度
free water surface	地下水位；自由水位
free water table	自由水面；潜水面
free wave	自由波
free way	高速公路；快速道
free-air correction	自由空气校正；空间改正
free-air gravity anomaly	空间重力异常
free-air space	自由气隙
freeboard	超高；干舷高度；出水高度干舷；净空
freeboard section	超高段
freeboard storage	超高库容
free-body diagram	自由体受力图
free-caving	容易塌落的；容易自然崩落的
free-clamped end condition	一端自由一端固定的条件
free-correction	自由空间校正；海平重力校正
freed ore	采落的矿石
freedom	自由度；自由；间隙
free-drainage level	排水平硐；排水平巷
free-draining	自由排水的；自由脱水的；自流排水；天然排水
free-draining electron	自由电子
free-draining energy	自由能
free-draining energy difference	自由能差
free-draining face	临空面；自由排水面
free-draining field	自由场
free-draining material	自由排水物质；自由排水层
free-draining-energy change	自由能变化
free-draining-fall	自由落下
free-draining-fall boring	重力冲击钻进
free-draining-fall core cutter	自落岩心钻头
free-earth support	自由端支承；无土支承
free-end piles	桩顶自由的桩
free-end triaxial test	自由端三轴试验
free-fall boring	冲击钻进
free-fall core cutter	冲击式岩芯钻头
free-fall corer	冲击取芯器
free-fall gravity	自由落体重力
free-falling	自由下沉的；自由沉降的
free-falling bit	冲击式钻头
free-falling device	自由降落装置
free-falling grab	自返式采样器；无缆采样器
free-field stress	自由场应力
free-flow channel	阴渠；自流渠；明流渠道
free-flow jet	自由射流
free-flow tunnel	明流隧洞；无压隧洞
free-flow weir	畅流堰；溢流堰
free-flowing	自由流动的
free-flowing aqueous coal	易流动水煤浆
free-flowing material	自由流动物料；松散材料
free-flowing well	自喷井；自流井
free-fluid index	自由流体指数
free-fluid index log	自由流体指数测井
free-fluid index logging	自由流体指数测井
free-fluid log	自由流体测井
free-gas breakthrough	游离气突破
free-growth crystal	自由生长的晶体
free-hanging geophone	自由悬挂检波器
freeing port	（船）排水口
freely falling body	自由落体
freely movable bearing	活动支座
freely supported	简支的
freely supported beam	简支梁
freely supported end	自由端
freely supported structure	简支结构
freely-rising convective stream	自由上升对流
free-masonry	毛石圬工
free-oscillation	自由振荡
free-pouring	容易倾出的；自由流动的
free-running frequency	固有频率；自然频率
free-settling zone	自由沉降带
freespace	自由空间
free-space coupling	自由空间耦合
free-space field	自由空间场
free-standing pile	自由支承桩
freestone	毛石；乱石；砂石；易切砂岩；易劈石
free-stone masonry	毛石砌体
freestream turbulence	自由流湍流
free-streamline	自由流线
free-surface energy	自由表面能
free-surface flow	自由面流动
free-surface reflection	自由表面反射
free-swelling index	自由膨胀系数
free-swelling test	自由膨胀试验
free-swing period	自由摆动期
free-water layer	自由水层；潜水层
free-water level	气水接触面；自由水面
free-water table	自由水位；潜水面
freeway	高速道路；高速公路；快速路
freeway overpass	高速公路立交桥；立体交叉桥
freeway system	高速公路系统
free-working	易采的；可采的；自由开采；自由作业
freeze	冻，冻结，冰冻；结冰；卡住；卡钻；凝固
freeze box	防冻套管
freeze deep	冻结深度
freeze expansion	冻胀
freeze hole	冻结钻孔，冻结孔
freeze in	卡钻；冻住
freeze index	冰冻指数；冻结指数
freeze out lake	深冻湖；冻透湖
freeze pipe	冷冻管
freeze plant	冻结站；冷冻厂
freeze point	冰点；凝固点
freeze resistance	抗冻性
freeze road	冻板结的道路
freeze sinking	冻结法凿井
freeze sinking method	冻结凿井法
freeze thaw	冻融
freeze thaw collapse	冻融滑塌
freeze thaw debris	冻融岩屑
freeze thaw resistance	抗冻融性
freeze up	冻结
freeze-and-thaw action	冻融作用
freezed rock mass	冻结岩体
freeze-dried sediment	冷冻干燥的沉淀物
freeze-free period	不冻期
freeze-proof	抗冻的；抗冻性；耐冻的；防冻的
freeze-proofing	防冻处理
freeze-proofing coal	防冻煤
freeze-thaw	冻融
freeze-thaw action	冻融作用；冰融作用

frequency stability 频率稳定性
frequency standard 频率标准
frequency sweeping 频率扫描
frequency synthesis 频率合成
frequency table 频率表
frequency teller 频率计
frequency tolerance 频率容差；频差容限
frequency tracking switching mode 频率跟踪切换方式
frequency transformation 频率变换
frequency transformer 频率转换器；变频器
frequency trim 频率微调
frequency variation method 变频法；变频激发极化法
frequency variation rate 频率变化速率
frequency vector 频率向量
frequency window function 频率窗函数
frequency-amplitude curve 频率-振幅曲线
frequency-dependent attenuation 频率相关衰减
frequency-dependent duration characteristic 与频率有关的持续时间特征；频率相关的时程特性
frequency-dependent energy loss 频率相关能量损耗
frequency-dependent phase delay 频率相关相位延迟
frequency-dependent signal-to-noise ratio 频率有关的信噪比
frequency-difference cycle 频率差周期
frequency-distance space 频率-距离空间
frequency-domain electromagnetic sounding 频率域电磁测深
frequency-domain seismic data analysis 频率域地震资料分析
frequency-energy distribution 频率-能量分布
frequency-magnitude coefficient 频度-震级系数
frequency-magnitude parameter 频度-震级参数
frequency-modulated 调频的
frequency-modulated wave 调频波
frequency-modulation index 调频指数
frequency-phase modulation 频率-相位调制
frequency-response analysis 频率特性分析；频响应分析
frequency-response characteristic 频率响应特性
frequency-response curve 频率特性曲线；频率响应曲线
frequency-selecting amplifier 选频放大器
frequency-sensitive load 频敏负载
frequency-time record 频时图
frequency-time spectral density function 频率-时间谱密度函数
frequency-wave number analysis 频率波数分析
frequency-wavenumber domain 频率波数域
frequency-wavenumber migration 频率波数域偏移；频率波数偏移
frequency-wavenumber space 频率-波数空间
fresh 新鲜的；淡的；急流，泛滥，暴涨；新鲜；淡水河，淡水水流，小河；新调制的
fresh bit 新修复的钻头
fresh coal 新采出的煤
fresh concrete 新浇混凝土，新浇筑的混凝土；新拌混凝土；生混凝土
fresh core 新鲜岩心
fresh core sample 新鲜岩样
fresh excavation 新鲜开挖；新近挖方
fresh exposure 新鲜露头
fresh flue 新风道；进风道
fresh fracture 新断口；新破裂
fresh groundwater 地下淡水

fresh lagoon facies 淡化泻湖相
fresh lake 淡水湖
fresh mineral 新采矿物；未风化矿物；新鲜矿物
fresh mud 淡水泥浆；清水泥浆
fresh oil 新采石油
fresh rock 新鲜岩石；未风化岩石
fresh snow layer 新雪层
fresh surface 新表面；新生面
fresh volcanic rock 新鲜火山岩
fresh warm water 淡温水
fresh water 淡水，淡水的；新鲜水；净水；食水；清水；甜水
fresh water alluvium 淡水冲积层
fresh water aquifer 淡水层
fresh water barrier 淡水障壁；淡水防渗层；淡水帷幕
fresh water deposit 淡水沉积；淡水沉积物
fresh water facies 淡水相
fresh water formation 淡水建造
fresh water gel mud system 淡水泥浆体系
fresh water lake 淡水湖；淡湖
fresh water lake deposits 淡水湖沉积物
fresh water lens 淡水透镜体
fresh water line 清水管道
fresh water marsh 淡水沼地
fresh water pond 清水池；淡水池
fresh water reservoir 淡水库；淡水储层
fresh water resource 淡水资源
fresh water tank 淡水缸；食水缸；淡水箱
fresh-air raise 进风天井
fresh-deposit 淡水沉积物
freshening sea 淡化海
fresh-estuary 淡水河口；湖泊浸没的河口
freshet 融雪汛，春汛；淡水流，入海的淡水河流；洪水；山洪暴涨，山洪；淡水河流
freshet period 春汛期；汛期
fresh-lake 淡水湖
fresh-lens 淡水透镜体
freshly broken 新断裂的
freshly exposed roof 新露出顶板
freshly exposed surface 新鲜露头
freshly mixed bentonite suspension 新拌膨润土悬浮液
freshly placed concrete 新浇混凝土
freshly set mortar 新凝砂浆
freshness 淡水度；淡化度；新鲜度
fresh-opened sample 剥露面样品；新鲜面样品
fresh-phreatic 淡水潜流
fresh-salt water interface 淡-咸水界面
fresh-vadose 淡水渗流
fresh-water 淡水
freshwater alluvium 淡水冲积层
freshwater aquifer 淡水含水层
freshwater barrier 淡水屏障；淡水堤
freshwater blanket 淡水覆盖层
freshwater body 淡水体
freshwater clay 淡水黏土
freshwater degradation 淡水恶化
fresh-water deposit 淡水沉积；淡水地层建造
freshwater ecosystem 淡水生态系统
freshwater facies 淡水相
freshwater fish 淡水鱼

fresh-water flow	淡水流量
freshwater formation	淡水层；淡水生成
freshwater geomagnetic electrokinetograph	淡水地磁动电测流器
freshwater habitat	淡水生境
freshwater injection	注淡水
freshwater lake	淡水湖
freshwater lens	淡水透镜体
freshwater limestone	淡水灰岩；淡水石灰岩；内陆灰岩
freshwater marsh	淡水草本沼泽；淡水沼泽；内陆湿地
freshwater mire	淡水沼泽；淡水泥潭
freshwater mud	清水泥浆
fresh-water mud filtrate	淡水泥浆滤液
fresh-water origin	淡水成因
freshwater peat	淡水泥炭
fresh-water pool	淡水池
freshwater preflush	淡水预冲洗
freshwater resources	淡水资源
freshwater sediment	淡水沉积
freshwater source	淡水源
freshwater supply	淡水供水
freshwater swamp	淡水沼泽
freshwater-saltwater interface	淡-咸水界面
freshwater-sensitive clay	淡水膨胀黏土
Fresnel lens	菲涅尔透镜
Fresnel zone	菲涅尔带
fret	侵蚀；使粗糙
frettage	摩擦腐蚀
fretted upland	磨蚀高地
fretting	路面毁坏；侵蚀，剥蚀，蜂窝状风化；表面机械损伤，表面侵蚀；磨损；微振磨损；流水磨蚀，侵蚀，风力磨蚀
fretting corrosion	磨损腐蚀
fretting wear	侵蚀磨损
fretum	海峡；海湾
fretwork	粒岩类的风化；粒状岩石风化；蜂窝式风化
fretwork weathering	蜂窝状风化
freyalite	硬硅铈钍矿；硬铈钍石
Freyssinet jack	压力枕；扁平千斤顶；弗雷西奈式双动千斤顶
Freyssinet system of prestressing	弗雷西奈预应力法
friability	脆性；脆度；易碎性；脆弱
friability index	脆性指数
friability of coal	煤的脆性；煤的易碎性
friability test	脆性试验
friable	脆的；易碎的；松散的，散粒的
friable coal	脆煤
friable core	易碎岩心
friable formation	易碎地层；破碎地层
friable ground	脆性地层；脆性岩石
friable gypsum	脆性石膏
friable immediate roof	伪顶；易碎直接顶
friable lignite	脆性褐煤
friable material	脆性材料；易碎材料
friable rock	松散岩石；脆性岩石；易碎岩石
friable roof	破碎顶板；易碎顶板
friable sand	脆砂；散粒砂
friable sandstone	脆性砂岩
friable seam	易碎煤层
friable soil	松散土；酥性土
friable state	适碎性
friable structure	脆性结构
friable top	脆性顶板；易碎顶板
fricative	摩擦的；摩擦的
friction	摩擦，摩阻；摩擦力；阻力；摩擦离合器
friction absorber	摩擦减振器
friction against itself	自摩擦
friction anchor	摩擦型锚杆
friction anchor bolt	摩擦型锚杆
friction anchorage	摩阻锚固
friction anchored rock bolt	摩擦锚固锚杆
friction angle	摩擦角
friction angle for discontinuity	不连续面的摩擦角
friction angle of discontinuity	岩体弱面内摩擦角
friction band	摩擦带；摩阻带
friction band clutch	摩擦带离合器
friction base isolation	摩擦基底隔振
friction bearing	普通轴承；滑动轴承；摩擦轴承
friction block	摩擦块；摩擦闸瓦
friction brake	摩擦制动器；摩擦刹车
friction braking	摩擦制动
friction breccia	擦碎角砾岩；摩擦角砾岩
friction calendaring	擦胶压延
friction cap	摩擦柱帽
friction centre	摩擦中心
friction circle	摩擦圆
friction circle analysis	摩擦圆分析法
friction circle concept	摩擦圆的概念
friction circle method	摩擦圆法
friction cleaning	摩擦选煤；摩擦法清理
friction clutch	摩擦离合器；摩擦连接器
friction clutch coupling	摩擦离合器
friction coefficient	摩擦系数；摩阻系数
friction coefficient of pavement	路面摩擦系数
friction cone	摩擦锥；摩擦轮
friction conglomerate	擦碎砾岩
friction control device	摩擦控制装置
friction coupling	摩擦联轴节
friction course	防滑面层
friction crack	摩擦裂隙；摩擦破裂隙；冰川擦口
friction criterion of fault	断层摩擦定律
friction current	摩擦电流；摩擦流
friction damper	摩擦阻尼器；摩擦减振器
friction damping	摩擦阻尼
friction disc	摩擦圆碟；摩擦片
friction disk	摩擦轮；摩擦盘
friction dissipation	摩擦损耗
friction drag	摩擦阻力；摩阻
friction drive	摩擦传动
friction drive hoist	摩擦传动提升机
friction dynamometer	摩擦测力计；摩擦功率计
friction effect	摩擦效应；摩擦作用
friction factor	摩擦系数；摩擦因数；摩擦因子；摩阻因子
friction feed	摩擦进给
friction flow	摩擦流
friction for mine fill	矿石充填的摩擦
friction force	摩擦力
friction formula	纯摩公式
friction gear	摩擦轮

friction gearing 摩擦传动装置；摩擦传动
friction grip 摩擦咬合；摩擦柱锁；摩擦握力
friction grip bolt 摩擦紧固螺栓
friction head 摩擦水头；摩擦位差；摩擦损失；摩擦压头
friction head loss 摩擦水头损失；沿程水头损失
friction hoisting 摩擦式提升；摩擦提升
friction index 摩擦指数
friction isolation 摩擦隔振
friction joint 摩擦缝；摩擦节点
friction lag 摩擦滞后
friction lap 摩擦圈
friction layer 摩擦层
friction less fluid 无摩擦流体
friction lining 耐摩擦的衬里；摩擦垫；摩擦片；摩擦衬层
friction loss 摩阻损失；摩擦损失
friction loss in transition section 过渡段摩擦损失
friction mandrel 摩擦轴
friction mark 擦痕；波痕
friction material 摩擦材料；产生摩擦作用的物料
friction measuring weir 摩阻量水堰
friction of motion 动摩擦；动摩擦
friction of rest 静摩擦
friction of rolling 滚动摩擦
friction path 摩擦路径
friction pawl 摩擦爪
friction pendulum 摩擦摆
friction pile 摩擦桩；摩阻桩；抗滑桩
friction plate 摩擦板
friction pressure 摩擦压力
friction pressure loss 摩阻压力损失
friction prop 摩擦支柱
friction pulley 摩擦轮
friction pump 摩擦泵
friction reinforcing effect 摩擦加固效应
friction resistance 摩擦阻力，摩阻力
friction resistance of airflow 风流摩擦阻力
friction resistance to sliding 抗滑摩擦
friction ring 摩擦环
friction rock 擦碎岩
friction rock stabilizer 摩擦铆管
friction roller drive 摩擦滚柱传动
friction routeds 摩擦圈
friction sensitivity 摩擦灵敏度
friction separation 摩擦选；摩擦选矿；摩擦分离
friction shock absorber 摩擦式减振器
friction shoe 摩擦瓦
friction sleeve 摩擦套；摩擦套管
friction slip 摩擦滑块
friction slope 摩擦比降；摩擦坡角；摩擦坡度
friction socket 摩擦打捞器
friction spring 摩擦弹簧
friction strength 摩阻强度；摩擦强度
friction stress 摩擦应力
friction support 摩擦支座
friction surface 摩擦面；摩擦表面
friction tangent matrix 摩擦切向矩阵
friction tape 摩擦带
friction test 摩擦试验
friction torque 摩擦转矩

friction type bolt 摩擦型锚杆
friction velocity 摩擦速度
friction washer 摩擦垫圈
friction welding 摩擦焊接；摩擦焊
friction welding machine 摩擦焊机
friction wheel 摩擦轮，摩擦盘
friction yielding prop 让压摩擦支柱
frictional 有内摩擦的，黏性的；摩擦的
frictional acceleration response spectrum 摩擦加速度反应谱
frictional anchorage 摩擦锚碇；摩擦锚具；摩擦锚地
frictional angle 摩擦角
frictional attenuation 摩擦衰减
frictional behaviour 摩擦性能；摩擦特性
frictional binding 摩擦结合；摩擦约束；条纹图形法；摩擦咬合
frictional brake 摩擦制动器
frictional characteristics of lubricant 润滑剂的摩擦特性
frictional clutch 摩擦联结器
frictional coefficient 摩擦系数；摩阻系数
frictional coefficient of foundation bottom 基底摩擦系数
frictional concentration 摩擦选矿
frictional constraint 摩擦约束
frictional contact 接触摩擦
frictional current 摩擦海流；摩擦流
frictional damping 摩擦阻尼
frictional dissipation 摩擦损耗；摩擦耗散
frictional drag 摩擦阻力
frictional energy dissipation brace 摩擦耗能支撑
frictional flow 摩擦流
frictional force 摩擦力
frictional fusion 摩擦熔融
frictional head loss 摩阻水头损失
frictional heat 摩擦热
frictional heating 摩擦生热
frictional index 摩擦指数
frictional layer 摩阻层
frictional loss 摩擦损失；摩擦损失
frictional loss of head 水头摩擦损失
frictional melting 摩擦熔融
frictional metal prop 摩擦式金属支柱
frictional power loss 摩擦功率损耗
frictional pressure drop 摩擦压降
frictional pressure loss 摩擦压力损失；沿程压力损失
frictional prop 摩擦支柱
frictional resistance 摩擦阻力；摩擦抗力；摩阻力；沿程阻力；摩阻
frictional resistance coefficient 摩擦阻力系数；沿程阻力系数
frictional resistance ratio 摩阻比
frictional restriction 摩擦限制
frictional sliding 摩擦滑动
frictional soil 无黏性土；摩擦土；内摩阻力大的土；内摩擦系数高的土壤
frictional spark 摩擦火花
frictional sparking 摩擦发火花；摩擦起火
frictional state 摩擦状态
frictional strength 摩擦强度
frictional stress 摩擦应力
frictional testing machine 摩擦试验机
frictional wear 摩擦磨损

frictional welding 摩擦焊
frictional work 摩擦功
friction-ball-friction 摩擦-滚珠-摩擦
friction-breccia 擦碎角砾岩
friction-free sensor 无摩擦传感器
friction-gear 摩擦轮；摩擦传动装置
frictionless 无摩擦的
frictionless flow 无摩擦流动
friction-reducing agent 摩阻减低剂；减摩剂
friction-resistance ratio 摩阻比
friction-resistant 耐磨的
friction-type unit 摩擦型装置
friedelite 红锰铁矿
Friedel's reflection law 弗里德尔反射定律
frieseite 杂硫银铁矿；富硫银铁矿
frieslebenite 硫锑铅银矿
frigid 寒冷的
frigid climate 寒带气候
frigid karst 寒带喀斯特；寒带岩溶
frigid zone 寒带
frill 褶边；起褶边；壳皱
fringe 边，边缘，条纹，干涉条纹；尖灭带；零散漂砾；带，干扰带；端
fringe alteration 边缘蚀变
fringe area 边缘区
fringe coal 煤层露头煤；风化煤
fringe drift 边缘平巷；脉外平巷；岩石平巷
fringe effect 边缘效应；边际效应
fringe joint 边缘节理
fringe margin 皱边；外围边缘；次要边缘；附加边缘
fringe of sea 海岸；海滨
fringe order measurement 光弹条纹级测定
fringe pattern 应力光图；等色线条纹，色线条纹；干涉条纹图形；条纹图形；干涉图样
fringe reef 岸礁
fringe water 毛细端带水；边水；毛细带水；毛细边缘水；边缘水
fringedrift 脉外平巷
fringing 边缘现象
fringing beach 近岸海滩
fringing coast 边缘海岸
fringing effect 边缘效应
fringing reef 岸礁；裾礁；岩礁
fringing sea 边缘海
frith 三角港；河口湾，河口，海口；峡海湾；峡湾；海湾
fritted rock 烧变岩；熔合岩
fritting 烧结；熔化；加热试料接近熔点或成糊状
fritzcheite 磷锰铀矿
fritzscheite 磷锰铀矿
friz 卷曲
frog angle 辙叉角；道岔辙角
frog area 蛙夯区；辙叉区
frog clamp 拉钳
frog crossing 弯轨交道叉；辙叉道岔
frog heel 辙叉跟端
frog number 辙叉号数
frog number of points 道岔号数
frog point 岔心尖；辙叉尖
frog rammer 蛙式打夯机；跳跃式打夯机；蛤蟆夯

frog tamper 蛙夯
frog tamping machine 蛙式夯
frog-rammer 蛙式打夯机
frohbergite 斜方碲铁矿
frolovite 水硼钙石
frond 藻体；叶状体
frondelite 锰绿铁矿
frondescent cast 菜叶状铸模；三角洲状铸型
frondescent mark 菜叶状印痕
frondescent structure 菜叶状构造
front 工作面，工作线，工组，开挖面；钻孔口；前面的；锋，前锋，前沿，前缘，前面，前部，前方，锋面；正面；波前
front abutment 前支座；前支承点；工作面前方支承压力带
front abutment pressure 前支撑压力；前支承压力
front alignment 前轮定位
front anomaly 前缘异常
front bank 开挖面；开采面
front bay 前湾
front bricking 前部砌砖；正面砌砖
front bridge 前过桥
front canopy 前顶梁；前顶盖
front chamber 前殿；前室
front coil 前圈
front connecting rod 前连杆
front contact 前接点
front cover 前盖
front cross member 前横挡；前过梁；前横梁
front draft lug 前从板座
front draft stop 前从板座
front dumper 前自卸车
front dumping 前端翻卸
front elevation 正面图；正视图；前视图
front elevation drawing 正面图；立面图
front embankment 前堤
front end 前端；高频端；前端处理机
front end mute 前端切除
front entry 纸巷；主平巷；大巷
front facade 前视图
front fitting radius 前回转半径
front focal distance 前焦距
front focal point 前焦点
front haulage 前牵引
front head 前头部
front interface 前缘界面
front intermontane basin 前山间盆地
front land 前陆；山前地带
front leg 前支柱
front lens 前锋透镜体
front line 前线；大门绷绳
front loaded 前载
front loaded wavelet 前载子波
front of fold 褶皱前锋
front of over-riding plates 上伏板块前缘
front of sliding mass 滑体前缘
front of thrust 冲断层前锋；逆掩断层前缘部分
front part theoretical length of turnout 道岔前部理论长度
front pinacoid 前平行双面；前轴面
front pitch 前节距

front plan	正面图
front range	前山
front resistance	前沿阻力
front rod of a point	尖端杆
front row	前排
front sectional elevation	前视剖面图
front sectional view	前视图
front shield	前护盾
front shot	前视
front side	正视图；正面；进料侧
front side support	前侧支撑
front slope	单面山的陡坡；前坡；外坡；迎坡；正面坡
front support	前支撑
front target	前靶
front tipper	前翻式矿车；前转翻笼；前翻斗车
front view	正面图；前视图；对景图；正视图；前景
front view of gate of ventilation	风道门正面
front wall	前墙
front yard	前庭
frontage	临海或临路部分的土地；楼面阔度；临街地界；屋前空地；临岸线；滩岸
frontal	前面的；正面的
frontal advance	前缘推进
frontal advance displacement method	前缘推进驱替法
frontal advance rate	前缘推进速率
frontal analysis	锋前分析，前沿分析
frontal apron	冰川前堆积层；冰水沉积平原
frontal apron-plain	山前冲积平原
frontal aprons	冰前沉积
frontal arc	前弧；正面弧
frontal arc terrane	前缘弧区
frontal displacement	前缘驱替
frontal drive	前缘驱动；前轴驱动
frontal dumping	前缘倾倒
frontal erosion	前缘侵蚀
frontal fault	前缘断层
frontal faulting	前缘断层作用
frontal furrow	前缘沟
frontal geosyncline	前缘地槽
frontal guide	纵向罐道；前方导引
frontal kame	前缘冰砾阜；冰前阜；前缘沙丘
frontal moraine	前碛；终碛
frontal moraine bar	前碛堤；前冰碛沙洲
frontal movement	前缘推进
frontal region	前缘区
frontal saturation	前缘带饱和度
frontal section	纵切
frontal seepage	坝前渗流
frontal side	前端；前面
frontal slope	锋面坡度
frontal surface	锋面
frontal terrace	浆砾阜阶地；冰前阶地；冰水阶地；冰水沉积阶地；山前阶地
frontal velocity	前缘速度
frontal view	正视图；前视图
frontal zone	前缘带；锋带；前缘地带；锋区
front-end effect	前缘影响段
fronthead	前端
front-head air release	前端排气孔
front-head release port	前端泄气口
frontier	限界，境界，边界；新领域，未勘探地区，未开发地区；边疆，边境；尖端；尖端领域；前沿；前缘
frontier area	边缘地区
frontier basin	边缘盆地
frontiers	拓宽
frontland	岬；前陆；山前地带；前沿地；海角
front-limb thrust fault	前翼冲断层
front-loaded	前载，前缘加载
frontogenesis	锋面生成；锋生
frontology	锋面学
frontolysis	锋面消灭；锋消；锋面消失作用
frontpanel	面板
front-to-back ratio	前后比
front-to-back slope	总坡度；前后坡度
front-tracking technique	前缘跟踪法
froodite	斜铋钯矿
frost	霜，霜冻，冰冻，严寒；冰点以下温度；冻结；霜面
frost action	冻结作用；寒冻作用；冻裂；冰冻作用
frost active soil	冰冻作用土层
frost area	冰冻区
frost belt	冰冻带
frost blister	冻土丘
frost boil	冻融；翻浆；冰沸现象；道路翻浆
frost boil soil	冻胀土
frost boiling	翻浆
frost cleft	冻裂隙
frost climate	冰冻气候
frost crack	冻裂缝，冰冻裂隙，冻裂
frost cracking	冻裂作用
frost damage	冻坏；冻裂；冻害
frost damage in tunnel	隧道冰害
frost depth	冻结深度
frost effect	冻结作用；冻结效应
frost fissure	冰冻裂隙
frost force	冻胀力
frost fracture	冻结裂缝；冰冻裂缝
frost front	冻结前缘；冻结深度
Frost gravimeter	弗罗斯特重力仪
frost heave	冻胀；冻胀性；冰冻隆胀
frost heave bard	冻害垫板
frost heave board	冻害垫板
frost heave capacity	冻胀量
frost heave force	冻胀力
frost heave ratio	冻胀率
frost heave test	冻胀测试
frost heaving	冻胀作用；冻害；解冻后隆起；鼓胀
frost heaving effect	冻胀效应
frost heaving force	冻胀力
frost heaving mound	冻胀丘
frost heaving pressure	冻胀力；冻胀压力
frost hillock	冻丘
frost in the soil	土壤结冰
frost injury	霜害；霜冻
frost line	冰冻线；冻结线；地冻深度；冻深线
frost mark	冻痕
frost mound	冻土丘
frost penetration	冻结深度；冰冻深度；冻土深度；冻透
frost pressure	冻胀压力
frost proof	防冻；防冻的

frost protection	防霜
frost protection course	防冻保护层
frost protection measure	防冻保护措施
frost resistance	抗冻力；抗冻性；冰霜抗性
frost resistant material	防冻材料；抗冻材料
frost resisting	抗冻
frost resistivity of rock	岩石的抗冻性，岩石抗冻性
frost riving	冻裂作用；融冻作用
frost shattering	霜冻损坏；冻融崩解
frost shifting	冰冻转移
frost shim	冻害垫板
frost slabs	冻土层；冻顶；冻盖
frost smoke	冻雾；冻烟
frost soil	冻土
frost splitting	冻裂作用
frost strength	冻结强度
frost susceptibility	易冻性；冻敏性
frost susceptibility classification	易冻性分类
frost susceptibility test	易冻性试验
frost table	冰冻面；冻结面；冻土面
frost thaw resistance	冻融抗力；抗冻融力
frost upheaval	冰冻隆起；冻胀
frost wall	冻结壁
frost weathering	冰冻风化；冰冻解裂；寒冻风化
frost wedging	冰冻解裂；冰楔作用
frost work	冻裂；冻裂
frost zone	冻结区，冻结带，冰冻区；冻土层，冻土带；季节冻土带
frost-active soil	冻结土
frostbite	霜冻；霜害；冻伤
frost-bound	封冻的
frost-cracking	冻裂作用
frosted glass	毛玻璃；磨砂玻璃
frosted slab	拉毛板
frost-free days	无霜期
frost-free growing season	无霜生长期；无霜期
frost-free period	无霜期
frost-gritting	用砂砾防冻
frost-heave capacity	冻胀量
frost-heave force	冻胀力
frost-heaving pressure	冻胀压力
frost-heaving soil	冻胀土
frostless zone	无霜带
frostproof	防冻的；防冻
frost-proof depth	防冻深度；防冻厚度
frost-resistance mark	抗冻标号
frost-resistant concrete	防冻混凝土；抗冻混凝土
frost-resisting	抗冻的；耐冻的
frost-resisting property	耐寒性；抗冻性
frost-resistivity	抗冻性
frost-susceptible soil	易冻土；冻结敏感土
frost-thawing test	冻融试验
frost-weathering	冰冻风化作用；寒冻风化作用；冰冻风化
frost-wedging	寒冻风化作用；冰劈作用；冰楔作用
frost-zone	冻结区；冻结带
froth	泡沫；起泡；浮渣
froth stability	泡沫稳定性指数
frothability index	起泡性指数
frothed coal	浮选精煤
frother performance	起泡性能
frothing	起沫，产生泡沫；翻浆；起泡；翻浆
frothy	有泡沫的；泡沫多的；起泡的
frothy crude oil	泡沫原油
frothy texture	泡沫状结构
Froude's curve	弗罗德曲线
frozen	冻结的；凝固的；冰冻的；卡住的；卡钻；黏着的
frozen blocking water in well	井筒冷冻法堵水
frozen bulge	冻胀
frozen casing	被卡套管
frozen coal	贴顶煤；黏顶煤；冻结煤，冻煤
frozen crust	冻结壳，冻壳；冻结层
frozen depth	冰冻深度；冰冻厚度
frozen dowel bar	冻结传力杆
frozen drill pipe	卡住的钻杆
frozen earth	冻土
frozen earth process	冻土法
frozen earth reservoir	冻土储罐
frozen earth storage	冻土储存
frozen expansion	冻结膨胀
frozen glass	毛玻璃
frozen ground	冻地；冻结地层；冻土；永冻层；水冻层；永冻土
frozen heave force	冻胀力
frozen heave test apparatus	冻胀仪
frozen intensity	冰冻强度
frozen rock mantle	冻结岩石地幔
frozen sand	冻砂
frozen shaft wall	冻结井壁
frozen soil	冻结土壤；冻土
frozen soil engineering disaster	冻土工程地质灾害
frozen soil layer	冻土层
frozen soil strength	冻土强度
frozen soil with rich ice	饱冰冻土
frozen soil with rich water	富水冻土
frozen state	冻结状态
frozen strain	冻结应变
frozen stress	冻结应力
frozen water	冻结水；固态水
frozen zone	冰冻带；冻结带；永冻区
frozen-ground phenomenon	地冻现象
frozen-heave factor	冻胀率
frozen-heave force	冻胀力
frozen-in ratio	冻结比
frozen-soil layer	冻土层
frozen-soil structure	冻土结构；冻土构造
frozen-thawed soil system	冻融土系统
fruchtschiefer	粒斑板岩；瘤状板岩
fruchtschieste	粒斑片岩
frugardite	镁符山石
frustose	龟裂的
frustule	硅藻骸
frustulent	碎屑的；硅藻壳的
frustum	截头锥体
frustum embankment	截锥形护坡
fry-dry method	炒干法
Fuchashan Formation	浮槎山组
Fuchikou Formation	富池口组
fuchsite	铬云母
fucoid marking	树枝状印痕；海藻状印痕

fucosite	岩藻糖石
Fudashan conglomerate	付大山砾岩
Fudian Formation	付店组
fuel ash	燃料灰分；燃料灰；煤灰
fuel ash corrosion	燃灰腐蚀
fuel ash deposition	燃料灰分沉积，燃料灰沉积；燃料灰沉积
fuel ash lagoon	煤灰湖
fuel bed	燃料层
fuel gas	可燃气体；气体燃料，气态燃料；燃料气体
fuel injection	燃油喷射；燃油注射；燃料喷射
fuel line	燃油管道
fuel pipe	燃油管道
fuel space	煤槽
fugacity	逸度；逸散能；挥发性
fugacity coefficient	逸度系数
Fugichnia	逃逸迹
fugitive	逃逸的；易挥发的；不稳定的
fugitive constituent	挥发分；挥发性组分
fugitive water	渗漏水；下渗水；漏失水
fugitiveness	不稳定性；不耐久性
Fuji sandstone	福基砂岩
Fujin Limestone	福金石灰岩
Fujinshan Formation	福金山组
Fujunshan Formation	府君山组
fulcrum	转轴；转动中心；支点；支轴；支柱
fulcrum bar	支杆
fulcrum bearing	支承；支点承座
fulcrum effect	支轴效应
fulcrum jack	支点千斤顶
fulcrum lever	支杆
fulcrum pin	支销；支轴销；支销
fulgurite	闪电管石；硅管石；电焦石英；雷击石；熔岩
fulje	沙丘间洼地
full 3-D migration	全三维全偏移
full advance	全截面掘进
full advance system	全面前进式系统；全断面推进系统
full bonded bolt	黏接型锚杆
full bore	全径；贯眼；与管子等径的
full capacity	全容量
full capacity relief valve	全流量安全阀
full car vibration test rig	全车振动试验台
full caving	完全放顶；全面陷落
full caving method	全部陷落法；全放顶开采法
full cavitation	全空化作用
full coal mining	完整的煤炭开采
full coefficient	充盈系数
full communicating	完全连通的
full completed drifting	一次成巷
full completed shafting	一次成井
full cover cementation	全面灌浆
full cover stabilization	全覆盖式稳定
full cross-sectional pumping	全断面取水
full cut-off	全闭；全截止
full depth	最大深度；大吃刀
full depth asphalt pavement	全厚式沥青路面
full depth migration	全深度偏移
full diameter	最大直径；外径
full diameter core	全直径岩心
full diameter radial permeameter	大直径径向渗透率仪
full diamond drill head drilling	全断面金刚石钻头钻进
full dip	真倾斜；微倾斜；真倾角
full double tracking	在全长铺双轨
full dynamic range	全动态范围
full elastic migration theory	完全弹性偏移理论
full elastic wave equation	完全弹性波动方程
full extraction	全面回采
full face	全工作面
full face boring machine	全断面掘进机
full face cutter head	全断面式刀盘
full face drilling	全面钻进；全断面钻进
full face driving	全断面掘进
full face driving method	全断面掘进法
full face excavation	全断面开挖
full face excavation method	全断面开挖法；全断面掘进法
full face firing	全断面爆破
full face method	全段面法
full face rock TBM with reaming type	扩孔式全断面岩石隧道掘进机
full face rock tunnel boring machine	全断面岩石隧道掘进机
full face round	全断面炮孔组，全断面掘进炮眼组，全断面掘进炮孔组
full face routed	全断面掘进炮眼组
full face tunneling	全断面隧道开挖；全断面隧道掘进
full face tunneller	全断面隧道掘进机，隧道掘进机
full face tunnelling	全断面隧道掘进
full facer	全断面掘进机
full fillet weld	大填角焊缝；全角焊；满角焊
full flighted screw	全程螺杆
full gate opening	门孔全开度
full gauge	全轨距；标准轨距；足尺寸；全孔径；整个范围；全尺寸的
full gauge bit	全径钻头
full gauge deflecting bit	全径造斜钻头
full gauge hole	足尺寸井眼
full gauge stabilizer	全尺寸稳定器
full gauge whipstock	全尺寸造斜器
full group sampling	整群抽样
full hardening	全硬化，穿透硬化；淬透；全淬硬
full head	全水头
full height	全高度；由楼面至天花板
full hole	贯眼；全井眼；全孔
full hole casing	全井眼套管柱
full hole cementing	全井眼注水泥
full hole drilling	全孔钻进；无岩芯钻进；孔底全部钻井；满眼钻井
full hole joint	贯眼接头
full hole testing	全眼地层测验
full hole tool joint	贯眼式接头
full hydraulic core drill	全液压岩芯钻
full initial excavation	全断面掘进
full laden	满载的
full length of station track	站线全长
full load	满载；全负荷；满负荷
full load characteristic	满载特性
full load run	满载运转；满载工作
full load running	满载运转
full load test	满载试验；全负荷试验；最高载重测试
full load torque	满载转矩；满载扭矩

full lock shaft guide rope　全密封罐道钢丝绳
full locked coil rope　全密封式钢丝绳
full longwall　全面长壁开采法
full mature relief　成熟地形；盛壮年地形
full migration　全偏移
full open position of coupler　连接器全开位置
full open well bore　全开井筒
full pack　完全充填；满眼
full packing　全部充填
full penetrating well　全透水井
full pillar extraction　全柱开采
full pitch　正常节距
full pitch winding　整节距绕组
full plain bearing　整体式滑动轴承
full pool level　满库水位
full pore-pressure ratio　全孔隙压力比
full prestressing　全预应力
full probability　全概率
full recovery　全部回收；完全复原；完全恢复
full reservoir level　满库水位
full retreat　全面后退开采
full retreat mining　全面后退式开采法
full retreat system　全面后退式系统
full rheological curve　全流变曲线
full rotating derrick　全转式起重机
full saturation　全饱和
full saturation diving procedure　全饱和潜水法
full scale　实尺；全面；足尺的；原尺寸的；真实尺寸；足尺
full scale clearance　全部清除
full scale construction　全面施工
full scale experiment　足尺实验
full scale model　实尺模型；原尺寸模型；足尺模型
full scale test　足尺试验；全尺寸试验；实物试验；工业性试验
full screw earth auger　长螺旋钻孔机
full sea　满潮
full seam cutting　一次采全高；全层厚度采割
full seam extraction　整层开采
full section ballast consolidating machine　全断面道床夯实机
full section method　全断面法
full section view　全剖视图
full size　原大，原尺寸，原型；实际大小，实际尺寸；足尺寸；真实尺寸
full size design　足尺设计
full size drawing　原尺寸图；按实物真实尺寸绘制的图
full size field test　实际现场试验
full size foundation　足尺基础
full size model　足尺模型；实尺模型；全尺寸模型
full sized　与实际尺寸相同的；足尺的
full spectrum seismograph　全频带地震仪；宽频地震仪
full speed　全速；开足马力速度
full stream sampler　自动顺向全宽截流取样机
full subsidence　完全下沉；最大下沉；充分采动；充分沉陷，完全沉陷
full teetering　完全摇摆流化态
full thickness　全厚度
full tide　满潮
full timbering　完全棚子；完全支护；密集支护；全木支架；全木材结构
full tree-dimensional migration　全三维偏移
full tubular rivet　空心铆钉
full universal drill　万能钻床
full view　全视图；全景
full viscosity-temperature curve　全黏-温曲线
full wave equation　全波动方程
full wave theory　全波理论
full wave train acoustic logging　全波列声波测井
full wave train log　全波列测井
full waveform　全波形
full waveform log　全波形测井
full waveform recording　全波形记录
full waveform tomographic reconstruction　全波形层析成像
full web cut　全截深割煤
full weight　满载重量；毛重；全重
full weight load　满重负载
full zone development pattern　合采开发井网；全区域开发格局
full zone injector　合注井
full zone producing well　合采井
full-advance method　全面前进式开采法；前进式房柱开采法
full-advance panel　全面前进式盘区
full-automatic　完全自动的；自动化的
full-bee tunneling　全断面隧道掘进
full-bore spinner flowmeter　全井眼涡轮流量计
full-bottom advance　井筒全断面掘进；全断面掘进
full-car haulage drift　重车平巷
full-circle-swinging　全转式的
full-column resin grouted bolt　全孔树脂锚杆
full-column resin grouting of bolts　锚杆全长树脂锚固
full-continuous transportation of open-pit　露天矿连续运输开采工艺
full-cut brilliant　完整多面形；标准多面形
full-depth asphalt pavement　全厚度沥青铺面
full-depth grouting　全深灌浆
full-diameter core　大直径岩芯
full-disc cloud picture　全景圆盘云图
fuller's clay　漂白土
fuller's earth　漂白土；埃洛石；高岭石；蒙脱石；漂泥
full-face　全断面的
full-face advancing　全断面掘进
full-face attack　全断面掘进法；全断面掘进
full-face bit　全面钻头；不取岩芯钻头
full-face blast　全断面爆破
full-face blasting　全工作面一次爆破；全断面一次爆破
full-face blasting method　全断面一次爆破
full-face cutting machine　全断面掘进机
full-face diamond bit drilling　全断面金刚石钻头钻进
full-face digging　全断面开挖
full-face drilling　全断面钻进
full-face driving　全断面掘进
full-face excavating method　全断面掘进法
full-face excavation　全断面掘进
full-face excavation method　全断面开挖法
full-face inclined sumping　留三角煤斜切进刀方式
full-face triangle inclined sumping　斜切进刀方式
full-face tunnel boring machine　全断面隧道掘进机；全断面TBM
full-face tunneling　全断面掘进

full-face tunnelling method	全断面隧道开挖法
full-flow valve	全流阀
full-gauge deflecting bit	全径偏转钻头；全径造斜钻头
full-hole bit	全面钻进钻头；不取芯钻头；无岩芯钻头
full-hole casing	全孔下套管
full-hole core	全直径岩芯；全孔岩芯
full-hole drilling	全孔钻进；无岩芯钻进；全面钻进；全尺寸钻进
full-hole rock bit	全断面钻头
fulling clay	漂土，漂泥
full-laden	满载的
full-length anchored bolt	全长锚固式锚杆
full-length grouted anchor	全长注浆锚杆
full-load	满载；满荷
full-load character	满载特性
full-load condition	最大负载条件；满载状态；极限负载状态
full-load loss	满载损耗，满载损失
full-load running	满载运行
full-mechanized mining face	综采工作面
fullness coefficient	充盈系数；满蓄率
fullness coefficient of indicator diagram	示功图丰满系数
fullness coefficient of pool	油气藏充满系数
fulloppite	柱硫锑铅矿
full-oven test	全炉试验；大炉试验
full-packed assembly	满眼组合钻具
full-packed bottom-hole assembly	满眼底部钻具组合
full-pillar extraction	全柱开采
full-pool level	满塘水位
full-retreat panel	全面后退式盘区
full-revolving	全转式的；全循环式的
full-round placing system	全断面衬砌系统
full-scale bit	足尺寸钻头
full-scale deflection	足尺挠度
full-scale dual operation	全面进行生产操作
full-scale gravel packing model	实际尺寸的砾石充填模型
full-scale measurement	自然测量；全尺寸测量
full-scale model	实比模型；与原物大小相同的模型；足尺模型；原型模型
full-scale model test	足尺寸模型试验；全尺寸模型试验
full-scale structure	原型结构
full-scale test	原型试验；足尺试验；全尺寸试验
full-scale value	满标值；满量程
full-scale water injection	大规模注水
full-scale waterflood	全面注水开发（采油）
full-scale well pattern	全面开发井网
full-seam coal	全厚采出的煤
full-seam extraction	整层开采；全厚开采
full-seam mining	全煤厚开采；整层开采
full-size	足尺的；原尺寸的
full-size tunneling machine	全断面岩巷掘进机
full-size tunneling shot	全断面隧道爆破
full-sized bit	原尺寸钻头；全径钻头
full-station point	基点
full-tide	满潮；高潮
full-tide cofferdam	围护墙；满潮围堰
full-wave	全波
full-wave acoustic logging	全波声波测井
full-wave rectification	全波整流
full-wave theory	全波理论
full-waveform log	全波形测井
full-web inclined sumping	全截深斜切进刀方式
full-web section switch rail	特种断面尖轨
full-width weir	满宽堰
fully aerated flow	充分掺气水流
fully automatic	全自动的
fully automatic cement mix plant	全自动水泥拌合站
fully automatic computer controlled mixing and grouting unit	全自动计算机控制拌合与灌浆机
fully automatic control	全自动控制
fully bonded bolt	全长黏接型锚杆
fully bounded reinforced bar	全长灌注的加固锚杆
fully circular length	满圆长度
fully closed	完全闭合的
fully closed waterproofing	全封闭式防水
fully compacted gravel	完全密实的砾石
fully compensated foundation	全补偿式基础
fully compositional simulation	全组分成模拟
fully coupled simulation	完全耦合模拟
fully developed mine	开拓工程全部完成的矿山
fully developed secondary scheme	全面开发的二次开采方案
fully digital mapping	全数字化制图
fully drawn yarn	全拉伸丝
fully dredged reclamation	全面挖泥填海工程；全面疏浚填海
fully enclosed	全封闭的；防爆的
fully extended length	完全伸展长度；总伸展长度
fully flexible adjacent control system	完全灵活邻架控制系统
fully flexible footing	纯柔性基础
fully floating foundation	全浮筏基础
fully grouted bolt	全孔注浆锚杆；充分灌浆锚杆
fully grouted reinforcing bar	全长锚固钢筋；全长锚固锚杆
fully grouted rockbolts	全长灌浆锚杆
fully heat treated rail	全面热处理钢轨
fully hydro-mechanized mine	全部水力机械化的矿山；全部水力机械化的煤矿
fully implicit method	全隐式方法
fully mechanical support	全机械的支护
fully mechanized bolting rig	综合机械化锚杆安装机
fully mechanized caving face	综放工作面
fully mechanized coal face	综合机械化采煤工作面
fully mechanized coal mining	综合机械化采煤；综采
fully mechanized coal mining face	综采煤矿工作面
fully mechanized longwall	全机械化长壁
fully mechanized mining	综采
fully mechanized mining face	综采工作面
fully mechanized mining with sublevel caving	综采综放开采
fully mechanized top coal caving	综放顶煤开采
fully mechanized top coal caving technology	综采放顶煤开采技术
fully mechanized top-coal caving face	综采放顶煤工作面
fully mobile dual operation	全移动式作业方式
fully open position	全开位置
fully penetrating well	完整井
fully portable sprinkler system	全移动喷灌系统
fully resin anchored roof bolt	全孔树脂锚固顶板锚杆
fully resin grouted rock bolt	全长锚固树脂锚杆
fully rigid footing	全刚性基础
fully saturated soil	完全饱和土

fully seam cutting 一次采全高
fully softened strength 全软化的强度
fully softened strength of clay 黏土完全软化强度；黏土全软化强度
fully stabilized rotary assembly 全稳定的旋转钻具组合
fully stressed design 满应力设计
fully-automatic 完全自动化的
fully-built structure 全建成搭建物；完建的结构
fully-closed 完全降柱的；完全缩回的；完全关闭的
fully-extended height 最大高度
fully-extended length 最大拉伸长度
fully-mechanized coal face with great mining height 大采高综采工作面
fully-mechanized coal mining machine complement 综合机械化采煤机组
fully-mechanized coal winning face unit 综合机械化采煤机组
fully-mechanized coal winning technology 综合机械化采煤工艺；综采
fully-mechanized longwall face 综采长壁工作面
fully-mechanized mining 综合机械化开采，综采
fully-mechanized mining with large mining height 大采高综采
fully-mechanized sublevel caving mining 完全综放开采
fully-opened 完全升柱的；完全伸出的；完全打开的
fully-seam mining 整层开采；即一次采全高
fulminic 爆炸性的；爆发性的
Fulongao Formation 扶隆坳组
Fulongquan Formation 伏龙泉组
Fulushan conglomerate 福禄山砾岩
fulvous 黄褐色的
fulvurite 褐煤
fumarole 喷气孔，气孔；火山喷气孔；硫磺气喷出孔
fumarole condensate 喷气孔凝结水
fumarole stage 喷气期
fumarolic effusion structure 喷气溢流构造
fumarolic field 喷气孔田；冒气地面
fumarolic gas 火山气
fumarolic spring 气泉；喷气孔
fumarolic stage 喷气阶段
fumarolic sublimate 喷气孔升华物；喷口升华物
fume character 瓦斯生成性；产瓦斯性
fume classification 炮烟分级
fumehood 通风柜；排风柜；排烟罩
fume-off 冒烟；排烟
fumes 烟气；烟雾
fuming 冒烟；烟化；发烟
Funan sandstone 福南砂岩
function 作用.功能，用途，功用；函数；官能职责任务；机构，机能；起作用
function of bounded variation 有界变差函数
function of fill mass 充填体作用
function of functions 复合函数；叠函数
function of limited variation 有界变分函数
function regression 函数回归
functional analysis 泛函分析；函数解析；函数解析学
functional architecture 功能结构；功能建筑
functional diagram 功能图；方块图；工作原理图；作业图；施工图；工作原理图
functional equation 泛函方程；函数方程

functional gravel pack 有效的砾石充填
functional joint 构造缝；工作缝
functional relationship 函数关系
functional requirement 功能要求
functional viscosity 实用黏度
function-fitting method 函数拟合法
fundamental assumption 基本假定
fundamental bench mark 基本水准点
fundamental circle 基准圆
fundamental combination for action effects 作用效应基本组合
fundamental complement 基底杂岩
fundamental complex 基底杂岩；基底
fundamental construction 基本建设
fundamental construction clearance 基本建筑限界
fundamental crystalline formations 基底结晶建造；结晶地盾
fundamental differential equation of geodetic gravimetry 大地重力学测量基本微分方程
fundamental fault 基底断裂
fundamental fracture 基底断裂
fundamental frequency 基础频率，基频，基本频率；主频
fundamental frequency band 基谐频带
fundamental frequency of dam 坝的基本频率
fundamental function 基本功能；特征函数
fundamental geographic information 基础地理信息
fundamental geotectonics 基础大地构造学
fundamental harmonic 基本谐波，基波
fundamental inequality 基本不等式
fundamental law of disintegration 衰变基本规律
fundamental mining survey map 基本矿山测量图
fundamental mode of vibration 基本振型
fundamental norms 基本规范
fundamental oscillation 基本振荡；基波振荡
fundamental parameter 基本参数；基波参数
fundamental particle 基本粒子；基本质点
fundamental period 基本周期
fundamental period of vibration 基本振动周期
fundamental plane 基础面；基面
fundamental point 基本点；基点
fundamental reverberation frequency 基本混响频率
fundamental rheological curve 基础流变曲线
fundamental rock 基岩；主要岩石；基本岩石
fundamental seismic degree 基本地震烈度
fundamental series 基底岩系；基系
fundamental solution 基本解
fundamental strength 极限应力；屈服点；基本强度；基底强度
fundamental structural unit 基本结构单元
fundamental structure gauge 基本建筑限界
fundamental survey 基本测量
fundamental unit 基本单位；普通单位
fundamental wave 基波；主波
fundamental wavelength 基本波长，基波长
fungotelinite 真菌质结构镜质体；真菌类结构凝胶体
fungus resistance 耐霉性
funicular 纤维的；索状的；索的
funicular bridge 悬索桥
funicular flow 索状流动

funicular line　张力线；索线
funicular polygon　索多边形，索线多边形
Funing Formation　阜宁组
funnel　漏斗；烟囱
funnel coast　漏斗形海岸
funnel erosion　漏斗状侵蚀
funnel intrusion　漏斗状侵入体
funnel mining　漏斗式采矿
funnel mining method　漏斗式采煤法
funnel pit　漏斗状坑
funnel rail　护轮轨；槽形轨
funnel shape　漏斗形
funnel shaped　漏斗形的
funnel timbering　漏斗状支护法；漏斗状木桁架结构
funnel viscosimeter　漏斗黏度计
funnel viscosity　漏斗黏度；漏斗黏度计测定的黏度
funneled　漏斗形的
funneling　漏斗缩分法；圆锥缩分器缩分法；形成圆锥堆
funnel-shaped　漏斗形的
funnel-shaped bay　漏斗状海湾；漏斗湾
funnel-shaped opening　漏斗形放矿口
Fuping Formation　伏平组
Fuping Group　阜平群
furcap rock　蘑菇石
furiotile lake　河成湖
furious cross-lamination　双重交错纹理
furious cross-stratification　双重交错层理
furnace coal　碎块级无烟煤；冶金煤
furnace shaft　火炉回风井；上风坚井
Furongite　芙蓉铀矿
Furongshan Formation　芙蓉山组
Furongshan gneiss　芙蓉山片麻岩
furrow　沟槽，槽沟，潮沟，沟，水沟，槽，渠，滩槽；做沟槽；断层面与地面的交格，断层迹；海渠，海谷；拗陷带，纵长拗陷带
furrow cast　沟状铸型
furrow dam　沟堤
furrow damming　灌水垄沟堵塞
furrow drain　排水垄沟；排水毛沟
furrowed aa flow　沟蚀熔岩流
furry sinter　毛皮状泉华
further deformation　进一步变形
fusain　丝炭；丝煤
fusain groundmass　丝炭化基质
fusainisation　丝炭化
fuscous soil　暗色土
fuse　引线；导火索，导火线，引信，导爆线；熔化，熔解；保险管；熔断器，熔线，熔断丝，保险丝；火药线
fuse blasting　点火放炮；导火线爆破，导火线起爆；点火爆破
fuse cutter　导火线切割器；导火线切割工
fuse initiation　导火线起爆
fuse point　熔点
fused basalt　熔玄武岩；熔化玄武岩
fused cast basalt　熔铸玄武岩
fused quartz　熔凝石英
fused quartz envelope　熔融石英外壳
fused rock glass　熔石玻璃

fused salts　熔盐
fuselage stress diagram　机身应力图
fuselage strut　机身支柱
Fushan sandstone　浮山砂岩
Fushi limestone　浮石石灰岩
Fushun Formation　抚顺组
fusibility　可熔性；熔度；熔性
fusible　可熔的，可熔化的；易熔的
fusible link　保险连杆；熔断保险器
fusible plug　易熔塞栓，易熔塞；安全塞；保险丝插塞
fusible point　熔点
fusible salt　易熔盐
fusibleness　熔性；熔度
fusiform　两端尖的；流线型的；纺锤状；梭形
fusil　旋转式结核
fusing　熔断；烧熔；熔融；熔化
fusing point　熔点
fusinite　丝质体；丝质组
fusinite coal　丝质煤
fusinite-collinite coal　丝质无结构镜质煤
fusinite-posttelinite coal　丝质亚结构镜质煤
fusinite-precollinite coal　丝质似结构镜质煤
fusinite-telinite coal　丝质结构镜质煤
fusinitic coal　丝质煤
fusinitization　丝炭化作用
fusinito-collinite　丝质无结构镜质煤
fusinito-posttelinite　丝质亚结构镜质煤
fusinito-precollinite　丝质似结构镜质煤
fusinito-telinite　丝质结构镜质煤
fusinization　丝炭化作用
fusinized　丝炭化的
fusinized groundmass　丝炭化基质
fusinized microcomponent　丝炭化微量组分
fusinoid　丝质组；丝质类
fusinoid group　丝质组；丝炭化组
fusinte needles　针状丝质体
fusion　熔化，熔合；熔解；熔融；消融；合成；聚变
fusion crust　熔壳
fusion cutting　熔化切割
fusion depth　熔化深度
fusion drilling　熔化钻眼法
fusion face　（焊接）坡口面
fusion heat　熔化热
fusion index　熔融指数
fusion line　熔合线
fusion penetration　熔融深度
fusion piercing　火力钻进；熔化穿孔；热力法钻进
fusion piercing drill　熔化钻机
fusion point　熔点
fusion tectonic　熔融构造
fusion tectonite　熔融构造岩；原生构造岩
fusion temperature　熔化温度；熔点
fusion zone　熔融带；母材熔合区
fusion-breccia　熔融角砾岩
fusion-piercing drill　热熔钻头
fusion-piercing equipment　热力穿孔设备
fusi-resinite　丝炭化树脂体，丝炭树脂体
fusite　丝炭，丝炭型；丝煤，乌煤；微丝炭；微丝煤
fusitoid　丝煤类
fusoclarain　丝质亮煤；丝亮煤

fusoclarite	丝质亮煤；微丝质亮煤	fuzzy image recognition	模糊图像识别
fusodurain	丝暗煤；丝质暗煤	fuzzy integrated evaluation	模糊综合评价
fusotelain	丝质结构镜煤	fuzzy language	模糊语言
fusovitrain	丝质镜煤；丝质结构镜煤	fuzzy linear programming	模糊线性规划
fusovitrite	微丝质镜煤；丝质镜煤；半丝质体	fuzzy mathematics	模糊数学
future reserves	远景储量	fuzzy matrix	模糊矩阵
future reservoir pressure performance	未来油层压力动态	fuzzy measure	模糊测度
Fuxian Formation	鄜县组	fuzzy model	模糊模型
Fuxianling Formation	幅先岭组	fuzzy optimization	模糊优化；动态优化
Fuxin Formation	阜新组	fuzzy pattern distinction	模糊模式识别
Fuzhou Formation	复州组	fuzzy phenomenon	模糊现象
Fuziya limestone	夫子垭石灰岩	fuzzy region	模糊区域
fuzzy clustering	模糊聚类	fuzzy reliability	模糊可靠度
fuzzy decision-making	模糊决策	fuzzy risk	模糊风险
fuzzy equation	模糊方程	fuzzy stochastic reliability	模糊随机可靠度
fuzzy estimation	模糊评估	fuzzy synthetic evaluation	模糊综合评价
fuzzy evaluation	模糊综合评价	fuzzy vibration	模糊振动
fuzzy function	模糊函数	fynchenite	凤凰石

G

G criterion　G 判据；能量释放率判据
gabbride　辉长岩类
gabbrite　淡辉长细晶岩；微晶辉长岩
gabbro　辉长岩
gabbro aplite　辉长细晶岩
gabbro diorite　辉长闪长岩
gabbro felsite　辉长霏细岩
gabbro granite　辉长花岗岩
gabbro nelsonite　辉长钛铁磷灰岩
gabbro norite　辉长苏长岩
gabbro pegmatite　辉长伟晶岩
gabbro porphyrite　辉长玢岩
gabbro porphyry　辉长斑岩
gabbro syenite　辉长正长岩
gabbro texture　辉长结构
gabbroic layer　辉长岩层
gabbroid　辉长岩状的；似辉长岩的
gabbroite　辉长岩类
gabbroitic　辉长岩类的
gabbroization　辉长岩化
gabbrophyre　辉长斑岩；辉长煌斑岩
gabbrophyrite　斑状辉长细晶岩；辉长玢岩
gabion basket　铁丝石笼；石筐
gabion system　笼石系统
gabion wall　框式挡土墙；石笼墙
gabion works　铁丝石笼护岸工程
gabionade　土石笼；石笼坝
gadiometer　磁强梯度计
gadolinite　硅铍钇矿
gadolinium gallium garnet　钆镓榴石
gagate　煤精；煤玉
gagatite　煤玉；煤精；碳化木
gagatization　煤玉化作用
gage　量视；计量测量仪器；规格；标准尺度；量规，量计
gage glass　量水玻璃管
gage table　计量图表；校正表
gage tank　量水桶；量水槽；量测箱
gage tolerance　量规公差
gage zero　水文零位；水尺零点
gaged　校准的；校正的；计量的；标记的
gageite　水硅锰镁锌矿
gager　计量器；计量者
gaging　检验；校准；调整；校正；量测
gaging station　观测站；水文测量站，水位站
gaging tank　量水箱
gaging weir　量水堰
gagitite　煤玉；煤精；炭化木
Gaia hypothesis　盖亚假说；生物对地球作用说
gain factor　增益因数；增益因子；增益系数
gain reduction　增益衰减
gaining river　盈水河；地下水补给河
gaining stream　盈水河；地下水补给河流
gain-ranging analog　增益-变化模拟

gal　伽
galactic year　银河年
galactite　针钠沸石；乳石
galantine powder　明胶炸药
Galapagos rise　加拉帕戈斯洋脊
Galapagos shield volcano　加拉帕戈斯盾形火山
galapectite　埃洛石；蒙脱石
galapektite　埃洛石；蒙脱石
galaxical physics　星系物理学
galaxite　锰尖晶石；铁尖晶石
galaxy　星系；银河系；银河
galeite　氟钠矾，菱钠矾
galena　方铅矿；硫化铅
galenite　方铅矿
galenobismutite　辉铅铋矿
galenoid　方铅矿类
Galerkin approximation　加勒金近似
Galerkin finite element method　加勒金有限元法
Galerkin finite element model　加勒金有限元模型
Galerkin form　加勒金形式
Galerkin least square　加勒金最小二乘法
Galerkin method　加勒金法
Galerkin solution method　加勒金解法
Galerkin weighted residual method　加勒金加权残值法
Galerkin weighting　加勒金加权
Galerkin's Finite Element Method　加勒金有限元法
Galerkin's Method　加勒金法
Galilean frame of reference　伽利略参考格网；伽利略参考系
Galilean satellite　伽利略卫星
Galitzin seismograph　加利津地震仪
Galitzin seismometer　加利津地震仪
Gall projection　戈尔投影
galleried cave　有廊道的洞穴；阶状岩洞
gallery　平硐；坑道，地道，廊道；集水沟，排水沟
gallery gravity survey　坑道重力测量
gallery heading　主平巷
gallery testing of explosive　爆破巷道试验
gallery ventilation　坑道通风；巷道通风
gallet　石片；碎石；石屑
galleting　碎石填缝；碎石片嵌灰缝
gallium albite　镓钠长石
gallon　加仑
gallows frame　门式吊架；龙门起重架；井架
gallows frame derrick　龙门架起重机；门式起重架；门式井架
galmey　异极矿；含二氧化硅锌矿
game theory　对策论；博弈论
gaming simulation　对策模拟；博弈模拟
gamma　伽马；伽马射线；反差系数；灰度系数
gamma absorptiometry　伽马射线吸收的测定；吸光测定法
gamma activity　放射性
gamma background　伽马本底；伽马射线本底

gamma constant 伽马常数
gamma contour map 伽马等值线图
gamma decay 伽马衰变
gamma distribution 伽马分布
gamma documentation 伽马编录
gamma energy 伽马-能；伽马射线能量
gamma log 伽马-射线测井记录；伽马射线测井；伽马测井剖面
gamma logging 伽马-射线测井；伽马测井
gamma neutron log 伽马中子测井剖面
gamma neutron logging 伽马中子测井
gamma neutron method 伽马中子法
gamma probability distribution 伽马概率分布
gamma ray 伽马射线
gamma ray amplitude 伽马射线射程
gamma ray curve 自然伽马曲线
gamma ray log 伽马射线测井剖面；伽马射线测井
gamma ray logging 伽马线测井
gamma ray sonde 伽马线探头；伽马射线测井电极系
gamma ray spectroscopy 伽马线能谱学
gamma spectrometer 伽马能谱仪；伽马分光仪
gamma spectrometry 伽马能谱测量法
gamma spectrum 伽马射线能谱
gamma survey 伽马测量；伽马探矿法
gamma-contamination radiator 伽马辐射污染源
gamma-contamination ray 伽马射线
gamma-contamination-emitting radioisotope 伽马射线放射性同位素
gamma-gamma logging 伽马-伽马测井；伽马-伽马射线测井
gamma-neutron log 伽马-中子测井记录；伽马-中子测井图
gamma-ray correction 自然伽马测井曲线校正
gamma-ray curve 自然伽马曲线
gamma-ray density gauge 伽马射线密度仪
gamma-ray exploration 伽马射线勘探
gamma-ray log 伽马射线测井记录
gamma-ray logging 伽马射线测井；自然伽马测井
gamma-ray neutron tool 自然伽马-中子测井仪
gamma-ray radiation 伽马射线辐射
gamma-ray source 伽马源
gamma-ray tool 伽马测井仪
gamma-ray transmission method 伽马射线透射法
gang of wells 井群组；组合井
gang of wells pumping 抽水井群
gangmylonite 侵入糜棱岩
gangue 脉石；矿石中杂质；尾矿；矸石；煤矸石
gangue froth 脉石泡沫；尾矿泡沫
gangue mineral 脉石矿物，矿物矸子
gangue rock 脉石岩；矸子石
gangway 主运输平巷；木桥；矿山进出通路；通路；主巷道；内走廊
gangway conveyor 主平巷输送机
gannister 细粒石英砂岩；细粒石英岩；致密硅岩
gantry 起重机，起重台架，桥式吊车；铁路信号架；跨线桥，栈桥；地下支架；门架，吊架
gantry crane 高架起重机；桥式起重机；龙门起重机；龙门吊；门架吊车
gantry-like platform 栈桥式平台
gap fault 张开断层；开口断层
gap gradation 不连续级配；间断级配

gap graded soil 不连续级配土；间断级配土
gap grading 间断级配
gap in the succession of strata 地层间断
gap observation 裂缝观测
gap of outcrop of beds 岩层露头缺失
gap test 殉爆试验；空隙检验
gap zone 间断带
gap-graded soil 不连续级配土
gaping 缝隙；张口的
gaping fault 张开断层
gaping fissure 张开裂缝
gaping hole 不渗透层的孔
gaping joint-fissure 张开裂缝
gapping 裂隙；裂缝；间隙
gapping fault 张开断层
gapping fissure 张开裂缝
gapping joint-fissure 张开裂缝
gappy 有裂口的；破裂的；裂缝多的
gap-sized grading 间断级配
garbage disposal 垃圾处理
garbage dump 垃圾填土；垃圾堆填
garganite 闪辉煌斑岩
garland 井筒壁的集水圈；集水圈；流水环沟；金属制花环
garnet 石榴石
garnet gneiss 石榴子石片麻岩
garnet jade 钙铝榴石
garnet rock 石榴石岩；石榴子石岩
garnet sand 石榴石砂
garnet-gneiss 石榴片麻岩
garnetiferous 含石榴子石的；含榴石的
garnetite 石榴石岩
garnetization 石榴石化；石榴子石化
garnet-muscovite granite 石榴石白云母花岗岩
garnetoid 似石榴石；似石榴子石类
garnet-rock 石榴石岩
garnierite 硅镁镍矿；硅镍矿
garrelsite 硅硼镁石；硅硼钠钡石
garret reel 加勒特式小型钢卷取机
garronite 十字沸石
gas 气体；瓦斯；天然气；瓦斯，煤气，沼气，毒气；气态
gas absorption 瓦斯吸附；气体吸附；气体吸收
gas accumulation 气藏；天然气聚集；瓦斯聚集
gas adsorption method 气体吸附法
gas analytical apparatus 气体分析器
gas aspiration 瓦斯抽放
gas atomization 气体雾化
gas bearing basin 含气盆地
gas bearing block 含气断块
gas bearing formation 含气层；气层；含气地层
gas bearing interval 含气层段
gas bearing reservoir 储气层
gas bearing shale 含气页岩
gas black 天然气炭黑
gas blanket 气体覆盖层
gas blanketing 气体覆盖层
gas blowout 瓦斯突出
gas booster 气体压缩机；压气机
gas boriding 气体渗硼

gas burner	煤气喷灯；煤气燃烧器；煤气炉
gas burst	瓦斯突出；瓦斯爆炸
gas burst disaster	瓦斯突出灾害
gas calorimeter	煤气量热器；气体量热器
gas cap	气顶；气帽
gas capillary seal	气体毛细管密封
gas carbon	气碳；炭黑
gas carbonitriding	气体碳氮共渗
gas cavity	气孔
gas chromatographic analysis	气相色谱分析
gas chromatography-mass spectrography	气相色谱-质谱法
gas chromatography-mass spectrometer	气相色谱-质谱仪
gas colliery	瓦斯煤矿
gas column	气柱；含气层
gas column height	含气高度
gas combustion	瓦斯燃烧；气体燃烧
gas compressor	空压机；空气压缩机；压气机
gas concentration	瓦斯浓度；气体浓度；气含量
gas concrete	加气混凝土；气孔混凝土
gas condensate	凝析油；气体冷凝物
gas condensate field	凝析气田
gas condensate ratio	气体凝析率
gas condensate reserve	凝析油储量
gas condensate reservoir	凝析气藏
gas condensate surface	凝析油形成深度
gas conduction	气体导电
gas conduit	煤气管道
gas consistency	瓦斯浓度
gas constant	气体常数
gas consumption	煤气耗用量；气体耗量
gas content	含气量；气体含量；瓦斯含量
gas content in coal seam	煤层瓦斯含量
gas control	瓦斯泄出控制；瓦斯治理；气控
gas densitometer	气体密度计
gas deposit	气藏；气体矿床
gas desorption	瓦斯解吸；气体解吸
gas desorption law	气体解吸法
gas detection	瓦斯检查；沼气检查；气体检测
gas detector	瓦斯测定器；瓦斯检查器；沼气检定器；可燃气体探测器
gas detector for mine gas	井下瓦斯检测仪
gas detonation	气相起爆
gas detonator	气相雷管
gas deviation factor	天然气压缩因数
gas diffusing exponent	瓦斯扩散指数；气体扩散指数
gas diffusion	气体扩散
gas dilution	冲淡瓦斯
gas disaster	瓦斯灾害
gas drainage	瓦斯抽放；气体抽采；排瓦斯设备
gas drainage hole	瓦斯排放钻孔
gas drainage method	瓦斯抽放方法
gas drainage plant	瓦斯抽放站
gas drainage technology	瓦斯抽放技术
gas drainage under suction	抽放瓦斯
gas draining	排出瓦斯法
gas draining method	瓦斯排放法
gas dynamics	气体动力学
gas emission	瓦斯涌出；瓦斯泄出；气体排放
gas emission irregularity	瓦斯涌出不均匀性
gas emission isovol	瓦斯等量线
gas emission quantity	瓦斯涌出量
gas enclosure	气包体
gas engine	煤气内燃机；煤气机；燃气发动机
gas equation	气体方程式
gas equilibrium	气体平衡
gas expansion method	气体膨胀法
gas explosion accident	瓦斯爆炸事故
gas fat coal	气肥煤
gas flow velocity	气流速度
gas geochemical survey	气体地球化学测量
gas geology theory	瓦斯地质理论
gas gravimeter	气体重差计
gas gun	气枪；燃气气枪
gas head	气体压头
gas heave structure	气鼓构造
gas high mine	高瓦斯煤矿区
gas holder	装气罐；储气罐；煤气罐；气柜；气体贮存缸
gas hydrate	天然气水合物，天然气水化物；可燃冰
gas hydrate phase diagram	天然气水合物相图
gas hydrate reservoir	天然气水合物储层
gas hydrate stability zone	天然气水合物稳定带
gas ignition	瓦斯燃烧；气体着火
gas impermeability	不透气性
gas inclusion	气体包裹体
gas indicator	瓦斯检定器；气体指示物
gas input well	注气井
gas inrush	瓦斯喷出；喷气
gas instrument	气测仪
gas intake	进气
gas interface	油气界面；气-水界面
gas interference	气体干扰
gas invaded zone	气侵带
gas issue	瓦斯突出
gas law	气体定律
gas law constant	气体常数
gas leakage manifestation	气体漏泄显示
gas leakage test	漏气试验
gas line	输气管道；煤气管
gas line network	输气管网
gas liquid chromatography	气液色层分离法
gas liquid exchanger	气-液换热器
gas liquid inclusion	气液包裹体
gas liquid ratio	气液比
gas liquids	液化石油气
gas manometer	气压表；气体压差计
gas nitrocarburizing	气体共渗氮碳
gas oil contact	油气接触面；油气界面
gas oil reservoir	气饱和油藏
gas outburst potential	瓦斯突出的可能性
gas permeability	气体渗透率；透气性；透气率
gas permeation	瓦斯渗透；气体渗透
gas pipe	煤气喉管；气体喉管；煤气管；瓦斯管
gas pipeline	煤气管道；气体管道；燃气管道
gas pore	气孔；气隙
gas pressure	气体压力；燃气压力
gas pressurization	气体加压
gas radon	氡气
gas ratio	气油比
gas reservoir rock	储气岩层

gas sand	含气油砂；含气砂岩；天然气砂；天然气砂岩	gasification zone	气化区；气化带
gas seam	瓦斯层	gasiform	气态的
gas segregation	气体偏析	gasket	衬垫，垫圈，垫片；密封垫，垫密片；填料，接合垫料；胶边
gas slippage effect	气体滑脱效应		
gas sorption	气体吸附	gasket cement	密封胶；衬片黏胶
gas storage cavern	贮气洞室；储气洞库	gasket factor	密封系数
gas storage in salt caverns	岩盐储气库	gasket packing	板式填料
gas stripping	气洗作用；气提	gasket ring	填密环；垫环
gas surveying	天然气测量；瓦斯量测量；气体测量	gasket seat	垫圈座
gas vent	气孔；通气孔；排气口；喷气孔；喷气口	gasketed joint	有衬垫接合；垫片接合；填实接缝
		gasket-sealed joint	衬垫密封连接
gas volcano	喷气火山	gas-law constant	气体定律常数
gas welding machine	气焊机	gasless	无气的；无瓦斯的
gas well	气井，天然气井	gasless delay composition	无气体延期混合炸药
gas zoning	瓦斯分带	gas-lift flowing well	空压机抽水井；气举出水井
gas-bearing area	含气面积；含气区	gas-mask	防毒面具
gas-cap drive	气顶驱动	gasogenetic anomaly	气成异常
gaseogenic anomaly	气成异常	gas-oil ratio	气油比
gaseous anomaly	气体异常	gasoline breaker	燃汽油混凝土打碎机
gaseous halo	气成晕；气体晕	gasometric	气量的
gaseous liquid inclusion	气液包裹体	gasometric geochemical prospecting	气量法化探
gaseous phase	气相	gasometric study	气量研究
gaseous state	气态	gas-producing agent	产气催化剂
gaseous transfer	气体迁移；挥发分迁移	gas-prone	蕴气性
gaseous transfer differentiation	气体搬运分异作用；气体分异作用	gasproof shelter	瓦斯躲避峒；防毒掩蔽部
		gas-saturated crude oil	饱和天然气原油
gaseous volcano	气体火山	gasser	喷气井；天然气井
gaseous water	含气水；气态水	gasser show	气苗；天然气露头；气显示
gases geothermometer	气体地热测试仪；气体地温表	gassi	沙丘沟
gas-exhalation	气孔；喷气孔	Gassman equation	加斯曼方程
gas-expansion factor	天然气膨胀系数	gassy sediment	含气沉积物
gas-field water	气田水	gastaldite	蓝闪石
gas-fluid-rock interaction	气体-流体-岩石相互作用	gas-tidal hypothesis	气体潮说
gas-fuel heater	瓦斯点火器；燃气加热器	gate-road locomotive	运输平巷电机车；调车机车
gas-gravity differentiation	气体重力分异	gather	聚集，道集，集合，收集；增长；选排
gash breccia	崩塌角砾岩；裂隙角砾岩	gather stack	选排叠加
gash fracture	张口破裂；张裂缝	gather up	收集
gash joint	张节理；逐渐尖灭张裂隙	gathering arm loader	蟹爪式装载机；聚臂式装载机；集爪式装载机
gash vein	裂缝矿脉；裂伤矿脉		
gasification by bore-hole method	钻孔地下气化法	gathering barrel	收集桶
gasification by fire percolation method	火力渗透地下气化法	gathering center	集油中心
		gathering conveyor	汇集输送机；集料输送机
gasification by hydraulic fracturing method	水力破煤地下气化法	gathering facility	油气集输设施
		gathering flow	聚流
gasification by hydro-linking method	钻孔水力贯通地下气化法	gathering ground	集水区；流域；储油面积；汇水面积
		gathering locomotive	调车机车
gasification by nuclear fracturing method	核爆裂煤层地下气化法	gathering network	集油管网
		gathering point	集油站
gasification by stream method	倾斜煤层底部建立燃烧槽的地下气化法	gathering process	选排处理
		gathering tank	集油罐；收集桶
gasification chamber	气化室；气化燃烧区	gathering zone	积聚带；集水带
gasification channel	气化槽；气化通道	gather-scatter instruction	集散指令
gasification coal	气化煤	gather-scatter operation	集散操作
gasification desulfurization	气化脱硫	gating pulse	控制脉冲
gasification of coal	煤的气化	gating system	选通系统；浇注系统
gasification of coal in place	煤的地下气化	gating time function	选通时间函数
gasification of oil in place	地下原油气化	gatumbaite	水羟磷铝钙石
gasification panel	气化盘区，气化燃烧盘区	gauffer	皱褶；压制波纹
gasification path	地下气化燃烧通路	gaufrage	微褶皱；皱纹
gasification strength	气化强度	gauge compact	背锥齿
gasification system	气化法；气化系统		

gauge diamond 侧刃金刚石
gauge edge 刮刀钻头外缘刃
gauge factor 仪器常数；仪表灵敏度；应变系数
gauge feeler 塞规；探规；触探仪
gauge hand 压力表指针
gauge hatch 量油口
gauge height 水位高度；水位；量高
gauge hole 量孔
gauge joint 限径管
gauge length 标距长度
gauge line 卷尺；规线；应变电阻丝
gauge loss 直径磨损
gauge pile 定位桩
gauge pressure 表示压力；计示压力；表压
gauge ring 内径规
gauge rod 探测杆；表尺；计量杆；量油尺
gauge row 保径齿圈
gauge saver 压力表机构的减震器
gauge seal 量孔盖
gauge size 计量尺寸
gauge staff 测量标杆
gauge station 水文站；水文测量站
gauge stick 标杆；量标
gauge stone 侧面金刚石
gauge table 油罐计量表；规格尺寸表
gauge tank 计量罐
gauge tape 卷尺
gauge tie rod 轨距拉杆
gauge tube 测流管
gauge wear 直径磨损；径向磨损
gauge weight 检尺重锤
gauge well 观测井；水文测量井
gauge zero 表零点
gauged 标准的；校核过的
gauged mortar 混合砂浆；速凝石膏砂浆；标准砂浆
gauge-discharge curve 水位-流量关系曲线
gauge-pole 量油杆
gauger 计量器；计量员
gauging line 水位线
gauging station 空气流测量站；气象测量站；水文测量站；水位站
gauging water 调和水
gauging well 水位井；水尺井；验潮井
gault 压实黏土；泥灰质黏土；重硬黏土
gault clay 重黏土；泥灰质黏土
gauss 高斯
Gauss algorithm 高斯算法
Gauss chron 高斯年代
Gauss conformal projection 高斯正形投影
Gauss constant 高斯常数
Gauss coordinate system 高斯坐标系
Gauss differential equation 高斯微分方程
Gauss distribution 高斯分布
Gauss Divergence Theorem 高斯散度定理
Gauss elimination 高斯估计；高斯消去法
Gauss epoch 高斯正极性期；高斯期；高斯正向期
Gauss equation 高斯方程
Gauss error function 高斯误差函数
Gauss excitation 高斯激励
Gauss eyepiece 高斯目镜

Gauss function 高斯函数
Gauss integral 高斯积分
Gauss integration 高斯积分
Gauss interpolation formula 高斯插值公式
Gauss Law 高斯定理
Gauss map projection 高斯地图投影
Gauss method of reduction 高斯约化法
Gauss midlatitude formula 高斯中纬度公式
Gauss normal distribution curve 高斯正态分布曲线
Gauss normal epoch 高斯正常期；高斯正向期
Gauss normal polarity chron 高斯正向极性时
Gauss number 高斯随机数
Gauss plane coordinate 高斯平面坐标
Gauss plane coordinate system 高斯平面坐标系
Gauss plume model 高斯烟流模式
Gauss point 高斯点
Gauss probability integral 高斯概率积分
Gauss projection 高斯投影
Gauss pulse 高斯脉冲
Gauss quadrature 高斯正交；高斯积分
Gauss stationary process 高斯平稳过程
Gauss Theorem 高斯定理
gaussage 高斯磁感应强度
gaussbergite 橄辉白榴玻斑岩；橄闪白玻斑岩
Gauss-Boag map projection 高斯-博格地图投影
Gaussian curvature 高斯曲率
Gaussian curve 高斯曲线
Gaussian distribution 高斯分布
Gaussian elimination 高斯消元
Gaussian elimination technique 高斯消元法
Gaussian error curve 高斯误差曲线
Gaussian filter 高斯滤波器
Gaussian function 高斯函数
Gaussian integration 高斯积分
Gaussian kernel 高斯核函数
Gaussian law of error 高斯误差定律
Gaussian number 高斯随机数
Gaussian permeability distribution 高斯渗透率分布
Gaussian probability curve 高斯概率曲线
Gaussian probability distribution 高斯概率分布
Gaussian process 高斯过程
Gaussian pulse 高斯脉冲
Gaussian quadrature 高斯积分
Gaussian random process 高斯随机过程
Gaussian white noise 高斯白噪声
Gaussian window 高斯窗
gaussistor 磁阻放大器
Gauss-Jordan elimination method 高斯-约当消去法
Gauss-Kruger coordinate 高斯-克吕格坐标
Gauss-Kruger plane rectangular 高斯-克吕格平面直角
Gauss-Kruger projection 高斯-克吕格投影
Gauss-Laborde map projection 高斯-拉勃德地图投影
Gauss-Laborde quadrature 高斯-拉勃德正交
Gauss-Legendre point 高斯-勒点
gaussmeter 高斯计；磁强计；磁通计
Gauss-Newton method 高斯-牛顿算法
Gauss's Law 高斯定律
Gauss's system 高斯单位制
Gauss-Seidel method 高斯-赛德尔迭代法
Gauss-Seidel procedure 高斯-赛德尔程序

gauze filter	滤网；网状过滤器
gauze strainer	网状滤器；网式过滤器
gauze suction strainer	网状吸滤器
gavite	水滑石
gaw	狭窄岩浆岩脉；沟，水沟
gaylussite	单斜钠钙石；斜碳钠钙石
geanticlinal	地背斜的
geanticlinal arch	地背斜拱起
geanticlinal axis	地背斜轴
geanticlinal belt	地背斜带
geanticlinal crest	地背斜脊
geanticlinal ridge	地背斜岭
geanticlinal welt	地背斜带
geanticline	地背斜；地穹
geburite dacite	紫苏英安岩
gedanite	脂状琥珀；软琥珀
gedrite	铝直闪石
gegenion	反离子
geierite	硫砷铁矿
Geiger vibrograph	盖革振动器
Geiger-Muller counter	盖革-缪勒型计数器
Geiringer velocity equation	盖林格速度方程
gel filtration	凝胶过滤法
gel lignite	凝胶状褐煤
gel rate of mud	泥浆胶体率
gel strength	凝胶强度
gel structure	凝胶结构；凝胶构造
gel texture	胶状结构
gel treatment	胶结处理
gelamite	胶凝硝化甘油炸药；吉拉买特半胶质炸药
gelate	胶凝
gelatification	凝胶化作用；胶凝
gelatin	明胶，动物胶；凝体；胶质炸药
gelatin borehole commutator	凝胶管
gelatin dynamite	明胶炸药；胶质硝酸甘油炸药
gelatin explosive	胶质炸药；黄色炸药
gelatination	胶凝作用；凝胶化；凝胶化作用
gelatine	明胶，动物胶；凝体；胶质炸药
gelatinization	凝胶化作用；胶质化作用；胶化作用
gelatinize	胶凝；胶化；涂胶
gelatinizer	胶化剂；胶凝剂；软胶化剂
gelatinizing agent	胶凝剂
gelatinous	凝胶的；胶状的；胶质的
gelatinous bed	胶质层
gelatinous explosive	胶质炸药；胶炸药
gelatinous permissible	凝胺安全炸药
gelation	胶凝；凝胶化；冻结
gelex	吉莱克斯炸药
gelid	冰冷的
gelification	胶凝；凝胶化作用
gelification character	胶凝特性
gelified	凝胶化的
gelified circleinite	凝胶化浑圆体
gelified groundmass	凝胶化基质
gelified group	凝胶化组
gelified microcomponent	凝胶化微组分
gelifluction	冰冻泥流
gelifraction	冰冻崩解
gelifusinite	胶丝质体
gelifusinite coal	凝胶化丝质煤；胶丝质煤
gelifusinization	凝胶丝炭化
gelignite	葛里炸药；吉利那特；凝胶体；凝胶状褐煤；凝胶体
gelinite coal	凝胶煤
gelinite group	凝胶化组
gelinito	胶质
gelinito-coal	胶质结构镜质煤
gelinito-collinite	胶质无结构镜质体
gelinito-posttelinite	胶质次结构镜质煤
gelinito-posttelinite coal	胶质无结构镜质煤
gelinito-precollinite	胶质似无结构镜质体
gelinito-telinite	胶质结构镜质煤
gelisol	冻土
gelite	蛋白石
geliturbate	融冻扰动
geliturbation	融冻扰动作用
gelivation	冰冻作用；冻劈作用
gelled slurry	胶状浆液
gelled water	胶凝水；稠化水
gelling	胶凝；凝胶化作用
gelling agent	增稠剂
gelling property	胶凝性
gelmagnesite	胶菱镁矿
Gelobel	吉罗拜尔安全炸药
gelocollinite	胶质镜质体
Gelodyn	吉罗达因硝化甘油炸药
gelose	棕腐质
gelosic coal	藻煤；胶质煤
gelosite	芽孢油页岩；藻烛煤
gel-pyrite	胶黄铁矿
gel-thorite	胶钍石
gem	宝石
gem geology	宝石地质学
gem grade	宝石品级
gem grade diamond	优质金刚石；宝石级金刚石
gem gravel	含重硬矿物的砾石层
gem mineralogy	宝石矿物学
gem treating	宝石处理
gem washing	砂金洗选；宝石洗选
gem washings	冲选宝石后的废液
gemel	铰链；成对的；双生的
gemel arch	铰链拱
gemmary	宝石学；宝石雕刻师；宝石类；宝石贮藏室；宝石贮藏器
gemologist	宝石学家
gem-quality crystal	优质晶体
gem-quality diamond	宝石级金刚石
gemstone	宝石
gemstone appraisal	宝石估价
gemstone enhancement	宝石改善；宝石优化
gendarme	岩柱
genealogy	系统学；族谱
general astronomy	普通天文学
general atlas	普通地图集
general atmospheric circulation	大气环流
general base level	总基准面
general base level of erosion	总侵蚀基准面
general cartography	普通地图学
general chart	一览图；总图；普通海图
general chemistry	普通化学

general circulation 大气环流
general circulation of atmosphere 大气环流理论
general collapse 大量岩石崩落；整体坍塌
general computer 通用计算机
general contingency reserve 一般意外损失准备金
general data 主要数据
general description of construction 施工说明；构造说明
general ellipsoid 通用椭球
general equation 一般方程；通式
general equilibrium theory 综合平衡理论
general exhaust ventilation 全面排风；全面通风
general expression 通式
general extension 均匀伸长
general extrusion 广义拉伸
general facies model 综合相模式
general flowchart 综合流程图
general formula 通式
general heading 通用标题
general instability 总体失稳；总体不稳定
general instrument rack 通用机柜；通用计测器支架
general layout 总平面图；总体布置；总体计划，总体设计
general layout drawing 总布置图
general layout plan 总平面图；平面布置总图
general leveling 普通水准测量
general limit equilibrium method 总体极限平衡法
general linear model 一般线性模型
general location balance 总位置图
general map 普通地图；一览图，总图
general mass matrix 一般质量矩阵
general neutron log 普通中子测井
general neutron logging tool 普通中子测井仪
general of stress 空间应力状态
general plan 总平面图；总图；计划概要，总体计划
general planning 总体规划
general pressure 总压力
general purpose motor 通用电动机
general purpose Portland cement 普通硅酸盐水泥
general rating of maintenance quality 养护质量综合值
general recursive function 一般递归函数
general register 通用寄存器
general risk 一般风险
general scale 基本比例尺
general scour 总体剥蚀；总体冲刷；一般冲刷
general sketch 概略图，示意图
general solution 通解
general specification 一般规格；总体规定；总体技术规范书，通用说明书
general stability analysis 整体稳定分析
general standard 通用标准；基础标准
general state of strain 空间应变状态；一般应变状态
general strain 总应变
general stratigraphic column 综合地层柱状剖面图
general strike 一般走向
general survey 总体勘测；普查
general symbol 通用符号
general variational principle 广义变分原理
general view 全视图；概貌图，鸟瞰图
generalised finite element method 广义有限元法
generalised patch test 广义小片实验

generalized acceleration 广义加速度
generalized adjustment 广义平差
generalized binomial distribution 广义二项分布
generalized configuration 广义排列
generalized contour 简化等高线；综合等高线
generalized convolution 广义褶合式
generalized Coulomb's equation 广义库仑公式
generalized damping 广义阻尼
generalized Darcy's law 广义达西定律
generalized data management system 通用资料管理系统
generalized discriminant analysis 广义判别分析
generalized displacement 广义位移
generalized distance 广义距离
generalized exciting force 广义激振力
generalized experience 综合经验
generalized expression 广义表达式
generalized Fick's law 广义菲克定律
generalized force 广义力
generalized forcing function 广义力函数
generalized Fourier analysis 广义傅里叶分析
generalized Fourier transform 广义傅里叶变换
generalized function 广义函数
generalized geologic map 综合地质图；地质简图
generalized geological section 综合地质柱状图；概略地质剖面；简化地质剖面
generalized geometric stiffness 广义几何刚度
generalized Green's function 广义格林函数
generalized Hook's law 广义虎克定律
generalized hydrostatic equation 广义流体静力学方程
generalized inertia force 广义惯性力
generalized integral 广义积分
generalized inverse matrix 广义逆矩阵
generalized isoreflectance contour map 广义反射率等值线图
generalized Lagrange multipliers 广义拉格朗日乘子
generalized least squares method 广义最小二乘法
generalized limit 广义极限
generalized linear inversion 广义线性反演
generalized load 广义载荷
generalized mass 广义质量
generalized Maxwell model 广义麦克斯韦模型
generalized minimal residual method 广义极小残值法；一种线性方程组的迭代解法
generalized Newmark algorithm 广义纽马克算法
generalized Newtonian fluid 广义牛顿流体
generalized Ohm's law 广义欧姆定律
generalized procedure of slices 广义系法；广义条分法
generalized property 广义特性
generalized pulse spectrum technique 广义脉冲谱技术
generalized quasilinearization theory 广义拟线性理论
generalized ray 广义射线
generalized ray theory 广义射线理论
generalized reduced gradient search 广义简化梯度搜索
generalized section 综合剖面；简化剖面，概化剖面；剖面示意
generalized spectral density 广义谱密度
generalized stiffness 广义刚度
generalized strain 广义应变
generalized stratigraphic chart 综合地层图
generalized stratigraphy 综合地层；综合土层

generalized stress　广义应力
generalized terrain correction　广义地形改正
generalized theorem　推广定理
generalized upper bound constraint　广义上界约束
generalized variance　广义方差
generalized variational principle　广义变分原理
generalized velocity　广义速度
generalized vibration equation　广义振动方程
generalized virtual displacement　广义虚位移
generalized Voigt model　广义伏格特模型
general-purpose simulation system　通用模拟系统
general-shear failure　整体剪切破坏
general-survey drill hole　普查孔
generated crack surface　新产生的裂纹面
generating area　风浪区；风波区；发源区域
generating function　母函数，生成函数
generating rock　生油岩
generation of fold　褶皱期；褶皱世代
generation of mineral　矿物世代
generation of mineralization　矿化世代
generation of opening　裂隙的生成
generation of structure　构造世代
generative basin　生油盆地
generative fold　增长褶皱
generative window　生油窗
generator furnace　煤气发生炉
generator gate　脉冲发生器
generator matrix　生成矩阵
generator program　生成程序
generic　一般的；属的；类的
genescope　频率特性观测仪
genesis　成因；起源；创始；发生
genesis of oil　石油的成因
genesis of sediment　沉积物的成因
genetic agent　遗传因子
genetic algorithm　遗传算法
genetic classification　成因分类
genetic classification system　成因分类系统
genetic code　遗传密码
genetic composition　原生成分
genetic condition　生成条件
genetic damage　遗传损害
genetic development　成因演化过程
genetic engineering　基因工程；遗传工程学
genetic evolutionary algorithm　遗传进化算法
genetic factor　遗传因子
genetic feature　成因特征
genetic increment of strata　地层的成因增量
genetic information　遗传信息
genetic interval of strata　地层成因段
genetic learning　遗传学习法
genetic mechanism　成因机制
genetic mineralogy　成因矿物学
genetic model　成因模型；成因模式；遗传模型
genetic morphology　地貌成因学
genetic porosity　原生孔隙度
genetic potential　生油潜力
genetic relation　成生联系
genetic relationship　成因关系；亲缘关系
genetic separation　成因划分

genetic sequence of strata　地层的成因序列
genetic series　成因系列
genetic soil　原生土
genetic stratigraphic sequence　成因地层层序；成因层序
genetic type　成因类型
genetic type of mineral deposit　矿床成因类型
genetic type of ore deposit　矿床成因类型
genetical　成因的；发生的
genetical association　成因组合
genetical marking　成因标志
genetical mineralogy　成因矿物学
genetical morphology　地貌成因学
genetics　遗传学；发生学
genetics-based technique　基于遗传学的技术
geniculate twin　膝状双晶
geniculated twin　膝状双晶
genthite　镍叶蛇纹石；水硅镁镍矿
gentle anticline　平缓背斜
gentle ascent　缓慢上坡
gentle asymmetric fold　平缓不对称褶皱
gentle dip　微倾斜；平缓倾斜
gentle flank dip　平缓侧翼倾斜
gentle folding　平缓褶皱作用
gentle homocline　平缓同斜层
gentle incline　缓坡度；缓坡斜井
gentle regional uplift　平缓区域隆起
gentle slope　平缓坡度；平缓边坡；缓坡；平坡
gentle structure　平缓构造
gently dipping　缓倾；缓倾斜
gently dipping fault　缓倾断层
gently dipping joint　缓倾节理
gently inclined　缓倾斜的
gently inclined joint　缓倾角裂隙
gently inclined seam　缓斜煤层；缓斜矿层
gently rolling country　缓丘陵地带
gently rolling topography　平缓起伏地形
gently sloping　缓斜的
gently sloping coast　缓斜岸
gently-dipping domal structure　平缓穹窿构造
gently-dipping monoclinal slope　平缓单斜坡
genuine shale　纯页岩
genuine vitrinite　真正镜质体；纯镜质体
geoacoustic model　地声学模型
geoacoustic spectrum　地声谱
geoacoustics　地声学
geoanalysis　地理学分析
geoanalysis system　地质分析系统
geoanalytical technology　地质分析技术
geoanticlinal axes　地背斜轴
geoanticline　地背斜
geoarchaeological　地质考古的；地质考古学的
geoarchaeology　地质考古学；地质考古
geoastronomy　宇宙地质学；天体地质学
geoastrophysics　天文地球物理学
geobarometer　地质压力计；地球压力计
geobarometry　地压测量法；地压测量学
geo-based file　大地原始编码文件
geo-based information system　大地信息系统
geo-based observation　大地原始观测
geobasin　深厚水平沉积盆地；盆地

geobelt 土工带
geobiochemistry 地球生物化学
geobiology 地质生物学；地球生物学；地生物学
geoblock 大地断块；地断块
geobody shadow feature of photograph 像片地物阴影特征
geocartographical unit 地质制图单位
geocell 土压力盒
geocenter 地心
geocentral longitude and latitude 地心经纬度
geocentric 以地球为中心的；地球中心说的；地心的
geocentric angle 地心角
geocentric coordinate 地心坐标
geocentric coordinate system 地心坐标系
geocentric datum 地心基准
geocentric gravitational constant 地心引力常数
geocentric horizon 地心地平面
geocentric latitude 地心纬度
geocentric longitude 地心经度
geocentric origin 地心原点
geocentric rectangular coordinate 地心直角坐标
geocentric theory 地球中心说；地心说
geocentricism 地心说
geocerite 硬蜡
geochemical 地球化学的
geochemical abundance 地球化学丰度
geochemical activity 地球化学活性
geochemical analyse 地球化学分析
geochemical analysis 地球化学分析
geochemical anomaly 地球化学异常
geochemical balance 地球化学平衡
geochemical block 地球化学块体
geochemical classification 地球化学分类
geochemical column 地球化学柱状图；地球化学柱剖面
geochemical composite sample 化探组合样
geochemical continent 地球化学洲
geochemical criterion 地球化学判据；地球化学标准
geochemical cycle 地球化学旋回；地球化学循环
geochemical data 地球化学资料
geochemical data processing 地球化学数据处理
geochemical degradation 地球化学降解
geochemical differentiation 地球化学分异作用
geochemical dispersion halo 地球化学分散晕
geochemical domain 地球化学域
geochemical drainage survey 地球化学水系测量；地球化学水系调查；水地球化学测量
geochemical ecology 地球化学生态学
geochemical engineering 地球化学工程学
geochemical exploration 地球化学勘查；地球化学探矿；化探
geochemical fence 地球化学边界
geochemical gas survey 气体地球化学测量
geochemical index 地球化学指数
geochemical indicator 地球化学指示物；地球化学标志
geochemical logging 地球化学测井
geochemical megaprovince 地球化学巨省
geochemical modeling 地球化学建模
geochemical parameter 地球化学参数
geochemical plane map 地球化学平面图
geochemical process 地球化学作用；地球化学过程
geochemical profile 地球化学剖面

geochemical prospecting 地球化学探矿，地球化学勘探
geochemical research 地球化学研究
geochemical rock survey 地球化学岩石测量
geochemical sample 地球化学样品
geochemical sampler 地球化学取样器；地球化学采样工
geochemical sampling 地球化学采样
geochemical section 地球化学剖面
geochemical soil survey 土壤地球化学测量
geochemical specialization 地球化学专属性
geochemical standard sample 地球化学标准样
geochemical stream sediment survey 地球化学河流沉积物测量
geochemical structure 地球化学结构
geochemical subblock 地球化学子块体
geochemical system 地球化学体系
geochemical table 地球化学表
geochemical threshold 地球化学阈值
geochemical tracer 地球化学示踪剂；地球化学示踪物
geochemical zone 地球化学区
geochemist 地球化学家
geochemistry 地球化学
geochemistry of individual element 个别元素地球化学
geochemistry of landscape 景观地球化学
geochemistry of lithosphere 岩石圈地球化学
geochemistry of mineral deposits 矿床地球化学
geochemistry of modern sediments 现代沉积物地球化学
geochemistry of ore deposit 矿床地球化学
geochemistry of ore formation 成矿作用地球化学
geochemistry of organic matter 有机质地球化学
geochemistry of stable isotope 稳定同位素地球化学
geochemistry of the heavy element 重元素地球化学
geochemistry of the lithosphere 岩石圈地球化学
geochemistry of trace elements 痕量元素地球化学；微量元素地球化学
geochemistry of volcanics 火山岩地球化学
geochromatography 地质色层
geochron 地质年代；岩石地层时代
geochronic 地质年代的
geochronic geology 地史学；历史地质学
geochronical 地质年代的
geochronologic 地质年代的
geochronologic chart 地质年代表
geochronologic data 地史资料
geochronologic information 地质年代信息
geochronologic interval 地质年代间隔
geochronologic sequence 地质年代顺序
geochronologic study 地质年代学研究
geochronologic unit 地质年代单位
geochronological 地质年代的
geochronological interval 地质年代间隔
geochronological mark 地质年代单位代号
geochronological scale 地质年代表
geochronological system 地质年代学体系
geochronological unit 地质年代单位
geochronology 地质年代学；地质纪年学
geochronometer 地质年代计；地质时标
geochronometric scale 地质编年表；地质年表
geochronometry 地质年代测定法；地质年代测定学
geochrony 地质年代学
geocinetics 地壳运动学

geoclinal 地斜
geoclinal valley 大向斜谷
geocline 地斜；地形差异类型；地理梯度
geocode 地理编码
geocoding system 地理编码系统
geocomposite 土工复合材料
geocomposite drain 土工复合排水材料
geocorona 地冕
geocosmogony 地球宇宙进化论；地球演化学
geocosmology 天文地质学；宇宙地质学
geocratic motion 造陆运动
geocratic movement 造陆运动；大陆扩展运动
geocratic phase 造陆期
geocronite 斜方硫锑铅矿；砷硫锑铅矿
geocryological 地球冰雪学的；多年冻土学的
geocryology 冰冻地质学；地球冰雪学；多年冻土学；地质低温学
geocushion 土工垫
geocycle 地质旋回
geocyclic 地质旋回的；地球旋转的
geocyclicity 地质旋回性
geodata 地质数据；地理数据
geode 空心石核；晶洞，晶孔，晶球；异质晶旋；晶脉
geodepression 地洼；大地凹陷
geodepression theory 地洼学说
geodepressional region 地洼区
geodesic 大地线；测地线；测地学的，大地测量学的；短程线
geodesic base line 大地测量基线
geodesic coordinate 大地坐标；测地坐标
geodesic coordinate system 大地测量坐标系统；测地坐标系统
geodesic curvature 大地曲率
geodesic datum 大地基准点
geodesic engineering 大地测量工程
geodesic latitude 测地纬度
geodesic levelling 大地水准测量
geodesic line 大地线；测地线
geodesic longitude 测地经度
geodesic measurement 大地测量
geodesic meridian 大地子午线
geodesic network 大地测量网
geodesic parallel 大地纬圈；测地平行线
geodesic position 大地定位
geodesic satellite 测地卫星
geodesy 大地测量学
geodesy and cartography 测绘学
geodetic astronomy 大地天文学
geodetic azimuth 大地方位角
geodetic base line 大地测量基线
geodetic boundary value problem 大地测量边值问题
geodetic computer 大地测量计算机
geodetic conjunction 大地联测
geodetic construction 轻型受拉杆系结构
geodetic control 大地测量控制
geodetic control point 大地控制点
geodetic coordinate 大地坐标
geodetic coordinate system 大地测量坐标系
geodetic data 大地测量资料
geodetic database 大地测量数据库

geodetic datum 测地基准点；大地基准点
geodetic dial 大地测量度盘
geodetic earth orbiting satellite 大地测量轨道卫星
geodetic equator 大地赤道
geodetic gravimeter 大地型重力仪
geodetic gravimetry 大地重力学；物理大地测量学；重力大地测量学
geodetic gravity meter 大地型重力仪
geodetic head 静压力水头
geodetic height 大地高程
geodetic information system 大地信息系统
geodetic instrument 大地测量仪器
geodetic inversion 大地测量反演
geodetic levelling 大地水准测量
geodetic line 大地线；测地线
geodetic measurement 大地测量
geodetic meridian 大地子午线；大地子午圈
geodetic network 大地网；大地测量网
geodetic origin 大地原点；大地测量原点
geodetic parallel 大地纬圈
geodetic point 大地测量点
geodetic reference ellipsoid 大地参考椭球
geodetic reference system 大地测量参考系；测量基准；测地参考系统
geodetic sea level 大地测量海平面
geodetic station 大地测量点
geodetic survey 大地测量
geodetic tower 大地觇标
geodetic triangulation 大地三角测量
geodetic vertical network 大地高程网
geodetic zenith 大地测量天顶
geodetics 大地测量学
geodic 晶洞的
geodiferous 含晶洞的
geodimeter 光电测距仪；光速测距仪
geodome 地穹
geodome series 地穹列
geodrain 排水板；地内排水板
geo-drilling 地质钻探
geodynamic 地球动力学的
geodynamic evolution 地球动力学演化
geodynamic method 地球动力学法
geodynamic model 地球动力学模型
geodynamic pressure 地动压力
geodynamical 地球动力学的
geodynamics 地球动力学；大地动力学
geodynamics experimental ocean satellite 地球动力学实验海洋卫星
geoecology 地质生态学；环境地质学；地生态学
geoeconomy 地质经济；地质经济学
geoecotype 地理生态型
geoelectric 地电的；大地电流的
geoelectric field 地电场
geoelectric section 地电剖面
geoelectric sounding method 地电测深法
geoelectric survey 地电测量
geoelectrical anomaly 地电异常
geoelectrical resistivity 地电阻率
geoelectrical survey 地电测量
geoelectrical work 地电勘探法

geoelectricity	地电场；大地电，地电
geoelectrics	地电学
geoengineer	地质工程师
geoengineering	地质工程
geo-environment	地质环境
geo-environmental problem	地质环境问题
geoevolution	地球演化
geoevolutionism	地球演化论
geoexploration	地质勘探
geoexplorationist	地质勘查工作者
geofabric	土工布；土工织物
geofabricform	土工模袋
geofacies	地质相
geofact	地质制品
geofactor	地质因素
geofault	地壳断裂；地质断层
geofissure	地裂缝
geoflex	弯曲造山带；爆炸索，导爆线
geofluid	地质流体；地球流体；岩石孔隙中的各种流体
geofracture	区域性大断裂；地裂缝，地断裂；地缝合线；断裂地貌
geo-gas method	地气法
geogeneration	地质建造
geogenesis	地球成因论
geogeny	地原学；地球成因学
geoglyphics	化石遗迹
geognosy	描述地质学，记录地质学；构造地质学；地球构造学
geogram	地质环境制图；综合地质柱状图；地层岩柱图
geographic base file	地理基文件
geographic base map	地理底图
geographic base system	地理基础系统
geographic center	地理中心
geographic coordinate	地理坐标
geographic cycle	地理旋回；侵蚀旋回
geographic distribution	地理分布
geographic entity	地理特征；地理实体
geographic equator	赤道
geographic geomorphology	地理地貌学
geographic horizon	地平线
geographic identifier	地理标志符
geographic information	地理信息
geographic information science	地理信息科学
geographic information system	地理信息系统
geographic isolation	地理隔离
geographic landscape	地理景观
geographic latitude	地理纬度
geographic location	地理位置
geographic longitude	地理经度
geographic map	地形图；地理图
geographic meridian	地理子午线
geographic mesh	经纬线网
geographic mile	地理英里
geographic net	地理坐标格网
geographic north	地理北
geographic orientation	地理方位
geographic position	地理位置
geographic proximity	地理相邻性
geographic setting	地理位置；地理环境
geographic value	地理值
geographical azimuth	地理方位角
geographical base map	地理底图
geographical coordinate	地理坐标
geographical cycle	地理旋回；侵蚀旋回
geographical geomorphology	地理地貌学
geographical information	地理信息
geographical information system	地理信息系统
geographical isolation	地理隔离
geographical landscape	地理景观
geographical latitude	地理纬度
geographical location	地理位置
geographical longitude	地理经度
geographical meridian	地理子午线
geographical network	地理格网
geographical orientation	地理定向
geographical pole	地极；南北极
geographical pole coordinate system	地极坐标系
geographical space	地理空间
geographical timescale	地层年表
geographical unconformity	地理不整合
geographical zone	地理分区
geogrid	土工格栅；网格形土工织物；土工网格
geogrid reinforced	土工网格加固
geohazard	地质灾害
geoheat	地热，地下热；地球热能
geoheritage	地质遗迹
geo-historic analysis	地质历史分析法
geohistory	地史学；地史
geohydrochemistry	水文地球化学
geohydrologic	地下水文的；水文地质的
geohydrologic condition	水文地质条件
geohydrologic data	地下水文数据；地下水文资料；水文地质数据
geohydrologic environment	水文地质环境
geohydrologic model	地下水文模拟；水文地质模拟；地下水文模型
geohydrologic study	水文地质研究
geohydrologic unit	水文地质单元
geohydrological	水文地质学的
geohydrological condition	水文地质条件
geohydrological constant	地下水文常数；水文地质常数
geohydrological environment	水文地质环境
geohydrological map	地下水文图；水文地质图
geohydrological reconnaissance	地下水文勘测；水文地质普查
geohydrological zone	地下水文带；水文地质带
geohydrology	水文地质学；地下水水文学
geohydrometry	水文地质测量
geoid	大地水准面
geoid anomaly	大地水准异常
geoid contour	地球体等高线
geoid determination	大地水准面测量
geoid height	大地水准面高
geoid separation	大地水准面差距
geoid undulation	大地水准面波动
geoid warping	大地水准面翘曲
geoidal azimuth	大地水准方位
geoidal body	大地体
geoidal coordinate system	大地坐标系
geoidal horizon	大地水准面地平圈

geoidal profile 大地水准剖面
geoidal surface 大地水准面
geoidal undulation 大地水准面高
geoinduction 大地感应
geoinformatics 地理信息科学；地球空间信息科学
geoinformation system 地学信息系统
geoisotherm 等地温线；等地温面
geoisothermal 等地温线的
geoisothermal surface 地热等温面
geolipid 地质类脂
geolith 岩石地层单位；岩石地层单元
geologic 地质的
geologic age 地质时代；地质年龄
geologic agent 地质因素；地质作用力；地质营力
geologic analogy 地质类比；地质模拟
geologic analogy method 地质类比法；地质学的类比法
geologic analysis 地质分析
geologic anomaly 地质异常
geologic aspect 地质概况
geologic background 地质背景
geologic barometer 地质压力计
geologic basin 地质盆地
geologic body 地质体
geologic boundary 地质界线
geologic calamity 地质灾害
geologic camp 地质野营
geologic carbon dioxide storage 二氧化碳地质封存
geologic carbon sequestration 碳地质封存
geologic carbon storage 碳地质封存
geologic characteristic 地质特征
geologic chemistry 地质化学
geologic chronology 地质年代学
geologic climate 地质气候；古气候
geologic climate unit 气候-地层单元
geologic clock 地质年代表
geologic CO$_2$ sequestration 二氧化碳地质隔离
geologic column 地质柱状剖面；地层柱状图
geologic column section 地质柱状剖面图
geologic columnar section 地质柱状剖面
geologic compass 地质罗盘
geologic complexity 地质复杂程度；地质复杂性
geologic condition 地质条件
geologic correlation 地层对比
geologic cross section 地质横剖面
geologic cycle 地质旋回
geologic data 地质资料
geologic dating 地质断代；地质年龄测定
geologic defect 地质缺陷
geologic description 地质描述
geologic disaster 地质灾害
geologic disaster monitoring 地质灾害监测
geologic disaster prediction 地质灾害预测
geologic disaster prevention and cure 地质灾害防治
geologic discontinuity 地质不连续面
geologic drill 地质钻机
geologic drilling 地质钻探
geologic dynamic simulation 地质动态模拟
geologic element 地质单元；地质单位
geologic engineering 地质工程；地质工程学
geologic entity 地质特征

geologic environment 地质环境；周围地质情况
geologic environment capacity 地质环境容量
geologic environment element 地质环境要素
geologic environment evaluation 地质环境评价
geologic epoch 地质时期
geologic era 地质时代
geologic erosion 地质侵蚀作用
geologic error 地质误差
geologic event 地质事件
geologic examination 地质调查
geologic exploration 地质勘探
geologic facies 地质相
geologic feature 地质特征
geologic field method 野外地质方法
geologic formation 地质建造；地质岩系；地质层组
geologic framework 地质结构；地质格架
geologic function 地质作用
geologic geomorphology 地质地貌学
geologic hammer 地质锤；手锤
geologic hazard 地质灾害
geologic hazard prone zone 地质灾害易发区
geologic heritage 地质遗迹
geologic high 地质上的隆起；地质高区域
geologic history 地质历史；地史；地史学
geologic horizon 地质层位
geologic horizontal plan 水平地质断面图；水平地质剖面图
geologic information 地质资料；地质信息
geologic interpretation 地质解释
geologic investigation 地质勘察
geologic landform 地质景观
geologic legend 地质图例
geologic log 地质柱状图；地质测井记录；地质编录
geologic logging 地质编录；地质录井
geologic low 地质上的凹陷；下部地质建造
geologic map 地质图
geologic map sheet 地质图幅
geologic mapping 地质测绘；地质制图；地质填图
geologic marker 地质标志物
geologic media 地质环境
geologic model 地质模型
geologic modeling 地质模拟
geologic noise 地质噪声
geologic norm 天然地质环境
geologic observation point 地质观察点；地质点
geologic observation route 地质观察路线；地质考察路线
geologic observation spot 地质观察点
geologic oceanography 海洋地质学
geologic ore resource 推测的矿产资源量
geologic origin 地质成因
geologic painting 地质色标
geologic paleontologic method 地质古生物学方法
geologic parameter 地质参数
geologic period 地质时代
geologic photography 地质摄影学
geologic point 地质点
geologic position 地层层位；地质位置
geologic premise 地质前提
geologic process 地质作用
geologic profile 地质剖面

geologic profile survey 地质剖面测量
geologic property 地质性质
geologic prospecting 地质勘探
geologic province 地质区域
geologic quadrangle map 地质标准地形图图幅
geologic range 地质延续时限
geologic reconnaissance 概略地质调查；地质踏勘
geologic record 地质记录；地质编录；地史
geologic reef 地质礁
geologic remote sensing 地质遥感
geologic report 地质报告
geologic repository 地质处置库
geologic reserve 地质储量
geologic reservoir description 油藏地质描述
geologic resources 地质资源量
geologic risk 地质风险
geologic route map 路线地质图
geologic satellite 地质卫星
geologic scheme 地质略图
geologic section 地质切图；地质剖面
geologic sedimentation 地质沉积
geologic setting 地质背景；地质条件；地质位置；地质环境；地质构造
geologic site condition 场地地质条件
geologic sketch 地质素描
geologic sketch map 地质草图；地质素描图
geologic static simulation 地质静态模拟
geologic steering drilling 地质导向钻井
geologic stereometer 地质立体量测仪
geologic storage 地质存储；地质封存
geologic structure 地质构造
geologic study 地质研究
geologic substratum 地质基底
geologic succession 地层系统
geologic survey 地质勘测；地质调查
geologic survey certificate 地质调查执照
geologic symbols 地质符号；地质图图例
geologic synchronism 地质同期性
geologic thermometer 地质温度计；地质温度表
geologic time 地质时期；地质年代
geologic time scale 地质年代表
geologic time unit 地质时间单位
geologic tracer 地质示踪剂
geologic trap 地质圈闭
geologic trap configuration 地质圈闭构造
geologic trend 地质构造方向
geologic unit 地质单元
geologic well log 钻井地质柱状图
geologic window 构造窗；构造内围层
geologic zonation 地质分带
geological 地质学的
geological action 地质作用
geological aerosurveying 地质航空测量
geological age 地质年代；地质年龄
geological agent 地质营力；地质作用力
geological analogue 地质类比法
geological analysis 地质分析
geological and topographic map 地质地形图
geological application 地质应用
geological aspects 地质概况

geological base map 地质地图；地质工作草图
geological body 地质体
geological chronology 地质年代表；地质年代学
geological classification of soil 土壤的地质分类
geological climate 古气候；地质气候
geological column 地层柱状图；地质柱状图
geological columnar section 地质柱状剖面
geological compass 地质罗盘；矿用罗盘
geological condition 地质条件；地质状况；地理条件
geological configuration 地质结构；地质形态
geological control 地质管理；地质检查；地质控制
geological conversion chart 地质换算表
geological correlation 地质条件对比；地层相关；地质学的对比；地质相关
geological cross-section 地质剖面
geological data 地质资料；地质数据
geological data base 地质数据库
geological data computer system 地质资料电脑系统
geological data file record 地质资料文件记录
geological dating 地质年龄测定
geological defect 地质缺陷
geological deformation 地质变形
geological descriptive book 地质说明书
geological diagnostics of rock grade 岩石完好度等级的地质鉴定
geological disaster 地质灾害
geological discontinuity 地质不连续性；地质不连续面；地质结构面
geological disposal of nuclear waste 核废料地质处置
geological distribution 地质分布
geological drill 地质钻机
geological drilling 地质钻探
geological element 地质单位；地质单元
geological engineer 地质工程师
geological engineering 地质工程；地质工程学
geological environment 地质环境
geological era 地质时代
geological erosion 地质侵蚀
geological event 地质事件
geological examination 地质勘测；地质勘察；地质调查
geological exploration 地质勘探
geological factor 地质因素
geological fault 地质断层
geological feature 地质特征；地质要素
geological field data 野外地质数据；现场地质资料
geological field method 野外地质方法
geological field party 野外地质队
geological field record 野外地质记录
geological fold 地质褶皱
geological formation 地层建造；地质岩系；地质层组
geological fracture 地质断裂
geological framework 地质框架
geological function 地质作用
geological ground survey 地面地质勘查
geological group 地质年代，地质时代
geological hammer 地质锤
geological hazard 地质灾害
geological hazard assessment 地质灾害评价
geological hazard factor 地质灾害危险因子
geological heritage 地质遗迹

geological high	上部地质建造；晚期地质建造；地质高部位，地质隆起
geological history	地质历史
geological horizon	地质层位
geological identification	地质鉴定；地质断定；地质鉴别
geological information	地质信息；地质资料
geological information recognition	地质信息识别
geological interpretation	地质判读
geological interpretation of aerial photo	航空照片地质判读
geological interpretation of aerial photograph	航片地质判读
geological interpretation of photo	像片地质判读
geological interpretation of photograph	像片地质判读；像片地质解译
geological interpretation of satellite	卫星照片地质判读
geological legend	地质图例
geological line	地质线
geological log	地质记录；地质柱状图
geological log of drill-hole	钻孔地质柱状图；钻孔柱状图
geological logging	地质测井
geological loss	地质损失
geological low	下部地质建造；相对老地层；地质凹陷
geological map	地质图
geological map of interpretation	判读地质图
geological map of reservoir	水库地质图；储层地质图
geological mapping	地质测绘；地质勘察；地质填图；地质制图
geological mechanisms	地质机制
geological memoir	地质纪要；地质报告
geological model	地质模型
geological observation point	地质点；地质观测点
geological observation spot	地质观察点
geological oceanography	海洋地质学
geological origin	地质成因
geological period	地质年代；地质时代；地质时期
geological photomap	影像地质图
geological plan	地质平面图；水平地质剖面
geological plan of rock quarry	石料场地质平面图
geological plan of soil borrow area	土料场地质平面图
geological point survey	地质点测量；地质点调查
geological precondition	地质条件
geological principle	地质原理
geological problem index	地层问题指数
geological process	地质作用
geological process model	地质过程模型
geological process of sea and lake	海湖地质作用
geological profile	地质纵剖面图；地质剖面
geological profile of exploratory line	勘探线地质剖面图
geological profile survey	地质剖面测量
geological prospecting	地质勘探；地质找矿
geological province	地质区域
geological radar	地质雷达
geological reasoning	地质推论；地质推断
geological reconnaissance	地质调查；地质勘探；地质勘测；地质踏勘
geological record	地质记录；地质图；地质编录
geological reference file	地质参考资料
geological remote sensing	地质遥感
geological report	地质报告
geological research	地质调查；地质研究
geological resources	地质资源
geological retrieval and storage program	地质检索和存储程序
geological retrieval and synopsis program	地质检索和摘要程序
geological risk	地质风险
geological risk factor	地质风险因素
geological route map	路线地质图
geological sampling	地质采样
geological satellite	地质卫星
geological scheme	地质略图
geological scouting	初步普查
geological section	地质剖面
geological section line and number	地质剖面线及编号
geological section map	地质剖面图
geological sequestration	地质封存；地质隔离
geological setting	地质背景；地质条件；地质位置；地质环境
geological sketch	地质素描
geological sketch map	地质草图；地质素描图
geological specimen	地质标本
geological stereometer	地质立体量测仪
geological strain	地质应变
geological strength index	地质强度指标
geological stress	地质应力
geological structure	地质构造
geological structure map	构造地质图
geological structure of ocean bottom	海底地质结构
geological succession	地质演变；地质系统；地质次序；地质层次
geological suitability of site	场址地质适宜性
geological survey	地质踏勘；地质调查；地质测量，地质测绘
geological survey memoir	地质调查报告
geological survey plan	地质调查图
geological symbol	地质符号；地质图图例
geological system	地质系统
geological technique	地质技术手段
geological tectonic	地质构造
geological texture	地质构造
geological thermometer	地质温度计；地温表
geological thermometry	地质计温学；地温测定法
geological time	地质时期；地质时代
geological time-scale	地质时代表
geological tracer	地质示踪剂；地质标志物
geological tracing of photograph	像片地质追索
geological transportation	地质搬运
geological trend	地质构造方向
geological unconformity	地质不整合
geological unit	地质单元；地质单位
geological variable	地质变量；地质的变化因素
geological well log	钻孔地质柱状图
geological-engineering investigation	地质工程调查；地质工程研究
geological-geophysical coordination	地质-物探工作配合
geologically ore blocking method	地质块段法
geologically stable area	地质稳定区
geological-survey credence	地质调查证

geologic-mineralogical factor	地质-矿物因素
geologic-paleontologic method	地质古生物学方法
geologic-time unit	地质时代单位
geologist	地质学家，地质学者；地质师；地质工程师
geolograph	钻速记录仪
geolograph chart	地质编录图
geology	地质学；地质；地质状况
geology engineering	地质工程
geology of celestial body	天体地质学
geology of coal-bed gas	煤层气地质学
geology of earthquake	地震地质学
geology of loess	黄土地质学
geology of mineral deposit	矿床地质学
geology of oil and gas	油气地质学
geology radar	地质雷达仪
geology-stratigraphy	地质-地层学
geolyte	光沸石
geomagnetic	地磁的
geomagnetic activity	地磁活动；地磁活动性
geomagnetic anomaly	地磁异常
geomagnetic axis	地磁轴
geomagnetic bay	磁湾；地磁湾扰
geomagnetic chart	地磁图
geomagnetic chronology	地磁年代学
geomagnetic coordinate	地磁坐标
geomagnetic coordinate system	地磁坐标系
geomagnetic cut-off momentum	地磁静止动量
geomagnetic daily quiet variation	地磁静日变化
geomagnetic daily variation	地磁日变化
geomagnetic data	地磁资料
geomagnetic declination	磁偏角
geomagnetic dipole field	地磁偶极子场
geomagnetic disturbance	地磁扰动
geomagnetic disturbance activity	地磁扰动活动性
geomagnetic diurnal change	地磁日变
geomagnetic effect	地磁效应
geomagnetic element	地磁要素
geomagnetic equator	地磁赤道
geomagnetic equator zone	磁赤道带
geomagnetic excursion	地磁漂移
geomagnetic field	地磁场；地球磁场
geomagnetic field reversal	地磁场倒转；地磁场反转
geomagnetic inclination	磁倾角
geomagnetic index	地磁指数
geomagnetic latitude	地磁纬度
geomagnetic longitude	地磁经度
geomagnetic map	地磁图
geomagnetic measurement	地磁测量
geomagnetic meridian	地磁子午线
geomagnetic micropulsation variation	地磁微脉动；微磁变
geomagnetic north pole	地磁北极
geomagnetic observation	地磁观测
geomagnetic observatory	地磁台
geomagnetic physics	地磁物理学
geomagnetic polarity chron	地磁极性时
geomagnetic polarity excursion	地磁极性漂移
geomagnetic polarity reversal	地磁极性反向；地磁极性倒转
geomagnetic polarity reversal time	地磁极性倒转时间表；地磁极性倒转时标
geomagnetic polarity scale	地磁极性年表；地磁极性时间表
geomagnetic polarity time scale	地磁极性年表；古地磁极性年代表
geomagnetic polarity zone	地磁极性带
geomagnetic pole	地磁极
geomagnetic pulsation	地磁脉动
geomagnetic response	地磁响应
geomagnetic reversal	地磁反转；地磁反向；地磁极性倒转
geomagnetic secular variation	地磁长期变化
geomagnetic south pole	地磁南极
geomagnetic station	地磁台
geomagnetic storm	地磁暴；磁暴
geomagnetic survey	地磁测量；磁测
geomagnetic time scale	地磁年代表
geomagnetic variation	地磁变化
geomagnetic variation method	大地磁变法；地磁变异法
geomagnetic variometer	磁变仪
geomagnetism	地磁；地磁学
geomagnetization	地磁化作用
geomagnetochronology	地磁年代学
geomancer	堪舆师；地占者，地卜者
geomat	土工网垫；土工垫
geomathematics	地质数学；数学地质学
geomatic engineering	地理信息工程；测绘工程
geomatics	地理空间信息；测绘学
geomatrix	网格草皮；土工织物护坡
geomechanical	地质力学的
geomechanical classification	地质力学分类
geomechanical measurement	地质力学测量
geomechanical model test	地质力学模型试验
geomechanical property	岩土特性；工程地质特性；地质力学性质
geomechanical property of discontinuity	不连续面的岩土力学性质
geomechanical stability	地质力学稳定性
geomechanics	地质力学，地质学；岩石力学；岩石力学
geomechanics classification	地质力学岩石分类法
geomechanics classification scheme	岩土力学分类法
geomedicine	地理医学；环境医学
geomenbrane	土工膜；土工薄膜；隔泥网膜；地质处理用膜
geometer	几何学家；地形测量家
geometric	几何学的；几何图形的
geometric aberration	几何像差
geometric accuracy	几何精度
geometric acoustics	几何声学；射线声学
geometric analysis	几何分析
geometric and physical analysis	几何物理分析
geometric angle	几何角
geometric anisotropy	几何异向性
geometric arrangement of well	井位布置
geometric asymmetry	几何不对称性
geometric average	几何平均
geometric brightness	几何亮度
geometric condition	几何条件
geometric condition of rectification	纠正几何条件
geometric configuration	几何形状

geometric construction	几何作图；几何构造
geometric correction	几何校正
geometric covariogram	几何协方差图
geometric cross-section	几何横断面
geometric crystallography	几何结晶学
geometric damage theory	几何损伤理论
geometric damping	几何阻尼
geometric diagram	几何图解
geometric distortion correction	几何畸变校正
geometric distortion in hydraulic model	水工模型几何变态
geometric distribution	几何分布
geometric divergence	几何发散
geometric ellipsoid	几何椭球
geometric error	几何误差
geometric factor	几何因素；几何因子
geometric fidelity	几何保真度
geometric figure	几何图形
geometric form of ground fissures	地裂缝几何形态
geometric formula	几何公式
geometric frequency function	几何频率函数
geometric geodesy	几何大地测量学；几何测地学
geometric grade scale	几何粒级表；几何粒级标准
geometric head	几何学水头
geometric identification algorithm	几何识别算法
geometric identity	几何尺寸一致
geometric image characteristics	几何图像特征
geometric isomer	几何异构体
geometric isomeride	几何异构体
geometric isomerism	几何异构性
geometric leveling	几何水准测量
geometric locus	几何轨迹
geometric matrix	几何矩阵
geometric mean	几何平均数；等比平均数；几何均值
geometric mean diameter	几何平均直径
geometric mean size	平均几何尺度
geometric metamerism	几何位变异构
geometric method	几何法
geometric model	几何模型
geometric non-linearity	几何非线性
geometric optics	几何光学
geometric output flow	理论输出流量
geometric parameter	几何参数
geometric pattern	几何形状；几何图形
geometric permeability model	渗透率几何模型
geometric point	图解点
geometric position	几何位置
geometric power diagram	矢量功率图；复功率图
geometric primitive	几何元素
geometric probability	几何概率
geometric progression	几何级数；等比级数
geometric projection	几何投影
geometric rate allocation factor	几何速率分配因子
geometric ray tracing	几何射线追踪
geometric rectification	几何校正
geometric rectification of imagery	图像几何纠正
geometric registration	几何配准
geometric registration of imagery	图像几何配准
geometric relationship	几何关系
geometric resolution	几何分辨力
geometric response	几何响应
geometric scheme	几何图案
geometric seismology	几何地震学
geometric series	几何级数；等比级数
geometric shadow	几何影区
geometric shape	几何形态
geometric similarity	几何相似性；几何相似
geometric sounding	几何测深法
geometric stereoisomer	立体几何异构体
geometric stereoisomeride	立体几何异构体
geometric stiffness matrix	几何刚度矩阵
geometric structural plane	几何性结构面
geometric symmetry	几何对称
geometric warping	几何扭曲
geometric well pattern	面积井网
geometrical	几何学的
geometrical anisotropy	几何各向异性
geometrical approximation	几何近似
geometrical center	几何中心
geometrical damping	几何阻尼
geometrical drawing	几何图
geometrical factor	几何系数，几何因子；几何因素
geometrical isotropy	几何各向同性
geometrical leveling	几何水准测量
geometrical mean	几何平均
geometrical measurement	几何测量
geometrical model	几何模型
geometrical optics	几何光学
geometrical orientation	几何定向
geometrical pattern	几何图案
geometrical projection	几何投影
geometrical seismology	几何地震学
geometrical shape	几何形状
geometrical similarity	几何相似性
geometrical similarity method	几何相似法
geometrical stiffness matrix	几何刚度矩阵
geometrical structure factor	几何构造因素
geometrical wave propagation	波的几何传播
geometrician	几何学家
geometrics	大地测量；测地学
geometrization	几何制图
geometrization of coal seam fault	煤层断裂几何制图
geometrization of minerals characteristic	矿产特性几何制图
geometrization of ore body	矿体几何制图
geometrization of ore deposit	矿体几何制图
geometrodynamics	几何动力学
geometry	几何学；几何条件，几何形状；几何结构
geometry correction	几何校正
geometry discontinuity	几何不连续性
geometry factor	几何因数
geometry number	几何数
geometry of ore deposit	矿体几何学
geometry of packing	填料的几何排列
geometry of plate motion	板块运动几何学
geometry sum	几何和
geomicrobiology	地质微生物学
geomonoclinal	地单斜的
geomonocline	地单斜；地槽边缘单斜沉积
geomonomer	地质有机单体

geomorphic	地表形态的；地形的；地貌的；地貌学的
geomorphic accident	地貌突变
geomorphic anomaly	地貌异常
geomorphic association	地貌组合
geomorphic composite profile	地貌综合剖面图
geomorphic cycle	地貌旋回
geomorphic development	地貌发育
geomorphic disruption	地貌破坏
geomorphic feature	地貌特征；地貌要素
geomorphic geology	地貌学
geomorphic glaciology	地貌冰川学
geomorphic interpretability	地貌解译能力
geomorphic interpretation	地貌解译
geomorphic process	地貌形成过程
geomorphic profile	地貌剖面图
geomorphic province	地貌区域
geomorphic snow line	地形雪线
geomorphic structure	地貌构造；地貌结构
geomorphic survey	地貌调查
geomorphic trap	地貌圈闭
geomorphic type	地貌类型
geomorphic unit	地貌单元
geomorphic-structural relation	地貌-构造关系
geomorphochronology	地貌年代学
geomorphogeny	地貌成因学；地形发生学
geomorphography	地貌叙述学；地貌描述学
geomorphologic	地貌的；地形的
geomorphologic agent	地貌营力
geomorphologic analysis	地貌分析
geomorphologic anomaly	地貌异常
geomorphologic division	地貌区划
geomorphologic landscape	地貌景观
geomorphologic map	地貌图
geomorphologic method	地貌法
geomorphologic principle	地貌原理
geomorphologic profile	地貌剖面图
geomorphologic sequence	地貌序列
geomorphologic structural interpretation	地貌构造解译
geomorphologic survey	地貌调查
geomorphological	地貌的
geomorphological analysis	地貌分析
geomorphological division	地貌区划
geomorphological landscape	地貌景观
geomorphological map	地形图；地貌图
geomorphological model	地貌模式
geomorphological structure	地貌构造
geomorphological survey	地貌测量
geomorphologic-seismic zonation	地貌地震分带性
geomorphology	地形学；地貌学；地貌
geomorphology division	地貌划分；地貌区分
geomorphology method	地貌法
geomorphy	地貌学，地貌；地形学
geo-navigation	地文航海法；地标航行
geonet	土工网
geonomy	动力地球学；地学
geooptimal	最佳地质的
geopark	地质公园
geopedology	地质土壤学
geopetal	向地性；示顶底的
geopetal criterion	示顶底标志
geopetal fabric	示顶底组构
geopetal structure	示顶底构造；示序构造
geopetality	觅序性
geophone	地震检波器；地下听音器；地音探测器；小型地震仪
geophone assembly	检波器组件
geophone break aligned plot	检波器波跳校直图
geophone cable	检波器电缆
geophone calibration	检波器标定
geophone characteristic	检波器特性
geophone offset	炮检距
geophone station	检波站
geophone time starting point	检波器时间起点
geophoto	地质摄影；地质照片
geophysical	地球物理的；地球物理学的
geophysical airborne prospecting	航空地球物理探矿
geophysical anomaly	地球物理异常；地理异常
geophysical computerized tomography	地球物理层析成像
geophysical data	物理勘探资料；地球物理数据
geophysical exploration	地球物理勘探；物探；地球物理探矿
geophysical forward calculation	地球物理正演
geophysical gravity method	地球物理重力法
geophysical indicator horizon	地球物理标准层
geophysical instrument	地球物理勘探仪
geophysical inversion	地球物理反演
geophysical investigation	地球物理调查
geophysical log	地球物理测井；地球物理记录
geophysical logging	地球物理测井
geophysical magnetic method	地球物理磁法
geophysical marine exploration	地球物理海洋勘探
geophysical medium	地球物理介质
geophysical method	地球物理勘探法
geophysical monitoring	地球物理监测
geophysical observation	地球物理场观测
geophysical profiling	地球物理剖面法
geophysical prospecting	地球物理勘探；地球物理探矿；物探
geophysical prospecting work quantity	物探工作量
geophysical research	地球物理研究
geophysical resistivity method	地球物理电阻率法
geophysical search equipment	地球物理勘探设备
geophysical seismic refraction method	地球物理地震折射法
geophysical sensing	地球物理探测
Geophysical Service International (GSI)	国际地球物理服务
geophysical shielding effect	地球物理屏蔽效应
geophysical signature	地球物理特征
geophysical site investigation	现场物探
geophysical software library	地球物理程序库
geophysical survey	地球物理勘探；地球物理测量；地球物理探测；地球物理调查
geophysical technique	地球物理方法；地球物理技术
geophysical test	地球物理试验
geophysical three dimensional structure	地球物理三维结构
geophysical tomography	地球物理层析成像
geophysical well logging	地球物理测井
geophysical-geological interpretation	地球物理-地质解释
geophysicist	地球物理学家

geophysics 地球物理学；地球物理	geostationary satellite 通信卫星；地球同步卫星
geophysiography 地球学	geostatistic transitive theory 地质统计传递论
geoplain region 地原区	geostatistical analysis 地质统计分析
geoplanetology 行星地质学	geostatistical software library 地质统计学程序库
geopolar 地极的	geostatistician 统计地质师
geopolarity of island arc 岛弧成分极性	geostatistics 地质统计学
geopolymer 土工聚合物；地质聚合物	geosteam 地热蒸气；天然蒸气
geopotential 地球重力势；地球位；大地位；位势	geosteering 地质导向
geopotential field 地球重力位势场	geostenogram 地理速测图
geopotential number 地球位数	geostenography 地理速测
geopressure geothermal resources 高压地热资源	geostratigraphic 全球地层学的
geopressure gradient 高压梯度	geostratigraphic scale 全球地层表
geopressured aquifer 地压含水层；高压含水层	geostratigraphic standard 全球地层标准
geopressured basin 地质压力盆地；高压盆地	geostratigraphy 全球地层学
geopressured deposit 高压型沉积	geostress 地应力
geopressured energy well 高压能源井	geostress coefficient 地应力系数
geopressured gas 地压气	geostress field 地应力场
geopressured geothermal area 高压地热区	geostress measurement technique 地应力测量技术
geopressured geothermal energy 高压地热能；地压型地热能	geostress survey 地应力测量
geopressured geothermal resources 高压地热资源；地压地热资源	geostrome 洲际地层
	geostrophic 地转的
geopressured reservoir 高压型储层	geostrophic current 地转流；地转水流
geopressured sequence 高压层序	geostrophic cycle 地转旋回
geopressured well 高压井	geostrophic motion 地转运动
geoprobe 地球探测火箭	geosurvey 大地测量
geoproduct 土工产品	geosuture 地裂缝，地断裂带；地缝合线；大地缝
georadar 地质雷达；探地雷达	geosynchronous 对地静止的；对地同步的
georadar exploration 地质雷达勘探	geosynchronous altitude 同步高度
geo-referenced data 大地编码资料	geosynchronous orbit 对地同步轨道
GEOSAT 测地卫星；地质卫星	geosynchronous satellite 地球同步卫星
geoscience 地质科学；地球科学	geosynchronous satellite sensing 对地同步卫星传感
geoscience database 地质科学资料库；地球科学数据库	geosynclinal 地槽的；地向斜的
geoscience information network 地球科学信息网；地学信息网	geosynclinal anticlinorium 地槽内复背斜；槽背斜
	geosynclinal area 地槽区
Geoscience Information Society 地球科学信息学会；地学信息学会	geosynclinal axis 地槽轴
	geosynclinal belt 地槽带
geoscience information system 地球科学信息系统	geosynclinal bicouple 地槽双联体；地槽双对偶
geoscientist 地学科学家	geosynclinal block 地槽块；槽块
geoseismic modeling 地质地震模拟	geosynclinal concave 地槽凹
geospace 地球空间	geosynclinal convex 地槽凸
geospatial information science 地球空间信息科学	geosynclinal couple 地槽偶；地槽对偶
geosphere 地圈，岩石圈；地质圈；壳圈	geosynclinal cycle 地槽式旋回；构造旋回
geostandard 地质标准样	geosynclinal deposit 地槽沉积
geostatic 地压的；耐地压的；耐土压的；土静力的	geosynclinal facies 地槽相
	geosynclinal folded system 地槽褶皱系
geostatic arch 地压拱，土拱	geosynclinal folding 地槽褶皱作用
geostatic curve 地压曲线	geosynclinal migration 地槽迁移
geostatic gradient 地静压力梯度	geosynclinal orogenesis 地槽造山作用
geostatic load 地静压负荷	geosynclinal pile 地槽堆积
geostatic pressure 地压力；自重压力；自重应力；静地压力，地静压力	geosynclinal polarity 地槽极性
	geosynclinal prism 地槽堆积柱
geostatic ratio 地静压比	geosynclinal region 地槽区
geostatic stress 自重应力；地应力	geosynclinal sedimentation 地槽沉积作用
geostatic stress field 地应力场；原岩应力场	geosynclinal subside 地槽陷
geostatics 地质静力学；刚体动力学	geosynclinal subsidence 地槽沉陷
geostationary 对地静止的；对地同步的	geosynclinal succession 地槽岩系
Geostationary Operational Environmental Satellite 地球同步轨道环境卫星	geosynclinal synclinorium 地槽内复向斜；槽向斜
	geosynclinal system 地槽系
geostationary orbiting environmental satellite 地球同步运转环境卫星	geosynclinal trough 地槽式槽地
	geosynclinal uprise 地槽隆

geosynclinal zone 地槽带
geosyncline 大地槽，地槽；大向斜；地向斜
geosyncline close 地槽封闭
geosyncline fold buildings 地槽褶皱组合体
geosyncline province 地槽区
geosyncline-platform theory 地槽-地台说
geosyncline-type metallogenic formation 地槽型成矿建造
geosynclinic 地槽的
geosynclinic sedimentation 地槽沉积作用
geosynthetic clay liner 土工合成纤维膨润土垫
geosynthetic fiber mattress 土工合成纤维网垫
geosynthetics 土工合成材料；土工合成纤维织物；土工聚合物
Geosynthetics International 国际土工合成材料学报
geotechnic synthetic material 土工合成材料
geotechnical 岩土工程的；大地工程的；地工的；岩土工程的，土力工程的
geotechnical analysis and evaluation 岩土工程分析评价
geotechnical appraisal 岩土工程评估；土力评估
geotechnical approval 岩土工程审批
geotechnical assessment 岩土评估；土力评估
geotechnical assessment study 岩土评估研究；土力评估研究
geotechnical belt 地槽带；土工带
geotechnical centrifugal model 土工离心模型
geotechnical centrifugal model test 土工离心模型试验
geotechnical classification 岩土分类
geotechnical consultant 岩土工程顾问；土力工程顾问
geotechnical control 岩土工程控制，岩土工程管制
geotechnical data 岩土数据
geotechnical design 岩土工程设计
geotechnical design assumption 岩土设计假定
geotechnical engineer 岩土工程师；土木工程师；土力工程师；地质技术工程师
geotechnical engineering 岩土工程；土木工程；大地工程，地工；岩土工程，土力工程
geotechnical engineering and underground engineering 岩土与地下工程
geotechnical engineering investigation 岩土工程勘察
geotechnical engineering test 岩土工程测试
geotechnical exploration 岩土工程勘探
geotechnical exploration report 岩土工程勘探报告
geotechnical fabrics 土工布；土工织物
geotechnical factor 工程地质因素；岩土工程因素
geotechnical geology 岩土地质学
geotechnical grid 土工格栅
geotechnical grouting 岩土灌浆
geotechnical index property test 岩土特性指标试验
geotechnical investigation 岩土工程勘察
geotechnical investigation report 岩土工程勘察报告
geotechnical land use map 岩土分析土地用途图
geotechnical log 钻孔柱状图；土工柱状图
geotechnical map 土工图；岩土工程图件
geotechnical material 土工材料
geotechnical membrane 土工膜
geotechnical model test 土工模型试验
geotechnical monitoring 岩土工程监测
geotechnical parameter 岩土参数；土工技术参数
geotechnical process 岩土工程方法；土工处理方法
geotechnical property 地质力学特性；岩土力学性质

geotechnical recommendation 岩土工程建议
geotechnical record 岩土记录
geotechnical specialist 岩土工程专家
geotechnical standard 岩土工程标准
geotechnical study 岩土工程研究
geotechnical study report 岩土工程勘察报告
geotechnical supervision 岩土工程监督；岩土工程监理
geotechnical survey 土力测量
geotechnical test 土工试验
Geotechnical Testing Journal 岩土试验学报
geotechnical textile 土工织物
geotechnical tool 土工器具
Geotechnical Engineering, Proceedings of ICE 岩土工程-英国土木工程协会进展
geotechnician 岩土工程师；岩土工程技术人员
geotechnics 土工学；岩土工程学
geotechnique 岩土工程；土工学；土工技术
geotechnology 岩土工程；土工学；土工技术
geotectocline 大地构造槽；地槽沉积
geotectogene 坳陷带；沉降带；深坳槽；构造槽
geotectogenesis 大地构造成因
geotectology 大地构造学
geotectonic 大地构造的
geotectonic areal division 大地构造分区
geotectonic component 大地构造组成
geotectonic cycle 地质构造旋回；大地构造旋回
geotectonic division 大地构造区划
geotectonic evolution 大地构造演化
geotectonic framework 大地构造构架
geotectonic geology 大地构造学；大地构造地质学
geotectonic hypothesis 大地构造假说
geotectonic map 大地构造图
geotectonic movement 地壳构造运动
geotectonic stress 构造地应力
geotectonic stress field 构造应力场
geotectonic system 大地构造体系
geotectonic unit 大地构造单元
geotectonic valley 大地构造谷
geotectonics 大地构造学；地球构造学
geotector 地震检波器
geotemperature 地热温度；地下温度；地温
geotextile 土工织物
geotextile filter 土工织物滤层
geotextile interlayer 土工织物夹层
geotextile polymer 土工布聚合物
geotextile reinforcement 土工织物加固层
geotextile-encapsulated soil 土工织物封装的土
Geotextiles and Geomembranes journal 土工织物和土工膜学报
geotexture 地表结构；地体结构
geotherm 地热；地温；等地温线；等地温面
geothermal 地热的；地热学的；地温的；地温梯度
geothermal active area 地热显示区
geothermal activity 地热活动
geothermal anomalous area 地热异常区
geothermal anomaly 地热异常
geothermal aquifer 地热含水层；地下热水含水层
geothermal area 地热区
geothermal atlas 地热图集
geothermal behavior 地热活动；地热显示

geothermal borehole	地热钻孔
geothermal brine	地热卤水；地热盐水
geothermal capacity	地热能容量
geothermal convection system	地热对流系统
geothermal corrosion	地热腐蚀
geothermal damage	地热损伤
geothermal data	地热资料
geothermal degree	地热增温梯度；地热增温级
geothermal deposit	地热储；地热储集层
geothermal depth	地温级；增温深度
geothermal development	地热开发
geothermal discharge	地热显示
geothermal district heating	地热区域供热
geothermal drillhole	地热钻孔
geothermal drilling	地热钻进；地热钻探
geothermal dry steam	地热干蒸气
geothermal drying	地热烘干
geothermal earthquake	地热地震
geothermal effluent	地热排污
geothermal electric power generation	地热发电
geothermal electricity production	地热发电
geothermal energy resource	地热能源；地热资源；地热能
geothermal energy stock	地热储；地热能储
geothermal energy system	地热能系统；地热能利用系统
geothermal equilibrium	地温平衡
geothermal exchange system	地热交换系统
geothermal exploration	地热勘探；地热调查
geothermal exploration survey	地热勘察
geothermal extraction	地热资源开发；地热流体抽采
geothermal extraction technology	地热开采技术
geothermal feature	地热显示
geothermal field	地热田；地热场；地热区
geothermal field of convective type	对流型地热田
geothermal flow	地热流体
geothermal fluid	地热流体；地下热液
geothermal fluid field	地热流体场
geothermal formation	地热储集层
geothermal framework	地热格局；地热背景
geothermal gas	地热气
geothermal geochemical exploration	地热地球化学勘查
geothermal geochemistry	地热地球化学
geothermal geologic map	地热地质图
geothermal geology	地热地质学
geothermal geophysical exploration	地热地球物理勘查
geothermal geophysics	地热地球物理学
geothermal girdle	地热带；地热活动带
geothermal gradient	地热增温率；地热梯度；地温梯度
geothermal gradient map	地热梯度图
geothermal greenhouse	地热温室
geothermal halo	地热晕
geothermal heat	地热
geothermal heat flow	地热流；大地热流；大地热流量
geothermal heat pump	地热泵；地源热泵
geothermal heating	地热供热；地热采暖
geothermal history	地热史
geothermal hot water	地下热水
geothermal implementation	地热利用
geothermal increment rate	地热增温率
geothermal infrared anomaly	地热红外异常
geothermal isogradient line	地热等梯度线
geothermal leakage manifestation	地热漏泄显示
geothermal logging	地热测井
geothermal manifestation	地热显露，地热显示；热泉
geothermal measurement	地热测量
geothermal metamorphism	地热变质作用
geothermal methane	地质变甲烷；深源甲烷
geothermal method	地热法
geothermal observation	地热观测，地热测量
geothermal phenomenon	地热现象
geothermal pollutant	地热污染物
geothermal pollution	地热污染
geothermal pool	地热储；地热田
geothermal power	地热电力；地热发电
geothermal power generation	地热电站；地热发电
geothermal power plant	地热电厂；地热电站
geothermal power production	地热发电
geothermal power system	地热发电系统
geothermal process	地热过程
geothermal producing zone	地热生产层
geothermal profile	地温剖面
geothermal prospecting	地热普查；地热勘测；地热勘探
geothermal ranking	地热分级
geothermal region	地热区
geothermal reinjection	地热回灌
geothermal reinjection pump	地热回灌泵
geothermal remote sensing	地热遥感
geothermal reservoir	地热储；地热储集层；地热库
geothermal reservoir engineering	地热储工程
geothermal reservoir evaluation	地热储评估
geothermal reservoir modelling	地热建模
geothermal resource	地热资源
geothermal resource base	地热资源基数
geothermal resource of hydrothermal type	水热型地热资源
geothermal saturated steam	地热饱和蒸汽
geothermal sediment	地热沉积
geothermal simulation	地热模拟
geothermal site	地热场地
geothermal source	地热源
geothermal spring	地热泉
geothermal steam	地热蒸汽
geothermal stress	地热应力
geothermal survey	地热调查；地温测量
geothermal system	地热系统
geothermal technology	地热技术
geothermal temperature	地热温度
geothermal turbodrill	地热涡轮钻具
geothermal turbodrilling	地热涡轮钻井
geothermal wastewater	地热废水
geothermal water	地下热水；地热水
geothermal water-rock interaction	地热水-岩相互作用
geothermal well	地热井
geothermal well completion	地热井成井；地热井完井
geothermal well log	地热测井
geothermal well test	地热试井
geothermal zoning	地热分带
geothermal-gradient temperature	地热梯度推算的温度
geothermal-heating project	地热供热工程
geothermally-anomalous area	地热异常区

geothermally-induced seismicity	地热诱发的地震活动	giant	水枪；巨大的
geothermic	地热的；地温的	giant accumulation	巨型油气藏
geothermic degree	地温梯度；地热增温率	giant crystal	伟晶；巨晶
geothermic energy	地热能	giant gelatin	烈性硝甘炸药
geothermic gradient	地温梯度；地热增温率	giant granite	伟晶花岗岩
geothermics	地热学	giant jet	气枪射流
geothermobarometry	地温压力测定；地质温压计	giant mineral deposit	巨型矿床
geothermodynamics	地球热力学	giant nozzle	水枪喷嘴
geothermograph	地温记录仪；地热自记测温仪	giant planet	巨行星
geothermometer	地质温度计；地球温度计；地温计；地热测温仪	giant powder	杰恩特炸药
geothermometry	地温学；地球温度测量学；地热学	giant ripple	大沙波；巨波痕
geothermosiphon	地热虹吸器	giant soil	巨粒土
geothermy	地热学	giant-grained	巨粒的
geotomography	地学层析成像技术	gib arm	斜撑；悬臂；吊杆
geotraverse	综合地质调查；导线式地质调查；地学大断面	gib crane	扒杆起重机
		gib hoist	已利用河段
geotropism	向地性；趋地性	gibber	三棱石；风刻石
geotumour	地瘤	gibbet	吊臂；起重杆；吊机臂
geoundation	大地波动运动；地壳升降运动；造陆运动	gibbous	凸圆的；凸状的
German cut	德国式掏槽；锥形掏槽	Gibbs adsorption theorem	吉布斯吸附定理
German method	德国掏槽法	Gibbs free energy	吉布斯自由能
German resin capsule	德国树脂锚固剂囊	Gibbs function	吉布斯函数
German timber set	德式木框支架	Gibbs phase rule	吉布斯相律
germanotype orogenesis	日耳曼型造山运动；日耳曼型造山作用	Gibbs' phenomenon	吉布斯现象
		Gibbs thermodynamic potential	等压等温位；吉布斯热力势
germanotype orogeny	日耳曼型造山作用	Gibbs-Helmholtz constrained equation of state	吉布斯-亥姆霍兹约束状态方程
germanotype structure	日耳曼型构造		
germanotype tectonics	日耳曼型大地构造	gibbsite	三水铝石，三水铝矿
gerontic	老年期，老年期的；衰老期	gibbsite sheet	水铝石层
gersdorffite	辉砷镍矿；砷硫镍矿	gibbsitite	三水铝岩
gesso	石膏粉	Gibbs's distribution	吉布斯分布
get overpull	起钻遇卡	gibelite	歪长粗面岩
Getty Conservation Institute（GCI）	盖蒂保护研究所	Gibraltar stone	直布罗陀石；条纹大理岩
geyerite	硫砷铁矿	Gieseler fluidity	吉泽勒流动度；吉氏流动度
geyser	间歇泉；间歇喷井；喷泉；热水锅炉；热水器	giga year	十亿年
		gigacycle	千兆周
geyser activity	间歇泉活动；间歇泉喷发	giga-electron-volt	千兆电子伏
geyser basin	间歇泉盆地；喷泉盆地区	gigantic landslide	巨型滑坡
geyser crater	喷泉口	gigantolite	堇青云母
geyser field	间歇泉区；间歇泉田	gigasequence	巨层序
geyser jet	喷泉射流；喷泉水柱	gilbert	吉伯
geyser pool	间歇泉池，喷泉池	Gilbert epoch	吉尔伯特期
geyser water	喷泉水	Gilbert reversed epoch	吉尔伯特反向期
geysering	间歇性活动；间歇性涌水；间歇式喷出	Gilbert reversed polarity chron	吉尔伯特反极性时
geyserite	硅华；间歇泉周围的沉积物	Gilbert reversed polarity epoch	吉尔伯特反极性期
geyserland	间歇泉区	Gilbert-type delta	吉尔伯特型三角洲
geyser-like explosive eruption	间歇泉式爆喷	gillespite	硅铁钡矿
geyser-like spring	类间歇泉	Gilliland diagram	吉利兰线图
ghaut	码头；石堤；山路	gillingite	水硅铁矿
GHG	温室气体	gilpinite	铁铀铜矾；硫酸铜铀矿
GHG emission	温室气体排放	Gilsa event	吉尔萨正向事件；吉尔萨期
GHG emission reduction	温室气体减排	gilsonite	硬沥青；天然沥青
ghizite	云沸橄玄岩	gilsonite cement	硬沥青水泥
ghost	次反射；虚反射；反常回波	gilt	镀金材料；镀金的
ghost arrival	虚反射初至	gimbal	万向接头；常平架；万向支架；平衡环，平衡架
ghost crater	残火山口		
ghost migmatite	阴影混合岩	gimbal correction	常平架校正
ghost stratigraphy	残迹地层	gimbal error	常平架误差
ghost suppression	虚反射压制；重影抑制	gimbal frame	罗盘架；常平架
ghyll	涧流；沟壑；峡谷		

gimbal lock	常平架锁定	glacial carved valley	冰蚀谷
gimbaled base	平衡基座	glacial cascade	冰川急流；冰瀑
gimbaled geophone	万向架固定式检波器	glacial chute	冰流槽
gimbal-mounted	用万向架固定的	glacial cirque	冰斗
gimlet bit	螺旋钻头	glacial clay	冰碛黏土；冰川黏土
GIN	地球科学信息网；地学信息网	glacial cliff	冰川崖
ginney	自重滑行坡；轨道上山	glacial cone	冰川碎屑锥
ginorite	基性硼钙石	glacial conglomerate	冰碛砾石；冰川砾岩
gioberite	菱镁矿	glacial control theory	冰川控制学说
gipsmantle	石膏盖层	glacial cycle	冰川周期；冰川旋回
girder	纵梁；大梁；桁架	glacial debris	冰川碎屑；冰碛物
girder and beam connection	梁节点，纵梁和横梁连接处	glacial delta	冰川三角洲
girder construction	桁架结构	glacial deposit	冰川沉积；冰积土；冰碛物
girder foundation	梁式基础	glacial diffluence	冰川越流；冰川分流
girder longitudinal	纵梁	glacial dispersion	冰川分散
girder of shield	盾构支承环	glacial diversion	冰川转移；冰川转向
girder pole	桁架式杆，梁杆	glacial drainage	冰川水系；冰川流系
girder system	梁结构体系	glacial drainage channel	冰川溢流道；冰川溢水道
girder truss	梁式桁架	glacial drift	冰碛；冰川漂移；冰川沉积物
girder without ballast and sleeper	无碴无枕梁	glacial epoch	冰河时代；冰河期；冰川期
girderless floor	无梁楼板；自动关闭装置	glacial epoch origin theory	冰期起源假说
girder-section bar	型钢横梁	glacial erosion	冰蚀作用
girdle axis	环带轴	glacial erosion cycle	冰蚀旋回；冰蚀轮回
girdle band	环带	glacial erosion lake	冰蚀湖
girdle beam	圈梁	glacial erratic	冰川漂砾
girdle of fire	环太平洋火山带	glacial erratic boulder	冰川漂砾
girdle of sediments	沉积带	glacial eustasy	冰期海面升降；冰川性海面升降
girdle pattern	环带状模式	glacial eustatic change	冰川型海平面升降
girdle plane	环带面	glacial facies	冰川相
girdle zone	环带	glacial float	冰川漂流物
girt	围梁；小梁；方框支架的横撑；柱间连续梁	glacial flood	冰川洪水
girt rail	横铁；横轨	glacial flow	冰川水流
girth	横梁；围梁；圈板；周长	glacial fluctuation	冰川变化
girth joint	环形焊缝	glacial geology	冰川地质学
girth welding	环形焊接；焊圆焊缝；周围焊接	glacial gravel	冰川砾石
girtwise	沿走向的	glacial groove	冰川刻槽；冰川沟
GIS	地理信息系统	glacial hanging valley	冰成悬谷；冰蚀悬谷
gisement	坐标纵线偏角；坐标偏角	glacial horn	冰川角峰
Gismo method of mining	吉斯莫型采掘机组开采法	glacial ice drilling	冰上钻进
Gismo-jumbo	吉斯莫钻车	glacial karst	冰川岩溶
gismondine	水钙沸石；斜方钙沸石	glacial lakebed	冰川湖床
gismondite	多水高岭土	glacial landform	冰川地质作用景观；冰川地貌；冰川地貌学；冰川地形
glacial	冰的；冰川的；冰成的	glacial limit	冰川界线
glacial ablation	冰川消融	glacial lobe	冰川舌
glacial abrasion	锉磨作用；冰川磨蚀作用；冰蚀	glacial loess	冰川黄土
glacial accumulation	冰川堆积	glacial marginal lake	冰缘湖；冰前湖
glacial action	冰川作用	glacial marine sediment	冰川海相沉积物
glacial age	冰期；冰川时代	glacial material	冰碛物
glacial alluvium	冰川冲积层；冰迹层	glacial maximum	冰川作用极盛期
glacial amphitheater	冰斗	glacial minimum	冰川作用极弱期
glacial aquatic	冰川水的	glacial moraine	冰碛
glacial aquifer	冰川含水层	glacial motion	冰川运动
glacial basin	冰川盆地	glacial outwash	冰水沉积；冰碛
glacial berg	冰川冰山；冰山	glacial overburden	冰碛；冰川沉积
glacial block	冰川岩块	glacial pavement	冰川削平作用；冰溜面
glacial boulder	冰川漂砾	glacial period	冰期；冰河时期
glacial breccia	冰川角砾；冰川角砾岩	glacial phase	冰川相；冰川期
glacial burst	冰川爆发	glacial plain	冰成平原
glacial canyon	冰川峡谷；冰川谷	glacial planation	冰川平夷作用
glacial cap	冰帽		

glacial ploughing	冰川刨蚀作用
glacial plucking	冰川挖掘作用；冰川剥蚀作用
glacial polish	冰川磨光面
glacial pot-hole	冰川水穴
glacial retreat	冰川后退
glacial rock	冰成岩；冰碛岩
glacial sand	冰碛砂
glacial sapping	冰川挖掘作用
glacial score	冰川擦痕；冰擦痕
glacial scoring	冰川擦痕作用
glacial scour	冰川冲蚀作用；冰川刨蚀状
glacial scratch	冰川擦痕；冰川刮痕
glacial sediment	冰川沉积；冰川沉积物
glacial sedimentary facies	冰川沉积相
glacial series	冰碛层
glacial sheet	冰川覆盖层；冰盖
glacial soil	冰积土；冰碛土；冰川土
glacial spill-way	冰流溢口
glacial stadial phase	冰川阶段
glacial stage	冰期；冰川阶段
glacial staircase	冰阶
glacial stratigraphy	冰川地层
glacial striation	冰川擦痕
glacial table	冰桌；冰台
glacial terrace	冰川阶地；冰成阶地
glacial terrain	冰川地带；冰川地形
glacial theory	冰川学说
glacial till	冰碛物；冰碛土；冰川泥砾
glacial topography	冰川地形
glacial transfluence	冰川越流
glacial transport	冰川搬运
glacial valley	冰蚀谷；冰蚀槽
glacial varve	冰川纹泥
glacial vestige	冰川遗迹
glacial-fluvial soil	冰水沉积土
glacial-interglacial cycle	冰期-间冰期旋回
glacialism	冰川作用；冰川学说
glacialite	白蒙脱石
glacial-lake deposit	冰湖沉积
glacially disturbed	冰川扰动的
glacially-induced	冰川诱发的
glaciated	受冰川作用的；冻结成冰的
glaciated coast	冰蚀海岸
glaciated landform	冰川地形；冰川地貌；冰蚀地貌
glaciated rock	冰蚀岩
glaciated rock knob	冰蚀岩丘
glaciated valley in glaciated valley	冰川套谷
glaciated zone	冰冻地带
glaciation	冰川作用；冰蚀现象；冻结成冰
glaciation maximum	冰期最盛期
glaciation minimum	冰川退缩极限
glacier	冰川的；冰川；冰河
glacier avalanche	冰崩
glacier bed	冰川床
glacier cap	冰帽；冰冠
glacier cascade	冰川瀑布
glacier debris flow	冰川泥石流
glacier drift	冰碛物
glacier erosion	冰川侵蚀
glacier fall	冰川瀑布；冰瀑
glacier ice	冰川冰
glacier lake	冰湖；冰川湖
glacier meal	冰川岩粉
glacier milk	冰川乳；冰川乳浆；白浆冰水
glacier plain	冰川平原
glacier rubbish	冰川岩石碎块；冰川面岩屑
glacier silt	冰碛砂泥；冰碛粉土；冰川粉
glacier sounding	冰川探测
glacier table	冰碛台地
glacier theory	冰川学说
glacier till	冰碛物
glacieret	小冰川；新月形沙丘
glacierization	冰川作用；冰川化
glacierized	冰川覆盖的；受冰川作用的
glacigene	冰成的
glacigenous	冰成的
glacioaqueous	冰水形成的
glacioclimatology	冰川气候学
glacio-eustatism	冰川型海面升降
glaciofluvial	冰水的，冰水沉积的；冰川洪积
glaciofluvial deposit	冰水沉积
glaciofluvial fan	冰水扇
glaciofluvial laccolith	冰水岩盘；冰水岩盖
glacio-fluvial landform	冰川河流地貌
glacio-fluviatile placer	冰河沉积砂矿
glaciogeology	冰川地质学
glacioisostasy	冰川地壳均衡作用；冰川地壳均衡
glacio-lacustrine	冰湖的
glacio-lacustrine deposit	冰湖沉积
glacio-lacustrine landform	冰湖地形
glaciology	冰川学；冰河学；冰川特征
glaciomarine	冰海的
glaciometer	冰川仪；测冰计
glaciospeleology	冰川洞穴学
glaciotectonic	冰川构造学的
glaciotectonic structure	冰川地质构造
glacis	缓斜坡；斜堤
glade	林中空地；湿地；沼泽地
gladite	柱硫铋铜铅矿
gladkaite	英斜煌岩；石英斜煌岩
Gladstone constant	格拉斯顿常数
glance coal	沥青煤；辉褐煤；光亮煤；辉煤
glance pitch	辉沥青
glancing angle	掠射角
glancing impact	偏斜冲击
glaserite	钾芒硝
glass	玻璃；熔岩速凝体
glass augite	玻璃辉石
glass basalt	玻璃玄武岩
glass bead	玻璃珠
glass bead pack	玻璃珠人造岩心
glass block screen wall	玻璃幕墙
glass capillary	玻璃毛细管
glass capillary gas chromatography	玻璃毛细管气相色谱
glass capillary tube viscometer	玻璃毛细管黏度计
glass circle	玻璃分度盘
glass extrusion	玻璃润滑热挤压
glass fiber	玻璃丝；玻璃纤维
glass fibre insulator	玻璃纤维绝缘器
glass geogrid	玻纤网

glass grid	玻纤网
glass hemisphere	玻璃半球
glass inclusion	玻璃包裹体
glass jet	玻璃筒聚能装药
glass jet perforator	玻璃管冲孔器
glass matrix	玻璃基质
glass porphyry	玻基斑岩
glass reinforced concrete	玻璃纤维混凝土
glass reinforced plastic pipe	玻璃强化胶管
glass shard	玻屑，火山玻屑；玻质碎片
glass slide	载玻片
glass stone	玻璃岩；透明彩石
glass texture	玻璃结构
glass tile	玻璃砖
glass water gauge	玻璃水位表
glass wrap	玻璃包封
glass-covered	玻璃敷层的；玻璃覆盖的；玻璃绝缘的
glass-fiber board	玻璃纤维板
glass-fiber rod with polyester resin	聚酯树脂加固的玻璃纤维钢杆
glass-fibered-reinforced rod	玻璃纤维加固锚杆
glass-lined pipe	玻璃衬管
glasspaper	砂纸
glass-pot clay	陶瓷耐火黏土；陶土
glass-section model	矿层剖面玻璃模型
glassy	玻璃质；玻璃状的
glassy feldspar	透长石；玻璃长石
glassy fracture	玻璃状断口
glassy luster	玻璃光泽
glassy mass	玻璃状体
glassy rock	玻璃质岩
glauberite	钙芒硝
Glauber's salt	芒硝；硫酸钠
glaucamphibole	蓝角闪石类；蓝闪石类
glaucocerinite	锌铜铝矾；锌铜矾
glaucochroite	绿粒橄榄石；钙锰橄榄石
glaucodot	铁硫砷钴矿
glaucolite	海蓝柱石
glauconarenite	海绿石砂屑岩
glauconite	海绿石
glauconite sandstone	海绿石砂岩；海绿砂岩
glauconitic	海绿石的
glauconitic limestone	海绿石灰岩
glauconitic sandstone	海绿石砂岩
glauconization	海绿石化
glaucony	海绿石相
glaucophane	蓝闪石
glaucophane schist	蓝闪石片岩
glaucophane schist zone	蓝闪石片岩带
glaucophane-greenschist facies	蓝闪绿片岩相
glaucophane-lawsonite schist facies	蓝闪石-硬柱石片岩相
glaucophanite	蓝闪岩
glaucophanitic metamorphism	蓝闪石型变质作用
glaucophanization	蓝闪石化
glaucopyrite	钴砷铁矿
glaucoquartzite	海绿石英岩
glaze wheel	研磨轮
glazed ceramic tile	釉面瓷砖
glazed earthenware	釉面陶土
glazed earthenware pipe	上釉陶管
glazed frost	冰雨；雨冰膜
glazing	装配玻璃；上釉；磨光；釉料；光泽
glebe	成矿区；含矿地带
glei soil	潜育土壤；潜育层
gleitbrett fold	滑褶皱
gleitbretter	劈理片；滑板，滑片
gleization	潜育作用；潜育化
gleization bedding	潜育化层理
Glen-Coe-type caldera	格伦考型破火山口
glessite	圆粒树脂石；棕琥珀；褐色琥珀
glesum	琥珀
gley	潜育土
gley horizon	潜育层
gley process	潜育作用
gley soil	潜育土壤
gley water	潜育水
gleying	潜育作用
gleying mire	潜育沼泽
glide	滑动；滑移；滑裂带
glide band	滑移带
glide bedding	滑移层理；滑动层面
glide breccia	滑动角砾岩
glide contact	滑动接触
glide direction	滑动方向
glide fault	滑动断层；滑断层
glide fold	滑褶皱；剪褶皱
glide lamella	滑移层
glide line	滑移线
glide motion	滑移运动
glide reflection plane	滑移反射面
glide slope	下滑斜度
glide tectonics	滑动构造
glide twin	滑移双晶
glide-plane of symmetry	对称滑动面
gliding	滑移；滑动
gliding block	滑块
gliding flow	滑移流动
gliding fracture	滑移断裂
gliding mass	滑体
gliding nappe	滑动推覆体；滑覆
gliding plane	滑移面；滑面
gliding slab	滑片
gliding system	滑动系统
gliding tectonics	滑动构造；滑动构造学
glimmering luster	微光泽
glimmerite	云母岩
glimmerton	伊利石
glinkite	绿橄榄石
glint	闪烁；反射；回波起伏
glissade	侧滑；滑降
glissile dislocation	滑动位错
glist	云母；闪光
gloap	天然风洞；吹蚀穴
global	球形的；全世界的，全球的；总的，整体的
global air pollution	全球大气污染
global approval	全世界承认
global area database	全球数据库
global array	全局数组
global asymptotic stability	全局渐近稳定性
global atmosphere research programme	全球大气研究

计划
global atmospheric circulation 全球大气环流
global carbon cycle 全球碳循环
global chronostratigraphic unit 国际性年代地层单位；全球年代地层单位
global circulation 全球性环流；全球环流
global coefficient matrix 全局系数矩阵
global collapse 整体崩塌
global coordinate system 整体坐标系
global coordinates 总体坐标
global coverage 全球摄影范围；全球覆盖
global cycle 全球性周期
global cyclicity 全球旋回性
global definition 全局定义
global density 整体密度
global digital seismograph network 全球数字地震台网
global ecology 全球生态学
global emission 全球排放量
global energy norm 全局能量范数
global energy norm error 全局能量范数误差
global environment 全球环境；地球环境
global environmental monitoring system 全球环境监测系统
global error 全局误差；全域误差
global error estimator 全局误差估计
global eustasy 全球海平面升降
global event 全球事件
global expansion hypothesis 全球膨胀假说；全球扩张说
global factor of safety 综合安全系数
global faulted network 全球性断裂网格
global finite element approximation 全局有限元近似
global fit 总体拟合
global function 全局函数
global geomorphology 环球地貌学
global geoscience transect 全球地学断面
global gravity anomaly 全球重力异常
global gravity measurement 全球重力测量
global heat balance 全球热平衡
global heat budget 全球热量收支
global mass matrix 整体质量矩阵
global matrix 总体矩阵；全局矩阵
global maximum 全局极大值；总体极大值
global meteorology 全球气象学；全球气象状态
global minimization 整体最小化
global minimum 全局最小值；总体极小值
global navigation and positioning system 全球导航定位系统
global navigation satellite system 全球导航卫星系统
global oil movement 油动；总体石油运移
global optimization 全局优化
global optimum 全局最优值
global paleoenvironmental variation 全球古环境变化
global picture 全球图；架构图
global plate tectonics 全球板块构造
global position survey 大地定位测量
global position system 全球定位系统
global positioning 全球定位
global property 整体性质
global remote sensing 全球遥感
global rifting system 全球裂谷系
global satellite communication 全球卫星通信

global scale 全球规模
global sea level observing system 全球海平面观测系统
global search 全局搜索；全球搜索
global sedimentology 全球沉积学
global seismic activity 全球地震活动性
global seismicity map 全球地震活动性图
global shape function 全局形函数
global spatial data infrastructure 全球空间数据基础设施
global stability 整体稳定性；全局稳定性
global stiffness 整体刚度
global stiffness matrix 整体刚度矩阵；全局刚度矩阵
global symbol table 全局符号表
global tangent matrix 全局切线矩阵
global tectonic regime 全球构造体制
global tectonics 全球性大地构造学；全球构造学；全球构造
global transgression and regression 全球性海水进退
global truncation error 全域截断误差；全局截断误差
global variable 全程变量；全局变量
global variogram 整体方差图
global warming 全球变暖；全球增温；地球气温上升；地球温室效应
global warming potential 全球变暖潜力；全球变暖潜能
global weather reconnaissance 全球天气侦察
global wind system 全球风系
globalization 全球化
globar 碳化硅棒
globate 球形的
globe mill 球磨机
globigerina 海底软泥；球房虫
globigerina and coralline limestone 球房虫和珊瑚灰岩
globigerina marl 球房虫泥灰岩
globigerina mud 球房虫泥
globigerina ooze 球房虫软泥
globoid 球状体；球状的
globose 球形
globosity 球状；球形结构
globospherite 团状球锥晶
globular 球形，球形的；球状的
globular cementite 粒状渗碳体；球状碳化铁
globular flow 球滴状流动
globular inclusion 球状夹杂物
globular jointing 球状节理
globular perlite 球状珍珠岩
globular structure 球状构造
globularity 球状；球形
globulation 成球作用
globulite 球锥晶；滴晶
globulith 球状侵入体
globulitic limestone 鲕状灰岩
glockerite 纤水绿矾
glomerate 团块；团聚的；成球状的
glomeroblast 聚合变晶
glomeroblastic structure 聚合变晶状构造
glomeroblastic texture 聚变晶结构
glomeroclastic 聚合碎屑状
glomerocryst 聚晶
glomerocrystalline 聚合晶簇的
glomerogranular 聚合微粒状
glomerogranular texture 聚粒结构

glomerolepidoblastic 聚合鳞片状变晶的
glomerophitic 聚合辉玄岩状的
glomerophyric 聚合斑状的
glomerophyric texture 聚斑状结构
glomeroplasmatic 聚束状的
glomeroplasmatic blastic texture 束状变晶结构
glomeroplasmatic texture 聚束状结构
glomeroporphyric 聚斑状结构；聚合斑状的
glomeroporphyritic texture 聚斑状结构
glomerospheric 球粒状的
glory hole 露天放矿漏斗；入井漏斗；落碴孔，漏碴孔
glory hole spillway 竖井式溢洪道；漏斗式溢洪道
glory-hole method 漏斗采矿法
glory-hole system 漏斗采矿法
gloss 光泽；珐琅
gloss coal 亮褐煤
gloss finish 抛光；光泽整理
gloss index 光泽度
glossiness 光泽性；光泽度
glossmeter 光泽计
glossy 光泽的；光面的；光亮的；光滑的
glossy coal 光亮型煤；辉煤
Gloster getter 格劳斯特型采煤机；一种连续采煤机
gloup 天然风洞；吹蚀穴；海蚀洞顶开口
glow discharge 辉光放电
glow discharge boriding 辉光放电渗硼
glow discharge carbonitriding 辉光放电碳氮化
glow discharge lubricant 辉光放电合成润滑剂
glowing avalanche 火山灰流
glowing cloud 炽热灰云；火山发光云
glucinum 铍
glue 胶；胶水
glue bar 胶棒
glued belt joint 胶带胶接
glued board 胶合板
gluing 胶合
gluing rock 含铁黏土；黏合土
glutenite 砂砾岩
glyptolith 风棱石；风刻石
gmelinite 钠菱沸石
gnarly bedding 扭曲层理；卷曲层理
gneiss 片麻岩
gneiss complex 片麻杂岩
gneiss dome 片麻岩穹隆
gneiss granite 片麻花岗岩
gneiss suite 片麻岩套
gneissic 片麻岩的；片麻状的
gneissic granite 片麻状花岗岩；片麻花岗岩
gneissic sand 片麻岩砂
gneissification 片麻岩化
gneissoid 片麻状的；片麻岩状的；似片麻岩状
gneissoid structure 片麻状构造
gneissose 片麻岩的；片麻岩状
gneissose banding 片麻状条带
gneissose granite 片麻状花岗岩
gneissose texture 片麻状构造
gneissosity 片麻理；片麻状构造
gnomonic projection 极射赤平投影；心射极平投影；球心投影；日晷投影；大环投影
goaf 采空区

goaf area 采空区
goaf edge 采空区边界；冒顶区边界
goaf filling 采空区充填
goaf handling 采空区处理
goaf pack building 充填采空区
goaf packing 采空区充填
goaf stowing 采空区充填；废石充填
goaf water 采空区水；老塘水
goafed 采空的；冒落的
goaf-side 采空区侧；靠采空区侧的；采空区旁边
goal canopy 采空区侧护板
goal post support 球门式棚架
goal-directed inference 目标引导推理
goal-directed reasoning 目标引导推理
go-and-return test 回线测试
gob 矿内废石，充填废石；采空区；空岩，填充用的杂石
gob area 采空区；老塘；塌落区
gob atmosphere 采空区空气；老塘空气
gob bin 废石仓
gob bleeder 采空区排气孔
gob caving 采空区落顶；采空区崩落
gob entry 在采空区中维护的巷道；沿空留巷
gob fire 采空区火灾；老塘火灾
gob floor 木板假顶
gob flushing 采空区水砂充填
gob heading 采空区石墙平巷
gob hydraulic fill 采空区水砂充填
gob lagging 废石隔板；充填物隔板
gob line 采空区边线
gob loading chute 充填天井
gob pack 废石充填带
gob pile 废石堆
gob roadway 充填带侧巷；采空区内平巷；采空区巷道
gob rock 采空区塌落岩石
gob run 采空区落石移动
gob shield 掩护梁
gob side 采空区侧
gob silting 采空区充填
gob slushing 采空区水力充填
gob stink 采空区燃臭味
gob stopping 采空区隔墙；用废石砌的隔墙
gob stower 废石充填机；投掷式充填机
gob up 充填
gob water 老塘水
gob way 采空区平巷
gobbed material 采空区遗留物料；采空区充填料
gobber 废料充填装置
gobbet 石块
gobbing 充填；将劣质煤弃在矿内
gobbing of middleman 用夹矸充填
gobbing-up 废石充填
gobi 戈壁
gobi stone 戈壁石
Goble pile-driving analyzer 戈布尔打桩分析器
gob-road system 采空区巷道开采法；前进式长壁开采法
gob-side entry 沿空巷道
gob-stowing machine 采空区充填机
GOES 地球同步轨道环境卫星
goeschwitzite 伊利石

goethite	针铁矿
going bord	运煤支巷
going down	下钻
going headway	运输平巷；铺轨平巷
going in	下钻
gold amalgamation	混汞法提金
gold beating	金箔制造；含金的
gold coating	镀金
gold diggings	金矿区；金砂矿
gold dredge	采金船
gold dredging	挖掘船采金
gold dust	金粉，金末
gold electroplate	镀金
gold field	产金矿区；金矿区
gold ore	金矿
gold pan	淘金盘；精选盘
gold panning	淘金
gold placer	含金砂矿；砂金矿床
gold plate	镀金；包金
gold quartz	含金石英；金丝水晶
gold quartz vein	含金石英脉
gold reserve	黄金储备
gold spine	金钉标志
gold stockwork	网状金矿脉
gold trap	金粒捕集器
gold veinlet	含金细脉
gold-bearing conglomerate	含金砾岩
gold-bearing gravel	含金砾石
golden section	黄金分割
goldfield	金矿田；金矿区
gold-mining	采金
gold-mining district	金矿区；采金区
gold-quartz vein	含金石英脉
gold-saving device	捕金装置
gold-saving table	选金摇床
Goldschmidt's law	戈尔德施米特定律
Goldschmidt's phase rule	戈尔德施米特相律
Goldschmidt's process	戈尔德施米特铝热焊接法
goliath	移动式大型起重机
goliath take-up	千斤顶拉紧装置
Gomperz law	冈珀兹定律法
gompholite	钉头砾岩；泥砾岩
gon grade	百分度
gondite	锰榴石英岩
Gondwana	冈瓦纳古陆；南方古陆
Gondwana coal	冈瓦纳型煤层
Gondwana craton	冈瓦纳克拉通
Gondwana flora province	冈瓦纳植物区
Gondwana land	冈瓦纳古陆
Gondwana land bridge	冈瓦纳陆桥
Gondwanaland	冈瓦纳古陆；冈瓦纳大陆
gone-off hole	偏斜钻孔
gongylite	黄块云母
goniasmometer	罗盘测角器；量角器
goniometer	测角仪，角度仪
gonnardite	纤沸石
gonyerite	富锰绿泥石
good ground	稳定地层；坚实地层
good permeability	良好透水性
good resolution	良好分辨率
good running	正常运转
gooderite	含霞钠长岩
Goodman miner	古德曼型采煤机
goodness of fit	拟合优度
good-quality sand	高质量砂
Google Earth	谷歌地球
Google Scholar	谷歌学术搜索
googol	10 的 100 次方；巨大的数字，天文数字
gopher	浅井小钻机；地鼠
gopher rig	浅井小钻机
gopherhole charge	洞室装药
gophering	平硐勘探，浅井勘探，巷探；滥采；中硬地层掏药壶法
gorceixite	磷钡铝石；磷钡铝矿
gordian technique	关键技术
gordunite	榴辉橄榄岩
gorge	峡；峡谷；凹槽
goshenite	透绿柱石；白柱石
goslarite	皓矾
goslin	小型轻便泵
gossan	铁帽
gossan-type deposit	铁帽型矿床
gosseletite	锰红柱石
gothardite	硫砷铅铁矿
Gothic architecture	哥特式建筑
gothic groove	弧边菱形孔型
gothic pass	弧边菱形孔型
gotten	采完报废的煤矿；采出的；采空的
gottenburg	戈德堡正极性期
gouge angle	铲凿角
gouge auger	匙形螺钻
gouge bit	空心钻头；弧口凿；勺形钻头；弧刃钻
gouge channel	槽水道；大槽流痕
gouge-filled fissure	充泥裂隙
gouger	采富留贫的采矿者；掠夺式采矿者
gouging	地沟洗矿；只采富矿；不采边缘和贫矿；刮削
gouging shot	掏槽炮眼，开槽炮眼，开口炮眼
gouging-scrapping action	铲凿刮削作用
Gould plotter	戈得绘图仪
Goupillaud layer-earth model	古皮劳德层状地层模型
goutwater	含硫化氢的矿水
governing motion	调节装置
governing principle	指导原则
government assurance	政府担保
governor valve	调速阀；调节阀；节流阀
gow caisson	多级套筒式沉井；高氏沉箱
gow caisson pile	高氏沉箱桩
goyazite	磷铝钙石
goz	沙丘状积砂
GPR	探地雷达，地质雷达；通用寄存器
GPS	全球定位系统
GPU	图形处理器，图形处理单元；地面动力装置
grab	抓斗；挖掘机；抓岩机；钻具打捞器
grab bucket	抓斗；抓岩机的抓斗
grab bucket dredger	抓斗式挖泥器
grab crane	抓岩机吊车
grab dredge	抓斗式挖泥船；岩铲
grab dredging	抓斗挖掘船采矿法
grab driver	抓岩机驾驶员
grab excavator	抓斗式挖掘机

grab hoist 抓斗绞车
grab hook 起重抓钩
grab iron 铁撬棍；钻具捞取器；扶手
grab jaw 抓铲
grab loader 抓岩机
grab pipe machine 抓管机
grab platform 抓斗平台
grab rope 系索；系绳
grab sample 抓斗样本；随机采集的试样；手工取样
grab sampler 抓式取样器；海底取样器；简单取样
grab sampling 刻槽取样；抓取法取样；海底取样；瞬时采样
grab tensile strength 抗抓拉强度；握持抗拉强度
grab tensile test 抓拉试验；握持拉伸试验
grab type dredge 抓斗挖掘船
grab type drill bit 抓式钻头
grabber 井架工
grabbing crane 抓岩机吊车
grabbing excavator 抓岩机；抓斗式挖掘机；抓斗挖土机
graben 地堑，地堑带；断层槽，槽形断层
graben block 地堑断块
graben cycle 地堑旋回
graben deep 地堑深部
graben fault 地堑断层
graben faulting 地堑断陷作用
graben lake 地堑湖
graben segment 地堑截体
graben shoulder 地堑肩
graben-fill basin 地堑填充盆地
graben-like basin 似地堑盆地；地堑式盆地
graben-shaped basin 地堑形盆地
graben-type basin 地堑型盆地
grabenward inclination 向地堑侧倾斜
graben-wide upward 地堑宽隆起
grabhook 抓爪
grab-sample assay 铲勺采样分析
grab-type dredge 抓斗挖掘船
grace of payment 付款宽限
grace period 宽限期
gradability 可分级性，可分等级性；爬坡能力
gradated filter 级配反滤层
gradated soil 级配土
gradation analysis test 粒度试验
gradation coefficient 级配系数
gradation composition 粒度组成；级配组成
gradation factor 粒度因数
gradation filter 级配反滤层
gradation of stone 石料等级
gradation period 均夷相；均夷期
gradation test 级配筛分试验；粒度分级试验；粒径分析
gradation unit 配料装置
gradation zone 均变带；渐变带
gradational boundary 分级界限
gradational contact 渐变界面
gradational multiple basin theory 多级盆地说
gradational relation 逐变关系
gradational stratification 渐变成层
gradational transition 逐渐变化
gradational type 过渡类型

gradational zone 序粒带
grade 等级，分类，分级；粒径，粒级；坡度，梯度；矿产品位
grade ability 爬坡性能；爬坡能力
grade against the load 逆坡；不利重车运行的坡度
grade beam 斜坡梁，基础梁
grade bedding 粒序层理；递变层理
grade board 坡度规
grade change 坡度变化
grade change point 变坡点
grade climbing 沿坡上爬
grade compensation 纵坡折减；坡度折减
grade computation 坡度计算
grade control 坡度控制
grade correction 坡度校正，倾斜改正
grade crossing 铁路交叉道口
grade efficiency 分离率；选矿效率；分级除尘效率
grade elimination 减缓坡度；缓和坡度
grade estimation 质量评定；等级评定
grade facies 序粒相；粒级递变相
grade factor 坡度修正系数
grade length limitation 坡长限制
grade level 基准面；底面
grade limit 限制坡度；坡度限制
grade line 腰线；坡度线；基准线
grade mining 阶段采矿法
grade of accuracy 精度等级
grade of bench 阶段水平
grade of coal 煤炭品质分类；煤的等级
grade of concrete 混凝土等级
grade of diamond 金刚石等级
grade of discharge 泄水效度
grade of fit 配合等级；吻合度
grade of floatability 可浮性等级
grade of heat 热能等级
grade of metamorphism 变质程度
grade of ore 矿石品级；矿石品位
grade of side slope 边坡坡度
grade of slope 坡度
grade of soil liquefaction 土壤液化等级
grade of washability 可洗选性等级
grade pan 分段溜槽；分级溜槽
grade peg 坡度标桩；坡度标志
grade plain 均夷平原
grade post 坡度标
grade rate 坡度率
grade rod 水准标尺；坡度尺
grade scale 分级标准；粒级标准；粒级表
grade separation 分级；等级分类；邻接坡面分离台
grade size 粒级
grade slope 均衡坡度
grade stake 平整标桩；坡度桩；填挖土方高程标准桩
grade surface 均夷面；坡面
grade texture 粒级结构
grade turbidite 序粒浊积岩
grade yield 粒级产量
gradeability 拖曳力；爬坡能力
gradebuilder 斜板推土机
graded 分级的，分类的；级配的；分度的，刻度的；校准坡度的，平整过的；有坡度的；递级的

graded aggregate 颗粒级配；级配骨料
graded aggregate mixture 分级集料混合物
graded aggregate pavement 级配路面
graded bed 递变层；粒序层；分级层
graded bedding 粒序层理；递变层理；粒级层；序粒层；渐变层
graded broken stone 分级碎石；过筛碎石；级配碎石
graded coal 过筛煤
graded commutator 刻度管
graded crushing 分级压碎；分级破碎
graded earth road 整平土路
graded facies 粒级递变相
graded filter 级配反滤层；级配滤器
graded gravel 按尺寸分级的砾石
graded hardening 分级淬火
graded hydrolysis 分段水解
graded material 级配材料
graded plain 均夷平原
graded product 分级产品；过筛产品
graded profile 阶梯形剖面；均衡剖面
graded rhythmite 粒级韵律层；递变韵律层
graded sand 分选砂；分级砂；级配砂；粒级砂；过筛砂
graded sandstone 粒级砂岩
graded sediment 分选沉积物；分粒沉积；粒度递变沉积
graded siltstone turbidite 序粒粉砂浊积岩
graded size 过筛粒度；分级粒度
graded slope 阶梯形斜坡
graded soil 级配土
graded standard sand 标准级配砂
graded suspension 递变悬浮
graded terrace 微斜阶地
graded test 粒径测试
graded texture 粒级结构
graded unconformity 均夷不整合，渐变不整合
graded-bed sequences 粒级层序
graded-channel type terrace 斜沟式阶地
graded-density skin 变密度表层
graded-stratified bed 层理状递变层
gradeline 纵坡线；坡度线
gradeline of road 巷道腰线
grader 平土机，推土机，平路机，分选机，分类器
grader blade 平路机铲刀；平路机刮板；推土机推板
grader elevator 高架平路机；高架推土机
grader man 推土机手
grade-separated junction 立体交叉
grade-speed ability 一定速率的爬坡能力
grade-stabilizing structure 固坡结构
grade-up 上坡
gradient 梯度，陡度；斜坡，坡道，倾斜度，倾斜率；比降
gradient angle 坡度角；倾角
gradient arrangement 梯度排列
gradient array 梯度排列；梯度电极系
gradient break 坡度转折点；坡折
gradient compensation 坡度折减
gradient correction factor 梯度校正因子
gradient current 斜坡流；梯度流
gradient formula 水力坡度公式
gradient measurement 梯度测量；坡度测量

gradient meter 测坡仪，测斜仪，陡度计
gradient method 梯度法
gradient of anomaly 异常梯度
gradient of equal traction 等牵引力坡度；等阻坡度
gradient of gravity 重力梯度
gradient of head 水头梯度
gradient of metamorphism 变质梯度
gradient of river bed 河床比降
gradient of slope 坡度；坡度角
gradient of temperature 温度梯度
gradient of water table 水位坡降；地下水位坡降
gradient operator 梯度算子
gradient peg 坡度桩
gradient plane 梯度平面
gradient ratio 梯度比；坡降；比降
gradient ratio test 梯度比试验
gradient resistance 坡道阻力
gradient search 梯度搜索；梯度寻优法
gradient search procedure 梯度搜索法
gradient singularity 梯度奇异
gradient slope 梯度坡度
gradient solvent 梯度溶剂
gradient sonde 梯度测井电极系；梯度探头；梯度探测器
gradient start 梯度起点
gradient stop 测压力梯度停点
gradient trend line 递减率趋势线
gradient vector 梯度向量
gradient wind 梯度风
gradient work 斜坡工作
gradient zone 梯度带
gradient-related method 梯度相关法
grading 粒度分级，级配，粒级成层；分选，筛选，筛分；定坡降线；修坡
grading analysis 粒度分析，级配分析；筛析
grading analysis of soil 土的颗粒分析
grading collapse settlement 分级湿陷量
grading curve 级配曲线；粒径分布曲线；粒度曲线
grading dual operation 土方修整工作
grading elevation 路基标高
grading envelope 级配包络线
grading factor 分选系数；粒度比；分级因子；级配系数
grading fraction 粒径分级
grading instrument 坡度测定器；测定坡度水准仪
grading job 土方修整工作；路面整坡工作
grading limit 级配范围；粒级界限
grading loading 级配装填
grading of aggregate 集料级配
grading of gravel 砾石颗粒级配
grading of soil 土粒级配；土壤级配
grading plan 地面平整图
grading range 级配范围
grading requirement 级配标准；场地路基整平标准
grading screen 分级筛；分级筛组
grading structure 倾斜结构，倾斜段
grading test 级配试验；级配筛分试验
grading work 土方修整工作；地面平整工作
grading zone 分选区；级配区
gradiolog 压力梯度测井

gradiometer 重力梯度仪
gradiometry 重力梯度测量
gradual commencement magnetic storm 缓始磁暴
gradual constraint removal method 逐步去除约束法
gradual declining 逐渐下降的
gradual evolution 渐变演化
gradual expansion 渐进式扩张
gradual increment period 逐渐递增时期
gradual load 逐渐加载法
gradual loss 渐次损失
gradual modification 逐渐改变
gradual reduction period 逐渐递减时期
gradual sagging method 缓慢下沉法
gradual sagging roof 缓慢下沉顶板
gradual sagging zone 缓慢下沉带
gradual transformation 渐变式转化
gradual transition 渐变过程
gradualism 渐变论
gradualistic model 渐变模式
gradually applied load 渐加荷重；渐加负载
gradually generated hazard 渐发性灾害
gradual-pressure crusher 逐渐增压式破碎机
graduate 量筒，量杯；刻度，分度
graduated arc 分度弧；刻度弧；量角器
graduated circle 刻度盘；分度弧；刻度圈
graduated collar 分度环；环形刻度盘
graduated cylinder 量筒
graduated disk 刻度盘
graduated glass 量杯
graduated horizontal circle 水平刻度盘
graduated increasing 阶段提升
graduated micrometer collar 测微环；微动环
graduated pipette 刻度移液管
graduated plate 刻度板
graduated reference plug 有刻度的基准塞
graduated release 阶段缓解；分级释放
graduated ring 分度圈；刻度圈
graduated scale 分度标；分度尺；刻度尺；比例尺
graduating valve 递开阀；调节阀
graduation of curve 曲线校准；曲线修匀
Graefenberg array 格拉芬堡台阵
Graf sea gravimeter 格拉夫海洋重力仪
graftonite 磷铁锰矿
Graham magnetic interval 格拉姆磁间段
Graham ratio 格雷姆系数
Graham's law 格雷姆定律
grain 粒，颗粒；纹理；砾石，砂，碎屑
grain accumulation curve 颗粒累积曲线
grain analysis 颗粒分析
grain boundary 粒径界限；晶粒界限，晶缘，晶界；颗粒边界
grain boundary sliding 晶界滑动
grain boundary surface energy 晶界面能
grain breakage 颗粒破碎率
grain complex 土粒复合体
grain composition 颗粒级配；颗粒组成；颗粒级配百分含量统计表
grain corrosion 结晶类腐蚀；颗粒腐蚀
grain crushing 颗粒压碎
grain density 颗粒比重；颗粒密度

grain diameter 粒径；颗粒直径
grain diminution 微粒化；退变重结晶
grain distribution 粒度分布
grain fineness 颗粒细度
grain fineness number 砂粒细度
grain flow 颗粒流；沙流
grain fraction 颗粒组分
grain fragmentation 晶粒碎裂
grain grade 粒度
grain grade scale 粒径标准；粒级标准；粒度表
grain growth 晶体长大；晶粒生长
grain limestone 粒屑石灰岩
grain lineation 粒线理
grain orientation 颗粒定向
grain packing 颗粒充填；颗粒填塞；颗粒排列
grain powder 粒状火药
grain pressure 粒间压力；结晶颗粒压力
grain pump 颗粒物料泵
grain roughness 颗粒粗糙度；颗粒圆度
grain separation 土粒分离；粒选
grain shape of rock 岩石颗粒形状
grain size 粒度，粒径，粒级；颗粒尺寸；晶粒度，结晶大小，晶粒大小
grain size accumulation curve 粒径累积曲线
grain size accumulation test 粒径累积试验
grain size analysis 粒度分析
grain size analysis of sediment 泥砂粒径分析
grain size category 粒度等级
grain size characteristic curve 粒度特性曲线
grain size composition 粒径组成
grain size curve 粒度曲线
grain size distribution 粒度分布；粒级配；粒径分布
grain size distribution curve 粒径分布曲线
grain size fraction 粒级
grain size frequency curve 粒径频率曲线
grain size grade 粒级；粒度等级
grain size measurement 颗粒度分析
grain size of rock 岩石颗粒大小
grain size parameter 粒度参数
grain size probability plot 粒度概率图
grain size range 粒度范围；粒度上下限
grain size segregation 粒度分级
grain sizing 粒径分级
grain skeleton 颗粒骨架
grain sphericity 颗粒球度
grain structure 粒状结构；粒状构造
grain structure analysis 颗粒结构分析；粒度分析
grain surface structure 颗粒表面结构
grain texture 粒状结构
grain tin 粗粒锡石
grain type coefficient 粒型系数
grainage 粒度
grain-by-grain settling 分粒沉降
grained 粒状的；成粒的；有纹理的
grained clast 粒屑
grained formation sand 粒状地层砂
grained rock 粒状岩
grained soil 粒状土
grain-flow sandstone 颗粒流砂岩
graininess 粒度；粒状；多粒；微粒状态

graining sand	细砂
grain-layer filter	颗粒层过滤器
grain-size accumulation curve	粒度累计曲线
grain-size analysis	粒度分析；粒径分析；筛分分析
grain-size analysis curve	颗粒分析曲线
grain-size characteristic curve	粒度特性曲线
grain-size classification	粒径分级；粒度分级
grain-size comparator	粒度比测器
grain-size curve	粒度曲线；筛分曲线
grain-size distribution	粒度分布
grain-size distribution curve	粒度分布曲线；粒径分布曲线
grain-size fraction	粒径组合
grain-size frequency curve	粒径频率曲线
grain-size frequency diagram	粒度频率图
grain-size frequency spectrum	粒度频谱图
grain-size grade	粒径等级
grain-size graph	粒径图
grain-size number	粒度指数
grain-size parameter	粒度参数
grain-size parameter plot	粒度参数散点图
grain-size scale	粒径标准；粒度表
grain-size sorting	粒度分选
grain-size statistic	粒度统计量
grainstone	粒状灰岩；颗粒灰岩
grain-supported	颗粒支撑的
grain-to-grain boundary	粒间边界
grain-to-grain cementation	粒间胶结
grain-to-grain cohesion	粒间黏聚
grain-to-grain contact	粒间接触
grain-to-grain loading	粒间载荷
grain-to-grain pressure	粒间压力
grain-to-grain shear stress	粒间切应力
grain-to-grain stress	粒间接触应力；粒向应力
grainy	多粒的；粒状的
grait	砂；卵石
gram	克
gram atom	克原子
gram atomic weight	克原子量；克原子重量
gram calorie	克卡
gram centimeter	克厘米
gram equivalent	克当量
gram force	克力
gram formula weight	克式量；克分子量
gram ion	克离子
gram mass	克质量
gram mol	克分子
gram molecular solution	克分子溶液
gram molecular volume	克分子体积
gram molecular weight	克分子量
gram molecule	克分子
gram particle	克粒
gram particle weight	克粒量
gram weight	克重量
gram-atom	克原子
gram-element specific activity	克元素比放射性；克元素比活度
gram-equivalent	克当量
gram-equivalent weight	克当量
grammage	克重
gramme	克
gramme calorie	克卡；小卡
gramme-atom	克原子
gramme-equivalent	克当量
gramme-mol	克分子
gramme-molecule	克分子
grammite	硅灰石
grammol	克分子
gram-molecular weight	克分子量
granby car	格兰贝型矿车；侧卸式矿车
grand calorie	大卡；千卡
Grand Canal	大运河
Grand Canyon	大峡谷
grand granolithic screed	细石混凝土地面
grand total	总数；总和；总计
grandfather cycle	磁带原始周期；存档周期
grandfather file	原始文件
grandidierite	复合矿
grandtotal	总计
granide	花岗岩类岩石
graniphyric	花斑状；文象斑状
granite	花岗岩；花岗石
granite block	花岗石块；花岗岩砌块
granite chip	花岗岩碎屑
granite chipping	花岗岩屑
granite concrete	花岗石碴混凝土
granite face	花岗岩面层；花岗岩砌面
granite facing	花岗岩面层；花岗石砌面
granite greenstone belt	花岗岩-绿岩带
granite gruss	花岗岩砂砾
granite minimum	花岗岩最低温度点
granite pegmatite	花岗伟晶岩
granite porphyrite	花岗玢岩
granite porphyry	花岗斑岩；石英斑岩
granite sand	花岗岩砂；花岗岩砂粒
granite sett	小方块花岗岩
granite tectonics	花岗岩构造学
granite wash	花岗岩冲积页岩；花岗岩冲积物
granite-aplite	花岗细晶岩
granite-gneiss	花岗片麻岩
granite-greenstone terrain	花岗岩-绿岩地区；花岗-绿岩地体
granite-greisen	花岗云英岩
granitelle	二元花岗岩；二云花岗岩；辉石花岗岩
granite-pegmatite	花岗伟晶岩
granite-porphyry	花岗斑岩
granitic	花岗状，花岗状的；花岗质的
granitic aplite	花岗细晶岩
granitic gneiss	花岗片麻岩；花岗质片麻岩
granitic intrusion	花岗岩侵入
granitic layer	花岗岩层；花岗质岩层
granitic mass	花岗岩岩体
granitic microfeature	花岗状微地貌
granitic migmatic gneiss	花岗质混合片麻岩
granitic pegmatite	花岗伟晶岩
granitic pluton	花岗岩深成体
granitic porphyry	花岗斑岩
granitic rock	花岗质岩石；花岗岩
granitic saprolite	花岗岩残积土；花岗岩腐泥土
granitic texture	花岗结构

granitic vein 花岗岩岩脉
granitification 花岗岩化；花岗岩化作用
granitine 细晶岩；非花岗岩质结晶岩
granitite 黑云花岗岩
granitization 花岗岩化
granitization granite 混合花岗岩
granitization metamorphism 花岗岩化变质
granitize 花岗岩化
granitoid 花岗岩类，花岗岩；花岗质的；花岗状，似花岗岩状；人造花岗石面
granitoid texture 花岗状结构
granitoidite 花岗状变质岩
granitophile element 亲花岗岩元素
granitophyre 花斑岩
granitotrachytic 花岗粗面状
granitotrachytic texture 花岗粗面结构
granoblastic 花岗变晶状；花岗变晶
granoblastic texture 花岗变晶状结构
granoblastite 花岗变晶岩
granoclastic 花岗碎裂
granodiorite 花岗闪长岩
granodiorite-aplite 花岗闪长细晶岩
granodiorite-granophyre 花岗闪长花斑岩；花岗闪长文象斑岩
granodiorite-porphyry 花岗闪长斑岩
granodioritic basement 花岗闪长岩质基底
granodioritization 花岗闪长岩化
granodolerite 花岗粒玄岩
granofelsophyre 花岗霏细斑岩
granogabbro 花岗辉长岩
granolepidoblastic texture 花岗鳞片变晶结构
granoliparite 花岗流纹岩
granolite 花岗状火成岩；花岗深变岩；麻粒岩
granolith 碎花岗岩混凝土铺面；人造铺地石
granolithic 人造石铺面的
granomasanite 斜斑花岗岩
granophyre 花斑岩；文象斑岩
granophyric 花斑状的
granophyric structure 文象斑状构造
granosphaerite 放射聚心球粒
granospherite 放射球粒
granosyenite 花岗正长岩
granular activated carbon 粒状活性炭
granular base 粒料基层；粒料底层
granular bearing skeleton 粒状支承骨架
granular Binghamian grout 粒状宾汉体浆液
granular blanket 粒料覆盖层
granular carbon 碳粒
granular cementite 粒状渗碳体
granular composition 颗粒结构；颗粒组分；颗粒级配
granular concentrate 粒状精煤
granular deposit 碎屑沉积；颗粒沉积
granular disintegration 粒状崩解
granular exfoliation 粒状剥落；粒状剥落物
granular fill 粒料填方
granular flow 颗粒流
granular form 颗粒形
granular fracture 粒状断口
granular frost blanket 粒料防冻层
granular ice 晶状冰；粒状冰
granular interlocking texture 粒状连生结构
granular iron formation 粒状铁建造
granular limestone 粒状灰岩，粒状石灰岩；大理石
granular loss circulation material 粒状防漏剂
granular material 粒料，粒状物料，粒状材料
granular material arching 颗粒材料造拱
granular membrane 筛网过滤器
granular moulding compound 粒状模塑化合物
granular particle 颗粒
granular polymer 粒状聚合物
granular porosity 粒间孔隙性
granular powder 粒状粉
granular rock 粒状岩石，粒状岩
granular snow 粒状雪
granular soil 粒状土；粗粒土；颗粒土壤
granular stabilized soil mixture 粒状稳定土混合料
granular structure 粒状结构；团粒结构
granular texture 粒状结构；团粒结构
granular type road 颗粒石面道路
granular variation 粒度变异
granular-cleavable structure 片粒状构造
granular-cohesive soil 粒状黏性土
granularity 粒度；颗粒度；间隔尺寸
granularity value 粒度值
granularmetric analysis 粒径分析
granulate 粒化；粒状；使成粒状
granulated 成粒的；粒状的
granulated coal 碎粒煤
granulated powder 粒状粉；粒状火药
granulated soil 团粒土壤
granulating 粒化；成粒
granulating mill 球磨机
granulating plant 粒化装置
granulation 粒化；成粒作用，团粒作用；造粒
granulation tower 造粒塔
granule 颗粒；颗粒料
granule composition 粒度成分
granule gravel 砂砾；细砾
granule roundstone 细砾；粒状圆石
granule stone 细粒岩；细砾岩
granule texture 粒状结构
granuliform 细粒状的
granulite 麻粒岩；粒变岩；粒状深变岩
granulite belt 麻粒岩带
granulite facies 麻粒岩相；粒变岩相
granulite facies metamorphic rock 麻粒岩相变质岩
granulite terrane 麻粒岩地体
granulitic 麻粒岩的
granulitic facies 麻粒岩相
granulitic structure 碎粒结构；粒化结构
granulitic texture 粒状结构；等粒结构；麻粒结构
granulitization 麻粒岩化；压碎作用
granulo ophitic 粒状辉绿岩状的
granulometer 颗粒测量仪；粒度计
granulometric 颗粒的；测粒度的；粒度的
granulometric analysis 粒径分析；粒度分析
granulometric composition 粒径组成；粒度成分
granulometric distribution 颗粒级配；颗粒分布
granulometric facies 粒度相
granulometric principle 颗分原理

granulometry 粒度测量；粒度分析；颗粒测定法
granulophyre 微花斑岩；微文象斑岩
granulose structure 麻粒构造
granulous 粒状的
grapestone 葡萄状灰岩；葡萄石
grapestone facies 葡萄状灰岩相
graph data 图表资料；图形数据
graph display 图形显示
graph follower 读图器
graph of development 展开图
graph of error 误差曲线
graph paper 自动记录仪用描线纸；毫米纸；方格纸
graph plotter 绘图仪；制图仪；图表复制机
graph search control 图搜索控制
graph theory 图论，图形理论
graph user interface 图形用户界面
graph-densimetrical method 图示密度法
grapher 自动记录器；记录仪
graphic chart of magnetization direction 磁化方向图示法
graphic coordinate 曲线坐标
graphic correlation 图形对比
graphic data processing 图形处理
graphic data system 图形数据系统
graphic display unit 图形显示装置
graphic fabric 文象组构
graphic formula 图解式；结构式
graphic granite 文象花岗岩
graphic instrument 自动记录仪器；图示器
graphic intergrowth 文象交生；文象共生
graphic log 岩性测井；图示录井图
graphic log of strata 地层柱状图
graphic mechanics 图解力学
graphic method 图解法
graphic processing unit 图形处理器；图形处理单元
graphic rating scale 图尺度评价法；图示评定量表
graphic scale 图示比例尺；图解比例尺
graphic section 截面图
graphic statics 图解静力学
graphic structure 文象构造
graphic texture 文象结构
graphic transmission 图像传输；图形传输
graphic triangulation 图解三角测量
graphic well log 钻井剖面图
graphic workstation 绘图工作站
graphical 文象的；图像的；图示的，图解的
graphical accuracy 图解精度
graphical adjustment 图解平差法
graphical analysis 图解；图形分析
graphical analytic method 图解分析法
graphical art 图解法；图示法
graphical calculation 图算；图解计算
graphical chart 图表；图解表；曲线图
graphical computation 图解计算；图算
graphical data 图表资料
graphical data processing 图解数据处理
graphical depiction 图解法
graphical derivation 图解推导法
graphical determination 图表测定法
graphical differentiation 图解微分法
graphical display 曲线显示；图形显示
graphical input device 图形输入装置
graphical integration 图解积分法
graphical interpolation 图解内插法
graphical log 钻孔岩性记录图；柱状剖面图
graphical mean 图解平均值
graphical measurement 图解量测
graphical method 图解法；图示法
graphical method for differentiation 微分的图解法
graphical model 图形模型
graphical model fitting 图形模型拟合
graphical output device 图形输出装置
graphical parameter 图形参数
graphical performance 图解性能；图解特性
graphical plotter 绘图仪
graphical recorder 图表记录器；图形记录仪
graphical representation 图示法；图解表示法
graphical representation of biaxial stress 双轴应力图示法
graphical representation of stress 应力图示法
graphical scale 图示比例尺
graphical solution 图解法
graphical solution in hodograph plane 时距曲线平面图解法
graphical statics 图解静力学
graphical structure 文象构造
graphical symbol 图例
graphical texture 文象结构
graphical transform 图形变换法
graphical treatment 图解法
graphical water-hammer 水击图解法
graphical-analytical treatment 图解分析处理
graphical-extrapolation method 图解外推法；图解外插法
graphics device 图形显示装置
graphics display 图表显示；绘图显示
graphics library 图形软件库；图形库
graphiphyre 微文象斑岩
graphite 石墨；方晶石墨
graphite composite material 石墨复合材料
graphite lubricant 石墨润滑剂
graphite paint 石墨涂料
graphite schist 石墨片岩
graphite-gneiss 石墨片麻岩
graphite-mica schist 石墨云母片岩
graphitic 石墨的
graphitic bitumen 含石墨沥青
graphitic chert 含石墨燧石
graphitic corrosion 石墨腐蚀
graphitic lubricant 石墨润滑剂
graphitic structure 石墨结构
graphitic zone 石墨带
graphitiferous 含石墨的
graphitite 不纯石墨
graphitization 石墨化作用
graphitized 石墨化的
graphitizer 石墨化剂；石墨化促进剂
graphitizing 石墨化
graphitizing annealing 石墨化退火
graphitizing carbon 石墨化碳
graphitizing of diamond 金刚石石墨化
graphitoid 次石墨，不纯石墨；似石墨的，石墨状的

graphocite	石墨岩；石墨质岩；石墨片岩
grapholith	黏土质页岩
graphology	图解法；图表法
graphophyre	粗文象斑岩
graphophyric mudstone	花斑泥岩
grapple	抓斗，吊车；钻孔打捞器；岩心提取器；抓住，钩住
grapple dredge	卸料塔
grapple travelling crane	抓斗行走吊车
grappling bucket	抓斗
grappling hook	抓升钩
grappling method	拣块法；攫取法
graptolite	笔石；笔石类
graptolite facies	笔石相
grass planting by hydraulic jetting	液压喷播植草
grass-bog peat	草本沼泽泥炭；草甸泥炭
grasshopper approach to exploration	"蚱蜢"式勘探找矿法
grassland degeneration	草原退化
grassland ecosystem	草原生态系统
grassroot exploration	浅层普查；浅层找矿
grass-root mining	投资不足的采矿；露天开采
grate mill	格子排料式磨机；栅条磨
grate surface	炉栅表面；炉床面
grater	粗齿木锉；磨光机
graticule line	方格线；格网线
graticule mesh	格网
graticule tick	坐标网延长线；经纬网延伸短线
grating	光栅；栅格；点阵晶格；筛条，格子，炉排
grating analysis	筛析；颗粒分析
grating constant	光栅常数；晶格常数
grating texture	格状结构
gravel	砾石；砂砾；碎石，角砾
gravel aggregate	砾石骨料；卵石骨料
gravel angularity	砾石棱角度
gravel aquifer	砾石含水层
gravel arrangement	砾石颗料排列
gravel auger	砾石螺旋钻
gravel backfilling	砾石充填；砾石回填
gravel ballast	砾石道碴
gravel bank	砾石岸；砾石滩
gravel bar	砾石洲；砾石滩
gravel bed	砾石河床；砾石层
gravel box	砾石笼
gravel coating	砾石外壳；砾石面
gravel cobble	卵石
gravel conglomerate	砾岩
gravel deposit	砾石沉积物
gravel desert	砾漠
gravel drain	卵石排水沟；砾石排水沟
gravel envelope	砾石包层；砾石围填；砂砾封皮
gravel face	砂矿工作面
gravel fill	填砾；砾石充填层；砂砾淤积
gravel filter	砂砾滤层；砾石过滤器；卵石过滤器
gravel foundation	砾石基础；砾石地基
gravel fraction	砾石粒级；砾石成分
gravel grain	砾粒
gravel ground	砾质土
gravel layer	砾石层
gravel method	砾石找矿法
gravel mine	砂金矿
gravel mulch	砾石覆盖层
gravel pack	砾石填充层，填砾；砾石填料
gravel packing	砾石充填；砾石填塞；砾石衬垫
gravel pebble	小卵石；小圆石；砾
gravel pile	碎石桩
gravel pit	砾石坑；采石场；砾石料坑
gravel plain	砾石平原；砂砾平原
gravel plain placer	砾石平原砂矿；砾石平原冲积矿
gravel plant	砾石筛选厂；碎石厂
gravel pocket	砾石填坑
gravel pump	砂砾泵
gravel pump mining	砂泵开采
gravel ridge	砾石堤
gravel road	砾石路
gravel sand	砾石砂
gravel soil	砾石土；砂砾土；碎石土
gravel spread	砾石覆盖层；卵石覆盖层
gravel stone	砾岩
gravel stratum	砾层；砂砾层
gravel terrace	砾石阶地
gravel trap	砾石拦截坑
gravel wall	砾石围填层；砾石墙
gravel wash	冲积砾石
gravel washer	洗砾机
gravel washing screen	洗砾筛
gravel well	砾壁井
gravel-filled drain trench	砾石排水沟
gravel-filled trench	砾石排水沟
gravelling	砾石铺道；铺砾石
gravelly clay	砾质黏土
gravelly ground	砾石土壤
gravelly loam	砾质壤土
gravelly sand	砾砂；砾石砂
gravelly soil	砾质土；砾土
gravelly soil foundation	碎石土地基；砾石土地基
gravel-packed liner	砾石过滤层
gravel-packed screen	填砾过滤器
gravel-packed well	填砾井；砾石壁井
gravel-packing water well	砂砾填料水井
gravel-rock contact	砾-岩接触面
gravel-sand cushion	砂砾垫层
gravel-to-sand ratio	砾砂比
gravel-wall well	砾石围填井；砾壁井
graver-filled drain trench	砾石排水沟
gravics	重力场学
gravimeter	比重计；重力仪
gravimeter method	重力勘探法；重力探矿法
gravimeter survey	重力计勘探
gravimetric	重力测量的；重量分析的
gravimetric altimeter	涡流层
gravimetric analysis	重量分析法
gravimetric concentration	重力选矿
gravimetric density	重力密度
gravimetric determination	重力测定
gravimetric dust sampler	重力粉尘采样器；矿尘重量取样
gravimetric dust sampling instrument	重量尘样采取器
gravimetric geodesy	重力大地测量
gravimetric investigation	重力分析

gravimetric loading	自溜装载；自重装载
gravimetric measurement	重力测量
gravimetric method	重量分析法
gravimetric method of dust measurement	矿尘重量分析法
gravimetric network	重力测网
gravimetric observation	重力观测
gravimetric point	重力点
gravimetric prospecting	重力勘探
gravimetric prover	重力校准仪
gravimetric proving	重力校正
gravimetric sample	重量分析试样
gravimetric survey	重力测量
gravimetric water content	重力含水量
gravimetrical	测定重量的；重量分析的；重力的
gravimetrical analysis	重量分析
gravimetrical prospecting	重力勘探
gravimetry	重力测量学；重量分析法；重力测定法
gravipause	重力边缘
graviplanation	重力夷平
gravireceptor	重力感受器
gravisphere	引力范围，引力圈
gravitate	重力吸引；自重下落；下沉
gravitate pressure	重力压力
gravitation	重力；地心引力；万有引力；引力
gravitation constant	重力常数；万有引力常数
gravitation field	引力场；重力场
gravitation gliding tectonics	重力滑动构造
gravitation of ore	矿石自重下放
gravitation pore space	重力孔隙
gravitation settler	重力沉淀槽
gravitation transporter	重力运输装置
gravitational	引力的；重力的
gravitational acceleration	重力加速度
gravitational anomaly	重力异常
gravitational attraction	地球引力；万有引力；重力
gravitational compaction	重力压实
gravitational constant	重力常数；万有引力常数；引力常数
gravitational convection	重力对流；自然对流
gravitational conveying	重力运输
gravitational creep	重力蠕动
gravitational deposition	重力沉降
gravitational differential	重力弥散；重力分异
gravitational differentiation	重力分异
gravitational differentiation hypothesis	重力分异假说
gravitational disturbance	引力扰动
gravitational equilibrium	重力平衡
gravitational equipotential surface	重力等位面；重力等势面
gravitational erosion	重力侵蚀
gravitational exploration	重力勘探
gravitational field	重力场；引力场，万有引力场
gravitational flow	重力水流；自由水流；自流
gravitational flowage	重力流动
gravitational force	万有引力，地心引力；重力
gravitational head	重力水头
gravitational instability	重力不稳定性
gravitational intensity	重力强度；引力强度
gravitational interaction	引力相互作用
gravitational loading	重力负荷；重力荷载
gravitational method	重力法；引力法
gravitational method of exploration	重力勘探法
gravitational potential coefficient	重力势系数
gravitational potential energy	重力势能
gravitational potential gradient	重力势梯度
gravitational prospecting	重力勘探；重力找矿
gravitational pull	万有引力；重力牵引力
gravitational radiation	引力辐射；重力辐射
gravitational red shift	引力红移
gravitational screening	重力筛分
gravitational sedimentation	重力沉淀
gravitational segregation	重力分离；重力分选；重力偏析
gravitational selection	重力分选
gravitational separation	重力分离
gravitational settling	重力沉降
gravitational signature	重力特征；引力特征
gravitational sliding	重力滑动
gravitational sliding plate	重力滑动板块
gravitational slip	重力滑坡
gravitational spreading	重力扩张
gravitational stratification	重力分层
gravitational stress	自重应力
gravitational survey	重力测量
gravitational system of units	重力单位制
gravitational tectogenesis	重力构造作用
gravitational tectonics	重力构造；重力构造地质学
gravitational test	重力试验
gravitational tide	引力潮；天文潮
gravitational torque	重力矩
gravitational water	重力水；自流水
gravitational wave	重力波
gravitational-inertial classifier	重力惯性分级机；重力惯性分类器
gravitationally induced stress field	重力引起的应力场
gravitation-capillary pore space	重力-毛管孔隙
gravitative adjustment	重力调节；重力析离
gravitative arrangement	按比重分布，重力分离
gravitative deposit	重力堆积物；重力沉积
gravitative differentiation	重力分异；重力差异
gravitative faulting	重力断层
gravitative pressure	重力压力
gravitative separation	重力分离
gravitative settling	重力沉降
gravitative stratification	重力分层
gravitative transfer	重力移动
gravitite	重力碎屑沉积层；重力堆积物
gravitometer	比重计；重力仪
graviton	重子，重力子；引力子
gravity	重力；引力，地心吸力；重量
gravity abutment	重力墩；重力式桥台
gravity acceleration	重力加速度
gravity accumulation	重力积聚
gravity action	重力作用
gravity alignment	重力校准
gravity anchor	重力锚定；重力式锚定；重力锚
gravity anomaly	重力异常
gravity anomaly inversion	重力异常反演
gravity anomaly simulator	重力异常模拟器；重力反常模拟器

gravity apparatus	重力仪
gravity arch dam	重力拱坝
gravity axis	重力轴；重心轴
gravity balance	比重秤；重力平衡；重力秤
gravity base	重力基座；重力测量基点
gravity base station	重力基准点站
gravity based platform	重力基座平台
gravity basement	重力基底
gravity bottle	比重瓶
gravity breakwater	重力式防波堤
gravity bucket conveyor	重力斗式输送机
gravity bulkhead	重力式挡土墙；重力式护堤岸
gravity carrier	重力投料设备
gravity casting	重力浇注；重力式铸造法
gravity center	重心
gravity classification	按比重分级
gravity coal washing	重力洗煤法
gravity cofferdam	重力围堰；重力围堰坝
gravity collapse	重力坍塌
gravity collapse structure	重力塌陷构造
gravity compaction	重力压实；重力压缩作用
gravity component	重力分量；重力组分
gravity concentration	重力选矿；重力选；重力选煤
gravity concentration shop	重选车间
gravity concentrator	重力选矿机；重力选煤机
gravity constant	重力常数
gravity contour	重力等值线；重力等高线
gravity conveyor	重力输送器；自重输送器
gravity core	重力取芯；重力取样
gravity core sampler	重锤岩芯取样器；重锤取芯器
gravity core sampling	重锤岩芯取样
gravity corer	重力取心器
gravity correction	重力校正；重力换算
gravity cover	重力覆盖
gravity coverage	重力测区；重力测网
gravity culvert	重力排水涵洞
gravity curvature	重力曲率
gravity dam	重力坝
gravity dam of triangular section	三角形断面重力坝
gravity damped	重力摆
gravity datum	重力基准
gravity deformation	重力变形
gravity density	重力密度；重度
gravity deposit	重力沉积
gravity derived water	重力导出水
gravity destabilization hypothesis	重力失稳说
gravity determination	重力测量
gravity differentiation	重力分异
gravity distribution	重力分布
gravity disturbance	重力扰动
gravity drain	重力排水；自重泄油
gravity drainage	重力流泄；重力排水；重力疏干；自流排水
gravity dressing	重力选矿
gravity drive	重力驱动
gravity effect	重力效应
gravity elevation correction	重力高度校正；重力高程改正
gravity equipotential surface	重力等势面；重力等位面
gravity erosion	重力侵蚀
gravity exploration	重力勘探
gravity fault	重力断层
gravity field	重力场；引力场，万有引力场
gravity flow	重力流；自流；重力流动
gravity flow folding	重力流动褶皱
gravity flow pipe line	自流管线
gravity flow sewer	自流污水渠；无压污水渠
gravity flushing	自重水力运输；自重冲洗
gravity fold	重力褶皱
gravity force	地球引力；重力
gravity force capillary viscometer	重力毛细管黏度计
gravity fractionation	重力分离作用
gravity gliding	重力滑移
gravity gliding tectonic	重力滑动构造
gravity graben	重力地堑
gravity gradient	重力梯度
gravity gradient anomaly	重力梯度异常
gravity gradient measurement	重力梯度测量
gravity gradient survey	重力梯度测量
gravity gradient zone	重力梯度带
gravity gradiometer	重力梯度仪；重力倾斜仪；重力坡度测定仪
gravity groundwater	重力地下水
gravity hammer	重力锤；打桩落锤
gravity haulage	自重运输；重力运输
gravity head	压力差；静压头；重力落差；重力压头
gravity high	重力高
gravity hopper	重力自流斗；重力送料斗仓
gravity horizontal gradient	重力水平梯度
gravity hydraulic transport	自重水力运输
gravity in bulk	容重
gravity incline	轮子坡；重力坡；自重滑行坡
gravity injection	重力注水；重力进样；重力流入
gravity inversion method	重力反演法
gravity jigging	重介质跳汰选矿；重力簸选
gravity landform	重力地貌
gravity load	自重；重力荷载；重力负载
gravity loading	重力荷载
gravity loading conveyor	自重装载输送机
gravity lock wall	重力式闸墙
gravity logging	重力测井
gravity loss	重力损失；失重
gravity low	重力低
gravity main	重力泄水管；重力输水管道
gravity maximum	重力高；重力最大值；重力最大变异
gravity measurement	重力测量
gravity measurement at sea	海洋重力测量
gravity method	重力法
gravity mill	重力选矿厂；重力选煤厂
gravity minimum	重力低；重力最小异常
gravity negation	重力低
gravity observation of earth tide	重力固体潮观测
gravity ore pass	重力放矿溜道；重力放矿溜井
gravity orogenesis	重力造山运动；重力造山作用
gravity pier abutment	重力式墩，重力式桥墩
gravity plane	重心平面；滑动面；轮子坡，自重滑行坡
gravity platform	重力式平台；重力基座平台
gravity point	重力点
gravity potential	重力位；重力势
gravity potential energy	重力势能

gravity precursor anomaly	重力前兆异常
gravity preparation	重力选
gravity prospecting	重力勘探
gravity railroad	重力轨道；重力铁道；重力缆道
gravity reconnaissance survey	重力普查勘探；重力踏勘测量
gravity reduction	重力校正；重力改正；重力换算
gravity restoring moment	重力恢复力矩；重力复原力矩
gravity retaining wall	重力式挡土墙；重力支承墙
gravity road	自重下行巷道
gravity rope way	重力索道
gravity runway	轮子坡；自重运输斜坡道；轨道上山；自重跑道
gravity segregation	比重分聚；比重析离；重力分异
gravity separation	重力分离；重力选；沉淀分离
gravity settled shale	重力沉淀页岩
gravity settling	重力沉降
gravity sewer system	引力污水渠系统；落水式下水道
gravity slide	重力滑动
gravity sliding	重力滑动
gravity slipping	重力滑动
gravity solution	重液；重介悬浮液
gravity sorting	重力分选
gravity speed	重力沉降速率
gravity spillway dam	重力式溢流坝
gravity spreading	重力扩张
gravity spring	下降泉
gravity stamp	重力捣矿机；落锤捣矿机
gravity stowing	重力充填；自重充填
gravity stress	自重应力
gravity structure	重力式结构；重力基座平台
gravity survey	重力调查；重力测量；重力勘探
gravity surveying	重力测量
gravity suspended water	重力悬着水
gravity take-up	重力拉紧装置
gravity tectonics	重力构造；重力构造地质学
gravity tensioner	张力锤；重力拉紧装置
gravity tensioning	重力拉紧
gravity tongue	重力舌
gravity track	自重滑行道；重力坡
gravity transportation cableway	重力输送索道
gravity tube	比重计
gravity type quay	重力式码头
gravity unit	重力单位
gravity variometer	重力变感器
gravity vertical gradient	重力垂直梯度
gravity wall	重力式挡土墙
gravity water	重力水
gravity wave	重力波；引力波
gravity wharf	重力式码头
gravity-arch dam	重力拱坝
gravity-artesian well	重力自流井
gravity-aspirator bit	重力吸气钻头
gravity-bar screen	自重流动棒条筛；固定棒条筛
gravity-chute ore collecting	用重力溜槽收集矿石
gravity-chute ore-collection method	溜井集矿法
gravity-collapse structure	重力崩塌构造
gravity-controlled instrument	重力控制仪器
gravity-displaced deposit	重力移位沉积
gravity-displacement product	重力位移产物
gravity-flow screen	自重流动筛；固定筛
gravity-focusing	重力聚集
gravity-glide fault	重力滑动断层
gravity-gliding mechanism	重力滑动机制
gravity-gliding tectonics	重力滑动构造
gravity-induced tectonics	重力引起的构造；重力诱导构造
gravity-operated tilt sensor	重力式测斜传感器
gray antimony	辉锑矿
gray antimony ore	辉锑矿石
gray bed	页岩砂岩交错层
gray brown podzolic soil	灰棕色灰化土
gray cobalt	砷钴矿
gray cobalt ore	砷钴矿石
gray correlation	灰色关联
gray correlation degree analysis	灰色关联度分析
gray desert soil	灰漠土；灰钙土
gray ferruginous soil	灰色铁质土
gray forest soil	灰色森林土
gray level	灰阶；灰级
gray level transformation	灰度变换
gray mud	灰色泥；灰软泥
gray relationship analysis	灰色关联分析
gray shade scale	灰度等级
gray system	灰色系统
gray theory	灰色系统理论
gray-cinnamonic soil	灰褐土
gray-desert steppe soil	灰漠草原土
Gray-King assay	葛-金干馏试验
Gray-King carbonization assay	葛-金碳化试验
Gray-King coke type	葛-金氏焦型
Gray-King test	葛-金结焦性试验
gray-scale image	灰度图；灰阶图
graystone	中粒玄武岩；灰色岩
graywacke	杂砂岩；硬砂岩
grazinite	响霞岩
grease-proof	防油的；耐油的
greasing	润滑；加油；涂油脂
greasing apparatus	润滑器
greasing point	注润滑脂油点
greasy	油脂的；油污的；油性的
greasy blae	光滑灰青碳质页岩
greasy gold	纯金
greasy luster	油脂光泽；脂肪光泽
greasy quartz	乳石英
great calorie	大卡；千卡
great circle projection	大圆投影
great circle tectonic belt	大圆构造带
great circum-Pacific fault	环太平洋大断裂
great divide	大分水岭
great earthquake	特大地震
great glacial epoch	大冰期
great ice age	大冰期
great igneous province	大火成岩区
great interglacial epoch	大间冰期
Great Rift Valley	东非大裂谷
great soil group	大土类
greatest common measure	最大公约数；最大公测度
greatest diurnal tidal range	最大日潮差
greatest peak	最大峰值

green beryl 绿宝石；绿柱石
green briquette 生型煤；生煤球
green coal 原煤；新采出的煤
green concrete 新浇混凝土；新拌混凝土；生混凝土；未硬混凝土
green copper ore 硅孔雀石
green earth 海绿石；绿鳞石；绿土
green house gas 温室气体
green iron ore 绿磷铁矿；绿铁矿
green karst 绿岩溶
green lead ore 磷氯铅矿
green mining 绿色开采；绿色矿业
green mining industry 绿色矿业
green mud 海绿泥，绿泥
green ore 原矿
green permeability 湿砂透气性；生砂通气度
green retaining wall 重力式挡土墙
green rock 绿色岩；碧岩
green sand 海绿砂；湿砂
green sandstone 绿砂岩；青砂岩
green schist 绿片岩；绿色片岩
green shotcrete 未固化喷射混凝土；湿喷混凝土
green tuff formation 绿色凝灰岩组
green vitriol 绿矾；水绿矾
greenalite 铁蛇纹石；土状硅铁矿
greenalite rock 硅铁岩
greenhouse 温室
greenhouse effect 温室效应
greenhouse effect gas 温室效应气体
greenhouse gas 温室气体
greenhouse gas abatement 温室气体减排
greenhouse gas accounting 温室气体核算
greenhouse gas balance 温室气体平衡
greenhouse gas emission 温室气体排放；温室气体排放量
greenhouse gas reduction 温室气体减排
greenhouse warming potential 温室增温，温室加温；温室变暖趋势；温室升高潜能
greenite 绿泥石
Greenland ice sheet 格陵兰冰盖
greenlandite 铁铝榴石；镁铝榴石；铌铁矿
greenockite 硫镉矿
greenovite 红榍石
Green's equivalent layer 格林等效层
Green's formula 格林公式
Green's function 格林函数
Green's law 格林定律
Green's theorem 格林定理
greensand 海绿砂；湿砂；黏性模砂；绿砂
greenschist 绿片岩
greenschist facies 绿片岩相
greenstone belt 绿岩带
greenstone slate 绿石板岩
green-weight 湿重
Greenwell formula 格林韦尔公式
Greenwich 格林尼治
Greenwich civil time 格林尼治民用时
Greenwich mean time 格林尼治标准时间；格林尼治平均时
Greenwich meridian 格林尼治子午线；本初子午线

Greenwich time 格林尼治时间
gregale 格雷大风
gregaritic texture 辉石聚斑状结构
Gregorian calendar 格雷戈里历；公历
Gregory-Newton formula 格雷戈里-牛顿公式
greigite 硫黄铁矿
greisen 云英岩
greisenization 云英岩化
grena 原矿；脏矿石
grenatite 十字石；白榴石；微霞正长岩
grey absolute correlation degree 灰色绝对关联度
grey antimony 辉锑矿
grey body 灰体
grey brown soil 灰棕壤
grey clustering 灰色聚类分析
grey cobalt ore 砷钴矿
grey correlation analysis 灰色关联度分析
grey desert soil 灰钙土
grey durain 灰色暗煤；贫孢子微暗煤
grey durite 灰色微暗煤；惰性微暗煤
grey entropy 灰熵
grey forecast 灰色预测
grey forest soil 灰色森林土
grey gneiss 灰色片麻岩
grey interrelated analysis 灰色关联分析
grey level 灰阶；灰度级；灰度水平
grey model 灰色模型
grey neural network 灰色神经网络
grey prediction 灰色预测
grey relation 灰色关联
grey relation model 灰色关联模型
grey relation projection model 灰色关联投影模型
grey relation projection value 灰色关联投影值
grey relational analysis 灰色关联分析
grey synthetically relational degree 灰色综合关联度
grey system analysis 灰色系统分析
grey system modelling 灰色系统建模
grey system theory 灰色系统理论
grey theory 灰色系统理论
greywacke 杂砂岩；硬砂岩；灰瓦克
grid average 格子间平均值
grid azimuth 坐标方位角
grid chart 方格图
grid column 格栅柱
grid coordinate 平面直角坐标；方格坐标
grid data 网格化数据；坐标网数据
grid foundation 格栅式基础；交叉梁基础
grid leveling 面积水准测量；方格水准测量
grid line 坐标线，坐标格网线；分隔条；栅格线
grid magnetic azimuth 栅磁方位角
grid matrix 网格矩阵
grid origin 坐标原点
grid pattern 网格图形
grid plate 跳汰机筛板；坐标网板；格网板；格栅盖板
grid point 网络点；格点
grid principle 网格法；网格原则
grid residual 网点剩余
grid residual method 网格剩余法；网点剩余值法
grid sampling 网格采样
grid scale 坐标比例尺；网格比例尺；方眼尺

grid stiffened plate	网格加筋板
grid variation	磁偏角
gridded data	网格化数据
gridding	网格内插法；网格化；绘格线
griddle	筛子；大孔筛
gridiron	格状物；方格形；梁格结构
gridiron pattern	格子形
gridiron structure	格子状构造
gridiron twinning	格子双晶
grid-leak resistance	栅漏电阻
gridline	方格坐标线
grid-mat foundation	格筏基础
grid-point approximation	网格点近似法
grid-reference	参考坐标网
grief kelly	方钻杆
grief stem	方钻杆；主动钻杆
Griffin mill	格里芬型磨机
Griffith crack	格里菲斯裂纹
Griffith crack theory	格里菲斯破裂理论；格里菲斯裂缝理论；格里菲斯微裂纹理论
Griffith cracks	格里菲斯裂缝
Griffith criterion of brittle	格里菲斯脆性破坏准则
Griffith failure criterion	格里菲斯破坏准则
Griffith fracture	格里菲斯破裂
Griffith-failure criterion	格里菲斯破坏准则
griffithite	绿水金云母；水绿皂石
Griffith-Murrell failure	格里菲斯-穆雷尔破坏准则
Griffith's crack theory	格里菲斯破裂理论
Griffith's strength criterion	格里菲斯强度准则
Griffith's strength theory	格里菲斯强度理论
Griffith's theory	格里菲斯理论
grike	溶沟；岩溶沟；竖溶隙
grill footing	格栅基础
grillage column base	格排柱基
grillage foundation	格排基础；格排地基
grillage raft	格排基础
grind	磨，研磨，磨碎；磨矿；磨光
grind ability rating	可磨性指标
grind per pass	一次通过的研磨物料量
grindability	可磨性；可磨碎性；易磨性
grindability rating	可磨碎性指数
grindability test	可磨性试验
grinding additive	研磨添加剂；助磨剂
grinding attachment	钻头磨刃装置；磨钎装置
grinding ball	磨球
grinding circulating load	磨矿循环负荷
grinding cycle	研磨循环
grinding density	磨矿浓度
grinding disk	砂轮；磨光砂轮
grinding dust	磨屑
grinding efficiency	磨矿效率
grinding flowsheet	磨矿流程；磨矿流程图
grinding hardness	研磨硬度；耐磨硬度
grinding machine head stock	磨床头
grinding machinery	磨矿机械
grinding medium	磨料；研磨剂
grinding mill	研磨机；细磨机；磨机，磨碎机
grinding out	磨去
grinding pan	磨盘；混汞磨盘
grinding paste	研磨剂；研磨膏
grinding plate	磨板
grinding rate	研磨速度
grinding ring	磨环
grinding saw	砂轮锯
grinding stone	磨石
grinding test	研磨试验
grinding wheel	磨轮
grinding wheel machine	砂轮锯；砂轮机
grinding-concentration unit	磨矿-精选装置
grindstone	砂轮；磨石
Gringarten type curve	格林加顿标准曲线
griotte marble	棱角石大理石；红纹大理岩
grip anchorage	握裹锚固
grip block	断绳防坠器；罐座；夹块
grip device	夹具；夹紧装置
grip end	抓物端
grip gear	抓取机构；夹紧装置；罐笼防坠器
grip length	握裹长度，握固长度；锚固长度
grip of concrete	混凝土握固力；混凝土黏着力
grip of hole	炮眼斜角
grip of rib	工作区的侧向推进
grip pawl	夹爪
grip ring	夹圈；锁紧环；压环；夹环
grip socket	卡盘，夹盘；夹头；夹圈，锁紧环，压环
grip surface	夹持面
griping hand system	锚爪系统
gripper	抓持器，抓器；夹具，钳子；掘进机掘进装置；支撑
gripper ring	支承圈
gripper shield	后护盾；支撑护盾
gripper shoe	支撑板；支撑靴；夹紧支座
gripper type full face rock TBM	支撑式全断面岩石隧道掘进机
gripper-type tunnel boring machine	抓爪型钻巷机；抓爪型全隧道掘进机
gripping	夹住
gripping angle	咬入角
gripping apparatus	抓取器；扣紧工具；固定器
gripping capability	握裹能力
gripping device	夹具；固定器；抓取器
gripping head	锚具，锚头；夹头
gripping of screen surface	筛面夹紧
grit	砂粒，砂砾；粗砂，粗砂岩；磨料，人造磨石
grit blast	喷砂装置
grit blasting	喷砂处理
grit catcher	沉砂池；集砂器
grit cement	粗粒水泥
grit chamber	沉砂池；沉砂槽
grit gravel	细粒砾石；砂砾
grit sandstone	粗砂岩
grit separator	煤灰分离器
grit stone	粗砂岩
grit tank	沉砂槽；砂砾池
grit-proof	防岩尘的
gritrock	粗砂岩
gritstone	粗砂岩
gritstone dust	粗砂岩岩尘
gritter	铺砂机；撒尘粉器
gritting machine	撒砂机，撒碎石机
gritting material	砂砾材料；粗砂，粗砂石

gritty	粗砂质的；砂质	grossular	钙铝榴石
gritty consistence	含砂砾度；砂砾稠度	grossularite	钙铝榴石
gritty coverstone	砂砾质面层；粗砂质面层	grothine	钙铍柱石；铁橄石
gritty dust	砂砾屑；石粉	grotto	洞穴；岩穴；人工洞室；石窟
gritty soil	粗砂砾质土；砂砾质土	grotto cultural relics	石窟文物；石窟文化遗迹；石窟文化遗址
gritty surface dressing	路面铺石碴		
gritty water	含砂水；磨蚀水	ground absorption	地面吸收
grivation	磁斜坐标纵线偏角，栅格差，格网磁差；方格角，方格磁角	ground acceleration	地面加速度
		ground air	土中空气；亚土层上层的气体
grizzle	高硫低级煤	ground altitude	地面高度
grizzly bar	格筛条；棒条筛	ground amplitude	地面振幅
grizzly blasting	溜井爆破；格筛爆破	ground anchor	地基锚杆；地锚；锚钉
grizzly chamber	格筛室；棒条筛室	ground anisotropy	大地各向异性
grizzly chute	格筛溜槽	ground arch	塌落拱；平衡拱
grizzly drift	格筛平巷	ground argillization	岩石黏土灌浆
grizzly feeder	棒条给料机	ground auger	土螺旋钻；螺旋钻；土钻
grizzly fringedrift	脉外格筛平巷	ground barium sulfate	重晶石粉
grizzly screen	格筛	ground base	地基
grizzly station	格筛车间	ground beacon antenna	地面信标天线
grobaite	淡辉二长岩	ground beam	地梁，地基梁；基础梁
grochauite	镁蠕绿泥石	ground bearing pressure	地基承压力
groining	交叉拱；丁坝	ground bearing test	地基承载试验
grommet type seal	环形密封圈	ground block	地面滑轮
Groningen effect	格罗宁根效应	ground bolt	地脚螺栓；锚栓；板座栓；锚固螺栓
gronlandite	紫苏闪石岩	ground calcium carbonate	重质碳酸钙
Gronlund type cut	格朗隆德式掏槽	ground calibration	地面校准
groove	槽，凹槽，槽沟；孔型，纹道，轧槽；沟	ground cementation	土层灌浆，土壤灌浆；岩层灌浆
groove sample	刻槽试样；刻槽煤样	ground characteristic line	地层特征线
grooved lava	槽状熔岩；沟切熔岩；槽面熔岩	ground check	地面校正；地面核查；基线校正
grooved rail	槽形轨道	ground coal	底煤，煤层底部煤
groover	矿工；挖槽机	ground collapse	地面塌陷
grooving	挖槽；凹凸榫接；割槽处理	ground compaction	地基压实；地基压缩；压实土
grooving tool	挖槽机，切缝机	ground condition factor	场地条件系数
grospydite	榴辉蓝晶岩	ground consolidation	地基固结
gross amount	总额；总量	ground control	地面控制；地层控制；岩层控制
gross analysis	全量分析；总量分析	ground control point	地面控制点
gross calorific value	总热量；总热值	ground control-point survey	地面控制点测量
gross count gamma ray	总自然伽马计数率	ground cover	地面覆盖物；植被；地表盖层
gross cut	总挖方量	ground coverage	地面覆盖范围
gross error	总误差；显著误差；粗差，粗误差	ground crack	地裂缝
gross exploration	总体勘探；大面积普查；普查	ground data	综合地面资料；地面数据
gross head	总水头；毛水头	ground data-collection platform	地面数据收集站；地面数据自动遥测站
gross heat	总热量		
gross heat value	高位热值；总热值	ground deformation	地基变形；地面变形
gross heating value	总热值	ground deposition	地面沉积
gross load	总荷载；毛重	ground displacement	地面位移；地动位移
gross loading intensity	总荷载强度；总荷载	ground displacement monitoring	地面位移监测
gross operating head	总运行水头	ground drill	地钻；土钻；钻土机
gross output	总产量；总产值	ground effect	地面效应
gross porosity	总孔隙度	ground elevation	高程点；地面高程，地面标高
gross pressure	总压力	ground engineering	地基工程学；地面工程
gross production	毛产量；总开采量；总产量	ground excavation	地基开挖
gross record	原始记录	ground failure	地面破坏；地基失效
gross recovery	总开采量；总回收率	ground failure opportunity map	地面破坏可能图
gross sample	总样；总样品；总试样	ground feature	地面特征；地物；地貌
gross section	全断面，总截面；毛截面	ground filter	土层过滤；土层过滤器
gross settlement	总沉降量	ground fissuration	地裂
gross strain	总应变	ground fissure	地裂缝
gross structure	总体构造	ground fissuring	地面开裂
gross weight	总重，总重量；毛重；全重	ground floor	底层，基层；地面层

ground floor construction 底层地面构造
ground flow 岩石流动；地层移动
ground fluid 地下流体
ground follow-up 地面异常检查；地面检查，地面查证；地面追踪
ground fracture 地裂缝
ground fracturing 地裂
ground freezing 地基冻结法；地下冻结
ground frost 永久冻土；地面霜
ground gas 地气
ground geology 地表地质
ground geophysics 地面地球物理学
ground heat exchange system 地埋管换热系统
ground heat exchanger 地埋管换热器
ground height 地面标高
ground ice 底冰；锚冰；潜冰
ground improvement 地基改良；地基加固
ground installation 地面装置；地面设备
ground investigation 场址勘探，场地勘探，土地勘测，探土；地面调查
ground investigation field work 现场勘探工程
ground isotherm 地下等温线
ground joist 地隔栅
ground layer 底土层
ground level 地面，地平面；地面标高，地面高程；基态能级，基级；井口水平，井口平台
ground level winder 地面安装的绞车
ground leveling 地面水准测量
ground line 基线；地面线；地平线
ground line gradient 地面坡度；自然坡度；地平线坡度
ground loading intensity 基底压力；总荷载强度
ground location 地面定位；地面位置
ground loss 地面下陷；塌方；矿柱中矿物的损失
ground loss in effectiveness 地基失效
ground magnetic prospecting 地面磁法勘探
ground man 掘土工
ground map 地形图
ground marker navigation 地标导航
ground mass 基质；金属基体；基体
ground measurement around tunnel 隧道周围的地表测量
ground mica 云母
ground mineralization 地下水矿化度；地下水矿化作用
ground mining 地下水开采
ground mixing 地面混凝
ground mixing effect 地下水混合作用
ground modification technique 土的加固技术
ground monitoring 地面监测
ground moraine 底碛；地面冰碛层；基碛
ground motion 地面运动；地震动
ground motion array 地面震动观测台阵；地面运动观测台阵
ground motion attenuation 地震动衰减
ground motion component 地震动分量
ground motion duration 地震动持时
ground motion intensity 地震震动强度
ground motion parameter 地面运动参数；地震动参数
ground motion record 地震动记录
ground motion zoning 地震动区划
ground movement 地面移动，地面运动；地层移动；地面塌陷；围岩移动

ground movement and deformation 地表移动变形
ground nadir 地面天底点；地底点
ground nadir point 地底点
ground navigation 地面导航
ground network 地下水网络
ground object 地物；地面地标
ground peg 消波器；量地木桩
ground penetrating radar 探地雷达；地质雷达；地下透视雷达
ground photogrammetric survey 地面摄影测量
ground photogrammetry 地面摄影测量；地面摄影测量学
ground pipe 地下管道
ground plan 地面图；平面图；水平投影图；底面图
ground plan of mine workings 矿山巷道底面图
ground plate 底梁；枕木支承板；埋板
ground plot 地基图；底面图
ground plumb point 地面铅锤校准点
ground position 地面位置
ground power unit 地面发电机；地面动力装置
ground pressure 接地压力；地面压强；地面压力；地压，地层压力，土压；山体压力，围岩压力；地基应力
ground pressure activity 地压活动
ground pressure control in stope 采场地压控制
ground pressure in stope 采场地压
ground pressure measurement 地压测量；地压观测
ground pressure phenomena 地压现象
ground pressure show 地压显现
ground probing radar 地质雷达；探地雷达
ground processing 地面处理
ground prop 地下支撑
ground quartz 石英粉
ground range 地面距离；水平距离
ground reaction curve 岩体变形特性曲线；地层反作用曲线
ground receiving station 地面接收站
ground receiving telemetering station 地面接收遥测站
ground reconnaissance 初步勘察；初步踏勘
ground resistivity 地电阻率
ground response 地面反应
ground rock 石屑；碎磨岩石
ground rock powder 石粉
ground roll 地表空气界面波；地滚波
ground run 地面试验
ground rupture 地裂；地面破裂
ground sensitivity 地面灵敏度
ground settlement 地表塌陷
ground settlement observation 地表沉降观测
ground shaking 地面振动
ground shaping 地面整平
ground shock 地震动
ground signature 地面特征
ground sill 底板；基础板；底梁
ground slab 地板
ground slag grouting 地下矿渣灌浆
ground slope 地面坡度；地面倾斜
ground slope factor 地面坡度因子；地面倾斜系数
ground sluice 地沟；地内流矿槽
ground soil 地基土

ground source field	地下水水源地
ground stability	地基稳定性
ground stabilization	地基加固
ground stereo camera	地面立体摄影机
ground stereogram	地面立体图
ground stereophotography	地面立体摄影学
ground stiffness	地基刚度
ground storage	地下储藏
ground strain	地基应变；地应变
ground strength	地基强度；土壤强度
ground strengthening	地基加固
ground stress	地应力
ground stress field	地应力场
ground stress relief	地应力解除
ground stress restoration	地应力恢复
ground subsidence	地层沉陷，地层陷落；地面沉降，地面沉陷
ground subsidence control	地面沉降控制
ground support	地面支护；矿山支架
ground surface	地表；地面
ground surface acceleration	地面加速度
ground surface deformation	地面变形
ground surface elevation	地面标高
ground surface heave	地面隆起
ground surface settlement	地表沉降
ground surface settlement measure	地表下沉测量
ground surface survey	地面测量
ground surge	地面涌流
ground survey	地面测量；地形测量；地面调查
ground swell	地面隆起
ground swing	地面反射变化
ground swing error	地面反射误差
ground table	地面标高；地平高程
ground temperature	地温；地面温度；井上温度
ground temperature-depth curve	地温-深度曲线
ground thermal resistivity	土壤热阻系数
ground tilt	地倾斜
ground treatment	地基处理；地基加固处理
ground treatment work	土地处理工程；地层处理工程
ground type	岩层类型
ground unrest	背景噪声；外界噪声
ground vibration	地表振动
ground wall	地墙，基础墙
ground warm water	地下热水
ground water	地下水；潜水；雨水；来自大气中的水；矿井水
ground water accumulation	地下水堆积
ground water anomaly	地下水异常
ground water balance	地下水均衡
ground water basin	地下水盆地
ground water boring	地下水钻探
ground water budget	地下水平衡；地下水均衡
ground water circulation	地下水循环
ground water contamination	地下水污染
ground water data bank	地下水数据库
ground water discharge area	地下水排泄区
ground water divide	地下水分水岭
ground water drawdown	地下水降深；地下水降落
ground water dynamics	地下水动力学
ground water elevation	地下水位；潜水位
ground water flow	地下水径流；地下水流动；潜流
ground water geochemistry	地下水地球化学
ground water geology	地下水地质学；水文地质学
ground water gushing disaster	地下水突水灾害
ground water hardness	地下水硬度
ground water hydrology	地下水水文学
ground water inflow	地下水径流流入量
ground water level	地下水位；地下水面；潜水面；潜水位
ground water level lowering	地下水位的降低
ground water lowering	地下水位的降低
ground water monitoring	地下水监测
ground water network	地下水网络
ground water occurrence	地下水赋存条件
ground water of aeration zone	包气带水
ground water origin	地下水起源
ground water outflow	地下水流出量
ground water overdraft	过度抽取地下水；过度开采地下水
ground water recharge	地下水补给
ground water regime	地下水动态
ground water regime monitoring system	地下水动态监测系统
ground water reserve	地下水储量
ground water reservoir	地下水储层；地下水库
ground water resource	地下水资源
ground water resource exhaustion	地下水资源枯竭
ground water resource potential	地下水资源潜力
ground water runoff	地下水径流；地下水径流量
ground water sediment	地下水沉积物
ground water seepage	地下水渗漏；地下水渗透
ground water station	地下水站
ground water storage	地下水储量
ground water system	地下水系统
ground water table	地下水位；潜水位；潜水面
ground water table level	地下水位
ground water tracer	地下水流示踪物
ground water vertical zoning	地下水垂直分带
ground water vulnerability	地下水脆弱性；地下水易损性
ground water withdrawal	地下水开采量
ground wave	地波；地面波；地面电波
ground work	基础；路基；土方工程
Ground Improvement	地基加固学报
ground-anchorage tendon	地锚预应力钢丝束
ground-batching plant	泥浆拌和机
ground-breaking ceremony	动土礼；破土动工仪式
ground-checking	地面检查；地面验证
groundchute	底板溜槽；溜煤板
ground-data processing system	地面数据处理系统
groundhog	挖土机
ground-installed	装在地面上的
ground-line	基线；地平线
groundman	掘土工
groundmass	基质；石基
ground-moraine shoreline	底碛滨线
ground-mounted	装在地面上的
ground-particle velocity	地面质点速度
ground-probing radar	地质探测雷达；顶板岩层雷达
ground-range image	地面距离图像

ground-range scale 地面距离比例尺
ground-source heat pump system 地源热泵系统
ground-stone 基石
ground-surface movement 地表移动
ground-thermometer 地温计；地温表
ground-type friction winder 落地式摩擦提升机
ground-type Koepe winder 落地式戈培式提升机
ground-up 地面上拱
groundwater 地下水；潜水；自由水
groundwater abstraction 抽取地下水；汲取地下水
groundwater aggressiveness 地下水侵蚀性
groundwater anomaly 地下水异常
groundwater aquifer 地下水含水层
groundwater available velocity 地下水实际平均速度
groundwater balance 地下水均衡
groundwater balance element 地下水平衡要素
groundwater barrier 地下水屏障；地下水阻体；地下水坝；地下水堡坝
groundwater basin 地下水盆地；地下水流域
groundwater body 地下水体
groundwater boring 地下水钻探
groundwater capture 地下水袭夺；地下水截取；截潜流
groundwater characteristics 地下水特征
groundwater chemical quality 地下水化学质量
groundwater chemistry 地下水化学
groundwater circulation 地下水循环
groundwater compartment 潜水含水层
groundwater condition 地下水条件；地下水情况
groundwater conditioning 地下水调节
groundwater conditions 地下水条件；地下水情况；地下水状况
groundwater consumption 地下水消耗量，地下水使用量
groundwater contamination 地下水污染
groundwater contour 地下水等水位线；地下水位等高线
groundwater control 地下水控制
groundwater corrosion 地下水腐蚀作用
groundwater corrosivity 地下水侵蚀性
groundwater dam 地下水阻挡层；地下水坝
groundwater decline 降低地下水位
groundwater decrement 地下水消耗；地下水减少
groundwater degradation 地下水恶化
groundwater depletion 地下水量枯竭；地下水疏干；地下水消落
groundwater depletion curve 地下水消退曲线；地下水损耗曲线
groundwater deposit 地下水沉积
groundwater depression cone 地下水降落漏斗
groundwater depth of occurrence 地下水埋藏深度
groundwater development 地下水开发
groundwater discharge 地下水排泄；地下水流量；地下水溢出量；地下水出流
groundwater discharge area 地下水排泄区；地下水溢出区
groundwater divide 地下水分水岭；地下水位界面；地下分水界
groundwater drawdown 地下水降深；地下水降落；地下水位下降
groundwater dynamics 地下水动力学
groundwater economics 地下水经济学
groundwater electro-osmosis 地下水电渗

groundwater elevation 地下水位；地下水位标高；潜水位
groundwater engineering 地下水工程
groundwater equation 地下水方程
groundwater evaluation 地下水评价；地下水估算
groundwater exploitation 地下水开发；地下水开采
groundwater exploration 地下水勘探
groundwater extraction 地下水抽取；地下水开采
groundwater fall line 地下水陡降线
groundwater flow 地下水径流；地下水流动；潜流；地下水流
groundwater flow dynamic 地下水流动态
groundwater flow system 地下水流系统；地下水系
groundwater flow velocity 地下水流速
groundwater fluctuation 地下水波动；地下水变幅
groundwater forecast 地下水预报
groundwater geochemistry 地下水地球化学
groundwater geology 地下水地质学；水文地质学
groundwater geophysics 地下水地球物理学
groundwater gradient 地下水坡度；地下水梯度
groundwater hazard 地下水灾害
groundwater head level 地下水头高程
groundwater hill 地下水丘；潜水丘
groundwater horizon 地下水层；含水层
groundwater hydraulics 地下水水力学
groundwater hydrological factor 地下水水文因素
groundwater hydrology 地下水水文学；地下水文学
groundwater increment 地下水补给，地下水补给量；地下水增量
groundwater inflow 地下水涌入量；地下水径流流入量
groundwater information system 地下水信息系统
groundwater injection 地下水回灌
groundwater inventory 地下水资源平衡表；地下水资源总量表
groundwater investigation 地下水调查；地下水勘察
groundwater isopiestic line 地下水等位线
groundwater isopotential line 地下水等位线
groundwater laterite 潜水砖红壤
ground-water lateritic soil 潜水砖红壤性土
groundwater law 地下水定律
groundwater leaching 地下水溶滤
groundwater leakage 地下水越流；地下水渗漏
groundwater lens 地下水透镜体
groundwater level 地下水位，地下水位线；潜水位；地下水面；潜水面
groundwater level contour map 地下水等水位线图
groundwater level in drill hole 钻孔地下水位
groundwater level network 地下水位观测网
groundwater line 地下水位；地下水面；潜水面
groundwater lowering 地下水下降；降低地下水；降低地下水位
groundwater management 地下水管理
groundwater map 地下水分布图
groundwater mapping 地下水测绘；地下水制图；地下水填图
groundwater mass 地下水体
groundwater mathematical modelling 地下水数学模型
groundwater mineralization 地下水矿化度；地下水矿化作用
groundwater mining 地下水开采

groundwater mode of discharge 地下水排泄方式
groundwater mode of occurrence 地下水埋藏方式
groundwater mode of recharge 地下水补给方式
groundwater model 地下水模型
groundwater modeling 地下水模拟
groundwater monitoring 地下水监测；地下水监测系统
groundwater monitoring network 地下水监测网
groundwater monitoring system 地下水监测系统
groundwater motion 地下水活动；地下水运动；地下水移动
groundwater mound 潜水丘；地下水丘；地下水隆起
groundwater movement 地下水运动
groundwater observation 地下水动态观测
groundwater observation station 地下水观测站
groundwater occurrence 地下水产状；地下水赋存；地下水埋藏；地下水赋存条件；地下水分布
groundwater origin type 地下水成因类型
groundwater outflow 地下水出流量；地下水流量；地下水溢出
groundwater overdraft 过度抽取地下水；过度开采地下水
groundwater physical property 地下水物理性质
ground-water plane 地下水面；潜水面
groundwater pollutant 地下水污染物
ground-water pollution 地下水污染
groundwater potential 地下水势；地下水潜力
ground-water pressure 地下水压力
groundwater pressure head 地下水压头
groundwater pressure measurement 地下水压力测量；地下水水压测试
groundwater prospecting 地下水勘探
groundwater pumpage 地下水抽取
groundwater pumping plant 地下水抽水站
groundwater quality 地下水水质；地下水质量
groundwater quality deterioration 地下水质恶化
groundwater quality map 地下水水质图
groundwater quality model 地下水质模型；地下水水质模型
groundwater quality monitoring 地下水水质监测
groundwater recession 地下水衰退；地下水枯竭；地下水后退；地下水消落；地下水下降
groundwater recession curve 地下水消耗曲线；地下水下降曲线；地下水后退曲线
groundwater recharge 地下水补给；地下水回灌
groundwater recharge rate 地下水补给率；地下水补给量
groundwater recharge station 地下水补给站
groundwater recharge trench 地下水补给槽
groundwater recovery 地下水回收；地下水开采
groundwater recreation curve 地下水消耗曲线
groundwater regime 地下水动态；地下水体系
groundwater regime element 地下水动态要素
groundwater regime regularity 地下水动态规律
groundwater regression 地下水消落
groundwater regulation 地下水调节
groundwater remediation strategy 地下水治理对策；地下水修复策略
groundwater replenishment 地下水补给
groundwater research 地下水调查；地下水研究
groundwater reserve 地下水储量
groundwater reservoir 地下储水层；地下水储水体；地下水库
groundwater resource 地下水资源
groundwater resource assessment 地下水资源估计；地下水资源评价
groundwater resource evaluation 地下水资源评价
groundwater resources protection 地下水资源保护
groundwater ridge 地下水分水线；地下水脊；地下水分水岭
groundwater right 地下水权
groundwater rise 地下水上升
groundwater runoff 地下水径流；地下水出流
groundwater sampling 地下水取样
groundwater seepage 地下水渗流，地下水渗漏
groundwater seepage velocity 地下水渗透速度
groundwater shortage 地下水缺乏
groundwater simulation 地下水模拟
groundwater source 地下水水源
groundwater source field 地下水源地
groundwater stage 地下水位
groundwater stage graph 地下水位曲线；地下水位图
groundwater storage 地下水储存；地下水储量；地下水储存资源
groundwater storage capacity 地下水储存能力；地下水储存量
ground-water supply 地下水供水；地下水补给
ground-water surface 地下水面，地下水位；潜水面
groundwater table 地下水位；潜水位；地下水面；潜水面
groundwater table fluctuation 地下水位变动
ground-water tapping 引取地下水
groundwater total solid 地下水总固化物
groundwater tracer test 地下水示踪剂试验；地下水跟踪试验
groundwater transient flow 地下水非稳定流；地下水暂态流
groundwater trench 地下水谷；地下水沟；地下水槽形水面；地下水槽
groundwater turbidity 地下水浊度；地下水浑浊性；地下水混浊度
groundwater unsteady flow 地下水非稳定流
groundwater utilization 地下水利用
groundwater velocity 地下水流速
groundwater well 地下水水井；潜水井
groundwater withdrawal 地下水抽取；地下水开采
groundwater work 地下水工程
groundwater yield 地下水涌水量
groundwater zone 地下水带；潜水带
groundwater-related hazard 与地下有关的水灾害；地下水公害
group 组，队，群，界；类，族，炮眼组
group action of piles 桩群作用
group control 群控；组控
group data 归组数据
group delay 群延迟；组延迟
group demodulator 群解调器
group deviation method 群偏差法
group diffusion 群扩散
group drilling 丛式钻进；多孔底钻进
group drive 成组传动
group driving 屏风式打桩法

group effect of anchor	群锚效应
group efficiency of piles	群桩效率
group element	同组元素
group filter	群滤波器
group frequency	群频率
group frequency selector	频组选择器
group index	分组指数
group iterative method	群迭代法
group level	集中平巷水平
group mother entry	分组集中平巷
group motion	群动
group of bars	钢筋束
group of beds	岩层群
group of building	建筑群
group of faults	断层组
group of gentle dipping	平缓倾斜组
group of low dipping	低倾斜组
group of medium dipping	中倾斜组
group of mountains	山群
group of piles	桩群，桩组，群桩
group of seams	煤组；煤层组；矿层组
group of steep dipping	陡倾斜组
group of volcanoes	火山群
group of waves	波群
group piece work contract	小包工计件合同
group pile	群桩
group precipitation	分组沉淀
group reduction factor of piles	桩组折减系数
group relaxation	群松弛
group repetition interval	组重复周期
group retention index	基团保留指数
group ring	群环
group sampling	分层抽样；分组抽样；分层采样
group selector	群选择器，选组器
group sequence control	分组程序控制
group settlement	群体沉陷
group slowness	群慢度
group speed	群速度
group technology	成组技术
group theory	群论
group transfer polymerization	基团转移聚合
group transfer reaction	基团转移反应
group valve	组合阀，阀组
group velocity	群速度
group wave	群波
grouped array	组排列
grouped column	群柱
grouped data	分类数据
grouping	归组；集合；组合
grouping contactor	组合接触器
grouping error	分组误差
grouping process	分组过程
group-velocity dispersion value	群速度频散值
group-velocity method	群速度法
grout absorption	吸浆量；吃浆量
grout acceptance	吸浆量；吃浆量；浆液注入量
grout agent	注浆材料
grout agitator	浆液搅拌器
grout box	灌浆槽
grout cohesion	浆液凝聚力；浆液黏聚力
grout concrete	灌浆混凝土
grout consistency	泥浆稠度
grout consumption	耗浆量
grout coordinate ratio	浆液配比
grout core	水泥黏结的岩心
grout curtain	灌浆帷幕；注浆帷幕；薄浆隔墙
grout delivery	供浆量
grout discharge valve	浆液排放阀
grout distribution line	配浆管路
grout effectiveness	灌浆效果
grout fabrication	浆液的配置
grout filler	灌缝料；灌缝砂浆
grout fluid	未凝固的注浆液
grout fluidifier	灌浆流化剂
grout formulation	浆液配比
grout gallery	灌浆廊道
grout head	灌浆头
grout header	灌浆集管
grout hole	灌浆孔
grout hole sealing	灌浆封孔
grout hole spacing	灌浆孔孔距
grout injection apparatus	注浆设备
grout injection parameter	灌浆参数
grout injection pipe	灌浆管
grout injection pressure	灌浆压力
grout injector method of cementing	注浆法
grout joint	灌浆缝
grout leaking	串浆
grout level	灌浆水平
grout line	灌浆管道
grout line guide	注浆管导管
grout loss	漏浆
grout machine	灌浆机
grout mix computation chart	水泥浆拌合计算图表
grout mix design	浆液配合比设计
grout mix proportion	浆液配合比
grout mixer	水泥浆搅拌机；薄浆搅拌机；拌浆机
grout mixer and placer	砂浆搅拌喷射器
grout mixing plant	制浆站；浆液搅拌厂
grout mixture	薄浆混合物
grout oozing out	冒浆
grout outlet	出浆口
grout packer	灌浆塞
grout permeation zone	浆液扩散区
grout pipe system	灌浆管系统
grout placement	注浆工序
grout plant	制浆站
grout proportion	浆液配比
grout pump pressure	灌浆泵压力
grout return pipe	回浆管
grout rock bolt	灌浆岩石锚杆
grout rock cap	注浆岩帽；注浆盖层
grout setting time	浆液凝固时间
grout spreading range	灌浆扩散半径
grout stone	浆液结石
grout stop	止浆片
grout supply pipe	供浆管
grout system	注浆方法
grout take	吃浆量；灌浆量；吸浆量
grout travel	浆液渗透距离；泥浆散布距离

grout treatment	灌浆处理	grouting plan	灌浆计划
grout vent	溢浆孔；灌浆孔	grouting plant	注浆站
grout volume injected	灌浆量	grouting port	灌浆口；出浆口
groutability ratio	可灌比	grouting pressure	灌浆压力；注浆压力
grout-bleeding	薄浆渗出	grouting procedure	灌浆工序；灌浆法；灌浆过程
grouted anchor	灌浆锚杆；灌注锚定；灌浆地锚	grouting profile	灌浆剖面图
grouted area	灌浆区	grouting programme	灌浆方案
grouted base pile	基底灌浆桩	grouting project	灌浆工程
grouted bolt	混凝土锚杆；灌浆固定的锚杆	grouting pump	灌浆泵；压浆泵
grouted concrete	灌浆混凝土；预填集料混凝土；压浆混凝土	grouting quantity	灌浆量
		grouting radius	灌浆半径
grouted cut-off wall	灌浆截水墙；灌浆帷幕	grouting rate	注浆率；注浆速度
grouted joint	灌浆接缝	grouting reinforcement	注浆加固
grouted pile	灌浆桩；注浆桩	grouting result	灌浆效果
grouted resin bolt	灌浆树脂锚杆	grouting screen	灌浆帷幕
grouted riprap	浆砌乱石；乱石灌浆；灌浆乱石	grouting section	灌浆段
grouted rock bolt	灌浆岩石锚杆；灌浆锚杆	grouting seepage force	灌浆渗流力
grouted tubing bolt	管式注浆锚杆	grouting sequence	灌浆程序；灌浆分序；灌浆顺序
grouted vein	灌浆脉	grouting sinking method	注浆凿井法；注浆凿井法施工测量
grouted-aggregate concrete	骨料灌浆混凝土		
grouter	灌浆管；灌浆头；灌浆泵；注浆机	grouting specification	灌浆技术规程
grout-filled fabric mat	灌浆土工布护排	grouting stage	灌浆阶段；灌浆段；灌段
grout-flow cone efflux time	浆液流度锥流出时间	grouting support	注浆支护
grout-hole drilling	灌浆孔钻进	grouting system	注浆法；注浆系统；灌浆法
grouting admixture	灌浆外加剂	grouting technique	灌浆技术
grouting agent	胶结剂；灌浆助剂；灌浆材料	grouting technology	灌浆工艺
grouting ahead of tunnel	隧道超前灌浆	grouting test	灌浆试验
grouting and sealing technique	注浆堵水技术	grouting-in	套管放入钻孔后灌浆
grouting blanket	灌浆铺盖	grout-injected volume	灌浆量
grouting control	灌浆控制	grout-injection method	灌浆法；注浆法
grouting cost	灌浆费	grout-stone	浆体结石
grouting culvert	灌浆廊道	growth anisotropy	生长各向异性
grouting cup	灌浆漏斗	growth anticline	生长背斜
grouting data recording system	灌浆数据记录仪	growth bedding	生长层理
grouting depth	灌浆深度	growth diapir	生长底辟；生长刺穿构造
grouting design	灌浆设计	growth fabric	生长组构
grouting efficiency	灌浆效率；灌浆效果	growth factor	生长因素；增长系数；成长因素
grouting engineer	灌浆工程师	growth fault	生长断层；沉积断层；同生断层；新生断层
grouting engineering	灌浆工程		
grouting equipment	灌浆设备	growth fault trap	同生断层圈闭
grouting equipment supplier	灌浆设备供应商	growth fibre	生长纤维
grouting gallery	灌浆廊道	growth front	生长阵面；生长前沿
grouting gun	喷浆器；水泥浆喷枪	growth lattice	生长格子；生长格架
grouting hole	灌浆孔；注浆孔	growth layer	生长层
grouting hose	灌浆软管	growth line	生长线
grouting in separated-bed	离层带注浆充填	growth matrix	生长基质
grouting inspector	灌浆工程检查员	growth monocline	生长单斜
grouting intensity number	灌浆强度值；灌浆强度法	growth of concrete	混凝土膨胀
grouting job	灌浆工作	growth of continent	大陆增长
grouting log	灌浆记录	growth of crystal	晶体生长
grouting machine house	灌浆机房	growth of rock crack	岩石裂纹增长
grouting material	灌浆材料	growth potential	增产潜力
grouting material in urgent need	应急注浆材料	growth striation	生长条纹
grouting mortar	灌浆灰浆	growth syncline	生长向斜
grouting nipple	灌浆管接头	growth transcurrent fault	生长横推断层；生长平移断层
grouting of subsoil	地基土石灌浆	growth twin crystal	生长双晶
grouting off	注水泥止水；水泥封堵	growth-fault movement	生长断裂运动
grouting operation	灌浆作业；灌浆施工	growth-faulted region	生长断层区
grouting parameter	灌浆参数	growth-framework porosity	生长格架孔隙性
grouting path	灌浆途径；灌浆路线	growth-in-place	原地生长；原地生成

groyne	拦沙坝；海堤；防波堤
GRS	陀螺基准系统，陀螺参考系；大地测量参考系，测地参考系统；图尺度评价法，图示评定量表
grueling test	疲劳试验
Grumbrecht number series	顾氏数系
grunauite	辉铋镍矿
grunerite	铁闪石
GSIS	地球科学信息学会；地学信息学会
GTJ	岩土试验学报
guanajuatite	硒铋矿
guanite	鸟粪石
guano bed	鸟粪层
guano-phosphatic deposit	鸟粪磷矿
guanophosphorite	鸟粪磷灰岩
guanovulite	填卵石
guaranteed stage of water level	保证水位
guard pile	护桩
guard stage of water level	警戒水位
guard tube	套管；防护管；保护管
guard wall	护墙
guardiaite	霞长粗安岩
guarinite	片楣石
gudmundite	硫锑铁矿
guest mineral	后成矿物；交代矿物；充填矿物
guestimate	概算，粗略估计；推测
guide bed	标志层；标准层
guide commutator	导管
guide drill rod	导向钻杆
guide drum	导向滚筒
guide element	标型元素；指示元素
guide frame	导架；导向架
guide head	导向头
guide hole	导向钻孔；先导孔，超前孔，超前钻孔
guide horizon	测量标志水平面
guide idler	导辊
guide installation	导向装置；安装罐道
guide key	导向键
guide ledge	导板，导向尺；导杆
guide lever	导向杆；基准杠杆
guide line	导线；导引线；视线
guide member	导引构件；导轨
guide mineral	指示矿物；主导矿物；标型矿物
guide pad	导向垫；导向块
guide passage	导路
guide pile	导桩；定位桩
guide pilot-hole	超前探孔
guide pipe	导管
guide plate	导向板；导板；导向隔板
guide post	导向支柱
guide rod	导杆；导向钻杆
guide roller	横向导向轮；导滚
guide tracer	导向性示踪剂；指示性示踪剂
guide trench	导沟；导槽
guide tube	钻探导管；套管
guidebook	参考手册；指南
guided drilling	定向钻进
guideline	指南；导向图表；方针，指标；准则
guideseam	标志层，标准层
guide-wall	导水墙；护井
guiding bed	标准层；标志层；标志煤层
guiding collar	导向垫圈；导向轴环
guiding design for construction scheme	指导性施工方案设计
guiding device	导向装置
guiding edge	导向边；导杆
guiding hole	导向孔
guiding landmark	方向标
guiding laser	激光导向器；导向仪
guiding mark	基点；控制点
guiding member	导向构件
guiding ramp	滑架；导向架
guiding rule	规准；量规，准则
guildite	多水铜铁矾
guillotine shear	闸刀式剪切机
guillotine-type gate	吊升垂直闸门；起落式闸门
guimaraesite	钽铌钛铀矿
Guizhou-Guangxi oldland	黔桂古陆
gulch	冲沟；干谷；峡谷
gulch placer	溪沟砂矿；冲沟砂矿
gulching	顶板下沉的碎裂声
gulf	海湾；地下水系；深峡谷，深坑；封闭洼地；落水洞
gulf coast-type fault	海湾沿岸式断层；生长断层
gulf ice	港湾冰块
Gulf Oil Corporation	海湾石油公司
gulley	集水沟；排水沟
gulley trap	水沟排污阱
Gullick Dobson self-advancing support	伽立克-道布森自移式支架
gully	山沟；冲沟，水沟，集水沟
gully correction	沟壑整治
gully cover	污水井盖
gully drain	排水渠；下水道；雨水口连接管
gully gravure	沟蚀后退；沟谷溯源侵蚀
gully hole	沟洞
gully pot	雨水井；排水井；集水坑
gully stoping	贮矿堑沟回采法
gully trap	集水沟气隔；集水沟气隔弯管
gully-control dam	水土保持坝
gullying	沟蚀作用；沟状冲刷；冲沟作用；沟道下切
gully-stopping method	贮矿沟开采法
gum bed	地蜡
gum dirt flinger	截粉清除器
gum dynamite	胶质硝甘炸药
gum formation	胶质形成；黏胶形成作用
gumbelite	镁水白云母
gumbo	肥黏土；强黏土；坚硬黏土；黏泥；碱性黏土
gumbo clay	碱性黏土；重黏土
gumbo soil	强黏土
gumbotil	硬黏土
gum-forming hydrocarbon	生成胶质的烃
gum-forming tendency	生胶倾向
gumlike material	类胶物质
gummer	除截粉器
gumminess	胶黏性；树胶状
gumming	清除截粉，除粉；胶结；浸液，涂液
gummite	脂铅铀矿
gummosity	黏着性；胶黏性
gummy	黏而无润滑能力的
gummy formation	黏性地层

gummy oil 含大量胶质的油
gun 喷镀枪，注油枪；空气枪，气爆震源；井壁取心装置；混凝土喷射机
gun boat 自卸式箕斗
gun finish 喷枪修整
gun perforation 射孔；射击冲孔
gun powder 火药
gun sampler 喷射式取样器；管式采样器
gunboat 自动卸载车；翻斗车；斜井箕斗
gunboats skip 提煤箕斗
gunite 喷射水泥浆；喷枪；喷射混凝土，喷浆；喷射砂浆；压力喷浆；喷射灌浆
gunite coat 喷浆面层
gunite coating 喷浆涂层；喷浆表面
gunite concrete 喷浆混凝土
gunite concrete lining 混凝土喷浆衬砌
gunite covering 喷浆面，喷浆面层；喷浆；喷射水泥
gunite layer 喷浆层
gunite lining 喷混凝土衬砌；喷浆衬砌；水泥喷射灌浆衬砌
gunite material 喷浆材料
gunite method 压力喷浆法
gunite work 喷浆工作，混凝土喷灌
gunited 喷浆的
gunited concrete 喷浆混凝土
gunite-lined shaft 喷射混凝土支护的井筒
guniting 喷浆支护
gunjet 喷枪
gunk 方钻杆防磨用油；油泥
gunk plug 柴油泥塞
gunning mix 喷补料；喷补混合物
gun-perforated completion 射孔完井
gunpowder 火药，黑色火药
gunpowder core 火药芯
gunpowder explosive 火药，黑火药
gunsight 瞄准器；标尺
gun-to-casing clearance 射孔器-套管间隙
gun-type coring machine 枪式取土器；枪式取芯器
Günz glacial stage 贡兹冰期
Günz-Mindel interglacial 贡兹-民德间冰期
Günz-Mindel interglacial stage 贡兹-民德间冰期
Günz-Mindelian interglacial stage 贡兹-民德间冰期
gurgling well 脉动喷井
gurmy 矿井水平
gush out water 突水
gusher hole 喷泉口
gusher sand 高压油砂层
gusher type well 自流井；自喷井
gusher well 自喷井
gushing spring 喷泉；火山泉
gushing water 涌水；突水
gushing water hole 涌水孔
gushing well 自喷井
Gusto multiple plough 格斯托型复式刨煤机
Gutenberg and Richter's law of magnitude 古登堡-里克特震级定律
Gutenberg discontinuity 古登堡不连续面；古登堡界面；古登堡间断面
Gutenberg energy-magnitude relation 古登堡能量-震级关系式
Gutenberg low-velocity zone 古登堡低速带
Gutenberg velocity-depth profile 古登堡速度-深度剖面图
Gutenberg wave 古登堡波
Gutenberg-Richter energy magnitude relation 古登堡-里克特能量-震级关系
Gutenberg-Richter relation 古登堡-里克特关系式
Gutenberg-Richter table 古登堡-里克特走时表
Gutenberg-Richter travel time 古登堡-里克特走时
Gutenberg's curve 古登堡曲线
Guthrie-Greenberger method 格思里-格林伯格法
guttameter 滴法张力计
gutter bed 天沟挡水板
gutter grade 沟底坡度
gutter grade line 水沟纵坡线；水沟坡度线
gutter plough 犁式挖沟机
guttering 井筒周边的引水槽
guttering corrosion 沟槽腐蚀
gutter-plough 犁式挖沟机
gutterway 水沟；排水沟
Gutzeit method 古特蔡特法；砷斑法
guy anchor 拉索锚，绷绳锚，拉线地锚；拉线桩，拉索桩
guy cable 拉索；牵索
guy derrick 牵索起重机
guy hook 拉线钩
guy rod 拉线桩
guy wire 牵索导线；绷绳；拉线
guy-cable tower 拉索桥塔
guyed 牵索加固的
guyed tower structure 锚索塔架结构
guyot 海台；海底高原；平顶海山，海底平顶山
gymnite 水蛇纹石
gyprock 石膏岩；易坍塌岩层
gyps 石膏
gypseous 石膏状的；含石膏的
gypsic horizon 石膏层
gypsiferous 含石膏的
gypsiferous marl 石膏泥灰岩
gypsification 石膏化
gypsodolomite 含石膏白云岩
gypsolith 石膏岩
gypsolith-anhydrite 石膏-硬石膏岩
gypsolyte 石膏岩
gypsum 硫酸钙；石膏；灰泥板
gypsum aluminate expansive cement 石膏矾土膨胀水泥
gypsum anhydrite 无水石膏
gypsum block 石膏块材；石膏块
gypsum caprock 膏盐类盖层
gypsum cave 石膏洞穴
gypsum cement 石膏水泥
gypsum flower 石膏花
gypsum karst breccia 膏溶角砾岩
gypsum mud 石膏泥浆
gypsum plate 石膏试板；石膏板；石膏片
gypsum rock 石膏岩
gypsum twin 石膏双晶
gypsum-anhydrite rock 石膏-硬石膏岩
gypsum-free layer 不含石膏层
gypsum-slag cement 石膏矿渣水泥
gypsy wheel 锚链轮

gyradisc crusher　埋顶曲线；旋回圆锥破碎机
gyral　旋转的；回旋的；环行洋流；涡流
gyrasphere crusher　旋回球面破碎机
gyrate　旋转的；螺旋状的；回转
gyrating crusher　回转压碎机
gyrating vanner　回转带式流槽
gyration moment　压差式取样器
gyration precision　旋转精度
gyrator　方向性移向器；回转器
gyratory　回转的
gyratory breaker　回转式破碎机；旋回破碎机
gyratory centrifuge　振动式离心机；旋回离心机
gyratory compactor　旋转式压实机
gyratory cone breaker　回转式锥形破碎机
gyratory cone crusher　圆锥破碎机
gyratory screen　回转筛
gyredozer　回转式推土机
Gyrex screen　杰瑞克斯型振动筛
gyro and flume stabilizer　陀螺仪和防摇水槽稳定器
gyro dozer　回转式推土机
gyro flux-gate compass　陀螺感应同步罗盘
gyro frequency　旋转频率
gyro horizon　陀螺地平仪
gyro horizon indicator　回转地平仪
gyro integrating accelerometer　陀螺积分加速度计
gyro magnetic compass　陀螺磁罗盘
gyro magnetic ratio　回转磁比率；磁矩与角动量之比
gyro motor　陀螺马达
gyro orientor　陀螺定向器
gyro reference system　陀螺基准系统；陀螺参考系
gyro remanent magnetization　旋转剩余磁化作用

gyro shaft　回转轴
gyro shock absorber　陀螺仪减震器
gyro soil　腐泥土
gyro speed　旋转速度
gyro stabilization unit　陀螺稳定平台
gyro theodolite　陀螺经纬仪；回转经纬仪
gyro transit　陀螺经纬仪
gyro traverse moment　回转力矩
gyrocompass　陀螺仪罗盘；回转罗盘；陀螺罗盘
gyrodynamics　陀螺动力学
gyro-flywheel　回转式飞轮
gyrograph　记转器；旋转测度器；转数记录器
gyro-kinetics　陀螺动力学
gyroradius　回转半径
gyroscope　陀螺仪；回转仪；回转器回旋器
gyroscopic action　回转作用；陀螺作用；回转效应
gyroscopic azimuth　陀螺方位角
gyroscopic compass　陀螺磁罗盘；回转罗盘
gyroscopic couple　回转力矩；陀螺力矩
gyroscopic horizon　回转水平仪；陀螺地平仪；人工地平仪
gyroscopic inclinometer　回转倾斜计；陀螺式倾斜计
gyroscopic moment　回转力矩
gyroscopic motion　回转运动
gyroscopic orientation　陀螺定向
gyroscopic transit　陀螺经纬仪
gyro-sextant　回转式六分仪
gyrostabilization unit　陀螺稳定装置；回旋稳定器
gyrostatic core orientator　回转轮式岩心定位仪
gyrostatic orientation　陀螺定向测量

H

Haalck gas gravimeter　哈氏气体重力仪
haapalaite　叠羟镁硫镍矿；红磷铁镍矿
Haase system　海斯铁管桩法凿井
habitat　海底探测船，居留仓；栖身地，生长环境，生境；自然环境；产地；聚居地
habitat of oil　石油产地
hachure　晕滃；影线
hachure map　用袭状线表示山岳的地图
hachure method　晕法；晕滃线法
hachuring　晕滃法；地貌晕线法
hack　刻痕，劈砍；格架，晒架，堆晒；碎土；鹤嘴锄
Hackberry facies　浊积扇相
hacking　刻痕
hackly　锯齿状；粗糙破裂面
hackly fracture　锯齿状断口
hackly surface　粗糙不平表面；锯齿状表面
hackmanite　紫方钠石
hacksaw structure　锯齿状构造
hadal　超深渊；深海的，海沟的
hadal depth　超深深度
hadal marine zone　超深海渊带
hadal sedimentation　超深渊沉积
hadal zone　超深渊带
hade　断层余角，伸角；倾斜，坡度，倾斜余角；偃角
hade angle　伸角；伸向
hade of fault　断层余角
hade slip fault　倾向滑断层
hadean　冥古宙；冥古代
hade-slip fault　倾向滑动断层
hading　倾斜的
hading ram　倾斜千斤顶
hading side　下盘
hadopelagic zone　远洋海底带
haematite　赤铁矿，红铁矿
haematization　赤铁矿化
haematocinite　赤石灰岩
haematogelite　胶赤铁矿
haff　泻湖
hafnefjordite　拉长石
hagatalite　波方石；钇铈锆石
hagemannite　铁方霜晶石
hagendorfite　黑磷铁钠石
haggite　黑斜钒矿；三水钒石
Hague anchor　海牙型气锚
haidingerite　砷钙石
hail imprint　雹痕
hainite　铈片楣石
hair checking　发状辐裂；发丝状裂缝，细裂缝
hair crack　毛细裂缝，细裂纹；发状裂缝
hair crystal　发晶
hair fibered plaster　麻刀灰泥
hair line crack　下落断层；细裂纹，发状裂纹
hair mortar　麻刀灰泥
hair seam　发缝，细裂缝

hairline cracking　细裂缝，毛细裂缝；细微裂缝
hairline cracks　细裂缝，毛细裂缝
hairline finish　幼纹面
hairline fracture　发丝状裂缝
hairpin　马蹄形的；U形的
hairpin dune　马蹄形沙丘；发针形沙丘
hairpin heat exchanger　回弯管壳式换热器
hairsalt　泻利盐
hairstone　毛发水晶
haiweeite　水硅钙铀矿；多硅钙铀矿
hakutoite　白头岩；英钠粗面岩
halation　晕光，成晕现象；晕影
Hales refraction method　哈利斯折射法
half angle　半角
half attenuation thickness　半衰减厚度
half basin　箕状盆地；半盆地
half bog soil　半沼泽土
half course　与倾斜交成45°的斜巷
half cycle　半周；半衰期
half duplex channel　半双工信道；半双向通道
half ebb　半落潮，半退潮
half end　与主解理成45°推进的采煤工作面
half graben　箕状地堑；半地堑
half graben-like basin　箕状断陷盆地；半地堑式盆地
half hole rate　半孔率
half horst　半地垒
half infinite extraction　半无限开采
half life　半衰期；半寿期
half load　半载
half lunarscape pebble　半月湾球石
half period　半衰期，半周期
half plane　半平面
half saturated　半饱和的
half saturation constant　半饱和常数
half scale　二分之一比例
half section　半断面
half section view　半剖视图
half set shotcreting　潮喷混凝土，半湿喷混凝土
half shear　半剪切
half subsidence　半沉陷
half subsidence point　沉陷量为最大值一半的点；半沉陷点
half tide　半潮
half tide level　半潮面，平均潮面；半潮水位
half timber construction　砖木结构
half V system　半V形工作面开采法
half value layer (HVL)　半值层
half wave　半波
half wet shotcreting　潮喷混凝土
half wet shotcreting machine　半湿喷混凝土机
half-absorption value　半吸收值
half-consolidated formation　半固结岩层；半坚硬岩层
half-decay time　半衰期
half-desert　半荒漠

half-dressed quarry stone	粗琢原石
half-duplex	半双向；半双工
half-edge layer	陡倾斜地层；倾角为45°的岩层
half-edge seam	急倾斜煤层；急倾斜矿层
half-flood	半涨潮，半潮汛
half-graben structure	半地堑构造
half-infinite	半无限的
half-infinite earth	半无限地层
half-interval contour	间曲线；半距等高线
half-life period	半衰期；半寿期
half-mark	实测标
half-maximum distance	半极大值距离
half-maximum line	半峰线
half-mechanized shield	半机械开挖盾构
half-obliterated texture	残余结构
half-peak width	半峰宽度
half-quad	半象限
half-round floor	半圆形地沟
half-sine pulse	半正弦脉冲
half-sine shape wave	半正弦波
half-size ratio	筛下粒度比
half-slope method	半坡度法
half-tide basin	半潮港地
half-V	半V形工作面开采法
half-V method	半V形开采法；长柱对角工作面开采法
half-width	半宽度，半值宽度
halide	卤化物；卤化物的
halilith	石盐岩
halilyte	石盐岩
halite	石盐；岩盐
halitic	石盐的；岩盐的；含石盐的
Hall coefficient	霍尔系数
Hall conductivity	霍尔电导率
Hall constant	霍尔常数
Hall effect	霍尔效应
halleflint gneiss	长英麻粒岩
Hallimond commutator	海力蒙浮选试验管
halloysite	多水高岭石，埃洛石，变埃洛石，叙永石
Hall-Row wedging method	霍尔-罗型偏向楔定向钻进法；霍尔-罗人工造斜方法
halmeic	海水生成的，深海溶液生成的
halmeic deposit	深海溶液沉积物，海成沉积物
halmirolysis	海底分解，海底风化；海解作用，海底风化作用
halmyrolysis	海解作用，海底风化作用
halo	晕圈；矿物晕圈；光晕
halo anomaly	晕状异常
halo burrow	潜穴晕
halo effect	光环效应，晕圈效应，环晕效应
halo of dispersion	分散晕
halo ore	边缘矿石
halocarbon	卤化碳
halocline	盐度跃层；盐跃层；用水效率
halogen rock	盐类岩石
halogenation	卤化
halogenesis	卤化酌；成盐作用
halogenic	海水生成的，深海溶液生成的
halogenic deposit	海盐沉积
halogenic soil	盐成土
haloid	卤的；卤化物；卤族的
haloid sediments	卤化沉积物
halokainite	钾镁盐岩；钾盐钾镁矾
halokinesis	盐构造学；盐类构造作用
halokinetic	盐体动力的
halokinetic deformation	盐体动力形变
halokinetics	海洋动力学；盐体动力学
halomorphic	盐生的，盐成的
halomorphic soil	盐渍土，盐碱土，盐成土
halopelite	盐质泥岩
halophyllite	适盐矿物
halosere	盐生演替系列；海洋演替系列
halo-sylvite	钾石岩；钾石盐
halotolerancy	耐盐性
halotrichite	铁芒
halvans	杂质多的矿石；贫矿
halvings	贫矿
hamada	石漠，岩漠
hamartite	重硫铋铜铅矿；硫铜铅铋矿
hamatophanite	铁氯铅矿
hambergite	硼铍石
hamelite	水硅铝镁石
Hamilton beds	汉密尔顿层
Hamiltonian expression	哈密顿式
Hamiltonian function	哈密顿函数
Hamiltonian operator	哈密顿算子
Hamilton's principle	汉密尔顿原理
Hamilton's variational principle	哈密顿变分原理
hammer cushion	桩锤垫
hammer dressed slab	锤击板
hammer drill stopper	伸缩式凿岩机
hammer drilling	冲击钻眼，冲击钻进，锤击钻进
hammer hand drill	手持式冲击钻机，手持式凿岩机
hammer rock drill	凿岩机
hammer seismic survey	锤击震波测试
hammer seismograph	锤击地震仪
hammer sinker	凿井用凿岩机，手持式凿岩机
hammer stroke	微压；锤冲程
hammer tamping	锤夯
hammer test	锤击试验
hammer throwing zone	铅球投掷区
hammer tool	震击钻具
Hammer-Aitoff projection	哈麦-澳托夫投影
hammer-blow tamper	锤击夯
Hammer's method	哈默法
hammer-type vibrator	锤式振动器
Hammite	鲕状岩
hammock	沼泽小岛；残丘；吊铺
hamonic constant of tide	潮汐调和常数
hampdenite	硬蛇纹石
hamper	阻碍物
hampshirite	滑蛇纹石
hamrongite	英云斜煌岩
hancockite	锶帘石，铅黝帘石
hand auger	手钻，手动螺旋钻
hand auger boring	手动钻探，手动螺旋钻探
hand augering	手摇钻
hand boring	人工打钻，手工钻探，手摇钻探
hand brace	手摇钻
hand carved sample	手切试样
hand charging	人工装料

hand compacted	手工捣实的
hand displacement meter	指针位移计
hand drilling	人工打钻
hand dug well	人工开掘井，人力钻井，人力挖井，手工挖掘井
hand excavation	人工挖方
hand floating dry concrete	手工浇筑的干混凝土
hand gobbing	人工充填
hand goniometer	手持测角仪
hand gritter	手持式撒岩粉器
hand hammer	手持式凿岩机；轻型凿岩机
hand lens	手持放大镜
hand level	手用水平仪，手持水准仪；辅助平巷
hand mix procedure	手拌法，人工拌合
hand packing	人工充填
hand palm broom	手棕扫
hand pile driver	人工打桩机
hand pit	人工掘井；人工挖掘坑
hand placed riprap	人工抛石护坡
hand puddling	人工捣实
hand punning	手丁捣实
hand ram	手夯，人工夯
hand rammer	手夯锤；人力夯，手动夯
hand reset	人工重调，人工复原
hand sampler	手工取样；人工取样器
hand sampling	手工取样，人工取样
hand signal	手示信号
hand sinking	人工凿井
hand specimen	手标本，手工样品；手标本
hand spreading	手工浇注
hand stowing	手工充填；人工充填
hand tamper	手工捣锤，手夯锤；手工夯
hand tamping	手工夯实，人工捣实
hand vane	手动十字板
hand winch	手动绞车；手摇绞车
hand-compacted concrete	人工捣实混凝土
hand-dug	手掘的；人工开挖
hand-dug caisson	人工挖掘沉箱
hand-feed drifter	手摇推进架式钻机
hand-feed drill	手摇推进钻机
hand-filled	手工装载的；手工充填的
hand-got	手工采掘的
hand-hammer drilling	手锤钻眼；手锤凿岩
hand-held grinder	手持研磨机
hand-held pneumatic	手风钻
handite	锰砂岩
handle	手把；手柄；处理，操作，管理；运输，搬运
handle bar	把手
handle hammer	手锤
handlead sounding	手锤测深
handling of broken ore	矿石运搬
handling pump	手摇泵
handling radius	作用半径；悬臂半径
handling stress	起吊应力
hand-operated	手动的，手工操纵的，人工的
hand-operated auger	手动螺旋钻
hand-packing strip	手工充填带
hand-pitched broken stone	手铺碎石
hand-positioned drill	手持式凿岩机，人工定位凿岩机
handpump	手摇泵；唧水筒
hand-rodding	手工捣固，人工捣实
hand-set bit	手工镶嵌钻头
handstone	卵石；肘节杆
hand-trimming method	手修法
hand-vice	手钳
hand-won face	手工开采工作面
Hanfushan Sandstone	韩府山砂岩
hang	悬挂，垂，吊，倾斜；斜坡
hang wall raise	上盘天井
hang weight	悬重
hanger	挂钩，挂孔，吊架，悬吊装置，悬挂器，悬挂舵，轴承架，吊架，吊杆，吊环；上盘，顶盘，顶板，绝壁，陡坡林地，悬崖；吊螺栓
hanger-supported shaking screen	悬挂式振动筛；吊架摇动筛
hangfire	迟爆；延时爆炸，延时发火
hanging	陡的，悬的，悬空，悬吊的，居上方的；上盘，上盘的，顶盘，顶板；吊挂
hanging adit	上盘平硐，悬帮平硐
hanging arch	悬拱
hanging buttress	悬挂支墩；悬扶垛
hanging cirque	悬浮式冰斗
hanging clinometer	悬挂式倾斜仪；悬挂式半圆仪；最大体积变化
hanging compass	悬挂罗盘
hanging curtain	悬挂式帷幕
hanging dam	浮冰坝
hanging gallery	上盘平巷
hanging glacier	悬浮式冰河；悬冰川
hanging gutter	悬沟
hanging ice	悬冰
hanging junction	悬垂合流悬垂汇流；交叉路口
hanging layer	顶板；上盘
hanging leveling rod	悬挂水准尺
hanging load	井架极限载荷；悬重
hanging pendulum	悬式摆
hanging river	悬河
hanging rock	危岩
hanging scaffolding	悬臂脚手架
hanging side	顶板悬壁，悬崖，顶壁；上盘；顶板
hanging stairs	温跃层
hanging theodolite	悬式经纬仪
hanging tributary	悬谷支流
hanging tributary valley	悬支谷
hanging valley	悬沟，悬谷；消能建筑物
hanging wall	悬帮，顶壁；上盘，顶盘，顶盘
hanging wall block	断层上盘；顶壁断块
hanging wall cutoff	上盘断壁
hanging wall drift	上盘平巷
hanging wall imbrication	上盘叠瓦构造
hanging wall ramp	上盘陡冲断层
hanging wall waste	上盘废石
hanging-wall alteration	顶壁蚀变
hanging-wall zone	顶盘带
hanging-water	悬着水
hanksite	碳酸芒硝；碳钾钠矾
hanlance	手压泵
hanleite	铬镁榴石
hannayite	水磷铵镁石
Hanning function	汉宁函数

Hannuoba Basalt	汉诺坝玄武岩
Hansen's bearing capacity formula	汉森承载力公式
hanusite	针镁石
haphazard	偶然；无计划的；随意的
haphazard distribution	无规律分布
haplite	细晶岩；简单花岗岩
haplobasalt	人造玄武岩
haplodiorite	人造闪长岩
haplogabbro	人造辉长岩
haplogranite	人造花岗岩；细岗岩
haplome	粒榴石
haplophyre	碎斑花岗岩
haplo-pitchstone	人造松脂岩
haplosyenite	人造正长岩
Harang discontinuity	哈朗间断
harbolite	硬辉沥青，硫氢氮沥青
harbor bar	港口沙坝
harbor engineering survey	港口工程测量
harbor entrance	海港入口
harbor plan	港湾平面图
harbor reach	港区
harbor work	海港工程
harbour	海港，港湾，码头，港口
harbour engineering	港口工程，港湾工程
harbour engineering geological mapping	海港工程地质测绘
harbour limit	海港界线
harbour salutation	海港淤积
harbour siltation	港口淤积，海港淤积
hard base rock	硬基岩
hard bending	硬弯
hard bottom	坚固底板，硬底，坚固的巷道底
hard brittle material	脆性材料
hard broken ground	硬碎地层
hard cemented soil	硬结土
hard clay	硬质黏土，硬黏土
hard clip limits	硬剪切限
hard compact clod	坚硬土块
hard compact soil	压实土，夯实土；稳定土
hard concrete	干硬性混凝土
hard condition	刚性条件
hard core	岩石碎块；碎砖垫层，矿渣碎块垫层
hard crust	硬壳层
hard cutting	硬岩掏槽
hard drawing	冷拉
hard drawn	冷拉的
hard drilling	坚硬地层钻井
hard facing	镀硬面；表面硬化，表面淬火
hard floor	坚硬底板；老底
hard formation	坚硬岩层；岩石圈
hard formation force	硬地层力
hard ground	石质河床；基岩面，不整合面，无沉积底，硬底，坚硬地层，难采掘地层
hard heading	岩石平硐，岩巷
hard heading work	石硐掘进工作；硬岩掘进工作
hard impervious material	坚硬不透水材料
hard limestone	硬石灰岩
hard magnetic material	硬磁性材料
hard magnetization	硬磁化
hard mineral	硬矿物；硬矿石
hard mortar	干硬性砂浆
hard oil sand	坚硬含油砂层
hard ore	硬矿石
hard pitch	硬质沥青
hard point	硬化点
hard rock	硬岩，岩浆岩，变质岩；坚石
hard rock concrete	硬石混凝土
hard rock coring tool	坚硬岩石井壁取芯器
hard rock country	坚硬岩石地区
hard rock geology	结晶岩地质学
hard rock mines	硬岩矿
hard roof	坚硬顶板，坚固顶板，老顶
hard sand	硬砂岩
hard shore	基岩岸
hard shotcrete	已固化喷射混凝土
hard site	坚固发射场
hard slab avalanche	硬块崩塌
hard spar	刚玉，硬玉，红柱石
hard spot	硬点
hard spot focal point of stress	应力集中点
hard stone	硬石；硬岩
Hard strata	坚硬岩层
hard streak	硬条带夹层
hard top	硬顶板，坚固顶板
hard wall	坚实围岩
hard water	硬水
hardbanding	环形加硬层
hard-bed stream	石底河流
hard-drawn wire reinforcement	冷拉钢筋
harden	硬化，凝固，变硬；淬火
harden ability	可淬性，淬透性，可淬硬性；硬化程度；脆裂性
harden ability characteristics	可淬硬特性
harden ability test	淬透性试验
harden grout film	水泥结石
harden modulus	硬化模量
hardened	淬硬的
hardened adhesive	固化黏结剂
hardened concrete	硬化混凝土
hardener	硬化剂，固化剂
hardening	硬化，变硬；硬化剂；淬火
hardening accelerator	促硬剂，早强剂
hardening agent	固化剂，增硬剂，硬化剂；淬火剂
hardening behavior	硬化行为
hardening capacity	硬化度；可淬性
hardening flaw	淬火缺陷，淬火裂纹
hardening heat	硬化热；淬火加热
hardening liquid	硬化液，淬火液
hardening of cement	水泥硬化
hardening of concrete	混凝土硬化
hardening parameter	硬化参数
hardening period	硬化期
hardening point of asphalt	沥青硬化点
hardening process	硬化过程；硬化法
hardening rule	硬化规则，硬化法则，强化法则
hardening strain	淬火变形；硬化应变
hardening strength	硬化强度
hardening stress	淬火应力
harder spectrum	较硬能谱
hardest water	极硬水

hard-faced 表面硬化的
hardfacing 表面硬化；硬质焊敷层；表面耐磨堆焊；表面堆焊硬合金
hard-frozen soil 硬冻土
hard-grained 粗粒的
hardground 硬底化；硬灰岩层
hardground structure 硬底构造
Hardgrove grindability index 哈氏可磨性指数；哈德格鲁夫可磨性指数
Hardgrove number 哈德格鲁夫可磨性指数
hardness 硬度，刚度，坚固
hardness curve 硬度曲线
hardness degree 硬度标度
hardness degree of rock 岩石坚硬程度
hardness gage 硬度计
hardness gauge 硬度计；硬度标
hardness impression 硬度试验压痕
hardness indentation 硬度试验压痕
hardness index 硬度指数
hardness number 硬度值，硬度指数
hardness of crystal 晶体硬度
hardness of joint surfaces 节理表面的硬度
hardness of rock 岩石硬度
hardness of the spectrum 能谱硬性
hardness of water 水硬度
hardness on Rockwell 洛氏硬度
hardness penetration 淬硬深度；淬火深度；硬化深度
hardness point 硬度标
hardness points 坚硬角
hardness profile 硬度剖面
hardness reducer 硬度软化剂
hardness scale 硬率；硬度计；硬度分级，硬度标
hardness test 硬度试验
hardness tester 硬度仪，硬度试验机
hardness testing device 硬度计，硬度试验机
hardness testing machine 硬度试验机；硬度计
hardness value 硬度值
hardness-upward 上硬型；上硬的
hardometer 硬度计；回跳硬度计
hard-packed 结构紧密的
hardpan 硬土层，硬质地层；铁磐；灰质壳，硬盘；铁盘；不透水硬黏土层
hard-rock 硬岩
hard-rock aquifer 硬岩含水层
hard-rock geology 硬岩地质学；岩浆岩和变质岩地质
hard-rock mine 硬岩矿
hard-shell 硬壳的
hard-shot ground 不易爆破地层
hardstem 速凝石膏炮泥
hard-to-cut rock 难截割的岩石
hard-to-drill 难钻进的
hardwood forest 阔叶林
hardystonite 锌黄长石
harfour deposit 湖成堆积，湖相砂矿
Harker diagram 哈克图解
harkerite 碳硼硅镁钙石
harlequinopal 蜡蛋白石
harmful geologic phenomena 不良地质现象
harmonic amplitude 谐振幅
harmonic analysis 谐波分析；调和分析，傅里叶分析

harmonic attenuation 谐波衰减
harmonic characteristic 谐波特性
harmonic coefficient 谐波系数
harmonic component 谐波分量
harmonic content 谐波含量
harmonic curve 谐波曲线，正弦曲线，简谐曲线，调和曲线
harmonic decline 调和递减
harmonic decline law 调和递减规律
harmonic detector 谐波检波器
harmonic differential 调和微分
harmonic distortion 谐波畸变；谐波失真；谐畸变
harmonic disturbance 谐波干扰
harmonic elastic wave 谐弹性波；弹性谐波
harmonic elimination 谐波消遏器
harmonic equation 调和方程
harmonic excitation 简谐激振；谐波激励
harmonic field 谐波场
harmonic filter 谐波滤除器
harmonic filtration 谐波过滤；谐波滤除
harmonic fold 谐和褶皱；协调褶皱
harmonic folding 谐和褶皱
harmonic force 简谐力
harmonic frequency 谐频
harmonic frequency component 谐频分量
harmonic function 调和函数
harmonic generator 谐波发生器
harmonic interference 谐波干扰
harmonic ions 谐振离子
harmonic lake type 谐和湖型
harmonic loading 谐振荷载，简谐荷载
harmonic mean 调和平均；调和中项
harmonic mining 中和地压采煤法；协调开采
harmonic motion 谐运动；谐运动
harmonic mountains 调和山脉
harmonic oscillation 谐和摆动；简谐振荡
harmonic oscillator 谐波发生器；谐波振荡器
harmonic peak 谐振峰
harmonic plane wave 平面谐波
harmonic producer 谐波发生器
harmonic progression 调和级数；调和数列
harmonic pulsation 谐波脉动
harmonic ratio 谐比例
harmonic relation 谐和关系
harmonic response 简谐反应
harmonic series 调和级数
harmonic stress situation 谐振应力状态
harmonic synthesis 谐波合成
harmonic system of working 协调开采
harmonic test 谐振试验
harmonic tremor 长期火山地震；驻波振荡
harmonic trend analysis 调和趋势分析
harmonic trend surface analysis 调和趋势面分析
harmonic vibration 简谐振动，谐振；简谐振动试验
harmonic wave 谐波；调和波
harmonic wave analyser 调和波分析器；谐波分析仪
harmonic wave function 谐波函数
harmonically inhomogeneous medium 调和非均质介质
harmonic-regression analysis 调和回归分析
harmonic-restraint relay 谐波抑制继电器

harmonics 谐；谐波；调和函数，谐函数；谐音
harmonization 谐和；谐波；调谐
harmonograph 谐振记录器
harmony 谐和性；谐和
harmophane 刚玉，金刚砂
harmotome 交沸石
harness 马具，挽具，装置；装具线束，导线；开发治理
harnessing 利用水力、风力等做动力
harpolith 岩镰
Harris' geologic occurrence model 哈里斯地质产状模型
harrisite 方辉铜矿，方铜矿；钙长橄榄岩
harrow mark 耙痕
harsh 粗糙的
harsh concrete 干硬性混凝土
harsh desert 荒漠
harsh grain 粗粒
harsh mix 粗糙搅拌；干硬性混合料
harsh mixture 粗糙搅拌混凝土；干硬性混合料；劣拌混合料
harsh sand 棱角砂；粗砂
harshness 粗糙；刚性
harshness of concrete 混凝土粗糙度
harstigite 镁柱石
hartite 晶蜡石
hartleyite 碳页岩
Hartmann line 哈特曼拉伸应变痕迹线
hartmannite 红锑镍矿
hartsalz 硬盐
hartschiefer 硬片岩
hartungite 暗霓霞质岩
harzburgite 斜方辉石橄榄岩；斜方辉橄岩；方辉橄榄岩
hashed 散列的
hashing 散列法
Hassler chamber 哈斯勒岩心盒
Hassler type permeameter 哈斯勒式渗透率仪
hassock 软钙质砂岩；草垫，草丛
hassock structure 丛状构造；蒲团构造
hastingsite 绿钙闪石；富铁钠闪石
hasty soil survey 快速土质调查
hatch box 检查箱
hatchettite 伟晶蜡石
hatchettolite 铀烧绿石；铀钽铌矿
hatching 阴影线；剖面线
hatchite 细硫砷铅矿
hatherlite 歪闪正长岩
hatterikite 钇钽矿
hauchecornite 硫镍铋锑矿
hauerite 褐硫锰矿
haugh 河岸平台，洪泛平原；低丘陵
haughland 河岸平台
haulage adit 运输平硐
haulage berm 运输平台，运输平盘
haulage force 牵引力
haulage gateway headgate belt gate 工作面运输巷；运输顺槽，下顺槽
haulage in country rock 脉外运输；围岩巷道运输
haulage road 运输，运输道
haulage roadway 运输平巷，运输道

haulage shaft 提升井
haulage tunnel 运输平硐
haulage way 运输道；出土，出渣
haulage winch 运输绞车；曳引绞车，拖运绞车
hauling engine 牵引机，卷扬机；运料机
hauling shaft 提升井
haulroad 进出道；运输巷道
haulroad dewatering 道路排水
haulway 运输道；拖运道
haunch up 拱起
haunched arch 加腋拱，护拱
haunched roof 弓状顶
haunching 护拱
hauptsalz 硅杂盐
hausmannite 黑锰矿
Hauy law 阿羽依定律
hauyne 蓝方石；层状构造
hauyne basalt 蓝方玄武岩
hauyne basanite 蓝方碧玄岩
hauyne phonolite 蓝方响岩
hauynite 蓝方石；蓝方玄武岩
hauynitite 钛辉蓝方岩
hauynolite 蓝方石岩
hauynophyre 蓝方岩；蓝方斑岩
Hauy's law 郝欧定律
haven ore 多杂质矿石
haversine 半正矢
Hawaiian activity 夏威夷式活动
Hawaiian shield 夏威夷型盾形火山
Hawaiian type eruption 夏威夷型火山喷发
Hawaiian-type caldera 夏威夷型破火山口
Hawaiian-type volcano 夏威夷式火山
hawaiite 夏威夷岩；深绿橄榄岩
hawiite 中长玄武岩
hawleyite 方硫镉矿
haydite concrete 陶粒混凝土
hayesine 水硼钙石，硼酸方解石
Hayford ellipsoid 海福德椭球
haystack 锥形溶蚀丘；干草堆
haystack hill 溶蚀丘；灰岩残丘
haytorite 硼石髓；硅硼钙状玉髓
hazard assessment 危险性评估，灾害评估
hazard assessment study 危险程度评估研究
hazard detection 隐患探测
hazard effect 灾害效应
hazard evaluation 危险估计；灾害评估
hazard geology 灾害地质学
hazard geoscience 灾害地学
hazard intensity 灾害烈度
hazard level 危险程度
hazard mitigation 减灾
hazard model 灾害模式
hazard potential 危险可能性，危险势；潜在危险性
hazard rate 危险率；冒险率
hazard reduction 减灾
Hazard risk prediction 灾害风险预测风险；灾害风险预测
hazard sign 危险标志
hazard standard 危险标准
hazardous 有害的；危险的

hazardous condition	围筑畦埂
hazardous earthquake	危险性地震；灾害性地震
hazardous rock	软岩石；危险性岩石
hazardous slope	危险斜坡
hazardous structure	危险结构；危险构造
hazardous top	危险顶板
hazardous waste control systems	有害废料控制系统
hazardous waste management	有害废料处理
hazardous waste site	有害废料场
hazards	危害，危险
hazard's monitoring	危险监测
Hazen and Williams formula	黑增-威廉斯公式
Hazens equation	哈森公式
Hazhihai Volcanics	哈只海火山岩
haziness	浑浊性
H-component	水平分量
HDR geothermal reservoir	干热岩热储
head	水头，头；头山甲；海角，压头，工作面长度掘进水头压头；落差，高差，深度，渠首，坡顶，巷道，头上部头部顶部盖开拓煤巷工作面近顶板部分；闸首；上游；河源；残积物
head amplifier	前置放大器；前级视频放大器
head bay	小褶皱；上游池，上游河段；上闸首
head boundary condition	水头边界条件
head conduit	压力输送管道；有水头管道；压力水管
head current	急流头部
head cut	向源切割
head decline	水位下降
head deposit	岩屑堆积；残积物；前端堆积，前碛
head difference	水头差
head ditch inlet channel	进水渠
head drop	落差；水头降落；水头下降
head duration curve	水头历时曲线
head elevation	水头高程
head erosion	向源侵蚀；溯源侵蚀
head fall	水头差；落差；压力降落；纵向坡度；跌水
head flow	脉动式喷流；压差流动
head flume	头部水槽
head gradient	水头梯度
head line	水头线
head lose	水头损失，压头损失，落差损失
head losing	水头损失
head loss due to contraction	收缩段水头损失
head loss due to enlargement	扩大段水头损失
head loss of pipe line	管道水头损失
head lost	水头损失
head measurement tube	压头测量管
head of beach	滩头；滩顶
head of breakwater	防波堤堤头
head of landslide	滑坡顶端；山泥倾泻源头
head of pile	桩头，桩顶，桩帽
head of rip current	离岸流头，离岸裂流头
head of tide water	感潮区上游段
head of water	水头；水位差；水柱高度；扬程
head oscillation curve	水头摆动曲线
head plate	顶盖板
head pressure	水头压力；压位差；承压水头；压头
head race	引水隧洞，引水渠，引水槽；水工建筑上游，上游水道，前渠；预报高水位
head range	水头范围
head ratio	水位比，不同水位比，水头比
head reservoir	压力水池；上水库
head resistance	正面阻力，迎面阻力；有害阻力，废阻力，
head river	河源段
head room under a bridge	桥梁出水净高
head room under bridge	桥下净高
head sample	原矿试样；原煤试样
head scarp	滑坡后壁；山头陡坡；滑坡陡坡
head section of breakwater	防波堤头断面
head surge basin	上游调压池
head surge chamber	上游调压室
head tide	逆潮流；落潮流
head water	上层水，源头水；水源；上游
head water tank	高位水箱
head wave	头波，首波，前波
head-capacity curve	水头容量曲线；压头-扬量曲线
head-center	破碎头中轴
head-discharge relation	水头流量关系
headed bolt	撑帽式杆柱；有膨胀头的锚杆
headentry	运输平巷；下顺槽
header	丁面，顶梁，丁砌；集水管，分水器，集水器，水箱，记录表头；平巷掘进机；露头石，露头砖；掘进时遇到的硬岩高过人头的炮眼集管联管箱
header box	联管箱；汇流箱
header course	露头层；单砖砌层
header pipe	总管；集管；接头管
header tank	定压池；配水池；高架罐
head-gate	首部闸门；渠首闸门
head-gate duty	渠道用水率
headgate duty of water	首部闸门灌水率；毛灌水率
heading	导洞，导坑，隧道导坑，掌子面；掘进工作面，大断面掘进的顶或底导巷，平巷端，平巷掘进，精矿选矿所得重质部分，水平巷道上方的矿脉巷端煤层巷道平巷风巷
heading advance	工作面推进，作业面推进，工祖推进；巷道掘进
heading and bench method	短台阶法
heading and cut method	导洞掘进法
heading blast	阶段爆破；超前工作面爆破；硐室爆破，坑道爆破
heading blast charge	硐室装药
heading blasting	掘进爆破；硐室爆破，坑道爆破
heading face	掌子面，平巷掘进工作面，工作面前壁
heading machine	掘进机，平巷掘进机，导坑掘进机
heading method	上部超前工作面开采法；导硐法
heading shop	回采工作面；采矿工作面
heading side	底板，下盘
heading system	平巷下向采矿法
heading through	贯通，贯穿；掘通巷道
heading up	壅水，抬高水位
heading wall	断层下盘
heading work	平巷掘进工作；拱顶石饰
heading-and-cut	上部超前平巷梯段掘进；正梯段掘进
heading-and-overhand	下部超前平巷梯段掘进；倒梯段掘进
heading-and-stall	房柱式开采法
headland	海角，陆角；岬角，河源；岬，陆岬
headland beach	岬滩；岬滩
head-migrating erosion	向源侵蚀

head-of-hollow fill method 填谷排土法
head-on sensitivity 正端灵敏度
headpool 河源地
head-race bay 前池
headrace conduit 上游进水道，上游输水道
headrace surge tank 上游调压井，首部调压井
headrace tunnel 上游隧洞
headroom 巷道高度，净空，净空高度，通行高度，钻机平台至钻机绳轮的高度
head-size stone 人头大小的石块
headspan 横跨
headstream 河源；源头，发源地，源流
headwall 端墙；山墙
headward 溯源的；向源的
headward erosion 向源侵蚀，溯源侵蚀；溯源；圆锥贯入试验
headwater area 上游地区
headwater channel 引水渠
headwater control 上源控制；水头调节；水压控制
Headwater depth 上水深
headwater elevation 上游水位；水源标高，水头高程
headwater erosion 溯源侵蚀，向源侵蚀；溯源
headwater level 上游水位，上游水面
headwater region 上游区，源头区
headwater reservoir 上游水库
headwater survey 水源调查；河源调查
headwater tributary 上游支流
headwaters 河源；上游
headwaters region 河源区
headwaters river source 河源
headway 进尺，房柱法的二次回采，主割理，割理的方向；钻进，横巷平行于主割理的平巷；时间车距；班距；行进间距
headway per drill bit 钻头进尺
headwind 逆风；渠首工程，渠头控水建筑物；进水口工程；掘进工程；平巷掘进工作，拱顶石饰；井架
headworks 进水口工程
heal 合拢
healed cracks 愈合裂纹
healed fault 愈合断层
healed fissure 愈合裂隙
healed fracture 弥合裂缝
healed structure 覆瓦状构造
healing agent 修补剂
healing front 愈合前沿
healthy stream 有效发电水头
heap 矸石堆，堆，废石堆，土堆，堆积；大量
heap capacity 装载容量；加载重
heap leaching pollution 堆浸污染
heap leaching site 堆浸场
heaped concrete 堆浇的混凝土；未捣固的混凝土
heaped load 堆尖装料；堆积荷载
heap-water platform 高潮浪蚀台
heap-wax crude oil 高蜡原油
heart-break rock 低渗透沉积岩
heat 热，加热，热学；热量
heat absorbing reaction 吸热反应
heat absorption 热吸收，吸热量；吸热
heat absorption capacity 吸热能力
heat absorptivity 吸热性

heat anomaly 热异常
heat balance 热平衡，热量平衡
heat balance of ocean 海洋热平衡
heat balance test 热平衡试验
heat basin 热盆
heat bonding 热黏合
heat budget 热总量；热量收支
heat budget method 热量平衡法
heat calculation 热平衡计算；热力计算
heat capacitance 热容，热容量
heat capacitivity 热容率
heat capacity 热容，热容量
heat capacity at constant pressure 等压热容
heat capacity at constant volume 等容热容
heat change 热变化；热交换
heat checked 热稳定的
heat condensation 凝结热
heat conductance 热导性
heat conducting bar 导热杆
heat conduction 导热；热传导；导热率
heat conduction analogy 热传导模拟
heat conduction coefficient 导热系数
heat conduction equation 热传导方程
heat conductivility 导热性
heat conductivity 导热性；导热系数；热导率
heat conductor 导热体
heat conservation 热量守恒
heat consumption 耗热量
heat consumption rate 耗热率
heat content 热容量；热含量；焓；热函
heat control 热控制；温度调节
heat convection 热对流；热运流；对流传热，热转换
heat conversion factors 热换算因数
heat crack 热裂纹
heat curing 热固化；热养护
heat current 热流
heat cycle 热循环
heat cycle effect 热循环效应
heat death 热寂
heat decomposition 热分解
heat degree 热度
heat delivery surface 散热表面，放热表面
heat density 热能密度
heat detector 热源探测器；热量探测器
heat diffusion 传热，导热；热扩散
heat diffusivity 热扩散率
heat disaster 矿井热害
heat discharge rate 放热率
heat dispersion 热分散；热散射
heat dissipating capacity 散热量
heat dissipation 散热损失；热扩散，热散逸；热消耗；散热作用
heat dissipation factor 热消散系数
heat dissipation rate 热耗散率
heat distortion temperature 热变形温度
heat distribution 热量分布；配热
heat distribution profile 热量分布剖面
heat drop 热降
heat duty 热负荷
heat effect 加热效应，热效应

heat efficiency 热效率
heat eliminating medium 冷却介质
heat elimination 热气排除，散热，冷却
heat emission 热量发射，热辐射，放热
heat emission rate 热量排放率
heat emissivity 热发射率
heat emissivity coefficient 热发射系数；热辐射系数
heat endurance 热稳定性；耐热性
heat energy 热能
heat engineering 热力工程；热工学
heat equation 热方程
heat equator 热赤道
heat equilibrium 热平衡
heat equivalent 热当量
heat equivalent of work 热功当量
heat erosion 热侵蚀
heat escape 热逸散，热消失
heat evolution 放热
heat exchange 热交换；热平衡
heat exchange area 热交换面积
heat exchanger 换热器，热交换器
heat expansibility of rock 岩石热胀性
heat expansion 热膨胀
Heat extraction 取热；白冰自定名
heat flow 热流，热流动；热流量，热量
heat flow anomaly 热流异常
heat flow component 热流分量
heat flow density 热流密度；热流量
heat flow equation 热流方程
heat flow intensity 热流强度
heat flow measurement 热流量测量；热流量实测数据
heat flow meter 热流计
Heat flow modelling 热流模型
heat flow of the crust 地壳热流
heat flow province 热流区；热流省
heat flow subprovince 热流亚区
heat flow transducer 热流传感器
heat flow unit 热流单位
heat flux 热通量，热流密度；热流
heat flux density 热通量密度
heat flux probe 热流计；海底温差计；热流枪
heat flux sensor 热流传感器
heat from strata 地热
heat front 热前缘
heat function 热函；焓
heat gain 热增量；热增加
heat gauge 热传感器；热量计；量热规
heat generating reaction 放热反应
heat generation 生热，发热，热生成量
heat generation balance 热平衡
heat generation calefaction 发热
heat generation unit 生热率单位
heat gradient 热梯度
heat heave 热胀性
heat impedance 抗热性
heat index 热指数
heat indicator 温度指示器；温度计
heat influx 热补给；热补给量
heat injection 注热；热注入
heat input 热输入，热补给

heat insulated 隔热，绝热
heat insulating 绝热的
heat insulation 热绝缘，隔热；保温
heat insulator 热绝缘体
heat interchange 热交换
heat interchanger 热交换器，换热器；热交换器
heat interchanging area 换热面积
heat island effect 热岛效应
heat leak 热量漏泄，热渗透；热显示；放热；漏泄热显示
heat liberation 放热
heat load 热负荷
heat loss 热量耗损，热耗，热耗量，散失的热量
heat loss by transmission 传热损失
heat loss survey 热耗测量
heat migration 热传递
heat mining 采热，热开采
heat of adsorption 吸附热
heat of coagulation 凝结热
heat of condensation 凝结热；冷凝热
heat of cooling 冷却热
heat of crystallization 结晶热
heat of decomposition 分解热
heat of desorption 除去吸收的热；解吸热
heat of dissociation 离解热；分解热
heat of explosion 爆炸热
heat of formation 生成热
heat of hardening 凝固热
heat of hydration 水化热；水合热
heat of precipitation 沉淀热
heat of solidification 固化热
heat of solution 溶解热
heat of sublimation 升华热
heat of transformation 转变热
heat of vaporization 汽化热；蒸发热
heat of wetting 湿化热；润湿热
heat of wetting method 湿热法
heat ore deposit 热矿床
heat outflow 热流出；热流
heat passage 传热；热传递
heat path 传热孔道；传热途径
heat penetration 热透入；透热厚度
heat plume 热柱
heat pollution 热污染
heat production 热能生产；热能产量
heat productivity 发热量；生热率
heat productivity of rock 岩石的导热性
heat proof 耐热，抗热，隔热
heat proof quality 耐热性
heat property of concrete 混凝土热学性能
heat quality 热源品质
heat quantity 热量
heat radiation 热辐射；散热
heat rate 热耗率；热耗，发热量
heat ratio 热力系数
heat rays 热线
heat recharge 热补给
heat recovery 热回收
heat reinjection system 热能回注系统
heat rejection 热消耗，散热；废热排放

heat release	热释放，放热
heat removal	热量排除；排热；取热
heat requirement	热需要量；需热量
heat reserve	热储量
heat reservoir	热储，热库；储热器
heat residual	剩余热量
heat resistance	耐热性，耐热度，抗热；热稳定性
heat resistant	耐热的
heat resource	热资源
heat return	热量补给；热量回流率
heat rise in mass concrete	大体积混凝土内热量升高
heat run	加热试验；加热运行；发热试验
heat scavenging method	余热利用法
heat sensing cable	热敏电缆
heat sensor	热传感器
heat shield	隔热屏障；热屏障
heat shrinkable tubing	遇热收缩软管
heat sink	热壑，热阱；热汇，热吸收
heat sink device	散热设备
heat soil	受热土层
heat source	热源
heat source well	热源井
heat sources of the geothermal resource	地热资源热源
heat stowing	热量积累，热量储存
heat strength	热强度
heat stress	温度应力，热应力；热负荷
heat stress index	热应力指标
heat summation	热量总和
heat supply	热补给，供热
heat surge	温度突然上升
heat test	加热试验；耐热试验
heat transfer	热传导，传热，热能量输送
heat transfer by convection	对流传热
heat transfer capacity	传热量
heat transfer characteristic	传热特性
heat transfer model	热量传输模型
heat transference	传热
heat transmissibility	传热性
heat transmission	传热，热传输；热透射
heat treated corundum	热处理刚玉
heat treated gemstone	热处理宝石
heat treatment	热处理
heat treatment crack	热处理裂纹
heat treatment of clay minerals	黏土矿物的热处理
heat value	热值，发热量
heat withdrawal	热能汲取
heat yield	热能产量
heat-absorbing	吸热的
heat-affected zone	稀释剂；热影响区
heat-balance of the earth	地球的热平衡
heat-carrying capacity	载热能力
heat-collecting zone	热量积聚带；储热层
heat-concentrating mechanism	热能富集机制
heat-conducting	导热的，热导的
heat-conducting media	热导介质；导热介质
heat-dry	烘干的
heat-drying	加热干燥
heated soil	加热土
heated stone	热变石
heat-eliminating medium	冷却介质
heater	加热器，发热器；热水炉
heat-flow analysis	热流量分析
heat-flow hole	热流量测孔
heat-flow probe	热流量探头
heat-flow rate	热流率
heat-flow vector	热流向量
heat-flux region	热通量区
heating	生热，加热，供热；供暖，暖气；保温；白热灼热
heating and bench method	台阶掘进法
heating capacity	热容量；热值；供暖能力，给热能力
heating effect	热效应
heating efficiency	热效率
heating load	供热量；热负荷
heating microscope	高温显微镜
heating plant	供热装置；供热站
heating power	发热量；热值；卡值；热力；燃烧热
heating rate	升温速率；加热速率；发热率
heating source experiment	热源实验
heating stage	显微镜热台，热台
heating strip	加热带；加热条
heating test	加热试验；耐热试验
heating value	热值
heatless	无热的
heat-power engineering	热工学；热力工程
heat-proof aging	耐热老化
heat-protection	防热
heat-released mercury survey	热释汞测量
heat-resistant concrete	抗热性混凝土
heat-resistant current metre	耐热流速计
heat-resisting	耐热
heat-resisting quality	耐热性
heat-retaining capacity	保热容量
heat-seeking device	探热装置；探热部件
heat-seeking head	测热能量探头；探热头
heat-sensitive	热敏的
heat-sensitive material	热敏材料
heat-sensitive packer	热敏性封隔器
heat-sensitive sensor	热敏传感器
heat-shielding performance	隔热性能
heat-shrink property	热收缩特性
heat-source characteristics	热源特性
heat-storage capacity	储热能力；热容量
heat-storage reservoir	热储；储热层；储热库
heat-storage well	储热井
heat-transfer capability	传热能力
heat-transfer coefficient	传热系数，放热系数；导热率
heat-transfer commutator	传热管
heat-transfer fluid	传热流体
heat-transfer medium	传热介质
heat-transfer process	传热过程
heat-transfer system	传热系统
heat-transport equation	热传导方程
heat-transporting fluid	传热流体
heat-up	加热
heave	平错，断层，水平移动，水平断距；隆起，鼓胀，冻胀，隆胀；抬动，上升
heave amplitude	升降幅度
heave and set	起伏
heave calculations	隆起计算

heave compensation 隆起补偿；波浪补偿；升沉补偿；升沉补偿器；涌浪滤波器；浮升补偿器；升沉补偿装置	heavy cover 厚覆盖岩层，厚层覆盖岩层
heave fault 横推断层，捩转断层；平推断层；平移断层	heavy crude oil 重质原油；重油
heave force 隆胀力	heavy current 强电流，大电流
heave gauge 隆胀计	heavy cut 深开采；深掏槽；深挖掘
heave height 升沉高度	heavy cutting 深槽；深采掘；深槽
heave meter 升沉仪	heavy damage 严重破坏
heave movement 升沉运动	heavy damping 强阻尼
heave of base 基底隆起	Heavy Density Agglomerated Cork Sheet 黄水松机毡
heave of fault 断层平错，断层水平移位	heavy duty 重型；重型的，重级的；永久封冻湖
heave rate 隆起速率	Heavy Duty Bull Point Chisel 大尖凿
heave ratio 冻胀比	heavy duty carrier 重负荷载体
heave stake 隆起标桩；鼓胀标桩	heavy duty crusher 大型破碎机
heave up 隆起；鼓胀	heavy duty electrical conductor 大功率导电体
heaved block 脊状断块	heavy duty road header 重型巷道掘进机
heaved pile 挤升桩	heavy duty service 深井作业
heaved side 上升盘	heavy dynamic sounding 重型动力触探
heaver 杠杆；起重具；举起重物的人	heavy earth 重晶土；重土
heave-resistant 抗隆起	heavy earthwork 大量土方工程
heavily shaken area 严重受震区	heavy excavation 大开挖
heavily watered 涌水量大的	heavy flow of water 急流
heavily watered ground 富水地层	heavy fluid 重液
heavily-watered area 大量充水区；大量浸水区	heavy fluid separation 重液分选；重液体分离
heavily-weathered rock layer 风化严重岩层	heavy foundation 重型基础；重型地基
heaving 隆起，鼓胀，冻胀；隆起的，鼓胀的；升沉	heavy gauge 大尺寸的；大口径的
heaving action 隆起作用	heavy gold 金块；狗头金
heaving bottom 隆起底板；冻胀土基；隆起的巷道底	heavy gradient 大坡度，陡度
heaving deformation 坑道底部变形；底鼓	heavy grading 大量平整土地
heaving effect 膨胀效应	heavy grease 稠润滑脂
heaving floor 隆起的底板	heavy ground 不稳定地层；不易掘进的地层；危险上盘
heaving ground 起伏地面	heavy hammer 重锤；大锤
heaving knob 冻胀丘	heavy hand-held rockdrill 重型手持式凿岩机
heaving of floor 底板鼓起	heavy hanging wall 易冒落上盘；不稳固上盘
heaving of the bottom 基坑底隆胀	heavy hilly area 重山丘区
heaving of the tunnel 底臌	heavy intermittent test 大负载断续试验
heaving pressure 隆起压力	heavy isotope 重同位素
heaving rock 隆起的岩石	heavy joist 重型托梁；吸气作用
heaving sand 流沙；流砂；冒砂	heavy layer 巨厚岩层；厚层
Heaviside load 阶跃荷载	heavy liquid 重质液体
heavy asphalt 重质沥青；黏质沥青	heavy liquid separation 重液选矿；未处理污水
heavy asphalt crude 沥青基重质原油	heavy load 重负载
heavy asphalt residues 重质沥青残渣	heavy loam 重壤土
heavy bitumen 重质沥青	heavy media process 重介质选矿法
heavy blasting 强力爆破	heavy media separation 重介质选矿
heavy bottoms 渣油	heavy medium 重悬浮液；重液
heavy brine 超浓卤水；高比重卤水	heavy medium separation 重介质选矿
heavy case 深渗碳层	heavy medium suspension 重悬浮液
heavy casing 厚壁套管	heavy metal mud 多金属泥；重金属泥
heavy clamshell 重型抓斗	heavy mineral 重矿物；重砂矿物
heavy clay 重黏土；致密黏土；压气槽	heavy mineral analysis 重矿物分析
heavy clay loam 重黏壤土；重亚黏土	Heavy mineral sands 重矿物砂
heavy clay soil 重黏土	heavy minerals 重矿类；重矿物
heavy clay suspension 重黏土悬浮体；重黏土悬浮液	heavy mortar 稠灰浆；浓厚灰浆
heavy compaction plant 重型固土机	heavy mud 重泥浆
heavy component 较重组分	heavy neutral oil 重质中性油
heavy concentrate 重砂	heavy oil recovery 重油开采
heavy concrete 重混凝土；细砂壤土	heavy oil sand 重油砂层
heavy condensation 高度冷凝	heavy oil-bearing formation 重油油层
heavy construction 大型工程	heavy overburden 厚覆盖
	heavy pitch 陡倾；急倾斜
	heavy placer mineral 重砂矿物

heavy pneumatic hammer　重型风动锤
heavy pressure　重压力；大压力
heavy product　重介质材料；加重剂；重悬质重浮质；比重大的产品
heavy pumping　剧烈抽水
heavy rain　暴雨
heavy rain downpour　大雨
heavy rainstorm　暴雨
heavy rare earth elements　重稀土元素
heavy sand　重质油砂层；重砂
heavy separation　重液分离
heavy silty soil　重粉砂土
heavy soil　重质土；致密土；重湿黏土；下滑漂移
heavy solids　机械杂质粗粒
heavy solution　重液，重溶液
heavy spar　重晶石
heavy suite　重颗粒，重组分
heavy tamping　重锤夯实；强夯法
heavy tamping method　重锤夯实法
heavy test　重载试验
heavy textured soil　重黏结土；重黏土质
heavy timber construction　重型木结构
heavy toe　阻力很大的难爆下盘
heavy toe hole　底部负载重的炮眼
heavy viscous crude oil　重稠原油
heavy viscous mud　高黏度泥浆
heavy wall drillpipe　厚壁钻杆
heavy wall tubing　厚壁油管
heavy walled　厚壁的
heavy water　重水；大涌水量；最大洪峰流量
heavy water clean-up　重水净化
heavy water loss　重水损失
heavy water purification　重水净化
heavy water reconcentration　重水再浓缩
heavy waxy crude　重质含蜡原油
heavy weathering　严重风化，强风化
heavy weight concrete　重质混凝土，特重混凝土
heavy weight drill pipe　加重钻杆
heavy weight solution　高比重溶液
heavy weight tubing　加厚油管
heavy-bedded　厚层的
heavy-bodied　很黏的；很稠的
Heavy-Duty Caster　生铁辘
heavy-duty crane　重型起重机
heavy-duty drill　重型钻机
heavy-duty drill rig　重型钻塔
Heavy-Duty Electric Blower　大型吹风机；大电动锤
heavy-duty equipment　重型设备
heavy-duty hex auger　重型六角螺旋钻
heavy-duty hydraulic breaker　重型液压破碎机
Heavy-Duty Impact Wrench　重型气动扳手
Heavy-Duty Pneumatic Hammer　大风炮
heavy-duty pressure tunnel　高压隧洞
Heavy-Duty PVC Pump　重庄胶泵
heavy-duty tyre　无孔隙体
heavy-lift equipment　悬移质
heavy-media method　重介法
heavy-mineral prospecting　重矿物勘探
heavy-mineral province　重矿物区
heavy-oil reservoir　重质油油藏

heavy-type hydraulic ram　重型液压千斤顶
heavy-wall pipe　厚壁管
heavy-water reactor　重水反应堆
Heavy-Weight Aggregate　重质粒料
heavy-weight drilling　重泥浆钻井；电泥浆钻井
heavy-weight drilling fluid　重质钻探用泥浆
heazlewoodite　赫硫镍矿，希兹硫镍矿
hebetine　硅锌矿
hebraic granite　文像花岗岩
hebronite　锂磷铝石
hecatolite　月长石
hectocotylus　交接腕；化茎腕
hectometer stake　百米桩
hectorite　锂蒙脱石；水辉石；锂皂石；富镁皂石；蚀变辉石
hedenbergite　钙铁辉石
hedgehog spine anchorage　刺猬式锚固
hedrumite　霞碱正长斑岩，霞碱正长岩
hedyphane　钙砷铅矿
heel　炮眼口，钻孔口；挡土墙底部；踵，后跟，坝踵；柱脚，棱；底垫
heel of dam　坝踵
heel over　倾斜
heel slab　踵板
heidarnite　枪硼钙钠矾
height　高度，厚度；地平纬度；顶点
height above datum　基面以上高程，高于基准面的高度；水尺零点以上高程
height above ground level　地面以上高度
height above mean sea level　平均海拔高度
height accuracy　高程精度
height adjustment　高度调整
height adjustment knob　调高旋钮
height altitude　高度
height anomaly　高程异常
height compensator for mine car　高度补偿器
height computation　高程计算
height control　高度限制
height datum　高程基准
height deviation　高程差
height difference　高差；高低潮；高度订正
height displacement　投影差
height finder　测高仪；高度计
height finding　测高
height finding instrument　高度计
height gain　高度增量
height gauge　高度计；测高规
height index　高度标志
Height indicator　高度表；高度指示器
height interval　高度间隔
height loss　高度损失；压头损失
height mark　高度标记
height mark spot level　标高
height measurement　高程测量；测高法
height measurer　测高仪；高度计
height of arch　拱高
height of aurora　极光高度
height of capillary rise　毛细上升高度
height of capillary water　毛管水高度
height of caving　落顶高度；冒落高度

height of collimation 视准高度
height of dam 坝高
height of damming 回水高度
height of delivery 输送高度；排送高度；扬程
height of drop 落距，落差
height of earth column 土柱高度
height of floor heave 底臌量
height of hydraulic jump 水跃高度
height of instrument 仪器高度
height of level 水平的高度；水准仪器高度
height of lift 扬程；提升高度
height of lifting 拔起高度；克服高度
height of loosen bedrock 松弛荷载高度
height of low water 低潮高；低水位高程；枯水位高程
height of overflow 溢流水头；溢流高度
height of precipitation 降水量
height of reserve layer 储备层高度
height of side slope 边坡高度
height of site 测站高程
height of strata requiring support 需要支承的岩层厚度
height of swell 涌浪高度；涌高；涨水高度
height of tide 潮高；潮位
height of water 水位；水柱高，水头
height point 高程点
height ratio 高度比；潮高比
height reduction 高度改正；高度修正；自由空间改正；高程归算
height restriction 高度限制
height sensor 高度检测器
height survey 高程测量
height system 高程系统
height transmission 高程传递
height traverse 高程导线
height traversing 高程导线测量
height-area curve 高度面积曲线
height-diameter ratio 高径比
heighting 测高差；水准测量
height-of-eye correction 视高改正
height-of-image adjustment 象高修正
height-to-width ratio 高宽比；高度-宽度比
height-volume curve 高度体积曲线
heikolite 平康石
Heim's hypothesis 海姆假说
Heim's rule 海姆公式；海姆法则
heinrichite 砷钡铀矿；钡铀钙云母
heintzite 硼钾镁石
Heishan Conglomerate 黑山砾岩
Heishan Limestone 黑山石灰岩
Heishui Conglomerate 黑水砾岩
Heiskannen modification 海斯卡纳修正
Heizuojiang Conglomerate 黑座江砾岩
held water 毛细水；结合水；吸着水
helenite 弹性地蜡
helical 螺旋形的；螺线
helical auger 螺旋钻
helical compression spring 螺旋压缩弹簧
helical configuration 螺旋构型
helical curve 螺旋曲线
helical dislocation 螺位错；螺型位错
helical drill 螺旋凿岩机；螺旋钻

helical flow 螺旋状流；弯道环流
helical line 螺旋线
helical motion 螺旋运动
helical pump 螺旋泵
helical screw pump 螺杆泵
helical sounding bore 螺旋形触探钻
helical spline type actuator 带螺旋活塞杆的摆动油缸；螺旋棒式摆动油缸
helical spring 盘簧；螺旋弹簧
helical-gear pump 螺旋齿轮泵
helicitic 残缕状的
helicitic structure 残缕构造
helicitic texture 残缕结构；螺纹结构
helicity 螺旋形；螺旋性
helicoid hydraulic motor 螺杆钻具
helicoidal 螺旋形的
helicoidal axis 圆偏光轴；螺旋轴
helicoidal flow 螺旋流
helicopter electromagnetics 直升机电磁法；水平线圈电磁法
helicopter rig 直升机钻机
helicopter survey 直升机航空勘测
helicotron 螺线质谱计
helictite 石枝；不规则石钟乳
heligmite 螺旋状石笋；不规则石笋
heliocentric coordinate 日心坐标
heliocentric distance 日心距离
heliochromy 天然色摄影术
heliodor 日长石；金绿柱石
heliogram 回光信号
heliographic coordinate 日面坐标
heliographic latitude 日面纬度
heliographic longitude 日面经度
heliolite 日长石；日叶石
heliophyllite 日叶石；直氯砷铅矿
heliotrope 血玉髓；鸡血石；日光回照器；血滴石；红斑绿石髓；日光测距仪
heliotropic stalactite 向光钟乳石
heliotropic wind 日成风；日转风
helitron 电子螺线管
helium-helium method 氦-氦测年法
helix 螺旋管；螺旋弹簧；螺旋线；螺旋形
helix angle 螺旋角
helk 溶蚀深槽
hellandite 钙铒钇矿
helldiver carriage 掘进机组架
helluhraun lava 波状熔岩
helminthe 蠕绿泥石
helminthorhaphe 蠕缝迹
helsinkite 绿帘钠长岩；钠长绿帘岩
helvine 日光榴石
hemachate 血点玛瑙
hemafibrite 红纤维石
hematite 赤铁矿；赤血石；赤铁矿粉
hematite concentrate 赤铁矿精矿
hematite rock 赤铁岩
hematitic 赤铁矿的
hematitization 赤铁矿化
hematolite 红砷锰矿
hematophanite 绿铁铅矿

hematostiblite	红锑锰矿
hemeroecology	人工环境生态学；栽培生态学；人为地面形态
hemi spherical shell	半球壳
hemi-arid fan	半干旱地区冲积扇
hemichalcite	硫铜铋矿
hemicone	半锥体；冲积锥
hemicrystalline	半晶质；半晶质的
hemicycle	半旋回；半圆形
hemicycle half cycle	半轮回；半循环
hemicyclothem	半旋回层；半韵律层
hemidome	半穹窿；半坡面；半圆屋顶
hemielytra	半翅鞘
hemihedral crystal	半面形晶体
hemihedral form	半面形
hemihedrism	半面象；半对称
hemihedron	半面晶形；半面体；半面
hemihedry	半面形；半对称
hemihydrate	半水化合物
hemihydrate plaster	半水化热石膏；熟石膏灰泥
hemimorph	异极象；半形
hemimorphic	半形态的；半自形的；异极的
hemimorphic form	异极象
hemimorphic hemihedry	异极半面象
hemimorphism	半对称形；异极性；异极象
hemimorphite	异极矿
hemimorphy	异极象
Hemingfordian	赫明福德阶
hemipelagic	近海的；半远洋的
hemipelagic deposit	半远洋沉积；半深海沉积
hemipelagic environment	半远洋环境
hemipelagic intercalation	半深海沉积夹层
hemipelagic mud	半远洋软泥
hemipelagic sediment	半远洋沉积物；近海沉积物
hemipelagic-abyssal	半远洋-深海的
hemipelagite	半远洋岩；近海岩
hemiprism	半柱；半棱晶
hemispheric	半球形
hemispherical flow	半球形流
hemispherical probe	半球型探头
hemispherical projection	赤平极射投影
hemispherical projection contouring	赤平极射投影轮廓线
hemispherical projection counting net	赤平极射投影的计算网
hemispherical wave front	半球状波前
hemispheroid	半扁球体
hemist	半分解有机土
hemisymmetric	半对称的；半面像的
hemisymmetrical	半对称的；半面像的
hemitrope	半体双晶
Hengduanshan geosyncline	横断山地槽
hengleinite	钴镍黄铁矿
Hengludongmen Sandstone	横路硐门砂岩
Hengshuitang Limestone	横水塘石灰岩
henkelite	辉银矿
Henry's law	亨利定律
Henry's law constant	亨利定律常数
henwoodite	磷铝铜矿
hepatic cinnabar	肝辰砂；朱砂-硫化汞
hepatinerz	肝赤铜矿
heptorite	蓝方煌沸岩；蓝方碧玄岩
herbaceous cover	草皮护面
herbaceous soil covering	草皮覆盖层
Hercules piling system	大力士打桩装置
hercynite	铁尖晶石
herderite	磷铍钙石
hermannite	蔷薇辉石
hermarmolite	铌铁矿
hermatobiolith	生物礁岩
hermatolith	礁岩；生物礁岩
hermatopelage	沉没礁群
hermatype	造礁珊瑚
hermatypic	造礁的；同性型的
hermatypic coral	造礁珊瑚
hermetic	密封的，气密的，不透气的
hermetic seal	密封止水；密封
hermetic sealing	密封；气密封
hermetically sealed	密封的
hermetically sealed enclosure	气密外壳
hermetically sealed thermometer	密闭温度计
Hermite interpolation	埃尔米特插值
Hermite polynomial	埃尔米特多项式
Hermitian interpolation function	埃尔米特插值函数
Hermitian matrix	埃尔米特阵；埃尔米特矩阵
Hermitian operator	埃尔米特算子
heronite	浅沸绿岩；正长球粒霞霓岩
herrengrundite	钙铜矾
herrerite	铜菱锌矿
herringbone	鱼骨；人字形，人字形的；鲱骨；人字形支架
herringbone bedding	鱼骨状层理
herringbone bridging	人字撑
herringbone cleavage	人字形劈理
herringbone cross bedding	羽状交错层理；人字形交错层理；鱼骨状交错层理
herringbone cross lamination	人字形交错层理
herringbone drain	梳式排水渠，人字形排水渠；鱼骨式排水渠；人字形排水
herringbone drainage	人字形管道排水
herringbone drainage ditch	人字形排水沟
herringbone drainage system	人字形排水系统
herringbone fashion	人字形
herringous mark	人字痕；鱼骨状流痕
herringeous marking	人字形流痕
herringbone method	人字形矿房开采法
herringbone stoping	人字形回采法
herringbone structure	鱼骨状构造
herringbone system	人字形排水系统；入字形系统
herringbone texture	人字形结构；鱼骨结构
herringbone timbering	人字形支撑；人字形支架
herringbone tooth	人字齿
herringboning	人字形采矿法
herschelite	碱菱沸石
herst	沙洲
Hertzian contact	赫兹接触
Hertzian oscillator	赫兹振荡器
Hertz-Mindlin theory	Hertz-Mindlin 理论
hervidero	泥火山
herzenbergite	硫锡矿
hesitation set	假凝；暂凝

hesitation squeeze 间歇挤水泥
Hess brisance test 赫氏猛度试验
Hessenberg form of a matrix 矩阵的 Hessenberg 形式
hessite 天然碲化银
hessonite 桂榴石，钙铝榴石
hetaerolite 锌黑锰矿
hetairite 锌黑锰矿
heteroaxial 不同轴的
heteroaxial fold 异轴褶皱
heteroblastic 异变晶的；不等粒变晶的
heteroblastic texture 异变晶结构；不等粒变晶结构
heterochromatic 异色的，多色的
heterochronism 异时性
heterochronogenous soil 杂生土；次生土
heterochronous homeomorph 后成异质同形体
heterochrony 异时发生
heterochthonous 异地的
heteroclinal 不对称的
heterocline 杂褐锰矿
heterococcolith 异颗石
heterocollinite 不均匀镜质体，基质镜质体
heterocrystal 异质晶体
heterodesmic 杂键型的；多键型的
heterodesmic bond 多型键
heterodesmic structure 杂键型结构；多键型结构；杂键结构
heterodisperse 非均相分散
heterodyne 外差；外差法，外差作用；外差振荡器
heterodyne frequency 外差频率
heterodyne frequency meter 外差式频率计
heterodyne method 外差法
heterodyne slave filter 外差式从动滤波器
heterodyne wave analyzer 外差式谐波分析器
heterogene structure 非均匀组织
heterogeneity 异质性；不均匀性，多相的，异质性；非均质性；不统一性
heterogeneity coefficient 非均质系数
Heterogeneity factor 不均匀系数
heterogeneity non-homogeneity 非均匀性
heterogeneity of backfill 充填料的非均匀性
heterogeneity test 非均匀性试验
heterogeneous 不均匀的；非均质的；多相的；复杂的；非齐次的；异质的；不同种类的
heterogeneous accumulation 多相堆积
heterogeneous anisotropic 非均质各向异性
heterogeneous anisotropic medium 非均质各向异性介质
heterogeneous aqueous system 不均匀含水体系；不均匀水系
heterogeneous aquifer 非均质含水层
heterogeneous azeotrope 非均相共沸物
heterogeneous body 非均质体
heterogeneous carbonate 非均质碳酸盐岩
heterogeneous catalysis 多相催化
heterogeneous catalyst 多相催化剂
heterogeneous chain compound 杂链化合物
heterogeneous coal 不均匀煤；不均质煤
heterogeneous crystallization 多相结晶；非均匀结晶
heterogeneous deformation 非均匀变形
heterogeneous deposit 非均匀沉积层
heterogeneous distortion 不均匀破坏

heterogeneous distribution 不均匀分布
heterogeneous effect 非均匀效应
heterogeneous elastic medium 非均匀弹性介质
heterogeneous equilibrium 非均质平衡；异质平衡；多相平衡；复杂平衡
heterogeneous flow 非均质流
heterogeneous fluid 非均匀流体；非均质流体
heterogeneous foundation 非均匀地基
heterogeneous ground 非均匀土
heterogeneous isotropic 非均质各向同性
heterogeneous isotropic media 非均质各向同性介质
heterogeneous kinetics 多相动力学
heterogeneous layer 非均质层
heterogeneous material 不均匀物质，非均质材料
heterogeneous medium 非均匀介质
heterogeneous mixed-layer clays 异构混合层黏土
heterogeneous mixture 不均匀混合物
heterogeneous organic matter 多种有机物
heterogeneous organic waste 不均匀的有机废物
heterogeneous pack 不均匀填砂模型
heterogeneous polymerization 多相聚合
heterogeneous porous media 非均质多孔介质
heterogeneous reaction 多相反应；非均相反应
heterogeneous reservoir 非均质储集层
heterogeneous reservoir sand 非均质储油砂层
heterogeneous rock mass 非均质岩体
heterogeneous sample 不均匀试样
heterogeneous seam 不均匀土层
heterogeneous sediment 异构沉积物
heterogeneous soil 非均质土
heterogeneous strain 非均匀应变，不均匀应变
heterogeneous stratum 非均匀地层
heterogeneous stress 非均匀应力
heterogeneous stress field 不均匀应力场
heterogeneous system 多相体系；不均匀体系
heterogeneous texture 非均匀结构
heterogenetic 异质的；非均匀的，不均一的
heterogenite 水钴矿
heterogenous 非均质
heterogenous nucleation 不均质核形成
heterogranular 不等粒的
heterogranular porphyritic texture 不等粒斑状结构
heterogranular texture 不等粒状结构
heterolite 锌锰矿
heterolithic facies 异粒岩相
heterolithic unconformity 异岩不整合；岩性不整合
heterolysis 异种溶解；外力溶解
heteromerite 符山石
heteromesic facies 异媒相
heterometrically grained 不均粒的
heteromorphic 异形的；异态的；多晶的；复形性的；异象的
heteromorphism 异形性；多晶现象；复形性；同质异象
heteromorphite 异硫锑铅矿；脆硫锑铅矿
heteropic 非均性的；不均匀的；异相的
heteropic deposit 异相沉积
heteropic facies 异相
heteropic unconformity 异岩不整合
heteropical 异相的；非均匀性的
heteropical deposit 异相沉积

heteropical facies 异相
heteropolar 异极的
heteropolar colloid 有极胶体
heteropolar compound 异极性化合物
heteropolar crystal 有极结晶
heteropolar structure 异序构造
heteropolarity 异极性
heteroscedasticity 异方差性
heterosite 异磷铁锰矿；磷铁锰矿
heterosphere 非均质层；非均匀气层
heterostrophism convergence 斜交汇聚
heterotactic 异列的；异变的
heterotaxial 异列的
heterotaxial deposit 异列沉积
heterotaxy 异列；异常排列
heterothermozone 变温层
hettonite 硅钍石
heubachite 水钴镍矿；片沸石；黄束沸石
heumite 钠霞正煌岩；棕闪碱长岩
heuristic 启发式算法；探试的；探索的，探索式的；启发的,启发式的；试探法
hew 开凿；人工采掘；砍；劈；伐
hewettite 针钒钙石；薄晶钒钙石
Hex Key Set 套庄六角匙
Hex Let sampler 赫克斯莱特型矿尘取样器
hexabolite 玄闪石
Hexadecimal 十六进制；十六进制的
hexagon integral drill rod 六角整体钻杆
hexagonal axis 六次对称轴
hexagonal close packed lattice 密排六方晶格
hexagonal close packing 六方紧密堆积
hexagonal close-packed structure 六方紧密堆积结构
hexagonal closest packing 六方紧密填集；六方最紧密堆积
hexagonal cross ripple mark 六边形交错波纹
hexagonal crystal system 六方晶系
hexagonal dipyramid 六方双锥
hexagonal dipyramidal class 六方双锥类
hexagonal drill collar 六方钻铤
hexagonal enatiomorphic hemihedral class 六方对映半面象晶族
hexagonal Kelly 六角方钻杆
hexagonal pack 六方充填
hexagonal packing 六方堆积
hexagonal prism 六方柱
hexagonal pyramid 六方锥
hexagonal pyramidal class 六方锥类
hexagonal system 六方晶系
hexagonal trapezohedral class 六方偏方面体类
hexagonal trapezohedron 六方偏方面体
hexagonal trommel 六面滚筒筛
hexagonite 含锰透闪石；浅紫透闪石
hexahedra 六面体
hexahedral 六面体的
hexahedral mesh 六面体网格
hexahedral mesh generator 六面体网格生成
hexahedrite 方陨铁类；六面体铁陨石
hexahedron 六面体；立方体
hexahydrite 六水泻盐；六水镁矾
hexakisoctahedron 六八面体

hexaleg block 六脚块体
hexastannite 六方黄锡矿
hexatetrahedral class 六四面体类
hexatetrahedron 六四面体
Hexi Conglomerate 河西砾岩
hexoctahedral class 六八面体晶组；六八面体晶类
hexoctahedron 六八面体
hextetrahedral class 六四面体晶组；六四面体晶类
hextetrahedron 六四面体
hiatal 越级结构；间断的；超级不等粒状
hiatal fabric 超级组构
hiatal fault 开断层
hiatal texture 裂孔构造
hiatus 沉积间断，间断，缺失；裂隙
hiatus abundance 间断丰度
hiatus concretion 沉积间断结核
hiatus stratigraphy 间断地层学
hibbenite 杂磷锌矿；板状磷锌矿
hibbertite 水菱钙镁矿
Hibernian orogeny 希伯尼造山运动
hibinite 胶绿硅碲钛矿；黑复铝钛石
hibschite 水榴石；水钙铝榴石
Hida sangun paired metamorphic belt 飞蝉-三郡双变质带
hidalgoite 砷铝铅矾
hidden abutment 埋置式桥台
hidden anomaly 隐伏异常
hidden danger 隐患；潜在事故
hidden danger in embankment 堤身隐患
hidden deposit 潜伏矿床；隐伏矿床
hidden ditch 暗渠
hidden fault 隐伏断层；潜伏断层
hidden karst 隐伏岩溶；潜伏喀斯特
hidden landform 隐地形
hidden layer 掩盖层；盲层；隐蔽层；隐含层
hidden mineral deposit 掩蔽矿床
Hidden nodes 隐含节点
hidden ore 盲矿；隐伏矿
hidden outcrop 隐蔽露头；掩蔽露头
hidden peril 潜在事故
hidden project 隐蔽工程
hidden relief 隐地形
hidden reservoir 地表无显示热储
hidden resources 地下资源
hidden rock 暗礁
hidden spring 潜流；隐伏泉水
hidden structure 掩蔽构造
hidden surface 隐面
hidden suture zone 隐伏缝合带
hidden trouble 潜在事故
hidden water 隐藏水；潜藏水
hidden work acceptance 隐蔽工程验收
hidden zone 隐蔽带
hiddenbog 潜育沼泽
Hidden-Dip 躲坑
hiddenite gem 翠绿锂辉石
hi-density gravel pack 高密度砾石充填
hiding in earthquake 潜伏地震
hiding ram 倾斜千斤顶
hiding-in earthquake 潜伏地震
hiduminium 希杜铝合金；铝铜镍合金；海度铝合金；

镍铜铝合金
hielmite　钙铌钽矿
hierarchical　分层的；分级的
hierarchical analysis method　层次分析法
hierarchical brick element　分层块单元
hierarchical bubble function　分层气泡函数
hierarchical diagram　谱系图；枝状图
hierarchical file system　分级文件系统
hierarchical graph　谱系图
hierarchical image understanding model　逐级图像理解模型
hierarchical interpolation　分层插值
hierarchical method　分层法
hierarchical model　分层模型
hierarchical moving least square　分层移动最小二乘
hierarchical organization　等级结构
hierarchical path　层次路径
hierarchical pointer　层次指针
hierarchical system　等级体制
hierarchical tetrahedron element　分层四面体单元
hierarchical triangle element　分层三角形单元
hierarchical variable　分层变量
hierarchization　分级
hierarchy　分级结构；层次结构，结构；阶层层次，分层，层次；谱系；等级、体系；结构
hierarchy analysis　层次分析
hierarchy of geochemical patterns　地球化学模式谱系
hierarchy of memory　分级存储器系统
hieratite　方氟硅钾石；氟硅钾石
hieroglyph　似文象结构；虫迹；象形印痕
higginsite　砷钙铜矿；绿水砷钙铜矿
Higgins-Leighton geometric factor　赫金斯-莱顿几何因子
Higgins-Leighton shape factor　赫金斯-莱顿形状因子
high accuracy　高精度
high activity detection loop　高放射性探测回路
high air entry piezometer tip　高进气孔隙水压力测头
high airproof　高密封性
high albite　高温钠长石
high alumina basalt　高铝玄武岩
high alumina cement　高铝水泥；矾土水泥；高矾土水泥
high alumina clay　高铝黏土
high amphibolite facies　高角闪岩相
high amplitude down going event　强振幅下行波
high angle　高角度，陡角
high angle boundary　大角晶粒间界
high angle conveyor　大倾角胶带运输机
high angle cross-bedding　陡角交错层理
high angle deviated hole　大斜度井
high angle fault　陡角断层；高角度断层
high angle fracture　陡角裂隙；高角度裂隙
high angle fracture plane　陡角破裂面
high angle gravel packed well　砾石充填大斜度井
high angle hole　大角度井
high angle planar normal fault　陡角面状正断层
high angle reverse fault　逆断层；高角度逆断层
high angle thrust　陡冲断层；高角度冲断层
high angle well drilling　大斜度井钻进
high arch dam　高拱坝
high attenuation zone　高衰减带
high bank　高地河岸；高沙洲；高堤

high bending strength of　高抗弯强度
high bottom phase　高滩地
high calcium lime　高钙石灰
high capacity reservoir　高产油层
high capacity well　高产能井
high clay content　高黏土含量
high coal　厚煤层
high coast　高海岸；陡崖岸
high concentration　高浓度
high concentration dispersion　高密度悬浮液
high condensation and viscosity　高凝高黏度
high conductive layer　高传导层
high conductivity copper　高导电铜
high conductivity formation　高电导率地层
high consequence　严重后果
high consistency　高浓度
high constructional delta deposit　高建设性三角洲沉积
high constructive delta　高建设性三角洲
high containment　强阻挡层
high contrast　强反差
high craton　高位稳定地块
high curvature　大曲率
high curve　陡升曲线
high cut　上挖；向上切割
high cutting slope　路堑高边坡
high dam　高坝；坝高
high definition　高清晰度
high degree of drilling accuracy　高精度钻进
high density　高密度
high density concrete　高密度混凝土
high density current　高密度流
high density electronic method（HDEM）　高密度电法；高密度电阻率法
high destructional delta deposit　高破坏性三角洲沉积
high destructive delta　高破坏性三角洲
high dip　高角度倾斜，高倾斜；陡倾斜；陡坡
high dip angle　陡倾角；高倾角
high dip angle steep dip　陡倾角
high dipping　高倾斜；陡倾斜的
high drift angle　高偏角
high early cement　早强水泥
high early strength　早期强度；早高强
high early strength agent　早强剂
high early strength cement　早强水泥；快硬水泥；速凝水泥
high early strength Portland cement　早强硅酸盐水泥
high efficiency　高效率
high efficiency filtration　高效率隔滤设施
high efficiency frac fluid　高效压裂液
high efficiency water reducing agent　高效减水剂
high emission　高速度喷出
high energy electromagnetic induction　高能电磁感应
high energy environment　高能环境
high energy gamma activation analysis　高能伽马活化分析
high energy gas fracturing　高能气体压裂
high energy marine environment　高能海洋环境
high energy particle　高能粒子
high entry porous stone　高进气透水石
high face　上挖工作面；高阶段工作面
High field stress areas　高地应力区

high field-strength element	高场强元素
high fill	高填方；高填土
high fill and deep cut	高填深挖
high flood level	高洪水位
high flow	洪流；洪水流量；高流量；高流动性
high flow regime	高流态
high fluid potential belt	高流体势能带
high fluid pressure	高液压
high fluid state concrete	高流态混凝土
high fluidity concrete	高流动性混凝土
high fluoride-bearing water	高含氟水
high fluorine water	高氟水
high frequency	高频；高频率的
high frequency amplitude	高频振幅
high frequency compensation	高频补偿
high frequency generator	高频发生器；高频发电机
high frequency geophone	高频检波器
high frequency ground wave radar	高频地波雷达
high frequency oscillation	高频振荡
high frequency oscillator	高频振荡器
high frequency residual statics	高频剩余静校正
high frequency seism	高频地震
high frequency wave	高频波
high gas rate well	高产气井
high gaseous mine	高瓦斯矿井
high geostress	高地应力
high grade cement	高标号水泥；优质水泥
high grade coal	高品位煤，按国际煤层煤分类；干基灰分小于10%的煤
high grade concrete	高标号混凝土
high grade metamorphism	高级变质作用
high grade mining	选择性开采；选优开采
high grade oil	高级油
high grade ore	高品位矿；富矿
high grade product	高品位产品
high grade project	优质工程；优良工程
high grade reserve	高级储量
high greenschist facies	高绿片岩相
high head	高压头；高水头；高压头的；高压的
high humidity	高湿度
High in situ stress breakage	高原位应力破坏
high inclination-angle	大倾角
high initial stress	高初应力
high intensity manifestation	高强度显示
high intensity pumping	高强度抽水
high island	高岛
high land	高原；高地
high land peat	高位泥炭
high latitude	高纬度
high layer	高帮；高位砂层
high level crossings	立体交叉；高架跨越
high level intake	浅孔式进水口
high level of production	高速开采
high level radioactive waste	高水平放射性废物
high level reservoir	高地水库
high level tank	高位罐
high level waste	高放废物
high lift pump	高扬程水泵
high lime mud	高钙泥浆
high limit	最大限度；上限
high line	索道；高压线
high liquid limit soil	高液限土
high load melt index	高负荷熔融指数
high log rate	高对数变化率
high magnesian calcite	高镁方解石
high mark	高潮线
high melting point	高熔点
high moisture content	高水分含量
high molecular	高分子的
high molecular compound	高分子化合物
high moor	高位沼泽
high moor bog	高位沼泽
high moorland	高地泥炭田
high mountain karst	高山岩溶
high oblique	高度倾斜的
high oblique photograph	高斜度影像
high observing tower	高测标；观测高塔
high order derivative	高阶导数；高阶微商
high order mode	高阶振型
high order structure	高级构造
high pass	高通
high pass filter	高通滤波器；低阻滤波器
high pass filtering	高通滤波
high penetration bitumen	高渗透性沥青
high performance	优越性能；高准确度
high permeability	强渗透性；强透水性
high permeability channel	高渗透孔道
high permeability formation	高渗透率地层
high permeability streak	高渗透夹层
high permeability zone	高渗透带
high pitch	陡坡
high plain	高地平原
high plastic limit	高塑限
high plasticity soil	高塑性土
high Poisson's ratio	高泊松比
high polymer	高分子聚合物，高聚物
high pour-point crude	高倾点原油
high power pulse	强脉冲
high precision	高精度；高精度的
high precision computation	高精度计算
high pressure	高压；高压
high pressure barrier	高压坝
high pressure belt	高压带
high pressure burner	高压炉头
high pressure capillary	高压毛细管
high pressure capillary rheometer	高压毛细管流变仪
high pressure cell	高压腔，高压室；高压传感器
high pressure charge	孔内爆破药包
high pressure circulating pump	高压循环泵
high pressure cleaner	高压清洁器
high pressure completion	高压完井
high pressure draught	高压气流；高压通风
high pressure drilling	高压钻进
high pressure facies series	高压相系
high pressure field	高压油田
high pressure formation	高压地层
high pressure gauge	高压力计
high pressure greenschist facies	高压绿片岩相
high pressure grouting	高压灌浆；高压注浆
high pressure grouting pump	高压灌浆泵

high pressure hose	高压胶管；水龙带
high pressure intake	高压进水口
high pressure jet drill	高压射流钻
high pressure jet grouting	高压喷射灌浆
high pressure jet grouting pile	高压旋喷桩
high pressure jetting	高压射水
high pressure large diameter jet	高压大射流
high pressure limiter	高压限压器
high pressure mercury lamp	高压水银灯
high pressure metamorphic facies	高压变质相
high pressure metamorphism	高压变质作用
high pressure oil	高压油
high pressure oil pump	高压油泵
high pressure phase	高压相
high pressure pump	高压泵
high pressure tunnel	高压隧洞
high pressure type metamorphism	高压变质酌
high pressure water blast	高压水爆破
high pressure water infusion	高压注水
high pressure water injection	高压注水
high pressure water jet	高压水柱，高压水射流；高压喷水器；高压水枪
high pressure water jet drilling system	高压水射流钻井系统
high pressure water jet drivage	高压水射流掘进
high pressure water pipe	高压冲水管
high pressure water propagating	高压水推送清洗器
high pressure water pump	高压水泵
high pressure well	高压井
high pressure well completion	高压井完井
high pressure zone	高压带
high pressure-low temperature metamorphism	高压-低温变质作用
high productivity layer	高产层
high pulling torque	大扭矩
high purity germanium detector	高纯锗探测器
high pyrite	高硫黄铁矿
high quality cement	优质水泥；高标号水泥
high quartz	高温石英；石英
high range	大比例，大刻度；高山脉
High Range Water-Reducing Admixture	高性能减水剂
high rank metamorphism	高级变质
high rate	高速
high rate activated sludge process	高速活性污泥法
high rate filter	高效过滤；高效率滤池
high rate filtration	高速过滤
high rate producing well	高产井
high rate rapid filter	高速快滤池
high reference point	高基准点
high reliability	可靠性高；高可靠度
high relief	高峻地形
High Relief Carbonate Mud Mound	高起伏碳酸盐泥丘
high relief reservoir	高闭合度储层
high residue level	高残留度
high resistivity shielding	高阻屏
high resolution	高清晰度；高分辨率；高清晰；高精度
high resolution analysis	高清晰度分析
high resolution boomer	高解像度探测器
high resolution capacity	高分辨能力
high resolution climatostratigraphy	高分辨率气候地层
high resolution column	高分辨柱
high resolution detector	高分辨率检波器
high resolution dipmeter	高分辨率地层倾角仪
high resolution pressure gauge	高分辨率压力计
high resolution scheme	高分辨率格式
high resolution seismic survey	高分辨率地震勘探
high resolution seismic work	高分辨率地震测量
high resolution sequence stratigraphy	高分辨率层序地层学
high resolution spectrum	高分辨谱
high resolution thermometer	高分辨率井温仪；高分辨力温度计
high resolving power	高分辨率
high resonance capture	强共振俘获
high rise cap pile	高承台桩
high risk	高风险
high risk area	高风险区
high risk structure	高风险结构
high risk tunnels	高风险隧道
high salt content	高含盐量
high sample rate instrumentation	高速取样装置
high sand content production	高含砂采油
high sanidine	高温型透长石
high seam	厚煤层；厚矿层；厚层
high seismic risk	高地震危险性
high sensitive	高灵敏度的
high sensitivity detector	高灵敏度检测器
high sequence	高频层序
high shrinkage oil	高收缩率原油
high slope	高边坡；陡坡
high solid content fluid	高固相含量流体
high solid loading	高固相含量
high solid mud	高固相泥浆
high specific activity	高比放射性
high specific gravity medium	高比重介质
high specific thrust	最大比推力
high speed	高速度
high speed bit	高速钻头
high speed cement	速凝水泥
high speed colloidal grout mixer	高速胶体浆液搅拌机
high speed concrete	速凝混凝土
high speed decline of water level	水位快速消落
High speed drill	锋钢钻嘴；高速钻机
high speed earth compactor	高速土方压实机
high speed grout mixer	高速浆液搅拌机
high speed jet	高速射流
high speed landslide	高速滑坡
high speed mixer	高速搅拌机
high speed motorway/road	快速道路；快速公路
High speed steel hole saw	锋钢令梳
high spin state	高自旋状态
high spud angle	大倾角
high stability	高稳定性
high stand systems tract	高位体系域
high steel	高碳钢；硬钢
high steep slope	高陡边坡
high strength	高强度；高浓度
high strength cement	高强水泥；高标号水泥
high strength concrete	高标号混凝土；高强混凝土
high strength expanding grout	高强膨胀性浆液
high stress	高应力

high stress layer	高应力层
high stress zone	高应力区
high temperature bending strength	高温弯曲强度
high temperature catalytic gas	高温裂解气
high temperature cement	耐高温水泥
high temperature completion	高温完井
high temperature form	高温型
High temperature high pressure	高温高压
high temperature hydrothermal deposit	高温热液矿床
high temperature isostatic pressing	高温等静压
high temperature plagioclase	高温型斜长石
high temperature shift	高温变换
high temperature stabilizer	高温稳定剂
high temperature test	高温试验
high temperature torsional strength	高温扭转强度
high temperature viscometer	高温黏度计
high temperature volume stability	高温体积稳定性
high temperatures bending strength	高温抗折强度；高温弯曲强度
high temperatures crushing strength	高温耐压强度
high tensile steel	高抗拉纲；高强度钢
high tensile steel tendon	高抗拉钢缆
High Tensile Steel Welding Electrode	高拉力焊支
high tensile steel wire	高强钢丝
high tensile strength	高抗拉强度；高抗拉应力
high tension	高压力；高拉力
high tension steel wire	高强钢丝
high test	高质的
high tidal level	高潮面
high tidal terrace	高潮阶地
high tide	高潮；满潮
high tide level	高潮水位；高潮位
high tide line high water line	高潮线
high tide shoreline	高潮海岸线
high vacuum pump	高真空泵
high value	高金属矿石；高值
high velocity ground roll	高速地滚波
high velocity impact	高速冲击
high viscosity	高黏性
high viscosity fracturing fluid	高黏压裂液
high viscosity gelled carrier fluid	高黏稠化携砂液
high volt starter	高压起动器
high voltage cable trough	高压缆槽
high voltage electricity	高压电力
high voltage injection	高压输入
high voltage insulating oil	高压绝缘油
high voltage probe	高压探针
high voltage tester	高压测试器
high voltage wire	高压线
high volume hole	大容积药壶钻孔
high volume sampler	高流量采样仪器；大容量采样器
high wall	顶帮边坡；边坡；阶段边坡；工作面
high wall ramps	顶帮沟道
high water	高水，高水位；洪水；高潮，高潮位；满潮
high water bed	洪水期河床；高水位河床
high water dam	高水坝
high water full and change	朔望高潮
high water interval	高水位间隔
high water level	高潮面；高水位；高水线
high water line	高潮线；高潮水位线
high water mark	高潮标，高水位，高潮线；高水位；高潮标记，高潮痕
high water of ordinary spring tide	淤泥土
high water river bed	洪水河床
high water spring	大潮高潮
high water stand	高潮停潮
high water terrace	高潮阶地
high water-content crude oil	高含水原油
high wax content oil	高含蜡原油
high workability	高和易性
high yield bentonite	高屈服值的膨润土
high-accuracy survey	高精度测量
high-alkali cement	高碱水泥
high-altitude and cold region	高寒高海拔地区
high-altitude magnetic field	高空磁场
high-altitude photo	高空摄影
high-alumina	富钒土的；高铝的
high-alumina basalt	高铝玄武岩
high-amplitude	强振幅
high-angle basement fault	陡倾基底断层；高角度基底断层
high-angle tubing string	大斜度井油管柱
high-angled	急倾斜的，陡的
high-boiling oil	高沸点油
high-boost filter	高增益滤波
highboy	撬装式或拖车式手摇泵储罐加油装置
high-calcium limestone	富钙灰岩；高钙质灰岩
high-capacity channel	高渗透孔道
high-capacity prop	高负载顶柱
high-carbon	高碳
high-centre bit	中部凸出的钻头；中部凸出的钎头
high-coal bolting pattern	厚煤层锚杆布置方式
high-coal deposit	厚煤层
high-consistency viscometer	高稠度黏度计
high-contrast anomalies	高对比度异常
high-crowned	高度凸出的；高度隆起的
high-cut filter	高阻滤波器
high-definition lidar	高分辨率激光雷达
high-deformation regime	高变形状态
high-discharge mill	高排料水平磨机
high-duty	重载的；重型的；大功率的；大型的；高生产能力的；高产的
high-duty pump	大排量泵
high-early strength concrete	快硬混凝土；早强混凝土；高早强混凝土
high-enthalpy thermal water	高焓热水
high-enthalpy well	高焓井
higher critical velocity	超临界速度；高临界流速
higher difference	高阶差分
higher frequency component	高频分量
higher geodesy	高等大地测量学
higher harmonic	高次谐波
higher high tide	高高潮
higher high water	高高潮；高高水
higher high water interval	高高潮间隙
higher level	上水平；上平巷；较高水位；上部层位
higher low tide	高低潮
higher low water	高低潮
higher low water interval	高低潮间隙
higher mode	高阶振型；高振型

higher mode surface wave	高振型表面波
higher order patch test	高阶小片实验
higher order stability	高阶稳定
higher order structure	高级构造
higher order surface wave	高阶表面波
higher order wave	高阶波
higher slice	上分层
higher surveying	高等测量学
higher white interference colour	高级白干涉色
higher-gain long-period seismograph	高增益长周期地震仪
higher-mode interference	高阶干扰
higher-order attenuation	高次衰减
highest beach ridge	风暴海滩脊
highest frequency	最高频率
highest groundwater level	地下水最高水位
highest high water	最高洪水位；最高高水位；最高高潮；向心排水
highest low water	最高低水位；最高低潮
highest mode	最高振型
highest navigable flood level	最高通航洪水位
highest navigable flood stage	最高通航洪水位
highest normal high water	理论最高高潮面
highest order	最高位；最高阶
highest quotation	顶盘
highest record stage	永久潜水位
highest recorded stage	最高记录水位；实测最高水位
highest stage	最高水位；淹没平面
highest tide	最高潮
highest tide level	最高潮位
highest upsurge	最高上涌浪
highest water level	最高水位
highest wave	最高波
high-expansion cement	高膨胀水泥
high-frequency alternator	高频振荡器；高频发生器
high-frequency anomaly	高频异常
high-frequency error	高频误差
high-frequency induction	高频感应
high-frequency microtremor	高频脉动
high-frequency percussion	高频冲击
high-frequency vibration	高频振动
high-gel salt cement	高胶质含盐水泥
high-grade assemblage	高级变质组合
high-gradient magnetic separation	高梯度磁选
high-gravity	高比重的
high-gravity oil	轻质油；高比重油
high-head dam	高水头坝
high-head pump	高压泵；高扬程泵
high-inclination well	大斜度井
high-intensity	高烈度
high-intensity oscillation	高强度振荡；高强度振动
high-iron portland cement	高铁硅酸盐水泥
high-lake phase	高位湖相
highland grade compensation	高原纵坡折减
highland ice cap	高地冰帽
highland moor	高地湖沼
highland plagioclasite	高地斜长岩
high-latitude glacier	高纬冰川
high-latitude precipitation region	高纬降雨雪区；高纬降水区
high-latitude spot	高纬黑子
high-lift concrete construction method	混凝土高块浇筑法
high-lift construction	高浇注块施工
high-line conduit	引水渠
high-liquid-ratio gas well	高含液气井
high-low bias test	极限状态测试
high-low piles	多层桩；高低桩
high-low pump	双压泵
high-low sensing device	高低压传感装置
highly acidic	高酸性的
highly compressed clay	高压缩黏土
highly compressible	可高度压缩的
highly compressible soil	高压缩性土
highly conductive fluid	高导电流体
highly decomposed granite	高度风化花岗岩，高度风化花岗岩层；高度破碎花岗岩；强风化花岗岩
highly decomposed rock	高度风化岩
highly decomposed volcanic rock	高度风化火山岩；高度风化火山岩层
highly decomposed volcanics	强风化火山岩
highly deformed area	强烈变形地区
highly dissected terrain	强切割地区
highly disturbed	剧烈变动的；强烈破坏的
highly efficient regeneration	高效再生
highly folded structure	强烈褶皱的构造
highly fractured linear belt	强烈破碎线形带
highly fractured region	强破裂区
highly gassy mine	高瓦斯矿井
highly jointed rock	节理多的岩石
highly magnetic	强磁性的
highly metamorphosed	高度变质的
highly mineralized water	高矿化水
highly oxidizing system	强氧化系统
highly permeable	强透水的；透水良好的
highly permeable sublayer	高渗透夹层
highly pervious soil	高透水性土壤
highly plastic clay	高塑性黏土
highly plastic soil	高塑性土
highly porous	高孔隙度的
highly resistant mineral	强稳定矿物
highly saline aquifer	高盐水含水层
highly saline drilling fluid	高矿化度钻井液
highly saline oilfield brine	高含盐油田卤水
highly seismic region	高烈度区；强震区
highly sensitive	高灵敏度的
highly sensitive soil	高敏感土
highly strained area	强烈应变地区
highly stressed	高度受力
highly stressed rock	高应力区岩石
highly stressed zone	高应力区
highly viscous crude	高黏原油
highly weathered	强风化的；强风化
highly weathered granite	高度风化花岗岩；高度风化花岗岩层
highly weathered volcanic rock	高度风化火山岩；高度风化火山岩层
highly weathered volcanics	强风化火山岩
highly-deviated well	大斜度井
high-lying deposit	高地矿床
high-magnification seismograph	高倍率地震计

high-melting 高熔点的
high-Mg andesite 高镁安山岩
high-mountain station 高山测站
high-order bit 高位
high-order elastic model 高阶弹性模型
high-order elastic modulus 高阶弹性模量
high-order elasticity 高阶弹性力学
high-order polynomial 高次多项式
high-output geophone 高输出检波器
high-oxide ore 高氧化矿石
high-pass 高通的
high-performance concrete 高性能混凝土
high-performing 高性能的
high-polar glacier 高极地冰川
high-portland 含大量波特兰水泥的
high-power 大功率
high-power field 高倍放大视域
high-power load 大功率负载；大功率负荷
high-powered 高能的
high-precision 高精度的；精密的
high-precision depth recorder 高精度深度记录仪
high-precision leveling 高精度水准测量
high-precision standard aneroid 高精度标准无液气压计
high-pressure 高压的
high-pressure air and water jet 高压气水射流
high-pressure belt 高压带
high-pressure bit 高压射流钻头
high-pressure desander 高压除砂器
high-pressure erosion bit 高压冲蚀钻头
high-pressure erosion drill 高压射流成孔钻机
high-pressure fluid 高压液
high-pressure fumarole 高压喷气孔
high-pressure grouting 高压灌浆
high-pressure hole 高压孔
high-pressure hose 高压软管
high-pressure liquid system 高压水系统
high-pressure metamorphic type 高压变质型
high-pressure metamorphism 高压变质
high-pressure phase 高压相
high-pressure rocks 高压岩石
high-pressure spray 高压射流
high-pressure sprinkler 高压喷灌机
high-pressure storage 高压储存
high-pressure water 高压水
high-pressure water derusting 高压水除锈
high-quality 优质
high-quality concrete 高级混凝土
high-quality crude 优质原油
high-rank greywacke 高级杂砂岩
high-rank metamorphism 高级变质
high-relief mountains 强烈起伏的山脉
high-resistance 高阻
high-resistance load 高阻负载
high-resistance voltmeter 高阻伏特计
high-resistant cement 高强度水泥
high-resolution bathymetry 高分辨率的测深仪
high-resolution interferometer 高分辨率干涉仪
high-resolution magnetic survey 高分辨率磁力测量
high-resolution sedimentary record 高分辨率沉积记录
high-resolution seismic 高分辨率地震

high-resolution seismic reflection 高分辨率地震反射
high-resolution seismic stratigraphy 高分辨率地震地层学
high-resolution vibroseis 高分辨率可控震源
high-rise pile cap 高桩承台
high-rise platform pile foundation 高承台桩基础
high-roof 顶板高的
high-saline 高盐分的
high-salinity brine 高矿化度盐水；高矿化度；高卤水
high-sensitivity 高灵敏度
high-sensitivity flowmeter 高灵敏度流量计
high-side coal bump 煤层上帮突出
high-side pack 上部充填带
high-silica cement 高硅水泥
high-silica dust 高硅尘末
high-silica rock 高硅岩
high-silicon chromium iron 高硅铬铁
high-silicon iron 高硅铸铁
high-sill spillway 高槛溢洪道
high-span 大跨度
high-speed 高速的
high-speed agitator 高速搅拌机
high-speed bed 高速层
high-speed belt stowing 快速胶带充填
high-speed bus 高速母线
high-speed camera 高速摄影机
high-speed carry 高速进位
high-speed centrifuge 高速离心机
high-speed channel 高速通道
high-speed data acquisition 高速数据采集
high-speed differential protection 高速差动保护
high-speed drivage 高速掘进
high-speed layer 高速层
high-speed motion-picture camera 高速摄影机
high-speed oscilloscope 快速扫描示波器
high-speed photograph 高速照片
high-speed photographic method 高速照相术
high-speed photography 高速摄影
high-speed pictorial presentation 高速显图
high-speed plotting 高速绘图
high-speed pump 高速泵
high-speed radial drill 高速摇臂钻床
high-speed rolls 高速辊碎机
high-speed tunneling 隧洞快速掘进法
high-speed vibrating screen 高速振动筛
highstand 高水位期
high-stand deposit 高水位期沉积
high-strength inorganic grout 高强度无机浆液
high-strength reinforcement 高强钢筋
high-strength steel 高强度钢；强力钢
high-strength steel reinforcing 高强度钢筋
high-stress mining 高应力采矿
high-temperature corrosion 高温腐蚀
high-temperature insulation 高温绝缘
high-temperature stability 高温稳定性
high-temperature tar 高温焦油
high-tensile reinforcement 高强钢筋
high-tensile steel bar 高强度钢筋
high-tensile wire 高强钢丝
high-tension 高压的；高拉力
high-test cast iron 高级铸铁；优质铸铁

high-test cement	高级水泥
high-thrust drill	高推力钻机；高推力凿岩机
high-tide line	高潮线；高水位
high-tide period	高潮期
high-tide slackwater	高潮憩流
high-torque	大扭矩
high-torque rotary drill	高扭矩旋转钻机
high-torque screwdriver	大扭矩螺丝刀
high-transmissibility path	高渗透流道
high-type highway	高等级公路
high-type road	高等级道路
high-velocity anomaly	高速异常
high-velocity area	高速区
high-velocity bed	高速层
high-velocity channel	高速渠道
high-velocity detonation	高速爆轰
high-velocity flow	高速水流
high-velocity gunite	高速喷浆
high-velocity jet	高速射流；高速喷射
high-velocity layer	高速层
high-velocity zone	高速带
high-viscosity fluid	高黏性流体
high-viscosity oil	高黏度油
High-voltage bushings	高压套管
high-voltage cable tunnel	高压电缆洞
highwall augering	边坡螺旋钻开采
highwall boxcut	顶帮拉沟
highwall drill	台阶钻机
highwall miner	边坡开采系统
highwall slope	高边坡
highwall tunnel	顶帮运煤巷
highwall-drilling machine	立式钻机；边坡钻机
high-water bed	洪水河床；洪水河槽；河漫滩
high-water control	高水位控制
high-water elevation	高水位
high-water inequality	高潮不等
high-water lunitidal interval	月潮高潮间隙；高潮月潮间隙
high-water platform	高潮浪蚀台
high-water river bed	洪水期河床
high-water season	汛期
high-water slack	高水平潮
high-water training	洪水治导
high-water-cut stage	高含水期
highway and rail transit bridge	公铁两用桥
highway clearance limit	道路净空界限
highway cut	公路挖方
highway drainage	公路排水
highway embankment	公路路堤
highway engineering	公路工程
highway geology	公路地质学
highway layout	道路设计
highway loading	道路荷载
highway pavement	公路路面
highway slopes	公路边坡
highway subgrade	公路路基
high-wearing feature	高耐磨性
high-withdrawal area	高产区
high-yield clay	高造浆黏土；高膨胀黏土
high-yield nuclear explosion	高效核能爆炸
high-yield well	高产井；高出水量井；高流量井
high-yielding well	高产井
hilairite	钠长正霞正长岩；方钠霞石正长斑岩；三水钠锆石
Hilbert determinant	希尔伯特行列式
Hilbert space	希尔伯特空间
Hilbert transform	希尔伯特变换
Hiley formula	希利公式
Hileys formula	黑莱公式
hilgardite	水氯硼钙石
hilgenstockite	板磷钙石
hill	丘陵，小山，岗，土堆；矿内高地，矿山地面；轨道上山
hill country	丘陵地区
hill creep	滑坡；顺坡潜移，山坡蠕动
hill crest	山顶
hill depression	溶丘洼地
hill of circumdenudation	环蚀丘陵
hill of planation	基岩侵蚀平面
hill of upheaval	隆起丘陵
hill or moderately rolling terrain	丘陵地形
hill peat	高地泥炭
hill shading method	晕渲法；阴影法；光影法
hill slope debris	坡地沉积物
hill torrent	涧；山区急流，山溪
hill wash	坡地坍滑
hillebrandite	针硅钙石
hillhouse basalt	苦橄玄武岩
hill-island	冰碛丘
hillock	废石堆，矸石堆；小丘，土坡
hillock moraine	冰碛丘陵
hillock-top elevation	丘顶高程
hillocky	丘陵地带的
hillside	山腰，山麓，坡脚，丘陵侧面；山坡，山腰
hillside canal	环山渠道
hillside covering works	山坡覆盖工程
hillside creep	山坡物质蠕动；滑坡
hillside cut and fill	半填半挖
hillside dam	山麓小坝；山坡挡水坝
hillside ditch	单侧沟，山脚沟
hillside flanking	护坡；加固山坡
hillside gravel	坡地砂矿
hill-side line	山坡线
hillside open cut	山坡露天矿
hillside open pit	山坡露天矿
hillside placer	山腰砂矿
hillside sand	仅磨钻头一侧的砂岩
hillside spring	山腰泉
hillside swamp	山腰沼泽
hillside waste	山腰岩屑；坡积层
hillslope	山坡
hillslope erosion	坡面侵蚀；山坡冲刷；山坡侵蚀
hillslope run-off	山坡径流
hill-stream	丘陵区河流
hilltop	小山顶；丘顶
hilltop reservoir	高山水库；山顶水库
hilltop surface	丘顶面
hill-torrent	山区急流
hillwash	坡地侵蚀堆积物；坡地冲刷作用；坡地片冲物，坡地坍滑；坡滑

hillwork 山影法；地貌晕渲法
hilly 有斜坡的，陡的；岗陵起伏的；丘陵多的，丘陵的
hilly area 丘陵区
hilly coast 丘陵海岸
hilly country 丘陵地带
hilly ground 丘陵地
hilly land 丘陵地
hilly land location 丘陵地段选线
hilly land terrain 丘陵
hilly landscape 丘陵景观
hilly region 山区
hilly terrain 重丘区
Hilt's law 希尔特法则；希尔特规律
Hilt's rule 希尔特法则
Himalaya geosyncline 喜马拉雅地槽
Himalayas mountains 原沟
hinder land 腹地
hindered falling 干扰沉降
hindered settling 干涉沉降；干扰下沉；干扰沉降；沉降受阻，受阻沉降
hindered-settling column 干扰下沉柱；干扰下沉室
hindered-settling commutator 干扰沉落管；干扰下沉管
hindered-settling hydraulic classifier 干扰下沉式水力分级机
hindered-settling ratio 干扰下沉比；干涉沉降比
hinge 铰，铰片，平面铰链，铰链，铰合线；折叶；转枢，转枢；脊线
hinge axis 枢纽轴
hinge bar 铰链杆
hinge belt 枢纽带
hinge fault 捩转断层；枢纽断层
hinge line 枢线；脊线；铰合线；枢纽线；构造转折线
hinge of anticline 背斜枢
hinge of fold 褶皱枢纽
hinge of syncline 向斜枢
hinge pin 铰链销；铰销；胶带铰销
hinge region 枢纽区
hinge surface 枢纽面
hinge zone of earth 地震带
hinged scraper 绞式刮土机
hingeless arch 无铰拱
hinge-line fault 转枢线断层
hinge-line plunge isogons 枢纽线等倾伏线
hinge-type faulting 捩转断层作用；枢纽断层作用
hinsdalite 磷硫铅铝矿
hinter basin 后陆盆地
hinter deep 后海渊；后海沟
hinterland 海岸或河岸后部地方；后置地；腹地；后陆
hinterland basin 内陆盆地
hinterland dipping duplex 倾向腹陆式双冲构造
hinterland eugeosyncline 腹地优地槽
hinterland sequence 内陆层序；腹地层序
hintzeite 硼钾镁石
hiortdahlite 片楣石
hip 脊；斜脊；角梁；重力式引水堰
hip and gable roof 人字坡山墙
hirnantite 绿钠角斑岩；绿泥钠长角斑岩；绿泥角斑岩
hirst 沙堆，沙丘；沙洲；砂洲；沙滩；沙坝；河岸砂堆积

hi-saltstable polymer 高抗盐聚合物
hisingerite 硅铁石；硅铁土；黑高岭土
hislopite 海绿方解石
histic epipedon 泥炭表层
histogram 直方图，柱状图，直方统计图；频率分布器；频率曲线，阶梯频率分布图；柱式图解
histogram adjust 直方图调整
histogram adjustment 直方图调整
histogram equalization 直方图均衡；直方图均衡化
histogram frequency distribution diagram 柱状图解频率分布图
histogram linearization 直方图线性化
histogram match 直方图匹配
histogram method 直方图法
histogram modification 直方图修改
histogram normalization 直方图正态化；直方图标准化
histogram of pixel values 像元值直方图；像素值直方图
histogram specification 直方图规格化
histogrammic 直方图的
histogrammic paleostructural analysis 柱状古构造分析
histograph 输水时间曲线图；频率分布图
histometabasis 保存构造
historic flood level 历史洪水位
historic geomorphology 古地貌学
historic period 历史时代
historic site archaeology 历史遗址考古学
historic spot 古迹
historical baseline 历史基线
historical earthquake 历史地震
historical event 历史事件
historical flood 历史洪水
historical flood damage 历史洪水灾害
historical flood investigation 历史洪水调查
historical geochemistry 历史地球化学
historical geography 历史地理学
historical geology 历史地质学；地史学
historical geomorphology 历史地貌学；地貌史学
historical production performance 历史开采动态
historical seismicity 历史地震活动性
historical seismology 历史地震学；地震史学
historical well analysis 老井分析；已钻井分析
historically active fault zone 历史上活动断层区
history match 历史拟合；历史匹配
history of earthquake 地震史
history of instability 过去不稳定记录；过去不稳定历史
histosol 有机土；有机质土
hi-strength 高强度
histrixite 硫铋锑铜矿
hitch 小断层；梁窝柱窝；挂钩套拴；障碍索结，牵引装置；故障，钩住；急拉
hitch a cut 挖窝；掏凹槽
hitchcockite 水磷铝铅矿
hoarding 临时围篱；围板；悬臂式模板
Hoarding Sheet 围街板
hoarstone 界石
hobiquandorthite 闪云二长岩
hochkraton 陆间克拉通
Hock triaxial cell 霍克三轴仪压力室
Hock-Brown empirical strength criterion 霍克布朗经验强度准则

hod	灰砂斗；砂浆桶；煤斗；灰浆桶
hodgkinsonite	褐锌锰矿
hodograph	速度图；速矢端线，速矢端迹；时距曲线；高空风速分析图；耐距曲线
hodograph method	速端曲线法；时距曲线法
hodograph plane	速端平面；速度图面
hodograph transformation	速度矢端曲线变换
hodometer	里程计；车程计；计距器；测距器；轮转计
hodoscope	描迹仪；辐射计数器
hoek	隐谷
Hoek-Brown failure criterion	Hoek-Brown破坏准则
Hoeppler rolling ball viscosimeter	郝普勒落球黏度计
hoevillite	钾盐
Hofmann degradation	霍夫曼降解
hofmannite	晶蜡石
hog back	煤层底部突起；豚脊；猪背岭
hog bowl	沉淀池；沉淀箱
hogauite	钠沸石
hogback strike coast	猪背脊走向海岸
hogbacked	豚脊形的；猪背状的
hogbacked bottom	突起底板
hogbacked floor	突起的底板
hog-frame	弓背构架
hoggin	夹砂砾石；道碴；夹砂砾石结合料
hogging	挠度；弯曲，翘曲，拱曲；船身弯曲
hogging moment	负弯矩；中拱力矩
hoist	提升机，绞车，卷扬机；起重机，吊重机；提升；每班提升量；升降机，启闭机
hoist effort	提升力
hoist engine	提升机，绞车；绞车发动机，起重发动机
hoist in shaft sinking	竖井掘进提升
hoist track way	吊车滑道
hoist winch	卷扬机
hoister	提升机，绞车；起重机；提升机司机，起重机司机
hoisting	提升，起重；提升的；起重的；卷扬机；吊装
hoisting crab	提升绞车
hoisting crane	起重机；升降起重机
hoisting cycle	提升周期
hoisting cycle time	提升循环时间
hoisting height	提升高度
hoisting hook	起重吊钩，起重吊钩，提升钩；提升绳钩；起重机钩
hoisting motor	起重马达
hoisting rope	起重绳；提升钢丝绳
hoisting stopper	起吊止挡
hoisting swivel elevator	提引器
hoisting winch	卷扬机；绞车；绞盘；提升绞车
hoisting wire	起重钢索
hokutolite	北投石；铅重晶石
holard	土壤水；土壤总含水量
hold of pile	桩入深度
hold tank	贮槽
holdenite	红砷锌锰矿
holder base plate	支座垫板
holdfast	固定器；锚钉；夹钳；"磨耳"；支架
holding bolt	固定螺栓
holding down bolt	地脚螺栓；定位螺栓；压紧螺栓；锚杆；压栓
holding force	夹持力
holding ground	持力层
hold-up	滞留量；停止；阻塞；阻碍；举起
hold-up capacity	滞留容量
holdup factor	悬持系数
hole	洞穴，孔；炮眼，井眼；钻孔，弯曲；通道；井筒，拉槽；内多边形
hole alignment	钻孔定向
hole angle	井斜角
hole annulus	井眼环形空间
hole anomaly	井眼不规则
hole aperture	孔径
hole axis	井眼中心线；钻孔轴线
hole azimuth angle	井眼方位角
hole back	炮眼底；钻孔底
hole backfilling	钻孔回填；封孔
hole bit	钻头；钎头
hole blasting in open-pit	露天矿深孔爆破
hole block	空心块
hole blow	井喷，井喷干扰
hole boring	井孔钻进
hole bottom	钻孔底，炮眼底
hole bottom pressure gauge	孔底压力计
hole bottom region	钻孔底
hole caliper	井径规；孔眼测径器
hole capacity	井眼容积
hole cave-in	井孔坍塌
hole caving	孔塌
hole cementation	钻孔灌浆
hole cleaning	井眼净化
hole clearance	套眼与井眼的间隙；套管与孔壁间隙
hole collapse	井壁坍塌
hole collar	炮眼口；钻孔口
hole condition	井眼状况
hole configuration	井孔内廓；井身结构
hole control	井眼控制
hole coordinates	钻孔坐标
hole correction	井眼校正
hole cover	探区井网；井口盖；钻孔盖
hole coverage	钻头牙齿与井底接触面积
hole curvature	井身斜度；钻孔弯曲度；井身曲率
hole deflection	钻孔偏斜度；孔斜；钻孔弯曲；钻孔偏斜；井斜
hole density	钻孔密度
hole depth	炮孔深度；孔深
hole deviation	钻孔偏斜；孔斜；井斜
hole deviation angle	钻孔倾角；井斜角
hole deviation prevention	钻孔防斜
hole diameter	钻孔直径；孔径
hole diameter measurement	井径测量
hole dip angle	钻孔倾角
hole direction	井眼方位角
hole distance	孔距；钻孔间距
hole drag	井眼阻力
hole drift	支巷
hole drift angle	钻孔顶角；井斜角
hole drilling	井孔钻进
hole drilling machine	钻眼机
hole effect	空穴效应；孔洞效应
hole effective mass	空穴有效质量
hole enlargement	扩径

hole enlarger 扩孔器
hole erosion 井眼冲蚀
hole flushing 钻孔冲洗；洗井
hole for measuring temperature 测温孔
hole gauge 验孔规；钻孔大小；内口量具
hole inclination 钻孔倾斜度，钻孔角度
hole inflow 孔内涌水
hole instability 井壁不稳定
hole irrigation 点浇；穴灌
hole load 炮眼装药
hole loading 炮眼装药；测井
hole method 打眼法
hole micrometer 测孔千分尺
hole migration 空穴迁移
hole mobility 空穴迁移率
hole mouth 井口，钻孔口
hole opener 扩眼器
hole opening device 扩眼设备
hole overflow of water 钻孔涌水
hole pattern 布井方式；井网；钻孔布置方式
hole pitch 孔间距
hole placement 钻孔布置；布孔；炮眼的布置
hole placing 炮眼布置；炮眼排列
hole plug 井塞
hole pressure 孔压力；孔压强
hole probe 测井；电测井
hole probing 测井
hole problems 钻孔复杂情况
hole prop 横撑
hole protection 孔壁保护
hole punching 穿孔；冲孔
hole reaming 剧大钻孔；扩眼
hole resistance 钻孔阻力
hole rugosity 井壁不平度
hole scattering 钻眼偏斜
hole sealing 封孔
hole shooting 井中爆炸
hole site 孔位
hole siting 布孔；确定井位
hole size 钻孔尺寸，钻孔规格，钻孔直径；井眼尺寸
hole sloughing 井塌
hole spacing 孔距；炮眼间距
hole spotting 炮眼定点；钻孔定点
hole spread 炮眼分布；钻孔分布
hole springing 钻孔底扩大
hole stability 井眼稳定性
hole stemming 钻孔堵塞物
hole straightening 校直井眼；钻孔矫直
hole stripping 钻孔；钻进
hole structure 钻孔结构；井身结构，井身结构
hole sweep 洗井
hole tester 钻孔试验器；钻孔压水试验用栓塞
hole through 贯通，贯穿，打通；掘通
hole through on line 钻孔保持在勘探线上穿透
hole toe 井底；孔底；炮眼底；孔口，钻孔上部；井口
hole trap 空穴陷阱
hole volume 井筒容积；炮眼容积
hole walk 井眼弯曲方向
hole wall 孔壁；井壁
hole wall core sampling 孔壁取心样

hole wall stability 孔壁稳定
hole washout 井眼冲蚀
hole water level measurement 孔内水位测定
hole-bottom power drilling 井下动力钻井
hole-bottom working characteristic 井底工作特性
hole-closure rams 封井眼闸板
holed 掘通的；凿通的；穿眼的；掏槽的
hole-deviation correction 井斜校正
hole-digger 掘孔机
hole-injection 空穴注入
hole-injection rate 空穴注入速度
hole-per-bit 钻头进尺数
holes pitch 孔距
hole-size elongation 井眼变形
hole-wall collapsing 井壁坍塌
hole-wall-protecting mud 护孔壁泥浆
holiday-free gravel pack 无空穴的砾石充填
holing 钻孔；打洞；掏槽落煤法；打眼，穿孔于；钻孔打眼底部掏槽；巷道交叉贯通
holing a seam 掏槽
holing blast 联络巷道贯通爆破
holing through 掘通
holing through survey 贯通测量
holing-through 联络小巷
holing-through rotary drill 联络小巷钻机
holistic approach 整体分析
hollaite 方解霞辉脉岩
hollandite 锰钡矿
hollow 凹地，浅凹地；孔，穴；洼地；山谷；中空的，空心的
hollow abutment 空心桥台
hollow area 空心区域
hollow axle 空心轴
hollow bit 空心钻头
hollow block 空心砖；空心砌块
hollow buttress 空腹支墩
hollow casing spear 空心套管打捞矛
hollow center load cell 空心载荷传感器
hollow chamfer 凹斜面
hollow clay tile 空心黏土砖
hollow concrete 空心混凝土；大孔隙混凝土
hollow concrete block 空心混凝土砌块；空心混凝土砖
hollow concrete unit 空心混凝土砌块
hollow cone 空心圆锥体
hollow core rock bolt 空心岩栓
hollow cylinder apparatus 空心圆柱三轴试验机
hollow cylinder inner axial stress 空心圆柱体内轴向应力
hollow cylinder inner radial stress 空心圆柱体内径向应力
hollow cylinder test 空心圆筒体测试；空心圆柱体试验
hollow dam 空心坝
hollow drill 空心钻杆
hollow drill bit 空心钻头
hollow drill rod 空心钻杆
hollow drill steel 空心钻杆
hollow drive shaft 空心驱动轴
hollow flight auger 空心管螺纹钻头
hollow foundation 空心基础
hollow gravity arch dam 空腹重力拱坝
hollow gravity dam 空腹重力坝；空心重力坝

hollow lode	洞穴脉；晶洞脉
hollow masonry wall	虚圬墙
hollow pile	空心桩
hollow preformed pile	预制空心桩
hollow prop	空心支柱
hollow ridding	钻管柱
hollow rod	空心钻杆；空心抽油杆
hollow rope	空心钢索
hollow spar	空晶石；红柱石
hollow square	空心四方块体；消波混凝土块体
hollow stem	空心钻杆
hollow stem auger	空心钻杆螺旋钻；空心钻杆螺旋钻头
hollow stone	空心石
hollow tube pile foundation	管柱桩基础
hollow tube sampler	空心管取样器
hollow way	谷道
hollow weir	空心堰
hollow zone	凹陷带
hollow-bored	钻成空心的
hollowed-out	冲蚀空的
hollowness	空洞；多孔性；空心度
hollow-plunger pump	空心柱塞泵
hollow-rod churn drill	空心钻杆钢丝绳冲击式钻机；空心钻杆蹾钻
hollow-tube pile	管柱桩
holme	小岛；滩地
holmite	绿脆云母；云辉黄煌岩；不纯灰岩
holoaxial	全对称的；全轴的
holoblast	全变晶的
holoclastic	全碎屑的
holoclastic rock	全碎屑岩
holococcolith	全颗石
holocrystalline	全晶质；全结晶的；全晶质的
holocrystalline interstitial	全晶间片状的
holocrystalline porphyritic	全晶斑状的
holocrystalline rock	全晶质岩
holocrystalline texture	全晶质结构
holocrystalline-porphyritic	全晶质斑状的；似斑状
hologram	全像图；全息照相；全息照片
hologram image	全息图像
hologram photography	全息摄影
hologram view	全息图
hologrammetry	全息摄影测量
holograph	全息照相；全息摄影；手书；亲笔写的
holograph photogrammetry	全息摄影测量
holographic	全息的；全息摄影的
holographic color map	彩色全息地图
holographic filter	全息滤波器
holographic grating	全息光栅
holographic imaging	全息成像
holographic interferometer	全息干涉测量；全息照相干涉仪
holographic interferometry	全息干涉测量；全息摄影干涉测量学；全息干涉测量法
holographic memory	全息照相存储器
holographic radar	全息雷达
holographic recording	全息照相记录
holographic storage	全息存储器
holographic tomography	全息层析术
holographic virtual	全息虚像
holography	全息摄影；全息摄影学；全息照相术
holography principle	全息摄影原理
holohedral	全面象的；全对称形的；全面体的
holohedric	全对称的
holohedrism	全对称性
holohedron	全面体；全面形
holohedry	全面象；全对称
holohyaline	全玻质；全透明
holohyaline texture	全玻质结构
holokarst	发育完全岩溶；全喀斯特；高度发育喀斯特
holomafic	全镁铁质的
holomelanocratic	全黑色的
holometer	测高仪；测高计
holomictic lake	全循环湖
holomixed lake	完全混合湖
holomorphic function	全纯函数；解析函数
holoorogenic phase	全造山幕
holoseismic	全息地震的；全息地震
holoseismic method	全息地震；全像震测法；全息地震法
holostratigraphy	全息地层学
holostratotype	正层型
holostrome	全层系
holosymmetry	全对称
holotype	正模；正型；全型
Holsteinian interglacial stage	荷斯廷间冰期
holtonite	小藤石；镁硼石
holyokeite	辉绿钠长岩；钠长辉绿岩
holystone	软砂岩；浮石；磨石；用磨石磨
Homalographic projection	等积投影；古德投影
homate	低矮火山；火山碎屑丘
home	停击阻力；后退式的朝井底方向内部的；内地的；国内的
home waters	国内水域
home-driven pile	打到尽头的桩
homeland rock	原地岩石
homeoblastic	等粒变晶状的
homeocrystalline	等粒的；等晶粒的；等粒晶质的
homeogranular	等粒的
homeomorph	异物同形；异种同态；异质同形物；异质同形晶体
homeomorphic	异质同形的
homeomorphism	异质同形现象；同胚；异种同态；异质同晶
homeomorphous	异质同形的
homeopolar compound	无极化合物
homeostat	同态调节器；动态平衡系统
homeostatic model	同态调节模型
homeotheric map	组合地图
homestead	小岛
homilite	硅硼钙铁矿
homoaxial	同轴向的；平行轴的
homoaxial fold	同轴褶皱
homoaxial folding	同轴褶皱
homocentric	同中心的
homochronous	同时的
homoclinal	同斜层的；单斜的，同斜的
homoclinal dip	同斜倾斜
homoclinal dipping	同斜倾斜
homoclinal mountain	单斜山

homoclinal ramp facies	等斜缓坡相
homoclinal ridge	单斜脊，同斜脊
homoclinal river	单斜河
homoclinal shifting	同斜迁移；同斜变位
homoclinal structure	均斜构造；同斜构造
homoclinal valley	单斜谷
homocline	均斜；同斜层；单斜层
homocoagulation	同凝结；同粒凝结（作用）
homoeoblastic	等粒变晶状的
homoeoblastic structure	均质构造；等粒变晶结构
homoeocrystalline texture	等晶粒结构
homoeogene enclave	同源包体
homogeneity	均匀性；同质性，均质性；同种；同质性试验；均一性；齐性
homogeneity coefficient	均质系数
homogeneity index	均匀度指数
homogeneity nucleation	均匀成核作用
homogeneity test	均匀性检验
homogeneization	均化作用
homogeneous	均质的，均一的，各向同性；均匀的，单向的，均相的；同类的；同质的；齐次的
homogeneous accretion models	均匀聚积模型
homogeneous alloy	均质合金；单相合金
homogeneous anisotropic	均质各向异性的
homogeneous anisotropy	均质各向异性；均匀各向异性
homogeneous batholith	均匀基岩
homogeneous bedding	均匀层理
homogeneous body	均质体
homogeneous boundary condition	齐次边界条件
homogeneous carbonate	均质碳酸盐岩
homogeneous chemical reaction	均匀化学反应
homogeneous coefficient	齐次系数
homogeneous compaction of fill slope	填土斜坡的均等压实度
homogeneous concrete	均质混凝土
homogeneous crystallization	均匀结晶
homogeneous dam	均质坝
homogeneous deformation	均匀变形
homogeneous differential equation	齐次微分方程
homogeneous dike	均匀岩墙；均匀岩脉
homogeneous earth dam	均质土坝
homogeneous earth fill dam	均质坝
homogeneous elastic solid	均质弹性体
homogeneous elastomer	均质弹性体；均质高弹体
homogeneous embankment dam	均质填筑坝；均质土坝
homogeneous equation	齐次方程
homogeneous equilibrium	均匀平衡；单相平衡
homogeneous flow	均质流
homogeneous fluid	均质流体
homogeneous formation	均质地层
homogeneous foundation	均质地基
homogeneous fractionation	均匀分离；均匀分化
homogeneous function	齐次函数
homogeneous half-space	均匀的半空间
homogeneous invariant	齐次不变式
homogeneous isotropic	均质各向同性的
homogeneous isotropic elastic body	均质各向同性弹性体
homogeneous isotropic medium	均质各向同性介质
homogeneous isotropy	均质各向同性
homogeneous layer	均质层
homogeneous linear differential equation	齐次线性微分方程
homogeneous material	均质材料
homogeneous medium	均匀介质；均匀媒质
homogeneous mixture	均匀混合物
homogeneous neutron transport equation	齐次中子迁移方程
homogeneous pay zone	均质产层
homogeneous permeable bed	均质渗透层
homogeneous phase	均相
homogeneous plasma	均匀等离子体
homogeneous polynomial	齐次多项式
homogeneous porous medium	均质多孔介质；均匀多孔介质
homogeneous precipitation	均匀沉淀
homogeneous reaction	均相反应
homogeneous reservoir	均质储层
homogeneous rock	均匀岩石；均质岩石
homogeneous soil	均质土
homogeneous solution polymerization	均相溶液聚合
homogeneous state of strain	均匀应变状态
homogeneous state of stress	均匀应力状态
homogeneous strain	均匀应变
homogeneous stratum	均质层；均质地层
homogeneous stress	均匀应力
homogeneous stress field	均匀应力场
homogeneous structure	均一构造
homogeneous substance	均质物质
homogeneous system	均匀系
homogeneous texture	均匀结构；均质结构
homogeneous transformation	齐次变换
homogeneous turbid medium	均匀浊积介质
homogeneous volcano	均火山
homogeneous wave equation	齐次波动方程
homogeneous wettability	均匀润湿
homogeneous Young's modulus	均质杨氏模量
homogeneous zones	均质区
homogenetic	均匀的；均质的
homogenic migmatite	均质混合岩
homogenisation	均化
homogenization	均质化；均一化，均一化作用；等质化；均化；概化方法
homogenization method	均一法
homogenization temperature	均一化温度
homogenization transient pressure	均一瞬间压力
homogenize	搅匀；均化
homogenized	均匀化的
homogenizing	均化
homogenizing annealing	均质化退火
homogenous material	单一材料
homogenous slope	均质边坡
homoiolithic	同源岩屑的
homoionic soil materials	同离子土
homoiosmotic	同渗透的；等渗压的
homoisothermal	等温的
homojunction	同质结
homolog	对等质；同系物
homologous	相应的；类似的；同系的；同源的
homologous hypothesis	同源假说
homologous pair	对应线对

homologous series 同源系列；同系列
homolographic projection 等面积投影
homologue 同系物；同源包体；同源生物，相同；相似物；同源，同素，同调；透射
homolosine map projection 等积地图投影
homolosine projection 等积投影
homomorphic deconvolution 同态反褶积
homomorphic filter 同态滤波器
homomorphic function 同态函数
homomorphic inverse filtering 同态反滤波
homomorphic mapping 同态映射
homomorphism 异质同形；同态
homomorphism effect 异质同晶效应
homomorphosis 同形性
homomorphs 异质同形
homomorphy 同形；同系
homonym 异物同名
homopause 均质层顶；均匀层顶；匀气层顶层
homophaneous structure 均质构造
homophase 同相
homoplasy 相似；同形
homoplasy-non 不相似
homopolar crystal 共价结晶；同极晶体
homopycnal flow 等密度流
homoscedastic 同方差的
homoscedasticity 同方差性
homoseism 同震线；共震线
homoseismal line 同震线；共震线
homosphere 均匀层；均质层
homostasis 同态；物理参数
homotactic 排列类似的
homotaxial 等列的；等列性
homotaxic arrangement 相似排列；一次调节
homotaxis 排列类似
homotaxy 排列相似；层序类似性
homothermal 同温的；等温的
homothermal condition 同温条件
homothermal water 等温水；恒温水
homothetic centre 相似中心
homothetic fault 同位断层
homotypic 同型的
homotypic synonym 同模异名
homotypism 似型；同型
hondrometer 粒度计
hone 磨刀石；极细砂岩；未夯实混凝土；磨
honessite 铁镍矾
honestone 均密砂岩；磨刀石；极细砂岩磨石
honeycomb 蜂窝，蜂窝状的，麻面；蜂窝；蜂窝状物；蜂窝结构；整流格，蜂窝器；整流器
honeycomb check 蜂窝裂缝
honeycomb coil 蜂巢线圈
honeycomb crack 蜂窝状裂缝；龟裂；网状裂缝
honeycomb dune 蜂窝状沙丘
honeycomb method 蜂房式开采法；短柱式开采法
honeycomb structure 蜂窝状结构；蜂窝状构造
honeycomb system 蜂窝式开采法；短柱式开采法
honeycomb texture 蜂窝结构
honeycomb weathering 蜂窝状风化
honeycombed 蜂窝状的；多孔的
honeycombed catalyst 蜂窝状催化剂

honeycombed cement 多孔水泥
honeycombed check 蜂窝状裂纹
honeycombed concrete 蜂窝混凝土；"黄蜂窦"
honeycombed dune 蜂窝状沙丘
honeycombed grain 多孔隙颗粒
honeycombed rock 蜂窝岩
honeycombed wall 蜂窝式墙
honeycombing 短柱开采法；形成蜂窝状；蜂窝
honeystone 蜜蜡石
Hong Kong 1980 Geodetic Datum 香港1980大地基准
Hong Kong 1980 Grid 香港1980方格网
Hong Kong 80 Geodetic Datum 1980年香港大地测量基准
Hong Kong Principal Datum 香港主水平基准
Hong Kong Principal Datum HKPD 香港高程基准面
Hongchunping Limestone 洪椿坪石灰岩
Honigmann drop-shaft method 霍尼曼沉井法
Honigmann method 霍尼曼钻井法
hoo cannel 土状烛煤；混有黏土的烛煤
hood pile cap 套式桩帽
hooded dry drilling 加罩干式钻眼；加罩干式凿岩
hoodoo 石林；怪岩柱；峰林；侵蚀柱
hoodoo rock 石林地形
hooibergite 闪正辉长岩
hook 钩，沙钩；弯钩，挂钩；河湾，钩形沙嘴；钩状物；线路中断联播；箍圈
hook air-door 斜井挂钩工；地面挂钩工
hook and eye 风钩
hook bac 钩形沙嘴
hook faulting 钩状断裂作用
hook fold 钩状褶皱
hook for jaw-crusher feeding 颚式破碎机给料钩
hook gauge 钩尺，钩式量尺；钩式水位计；钩形测针；钩式精密液位计；弯管压力计
hook groin 钩状丁坝
hook groyne 钩形丁坝
hook height 大钩悬挂高度；大钩高度
hook load 大钩负荷
hook valley 逆向支流
hook wall flooding packer 注水井悬挂式封隔器
hook wall packer 悬挂式井壁封隔器；挂壁式封隔器
hook wall pumping packer 抽油井悬挂式封隔器
hook wall testing 悬挂式测试
hook weigh 大钩悬重
hook wrench 钩形扳手；钩头扳手
hooke joint 万向接头；万向联轴节
Hookean body 虎克体；虎克模型
Hookean effect 虎克效应
Hookean elastic body 虎克弹性体
Hookean elasticity 虎克弹性
Hookean model 虎克模型
Hookean solid 虎克体；虎克刚体；虎克固体
hooked bar 带钩钢筋；有钩的钢筋条；钩筋；钩型钢筋
hooked bay 钩状湾
hooked dune 钩状沙丘
hooked nail 钩头钉
hooked spit 钩状沙嘴
hooked-groin contraction works 钩形丁坝束水工程
Hooke's law 胡克定律；虎克定律
hook-like structure 似钩状构造

hook-shaped structure 钩曲构造
hoop load 环向荷载
hoop pressure stress 周压应力
hoop reinforcement 箍筋；环状加固
hoop reinforcement stirrup 箍筋
hoop strain 环应变；环变形
hoop stress 周应力；环向应力
hoop tension 箍拉力；环向拉力
hoop type heater 环形管加热炉
hooped bar 环筋；箍筋
hooped column 环柱
hooped concrete 环筋混凝土；配箍筋混凝土
hooped penstock banded penstock 加箍式压力水管
hooped pile 配箍筋桩
hooped reinforcement 环框锚杆；网状钢筋；环箍钢筋；环筋
hooping 箍筋；环筋；螺旋钢箍；环筋；加箍筋
hoop-iron 箍铁；箍钢
hoop-stress loading 周应力加载
Hoover's flotation 胡佛式浮选法
hoove-up 隆起
hope 小盲谷；沼泽中干地；封闭谷；盲谷
hopeite 磷锌矿
Hopkinson effect 霍普金森效应
hopper 漏斗；给料斗；入料斗；底开式泥舱；贮料斗；仓，有效空间，泥舱，斗舱；底卸式车，漏斗车
hopper of recovery 开采漏斗
hopper structure 漏斗构造
horadiam 水平辐射钻孔
horary angle 时角
horicycle 极限圆
horizon 层位，范围，地层；地平线，水平线，视界，地平圈，母质层；地平仪，水平仪；水平，阶段；垂直于铅垂线的平面；反射界面
horizon A A 层；淋溶层
horizon B B 层；淀积层
horizon C C 层；母质层
horizon closure 全圆闭合差；水平闭合差
horizon D D 层；基岩层
horizon flattening 层位拉平
horizon glass 地平镜；水平玻璃观察孔
horizon interval 阶段垂高；水平垂高
horizon line 地平线
horizon lowering 水位降低
horizon marker 标准层位；指示层；标志层
horizon mining 多水平采矿；多水平采矿法；卧式；阶段开采
horizon mining system 多水平开采系统
horizon of soil 土层
horizon picking 层位拾取
horizon plain 地平面
horizon refraction 水平折射
horizon sample 水平试样；阶段试样
horizon seiscrop section 层位振幅水平切片；水平层位切片剖面
horizon smoothing 地层平滑
horizon system of coordinates 地平坐标系
horizon trace 像地平线
horizon track 层位踪迹
horizon velocity analysis 层速度分析

horizon velocity seiscrop section 层位速度水平切片
horizontal fault 水平断层
horizontal joint 水平节理
horizontal pendulum tiltmeter 水平振子倾斜计
horizonntlly polarized S waves 水平偏振横波
horizon-slice map 水平切面图
horizontal 水平，水平位置；横向的；横平的；卧式的；地平线
horizontal acceleration 水平加速度
horizontal accelerometer 水平加速度计
horizontal accuracy 平面位置精度
horizontal adit 横通道；横洞
horizontal adjustment 水平校正
horizontal agreement 横向协议
horizontal alignment 平面定线；水平校准；水平线向；平面线形；水平线路；成水平直线；水平校准
Horizontal and inclined structure 水平构造和单斜构造
horizontal and vertical reference points 平面及高程参考点
horizontal angle 水平角；平面角
horizontal angle of inclination 水平角
Horizontal angle of the cone of vision 水平视锥角
horizontal aquifer 水平含水层
horizontal arch element 水平拱单元；拱环
horizontal area 水平面积
horizontal auger 水平螺钻；卧式土钻机
horizontal axis 水平轴；平面轴
horizontal bar 水平杆；横梁；横杆；单杠；水平筋
horizontal bedding 水平层理
horizontal belt filter 水平带式过滤机
horizontal blasting 水平爆破
horizontal bolt retrieval 螺栓回复器
horizontal bore hole drain 水平钻孔排水
horizontal borehole 水平钻孔
horizontal borehole drain 水平钻孔排水
horizontal borehole pressure recorder 水平钻孔压力测定器
horizontal borehole profile 水平钻孔断面
horizontal borer 水平钻孔钻机
horizontal boring machine 水平钻探机
horizontal boundary 水平界面；水平边界
horizontal brace 水平横撑；平台横支架
horizontal breadth 水平厚度
horizontal bridging 水平格栅撑；水平支杆
horizontal centrifuge 卧式离心机；水平离心机
horizontal change 横向变化
horizontal circle 水平度盘，水平分度盘；地平圈，水平圆
horizontal cleat 平割理
horizontal closure error 平面闭合差
horizontal coherence 水平相干性
horizontal coil system 水平线圈系统
horizontal comparator 水平比较器
horizontal component 水平部分；水平分量；水平分力；水平线圈
horizontal component of force 力的水平分量
horizontal component of the earth's magnetic field 地磁水平分力
horizontal component seismograph 水平向地震仪；水平地震仪；水平动跳计
horizontal compression 横向压缩
horizontal compressional effect 水平压缩效应

horizontal compressional stress	水平压缩应力
horizontal confined flow	水平承压水流；水平承压流
horizontal contraction	水平收敛
horizontal control	平面控制；水平控制
horizontal control network	平面控制网；水平控制网
horizontal control point	平面控制点
horizontal control station	平面控制点
horizontal control survey	平面控制测量
horizontal control survey network	平面控制测量网
horizontal convergency measurement	水平收敛位移量测
horizontal coordinate	平面坐标；水平坐标；横坐标
horizontal correction	水平校正
horizontal couple	水平力偶
horizontal cross section	水平断面；水平剖面
horizontal crustal deformation measurement	水平地壳形变测量
horizontal current	水平流
horizontal curve	平面曲线；横向曲线；水平曲线
horizontal cut-and fill stoping	水平分层充填采矿法
horizontal cylinder	水平圆柱
horizontal damped	水平阻尼的
horizontal damper	水平减震器
horizontal datum	平面基准
horizontal datum plane	水平基准面
horizontal definition	水平清晰度
horizontal deflecting electrode	水平偏转电极
horizontal deflection	水平挠度
horizontal departure	水平偏距；水平位移
horizontal detachment	水平挤离
horizontal detector	水平检波器
horizontal deviation	水平位移
horizontal dial	水平度盘
horizontal diameter	横向直径
horizontal diffuser	水平扩散器
horizontal diffusion	水平扩散；水平尺寸
horizontal dip slip	水平倾向滑动；水平滑动
horizontal directional drilling（Hdd）	水平定向钻进
horizontal directive tendency	横向度；水平方向性
horizontal discharge	水平排泄
horizontal disk crusher	自动化电站
horizontal dislocation	水平位错；水平断错
horizontal dispersion	水平色散
horizontal displacement	水平位移；水平变位
horizontal displacement measurement	水平位移测量
horizontal distance	平距；水平距离；水平间距
horizontal distribution	水平分布
horizontal ditch	水平沟
horizontal divergence	水平散度
horizontal drain	水平排水
horizontal drain hole	水平井；水平排泄孔
horizontal drainage	水平排水
horizontal drainage channel	水平排水孔
horizontal drilling	水平钻井；水平钻探
horizontal drilling machine	卧式钻床
horizontal drilling process	水平钻孔法
horizontal drilling rig	水平钻机；卧式钻机
horizontal drive	平巷掘进
horizontal earth movement	地壳水平运动
horizontal earth pressure	水平土压力
horizontal earth pressure stress	水平土压应力
horizontal earthboring machine	横向钻孔机
horizontal equalization	水平均衡
horizontal equivalent	水平等量值；水平距离等量值
horizontal exaggeration	水平放大
horizontal expansion	横向延伸
horizontal extent	水平范围；水平引伸
horizontal extruder	卧式挤压机
horizontal fault	水平断层；平移断层
horizontal field balance	水平磁力仪；水平分量的磁力仪；水平磁强计
horizontal filter	水平反滤层；卧盘式真空过滤机
horizontal filter well	水平滤井；导滤孔
horizontal flood	水平注水
horizontal flow	水平流；水平流动
horizontal fluid layer	水平流层
horizontal fold	水平褶皱
horizontal foliation	水平叶理
horizontal force	水平力
horizontal force of roof	顶板水平作用力
horizontal form index	水平形态指数
horizontal gallery	水平集水廊道，平巷；水平通道
horizontal geologic profile	水平地质断面图
horizontal geophone	水平检波器
horizontal gradient	水平坡度，水平梯度
horizontal gradient of gravity	重力水平梯度
horizontal gravity gradient	水平重力梯度
horizontal ground heat exchanger	水平地埋管换热器；水平地热交换器
horizontal ground motion	水平向地面运动
horizontal ground stress	水平地应力
horizontal grout blanket	水平灌浆铺盖
horizontal guide face	水平导面
horizontal heave force	水平冻胀力
horizontal hole	水平炮眼，水平钻孔
horizontal hole drainage	平孔排水
horizontal hole drilling	水平孔钻进
horizontal homogeneity	横向均质性
horizontal hydraulic fractures	水平水力压裂缝
horizontal inclined fold	水平不对称褶皱；水平倾斜褶皱
horizontal inclinometer	水平测斜仪
horizontal inertia bar	水平惯性摆
horizontal instability	水平不稳定
horizontal intensity	水平强度
horizontal interface	水平分界面
horizontal jet grouting	水平喷射灌浆；平喷灌浆
horizontal joint	水平节理；水平接缝
horizontal joint system	水平节理系
horizontal lamination	水平层理；水平叠合
horizontal lateral stresses	水平侧向应力
horizontal layer	水平层
horizontal line	水平线
horizontal load	水平荷载
horizontal load carrying member	水平受力构件
horizontal load test	水平荷载试验
horizontal loop method	水平回线法；水平线圈电磁法；水平回路法
horizontal magnetic anomaly	水平磁异常
horizontal magnetic intensity	水平磁强
horizontal magnetization	水平磁化
horizontal magnetometer	水平地磁仪；水平磁强仪

horizontal mantle	水平覆盖层
horizontal map of heights	高程平面图；等高线图
Horizontal marking	横向标线
horizontal matrix permeability	基质的水平渗透率
horizontal matrix stress	基岩水平应力
horizontal member	水平构件
horizontal mixing	水平混合；水平混波
horizontal monitoring control network	平面监测网
horizontal motion	水平运动
horizontal motion seismograph	水平地震仪
horizontal movement	水平位移；水平运动；横向移动
horizontal movement gauge	水平位移计
horizontal normal fold	水平正常褶皱
horizontal normal offset	水平正错距
horizontal offset	水平偏离距；水平偏移距
horizontal opening	平洞
horizontal overlap	水平掩覆
horizontal peak acceleration	水平加速度峰值
horizontal pendulum	水平摆
horizontal permeability	水平向渗透率；横向渗透；水平渗透性
horizontal permeability barrier	横向阻渗层
horizontal pipe rack	水平排管架
horizontal plan	平面图，水平投影
horizontal plan drawing	平面图则
horizontal plane	水平面；地平面
horizontal plate	水平板；水平反射板
horizontal plate gauge	板式水平位移计
horizontal point of tunnel portal	隧道洞口投点
horizontal position	水平位置
horizontal pressure	水平压力
horizontal profile	水平钻孔断面
horizontal profiling	水平剖面
horizontal projection	水平投影；示准面
horizontal projection plane	水平投影面
horizontal pull	水平拉力
horizontal pump	卧式泵；沿岸洲
horizontal pushing force	水平推力
horizontal radial motion geophone	水平径向运动检波器
horizontal range	水平距离；广度
horizontal reference plane	基准水平面
horizontal refraction	水平折光；地平折射
horizontal refraction error	水平折光差；旁折光差
Horizontal refractor	水平折射面
horizontal resolution	水平分辨能力；水平分辨率
horizontal response	水平响应
horizontal restrain	横向约束
horizontal ring	水平炮眼组圈
horizontal road alignment	道路水平线形
horizontal roof	水平顶板
horizontal rotary cylinders carbonizer	水平回转筒式焦化器
horizontal scale	水平比例尺；横坐标；水平尺度
horizontal scanning	水平扫描
horizontal screen	平筛
horizontal seam	水平煤层，水平矿层，水平岩层，水平层；水平焊缝
horizontal section	水平截面；水平切面；水平剖面；水平段；水平断错
horizontal section method	水平断面法
horizontal seismic coefficient	水平地震系数
horizontal seismic section	水平地震剖面
horizontal seismograms	水平地震动图
horizontal seismograph	水平地震仪
horizontal separation	水平离距；水平继距
horizontal separator	卧式分离器
horizontal shaft	水平轴
horizontal shear machine	水平剪力仪
horizontal shear method	水平剪切法
horizontal shear stress	水平剪应力
horizontal shift	水平位移
horizontal shore	横撑；水平撑；水平支承架；跨间支撑；水平支撑梁
horizontal shoring	水平支撑
horizontal skew	水平偏斜
horizontal slice	水平分层
horizontal slicing	水平分层开采法；水平分层采煤法
horizontal sliding	水平滑移
horizontal sliding gate	横向滑动闸
horizontal slip	水平滑动
horizontal slip fault	水平滑动断层
horizontal slope	平坡
horizontal slowness	水平慢度
horizontal split-case centrifugal pump	水平中开式离心泵
horizontal stacking	水平迭加；水平叠加
horizontal strain	水平应变
horizontal strata-bound mineral deposit	深藏水平矿床
horizontal stratification	水平分层；水平层理
horizontal stratum	水平岩层
horizontal stress	水平应力
horizontal structure	水平构造
horizontal subgrade reaction	水平基床反力
horizontal surface	水平表面
horizontal survey	水平测量
horizontal sway	水平侧倾
horizontal system of coordinates	地平坐标系
horizontal tectonics	水平构造
horizontal temperature gradient	水平地温梯度
horizontal tension	横张力
horizontal thermal gradient	水平热梯度
horizontal throw	水平距离；平错
horizontal thrust	水平推力
horizontal thrust fault	平冲断层
horizontal thrust plane	水平冲断面
horizontal time scale	横向时间比例尺
horizontal time slice	水平时间切片
horizontal tipping method	平堵
horizontal topslicing	下行水平分层采矿法
horizontal transmit horizontal return（HH）	水平发射水平回波
Horizontal transmit vertical return（HV）	水平发射垂直回波
horizontal travel	水平行程
horizontal turbine	游荡河槽
horizontal unsaturated flow	非饱和水平流
horizontal variometer	水平磁力变感器；水平磁变记录仪；水平磁秤
horizontal velocity	水平速度；水平流速
horizontal velocity distribution	水平速度分布；水平流速分布

horizontal vibrating centrifuge	卧式振动离心机
horizontal vibrating screen	水平振动筛
horizontal vibrating screening centrifuge	卧式振动离心脱水机
horizontal vibration	水平振动
horizontal vibrator	水平振动器
horizontal viewing angle	水平视角
horizontal visibility	水平能见度
horizontal visual angle	水平视角
horizontal water-collecting layout	水平集水布置
horizontal waterflood	水平注水
horizontal wave	水平波
horizontal well	水平井
horizontal wellbore logging	水平井测井
horizontal workings	平巷
horizontal zonality	水平分带；水平地带性
horizontal zoning	水平分带
horizontal-dipole sounding	水平偶极测深
horizontale	水平轴
horizontality	水平度；水平状态
horizontalization	置平
horizontally bedded sandstone	水平层状砂岩
horizontally fractured well	水平裂缝井
horizontally layered elastic medium	水平层状弹性介质
horizontally layered medium	水平层状介质
horizontally loaded piles	承受水平面荷载的桩
horizontally stratified sandstone	水平成层砂岩
horizontally-laminated bed	水平纹层
horizontally-projected area	水平投影面积
horizontal-pitman crusher	水平摇杆破碎机
horizontal-projected area	水平投影范围
horizontal-push landslide	平推式滑坡
horizontal-radial diamond drilling	水平辐射状金刚石钻孔
horizontal-radial drilling	水平辐射状钻进
horizontal-shaft arrangement	横轴式布置
horizontal-shaft disk crusher	平轴盘式碎矿机
horizontal-type pump	卧式泵
horizon-to-horizon camera	三镜头航空摄影机
hormannsite	碳酸辉石二长岩
horn	悬臂，角；角峰主解理与采煤工作面成45°的线；牛角制验金淘洗器；角喇叭操纵杆；导绳棱；喷枪；喇叭
horn lead	角铅矿
hornberg	角页岩；角山
hornblende	角闪石；角闪岩
hornblende andesite	角闪安山岩
hornblende diorite	角闪闪长岩
hornblende gabbro	角闪辉长岩
hornblende hornfels facies	角闪石角岩相
hornblende hornstone	角闪石
hornblende peridotite	角闪橄榄岩
hornblende pyroxenite	角闪辉石岩
hornblende rock	普通角闪石岩
hornblende schist	角闪片岩
hornblende-porphyry	角闪斑岩
hornblende-schist facies	角闪石片岩相
hornblendic schist	角闪片岩
hornblendite	角闪石岩
Horner analysis	赫诺分析
Horner extrapolation method	赫诺外推法
Horner slope	赫诺斜率
Horner straight line	赫诺曲线的直线段
Horner time	赫诺时间
Horner type-curve	赫诺样板曲线
hornesite	砷镁石
hornfels	角页岩；角岩
hornfels facies	角页岩相
hornfels texture	角岩结构；角页岩构造
hornfelsization	角闪岩化
hornito	溶岩滴丘
hornstone	角岩；角状岩；燧石
horny	劣质气煤；角制的；角质的；角的
horny layer	角质层
horny scale	角质鳞
horocycle	极限圆
horosphere	极限球面
horse	夹层，夹岩，夹块；台架；底板隆起；夹石顶板或底板凸出处断层壁间巨石块；矿体中不规则的掘进井筒中吊座；有脚架马
horse back	马脊岭；底板隆起；煤层冲蚀层
horse cock	钻头；泥浆软管
horse dam	临时坝
horse head	前探双纵梁
horse of barren rock	脉石夹层
horse saddle	马鞍石；马厩
horseback	底板隆起；马脊岭；顶板或底板凸出处脉石夹层；扁豆形板岩
horses	断片
horseshoe	马蹄形；马蹄铁
horse-shoe arch	马蹄形拱；马蹄券
horseshoe atoll	马蹄形环礁
horseshoe bend	马蹄形弯道
horseshoe conduit	马蹄形排水管
horse-shoe curve	马蹄形弯；马蹄形曲线
horseshoe lake	弓形湖
horse-shoe mining method	马蹄形开采法
Horseshoe Reef	马蹄形礁
horseshoe shape	马蹄形
horseshoe shaped betwixt land	马蹄形盾地
horseshoe tunnel	马蹄形隧洞
horseshoed section	马蹄形断面
horseshoe-shaped	马蹄形的
horseshoe-shaped shield	马蹄形盾构
horseshoe-shaped tunnel	马蹄形隧道
horsestone	夹石
horsetail fault	马尾状断层
horsetail fissure	马尾状裂隙；马尾状破裂；马尾构造
horsetail structure	马尾丝状构造；马尾构造
horsetailing	马尾形裂楔效应
horsfordite	锑铜矿
horsing	临时挑出支撑
horst	地垒；悬垂体
horst block	地垒断块
horst block mountain	地垒式断块山
horst fault	地垒断层；地垒状断层
horst mountain	地垒山；垒断山
horst ridge	山脊；脊梁
horst structure	地垒构造
horst-anticline	地垒背斜
horticulture planning	细部结构

hortite	方解辉长混染岩
horton sphere	哈通球形储罐
hortonolite	镁铁橄榄石
hortonolitite	镁锰铁橄榄岩
hose	软管，软管体，软喉，水龙带；钩环，绳钩环；制动软管
hose adapter	软管接头
Hose Clips	喉码
hose leveling instrument	软管水准台；软管水准仪；软管水准器；软管水平尺
hose pump	软管式泵
hose reel	消防喉辘；软喉辘；消防水喉；软管卷筒
hose tube	软管
hose wrench	软管扳手
hose-type detector	软管式检波器
Hosi tectonic system	河西构造体系
host	主体；基质材料；主岩；主矿物；基质；容矿岩
host crystal	主体晶；寄主晶
host formation	宿主地层；主体地层
host layer	主层
host lithology	主岩岩性
host material	基质材料
host mineral	主矿物；原生矿物；主矿物
host rock	主岩，围岩；母岩；容矿岩；主体岩石
host series	主岩系
host structure	容矿构造
host volcanic liquid	主火山熔体
hostile environment logging	恶劣环境测井
hostile water	险恶水域
host-rock	主岩；围岩
hot acid treatment	热酸处理
hot and cold steeping	冷热浸渍
hot and pressurized state	热备用状态；高温高压状态
hot area	热区
hot bearing alarm	热轴承警报器
hot brine field	热海水地；热盐水场
hot brine in the Red Sea	红海热卤
hot brine metallogenesis	热卤水成矿作用；热卤水成矿酌
hot briquetting	热成型
hot brittleness	热脆性
hot cement	热水泥
hot channel	热管；热通道
hot compressive strength	热抗压强度
hot corrosion	热腐蚀
hot creek	热水溪
hot cross-section	热截面
hot desert	热沙漠区
hot drawn	热拉
hot drilling water	加热钻孔冲洗水
hot dry rock	干热岩
hot dry rock cranny	干热岩裂缝
hot dry rock formation	干热岩体；干热岩地层
Hot dry rock potential	干热岩潜力
hot extrusion pressing	热挤压
hot face	受热面；加热面
hot fluid	热流体
hot fractured rock	热破裂岩体
hot gas	热气
hot gaseous spring	含气热泉
hot geothermal brine	地下热盐水
hot ground	放热地；放热地面
hot groundwater	地下热水；热地下水
hot hardness	热硬性
hot hole	高温井
hot hydrothermal brine	热卤水
hot interwell communication	井间热传递通道
hot karst spring	热岩溶泉；岩溶温泉；热喀斯特泉
hot lahar	热泥流
hot lake	热水湖
hot lineation	热异常构造线；热泉集中出露构造线
hot liner completion	热衬管完井法
hot load test	热荷载试验
hot marsh	热水沼泽
hot metal	液态金属；熔融金属
hot metalliferous brine	多金属热卤水；含矿热卤水；热矿水
hot mine	高温矿井；瓦斯矿
hot mix asphalt overlay	热拌沥青混凝土加铺
hot mix asphalt（HMA）	热拌沥青混凝土
hot mud	热泥；热泥塘
hot mud flow	热泥流
hot peening	热珠击法
hot penetration method	热灌法
hot performance test（HPT）	热态性能试验
hot plume	热焰；热地幔柱
hot point	热点
hot point test	热针法
hot pool	热水塘；热水储；下部地层
hot pressurized well	高温耐压井
hot reflux	热回流
hot reservoir	热储；热储层；热库
hot reservoir parameters	高温热储参数
hot resistance	耐热性；热电阻
hot river	热水河
hot saline reservoir	热咸水库；热盐湖
hot sand avalanche	热沙崩
hot sand flow	热沙流
hot seepage	热水渗出点；热水渗眼；热渗流
hot shale	富含有机物的页岩
hot shortness	热脆性
hot shot	腐蚀点；热点；高速飞机
hot shutdown	热停堆
hot sink	热阱
hot spot	热点；热柱
hot spot corrosion	点腐蚀
hot spot hypothesis	热点说
hot spot reference frame	热点参照系
hot spray coating	热喷涂涂层
hot spray painting	热喷涂
hot spraying	热喷涂
hot spring	温泉；热泉；下垫面地形
hot spring area	温泉区；热泉区
hot springs	热泉群；热泉区
hot stability	热稳性
hot stability of bitumen	沥青热稳性
hot stream	热水溪；热水河
hot strength	拉伸强度；热强度
hot tensile strength	热态拉伸强度
hot tunnel climatic test	热隧道气候试验

hot volcano 热火山
hot waste 强放射性废物
hot water bank 热水带
hot water circulation 热水循环
hot water circulation pump 热水循环泵
hot water circulation system 热水循环系统
hot water corrosion 热水溶蚀
hot water deposit 热水沉积；热水矿床
hot water field 热水田
hot water metamorphism 热水变质作用
hot water process 热水溶解法
hot water well 热水井
hot wave 热浪
hot well 热水井
hot wet rock 湿热岩
hot zone 热区；高温带
Hotchkiss super-dip 赫奇基斯超倾磁力仪
hot-dip galvanizing 热浸镀锌
hot-laid asphaltic concrete 热铺沥青混凝土
hotness 热度
hot-short 不耐热的
hot-spot plume 热点地幔柱
hot-spring apron 泉华裙；泉华台；热泉台
hot-spring bowl 热泉塘；热水塘
hot-spring chamber 热泉塘；热泉室
hot-spring complex 热泉群
hot-spring effluent 热泉水
hot-spring sediment 热泉沉积物；泉华
hot-spring silica 热泉氧化硅；硅华
hot-spring species 热泉种类
hot-spring terrace 泉华阶地；泉华台地
hot-spring vapor 热泉汽雾；热泉上方水雾
hot-spring water 温泉水；热泉水
hot-stretched cord 热拉伸合成帘子线
hot-tank process 热桶处理法
hot-tapping 热分接
hot-water aquifer 热水含水层
hot-water cup 热水洼
hot-water displacement 热水驱替
hot-water geothermal system 热水地热系统
hot-water growths 热水生物
hot-water manifestation 热水显示；热泉
hot-water occurrence 热泉出现；热泉显现
hot-water rate 热水耗率；热水流量
hot-water reservoir 热水储；热水储集层
hot-water spring 热泉
hot-water supply 热水供应；热水补给
hot-wire ammeter 热线电流表
hot-wire anemometer 热线风速计
hot-wire detector 热丝检测器；热线地震仪
hot-wire instrument 热线仪表
hot-wire pressure gauge 热线压力计
hot-wire voltmeter 热线式电压表
hot-work 热加工；高温加工
hotworking polygonization 热加工多边形化
hot-zone 热区
Hough function 霍夫函数
hour angle 时角
hour circle 时圈
hour glass structure 砂钟构造
hour-angle coordinate system 时角坐标系
hourglass mode 沙漏模式
hourly load 时负荷
hourly observation 逐时观测；物理环境
housed joint 套入接合；封装接头
housed pump 箱式泵
hover ground 松散地基；不坚固土壤；松软地；松软土壤
hovlandite 橄云二长辉长岩
howardite 钙长紫苏辉石无球粒陨石
howe 洞；孔
howieite 硅铁锰钠石
howlite 硅硼钙石
Hoyt gravimeter 哈脱重力仪
H-shaped antisliding pile H形抗滑桩
hsianghualite 香花石
hsihutsunite 西湖村石；镁蔷薇辉石
H-type-section H型断面
Huanghai Coastal Current 黄海沿岸流
Huanghai mean sea level 黄海平均海水面
Huanghe River Delta 黄河三角洲
huanghoite 黄河矿
Huanghua transgression 黄骅海侵
Huanghuashan Breccia 黄花山角砾岩
hub 衬套；轴，柄；中心，中枢；凸起，突口，插孔；集线器；毂套
hubnerite 钨锰矿
huckle 背斜顶部；背斜鞍
hudsonite 角闪橄榄岩
huebnerite 钨锰矿
huegelite 水钒锌铅矿
huelvite 杂锰矿
huerfano 老岩丘
huge 巨大的
huge concrete block 大型混凝土砌块
huge strike fault 大型走向断层
hugelite 钒锌铅矿
Hughes tunnel-boring machine 休斯型隧道掘进机
huhnerkobelite 钙磷铁锰矿
Huicaozi Quartzite 灰槽子石英岩
hum 孤峰；溶蚀残丘；交流声；杂音；电干扰
Hum mogote karst tower 灰岩残丘
human ecological environment 人类生态环境
human erosion 人为侵蚀
human perception vibration 人感振动
Humangdong Gneiss 虎忙岽片麻岩
humate 腐殖酸盐；腐殖酸
humatite 有机矿物
Humble formula 汉布尔公式
humboldtilite 硅黄长石
humboldtine 草酸铁矿
humboldtite 硅硼钙石；草酸铁矿
humic 腐殖的
humic alkali soil 腐殖质碱土
humic gley soil 腐殖质潜育土
humic latosol 腐殖质砖红壤
humic mud 腐殖泥
humic mulch 腐殖覆盖物
humic organic matter 腐殖有机质
humic soil 腐殖土

humic substance 腐殖质
humid 有湿气的，潮湿的
humid analysis 湿法分析
humid and arid lithogenesis 潮湿和干旱的成岩
humid area 湿润区；潮湿区
humid fan 湿地扇
humid gas 湿气
humid glaciated area 湿润冰川作用地区
humid heat test 湿热试验
humid land 湿地
humid lithogenesis 湿成岩石成因说
humid rock 湿成岩
humid soil 湿境土；腐殖土；湿润土
humid test 湿度试验
humidification 湿润；浸湿；增湿；润湿化
humidifier 喷雾机；加湿器
humidifying 增湿；使湿
humidiometer 湿度计
humidistat 恒湿器
humidity 湿度；湿气
humidity coefficient 湿度系数
humidity collector 水分收集器
humidity control 湿度控制
humidity controller 湿度控制器
humidity correction factor 湿度校正系数
humidity corrosion test of grease 润滑脂潮湿腐蚀试验
humidity deficit 湿度不足
humidity density metre 湿度密度计
humidity factor 湿度因子；湿润系数
humidity field 湿度场
humidity of rock 岩石湿度
humidity probe 湿度探头
humidity ratio 湿度比
humidity sensor 湿度传感器
humidity-controlled oven 湿度控制炉
humidity-sensitive element 湿敏元件
humidizer 增湿剂
humidostat 湿度调节仪；恒温仪
humification 腐殖质形成；腐殖化；腐殖化作用
humin 胡敏素；腐殖质
humite 硅镁石
hummerite 水钒镁矿
hummock 波状地；小丘；堆积冰；冰丘沼泽中的高地
hummock-and-hollow topography 丘洼地形
hummocked ice 冰丘；冰丘冰
hummocky 小圆丘的
hummocky clinoform reflection configuration 乱岗状斜坡反射结构
hummocky cross-bedding 丘状交错层理
hummocky surface 丘形地面
hummocky terrain 丘形地带
humod 腐殖质灰土
humoferrite 土赭石
humonigritite 沉积煤化沥青
humo-silicate 有机硅酸盐矿物
humox 腐殖质氧化土
hump 小丘，丘陵；凸起处，隆起；驼峰，峰，巅峰值；山岗；曲线上的高峰值
hump bridge 拱桥
hump crest 峰顶

hump joint 驼峰节理
humpback 弓背
humulite 腐殖岩；腐殖煤
humult 腐殖质老成土
humus 腐殖质；腐殖土
humus alkali soil 腐殖质碱土
humus analog 腐殖质类似物
humus carbonate soil 腐殖质碳酸盐土
humus carbonatic soil 腐殖质碳酸盐土
humus fen soil 腐殖质沼泽土
humus formation 腐殖质形成
humus horizon 腐殖质层
humus layer 腐殖质层
humus layer humus stratum 腐殖层
humus loam 腐殖壤土
humus podzol 腐殖质灰壤
humus sapropel 腐殖质
humus sludge 腐殖土；腐殖污泥
humus soil 腐殖土
humus tank 腐殖质沉淀池
hunch 瘤；圆形隆起物；厚片；弯成弓状
hungarite 角闪安山岩
hungry 无价值的；低质的
hungry-looking 看来无价值的
hunji rock 混积岩
huntilite 杂砷银矿；砷银矿
hunting oscillation 长周期振动；长周期纵向振动
huntite 碳酸钙镁矿；碳钙镁石
hurdled ore 粗筛矿石
hureaulite 红磷锰矿
hurl 拖运；投，掷
Hurley 矿槽；跳汰机；木制矿车
hurling pump 旋转泵
huronite 钙长石
hurst 岩溶泉；沙洲；沙岸
Hurst method 赫斯特法
hurter 防护短柱
hurumite 钾英辉正长岩
husebyite 斜霞正长岩
hussakite 磷钇矿
hutch road 井下主运输平巷
hutchinsonite 硫砷铊铅矿
huttonite 硅钍石；斜钍石
huyssenite 铁纤硼石
H-variometer 水平磁力变感器；平磁变记录仪
Hvorslev parameter 伏斯列夫参数
Hvorslev soil model 伏斯列夫模型
Hvorslev surface 伏斯列夫面
hyacinte 红锆石；钙柱石；符山石；交沸石
hyacinth 红锆石
hyaline 玻璃状的；明的；玻璃质的；透明的
hyalite 玻璃蛋白石；玉滴石
hyalobasalt 玻质玄武岩
hyaloclastite 玻屑岩
hyalocrystalline 玻晶质；玻晶质的
hyalodacite 玻质英安岩
hyaloid 玻璃状的；明的
hyalomelane 玄武斑状玻璃
hyalomicte 云英岩
hyalo-mylonite 玻璃摩棱岩

hyaloophitic	玻质辉绿岩的	hydrant	消防龙头；消火栓，水龙头
hyalo-ophitic texture	玻质辉绿结构	hydrant outlet	消防龙头出口
hyalopantellerite	玻璃碱流岩	hydrarenite	水成杂砂岩
hyalophane	钡冰长石	hydrargillite	水铅矿
hyalopilitic	玻晶交织	hydra-shock absorber	液压式减震器
hyalopilitic texture	玻晶交织结构	hydratability	水化能力
hyalopsite	黑曜岩；玻质岩	hydratable clay	水合性黏土
hyalosiderite	透铁橄榄石	hydratable shale	易水化页岩
hyalotekite	硼硅钡铅矿	hydrate	水合物；水合
hybrid	混合；混合网络；混合的；杂化的	hydrate covering stratum	水合物盖层
hybrid analysis	混合分析；杂交分析	hydrate deposit	水合物沉积
hybrid arenites	混合砂屑岩	hydrate of lime	熟石灰
hybrid bit	混合式钻头	hydrate pingo	水合物冰丘
hybrid coordinate	混合坐标	hydrate reflection	水化层反射
hybrid coring system	组合取心系统	hydrate water	水合水；结合水
hybrid curve	混合比例尺曲线	hydrated	水化的
hybrid electromagnetic wave	混合电磁波	hydrated alumina	水合氧化铝
hybrid element	杂交单元	hydrated aluminium silicate	水化钙硅酸盐
hybrid fluidized bed gasification	混合流化床气化	hydrated bentonite	水化膨润土
hybrid granite	混杂花岗岩	hydrated cement	水化水泥
hybrid gravity type platform	混合式重力平台	hydrated chlorite	水化型绿泥石
hybrid isolation system	混合隔震系统	hydrated clay	水合黏土
hybrid linear-exponential scale	线性-指数混合比例尺	hydrated ion	氢离子；水合离子；氢氧离子
hybrid measurement	混合测量	hydrated lime	熟石灰；水化石灰；消石灰
hybrid migration	混合偏移	hydrated lime stabilisation	熟石灰加固法
hybrid ranging	混合测距	hydrated mineral	含水矿物；水化矿物
hybrid resonance	混合共振	hydrated oxide	氢氧化物
hybrid rock	混杂岩；混浆岩	hydrated peroxide	水合过氧化物
hybrid sandstone	混合砂岩	hydrated silica	硅的水化物
hybrid scale	混合比例尺	hydrated surface	水化表面
hybrid sedimentary rock	混积岩	hydrates of natural gas	天然气水合物
hybrid sedimentation	混杂沉积	hydrating polymer	水解聚合物
hybrid stress	混合应力	hydrating solution	水解溶液
hybrid wave	混合波	hydration	水化；水合；水合酌；水化作用
hybridization	黄铁矿化；混合岩化作用；混染作用，杂种型；杂拼，杂交，配种；杂化，杂化作用；混合酌	hydration catalyst	水合催化剂
		hydration heat	水化热；水合热
hybridized rock	混杂岩	hydration number	水化数
hybrid-stress element	混合应力单元	hydration of cement	水泥水化作用
hydatogen sediment	水成沉积物；液成沉积物	hydration of colloidal particles	胶粒的水化酌
hydatogenesis	水成作用；热液成矿酌	hydration product	水化物
hydatogenetic	岩浆水溶液沉积的；液成矿的；成的；水成的	hydration reaction	水化反应
		hydration shattering	水化崩解作用
hydatogenic	热液成的；水成的	hydration structure	水化构造
hydatogenic rock	水成岩；蒸发岩	hydration variation	水化变异
hydatogenous	水成的；液成的；液成矿的	hydration water	吸附水；结合水
hydatogenous mineral	水成矿物；液成矿物	hydratisomery	水合同分异构
hydatogenous rock	水成岩；液成岩；沉积岩	hydrauger	水力螺旋钻
hydatometamorphic rock	水成变质岩	hydrauger hole	水冲钻孔；水螺旋钻孔
hydatomorphic	热液成的；水成的	hydraulic accumulator	预制混凝土薄壳；蓄水池；液压蓄能器
hydatomorphism	水变质酌；水成作用		
hydatopneumatogenic	气水成的；气液成的	hydraulic action	水力作用
hydatopyrogenic	火水成的	hydraulic admixture	水硬性掺料
hydatothermal	气成的；热液的	hydraulic analogy	水力模拟；水力相似
hydatothermal phase	气成相	hydraulic analysis	水力分析
hydrability	水化性	hydraulic anchor	水力锚
hydra-cushion shock tool	液压减震工具	hydraulic anchorage	液压锚固；锚固式液压支架
hydrafrac	水力压裂	hydraulic and harbor engineering	水利及港湾工程
hydralime	熟石灰	hydraulic and hydroelectric engineering	水利水电工程
hydralsite	水硅铝石	hydraulic approach	水力学方法
hydranlic tubing holddown	油管水力锚	hydraulic architecture	水利工程；水工建筑

hydraulic axial thruster 轴向加压器
hydraulic backfilling 水力充填；水砂充填
hydraulic backhoe 型板；液压反铲
hydraulic back-pressure valve 水压逆止阀
hydraulic bender 沿岸漂移；液压折弯机；锥摆调速机
hydraulic binder 水硬性胶结剂；水硬性胶凝材料
hydraulic binding agent 水硬性胶结剂；水硬性胶凝材料
hydraulic bladder 水胞；软壳水圈
hydraulic blast 水力清砂
hydraulic blasting 水力爆破
hydraulic bolt tensioner 液压锚杆拉紧器
hydraulic bolt-hole drilling machine 液压锚孔钻机
hydraulic bore pororoca 河口涌潮
hydraulic borehole gauge 挖掘机械
hydraulic borehole mining 钻孔水力开采
hydraulic boring unit 液压钻孔机
hydraulic boundary condition 水力边界条件
hydraulic brake 液力制动器，液压制动器；水力制动器；镶嵌板；闸式水力测功器
hydraulic brake fluid 液压制动液
hydraulic break out cylinder 卸扣液缸
hydraulic breaker 液压破碎机
hydraulic bucket wheel drive 轮斗液压驱动
hydraulic buffer 水力缓冲器
hydraulic bumper 液压缓冲器
hydraulic burster 水力爆破筒
hydraulic calculating 水力计算；水利计算
hydraulic capacity 水流容量
hydraulic capstan 水力绞盘；水力起锚机；液压绞盘
hydraulic capsule 液压舱；液压容器；压力抛物线分布；野外三角测量；测压计；液压传感器
hydraulic cartridge 水力爆破筒；外部设备
hydraulic cell 液力测力计
hydraulic cement 水硬性水泥；水硬性胶凝材料；水凝水泥
hydraulic cementing 水压胶接
hydraulic cementing agent 水硬性胶结料
hydraulic cementitious material 水硬性胶凝材料
hydraulic centrifuge 水力离心机
hydraulic character 水力特性
hydraulic characteristic curve 水力特性曲线
hydraulic characteristics 水力特性
hydraulic chart 水力图板
hydraulic chock 垛式液压支架
hydraulic circuit diagram 液压系统图
hydraulic circulating head 水力循环给水器；循环头
hydraulic circulating system 正循环钻进方法；液压循环系统；水力循环系统
hydraulic classification 水力分级
hydraulic closed packer 水力密闭式封隔器
hydraulic clutch 液压离合器
hydraulic coalburster 水压爆煤筒
hydraulic complex 水利枢纽
hydraulic computation 水力计算
hydraulic condition 水力条件
hydraulic conductivity 水力传导率，渗透系数；导水率；导水性；水力传导系数；透水性
hydraulic conductivity barrier 水力传导屏障
hydraulic connection 水力联系；液压连接
hydraulic connector 水力操纵的连接器

hydraulic constituent 水力要素
hydraulic control 水力控制；液压控制
hydraulic control plan 液控平面图；液压系统图
hydraulic controlled scraper 液压式铲运机
hydraulic controller（HCR） 液压控制阀
hydraulic conveying 水力运输；水力输送
hydraulic core extractor 液压岩心推取器；水力岩心提取器
hydraulic core pump 岩心推取泵
hydraulic current 落差流
hydraulic cut method 水力削减法
hydraulic cutter 水力割刀；水力清焦器
hydraulic cutter drive 截割部液压传动头
hydraulic cutting 液压切割
hydraulic cutting coal 水力采煤
hydraulic cyclone 水力旋流器；水力旋流；旋液分离器；水力旋风器
hydraulic cylinder feed 液压给进
hydraulic cylinder hoist 液压圆筒式启闭机
hydraulic dam 水压坝；水力冲积堤
hydraulic damper 液压减震器；液压阻尼器
hydraulic design 液压设计；水力工程设计
hydraulic development 水力开发；水力枢纽
hydraulic device 液压装置
hydraulic diameter 水力直径
hydraulic diaphragm transducers 液压膜片式传感器
hydraulic diffusivity 压力传导系数；水力扩散系数；水力扩散；水力传导系数
hydraulic discharge 涌水量；出水量；流量；地下水出流
hydraulic disorder 水力失调
hydraulic dispersion 水力弥散
hydraulic drag 水力牵引；水阻力
hydraulic dredger 水力疏浚机；水力挖泥机
hydraulic drifter 液压架式钻床；液压架式凿岩机
hydraulic drill 水力钻机；水力凿岩机；液压凿岩机
hydraulic drill rig 液压钻车
hydraulic drilling 液压钻进；水力凿岩；水力钻眼；水力钻探
hydraulic drilling jar 液压式钻井震击器
hydraulic drive 水力传动；转换阀；液压传动
hydraulic drive rig 液压钻机
hydraulic drop 水力落差
hydraulic dump 水力排土场
hydraulic dynamometer 液压测力器；洋流；液压测力计
hydraulic effect 水力效应
hydraulic efficiency 水力效率
hydraulic ejector 排沙泵；水力排泥管
hydraulic element 液压元件；水力单元；水力因素
hydraulic elements 水力要素
hydraulic energy 水能
hydraulic engine 水力发动机；水力机
hydraulic engineer 水力工程师
hydraulic engineering 水利工程；水工学；液压工程
hydraulic equivalent 水压等效；水力当量
hydraulic evapotranspirometer 水力式蒸腾计
hydraulic excavation 水力开采；水力开挖；水力掘进
hydraulic excavator 水力冲采机；水力挖掘机；液压挖掘机；吸泥泵；液压式挖土机
hydraulic excavator loading 液压挖掘机采装
hydraulic excitation test 液压激振试验

hydraulic expansion wall scraper	液压扩张的井壁刮刀	hydraulic haulage unit	液压牵引部件
hydraulic exponent	水力指数	hydraulic head	水头；扬程；水位差
hydraulic extension test	液压拉伸试验	hydraulic head gradient	水头梯度
hydraulic extraction	水力开采；水采；水力落煤和运输；水力采掘相运输	hydraulic head surface	水头面
hydraulic extruder	液压挤出器	hydraulic heading through	水力贯通
hydraulic facilities	水力设施	hydraulic hoisting	水力提升；水力提升采矿法
hydraulic factor	水力因素	hydraulic hoisting equipment	水力提升设备
hydraulic failure	水力破坏；液压故障	hydraulic hold-down	水力锚
hydraulic feed	液压推进，液压给进；水力传送；重水；水力给料	hydraulic hole	水冲钻孔
hydraulic feed drill	液压给进钻进；油压钻机	hydraulic hole caliper	水力井径仪
hydraulic feedback	水力反馈	hydraulic hole calipers	水力孔眼测径器
hydraulic feeding equipment	液压给进设备	hydraulic hopper	液压漏斗
hydraulic fill	水力冲填；吹填，冲填土；吹填土；水砂充填	hydraulic horsepower	水马力
hydraulic fill dam	水力冲填坝；冲垒坝	hydraulic hose	水力软管；液压软管
hydraulic filler	水砂充填浆	hydraulic HP output	水马力输出
hydraulic filling	水力充填	hydraulic impact	水力冲击
hydraulic filling system	水砂充填系统	hydraulic impact unit	液压冲击装置；液压冲击钻机
hydraulic flounder	钻屑阻进效应	hydraulic impulse	液压冲击；液压脉冲
hydraulic flow	水力的渗流；湍流	hydraulic impulse turbine	水力冲击式透平
hydraulic flow net	水力流网	hydraulic index	水硬率；水硬度
hydraulic fluid	液压机液体；液压液；锥形锚塞	hydraulic indicator	水压计
hydraulic flume	水力试验槽	hydraulic instability	水力不稳定性
hydraulic flushing	水砂充填；湿法充填	hydraulic instrument	水力学仪器
hydraulic flushing in hole	水力冲孔	hydraulic integraph	水力积分仪
hydraulic forging	水压机；锻造	hydraulic integrator	水力积分仪
hydraulic fracture	水力劈裂；水力破坏；圆形断面；水力致裂法	hydraulic intensifier	液压增强器；余波龄
hydraulic fracture technique	水压破裂法；水力压裂法	hydraulic interrelation	水力联系
hydraulic fracture test	水力压裂试验；水力破裂试验；水力劈裂试验	hydraulic jack	液压千斤顶；液压起重器
hydraulic fracturing	水力劈裂；水力压裂；水压破裂，水压破裂作用；水压胀裂法；压裂；水力碎裂	hydraulic jam	水锤扬水机
Hydraulic Fracturing Method	水压致裂法	hydraulic jar	水力震击器
Hydraulic Fracturing of Rock Mass	岩体水力压裂	hydraulic jar penetration	水力振击钻孔
hydraulic fracturing stimulation	水力压裂增产措施	hydraulic jar perforation	水力振击打孔
hydraulic fracturing technique	水力劈裂法；水力压裂法	hydraulic jar-down	水力振击打下去
hydraulic fracturing test	液力压裂试验	hydraulic jet	水力射流；水力喷射
hydraulic friction	水力摩阻	hydraulic jet cutting	水力喷射挖方；水射流落煤
hydraulic gauge	液压计	hydraulic jet mining	水力喷射开采；水力开采；水采
hydraulic giant	水采水枪	hydraulic jet mixer	水力喷射式混合器；水力喷射搅拌器
hydraulic gobbing	水力充填；水砂充填	hydraulic jet pump	水力喷射泵
hydraulic governing	液压调节	hydraulic jetting	水力冲洗；水射流冲洗；水力钻探；水力喷射
hydraulic grade line	水力坡降线；水压线；淤沙地；流水压力线	hydraulic jig	水力跳汰机
hydraulic gradient	水力坡度；水力坡降；液压梯度；液位梯度；水力梯度	hydraulic jump	水跃；液压跳
hydraulic gradient line	水力比降线；水力坡降线；锥形铰刀；静水头线	hydraulic jump value	水跃值
hydraulic gradient slope	水力坡降	hydraulic jump wave	水跃波
hydraulic gravel	水力冲刷砾石	hydraulic laboratory	水工试验室；水力实验室
hydraulic gravel fill	水力砾石充填	hydraulic leg	液压腿；液压柱
hydraulic gun	水力冲射器；水枪；水力冲采机	hydraulic levee	水力冲填土坝；淤填土坝
hydraulic gun release sub	射孔器水力丢手接头	hydraulic leveling device	流体测沉计；液压调平装置
hydraulic handling	水力运搬	hydraulic lift force	水力上顶力
hydraulic hardening	水硬性作用	hydraulic light-weight jack	轻型液压千斤顶
hydraulic haulage	水力运输；液压牵引	hydraulic lime	水硬性石灰；水硬石灰
hydraulic haulage drive	液压拖运传动装置	hydraulic limestone	水硬石灰石
		hydraulic load cell	液压-电子测力计
		hydraulic loading	水力装料
		hydraulic loosening	水力崩落
		hydraulic loss	水力损失；液压损失
		hydraulic magnitude	液压幅度；水力幅度
		hydraulic main	水压主管；总水管
		hydraulic mean depth	水力平均深度

hydraulic mean diameter	均水力直径
hydraulic mean path	水力均匀路径
hydraulic measurement	水力测量
hydraulic mine	水力开采矿井；水力机械化矿井
hydraulic mining	水力采矿；水力开采；水采
hydraulic mining dual operation	水采作业
hydraulic mining method	水力开采
hydraulic model	水力模型；水工模型
hydraulic model test	水力模型试验
hydraulic modeling	水力模型
hydraulic modulus	水硬模数；高压水枪
hydraulic monitor	高压水枪
hydraulic mortar	水硬性砂浆；水下速凝砂浆；液压马达；水力发动机
hydraulic mucking	水力出渣
hydraulic mud fill	吹填淤泥
hydraulic oil jar	油压震击器
hydraulic overflow settlement cell	溢水式沉降计
hydraulic packer	水力封隔器
hydraulic packing	液压填密；水力充填
hydraulic parameter	水力参数
hydraulic parting	水力压裂
hydraulic percussion drilling	液压冲击钻井
hydraulic percussion method	水力冲击钻探法
hydraulic percussive boring	水力冲魂井
hydraulic percussive drill	液压冲击式凿岩机
hydraulic perforating gun	液压射孔器
hydraulic perforation	水力喷射射孔
hydraulic performance	水力特性
hydraulic permeability	水力透水性；液体渗透性
hydraulic physical and mathematical models	水力实体及数学模型
hydraulic piezometer	水压力计；水力式测压计；液态孔隙水压计
hydraulic piezometer tip	液压孔隙水压力计测头
hydraulic pile driver	液压打桩机
hydraulic pile driving	液压打桩；吸入管
hydraulic pipe bending equipment	液压弯管设备
hydraulic pipe catcher	水压捞管器
hydraulic pipe cutter	水压割管器；水压切管器；液压管道切割机
hydraulic pipe transport	管式水力运输
hydraulic pipeline	水力管道
hydraulic pipeline dredge	水力排管式挖掘船
hydraulic piston corer	液压活塞取样管
hydraulic piston pump	水力活塞泵
hydraulic piston sampler	水压活塞取土器；液压活塞取样器
hydraulic pit prop	液压矿柱
hydraulic placement	水力浅筑法
hydraulic potential	水力位能；水力势能
hydraulic power	液压动力；水力；水能
hydraulic power feed drill	液压进给钻
hydraulic power station	水电站
hydraulic press	液压机；水压机
hydraulic pressure	水压，液压；自由降落
hydraulic pressure cells	液压盒；液压压力盒
hydraulic pressure head	水压压头；水压压头
hydraulic pressure percussion drill	液压冲击钻
hydraulic pressure test	液压试验；水压试验；压水试验
hydraulic pressure test machine	液压式压力试验机
hydraulic pressure transducer	液压传感器
hydraulic prime mover	水力原动机
hydraulic profile	水压断面；水力学剖面；蓄水层水压断面；水力断面
hydraulic prop	液压支柱；液压支柱
hydraulic propeller	喷水推进机；水力推进机
hydraulic property	水力特性
hydraulic props	液压立柱
hydraulic prospecting	水力勘探
hydraulic pull tester	液力拉力检定器
hydraulic puller	液压拆卸器
hydraulic pump	液压泵
hydraulic punching machine	水力冲孔机
hydraulic radius	水力半径
hydraulic rail setter	液压匀轨机
hydraulic ram	水锤泵，水力夯锤，液力冲头；液压泵，液压防喷器闸板；液压油缸；液压推车机；液压千斤顶液压压头液压撞击机，液压支腿；水压扬汲机
hydraulic ratio	水力比；水力比率
hydraulic reamer	水压扩孔器
hydraulic reclamation	吹填造陆
hydraulic regime	水力状态；水力工况
hydraulic regulator	水力调节器；液体测压计
hydraulic relation	水力联系
hydraulic relay	液压继电器
hydraulic relief valve	液压减压阀
hydraulic resistance	水力阻力；水力摩阻；水力阻抗
hydraulic resistivity	水力阻力
hydraulic resources	水利资源
hydraulic response time	水力反应时间
hydraulic retaining structure	挡水水工建筑物
hydraulic ridge	水力分水岭
hydraulic rig	液压钻机
hydraulic rock cutting	水力凿岩
hydraulic rock drill	液压凿岩机
hydraulic rolling compaction	洒水压实
hydraulic roof bolting machine	液压式锚杆安装机
hydraulic rotary	水力转盘
hydraulic rotary drill	液压旋转式钻机
hydraulic rotary drilling	水力旋转钻探；水冲回转钻探
hydraulic rotary rig	液压回转钻机
hydraulic sampling machine	液压采样机
hydraulic sand filling	水砂充填
hydraulic sand sizer	水力分砂器
hydraulic scale model	水工缩尺模型
hydraulic scheme	液压系统图
hydraulic scouring	水力冲刷
hydraulic seal	水封；防水层；液体密封
hydraulic section	水力断面
hydraulic self-sealing packer	水力自封式封隔器
hydraulic sensing pad	液压传感垫
hydraulic set	水凝；水硬；液压装配
hydraulic setting	水凝；水硬
hydraulic setting mortar	水凝灰浆
hydraulic settlement gauge	水压沉降计
hydraulic shaft jumbo	液压凿井钻机
hydraulic shock absorber	液压减震器；液压减震器油压缓冲装置；外港；水力减振器；水力减震器
hydraulic shock eliminator	纵向收缩；水力减振器

hydraulic side-wall coring tool 水力井壁取心器
hydraulic similarity 水力相似；水力相似性
hydraulic simulation 水力模拟；水力仿真
hydraulic sliding form 液压滑动模板
hydraulic slip 液压损失；液体漏失
hydraulic slip coupling 液压滑动联轴节
hydraulic slope 水力坡度；水力坡降；水力比降
hydraulic sluicing 水力冲沙法；水力冲采
hydraulic slushing 水力充填
hydraulic snubber 水力减震器
hydraulic spear 水力打捞矛
hydraulic spray-seeding 液力喷播
hydraulic stability 水力稳定性
hydraulic static cone penetrometer 液压静力触探仪
hydraulic steering 水力操纵，液压操纵；液压控制
hydraulic steering gear 液压操舵装置
hydraulic steering system 液压转向系统
hydraulic stepless transmission 液压无级变速机
Hydraulic stimulation 水压激励；水力压裂
hydraulic stowing 水砂充填；水力充填
hydraulic stowing floor mat 水砂充填铺底
hydraulic stowing method 水砂充填采煤法
hydraulic stowing partition 挡砂帘；砂门子
hydraulic strength 水硬强度
hydraulic strengthener 液压助力器
hydraulic stress-path triaxial cell 液控应力路径三轴室
hydraulic stripping 水力剥离，水力开采；水力清除
hydraulic structure 水工建筑；水工结构；水力结构
hydraulic subsiding value 水力沉降值；水力下沉值
hydraulic supply line 液压供给管线
hydraulic supply unit 液压供给装置
hydraulic support leg 液压千斤支腿
hydraulic surface loading 表面水力负荷
hydraulic system 液压系统；水力循环系统
hydraulic tamping machine 液压捣固机；液压捣固车
hydraulic tensioning 水力张拉
hydraulic test 水力学试验；液压检验；水压试验；水工试验
hydraulic testing 水压试验
hydraulic theory 水动力学理论；水力学理论
hydraulic thrust boring machine 液压冲钻机；水力钻机；涡轮钻；涡轮钻具
hydraulic top hammer drill 液压冲击钻机
hydraulic torque 水力转矩；转盘扭矩
hydraulic torque converter 液力变扭器；液压变矩器
hydraulic tortuosity factor 水力弯曲系数
hydraulic traction 水力曳引
hydraulic transient 水力瞬变过程
hydraulic transmission 液力传动；液压传动
hydraulic transmission control 液压传动控制系统
hydraulic transmission fluid 水力传动系统用的液体
hydraulic transmission gear 水力传动装置；溢洪道下消能
hydraulic transmission gear box 液力传动箱
hydraulic transmission system 液力传动系统
hydraulic transport 水力运输
hydraulic transport core drilling 水力输岩钻井；水力反循环连续取心钻进
hydraulic transport of placer 砂矿水力输送
hydraulic transport system 水力运输系统

hydraulic transportation 水力输送
hydraulic transporting device 水力运输设备
hydraulic travel 液压行程
hydraulic treble-ram pump 三柱塞液压泵
hydraulic tripping device 液压脱扣器
hydraulic tunnel 水工隧道
hydraulic turbine 水轮机；液力涡轮
hydraulic turbine generator 水轮发电机
hydraulic turbine type pump 水力涡轮泵
hydraulic underreamer 液压管下扩眼器
hydraulic uplift pressure 水力扬压力；静水浮托力
hydraulic vibrator 液压振动器
hydraulic wall sampler 液压井壁取样器
hydraulic wall scraper 液压井壁泥饼清除器
hydraulic washing pump 液压清洗泵
hydraulic water 加压水；水压水
hydraulic water resource 水力资源；水利资源
hydraulic whipstock 水力造斜器
hydraulic work 水工结构；木力开采；水利工程
hydraulically balanced 水力平衡的
hydraulically created fracture 水力压裂裂缝
hydraulically driven crawler-mounted jumbo 液压驱动履带式钻车
hydraulically operated 液压操纵的；水力操纵的
hydraulically operated rear dump truck 滞阻
hydraulically powered underground drill 井下液压钻机
Hydraulically stabilized layer 液压稳定层
hydraulically-operated tele-scopic action 液控伸缩操作
hydraulically-powered winch 液动绞车
hydraulic-feed drill rig 液压给进钻机
hydraulic-fracturing program 水力压裂方案
hydraulician 水力学家；水力工程师
hydraulicity 水硬性；水泥性；水凝性
hydraulicking 水力掘进；水力冲刷；水力开采，水力冲挖；水力挖土
hydraulicking stripping 水力剥离
hydraulic-pump placer 液压泵浇注装置
hydraulics 水力学；液压系统；水力条件
hydraulics of groundwater 地下水水力学
hydraulics optimization 水力优化
hydride 氢化物
hydrion 氢离子；质子
Hydrion pH dispenser pH试纸
hydrite 微亮煤；沸石
hydro geochemical prospecting 水文地球化学勘探；水文物探
hydro orientator 水压定向器
hydro percussion hammer 液动冲击器；液动冲击
hydro power tunnels 水力隧洞
hydro project 水利工程；水利枢纽
hydro seal pump 水封式泵
hydroacoustic 水声的
hydroacoustic bearing indicator 水声测向仪
hydroacoustic noise 水下噪声
hydroacoustic positioning 水声定位
hydroacoustic range finder 水声测距仪
hydroamphibole 水闪石
hydroastrophyllite 水星叶石
hydrobarometer 水深计
hydrobiolite 生物堆积岩

hydrobiotite 水黑云母
hydroblast 水力清理；水力清砂；喷水冲洗
hydro-boom jumbo 带液压钻架的钻车
hydroboracite 水方硼石
hydroborer 水力钻头
hydro-boring pump 钻进用水力冲洗泵
hydrocalcilutite 水成泥屑灰岩
hydrocalcite 水方解石
hydrocalumite 水铝钙石
hydrocarbon black 石油炭黑
hydrocarbon detector 碳氢化合物探测仪
hydrocarbon entrapment 油气滞留；油气滞集
hydrocarbon flow 油气流
hydrocarbon impregnation 油气浸染
hydrocarbon movability 油气迁移率
hydrocarbon potential 油气潜力
hydrocarbon production 油气生产
hydrocarbon ratio 烃比值；碳氢比
hydrocarbon reservoir 油气层
hydrocarbon reservoir assemblage 含油层系
hydrocarbon reservoirs 油气藏
hydrocarbon resources 油气资源
hydrocarbon saturation 烃饱和度
hydrocarbon seep-carbonate 冷泉碳酸盐岩烃
hydrocarbon seeps 碳氢化合物渗漏
hydrocarbon source rocks 烃源岩
hydrocarbon sub-marine field 海底油气田
hydrocarbon trap 油气圈闭
hydrocarbon zone 油气层
hydrocarbonate 重碳酸盐；碳酸氢盐
hydrocarbonate calcium water 重碳酸钙水
hydrocarbonate water 重碳酸水
hydrocarbonate-calcium water 重碳酸钙水
hydrocarbon-bearing pool 油气藏
hydrocarbon-filled porosity 含烃有效孔隙度
hydrocarbon-water contact 油-水接触面
hydrochemical 水化学的；水文化学的
hydrochemical analysis 水化学分析
hydrochemical anomaly 水化学异常
hydrochemical characteristics 水化学特性；水文化学特性
hydrochemical chart 水化学图；水文化学图
hydrochemical facies 水化学相
hydrochemical geothermometer 水化学地热计；水化学地热温标
hydrochemical geothermometry 水化学地温测量；水化学计温
hydrochemical horizontal zoning 水化学水平分带
hydrochemical index 水化学指标
hydrochemical inversion 水化学倒置
hydrochemical map 水化学图
hydrochemical map of groundwater 地下水水化学图
hydrochemical metamorphism 水化学变质作用；水化变质
hydrochemical modeling 水化学建模
hydrochemical network 水化学网
hydrochemical prospecting 水化学勘探
hydrochemical prospecting by pumping water from bore 钻孔抽水水化学找矿
hydrochemical prospecting in adit exploration engineerings 硐探工程水化学找矿
hydrochemical prospective mineralization area 水化学成矿远景区
hydrochemical regime 水化学动态
hydrochemical survey 水化学调查
hydrochemical vertical zoning 水化学垂直分带
hydrochemical zonality 水化学分带性
hydrochemist 水化学工作者
hydrochemistry 水文化学；水质化学
hydrochlorborite 多水氯硼钙石
hydrochloric acid 盐酸
hydroclassification 水力分级
hydroclast 水成碎屑；水成碎屑岩
hydroclastic 水力沉积的石屑；水成碎屑的；水成碎屑岩
hydroclastic rock 水成碎屑岩
hydroclastics 水成碎屑岩
hydroclimate 水文气候；水面气候
hydroclimatic factor 水候因素；水文气候因素
hydroclimatology 水文气候学；水候学
hydrocompaction 湿陷；水力压实
hydrocomplex 水利枢纽
hydrocone type 虹吸式
hydrocone-crusher 液压调节的圆锥破碎机
hydroconsolidation 水力固结作用；水力固结；湿陷
hydrocracking 加氢裂解；加氢裂化
hydrocrane 液压起重机；水力起重机
hydrocratic motion 水动型运动
hydrocyclonic desander 旋流除砂器
hydro-development 水力开发
hydrodialeima 水下不整合
hydro-disperser 水力分散器
hydrodolomite 水白云石
hydrodrill 液压钻机
hydro-drill rig 液压钻架；液压架式钻机
hydrodynamic 流体动力的；水力的；水动力的；流体的；流体动力的；动水压的
hydrodynamic analog 流体动力学模拟；流体动力学模型
hydrodynamic analogy 水动力模拟
hydrodynamic and advection-diffusion model 水力及对流——扩散模型
hydrodynamic brake valve 液力制动阀
hydrodynamic closure 水动力封闭
hydrodynamic codes 流体动力规范
hydrodynamic computation 流体动力计算
hydrodynamic conductibility 水动力传导性
hydrodynamic criterion 水动力指标
hydrodynamic damping 水动力阻尼
hydrodynamic derivative 水动力导数；流体动力导数
hydrodynamic differential equation 流体动力微分方程
hydrodynamic dispersion 水动力弥散；水力扩散；流体动力弥散
hydrodynamic drag 水动力阻力；动水拖曳力
hydrodynamic drive 液力传动
hydrodynamic effect 流体动力效应；水动力效应
hydrodynamic entrapment 水动力圈闭
hydrodynamic environment 水动力环境
hydrodynamic equation 流体动力学方程；水动力方程
hydrodynamic force 水动力；流体动力
hydrodynamic forcing 水动力强迫
hydrodynamic framework 水动力结构
hydrodynamic gauge 动水压力计

hydrodynamic governor	液压动力调速器	hydroelectric generator	水力发电机
hydrodynamic gradient	流体动力梯度；水动力梯度	hydroelectric potentiality	水电蕴藏量
hydrodynamic head	动水头	hydroelectric power development	水力发电开发
hydrodynamic impact	水动力冲击	hydroelectric power planning	水力发电规划
hydrodynamic intensity	水动力强度	hydroelectric power plant	水力发电厂
hydrodynamic interaction	流体动力相互作用；流体动力干扰	hydroelectric power station	水电站
hydrodynamic lag	水动力滞后；水力延滞	hydroelectric power (HEP)	水力发电
hydrodynamic load	流体动力荷载；动水荷载	hydroelectric project	水电工程
hydrodynamic magnification factor	动水放大因数	hydroelectric resource	水电资源；水力资源
hydrodynamic map	水动力图	hydroelectric structure	水电建筑；水电结构物
hydrodynamic mass effect	动水质量效应	hydroelectricity	水电；水力发电；热电
hydrodynamic mechanical seals	液压机械密封	hydroelectro dynamic analogy	水电比拟
hydrodynamic model	水动力模型	hydroenergetics	水能学
hydrodynamic net	水动力网	hydroenergy	水能
hydrodynamic noise	水动力噪声	hydro-energy gradient	水能梯度
hydrodynamic piezometer	水动力测压计	hydroenvironment	水环境
hydrodynamic pressure	流体动压力；动水压力	hydro-environmental capacity	水环境容量
hydrodynamic pressure gradient	水动压梯度	hydro-extracting cage	甩水机；脱水机
hydrodynamic process	流体动力过程	hydroextraction	水力提取；水力采矿；水力采煤；水采
hydrodynamic profile	水动力剖面	hydroextractor	脱水器；水力离心机；水力挤压机
hydrodynamic property	水动力特性	hydrofacies	水相
hydrodynamic regime	水动力动态	hydroferrite	含水碳酸盐
hydrodynamic research	流体动力研究	hydrofixation	湿固着
hydrodynamic retention	水动力滞留	hydrofoil weir	水翼式堰
hydrodynamic reverser	液力换向传动箱	hydroform	液压成形
hydrodynamic sieving	水动力筛选法	hydrofrac	水力压裂
hydrodynamic spectrum	流体动力谱	hydrofrac fluid	水力压裂液
hydrodynamic stability	动水稳定性	hydrofracking	水压致裂；水力压裂
hydrodynamic stress	液体动应力	hydrofraction	水分离作用；水热破碎
hydrodynamic synthesis	水动力综合法	hydrofracture	水力劈裂；水力破碎；水压破碎；水力裂缝
hydrodynamic system	水动力系统	hydrofracture grouting	劈裂灌浆
hydrodynamic texture	水动力结构	hydrofracturing	水破碎；水压致裂；水力劈裂；液力加压开裂测试
hydrodynamic theory	水动力理论		
hydrodynamic tilted fluid contact	水动力倾斜液面	hydrofracturing disposal	水力压裂处置
hydrodynamic trap	水动力圈闭	hydrofracturing method	水力压裂法
hydrodynamic trapping	水动力圈闭作用	hydrofuge	不透水的；防湿的
hydrodynamic vertical zoning	水动力垂直分带	hydrogen bond	氢键
hydrodynamic wave	水动力波	hydrogen bonding	氢结合
hydrodynamic zone	水动力带	hydrogen geothermometer	氢地热温标
hydrodynamical differential equation	流体动力学微分方程	hydrogen halation	氢晕
hydrodynamical interaction	流体动力学的相互作用	hydrogen index	含氢指数
hydrodynamically rough surface	水动力粗糙面	hydrogen induced cracking	氢致开裂
hydrodynamically smooth surface	水动力光滑面	hydrogen peroxide	过氧化氢
hydrodynamics	流体动力学；水动力学；液体动力学	hydrogenation	氢化；加氢
hydrodynamics conductibility	水动力传导性	hydrogen-bearing index	含氢指数
hydrodynamics exploration analysis	水动力学勘探分析	hydro-generator	水力发电机；水轮发电机
hydrodynamics modelling methodology	水动力学的模拟方法学	hydrogenesis	氢解作用；水分子凝结作用；水成作用
hydrodynamism	水动力作用	hydrogenic	水成的；水生的；含氢的
hydrodynamometer	水力测功器；流速器；流速计	hydrogenic rock	水生岩；水成岩
hydroecology	水文生态学；水域生态学	hydrogenic soil	水生土
hydroeconomics	水力经济学	hydrogen-ion concentration	氢离子浓度
hydroelastic vibration model	流体弹性振动模型	hydrogen-ionized	氢离子化
hydroelasticity	水弹性；流体弹性理论；水弹性力学	hydrogenite	氢发生剂
hydro-elastoplastic medium	流体弹塑性体	hydrogenium	金属氢
hydroelectric	水电的；水力发电的	hydrogenization	氢化作用；氢化
hydroelectric dam	水力发电坝	hydrogenolysis	加氢作用；加氢分解；氢解
hydroelectric development	水电开发	hydrogenous	水成的；水生的，水成的；含氢的；火成的
hydroelectric generation	水力发电	hydrogenous phase	水成相

hydrogenous sediment 水成沉积；水生沉积
hydrogeochemical 水文地球化学的
hydrogeochemical anomalous value 水文地球化学异常值
hydrogeochemical anomaly 水文地球化学异常
hydrogeochemical background value 水文地球化学背景值
hydrogeochemical barrier 水文地球化学障
hydrogeochemical environment 水文地球化学环境
hydrogeochemical environment zonation 水文地球化学环境分带性
hydrogeochemical exploration 水文地球化学勘探；水文地球化学探矿
hydrogeochemical field 水文地球化学场
hydrogeochemical halo 水文地球化学分散晕
hydrogeochemical indication 水文地球化学标志
hydrogeochemical map 水文地球化学图
hydrogeochemical map of groundwater 地下水水化学图
hydrogeochemical ore-prospecting 水文地球化学找矿
hydrogeochemical ore-prospecting indicator 水文地球化学找矿标志
hydrogeochemical process 水文地球化学作用
hydrogeochemical prospecting 水文地球化学勘探
hydrogeochemical province 水文地球化学省；水文地球化学区
hydrogeochemical reconnaissance 水文地球化学勘测
hydrogeochemical survey 水文地球化学调查
hydrogeochemical trace 水文地球化学形迹
hydrogeochemical zonality 水文地球化学分带
hydrogeochemistry 水文地球化学
hydrogeochemistry cycling 水文地球化学循环
hydrogeography 水文地理学
hydrogeologic 水文地质的
hydrogeologic analogue 水文地质模拟
hydrogeologic analogue method 水文地质比拟法
hydrogeologic analogy method 水文地质比拟法
hydrogeologic appraisal 水文地质评价
hydrogeologic assessment 水文地质评价
hydrogeologic characteristics 水文地质特性
hydrogeologic chart 水文地质图
hydrogeologic complex 水文地质综合体
hydrogeologic condition 水文地质条件
hydrogeologic cycle 水文地质循环
hydrogeologic deciphering 水文地质解释
hydrogeologic division 水文地质分区
hydrogeologic drill hole 水文地质钻孔
hydrogeologic drilling 水文地质钻探
hydrogeologic drilling rig 水文水井钻机
hydrogeologic element 水文地质单元
hydrogeologic evaluation 水文地质评价
hydrogeologic factor 水文地质因素
hydrogeologic feature 水文地质特征
hydrogeologic framework 水文地质结构
hydrogeologic log 水文测井
hydrogeologic map 水文地质图
hydrogeologic map of pits 坑道水文地质展示图
hydrogeologic massif 水文地质地块
hydrogeologic mechanics 水文地质力学
hydrogeologic model 水文地质模型
hydrogeologic nature 水文地质特性
hydrogeologic network 水文地质网
hydrogeologic parameter 水文地质参数
hydrogeologic profile 水文地质剖面
hydrogeologic reconnaissance 水文地质普查；水文地质踏勘
hydrogeologic section 水文地质剖面
hydrogeologic setting 水文地质背景水文地质环境
hydrogeologic station 水文地质站
hydrogeologic subdivision 水文地质分区；水文地质区划
hydrogeologic survey 水文地质调查
hydrogeologic team 水文地质队
hydrogeologic test 水文地质试验
hydrogeologic unit 水文地质单元；水文地质单位
hydrogeological analysis 水文地质分析
hydrogeological condition 水文地质条件
hydrogeological division 水文地质部门
hydrogeological drilling 水文地质钻探
hydrogeological environment 水文地质环境
hydrogeological evaluation of ore deposit 矿床水文地质评价
hydrogeological exploration 水文地质勘探
hydrogeological exploration borehole 水文地质勘探孔
hydrogeological interpretation 水文地质解释
hydrogeological interpretation of photo 像片水文地质解释
hydrogeological interpretation of remote sensing images 遥感图像水文地质解译
hydrogeological investigation 水文地质勘探；水文地质勘察
hydrogeological investigation for water supply 供水水文地质勘查
hydrogeological investigation stage 水文地质勘查阶段
hydrogeological logging 水文测井
hydrogeological map 水文地质图
hydrogeological map of ore field 矿区水文地质图
hydrogeological mapping 水文地质测绘；水文地质编图
hydrogeological observation point 水文地质观测点
hydrogeological parameter 水文地质参数
hydrogeological profile 水文地质剖面图
hydrogeological section 水文地质剖面
hydrogeological structure 水文地质构造
hydrogeological survey 水文地质测绘；水文地质调查
hydrogeological test 水文地质试验
hydrogeological test borehole 水文地质试验孔
hydrogeological unit 水文地质单元
hydrogeological well logging 水文物探测井
hydrogeological work in mine 矿山水文地质工作
hydrogeological zoning 水文地质分区
hydrogeologist 水文地质工作者；水文地质学家；水文地质师
hydrogeology 水文地质学
hydrogeophysical property 水文地球物理性质
hydrogeophysical prospecting 水文地质地球物理勘探
hydrogeophysics 水文地球物理
hydrogeothermal resource 水文地热资源；水热型地热资源
hydrogeothermics 水文地热学
hydrogeothermometer 水文地热计
hydroglauberite 水钙芒硝
hydroglimmer 水云母类
hydrogoethite 含水针铁矿，水纤铁矿
hydrograph 水文曲线；水文过程曲线；水文图；水流测量图；自记水位计
hydrograph cutting method 流量过程线切割法
hydrograph of groundwater level 地下水水位动态曲线图

hydrograph of spring discharge　泉水流量过程曲线
hydrograph of surface runoff　地面径流过程线
hydrograph separation　水文曲线分割；水文曲线分离；水文图分割；水文过程线分割；水文分析
hydrograph separation method　水文过程线分割法
hydrographer　水文地理工作者；水道测量人员；水文地理学家
hydrographic　水文的；水道测量的
hydrographic basin　河流流域；湖泊流域
hydrographic cast　测深锤
hydrographic chart　水道图；河道图；水文图；水文曲线图；水系图，水路图
hydrographic control point　海控点
hydrographic curve　水文曲线
hydrographic data　水道测量资料；水文地理资料；水文资料；水文数据
hydrographic datum　水文测量基准面；水深基面；水位基面；水位过程线零点；海图基面
hydrographic engineering survey　水利工程测量
hydrographic feature　水文特点；水文性质；水文要素
hydrographic information system　水文信息系统
hydrographic map　水文图，水道图；水文网；水系；河网
hydrographic net　水文网
hydrographic net area height survey　水网区测高
hydrographic network map　水系图
hydrographic signal　海道测量标志
hydrographic station　水文站；定点观测站
hydrographic survey　海道测量；水路测量；水文测量；河海测量
hydrographic surveying and charting　海洋测绘
hydrographic system　水系
hydrographical feature　水文地理特征
hydrographical ship　水道测量船
hydrographical table　水道测量表
hydrography　水文地理学；水文学；水道测量学；海道测量学，水道测量术；水文曲线；水文图
hydrogrossular　水钙铝榴石；水绿榴石
hydrogrossularite　水绿榴石岩
hydrogyro　流体浮悬陀螺仪
hydrohalite　水石盐
hydrohalloysite　氢化多水高岭土；水埃洛石
hydrohematite　水赤铁矿
hydrohetaerolite　水锌锰矿
hydrointegrator　水力求积仪
hydro-isobaric line　等水压线
hydro-isobath　地下水等水深线
hydroisohypse　地下水等高线；等深线
hydroisomerisation　氢化异构现象
hydroisomerization process　临氢异构化过程
hydroisopiestic line　等水压线；等测压水头线
hydroisopleth　水位等值线
hydroisopleth map　水文等值线图
hydro-isostasy　水均衡作用
hydrojacking force　水力压裂力
hydrojet　水力喷射
hydro-jet dredge　高压水喷射式挖掘船
hydro-junction　水利枢纽
hydrokinematics　水运动学；流体运动学
hydrokinetic　流体动力的；流体动力学的

hydrokinetic crushing　流体动力破碎
hydrokinetic power transmission　流体动力传送
hydrokinetic symmetry　流体动力对称
hydrokinetics　流体动力学；水动力学
hydrolaccolith　水成岩盖
hydro-linking　用高压水枪冲掘贯通
hydroliquefaction　加氢液化
hydrolite　硅华；钠菱沸石
hydrolith　水成沉淀岩；水生岩；氢化钙；水成碳酸盐碎屑岩
hydrologic　水文的
hydrologic analogy　水文比拟；水文模拟
hydrologic analysis　水文分析
hydrologic atlas　水文图集
hydrologic balance　水文平衡；水文过程量平衡
hydrologic barrier　地下水屏障；水文阻体；阻水障壁；水文堰洲
hydrologic basin　水文流域；水文盆地；地下水流域
hydrologic benchmark　水文用水准点
hydrologic budget　水文平衡计算；水均衡预测；水均衡
hydrologic characteristics　水文特征；水文特性
hydrologic computation　水文计算
hydrologic condition　水文条件
hydrologic cross-section　水文断面
hydrologic cycle　水文周期；水文循环；水循环
hydrologic data　水文资料
hydrologic design　水文设计
hydrologic discharge network　流量站网
hydrologic divide　水文分水界
hydrologic element　水文要素
hydrologic engineering geological profile along impervious curtain on dam axis　坝轴线防渗帷幕水文地质剖面图
hydrologic engineering method　水文工程方法
hydrologic equation　水量平衡方程；水文学方程
hydrologic evaluation　水文评价
hydrologic experiment　水文试验
hydrologic exploration　水文查勘
hydrologic factor　水文要素；水文因素
hydrologic feature　水文要素
hydrologic forecasting　水文预报
hydrologic frequency analysis　水文频率分析
hydrologic gradient　水文梯度
hydrologic handbook　水文手册
hydrologic investigation　水文研究；水文查勘
hydrologic manual　水文手册
hydrologic map　水文图
hydrologic mass-balance equation　水文质量平衡方程
hydrologic meteorologic survey　水文与气象测验
hydrologic model　水文模型
hydrologic monitoring　水文地质监测
hydrologic mud and sand network　泥沙站网
hydrologic network　水文网；水文站网
hydrologic observation station　水文测站
hydrologic passage　水力通道
hydrologic prediction　水文预测
hydrologic process　水文过程
hydrologic prognosis　水文预测
hydrologic property　水文特性；水文属性
hydrologic regime　水文状况；水文动态；水情
hydrologic region　水文区域

hydrologic research 水文研究
hydrologic routing 水文路由；水文路线选择
hydrologic section 水文断面
hydrologic sectional drawing 水文断面
hydrologic sensor 水文传感器
hydrologic station 水文测站；水文观测站；水文站
hydrologic storm 水文暴雨
hydrologic survey 水文调查
hydrologic survey of glaciers 冰川水文调查
hydrologic survey of lake 湖泊水文调查
hydrologic survey of river basin 流域水文调查
hydrologic survey of watershed 流域水文调查
hydrologic water level network 水位站网
hydrologic year 水文年
hydrologic yearbook 水文年鉴
hydrological 水文的
hydrological and geological data 水文地质资料
hydrological geologic condition 水文地质条件
hydrological geology 水文地质
hydrological maps 水文图
hydrological marks 水文标志
hydrological observation 水文观测
hydrological stratigraphy 水文地层学
hydrologist 水文工作者；水文学家
hydrology 水文；水文学；水文地理学
hydrolysate 水解产物
hydrolysis 水解；加水分解；热解；高温分解
hydrolyte 水解质
hydrolytic 水解的
hydrolytic acidity 水解酸度；水解性酸度
hydrolytic action 水解作用
hydrolytic dissociation 水解作用；水离解
hydrolyzable 可水解的
hydrolyzate 水解产物
hydrolyzing 水解的；水解
hydromachine 水力机械
hydromagmatic eruption 潜水水汽喷发；岩浆水汽喷发
hydromagnesite 水碳镁石；水菱镁石
hydromagnetic effect 磁性流体效应
hydromagnetic oscillation 磁流体振荡
hydromagnetics 磁流体力学
hydromanganite 水锰矿
hydromanganosite 水方锰矿
hydromanometer 流体压力计；测压计
hydromatic 液压自动传动
hydromatic brake 水刹车；水力刹车
hydromechanical 水力机械的；流体力学的
hydromechanical coupling 水力耦合
hydromechanical effects 水力学效应
hydromechanical modelling 热力耦合模型
hydromechanics 流体力学，液体力学；水力学
hydromechanization 水力机械化；液压机械化
hydro-mechanization 水力机械化
hydromelioration 水利土壤改良
hydrometallurgical purification 湿法净化
hydrometallurgy 湿法冶金，湿法冶金学；水法冶金；火冶学；热冶学
hydrometamorphic rock 水变质岩
hydrometamorphism 水变质作用；水热变质；水交代作用
hydrometasomatose 热液交替酌

hydrometeor 水汽凝结体；降水；水文气象；空中水分凝结物
hydrometeorological 水文气象学的
hydrometeorological condition 水文气象条件
hydrometeorologist 水文气象学家
hydrometeorology 水文气象学
hydrometer 比重计；流速仪
hydrometer analysis 比重计法
hydrometer condition 空气湿度
hydrometric 液体比重测定的
hydrometric basic station 水文基本测站
hydrometric float 水文测验浮标；测流浮标
hydrometric measurement 水文测验
hydrometric propeller 水速计
hydrometric section 水文测验断面
hydrometric section line 水文测验断面线
hydrometric station 水文观测站
hydrometric station control 水文测站控制
hydrometric survey 水文勘测
hydrometrist 水文测验工作者
hydrometry 水文测量学；水文测验；比重测定；流量测定法，流速测定
hydromica 水云母
hydromica clay 水云母黏土
hydromicazation 水云母化
hydromodulus 流量模数
hydromorphic 水成的
hydromorphic anomaly 水成异常
hydromorphic soil 水成土
hydromorphic zone 水成土带
hydromorphous process 水成过程；水渍过程
hydromuscovite 水白云母
hydronautics 海洋工程学
hydronepheline 水霞石
hydro-osmosis 水渗滤；水渗透
hydropathy 水疗法
hydroperiod 水文周期；洪水淹没期
hydrophane 昙白欧泊；水蛋白石
hydrophile 亲水物；亲水胶体；亲水性
hydrophilic 亲水性；吸水的；亲水的
hydrophilic colloid 亲水胶体
hydrophilic emulsion 亲水乳状液
hydrophilic mineral 亲水性矿物
hydrophilic particle 亲水性颗粒
hydrophilicity 亲水性
hydrophilic-lipophilic balance 亲水亲油平衡
hydrophilite 氯钙石
hydrophilous nature 亲水性质
hydrophobe 疏水物；疏水胶体；憎水物
hydrophobic 憎水的；疏水的；疏水性；厌水性；引火的；生火花的
hydrophobic colloid 憎水胶体；疏水胶体
hydrophobic mineral 疏水性矿物；憎水性矿物
hydrophobic nature 疏水性；憎水性
hydrophobic particle 疏水性颗粒
hydrophobic pores 疏水孔隙
hydrophobic property 疏水性
hydrophobicity 疏水性；憎水性
hydrophobicity index 疏水性指数
hydrophone 水下测听器；水中地震检波器；漏水检查

器；水听器，水声仪；水下检波器
hydrophonic detector 水声探测器
hydrophore 采水器；海水取样器
hydrophysical law 水文物理定律
hydrophysical of rock 岩石水理性
hydrophysical process 水文物理过程
hydrophysical property of soil 土的水理性质；土的水理特性
hydrophysics 水文物理学
hydrophyte 水生植物；水铁蛇纹石；含水石
hydroplant 水电站
hydroplastic 水塑性的
hydroplastic deformation 水塑性变形
hydroplastic fold 水塑性褶皱
hydroplastic sediment 含水塑性沉积物
hydroplasticity 水塑性
hydropneumatic 液压气动的；气水成的；液气的
hydro-pneumatic settlement cell 水气式沉降盒
hydro-pneumatic system 水压气动系统
hydropneumatics 液压气动学；雨成的
hydropower 水能；水力；水电
hydropower development 水电开发
hydropower plant 水力发电厂
hydropower project 水电工程；水电枢纽
hydropower station 水电站
hydropower tunnel 水力发电隧洞
hydroscience 水科学；水利科学；水文科学
hydroscope 水气计
hydroscopic 收湿的；吸湿的
hydroscopic capacity 吸湿容量
hydroscopic coefficient 吸湿系数
hydroscopic moisture 吸湿性；吸着水分，吸湿含水量
hydroscopic water 吸着水；吸湿水；易风化性；阴沟头
hydroscopic water coefficient 钻塔
hydroscopic water content 吸着水含水量；吸湿含水量
hydroscopicity 吸湿性
hydroseal 水封；液封
hydroseismic 海洋地震
hydroseparation 水力分级
hydrosequence 水文系列；水文序列
hydrosiderite 褐铁矿
hydrosilicarenite 水成硅质砂屑岩
hydrosilicate 含水硅酸盐；水化硅酸盐
Hydrosilicirudyte 石英砾岩；含水硅质砾岩
hydrosol 水溶胶体；水质土；脱水溶胶；液悬体；水溶胶
hydrosol of silicic acid 硅酸水溶胶
hydrosolvent 水溶剂
hydrosonde 水下探测器；海洋地震剖面仪；海底地貌测定仪
hydrospace detection 海洋探测
hydrosphere 水圈；水界；水气
hydrosphere geophysics 水圈地球物理
hydrosphere-lithosphere relationship 水圈-岩石圈关系
hydrospheric geochemistry 水圈地球化学
hydrosplitting 水力劈裂；水力撕裂
hydrostability 抗水稳定性
hydrostat 水压调节器；恒湿仪；锅炉防爆装置
hydrostatic 静水的；流体静力的；静水压力的；水压式的

hydrostatic accelerator 流体静压加速器
hydrostatic accelerometer 流体静力加速仪
hydrostatic axis 静水压力轴
hydrostatic brine gas 静压卤水气
hydrostatic column pressure 静液压力
hydrostatic components of stress 静水应力分量
hydrostatic condition 静水压条件
hydrostatic confining pressure 流体静力围压；静水围压
hydrostatic distribution 静水压力分布
hydrostatic equation 流体静力方程
hydrostatic equilibrium 液体静力平衡；流体静力平衡；静水压平衡
hydrostatic excess pressure 超静水压力；超流体静压力
hydrostatic fluid column pressure 静液柱压力
hydrostatic force 流体静力；静水压力
hydrostatic gradient 静水压力梯度；流体静力梯度
hydrostatic head 流体静压头；静水头
hydrostatic head method 常水头法
hydrostatic hysteresis 静水力滞后现象
hydrostatic instability 流体静力不稳定；流体静力不稳度
hydrostatic level 静压水平；水平面；静水位
hydrostatic leveling 流体静力水准测量
hydrostatic line 静水压力线
hydrostatic load 静水荷载；静液压负载
hydrostatic mud pressure 泥浆柱静水压力
hydrostatic pressing 液静压制
hydrostatic pressure 静水压力；液体静压；流体静压力；静压
hydrostatic pressure chamber 静水压力室
hydrostatic pressure hypothesis 岩体中的静水压力状态
hydrostatic pressure intensity 静水压强
hydrostatic profile gauge 静水位位移监测计；静水压力沉降计
hydrostatic pull 静水拉力
hydrostatic readjustment 静压再调整
hydrostatic restoring force 流体静恢复力
hydrostatic stability 流体静力稳定
hydrostatic state of stress 静水应力状态
hydrostatic strain 流体静应变
hydrostatic strength 流体静力强度
hydrostatic stress 静水应力；各向等应力；流体静应力
hydrostatic stress field 静水应力场
hydrostatic test 水压试验；静水压试验；流体静压试验
hydrostatic uplift 静水上托力；静水浮力
hydrostatics 流体静力学；静水压力的；水静力学
hydrostratigraphic unit 水文地层单元；含水地层单位
hydro-structure 水工结构
hydrosystem 水文系统
hydrotalcite 水滑石
hydrotaxis 趋水性
hydrotechnic amelioration 水工土壤改良
hydrotechnic construction 水工建筑
hydrotechnical construction 水工建筑；水工构筑物
hydrotechnics 水利技术；水力工程学；水工学
hydrotechnique 水利技术；水利工程；水工学
hydrotension 流体张力
hydrotest 水压试验；液压试验
hydrothermal 热液；热液的；水热的
hydrothermal activity 热液活动；水热活动

hydrothermal alteration-type deposit	热液蚀变型矿床
hydrothermal alternation	热液蚀变；水热蚀变
hydrothermal area	水热活动区；水热区
hydrothermal brine	热卤
hydrothermal carbon dioxide	水热二氧化碳
hydrothermal cement	热井水泥
hydrothermal cemented rock	泉胶岩石
hydrothermal circulation	热液循环；水热循环
hydrothermal clayrock	热液黏土岩
hydrothermal coastal vents	沿海热液喷口
hydrothermal convection	热液对流
hydrothermal convection system	水热对流系统
hydrothermal convective movement	水热对流运动
hydrothermal deposit	热液矿床
hydrothermal deposit of ocean floor	洋底热液矿床
hydrothermic eruption	水热喷发
hydrothermal exchange	热液交换
hydrothermal experiment	热水实验
hydrothermal explosion	水热爆炸
hydrothermal field	水热田；水热型热地；热水地
hydrothermal filling	热液充填
hydrothermal fluid	水热流体；地热流体；热液流体；热液
hydrothermal fluid-rock interaction	热液流体岩石相互作用
hydrothermal gas	水热成因气
hydrothermal gas geochemistry	热液气体地球化学
hydrothermal geothermal field	水热型地热地；热水地
hydrothermal glaciology	水热冰川学
hydrothermal metamorphism	热液变质；热液交代作用
Hydrothermal method	水热方法
hydrothermal mineral	热液矿物
hydrothermal mineralization	水热矿化作用；热液矿化作用；水热矿化
hydrothermal mound	热液丘
hydrothermal nontronite	热液自生绿脱石
hydrothermal ore	热液矿石
hydrothermal ore deposit	热液矿床
hydrothermal origin	热液成因
hydrothermal plume	热液羽状流；热液柱；热液流
hydrothermal process	热液作用；水热法；热液酐；水热过程
hydrothermal reaction	水热反应
hydrothermal reservoir	水热型热储；热水储集层
hydrothermal resources	水热资源
hydrothermal sediment	热液沉积物
Hydrothermal sedimentation	热液沉淀；热液沉积
hydrothermal solution	热水溶液
hydrothermal spring	热水泉
hydrothermal stability	水热稳定性
hydrothermal stage	热液期；热液阶段
hydrothermal synthesis	热液合成
hydrothermal theory	热液说
hydrothermal vent	热液喷口；热水泉口；喷汽孔
hydrothermalism	水热活动；水热过程；热液作用；热液论
hydrothermic coefficient	水热系数
hydrothermic factor	水热因子
hydrotimeter	水硬度计
hydrotropic action	促水化作用
hydrotropism	向水性
hydrous	含水的；水化的；水状的；含氢氧根的；水合的
hydrous complexes	含水配合物；含水络合物
hydrous facies	含水相
hydrous lime	熟石灰
hydrous manganese oxide	含水氧化锰
hydrous mantle melting	含水地幔熔融
hydrous mica	水云母
hydrous pyrolysis	有机物高温水解作用；加水热解
hydrous salt	含水盐
hydrous sulphate	水合硫酸盐
hydrous sulphate of lime	石膏；硫酸钙
hydrovalve	液压阀
hydrowave	水波
hydroxide ion	氢氧离子；氢氧根离子
hydrozincite	水锌矿
hyetal	雨的；降雨的
hyetal coefficient	降雨系数；雨量系数
hyetology	降水学；降雨量学；降雨学
hyetometer	雨量计
hygral expansion	湿膨胀
hygral shrinkage	润湿收缩
hygrograph	湿度计；湿度记录表；湿度记录仪
hygrology	湿度学
hygromagmatophile	湿亲岩浆的
hygromagmatophile element	湿亲岩浆元素
hygromagmatophile water	吸湿水；吸湿度
hygrometer	湿度计，湿度表；测湿器
hygronom	湿度计
hygrophilous	适湿的
hygroscope	湿度计，测湿器；验湿器；吸水
hygroscopic	吸湿的；收湿的
hygroscopic capacity	吸湿容量
hygroscopic coefficient	吸湿系数
hygroscopic effect	吸湿作用
hygroscopic moisture	吸湿度；吸湿水，吸着水，湿存水；有效容量；预涂
hygroscopic nucleus	吸湿核；吸湿核心
hygroscopic property	吸水性；收湿性
hygroscopic soil water	吸湿土壤水分
hygroscopic water	吸湿水；吸着水；吸湿器可计量的水
hygroscopic water content	吸湿含水量
hygroscopicity	吸湿性，吸水性；吸湿度；润湿度
hygroscopy	湿度测定
hygrothermograph	温湿计
hylotropy	恒溶性；恒沸性
hypabyssal	半深成的；浅成的；浅成岩
hypabyssal intrusive	浅成侵入体；半深成侵入体
hypabyssal rock	浅成岩；半深成岩
hypautochthonous granite	半原地花岗岩
hypautochthony	亚原地生成；微异地生成；微异地生成煤
hypautomorphic	半自形的
hypautomorphic crystal	半自形晶
hypautomorphic granular	半自形粒状
hypautomorphic granular texture	半自形粒状结构
hypautomorphic-granular fabric	半自形粒状组构
hyper normal distribution	超正态分布
hyperacidite	超酸性岩
hyperalkaline	富碱性的；超碱性的；超碱性
hyperbaric	高压的

hyperbasite	超基性岩
hyperbola	双曲线
hyperbola equation	双曲线方程
hyperbola function	双曲线函数
hyperbolic and eccentric arc structure	双曲偏心拱结构
hyperbolic decline law	双曲线递减规律
hyperbolic model	双曲线模型
hyperbolic paraboloid	双曲抛物面
hyperbolic positioning	双曲线定位；距离差定位
hyperbolic stress distribution	双曲线应力分布
hyperboloid of revolution	旋转双曲面
hyperborean	极北的；北极的
hyperconcentrated flows	高含沙水流
hyperelastic law	超弹性定律
hyperelastic material	超弹性材料
hyperelastic model	超弹性模型
hypereutectic	过共晶体
hypereutectoid	过共析体
hyper-finite element	超有限元
hyperfrequency waves	微波；超高频波
hyperfusible component	超熔组分
hypergene	表生；表生的；浅成的；下降的；下降溶液生成的
hypergene mineral	表生矿物
hypergene mobility	浅成活动性
hypergene structure	表生构造
hypergene zone	浅成带
hypergenesis	表生作用；深埋成岩作用；表面蚀变
hypergeometric differential equation	超几何微分方程
hyperglyph	风化印痕
hyperhaline	高盐的
hyperite	紫苏辉石；橄榄苏长岩；辉长苏长岩
hyperitite	无橄紫苏岩
hypermelanic	深暗色的
hypermelanic rock	深暗色岩；超暗色岩
hypermetamorphism	超变质作用
hyperpiestic water	超承压水；自流水
hyperplasticity	超塑性
hyperpressure	超压
hyperpycnal flow	异重流；高密度流
hyperpycnal inflow	重入流
hypersaline	超盐度的
hypersaline brine	超浓卤水；高盐度卤水
hypersaline fluid	高咸度流体
hypersaline geothermal brine	超浓地热卤水
hypersaline water	超咸水
hypersalinity	高矿化；高盐度
hypersolvus granite	超熔花岗岩
hyperstatic	超静定的
hyperstatical	超静定的
hypersthene	紫苏辉石
hypersthene chondrite	紫苏辉石球粒陨石
hypersthene granite	紫苏花岗岩
hypersthenfels	苏长岩类
hypersthenite	紫苏岩；苏长岩；紫苏辉石岩
hypersubsidence	超沉降
hyperthermal spring	高温热泉
hyperthermal stage	高温期
hyperthermal water	高温水
hyperthermic	超热状况
hypertonicity	高渗性
hypertron	超小型电子射线加速器
hypertrophy	过渡生长
hypervelocity	超高速
hypervelocity free flow	超高速自由流
hypichnia	底迹；底生迹
hypidioblast	半自形变晶
hypidioblastic	半自形变晶的
hypidiomorphic	半自形的
hypidiomorphic crystal	半自形晶
hypidiomorphic granular	半自形粒状的；半自形粒状
hypidiomorphic granular texture	半自形晶粒状结构；半自形粒状结构
hypidiomorphic structure	半自形构造
hypidiomorphic texture	半自形结构
hypidiomorphic-granular	半自形粒状
hypidiotopic	半自形的；半自形组构
hypidiotopic crystal	半自形晶
hypo eutectic	亚共晶
hypo gene rock	深成岩
hypobasite	深成超基性岩
hypobatholithic	深岩基的
hypobatholithic zone	深成岩基带
hypocenter computing process	震源计算方法
hypocenter data file	震源资料档案
hypocenter distance	震源距
hypocenter location	震源定位
hypocenter of the explosion	爆炸源
hypocenter parameter	震源参数
hypocentral distance	震源距离；震源距
hypocentral location	震源定位；震源位置
hypocentral plot	震源图
hypocentral region	震源区
hypocentral shift	震源迁移
hypocentre	震源
hypocentre parameter	震源参数
hypocentrum	震源
hypocrystalline	半晶质的；半晶质
hypocrystalline porphyric	半晶质斑状的
hypocrystalline porphyritic	半晶质斑状的
hypocrystalline texture	半晶质结构
hypo-differentiation	深成分异作用
hypodispersion	平均分布
hypoelastic law	次弹性定律
hypoelastic material	次弹性材料
hypoelastic model of soil	土的次弹性模型
hypoelastic yield	次弹性屈服
hypoelasticity	次弹性
hypoeutectic	亚共晶；亚共晶的
hypofiltration	深成渗透作用
hypofocus	震源
hypogeal	地下的
hypogee	岩洞建筑，地下建筑；循环回水管；地下建筑
hypogeic	地下的
hypogene	上升；深成；上升溶液生成的；内力的；内生的；上升的；深生的
hypogene action	深成作用
hypogene deposit	深成矿床；内生矿床
hypogene fluid	深成流体；上升流体
hypogene magma	深成岩浆

hypogene mineral 深成矿物
hypogene mineral deposit 深成矿床
hypogene mobility 深成活动性
hypogene ore 深成矿石
hypogene ore mineral 深成矿物；原生矿物
hypogene rock 深成岩
hypogene spring 深部泉；内生泉；上升泉
hypogene vein deposit 上升脉状矿床
hypogene water 上升水；深成水
hypogenic 上升生成；深成的；上升的；内力的
hypogenic action 深成作用
hypogenic alteration 深成蚀变
hypogenic fluid 深成流体
hypogenic ore deposit 深成金属矿床
hypohyaline 半玻质的
hypolimnion 均温层；下层滞水带；冷水层；深水层；深层层状水；湖底静水层
hypomagma 深部岩浆
hypomagmatic 深部岩浆的
hypomagmatic dyke 深成岩脉；深成岩墙
hypometamorphic 深带变质的；深变质的
hypometamorphic rock 深变质岩
hypometamorphism 深变质作用；深带变质作用
hypomigmatization 深混熔作用
hypophtanite 次致密硅质岩
hypopiestic water 次承压水；半自流水
hypopycnal flow 低密度流
hypopycnal inflow 轻入流
hyporelief 底浮痕；下浮雕；底面起伏
hyposeismic 深震
hypostasis 沉渣
hypostratotype 次层型；亚地层型
hypostratum 下壳层
hypotaxic deposit 地面矿床
hypothermal 深成高温热液的；深热液的；低体温的
hypothermal deposit 高热液矿床；深成热液矿床
hypothermal fluid 低温流体
hypothermal process 深成高温热液作用
hypothermal rock 高温岩石
hypothermal spring 低温热泉
hypothermal vein 高温热液矿脉；深成热液矿脉
hypothermal water 低温热水
hypothermophilous 适低温的
hypothesis of bench subsidence 台阶下沉假说
hypothesis of horizontal movement 水平论；活动论
hypothesis of isostasy 地壳均衡假说
hypothesis of mantle creep 地幔蠕动说
hypothesis of meteorite impact mineralization 陨石冲击成矿说
hypothesis of radioactive cycle 放射性旋回说
hypothesis of vertical movement 垂直论
hypothetic resource 假定资源；推测资源
hypothetical anomaly 假说异常
hypothetical aquifer 假设的含水层；推定的含水层
hypothetical hydrograph 设计水文过程线

hypothetical reserves 推测储量；假定储量；可能储量
hypothetical resources 推测资源；假定的资源
hypothetical roof block 假设的顶板岩块
hypozoic 深生
hypozonal metamorphism 深成变质；深成变质作用
hypsograph 测高仪
hypsographic curve 陆高海深曲线；等高线；高度深度曲线；等深线
hypsographic feature 地形特征；地势起伏
hypsographic map 地势图；分层设色图；等高线地形图
hypsography 地貌表示法；测高学；等高线法；地形测绘学；测高仪
hypsometer 测高计；沸点气压计；沸点测高计；沸点测定器
hypsometric 测高的；有标高的
hypsometric analysis 测高线分析，等高线分析；面积高程分析
hypsometric chart 等高线地形图
hypsometric curve 陆高海深曲线，湖海等深线；高度深度曲线；测高线
hypsometric formula 测高公式
hypsometric map 分层设色图；等值线图
hypsometric method 分层设色法
hypsometric tending method 分层设色法
hypsometric unit 分层设色单位
hypsometry 测高术；测高法
hypsometry altimetry 测高术
hysteresis 滞后；滞后；滞变；滞回
hysteresis characteristic 滞变特性
hysteresis coefficient 滞后系数；磁滞因数
hysteresis compaction 滞后压实；滞回压实
hysteresis diagram 滞变图
hysteresis effect 滞后效应；磁滞效应
hysteresis loop 滞后回线，滞后环
hysteresis modulus 回滞模量；滞回模量
hysteresis of soil water 土壤水的滞后
hysteresis phenomenon 滞后现象
hysteresis loop 滞后回线，滞后环
hysteretic 滞变；滞后
hysteretic angle 滞后角
hysteretic area 滞变面积
hysteretic behaviour 滞变性能
hysteretic curve 滞变曲线
hysteretic effect 滞变效应
hysteretic elastic model of soil 土的滞弹性模型
hysteretic envelope 滞变包线
hysteretic strain energy 滞变应变能
hysterocrystalline 次生结晶作用
hysterocrystallization 次生结晶作用
hysterogenetic 岩浆末期的
hysterogenic 后生的
hysterogenous 表成次生沉积的；后生的
hysteromorphous 表成次生沉积的；后生的
hysteromorphous deposit 表成次生沉积矿床
hythergraph 温湿图

I

IAEG 国际工程地质协会
ianthinite 水斑铀矿
ianthite 水铀矿
I-arch 工字钢拱
I-beam 工字钢
IBRD 国际复兴开发银行；世界银行
ICE 英国土木工程师学会
ice accretion 结冰；积冰；覆裹冰层
ice accumulation 积冰
ice action 冰的作用
ice age 冰河时代；冰期
ice apparatus 测冰器
ice apron 挡冰板
ice atlas 冰图
ice auger 冰钻
ice avalanche 冰崩
ice band 冰层
ice bar 冰坝
ice barrier 冰障；冰堤
ice barrier basin 冰围盆地
ice blister 冰堆
ice block 冰块
ice boat 破冰船
ice bond 冰连接；冰胶结
ice bore 冰钻孔
ice boulder 冰川巨砾
ice breaking 破冰
ice breccia 冰角砾
ice bridge 冰桥
ice cake 冰块
ice calorimeter 冰量热计
ice cap 冰帽；冰盖；冰冠
ice carried placer 冰积砂矿
ice cascade 冰瀑布
ice cave 冰穴
ice cavern 冰洞
ice chart 冰情图；冰区图；海冰图
ice chisel 冰凿
ice chute 排冰道
ice climate 冰雪气候
ice clogging 冰塞
ice cloud 冰云
ice concrete 冰混凝体；冰混凝土
ice condition 结冰条件
ice cone 冰锥
ice contact deposit 冰川接触沉积；冰界沉积
ice contact sediment 冰界沉积
ice content 含冰率
ice control 防冰措施
ice control structure 防冰构筑物
ice core 冰芯
ice core drilling 冰芯钻探
ice cover 冰层
ice crack 冰裂；冰冻裂隙

ice creep 冰蠕动；冰蠕变
ice crevasse 冰隙
ice crust 冰壳
ice crystal 冰晶体；冰晶
ice cylinder 冰柱
ice dam 冰坝
ice dammed glacial 冰川冰堰塞湖
ice discharge 冰流量
ice divide 分冰岭
ice dome 冰穹
ice doubling 抗冰衬板
ice drift 流冰
ice engineering 冰工程学
ice epoch 冰期
ice erosion 冰蚀，冰侵蚀
ice escape channel 泄冰渠
ice fall 冰瀑；冰崩
ice field 冰原
ice float 浮冰
ice float velocity 浮冰流速
ice floe 浮冰
ice floe survey 冰凌调查
ice flood 凌汛
ice flow 冰流；冰水径流
ice flow control 冰流控制
ice flower 冰花
ice fog 冰雾
ice foot 冰壁；冰棚
ice forecast 冰情预报
ice forecasting 冰情预报
ice formation 结冰；成冰作用
ice free port 不冻港
ice fringe 冰条带；冰条纹
ice gage 测冰尺
ice gang 融冰流
ice gate 排冰门
ice gauge 测冰尺
ice gland 冰腺
ice gorge 冰峡
ice growth rate 积冰率；冰增长率
ice guard 拦冰栅
ice gush 冰涌
ice hail 冰雹
ice harbour 冻港
ice hummock 冰堆
ice in soil 土中冰
ice island 浮冰冰岛
ice jam 冰塞；流冰壅塞；冰障；冰坝
ice jam blasting 冰凌爆破
ice jam flood 冰壅凌汛
ice jam stage 冰坝水位
ice laid drift 冰碛
ice lake 冰川湖
ice layer 冰层

ice layer with soil	含土冰层
ice ledge	冰壁；冰棚
ice lens	底冰；冰透镜体
ice load	覆冰荷载
ice mantle	冰盖；冰原
ice margin	冰边；冰缘
ice margin delta	冰水三角洲
ice mark	冰川标志
ice mechanics	冰力学
ice mill	冰川磨蚀
ice mound	冰堆
ice mountain	冰山
ice movement	冰川运动
ice needle	冰针
ice nucleus	冰核
ice pack	积冰；大片浮冰
ice patch	流冰区
ice pavement	冰川磨蚀面
ice pellet	冰丸
ice period	冰冻期
ice pillar	冰柱
ice platform	冰质平台
ice plug	冰塞
ice point	冰点
ice pressed form	冰压地形
ice pressure	冰压力
ice prevention	防冰凌
ice push	冰推力
ice pushed terrace	冰推阶地；冰成阶地
ice quake	冰崩
ice raft	浮冰排
ice rafting	冰筏
ice rain	冰雨
ice rampart	冰河堤
ice ran	浮冰流；融冰流；淌凌
ice reconnaissance	冰情勘察
ice reef	冰礁
ice regime	冰情
ice regime forecast	冰情预报
ice regime observation	冰情观测
ice reinforced drillship	抗冰钻井船
ice rind	冰壳；冰皮
ice rise	冰隆
ice river	冰川；冰河
ice run	流冰；漂冰；冰凌
ice runoff	冰水径流；融冰径流
ice sample	冰样
ice segregation	冰隔离
ice sheet	冰盖；冰原
ice shelf	冰架；冰棚
ice shelf fluctuation	冰架波动
ice shock	冰震
ice situation	冰情
ice slope	冰坡
ice slush	冰凌
ice spar	透长石；冰晶石
ice stone	冰晶石
ice storm	冰暴
ice stream	冰川；冰河；冰川水流
ice thrust	冰川推力
ice tongue	冰舌
ice tower	冰塔
ice undercurrent	冰川潜流
ice vein	冰脉
ice wall	冰壁；冻结壁
ice warning	冰情警报
ice water	冰水
ice wedge	冰楔
ice wedging	冰楔作用
ice wind	冰川风
ice-anchor	冰锚
ice-bar	冰塞
ice-bearing current	挟冰水流
iceberg	冰山
iceblink	冰映光
ice-border drainage	冰川边流
ice-borne sediment	冰积物
ice-bound	被冰堵塞的
ice-breaker	破冰船
ice-breaking cutwater	破冰体
icebreaking service	破冰作业
ice-cap climate	永冻气候
ice-cold	冰冷的
ice-contact deposit	冰界沉积
ice-contact end moraine	终碛冰界
ice-contact fan	冰水扇
ice-contact feature	融冰沉积地形
ice-contact forms	融冰沉积地形
ice-contact plain	冰砾平原
ice-contact stratified drift	融冰成层沉积
ice-cored moraine	芯冰碛；芯状冰碛石
ice-cover water rating-curve	冰盖水位流量曲线
ice-covered	冰川覆盖的
ice-covered channel	冰封的渠道
ice-crystal cast	冰晶铸形
ice-crystal mark	冰晶痕
ice-crystal theory	冰晶学说
iced mixing	加冰拌合
ice-dam lake	冰坝湖，冰塞湖
ice-dammed lake	冰坝湖，冰塞湖
ice-erosion	冰蚀；冰川侵蚀
ice-float	浮冰
ice-free	无冰的；不冻的
ice-free area	无冰区；不冻区
ice-free harbour	不冻港
ice-free period	解冻期；不冻期
ice-free port	不冻港
ice-free waterway	无冰水道；不冻水道
ice-laid deposit	冰川沉积
ice-laid drift	冰碛；冰碛物
Iceland	冰岛
Iceland crystal	方解石；冰洲石
Iceland plume	冰岛地幔柱
Iceland spar	方解石；冰洲石
Icelandic low	冰岛低压
Icelandic shield volcano	冰岛型盾形火山
Icelandic-type eruption	冰岛型喷发
icelandite	冰岛岩
ice-loading	冰荷载
ice-marginal lake	冰缘湖；冰前湖

ice-melt water	融冰水
ice-plagued area	冰患海域
ice-pushed ridge	冰脊
ice-pushed terrace	冰壅阶地
ice-removal salt	防冻盐
ice-scarp	冰崖
ice-scoured plain	冰蚀平原
ice-shelf	冰架
ice-strengthened tanker	抗冰油轮
ice-surface stream	冰面水流
ice-wall forming	形成冰墙
ice-wedge polygon	冰楔多边形
ice-worn	冰蚀的
ichnite	足迹化石；遗迹；印痕
ichnite footprint	动物足迹
ichnocoenose	足迹化石群
ichnocoenosis	遗迹群落；痕迹化石群体
ichnofacies	遗迹相；虫迹相
ichnoflora	植物遗迹群
ichnofossil	遗迹化石；足迹化石
ichnograph	平面图；平面绘图法
ichnography	平面图；平面绘图法
ichnolite	足迹化石；化石足迹
ichnolithology	遗迹岩石学
ichnological	足迹化石的
ichnology	遗迹学
ichor	岩汁；残余岩浆
ichthyolite	鱼的化石
ichthyoliths	鱼石类
icicle	冰柱；垂冰
icicle prevention device	防冻装置
icing	冰川；冰原；冰积
icing load	结冰载荷
icing system	冰冻系统
ICOLD	国际大坝委员会
icon	图像；图标
iconography	插图
iconometer	测距镜；光像测定器；量影仪
icosahedron	二十面体
icosinene	液态碳化氢
icy region	冰覆盖区
icy shower	冰雨
icy soil	冻土
iddingsite	伊丁石；褐绿泥石
ideal adsorbed solution theory	理想吸附溶液理论
ideal average displacement	理论平均排量
ideal behavior	理想特性
ideal black-body	理想黑体
ideal body	理想体
ideal boundary	理想边界
ideal brittle material	理想脆性材料
ideal chain	理想链
ideal climate	理想气候
ideal condition	理想状态；理想条件
ideal constraint	理想约束
ideal coordinate	理想坐标
ideal crystal	理想晶体
ideal crystalline form	理想晶形
ideal cycle	理想循环
ideal density	理想密度
ideal diagram	理想图
ideal efficiency	理想效率
ideal elastic medium	理想弹性介质
ideal elasticity	理想弹性
ideal elastoplastic	理想弹塑性
ideal elastoplastic model	理想弹塑性模型
ideal energy grade line	理想能量梯度线；理想能量均衡线
ideal filter	理想滤波器
ideal fluid	理想流体
ideal form	理想晶形
ideal function	理想函数；广义函数
ideal gas	理想气体
ideal gas law	理想气体定律
ideal grading curve	理想颗粒级配曲线
ideal lattice	理想晶格
ideal liquid	理想液体
ideal lubricant	理想润滑剂
ideal model	理想模型；理想模式
ideal network	理想网络
ideal orientation	理想取向；理想方位
ideal pendulum method	理想钟摆法
ideal plastic body	理想塑性体
ideal plastic solid	理想塑性固体
ideal polarized electrode	理想极化电极
ideal productivity index	理想采油指数；理想生产指数
ideal profile	理想剖面
ideal ray	理想射线
ideal recovery	理想采收率
ideal rheological body	理想流变体
ideal sea level	理想海水面
ideal section	理想剖面
ideal solid	理想固体
ideal solution	理想溶液
ideal time domain filter	理想时域滤波器
ideal tracer	理想示踪剂
ideal value	理想值
ideal viscous fluid	理想黏性流体
ideal wave	理想波
ideal well	理想井
ideal white noise	理想白噪声
idealized character	理想特性
idealized characteristic	理想特性
idealized earthquake ground motion	理想化地震动
idealized elastic continual mass	理想弹性连续体
idealized fragmental mass	理想碎屑体
idealized model	理想模型
idealized plan	理想平面图
idealized pressure distribution	理想化的压力分布
idealized reservoir	理想化储层
idealized section	理想剖面
idealized source model	理想震源模型
idealized structure	理想化结构
ideally rigid body	理想刚体
ideally-elastic member	理想弹性构件
idemfactor	等幂矩阵；归本因素
idempotence	幂等性
idempotent matrix	幂等矩阵
identical equation	恒等式
identical map	等角投影地图

identical quantity	恒等量
identical relation	恒等式
identical strata	相同地层
identical substitution	恒等代换
identical transformation	恒等变换
identifiability	可辨识性
identifiable factor	鉴定因素
identifiable object	可辨认地物
identifiable point	可辨认点
identification	鉴别；确定；识别
identification and investigation	鉴定与调查
identification code	识别码
identification key	辨认标志
identification label	识别标签
identification mark	识别证；识别符号
identification marking	认识标记；标志
identification method	识别方法
identification number	标志号；识别号；设备编号
identification of input	输入反演
identification of risk	风险辨识
identification of seam	岩层鉴别
identification of soil	土的识别
identification of strata	地层鉴别
identification system	识别系统
identification technique	识别技术
identification test	鉴别试验
identification threshold	鉴定阈值
identified	已证实的；已鉴别的；已查明的
identified mineral resource	查明矿产资源
identified resource	已查明资源
identifier	鉴定剂
identify	辨认；鉴定；测定
identifying feature	识别要素
identifying information	识别信息
identifying signature	识别特征信息
identity element	单位元素
identity matrix	单位矩阵
identity operator	恒等算子
identity period	恒等周期
identity relation	恒等关系式
ideograph	模式图；表意符号
ideologist	思想家；理论家
ideology	思想意识；思想方式；意识形态；空想
idioblast	自形变晶
idioblastic	自形变晶的
idioblastic texture	自形变晶结构
idiochromatic	本色的，白色的；本质的
idiogenous	同成的；同生的
idiogeosyncline	山间地槽；独地槽；晚期地槽
idiomorphic	自形的
idiomorphic crystal	自形晶
idiomorphic granular texture	自形粒状结构
idiomorphic soil	自型土
idiophanous	自现干涉图的
idiostatic	等位差的
idiot stick	铲，锹；便携式螺旋钻；挖土器
idiotopic	自形的
idiotopic crystal	自形晶
idle drum	随转滚筒；空转滚筒；惰筒
idle equipment	闲置设备
idle hour	停工时间
idle mine	停产矿井
idle test	空载试验
idle time	停机时间
idle work	虚功；无用功
idocrase	符山石
idrialite	绿地蜡
igalikite	伊加利科石
igmerald	人造绿宝石
igneous	火成的；熔融的；岩浆的
igneous accumulate	火成堆积
igneous accumulation	火成堆积
igneous activity	火成活动；岩浆活动
igneous agglomerate	火成集块岩
igneous autobreccia	自碎火成角砾岩
igneous body	火成岩体
igneous breccia	火成角砾岩
igneous complex	火成杂岩
igneous cycle	火成旋回
igneous drilling	火力钻眼；热力钻眼
igneous emanation	岩浆喷射
igneous exhalation	岩浆喷气
igneous geothermal well	火成岩地热井；岩浆岩地热井
igneous intrusion	火成岩侵入
igneous lamination	火成纹理
igneous magma	岩浆
igneous metamorphism	火成变质作用
igneous petrology	火成岩石学
igneous plug	岩栓；火成岩栓，火山颈
igneous province	火成岩区
igneous rock	火成岩；岩浆岩
igneous rock series	火成岩系
igneous stock	岩浆岩干
igneous structure	火成岩构造
igneous structure surface	火成结构面
igneous substratum	火成底层；火成岩基
igneous tectonic assemblage	火成构造组合
ignimbrite	熔结凝灰岩；熔灰岩
ignimbrite eruption	熔结凝灰岩喷发
ignimbrite texture	熔结凝灰结构
ignitability	可燃性；着火性
ignitable coal	自燃煤层
ignitable dust	可燃粉尘
ignitacord	点火线
igniter	点火器；打火机；引火剂
igniter cord	点火绳；点火索
igniter fuse	导火线
igniter gas	点火气
igniter pellet	点火雷管
igniter plug	点火塞
igniting circuit	点火电路
igniting composition	起火剂
igniting layer	点火层
igniting material	点火料
igniting primer	起爆药包；雷管
ignition	点火；灼烧；发火；燃烧
ignition alloy	点火合金；发火合金
ignition angle	点火角
ignition cap	雷管
ignition cause	着火原因；火源

ignition center	发火中心
ignition charge	导火炸药
ignition compound	发火剂
ignition control compound	易燃化学品
ignition delay	着火迟后
ignition delay time	点火延迟时间
ignition device	点火装置；起爆装置
ignition explosive	导火炸药
ignition fuse	导火线
ignition gallery	点火平巷
ignition hazard	发火危险；燃烧危险
ignition heat	着火热
ignition indication	着火显示
ignition inhibitor	爆炸抑制剂
ignition lag	点火滞后
ignition location	点火位置
ignition loss	灼失量；烧失量
ignition order	点火次序；起爆次序；放炮次序
ignition pattern	点火次序，放炮程序
ignition period	发火期
ignition plug	火花塞
ignition point	燃点；着火点
ignition powder	点火药；引燃剂
ignition property	着火特性
ignition rate	着火率
ignition rectifier	灼管；引燃管
ignition residue	灼烧残渣
ignition resistance	起爆电阻
ignition scattering	起爆时值分散
ignition scope	点火检查示波器
ignition source	火源
ignition temperature	点火温度；着火温度；燃点
ignition temperature of methane	沼气着火温度
ignition time	点火时间
ignition voltage	点火电压；起爆电压
ignition zone	点火带
ignition-first method	预先灼烧法
ignition-first treatment	预烧处理
ignitor	点火器，打火机；引火剂
ignitor pad	点火药包
ignitron	点火管，引燃管
IGS	国际土工合成材料学会；惯性制导系统
ihleite	黄铁矾；叶绿矾
I-iron	工字铁
IJCG	国际煤田地质学报
IJG	国际地质力学学报
IJNAMG	国际地质力学数值和解析方法学报
IJNME	国际工程数值方法学报
I-joist	工字形小梁
IJRMMS	国际岩石力学与采矿科学学报
ijussite	棕闪钛辉岩
ill air	不洁空气
ill conditioning	病态条件
ill effect	有害作用
ill use	不正确使用
ill-condition matrix	病态矩阵
ill-defined boundary	不清晰的边界
ill-defined divide	不明显的分水岭
ill-defined problem	不完善的问题
illegal	不合法的；非法的
illegal character	非法字符
illegal code	非法代码
illegal command	非法指令，非法信号
illegal contract	非法合同
illegal dual operation	非法操作
illegal encroachment	非法据用
illegal excavation	非法挖掘
illegal extension	非法扩建
illegal operation	非法操作
illegal product	非法产品
illegal production	非法生产
illegal profit	非法利润
illegal structure	违例构筑物；违例搭建物
illegal-command check	非法字符校验
illegitimate error	不合理误差
illimerization	黏粒的移动
Illinoian Glacial Stage	伊利诺伊冰期
Illinoian phase	伊利诺伊冰期
illite	伊利石；水白云母
illite crystallinity	伊利石结晶度
illite glimmerton	伊利石
illogic	不合逻辑
illogical	非逻辑的
illuminance level	照明度
illuminance of ground	地面照度
illuminance uniformity	照度均匀度
illuminated display	灯光显示
illuminated panel system	信号灯指示器
illuminated sign	照明标志
illuminating effect	照明作用；照明效果
illuminating engineering	照明工程
illuminating gas	可燃气体
illuminating lamp	照明灯
illuminating mark	光点测标
illuminating power	亮度
illumination	照明，照度；电光饰，灯饰；阐明；启发
illumination design	照明设计
illumination equipment	照明装置
illumination intensity	照度
illumination lamp	照明灯具
illumination length	光照长度
illumination level	照明级；照明程度；照度级
illumination measurement	照度测定
illumination meter	照度计
illumination photometer	照度计
illumination source	光源
illumination voltage stabilizer	照明稳压管
illuminator	发光器；照灯，照明装置
illuminometer	照度计；照度仪
illustrate	插图，证解，例证
illustration	说明；图解；例释；插图
illustrative	图解的；例释的；插图的
illuvial	淀积的
illuvial clay	淀积黏粒；淀积黏土
illuvial deposit	淋滤矿床
illuvial gravel	残积砾；淋滤砾；淀积砾
illuvial horizon	淋积层；淀积层
illuvial humus	淀积腐殖质
illuvial iron	淀积铁
illuvial layer	淋积层；淀积层

illuvial soil 淀积土
illuviation 淀积作用；淋积作用
illuvium 淋滤层；淋积层；淀积层
ilmenite 钛铁矿
ilmenite cement 钛铁矿粉加重水泥
ilmenite sand 钛铁矿砂
ilmenitite 钛铁岩
ilmenomagnetite 钛铁磁铁矿
ilmenorutile 黑金红石
ilsemannite 蓝钼矿
ilvaite 黑柱石
ilyogenous 泥质的
ilzite 英云微闪长岩
image 像，图像
image acquisition 图像搜索；图像采集
image acquisition system 图像采集系统
image activator 映像激活程序
image amplifier 图像放大器；影像放大器
image analysis 图像分析
image angle 像角，图像倾斜角
image annotation 图像注释；影像批注
image annotation record 图像注释记录；影像批注记录
image annotation tape 图像注释带
image at night 夜间影像
image attenuation constant 影像衰减常数
image averaging 图像取均值
image binary 图像二值化
image boundary 图像边缘；影像边缘
image brightness 图像亮度；影像亮度
image carrying fibre bundle 导像纤维束
image circle 成像圈；像圈
image classification 图像分类法；影像分类法
image coding 图像编码
image composite 图像合成
image compression encoding 图像压缩编码
image confusion 图像弥散
image contract 图像缩小
image conversion system 图像转换系统
image converter 光电图像变换管
image coordinate 影像坐标
image correcting 图像校正
image corrector 图像改正器
image correlation 影像相关
image curvature 图像弯曲
image data 影像数据；图像数据
image data acquisition 图像数据采集
image data base 图像数据库
image data compression 图像数据压缩
image data tablet 图像数据输入板；影像数据输入板
image database 影像数据库
image defect 图像失真
image definition 图像清晰度；影像清晰度
image description 影像描绘
image detail 图像细节；影像细节
image diagnostic system 图像诊断系统；影像诊断系统
image diffusion 图像模糊
image digitization 图像数字化
image digitizer 图像数字化仪；图像数字转换器
image displacement 图像位移；影像位移
image display 图像显示
image distortion 图像畸变
Image element 像素；像元；像点
image encoding 图像编码；影像编码
image enhancement 图像增强
image enhancement technology 图像增强技术
image enlargement 影像放大
image fault 影像失真
image feature 影像特征
image fidelity 图像保真度；影像保真度
image file 图像文件
image filter matrix 图像滤波器矩阵
image filtering 滤像
image flicker 图像闪烁
image format 影像格式
image formation 成像
image frequency 像频；帧频
image fusion 影像融合
image geometry 影像几何
image grayscale 影像灰度
image hologram 影像全息图
image horizon 图像地平线
image identification 图像识别
image impedance 影像阻抗
image input device 图像输入设备；影像输入设备
image intensification 图像增强
image intensifier 图像增强器；变像管
image intensity 图像强度；影像强度
image interpretability 影像可判性
image interpretation 图像解泽；影像判读
image interpretation method 图像解译方法
image library 图像库，图片库
image log 成像测井
image magnification 图像放大；影像放大
image map 影像地图
image masking 影像掩蔽
image matching 图像匹配；影像匹配
image method 镜像法；映射法；图像法
image method of well 井的镜像法
image mixing 图像混合
image mosaic 影像镶嵌；图像镶嵌
image motion compensation 像移补偿
image multiplier 图像放大器
image output 图像输出；影像输出
image output device 图像输出设备；影像输出设备
image output transformer 图像信号输出变压器
image overlaying 图像复合
image pattern 像图
image plane 像平面
image point 像点
image processing 图像处理
image processing facility 图像处理设备；影像处理设备
image projector 图像投影仪；影像投影仪
image pulse 映像脉冲
image pyramid 影像金字塔
image quality 影像质量；图像质量
image recognition 图像识别
image reconstruction 图像重建
image registration 影像配准
image representation 图像显示；影像显示
image reproducer 影像重现装置

image reproduction	影像重现
image resolution	影像分辨力；像元地面分辨力
image restoration	影像复原；图像复原
image rotation	图像旋转；影像旋转
image scale	图像比例尺；影像比例尺
image scanner	图像扫描器
image segmentation	图像分割；影像分割
image sensing	图像传感
image sensor	图像传感器
image sharpness	图像清晰度；影像清晰度
image signal	图像信号
image size	图像尺寸；影像尺寸
image smoothing	图像平滑；影像平滑
image sort	映像分类
image source	虚震源；像源
image space coordinate system	像空间坐标系
image storage	影像存储；图像存储
image storage device	录像装置；图像存储设备
image subtraction	图像相减；影像相减
image theory	镜像原理
image transfer	图像转绘；影像转绘
image transfer technique	影像转绘技术
image transform	图像变换；影像变换
image transformation	图像变换
image transformation matrix	影像转换矩阵
image translation vector	影像转移向量
image translator	图像转换器；影像转换器
image transmission	传真；图像传输
image type	影像型式
image understanding	图像理解；图像识别
image well	映像井；映射井；推测井
image zooming	图像变焦
imaged nucleus	成像基点
imagery	影像；造像术，成像；雕像
imagery digitisation	图像数字化
image-superimposable molecule	镜像重合分子
image-transfer	图像传输
image-well method	映像井法；设想井法
image-well theory	映像井理论；映射井理论
imaginal well	镜像井
imaginary	虚的；想象的；假想的
imaginary accumulator	虚数累加器
imaginary axis	虚轴
imaginary boundary surface	虚边界面
imaginary circle	虚圆
imaginary component	虚分量；无功分量
imaginary component method	虚分量法
imaginary demand	虚假需求
imaginary exponent	虚指数
imaginary geomagnetic pole	虚地磁极
imaginary intersection point	虚交点
imaginary layer	假想层
imaginary line	虚设线；设想线
imaginary load	虚荷载
imaginary loading	虚负荷
imaginary number	虚数
imaginary part	虚部
imaginary quantity	虚量
imaginary root	虚根
imaginary source principle	虚震源原理
imaginary term	虚数项
imaginary trace	虚道
imaginary unit	虚数单位
imaginary wavefront	虚像波前
imaginary well	假想井；虚井
imaginary-real component method	虚实分量法
imaginative geomorphologic figuration	虚拟造型地貌
imaging	成像；影像学
imaging agent	显像剂；显影剂
imaging condition	成像条件
imaging device	成像装置
imaging equation	构像方程
imaging log	成像测井
imaging radar	成像雷达
imaging radar system	成像雷达系统
imaging sensor	成像传感器
imaging sonar	成像声呐
imaging spectrometer	成像光谱仪
imaging technique	成像技术
imaging theory	成像理论
imandrite	石英钠长岩
imbalance	不平衡；不稳定
imbedded	嵌镶的；夹杂的；埋藏在层间的
imbedment	埋置；包埋
imbibe	渗入；吸收；吸入
imbibition	吸液；加水
imbibition capillary pressure curve	自吸毛细管压力曲线
imbibition channel	自吸孔道
imbibition contact angle	自吸接触角
imbibition curvature	自吸曲率
imbibition cycle	吸入周期
imbibition displacement	自吸驱替
imbibition hysteresis	自吸滞后
imbibition oil recovery	自吸采油
imbibition relative permeability	自动吸入相对渗透率
imbibition water	渗吸水
imbibitional moisture	吸入水
imbibometry	自吸测定法
imbricate	叠瓦状的；重叠的；叠盖
imbricate arrangement	叠瓦状排列
imbricate fault	叠瓦断层
imbricate structure	叠瓦构造
imbricate thrusting	叠瓦式冲断裂
imbricate zone	叠瓦带
imbricated anticline	叠瓦状背斜
imbricated block structure	叠瓦块状构造
imbricated fabric	叠瓦状组构
imbricated fault	叠瓦状断层
imbricated fault zone	叠瓦断层带
imbricated overlapping	叠瓦状超覆
imbricated relief	叠瓦状地形
imbricated structure	叠瓦构造
imbricated texture	叠瓦结构
imbricated thrust zone	叠瓦冲断层带
imbricated vein	叠瓦状矿脉
imbricating delta	叠瓦三角洲
imbrication	叠瓦作用；叠瓦构造
IMF	国际货币基金组织
imitation marble	人造大理石；仿云石
imitative	仿制的；伪造的

imitative stable adjustment 拟稳平差
imitator 模仿者；模拟器，仿真器
immature 幼年的；未成熟的
immature oil 低熟油
immature profile 不完整剖面
immature region 幼年地区
immature residual soil 新残积土
immature soil 未成熟土；原状土，未扰动土
immature source rock 未成熟烃源岩
immature stage 未成熟阶段
immature technology 不成熟技术
immaturity 不成熟性；幼年期
immeasurability 不可计量性
immediate bottom 直接底板
immediate compensation 立即赔偿
immediate data 直接数据
immediate deformation 瞬时变形
immediate delivery 立即交货
immediate display 直接显示
immediate entry 采区平巷
immediate fill 立即充填；及时充填
immediate floor 直接底板
immediate forward support 即时超前支架；立即支护
immediate hanging wall 直接上盘
immediate landlord 直接业主
immediate oxygen demand 直接需氧量
immediate power amplifier 直接功率放大器
immediate prediction 即时预报
immediate report 速报
immediate rockburst 即时型岩爆
immediate roof 直接顶板
immediate roof bed 直接顶板层
immediate runoff 直接径流；瞬时径流
immediate settlement 初始沉降；瞬时沉降
immediate stability analysis 瞬时稳定分析
immediate strain 瞬时应变
immediate support action 直接支护作用
immediate surface 接触面
immediate transportation entry 即时报关进口手续
immediate vicinity 紧邻
immediate vicinity of wellbore 近井地带
immediate-bearing prop 及时承载支柱
immense 无限的；广大的；巨大的
immensity 广大；无限；无限的空间
immerse 浸渍
immerseable embankment 渗水路堤
immersed 潜没的；沉没的；浸入的；水下的
immersed belt 浸渍胶带
immersed body 潜没体
immersed bog 淹没沼泽
immersed density 潜容重
immersed depth 浸水深度
immersed method 浸入法
immersed section 浸水剖面；浸水断面
immersed structure 浸没结构
immersed tube 沉管；沉埋管道；沉管式隧道
immersed tube method 沉管法
immersed tube tunnel 沉管隧道
immersed tunnel 沉埋式隧道
immersed tunnel trench method 沉埋法，沉管法

immersed tunneling method 沉埋法，沉管法
immersed tunnelling 沉埋法隧道掘进
immersed tunnelling method 沉埋法，沉管法；沉埋法隧道掘进；沉管法隧道掘进
immersible 可淹没的；可浸入的
immersible pump 潜水泵
immersing looseness 浸散性
immersion 浸没；浸入；沉入
immersion curing 浸水养护
immersion depth 浸水深度
immersion fluid 浸液
immersion lens 油浸透镜
immersion liquid 浸液
immersion medium 浸没介质；浸液
immersion method 油浸法
immersion objective 浸没物镜
immersion oil 浸油
immersion plating 浸镀；浸渍涂敷
immersion proof 抗浸水性
immersion surface 浸流面
immersion test 浸水试验
immersion vibrator 插入式振捣器
imminent and short-term earthquake prediction 短临地震预报
imminent danger 迫切危险
imminent earthquake emergency management 临震应急管理
imminent earthquake forecast 临震预报
imminent earthquake forecast scheme 临震预报方案
imminent earthquake prediction 临震预测
imminent failure 即时毁坏；即时损毁
imminent-term earthquake precursor 临震地震前兆
immiscibility 不溶混性；不可混合性
immiscible 不溶混的；不混合的
immiscible displacement 非混相驱；不混溶驱替
immiscible displacement apparatus 非混相驱替仪
immiscible displacement mechanism 非混相驱替机理
immiscible displacing fluid 非混相驱替液
immiscible drive-gas displacement 非混相气驱
immiscible equation 不溶混方程
immiscible flow 非混相流动
immiscible fluid 不混溶液体
immiscible fluid flow 不混溶流体流动
immiscible gas drive 非混相气驱
immiscible inclusion 不混溶包裹体
immiscible liquid 非混相液体
immiscible magma 不混溶岩浆
immiscible mixture 不溶混的混合物
immiscible performance 非混相动态
immiscible phase 不溶混相
immiscible pump 潜水泵
immiscible region 非混相区
immiscible solution 不溶混溶液
immiscible solvent 不溶混溶剂
immobile 不能移动的；固定的
immobile interface 固定界面
immobile liquid 固定液
immobile oil 不动油
immobile phase 固定相
immobile water 束缚水

immobile water saturation	束缚水饱和度	impact idler	缓冲托辊；防冲托辊
immobility	固定性	impact indentation apparatus	冲击压痕试验装置
immobilization	固定；定位；不能移动	impact ionization	碰撞电离
immobilizing phase	不动相	impact jet flow	冲击射流
immovability	流动性；可动性	impact kinetic energy	碰撞动能
immovable property	不动产	impact load	冲击荷载
immunity	免疫性；抗扰性；不敏感性	impact loading condition	冲击加荷状态
immunity limit	抗扰性限值	impact machine	冲击试验机
immunization	免疫；免除	impact metamorphism	冲击变质作用
impact	冲击；碰撞	impact meter	冲击计
impact absorber	缓冲器；减震器	impact method	冲击法
impact absorption device	冲击吸收装置	impact mill	冲击式磨机；冲击式粉碎机
impact acceleration	冲击加速度	impact mineral	冲击矿物
impact action	碰撞作用；冲击作用	impact mixer	冲击式混合机
impact adhesive	冲击黏合剂	impact modifier	耐冲击性改进剂
impact allowance	冲击容许量；容许冲击荷载	impact modulator	冲击型调节器
impact bend test	冲弯试验；冲击弯曲试验	impact modulus	冲击模量
impact bending test	冲击弯曲试验	impact noise	冲击噪声
impact block	冲击块	impact of earthquake on society	地震社会影响
impact blow	冲击	impact of recoil	反跳冲击
impact breaker	反击式破碎机；冲击式破碎机	impact of rolling	机车车辆冲击
impact breccia	撞击角砾岩	impact of rolling stock	机车车辆冲击
impact briquetting	冲压成型	impact penetrometer	锤击贯入仪
impact briquetting machine	冲压式成型机	impact pile driving	锤击打桩
impact brittleness	冲击脆性	impact piston	撞击活塞
impact center	冲击中心	impact plate	冲击板
impact coefficient	冲击系数	impact plough	冲击式刨煤机
impact compactor	冲击式压实机	impact prediction	影响预测
impact crater	陨石坑；撞击坑	impact pressure	冲击压力
impact cratering	撞击成坑作用	impact property	冲击性能
impact crusher	冲击破碎机	impact pulse	冲击脉冲
impact crushing	冲击破碎；碰撞破碎	impact rate	冲击速率
impact crushing value test	冲击压碎值试验	impact relict	撞击残留物
impact damage	撞击破坏	impact resilience	冲击回弹性
impact damper	冲击阻尼器	impact resistance	冲击阻力；撞击阻力
impact damping device	冲击阻尼装置	impact ripper	冲击式挑顶机；冲击式平巷掘进机
impact detonation	冲击爆破	impact ripping machine	冲击式挑顶机
impact drill bit	冲击钻头	impact rock breaking	冲击凿岩
impact drilling	冲击钻进	impact rock drilling machinery	冲击式凿岩机
impact ductility	冲击韧性	impact rock pressure	冲击岩压
impact effect	冲击效应	impact screen	振动筛；冲击式振动筛
impact elasticity	冲击弹性	impact sensitivity	冲击敏感度；撞击敏感度
impact endurance test	耐冲试验	impact strength	冲击强度
impact energy	冲击能	impact strength index	冲击强度指数
impact excavation	冲击开挖	impact strength test	冲击强度试验
impact factor	冲击因数；冲击荷载系数	impact stress	冲击应力
impact failure	冲击破坏	impact tension test	冲击抗张试验
impact fatigue	冲击疲劳；碰撞疲劳	impact test	冲击试验；碰撞试验
impact flexural strength	冲击抗弯强度；冲击弯曲强度	impact testing machine	冲击试验机
impact force	撞击力；冲击力	impact time	冲击时间
impact force of falling stone	落石冲击力	impact toughness	冲击韧性
impact force of train	列车冲击力；列车冲击荷载	impact toughness machine	撞击韧度试验机
impact fracture	冲击断裂	impact toughness of matrix	基质抗冲击韧性；脉石抗冲击韧性
impact frequency	冲击频率	impact value	冲击值
impact frequency rock drill	冲击频率式凿岩机	impact velocity	冲击速度
impact fuse	碰炸导火线	impact vibration	冲击振动
impact grinding	冲击研磨；冲击磨矿	impact viscosity	冲击黏度
impact hammer	冲锤	impact wave	冲击波
impact hardness	冲击硬度	impact wrench	气动扳手；冲击扳钳
impact head loss	冲击水头损失		

impact zone	冲击带
impact-allowance load	容许冲击荷载
impact-endurance test	重复冲击试验；冲击耐力试验；冲击疲劳试验
impacter	冲击器
impaction	碰撞；压紧；嵌塞
impactite	冲击岩；击变岩
impactogen	冲击裂谷；内陆裂谷
impactor	冲击机
impact-pressure velocity meter	冲压式流速仪
impact-rotary drilling	冲击回转式凿岩
impact-type hammer crusher	冲击锤式破碎机
impact-vane anemometer	冲击叶式风速计
impair	损害；损伤；减少
impairment	损伤
impalpable	极细的；微粒的
impalpable structure	极细粒构造
imparity	不均等，差异；不同位；非对称性；掺杂物质
impedance	阻抗
impedance character	阻抗特性
impedance conversion	阻抗变换
impedance corrector	阻抗校正器
impedance coupled amplifier	阻抗耦合放大器
impedance coupling	阻抗耦合
impedance factor	阻抗系数
impedance function	阻抗函数
impedance interface	阻抗界面
impedance loss	磁漏损失；阻抗损失
impedance matching	阻抗匹配
impedance measurement	阻抗测量
impedance meter	阻抗计
impedance method	阻抗法
impedance muffler	阻抗复合消声器
impedance of a rock	岩石阻抗
impedance of an explosive	炸药阻抗
impedance of slot	隙缝阻抗；槽阻抗
impedance probe	阻抗探针
impedance protection	阻抗保护
impedance ratio	阻抗比
impeded drainage	不畅排水；受阻的排水
impeded flow	受阻抗水流
impediment	阻碍
impeller	叶轮；水泵转子；推进器
impeller pump	转子泵；叶轮泵
impending cliff	悬崖
impending earthquake	临震；即将发生的地震
impending failure	紧急故障
impending landslide moment	临滑时刻
impenetrability	不渗透性；不可穿入性
imperfect cleavage	不完全解理
imperfect combustion	不完全燃烧
imperfect contact	不良接触
imperfect crystal	不完美晶体
imperfect crystal formation	不无定型结晶
imperfect dislocation	不完全位错
imperfect drainage	不完善排水
imperfect elastic collision	非完全弹性碰撞
imperfect elasticity	不完全弹性
imperfect fluid	不完全流体
imperfect frame	静不定结构
imperfect gas	非理想气体
imperfect geosyncline	不完整式地槽
imperfect pile	缺陷桩
imperfection cross coupling	不完善交叉耦合
imperfection of elasticity	不完全弹性
imperfectly drained	排水不良的
imperfectly elastic	非完全弹性的
imperfectly elastic body	非完全弹性体；塑性体
imperfectly elastic media	非完全弹性介质
imperforate	无孔的
imperiousness	不透水性
impermeability	不透水性；抗渗性
impermeability factor	不渗透系数
impermeability pressure	抗渗压力
impermeability test	抗渗试验
impermeability to water	防水性，抗渗水性
impermeable	不透水的；不渗透的
impermeable barrier	不透水阻体；隔水屏障；不透水层
impermeable bed	不透水层；隔水层
impermeable boundary	不透水边界
impermeable break	不渗透夹层
impermeable breakwater	不透浪防波堤
impermeable cap formation	隔水盖层；隔水顶板
impermeable confining bed	不透水隔层
impermeable cover	不透水护面
impermeable formation	不透水岩层
impermeable ground	不透水土
impermeable layer	不透水层；隔水层
impermeable lithologic horizon	隔水岩土层
impermeable material	防渗材料；不透水体
impermeable membrane	不透膜
impermeable river bed	不透水河床；不渗水河床
impermeable rock	不渗透岩石
impermeable rock bed	不透水岩层
impermeable rock stratum	不透水岩层
impermeable seam	不透水层
impermeable section	不渗透层段
impermeable soil	不渗透土；重黏土
impermeable spur dike	不透水折流坝
impermeable strata	不透水层
impermeable system	不透水岩系；隔水岩系
impermeable to water	不透水的
impermeable wall	不透水墙
impermeable zone	不透水层；不透水区
impersonal entity	法人单位
impertinent	不恰当的，不适合的；不相干的
impervious	非渗透性的；不可渗透的
impervious area	不渗水面积；不透水区
impervious barrier	防渗层
impervious base	不渗透底层
impervious bed	不透水层
impervious blanket	不透水面层；防渗铺盖
impervious boundary	不透水边界
impervious cap rock	不透水盖层
impervious capping	不透水盖层；隔水盖层；隔水顶板
impervious clay blanket	不透水黏土层
impervious core	不透水心墙
impervious curtain	防渗帷幕；不透水帷幕
impervious dam	不透水坝
impervious element	防渗设施；防渗体

impervious embankment	非渗水土路基	import	输入；进口
impervious facing	不透水盖面	import and export license	进出口许可证
impervious factor	防渗参数；抗渗系数	import licensing	进口许可制
impervious fill	输入填料	import material	进口材料
impervious foundation	不透水地基	import of advanced technology	引进先进技术
impervious ground	不透水围岩	import quota	输入限额；进口限额
impervious horizon	不透水层	importance function	价值函数
impervious layer	不透水层	importance sampler	重点抽样
impervious liner	不透水衬层	important imposed load	重要外加荷载
impervious material	不透水材料	important structure	重要结构
impervious rock	不透水岩石	importation	输入；进口
impervious seal	不透水层；不透水密封；不透水盖层	imported fill	场外运入填料；外来填料
impervious skin	不透水膜；不透水层	imported stowing	外来材料充填
impervious spur dike	不透水折流坝；不透水丁坝；不透水挑水坝	imported subsoil	外来底土
		imported topsoil	外来表土
impervious stratum	不透水地层	imported water	输入水；引进水；外来水
impervious structure	不可透结构	imposed load	外加荷载；作用荷载
impervious to moisture	不防潮的	impost	进口税；关税
impervious wall	防渗墙	impost capital	拱墩帽
impervious zone	不透水区	impost springer	拱脚石
impervious-area runoff	不透水区径流	impotable water	非饮用水
imperviousness coefficient	不透水系数	impound	筑堤堵水；集水
impetuous torrent	湍急水流；急流	impound water	圈闭水；静水
impetus	动力；冲力；促进	impounded area	蓄水面积；围堤面积
impingement	碰撞；冲击；动力附着	impounded basin	集水盆地；蓄水池
impingement attack	冲击腐蚀	impounded body	静止水体
impingement black	烟道炭黑	impounded drainage	淤塞的流道；淤塞的排水系统
impingement corrosion	冲刷腐蚀	impounded regulating reservoir	蓄水调节水库
impingement curtain	动力附着型风帘	impounded water	拦蓄水
impingement plate	防冲挡板	impounding basin	蓄水池；蓄水库
impinging sand	撞击砂	impounding dam	蓄水坝
implantation technique	注入技术	impounding of reservoir	水库蓄水
implanted gauge	永置式压力计	impounding reservoir	蓄水池；蓄水库
implement	工具，用具；履行，执行	impoundment	蓄水；人工湖；蓄水池
implementation	实施；执行；履行	impoverishment	贫化；贫乏
implementation plan	执行计划	impracticability	不实用
implementation schedule	执行进度	impractical	不实用的
implicated structure	显微共生构造	imprecision	不精密；不精确
implicit algorithm	隐式算法	impregnable	坚固的
implicit and explicit methods	隐式和显式算法	impregnant	浸渍剂；溶剂
implicit difference equation	隐式差分方程	impregnated	浸染浸渍的；渗透的
implicit equation	隐式方程	impregnated prop	防腐剂浸渍支柱
implicit Euler method	隐式欧拉法	impregnated shale rock	浸染页岩
implicit function	隐函数	impregnating apparatus	浸染器；浸渍机
implicit method	隐式法	impregnating compound	浸渍剂；浸渍化合物
implicit parallelism	隐式并行性	impregnating crack detector	浸透裂纹探伤器；浸透裂纹探测仪
implicit scheme	隐格式		
implicit-explicit dynamic analysis	隐式-显式动力分析	impregnating metal	浸渍金属
implicit-explicit method	隐-显混合法；隐-显格式	impregnating solution	浸渍溶液
implicit-explicit partition	隐式-显式分区	impregnating tank	浸渍槽
implicit-explicit scheme	隐式-显式方案	impregnating varnish	涡旋压缩机
implicit-explicit solution	隐式-显式解	impregnation	浸渍，浸染；灌注；防渗处理
implied addressing	隐含寻址	impregnation deposit	浸染矿床
implosion	内爆，聚爆；压破；突然倒塌	impregnation depth	浸渗深度
implosive	内爆，聚爆；挤压震源	impregnation layer	浸染层
implosive source	挤压震源	impregnation method	浸渍法
implosive treatment	内向爆炸处理	impregnation of zone	区段灌注
imponderability	无重量；失重；不可称量	impregnation structure	浸染状构造
imporosity	不透气性；无孔隙性	impress	刻痕；压痕
imporous	无孔隙的	impression	模槽；刻痕；压痕

imprest	预付款；垫付款
imprint	刻印；痕迹
improper	不合适的；不恰当的
improper fraction	可约分数；假分数
improper symbol	非正常符号
improper works	不合格工程
impropriety	不适当；不合宜；不正当行为
improved foundation	加固地基
improved ground	加固地基
improved longwall system	改进的长壁开采法
improved plow-steel-grade rope	改进级高强钢丝绳
improved polygon method	改进的多边形法
improved recovery method	提高采收率方法
improved soil	改良土
improvement	改善；改进
improvement area	改建区
improvement line	道路改建界线，道路扩建用地线
improvement of alkaline soil	碱性土改良
improvement of liquefiable soil	液化地基加固
improvement of river	河道整治；治河
improvement of river bed	河床改善
improvement of soil	土壤改良
improvement works	改善工程
improver	改良剂；改进剂
impsonite	脆沥青岩；焦性沥青
impulse	脉冲；冲量；冲力
impulse amplitude	脉冲幅度
impulse correction	校正脉冲
impulse deviser	脉冲发生器
impulse duration	脉冲宽度
impulse effect	弹冲效应
impulse excitation	脉冲激发
impulse firing	脉动水力爆破；脉动水力落煤
impulse firing installation	水力脉冲爆破装置
impulse force	冲力
impulse frequency	脉冲频率
impulse function	脉冲函数
impulse generator	脉冲发生器
impulse hardening	冲击淬火
impulse heating	脉冲加热
impulse inclinometer	脉冲测斜仪
impulse load	脉冲荷载
impulse meter	脉冲计数器
impulse noise	脉冲噪声
impulse onset	脉冲初动
impulse oscillograph	脉冲示波器
impulse pump	脉冲泵
impulse radiation	脉冲辐射
impulse ratio	脉冲比；脉冲系数
impulse response	脉冲响应
impulse response function	脉冲响应函数
impulse scalar	脉冲计数器
impulse seismic device	激震装置
impulse seismogram	脉冲地震记录
impulse source	脉冲源
impulse starting	脉冲启动
impulse test	冲击试验
impulse valve	脉冲阀
impulse voltage	冲击电压；脉冲电压
impulse voltage withstand level	耐冲击电压水平
impulse water wheel	冲击式水轮
impulse wave	脉冲波；冲击波
impulse waveform	脉冲波形
impulse-length ratio	冲击长度比
impulse-momentum equation	脉冲-动量方程
impulse-momentum relationship	脉冲-动量关系
impulser	脉冲传感器
impulsion	脉冲；推力；冲击；冲量
impulsive	冲击的；碰撞的
impulsive force	冲力
impulsive load	冲击荷载；脉冲荷载
impulsive motion	脉动；脉冲运动
impulsive noise	冲击噪声
impulsive pressure	冲击压力
impulsive release	突然释放；瞬时释放
impulsive response	脉冲响应
impulsive source	脉冲源
impulsive surface seismic source	地面脉冲地震震源
impulsiveness	脉冲性
impure arkose	杂长石砂岩
impurities handling	杂质处理
impurity	污染物；杂质
impurity absorption	杂质吸收
impurity and greasy dirt	杂物及油污
impurity concentration	杂质浓度
impurity content	杂质含量
impurity picking	拣矸
impurity yield	矸石回收率
impurity-profile measurement	杂质分布测量
in bad repair	维修不良
in compression	受压缩的；在压力下的
in favour of	付款给…人；有利于
in force	有效
in layers	成层；层状
in phase	同相
in place density	原地密度；原状密实度
in place test	实地试验；原位试验
in plane load	平面内荷载
in position	就位；在原位
in pressure balance condition	平压条件下；在压力平衡条件下
in quincunx	五点形；五点形排法
in real time	实时
in series	串联的；连续的；按顺序排列
in situ	原位；原地；就地
in situ actual strength of concrete	现场混凝土实际强度
in situ analysis	现场分析技术
in situ California bearing ratio test	原位加州承载比试验
in situ CBR test	原位加州承载比试验
in situ combustion	地下燃烧；原地燃烧
in situ compression curve	现场压缩曲线
in situ concrete	现场浇筑混凝土
in situ concrete lining	现浇混凝土衬层
in situ concrete pile	现浇混凝土桩
in situ concrete strength	现场混凝土强度
in situ condition	天然条件
in situ conversion	原址改建
in situ cube strength	现场立方体强度
in situ deformation test	原位变形试验
in situ density	原地密度；原位密度

in situ density test 原位密度试验
in situ determination 原地测定
in situ EIS measurements 现场环境影响测量
in situ environmental improvement 原地进行环境改善工程
in situ exchange 原址换地
in situ field testing 现场野外试验
in situ geochemical analysis 现场地球化学分析
in situ horizontal stress 原位水平地应力
in situ improvement works 原址改善工程
in situ impulse test 原地激振试验
in situ inspection 现场检验
in situ leach 原地浸出；就地浸出
in situ material testing 物料现场测试；物料就地测试
in situ matrix suction 原位基质吸力
in situ measurement 现场测量；原位测试
in situ metamorphism 原地变质
in situ migmatite 原地混合岩
in situ mining 原地开矿；就地开矿
in situ modulus 现场模量
in situ moisture content 现场含水量
in situ monitoring 现场监测
in situ observation 原位观测
in situ origin 原地成因
in situ permeability 现场渗透性；现场渗透率
in situ permeability test 原位渗透试验
in situ pile 现场灌注桩；就地灌注桩
in situ porosity 地层原始孔隙度
in situ preservation 原址保存
in situ principal stresses 原地主应力
in situ probe 原地探针
in situ processing 现场加工
in situ rock triaxial compression test 现场岩石三轴压缩试验
in situ shear test 现场剪切试验
in situ soil 原地土
in situ soil test 原位土工试验
in situ soil testing 土的原位试验；原位土工试验
in situ state of rock stress 岩石应力的原始状态
in situ statistical sampling 实测统计法取样
in situ strength 原位强度
in situ stress 现场应力；原位应力
in situ stress field 地应力场
in situ stress measurement 现场应力测量
in situ stress state 原位应力状态
in situ temperature 原地温度
in situ test 原位测试；现场试验；实地试验
in situ testing 原位试验
in situ texture 原地结构
in situ thrust test 原位推力试验
in situ treatment 就地处理；现场处理
in situ triaxial test 现场三轴试验
in situ upgrading 就地升级
in solid 在实体中
in staggered pattern 左右交错
in step 同步的
in tandem with 同…串联
in the large 全局的；大规模的
in the medium 平均来说
in the seam mining 层内开采法；单一煤层开采法
in the solid 固体；立方体
in the strike 平行于走向；沿走向
in tunnel control survey 隧道洞内控制测量
in tunnel installation 隧道敷设
in vacuo 在真空中
in vacuum 在真空中
inaccessibility 难接近性；不可存取性
inaccessible 难接近的；不可存取的
inaccessible area 难接近区；难到区
inaccessible distance 不能直接用尺量的距离
inaccessible pore 不可进入的孔隙
inaccessible region 难接近地区
inaccessible site 不能达到的地点
inaccuracy 不精确；误差；偏差
inaccurate 不精确的
inaction 不活动；钝化；不旋光
inaction period 钝化周期
inactivation 钝化；失活
inactive 不活动的；钝化的
inactive basin 稳定海盆
inactive face 未生产的工作面；停产的工作面
inactive fault 不活动断层
inactive period 停产时间
inactive plate 静止板块；不活动板块
inactive porosity 无效孔隙度
inactive reserve 呆滞储量；难开采的储量
inactive source rock 非活性生油岩
inactive spring 死泉
inactive storage 死库容；垫底库容；无效库容
inactive substance 非活性物质
inactive volcano 休眠火山；静火山
inactive well 不出油的井；干井
inactive workings 停工工作区；废巷道
inactivity 不活动；钝化；不放射性；不旋光性
inadaptability 不适应性
inadaptation 不适应
inadapted river 不相称河
inadequate source rock 不充分烃源岩
inalterability 不变性
in-and-out movement 往复运动；出入运动
inane 空洞的；无限空间
inapplicability 不适用性；不适宜性
inapplicable 不适用的
inbreak 崩落；陷落
inbreak angle 崩落角；陷落角
inbye 向矿井内部的；向工作区的；地下的；巷道内的
inbye compression 井下空气压缩机
inbye end 矿井内端；给料端
inbye opening 内向巷道
inbye system 顺槽系统；向工作区系统
inbye transformer 矿井内部变压器；采区变压器
inbye transport 内向运输；采区运输；顺槽运输
inbye unit 矿井内端工区
inbye workings 矿井内部巷道；通往工作面的巷道
incandescent lighting 白炽灯照明
incarbonization 煤化作用；成碳作用
incaving 淘刷
incendiary 燃烧的
incendiary agent 燃烧剂
incendiary composition 发火剂

incendive 引火的；可燃的
incendivity 燃烧性；引火性
incentive sparking 引燃火花
inceptisol 始成土
inch 英寸
inch of water 英寸水柱；水柱英寸数
inches mercury 英寸水银柱；水银柱英寸数
inches water gauge 英寸水柱；水柱英寸数
inching valve 微动阀
inch-ounce 英寸-盎司
incidence 影响范围，入射；发生率
incidence angle 入射角
incidence indicator 倾角指示器；入射指示器
incidence matrix 关联矩阵
incidence of taxation 征税范围
incidence plane 入射面
incidence point 入射点
incidence wave 入射波
incidence wire 倾角线
incident 入射；发生范围，发生次数
incident angle 入射角
incident bar 入射杆
incident beam 入射束
incident compressional angle 纵波入射角
incident direction 入射方向
incident earthquake 入射地震
incident energy 入射能
incident field 入射场
incident flow 来流
incident intensity 入射强度
incident light 入射光
incident management 事件管理
incident plane 入射面
incident power 入射功率
incident pulse 入射脉冲
incident radiation 阳光辐射能
incident ray 入射线，入射光线
incident wave 入射波
incidental observation station 临时性观测站
incinerator 垃圾焚化炉；煅烧炉
incipient arc volcanism 初始岛弧火山作用
incipient bar 雏形沙洲
incipient cavitation 初始空蚀
incipient cavitation number 初生空穴数
incipient crack 原始裂缝；起始裂缝
incipient detonation 初爆
incipient failure 初始破坏
incipient fire 初发火灾
incipient flood plain 原始河漫滩
incipient fluidization 初步流态化
incipient fold 初始褶皱
incipient fracture 早期断裂
incipient fracture porosity 原始裂缝孔隙度
incipient joint 萌芽节理
incipient metamorphism 初级变质作用
incipient peneplain 初期准平原
incipient piping 初始管涌
incipient sediment motion 初始泥沙运动
incipient stability 初始稳定性
incipient stage 初始阶段

incise 下切；深切；刻切；雕刻
incised meander 深切曲流
incised river 嵌入河；深切河
incised stream 深切河流
incising erosion 下切作用
incising of wooden tie 木枕刻痕
incision 深切作用；下切作用
incisive 切入的
inclement weather 恶劣天气
inclination 坡度；斜度；倾角；倾斜度
inclination angle 倾角
inclination compass 磁倾角罗盘
inclination correction factors 荷载倾斜因子
inclination curvature 井斜曲率
inclination drilling 倾斜钻进
inclination factor 倾斜因子
inclination logging 测井斜；井斜测井
inclination maximum 最大倾角
inclination measurement 倾角测定，测斜
inclination method 倾角计算法
inclination minimum 最小倾角
inclination observation 倾斜观测
inclination of ground 地面坡度
inclination of hole 钻孔倾角；孔斜
inclination of seam 地层倾斜
inclination pitch 倾斜度
inclinator 倾斜仪；倾角仪
incline 斜井；斜面；斜坡
incline bottom station 斜井井底车场
incline grade 倾度
incline hoist 斜井提升机
incline impact test 斜面冲击测试
incline inset 斜井井底车场
incline stop 斜井挡车器
inclined 倾斜的
inclined adit 斜硐
inclined angle 倾斜角
inclined anticline 倾斜背斜
inclined arch 倾斜拱
inclined bedding 倾斜层理
inclined bedding slope 倾斜山体
inclined belt conveyor 倾斜胶带输送机
inclined body wave 斜体波
inclined bore 偏斜钻孔
inclined coast 倾斜海岸
inclined conveyor 倾斜输送机
inclined crack 倾斜裂纹；斜向裂缝
inclined current 倾斜流
inclined cut 倾斜工作面
inclined cut-and-fill method 倾斜分层充填开采法
inclined cut-and-fill stoping 斜梯段随采随填方框支架回采；V形梯段随采随填回采
inclined cut-and-fill system 上向梯段充填采矿法
inclined dip-slip fault 斜倾滑断层；斜滑断层
inclined drift 倾斜巷道
inclined drilling 倾斜钻进
inclined extinction 斜消光
inclined fault 倾斜断层
inclined fill 倾斜分层充填
inclined fold 倾斜褶皱

inclined force	倾斜力
inclined gauge	倾斜计，倾斜高度计
inclined grizzly	倾斜格筛
inclined haulageway	倾斜运输巷道
inclined height	斜高
inclined hoist	斜井提升机
inclined hole	斜井
inclined layer	倾斜地层
inclined load	斜向荷载
inclined oil water contact	倾斜油水界面
inclined paralled plates	倾斜平行板
inclined pile	斜桩
inclined plane	斜面
inclined plane tester	斜面应力试验仪
inclined plane viscometer	斜面黏度仪
inclined ramp	倾斜滑道
inclined reaction	斜向反力
inclined rig	倾斜式钻机
inclined rill method	倾斜分层充填开采法；倒角锥形矿房开采法
inclined road	倾斜巷道
inclined roadway	斜巷
inclined rock slope	岩石斜坡
inclined seam	倾斜矿层；倾斜煤层
inclined seismic zone	倾斜地震带
inclined shaft	斜井
inclined shaft development	斜井开拓
inclined shaft driven in country rock	围岩斜井
inclined shaft hoisting	斜井提升
inclined shaft mucking apparatus	斜井装岩机
inclined shaft sinking	斜井掘进
inclined shaft sinking in rock	斜井基岩掘进
inclined shaft sinking in surface soil	斜井表土施工
inclined slice	倾斜分层
inclined slicing	倾斜分层崩落回采法；倾斜分层开采
inclined slicing with caving	倾斜分层下行陷落采煤法
inclined sliding surface	斜滑移面
inclined stone drift	倾斜岩石巷道
inclined stratum	倾斜地层
inclined strut	斜撑；斜支柱
inclined undercut gate	倾斜下开式闸门
inclined upstream face	倾斜上游面
inclined vibrating screen	倾斜振动筛
inclined water gauge	斜管液面压力计；微压计
inclined winze	暗斜井
inclined-directional drilling	倾斜定向钻进
inclining experiment	倾斜实验
inclinograph	偏角记录器
inclinometer	测斜仪；磁倾计
inclosed meander	环形河湾；封闭曲流
inclosure	包围；包裹体；包含物
included angle	内角；夹角
inclusion	包体；包含；包留
inclusion geochemical survey	包裹体地球化学测量
inclusion relation	包含关系
inclusion texture	包体结构
inclusion water	包裹体水
incoalation	成煤作用；煤化作用
incoherent	松散的；未胶结的
incoherent material	松散碎屑物；未胶结碎屑物
incoherent rock	不黏结岩石
incoherent scattering	非相干散射
incoherent scattering radar	非相干散射雷达
incohesion	无内聚性；无胶黏性
incombustibility	不可燃性
incombustible construction	耐火结构
incombustible dust	不可燃尘末；不可燃岩粉
incombustible lining	耐火支架；不燃支架
incombustible material	不燃材料
incombustible organic rock	非可燃有机岩
incoming air	外部进入的空气
incoming flow	入流，来流
incoming fluid	流入液
incoming signal	输入信号
incoming wave	入射波；来波
incompact	不紧密的；松散的；不结实的
incompatibility	不相容性；不共存性；不协调性
incompatible development	不协调的发展；互不配合的发展
incompatible displacement function	非协调位移函数
incompatible element	不协调单元
incompatible equation	不相容方程
incompatible event	不相容事件
incompatible formulation	不协调组合
incompatible mineral	不相容矿物；不共存矿物
incompatible mode	非协调模式
incompatible parameter	不兼容参数
incompetence	无能力；不合格
incompetent	软的，弱的；不适当的；无能力的
incompetent bed	弱岩层
incompetent folding	软弱褶皱
incompetent formation	胶结程度差的地层；弱岩层；软岩层
incompetent rock	弱胶结岩层
incompetent strain domain	弱应变域
incomplete combustion	不完全燃烧
incomplete convergence	不完全收敛；不完全会聚
incomplete coupling	不完全耦合
incomplete data	不完全资料
incomplete detonation	不完全起爆；熄爆
incomplete explosion	不完全爆炸
incomplete field	不完全域
incomplete filled groove	坡口未填满
incomplete filling	不完全充填
incomplete fusion	未熔合；未焊透，不完全焊透；未完全熔合
incomplete hole	残孔
incomplete rarefaction wave	不完全稀疏波；不完全余波；不完全膨胀波
incomplete reaction	不完全反应
incomplete recombination	不完全的复合
incomplete reduction	不完全还原
incomplete ripple	不完全波痕
incomplete vibration	不完全振捣；漏振
incompletely saturated	不完全饱和的
incompletely saturated soil	不完全饱和土
incompressibility	不可压缩性
incompressibility modulus	不可压缩模量
incompressibility patch test	不可压斑片试验
incompressible	不可压缩的

incompressible boundary layer	不可压缩边界层
incompressible constraint	不可压缩约束
incompressible elasticity	不可压缩弹性
incompressible energy equation	不可压缩能量方程
incompressible field	不可压缩场
incompressible filter cake	不可压缩滤饼
incompressible flow	不可压缩水流
incompressible fluid	不可压缩流体；非压缩性流体
incompressible material	不可压缩材料
incompressible medium	不可压缩介质
incompressible stratum	不可压缩层
incompressible viscous fluid	不可压缩黏性流体
inconformity	不一致
incongruity	不适合；不一致；不相称
incongruous	不适合的；异向的
inconsequence	不连贯；不一致；不重要
inconsequent stream	非顺向河；无定向河
inconsistency	不相容性；不一致性
inconstancy	易变；无常；不定
inconvertibility	不可逆性；不可交换性
incoordination load	不匹配荷载；不协调荷载
incorrect	不正确的；不恰当的
incorrelate	不相关的
incorruptible	不易腐蚀的东西；不易腐蚀的
increase by degree	递增
increase in cost	费用的增加；成本增加
increasing function	递增函数
increasing sequence	递增序列
increasing value	增值
increasing velocity	递增速度
increasing zone of subsurface temperature	地下温度增温层
increment	子样，小样；增量，递增量；增大，增加
increment adding	增量加载
increment cost	增量成本
increment depth	增量深度
increment function	增量函数
increment load method	递增荷载法
increment of a function	函数增量
increment of aquifer	含水层补给量
increment of coordinate	坐标增量
increment of face advance	工作面每循环推进进度
increment of fines	细粒增量
increment of load	荷载增量
increment of plastic strain	塑性应变增量
increment of time	时间增量
increment oil produced	增产油量
increment pressure ratio	压力增量比
increment rate	增加率
incremental capital-output ratio	资本-产出增量比
incremental compilation	增量编译程序
incremental deformation	前进变形
incremental displacement	增量位移
incremental dynamic analysis	增量动力分析
incremental elastoplastic theory	弹塑性增量理论
incremental energy	能量增量
incremental equation of motion	增量运动方程
incremental equilibrium equation	增量平衡方程
incremental finite element	增量有限元
incremental hysteresis loop	增量磁滞回线
incremental initial stress method	增量初应力法
incremental integration	增量积分
incremental investment	增量投资
incremental launching	分段曳进；顶推
incremental launching method	分段曳进法；顶推法
incremental load method	荷载增量法
incremental loading	递增加荷
incremental loading test	增量荷载试验
incremental manner	递增方式
incremental method	增量法
incremental movement	增量运动
incremental oil	增产油量
incremental oil production	增产的采油量
incremental pressure	增加的压力
incremental productive capacity	新增生产能力
incremental project oil	方案增产油量
incremental ratio	增量比
incremental recorder	步进记录器
incremental stiffness	增量刚度
incremental stiffness matrix	增量刚度矩阵
incremental stress	增量应力
incremental tertiary oil	三次采油增加的油量
incremental theory of plasticity	塑性增量理论
incremental time	时间增量
incremental value	增值
incremental velocity	增量速度
incremental water injection volume	增加注水量
incremental-iterative approach	增量迭代法
incriminatory ratio	增值比；增量比
incrop	隐蔽露头；掩盖露头
incrustation	结壳；外壳；渣壳；水锈
incrusted	结泥饼的；结水垢的；结锈的
incrusted soil	结皮土；结壳土壤
incubation	培养；培育；保温
incubation period	潜伏期；孵化期；培育期
incubation period of crack nucleation	裂纹形核孕育期；形成裂缝的孕育期
incubative stage	潜伏期
incursion	进入；流入；侵入
incurvation	向内弯曲；内曲；挠度
indefinite equation	不定方程
indefinite integral	不定积分
indefinite scale	任意比例尺
indeformable	不变形的
indehiscent	不开裂的
indemnification	赔偿，补偿；保障
indemnify	赔偿，补偿；保障
indemnity for risk	风险赔偿
indemnity form	保证书
indemnity liability	赔偿责任；损害赔偿
indemnity of insurance	保险赔偿
indent	刻痕；凹槽；锯齿形
indentation	缺刻；凹入；成穴
indentation fracture toughness	压痕断裂韧度
indentation hardness	压痕硬度
indentation machine	压痕硬度测定仪
indentation method	压痕法；刻痕法
indentation test	压入硬度试验；压痕试验
indented bar	螺纹钢筋
indented bolt	带纹螺栓；齿纹螺栓
indented coast	港湾海岸；弯曲海岸

indented coast line 曲折海岸线
indented coast of erosion 海蚀港湾岸
indented coast-line 曲折岸线；锯齿形岸线；海湾岸线
indented wire 齿痕钢丝
indenture 合同，契约
independent assessment 独立评价
independent audit 独立审核
independent component 独立组分
independent condition 独立条件
independent control 独立控制；单架控制
independent coordinate system 独立坐标系
independent criterion 非相关性指标；无关的准数
independent density source 独立密度源
independent drive 单独驱动
independent equipment 独立设备
independent event 独立事件
independent foundation 独立基础
independent function 独立函数
independent haulage 独立牵引；外牵引
independent increment 独立增量
independent inspection authority 独立检查机构
independent insurer 独立保险人
independent jobber 独立运销商
independent liability 单独的责任
independent linear relation 独立线性关系
independent model aerial triangulation 独立模型法空中三角测量
independent network 独立网
independent observation 独立观测
independent observation loop 独立观测环
independent rotation 自转；独立转动
independent side band 独立边带；独立边频带
independent strain relation 独立应变关系
independent subsidence 单独下沉；自由下沉
independent variable 自变量；独立变量
in-depth filtration 深层过滤
in-depth formation damage 深部地层损害；深部油层损害；深部储层损害
in-depth study 深入研究
inderite 多水硼镁石
indeterminacy 不确定性
indeterminacy degree 不定度
indeterminancy principle 测不准原理；不确定性原理
indeterminate 不确定的；模糊的
indeterminate coefficient 待定系数
indeterminate constant 不定常数
indeterminate equation 不定方程
indeterminate error 不定误差
indeterminate form 不定形；未定形
indeterminate principle 测不准原理；不确定性原理
indeterminate structure 超静定结构
indeterminateness 不定性；超静定性
index 指数索引；指标；变址；下标
index arm 指示杆；分度臂；游标
index bed 标准层
index contour 计曲线；示量等高线，注数字等高线
index correction 仪表误差校正；指标订正；刻度改正
index curve 指数曲线
index diagram 图幅接合表；接图表
index dial 刻度盘
index error 指数误差；指标误差；指差
index error of vertical circle 竖盘指标差；垂直度盘指标差
index factor 指标因子
index for selection 选取指标
index form 指标形式
index formation 标准建造
index fossil 标准化石
index law 指数律
index liquid 浸液；折射率液
index map 指示图；索引图
index mark 指数；索引
index mineral 指示矿物；标志矿物
index number 指数
index of computation 计算指标
index of correlation 相关指数
index of estimate 估算指标
index of heat stress 热应力指数
index of hydrogeochemical environment 水文地球化学环境指标
index of inertia 惯性指数
index of oscillation 振荡指数
index of overall concentration 总集中指数；总浓度指数
index of plasticity 塑性指数
index of porosity 孔隙指数
index of refraction 折射率；折光指数
index of reliability 可靠性指数
index of ripple 波痕指数
index of size distribution 粒度分布指数
index of size reduction 破碎度指数
index of spaciousness 宽敞度指数
index of stability 稳定性指数
index of stabilization 稳定性指数
index pin 指度针，指针
index plane 标准面
index plate 分度盘；指示牌；说明牌
index plunger 分度销
index property 指标特性
index property test 指标特性试验
index stress 特征应力
index zone 标志带；标准带；标准层
indexing 编索引，标引；指标化；分度法
indexing error 分度误差
indexing fixture 分度夹具
indexing language 标引语言
indexing mechanism 转位机构；索引机制
indexing plate 标度盘
indexing ring 刻度环
indexing section 指标部分
Indian Ocean 印度洋
indianite 粒钙长石；埃洛石
indicated altitude 指示高度
indicated attractive power 指示牵引力
indicated efficiency 指示效率
indicated object 表示对象
indicated ore 推定储量
indicated output 指示输出；指示产量
indicated power 指示功率
indicated pressure 指示压力
indicated reserve 推定储量

indicated work	指示功
indicating control	指示控制器
indicating controller	指示控制器；刻度调节器
indicating lamp	指示灯；信号灯
indicating length	信示长度
indicating light	指示灯
indicating meter	指示计；指示仪表
indicating needle	指针
indicating panel	仪表盘
indicating pointer	指针
indicating pressure gauge	指示压力计
indication	指示，指出，表示；迹象；读数；显示，指标
indication cycle	表示周期
indication of fracture	裂隙迹象
indicative feature	指示要素
indicative grain size	特征颗粒
indicative list	指示性名单
indicative mark	指示性标志
indicative pore size	特征孔径
indicative sand	标志砂
indicator	指示器；指示剂；示踪剂；指示符，压力计
indicator board	指示板
indicator breakdown	指标分解
indicator constant	指示剂常数
indicator diagram	指示图；示功图；器示压容图
indicator dial	指示盘
indicator element	示踪元素；指示元素
indicator hand	指针
indicator hole	指示钻孔
indicator horizon	标准层位；指示层
indicator light	指示灯；信号灯
indicator of output	产量指标
indicator of plan	计划指标
indicator of product variety	品种指标
indicator pointer	指针；指示器
indicator range	指示幅度
indicator ratio	指标比值
indicator seam	指示层；标志层
indicator signal	指示信号
indicator stem	指示杆
indicator sub	指示接头
indicator test	绘示功图；指示剂试验
indicatrix	指标；指示量；指示线
indicial equation	指数方程
indifferent gas	惰性气体
indifferent oxide	惰性氧化物
indifferent solvent	惰性溶剂
indiffusible	不扩散的
indigenous	本地的；原地的；固有的；自在的
indigenous coal	原地生成煤
indigenous construction material	地方建筑材料
indigenous forest	原始森林；本地森林
indigenous method	土法
indigo copper	铜蓝；靛铜矿
indigolite	蓝电气石
indirect adjustment	间接平差
indirect aerological analysis	间接高空分析
indirect analysis	间接分析
indirect approach	间接法
indirect bank protection	间接护岸
indirect benefit	间接受益
indirect catchment area	间接集水区
indirect charge	间接费
indirect circulation	逆环流
indirect cost	间接成本；间接费用
indirect cost account	间接成本账；间接成本表
indirect cost of project	工程间接费
indirect demonstration	间接证法；间接证明
indirect distribution	间接分配
indirect drainage	间接排水
indirect economic loss	间接经济损失
indirect effect	间接效应
indirect eruption	间接喷发
indirect excitation	间接激励
indirect flood damage	洪水间接灾害
indirect flushing	反循环冲洗；反向洗孔
indirect heating system	间接供暖系统
indirect holding fastening	分开式扣件
indirect input	间接输入
indirect interpretation key	间接解译标志
indirect liability	间接责任
indirect liquefaction	间接液化
indirect load	间接荷载
indirect measurement	间接量度；间接观测
indirect method	间接法
indirect observation	间接观测
indirect orientation method	间接定向法
indirect output	间接输出
indirect oxidation	间接氧化
indirect paper	间接票据
indirect participation	间接参股
indirect pollution	间接污染
indirect priming	底部点火；反向起爆
indirect probability	间接概率
indirect production cost	间接生产成本
indirect reduction	间接还原
indirect refrigerating system	间接制冷系统
indirect refrigeration	间接冷冻
indirect scanning	间接扫描
indirect scheme of digital rectification	间接法纠正
indirect sintering	间接烧结
indirect stratification	次生层理
indirect stress	间接应力；合成应力
indirect synthesis technique	间接合成法
indirect tensile test	间接拉伸试验
indistinct boundary	不清晰的边界
indistinct cleavage	不清晰解理
indistinction	无区别；不清楚；难辨认；不确定
indistinguishable	难分辨的
inditron	指示管
individual	个人，单一体；个别的，单独的；独特的
individual control	个别控制；单独控制
individual footing	单独基础；独立基础
individual hole	单孔
individual project	单项工程
individual project estimate	单项工程概算
individual prop	单体支柱
individual shot	单孔爆破
individual stone column	单根碎石桩
indoor air pollution	室内空气污染

indoor air velocity 室内空气流速
indoor climate 室内气候
indoor fire extinguishing system 室内消防系统
indoor fire hydrant 室内消火栓
indoor hydroelectric station 室内式水电站
indoor moisture 室内湿度
indoor rainwater system 雨水内排水系统
indoor reference for air temperature and relative humidity 室内温湿度基数
indoor steel-tape length checking device 室内钢尺比长器
indoor substation 室内变电站
indoor temperature 室内温度
indoor test 室内测试
indoor test of geotechnique 岩土室内试验
indoor thermal environment 室内热环境
indoor transportation system 室内运输系统
indoor ventilation system 室内通风系统
indoor water supply system 室内供水系统
indoor work 室内工作；内业工作
Indosinian cycle 印支旋回；印支期
Indosinian movement 印支运动
Indosinides 印支造山带
indraft 吸入；流入；向岸流
induce 感应；诱导
induce stress 诱发应力
induced 诱发的
induced air 抽入空气；引入空气
induced air flotation 加气浮选
induced artificial recharge 人工引导补给
induced block caving 阶段人工崩落开采法；分段强制崩落开采法
induced break 诱导裂缝
induced catalysis 诱导催化
induced caving 人工崩落；人工放顶
induced charge 感生电荷；感应电荷
induced cleavage 诱导断裂，诱导裂缝；采动劈理；压力裂隙
induced current 感应电流
induced current log 感应测井
induced draft 诱导通风；诱导气流
induced earthquake 诱发地震
induced effect 引发影响；诱导效应
induced fracture 采动裂隙；诱发破裂；派生裂隙
induced gamma-ray spectroscopy log 感应伽马能谱测井
induced geological hazard 诱发地质灾害
induced infiltration 诱导渗透
induced landslide 诱发滑坡
induced leak 诱导泄漏；人为泄漏；引起泄漏
induced leakage 诱发渗漏；诱发泄漏
induced magnetization 感应磁化；感应磁化强度
induced microseismicity 诱发微震
induced mutagenesis 诱变，诱发突变
induced phenomenon 激发现象
induced polarization 感应极化；激发极化
induced polarization method 激发极化法
induced porosity 次生孔隙度；感生孔隙率
induced radioactivity 感应放射性；人工放射性
induced seismicity 诱发地震
induced stress 感生应力；诱导应力
induced stress in rock mass 岩体二次应力；岩体诱导应力
induced structure 派生构造
induced surface runoff 诱导地面径流
induced surface water recharge 地表水诱导补给
induced vibration 诱发振动；诱振
induced voltage 感应电压
inducer 诱导物；电感器
inducer shot-firing 感应爆破
inductance 电感；感应系数
inductance amplifier 电感耦合放大器
inductance coil 电感线圈
inductance figure 电感系数
inductance meter 电感计；电感测定器
inductance pick-up 感应传感器
inductance-capacitance coupling 电感电容耦合
induction 感应；诱导；归纳法
induction balance 电感平衡
induction blasting 震动爆破；松动放炮；诱导爆破
induction coil 感应线圈
induction compass 感应式罗盘
induction conductivity 感应电导率
induction current 感应电流
induction curve 感应曲线
induction effect 感应作用；诱导效应
induction electrical module 感应电测井模块
induction electrical survey 感应电测井
induction electrolog 感应电测井
induction field 感应场
induction force 感应力；诱导力
induction generator 感应发电机；异步发电机
induction hardening 感应硬化；感应淬火
induction heating 感应加热
induction height survey 导入高程测量
induction height survey through shaft 立井导入高程测量
induction load 感应负载
induction log 感应测井
induction log resistivity 感应测井电阻率
induction logging 感应测井
induction machine 感应电机；起电机
induction melting 感应熔化
induction meter 感应式仪表；感应式电度表
induction motor 感应电动机
induction off-machine control equipment 感应式离机控制装置
induction period 诱导期；感应期
induction pick-up 感应传感器
induction pipe 进口管
induction port 进口
induction pulse transient method 感应脉冲瞬变法
induction quadrature 感应直角相移；感应求积法
induction reactance 感抗
induction resistivity 感应电阻率
induction sheath 感应屏蔽；磁屏蔽
induction time 引爆时间；感应时间
induction torque 异步力矩；感应力矩
induction type alternator 感应式交流发电机
induction unit 诱导器
induction valve 吸入阀
inductive 电感的；感应的
inductive action 感应作用

inductive capacity	感应能力；电感容量
inductive coupling	电感耦合
inductive displacement pickoff	感应式位移传感器
inductive disturbance	感应干扰
inductive electromagnetic method	电磁感应法
inductive electromagnetic survey	电磁感应测量
inductive inference	归纳推理
inductive interference	感应干扰
inductive leak	电感漏泄；诱导漏泄
inductive method	感应电磁勘探法
inductive resistor	有感电阻器
inductivity	诱导性；感应率
inductometer	电感计
inductor	感应物；电感线圈；磁石发电机
indurated	固结的；硬化的
indurated clay	硬化黏土
indurated mudstone	坚硬泥岩；泥板岩
indurated red earth	坚硬红壤
indurated rock	硬化岩
indurated shale	硬化页岩
induration	坚固；变硬
induration degree	硬化程度
induration temperature	硬化温度；固化温度
industrial adjustment	产业调整
industrial air pollution	工业空气污染
industrial architecture	工业建筑
industrial area	工业区
industrial briquette	工业型煤
industrial building	工业建筑物
industrial climate	工业气候
industrial computed tomography scanning	工业CT扫描
industrial contaminant	工业污染
industrial CT scanning	工业CT扫描
industrial deposit	工业用矿物矿床
industrial dust	工业粉尘
industrial economics	工业经济；工业经济学
industrial facility	工业设备
industrial fatigue	工业疲劳
industrial interference	工业干扰
industrial management	工业管理
industrial measuring system	工业测量系统
industrial mineral	工业矿物
industrial organization	工业企业
industrial output	工业产值
industrial park	工业区
industrial photogrammetry	工业摄影测量
industrial pollution	工业污染
industrial property right	工业产权；工业所有权
industrial property system	工业产权制度
industrial psychology	工业心理学
industrial railway	铁路专用线；厂内窄轨铁路
industrial rationalization	工业合理化
industrial refuse	工业垃圾
industrial reserves	工业储量
industrial salt	工业盐
industrial scale	工业规模
Industrial Security Commission	工业安全委员会
industrial sewage	工业废水
industrial standard	工业标准
industrial station	工业站
industrial structure	工业结构
industrial system	工业体系
industrial use	工业用途
industrial waste	工业废水；工业污水；工业废弃物
industrial waste base course	工业废渣基层
industrial waste discharge	工业污水
industrial wastewater	工业废水
industrial water	工业用水
industrial water need	工业需水量
industrial water pollution	工业水污染
industrial water quality	工业水质
industrial water service	工业供水
industrial water supply	工业供水
industrial water use	工业用水
industrial zone	工业分区
industrialization	工业化
industrialization of disaster reduction	减灾产业化
industrialized building system	工业化建筑体系
industry railway	工业企业铁路
industry tax	产业税
inefficient	无效的；效率不高的
inefficient well	无效井；不成功井；低效井
inelastic	无弹力的；无弹性的
inelastic activity	非弹性性状
inelastic analysis	非弹性分析
inelastic attenuation	非弹性衰减
Inelastic attenuation factor	非弹性衰减因子
inelastic behaviour	非弹性性状
inelastic collision	非弹性碰撞
inelastic deformation	非弹性变形
inelastic densification	非弹性致密
inelastic design response spectrum	非弹性设计反应谱
inelastic displacement response	非弹性位移响应
inelastic drift	非弹性侧移
inelastic dynamic analysis	非弹性动力分析
inelastic earthquake design spectrum	非弹性地震设计谱
inelastic earthquake response	非弹性地震反应
inelastic finite element	非弹性有限元
inelastic frame structure	非弹性框架结构
inelastic lateral buckling	非弹性侧向压屈
inelastic material	非弹性材料
inelastic range	非弹性范围
inelastic rebound	非弹性回弹
inelastic reinforced concrete frame	非弹性钢筋混凝土框架
inelastic resistance	非弹性抗力
inelastic response spectrum	非弹性反应谱
inelastic restitution	非弹性恢复
inelastic scattering	非弹性散射
inelastic state	非弹性状态
inelastic strain	非弹性应变
inelastic structure	非弹性结构
inelastic system	非弹性系统
inelastic vibration	非弹性振动
inelastic volumetric strain	非弹性体积应变
inelastic wave	非弹性波
inelasticity	无弹性，非弹性
inequigranular	不等粒的；不等粒状
inequigranular texture	不等粒状结构
inequilateral	不等边的
inequilibrium	不平衡

inequilibrium stage 不平衡阶段
inequitable exchange 不等价交换
inequivalent site 不等效晶位
inert 惰性的；不活泼的
inert additive 惰性添加剂
inert aggregate 惰性骨料
inert component 惰性组分
inert dust 惰性岩粉；惰性粉尘
inert electrode 惰性电极
inert element 惰性元素
inert gas 惰性气体
inert material 惰性物料
inertia 惯性；惰性
inertia axis 惯性轴
inertia balance 惯性平衡
inertia coefficient 惯性系数
inertia constant 惯性常数
inertia couple 惯性矩；惯性力偶矩
inertia criterion 惯性标准
inertia current 惯性流
inertia damping 惯性阻尼
inertia displacement 惯性位移
inertia drift 惯性漂移
inertia effect 惯性效应
inertia flow 惯性流
inertia force 惯性力
inertia governor 惯性调节器
inertia grade 惯性坡度
inertia head 惯性水头
inertia law 惯性定律
inertia loads 惯性荷载
inertia mass 惯性质量
inertia moment 惯性矩；转动惯量
inertia navigation system 惯性导航系统
inertia of flowing water 流水的惯性
inertia oscillation 惯性振动
inertia pressure 惯性压力；惯性压力
inertia principal axis of area 截面主惯性轴
inertia ratio 惯性矩比
inertia reaction 惯性反作用力
inertia resistance 惯性阻力
inertia tensor 惯性张量
inertia test 惯性试验
inertia vibrating screen 惯性振动筛
inertia wave 惯性波
inertia-gravity wave 惯性重力波
inertial 惯性的
inertial axis 惯性轴
inertial centrifugal force 惯性离心力
inertial coefficient 惯性系数
inertial control 惯性控制
inertial coordinate system 惯性坐标系
inertial coupling 惯性耦联
inertial discharge centrifuge 惯性卸料离心脱水机
inertial dust separator 惯性除尘器
inertial effect 惯性效应
Inertial Guidance System 惯性制导系统
inertial hydraulic head 惯性水头
inertial instability 惯性不稳定
inertial interaction analysis 惯性相互作用分析

inertial load 惯性载荷
inertial mass 惯性质量
inertial navigation 惯性导航
inertial oscillation 惯性振动
inertial resistance 惯性阻力
inertial seismometer 惯性地震检波器
inertial sensor 惯性传感器
inertial surveying system 惯性测量系统
inertial system 惯性系统
inertialess 无惯性的
inertodetrinite 碎屑惰性体
inertodetrite 微碎屑惰性煤
inerts 惰性组分；惰性成分；惰性气体
inesite 红硅钙锰矿
inexact data 不准确数据
inexact integration 非精确积分
inexperience 无经验；不熟练；无经验的；不熟练的
inference 推断；推论
inferred fault 推断的断层
inferred reserves 推算储量
inferred resources 推断的资源
infill 填实空隙；填料
infill material 填充材料
infill well 加密井
infill well-drilling reserves 加密钻井储量
infill wellsite 加密井位
infilled frame 充填框架
infilled joint 充填节理
infilled structure 充填结构
infilled well pattern 加密井网
infiller 加密井；插补井
infilling 充填
infilling material 充填物
infilling of discontinuity 不连续面的充填物
infilling percentage 充填度
infilling vein 充填脉
infilling well 井网加密井
infiltrability 可渗入性；可渗透性
infiltrate 渗入；渗透；渗入物
infiltrated water 入渗水
infiltrated zone 泄油带
infiltrating water 入渗水；渗透水
infiltration 渗滤；渗透；渗入；渗填作用
infiltration action 渗入作用
infiltration anomaly 渗滤异常
infiltration area 入渗区；入渗面积；入渗范围；地下水补给区
infiltration basin 渗透盆地；渗水池
infiltration capacity 下渗容量；下渗能力；入渗容量；入渗能力
infiltration capacity curve 下渗容量曲线；入渗容量曲线
infiltration cell 渗透池
infiltration center 入渗中心
infiltration channel 渗流集水渠
infiltration coefficient 入渗系数；浸润系数；渗透系数
infiltration coefficient of precipitation 降水入渗系数
infiltration criteria 入渗标准
infiltration deformation 渗透变形
infiltration deposit 淋积矿床
infiltration depression 渗水洼地

infiltration ditch	入渗渠；引渗渠
infiltration diversion	渗水管引水
infiltration experiment	渗透试验
infiltration feature	渗漏地形
infiltration flow	渗入流量
infiltration front	入渗前锋；渗透面
infiltration funnel	入渗漏斗
infiltration gallery	集水廊道
infiltration index	入渗指数；渗透指数
infiltration intensity	渗入强度
infiltration line	浸润线
infiltration loss	入渗损失；渗失量
infiltration of water	水渗透
infiltration path	入渗路径
infiltration pit	渗水坑
infiltration point	入渗点
infiltration pond	入渗池；渗透池
infiltration pond recharge	渗透池补给
infiltration pressure coefficient	渗入压力系数
infiltration process with cold pressing	冷压浸渍法
infiltration process with vibration	无压浸渍法
infiltration rate	入渗速率；入渗率；渗透率
infiltration ratio	入渗比
infiltration recharge	入渗补给
infiltration recharge from land surface	地面引渗回灌
infiltration recharge of irrigation water	灌溉水回渗补给
infiltration routing	下渗演算；下渗推算；入渗演算；入渗推算
infiltration slit	渗水缝
infiltration stress	渗透应力
infiltration system	渗透系统；渗滤系统
infiltration test	渗透试验；渗水试验
infiltration tunnel	渗水隧洞
infiltration velocity	入渗速度；渗透速度
infiltration water	渗入水；下渗水；过滤水
infiltration well	渗水井；滤水井
infiltrometer	渗透仪；渗透计
infinite	无限物；无穷的，无尽的；无限的，无数的；非限定的
infinite aquifer	无限含水层
infinite boundary	无穷远边界
infinite conducting medium	无限传导介质
infinite confined aquifer	无界承压含水层
infinite decimal	无穷小数
infinite dimension	无限尺寸；无限维
infinite distance	无穷远
infinite domain	无限域
infinite duration	无限持续时间
infinite elastic solid	无限弹性体
infinite elastic strip	无限弹性带
infinite elastic wedge	无限弹性楔形体
infinite electrode	无穷远电极
infinite element	无限单元
infinite element method	无限元法
infinite expansion	无限扩展
infinite extrapolation	无限外推
infinite flooding pattern	无限注水井网
infinite half space	半无限空间
infinite homogeneous medium	无限均匀介质
infinite impulse response	无限冲激响应
infinite integral	无穷积分
infinite media	无线介质
infinite series	无穷级数
infinite slope	无限界斜坡；无限边坡
infinite solution	无限解
infinite space	无限空间
infinite strip aquifer	无限条带状含水层
infinite system	无限系统
infinite wave train	无限波列
infinite-acting	无限作用
infinite-acting decline	自然递减
infinite-acting period	无限作用阶段
infinite-acting regime	初始生产时期
infinite-acting system	无限作用流动系统
infinitely great	无穷大；无限大
infinitely rigid	无限刚性的
infinitely small	无穷小
infinitely stiff	无限刚性
infinitesimal	无限小；无穷小
infinitesimal analysis	微积分
infinitesimal area	无穷小面积
infinitesimal calculus	微积分
infinitesimal geometry	微分几何
infinitesimal increment	无穷小增量
infinitesimal method	无穷小法
infinitesimal patch	无穷小片
infinitesimal strain	微应变
infinitesimal strain theory	无穷小应变理论；微应变理论
infinitive	不定式；不定的
infinity	无穷远
infinity point	无限大点；无穷远点
inflammability	可燃性；易燃性
inflammable	易燃
inflammable gas	可燃气体；易燃气体
inflammable material	易燃材料
inflammable minerals	可燃矿物
inflammable mixture of gases	易燃的混合气体
inflammable seam	易燃煤层
inflammableness	易燃性
inflatable seal	膨胀性密封
inflatable straddle testing	膨胀跨隔测试
inflatable structure	充气结构；可膨胀结构
inflation pressure	充气压力；充胀压力
inflator	增压泵；压送泵；打气筒
inflection	反弯曲；反挠曲；拐折
inflection point	反弯点；拐点
inflection point of subsidence profile	完全沉陷区剖面拐点
inflectional elastic curve	反弯曲弹性线
inflection-point method	反弯点法
inflexibility	不弯曲性；非挠性
inflexible	不可弯曲的；非挠性的
inflexion	弯曲，向内弯曲；反挠；屈曲，拐折，拐点；偏转
inflexion point	拐点
inflow	流入；流入物
inflow barrier	入流障碍
inflow capacity	流入能力
inflow cave	入流洞；流入型洞穴
inflow channel	进水渠道
inflow curve	入流曲线

inflow device	流入装置
inflow discharge curve	来水流量曲线
inflow duration	入流历时
inflow face	岩心筒入口截面
inflow forecast	来水预报；入库径流预报
inflow hydrograph	进水过程线；入流过程线
inflow into river system	河网入流；河系入流
inflow lake	内流湖
inflow of water	涌水量
inflow performance curve	向井流动动态曲线
inflow performance relationship	向井流动动态
inflow point	进水点
inflow production relation	井底流压-产量关系
inflow rate	涌水量；涌水率
inflow storage out-flow curve	入流储量出流曲线
inflow stream	流入河；流入水系；地下入流河
inflow to reach	河段来水
inflow velocity	流入速度
inflow water	流入水
inflowing stream	入流河；支流
inflow-outflow method	进出流法
influence area	受影响的面积；作用区
influence basin	影响洼地
influence chart	影响图；感应图
influence coefficient	影响系数
influence coefficient for spacial action	空间性能影响系数
influence depth	影响深度
influence diagram	影响图；感应图
influence factor of strain	应变感应系数
influence function	影响函数
influence line	影响线
influence line for reaction	反力影响线
influence matrix	影响矩阵
influence of gravity	重力效应；重力影响
influence of ground	地面效应
influence of ground current	地电流影响
influence of local geology	局部地质影响
influence radius	影响半径
influence region	影响区域
influence value	影响值；感应值
influencing characteristic	影响特性
influencing current	影响电流
influent	液体；流入，支流；流入的；诱导油流
influent action	渗水作用
influent flow	渗流
influent impounded body	补给地下水的水体
influent seepage	入渗；渗入补给
influent stream	潜水注入河
influent water	渗漏水
influx	流入；灌注；汇流；河口
informatics	信息学；信息科学；情报学
information	情报；资料；信息
information acquisition	信息采集
information analysis	信息分析
information attribute	信息属性
information bank	现况调查
information base	信息库
information bit	信息位
information capacity	信息容量
information circular	通告；资料通报
information code	信息码
information collection	信息采集
information consultant	信息咨询
information content	信息量
information engineering	信息工程
information exchange	情报交流
information extraction	信息提取；信息抽取
information extraction process	信息提取法
information feedback	周期性变形
information processing	信息处理；情报处理
information processing code	信息处理编码
information processing system	信息处理系统
information processor	最高水位
information protection system	信息保护系统
information pulse	信息脉冲
information quantity	信息量
information radius	信息半径
information rate	信息率
information register	信息寄存器
information reply	信息应答
information request	信息请求
information retrieval	信息检索；情报检索
information retrieval system	信息检索系统；情报检索系统
information science	信息科学
information service	询问处，查询台；情报所；讯息服务，情报服务
information sharing	信息共享
information sign	指路标志；信息标志
information society	信息社会
information source	信息来源
information storage	信息存储；信息库
information storage and retrieval system	信息存储和检索系统
information storage density	信息存储密度
information storage means	信息存储方法
information stream	信息流
information structure	信息结构
information superhighway	信息高速公路
information synergism	信息协同
information system	信息系统
information technology	信息技术
information theory	信息论
information transmission	信息传输
information utility	信息效用
information utilization	信息利用
information vector	信息矢量
informative abstract	内容提要
informative pattern	信息模式；可提供数据的图形
informative sign	说明标记；说明符号
informative weight	信息权
infracraton	深克拉通
infracrustal	内地壳的
infracrustal rock	深火成岩
infracrustal structure	地壳下部构造
infraglacial deposit	底碛
infralittoral deposit	远岸沉积
infraneritic	浅海的
infraneritic environment	浅海外环境
infranics	红外线电子学
infrapermafrost water	冻结层下水；冰冻层下水

infrared	红外线；红外线的	infrared thermometer	红外线测温仪
infrared absorption	红外吸收	infrared thermometry	红外线测温
infrared absorption anisotropy	红外吸收各向异性	infrasalt water	盐下水
infrared absorption method	红外线吸收分析法	infrastructural	构造基底的
infrared absorption spectrometry	红外吸收光谱法	infrastructural belt	深构造带
infrared absorption spectrum	红外吸收光谱	infrastructural project	基础建设工程；基建工程
infrared aid	红外装置	infrastructural upwelling	下部构造层物质上涌
infrared analysis	红外光分析	infrastructure	基础建设；基础设施；基本建设
infrared astronomy	红外天文学	infrastructure gauge	基本限界
infrared band	红外频带；红外波段	infrastructure project	基础设施项目
infrared blasting	红外引爆；红外爆破	infratid zone	潮下带
infrared detection	红外探测	infrequent	不常见的
infrared detection device	红外探测装置	infrequent occurrence	偶尔发生
infrared detection system	红外探测系统	infringement	违约
infrared detector	红外线检测器	infuse	注入；灌注；泡制，浸制
infrared detector package	红外探测器装置	infused seam	注水煤层
infrared device	红外装置	infusibility	难熔性；不溶性
infrared diffusion	红外线散射	infusion in seam	向煤层内注水
infrared electronic distance meter	红外线电子测距仪	infusion method	注水法；浸渍法
infrared emission	红外发射	infusion shot firing	水封爆破
infrared filter	红外线滤光器	infusion time	灌注时间
infrared flux	红外辐射通量	infusorial earth	硅藻土
infrared illuminator	红外发光器	infusorial silica	硅藻土
infrared image	红外线图像	infusoriolite	硅藻土
infrared imaging	红外线成像	ingate	马头门；入口孔
infrared inspection	红外检验；红外探伤	ingoing flood	进潮流
infrared light	红外线；红外光	ingoing side	进入的一侧
infrared line scanner	红外线扫描器	ingoing stream	进潮流
infrared linear scan device	红外线扫描仪	ingredient	成分；拌分；拌料
infrared locator	红外定位器；红外探测器；热探测器	ingredient of concrete	混凝土成分
infrared maser	红外激射器	ingredient	成分，材料
infrared measurement	红外测量	ingress	进口；入口；进入
infrared methanometer	红外线沼气测定仪	ingress of groundwater	地下水浸入
infrared microscope	红外显微镜	ingression	海进；海侵
infrared photography	红外摄影；红外照相	ingression coast	海侵岸
infrared pickoff	红外传感器	ingression of groundwater	地下水侵入
infrared picture	红外图像	ingression sea	进浸海
infrared region	红外光域	ingrown meander	内生曲流
infrared remote sensing	红外遥感	ingrown meander valley	内生曲流谷
infrared remote sensor	红外遥感器；红外遥感传感器	ingrown valley meander	深切河谷曲流
infrared scanner	红外扫描仪	inhabited area	居民区
infrared scanner imagery	红外扫描成像	inhalant canal	进水沟
infrared scanning	红外扫描	inhalant orifice	入水孔
infrared scanning imaging	红外扫描成像	inhalant pore	入水孔
infrared scanning radiometer	红外扫描辐射计	inhalant siphon	进水管
infrared scanning test	红外线扫描法检漏	inherence	内在；固有；基本属性
infrared sensitivity	红外灵敏度	inherent	固有的；特有的；内在的
infrared sensor	红外传感器	inherent error	固有误差
infrared signal	红外信号	inherent feedback	固有反馈；内反馈
infrared sounder	红外探测仪	inherent frequency	固有频率
infrared source	红外线源	inherent gas	赋存瓦斯；赋存气体
infrared spectrometry	红外光谱法	inherent geometrical distortion	固有几何失真
infrared spectrophotometer	红外分光光度计	inherent impurity	固有杂质；内在杂质
infrared spectroscopy	红外线光谱学	inherent matter	固有物质；内在物质
infrared spectrum	红外线光谱	inherent mineral matter	固有矿物质
infrared television	红外电视	inherent mode	固有振型
infrared television camera	红外电视摄像机	inherent moisture	内在水分
infrared thermogram	红外热异常图	inherent regulation	固有调节
infrared thermography	红外线温度记录法	inherent settlement	固有沉降
infrared thermography survey	红外线热像图法检漏	inherent shear strength	固有抗剪强度

inherent stability	固有稳定性	initial connate water saturation	原始原生水饱和度
inherent stress	固有应力；初应力；内在应力	initial consequent river	原始顺向河
inherent structure	固有结构	initial consolidation	初始固结
inherent vibration	固有振动	initial consolidation pressure	初始固结压力
inherent viscosity	特性黏度	initial convergence	初始下沉
inherent-residual stress	固有残余应力	initial cost	生产成本；原价；初期成本
inheritable tenancy	可继承的租住权	initial course	起程航路
inheritance	遗传	initial crack	初裂缝；初始裂纹
inherited mineral	继承矿物	initial crack length	初始裂纹长度
inherited property	继承的财产	initial creep	初蠕变；初始蠕变
inherited river	遗留河	initial data	原始数据；原始资料
inherited stream	遗留河	initial data error	起始数据误差
inherited structure	继承性构造	initial deflection	初始挠度
inherited tectonics	继承性构造	initial deformation	原始变形
inhibit	抑制；阻止；禁止	initial delay	起始延迟
inhibition	禁止；抑制；抑制物；阻化	initial density	初始密度
inhibitor	抑制剂；防腐蚀剂；阻化剂	initial depth	初始水深；跃前水深
inhibitory coating	保护层	initial detention	初始持水量
inhomogeneity	不均匀性；多相性；异质性	initial development phase	初期开发阶段
inhomogeneity breccia	非均一角砾岩	initial development zone	初步发展区
inhomogeneous	不均匀的；多相的；非均质的；不同类的	initial dip	原始倾斜
inhomogeneous accretion model	非均匀聚积模型	initial displacement	初始位移
inhomogeneous anisotropic medium	不均匀各向异性介质	initial earth stress	初始地应力
inhomogeneous boundary condition	非齐次边界条件	initial filling	早期注水
inhomogeneous deformation	非均匀变形	initial fissure	原生裂隙
inhomogeneous differential equation	非齐次微分方程	initial flow	初始流量；起始流量
inhomogeneous equation	非齐次方程	initial flow period	初流动阶段
inhomogeneous flow	不均匀流	initial flow potential	初始最大产能；初始最大流量；初始流动势
inhomogeneous function	非齐次函数		
inhomogeneous glide	不均匀滑移	initial flowing pressure	初始流压
inhomogeneous linear difference equation	非齐次线性差分方程	initial fluid saturation	原始流体饱和度
		initial force	初始压力；起始力
inhomogeneous material	非均质材料	initial form	原始地形；原始类型；早期阶段
inhomogeneous medium	不均匀介质	initial formation pressure	原始地层压力
inhomogeneous plane wave	不均匀平面波	initial fracturing	初步破碎
inhomogeneous rock	非均质岩石	initial fragmentation	初次爆破
inhomogeneous soil	非均质土	initial geodetic data	起始大地数据
inhomogeneous strain	不均匀应变	initial geodetic point	起始大地测量点
inhomogeneous wave	非均相波；非齐次波	initial geological logging	原始地质编录；原始地质录井
initial abstraction	初始吸收量；初始渗失量	initial gradient	初始梯度
initial acceleration	初始加速度	initial gravimetric point	起始重力点
initial activity	初活性；初始放射性	initial gross separation	初步分离
initial address	起始地址	initial ground-stress field	初始地应力场
initial air blow	初始气喷	initial hardness	起始硬度；初始硬度
initial alignment	起点定向	initial heat of hydration	初始水化热
initial amplitude	初振幅	initial hole diameter	开孔直径
initial approximation	初始近似值	initial hydrostatic pressure	起始静液压力
initial blast	初爆破	initial image	初始图像
initial blasting	初次爆破	initial imperfection	初始缺陷
initial boiling point	初沸点	initial impulse	初始脉冲
initial capacity	最初产量	initial incidence angle	初始入射角
initial capacity of well	井的最初产量	initial index	初始变址
initial capital investment	初期基建投资	initial infiltration capacity	初始入渗能力；初始入渗量
initial caving	初次放顶	initial infiltration rate	初期浸透强度；初始入渗率
initial collapse pressure	湿陷起始压力；坍塌起始压力	initial investment	初期投资
initial compression	初始压缩	initial irreducible water saturation	原始束缚水饱和度
initial concentration	初浓度；起始浓度	initial isotopic composition	原始同位素组成
initial condition	初始条件；初值条件	initial kinetic energy	初动能
initial condition of seepage	渗流的初始条件	initial laboratory phase	实验室初期试验阶段
initial confining pressure	初始围压；初始封闭压力	initial landform	原始地形

initial level 起始水准面
initial leveling 起始标高
initial liquefaction 初始液化
initial loading stage 初始荷载阶段
initial looseness coefficient 初始松散性系数；初始松散度系数
initial loss 初始损失
initial measurement 初始量测
initial migration 初始运移
initial mixing 初期搅拌
initial mobility period 初始运动期
initial model 原始模型；初始化模型
initial modulus 起始模数
initial moisture content 初始含水量
initial numerical data 原始数据；起算数据
initial oil in place 原始石油地质储量
initial oil saturation 原始含油饱和度
initial operation 初始运行；启动操作
initial order 初始指令
initial outlay 基建费用；初始投资
initial output 初产量
initial overburden pressure 初始超载压力；初始覆盖地层压力；初始积土压力
initial P wave 初至P波；初至纵波
initial pack 新井砾石充填防砂
initial parameter method 初值参数法
initial pattern 起始井网
initial payment 首期付款
initial period 开始时期；初期
initial period of construction 施工初期阶段
initial period of operation 运行初期，经营初期
initial permeability 起始磁导率；初始渗透率
initial phase 初相
initial phase angle 初相角
initial piezometric head 初始承压水头；初始测压管水头
initial pit 初始坑
initial placement condition 初始填筑条件
initial plastic flow 初始塑性流动
initial plastic state 初始塑性状态
initial point 起点；原点；始点
initial pore volume 原始孔隙体积
initial pore water pressure 初始孔隙水压力
initial position 原始位置；初始位置
initial potential 初期势能；初期潜能
initial potential flowing 初期自喷产量
initial precipitation 初期降水
initial pressure of collapsing 湿陷起始压力；坍塌起始压力
initial prestress 初预应力
initial production 初期产量；最初产量
initial protection 初期支护
initial rain 初始雨量
initial rainfall 初期降雨
initial rate of penetration 初始钻速
initial reading 初始读数
initial reserves 原始储量
initial rock stress 原岩应力
initial rock stress field 原岩应力场
initial rolling 初次碾压
initial roof pressure 顶板初次来压

initial S wave 初至S波；初至横波
initial sand control 先期防砂
initial saturated thickness 原始饱和厚度
initial saturation 原始饱和度
initial screening of slopes 斜坡初步筛选
initial section 起始段
initial segment 初始段；前节
initial separation 初步分选
initial set 初始设置；初凝；始集
initial setting 初步设定；初凝
initial setting strength 初凝强度
initial setting time 初凝时间；初凝期
initial settlement 初始沉降；原有沉降
initial shear 初始剪切
initial shear modulus 初始剪切模量
initial shear resistance 初始剪切阻力
initial shear stress ratio 初始剪应力比
initial signal 起始信号
initial sliding direction 初始滑移方向
initial slope 初始斜率
initial soil 原始土壤
initial soil moisture 初始土壤水分
initial soil storage 初期土壤储水量
initial solution 初始解
initial sounding survey 首次海洋回声测深
initial speed 初速，初速度
initial stability 初始稳定性
initial stage 初阶段；初期
initial state 初始状态
initial steady level 起始稳定水位
initial stiffness 初始刚度
initial storage 初始库容；初始容量
initial strain 初始应变
initial strength 早期强度；初始强度
initial stress 初始应力；预应力
initial stress condition 初始应力状态
initial stress field 初始应力场
initial stress matrix 起始应力矩阵
initial stress measurement 初始应力测量
initial stress method 初始应力法
initial stress of rock mass 岩体初始应力
initial stress state 初始应力状态
initial support 初期支护
initial suppression 初始压制；预压制
initial surface 初始面；初始曲面
initial surface absorption test 初始表面吸水测试
initial survey 施工前测量；初步检测；初级检验
initial tangent modulus 初始正切模量
initial temperature 原始温度
initial tension 初张力；起始拉力
initial thrust 起始推力；初推力
initial transient creep 初始不稳定蠕变；初始瞬时蠕变
initial value 初始值；基础资料
initial value of groundwater pollution 地下水污染起始值；地下水污染基础资料
initial value of isotopic ratio 同位素比初始值
initial value of parameter 参数初值
initial value problem 初值问题
initial volume 初体积；初容积
initial water breakthrough 开始水窜

initial water deficiency	初始水分不足量；初始缺水量	injection into an aquifer	边外注水；向含水层注入
initial water front	初始水前缘	injection material	灌浆材料
initial water level	初始水位	injection pipe	灌浆管；注浆管
initial water table	初始水位	injection pore volume	注入孔隙体积
initial wave	初生波；初始波	injection port	进样口；注入口
initial working resistance	初始工作阻力	injection pressure	灌注压力；注入压力
initial yield load	初始屈服荷载	injection pressure buildup	注入井压力上升
initial yield surface	初始屈服面	injection process	灌浆方法；注入工艺
initial zone	零时区	injection profile	注水断面；注入剖面
initialization	初始化	injection profile height	注入剖面高度
initializing constraint	初始约束条件	injection profile log	注入剖面测井
initializing data	起始数据	injection profile modification	注入剖面调整
initiate	开始；起爆	injection proration	配注
initiating ability	起爆能力	injection pump	注入泵；喷射泵；高压油泵
initiating agent	引发剂	injection rate	注入速度；注水量
initiating charge	起爆剂	injection recharge	注水补给
initiating device	起爆装置	injection system	注入系统；喷射系统
initiating explosive	起爆剂；引爆药	injection technique	灌浆技术；注入技术
initiating signal	起始信号；启动信号	injection well	注入井；注水井；补给井；回灌井
initiating trigger	启动触发；启动触发器	injection well barrier	注入井屏障
initiation effect	起爆能力	injection well conversion	注入井转换
initiation interval	起爆间隔	injection well head assembly	注水井井口装置
initiation point of slip surface	滑面起点	injection well pattern	注入井井网
initiation pressure	引发压力	injection well performance	注入井动态
initiation reaction	引发反应	injection well pressure	注入井压力
initiation sensitivity	起爆感度	injection well row	注入井排
initiation sequence	起爆顺序	injection well spacing	注入井控制面积；注入井井距
initiation strength	起爆强度	injection well testing	注入井试验
initiation time	起爆时间	injection withdrawal balance	注采平衡
initiation-timing system	定时起爆方法	injection-earthquake relationship	注入液-地震关系
injected body	贯入体	injection-fluid front	注入液前缘
injected fluid front	注入流体前缘	injection-gas-liquid ratio	注入气-液比
injected fluid viscosity	注入流体黏度	injection-pore volume ratio	注入孔隙体积比
injected water	注入水	injection-production cycle	注采周期
injected water volume	注入水量	injection-production pattern	注采井网
injected zone	注水区	injection-production ratio	注采比
injecting additive	注添加剂	injection-production strategy	注采对策；注采方案
injecting grout	注浆；注入灌浆	injection-production voidage balancing	注采平衡
injecting interval	注水井段	injection-production well arrangement	注采井布置
injecting material	注入材料	injection-production well conversion	注采井转换
injection aquifer	注水含水层；注入含水层	injectivity	注入性；注入能力；可灌性
injection area	注水面积；注入区	injectivity behavior	注入动态
injection breccia	贯入角砾岩	injectivity coefficient	注入系数；吸水指数
injection commutator	注射管	injectivity enhancement	注入性增强
injection complex	贯入杂岩	injectivity index	注入指数；吸水指数
injection dyke	贯入岩墙	ink bottle effect	颈缩效应
injection efficiency	注入系数；注入效率	inkstone	砚石；水绿矾
injection end	注入端	inland coal-bearing formation	内陆型含煤建造
injection face	注入面	inland delta	内陆三角洲
injection flow	注入流量	inland desert	内陆沙漠
injection flow rate	注入速率；注入流量	inland drainage	内陆水系
injection fluid	注入流体	inland dune	内陆沙丘
injection folding	贯入褶皱作用	inland earthquake	内陆地震
injection grout	压力灌浆；高压注浆	inland lake	内陆湖
injection hole	注入钻孔	inland plain	内陆平原
injection horst	贯入地垒	inland rig	陆上钻机
injection imbalance	注入不平衡	inland river	内河，内陆河
injection index	注入指数	inland salinization	内陆盐渍化；内陆盐化作用
injection induced earthquake	注入流体诱发的地震	inland sea	内海
injection interval	注入层段	inland sea dike	内陆海堤

inland stream	内陆河流
inland transportation	内陆运输
inland water grouping	内陆水域组别
inland water transportation	内河航运
inland watercourse	内陆水道
inland waters	内陆水域
inland waterway	内地水路；内河航路
inleakage	漏泄；渗入；吸入
inlet	入口；进入孔；入口管
inlet branch	输入支管；进水管；进气管
inlet channel	进水沟；进气道
inlet for storm water	雨水口
inlet manifold	吸入管；吸水管；进水管
inlet opening	进风口；进水口；进口
inlet shaft	通风井
inlet side	进口侧
inlet temperature	进口温度
inlet unsubmerged culvert	无压力式涵洞
inlet velocity	进口速度；进入速度
inlet well	集水井；注浆钻孔
inmost depth	最大深度；到底深度
inner barrel	装岩芯内管
inner bay	内湾
inner border	内图廓
inner casing	内套管
inner clearance ratio	内间隙比
inner continental shelf	内陆架
inner core	内核
inner cross-section	内断面；净空断面
inner diameter	内径
inner dimension	内尺寸
inner face	内表面
inner facies association	内相组合
inner force	内力
inner orientation	内方位；内部定向
inner product	内积；标量积
inner section	内断面；净空断面
inner stope	内回采工作面
inner viscosity	结构黏度
inner wall	内壁
innocuous	无害的；无毒的
innoxious	无害的；无毒的
inoperative	不工作的；不起作用的；无效的
inoperative period	非运行期
inorganic acid	无机酸
inorganic architectural coating	无机建筑涂料
inorganic base	无机碱
inorganic binder	无机结合料；无机黏结剂
inorganic chemistry	无机化学
inorganic clay	无机黏土
inorganic coating	无机涂层
inorganic component	无机组分
inorganic constituent	无机成分
inorganic contaminant	无机污染物
inorganic environment	无机环境
inorganic flocculant	无机絮凝剂
inorganic geochemistry	无机地球化学
inorganic liquid	无机液体
inorganic material	无机材料
inorganic matter	无机物
inorganic nitrogen	无机氮
inorganic sediment	无机沉积物
inorganic silt	无机泥沙
inorganic soil	无机土
inoxidizability	抗氧化性
inoxidizing coating	防护层；防蚀层
in-place	原位；原地；就地；现场
in-place investigation	实地调查；现场调查
in-place measurement	原位测定
in-place nonwetting phase	原地非湿相，现场非湿相
in-place permeability	地层渗透率；原位渗透率
in-place soil-shearing test	现场土壤剪力试验
in-place stability	原地稳定性；现场稳定性
in-place test	现场试验
in-place void ratio	原位孔隙比
in-plane	平面内的
in-plane bending	平面内弯曲
in-plane displacement	平面内位移
in-plane force	平面内力
in-plane loading	平面内加载
in-plane mode	平面模态；面内模态
in-plane phasing	平面相
in-plane rigidity	平面内刚度
in-plane shear	面内剪切
in-plane shear crack	平面剪切裂纹；面内剪切裂纹
in-plane shear modulus	平面内剪切模量
in-plane shear strength	平面内剪切强度
in-plane stiffness	平面内刚度
input block	输入信息组；输入模块
input character	输入特性
input data	输入数据；原始数据
input flow	输入流量
input format	输入格式
input function	输入功能
input gas	注入气
input gas-oil ratio	注气量与采油量之比
input horizon	注入层
input image	输入图像；输入影像
input imagery	输入成像
input information	输入信息
input instruction code	输入指令码
input inversion	输入反演
input layer	输入层
input port	输入端；输入口
input power	输入功率；输入电源
input power control	输入功率控制；输入电源控制
input pressure	进口压力
input process	输入过程
input profile	注入剖面
input program	输入程序
input range	输入范围
input rate	注入率
input routine	输入程序
input selector	输入选择钮
input sensitivity	输入灵敏度
input signal	输入信号
input simulation network	输入模拟网络
input speed	输入速度；注入速度；进料速度
input unit	输入装置；输入器；输入设备
input variable	输入变量

input well	注入井；注水井；补给井
input-output ratio	投入产出比率
inrush	涌入；流入；浸入；喷出
insensitive	不敏感的；低敏感度的
insert depth	嵌入深度
insertion	插入；嵌入；插入物；嵌入物
inshore	向岸的；近岸的；外滩；近海
inshore current	近海流；近滨流
inshore engineering	近海工程；海岸工程
inshore hydrographic survey	近岸水文测量
inshore region	近岸地区
inshore subtidal zone	内滨潮下带
inshore waters	近岸水域；沿岸水域
inshore waters development	近岸水域开发
inside diameter	内径；内直径
inside lining	内衬；内衬砌
inside slope	井下斜井；暗斜井
insignificant error	不显著误差
in-situ activation analysis	现场活化分析；现场放射性分析
in-situ and laboratory test	现场试验与室内试验
in-situ broken-ore leaching mining	原地破碎浸出开采
in-situ concrete	原地浇筑混凝土；整体混凝土
in-situ concrete lining	现场浇筑混凝土衬砌
in-situ concrete pile	现场浇筑混凝土桩
in-situ concrete strength	现场浇筑混凝土强度
in-situ concreting	现场浇筑混凝土
in-situ condition	天然条件；现场条件
in-situ cube strength	现场立方体强度
in-situ deformation test	原位变形测试
in-situ density	原地密度
in-situ direct shear test	原位直剪试验
in-situ experiment	原位试验；现场试验
in-situ explosive publication	原地成块爆破；原地块石化爆破
in-situ field test	现场试验
in-situ fracture toughness	原位断裂韧度
in-situ gasification	地下气化
in-situ heat system	井下热液
in-situ inspection	现场检验
in-situ measurement	原位测试；现场量测
in-situ mining	原地开采
in-situ modelling	现场模拟
in-situ monitoring	现场监测
in-situ permeability test	原位渗透试验
in-situ porosity	原位孔隙度
in-situ rock stress	原地岩石应力
in-situ sediment velometer	原地沉积速度计
in-situ shear test	原地抗剪试验
in-situ soil test	原地土壤试验；现场土工试验
in-situ strength	现场强度
in-situ stress azimuth	地应力方位角
in-situ stress gradient	地应力梯度
in-situ stress magnitude	地应力大小
in-situ stress orientation	地应力方向
in-situ test	原位测试；现场试验
in-situ thrust test	原位推力试验
in-situ wave velocity	原位波速
insoluble fraction	不溶组分
insoluble material	不可溶物质
insoluble precipitate	不溶性沉淀物
insoluble residue	不溶残渣
insoluble sludge	水不溶物
insoluble solid	不溶性固体
insoluble substance	不溶物质
inspection balance	检验单；检验卡片
inspection bench	检验台
inspection borehole	检查孔；观测孔
inspection chamber	检查井
inspection department	检验部门
inspection gallery	检查廊道
inspection hole	检查孔
inspection of site	现场视察
inspection schedule	检查图表；监控调度
inspection screen	检查筛分
inspection shaft	检查井
inspection tunnel	检查隧洞
inspection well	检查井
inspectional analysis	检验分析
instability	不稳定；不安定
instability constant	不稳定常数
instability criterion	失稳准则
instability effect	非稳定效应
instability in algorithm	算法的不稳定性
instability in tension	拉伸失稳
instability index	不稳定指数
instability mechanism	失稳机制
instability of first kind	第一类失稳
instability of second kind	第二类失稳
instability of slope	边坡的不稳定性
instability phenomenon	不稳定现象
instability point	不稳定点
instability ratio of spring discharge	泉水泄量不稳定系数
instability theory	不稳定性理论
instability threshold	不稳定性阈值
instable equilibrium	不稳定平衡
installation diagram	装配图，安装图
installation drawing	装置图；施工图
installation engineering	安装工程
installation error	安装误差
installation expenditure	安装费
installation fee	安装费
installation fundamental circle	安装基准圆
installation guide	安装指导
installation procedure	安装程序
installation rate	安装速度
installation site	安装现场
installation specification	安装规范
installation survey	安装测量
installation technique	安装技术
installation time	安装时间
installation tool	安装工具
installed cost	安装费
instant grouting	即时灌浆
instant time line	瞬时压缩线
instantaneous	瞬时的；即时的；同时的；同时发生的
instantaneous action	瞬时作用；即时作用
instantaneous amplitude	瞬时振幅
instantaneous corrosion-rate	瞬时腐蚀率
instantaneous couple	瞬时力偶
instantaneous deformation	瞬时变形

instantaneous delivery	瞬时流量
instantaneous elastic strain	瞬时弹性应变
instantaneous elasticity	瞬时弹性
instantaneous flow	瞬时流量
instantaneous frequency	瞬间频率
instantaneous frequency stability	瞬时频率稳定度
instantaneous fuse	瞬发导火线；瞬发引信
instantaneous load	瞬时荷载
instantaneous modulus	瞬时模量
instantaneous modulus of deformation	瞬时变形模量
instantaneous modulus of elasticity	瞬时弹性模量
instantaneous molal concentration	瞬时摩尔浓度；瞬时克分子浓度
instantaneous optimal control	瞬时最优控制
instantaneous outburst mine	瞬时突出矿
instantaneous outburst of coal and gas	煤与瓦斯瞬时突出
instantaneous overload	瞬时过载
instantaneous peak discharge	瞬时洪峰流量
instantaneous peak flow	瞬时最高流量；瞬时顶峰流量
instantaneous penetration rate	瞬时钻速
instantaneous phase	瞬时相位
instantaneous pore pressure	瞬时孔隙压力
instantaneous profile method	瞬时剖面方法
instantaneous rate of flow	瞬时流量
instantaneous rupture	瞬时破坏
instantaneous sampling	瞬间采样
instantaneous settlement	瞬时沉降
instantaneous stability of slope	边坡瞬时稳定
instantaneous state	瞬时状态
instantaneous strain	瞬时应变
instantaneous strength	瞬时强度
instantaneous stress	瞬时应力
instantaneous stress-strain curve	瞬时应力-应变曲线
instantaneous stress-strain response	瞬时应力-应变响应
instantaneous stripping ratio	瞬时剥采比
instantaneous translation	瞬时平移
instantaneous unit hydrograph	瞬时单位过程线
instantaneous value	瞬时值
instantaneous velocity	瞬时速度
instantaneous vibration	瞬时振动
Institution of Civil Engineers	英国土木工程师学会
Instron testing machine	英斯特朗试验机
instruction control	指令控制
instruction counter	指令计数器
instruction cycle	指令周期
instruction manual	使用手册
instruction register	指令寄存器
instruction unit	指令部件
instrument adjustment	仪器校正；设备调试
instrument board	仪表盘；仪表板；配电板；分配板
instrument effect	仪器效应
instrument epicenter	仪器震中
instrument error	仪器误差
instrument factor	仪器因数
instrument for aerial photographic record	航摄记录仪器
instrument for water quality analysis	水质分析仪器
instrument gallery	观测廊道；仪表廊
instrument of flow velocity measurement	流速量测仪器
instrument of surveying and mapping	测绘仪器
instrument precision	仪表精度
instrument response	仪器响应
instrument station	测站；仪器站
instrumental analysis	仪器分析
instrumental observation	仪器观测
instrumentation engineering	仪表工程
instrumentation imbalance	仪器不平衡
instrumentation program	仪器测试计划
insulated stream	隔底河流；隔绝河
insulating bushing	绝缘衬套
insulating capacity	绝缘能力
insulating caprock	隔水盖层；隔热盖层
insulating cement	保温水泥
insulating coating	绝缘涂层
insulating course	绝缘层
insulating fire brick	绝热耐火砖
insulating layer	隔绝层；保温层
insulating lining	绝缘里衬
insulating material	绝缘材料；隔热材料
insulating oil	绝缘油
insulating packing	绝缘材料
insulating rod	绝缘棒
insulating shield	绝缘护片
insulating sleeve	绝缘套管
insulating socket	绝缘承口；绝缘套接口
insulation course	绝缘层；隔热层
insulation gap	绝缘间隙
insulation grade	绝缘等级
insulation lagging	绝缘层
insulation paste	绝缘胶；绝缘膏
insulation resistance	绝缘电阻
insulation sleeve	绝缘护套；绝缘套管
insulation slit	绝缘缝
insulation test	绝缘试验
insulation works	隔音工程
insulator	绝缘体；绝缘子；隔电子
insurance	保险；保险额；保险费；保障
insurance against accident to workmen	工伤事故保险
insurance against nonperformance	对方不履约保险
insurance amount	保险金额
insurance and freight cost	保险费和运费
insurance appraiser	保险鉴定人
insurance assessment	保险估价
insurance binder	保险承诺书
insurance certificate	保险单，保单；保险凭证
insurance charge	保险费
insurance claim	保险索赔
insurance clause	保险条款
insurance contract	保险合同
insurance coverage	保险范围
insurance device	保险装置；保险器
insurance division	保险部
insurance document	保险单据
insurance expense	保险费
insurance for liability	责任保险
insurance fund	保险费
insurance group	保险集团
insurance in force	有效保险
insurance incident	保险事件
insurance indemnity	保险赔偿
insurance interest	保险权益

insurance law	保险法
insurance liability	保险责任
insurance market	保险市场
insurance of client's property	业主财产保险
insurance of goods	货物保险
insurance of workmen	劳工保险
insurance of works	工程保险
insurance of works and contractor's equipment	工程及承建商设备的保险；工程及承办商设备的保险
insurance period	保险期限
insurance policy	保险单；保单；保险合同
insurance policy in force	生效的保险单
insurance premium	保险金；保险费
insurance rebate	保险回扣
insurance reserve fund	保险准备基金
insurance slip	保险单
insurance subject	保险标的
insurance tax	保险税
insurance value	保险价值；保险额
insurant	被保险人；受保人；投保人
intact clay	原状黏土
intact displaced material	无扰动位移物质
intact displacement mass	无扰动位移块体
intact displacement material	无扰动位移物质
intact ground	未扰动地基
intact loess	原状黄土
intact modulus	完整系数；完整性模数
intact rock	原状岩石
intact rock mass	未采动的岩体；原岩体
intact rock sample	原状岩心样品
intact rock strength	未扰动岩石的强度
intact roof	完整顶板；未采动的顶板
intact sliding zone soil	原状滑带土
intact specimen	原状试验
intactness index	完整性指数
intactness index of rock mass	岩体完整性指数
intake	进口巷道；进风，入风道；进水口；吸水管；
intake air	进风；进风流
intake air course	进风道
intake air way	进风道
intake airflow	进风
intake airway	进风道；进风巷
intake area	受水区；补给区；渗入区
intake area of alluvial cone	冲积锥进水区
intake area of aquifer	含水层来水区
intake brine	外来盐水
intake bulk station	收货油库
intake capacity	吸入能力
intake chamber	进水室；进气室
intake channel	进水槽
intake conduit	进水管道
intake cone	进水曲面
intake culvert	进水涵洞
intake duct	进气道
intake end	终点
intake entry	进风平巷
intake facility	接油设施；取水设备
intake fan	进气风扇
intake fish baffle	进水口挡鱼结构
intake flow	吸入液流

intake gallery	进风平巷
intake gate	进水闸门；引水闸门
intake head	进风压头
intake horizon	进风水平
intake line	进风路线；入风管道
intake lip	进口边缘
intake manifold	进口管汇
intake of groundwater	地下水补给
intake of the hole	钻孔吸收泥浆量
intake pipe	取水管；进气管
intake pipeline	进水管道；进气管道
intake place	给水区；流域
intake pressure	进气压力；进水压力；入口压力
intake profile	吸水剖面
intake quantity	进风量；进水量
intake rate	吸入速度
intake resistance	进气阻力；进水阻力
intake screen	进口拦污栅
intake shaft	进风井
intake slot	进水孔
intake specimen	原状试件
intake stroke	进气行程
intake structure	进水建筑物
intake swirl	进气涡流
intake temperature	进气温度
intake tower	进水塔
intake tunnel	引水隧洞
intake valve	进气阀
intake velocity	进口速度；进风速度
intake volume	注入量
intake well	注入井
intake well head pressure	注入井井口压力
intake workings	进风巷道
intake works	进水工程；取水工程；进水建筑物；取水建筑物
intake yield	进入水量；入流水量
integral calculus	积分学
integral circuit	积分回路
integral coring method	整体取芯法
integral differential equation	积分微分方程
integral distribution curve	累积分布曲线
integral divisor	整因子
integral dose	累积剂量
integral effective spectrum	有效积分谱
integral electron concentration	积分电子浓度
integral enclosure	整体密闭罩
integral equation	积分方程
integral exponent	整指数
integral expression	积分表达式
integral factor	整因子
integral finite difference	积分有限差分
integral flow curve	累积流量曲线；径流累积曲线
integral flux	积分通量
integral formula	积分公式
integral geometry	积分几何学
integral invariant	积分不变量
integral matrix	整数矩阵
integral mean	积分平均值
integral mean-value theorem	积分中值定理
integral method	积分法

integral multiple　整倍数
integral number　整数
integral part　主要部分；积分部分；整体部分
integral quantity　积分值
integral rational invariant　整有理不变式
integral rigidity　整体刚度
integral sampling method　整体采样方法
integral spectrum　积分谱
integral square error　积分平方误差
integral stowing　整体充填
integral structure　整体结构
integral surface　积分曲面
integral transform　积分变换
integral transform method　积分变换法
integral type　整体式
integral value　整数值
integrate　求积；综合；综合的；完整的
integrated　综合的；集成的，完整的，完全的；全盘的；积分的
integrated analysis　综合分析；整体分析
integrated anomaly map　综合异常图
integrated approach　综合方法
integrated assemblage-zone　交互组合带
integrated assessment　综合评价
integrated automatic test system　综合自动测试系统
integrated ballast bed　整体道床
integrated budget formula　综合预算方案
integrated cabinet　综合柜
integrated circuit computer　集成电路计算机
integrated coastal zone management　海岸带综合管理
integrated computer aided manufacture　综合计算机辅助制造
integrated conservation　整体维护
integrated conservation and development project　整体维护和开发项目
integrated control　综合控制；综合防治
integrated control system　综合控制系统
integrated curve　综合曲线
integrated data　综合数据
integrated data base　综合数据库
integrated data processing　综合数据处理
integrated data store　集中数据库；集成数据存储系统
integrated decision making　集成决策
integrated deep water navigation system　深水综合导航系统
integrated development program of oil field　油田开发总体规划
integrated digital network　综合数字网
integrated dispatching center　综合调度中心
integrated drainage　合成水系
integrated dual operation　整体操作；连续顺序操作
integrated earthing wire　综合接地线
integrated electronics　集成电子学
integrated element　集成元件
integrated equipment　整机，整件；综合设备，成套装置
integrated expert system　集成的专家系统
integrated exploration　综合勘探
integrated external dose of radiation　外来辐射的累积剂量
integrated filter　整合滤波器；集成滤波器
integrated flow　累计流量
integrated gasification combined cycle　煤气化联合循环

integrated geodesy　整体大地测量
integrated geophysical interpretation　综合地球物理解释
integrated geophysical system　综合物探系统
integrated geophysics　综合地球物理学
integrated geothermal exploration program　综合地热勘探计划
integrated industrial system　综合工业体系
integrated information system　综合信息系统
integrated interpretation　综合解释
integrated leak rate test　整体泄漏率试验
integrated load curve　荷载累积曲线
integrated logging　综合测井
integrated magnetic memory　集成磁存储器
integrated management　综合管理
integrated management approach　综合管理方法
integrated management function　综合管理任务
integrated manufacturing system　综合生产系统
integrated marketing system　综合销售系统
integrated mechanized coal getting system　综合机械化采煤系统
integrated mill　大型工厂；联合工厂
integrated monitor panel　综合监控台
integrated navigation　组合导航
integrated navigation system　综合导航系统；组合导航系统
integrated network　集成管网；综合网络
integrated neutral point　集成中性点
integrated neutron flux　中子积分通量
integrated package　完整的软件包
integrated planning　整体规划
integrated plant evaluation　电厂综合评估
integrated positioning　组合定位
integrated power grids　综合电力网；完整电力网
integrated production line　综合生产线
integrated project　综合计划
integrated pseudo geometrical factor　近似积分几何因子
integrated pulse　积分脉冲
integrated rack　综合架
integrated radial geometrical factor　积分径向几何因子
integrated radiation　累积辐射
integrated research vessel　综合调查船
integrated river basin development　江河流域综合开发
integrated road network　综合式路网
integrated service　整套服务
integrated services digital network　综合业务数字网
integrated simulator　集成模拟器
integrated solids control　成套固相控制
integrated sonar system　综合声呐系统
integrated spectrum　积分谱
integrated square error　积分平方误差
integrated storage　综合库存；累积存储器
integrated study　综合研究
integrated survey　综合勘测；综合调查
integrated tester　综合测试仪
integrated travel time　累计传播时间
integrated unit　联合装置
integrated unit substructure　拼装式钻机底座
integrated utilization　综合利用
integrated vertical geometrical factor　纵向积分几何因子
integrated vertical response　积分垂向响应
integrated waste reduction plan　综合减废计划

integrated waste treatment service 废物处理综合服务
integrated wastewater treatment plant 污水综合处理厂
integrated water resource planning 水资源综合利用规划
integrating 积分
integrating accelerometer 综合加速度计
integrating amplifier 积分放大器
integrating analogue digital converter 积分型模数转换器
integrating depth recorder 积分式自记深度计
integrating device 积分器
integrating digital voltmeter 积分数字电压表
integrating dosemeter 积分剂量计
integrating factor 积分因子
integrating instrument 积分仪；累计仪
integrating inverse of gravity anomaly 重力异常的积分反演
integrating mechanism 积分机构
integrating meter 积分仪；积量计
integrating orifice meter 积分式锐孔流量计
integrating point 积分点
integrating sound level meter 积分声级计
integrating sphere 积分球
integrating watermeter 累计式水表
integration 积分；集成；综合；积累
integration by parts 分部积分法
integration constant 积分常数
integration filter 积分滤波器
integration formulae 积分公式
integration formulae for tetrahedron 四面体的积分公式
integration formulae for triangle 三角形的积分公式
integration limit 积分限
integration method 积分法
integration method of velocity measurement 流速积测法
integration of chargeability 积分荷电力
integration order 积分顺序
integration procedure 积分程序
integration sphere 积分球
integration step 积分步骤
integration system for mine dynamic disaster monitoring 采矿灾害动态监测集成系统
integration theorem 积分定理
integration time constant 积分时间常数
integrator 积分仪；积分器；积分电路
integrity 完整性；强韧性；牢固性
integrity test 无损试验；完整性测试
integro-differential equation 微积分方程
intellectual property right 知识产权
intellectualization 智能化
intellectualized geographic information system 智能化地理信息系统
intelligent console 智能控制台
intelligent control 智能控制
intelligent device 智能装置
intelligent digitizer 智能数字化仪
intelligent editing system 智能编辑系统
intelligent editor 智能编辑程序
intelligent energy dissipation device 智能耗能装置；智能消能设施
intelligent maintenance 合理维修；合理保养
intelligent material 智能材料
intelligent mining 智能采矿
intelligent monitoring instrument 智能监控仪器
intelligent monitoring system 智能监测系统
intelligent numerical method 智能数值方法
intelligent optimum design 智能优化设计
intelligent rock engineering 智能岩石工程
intelligent rock mechanics 智能岩石力学
intelligent semi-active seismic isolation 智能半主动隔震
intelligent structure 智能结构
intelligent system 智能系统
intelligent terminal 智能终端；智能型终端设备
intelligent use 合理使用
intelligent vehicle 智能车
intelligent well system 智能井控系统
intelligent workstation 智能工作站
intelligibility 可懂度，可理解性；清晰度
intelligible 可理解的
intended size 公称尺寸；给定尺寸
intended target 预定目标
intended zone 目的层
intense absorption 强吸收
intense anomaly 强异常
intense burst 强爆发
intense earthquake 强烈地震
intense earthquake motion 强烈地震运动
intense erosion 强烈侵蚀
intense ground shaking 强烈地震动
intense water exchange 强烈水交替
intensely weathered 强烈风化的
intensely weathered zone 强风化带
intensification 增强；加厚；加强明暗度
intensifier 增强器；增加剂
intension 加强，加剧；强度，烈度；紧张
intensity abnormal region 烈度异常区
intensity adjustment 烈度调整
intensity assessment 烈度评定
intensity attenuation 烈度衰减；强度衰减
intensity calibration 强度校准
intensity characteristic 烈度特征；强度特性
intensity control 亮度调整；强度控制
intensity determination 烈度测定；强度测定
intensity distribution 烈度分布
intensity envelope function 强度包线函数
intensity exceedance probability 烈度超越概率
intensity exceedance rate 烈度超过率
intensity factor 强度因素
intensity function 强度函数
intensity level 亮度；强度级
intensity of anomaly 异常强度，异常程度
intensity of burst 瓦斯突出强度
intensity of compression 压缩度
intensity of deformation 形变强度
intensity of earthquake 地震烈度
intensity of excavation 开挖强度
intensity of field 场强
intensity of frequently occurred earthquake 多遇地震烈度
intensity of gamma radiation 伽马辐射强度
intensity of gravity 重力强度
intensity of ground shaking 地震动强度
intensity of hydrostatic pressure 静水压强
intensity of joint 节理密度
intensity of magnetic field 磁场强度

intensity of outburst	突出强度；火山爆发强度；破裂强度
intensity of precipitation	降水强度
intensity of pressure	压力强度；压强
intensity of radiation	辐射强度
intensity of rain	降雨强度
intensity of rainfall	降雨强度
intensity of rainstorm	暴雨强度
intensity of rockburst	岩爆等级；岩爆烈度
intensity of seldom occurred earthquake	罕遇地震烈度
intensity of shaking	震动强度
intensity of stress	应力强度
intensity of vibration	振动强度
intensity of wave pressure	波压强度
intensity of weathering	风化强度；风化程度
intensity rating	烈度值；强度率
intensity ratio	强度比
intensity recurrence period	烈度重现周期
intensity scale	烈度表
intensity statistics	烈度统计量
intensity target method	强度指标法
intensity-attenuation	烈度衰减
intensity-duration curve	强度-历时曲线
intensity-duration formula	强度-历时公式
intensive mixing	充分搅拌；强力混合
intensive parameter	强度参数；烈度参数
intensive quantity	强度量
intensive support	加强支架；密集支架
intensive variable	强度变量
intensive water exchange	强烈水交换
intention agreement	意向协定
intention to claim	索赔意向
inter isochron area	等时线间面积
interactant	相互作用物；反应物
interacting element	相关要素
interacting natural system	相互作用自然系统
interaction effect	相互作用
interaction energy	相互作用能
interaction factor	相互作用因素
interaction force	相互作用力
interaction of footing	基础的相互影响
interaction of shear walls	剪力墙相互作用
interaction parameter	相互作用参数
interaction sequence	相互作用顺序
interaction space	相互作用空间
interaction stress	相互作用应力
interactive	交互式的；相互作用的
interactive analysis	人机对话分析
interactive environment	交互环境
interactive graphic design system	交互式绘图设计系统
interactive graphic interpretation system	交互图示解释系统
interactive graphic system	交互式制图系统
interactive interpretation	交互解释
interactive model	交互模式
interactive pattern	交互模式
interactive pattern recognition	交互式模式识别
interactive plotting	人机对话绘图
interactive plotting system	交互绘图系统
interactive processing	人机交互处理
interactive simulator	人机模拟器
interaquifer flow	层间越流；层间流
interarc basin	弧内岛地
interarc marine basin	弧间海盆
interarc oceanic basin	弧间洋盆
interarc sea	弧间海
interatomic force	原子间力
interaxial angle	轴间角
interbanded coal	夹矸煤
interbasin area	流域间地区；流域间面积
interbasin diversion	流域间分洪；跨流域调水
interbasin flow	流域间流动
interbasin length	流域间长度
interbasin transfer of water	流域间水迁移
interbasin water diversion	跨流域引水
interbasin water diversion project	跨流域引水工程
interbasin water transfer	跨流域调水；跨流域引水
interbed	互层；夹层
interbed multiple reflection	层间多次反射
interbed multiple wave	层间多次波
interbedded	夹层的；互层的
interbedded clay	黏土夹层
interbedded formation	交互层
interbedded gas	夹层气
interbedded layer	夹层
interbedded sand	砂岩夹层；层状砂；含夹层砂
interbedded sedimentation	层间沉积作用
interbedded source rock	互层状生油气岩；互层状烃源岩
interbedded strata	互层；夹层；间层
interbedded water	层间水
interbedding	互层；中间层
interborehole measurement	井间测量
intercalate	夹入；插入；添入
intercalated	夹层的；间层的；插入的
intercalated coal seam	含夹石层的煤层
intercalated contact	穿插接触
intercalated layer	夹层
intercalated shale	页岩夹层；含夹层页岩，层状页岩
intercalated texture	夹层结构；插层型构造
intercalated zone	夹层带
intercalation	夹入，插入；夹层
intercellular liquid	晶堆间液体
intercept	截断，截捕；阻断，遮断；截距
intercept method	截距法
intercept time	延迟时间
intercepted water	拦蓄水；蓄积水；截留水
intercepting drain	截水沟；截水管
interception	相交；拦截
interception factor	拦截因子；截流倍数
interception loss	截留损失
interception of water	截断水流；截流
interceptor	拦截物；截水器；截流管
interceptor drain	截水沟
interceptor sewer	截流污水渠
interchange instability	交换不稳定性
interchange method	替换法，互换法
interchange of energy	能量交换
interchange of sea and land	沧海桑田；海陆变迁
interchangeability	互换性；可交换性
interchangeable bit	活钻头；可替换的钻头
interchannel	河道间；信道间

interchannel area	河道间地区；河道间面积	interface level	界面液面；界面水位，界面标高
interclass correlation coefficient	组内相关系数	interface limit	界面限度
interclass variance	组内方差；类间方差	interface location	界面位置
intercloud medium	互联云媒介；网络云际媒介	interface locator	界面探测器
intercluster bus	集群间互连总线	interface reaction in seawater	海洋界面反应；海洋界面作用
intercoagulation	相互凝聚	interface routine	接口程序
intercoastal waterway	岸间水道	interface standard	接口标准；接口规范
intercolumnar	柱间的	interface strength	接触面强度；界面强度
intercolumniation	柱间距间定比；分柱法	interface velocity	界面速度
intercomparison	相互比较	interfacial	界面的；面间的
interconnected control	互联控制	interfacial adhesion failure	界面黏结破坏
interconnected estuarine channels	河口河网	interfacial adsorption	界面吸附作用
interconnected fracture	连通裂缝	interfacial angle	界面角
interconnected hydrodynamic system	连通的水动力系统	interfacial area	界面面积
interconnected pores	连通孔隙	interfacial bond strength	界面黏结强度
interconnection	互连；互相连接	interfacial boundary	界面边界
intercontinental basin	内陆盆地；陆间盆地	interfacing concentration	界面浓度
intercontinental connection	洲际联测	interfacial configuration	界面形态
intercontinental geosyncline	内地槽；陆间地槽	interfacial contact	界面接触
intercontinental railway	大陆桥；洲际铁路	interfacial corrosion	界面腐蚀
intercontinental rift system	陆间裂谷系	interfacial coupling	界面耦合
intercontinental sea	陆间海	interfacial curvature	界面曲率
intercontinental subduction	陆间俯冲	interfacial dilation	界面膨胀
intercontinental superregion	陆间区	interfacial effect	界面效应
interconversion	互换；互变	interfacial energy	界面能；界面张力
interconvertibility	可相互转换	interfacial film	界面薄膜
intercrystal porosity	晶间孔隙度	interfacial fluid	界面流体
intercrystalline brittleness	晶粒间脆性	interfacial force	面际力，界面力；界面张力
intercrystalline corrosion	晶间腐蚀	interfacial fracture mechanics	界面断裂力学
intercrystalline crack	晶间裂纹	interfacial fracture toughness	界面断裂韧性，界面断裂韧度；界面结合强度
intercrystalline fluid	晶间流体	interfacial free energy	界面自由能；界面能
intercrystalline fracture	晶间裂缝；晶间断裂	interfacial friction	层间摩擦；界面摩擦
intercrystalline pore	晶粒间孔隙	interfacial grain boundary	颗粒界面边界
intercrystalline porosity	晶间孔隙度	interfacial molecular force	界面分子力
intercrystalline swelling	晶粒间膨胀	interfacial potential	界面势；界面位能
interdeep	山间坳陷；岛弧间海渊	interfacial pressure	界面压力
interdeltaic	三角洲间的	interfacial reaction	界面反应
interdeltaic shoreline	三角洲间滨线；三角洲间海岸线	interfacial rheological effect	界面流变效应
interdependence	相互依赖；相互关联	interfacial rheological resistance	界面流变阻力
interdistributary area	支流区间；分流河道间区域	interfacial rheology	界面流变学
interdistributary shell bank	分流河道间介壳滩	interfacial rheometer	界面流变仪
interdiurnal change	日际变化	interfacial shear strength	界面剪切强度
interdiurnal variation	日变度化	interfacial shear stress	界面剪切应力
interface between horizons	地层界面	interfacial spreading agent	界面分散剂
interface crack	界面裂纹	interfacial spreading coefficient	界面扩散系数
interface curvature	界面曲率	interfacial state	界面状态
interface cut	界面切割	interfacial tensiometer	界面张力计
interface detector	界面探测器	interfacial tensiometry	界面张力测定法
interface determination	界面确定	interfacial tension	界面张力
interface device	接口设备	interfacial tension force	界面张力
interface dip	界面倾角	interfacial tension gradient	界面张力梯度
interface displacement frame	界面位移框架	interfacial velocity	界面速度
interface equipment	接口设备；转换装置	interfacial viscoelasticity	界面黏弹性
interface fracture model	界面断裂模型	interfacial viscosity ratio	界面黏度比
interface geometry	界面几何形状	interfacial wave	层面波
interface gradient	界面梯度	interfacial zone	界面区
interface identifier	接口标识	interfacing equipment	接口设备
interface in seawater	海洋界面	interference	干涉；干扰；相互影响
interface joint meter	界面变位计；界面测缝计	interference among wells	井间干扰
interface layer	中间层		

interference analysis	干扰分析	interformational foliation	层间叶理，层间片理
interference band	干扰光带	interformational inlayer	层间夹层
interference coefficient	干扰系数；干涉系数	interformational laccolith	层间岩盖
interference color	干涉色	interformational multiples	层间多次波
interference colour chart	干涉色图；干涉色图谱	interformational sheet	层间岩席
interference current	干扰电流	interformational sliding	层间滑动
interference effect	干扰作用	interformational unconformity	层间不整合
interference field	干扰场	interfragmental	碎片间的
interference field strength	干扰场强	interglacial	间冰期的
interference figure	干涉图	interglacial deposit	间冰期沉积
interference filter	干扰滤波器	interglacial epoch	间冰期
interference frequency	干扰频率	interglacial feature	间冰期地形
interference fringe	干涉条纹	interglacial period	间冰期
interference of wells	井间干扰	interglacial sea level	间冰期海平面
interference parameter	干扰参数	interglacial stadial phase	间冰期冰退阶段
interference pattern	干涉条纹；干涉图样	interglacial stage	间冰期
interference range	干扰区；干扰范围	interglaciation	间冰期；间冰期作用
interference refractometer	干涉折射仪	Intergovernmental Panel on Climate Change	政府间气候变化专门委员会
interference region	干涉区	intergrade	中间级配；过渡性
interference response	干涉响应	intergrade layer silicate	中间型层状硅酸盐
interference ripple	干涉波痕	intergrade soil	过渡性土壤
interference ripple mark	扰动波痕；交错波痕；干涉波痕	intergrain boundary	粒间界面
interference settlement	干扰性沉陷	intergrain loss	粒间损耗
interference source	干扰源	intergranular	晶粒间的；粒间的
interference spectrum	干涉谱	intergranular cement	粒间胶结物
interference structure	干涉构造	intergranular cracking	晶间开裂；沿晶破裂
interference test	干扰试验	intergranular diffusion	粒间扩散
interference wave	干扰波	intergranular effective stress	粒间有效应力
interference well pumping test	干扰抽水试验	intergranular film	粒间薄膜
interference zone	干扰区；干扰范围	intergranular fracture	粒间断裂
interfering component	干扰组分	intergranular friction	颗粒间摩擦
interfering impulse	干扰脉冲	intergranular movement	粒间运动
interfering noise	干扰噪声	intergranular opening	粒间空隙
interfering resource	干扰源	intergranular permeability	粒间渗透性；粒间透水性
interfering signal	干扰信号	intergranular pore-space	粒间孔隙
interfering wells	干扰井；互扰井	intergranular porosity	粒间孔隙度
interfering wells pumping	干扰孔抽水	intergranular pressure	粒间压力
interferogram	干涉图	intergranular skin	粒间薄膜
interferometer	干扰仪；干涉仪	intergranular slip	粒间滑移
interferometric sea-bed inspection sonar	相干海底探测声呐系统	intergranular slip-plane	粒间滑动面
interferometric spectrometer	干涉分光计	intergranular stress	粒间应力
interferometric spectroscopy	干涉光谱学	intergranular stress corrosion cracking	粒间应力腐蚀开裂
interferometric synthetic aperture radar	合成孔径雷达干涉测量	intergranular structure	晶间组织
interferometry	干涉量度学；干扰测量法	intergranular texture	间粒结构；间粒组织
interferometry synthetic aperture radar	干涉合成孔径雷达	intergranular water	粒间水；孔隙水
interfibrillar friction	纤维间摩擦	intergrown	共生的；交互生长
interfingering	相互穿插；指状交错	intergrowth	互结生长；共生；连晶
interflow	壤中流，土内水流；层间流；过渡流量	interim	中间；暂时的，临时的
interflowing	层间流动的；交流的；交错的	interim completion date	中间完工日期
interfluent	交流的，合流的；流入中间的	interim criterion	暂行准则
interfluent flow	层间水流	interim determination of extension	临时延期的决定
interfluve	河间地；泉间地	interim measurement	临时测量
interfluvial slope	河间分水坡	interim meeting of the board	临时董事会议
interfolding	混合褶皱；交叉褶皱	interim payment	期中付款；临时付款
interfoliated vein	交织裂缝岩脉	interim provisions	临时管理办法；暂行规定
interformational	层间的	interim regulations	暂行条例；暂行管理办法
interformational conglomerate	层组间砾岩	interim report	阶段报告；临时报告
interformational fold	多层褶曲；层间褶皱	interim results	中期业绩
		interim survey	中期测量

interionic attraction	离子互吸；离子间引力	interlinkage	互连
interior angle	内角	interlock	联锁装置；联锁；结合；连结
interior basin	内陆盆地	interlocked area	联锁区
interior decoration	内部装修；室内装修	interlocked pile	搭接桩
interior drainage	内陆水系；内部排水	interlocked structure	镶嵌结构
interior lake	内陆湖	interlocking	咬合作用
interior land basin	内陆盆地	interlocking bond	嵌合联结
interior layer	内部层	interlocking device	联锁装置
interior lowland	内陆低地	interlocking pore	连通孔隙
interior measure	内测度	interlocking property	咬合性质
interior orientation	内部定向	interlocking recording	联锁记录
interior plain	内陆平原	interlocking switch	联锁开关
interior sea	内陆海	interlocking texture	交错结构
interior shaft	暗竖井；盲竖井	interlude	阻隔，使中断；插算
interior stress	内应力	interm slope	中期边坡
interior structure	内部结构	intermap relationship	图幅接合表
interior wall	内墙	intermediary position	中间位置；媒介位置
interior wall coating material	内墙涂料	intermediary slope	采区中间斜坡
interior waterway	内陆水道；内河水道；内河航道	intermediate	中性的；中间的；过渡的
interior well	内部井	intermediate belt	中间带；过渡带
interirrigation	灌溉间期	intermediate compound	中间化合物
interjacent	在中间的；居间的	intermediate contour	首曲线；基本等高线
interlace	交织；交错	intermediate contour line	首曲线；基本等高线
interlaced scanning	隔行扫描	intermediate data	中间数据
interlaced structure	交织构造；交错构造	intermediate data set	中间数据集
interlacing	隔行；隔行扫描；交错	intermediate density	中等密度
interlacing drainage	游荡水系；辫状水系	intermediate deposit	中段沉积物
interlacing drainage pattern	游荡水系；辫状水系	intermediate drivage	区段平巷；区段石门
interlaid	互层的；夹层的	intermediate entry	中间平巷；中巷
interlaminar	层间的	intermediate focus shock	中源地震
interlaminar shear strength	层间剪切强度	intermediate footbridge support	中间行人天桥支座
interlaminar strength	层间强度	intermediate grinding	中级研磨；中磨，中碎
interlaminar stress analysis	层间应力分析	intermediate groin	中间丁坝
interlaminate	薄层交替	intermediate haulage	中间运输；中间拖运
interlaminated	交替纹层状的	intermediate head	中间接头
interlayer	间层；夹层；层间的	intermediate heading	中间平巷
interlayer bonding	层间联结	intermediate index	中间指数；中间折光率
interlayer cation	层间阳离子	intermediate inflow	区间来水
interlayer compression	层间挤压	intermediate information	中间信息
interlayer contact	层间接触	intermediate layer	中间层；夹层
interlayer continuity	层间连续性	intermediate level	中间平巷；中巷
interlayer crossflow	层间对流	intermediate loading section	中间转载装置；中间转载节段
interlayer distorted bedding	层间扭曲层理	intermediate magnitude earthquake	中强地震；中等地震
interlayer gangue parting	层间夹石层	intermediate observation	中间观测
interlayer interference	层间干扰	intermediate oxide	两性氧化物
interlayer oxidation zone	层间氧化带	intermediate phase	中间相
interlayer pressure	层间压力	intermediate pillar	中间煤柱；中间矿柱
interlayer rejected thickness	夹石剔除厚度	intermediate plate	中间板块
interlayer slide	层间滑动	intermediate principal plane	中间主平面
interlayer spacing	层间距	intermediate principal strain	中间主应变
interlayer stripping material	层间剥离物	intermediate principal stress	中间主应力
interlayer water	层间水	intermediate product	中间产品；半成品
interlayer-gliding fault	层间滑动断裂	intermediate pump station	中间泵站
interlayering	间层理；互层理	intermediate pumping station	中间泵站
interleaf	中间层	intermediate rock	过渡性岩石；中间岩石
interleave	交叉；交织；交错	intermediate section	中间部分；过渡层
interlensing	透镜状夹层	intermediate shaft	中间轴
inter-level correlation	层间对比	intermediate shock	中强震动
inter-limb angle	翼间角	intermediate sight	插视；中间观测
interlink	链接；互连	intermediate station	中间站

intermediate straight line	夹直线
intermediate stress	中间应力
intermediate strut	中间横撑
intermediate studdle	中间支柱
intermediate support stool	中间支撑座
intermediate surge drum	中间平衡筒
intermediate test	中间试验
intermediate truss	中间桁架
intermediate type pavement	中级路面
intermediate vadose water	中间带渗流水
intermediate value	中间值
intermediate value theorem	中值定理
intermediate viscosity carrier fluid	中黏携砂液
intermediate water	层间水；过渡带水
intermediate water layer	过渡带水层
intermediate wave	中波
intermediate well	中间井；加密井
intermediate zone	中间带；过渡带
intermediate-depth shock	中源地震
intermediate-period seismograph	中周期地震仪
intermediate-scale map	中比例尺图
intermediate-strength proppant	中等强度支撑剂
intermediate-term earthquake forecast	地震中期预报
intermediate-wet reservoir	中性润湿储层
intermediation	中介；媒介；调解
intermedium	中间物；媒介物
intermeshed	相互交错的
intermeshed structure	相互交错结构
intermetallic	金属间的
intermittence	间歇性；周期性；脉冲性；中断
intermittency correction	断续修正系数
intermittency factor	断续因素
intermittent	暂歇的；间歇的；断续的
intermittent dual operation	间歇操作
intermittent dust removal	定期除灰
intermittent ejection	间歇喷发
intermittent error	间歇误差
intermittent eruption	间歇喷发
intermittent exposure	断续曝光
intermittent extrusion	间歇喷出
intermittent failure	间歇失效；间发故障
intermittent faulting	间歇性断裂活动
intermittent flow	间歇性水流
intermittent gradient	断续坡度；断续粒度组成
intermittent grading	间断级配
intermittent grinding	间歇研磨
intermittent heating	间歇加热
intermittent horizontal cut-and-fill stoping	间断式水平分层充填采矿法
intermittent injection	间歇注入
intermittent irrigation	间歇灌溉
intermittent issuing	间歇性涌水
intermittent lake	间歇湖
intermittent layering	间歇成层作用
intermittent load	间歇荷载
intermittent lubrication	间歇润滑；定期润滑；周期润滑
intermittent mining method	间接开采法
intermittent mixing	间歇搅拌
intermittent operation	间歇操作
intermittent output	间歇产量；间歇输出
intermittent recharge	间歇补给
intermittent river	季节河
intermittent sampling	间歇采样
intermittent sand filter	间歇砂滤器
intermittent sedimentation	不连续沉积作用；断续沉积
intermittent semi-long wall	间隔留柱半长壁开采法
intermittent service	间歇工作；断续使用
intermittent spring	间歇泉
intermittent stream	临时性河流
intermittent stream sediment	间歇性河流沉积物
intermittent streamflow	间歇河流；间歇水流；间歇水道
intermittent surge	间断涌浪
intermittent suspended load	间歇悬移质；间歇悬浮载荷；间歇悬载重
intermittent take-up motion	间歇式卷取装置
intermittent thickener	间歇式浓缩机；间歇式增稠剂
intermittent thickening	间歇浓缩过程
intermittent wet fumarole	间歇湿喷孔；间歇喷气孔
intermittent zoning	间歇分带
intermittently flowing well	间歇自喷井；间歇自流井
intermittently overflowing spring	间歇性溢流泉
intermittently rainy climate	间歇雨季气候
intermittently submerged sill	间断式潜槛
intermitting spring	间歇泉；间断泉
intermitting system	间歇控制系统；间歇系统
intermix	混合
intermixing	混合；搅拌
intermixture	混合物
intermodulation	互调；相互调制
intermodulation interference	互调干扰
intermolecular attraction	分子间引力
intermolecular bonding energy	分子间键能
intermolecular bonding force	分子键力
intermolecular cohesion	分子间内聚力
intermolecular condensation	分子间缩合
intermolecular migration	分子间转移
intermolecular ordering	分子间有序化
intermolecular potential	分子间势能
intermont	山间的；山间凹地
intermont basin	山间盆地
intermont glacier	谷冰川
intermontane	山间的
intermontane basin	山间盆地
intermontane depression	山间洼地；山间凹地
intermontane plain	山间平原
intermontane space	山间地带
intermontane trough	山间槽地
intermountain area	山间地区
intermountain basin	山间盆地
intermountain seismic belt	山间地震带
intermountainous plain	山间平原
internal	内的；内部的
internal absorption	内吸收
internal accelerator	内部催化剂
internal access	内部存取
internal action	内部作用
internal activator	内部催化剂；内部活化剂
internal agency	内力作用；内营力；内因；内部作用
internal air circulation	内部空气循环
internal and external forces	内力与外力

internal and external partition	内外间隔
internal angle	内角
internal antistat	内用抗静电剂
internal area	内部面积
internal calibration	内部校准
internal classification	内部分级
internal cohesion	内黏聚力；内凝聚力
internal component	内部组件
internal control	自动控制；自律控制
internal conversion	内转换
internal cooling	内部冷却；内冷法
internal crack	内部裂纹；内部裂缝
internal crystal framework	内部晶格
internal curing agent	内部养护剂
internal cutter	内割刀
internal cycle time	内部存取周期；内部循环时间
internal damping	内阻尼
internal displacement	内部位移
internal dissipation	内耗散
internal drainage	内陆水系
internal drainage layer	内部排水层
internal drainage pattern	内陆水系
internal element	内部单元
internal element boundary	内部单元边界
internal emission	内发射
internal energy	内能
internal environment	内部环境
internal erosion	层内侵蚀；内侵蚀作用
internal esterification	内酯化
internal face	内面；内壁面
internal field	内源场
internal field distribution	田间配水
internal filter	内滤器
internal flow	内流，内部流动
internal flush	内平式
internal flush drill pipe	内平钻杆
internal force	内力
internal force redistribution	内力重分布
internal format	内部格式
internal friction	内摩擦；内耗
internal friction angle	内摩擦角
internal friction coefficient	内摩擦系数
internal friction resistance	内摩擦阻力
internal frictional resistance	内摩擦阻力
internal gas drive	溶解气驱；内部气驱
internal heat	内热
internal hydraulic pressure	内部液压
internal ignition	内燃，内部燃烧
internal inclined shaft	井下斜井；暗斜井
internal injection line	内部注水井排；内部注水管线
internal inland drainage	内陆排水系统
internal inserts	内部填充物
internal instrument installation	内部仪器埋设
internal layout	内部间隔；内部布局
internal leak	内部泄漏
internal leakage	内部泄漏
internal lubrication	内部润滑
internal memory	内存储器
internal migration	层内运移
internal model	内部模型
internal moisture	内在水分
internal node	内节点，内部节点
internal partition	内部间隔
internal phase	内相
internal pressure	内压力
internal pressure strength	抗内压强度
internal pressure stress	内压应力
internal pressure test	内压试验
internal preventer	钻杆内防喷器；钻柱防喷回压阀
internal primitive water	深层原生水
internal radius	内半径
internal ramp	内部沟；内坡道
internal rate of return	内部收益率
internal rate of return method	内部回收率法
internal reaction	内部反应
internal reflection	内反射
internal reflux	内回流
internal relation	内部关系
internal relief valve	内置泄压阀
internal resistance	内阻
internal risk	内部风险；内在风险
internal road	区内道路；厂内公路
internal rock-mechanics instrumentation	岩石力学性质测试仪器
internal schema	内模式；存储模式
internal scour	潜蚀
internal scouring	内部冲刷
internal sea	内海
internal sedimentation	层内沉积作用
internal settlement measurement	内部沉陷量测
internal shaft	暗竖井；盲竖井
internal stagnant water	内滞水
internal standard	内标；内标物；内标准
internal standard element	内标元素
internal standard method	内标法
internal storage	内存储器
internal strain	内应变
internal stress	内应力
internal structure	内部结构；内部构造
internal support	内支护；内支撑
internal surface	内表面；内层
internal surface area	内表面积
internal surge	内涌浪
internal survey	内区调查
internal transport	矿内运输；内部运输
internal trench	内部沟
internal vibrator	插入式振捣器
internal view	内视图
internal viscosity	内黏滞性
internal viscous damping	内黏滞阻尼
internal voltage	内电压
internal water	深层水；内部水
internal water pressure	内水压力
internal wave	内波
internal wave of ocean	海洋内波
internal work	内功
internally indeterminate construction	内部超静定结构
international assistance	国际援助
International Association of Engineering Geology	国际工程地质协会

international atomic time	国际原子时
International Bank for Reconstruction and Development	国际复兴开发银行；世界银行
International Commission on Large Dam	国际大坝委员会
international convention	国际惯例
international cooperation	国际合作
international custom	国际惯例
International Journal for Numerical and Analytical Methods in Geomechanics	国际地质力学数值和解析方法学报
International Journal for Numerical Methods in Engineering	国际工程数值方法学报
International Journal of Coal Geology	国际煤田地质学报
International Journal of Rock Mechanics and Mining Sciences	国际岩石力学与采矿科学学报
International Monetary Fund	国际货币基金组织
international practice	国际惯例
international price	国际价格
International Science Organization	国际科学组织
International Society for Rock Mechanics	国际岩石力学学会
International Society of Environmental Geotechnics	国际环境岩土工程学会
International Space Congress	国际宇宙空间大会
International Standardization Organization	国际标准化组织
international symbol	国际符号
international system	国际制单位
international system of units	国际单位制
International Tunneling Association	国际隧道协会
International Union of Geodesy and Geophysics	国际大地测量与地球物理联合会
international unit	国际单位
International Geosynthetics Society	国际土工合成材料学会
International Journal of Geomechanics	国际地质力学学报
internet of things	物联网
interoperability	互通性
interparticle attraction	颗粒间的引力
interparticle attractive force	粒间引力
interparticle dispersion	粒间分散作用
interparticle dispersion force	粒间弥散力；粒间分散力
interparticle force	粒间作用力
interparticle friction	粒间摩擦
interparticle pore	粒间孔隙
interparticle porosity	粒间孔隙度
interparticle repulsion	粒间斥力
interparticle spacing	粒子间距
interparticle stress	粒间应力
interpenetration	互相渗透
interpermafrost water	冻结层间水；永冻层间水
interphase interface	中间相界面；相间界面
interphase mass transfer	相间传质
interphase transformer	相间变换器
interplain channel	平原间峡谷
interplanar spacing	面间距；面网间距；晶面间距
interplanetary discontinuity	行星际间断
interplay	相互影响；相互作用
interpolated contour	内插等高线
interpolated map	插图
interpolating function	插值函数
interpolating spline	插值样条
interpolation	内插法；内推法；插值法
interpolation error	内插误差
interpolation function	内插函数
interpolation method	内插法
interpolation table	插值表
interpolation weight	插值权
interpret	解释；诠释；说明；翻译
interpretability	可判读性；判读能力；可解释性
interpretable	可解释的
interpretable anomaly	可解释异常
interpretable limit	解译范围
interpretation	解释；解译；判读；翻译
interpretation algorithm	解释算法
interpretation ambiguity	解释的多解性
interpretation element	判读要素
interpretation key	判读标志
interpretation map	解释推断图
interpretation mark	判读标志
interpretation model	解释模型
interpretation of air photograph	航空照片判读
interpretation of data	数据判读；数据解释
interpretation of geological map	地质图判读；地质图解释
interpretation package	解释程序包
interpretation procedure	解释方法
interpretation work	解释工作
interpretational criteria	评价准则
interpretative factor	解译因素
interpretative geologic map	推断性地质图；解译地质图
interpretative log	钻探解释剖面；可供解释录井
interpretative stratigraphy	解释地层学；解译地层学
interpretative tool	解释工具；解译工具
interpretive availability	可判读程度；解译能力
interpretive classification	解释分类
interpretive code	解释代码
interpretive geologic map	推断性地质图；解译地质图
interpretive language	翻译语言
interpretive log	解译测井；可供解释录井
interpretive order	解释指令
interpretive program	解释程序
interpretive programming	解译程序设计
interpretive routine	解释程序
interpretoscope	判读仪；解译镜
interprocessor communication	处理机间的通信
interreaction	相互反应；相互作用
interregional unconformity	区域不整合面
interrelationship constraint	相互关系约束条件
interrupted fault	不连续断层
interrupted fold	不连续褶皱
interrupted profile	不连续剖面
interrupted stream	中断河；间断河；间断水道
interrupted water table	中断地下水面；阻降水面；阻变水面
interrupted wave	断续波
interrupter	阻碍者；障碍物；断续器
interrupting drainage well	间歇排水井
interruption	中断；停止；断路
interruption in deposition	沉积中断
intersect	交切；交错；相交；交叉
intersected country	交错地带
intersected crack	交错裂缝
intersected crack and seam	交叉裂缝和夹层

intersected peneplain 交切准平原
intersected terrace 交切阶地
intersecting fault 交错断层
intersecting line 交线
intersecting peneplain 交错准平原
intersecting perforation 垂直于裂缝面射孔
intersecting point 交点
intersection 交切交叉；交点交叉线；前方交会；交集
intersection angle 交叉角；转角
intersection chart 网络图
intersection classification 交叉分类
intersection cleavage 交叉劈理；交叉解理
intersection entrance 交叉口进口
intersection error 交叉点误差
intersection line 交线
intersection line of two planes 两个平面的交线
intersection method 交会法
intersection of network 测网交点
intersection of shock waves 冲击波交会
intersection of two planes 两平面相交
intersection plan 交叉口平面图
intersection plane 交面
intersection point 交点
intersection station 交会点；交会测站
intersheet bonding 片间黏结
interslice force 条间力
interslice force function 条间力函数
interslice normal force 条块间的法向力
interstice 空隙；孔隙；间隙；缝隙；
interstice of soil 土壤空隙；土壤裂缝
interstitial compound 间充化合物；填隙化合物
interstitial connate water 间隙原生水
interstitial deposit 间隙沉积
interstitial diffusion 间隙扩散
interstitial element 间隙元素
interstitial filling 间隙填充作用
interstitial flow 间隙渗流
interstitial heterogeneity 孔隙不均质性
interstitial ice 间隙冰；空隙冰
interstitial liquid 间隙液体；孔隙流体
interstitial material 填隙物质；填隙物
interstitial pore 隙溜微孔；间隙微孔
interstitial pore space 孔隙空间
interstitial pressure 孔隙压力
interstitial ratio 隙溜比
interstitial sedimentary water 隙间沉积水
interstitial site 间隙位置
interstitial solid solution 填隙式固溶体；间充固溶体
interstitial space 孔隙空间
interstitial surface area 孔隙面积
interstitial surface of soil 土壤间隙面
interstitial tension 面间张力
interstitial velocity 隙间流速
interstitial water 间隙水；填隙水；孔隙水
interstitial water pressure 孔隙水压力
interstitial water saturation 隙间水饱和度
interstitial wire 填充线
interstratal brine 层间卤水；层间盐水
interstratal karst 层间岩溶
interstratal water 层间水

interstratification 间层作用；互层作用
interstratified bed 互层；间层
interstratified coal sample 分层煤样
interstratified coal seam sample 分层煤样；分层煤矿层样品
interstratified layer 间层
interstratified mineral 间层矿物
interstratified rock 层间岩
interstratified seam 间层煤层
interstratified seam sample 矿层分层试样
interstratified slate 层间板岩；夹层板岩
interstratified water 层间水
interstratify 间层
interstream area 河间地
interstream groundwater ridge 河间地下分水岭；河间地下分水线
intertextic fabric 交织状构造
intertidal 潮间带
intertidal abrasion platform 潮间带海蚀平台
intertidal communities 潮间带群落
intertidal ecology 潮间带生态学
intertidal facies 潮间相
intertidal fossil 潮间带化石
intertidal marsh 潮间沼泽地
intertidal mudflat 潮间带泥滩
intertidal realm 潮间域
intertidal reef 潮间带珊瑚礁；潮间带矿脉
intertidal regime 潮间带动态
intertidal region 潮水浸淹带
intertidal zone 潮位变动区；潮漫区；水位变动区；潮位变化段
intertongued lithofacies 舌形交错岩相
intertonguing 交错变相；交错沉积
interval 间隔间歇；时间；时限；区间
interval between fissures 裂隙间距
interval correlation 间隔对比
interval density 区间密度；间隔密度
interval estimator 区间估计
interval from high water 高潮间隙
interval from low water 低潮间隙
interval function 区间函数
interval of isoline 等值距
interval of measuring profiles 量测断面间距
interval of sampling 取样间距
interval scaling 等距量表
interval slowness model 层慢度模型
interval thickness 层段厚度
interval time 间隔时间
interval transit time 声波时差
interval transit time curve 声波时差曲线
interval velocity 间隔速度；层速度
interval zone 间隔带
intervening pillar 矿房间矿柱；煤房间煤柱
intervening soft soil layer 软土夹层
intervening stratum 夹层
intervening well 插入井
interventilation 联合通风
intervisible 可互见的
interwell permeability 井间渗透率
interwell spacing 井距

interwell structure	井间构造	intrinsic contact angle	固有接触角
interwell sweep efficiency	井间波及效率；井间波及系数	intrinsic contaminant	固有的杂质
interwell tracer test	井间示踪试验	intrinsic corrosiveness	内在的腐蚀性；固有的侵蚀作用
interzonal cross-flow	层间窜流	intrinsic energy	内能
interzonal isolation	层间隔离	intrinsic fracture	天然裂缝
interzonal soil	未成熟土；带间土	intrinsic friction angle	固有摩擦角
interzonal time	带间时；地层间断时间	intrinsic function	内部函数；本征函数
intraclass variance	组合方差	intrinsic heat	内热
intraclast	内碎屑；内成碎屑灰岩	intrinsic hypothesis	内蕴假设；本征假设
intracontinental deformation	陆内变形作用	intrinsic impedance	固有阻抗
intracontinental geosyncline	内地槽；陆内地槽；内地向斜	intrinsic instability	固有不稳定性
intracratonic geosyncline	克拉通内地槽；陆间地槽	intrinsic internal angle of friction	固有内摩擦角
intracrystalline penetration	晶内渗透	intrinsic permeability	内在渗透性；固有渗透性
intracrystalline plasticity	晶内塑性	intrinsic property	固有性质；本征性质
intracrystalline pore	晶内孔隙	intrinsic safety	内在安全
intracrystalline porosity	晶内孔隙度	intrinsic shear strength	固有抗剪强度
intracrystalline rupture	晶内断裂	intrinsic shear strength curve	固有抗剪强度包线
intracrystalline slip	晶内滑动	intrinsic shrinkage	内在收缩；内部收缩
intracrystalline swelling	晶内膨胀	intrinsic spin	固有自旋；特征自旋
intrafolial fold	层内褶皱	intrinsic strength anisotropy	内在强度各向异性
intraformational bed	层内夹层	intrinsic thermal neutron decay time	固有热中子衰减时间
intraformational breccia	层内角砾岩	intrinsic value	本征值
intraformational conglomerate	层内砾岩	intrinsic viscosity	特性黏度；固有黏度；本征黏度
intraformational contortion	层内扭曲；层间扭曲	intrinsical	固有的；内在的；本质的
intraformational corrugation	层内揉褶	intrinsically	真正地；实在地；本质上地
intraformational fold	层内褶皱	intrusion	进入；侵入
intraformational folding	层内褶皱	intrusion displacement	侵入位移；侵入变位
intraformational recumbent fold	层内伏卧褶皱；层内平卧褶皱	intrusion of seawater	海水入侵
		intrusion pipe	灌注管
intrageoanticline	内地背斜	intrusion rock	侵入岩
intrageosyncline	内地槽；陆内地槽；内地向斜	intrusion tectonics	侵入构造
intraglacial	冰川内的	intrusive breccia	侵入角砾岩
intragranular movement	粒内运动；粒内滑动	intrusive contact	侵入接触
intragranular pore	粒内孔隙	intrusive dike	侵入岩墙
intragranular porosity	粒间孔隙度	intrusive facies	侵入岩相
intramassif	地块内的	intrusive force	侵入力
intramassif basin	断块盆地	intrusive granite	侵入花岗岩
intramicrite	内碎屑微晶灰岩	intrusive igneous mass	侵入火成岩体
intraplate	板块内的；内陆板块	intrusive igneous rock	侵入火成岩；侵入岩
intraplate earthquake	板内地震	intrusive mass	侵入体
intraplate tectonics	板块内部构造；板内构造地质学	intrusive rock	侵入岩
intraplate volcanism	板内火山活动	intrusive sheet	侵入岩床
intraplate volcano	板内火山	intrusive sill	侵入岩床
intra-sand shale	砂层内页岩	intrusive stock	侵入岩株
intraseam	煤层内部的	intrusive tuff	侵入凝灰岩
intrasparrudite	内碎屑亮晶砾屑灰岩；粗砾石灰岩	intrusive vein	侵入脉，侵入岩脉
intrastratal	层内的	inundated area	泛滥区；淹没面积
intrastratal contortion	层内扭曲	inundated district	泛滥地区；淹水地区
intrastratal flowage	层内流动	inundation	洪水；泛滥；淹没
intrastratal solution	层内溶液；层间溶液；层内深溶解作用	inundation method	漫灌法
		inundation phase	海侵相；淹水相
intrazonal	带内的；隐域的	invaded sediment	受侵沉积物
intrazonal soil	隐域土	invaded zone	侵入带
intrinsic	内在的；固有的；本质的	invading water	侵蚀水；侵入水
intrinsic absorption	内部吸收	invading wave	入侵波
intrinsic accuracy	固有精度	invading wave in estuary	河口入侵波
intrinsic anisotropy	内在非均质性；固有非均质性	invariable	不变的；固定的
intrinsic brightness	本身亮度	invariance principle	不变性原理
intrinsic conductivity	本征电导率；固有导电性；固有传导率	invariant	不变量；不变的；不变式
		invariant factor	不变因子；不变因素

invariant geomagnetic coordinate	不变地磁坐标
invariant integral	不变积分
invariant integration	不变积分
invariant latitude	不变纬度
invariant of stress	应力不变量
invariant of stress tensor	应力张量不变量
invariant point	不变点
invariant relation	不变关系
invasion depth	侵入深度
invasion diameter	侵入带直径
invasion efficiency	侵入效率
invasion interface	侵入界面
invasion of the sea	海侵
invasion parameter	侵入参数
invasion profile	侵入带剖面
invasion radius	侵入半径
invasion response	侵入响应
invasion water	侵入水
inventory record	盘存记录；盘存清单
inventory reserve	地质储量；库存储备
invernite	正斑花岗岩；闪云斑状花岗岩
inverse analysis	反演分析
inverse analysis of parameter	参数反分析
inverse calculation	反算；逆运算，逆计算；回代
inverse chart	横轴投影地图
inverse computation	反算
inverse configuration	反电极系
inverse contact metamorphism	逆接触变质作用
inverse convolution	逆褶积
inverse correlation	逆相关
inverse curve	反曲线；逆曲线
inverse direction	反向
inverse discharge	反排矸
inverse dual operation	反对偶运算
inverse dynamic correction	反动态校正
inverse electrode current	反向电极电流
inverse filter	逆滤波器；反滤波器
inverse function	反函数
inverse gas liquid chromatography	反相气液色谱法
inverse gravity and magnetic problem	重磁反演问题
inverse gravity number	逆重力数
inverse hyperbolic function	反双曲函数
inverse image	逆像；反图像
inverse interpolation	逆插法；反插值法
inverse isochron	反等时线
inverse iteration	反迭代；逆迭代
inverse magnetostriction effect	反磁致伸缩效应
inverse mapping	逆映射；反向映射
inverse matrix	逆矩阵
inverse metamorphism	逆变质
inverse method	反演法
inverse migration	反偏移
inverse model	反演模型
inverse modeling	反演模拟；反演解
inverse of matrix	矩阵的逆
inverse of weight matrix	权逆矩阵
inverse operation	逆运算
inverse operator	反算子
inverse piezoelectric effect	反压电效应
inverse plummet observation	倒锤线观测
inverse pore compressibility	逆孔隙压缩系数；逆孔隙压缩性
inverse position computation	后方交会计算
inverse power law	逆幂法则
inverse problem	逆问题；反演问题
inverse proportion	反比例
inverse ratio	反比
inverse ray theory	逆射线原理
inverse reaction	逆反应
inverse relation	反比关系
inverse relation of elasticity	逆弹性关系
inverse scattering approach	逆散射法
inverse scattering method	逆散射法
inverse sine	反正弦
inverse solution of geodetic problem	大地测量问题反解
inverse square	负二次方；平方反比
inverse square law	平方反比律
inverse square matrix	逆方阵；方阵求逆
inverse substitution	逆代换
inverse system	逆向系统
inverse tangent	反正切
inverse technique	反演法
inverse theorem	逆定理
inverse transformation	逆变换；反变换
inverse trigonometric function	反三角函数
inversion	倒转，倒置；转变；转化；反演变换；
inversion axis	反演轴；倒转轴
inversion center	反演中心
inversion formula	反演公式
inversion layer	逆温层
inversion of rainfall	雨量逆增
inversion of relief	地形倒置；地形倒转；反立体感
inversion point	转变点；倒转点
inversion tectonics	反演构造学
inversion temperature	转化温度
inversion theorem	反演定理
inversion with artificial intelligence	人工智能反演
invert elevation	底拱高程
invert structure	仰拱结构
inverted arch	底拱；反拱；仰拱
inverted arch foundation	仰拱基础
inverted arch set	底拱；反拱；仰拱
inverted canoe fold	覆舟形褶皱；舟状褶皱；紧闭向斜
inverted capacity	吸收容量
inverted capacity of well	井的吸收容量
inverted drainage well	反渗排水井；逆向排水井
inverted draw cut	上部掏槽；顶槽
inverted filter	反滤层
inverted fold	倒转褶皱
inverted order	反逆次序；逆序
inverted position	倒转层位
inverted relief	地形倒置
inverted river	逆流河
inverted siphon	倒虹吸管
inverted strata	倒转岩层；倒转地层
inverted stratum	倒转岩层；倒转地层
inverted unconformity	倒转不整合
inverted valve	止回阀；逆止阀；单向阀
inverted well	回灌井；回渗井；注水井；吸水井；
investigation	勘测；勘察；调查研究

investigation assignment	勘测任务；调查任务
investigation method	勘测方法
investigation of groundwater	地下水调查
investigation of groundwater withdrawal	地下水开采量调查
investigation of selected places	复查
investigation of tunnel	隧道调查
investigation program	勘察计划
investigation stage	勘察阶段
investigation survey	调查测绘
investigation surveying and sketching	调查测绘
investigation test	研究性试验
investment analysis	投资分析
investment margin	投资差额
investment method of valuation	投资估价法
investment project	投资项目
investment rating	投资分级
investment recovery	投资回收
investment venture	投资风险
inviscid fluid	理想液体；非黏性液体
involute	内卷；内旋式；包旋式
involute function	渐开线函数
involute surface	渐开线表面
involution	乘方；回旋；内卷
involvement	卷入；包含
inward flow	内向水流
inward pressure	内向压力
inwash	冲积；淤积；水砂充填
ion activity product	离子活度积
ion adsorption	离子吸附
ion chemistry	离子化学
ion chromatography	离子色层分析；离子色谱法
ion concentration	离子浓度
ion diffusion	离子扩散
ion exchange	离子交换
ion exchange adsorption	离子交换吸附
ion exchange capacity	离子交换量
ion exchange column	离子交换柱
ion hydrate	离子水合物
ion hydration	离子水化作用
ion migration ratio	离子迁移率
ion probe	离子探针
ion probe analyzer	离子探针分析仪
ion sieve separation	离子筛选
ion sputtering mass analyzer	离子溅射质量分析仪
ionic activity	离子活度
ionic activity measurement	离子活度测量
ionic addition reaction	离子加成反应
ionic adsorption	离子性吸附
ionic bond	离子键；异极键
ionic charge	离子电荷
ionic conductivity	离子导电性；离子电导率
ionic crystal	离子晶体
ionic crystal lattice	离子晶格；离子晶体点阵
ionic derivative	离子衍生物
ionic diffusion	离子扩散
ionic diffusion coefficient	离子扩散系数
ionic equilibrium	离子平衡
ionic exchange	离子交换
ionic exchange method	离子交换法
ionic migration	离子迁移
ionic potential	离子电位
ionic radius	离子半径
ionic strength	离子强度
ionic structure	离子结构
ionic substitution	离子替换
ionization effect	电离效应；电离作用
ionization energy	电离能
ionization heat	电离热
ionization layer	电离层
ionization measurement	电离测量
ionization potential	电离电位
ionizing radiation	电离辐射；游离辐射
ionizing solvent	离子化溶剂
ionosphere	电离层
ionosphere absorption	电离层吸收
ionosphere disturbance	电离层干扰
ionosphere propagation	电离层传播
ionosphere reflection	电离层反射
ionosphere reflection propagation	电离层反射传播
ionosphere refraction	电离层折射
ionosphere scatter propagation	电离层散射传播
ionospheric disturbance	电离层扰动
ionospheric refraction	电离层折射
ionospheric refraction correction	电离层折射改正
IPCC	政府间气候变化专门委员会
iron clay	铁泥矿；铁质黏土
iron meteorite	陨铁；铁陨石
iron reinforcement	钢筋
iron-ore cement	铁矿渣水泥
iron-oxide cement	氧化铁水泥
iron-portland cement	含铁波特兰水泥；含铁硅酸盐水泥
irradiation	光渗；照光；照射；辐照
irradiation embrittlement	辐照脆化
irradiation hardening	辐照硬化
irradiation polymerisation	辐射聚合作用
irradiation treatment	辐射处理
irradiator	辐射源；辐射器
irrational equation	无理方程
irrational expression	无理式
irrational flow	非旋流
irrational function	无理函数
irrational number	无理数
irrational orientation	不规则排列
irreducible	不能分解的；不可约的；不能削减的；残余的
irreducible matrix	不可约矩阵
irreducible minimum saturation	最小限饱和度
irreducible saturation	残余饱和率；束缚饱和度
irregular	不规则的；非正规的
irregular bedding	不规则层理
irregular boundary	不规则边界
irregular cross section	不均匀横断面
irregular crystal	不规则晶体
irregular curve	不规则曲线
irregular curve fracture	不规则弯曲裂缝
irregular deformation	不规则变形
irregular dip	不规则倾斜
irregular distribution	不规则分布
irregular error	偶然误差；不规则误差
irregular excavation	不规则挖掘
irregular fracture	不规则断口

irregular frame	不规则框架
irregular geomagnetic field	不规则地磁场
irregular settlement	不规则沉陷；不均匀沉陷
irregular slab	异型板
irregular slip surface	不规则滑面
irregular specimen for rocks	不规则岩石试件
irregular stone	异型石材
irregular strike	不规则走向
irregular terrain	不规则地形
irregular topography	不规则地形
irregular wave	不规则波
irregular workings	不规则采区
irregularity	不规则性；无规律；参差不齐
irregularity coefficient	不平整系数
irregularity degree	不平整度
irregularity index	不规则指数
irregularly caving zone	不规则垮落带
irregularly distributed load	不均匀分布荷载
irrelevant variable	无关变量
irremediable defect	永久性损坏；不可弥补的缺陷
irreproducible	不能再得的；不能复现的
irresistible force	不可抗力
irreversibility	不可逆性
irreversible	不可逆的；不可反转的；不能倒置的
irreversible deformation	不可恢复变形；永久变形
irreversible operation	不可逆运算
irreversible parity error	不可逆奇偶误差
irreversible permeability	不可逆磁导率；不可逆渗透性
irreversible process	不可逆过程
irreversible reaction	不可逆反应
irreversible resistance	不可逆阻力
irreversible shrinkage	不可逆收缩
irreversible system	不可逆系统
irrigable area	可灌面积
irrigated area	灌区；灌溉面积
irrigated plot	灌溉地区
irrigating water	灌溉水
irrigation application efficiency	田间灌水效率
irrigation basin	灌溉盆地
irrigation canal	灌溉渠
irrigation canal network	灌溉渠道网
irrigation canal system	灌渠系统
irrigation channel	灌溉渠
irrigation coefficient	灌溉系数
irrigation cycle	灌溉周期
irrigation demand	灌溉需水量
irrigation design	灌溉设计
irrigation district	灌区
irrigation ditch	灌溉渠
irrigation ditch output	灌渠出水口
irrigation draught	灌溉取水
irrigation efficiency	灌溉效率
irrigation farming	灌溉农业
irrigation for artificial recharge	人工补给灌溉
irrigation frequency	灌水频率
irrigation head	灌溉水头
irrigation interval	灌溉周期；灌水时间间隔
irrigation lateral	灌溉支渠
irrigation layout plan	灌区布置规划；灌区平面布置图
irrigation level	灌溉水位
irrigation method	灌水方法；灌注法
irrigation network	灌溉网；灌溉渠网
irrigation norm	灌溉标准
irrigation project	灌溉工程
irrigation requirement	灌溉需要量；灌溉需水量
irrigation reservoir	灌溉水库
irrigation return flow	灌溉回归水
irrigation return flow rate	灌溉回归系数
irrigation system	灌溉系统
irrigation tunnel	灌溉隧洞
irrigation water	灌溉水
irrigation water quality	灌溉水质
irrigation water requirement	灌溉需水量
irrigation well	灌溉井
irrotational	不旋转的；无旋的
irrotational deformation	非旋转变形；无旋变形
irrotational disturbance	无旋扰动
irrotational field	非旋场
irrotational flow	旋流；非旋转流；无涡流
irrotational wave	无旋波
irruption	突入；侵入
irruptive rock	侵入岩
Irwin's expression	埃尔文表达式；埃尔文方程
isallobar	等变压线；等气压图
isallo-stress line	等变应力线
isallotherm	等变温线
isanomalic line	等异常线
isarithm	等值线
ISC	工业安全委员会；国际宇宙空间大会
ISEG	国际环境岩土工程学会
isentropic change	等熵变化
isentropic chart	等熵图
isentropic expansion	等熵膨胀
isentropic flow	等熵流动
isentropic index	等熵指数
isentropic process	等熵过程
isentropic surface	等熵面
I-shaped fracture system	I型断裂系统；I型裂缝系统
iskymeter	现场剪切触探仪
island	岛，岛屿；安全岛
island arc	岛弧；岛链；弧形列岛
island arc geothermal belt	岛弧型地热带
island arc geothermal zone	岛弧地热带
island arc igneous rock	岛弧火成岩
island arc island chain	岛弧；岛链
island arc magmatic association	岛弧岩浆共生组合
island arc sea	岛弧海
island arc structure	岛弧结构
island arc tholeiite	岛弧型拉斑玄武岩
island arc type geosyncline	岛弧型地槽
island arc-bearing plate	含岛弧板块
island arc-island arc collision	岛弧对岛弧碰撞作用
island arc-marginal basin system	岛弧-边缘盆地体系
island arc-trench system	岛弧-海沟体系
island barrier	屏障岛
island mining	小规模建井开采
island survey	岛屿测量
island-arc trench belt	岛弧海沟带
island-arc type	岛弧型
island-arc type geosyncline	岛弧型地槽

island-arc-inner arc basin zone	岛弧内弧盆地带	isochemical metamorphism	等化学变质
island-geosyncline	岛弧型地槽	isochemical series	等化学系列
island-mainland connection survey	岛陆联测	isochlorosity	等含氯量线
island-reef fishing ground	岛礁渔场	isochore	等容线；等体积线；等层厚线
islands	群岛；列岛	isochore map	等容线图；等体积图；等层厚图
islands arc	复岛弧	isochoric	等层厚的
island-trench type geosyncline	岛弧-海沟型地槽	isochoric flow	等容流动
island-type	岛弧型	isochoric process	等容过程
island-type construction	独立式建筑	isochromatic	等色的；等色线
island-type geosyncline	岛弧型地槽	isochromatic line	等色线
isle	岛；小岛	isochromatic pattern	等色线图
ISO	国际标准化组织；国际科学组织	isochromatic photograph	等色线照片
isoabnormal	等异常线	isochrome map	等色线图
isoabsorption line	等吸收线	isochron layer	同期层
isoacceleration contour	等加速度线	isochrone	等时线；等反射时间线；等时水位线
isoacceleration line	等加速度线	isochrone chart	等时线图
isoanomalic contour line	等异常线	isochrone map	等时差线图
isoanomalous line	等异常线	isochrone method	等时线法
isoanomaly chart	等异常曲线图	isochrone of ocean floor	海底等时线；洋底等时线
isoanomaly curve	等异常曲线	isochroneity	等时性
isoanomaly map	等异常曲线图	isochronic	等时线的；等时的
isobaric change	等压变化	isochronic flame temperature	等时火焰温度
isobaric chart	等压线图	isochronic surface	等时面
isobaric surface	等压面	isochronism	等时性；同步
isobase	等基线	isochronograph	等时计
isobase contour	等基线图	isochronous	等时的；等时线的
isobath	等深线；等高程线	isochronous diastem	同时小间断；同时沉积暂停
isobath curve	等深线	isochronous governor	同步调节器
isobath interval	等深距	isochronous oscillation	等时振荡
isobath map	等深图	isochronous stratigraphic unit	等时性地层单位
isobath map of coal seam	煤层等深线图	isochronous surface	等时面
isobath of water table	潜水位等深线	isochronous vibration	等时振动
isobathic line	等深线	isoclinal	等倾斜的；等伏角的
isobaths map of aquifer	含水层等埋深图	isoclinal anticline	等斜背斜
isobathy submarine contour	等深线	isoclinal fault	等斜断层
isobathye	等深的	isoclinal fold	等斜褶皱
isobathye line	等深线	isoclinal line	等斜线；等倾线
isobathytherm	等温深度线	isoclinal method	等倾线法
isobed map	岩层等厚线图	isoclinal ridge	等斜脊
isobiolith	等化石年代岩石单位	isoclinal syncline	等斜向斜
isobutane	异丁烷	isoclinal valley	等斜谷
isobutanol	异丁醇	isocline	等倾线；等斜褶皱
isobutene	异丁烯	isoclinic line	等磁倾线；等斜线
isobutyl	异丁基	isoconcentration	等浓度
isobutyl alcohol	异丁醇	isoconcentration contour map	等浓度线图
isobutyl isobutyrate	异丁酸异丁酯	isoconcentration line	等浓度线
isobutyl methacrylate	甲基丙烯酸异丁酯	isoconcentration map	等浓度图
isobutyl phenylacetate	苯乙酸异丁酯	isocorrelation	等相关线
isobutyl salicylate	水杨酸异丁酯	isocratic movement	均衡运动
isobutylaldehyde	异丁醛	isocurlus	等旋涡强度线
isobutylene	异丁烯	isodense	等密度线
iso-butyraldehyde	异丁醛	isodensity	等密度线
isobutyric acid	异丁酸	isodepth contour line	等深线
isobutyronotrile	异丁腈	isodepth map of coal seam	煤层等深线图
iso-capacity map	等地层系数图	isodesmic structure	等键结构
isocarb	等含碳线；等碳线；等碳比线	isodiametric	等直径的
isocenter	等角点；等深点	isodip line	等倾线
isocenter of photograph	像等角点	isodynamic line	等磁线；等力线
isochasm	等峡；等裂口；极光等频线	isofacial map	等相图
isocheim	等冬温线	isofacial metamorphism	等相级变质

isofacial rock	等相岩
isofacies	等相
isofacies map	等相图
isofactor model	等因子模型
isoflux	等通量
iso-flux contour	等通量线
isofrequency map	等频率图
isofrigid	常年冻温的；等冻温的；寒冷等温线
isogal	等重力线
isogenetic	同源的；同成因的；同期成的
isogeotherm	等地温线；地下等温线
isogeothermal contour	等地温线；地热等值线
isogeothermal line	等地温线
isogeothermal surface	等地温面；地热等温面
isogonal	等角的；等方位线的；等磁偏线的
isogonal line	等磁偏线；等偏角线；等方位线
isogonality	等角性；等角变换
isogonic	等偏线；等角的；等偏角的
isogonic chart	等磁差图；等磁偏线图
isogonic curve	等磁偏线；等方位线
isogonic line	等磁偏线
isograd	等变线；等变质级；等变度
isograd rock	等变岩；等变质岩
isogradal	等变质级的
isograde	等坡的；等变质级
isogradient	等梯度线
isogradient contour	等梯度线
isogradient map	等梯度图
isogram	等值线
isogram map	等值线图
isogram of ash content	灰分等值线图
isogram of coal ash content	煤层灰分等值线图
isogram of coal linear reserves	煤层线储量等值线图
isogram of rock fissure density	岩石裂缝密度等值线图
isogram of stripping ratio	剥离系数等值线图；剥采比等值线图
isogranular	等粒的，等粒度的；等粒状的
isograph	求根仪；等线图
isogriv	地磁等偏线；等磁针坐标偏角线
isogyre	消光影；同消色线
isohaline	等盐度线；等含盐量
isohaline map	等盐度线图；等含盐量图
isoheight	等高线
isohel	等日照线
isohume	等水分线；等湿度线
isohydric concentration	等氢离子浓度
isohyet	等降雨量线
isohyetal line	等雨量线；等降水量线
isohyetal line chart	等雨量线图
isohyetal map	等雨量线图；降水量分布图
isohypse	等高线；等深线
isohypsometric line	等高线
isoinclination line	等斜线
iso-intensity curve	等强度曲线
isokinetic condition	等动力条件
isokinetic suction	等动力空吸
isolated area	隔离区；通风不良区段
isolated basin	孤盆地
isolated beam	独立梁
isolated blob	孤立液滴
isolated body	隔离体
isolated breakwater	岛式防波堤
isolated cell	孤立单体
isolated chimney	独立烟囱
isolated column	独立柱
isolated component	孤立成分
isolated error	孤立错误
isolated event	孤立事件
isolated feature	独立地物
isolated footing	独立底脚；独立基础
isolated foundation	隔离基础；独立基础
isolated interstice	隔离空隙
isolated interval	封隔层段
isolated network	独立网络
isolated opening	孤立或隔离的矿房；孤立或隔离的井下巷道
isolated peak	孤峰
isolated pier	独立桥墩
isolated pipe	孤立管道；绝缘管道
isolated plant	单独运转电站
isolated platform facies	孤立台地相
isolated plot	隔离区
isolated point	孤点
isolated pore	孤立孔隙
isolated position	隔离位置
isolated power system	独立电力系统
isolated sample	孤立样点
isolated shear wall system	独立抗剪墙系统
isolated single pile	孤立单桩
isolated stacking	孤立堆积
isolated storm	孤立暴雨
isolated support type	单支撑式
isolated system	孤立系统
isolating partition	隔离间壁；绝缘隔板
isolating ring	隔离环
isolating switch	断路器；切断开关
isolating switch assembly	隔离开关
isolating switch for control supply	控制电源隔离开关
isolating switch for main silicon rectifier cubicle	主整流柜隔离开关
isolating valve	隔离阀；隔断阀
isolation	隔离，隔开；绝缘；离析
isolation board	冻结板
isolation effectiveness	隔振效果
isolation layer	隔离层
isolation method	等值线法
isolation partition	隔实隔墙
isolation room	隔离室
isolation system	隔振系统
isolation technique	分离技术
isolation tester	绝缘试验仪
isolation trench	隔振沟
isolation valve	隔离阀
isoline map	等值线地图
isoline method	等值线法
isolith map	等岩性图
isomagnetic	等磁线；等磁力的
isomagnetic chart	等磁图
isomagnetic line	等磁线
iso-maturity line	等成熟度线

isomer	同分异构物；同质异能素；异构体	isoreliable seismic design	等可靠性抗震设计
isomeric	同质异构的	isoresistivity map	等电阻图
isomeric alkane	异构烷	isorheic	等黏的；等黏液
isomeric compound	异构化合物	isosaturation surface	等饱和度面
isomeric hydrocarbon	异构烃	isosceles	二等边的；等腰的
isomeric state	同分异构态	isoseism	等震线；等震
isomeric transition	同核异能跃迁	isoseismal	等震线
isomeride	异构体	isoseismal curve	等震线
isomerism	同分异构现象	isoseismal line	等震线
isomerization	异构化；异构化作用	isoseismal map	等震线图
isomerization equilibrium	异构化平衡	isoseismic line	等震线
isomerization reaction	异构化反应	isosinal map	等坡角正弦图
isomerized	异构化的	isosmontic pressure	等渗压
isometamorphic	等变质的	isosmotic	等渗压的
isometamorphic coal	等变质煤	isosmotic pressure	等渗压
isometamorphous vitrinite	等变质镜质体	isosmotic pressure line	等渗压线
isometric	等轴的；等体积的；等容的；同质异能的	isostatic	地壳均衡的；等压的；均衡的
isometric block diagram	等体积的块状图	isostatic adjustment	均衡调整；均衡调节
isometric chart	等距图	isostatic anomaly	均衡异常
isometric correspondence	等距对应	isostatic balance	均衡；平衡
isometric crystal system	等轴晶系	isostatic correction	均衡校正；均衡改正
isometric drawing	等视图	isostatic curve	等压曲线
isometric latitude	等量纬度；等角纬度	isostatic gravity anomaly	均衡重力异常
isometric line	等容线；等值线	isostatic hypothesis	均衡假说
isometric projection	等角投影	isostatic line	主应力迹线；等应力线；等压线；均衡线
isometric system	等轴晶系	isostatic line of aquifer	含水层等压线
isometric view	等距投影图	isostatic movement	均衡运动
isometry	等距；等轴	isostatic readjustment	均衡调整
isomorphic space	同构空间	isostatic reduction	均衡修正；地壳均衡改正
isomorphous replacement	同晶替代；同晶置换	isostatic settling	均衡下沉；均衡沉降
isomorphous series	类质同象系列	isostatic structure	等静力结构
isomorphous substitution	同晶置换	isostatic subsidence	均衡下沉
isopach deposition	等厚沉积作用	isostatic surface	均衡面
isopach interval	等厚沉积间断	isostatic system	均衡系统
isopach map	等厚图	isostatic theory	地壳均衡理论
isopach strike	等厚线走向	isostatic uplift and depression	均衡升降
isopachous line	等厚线	isostatic warping	地壳均衡挠曲
isopachyte	等厚线；等厚图	isoster	等容线
isoparametric	等参数的	isosteric	等比容线
isoparametric transformation	等参变换	isosteric surface	等比容面
isoparametric type	等参数型	isostratification map	等层厚图；岩层等厚图
isoperimetric line	无变形线；等周线	isostructure	等构造；等结构
isopic facies	同相	isotach	等速线；等风速线
isopic zone	同相带	isotherm	等温线；等温过程
isopical deposit	同类沉积	isothermal adsorption	等温吸附
isopiestic surface	等压面	isothermal analysis	等温分析
isopleth of thickness	等厚度线	isothermal annealing	等温退火
isopluvial chart	等雨量图	isothermal atmosphere	等温大气
isopluvial line	等雨量线	isothermal change	等温变化
isopluvial map	等雨量线图	isothermal compressibility	等温压缩性
isoporic chart	等磁变线图	isothermal compression	等温压缩
isoporic line	等磁变线；地磁等年变线	isothermal condition	等温情况；等温条件
isopotential	等势的；等位线	isothermal contour	等温线；等温图
isopotential map	等位能图；等产量图；等势图	isothermal curve	等温曲线
isopotential surface	等势面	isothermal efficiency	等温效率
isopressure surface	等压面	isothermal expansion	等温膨胀
isopycnic	等密度线	isothermal explosion	等温爆炸
isorank line	等类线；等变质线；等煤级线	isothermal extrusion	等温挤压
isoreaction grade	等反应线；等反应度；等变质反应级	isothermal flame ionization chromatography	等温火焰电离色谱
isoreflectance	等反射率		

isothermal flow	等温流	isotopic age determination	用同位素的时代测定
isothermal process	等温过程	isotopic analysis	同位素分析
isothermal quenching	等温淬火	isotopic composition	同位素成分
isothermal reactor	等温反应器	isotopic dating	同位素测年
isothermal remanent magnetization	等温剩余磁化强度；等温剩磁	isotopic deposit	同环境沉积
isothermal section	等温截面	isotopic difference	同位素差异
isothermal sintering	等温烧结	isotopic differentiation	同位素差异
isothermal strain	等温应变	isotopic dilution	同位素稀释
isothermal surface	等温面	isotopic distribution	同位素分布
isothermal zone	恒温带	isotopic effect	同位素效应
isothermic gradient	等温梯度	isotopic element	同位素
isothermic surface	等温面	isotopic equilibrium	同位素平衡
isotime	等时线	isotopic exchange	同位素交换
isotime line	等时线	isotopic exchange effect	同位素交换效应
isotope	同位素	isotopic exchange equilibrium	同位素交换平衡
isotope abundance	同位素丰度	isotopic exchange reaction	同位素交换反应
isotope age	同位素年龄	isotopic fractionation	同位素分馏
isotope analysis	同位素分析	isotopic geochronology	同位素地质年代学
isotope assay	同位素测定	isotopic geology	同位素地质学
isotope assaying of groundwater	地下水同位素测定	isotopic geothermometer	同位素地热温标；同位素地质温度计
isotope buildup	同位素积累	isotopic hydrogeology	同位素水文地质学
isotope carrier	同位素载体	isotopic indicator	同位素指示剂
isotope chemistry	同位素化学	isotopic measurement	同位素测定法
isotope chronology	同位素年代学	isotopic moisture gauge	同位素湿度表
isotope composition	同位素成分	isotopic pattern	同位素模式
isotope composition contour	同位素组成等值线	isotopic power	同位素动力
isotope contour	同位素等值线	isotopic spike	同位素稀释剂
isotope dating	同位素年代测定	isotopic standard	同位素标准
isotope dilution analysis	同位素稀释分析	isotopic symbol	同位素数字符号
isotope dilution mass spectrometry	同位素稀释质谱法	isotopic thermo hydrology	同位素热水文学
isotope dilution method	同位素稀释法	isotopic thermometer	同位素温度计
isotope effect	同位素效应	isotopic thickness gauge	同位素厚度计
isotope exchange reaction	同位素交换反应	isotopic tracer	示踪同位素
isotope fractionation	同位素分离	isotopic variable	同位素变数
isotope fractionation factor	同位素分馏系数	isotopic weight	同位素量；同位素的原子量
isotope geochemical prospecting	同位素地球化学勘探	isotopical deposit	同环境沉积
isotope geochemistry	同位素地球化学	isotrope	均质；均质体；各向同性
isotope geology	同位素地质学	isotropic aquifer	各向同性含水层
isotope geothermometer	同位素地质温度计	isotropic bedding	均质层理
isotope labeling	同位素标记	isotropic behavior	各向同性行为
isotope line	同位素线	isotropic body	均质体；各向同性体；等向性体
isotope method	同位素法	isotropic compression	各向等压压缩
isotope mineralogy	同位素矿物学	isotropic consolidation	各向等压固结
isotope ratio	同位素比值	isotropic consolidation state	等向固结状态
isotope ratio mass spectrometer	同位素比质谱计	isotropic creep compliance	各向同性蠕变柔量
isotope ratio tracer method	同位素比示踪剂法	isotropic deep water	各向同性的深层水
isotope scanner	同位素扫描仪	isotropic deformation	各向同性变形
isotope separation	同位素分离	isotropic deposit	各向同性沉积；均匀沉积
isotope source	同位素源	isotropic fabric	各向同性组构
isotope specific activity	同位素比放射性	isotropic fluid	各向同性流体
isotope standard	同位素标样	isotropic flux	各向同性通量
isotope stratigraphy	同位素地层学	isotropic formation	各向同性地层；均质地层；均质建造
isotope structure	同位素结构	isotropic hardening	等向硬化；各向同性硬化
isotope table	同位素表	isotropic hardening theory	等向硬化理论；各向同性硬化理论
isotope temperature scale	同位素温标	isotropic homogeneity	各向同性
isotopic	同位素的；同沉积区的	isotropic homogeneous formation	各向同性均质地层
isotopic abundance	同位素丰度	isotropic inversion	各向同性反演
isotopic abundance measurement	同位素丰度测量	isotropic layer	各向同性层
isotopic age	同位素年龄		

isotropic line	空间对角线
isotropic loading	各向等压加载
isotropic material	各向同性材料
isotropic medium	各向同性介质；均质体介质
isotropic mineral	均质矿物
isotropic rock	各向同性岩石
isotropic semi-infinite solid	各向同性半无限体
isotropic soil	各向同性土；均质土
isotropic stress	各向等应力；各向同性应力
isotropic stress line	各向等应力线；各向同性应力线
isotropic stress loading	各向等应力线加载；各向同性应力加载
isotropic structure	各向同性结构
isotropic substance	各向同性物质
isotropic surface	等值面；各向同性面
isotropic symmetry	均质对称；等向对称
isotropic thermal expansion	各向同性热扩散
isotropic thickness gauge	同位素测厚仪
isotropic transparent material	各向同性透明材料
isotropic turbulence	各向同性湍流；各向同性紊流度
isotropism	各向同性现象
isotropization	均质化；各向同性化
isotropy	各向同性
isotropy of mine product	等品位线图
isotropy of mine product layout	图面配置
isotropy of mine product map	矿产等品位线图
isotropy of mine product out	绘图；绘制
isotype	同型；等型；同结构
isotypic	等型的；同型的；等结构的
isotypic crystal	等型晶体
isotypic specificity	同型特异性
isotypic variation	同型变异
iso-variable velocity	等变速
iso-velocity contour	速度值线图
isovelocity cross section	等速剖面
isovelocity interface	等速界面
isovelocity layer model	等速层模型
isovelocity surface	等速度面
isoviscous temperature	等黏温度
isovois	等体积线；等容线
isovol	等挥发分线
isovol map	等体积图
isovolumetric wave	等容波
isowarping	等挠曲线；等挠曲的
ISRM	国际岩石力学学会
ISRM Secretariat	国际岩石力学学会秘书处
issite	暗闪辉长岩
issuance of methane	沼气泄放
issue of firedamp	沼气泄出
issue of license	发给执照
issuing velocity	射出速度
issuing water	涌出水；冒出水；泉水
isthmus	地峡
ITA	国际隧道协会
itabirite	铁英岩
itacolumite	可弯砂岩
italite	粗白榴岩
item analysis	项目分析
item design	项目设计；课题设计
item-by-item inspection	逐项检查
itemized record	明细记录
itemized schedule	项目一览表
iterate	迭代；重复
iterated interpolation	迭代插值
iterated interpolation method	迭代插值法
iteration	迭代；逼近
iteration chromatography	循环色谱法
iteration factor	迭代因子
iteration index	迭代指数
iteration loop	迭代循环
iteration method	迭代法
iteration method with variable weight	选权迭代法
iteration number	迭代次数
iteration parameter	迭代参数
iteration point	迭代点
iteration procedure	迭代程序
iteration solution	迭代解
iteration system	迭代系统
iterative	重复的；迭代的
iterative addition	迭代加法
iterative algorithm	迭代算法
iterative analysis	迭代分析
iterative approximation	迭代近似
iterative balance	迭代平衡
iterative computation	迭代计算
iterative computing method	迭代计算法
iterative dominance	迭代优势法
iterative instruction	迭代指令
iterative inversion	迭代反演
iterative least square	迭代最小平方
iterative least square inversion	迭代最小平方反演
iterative least square method	逐次最小二乘法；迭代最小平方法
iterative loop	迭代循环
iterative method	迭代法
iterative modeling	迭代模拟
iterative nature	迭代性质
iterative operation	迭代操作
iterative optimizing technique	迭代最优法
iterative procedure	迭代程序
iterative process	迭代过程
iterative program	迭代程序
iterative reconstruction algorithm	迭代重建算法
iterative refinement	迭代求精
iterative reweighted least square	迭代再加权最小平方
iterative shooting method	反复试凑法
iterative solution	迭代解
iterative solution method	迭代解法
iterative step	迭代步骤
iterative technique	迭代方法
iteratively faster	快速迭代
itinerary	航海日程表；旅行指南；旅程；旅程的
itinerary map	路线图
I-type granite	Ⅰ型花岗岩
IUGG	国际大地测量与地球物理联合会
ivernite	二长斑岩
iverse layer landslide	逆层滑坡
ivoirite	单斜辉石苏长岩
ivory coast tektite	象牙海岸石；象牙海岸玻陨石
ixiolite	锰钽酸铁矿；锰钽矿

J

J　焦，焦耳
J Integral　J积分
jack　千斤顶，起重器；插口，插座
jack arch　平拱，单砖拱，等厚拱
jack arm　起重平衡臂
jack bar　千斤顶垫木
jack bit　钻头，岩芯钻头；可折式钻头
jack blocking　千斤顶垫板
jack board　插孔板
jack bolt　定位螺栓；起重螺栓
jack calibration　千斤顶标定
jack column　支柱，撑杆，可伸缩立柱
jack crusher　碎石机
jack cylinder　千斤顶油缸；起重油缸
jack drill　风镐；凿岩机
jack frame　起重架；液压反力器
jack hammer　手持式凿岩机；凿岩锤；气锤
jack hammer drill　手持式钻机；手持式凿岩机
jack hole　石门，横巷，联络小巷
jack horse　脚手架；台架
jack lagging　坑壁承重填木；承重木构件
jack latch　打捞工具；带闩打捞钩
jack lever　起重杠杆
jack lift　起重吊车
jack loading test　千斤顶加载试验
jack panel　塞孔盘，接线板
jack pile　千斤顶压入桩
jack pile puller　千斤顶式拔桩机；液压拔桩机
jack position transducer　千斤顶位置转换器；千斤顶位置传感器
jack pressure　千斤顶压力
jack rack　千斤顶齿条
jack screw analogy　起重螺旋模拟
jack shaft　浅井，暗竖井，溜眼；起重轴；变速箱传动轴；钻机支柱
jack shoe　撑靴
jack spring　塞孔弹簧
jack test　千斤顶试验
jack timber　短撑木，顶木
jack well　抽水井，抽油井
jacked pile　顶压桩，压入桩
jacked pile underpinning　压入桩托换
jacked-in sampler　压入式取样器
jacket　套，罩，外壳；导管架；加套，加罩
jacket collar　衬套接箍
jacket cooling　套式冷却，壳式冷却
jacket handling　导管架扶正
jacket heater　套式加热器
jacket launching　导管架下水
jacket launching barge　导管架下水驳
jacket leak　护套渗漏
jacket leg　支撑腿，导管架腿柱；支撑桁架
jacket lifting eye　导管架吊耳
jacket panel　导管架组片
jacket panel assembly　导管架组对
jacket positioning　导管架定位
jacket securing　导管架固定
jacket upending　导管架竖立
jacket valve　缸套阀
jacket walkway　导管架走道
jacket water　水套冷却水
jacket-cooled reactor　套式冷却反应器
jacketed cable　包皮电缆
jacketed cooler　套式冷却器
jacketed evaporator　套式蒸发器
jacketed heat exchanger　套式换热器
jacketed reactor　包壳式反应器
jacketed specimen　加套样品，封装样
jacketed-pipe high-pressure steam preheater　包套管式高压蒸汽预热器
jacketing　套式冷却，套式加热
jacketing machine　包套机，包封机
jack-hammer man　打眼工，凿岩工
jack-hammer with air leg　气腿式轻型凿岩机
jack-in method　顶进法
jacking　抬起；上抬；顶起；张拉
jacking beam　顶梁，反力梁
jacking block　千斤顶垫块
jacking bracket　千斤顶支座
jacking conditions　升降条件
jacking cylinder　小型液压起重设备
jacking device　升降设备
jacking force　顶托力，千斤顶举升力
jacking gear　升降齿轮
jacking header　千斤顶顶头
jacking method　压入法，顶升法；推进工法
jacking motor　升降电机
jacking of foundation　基础抬升，基础顶托
jacking of pile　桩支撑
jacking of roof　顶板支撑
jacking oil　顶轴油
jacking oil pump　顶轴油泵
jacking oil system　顶轴油系统
jacking pad　千斤顶垫
jacking pile　顶压桩，压入桩，顶进桩
jacking plate　千斤顶垫板
jacking pressure　千斤顶起重压力
jacking pump　顶轴泵
jacking state　升降状态
jacking stress　顶托应力
jacking system　升降系统
jacking test　顶托试验，千斤顶试验，千斤顶法试验
jacking through anchorage　锚式顶托
jack-knife derrick　折叠式钻塔，悬臂式井架
jack-knife drilling mast　折叠式井架；折叠式桅杆
jackknife rig　折叠式钻机
jackleg　轻型钻架，钻机腿
jackleg drill　气腿式凿岩机

jack-loading method 千斤顶加载法
jackmill 修钎机
jack-prop 千斤顶支柱
jackrod 钻杆，钎杆
jackrod furnace 制钎炉
jackscrew 螺旋千斤顶，起重螺旋
jack-squib 井下爆炸器
jackstay 撑杆；支索
jack-type bar shear 手动液压钢筋剪切机
jack-up 顶高，起重；海上钻井
jack-up drilling platform 自升式钻探平台，自升式钻井船，桩脚式钻探平台
jack-up drilling rig 自升式钻机
jack-up pile driver 自升式打桩机
jack-up rig 自升式钻机；自升式钻井平台
Jacobi formula 雅可比公式
Jacobi iteration method 雅可比迭代法
Jacobi matrix 雅可比矩阵，导数矩阵
Jacobi matrix method 雅可比矩阵法
Jacobian determinant 雅可比行列式，导数行列式
Jacobian transformation 雅可比变换
Jacobian variety 雅可比簇
Jacobi's linear analysis 雅可比线性分析
Jacobi's staff 罗盘支杆；连杆
jadder 割石机，截石机
jade 玉，硬玉
jadeite 硬玉
jadeitite 硬玉岩
jaffaite 树脂
jag 锯齿状缺口
jagged 锯齿状的，参差不齐的
jagged core bit 牙轮岩心钻头
jagged edge 锯齿状边缘
jam 卡住，阻塞，挤塞；矿柱
jam nut 防松螺母，锁紧螺母
jam release 解卡
jam signal 干扰信号
jam welding 对接焊，对头焊接
jamb 矸石；斜坡，陡坡；巨砾；矿柱；边框
jamb post 门窗侧柱
jamb shaft 门窗立柱
jamb wall 门窗侧墙
jambo 钻车，凿岩机车
Jamin effect 贾敏效应；油阻效应
Jamin-Lebedeff interference equipment 贾敏-列别捷夫干涉装置
jammed chute 堵塞的溜槽；堵塞的溜眼，堵塞的溜口
jammer 干扰发射器，干扰器；支架，支柱，支座
jamming 干扰；阻塞；挤压；卡住；夹紧
jamming intensity 干扰强度
jamming margin 干扰容限
jamming of drill 卡钻
jamming-nut 锁紧螺母
jam-on packer 卡压式封隔器
jam-packed 卡紧的
jamproof 抗干扰的
jam-to-signal ratio 噪信比
Janbu method of slope stability analysis 简布斜坡稳定性分析法；简布边坡稳定性分析法
Janbu simplified method 简布简化法

Janka indentation test 詹卡刻痕测试；压痕测试
Janus configuration 詹纳斯配置
jap 手持式凿岩机
Japan Nuclear Cycle Development Institute 日本核循环发展研究所
Japan Society of Civil Engineering 日本土木工程学会
Japan Tunneling Association 日本隧道学会
Japanese Committee for Rock Mechanics 日本岩石力学委员会
Japanese excavation method 日本挖掘法；日本开挖法
Japanese Geotechnical Society 日本岩土工程学会
Japanese Society for Rock Mechanics 日本岩石力学学会
Japanese Society of Material Science 日本材料科学学会
Japanese Society of Soil Mechanics and Foundation Engineering 日本土力学与地基工程学会
jar 震动，震击；冲击钻进；震击器
jar accelerator 震击加速器
jar block 冲击锤，吊锤
jar booster 震击助力器
jar bumper 震击缓冲器；钻杆夹
jar collar 震击环
jar coupling 震击连接器
jar crotch socket 震击打捞筒
jar down spear 下冲打捞矛
jar intensifier 震击助力器
jar knocker 震敲器
jar latch 碰门，碰钩
jar mill 罐形磨机
jar ram moulding machine 震动制模机
jar rein socket 绳式震击打捞器
jar rod 加重冲击钻杆
jar socket 震击打捞筒
jar stem 震击杆
jar tester 震动检测仪
jar tong socket 杆式震动打捞器
jar tongue socket 舌簧式震击打捞筒
jar weight 冲击锤
jar-proof 防震的
jarring 震动，震击
jarring effect 震击效应
Jarva tunnel-boring machine 贾瓦全断面隧道掘进机
jasper 碧玉；墨绿色
jasper rock 碧玉岩
jaw 爪，钳，夹具，卡盘；颚，颔；颚板
jaw bolt 颚头螺栓
jaw brake 爪形制动器，爪闸
jaw breaker 颚式破碎机
jaw chuck 爪式卡盘，孔口夹持器，孔口卡盘
jaw clip 颚式夹，爪形夹箍
jaw clutch 颚式离合器，牙嵌式离合器
jaw coupling 爪盘连轴节/爪盘联轴节
jaw crusher 颚式破碎机，颚式碎石机，老虎口
jaw of pile 桩靴，桩爪
jaw of spanner 扳手钳口
jaw plate 颚板
jaw socket 虎钳口
jaw vice 虎钳
jawab 对称房屋
jaw-clutch differential 爪形离合器差动器
jazz rail 弯轨

JCRM 日本岩石力学委员会
jealous glass 不透明玻璃
Jeans instability criterion 琼斯不稳定判据
Jeans length 琼斯长度
Jeans-Jeffreys hypothesis 金斯-杰弗里斯假说
jeeper 涂层检漏器
jeeping 涂层检漏
jefferisite 水蛭石
Jeffrey diaphragm jig 杰弗雷隔膜跳汰机
Jeffreys-Bullen travel time table 杰佛利斯-布伦走时表
Jeffrey-Traylor feeder 杰弗里-特雷勒给料机，磁力振动槽式给料机
Jeffrey-Traylor screen 杰弗雷-特赖勒筛，电磁振动筛
jel 凝胶，冻胶
jellied gasoline 凝固汽油
jellification 胶凝作用
jelling agent 胶凝剂
jellous 凝胶状的，胶状的
jelly 胶质，胶状物；凝胶状
jelly filled-type optical fiber cable 充油型光缆
jemmy 短撬棍
jenny 单轮滑车，移动吊机，桥式起重机
jenny roadway 自重滑行巷道，自重滑行坡；轨道上山
jenny scaffold 活动脚手架
jerk 急引，急牵，急撞
jerking 溜放调车；急动的
jerking table 摇床，摇动式淘汰盘
jerk-line swing 变向肘节
jerkmeter 加速度计
jerry 碳质页岩；草率了事的，偷工减料的
jerry building 偷工减料的建筑
jerry man 矸石清理工，坍塌清理工
jerry-built 匆忙建筑的，草率建造的，偷工减料建造的
jerry-built construction 偷工减料的施工
jet 射流；喷注，喷射；喷口，喷嘴，喷射钻井；
jet acidizing 射流酸化
jet action 喷射作用
jet agitation 喷射搅拌
jet aircraft 喷气式飞机
jet aircraft fuel 喷气式飞机燃料
jet alloy 火箭合金
jet ammonia refrigerating machine 喷射式氨制冷机
jet analyzer 喷射式分析器
jet axis 射流轴
jet axis dynamical pressure 射流轴动压力
jet basket 喷射式打捞篮
jet bit 喷射钻头
jet bit drilling 喷射钻头钻进
jet blender 喷射搅拌机；喷射混合器；喷射漏斗
jet blocking coefficient 射阻系数
jet blocking length 射阻长度
jet blower 喷射式通风机；射流风机；喷气鼓风机
jet boring 喷射钻探
jet boring machine 喷射式钻孔机
jet boundary 射流边界
jet brake 射流制动器
jet car 水枪车，喷射车
jet carburetor 喷射式汽化器
jet casing cutter 射流切管器
jet cement mixer 喷射式水泥混合器
jet cementing 喷浆；喷浆法
jet centrifugal-pump 喷射式离心泵
jet chamber 喷雾室
jet charge 聚能射孔弹
jet circulation 喷射环流
jet coefficient 喷射系数
jet column 射流柱
jet compression 喷射压缩
jet compressor 喷射式压缩机
jet concrete 喷射混凝土
jet condensation 喷水冷凝；喷水凝汽
jet condenser 喷水凝汽器；喷水冷凝器
jet condenser pump 喷射凝气泵；喷水冷凝泵
jet contraction 喷射收缩
jet control system 射流控制系统
jet conveyor 喷射输送器
jet cooling 喷射冷却
jet current 喷射流
jet cutter 喷射式切割器，聚能切割器，聚能切割弹
jet cutting 射流切割，水力开凿
jet cutting shearer 射流切割机
jet deep-well pump 喷射式深井泵
jet deflection 喷射造斜
jet deflector 射流转向器
jet diffuser 射流扩散器
jet dispersion 射流扩散
jet divergence angle 射流扩散角
jet dredge 喷射式挖泥船
jet drill 喷射式钻机
jet drill technique 射流成井技术；射流钻井技术；射流钻探技术
jet drilling 射流钻进
jet drilling mud 射流钻井液
jet dust counter 喷射测尘计
jet eductor 射流器，喷射器
jet efflux 喷射外排
jet ejector 射流引射器
jet engine 喷气发动机
jet engine fuel 喷气发动机燃料
jet erosion drill 射流冲蚀钻机
jet fan 射流风机
jet flotation machine 喷射式浮选机
jet flow 射流，喷流，喷射流
jet fuel 喷气机燃料；喷气发动机燃料
jet fuel oil 航空煤油；喷气发动机燃料油
jet grouting 喷射灌浆，喷射注浆
jet grouting column 喷注桩柱
jet grouting stem 喷射灌浆钻杆
jet gun 聚能射孔器
jet hole 射流孔，水力喷射眼；喷嘴口
jet hopper 喷射漏斗
jet hose 射流软管
jet humidifier 喷射加湿器
jet hydraulic horse-power 射流水马力
jet impact area 射流冲击区
jet impact force 射流冲击力
jet impact scrubbing 射流冲击擦洗
jet impulse 射流脉冲
jet leaching 喷射淋滤，喷射浸出
jet lift dredger 喷射泵挖泥船

jet method	冲采法
jet mill	气流磨，喷射碾磨机
jet mixer	喷射式混合器，射流搅拌机
jet mixing flow	喷射混合流，混射流
jet molding	注塑，喷射模塑
jet noise	喷注噪声
jet nozzle	射流喷嘴
jet orifice	喷射口，喷嘴
jet outlet velocity	射流出口速度
jet path	射流路径
jet perforating process	聚能射孔作业，聚能炸药射孔作业
jet perforation	聚能射孔，聚能炸药射孔
jet perforation circulation washer	聚能循环洗孔工具
jet perforator	聚能射孔器
jet perforator charge	聚能射孔炸药
jet pierce drill	热力喷射钻机
jet pierce machine	热力喷射钻机
jet piercer	热力喷射钻机
jet piercing	喷焰钻进，热力喷射钻进
jet pile	射水桩
jet pipe	射流管，喷管
jet pipe servomotor	射流管伺服马达
jet pipe type preamplifier	喷射管式前置放大器
jet pressure	射流压力，喷射压力
jet priming	射流引发，射流启动
jet propeller	喷气推进器
jet propulsion	喷气推进
jet pulverizer	喷射式粉磨机
jet pump	射流泵，喷射泵
jet range	射流射程
jet ratio	射流比
jet reverse circulation drilling	射流反循环钻进
jet rock	煤玉岩
jet runway	喷气式飞机跑道
jet scrubbing	水力喷射擦洗
jet separation	射流分离，喷射分离
jet separator	射流分离器，喷射分离器
jet shale	煤玉页岩，褐煤页岩
jet sieve	射流筛，喷射筛
jet splitter	射流分束器
jet spread coefficient	射流扩散系数
jet spreading	射流扩散
jet spud bit	喷射开眼钻头
jet stone	黑色电气石；黑碧玺
jet stower	风力喷射充填机
jet stream	射流
jet stream velocity	射流速度
jet theory	射流理论
jet torch	喷灯
jet trajectory	射流轨迹
jet trajectory length	射流挑出长度
jet tray	射流塔盘
jet tube	喷管
jet tube pressure relief outlet	喷管减压出口
jet vacuum mixer	喷射真空搅拌器
jet velocity	喷射速度
jet viscometer	喷射黏度计
jet water course	喷射水流
jet-air pump	喷气泵
jetboat	喷气快艇
jet-circulation type rolling cutter bit	喷射环流型牙轮钻头
jetcreting	混凝土喷射浇筑
jet-cyclo device	旋喷装置
jet-cyclo flotation cell	旋喷浮选机
jet-cyclo flotator	喷射旋流式浮选机
jet-cyclone floatation machine	旋喷浮选机
jet-fill tensiometer	喷射注入式张力计
jet-flame drill	喷焰钻机
jet-grout underpinning	喷浆托换
jet-grouted pile	喷注桩
jetliner	喷气式客机
jet-nozzled rock bit	喷射式牙轮钻头
jetometer	润滑油腐蚀性测定仪
jet-piercing burner	热力钻进喷热器
jet-piercing drill	火焰喷射钻机
jet-piercing method	热焰喷射钻孔法
jet-powered junk retriever	反冲式废弃物打捞篮
jet-propelled carrier	喷气推进的运载工具
jet-propelled sprayer	喷气推进的喷雾器
jet-pump pellet impact drill bit	喷射泵钻粒冲击钻头，霰弹冲击钻头
jetrig	射水枪，喷射架
jet-rocket combination	喷气火箭发动机组合
jetted particle drilling	喷射钢粒钻进，喷砂钻进
jetted pile	水冲桩，射水沉桩
jetted well	水力钻井，喷成井
jetting	喷射，灌注；冲孔，射流洗孔；水力喷射钻进
jetting drilling	水力钻进，喷射钻进
jetting equipment	水力喷射开沟设备
jetting fill	水冲法填土
jetting fluid	喷射液
jetting gun	喷射枪
jetting height	喷射高度
jetting nozzle	喷嘴
jetting piling	射水打桩，水力沉桩
jetting process	水冲法
jetting scraper	喷射刮管器
jetting stream	射流
jetting type well-point	喷射型过滤器；喷射型有孔管；降低地下水位用的喷射型穿孔管
jetting velocity	喷射速度
jet-trajectory height	射流挑射高度
jet-turbine engine	喷气涡轮机
jetty	导流堤，突堤，防波堤；突堤式码头，栈桥
jetty bent pile	码头排架桩
jetty dock	堤式船坞
jetty harbor	突堤港
jetty head	突堤堤头，导流堤头头，防波堤头部
jetty pier	栈桥码头，突堤码头
jetty-mounted radar	码头雷达
jetty-mounted sonar	码头声呐
jet-type carburetor	喷射式汽化器
jet-type deaerating heater	喷射式除氧加热器
jet-type deaerator	喷射式除气器
jet-type humidifier	喷射式加湿器
jet-type mill	喷射式磨机
jet-type reverse circulation tool	喷射式反循环钻具
jet-type tricone bit	喷射式三牙轮钻头
jet-type washer	喷射式清洗机

jet-type water course 喷射式水流
jetty-type wharf 栈桥式码头，突堤式码头
jewel 宝石
jewel bearing 宝石轴承
JGR 地球物理学研究
JGS 日本岩土工程学会
Jiangnan old land 江南古陆
jib 起重臂，起重杆，吊机臂；吊杆，挺杆，悬臂；突梁
jib boom 起重臂，起重杆
jib crane 悬臂起重机，挺杆起重机，起重吊机，旋臂吊机
jib head radius 悬臂作用半径
jib hoist 悬臂起重机
jib loader 旋臂装载机
jib motor 悬臂电动机
jib-head pulley 悬臂主滑轮
jib-length 悬臂长度
jib-type scraper loader 悬臂式扒斗装载机
jib-type stationary hoist 悬臂式固定起重机，悬臂式固定卷扬机，悬臂式固定升降机
jig 夹具；跳汰机；钻模
jig amplitude 跳汰振幅
jig bed 跳汰床
jig bush 钻套
jig compartment 跳汰分段
jig concentration 跳汰选别
jig cycle 跳汰周期
jig ejector 跳汰排渣器
jig force 跳汰力
jig frequency 跳汰频率
jig haulage 自重运输，重力运输
jig refuse 跳汰矸石，跳汰废渣
jig stroke 跳汰冲程
jig suction 跳汰吸啜
jig transit 支架式经纬仪；定向经纬仪
jig washer 跳汰洗选机
jig washing 跳汰洗选
jigger 跳汰机，振动筛；盘车，辘轳；可变耦合变压器
jigger conveyor 振动式输送机
jigger coupling 电感耦合
jigger rotor 盘车转子
jigger without piston 无活塞跳汰机
jigging 跳汰，跳汰选矿
jigging area 跳汰面积
jigging conveyor 振动式输送机
jigging cycle curve 跳汰周期曲线
jigging machine 跳汰机
jigging motion 跳汰运动，簸动
jigging platform 振动台
jigging ratio 跳汰比
jigging refuse 跳汰废渣
jigging screen 振动筛
jigging sieve 振动筛
jigsaw 线锯，竖锯；
jigtank 跳汰箱
jim crow 弯轨器
jimmy 料车；短撬棍
jinny 矿车滑行斜坡；固定绞车
jinny roadway 矿车滑行斜坡
J-integral J积分
jitter 颤动，抖动，振动；不稳定；偏差
jitterbug 混凝土表面压平器；捣实机；图像跳动
jitty 联络巷道
Jixian Group 蓟县群
Jixianian 蓟县纪；蓟县系
JM type anchor device JM型锚具
JM type anchorage JM型锚具
JNC 日本核循环发展研究所
job 作业，工作；零工；工地
job analysis 工作分析，职务分析，岗位分析
job classification 工作分类
job classification method 工作分类法
job cleanup 工地清理
job completion 完工
job control communication 作业控制通信
job cost 工程成本
job costing method 成本计算分批法
job description 任务书，任务单
job design 施工设计
job engineering 作业工程
job entry system 作业调入系统
job environment 工作环境
job evaluation 工作评价
job execution 作业执行
job facilities 工地设施
job flow 作业流程
job flow control 作业流程控制
job handling routine 作业处置常规
job implementation 作业实施，施工
job initiation 开工
job instruction manual 工作说明手册
job laboratory 工地实验室
job layout 作业放线，作业布置
job library 作业库
job location 施工地点
job management 作业管理
job materials 工料
job mixed concrete 现场搅拌的混凝土
job note 工作笔记
job number 工作号
job opening 招工
job order 工作通知单
job organization 作业组织
job organization language 作业组织语言
job out 外包，分包
job overhead cost 作业管理费
job plan 工作计划；工地平面图
job practice 施工方法；施工实践
job procedure 作业程序
job production 零星生产
job program 作业程序，加工程序
job progress 作业进度
job ranking method 工作评定法
job rate 工作定额
job record 工作记录
job report 工作报告
job safety analysis 工作安全分析
job scheduler 作业调度程序；作业调度计划员；作业调度程序机

job scheduling 作业调度
job scheduling routine 作业调度常规
job sequence 加工程序；施工程序
job site 施工现场，工地
job site facilities 工地设施
job site story 施工现场描述
job specification 工作说明手册；施工规范，工作规范，操作规程；任务说明书
job specification balance 工作规范明细表
job specification language 作业说明语言
job stacking 成批生产
job step 作业段
job supervisor 监工
job ticket 记工分签条
job title 职称
job work 包工；计件工
jobber 散工，包工
jobbing 计件工作；零修
jobbing mill 零批轧机
jobbing pulley 导轮；张紧轮
jobbing sheet rolling mill 零批薄板轧机
jobbing shop 修理车间
job-control language 工作控制语言
job-poured concrete 现场浇筑混凝土
job-proved 生产经验证实的，实践证实的
job-splitting 分工
jock 斜井矿车防跑车叉
jockey 矿车无极绳抓叉；自动释车器
jockey chute 放矿溜槽
jockey pulley 导轮，导向滑车
jockey pump 稳压泵，管道补压泵；补给水泵
jockeying back-and-forth 来回调动
jog 凹进或凸出，粗糙面；突然转向
joggle 啮合扣，接榫；套索钉
johannite 铀铜矾，硫铜铀矿
John's prop intercalation press 约翰支柱压入机
Johnson counter 约翰逊计数器
Johnson noise 约翰逊噪声
Johnson separator 约翰逊多辊静电分选机
join 连接，接合；连接线，连接面，连线，共轭线
joining balk 联系木梁，系梁
joining beam 联系横梁，系梁
joining by mortise and tenon 榫槽接合
joining by screw 螺丝接合
joining nipple 接合螺管
joining on butt 对头接合
joining up in parallel 并接
joining up in series 串接
joining with swelled tenon 扩榫接合
joint 节理；接头，接缝，接合；关节，节
joint adventure 合营企业，合资企业，联营体
joint alteration 节理蚀变
joint alteration number 节理蚀变系数
joint alternation factor 节理蚀变因素
joint angle 接头角度
joint application 联合应用；综合利用
joint bar 接头夹板，鱼尾板
joint bid 联合投标
joint block 接头凸爪；节理块，节理岩块
joint bond 接头跨接线；共同债券
joint bowl 电缆接线盒
joint box 电缆接线箱；连接套筒
joint branch 分岔接头
joint breakage index 节理破碎指数
joint business operation 合作经营
joint capital 合资
joint cathodic protection system 联合阴极保护系统
joint cavity 节理溶穴
joint cement 填缝水泥
joint chair 接座；接轨座板
joint closure 封缝；封拱
joint coating 接头涂覆，补口
joint compound 镶缝胶
joint compressive strength 节理抗压强度
joint construction 衔接施工
joint construction method 衔接施工方法
joint coupling 活节联接；活节联接器
joint crack 节理缝
joint creditor 共同债权人
joint cross 十字接头；十字轴
joint damper 节点阻尼器
joint denial gate 或非门
joint dense zone 节理密集带
joint density 节理密度
joint diagram 节理图解
joint dispersion index 节理扩散指数
joint distribution function 联合分布函数
joint elbow 弯头
joint element stiffness matrix 节理单元刚度矩阵
joint enterprise 联合企业
joint entropy 联合熵，联合平均信息量
joint epicenter method 联合震中法
joint face 接缝面，接合面；节理面
joint filler 填缝料，接缝压条；节理充填物
joint financing 联合筹资
joint fissure 节理裂缝
joint force 结点力；节点力
joint frequency 节理频率
joint gauge 测缝计
joint grouting 接缝灌浆
joint guarantee 共同担保
joint halved 两块搭接的，对搭的
joint hinge 接缝铰接
joint hoisting shaft 共用提升井
joint hoop 接缝箍筋
joint hydraulics 裂隙水力学
joint hypocentral determination 联合震源定位
joint inspection 联合检查，共同检验
joint integrity 连接完整性
joint intensity 节理密度
joint intersection 节理交切
joint inversion 联合反演
joint investment 联合投资
joint juncture seam 裂隙接合痕
joint leakage 接缝渗漏
joint leasehold 联合租用
joint liability 共同义务，连带责任
joint line 连线；关节线；节理线
joint loan 联合贷款
joint measurement 节理测量，缝隙测量

joint meter	测缝仪；节点计
joint mortgage	联合抵押
joint nodal load	节点荷载
joint nut	铰接螺母
joint observation	接缝观测；联合观测
joint of framework	构架接头
joint of trigonal frame	三脚架接头
joint offset	接缝错开；节理错开
joint opening	缝隙，缝口；裂缝开度；节理张开
joint operation	联合经营；联合作业
joint owner	共同业主，共同拥有人
joint ownership	共同所有权
joint packing	接头垫圈
joint pattern	节理型式，节理形态组合
joint pin	铰链销，连接销
joint plane	节理面
joint probability	联合概率
joint probability density function	联合概率密度函数
joint probability distribution	联合概率分布
joint production	合作生产
joint property	共有财产
joint random variable	联合随机变数
joint reinforcement	连接钢筋
joint research center	联合研究中心
joint rose	节理玫瑰图
joint rose diagram	节理玫瑰图解
joint rosette	节理玫瑰图
joint roughness	节理面粗糙度
joint roughness coefficient	节理粗糙度系数；裂隙粗糙系数
joint roughness factor	裂隙面粗糙因子
joint roughness number	节理糙度数
joint seal	接缝密封；接缝止水
joint sealant	填缝料，夹口胶；接缝止水剂
joint sealing	接缝密封；接头防水
joint sealing material	填缝料，接头密封料；接缝止水材料
joint seam	节理缝；接缝缝
joint set	节理组；裂隙组
joint set factor	节理组系数；裂隙组系数
joint set strike	节理组走向
joint sheet packing	接缝密封片
joint shield	接缝盖面
joint shim	接缝垫片
joint spacer	接缝分隔物，连缝隔片
joint spacing	节理间距；接缝间距
joint spring	节理泉，裂隙泉
joint staggering	接缝错开，接缝交错
joint stitching	接缝缝合
joint surface	节理面
joint survey	节理调查
joint system	节理系统，裂缝系统
Joint Technical Committee	联合技术委员会
joint tendering	联合投标
joint tongue	舌形接榫
joint tongue and groove	舌槽接榫，雌雄接榫
joint valley	节理谷
joint ventilation shaft	联合风井，共用风井
joint venture	合资企业
joint venture bank	合资银行
joint venture company	合资公司，联营公司
joint venture contract	合资合同
joint venture for project	项目联营体
joint venture using Chinese and foreign investment	中外合资企业
joint wall	节理面
joint washer	密封垫圈
joint water reduction factor	裂隙水折减系数
joint waviness	节理面起伏度
joint weld	接口焊接
joint with dovetail groove	燕尾槽接合
joint with hinge	铰接
joint-controlled cave	受节理控制的洞穴
joint-cutting	切缝
jointed brake	活节闸瓦，活节制动器
jointed concrete pavement	接缝式混凝土铺面
jointed coupling	活节联接，铰接连接；活节联接器
jointed loading stick	组合式装药棒
jointed reinforced concrete pavement	接缝式钢筋混凝土铺面
jointed rock	多节理岩石
jointed rock mass	多节理岩体，多裂缝岩体
jointed rod	组合钻杆；接杆钎子
jointed shaft	铰接轴
jointed track	有缝线路；有缝轨道
jointed-arm robot	多手臂机器人
jointer	连接工具，接缝器，接缝刨；连接件，联结件；电缆焊接工
jointing	成缝；连接，黏结，胶结，焊接；填料连接，填缝
jointing by cooling	冷却成缝
jointing cable	连接电缆
jointing clamp	压接线夹，接线线夹
jointing compound	接合剂，密封剂
jointing index	成缝指数
jointing material	胶结材料，黏结材料
jointing nail	接合钉
jointing paste	填封腻子，连接膏，接合涂料
jointing plane	接合面
jointing shoe	接合靴
jointing sleeve	连接套管；雷管引线绝缘套
jointing-rule	接榫量规
jointless conveyer belt	无接缝输送带
jointless line	无接缝线路
jointless track	无接缝轨道
jointless track circuit	无接缝轨道电路，无绝缘轨道电路
jointly ergodic random process	联合遍历随机过程
jointly owned railway	合资铁路；合营铁路
jointly stationary random process	联合平稳随机过程
joint-normal stiffness	节理法向刚度
joint-shaped support	铰接形支架
joint-shear stiffness	节理剪切刚度
joint-stock company	股份公司
joint-type fracture	节理型裂缝
jointy	多细裂缝的，多节理的
joist	横梁，托梁，小梁；格栅；桁条；工字钢；子程序
joist anchor	格栅锚栓
joist mill	钢梁轧机，工字钢轧机

joist section 双丁字型断面，双 T 型断面，H 型断面
joist steel 梁钢，工字钢
joker chute 紧急溜槽，紧急溜井
Jolly balance 育氏天平，弹簧比重天平，测密实度天平
Jolmo lungma 珠穆朗玛
jolt ramming 振捣，振夯，震动捣井
jolter 震实机；震动造模机
jolt-packing 震动充填，振动填料
jolt-squeeze 振实挤压
Jominy test 乔米尼试验，淬透性试验，顶端淬火试验
Jones riffle sampler 琼斯分格槽取样器
Jones splitter 琼斯分样器
jonnies 不纯烛煤
Joosten process 乔斯登法，硅化加固法
Jora lift 乔拉升降机
Joule 焦，焦耳
Joule cycle 焦耳循环
Joule dissipation 焦耳耗散
Joule magnetostriction expansion 焦耳磁致伸缩性膨胀
Joule-Kelvin effect 焦耳-开尔文效应
Joule-Lenz effect 焦耳-楞次效应
joulemeter 焦耳计
Joule's equivalent 焦耳热功当量
Joule's law 焦耳定律
Joule-Thomson coefficient 焦耳-汤姆逊绝热节流系数
Joule-Thomson cooler 焦耳-汤姆逊冷却器
Joule-Thomson cooling 焦耳-汤姆逊冷却
Joule-Thomson effect 焦耳-汤姆逊效应
journal 轴颈；钻井记录；杂志，期刊
journal angle 牙轮轴颈角
journal bearing 轴颈轴承，滑动轴承
journal bearing bit 滑动轴承钻头
journal bearing button bit 轴颈轴承球齿钻头
journal bearing insert bit 轴颈轴承镶齿钻头
journal bearing wedge 轴颈轴承楔
journal book 日记账，流水账
journal brass 轴颈铜套
Journal of Geophysical Research 地球物理学研究
Journal of Performance of Constructed Facilities 建筑设施性能学报
Journal of Rock Mechanics and Geotechnical Engineering 岩石力学与岩土工程学报
Journal of Structural Geology 构造地质学报
journal with collars 颈轴带环
Journal of Earthquake Engineering 地震工程学报
Journal of Geotechnical and Geoenvironmental Engineering 岩土与环境岩土工程学报
journey 旅行，旅程，路程；历程
journeyman 雇工，短工，计日工；熟练工人
journey-work 短工性工作；雇用性工作
Jovian planet 类木行星
jowling 敲顶，问顶，叫顶，敲帮信号
Joy continuous miner 乔伊连续采煤机
Joy double-ended miner 乔伊双端滚筒采煤机
Joy extensible conveyor 乔伊伸缩式输送机
Joy loader 乔伊装载机
Joy pushbutton miner 乔伊按钮自动式采煤机
Joy walking miner 乔伊迈步式采煤机
joystick 操纵杆，控制杆，控制手柄
Joy-Sullivan hydrodrill rig 乔伊-萨利文液压钻机

JPCF 建筑设施性能学报
JRC 节理粗糙度系数；联合研究中心
JRMGE 岩石力学与岩土工程学报
JSCE 日本土木工程学会
JSG 构造地质学报
JSMS 日本材料科学学会
JSRM 日本岩石力学学会
JSSMFE 日本土力学与地基工程学会
JTA 日本隧道学会
JTC 联合技术委员会
jubilee truck 小型侧卸式货车；矿车
jubilee wagon 小型货车；矿车
judge 井下量尺；判断，审判；法官
judge distance 目测距离
judgment sample 鉴定样本
judgment test 鉴定试验
judicial person 法人
jug 地震检波器，地音探测器；水罐
jug hustler 放线员
jug line 大线；排列
juggie 放线工
juggler 斜柱
jumbled rock 混杂岩
jumbo 凿岩台车，大型钻车；隧道盾构机；大型喷气式客机
jumbo rig 多机台车
jumbo-type tunnel-boring machine 大型隧道掘进机
jump 跳变，跃迁；冲击打眼，手动式给进；出轨，脱轨；半截风障；断层
jump cloth 半截风障
jump condition 跳变条件
jump correlation 跳点对比
jump counter 跳跃式脉冲计数器
jump coupling 跳合联轴节
jump discontinuity 跳跃间断性
jump drilling 钢绳冲击钻进
jump factor 跃迁因子
jump function 跃变函数
jump grading 间断级配
jump if not 条件转移
jump in brightness 亮度跃变
jump in temperature 温度跃变
jump instruction 跳变指令
jump joint 对接，对头接，平接
jump on rail 跳轨
jump operator 跳跃算子
jump path 跳程
jump ratio 跃迁比
jump spark ignition 跳火花点火
jump zone 跃变带
jumper 冲击钻杆；撬棍；跨接线；非法占用矿地者
jumper bar 凿岩钎
jumper bit 冲击锤；冲击钻头；钎子
jumper cable 跨接电缆
jumper detacher 冲击钻杆拆装机
jumper extractor 冲击钻杆提取器
jumper pipe 跨接管
jumper plug 跨接线插头
jumper stem 手钻钎子；冲击钻钻杆
jumper wire 跨接线

jumper-boring bar 冲击钻杆
jumping 跳动，跃变
jumping bar 探眼工具
jump-over connection 越管连接
jump-over distance 跳跃距离
junction 接合，接触；汇合处，汇流；接边；交叉，枢纽
junction arch 券拱，碹岔
junction arch lining 券拱衬砌，碹岔衬砌
junction board 连接台
junction bowl 接线盒，接线箱，分线盒；连轴器；套管
junction box 分线盒；集管箱
junction cable 中继电缆
junction call 区间呼叫
junction center 中心站
junction laser 面结型激光器
junction line 连接管线；中继线；转辙接合线
junction manhole 管路交汇处进人检查井
junction of haulages 运输巷道连接点
junction piece 连接件，连接短管
junction pile 联结桩
junction point 结点，结合点，联结点，接合点
junction point of traverse 导线连接点
junction potential 接界电位
junction signal 交叉点信号
junction surface 接合面
juncture 接合，接合点，接合处；时机，关头
juncture plane 接触面，接合面
jungle 热带丛林，密林；稠密居住区

junior beam 轻型梁，次梁
junk 废物，废料，废缆绳，废料堆
junk bit 磨铣钻头
junk ring 填料函压盖；活塞顶密封环
Jupiter 木星
Jura 侏罗纪；侏罗系
Jurassic 侏罗纪；侏罗系
Jurassic period 侏罗纪
Jurassic system 侏罗系
Jura-Trias 侏罗-三叠纪
jury pump 备用泵
justifiable 可证明正当的，有理由的
justify 证明…是正确的或合理的
jut 突出部，尖端；悬臂
jute insulated cable 黄麻绝缘电缆
juvenile 原生的，初生的；岩浆源的
juvenile basalt 初生玄武岩
juvenile drainage 初生水系
juvenile gas 岩浆气体；原生气体
juvenile hydrogen sulfide 深源硫化氢
juvenile mantle 原生地幔
juvenile oxygen 初生氧
juvenile soil 原生土壤
juvenile topography 幼年地形
juvenile water 原生水，初生水，岩浆水
juxtaposed 并列的，并置的
juxtaposed reservoir 并置储层
juxtaposition 并置，并列；斜接；毗连

K

K coefficient method　K 法
K wave　K 波，地球外壳的纵波
K0 consolidated drained compression test　K0 固结排水压缩试验
K0 consolidated undrained compression test　K0 固结不排水压缩试验
K0-consolidation　K0 固结
kabaite　陨地蜡
kachchi　泛滥平原
Kaena event　凯纳反向事件
Kaena polarity subchron　凯纳亚磁极期
kahlerite　黄砷铀铁矿，铁砷铀云母
kahruba　琥珀
kainovolcanic rock　新火山岩
Kainozoic　新生代；新生界
Kainozoic era　新生代
Kaiser effect　凯塞尔效应
kalema　激浪
kali　苛性钾，氧化钾
kali salt　钾盐
kalisaltpeter　钾硝石，硝石
kalisyenite　钾质正长岩
kalium base mud　钾基泥浆
kalkspath　方解石
kalktalkspath　白云岩
kalk-uran-carbonate　铀碳钙石
Kalman filter　卡尔曼滤波器
kaluszite　钾石膏
Kama plough　卡马型刮斗刨煤机
Kamal-Brigham analysis method　卡玛尔-布里格姆分析法
kame-and-kettle topography　凹凸地形
kam-tap screen　凸轮垂直振动筛
kanat　暗渠，地下水道
kanch　挑顶卧底，扩大巷道
kandite　高岭石族
K-and-K machine　KK 式浮选机，长缝滚筒式浮选机
kanigen process　镍合金镀覆法
kankar　钙结核；结核灰岩
kann　荧石
Kant-Laplace nebular theory　康德-拉普拉斯星云说
Kanzashi-beam method　托梁法
kaolin　高岭土；高岭石
kaolin catalyst　高岭土催化剂
kaolin clay　高岭石黏土；陶土
kaoline　高岭土
kaoline powder　高岭土粉
kaolinic porridge　粥状风化土，饱水风化土
kaolinic shale　高岭石页岩
kaolinite　高岭石
kaolinite-bearing mining waste　含高岭石矿渣
kaolinitic clay　高岭石黏土
kaolinitic laterite　高岭砖红壤
kaolinization　高岭石化作用；高岭土化
kaolinized zone　高岭石化带

kaolin-sandstone　高岭土砂岩
kaolinton　高岭黏土
kaolin-type residuum　高岭土型风化壳
kaolisol　高岭化土
Kaplan runner　卡普兰转轮
Kaplan turbine　卡普兰水轮机
Kaplan wheel　卡普兰水轮
kapok　木棉
kappameter　卡帕仪，磁化率仪
kar　冰斗
K-Ar dating　钾-氩法定年
kara　黑碱土，黑色盐碱土
karabe　琥珀
karaburan　风沙尘，黑风暴
Karakorum-Nyainqentanglha geosyncline　喀喇昆仑-念青唐古拉地槽
kararfveite　独居石
karat（K）　克拉
karez　坎儿井，灌溉暗渠
Karhunen-Loeve transformation　卡洛变换
karlsteinite　富钾花岗岩
karma　卡马镍铬高电阻合金
Karman constant　卡曼常数
Karman vortex street　卡曼涡列
Karnaugh map　卡诺图
karoo　湿季草原，干燥台地高原，阶地形荒原
karpathite　黄地蜡
karren　溶沟，石牙，石灰岩沟
karst　岩溶，喀斯特
karst aquifer　喀斯特含水层，岩溶含水层
karst base level　岩溶侵蚀基准面，喀斯特侵蚀基准面
karst basin　岩溶盆地，喀斯特盆地
karst box canyon　岩溶箱状谷，喀斯特箱状谷
karst breccia　岩溶角砾岩，喀斯特角砾岩
karst bridge　喀斯特桥，天生桥
karst calcareous tufa　岩溶石灰华，喀斯特石灰华
karst cave　溶洞，喀斯特洞穴
karst channel　岩溶通道，喀斯特通道
karst coast　喀斯特海岸，岩溶海岸
karst collapse　喀斯特塌陷，岩溶塌陷
karst collapse column　喀斯特陷落柱，岩溶陷落柱
karst cone　岩溶锥形丘，喀斯特锥形丘
karst cycle　岩溶旋回，喀斯特旋回
karst denudation plane　岩溶剥蚀面，喀斯特剥蚀面
karst deposit　岩溶沉积，喀斯特沉积
karst depression　喀斯特洼地，岩溶洼地
karst ditch　岩溶沟，喀斯特沟
karst drainage base level　岩溶排水基准面，喀斯特排水基准面
karst dynamics　喀斯特动力学，岩溶动力学
karst ecosystem　喀斯特生态系统，岩溶生态系统
karst engineering geology　喀斯特工程地质学，岩溶工程地质学
karst environment　喀斯特环境，岩溶环境

karst erosion 喀斯特侵蚀，岩溶侵蚀
karst fault block mountain 喀斯特断块山，岩溶块状山
karst feature 岩溶地形，喀斯特地形
karst fen 岩溶沼泽，喀斯特沼泽
karst fenster 岩溶天窗，喀斯特天窗
karst filling 喀斯特充填，岩溶充填
karst fissure 喀斯特裂隙，岩溶裂隙
karst flow passage 岩溶落道，喀斯特落道；岩溶流动通道，喀斯特流动通道
karst formation 喀斯特堆积物，岩溶堆积物
karst funnel 喀斯特漏斗，岩溶漏斗
karst gas explosion 喀斯特气爆，岩溶气爆
karst geochemistry 喀斯特地球化学，岩溶地球化学
karst geomorphology 喀斯特地貌学，岩溶地貌学
karst groundwater 喀斯特岩溶地下水，岩溶地下水
karst groundwater flow 岩溶地下水流，喀斯特地下水流
Karst groundwater system 喀斯特岩溶地下水系，岩溶地下水系统
karst hanging valley 喀斯特悬谷，岩溶悬谷
karst hill 喀斯特丘陵，岩溶丘陵；喀斯特山丘，岩溶山丘
karst hydrodynamic unit 喀斯特水动力单元，岩溶水动力单元
karst hydrogeology 喀斯特水文地质学，岩溶水文地质学
karst hydrology 喀斯特水文学，岩溶水文学
karst lake 喀斯特湖，岩溶湖
karst landform 喀斯特地形，岩溶地形
karst landscape 喀斯特景观，岩溶景观
karst medium 喀斯特介质，岩溶介质
karst mineral deposit 喀斯特矿床，岩溶矿床
karst peneplain 喀斯特准平原，岩溶准平原
karst percentage 岩溶率，喀斯特率
karst phreatic zone 喀斯特潜水带，岩溶潜水带
karst pit 喀斯特坑，岩溶坑
karst plain 喀斯特平原，岩溶平原
karst planation surface 喀斯特夷平面，岩溶夷平面
karst plateau 喀斯特高原，岩溶高原
karst pulsating spring 岩溶脉动泉，喀斯特脉动泉
karst ratio 岩溶率，喀斯特率
karst red earth 岩溶红土，喀斯特红土
karst region 喀斯特区，岩溶区
karst reservoir 喀斯特油气藏，岩溶油气藏；喀斯特水储，岩溶水储；喀斯特储层，岩溶储层
karst river 喀斯特河，岩溶河
karst rocky desertification 岩溶石漠化，喀斯特石漠化
karst run-off 喀斯特径流，岩溶径流
karst saturation zone 岩溶饱水带，喀斯特饱水带
karst sedimentation 岩溶沉积作用，喀斯特沉积作用
karst source 喀斯特水源区，岩溶水源区
karst spring 喀斯特泉，岩溶泉
karst stone column 岩溶石柱，喀斯特石柱
karst stream 喀斯特河，岩溶河
karst street 喀斯特通道，溶蚀通道；岩溶通道
karst subgrade 喀斯特地基，岩溶地基
karst tectonic basin 岩溶断陷盆地，喀斯特断陷盆地
karst terrain 喀斯特地区，岩溶地区
karst thermal water 喀斯特热水，岩溶热水
karst topography 喀斯特地形，岩溶地形
karst trench 岩溶槽，喀斯特槽
karst type residuum 岩溶型风化壳，喀斯特型风化壳
karst underground water flow 岩溶地下水流，喀斯特地下水流
karst valley 岩溶谷，喀斯特谷
karst vegetation 岩溶植被，喀斯特植被
karst water 喀斯特水，岩溶水
karst water aeration zone 岩溶水包气带，喀斯特水包气带
karst water bursting 岩溶突水，喀斯特突水
karst water filling deposit 岩溶充水矿床，喀斯特充水矿床
karst water inflow 岩溶涌水，喀斯特涌水
karst water level 喀斯特水位，岩溶水位
karst water resource 喀斯特水资源，岩溶水资源
karst water vadose zone 喀斯特水包气带，岩溶水包气带
karst water-bearing massif 岩溶含水地块，喀斯特含水地块
karst water-bearing system 岩溶含水系统，喀斯特含水系统
karst well 喀斯特井，岩溶井
karst window 喀斯特天窗，岩溶天窗
karst zonality 岩溶地带性，喀斯特地带性
karst-cave connecting test 溶洞连通试验，喀斯特洞穴连通试验
karsten 岩溶沟，喀斯特沟
karstenite 硬石膏
karstic conduit 岩溶通道，喀斯特通道
karstic discharge 岩溶水流量，喀斯特水流量
karstic earth cave 岩溶土洞，喀斯特土洞
karstic emergence 岩溶泉，喀斯特泉
karstic filling 岩溶充填，喀斯特充填
karstic formation 喀斯特地层，岩溶地层
karstic groundwater 岩溶地下水，喀斯特地下水
karstic loss 岩溶漏水，喀斯特漏水
karstic network 岩溶网，喀斯特网
karstic spring 喀斯特泉，岩溶泉
karstic water table 喀斯特水水位，岩溶水水位
karstification 喀斯特化，岩溶化
karstification zone 喀斯特作用带，岩溶作用带
karstified limestone 岩溶石灰岩，喀斯特石灰岩
karstology 喀斯特学，岩溶学
kasolite 硅铅铀矿
kassiterite 锡石
Kasten box cover 开斯顿取样器，盒式取样器
Kataarchean 远太古代的
Kataarcheozoic 远太古代
katabolism 新陈代谢，分解代谢
kataclasite 碎裂岩，破碎岩
kataclastic breccia 压碎角砾岩，碎裂角砾岩
kataclastic structure 破碎构造
kataclastics 压碎岩，碎裂岩，碎屑岩
katagenesis 碎裂作用，后生作用；退化
katagenetic 分解的；后退演化的
katagenetic metamorphism 分解变质
kata-metamorphism 深变质
katamorphic 碎裂变质的，地壳浅部变质的
katamorphic zone 碎裂变质带，地壳浅部变质带
katamorphism 碎裂变质，地壳浅部变质，风化变质
Katarchean 远太古的
katarock 破碎岩；深变质带岩石
kataseism 向震源地壳运动，向震中地壳运动
kataseismic onset 向源初动
katatectic layer 深积层，溶蚀残余层

katatectic surface 溶积层面
katathermal ore deposit 深成热液矿床
kata-zonal metamorphism 深带变质
katazone 深变质带
katergol 液体火箭燃料
kathion 阳离子，正离子
kathode 阴极，负极
kation 阳离子，正离子
katogene rock 水成岩，沉积岩
katohaline 盐度随深度增大
katothermal lake 逆温湖
Kaufmann arch saver 科夫曼拱形回收机
kaustobiolite 可燃生物岩，可燃有机岩
kautschuk fossiles 弹性地蜡
kavir 盐沼；盐漠
kawk 荧石
K-band radar imagery K波段雷达成像
kbar 千巴
K-braced frame K形支架
kedge anchor 小锚
keel 龙骨；平底船；脊棱；煤的重量单位
keel line 龙骨线，首尾线
keel piece 龙骨构件
keelage 停泊费
keel-and-bilge block 垫船木块
keelblock 龙骨墩；铸锭
keel-block capping timber 龙骨墩顶木
Keene's cement 金氏水泥
keeper 夹子，卡箍，衔铁；保管人
keeping-to-lining grab 靠壁式抓岩机
keeve 洗矿桶
keffekilite 漂布土
keffekill 海泡石
keg buoy 桶形小浮标
keld 岩溶河外流处，喷泉，大泉；水体静止部分，水流静止区
Kelly ball test 凯利球体贯入试验
kelly bar 方钻杆
kelly bar sub 方钻杆接头
kelly bushing 方钻杆补芯
kelly bushing guide skirt 方钻杆补芯导盘
kelly cock 方钻杆旋塞阀
kelly drive rod 方钻杆，主动钻杆
kelly hole 方钻杆用鼠洞
kelly joint 方钻杆接头
kelly rod 方钻杆
kelly safety valve 方钻杆安全阀
kelly spinner 方钻杆旋扣器
kelly stem 方钻杆
kelly stop cock 方钻杆回流阀
kelly straightener 方钻杆矫直器
kelly wiper 方钻杆刮泥器
kelly-in 方入，转盘补芯面以下方钻杆长度
kelly's rat hole 方钻杆用鼠洞
kelly-saver sub 方钻杆保护接头
kelly-up 方余
kelmet 油膜轴承；油膜轴承合金
kelp bed 海藻床
kelve 碳质页岩；荧石
Kelvin body 开尔文体

Kelvin bridge 开尔文电桥
Kelvin degree 开氏度
Kelvin effect 开氏效应
Kelvin equation 开尔文方程
Kelvin material 开尔文物质
Kelvin model 开尔文模型
Kelvin scale 绝对温标，开氏温标
Kelvin substance 开尔文物质
Kelvin temperature 绝对温度，开氏温度
Kelvin temperature scale 开氏温标，绝对温标
Kelvin wave 开尔文波
kelvin（K） 开，开尔文
Kelvin-Voight model 开尔文-沃伊特模型
kemetine 石棉
Kendall coefficient of rank correlation 肯德尔秩相关系数
kendallite 陨铁
Kennedy Space Center（KSC） 肯尼迪空间中心
kennel coal 烛煤；长焰煤
kenotron 高压二极整流管
kentledge 压重，压载块，压舱生铁
Kent-Maxecon mill 肯特-马克松型磨机
Kepes model 开派斯模型
Kepler coordinates 开普勒坐标
Kepler's equation 开普勒方程
Kepler's planetary law 开普勒行星定律
K-Epsilon turbulence model K-ε湍流模型
kerabitumen 油母沥青，干酪根
keramsite 轻膨土，有孔黏土
keratin 石膏灰缓凝剂；角朊，角质
keratinization 角质化
keratophyre 角斑岩
kerb 道路镶边石，路缘石
kerb line 路边线，路缘线
kerb shoe 路边脚；刃脚
kerbside lane 路旁行车线
kerbstone 路缘石
kerf 切缝，切口，锯痕，劈痕，掏槽，截槽，拉槽
kerf blasting 拉槽爆破
kerite 沥青类，沥青岩类；硫沥青
kern 仁，核；颗粒；古地块
kernbut 断层外侧丘
kerncol 断层外侧洼
kernedge 核心边
kernel 核，中心
kernel function 核函数
kernel method 核方法
kernel roasting 中心焙烧
kernstone 粗粒砂岩
kerobitumen 油母沥青
kerogen 干酪根，油母岩，油母沥青
kerogen rock 油母岩
kerogen sapropelite 干酪根腐泥岩
kerogen shale 油页岩，干酪根页岩
kerogenetic coal 油母煤
kerogenite 油母岩
kerogenous clay 含干酪根黏土
kerogenous shale 含干酪根页岩
keronigritite 油母煤化沥青，干酪根煤化沥青
kerosene 煤油
kerosene distillate 煤油馏出物

kerosene engine	煤油发动机
kerosene floatation	煤油浮选
kerosene oil gas	煤油气
kerosene propellant	煤油推进剂
kerosene shale	油页岩
kerosene sulphur test	煤油硫测定
kerosene-type jet fuel	煤油型喷气燃料
kerosene-type turbine fuel	煤油型燃气轮机燃料
kerosine	煤油
ketone	酮
kettle	锅状穴；吊桶
kettle basin	锅穴盆地，锅形洼地，锚洼地；融冰洼地
kettle depression	锅穴陷落
kettle giant	瓯穴，锅状巨穴
kettle hole	锅穴，壶穴
kettle-type reboiler	釜式重沸器
kevel	盘绳栓，系索耳
kewal	冲积土
key	键，电键，开关，按钮；拱顶石，封闭块，键板；扳手，钻杆扳手；钥匙；关键，重点；检索表
key aggregate	嵌缝集料
key base	键座
key bed	标志层
key block	刹尖，封顶块，拱顶石；关键块体
key block theory	关键块体理论
key borehole	基准钻孔
key button	键钮
key city	中心城市
key controlling factor	主控因素
key drawing	索引图；解释图
key engineering project	关键工程项目
key entry	密钥输入
key equipment	关键设备
key fossil	标准化石
key fraction	关键馏分
key groove	键槽
key horizon	标志层，标准层，基准层；主要土层
Key Laboratory of Ministry of Education for Geomechanics and Embarkment Engineering 岩土力学与堤坝工程教育部重点实验室	
key map	总图；解释图；索引图；映射图
key observation well	基准观测井
key off	切断
key on	接通
key operated switch	按键开关
key panel	按键板
key parameter	关键参数
key petroleum source bed	标准生油层
key pile	主桩，枢桩
key plan	总平面图；要览图；索引图；总图
key pollutant	主要污染
Key Scientific Research Base for Ancient Wall Paintings Conservation 古代壁画重点科研基地	
key seated hole	钻具卡槽钻孔
key section	标准剖面
key segment	封顶块
key stone	拱顶石
key stratum	关键层
key support	基准支架
key switch	按键开关
key technology	关键技术
key trench	键槽；齿墙槽
key water-control project	水利枢纽；重点治水工程
key well	基准井，基准钻孔，控制钻孔
key word	关键词，检索词，索引词
key wrench	套筒扳手
keyboard	键盘
keyboard encoder	键盘编码器
keyboard entry	键盘输入
keyboard perforator	键盘穿孔器
keyboard printer	键盘打印机
keyed beam	键接梁，加键梁
keyer	键控器
key-in	键入
keying device	键控装置
keying effect	锁固效应
keying pulse	键控脉冲
keying relay	键控继电器
keying strength	咬合强度
keylock	锁定
keylock switch	锁定开关
keynote	主旨，要点
keypad	便携式键盘，手提键盘，袖珍键盘
keypoint	要点，关键
keypunch	键控穿孔；键控穿孔机
key-punched data	键控穿孔数据
keyset	键盘；配电板
keystone effect	楔效应
keystone plate	瓦垄钢板，波纹钢板
keystone shape wire	梯形断面钢丝
keyway	键槽，接榫
K-feldspar	钾长石
K-G model	K-G 模型
khud	悬崖；峡谷
kibble	矿石吊桶；碎矿
kibble winder	吊桶提升机
kibbler	破碎机，粗磨机；
kick	跳动；到达；井涌，轻微井喷；泥浆流失；折断；偏斜
kick board	跳板
kick control	井涌控制
kick sign	溢流征兆，井涌标志
kick sub	造斜工具接头
kickback	车辆重力反行装置；反冲，逆转；返程；酬金，回扣
kickdown	下弯；弹落；调低速；调速器；移位装置
kicker	喷射器；推出器，抛掷器，推杆；船用小型内燃机；侧刃；加固木块
kicker stone	侧刃金刚石
kicking	微井喷，微气侵
kicking coil	反作用线圈
kicking down	开钻；钢绳冲击钻进
kicking off	开始造斜
kicking piece	垫木
kick-killing	压井
kick-off	启动压力；始造斜位置；开始出油
kick-off assembly	造斜钻具组合；启动装置
kick-off casing pressure	启动套压
kick-off depth	造斜井深
kick-off gas-lift well	带启动阀的气举井

kick-off point 造斜点
kick-off pressure 启动压力
kick-off section 造斜井段
kick-off spring 跳闸弹簧
kick-off valve 反冲阀；启动阀
kickplate 脉冲板
kickpoint 转折点
Kick's law 基克定律
Kick's law deviation exponent 基克定律偏差指数
kicksort 脉冲振幅分析
kicksorter 脉冲振幅分析器
kick-up 预留空余，向上弯曲；封顶；翻车器，翻罐笼
kick-up block 阻车器，挡车器，道栏，阻块
kick-up rail stop block 折合轨挡车器
kick-well-off operation 诱喷
kid 柴捆，柴排，埽
kidding 打捆；填梢护岸，埽工护岸
kidney 腰形阀；结块；肾状卵石；肾状矿石
kidney joint 挠性接头
kieselton 硅质黏土
kietyoite 磷灰石
Kilburn method 吉尔伯恩法，矿业权评估地质工程法
kilkenny coal 无烟煤
kill job 压井作业
kill mud 压井泥浆
kill pump 压井泵
kill valve 压井阀
killed lime 熟石灰，消石灰，失效石灰
killed steel 镇静钢，脱氧钢
killer 限制器；断路器；瞄准器；消色器
killesse 沟渠，渠道，水沟
killick 小锚；大石块
killing agent 脱氧剂；镇静剂
killing fluid 压井液
killing operation 压井操作
killing well 压井
kill-weight brine 压井重卤水，压井加重盐水
kill-weight fluid 加重压井液
kiln 窑，烘干炉
kiln dust 窑灰
kiln-gas gasifier 回转窑气化炉
kiln-stack gas sampler 窑炉烟囱气采样器
kilo-（k） 千
kiloampere 千安
kilobar 千巴
kilobit 千位，千比特
kilobyte 千字节
kilocalorie（Kcal） 千卡，大卡
kilocoulomb 千库仑
kilocurie 千居里
kilocycle 千周，千赫
kilocycle electromagnetics 千周波电磁勘探法
kilocycle（Kc） 千周
kilocycles per second（kHz, kc/s） 千赫；千周/秒
kilodyne 千达因
kilo-electron-volt（KeV） 千电子伏
kilogamma 千微克
kilogauss 千高斯
kilogram（kg） 千克，公斤
kilohertz 千赫

kilohm 千欧
kilojoule 千焦
kiloline 千磁力线
kiloliter 千升
kilolumen 千流明
Kilomegabit 千兆位
kilomegacycle 千兆周
kilometer grid 方里网；公里网，千米网
kilometer post 公里标；千米标
kilometer scale 千米尺；千米标度
kilometer stone 里程桩，里程碑
kilometer（km） 公里，千米
kilometre 千米，公里
kilonewton 千牛
kilopascal（kPa） 千帕，千帕斯卡
kilopost 公里标
kilopound 千磅
kilorad 千拉德
kilorentgen 千伦琴
kilostere 千立方米，千立方公尺
kiloton 千吨
kilovar 千乏；无功千伏安
kilovolt 千伏
kilovoltampere 千伏安
kilovolt-ampere-hour 千伏安小时
kilovoltmeter 千伏表
kilowatt 千瓦
kilowatt hour 千瓦时；度
kilowatt hour metre 千瓦小时计，电度表
kilowatt meter 千瓦计
kim shale 油页岩
Kimberley method 金勃雷开采法，下行水平分层临时留矿开采法
Kimberley-type skip 金勃雷型箕斗
kimberlite 金伯利岩，角砾云母橄榄岩
kimolite 水磨土
Kind-Chaudron method 金乔德隆钻井法
K-index K指数，磁扰强度量
kindling point 着火点
kindling temperature 着火温度
kine 凯恩，基尼
kinematic 运动学的，运动的
kinematic analysis 运动学分析
kinematic constraint 运动学约束
kinematic ductility 动态延性，动态韧性
kinematic eddy viscosity 动涡黏滞度
kinematic energy 动能
kinematic equation 运动方程
kinematic friction 动摩擦
kinematic hardening 运动硬化
kinematic interaction 运动相互作用
kinematic link 传动连杆
kinematic mode 动态模式
kinematic model 运动学模型
kinematic parameter 运动学参数
kinematic positioning 动态定位
kinematic similarity 运动学相似性
kinematic survey 动态测量
kinematic system 运动系统
kinematic train 传动系统

kinematic viscosity 动黏度，动黏滞系数
kinematical boundary condition 运动边界条件
kinematical equilibrium 运动平衡
kinematics 运动学
kinemometer 转速计；流速计
kinescope 显像管
kinesis 动态技术；应激运动
kinetheodolite 机动式经纬仪
kinetic 动力学的，动力的；运动的，能动的
kinetic analysis 动力学分析，动力分析
kinetic control 动态控制
kinetic effect 动力学效应
kinetic energy 动能
kinetic energy density 动能密度
kinetic energy transformation 动能转换
kinetic equilibrium 动态平衡
kinetic expression 动力学表达式
kinetic factor 动力学因素
kinetic filter 动态滤波器
kinetic flow factor 流动因数
kinetic friction 动摩擦
kinetic friction coefficient 动摩擦系数
kinetic frictional stress 动摩擦应力
kinetic hardening 随动硬化；运动硬化
kinetic head 动压头，动力水头，动能头，流速水头
kinetic metamorphism 动力变质
kinetic moment 动力矩
kinetic momentum 动力动量
kinetic oscillogram 动态波形图
kinetic parameter 动力学参数
kinetic potential 动力势
kinetic pressure 动压力
kinetic property 动力学性质
Kinetic reaction 动力学反应
kinetic simulator 动态模拟器，动态仿真器
kinetic stability 动力学稳定性
kinetic state 动态
kinetic theory 动力说；分子动力说
kinetic wave 动力波
kinetic Young's modulus 动态杨氏模量
kinetic-control 动态控制
kinetic-friction coefficient 动摩擦系数
kinetics 动力学
king and queen post truss 十字双柱桁架
king bolt 主栓，中心立轴，中枢销，大螺栓
king journal 主轴颈
king oscillator 主振荡器
king pile 定位桩，主桩，中心桩
king pin 关节销，中心销，大王销；中心立轴
king post 主柱，桁架中柱，主杆；主桩
king post girder 单柱大梁
king post truss 单柱桁架
king rod 桁架中吊杆
king truss 主构架
king valve 主阀，总阀
king wire 主钢丝
kingle 硬岩石，坚硬砂岩
King-Maries-Crossley formula 金-马里斯-克罗斯莱公式
king-post spring 桁架中柱弹簧
Kingsbury thrust bearing 金斯勃雷型止推轴承，倾斜板式滑动止推轴承
king-size cartridge 长药筒，长药卷
kink 膝折，扭折，纽结，弯曲，急曲
kink angle 扭折角
kink band 扭折带
kink band boundary 扭折带边界
kink fold 膝折褶皱，尖角褶皱
kink instability 扭曲不稳定性
kinkle 小纽结，卷曲；
kinship 同源关系，血缘关系
kinsite 海泡石
kinzel test 焊接弯曲试验
Kipp's apparatus 基普发生器
kir 含沥青岩，油砂，硬化石油
Kirchhoff integration migration 基尔霍夫积分偏移
Kirchhoff's law 基尔霍夫定律
Kirchhoff-summation method 基尔霍夫求和偏移法
Kirchoff algorithm 基尔霍夫算法
Kirchoff diffraction equation 基尔霍夫绕射方程
Kirchoff forward modeling 基尔霍夫正演模拟
Kirishima geothermal system 基里希玛地热系统
kirn 人工打眼，手工钻进
kirner 手持冲击钻
Kirsch equations 吉尔西方程
Kiruna method 基鲁纳钻孔偏斜测量法
kish 片状石墨
Kiso-Jiban Consultants. Co. Ltd. KJ基础地盘株式会社
kit 成套工具，工具箱，用具包；配套元件
kitten 小型履带拖拉机
kivuite 水磷钍铀矿
Kjeldahl method for nitrogen determination 克氏定氮法
Kjellin furnace 凯林感应电炉
Klauder wavelet 克劳德子波
klaxon 电喇叭，电警笛，电笛
Klein-Fishtine equation 克莱恩-菲希丁方程
kleit 高岭石
klieg-light 溢光灯，弧光灯
Kling test 克林试验
Klinkenberg effect 克林肯伯格效应
Klinkenberg permeability 克林肯伯格渗透率
klinkstone 响岩
klint 陡崖
klip 断崖，悬崖；陡岩岸；浪蚀阶地
klippe 飞来峰，孤残层
klirr 波形失真，非线性失真
klirr factor 波形失真系数
klirr-attenuation 失真衰减量
klizoglyph 干裂
Kloeckner-Ferromatik powered roof support 克洛克纳-弗罗马蒂克液压支架
klong 水道；运河
kloof 冲沟，峡谷
kluf 沟，峡谷
klydonogram 脉冲电压记录图
klydonograph 脉冲电压记录器
knack 技巧，窍门，诀窍
knaggy wood 多节木材，有瘤木材
knallgas 氢氧混合气
knap 打碎，砸碎；小山，丘顶
knap hammer 碎石锤

knapper	碎石锤，击碎器，碎石机；碎石工	knock out sub	撞击接头
knapping machine	碎石机	knock suppressor	抑爆器
kneader	捏和机，搓揉机	knock wave	冲击波，爆振波
kneading	搓揉；搅拌，混合	knock-boring machine	冲击钻机
kneading compaction	搓揉压实	knocker	冲击架；冲击器；门环；巨石
kneading machine	捏和机，揉捏机，搅拌机	knocking combustion	爆振燃烧
knee	弯头；弯管；拐点；膝形杆；膝	knocking explosion	爆振爆炸
knee bend	直角弯头	knocking noise	敲击噪声
knee brace	斜撑，隅撑，角撑	knocking test	爆振试验
knee brake	弯闸	knock-off block	停泵装置
knee head loss	弯管水头损失	knock-off hook	敲落式挂车钩
knee hole	弯曲钻孔	knock-off joint	脱钩装置；脱钩铰
knee joint	膝状接头	knockout bleeder collar	泄阀接头
knee of curve	曲线转折点	knockout drum	气液分离罐
knee pipe	弯管，曲管	knock-reducer	减振器，消振器；抗振剂
knee plate	弯板	knock-sedative	抗振的，防爆的
knee point	拐点，曲线弯曲点	knock-sedative dope	防爆剂；抗振剂
knee strut	抗压柱	knock-test	抗振试验，抗爆试验
knee timber	多节材	knock-test engine	爆振试验机
knee twin	膝状双晶，肘状双晶	knoll	圆丘，小山，墩；圆丘礁；海底丘
knee-iron	隅铁，曲铁	Knoop hardness	努普硬度，努氏硬度
kneepiece	三角肘板	Knoop microhardness test	努氏显微硬度试验
knee-type structure	膝型构造	knot	包裹体；矿结；瘤；海里；节
knick	裂点；扭折	knotlike	节状，瘤状
knick band	扭折带	knotmeter	测速仪
knick in slope	斜坡裂点，坡折	Knott's equation	诺特方程
knick point	裂点，急折点，坡折点	knotty ore	瘤状矿
knick zone	扭折带	knotty structure	瘤状构造
knicking	膝曲	know-how	实践知识；技能；诀窍
knickline	坡折线，膝折线	know-how license	专有技术许可证
knickplane	膝曲面	knowledge base	知识库
knife cutter	锋刃刮蜡器	knowledge base management system	知识库管理系统
knife dog	钻杆升吊器，楔形钻杆夹持器	knowledge concentrated industry	知识密集型产业
knife edge	刀口，刃	knowledge engineering	知识工程
knife edge bearing	刀口支承	knowledge information	知识信息
knife stone	磨刀石	knowledge intensive enterprise	知识密集型企业
knife structure electrode	刀形电极	knowledge intensive industry	知识密集工业
knife-edge crest	刃脊；刀状山脊	knowledge representation	知识表达
knife-edge surface viscosimeter	刀刃面黏度计	knowledge requisition	知识获取
knife-edge valve	刀形阀	knowledge source	知识源
knife-edge weir	锐缘堰	Knowles-Curran process	诺尔斯-克兰低温碳化法
knife-edged	极锋利的；极精密的	known danger	已知危险
knife-like demarcation	截然分界	known function	已知函数
knife-switch	刀形开关	known loss	已知损失
knife-type scraper	刀型刮蜡器	known number	已知数
knitted fabric	编织物	known quantity	已知量
knitted geotextile	土工编织物，土工编织布	known reserves	已知储量，探明储量
knob	小丘，圆丘；钮，旋钮，按钮；圆形把手	known slip surface	已知滑面
knob and trail	丘尾地形	knuckle	斜井坡度突变处；抓钩；转向节，铰链，万向接头；关节；肘节
knob dial	鼓形刻度盘	knuckle and socket joint	铰链活节连接，链球形联结
knob knot	节疤	knuckle bearing	关节轴承；铰式支座
knob-and-basin topography	盆丘地形，凹凸地形	knuckle centre	万向节十字轴
knob-and-kettle topography	凹凸地形	knuckle drive flange	关节传动凸缘
knob-and-tube wiring	穿墙布线	knuckle gear	圆齿齿轮
knock	冲击，敲击，撞击；振动，爆振；敲击信号；小丘	knuckle guide	铰式导向装置
knock burst	爆发	knuckle joint	万向接头，万向节；铰接，铰链接合
knock inception	起爆	knuckle pin	肘销，关节销
knock indicator	爆振指示器	knuckle pivot	关节枢
knock intensity	爆振强度	knuckle plug	关节销
knock meter	爆振仪		

knuckle spindle 转向节轴
knuckle thrust bearing 关节止推轴承
knuckle-joint press 曲柄压力机
knuckle-joint shoe 铰链接合靴
knuckle-type coupler 关节式车钩
knurl 滚花旋钮
knurl screw 滚花螺杆
knurled knob 滚花捏手，滚花旋钮
knurled nut 滚花螺母
knurled roll 网纹轧辊，槽纹轧辊
knurling tool 滚花工具
koalmobile 自行井下无轨矿车
kobelite diamond bit 细粒金刚石镶条硬焊钻头
Koch freezing process 科赫冻结法
koehler 石脑油火焰安全矿灯
Koehler lamp 科奇勒火焰安全灯
Koenigsberger ratio 科尼斯贝格比，剩磁强度与地磁感应强度比
koenlinite 重碳地蜡
Koepe hoist 戈培提升机，单绳摩擦提升机
Koepe hoisting 戈培提升，单绳摩擦提升；摩擦轮提升机
Koepe pulley 戈培轮，单绳摩擦轮
Koepe winder 单绳摩擦轮提升机，戈培提升机
Koepe winder brake 戈培提升机制动器，摩擦轮提升机制动器
Kohlrausch magnetometer 克尔劳奇型地磁仪
kokowai 红赭土，新西兰赭石
kolm 铀煤；含铀煤结核；柯姆煤
kolm shale 铀煤页岩
Kolmogorov differential equation 科尔莫格洛夫微分方程
Kolmogorov entropy 科尔莫格洛夫熵，测度熵
Kolthoff buffer solution 考尔索夫缓冲溶液
Komata lining 柯马塔衬里
kominuter 磨矿机
kona 背风面
kone shot 聚能射孔弹
konimeter 测尘器，尘量计
konimetric method 计尘法
konimetry 空气浮尘计量学，空气浮尘计量法
koniogravimeter 微尘重力仪
koniology 微尘学
koniosis 尘肺，尘埃沉着病
koniscope 检尘器；尘粒镜
konisphere 尘圈，尘层
konstantan 康铜
Koppers Battelle launder 科珀斯-巴特尔型选煤槽
Koppers-Totzek gasifier K-T 气化炉
koppie 独山，孤山，残丘
Koran 古兰经
Korb-Pettit screen 考布派蒂特筛，不平衡轮式振动筛
Korea Atomic Energy Research Institute（KAERI） 韩国原子能研究院
Korea Institute of Geoscience and Mineral Resources（KIGMR） 韩国地质科学与矿产资源研究院
Korean Geotechnical Society（KGS） 韩国岩土工程学会
Korean Institute of Energy and Resources（KIER） 韩国能源与资源研究院
Korean Society for Rock Mechanics（KSRM） 韩国岩石力学学会

koreite 寿山石
Korfmann power loader 科夫曼型机动装载机
koroseal 氯乙烯树脂
koum 砂质沙漠，纯沙沙漠
koup 最干旱区
Kovats retention index 科法兹保留指数
Kozeny-Carman equation 科泽尼-卡尔曼方程
K-point K 点
kraft board 牛皮纸板
kraft paper 牛皮纸
Krasnopolsky's law 克拉斯诺波尔斯基定律
Krasovsky ellipsoid 克拉索夫斯基椭球
kraton 克拉通，稳定地块
kremastic water 包气带水，渗流水
Krey wave 克雷波
krezp 克里兹普岩
Kriging method 克里格储量计算法
Kriging variance 克里格方差
Kroll process 克罗尔法
Kronecker delta function 克罗内克 δ 函数
Krupp ball mill 克鲁普筛筒球磨机
Krupp tunnel boring machine 克鲁普隧道钻进机；克鲁普隧道掘进机
Krupp universal screen 克鲁普万能筛
kryometer 低温计
kryoscope 凝固点测定计
kryptoseismic 隐式地震的
krystic 冰的，冰雪的
krystic geology 冰雪地质学
krytron 弧光放电充气管
ksimoglyph 拖迹，拖曳痕
K-support K 形支架
K-truss K 形桁架
K-type anchorage K 型锚
K-type section K 型断面
Kucha Grottoes 龟兹石窟
Kula plate 库拉板块
Kullenberg sampler 库伦堡取样器，海底土取样器
kundaite 脆沥青煤，脆沥青
kunkar 钙结核；结核灰岩
kunkur 钙结核；结核灰岩
kupholite 蛇纹石
Kurllov's formula 库尔洛夫公式
kuroko deposits 黑矿型矿床
kuroko ore 黑矿矿石
kuroshio 黑潮
kuroshio countercurrent 黑潮逆流
kurtosis 峰态，峭度，尖锋值
Kuster gauge 库斯特井下压力计
kwarc 石英
kyanite 蓝晶石
kyanizing 升汞防腐
kybernetics 控制论
kyle 海峡
kymatology 波浪学
kymogram 记波图
Kyoto Protocol 京都议定书
Kyoto University 京都大学
kyr 高地，小山，山顶
kyrock 沥青质岩，沥青砂岩

L

laachite 黑云透长岩
laangbanite 硅锑锰矿
laanilite 榴铁伟晶岩
lab 实验室，研究室
lab analysis 实验室分析
lab core test 实验室岩心试验
lab flotation machine 实验室用浮选机
lab investigation 实验室测试
lab screening 室内筛选
lab setup 实验室装置
lab simulation 室内模拟
lab test 室内试验
lab test procedure 室内试验程序
label check 标记检验
label coding 标记编码
label grade 标号
label insurance 标签保险
label mark 标记
label processing 标记处理
label record 标号记录
labelled 标记的
labelled atom 标记的原子
labile zone of crust 地壳不稳定带
labilizing force 不稳定力
laboratory coal crusher 实验室型碎煤机
laboratory coefficient of permeability 实验室渗透系数
laboratory flotation experiment 单元浮选试验
laboratory flotation machine 实验室用浮选机
laboratory furnace 试验炉；实验室窑炉
laboratory pack 实验室填砂模型
laboratory roasting test 简易焙烧试验
laboratory sample divider 实验室分样机
laboratory sampling error 实验室抽样误差
laboratory simulation test 实验室模拟试验
laboratory soil mixer 土样拌和机
laboratory soil test 室内土工试验
laboratory soil testing 土质化验
laboratory vane test 室内十字板剪切试验
labradite 拉长岩；钙钠斜长石
labradophyre 拉长斑岩
labradorite 拉长石；富拉玄武岩；闪光拉长石
labradorite anorthosite 拉长斜长岩
labradoritite 拉长岩
laccolite 岩盖；岩盘；岩株
laccolithic dome 菌状穹隆
laccolithic mountain 岩盘山
lace 缝合物；断面擦痕上的刻槽；束带
laced beam 空腹梁
lacerating machine 拉力试验机；切碎装置
lacet road 盘山道；回旋道路
lacework 岩石中网状脉
lacquer deposit 漆状沉积
lacroixite 锥晶石；晶锥石
lacullan 黑沥青灰岩

lacustrine 湖泊的；栖湖的；湖上的；湖成的
lacustrine basin 湖成盆地
lacustrine carbonate 湖相碳酸盐
lacustrine clay 湖积黏土
lacustrine deposit 湖泊沉积
lacustrine facies 湖相
lacustrine formation 湖泊建造
lacustrine landform 湖成地形
lacustrine muck 湖积腐泥
lacustrine peat 湖积泥炭
lacustrine sedimentation 湖相沉积
lacustrine sedimentology 湖泊沉积学
lacustrine soil 湖积土
lacustrine source rock 湖相源岩
lacustrine terrace 湖成阶地
lacustrine water balance 湖泊水平衡
ladar 激光雷达
ladder drilling 梯式钻眼
ladder drilling method 架梯钻眼法
ladder excavator 链斗式挖土机
ladder fault 阶梯断层
ladder lode 梯状矿脉
ladder scraper 皮带装料铲运机
ladder sollar 梯子平台
ladder suspension system 挖掘架吊挂系统
ladder trencher 多斗式挖沟机
ladder type trencher 梯式挖沟机
ladder vein 梯状矿脉
ladder wall 梯式加筋锚定墙
lade hole 底板上浅集水洞
lag 槽面焊板；落后，延迟，移后，滞后；板桩；外套
lag breccia 残留角砾岩
lag fault 残留断层
lag gravel 滞留砾石
lag pile 套桩
lag time of cutting 岩屑滞后时间
lagengneiss 层状片麻岩
lagged 有背板的；绝热的；滞后的，延迟的
lagging board 背板
lagging edge 后沿；下降边
lagging of pile 桩箍
lagging plank 背板
lagging slab 背板
lagging water 滞留水
lagonite 硼铁矿
lagoon 潟湖；环礁湖
lagoon deposit 潟湖沉积
lagoon facies 潟湖相
lagoon flat 潟湖礁坪
lagoon floor 潟湖底
lagoon margin 潟湖边缘
lagoon phase 潟湖相
lagoon plain 潟湖平原
lagoon reef margin 潟湖礁边缘

lagoon sediment	泻湖沉积物
lagoon shelf	泻湖礁架
lagoon slope	泻湖坡
lagoonal mud	泻湖软泥
lagooning	污水储存；污泥储存
lagoonlet	小泻湖
lagoonside	泻湖陆侧地
lagoriolite	钠榴石
laharic breccia	火山泥流角砾岩
lahnporphyry	霓角斑岩
laid length	敷设长度
laihunite	莱河矿；高铁橄榄石
lair	洞穴；泥潭
laired	泥土堵塞的
laitance	水泥浮浆
laitance coating	浮浆层
laitance layer	翻沫层
lakarpite	含霞钠闪正长岩
lake asphalt	湖沥青
lake asphaltite	湖沥青岩
lake bank slope	湖岸岸坡
lake bank terrace	湖岸阶地
lake basin	湖盆
lake basin deposit	湖盆沉积
lake bed	湖成层；湖沉积
lake biscuit	湖积藻饼
lake bottom	湖底
lake clay	湖成土
lake cliff	湖蚀崖
lake dammed by rockfall	岩崩湖
lake delta	湖泊三角洲
lake deposit	湖沉积
lake drainage	湖泊流域
lake due to landslide	滑坡湖
lake eroded cave	湖蚀洞穴
lake floor	湖底
lake loess	湖积黄土
lake marl	湖底泥灰岩
lake marsh	湖沼
lake mud	湖泥
lake of marine origin	海源湖
lake ore	沼铁矿
lake peat	湖成泥炭
lake pitch	湖沥青
lake sediment	湖沉积物
lake seiche	湖震；湖面波动；假潮
lake shoreline	湖岸线
lake side	湖滨
lake stage	湖水位
lake strand zone	湖岸带
lake strandline	湖岸线
lake succession	湖成层序
lake survey	湖泊测量
lake terrace	湖阶地
lake valley	湖谷
lakebed	湖积层；湖底
lamb and slack	含泥末煤；废煤
lambertite	硅钙铀矿；斜硅钙铀矿
lamboanite	暗色混合岩
lambskin	劣质无烟煤
lamellar	纹层状；薄层状
lamellar coal	页状煤；片状煤
lamellar eutectic	层状共晶
lamellar fracture	叶片状断口
lamellar intergrowth	片晶连生
lamellar layer	层纹状岩层
lamellar mesh texture	片状网眼结构
lamellar plane	层面
lamellar pore	叶片状孔隙
lamellar prop	薄层状支柱；薄板状；多咬合面摩擦支柱
lamellar roof	叠层式屋顶
lamellar structure	层状结构
lamellar tearing	层状拉裂
lamellated	层状的
lamellated material	片状材料
lamellation	叶理；薄层状
Lame's constant	拉梅常数；λ常系数
lamina	层状体；叶片，薄层，薄板
lamina form network	平行细脉带
lamina not parallel to the roof	与巷道顶成倾角的直接顶
lamina parallel to roof	与巷道顶平行的直接顶
laminar	薄片状的，层状的，板状的
laminar bedding	纹层；层状矿床
laminar cavitation	片状空泡
laminar displacement	层状驱替；层状位移
laminar flow	层流
laminar flow structure	层状流动构造
laminar formation	层状地层
laminar fracture	成层断裂
laminar fusain	薄片状丝煤
laminar layer	层流层；薄层
laminar model	层流模型
laminar motion	层流；层状运动
laminar region	层流区
laminar separation	层流分离
laminar skin friction	层流表面摩擦
laminar structure	层状结构；薄片状结构；纹层状构造；层压结构
laminarization	层化；层状
laminate	将锻压成薄片，分成薄片；薄片制品，层压制件
laminate material	层压材料
laminate structure	片状结构
laminated	分层的；叠层的
laminated arch	叠层拱
laminated beam	叠层梁
laminated bearing	层压支承
laminated briquette	分体式型煤；层式型煤
laminated brush	叠片电刷
laminated clay	层状黏土；纹层状黏土；纹泥
laminated coal	叠层煤；层状煤
laminated construction	叠层结构
laminated core	叠片铁心；层状岩心
laminated fabric	层压织物；叠层织物，胶合织物；层状结构
laminated fiber board	层压纤维板
laminated fiber wallboard	层压纤维墙板
laminated form	片形
laminated fracture	层状断裂；片状断口
laminated glass	夹层玻璃

laminated ground	纹层地层
laminated guide	叠层罐道
laminated layer	薄片层
laminated limestone	层纹状石灰岩；叠层石灰岩
laminated magnet	叠片磁铁
laminated material	层压材料
laminated metal	叠层金属；层状金属
laminated moor	层状泥炭
laminated plastic	层压塑料
laminated pole	叠片极；薄片极
laminated rock	层状岩；纹层岩；页岩
laminated sands	层状砂质夹层
laminated slate	层状板岩
laminated soil	层状土
laminated spring	叠板弹簧；板弹簧；板簧
laminated structure	纹理构造；纹层构造；片状结构；层状结构
laminated subbituminous coal	层状半烟煤
laminated texture	纹层结构
lamination	成层，分层；纹理；薄层，薄片
laming	薄层；薄板
laminite	细复理岩
laminogram	X射线分层照片
laminography	层析X射线照相法；X射线分层摄影法
lamosite	湖成油页岩
lamp battery	矿灯蓄电池
lamp bonnet	火焰安全灯网上覆罩
lamp bracket	壁灯架
lamp cabin	矿灯房
lamp cup	三脚架承灯器
lamp gauze cap	火焰安全灯的网罩
lamp house	矿灯房
lamp key	安全矿灯保险闩
lamp key lock	矿灯锁闩
lamp picker	安全矿灯调焰器
lamp pole	灯柱；灯杆
lamp post	路灯柱
lamp room	矿灯房
lamp signal	信号灯；灯光信号
lamp station	灯房
lamp tender	矿灯工
lamp test	灯法定硫试验；灯泡试验
lamphole	灯孔
lamping	用紫外线灯勘探
lampitude	同态振幅谱
lampman	矿灯管理员
lamppost	灯柱
lamprobolite	玄武角闪石
lamproite	钾镁煌斑岩类；钾镁煌斑岩
lamproom	矿灯房
lamprophyllite	闪叶石
lamprophyre	煌斑岩
lamprophyric	煌斑岩状的
lamproschist	煌斑片岩
lamprosyenite	煌斑正长岩
lanarkite	黄铅矿
lance	固定泥心的铁杆
lance boom	杆式钻臂
lance cutting	氧炬切割
lance pipe	喷管；喷枪杆
lance point	钎子尖
lance port	吹风孔；吸尘孔
lance system	矛式系统
lancet arch	尖拱
land abutment	桥台；岸墩
land accretion	填筑土地
land aggradation	地面增高；地面加积
land allocation	拨地
land amelioration	土壤改良
land application	土地利用
land appraisal	土地估价；土地鉴定
land asphalt	劣质沥青；地沥青
land bank	土地备用区；土地储备；预留地带
land barrier	陆障
land bedding	土地分层耕作
land betterment	土壤改良
land block	陆块；地块
land board	土地局，国土局
land boundary	地界
land boundary map	地类界图
land boundary plan	土地界线图
land boundary record	土地界线记录
land boundary survey	土地界线测量
land bridge	陆桥；旱桥
land caisson	陆地沉井；陆上沉井
land carrying capacity	土地负荷能力
land classification	土地分等
land clearance	清理场地；房屋拆除；清除地面树木
land clearing	清理土地
land collapse	地陷；地面塌陷
land compass	测量罗盘
land configuration	地形
land conservation area	陆上保护区
land contract	土地契约
land cover mapping	大地覆盖显示图
land crack	地裂缝
land creep	地表蠕动
land crossing	陆上横道
land deformation	地面形变
land deterioration in mining area	矿区土地破坏
land drain	地面排水沟；土地排水沟
land drainage	土地疏干；地面排水
land drainage network	土地排水网
land dredge	陆地挖掘机
land dredger	挖土机；挖泥机
land driver	陆上打桩机
land element	地面单元
land erosion	土地侵蚀
Land evaluation	土地评价
land facies	陆相
land fall	地崩
land feature	地物；地貌
land fill	土地填筑；填土；填地
land filtration	土地浸润
land filtration irrigation	土地浸润灌溉
land for crosshead	十字头填料压盖
land formation	陆地建造；开拓土地；平整土地
land geophone	陆上检波器；地震检波器
land gobi desertization	戈壁化
land grader	平地机；推土机

land grading	平整土地
land index	土地指数
Land Information System	土地信息系统
land levelling	土地平整；陆地水准测量
land liable to flood	漫滩；易淹地区
land line	土地界线；陆线
land main	陆上水管
land mark	界标；陆标
land movement	土体移动
land operation	陆上作业
land pebble	陆地磷灰岩砾
land pier	海岸防波堤；陆堤
land pipeline	陆地管线
land pitch	劣质沥青；地沥青
land planning	土地规划
land planning survey	土地规划测量
land plat	土地图；地籍图
land pollution	陆地污染；土壤污染
land preparation	整地
land processing	陆上处理
land production	土地开发
land production programme	土地开发计划
land productivity	土地生产力
land qualification	土地定级
land quota	土地配额
land reclamation	土地开垦；采区复田
land record	土地记录
land reserve zone	土地预留区；土地储备；专用土地
land resistance	土壤阻力
land resource	土地资源
land resource survey	土地资源调查
Land Resources Information System	土地资源信息系统
land restoration	土地恢复
land rig	陆上钻机
land rock	天然磷钙石
land sale colliery	供应当地的煤矿；本地销售煤矿
land sale siding	当地售煤场岔线
land sediment	陆地沉积物
land seismic source	陆上地震震源
land shooting	陆上放炮
land side	背水侧
land side slope	背水坡；内坡
land slide	滑坡；地崩
land slide mass	滑坡体
land slide survey	滑坡测量；滑坡调查
land slide warning device	滑坡报警器
land slope	土地坡度；斜坡
land smoothing	土地细整
land status	土地类别；土地性质
land strip	剥除表土
land subsidence	地面沉降；地层陷落；地基下沉；地表下沉
land survey	土地调查
land surveyor	陆地测量员；土地测量师
land swamping	土地沼泽化
land the casing	下放套管到孔底
land tie	地锚拉杆
land upheaval	地面隆起
land weight	地压；覆盖层压力
land yielding	地面塌陷
landbased	陆基的
landcreep	土地蠕变；滑坡；滑塌
landerite	蔷薇榴石；蔷薇钙铝榴石
landfall	山崩；滑坡
landform	地形；地貌
landform analysis	地形分析；地表形态分析
landform element	地形要素
landform evolution	地貌演化
landform of lake	湖泊地貌
landform unit	地貌单元
landforms of glacial erosion	冰蚀地貌
landholding	拥有土地，占有土地；土地所有的，占用土地的
landing	罐笼装卸台；斜井口地面平台井口井底；楼梯平台罐笼座
landing block	井底罐笼承接梁
landing bottom	井筒出车场
landing depth	套管；下放深度；提升深度
landing head	油管头
landing joint	联顶节
landing nipple	坐放短节
landing pier	栈桥码头；靠岸码头；桥台；起货码头
landing pitch	路面
landing platform	装卸平台
landing point	罐笼装卸点
landing quay	靠岸码头
landing ring	联顶环
landing shaft	提煤井；出煤井
landing stage	井筒装卸台；井筒出车场
landing stage of scaffold	脚手架平台
landings	出煤量；上市量
landmark feature	方位物
landmark image	地标像
landsat	陆地卫星；地球资源探测卫星；地球遥感卫星
landsat band	卫星频带
landsat catalog microfiche	陆地卫星缩微胶片目录
landsat colour ratio composite image	陆地卫星彩色比值合成图像
landsat coregistered principal component image	陆地卫星等注记主分量图像
landsat coverage	陆地卫星视界
landsat data	陆地卫星数据
landsat data reception	陆地卫星数据接收
landsat discriminant analysis image	陆地卫星判别分析图像
landsat image	陆地卫星图像
landsat imagery	陆地卫星影像
landsat mosaic	陆地卫星影像镶嵌图
landsat multispectral scanner	陆地卫星多波段扫描仪
landsat orbit	陆地卫星轨道
landsat principal component image	陆地卫星主分量图像
landsat sensor	陆地卫星传感器
landsat solar azimuth	陆地卫星太阳方位角
landsat system	陆地卫星系统
landseismic crew	陆上地震队
landslide	滑坡；山崩；塌方
landslide analysis	滑坡分析
landslide and avalanche	崩滑
landslide area	塌方区；滑坡区
landslide body	滑坡体

landslide breccia	滑坡角砾岩；山崩角砾岩
landslide classification	滑坡分类
landslide coast	地滑海岸
landslide control	滑坡控制；塌方控制
landslide criterion	滑坡判据
landslide dam	堰塞坝
landslide debris	滑坡泥石；滑坡碎屑；山泥倾泻的泥石
landslide deformation stage	滑坡变形阶段
landslide development	滑坡发育
landslide disaster	滑坡灾害
landslide disaster prevention	滑坡灾害防治；山泥倾泻灾害防治
landslide distribution	滑坡分布
landslide drumlin	滑坡鼓丘
landslide due to engineering	工程滑坡
landslide effect	滑坡影响
landslide erosion	滑坡冲蚀
landslide evaluation	滑坡评价
landslide fissures development process	滑坡演化过程
landslide forecast	滑坡预测
landslide from slide accumulation	滑坡堆积体滑坡
landslide from talus slides	崩塌堆积体滑坡
landslide hazard	滑坡危险；滑坡灾害
landslide hazard assessment	滑坡危害性评估
landslide in flat rock bed	近水平岩层滑坡
landslide investigation	滑坡调查；滑坡勘查
landslide lake	山崩湖；崩滑堰塞湖
landslide mass	滑坡体
landslide modelling	滑坡建模
landslide monitoring	滑坡监测
landslide premonition	滑坡前兆
landslide prevention skirting	护脚
landslide preventive treatment	滑坡防治
landslide protection wall	抗滑挡墙
landslide protection works	滑坡防护工程
landslide revival	滑坡复活
landslide risk management	滑坡风险管理
landslide scar	崩塌残痕；滑坡残壁；滑坡断崖
landslide shape	斜坡形态
landslide shear surface	滑坡剪切面
landslide spatial prediction	滑坡空间预测
landslide spring	滑坡泉
landslide stability	滑坡稳定性
landslide surface	滑坡面；山泥倾泻面
landslide surge	山崩涌浪
landslide surveillance	滑坡监视
landslide susceptibility index	滑坡敏感性指数
landslide temporal prediction	滑坡时间预测
landslide terrace	坍塌阶地
landslide topography	滑坡地形
landslide traces landform	滑坡遗迹景观；滑坡遗迹地形
landslide track	坍塌径迹
landslide valley	滑坡谷
landslide volume	滑坡体积
landslide zone of slope	边坡塌滑区
landward	向陆的；朝陆的
landward side	向陆地一侧
landwaste	岩屑；风化石；砂砾
langbanite	硅锑锰矿
langbeinite	无水钾镁矾
langer bridge	椰格尔式桥；刚性系杆柔性拱桥
langite	蓝铜矾
Langmuir constant	兰米尔常数
lansfordite	五水碳镁石；多水菱镁矿
lantern coal	烛煤
lap	余面；搭接，互搭；提升机滚筒上的绳圈；盖板，盖片；研磨
lap seam	重叠岩层；搭接缝
lap winding	叠绕法
lapiaz	喀斯特沟；溶槽；岩溶沟
lapidification	石化；岩化
lapidify	成岩；石化
lapidite	块熔凝灰岩
lapiesation	溶沟化作用
lapillus	火山砾
lapis lazuli	天青石；青金石
lapout	超覆尖灭；沉积尖灭
lapped type	重叠式；折叠，围绕
lapper	研磨机
lapping machine	精研机
Laray viscometer	莱雷黏度计
larboard	左舷
lardalite	歪霞正长岩
larderellite	硼铵石
lardite	冻石
large adit	主平硐
large amplitude theory	大振幅理论
large arm caliper	大臂井径仪
large ballast undercutting cleaners	大型全断面清筛机
large bore	大直径钻孔
large bored shaft	大口径钻井
large borehole	大孔径钻孔
large capacity well	高流量井；出水量大的井
large clod	大土块
large coal	大于规定粒度的煤；大块级煤
large deflection	大挠度；大垂度
large deflection theory	大挠度理论
large deformation	大变形
large deformation of surrounding rock	围岩大变形
large destructive earthquake	破坏性大地震
large diameter bored pile	大直径螺旋钻孔桩
large diameter borehole	大口径钻孔
large diameter boring	大口径钻探
large diameter evaporation pan	大口径蒸发皿
large diameter hole drilling machine	大直径钻孔机
large diameter pipe	大直径管
large displacement analysis	大位移分析
large displacement matrix	大位移矩阵
large drill	重型凿岩机
large earthquake	大地震
large elastic-plastic deformation	弹塑性大变形
large groundwater project	大型地下水工程
large groundwater supply	大型地下水供水
large high velocity landslide	大型高速滑坡
large hole diameter perforation	大直径炮眼
large hole drilling	大井眼钻井
large igneous province	大火成岩区
large low angle fold	大型平缓褶皱
large muck	大块崩落岩石
large navigable river	大型通航河流

large ore	大块矿石
large panel building	大板建筑物
large panel construction	大墙板建造法；大模板建造法
large platform elevator	大平台式升降机
large producer	高产井
large radius elbow	大弧弯头
large region yield	大面积屈服
large riprap	大块乱石护面
large scale blast	大爆破
large scale earthquake	大规模地震
large scale field test	大规模野外试验；大型现场试验
large scale map	大比例尺图
large scale mining	大规模开采
large scale ore car	大型矿车
large scale pumping	大规模抽水
large scale soil test	大尺寸土工试验，大规模土工试验
large scale system theory	大系统理论
large scale topographic mapping	大比例尺地形测图
large scale yielding	大范围屈服
large shock	大地震
large stove	大炉级无烟煤
large strain	大应变
large strain amplitude	大应变振幅
large structure system	大型结构体系
large structure testing laboratory	大型结构试验室
large temporary engineering	大型临时工程
large twin compressor	大型双联压缩机
large wall panel	大型墙板
large wave tank	大型模拟波浪水池
larrying	灌浆；薄浆砌筑法
larsenite	硅铅锌矿
larvikite	歪碱正长岩
lasecon	激光转换器
laser acquisition	激光探测
laser aligner	激光准直仪
laser alignment	激光定向；激光校直
laser alignment system	激光准直系统
laser alignment with zone plate	波带板激光找正
laser altimeter	激光测高仪
laser and flat bed plotter	激光绘图台
laser bathymetric survey	激光测深
laser beam drilling	激光束钻井
laser beam effect	激光束效应
laser beam image reproducer	激光束图像重现器
laser beam microprobe	激光探针
laser beam setup	激光照准器
laser blasting	激光引爆
laser bombardment	激光照射；激光轰击
laser carrier	激光载波
laser ceilometer	激光云幂仪
laser channel marker	激光航道标
laser collimator	激光准直仪
laser depth sounder	激光测深仪
laser detection	激光探测
laser diagnostic	激光诊断
laser diode	激光二极管
laser direction indicator	激光指向器
laser distance measurement	激光测距
laser distance measuring instrument	激光测距仪
laser distance meter for observing satellites	人卫激光测距仪
laser doppler velocimeter	激光多普勒测速仪
laser drill	激光钻孔；激光打孔
laser drilling	激光钻孔；激光打孔
laser dual operation	激光器运转
laser geodimeter	激光测距仪
laser gravimeter	激光重力仪
laser guidance unit	激光导向机构；激光导向装置
laser guide	激光导向；激光指向
laser guide instrument	激光导向仪
laser guide of vertical shaft	立井激光指向
laser guiding system	激光导向系统
laser gyroscope	激光陀螺仪
laser heater	激光加热器
laser height accuracy	激光测高精度
laser hologram	激光全息图
laser hologram photography	激光全息照相
laser holography	激光全息术
laser illuminated imaging	激光照明成像
laser illuminator	激光照明器
laser image	激光图像
laser induced breakdown spectroscopy	激光诱导击穿光谱
laser instrument	激光测定仪
laser interferometer	激光干涉仪
laser interferometry	激光干涉测量
laser level	激光水准仪
laser leveling instrument	激光水准仪
laser light vibration sensor	激光振动传感器
laser line control	激光定向控制
laser line follower	激光线跟踪器
laser local oscillator	激光本机振荡器
laser locator	激光定位器
laser mapping	激光测绘
laser mapping system	激光测图系统
laser methanometer	激光沼气测定仪
laser microanalysis	激光显微分析
laser microprobe	激光探针
laser microprobe emission spectrometer	激光探针发射光谱术
laser microprobe mass spectrometer	激光探针质谱测定
laser microprobe spectrochemical analysis	激光显微探针光谱化学分析
laser microspectral analyzer	激光显微光谱分析仪
laser microspectrography	激光显微光谱分析
laser navigation	激光导航
laser oscillation	激光振荡
laser oscillator	激光振荡器
laser photoreconnaissance package	激光侦察装置
laser pickoff	激光传感器
laser pickoff unit	激光接收组件
laser plasma	激光等离子体
laser plotter	激光绘图仪
laser plumbing	激光投点
laser plummet apparatus	激光铅垂仪
laser positioning	激光定位
laser probe	激光探针
laser radar	激光雷达
laser radar altimeter	激光雷达测高仪
laser range finder	激光测距仪
laser range finding	激光测距

laser rangefinder	激光测距仪；激光定位仪
laser ranger	激光测距仪
laser ranging	激光测距
laser ranging equipment	激光测距仪
laser ranging sensor	激光测距传感器
laser reconnaissance	激光勘测
laser reconnaissance system	激光勘测系统
laser recording	激光记录
laser remote sensing	激光遥感
laser rotational sensor	激光转动传感器
laser scan	激光扫描
laser scanner	激光扫描仪
laser scanning	激光扫描
laser scattering	激光散射
laser scatterometer	激光散射计
laser sensor	激光传感器
laser sounder	激光测深仪
laser strain seismometer	激光应变地震计
laser strainmeter	激光应变计
laser swinger	激光扫平仪
laser theodolite	激光经纬仪
laser topographic position finder	激光地形仪
laser tracking	激光跟踪
laser transit	激光经纬仪
laser tripler monochrometer	激光三联单色光度计
laser velocimeter	激光测速计
laser vibrometry	激光振动测量技术
laserlog	激光记录
lash	鞭打冲击拉紧束缚；清除岩石；清除矿石
lash back	清除岩石；清除矿石
lashing	清除岩石；清除矿石；向放矿道放石
lashing equipment	装岩设备
lashing gang	装载组
lashing shift	矿石清理班；岩石清理班
lassenite	新鲜粗玻岩；英安玻璃
last appearance datum plane	末现基准面
last casing	末层套管
last climate cycle	过去气候周期
last deglacial	末次冰期的；冰消期的
last eustatic cycle	上个海平面升降周期
last glaciation	最后冰期
last interglacial	末次间冰期；上个间冰期
last interglacial epoch	末次间冰期；上个间冰期
last lift	最后采掘层
last pour	上一浇注段高
lasurapatite	青磷灰石
lasurite	青金石
latch	闩，弹键，弹簧锁；上插锁，销住
latch jack	打捞器
latching	闩住；锁住；系住
late Albian tectonism	晚阿尔布期构造运动
late Archean	晚太古代
late bearing prop	缓增阻支柱
late Cenozoic uplift	晚新生代隆升
late diagenesis	晚期成岩作用
late formation	晚期形成
late geological time	晚近地质时期
late glacial	晚冰期
late glacial deposit	后期冰川沉积
late Jurassic	晚侏罗纪
late magmatic ore deposit	晚期岩浆矿床
late mature valley	晚成年期谷
late Miocene	晚中新世
late Neocathaysian system	新华夏构造体系
late optimum	晚气候适宜期
late origin theory	晚期成油说
late Paleozoic	晚古生代
late Permian	晚二叠纪
late Pleistocene	晚更新世
late Quaternary	晚第四纪
late stage diversion	后期导流
late Tertiary	晚第三纪
late Triassic	晚三叠纪
lateglacial	后冰期
latent defect	潜在的建筑缺陷；潜在的建筑瑕疵
latent geothermal site	隐伏地热区；潜在地热区
latent hardening	潜在硬化
latent instability	潜在不稳定性
latent magma	潜在岩浆
latent magnetization	潜磁化力
latent solvent	潜溶剂；惰性溶剂
latent source rock	潜在生油气岩
latent strain	潜应变
latent stress	潜应力
later arrival	续至波
later modules of subgrade reaction	侧向基床反力模量
later thrust of roof	顶板倾斜作用力
later waterflood	晚期注水；晚期水驱采油
lateral	横的；侧向的；走向的
lateral abrasion	侧向磨蚀
lateral accretion	侧向增生；侧向加积
lateral anomaly	侧移异常
lateral attraction	侧向引力
lateral bar	侧向沙坝
lateral bearing capacity of piles	桩的横向承载力
lateral cofferdam	侧向围堰
lateral compression	侧向挤压；横向挤压
lateral compression coefficient	侧压系数
lateral compression test	侧压试验
lateral compressive force	侧向压力
lateral cone	侧火山锥；寄生火山锥
lateral confinement	侧限
lateral confining pressure	侧限压力
lateral conveyor	平巷输送机
lateral core	井壁取芯
lateral corer	井壁取芯器
lateral coring	井壁取样；侧向取样；侧向取岩芯
lateral coring gun	井壁取芯器
lateral corner	侧角
lateral corrosion	侧向磨蚀；河岸磨蚀
lateral coverage	侧向覆盖范围；侧向覆盖率
lateral crater	侧火山口
lateral creep	侧向蠕动
lateral crevasse	侧向裂缝
lateral curve	梯度测井曲线
lateral cutting	侧向开挖
lateral cutting ability	侧向切削能力
lateral cutting head	横轴式截割头
lateral daylighting	侧面采光
lateral deflection	侧向挠曲；横向变位

lateral deformation 横向形变
lateral deposit 侧边沉积物
lateral deposition 侧向沉积作用
lateral developing 水平开拓；巷道开拓
lateral development 水平开拓
lateral deviation 水平偏斜
lateral diagonal 横向斜撑
lateral digging force 横向切割力
lateral dimension 水平尺寸；横向尺寸
lateral dipmeter 梯度电极地层倾角仪
lateral direction 水平方向
lateral dispersion 横向弥散；侧向分散
lateral dispersion coefficient 横向弥散系数
lateral displacement 水平位移，横向位移
lateral ditch 边沟
lateral divergence 横向发散
lateral drainage 横向排水；侧向排水
lateral drainage ability 侧向排水能力
lateral drift 横向漂移；侧平巷；石巷
lateral drill 水平钻
lateral drilling 侧向钻眼；侧向凿岩；侧向钻井
lateral dynamic response 横向动力响应
lateral earth movement 横向地面运动
lateral earth pressure 侧向土压力
lateral earth pressure coefficient 侧向土压力系数
lateral edge of sliding 滑坡侧边
lateral electric sounding 侧向电测深
lateral en echelon arrangement 侧向雁行排列
lateral enrichment 侧富集
lateral entrainment 侧向掺混
lateral erosion 侧蚀作用；侧蚀，旁蚀，侧向侵蚀
lateral error 横向误差；水平误差
lateral error of traverse 导线横向误差
lateral eruption 侧喷发
lateral excursion 横向偏移
lateral expansion 横向膨胀；侧向膨胀
lateral exploration 侧向探勘
lateral extension 横向伸长
lateral extension fracture 侧向延伸破裂
lateral facies variation 横向相变
lateral fault 横移断层；走向分离断层；侧向断层
lateral flexure 横向弯曲
lateral flow 横流；旁流
lateral flushing hole 侧冲水孔
lateral force 横向力；侧向力
lateral force coefficient 侧向力系数
lateral friction 侧面摩擦；侧向摩阻力
lateral glacial stream 侧面冰川河
lateral gradation 侧向过渡；侧向渐变；横向分级作用
lateral gradient 横向坡降
lateral growth 横向生长
lateral guide 侧向导向
lateral hanging valley 侧悬谷
lateral heterogeneity 横向不均匀性
lateral impact 横向冲击
lateral in situ stress 原地侧应力
lateral incaving 侧向淘刷
lateral inclinometer 横向倾斜计
lateral inertia 横向惯量
lateral inertia force 横向惯性力

lateral inflow 侧向来水；旁侧入流
lateral inhomogeneity 横向不均匀性
lateral integration 横向结合
lateral intergradation 横向渐变
lateral intersection 侧方交会
lateral isotropy 横观各向同性体
lateral kink plane 侧向扭折面
lateral lake 支流湖
lateral lap 横向重叠
lateral layer 两侧地层
lateral leaching 侧向淋滤
lateral levee lake 天然堤后湖
lateral line 支线；侧线
lateral load 侧向荷载；横向荷载
lateral load equivalent frame 侧向荷载等效框架
lateral load requirement 侧向荷载要求；侧向荷载需要量
lateral load resisting member 抗侧力构件
lateral load test 侧向载荷；侧向荷载试验
lateral loaded pile 水平荷载桩
lateral loading test of pile 桩的横向载荷试验
lateral log 侧向测井；梯度电极系测井
lateral logging 侧向测井；聚焦测井
lateral migration 侧向迁移；横向运移
lateral mixing 侧向混合
lateral modulus of subgrade reaction 侧向地基反力系数；侧向基床反力模量
lateral moraine 侧碛；侧斜冰碛层
lateral motion 侧向运动；横向运动
lateral movement 横向移动；沿层错动；平行层理错动；横向位移
lateral opening 走向平巷；边孔
lateral oscillation 侧振荡；横向摆动
lateral overlap 横向重叠；旁向重叠
lateral parametric excitation 侧向参数激振
lateral pattern 横移模式；侧生式
lateral permeability 水平方向渗透率
lateral pile load test 桩的侧向荷载试验
lateral plan 侧视图
lateral planation 旁夷作用；侧夷作用
lateral plate 侧板
lateral play of wheel set 轮对横动量
lateral positioning groove 横向定位凹槽
lateral pressure 侧压力；横向压力
lateral pressure apparatus 侧压仪
lateral pressure coefficient 侧压力系数
lateral pressure difference 横向压差
lateral pressure ratio 侧压系数
lateral profile 横断面
lateral ramp 侧断坡
lateral reaction 侧反力
lateral recharge increment 侧向补给增量
lateral reflection 侧反射；侧波
lateral refraction 旁折光；侧向折射
lateral reinforcement 箍筋；横向钢筋；横向加固
lateral resistance 侧向抗力；横向弹性
lateral resisting capacity 抗侧力能力
lateral resisting structure 抗侧力结构
lateral resisting system 抗侧力体系
lateral resistivity tool 横向电阻率下井仪
lateral resolution 横向分辨率

lateral response	侧向反应	lateral wave	侧面波；横向波
lateral restrain	侧向约束	lateral wedge	侧向楔体
lateral restraining pressure	侧限压力	lateral wing	横巷
lateral restraint	侧向约束；横向约束	lateral yield	侧向屈服
lateral restraint reinforcement	侧限加固	lateral zonation	侧向分带性；横向分层性
lateral restriction	侧向约束力	lateral zoning	侧向分带
lateral ridge	侧脊	laterally composite basin	横向复合盆地
lateral rigidity	侧向刚度；侧向刚性	laterally confined specimen	受侧限的试样
lateral road	走向平巷；大巷	laterally loaded pile	侧向受荷桩
lateral roadway	走向平巷；运输巷道	laterite	红土，铁矾土；砖红壤性土
lateral roof movement	顶板横向移动	laterite cement	红土胶结物
lateral runout	侧向运动；横向偏离	lateritic	红土化的；风化的
lateral scouring	侧向冲刷	lateritic cement	红土胶结物
lateral seal	侧向封闭	lateritic clay	红黏土
lateral section	横截面；横断面；侧截面；横剖面	lateritic gravel	红土砾石
lateral seepage	侧向渗流	lateritic ore	红土矿石
lateral seepage recharge	侧向渗透补给	lateritic soil	铝红壤；红土；砖红壤性土
lateral segment	水平井段	lateritite	红土碎屑岩；次生红土
lateral seismic coverage	横向地震覆盖	lateritization	红土化
lateral sewer	侧向暗沟	lateritoid	砖红壤状土壤；似红土
lateral shear	横向剪切	laterization	红土化作用；红土，砖红壤
lateral shear factor	侧向剪力系数	laterlog	侧向测井
lateral shift	侧向变位；侧向位移	laterlogging	侧向测井
lateral shrinkage	侧向收缩；横向收缩；侧向收缩率	laterolog resistivity	侧向测井视电阻率
lateral slide	侧向滑坡；侧向滑动	laterology	钻孔电阻记录；侧向测井记录
lateral soil pressure	侧向土压力	lath	板条；护井板桩
lateral spread	侧向扩张；横向扩张	lath and plaster	板条抹灰
lateral springing of pile	桩侧向弹动	lath and plaster ceiling	板条抹灰吊顶
lateral stability	横向稳定性，侧向稳定性	lath ceiling	板条顶棚
lateral stability assessment	侧向稳定性评估；横向稳定性评估	lath crib	基环的背板框架
lateral stabilizer	侧向稳定器	latialite	蓝方石
lateral stacking	侧向堆积	latite	安粗岩；二长安山岩
lateral stiffness	侧向刚度	latite andesite	安粗安山岩
lateral stopper	横向限动块；侧向阻挡器	latite basalt	安粗玄武岩
lateral strain	侧向应变；横向应变	latitude	纬度；黄纬；范围；幅度
lateral strain indicator	侧向应变指示器	latitude and longitude	经纬度
lateral stream	侧向水流	latitude correction	纬度改正；纬度校正
lateral stress	侧应力；横向应力	latitude determination	纬度测定
lateral strut	横撑；侧向支撑	latitude difference	纬度差；纵距
lateral support	侧向支撑；横向承托	latitude effect	纬度效应
lateral support system	横向承托系统	latitude line	纬度线
lateral support works	横向承托工程	latitude migration	纬向迁移
lateral supported point	侧向支撑点	latitude of epicenter	震中纬度
lateral surrounding rock pressure	侧向围岩压力	latitude of pedal	底点纬度
lateral tensile stress	侧向张应力	latitude of reference	基准纬度
lateral terrace	侧向阶地	latitude parallel	黄纬圈
lateral thinning	侧向变薄	latitude shift	纬向移动
lateral thrust	侧压力；横向推力；横平推断层	latitude small circle	纬线小圆
lateral tie	横向拉件；侧面拉条	latitudinal direction	纬度方向
lateral ties of column	横向柱箍	latitudinal drain	横向排水槽
lateral tilt	旁向倾角；侧向倾斜	latitudinal grade	横向坡度
lateral torsional buckling	横向扭曲；侧扭屈曲	latitudinal inertia force	纬向惯性力
lateral turnout	单开道岔	latitudinal motion	纬向运动
lateral underpinning	侧向托换	latitudinal tectonic subzone	东西构造亚带
lateral valley	纵向谷	latitudinal zoning	横向分带
lateral vibration	横向振动	latiumite	硫硅碱钙石；硫硅石
lateral view	侧视图	latosol	红化土类；砖红土；砖红壤
lateral volcanic cone	侧火山锥	latosolization	砖红壤化
lateral waterflood	侧向注水	latrobite	淡红钙长石
		lattic boom	桁架式悬臂

lattice	格子；晶格；点阵	laurite	硫钌锇矿；硫钌矿
lattice anchor	格构锚固	laurvikite	歪碱正长岩
lattice arch	格构拱；钢花拱；格棚拱	lautarite	碘钙石
lattice bar	格条	lautite	辉砷铜矿
lattice base	点阵基底	lava	熔岩
lattice beam	桁架横梁	lava ash	熔岩灰；火山灰
lattice clay	网纹黏土	lava ball	熔岩球
lattice column	格构柱	lava banding	熔岩条带
lattice column mast	栅格柱轻便架	lava bed	熔岩层
lattice drainage	格状排水系统	lava blister	熔岩泡
lattice frame	格构框架	lava bomb	火山弹
lattice framing	格式框架	lava breccia	熔岩角砾岩；火山角砾岩
lattice girder	格子梁；格构梁；钢桁架	lava cascade	熔岩瀑布
lattice grid structure	网架结构，格构梁结构	lava cauldron	熔岩凹集地
lattice reinforcement	格构配筋；格构加固	lava cavern	熔岩洞
lattice retaining wall	网格式挡墙；网格式护岸	lava column	熔岩柱
lattice scattering	晶格散射	lava cone	熔岩锥
lattice spacing	格子间距	lava crater	熔岩火口
lattice steel pole	桁架钢结构支柱	lava desert	熔岩荒地
lattice steel tower	格式结构钢塔	lava dome	熔岩丘；熔岩穹丘
lattice strain	晶格应变	lava eruption	熔岩喷溢
lattice strut	格构支撑	lava extrusion	熔岩喷发
lattice texture	格子结构	lava field	熔岩地；熔岩荒野
lattice theory	晶格理论	lava flood	熔岩泛流
lattice truss	格构桁架	lava flow	熔岩流
lattice type reinforcement	桁架型钢筋	lava flow plateau	熔岩流高原
lattice vibration	晶格振动	lava fluidity	熔岩流动性
laubanite	白沸石	lava fountain	熔岩泉
laueite	劳埃石；黄磷锰铁矿	lava hole	熔岩穴
laugenite	奥长闪长岩	lava lake	熔岩湖
laumontite	浊沸石	lava landscape	熔岩景观
laumontite facies	浊沸石相	lava levee	熔岩堤
launching chamber	发送筒	lava morphology	熔岩形态
launching cradle	下水架	lava neck	火山颈；熔岩颈
launching dolly	浮筒	lava pillow	熔岩枕；枕状熔岩
launching equipment	下水设备	lava pit	熔岩坑
launching erection	曳进架设法	lava plain	熔岩平原
launching girder	曳进吊梁机	lava plateau	熔岩高原
launching lug	发射环	lava plug	熔岩栓；熔岩塞；熔岩颈
launching method	拖拉架设法	lava rheology and morphology	火山熔岩流变性和形态
launching nose	曳进导梁	lava sack	熔岩囊
launching ramp	管线下水坡道	lava scarp	熔岩崖
launching stand	管线下水坡道	lava scum	熔岩沫
launching trap	投放井；发送筒	lava sea	熔岩海
launching valve	发送阀	lava sheet	熔岩被；熔岩席
launder	流水槽；流槽；洗涤，洗熨，耐洗	lava shield	熔岩盾；盾状火山
launder classifier	流槽分级机	lava slag	熔岩渣
launder overflow	洗煤槽溢流	lava soil	熔岩土壤
launder powder	刚铝石粉；铝氧粉	lava spine	熔岩刺；熔岩脊
launder recirculation	洗煤槽循环	lava spring	熔岩泉
launder refining	流槽精炼	lava stalactite	熔岩钟乳石
launder refuse	洗煤槽废渣	lava stalagmite	熔岩石笋
launder screen	洗槽筛；流槽筛	lava streak	熔岩脉
launder separation process	流水槽比重分选法	lava stream	熔岩流
launder splitting	流槽分支	lava structure	熔岩构造
laundering	槽洗；槽选	lava tear	熔岩泪
laundry bowl	洗煤槽；流洗槽	lava terrace	熔岩阶地
laundry box	洗煤槽	lava tree mould	熔岩释
laurdalite	歪霞正长岩	lava tube	熔岩管；熔岩通道
laurionite	羟氯铅矿	lava tumulus	熔岩鼓包；熔岩瘤；熔岩丘

lava tunnel 熔岩隧道；熔岩洞
lava volcano 熔岩火山
lava wedge 熔岩楔
lava well 熔岩井
laval 熔岩的
lavatory 金矿洗选厂；盥洗室
lavendulan 铜钴华
lavenite 褐锰锆矿；锆钽矿；钠钙锆石
lavialite 残拉角闪岩
lavic 熔岩的
lavrovite 铬透辉石；钒辉石
law of correlation of facies 岩相相关律
law of fossil sequence 化石层序律
law of initial horizontality 原始水平定律
law of injection 注入规律
law of laminar filtration 层流渗透定律
law of mobile equilibrium 流动平衡定律
law of molecular concentration 分子浓度定律
law of strata superposition 地层层序律
law of stream gradient 河流坡降定律；河流比降定律
law of stream length 河长定律
law of stream slope 河流坡降定律；河流比降定律
law of superposition 地层层序律；叠覆律
law of surface relationships 地层分布关系律
law of symmetry 对称定律
law of tectonic layer superposition 构造层叠置定律
law of unequal slope 山坡不对称定律
lawrowite 钒辉石
lawsonite 硬柱石
laxite 机械碎屑岩
laxmannite 磷铬铜铅矿
laxolithus 宽轭石
lay fault 滞后断层
lay of land 地形；地貌
lay out 绘样设计；布置
lay out of round 成组炮眼的排列
lay out of route 放线
lay ramp 敷管滑道
lay system 金属矿留矿回采法
lay the line 延伸输送管；延伸线路
laydown area 下管区
laydown drill pipe 甩钻具
laydown machine 铺设机
layer 层，层次；地层，岩层；夹层
layer aluminosilicate 层状铝硅酸盐
layer anisotropic media 层状各向异性介质
layer bound vein 层控矿脉
layer breccia 层状角砾岩
layer construction 分层铺筑
layer corrosion 层状腐蚀
layer cross flow 层间窜流
layer cut 分层拉沟
layer flow 层流
layer impedance 层阻抗
layer in layer 层内层
layer loading 分层装车法
layer mining 分层开采
layer of coal 煤层
layer of overburden 覆盖层
layer of shale 夹层

layer of surface detention 地表滞流层
layer of wooden blocks 方木层；木块层
layer replacement 分层替代
layer silicate 层状硅酸盐
layer stripping 剥层法
layer structure 层状结构；层状构造；分层构造
layer structure factor 成层构造因子
layer summation method 分层总和法
layer system 层系
layer trenching 分层掘沟法
layered and blocked excavation 分层分区开挖
layered aquifer 层状含水层
layered clay 层状黏土
layered column chromatography 层叠式柱型色谱法
layered crust structure 层状地壳构造
layered earth model 层状地球模型
layered echo 层状回波
layered elastic medium 层状弹性介质
layered gneiss 层状片麻岩
layered half space 层状半空间
layered heading 分层巷道
layered hydrate 层状水合物
layered hydrous aluminum silicate 含水的层状硅酸铝
layered intrusion 层状侵入体
layered map 分层着色地形图
layered media 层状介质
layered medium 层状介质
layered periodic medium 成层周期性介质
layered permafrost 层状永冻土
layered plate 层状板
layered reservoir 层状储集层
layered rock 层状岩石
layered sand 层状砂
layered sequence 层状岩序
layered series 成层岩系
layered soil 成层土
layered strata 层状地层
layered structure 层状构造
layered system 层状体系
layering 分层；层理；似层理；层状构造；成层作用
layering effect 分层效应
layering number 成层数
layering of firedamp 瓦斯成层；沼气分层
layers of shale 页岩层
layerwise adsorption 层状吸附
layerwise summation method 分层总和法
laying 布置；铺筑；敷设；建筑
laying barge 铺管驳船
laying buoy 铺管浮筒
laying caterpillar 铺管用的履带拖拉机
laying crew 铺管班
laying depth 埋置深度
laying gang 铺管队
laying in one ditch 同沟敷设
laying mode of route 线路敷设方式
laying of a borehole 钻孔布置
laying off 放样
laying off curve 曲线测设
laying out 定线；放样；布置；规划
laying pipe 管道敷设

laying point	照准点
laying progress chart	铺管进度表
laying rate	铺管速度
laying ship	铺管船
laying speed	铺管速度
laying the foundation stone	奠基
laying tractor	履带式吊管机
laying winch	铺管绞车
layout area	蓝图区；详细设计区
layout balance of exploratory engineering	勘探工程分布图
layout chart	观测系统图；布置图；线路图；施工流程图
layout diagram	布置图；配置图
layout drawing	草图；配置图
layout machine	测绘缩放仪
layout of blast holes	炮眼布置
layout of construction site	施工场地布置
layout of construction work	施工总平面图
layout of control network	控制网布设方案
layout of geophysical prospecting work	物探工作布置图
layout of hydroproject	水利枢纽布置；水利工程布局
layout of level	阶段水平布置
layout of plan	编制设计书；设计方案
layout of power station	电站布置
layout of reinforcement	加固方案
layout of the shaft	井筒布置方式
layout sheet	总布置图；总规划图；总平面图
layout survey	施工放样测量；定线测量
lazurite	天青石；青金石；琉璃
lazy board	带柄管端支撑板
lazy chain	吊桶卸载钩链
lazy rope	有钩的翻卸吊桶绳
leach	沥滤；浸滤；溶滤
leach hole	溶洞；落水洞
leach material	沥滤物料
leach mineral	淋溶矿物
leach out	渗漏；浸出；溶滤
leach water	溶渗水；淋滤水
leachability	可浸出性；淋溶力；可滤取
leachable	可滤取的，可滤去的
leachate	渗滤污水；淋滤液；淋滤物
leachate collection drain	渗滤污水渠；渗滤污水收集喉管
leachate collection layer	渗滤污水收集层
leachate collection well	积液井
leachate contamination	溶滤污染；淋滤污染；渗滤污染
leachate drainage layer	渗滤污水排水层
leachate level	渗滤液水位
leachate recirculation	渗滤液回灌
leachate treatment	渗滤液治理
leached alkali soil	淋溶碱土
leached brine	淋滤卤水
leached capping	淋滤覆盖层；淋滤露头
leached chernozem	淋溶黑钙土
leached horizon	淋滤层；淋滤层
leached layer	淋滤层；淋滤层
leached outcrop	淋滤露头
leached profile	淋溶剖面
leached saline soil	淋溶盐土
leached soil	淋溶土
leached surface	溶滤面；溶滤带
leached zone	溶滤带；淋滤带
leaching	沥滤，浸出；淋滤作用；溶滤作用；淋溶作用
leaching agent	浸出剂；溶浸剂
leaching aquifer	渗漏含水层
leaching cavity	浸析腔；浸出洞
leaching channel	沥滤槽
leaching concentrate	浸出精矿
leaching effect	淋溶效应
leaching factor	淋洗因数
leaching horizon	淋溶层
leaching layer	淋溶层
leaching method	淋洗法
leaching mineral deposit	淋滤矿床；淋积矿床
leaching mining	溶浸开采
leaching operation	淋洗作业
leaching out	淋失；滤失；浸出
leaching pit	渗水坑；污水坑
leaching plant	沥滤厂；沥滤车间
leaching process	沥滤溶解采矿法；淋溶过程
leaching process to desulfurize coal	煤的浸出脱硫
leaching rate	浸出率
leaching requirement	冲洗需水量
leaching residue	溶滤残渣
leaching solution	溶浸液
leaching tank	浸出槽；沥滤槽
leaching test	淋溶试验
leaching vat	沥滤瓮；沥滤桶
leaching well	渗水井；污水渗井
leaching zone	淋滤带
lead	导火线的超前长度；移前，超前，导向；导线，引线；铅
lead amalgam	铅汞合金
lead angle	超前角；导程角
lead annealing	铅退火
lead antimony alloy	铅锑合金
lead ash	铅灰
lead azide	叠氮化铅
lead ball	铅球；铅锤
lead base alloy	铅基合金
lead bath furnace	铅浴炉
lead bit	导向钻头
lead blade	导向刀翼
lead block	导块
lead block compression test	铅块压缩试验
lead block crushing test	铅块挤压试验
lead block expansion	铅块膨胀
lead block expansion test	铅块膨胀试验；铅块延伸度试验
lead block test	铅铸试验；铅块试验
lead cement slurry	先行水泥浆；超前水泥灌浆
lead collar	导向钻铤
lead concentrate	铅精矿；精铅
lead curve	导曲线；测深曲线
lead cylinder compression test	铅柱压缩试验
lead damp course	铅皮防水层；防潮层
lead damper	铅阻尼器
lead drilling hole	引线孔
lead drum	前导滚筒
lead error	导程误差

lead extrusion device	铅挤出装置
lead gasket	铅垫片
lead glance	方铅矿
lead gutter	点火导火线
lead hole	导孔
lead hydronitride	叠氮化铅
lead joint	填铅接合
lead line	测深绳；铅锤线
lead method	铅法年龄测定
lead plate test	铅板试验
lead rail	导轨
lead sheath cable	铅包电缆
lead sinker	铅锤
lead slime	铅泥
lead slurry	前导浆
lead solder	铅焊料
lead sounding	水铊测深
lead splitter	一种安全导火线点火器
lead stabilizer	导向稳定器；铅稳定剂
lead stamp	铅模
lead stope	上向梯段第一分层回采工作面
lead sulfide	硫化铅
lead survey plug	铅熔塞；铅水准标石
lead water structure	导水结构
lead weight	吊锤；铅锤
lead zinc ore	铅锌矿
leaded gasoline	含铅汽油
leader pin	导杆
leader pipe	水落管
leadhillite	硫碳酸铅矿
leading band	仰斜巷道
leading beacon	主指向标，定向标；引导示标；铅锤测深法
leading bogie	前转向架
leading coefficient	首项系数
leading displacement edge	驱替前缘
leading edge of slope	坡体前缘
leading effect	超前效应
leading face	超前工作面
leading fossil	标准化石
leading frame	主构架
leading heading	超前平巷
leading hole	导孔；起爆孔
leading hooked fibre	前弯钩纤维
leading imbricate fan	前导叠瓦扇
leading light	导灯；导航标灯
leading pile	导桩；定位桩
leading place	超前工作区；超前巷道
leading stope	超前工作面
leading wall	导墙
leading wave	先导波；先行波
leading wire	引线；电雷管引线；爆破母线；脚线
leady matte	含铅锍
leaf by leaf injection	层层注入
leaf clay	叶状黏土；页状黏土
leaf coal	页状煤；纸煤
leaf gold	叶片状自然金
leaf injection	间层注入；分层注入
leaf like texture	叶片状结构
leaf mining	分层开采
leaf peat	页状泥炭；页泥炭
leaf physiognomy	叶相分析
leaf shale	片状页岩
leaf spring	钢板弹簧；叶片弹簧
leafing	脱离；剥离；分层
leak	泄漏；漏洞；漏出量
leak check	检查泄漏；检查漏气
leak checking collar	检漏套管
leak checking well	检漏井
leak clamp	堵漏卡箍
leak detection	泄漏检查
leak detection method	探漏法
leak detector	检漏器，查漏仪；漏失检验器；渗漏探测仪
leak finding	查漏
leak gas	漏气
leak in the casing	套管渗漏
leak locater	测漏器；泄漏定位器；检漏仪
leak oil line	泄油管
leak out	漏失，泄漏
leak path	渗漏通道
leak piping	渗漏管涌
leak scene	泄漏现场
leak sealing	封漏
leak test	密封性试验；泄漏检验
leak through	渗漏
leak to ground	漏入地下
leak tolerance	漏损的允许限度
leak tunnel	渗水隧洞
leakage	泄漏；漏出量；渗漏
leakage anomaly	渗漏异常
leakage around dam abutment	绕坝渗漏；坝肩周围渗漏
leakage boundary	渗漏边界
leakage coefficient	泄漏系数
leakage control	防漏；泄漏控制
leakage detection device	检漏装置
leakage detector	漏水探测仪
leakage distance	泄漏距离
leakage factor	渗漏系数；越流因数；泄漏因数；漏磁系数；漏水系数
leakage flow	泄漏流；漏流
leakage halo	渗漏晕
leakage impedance	漏泄阻抗
leakage loss	漏失量；渗漏损失
leakage noise	直达干扰；泄漏噪声
leakage of air	漏气
leakage of dam foundation	坝基渗漏
leakage of grout	冒浆；薄浆流失；薄浆漏出
leakage of mortar	漏浆
leakage path	漏电路径；渗漏途径；漏水路线
leakage percentage	漏损率
leakage protection	防渗；堵漏；漏电保护
leakage rate	漏水率；泄漏率
leakage reactance	漏磁电抗；漏抗
leakage recharge	越流补给
leakage reduction	泄漏抑制
leakage resistance	泄漏电阻；泄漏阻力
leakage survey	检漏
leakage test	漏水试验；越流试验；渗漏试验
leakage water	渗透水；越流水；渗漏水

leakage zone 漏泄带；漏失层
leaker 漏失井；泄漏构件；漏失孔
leakiness 泄漏；易漏
leaking 渗漏的；漏出的
leaking coefficient 漏损系数
leaking mode 波能漏失；泄漏模式
leaking recharge 越流补给
leaking stoppage 堵漏
leaking system 越流系统
leaking well 渗漏井
leakless 不漏的
leak-off pressure 滤失压力；裂开压力
leakproof 密封的；防漏的
leakproof of cofferdam 围堰防渗
leakproof seal 防漏密封
leakproofness 密封性；密闭性；气密性
leaky 越流的；渗漏的
leaky aquifer 越流含水层；渗漏含水层
leaky aquifer system 越流系统
leaky artesian aquifer 越流性自流含水层
leaky coefficient 越流系数
leaky confined aquifer 越流性承压含水层
leaky confining bed 越流性隔水基底
leaky confining stratum 越流性隔水层
leaky duct 泄漏管道
leaky duplex 漏顶式双冲构造
leaky effect 越流效应
leaky factor 越流因数；阻越流系数
leaky formation 越流地层
leaky foundation 漏水地基；渗漏地基
leaky joint 漏接；不紧密接合
leaky layer 渗漏层；越流层
leaky mode 泄漏振型；泄漏模式
leaky phreatic aquifer 越流性潜水含水层
leaky pipe 渗漏管
leaky reservoir 泄漏水库；泄漏储层
leaky seam 松接缝；漏水缝
leaky transform fault 渗透型转换断层；裂缝型转换断层
leaky wave 泄能波
lean 倾斜，偏向；瘦的；贫瘠的，少的
lean cannel coal 瘦烛煤
lean cement 水泥含量少的水泥浆
lean clay 瘦黏土；贫黏土；低塑性黏土
lean coal 瘦煤；变质程度高的烟煤
lean concrete 少灰混凝土；低强度混凝土；低标号混凝土
lean gas 贫气；贫煤气；低热量燃气
lean halvings 贫矿
lean lime 贫石灰
lean material 选矿后的废石
lean mix concrete 贫搅拌混凝土；少灰混凝土
lean mixture 贫燃料混合物
lean mortar 低标号砂浆
lean oil 贫油
lean oil shale 贫油页岩
lean ore 低品位矿石；贫矿石
lean over 倾斜
lean protore 贫矿胎；贫瘠胚胎矿；贫原生矿
lean residue gas 贫残气

lean soil 瘦土
lean solution 稀溶液
leaning instability 倾斜不稳定性
leaning tower 斜塔
leaning wheel grader 斜轮式平路机
leap 跳跃；错位；突变；断层
leapfrog 蛙式夯土机；动力打夯机
leapfrogging 跳点钻探；跳步法；间断勘探
leaping divide 变动的分水岭
leaping frog 蛙式打夯机
learies 采空区
lease 租地；租约；租借权；采油租地
lease boss 矿区采油领班
lease condensate 伴生气凝析油
lease conditions 租契条件；批租条件
lease crude 矿产原油；当地原油
lease for special purpose 特殊用途土地契约；特殊用途楼宇契约
lease in perpetuity 永租权
lease line 矿场集输管线
lease modification 契约修订
lease pipeline 租借管道
lease tank 油田储罐
lease water 矿产水
leased equipment 租用的设备
leased interest 租赁权益
leased land 已批租土地
leasehold of land 土地批租
leasehold system 批租制度；租地制度
least permeability system 低渗透系统
least permeable medium 低渗透介质
least radius of gyration 最小回转半径
least significant bit 最低有效位
least significant character 最低有效位组
least significant difference 最低有效位差
least strain axis 最小应变轴
least waterholding capacity 最小挡水能力；最小持水容量
leastone 层状砂岩
leat 沟渠；水道；露天水渠
leather packing 皮垫
leather seal 皮密封垫圈
leather washer 皮衬；皮密封垫片
leathering 皮制密封填料
leavings 残渣，剩余物，残留物
lechatelierite 焦石英
leck 硬黏土；石状黏土；致密黏土
leckstone 沸绿岩
lecontite 钠铵矾
lectostratotype 选定标准地层剖面；选层型
lectotype 选模式；选型
ledeburite 莱氏体；杂砷锌钙铁矿
ledeburitic steel 莱氏体钢
ledge blasting 突岩爆破
ledge excavation 岩面开挖
ledge matter 脉质
ledge rock 基岩；坚硬岩盘；突出岩架
ledge wall 底板，下盘
ledger wall 下盘；底帮
ledikite 伊利石

ledmorite 榴霞正长岩
lee 背风面
lee breakwater 背风防波堤
lee coast 背风岸
lee face 背风面
lee side 冰川岩石的前面；背风面
lee slope 背风坡
lee trough 背风槽
leelite 正长石；肉红正长岩
lees 沉积物；沉淀物；残渣
leeside 背风面
leeuwfonteinite 歪闪正长岩
leeward side 背风面
leeward slope 背风坡
LEFM 线弹性断裂力学
left elevation 左视图；左侧立面图
left hand turnout 左开道岔
left lane 左边道；左通路
left margin 左侧刃
left marginal bank 左侧边滩；左边岸
left toe 遗留岩坎
left-hand crusher 左式破碎机
left-hand drill pipe 左螺纹钻杆
left-hand safety connector 左旋安全接头
left-hand square thread 左旋方形螺纹
left-hand string 左扣钻柱
left-hand tool joint thread 左旋钻杆接头螺纹
left-hand turning 左向旋转
left-hand turnout 左侧岔道
left-handed crystal 左晶
left-handed direction "左旋方向"
left-handed polarization 左旋偏光；左旋极化
left-handed rotation 左向旋转
left-lateral displacement 左旋位移
left-lateral fault 左旋断层
left-lateral motion 左旋运动
left-lateral movement 左旋运动
left-lateral sense 左行
left-lateral separation 左行离距；左移间距
left-lateral shear 左旋剪切
left-lateral shear movement 左旋剪切运动
left-lateral slip 左旋滑动
left-lateral slip fault 左旋滑动断层
left-lateral strike-slip fault 左旋走向滑动断层；左旋平移断层
left-lateral transcurrent fault 左旋横推断层
left-oblique 左倾斜的
left-oblique slip fault 左斜滑断层
left-reverse-slip fault 左旋逆滑断层
leg base anchorage 井架支腿底脚的固定
leg drill 支腿钻机
leg load gauge 立柱载荷计
leg load testing adaptor 液压支柱载荷试验接头
leg member 支柱
leg nipple 支柱铜头
leg piece 柱
leg pipe 大气冷凝管；冷凝器气压管
leg resistance 支柱阻力
leg restoration 支柱复位
leg seal 支柱密封

leg tie bracket 立柱固定托架
leg well cluster 腿柱井组
leg wire 分支导线；电雷管导线；脚线
leg yield pressure 支柱屈服压力；支柱额定工作阻力
legrandite 基性砷锌石
leguminous structure 豆荚状构造
lehiite 白磷碱铝钙石；白磷碱铝石
lehmanite 长英火成岩；糟化石
lehrzolite 两辉橄榄岩
leidleite 玻质英安岩；玻质流纹英安岩；微斑安山岩；流安松脂岩；安山质松脂岩
leifite 白针柱石
lembergite 绿蒙脱石；绿胶岭石
lemniscate shield support 双纽线式掩护支架；四连杆式掩护支架
lencostite 中碱性火山岩
lengaite 碳酸熔岩；钠质碳酸熔岩
length of a working section 采区长度
length of cantilever 悬臂长度
length of connecting rod 连杆长度
length of dam 坝长
length of draw of subsidence 地面塌陷区与井下相应采空区边点的水平距离
length of drill rod 钻杆长度
length of ducting 风管节段长度；管道节段长度
length of embedment 埋入长度
length of faulting 断层作用长度
length of flame proof joint 隔爆接头长度；防火接头长度
length of grade 坡长
length of grade section 坡段长度
length of lay 捻距；节距
length of over end sill 底架长度
length of overland flow 地面漫流长度；地表径流长度；坡面流长度
length of penetration 贯入长度
length of restraint 嵌固长度
length of run 滑走长度
length of screen 过滤器长度
length of shear plane 剪断面长度；剪切面长度
length of sheet flow 漫流长度；层流距离
length of stable fissure 稳定裂纹长度
length of switch rail 尖轨长度
length of tank 罐体长度
length of tracklaying 铺轨长度
length of underframe 底架长度
lengthening 拉长；延伸
lengthening piece 延接段
lengthening zone 拉伸带
lengthways 纵向的，沿长度方向的
lengthwise section 纵剖面；纵向截面
lengthwise with the strike 沿走向的
lenne porphyry 角斑岩
lenrodiscontinuity 光面不整合
lenrohydrodialeima 光面水下不整合
lens 透镜；扁豆状矿区；沉积透镜体
lens of clay 黏土透镜体
lens of gravel 凸体镜砾石层
lens out 尖灭
lensing 透镜状
lens-like structure 透镜状构造

lensoid	透镜状；似透镜状
lensoid body	扁豆状体
lensoid ore body	透镜状矿体
lensometer	焦度计
lenticle	扁豆体；透镜体
lenticular	透镜状的；凸镜状的；扁豆状的；饼状的
lenticular arch	双叶拱
lenticular beam	组合梁；鱼形梁
lenticular bedding	透镜状层理
lenticular body	透镜体
lenticular deposit	透镜状油藏；透镜状矿床
lenticular girder bridge	鱼腹式梁桥
lenticular intercalation	扁豆状夹层
lenticular lithologic reservoir	透镜状岩性油气藏
lenticular mass	扁豆体；透镜体
lenticular oil pool	透镜状油藏
lenticular ore body	透镜状矿体
lenticular orebody	透镜状矿体
lenticular pyrite	透镜状黄铁矿；扁豆状黄铁矿
lenticular reservoir	透镜状储集层
lenticular sand	透镜状砂层
lenticular structure	透镜状构造；扁豆状构造
lenticular trap	透镜状圈闭
lenticular truss	鱼腹式桁架
lenticular twin	扁豆状双晶
lenticular vein	扁豆状矿脉；透镜状矿脉
lenticularity	透镜状产状；扁豆状产状
lenticule	小透镜状层
lenticulite	豆状岩；扁豆熔灰岩
lentiform	透镜状；扁豆状
lentiform beam	组合梁；鱼形梁；鱼腹式梁
lentil	小扁豆层；小扁豆体；扁豆状矿体
leonardite	风化褐煤
leonhardite	黄浊沸石
leonite	钾镁矾
leopard limestone	豹皮灰岩
leopard rock	豹斑岩
leopardite	豹斑石英斑岩
lepidoblastic	鳞片变晶状的
lepidoblastic texture	鳞片变晶状结构
lepidochlorite	鳞绿泥石
lepidocrocite	纤铁矿
lepidogranoblastic	鳞片花岗变晶的
lepidogranoblastic texture	鳞片粒状变晶结构
lepidolite	锂云母
lepidomelane	铁黑云母；铁锂云母
lepidometer	纤维间摩擦力测定仪
lepidophytotelinite	鳞木结构镜质体；鳞木结构凝胶体
leptite	浅粒岩；长英质片麻岩；长英质麻粒岩
leptochlorite	鳞绿泥石
leptogeosynclinal facies	薄地槽相
leptogeosyncline	薄地槽
leptokurtic distribution	尖峰态分布
leptokurtosis	峰态
leptometer	比黏计
lepton	轻子；轻粒子
leptonematite	硬锰矿
leptothermal	中热带与浅热带之间的矿床；次中深热液的
leptothermal deposit	亚中温热液矿床
leptynite	变粒岩；长英粒变岩；长石白粒岩
lerbachite	杂硒铅汞矿
lermontovite	水锆铀磷钙石
less active fault	活动不明显断层；弱活动性断层
less permeable	弱透水的；低渗透性的
less permeable layer	弱透水层；低渗透层
lesser ebb	小落潮流
lessingite	钙硅铈石
lessivation	淋洗作用；淋洗过程
lessive	淋洗的；洗涤液的
lessive pseudogley	淋洗假潜育土
lestiwarite	辉闪正长细晶岩
letdown	排出；松弛；减低
lethal	致命的；致死的
lethal concentration	致死浓度
lethal concentration low	最低致死浓度
lethal dose	致死剂量
lethal gas	致死毒气
lethal percentage	致死气体百分含量
lethal quantity	致死量
lethality	致死性；死亡率
letter	证书，许可证
letter of acceptance	中标通知书；同意书；接纳书
letter of advise	通知书
letter of application	申请书
letter of appointment	委任书
letter of assignment	转让书
letter of assurance	保证书
letter of attorney	委任书；授权书
letter of authorization	授权书
letter of award	授标函
letter of commitment	承诺书
letter of compliance	合约完成证明书；完工证；完成规定事项证明书
letter of confirmation	确认书
letter of exchange	换地证
letter of guarantee	保函；保证书
letter of hypothecation	抵押证书
letter of indemnity	赔偿保证书
letter of instruction	通知信
letter of intent	意向书
letter of invitation to tender	招标书
letter of lien	留置权书
letter of modification	建筑牌照规约修订书；批约修订书；批地条款修订书
letter of no objection to occupy	不反对入住书
letter of notice	通知书
letter of patent	专利证书；特许证书
letter of recommendation	推荐信
lettering of chart	海图注记；图表注释
leuchtenbergite	淡绿绿泥石
leucite	白榴石
leucite basalt	白榴玄武岩
leucite phonolite	白榴响岩
leucite syenite	白榴正长岩
leucite tephrite	白榴碱玄岩
leucite trachyte	白榴粗面岩
leucitite	白榴石岩
leucitite basanite	白榴碧玄岩
leucitite tephrite	白榴碱玄岩
leucitoid	似白榴岩；白榴石状的

leucitolith 白榴石岩；纯白榴岩
leucitonephelinite 白榴霞岩
leucitophyre 白榴斑岩
leucobasalt 淡色玄武岩
leucocrate 淡色岩；浅色岩
leucocratic 淡色的
leucocratic dyke 淡色岩脉
leucocratic rock 浅色岩
leucodiorite 淡色闪长岩
leucogranite 淡色花岗岩
leucolite 淡色岩
leucopetrite 蜡褐煤
leucophanite 白铍石
leucophoenicite 水硅锰矿
leucophyre 浅色斑岩；糖化辉绿岩
leucopyrite 低砷铁矿；斜方砷铁矿
leucoscope 光学高温计；感光计
leucosome 浅色体
leucosphenite 淡钡钛石
leucotephrite 淡灰玄岩
leucoxene 白钛石；金红石
leurodiscontinuity 具规则面的不整合
levecon 信号电平控制；液面控制
levee 天然堤；冲积堤
levee back 堤背
levee base 堤底
levee body 堤体
levee breach 土堤决口；堤坝裂口
levee building machine 堤坝建筑机械
levee core wall 堤心墙
levee crest 堤顶
levee delta 堤状三角洲
levee drain valve 防火堤排水阀
levee foundation 堤基
levee front 堤防临河面
levee gate 堤坝闸门
levee grade 堤顶纵坡
levee lake 天然堤湖
levee maintenance 堤防维护
levee protection 堤防保护
levee raising 堤坝加高
levee ramp 堤上坡道
levee revetment 堤防护岸
levee ridge 天然堤河床
levee safety 堤防安全
levee slide 堤身滑坡
levee slope 堤坡
levee sloughing 堤身崩坍
levee sluice 堤上泄水闸
levee spacing 堤间距
levee system 堤防系统
levee toe 堤脚
levee undermining 堤底淘刷
levee widening 堤坝培厚；堤坝加宽
leveed 天然成堤的，堤成的；淤填的；冲填的
leveed bank 筑堤护岸；天然堤岸；淤积沙滩
leveed channel 堤成渠
leveed pond 堤成池
level 水平面；级，能级，层；水准
level axis 水平轴

level bar 水平尺
level bearing 走向线
level bubble 水准器；水准器气泡
level circuit 水准闭合环
level coal 平煤层
level commutator 水准管
level conductivity coefficient 水位传导系数
level constant 水准仪常数
level control 水准控制；水位控制
level control valve 液位控制阀
level controller 水平控制器；自动液面调节器
level correction 水准器改正；海平面修正
level course 沿走向方向；水平巷道；水平层；煤层等高线
level crossing 平面交叉
level crossing watchman 道口看守工
level cut slice 水平分层
level cutting 水平掘槽；水平采掘；平巷掘进
level datum 水平基准面
level density 能级密度；电平密度
level detector 水位计
level development 水平开采
level deviation 水准器气泡偏差
level device 定平装置
level diagram 电平图
level difference 标高差距，高差；电平差
level distribution 能级分布；水平分布
level drift 水平巷道
level drive 沿走向掘进；平巷掘进
level ellipsoid 水准椭球
level error 水准误差
level fall 水位下降
level fold 水平褶皱
level gaging 液面测定
level gauge 水准仪，水平规，水准指针；液面指示器，液位计
level grade 平坡
level ground 平地
level height 采矿阶段高度
level I fracture toughness test 水平Ⅰ断裂韧度测试
level II fracture toughness test 水平Ⅱ断裂韧度测试
level indicating lamp 水平指示灯
level indicator 水平指示器，水平规，水准指针；液面指示器，液位计
level instrument 测量水平仪；水准仪；液面测量仪
level interval 水平间距水平垂高
level life 水平开采年限；阶段性开采年限
level line 水准线
level luffing crane 平臂起重机
level mark 水准面标记；基准面标记
level mark on column 柱上的水准点标志
level measurement 电平测量；标高测量；液位测量，水平面测量
level meter 电平表；水准测量仪
level mining 中段开采；水平开采
level note 水准测量记录
level of addressing 定址级数
level of compensation 补偿深度
level of foundation 基础标高
level of ground water 地下水水位

level of hydraulic pressure	承压水水位
level of no strain	无应变层位
level of optimization	最优化水平
level of oxygen	含氧量
level of risk	风险程度
level of safety	安全度
level of saturation	饱和水面；潜水面
level of significance	显著性水平
level of stretch	拉伸程度
level of subsoil water	地下水位；底土水位
level of the hydraulic pressure	承压水位；液压压力水平
level of uncertainty	不确定性程度
level off	夷平
level pad	平衡缓冲液
level pillar	水平煤柱；水平矿柱
level plan	水平平面图；主要巷道平面图
level plane	水准面
level position	水平位置
level pressure	定压；恒压
level pressure control	基准压力调节
level recession	水位下降
level recorder	液面记录器；水位记录仪
level reduction	水准折算
level regulator	液面调节器
level road	主平巷；阶段平巷
level rod	水准标尺
level scheme	能级图
level seam	水平煤层；平矿层
level sensitivity	气泡灵敏度
level sensor	液面传感器
level service	水平运输；阶段运输
level shoe	水准尺垫
level slicing	分级，分度；灰度分割；密度分割；水平分割
level spacing	水平间距，水平高度
level staff	水准标尺
level stretch of grade crossing	道口平台
level string	水准线
level surface	等势面；水准面；水平面；液面
level survey	水准测量
level switch	液位控制开关
level testing instrument	水准检定器
level theodolite	水准经纬仪
level theory	平衡理论
level thrust	水平推力
level track	水平轨道
level transmitter	液面发送器
level trier	水准检定器
level tube	水准管
level tube axis	水准管轴
level tube bubble	水准管气泡
level up	平衡
level vial	水平仪气泡
level workings	水平巷道
leveler	水准测量员；校平器；平地机
leveling	水准测量
leveling adjustment	水准调整；电平调整；调平
leveling along the line	沿线整平；沿纵向校平
leveling base	基准面；水准面
leveling board	水平板
leveling branch	支水准路线
leveling bubble	水准气泡
leveling closure	水准测量闭合差
leveling correction	水准测量改正
leveling course	平整层
leveling cylinder	调平液缸
leveling datum	水准测量基准；水准测量数据
leveling error	调平误差
leveling grade	平整度
leveling instrument	水准测量仪
leveling jack	调平用千斤顶
leveling line	水准线
leveling machine	薄板矫平机
leveling method	平衡法
leveling motor	调平马达
leveling network	水准网
leveling of high precision	高精度水准测量
leveling of model	模型置平
leveling origin	水准原点
leveling peg	水平桩；水准小桩
leveling point	水准点
leveling pole	水准标尺
leveling process	水准测量
leveling rod	水准尺
leveling staff	水准标尺
leveling survey	水准测量
leveling system	调平系统
leveling valve	高度调整阀
leveling work	水准测量工作；找平；填平地坪
leveller	水平测量员；校平机；墨斗
levelling	水准测量；高程测量；校平，调平
levelling adjustment	水准调整；电平调整
levelling bar	平煤杆；校平杆
levelling box	水准仪
levelling course	整平层；水平层
levelling instrument	水准仪；水平尺
levelling member	平衡杆；平衡构件
levelling peg	水准标桩
levelling pole	水准尺；水准标杆
levelling process	水准测量
levelling roll	矫直辊；平整辊
levelling route	水准线路
levelling screw	校平螺旋；水准螺旋
levelling staff	水准标尺
levelling stone	整平石
levelling survey	水准测量；水平测量
levelling tape	水准卷尺
levelman	水准测量员
levelness	水平度
lever	杠杆，杠，柄，操纵杆
lever and weight safety valve	杠杆重锤式安全阀
lever arm	杠杆臂
lever connecting rod	杠杆上拉杆；制动拉杆
lever control	杠杆操纵；手柄操纵
lever feeler	杆臂感应器
lever fulcrum	杠杆支点
lever jack	杠杆起重器，杠杆千斤顶
lever lift	杠杆起重机
lever lock	杠杆锁
lever of stability	稳性力臂；回复力臂；扶正力臂

lever shaft	曲轴；曲柄轴；杠杆轴
lever slide gate	杆式滑板闸门
lever switch	杠杆式开关
lever test bar	杠杆试棒
lever type gate lifting device	杠杆式启门设备
lever valve	杠杆阀
leverage	杠杆率；杠杆作用；杠杆机构
leverrierite	晶蛭石
levigation	水中细磨；水磨；研磨
levyne	插晶菱沸石
levynite	插晶菱沸石
lewisite	锑钛烧绿石
leyteite	沥青
lherzite	黑云褐闪岩
lherzolite	铬尖晶石；二辉橄榄岩；二辉橄家
liability clause	责任条款
liability for acceptance	承兑责任
liability for breach of contract	违约责任
liability for damage	赔偿责任
liability for delay	误期责任
liability for endorsement	背书责任
liability for fault	过失责任
liability for pollution	污染责任
liability of cracking	易裂性
liability of the client	委托人的责任；业主的责任
liability of the consultant	咨询专家的责任
liability to cracking	易裂性
liability to frost damage	易冻性
liability to weathering	易风化性
liandratite	铌钽铀矿
liberate water	析水
liberated gas	释放气
liberated mineral	单体矿物
liberating size	解离粒度
liberation	矿物释出；解离；释放
liberation mesh	解离网目
liberation of gas	气体释出；瓦斯放出
liberation of intergrown constituents	连生组分的解离
liberation of methane	沼气放出
liberite	锂铍石
libethenite	磷铜矿
libollite	暗沥青
licence block	许可证区块
licence to discharge	排放许可
licencee	许可证接受人；领有执照者
license	许可证；执照
licensed architect	注册建筑师
licensed contractor	注册承包商
licensed engineer	有开业执照的工程师
licensee	领有许可证者
licenser	许可证颁发人；出证方
licensing agreement	专利使用许可
licensing basis	审批基准
lick	盐沼地
lid	盖；柱帽；短顶梁
lidar	激光雷达；激光探测；激光测距；激光定位；激光定向
lidar altimeter	激光雷达测高计
lidar laser radar	激光雷达
lidar meteorology	激光雷达气象学

lie	旁道，岔道；矿脉方向
liebenerite	白霞石
liebenerite porphyry	白假霞石
liebigite	铀钙石
liebnerite	蚀霞石块云母
lierne vaulting	扇形肋穹顶
lievrite	黑柱石
life expectancy	预期寿命；预计期限；使用年限
life factor	寿命系数
life length	寿命；使用年限；使用时期
life line	生命线；安全线
life loss	建筑物寿命损失
life of a building	建筑物存在期
life of face	作业面寿命
life of lease contract	土地租约寿命
life of product	产品寿命
life of project	项目寿命期
life of well	井开采期限
life parameter estimate	寿命参数估计
life prediction	寿命预测
life preserver	救生设备
life repair cost	全使用期内修理费
life span	使用期限
life support	生命保障
life test	寿命试验
life time	寿命；操作年限；使用期限
lifeline earthquake engineering	生命线地震工程
lifeline engineering	生命线工程
lifeline network	生命线网
lifeline system	生命线系统
lifesaver	井架工安全带
lifesaving appliance	救生设施
lifesaving equipment	救生设备
lift	扬程；上升高度；电梯；升降机
lift arm	起重臂
lift between two levels	水平间垂距
lift bracing	升力拉条
lift bridge	升降桥
lift car	升降机厢
lift check valve	升举式止回阀
lift coefficient	升力系数
lift column	空气升液器压力管
lift dobby	多臂机
lift efficiency	升举性能；升力特性
lift engager	提升装置接合器
lift force	悬浮力
lift gate	升降式闸门
lift hammer	起落锤
lift head	扬程
lift height	提升高度；扬程，扬高；分段厚度
lift into position	将起吊到位
lift joint	水平张节理
lift level interval	开采水平垂高；开采层段间距
lift of pump	泵的扬程
lift platform	上升平台；平台升降机
lift plug	提升旋塞
lift pump	抽扬泵；升液泵
lift shaft	升降机井；电梯井
lift slab construction	升板法施工
lift station	扬水站

lift the floor	开凿底板	light ends fractionation	轻烃分馏
lift thickness	层厚	light ends unit	轻烃回收装置
lift tower	升降机系统；升降机塔	light fill	低填土；矮路堤
lift truck	升降台车	light grade	缓坡，小坡度
lift tube	升水管	light hammer drill	轻型钻机；轻型凿岩机
lift valve	升阀	light hilly area	微丘区
lift way	升降通道	light hole	地面塌陷洞
lift well	电梯井道；升降机竖井	light hydraulic jack	轻型液压顶柱
lift wire	提升钢丝绳	light hydraulic prop	轻型液压支柱
liftdrag ratio	升力阻力比	light hydraulic tamping machine	小型液压捣固机
lifted block	地垒；隆起断块	light line	轻质油品输送管
lifter	升降机，举扬器，起重机	light liquid hoisting	轻液提升
lifter cut	底板掏槽	light loam	轻质垆坶；轻质壤土
lifter hole	底部炮眼	light meter	照度计；光度计
lifting	提升；举起；升起	light mineral	轻矿物；浅色矿物
lifting appliance	起重机械；起重设备；吊具	light oil	轻油；低黏度油
lifting arm	起吊杆	light pitch	缓倾斜
lifting beam	起吊梁；起重臂	light pitch seam	缓倾斜煤层
lifting brake	起重制动器	light pitching	缓倾斜
lifting bridge	吊桥	light pneumatic drilling machine	轻型风动式钻机
lifting bridge for inclined shaft	斜井吊桥	light rig	轻便钻探设备
lifting cap	提升钻头的螺丝帽	light sand	松砂
lifting capacity	提升能力，起重量	light section	小型型钢
lifting case shaft	吊箱井	light service rig	清孔用轻便钻机；修井用轻便钻机
lifting casing	吊销套	light shaft	采光井
lifting chute	闸板溜槽；闸板溜口	light shale	软页岩；轻页岩
lifting condensation level	抬升凝结高度	light silty loam	轻粉壤土
lifting crane	提升起重机	light soil	轻质土
lifting cylinder	起卸液压伸缩筒	light solid	轻固相
lifting device	提升设备	light sounding test	轻便触探试验
lifting dog	钻杆升器器	light source in mine	井下光源
lifting height	拔起高度	light surface mulch	表面薄松土层
lifting hook	吊钩	light viscosity oil	低黏度油
lifting platform	升降平台；举重台	light volatile fuel	轻质易挥发燃料
lifting ram	调向油缸	light weight cement	低密度水泥
lifting table	升降台	light weight diverter	低比重转向剂
lifting tackle	起重滑车；提升滑轮	light weight fluid	轻质液体
lifting tower	竖井井架	light weight mud	低比重泥浆
lifting valve	提升式阀	light well	光井
lifting winch	提升绞车	light well point	轻型井点；真空井点
ligament	韧带	light year	光年
ligamented charge holder	带箍的炸药容器	lightening admixture	减轻剂
light blasting	轻爆破，松动爆破	lightening hole	减轻孔
light brick	轻质砖	lighter	发光器；点火器；打火器
light charge	少量装药	lightermeter	照度计
light chestnut earth	淡栗钙土	lighting arrester	避雷器
light clay	轻黏土	lighting facilities of road	道路照明设备
light coloured soil	淡色土	lighting of the site	工地照明
light component	轻组分	lighting order	点火次序；放炮次序
light concrete	轻质混凝土	lighting protection	防雷保护
light constituent	轻组分	lighting protection conductor	防雷线
light crude	轻质原油；石蜡基原油	lighting protection device	防雷装置
light cut	轻馏分	lighting switch	避雷开关
light cycle oil	轻循环油	lighting torch	点火棒
light detection and ranging	光探测和测距	lightning arrester	避雷器，避雷针
light dirt	轻矸石；小块矸石	lightning conduction	避雷网
light drifter	轻型凿岩机	lightning conductor	避雷针；避雷导线
light drill	轻型钻机	lightning earth	防雷接地
light drilling	轻型钻机钻眼；轻型凿岩机凿岩	lightning prevention design	避雷设计
light ends	轻质烃；轻馏分	lightning protection	避雷装置；防雷接地；防雷保护

lightning protection equipment	雷电防护设备
lightning protection in open pit	露天采场防雷
lightning protector plate	避雷板
lightning rod	避雷针；避雷器
lightweight aggregate	轻砂石；轻骨料
lightweight aggregate concrete	轻骨料混凝土
lightweight alumina brick	轻质高铝砖
lightweight board	轻质板材
lightweight concrete	轻混凝土
lightweight concrete block	轻质混凝土块
lightweight construction	轻质构造
lightweight cover	轻型上盖
lightweight filler	轻填料；轻质掺合料
lightweight lime concrete	轻石灰混凝土
lightweight matter in aggregate	轻物质含量
lightweight oil	轻质油
lightweight rod string	轻型抽油杆
lightweight screed	轻质砂浆
lightweight soil	轻质混合泥
lightweight steady arm	轻型稳定臂
lightweight stopper	轻型伸缩式凿岩机
ligniferous	褐煤质的
ligniferous shale	褐煤质页岩
lignification	木质化；褐煤化
lignified	褐煤化的
lignin grout	木质素浆
lignite	褐煤
lignite alkali reagent	煤碱剂
lignite shale	褐煤页岩
lignite uranium deposit	褐煤铀矿床
lignitic	褐煤的；含褐煤的
lignitic and bituminous oil shale	褐煤和沥青质油页岩
lignitic clay	褐煤黏土
lignitic coal	褐煤
lignitic material	褐煤
lignitiferous	褐煤化的；褐煤质的
lignitoid	木质状；褐煤状
lignosulfonate mud	磺化木质素泥浆
ligroin	轻石油，石油英
likelihood of failure	破坏可能性
lillianite	硫铋铅矿
liman	泻湖；溺谷；泥湾；河口
limb fault	翼部断层
limb of fold	褶皱翼
limber string	柔性钻具
limbo	中间过渡地带；安置遗弃物的地方；垃圾场
limburgite	玻基辉橄岩
lime	石灰；石灰质；氧化钙；钙质
lime alkali rock	钙碱性岩
lime and cement mortar	石灰水泥砂浆
lime base mud	灰基泥浆
lime bloom	混凝土表面起霜；石灰花
lime brick	石灰砖
lime burning	煅烧石灰
lime carbonate	碳酸钙
lime cartridge	生石灰卷筒
lime cave	灰岩洞
lime cavitation	石灰岩洞穴化
lime cement	石灰胶结料；石灰水泥
lime chloride	氯化钙
lime clay	灰质黏土
lime coal	烧石灰用煤
lime coating	石灰处理
lime column	石灰桩；石灰柱
lime compaction pile	石灰桩挤密；石灰压实桩
lime concrete	石灰混凝土
lime concretion	石灰结核；钙质结核
lime content	石灰含量
lime cream	石灰乳
lime deep mixing method	石灰系深层搅拌法
lime deposit	石灰质沉积物
lime dust	石灰粉
lime earth concrete wall	三合土墙
lime earth rammed	灰土夯实
lime earth surface	灰土地面
lime feldspar	钙长石；石灰长石
lime grout	石灰灌浆
lime gyttja	灰质腐殖黑泥；灰质湖底软泥
lime hardpan	石灰硬磐；石灰硬质地层
lime hydrate	消石灰；熟石灰
lime injection	石灰粉喷射；打石灰桩
lime iron concretion	石灰铁质结核
lime kiln	石灰窑
lime marl	灰质泥灰岩
lime metasomatism	石灰交代作用；石灰交代变质
lime mica	珍珠云母
lime milk	石灰乳
lime mixer	石灰拌和机
lime mortar	石灰砂浆
lime mud	灰泥；石灰泥
lime mudrock	灰泥岩
lime mudstone	灰泥岩
lime nanoparticle	石灰纳米颗粒
lime nodule	钙质结核
lime pan	石灰硬磐
lime phyllite	灰质千枚岩；灰质硬绿泥石
lime pile	石灰桩
lime pile method	石灰桩法
lime pit	石灰池
lime plaster	石膏；石灰泥，石灰粉饰
lime powder	石灰粉
lime pozzolanic cement	石灰火山灰水泥
lime producer	石灰岩储层生产井
lime putty	石灰膏
lime quarry	石灰石采石场；石灰石露天矿坑
lime reactivity	石灰活化性
lime requirement	石灰需要量
lime rock	石灰岩
lime rubble rock	灰质角砾岩
lime sand	石灰砂
lime sand brick	石灰砂砖
lime sand pile	石灰砂桩
lime sandstone	石灰砂岩
lime saturation degree	石灰饱和系数；石灰饱和度
lime shale	钙质页岩
lime sink	落水洞
lime slaking	石灰熟化；石灰消化
lime slurry	石灰浆
lime soda feldspar	钙钠长石
lime softening	石灰软化

lime soil	石灰土；钙质土
lime soil pile	灰土桩
lime spraying	喷刷石灰；刷白
lime stabilization	石灰加固
lime stabilized soil	石灰稳定土；石灰加固土
lime treated mud	石灰处理泥浆
lime treated soil	灰土
lime treatment	石灰处理
lime water	钙质水；石灰水
lime white	熟石灰；石灰浆
limeclast	灰岩屑；灰屑
limed	石灰处理的
limerock	灰质岩；未固结石灰岩；未固结介壳灰岩
limespar	方解石
limestone	石灰岩；石灰石；灰岩
limestone aquifer	石灰岩含水层
limestone cave	溶洞；灰岩洞
limestone consolidation	石灰岩加固
limestone decay	石灰岩衰变
limestone deterioration	石灰岩劣化
limestone diagenesis	石灰石成岩作用
limestone karst	灰岩喀斯特；灰岩岩溶
limestone log	石灰岩电极系测井
limestone pillar	灰岩柱
limestone quarry	石灰石采石场
limestone red loam	石灰岩红色土；红色石灰土
limestone reef reservoir	礁灰岩储集层
limestone reservoir	灰岩储层
limestone sand	石灰岩砂
limestone sink	落水洞；石灰岩陷穴，石灰岩阱
limestone sink hole	石灰岩落水洞；灰岩坑
limestone slate	灰质板岩
limestone soil	石灰岩土；石灰土
limestone sonde	石灰岩电极系
limestone terrain	石灰岩地区；石灰岩地形
limestone weathering	石灰岩风化
limewash	石灰水；用稀石灰粉刷，刷灰；粉刷墙壁用的稀石灰粉
limewater	石灰水
limewhiting	石灰刷白
limey clay	灰质黏土
liminal	阈限的
laminated structure	片状结构
liming	加石灰
liming tank	石灰槽；加灰槽
liminite	褐铁矿
limit	限；极限；界限；限度；范围
limit analysis method	极限分析法；极限解析法
limit analysis solution	极限分析解；极限解析解
limit angle	边界角；极限角；塌陷角
limit aquifer	有限含水层
limit bearing capacity	极限承载力
limit charge	装药量限度
limit design	极限设计
limit equilibrium	极限平衡
limit equilibrium analysis	极限平衡分析
limit equilibrium area	极限平衡区；极限平衡面积
limit equilibrium condition	极限平衡条件
limit equilibrium method	极限平衡法
limit equilibrium theory	极限平衡理论
limit grade	极限坡度；最大坡度
limit line	极限线；界线；矿区边界
limit load	极限荷载
limit moisture content	界限含水率
limit moisture content test	界限含水率试验
limit of acceptance	验收界限
limit of accuracy	精确限度
limit of bearing capacity	承载能力极限
limit of bearing power	极限承载力
limit of compression	压缩极限
limit of consistency	稠度界限；相容性界限；一致性界限
limit of creep	蠕变极限；蠕动极限
limit of deformation	变形极限
limit of detectability	探测极限；检测能力极限；可检测性极限
limit of draw	地表塌陷极限
limit of drift ice	浮冰界，浮冰边界
limit of ductility	延性极限
limit of elasticity	弹性极限
limit of endurance	耐久极限
limit of equilibrium	平衡极限
limit of error	误差限度
limit of excavation	开挖范围界线
limit of flow regime	流态界限
limit of friction	摩擦限度；摩擦极限
limit of identification	鉴定限度
limit of interference	干扰限值；干扰极限
limit of plastic flow	塑流极限；塑性流动极限
limit of proportionality	比例极限；均衡极限
limit of recharge zone	补给带宽度
limit of reclamation	填海界限
limit of release	排放极限
limit of resolution	分辨极限
limit of secondary gas cap	次生气顶边界
limit of solution	混溶极限
limit of stability	稳定极限
limit of strength	强度极限
limit of works area	施工范围；工地范围
limit of yielding	屈服极限
limit point of subsidence basin	塌陷盆地边界点；沉陷槽边界点
limit slope	极限坡度
limit state	极限状态
limit state design	极限状态设计
limit state design method	极限状态设计法
limit state method	极限状态法
limit state of structure	结构极限状态
limit stress	极限应力
limit switch	极限开关；行程开关；限制开关；限位开关
limit water surface	极限水面；极限油水界面
limitation of output	限制产量；产量限度
limited aquifer	有限含水层
limited cracking	有限裂缝
limited deformation	有限变形
limited dependent variable	有限因变量；受限因变量
limited duration of orogeny	短期造山运动
limited end float coupling	限流浮箍
limited entry fracturing	限流压裂
limited entry treatment	限流法压裂

limited flow deformation	有限流动变形
limited flow strain	有限流动应变
limited gravel reserve	有限砾石储留量
limited life structure	寿命不长的建筑物
limited oxygen index	极限氧指数
limited penetration perforator	有限穿深射孔器
limited prestressing	有限预应力
limited production	限定的产量
limited proportionality region	有限正比区
limited thickness extraction	限厚开采
limited thickness of horse stone	夹石剔除厚度
limited vertical coverage	有限垂向视域；有限垂向覆盖范围
limiting depth	极限深度
limiting design value	设计限值
limiting gas-oil ratio	极限气油比
limiting grade	限制坡度；极限品位
limiting grade of ore	矿石极限品位，矿石最低品位
limiting gradient	极限斜度；极限坡度；最大陡度
limiting grain size	极限粒度；极限晶粒尺寸
limiting holing	限定掘进，限制钻进；限定掏槽落煤法
limiting intensity	极限强度
limiting load	极限荷载
limiting mesh	极限网目
limiting point	极限点
limiting point instability	极值点失稳
limiting pressure	极限压力
limiting producing water cut	极限生产含水量；极限产液含水量
limiting range of stress	极限应力范围
limiting rate of settlement	极限沉降速率
limiting reduction ratio	极限破碎比；极限压缩比；极限减速比；极限折减率
limiting safety system setting	极限安全系统设定
limiting screen	极限筛；限粒筛
limiting state design	极限状态设计法
limiting stratum	界限层
limiting stress	极限应力
limiting stress circle	极限应力圆
limiting surface	界面
limiting tangential stress	极限切向应力
limiting tractive force	极限牵引力
limiting transition probability	极限转移概率
limiting value for sectional dimension	截面尺寸限值
limiting value for supporting length	支承长度限值
limiting value of friction	摩擦力极限值
limiting viscosity	特性黏度，本征黏度
limiting viscosity number	本征黏度值
limitrophe well	位于边界上的井；邻接井
limnetic	湖沼的；淡水的
limnetic basin	湖沼盆地；淡水盆地；陆相盆地
limnetic coal	沼煤；淡水煤；陆相煤
limnetic coal basin	陆相煤田；淡水生成的内湖煤田
limnetic facies	湖沼相
limnic basin	淡水盆地
limnic coal deposit	内陆相含煤沉积；湖沼相煤炭沉积
limnic coal measures	内陆相含煤岩系；湖沼相煤层
limnic peat	湖成泥炭；淡水泥炭
limnocalcite	淡水泥质灰岩
limnogenic rock	淡水沉积岩
limnogeology	湖沼地质学
limnoquartzite	淡水石英岩
limonite	褐铁矿
limonite nodule	褐铁矿结核
limonitization	褐铁矿化
limpid dolomite	宝光白云石；透明白云石
limpidity	透明度；清澈度
limy	含石灰的；石灰质的
limy dolomite	灰质白云岩
limy soil	石灰性土壤；灰质土壤
limy streak	含灰质夹层；石灰质条纹
linarite	青铅矿
lindgrenite	钼铜矿
lindinosite	富钠闪花岗岩
lindoite	钠闪正长细晶岩
lindstromite	硬硫铋铅铜矿；辉铋铜铅矿
line blasting	单排孔爆破；线爆破
line booster pump station	输油管道加压泵站
line boring	沿同一勘探线钻孔
line brattice	纵向风幛
line brattice ventilation	纵向风幛通风
line break detection	管线破裂检测
line clinometer	线性测斜仪
line clogging	管路堵塞
line cut	缝形掏槽，直线掏槽；线切割
line cutting	线刻技术；线切割；直线掏槽
line cutting waterflooding	行列注水
line depot	管路沿线的油库
line development	展线
line drawing	线提取
line drawn map	线划图
line drilling	排钻；成行钻孔；线形钻孔
line drive flood	行列注水驱
line drive pattern	直线驱井网；行列注水驱井网
line drive water injection	行列注水
line equipment	线路设备；输送管线设备
line expansion	线膨胀，线性膨胀，线性伸长
line fault	线路故障；线状断层
line filter	线路滤波器
line fitting	线性拟合；直线拟合
line hole	周边炮眼；圈定炮眼
line hose	主水力分配软管
line in tunnel	隧道内线路
line inclinometer	线性测斜仪
line inclusion	线状夹渣；线状包裹体
line layout	测线布置
line level	线水准仪；气泡水准器；测定水平高程
line list	管道沿线概况表；管线表
line locator	管线位置探测仪；管线定位器
line map	路线图；平面图
line marking	标线
line mixer	管道混合器
line occupation for works	施工封闭线路
line of bearing	走向线；方位角线
line of break	截断线
line of collimation	瞄准轴线；准直线；平行光线
line of correlation	剖面对比线；相互关系线
line of creep	渗流路径；蠕动线；蠕变线
line of dislocation	断裂线；错位线
line of disturbance	地壳破坏带；干扰带

line of equal shear	等剪力线	line stake	定线标桩
line of equal subsidence	等下沉值曲线	line stopper	管塞
line of excavation	开挖线	line strainer	管道过滤器
line of face of coal	煤的主解理	line stretcher	拉线器；紧线器
line of fracture	破裂线	line structure	缆索结构
line of geological limitation	地质界线	line surcharge load	线超载
line of hole	排孔线	line surge	管路水击
line of injecting wells	注入井排	line survey	线路测量
line of intersection	交切线	line swelling	线膨胀
line of least resistance	最小抵抗线；最小阻力线	line tank farm	管路沿线的油库
line of level	水准线路	line tape	卷尺
line of leveling	水准测量路线	line tester	线路试验器
line of lode	矿脉走向线	line the hole	下套管
line of maximum velocity	最大速度线	line ties	测线闭合
line of outcrop	露头线	line type filter	管路型过滤器
line of percolation	渗流线；渗漏线	line vortex	线涡流
line of position	定位等值线	line walker	管道巡查工
line of precise levels	精确水准线	line walking	巡线
line of pressure	压力线	line water drive	行列注水
line of principal stress	主应力迹线	line well	边界井；生产线井
line of producing wells	生产井排	line well drilling	沿油藏边界钻井
line of resistance	抗力线	line well pattern	行列井网
line of rods	杆柱；钻杆柱	line with casing	用套管护孔
line of rupture	破损线；破裂线	lineament	线性体；线性影像；线形构造
line of saturation	浸润线；饱和线	lineament interpretation	线性要素解译
line of section	剖面线	lineament map	线性要素图；区域断裂构造图
line of seepage	浸润线；渗流线	lineament of regional structure	区域构造线
line of slide	滑裂线；滑动线；塌方线	lineament structure	区域线形构造；棋盘格式构造
line of sliding	滑线	lineament system	地质构造体系
line of slope	坡面线	linear arch	线性拱
line of sounding	测深线	linear arch mechanism	线性拱理论
line of stratification	层线；层理线	linear area of subsidence	线状沉降带
line of strike	走向线；走向	linear arrangement	直线布井；直线排列
line of thrust	推力线；逆断层线	linear bar	线形沙洲
line of tunnel	隧道的边线；平硐的边线	linear channel	线形河槽
line of vision	照准线	linear coordinates	线坐标
line of wave	波线	linear core flood	线性岩心的驱替
line of weakness	最小抵抗线	linear deformation	线性变形
line of zero moment	零矩线；零力矩线	linear denudation	线状剥蚀
line oiler	压气管加油器	linear detection	线性检波
line oscillator	行扫描振荡器；行扫描信号发生器	linear development	线状发展；线状开发
line pack	管线充填量	linear displacement	线性位移
line pack storage	管道中储存	linear displacement efficiency	线性驱替效率
line pan	机槽；中部槽；溜槽；输送机链槽	linear distortion	线性失真；线性畸变
line pattern	直线布井；线性井网	linear drainage	线性水系；线状排水系统
line pipe	干线用管	linear dune	线状沙丘
line plan	线路平面图	linear earthquake	线状地震
line plot survey	网状沙洲	linear elastic body	线弹性体
line plunger pump	往复柱塞泵	linear elastic fracture	线弹性断裂
line pressure	主压力；管路压力，输送管压力	linear elastic fracture mechanics	线弹性断裂力学
line profile	线路纵断面图	linear elastic material	线弹性材料
line reflection	直线反射	linear elastic model	线弹性模型
line rod	标杆	linear elastic model of soil	土的线弹性模型
line scan imagery	行扫描图像	linear elastic supposition	线弹性假定
line scanning	行扫描；线扫描	linear elasticity	线弹性
line search	线搜索	linear elastodynamics	线弹性动力学
line shooting	纵测线观测	linear energy transfer	传能线密度；线性能量传递
line source	线震源	linear eruption	裂隙式喷发；线状喷发
line source model	线震源模型	linear fault	直线断层
line spread	地震仪直线排列法	linear filter	线性滤波

linear filtering	线性滤波
linear filtering flow	线性渗流
linear flow	线性流；直线流
linear flow characteristic	线性流量特性；线性流动特征
linear flow leak-off mechanism	线性滤失机理
linear flow period	线性流动阶段
linear flow structure	线状流动构造
linear fold	线状褶皱
linear foliation	线状叶理；线状剥理
linear fracture zone	线状破裂带
linear frontal displacement	线性前缘驱替
linear interpolation	线性插值
linear intersection	边交会法；测边交会
linear inversion	线性反演
linear load	线性荷载；单位长度线荷载
linear magnetic charge	线磁荷
linear mass	线质量
linear measurement	长度测量；直线测量
linear migration of large earthquake	大地震直线迁移
linear multiplex	线性多路传输
linear multiplexing temperature tool	线性多路传输井温仪
linear optimization	线性优化；线性规划
linear parting	线状裂隙；线状节理
linear pattern shooting	直线型爆炸
linear pipeline	线性流水线；直线管线
linear plate	衬砌板
linear porosity	线状孔隙
linear pressure	线向压力
linear prestressing	线性预加应力法
linear profiling	线性断面
linear program	线性规划
linear program method	线性规划法；线性编程法
linear programming	线性规划
linear programming method	线性规划法
linear programming model	线性规划模型
linear projection	直线投影
linear proximity	线性逼近
linear ramp	线性斜坡
linear range	线性范围
linear rate of flow	流动线速度
linear recombination	线性复合
linear recurrence	线性递归关系
linear referencing system	线性参考系统
linear reserve method	线储量法
linear reservoir model	线性储层模型
linear residuum	线状风化壳
linear rift valley	线状裂谷；线状地堑
linear robust control theory	线性鲁棒控制论；线性强健控制论
linear sampled-data control system	线性采样数据控制系统
linear scanner	线性扫描仪
linear scanning	线性扫描
linear schistosity	线状片理
linear screen	直线往复筛
linear seepage	线性渗流
linear seepage law	线性渗透定律
linear strain	线性应变
linear strain rate	线应变率
linear strain seismograph	线性应变地震仪
linear strain strength theory	线应变强度理论
linear stratification	线性层理
linear stress	线性应力
linear stress state	单向应力状态；线性应力状态
linear stretch	线性拉伸；线性伸展
linear structure	线状构造；线性结构
linear synchronous motor	直线同步电动机
linear system	线性系统
linear transformation	线性变形；线性转换
linear transient analysis	线性瞬态分析
linear transition	线性转移
linear transponder	线性转发器；线性变换器
linear trend	线性趋势
linear trend surface	一次趋势面
linear triangulation	线形三角测量
linear triangulation chain	线形三角锁；线形三角链
linear triangulation network	线形三角网
linear trough	线状海槽；线状地堑
linear uplift	线状隆起
linear valley	线状谷
linear variable differential transformer	线性差动变压器；线性差动位移传感器
lineation	线理；线状构造；区域构造线；轮廓
lineation plunging	线理倾伏
lineation structure	线理构造；线状构造
lined	有衬里的；有护板的；有壁的
lined borehole	加衬钻孔；护壁钻孔
lined canal	衬砌的渠道
lined shaft	砌壁竖井
lined tunnel	衬砌隧洞，衬砌隧道
lineman	线务员；通信员；路线看守人
lineoid	超平面
liner	衬里，衬垫，衬圈；两梁间的横梁；衬砌
liner adapter	衬管接头
liner assembly	衬管管柱组合
liner barrel	衬套
liner bushing	衬套
liner centralizer	衬管扶正器；衬管定心夹具
liner clamp	衬套座
liner completion	衬管完井
liner cup packer	碗状衬管封隔器
liner extension	衬管加长短节
liner flow capacity	衬管流通能力；衬管流量
liner foundation	基础层；衬管的基础
liner gravel packing	衬管砾石充填
liner hanger	尾管悬挂器
liner jacket	泵筒
liner mill	衬管磨铣器
liner packer	衬管封隔器
liner pipe	衬管
liner plate	衬板
liner puller	缸套拉拔器
liner pump	衬管泵
liner releasing tool	衬管解卡工具；衬管释放工具
liner setting tool	尾管安放器
liner shoe	衬管鞋
liner slot	衬管割缝
liner squeeze	下衬管挤压
liner stab	衬管插入接头
liner swivel	尾管活动短节

liner system	衬里系统
liner top sub	衬管顶部接头
liner tube	衬管
liner vibration flow pack	衬管振动循环充填
liner vibration pack	衬管振动充填
liner vibration tool	衬管振动工具
liner washing	衬管冲洗
liner wiper plug	衬管刮塞；衬管封隔塞
liner wire spacing	筛管绕丝间隙
linesman	巡线工人；线务员，架线工；通信员；路线看守人
lingual bar	舌形沙坝
linguoid bar	舌形沙坝
linguoid ripple mark	舌形波痕
lining	衬里，衬套
lining arch	衬砌拱
lining area	衬里面积
lining board	衬板
lining brick	砌壁砖；面砖
lining commutator	套管；衬里换向器
lining cracking	衬砌裂损
lining material	支护材料；衬砌材料
lining of arch	拱形衬砌
lining of slope	坡面铺砌
lining of tunnel	隧道衬砌
lining of tunnel portal section	洞口段衬砌
lining peg	测桩；标桩
lining pipe	衬里管
lining plate	衬板
lining platform	衬砌模板台车
lining pole	衬砌杆
lining pouring sequence	衬砌浇注顺序
lining resistance	支护抗力
lining sight	简单瞄准器；方向照准器
lining split	衬砌裂损
lining thickness	衬砌厚度
lining tube	套管
lining with bricks	砖石衬砌
lining with column-typed sidewalls	柱式边墙衬砌
lining with continuous-arched sidewalls	连拱墙衬砌
link	连接；测链环长；链指针；链路
link arm	连杆臂
link bar	联杆梁；铰接梁
link road	连接路；接驳道路
linkage	链系，联动装置；联络巷道，连锁
linkage assembly	连杆机构
linkage bore hole	贯通孔
linked vein	链状矿脉
linking beam of foundation	基础连系梁
linksland	海岸沉积沙带；海边沙丘地带
linnaeite	硫钴矿
linoleum surface	油地毡地面
linophyre	线状斑岩
linophyric texture	线斑状结构
linosaite	含霞碱玄岩
liottite	硫碳钙霞石
lip	溜槽延伸部分；顶板边缘
lip curb	边石，路缘石
lip elevation	坎唇高程
lip of a crater	火山口边缘
lip packing	法兰密封；带唇密封；带边密封
lip screen	分级筛；唇筛
lip seal	唇形密封；边缘密封
lip seal arrangement	唇形密封结构
liparite	流纹岩；萤石；硅孔雀石
liparite-dacite	流纹英安岩
liparite-tuff	流纹凝灰岩
liparophyre	流纹斑岩
lipkin bicapillary pycnometer	利普金双毛细管比重计
liptinite	类脂质；类脂组；壳质组
liptinite coal	类脂煤；稳定煤；壳质煤
liptinite group	稳定组；类脂组
liptite	微稳定煤；微壳煤
liptobication	残植化作用
liptobiolite	残植煤；残植质
liptobiolith	残植煤；残植岩；残留生物岩
liptodetrinite	碎屑类脂体；碎屑稳定体；碎屑壳质体
liquation	熔融；分熔；熔离作用
liquefaction	液化；液化过程
liquefaction coal	液化用煤
liquefaction counter-measures	抗液化措施
liquefaction during earthquake	地震液化
liquefaction grade	液化等级
liquefaction index	液化指数
liquefaction index of foundation soil	地基土液化指数
liquefaction of coal	煤的液化
liquefaction of fine sand layer	细砂层液化
liquefaction of sand	砂土液化
liquefaction of saturated soil	饱和土液化
liquefaction of silt	粉砂液化
liquefaction of soil	土壤液化
liquefaction point	液化点
liquefaction potential	液化势；液化潜力；液化可能性
liquefaction resistance factor	抗液化系数
liquefaction safety coefficient	液化安全系数
liquefaction slide	液化滑移
liquefaction strength	抗液化强度；液化强度
liquefaction stress ratio	液化应力比
liquefaction susceptibility	液化灵敏度
liquefaction with limited strain potential	具有有限应变势的液化
liquefaction zone	液化区
liquefiable	可液化的
liquefied breccia	液化角砾岩
liquefied cohesionless-particle flow	液化松散颗粒流
liquefied crinkled deformation	液化卷曲变形
liquefied foundation	液化地基
liquefied natural gas	液化天然气
liquefied natural gas tank	液化天然气罐
liquefied particles	液化颗粒集合体
liquefied petroleum gas	液化石油气
liquefied petroleum gas installation	石油气装置
liquefied petroleum gas sphere	球形石油气缸
liquefied region	液化区
liquefied soil	液化土
liquefied soil layer	液化土层
liquefied water escape vein	液化泄水岩脉
liquid absorption vapour recovery	液体吸收法蒸气回收
liquid adsorption method	液体吸附法
liquid asphalt	液态沥青

liquid bitumen	液态沥青	liquidize	使液化
liquid cement	水泥浆	liquid-jet pump	液体射流泵
liquid charge pump	注液泵	liquid-junction potential	液体接触电位
liquid chiller	液体冷冻机	liquid-knockout surge drum	液体分离缓冲罐
liquid chromatogram	液相色谱	liquid-level controller	液位控制器；液面控制器
liquid chromatography	液相色谱分析法；液相层析法	liquid-level gauge	液位压力计；液面式压力计
liquid clutch	液体联轴节	liquid-level manometer	液位压力计；液面式压力计
liquid column	液柱	liquid-liquid chromatography	液-液色谱分析法
liquid column hydrostatic pressure	静液柱压力	liquid-liquid displacement	液-液驱替
liquid column manometer	液柱式压力计	liquid-liquid equilibrium	液-液平衡
liquid compressibility	液体可压缩性	liquid-liquid extraction	液-液萃取
liquid concrete	流态混凝土	liquid-liquid interface	液-液界面
liquid control valve	液体控制阀	liquid-liquid miscibility	液-液互混性；液-液混相性
liquid conveyance	液体输送	liquid-liquid transfer loop	转环联轴节
liquid coolant	冷却液	liquid-oxygen cartridge	液氧爆破筒
liquid cooling	液体冷却	liquid-phase oxidation catalyst	液相氧化催化剂
liquid core	液核	liquid-phase separation	液相分离
liquid counter	液压计数器	liquid-phase suspension process	液相悬浮过程
liquid damper	液压减震器；油液缓冲器	liquid-plastic limit tester	液塑限联合测定仪
liquid explosive	液体炸药	liquid-pressure transducer	液压传感器
liquid field	液相区	liquid-producing capacity	产液能力
liquid film	液膜	liquid-retaining structure	挡水构筑物
liquid film lubrication	液膜润滑	liquid-ring vacuum pump	液环真空泵
liquid gas	液化气体	liquid-solid adsorption chromatography	液-固吸附色谱
liquid gauge	液体负压计；水柱负压计	liquid-solid interface	液-固相界面
liquid gum	液态胶质	liquid-solid ratio	液固比
liquid hang-up effect	悬液效应	liquid-solid separation	液-固相分离
liquid head	液柱压力；液头	liquid-state diffusion	液态扩散
liquid level test	液面测试	liquidus	液线，液相线；液相的，液态的
liquid limit	液限；液体极限	liquidus boundary	液相线边界
liquid limit device	液限仪	liquidus curve	液相线
liquid lubrication	液体润滑	liquidus mineral	液相线矿物
liquid magma	液态岩浆	liquidus temperature	液相线温度
liquid manometer	液柱压力表	liquified natural gas	液化天然气
liquid oxygen explosive	液氧炸药	liroconite	水砷铝铜矿；豆铜矿
liquid petroleum gas system	液化石油气系统	liskeardite	砷铁铝矿
liquid phase	液相	listening depth	测声深度
liquid press	油压机	listric	铲状的；上凹形的
liquid pressure	液体压力	listric fan	犁式扇，铲状扇
liquid resistance	液流体变阻器；液体电阻	listric fault	铲状断层；上凹断层；犁式断层
liquid ring	液环	listric normal faulting	铲状正断层作用
liquid saturation	含液饱和率	listric surface	铲状破裂面；上凹形面
liquid scintillation	液体闪烁法	listvenite	滑石菱镁片岩
liquid seal	液封；水封；油封	listvenitization	滑石菱镁片岩化
liquid unmixing deposit	熔离矿床	listwanite	滑石菱镁岩
liquid viscous-gas turbulent	液相层流-气相紊流	lit by lit injection	层层贯入，层层注入
liquid withdrawal rate	采液速率	lit par lit gneiss	间层片麻岩
liquidation of hole	封孔	lit par lit intrusion	间层侵入
liquid-carburizing	液体渗碳	lit schistosity	间层片理
liquid-crystal display	液晶显示	litchfieldite	霞云钠长岩
liquidensitometer	液体密度计	lithia mica	锂云母
liquid-filled porosity	充满液体的孔隙；含液孔隙	lithia spring	含锂矿泉
liquid-gas boundary	液-气边界	lithia water	锂盐矿水
liquid-gas interface	液-气相界面	lithian muscovite	锂白云母
liquidifiable	可液化的	lithic	石质的；岩屑的
liquid-immersed reactor	油浸电抗器	lithic arenite	岩屑砂屑岩
liquidity	流动性	lithic arkose	岩屑长石砂岩
liquidity factor	液性指数；流动性指数	lithic arkosic wacke	岩屑长石砂岩质杂砂岩
liquidity index	液性指数；流动性指数	lithic contact	母质层；土壤岩石接触层
liquidity rate	流动比率	lithic drainage	地下水系

lithic facies	岩相
lithic fragment	岩屑
lithic graywacke	岩屑杂砂岩
lithic percentage map	岩性百分比图
lithic sandstone	岩屑砂岩
lithic subarkosic wacke	岩屑亚长石砂岩杂砂岩
lithic tuff	岩屑凝灰岩
lithic wacke	岩屑瓦克岩
lithical	石质的
lithiclast	岩屑
lithification	岩化作用；成岩作用
lithified	岩化的
lithify	岩化作用；成岩作用
lithionite	锂云母
lithiophilite	磷锰锂矿
lithiophorite	锂硬锰矿
lithium drifted detector	锂漂移探测器
lithium hydroxy-stearate grease	羟基硬脂酸锂润滑脂
lithium minerals	锂矿类
lithium oxide	氧化锂
lithium phosphate catalyst	磷酸锂催化剂
lithizone	岩性带；岩性地层带
lithocalcarenite	灰岩屑砂岩
lithocalcilutite	灰岩屑泥岩
lithocalcirudite	灰岩屑砾岩
lithocalcisiltite	灰岩屑粉砂岩
lithochemical	岩石化学的
lithochemistry	岩石化学
lithoclase	岩石裂隙；破裂面
lithoclast	岩屑
lithoclastic	灰质碎屑岩
lithodemic unit	岩石谱系单位
litho-density logging	岩性密度测井
litho-density quicklook	岩性密度测井快速直观解释
litho-density tool	岩性密度测井仪
lithofacies	岩相
lithofacies association	岩相协会；岩相组合
lithofacies map	岩相图
lithofacies model	岩相模式
lithofacies paleogeographic map	岩相古地理图
lithofraction	岩裂作用
lithogeneous component	造岩组分
lithogenesis	岩石成因；岩石成因论；成岩作用
lithogenesy	岩石成因；岩石成因论
lithogenetic evidence of climate	成岩气候证据
lithogenetic unit	岩层单位
lithogenic	成岩的；造岩的；岩生的
lithogenous	造岩的；石质的；岩生的；岩石成因的；成岩的
lithogenous component	造岩组分；成岩组分
lithogenous constituent	造岩组分；成岩组分
lithogenous material	成岩物质
lithogenous mineral	造岩矿
lithogenous phase	成岩相
lithogenous process	成岩作用
lithogeny	岩石成因；岩石成因论，岩石成因学；成岩作用
lithogeochemical	岩石地球化学的
lithogeochemical method	岩石地球化学方法
lithogeochemical survey	岩石地球化学测量
lithogeochemistry	岩石地球化学
lithogram	岩谱图
lithographic	石印的；石板的；平板的，平板印刷的
lithographic limestone	石印灰岩
lithographic stone	石印石；印版石
lithoherm	岩礁或岩丘
lithohorizon	岩石层位；岩性层位
lithoid	岩状的；像石头的
lithoidal	石质的；细密晶质
lithoidal tufa	石质石灰华
lithoidite	隐晶流纹岩
lithoidite texture	石质结构
lithoidtic	显微霏细质的
lithologic affiliation	岩性亲缘关系
lithologic boundary	岩性界线
lithologic break	岩性间断
lithologic change	岩相变化；岩性变化
lithologic character	岩性特征
lithologic characteristic	岩性特征
lithologic classification	岩性分类
lithologic composition	岩石成分
lithologic control	岩性控制
lithologic correlation	岩性对比；岩相对比
lithologic criteria	岩性标志；岩性标准
lithologic criterion	岩性标志；岩性标准
lithologic deposit	岩性油藏
lithologic discrimination	岩性识别；岩性判定
lithologic factor	岩性要素
lithologic feature	岩性特征
lithologic gradient	岩性梯度
lithologic homogeneity	岩性均一性
lithologic identification	岩性鉴别
lithologic identity	岩性鉴定
lithologic information	岩性资料
lithologic interface	岩性界面
lithologic interpretation	岩性解释
lithologic layering	岩性层理；岩性分层
lithologic log	岩性测井，岩性录井；岩性柱状图
lithologic map	岩性图
lithologic oil and gas pool	岩性油气藏
lithologic oil pool	岩性油藏
lithologic parameter	岩性参数
lithologic pinch out reservoir	岩性尖灭油气藏
lithologic reservoir	岩性油气藏
lithologic section	岩性剖面
lithologic sequence	层序
lithologic series	岩石系列
lithologic similarity	岩性相似
lithologic symbol	岩性符号
lithologic transition	岩性转换
lithologic trap	岩性圈闭
lithologic triangle	岩性三角图解
lithologic unit	岩性单位；岩石地层单位
lithological analysis	岩性分析
lithological anisotropy	岩石的各向异性
lithological association	岩性组合
lithological boundary	岩性界线
lithological change	岩性变化
lithological character	岩性特征
lithological classification	岩石分类

lithological column 岩性柱状图
lithological combination 岩性组合
lithological composition 岩石成分
lithological context 岩性范围
lithological contrast 岩性差异
lithological correlation 岩性对比
lithological criteria 岩性标志
lithological discrimination 岩性识别
lithological factor 岩性要素
lithological feature 岩性特征
lithological identification 岩性识别
lithological information 岩性资料
lithological interpretation of photo 像片岩性判读；像片岩性解译
lithological layering 岩性分层；岩性层理
lithological map 岩性图
lithological prediction 岩性预测
lithological profile 岩性剖面
lithological response 岩性响应
lithological sequence 岩性序列
lithological settings 岩性环境
lithological triangle 岩性三角形图解
lithological unit 岩性地层单位
lithology 岩性；岩石学
lithology analysis package 岩性分析程序包
lithology and geochemistry of sediments 沉积物岩性和地球化学
lithology breakdown 岩性分类
lithology complex reservoir analysis 复杂岩性储集层测井分析
lithology factor 岩性系数；岩性要素
lithology identification 岩性鉴别
lithology indicator 岩性指示
lithology information 岩性资料
lithology line 岩性线
lithology logging 岩性录井
lithology of lithospheric plate 岩石圈板块的岩性
lithology parameter 岩性参数
lithology triangle 岩性三角形图解
lithology window 岩性窗
lithomarge 密高岭土；不纯高岭土
lithomechanics 岩矿力学；岩石力学
lithometeor 大气尘粒
lithomorphic soil 岩成土；石质土
lithophile 亲岩的；亲石的
lithophile affinity 亲石性
lithophile element 亲石元素；亲岩元素
lithophotography 光刻照相术
lithophysa 石核桃；石泡
lithophysa structure 石泡构造
lithoplate 岩石板块
lithopone 锌钡白
litho-porosity crossplot 岩性-孔隙度交会图；岩性-孔隙度对比图
lithorelict 岩石风化残余物
lithorizing 防蚀处理
lithosiderite 石铁陨石
lithosol 石质土
lithosome 岩体，岩石体
lithosphere 岩石圈；构造圈；岩界；陆界

lithosphere plate 岩石圈板块
lithosphere root 岩石圈根
lithosphere slab 岩石圈板片；岩石圈板块
lithospheric 岩石圈的
lithospheric block 岩石圈断块；岩石圈地块
lithospheric bulge 岩石圈膨胀
lithospheric earthquake 岩石圈地震
lithospheric fault 岩圈断裂
lithospheric flexure 岩石圈弯曲
lithospheric isostasy 岩石圈均衡；地壳均衡
lithospheric motion 岩石圈运动
lithospheric plate 岩石圈板块
lithospheric pressure 岩石圈压力
lithospheric profile 岩石圈剖面
lithospheric splitting 岩石圈分裂
lithospheric stress 岩石圈应力
lithospheric subplate 岩石圈下板块；岩石圈小板块，岩石圈次板块
lithospheric velocity structure 岩石圈速度结构
lithosporic 石斑
lithostatic 地压的；静岩的
lithostatic fluid pressure 静岩流体压力
lithostatic gradient 岩层静压力梯度；地压梯度
lithostatic load 岩石静载荷；静岩荷载
lithostatic pressure 岩石静压力；静岩压力
lithostatic state of stress 岩石应力状态；静岩应力状态
lithostatic stress 岩层静应力；静岩应力
lithostratic unit 岩性地层单位
lithostratigraphic boundary 岩性地层界线
lithostratigraphic classification 岩性地层划分
lithostratigraphic column 岩性地层柱
lithostratigraphic correlation 岩性地层对比
lithostratigraphic stratotype 岩性地层的层型
lithostratigraphic unit 岩性地层单位
lithostratigraphic zone 岩性地层带
lithostratigraphy 岩石地层学
lithostratigraphy boundary 岩性地层界线
lithostrome 均质岩性体；均质岩层
lithostructural contact 岩性构造接触带
lithotype 岩性类型；煤岩类型
lithozone 岩性带；岩性地层带
lit-par-lit 间层的
lit-par-lit gneiss 间层状片麻岩；夹层状片麻岩
lit-par-lit intrusion 顺层侵入；间层侵入
lit-par-lit structure 间层构造；成层构造
little ice age 小冰期
little opened void 细小开启裂隙
little vehicle 斗车
littoral accumulation 沿岸堆积；沿岸淤积
littoral area 海岸区；潮汐区
littoral current 沿岸流
littoral deposit 滨海沉积；沿岸沉积；海滩沉积；湖滩沉积
littoral drift 沿岸沉积物流；沿岸漂移
littoral dune 沿岸沙丘；滨海沙丘
littoral facies 潮岸相；滨海相
littoral flat 滨海坪
littoral flow 沿岸流
littoral fringe 海岸带边界
littoral landform 海岸地形

littoral nourishment 沿岸淤滩
littoral placer 海滨砂矿床
littoral plain 沿岸平原
littoral prism 沿海棱镜
littoral region 海岸区；沿岸区；滨海区；潮汐区
littoral ridge 滨岸堤
littoral sediment 滨海沉积物；沿岸沉积物
littoral shelf 浪成湖滨台地；湖棚；沿岸台地
littoral terrace 海岸阶地
littoral topography 沿岸地形
littoral zone 滨海带；潮汐带；潮滩带；潮间带
live coal 易采掘煤
live crack 活裂缝
live crude 含气原油
live fossil 活化石
live geophone 有效地震检波器
live glacier 活冰川；流动冰川
live gravity load 重力活荷载
live ice 活动冰；流动冰
live landform 现代地形
live line 装满油的管线
live load 活荷载，动荷载；活重；有效负荷；交变负荷
live load moment 活荷载弯矩；活荷载力矩
live load of train 列车活载
live lode 可采矿脉
live oil 含气石油；充气石油
live primer 活性起爆药包
live pulley 动滑轮
live quartz 含矿石英
live roller 传动辊道；辊式输送机
live spindle 回转轴；旋转轴；动轴
live stress 活荷载应力
live well 充气油井
live working 生产工作区；生产巷道
lively coal 易碎煤
liver ore 肝色赤铜矿；一种辰砂
liver rock 易裂岩石
live-roller conveyor 辊道输送机，滚柱式输送机
livestock reservoir 有调节池的径流式电站
living fossil 活化石
living soil 地表土壤
living substance 活质
living volcano 活火山
livingstonite 硫汞锑矿
lixivial 淋滤的
lixiviant 浸滤剂
lixiviate 浸出
lixiviation 淋蚀；溶滤；浸出
lixiviation process 浸出法；溶滤作用
lixiviation screen 浸滤筛
lixiviation water 溶滤水
lixivium 浸出物；浸滤液；淋余土
lizardite 利蛇纹石
load 荷载；载重，负荷，负载，装载，装填
load and strength histogram 荷载-强度直方图
load and stress in pillar 矿柱荷载和应力
load and unload line 装卸线
load application 加重；施加荷载
load array 荷载数组

load at certain elongation 定伸长载荷
load at failure 破坏荷载
load at first crack 初裂荷载
load at rupture 破坏荷载；断裂荷载
load at specified elongation 定伸长载荷
load ball 负载球
load bearing 承重；承载
load bearing capacity 承载力
load bearing characteristic 负荷特性
load bearing grain 承载颗粒
load bearing layer 持力层
load bearing masonry building 承重砖石房屋
load bearing solids 骨架砂；承载固体颗粒
load bearing structure 承载构筑物；荷载支承结构
load binder 扎紧管子的装置；锁紧器
load brake 重锤闸
load button 输入按钮；加载键
load cable 牵引钢丝绳
load calibrating device 负载检验器；负载校准装置
load capacity 载重量；载重能力；承载能力；负载容量
load capacity of lubricant 润滑剂负荷能力
load car 空中吊运车
load carrying arch 承载拱；压力拱
load carrying capacity 负荷能力；承载能力；承载量；支承荷载能力
load cell 拉压传感器，荷载传感器；压力盒，加载盒，测力仪，测压仪
load center 负荷中心；加载中心
load change 负载变化；载荷变化
load change test 变荷载试验
load characteristic 负荷特性；负载特性
load classification number 荷载分类指数
load coefficient 荷载系数
load combination 荷载组合
load condition 承载状态；负荷条件
load consolidation curve 载荷固结曲线
load control 负载控制；载荷控制
load controller 载荷控制器
load criterion 荷载准则
load cross line 荷载交叉线
load curve 荷载曲线；负荷曲线；钻压曲线
load curve during a day 日负荷曲线
load deflection curve 荷重挠度曲线
load deflexion 载荷挠度
load deformation 负载变形
load deformation characteristics 荷载-变形特征
load deformation diagram 荷载-变形图；荷载与变形量关系图
load density 荷载密集度；荷载强度
load diagram 负载曲线；负荷图
load difference 荷载差
load discontinuity 荷载非连续；不连续荷载
load distribution 载荷分布
load distribution angle 荷载分布的角度
load distribution curve 荷载分布曲线
load distribution factor 荷载分布系数
load down 降负荷，降载荷
load duration curve 荷载历时曲线
load effect combination 荷载效应组合
load elongation curve 荷载伸长曲线

load equivalency　荷载当量
load estimate　荷载估算
load factor　荷载因子；荷载系数
load factor design　荷载系数设计
load factor for cohesion　黏聚力荷载因子
load factor method　荷载因数法；荷载系数法
load factor rating　承载系数检定
load fluctuation　载荷波动；负荷波动
load fluid　携砂液
load fold　负载褶皱
load frequency relationship　荷载频率关系
load gauge　测力计；压力盒
load geologic process　负荷地质作用
load grade　负荷等级
load growth　负荷增长
load haul dumper　铲运机
load history　荷载历史；加载历史
load impact allowance　容许冲击荷载
load impedance　负载阻抗
load inclination　荷载倾斜角
load increment　荷载增量
load increment ratio　荷载增量比
load indicator　指重表；载荷指示器；测力计
load influence zone　荷载影响区
load input　加荷；荷载输入
load input tensile tester　载荷输入拉力试验机
load intensity　荷载强度
load length　负荷长度
load limit　载荷极限；载重限制
load limitation　负载限度
load limiter　载荷限制器
load maintainer　试验机加载稳定器；荷载稳定器
load matrix　荷载矩阵
load measuring recorder　荷载测录器
load metamorphism　负载变质作用
load module　输入模块
load module library　输入模块库
load mold　压模
load moment　负载力矩
load of earth pressure　土压力荷载
load of loosen bedrock　松弛基岩荷载
load of river　河流输砂量
load of roof　顶部负荷；顶板荷载
load of support　支架荷载
load oil　压裂工作油；油井起剂油
load on aquifer　含水层的负荷
load on axle　轴载
load on base　底座负荷；基础负荷
load on bit　钻头负荷
load on side bearing　旁承载荷
load pattern　负载曲线图
load peak　负载峰值；最大负载
load per leg　每根支柱的工作阻力
load per unit run of face　工作面单位长度的支架阻力
load pocket　负载囊
load point　加载点；装入点；信息起止点
load pouch　负载囊
load power　负载功率
load prediction　负荷估计
load prediction curve　负荷预测曲线

load pressure　负荷压力；荷载压力
load pressure feedback type servovalve　荷载压力反馈型伺服阀
load proportional valve　载荷比例阀
load range　荷载幅度；荷载范围
load rate　荷载率
load rating　荷载率定；荷载额定
load reading　荷载读数
load reapplication　重新加载
load record　载荷记录
load redistribution　荷载重分布
load regulation　载荷调整
load regulator　载荷调节器
load removal　卸载；卸去负荷
load resistance　负载电阻
load reversal　荷载反向
load ring　载重环；量力环
load rope　吊重绳
load screen　筛分装载机
load sensor　荷载传感器
load sensor valve　载荷传感阀
load setter　负载定值器
load setting unit　负载定值装置
load settlement curve　荷载沉降曲线
load sharing equipment　负荷分担装置
load shedding　支架突然丧失支承能力
load spectrum　荷载谱
load speed　加载速度；受压速度
load spread　荷载分布
load spreading property　荷载分布特性
load stage　加载阶段
load strain diagram　载荷应变曲线图
load stress　荷载应力
load striation structure　负载擦痕构造
load structure　负荷构造
load supporting capability　承载能力
load switch　加载开关
load tension　负载张力
load test　荷载测试；负载测试；承重测试；加载试验
load test on pile　桩的载荷试验
load testing　负载测试；荷载测试；承重测试
load testing of structure　结构载荷试验；建筑物载荷试验
load thrown off　负荷卸除；卸载
load tide　载荷潮；负荷潮
load torque　负载转矩
load transducer　荷载传感器
load transfer　负载转移；荷载转移
load transfer device　荷重传递设施
load transfer efficiency　荷重传递效率
load transfer function method of pile　桩的荷载传递函数法
load transfer law　荷载传递规律
load transfer mechanism　荷载传递机理
load transfer method　载重转移法；荷载传递法
load vector　荷载向量
load water　压裂工作水；积水
load wave　负载波纹；重荷波痕
load weight　载重；负载重量
load yield curve　载荷屈服曲线
loadability　承载能力；载荷能力

loadable coal 适于机装的煤；易装煤
loadage 装载量
loadamatic control 负载变化自动控制
load-back method 负载反馈法
load-bearing aggregate 承载集料
load-bearing brick wall 承重砖墙
load-bearing capacity 支承重量；支承荷载；支承力；载重力；支承能力；载重能力
load-bearing cross wall 承重横墙
load-bearing formation sand 承载地层砂
load-bearing frame 承重构架
load-bearing reinforced concrete wall 钢筋混凝土承重墙
load-bearing ring 承载环
load-bearing shear wall 承重剪力墙
load-bearing skeleton 承重骨架；承重构架
load-bearing soil 承重土层；土持力层
load-bearing stratum 持力层
load-bearing strength 承载强度
load-bearing structure 承重构筑物
load-bearing test 承载试验；载荷试验
load-bearing wall 承重墙
load-bearing wall panel 承重墙板
load-carrying 载重的
load-carrying ability 承载能力；载重量
load-carrying area 加载面积；承载面积
load-carrying capacity 载重量；支承力；承压力；支承荷载能力；承重能力
load-carrying capacity per bolt 单个螺栓承载能力
load-carrying capacity per rivet 单个铆钉承载能力
load-carrying member 承重构件
load-carrying property 承载性能
load-carrying swivel joint 承载旋转接头
load-cell 负载管；测压元件；测力传感器
load-characteristic test 负荷特性试验
load-compression diagram 荷载压缩图
load-current meter 负载电流表
load-deflection curve 荷载-挠度曲线
load-deflection relation 荷载-挠度关系
load-deformation character 荷载-变形特性
load-deformation curve 荷载-变形曲线
load-deformation relationship 荷载-变形关系
load-deformed mark 荷载变形痕迹
load-dispersive anchor 荷载分散型锚杆
load-displacement curve 荷载-位移曲线
load-displacement relation 荷载-位移关系
load-dividing valve 负载分配阀
loaded 有负荷的，加载的，承载的；装药的；填料的；加感的
loaded area 承载面积
loaded beam 承重梁
loaded cable 加感电缆
loaded dip needle 加感磁倾仪
loaded filter 压载滤水体；被阻塞的滤水层；反滤层
loaded impedance 负载阻抗
loaded length 炮眼装药长度
loaded line 载荷线
loaded matching 载荷匹配
loaded mud 高比重钻泥
loaded region 负荷区
loaded speed 装载速度；加载速度
loaded stock 填料
loaded strand 负载段；工作段
loaded stream 挟沙河流；含沙河流
loaded string 充液钻柱
loaded test 负载试验；加载试验
loader 装载机，炮眼装药车；装载设备
loader belt 装载机皮带
loader chassis 装载机底架
loader conveyor 装运机
loader discharge conveyor 装载机的卸载输送机
loader drift 装载平巷
loader for shaft 竖井用装运机
loader front conveyor 装载机的前端输送机
loader gate 采区装运顺槽
loader gathering arm 装载机扒爪
loader gathering chain 装载机集料链
loader loading head lift 装载机装料斗的高程
loader loading range 装载机装载范围
loader main motor 装载机的主电动机
loader operator 装载机司机
loader rear conveyor motor 装载机尾输送机的电动机
loader routine 输入程序
loader shovel 装载机铲斗
loader snubbing arm 装载机耙臂
loader stripping 装载机倒堆
loader team 装车队
loader tramming 装载机调动
loader tramming speed 装载机的调动速度
loader truck stripping 单斗铲-卡车剥离
loader vertical range 装载机铲斗的扬起高度
loader-digger 装载挖掘两用机
loading 加载；装载；荷载，载重；填塞
loading agent 增重剂
loading analysis 荷载分析
loading and hauling muck 装拉碴
loading and unloading 加载与卸荷
loading and unloading accident 装卸事故
loading and unloading cycle 加载卸载循环
loading and unloading operation 装卸作业
loading and unloading track 装卸线
loading apparatus 装药器
loading area 装载面积；负载面积；装载场
loading arm 装载刮板；装载耙臂
loading assumption 荷载的确定
loading bay 进料台；装货场；安装现场
loading belt 装载胶带
loading berm 加载台肩；反压平台，反压护道
loading berth 装货泊位
loading bin 装载仓
loading board 加料板
loading boom 吊货杆
loading bowl 装载仓
loading box 装载仓
loading box steelwork 装载设施
loading bridge 装载桥
loading bucket 装料斗；装载桶
loading bunker 装车仓
loading bunker with discharge point 单点装车仓
loading bunker with several discharge points 多点装车仓
loading buoy 装油浮筒

loading cable 加感电缆
loading capacity 装载能力；载荷能力，加载能力
loading cartridge 限量装置；计量给料器
loading chain 装载链
loading chute 装载槽
loading clearance limit 装载限界
loading coal by explosive force 爆破装煤
loading code 荷载规范
loading coefficient 荷载系数；加载系数
loading coil 加感线圈
loading coil box 加感箱
loading coil spacing 加感节距
loading combination 荷载组合
loading condition 荷载条件
loading conveyor 装载运输机
loading crane 装载起重机
loading curve 加载曲线
loading cycle 加载周期
loading density 装药密度；装填密度；装药体积系数
loading depot 装货站
loading device 装载装置
loading digger 装载铲斗
loading disc 盘式装载机构；装载盘
loading dock 灌油码头
loading drift 装载平巷
loading dual operation 装车作业
loading duration 加载时间；荷载历时
loading edge 装岩斗
loading effect 负载效应
loading elevator 装载提升机
loading embankment 加荷填方；加载路堤
loading equipment 加载设备
loading error 装入误差；加载误差
loading facility 装油设施；加载设备
loading factor 负载系数
loading filter 压载滤水体；压载过滤器
loading fixture 加载夹具
loading flexible conduit 装油软管
loading flight 装煤刮板
loading floor 装车台
loading force 加载力
loading frame 载荷架；加载架
loading function 加载函数
loading gate 装载平巷
loading gauge 装载限界
loading grab 抓岩机
loading head 装载头；荷载压头
loading height 装载高度
loading history 荷载历史
loading hopper 装料漏斗；装载漏斗
loading hose 装油软管
loading impact 载荷冲击；加载效果，加载影响
loading in increment 分级加载
loading index 承载指数；荷载指数
loading induced stress 载荷引起的应力
loading installation 装油设施；加载装置
loading intensity 载荷强度
loading jib 装载盘；装载臂
loading level 装载水平；装载平巷；加载水平
loading lift 装载高度；装料斗的高程

loading line 灌注线；充气管线；加载线
loading machine 装载机；装药机
loading machine man 装载机司机
loading machine with chain buckets 链斗装车机
loading material 填料
loading measuring device 负荷测量装置
loading mechanism 加载机械装置；加载原理
loading method 加载方式
loading method of initial velocity 初速度加载法
loading of forked coal 块煤装载
loading of water 水体载荷
loading on board 装船
loading operation 装载作业
loading ore 装矿
loading ore through gate chute 闸门漏斗装矿
loading ore through vibrating chute 振动漏斗装矿
loading pad 加载垫板
loading pan 吊桶装载斗
loading parameter 荷载参数
loading path 加载途径
loading piston 承载活塞
loading plant 装载设备；装料设备
loading plate 加载板；载荷板；承载板；载重指示牌
loading plate form 载荷台
loading plate test 载荷实验
loading plough 装载刨煤机
loading pocket 装载仓；装载硐室
loading point 装载点；加载点
loading point displacement 施力点位移；施加载荷作用点处的位移
loading point man 装载点工人
loading pole 炮棍；爆炸填药杆
loading pool 装料池
loading port 装货港口
loading program 装油计划；加载程序
loading pump 装油泵
loading quartz sand transformer 充石英砂型变压器
loading rack 灌油桥台
loading ramp 装煤板；自动滑行装车台
loading range 装载范围；加载幅度
loading rate 加载速率
loading ratio 装炸药比，装药率；荷载比；充填比
loading resistance 加载阻力；负载电阻；抗增重
loading reversal 反复荷载
loading room 装矿硐室
loading routine 加载程序
loading run 装载循环；加载循环
loading section 承载段；装载部分
loading sensing cell 荷载传感器
loading shift 装载班；装煤班
loading shovel 推土机；装载机铲斗
loading slab 重压板；加荷板
loading soil 承载土层
loading space 装载空间；装料室，装料仓
loading speed 加载速率
loading spring 负载弹簧
loading standard for design vehicle 车辆负载标准
loading station 装载站
loading stick 装药棍；塞药棒
loading strength 承载强度

loading stress	荷载应力
loading structure	装载建筑
loading surface	加载面；承载面
loading survey	荷载勘测；载重勘测；装舱检查
loading system	装车系统；加载系统
loading table	容积表；装车台
loading terminal	装车终点站；装货码头；装料站
loading test	加载试验；载重试验
loading test for pile	桩的载荷试验
loading time	装载时间；加载时间
loading to failure	加载至破坏
loading track	装载轨道
loading transfer	荷载传递
loading tray	装车槽；给矿槽；给料盘
loading trough	装载槽
loading truck	装药车
loading type	荷载类型
loading unit	装载机构；加载装置
loading valve	装卸阀；加载阀；装气门
loading vessel	装药器；装卸船；加载容器
loading voucher	装货单
loading weight	载重
loading winch	起重绞车
loading zone	荷载区；加载区域
loading-unloading cycle	加卸载循环
loading-unloading loop	加卸载环
loading-unloading rate	装卸率；加卸荷速度
loadometer	荷载计；测荷仪；轮载测定器；落地磅
loadstone	极磁铁矿；天然磁石
loam	亚黏土；壤土，肥土
loam brick	黏土砖
loam clay	壤黏土
loam moulding	黏土造型
loam pit	采黏土场
loam sand mould	黏土沙泥型
loam soil	壤土
loamification	壤质化
loamy	壤土质的
loamy clay	壤质黏土
loamy coarse sand	壤质粗砂
loamy fine sand	壤质细砂
loamy gravel	含粉质黏土的砾石；壤质砾石
loamy ground	亚黏土质地；泥质土
loamy light sand	壤质松砂
loamy sand	壤质砂土
loamy soil	壤质土
loamy texture	壤质
lobate bedding	叶状层理
lobate delta	舌状三角洲；朵状三角洲；浆叶状三角洲
lobate rill mark	舌状流痕
lobe plough	罗勃型高速刨煤机
lobed	浅裂的
lobulate	具小裂片的
lobus	裂片
local activity	局部活动性
local base level	局部基准面
local bearing area	局部承压面积
local bearing capacity	局部受压承载力
local bending	局部弯曲
local borrow pit	就地取土坑

local break	局部开裂
local buckling	局部失稳；局部屈曲
local building code	地方建筑规范
local building regulation	地方建筑规程
local cleavage	局部劈理
local datum	地方基准面
local deformation	局部变形
local depression	局部陷落；局部下沉，陷落漏斗
local distortion	局部变形；局部畸变；局部失真
local disturbance	局部扰动
local drainage spring	部分排泄型泉
local earthquake	区域地震；地方震
local effect array	场地影响台阵
local energy release rate	局部能量释放率
local erosion base level	地方侵蚀基准面
local excavation	局部开挖
local factor	局部因素；地方条件
local fan	局部扇
local fan ventilation	局扇通风
local field	局部场
local flexure	局部弯曲
local flow system	局部水流系；局部流动系统
local fluctuation	局部涨落；局地波动；局地起伏
local forecast	局部地区预报
local freezing	局部冻结
local geology	地区地质学
local gravity	局部重力值
local gravity anomaly	局部重力异常
local gravity map	局部重力图
local grid refinement	局部网格加密
local ground water	区域地下水
local horizon	视地平
local inclination	局部倾斜
local influence	地方影响；局部影响
local instability	局部不稳定；局部失稳
local irrigation works	地方灌溉工程
local isostatic anomaly	局部均衡异常
local laboratory array	本地实验室台阵
local loading and unloading rate	管内卸率
local locked-up stress	局部紧锁应力
local loose	局部松裂
local magnetic anomaly	局部磁异常
local magnetic disturbance	地区磁扰
local magnitude	地方震级
local material	当地材料；地方材料
local mean sea level	当地平均海面
local mean time	地方平均时
local meridian	地方子午线
local metamorphism	局部变质
local microseismic background level	局部微震背景值
local minimization problem	局部极小化问题
local natural resources	地方自然资源
local network of stations	地方台网
local orientation	局部定向
local peat	潜育泥炭
local pneumatic process	局部气压法
local pore pressure	局部孔隙压力
local power pack	局部液压设备；局部动力源组；局部动力泵站；移动式动力泵站
local precipitation	当地降水量；局部降水量

local pressure	局部压力
local pressure system	局部压力系统
local project	地方项目
local public works	乡村工程；地区小型工程
local public works programme	乡村工程计划；地区小型工程计划
local quenching	局部淬火
local railway	地方铁路
local rehousing	原区安置
local relief	局部地形起伏
local river basin	局部河流流域
local rounding error	局部舍入误差；局部化整误差
local saddle	局部鞍状构造
local scale	局部比例尺
local scour	局部冲刷
local scour near pier	桥墩局部冲刷
local seismic network	地区地震台网
local seismic requirement	地方性抗震要求
local seismicity	地区地震活动性
local shear	局部剪力
local shear failure	局部剪切破坏
local side friction	局部侧向摩擦；局部面摩擦
local sidereal time	地方恒星时
local similarity	局域相似
local site condition	局部场地条件
local site geology	局部场地地质情况
local slope failure	局部边坡破坏
local soil type	土组
local solar time	地方太阳时
local stability	局部稳定性
local stowing	局部充填
local strain	局部应变
local stratigraphic range	局部地层延续；地层延续时限
local stratigraphic unit	地方性地层单位
local stress	局部应力
local stress concentration	局部应力集中
local stress error	局部应力误差
local stress field	局部应力场
local stress raiser	局部应力集中；局部应力集中器
local survey	局部测量；地方测量
local tangent modulus	局部切线模量
local tectonics	局部地质构造
local terrace	局部阶地
local thickness	局部厚度
local topography	局部地形
local trap	局部圈闭
local triangulation	局部三角测量
local triangulation networks	局部三角网
local truncation error	局部截断误差；局部舍位误差
local unconformity	局部不整合
local value	局部值
local variable	局部变量
local variometer	局部磁性变感器；局部磁变记录仪
local velocity	质点速度；局部流速
local ventilation	局部通风
local water discharge pump room	局部排水泵房
local waviness	局部波纹度；局部波痕
local yielding	局部屈服
locale	场所，现场；地点
localised landslip	局部滑坡；局部塌方
locality	位置；地点
locality map	位置图
localizability	可定位性，可定域性；可局限性；本地化
localizable	可定位的
localization	定位，局部化
localization deformation	局部化变形
localization method	定位法
localization of fault	故障定位
localization shear zone	局部化剪切带
localizator	定位器
localize	局部化
localized attack	局部侵蚀
localized control	就地控制；局部控制
localized cooling	局部冷却
localized damage	局部损害；局部破坏
localized deposition	局部沉积
localized ducting	集中管道
localized landslip	局部场方；局部滑坡
localized lighting	分区照明
localized oxide	局部氧化层
localized repair	局部修复
localized scour	局部冲刷
localized seepage	局部渗漏
localizer	定位器；探测器
locally available material	当地可获取的材料；当地可用材料
locally controlled power station	当地控制电站
locate	定位；查出；探明；位于；设置
locate mode	定位方式
located	定位的；设置的
located object	定线目标
locating	划定矿区；设置；找出；探出
locating arrangement	定位装置
locating bar	定位横栏
locating bearing	定位轴承
locating bush	定位套
locating device	定位装置
locating earthquake	地震定位
locating engineer	地震勘探工程师
locating pin	定位针；定位销
locating plate	定位片
locating plunger	定位销
locating point	定点
locating position	定位
locating ring	定位环
locating roller	定位滚柱
locating shoulder	定位台肩
locating sleeve	定位套管
location	定位
location area	位置区
location area code	位置区码
location area identification	位置区识别
location area identity	位置区号；位置区标识
location beacon	位置信标
location bias	定位偏差
location cancellation procedure	位置取消规程
location capability	定位能力
location center	定位中心
location coefficient	位置系数
location condition	位置条件；选点条件

location correction	位置校正
location correction factor	方位修正系数
location crosshair	定位准星
location detection	位置探测
location diagram	图幅位置示意图
location dimension	定位尺寸
location factor	地点因素；选址因素
location file	地点档案
location from traverses	导线测量定位
location in plain region	平原地区选线
location index plan	位置索引图
location indicator	定位指示器
location information	位置信息
location jump of ridge	洋脊跳位
location map	定位图
location map of structure	建筑物位置图
location mark	位置标记
location marker	位置标识器
location method	定位法
location model	选址模型
location monument	定位标石
location of anchorage	锚座的位置
location of drill hole	钻孔定位；钻孔位置
location of earthquake	地震地点
location of epicenter	震中位置
location of groundwater	地下水位置；圈定地下水
location of hydraulic jump	水跃位置
location of leak	泄漏地点
location of line	定线
location of line in mountain region	越岭选线；山岭地区选线
location of line on hilly land	丘陵地段选线
location of malfunction	事故位置测定
location of mistake	错误勘定
location of mountain line	越岭选线
location of pier	桥墩定位
location of railway route selection	铁路选线；铁道选线
location of route	线路中线测量；选线
location of spoil	弃碴位置
location of stop	站台位置
location of structure axis	构造轴部
location of well	定井位；钻孔布置
location over mountain	越岭选线
location plan	位置图
location plate	定位板
location reference	方位数值；参照方位
location register	位置登记器
location registration zone	位置区
location seat	场所
location selection	选址
location sketch	现场草图
location survey	定线测量
location technique	定位方法
location tolerance	位置公差
location tracking	位置跟踪；位置追踪
location update	位置更新
location updating procedure	位置更新规程
location verification	位置确认；位置验证
location with cross-section method	横断面选线
locational accuracy	定位精度
locational precision	定位精度
locational requirement	定位需求
locator	探测器；测位器；位置测定仪；定位程序
locator mandrel	定位心轴
locator means	定位方式，定位方法
locator of pipe	探管器
locator tubing seal nipple	油管定位密封接头
loch	滨海湖；海湾
lock bar pipe	加箍管
lock bar steel pipe	箍锚钢管
lock bolt support with shotcrete	喷锚支护
lock face	锁扣面
lock hopper	闸斗仓
lock in gear	闭锁机构
lock piece	坑木；横梁
locked coil construction	封闭式线股结构
locked coil rope	封闭式钢丝绳
locked coil wire	密封钢丝绳
locked fault	闭锁断层
locked joint	连锁接头；咬口接缝；闭锁节理
locked middling	夹矸中煤；连生中矿
locked position of coupler	闭锁位置
locked sand	嵌固砂
locked segment	锁固段
locking in geologic fault	断层闭锁作用
locomorphic stage	胶结阶段；硬结期
locomotive roadway	机车运输巷道
locomotive routing	机车交路
locus	轨迹；场所；位置
locus of concentration	富集中心
locus of foundering	塌陷中心；沉降中心
locus of principal stress	主应力迹线
locus trace	轨迹
lode	矿脉
lode angle	矿脉角
lode chamber	矿脉变厚处
lode country	矿域
lode formation	矿脉层
lode mining	矿脉开采
lode ore	脉矿石
lode rock	脉岩
lodestone	磁铁矿；天然磁石
lodestuff	矿脉内矿物
lodgement till	底碛，底积冰碛；底积层
lodranite	橄榄古铜陨铁
loellingite	斜方砷铁矿
loess	黄土
loess area	黄土区
loess bridge	黄土梁
loess caliche	黄土砂礓；黄土钙质层
loess child	黄土结核
loess collapsibility test	黄土湿陷性试验
loess collapsing	黄土塌陷
loess column	黄土柱
loess concretion	黄土结核；礓石
loess deposit	黄土沉积
loess doll	黄土结核；礓石
loess earthflow	黄土泥流
loess erosion	黄土侵蚀
loess fissure	黄土裂隙

loess flow 黄土流
loess formation 黄土层
loess foundation 黄土地基
loess fracture 黄土裂缝
loess funnel 黄土漏斗
loess genesis 黄土成因
loess groove 黄土细沟
loess gully 黄土冲沟
loess highland 黄土高原；黄土塬
loess hill 黄土丘陵；黄土峁
loess hilly region 黄土丘陵区
loess joint 黄土节理
loess karst 黄土岩溶
loess kindchen 黄土结核
loess landform 黄土地貌
loess landslide 黄土滑坡
loess like rock 黄土状岩
loess like soil 黄土状土
loess material 黄土层
loess nodule 黄土钙结核
loess origin 黄土成因
loess plain 黄土平原
loess plateau 黄土高原
loess process 黄土重介选法
loess ridge 黄土梁
loess sinkhole 黄土陷穴；黄土落水洞
loess soil 黄土质土壤
loess table 黄土塬
loess tableland 黄土塬；黄土台地
loess terrace 黄土阶地
loess terrace-like plain 黄土台塬
loess wall 黄土墙
loessal 黄土的
loessal soil 黄土性土；黄土类土
loessial 黄土的；黄土类的
loessial plateau 黄土高原
loessial soil 黄土类土
loessial subgrade 黄土路基
loessic stage 黄土期
loessification 黄土化
loessite 黄土岩；古黄土
loessland 黄土地区
lofting 放样；划线
loftsman 放样工
log 钻井柱状图；测井记录，测井；钻孔记录
log analysis center 测井分析中心
log analyst 测井分析员
log book 记录簿
log calibration 测井曲线标定
log computing center 测井计算中心
log dam 堆木坝
log data 测井数据
log data base 测井数据库
log deflection opposite 测井曲线偏转幅度
log editing 测井曲线编辑
log error 测录误差
log file editing 测井文件编辑
log format 测井记录格式
log heading 测井图头
log information 测井资料
log information standard format 测井信息标准记录格式
log interpretation 测井解释
log measurement 测井测量
log of bore hole 钻孔记录；钻孔柱状图
log operation system 测井操作系统
log parameter 测井参数
log presentation 测井显示
log processing center 测井资料处理中心
log quality control 测井质量控制
log reading 测井读数
log repeatability 测井重复性
log resolution 记录分辨率；测井分辨率
log response 测井响应
log seal 木条止水
log sheet 钻井地质柱状图
log temperature 测井温度；记录温度
log training wall 原木导流墙
log value 测井值
log washer 洗矿槽
log well 观测井
log zero 测井零线
logarithmic strain 对数应变
logger 记录器；测井仪
logging 伐木；测井；记录
logging cable 测井电缆
logging caliper 井径仪
logging correction 测井校正
logging current 测井电流
logging curve 测井曲线
logging data 测井数据
logging device 测井仪器
logging down 下行测井
logging drill holes and core 钻孔与岩心的记录
logging equipment 测井装备
logging evaluation 测井评价
logging excavation and test pit 开挖与探坑的记录
logging head 测井仪器接头
logging instrument 测井下井仪器
logging instrumentation 测井仪器
logging interpretation 测井解释
logging interval 记录间隔；测井区间；测井间距
logging method 测井方法
logging mode 测井方式
logging motion arrestor 测井升沉制动器
logging multiple regression analysis 测井多元回归分析
logging of core 岩心编录
logging oscillograph 测井示波仪
logging point 测点
logging program 测井程序
logging response 测井响应
logging sensitivity 测井灵敏度
logging speed 测井速度
logging suite 测井系列
logging technology 测井技术
logging tool 测井仪器
logging tool response 测井仪器响应
logging trace 测井曲线，测井迹线
logging truck 测井车，测井工程车
logging unit 测井仪；测井装置
logging while drilling 随钻测井

logging while drilling apparatus 随钻测井仪
logging wire selection 测井缆芯选择
logistic 逻辑的
logy drill column 不易起下的钻柱
long bar 沿岸沙坝
long baseline interferometer 长基线干涉仪
long baseline positioning 长基线定位
long blast-hole work 深炮眼崩落开采
long boom dragline 长悬臂索斗铲
long borehole 深孔
long clay 高塑性黏土
long distance recorder 远程记录仪；遥测仪
long distance water level recorder 远程水位记录仪；水位遥测仪
long drivage 长距离巷道掘进
long eye auger 深眼木钻；深眼螺旋钻
long face 长壁工作面
long face mining 长壁工作面开采
long face place 长壁工作面
long face system 长壁工作面开采法
long fishing neck 长打捞颈
long gravity wave 长重力波
long haulage 长距离运输
long heavy down grade 长大下坡道
long heavy grade 长大坡度
long hole 深炮眼
long hole benching 深孔台阶式开采
long hole blasting 深孔爆破
long hole drilling 深孔钻进
long hole grouting 深孔灌浆
long hole infusion 深孔注水
long hole rod 深孔钻杆
long hole surveying 深孔测量
long holing 打深孔炮眼
long hollow barrel 长空心筒
long horizontal traverse well 大斜度井
long internal wave 长内波
long interval 长层段；长间距，长间隔
long interval cementing 长封固段固井
long interval method 长间隔法
long lasting sand control treatment 长效防砂作业
long lateral logging 长梯度电极系测井；长梯度测井曲线
long lived fault 长期活动断层
long luminous-flaming coal 长亮焰煤
long migration 长距运移；长距离迁移
long mining face 长工作面；长回采工作面
long period earthquake 长时间地震
long period ventless delay 长时无孔迟发
long period vertical seismograph 长周期垂直地震仪
long pile 长桩
long pillar method 长柱分段开采法
long pillar mining 长柱分段开采
long pillar work 长柱分段开采
long range stress field 长程应力场
long reach outboard arm boom 外侧长摇臂
long rectangular shaft 长矩形立井
long ridged dune 长脊沙丘
long rigid logging tool 长刚性测井仪器
long routed 深炮孔组
long shank drifter 长钎架式钻机
long shore bar 沿岸堤
long spaced 长源距的
long spaced acoustilog 长源距声波测井
long spaced densilog 长源距密度测井
long spaced detector 长源距探测器
long spacing count rate 长源距计数率
long spacing curve 长电极距测井曲线
long spacing detector 长源距探测器
long span 大跨度；长跨度
long span bridge 大跨度桥梁
long span structure 大跨度结构
long span tunnel 大跨度隧道
long spread of detector 长检波器排列
long squeeze 长程挤注
long steep grade 长大坡道；长陡坡
long string 长管柱
long string tailpipe 长管柱尾管
long strip footing 长条形基础
long stroke bumper sub 长冲程震击器
long term integrity of the structure 结构长期完整性
long term loading 长期荷载
long term loading test 长期荷载试验
long term monitoring 长期监测
long term observation 长期观测
long term predictive deconvolution 长期预测反褶积
long term rupture strength 持久断裂强度
long term settlement 长期沉降
long term stability 长期稳定性
long term stability of slope 边坡长期稳定
long toothed bit 长齿钻头
long tunnel 长隧道
long tunnel system 长隧道系统；长隧道掘进法
long wall 长壁
long wall caving 长壁落顶采煤法；长壁崩落法
long wall method 长壁开采法
long wall mining 长壁开采法
long wall system 长壁开采法
long-base method 长基线法
long-core test 长岩心驱替试验
long-crested wave 长峰波；长脊波
long-crested weir 长顶堰
long-distance conveyance 远距离运输
long-distance conveyor 长距离输送机
long-duration static test 长期静载试验；蠕变或疲劳试验
long-duration test 长期试验
long-established station 长期观测站
long-face method 长壁工作面开采法；长壁开采法
long-face place 长壁工作面
long-face system 长壁工作面开采法；长壁开采法
long-feed drill jumbo 长给进钻车
long-flame coal 长焰煤
long-half-life material 长半衰期物质
long-haul 远程的，远距离的
long-hole benching 深孔台阶式开采
long-hole blasting 深孔爆破
long-hole blasting method 深孔爆破法
long-hole grouting 深孔灌浆
long-hole infusion blasting mining 深孔水封爆破采煤

long-hole jetting	深孔水力冲采；深孔水力喷射
long-hole percussion machine	深孔冲击钻机
long-hole pulsed infusion shot-firing method	深孔脉动注水爆破采煤法
long-hole rod	深孔钻杆
long-hole stoping	深孔崩落回采法；深孔回采
long-hole stoping with delay fill	深孔爆破延迟充填采矿法
long-hole system	深孔崩落采矿法；深孔爆破采煤法
long-hole type blasting	深孔爆破
long-holed pillar	钻有深孔的矿柱
long-horn room	煤面与主解理面小于45°角的煤房
longitude horizontal bracing	纵向水平支撑
longitude of epicenter	震中经度
longitude strut	纵向联系梁
longitudinal	纵向的
longitudinal aberration	纵向象差
longitudinal arrangement for mine industrial site	矿山工业场地竖向布置
longitudinal axial force	纵向轴力
longitudinal backstoping and filling	沿走向上向梯段充填回采
longitudinal baffle	纵向折流板
longitudinal bar	纵向沙坝；纵筋，纵钢筋
longitudinal beam	纵梁
longitudinal bending	纵向弯曲
longitudinal bending factor	纵向弯曲系数
longitudinal bent	纵向弯曲的
longitudinal broom	纵向扫纹
longitudinal buckling	纵向弯曲
longitudinal bulkhead	纵舱壁，纵向隔板
longitudinal cap	纵向顶梁
longitudinal carrier cable	纵向承力索
longitudinal center of buoyancy	浮心纵向位置
longitudinal center of gravity	重心纵向位置
longitudinal cofferdam	纵向围堰
longitudinal collision avoidance	纵向防撞
longitudinal compression	纵向压力；纵向压缩
longitudinal consequent stream	纵顺向河
longitudinal contraction joint	纵向收缩缝；纵向收缩节
longitudinal crack	纵裂隙
longitudinal crevasse	纵裂隙；纵向冰隙
longitudinal crown stay	纵向顶撑
longitudinal curl	纵向卷曲
longitudinal cut mining method	纵向工作线采矿法
longitudinal dam	顺坝
longitudinal defect	纵向缺陷
longitudinal deformation	纵向变形
longitudinal dehiscence	纵向开裂
longitudinal dextral shear	纵向右旋剪切
longitudinal diagram	纵剖面图
longitudinal diffusion	纵向扩散
longitudinal diffusivity	纵向扩散系数
longitudinal dike	顺坝；纵向岩脉；纵向岩墙
longitudinal dispersion	纵向弥散
longitudinal dispersion coefficient	纵向弥散系数
longitudinal dispersivity	纵向弥散性
longitudinal displacement	纵向位移
longitudinal disruption	纵断裂
longitudinal ditch	纵向沟
longitudinal divide	纵向分水岭
longitudinal division	纵裂
longitudinal earthquake	纵向地震
longitudinal efflux	顺向冲采；纵向外流
longitudinal expansion joint	纵向伸缩缝；纵向膨胀节
longitudinal extension	纵向伸长
longitudinal fault	纵断层
longitudinal fold	纵褶皱
longitudinal force	纵向力
longitudinal frequency	纵波频率
longitudinal frictional force	纵向摩擦力
longitudinal furrow	纵沟
longitudinal girder	纵梁
longitudinal grade	纵坡；纵向坡度
longitudinal gradient	纵向坡度；纵坡；纵向梯度
longitudinal head	纵轴式截割头
longitudinal impact	纵向碰撞；纵向冲击
longitudinal inclinometer	纵向倾斜计，纵向测斜仪；纵向井斜仪
longitudinal joint	纵节理
longitudinal line	纵线
longitudinal load factor	纵向荷载因子
longitudinal magnetization	纵向磁化
longitudinal member	纵向构件
longitudinal migration	纵向运移
longitudinal moraine	纵向冰碛
longitudinal motion	纵向运动
longitudinal plan	纵剖面图
longitudinal pressure	纵向压力
longitudinal prestress	纵向预应力
longitudinal profile	纵剖面
longitudinal projection map	纵投影图
longitudinal ridge	纵向山脊
longitudinal rift	纵向裂谷
longitudinal rift motion	纵向裂谷运动
longitudinal seam	纵向缝
longitudinal section	纵剖面，纵断面
longitudinal section design	纵断面设计
longitudinal separation	纵向间隔
longitudinal shear	纵向剪切
longitudinal shear crack	纵剪切裂隙；纵剪切裂缝；纵剪切裂纹
longitudinal shear modulus	纵向剪切模量
longitudinal shift	纵向移动
longitudinal shrinkage stoping	纵向留矿回采法
longitudinal slope	纵向坡度
longitudinal slope of the bridge surface	桥面纵坡
longitudinal slot	纵向割缝
longitudinal span length	纵向跨距
longitudinal stability	纵向稳定性
longitudinal stability test	纵向稳定性试验
longitudinal strain	纵向应变
longitudinal strength	纵向强度；拉伸强度
longitudinal stress	纵向应力
longitudinal striation	纵向条纹状
longitudinal structural transposition	纵向构造置换
longitudinal sublevel caving	纵向分段崩落
longitudinal subsidence profile	地表沉陷纵剖面
longitudinal survey	线路测量；纵向调查
longitudinal tectonic system	纵向构造体系

longitudinal thrust 纵向推力
longitudinal thrust fault 纵向冲断层
longitudinal tie 纵向轨枕
longitudinal timber 木纵梁
longitudinal track 长枕埋入式无碴轨道
longitudinal type district station 纵列式区段站
longitudinal valley 纵向谷
longitudinal wave in a bar 杆中的纵波
longitudinal wave velocity 纵波波速
longshore drift 沿岸泥沙流；沿岸沉积物流；沿岸漂移
longshore sediment transport 沿岸输沙；沿岸沉积物运移；沿岸泥沙流移
longulite 长联雏晶
longwall 长壁开采法的
longwall advanced mining method 前进式长壁开采法
longwall advancing 长壁前进式回采
longwall advancing mining 前进式长壁采
longwall advancing on the strike 前进式走向长壁开采法
longwall advancing to the dip 前进式俯斜长壁开采法
longwall advancing to the rise 前进式仰斜长壁开采法
longwall alternating retreat 交替后退式长壁开采法
longwall coal plough giant 长壁刨煤机式水枪
longwall coal-mining method 长壁采煤法
longwall coal-mining system 长壁采煤系统；长壁采煤法
longwall face 长壁工作面
longwall face development 长壁工作面升采
longwall geometry 长壁开采的几何结构
longwall method 长壁开采法
longwall mining 长壁采矿，长壁开采
longwall mining method 长壁采矿法
longwall mining method in coal 煤矿长壁采矿法
longwall mining method in hard rock 硬岩长壁采矿法
longwall mining on the strike 走向长壁采煤法
longwall mining to the dip 倾斜长壁采煤法
longwall mining to the rise 仰斜长壁采煤法
longwall mining with sublevel caving 长壁放顶煤采煤法
longwall mining with top-coal drawing 长壁放顶采煤法
longwall peak method V形工作面全面采矿法
longwall peak stoping V形长壁回采法
longwall peak system V形长壁开采法
longwall piggyback conveyor 长壁工作面转载输送机
longwall pillar method 长壁回采煤柱法
longwall plough installation 长壁工作面刨煤机组
longwall plough machine 长壁刨煤机
longwall retreat method 后退式长壁开采法
longwall retreating 后退式长壁开采
longwall retreating mining 后退式长壁采
longwall retreating on the strike 走向长壁后退开采法
longwall retreating to the dip 俯斜长壁后退开采法
longwall retreating to the rise 后退式倾斜长壁开采法；仰斜长壁后退开采法
longwall shearer installation 长壁工作面滚筒式采煤机组
longwall shearer machine 长壁滚筒采煤机
longwall slicing 长壁分层开采法
longwall slicing along the dip 倾斜分层开采法
longwall stall system 长壁开采法
longwall stoping 上向长壁回采法
longwall strip mining 长壁条带采煤法
longwall system 长壁开采法
longwall tophole system 上向全长长壁开采法

longwall with caving 崩落式长壁开采法
longwall work 长壁开采法
longwall working 长壁开采法
longwall working with caving 崩落式长壁开采法
longwalling 长壁开采法
look for bottom 下放钻柱测井底；放钻具探测井底
looking-glass ore 镜铁矿；光泽赤铁矿
loop lake 牛轭湖
loop pit 环形矿井
loop shaft bottom 环形井底车场；活套坑
looped bar 环形沙洲；环形沙坝
looped barrier 环形沙洲；环形沙坝
looping 构成环形；闭合导线观测
loop-shaped sounding 环形测探；环形测深
loop-type shaft bottom 环形井底车场
loose aggregate 松散集料，松散骨料
loose alluvium 松散冲积层；松散冲击土
loose base 松散基底；松散地基
loose bed 松散岩层
loose blasting 松动爆破
loose block 松散块体
loose cement 松散水泥
loose coal 疏松煤；松散煤；易破碎的煤
loose coefficient of soil 土壤松散系数
loose cutting 从原矿体中切开
loose debris 疏松碎屑物
loose earth 松土
loose earth layer 虚土层；松散土层
loose end 已掘槽采煤工作面；爆破松动了的采煤工作面
loose explosive 松装炸药
loose fill 松散填土
loose foundation 松散地基
loose ground 松散岩体；松散地面
loose pressure of surrounding rock 围岩松动压力
loose rock 松散岩石
loose rock check dam 碎石拦沙坝
loose rock cutting 松散岩质路堑
loose rock dam 碎石坝
loose rock fill 抛石填充
loose rock mass 松散岩体
loose roof 软弱顶板
loose sand 松砂
loose sediment 松散沉积物
loose soil 松散土
loose soil layer 松散土层
loose state 松散状态
loose stemming 松炮泥；松散填塞物
loose structure 散体结构
loose stuff 松散填充物
loose subgrade 路基松软
loose suture 松散缝合线
loose zone 松动区
loose-coupled type pipe 松接式管道
loosely aggregated structure 松聚集结构
loosely bound water 弱结合水
loosely compacted slope 夯压不足的山坡；欠压实的山坡
loosely grained 松散粒状
loosened rock 松散岩石
loosened rock mass 松动岩体
loosened zone 松动区

looseness	松散度	lost river	干河；隐入河
looseness of soil	土壤松散度	lost routed	失效的炮眼组
looseness of structure	结构松散	lost strata	层序间断
loosening	崩松，崩落	lost stream	隐入河；干河
loosening blasting	松动爆破	lost synchronization coefficient	失步系数
loosening ground pressure	散体地压	lost tool	落在井内的工具
loosening junk	松散杂物	lost velocity head	速度头损失
loosening pressure	松动压力	lost volcano	熄火山
loosening resistance	松碎阻力	lot boundary	地段界线
loosening zone of surrounding rock	围岩松动区	lot division	工区划分
loosing pressure	松散压力；松动压力	lot identification	地段鉴辨
loparite	铈铌钙钛矿	lot identification plan	地段鉴辨图
lopezite	铬钾矿	lot index plan	地段索引图
lopolith	岩盆	lot number	地段编号
lopolithic sill	岩盆式岩床	lot owner	地段业权人
lorandite	红铊矿	lot plan	地段图
loranskite	钇钽矿	lot planning	基地规划
lorenzenite	硅钠钛矿	lot size	批量；地块尺寸
lorettoite	黄氯铅矿	lot striping	地块划线
lose circulation	漏失循环液	lotus form structure	莲花状构造
loseyite	蓝锌锰矿	loud tapping	瓦斯啸叫声
losing return	循环液漏失	loudspeaker face telephone	回采工作面扬声电话
losing stream	渗失河	lough	湖；港湾
loss and dilution	损失与贫化	louver dryer	百叶式干燥机
loss by erosion	侵蚀间断；侵蚀损失	louver liner	百叶窗式割缝衬管
loss by percolation	渗漏损失	louver slot	百叶窗式割缝
loss by solution	溶失量	louver vane	百叶窗式导流栅
loss by vapour emission	蒸气放出损耗	lovenite	锆钽矿
loss circulation material	防漏剂	loveringite	钛铈钙矿
loss due to anchorage take-up	锚固损失	low alloy steel	低合金钢
loss due to earthquake	地震灾害损失	low amphibolite facies	低角闪岩相
loss of gauge	量规损失；容量规格损失	low amplitude vibration	低幅振动；低振幅
loss of grout	薄浆流失；薄浆漏出	low and ball	槽脊相间地形
loss of petroleum product	石油产品损耗	low angle	低角度
loss of prestress	预应力损失	low angle boundary	小角晶界
loss of return	井口无返；返排损失	low angle cross-bedding	缓角交错层理
loss of slump	坍落度损失	low angle extensional detachment	缓角延伸脱顶作用
loss of water from tunnel	隧道渗水	low angle fault	缓角断层
loss of well control	井失控	low angle faulting	缓角断层作用
loss of working diameter	井径缩小	low angle hole	低倾角井
loss on ignition	烧失量	low angle normal detachment fault	低角度滑脱断层；低角度拆离断层
loss rate of circulation	井漏速率	low angle reverse fault	低角度逆冲断层
loss through breathing	蒸发损耗	low angle scattering	小角散射，低角散射
lossening force	松碎力	low angle thrust	低角冲断层
lossless filter	无损滤波器	low ash coal	低灰分煤
lossless formation	无损耗地层	low assay	低指标；低含量
lossy formation	有损耗地层	low bog	低地沼泽
lost circulation	损失循环液；井漏	low bush tundra	灌木冻原；灌木冻土层；灌木苔原
lost circulation additive	堵漏剂	low byte strobe	低位组选通脉冲
lost circulation alarm	漏失报警器	low carbon ferrochrome	低碳铬铁
lost circulation interval	漏失段，井漏段	low carbon pig iron	低碳生铁
lost circulation material	堵漏材料	low carbon steel	低碳钢
lost circulation plug	堵漏塞	low cement content	低水泥用量
lost circulation zone	循环漏失区	low channel seismogram	低信道地震图
lost cone	掉牙轮	low coal	薄煤层
lost head	压头损失	low coal rank	低煤阶，低煤级
lost hole	废孔	low country	低洼地区
lost level	定线不准确的地下平巷	low cover tunnel	浅埋隧道
lost record	地层缺失	low current velocity method	烟雾测量低速风流法；低流
lost return	循环液漏失		

速法
low cut filter 低阻滤波器
low dam 低坝
low density concrete 低密度混凝土
low density explosive 低密度炸药
low density foam pig 低密度泡沫清管器
low density gravity flow 低密度重力流
low dip 缓倾斜
low dip angle 缓倾角
low discharge 低流量；小流量
low down pump 浅井潜水泵
low earth barrier 低土围堤；低土围墙
low embankment 低路堤；低堤岸
low energy well 低能量井
low excess air 低过剩空气
low excess air burner 低过剩空气火嘴
low excess air operation 低过剩空气操作
low explosive 低效炸药
low explosive limit 爆炸下限
low fill and shallow cut 低填浅挖
low fine gravel 细砂含量低的砾石
low flow forecast 低流量预报
low flowing pressure 低流动压力；低动水压力
low fluid level well 低液面井
low fluid-loss cement 低失水水泥
low freezing dynamite 难冻硝甘炸药
low frequency 低频率
low frequency earthquake 低频率地震
low frequency electrified railway 低频电气化铁道；工频电气化铁路
low frequency filter 低频滤波器
low frequency rock drill 低频率冲击式凿岩机
low friction compound 减摩剂
low gaseous mine 低瓦斯矿井
low grade coal 低品位煤；劣质煤
low grade metamorphism 低级变质作用
low grade ore 贫矿；低品位矿石
low gradient 平缓坡度
low gravity 低比重
low gravity crude 低比重原油
low gravity float 低比重浮选
low greenschist facies 低绿片岩相
low greywacke 低杂砂岩
low head 低水头
low impact sensitivity 轻冲击敏感性；低撞击感度
low impedance interval 低阻抗层段
low impedance sampling oscilloscope 低阻抗采样示波器
low injection pressure 低注入压力
low interfacial tension displacement 低界面张力驱替
low land peat 低位泥炭
low level bog 低沼泽地
low level laterite 低地砖红壤
low liquid limit silt 低液限粉土
low loss 低损耗
low magnitude earthquake 低震级地震
low marsh 低平沼泽
low moor 低位沼泽
low moor peat 低位泥炭
low moor soil 低位沼泽土
low moor wood peat 低位沼泽森林泥炭

low mountain 低山
low pass 低通，低通量
low pass filter 低通滤波器
low pass filtering 低通滤波
low permeability 低渗透性；低渗透率
low permeability coal seam 低渗透煤层
low permeability contrast reservoir 渗透率差异小的储层
low permeability gas reservoir 低渗透气藏
low permeability layer 低渗透层
low permeability mud cake 低渗透性泥饼
low permeable sublayer 低渗透夹层
low perviousness 低透水性；低渗透性
low phosphorus iron 低磷生铁
low pile cap 低桩承台
low pitch 缓倾斜；低螺距
low pitch cone roof 小坡度锥形顶盖
low pitched roof 缓坡屋顶
low place 低工作区；薄煤层工作面
low plasticity clay 低塑性黏土
low plasticity soil 低塑性土
low plunge 平缓倾伏
low pressure grout hole 低压灌浆孔
low producing well 低产井
low profile crusher 低机身破碎机
low quartz 低温石英
low rank bituminous coal 低级无烟煤
low rank coal 低煤阶煤；低级煤；劣质煤
low red heat 低红热
low relief 低地势；浅浮雕
low residue fracturing fluid 低残渣压裂液
low resistivity 低阻
low resistivity pay zone 低电阻率产油层
low resistivity zone 低阻层
low Reynolds number flow 低雷诺数流体
low ring lock 下环梁锁销
low river basin 低洼盆地；低河盆；低流域
low runoff 枯水径流；枯水流量
low salinity water 低矿化度水；低盐度水
low seam 薄煤层；薄矿层
low seam conveyor 薄煤层输送机
low selector 低压选择阀
low setting load 低初撑力；低预压负荷
low shaft furnace 矮高炉
low shrink resin 低收缩树脂
low shut pressure 低关井压力
low silicon iron 低硅生铁
low solid content 低固相含量
low solid mud 低固相泥浆
low specific speed 低比速；低比转数
low steel proportion 低含钢率
low strain 低应变；弱应变
low strength 低强度
low strength cement 低强度水泥
low strength concrete 低强度混凝土
low sulfidation 低硫化
low sulfur waxy residue 低硫含蜡渣油
low swampy land 低湿地
low temperature annealing 低温退火
low temperature briquette 低温炭化煤砖；低温半焦化煤球

low temperature carbonization	低温碳化；半焦化
low temperature chromatography	低温色谱分析法；低温层析法
low temperature distillation	低温干馏
low temperature drying	低温干燥
low temperature gas separation process	深冷气体分离过程
low temperature hydrothermal process	低温热液过程；低温热液作用
low temperature resistance	耐低温性
low temperature storage tank	低温储罐
low to moderate consequence	轻微至中等危害后果
low to moderate risk	低至中等风险
low truss	低架式；低桁架
low truss bridge	低桁架桥
low type pavement	低级路面
low vacuum	低真空
low vein	薄矿脉；贫矿脉
low velocity explosive	低速炸药
low velocity layer	低速层
low velocity region	低速带
low velocity zone	低速带
low viscosity flow packing	低黏液循环充填
low wall	下盘；底帮
low wall ramp	底帮坡道
low water cut well	低含水井
low water ratio cement	低水灰比水泥
low water-bearing formation	弱含水层
low weir	低堰
low yield clay	低造浆黏土
low yield explosion	低量级爆炸
low-alkali cement	低碱水泥
low-amplitude paleofold	小幅度古褶皱
low-amplitude paleostructure	小幅度古构造
low-angle cross-bedding	低角度交错层理
low-angle dip	缓倾角
low-angle fault	缓倾角断层
low-angle growth fault	低角生长断层
low-angle negative rake	小负角倾斜定向
low-angle overthrust	低角度逆掩断层；低角度上冲断层
low-angle overthrust fault	低角度逆掩断层；低角度上冲断层
low-angle shear plane	低角度剪切面；低角度剪断面
low-arched	低拱的；缓拱顶的；微隆起的
low-ash coal	低灰分煤
low-ash material	低灰分物料
low-attenuation sinking plate	低衰减下沉板块
low-capacity pump	低流量泵
low-capacity well	低出水量井
low-coal bolting pattern	薄煤层锚杆布置方式
low-coal deposit	薄煤层
low-coal mine	薄煤层矿
low-coal seam	薄煤层
low-curing cutback asphalt	慢干稀释沥青
low-deformation zone	弱变形带
low-density layer	低密度层
low-density material	低比重物料
low-density oil	轻质石油
low-density slurry	低密度水泥浆
low-density solid	低密度固体颗粒
low-dipping	低倾角的；缓倾角的
low-dipping fault	缓斜断层；缓倾向断层
low-discharge mill	低排料磨机
low-dispersed mud	粗分散泥浆
low-energy seismic-source system	低能地震源系统
low-energy source	低能量震源
low-energy state	低能态
low-energy zone	低能带
low-enrichment uranium	低浓缩铀
low-enthalpy water	低焓热水
lower adaptor	下接头
lower and middle Jurassic	下-中侏罗统；早-中侏罗世
lower apex of fold	向斜顶点；褶皱底，褶皱下顶
lower arch bar	下拱杆；下拱板
lower bearing spider	下轴承支座
lower bed	低河床；下层河床
lower bench	下台阶；下梯段
lower bend	凹槽
lower block	下盘
lower bound solution	下限解
lower bound theorem	下限定理
lower boundary	下边界
lower break	褶皱的下弯折部分
lower catchment area	下段集水区；低汇水面积
lower coal measures	下部煤系；低煤层
lower confining bed	隔水层底板
lower core	下地核
lower course of river	河流下游
lower Cretaceous	早白垩世；白垩纪早期；下白垩系
lower critical velocity	低临界流速；低临界速度
lower crust	下地壳
lower crust seismic structure	下地壳地震结构
lower crustal composition	下地壳组成；下地壳成分
lower crustal layer	下地壳层；康拉德层
lower curved section	斜井段
lower deviation	下偏差
lower discharge	小流量
lower drilling limit	低钻井范围
lower elastic limit	弹性下限
lower ever known discharge	迄今最小流量
lower ever known water level	迄今最低水位
lower expansion chamber	钟形进水口
lower explosive limit	爆炸下限
lower extreme	下端；下限
lower fan	冲积扇下段
lower flange of girder	梁下翼
lower flow regime	下部水流动态
lower foot slope	坡脚；底帮
lower frame	底座；底部框架
lower gangway	下运输道；下平巷
lower gravel pack	下层砾石充填
lower guard sill	下游防护栏
lower kelly cock	方钻杆下旋塞阀
lower lateral bracing	下弦横向支撑；下弦水平支撑
lower layer	底层
lower leaf	下分层回采
lower limit magnitude	起算震级；震级下限
lower limit of plasticity	塑性下限
lower limit of separation	分选下限
lower limit of yield	屈服下限

lower lip 出口戽斗
lower mantle 下地幔
lower member 下部构件；底部构件
lower nappe profile 水舌下缘线
lower plane 底面
lower plastic limit 塑性下限
lower plasticity clay 低塑性黏土
lower plate 断层下盘
lower polar stratosphere 极地平流层下部
lower power station 下游电站
lower prop 下支柱；下部支撑
lower punch 下冲头；下冲杆
lower rank coal 低级煤；低煤化程度的煤
lower rank coking gas coal 低级焦化气煤
lower relief 平缓起伏；低地势
lower reservoir 下游水库；储层底部
lower ring 下环梁
lower roll 下辊；底辊
lower sample 下部取的样品；下层位样品
lower series 下统
lower side wall 下边墙；下侧壁
lower sideband 下边带
lower size limit of feed 入洗下限粒度
lower sliding part 下部滑动段
lower sublittoral zone 亚滨海带底部；低亚滨海带
lower subsoil 心土下层；下层底土
lower surface of permafrost 永冻层底面
lower surge basin 低冲击盆地；低浪涌盆地
lower surge chamber 下调压室；下缓冲室
lower tension carriage 下部拉紧装置；下部拉紧拖架
lower the leg 液压支架降柱
lower thermosphere 低热层；低热电离层
lower transverse strut 下横撑；下横梁
lower wall 下盘
lower-bound solution 下界解
lower-hemisphere equal-area diagram 下半球等面积图解
lower-hemisphere projection 下半球投影
lower-horizon soil 下层土
lowering of ground water level 地下水位下降
lowering of labor force 工人下井
lowering of roadbed 落底
lowering of track 落道
lowering of water table 潜水面下降
lowering the water table by well-point 井点降水法
lowermost mantle 地幔最下部
lowermost packer 最下部封隔器
lowest groundwater level 地下水最低水位
lowest water level 最低水位；最低潮位
low-fluid loss mud 低失水泥浆
low-fluid well 低产井
low-freezing 低凝固点的
low-freezing explosive 低凝固点炸药；低冻炸药
low-grade area 贫矿区；低品位区
low-grade cement 低标号水泥
low-grade concrete 低标号混凝土
low-grade deposit 贫矿床；低品位矿床
low-grade gas 低级煤气；贫煤气
low-grade oil 低级油
low-grade phosphate rock 低品位磷矿
low-grade reserves 低级储量

low-hardness water 低硬度水
low-heat cement 低热水泥
low-heat portland cement 低热硅酸盐水泥
low-heat value gas 低热值气体；低热值煤气
lowland bog 低位沼泽；低位泥塘
lowland catchment 低位集水区；低流域
lowland coast 低地海岸；低海岸
lowland deposit 低地沉积
low-level groin 河底潜坝
low-level groyne 河底潜坝；低位防波堤
low-level switch 最低充液面开关；下限油位开关
low-lift centrifugal pump 低扬程离心泵
low-lift construction for mass concrete 大体积混凝土薄层施工
low-lying deposit 谷地矿床；低地矿床
low-lying ground 低洼地
low-lying land 低洼地
low-lying sand 深层砂
low-magnesian calcite 低镁方解石
low-magnification seismograph 低倍率地震仪；低放大率地震仪
low-moor peat 低沼泥炭
low-noise ventilator 低噪声通风器
low-oil 轻质石油
low-permeability gas reservoir 低渗气藏；低渗天然气储层
low-permeability strata 低渗地层
low-porosity concrete 低孔隙度混凝土
low-porosity rock 低孔隙度岩石
low-potassium tholeiite 低钾拉斑玄武岩
low-pressure filter press 低压压滤机
low-pressure groundwater 低压地下水
low-pressure grouting 低压灌浆
low-pressure tunnel 低压隧洞
low-pressure well 低压井
low-productivity layer 低产层
low-productivity zone 低产层
low-rank graywacke 低级杂砂岩
low-ranking heavy oil 低级重油
low-seam conveyor 薄煤层输送机
low-seismicity macrozone 低地震活动性大区
low-seismicity region 低地震活动性区
low-side coal bump 煤层下帮突出
low-slump concrete 低坍落度混凝土；干硬混凝土
low-soild 低固相的
low-solid system 低固相体系
low-speed jet 低速射流
low-speed landslide 低速滑坡
low-speed layer 低速层
low-speed mill 低速辊碾机
low-sulfur coke 低硫焦
low-sulfur crude oil 低硫原油
low-sulphur coal 低硫煤
low-sulphur crude oil 低硫原油
low-sulphur fuel 低硫燃料
low-temperature ashing of coal 煤的低温灰化
low-temperature coke 低温焦炭
low-temperature steel 低温钢
low-temperature tar 低温焦油
low-type machine 低型机；薄煤层用的机器

low-type ranging shearer	低型可调高滚筒采煤机
low-viscosity layer	低黏度层
low-volatile bituminous coal	低挥发分烟煤
low-volatile coal	低挥发分煤
low-volume producer	低产井
low-volume well	低产井
low-yield clay	低膨胀黏土；低造浆黏土
lozenge-shaped bar	菱形砂坝
lozenge-shaped block	菱形断块
LPD	施力点位移；施加载荷作用点处的位移
LSY	大范围屈服
lubeckite	铜钴锰土
lubumbashite	水钴矿
lucianite	富镁皂石
luciite	细粒闪长岩；细闪长岩
lucinite	磷铝石
luckite	绿锰铁矾
lucullan	碳大理石
ludlamite	板磷铁矿
ludwigite	硼镁铁矿
luneburgite	硼磷镁石
lueshite	钠铌矿
lugarite	沸基辉闪斑岩；闪辉沸霞斜岩
Lugeon	吕容
Lugeon test	钻孔压水试验；吕容试验
Lugeon unit	吕容单位
luhite	蓝方黄长霞煌岩
luigite	结灰石
lujavrite	异霞正长岩
lujavritite	长辉霞石岩；辉霞正长岩
Luliang movement	吕梁运动
lumachelle	贝壳大理岩；介壳层；介壳砾岩
lumbar region	腰区，腰部
Lumb's index	伦布指数
luminescence mineral	发光矿物
luminite cement	矾土水泥
luminous mineral	发光矿物
lump breaker	块煤破碎装置
lump coal	块煤
lump coal yield	块煤产量
lump coke	块焦炭
lump lime	生石灰
lump limestone	团块灰岩
lump middling	块中煤；块中矿
lump ore	大块矿石；矿块
lump pumice flow	块状浮石流
lump screen	块煤筛
lump soil	块土
lump sum contract	定额工程合约
lump sum item	包干项目
lumped mass	集中质量
lumped mass element	集中质量单元
lumped mass idealization	集中质量理想化
lumped mass matrix	集中质量矩阵
lumped mass method	集中质量法
lumped mass model	集中质量模型
lumped mass system	集中质量体系
lumped model	集总模型
lumped parameter	集总参数
lumped parameter method	集总参数法
lumped parameter model	集中参数模型
lumped parameter system	集总参数系统
lumped reactance	集总电抗
lumped seismometer	集总地震检波器
lumped voltage	集总电压
lumped-coefficient system	集总参数体系
lumped-constant simulation	集总常数模拟
lumped-parameter model	集总参数模型
lumping	成块，成团；换装整体铁轨枕木法
lumpy cement	块状水泥
lumpy clay	块状黏土
lumpy groundmass	团粒结构基质
lumpy soil	碎块土；土块
lumpy structure	碎块结构
lunabase	月岩；月球上低平面的；月海的
lunar anorthosite	月球斜长岩；月球钙长岩
lunar bar	新月形沙洲；新月形沙丘；新月形沙坝
lunar basalt	月球玄武岩
lunar basin	月盆
lunar celestial equator	月球赤道
lunar chronology	月面年代学
lunar core	月核
lunar corona	月华；月晕
lunar crack	新月形裂缝
lunar crater	月坑；月球陨石坑；月球火山口
lunar crust	月壳
lunar diurnal component	月球全日分潮
lunar diurnal variation	月球日际变化
lunar drill	月球钻
lunar exploration	月球探索；月球探测
lunar fortnightly tide	月球二周潮；月球半月潮
lunar geodesy	月面测量学；月球测绘
lunar geology	月球地质学
lunar gravimetry field	月球重力场
lunar halo	月晕
lunar highland	月球高地
lunar hour	太阴时；月球时
lunar igneous rock	月球火成岩
lunar oscillation	月球振荡
lunar parallax	月球视差
lunar parallax inequality	月球视差不均衡
lunar petrology	月岩学；月球岩石学
lunar ranging retroreflector	月球测距后向反射器
lunar regolith	月土；月壤
lunar ring	月球环形山
lunar rock	月岩
lunar rover	月球车
lunar seismogram	月震图
lunar seismology	月震学
lunar semidiurnal tide	太阴半日潮
lunar soil	月壤
lunar soil mechanics	月球土力学
lunar stratigraphy	月球地层学
lunar tectonism	月球构造作用
lunar tidal component	太阴潮汐分量
lunar tide	太阴潮
lunar time	太阴时
lunar volcanism	月球火山作用
lunar year	太阴年，农历年
lunarite	月陆，月面；酸性月岩；月球高地表

lunate bar	新月形沙坝；新月形沙洲
lunate fracture	新月形破裂
lundyite	红钠闪正长斑岩
lunette	风成新月形脊；月牙形低丘；弦月窗，弧形窗
lunisolar attraction	日月引力
lunisolar diurnal tide	日月潮汐
lunisolar gravitational perturbation	日月引力摄动
lunisolar precession	日月岁差
lunule	前月面；半月形
lupatite	霞长斑岩
lurching coal seam	杂乱煤层；突然倾斜煤层
lurching seam	杂乱煤层；突然倾斜煤层
lurescope	月球测距镜
lusakite	钴十字石
luscladite	橄辉斜霞岩；橄榄霞斜岩
lusec	流西克
lusitanite	霓闪钠长正长岩；斜磷锌矿
lussatine	负绿方石英
lustrous coal	镜煤，闪光煤
lusungite	水磷铁锶矿
lutaceous	黏土质的；泥质的
lutalite	暗碱玄白霞岩
lute	密封胶泥；封泥
luting paste	封填膏
lutite	泥质岩
lutyte	泥岩
luvisol	淋溶土
luxmass	活性黄土
luxullianite	电气花岗岩
luzonite	四方硫砷铜矿；锑硫砷铜矿
LVDT	线性差动变压器；线性差动位移传感器
lydite	燧石板岩；试金石；碧玄岩
lye	碱液；用碱液洗涤
lying back cleat	后倾节理
lying fold	伏卧褶皱；伏褶皱
lying forward cleat	前倾节理
lying overfold	伏卧倒转褶皱
lying tube well	卧管井
lyndochite	钙钛黑稀金矿
lynx stone	琥珀；山猫石
lyosorption	吸附溶解作用；吸收溶剂
lype	有擦痕光面的断层；顶板易脱落岩石；破裂易落顶板
lysimeter	测渗计；溶度计
lysimeter pit	渗透仪坑；溶度计坑
lysimetric experiment	渗漏试验
lysimetric method	土中渗透仪法
lysimetry	渗透测定计
lysocline	溶跃面；溶跃层
lythrodes	白斑霞石
lytomorphic	热液溶解变形的
lytomorphic crystal	溶解变形晶体

M

Ma'anshan beds 马鞍山层
Ma'anshan clay 马鞍山黏土
Ma'anshan formation 马鞍山组
Ma'anshan granite 马鞍山花岗岩
Ma'anshan limestone 马鞍山石灰岩
Ma'ao formation 马凹组
maar 马尔式火山；低平火山口；火山口湖
Maas compass 马氏罗盘
Maas survey instrument 马氏测斜仪
MAASP 套管环隙最大容许压力
Maastrichtian 马斯特里赫特阶
Maba limestone 马坝石灰岩
macadam 碎石；碎石路
macadam aggregate 粗粒掺合料
macadam base 碎石基层；碎石底层
macadam effect 自胶结作用
macadam foundation 碎石地基；碎石基层
macadam pavement 碎石路面
macadam road 碎石路
macadamization 碎石路修筑法；碎石铺路法；碎石铺路
macallisterite 三方水硼镁石
macaluba 泥火山
macaroni 小直径管
macaroni pipe 细管
macaroni rig 小直径油管修井机
macaroni string 小直径油管柱
macaroni tubing string 小直径油管柱
macasphalt 碎石沥青混合料
macasphalt type pavement 碎石柏油路面
maccaboy 一种雷达干扰寻觅器
maccaluba 泥丘；泥火山
macedonite 云橄粗面岩；云橄粗安岩
maceral 显微组分；煤显微组分
maceral suite 显微组分组
maceral variety 显微组分变种
maceration 浸渍作用；浸解；浸软；浸化；离析
macerator 碎渣机
macfallite 锰绿帘石
macfarlanite 杂砷钴镍矿
MacGeorge's method 麦克乔治法
mach 马赫
mach number 马赫数
mach principle 马赫原理
mach reflection 马赫反射
mach wave 马赫波
machanical amplitude 机械振幅
machanical impedance 机械阻抗
machanism of rock blasting 岩爆机理
machinability 切削加工性；切削性
machinable 可切削的；可用机器加工的
machine 机械加工；切削加工；机器；机械
machine aided cognition 计算机辅助识别
machine arithmetic 机器运算
machine attendance 机器保养

machine attention 机器保养
machine availability 机械利用率
machine bay 机械工段
machine bit 采煤机截齿
machine boreability 机械钻眼性；机械可钻性
machine check 自动检验；程序检验；机器校验
machine check interruption 计算机检查中断
machine code 机器代码；指令表
machine cognition 机器识别
machine computation 机器计算
machine controlled analysis 计算机控制分析
machine cut 机械切削；机械切割
machine cycle 计算机周期；机器工作周期
machine design 机械设计
machine drawing 机械制图；机械设计图
machine drill 钻机；凿岩机
machine driller 钻眼工；凿岩工；钻探工
machine drilling 机械钻眼；机械钻进；机械凿岩；机械穿孔
machine drive 机械传动
machine efficiency 机器效率
machine engineer 机械工程师；机械制造工程师
machine finishing 机械精加工
machine for roadway work 路基机械
machine hand 机械手
machine height 机身高度；自采煤工作面底板至采煤机机身上表面的高度
machine holing 机械成孔
machine hours 机器运转时间
machine idle time 设备闲置时间
machine instruction 机器指令
machine instruction code 机器指令码
machine intelligence 机器智能
machine kibble 小吊桶
machine language 机器语言；计算机语言
machine language program 机器语言程序
machine learning 机器学习
machine loader 装载机
machine loading 机械装载；机械加载
machine loading ore 机械装矿
machine mining 机器开采；机械化开采
machine mixer 搅拌机
machine mixing 机械搅拌；机械拌合
machine oil 机油
machine operation 机器操作
machine operator 机器操作员
machine oriented 面向机器的
machine processing 计算机处理
machine program 机器程序
machine raising 机械升举
machine rammer 动力捣固锤；打夯机，夯击机；机械夯
machine readable 机器可读的
machine recognizable 机器可识别的

machine reliability	机器的可靠性
machine rock drill	岩石钻机
machine room	机房
machine run	机器运行
machine runway	机道
machine sampler	取样机
machine script	机器可读数据
machine sensible	机器可感的；机器可读的；机器可感知的
machine set	机组
machine set bit	机镶细粒金刚石钻头
machine setting	机床安装；机械设置
machine shoe	滑靴
machine shop	金工车间；机械加工车间
machine starting	机器启动
machine stroke	机械冲程
machine time	机械开动时间
machine tool	机床；工作母机；工具机
machine translation	机器翻译
machine trapping rail	采煤机导向轨
machine tunnel boring	隧洞全断面掘进机
machine tunneling	机械开挖隧洞
machine unit	机组；运算装置
machine variable	计算机变量
machine walking fixture	掘进机步进装置
machine wall	机械化作业面
machine with external blower	外部吹风机式浮选机
machine with internal air-pump	内部空气泵式浮选机
machine work	机械加工；切削加工
machine worked mine	机械化煤矿；机械化矿山
machine-assembled	机器装配的
machine-assembly department	机械装配车间
machine-controlled analysis	计算机控制分析
machine-cut	机械掘进的
machine-cut peat	机采泥炭
machine-cutting shop	金工车间
machined	机械加工的
machined gate seat	机加工闸门支座
machined parts	已加工部件
machine-gun hydraulic jet	高压水枪射流
machine-hours	机器工时；机器的运转时间
machine-mixed concrete	机拌混凝土
machine-oriented language	面向计算机的语言
machine-processable form	机器可处理形式
machine-readable data	机器可读数据
machinery	机械设备；机器制造；机构；机械
machinery and equipment	机器设备
machinery cost	施工机械使用费
machinery duty	机械的功能
machinery hall	机器房；机器间
machinery noise	机械噪声
machinery parts	机械部件；机械零件；机械零部件
machinery vibration	机器振动
machine-sharpened bit	机磨钎头；机磨截齿
machine-spoiled time	机器故障时间
machine-tool analyser	机床测定装置
machinework	机加工；机械制品
machine-worked mine	机械化矿；机器开采矿
machining	加工；机械加工
machining precision	加工精度
machining property	机械加工性能
machining quality	可加工性；可切削性
machmeter	马赫表
mackayite	水碲铁矿
mackensite	铝绿泥石
macker	碳质页岩
mackerel sky	预压法
mackintoshite	黑铀钍石；黑铀钍矿
Mackower gas-adsorption apparatus	麦柯沃型气体吸附仪
Maclaurin series	麦克劳林级数
Maclaurin series expansion	麦克劳林展开
Maclaurin's formula	麦克劳林公式
macle	矿石暗斑；短空晶石
macrinite	粗粒体
macro	巨大的；大量的；宏观的；宏指令
macro analysis	总体分析；宏观分析
macro assembler	宏汇编程序
macro assembly language	宏汇编语言
macro axis	长轴
macro call	宏调用
macro check	宏观分析；宏观检查
macro crack	大裂隙；宽裂隙
macro definition	宏定义
macro element	大量元素；宏量元素
macro epicenter	宏观震中
macro etch	宏观腐蚀
macro etching	宏观侵蚀
macro generator	宏功能生成程序
macro instruction	宏指令
macro irregularity	宏观不均质性
macro parameter	宏参量；宏参数
macro qualitative analysis	常量定性分析
macro subsurface model	宏观地层模型
macroalgae	大海藻
macroanalysis	宏观分析；常量分析
macro-anisotropy	宏观各向异性
macroassembler	宏汇编程序
macroaxis	长轴
macroburrowed strata	巨潜穴地层
macrocale	大比例尺；宏观尺度
macroclastic rock	粗屑岩
macrocleavage	粗劈理
macroclimate	大气候
macroclimatology	大气候学
macrocode	宏代码
macrocoding	宏编码
macroconcentration	常量浓度
macroconstituent	常量成分
macrocrack	宏观裂纹；宏观裂缝；宽裂缝；大裂隙
macrocrystalline	粗晶的；粗晶质；宏晶
macrocycle	大旋回
macrodcopic displacement efficiency	宏观驱替效率
macro-diagonal	长对角线
macrodiscontinuity	宏观非连续性
macrodispersoid	粗粒分散胶体
macrodome	长轴坡面
macroeffect	宏观效应
macroelement	常量元素；宏组件；大量元素
macro-epicenter	宏观震中
macroetching	粗形侵蚀；宏观侵蚀

macroevolution	宏观演化；大阶段演化；长期演化
macroexamination	宏观分析；宏观检验
macrofabric	宏观组构；大型组构；粗组构
macroface	宏观相
macrofeature	宏观特征
macrofluid	宏观流体
macrofold	宏观褶皱；大褶曲；大褶皱
macrofossil	巨体化石；大化石
macrofractography	宏观断口分析
macrofracture	大破裂；宏观裂缝
macrofracture fabric	大裂缝组构
macrofragment	粗碎屑
macrofragmental coal	显组分煤
macrogeneration	宏生成
macrogeography	宏观地理学
macrogeological cycle	地质大旋回
macrogphysics	宏观物理学
macrograin	粗粒；粗晶粒
macrograined	粗粒的
macrogranular structure	粗粒组构
macrograph	肉眼图；宏观图
macrography	宏观摄影术；肉眼检查
macroinstruction	宏指令
macroite	粗粒惰性体
macrokinetics	宏观动力学
macrolibrary	宏库；宏程序库
macrolithic	粗大石器
macrolithology	宏观岩性特征；宏观岩性学
macrolithotype	宏观煤岩类型
macromeritic	粗晶粒状
macrometeorology	宏观气象学；大气象学
macrometer	测距器；测远计
macromethod	常量法；宏观法
macro-modular computer	宏模组件计算机
macromodule	大模块
macromolecular	大分子的
macromolecular chemistry	大分子化学
macromolecular compound	大分子化合物
macromolecular dispersion	大分子分散体系
macromolecular fluid	大分子流体
macromolecular network structure	大分子网状结构
macromolecule	大分子
macromosaic structure	宏观镶嵌结构
macronucleus	巨核；大核
macrooscillograph	常用示波器
macroparameter	宏参数
macropetrographical section of a coal seam	宏观煤层柱状剖面图
macrophotogrammetry	超近摄影测量
macrophyric	粗斑状的
macropinacoid	长轴面
macroplate	大板块
macropore	大孔隙
macropore coefficient	大孔隙系数
macroporosity	大孔隙度；宏观孔隙率
macroporous	大孔隙的；大孔的
macroporous coefficient	大孔隙系数
macroporous soil	大孔性土；大孔土
macroporous structure	大孔结构
macroporphyritic	大斑晶的；粗斑状的
macroprocessor	宏处理程序
macroprogram	宏程序
macroquake	巨大地震；大震
macroreef	宏观礁；大礁
macro-regionalization	大区域划分
macrorelief	宏观地形；大起伏
macro-rheology	宏观流变学
macrosamlpe	大试样
macroscale	宏观尺度；大规模
macroscopic	巨型的；宏观的；粗大的；肉眼可见的
macroscopic absorption cross section	宏观吸收截面
macroscopic analysis	宏观分析
macroscopic anisotropy	宏观各向异性
macroscopic boundary condition	宏观边界条件
macroscopic capture cross section	宏观俘获截面
macroscopic cavitation	大空穴；大洞穴
macroscopic constitutive behavior	宏观本构特性
macroscopic criterion	宏观判据
macroscopic cross-section	宏观截面
macroscopic damage	宏观损伤
macroscopic deformation property	宏观形变性质
macroscopic description	宏观描述
macroscopic earthquake precursor	宏观地震前兆
macroscopic effective cross section	宏观有效截面
macroscopic epicenter	宏观震中
macroscopic examination	肉眼检查
macroscopic failure localization	宏观破坏局部化
macroscopic fault motion	宏观断层运动
macroscopic flow direction	宏观流向
macroscopic flow velocity	宏观渗流速度
macroscopic flux	宏观流量
macroscopic fraction	宏观的粒级
macroscopic fracture	宏观裂缝
macroscopic granular	粗颗粒的
macroscopic heterogeneity	宏观非均质性
macroscopic homogeneity	宏观均质性
macroscopic instability	宏观不稳定性
macroscopic interface	宏观界面
macroscopic level	宏观水平
macroscopic lithotypes	宏观煤岩类型
macroscopic method	宏观法
macroscopic observation	宏观观测
macroscopic permeability	宏观渗透率
macroscopic pore structure	宏观孔隙结构
macroscopic pre-earthquake anomaly	地震宏观异常
macroscopic property	宏观特性
macroscopic scale	宏观规模
macroscopic seismic phenomenon	宏观地震现象
macroscopic streamline	宏观流线
macroscopic structure	大型构造；宏观构造
macroscopic sweep efficiency	宏观波及效率
macroscopic test	宏观检验
macroscopic velocity	宏观速度
macroscopic viscosity	宏观黏度
macroscopic void	大孔洞
macroscopic water-oil interface	宏观油水界面
macroscopy	宏观
macrosection	宏观断面；宏观截面
macrosegregation	宏观偏析
macroseism	强震

macroseism seismology	强震地震学
macroseismic	强震的
macroseismic area	强震区
macroseismic data	强震资料
macroseismic effect	强震效应
macroseismic epicenter	强震震中
macroseismic epicentre	强震震中
macroseismic field	强震场
macroseismic intensity	强震烈度
macroseismic magnitude estimation	强震震级估计
macroseismic observation	强震观测
macroseismic survey	强震考察
macroseismicity	强震活动性
macroseismograph	强震仪
macroseisms seismology	强震地震学
macroshrinkage	宏观缩孔
macrosolifluction	大型泥石流
macrostatistical approach	宏观统计法
macrostrain	大应变
macrostratigraphy	宏观地层学
macrostress	宏观应力
macrostructure	宏观结构；粗显结构；大型构造
macrosymbiont	大共生体
macro-system	宏观系统；大规模
macrotectonic	大构造；巨型构造
macrotectonics	大地构造学
macrotidal estuary	强潮河口
macrotidal range	大潮差
macrotide	大潮
macroturbulence	大尺度紊动；宏观湍流
macro-type	大型
macrovariolitic	粗球粒玄武岩的
macrovisual study	宏观研究
macrovoid ratio	大孔隙比
macro-zone	大区域
macula	次生岩浆囊；暗斑；斑点
maculose	斑点状的；瘤状矿石
maculose rock	斑结状岩
macusanite	酸性火山玻质岩
made ground	人造陆地；填土；翻填地
made land	人造陆地，人造土地；填土，填地；造地；翻填地
madeirite	钛辉苦橄斑岩
madupite	透辉金云斑岩；透辉金云白榴岩
madupitic lamproite	金云嵌晶钾镁煌斑岩
maenaite	富钙淡歪细晶岩；富闪二长霏细岩
Ma'er limestone	马尔石灰岩
Ma'ershan Formation	马尔山组
mafelsic	镁铁硅质的；镁铁长英质
mafic	镁铁质的
mafic front	铁镁质前缘；基性前缘
mafic gneiss	镁铁质片麻岩
mafic hornfels	镁铁质角岩
mafic index (MI)	镁铁指数
mafic margin	镁铁质边缘
mafic mineral	镁铁质矿物
mafic rock	镁铁质岩
mafite	镁铁矿物；暗色造岩矿物；暗色隐晶岩
mafitite	超镁铁质岩
mafraite	钠闪辉长岩
mafurite	橄榄单辉钾霞岩；橄辉钾霞斑岩；马浮石
magacycle	巨旋回
magallanite	沥青砾石
magaluma	铝镁合金
magamp	磁放大器
magaseism	剧震；特大地震
magator	马伽脱泵
magazine mining	金属矿留矿采矿法
magazine type drill rods carrier	桶式钻杆箱
magcogel	膨润土粉；泥浆状土
magdolite	二次煅烧的白云石
maggie	砂质劣铁石；劣质煤
maghemite	磁赤铁矿
magic chuck	快换夹具
magic guide bush	变径导套；张缩导套
magistoseismic area	极震区
magistral	焙烧黄铜矿
maglev	磁悬浮；磁悬浮火车
maglev vehicle	磁悬浮车辆
magma	岩浆
magma association	岩浆共生组合
magma basalt	玻璃玄武岩；岩浆玄武岩
magma basin	岩浆储源
magma chamber	岩浆库；岩浆房；岩浆储源
magma consolidation	岩浆固结
magma contamination	岩浆混染作用
magma distillation	岩浆蒸馏作用
magma evolution	岩浆演化
magma formation	岩浆建造
magma fragment	浆屑
magma genesis	岩浆成因；岩浆起源
magma heat source	岩浆热源
magma hydrothermalism	岩浆热液作用
magma impact hypothesis	岩浆冲击说
magma intrusion	岩浆侵入
magma mingling	岩浆混合
magma mixing	岩浆混合作用
magma pocket	岩浆储源
magma province	岩浆区
magma reservoir	岩浆库；岩浆房；岩浆储源
magma sediment mingling	岩浆泥沙混杂
magma series	岩浆系列
magma splitting	岩浆分异；岩浆分裂
magma tapping	岩浆热开发
magma thermodynamic structure	岩浆热动力构造
magma transport	岩浆运输
magma trap	岩浆囊
magma type	岩浆类型
magmabasalt	玻璃玄武岩；岩浆玄武岩
magma-coalescence process	岩浆结合作用；岩浆合并过程
magmacyclothem	岩浆旋回
magmagene	火成的；岩浆派生的；岩浆成因的
magmagranite	岩浆花岗岩
magmametamorphism	岩浆变质作用
magmatectonic classification	构造岩浆分类
magmatic	岩浆的
magmatic activity	岩浆活动；岩浆作用
magmatic affiliation	浆岩亲缘
magmatic and tectonic accretion	岩浆岩和构造增生

magmatic arc	岩浆弧；火山弧
magmatic assimilation	岩浆同化
magmatic autocatalysis	岩浆自催化作用
magmatic breccia	岩浆角砾岩
magmatic chamber	岩浆房
magmatic circulation	岩浆环流
magmatic column	岩浆柱
magmatic complex	岩浆杂岩
magmatic concentration	岩浆富集
magmatic corrosion	岩浆熔蚀
magmatic crystallization	岩浆结晶作用
magmatic cycle	岩浆旋回
magmatic deposit	岩浆矿床
magmatic diapirism	岩浆底辟作用
magmatic differential	岩浆分异
magmatic differentiation	岩浆分异
magmatic digestion	岩浆同化
magmatic ejecta	岩浆抛出物；岩浆喷出物
magmatic emanation	岩浆喷发；岩浆喷气
magmatic emplacement	岩浆侵位；岩浆贯入
magmatic eruption	岩浆喷溢；岩浆喷发
magmatic evolution	岩浆演化
magmatic exhalation	岩浆喷发
magmatic explosion	岩浆爆发
magmatic flow	岩浆流
magmatic focus	岩浆源
magmatic formation	岩浆建造
magmatic fractionation	岩浆分离
magmatic gas	岩浆气
magmatic gradient	岩浆活动梯度
magmatic hearth	岩浆源
magmatic hydrothermal replacement	岩浆热液交代
magmatic hydrothermalism	岩浆热液作用
magmatic inclusion	岩浆包裹体
magmatic inflow	岩浆流入
magmatic injection	岩浆贯入
magmatic injection deposit	岩浆贯入矿床
magmatic intrusion	岩浆侵入
magmatic melt	岩浆；熔浆
magmatic mixing	岩浆混合作用
magmatic ore	岩浆矿石；岩浆矿
magmatic ore deposit	岩浆矿床
magmatic oreforming process	岩浆成矿作用
magmatic origin	岩浆成因
magmatic petrology	岩浆岩岩石学
magmatic plug	火山塞
magmatic pneumatolysis	岩浆气化作用
magmatic pocket	岩浆房
magmatic process	岩浆过程
magmatic reemplacement	岩浆再侵位
magmatic residual phase	岩浆残余相
magmatic resorption	岩浆熔蚀
magmatic rock	岩浆岩
magmatic segregation	岩浆分结作用；岩浆分凝作用
magmatic segregation deposit	岩浆分凝矿床；岩浆析离矿床
magmatic solution	岩浆溶液
magmatic stoping	岩浆升蚀；岩浆顶蚀
magmatic subsidence	岩浆沉降
magmatic succession	岩浆序列
magmatic suite	岩浆岩套
magmatic sulfides	岩浆硫化物
magmatic vent	岩浆通道
magmatic water	岩浆水
magmatic wedging	岩浆楔入
magmatic withdrawal	岩浆沉淀
magmatism	岩浆作用；岩浆活动
magmatist	岩浆论者；火成论者
magmatite	岩浆岩
magmatogene	岩浆成因的；岩浆派生的
magmatogene rock	岩浆生成岩
magmatogenic	岩浆成因的
magmatology	岩浆学
magnacycle	巨旋回
magnafacies	大相；主相；同性相；相体
magnaflux	磁铁粉检查法；电磁探矿法；磁粉探伤机；磁通量；磁力探伤
magnaflux examination	磁粉检验
magnaflux inspection	磁粉探伤检验
magnaflux method	裂缝及缺陷的磁通量检测法
magnalite	绿玄武土
magnalium	铝镁铜合金
magnascope	放像镜；扩大镜
magnatector	测卡点仪
magnechuck	电磁吸卡盘
magneic hysteresis	磁滞
Magnel system	马氏预应力张拉系统
magnelog	磁测井
magner	无功功率；无效功率
magnesia	氧化镁；镁云母
magnesia binder	氧化镁黏结剂
magnesia carbon refractory	镁碳质耐火材料
magnesia cement	菱镁土水泥；镁石水泥；高镁水泥
magnesia ferro ratio	镁铁比值
magnesia mica	金云母；镁云母
magnesia spinel	镁氧尖晶石
magnesia-glimmer	金云母
magnesial	镁质的
magnesian	含镁的；镁质的
magnesian chalk	镁质白垩；含镁石灰岩
magnesian lime	镁石灰；含镁石灰岩
magnesian limestone	镁质灰岩；白云质灰岩
magnesian schist	镁质片岩
magnesian siderite	镁菱铁矿
magnesian slate	镁质板岩
magnesio orthite	镁褐帘石
magnesioastrophyllite	镁星叶石
magnesiocarbonatite	镁质碳酸岩
magnesiochromite	镁铬铁矿
magnesioferrite	镁铁矿
magnesiohulsite	镁硼锡铁矿
magnesioludwigite	硼镁铁矿；富镁硼铁矿
magnesioriebeckite	镁钠闪石
magnesiosussexite	镁硼锰石
magnesite	菱镁矿；菱镁石；菱镁土
magnesite cement	菱镁土水泥
magnesite dolomite refractory	镁质白云石耐火材料
magnesite refractory	镁质耐火材料
magnesitization	菱镁矿化
magnesium	镁

magnesium acetate	醋酸镁
magnesium alloy	镁合金
magnesium alloy diving suit	镁合金潜水服
magnesium anode protection	镁阳极防蚀
magnesium base alloy	镁基合金
magnesium chloride	氯化镁
magnesium compounds	镁化合物
magnesium hydroxide	氢氧化镁
magnesium limestone	含镁石灰岩
magnesium minerals	镁矿类
magnesium nitrate	硝酸镁
magnesium nitrite	亚硝酸镁
magnesium octahedron	镁氧八面体
magnesium oxide	氧化镁
magnesium oxychloride cement	氯镁氧水泥
magnesium silicate	硅酸镁
magnesium skarn	镁质矽卡岩
magnesium soddyite	镁水硅铀矿
magnesium subgroup	含镁水亚组
magnesium sulfate	硫酸镁
magnesium sulphate	硫酸镁
magnesium sulphate salts	硫酸镁盐
magnesium-zirconium drill rod	镁锆钻杆
magnestat	磁调节器
magnet	磁体；磁铁；磁石
magnet band	磁带
magnet bank	磁铁组
magnet check	磁力探伤；磁力探伤器
magnet coil	电磁铁线圈
magnet contactor	磁开关
magnet core	磁芯
magnet coupling	磁力联轴器
magnet crane	磁力起重机
magnet drilling machine	磁力钻床；磁力钻孔机
magnet driver	电磁铁驱动器
magnet feed	磁性传动
magnet insert	磁心棒
magnet junk retriever	磁力碎屑打捞工具
magnet keeper	永久磁铁衔铁
magnet meter	磁通量表
magnet pole	磁极
magnet ring	环形磁铁
magnet spool	电磁铁线圈
magnet starter	磁力起动器
magnet wheel	磁轮
magnet winding	电磁铁绕组
magnet yoke	磁轭
magnetic	磁性的；地磁的
magnetic action	磁效应；磁性作用
magnetic activity	地磁活动性
magnetic after effect	磁后效；剩磁效应
magnetic aftereffect	磁后效
magnetic aging	磁时效；磁老化
magnetic airborne surveys	航磁测量
magnetic alignment	磁力校准
magnetic alloy	磁性合金
magnetic amplifier	磁放大器
magnetic amplitude	磁化曲线振幅
magnetic analysis	磁力分析；磁性分析
magnetic anisotropy	磁各向异性
magnetic annealing	磁性退火
magnetic anomaly	磁力异常；磁异常
magnetic anomaly area	磁力异常区
magnetic anomaly isotropy of mine product	地磁异常图
magnetic anomaly offset	磁异常位移
magnetic area moment	磁矩
magnetic artifact	人工磁效应
magnetic attitude control system	磁力姿态控制系统
magnetic attraction	磁吸引；磁引力
magnetic axis	磁轴；地磁轴
magnetic azimuth	磁方位角
magnetic basement	磁性基底
magnetic bay	磁湾扰；磁湾
magnetic bearing	磁方位角；磁象限角
magnetic biasing	磁偏
magnetic bit extractor	磁力钻头打捞器
magnetic blow	磁偏吹
magnetic blowout	磁吹熄弧；磁性熄弧
magnetic body	磁性体
magnetic brake	磁力制动器
magnetic bubble memory	磁泡存储器
magnetic calibrating device	磁标定装置
magnetic capacity	磁化率
magnetic card	磁性卡
magnetic catcher	磁力吸盘
magnetic cell	磁元件
magnetic characteristics	磁性
magnetic charge	磁荷
magnetic chart	地磁图；磁力线图
magnetic chute	磁力放矿槽
magnetic circuit	磁路
magnetic cleaning	磁清洗
magnetic cobber	磁选机
magnetic cobbing	磁选
magnetic cobbling machine	磁性分离器；磁选机
magnetic coil	电磁线圈
magnetic colatitude	磁余纬
magnetic collar locator	磁性定位接箍
magnetic compass	罗盘仪；磁针罗盘
magnetic concentrate	磁选精矿；磁铁精矿
magnetic concentration	磁力选矿；磁选
magnetic concentrator	磁选机；磁力选矿机
magnetic conductance	磁导
magnetic conductivity	导磁性；磁导
magnetic configuration	磁位形
magnetic conjugate point	磁共轭点
magnetic contactor	磁接触器
magnetic control relay	磁控继电器
magnetic coordinate	磁坐标
magnetic core	磁心
magnetic core matrix	磁心矩阵
magnetic core storage	磁心存储器
magnetic core-orientation test	磁法岩芯定向测定
magnetic correction	磁力校正
magnetic correlation	磁定向
magnetic Coulomb's law	磁库仑定律
magnetic coupling	磁耦合
magnetic crack detection	磁力探伤；磁力裂缝检查
magnetic crack detector	磁性探伤器；磁性裂隙检验器
magnetic crane	磁力起重机；磁力吊机

magnetic creeping	磁滞
magnetic crochet	磁钩扰
magnetic current	磁流
magnetic current line source	磁流线源
magnetic currentmeter	磁力流速仪
magnetic curve	磁化曲线
magnetic cycle	磁化循环
magnetic damper	磁铁阻尼器
magnetic damping	磁阻尼
magnetic data	磁力资料
magnetic dating	磁年龄测定
magnetic declination	磁偏角；磁差
magnetic deflection	磁场偏转；磁偏角
magnetic deflection sensitivity	磁偏转灵敏度
magnetic deformation	磁致变形
magnetic delay-line	磁延迟线
magnetic density	磁感应密度；磁场强度
magnetic detection	磁性深测
magnetic detector	磁场探测器；磁性检波器
magnetic deviation	磁差；磁力差
magnetic dial gauge	磁性指示表
magnetic differential flow recorder	磁性压差流量记录仪
magnetic dip	磁垂直偏角；磁倾斜；磁倾角
magnetic dip angle	磁倾角
magnetic dip pole	磁倾极
magnetic dipole	磁偶极子；磁偶极
magnetic dipole moment	磁偶极矩
magnetic dipole time	磁偶极时
magnetic directional clinograph	磁针式测斜仪
magnetic disc	磁盘
magnetic disc capacity	磁盘容量
magnetic disc head	磁盘磁头
magnetic disc storage	磁盘存储器
magnetic disk	磁盘
magnetic disk file	磁盘文件
magnetic disk memory	磁盘存储器
magnetic disk unit	磁盘机
magnetic dispersion	磁漏
magnetic displacement	磁位移
magnetic disturbance	磁扰；磁干扰
magnetic diurnal variation	磁周日变化；地磁日变
magnetic domain	磁畴；磁域
magnetic doublet	磁偶极子
magnetic drag	磁吸力
magnetic dressing	磁选
magnetic drill press	电磁钻床
magnetic drop-type survey	磁性投入式测斜仪
magnetic drum	磁鼓；磁转筒
magnetic drum memory	磁鼓存储器
magnetic drum storage	磁鼓存储器
magnetic effect	磁化；磁效应
magnetic elongation	磁致伸长
magnetic equator	地磁赤道；磁赤道
magnetic equipotential surface	磁等势面
magnetic exploration	磁法勘探
magnetic extensometer	磁性伸长计
magnetic fabric	磁组构
magnetic fabrics and anisotropy	磁组构和各向异性
magnetic field	磁场
magnetic field balance	磁力仪
magnetic field change	磁场变化
magnetic field density	磁场强度
magnetic field force	磁场力
magnetic field gradient	磁场梯度
magnetic field intensity	磁场强度
magnetic field leveling	磁场水平调整
magnetic field map	磁场图
magnetic field polarity	磁场极性
magnetic field reversal	磁场倒转
magnetic field strength	磁场强度
magnetic field-balance	磁力仪
magnetic fieldforce	磁场力
magnetic figure	磁力线图
magnetic filter	磁性过滤器
magnetic finisher	最后磁选机
magnetic fishing tool	磁力打捞工具；磁力打捞器
magnetic flaw detection	磁力探伤
magnetic flaw detector	磁力探伤器；磁力探伤仪
magnetic flocculation	磁力絮凝；磁团聚
magnetic flow	磁通量；磁铁矿流
magnetic flow meter	磁性流量计
magnetic flowmeter	磁流量计；磁力流速计
magnetic fluid	磁流体；磁流
magnetic fluid separation	磁流体选矿
magnetic flux	磁通量
magnetic flux density	磁感应强度；磁通密度
magnetic flux leakage	磁漏
magnetic flux line	磁通线
magnetic flux test	磁力线检验
magnetic force	磁力
magnetic fossil	磁性化石
magnetic gap	磁隙
magnetic gauge	磁性测微计
magnetic geophysical method	地球物理磁测法；磁法勘探
magnetic gradient	磁场梯度
magnetic gradient drift	磁场梯度漂移
magnetic gradient survey	磁场梯度测量
magnetic gradiometer	磁力梯度仪
magnetic hammer	磁性锤
magnetic head	磁头
magnetic heading	磁航向
magnetic high	磁力高
magnetic hot spot	磁热点
magnetic hysteresis	磁滞
magnetic hysteresis curve	磁滞曲线
magnetic hysteresis loop	磁滞回线
magnetic hysteresis loss	磁滞损耗
magnetic inclination	磁偏角；磁倾角；磁力线偏转
magnetic induced polarization method	磁激发极化法
magnetic induction	磁感应
magnetic induction density	磁感应强度
magnetic induction flowmeter	磁感应式流量计
magnetic induction loop	磁感线圈
magnetic inductive capacity	导磁率
magnetic inductivity	导磁率
magnetic inspection	磁性探伤；磁力探伤
magnetic instability	磁力不稳定性
magnetic instrumentation	磁法勘探仪器装备；磁法仪表化

magnetic intensity	磁场强度；磁化强度	magnetic perturbation	磁扰
magnetic interface inversion	磁性界面反演	magnetic pick up	磁性传感器
magnetic interference	磁干扰	magnetic pickup	电磁式拾波器
magnetic interpretation method	磁测资料解释法	magnetic plant	磁选厂；磁选车间
magnetic iron ore	磁铁矿	magnetic polarisation	磁极化
magnetic isoanomalous line	等磁异常线	magnetic polarity	磁极性
magnetic isoclinic line	等磁倾线	magnetic polarity reversal	磁极性转向；磁极反转
magnetic lag	磁滞	magnetic polarity stratigraphic classification	磁极性地层划分
magnetic latitude	地磁纬度；磁纬	magnetic polarity stratigraphy	磁极地层
magnetic leakage	磁漏	magnetic polarity time scale	磁极性年代表
magnetic leakage factor	漏磁系数	magnetic polarization	磁极化；磁极化强度
magnetic leakage flux	漏磁通量	magnetic pole	磁极
magnetic levitation	磁悬浮	magnetic pole strength	磁极强度
magnetic levitation railway	磁浮铁路	magnetic pole wandering	磁极迁移
magnetic line	磁力线	magnetic potential	磁位；磁势
magnetic line of force	磁力线	magnetic powder	磁铁粉
magnetic lineation	磁条带	magnetic power	磁场功率；磁力
magnetic linkage	磁链；磁通连匝数	magnetic pressure	磁压
magnetic loading	磁负载；全磁通	magnetic probe extensometer	磁性沉降计
magnetic locator	磁性探测器；磁定位器	magnetic profile	磁测剖面；磁力剖面
magnetic log	磁测记录；磁测井	magnetic property	磁性
magnetic logging	磁法测井	magnetic prospecting	磁性勘探
magnetic low	磁力低	magnetic proximity logging tool	磁性邻近径向测井仪
magnetic map	地磁图	magnetic pull	磁引力
magnetic map analysis	磁力图分析	magnetic pumping process	磁泵过程
magnetic material	磁性材料	magnetic pyrite	磁黄铁矿
magnetic maximum	磁力高	magnetic quantum number	磁量子数
magnetic measurement	磁测；磁力测量	magnetic quiet zone	磁静带
magnetic medium	磁性介质	magnetic reactance	磁抗
magnetic memory	磁存储器	magnetic receptivity	磁化系数
magnetic meridian	磁子午线	magnetic recording	磁记录
magnetic method	磁法；磁测法	magnetic recording system	磁记录系统
magnetic micropulsation	地磁微脉动	magnetic reflectivity	磁阻率
magnetic microscope	磁显微镜	magnetic relaxation	磁性弛豫
magnetic mill	磁力选矿厂	magnetic reluctance	磁阻
magnetic mineral	磁性矿物	magnetic reluctivity	磁阻率
magnetic mineralogy and petrology	磁矿物学和岩石学	magnetic remanence	剩磁；剩余磁感应；剩余磁通密度
magnetic minimum	磁力低	magnetic resistance	磁阻
magnetic moment	磁矩	magnetic resolution	磁性分离
magnetic multishots	磁力多点测斜仪	magnetic resonance	磁共振
magnetic MWD data	磁力随钻测量数据	magnetic resonance imaging logging	核磁共振成像测井
magnetic needle	磁针	magnetic retardation	磁滞
magnetic needle clinograph	磁针式测斜仪	magnetic retentivity	顽磁性
magnetic needle inclination	磁针倾角	magnetic return path	磁通量回路
magnetic neutral state	磁中性状态	magnetic reversals	地磁倒转；地磁反向
magnetic north	磁北	magnetic rigidity	磁刚性；磁刚度
magnetic north pole	磁北极	magnetic roasting	磁化焙烧
magnetic observatory	地磁观察台	magnetic rock	磁性岩石
magnetic ore	磁性矿石	magnetic rotation	磁旋
magnetic orientation	磁定向	magnetic rougher	磁力粗选机
magnetic overprinting	磁叠印	magnetic sand	磁性砂
magnetic paleointensity	磁场古强度	magnetic saturation	磁饱和
magnetic particle examination	磁粉检验	magnetic scalar potential	磁标量位
magnetic particle inspection	磁粉检验；磁粉探伤	magnetic scanning	磁扫描
magnetic particle testing	磁粉检验	magnetic scattering	磁散射
magnetic path	磁路；磁通路径	magnetic screen	磁屏蔽
magnetic permeability	磁导率	magnetic sector structure	磁扇形结构；扇形磁场
magnetic permeability curve	磁导率曲线	magnetic sensor	磁性传感器
magnetic permeameter	磁导计	magnetic separation	磁力分离；磁选
magnetic permeance	磁导		

magnetic separator	磁选机；磁力除铁器；磁力分离器
magnetic shell	磁壳
magnetic shield	磁屏蔽
magnetic shielding	磁屏蔽
magnetic shift register	磁移位寄存器
magnetic signature	磁异常特征
magnetic single-shot tool	磁性单点测斜仪
magnetic smooth zone	磁平带
magnetic sorting	磁力选矿；磁选
magnetic sounder	磁测深仪
magnetic sounding	磁测深
magnetic south	地磁南极
magnetic south pole	地磁南极
magnetic spectrograph	磁谱仪
magnetic spin	磁偶自旋
magnetic spin quantum number	磁自旋量子数
magnetic spirit level	磁力水平尺；磁力水准仪
magnetic stability	磁稳定性
magnetic starter	磁力起动器
magnetic storage	磁存储器
magnetic storage drum	存储磁鼓
magnetic store	磁存储器
magnetic storm	磁暴
magnetic storm phase	磁暴相
magnetic stratigraphic chassification	磁性地层划分
magnetic stratigraphy	古地磁地层学；磁性地层学
magnetic stratum	磁性地层
magnetic stress	磁应力
magnetic substance	磁体
magnetic substorm	磁亚暴
magnetic survey	地磁测量；地磁查勘图；磁法勘探
magnetic susceptibility	磁化率
magnetic susceptibility logging	磁化率测井
magnetic susceptibility meter	磁化率计
magnetic sweeping	磁力扫海测量
magnetic systems	磁系
magnetic tape plotting system	磁带绘图系统
magnetic testing	磁力探伤
magnetic theodolite	磁经纬仪；罗盘经纬仪
magnetic thickness log	套管壁厚磁测井
magnetic thickness tester	磁性测厚仪
magnetic thin film memory	磁膜存储器
magnetic torque	磁矩；磁转矩
magnetic torsion balance	磁扭秤
magnetic track	磁道；磁通
magnetic transformation	磁性变化；磁性转变
magnetic trough separator	槽式磁选机
magnetic value	磁值
magnetic variable-reluctance system	变磁阻系统
magnetic variation	磁偏角；磁变
magnetic vertical component	地磁垂直分量
magnetic vertical intensity	地磁垂直强度
magnetic viscosity	磁黏滞性
magnetic viscosity decay	磁黏滞衰减
magnetic washing	磁清洗
magnetic wave detector	磁性振荡指示器；磁性检波器
magnetic well logging	磁法测井
magnetic work	磁法勘探
magnetically disturbed day	磁扰日
magnetically hard alloy	硬磁合金
magnetically hard material	磁性硬物
magnetically quiet day	磁静日
magnetically saturated	磁饱和的
magnetically soft alloy	软磁合金
magnetically soft material	磁性软物
magnetically vibrated screen	磁力振动筛
magnetic-coupling flowmeter	磁感应流量计；磁耦合流量计
magnetic-dise memory	磁盘记忆；磁盘存储器
magnetic-drum computer	磁鼓计算机
magnetic-drum memory	磁鼓存储器
magnetic-drum reader	磁鼓读出器
magnetic-field test	磁力探伤
magnetic-film	磁膜
magnetic-film memory	磁膜存储器
magnetic-particle method	磁力探伤法
magnetic-polarity sequence	磁极层序
magnetic-pulse	磁脉冲
magnetics	磁学
magnetic-tape recorder	磁带记录器
magnetic-tape recording	磁带记录
magnetic-tape recording accelerograph	磁带记录加速度仪
magnetic-tape recording seismic system	磁带式地震记录系统
magnetic-tube concentrator	磁管磁选机
magnetisablilty	磁化能力
magnetisation	磁化
magnetisation curve	磁化曲线
magnetism	磁学；磁性；磁力
magnetism theodolite	地磁经纬仪
magnetite	磁铁矿
magnetite concentrate	磁铁精矿
magnetite consumption	磁铁矿耗量
magnetite dense medium	磁铁矿悬浮液
magnetite feeder	磁铁矿给料器
magnetite loss	磁铁矿损失
magnetite recovery	磁铁矿回收率
magnetite spinellite	磁铁尖晶石岩
magnetite-rich rock	富磁铁岩
magnetitite	磁铁岩
magnetitization	磁铁矿化
magnetizability	磁化能力；磁化性
magnetization	磁化作用；起磁；磁化强度
magnetization characteristic	磁化特性
magnetization coefficient	磁化系数
magnetization curve	磁化曲线
magnetization cycle	磁化回线；磁化循环
magnetization direction	磁化方向
magnetization distribution	磁化分布
magnetization error	磁化误差
magnetization intensity	磁化强度
magnetization mapping	磁化制图
magnetization of rocks	岩石磁化
magnetization ratio	磁性率
magnetization roasting	磁化焙烧
magnetize	磁化
magnetized area	磁化区域
magnetized bit	磁化位
magnetized drilling assembly	磁化钻具
magnetized layer	磁化层

magnetized magnetic field　磁化磁场
magnetized spot　磁化点
magnetized water　磁化水
magnetizer　感磁物；磁化器；导磁体
magnetizing　充磁
magnetizing along stratum　顺层磁化
magnetizing block　磁化块；起磁块
magnetizing detector of drilling tools　钻具磁性探伤
magnetizing field　磁化场
magnetizing force　磁化力
magnetizing roast　磁化焙烧
magneto detector　磁力检波器
magneto field scope　磁场示波器
magneto fluid mechanics　磁流体力学
magneto gyrocompass　磁力回转罗盘
magnetobiology　磁生物学
magnetochemistry　磁化学
magnetochronology　地磁年代学
magnetoconductivity　磁导率
magnetocrystalline anisotropy　磁晶各向异性
magneto-dipole　磁偶极子
magnetoelastic device　磁弹性装置
magnetoelastic dynamometer　磁弹式测力计
magnetoelastic meter　磁弹性计
magnetoelasticity　磁致弹性
magnetoelectric　磁电的
magnetoelectric effect　磁电效应
magneto-electric induction　磁电感应
magnetoelectric pile driver　电磁打桩机
magnetoelectric pile hammer　电磁桩锤
magneto-electrotelluric　大地电磁的
magneto-electrotelluric exploration　大地电磁勘探
magnetofluid mechanics　磁流体力学
magnetogasdynamics　磁性气体动力学
magnetogram　地磁自记图；磁力图；磁强记录图
magnetograph　地磁记录仪；磁强记录仪；磁力记录仪
magnetography　磁记录法
magnetogravimetric separation　磁重分离
magnetogyric ratio　磁旋比
magnetohydrodynamic　磁流体动力的
magnetohydrodynamic flow　磁流体流
magnetohydrodynamics　磁流体力学；磁动流体力学
magneto-ionic theory　磁离子理论
magnetometer　磁强计；地磁仪；磁力仪
magnetometer method　地磁仪勘探法
magnetometer sensor　磁强仪传感器
magnetometer survey　磁法勘探；磁法测量；磁力仪测量
magnetometric　磁力的
magnetometric analysis　测磁分析
magnetometric induced polarization method　磁激发极化法；磁感应极化法
magnetometric offshore electric sounding　近海磁电测深
magnetometric offshore electrical sounding　磁测近岸电测深
magnetometric resistivity method　磁电阻率法
magnetometry　地磁测量；磁力测定；测磁强术；测磁学
magnetomotive　磁力作用的
magnetomotive force　磁通势；磁动势
magneton　磁子

magnetooptics　磁光学
magnetopause　磁层顶
magnetoplasmodynamics　磁等离子动力学
magnetoplumbite　磁铅矿
magnetoresistance　磁阻；磁致电阻
magnetoresistivity　磁致电阻率
magnetoresistor　磁控电阻
magnetorheological　磁流变的
magnetorheological damper　磁流变阻尼器
magnetorheological fluid　磁流变液
magnetoscope　验磁器
magnetosheath　磁鞘；磁套
magnetosheath region　磁鞘区
magnetosphere　磁层；磁圈
magnetospheric　磁性层的
magnetospheric convection　磁层对流
magnetospheric dumping　磁层倾泻
magnetospheric plasma　磁层等离子体
magnetospheric response　磁层响应
magnetospheric storm　磁层暴
magnetospheric substorm　磁层亚暴
magnetostatic field　静磁场
magnetostatic well-tracking　静磁井迹跟踪
magnetostatics　静磁学
magnetostratigraphic　磁性地层的
magnetostratigraphic classification　磁性地层划分
magnetostratigraphic polarity unit　磁性地层极性单位
magnetostratigraphic polarity zone　磁性地层极性带
magnetostratigraphic polarity-reversal horizon　磁性地层极性划分
magnetostratigraphic unit　磁性地层单位；地磁地层单位
magnetostratigraphy　地磁地层学
magnetostriction　磁致伸缩
magnetostriction constant　磁致伸缩常数
magneto-striction oscillation　磁致伸缩振荡
magneto-striction phenomenon　磁致伸缩现象
magnetostriction transducer　磁致伸缩换能器
magnetostriction vibrator　磁致伸缩振动器
magnetostrictive　磁致伸缩的
magnetostrictive actuator　磁致伸缩作动器
magnetostrictive drill　磁致伸缩钻具
magnetostrictive oscillator　磁致伸缩振荡器
magnetostrictive transducer　磁致伸缩换能器
magnetostrictor　磁致伸缩体
magnetotactic bacteria　趋磁细菌
magnetotail　磁尾
magnetotelluric　地球磁场的；大地电磁的
magnetotelluric curve　大地电磁测深曲线
magnetotelluric effect of ocean bottom terrain　海底地形电磁效应
magnetotelluric field　大地电磁场
magnetotelluric impedance　大地电磁阻抗
magnetotelluric method　磁大地电流法；大地电磁法
magnetotelluric noise　大地电磁噪声
magnetotelluric profiling method　大地电磁剖面法
magneto-telluric prospecting method　大地电磁勘探法
magnetotelluric sounding　大地电磁测深法；大地电磁探测
magnetotelluric variation method　大地磁变法
magnetotellurics　大地电磁学

magnetpmeter	磁力仪
magnetrol	磁放大器
magnification	放大；放大倍数；放大率
magnification coefficient	放大系数
magnification constant	放大常数
magnification curve	放大曲线
magnification factor	放大系数；放大因子
magnification for rapid ground movement	快速地动放大
magnification for rapid waves	速波放大；快波放大
magnification of image	影像放大
magnification ratio	伸缩比；放大比率
magnifier	放大镜；放大器
magnifier for reading	读数放大镜
magnify	放大
magnifying glass	放大镜
magnifying lens	放大镜
magnifying power	放大率
magnioborite	硼镁石；遂安石
magniophilite	磷镁铝矿
magniphyric	微粗斑状
magni-scale	放大比例尺
magnitude	地震震级；量；大小；强度；尺寸
magnitude and direction	大小和方向
magnitude anomaly	震级异常
magnitude chart	震级图
magnitude determination	震级测定
magnitude distribution	震级分布
magnitude estimation	震级估算
magnitude for local shock	局部地震震级
magnitude formula	震级公式
magnitude interval	震级区间；震级档
magnitude of body wave	体波震级
magnitude of degradation	刷深程度；冲刷程度
magnitude of discharge	流量大小；排水量大小
magnitude of down-throw	断层下落幅度
magnitude of earthquake	地震震级
magnitude of fault	断距
magnitude of fault displacement	断层位移量；断距
magnitude of inclination	倾斜幅度
magnitude of load	荷载大小
magnitude of maximum shearing stress	最大剪应力值
magnitude of stresses	应力值
magnitude of surface wave	面波震级
magnitude of water carrier	含水层厚度
magnitude recurrence curve	震级重现曲线
magnitude residual	震级残差
magnitude scale	震级表；震级表
magnitude size	震级大小
magnitude statistics	震级统计
magnitude threshold	震级限值
magnitude time curve	震级时间曲线
magnitude-dependent peak acceleration	依赖于震级的峰值加速度
magnitude-frequency equation	震级频度方程
magnitude-frequency law	震级频度法则
magnitude-frequency relation	震级-频度关系
magnitude-intensity correlation	震级-烈度相关关系
Magno	镍锰合金
magnochromite	镁铬铁矿
magnoferrite	镁铁尖晶石
magnophorite	含钛钾钠透闪石
magnophyric	粗斑状
magsat	磁场测量卫星
magsat surveying	卫星地磁测量；卫星磁测
maiden	新的；未开拓的
maiden coal	未采煤层
maiden field	未采的矿区；未开发油、气田
Maijishan grottoes	麦积山石窟
main access ramp	露天矿出入沟；露天矿基本沟；露天矿主沟
main active fault	主要活动断层
main adit	主平硐
main airway	主风道；主风巷
main anchor	主锚
main angle	主交叉巷
main anti-siphonage pipe	总反虹吸管
main aperture	主孔径
main aquifer	主含水层
main arch ring	主拱圈
main axis	主轴
main bar	主筋
main base	控制点；主基点
main beam	主梁
main bearing	主轴承
main bed	主河床；主河槽
main belt	干线皮带运输机
main blasting lead	主爆破导线
main block valve	总闸门
main bottom	基岩；基底
main boundary thrust	主边界逆断层
main branch	主汊道
main breakwater	主防波堤
main bridge	正桥；主桥
main buoy	主浮筒
main canal	总渠；干渠
main canopy	主顶梁
main centenary wire	主承力索
main channel	主河槽；主河道；主航道
main channel flow	干流；主水道流量；干渠
main cleat	主解理
main coking coal	主焦煤
main cokingcoal	主焦煤
main collector	总干管；主干管；集水总管；集水干管
main column	主柱
main connecting rod	主连杆
main connection	主竖井
main consolidation settlement	主固结沉降
main control console	主控制台
main control equipment	主控设备
main control head	总闸门
main conveyor	主输送机
main conveyor roadway	主输送机巷道
main coordinate	主坐标
main course	主航线
main crosscut	主要石门
main cross-section	主横断面
main current	主流；干流
main current trend line	主流倾向线
main cycle	主旋回
main dam	主坝

main deck 主桥面
main delta arm 三角洲主汊道
main deviator 主偏应力
main diagonal 主斜杆；主对角线
main dike 干堤
main dimensions 主要尺寸
main dip 真倾角
main dip haulage 主下山运输
main distributing conveyor 主分卸输送机
main divide 主分水岭；主分水界；主分水线
main dome 主穹顶
main drain 排水干管；总排水管
main drainage channel 主要集水区排水道；主排水渠
main drift 主平巷；主水平
main drilling packer 主封隔器
main drive 主传动
main drive shaft 主动轴
main dyke 干堤
main earthquake 主震
main effect model 主效应模型
main elevation 主要立视面
main end 主要采准巷道掘进端
main endings 煤巷间每隔一定距离的横巷
main entry 大巷；主平巷
main fault 主断层
main fault system 主断层系
main fault zone 主断层带
main flow 主流
main flow direction 主流向
main flux 主磁通
main fracture 主破裂
main fringedrift 主要脉外平巷
main gallery 主平巷
main gangway 主运输平巷
main gathering station 总集油站
main generator 主发电机
main geosynclinal stage 主地槽阶段
main girder 主梁
main haulage level 主运输道；主运输水平
main haulage roadway 运输大巷
main haulage service 主要巷道运输
main haulageway 运输大巷；主运输巷道
main header 总集气管
main hoist 主提升机
main hole 主孔
main ice stream 主要冰流
main incline 主斜井
main inclined shaft 主斜井
main influence radius 主要影响半径
main intake 主进风道；主进气口
main intake airway 主进风巷道
main irrigation canal 灌溉总渠
main jack 盾构千斤顶
main jet 主射流；主喷嘴；工作喷嘴
main joint 主节理
main lateral 主平巷；主运输道
main levee 主堤；干堤
main line 干线
main line belt conveyor 主要运输平巷胶带输送机；干线胶带输送机

main line of drainage 排水干线
main line of turnout 道岔主线
main line railway 干线铁路
main line ramp 坡道主干线
main line track 干线
main loading level 主装载水平；主装载平巷
main lock 主水闸
main longitudinal bar 纵向主筋；纵向沙坝
main maximum 主峰；主最大值
main memory 主存储器
main module 主模块
main opening 主井；主要巷道
main orebody 主矿体
main pay 主要生产矿层
main peak 主峰
main phase 主相；主要阶段
main pile 主桩
main pipe 排水干管；总排水管
main pipeline 干线
main piping 主要管线
main pocket 主矿仓；主仓
main points of control 防治要点
main pole 主磁极
main process 主要作业
main producing horizon 主力生产层
main productive zone 主力生产层
main profile 主测线
main program 主程序
main program sequence 主程序序列
main pull rod 主拉杆
main pulse 主脉冲
main pump room 主排水泵硐室；中央水泵房
main pumping station 总水泵站
main ram 主千斤顶
main ram valve block assembly 主千斤顶阀体总成
main reaction 主反应
main reef 主矿层；主脉
main reinforcement 主筋；主钢筋；受力筋
main return 主回风道；总回风道
main return airway 主回风巷道
main river 干流；主河道
main river bed 主河床
main river channel 主河槽
main road 干路；干线
main roadway 主要巷道；大巷
main roadway for single seam 单煤层大巷
main roadway in seam 煤层大巷
main roof 老顶
main roof breaks 老顶破裂
main roof caving 老顶垮落
main roof collapse 老顶垮塌；老顶垮落
main rope 主绳；牵引绳
main route 主要线路；干线
main routine 主程序
main scarp 主破裂壁；主断崖
main sea 外海；开阔海面
main seam 主煤层；主矿层
main sectional plane 主断面
main servomotor 主伺服马达
main sewer 总排水管

main shaft 主井；竖井；主轴	main-roof rock 老顶岩石
main shaft hoist 主井提升机	mains firing 电力线引电爆破
main shelf 主大陆架	mainshock 主震
main ship channel 斜坡推进灌浆	maintain angle 稳斜
main shock 主震	maintain at grade 保持坡度
main shock type 主震型	maintained load pile test 桩的维持荷载试验
main shock-after shock type 主震-余震型	maintained load test 维持荷载法
main sill 主底梁；钻台主基木	maintenance 维修；维护；保养
main slant 主斜井	maintenance of bridge and tunnel 桥隧养护
main slide mass 主滑动体	maintenance of estuarine channel 河口航道的维护
main slide section 主滑段	maintenance of mud 泥浆性能保持；泥浆的维护
main slope 主坡	maintenance of plant 设备的维修保养
main soil group 主要土类	maintenance of reservoir pressure 油层压力保持
main span 主跨	maintenance of the strata in undisturbed condition 保持地层原位状态；保持地层不受扰动
main specifications 主要技术参数	
main spillway 正常溢洪道	maitlandite 钍脂铅铀矿
main spindle 主轴	major aftershock 主余震
main station 总站；主井底车场	major axis 长轴；主轴
main station peg 主测站标桩；基准点标桩	major bed 主河床
main stem 主河道；干流	major blast 大爆破
main storage 主存储器	major bridge 大桥
main store 主存储器	major component 主要组分
main stream 主流；主河道	major constituent 主要成分
main stream channel 主水道；主河道	major construction project 大型建筑工程计划；重点建筑工程计划
main streamline 主流线	
main stress line 主应力迹线	major cross-section of subsidence trough 移动盆地主断面
main structure 主体结构	major dam 主坝
main supply conduit 给水干管	major defect 主要缺陷
main technical requirement of railway 铁路主要技术条件	major deposit 主矿床；主矿体
main technical standard of railway 铁路主要技术标准	major diameter 外径
main tilting axis 主倾斜轴	major disturbance 大扰动
main timber 底梁	major divide 主分水界
main track 正线；主轨	major drainage basin 大流域
main track of turnout 道岔主线	major earthquake 主震；大地震
main trunk highway 主干公路	major earthwork 大规模土方工程；主体土方工程
main tunnel 正洞；主隧道；主巷	major element 主要元素；大量元素；常量元素
main vein 主脉	major emitter 主要排放源
main watercourse 主要水道	major equipment 主要设备
main waterway 主航道	major event 重大事件
main way 主巷	major fault 主干断裂；主断层；大断裂；大断层
main winding shaft 主提升井	major field 大油气田
main working 永久巷道	major flood 大洪水
main working shaft 主生产井	major flood flow 大洪水流量
main workings plan 主要巷道平面图	major fold 大褶皱；主褶皱
main works 主体工程；主要工程	major folded zone 大褶皱带
main zone 主带	major fracture 主裂缝
mainbar 钢筋混凝土的主钢筋	major function 主要作用；主要功能；优函数
mainbody 主体	major highway 主要公路
main-entry pillar 主平巷煤柱	major horizontal principal stress 最大水平主应力
mainframe computer 主计算机	major industry 主要工业；重工业
mainframe memory 主体存储器	major influence radius 主要影响半径
mainframe network 主机网络	major infrastructure project 大型基建工程
mainland 本土；大陆	major joint 主节理
mainland climate 大陆气候	major landslide 大型滑坡
mainline 干线	major landslip 大型山泥倾泻；大型滑坡
main-line conveyor 干线输送机；主输送机	major opening 主要巷道
mainline haulage 干线运输；干线拖运	major overhaul 大修
main-line track 干线	major parameter 主要参数
mainpole coil 主极线圈	major phase 主相
mainpole core 主极铁芯	major planet 大行星

major plate	大板块
major principal plane	大主平面
major principal strain	大主应变；最大主应变
major principal stress	最大主应力；大主应力
major project	大型工程；重点工程
major radius of ellipsoid	椭球长半径
major repair	大修
major repair depreciation rate	大修折旧率
major reservoir	主力油层；主力储层
major restoration	重点修复
major rig repair	钻机大修
major river bed	主河床
major road	主要道路
major semi-axis	长半轴
major semi-axis of the earth	地球长半径
major shock	主震
major sources of pollution and pollutants	主要污染源和主要污染物
major stress	主应力
major thrust plane	主冲断面
major tremor	主震
major wave	主波
major works	大型工程
majorant	强函数；控制函数
majorant series	强级数
majorizing sequence	优化序列
Majuanzi Formation	马圈子组
Makan Formation	马坎组
make a pull	起钻
make a trip	起下钻
make a well	钻一口井
make and break rotary	上卸扣旋转工具
make down	拆卸
make footage	钻进；钻探进尺
make hole	钻孔；打井；挖洞；打孔；成孔
make location	定井位
make of casing	套管下放；下套管
make of ore	脉状矿体；矿石开采
make of refuse	尾矿产量
make of water	矿井涌水率
make the gas	泄出气体；放出沼气
make the rounds	检查油井或机器
make time	匆忙完成作业
make up	上扣
make-and-break	断续；开闭；拧上拆下
make-and-break coupling	快速接卸接头
make-and-break rotor	钻管装卸转动器
make-and-break test	上卸扣试验
makensite	铝绿泥石
makeup and breakout	上紧-卸开螺纹
make-up chuck	钻杆旋接夹盘
make-up fluid	补充液
make-up gas	补给气
make-up gun	上紧-卸开螺纹装置
make-up hydrogen compressor	新氢压缩机
make-up pump	补给水泵；供水泵
make-up ring	金属井旋模板接合圈
make-up suspension	补充悬浮液
make-up torque	上紧力矩
make-up valve	备用阀
make-up water	补给水
making hole	钻井；日钻进深度
malachite	孔雀石
malacolite	透辉石；浅色透辉石
malacon	水锆石；变水锆石
malakon	水锆石；变水锆石
Malakovian	马拉科夫阶
malalignment	轴线不对准；不成直线；相对位偏；偏心率
Malan formation	马兰砾石
Malan loess	马兰黄土
Malanguan Formation	马兰关组
malanite	马兰矿
Malanloess	马兰黄土
Malanloess Stage	马兰期
malaspina glacier	山麓冰川
malaxator	捏土机；碾泥机；黏土拌合机
malchite	微闪长岩
malconformation	不均衡性；畸形
maldistribution	分布不均
maldonite	黑铋金矿
male type bit	公扣钻头
malformation	畸形
malfunction	失效；故障；不正常工作
malfunction probability	工作失常；故障概率
malfunction rate	故障出现率；故障概率
Malianjing Formation	马莲井组
Malibuer Formation	马莉卜尔组
malignite	暗霞正长岩
Maliu limestone	麻柳石灰岩
malladrite	氟硅钠石
malleability	可塑性；展延性；韧性
malloseismic region	破坏性频震区
malm	泥灰岩；钙质砂土；白垩黏土
malm rock	泥灰岩；泥质砂岩；白垩土
Malm Series	麻姆统
malmrock	黏砂岩
malmstone	燧石质砂岩；泥灰质岩；白垩质岩
malobservation	观测误差
Malong Formation	马龙组
maloperation	误操作
malpais	熔岩区
malposition	错位
maltha	软沥青
malthacite	水铝英石
malthene	软沥青质
malthenes	石油脂
malthite	软沥青
malting coal	无烟煤
Malukou Shale	马路口页岩
Malutang Formation	马鹿塘组
Malvernian	莫尔文阶
mamelon	圆丘
mamilite	镁铁白榴金云火山岩
mammal fossil	哺乳动物化石
mammillary hill	鼓丘
mammillated surface	丘形面
mammillated topography	乳房状地形
Mammoth	美国无烟煤区厚煤层名
mammoth blast	大爆破

mammoth blasting 大爆破
Mammoth event 马默思地磁反向事件
mammoth mill 大型选矿厂
Mammoth polarity subchron 马默思亚时
mammoth pool 大油藏
mammoth pump 大型泵
mammoth structure 巨型结构
mammoth tanker 巨型油轮
man access 出入沟
Man and Biosphere Programme 人类与生物圈计划
man cage 升降人员用的罐笼
man cage compartment 乘人罐笼隔间
man car 运送人员小车
man carrier 载人车
man door 人行门
man efficiency 人工效率
man engine 老式竖井人员升降机
man hoist 人员提升机
man hole 人孔
man hours 工时
man hudge 提人吊桶
man load 人员载重；人员重量
man machine 人员升降机
man made accumulation 人工堆积物
man made deposits 人工堆积物
man made earthquake 人工地震
man made geologic hazard 人为地质灾害
man made geothermal field 人工地热田
man made restrictions 人为限制
man power 人力
man processing 人工处理
man rack 管道工用平板篷车
man riding 人具运输
man sledging 人工锤碎大石块
man tramming 人力运输
man trip 人员运送
man way 人行道；梯子间
man way shaft 人行竖井
management information system 管理信息系统
management information system in mine 矿山管理信息系统
management model of groundwater resources 地下水资源管理模型
management of earthquake data 地震数据管理
management of earthquake-related data 地震数据管理
management of hazard risk 灾害风险管理
management of land resources 国土资源管理
management of marine disaster effect 海洋灾害管理
management of marine disaster reduction and relief 海洋减灾救灾管理
management of marine resources 海洋资源管理
management of ore-pass in open pit 露天矿溜井生产管理
management software 管理软件
managerial data 管理数据；管理资料
managerial mathematics 管理数学
managerial organization 管理组织；管理机构
man-and-material hoist 人员材料提升机；副井提升机
man-basket 人员吊桶
mancage 乘人罐笼
mancar 乘人矿车

man-carried 手提的；便携的
man-computer communication 人-机通信
man-computer interaction 人-机交互；人机对话
man-computer interactive processing 人机交互处理
Mancos shale 曼柯斯页岩
mandatory seismic standard 强制性地震标准
Mandel-Cryer effect 曼代尔-克雷尔效应
mandelstone 杏仁状熔岩
mandoor 进人门
mandrel socket 梨形套管打捞器；捞管器
mandrel substitute 异径接头
mandrel type tool 心轴型下井仪
mandschurite 玻霞碧玄岩
manebach twin 底面双晶
Manebach twin law 底面双晶律；曼尼巴双晶律
man-excited vibration 人工激振
mangan blende 硫锰矿
mangan ludwigite 硼镁锰矿
manganalmandine 锰铁铝榴石
manganapatite 锰磷灰石
manganate 锰酸盐；锰酸盐类
manganaxinite 锰斧石
manganblende 硫锰矿
manganchlorite 锰绿泥石
mangandolomite 锰白云石
manganese 锰
manganese agglutination 锰凝结物
manganese alloy 锰合金
manganese brass 锰黄铜
manganese bronze 锰青铜
manganese chloride 氯化锰
manganese crust 锰结壳；铁锰结壳
manganese deposits 锰矿
manganese dioxide 二氧化锰；软锰矿
manganese green 绿锰矿
manganese hat 锰帽
manganese iron 锰铁
manganese minerals 锰矿类
manganese nodule 锰结核；锰团块；锰矿球
manganese ore 锰矿
manganese oxide 氧化锰
manganese polymetallic nodule 锰-多金属结核
manganese rock 锰质岩
manganese spar 蔷薇辉石；菱锰矿
manganese stain 锰结斑
manganese steel 锰钢
manganese superoxide 过氧化锰
manganese wad 锰土；石墨
manganese yield 锰收得率
manganese-spar 蔷薇辉石
manganhedenbergite 锰钙铁辉石
manganic oxide 三氧化二锰
manganiferous 含锰的
manganiferous iron ore 含锰铁矿
manganiferous zinc ore 含锰锌矿
manganin 锰铜
manganite 水锰矿
mangankoninckite 锰针磷铁矿
manganmonticellite 锰钙橄榄石
manganmuscovite 锰白云母

mangan-neptunite	锰柱星叶石
manganobrucite	锰水镁石
manganocalcite	含钙菱锰矿；锰方解石
manganolite	锰矿岩；蔷薇辉石
manganophyllite	锰黑云母
manganosiderite	锰菱铁矿
manganosite	方锰矿
manganostibiite	锰砷锑矿
manganostibite	锑砷锰矿
manganotantalite	钽锰矿；锰钽铁矿
manganous oxide	氧化亚锰；一氧化锰
mangerite	纹长二长岩
mangrove peat	红树泥炭
mangrove peat mire	红树林泥炭沼泽
mangrove swamp	红树林沼泽
man-handled boring	人力钻进
manhead	罐底观测孔
manheim gold	铜锌锡合金
manhole	巷道侧躲避洞；探井探孔；进人孔口；检修井
manhole cover	探井盖；人孔盖；检查井盖
manhole frame	人孔颈口
manhole frame and cover	检查井盖架；进入孔盖架
manhole ladder	内梯
manhole sump	检查井沉泥槽；检查井落底
manifold flow	簇流；分叉管水流
manifold gasket	歧管垫密片
manifold gauge	歧管仪表
manifold header	集管头
manifold method	流形元法
manifold penstock	多叉压力水管
manifold platform	多层吊盘
manifold pressure	管汇压力
manifold pressure gauge	歧管压力计
manifold seismic wave	多波地震勘探
manifold system	多头管系
manifold tunnel	多岔隧洞
manifold valve	管汇阀组；多支管阀
manifolding	多叉管装置；总管装置
Manihiki Plateau	马尼希基海台
man-induced	人为的，人工诱发的
man-induced contamination	人为污染
man-induced earthquake	人工诱发地震
man-induced erosion	人为的冲蚀
man-intiated earthquake	人为诱发地震
manipulated curve	控制曲线
manipulated image	处理后影像
manipulated soil	重塑土
manipulated variable	控制变量
manipulator	控制器；操纵器；操纵者；机械手；操纵型机器人
manipulator arm	操纵杆；控制杆
manipulator device	机械手
manipulator finger	机械手抓手
manipulator-type robot	操作手型机器人
manjak	纯沥青；硬化沥青
manless coal mining	无人采煤法；遥控采煤法
manless face	无人操作工作面；自动化工作面；遥控工作面
manless longwall face	无人操作长壁工作面；自动化长壁工作面；遥控长壁工作面
manless mining	无人开采
man-machine chart	人机作业图
man-machine communicate	人机交换
man-machine communication	人机联系；人机通信
man-machine conversation	人机对话
man-machine dialog	人机对话
man-machine input-output system	人机联用输入-输出系统
man-machine interaction	人机互动；人机联作
man-machine interactive	人机交互
man-machine interactive processing system	人机交互处理系统；人机对话处理系统
man-machine interactive system	人-机交互系统
man-machine interface	人机接口；人机界面
man-machine simulation	人机模拟
man-machine system	人机系统
man-made cavern	人工开凿山洞
man-made diamond	人造金刚石；合成金刚石
man-made disturbance	人为扰动
man-made dust	人工尘埃
man-made earth satellite	人造地球卫星
man-made earthquake	人工地震；人为的地震
man-made earthwork	人工土石方工程
man-made effect	人为效应
man-made erosion	人为侵蚀
man-made explosion	人工震源
man-made factor	人为因素
man-made feature	人工地物；人文要素
man-made fiber	人造纤维
man-made fill	人工回填土
man-made fracture	人工裂缝
man-made geologic hazard	人为地质灾害
man-made geothermal well	人工地热井
man-made influence	人为影响
man-made injection	人工注水
man-made interference	人为干扰
man-made island	人工岛
man-made isotope	人造同位素
man-made lake	人工湖；人造湖
man-made land	人造地；人工建筑的土地
man-made noise	工业噪声；人为噪声
man-made object	人工地物
man-made pollution	人为污染
man-made pollution source	人为污染源
man-made polycrystalline diamond	人造聚晶金刚石
man-made recovery	人工开采
man-made reservoir	人工水库
man-made satellite	人造卫星
man-made seismic hazards	人为震害
manmade slope	人工边坡；人造斜坡
man-made slope	人工边坡；人造斜坡
man-made soil	人造土
man-made source	人工放射源
man-made stones	人造石
man-made vibration	人为振动
manned diving	载人潜水
manned platform	有人平台；驻人
manned pressure hull	载人的承压舱体
manned space station	载人空间站；载人航天站
manned space surveillance system	载人空间监视系统

manned spacecraft 载人飞船；载人航天器
manned spaceship 载人宇宙飞船
manned submersible 载人潜水的
manned underwater vehicle 载人潜水器
manned work enclosure 载人密闭工作舱
manner of occurrence 埋藏状态；生成状态
manner of origin 起源形式
Mannesmann mill 曼内斯曼斜轧机
Mannesmann piercer 曼内斯曼式穿孔机
Mannesmann process 曼内斯曼轧管法
Manning formula 曼宁公式
Manning roughness factor 曼宁糙率系数
manocryometer 加压溶点计
manograph 压力记录器
manometer 流体压力计；测压管
manometer method 测压管法
manometer piezometer pressure gauge 压力计
manometer pressure 压力计压力
manometer regulator 风压调整器
manometer tube 压力表的边接管
manometer-thermometer sonde 压力计-温度计探头
manometer-type tube 测压管
manometric 压力的；压差的；测压的
manometric efficiency 负压效率；有效压头比
manometric head 测压水头；压力头
manometric liquid 测压液
manometric setup 压力法测量装置
manometrical 压力计的；测压的
manometry 测压法
manoscope 流压计；气体密度测定仪
manoscopy 气体密度测定
manostat 稳压器；恒压器；压力调整器
MANOVA-biplot 多元方差分析-双标图
man-pollution sources 人为污染源
mantle 地幔；覆盖层；表层
mantle array 地幔趋向线
mantle bulge 地幔鼓胀；地幔隆起
mantle composition 地幔成分
mantle convection 地幔对流
mantle convection cell 地幔对流环
mantle convection circle 地幔对流环
mantle convection hypothesis 地幔对流说
mantle convection plume 地幔对流柱
mantle core boundary 地幔-地核界面；核幔边界
mantle crust mix 幔壳岩；地幔地壳混杂体
mantle current 地幔流
mantle diapir 地幔底辟；地幔挤入构造
mantle diapirism 地幔底辟作用；地幔挤入褶皱
Mantle drag force 地幔拖曳力
mantle dynamics 地幔动力学
mantle earthquake 地幔地震
mantle fold 盖层褶皱
mantle folding 盖层褶皱
mantle friction 表面摩擦；表面摩阻力
mantle heat flow 地幔热流
mantle hotspot 地幔热点
mantle isochron 地幔等时线
mantle metasomatism 地幔交代作用
mantle of ash 火山灰盖层
mantle of rock 风化层；表土层

mantle of soil 土的表层；表皮土；风化层；土壤覆盖层
mantle of the Earth 地幔
mantle of vegetation 植被；绿化覆盖
mantle of waste 风化壳；风化残积层
mantle origin 幔源；地幔起源
mantle origin granite 幔源型花岗岩
mantle petrology 地幔岩石学
mantle pipe 套管
mantle plume 地幔柱
mantle plume theory 地幔柱说
mantle refraction 地幔折射
mantle reservoir 地幔层
mantle rheology 地幔流变学
mantle rock 风化层；覆盖层；表岩层
mantle soil 表土层
mantle source composition 地幔源区组成
mantle source gas 幔源气
mantle source rock 幔源岩
mantle source volume 幔源区
mantle surface wave 地幔面波
mantle tomour 地幔鼓胀
mantle transform fault 地幔转换断层
mantle water 地幔水
mantle wave 地幔波
mantle wedge 地幔楔；地幔楔体
mantled 覆盖的
mantled gneiss dome 沉积岩覆盖的片麻岩穹丘
mantle-derived rock 幔源岩石
mantle's convection 地幔对流
manto 席状交代矿床；管状矿床；平卧层控矿床
mantolike 席状的；管状的
mantomarine platform 海蚀台地
Mantou Formation 馒头组
manual 手工的；手动的；指南；手册
manual auger 手摇钻
manual batcher 人工配料拌合机
manual bore 人工钻孔；人工成孔
manual boring 人力钻孔；人工成孔
manual call point 手控火警警报装置
manual choke 手动节流阀
manual consolidation 手工捣实；人工固结
manual control 手动操纵；手动控制
manual control system 手动控制系统
manual counts 人工调查法
manual crane 手动起重机
manual cutting 手工切割；人工切割
manual decode 人工译码
manual design 手工设计；人工设计
manual digitizing 人工数字化
manual dual operation 人工操纵；人力操作
manual encoding 人工编码
manual excavation 人工挖土方；人工开挖
manual excavation shield 人工开挖盾构
manual feed 人力进给；手进给；人力推进；人力给料；手操纵进刀
manual fire alarm call point 手动火灾报警按钮
manual fire alarm system 手控火警警报系统
manual firing 人工点火
manual fitting 手工组装
manual gaging 人工量油

manual handling	人工操作
manual index test	手工指标试验
manual input unit	人工输入设备
manual interpretation	人工判读
manual interpretation technique	人工判读技术
manual knob	手动旋钮
manual loading	人工装载；人工加载
manual mode controller	手动控制器
manual mode operation	手控操作
manual of maintenance	维修手册
manual of photogrammetry	摄影测量手册
manual of seismology	地震学指南
manual operation	单独操纵作业；手动作业
manual part programming	手工编制零件加工程序
manual record	人工记录
manual recording techniques	手工记录技术
manual regulation	手动调整；人工调节
manual release	人工解锁
manual release of switch	道岔人工解锁
manual reset	人工重调；人工重安装；手动复位；手工镶嵌
manual sampler	手工取样器
manual sampling	手工取样；人工取样
manual selectivity control	手动选择性调整；人工选择性控制
manual setter	手摇装置
manual setting	手工装配
manual shield	手掘式盾构
manual shift	人工调节
manual shunt	人工分路
manual steering	人工操작
manual switchboard	人工交换台
manual valve	手动阀
manual voltage regulation	手动调压
manual welding	手工焊接
manual winding control	提升机手动控制
manual zoning	人工分层
manual-acting	手动操作的；手动的；手控的
manualauto switch	手动-自动开关
manual-automatic	半自动的；人工-自动的
manual-automatic switch	手动-自动开关
manual-lock	手动关闭
manually actuated control valve	手动控制阀
manually assisted digitizer	手工辅助数字化仪
manually controlled	手控的
manually controlled work cycle	手控工作周期
manually excavated cast-in-place pile	挖孔灌注桩
manually operated	手控；人工操纵
manually operated control valve	手动控制阀
manually-operated gate	手动闸门
manual-operated pumping well	手压井
manual-return unidirectional prover	手动返回单向标准体积管
manufactured mud	特制泥浆
manufactured rich ore	人造富矿
manufactured sand	人造砂
manufactured soil conditioner	人造土壤改良剂
manu-marble	人造大理石
manus tester	石油闪点测定器
manuscript base	原稿
manuscript map	稿图
manway	人行走道；人孔
manway compartment	人行道隔间
manway landing	人行道平台；梯子间平台
man-way shaft	人行井
many-degree-of-freedom system	多自由度系统
many-group model	多群模型
many-stage	多级的；多段的
many-staged	多级的
many-staged cementing	分级注水泥；多级注水泥
many-valued function	多值函数
mao loess hill lock	黄土峁
Maokouan	茅口阶
Maozhuangian	毛庄阶
map	图；地图；绘图；映射
map accuracy	填图精度；测图精度；地图精度
map adjustment	地图接边；图幅接边
MAP algorithm	后验概率极大化算法
map analysis	映射分析
Map and Air-photo Library	地图及空摄图片资料室
map annotation	地图注记
map appearance	图示特征
map application	地图利用
map archive	地图档案
map binding	地图装帧
map block address assignment	变换块地址分配
map board	图板
map border	图廓
map border decoration	图廓整饰
map cartometry	地图量算
map clarity	地图清晰性
map code	地图符号
map collar	图廓；图边
map colour theorem	地图着色定理
map compilation	地图编制；地图编绘；编图
map complexity	地图复杂性
map composition	地图编制；地图编绘
map construction	图件编绘
map convolution	平面图褶积
map coordinate system	地图坐标系
map coordinates	地图坐标
map coordinates system	地图坐标系
map coverage	有图地区
map cracking	网状裂缝；龟裂；将岩层裂缝标在地图上
map data base management system	地图数据库管理系统
map data structure	地图数据结构
map database	地图数据库
map delineation	地图清绘
map design	地图设计
map development	构图
map digitization	地图数字化
map digitizing	地图数字化
map dimension	图幅尺寸；图幅大小
map display	地图显示
map distance	图上距离
map drawing	绘图
map duplicate	地图复制
map duplication	地图复制
map editing	地图编辑；图件编辑
map editorial policy	地图编辑大纲

map element 地图要素
map face 图面；图幅面积；图幅尺寸
map fair drawing 地图清绘
map for prediction 预测图
map form 图式
map format 地图开幅
map grid 地图网格
map inset 插图
map interlinking 接图
map interpretation 地图判读
map join 图幅连接
map joining 接图
map layer 图层
map layout 图面配置
map legend 地图符号；图例
map legibility 地图易读性
map lettering 地图注记
map library 图库
map load 地图负载量
map making 制图
map making application software 制图应用软件
map manoeuvre 图上作业
map margin 图廓；图幅边缘
map matching method 地图配对法
map measure 量图规
map measurer 测图器；量图仪
map microphotography 地图缩微摄影
map migration 图面空间校正
map model 地图模型
map nadir 图幅天底点
map name 地图名
map of age data 年龄数据图
map of burial depth of bedrock 基岩埋深图
map of buried depth groundwater 地下水埋藏深度图
map of energy distribution 能量分布图
map of frequency distribution 频率分布图
map of geomorphological type 地貌类型图
map of groundwater resources 地下水资源分布图
map of groundwater table 地下水分布图
map of isopiestic level of confined water 地下水等水头线图
map of mine 矿山工程图
map of mine working 采掘工程平面图
map of mineral deposits 矿产图
map of mining subsidence 开采沉陷图
map of primitive data 实际数据图
map of seismicity 地震活动分布图
map of water table 水位地图
map out 绘制；在图上标出
map overlay 地图着色；图覆盖法
map overlay analysis 地图叠置分析
map perception 地图感受
map photograph 航空制图像片
map photography 地图复照
map platemaking 地图制版
map plotting 填图
map plumb point 图面天底点
map positioned 地图展视器
map printing 地图印刷
map printing plate making 地图制版
map process and printing 制印

map production 地图制印
map projection 地图投影
map projection transformation 地图投影转换
map projector 地图投影仪
map range 地图水平距离；图上距离
map reading 读图
map reference 地图索引
map region 已填图地区
map relationship 图幅接合表
map reproduction 地图复制
map reprography 复照
map revision 地图修测；地图修订；地图更新
map scale 图比例尺；地图比例尺
map series 图辑；成套地图；地图类型；地图组别
map sheet 单张地图；一幅图；图幅
map sheet separation 地图分幅
map showing the degree of geological studiedness 地质研究程度图
map spotting 填图
map symbols 图例；地图符号
map symbols bank 地图符号库
map technique 制图技术；制图方法
map title 图名
map unit 填图单位；图上单元
map updating 地图更新
map use 地图利用
map-area 图幅
mapland 制图地区
map-making 制图
map-making satellite 测图卫星
mappable horizon 制图层位
mapped area 已填图地区
mapped element 映射单元
mapped mesh 映射网格
mapped mesh generation 映射网格生成
mapped shape function 映射形函数
mapper 测绘仪；制图员
mapping 绘图；制图；测图；地质填图；地质素描；巷道布置；映射
mapping accuracy 测图精度；填图精度；制图精度
Mapping Advisory Committee 绘制地图咨询委员会
mapping area 测绘范围
mapping by survey and record method 测记法成图
mapping control 测图控制；图根控制
mapping control leveling 图根水准测量
mapping control point 图根控制点
mapping control survey 图根控制测量
mapping exposures 露头测绘
mapping fault 变换误差；素描断层；测绘断层
mapping from photograph 相片测图
mapping function 映射函数
mapping height survey 图根高程测量
mapping instrumentation 绘图仪器
mapping intelligence 地图资讯
mapping intelligence system 地图资讯系统
mapping lineament 映象线性体
mapping method with transit 经纬仪测绘法
mapping mode 变换方式；映射模式
mapping of deterioration 病害绘图
mapping package 绘图程序包

mapping photograph	航空制图相片
mapping photography	航空制图摄影
mapping program	绘图程序
mapping programme	地图测绘计划
mapping radar	成像雷达
mapping recorded file	图历簿
mapping satellite	测图卫星
mapping scale	测绘比例尺
mapping space	映射空间
mapping strip	制图航线
mapping survey	地图制作测量
mapping title	图名
mapping traversing	图根导线测量
mapping triangulation	图根三角测量
mapping unit	填图单位；图上单元
mapsat	制图卫星
mapsheet	图幅
maraging	高强度热处理；马氏体时效处理
marahuite	土状沥青褐煤
marble	大理岩；大理石；云石
marble cavern	大理岩溶洞
marble coating	大理石墙面
marble conservation	大理石保护
marble degradation	大理石劣化
marble deterioration	大理石病害
marble division	大理岩分层
marble factory	大理石厂
marble floor	大理石地板
marble fracture	大理石状断口
marble limestone	大理岩化灰岩
marble patina	大理石光泽
marble quarry	大理石采场
marble samples	大理石样品
marble sand	大理石砂
marble sculptures	大理石雕塑
marble slab	大理石板
marble stone	大理石
marble surface	大理石面层
marble wax	地蜡
Marble weathering	大理石风化
marble worker	大理石开采工；大理石石工
marble-cement interface	大理石-水泥界面
marbled	大理石的；用大理石建造的
marbled glass	大理石纹玻璃
marbled limestone	大理岩状灰岩
marbleization	大理岩化；仿大理石
marcasite	白铁矿；具金属光泽的宝石
marcasitization	白铁矿化
marceline	杂褐锰矿
Marcellus shale	马塞勒斯页岩
march place	开到矿界的巷道；沿矿界巷道；井田边界巷道
march stone	标界石
marches	煤矿边界；界
marchetti	扁式松胀仪
Marchetti's flat dilatometer	马氏扁式松胀仪
marching	沿煤矿边界采掘
marchite	顽火透辉岩
marcofabric	粗组构
mare basin	月海盆地
mare liberum	公海
mare material	月海物质
mare platform	海上工作平台
marebase	月海基性岩
marekanite	珍珠岩；贫水黑曜岩
maremma	沿岸低沼泽区；近海岸沼泽地
mareogram	潮汐涨落曲线
mareograph	自动记潮仪；自记水位计
mareugite	蓝方辉长岩
marezzo	人造大理石
marforming	形变热处理；橡皮垫深拉法
marga porcellana	高岭石
margarite	珍珠云母；串珠雏晶
margarosanite	针硅钙铅矿
margianl-ocean basin	边缘洋盆
margin	安全系数；边缘；边界；界限；储备
margin capacity	备用容量
margin check earthquake	安全极限校核地震
margin design	边限设计
margin for contingencies	发生意外预留金
margin knot	边节
margin line	界限线
margin of continental shelf	大陆架边缘
margin of drill	钻头切削边缘；钻锋圆边
margin of energy	能量储备
margin of error	误差界限；误差限度
margin of lift	升举限度
margin of safety	强度储备〔混凝土强度〕；安全度；安全储备；安全系数
margin of stability	稳定系数
margin plate	缘板
margin reefs	缘礁
margin tolerance	公差范围
marginal	边际的；边缘的；最低限度的；微小的；少量的
marginal accretion	边缘增生
marginal adjustment	边际调整
marginal analysis	边际分析
marginal arc	边缘弧
marginal assimilation	边缘同化
marginal aulacogen	边缘坳拉槽
marginal bank	边滩
marginal barrier	边缘堤
marginal basin	弧后盆地；边缘海盆地
marginal beam	边缘梁
marginal bund	溢流闸室
marginal channel	边缘海峡；陆缘海峡
marginal checking	边缘校验
marginal compression	边缘压缩
marginal crevasse	边冰隙；冰缘裂隙
marginal crude oil production	边际原油生产
marginal data	临界数据；图例说明
marginal deep	边渊
marginal deep sea	陆缘深海
marginal deposit	开采赢利最低的矿床；边界堆积；边缘沉积
marginal depression	边缘坳陷
marginal development	从油田边缘向中间开发
marginal distribution	边缘分布
marginal ditch	边缘沟

marginal drainage 边界排水
marginal earthquake 边缘地震
marginal effect 边际效应
marginal facies 边缘相
marginal facies association 边缘相组合
marginal fault 边缘断层
marginal field 边际油气田
marginal flood 边缘注水
marginal fold 边缘褶皱；附加褶皱
marginal furrow 外缘沟
marginal geosyncline 陆缘地槽
marginal grade 边缘品位；边际品位；可采矿产的最低品位
marginal indifference curve 边际无差别曲线
marginal information 图廓注记
marginal joint 边缘节理
marginal lagoon 边缘泻湖
marginal land hypothesis 边缘陆块说
marginal layer 边际油气层
marginal metasomatic texture 边缘交代结构
marginal migmatization 边缘混合岩化作用
marginal mine 刚够维持成本的矿山
marginal mining place 边缘采矿场
marginal moraine 边缘冰碛
marginal note 旁注；附注
marginal offset 边缘断错
marginal oil field 边际油田
marginal ore 边缘矿石；极限矿石
marginal output curve 边际产量曲线
marginal pacific fault system 滨太平洋断裂系
marginal permafrost 边界永冻土
marginal plain 边界平原
marginal plateau 边缘台地；陆缘海台
marginal principle 边际原理
marginal probability 边际概率
marginal probability distribution 边缘概率分布
marginal producer 边际生产井
marginal production rate 边际产量
marginal productivity 边际生产力
marginal prospects 边际矿区
marginal quay 顺岸式码头；顺岸码头
marginal reserves 边界储量
marginal ridge 围脊
marginal ring depression 陆缘环形盆地；边缘环形坳陷
marginal rise 边缘隆起
marginal schistosity 边缘片理
marginal sea 边缘海；陆缘海
marginal sea basin 边缘海盆地
marginal sea basin magnetic association 边缘海盆地岩浆共生组合
marginal sea level movement 边缘海平面运动
marginal sharpness 边缘清晰度
marginal stability 临界稳定性
marginal stream 边缘河流；冰缘河流
marginal strike 边际产量发现井
marginal subsidence 边缘沉降；边缘坳陷；边缘沉陷；临界坳陷
marginal swell 边缘隆起
marginal syncline 边缘向斜
marginal terrace 边缘阶地；冰缘阶地

marginal texture 边缘结构
marginal thrust 边缘冲断层；临界推力
marginal time 临界时间
marginal trench 边缘海沟；陆缘海沟
marginal trough 边缘坳陷；边缘凹陷；边缘地槽
marginal type wharf 顺岸码头
marginal unconformity 边缘不整合
marginal upthrust 边缘上冲断层
marginal utility 边际效用
marginal value 临界值；边际值
marginal well 接近采完的井；枯竭井
marginal wharf 堤岸码头
marginal zone 边缘带
marginalia 图廓；旁注；页边说明
marginalize 忽略；边缘化
marginal-marine 滨海的
marginal-marine formation 海岸沉积；海岸层
marginal-slope accumulation 边坡堆积作用；临界边坡堆积
marginal-type wharf 顺岸码头
marginarium 边缘区
marialite 钠柱石
marialitization 钠柱石化
Mariana arc 马里亚纳弧
Mariana trench 马里亚纳海沟
Mariana trench type subduction zone 马里亚纳型俯冲带
Mariana-type subduction zone 马里亚纳型俯冲带；马里亚纳型隐没带
marienbergite 钠沸响岩
Marietta continuous miner 玛丽埃塔型连续巷道掘进机；玛丽埃塔重型履带式连续采煤机
marigram 海潮图；潮汐图
marigraph 验潮计
marine 海洋的；海成的；海相的；海积的；海蚀的；海事的；海运的
marine 3-D survey 海上三维勘探
marine abrasion 海蚀；浪蚀
marine abrasion topography 海蚀地形；浪蚀地形
marine accumulation 海洋堆积；海积
marine acoustics 水声学
marine aerosol 海洋气溶胶
marine aggregate 海砂粒；海骨料
marine air lift mining dredger 海上空气提升式采矿船
marine algae 海藻
marine alluvium 海洋冲积层；海成冲积层
marine and airborne magnetometer 海空磁力仪
marine and non-marine stratigraphic correlation 海相和非海相地层对比
marine arch 海蚀拱桥；海蚀拱
marine arch deposit 海洋火山灰沉积
marine artificial port 海上港口
marine atlas 海洋图集
marine authigenic mineral 海相自生矿物
marine band 海相夹层
marine bar 海相砂坝
marine barium cycle 海洋钡周期
marine barometer 船用气压计
marine basin 海盆
marine beach 海滩
marine bed 海相层；海底；海床

marine belt 领海
marine bench 海蚀台；浪蚀台
marine biocycle 海洋生物带
marine biological chart 海洋生物图
marine biology 海洋生物学
marine biosphere 海洋生物圈
marine borrow area 海洋采泥区；海洋取土区；海洋采料区；海上采砂区
marine borrowing 海洋取土
marine bottom community 海底群落
marine bottom material survey 海底物质调查；洋底物质调查
marine bottom proton sampler 海洋质子采样器
marine bottom sediment 海底沉积物
marine brine 海相卤水
marine cable 船用电缆；海底电缆
marine cave 海蚀洞；海蚀穴
marine chain-bucket mining dredger 海上链斗式采矿船
marine chart datum 海图基准面
marine charting 海洋测绘
marine charting database 海洋测绘数据库
marine chemistry 海洋化学
marine clastics 海上碎屑岩
marine clay 海成黏土；海相黏土
marine cliff 海蚀崖
marine climate 海洋性气候
marine climatology 海洋气候学
marine communications and transportation industry 海洋交通运输业
marine conductor 海上钻井导管
marine conservation 保护海洋环境；海洋保护
marine conservation area 海洋保育区
marine construction 海洋构筑物
marine construction work 海事建筑工程
marine contained disposal 密封式海上卸置
marine contamination 海洋污染
marine continental alternative facies 海陆交互相；海陆过渡相
marine corrosion 海水腐蚀；海洋腐蚀
marine crossing 水下穿越
marine cryology 海冰学
marine cut terrace 海蚀阶地
marine cycle 海蚀旋回
marine data 海洋水文注记
marine datum 海洋基准面
marine debris 海洋碎屑物
marine delta 海洋三角洲
marine delta plain 海洋三角洲平原
marine demarcation survey 海洋划界测量
marine denudation 海水侵蚀；海水剥蚀作用
marine deposit 海相沉积；海积物
marine deposit displacement 海相沉积土排移法
marine deposit removal 海相沉积土清除法
marine deposition 海积作用；海洋沉积作用
marine deposition coast 海积海岸
marine deposition graded coast 海积夷平海岸
marine deposition plain 海积平原
marine deposition terrace 海积阶地
marine deposition topography 海积地形；浪积地形
marine detector 海上检波器

marine detector cable 海上检波电缆
marine detritus 海洋碎屑
marine development 海洋开发
marine development planning 海洋开发规划
marine Devonian 海洋泥盆纪
marine directional well 海上定向井
marine disaster 海洋灾害
marine disaster forecasting and warning 海洋灾害预报和警报
marine disaster prevention 海洋防灾
marine disaster reduction 海洋减灾
marine disaster reduction engineering 海洋减灾工程
marine disposal scheme 海上倾卸计划
marine dolomitization 海洋白云岩化
marine drift 海流
marine drilling 海洋钻探；海上钻探；海上钻井
marine drilling catamaran 筏式海洋钻机
marine drilling platform 海上钻井平台
marine drilling rig 海上钻机
marine drilling ship 海上钻井船
marine dumping 倾物入海；海上倾倒物料
Marine Dumping Action Plan 倾物入海行动计划
marine dumping area 海洋倾倒物料区
marine dumping permit 倾物入海许可证
marine dumping site 海上卸泥区
marine ecology 海洋生态；海洋生态学
marine ecosystem 海洋生态系统
marine electrical prospecting 海洋电法勘探
marine electromagnetics 海洋电磁
marine energy sources 海洋能源
marine engineering 海事工程
marine engineering dynamic geology 海洋工程动力地质学
marine engineering geologic investigation 海洋工程地质勘察
marine engineering geologic map 海洋工程地质图
marine engineering geological analysis map 海洋工程地质分析图
marine engineering geological condition 海洋工程地质条件
marine engineering geological drilling 海洋工程地质钻探
marine engineering geological investigation 海洋工程地质勘察
marine engineering geological long-term survey 海洋工程地质长期观测
marine engineering geological phenomenon 海洋工程地质现象
marine engineering geological prospecting 海洋工程地质勘探
marine engineering geological prospecting point 海洋工程地质勘察点
marine engineering geological sampling 海洋工程地质取样
marine engineering geological sruvey line 海洋工程地质勘察线
marine engineering geological survey 海洋工程地质测绘
marine engineering geological test 海洋工程地质试验
marine engineering geology 海洋工程地质学
marine engineering survey 海洋工程测量
marine environment 海洋环境
marine environment monitoring 海洋环境监测

marine environment protection	海洋环境保护
marine environmental assessment	海洋环境评价
marine environmental capacity	海洋环境容量
marine environmental chart	海洋环境图
marine environmental data and information referral system (MEDI)	海洋环境数据和资料检索系统
marine environmental forecasting and prediction	海洋环境预报预测
marine erosion	海蚀；浪蚀
marine erosion coast	海蚀岸
marine erosion plain	海蚀平原
marine erosional unconformity	海蚀不整合
marine erratics	海相漂砾
marine evaluation well	海上评价井
marine evaporite	海相蒸发岩
marine exploration	海洋勘探
marine exploration well	海上勘探井
marine explosive source	海洋炸药震源
marine exposure test	海洋环境腐蚀试验
marine facies	海相
marine facies source rock	海相烃源岩
marine facilities	海洋设施
marine fill	海相填料；海沙填料
Marine Fill Committee	海洋填料委员会
marine flooding surface	海泛面
marine foreland	海岬
marine foundation	海工基础
marine gas hydrate	海洋天然气水合物
marine gathering line	海上集输管线
marine geochemical cycling	海洋地球化学循环
marine geochemical exploration	海洋地球化学勘探；海洋化探
marine geochemical facies	海洋地球化学相
marine geochemical prospecting	海洋地球化学探矿；海洋化探
marine geochemistry	海洋地球化学
marine geodesy	海洋大地测量学；海洋测量学
marine geodetic network	海洋大地控制网
marine geodetic survey	海洋大地测量
marine geographic information system (MGIS)	海洋地理信息系统
marine geography	海洋地理学
marine geohazard	海洋地质灾害
marine geoid	海洋大地水准面
marine geologic investigations	海洋地质调查
marine geological process	海洋地质作用
marine geological research vessel	海洋地质调查船
marine geological storage	海洋地质储存；海洋地质封存
marine geological survey	海洋地质调查
marine geology	海洋地质学
marine geology and geophysics	海洋地质与地球物理
marine geology of the Pacific	太平洋海洋地质
marine geomagnetic anomaly	海洋地磁异常
marine geomagnetic survey	海洋磁测；海洋地磁调查
marine geomorphology	海洋地貌；海洋地貌学
marine geophysical prospecting	海洋地球物理勘探；海洋物探
marine geophysical survey	海洋地球物理测量
marine geophysics	海洋地球物理学
marine geoscience	海洋地球科学
marine geotechnical exploration	海洋工程地质勘探；海洋岩土工程勘探
marine geotechnical test	海洋土工试验
marine geotechnique	海洋土工学
marine geotechnology	海洋地质工程学；海洋岩土工程学
marine GIS database	海洋地理信息系统数据库
marine glaciation	海洋冰川
marine gold dredger	海上采金船
marine gravel	海相砾石
marine gravimeter	海洋重力仪
marine gravimetric survey	海洋重力测量
marine gravimetry	海洋重力学；海洋重力测量
marine gravity	海洋重力
marine gravity anomaly	海洋重力异常
marine gravity meter	海洋重力仪
marine gravity prospecting	海洋重力勘探
marine gravity survey	海洋重力勘探；海洋重力调查
marine ground investigation	海上勘探
marine hardground	海洋硬底
marine heat flow	海洋热流
marine heat flow survey	海洋地热流调查
marine humus	海洋腐殖质
marine hydrographic survey	海洋水文测量
marine hydrography	海洋水文学
marine hydrological chart	海洋水文图
marine hydrology	海洋水文学
marine influence	海洋的影响
marine ingression	海进
marine invasion	海侵
marine invertebrate	海洋无脊椎动物
marine iron ore sand dredger	海上铁矿砂开采船
marine Jurassic	海洋侏罗纪
marine laboratory	海洋实验室
marine landform	海成地形；海洋地貌
marine leveling	海洋水准测量
marine limestone	海相石灰岩
marine limit	海岸线
marine linear magnetic anomaly	海洋条带状磁异常
marine magnetic anomaly	海洋磁力异常
marine magnetic chart	海洋磁力图
marine magnetic prospecting	海洋磁法勘探
marine magnetic survey	海洋磁力测量
marine magnetometer	海洋磁力仪
marine magnetometry	海洋磁力测量
marine magnetotelluric	海洋大地电磁
marine magnetotelluric sounding	海洋大地电磁测深
Marine Managed Area	海洋管理区
marine map	海图
marine mapping	海洋测绘
marine marl	海成泥灰岩；海成灰泥
marine marsh	沿海沼泽
marine massive sulfide	海底块状硫化物
marine meteorological chart	海洋气象图
marine meteorology	海洋气象学
marine microbiology	海洋微生物学
marine mineral	海洋矿物
marine mineral deposits	海洋矿藏
marine mineral resource	海洋矿产资源
marine mining	海洋采矿

marine mining dredger	海上采矿船
marine mud	海积淤泥；海泥
marine natural resources	海洋自然资源
marine navigation	海上导航
marine observational section	海洋观测断面
marine offlap	海上退覆
marine oil	海洋石油
marine oil industry	海洋石油工业
marine oil production well	海上采油井
marine oil shale	海洋油页岩
marine oil spill	海上漏油
marine onlap	海相超覆；海进
marine operation	海上作业
marine organic matter	海洋有机物
marine organism	海洋生物
marine origin	海相成因；海成
marine origin of petroleum	海相生油
marine originated saline lake	海源咸水湖
marine originated salt lake	海源盐湖
marine originated salt water lake	海源咸水湖
marine outfall	出海排水口
marine paleoenvironment	海洋古环境
marine peat	海相泥炭；咸水泥炭
marine pendulum	海洋摆仪
marine peneplain	海成准平原；海蚀准平原
marine peneplanation	海洋准平原作用
marine petroleum development	海洋石油开发
marine petroleum exploitation	海洋石油开采
marine petroleum pollution	海洋石油污染
marine phase carbonate rocks	海相碳酸盐岩
marine phreatic environment	海水潜流环境
marine physical chemistry	海洋物理化学
marine physics	海洋物理学
marine pipe-laying	海上铺管
marine placer	海砂矿；海成砂矿
marine plain	海蚀平原；海积平原
marine planation	浪蚀夷平作用；浪削
marine plane	浪蚀平面
Marine Planning Areas	海洋规划区
marine platform	海蚀台地；海上平台
marine pneumatic seismic source	海洋压缩空气震源
marine pollutant	海洋污染物
marine pollution	海洋污染
marine pollution prevention law	海洋防污法规
marine position recorder	海上位置记录仪
marine positioning	海洋定位
marine processing technique	海上处理技术
marine production well	海上生产井
marine progressive overlap	海侵超覆
marine prospecting	海上勘探
Marine Protected Area	海洋保护区
marine protection	海洋保护
marine proton magnetometer	海洋质子磁力仪
marine reconnaissance	海上普查
marine reflection seismic survey	海洋反射地震调查
marine reflection seismics	海洋反射地震
marine reflection survey	海洋反射测量
marine refraction seismic survey	海洋折射地震调查
marine refraction survey	海洋折射测量
marine regression	海退
marine research ship	海洋科研船
marine reserve	海洋保护区
marine reservoir rock	海相储集岩
Marine Resource Areas	海洋资源区
marine resource conservation	海洋资源保护
marine resource exploitation	海洋资源开发
marine resource utilization	海洋资源利用
marine resources	海洋资源
marine resources chart	海洋资源图
Marine Resources Conservation Working Group	海洋资源保护工作小组
marine riser	海上隔水导管；海底取油管；海水隔管
marine rock	海成岩石
marine salina	海边盐沼；盐滩
marine salt	海盐
marine sampling	海洋取样
marine sampling station	海水抽样站；海上取样站
marine sand	海积砂；海砂
marine sand borrow area	海沙采挖区
marine sand borrow pit	采海沙坑
marine science	海洋科学
marine science and technology	海洋科学和技术
marine sdeimentation	海洋沉积作用
marine section	海洋断面
marine sediment	海洋沉积物；海相沉积物；海床底泥；海泥
marine sediment acoustics	海洋沉积声学
marine sediment extract	海洋沉积物中提取物
marine sediment mineralogy	海洋沉积物矿物学
marine sediment quality	海洋沉积物质量
marine sedimentary sequence	海相沉积序列
marine sedimentation	海洋沉积作用
marine sedimentology	海洋沉积学
marine sedimentophysics	海洋沉积物理学
marine sediments	海洋沉积物；海积物
marine seepage	海底渗流；海底渗漏
marine seismic acquisition	海上地震采集
marine seismic cable	海洋地震电缆
marine seismic crew	海上地震队
marine seismic exploration	海洋地震勘探
marine seismic generator	海洋地震信号发生器
marine seismic operation	海上地震作业
marine seismic profiler	海洋地震剖面仪
marine seismic reflection data	海洋反射地震数据
Marine seismic source	海上震源
marine seismic streamer	海洋地震拖缆；海洋地震漂浮电缆
marine seismic survey	海洋地震勘探；海洋地震观测；海洋地震调查
marine seismics	海洋地震学
marine seismograph	海洋地震仪
marine seismometer	海洋地震计
marine self purification	海洋自净能力
marine self-potential method	海洋自然电位法
marine sensing	海洋遥感
marine sequences	海相层序
marine services support area	海事服务支援区；海事服务后勤用地
marine sewer	海底污水渠
marine shale	海相页岩

marine snow	海雪
marine soil	海积土；海相土壤；海泥
marine soil mechanics	海洋土力学
marine soil science	海洋土质学
marine sonoprobe	海洋声呐测深仪；海洋声波测量仪
marine sour natural gas	海洋酸性天然气
marine source	海上震源
marine source of pollution	海洋污染源
marine source rock	海相烃源岩
marine sources of fill	海上填料来源
marine spoil ground	海上废土场；海上弃土区；海上卸泥场
marine strata	海相地层
marine stratigraphy	海洋地层学
marine stratum	海相地层
marine streamer	海洋拖缆
marine stromatolites	海洋叠层石
marine structural dynamics	海洋结构动力学
marine structure	海事结构；海工结构；海洋建筑物；海上建筑物
marine succession	海洋生态演替
marine suction mining dredger	海上吸扬式采矿船
marine survey	海洋测量
marine survey positioning	海洋测量定位
marine survey vessel	海上监察船
marine surveying	海洋调查；海上测量；海洋测量
marine surveying and mapping	海洋测绘
marine swamp	海湿地；沿海木本沼泽
marine talus	海下岩屑堆积
marine tankage	油轮载量
marine technology	海底矿技术工艺；海洋技术
marine tectonic settings	海洋构造环境
marine terminal	海洋油库；海运末站；海运终点站
marine terrace	海阶；海成阶地；海洋阶地
marine thematic survey	海洋专题测量
marine time	海洋地质时代
marine traffic	海路交通
marine transect	海洋断面
marine transgression	海侵
marine transport	海上运输
marine transportation	海洋搬运作用
marine trap	海相圈闭
marine unconformity	海成不整合
marine vertical seismic profiling	海洋垂直地震剖面
marine warp soil	海相淤泥土
marine water	海水
marine water quality monitoring station	海上水质监测站
marine waters	海洋水域
marine weather forecast	海洋天气预报
marine wide-angle reflection seismic survey	海洋广角反射地震调查
marine works	海上工程；海洋工程；海事工程
marine-based investigation	海上调查
marine-based port facility	海上港口设施
marine-built	海积的
marine-built platform	海成台地
marine-built terrace	海积阶地；海成阶地
marine-continental correlations	海陆相关；海相陆相对比
marine-cut	海蚀的
marine-cut bench	海蚀阶地；海台
marine-cut platform	海蚀台地
marine-cut terrace	海蚀阶地；浪蚀阶地
marine-erosion coast	海蚀海岸；海蚀岸
marine-erosion plain	海蚀平原
marine-erosion platform	海蚀平台
marineline	海底管线
mariner's compass	船用罗盘
marine-terrestrial correlations	地球海洋的相关性；海相陆相对比
marining	海侵
mariograph	自记验潮仪
marionite	水锌矿
mariposite	铬硅云母
mariposition	海洋沉积
marisat	海上卫星通信系统
maritime	海的；海上的；海运的
maritime affair	海运事务
maritime air mass	海洋气团
Maritime Arbitration Commission	海事仲裁委员会
maritime archaeology	海洋考古学
maritime bridge	海上桥梁
maritime canal	通海运河
maritime city	海上城市
maritime civilization	海洋文明
maritime claim	海事索赔
maritime climate	海洋性气候
maritime consciousness	海洋意识；海洋观念
maritime court	海事法庭
maritime customs	海关
maritime distress	海难
maritime engineering	海事工程；海洋工程学
maritime exploration	海洋探险
maritime factory	海上工厂
maritime glacier	海洋性冰川
maritime law	海洋法
maritime meteorological observation	海洋气象观测
maritime meteorology	海洋气象学
maritime peril	海上遇险
maritime port	海港
maritime province	沿海区域
maritime resources	海洋资源
maritime sense	海洋意识；海洋观念
maritime sequence	海域层序；海岸层序；海洋层序
maritime territory	领海
maritime transportation	海运
maritime waste	海上废物
maritime work	海工；海事工程；海塘工程
mariupolite	淡霞钠长岩；钠霞正长岩
mark at or below ground level	埋石
mark buoy	标识浮标
mark card	标记卡片
mark detail	标志大样
mark for geological interpretation	像片地质解释标志
mark for measuring velocity	测速标
mark number	标号；编号
mark of seismic phase	震相标志
mark off	划分出；给…划界；区分
mark post	标柱；标桩
mark pulse	标志脉冲
mark reader	标记读出器

mark scanning	标记扫描；特征扫描	markstone	标石
mark scraper	划线器	marl	泥灰岩
mark sensed card	标记读出卡片	marl ball	泥灰球
mark sensing	符号识别	marl biscuit	泥灰饼
mark stone	标石	marl brick	泥灰岩砖
marked	有标记的；定线的	marl clay	泥灰质黏土；泥灰土
marked bed	标志层；标准层	marl earth	泥灰土
marked compound	标记化合物	marl lake	泥灰湖
marked depth line	标定深度线	marl loam	泥灰质壤土
marked end	磁针的指北端	marl sandstone	泥灰砂岩
marked flask	带刻度烧瓶	marl shale	泥灰页岩
marked point	觇标点	marl slate	泥灰板岩
marked prescription bottle	有标记的样品瓶	marl soil	泥灰土；泥灰质土壤
marked relief	显著地形起伏	marl stone	硬泥灰岩
markedness	显著	marlaceous	泥灰岩的；泥灰质的
marker	标准层；指示层；指示器；标记器；划线盘	marlaceous chalk	泥灰质白垩
marker and cell method	标记网格法	marlaceous face	泥灰岩相
marker band	标志带；指示层	marlaceous facies	泥灰岩相
marker bed	标志层；标准层；指示层位	marlaceous loam	泥灰质壤土
marker block	岩心箱标志牌；标志块	marlekor	冰泥钙结核
marker bolt	标桩锚杆；标志锚杆	marline sand	海砂
marker buoy	标志浮筒	marlite	硬泥灰岩；泥灰岩
marker crude	标记原油	marlne mining	海底采矿法；海洋采矿法
marker crude price	基准油价	marloesite	橄榄钠长斑岩；钠长橄斑岩
marker dolphin	标志柱；标志墩	marl-pit	泥灰岩坑
marker formation	指示层；标准层；标准层位；标准建造	marls and marly-limestones	泥灰岩和泥灰质灰岩
marker horizon	标志面；标志层；指示层；标准层；标准层位	marlstone	泥灰岩；铁质鲕状泥灰岩；硬泥灰岩
		marly	泥灰岩的；泥灰质的
marker lamination	标准层；指示层	marly bituminous shale	泥灰沥青质页岩
marker post	标杆	marly clay	泥灰质黏土
marker register	时标寄存器	marly facies	泥灰岩相
marker seam	标志煤层；标记煤层	marly limestone	泥灰质灰岩；泥灰质石灰岩
marker spacing	标志间距	marly sandstone	泥质砂岩；泥灰质砂岩
marker velocity	标准层速度	marly sandy loam	泥灰砂质壤土
marker-and-cell method	标记点和格子法	marly shale	泥灰质页岩
marker-bed	标志层	marly soil	泥灰质土壤；泥灰土
marker-defined unit	标志明确的单位	Marmara Sea	马尔马拉海
marketable coal	旺销煤	marmatite	铁闪锌矿
marketable gas	商品气	Marmo Stage	马莫阶
marketable natural gas	商品天然气	marmolie	白蛇纹石
marketable ore	商品矿石	marmolite	白叶蛇纹岩
markfieldite	斜长花斑岩；闪长花斑岩	Marmor	马莫阶
marking	印痕；痕迹；标志；标记	marmoraceous	大理石状的
marking of points	测点的标定	Marmorian	马莫阶
marking out of foundation trench	基坑划线	marmorization	大理岩化
marking pin	测钎	marmorize	大理岩化
marking point	标志点	marmorosis	大理岩化
Markov analysis	马尔可夫分析	marosion	海洋侵蚀
Markov chain	马尔可夫链	marosite	云辉等色岩
Markov chain model	马尔柯夫链模型	Marquesas archipelago	马克萨斯群岛
Markov model	马尔可夫模型	Marquesas fracture zone	马克萨斯断裂带
Markov operator	马尔科夫算子	Mars	火星
Markov sequence	马尔科夫序列	Mars dust storm	火星尘暴
Markov stochastic field	马尔可夫随机场	Mars escape velocity	火星逃逸速度
Markovian decision	马尔科夫决策	Mars ionosphere	火星电离层
Markovian variable	马尔科夫变量	Mars polar cap	火星极盖
marks for measuring velocity	测速标	marscoite	花岗辉长混染岩
marks of interpretation directly	直接判读标志	Marsdenian	马斯登阶
marks of interpretation indirectly	间接判读标志	marsh	沼泽；沼地；湿地
		marsh bank	沼泽滩

marsh basin 沼泽盆地
marsh bog 沼泽地
marsh coast 沼泽海岸
marsh deposit 沼泽沉积
marsh funnel 钻泥黏度测量漏斗；马氏漏斗
Marsh funnel viscosimeter 马氏漏斗泥浆黏度计
Marsh funnel viscosity 马氏漏斗黏度
marsh gas 沼气；甲烷
marsh geophone 沼泽检波器
marsh ground 沼泽地
marsh lake 沼泽湖
marsh land 沼泽地
marsh land coast 沼泽海岸
marsh muck 沼泽腐殖土
marsh ore 沼铁矿；褐铁矿
marsh peat 沼泽泥炭
marsh pipeline 沼泽地管线
marsh soil 沼泽土
Marsh viscometer 马氏黏度计
Marsh viscosity 马氏黏度
Marshall line 马歇尔线
Marshall mini booster 马歇尔小型增压器
Marshall stability apparatus 马歇尔稳定度仪
Marshall stability test 马歇尔试验
Marshall test 马歇尔试验
marshing of soil 土壤沼泽化
marshite 碘铜矿
marshland 沼泽地；泥沼地；湿地
marshland coast 湿地海岸
marshy 沼泽地的
marshy area 沼泽地区；潮湿地区
marshy grassland 草甸子
marshy ground 沼泽地
marshy soil 沼泽土
marshy terrain 沼泽地带
Marsquake 火星地震
Martens hardness 马氏硬度
marthozite 硒铜铀矿
Martian 火星的
martinite 响白碱玄岩；板磷钙石
martite 假象赤铁矿
martite concentrate 假象赤铁矿精矿
marundite 珠云刚玉岩
masafuerite 多橄玄武岩
masanite 马山岩
masanophyre 马山斑岩
Masaya type caldera 玛萨亚型火山口
mascagnite 铵矾；硫铵矿
mascareignite 植物蛋白石
maser 脉塞；微波激射器
maser action 微波激射作用
mash 矿浆
mashroom rock 蘑菇状岩
maskelynite 斜长玻璃；熔料长石
masonite 硬绿泥石
masonry beam 砖石梁
masonry bearing wall 砖石承重墙
masonry block 砌石块；砌墙块
Masonry bridges 石板桥
masonry cement 砌筑用水泥

masonry construction 砖石结构
masonry dam 圬工坝；砌石坝
masonry foundation 砌筑基础
masonry gravity dam 浆砌石重力坝
masonry lining 砖石衬砌；砌石支架
masonry pier 砖石墩
masonry pitching of slope 砌石护坡
masonry plate 支承垫石
masonry reservoir 围海造田
masonry retaining wall 砖石挡土墙；砌石挡土墙
masonry shaft 砖砌立井；石砌立井
masonry shaft lining 砌筑式井壁
masonry shear wall 砌体剪力墙
masonry skinwall 砖石薄墙；砌石薄墙
masonry structure 砌体结构；砖石结构
masonry unit 砌块
masonry wall 砖石墙；砌石墙
masonry work 石工
masonry-lined 圬工衬砌的
masonry-lined tunnel 圬工衬砌隧道
mass absorption coefficient 质量吸收系数
mass accumulation 质量累积
mass action 质量作用；浓度作用
mass action effect 质量作用效应
mass action law 质量作用定律
mass analysis 质量分析
mass analyzer 质谱仪；质谱分析仪
mass attenuation coefficient 质量衰减系数
mass avalanche 大规模崩坍
mass average flux 质量平均通量
mass balance 质量平衡
mass balance approach 质量平衡方法
mass balance equation 质量平衡方程
mass balance modelling 质量平衡模式
mass breaking 大崩矿；大量采矿；大爆破
mass building earthquake disaster prediction 房屋群体震害预测
mass caving 大冒顶；大崩落
mass cement 普通水泥
mass center 质心
mass characteristic 质量特性
mass chromatogram 质量色谱图
mass chromatography 质量色谱
mass circuit 物质循环
mass coefficient 质量系数
mass collection 大量采集
Mass compass 马氏罗盘
mass concentration 质量浓度；质量瘤
mass concrete 无钢筋混凝土；大块混凝土；大体积混凝土
mass concrete dam 大体积混凝土坝
mass concrete retaining wall 大体混凝土挡土墙
mass concrete wall 无钢筋混凝土墙；大块混凝土墙
mass conservation 质量守恒
mass conservation law 质量守恒定律
mass control 质量控制
mass coupling 质量耦合
mass coupling coefficient 质量耦合系数
mass curve 累积曲线
mass curve of water 累计水量曲线

mass data 大批数据；大量数据
mass defect 质量亏损
mass density 质量密度
mass depletion 质量亏空
mass detector 质量检测器
mass diagram 积分曲线图；土方累积图；土方叠积图
mass diagram of earthwork 土方累积曲线图
mass diagram of runoff 径流累积曲线图
mass difference 质量差
mass diffusion 质扩散
mass distribution 质量分布
mass effect 质量效应
mass energy absorption coefficient 质能吸收系数
mass erosion 重力侵蚀
mass estimation 质量测定
mass excavation 大面积开挖
mass excess 质量剩余
mass extinction 集群灭绝；大量灭亡
mass float 惯性浮标；集束浮标
mass flow 块状流；重力流；质量流
mass flow imbalance alarm 质量流量不平衡报警器
mass flow meter 质量流量计
mass flow rate 质量流量
mass flow rate sensitive detector 质量流量敏感型检测器
mass flowmeter 流量计；质量流量计
mass flux 质通量；质量通量
mass for stopping grout 止浆岩盘
mass force 体积力
mass foundation 大体积基础
mass fraction 质量分数
mass function 质量函数
mass haul curve 开挖运输曲线
mass haul diagram 土方掘填搬运图
mass heaving 大量冻胀；大面积冻胀；总体冻胀
mass influence coefficient 质量影响系数
mass law 质量定律
mass load 惯性力；惯性负载
mass loading 大量负荷；惯性负荷
mass loss 质量损失
mass magnetic moment 质量磁矩
mass magnetic susceptibility 质量磁化率
mass matrix 质量矩阵
mass memory 大容量存储器
mass metering 测质量
mass moment of inertia 质量惯矩；转动惯量
mass motion 整体运动
mass movement 重力迁移；整体移动；块体运动
mass movement of rocks 大量岩层移动
mass observation 大量观测
mass of earth 地球质量
mass of igneous rock 火成岩体
mass of inertia 惯性质量
mass of moisture 水分质量
mass of ore 矿体
mass of rock 岩体
mass overthrust 推覆大断层
mass peak 质量峰
mass per unit area 单位面积质量
mass percent 质量百分比
mass permeability 总渗透性；体渗透性；大体积渗透性

mass pile cap 大型桩帽
mass point 质点
mass polymerization 本体聚合法；整体聚合法
mass property 块体性质；整体特性
mass ratio 质量比
mass reservoir supply line 总体水库供水线
mass resistivity 电阻率；比电阻
mass resolution 质量分辨率
mass rock and soil accumulation body 块石土堆积体
mass runoff curve 径流累积曲线
mass sampling 群集抽样
mass scale 质量标度
mass separation 同位素分离
mass separator 质量分离器
mass shooting 大爆破；齐发爆破
mass spectra 质谱
mass spectrogram 质谱
mass spectrograph 质谱仪
mass spectrographic analysis 质谱分析
mass spectrographic method 质谱法
mass spectrography 质谱分析
mass spectrometer 质谱仪；质量分光仪
mass spectrometric analysis 质谱分析
mass spectrometric differential thermal analysis 质谱差热分析
mass spectrometry 质谱分析；质谱法；质谱学
mass spectrophotometer 质谱分光光度计
mass spectrophotometry 质谱分析
mass spectroscope 质谱仪
mass spectroscopy 质谱学；质谱分析；质谱法
mass spectrum 质谱
mass storage 大容量存储器
mass storage facility 大容量存储器设备
mass storage system 大容量存储系统
mass stratum 主要地层
mass strength 岩土体强度
mass synchrometer 同步质谱计
mass thickness 质量厚度
mass transfer 质量转移；质量传递；传质
mass transfer coefficient 传质系数；质量传播系数
mass transit line 集体运输路线
Mass Transit Railway 城市轻轨地铁工程
mass transit system 集体运输系统
mass transport 重力迁移
mass unit 质量单位
mass value 质量值
mass velocity 水体流速；土体流速
mass wasting 块体坡移
mass weighted mean value 质量加权平均值
mass-calculation 土方计算
mass-concrete wall 大体积混凝土岸壁
mass-conservation equation 质量守恒方程
mass-damper-spring system 质量-阻尼器-弹簧系统
mass-energy 质能关系
mass-energy conversion 质-能转换
Mass-energy equation 质量能量方程
mass-energy equivalence 质能等价性
mass-energy relation 质量-能量关系
mass-flow deposit 块体流沉积
mass-flow gas meter 气体质量流量计

mass-gravity dam	重力坝
mass-gravity transport	块体重力搬运
massif	地块；断层块
massif central	中央地块
massif landslide	山体滑坡
massing of buildings	建筑物的密集程度
massive	大的；重的；块状的；大块的；均匀构造的；非晶质的
massive abutment	重力式桥台
massive and rigid structure	大体积刚性结构
massive arch dam	重力拱坝
massive bedding	块状层理
massive clay	块状黏土
massive coal	煤柱中的煤；煤层中未采动部分；块煤
massive concrete	大体积混凝土
massive conglomeratic reservoir	块状砾岩储集层
massive deposit	块状矿床
massive foundation	大块式基础；实体式基础
massive geologic body	块状地质体
massive gold	自然金
massive head dam	大头坝
massive hydrate	块状天然气水合物
massive hydraulic fracturing	大型水力压裂
massive layer	大体积浇筑层
massive mountain	块状山
massive movement	块状运动
massive oil pool	块状油藏
massive ore	块状矿
massive ore deposit	块状矿床
massive orebodies	巨大的矿体
massive quay wall	大体积岸壁
massive reservoir	大型油气藏
massive road network	大型公路网
massive rock	块状岩石；整体岩石；硬岩；岩体
massive rock mass	块状岩体
massive sandstone	块状砂岩
massive sandstone facies	块状砂岩相
massive slump	块状滑动
massive structure	块状构造；整体构造；大体积结构
massive sulfide	块状硫化物
massive sulfide deposit	块状硫化物矿床
massive sulphides	块状硫化物
massive texture	块状结构
massive-head dam	大头坝
massively	整体地，大块地
massiveness	厚实；块状；非晶质；均匀构造
massivity	整体性；巨块结构
massless	无质量
mass-proportional damping	质量比例阻尼
mass-spectrometer	质谱仪
mass-spectrometer measurement	质谱测量
mass-spectrometer tube	质谱仪管
mass-spectrometric analysis	质谱分析
mass-spectrometric technique	质谱法
mass-spectrometry	质谱学
mass-spectrum	质谱
mass-spring system	质量弹簧系统
mass-spring-dashpot system	质量-弹簧-阻尼器体系
mass-synchrometer	同步质谱仪
mass-to-charge ratio	质荷比
mass-transfer control	质量传输控制
mass-translation	整体平移
mass-transport mechanism	质量转移机制
mass-wasting	块体坡移
mass-wasting deposition	块体坡移沉积作用
master and slave flip	主从触发器
master check	校正
master clock	母钟；时钟脉冲；主脉冲
master clock-pulse generator	母时钟脉冲发生器
master computer	主计算机；主机
master console	主控制台
master control	主控制；中心控制站
master control board	主控制台；中心控制台
master control centre	主控制中心
master control code	主控制代码
master control gate	总闸门；主控制栅；主控制门
master control integrated	主控制台；主操纵台
master control panel	主控制盘；主操纵盘
master control program	主控程序
master control routine	主控例行程序
master control system	中央控制系统；中央控制台
master controller	主控制器
master curve	理论曲线；标准曲线
master data	基本数据；主要数据
master design	总体设计
master development plan	总发展计划图
master drawing	主图；原图；布设总图
master dual operation	主要工序
master earthquake	主导地震
master end	主动侧
master equation	主导方程
master event	主导事件
master factor	决定因素
master fault	主断层
master format	标准格式
master fracture	主裂隙；主裂缝
master frequency	主频
master fuse lighter	主导火线点火器
master gauge	标准规
master geological observation route	主干地质路线
master hydraulic control manifold	总液压控制管汇
master joint	主节理
master layout plan	总纲发展蓝图
master lever	主操作杆
master link	主连杆
master lode	主矿脉；巨矿脉
master map	原始资料图；底图
master menu	主菜单
master meter calibration unit	标准流量计标定装置
master meter method	标准仪表检验法
master meter prover	标准流量计检定装置
master mode	主控式；主控方式
master oscillator	主控振荡器
master pattern	原始模型；母模型
master phase	基本相位
master pin	主销；中心立轴
master plan	总纲计划；总纲图则；总图；总平面图
master profile	标准剖面图
master program	主程序
master programme	总纲计划

master ranking list	总评级次序表	material universal test machine	万能材料试验机
master record	主数据；基本数据	material with memory	具有记忆的材料
master river	干流；主要河流	material yielding	材料屈服
master route sheet	总路线表	material-mud	泥浆料
master routine	主程序	materials and techniques of rock support and reinforcement	岩石支护与加固的材料和技术
master sample	标准样品		
master scale	标准刻度；标准秤	materials behavior	材料性质
master schedule	总进度计划	materials characterization	材料表征
master scheduling group	主要工程项目计划小组	materials evaluation test	材料评价试验
master seismic event	主导地震事件	materials fatigue	材料疲劳
master selector	主选择器；主选波器	materials science	材料科学
master set	校对调整	materials testing	材料试验
master slave system	主从系统	mathematic analysis	数学分析
master station	主站；总站；主控站	mathematic statistics	数理统计
master station amplifier	主台放大器	mathematical algorithm	数学算法
master station service	主台；主控台	mathematical analysis	数学分析
master stratum	主要地层	mathematical and physical modelling	数学物理建模
master stream	主流；干流	mathematical approximation	数学近似法
master switch	总开关；主控开关	mathematical blocking routine	数学分层程序
master terminal unit	主控终端	mathematical cartography	数学地图学
master valley	主谷	mathematical characterization	数学表征
master valve	总阀	mathematical computation	数学计算
master well course map	井筒走向图	mathematical correlation	数学相关关系
master-slave computer	主从计算机	mathematical dispersion model	数学散布模拟系统
master-slave configuration	主从配置	mathematical equation	数学方程
master-slave flip-flop	主从触发器	mathematical expectation	数学期望
mastic cement	胶脂水泥	mathematical expectation value	数学期望值
mastic pointing	沥青填缝	mathematical filtering	数学滤波
mastic sealant	胶泥密封剂	mathematical forecast	数值预报
mat etching	变暗浸蚀	mathematical formula	数学公式
mat footing	平板底脚；板式基础；席式基脚	mathematical formulation	数学描述
mat foundation	席形基础	mathematical geodesy	数学大地测量学
mat lignite	暗褐煤	mathematical geography	数学地理学
mat support jack-up rig	沉垫自升式钻井平台	mathematical geology	数学地质学
matched generator	匹配振荡器	mathematical geology software library	数学地质程序库
matched image	匹配图像	mathematical geomorphology	数学地貌学
matched impedance	匹配阻抗	mathematical horizon	理想地平；数理地平
matched terrace	对应阶地	mathematical idealization	数学理想化
matching construction	配套工程	mathematical induction	数学归纳法
matching difference of water level	水位拟合误差	mathematical justification	数学证明
matching error	匹配误差	mathematical logic	数理逻辑
matching of marker beds	标准层对比；标准层配位	mathematical manipulation	数字变换
matching of succession	层序对比；层系配位	mathematical method	数学方法
matching principle	匹配原理	mathematical mode	数学模式
material damping	材料阻尼	mathematical model	数学模拟；数学模型
material ductility	材料延性	mathematical model of a coal preparation flow sheet	选煤厂工艺流程的数学模型
material dynamic property	材料动力性能		
material fabric	物质组构	mathematical model of drilling	钻进数学模型
material flaws	材料缺陷	mathematical model of groundwater	地下水数学模型
material mechanics	材料力学	mathematical modeling	数学模拟
material modulus matrix	材料模量矩阵	mathematical modelling of sulphur dioxide dispersion	二氧化硫散布的数学模拟系统
material nonlinearity	材料非线性		
material of construction	建筑材料	mathematical modelling technique	数学模拟技术
material parameter	材料参数	mathematical operation	数学运算
material properties	材料特性	mathematical optimization	数学最优化
material specification	材料规格；材料说明书	mathematical parameterization	数学参数化法
material specification manual	材料规格手册	mathematical pendulum	数学摆；单摆
material stiffness property	材料刚度特性	mathematical physics	数学物理学
material strength	材料强度	mathematical proceedings	数学计算过程
material testing	材料试验	mathematical processing	数学处理

mathematical programming	线性规划；最优化理论	matrix function	矩阵函数
mathematical programming method	数学程序设计法	matrix invariants	矩阵不变量
mathematical regression	数学回归	matrix inverse	矩阵反演
mathematical relation	数学关系	matrix inversion	矩阵求逆
mathematical routine	数学程序	matrix iteration	矩阵迭代法
mathematical search procedure	数学寻优过程	matrix limestone	微晶石灰岩
mathematical sedimentology	数学沉积学	matrix list	矩阵表
mathematical simplification	数学简化	matrix matching	阵列匹配
mathematical simulation	数学模拟；数字模拟	matrix material	基质材料
mathematical stability	数学稳定性	matrix mechanics	矩阵力学
mathematical statistical method	数理统计分析法	matrix memory	矩阵存储器
mathematical statistics	数理统计；数理统计学	matrix method	矩阵法
mathematical statistics of hydrology	水文数理统计	matrix method of inversion	矩阵反演法
mathematical treatment	数学处理	matrix mineral	骨架矿物
mathematically transform diagram	数学变换图表	matrix multiplication	矩阵乘法
mathematical-statistical analysis	数理统计分析	matrix notation	矩阵符号
mathematician	数学家	matrix of a liner transformation	线性交换矩阵
mathematics	数学；运算	matrix of coal	煤的基质
mathematics manipulation	数学处理	matrix of coefficient	系数矩阵
mathematics planning model	数学规划模型	matrix of derivatives of shape functions	形函数导数矩阵
Matheron geostatistical theory	马特隆地质统计学理论	matrix of elasticity	弹性矩阵
Matin heat resistance test	马丁耐热试验	matrix of linear transformation	线性变换矩阵
mating	配合	matrix of strain	应变矩阵
mating endcap	配套堵头	matrix of stress	应力矩阵
mating face	接触面；接合面	matrix operation	矩阵运算
mating flange	法兰组；配对法兰	matrix orthogonal	正交矩阵
matraite	角闪钙长岩	matrix pair	骨架对
matri displacement method	矩阵位移法	matrix palaeosome	基体；古成体
matric algebra	矩阵代数	matrix permeability	基岩渗透率；原生渗透率；基质渗透率
matric suction	基质吸力		
matric suction envelope	基质吸力包线	matrix porosity	基质孔隙度；基岩孔隙度
matric suction gradient	基质吸力梯度	matrix progression method	矩阵级数法
matric suction head	基质吸力水头	matrix propagator method	矩阵传播函数法
matric suction profile	基质吸力分布	matrix representation	矩阵表示
matrice	矩阵	matrix shear strength	基体剪切强度
matrix	基质；填质；模型；矩阵；母岩	matrix shrinkage	基质收缩
matrix acidizing	基岩酸化；基质酸化	matrix singularity	矩阵奇异性
matrix acoustic velocity	基岩中声速	matrix solid material	造岩物质
matrix addition	矩阵加法	matrix store	矩阵存储器
matrix algebra	矩阵代数	matrix stress	基岩应力；基体应力
matrix algorithms	矩阵算法	matrix stress calculation	应力矩阵计算
matrix analysis	基质分析；矩阵分析	matrix structural analysis	矩阵结构分析
matrix arithmetic processor	矩阵运算处理器	matrix structure	基质状结构；基质状构造；矩阵结构
matrix array algorithm	矩阵数列算法	matrix subtraction	矩阵减法
matrix calculus	矩阵计算	matrix suction	基质吸力
matrix compilation	矩阵编译程序	matrix support	基质支撑
matrix compiler	矩阵编译程序	matrix swelling	基质膨胀
matrix cracking	基质裂化基质开裂	matrix symmetric	对称矩阵
matrix cutting flood	行列式切割注水	matrix topology	骨架拓扑结构
matrix density	岩石骨架密度	matrix trace	矩阵的迹
matrix diagonalization	矩阵对角化	matrix transfer method	矩阵传递法
matrix differential equation	矩阵微分方程	matrix transpose	矩阵转置
matrix diffusion	基质扩散	matrix type structure	矩阵式结构
matrix dipole	基质偶极子	matrix velocity	骨架速度
matrix effect	基体效应	matrix-banding scheme	矩阵列图
matrix eigenvalue	矩阵特征值	matrix-supported reefs	基质支撑的生物礁
matrix element	矩阵元；基体元素；母质元素	matrix-whole rock	基质全岩
matrix encoder	矩阵编码器	matrosite	孢芽油页岩
matrix equation	矩阵方程	mat-supported drilling platform	刚性腿座架支承的钻井平台
matrix exponential	矩阵指数		

Matsushiro area earthquake	松代地震群地震
Matsuyama reversed epoch	松山反向期；松山反向极性时
matt	无光泽的
matt surface	粗面
matter wave	物质波
mattress slope protection	埽褥护岸
maturation	煤化作用；熟化作用
maturation factor	成熟因子
maturation gradient	熟化梯度
maturation index	成熟指数
maturation period	熟化期
mature coast	壮年期海岸；成熟海岸
mature conglomerate	分选良好的砾岩；成熟砾岩
mature form	壮年期地形
mature gas	成熟气
mature index	成熟度指数
mature karst	壮年岩溶；发育的岩溶
mature mountain	壮年山
mature mud	稳定泥浆；充分水化的泥浆
mature period	壮年期
mature plain	壮年切割平原
mature river	壮年河
mature sandstone	成熟砂岩
mature shore line	成熟海岸线；成熟滨海线
mature soil	成熟土壤；熟土
mature stage	壮年期；成熟阶段
mature stage of a stream	河流的壮年期
mature stream	壮年河流
mature topography	壮年地形
mature turbidite	成熟浊积岩
mature valley	壮年谷
mature waterflood	成熟注水
matured concrete	经养护硬化的混凝土；足龄混凝土
maturing	硬化；成熟
maturing agent	催熟剂；早熟剂
maturing field	老油田
maturing period	成熟时期
maturing well	老井
maturity	成熟度
maturity index	成熟度指数
maucherite	砷镍矿
maucherite molecular moisture	最大分子含水率
max	最大值；最高额；极大值
max density	最大密度
max permissible concentration	最大允许浓度
max value	极大值；最大值
maxim	原理；准则
maxima	最大值；极大
maxima of regular waves in the principal phase	主震最大波
maxima of regular waves occurring in the preliminary tremor	初期微震最大波
maximal	极大；最大
maximal contact angle	最大接触角
maximal draw ratio	最大拉伸比
maximal head	最大水头
maximal invariant	极大不变量
maximal of waves	波的极大值
maximal value	最大值
maximal-flow algorithm	最大流量算法
maximax criterion	极大准则
maximised locking force	最大闭锁力
maximization	增加到最大值；极大化
maximize	增到最大程度；最大化
maximizing	达到最大值
maximum	极密区；最大值；最大量；最多的；最高的
maximum a posteriori method	最大后验法
maximum acceleration	最大加速度
maximum acceleration of vibration	振动最大加速度
maximum acceleration response	最大加速度反应
maximum acceptable concentration	最大容许浓度
maximum aggregate size	集料最大粒径
maximum allowable concentration	最高容许浓度
maximum allowable deflection	最大容许挠度
maximum allowable emission quantity	最高容许排放量
maximum allowable gradient	最大容许坡度
maximum allowable noise level	最高许可噪声声级
maximum allowable operating pressure (MAOP)	最大允许运行压力
maximum allowable pressure	最大容许压力
maximum allowable side friction factor	最大容许侧向摩擦系数
maximum allowable velocity	最大容许速度；最大容许流速
maximum allowable velocity of flow	最大允许流速
maximum allowable weight on bit	最大允许钻压
maximum allowance working pressure	最大容许压力
maximum allowed casing pressure (MACP)	最大允许套压
maximum allowed pressure drop	最大允许压力降
maximum amount	最大量；最大金额
maximum amplitude	最大振幅
maximum amplitude of subsidence	最大沉陷幅度
maximum amplitude pulse	最大振幅脉冲
maximum and minimum thermometer	最高最低温度计
maximum angle of incidence	最大入射角
maximum annual flood	年最大洪水
maximum annual hourly volume	年最大小时交通量
maximum arterial flow method	最大干道流量法
maximum attainable growth rate	可达到的最大增长率
maximum available storage	最大有效库容；最大有效封存容量
maximum base shear	最大基底剪力
maximum basin relief	流域最大高差；盆地最大起伏
maximum bearing	最大承压力
maximum bending moment diagram	最大弯矩图
maximum boring grade	最大掘进坡度
maximum breaking load	最大断裂载荷；最大破坏载荷
maximum burst	极限爆裂
maximum capable earthquake	最大可能地震
maximum capacity	最大容量
maximum capacity of well	井的最大出水量
maximum capillarity	最大毛细管力
maximum capillary capacity	最大毛细管持水量；最大毛细管封闭能力
maximum carrying capacity	最大承载能力
Maximum circumferential tensile stress criterion	最大周向拉应力准则
maximum clearance	最大间隙
maximum coefficient of heat transfer	最大传热系数

maximum coherency filtering 最大相干滤波
maximum combustion temperature 最高燃烧温度
maximum compression 最大压力
maximum computed flood 最大估计流量；最大估计洪水流量；最大计算洪水流量
maximum consolidation pressure 最大固结压力
maximum constructive height 最大结构高度
maximum contaminant level 最大污染级
maximum continuous load 最大连续负载
maximum controllable level 最高控制水位；最高可控水位
maximum converted bulk density 最大转换体积密度
maximum convexity 最大凸率
maximum convexity migration 最大凸点偏移
maximum coordinate value 最大坐标值
maximum corona 极大日晕
maximum credible earthquake 最大可信地震
maximum credible ground motion 最大可信地面运动
maximum credible intensity 最大可信烈度
maximum crest 最大峰值
maximum critical void ratio 最大临界孔隙率；最大临界孔隙比
maximum curvature 最大曲率
maximum daily consumption of water 最高日用水量
maximum daily discharge 日最大流量
maximum daily footage 最大日进尺
maximum daily out-put 最高日供水量
maximum daily water consumption 最大日用水量
maximum decline of water level 最大水位降深
maximum delivery 最大泵量；最大流量；最大排量
maximum density 最大密度
maximum density grading 最大密实度级配
maximum dependable gross capacity 最大可靠总容量
maximum dependable net capacity 最大可靠净容量
maximum depth 最大深度；最大水深
maximum depth of frozen ground 最大冻土深度
maximum depth of plastic zone 塑性区最大深度
maximum depth of scour 最大冲刷深度
maximum depth of subsidence trough 沉陷盆地最大深度
maximum depth-area-duration data 最大深-面-时资料
maximum design condition 最大设计条件
maximum design pressure 最高设计压力
maximum design wave height 最大设计波浪高度
maximum design wind speed 最大设计平均风速
maximum detectable range 最大探测距离
maximum deviation 最大偏差
maximum deviator stress failure criteria 最大偏应力破坏准则
maximum diameter of hole 最大孔径
maximum dip 最大倾角
maximum discharge 最大流量；极大流量；最大排放量
maximum discharge loading 最高污水排放量
maximum discharge pressure 最大排放压力
maximum discharge principle 最大流量原理；最大流量原则
maximum disk capacity 最大磁盘容量
maximum displacement 最大位移
maximum displacement response 最大位移反应
maximum distant earthquake 极远地震
maximum distortion strain energy theory 最大剪应变能学说
maximum drawdown 最大降深；最大跌幅
maximum drilling depth 最大成孔深度
maximum drilling string load 最大钻柱载荷
maximum dry density 最大干密度
maximum dry unit weight 最大干重度；最大干容重
maximum duty of water 最大灌水定额；最大灌水率
maximum earthquake 极大地震
maximum earthquake deflection 最大地震变位
maximum earthquake force 最大地震力
maximum ebb 最大落潮流速
maximum effect force 最大有效力
maximum effective moment criterion 最大有效力矩准则
maximum efficiency 最优逼近
maximum elevation 最大高程；最大仰角
maximum elongation 最大伸长度；扯断伸长率
maximum emission concentration 最大排放浓度
maximum energy release rate criterion 最大能量释放率准则
maximum enhancement 最大增强
maximum entropy filtering 最大熵滤波
maximum entropy method 最大熵法
maximum entropy spectral analysis 最大熵光谱分析
maximum entropy spectrum 最大熵谱
maximum envelope curve holding capacity 吸振基础
maximum erosion depth 最大冲刷深度
maximum error 最大误差；极限误差
maximum event 最大现象；极大事件
maximum expectable earthquake 最大可预期地震
maximum experienced flood 实测最大洪水；经历最大洪水
maximum experimental safe gap 最大试验安全间隙
maximum explosive point of methane 沼气-空气混合体最高爆炸点
maximum explosive pressure 最高爆发压力
maximum extension 最大扩张
maximum extracting force on pile 最大拔桩力
maximum feasible intensity 最大可能烈度
maximum feasible single well capacity 单井最大可用出水量
maximum field 最大磁场；最大视场
maximum field moisture capacity 最大田间持水量
maximum filling working temperature 最高充填工作温度
maximum firing pressure 最高爆发压力；最高燃烧压力
maximum flexural response 最大弯曲反应
maximum flood 最大洪水；最大涨潮流速
maximum flood discharge 最大洪水流量
maximum flood level 最大洪水位
maximum flooding surface 最大海泛面
maximum flow 最大流量；最大径流
maximum flow capacity 最大过水能力
maximum flow rate 最大流率；最高流量率
maximum fluidity 最大流动度
maximum frequency 最大频率
maximum gain 最大增益
maximum glaciation 最大冰期
maximum gradability 最大爬坡坡度
maximum grade 最大坡度
maximum gradient 最大坡度；最大梯度
maximum gross 最大载重量
maximum gust 最高阵风；最高阵风风速

maximum head	最高水头
maximum headwater	上游最高水位
maximum heat transfer coefficient	最大传热系数
maximum heating load	最大热负荷
maximum high water	最高水位
maximum hole inclination	最大井斜角
maximum horizontal compression stress	水平最大主压应力
maximum horizontal displacement of contact wire	接触线最大水平偏移值
maximum horizontal stress	最大水平应力
maximum hourly rainfall	每小时最高雨量
maximum hourly traffic volume	最高小时交通量
maximum humidity	最大湿度
maximum hydraulic horsepower	最大水马力
maximum hygroscopicity	最大吸湿着水容度；最大吸湿量
maximum image visibility	最大影像可见度
maximum inclining position	最大倾斜位置
maximum initial field pressure	最大初始地层压力；最大原始油田压力
maximum inlet pressure	最大进口压力；最大入口压力
maximum instantaneous discharge	最大瞬间流量
maximum instantaneous wind speed	最大瞬时风速
maximum intensity	最great烈度
maximum intensity map	最大烈度图
maximum joint closure	节理的最大闭合度
maximum kinetic energy	最大动能
maximum known flood	历史最大洪水
maximum lateral shrinkage	最大侧向收缩
maximum lift	最大举升力；最高升程
maximum likelihood	最大似然率；最大可能性
maximum likelihood classification	最大似然分类
maximum likelihood classifier	最大似然分级器
maximum likelihood criterion	最大似然率准则
maximum likelihood deconvolution	最大可能重叠度；最大似然反褶积
maximum likelihood estimate	极大似然估计
maximum likelihood estimator	最大似然估计量
maximum likelihood function	最大概似方程
maximum likelihood method	极大似然方法
maximum likelihood ratio	最大似然率
maximum limit	最大极限
maximum limit of size	最大极限尺寸
maximum linear strain theory	最大线应变理论
maximum load	最大荷载；最大荷载量
maximum longitudinal gradient	最大纵坡；最大纵向坡度
maximum looseness	最大分散度
maximum magnitude	最大震级
maximum maximum-shearing	最大剪切力
maximum mean temperature	最高平均温度
maximum membership principle	最大隶属度原则
maximum microcline	最大微斜长石
maximum minimum coupling method	极大极小耦合法
maximum mining yield	最大开采水量；最大采矿量
maximum mixedness	最大混合度
maximum mixing depth	最大混合深度
maximum mobility period	激烈期；最大移动期
maximum moisture capacity	最大持水量
maximum molecular moisture capacity	最大分子水容量；最大分子吸湿度
maximum molecular moisture content	最大分子水容量；最大分子吸水量
maximum molecular water content	最大分子含水量；最大分子吸水量
maximum moment	最大力矩
maximum normal stress	最大正应力
maximum observed precipitation	实测最大降水量
maximum of waves	波的极大值
maximum offset	最大偏移距；最大位移；最大偏差
maximum one-day rainfall	最大日降水量
maximum operation frequency	最高工作频率
maximum outlet pressure	最大出口压力；最大排出压力
maximum output	最大产量；最大输出；峰值输出功率
maximum outreach	最大外伸范围；最大跨距
maximum outward position	最大外移位置
maximum overall efficiency	最高总效率
maximum parallelism degree	最大并行度
maximum particle size	最大粒径
maximum past pressure	最大历史压力
maximum past stress	最大历史应力
maximum peak	最大峰值
maximum penetration	最大穿透深度
maximum penetration rate	最高钻速
maximum permeability	最大磁导率；最大渗透率
maximum permissible concentration	最大允许浓度
maximum permissible dustiness	最大允许含尘量
maximum permissible load	最大容许荷载
maximum permissible loading	最大容许荷载
maximum permissible pressure	最大允许压力
maximum permissible rate	最大允许产量
maximum permissible service temperature	最高允许使用温度
maximum permissible velocity	最大允许速度
maximum permitted charge	最大允许装药
maximum permitted gap	最大许可间除
maximum permitted load	最大容许荷载；最大容许荷载量
maximum phase	最大相位；最高期
maximum phase wavelet	最大相位子波
maximum plot ratio	最高地积比率；最高建造比率
maximum pool level	最高库水位；最高壅水位
maximum pore ratio	最大孔隙比
maximum pore-volume	最大孔隙体积
maximum possible earthquake	最大可能地震
maximum possible flood	最大可能洪水
maximum possible loss	最大可能损失
maximum possible magnitude	最大可能震级
maximum possible rate	最大可能产量
maximum precipitation	最大降水量
maximum pressure	最大压力
maximum pressure arch spacing	压力拱最大跨距
maximum pressure on pile	最大压桩力
maximum pressure rating	最大压力额定值
maximum primary variable	最大主变量
Maximum principal effective stress	最大有效主应力
maximum principal energy	最大主能量
maximum principal stress	最大主应力
maximum principal stress difference	最大主应力差
maximum principal stress ratio	最大主应力比

maximum principal stress ratio failure criteria 最大主应力比破坏准则
maximum principle 极大值原理；最大值原理
maximum probable flood 最大可能洪水
maximum probable precipitation 最大可能降水量
maximum production capacity 最大生产能力
maximum pulling capacity 最大提升能力；最大拉力
maximum rainfall 最大降雨量
maximum rainfall intensity 最大降雨强度
maximum range of EDM 电磁波测距最大测程
maximum reading thermometer 最大读数温度计
maximum recorded 实测最大
maximum recorded flood peak 实测最大洪峰流量
maximum recording thermometer 最高读数温度计
maximum recovery rate 最大开采速度
maximum reflectance 最大反射率
maximum reinforcement force criterion 最大加固力准则
maximum resultant ground displacement 最大合成地面位移
maximum retention level 最大蓄水位
maximum rolling rainfall 最大累积降雨量
maximum runoff 最大径流
maximum safe capacity 最大安全容量
maximum safe concentration 最大安全浓度
maximum safe density 最大安全密度
maximum safe earthquake intensity 最大安全地震烈度
maximum safe limit 安全极限
maximum safe speed 最高安全速度
maximum safe working load 安全操作最高负荷；安全操作最高负荷量
maximum safe working pressure 最高安全工作压力
maximum safety charge 最大的安全装药量；最高安全容量
maximum salinity 最大盐度
maximum salinity layer 最大盐度层
maximum sand free production 最高无砂产量
maximum saturation 最大饱和度
maximum saturation condition 最大饱和条件
maximum scour depth 最大冲刷深度
maximum sediment concentration 最大沉积物浓度；最大含沙量
maximum seismic intensity 最大地震烈度
maximum service life 最大使用期限；最大工作寿命
maximum shear 最大剪力；最大剪应力
maximum shear strain 最大剪应变
maximum shear stress 最大剪应力
maximum shear stress failure criteria 最大剪应力破坏准则
maximum shear stress theory 最大剪应力理论；第三强度理论
maximum shear theory 最大剪应力理论
maximum shearing stress 最大剪应力
maximum size 最大粒径；最大尺寸
maximum slip 最大滑动
maximum slope 最大坡度；最大斜率
maximum span 最大跨度
maximum speed 最高速度
maximum stability period 和缓期；最大稳定期
maximum stacking velocity 最大叠加速度
maximum stage 最高水位；最高潮位

maximum static pressure 最高静水压力
maximum stiffness 最大刚度
maximum strain 最大应变
maximum strain energy 最大应变能
maximum strain failure criterion 最大应变破坏准则
maximum strain theory 最大应变理论；第二强度理论
maximum streamflow 最大流量；最大河川径流
maximum strength 极限强度
maximum stress 最大应力
maximum stress failure criteria 最大剪应力破坏准则
maximum stress failure criterion 最大应力破坏准则
maximum stress theory 最大应力理论；第一强度理论
maximum subsidence 最大沉陷；最大沉降量
maximum subsidence depth to seam thickness ratio 最大沉陷深与煤层厚之比
maximum surge 最大涌浪
maximum surge pressure 最大水击压力
maximum sustained yield 最大持续开采水量；最大连续出水量
maximum tensile to shear stress ratio 最大张剪应力比
maximum tension 最大张力；最大拉力
maximum term 极大项
maximum test pressure 最大试验压力
maximum theoretical specific gravity 最大理论比重
maximum thermometer 最高温度计
maximum thrust 最大推力
maximum tip load at setting 初撑时支架顶梁端最大载荷
maximum total energy filter 最大总能量过滤器；最大总能量滤波器
maximum turnover 最高周转率
maximum unambiguous range 最大不模糊距离
maximum uncertainty 最大不确定性
maximum upstream water level 最大上游水位；最大上层带水位
maximum upsurge 最大上涌浪
maximum usable amplitude deviation 最大可用幅偏
maximum usable frequency 最大可用频率
maximum value 极大值
maximum variance norm deconvolution 最大方差模反褶积
maximum velocity 最大速度；最大流速
maximum vertical shrinkage 最大垂向收缩率
maximum void ratio 最大孔隙比
maximum volumetric shrinkage 最大体积收缩率
maximum water 油井固井中最高水灰比
maximum water capacity 最大水容量；最大含水容量
maximum water consumption 最高耗水量
maximum water content 最大含水量
maximum water demand 最大需水量
maximum water level 最高水位；最高潮位；最高壅水位
maximum water table boundaries 最大水头边界；最大潜水面边界
maximum water yield 最大涌水量
maximum water-holding capacity 最大持水量
maximum well discharge 井的最大出水量
maximum well yield 井孔最大涌水量
maximum wind pressure 最大风压
maximum wind speed 最大风速
maximum working flow 最大实际流量；最大驱动油量

maximum working height 最大工作高度
maximum yield 最大出水量；最高产量
maximum yield of drillhole 钻孔最大出水量
maximum yield of water well 水井最大出水量
maximum, intermediate and minor principal stresses 最大、中间和最小主应力
maximum-delay 最大延迟
maximum-entropy estimate 最大熵估计
maximum-likelihood criterion 最大似然准则
maximum-likelihood deconvolution 最大似然反褶积
maximum-minimum theory 最大-最小理论
maximum-phase 最大相位
maximum-pressure arch 极限压力拱
maximum-reading thermometer 最高读数温度计
maxipulse 最大脉冲法
MAXIS 多功能数据采集和成像系统
max-operator 取大算子
Maxton screen 麦克斯登型筛
max-value function 极大值函数
Maxwell diagram 麦克斯韦图
Maxwell distribution 麦克斯韦分布
Maxwell elastoviscous model 麦克斯韦弹黏性模型
Maxwell equation 麦克斯韦方程式
Maxwell material 麦克斯韦体
Maxwell model 麦克斯韦模型
Maxwell-Betti reciprocal theorem 麦克斯韦-贝蒂互等定理
Maxwell-Boltzmann distribution 麦克斯韦-玻尔兹曼分布律
Maxwell-Boltzmann distribution law 麦克斯韦-玻尔兹曼分布定律
maxwellmeter 磁通计；磁通量测定计
Maxwell's equation 麦克斯韦方程
Maxwell's law of reciprocal deflection 麦克斯韦位移互等定律
Maxwell's model 麦克斯韦尔模型
mayberyite 富硫石油
Mayer curve 迈耶尔曲线；矢量可选性曲线
Maysvillian 迈斯维尔阶
Maysvillian Stage 迈斯维尔阶
mazapilite 黑砷铁矿
mazier barrel 麦氏取样器
mazout 重油
M-boundary 莫霍不连续面
mcgovernite 红砷硅锰矿
mckittinite 地沥青
mdarcy 毫达西
MDH analysis 米勒-戴斯-赫钦森试井分析法
MDH graphs 米勒型处理压力恢复资料的图解法
MDH-type curve MDH 型曲线
M-discontinuity 莫霍不连续面
meadow black soil 草原黑土
meadow bog 草甸沼泽
meadow burozem 草甸棕壤
meadow chernozem 草原黑钙土
meadow chernozemic soil 草甸黑土
meadow chestnut soil 草甸栗钙土
meadow coal 贫煤
meadow marsh 草甸沼泽
meadow moor 草地沼泽
meadow ore 沼铁矿

meadow peat 草甸泥炭
meadow podzolic soil 草甸灰化土
meadow red earth 草甸红壤
meadow soil 草地土壤；草甸土
meadow-bog 水湿地
meadow-boggy soil 草甸沼泽土
meadstream 河流航道；河流中泓
meager clay 瘦黏土
meager coal 贫煤
meager lean coal 贫瘦煤
meagre coal 贫煤
meagre lean coal 贫瘦煤
meagre lime 贫石灰
meal 细磨石料；岩粉；粉
meal bore 钻粉；钻屑
mealy 粉状的
mealy sand 粉砂
mealy structure 粉状结构；粉状构造
mean 平均数；平均值；中间的；中数
mean absolute deviation 平均绝对偏差
mean absolute error 平均偏差；平均绝对误差
mean accumulation 平均积累
mean accuracy 平均精度
mean activity 平均活度
mean advance 平均进尺
mean annual discharge 年平均流量
mean annual efficiency 全年平均效率
mean annual flood 平均年洪水流量
mean annual precipitation 年平均降水量
mean annual rainfall 平均年降水量
mean annual range 多年平均变幅
mean annual runoff 年平均径流
mean annual temperature 年平均温度
mean annual temperature range 平均年温差
mean annual water level 年平均水位
mean anomaly 平均异常
mean approximation 平均近似
mean arrival rate 平均到达率
mean asperity height of fracture face 裂缝面平均粗糙度
mean axis 中轴
mean basin height 流域平均高度；流域平均高程
mean basin slope 流域平均坡降；流域平均坡度
mean basin width 流域平均宽度
mean calorie 平均热值
mean channel height 平均河床高度；平均河床高程
mean collision time 平均碰撞时间
mean concentration 平均浓度
mean current 平均流
mean curvature 平均曲率
mean daily discharge 日平均流量
mean daily temperature 日平均温度
mean damage index 平均震害指数
mean density 平均密度
mean depth 平均深度
mean deviation 平均偏差；平均偏斜度
mean diameter 平均直径；平均粒径
mean difference 均差
mean discharge 平均流量
mean displacement rate 平均驱替速度
mean distance 平均距离

mean diurnal low-water inequality	平均日潮低潮不等
mean down time	平均停工时间；平均停机时间
mean dynamic amplification factor	平均动力放大系数
mean earth ellipsoid	平均地球椭球
mean effective diameter	平均有效直径
mean effective horsepower	平均有效马力；平均有效功率
mean effective pore diameter	微孔平均有效直径
mean effective pressure	平均有效压力
mean effective stress	平均有效应力
mean effective value	平均有效值；均方根值
mean elevation of basin	流域平均高程
mean error	平均误差；标准误差
mean error ellipsoid	平均误差椭圆
mean extreme high-water springs	平均大潮极高潮
mean extreme low-water springs	平均大潮极低潮
mean extreme-value	平均极值
mean flow	平均流量
mean formation level	平均地基水平线；平均地基面
mean free path	平均自由程
mean freeze-up date	平均封冻日期
mean geometrical distance	几何平均距离
mean grain size	平均粒径
mean gravity measurement	均匀重力测量
mean gross head	平均毛水头
mean ground elevation	平均地面海拔高度
mean ground pressure	平均地压
mean heat capacity	平均热容量
mean height	平均高度
mean height difference	平均高差；平均潮高差
mean height of apparent sea	平均视浪高
mean height of survey area	测区平均高程面
mean high tide	平均高潮面
mean high water	平均高潮面；平均高水位
mean high water neap	平均小潮高潮面
mean high water spring	平均大潮高潮面
mean higher high water	平均最高高水位；平均较高潮
mean higher high-water springs	平均大潮高高潮
mean higher low water	平均高低潮
mean highest discharge	平均最大流量
mean high-water interval	平均高潮间隙
mean high-water level	平均高水位
mean high-water line	平均高潮线
mean high-water neaps	平均小潮高潮
mean high-water springs	平均大潮高潮
mean hourly wind speed	每小时平均风速
mean hydraulic depth	平均水力深度
mean hydraulic radius	平均水力半径
mean indicated pressure	平均指示压力
mean indication pressure	平均指示压力
mean intensity	平均强度
mean interdiurnal variation	平均日际变化
mean land level	大陆平均高度
mean level	平均海面；平均水面
mean level of wave	波浪中线
mean lifetime	平均寿命
mean line	等分线
mean linear range	平均直线射程
mean load density	平均装填密度；平均支护强度
mean load per unit cycle	循环平均阻力
mean loss ratio	平均损失比
mean low level	平均低潮面
mean low tide	平均低潮
mean low water	平均低潮水位；平均低潮面
mean low water level	平均低水位
mean low water neap	平均小潮低潮面
mean lower low water	平均最低低水位
mean lowest discharge	平均最小流量
mean lowest water level	平均最低水位；平均最低潮位
mean low-water level	平均低水位
mean low-water neaps	平均小潮低潮
mean low-water springs	平均大潮低潮
mean map	平均值图
mean matrix	平均矩阵
mean meridional circulation	平均径向环流
mean minimum temperature	平均最低温度
mean molar quantity	平均摩尔数量
mean molecular weight	平均分子量
mean monthly air temperature of yearly hottest month	历年最热月平均温度
mean monthly discharge	月平均流量
mean monthly highest water level	月平均最高水位；月平均最高潮位
mean monthly lowest water level	月平均最低水位；月平均最低潮位
mean monthly maximum temperature	月平均最高温度
mean monthly minimum temperature	月平均最低温度
mean monthly stage	月平均水位
mean motion	平均运动
mean neap	平均小潮
mean neap range	平均小潮差
mean neap rise	平均小潮升
mean net head	平均净水头
mean neutron lifetime	中子平均寿命
mean normal intension	平均法向强度
mean normal stress	平均法向应力
mean occurrence rate	平均发生率
mean of high stages	平均高水位
mean of high water	平均高水位；平均高潮位
mean of low stages	平均低水位
mean of low water	平均低水位；平均低潮位
mean of the errors	误差平均值
mean parallax	平均视差
mean parameter	平均参数
mean particle diameter	颗粒平均粒径
mean particle size	平均粒度；平均颗粒大小
mean permeability	平均渗透率
mean pore pressure	平均孔隙压力
mean position	平均位置
mean potential water power	平均水力蕴藏量
mean power	平均功率
mean pressure	平均压力
mean principal directions of stress	平均主应力方向
mean principle stress	平均主应力
mean probable error	平均概差；平均可能误差
mean proportional	比例中项
mean radiant temperature of surrounding surfaces	周围表面平均辐射温度
mean radius	平均半径
mean radius of curvature	平均曲率半径
mean radius of the earth	地球平均半径

mean range	平均距离；平均射程；平均自由程；平均极差；平均潮差
mean range of tide	平均潮差
mean rate	平均速率
mean recession curve	平均退水曲线
mean recurrence time	平均重现时间；平均循环时间
mean reflectance	平均反射率
mean reflectivity	平均反射率
mean relative deviation	平均相对偏差
mean relief ratio of basin	流域平均起伏比
mean reservoir pressure	平均地层压力
mean response spectrum shape	平均反应谱曲线
mean return period	平均重现期；平均间隔期
mean return time	平均重现时间
mean rise	平均潮升
mean rise interval	平均潮升间隙
mean runoff	平均径流量
mean salinity	平均含盐度
mean scale	平均比例尺
mean sea depth	海平均深度；平均海水深度
mean sea level	平均海平面
mean sea level datum	平均海面基点
mean sea surface temperature	平均海面水温
mean sea-level	平均海平面
mean side length	平均边长
mean size	平均粒径；平均粒度
mean slope	平均坡度
mean specific gravity	平均比重
mean specific heat	平均比热
mean sphere level	地面平均高度
mean spring	平均大潮
mean spring low water	平均大潮低潮
mean spring range	平均大潮差
mean spring rise	平均大潮升
mean square	均方差
mean square bandwidth	均方带宽
mean square departure	均方偏差
mean square deviation	均方偏差；均方差
mean square dip	均方地层倾角
mean square displacement	均方位移
mean square error	均方误差；中误差
mean square error norm	均方误差范数
mean square error of a point	点位中误差
mean square error of angle observation	测角中误差
mean square error of azimuth	方位角中误差
mean square error of coordinate	坐标中误差
mean square error of distance measurenlent	测距中误差
mean square error of height	高程中误差
mean square error of prior weight	先权中误差
mean square error of side length	边长中误差
mean square error of weight unit	单位权中误差
mean square integral	均方积分
mean square modulus	均方模
mean square moment	均方弯矩
mean square power spectral density function	均方功率谱密度函数
mean square regression	均方回归
mean square response	均方反应
mean square root	均方根
mean square slowing-down length	均方慢化长度
mean square spectral density	均方谱密度
mean square value	均方值
mean square value of random variable	随机变量的均方值
mean square velocity	均方速度
mean squared error	均方误差
mean standard error	平均标准差
mean standard weighted deviation (MSWD)	平均标准加权偏差
mean steepness	平均陡度
mean strain	平均应变
mean strain axis	中等应变轴
mean strength	平均强度
mean stress	平均应力
mean surface	平均表面积
mean temperature difference	平均温差
mean test value	平均试验值
mean thermal transmittance	平均传热系数
mean tidal level	平均潮位
mean tidal range	平均潮差
mean tide level	平均潮位
mean tide rise	平均潮升
mean time	平均时；地方平时
mean value	平均值
mean variance	离均差
mean variogram	平均方差图
mean vector	均值向量
mean velocity	平均速度；平均流速
mean velocity curve	平均流速关系曲线
mean velocity in section	断面平均流速
mean velocity in vertical	垂线平均流速
mean velocity on a vertical	垂线平均流速
mean velocity point	平均流速点
mean water	平均水位
mean water level	平均水位；平均水平面
mean water stage	平均水位
mean water table	平均水头
mean wave	平均波
mean wave height	平均波高
mean winter temperature	冬季平均温度
mean yearly discharge	年平均流量
meand	老露天矿
meander	曲流；河曲；蛇曲
meander amplitude	蜿蜒幅度；曲流幅度
meander bar	河曲内侧坝；曲流边滩；曲流沙坝
meander belt	河道曲折带；河道蜿蜒带
meander bend	河曲
meander bluff	曲流陡岸
meander breadth	河曲宽度
meander core	离堆山；曲流环绕岛
meander curve	曲流
meander cut off	曲流裁弯取直
meander flow	曲流
meander lake	蛇曲湖
meander line	曲流线；折测线
meander loop	河曲的环状河道
meander migration	河曲迁移
meander neck	曲流颈
meander ratio	曲流比；蜿蜒比
meander reach	曲流段；曲流河段
meander river	曲流河

meander scroll	曲流湖；曲流内侧坝；河曲沙洲	measured geological map	测定地质图
meander terrace	曲流阶地；河曲阶地	measured geological section	实测地质剖面
meander trough	曲流槽	measured inclination	测量的井斜角
meander valley	曲流河谷	measured level	测量水准；测量水平面
meander width	河道曲折带宽度；曲流宽度；蜿蜒宽度	measured mineral resource	确定矿产资源
meanderiform	河曲状	measured ore	实测矿量；探明矿量
meandering	曲流，弯曲；蜿蜒	measured profile	实测断面；实测纵断面
meandering channel	游荡河道；弯曲河槽；蜿蜒河道；弯曲航道	measured reserve	实测储量；确定储量；开采储量
		measured resources	实测资源；查明资源
meandering coefficient of traverse	导线曲折系数	measured runoff	实测径流
meandering course	曲流	measured section	实测剖面；实测地层剖面
meandering movement	曲折移动	measured stage	实测水位
meandering reach	曲流段；蜿蜒段	measured strength	实测强度
meandering river	弯曲河；蛇曲河；游荡河流	measured temperature	实测温度
meandering river deposit	曲流河沉积；蛇曲河沉积	measured value	量测值；实测值
meandering stream	弯曲河；蛇曲河	measured variable	测定变量
meandering tidal channel	弯曲潮道	measurement	量度；量测；测定；计量；尺寸；分量
meandering valley	曲流河谷	measurement accuracy	量测精度
meaningless	无意义的；失效	measurement adjustment	测量平差
mean-square	均方	measurement bolt	测试锚杆；测量用锚栓
mean-square channel diameter	均方孔道直径	measurement circuit	测量回路；量测线路
mean-square deviation	均方差	measurement complexity	测量复杂度
mean-square error	均方差	measurement device	量测装置
mean-square root	均方根	measurement error	测量误差；量度误差
mean-value function	均值函数	measurement inaccuracy	测量误差
mean-value of random process	随机过程的平均值	measurement instrument	量测仪器；测量仪表
measurability	可测性	measurement matrix	测量矩阵
measurable	可计量的；可测的	measurement method	测量方法
measurable anomaly	可测异常	measurement noise	测量噪声
measurable property	可测特性	measurement of angle	角度测量
measurables	衡量标准	measurement of azimuth	方位角测量
measuration	测量；计量；量度	measurement of bearing	方位测量
measure	量器；测量；尺寸；分量；层组层系岩层	measurement of changes in stress	应力变化测量
measure analysis	体积分析；容积分析	measurement of coordinates	坐标测量
measure device	计量设备	measurement of dip angle	倾角测量
measure expansion	体积膨胀；体膨胀	measurement of drawdown	降落测量
measure matric suction	基质吸力测量	measurement of flow rate	流量量测
measure moisture	计量水分	measurement of Friction coefficients of power/particle and static wall	粉粒体静壁摩擦系数的测定
measure mud weight	测定泥浆比重；测泥浆重量		
measure of curvature	曲率测度	measurement of geodesy	大地测量；测地学；大地测量学
measure of discretization error	离散误差的测量	measurement of gravimetry	重力测量
measure of dispersion	离差的度量	measurement of in-situ state of stress	现场应力状态的测量
measure of fuzziness	模糊性测度		
measure of precision	精密程度；精确度	measurement of rock pressure	岩石压力测量
measure of skewness	偏度；偏斜度	measurement of strain	应变量测
measure out	起钻测算井深	measurement of stress of lining	衬砌应力测量
measure point	测量点	measurement of tension of bolts	锚杆轴力测量
measure specific gravity	测定比重	measurement of true dip	真倾角测定
measure total suction	总吸力测量	measurement point	量测点
measure water loss	测定失水	measurement precision	量测精度
measured	实测的；探明的	measurement procedure	计量程序
measured coal resources	煤炭资源测算	measurement range	测程；量程；量测范围
measured data	实测数据；测量数据	measurement reference	测量基准
measured depth	量测深度；量测井深	measurement report	测量报告
measured direction	实测方位	measurement result	测量结果
measured discharge	实测流量	measurement standard	计量标准
measured distance	实测距离	measurement station	测点
measured drilling depth	量测的井深	measurement stereoscope	量测立体镜
measured fall	实测落差	measurement system	测量系统；测量制
measured field settlement data	现场实测沉降资料	measurement technique	量测技术

measurement theory 测量理论
measurement transducer 测量传感器；测量换能器
measurement transformer 测量互感器
measurement unit 计量单位
measurement update 测量校正
measurement variable 测量变数
measurement well 测试井
measurement while drilling 随钻测量
measurements of blasting action and effect 爆破测试
measurements transmission 量测结果的传送
measurement-while-drilling 随钻测量
measurement-while-drilling log 随钻测井
measurer 测量员；量测仪；量测元件；量测器
measures 层系；层组
measures against tunnel fire 隧道防火措施
measures and weights 度量衡
measuring 测量；检测；测试
measuring accuracy 测量精度
measuring amplifier 测量放大器
measuring apparatus 测量仪器
measuring appliance 测量器具；测量设备；测量仪器
measuring bar 测杆
measuring basis 测量基准
measuring bellows 测量波纹管
measuring block 量块；块规
measuring bolt 测试锚杆
measuring bottle 量瓶
measuring buoy 测验浮标
measuring buret 量液滴定管
measuring by sight 目测
measuring cable 测量缆；测量电缆
measuring car for overhead contact line equipment 接触网测试车
measuring cell 测压计
measuring channel 量水槽
measuring circuit 测量线路
measuring coil 测量线圈
measuring column 汞柱；测量仪表柱
measuring control 计量管理
measuring cylinder 量筒
measuring dam 量水坝
measuring deformations of bore holes 钻孔变形测量
measuring device 量测装置；计量装置；量具；量测仪器
measuring duct 测量槽
measuring electrode 测量电极
measuring element 量测元件；计量元件
measuring equipment 量测装置；测量设备
measuring error 测量误差
measuring flask 量瓶
measuring fluid 测量液
measuring flume 量水槽；测量管渠
measuring frequency 量测时间间隔
measuring glass 玻璃量杯
measuring graduates 刻度量筒
measuring grid 量测格网
measuring head 测试头；测量头
measuring hopper 计量斗；计量仓；计量箱
measuring implement 量具；测量仪器
measuring index 度量指标

measuring installation 测站；测量装置；测量设备
measuring instrument 计量仪器；测量仪器
measuring laboratory 测量实验室
measuring lath 标杆；测量杆
measuring line 测线；测绳
measuring loop 测量用回线
measuring mark 测标；丈量标
measuring means 量测工具
measuring method 测量方法
measuring microscope 测微镜；测量显微镜
measuring orifice 流量孔板；测流口
measuring panel 测量仪表板
measuring platform 观测台
measuring plug 测量标志；测钉；测量标桩；测量塞
measuring point 测点；测量点
measuring pot 量杯
measuring principle 测量原理
measuring probe 探头
measuring pump 计量泵
measuring range 测量范围；测定范围；量程
measuring reel 卷尺；卷尺盘；测量绳卷筒
measuring reference surface 测量基准面
measuring repeatability 测量重复性
measuring resistance 量测用电阻；标准电阻
measuring rod 量杆
measuring rope 测索
measuring rule 量尺
measuring scale 量尺；测量用刻度
measuring screen plate 测微螺丝；微动螺丝
measuring screw 测量螺丝
measuring section 测验断面；量测断面
measuring sensitivity 测试灵敏度
measuring sensor 测量传感器
measuring set 测量装置；测试设备
measuring sonde 测量探头
measuring spring 测量弹簧
measuring stability 测试稳定性
measuring staff 测量杆
measuring stick 量尺；计量尺
measuring stress of steel support 钢支撑应力测量
measuring system 量度系统
measuring tank 计量罐；测箱
measuring tape 卷尺；量尺
measuring technique 计量技术；测量技术
measuring terminal 测量端子
measuring tour 野外测量；测量行程
measuring truck 仪表车
measuring tube 测量管；量筒
measuring uncertainty 观测不确定性
measuring unit 测量单位
measuring weir 量水堰；测水堰
measuring well 量测井；观测井
measuring wheel 测量轮
measuring-tape 测尺；皮尺；卷尺
mechanic jar 机械震击器
mechanical 机械的；力学的
mechanical accelerometer 机械加速计
mechanical accumulation 机械堆积
mechanical action 机械作用
mechanical activator 机械触发器

mechanical adjustment 机械调节
mechanical agitation flotation cell 机械搅拌式浮选机
mechanical agitator 机械搅拌器
mechanical alignment 机械校准
mechanical analysis 颗粒分析；粒度分析；粒径分析；力学分析
mechanical analysis curve 粒径分析曲线
mechanical analysis of soils 土壤粒度分析
mechanical anchor 机械锚固式锚杆；机械锚杆
mechanical and chemical indices 物理化学指标
mechanical anemometer 机械风速计
mechanical anisotropy 力学各向异性
mechanical arm 机械手
mechanical attack 机械冲击作用
mechanical balance 机械平衡；力学平衡
mechanical behavior 机械特性；力学特性
mechanical behavior of rock 岩石矿压显现
mechanical breaking 机械回采
mechanical burster 机械爆破筒
mechanical character of deposit 矿床物理力学特性
mechanical characteristics 力学特性
mechanical clay 机械黏土；岩屑土
mechanical cleaning 机械选矿；机械选煤
mechanical cleaning of broken ore 机械清理崩落矿石
mechanical coal mining 机械化采煤
mechanical competence 机械搬运力
mechanical composition 颗粒级别
mechanical compression 物理压缩
mechanical core taker 机械式取芯器
mechanical coring 机械式取芯
mechanical counter 机械计数器
mechanical coupler 机械套管
mechanical cover 机械的覆盖层
mechanical cutter 机械切割器
mechanical declinometer 机械倾斜仪
mechanical denudation 机械剥蚀
mechanical deposit 机械沉积
mechanical destruction of rock 岩石的力学破坏
mechanical device 机械装置
mechanical digger 挖掘机；机械铲
mechanical disintegration 机械崩解
mechanical drift indicator 机械式井斜仪
mechanical drill 机械钻机；机械钻孔
mechanical effect 机械作用
mechanical equation of state 力学状态方程
mechanical equilibrium 机械平衡；力学平衡
mechanical equipment 机械设备
mechanical equivalent of heat 热功当量
mechanical erosion 机械侵蚀
mechanical excavation 机械开挖
mechanical execution 机械施工
mechanical exhaust system 机械通风系统
mechanical extension bar 机械伸长顶梁
mechanical extension piece 机械接长件；机械内外螺纹管接头
mechanical extensometer 机械式伸长计
mechanical failure 力学破坏；机械故障
mechanical features 力学性能；机械性能
mechanical feed drill 机械推进的凿岩机
mechanical feeder 机械式给矿机

mechanical filling 机械充填；机械装载
mechanical force 机械力
mechanical friction 机械摩擦
mechanical friction coefficient 机械摩擦系数
mechanical gauge 机械式压力计
mechanical getting 机械化采煤
mechanical gradation 筛别分析
mechanical grading 粒度组成
mechanical hauling shearer 机械牵引采煤机
mechanical heading machine 平巷掘进机
mechanical hysteresis 力学滞后
mechanical impact screen 机械冲击筛
mechanical impedance 力阻抗
mechanical injection device 机械注入装置
mechanical jack 机械式千斤顶
mechanical logging 机械测井
mechanical magnetometer 机械式磁力仪
mechanical metamorphism 动力变质作用
mechanical mining 机械开采；机械化开采
mechanical model 力学模型
mechanical modelling 机械建模；力学建模
mechanical mole 掘进护盾；平巷联合掘进机
mechanical monitoring devices 机械式监测设备
mechanical moulding 机械成形
mechanical mucked 装岩机
mechanical mucking 机械装岩
mechanical oil production 机械采油
mechanical optical accelerograph 机械光学式加速度仪
mechanical ore breaking 机械落矿
mechanical orienting tool 机械式定向工具
mechanical oscillation 机械振动
mechanical packer 机械式封隔器；机械胀塞
mechanical packing 机械充填
mechanical parameter 力学参数
mechanical penetration rate 机械钻速
mechanical perforator 机械射孔器
mechanical performance 机械性能；力学性能
mechanical piping 机械潜蚀；机械缩孔
mechanical placer 液压泵浇注装置
mechanical plough 机械犁；刨煤机
mechanical press 机械压力机
mechanical pressure recorder 机械式压力记录仪
mechanical properties laboratory 力学性能试验室
mechanical properties of ground 岩土的力学性质
mechanical properties of materials 材料力学性能
mechanical properties of rock 岩石力学性质
mechanical property 力学性能
mechanical rammer 机械夯具；机动夯
mechanical reactance 机械反作用力；力抗
mechanical recovery 机械采油
mechanical resistance 力阻
mechanical rig 机械驱动钻机
mechanical sampler 机械取样器
mechanical sediment 碎屑沉积物；机械沉积
mechanical sedimentary deposit 机械沉积矿床
mechanical sedimentary differentiation 机械沉积分异作用
mechanical seismic source 机械地震源
mechanical seismograph 机械式地震计；机械地震仪
mechanical servo 机械伺服装置

mechanical set bit	机镶钻头
mechanical setting	机械沉积
mechanical sewage treatment	最大流速线
mechanical sharpener	磨钎机；钻头磨锐机
mechanical shield	机械开挖盾构
mechanical shovel	机械铲；挖土机
mechanical sieve analysis	机械筛分析
mechanical similarity	力学相似性
mechanical source	机械震源
mechanical sprag	机械煤面撑
mechanical stabilization	机械加固法；改善级配加固法
mechanical stowing	机械充填
mechanical strain	力学应变；机械应变
mechanical strain gauge	机械式应变计
mechanical strength	机械强度；力学强度
mechanical stress	机械应力；力学应力
mechanical stretching	机械张拉
mechanical swelling	力学膨胀
mechanical system	力学系统；机械系统
mechanical test	机械性能试验；力学试验
mechanical testing	力学试验；机械试验
mechanical tunneling machine	隧道掘进机
mechanical twin	机械双晶；形变孪晶
mechanical type manometer	机械式压力计
mechanical unconformity	机械不整合
mechanical unit	力学单位
mechanical ventilation	机械通风
mechanical ventilation equipment	机械通风装置
mechanical vibration	机械振动
mechanical wave	机械波
mechanical wear	机械磨损
mechanical weathering	机械风化；物理风化
mechanical-electronic load cell	机械-电子测力计
mechanically anchored rock bolt	机械锚固锚杆
mechanically anchored roof bolt	机械锚固顶板锚杆
mechanical-set packer	机械坐封封隔器
mechanics of blasting	爆炸力学；爆破力学
mechanics of continuous media	连续介质力学
mechanics of elasticity	弹性力学
mechanics of explosion	爆炸力学
mechanics of frozen soil	冻土力学
mechanics of granular bodies	粒状体力学
mechanics of granular media	散体力学
mechanics of jointed rock	节理岩石力学
mechanics of landslide	滑坡力学；滑坡的力学性质
mechanics of loose media	松散介质力学
mechanics of materials	材料力学
mechanics of mixture	混合体力学
mechanics of ploughing process	刨煤过程的力学
mechanics of quasi stochastic bodies	半随机体力学；半随机介质力学
mechanics of rock mass	岩体力学
mechanics of stick-slip	黏滑力学
mechanics of stochastic media	随机介质力学
mechanics of viscous fluids	黏性流体力学
mechanised drilling	机械钻眼
mechanised extraction	机械化采掘
mechanism at the source	震源机制
mechanism for soil liquefaction	土液化机理
mechanism of activation	活化作用机理
mechanism of assimilation	同化机理
mechanism of deformation	变形机理
mechanism of deformation and failure	变形破坏机理
mechanism of earthquake	地震机制
mechanism of landslide	滑坡机理
mechanism of mass sinking	块体沉陷机制
mechanism of oil displacement	驱油机理
mechanism of outburst	突出机理
mechanism of reaction	反应机理
mechanism of rock failure	岩石破坏机理
mechanism of rock mass deformation	岩体变形机制
mechanism of shotcrete-bolt supporting	喷锚支护原理
mechanism of slit deposition	泥沙沉积机理
mechanism of soil liquefaction	土壤液化机制
mechanism of the source	震源机制
mechanism of transport	搬运过程
mechanism of transport and deposition	搬运沉积过程
mechanisms of chimney caving	筒状崩落机制
mechanisms of sink hole formation	缩孔形成机制；落水洞形成机制
mechanized coal face	机械化采煤工作面
mechanized cut-and-fill method	机械化采掘充填开采法
mechanized drilling	机械钻眼；机械凿岩
mechanized extraction	机械化采掘
mechanized face	机械化工作面
mechanized mine	机械化矿井
mechanized mining	机械化开采
mechanized mining face	机采工作面
mechanized mucking	机械装岩
mechanized operating line for drifting	平巷掘进机械化作业线
mechanized operating line for shaft sinking	竖井掘进机械化作业线
mechanized output	机械化采煤量
mechanized packing	机械充填
mechanized post pulling	机械回柱
mechanized support	机械化支护
mechanogenesis	机械成岩作用
median depth	中值深度
median deviation	中位差
median diameter	中值粒径
median discharge	中流量
median filter	中值滤波器
median filtering	中值滤波
median fold	中隆
median formation sand grain size	地层砂粒度中值
median grain diameter	粒度中值
median gravel diameter	砾石直径中值
median gravel size	砾石粒度中值
median humidity	湿度中值
median mass	中间地块；中央地块
median massif	中央地块
median method	中值方法
median moraine	中碛
median of field strength	场强中值
median of signal level	信号电平中值
median pack-to-formation grainsize ratio	充填砾石与地层砂粒度中值比
median particle diameter	中值粒径
median ridge	洋中脊

median rift system 中央裂谷系；中央地堑系
median rift valley 洋中裂谷
median seam 中央矿层
median size 平均粒径；中值粒径；粒度中值
median valley 中央裂谷
median value 中值
median water level 中水位
media-structure interaction 介质-结构相互作用
medical geology 医疗地质学
medical mineralogy 药用矿物学
mediiphyric 显微斑晶的
Mediterranean 陆间地槽；地中地槽；地中海型
Mediterranean basin 地中海盆地
Mediterranean belt 地中海带
Mediterranean climate 地中海型气候
Mediterranean delta 地中海式三角洲
mediterranean petrographic province 地中海岩区
Mediterranean Ridge 地中海岭
Mediterranean Sea 地中海
Mediterranean suite 地中海岩套；地中海岩组
mediterranean-type margin 地中海型边缘地区
Medithermal 小冰期
Medittterian salt water 咸水
medium barrel 中硬地层取芯筒
medium bedded 中厚层的
medium caking coal 中黏煤
medium capacity 中等容量
medium coarse sand 中粗砂
medium compact sand 中等密度的砂
medium consistency 中等稠度
medium crooked hole 中度弯曲井眼
medium cross section 中断面
medium dense sand 中密砂
medium deposit 中厚矿床
medium depth bore 中深孔
medium diameter 中等粒径；中值粒径
medium dip 中倾斜
medium dipping bed 中倾斜岩层
medium drill 中型钻机；中型凿岩机
medium duty road header 中型巷道掘进机
medium dynamic sounding 中型动力触探
medium earthquake 中度地震
medium filtering 中值滤波器
medium frequency rock drill 中频率冲击式凿岩机
medium grade 中等坡度
medium grade coal 中品位煤
medium grade metamorphism 中级变质作用
medium gradient 中等梯度；中等坡度
medium grain 中粒
medium grained 中等粒度的
medium grained sand 中粒沙
mediugrained soil 中粒土
medium grained texture 中粒结构
medium granular sandstone 中粒砂岩
medium gravity 中等比重
medium ground 中硬地层
medium hard coal 中等硬度煤
medium hard coal seam 中硬煤层
medium hard rock 中硬岩石
medium height mountain 中高山

medium incline 中等坡度；中等倾斜
medium induction log 中探感应测井
medium investigation device 中等探测深度仪器
medium massif 中间地块
medium permeability 中等渗透率
medium pitch 中等倾斜
medium pressure metamorphic facies series 中压变质相系
medium pressure metamorphism 中压变质作用
medium pressure pump 中压泵
medium range 中距离；中程；中范围
medium relief 中等地势
medium resolution infrared radiometer 中分辨率红外辐射计
medium round nose 半圆端
medium salinity 中等盐度
medium sand 中砂；中粒砂
medium scale 中比例尺
medium scale regional geological reconnaissance 中比例尺区域地质调查
medium seism 中震
medium setting cement 中等凝结速度的水泥
medium shaft 中间竖井
medium short wave 中短波
medium silt 中粉砂；中粒粉砂
medium size 中等尺寸；中等大小
medium sized 中等大小的
medium sizing 中级筛分；中级分级
medium slope 中等坡
medium soft floor 中软底板
medium stable roof 中等稳定顶板
medium strain 中等应变
medium structure 中粒结构
medium term earthquake prediction 地震中期预报
medium thickness seam 中厚矿层
medium tunnel 中长隧道
medium value 中值；中间值；平均值
medium wave 中波
medium-angle thrust 中等角度的冲断层
medium-bedded 中厚层状的
medium-coarse sand 中粗砂
medium-deep well 中深井
medium-field earthquake motion 中场地震运动
medium-frequency positioning system 中频定位系统
medium-grained 中粒的
medium-grained clastics 中粒碎屑
medium-grained sand 中粒砂
medium-granular 中粒的
medium-granular rocks 中粒岩石
medium-gravity crude 中等比重原油
medium-hard 中等硬度的
medium-high frequency 中-高频
medium-inclined 中等倾斜的
medium-investigation induction log 中探感应测井
medium-longhole blasting in underground mine 地下矿中深孔爆破
medium-radius lateral drilling 中曲率半径侧向钻井
medium-scale geological surveying 中比例尺地质调查
medium-size computer 中型计算机
medium-size rig 中型钻机
medium-sized 中级尺寸的；中等粒度的

medium-soft sandstone 中软砂岩
medium-steep seam 倾斜煤层；倾斜矿层
medium-stone bit 中粒金刚石钻头
medium-support interaction 岩石介质与支架的相互作用
medium-term earthquake precursor 中期地震前兆
medium-term earthquake prediction 地震中期预报
medium-textured 中型颗粒结构的
medium-textured soil 中等质地土壤
medium-thick seam 中厚煤层
medium-to-fine-textured 中-细颗粒结构的
medium-to-short-range prediction 中短期预报
Medvedev intensity increment criterion 麦德维捷夫烈度增量准则
Medvedev scale 麦德维捷夫地震烈度表
Medvedev-Sponheuer-Karnik intensity scale 麦德韦杰夫-施蓬霍伊尔-卡尔尼克烈度表
mega reach well 特大位移井
megabar 兆巴
megabit 兆位；百万位
megablock 巨断块
megabreccia 巨型角砾岩；粗角砾岩
megaclast 巨碎屑；大碎屑
megacryst 大晶体；巨晶
megacrystalline 粗晶质的；大晶的
megacrystalline fabric 巨晶组构
megaculmination 巨型隆起
megacycle per second 兆赫
megacyclothem 巨旋回层；巨韵律层
megadelta 巨型三角洲
mega-depression tectonics 大坳陷构造
megaextended-reach well 大水平延伸井
megafabric 大型组构
megafacies 主相；异时同时相
megafan 巨型扇沉积；巨冲积扇
megafault 巨大断层
megafracture zone 巨型断裂带
megagea 巨陆
megageotectonics 大构造
megagrained 粗粒的
megahertz 兆赫
megajoule 兆焦耳
megalith 巨石
meganticlinorium 巨复背斜
megaphenocryst 巨斑晶
megaphyric 大斑晶的
megaphyric structure 大斑晶状构造
megaplate 大板块；巨板块
megaplume 巨型地幔喷流柱；巨地幔柱
megapore 大孔隙
megaporosity 巨孔隙；大孔隙
megaporphyric 大斑晶；粗斑的
megarelief 大地形；大起伏
megarhythm 巨韵律
megaripple 大波痕
megascopic 宏观的；肉眼识别的；粗大的
megascopic anthraxylon 肉眼镜煤
megascopic character 粗视特征
megascopic determination 肉眼鉴定
megascopic examination 肉眼鉴定
megascopic fabric 粗视组构

megascopic heterogeneity 肉眼可见的非均质性
megascopic method 肉眼识别法
megasedimentology 宏观沉积学；大沉积学
megaseism 大地震；剧震
megasequence 大层序
megashear 巨大剪切带；巨大切变带；大平移断层
megastructure 巨型构造
megasyncline 大向斜
megasynclinorium 大复向斜
megatectonic 巨型构造；大构造
megatectonics 宏观构造学；大构造；巨型构造
megatemperature 高温
megaterrane 巨地体
megatexture 大纹理；巨观构造
megathermal 高温型的
megathermal climate 高温气候；热带雨林气候
mega-thrust event 巨型逆冲事件
megathrust zone 大逆冲断层带
mega-tsunami 巨型海啸
megavarve 粗纹泥
meimechite 玻质纯橄岩；麦美奇岩
meionite 钙柱石
meizoseismal 极震的；最强地震的
meizoseismal area 强震区；极震区
meizoseismal region 强震区
meizoseismal zone 强震区；强震带
mekometer 光学测距仪
melabasalt 暗色玄武岩
melaconite 土状黑铜矿
meladiorite 暗色闪长岩
melagabbro 暗色辉长岩
melagranite 暗色花岗岩
melane 镁铁质矿物
melange 混杂堆积；混杂岩；滑塌岩
melanhydrite 黑色橙玄玻璃
melanite 含钛钙铁榴石
melanocerite 黑稀土矿
melanocrate 暗色岩
melanocratic mineral 暗色矿物
melanocratic rock 深色岩；暗色岩
melanolite 暗色岩
melanophyride 暗斑岩
melanotekite 硅铅铁矿
melanterite 水绿矾
melaphyre 暗玢岩；蚀变玄武岩
melaporphyre 暗长玢岩
melarhyodacite 暗色流纹英安岩
melasilexite 暗色英石岩
melasyenite 暗正长岩
melatonalite 暗色英闪岩；暗英云闪长岩
melfite 蓝方斑岩
melilite 黄长玄武岩；黄长石
melilitholith 纯黄长岩
melilitite 黄长岩
melilitolite 黄长石岩
melinite 软滑黏土；黏泥；红玄武土
melinophane 蜜黄长石
melioration 土壤改良
meliphanite 蜜黄长石
melite 水硅铝铁土

mellahite	杂海盐
mellilite	黄长石
mellite	密蜡石
mellow loam	松壤土
mellow soil	肥沃土；海绵土；熟土；松软土
mellow soil layer	熟土层
melnikovite	胶黄铁矿
melnikovitization	胶黄铁矿化
melnikovitolite	胶黄铁矿岩
melonite	碲镍矿
melt	熔化；熔融；熔解；熔体
melt band	融化带
melt basin	融水盆地
melt blending	熔融掺混
melt bonding	热熔黏合
melt conductivity	熔体热导率
melt cosmic dust	消融型宇宙尘
melt curve	熔融曲线
melt down	熔解
melt flow index	熔体流动指数
melt flow rate	熔体流动速率
melt generation and magma transport	熔体生成和岩浆运输
melt inclusion	熔融体包裹体
melt index	熔融指数
melt mass	熔体
melt percolation	熔渗
melt point	熔点
melt relaxation time	熔体弛豫时间
melt structure	熔体结构
melt temperature	熔化温度
melt viscosity	熔体黏滞性
melt water	融水；融雪水；融冰水
meltability	可熔性
meltableness	熔度
meltback	回熔
meltdown	熔化
melted	熔融的
melteigite	暗霓霞岩；霞霓钠辉岩
melting	熔融；熔化；融化
melting and refreezing	熔化与再凝
melting basin	融水盆地
melting curve	熔化曲线
melting furnace	熔炼炉
melting heat	熔化热
melting ice	融冰
melting inclusion	熔融包裹体
melting index	熔融指数
melting loss	熔化损；熔炼损耗
melting point	熔点；融解点
melting spot	熔融热点
melting temperature	融化温度
melt-mantle interaction	熔融幔相互作用
melt-out till	熔化冰碛土
meltwater	冰雪融水；冰河融水
meltwater channel	融水河道
meltwater deposit	融水沉积
meltwater lake	融水湖
meltwater stream	融河水道；融水河流
member	段；小层；分层；构件；成员
Member Hong Kong Institution of Engineers	香港工程师学会会员
member in bending	承曲构件；受弯构件
member in compression	受压构件；压杆
member in shear	承剪构件；受剪构件
member in tension	受拉构件；拉杆
member in torsion	受扭杆件
Member Institution of Civil Engineers	英国土木工程师学会会员
member structure	杆系结构
member support	构件支撑
Member, Institution of Professional Engineer, New Zealand	新西兰专业工程师学会会员
membrane	橡皮膜
membrane activity	膜渗作用
membrane analog	薄膜模拟
membrane analogy	薄膜模拟
membrane barrier	隔膜
membrane correction	橡皮膜校正
membrane curing	薄膜养护
membrane dam	薄膜坝
membrane efficiency	膜效率
membrane equilibrium	膜渗平衡
membrane filter	薄膜过滤器；过滤膜
membrane filtration	隔膜渗滤作用
membrane filtration factor	膜滤系数
membrane grouting process	薄膜灌浆法；帷幕灌浆法
membrane method	薄膜法
membrane method of waterproofing	薄膜防水法
membrane model	薄膜模型
membrane penetration effect	橡皮膜嵌入效应
membrane permeability	膜渗透性
membrane polarization	薄膜极化
membrane porosity	薄膜孔隙性
membrane pressure gauge	薄膜式压力计
membrane process	薄膜法
membrane pump	隔膜泵；膜式泵
membrane sealed	薄膜止水的
membrane separation	薄膜分离
membrane separator	薄膜分离器
membrane simulation	薄膜模拟
membrane stabilization	薄膜稳定法
membrane stress	薄膜应力；板块纬向运动挤压应力
membrane surface	薄膜面层
membrane theory	薄膜理论
membrane type fabric support	帷幕结构支撑
membrane type pressure gauges	薄膜式压力盒；薄膜式压力计
membrane-enveloped soil layer	薄膜封闭土层
membraneous waterproofing	薄膜防水
membranometer	薄膜式压力计
membranous waterproofing	薄膜防水
memnescope	瞬变示波器；存储管式示波器
memorize	内存化；内存储式
memorizer panel	存储器控制面板
memory	内存；存储；存储器
memory access control	存储器的存取控制
memory address register	存储地址寄存器
memory address register (MAR)	存储地址寄存器
memory address space	存储地址空间
memory allocation	存储器配置；存储器分配

memory area 存储区
memory array 存储器阵列
memory bank 存储体；内存条
memory block 存储区；存储块
memory board 存储板
memory buffer 缓冲存储器
memory buffer unit 存储缓冲器
memory capacity 储存容量
memory cell 存储单元
memory controller 存储控制器
memory cycle 存储周期
memory data bus 存储数据总线
memory decoder 存储器译码器
memory device 存储设备
memory dump 存储器转储
memory effect 记忆效应
memory element 存贮元件；记忆元件
memory function 存贮函数
memory interleaving 存储器交叉存取
memory location 存储器位置；存储单元
memory module 存储器模数；存储器模块
memory of digital computer 数字计算机存储器
memory parity interrupt 存储奇偶中断
memory port 存储器出入口
memory protection 存储保护
memory reference instruction 存储访问指令
memory reference order 存储访问指令
memory register 存储寄存器
memory row address 存储器行地址
memory shared computer 内存共享计算机
memory sharing 存储器共享
memory space 存储量
memory span 存储器容量
memory stick 内存条
memory storage 存储装置
memory system 存储系统
memory time 存储时间
memory unit 存储单元；存储器
memoscope 存储管式示波器
memotron 存储管
memphytic slate 辉绿板岩
menaccanite 钛铁矿；含钛火山灰砂岩；黑火山砂
Menap glacial stage 梅纳普冰期
Menard meter 旁压仪
Menard modulus 梅纳德模量
Menard pressure meter 梅纳德压力表；梅纳德旁压仪
Menard's method 梅纳德法
Mendeleev chart 门捷列夫周期表
Mendeleev group 门捷列夫族
Mendeleev law 门捷列夫周期律
mendeleevite 钙铌钛铀矿
Mendeleev's periodic system 门捷列夫周期系
Mendeleev's periodic table 门捷列夫周期表
Mendeleev's table 门捷列夫周期表
Mendeleyev Ridge 门捷列夫海岭
mendeleyevite 钙铌钛铀矿
mendipite 白氯铅矿
mendozite 钠明矾；水钠铝矾
meneghinite 斜方辉锑铅矿
menhir 石柱

menilite 肝蛋白石；硅乳石
meniscoid 新月形的
meniscus 新月形；凹凸透镜；弯液面
meniscus dune 新月形沙丘
meniscus shape 弯月面形状
mensuration 测量；测定；量度；求积法
Merapi type flow 默拉皮型火山碎屑流
Mercalli intensity 麦加利地震烈度
Mercalli intensity scale 麦加利地震烈度表
Mercalli scale 麦加利地震烈度表
Mercalli scale of intensity 麦加利地震烈度表
Mercalli-Cancani-Sieberg intensity scale 麦加利-肯肯尼-西贝尔格地震烈度表
mercallite 重钾矾
Mercaly-Cancany-Zeiberg scale 梅尔卡利-坎坎尼-蔡伯格烈度表
mercast 冰冻水银法
Mercator bearing 墨卡托方位角
Mercator chart 墨卡托海图
Mercator map projection 墨卡托地图投影
Mercator mosaics 墨卡托镶嵌图
Mercator projection 墨卡托投影；正轴墨卡托投影
Mercator track 墨卡托恒向线
Mercator's projection 墨卡托投影
merchantable bed 可采煤层；工业煤层
merchantable coal 可售煤；商品煤
merchantable coal bed 商品煤层；可采煤层
merchantable oil 商品原油
merchantable ore 可售矿石
mercurial gauge 水银压力计
mercurial thermometer 水银温度计
mercuric 汞的；二价汞的
mercuric pressure gauge 水银压力计
mercuric vacuum gauge 水银真空计
mercuride 汞化物
mercurimetry 汞量测量；汞液滴定法
mercurius 自然汞；汞；水银
mercurometric survey 汞量测量
mercury 汞；水银
mercury absolute pressure 水银柱压力
mercury analyzer 测汞仪
mercury barometer 水银气压计；水银晴雨表
mercury capillary-pressure curve 水银毛细管压力曲线
mercury column 水银柱；汞柱
mercury cracking 汞致晶界腐蚀开裂
mercury detector 水银检波器
mercury displacement pump 容积式水银泵
mercury ejection curve 退汞曲线
mercury ejection-injection ratio 退汞压汞比
mercury gage 水银压力计；水银气压计
mercury gauge 汞测压计
mercury injection apparatus 压汞仪
mercury injection capillary pressure 压汞毛管压力
mercury injection method 水银注入法
mercury intrusion 水银压入
mercury intrusion method 压汞法
mercury intrusion pore size distribution 压汞法孔隙大小分布
mercury intrusion porosimeter 压汞孔隙度仪
mercury intrusion porosimetry 压汞法；汞孔隙率法

mercury manometer　水银压力计
mercury minerals　汞矿类
mercury penetration　压汞
mercury permeameter　水银渗透率仪
mercury porometer　水银孔隙度仪
mercury porosimetry　水银孔率法；压汞分析仪
mercury porosimetry saturation　水银孔隙度仪测定的饱和度
mercury pressure　汞柱压力；水银柱压力
mercury pressure gauge　汞压力计
mercury pycnometer　水银比重计
mercury reservoir　水银容器
mercury seal　汞封口；水银封口
mercury sealing　汞密封；水银封口
mercury slug　水银柱
mercury thermal analysis curve　汞热释曲线
mercury thermometer　水银温度计
mercury thermostat　水银恒温计
mercury vacuum pump　水银真空泵
mercury vapor analyzer　测汞仪
mercury vapormeter　汞蒸汽仪
mercury withdrawal　退汞
mercury-injection method　压汞法
mercury-vapour method　汞汽勘探法
mere stone　界石
merenskyite　碲钯矿
merestone　界石
merged curve　组合测井曲线
merged log　合并测井
merging data　合并数据
meridian　子午线；经线
meridian angle　子午线角
meridian arc　子午弧
meridian circle　子午圈
meridian convergence　子午线收敛角
meridian distance　纬度；子午距
meridian ellipse　子午面
meridian indicator　回转方位测量罗盘
meridian line　子午线；经线
meridian of the ecliptic　黄道经线
meridian plane　子午面
meridian projection　子午投影
meridian spacing　经差
meridian stress　经向应力；经线应力
meridianal circulation　经向环流
meridianal tectonic system　经向构造体系
meridional　子午线的；经向的；向南的；南方的；偏南的
meridional central map projection　中心经线地图投影
meridional circulation　经向环流
meridional convergence　子午线收敛角
meridional part　渐长纬度；横轴正射投影
meridional projection　经向投影
meridional stress　纵向应力
meridional structural system　经向构造体系
meridional tectonic system　经向大地构造体系
merochrome　异色异构混晶；异色性结晶
merocrystalline　半晶质的
merokarst　不完全岩溶；半岩溶
meroleims　煤化残植屑

merolite　假碎屑岩
merolitic structure　假角砾构造
meromictic lake　不完全对流湖；半对流湖
meropelagic　半海洋性的
merosymmetry　非全对称；缺对称
merosyncline　局部地槽
merotropism　稳变异构
merotropy　稳变异构
meroxene　黑云母
MERP　最大有效钻进速度
merrillite　陨磷钙钠石
MERT　多电极电阻率测井仪
mertia moment　转动力矩；惯性矩
mertial oscillation　惯性振动
merwinite　镁硅钙石
merzlota　冻土
mesa　平顶山；方山；桌状山；台地
meseta　小方山；高原；陆台
mesh after straining　应变后的网格
mesh analysis　筛分分析；粒度分析
mesh aperture　筛孔尺寸
mesh before straining　应变前的网格
mesh boundary　网格边界
mesh column　网格列
mesh configuration　网格构形
mesh connection　网状接线；网格连接
mesh coordinate　网格坐标
mesh data checking　网格数据检查
mesh density　网格密度
mesh distortion　网格畸变
mesh division　网格划分
mesh enrichment　网格强化；密集网格
mesh equation　网格方程
mesh filter　滤网
mesh fraction　筛分粒度级；筛分粒组
mesh generation　网格生成
mesh index　网格指数
mesh input module　网格输入模块
mesh line　坐标格网线
mesh method　网格方法；阻抗法
mesh network　网形网络
mesh of separation　分离网目
mesh parameter　网格参数
mesh point　网格点；网点
mesh reef　网状礁
mesh refinement　网格细化；网格加密
mesh refinement discretization error　网格精化离散误差
mesh regeneration　网格重生成
mesh region　网格区域
mesh reinforcement　钢筋网；钢丝网配筋；网状钢筋
mesh row　网格行
mesh screen　网筛；滤网；网格过滤器
mesh sieve　网筛
mesh size　网目尺寸；筛孔尺寸
mesh space　网格空间
mesh spacing　网格步长；网眼间距
mesh structure　网状结构；格状构造
mesh texture　网状结构
mesh width　网格宽度；筛格尺寸
meshing and grouting　边坡挂网喷浆

meshless local boundary integral equation 无网格局部边界积分方程法
meshless local Petrov-Galerkin method 无网格局部彼得洛夫-伽辽金法
meshless method 无网格法
meshwork 筛；筛面；网状物
meshworm 网眼
meshy 筛的；筛孔的
meshy filter 网状过滤器
mesitite 菱铁镁矿
meso method 半微量法
meso-anthracite 中无烟煤
Mesoarchean 中太古代；中太古界
Mesoarchean Era 中太古界
meso-bituminous coal 中烟煤
Meso-cathaysian 中华夏系
mesoclimate 中气候；局地气候
mesoclimatology 中气候学
mesocrate 中色岩
mesocratic 中色的
mesocrystalline 中等结晶的；中晶质的；半晶质的
mesocrystallization 中等结晶作用
mesodiagenesis 中期成岩作用
mesodialyte 中异性石；中性石
mesofold 中型褶皱
mesogene 中深成矿作用；中深成岩作用
mesogenetic 中期形成的
mesogenetic porosity 中期形成的孔隙性
mesogenetic stage 中成岩阶段
mesogeosyncline 陆间地槽；中间地槽
mesograin 中粒
mesograined 中粒的
mesogranulite 中色麻粒岩
mesohaline 中盐的；中盐度的
mesoide 中生代褶皱带；中生代构造带
mesoiden 中生代山脉
mesokaite 褐煤
meso-lignite 中褐煤
mesolimnion 中间湖沼；湖水中层
mesolittoral zone 潮间带
mesomechanics 细观力学
mesometamorphic 中深变质的
mesometamorphism 中深变质作用
mesomorphic 介晶的
mesoneritic fascia 中浅海带；中近海带
mesonorm 中带变质标准矿物
mesopause 中间层顶；中气层顶
mesopelagic 中远洋的；海洋中层的；中深海
mesopelagic zone 海洋水中层带；中远洋带
mesoperthite 中条纹长石
mesoplate 中板块
mesopore 中孔隙；间隙孔
mesoporium 孔间面；孔间区
mesoporosity 中等孔隙性；中孔隙
mesoporous materials 介孔材料
Mesoproterozoic 中元古代；中元古界
Mesoproterozoic carbonate 中元古代碳酸盐
Mesoproterozoic Era 中元古代
Mesoproterozoic Erathem 中元古界
mesoreef 细观礁
mesorelief 中起伏；中地形
mesoscale 中尺度；中等规模的
mesoscale eddy 中尺度涡流
mesoscopic fracture 中等规模裂隙
mesoscopic structure 中型构造
mesosequence 中层序
mesosiderite 中铁陨石；中陨铁
mesosilicate 中性硅酸盐
mesoslope 中坡
mesosome 中色体
mesosphere 中圈；中间层；散逸层
mesostasis 最后充填物；间隙物质
mesostructure 中构造
mesotectonics 中等构造；中型构造
Meso-Tethys 中特提斯
mesothermal 中温的；中温热液的；中深热液的
mesothermal climate 中温气候
mesothermal deposit 中温热液矿床；中深热液矿床
mesothermal temperature 中温
mesothermophilous 适温带的
mesothorium 新钍
mesotidal range 中潮差
mesotrophic mire 中营养沼泽
mesotrophic peat 中营养泥炭
mesotype 中型；中色的；中沸石
mesotype rock 中性岩
Mesozoic 中生代；中生界
mesozoic epochs of compression 中生代挤压
Mesozoic era 中生代
Mesozoic erathem 中生界
Mesozoic group 中生界
Mesozoic plate tectonics 中生代板块构造
Mesozoic stratigraphy 中生代地层
mesozonal metamorphism 中深带变质作用
mesozone 中深带
message coding 信息编码
message decoder 信息解码器
message digit 信息数字
message exchange 信息交换
message processing program 信息处理程序
message-to-noise ratio 信噪比
messelite 次磷钙铁矿
messenger wire 承力索
messenger wire connection clamp 承力索接头线夹
Messer method 麦赛尔法
MEST 微电阻率扫描测井仪
mestigmerite 角闪辉霞岩
meta andesitization 变安山岩化
meta bituminous coal 高阶烟煤；变质烟煤
meta byte 兆字节
meta gabbro 准辉长岩
meta ripples 不对称波痕
metaamphibolite 变质角闪岩；变角闪岩
meta-andesite 变安山岩
meta-andesitization 变安山岩化
metaankoleite 变钾铀云母；准钾铀云母
meta-anthracite 碳化程度最高的无烟煤；变质无烟煤；准石墨；高煤化无烟煤
meta-anthracite mat 高阶无烟煤；高级碳化无烟煤
meta-arkose 变质长石砂岩

meta-autunite	变钙铀云母；准钙铀云母
metabasalt	变质玄武岩；变玄武岩
meta-basalt	变玄武岩
metabasaluminite	准基矾石
metabasite	变质基性岩；变基性岩；准基性岩
metabassetite	准磷铁铀矿
metabentonite	变蒙脱石；变斑脱岩
metabituminous coal	肥煤；中烟煤
metabitumite	准沥青岩
metablast	均匀变晶；变晶
metablastesis	均匀变晶作用
metabolism of rocks	岩石的再生作用
metabolite	蚀变粗玻岩；代谢产物
metabrushite	脂磷钙石
metacannel coal	高煤化烛煤
metacarbonatite	变质碳酸岩
metaceystalline	变质结晶岩
metachemical metamorphism	准化学变质作用
metachert	硅化灰岩；变燧石
metacinnabarite	黑辰砂矿
metaclase	劈理；次生劈理岩
metaclastic schistosity	次生片理
metaclastics	变碎屑岩
metacolloform structure	变胶状构造
metacolloid	偏胶质；结晶胶体；变胶体
metacolloidal structure	变胶状构造
metaconglomerate	变砾岩；变质砾岩
metaconite	重结晶灰岩
metacryst	次生晶；变晶
metacrystal	次生晶；变斑晶；变晶
metacrystal texture	变晶结构
metacrystalline rock	变晶岩
metadiabase	变辉绿岩
metadiagenesis	准成岩作用；后生作用
metadiorite	变质闪长岩
metadolerite	变粗玄岩
metadurain	变质暗煤
metadurite	高煤化微暗煤
meta-evaporite	准蒸发岩
metafluidal texture	动力流动结构
metagabbro	变质辉长岩
metagenesis	世代交替；后成岩作用；变生作用
metagenetic deposit	变质沉积物；复成沉积物
metagenetic gas	变生阶段成因气
metagenetic stage	变生作用阶段
metagenetics	变质
metaglyph	变质印痕
metagneiss	变片麻岩；准片麻岩
metagreywacke	变质杂砂岩
metahalloysite	准埃洛石；偏多水高岭土
metaharmosis	晚期成岩作用
metaheinrichite	变钡砷铀云母；准砷铀钡矿
metahewettite	变针钒钙石
metaigneous rock	变火成岩
metakaolin	偏高岭石
metal coal	含黄铁矿的煤
metal mine	金属矿；金属矿山
metal mining	金属矿开采
metal ridge	矿柱；隆起地层
metal winning	金属矿开采
meta-lignite	变质褐煤；高阶褐煤
metalignitous coal	高煤化褐煤；超褐煤
metalimnion	变温层；中层湖水
metallic mineral	金属矿物
metallic mineral deposit	金属矿床
metallic ore	金属矿石
metallic ore deposit	金属矿床
metallics	天然金属
metallics cement	矿渣水泥
metalliferous	含金属的
metalliferous brine	含金属卤水
metalliferous connate water	含金属原生水
metalliferous deposit	金属矿床
metalliferous sediment	含金属沉积物
metallization	矿物化；成矿作用
metallization phase	矿化期；矿化阶段
metallized lake	矿湖；矿化湖
metallogenetic	成矿的
metallogenetic element	成矿元素
metallogenetic epoch	成矿期；成矿时代
metallogenetic factor	成矿因素
metallogenetic formation	成矿建造
metallogenetic information	成矿信息
metallogenetic lineage	成矿谱系
metallogenetic megaprovince	成矿域
metallogenetic model	成矿模式
metallogenetic phase	矿化期
metallogenetic prognosis	矿产预测
metallogenetic prospective province	成矿远景区
metallogenetic province	成矿区；矿床区
metallogenetic specialization	成矿专属性
metallogenetic system	成矿系统
metallogenetic zoning	成矿分带
metallogenic	成矿的
metallogenic background and genesis	成矿背景与成因
metallogenic element	成矿元素
metallogenic epoch	成矿时代
metallogenic factors	成矿因素
metallogenic isotropy of mine product	成矿图；矿床成因图
metallogenic map	成矿图
metallogenic province	成矿区
metallogenic specialization of magma	成矿岩浆专属性
metallogeny	矿床成因论；金属成矿学
metallographer	金相学家
metallographic	金相的
metallographic examination	金相分析
metallographic laboratory	金相检验室
metallographic microscope	金相显微镜
metallographic microscopy	金相显微学；金相显微术
metallographic province	成矿区
metallographic section	金相磨片
metallographic specimen	金相试样
metallographic test	金相检验
metallographical microscope	金相学显微镜
metallographist	金相学家
metallography	金相学
metalloid	准金属的；类金属的；非金属的
metalloid luster	似金属光泽
metalloscope	金相显微镜

metalloscopy 金相显微检验法
metallostatics 金属静力学
metallotectonic 成矿构造的
metallotectonics 成矿构造；成矿构造学
metallurgical coal 冶金用煤；炼焦煤
metallurgical microscope 冶金显微镜；金相显微镜
metaloscope 金相显微镜
metal-rich geothermal system 富金属地热系统
metal-rich soil 含矿土壤；含金属矿的土壤
metals spring gravimeter 金属弹簧重力仪
metamarble 变质大理岩
metamathematics 元数学
metamerism 分节现象；位变异构性；位变异构现象
metameter 位变仪
metamict 变成非晶质的矿
metamict mineral 结晶变异矿物
metamictization 非晶化；蜕晶作用；似晶化
metamorphic 变质的
metamorphic age 变质年龄
metamorphic assemblage 变质杂岩；变质矿物组合
metamorphic aureole 接触变质圈；接触变质带
metamorphic bathograds 变质深度梯度；深变质级
metamorphic bathozone 变质深度带
metamorphic belt 变质带
metamorphic climax 变质作用峰期
metamorphic complex 变质杂岩体
metamorphic convergence 变质会聚作用；趋同变质作用
metamorphic core complexes 变质核杂岩
metamorphic cycle 变质旋回
metamorphic degassing 变质脱气作用
metamorphic degree 变质程度
metamorphic dehydration 变质脱水作用
metamorphic deposit 变质矿床
metamorphic differential 变质分异作用
metamorphic differentiation 变质分异作用
metamorphic diffusion 变质扩散作用
metamorphic event 变质事件
metamorphic fabric 变质组构
metamorphic facies 变质相
metamorphic facies belt 变质相带
metamorphic facies group 变质相组
metamorphic facies series 变质相系
metamorphic fluid 变质流体
metamorphic grade 变质级；变质程度
metamorphic gradient 变质梯度
metamorphic granite 变质花岗石
metamorphic homogenization 变质均匀化
metamorphic intensity 变质强度
metamorphic isograd 等变质线；等变线
metamorphic limestone 变质灰岩
metamorphic map 变质地质图；变质图
metamorphic mechanism 变质机理
metamorphic mineral 变质矿物
metamorphic mineral association 变质矿物共生组合
metamorphic mineralization 变质成矿作用
metamorphic overprint 变质叠加
metamorphic petrology 变质岩石学
metamorphic phase 变质幕
metamorphic polarity 变质极性
metamorphic process 变质作用；变质过程

metamorphic protolith recovery 变质原岩恢复
metamorphic reaction 变质反应
metamorphic reaction zone 变质反应带
metamorphic recrystallization 变质再结晶作用
metamorphic rock 变质岩
metamorphic rock seamounts 变质岩海山
metamorphic rock sequence 变质岩石序列
metamorphic schist 变质片岩
metamorphic secretion brine 变质分泌卤水
metamorphic segregation 变质析离作用
metamorphic sequence 变质顺序
metamorphic siltstone 变质粉砂岩
metamorphic soil 变质土
metamorphic structure 变成构造；变质构造
metamorphic structure surface 变质结构面
metamorphic subfacies 变质亚相
metamorphic syenite 变质正长岩
metamorphic tectonics 变质构造
metamorphic terrane 变质地体
metamorphic transformation 变质迁移作用
metamorphic water 变质成因水；变质水
metamorphic zone 变质带
metamorphic zoning 变质分带
metamorphics 变质学
metamorphide 变质褶皱带
metamorphism 变质；变质程度
metamorphism in lower crust 下地壳变质
metamorphism sequence 变质作用序列
metamorphism zone 变质带
metamorphogenetic 变质成因的
metamorphogenic 变质成因的
metamorphogenic deposit 变质生成矿床；变成矿床
metamorphose 变质
metamorphosed 变质的
metamorphosed basement 变质基底
metamorphosed basic dike swarm 变质基性岩墙
metamorphosed deposit 受变质矿床
metamorphosed inclusion 变生包裹体
metamorphosed ore deposit 变质矿床
metamorphosed plutons 变质深成侵入体；深成岩体
metamorphosed rock 变质岩
metamorphosed sedimentary deposit 受变质沉积矿床
metamorphosed sedimentary ore deposit 变质沉积矿床
metamorphosed supracrustal rock series 变质表壳岩系
metamorphosis 变质；变形
metamorphosize 使变质；使变化；改造；改变
metamorphous 变质的；变形的
metanovacekite 变镁砷铀云母；变水砷镁铀矿
metapegmatite 变质伟晶岩；准伟晶岩
metapelite 变泥质岩
metapepsis 水热变质
metaplatform 准地台
metaprogram 元程序
metaquartzite 变质石英岩
metargillitic 变黏土质的
metarhyolite 变流纹岩
metaripple 不对称波痕
metarossite 准水钒钙石；变水钒钙石
metasandstone 变质砂岩
metasapropel 压实腐泥

metascope	红外线显示器
metasediment	变质沉积物；变质沉积岩
metasedimentary	变质碎屑岩
metasedimentary province	变质沉积区
metasedimentary rock	变质沉积岩
metasedimentogenic	准沉积成因的
metaseds	变质沉积岩
metasilicate	硅酸盐
metasomaric texture	交代结构
metasomasis	交代
metasomatic	交代的
metasomatic alteration type deposit	交代蚀变型矿床
metasomatic antiperthitic texture	交代反条纹结构
metasomatic corrosion texture	交代侵蚀结构
metasomatic deposit	交代矿床
metasomatic edulcoration border texture	交代净边结构
metasomatic graphoid texture	交代似文象结构
metasomatic hydrothermal metallogenesis	交代热液成矿
metasomatic metamorphism	交代变质
metasomatic myrmekitic texture	交代蠕状结构
metasomatic ore	交代矿石
metasomatic perforation texture	交代穿孔结构
metasomatic perthitic texture	交代条纹结构
metasomatic porphyritic texture	交代斑状结构
metasomatic pseudomorph texture	交代假象结构
metasomatic relict texture	交代残余结构
metasomatic rock	交代岩
metasomatic texture	交代结构
metasomatic vein	交代脉
metasomatism	交代；交代变质
metasomatite	交代岩
metasomatose	交代作用
metasomatosis	交代作用
metasome	交代矿物；新成体
metaspilite	变质细碧岩
metastabilite	半晶矿物
metastability	亚稳度；亚稳性
metastable	准稳定的；亚稳定的
metastable composition	准稳成分
metastable compound	准稳定化合物
metastable equilibrium	亚稳定平衡；暂时稳定的平衡
metastable helium magnetometer	准稳态氦气磁力仪
metastable immiscibility surface	亚稳不混溶面
metastable ion	亚稳离子
metastable level	亚稳态能级
metastable mineral	准稳矿物
metastable phase	准稳定相；亚稳相
metastable range	亚稳区
metastable soils	亚稳土
metastable state	准稳态；亚稳态
metastable structure	亚稳结构
meta-stable structured soil	亚稳结构性土
metastasis	同质蜕变
metastasy	侧向均衡调整；水平均衡调整
metastibnite	胶辉锑矿；准辉锑矿
metastrengite	准红磷铁矿
metastructure	变质构造；次显微构造
meta-substitution	间位取代
metataxis	分异深熔作用
metatect	变熔体
metatectic transformation	包晶转变
metatectite	交代深熔岩；选择重熔混合岩
metatexis	分异深熔作用；带状混合作用；选择重熔作用
metatexite	交代深熔岩；变熔体
metathesis	复分解；置换；易位
metatorbernite	变铜铀云母；准铜铀云母
metatriplite	准磷铁锰矿
metatrophic bacteria	腐生细菌
metatropy	变性
metatyuyamunite	变钙钒铀矿；准钙钒铀矿
metauranocircite	变钡铀云母
metauranopilite	准硫铀钙矿
metauranospinite	变钙砷铀云母
metavariscite	准磷铝石
metavauxite	准蓝磷铝铁矿
metavolcanic	次火山活动的
metavolcanics	部分变质火山岩
metavolcanite	变质火山岩
metavoltite	黄磷铁矾
metazellerite	变碳钙铀矿
metazeunerite	变铜砷铜铀云母
metazoa fossil	后生动物化石
meteor	流星；陨星；大气现象
meteor crater	陨石坑
meteor dust	流星尘埃
meteor flare	流星爆发
meteor radar	流星雷达
meteor shower	流星雨；陨石雨
meteor trail	流星余迹
meteoric	大气的；流星的
meteoric dust	陨星尘
meteoric groundwater	大气成因地下水
meteoric ice	陨冰
meteoric iron	陨铁
meteoric shower	陨石雨；流星雨
meteoric stone	陨石
meteoric stream	流星群
meteoric water	大气降水
meteoric water ingress	雨水进入
meteoric water line	大气降水线
meteorite	陨石；陨星
meteorite age	陨石年龄
meteorite crater	陨石撞击坑；陨击坑
meteorite fall	陨石雨
meteorite impact	陨石撞击
meteorite impact instant rock	瞬变岩；陨击瞬变岩
meteorite shower	陨石雨
meteorite-impact feature	陨石撞击构造
meteoritic astronomy	陨星天文学
meteoritic dust	陨石尘
meteoritic hypothesis	陨星说
meteoritics	陨石学；陨星学
meteorogram	气象图
meteorograph	气象自记仪；气象计
meteoroid	陨星体；流星体；宇宙尘
meteorolite	陨石；陨星
meteorologic satellite	气象卫星
meteorological	气象的
meteorological acoustics	气象声学

meteorological chart	气象图
meteorological condition	气象条件
meteorological data	气象资料；气象数据
meteorological depression	气象低压区
meteorological dynamics	气象动力学
meteorological element	气象要素
meteorological factor	气象因素
meteorological front	气象锋
meteorological observation	气象观测
meteorological observation ship	气象观测船
meteorological phenomena	气象现象
meteorological problems	气象问题
meteorological radar	气象雷达
meteorological report	气象报告
meteorological representation error	气象代表误差
meteorological rocket	气象火箭
meteorological satellite	气象卫星
meteorological sensor	气象传感器
meteorological situation	天气形势
meteorological station	气象台；测候站
meteorological storm	气象风暴
meteorological survey	气象测量
meteorologist	气象学家
meteorology	气象学
meteotron	人造雨云设备
meter	米；计；表；仪表
meter accuracy factor	仪表准确度系数
meter board	仪表盘；量表板
meter characteristic	仪表特性
meter chart	仪表记录纸
meter constant	仪表常数
meter counter	仪表计数器
meter dial	仪表刻度；仪表度盘
meter distortion	仪表变形
meter error	仪表误差
meter factor	流量计校正系数
meter full scale	刻度范围；最大量程
meter proof	流量计检定
meter prover	仪表校准仪
meter rack	仪表架
meter reading	仪表读数
meter registration	流量计累积值
meter resistance	仪表电阻
meter scale	米尺；公制尺
meter sensitivity	仪表灵敏度
meterage	计量；测量；米数
metered inlet flow	计量的入口流量
metered outlet flow	计量的出口流量
metered system	计量系统
meter-in dual operation	进口节流调节
metering	计量；量测；记录；统计
metering circuit	测量电路
metering cylinder	计量筒
metering device	计量装置
metering equipment	测试仪器；测量装置
metering installation	计量装置
metering orifice	测流口；流量孔板；计量孔
metering outfit	计量设备
metering point	量测点
metering practice	流量测定工作
metering pump	计量泵
metering section	量测断面；计量测流断面
metering type screw	计量型螺杆
metering unit	计量装置
metering valve	计量阀
meter-kilogram-second system	米-千克-秒制
metermultiplier	仪表量程倍增器
meter-out dual operation	出口节流调节
meter-run data	表读数据
meter-ton-second system	米-吨-秒制
metewand	计量基准
methan explosive proportion	沼气爆炸点
methane	沼气；甲烷；瓦斯
methane accreting	甲烷增生作用
methane accumulation	沼气聚集；瓦斯聚集
methane adsorbing capacity	甲烷吸附容量
methane air mixture	沼气空气混合物
methane alarm	甲烷报警器
methane blower	沼气喷出口
methane concentration	瓦斯浓度；甲烷浓度
methane content	沼气含量
methane detection	沼气检查
methane detector	沼气检定器
methane determination	沼气测定
methane drainage	沼气抽放；沼气排放；排放瓦斯
methane drainage borehole	沼气排气钻孔
methane drainage efficiency	瓦斯抽放率
methane emanation	沼气泄出
methane emission	沼气泄出
methane emission gradient	瓦斯泄出梯度
methane eruption	沼气泄出
methane explosion	沼气爆炸
methane explosive proportion	沼气爆炸量
methane gas emission	瓦斯涌出
methane gas injection	注甲烷气
methane hydrate	甲烷水合物
methane hydrocarbon	甲烷系烃
methane ignition	沼气燃烧
methane indicator	沼气检定器
methane interferometer	沼气测定仪
methane liberation	放出沼气
methane limit	沼气极限
methane mine	瓦斯矿井
methane monitor	瓦斯监控器
methane monitoring	沼气监测
methane monitoring system	甲烷监测系统
methane oil	甲烷族石油
methane outburst	沼气突出；瓦斯突出
methane oxidation	甲烷氧化
methane pocket	沼气包
methane prediction	沼气预测；瓦斯预测
methane production	甲烷生成
methane production bacteria	产甲烷细菌
methane recorder	沼气记录器
methane recovery	沼气回收率
methane seeps	甲烷渗漏
methane series	甲烷系
methane sorption	甲烷吸附
methane tester	沼气检定器
methane-air mixture	沼气-空气混合物

methanogenesis 甲烷生成作用
methanometer 沼气测定器；沼气指示器
methanometry 沼气测量法；沼气检验法
methanophone 沼气信号器；瓦斯警报器
metharmosis 晚期成岩作用；后期成岩固结作用
method by geological blocks 地质地段法
method engineering 方法工程学
method for rock strength measuring 岩石强度测试方法
method for settlement calculation 沉降计算法
method in all combinations 全组合测角法
method of adjustment 平差法
method of aggregative indicator 综合指标法
method of analogue 类比法
method of application 使用方法
method of approach 渐近法
method of approximation 近似法
method of arithmetic average 算术平均法
method of average 统计法；平均法
method of bearings 坐标方位角法
method of bivariate least squares 二元最小二乘法
method of characteristic 特征线法
method of characteristic curves 特征曲线法
method of characteristic point 特征点法
method of characteristics 特征线法
method of charging 装药法；充电方法
method of chord deflection distance 弦线偏距法
method of class combination 类组合法
method of coal preparation 选煤方法
method of column analogy 柱比法
method of compensation 补偿法
method of conjugate gradient 共轭梯度法
method of construction 施工方法
method of continuous control 连续控制法
method of correction 校正法
method of correction coefficient 修正系数法
method of correlation analysis 相关分析法
method of curve-fitting 曲线拟合法
method of deflection angle 偏角法
method of descent 降维法；递降法
method of determining stress tensor from fault slip data 断层滑动反演构造应力张量方法
method of development 展开法
method of difference 差分法
method of digital image processing 数字图像处理方法
method of dimensions 因次法
method of direct interpretation 直接解译法
method of direction line intersection 方向线交会法
method of direction observation 方向观测法
method of direction observation in rounds 全圆方向法
method of discharge measurement 流量测验法；测流法
method of distance measurement by phase 相位法测距
method of division into groups 分组平差法
method of double refraction 双折射法
method of double sight 复觇法
method of drained sinking 排水下沉法
method of drilling bits 钻屑法
method of dynamic analysis 动态分析法
method of elimination 消元法
method of energy dissipation 能量耗散法
method of equal effect 等效法

method of equal-weight substitution 等权代替法
method of extrapolation 外插法
method of fictitious load 阻力曲线
method of finite difference 有限差分法
method of finite element 有限元法
method of finite increment 细粒冲积物
method of gamma-gamma 伽马-伽马法
method of grading mud-making clay 黏土分级评价法
method of identifying joint sets 鉴别节理组的方法
method of images 映像法；镜像法；映射法
method of impregnation 浸渍法
method of induction 归纳法
method of information design 动态设计法
method of initiation 起爆方法
method of inserting network 插网法
method of installation 安装方法
method of interior point 内分点法
method of interpolation 插值法
method of intersection 交点法；前方交会法
method of inversion 反演法
method of inversion points 逆转点法
method of isotope settlement 沉降核素法
method of iteration 迭代法
method of laminae 面元法；薄片法
method of laminar flow 线性风流法
method of laser alignment 激光准直法
method of leading variables 引入变量法；先导变量法
method of least squares 最小二乘法
method of lime-soil replacement and tamping 灰土换填夯实法
method of limit equilibrium 极限平衡法
method of line element 线元法
method of linearization 线性化方法
method of lithologic stratigraphy 岩石地层法
method of lowering the center of gravity of rocks 上轻下重法
method of making the front part of rock lighter than the back part 前轻后重法
method of making the rock equational and balanced 等分平衡法
method of mathematical model 数学模型法
method of maximum entropy 最大熵法
method of maximum likelihood 最大似然法
method of measurement 计量方法；测量方法
method of minimum squares 最小二乘法
method of mirror 镜像法
method of moment 矩量法；力矩法
method of neutron activation 中子活化法
method of observation set 测回法
method of optimization 最优化方法
method of palaeomagnetic stratigraphy 古地磁地层学方法
method of palaeontological stratigraphy 古生物地层法
method of paleomagnetic stratigraphy 古地磁地层法
method of perturbation 扰动法；摄动法
method of physical model 物理模型法
method of plotting position 定位法
method of point load 点荷法
method of preparation 连矿法
method of projection 投影法
method of redundant reaction 超静定反力法

method of reiteration　复测法
method of relaxation　松弛法
method of repetition measurement　复测法
method of resection　后方交会法
method of residual anomaly　剩余异常法；剩余场化
method of residue　剩余法
method of resistance proportion of mine roadways　井巷风阻比法
method of river diversion　导流方法
method of rounds　全圆法
method of sample taking　采样法；岩样分选法
method of sampling　取样法；采样法
method of scanning　扫描方法
method of sections　截面法
method of seismic prospecting　地震勘探法
method of selected points　选点法
method of sieving　筛分法
method of slice limit equilibrium　分块极限平衡法
method of slices　条分法
method of slime treatment　煤泥处理方法
method of slope deflection　角变法
method of slurry direct circulation　泥浆正循环法
method of slurry reverse circulation　泥浆反循环法
method of small angle measurement　小角法
method of soil and water conservation　水土保持法
method of spring flow attenuation　泉流量衰减方程法
method of stationary phase　驻相法
method of steepest descent　最速下降法
method of strength reduction　强度折减法
method of substitution　替代法；代换法
method of successive approximation　逐步逼近法
method of successive comparison　逐次比较法
method of successive correction　转镜照准仪
method of successive displacement　逐次置换法
method of successive iteration　逐次相迭代
method of summation　总和法
method of superposition　叠加法
method of tangent line　切线法
method of telemetering method　遥测方法
method of tension wire alignment　引张线法
method of time determination by star transit　恒星中天测时法
method of transit projection　经纬仪投点法
method of trial and error　试错法；逐步逼近法
method of undetermined coefficient　待定系数法
method of undrained sinking　不排水下沉法
method of unifying the rock materials　石料统一法
method of unifying the rock veins　纹理统一法
method of vector interpretation　矢量解释法
method of virtual displacement　虚位移法
method of volume element　体元法；点元法
method of water-softening　水软化法
method of weighted least squares　加权最小二乘法
method of weighted mean　加权平均法
method of weighted moving average　加权移动平均法
method of weighted residual　加权残值法
method of weighting　加权法
method of well uniform configuration　平均布井法
method of working　开采方法
method statement　施工方法纲领；施工说明书

method study　方法研究
method study engineer　研究工作方法的工程师
methodical　方法上的；有顺序的；有规律的
methodological　方法学的
methodology　方法论；方法学；分类研究法；方法系统；技术体系
methods of caving　崩落法
methods of groundwater resource evaluation　地下水资源评价方法
methods of non-equilibrium push　不平衡推力法
metric　度量标准；公制的；量度的
metric atmosphere　公制大气压
metric calibration　测量校正
metric camera　量测摄影机；测绘摄影机
metric carat　公制克拉
metric data　量测数据
metric density　公制密度
metric horse power　公制马力
metric horsepower　公制马力
metric module　公制模数
metric scale　米制比例尺
metric system　公制
metric ton　公吨；千公斤
metric ton standard fuel　吨标准燃料
metric ton（MT）　吨；千公斤
metric unit　公制单位；米制单位
metric wave　米波；超短波
metrical information　可度量信息
metrical photogrammetry　摄影测量学
metrical photography　测量摄影学
metrication　公制化
metro　地下铁道；地铁
metro engineering　地铁工程
metro network　轨道交通路网
metro station　地铁车站
metro tunnel　地铁隧道
metrological moisture　计量水分
metrologist　计量学家
metrology　计量学；度量衡学；计量制；度量衡制
metrophotography　测量摄影学
metropolitan area network　城域网
metropolitan railway　城市铁道；市内铁道
METT　多频电磁测厚仪
meuliere　磨石；硅化灰岩或砂岩
MeV　百万电子伏；兆电子伏
Mexican setup　墨西哥型凿岩机支架
mexphalt　沥青
Meyer hardness　迈耶耳硬度
meyerhofferite　三斜硼钙石
Meyerhof's bearing capacity formula　迈耶霍夫承载力公式
Meyerhof's formula　迈耶霍夫承载力公式
meymechite　麦美奇岩；玻质纯橄岩
mezogeosyncline　陆中地槽
mezokatagenesis　中后生作用
miagite　球状辉长岩
miargyrite　辉锑银矿
miarolite　晶洞花岗岩
miarolitic　晶洞状；洞隙
miarolitic cavity　晶洞

miarolitic granite	晶洞花岗岩
miarolitic structure	晶洞状构造
miarolitic texture	晶洞结构
miaskite	云霞正长岩
mica	云母
mica and silt content	砂云母含量及含泥量统计表
mica book	书页云母；云母层
mica clay mineral	云母黏土矿物
mica content	云母含量
mica gneiss	云母片麻岩
mica lamination	云母片
mica peridotite	云母橄榄岩
mica plate	云母板
mica powder	云母粉
mica schist	云母片岩
mica sheet	云母片
mica slate	云母板岩
mica-basalt	云母玄武岩
micaceous	云母的；含云母的；云母质的；云母状
micaceous hematite	云母形赤铁矿
micaceous iron ore	云母铁矿
micaceous mineral	含云母矿石
micaceous quartzite	云母石英岩
micaceous sand	云母砂；含云母砂
micaceous sandstone	云母砂岩
micaceous schist	云母片岩
micaceous shale	云母页岩
micaceous soil	云母土
micaceous structure	云母构造
micacite	云母片岩
micacization	云母化作用
mica-gneiss	云母片麻岩
micalamprophyre	云母煌斑岩
micalex	云母石；云母玻璃
micalike	云母状
mica-marble	云母大理石
micanite	人造云母；层合云母板
micaperidotite	云母橄榄岩
mica-schist	云母片岩
mica-slate	云母板岩
micatization	云母化
mica-type clay mineral	云母型黏土矿物
mica-type layer-silicate	云母型层状硅酸盐
Michelson interferometer	迈克尔逊干涉仪
mickle	软黏土；顶板黏土；夹石
micrinite	微粒组；不透明基质；微晶粒；碎片体
micrinoid	微粒体组
micrinoid group	细粒体组
micrite	泥晶；微晶灰质岩
micrite enlargement	微晶加大作用
micrite envelope	微晶套；微晶包壳；泥晶套
micritic	泥晶质；微晶的；微晶质
micritic aragonite	微晶文石
micritic components	微晶质组分
micritic envelope	泥晶套
micritic limestone	微晶质灰岩；泥晶灰岩
micritic rind	泥晶质壳层
micritization	微晶化；泥晶化
micro	微
micro adsorption detector	微吸附检测器
micro aggregate	微团粒
micro analysis	显微分析
micro analyzer	显微分析仪
micro balance	微量天平
micro computer	微型计算机
micro continental massif	微型陆块
micro crack	微裂纹
micro disseminated type deposit	细微浸染型矿床
micro electric log	微电极测井
micro FT-IR spectroscopy	显微傅里叶变换红外光谱
micro granite texture	微花岗岩结构
micro irregularity	微观不均质性
micro laterolog tool	微侧向测井仪
micro maceral	煤显微组分
micro or spot analysis	微观分析
micro paleopiezometer	显微古应力计
micro piezometer	微压计
micro pile	微形桩
micro processor	微型信息处理机
micro resistivity-induction-laterolog	微电阻-感应-侧向测井
micro seismic monitoring	微震监测
micro seismic precursor	微震预兆
micro spherically focused logging	微球形聚焦测井
micro structure	微观结构
micro texture	微观结构
micro tremor	微振
micro-activity	微反活性
microadjuster	微调装置
microadjustment	微调
microaftershock	微余震
microaggregate	微团聚体
microaggregate grain	微团聚粒
micro-altimeter	微高程计；微高度计
microanalysis	微量分析
microanalytical technique	微量分析技术
microanalyzer	显微分析器
microannulus path	微环隙通道
microaphanite structure	显微隐晶质构造
microaphanitic	显微隐晶质
micro-appendix	微突起
microarchitecture	显微原生构造
microarea	微区
microatoll	微环礁
microautograph	微射线自动照相
microautoradiogram	微射线自动照相
microautoradiograph	微射线自动照相机
microbalance	微量天平；测微天平
microbar	微巴
microbarogram	微气压记录图
microbarograph	微压计；微气压计
microbarometer	微气压计
microbe	微生物；细菌
microbending	局部严重挠曲
microbial	微生物的
microbial action	微生物作用
microbial activity	微生物作用
microbial agent	微生物催化剂
microbial alteration	微生物演化
microbial boundstone	微生物灰岩
microbial carbonates	微生物碳酸盐

microbial enhanced oil recovery 微生物强化采油
microbial geotechnology 微生物岩土工程
microbial laminite 微生物层纹状灰岩；叠层石灰岩
microbial metallogenesis 微生物成矿
microbial metallurgy 微生物冶金学
microbial method 微生物法
microbial monitoring 微生物监测
microbial prospecting 微生物勘探
microbial reefs 微生物礁
microbial weathering 微生物风化
microbialite 微生物岩
microbialite framework 微生物岩骨架
microbially induced carbonate precipitation 微生物诱导碳酸盐沉淀
microbially induced sedimentary structures 微生物引起的沉积构造
microbially rimmed platform margin 微生物镶边台地边缘
microbioclastic 微生物碎屑的
microbiodegradation 微生物降解作用
microbiofacies 微生物相
microbiogenic 微生物成因的
microbiological 微生物的
microbiological alteration 微生物蚀变
microbiological analysis 微生物分析
microbiological anomaly 微生物异常
microbiological assay 微生物测定
microbiological control 微生物控制
microbiological corrosion 微生物腐蚀
microbiological deterioration 微生物降解作用
microbiological exploration 微生物勘探
microbiological geochemical survey 微生物地球化学测量
microbiological indices 微生物学指标
microbiological manipulation 微生物控制
microbiological prospecting 微生物勘探
microbiological testing 微生物试验
microbiological tracer method 微生物示踪研究法
microbiological treatment 微生物处理
microbiology 微生物学
microbiostratigraphy 微体生物地层学
microbit 微型钻头；微比特
microbit drilling test 微钻头钻进试验
microbore 微孔
microbreccia 细角砾岩
microburette 微量滴定管
microburrowed strata 细潜穴地层
microcaliper 微井径仪
microcalipers 千分尺；测微器
microcalorimeter 微量热计
microcamera 微型照相机；显微照相机
microcator 指针测微计
microchannel 微孔道
microcharacter 显微划痕硬度计
micro-characteristic 微观特征；显微特征
microchemical 微量化学的
microchemical analysis 显微化学分析
microchemical test 微量化学试验
microchemical-analysis 微化分析
microchemistry 微量化学
microchipping 微碎裂
microclastic 细屑质；微细屑状

microclastic rock 微细屑岩
microclearance 微间隙
microclimate 小气候；局部气候
microclimate conditions 微气候条件
microclimate effects 微气候影响
microclimatic conditions 微气候条件
microclimatic investigation 微气候调查
microclimatology 小气候学
microcline 微斜长石；钾微斜长石
microcline perthite 微斜纹长石
microcline twin law 微斜长石双晶律
microcline-perthite 微斜纹长石
microclinite 微斜长石岩
microclinization 微斜长石化
microcode 微编码程序设计；微编码指令；微程序代码；微操作码
microcoded data processor 微码数据处理机
microcomponent 微量组分
micro-compression 微压缩
microcomputer 微量计算机
microcomputer program 微机程序
microcomputer software 微机软件
microcomputer-controlled 微型计算机控制的
microcomputer-controlled telemetry system 微型电脑控制的遥测系统
micro-concrete 微粒混凝土
microconductivity 微电导率
microconglomerate 细砾岩；微砾岩
microconstituent 显微组分
microcontinent 微型陆块；微大陆
microcontinental block 微型大陆断块
microcontrollers 微控制器
microcopy 缩微照片
Microcopying 缩微摄影
microcoquina 微贝壳灰岩；砂状贝壳灰岩
microcosm 微观世界
microcosmic 微观的
microcoulomb 微库仑
microcrack 小裂缝；微细裂纹；显微裂隙
microcrack fabric 微裂纹结构
microcracking 微裂缝；显微裂纹
microcreep 微观蠕变
micro-cross-bedding 微交错层理
microcryptocrystalline 微隐晶质的；显微隐晶质
microcrystal 微晶
microcrystalline 微晶质的；微晶的
microcrystalline allochemical rock 微晶异常化学岩
microcrystalline biogenic calcirudite 微晶生物砾屑灰岩
microcrystalline calcite ooze 微晶方解石软泥
microcrystalline fabric 微晶组构
microcrystalline limestone 微晶灰岩
microcrystalline ooze 微晶软泥
microcrystalline rock 微晶岩
microcrystalline texture 微晶结构
microcrystalloblastic 显微变晶的
microcrystallography 微观晶体学；微晶学
Microcurie 微居里
microdarcy 微达西
microdeformation 微小变形
microdensitometer 微密度计；测微密度计

microdensitometry 微显像密度计
microdetection 微量测定
microdetector 微量测定器
microdetermination 微量测定
microdiorite 微闪长岩
microdiscontinuity 微非连续性
microdistancer 微波测距仪
microdivision 测微分划
microdolomite 微白云岩
microdot strain gauge 微粒形应变片
microdrill 微型钻机
microdynamic 微动力的
microdynamometer 精测测力计；微功率计
microearthquake 微地震；微震
microearthquake analysis 微震分析
microearthquake array 微震台阵
microecology 微生态学
microeffect 微观效应；显微效应
microelectrical borehole wall imaging 微电极井壁成像
microelectrolytic determination 微量电解测定
micro-emission spectroscopy 微发射光谱学
microenvironment assessment 微环境评价
microequivalent 微当量
micro-erosion meter 微蚀侵蚀仪
microeutaxitic 微条纹斑状
microevolution 微演变；微观变化
micro-EXAFS and XANES 微延伸X射线吸收精细结构及X射线吸收近边结构
microexamination 微观检验；显微检验
microfabric 显微组构
microfacies 微相
microfault 微断层
microfaulting 微断层活动
microfeature 微地形特征
microfelsite 微霏细岩
microfelsitic 微霏细岩状
microfilm analyzer 显微照相分析仪
microfilm map 缩微地图
microfilm recorder 缩微胶片摄影机
microfilm-camera 缩微摄像机
microfilmer 缩微摄影机
microfilming 微缩照相法
microfine cement 超细水泥
micro-fisheye 微小聚合物团块
microfissure 微裂缝；微裂隙
microfissuring 微裂隙
micro-flaw 微裂纹
microflexing 微挠曲作用
micro-flotation cell 微量浮选试验机
microfluid 微观流体
microfluidal texture 微链结构
microfluorometer 测微荧光计
microfluorometric method 显微荧光测定法
microfluorometry 显微荧光测定法
microfolding 微褶皱作用
microfoliation 微叶理
microfossil 微化石
microfossiliferous limestone 微化石灰岩
microfossils 微化石
micro-frac 微压裂

micro-frac test 微型压裂实验
microfracture 微裂缝；显微裂隙
microfracture close down 微裂缝闭合
microfracture fabric 微裂缝组构
micro-fracture mechanics 微观断裂力学
microfracturing 微断裂；显微破裂作用
microfragmental 微碎屑的
microfragmental coal 微组分煤
microgabbro 微辉长岩
micro-gas analysis apparatus 微量气体分析器
micro-geodesy 微观大地测量学
microgeology 显微地质学；微观地质学
microgeometry 微观几何形态
microgeomorphology 微地形学；微地貌学
microgeophysics 微地球物理学
micrograined 微粒的
microgram 微克
microgranite 微花岗岩
microgranitic 微花岗状；微晶粒状
microgranitoid 微花岗状的
microgranodiorite 微花岗闪长岩
microgranular 微晶粒状的
microgranular texture 微粒结构
micrograph 显微照片；显微图
micrographic 微文象的；显微照相的
micrographic intergrowth 微文象共生
micrographic texture 微文象结构；显微文象结构
micrography 显微摄影；缩微摄影
microgravimetry 微重力测学；微重力测量
microgravity 微重力
microgravity exploration 微重力勘探
microgravity method 微重力方法
Microgravity survey 微重力测量
microhardness 显微硬度
microhardness of a rock 岩石显微硬度；岩石微观硬度
microhardness scale 显微硬度计
microhardness tester 显微硬度计
microhardness testing 显微硬度试验；微观硬度测量
microheterogeneity 微观不均一性
microholograph 微型全息照相
micro-image analysis 显微图像分析
micro-indentation hardness 显微压入硬度；微刻痕硬度
microinhomogeneity 微观不均匀性
micro-interferometer 显微干涉仪
microinterrupt 微中断
micro-IR 显微红外
microite 微惰性煤
microkarst 微喀斯特
microlaminae 微纹层
microlaminated bed 微层状岩层
microlamination 微纹理；微纹层
microlamp 显微镜用灯
microlaterolog 微侧向测井
microlayer 微层；细层
microlayered 微层理的
microlayered sandstone 微层理砂岩
microlayering 显微层理
microlepidoblastic 微鳞变晶的
microlevel gauge 精密水位计
microlite 微晶；细晶石

microliter	微升
microlith	显微煤岩；微晶；细石器
microlithic	微晶的
microlithofacies	显微岩相；微岩相
microlithology	显微岩性学
microlithon	微劈石；微解理片
microlithotype	显微煤岩类型；煤的显微岩性类型
microlithotype of coal	煤岩显微类型
microlitic	微晶的；微晶质的
microlitic structure	微晶结构
microlitic texture	微晶结构
microlock	卫星遥测系统；微波锁定
microlog	微电阻率测井；微电极测井
microlog continuous dipmeter	连续式微电极地层倾角测量仪
micrologging	微电极测井
microlog-proximity log	微电极-邻近侧向测井
micromachining	显微机械加工
micromagnet	微磁体
micromagnetic survey	微磁测量
micromagnetometer	微磁力仪；显微磁强计
micromanipulation	显微操纵；精密控制
micromanipulator	精密控制器；显微检验设备；小型机械手
micromanometer	微压计；精测流体压力计
micromap	缩微地图
micromatrix	微矩阵
micromechanics	微观力学
micromechanics base continuum model	微观力学连续介质模型
micromechanism	微观机构
micromerite	微晶粒状岩
micromeritic	微晶粒状
micromeritics	微晶学；微尘学；微粒学
micrometeorite	微陨石；陨尘
micrometeorite crater	微陨击坑
micrometeoroid	微陨星体
micro-meteorological system	微气象系
micrometeorology	微气象学
micrometer	千分尺；测微计
micrometer caliper gauge	千分测径规
micrometer depth gauge	深度千分尺
micrometer drum	测微鼓
micrometer eyepiece	测微目镜
micrometer gauge	测微规；千分尺
micrometer head	测微器转头；测微仪转盘
micrometer inside caliper	内径千分尺
micrometer instrument	微米经纬仪；光学经纬仪
micrometer jaw	千分尺量脚
micrometer microscope	微规显微镜；微米显微镜；测微显微镜
micrometer reading	测微器读数
micrometer run	测微器行差
micrometer screen plate	测微计螺旋
micrometer screw	测微螺丝
micrometer stop	千分尺定位器
micrometer theodolite	测微经纬仪
micrometering	微测
micrometering valve	微调阀
micrometers caliper	千分卡尺
micromethod	微量测定法；微量法
micrometre	微米
micrometric analysis	微量分析
micrometric depth gauge	深度千分尺
micrometric method of rock analysis	岩石微量分析法
micrometry	测微法
micromicro earthquake	超微地震
micromineralogy	显微矿物学
microminiature	超小型
microminiaturization	微型化
microminiaturize	微型化
micromodule	微型组件
micromonzonite	微二长岩
micromorphology	微形结构；微观形态学
micromotion	微动
micromotor	微电机
micronic dust particle	微尘粒；显微尘粒
micronic filter	微孔过滤器
micronite	微粒体
micronized salt system	微粉化盐完井液
micro-normal arrangement	微电位电极系
micro-normal curve	微电位曲线
micron-sized	微米级的
micro-ocean	微大洋
micro-oil	微量初生油
microophitic	微辉绿岩状的
microoptic level	光学测微水准器
micro-optics	微光学
microorganism	微生物
microoscillograph	显微示波器
microosmometer	微渗透压强计
micropedology	微土壤学
micropegmatic intergrowth	微文象共生
micropegmatite	微文象岩
micropelletoid	微球粒的
microperthite	微纹长石
microperthitite	微纹长岩
micropetrography	显微岩石学
micropetrological unit	显微组分
microphanerocrystalline	微显晶质的
microphenocryst	微斑晶
microphenomenon	微观现象
microphotogrammetry	显微摄影测量
microphotograph	显微照相；缩微照片
microphotographic apparatus	显微照相装置
microphotographic camera	显微照相机
microphotography	显微照相术；缩微摄影
microphotometer	显微光度计；测微光度计
microphyric	微斑状
microphysics	微观物理学
microphysiography	显微岩像学；显微地文学
microphytal deposit	微植物沉积
micro-pile	小型桩；微型桩
micro-pipette	微量吸移管
microplankton	微浮游生物；小型浮游生物
microplate	微板块；小板块
microplissement	微型褶皱
microplot	微缩测量图
micropluviometer	微雨量计
micropoikilitic	微嵌晶状的

micropoikilitic texture	微嵌晶结构
micropolarimeter	测微偏振计
micropolariscope	偏光显微镜；测微偏振镜
micropollutant	微量污染物；痕量污染物
micropopulation	微生物群
micropore	微孔隙
micropore space	微孔隙空间
microporosity	微观孔隙率
microporous	微孔隙的
microporphyritic	微斑状
micropressure	微压
microprobe	显微探针；电子探针
microprobe trace	电子探针扫描迹
microprocessor	微处理机
microprocessor controlled	微处理机控制的
microprocessor development system	微处理机开发系统
microprofile caliper log instrument	内径测微规
microprogram	微程序
micro-programmed computer	微程序控制计算机
microprogramming	微程序设计
microprojection apparatus	显微投影仪
microprojector	显微投影器；显微映像器
micropsammite	细粒砂岩
microptic theodolite	测微经纬仪
micropulsation	微脉动
micropulser	矩形脉冲发生器；微脉冲发生器
micropyrometer	精测高温计；微小发光体测温计
microquake	微地震；微震
microquantity	微量
microquartz	微石英
microradiograph	X射线显微照片；X光照相检验
microradiography	X射线显微照相术；X射线照相检验
microradiometer	显微辐射计
micro-Raman spectroscopy	显微拉曼光谱仪
microray	微射线；微波
microreef	微形礁
microrelief	微地形起伏；微地形；微地貌
microreservoir	小储集层
microresistivity	微电阻率
microresistivity anomaly	微电阻率异常
micro-resistivity curve	微电阻率曲线
microresistivity device	微电阻率测井仪
microresistivity log	微电极测井；微电阻率测井
microresistivity logging	微电极测井；微电阻率测井
microrig	微形钻机
microscale	微尺度
microscan	显微扫描
microscanning	显微扫描；细光栅扫描
microschist	微片石
microschist method	微片石法
microsclerometry	显微硬度测定
microscope	显微镜
microscope analysis	显微镜分析
microscope carrier	显微镜载物台
microscope condenser	显微镜聚光器
microscope field	显微镜的视域
microscope objective	显微镜物镜
microscope study	显微研究
microscopic	显微的
microscopic analysis	显微分析；显微镜分析
microscopic anisotropy	微观各向异性；显微异向性
microscopic behavior	微观动态
microscopic capillary	微毛细管
microscopic capture cross section	微观俘获截面
microscopic damage	微观损伤
microscopic damage mechanics	细观损伤力学
microscopic deformation	微观变形
microscopic displacement	微观位移
microscopic displacement efficiency	微观驱替效率
microscopic epicenter	微观震中
microscopic examination	显微镜检验；微观检验；金相试验
microscopic fabric	微细组构
microscopic feature	微观特征
microscopic flood front	微观注水前缘
microscopic geometry	微观几何形态
microscopic heterogeneity	微观非均质性
microscopic identification	显微镜鉴定
microscopic instability	微观不稳定性
microscopic level	微观水平
microscopic magnification	显微镜放大
microscopic mechanism	微观机理
microscopic method	显微镜法
microscopic model	微模型
microscopic objective	显微镜物镜
microscopic observation	微观观察
microscopic oil displacement efficiency	微观驱油效率
microscopic polarity spectrometer	显微偏振光谱仪
microscopic pore structure parameter	微观孔隙结构参数
microscopic porosity	显微孔隙率
microscopic pre-earthquake anomaly	地震微观异常
microscopic projector	显微投影仪
microscopic pyrite	显微黄铁矿；细微黄铁矿
microscopic roughness	微观粗糙度
microscopic sample	显微样品
microscopic seam analysis	显微煤层分析
microscopic section	显微切片；显微薄片
microscopic segregation	显微偏析
microscopic sizing	显微镜粒度法；显微分级；测微法
microscopic sizing analysis	显微镜微粒尺寸分析
microscopic strain	微观应变
microscopic stress	微观应力
microscopic structure	显微组织；微观结构；微型构造
microscopic surface texture	微观表面结构
microscopic topology	微观拓扑结构
microscopical earthquake precursor	微观地震前兆
microscopy	显微学；显微技术；显微镜检查法
microscratch	微痕
microsecond	微秒
microsecond photography	微秒照相术
microsection	显微薄片；磨片
microseepage	微油气苗
micro-segregation	显微偏析
microseism	微震；小震
microseism count	微震计数
microseism rate per minute	每分钟的微震率
microseism spectrum	微震谱
microseism storm	脉动暴
microseismic	微震的
microseismic activity	微地震活动性；微震活动

microseismic data	微震资料；微震数据
microseismic data acquisition	微震数据采集
microseismic effect	微观地震效应
microseismic event	微震事件
microseismic event monitoring	微震监测
microseismic field	微地震场
microseismic forecasting	微震预测
microseismic hazard prediction	微震风险预测
microseismic measurement	微震测量
microseismic method	微震测量法
microseismic monitoring	微震监测
microseismic monitoring system	微震监测系统
microseismic monitoring technique	微震监测技术
microseismic monitors	微地震监控器
microseismic movement	微震运动
microseismic noise	微震噪声
microseismic noise detection	微震噪声探测
microseismic observation	微震观测
microseismic peak	微震峰
microseismic region	微震区
microseismic section	微地震剖面
microseismic source location	微震源定位
microseismic storm	微地震扰动；脉动暴
microseismic zoning	微震区划
microseismicity	微震；微震活动性
microseismogram log	微地震测井
microseismograms	微地震记录
microseismograph	微震仪
microseismology	微地震学；地脉动学
microseismometer	微震计；地脉动计
microsequence	微层序
microshrinkage	显微缩孔
microsilica	微粒硅
micro-silica powder	细硅粉
microsize	微小尺寸；自动定寸
microslide	载玻片
microslip	微滑；微观滑移
microsolifluction	显微泥流
microsolubility	微溶性
microsome	微粒体
microsommite	微碱钙霞石；碱钾钙霞石
microsonic activity	微声活动性
microspar	微亮晶
microsparite	微亮晶灰岩
microsparry	微亮晶的
microsparry calcite	微亮晶方解石
microspectrofluorimeter	显微荧光分光计
microspectrofluorometry	微量荧光光度法
microspectrometer	显微分光计
microspectrophotometer	显微分光光度计
microspectrophotometry	显微分光光度法
microspectroscope	显微分光镜
microspectroscopy	显微光谱学
microsphere	微球体
microspherically focused conductivity	微球形聚焦电导率
microspherically focused log	微球形聚焦测井
microspherically focused tool	微球形聚焦测井仪
microspherulitic	微球粒状的
microspherulitic texture	微球粒结构
micro-spontaneous potential caliper	微自然电位井径仪
microspread	小排列
microstone	细粒度油石
microstrain	微应变
microstraining	微渗滤；微量应变
microstrata	微层
microstratification	微层理
microstratigraphy	微观地层学
microstratum	微层
microstrength	微强度
microstress	微应力
microstriation	微小擦痕
microstructural	显微结构的
microstructural damage	显微结构损害
microstructural evolution	微结构演变
microstructure	微结构；显微构造
microstructure analysis	显微结构分析
microstructure of coal	煤的显微结构
microstructure of soil	土的微结构；土的微观结构；土的显微结构
microstructures	显微构造学
microstylolite	微缝合线
microsyenite	微正长岩
microtaxitic	微斑杂状的
microtectonics	微观构造学
microtektite	微玻璃陨石；显微熔融石
microtest	精密试验
microtexture	显微结构；微观结构
microthermal	低温气候的
microthermal climate	低温气候
microthermometer	精密温度计；显微测温计
microtinite	透斜长石；浅粒二长岩
micro-tool	微测井下井仪
microtopography	微地形学；微地貌；微地貌学
microtrap	微圈闭
microtremor	微振动；脉动
microtremor behaviour	脉动性状
microtremor excited vibration	脉动激振
microtremor measurement	脉动测量
microtremor observation	微震观测
microtremor process	脉动过程
microtremor signal	脉动信号
microtremor spectrum	脉动谱
microtrilateration	小三边测量
micro-tsunami station	微海啸观测站
micro-tunnel boring machine	微型隧道掘进机
micro-tunneller	小型隧道掘进机
microtunnelling	微型隧道掘进
microturbulence	微弱紊动；微观湍流
microvarve clay	极细纹泥
microvesicular	微多孔状的
microvibrograph	微震仪
microviscosimeter	微量黏度计
microvision	微波观察仪
microvisual model	微观模型
microvoid	微空隙
microvolt	微伏
microvoltmeter	微伏计
micro-vuggy	微孔的
microvugular carbonate	微孔洞碳酸盐岩
microwatt	微瓦

microwave 微波
microwave acoustics 微波声学
microwave atmospheric sounding radiometer 微波大气探测辐射计
microwave attenuation 微波衰减
microwave band 微波频带
microwave beacon 微波信号柱
microwave communication 微波通讯
microwave communication equipment 微波通信机
microwave communication tower 微波通信塔
microwave communication vehicle 微波通信车
microwave computer 微波计算机
microwave control system 微波遥控系统
microwave distance measurement 微波测距
microwave distance measuring instrument 微波测距仪
microwave drill 微波钻井
microwave drying 微波干燥
microwave emission spectrum 微波发射光谱
microwave emissivity 微波发射率
microwave filter 微波滤波器
microwave frequency 微波频率
microwave heating 微波加热
microwave hologram 微波全息图
microwave hologram radar 微波全息雷达
microwave holography 微波全息术；微波全息摄影
microwave hops 微波射程
microwave image 微波图像
microwave imagery 微波成像
microwave interferometry 微波干涉量度学
microwave lens 微波透镜
microwave low noise amplifier 微波低噪声放大器
microwave moisture apparatus 微波含水量仪；微波水分仪
microwave moisture meter 微波水分分析仪
microwave observation 微波观测
microwave oven drying method 微波炉烘干法
microwave phase meter 微波相位计
microwave phase shifter 微波相移器
microwave power 微波动力
microwave power amplifier 微波功率放大器
microwave radiation 微波辐射
microwave radiometer 微波辐射计
microwave radiometric imagery 微波辐射测量成像
microwave radiometry 微波辐射测量学；微波辐射测量术
microwave ranger 微波测距仪
microwave ranging measurement 微波测距
microwave receiver 微波接收机
microwave region 微波波段
microwave relay station 微波中断站
microwave remote sensing 微波遥感
microwave remote sensor 微波遥感器
microwave satellite 微波卫星
microwave scanning radiometer 微波扫描辐射计
microwave scanning radiometry 微波扫描辐射测量
microwave shield door 微波屏蔽门
microwave spectrometer 微波分光仪
microwave spectroscopy 微波光谱学
microwave spectrum 微波波谱
microwave station 微波站
microwave terminal station 微波终端站
microwave tower 微波塔
microwave transmission 微波传输
microwave transmission tower 微波塔
microwave-navigation 微波导航
micro-waves 微波
micro-X-ray diffraction 微X射线衍射
microzoal deposit 微动物沉积
microzonation 小区划；地震小区域划分
microzonation parameter 地震小区划参数
microzoning 小区划；细分带
microzoning effect 地震小区划效应
mictite 混染岩
mid Arctic ridge 北冰洋中脊
mid Atlantic ridge 大西洋中脊
mid Atlantic rift valley 中大西洋裂谷；大西洋中央裂谷
mid Indian ridge 印度洋中脊；印度洋中央海岭
mid Indian rift system 印度洋中央裂谷系
mid ocean rift 大洋中央裂谷
mid ocean rise 大洋中隆
mid oceanic ridge 洋中脊
mid oceanic ridge basalt 中央海脊玄武岩
mid oceanic rift 大洋中谷
mid of year method 年中法
mid plate 中板块
mid plate volcano 板块内火山
mid point 中点；平均点
mid shaft 井筒中部
mid-Adriatic ridge 中亚得里亚海脊
midalkalite 霞石正长岩
mid-Arctic ridge 北冰洋中脊
mid-Atlantic ridge 大西洋中脊
mid-Atlantic rift valley 中大西洋裂谷
Midband 中频
mid-bay bar 海湾中间的沙洲
midchannel 河流中部；中央航道
mid-channel bar 河心沙洲
midchannel buoy 航道中央浮标
mid-circle 中点圆
mid-continent 中陆地区
mid-Cretaceous pulsation 白垩纪中期脉动
mid-depth 中深
mid-diameter 中央直径
middle 中央；中间的；中等的；中项
middle atmosphere 中层大气
middle band 夹矸层；夹石层
middle bench 中台阶；中梯段；中阶段
middle Cambrian 中寒武世
middle circle 中点圆
middle cutting 中部掏槽；掏腰槽
middle Devonian 中泥盆世
middle diagenesis 中期成岩作用
middle diameter 中径
middle epoch 中世
middle fan 海底扇中扇
middle flask 中间砂箱
middle girder 中主梁
middle ground 潮间浅滩
middle hydrothermal deposit 中深热液矿床
middle infrared 中红外

middle intertidal zone 中潮间带
middle Jurassic 中侏罗世
middle land peat 中位泥炭
middle landslide 中层滑坡；中型滑坡
middle latitude 中纬度
middle layer 中间层
middle magnetosphere 中磁层
middle man 岩层间的夹层；两个煤层间的夹石层
middle mineralized groundwater 地下咸水
middle Miocene 中中新世
middle Permian 中二叠世
middle Proterozoic 中元古界
middle Proterozoic subera 中元古代
middle rank coal 中等碳化度煤；中等品级煤
middle reaches 河道中游
middle seam 中间煤层；中间矿层
middle series 中统
middle span 中跨径
middle square method 平方取中法
middle surface 中层面
middle to late Triassic 中晚三叠世
middle Triassic 中三叠世
middle water 层间水
middle-depth earthquake 中等深度地震
middle-fan channel 中扇水道
middleman 夹石层
middle-upper Permian 中上二叠统
middling 中等的；中级品；中等
middling product 中矿
middlings 中煤
middlings crusher 中煤破碎机
middlings crushing 中煤破碎
middlings storage bin 中煤贮仓
middlings yield 中煤出量
midfan 中部洪积扇；中间扇形地；冲积扇中段
midfan mesa 冲积扇侵蚀残丘
midfeather 中间支护；中间细矿脉
mid-hard rock 次坚石
mid-Indian ridge 印度洋中脊
mid-Indian rift system 印度洋中央裂谷系
mid-infrared spectral 中红外光谱的
midland 内地；内陆
mid-latitude 中纬度；等比例纬线
mid-layer correction 中间层改正
midline 中线
midlittoral 潮间的
midocean 海洋中央
mid-ocean canyon 大洋中央峡谷；深海峡谷
mid-ocean plateau 洋中海台
mid-ocean ridge 大洋中脊
mid-ocean ridge basalt 洋中脊玄武岩
mid-ocean ridge crest 洋中脊峰线
midocean ridge processes 大洋中脊过程
mid-ocean rift 中央裂谷
mid-ocean valley 中央裂谷
midoceanic 大洋中央的
mid-oceanic carbonates 大洋中脊碳酸盐
mid-oceanic ridge 大洋中脊
mid-oceanic ridge basalt 大洋中脊拉斑玄武岩；深海拉斑玄武岩
mid-oceanic ridge belt 大洋中脊地带
mid-oceanic ridge processes 洋中脊的过程
mid-oceanic ridge tholeiite 海岭拉斑玄武岩；洋中脊拉斑玄武岩
midoceanic rift 洋中裂谷
midoceanic rise crest 洋中隆起顶部
mid-Pacific rise 太平洋中隆
mid-plate 中板块
mid-plate continental margin 板块中间大陆边缘
mid-plate melting anomaly 板内熔融异常
mid-point anchor in tunnel with anchor to the wall 固定于隧道壁的中心锚结
mid-point anchoring pole 中心锚结柱
mid-point distribution curve 中间分配曲线
midpoint of pay zone 油层中部深度
mid-point pump station 中间泵站
mid-point scattergram 中点数据离散图
mid-Proterozoic 中元古代的；中元古界的
midrange forecast 中期预报
mid-ridge spreading 洋中脊扩张型
midsection 中间截面
midshaft 井筒中部；中间水平
midshaft plug 井筒中间隔墙
mid-side node 边节点；边中节点
mid-side shape function 中侧形函数
midspan 最大横向收缩率；中跨
mid-span joint 中跨接缝
midspan load 跨中荷载
mid-span moment 跨中弯矩；跨中力矩
midst 正中；中间
midstream 中流；河道中游
midstream depth 中泓水深
mid-stream operation 中流作业
mid-structure 构造中部
midtrough 地堑中部
mid-value 中值
midwall 井筒木隔壁；井筒隔间木墙
midwall building 砌筑中间岩石带
midwall column 墙壁柱
midwinter 细土粒
miedziankite 锌黝铜矿
mienite 富玻流纹岩
miersite 黄碘银矿
mighty torrent process 洪流地质作用
migma 混合岩浆
migma pluton 混浆深成岩
migmatic pegmatite 混合伟晶岩
migmatite 混合岩
migmatite zone 混合岩带
migmatitic complex 混合杂岩
migmatitic gneiss 混合片麻岩
migmatitic granite 混合花岗岩
migmatitic ore deposit 混合岩化矿床
migmatization 混合岩化作用
migmatized magma 混合岩浆
migmatized metallogenic process 混合岩化成矿作用
migmatized metamorphic rock 混合岩化变质岩
migmatoblast 混合岩化变晶
migrated depth section 偏移深度剖面
migrated line 偏移测线

migrated seismicity	地震活动偏移
migrated time map	偏移时间图
migrated time section	偏移校正的时间剖面
migrating bar	移动沙洲
migrating contour map	偏移等值线图
migrating dune	移动沙丘
migrating oil stringer	运移油流
migrating wave	迁移波
migrating wave configuration	迁移波状结构
migrating wave seismic facies	迁移波状地震相
migrating weather system	迁移气候系统
migration	偏移；迁移
migration agent	运移动力
migration algorithm	偏移算法
migration aperture	偏移孔径
migration area	运移范围
migration before stack	叠前偏移
migration carrier	运移载体
migration circle	运移范围
migration coefficient of elements	元素迁移系数
migration conduit	运移通道
migration distance	运移距离
migration dynamics	运移动力
migration filling pathway	烃类运移途径
migration imaging	偏移成像
migration mechanism	运移机理
migration noise	偏移噪声
migration of continents	大陆漂移；地壳迁移
migration of divide	分水岭移动
migration of dunes	沙丘移动
migration of earthquake	地震迁移
migration of elements	元素的迁移
migration of epicenter	震中迁移；地震迁移
migration of gas	迁移瓦斯；运移瓦斯；采矿
migration of geosyncline	地槽迁移
migration of ions	离子迁移
migration of oil	石油运移
migration of petroleum	油气运移石油迁移
migration of stream channel	河道迁移；河流改道
migration of the plate boundaries	板块边界的迁移
migration of valley	河谷迁移；河道迁移
migration operator	偏移算子
migration parameter	偏移参数
migration passage	运移通道
migration path	运移途径
migration pathway	运移通道
migration pattern	运移形式；运移方式
migration phases	运移相态
migration potential	电泳电位
migration program	偏移程序
migration rate	移动速度；迁移速度
migration ratio	运移比
migration reconstruction	偏移重建
migration resistance	抗渗移性
migration route	迁移路线
migration stack	偏移叠加
migration stain	渗移斑
migration structure	迁移构造；同质假象
migration velocity	偏移速度；运移速度；驱进速度
migration velocity analysis	偏移速度分析
migration velocity determination	偏移速度确定
migrational mode	迁移模式
migratory	移动的
migratory concentration	运移富集
migratory direction	运移方向
migratory dune	移动沙丘
migratory fill	移动式充填
migratory oil	运移的石油
migratory pattern	迁移模式
migratory siliceous fine	流动硅质细砂
miharaite	英苏玄武岩
mihr	英里小时
mijakite	锰辉玄武岩
mikenite	斜长白榴岩
mikheevite	水钾钙矾
Milankovitch frequency	米兰科维奇频率
Milankovitch theory	米兰科维奇理论
milarite	整柱石
milcomsat	军用通信卫星
mild acid	弱酸
mild base	弱碱
mild carbon steel	低碳钢
mild channel	平缓河道
mild clay	亚黏土；软黏土；含游离硅石的黏土
mild compression	适度的压缩
mild crooked hole	轻度弯曲井眼；轻度弯曲的井身
mild slope	平缓边坡；缓坡
mild steel	软钢；低拉力钢；低碳钢
mild steel bar	低碳钢钢筋
mildly brackish water	微咸水；弱矿化水
mile	英里
mileage	英里里数；英里里程
milestone	里程碑
milestoning	测量打桩
miliolite	糜棱岩；风成细粒灰岩
military base at sea	海上军事基地
military base survey	军事基地测量
military bridge survey	军用桥梁测量
military chart	军用海图
military computer	军用计算机
military coordinate system	军用坐标系
military engineering	军事工程
military engineering survey	军事工程测量
military explosive	军用炸药
military geodesy and cartography	军事测绘
military geographic information	军事地理信息
military geography	军事地理学
military geology	军事地质学
military grid	军事坐标网格
military industrial target	军事工业目标
military interpretation	军事判读
military land	军事用地
military map	军用地图
military marine geodesy and cartography	军事海洋测绘学
military nautical meteorology	军事航海气象学
military navigation	军事航海
military ocean technology	军事海洋技术
military oceanology	军事海洋学
military site	军事用地
military surveying and mapping	军事测绘

military target 军事目标
military test area at sea 海上军事试验场
military topography 军事地形学
military tunnel 军用隧道
military tunnel survey 军用坑道测量
military use 军事用途
milk of lime 石灰浆;石灰乳
milk opal 乳蛋白石
milk quartz 乳石英
milky quartz 油脂状石英;乳石英
mill 磨机;选矿厂
mill barrel 磨矿机滚筒;磨碎机筒
mill chats 选矿厂废石
mill coal 非炼焦煤;工厂用煤
mill discharge 磨机排料
mill feed 磨机装料量;磨机给料
mill feeder 磨机给料器
mill fibre 普通石棉纤维
mill heads 入厂原矿
mill hole mining 漏斗采矿
mill liner 磨机衬里
mill load 磨机负载
mill method 漏斗采矿法
mill method along the dip 倾斜漏斗采煤法
mill ore 可选级矿石;有精选价值的矿石;需选矿石
mill overload 磨机过载
mill pebble 磨砾
mill raise 放矿漏斗天井
mill sampler 选矿厂取样
mill site 选矿厂厂址
mill tailings 矿山尾矿;选矿尾矿
mill tooth bit 磨牙钻头
mill water 选矿用水
mill-cut casing 铣割套管
mill-cut slot 铣切割缝
milled peat 铲采泥煤
milled tooth bit 铣齿钻头
milled twist drill 麻花钻
milled window in casing 套管磨出孔
Miller index 米勒结晶标志指数
Miller indices 米勒指数;结晶指数
Miller orientation 米勒定向
Miller projection 米勒投影
Miller symbol 米氏符号
Miller-Bravais indices 米勒布拉维指数
millerite 针镍矿;针硫镍矿
Miller's indices 密勒指数
Miller's notation 密勒记法
millet-seed sand 小米粒砂
millhole 放矿口;露天放矿漏斗
milliammeter 毫安计
milliamp 毫安
milliampere 毫安;毫安培
milliamperemeter 毫安表;毫安计
milliard 十亿
milliatom 毫克原子
millibar 毫巴
millibar-barometer 毫巴气压表
millicurie 毫居里
millidarcy 毫达西

milli-degree 千分之一度
milliequivalent 毫克当量
millier 公吨
millifarad 毫法拉
milligal 毫伽
milligram 毫克
milligram equivalent 毫克当量
milligram radium equivalent 毫克镭当量
milligramequivalent 毫克当量
millihenry 毫亨
millihertz 毫赫
milliliter 毫升
millimeter 毫米
millimeter wave 毫米波
millimeter-wave drilling 毫米波钻
millimetre-sized 毫米级的
millimetric 毫米的
millimho 毫姆欧
millimicro 毫微
millimicro-farad 毫微法
millimicron 毫微米
millimicrosecond 毫微秒
millimolar 毫摩尔的;毫克分子的
milling 选矿;研磨.磨碎;放矿漏斗开采法
milling bit 铣磨钻头
milling cutter 碎煤刀;铣刀;铣削工具
milling drum 破碎滚筒
milling dual operation 选矿作业
milling grade 值得精选的矿石
milling head 掘削头
milling hole 溜眼;溜道;溜井
milling jack 螺旋千斤顶
milling machine 铣床;磨床
milling method 选矿法;磨矿法;露天漏斗采矿法
milling operation 选矿作业;磨铣作业
milling ore 可选级矿石;二级矿石
milling pit 磨矿机;研磨井
milling slime 选矿厂矿泥
milling teeth 铣齿
milling tool 铣具
milling type of cutter 铣型切削钻机
milling type tunnel boring machine 铣式隧道钻巷机
milling up 磨碎
milling-grade ore 可选级矿石;有精选价值的矿石;需选矿石
milling-type continuous miner 研磨型连续采煤机
milling-type cutter 铣型切轧机
millisecond 毫秒
millisecond blasting 毫秒爆破;微差爆破
millisecond blasting cap 毫秒迟发电雷管
millisecond connector 毫秒继爆管;延期继爆管
millisecond delay 毫秒迟发
millisecond delay blasting 毫秒延期爆破
millisecond delay blasting 毫秒延时爆破
millisecond delay cap 毫秒延发电管
millisecond delay detonation 毫秒爆破
millisecond delay detonator 毫秒延时电雷管
millisecond detonator 毫秒迟发雷管
millisecond electric blasting cap 毫秒电爆雷管
millisecond routed 迟发毫秒爆破炮眼组

millisite 水磷铝碱石
millivolt 毫伏
millivoltpneumatic transducer 毫伏-气压变送器
milliwatt 毫瓦
millstone 磨石；燧石
millstone grit 磨石砂砾
Milne-Shaw seismograph 米尔恩-肖地震仪
miloschite 铬铝英石
Milun Conglomerate 米仑砾岩
Mimas 土卫一
mimesite 粗玄岩；粒玄岩
mimetesite 砷铅矿
mimetic crystal 拟晶
mimetic crystallization 拟晶结晶
mimetic tectonite 模拟构造岩；似构造岩；重结晶构造岩
mimetite 砷铅矿
mimophyre 似斑岩
mimosite 钛铁粒玄岩
minable 可开采的
minable boundary 可采边界
minable coal seams 可采煤层
minable deposit 可采矿床
minable limit 可采极限
minable removal increment 可采出的增量
minable reserves 可采储量
minable seam 可采煤层；可采矿层
minable seam sample 可采层试样
minable stratum 可采矿层
minable thickness 可开采厚度
minable width 可采宽度
Minamata disease 水俣病
minasragrite 钒矾
min-collapse pressure 最小挤毁压力；最小破坏压力
Mindel glaciation 民德冰期
Mindel-Riss interglacial stage 民德-里斯间冰期
mine 矿；矿山；矿井；开采；采矿
mine access 矿山通道
mine accident 矿山事故
mine accident prediction 矿山事故预测
mine adit 平硐
mine age 矿山寿命；矿山可采期
mine air 矿内空气；井下空气
mine air analysis 井下空气分析；井下气体分析
mine air analyzer 矿内空气分析器
mine air conditioning 矿井空气调节
mine air pollution control 矿山大气污染防治
mine air preheating 预热矿井空气
mine air refrigeration 矿井空气冷却
mine analyst 采矿专家；安全技术工程师；采矿工程师
mine annual output 矿山年产量
mine approved apparatus 矿用电气设备
mine automation 矿山自动化
mine bank 采矿台阶
mine block 采区
mine block water 矿井截流方法
mine cable 矿用电缆
mine cage 矿用罐笼
mine capacity 矿井生产能力；矿山储量
mine capital construction stage 矿山基建阶段

mine captain 矿山主任；采矿技师
mine car 矿车
mine car handling equipment 矿车运行设备；矿车处理设备
mine car locking device 矿车固锁装置
mine car runaway in inclined shaft 斜井跑车
mine characteristic and development principle 矿井特点及开发原则
mine climate 矿井气候
mine climatic condition 矿井气候条件
mine closing 闭坑；矿山关闭
mine coal 原煤
mine collapse 矿山塌陷
mine collapse disaster 采矿塌陷灾害
mine compound 矿场
mine concession 矿区；矿田
mine conditions 矿山条件
mine construction 矿井建设；矿山建设
mine construction design 矿井施工设计
mine construction volume 矿建工程量
mine contaminant 开采污染
mine conveying 矿山运输
mine dam 矿井防水墙
mine depletion 矿的储量递减
mine depression earthquake 矿山陷落地震
mine design 矿山设计
mine design technique 矿山设计工艺
mine developing 矿井开拓工程
mine development 矿山开发
mine development design 矿井开拓设计
mine dewatering 矿山排水
mine dial 矿用罗盘
mine district 采区采煤区；采矿区；采矿区域；采矿地带
mine drainage 矿山排水；涌水处理
mine drainage equipment 矿山排水设备
mine drainage gas 矿井气
mine drainage pollution 矿山排水污染
mine drainage system 矿山排水系统
mine dump 矿山废石堆；矿山矸石堆
mine dust 矿尘
mine dust monitor 矿尘监测器
mine dustiness 矿井含尘性；矿井含尘量
mine dynamic disaster 矿山动力灾害
mine earthquake 矿震
mine engineering 矿山工程
mine engineering geology 矿山工程地质学
mine engineering volume 矿山工程量
mine enterprise 矿山企业
mine environmental monitoring 煤矿环境监测
mine environmental engineering 矿山环境工程
mine equivalent orifice 矿井等积孔
mine evaluation 矿山评价
mine expert 采矿专家；采矿工程师
mine exploitation 矿井开发；矿井开拓
mine explosion 矿井爆炸
mine explosive 矿山炸药
mine explosive gas 矿井爆炸性气体
mine face 采矿工作面
mine feasibility study 矿井可行性研究

mine field	井田；矿田；油田；矿区
mine field developing	井田开拓
mine field development	井田开拓
mine fills setting	充填物沉沉
mine fills settlement	充填物沉实
mine fire	矿内火灾；矿火
mine fire accident	井下火灾事故
mine fire door	矿井防火门
mine fire truck	矿用消防车
mine flooding	矿井淹没；矿井涌水
mine flooding accident	矿井涌水事故
mine fracture forecast with satellite	卫星矿山断裂预测
mine gallery	水平巷道；廊道
mine gas	矿井瓦斯；井下气体；沼气；矿山有害气体
mine gas drainage plant	矿井瓦斯排泄装置
mine gel	矿用爆胶
mine geo-environment	矿山地质环境
mine geo-hazards	矿山地质灾害
mine geologic map	矿山地质图
mine geological isotropy of mine product	矿山地质图
mine geological map	矿山地质图
mine geologist	矿井地师；采矿地师；矿井地质工作者
mine goaf	采空区；老塘
mine haul	矿山运输；矿山拖运运输矿物
mine haulage	矿山运输；矿山拖运
mine haulage roadway	矿井运输大巷
mine hazard	矿山事故
mine head frame	井架
mine heat disaster	矿井热害
mine hoist	矿井提升机
mine hoisting	矿井提升
mine hoisting rope	矿井提升钢丝绳
mine hydrogeological investigation	矿区水文地质勘察
mine hydrogeological map	矿床水文地质图
mine hydrogeology	矿山水文地质学
mine hydrology	矿山水文学
mine industrial site	矿山工业场地
mine inflow	矿井涌水量
mine inflow growth	矿井涌水量
mine inspection office	矿山监察局
mine inspector	矿山监察员
mine interferometer	煤矿干涉仪
mine internal transport	矿山内部运输
mine inundation	矿山水灾；矿井淹没
mine jitney	轻便小型送人或材料车
mine jumbo	巨型矿车
mine land reclamation	矿山土地复垦
mine landscape	矿山景观
mine laser geodimeter	矿用激光测距仪
mine layout	矿山布置；矿山设计；井下巷道布置
mine layout and transport	矿山总图运输
mine life	矿山开采期限；矿山寿命
mine light	矿山照明；矿灯
mine lighting	矿山照明
mine lighting design	矿井照明设计
mine loco	矿用机车；矿山机车
mine locomotive	矿用机车
mine management	矿山管理
mine manager	矿长
mine map	采掘工程平面图；矿山图；矿山工程图
mine methane gas	矿井瓦斯
mine model	矿山模型
mine mouth	井口
mine mouth structure	井口建筑物
mine noise	矿山噪声
mine normal inflow	矿井正常涌水量
mine opening	矿山井巷；矿山巷道
mine orifice	等积孔
mine outside transport	矿山外部运输
mine overseer	矿山监工；采矿技术员
mine petering-out	矿山枯竭
mine pillar	矿柱
mine plan	矿区平面图；采掘工程平面图
mine plan engineering data	矿山规划工程数据
mine plant layout	矿山设备布置图
mine pollution	矿物污染
mine portal	平硐口
mine pressure	矿压；矿山压力
mine production curve	矿井生产曲线
mine property	矿山用地；矿山地产
mine pump	矿用水泵；矿用泵
mine pumping	矿山泵水；矿井排水
mine radioactive gas	矿井放射性气体
mine rail	矿用钢轨
mine railway	矿山铁路
mine recovery	矿井恢复
mine refuse	矿场废物
mine refuse impoundment	尾矿坝
mine rescue	矿山救护；采矿救援
mine rescue car	矿山救护车；井下救护车
mine rescue crew	矿山救护队
mine rescue equipment	矿山救护设备
mine rescue station	矿山救护站
mine rescue team	矿山救护队
mine resistance	矿井通风阻力；矿山阻力
mine respiration factor	矿井呼吸系数
mine return	矿山利润
mine rig	矿用钻车
mine roadway	矿山巷道
mine roadway rock	采出废石
mine rock	矿石
mine roof statistical observation	顶板统计观测法
mine roof supports	矿顶板支护
mine run	原煤；原矿
mine run coal	原煤；入厂原煤
mine safety	矿山安全；矿山保安；煤矿安全
mine safety appliance	矿山安全设备
mine safety car	井下救护车；矿山救护车
mine safety management	矿山安全管理
mine scheme	矿山规程
mine section	矿区；采区
mine seismology	矿震学
mine service life	矿井服务年限
mine shaft	矿井；井筒
mine shaft liner	矿井衬砌
mine shafts	矿井
mine site	厂址；场址
mine size	井型；矿山规模
mine skip	矿用箕斗

mine slope 斜井；倾斜巷道
mine smalls 粉矿
mine static head 矿井静压头
mine structure plan 矿山构造平面图
mine superintendent 矿长；坑长
mine surface arrangements 矿井地面布置
mine surface facilities 矿井地面设施
mine surface survey 矿山地面测量
mine surface transport 矿山地面运输
mine survey 矿山测量；矿山工程测量
mine survey isotropy of mine product 矿山测量图
mine survey map 矿图；矿山测量图
mine survey plug 矿山测量标桩
mine surveying 矿山测量
mine tailings 尾矿
mine tailings dam 尾矿坝
mine technical inspection 矿山技术检查；采矿技术监督
mine technique 采矿技术；采矿方法
mine terminology 采矿术语；采矿名词
mine thermal harm 矿井热害
mine timbering 井支架
mine track 矿山轨道
mine tractor 矿用拖拉机；矿用牵引机
mine transportation 矿山运输
mine travelling road 井下人行道
mine tremor 矿山地震
mine truck 矿车
mine tub 矿车
mine tunnelling method 矿山法；钻爆法
mine valuation 矿山评价；矿床评价
mine velocity head 矿井通风速度压头
mine ventilation 矿井通风
mine ventilation character 矿井通风阻力特性
mine ventilation diagram 矿井通风系统图
mine ventilation network 矿井通风网络
mine ventilation structure 矿井通风构筑物
mine ventilation survey 矿井通风检查
Mine ventilation system 矿井通风系统
mine vibration 矿山振动
mine wastewater 矿山废水
mine water 矿坑水；矿井水
mine water disaster 矿井水灾
mine water drainage 矿井水疏干
mine water management 矿井水防治
mine water pollution 矿水污染
mine winch 矿用绞车
mine wind blast 矿内暴风
mine winder 矿用绞车；矿井提升机
mine work schedule 矿山工程进度计划
mine workface 矿震；矿山地震
mine working 矿山巷道；矿井井巷；矿石
mine working longitudinal projection 巷道纵投影图
mine working stereogram 矿山巷道立体图
mine working vertical section 巷道剖面图
mine workings 采矿作业；矿山巷道
mine yard 矿场
mine yard plan 工业场地平面图
mineability 可开采性
mineable oil sand development 油砂矿坑开发法
mine-air refrigeration 矿井空气冷却

minecar clearance simulation 井下储运模拟
mine-concession 特许采矿地区；矿区
mined agricultural land 开采破坏的农用地
mined bed 开采层；已采层
mined bulk 采落的矿石
mined land reclamation 矿山土地复垦
mined lands 矿山开采破坏的土地；矿山采动的土地
mined ore grade 采矿品位
mined out area 井下采空区
mined out space 采空区
mined-out pit 采矿废坑；采空区
mined-out region 采空区
mine-drainage water 矿坑排出水
mine-dust sampler 矿尘取样
min-effective hole diameter 最小有效井眼直径
minefield development 井田开拓
minefield development and mining 井田开拓与开采
minefield structure and geological feature 井田构造及地质特征
mine-fill 矿山充填材料
mine-fills raise 下充填料溜井
mine-fills storing and mixing shaft 充填注砂井
mine-induced seismic events 采矿导致的地震事件；矿震
mine-lamp photometer 矿灯照度计
mine-openings 矿井巷道；矿内开掘的空间
miner 矿工；连续采煤机；联合采矿机
miner graphic method 矿相法
mineragenetic 成因矿物学的
mineragenetic epoch 成矿时代
mineragenetic province 成矿区
minerageny 成因矿物学；矿物成因论
mineragraphic analysis 矿相分析
mineragraphy 矿相学
mineral 矿物；矿物的
mineral aggregate 矿质集料；骨料；石料
mineral analysis 矿物分析
mineral and geochemical composition of reef carbonates 礁碳酸盐矿物和地球化学组成
mineral anomaly 矿异常
mineral assemblage 矿物聚合；矿物组合
mineral association 矿物组合；共生矿物
mineral bearing froth 含矿泡沫
mineral belt 矿化带；含矿带
mineral beneficiation 选矿
mineral black 石墨
mineral board 矿物纤维板
mineral burning oil 煤油
mineral caoutchouc 弹性沥青
mineral carbon 煤；石墨
mineral carbon sequestration 矿物固碳
mineral carbonation 矿物碳化
mineral cement 矿物胶结
mineral cementing agent 矿物胶结剂
mineral charcoal 天然木炭；丝炭；乌煤
mineral chemistry 矿物化学
mineral cleavage 矿物解理
mineral coal 煤；矿物煤
mineral coke 天然焦
mineral component 矿物成分
mineral composition 矿物成分；矿石成分

mineral composition determination of rock 岩石矿物成分分析
mineral composition determined by microscope 显微镜鉴定矿物组成结果
mineral compound 无机化合物
mineral concentration 矿物浓缩；选矿；矿物富集
mineral concentration district 矿化集中区
mineral conservation 矿物资源保护
mineral constituent 矿物组成；矿物成分
mineral content 矿物含量
mineral cotton 矿渣棉；矿棉；石棉
mineral crust 矿物壳
mineral crystal 矿物晶体
mineral density 矿物密度
mineral deposit 矿床
mineral deposit by chemical sedimentation 化学沉积矿床
mineral deposit by mechanical sedimentation 机械沉积矿床
mineral deposit geology 矿床地质学
mineral deposit modeling 矿床模型；矿床建模
mineral deposits geometry 矿体几何学
mineral detector 矿石晶体检波器
mineral diagenesis 成岩
mineral disintegration 矿物分解
mineral dissolution 矿物溶解
mineral dressing ability 矿石可选性
mineral dressing indices 选矿指标
mineral dust 矿物粉尘
mineral dust explosion 矿尘爆炸
mineral dust pollution 矿尘污染
mineral earth oil 石油
mineral economics 矿业经济学；矿产经济学
mineral engineer 选矿工程师
mineral engineering 矿物工程
mineral equilibrium 矿物平衡
mineral estate 矿地
mineral estimation by eye 目估定量
mineral exploitation 矿产开发
mineral exploration 矿产资源勘察；矿产勘探
mineral exploration license 矿产勘察许可证；探矿许可证
mineral exploration rights 探矿权
mineral extraction 矿物开采；采矿
mineral facies 矿物相
mineral fat 地蜡；矿脂
mineral fertilizer 矿质肥料
mineral field 矿田
mineral filler 矿物填料；矿物填充料
mineral fines 矿粉；岩粉
mineral flour 矿石粉
mineral forming 成矿作用
mineral formula 矿物化学式
mineral froth 矿物泡沫；矿化泡沫
mineral fuel 矿物燃料
mineral fuel deposit 燃料矿床
mineral generation 矿物世代
mineral genesis 矿物成因
mineral geothermometer 矿物地质温度计；矿物地温表
mineral grain 矿物颗粒
mineral granulity measure 矿物粒度测量
mineral hardness factor 矿石硬度系数
mineral hardness scale 矿物硬度计
mineral horizontal zoning 矿床水平分带
mineral hydration 矿物水合作用
mineral identification 矿物鉴别；矿物鉴定
mineral inclusions in coal 煤层内含有的矿物
mineral industry 矿业
mineral infrared spectroscopic analysis 矿物红外光谱分析
mineral interpretation of photo 相片矿产分析
mineral isotropy of mine product 矿物分布图；矿产图
mineral jelly 矿物冻，矿脂
mineral kingdom 矿物界
mineral laden bubble 矿化气泡
mineral land 含矿土地
mineral land patent 采矿用地执照
mineral layering 矿物成层作用；矿物浅层作用
mineral line 矿石运输线
mineral lineation 矿物生长线理；矿物线理
mineral logging 矿物测井
mineral manure 矿质肥料
mineral map 矿产地图
mineral material for metallurgical industry 冶金矿物原料
mineral matter 无机物质；矿物质；成灰物质
mineral matter free basis analysis 无矿物基分析
mineral maturity 矿物成熟度
mineral metabolism 无机代谢作用
mineral microscope 矿物学用显微镜
mineral mining 矿物开采；采矿
mineral mining registration system 矿产资源开采登记制度
mineral monument 矿用地永久界石
mineral nutrient 矿质养分
mineral occurrence 矿点
mineral oil 矿物油；石油
mineral oil spring 石油喷流；矿物油泉
mineral output 矿物产量
mineral pair 矿物对
mineral paper 页状煤；片状煤；纸煤
mineral paragenesis 矿物共生
mineral particle 矿物颗粒
mineral pharmacology 矿物药学
mineral phase 矿物相
mineral phase rule 矿物相律
mineral physics 矿物物理学
mineral pigment 矿物颜料；无机颜料
mineral pitch 地沥青；柏油
mineral pollution 矿物污染
mineral powder 矿粉
mineral process control 选矿工艺过程控制
mineral processing 矿产加工；选矿
mineral product 矿产；有用矿物
mineral prospecting 找矿；矿物勘探；矿产普查
mineral refuse 矿渣
mineral reserve 矿产储量
mineral reserve balance sheet 矿产储量平衡表
mineral reserve sheet 矿产储量表
mineral reserves database 矿产储量数据库
mineral reserves registration and control 矿产储量登记管理
mineral reserves review 矿产储量评审认定
mineral resin 硬沥青；矿质沥青；矿物树脂

mineral resources	矿物资源；矿产资源；矿石储量
mineral resources development program	矿产资源开发规划
mineral resources divisions	矿产资源区划
mineral resources economics	矿产资源经济学
mineral resources information management	矿产资源信息管理
mineral resources of the sea floor	海底矿产资源
mineral resources prognosis	矿产资源预测
mineral resources programming	矿产资源规划
Mineral resources reserves	矿产资源储量
mineral resources tax	矿产资源税
mineral right	采矿权
mineral rock	矿岩
mineral salt	天然盐；矿盐；岩盐
mineral saturation indices	矿物饱和指数
mineral scaling	矿物结垢
mineral separation	矿物分离；选矿
mineral separation plant	选矿厂
mineral separation process	选矿工艺
mineral sequence	成矿顺序；矿物共生顺序
mineral sequestration	矿物固碳
mineral skeleton	矿物骨架
mineral sludge	矿质污泥
mineral soap	膨润土
mineral soil	矿质土；矿质土壤
mineral species	矿物类别；矿物种类
mineral spot	矿物包体；矿物斑点
mineral spring	矿泉
mineral spron	精矿卸载槽
mineral stability	矿物稳定性
mineral stable texture	矿物稳定结构
mineral structure	矿物结构；矿物构造
mineral substance	矿物质
mineral surface	矿物表面
mineral tallow	地蜡；伟晶蜡石
mineral tar	矿质焦油；风化石油；软沥青
mineral time	矿物年代
mineral train	运矿列车
mineral trapping	矿物俘获；矿化圈闭
mineral vein	矿脉
mineral vertical zoning	矿物垂直带
mineral void	矿质填料孔隙
mineral waste residue	矿渣
mineral water	矿泉水；矿水；矿质水
mineral wax	地蜡
mineral wealth	矿产；矿产资源
mineral well	矿泉井
mineral wettability	矿物润湿性
mineral white	石膏
mineral wool	矿棉；石棉；石渣棉
mineral works	矿物工程
mineral zone	矿物带
mineral zoning	矿物分带
mineral-air interface	矿物-空气界面；固-气界面
mineral-air-water contact	矿物-空气-水接触；固-气-液接触
mineral-bearing froth	含矿泡沫
mineral-bituminous groundmass	矿物沥青基质
mineral-chemistry	矿物化学
mineral-durite	矿物暗煤
mineral-fuel deposit	矿物燃料矿床
mineralization	成矿作用；矿化作用
mineralization coefficient	矿化系数
mineralization degree	矿化度
mineralization epoch	成矿时代；矿化期
mineralization inventory	矿化藏量
mineralization liquor	成矿溶液
mineralization period	矿化期；成矿期
mineralization phase	矿化期
mineralization ratio	矿化系数
mineralization stage	矿化阶段
mineralize	矿化
mineralized	矿化的
mineralized anomaly	矿化异常
mineralized area	矿化地区
mineralized belt	矿化带
mineralized bubble	矿化气泡；带矿粒的气泡
mineralized degree of ground water	地下水矿化度
mineralized degree of groundwater	地下水矿化度
mineralized district	矿化区
mineralized froth	矿化泡沫
mineralized groundwater	矿化地下水
mineralized lake	矿湖；矿化湖
mineralized reef	矿脉
mineralized spot	矿化点
mineralized water	矿化水；矿泉水
mineralized zone	矿化带
mineralizer	造矿元素；矿化因素；矿化剂
mineralizing	矿化
mineralizing agent	矿化剂
mineralizing component	成矿组分
mineralizing epithermal system	浅低温成矿系统
mineralizing fault	成矿断层
mineralizing fluid	成矿流体
mineralizing solution	矿化溶液
mineralizing water	矿化水
mineral-laden water	含矿水；矿水
mineralogical	矿物学的
mineralogical analyses	矿物分析
mineralogical barometer	矿物压力计
mineralogical character	矿物特性
mineralogical chemistry	矿物化学
mineralogical component	矿物成分
mineralogical composition	矿物成分；矿物组成
mineralogical guide	找矿导引；探矿指南
mineralogical identification	矿物鉴定
mineralogical microscope	矿物显微镜
mineralogical oddity	矿物异态
mineralogical phase rule	矿物相律
mineralogical thermobarometer	矿物温压计
mineralogical thermometer	矿物温度计；地质温度计
mineralogist	矿物学家
mineralographic microscope	矿相显微镜
mineralography	矿相学
mineralogy	矿物学
mineralogy and petrography	矿物学与岩相学
mineralogy and petrology	矿物学和岩石学
mineralogy of drugs	药材矿物学
mineraloid	似矿物；准矿物；胶质矿物
mineral-processing engineer	选矿工程师；冶炼工程师
minerals availability system	矿产资源可供性系统

mineral-spring hydrogeological investigation	矿水水文地质勘查
mine-rescue	矿山救护
mine-rescue apparatus	矿山救护器械
mine-rescue crew	矿山救护队
mine-rescue equipment	矿山救护设备
mine-rescue man	矿山救护员
minerocoentology	矿物共生学
minerogenesis	成矿作用
minerogenetic	成矿的
minerogenetic epoch	成矿时代
minerogenetic province	成矿区
minerogenetic region	成矿区域
minerogenetic series	成矿系列
minerogenic rock	成矿岩石
miner's code of laws	矿山规程
miner's compass	矿用罗盘
miner's consumption	矿工矽肺病
miner's dial	矿用罗盘
miner's dip needle	矿用磁倾仪
miner's electric lamp charging rack	矿灯充电架
miner's lamp	矿工灯
miner's level	矿用水准仪
miner's light	矿灯；矿工用灯
miner's phthisis	矿工尘肺病
miner's pick	锤镐
miner's powder	黑火药
miner's rule	矿工规则
miner's tool	矿工用工具
mine-run	原矿；原煤
mine-run bin	原煤仓；原矿仓
minery	矿山群；采矿区；采石场
mines inspection	矿山视察；矿山检查
Mines Safety Regulations	矿场安全规例
mineshaft	井筒
mine-surveying instrument	矿山测量仪器
minette	云煌岩；鲕褐铁矿
minette ironstone	鲕状褐铁矿
minettefels	隐云煌岩
mine-water discharge	矿坑排水；矿坑疏干
minge	软煤；脆煤
mingle	混合
mingling	混合
Ming-Qing little ice age	明清小冰期
minguetite	黑硬绿泥石
mingy coal	软煤；脆煤
mini computer	小型计算机
mini cross section	特小断面
mini fracture test	小型压裂试验
mini fullfacer	小型全断面隧洞掘进机
mini hose	微型软管
mini pile	微型桩
miniature	小型；微型
miniature bit	小型钻头
miniature geophone	微型检波器
miniature hydrophone	微型海上检波器
miniature lamp	微型灯；指示灯
miniature radiography	微型射线照相术
miniature terracing	微形阶地
miniature whipstock theory	小型斜向器理论
miniaturization	小型化；微型化
miniborer	微型钻机；小型隧道掘进机
mini-cable	小型电缆
minicam	小型照相机
minicomputer	微型计算机；小型计算机
minidozer	小型液压推钻机
mini-driving apparatus	小型拖动装置；小型传动装置
minification	缩小尺寸
miniflexichoc	小型板式挤压震源
minifocused log	微聚焦测井
mini-frac	小型压裂；测试压裂
minifrac calibration treatment	小型标定压裂
mini-hydropower station	小型水电站
minilog	微电极测井
minilog electrode arrangement	微电极测井电极系
minilog instrument	微电极测井仪器
minilog pad	微电极测井极板
minilog-caliper	微电极测井卡尺测井中量测孔径的小型卡尺
minimal	极小的；最低的；最小的
minimal area	最小面积
minimal basis	最小基；极小基
minimal curve	极小曲线
minimal curve radius	最小转弯半径
minimal detectable activity	最低检出放射性强度
minimal head	最小水头
minimal safe distance	最小安全距离
minimal sandwich	最小多层结构
minimal value	最小值
minimal-path problem	最短路径问题
minimax	极小极大；最小最大
minimax approximation	极值逼近
minimax control	极小极大控制
minimax criteria	极小极大准则
minimax criterion	最小最大值准则；极小化极大准则
minimax dual method	极大极小对偶法
minimax estimate	极小化极大估计
minimax estimator	极小极大估计量
minimax method	极大极小法
minimax principle	最小最大值准则
minimax regret criterion	极小极大回归准则
miniaturization solution	极小极大解
minimax strategy	极小化极大策略
minimax technique	极小化极大技术
minimax test	极小极大检验；极小极大判别
minimax theorem	极小化极大定理
minimeter	指针测微计；千分比较仪
minimite	共结花岗岩
minimization	极小化；最小化；求最小值
minimization of bandwidth	带宽最小化
minimization process	求极小的方法
minimize	使减到最少；使缩到最小
minimized	减到最小的
minimizing chart	缩图
minimizing sequence	极小化序列
minimodel	小模型
minimum	最小的；最少的；最低的；最低量；最小值
minimum average payable grade	最低平均可采品位
minimum average value	小值平均值
minimum burial depth map of magnetic bodies	磁性体最

小埋深图；磁性体等深度图
minimum clearance 最小间隙
minimum closed-in pressure 最小关井压力
minimum concentration coefficient 最低浓度系数
minimum constant infiltration capacity 最小恒定入渗量
minimum constraint variable 最小约束变量
minimum constructive height 最小建造高度；最小施工高度
minimum content of ore 矿石的最低含量
minimum core size 最小岩芯直径
minimum cover of reinforcement 钢筋保护层的最小厚度
minimum curve radius 最小转弯半径
minimum daily discharge 最小日流量
minimum delay filter 最小延迟滤波器
minimum depth 最小深度
minimum design 极小设计
minimum detectable amount 最小检测量
minimum detectable concentration 最小检测浓度
minimum detectable quantity 最小检测量
minimum deviation 最小偏角；最小偏差
minimum deviation angle 误差曲线
minimum diameter 最小直径
minimum dip 最小倾角
minimum drawdown level 最低泄降水位
minimum earthquake 最小地震
minimum ebb 最小落潮流
minimum economic rate 最低经济产量
minimum effective hole diameter 最小有效孔径
minimum effective liquid rate 最小有效流量
minimum elevation 最小高程
minimum embedded depth of foundation 基础最小埋深
minimum energy ratio principle 最小能量比原理
minimum entropy 最小熵
minimum entropy deconvolution 最小熵反褶积
minimum entropy filtering 最小熵滤波
minimum excavation line 最小开挖线
minimum exploitable thickness 最小可采厚度
minimum exploitable width 最小可采宽度
minimum exploration input 最低勘察投入
minimum exploration work program 最低限度勘探工作量
minimum field 最小磁场
minimum flood 最小涨潮流
minimum flow 最小径流；最小流量
minimum fluidization velocity 最小流态化速度
minimum gap 最小间隙；最小间距
Minimum grade 最小纵断坡度；最小坡度
minimum grade of ore 可采矿石的最低品位
minimum gradient 最小坡度
minimum height of fill 最小填土高度
minimum hole diameter 最小孔径
Minimum horizontal compression stress 水平最小主压应力
minimum horizontal stress 最小水平应力
minimum industrial grade of ore 矿石最低工业品位
minimum industrial payable grade 最低工业可采品位
minimum initiating charge 极限起爆药量；最小起爆药量
minimum inlet pressure 最小进口压力
minimum interference 最小过盈；最小干涉
minimum lag 最小滞后
minimum level period 最低水位期

minimum longitudinal gradient 最小纵坡
minimum mapping unit 最小制图单元
minimum mean squared error estimation 最小均方误差估计
minimum mean squared error estimator 最小均方误差估计
minimum mineable thickness 最低可采厚度
minimum mining content 最小可采品位
minimum mining thickness 最小可采厚度
minimum mining thickness of orebody 矿体最小可采厚度
minimum mining width 最小可采宽度
minimum moisture content 最小含水量
minimum navigation depth 最低通航水深
minimum normal stress 最小正应力
minimum operating fluid level 最低工作液位
minimum outlet pressure 最小出口压力
minimum output 最低产量
minimum path tracing algorithm 最短路径跟踪算法
minimum penetration 最小贯入度
minimum permissible velocity 最小容许流速
minimum phase 最小相位
minimum phase filter 最小相位滤波器
minimum phase inverse filter 最小相位反滤波器
minimum phase operator 最小相位算子
minimum phase shift 最小相移
minimum phase wavelet 最小相位子波
minimum pile spacing 桩的最小间距
minimum pipe depth 管道的最小埋深
minimum point 最低点
minimum pool level 最低库水位
minimum pore ratio 最小孔隙比
minimum potential 最小势能
minimum potential energy 最小势能
minimum principal effective stress 最小有效主应力
minimum principal stress 最小主应力
minimum principle 最小原理
minimum radium of curve 最小曲线半径
minimum radius of curvature negotiable 通过最小曲线半径
minimum radius of curve 最小曲线半径
minimum radius of horizontal curve 最小平曲线半径
minimum receiving field strength 最小接收场强
minimum recorded 实测最小的
minimum recorded flood peak 实测最小洪峰
minimum reflectance 最小反射率
minimum relative entropy 最小相对熵
minimum residual method 最小剩余法
minimum risk function value criterion 最小风险函数值准则
minimum rock cover 最小岩石覆盖厚度
minimum runoff 最小径流
minimum safe height 最小安全高度
minimum safety factor for hoisting 提升钢丝绳的最小安全系数
minimum safety stability 最小安全稳定性
minimum sampling frequency 最小采样频率
minimum shut-in period 最短关井时间
minimum size of core 最小岩心直径
minimum slope 最小坡度
minimum solution 极小解
minimum span 最小跨度
minimum speed limit 最低速率
minimum square 最小二乘

minimum stacking velocity	最小叠加速度
minimum stage	最低水位
minimum storage	最小库容
minimum strain energy density criterion	最小应变能密度准则
minimum stress distribution	最小应力分布
minimum stress orientation	最小应力方向
minimum thermal resistance	最小热阻
minimum thickness of tunneling shield	最小衬砌厚度
minimum to maximum stress ratio	最小最大应力比
minimum torsional strength	最小抗扭强度
minimum total potential energy principle	最小势能原理
minimum turning circle of car	车辆的最小转弯半径
minimum turning radius	汽车最小转弯半径
minimum value	最小值
minimum value criterion	最小取值准则
minimum variance	最小方差
minimum variance deconvolution	最小方差反褶积
minimum variance estimate	最小方差估计
minimum velocity	最小速度
minimum void ratio	最小孔隙比
minimum water level	最低水位
minimum water saturation	最低含水饱和度
minimum water table boundaries	最小水头边界
minimum wave duration	最小波浪历时
minimum weight design	最小重量设计
minimum workable grade	最低可采品位
minimum workable thickness	最低可采厚度
minimum working height	最小工作高度
minimum yield required	最小需水量
minimum-curvature method	最小曲率法
minimum-delay	最小延迟
minimum-phase spectrum	最小相位谱
minimum-phase wavelet	最小相位涟波
minimum-void mixture	孔隙度最小的混合物
mining above aquifer	承压含水层上采煤
mining activity	采矿工作；采掘作业
mining and stripping capacity of surface mine	露天矿采剥能力
mining and stripping procedure of open-pit	露天矿采剥方法
mining area	开采区；采矿区
mining area capacity	矿区规模
mining area communication	矿区通信
mining area water pollution	矿区水污染；矿区水体污染
mining art	采矿工程
mining at high elevation	高山矿床开采；高海拔开采
mining block method	开采块段法
mining by areas	分区开采
mining by stages	分期开采
mining car	矿车
mining chart	采矿工程图表
mining city	矿业城市
mining claim	采矿执照；采矿许可证
mining coefficient	开采系数
mining compass	矿用罗盘；矿山罗盘仪
mining concentration index	采煤生产集中指数
mining concession	矿山开采权地区；矿山许可开采区域
mining condition	开采条件
mining conditions and hydrogeology conditions	开采技术条件及水文地质条件
mining consequences	回采后果
mining conservation	矿山环境保护
mining cut of open-pit	露天矿采掘带
mining damage	开采塌陷
mining depth	开采深度
mining depth for failure	破坏时开采深度
mining dirt	夹矸
mining district	采区；矿山管区
mining district survey	采区测量
mining drainage	矿山排水
mining dual operation	采矿作业；矿山经营
mining dust	矿尘；粉尘
mining effect	开采效果；采动效应
mining engineer	采矿工程师
mining engineering	采矿工程
mining engineering plan	采掘工程平面图
mining engineering qualitative management	矿山工程质量管理
mining enterprise	采矿企业
mining equipment	采矿设备；矿山设备
mining excavation	矿井挖掘；矿井采掘
mining explosive	矿用炸药
mining field	采矿区；采矿场
mining film	采矿科技影片
mining floor	方框支架最高采掘层；采场底板
mining geodesy	矿山大地测量学
mining geological logging work	矿山地质编录
mining geological sampling	矿山地质取样
mining geologist	采矿地质师；矿井地质师
mining geology	矿井地质；采矿地质学
mining geophysics	采矿地球物理学
mining groundwater	开采地下水
mining handbook	采矿手册
mining height	开采高度
mining hole	钻孔；炮眼
mining in bulk	大量回采
mining in installments	分期开采
mining in open pits	露天开采
mining in slices	分层开采
mining in strips	条带式开采
mining inclinometer	矿山测斜仪
mining induced cleavage	采动裂隙
mining induced seismicity	矿山诱发地震
mining industry	矿业；采矿业
mining influence	采动影响
mining inspection	采矿技术检查
mining intensity	开采强度；采矿强度
mining intensity method	开采强度法
mining law	采矿法；矿业法
mining lease	采矿租地；采矿用地
mining legislation	矿业法
mining level	开采水平；采掘水平
mining license	采矿许可证
mining limit	采区边界
mining locomotive	矿用机车；矿山机车
mining logging	矿业测井
mining loss	开采损失；采矿损失
mining machine	矿山机械；采掘机
mining machinery	矿山机械；采矿机械

mining map	矿山测量图
mining mechanical engineering	矿山机械工程
mining mechanics	矿山力学
mining method	采矿方法；开采方法
mining method employing fill	充填采矿法
mining method in tunnel construction	矿山法隧道施工
mining methods	采矿法
mining modulus analogy	开采模数比拟法
mining of groundwater	地下水开采
mining of pillars	矿柱回采
mining of self-flaming deposit	自燃矿床开采
mining operation	采矿作业；采矿工作
mining order	开采顺序
mining ore grade	采矿等级
mining ore losses	开采损失
mining organization	采矿机构；矿业机关；采矿工作组织
mining out region	采场；采矿地区
mining panel survey	采区测量
mining plan	矿山工作平面图；开采计划
mining plant design	矿场设计
mining plough	刨矿机；刨煤机
mining plough share	刨矿机犁板
mining ply	煤层中的软薄分层
mining policy	矿业政策
mining practice	采矿法；采矿操作
mining preparations	采准工作
mining procedure	采矿方法；开采方法；开采程序
mining progress	开采进度
mining pumping method	开采抽水法
mining rate	采出率；回收率；开采速度
mining raw materials	采矿原料
mining recovery	回采率
mining region	采矿区域；采掘区域
mining regions monitoring	矿区监测
mining regulation	采矿规程；矿山技术操作规程
mining research	采矿研究
mining reserve	开采储量
mining result	开采效果
mining right	采矿权；矿业权
mining right control	采矿权管理
mining right leasing	矿业权出租
mining right mortgage	矿业权抵押
mining right owner	采矿权人
mining rights nullification	采矿权注销
mining risk	矿业危险
mining rock mechanics	采矿岩石力学；矿山岩体力学
mining safety	采矿安全
mining science	采矿科学
mining section loss	采区损失
mining sequence	开采顺序
mining sequence of adjacent ore body	相邻矿体开采顺序
mining sequence of level	阶段开采顺序
mining sequence of ore block	矿块开采顺序
mining shield	采矿掩护支架；掘进掩护支架
mining shovel	采矿挖掘机；开采机铲
mining slope	开采坡面角；开采边坡
mining society	采矿界；采矿协会
mining specification	开采作业规程
mining stage of deposit	矿床开采步骤
mining starting line	始采线
mining sterile	贫矿
mining subsidence	开采沉陷；采空塌陷
mining subsidence observation	开采沉陷观测
mining survey	矿山测量
mining system	开采系统；开采方法
mining system engineering	采矿系统工程
mining technical economic index	矿山技术经济指标
mining technique	采矿技术；采矿工艺
mining technology	采矿工艺
mining theodolite	矿用经纬仪
mining title	开采权
mining to the boundary	前进式开采；向井田边界开采
mining tool	采矿工具
mining transit	矿用经纬仪
mining tunnel collapse	矿井巷道坍塌
mining under safe water pressure of aquifer	带压开采
mining venture	矿山投资
mining waste	矿场残渣；开采废渣
mining wastewater	采矿废水
mining water use	采矿用水
mining width	开采宽度
mining with filling	充填开采
mining with movable filling	放顶充填开采法
mining with self-filling	自动放顶充填开采
mining with stowing	充填开采
mining without destroying water resource	保水开采
mining yard plan	矿场平面图
mining yield	开采矿量；开采水量；疏干性开采量
mining-district design	采区设计
mining-employed reserves	动用储量
mining-induced damage	开采损害
mining-induced earthquake	矿山诱发地震；矿震
mining-induced fault	采矿引起的断层
mining-induced fault zone	采矿引起的断裂带
mining-induced fissure	采动裂隙
mining-induced fractures	采动裂隙
mining-induced landslide	采矿诱发滑坡
mining-induced stress	采动应力
mining-induced stress field	采动应力场；再生应力场
mini-permeameter	小型渗透率仪
miniphyric	微斑晶的
mini-pile	微型桩；小型桩
mini-pressuremeter	小型旁压仪
mini-roundabout	微形环交
minisat system	最低饱和系统
miniseis	微型检波器
mini-seismics	小地震法
mini-semi	浅海半潜式钻井平台
minitest	小型试验
mini-test injection	小型注水试验
minitexture meter	微型构造探测仪
minitrack system	干涉仪卫星跟踪系统
mini-tree	小型采油树
minnesotaite	铁滑石
minophyric	细斑晶的
minor angle method	小角度法
minor axis	短轴
minor crack	细微裂纹
minor crumple	次级褶皱

minor dam	副坝
minor damage	轻微破坏
minor determinant	子行列式
minor diagonal	次对角线
minor diameter	小直径
minor diameter of thread	螺纹内径
minor drainage	小型排水系统
minor earthquake	小震
minor element	少量元素；副元素；次要元素
minor element in seawater	海水微量元素
minor fault	小断层
minor fold	小褶皱
minor geologic details	次要地质细节
minor geosyncline	小地槽
minor horizontal principal stress	最小水平主应力
minor increment	最小距离
minor joint	小节理
minor karst feature	微喀斯特地形
minor landslide	小型滑坡
minor landslip	小型塌方；小型山泥倾泻；小型滑移
minor mineral	次要矿物
minor parameter	次要参数
minor partial fold	小型局部褶皱
minor phase	次生相
minor planet	小行星
minor principal plane	小主平面
minor principal strain	小主应变
minor principal stress	最小主应力
minor relief elements	微地貌因素
minor river bed	小河床
minor riverbed	枯水河床
minor road	次要道路；次级道路
minor road improvement work	小规模道路改善工程
minor seismic area	小震区
minor semi-axis	短半轴
minor shock	小震；微震
minor showing	次要油气显示
minor street	次要街道；次级街路
minor stress	小应力
minor structural works	小型结构工程
minor structure	小型构造
minor structure works	小型结构工程
minor tremor	微震
minor triangulation	小三角测量
minor works	小型工程
minorant	弱函数；劣函数
minor-scale strain	小规模应变
minor-scale structure	小型构造
minus angle	俯角
minus flower structure	负花状构造
minus grade	缓坡；下坡
minus minerals	负矿物
minus phase	负相位
minus-cement porosity	无胶结孔隙性
minute crack	细裂缝
minute extension	小体积伸长
minute fissure	微小裂隙
minute flake	微小薄片
minute folding	细褶皱
minute friction	毛细摩擦
minute opening	微小开口
minute pore space	微小孔隙空间
minute-pressure	微压
MINV	微梯度测井
minverite	钠长角闪辉绿岩
minyulite	水磷铝钾石
Miocene	中新世；中新统
Miocene epoch	中新世
Miocene Period	中新世
Miocene series	中新统
Miocene source rock	中新世源岩
miocrystalline	半晶质
miogeanticlinal	冒地背斜的
miogeanticline	冒地背斜
miogeo anticlinal ridge	冒地背斜脊
miogeocline	冒地斜
miogeosynclinal	冒地槽的
miogeosynclinal realm	冒地槽区
miogeosyncline	冒地槽；冒地向斜；副地槽
miohaline	中等咸度的
Mio-Oligocene age	中新-渐新世时期
mire	淤泥；泥沼地
mire black	极软的泥炭
mire climate function	沼泽气候功能
mire complex	沼泽复合体；复合沼泽
mire degradation	沼泽退化作用
mire ecosystem	沼泽生态系统
mire environment	沼泽环境
mire hydrology function	沼泽水文功能
mirror	反射体；反射界面
mirror reflection	镜面反射
mirror stone	白云母
mirror structure	镜面构造
mirror symmetry	镜面对称
miry ground	泥沼地
miry sand	泥质砂
miry soil	淤泥土
misalignment	未对准；有偏斜；不对准；角度误差；不符合；不一致；失调
miscellaneous fill	杂填土
miscellaneous geologic investigations	综合地质调查
miscellany rate	混杂率
miscibility	可混合性；溶混性
miscible	可混溶的；可混合的
miscible bank	混相带
miscible displacement	混合置换
miscible drive	混相驱动
miscible gushing flow	混相涌流
miscible mixture	可溶性混合物
misclosure	闭合导线误差
misclosure in azimuth	方位角闭合差
misclosure in leveling	水准测量闭合差
misclosure of angle	角度闭合差
misclosure of round	归零差
misclosure of traverse	导线闭合差
misclosure triangular	三角形闭合差
misdata	错误数据
misdirected hole	偏斜钻孔
miser	凿井机；钻孔机；管形提泥钻头
misering	管形提泥钻头钻进；钻探；凿井

miserite	钠硬硅钙石
Mises yield condition	米赛斯屈服条件
Mises yield criterion	米塞斯屈服准则
misfire hole	拒爆炮眼；拒爆炮孔
misfire removal	清除瞎炮
misfired cap	拒爆雷管
misfired charge	瞎炮；拒爆的装药
misfired hole	拒爆炮眼；炸药残孔；瞎炮
misfit river	不相称河；衰退河
misfit stream	不相称河流
misidentification	错误辨识
misinterpreting	曲解；误判
misoperation	操作失误
mispickel	砷黄铁矿
misplaced outcrop	失位露头
miss shot hole	拒爆炮孔
missed hole	拒爆炮孔
missed routed	拒爆炮眼组
miss-fire	瞎炮；拒爆
missfire shot	不爆发炮眼
missile orientation survey	导弹定向测量
Mississippi Valley type	密西西比河谷型矿床
Mississippi Valley type deposit	密西西比山谷型矿床；成层矿床
Mississippian System	密西西比系
Missourian	密苏里阶
missourite	白榴橄辉岩
mist blasting	喷水爆破
mist drilling	泡沫抑尘钻孔法；喷雾钻进
Mitchell slicing method	米切尔方框支架分层采矿法
Mitchell top slice	米切尔分层开采法的顶层
miter gauge	斜节规
miter lock gate	人字式船闸闸门
mitered inlet	水道斜坡进口
mitigating damage	减轻震害
mitigation	缓解；缓和；减轻
mitigation measure	减轻灾害措施
mitigation of seismic hazard	减轻地震灾害
mitigation works	防护工程
mix crystal	混晶；混合晶体
mix design	混合设计；配合比设计
mix design for cement grouting	水泥注浆的混合设计
mix design process for shotcrete	喷射混凝土混合设计过程
mix of concrete	混凝土拌合物
mix proportion	混合用料比例；混配比例
mix ratio of mortar	砂浆配合比
mix rock	混合岩
mix selector	混合配料选择器
mixability	可混溶性
mixable	可混合的
mix-crystal	混晶；混合晶体
mixed	混合的；拌合的
mixed age	混合年龄
mixed approximation	混合近似
mixed assemblage	混合化石群
mixed avalanche	混合崩坍
mixed base crude oil	混合基原油
mixed batholith	混合岩基
mixed bedding	复合层理
mixed boundary value problem	混合边界值问题
mixed carbonate siliciclastic	碳酸混合硅质碎屑
mixed cement	混合水泥
mixed CO_2-water injection	二氧化碳-水混注
mixed coal	混煤
mixed coal and rock heading	煤岩混合平巷
mixed concentrates	混合精矿
mixed concrete	搅拌好的混凝土
mixed construction	混合构造；混合结构
mixed cracking	混合破裂
mixed crystal	混合晶体；混晶；固溶体
mixed design	混合设计
mixed dislocation	混合型位错
mixed drive	混合驱动
mixed element	混合元
mixed explosion	混合爆炸
mixed explosive	混合炸药
mixed face	半软半硬掘进面
mixed fill	混合填土
mixed finite element formulation	混合有限元公式
mixed flow pump	混流式水泵
mixed fraction	带分数；混合分式
mixed gangues	混合脉石；混合矸石
mixed gases injection	混合气体注入
mixed gneiss	复片麻岩
mixed granular soil	混合粒状土
mixed ground	混合围岩
mixed laminar-turbulent flow	混合流
mixed layer mineral	互交层矿物
mixed method	混合法
mixed mining	混合开采；混采
mixed mode	复合型；混合型；混合模式
mixed model	混合模型
mixed ore	混合矿石
mixed phase	混相驱
mixed phase wavelet	混合相位子波
mixed pixel	混合像元
mixed power plant	混合式电站
mixed pumped-storage plant	混合式抽水蓄能电站
mixed pumping	混合抽水
mixed pumping test	混合抽水试验
mixed quantitative model	混合定量模型
mixed rock	混合岩
mixed sand	混合砂
mixed sedimentary rock	混合沉积岩
mixed siliciclastic-bioclastic segregated deposits	混合硅质碎屑生物碎屑隔离沉积
mixed siliciclastic-carbonate sediments	混合硅质碎屑
mixed sludge	混合软泥；混合矿泥
mixed soil	混合土
mixed spring	混合泉
mixed structure	混合结构
mixed-layer pumping test	混合抽水试验
mixed-grained	混粒的
mixed-grained sand	混粒砂；多种粒径混合砂
mixed-in-place	现场搅拌；就地搅拌
mixed-in-place pile	拌合桩；就地搅拌桩
mixed-in-place wall	就地搅拌连续墙
mixed-in-place-pile	就地搅凝桩；深层搅拌法
mixed-in-transit	运送搅拌；路拌
mixed-layer clay mineral	混层黏土矿物

mixed-layer mineral	混层矿物；间层矿物
mixed-layer pumping	混合取水
mixed-phase	混合相位
mixed-solvent ion exchange	混合溶剂离子交换
mixer	混合器；搅拌机；混合工；搅拌工
mixer arm	混合器桨臂；搅拌机桨叶
mixer granulator	混合制粒机
mixer lorry	汽车式混凝土搅拌机
mixer of continuous method	连续拌合机
mixer plant	拌合楼；搅拌站
mixer settler	混合-沉降器
mixer truck	搅拌运料车
mixer-placer machine	混凝土搅拌浇注机
mixer-settler	混合器-沉降器；混合器-沉淀器
mixes	拌合物
mixing	混合；搅拌；混频；混波
mixing analog	混合模拟
mixing at site	工地拌合
mixing bin	掺合仓；配煤仓
mixing bunker	混合仓
mixing chamber	搅拌室；混合室
mixing channel	混合槽
mixing coefficient	混合系数
mixing condenser	混合式冷凝器
mixing cycle	搅拌循环；混合期
mixing deposits	混合沉积
mixing drum	搅拌滚筒；搅拌鼓
mixing dual operation	搅拌工作
mixing effect	混合作用
mixing equipment	拌和设备
mixing hopper	搅拌斗
mixing hydrochemical reaction in groundwater	地下水混合化学作用
mixing in place	工地拌合
mixing layer	混合层
mixing length	混合长度
mixing machine	混合机；搅拌机
mixing manifold	混合集流管
mixing method	拌合法
mixing model	混合模式
mixing of concrete	混凝土搅拌
mixing of groundwater	地下水混合作用
mixing operation	搅拌工作
mixing parameter	拌合参数
mixing pit	搅拌池
mixing plant	混合车间；搅拌装置
mixing platform	拌合平台
mixing power	搅拌力
mixing process	混合过程
mixing proportion	混合比例；配合比
mixing pump	混合泵
mixing ratio	配合比；混合比
mixing rule	混合律
mixing temperature	混合温度
mixing test machine	搅拌试验机
mixing time	拌合时间
mixing velocity	混合速度
mixing water	混水；拌合用水
mixing zone	混合区
mixite	砷铋铜矿
mixometer	拌合计时器；搅拌计时计
mixtite	混杂沉积岩；混杂岩；冰碛岩
mixton	混杂沉积物
mixtum	混杂沉积物
mixture	混凝土拌合物；混合物
mixture corrosion	混合溶蚀作用
mixture design	配料设计
mixture dissolving	混合溶解
mixture ratio	配合比；混合比
mixture ratio controller	混合比控制器
mixture rule	混合律
mixture soil	混合土
mixture uniformity	拌合物的均匀性
mixture water	混合水
mizonite	针柱石
mizzonite	针柱石；中柱石；钙钠柱石
moat	冰川沟；海底环状洼地；海壕；牛轭湖
moatlike fault	渠状断层
mobile array	流动台阵
mobile bed	不稳定河床；动床
mobile belt	活动带
mobile channel	游荡性河槽；不稳定航道；变动的河道
mobile component	活动组分；活性组分
mobile concrete pump	移动式混凝土泵
mobile crustal deformation measurement	流动地壳形变测量
mobile drill	自行式钻机；轻便钻机
mobile drilling unit	移动式钻机
mobile dune	流动沙丘
mobile emergency winding equipment	移动式紧急提升设备
mobile ground	流沙层
mobile height finder	便移式测高计
mobile hydraulic jack	移动式液压千斤顶
mobile jumbo	自行钻车；机动钻车
mobile laboratory	流动实验室
mobile liquid	低黏度液体；流性液体
mobile lithosphere plate	活动岩石圈板块
mobile load	活动载荷；易变载荷
mobile loader	移动式装载机
mobile loading equipment	自行式装载设备；移动式装载设备
mobile loading machine	移动式装载机
mobile microwave communication	移动微波通信
mobile microwave communication equipment	移动微波机
mobile mill	可移式磨机
mobile mine drill	移动式矿用钻机；自行式矿用钻机
mobile miner	可移式采煤机
mobile moisture	流动水分；游离水分
mobile moving fault	活动断层
mobile observation	流动观测
mobile observation system	流动观测系统
mobile oil	机油
mobile phase	流动相；活动期
mobile piling rig	移动式打桩机
mobile platform	移动平台
mobile rig	移动式钻井机
mobile sand	流沙
mobile source of pollution	流动污染源
mobile stabilized soil mixing plant	移动式稳定土料拌合设备

mobile stacker 自行式排土机
mobile stage 活动平台
mobile stage loader 移动式装载机
mobile station 船台；移动台
mobilism 活动论
mobilist 活动论者
mobilistic approach 活动论观点
mobilistic theory 活动论
mobility 迁移率；流动性；活动性；滑动程度
mobility coefficient 迁移系数；迁移率
mobility in slope failures 滑坡泥石流动性；破坏斜坡移动性
mobility of concrete 混凝土流动性
mobility of debris 泥石的流动性
mobility of ion 离子迁移度
mobility of sand dune 沙丘流动性
mobilization 活动性；活化作用
mobilization degree 流动度
mobilized shear force 发挥出的剪力
mocha stone 藓纹玛瑙；苔石
mock lead 闪锌矿
mock ore 闪锌矿
mock up test 模型试验
mockup 模型
modal analysis 振型分析；模态分析
modal coefficient 模态系数
modal combination 振型组合
modal combination rule 振型组合规则
modal contribution 振型作用
modal contribution factor 振型贡献系数
modal cycle 模态旋回
modal damping 模态阻尼
modal damping matrix 振型阻尼矩阵
modal damping ratio 振型阻尼比
modal decomposition 模态分解
modal equation 振型方程
modal expansion method 振型展开法
modal force 模态力
modal identification 模态辨识
modal logic 模态逻辑
modal mass 振型质量
modal matrix 振型矩阵
modal orthogonality 模态正交
modal participation factor 振型参与系数
modal response 振型反应
modal response contribution 振型反应贡献
modal shape 振型形式
modal shear coefficient 振型剪力系数
modal superposition 模态叠加
modal truncation error 振型截断误差
mode analysis method 振型分析法；模态分解法
mode control algorithm 振型控制算法
mode conversion 模式变换；波形变换
mode coupling 振型耦合
mode field diameter 模场直径
mode of action of support 支护的作用方式
mode of arrangement 排列方式；结构；构造
mode of collapse 崩塌模式
mode of construction 建造方式
mode of decay 衰减方式
mode of deposition 沉积条件；沉积方式
mode of entry 井硐型式
mode of failure 破坏模式；破坏形式；崩塌形式
mode of fissuration 缝隙模式
mode of mining 开采法；采矿方法
mode of motion 运动方式
mode of movement 位移形式；位移模式
mode of occurrence 埋藏状态；存在形式；产状；产出条件
mode of opening 井硐型式
mode of origin 成因型式；成因方式
mode of oscillation 振荡方式
mode of slope failure 边坡破坏模式
mode of supply 补给方式；补给类型
mode of supporting wall 支撑方式
mode of transport 搬运方式；搬运条件
mode of vibration 振型
mode shape 振型
mode shape evaluation 振型计算
mode shape matrix 振型矩阵
mode shape vector 振型向量
mode superposition 振型叠加
mode superposition method 振型叠加法
mode-amplitude 振型幅值
model age 模式年龄
model algorithm 模型算法
model analysis 模型分析
model aquifer 模拟含水层
model bank 模型库
model base 模型库
model baseline 模型基线
model bit 标准钻头
model building code 典型建筑法规
model calculation 模型计算
model cell 模型单元
model construction 模型构造
model coordinate 模型坐标
model date 模式年龄
model deformation 模型变形
model dynamic test 模型动力测试
model earthquake motion 模型地震动
model efficiency 模型效率
model errors 模型误差
model experiment 模拟试验；模型试验
model experiment with clay 泥巴试验
model fabrication technique 模型制作技术
model flow test 流动模型试验
model following controller 跟随模型控制器
model footing test 基础模型试验
model for exploration 勘探模型
model formation 建立模型
model geyser 典型间歇泉
model guided identification 模型引导识别
model identifier 模型标识
model law 模型律
model library 模型库
model method 模型法；样板法；模拟法；式样法
model network 模拟网络
model of abrupt change of groundwater quality 水质突变模型

model of adverse response in an underground excavation due to release of energy 地下巷道中能量采放引起的不利响应模式
model of field source 场源模型
model of geological structure 地质构造模型
model of idealized elastic-plastic deformation 理想弹塑性变形模型
model of idealized rigid-plastic deformation 理想刚塑性变形模型
model of mechanics 力学模型
model of mineral deposits 矿床模型
model of mining system optimization 采矿系统优化模型
model parameter 模型参数
model pile group 模型桩群
model scale 模型缩尺；模型比例
model scope 模型范围
model seismology 模型地震学
model similarity 模型相似律；模型相似性
model simulation 模型仿真模拟
model specification 模范规格；模型规定
model stream 典型河流
model study 模型研究
model technique 模拟技术
model tectonics 模型大地构造学；模型地质构造学
model test 模型测试；模型试验
model testing basin 模型试验池
model uncertainty 模型不确定性
model verification 模型验证
model-aided decision system 模型辅助决策系统
modeling 模拟；模式化；模型化
modeling algorithm 模拟算法
modeling error 建模误差
modeling gradation 模拟级配
modeling process 建模过程
modeling time 建模期；建模时间
modelling 建模；模型化；模型试验
modelling exercise 模拟试验
modelling gradation 分级模拟；模拟级配
modelling of surfaces system 表面工程模型系统
modelling of the fracture 断裂建模
modelling research 模拟研究
modelling software 模拟软件；建模软件
modelling strategy 建模策略
modelling technique 模拟技术；建模技术；模拟试验技术
modelling verification 模型验证
model-prototype comparison test 模型与原型比较试验
model-prototype relationship 模型与原型关系
models for calculating age of groundwater 地下水年龄计算模型
model-scale measurement 比例模型测定
models-materials 模型材料
models-scaling laws 模型比例定律
model-stream 典型河流；模型河流
modem 调制解调器
mode-participation coefficient 振型参与系数
mode-participation factor 振型参与系数
moderate 中等的；适度的；温和的
moderate consequence 中等程度后果
moderate damage 中等破坏
moderate damping 中等阻尼
moderate deposit 半深海沉积
moderate dip 缓倾斜
moderate disturbance 中等扰动
moderate earthquake 中等地震；中强地震
moderate geomagnetic storm 中等磁暴
moderate ground shaking 中强地震动
moderate permeability 中等渗透性；中等渗透率
moderate pitch seam 中倾斜煤层；中倾斜矿层
moderate rainfall 中等降雨量
moderate relief 中等起状
moderate risk 中等风险
moderate sea deposit 半深海沉积
moderate seam 倾斜煤层
moderate shock 中等地震；中强地震
moderate size earthquake 中等地震
moderate strong earthquake 中强地震
moderate swell 中等涌浪
moderate tectonic activity 中等构造活动
moderate to high consequence 中等至严重危害后果
moderate to high risk 中等至高风险
moderate velocity air-flow measurement 中速风流测量
moderate weathered 中等风化的
moderately coarse 中间粒度的
moderately coarse texture 中等粗质土
moderately decomposed 中度风化
moderately drained 适当排水的；适度排水的
moderately firm ground 中等硬度土
moderately hard water 中等硬水
moderately inclined deposit 缓倾斜矿层
moderately mineralized 中等矿化的
moderately rapid permeability 常速渗透
moderately slow permeability 微慢渗透
moderately soluble salt test 中溶盐试验
moderately weathered 弱风化；中等风化的
moderately weathered zone 弱风化带；中等风化带
moderate-pitch mining 槽运倾斜煤层开采
moderate-resolution Imaging Spectroradiometer 中分辨率成像光谱仪
moderate-strong earthquake 中强地震
moderate-to-strong-motion 中强地震地面运动
moderate-weathering rock slope 弱风化岩石边坡
mode-ray duality 振型-射线双重性
modern continental margin sedimentation 现代欧陆缘沉积
modern crust movement 现代地壳运动
modern deposits 现代沉积
modern environment system 近代环境系统
modern geosyncline 现代地槽
modern geotechnical engineering 现代岩土工程
modern sand 现代砂
modern sediment 现代沉积物
modern stream sediment 现代河流沉积物
modern tectonic movement 现代构造运动
modern valley alluvium 现代河谷冲积物
modern wrench movement 现代旋扭运动
modes of failure for a roof beam in stratified rock 层状岩石中顶板梁的破坏模式
modes of fracture 破裂方式；破裂模式
modes of operation of monitoring devices 监测装置的操作模式

modes of pillar failure 矿柱破坏模式
modes of shear failure for subsoil 地基剪切破坏模式
modification factor 修正因数
modified 修正的；改性的；改良的
modified area mining 改进式倒推开采法
modified Atkinson formula 艾金森简化公式
modified bitumen 改性沥青
modified Bouguer anomaly 修正布格异常
modified briquette 改性型煤
modified cement 改性水泥
modified coefficient 修正系数
modified coefficient of seismic action effect 地震作用效应调整系数
modified coefficient of seismic bearing capacity 承载力抗震调整系数
modified compaction test 修正的击实试验
modified differential equation 修正微分方程
modified direct shear apparatus 改进的直剪仪
modified Euler method 改进的欧拉法
modified Galitzin seismograph 改进的伽利津地震仪
modified Griffith criterion 改进的格里菲斯准则
modified Griffith's criterion 修正的格里菲斯准则
modified Griffith's theory 修正的格里菲斯理论
modified groundwater 改质地下水
modified image description 修正影像描述
modified index 修正指数
modified longwall 变相长壁开采法；改进的长壁开采法
modified longwalling 改进的长壁开采法
modified longwalling method 改进型长壁开采法
modified mass 简化质量；修正质量
modified Mercalli intensity 修正麦氏烈度
Modified Mercalli intensity scale 修订的麦加利烈度表
modified Mercalli scale 修正麦加利烈度表
modified method of slices 改进的条分法
modified Parshall flume 改良巴歇尔量水槽
modified ripples 修饰波痕；改造波痕
modified Schmidt diagram 修订的施密特图
modified soil 改良土
modified spring constant 简化的弹簧常数
modified temperature difference factor 温差修正系数
modified triaxial apparatus 改进的三轴仪
modified value 修正值
modified variational principle 修正的变分原理
modified velocity 改正流速
modlibovite 云橄黄煌岩
modular 模的；模数的；系数的；模块
modular computer 模块化计算机
modular construction 单元结构
modular design 模块式设计；标准设计
modular design method 定型设计法
modular dimension 模数尺寸
modular function 模函数
modular programming 模块化程序设计
modular ratio 模量比
modular ratio effect 模拟比效应
modular ratio of elasticity 弹性模量比
modularity 模块化
modularization 模块化
modular-ratio design 模量比法设计
modulating frequency 调制频率
modulating oscillator 调制振荡器
modulation frequency 灯频率
modulation generator 调制振荡器
modulation measurement 调制度测量
modulation mode 调制方式
modulation monitor 调制监视器
modulation rate 调制速度
modulation transfer function 调制传递函数
modulation-demodulation 调制-解调
modulator 调制器
module 模数；模量；系数；模块
module coordination 模数协调
module declaration 模块说明
module design 模块设计
module of elasticity 弹性模数
module of resilience 回弹系数；回弹模数
module of rigidity 刚性模数；刚性模量
module programming 模块化程序设计
module test 模块测试
modulo 模数；残余数
modulo dual operation 模运算
modulus 模；模数；模量；系数
modulus in shear 剪切模量
modulus method of groundwater runoff 地下径流模数法
modulus of compressibility 压缩模量；压缩性系数
modulus of compression 压缩模量；体积弹性模量
modulus of deformation 变形模量
modulus of dilatancy 膨胀模量；剪胀模量
modulus of dilatation 膨胀系数；膨胀模量
modulus of dilation 膨胀模量
modulus of drainage 排涝模数
modulus of elastic aftereffect 后效弹性模量；滞后弹性模量
modulus of elastic compression 弹性压缩模量
modulus of elasticity 弹性模量；杨氏模量
modulus of elasticity in bending 弯曲弹性模量
modulus of elasticity in shear 剪切弹性模量；剪切模量
modulus of elasticity of bulk 体积弹性模量
modulus of elasticity of concrete 混凝土弹性模量
modulus of elasticity of rail support 钢轨支点弹性模量
modulus of elasticity of volume 体积弹性模量
modulus of elongation 伸长模量；拉伸模量
modulus of flow 径流模数；径流率
modulus of foundation 地基模量
modulus of groundwater flow 地下水流模数
modulus of groundwater runoff 地下水径流模数
modulus of incompressibility 非压缩性模数
modulus of instantaneous elasticity 瞬时弹性模量
modulus of linear deformation 线性变形模量
modulus of longitudinal elasticity 纵向弹性模量
modulus of plasticity 塑性模量
modulus of pressing 压缩模量
modulus of pressure meter 旁压仪模量
modulus of recharge 补给模数
modulus of resilience 回弹模量
modulus of resilience test 回弹模量试验
modulus of resistance 抗力模量
modulus of rigidity 刚性模量；刚性系数
modulus of rupture 断裂模量；挠折模量；破裂模数；破裂系数

modulus of rupture in torsion 扭转破坏模量
modulus of section 截面模量
modulus of shear deformation 剪切模量；剪切变形模量
modulus of shear resilience 剪切回弹模量
modulus of shearing 剪切模量
modulus of soil erosion 水土流失模数；土壤侵蚀模数
modulus of soil reaction 土的反力模量
modulus of subgrade reaction 地基反力模量
modulus of torsion 扭转模量；扭转弹性模数
modulus of torsional shear 扭切模量
modulus of transformation 变换系数
modulus of volume change 体积模数
modulus of volume elasticity 体积弹性模量；体积弹性系数
modulus ratio 模量比
modumite 碱辉斜长岩
mofette 碳酸喷气孔
Mogao Grottoes 莫高窟
mogote 灰岩残丘；单笔孤峰
Moho 莫霍界面；M界面
Moho boundary 莫霍面边界
Moho by regional gravity 区域重力莫霍面计算
Moho depth 莫霍面深度
Moho discontinuity 莫霍不连续面
Moho surface 莫霍面
Moho topography 莫霍面地形
Mohole 超深钻；莫霍钻；莫霍面钻探
Mohorovicic discontinuity 莫霍界面；M界面
Mohr circle 摩尔应力圆；摩尔圆
Mohr circle failure envelope 摩尔圆破坏包线
Mohr circle of stress 摩尔应力圆
Mohr circle pole point method 摩尔圆极点法
Mohr Coulomb yield criteria 摩尔-库伦屈服准则
Mohr criteria 摩尔准则
Mohr diagram 摩尔图
Mohr envelope 摩尔包络线
Mohr failure envelope 摩尔包络面；摩尔破坏包络线
Mohr scale of hardness 摩尔硬度计
Mohr strain circle 摩尔应变圆
Mohr strength criterion 摩尔强度准则
Mohr strength envelope 摩尔强度包线
Mohr-Coulomb criteria 摩尔-库仑准则
Mohr-Coulomb criterion 摩尔-库仑准则
Mohr-Coulomb envelope 摩尔-库仑包络线
Mohr-Coulomb failure cone 摩尔-库仑破坏锥
Mohr-coulomb failure criteria 摩尔-库仑破坏准则
Mohr-Coulomb failure criterion 摩尔-库仑破坏准则
Mohr-Coulomb law 摩尔-库仑定律
Mohr-Coulomb model 摩尔-库仑模型
Mohr-Coulomb soil model 摩尔-库仑土模型
Mohr-Coulomb strength envelope 摩尔-库仑强度包线
Mohr-Coulomb theory 摩尔-库仑理论
Mohr-Coulomb yield criterion 摩尔-库仑屈服准则
Mohr-Coulomb's theory 摩尔-库仑强度理论；摩尔-库仑理论
Mohr's circle 摩尔圆
Mohr's circle diagram 摩尔圆图
Mohr's circle of stress 摩尔应力圆
Mohr's criteria 摩尔准则
Mohr's criterion 摩尔准则

Mohr's diagram 摩尔图解；摩尔图
Mohr's dome 摩尔穹
Mohr's envelope 摩尔包络线；强度包线
Mohr's hypothesis 摩尔假说
Mohr's number 莫氏硬度值
Mohr's rupture diagram 摩尔破坏图解
Mohr's rupture envelope 摩尔破裂包络线
Mohr's scale 摩氏硬度表
Mohr's strength theory 摩尔强度理论
Mohr's stress circle 摩尔应力圆
Mohr's theory 摩尔强度理论；摩尔理论
Mohr's theory of failure 摩尔破坏理论
Mohs hardness 莫氏硬度
Mohs hardness range 莫氏硬度分级
Mohs hardness scale 莫氏硬度标；莫氏硬度表
Mohs number 莫氏硬度值
Mohs relative hardness scale 莫氏相对硬度计
Mohs scale 莫氏硬度计；莫氏硬度表
Mohs scale of hardness 莫氏硬度计
moire 龟纹；波纹
Moire fringe method 莫尔干涉法
moire pattern 波纹绢布状图样
moire texture 波纹绢布状结构
moire topography 叠栅条纹图；穆瓦条纹图
moissanite 碳硅石；碳化硅
moist ash-free basis 恒湿无灰基
moist curing 湿法养护
moist district 湿润地带
moist earth 湿土
moist ground 湿土；湿地
moist mineral-matter-free basis 恒湿无矿物质基
moist rodding 湿捣法
moist season 雨季
moist soil 湿土
moist tamping 湿击法
moistadiabatic process 湿绝热过程
moistening 变湿；浸湿
moistness 水分；湿气；湿度
moistograph 湿度仪
moisture 湿度；潮湿；水分
moisture absorbent 吸湿剂
moisture absorption 吸湿性
moisture absorption test 吸湿试验
moisture accumulation 水分聚集
moisture adjustment 水分调整；湿度调整
moisture allowance 允许水分留量
moisture and ash-free basis 减湿减灰计算法
moisture apparatus 湿度计；测湿器
moisture balance 水分平衡
moisture barrier 隔水体；防潮层
moisture bowl 测湿箱
moisture can 保湿土样盒；含水量盒
moisture capacity 湿度；容水度；持水量
moisture capacity method of thermal conductivity measurement 热导率测定含水量法
moisture circulation 水分循环
moisture circumstances 潮湿环境
moisture condition 湿度；含水状态
moisture conservation 水分保持
moisture content 含水量；湿度；含水率

moisture content analyses	水汽含量分析	moisture sensitivity	吸水灵敏度；湿度灵敏度
moisture content metre	含水量测定计	moisture source	水分来源
moisture content of aggregate	骨料含水量	moisture stabilization	湿度稳定
moisture content of coal	煤水分	moisture status	水分现状
moisture content of natural	天然含水量	moisture storage	水分储存量
moisture content test	含水率试验	moisture storage capacity	储水量
moisture control	含水量控制；水分控制；湿度控制	moisture stress	水分应力
moisture deficiency	湿度差；湿气缺乏	moisture tension	水分张力
moisture deficit	水分亏损；缺少水分；含水量不足	moisture test	湿度试验；含水量试验
moisture density test	湿度-密度试验	moisture tight	防潮的
moisture deposit	水分沉降	moisture trap	气水分离器
Moisture determination	含水量测定	moisture volume percentage	容计含水率
moisture entrance	透入湿气	moisture weight percentage	干重含水率
moisture equivalent	水分当量；持水当量	moisture-density curve	击实曲线；压实曲线；湿度-密度曲线
moisture excluding efficiency	水分排除效率	moisture-density nuclear gauge	同位素含水量-密度测定仪
moisture expansion	湿膨胀	moisture-density relationship	湿度-密度关系
moisture field	湿度场	moisture-density test	击实试验
moisture film	湿膜；水膜	moisture-film cohesion	薄膜水内聚力
moisture film cohesion	湿膜黏聚力	moisture-free	干的；无水分的
moisture flux	水汽通量	moisture-free basis	无水基
moisture free	不潮湿的；不含水分的；干的	moisture-holding capacity	持水量
moisture gage	含水量测定计	moisture-induced stress	水分诱导应力
moisture gain	湿度增加	moisture-laden	含水的；饱水的
moisture gradient	湿度梯度	moisture-laden aggregate	饱水骨料
moisture holding capacity	持水能力；最高内在水分	moisturemeter	湿度计；水分计
moisture in air-dried coal	内在水分；风干煤的水分	moisture-proof	防湿
moisture in air-dried sample	空气干燥煤样水分	moisture-proof adhesive	防水胶黏剂
moisture in the air-dried sample	风干试样的水分	moisture-repellent	防潮的；抗湿的
moisture index	含水指数；湿度指数	moisture-resistant	防潮的
moisture loss	水分损失	moisture-resisting	抗湿的
moisture meter	湿度计	moisture-retaining property	保水性
moisture migration	水分迁移	moisture-retentive	保湿的
moisture movement	水分移动	moisture-temperature index	温湿指数
moisture of absorption	吸湿性；吸收湿度；吸水量；含水量	moisturetight	耐湿的
moisture of blend	配料水分	moja	泥熔岩
moisture peak	水分峰值	molality	重量克分子浓度
moisture penetration	水分渗入深度	molar concentration	摩尔浓度
moisture permeability	透湿性	molar fraction	摩尔分数
moisture pick-up	吸水量	molar gas constant	摩尔气体常数
moisture prevention	防潮	molar ratio	摩尔比
moisture probe	同位素含水量探测仪	molar volume	摩尔体积
moisture proof	耐湿的	molarity	体积克分子浓度；容积摩尔数；摩尔浓度
moisture proofing	防湿；防潮	molasse	磨砾层；磨拉石；磨拉石建造；软砂岩沉积
moisture proofness	耐湿性；防湿性	molasse face	磨砾岩相
moisture protection	防潮	molasse formation	磨拉石建造
moisture ratio	含水比	molasse type sediment	磨拉石型沉积
moisture reduction	降低水分	moldavites	莫尔达维玻陨石；绿玻陨石
moisture regain	回潮	mole	防波堤；突堤
moisture regime	水分状况	mole channel	地下排水道
moisture resistance	抗潮湿性；抗湿性	mole coefficient	摩尔系数
moisture resistant	耐潮的	mole drain	暗沟；地下排水沟
moisture resistant grade	防潮等级	mole drainage	地下排水工程；开沟排水
moisture retardant	抗湿剂	mole head	突堤堤头
moisture retention	保水性；吸湿性	mole mining	用小型遥控连续采煤机采薄煤层
moisture retention capacity	持水量	mole-channel	鼠道；地下排水沟
moisture retention curve	持水曲线	molecular	分子的；克分子的
moisture room	保湿室	molecular absorption band	分子吸收带
moisture sample	全水分试样湿状样品	molecular adsorption	分子吸附
moisture sampler	水分取样；湿度取样湿态取样器		

molecular attraction	分子引力
molecular attraction force	分子引力
molecular bond	分子键；范德华键；分子联结
molecular cohesion	分子黏聚力；内聚力
molecular compound	分子化合物
molecular concentration	分子浓度
molecular constitution	分子结构
molecular crystal	分子晶体
molecular crystal lattice	分子晶格
molecular diffusion	分子扩散
molecular diffusion coefficient	分子扩散系数
molecular diffusivity	分子扩散率
molecular dispersion	分子分散
Molecular Dynamics	分子动力学
molecular energy level	分子能级
molecular force	分子力
molecular formula	分子式
molecular heat	分子热
molecular hydrogen	分子氢
molecular hydrogen chemistry	分子氢化学
molecular hydrogen spectrum	分子氢光谱
molecular level	分子水平
molecular markers	分子标志物
molecular mass	分子量
molecular modeling	分子模型
molecular moisture capacity	分子含水量；分子吸湿量
molecular movement	分子运动
molecular nitrogen	分子氮
molecular orbital method	分子道轨法
molecular orbital theory	分子轨道理论
molecular orientation	分子定向；分子定位
molecular oscillation	分子振动；分子振荡
molecular oxygen	分子氧
molecular oxygen emission	分子氧发射
molecular oxygen spectrum	分子氧光谱
molecular percentage	分子百分数
molecular physics	分子物理学
molecular polarity	分子极性
molecular polarizability	分子极化率
molecular proportion	分子比
molecular pump	分子泵
molecular scattering	分子散射
molecular sieve	分子筛
molecular sieve action	分子筛作用
molecular sieve effect	分子筛效应
molecular sieves gas purification	分子筛法煤气净化
molecular simulations	分子模拟
molecular solution	分子溶液
molecular spectroscopy	分子光谱学；分子光谱法
molecular spectrum	分子光谱
molecular structure	分子结构
molecular volume	克分子体积
molecular water	分子水
molecular weight	分子量
molecule	分子；克分子
mole-hole dual operation s	零星采矿；小量开采
mole-pipe drainage	暗沟管道排水
moler	硅藻土；硅藻泥岩
molisite	铁盐
mollic epipedon	松软表层土
mollification	软化；缓和；减轻
molling	开排水鼠洞
mollisol	冻土融化土层；冻融层；松软土；软土
mollition	冻土融化
Molodensky formula	莫洛坚斯基公式
Molodensky theory	莫洛坚斯基理论
Molokai Fracture Zone	莫洛凯断裂带
molten	熔融的
molten chamber	熔浆囊；熔浆房
molten condition	熔化状态
molten granite	熔融花岗岩
molten lava	熔浆；熔岩
molten magma	熔融岩浆；液态岩浆
molten mass	熔融体
molten rock	熔岩
moluranite	水钼铀矿
molybdate	钼酸盐
molybdenate	钼酸盐
molybdenite	辉钼矿
molybdenum	钼
molybdenum chloride	氯化钼
molybdenum disulfide	二硫化钼
molybdenum disulfide lubricant	二硫化钼润滑剂
molybdenum glance	辉钼矿
molybdenum isotopes	钼同位素
molybdenum minerals	钼矿类
molybdenum ore	钼矿
molybdenum oxide	氧化钼
molybdenum sulfide	硫化钼
molybdic	含钼的
molybdic ocher	钼华
molybdite	钼华
molybdophyllite	硅镁铅矿
molybdosodalite	钼方钠石
molysite	铁盐
moment	力矩；弯矩；挠矩
moment about point of support	支点弯矩
moment about the origin	原点力矩
moment area method	力矩面积法；弯矩面积法
moment arm	力矩臂
moment at support	支点弯矩
moment axis	弯矩轴
moment center	弯矩中心
moment centre	矩心
moment coefficient	力矩系数
moment compressing failure	弯压破坏
moment connection	抗弯联结
moment couple	偶矩
moment curve	弯矩曲线；弯矩图
moment diagram	力矩图；弯矩图
moment distribution	力矩分配；弯矩分配
moment distribution method	弯矩分配法；力矩分配法
moment equilibrium	力矩平衡
moment equilibrium factor of safety	力矩平衡安全系数
moment function	矩函数
moment gradient	弯矩梯度
moment load chart	弯矩荷载图
moment magnitude	矩震级
moment method	矩量法；力矩法
moment modified factor	弯矩调幅系数

moment of bending	弯矩
moment of couple	力偶矩
moment of force	力矩
moment of friction	摩擦力矩
moment of gyration	转动惯量；旋转力矩
moment of inertia	转动惯量；惯性矩
moment of inertia of a mass	物体的转动惯量；物体的转动力矩
moment of inertia of cross-section	截面惯性矩
moment of momentum	动量矩；角动量
moment of overturning	倾覆力矩
moment of resistance	阻力矩；抵抗力矩
moment of rotation	转矩
moment of rupture	破坏力矩；弯折力矩
moment of sliding force	滑动力矩
moment of sliding resistance	抗滑力矩
moment of spectral density function	谱密度函数矩
moment of stability	稳定矩
moment of statics	静力矩
moment of torsion	扭矩
moment redistribution	力矩再分配；弯矩重分布
moment tensor	力矩张量；矩张量
moment tensor analysis	矩张量分析
moment test	弯矩试验
moment-area method	力矩面积法
momentary fall of pressure	瞬时压降
momentary load	暂时负载；瞬时负载
momentary minimum discharge	瞬时最小流量
momentary peak discharge	瞬时洪峰流量
momentary rise of pressure	瞬时压力升高
momentary sea level	瞬时海面
momentary tractive effort	瞬时牵引力
momentary value	瞬时值
momentary water level	瞬时水位
moment-curvature relationship	弯矩-曲率关系
moment-density tensor	矩密度张量
moment-resisting connection	抗弯联结；抗弯节点
moment-resisting frame	抗弯框架
moment-resisting space frame	抗弯空间框架
moment-rotation relationship	弯矩-转角关系
moments equilibrium	力矩平衡
moments method	力矩法；矩量法
momentum	动量；冲量；冲力
momentum coefficient	动量系数
momentum correction factor	动量修正系数
momentum curve	动量曲线
momentum diffusion	动量扩散
momentum distribution	动量分布
momentum equation	动量方程
momentum flux	动量通量
momentum force coefficient	动量力系数
momentum grade	动能坡道；动力坡度
momentum gradient	动量梯度
momentum interchange	动量交换
momentum law	动量定律
momentum moment	动量矩
momentum principle	动量原理
momentum theorem	动量定理
momentum theory	动量理论
momentum thickness of boundary layer	边界层的动量厚度
momentum transfer	动量传递；动量交换
momentum-loaded block foundation	动量加载整体基础
monadnock	残山；残丘
monalbite	钠正长石
monazite	独居石
monazite sand	独居石矿砂
moncheite	碲铂矿
monchiquite	沸煌岩
mondhaldeite	闪辉二长煌斑岩；白榴闪辉斑岩
Mondi process	蒙特炼镍法
monetite	三斜磷钙石
monheimite	铁菱锌矿
monimolimnion	永滞层；深部滞水层
monimolite	绿锑铅矿
monite	黄胶磷矿；碳磷灰石
monitor	监察器；监控器；监测仪
monitor and control	监控
monitor device	监控设备
monitor manual	监测手册
monitor message	监控信息；监视信息
monitor program	监控程序
monitor signal	监测信号；监控信号
monitor station	监测台
monitor terminal	监视终端；监控终端
monitor well	监测井
monitor work	监控工作；监测工作
monitored drilling	仪表化钻进
monitoring	监测；施工监督
monitoring and control system	监控系统
monitoring and information system	监控显示屏
monitoring and prediction network	监测预报网络
monitoring apparatus	监控装置
monitoring borehole	监测孔
monitoring car	监测车
monitoring control facility	监控设备
monitoring device	观测装置；监测装置；监控装置
monitoring equipment	监测设备
monitoring feedback	监测反馈
monitoring for special term	专项监测
monitoring frequency	监测频次
monitoring gas	瓦斯监测
monitoring groundwater pollution	监测地下水污染
monitoring instrument	监控仪器；控制仪器；检验器
monitoring manual	监测手册
monitoring measurement	监控量测
monitoring methodology	监测方法
monitoring mine department	矿井监测部门
monitoring net	监测网
monitoring network	监测网
monitoring of building settlement and deformation	建筑物沉降变形观测
monitoring of landslide	滑坡监测
monitoring of microseismicity	微震监测
monitoring of pore-water pressure	孔隙水压力监测
monitoring of pressure in backfill	充填料中的压力监测
Monitoring of Radiation	辐射监测
monitoring of rock mass performance	岩体性能监测
monitoring of settlement and deformation	沉降变形监测
monitoring of slope deformation	斜坡变形监测

monitoring of support	井巷支护监测
monitoring of surrounding rock deformation of a tunnel	洞室围岩变形监测
monitoring of surrounding rock pressure of tunnel	洞室围岩压力观测
monitoring of the profile of the seabed	海床形状监测
monitoring of tunnel astringency	硐室收敛观测
monitoring performance of rock mass	岩体性能监测
monitoring point	监测点
monitoring point for level	水准监测点
monitoring program	监测项目
monitoring programme	监测计划
monitoring record	监测记录
monitoring recorder	监控记录器；监听记录器
monitoring relay	监控继电器
monitoring result	监察结果；监测结果
monitoring schedule	监测时间表；监测计划
monitoring scheme	监测布置
monitoring seismic activity in pillars	矿柱微震监测
monitoring site	监测点；监测场地
monitoring standard	监测标准
monitoring station	监察站；监测站
monitoring strategy	监测办法
monitoring survey	监测
monitoring surveyor	监察工程的测量师
monitoring system	监测系统
monitoring system analysis	监测系统分析
monitoring technique	监测技术
monitoring technology	监测技术
monitoring vessel	监察船
monitoring well	监测井
monitoring works	监测工程
monitors record	监视记录
monkey drift	通风联络斜巷；小探巷
monkey engine	打桩锤机；打桩机；锤式打桩机
monkey entry	小平巷；小巷道；急斜煤层的辅助平巷
monkey heading	小平巷；小巷
monkey hole	联络小巷
monkey rolls	小型辊轧机；小型辊碎机
monkey shaft	小暗竖井；溜井
monmouthite	闪霞岩
mono lake excursion	莫诺湖漂移
monoblock drill rod	整体铸造的钻杆
monobucket excavator	单斗挖土机
monochromatic wave	单色波
monochromator	单色仪；单色光镜
monochronogenous soil	原生土
monoclinal	单斜的
monoclinal block	单斜断块
monoclinal coast	单斜海岸
monoclinal dip	单斜的
monoclinal fault	单斜断层
monoclinal faulting	单斜断层作用
monoclinal flexure	单斜挠曲；单斜扰褶
monoclinal fold	单斜褶皱
monoclinal fold scarp	单斜褶皱崖
monoclinal mountain	单面山；单斜山
monoclinal ridge	单斜脊
monoclinal rising flood wave	单斜上升洪水波
monoclinal rising wave	单斜上升波
monoclinal river	单斜河
monoclinal shifting	单斜移动；单斜转移
monoclinal slope	单斜坡
monoclinal spring	单斜泉
monoclinal stratum	单斜地层；单斜层
monoclinal structure	单斜构造
monoclinal valley	单斜谷
monoclinal zone	单斜带
monocline	单斜；单斜层；单斜褶皱
monoclinic	单斜的；单结晶的；单斜的
monoclinic fabric	单斜组构
monoclinic form	单斜晶系
monoclinic lattice	单斜晶格；单斜点阵
monoclinic pyroxene	单斜辉石；斜辉石
monoclinic symmetry	单斜对称
monoclinic symmetry fabric	单斜对称组构
monoclinic syngony	单斜晶系
monoclinic system	单斜晶系
monoclonal	单斜的
monoclonal plunge	单斜插入
monoclonal ravine	单斜谷
monocomparator	单片坐标量测仪
monocrystal	单晶体
monocular hand level	手水准
monocular microscope	单目显微镜
monocular range finder	单目测距仪
monocycle	单旋回；单循环
monocyclic	单旋回的；单环的
monocyclic landform	单旋回地形
monocyclic orogenesis	单相造山运动；单旋回造山运动
monocyclic suturing	单旋回缝合
monodirectional	单向的
monodromic function	单值函数
monoelectrode sonde	单极探测器；单电极探针
monogene rock	单成岩
monogene volcano	单成火山
monogenetic	单成因的
monogenetic chain	单成山脉
monogenetic conglomerate	单成砾岩
monogenetic peneplain	单成准平原
monogenetic soil	单成因土
monogenetic volcano	单成因火山
monogenic rock	单矿岩；单成岩
monogeosyncline	单地槽；单主梁桥架
monohydrallite	铝红土；铝土矿
monohydrocalcite	单水方解石
monolayer	单层；单分子层
monolete	单裂缝
monolith	单块巨石；整块石料；原状土样；单成岩
monolith lysimeter	原状土中测渗计；原状土柱渗漏液采集器
monolithic	单一岩的；单成岩的；独石柱的；单片的
monolithic body	整体
monolithic concrete	整体混凝土
monolithic concrete bed	整体道床；整体混凝土底座
monolithic concrete lining	整体混凝土支架
monolithic concreting	整体混凝土浇筑
monolithic foundation	整体式基础
monolithic lining	整体浇筑衬砌
monolithic micro-computer	单片微型计算机

monolithic paving	整体浇筑地台
monolithic retaining wall	整体式挡土墙
monolithic structure	整体结构
monolithic track bed	整体道床
monolithic wall	整体墙
monolithically cast wall	整体浇灌墙
monometallic ore	单矿物矿石
monometamorphic diaphthoresis	单退化变质作用
monometamorphism	单变质作用
monomict breccia	单矿物碎屑角砾岩
monomict rock	单矿物碎屑岩
monomictic	单矿物碎屑的
monomineral ore	单矿物矿石
monomineralic	单矿物的
monomineralic rock	单矿物岩
monophasic soil	同相土壤；单相土
monophyletic	单系列的；单源的
monosymmetrical	单轴对称的
monotectic	偏晶体
monotectic point	偏晶点；独晶点
monotectic reaction	偏晶反应
monotectic transformation	偏晶转变
monotectoid reaction	偏析反应；独析反应
monothermite	单热石；高岭石
monotone difference scheme	单调差分格式
monotone function	单调函数
monotonic loading	单调加荷
monotonic static loading	单调静力荷载
monotonicity preserving difference scheme	保单调差分格式
monotonous loading curve	单调加载曲线
monotonous logic	单调逻辑
Monotron hardness test	莫诺硬度试验
monotropism	单变性
monotropy	单变性；单变现象
monotube pile	单管柱桩
monovalent ion	单价离子
monovariant system	单变体系
monoxide	一氧化物
monoxide detector	一氧化碳检定器
monstrometer	钻孔方向检验仪；坑道钻孔方向检测仪
Monte Carlo approach	蒙特卡罗方法
Monte Carlo calculation	蒙特卡罗计算
Monte Carlo debugging	蒙特卡罗排错技术
Monte Carlo method	蒙特卡罗方法；统计试验方法
Monte Carlo simulation	蒙特卡罗模拟
Monte Carlo simulation method	蒙特卡罗模拟法
Monte Carlo method	蒙特卡罗法
montebrasite	磷锂铝石
Monte-Carlo method	蒙特卡罗法
Monte-Carlo simulation	蒙特卡罗模拟
monteponite	方镉石
montesite	硫钨铅矿
monthly mean sea level	月平均海面
monticellite	钙镁橄榄石
monticule	小山岗；小丘；小火山丘
montiform	山形的；山状
montmorillonite	蒙脱土；蒙脱石；微晶高岭石
montmorillonite clay	蒙脱土；蒙脱石黏土
montmorillonitization	蒙脱石化；蒙脱石化作用
montmorillonoids	蒙脱石类
montomorillonite	蒙脱石
montoring	监测
montrealite	橄闪辉石岩
montroydite	橙红石；橙汞矿
monument protection	古迹保护
monument research	古迹研究
monument restoration	古迹修复
monument stone	界石
monument weathering	古迹风化
monumental ore	单矿物矿石
monumented bench mark	永久水准点
monuments	古迹
monzodiorite	二长闪长岩
monzogabbro	二长辉长岩
monzogranite	二长花岗岩
monzonite	二长岩
monzonitic	二长岩的
monzonitic granite	二长花岗岩
monzonitic structure	二长构造
monzonitic texture	二长结构
monzonorite	二长苏长岩
monzosyenite	二长正长岩
mooing regulations	采矿规程；矿山技术操作规程
moon basalt	月球玄武岩
moon probe	月球探测仪
moon seismograph	月震仪
moon slide	月球断层
moon tides	月球潮汐
moonland	高地；高沼草原
moonmilk	月奶石；月乳；岩乳
moonpool	钻探船操作平台
moonquake	月震
moon's orbit	月球轨道
moon's phase	月相
moonscape	月面景观
moonstone	月长石；月石
moor	沼泽；矿脉的富集部分；泥炭沼
moor coal	沼煤；松散褐煤；松散泥煤
moor land	沼泽地；荒野
moor peat	沼煤
moor rock	粗砂岩
moor soil	沼泽土；沼地土
moorcoal	沼煤
mooring pile	系船桩；锚定桩
moorland	沼地；沼煤；高沼草原
moorpeat	沼煤；高沼泥炭
moorstone	花岗岩
morainal	冰碛的
morainal apron	冰碛平原；冰水沉积平原
morainal dam	冰碛坝
morainal delta	冰碛三角洲
morainal deposit	冰碛沉积
morainal lake	冰碛湖
morainal plain	冰水平原；冰碛平原
morainal topography	冰碛地形
moraine	冰碛；冰碛层
moraine belt	冰碛带
moraine breccia	冰碛角砾岩
moraine clay	冰碛土；冰碛黏土
moraine deposit	冰碛；冰碛沉积

moraine fan	冰碛扇
moraine hill	冰碛丘陵
moraine kame	冰碛阜
moraine lake	冰碛湖
moraine land	冰碛地
moraine landscape	冰碛景观
moraine of advance	前进冰碛
moraine of recession	后退冰碛
moraine of retreat	后退冰碛
moraine plateau	冰碛台地；冰碛高原
moraine profonde	底碛
moraine soil	冰碛土
moraine terrace	冰碛阶地
moraine-breccia	冰碛角砾岩
moraines	冰碛
morainic	冰碛的
morainic apron	冰碛缘
morainic channel	冰碛河床
morainic clay	冰碛黏土
morainic dam	冰碛坝；碛坝
morainic debris	冰碛物；冰碛岩屑
morainic deposit	冰碛沉积
morainic fan	冰碛扇
morainic loop	冰碛环
morainic material	冰碛物
morainic topography	冰碛地貌
Morarian orogeny	莫拉尔造山运动
morass	沼泽；湿地；泥沼
morass iron	海绵状褐铁矿
morass ore	沼铁矿；褐铁矿
mordenite	丝光沸石；发光沸石
morenosite	碧矾
morinite	红磷钠矿
mormal force	法向力；正交力
mormal stress	正应力；垂直应力；法向应力
morphodynamics	地貌动力学
morphogenesis	地貌成因；地貌发生，地貌成因学；形态成因；形态发生
morphogenetic	地貌发生的；地貌成因的
morphogenetic classification of coast	海岸形态成因分类
morphogenetic force	地貌发生力
morphogenic environment	形态发生环境
morphogeny	地貌形成学；地貌形成作用
morphographic map	形态图；地势图
morphography	形态描绘学；描述地貌学
morphologic province	地貌区
morphologic region	地貌区
morphologic structural configuration	地形构造轮廓
morphologic unit	地貌单元
morphologic vertical zoning	地貌垂直分带
morphological	地貌的
morphological analysis	地貌分析；地貌形态分析
morphological association	形态缔合
morphological change	形态变异
morphological classification of coast	海岸形态分类
morphological crystallography	形态结晶学
morphological differentiation	形态分化
morphological evolution	形态演变
morphological filters	形态滤波器
morphological landscape unit	地貌景观单位
morphological parameters	形态参数
morphological property	形态特点
morphological region	地貌区
morphological unit	地貌单元；地貌地层单位
morphological variation	形态变异
morphology	形态学；地貌学
morphology of minerals	矿物形态
morphology of ore body	矿体形态
morphometric analysis	形态测定分析；地貌测量分析
morphometry	形态测定法；地貌测量
morphorogenic phase	造山幕
morphosculptures	形态风化
morphosequent	地表地貌
morphostratigraphic	地貌地层的
morphostratigraphic unit	地貌地层单位
morphostructure	构造地貌
morphotectonics	地貌构造学；地貌构造分析；构造地貌学
morphotectonics zones	地貌区
morphotropy	晶变
mortar	砂浆
mortar admixture	砂浆掺合料；灰浆混合料
mortar anchor	砂浆锚杆
mortar anchorage	砂浆锚固
mortar bed	砂浆层；灰砂层
mortar bed course	砂浆底层
mortar bolt	砂浆型锚杆
mortar bond	灰浆砌合
mortar bond strength device	砂浆黏结强度试验仪
mortar bound arch	浆砌拱圈
mortar bound surface	砂浆碎石面层
mortar characterization	灰泥表征
mortar consistency tester	砂浆稠度仪
mortar grouting	灌浆
mortar grouting method	注浆法；压浆法
mortar grouting pump	双液注浆泵
mortar injection	砂浆注射
mortar joint	砂浆接缝；灰缝
mortar mill	灰浆搅拌机
mortar mix	砂浆；灰浆
mortar mixer	灰浆拌合机
mortar mixing machine	灰浆搅拌机
mortar mixing plant	砂浆拌合设备；灰浆拌合厂
mortar mixture	胶砂混合料；砂浆混合料
mortar penetration tester	砂浆稠度仪
mortar penetrometer	水泥砂浆贯入仪
mortar pump	砂浆泵
mortar rubble	浆砌片石
mortar rubble foundation	浆砌片石基础
mortar rubble retaining wall	挡浆墙
mortar sand	灰浆砂
mortar setting	灰浆凝固
mortar spraying	洒浆
mortar spraying machine	喷浆机
mortar structure	碎斑结构
mortar texture	碎斑结构；碎斑组织
mortar top	砂浆抹面
mortar trough	灰浆槽
mortar void method	灰浆孔隙法
mortar void ratio	灰浆孔隙比

mortarless wall	干砌墙
mortar-mixing plant	灰浆搅拌机
mortar-void	砂浆空隙
mortar-voids ratio	灰浆空隙比
mortary	碎斑状的
morvan	交切侵蚀面；准平面交切
mosaic	镶嵌结构；马赛克
mosaic area	镶嵌面积
mosaic assembly	航空像片镶嵌图
mosaic block	镶嵌地块
mosaic breccia	镶嵌角砾岩
mosaic briquette	马赛克型煤
mosaic evolution	镶嵌式演化
mosaic granoblastic texture	镶嵌粒状变晶结构
mosaic image	镶嵌图像
mosaic index	镶嵌索引图；像片索引图
mosaic map	镶嵌图
mosaic of landslide	镶嵌式滑坡
mosaic soil	镶嵌土
mosaic structure	镶嵌构造
mosaic structure framework	镶嵌格局
mosaic texture	镶嵌结构
mosaics	镶嵌构造；拼图
mosandrite	层硅铈钛矿
mosesite	黄铵汞矿
mosor	溶蚀残丘；喀斯特残丘
moss	苔藓；沼泽；泥炭沼
moss land	泥炭苔地
moss peat	藓类泥炭；沼泽泥类
moss peat mire	藓类泥炭沼泽
moss-agate	苔纹玛瑙
Mossbauer absorption spectroscopy	穆斯堡尔吸收光谱法
Mossbauer effect	穆斯堡尔效应
Mossbauer spectroscopic analysis	穆斯堡尔谱学分析
Mossbauer spectrum	穆斯堡尔谱
mossite	重铌铁矿
mossy zinc	海绵状锌
most difficult coal	极难选煤
most frequent water level	最常见水位；常现水位
most probable extreme value	最可能极值
most probable value	最概然值；最可能值
most suitable bit speed	最佳钻头速度
mota	黏土
mota well	黏土层中水井
mote	微屑
mote comet tail	尘埃彗尾
mother coal	煤母；天然木炭
mother current	主流；本流；母流
mother entry	集中平巷；干线平巷
mother gate	集中运输平巷；公用主要运输平巷
mother liquid	母液
mother liquor	母液
mother lode	母脉；主矿脉
mother magma	母岩浆
mother of coal	煤母；丝煤
mother oil	原油；原生石油
mother rock	母岩；围岩；源岩
mother rock provenance	母岩区；母岩来源区
mother water	母液
mother well	母井
mother-lode	主矿体
moths scale of hardness	莫氏岩石硬度系数；莫氏岩石硬度标
motility	原动力；备用能力
motion capture	动态捕捉
motion equation	运动方程
motion of Earth poles	地极移动
motion of ground surface due to blasting	爆破引起的地标运动
motion of progressive wave	前进波运动
motion of the earth	地球运动
motion of the earth's pole	地极运动
motion of the ground	地面运动
motion of translation	平移运动
motive fluid	运动液体；运动流体
motive force	原动力；动力
motive power	原动力
motor drill	自备电动机式钻机
motor jack hammer	内燃凿岩机
motor octane number	马达法辛烷值
motor oil	马达油
motor pump	机动泵；马达泵
motor scraper	机动矿耙；机动铲运机
motorized scraper	摩托化矿耙；拖拉机式铲运机
motor-pump	电动泵
mottled limestone	斑状石灰岩
mottled sandstone	杂色砂岩
mottled soil	杂色土
mottled structure	斑杂状构造；斑点构造
mottling	斑点；成斑作用；斑点构造
mottling lustre	斑点光泽；斑点闪光
mottramite	矾铜铅矿
moulded coal	成型煤；煤砖
moulded lining	模筑衬砌
moulded specimen	模制试样；成型试样
moulded-in-place pile	就地灌注桩
moulding box	型箱；砂箱
moulding sand	型砂
mouldy peat	风化泥炭
mouldy substance	腐殖质
moulin	冰川锅穴
moult	松软土地
mound	筑堤；堆土；堆石堤；抛石堤
mound breakwater	堆石防波堤；斜坡式防波堤
mound crown	堆石堤顶
mound drain	堆石排水
mound seismic reflection configuration	丘形地震反射结构；丘形震波反射形态
mound-type breakwater	倾斜防波堤；墩式防波堤
mound-type reef	丘状礁
mount meal	硅藻土
mountain	山；山脉
mountain and valley region location	山区河谷选线
mountain apron	山麓冲积平原；山麓冲积裙
mountain belt	山岳地带
mountain body	山体
mountain bog	山地沼泽
mountain breeze	山风
mountain building	造山；造山运动；山脉形成
mountain chain	山脉

mountain climate	山地气候；山区气候；山岳气候
mountain coast	沿山海岸；陡海岸
mountain cork	石棉
mountain creep	山体滑坡；山体蠕动
mountain crest	山脊；山峰
mountain crystal	水晶
mountain disease therapy apparatus	高山疾病治疗仪
mountain flour	硅藻石
mountain foot	山麓
mountain forming	造山运动
mountain front	山系前线
mountain glacier	高山冰川；山地冰川；山岳冰川
mountain green	孔雀石；硅孔雀石
mountain group	山群
mountain ice cap	平顶冰川
mountain lake	山地湖
mountain land	山地；山区
mountain landform	山岳景观
mountain landscape	山地景观
mountain leather	紧密石棉
mountain limestone	坚硬石灰岩
mountain location	山地位置
mountain making force	造山力
mountain making movement	造山运动
mountain mass	山体
mountain mass stability	山体稳定性
mountain meadow soil	山地草甸土
mountain meadow steppe soil	山地草甸草原土
mountain meal	岩粉；硅藻土
mountain meteorology	高山气象学
mountain mire	山地沼泽
mountain of accumulation	堆积山
mountain of denudation	剥蚀山
mountain of dislocation	断层山
mountain peak	山峰
mountain peat	山泥炭；山地泥炭
mountain peat soil	山地泥炭土
mountain pediment	山麓侵蚀平原
mountain railway	山区铁路
mountain range	山脉
mountain region	山岳地区；山区
mountain reservoir	山区水库
mountain ridge	山脊；山岭
mountain ring	环形山
mountain river	山区河流；山流
mountain road	山区道路
mountain road location	山路定线
mountain root	山脚；山根
mountain scarp	陡壁；峭壁
mountain sickness	高山病；高原病
mountain side	山腰
mountain slide	山崩
mountain slip	地滑
mountain slope	山坡
mountain soil	山地土壤
mountain spur	山嘴
mountain station	山岳气象站
mountain steppe soil	山地草原土
mountain stream	山地河；山区河流
mountain surface mine	山坡露天采场；山坡露天矿
mountain surfacemine	山坡露天采场
mountain system	山系
mountain terrain	山岭区
mountain top	山顶
mountain top removal method	倒推开采法
mountain topography	山岳地形
mountain torrent	山溪；山洪
mountain tract	山地河段
mountain trench	山间洼地；山沟
mountain tundra	高山冻原；冻土地带；山地冻原
mountain tunnel	穿山隧道；山岭隧道
mountain uplift	山脉上升
mountain valley	山谷
mountain wall	陡峭山坡
mountain waste	山地岩屑
mountain zone	山地带
mountain-building activity	造山运动
mountain-building movement	造山运动
mountain-chain	山脉
mountained	多山的
mountain-making movement	造山运动；造山
mountainous area	高山区；山地；山区
mountainous belt	多山地带
mountainous country	山岳地带；山地
mountainous desert	山地沙漠
mountainous hazards	山区灾害
mountainous region	多山地区，山区
mountainous terrain	山岭区；山区
mountainous territory	多山地区
mountainous topography	山岭地形；山区地形
mountain-range	山脉
mountains of dislocation	错位的山脉；断层山
mountain-slide	山崩
mountanous area	多山地区
mountanous region	山地
mounted drill	架式钻机；架式凿岩机
mounted self-propelled coal drill	装在自进式车上的煤钻
mounting bar	钻架支杆
mounting column	安装柱；钻架
mounting rack	装置架；工作台
mounting shaft	安装井
mourite	紫钼铀矿
mouse ahead	小直径超前钻孔
mouth of river	河口
mouth of well	井口
movable array	流动台阵
movable bearing support	活动支座
movable bed	活动河床
movable dam	带闸坝；活动坝
movable deck	可动平台
movable fill	移动式充填
movable frame	活动构架
movable hydraulic pressure lift platform	移动式液压升降平台
movable lift and hydraulic pressure type platform	移动式液压升降平台
movable lift platform	移动式升降台
movable load	活动负载
movable oil plot	可动油图
movable part	活动部分

movable partition 移动式隔断
movable pore volume 可驱移的孔隙体积
movable river bed 移动河床
movable water method 可动水法
movable weir 活动堰；活动坝
movable working platform 活动工作台
movable-bed model test 动床模型试验
movable-bed routing in open channel 明渠动床演算
movable-bed stream 动床河流
move relative to each other 相对运动
move the earth 进行土方工作；挖掘
movement gage 测动仪
movement joint 伸缩缝；变形缝；移动接缝
movement meter 位移计
movement of dislocation 断错运动
movement of earth crust 地壳运动
movement of earth's crust 地壳运动
movement of floating debris 悬浮质运动
movement of groundwater 地下水运动
movement of ice 流冰；冰凌运动
movement of mass 块体运动
movement of river sediment 河流泥沙移动
movement of river silt 河流泥沙移动
movement of sand wave 沙波运动
movement of water vapor 水汽移动
movement picture 运动图像
movement rates 移动率
movement trace 运动轨迹
movement-oriented 流动导向
moving armature geophone 电动式地震探测器
moving average 移动平均；滑动平均
moving average method 移动平均法
moving average model 滑动平均模型
moving axle 动轴
moving barometer 移动式气压计
moving bed 移动床
moving bed filter 动床滤器
moving bed process 动态床法
moving boundary 移动边界；移动界面
moving compensation disturbance 动补干扰
moving conductor geophone 动导体式地音探测器；电动式地震检波器
moving dislocation 移动位错
moving dislocation model 移动断裂模型
moving dune 移动沙丘
moving fault block 移动断块
moving ground excitation 激发地面震动；地面激震
moving least square approximation 移动最小二乘近似
moving least square method 移动最小二乘法
moving load 活动荷载；移动载荷；动载
moving load test 移动荷载试验
moving machine sampler 移动式取样器
moving mechanical sampler 移动式机械取样机
moving moraine 移动碛；移动冰碛石
moving overburden 活动层
moving point load 移动点荷载
moving sand dune 流动沙丘
moving scale 活动刻度盘
moving screen 运动筛
moving source electromagnetic method 动源电磁法
moving source method 动源法
moving surge 移动涌浪
moving time 移动时间
moving uniform load 均布动荷载
moving upstream antidune 上行反沙丘
moving weight function 运动权函数
moving window Fourier spectrum 动窗傅里叶谱
moving window technique 动窗法
moving-coil geophone 动圈式地音探测器；动圈式检波器
moving-coil pickup 动圈式拾震器
moving-conductor electromagnetic seismograph 动导体电磁地震仪
moving-cone simulation 动锥模拟
moving-geophone 动圈式地震检波器
moving-seismograph 动圈式地震仪；动导体地震仪
moving-seismometer 动圈式地震计
moving-source method 动源式电磁法；动源法；震源移动法
moving-system 动源系统
moya 泥熔岩；火山泥
moyite 钾长花岗岩
Mozambique Channel 莫桑比克海峡
MS blasting 毫秒爆破
MS delay electric blasting cap 毫秒延期电雷管
MS delay electric detonator 毫秒延期电雷管
MSK scale of earthquake intensity MSK 地震烈度表
MSS 多谱段扫描仪
MST radar MST 雷达
Mt. Maiji Grottoes 麦积山石窟
M-type granite M 型花岗岩；类幔源型花岗岩
muck 覆盖层；腐殖土；淤泥；废土石；污泥；泥渣；岩渣
muck bank 弃碴场；弃土场
muck bin 储碴仓
muck bottom 软泥底；淤泥底
muck bucket 岩石吊桶；铲岩斗
muck cake 泥饼
muck car 运碴车；装岩车
muck chute 出渣槽
muck disposal 弃碴场；弃土场
muck disposal area 排矸区；排矸场
muck draw 放矿；排石
muck drawing 放碴
muck equipment 出碴装置
muck foundation 腐殖土基底；泥炭底；淤泥地基
muck handling 排碴
muck hopper 出碴漏斗
muck loader 装岩机
muck loading 装岩；装矸
muck loading control station 装碴控制台
muck out 装出矿石；装出岩石
muck pile 石堆；矿石堆
muck pile profile 爆堆形状
muck raise 岩石溜井
muck removal 出碴
muck rush 废石突然放落
muck sample 软泥样品
muck shifter 废碴装运机
muck shifting 土方搬运

muck slope 碎石堆坡度
muck soil 淤泥类土；腐泥土；黑色腐殖土
muck spoil 弃土；废石料
muck transfer equipment 出碴转载装置
muck truck 运碴卡车
muck-car 运矸石车
muck-ditch 污泥沟
mucking 出碴；排碴；清理岩石
mucking and haulage 出碴
mucking and removing 出碴
mucking apparatus 装岩器械
mucking bucket 出碴铲斗；排碴铲斗
mucking crew 装岩组；装矿组
mucking device 装岩机；抓岩机
mucking equipment 装岩设备；装矸设备
mucking machine 装岩机；抓岩机；出碴车
mucking method 出碴方法；挖土运装法
mucking of inclined shaft sinking 斜井掘进排矸
mucking pan 装岩槽
mucking plate 装岩用垫板
mucking rate 装岩速度
mucking ratio 排土率
mucking route 出碴路线
mucking system 出碴系统
mucking trains 土石碴运输车
muckle rig 大型钻机；大型凿岩机
muckpile 待装的岩石堆
muckpile profile 爆堆形状
muck-shifting 剥离工作
mucky soil 淤泥质土
mucous layer 黏液层
mud 淤泥；软泥；泥浆
mud accumulation 淤泥；灌浆
mud acid 无机泥酸
mud additive 泥浆附加物
mud agitator 泥浆搅拌器
mud analysis log 泥浆分析纪录
mud and microbiological accumulation 生物黏泥
mud and rock flow 泥石流
mud arrival 泥浆到达波
mud avalanche 泥流；泥石流；泥崩；土崩
mud avalanche aqueduct 泥石流渡槽
mud avalanche cave 泥石流明洞
mud avalanche ditch 泥石流排导沟
mud bailer 抽泥筒；泥浆泵
mud balance 泥浆天平；钻泥密度秤
mud ball 泥球
mud bank 泥滩；泥堆
mud bar 泥坝
mud belt 泥沉积带；泥带
mud bit 菱形尖钻头；菱形尖钎头
mud blasting 泥盖法爆破；泥封爆炸法；糊炮爆破
mud bowl 泥箱；泥渣捕集器
mud box 泥浆防喷盒；泥箱
mud breccia 泥角砾岩
mud cake 泥饼；泥皮
mud cap 泥浆帽
mud capping 糊炮；黏接爆破
mud chamber 泥浆池
mud chemist 泥浆或冲洗液化学师

mud circulation 泥浆循环
mud cleaning machine 除泥机；清道机
mud column 泥浆柱
mud cone 泥丘；泥火山锥
mud content 含泥量
mud crack 泥裂；干裂；龟裂
mud crack polygon 泥田龟裂
mud cracking 泥裂作用
mud crust 泥壳
mud cup 泥浆杯
mud density 泥浆密度
mud deposits 淤泥沉积物
mud desander 泥浆除砂器；泥浆筛
mud diapir 泥底辟；泥贯入体
mud disposal 泥土处置工程；废泥浆
mud disposal area 淤泥卸置区；卸泥场
mud ditch 泥浆槽
mud drag 疏浚机；刮泥机
mud drainage in underground mine 井下排泥
mud dredging 挖泥
mud drilling 钻探泥浆
mud dumping area 抛泥区
mud dumping foreshore 泥滩
mud dumping site 泥头倾卸区
mud engineer 泥浆工程师；冲洗液工程师
mud engineering 钻机冲洗液工程
mud filling 淤泥
mud filter 滤泥器
mud filtrate 泥浆滤出液
mud filtration 矿泥过滤
mud flat 泥质海滩；海滨泥地；淤泥滩；泥潮滩
mud flat coast 泥滩岸
mud flow 泥流；泥石流
mud flow deposit 泥流沉积
mud flow indicator 泥浆流量指示器
mud flow sensor 泥浆流量传感仪
mud fluid 钻探用泥浆；黏土浆
mud fluid displacement 换浆
mud flush 用泥浆冲洗；钻探用泥浆
mud flush drilling 泥浆钻探
mud foreshore 泥前滨；泥滩
mud forming clay 造浆黏土
mud formula 泥浆配方
mud gauge 泥浆计
mud geyser 泥喷泉
mud ground 淤泥地
mud grouting 泥浆灌浆；灌泥浆
mud guard 挡泥板
mud gun 泥浆枪
mud gusher 泥水喷泉
mud hole 出泥孔
mud in suspension 悬浮泥浆
mud injection 泥浆灌注
mud invasion 泥浆侵入
mud jack 压浆泵
mud jack method 压浆法
mud jacking 混凝土覆面层灌浆；喷射泥浆；压泥浆；压浆法
mud laden water 含泥水
mud land torrent 石洪

mud lava	泥熔岩	mud solid	泥浆颗粒；泥粒
mud layer	泥浆层	mud spout	泥浆喷出；喷水冒砂
mud lime	灰泥；泥质氧化钙	mud spring	泥泉
mud line	泥线	mud stone	泥岩；泥岩
mud lining	泥皮；结泥饼	mud stream	泥流；泥石流
mud lobe	泥舌	mud sump	沉泥池
mud logger	泥浆录井工	mud support	灰泥支撑
mud logging	泥浆测井；钻井液录井	mud tank	泥浆罐
mud loss	泥浆漏失	mud theology	泥浆流变学
mud lubrication	泥浆润滑；泥浆压井	mud thinner	泥浆减稠剂；泥浆稀释剂
mud lump	泥丘；泥火山；淤积池	mud tongue	泥舌
mud manifold	泥浆管线	mud transport	泄沙阀
mud mark	泥痕	mud trap	沉泥池；泥浆池
mud materials	泥浆材料	mud treating equipment	泥浆处理设备
mud mixer	拌泥机；泥浆搅拌机；泥浆拌合器	mud treatment	泥浆处理
mud mortar	泥灰浆	mud treatment plant	泥浆处理装置
mud motor	泥浆动力钻具	mud valve	排泥阀
mud mound	泥丘；泥岗	mud viscosity	泥浆黏度
mud mound complex	灰泥丘复合体	mud volcanic structure	泥火山构造
mud off	灌泥浆；泥浆护壁；形成泥封	mud volcano	泥火山
mud oil	钻井采油	mud volcano gas	泥火山气
mud peat	严重分解的泥炭；泥状泥炭；烂泥炭	mud volume meter	泥浆容积仪
mud pies	泥团	mud wall	土墙
mud pipe	排泥管	mud wave	泥浪；泥滑
mud pit	钻浆坑；泥浆坑	mud weight	泥浆比重
mud polygon	多边形泥裂；泥质多边形土	mud weight in	送入泥浆比重
mud port	泥浆排出口	mud weight out	排出泥浆比重
mud pot	热泥泉；泥喷泉；泥塘	mud yield	造浆率
mud program	泥浆标定	mud zone	泥带
mud pump	泥浆泵；钻进泵	mudball	泥球
mud pumping	翻浆冒泥；泥浆泵送	mud-bearing	含泥质的
mud reclamation	泥浆回收	mudcake	泥饼；泥皮
mud replacement by blasting	爆破排淤法	mudcake correction	泥饼校正
mud residue	泥渣	mudcake effect	泥饼影响
mud resistivity	泥浆电阻率；钻井液电阻率	mud-cake growth	形成泥皮
mud return line	泥浆返回管线	mudcap	封泥爆破法
mud ring	底圈；除泥孔环	mudcap blasting	裸露爆破；覆土爆破；外部装药爆破
mud rock	泥岩；弱凝固泥沉积物	mudcap process	泥封法
mud rock debris flow	泥石流	mudcapping	糊炮爆破；外部装药爆破
mud rock flow	泥石流	mudcapping method	泥盖法
mud rock flow sediments	泥石流沉积物	mud-capping without capping	不封泥的糊炮爆破；裸露装药爆破
mud run	泥砂突然涌入	mudcrack	泥裂
mud rush	泥砂涌入	mud-debris flow	泥石流
mud scale	泥浆比重计；泥浆密度计	mudded off	泥封的
mud scraper	刮泥机；刮泥板	mudding	泥浆护壁；泥封；黏土灌浆
mud screen	泥浆筛	mudding action	泥浆护壁作用；泥封作用
mud scum	翻浆冒泥；浮泥	mudding intercalation	泥化夹层
mud seam	泥夹层	mudding off	泥封；灌泥浆；造壁
mud settler	沉泥器	muddy	泥的；泥质的；淤泥的；泥浆的；混浊的
mud settling pit	沉淀池	muddy beach	泥滩
mud settling sump	泥浆沉淀池	muddy coast	泥质海岸；淤泥质海岸
mud shale	泥页岩	muddy creek	泥溪
mud sheath	泥浆造壁	muddy deposit	淤泥沉积
mud shield	泥水加压盾构	muddy gravel	泥砾；泥砾层；泥砾沉积
mud shovel	泥铲	muddy ground	淤泥地
mud sill	地梁	muddy intercalation	泥质夹层；泥化夹层
mud silting	淤积；泥淤积	muddy limestone	泥灰岩
mud snapper	抓泥器	muddy logging	泥浆录井
mud socket	钻井泥砂清除器；泥泵	muddy pool	泥塘；沸泥塘
mud soil	淤泥土		

muddy rip-up clast	泥浆撕裂碎屑
muddy sand	泥质砂；泥质砂层；泥砂沉积
muddy soil	未垦地；淤积池；淤泥土；淤泥质土
muddy spring	泥水泉；浊泉
muddy structure	泥状结构
muddy volcanic cone	泥火山锥
muddy water	泥水；泥浆水；浊水
mudflat	潮泥滩；泥坪；海滨泥地
mudflat polygon	泥滩龟裂土
mudflow	泥流；泥石流；火山泥流
mud-flow conglomerate	泥流砾岩
mudflow fan	泥石流扇
mud-flow gradient	泥流梯度
mud-flow soil	泥流土
mudflow terrace	泥流阶地
mud-flush	用泥浆冲洗的
mud-flush boring	泥浆冲洗钻进
mud-flush drilling	钻泥冲洗钻进
mud-flush method of sinking	泥浆冲洗钻井法
mud-flush of sinking	洗井
mud-hog pump	砂泵；泥浆泵
mudhole	排泥孔；除泥孔
mud-invaded zone	泥浆侵入带
mudjack	压浆
mud-jack method	水泥浆灌注修理法
mudline	泥线
mud-loss instrument	井漏探测器
mudlump	泥火山；泥丘
mudlump coast	泥丘海岸
mud-making capacity	造浆性能
mud-making formation	造浆岩层
mudprone	泥坡的
mudprone facies	泥坡相
mud-pulse	泥浆脉冲
mud-pulse transmitter	泥浆脉冲发射器
mudpump	抽泥；泥浆泵
mud-pumping	翻浆；冒泥
mudrock	泥岩；泥状岩
mudrock chemistry	泥岩化学
mudrock diagenesis	泥岩成岩作用
mudrock flow	泥石流
mudrock sedimentology	泥岩沉积学和动物
mud-rush	泥砂浆突然涌入；泥浆突然涌入
mud-sand flow	泥沙流
mud-scow method	泥船钻进去
mud-setting pit	泥浆沉淀池
mud-settling pit	沉泥池
mud-settling pool	沉淀池
mudsill	底梁；下槛；纵向平梁；泥岩嵌入层
mudslide	泥滑；泥塌；淤泥倾泻
mudslide platform	抗泥崩平台
mudslides	海底泥滑动
mudspate	泥流
mudstone	泥岩；泥状灰岩
mudstone caprock	泥质岩类盖岩；泥岩类冠岩
mudstone reservoir	泥质岩层
mudstream	泥流
mudsupported	灰泥支撑的
mud-supported biomicrite	灰泥支撑的生物微晶灰岩
muff	衬套；轴套；套筒
muff coupling	套筒联轴节；套接
muff joint	套管接头；套管连接
mugearite	橄榄粗安岩
muglalite	角闪硅酸岩
mulch	护根覆盖层；盖土
mulching soil	覆盖土
muldakaite	次闪辉绿岩
mulde	凹地；向斜层
mule foot a bit	钻头偏磨
mule shoe guide	斜口引鞋
mule shoe latch	斜口管鞋爪
mule-head hanger	驴头上挂抽油杆的装置
mule-shoe nipple	斜口管鞋短节
muleshoe orientation method	斜口管鞋定向法
muleshoe orienting device	斜孔造斜工具
muleshoe slinger lock	斜口管鞋投掷锁定器
muleshoe sub	斜口接头
mullanite	块硫锑铅矿
mullen burst test	耐破度试验
mullerite	软绿脱石
mullicite	蓝铁矿
mullion structure	窗棂构造；栅状构造
mullite	莫来石；多铝红柱石
mullitization	莫来石化作用
mullock	矿山废石堆
mullock chute	充填天井
mullock tip	废石堆；矸石山
mullocking	装运废石
multi beam echo sounder	多波束回声测深仪
multi channel logging truck	多线式测井站
multi cycle operation	对循环操作；多循环作业
multi directional shear	多向剪切
multi frequency exploration system	多频探测系统
multi grid	多层格栅；多重网格法
multi hole directional drilling	多开眼定向钻井
multi layer organization	多层结构
multi level hoisting	多水平提升
multi media extension	多媒体扩展软件
multi phase region	多相区
multi ring basin	多环盆地
multi service tunnel	合用沟
multi stage explosion effect	缓爆效果；迟爆效果
multi stage metamorphism	多阶段变质作用
multi step recurrence algorithm	多步递推算法
multiaddress	多地址
multi-address code	多地址码
multi-address computer	多地址计算机
multi-address imstruction	多地址指令
multi-address message	多地址信息
multi-address order code	多地址指令码
multi-amplifier	多级放大器
multi-analysis	多方面分析；全面分析；多重分析
multi-anchor anchorage	群锚锚固
multi-anchorage system	多层锚定系统
multi-anchored wall	梯式加筋锚定墙
multi-angle imager	多角度成像仪
multi-anomaly	多元素异常
multi-aperture	多孔的
multi-aperture device	多孔磁心；多孔器件
multiaquifer formation	多层含水层；多层含水建造

multiaquifer system	多含水层系
multiaquifer well	多含水层井
multi-arch dam	连拱坝；多拱坝
multi-arch tunnel	联拱隧道
multi-armed centralizer	多臂扶正器
multiaxial	多轴的
multi-axial compression	多轴压缩
multi-axial compression test	多轴抗压测试
multi-axial cyclic deformation	多轴向反复变形
multi-axial hardening coefficient	多轴向硬化系数
multiaxial tensile test	多轴拉伸试验
multi-azimuth	多方位；多仰角
multiband	多波段
multiband aerial camera	多波段航空摄影机
multiband color photography	多波段彩色摄影
multiband data	多波段资料
multiband image	多波段摄影图像；多波段扫描图像
multiband imagery	多波段成像
multiband imaging system	多波段成像系统
multiband information	多波段信息
multiband photograph	多波段摄影相片
multiband photographic image	多波段摄影图像
multiband photography	多波段照像；多波段摄影
multiband picture	多波段图像
multiband remote sensing	多波段遥感
multiband sensor	多波段传感器
multiband spectral analysis	多波段光谱分析
multi-band spectrum transformation	多波段频谱变换
multiband system	多波段系统
multi-band terminal	多频终端
multi-bank	多组的
multi-baseline solution	多基线解
multi-beam	多波束
multibeam antenna	多波束天线
multibeam bathymetric survey	多波束测深
multibeam bathymetry	多波束测深
multibeam data	多波束数据
multibeam echosounding	多波束回声测深
multi-beam holography	多光束全息术
multibeam mapping	多波束测绘
multibeam oscilloscope	多线示波器
multi-beam scan imaging method	多波束扫描成像法
multi-beam sonar	多波束声呐
multibeam sounding system	多波束测深系统
multi-beam survey	多波束调查
multi-beam swath bathymetry	多波束条带测深
multibeam ultrasonic meter	多束超声流量计
multi-bed	多床层的
multibench blasting	多阶爆破
multi-blade	多刃的
multi-blade bit	多刃式钻头
multiblast high-speed developing	多次爆破快速开拓
multi-body deformable contact	多体变形接触
multi-boom drill	多臂钻
multi-bore well	多底井
multi-bottom directional hole	多孔底定向孔；定向分支孔
multibranched drilling	多底井钻井
multi-bucket dredger	多斗挖泥机
multibucket excavator	多斗挖土机
multicarrier wave distancemeter	多载波测距仪
multicased deep well	多套管深井
multi-centered arch	多心拱
multi-central projection	多中心投影
multichannel	多道的；多路的；多信道的
multichannel access	多信道选取
multi-channel access system without central control	无中心多信道选址系统
multi-channel acquisition	多道采集
multichannel amplifier	多通道放大器
multichannel analyzer	多道分析器
multichannel coherence	多道相干
multichannel coherency filter	多道相干滤波器
multichannel colour sensor	多波道彩色传感器
multi-channel data acquisition system	多道数据采集系统
multichannel deconvolution	多道反褶积
multichannel discriminator	多道甄别器
multichannel electromagnetic oscillograph	多道电磁示波器
multichannel filtering	多道滤波
multichannel gamma ray spectrometer	多道伽马射线谱仪
multichannel image	多波道图像
multichannel magnetic tape	多道磁带
multi-channel memorizer	多道存储器
multi-channel modulator	多路调制器
multichannel multiplex	多道传输器；多道转换器
multichannel multiplier	多通道乘法器
multichannel ocean color sensor	多通道海洋彩色传感器
multichannel optimal filter	多道最佳滤波
multichannel oscillograph	多道振荡器；多道示波器
multichannel processing	多通道处理；多道处理
multichannel processor	多道信息处理机
multichannel record	多道记录
multichannel recorder	多道记录器
multichannel recording	多道记录
multichannel recording oscillograph	多路录波器
multichannel reflection seismology	多通道反射地震
multichannel seismic exploration	多道地震勘探
multichannel seismic instrument	多道地震仪
multichannel seismic profiler	多道地震剖面仪
multichannel seismic reflection	多道地震反射
multichannel seismograph	多道地震仪
multicarrier sensing capability	多道灵敏度
multichannel spectrometer	多道谱仪
multichannel system	多通道系统
multichannel tape recorder	多通道磁带记录器
multichannel technique	多道技术
multichannel telemetry seismic system	多道遥测地震系统
multicharge	混合装药
multicharge composite explosives	混合炸药
multichromatic spectrophotometry	多色分光光度计
multi-circuit hydraulic transmission	多循环液力传动
multicoil	多线圈
multicoil focused induction tool	多线圈聚焦感应测井仪
multicoil induction system	多线圈感应测井装置
multi-collar thrust bearing	多环式止推轴承
multicollisional	多次碰撞的；复碰撞的
multicolor	多色的
multicolor hologram	多色全息图
multicolor image	多色图像
multicolor isotropy of mine product	多色图

multicolor map 多色地图
multi-color spectrum 多色谱
multicolor three dimension image 多色三维图像
multi-column foundation 多柱基础
multi-column hydraulic press 多柱式液压计
multi-column pier foundation 群桩基础
multi-combination meter 多用途复合仪表
multicompartment shaft 多隔间立井
multicompleted well 多层完成的井
multicomponent 多分量的；多元的；多组分的
multicomponent brine 多组分盐水
multicomponent earthquake excitation 多分量地震激励
multicomponent effect 多分量效应
multicomponent mixture 多组分混合物
multi-component sorption 多组分吸附
multicompression 多级压缩
multicomputer 多机组；多计算机系统
multicomputing unit 多运算器处理机
multicontact miscibility 多次接触混相
multicontact miscible 多次接触混相的
multi-core magnetic memory 多磁心存储器
multicycle 多旋回；多周期；多循环
multi-cycle composite basin 多旋回复合盆地
multi-cycle dual operation 多循环作业
multicyclic 多旋回的；多循环的；多周期的
multi-data classification 多重数据分类
multideck screen 多层筛
multi-deck sinking platform 多层凿井吊盘
multi-decker cage 多层罐笼
multi-defective pile 多缺陷桩
multi-defence system of seismic engineering 多道抗震设防
multideformed terrain 多次变形区
multi-degree-of-freedom 多自由度
multi-degree-of-freedom system 多自由度系统
multidetector 多探测器
multidetector array 多检波器组合
multi-dimension 多维
multi-dimension structure 多维结构
multidimensional 多面的；多维的
multidimensional access 多维存取
multidimensional allocation 多维分布
multi-dimensional analysis 多维分析
multi-dimensional consolidation 多维固结
multidimensional convolution 多维褶积
multidimensional flow 多维流动
multidimensional Fourier trans form 多维傅里叶变换
multidimensional geometry 多维几何形态
multidimensional integrals 多维积分
multidimensional linearized inversion 多维线性反演
multidimensional model 多维模型
multi-dimensional multi-phase flow 多维多相流动
multidimensional normal distribution 多维正态分布
multidimensional objective function 多维目标函数
multidimensional optimization problem 多维最优化问题
multidimensional problem 多维问题
multi-dimensional scaled physical model 多维相似物理模型
multidimensional scaling 多维尺度；多维标度法
multi-dimensional search algorithm 多维检索算法
multi-dimensional searching 多维检索

multi-dimensional settlement 多维沉降
multi-dimensional signal 多维信号
multidimensional unconstrained 多维无约束
multidirectional block faulting 多向块断作用
multidirectional drilling 多筒钻井
multidirectional firing gun 多方向点火枪
multidirectional loading 多向加载
multidirectional normal fault 多向正断层
multidirectional photography 多向摄影
multidirectional shaking 多向振动
multidirectional shaking table test 多向振动台试验
multi-directional shear 多向剪切
multi-disc cutter 多刃盘形滚刀
multi-disciplinary 多学科的
multidisciplinary analysis 多学科分析；跨学科分析；综合分析
multidisciplinary approach 多学科研究方法
multidisciplinary diagnostic protocol 多学科诊断协议
multidisciplinary seafloor observatories 多学科海底观测站
multidisciplinary team work 跨学科团队协作
multi-disk flexible coupling 多盘弹性联轴节
multidomain 多磁畴；多区的
multi-domain approach 多域法
multi-domain F-K filtering 多域F-K滤波
multidomain grain 多畴颗粒
multi-domain technique 多域技术
multidomain thermal remanence 多畴热剩磁
multidrain well 多眼泄油井
multidraw 多点取样
multidrift 多导坑开挖法
multidrift construction 多巷掘进
multi-drift method 多导坑开挖法；多导坑法
multi-drill head machine 多轴钻床
multidrive 多点传动；复合传动
multi-drive conveyor 多传动输送机
multi-drum hoist for deep shaft 多滚筒深井提升机；布莱尔型提升机
multi-drum hoisting 多卷筒提升
multiecho 多次回声
multi-effect 多效的
multielement 多元素
multielement activation analysis 多元素活化分析
multi-element analysis 多元素分析
multielement bar 多元传感器
multi-element cooler 多元冷却器
multi-element filter unit 多芯过滤装置
multi-element geochemistry 多元素地球化学
multielement oscillograph 万用示波器；多元示波器
multielement parasitic array 多元无源天线阵
multielement seismoscope 多摆仪地震计
multi-expansion 多次膨胀
multifactor analysis 多元分析；多因素分析
multi-failure 多重失效
multi-featured mobile terminal 多功能移动终端
multifeeler casing caliper 多触点套管井径仪
multifid 多裂的
multi-finger caliper 多臂井径仪
multifinger contactor 多点接触器
multi-fissure 多裂隙

multiflow evaluator 多级流量地层测试器
multifold coverage 多次覆盖
multifold line 多次覆盖测线
multifold profiling 多次覆盖剖面法
multiform 多种的；多样的；多种形式的
multiform function 多值函数
multiformat output 多格式输出
multiformed 多次变形的
multiformity 多形
multifractals 多重分形
multi-fracture gas pool 多裂缝气藏
multifrequency 多频率
multi-frequency amplitude-phase method 多频振幅相位法
multi-frequency coding register 多频记发器
multifrequency exploration 多频勘探
multi-frequency horizontal loop electromagnet method 多频水平环路电磁波法
multifrequency signalling 多频信令
multifrequency transmission 多频发射
multifrequency vibration 复频振动
multifunction 多功能
multi-function hall 多功能厅
multi-function receiver 多功能接收器
multi-function vehicle 多功能车
multifunctional 多功能的
multifunctionality 多功能性
multigage 多用量测仪表；多用规；多用检测计
multigate decay-lithology tool 多门衰减岩性测井仪
multigelation 反复冻融作用
multigenerational dolomite 多代白云石
multigrade 多级的
multigrid method 多重网格法
multigrid preconditioning 多栅预处理
multigrid procedure 多重网格程序
multigroup approximation 多群近似
multigroup theory 多群理论
multi-harmonic function 多重调和函数
multi-harmonic resonance 多谐共振
multiharmonigraph 多谐记录仪
multi-hazard 多重灾害
multihazard mapping 多重灾害面分布图
multihole directional drilling 多井眼定向井
multihole drilling 多井眼钻井
multihole lance 多孔喷枪
multi-holed jet 多孔喷丝头
multihop 多跳
multihop path 多跳路径
multihypothesis test 多假设检验
multiimage 重复图像
multi-image enhancement 多重图像增强
multi-information back analysis 多元信息反分析
multi-input 多端输入
multi-input multi-output system 多输入多输出体系
multi-input system 多输入系统
multijaw 多颚式
multijaw grab 多瓣式抓岩机；多爪抓斗
multi-jet turbine 多喷嘴冲击式水轮机
multi-job 多道作业
multilateral sand 横向重叠砂体
multilateral well 多底井；分支井

multilateration 多边测量；多点定位
multilayer 多层的
multilayer adsorption 多层吸附
multilayer aquifer 多层含水层
multilayer feedforward neural network 多层前馈式神经网络
multilayer field 多油层油田
multilayer injector 合注井
multilayer karst cave 多层溶洞
multi-layer organization 多层结构
multilayer perceptron 多层感知器
multilayer producer 合采井
multilayer reservoir model 多油层油藏模型
multilayer semi-infinite solid 多层半无限体
multi-layered aquifer 多层含水层
multilayered elastic medium 多层弹性介质
multi-layered fluid 多层流体
multilayered fold 多层褶皱
multilayered karst cave 多层溶洞
multilayered medium 多层介质
multilayered model 多层模型
multilayered reservoir 多油层油藏
multi-leaf type cantilever 多层式悬臂梁
multi-legs intersection 复式交叉；多条道路交叉
multilength arithmetic 多倍长度运算
multilength working 多倍长度工作单元
multilens camera 多镜头摄影机
multilevel difference equation 多层差分方程
multilevel encoding 多电平编码
multilevel hoisting 多级提升；多水平提升
multi-level hoisting dual operations 多水平提升操作；多级提升操作
multilevel hoisting installation 多水平提升装置；多级提升装置
multilevel open cut pit 多水平露天矿
multi-level outlet shaft spillway 多层进水竖井式溢洪道
multilevel winding 多水平提升
multi-lift mining 多水平开采
multi-line 多线
multi-line acquisition 多线采集
multiline shooting 多线激发
multilinear failure 多线性破坏
multilinear function 多重线性函数
multilinker pump 多柱塞泵
multi-list processor 多道程序处理机
multilobate delta 多叶三角洲
multimagnet system 多磁铁系统
multimass system 多质点系统
multimedia 多媒体
multimedia database 多媒体数据库
multimedia map 多媒体地图
multi-metal mining 多金属开采
multimetal ore 多金属矿石
multimeter 万用表
multimetering 多点测量；多次计算
multimillion ton ore deposit 高储量矿床；几百万吨储量的矿床
multimineral model 多矿物模型
multimineral ore 多矿物矿石；复合矿石
multimodal distribution 多重模态分布；多峰分布

multimodal probability distribution	多重模态概率分布	multiphase flow rates	多相流量
multimodal sample	多峰态样本	multiphase fluid inclusion	多相气液包裹体
multimode	多方式；多模；多波型	multiphase pump	多相泵
multimode disturbance	多模干扰	multiphase system	复相体系；多相体系
multimode Kalman filter	多模卡尔曼滤波器	multiphase tectonic activity	多相构造活动
multimode optical cable	多模光缆	multiphone cable	多检波器电缆
multimode optical fibers	多模光纤	multiphysics	多物理场
multimode response	多振型反应	Multi-physics phenomena	多重物理量现象
multimode wave	多波型波	multiplanomural	多平壁的
multimoding	多型性；多波性	multiple	复合的；多重的；多倍的；多路的；多次波
multimolecular	多分子的	multiple access	多路存取
multimolecular adsorption	多分子吸附	Multiple Access Retrieval System (MARS)	多道存取检索系统
multimolecular film	多分子膜	multiple acting	多作用的
multimolecular reaction	多分子反应	multiple acting indicator	多功能指示器
multi-motor shearer	多电机采煤机	multiple action	多重作用；多倍作用
multinational company	多国公司；跨国公司	multiple analysis	复合分析
multinational corporation	多国公司；跨国公司	multiple aquifer	多层含水层
multinationalization	跨国化	multiple arch	连拱
multinomial	多项式；多项的	multiple arch bridge	连拱桥
multinomial coefficient	多项式系数	multiple arch dam	连拱坝
multinomial distribution	多项分布	multiple arrival	多次波波至
multinomial expansion	多项式展开	multiple attenuation	多次反射衰减
multinormal distribution	多维正态分布	multiple attenuator	多次反射衰减器
multinuclear imaging	多核成像	multiple auger continuous miner	多螺旋钻式连续采煤机
multi-objective decision making	多目标决策	multiple barrier	多边界
multiobjective decision making approach	多目标决策方法	multiple batholith	复合岩基
multi-objective optimization	多目标优化	multiple bed	叠层；复合层
multiobjective planning	多目标规划	multiple block slides	多块体滑坡
multiobjectives decision	多目标决策	multiple block slip	多块体滑动
multiparallel-layered medium	平行多层介质	multiple boss	复岩瘤
multiparameter	多参数	multiple bottom reflection	海底多次反射
multiparameter analyzer	多参数分析器	multiple branches	多值；多分支
multi-parameter detector	多参数侦测器	multiple cancellation	多次反射相消
multiparameter seismic inversion	多参数地震反演	multiple channel stripchart recorder	多道带状图记录器
multiparameter water quality instrument	多参数水质监测仪器	multiple charge breakage	多药包破碎
multiparticle spectrometer	多粒子光谱仪	multiple check	多重校验
multi-pass compiler	多遍编译程序	multiple choke	多层阻流器
multipass decomposition approach	多次分解法	multiple classification	多重分类
multipass exchanger	多程换热器	multiple classifying analysis	多重分类分析
multi-pass filter testing method	多通滤油器试验法	multiple collision	多次碰撞
multi-pass operations	多次操作	multiple completion	多层完井
multipass sort	多次扫描分类；多级分选	multiple computer terminal	多路计算机终端设备
multi-pass swivel	多路旋转接头	multiple condensed phases	多相凝析物
multipath delay	多路径延迟	multiple conductor cable	多芯电缆
multipath effect	多路径效应	multiple connection	并联连接；复接
multipath fading	多径衰落；多径衰减	multiple connection manifold	多路管汇
multipath interference	多道干涉	multiple correlation	多重相关；复相关
multipath propagation	多路径传播	multiple correlation coefficient	复相关系数；多重相关系数
multipath ray	多路径射线	multiple correlation method	多元相关法；复相关法
multipath signal propagation	多路径信号传播	multiple coverage	多次覆盖
multipath traffic assignment model	多重路线指派模式	multiple crack formation	多裂纹形成
multipath transmission	多径传播；多路传输	multiple criteria	多重准则
multipay field	多产层油气田	multiple cross-glide	多重交错滑动
multi-pen plotter	多笔绘图仪	multiple crystal	复晶体；多晶体
multiperiodic activity	多期活动	multiple curtain	多排帷幕
multipermeability	多重渗透率	multiple cut	多刀切削
multiphase	多相的；多期的	multiple deformation	多次变形
multiphase Darcy model	多相达西模型	multiple detection	组合检波；复合探测
multiphase flow	多相流		

multiple detection method 组合检波法
multiple detector 万能检查器；检波器组合
multiple diffraction 多重衍射
multiple diffusion domain model 多重扩散域模式
multiple dike 复合岩墙；复合岩脉
multiple discriminant analysis 多重判别分析
multiple disintegration 多分支衰变
multiple dome dam 双曲连拱坝
multiple drift 多层导坑；多巷掘进
multiple drill mounting 多钻机式钻车
multiple drill press 多轴钻床
multiple drilling 丛式钻井；多机式钻架；复式钻架
multiple dual operation 并联运行
multiple earthquake 多重地震；多发性地震
multiple echo 多重回声
multiple eigenvalue 多重特征值
multiple electrode system 多电极系
multiple equilibrium 多点平衡
multiple error 多级误差
multiple extraction 多次提取
multiple extreme 多重极值
multiple failure state test 混合破坏测试
multiple fan-shaped fold 复扇形褶皱
multiple fault 复断层；阶状断层；阶梯式断层
multiple faulting 多次断裂活动
multiple filter 复式滤波器
multiple flighted screw 多导程螺杆
multiple flow rate test 多级流量试井
multiple fluid grouting 多液注浆
multiple fluid sampler 多节式流体取样器
multiple fluvial sources 多河流源
multiple fold 复合褶皱
multiple fractionation 多次分馏；多次分级
multiple fracturing 多次压裂
multiple frame 多跨构架
multiple frequency 多频
multiple frequency amplitude-phase method 多频振幅相位法
multiple gated spillway 多闸门溢洪道
multiple genesis 多成因
multiple geophone 组合检波器；组合地震检波器
multiple glide 复滑移
multiple gravel pack 多层砾石充填
multiple grouting 混合式注浆
multiple heading 多条平巷掘进
multiple horizon mining 多水平井采
multiple hydraulic power pack 多泵液压泵站；组合液压泵站
multiple imaging 多重成像
multiple indeterminate 多次超静定
multiple in-hole motor 组合式钻孔马达
multiple injection recorder 多孔灌浆记录仪
multiple input 多点输入
multiple integral 多重积分
multiple integrator 多重积分仪
multiple interface 复接口
multiple interpolation 多重插值
multiple intrusion 重复侵入
multiple jack 复式塞孔
multiple jet flow 组合射流

multiple laccolith 叠岩盖；复合岩盖
multiple laminae 多层状体
multiple laminate horizontal roof 成多层的水平直接顶板
multiple laterals 多水平井段的井
multiple leaky aquifer 多层越流含水层
multiple level line 复测水准线
multiple line curtain 多排帷幕
multiple linear regression 多元线性回归；多重线性回归
multiple manometer 组合压力计；多量程压力表
multiple maximum 多极值
multiple measurement 组合测量
multiple melting 多重熔融
multiple metamorphism 多次变质作用；多期变质作用
multiple migration 多次波偏移
multiple navigation sensor 多路导航传感器
multiple neutron mass theory 多中子块理论
multiple nuclei theory 多核理论
multiple objective function 多目标函数；多重目标函数
multiple objective optimization algorithm 多目标优化算法
multiple open hole gravel pack 多层裸眼砾石充填
multiple order buckling 多次弯曲
multiple ore deposit 叠生矿床
multiple oxides 多氧化物类
multiple packer sleeved pipe 多栓塞袖套管
multiple pay bed 复油层
multiple periodicity 多重周期
multiple phase 多相
multiple phase flow 多相流
multiple physics coupling 多物理场耦合
multiple piezometer 多头测压管；多头水压计
multiple piles 群桩
multiple pipe perforation completion 多管射孔完井
multiple plough 复式刨煤机
multiple point borehole extensometer 多点钻孔伸长计
multiple point extensometer 多点应变计；多点位移计
multiple point loading 多点集中加载
multiple points measurement 多点量测
multiple position borehole extensometer 钻孔多点位移计
multiple position extensometer 多点位移计
multiple precision 多倍精度
multiple pressing 复式冲压
multiple pressure regime 多压力体系
multiple priority levels 多优先级
multiple process 多段法；多重处理
multiple projector 多倍投影仪
multiple projector system 多投影系统
multiple proportion 倍数比
multiple pulsation 复式脉动
multiple pumping 多泵抽油
multiple pumping log 多孔抽水记录
multiple punching machine 多头冲床
multiple purpose dam 多用途水坝
multiple purpose project 综合利用工程
multiple purpose reservoir 多用途水库；综合利用水库
multiple random variable 多元随机变量；多维随机变量
multiple rating curve 复式水位流量关系曲线
multiple ray path 多次反射射线路径
multiple reciprocity method 多重互易法
multiple recording device 多笔记录仪；复合记录仪

multiple recording group	检波器组合
multiple reflection	多次反射；多层面反射
multiple regression	多元回归；多重回归
multiple regression analysis	多元回归分析；多重回归分析
multiple reservoir completion	多产层完井
multiple route	多路径
multiple row method	多排炮眼爆破法
multiple row shot	多排列爆破
multiple rows	多排式；多行式
multiple runs	多次起下作业
multiple sampling	多次抽样
multiple sampling inspection plan	多次抽样检验方案
multiple sampling plan	多次抽样方案
multiple sand exploitation	多层开采
multiple scale problem	多重尺度问题
multiple scaled instrument	多量程仪器
multiple scan interferometer	多扫描干涉仪
multiple scattering	多次散射
multiple seam dual operation	多煤层作业
multiple seam operation	多煤层作业
multiple seams	多煤层；多矿层
multiple seismic event	多发性地震
multiple seismometer	组合地震计
multiple seismometer array	多检波器组合
multiple series	多重级数
multiple series connection	并串联；混联
multiple shock	多发地震
multiple shot	多孔爆破
multiple shot array	多炮组合
multiple shot points	多爆破点
multiple shot processing	多炮点处理
multiple shot tool	多点测斜仪
multiple shotfiring	多炮眼爆破
multiple shotholes	组合爆炸
multiple shotpoint	组合爆炸点
multiple shovel	多斗式挖掘机
multiple sidetracking	多层侧钻
multiple sill	复合岩床
multiple slide	多重滑移
multiple slip	多重滑移
multiple spindle perforating machine	多轴钻床
multiple splitting	劈裂煤层；多次分裂；多分支风流
multiple spot welding	多点焊
multiple squeeze	多次挤压
multiple squeeze slurry pack technique	多次挤压砂浆充填技术
multiple stable isotopes	多稳定同位素
multiple stage cementing	分级注水泥
multiple stage crushing	多级破碎
multiple stage fracturing	多级压裂
multiple stage nitriding	多段渗氮
multiple stage pumping	多级泵送
multiple stage sludge digestion	多级污泥消化
multiple statistical decision theory	多重统计决策理论
multiple step bit	多阶梯钻头；多台阶钻头
multiple step-drawdown test	多次降深试验；分层降深试验
multiple stress concentration	多重应力集中
multiple string completion	多管完井
multiple string small diameter well	小直径多管井
multiple structural sections	联合构造剖面；系列剖面图
multiple subsea well system	水下多井系统
multiple superimposed tectonic belt	多次叠加构造带
multiple supported structure	多重支撑结构
multiple survey	多点测斜
multiple textural elements	组合结构要素；复合结构要素
multiple thermocline	复斜温层
multiple track railway	多线铁路
multiple track tunnel	多线隧道
multiple transform	多重变换
multiple trap	复圈闭
multiple trip	多次起下管柱作业
multiple tropopause	复对流层顶
multiple truss	复式桁架
multiple tubing string	多油管柱的
multiple tubing wellhead	多油管井口装置
multiple twin	聚片双晶
multiple unconformity	重复不整合
multiple use	综合利用
multiple valued function	多值函数
multiple wave reflection	波的多次反射
multiple well	多层完成的井；多井的
multiple well platform	多井平台
multiple well pumping system	多井联动抽油系统
multiple well pumping test	多孔抽水试验
multiple well system	群井系统
multiple well test	多井联合试井；群井试验
multiple wells	群井；井群
multiple zone	多油层
multiple zone completion	多层完井
multiple zone open hole gravel pack	多层裸眼井砾石充填
multiple zone well	多层井
multiple-address instruction	多地址指令
multiple-address message	多地址信息
Multiple-Apertured Reluctance Switch (MARS)	多孔磁阻开关
multiple-arch bridge	连拱桥
multiple-arch dam	连拱坝
multiple-auger continuous miner thin seam mining	多螺旋钻式连续采煤机薄煤层开采
multiple-beam antenna	多波束天线
multiple-beam interferometer	多光束干涉仪
multiple-beam radar	多波束雷达
multiple-bedded	多层的
multiple-bench open-pit mine	多台阶露天矿
multiple-bowl overshot	多级打捞筒
multiple-bridge intersection	复式立交桥
multiple-channel mobile communication system with automatic dialing	多信道自动拨号移动通信系统
multiple-component system	多组分体系
multiple-contact	多次接触；多触点的
multiple-contact miscible displacement	多次接触混相驱
multiple-cut trench excavator	多斗式挖沟机
multiple-cycle	多旋回的
multiple-cycle mountain	多旋回山
multipled	并联的；复接的
multiple-detection method	多地震仪勘探法
multiple-drift method	多导坑开挖法
multiple-drill jumbo	多机钻车

multiple-drill shaft sinker 多钻凿井吊架
multiple-drilling machine 多头钻机
multiple-duct conduit 多孔管道；多管管道
multiple-exposure photography 多曝光照相术
multiple-fault trap 复断层圈闭
multiple-filter 复式过滤器
multiple-flow period DST 多级流量地层测试器试井
multiple-flow pump 多挡流量泵
multiple-frequency 多频
multiple-head auger machine 多钻头螺钻采煤机
multiple-heading method 多工作面法
multiple-hole 多孔的
multiple-hole blasting 成组爆破；多眼爆破；复爆破
multiple-hole recording equipment 多孔记录装置
multiple-horizon mining 多水平开采
multiple-horizon mining method 多水平开采法
multiple-lens camera 多镜头照相机
multiple-level mining 多水平开采
multiple-lift 多层的；复式的
multiple-machine-head 多切削加工头
multiple-order 多级
multiple-pass 多程；多路
multiple-pass rotary mixer 多行程转轴式拌合机
multiple-path 多路的
multiplepeaked hydrograph 多峰式水文过程线
multiple-peaked pore size distribution 多峰状孔隙大小分布
multiple-photograph orientation instrument 多次照相定向仪
multiple-pin 多刀；多测针的
multiple-point 多点的
multiple-point borehole extensometers 多点钻孔伸长计
multiple-point borehole rod extensometer 杆式钻孔多点位移计
multiplepoint displacement meter 多点位移计
multiple-point extensometers 多点伸长计
multiple-point gas lift 多层气举
multiple-position borehole extensometer 多点钻孔引伸计
multiple-programming 多程序
multiple-pulse 多脉冲
multiple-purpose project 综合利用工程
multiple-purpose reservoir 通用蓄水池
multiple-purpose tester 万能测试器
multiple-roller-cone machine 多牙轮式钻井机
multiple-row blasting 多行炮眼爆破
multiple-row method 多排炮眼爆破法
multiple-row shot 多排爆破
multiple-sand exploitation 多层砂开采
multiple-sand reservoir 多砂岩层油气藏
multiple-seam completions 多层完井
multiple-service 多功能随钻测量
multiple-shaft plumbing method 多垂线井筒定向法
multiple-shaft wall 复合井壁
multiple-short-delay blasting 成组瞬间迟发爆破
multiple-shot blasting 成组爆破；多眼爆破；复爆破
multiple-shot bullet gun 多点射孔器
multiple-shot contact watch 多点测斜仪钟表
multiple-shot film 多点测斜仪胶片
multiple-shot instrument 多点测试仪
multiple-shot survey 多点照相测斜

multiple-simultaneous blasting 成组同时爆破
multiple-speed gearbox 多级变速箱；多级减速器
multiple-spindle 多轴
multiple-spot 多点的
multiple-stable-state 多稳态的
multiple-stage 多级的；多段的
multiple-stage compressor 多级压缩机
multiple-stage crushing 多段破碎
multiple-stage drainage 多级排水
multiple-stage test 多阶段试验
multiple-stage triaxial test 多级三轴试验
multiple-string cementing 多管注水泥
multiple-string hookup 多管装置
multiple-string retrievable packer 多管可收回封隔器
multiple-structure 多层结构
multiplet 相重项；多重谱线；多重态
multiplet theory 多位理论
multiple-tool 多刀的
multiple-track 多道的
multiple-tuned mass damper 多调谐质量阻尼器
multiple-unit 多元的；活节的；联列的；复合的
multiple-unit controller 多元控制器
multiple-use reservoir 多目标水库；多用途水库
multiple-valued function 多值函数
multiple-valued logic 多值逻辑
multiple-well test 多井试井
multiple-well transient test 多井不稳定试井
multiple-wick oiler 多点润滑器
multiple-wire plumbing method 多垂线井筒定向法
multiplex 多倍投影测图仪；多倍仪；多路传输
multiplex aeroprojector 多倍投影测图仪
multiplex cementing collar 多级注水泥接箍
multiplex finite element 多重有限元
multiplex frequency 多路传输频率
multiplex heat treatment 复合热处理
multiplex transmitter 多路通信发射机
multiplex unit 多路调制装置
multiplexer 多路调制器；多路扫描器；倍增器；倍加器；乘数
multiplexer channel 多路转换通道
multiplexer delay 多路编排延迟
multiplexer holographic memory system 多路全息存储系统
multiplexing 多路复接；多道传输
multiple-zone field 多产层油、气田
multiple-zone production 多层开采
multiple-zone subsea completion 水下多层完井
multiplicand 被乘数
multiplication 倍增；相乘
multiplication algebra of matrix 矩阵的乘法代数
multiplication constant 乘常数
multiplication factor 放大系数
multiplication of dislocation 位错增殖
multiplication table 乘法表
multiplication theorem 乘法定理
multiplicative anomaly 累乘异常
multiplicative congruential method 剩同余法
multiplicative index 累乘指数
multiplicity 叠加次数；重复次数；多重性
multiplicity factor 倍增因子

multiplicity of hazards	灾害多重性
multiplicity process	倍增过程
multiplier	系数；乘数；倍率；乘子
multiplier coefficient unit	系数相乘器
multiplier detector	倍增检波器；倍增管探测器
multiplier digit	乘数的数
multiplier electrode	倍增电极
multiplier phototube	光电倍增管
multiplier principle	乘数原理
multiplier register	乘数寄存器
multiplier rule	乘数法则
multiplier traveling-wave photo diode	行波光电倍增二极管
multiplier-divider	乘除装置
multiplier-quotient register	乘-商寄存器
multiplough	复式刨煤机
multiplunger	多柱塞的
multiply	成倍增加；增殖；繁殖；乘
multiply-deformed terrane	多期变形区
multiplying arrangement	放大设备
multiplying factor	倍加系数；放大率；放大因子
multiplying lever	放大杠杆
multiplying power	倍率
multiplying system	放大系统
multipoint connection	多点连接
multipoint consistency curve	多点稠度曲线
multipoint deformeter	多点变位计
multi-point diffuser	多出口渗透器
multi-point displacement measurement	多点位移测量
multipoint electrode	多点电极
multipoint excitation	多点激振
multipoint method	多点测流法；多点法
multi-point open-flow potential test	多点无阻流量测试
multipoint priming	多点起爆；多点发动
multipoint recorder	多点记录仪
multipoint rotational viscometer	多点旋转黏度计
multi-point-scanner	多点扫描器
multipolar	多极的
multipolarity	多极性
multipolarization photography	多向偏振摄影
multipole	多极的
multipole acoustic log	多极声波测井
multipole array acoustic logging (MAC)	多极阵列声波测井
multipole transition	多极跃迁
multi-pore media	多重孔隙介质
multiporous	多孔的
multiport	多谐振荡器；多港埠的；多通道的；多口的
multiport memory	多端口存储器
multiport valve	多通阀；多进口阀
multi-position valve	多位阀
multiprecision	多倍精度
multiprocessing	多重处理
multiprocessor	多处理机；多处理器
multi-product line	多油品管线
multiproduct preparation	多产品选煤法
multiproduct separation	多产品分选
multiprogram operation	多道程序作业
multi-programmed computer	多道程序计算机
multiprogramming	多道程序设计
multipurpose	多种用途的；万能的；多效的
multipurpose dam	综合利用坝
multipurpose hydraulic project	水利枢纽
multipurpose instrument	全能仪器
multipurpose oscilloscope	多用途示波器
multipurpose photography	多用途摄影
multipurpose reservoir	多目标水库
multi-purpose semisubmersible	多功能半潜式装置
multi-purpose shotcreter	多用途混凝土喷射机
multipurpose submersible	多用途潜水器
multi-purpose tool	多功能工具
multipurpose triaxial test apparatus	多功能三轴仪
multipurpose water utilization	水利资源综合利用
multiquadric method	多元二次曲面法
multirange	多量程的；多刻度的；多波段的
multi-range meter	多量程仪表
multirate filtering	多率滤波
multi-redundant structure	多次超静定结构
multireflection	多重反射
multireflection theory	多次反射理论
multireflex	多次反射
multi-reinforcement	多层加强
multi-repeat station	多次重复测点
multireservoir	多油气层
multiresolution analysis	多分辨率分析
multi-resonant structure	多谐振荡结构
multirow and multi-interval millisecond blasting	多排多段毫秒爆破
multirow blasting	多行炮眼爆破
multirow firing	多排炮眼爆破
multirow grout curtain	多排灌浆帷幕
multisampling	多次取样
multiscale	多道定标；多尺度；多量程
multi-scale analysis	多尺度分析
multiscale database	多尺度数据库
multi-scale orebody controls	多尺度矿体控制
multiscope	简易测图仪
multi-screw pump	多螺杆泵
multi-seam coal mine	多煤层煤矿
multiseam coalmine	多煤层煤矿
multi-seam mining	多煤层开采
multi-seam stripping	多煤层露天开采
multi-seam working	多层开采；多接缝巷道
multi-seasonal storage reservoir	多季调节水库
multi-section pile	多节桩
multisensor MWD system	多传感器随钻测量系统
multisensor towfish	多传感器拖鱼
multi-serial-port card	多串口卡
multishock	激波系；多激波的
multishot camera	多点测斜照相机
multi-shot firing	多炮眼爆破；成组爆破
multi-shot gyroscopic instrument	多种数据钻孔测量回转仪
multishot instrument	多点测斜仪
multishot perforation penetration performance	多孔射孔穿透特性
multi-shot routed	多眼炮眼组
multishot survey	多点测斜
multishot test	多孔射孔试验
multi-shut-in tool	多次关井器

multi-source and multi-streamer off shore seismic acquisition
　多源多缆海上地震采集
multisource holography　多源全息术
multispaced neutron logging　多源距中子测井
multi-span　多跨的
multispan beam　多跨梁
multispan bridge　多跨桥
multi-span structure　多跨结构
multi-spans continuous　连续多跨
multispatial dimensioned source　多度空间源
multispectral　多光谱的；多波段的
multispectral additive color image　多光谱加色图像
multispectral analysis　多谱分析
multispectral band　多波段
multispectral camera　多波段摄影机；多光谱摄影机
multispectral classification　多波段分类；多光谱分类
multispectral image　多波段影像；多光谱影像
multispectral imagery　多波段成像；多光谱成像
multispectral imaging　多光谱成像
multispectral line scanner　多光谱行扫描仪
multispectral mapping system　多波段成像系统；多光谱成像系统
multispectral photographic facility　多光谱摄影装置
multispectral photography　多波段摄影；多光谱摄影
multispectral photometer　多谱光度
multispectral point scanner　多波段点扫描仪
multispectral radar　多光谱雷达
multispectral remote sensing　多波段遥感；多光谱遥感
multispectral remote sensing technique　多光谱遥测技术
multispectral scanner　多波段扫描仪；多光谱扫描仪
multispectral scanning　多波段扫描；多光谱扫描
multispectral scanning system　多波段扫描系统
multispectral sensor　多波段传感器；多光谱传感器
multispectral survey　多光谱测量
multispectral terrain photography　多光谱地面摄影
multispeed　多速的
multi-speed rheometer　变速流变计
multispindle drill　多轴钻床
multistabilizer　多稳定器底部钻具组合
multistabilizer configuration　多稳定器结构
multistabilizer holding assembly　多稳定器稳斜组合
multistabilizer hook up　多稳定器组合
multistage　多阶段的；多级的
multistage amplifier　多级放大器
multistage cementing　多级注水泥
multistage centrifugal booster pump　多级离心式增压泵
multistage centrifugal pump　多级离心泵
multistage cleaning　多段精选多阶段清洁
multi-stage compressor　多级压缩机
multi-stage decision procedure　多阶段决策程序
multi-stage decision processes　多期决策过程；多阶段决策过程
multistage deformation history　多级变形历史
multi-stage drainage　多级排水
multistage drawing　多阶段拉拔；多级拉拔
multistage expansion　多级膨胀
multistage extraction　多级抽提
multistage fractional crystallization　多阶段分离结晶作用
multistage fracturing　多次破裂；多级压裂
multistage gas lift　多级气举

multistage grinding　多段研磨；分段磨矿
multi-stage grouting　多次灌浆
multistage hoisting　多段提升
multistage interpretation　多阶段判读
multistage migration　多期运移
multi-stage mixedflow pump　多级混流式水泵
multi-stage motor　多级马达
multistage optimization　多级优选法
multi-stage optimization method　多阶段优选法
multistage photograph　多级摄影相片
multistage photography　多阶段摄影；多级摄影
multistage processes　多步工艺过程
multistage pump　多级泵
multi-stage pumping　多层抽水；多级排水
multistage rectification　多级纠正
multistage sampling scheme　多级摄影取样图
multistage sensing　多级传感
multistage separation　多级分离
multistage separator　多级分离器
multistage stabilization　多级稳定
multistage storage pump　多级蓄能水泵
multi-stage stressing　多阶段施加应力
multistage tests　多级试验
multistage torgue converter　多级变矩器
multi-stage triaxial compression test　多级三轴压缩试验
multi-stage triaxial test　多级应力三轴试验
multi-stage uniaxial orientation　多级单轴取向
multi-stage water power stations　多级水电站
multistage water seal　多级水封
multi-stage well point system　多级井点系统
multistation photograph　多站摄影相片
multistation photography　多站摄影
multistep　多步的；多级的
multistep method　多步法
multi-step polynomial　多步多项式
multi-step polynomial approximation　多步多项式近似
multistep process　多步过程
multistep recurrence　多步循环
multistep recurrence algorithm　多步递推算法
multi-step thread　多头螺纹
multistone bit　细粒金刚石钻头；多金刚石钻头
multi-storage system　多级蓄水系统
multistorey shear wall　多层剪力墙
multistorey structure　多层结构
multistory sand　重叠砂体
multistory sandbody　多层砂体；纵叠砂体
multistory sandstone body　多层砂岩体
multi-story structure　多层结构物
multistring hydraulic-set packer　多管水力坐封封隔器
multi-string target　多层管柱试验靶
multi-structure　复合结构
multi-structure system　多结构体系
multi-support beam　多支点梁
multi-support excitation　多点激励
multi-support seismic excitation　多点地震输入
multi-support structural system　多支承结构体系
multitemporal　多时态的；多时相的
multitemporal analysis　多时相分析
multitemporal image　多时相图像
multitemporal remote sensing　多时相遥感

multitemporal sensing	多瞬时传感
multi-tensioner	多向张力器
multiterminal	多端的；多接头的；多端网路
multi-terminal network	多端网络
multitester	万用表；多次地层测试器
multitrace equipment	多道设备
multitrace record	多道记录
multitrace seismogram	多道记录
multitrace seismograph	多扫描地震仪
multitracer	复式示踪剂
multitube manometer	多管式测压计
multi-use bit	多用钻头
multivalent	多价的
multivalent cation	多价阳离子
multivalent element	多价元素
multivalent function	多叶函数
multivalent ion	多价离子
multivalued amplitude	多值振幅
multivalued decision	多值判断
multivalued function	多值函数
multivariable	多变量的
multivariable analysis	多变量分析
multivariable control	多值调节；多参数控制
multivariable control system	多变量控制系统
multivariable system	多变数系统
multivariant analysis	多元分析
multivariate	多元的；多变量的
multivariate analysis	多变量分析；复变量分析；多元分析
multivariate correlation analysis	多元相关分析
multivariate data analysis	多元数据分析
multivariate discriminant analysis	多元判别分析
multivariate distribution	多元分布
multivariate equilibrium	多变平衡
multivariate function	多元函数
multivariate Gaussian distribution	多元高斯分布
multivariate general linear hypothesis	多元一般线性假设
multivariate histogram	多变量直方图
multivariate interpolation	多变量插值
multivariate linear regression	多元线性回归
multivariate model	多元模型
multivariate morphometrics	多元形态统计学；多变量形态度量学
multivariate normal distribution	多元正态分布
multivariate normality test	多元正态性试验
multivariate observation	多元观测值
multivariate probability density function	多元概率密度函数
multivariate random process	多变量随机过程
multivariate regression analysis	多元回归分析
multivariate regression model	多元回归模型
multivariate sample	多变量样本
multivariate scaling	多元标度法
multivariate statistical analysis	多元统计分析
multivariate statistics	多元统计学；多变量统计学
multivariate stochastic process model	多元随机过程模型
multivector	多重向量
multivelocity function	多速度函数
multivertor	复式变换器
multivibrator	多谐振荡器
multiwave	多波
multiwave seismic method	多波地震法
multi-wavelet	多小波
multi-well bounded reservoir	多井封闭油藏
multiwell cluster	井丛
multiwell development	多井开发
multiwell drilling	丛式井钻进
multiwell flooding pattern	多井注采井网
multi-well histogram	多井直方图
multiwell interference test	多井干扰试验
multi-well open caisson	多排孔沉井
multiwell pattern	多井井网
multiwell pilot test	多井先导性试验
multiwell platform	多井平台
multiwell producing pattern	多井生产井网
multi-well profile planning	丛式井剖面设计
multiwell subsea completion	水下多井完成
multi-well system	多井系统
multiwell template	多井底盘
multi-well transient test	多井不稳定试井
multiwell-gathering network	多井集输管网
multi-wheel bucket wheel excavator	多斗轮式挖掘机
multiyear ice	多年积冰；多年冰
multi-year mean sea level	多年平均海面
multiyear mechanical mucked	多瓣式抓岩机
multizone injection well	多层注入井
multizone limestone reservoir	多层灰岩油气藏
multizone reservoir	多层性油气藏
multizone stimulation	多层增产措施
multizone-flooding	多层注水驱油
municipal engineering	城市工程；市政工程
municipal highway	城市道路；都市公路
municipal landfill	城市填土
municipal planning act	城市规划条例
municipal pollution	城市污染
municipal port	城市港口
municipal program	城市规划
municipal rainfall pipe system	城市雨水管道系统
municipal refuse	城市垃圾
municipal road	城市道路
municipal rubbish landfill site	城市垃圾处理场
municipal sanitation	城市环境卫生
municipal sewage	城市污水
municipal sewage plant	城市污水处理厂
municipal sewerage	市政排水工程
municipal sewers	都市下水道
municipal sludge	城市污泥
municipal solid waste	都市固体废物
municipal transportation	城市交通运输
municipal waste	都市废物；城市废物
municipal waste material	城市废弃物
municipal waste pipe	市政污水管道
municipal wastewater	城市废水
municipal wastewater plant	城市废水处理工厂
municipal water	城市用水
municipal water distribution system	城市配水系统
municipal water facilities	城市供水设施
municipal water supply	城市给水
municipal water supply system	城市自来水
municipal water system	城市给水系统

municipal water use 城市用水
municipal works 市政工程
muniongite 碱长霞霓响岩
munjack 硬化沥青
munkrudite 磷铁钙矾
munroe effect 聚能效应
Munsell charts 孟塞尔氏图
Munsell color system 孟塞尔色系
Munsell colour 孟塞尔标准颜色判定法
Munsell colour system 孟塞尔色系
muntenite 琥珀；树脂石
Muntz brass 蒙次黄铜
Muntz metal 蒙次黄铜；熟铜
muon μ介子
muonium μ介子素
muptiple water sampler 复式取水样器
mural joint structure 垂直节理构造
murambite 暗白榴碧玄岩
murasakite 红帘石英片岩
murbruk structure 碎斑构造
murdochite 方铜铅矿；黑铜铅矿
muriacite 硬石膏
muriate 氯化物；氯化钾
muriatic acid 盐酸
murite 暗霞响岩
murmanite 水硅钛钠石；硅钛钠石
Murray fracture zone 默里断裂带
Murray meteorite 默里陨石
muschel sandstone 介壳砂岩
Muschelkalk 壳石灰岩；壳灰岩统
muschketowite 六方磁铁矿；穆磁铁矿
muscovadite 董云苏长岩；董青苏长角页岩
muscovite 白云母
muscovite granite 白云母花岗岩
muscovite schist 白云母片岩
muscovite-biotite gneiss 二云母片麻岩
muscovite-granite 白云母花岗岩
muscovite-schist 白云母片岩
muscovitization 白云母化
muscovy glass 白云母
mush 软泥质次煤；烟煤上的油质泥；噪声
mushketovite 假象磁铁矿
mushroom anchor 盘状锚
mushroom excavation 蘑菇形开挖法
mushroom fold 蘑菇式褶皱
mushroom head method 蘑菇头法
mushroom magnet 菌状起重磁铁
mushroom-type tunnelling method 蘑菇形开挖法
mushy 软的；多孔隙的；糊粉状的
mushy coal 泥炭；多孔煤
mushy consistency of concrete 混凝土流态稠度
mushy state 浆糊状态；粥状
Muskat intercept 麦斯盖特截距
Muskat method 麦斯盖特试井解释法
Muskat plot 麦斯盖特曲线
muskeg 厚苔泽；泥岩沼泽；水藓沼泽
muskeg terrain 沼泽地带
mutagenesis 致突变作用；突变发生
mutagenic 诱变的；致突变的
mutagism 诱变

mutamer 变构物
mutamerism 变构现象
mutant 突变体
mutant strain 突变株
mutarotation 变旋光
mutated microorganisms release 突变微生物释放
mutation 变化；变异；突变
mutation of water level 水位突变
mutation operator 变异算子
mutation pressure 突变压力
mutation probability 变异概率
mutational change 突变
mute 噪声抑制；切除
mute library 切除库
mute time 切除时间
mute tuning 静噪调谐
muthmannite 杂碲金银矿；板碲金银矿
mutilating 多次爆破
muti-line lock 多级船闸
muting sensitivity 低灵敏度
mutiple lock 多级船闸
mutiple-arch dam 连拱坝
mutton-fat jade 羊脂玉
mutual relation 相互关系
mutual 相互的；共同的
mutual action 相互作用
mutual affinity 相互亲和性
mutual anchorage 钢筋搭接；互锚
mutual assent 互相同意
mutual attraction 互相吸引；交互引力；相互引力
mutual contact zone 交互接触带
mutual correlation 互相关
mutual coupling factor 相互耦合系数
mutual effect 相互作用
mutual impedance 耦合阻抗；互阻抗
mutual indemnification 相互补偿
mutual inductance 互感
mutual inductance coefficient 互感系数；磁耦合系数
mutual induction 互感
mutual inductive impedance 互感阻抗
mutual inductor 互感器
mutual interference 互相干扰
mutual interference wells 互相干扰井
mutual inversion 联合反演
mutual reactance 互电抗；互抗
mutual relation 相互关系
mutual relationship 相互关系
mutual resistance 互电阻；互导的倒数
mutual solubility 互溶性；互溶度
mutual solvent 互溶剂
mutual solvent acidizing 互溶剂酸化
mutual spectral density 互谱密度
mutual synchronization 相互同步
mutual transfer function 相互传递函数
mutual viscosity 互黏性
mutual-boundary stratotype 共界层型
mutuality 互相关系；相关
mutually exclusive 互斥的；互不相容的
mutually exclusive alternatives 互斥方案
mutually exclusive events 互斥事件

mutually perpendicular joint sets	互相垂直节理组
mutually perpendicular planes	相互垂直平面
mutually soluble	互溶的
mycalex	云母块；云母玻璃
mylonite	糜棱岩；磨变岩
mylonite gneiss	糜棱片麻岩
mylonite schist	糜棱片岩
mylonite series	糜棱岩系列
mylonitic	糜棱状；糜棱质的
mylonitic coal	糜棱煤
mylonitic structure	糜棱状构造
mylonitic texture	糜棱结构；糜棱状结构
mylonitization	糜棱岩化
mylonitized zone	糜棱岩化带
mylonization	糜棱岩化作用
myriadoporous	多孔的
myriametric wave	超长波
myriametric wave communication	超长波通信
myrmekite	蠕状石
myrmekite texture	蠕英石结构；蠕状结构
myrmekitic structure	蠕状构造
myrmekitic texture	蠕状结构
myrmekitization	蠕状石化作用

N

N₂ enhanced coalbed methane　氮气驱替煤层气
N₂ flooding　氮气驱替
N₂ flooding test　氮气驱替实验
Na/K geothermometer　钠钾地热温标；钠钾地温表
Nabarro-Herring creep　纳巴罗-赫林蠕变，纳赫型蠕变；体积扩散蠕变
nabla　微分算符；倒三角算子
nablock　圆结核体岩石；铁质结核；圆结核
NaCl equivalent　氯化钠当量
nacre　珍珠层；珍珠母；珠母贝
nacreous layer　珍珠层
nacreous structure　珍珠构造
nacrite　珍珠石；珍珠陶土；珍珠高岭石
nadir　天底；镜头垂点，垂直点；最低点，最下点
nadir angle　天底角；最底角，最低点角度
nadir distance　天底距
nadir map　天底点图
nadir photograph　天底点像片；垂直摄影像片
nadir point　天底点；最低点
nadir point plot　天底点略图
nadir radial　天底点辐射线
nadir radial line　天底点辐射线
nadorite　氯锑铅矿
naegite　苗木石；锆铀矿；稀土锆石
Nafe-Drake relation　奈夫-迪瑞卡关系
nagatelite　磷褐帘石；长手石
nagelfluh　泥砾岩；钉头砾岩
nagelkalk　叠锥灰岩
nager　钎子；手摇钻
nagyagite　叶碲矿；叶碲金矿
nahcolite　苏打石；重碳钠盐
naibourne　间歇河；小河；溪
nail　钉；钉住；指甲
nail test　钉试验
nailable　可打钉的
nailable concrete　受钉混凝土
nailable floor　可钉底板
nailcrete　受钉混凝土
nailed joint　钉接，钉结合
nailhead scratch　钉头状擦痕
nail-holding power　吃钉力
nailing　打钉；土钉作业
nailing concrete　可钉混凝土
naked　裸露的
naked eye　肉眼；裸眼
naked flame　明火；活火焰；无遮火焰
naked hole　裸眼；未下套管的井
naked karst　裸露岩溶；露天喀斯特
naked light　明火灯
naked soil　残积土
naked-eye dustiness　肉眼可见含尘量
naked-eye observation　肉眼观测
naked-flame mine　明火灯矿井，无瓦斯矿
nakhlite　透辉橄无球粒陨石

Namurian stage　纳缪尔阶
Nanhua great ice age　南华大冰期
nano　毫微，纳；十亿分之一
nano fracture mechanics　纳米断裂力学
nanoammeter　毫微安计
nanoampere　毫微安培
nanocrystal　纳米晶体
nanocrystalline　超微晶质的；纳米晶体
nano-electro-mechanical system　纳电子机械系统
nanofossil　超微化石
nanofossil biostratigraphy　超微化石生物地层学
nanofossil group　超微化石类
nanofossil ooze　超微化石软泥
nano-indentation　纳米压痕
nano-indentation test　纳米压痕试验；纳米压痕硬度测试
nano-indentor　纳米压痕仪；纳米硬度计
nano-level indentation hardness test　纳米压痕试验；纳米压痕硬度测试
nanometer　纳米；毫微米
nanometer scale　纳米尺度；纳米级
nanomicropaleontology　超微古生物学
nanopaleontology　纳米古生物学；毫微古生物学
nanophotogrammetry　电子显微摄影测量
nanoscope　毫微秒示波器；纳秒示波器；超高频示波器
nanosecond　毫微秒；纳秒
nanostone　超微化石岩
nanotechnology　纳米技术
nanotesla　纳特斯拉；毫微特斯拉
nantokite　铜盐；氯化亚铜矿
napalite　蜡状烃类
napalm　凝固汽油；凝汽油剂，汽油胶化剂
naphthabitumen　石油沥青；石脑油沥青
naphthene base crude　环烷基原油；环烷基油
naphthene base crude oil　环烷基原油；环烷基油
naphthene hydrocarbon　芳香烃；环烷烃
naphthene index　环烷烃指标；环烷烃指数
naphthene oil　环烷油
naphthene series　环烷系
naphthenic acid　环烷酸；环酸；环己烷甲酸
naphthenic base　环烷基
naphthenic base oil　环烷基石油
naphthenic hydrocarbon　环烷烃
naphthenic oil　环烷基石油
naphthenic residual oil　环烷质残油
naphthenoaromatic hydrocarbon　环烷-芳香烃
naphthine　伟晶蜡石
naphtholite　沥青页岩
naphtholith　沥青页岩
naphthology　石油学
napierian logarithm　自然对数；纳氏对数，纳皮尔对数
napoleonite　球状辉长岩；球状闪长岩
napolite　蓝方石
nappe　推覆体；熔岩流；覆盖层；溢流水舌
nappe contraction　水舌收缩

nappe front	推覆体前锋
nappe inlier	构造窗
nappe outlier	飞来峰；残存推覆体
nappe profile	水舌截面；推覆体剖面
nappe root zone	推覆体山根带
nappe separation	射流分离；水舌分离
nappe sheet	推覆岩席；推覆岩片
nappe structure	推覆构造
nappe tectonics	推覆构造；推腹构造地质学
nappe-type oilfield	推覆体型油田
napping	推覆作用
narrow	狭窄的；变窄；峡谷，山峡，海峡；严密的；有限制的
narrow band channel	窄带通道；窄频带信道
narrow band filter	窄带滤波器
narrow beam projector	窄束投影仪
narrow capillary segment	毛细管颈
narrow defile	峡谷；窄路
narrow face	窄工作面
narrow fissure	狭窄裂隙；窄缝
narrow gorge	嶂谷；峡谷
narrow gradation	均匀级配
narrow joint	窄小节理；窄缝
narrow kerf core bit	薄壁取芯钻头；窄切口岩心钻头
narrow opening	窄巷道；窄口
narrow period band	窄周期带
narrow pillar	窄煤柱
narrow place	窄工作区；窄工作面；窄巷
narrow slot method	狭缝法；刻槽法
narrow spectrum	窄谱
narrow tube	小直径管
narrow valley	峡谷
narrow vein	窄矿脉；薄矿脉
narrow waters	窄狭水域；狭海峡
narrow web mining	浅截深采煤
narrow work	窄工作面作业
narrow working	窄工作面作业
narrow workings	窄工作面；窄巷道
narrow-angle aerial camera	窄角航摄仪
narrow-band	窄频带
narrow-band charactascope	窄频带特性观测设备
narrow-band characteristic	窄频带特征
narrow-band cross-correlation	窄带交叉相关；窄带互关联
narrowband device	窄带器件；窄带设备
narrow-band digital filtering	窄带数字滤波
narrowband disturbance	窄带干扰
narrowband emission	窄带发射
narrowband filter	窄带滤光片；窄频带滤波器
narrowband frequency	窄带频率
narrowband frequency modulation	窄带调频
narrowband interference	窄带干扰
narrow-band random process	窄带随机过程
narrow-band spectrum	窄带谱
narrow-band system	窄带体系
narrowband telemetry	窄带遥测
narrow-band vibration	窄带振动
narrow-band-pass filter	窄带通滤波器
narrow-beam	窄射线，窄束
narrow-beam absorption	窄束吸收
narrow-bore tunnel	小断面隧道

narrow-gauge	窄轨的；窄轨距
narrow-gauge railway	窄轨铁路
narrow-gauge surface haulage	地面窄轨运输
narrowing	变窄；收缩
narrow-mouthed	窄口的
narrow-necked	细颈的
narrowness	狭窄；严密
narrow-web shearer	浅截式采煤机
narsarsukite	短柱石
nascent	初生的；初期的
nascent continental margin	初生大陆边缘
nascent hydrogen	新生氢
nascent ocean basin	新生大洋盆地
nascent oil	初生石油
nascent oxygen	新生氧
nase	岬；海角
nasonite	氯硅钙铅矿
Na-spar	钠长石
nasturan	沥青铀矿；方铀矿
national atlas	国家地图集
National Basic Research Program of China	国家重点基础研究发展计划项目；973 项目
national benchmark	国家水准点；国家基准
national boundary	国界
national budget	国家预算
National Building Code	国家建筑规范
National Building Specification	国家建筑规范
National Bureau of Standards	国家标准局
National Coal Development Corporation	煤炭开发总公司
national coordinate system	国家坐标系
national degradation	自然降解，自然分解；自然退化
national emergency search and rescue team for earthquake disaster	国家地震灾害紧急救援队
National Emission Data System	全国排放数据系统
National Emission Standards for Hazardous Air Pollutants	国家危险性空气污染物排放标准
National Environmental Policy Act	国家环境政策法
National Environmental Protection Agency	国家环境保护局；环境保护署
national freeway	国道；国家高速公路
National Freeway Construction & Management Fund	国道高速公路建设管理基金
National freeway network	国道路网；国家高速公路网
National Fundamental Geographic Information System	国家基础地理信息系统
National Geodetic Coordinate System 1980	1980 国家大地坐标系；西安坐标系
National Geodetic Network	国家大地测网
National Geodetic Vertical Datum	国家大地高程基准
national geographic names database	国家地理名称数据库
national grid	国家测量坐标方格网
national grid coordinate	国家测量格网坐标
National Height Datum 1985	1985 国家高程基准
National High Technology Research and Development Program of China	国家高技术研究发展计划；863 项目
national highway	国道
national information infrastructure	国家信息基础设施
National Institute of Standards and Technology	国家标准和技术研究所

national land management 国土管理
national leveling network 国家水准网
national map projection 全国统一地图投影
National Marine Data and Information System 国家海洋信息系统；中国海洋信息网；中海网
National Natural Science Foundation of China 国家自然科学基金
national nature reserves 国家自然保护区
national network of seismograph 国家地震台网
National Oceanographical and Atmospheric Administration 国家海洋与大气管理局
National Oil Company 国家石油公司
National Oil Marketers Association 全国石油批发商协会
national park 国家公园
national petroleum reserve 国家石油储备
national railway 国有铁路
National Ready Mixed Concrete Association 美国国家预拌混凝土协会
national revenue 国家税收；国民收入；财政收入
National Science Foundation 国家科学基金
national soils data base 全国土壤数据库
National Spatial Data Infrastructure 国家空间数据基础设施
national standard 国家标准
national standard barometer 国家标准气压计
national trunk highway 国家干线公路；国道
native 本地；原生；天然的
native asphalt 天然沥青
native bitumen 天然沥青；原地沥青
native block 原地岩块
native block mass 自然块度
native clay 原生黏土
native coke 天然焦
native compound 天然化合物
native copper 自然铜
native core 新鲜岩心
native element 自然元素；天然元素
native formation fluid 原生地层流体
native gas 原生天然气
native gold 自然金
native groundwater 天然地下水；初始地下水
native iron 自然铁；天然铁
native lead 天然铅
native magnet 天然磁铁
native mercury 自然汞
native metal 自然金属；天然金属
native mud 天然泥浆；井内自造泥浆
native paper 土纸
native paraffin 天然石蜡
native permeability 原始渗透率
native platinum 自然铂；天然铂
native reservoir 原生油藏
native rock 原生岩；母岩
native silver 自然银
native soil 残积土；天然土
native state 天然状态
native stress 原地应力
native sulphate of barium 重晶石；天然硫酸钡
native sulphur 自然硫
native water 原生水；初生水；天然水；岩浆水

native well 天然井
native wettability 天然润湿性
native-state core 原状岩心
natrium 钠
natroalunite 钠明矾石，钠矾石
natrocalcite 硅硼钙石；钠方解石
natrocarbonatite 钠碳酸岩
natrochalcite 钠铜矾
natrojarosite 钠铁矾
natrolite 钠沸石；中柱石
natron 泡碱；天然碱；天然碳酸钠
natron lake 碱湖；苏打湖
natronbiotite 钠云母
natronjadeite 硬玉，翡翠
natronorthoclase 钠正长石
natronsanidine 钠透长石
natroopal 钠蛋白石
natrophilite 磷钠锰矿
natural 天然的；自然的；固有的
natural abundance 天然丰度
natural acceleration 自然加速度
natural accretion 自然增长；自然淤积
natural activity 天然活动性；天然放射性；天然活性
natural aeration 自然通风
natural aeroelectromagnetic field survey 天然航空电磁场测量
natural agent 自然力
natural aggregate 天然骨料
natural aging 自然时效；自然老化
natural air crossing 天然风桥
natural amalgam 天然汞膏
natural ambient level 自然环境的悬浮固体水平
natural analog 天然类比
natural analogue 自然类比；天然类似物
natural analogy 天然类比
natural angle of repose 天然休止角；自然休止角；天然静止角
natural angle of slope 天然坡角；边坡休止角；自然倾斜角
natural angular frequency 自然角频率；固有角频率
natural arch 自然拱，天然拱，天生桥；自然平衡拱；海蚀拱
natural asphalt 天然沥青
natural attenuation 自然衰减；固有衰减
natural background 自然本底；天然本底
natural background radiation 天然本底辐射
natural bar 浅滩；天然沙坝；沙洲
natural barrier 天然屏障；天然堤
natural base 天然地基
natural beach 天然海滩
natural bed 河床；自然层；平行成层
natural berm 天然护坡道
natural bitumen 天然沥青
natural bituminized sandstone 天然沥青砂岩
natural bonded sand 天然黏结砂
natural bottom hole temperature 原始井底温度
natural boundary 自然边界
natural boundary condition 自然边界条件
natural boundary element method 自然边界元法
natural boundary reduction 自然边界折减；自然边界归化

natural break 自然间断；自然间隙
natural bridge 天生桥，自然桥；喀斯特桥；曲流颈桥
natural brine 天然卤水；天然盐水
natural building material 天然建筑材料
natural bulk weight 天然容重
natural calamity 自然灾害
natural cavern 天然洞室；天然洞穴
natural cavity 天然洞穴
natural cement 天然水泥
natural cementing material 天然胶结材料
natural channel 天然河槽；自然航道；天然渠道
natural channel method 天然水道法
natural characteristic 自然特性；固有特性
natural circular frequency 自然圆频率；自然周频
natural circulation 自然环流；自然循环
natural classification 自然分类
natural classification system 自然分类系统
natural clay 天然黏土；自然黏土
natural coal fire 天然煤火
natural coke 天然焦炭
natural collapse 自然塌陷
natural color 天然色
natural completion 普通完井；正常完井
natural condition 自然条件；天然条件
natural condition test 自然条件试验
natural conservation 自然保护
natural conservation area 自然保护区
natural consistency 天然稠度
natural consistency test 天然稠度试验
natural consolidation 天然胶结；自然固结
natural consolidation pressure 天然固结压力；前期固结压力
natural convection 天然对流；自然对流
natural convection recirculation 自然对流再循环
natural cooling 自然冷却
natural corrosion test 自然腐蚀试验
natural course of a river 天然河道
natural cover 自然覆盖物；天然覆盖
natural crack 天然裂纹；天然裂隙
natural cracking catalyst 天然裂化催化剂
natural crude oil 天然原油
natural dam 天然坝
natural damping 固有阻尼；自然阻尼
natural decay 自然衰减
natural deceleration 自然减速度
natural decline 自然递减
natural defrosting 自然解冻
natural density 天然密度
natural depletion 自然递减；自然衰竭
natural depletion oil recovery 天然能量开采的原油采收率
natural depletion phase 天然能量开采阶段
natural deposit 天然沉积物
natural deterioration 自然恶化
natural detrition 自然风化破碎作用；自然损耗
natural diamond 天然金刚石；天然钻石
natural disaster 自然灾害
natural disaster mitigation 减少自然灾害，减灾
natural disaster reduction 减少自然灾害，减灾
Natural Disaster Traffic Management 自然灾害交通管理
natural discharge 天然流量；日常流量；天然排泄

natural disintegration 天然衰变
natural drainage 自然排水
natural drainage channel 自然排水沟；天然排水渠
natural drainage system 自然排水系统
natural draught 自然通风
natural draw ratio 自然拉伸比；固有延伸比
natural drift 自然偏斜距
natural drinking mineral water 天然饮用矿泉水
natural drive 天然能量驱动
natural dry 自然干燥
natural dry unit weight 天然干容重
natural drying 自然干燥
natural durability test 自然耐久性试验
natural earth 天然土
natural earth current 自然大地电流
natural earth potential 自然大地电势
natural earthquake 天然地震
natural ecology 自然生态学
natural ecosystem 自然生态系统
natural electric field 自然电场
natural electric field method 自然电场法
natural electromagnetic noise 天然电磁噪声
natural electromagnetic wave 天然电磁波
natural element 天然元素
natural elevation 自然标高；天然海拔
natural elimination 自然淘汰；自然消除
natural emission 自然放射；天然放射
natural encroachment 天然边水驱
natural energy 天然能量，天然能源
natural energy exploitation 天然能源开发
natural energy resource 自然能源
natural environment 自然环境
Natural Environment Conservation Act 自然环境保护法
natural environment deterioration 自然环境恶化
natural environment influence 自然环境影响
natural environment protection 自然环境保护
natural environment rehabilitation 自然环境复原；恢复自然环境
natural environmental abnormality 自然环境异常
natural environmental element 自然环境要素
natural environmental stress 自然环境压力
natural erosion 自然侵蚀；天然侵蚀
natural escape 自然流道；天然泄洪道
natural evaporation 自然蒸发
natural event 自然事件
natural exhaust system 自然排风系统
natural exposure 天然露头
natural fall 天然落差；自然下垂
natural fault 天然断层
natural feature 天然地形，天然地貌；自然要素；自然特征
natural field method 自然场法
natural flaw 天然裂纹；天然缺陷
natural flow 天然径流；未调节流量；自流；自喷
natural flow drain 自流暗渠
natural flow pipe 自流管
natural flow recharge through well 自流回灌
natural flowing well 自流井；自喷井
natural flux of heat 天然热流
natural forest 天然森林

natural form 原貌；天然形状
natural foundation 天然地基
natural fraction coal 自然级煤
natural fracture 天然裂缝；原生裂隙
natural fracture simulation 天然裂隙模拟
natural fracture system 天然裂缝系统
natural freezing method 自然冻结法；天然冻结法
natural frequency 固有频率；自振频率
natural frequency of foundation 基础自振频率
natural frequency of oscillation 自振频率
natural frequency of vibration 自振频率
natural gamma radiation 自然伽马辐射
natural gamma radioactivity 自然伽马放射性
natural gamma ray 自然伽马射线
natural gamma ray log 自然伽马测井
natural gamma ray spectrometry log 自然伽马能谱测井
natural gamma-ray logging 自然伽马射线测井
natural gamma-ray spectral log 自然伽马能谱测井
natural gamma-ray spectroscopy tool 自然伽马能谱测井仪
natural gamma-ray spectrum 自然伽马射线谱
natural gas 天然气
natural gas abundance 天然气丰度
natural gas conditioning 天然气净化
natural gas dehydrator 天然气脱水器
natural gas diverter 天然气导流器
natural gas facility 天然气设施
natural gas field 天然气田
natural gas gathering system 集气系统
natural gas geology 天然气地质学
natural gas hydrate 天然气水合物；可燃冰
natural gas leakage 天然气泄漏
natural gas liquid 天然气液；天然汽油；气体汽油
natural gas liquid recovery 天然气液回收
natural gas pool 气田；天然气藏
natural gas reserve 天然气储量
natural gas reservoir 气田；天然气层
natural gas source 天然气资源
natural gas system 天然气系统
natural gas treatment plant 天然气处理装置；天然气处理厂
natural gas well 天然气井
natural gas well completion 天然气井完井
natural gas well logging 天然气测井
natural gas well production 天然气井产量
natural gas-based fuel 天然气基燃料
natural gasoline 天然汽油
natural gasoline plant 天然汽油回收装置；天然汽油厂
natural gem 天然宝石
natural generation of electric power 天然发电
natural geography 自然地理
natural geologic hazard 自然地质灾害
natural geological disaster 自然地质灾害
natural geological process 自然地质作用
natural geomorphology 风景地貌
natural geothermal field 天然地热田；天然地温场
natural geothermal power station 天然地热电站
natural geothermal reservoir 天然地热层；天然热储
natural geothermal well 天然地热井
natural geyser 天然间歇喷泉
natural gold 自然金

natural grade 地表自然标高；自然坡度
natural graphite 天然石墨
natural gravel screen 天然砾石过滤器；天然砾石筛
natural ground 天然地基；原生岩，母岩；天然地面，原地面线
natural ground line 天然地面线
natural groundwater level 天然地下水位
natural groundwater outcrop 地下水天然露头
natural groundwater recharge 地下水天然补给
natural groundwater regime 地下水天然动态
natural groundwater replenishment 地下水天然补给
natural groundwater table 天然地下水位；天然地下水面；天然潜水面
natural hardness 自然硬度；固有硬度
natural hazard 自然灾害
natural hazard warning system 自然灾害预警系统
natural head 天然水头；天然落差
natural heat escape 天然热流量；天然地热出露
natural heat flow 天然热流
natural heat output 天然热流；自然热输出
natural heavy mineral 自然重矿物
natural heritage 自然遗产
natural high polymer 天然高聚合物
natural hillside 天然山坡
natural humidity 天然湿度；固有湿度；原湿度
natural hydrocarbon 天然碳氢化合物；天然烃
natural ilmenite 天然钛铁矿
natural impedance 固有阻抗；特性阻抗；波阻抗
natural impregnation 自然浸渍
natural impurity 天然杂质
natural inflow 天然进水量；天然入流量
natural in-place moisture 内在水分
natural interference 自然干扰
natural isotope 天然同位素
natural isotope variation 天然同位素变化
natural isotopic 天然同位素的
natural isotopic abundance 天然同位素丰度
natural land 天然土地
natural landscape 自然景观
natural landslide 自然滑坡
natural law 自然法则
natural levee 天然堤，天然冲积堤；自然堤
natural levee deposit 天然堤沉积物
natural levee sediment 天然堤沉积物
natural level 自然水平面
natural load 天然荷载；自然负荷；自然载重
natural load-transmitting arch 天然卸载拱
natural logarithm 自然对数
natural magnet 天然磁铁
natural magnetism 天然磁性
natural material 天然材料；天然物料
natural mineral 自然矿物
natural mode of vibration 固有振型；固有振动方式
natural moisture 天然湿度；天然含水量
natural moisture content 天然含水量；天然含水率
natural motion 固有振动；自动振荡；自然运动
natural mud 天然泥浆；井眼内自造泥浆
natural multiple 自然倍数
natural navigable waterway 天然通航水道
natural noise 自然噪声；天然噪声

natural noise barrier 天然噪声屏障；天然隔声屏障
natural number 自然数
natural number coding 自然数编码
natural occurrence 自然存在；天然产状
natural oil 天然石油
natural order 自然分类；自然顺序；自然秩序
natural ordering 自然排序
natural oscillation 固有振动；固有振荡；自然振荡
natural outcrop 天然露头
natural pack 天然堆积；天然填密
natural period 固有周期；自然周期
natural period of oscillation 振荡固有周期；自然回摆期
natural period of swing 摆动固有周期
natural period of vibration 自振周期
natural permafrost table 多年冻土天然上限；天然永冻土面
natural permeability 原始渗透率
natural phenomenon 自然现象
natural pollutant 天然污染物；自然污染物
natural pollution 天然污染；自然污染
natural pollution source 天然污染源
natural polymer 天然聚合物
natural pore ratio 天然孔隙比
natural porosity 原始孔隙度
natural porous media 天然多孔介质
natural potential 自然电位
natural process 自然过程；天然过程
natural producing energy 天然开采能量；天然地层能量
natural product 天然产物；自然产品
natural production decline 产量自然递减
natural profile 天然剖面；天然纵断面
natural protection area 自然保护区；自然保护面积
natural purification 自然净化；天然净化
natural quarrying 岩石自然崩落
natural quartz crystal 天然石英晶体
natural radiation 天然辐射
natural radiation dose 天然辐射剂量
natural radiator 自然辐射源
natural radioactive decay 放射性自然衰变
natural radioactive element 天然放射性元素
natural radioactive nuclide 天然放射性核素
natural radioactive series 天然放射系
natural radioactivity 天然放射性
natural radionuclide 天然放射性核素
natural recharge 天然补给；天然再充水
natural recharge rate 天然补给速率
natural recovery 天然开采
natural recovery mechanism 天然能量开采机理
natural recovery phase 天然能量开采阶段
natural recovery process 天然能量开采过程
natural regain 自然回潮率；自然吸湿率
natural regeneration 自然更新
natural regulation 天然调节；自然调节
natural remanence 天然剩磁
natural remanent magnetism 天然剩磁
natural remanent magnetization 固有剩余磁化；天然剩余磁化强度
natural replenishment 天然补给
natural repose angle 天然休止角
natural repose angle of sand 砂土天然坡角度

natural reserve 自然保护区；天然储量
natural reserve area 自然保护区
natural reservoir 天然水库；天然储库
natural reservoir drive 油层天然驱动
natural reservoir energy 天然油层能量
natural resonance 固有谐振；自然共振
natural resonant frequency 自然共振频率
natural resource 自然资源
natural resource distribution 自然资源分布
natural resource ecosystem 自然资源生态系统
natural resource evaluation 自然资源评价
natural resource management system 自然资源管理系统
natural resource of groundwater 天然地下水资源
natural risk 自然风险；正常风险
natural risk factor 自然风险因素；自然危险系数
natural risk zone 自然灾害带
natural river 天然河道
natural road 未加固土路；天然道路
natural rock 原生岩；天然岩
natural rock asphalt 天然沥青
natural rockfill dyke 天然堆石堤
natural rock-soil slope 岩土自然边坡，天然岩土边坡
natural sand 天然砂；天然骨料
natural sand and gravel 天然砂砾料
natural scale 自然比尺例；按实物大小的尺寸；自然数；自然量；固有量
natural scale model 天然比尺模型
natural scenic area 天然风景区
natural science 自然科学
natural scour 自然演变冲刷
natural sedimentation 自然沉淀；自然沉积；自然沉降
natural seismic wave 天然地震波
natural selection 自然选择
natural self-purification 天然自净化作用
natural self-purification capacity 天然自净能力
natural self-supporting arch 自然平衡拱；天然自稳拱
natural sinking 自然下沉；自然凹陷
natural size 真实尺寸，实际尺寸；原尺寸；天然粒径
natural slope 天然边坡，天然斜坡；天然坡面
natural slope angle of soil 土壤自然坡度角；土壤自然堆积角
natural sloping terrain 天然斜坡；天然倾斜地形
natural soft clay 天然软黏土
natural soft deposit 天然软土层；天然松软堆积物
natural soil 原土；底土；天然土
natural soil conditioner 天然土壤改良剂
natural soil deposit 天然土壤沉积
natural soil drainage 土壤自然排水
natural soil stratum 天然土层
natural solid bitumen 天然固体沥青
natural source 天然污染源
natural space radiation 自然空间辐射
natural stability limit 天然稳定极限
natural stability of side slope 自然稳定边坡
natural state of soil 土的天然状态
natural stone 天然宝石；天然石料
natural stone work 天然石方工程
natural stoping 自然崩落回采法；天然顶蚀作用
natural storage 天然容蓄量
natural strained well 天然滤井

natural strata	天然地层
natural stratum	天然地层；原始地层
natural stream	天然水道；天然河流
natural stream channel	天然河槽，天然河道；天然河床
natural strength	天然强度
natural strength of soil	土的天然强度
natural stress	自然应力；天然应力
natural stress field	天然应力场
natural stress of rock mass	岩体天然应力
natural stress state	天然应力状态
natural subgrade	天然路基
natural subirrigation	天然地下灌溉
natural subsoil	天然地基
natural succession	自然演替
natural synoptic period	自然天气周期
natural synoptic region	自然天气区
natural synoptic season	自然天气季节
natural synthesis	自然合成
natural system	自然系统；自然分类系统
natural tailwater	天然尾水
natural temperature field	天然温度场
natural tendency	自然趋势
natural terrain	天然地形
natural terrain feature	天然地形特征
natural terrain hazard mitigation measure	山体灾害缓减措施
natural terrain landslide	山体滑坡
Natural Terrain Landslide Inventory	山体滑坡目录
natural terrain landslide investigation	山体滑坡调查
natural terrain landslide risk management	山体滑坡风险管理
natural terrain landslide study	山体滑坡研究
natural tide	自然潮；真潮
natural tilt-angle test	天然坡角试验；休止角试验
natural time of oscillation	固有振动时间
natural tracer	天然示踪剂
natural tunnel	天然隧洞
natural tunnel wind	自然隧道风
natural type	自然类型
natural unit weight	天然容重；重度
natural uranium	天然铀
natural variational principle	自然变分原理
natural vegetation	天然植被
natural ventilating pressure	自然通风压力
natural ventilation	自然通风，自然换气
natural ventilation effect	自然通风效应
natural ventilation equipment	自然通风装置
natural ventilation pressure	自然风压
natural vibration	固有振动；自然振动
natural vibration frequency	固有振动频率
natural vibration mode	固有振型
natural vibration period of bridge	桥梁自振周期
natural void ratio	天然孔隙比
natural wastage	自然损耗
natural water	天然水
natural water content	天然含水量；天然含水率
natural water drive	天然水驱
natural water drive reservoir	天然水驱油藏
natural water dumping	自流注水
natural water encroachment	天然水侵
natural water escape	天然溢洪道
natural water flooding	天然水驱
natural water influx	天然水侵
natural water invasion	自然水侵
natural water level	自然水位
natural water resource	天然水资源；天然水利资源
natural water table	天然地下水位；天然地下水面；天然潜水面
natural watercourse	天然河道；天然水道
natural water-sensitive core	天然水敏性岩心
natural watershed	天然流域；大然分水岭
natural watertank	天然水槽；天然水柜
natural waterway	天然水道
natural wave	固有波；自然波
natural wavelength	固有波长；自然波长
natural wax	天然蜡
natural well	天然井；自然产油井
natural wettability	天然润湿性
natural whiteness	自然白度
natural year	回归年；太阳年；天文年
natural-color panchromatic photograph	天然色全色摄影相片
natural-color photography	天然色摄影术
natural-current method	自然电流勘探法
natural-draft pressure	自然通风压力
natural-gas field	天然气田
natural-gas injection	天然气注入
natural-gas-powered	天然气驱动的
naturalist	自然科学工作者；博物学家
naturally alloyed iron	天然铁合金
naturally alloyed steel	天然合金钢
naturally encroaching water	天然侵入水
naturally fractured reservoir	天然裂缝性储层
naturally impounded body	天然蓄水盆地；天然静止水体
naturally radioactive water	天然放射性水
naturally regulated river	自然调节河流
naturally supported mining method	天然支护采矿法
naturally-occurring earthquake	天然地震
natural-potential survey	自生电势勘探；等电势勘探；自然电位测量
natural-resource evaluation	自然资源评价
nature	自然界；性质；本性；特征
nature angle of repose	天然休止角
nature assemblage	自然群集
Nature Conservancy Council	自然保护委员会
nature conservation	自然保护
nature conservation area	自然保护区
nature earthquake	天然地震
nature evolution	自然演变
nature gas	天然气
nature node	自然节点
nature of coal	煤质
nature of ground	围岩的性质
nature of rock	岩石性质
nature of soil	土壤本质；土性质
nature of the coast	海岸性质
nature of works	工程性质
nature preservation	自然保护
nature protection	自然保护
nature reserve	自然保护区；自然保留地

nature slope 天然坡度
nature stress field 天然应力场
nature terrain landslide investigation 山体滑坡调查
nature terrain landslip hazard study 山体滑坡研究
naught 无；零
naujaite 方钠霞石正长岩
naujakasite 瑙云母
naumannite 硒银矿
nautical 航海的；海上的
nautical almanac 航海天文历；航海历
nautical archaeology 航海考古学
nautical astronomy 航海天文学
nautical chart 航海图
nautical ephemeris 航海历
nautical mile 海里
nautical receiving set 水听器；航海接收机
nautical scale 海图的经纬度网格；海图比例尺
navaid 助航装置；导航系统
navaid wind-finding 导航测风
naval brass 铜锡锌合金；海军黄铜
naval engineering technology 海军工程技术
naval light 航行信号灯
naval meteorology 海洋气象学
naval plotting chart 航线标绘图
naval port 军港
naval service survey 海军勤务测量
naval tank 船模试验池
naval target 海上目标
navascope 机载雷达显示器
navascreen 导航屏幕
navaspector 导航指示器
Navi-Drill 纳维钻具
Navier's equation 纳维尔方程
Navier's theory 纳维尔理论
Navier-Stokes equation 纳维尔-斯托克斯方程
navigability 适航性；航海性能
navigable 可航行的；适航的
navigable canal 通航运河
navigable channel 通航河道；航槽；航道
navigable database 导航数据库
navigable depth 通航水深
navigable power canal 通航发电渠道
navigable river 通航河流
navigable semicircle 可航半圆
navigable span 通航宽度
navigable water 航道；通航水域
navigable water level 通航水位
navigable waterway 通航水道
navigate 漫游导航；导航；航行
navigating instrument 导航仪；航海仪器
navigation 航行，航海，航运；导航，导向装置
navigation aid 助航设备；导航设施
navigation and positioning 导航定位
navigation beginning 通航河段起点
navigation by satellite timing and ranging-global positioning system 卫星导航全球定位系统
navigation canal 通航渠道；通航运河
navigation channel 航道；通航河道
navigation channel chart 航道图
navigation chart 导航图

navigation clearance 航道净空；通航净空
navigation control system 导航控制系统
navigation dam 通航坝
navigation dredging 航道开挖；航道清淤
navigation error 导航误差
navigation gyrocompass 导航回转罗盘
navigation head 水陆转运站；航运终点
navigation lock 船闸；通航用闸
navigation mark survey 航标测量
navigation obstruction 航行障碍物
navigation of aerial photography 航拍领航；空中摄影领航
navigation opening 港口；通航桥孔
navigation period 通航期
navigation positioning 导航定位
navigation positioning on sea surface 海面导航定位
navigation reservoir 通航水库
navigation satellite 导航卫星
navigation season 航海季节
navigation sonar 导航声呐
navigation span 通航宽度
navigation station location survey 导航台定位测量
navigation strip 航线
navigation structure 通航建筑物
navigation sub 导向接头
navigation system 导航系统
navigation tunnel 通航隧洞
navigation water level 通航水位
navigation waterway 航道
navigational aid 助航设备；导航辅助设备
navigational clearance 通航净空
navigational datum 航行基准面；航行数据
navigational mark 航标
navigational system 导航系统
navigational trace map 航迹图
navigational transmitter 导航发射机
navigational triangle 导航三角形
navigator 领航员；导航仪；导航系统；导航设备
navigraph 领航表
navsat 导航卫星
NAVSTAR global positioning system 导航星全球定位系统；天文导航全球定位系统
navvy 蒸汽挖土机；挖土工；矿工
navvy barrow 土方工程手推车；运土手推车
navvy excavator 挖掘机；挖土机
Navy Navigation Satellite System 海军导航卫星系统；子午卫星系统
naze 岬角；海角
n-dimensional normal distribution n维正态分布
n-dimensional vector space n维矢量空间
neap 小潮；低潮
neap range 小潮潮差
neap rise 小潮升
neap tide 小潮；低潮
neapite 霞磷岩；霞石磷灰岩
near accident 潜在事故；隐患；险些发生的事故
near balanced hydrostatic head 接近平衡的静水压头
near bottom boundary layer 近底边界层
near color infrared photography 近彩色红外线摄影
near desert 半沙漠区；似沙漠区

near detector 近检波器
near earth orbit 近地轨道
near earthquake 近震
near earthquake instrument 近震仪
near epithermal neutron count rate 近探测器超热中子计数率
near field 近场；近源场
near field area 近场区
near field domain of an excavation 开挖近场区
near field geophone 近源场检波器
near field measurement 近场测量
near field of an orebody 矿体近场
near incompressibility 几乎不可压缩
near infrared 近红外
near infrared absorption 近红外吸收
near infrared region 近红外区
near infrared spectroscopy 近红外光谱
near infrared window 近红外窗口
near isothermal decompression 近等温降压
near Level seam 近水平煤层
near lunar space 近月空间
near offset trace 近炮检距道
near point 近点
near polar orbit 近极地轨道
near real-time 近实时的；准实时的
near shaft control point 近井点
near shaft point 近井点
near shaft-point surveying 近井点测量
near shock 近震
near shore marine environment 近岸海洋环境
near shore zone 近岸带
near solar space 近太阳空间
near surface 近地表
near surface deposit 近地表矿床；近地表沉积物
near surface disposal 近地表处置
near surface foci earthquake 浅源地震
near trace 近道
near ultraviolet 近紫外
near ultraviolet excitation 近紫外激发
near ultraviolet radiation 近紫外线辐射
near vertical photograph 近似垂直摄影；近垂直摄影像片
near view 近景
near wellbore formation 近井地层
near wellbore formation damage 近井地层损害
near-bit reamer 近钻头扩孔器
near-bit stabilizer 近钻头稳定器
near-bit-influx detector 近钻头地层流体浸入检测器
near-bottom current 沿底流；底层流；近底流
near-bottom sonar 近底声呐；海底地貌仪
near-breaking wave 濒碎波；近碎波
near-by coal seam 近距煤层
nearby earthquake 近震
nearby field 卫星油田
nearby hole 相邻孔
nearby protection 近体防护
nearby view 近景
near-continental abyssal plain 近大陆深海平原
near-critical 近临界的
near-earth satellite 近地卫星

near-earth space 近地空间
near-end crosstalk 近端串音
near-end crosstalk attenuation 近端串音衰减
near-equilibrium state 近平衡状态
nearest distance 最近距离
nearest mean classification rule 最近均值分类法；最近平均分类规则
nearest neighbor 最近邻
nearest neighbor algorithm 最邻近法
nearest neighbor resampling 最近邻重复取样
nearest neighbor sequence 最近邻序列
nearest neighboring area method 最近邻区块储量计算法
nearest neighbour method 最近邻法
nearest neighbour sampling 最近邻采样
near-fault earthquake 近断层地震
near-fault site 近断层场地
near-field 近场；近区
near-field accelerogram 近场加速图；近场加速记录
near-field earthquake motion 近场地震运动
near-field effect 近场效应
near-field ground motion 近场地运动
near-field phenomenon 近场现象
near-field recording 近场记录
near-field seismology 近场地震学
near-field signature 近场特征
near-field stress 近场应力
near-field wavelet 近场子波
near-gauge hole 井径规则的井；直径与钻头相近的井孔
near-gravity material 邻近分选密度的物料；近比重物料；邻近比重物
near-gravity zone 近分选比重区
near-horizontal 近水平的
near-infrared 近红外
near-infrared band 近红外波段
near-infrared radiation 近红外辐射
near-infrared region 近红外区
near-infrared spectroscopy 近红外光谱学；近红外光谱法
near-infrared spectrum 近红外谱
near-infrared wave band 近红外波段
near-level grade 近水平坡；近水平坡度
near-level seam 近水平煤层
near-liquidus phase 近液相线相，近液相
near-lithostatic fluid pressure 近静岩流体压力
nearly consumed island 残岛
nearly flat-lying fault 近水平断层；近平伏断层
nearly horizontal layered landslide 近水平层状滑坡
nearly viscous flow 近黏滞流动；近黏性流
near-mesh bed 接近筛孔的物料床层；难筛粒床层
near-mesh grain 分界粒径颗粒
near-mesh material 接近筛孔尺寸的物料；难筛粒
near-mesh particle 接近筛孔尺寸的颗粒；难筛粒
near-mesh size 近筛孔尺寸粒度；难筛粒径，难筛粒度
near-perfect pack 接近理想的充填
near-plain of subaerial erosion 陆上侵蚀准平原
near-polar sun synchronous obit 近极地太阳同步轨道
near-quake magnitude 近震震级
near-range 近距离
near-real-time 近实时；准实时
nearshore 近岸；近滨；近海

nearshore bar 近滨沙坝
nearshore circulation 近滨环流
nearshore current 近岸流；沿岸流
nearshore current system 近岸流系
nearshore deposit 近岸沉积
nearshore environment 近岸环境；滨岸环境
Nearshore Environmental Analog Prediction System 近岸环境模拟预报系统
nearshore facies 近滨相
nearshore marine zone 近滨海相带
nearshore morphodynamics 近岸地形地貌
nearshore morphology 近岸形态
nearshore sandbank 近岸沙洲；近岸沙滩
nearshore sediment 近滨沉积物
nearshore sediment accumulation 近岸沉积物积累
nearshore sediment transport 近岸输沙
nearshore sedimentation 近岸沉积
nearshore structure 近海结构物；沿海结构物
nearshore wave action 近岸波浪作用
nearshore zone 近滨带；近滨区；近海区
nearshore-surface 近地表的
near-slot-size grain 尺寸与割缝宽度相近的颗粒
near-source 近震源
near-source effect 近震源效应
near-source ground motion 近源地震动
near-source response spectrum 近震源反应谱
near-source rock site 近震源岩石场地
near-source wave 近震源波
near-supercomputer 近似超级计算机
near-surface 近地表
near-surface anomaly 表层异常
near-surface correction 表层校正
near-surface damage 近地表损伤
near-surface effect 近地表效应
near-surface fault 近地表断层
near-surface foci earthquake 近地表地震；浅源地震
near-surface geophysical prospecting 近地表地球物理勘探；近地表物探
near-surface layer 表层
near-surface low-velocity layer 近地表低速层
near-surface multiples 表层多次波
near-surface probe data 浅孔探测资料
near-surface raypath 近地表射线路径
near-surface reflection 近地表反射
near-surface reflection anomaly 近地表反射异常
near-surface reflector 近地表反射层；近地表反射器
near-surface reverberation 近地表混响
near-surface rock 近地表岩石
near-surface sample 浅层样品
near-surface section 近地面井段；靠近地面的采区
near-surface seismic event 近地表地震事件
near-surface shooting 近地表激发
near-surface velocity anomaly 近地表速度异常
near-surface wave 表面波；近地表波
near-term 近期的
near-tip asymptotic field function 裂尖场函数
near-trace gateway 近道选排
near-trace gather 近道选排；近道道集
near-trace section 近道剖面
near-vertical 接近垂直

near-vertical aerial photography 近垂直航空摄影术
near-vertical well 近垂直井
near-well heat reservoir condition 井孔附近的热储条件
near-wellbore area 近井底地带
near-wellbore permeability 近井地带渗透率
near-wellbore scale 井筒附近地带
near-zone magnetic transient sounding 近区磁瞬变场测深法
neat area 净面积
neat cement 净水泥；净水泥浆
neat cement grout 纯水泥薄浆；净水泥浆
neat cement slurry 净水泥浆
neat line 内图廓线；准线；墙面交接线
neat plaster 净灰浆；纯灰浆
neat retarded cement 纯缓凝水泥
neat-bottom current 净底层流，纯底流；纯海底强流
neatline 图表边线；准线；内图廓线
Nebraska glacial stage 内布拉斯加冰期
nebular 星云的；星云状
nebular hypothesis 星云假说
nebulite 星云球雏晶；云染岩；星云岩
nebulite gneiss 云染片麻岩；星云片麻岩
nebulite migmatite 云染混合岩
nebulitic structure 云染状构造
necessary and sufficient condition 充要条件
necessary condition 必要条件
necessary living space 必要生活空间；必要居住面积
neck 颈；岩颈；矿筒；火山颈；地峡，海峡
neck bush 内衬套；轴颈套
neck channel 颈状海峡；颈状渠道
neck collar 轴承环；颈托，颈圈
neck current 束狭急流
neck cutoff 曲流裁直，曲流取直，截弯取直
neck diameter 孔颈直径
neck down 缩颈；颈状收缩
neck of land 地峡
neck of rip current 离岸流海峡；激流海峡
neck radius 孔颈半径；孔喉半径
neck-down 截面收缩
necked-in 向内弯曲
necked-out 向外弯曲
necking 颈缩；断面收缩
necking down 缩口；颈缩
necrolite 黑云安粗岩
needle 针，指针；磁针；针状物；针状晶体
needle beam 小托梁
needle beam method 针梁法；小托梁法
needle bearing 滚针轴承
needle coal 纤维状褐煤；纤维状木煤
needle coke 针状结晶石油焦；针状焦
needle crystal 针状结晶；针状晶体
needle deflection 指针偏转
needle density 针测密度
needle diabase 针状辉绿岩
needle fracture 针状断口
needle galvanometer 磁针电流计；指针检流计
needle hole test 针孔试验
needle ironstone 针铁矿；针铁石
needle mesh 针状筛孔
needle nozzle 针形阀；针形喷嘴

needle penetrometer 针式贯入仪；针穿硬度计；针入度仪
needle pile 针形桩
needle pointer 指针
needle probe 针形探头
needle punched process 针刺法
needle shaped and flaky particle 针状和片状颗粒
needle spar 霰石
needle stone 钠沸石
needle timber 柱脚撑木
needle traverse 罗盘施测导线；磁向导线
needle valve 针形阀；针状阀
needle valve seat 针阀座
needle vibrator 插入式振捣器；针状振捣器
needle-density 针入密度
needle-like 针状的；类似针形
needle-punched geotextile 针刺土工织物；针刺土工布
needle-punched process 针刺法
needle-shaped 针状
needle-shaped particle 针状颗粒
needle-shaped structure 针状构造
needling 针刺；横撑木，横撑杆
Neel point 奈尔点
Neel temperature 奈尔温度
nefedievite 红坚岭石
negative 负的；阴性的；否定的；负数；底片
negative acceleration 负加速度
negative adsorption 负吸着；负吸附作用
negative air pressure 负气压，空气负压力
negative allowance 负公差；负偏差
negative altitude 俯角
negative angle 俯角
negative anomaly 负异常
negative area 沉降区；负相区；负异常区
negative association zone 负共生带
negative balance 负均衡
negative benching tunnelling method 反台阶法
negative bias 负偏压
negative binomial distribution 负二项分布
negative breccia 负角砾岩
negative buoyancy 下沉力；负浮力
negative buoyancy of descending plate 俯冲板块负浮力；俯冲板块下沉力
negative carrier 负调制载波；负离子
negative catalyst 缓化剂，缓和剂；负催化剂
negative compression 负压缩
negative confining bed 隔水底板
negative correction 负修正
negative correlation 负相关
negative crystal 负晶；空晶；负光性晶体
negative curvature 负曲率
negative cutting angle 负切削角
negative damping 负阻尼
negative definite 负定的
negative definite matrix 负定矩阵
negative definite quadratic form 负定二次型
negative deflection 负挠度，负挠曲；负偏差，负偏转
negative delta 三角湾；喇叭口状三角洲
negative differential sticking 负差压卡钻
negative direction 负方向；反方向

negative electrode 负极；阴极
negative electron 阴电子；负电子
negative element 下降构造单元；负向构造单位
negative elongation 负延长
negative entropy 负熵
negative exponential distribution 负指数分布
negative factor 消极因素；不利因素
negative feedback 负反馈
negative film mosaic 航摄底片镶嵌图
negative form 负地形
negative frequency 负频率
negative friction 负摩擦；负摩擦力
negative friction pile 负摩擦桩
negative function 负函数
negative gravity anomaly belt 负重力异常带
negative growth 负增长
negative hardness 负硬度
negative head 负水头
negative incidence 负倾角
negative increment 负增量
negative induced polarization effect 负激发极化效应
negative interaction 负作用
negative inversion 负反转
negative ion 阴离子；负离子
negative ion generator 负离子发生器
negative landform 负地形
negative lens 负透镜
negative lumped mass 负集中质量
negative magnitude 负震级
negative mantle friction 负表面摩擦力
negative modulation 负极性调制
negative moment 负弯矩
negative moment reinforcement 负弯矩钢筋
negative mosaic 底片镶嵌图
negative movement 下降运动；负向运动；海平面下降
negative number 负数
negative output 逆功率
negative peak 最大负值；负峰值
negative phase contrast 负相位差
negative phase sequence 负相序；逆相序
negative plate 阴极板；负极板；底片
negative point of intersection 反交点
negative pole 阴极；负极
negative pore fluid pressure 负孔隙液体压力
negative pore pressure 负孔隙压力；负孔压
negative pore water pressure 负孔隙水压力；负毛细水压力
negative potential 负电位；负电势；负势；负位
negative power 负乘方；负幂
negative press 负压
negative pressure 负压，负压力；真空
negative pressure degasification 负压脱气
negative pressure of ventilation 通风负压
negative pressure perforating 负压射孔
negative pressure pulse system 负压脉冲系统
negative pressure stabilization tower 负压稳定塔
negative pressure wave 负压力波
negative pulse 负脉冲；负极性脉冲
negative quantity 负量
negative rake 负前角；负倾角

negative ray	阴极射线	negatron	阴电子；负电子
negative reaction	负反馈；负反力	negentropy	负熵；负平均信息量
negative reflection coefficient	负反射系数	negligible	可以忽略的
negative refraction	负折射	negligible amount	微量
negative reinforcement	负挠钢筋；负力矩钢筋	negligible deposition	微量沉积
negative relief	负地形	negligible probability	可忽略的概率
negative rotation	左旋；逆时针旋转	negligible risk	微风险；可忽略的风险
negative screening	负屏蔽现象	neighbor	邻居；邻近值；邻国
negative segregation	反偏析；负偏析	neighbor domain	邻域
negative semidefinite	半负定	neighborhood	邻近
negative separation	负幅度差	neighborhood analysis	相邻分析
negative sequence	逆序	neighborhood at infinity	无穷大邻域
negative sequence load	逆序负载	neighborhood method	邻元法
negative settlement	负沉降	neighboring anchor	相邻锚杆
negative shear	负剪力	neighboring line interference	邻线干扰
negative shoreline	上升岸线；负海滨线	neighboring pile	邻桩
negative side angle	负刃倾角	neighboring rock	围岩
negative side rake	负刃倾角	neighboring trace	邻道
negative sign	负号	neighboring well	邻井
negative signal	负信号	neighborite	氟镁钠石
negative skewness	负向偏态；负偏态；负偏斜度；负向偏斜度	neighbour	附近；邻近值
		neighbourhood	附近，邻近；周围；邻域
negative skin friction	负的表层摩擦；表面负摩擦力；负摩擦	neighbourhood analysis	邻域分析
		neighbourhood method	邻元法；邻域法
negative skin friction of pile	桩的负摩阻力	neighbourhood noise	环境噪声
negative slope	负斜率	neighbourhood sheet	邻接图幅
negative spike	负峰；负尖峰信号	neighbourhood tunnel	小净距隧道
negative stack power	负叠加能量	neighbouring	邻近的；接壤的
negative standard polarity	标准负极性	neighbouring area	邻域；附近地区；相邻地区
negative state water level	负静止水位	neighbouring building	邻近建筑物
negative stiffness	负刚度	neighbouring drainage basin	相邻流域；相邻流域盆地
negative strain	负应变	neighbouring node	邻近结点
negative stress concentration	负应力集中	neighbouring pile	邻桩
negative suction head	真空吸入压头	neighbouring region	邻域；附近区域
negative superelevation	反超高	neighbouring rock	围岩
negative superposition	负叠加	nek	垭口；山峡；鞍部
negative temperature	负温度；零下温度	nelsonite	钛铁磷灰岩
negative temperature anomaly	负温度异常	Nemagraptus	丝笔石属
negative temperature coefficient	负温度系数	nemalite	含铁水镁石；纤水镁石
negative temperature gradient	负温度梯度	nemaphyllite	绿蛇纹石
negative terminal	负极接线柱；阴极端子，负端子	nematoblastic	纤状变晶质
		nematoblastic texture	纤状变晶结构
negative test	负检验，负结果；反试验	nematomorphic	纤维状的
negative thixotropy	负触变性；负摇溶现象	nemite	暗白榴岩
negative utility	负效用	NEMS	纳电子机械系统
negative valence	负原子价	nenadkevite	硅钙铅铀矿
negative valency	负价	Neo-Alpine orogenesis	新阿尔卑斯造山运动
negative value	负值	Neoarchean	新太古代
negative vector	负矢量	Neoarchean carbonate rock	新太古代碳酸盐岩
negative well	吸水井；渗水井；排水井；注水井	neoautochthone	新原地岩体；新推覆基体
negative well bore pressure	井筒负压	Neo-Cambrian	晚寒武纪
negative-connected	接于负极的	neocatastrophic theory	新灾变说
negative-feedback amplifier	负反馈放大器	neocatastrophism	新灾变论
negative-going	负向的	Neocathaysian	新华夏式
negative-going zero	声波波形负向零点	Neocathaysian structural system	新华夏式构造体系
negative-positive procedure	负片-正片反转晒印法	Neocathaysian tectonic system	新华夏式构造体系
negative-positive process	负-正法	Neocene	晚第三纪；上第三系；新近纪
negative-positive reversal color film	正负反转彩色胶片	Neocene period	晚第三纪
negative-temperature-coefficient resistor	负温度系数电阻器	neochronology	新年代学
negativity	负性，阴性；否定性；负值性	neoclimatology	现代气候学；新气候学

neocolluvium 新崩积层；新风化壳
Neocretaceous 新白垩纪
neocryst 新生晶；次生单晶
neocrystallization 新结晶作用
neocrystic texture 次生重结晶结构；新生晶结构
neo-Darwinism 新达尔文理论；新达尔文主义
Neodevonian 新泥盆世；晚泥盆世
neodoxy 新学说，新观点
neodymium 钕
neodymium isotope 钕同位素
neodymium-samarium age method 钕-钐年龄法；钕-钐测年法
neoeffusive 新喷出岩
neoeluvium 新残积层
neoformation 新建造；新生作用；新形成作用
neofracture 新断裂
Neogene 上第三系；新近系；晚第三纪；新近纪
Neogene basin 晚第三纪盆地
Neogene climate 新近纪气候
Neogene fluvial deposit 新第三纪河流沉积
Neogene geodynamic evolution 新近纪的地球动力学演化
Neogene period 晚第三纪；新近纪
Neogene system 上第三系；新近系
neogenesis 新生成作用；新生作用
neogenic 新形成的；新生的
neogenic basin 新生盆地
neogenic crust movement 新近地壳运动；晚近地壳运动
neogenic Culture 新石器时代文化
neogenic mineral 新生矿物
neogenic mineral deposit 新生矿床
neogeosyncline 新地槽
neoglacial 新冰期的；新冰河作用的
neoglaciation 新冰期
Neohercynian 新海西的；新海西期的
neohexane 新己烷
neo-ice age 新冰期
neoichnology 新足迹化石学；新遗迹学；现代痕迹学
neoid 晚近
neoid crust movement 晚近地壳运动
neokaolin 新高岭土
neokerogen 新干酪根；新油母岩
neolenticular texture 次生透镜状结构
neolite 新火山喷出岩；新石
neolith 新石器
neolithic 新石器时代的
Neolithic Age 新石器时代
neolithic era 新石器时代
neo-loess 新黄土
neomagma 新生岩浆；新成岩浆
neometamorphism 初变质作用；新变质
neomineralization 新矿化作用；新成矿作用
neomorphic 新生变形
neomorphic fabric 新生变形组构；新生变形岩组
neomorphism 新生变形作用；新变成作用
Neon isotope anomaly 氖同位素异常
neoorogenic zone 新造山带
neopaleozoic 晚古生代；新古生代
neo-petroleum 新生石油
neoporphyrocrystic 次生斑状
neoporphyrocrystic texture 新生斑状结构

Neoproterozoic 新元古代
Neoproterozoic Era 新元古代
Neoproterozoic Erathem 新元古界
Neoproterozoic glaciation 新元古代冰期；新元古代冰川作用
neorift 新裂谷
neosome 新成体；新火岩体
neostratotype 新层型
neotantalite 黄钽铁矿；新钽铁矿
neotectonic 新构造的
neotectonic age 新构造年龄
neotectonic block motion 新构造断块运动
neotectonic element 新构造单元
neotectonic fluctuation 新构造运动波动性
neotectonic fracture zone 新构造破裂带
neotectonic framework 新构造构架
neotectonic generality 新构造运动普遍性
neotectonic inheritance 新构造运动继承性
neotectonic intermittence 新构造运动间歇性；新构造运动周期性
neotectonic joint 新构造节理
neotectonic map 新构造图
neotectonic movement 新构造运动
neotectonic oscillation 新构造运动振荡性
neotectonic period 新构造期
neotectonic type 新构造类型
neotectonic velocity field 新构造速度场
neotectonic zonation 新构造区带；新构造成带现象
neotectonics 新构造运动；新大地构造
neotectonism 新构造运动
neotethyan 新特提斯的
Neotethys 新特提斯海的；新特提斯洋
Neo-Tethys 新特提斯
Neo-Tethys ocean 新特提斯洋
neotocite 水锰辉石；硅锰矿
neovolcanic 新火山的
neo-volcanic rock 新火山岩
neovolcanism 新火山作用
neovolcanite 新火山岩
Neozoic 新生代的
Neozoic era 新生代
Neozoic Erathem 新生界
Neozoic group 新生界
nepheline 霞石
nepheline andesite 霞石安山岩
nepheline basalt 霞石玄武岩
nepheline basanite 霞石碧玄岩
nepheline basite 霞石基性岩
nepheline benmoreite 霞石歪长粗面岩
nepheline diorite 霞石闪长岩
nepheline gabbro 霞石辉长岩
nepheline hawaiite 霞石橄榄中长玄武岩；霞石深绿橄榄岩
nepheline latite 霞石安粗岩
nepheline monzogabbro 霞石二长辉长岩
nepheline monzonite 霞石二长岩
nepheline monzosyenite 霞石二长正长岩
nepheline mugearite 霞石橄榄粗安岩
nepheline plagisyenite 霞石斜长正长岩
nepheline porphyry 霞石斑岩

nepheline syenite	霞石正长岩
nepheline syenite-aplite	霞石正长细晶岩
nepheline syenite-pegmatite	霞石正长伟晶岩
nepheline tachylite	霞石玄武岩
nepheline tephrite	霞石碱玄岩
nepheline trachyandesite	霞石粗安岩
nepheline trachybasalt	霞石粗面玄武岩
nepheline trachyte	霞石粗面岩
nepheline tristanite	霞石碱长粗安岩
nepheline-basalt	霞石玄武岩
nepheline-porphyry	霞石斑岩
nephelinite	霞石岩
nephelinite basanite	霞石岩-碧玄岩
nephelinite tephrite	霞石岩-碱玄岩
nephelinitoid	霞石基质
nephelinitoid phonolite	霞响岩
nephelinization	霞石化
nephelinoith	霞石岩
nephelinolite	霞石岩
nephelite	霞石
nepheloid sediment	浑浊沉积物
nephelometric	浊度的
nephelometric analysis	浊度分析
nephelometry	浊度测定法
nephoscope	测云器；云观测器
nephrite	软玉；软透闪石
nephritic	软玉的；软玉状的
nephritoid	叶蛇纹石；纤蛇纹石；温蛇纹石
nepionic	幼年期
nepouite	镍蛇纹石；镍绿泥石
neptunian	水成论的；海王星
neptunian dyke	水成岩墙
neptunian theory	水成论
neptunianism	水成论
neptunic dyke	水成岩墙
neptunic rock	水成岩
neptunism	水成论
neptunist	水成论者
neptunite	星叶石；柱星叶石；海王石
neptunium	镎
neptunium series	镎系
nerchinskite	埃洛石
neritic	浅海的；近岸的
neritic area	浅海区
neritic coal measure	浅海型含煤岩系；浅海型煤系
neritic coal-bearing formation	浅海型含煤建造
neritic coal-bearing series	浅海型含煤岩系
neritic deposit	浅海沉积
neritic environment	浅海环境；大陆架环境；陆棚环境
neritic facies	浅海相；近岸相
neritic geology	浅海地质学
neritic marine basin	浅海盆地
neritic marine zone	浅海带
neritic region	浅海区
neritic sea	浅海
neritic sediment	浅海沉积物；浅海沉积
neritic sedimentation model	浅海沉积模式
neritic shelf	浅海陆架
neritic terrigenous sedimentation	浅海陆源沉积
neritic zone	浅海带；近海区；近岸带
neritopelagic	浅海的
nervous earth	地壳的震动部分；地震地带
nesosilicate	岛状硅酸盐
nesosilicate mineral	岛状硅酸盐矿物
nesquehonite	碳氢镁石，水碳镁石，碳酸镁石；三水菱镁矿
ness	海角；岬
nest like ore body	鸡窝状矿体
nest of minerals	矿巢
nest of screens	试验筛组；组合筛
nest outburst	瓦斯包突出；水包溃决；矿穴气体喷出
nested block	嵌套分程序
nested caldera	巢状破火山口
nested crater	巢状火山口
nested deposit	巢状矿床；窝子矿
nested loop	嵌套循环
nested observation wells	观测井群
nested subroutine	嵌套子程序
nested volcanic cone	巢状火山锥
nested-analysis of variance	方差套合分析
nesting	检波器组；嵌套
nesting grid	嵌套网格
nesting level	嵌套级；嵌套层次
nesting storage	后进先出储存器；嵌套存储
nest-like ore body	鸡窝状矿体
net	网，网络；网状物；净的
net added value	净增值
net amount	净量，净总值；净额
net amplitude	纯振幅；净振幅
net area	净面积
net back method	净回值法
net back value	净回值
net balance	净平衡
net bearing capacity	净承载力
net bearing pressure	净承载压力
net boundary	网格边界
net buoyancy	净浮力
net capacity	净容量，净容积，有效容量；净载货量
net cement grout	纯水泥浆
net confining pressure	纯封闭压力；净围压；净侧限压力
net contact foundation pressure	基底附加压力；净基底接触压力
net displacement porosity	驱替时的有效孔隙度
net drilling days	纯钻井日数
net drilling time	纯钻井时间
net effective pore	纯有效孔隙
net effective pore space	净有效孔隙
net efficiency	净效率
net energy gain	纯能获取；纯能量增益
net energy loss	纯能损失；纯能量耗散
net exchange capacity of soil	土壤净交换量
net fabric	网状组构
net force	净力
net foundation pressure	基底净压力
net function	网格函数
net gas	干气
net gravity force	净重力
net head	净水头；有效落差
net heat of combustion	净燃烧热；净热值

net heat value	净热值
net height	净高
net hoisting time	纯提升时间
net horizontal stress	净水平应力
net horsepower	有用功率
net hydrocarbon pore volume	有效含烃孔隙体积；净油气储集空间
net hydrocarbon sand	净含烃砂层
net impelling force	纯推动力
net irrigation requirement	田间灌水定额；净灌溉需水量
net jacket	滤网
net length of crest	堰顶长度
net lift	净举油高度
net like inclusion	网状包裹体
net like texture	网状结构
net load	净重；有效载重；净负载；净荷载
net load increment	净荷载增量
net loading intensity	基底净压力；净荷载强度
net loss	净损；净亏；全损耗
net make-up gas	净添加气
net mesh	网目
net meter registration	流量计净累积值
net moment	净弯矩
net normal stress	净法向应力
net of seepage lines	流线网络
net of triangulation	三角网
net of Wulff	吴氏网；极射赤平投影网
net oil	净采油量
net oil analyser	净油分析仪
net oil pay thickness	产油层有效厚度
net operation time	纯运转时间
net output	净产出；净功率
net overburden	超荷净重；净上覆荷载
net overburden stress	上覆岩层净应力
net pay	产层有效厚度；净产层
net pay thickness	有效含油层厚度；有效产层厚度
net pay volume	有效含油层体积
net peak flow	净峰流量
net point	网格点
net polymer	网状聚合物
net pore space	有效孔隙空间
net porosity	有效孔隙度
net positive suction head	净吸入扬程；剩余吸入绝对压头
net power	有效功率；净功率；净动力
net pressure	净压力
net pressure drop	净压力降
net pressure head	净压头；有效落差
net production	净产量
net production thickness	净生产厚度
net production time	净生产时间
net productive section	净生产层厚度；净生产层剖面
net prover volume	标准体积管的净体积
net pumping	净抽水量
net quantity of injected water	净灌量；净注水量
net radiation	净辐射
net recharge water	净补给水
net residual value	净残值；净剩余价值
net resource depletion	资源净损耗
net safe bearing capacity	净容许承载力
net safety distance	安全净距
net sand	净砂层
net sand isopach map	纯砂层等厚图
net sandface flowrate	油层表面净流量
net section	净断面；净截面；净剖面
net sectional area	有效截面积；净断面面积
net shaped	网状的
net shield	掩护网
net shift	总移距；净移距；总位移
net side load	侧向净荷载
net slip	总变位；总滑距；净滑距
net span	净跨；净跨度
net standard volume	净标准体积
net structure	网络构造；网状结构
net texture	网状结构
net thickness	有效厚度
net thrust	净推力
net time on bottom	纯钻井时间
net tolerance	净公差
net ton	短吨；美吨；净吨数
net tonnage	净吨位；净吨数
net tractive effort	净牵引力
net value	净值
net vertical stress	净垂直应力
net volume	净容积
net water cement ratio	净水灰比
net water requirement	净需水量
net water use	净用水量
net weight	净重
net weight filling	净重充填
nether	地下的；下方的
nether coal	厚煤层的底煤
nether roof	直接顶板
nether roof rock	直接顶板岩石
nethermost	最下面的，最低的
net-like stone soil	网状石质土
net-like texture	网状结构
net-oil computer	纯油自动测定仪
net-shaped	网状
net-shaped cracking	网状开裂；网裂
netted structure	网状构造
netted texture	网状结构
netting analysis	网格分析法
Nettleton's density profile method	奈特尔登密度剖面法
Nettleton's graphical method	奈特尔登图解法
net-to-gross ratio	有效厚度与总厚度的比值
Netwon-Raphson method	牛顿-拉弗森方法
network	网络；网系，网状物；网格结构；网状系统；配电网；配水管网；流线图，道路网
network access	进网，入网
network access license	进网许可证
network access of communications equipment	通信设备进网
network access point	网络访问点；网络接入点
network access server	网络接入服务器
network adapter	网卡；网络适配器
network address	网络地址
network address translation	网络地址解析；网络地址转换
network addressable unit	网络可寻址单元

network adjustment	网络平差；站网调整
network administration computer center	网络管理计算机中心
network administration system	网络管理系统
network analog	网络模拟
network analysis	网络分析；网络规划
network analysis system	网络分析系统
network analyzer	网络分析器；网络分析仪
network architecture	网络体系结构；网络结构
network basic input output system	网络基本输入输出系统
network cable	网线
network cable connector	网线接头；网线连接器
network calculator	网络计算机
network cementite	网状渗碳体；网状碳化铁
network characteristic	网状特征
network chart	网络图
network computer	网络计算机
network configuration	网络配置
network control protocol	网络控制协议
network database language	网络数据库语言
network densification	控制网加密
network deposit	网状矿床
network design	控制网设计；网络设计
network designer	网络设计工程师
network discharge simulative analysis	网流量模拟分析
network disk	网络硬盘；网盘
network driver	网络驱动器
network equipment	网络设备
network flow programming	网络流程规划；网络流程设计；网络流程程序编制
network flow routine	网络流程程序
network for the explicit method	显式网络法
network for the implicit method	隐式网络法
network function	网络函数
network geographic information system	网络地理信息系统
network ID	网络号；网络地址
network information center	网络信息中心
network information service	网络信息服务
network information system	网络信息系统
network infrastructure	网络基础设施
network interface unit	网络界面单元
network interworking	网络互联；网络交互
network intrusion detection system	网络入侵监测系统
network management	网络管理
network management center	网管中心
network management equipment	网管设备
network management requirement	网络管理要求
network management system	网管系统
network message	网络信息
network model	网状模型
network monitoring	网络监控
network node	网络节点
network of base points	基点网
network of capillary tube	毛细管网络
network of cracks	网状裂隙
network of drains	排水管网
network of fault	断层网；网状断层
network of flight strip	航线网；机场跑道网
network of fracture	裂缝网络
network of gravity-magnetic survey	重磁测网；重磁勘探网
network of marine conservation areas	海洋保护区网络
network of meteorological station	气象台站网
network of mine ventilation	矿井通风网络
network of multi-diagonal branches	多角联网路
network of pipelines	管道网
network of seismographic stations	地震台网
network of single diagonal branch	单角联网路
network of stations	台网
network of strong motion accelerograph	强震加速度仪台网
network of two diagonal branches	双角联网路
network of underground streams	暗河系；地下河系
network of veins	网状脉；脉网
network of waterways conflux	河网汇流
network operations center	网络操作中心
network operator	网络运行程序
network optimization	网络优化
network pattern	网格状；网络型式
network planning	网络规划；网络规划
network planning technique	网络计划技术；网络规划技术
network polymer	网状聚合物
network printer	网络打印机
network processing unit	网络处理器
network program	网络计划；网络程序
network protocol	网络协议
network protocol control information	网络协议控制信息
network resource	网络资源
network scheduling technique	网络进度计划法
network scheduling technique in mine construction	矿山建设施工的网络计划技术
network security	网络保密；网络安全
network service access point	网络服务访问点
network service provider	网络服务提供者；网络服务供应商
network size	网络大小
network solution	网络解决方案
network stream	河网；网状河流
network strengthening	控制网加强
network structure	网状构造；网脉状构造；网状结构
network switching center	网络交换中心
network terminal	网络终端
network terminal protocol	网络终端协议
network theory	网络理论
network topology	网络拓扑；逻辑结构图；网络布局
network training	网络训练
network transfer delay	网络传输时延
network vein	网状脉
network virtual terminal	网络虚拟终端
network wiring layout	网点布线
network with junction points	结点网
network-centric technology	以网络为中心的技术
network-flow method	网络流法
network-host protocol	网络主机协议
networking	联网
networking protocol	网络协议
Neumann bands	纽曼条纹；纽曼带
Neumann boundary condition	纽曼边界条件；第二类边界条件
Neumann triangle law	纽曼三角定律

neural computation 神经计算
neural computing 神经计算学
neural computing network 神经计算网络
neural dynamic modeling 神经网络动力学建模
neural net 神经网络
neural network 神经网络
neural network computer 神经网络计算机
neural network model 神经网络模型
neural network modeling 神经网络建模
neural network topology 神经网络拓扑结构
neuro-computing 神经计算技术
neurode 神经节点
neuro-fuzzy controller 神经模糊控制器
neuron 神经元
neuter 中性的
neutercane 中性热带气旋
neutral 中和的；中性的
neutral air 中性气体
neutral atmosphere 中性气氛；中性大气
neutral atom 中性原子
neutral axis 零轴；中性轴；中和轴
neutral coast 中性海岸
neutral complex 中性络合物
neutral compound 中性化合物
neutral condition 中和状态
neutral cross section 中性截面
neutral density filter 中性密度滤光片
neutral depth 中性深度；正常深度
neutral earthing 中性接地
neutral element 中性元素
neutral estuary 中性河口
neutral fluid 中性流体
neutral fold 中性褶皱；中性褶曲
neutral layer 中性层
neutral leaching 中性浸出
neutral level 中和界；中和面
neutral line 中和线；中性线
neutral mass spectrometer 中性质谱仪
neutral material 中性材料
neutral medium 中性介质
neutral micelle 中性胶粒
neutral mineral 中性矿物
neutral molecule 中性分子
neutral nitrogen 中性氮
neutral oil 中性油
neutral phase 中间相位
neutral plane 中和面；中性面
neutral plane height 中性面高度
neutral point 中性点
neutral point of pile 桩的中性点
neutral pressure 中性压力
neutral pressure level 中性压力水准；中性压力级
neutral reaction 中性反应
neutral reduction 中性还原
neutral resin 中性树脂
neutral rock 中性岩
neutral salt 中性盐
neutral salt effect 中性盐效应
neutral section 分相装置
neutral section insulator 分相绝缘器

neutral shaft 备用竖井
neutral shock wave 中性冲击波
neutral shoreline 中性滨线
neutral slag 中性炉渣
neutral soil 中性土壤
neutral solution 中性溶液
neutral stability 中性稳定
neutral stress 中性应力
neutral surface 中性面；中和面；无应变面
neutral water 中性水
neutral wave 稳定波；中性波
neutral wedge 中性楔形体
neutral wettability system 中性润湿体系
neutral zone 中性带；中性区域
neutralisation 中和
neutralisation heat 中和热
neutralism 中性关系
neutrality 中性
neutrality point 中和点；中性点
neutralization 中和；平衡；使失效
neutralization number 中和值；酸值
neutralization potential 酸中和潜力
neutralization potential ratio 酸中和比潜力
neutralization tank 中和池；中和罐
neutralize 中和
neutralizer 中和剂；中和器
neutralizing agent 中和剂
neutralizing tank 中和池
neutralness 中和
neutrino 中微子
neutrography 中子照相法
neutron 中子
neutron absorber 中子吸收剂；中子吸收体；中子吸收器
neutron absorption method 中子吸收法
neutron accelerator 中子加速器
neutron activation 中子激活
neutron activation analysis 中子活化分析
neutron activation log 中子活化测井
neutron activation logging 中子活化测井
neutron activation method 中子活化法
neutron beam collimation 中子束准直
neutron bombardment 中子轰击
neutron bombardment method 中子轰击法
neutron burst 中子猝发
neutron capture 中子俘获
neutron capture gammas 中子俘获伽马射线
neutron capture radiation 中子俘获辐射
neutron chain reaction 中子链式反应
neutron collimator 中子准直器
neutron collision 中子碰撞
neutron conservation principle 中子守恒原理
neutron count rate 中子计数率
neutron counter 中子计数器
neutron death 中子俘获
neutron decay 中子衰变
neutron density distribution 中子密度分布
neutron detector 中子探测器
neutron diffraction 中子衍射；中子绕射
neutron diffusion 中子扩散
neutron ejection 中子发射

neutron emitter 中子辐射源
neutron energy 中子能量
neutron energy group 中子能群
neutron energy spectrum 中子能谱
neutron epithermal neutron log 中子-超热中子测井
neutron exposure 中子照射
neutron gamma logging 中子伽马测井
neutron generation 中子产生；中子代
neutron generator 中子发生器
neutron geogas method 中子地气法
neutron howitzer 中子发射器；中子准直器
neutron induced process 中子诱发过程
neutron inelastic scattering 中子非弹性散射
neutron irradiation 中子照射；中子辐照
neutron leakage 中子泄漏
neutron lifetime log 中子寿命测井
neutron lifetime logging 中子寿命测井
neutron log 中子测井
neutron log porosity 中子测井孔隙度
neutron logging 中子测井
neutron moderator 中子慢化剂；中子缓和剂
neutron moisture gauge 中子测含水量仪；中子湿度计
neutron moisture meter 中子含水量测定仪；中子式湿度计
neutron moisture probe 中子水分仪；中子水分探头
neutron multiplication factor 中子的增殖系数；中子倍增因子
neutron neutron log 中子-中子测井
neutron neutron logging 中子-中子测井
neutron penetration depth 中子穿透深度
neutron physics 中子物理学
neutron population 中子密度
neutron porosity 中子测井孔隙度
neutron porosity index 中子孔隙度指数
neutron probe water log 中子水分测井；中子测水记录
neutron producer 中子发生器；中子源
neutron proton ratio 中子数与质子数比率
neutron quality 中子质量
neutron radiation analysis 中子放射分析
neutron radioactivation analysis 中子活化分析
neutron radiography 中子射线照相术；中子射线照相法；中子探伤法
neutron reflection 中子反射
neutron resonance absorption 中子共振吸收
neutron response 中子响应；中子灵敏度
neutron scattering 中子散射
neutron scattering method 中子扩散法
neutron shield 中子屏蔽
neutron soil moisture meter 中子土壤湿度计；中子土壤水分计
neutron source 中子源
neutron source calibration 中子源刻度；中子源校准；中子源标准化
neutron source holder 中子源夹持器
neutron spectrum 中子能谱
neutron spin 中子自旋
neutron star 中子星
neutron thermalization 中子热化
neutron transport 中子输运
neutron transport mean-free path 中子迁移平均自由程

neutron well logging 中子测井
neutron yield 中子产率
neutron-absorption cross-section 中子吸收截面
neutron-activation log 中子活化测井
neutron-capture cross section 中子俘获截面
neutron-capture gamma rays 中子俘获伽马辐射
neutron-capture gamma-ray logging 中子俘获伽马射线测井
neutron-counting equipment 中子计数设备
neutron-current density 中子流密度
neutron-densilog combination 中子-密度组合
neutron-density crossplot 中子-密度交会图
neutron-epithermal neutron logging 中子-超热中子测井
neutron-flux density 中子通量密度
neutron-flux distribution 中子通量分布
neutron-gamma log 中子伽马测井
neutron-gamma logging 中子伽马测井
neutronics 中子学；中子物理学
neutron-moisture-probe test 中子湿度探针试验
neutron-neutron logging 中子-中子测井
neutron-thermal neutron log 中子-热中子测井
neutron-thermal neutron logging 中子-热中子测井
neutron-γ logging 中子-γ测井
neutropause 中性层顶
neutrosphere 中性层；非电离层
Nevada system 内华达式方框支架法
Nevadan orogeny 内华达造山运动
Nevadian movement 内华达运动
nevadite 斑流岩
neve basin 粒雪盆地
neve line 永久雪线；雪线
never frozen soil 不冻土
never-slip 套管夹具
nevyanskite 铱锇矿
new archaeology 新考古学
New Austrian Method 新奥法
new Austrian tunneling method 新奥法；新奥地利隧道施工法
New Austrian Tunneling Method classification 新奥法岩石分类
New Austrian Tunnelling Method 新奥法；新奥地利隧道施工法
new basement tectonics 新基底构造
New Beijing Geodetic Coordinate System 1954 新1954年北京大地坐标系
New Britain Trench 新不列颠海沟
New British Standard 英国新工业标准
new generation fault 新生断层
new global tectonics 新全球构造；新全球构造说
New Guinea-type orogenic belt 新几内亚型造山带
New Hebrides plate 新赫布里底板块
new Laplace's hypothesis 新拉普拉斯假说
new level developing 矿井延深；开拓新水平
new level development 矿井延深；开拓新水平
new level preparation 新水平的准备工作；新矿井开采准备工作
new loess 新黄土
new material 新材料
new Mohs' scale 新莫氏硬度计
new polarography 新极谱法

new priority classification scheme	边坡工作排序新法；新优先等级分类法
new producer	新生产井
new project	新建项目
New Red Sandstone	新红砂岩统
new reserve	新储量
New Stone Age	新石器时代
new survey	重测；新测
New Zealand Plateau	新西兰海台
New Zealand sediment	新西兰沉积物
newberyite	镁磷石
newborn	新生的；再生的
new-field wildcat	新油田预探井
new-generated paleo reservoir	新生古储
newlandite	顽火榴辉岩
newly deposited soil	新近堆积土
newly drilled well	新钻井
newly formed ice	新冰
newly forming bank	新成岸
newly generating fault	新生断层
newly laid concrete	新浇灌的混凝土
newly laid track	新铺轨道
newly-built railway	新建铁路
newly-built railway construction	铁路新线建设
newly-laid	新铺的；新浇灌的
newly-located	新定位的
newly-placed concrete	新喷射的混凝土
new-made	新做的
Newmark algorithm	纽马克算法
Newmark beta method	纽马克类平均法
Newmark chart	纽马克图；纽马克感应图
Newmark inelastic response spectrum	纽马克非弹性反应谱
Newmark influence chart	纽马克感应图
Newmark method	纽马克法
Newmark-Hall	纽马克霍尔法
Newmark's chart	纽马克图；纽马克感应图
Newmark's influence chart	纽马克感应图
Newmark's method for fundamental mode	基本振型的纽马克分析法
Newmark's method for steady-state vibration	稳态振动的纽马克分析法
new-moon pool	月池；钻井平台上的月亮池
new-pool wildcat	新油藏预探井
new-replacement well	更新井
Newsom method	纽瑟姆钻粒钻井法
Newton	牛顿
Newton first law	牛顿第一定律
Newton fluid	牛顿流体
Newton interpolation	牛顿插值
Newton interpolation formula	牛顿内插公式
Newton iteration method	牛顿迭代法
Newton model	牛顿模型
Newton second law	牛顿第二定律
Newton third law	牛顿第三定律
Newton-Cotes integration formulas	牛顿-柯特斯积分法
Newton-Cotes quadrature	牛顿-柯特斯正交求积法
Newtonian behavior	牛顿特性
Newtonian damping	牛顿阻尼
Newtonian film	牛顿膜
Newtonian flow	牛顿流变；牛顿流
Newtonian fluid	牛顿流体
Newtonian fluid flow	牛顿型流体
Newtonian liquid	牛顿液体
Newtonian material	牛顿物质；黏滞物质
Newtonian mechanics	牛顿力学
Newtonian potential	牛顿位
Newtonian reference of frame	牛顿参考坐标系
Newtonian viscosity	牛顿黏滞性；牛顿黏度
Newtonite	斜方岭石；块矾石
Newton-Raphson algorithm	牛顿-拉夫申算法
Newton-Raphson formula	牛顿-拉夫申公式
Newton-Raphson iteration	牛顿-拉福松迭代法
Newton-Raphson method	牛顿-拉夫逊方法
Newton's approximation	牛顿近似法
Newton's divided-difference	牛顿差商
Newton's equation	牛顿方程
Newton's first law	牛顿第一定律
Newton's first law of motion	牛顿第一运动定律
Newton's law	牛顿定律
Newton's law of gravitation	牛顿万有引力律
Newton's law of motion	牛顿运动定律
Newton's law of universal gravitation	牛顿万有引力定律
Newton's law of viscosity	牛顿黏性定律
Newton's metal	牛顿易熔合金；牛顿合金
Newton's method	牛顿法
Newton's second law	牛顿第二定律
Newton's theory of collision	牛顿碰撞理论
Newton's three-eighth rule	牛顿八分之三法则
Newton's viscous law	牛顿内摩擦定律；牛顿黏滞定律
Ney's chart	奈伊投影地图
n-fold coverage	n 次覆盖
n-fold pole	n 阶极点
n-fold shooting	n 次激发
n-fold stack	n 次叠加
N-framed set	N 字形加固框架
N-girder	N 形梁；N 形支架
Nicalloy	镍锰铁合金
Nicaro process	尼加罗炼镍法
niccochromite	铬镍矿
niccolit	红砷镍矿
niccolite	红砷镍矿；红镍矿
nichrome	镍铬合金
nick	切口，切槽，缺口；刻痕
nick flaw	裂痕
nick point	裂点；侵蚀交叉点
nick-break test	切口试验；双缺口破断试验
Nickel	镍
nickel alloy	镍合金
nickel base alloy	镍基合金
nickel bloom	镍华
nickel brass	镍黄铜
nickel bronze	镍青铜
nickel chloride	氯化镍
nickel electrolyte	镍电解液
nickel electroplating	镀镍
nickel gymnite	镍叶蛇纹石
nickel hydroxide	氢氧化镍
nickel iron	镍纹石
nickel manganese steel	镍锰钢
nickel matte	镍冰铜；镍锍

nickel ocher	镍华	nip	浪蚀洞；小崖；夹；剪断；尖灭
nickel olivine	硅镍矿	nip angle	啮角；咬入角；轧入角
nickel ore	镍矿	nip-out	尖灭；变薄
nickel oxide	氧化镍	nipped	尖灭的；夹住的
nickel plated	镀镍的	niter	硝石；钾硝石
nickel plating	镀镍	niter spot	硝斑
nickel protoxide	氧化亚镍；一氧化镍	niton	氡
nickel pyrite	镍黄铁矿	nitosol	强风化黏磐土
nickel shot	镍粒；镍丸	nitramon	尼特拉芒炸药；硝胺火药
nickel silver	镍银	nitrate	硝酸盐；硝酸酯；硝化；用硝酸处理
nickel steel	镍钢	nitrate contamination	硝酸盐污染
nickelage	镀镍	nitrate cracking	热硝酸盐应力腐蚀开裂
nickel-base alloy	镍基合金	nitrate pollution	硝酸盐污染
nickel-chrome	镍铬合金	nitratine	钠硝石；智利硝石
nickel-chromium steel	镍铬钢	nitratite	钠硝石
nickel-gymnite	镍叶蛇纹石；水硅镁镍矿	nitre	硝石；钾硝石
nickeliferous	含镍的	nitric	氮的；含氮的；钾硝的
nickeline	红砷镍矿	nitric acid	硝酸
nickel-iron alloy	铁镍合金	nitric carbon oxide thermal water	含氮和二氧化碳的地下热水
nickelite	红砷镍矿	nitric oxide	一氧化氮
nickeloid	铜镍耐蚀合金	nitric water	氮水
nickel-pyrite	镍黄铁矿	nitridation	氮化
nickel-spinel	镍尖晶石	nitride	氮化物
nicking	煤壁剥落；片帮；煤屑，焦屑	nitride hardening	渗氮硬化；氮化
Nicol	偏光镜；尼科耳棱镜	nitride inclusion	氮化物夹杂
nicolayite	硅铅钍铀矿	nitride layer	氮化层；白亮层
nicomelane	黑镍矿	nitriding	渗氮；氮化
nicopyrite	镍黄铁矿	nitriding atmosphere	氮化气氛；渗氮气氛
nicrosilal	镍铬硅铸铁	nitriding steel	渗氮钢；氮化钢
niederigcraton	大洋克拉通	nitrification	氮化合；硝化作用
nife	镍铁地核	nitrite	亚硝酸盐
nifesima	镍铁硅镁带	nitro cementation	表面渗氮处理
nifesphere	镍铁地核；镍铁带	nitro compound	硝基化合物
nifontovite	粒水硼钙石	nitro elation	胶质炸药；明胶炸药
nigerite	锡铝矿	nitro meter	测氮管；氮量计
Niggli classification	尼格里岩石分类	nitroalloy	氮化合金；氮化钢
Niggli method	尼格里法	nitrobaryte	钡硝石
Niggli molecular norm	尼格里分子标准矿物	nitrocalcite	水钙硝石；钙硝石
Niggli number	尼格里值	nitrocarbonitrate	硝基碳酸铵
niggli value	尼格里值	nitrocarburizing	气体碳氮化；渗碳氮化
niggliite	碲铂矿	nitroexplosive	硝化炸药
nigrate	黑沥青	nitrogelatin	胶质炸药
nigrine	铁金红石	nitrogelatine	硝化明胶炸药
nigrite	沥青；氮沥青	nitrogen	氮；氮气
Ni-hard	硬镍合金	nitrogen atom	氮原子
Ni-hard grinding ball	硬镍合金磨球	nitrogen bottle	氮气罐
niklesite	三辉岩；透顽剥辉岩	nitrogen case-hardening	渗氮硬化
nine-point conic	九点二次曲线	nitrogen content	含氮量
nine-point sample	九点法试样；九点法取样	nitrogen content of coal	煤中氮
nine-spot well injection configuration	九点注入井网	nitrogen cycle	氮循环
nine-spot well network	九点井网	nitrogen deficiency	缺氮
niobite	铌铁矿；钶铁石	nitrogen dioxide	二氧化氮
niobium	铌	nitrogen displacement efficiency	氮气驱替效率
niobium carbide	碳化铌	nitrogen family explosives	氮族炸药
niobium minerals	铌矿类	nitrogen fixation	氮固定
niobium oxide	氧化铌	nitrogen flooding	氮气驱
nioboeschynite	铌易解石	nitrogen isotope	氮同位素
niobozirconolite	铌钙钛锆石	nitrogen isotope ratio	氮同位素比值
niobpyrochlore	烧绿石	nitrogen jar	气动震击器
niocalite	黄硅铌钙石		

nitrogen monoxide 一氧化氮
nitrogen oxide 氧化氮；氮氧化物
nitrogen pentoxide 五氧化二氮
nitrogen reinjection unit 回注氮气设备
nitrogen rejection facility 脱氮装置
nitrogen removal 除氮
nitrogen treatment 氮化处理
nitrogen water 含氮水
nitrogen workover 氮气修井法
nitrogen-bearing 含氮的
nitrogen-gas sand stimulation treatment 氮气加砂压裂
nitrogenized manganese 氮化锰
nitrogenous 含氮的
nitrogenous effluent 含氮废水
nitrogenous thermal water 含氮热水
nitrogen-reduction ratio 氮-还原系数比
nitroglauberite 钠矾硝石
nitroglycerin 硝酸甘油；硝化甘油；甘油三硝酸酯
nitroglycerin dynamite 硝甘炸药
nitroglycerin explosive 硝化甘油炸药
nitroglycerin leakage 硝化甘油渗出
nitroglycerine 硝化甘油；甘油三硝酸酯
nitroglycerine explosive 胶质炸药；硝化甘油炸药
nitroglycerine gelatine explosive 硝化甘油胶质炸药
nitro-glycerine low-density powder 硝化甘油低密度炸药
nitroglycerine safety explosive 硝甘安全炸药
nitro-glycerine-ammonium-nitrate powder 硝化甘油硝铵炸药
nitro-group 硝基
nitrohumic acid 硝基腐殖酸
nitrohydrochloric acid 王水
nitrokalite 钾硝石；印度硝石
nitrolite 特硝铵炸药
nitromagnesite 水镁硝石；镁硝石
nitrometer 氮量计
nitromethane 硝基甲烷
nitromuriatic acid 王水
nitronatrite 钠硝石
nitrosamine 亚硝胺
nitro-shooting 硝基炸药爆破工艺
nitroso compound 亚硝基化合物
nitrous acid 亚硝酸
nitrous oxide 氧化二氮；氧化亚氮
nival belt 雪带
nival climate 冰雪气候
nival erosion 冰雪侵蚀；雪水冲刷
nival karst 冰川喀斯特；高山喀斯特
nivation 雪蚀；雪蚀作用
nivation cirque 雪蚀冰斗；雪坑
nivation hollow 雪蚀凹地；积雪洼地
nivenite 钇铀矿；黑富铀矿
NMM 数值流形法
no borrow area 禁止采泥区；禁止取土区
no cohesive 不附着的；无黏结力的
no conformable 不整合的
no conservative force 非守恒力
no damage threshold 无危险临界值
no deformable 不变形的
no gassy 非气体的
no homogeneous 不均匀的

no leak gradient 无泄漏压降线
no leakage 不漏；无渗漏
no load 空载；无载荷的
no load device 空载装置
no load operation 无载运行
no pillar 无底柱；无矿柱
no shot 未放炮
no slump concrete 无坍落度混凝土；干硬混凝土
no support 无支撑开挖
no swell 零级涌浪
no yielding prop 非让压支柱；刚性支柱
noble element 贵重元素
noble gas 稀有气体；惰性气体
noble opal 贵蛋白石
noble point 结点；交点；节点
nocerite 针六方石
no-cut routed 不掏槽炮眼组
no-cut-hole blasting 不掏槽爆破
nodal analysis 节点分析法
nodal circle 节点循环
nodal coordinate 节点坐标
nodal cross section 节截面
nodal degree of freedom 节点自由度
nodal displacement 节点位移
nodal equation 节点方程
nodal force 节点力
nodal frequency 节点频率
nodal line 交点线；节线
nodal load 节点载荷
nodal load vector 节点荷载向量
nodal matrix 节点矩阵
nodal motion 交点位移；节点运动
nodal number 节点号；节点数
nodal parameter 节点参数
nodal period 节点周期
nodal plane 结点平面；节平面；节面
nodal point 节点；结点
nodal point of incidence 前节点；物方节点
nodal point of vibration 振动节点
nodal point quadrature 节点正交
nodal selection 结点选择
nodal surface 节面
nodal system graph 结点系统图
nodal well 结点井
nodality 波节数；节点数
noddle 矿瘤；岩球；矿结核
node 节点；交点；结点；波节
node diagram 节点图
node flow 节点流量
node number 节点号
node of a curve 曲线结点
node of the fault 断层节
node point 结点；节点；交点
node property 节点属性
nodeless internal parameter 无节点内部参数
nodeless variable 无节点变量
no-dig engineering 非明挖工程；暗挖工程
no-dig industry 非明挖产业；暗挖产业
nodular 结节的；结核状的；瘤状的；球状的；团块状的
nodular cast iron 球墨铸铁

nodular cementite 粒状渗碳体
nodular chert 燧石结核；结核状燧石
nodular fire clay 球状耐火黏土
nodular graphite cast iron 球墨铸铁
nodular material 瘤状物料
nodular ore 肾状矿石；结核状矿石；瘤状矿石
nodular pyrite 结核状黄铁矿
nodular structure 瘤状构造；结核状构造
nodular texture 结核结构
nodular troostite 细珠光体；团状屈氏体
nodulating process 烧结法
nodulation 结核作用
nodule 矿瘤；岩球；矿结核
nodule abundance 结核丰度
nodule coverage 结核覆盖率
nodule grade 结核品级
nodule hydrate 结核状水合物
nodulous limonite 结核状褐铁矿
no-fines concrete 无细料混凝土；无砂混凝土
no-flow boundary 不渗透边界
no-flow condition 不渗透条件；不渗透状态
nog 木垛；垛式支架；支柱垫楔
nog building 架设木垛
nogging 木架间水平短撑
nohlite 铌钇铀矿
noise 噪声；干扰
noise abatement 抑制噪声；减轻噪声；消声
noise abatement measure 减低噪声措施
noise absorbent 吸声设备；吸声物料
noise absorption 吸声料
noise amplitude 噪声振幅；噪声幅度
noise analysis 干扰波研究；噪声分析
noise assessment survey 噪声评估测量
noise attenuation 噪声衰减
noise attenuation measure 减低噪声措施；噪声衰减测量；噪声衰减程度
noise background 本底噪声；背景噪声
noise bandwidth 噪声带宽
noise barrier 隔声屏障；隔声围墙；隔声板；隔声屏障
noise burst 噪声猝发；突发噪声
noise cancellation 消除噪声
noise character 噪声特性
noise climate 环境噪声
noise coefficient 噪声系数；信息误检率
noise component 噪声分量
noise contaminant 噪声污染
noise contamination 噪声污染
noise contribution 噪声成分
noise control 噪声控制
noise control engineering 噪声控制工程
noise control equipment 噪声控制设备
noise control in mine 矿山噪声控制
noise control of main fan in mine 矿山主扇噪声治理
noise cover 噪声屏蔽器；噪声封闭器；隔声盖
noise criterion 噪声标准
noise elimination 噪声消除器；消除噪声
noise emission standard 噪声标准
noise emission test 噪声测试
noise enclosure 隔声罩
noise equivalent bandwidth 噪声等效带宽

noise equivalent power 噪声等效功率
noise error 噪声误差
noise exposure 噪声暴露量
noise exposure forecast contour 飞机噪声预测等量线
noise factor 噪声系数；噪声因数
noise field 噪声场
noise figure 噪声系数
noise filter 噪声滤波器
noise free synthetic seismogram 无噪声合成地震记录
noise fringe 噪扰带
noise generator 噪声发生器
noise grade 噪声等级
noise hazard 噪声危害
noise identification 噪声识别
noise immunity 抗干扰度
noise impact 噪声影响
noise impact assessment 噪声影响评估
noise improvement 噪声改善
noise improvement factor 噪声改善系数
noise in coal mine 煤矿噪声
noise insulation equipment 隔声设备
noise insulation work study 隔声工程研究
noise intensity 噪声强度
noise level 噪声声级；噪声级别
noise level meter 噪声级计；噪声电平表
noise log 噪声测井记录
noise logging 噪声测井
noise margin 噪声边限；噪声容限
noise measurement 噪声测量；噪声测定
noise meter 噪声计
noise mitigation 减低噪声
noise mitigation measure 噪声缓解措施
noise monitoring 噪声监测
noise muffler 噪声衰减器；消声器
noise of blasting 爆破噪声
noise pattern 噪声图；噪声频谱图；噪点分布
noise peak 噪声峰值
noise pollution 噪声污染；噪声污染
noise pollution level 噪声污染级
noise probability density 噪声概率密度
noise ranging sonar 噪声测距声呐
noise rate of rock 岩石信噪比
noise rating contour 噪声评价曲线
noise rating number 噪声评价数；噪声标定值；噪声额定值
noise ratio 噪声比
noise redress 减低噪声
noise reducing surface 吸声；减声道路面层
noise reduction 噪声降低；噪声衰减；减噪
noise reduction coefficient 噪声降低系数
noise reduction friction course 吸声道路面层；减声多孔道路面层
noise reduction surfacing material 低噪声铺路物料
noise remover 噪声抑制器
noise seismic channel 带噪声地震道
noise sensitivity 噪声灵敏度
noise signal 噪声信号
noise silencer 消声器
noise source 噪声源
noise spectra 噪声频谱

noise spectrum 噪声谱
noise spike 噪声脉冲；噪声尖峰
noise stability 噪声稳定度；噪声不变度
noise standard 噪声标准；噪声标准
noise suppressing case 灭声套
noise suppression 噪声抑制；抑制噪声
noise suppression coil 杂声抑制线圈
noise suppressor 杂声抑制器；噪声抑制器
noise survey 环境噪声测量
noise temperature 噪声温度
noise testing 噪声测试
noise testing report 噪声测试报告
noise thermometer 噪声探测器
noise tolerant use 能耐噪声的用途
noise trains 噪声波列
noise unit 噪声的量度单位
noise wavelength 噪声波长
noise-abatement equipment 消声装置
noise-attenuating hood 噪声衰减罩
noise-compensated deconvolution 噪声补偿反褶积
noise-free channel 无干扰通路；无噪声信道
noise-free seismogram 无干扰地震图
noiseless channel 无噪声道
noiseless piling technique 无噪声打桩技术
noise-like function 类噪声函数
noise-limited deconvolution 限带反褶积
noise-reducing road surface 低噪声路面
noise-reducing surface 吸声路面；减声路面
noise-reducing surfacing 铺设吸声路面层；吸声铺面料
noisiness 噪度；噪声特性
noisy reflection 干扰反射
noisy reproduction 干扰图像重现
no-lag cap 无滞后雷管；无时延雷管
no-lag seismograph cap 无滞后雷管；无时延雷管
nolanite 黑钒铁矿
no-linear 非线性的
no-load 空载；无负载
no-load character 空载特性
no-load characteristic 空载特性；无载特性曲线
no-load heat distortion 无载荷热变形
no-load resistance 无载阻力
no-load run 空转；无载运行
no-load running 空转
no-load saturation curve 无载饱和曲线
no-load test 无载试验；空载试验
no-loading 空载
nominal accuracy 标称精度
nominal accuracy of EDM 电磁波测距标称精度
nominal air delivery capacity 额定输气量
nominal air delivery pressure 额定输气压力
nominal area of screen 标称筛面面积
nominal bedrock surface 计算基岩面
nominal bore 标准孔径
nominal bucket capacity 额定铲斗能力
nominal capacity 额定功率；额定容量
nominal capacity pump 定量泵
nominal cement thickness 水泥环厚度名义值
nominal cross sectional area 标称横截面面积
nominal decline rate 标准递减率
nominal diameter 标称直径

nominal dimension 标称尺寸；通称尺寸
nominal ductility 额定延性值
nominal error 名义误差；标称误差
nominal frequency 标称频率；额定频率
nominal geographic code 名义地理编码
nominal horsepower 公称马力；额定功率
nominal load 额定荷载
nominal load-bearing capacity 额定承载力
nominal logging speed 额定测井速度
nominal maximum aggregate size 标称最大骨料尺寸
nominal maximum reduction ratio 公称最大破碎比
nominal outlet pressure 额定出口压力
nominal output 标称产量；标出量；名义产量
nominal overburden pressure 标称上覆压力
nominal pressure 公称压力；额定压力；标称压力
nominal range 额定使用范围
nominal reduction ratio 名义破碎比
nominal scaling 名义量表；名义尺度
nominal screen size 标称筛分粒度；通称筛分粒度；计划使用的筛分粒度
nominal shear stress 标称剪应力
nominal size 标称大小；标称尺寸
nominal speed 标称转速；额定速度
nominal static magnification 标称静态放大倍数
nominal strength 标称强度
nominal stress 名义应力；标称应力
nominal supporting force 标称支承力；额定支承力
nominal thickness 公称厚度；标称厚度
nominal throat diameter 喉部公称直径；喉径
nominal thrust 标准推力
nominal value 公称值；标称值；面值
nominal value of geometrical parameter 几何参数标准值
nominal width 公称宽度
nominal working condition 额定工作条件
nominal yield of support 支架可缩量
nomogram 列线图；诺模图
nomograph 计算图表；列线图；诺模图
nomographic chart 诺模图；列线图；计算图
nomography 图算法
non cohesive soil 无黏性土；砂性土
non commercial energy 非商业性能源
non condensable gas 非可凝性气体
non conducting fluid 非传导流体；非导电流体
non controllable drilling variable 不可控钻井参数
non destructive test 非破坏性试验
non destructive testing 无损检测
non fiery mine 无瓦斯矿
non gaseous mine 无瓦斯矿
non granular grout 非颗粒状材料浆液
non inflammability 不可燃性
non magnetic ore 非磁性矿物
non metallurgy ferrous metallurgy 非铁冶金；有色金属冶金
non Newtonian flow 非牛顿流
non Newtonian liquid 非牛顿流体
non plastic 无塑性
non plastic soil 无塑性土
non polar bond 非极性键
non polar linkage 非极性键
non real-time 非实时的

non real-time display	非实时显示	nonbedded	非成层的
non return valve	逆止阀；单向阀；止回阀	non-binary code	非二进制代码
non sheathed explosive	无被筒炸药	nonbiodegradable	非生物降解的；生物不可降解的
non SI unit	非国际单位	non-biodegradable	非生物降解的；生物不可降解的
non stable mining	无机窝采煤	non-biodegradable material	不可生物降解材料
non standard cable	非标准电缆	non-biogenic	非生物成因的
non stationary flow	不稳定流	nonbiological decomposition	非生物分解
non symmetry	不对称	non-bituminous coal	非黏结煤；瘦煤；非沥青煤；无烟煤
non tractor drill	非拖拉式钻机；非牵引式钻孔机	non-bleed character	不流淌特性
non valid trace	无效道	non-bonded tendon	无黏结的预应力筋
non working slope	非工作帮	non-brittle buffer	非脆性隔板
non working slope face	非工作帮坡面	non-building area	非建筑用地
non-absorbent material	非吸收性物料	non-building structure	非建筑结构
non-absorbing	不吸收的	non-built-up area	非建设区
nonadditivity	非叠加性	non-caking	不结块的
nonadiabatic change	非绝热变化	noncaking coal	非黏结煤；不黏煤
non-adiabatic well	非绝热井	non-calcareous alluvial soil	非石灰性冲积土；非钙质冲积土
nonaffine deformation	不均匀变形；非仿射形变	noncalcareous clay	非钙质黏土
non-agglomerating	不黏结的	noncalcic brown earth	非石灰性棕壤
non-aggressive	非侵蚀性的	noncalcic brown soil	非石灰性棕色土
non-airway inversed ventilation	无反风道反风	non-capillary pore	非毛管孔隙
nonaligned structure	非定向构造	noncapillary porosity	非毛细孔隙度
non-analytic function	非解析函数	noncarbonate	非碳酸盐
non-anchor length	无锚固长度；无锚固段	non-carbonate hardness	非碳酸盐硬度
nonangular unconformity	假整合；非角度不整合	non-carrot forming charge	无杵体射孔弹
non-antiparallel magnetic fields	非逆平行磁场	non-carrot forming type perforator	无杵体型射孔器
non-API tubular	非美国石油学会标准管材	non-causative fault	非发震断层
non-aqueous media	非水介质	noncaving mining method	非崩落采矿法
non-aqueous solution	非水溶液	non-cavitation flow	非空化水流
non-aqueous solvent	非水溶剂	non-centered difference scheme	非中心差分格式
non-aquifer area	无含水层地区	non-central load	偏心荷载
non-arithmetic shift	非算术移位	non-centralized interlocking	非集中集中联锁
non-artesian aquifer	非自流含水层；无压含水层	non-chain-pillar entry protection	无煤柱护巷
non-artesian ground water	非自流地下水	non-circular analysis	非圆弧分析法
nonartesian groundwater	非自流地下水	non-circularity of core	芯不圆度
nonartesian well	非自流井	non-circulating drilling	无循环钻进
non-artesian well	非自流井	noncirculating water	非循环水；静水
non-asphaltic base oil	非沥青基石油	non-circulating water	非循环水；静水
non-associated flow rule	不相适应性流动法则；非关联流动法则	non-classical damping	非典型阻尼
nonassociated gas	非伴生气；非伴生天然气	nonclastic	非碎屑的
non-associative	非伴生的	non-clastic brown soil	非碎屑褐土
non-associative flow rule	非关联流动法则；不相适应流动法则	nonclastic rock	非碎屑岩
non-associative theory of plasticity	非相关塑性理论	non-clastic sediment	非碎屑沉积
non-attenuating wave	等幅波	nonclastic texture	非碎屑结构
non-attraction injector	非吸上式注水器	non-clay fraction	非黏粒部分；非黏粒粒组
non-avoidable ore loss	不可避免的矿石损失	non-clay mineral	非黏土矿物
non-axial	非轴向的	non-closed geological feature	未闭合的地质构造
nonaxial load	偏心荷载	non-coal intra-seam band	煤层内非煤成分夹层
non-axial stress	非轴应力；偏应力	noncoaxial	非同轴的；非共轴的
nonaxiality	不同轴性	non-coccolithophore nannoliths	非颗石类球石
non-axisymmetric element	非轴对称单元	noncognate	非同源的
non-axisymmetric load	非轴对称荷载	noncoherent	无黏聚力的；松散的；非相干的
non-baking coal	不黏结煤；不结焦煤	noncoherent boundary	不连续边界；非共格界面
non-balance entropy	非平衡熵	non-coherent detection	非相干检测
non-balance model	非平衡模型	non-coherent echo	非相干回波
nonbanded coal	非条带煤；非带状煤	noncoherent exsolution	不连贯出溶作用
non-bearing structure	非承重结构	noncoherent imagery	非相干成像
non-bearing wail	非承重墙	non-coherent material	不黏附的材料；松散材料；无黏聚力的材料

noncoherent noise 非相干噪声
non-coherent scattering 非相干散射
noncoherent signal 非相干信号
noncohesive 无黏聚力的；松散的；非黏性的
noncohesive material 非黏性的材料；松散材料
non-cohesive soil 松散岩土；非黏结性土；无黏性土
noncoking coal 非黏结性煤；非结焦煤
non-colinear point 非共线点
non-collapsing 非湿陷性
non-collapsing soil 非湿陷性土
non-collection analysis 非聚集分析
noncollision plate boundary 非碰撞板块边界
non-colloidal particle 非胶质颗粒
non-color geological mass 消色地质体
noncombustibility 不燃性
non-combustibility test 不可燃性测试
noncombustible 不燃的；非燃性的；不燃物
noncombustible material 不燃物质
noncombustible solid waste 不可燃烧的固体废物
non-commercial pool 无开采价值的油藏
non-commercial producer 不经济的生产井
noncommercial reserve 表外储量；无开采价值的储量
non-commercial well 无开采价值的井
noncommunicating 不连通的
noncommunicating layers 互不连通的层
non-compliance with specifications 不符合技术规范
non-compressible 不可压缩的
non-compressible flow 不可压缩流动
noncompressible soil 不可压缩的土
non-condensable 不凝的
non-condensable gas 不凝性气体；不凝气；不可凝气体
non-condensing 不凝结的
non-conducting 不导电的
non-conducting dust 不导电尘末
non-conductive fibre 绝缘纤维
non-conductive oil base mud 非导电油基泥浆
nonconductive pore 不连通孔隙
non-conductor 非导体；绝缘体；电介质
non-confined compression test 无侧限压缩试验
nonconformable 不整合的
non-conforming element 非协调元；非相容元
nonconformity 不整合
nonconjuction gate 与非门
nonconservation 不守恒
nonconservation force 非保守力
nonconservative element 非保守元素
nonconservative motion 非守恒运动
nonconservative system 非守恒系统
non-consolidated sediment 未固结沉积物
nonconsolute 不混溶的
non-constancy 不稳定性
non-contact control system 无触点控制系统
non-contact filter paper 非接触滤纸
non-contact level sensing 非接触式液位传感
non-contact measurement 非接触测量
noncontact polarization curve method 非接触极化曲线法
noncontact sensor 非接触传感器
non-contact ultrasonic level measurement 非接触超声波测量
non-contactable region 不可接触区

non-contacting 非接触的
non-contacting analysis 不接触分析
noncontacting proximity sensor 非接触式近距离传感器
non-contacting radial deformation transducer 非接触径向变形传感器
non-contacting sensor 非接触式传感器
non-contacting system 非接触系统
non-contamination fuel 无污染燃料
noncontemporaneous deposit 非同期沉积
non-continental epicenter 非大陆震中
noncontinuous 不连续的
non-continuous transportation of open-pit 露天矿间断运输开采工艺
noncontinuous velocity function 非连续速度函数
non-contractual document 非契约性文件
non-contribution mode 无贡献振型
non-convective precipitation 非对流性降水
non-conventional bearing 非常规支承
non-conventional crude oil 非常规原油
non-conventional gas 非常规天然气
nonconventional investment project 非常规投资项目
non-conventional observation 非常规观测
non-conventional projection 非惯用投影
non-convergent 非收敛的
non-convergent series 非收敛级数；非收敛数列
non-convex fuzzy set 非凸模糊集合
non-convex trap 非背斜圈闭
non-copolymerizable monomeric additive 非共聚用单体添加剂
noncore drilling 无岩芯钻进；不取芯钻进
non-coring bit 不取芯岩芯钻头
non-coring drilling 无岩芯钻进；不取芯钻进
non-corrected image 未校正图像
non-correlated 不相关的
noncorrodibility 抗腐蚀能力；防腐性
noncorrodible 抗腐蚀的；非腐蚀的
noncorroding 抗腐蚀的
noncorroding metal 抗腐蚀金属
non-corrosive 抗腐蚀的；不锈的；无腐蚀性的
non-corrosive alloy 抗蚀合金
non-corrosive steel 不锈钢
non-corrosiveness 无腐蚀性
noncorrosivity 无腐蚀性
non-countable 不可数的
non-covalent bond 非极性共价键
noncrack-sensitive 不产生裂纹的
noncratonized 非克拉通化的
noncritical 非临界的；非关键的
noncriticality 非临界性
non-crystal body 非晶体
non-crystalline 非晶质；非结晶的；非晶性的
noncrystalline clay mineral 非晶质黏土矿物
non-crystalline clay mineral 非晶质黏土矿物
noncrystalline gel 非结晶胶体
noncyclic 非旋回的；非周期性的；不定期的
noncyclic basic-sequence 非旋回性基本层序
noncyclic variation 非周期性变化
non-damaged formation 未受损害地层
non-damaged well 未受污染井
non-damaging acid degradable polymer 不污染的酸降解

聚合物
non-damaging completion fluid 不污染的完井液
non-damaging drilling 无损害钻井
non-damaging earthquake 非破坏性地震
nondamaging gravel packing fluid 不污染的砾石充填液
non-Darcian flow 非达西水流
non-Darcy compressible flow 非达西可压缩流动
non-Darcy flow 非达西流动；非达西渗流；非达西流
non-Darcy flow coefficient 非达西流动系数
non-Darcy skin effect 非达西表皮效应
non-decimal base 非十进制基数
non-decimal system 非十进制数系统
nondeflecting 不变形的；不挠曲的
nondeformable 不可变形的；不变形的
non-deformable medium 不变形介质
nondeforming brittle propping agent 不变形脆性支撑剂
nondeforming steel 不变形钢
non-degradable 不可降解的
non-degradable waste 不可降解废料
non-delay 无迟发；不延发
non-deleterious 无害的
non-deltaic facies 非三角洲相
nondeposition 非沉积作用
nondepositional area 无沉积区
nondepositional hiatus 无沉积间断
nondepositional unconformity 非沉积不整合
non-destructive analysis 无破坏性分析；无损分析
nondestructive detection 无损检测；无损探伤
nondestructive earthquake 非破坏性地震
nondestructive examination 无损探伤；非破坏性检测；无损检验
nondestructive inspection 无损探伤；无损检验；非破坏性检查
non-destructive inspection of weld 焊缝无损检验
nondestructive material test 无损材料试验
nondestructive measurement 非破损测量
nondestructive measuring 不破坏测量法
nondestructive metallography 无损金相试验
nondestructive technique 不破坏样品的分析技术；无损探测技术
non-destructive test 非破坏性试验；探伤法试验；不伤测试；无损试验
non-detergent oil 无去垢性油
non-deteriorating restoring force 不退化恢复力
non-determinacy 不确定性
non-deterministic analysis 非定值分析；非确定性分析
non-deterministic excitation 非确定性激振；随机激振
nondeterministic information 非确定性信息
nondetrital 非碎屑的
non-detrital mineral 非碎屑矿物
non-deviative absorption 非偏移吸收
non-diastrophic 非地壳运动的；非构造的；非弯曲变形的
nondiastrophic strain 非构造应变
non-diastrophic structure 非造山构造；非地壳变形构造；非地壳运动构造
nondiastrophism 非构造变动
nondimensional 无量纲的；无因次的
non-dimensional coefficient 无量纲系数；无因次系数
nondimensional factor 无量纲因子
non-dimensional frequency 无量纲频率

nondimensional parameter 无量纲参数
non-dimensional parameter 无量纲参数
non-dimensional ratio 无量纲比
non-dipole magnetic field 非偶极子磁场；非偶极磁场
non-dipping formation 非倾斜岩层
non-directional 不定向的；无方向性的
non-directional counter 非定向计数器
non-directional detector 非定向探测器
non-directional point source 无定向点震源
non-discharging spring 无水流出的泉口
non-disjunction 或非
nondisjunction gate 或非门
non-dispersed beneficiated low solids mud 非分散增效低固相泥浆
non-dispersed system 非分散体系
non-dispersible 非扩散性
nondispersive 非弥散；非分散的
nondisplacement pile 不排土桩；钻孔桩
non-divergence level 无辐散层
nondivergent motion 无辐散运动
non-domestic accommodation 非住宅楼宇
non-domestic building 非住用建筑物
non-domestic car parking space 非住宅楼宇停车位
non-domestic equity 非住宅楼宇权益
non-domestic facility 非住宅设施
non-domestic premises 非住宅处所；非住宅楼宇；非住宅单位
non-domestic use 非住宅用途
nondrinkable water 非饮用水
non-drive end 非驱动端
non-driving end 非传动端
non-drying oil 非干性油
non-ductile 非延性的
non-ductile nature 非可延性质
nondurable 不耐用的；不耐久的
non-dusty mine 无尘矿
nondynamite seismic source 非炸药震源
non-earthquake-resistive building 非抗震房屋
noneconomic 不经济；无实用价值的
noneffective storage 无效储存
nonel 导爆管
nonel detonation 导爆管起爆
nonel detonator 导爆管雷管
nonel system 导爆管系统
non-elastic 无弹性的；非弹性的
non-elastic analysis 非弹性分析
non-elastic behaviour 非弹性性质；非弹性行为
non-elastic buckling 非弹性屈曲
nonelastic deformation 非弹性变形
non-elastic effect 非弹性效应
nonelastic scattering gamma-ray spectrometry logging 非弹性散射伽马能谱测井
non-elastic sediment 非弹性沉积物
non-elastic strain 非弹性应变
non-electric detonator 火雷管；非电雷管
non-electric heat detector 非电加热探测器
nonelectrolyte 非电解质
nonelectronic 非电子的
non-embedded 非埋藏的；露天的
non-emplaced 非侵位的；未就位的

non-emulsifier	防乳化剂
non-emulsifying acid	抗乳化酸
non-emulsifying agent	抗乳化剂
non-emulsifying surfactant	抗乳化活性剂
non-engineered structure	非正规设计的结构；非工程结构
nonentity	不存在；无足轻重
none-production well	停产井
nonequality	不等式
nonequality gate	异门
non-equant	非等径的
non-equilateral mine	不等翼开采的矿山
nonequilibrium	非平衡的；非平衡态；不平衡
non-equilibrium angle	不平衡角
non-equilibrium condensation model	非平衡凝聚模式；非平衡冷凝模式
non-equilibrium condition	非平衡条件
non-equilibrium confined aquifer	不平衡承压含水层
non-equilibrium equation	不均衡方程
non-equilibrium flow	非平衡流
nonequilibrium fractional crystallization	非平衡分离结晶作用
nonequilibrium state	非平衡状态
nonequilibrium thermodynamics	非平衡热力学
non-equilibrium well formula	非平衡井公式
non-equivalence gate	异门
nonequivalent	非等效的；非等价的
non-erasable storage	只读存储器
non-ergodic process	非各态历经过程
nonesite	顽拉玄斑岩
non-essential reservoir	非主要储层
non-essential resistance	外部电阻
non-essential stipulation	非主要条款
non-Euclidean finite element	非欧有限元
nonevident disconformity	不明显的不整合；掩蔽的不整合
nonevident unconformity	不明显的不整合；掩蔽的不整合
non-exchangeable ion	非可交换离子
nonexistent code	非法代码
nonexistent code check	非法码校验
nonexpansion	不膨胀
non-expansion soil	非湿胀土；非膨胀土
non-expansive condition	非膨胀状态
non-expansive fusion caking	不膨胀熔融黏结
nonexpansive soil	非膨胀性土
nonexplosible	不爆炸性的
non-explosion-proof	非防爆的
non-explosive	防爆的；不爆炸的
nonexplosive agent	无焰药包；安全炸药
non-explosive mining	不用炸药开采；无爆炸采矿法
non-explosive process	非爆炸法；不用炸药的破石法
non-explosive source	非炸药震源
nonexponential trend	非指数趋势
non-exposed area	摄影死区；非曝光区
non-extrusion ring	非挤压环
non-failure	无损坏
non-fatal accident	非死亡事故
nonfatal injury	非致命性的伤害
nonfault tension fissure	非断层张裂缝
non-ferrous	非铁的；有色金属的
nonferrous alloy	非铁合金；有色合金
nonferrous electrode	有色金属焊条
nonferrous metal	非铁金属；有色金属
non-ferrous metallic ores	有色金属矿产；有色金属矿石
nonferrous metallurgical works	非铁冶金工程
nonferrous metallurgy	非铁金属冶金；非铁金属冶炼术；非铁冶金学；有色冶金学
nonferrous welding	有色金属焊接
non-fines concrete	无细料混凝土；无砂混凝土
non-fireproof construction	不防火建筑；不耐火构造
non-fissile	不分裂的；非裂变的
nonfissured clay	无裂缝的黏土
nonflame blasting	无焰爆破
non-flame combustion nozzle	无焰燃烧火嘴
nonflame detector	无焰检定器
nonflammability	不易燃性
nonflammable	不易燃的；不燃的
non-flammable adhesive tape	阻燃胶带
non-flammable fluid	不燃液体
nonflammable gas	非易燃气体；不可燃气体
non-flanged	无法兰的
nonflexible	刚性的；非挠性的；不易弯的
nonfloat	浮选尾矿
nonfloating marine cable	非等浮式海上电缆
non-flooding feeder	防滥定量给料器
non-flowing	非自喷的
non-flowing artesian well	非自流承压井；非溢出自流井
nonflowing gas saturation	不流动气饱和度
nonflowing well	非自喷井；非自流井
non-fluctuating	非脉动的
non-fluid oil	厚质机油；润滑油
non-fluorescence	无荧光
non-fluvial	非河流的；非河流冲刷形成的
non-foaming	不起泡的
non-foliated rock	块状岩石；非叶理状岩石
non-fractured reservoir	非裂缝性储层
non-framework alumina	非骨架氧化铝
non-free jet	非自由射流
non-freezable oil	不冻油
non-freezing	不冻的
nonfreezing brine	不冻盐水
nonfreezing alloy	非冰冻土；不冻土
nonfrothing floatation	无泡沫浮选
nonfuel	非燃料的
nonfuel resource	非燃料资源
nonfuel use	非燃料用途
non-fulfilment	不履行
non-fusion caking	不熔融黏结
non-gas	非水气体；不凝气体
non-gaseous coal	瘦煤；无瓦斯煤
non-gaseous mine	无瓦斯矿
nongassy	无瓦斯的
non-Gauss noise	非高斯噪声
non-Gauss signal	非高斯信号
non-Gaussian linear process	非高斯线性过程
non-gear member	非齿轮构件
non-genetic	不涉及成因的；非遗传的
non-geogenic	非地质成因
nongeologic fracture	非地质裂隙

nongeosyncline	非地槽
nonglacial	非冰川的
non-governmental institution	非政府机构
Non-governmental Organization Environment Liaison Board	非政府组织环境联络会
non-gradational stratification	非递变成层
nongraded	不均粒的；分选差的
nongraded mix	非级配混合料
nongraded sediment	不均粒沉积
nongradient method	非梯度法
non-granular grout	非颗粒状材料浆液
nongravel-packed screen	非砾石预充填筛管
nongravitational water deposit	非重力水沉积
nongravity spring	升泉
non-harmonic response	非简谐反应
non-hazardous waste	非危险废物；无害垃圾
non-heaving soil	验潮杆
non-hermetic	不气密的
non-holonomic condition	不完整条件
non-holonomic constraint	非完整约束
nonhomogeneity	非均匀性；非均质性；多相性；非齐次
nonhomogeneity of rock	岩石非均质性
nonhomogeneity of soil layer	土层的非均匀性
nonhomogeneous	非均质的；不均匀的；非齐次的
nonhomogeneous boundary condition	非齐次边界条件
nonhomogeneous continuum structure	非均质连续介质构造
nonhomogeneous deformation	非均匀形变
nonhomogeneous differential equation	非齐次微分方程
nonhomogeneous linear differential equation	非齐次线性微分方程
nonhomogeneous linear equation	非齐次线性方程
nonhomogeneous media	非均匀介质
nonhomogeneous media medium	不均匀介质
nonhomogeneous medium	非均匀介质
non-homogeneous pay zone	不均质生产层；不均质产油层
non-homogeneous porous medium	非均质多孔介质
nonhomogeneous soil	非均质土
nonhomogeneous strain	不均匀应变
non-homologue	非同源特征
nonhomology	异源；非同源
nonhydrocarbon	非烃类；非碳氢化合物
non-hydrocarbon gas	非烃类气体
non-hydrodynamic condition	非水动力条件
nonhydrostatic pressure	非静水压力；非静压
nonhydrous	不含水的；无水的
non-hygroscopic	不吸湿的
non-hygroscopic insulation	防潮绝缘
non-hyperbolic moveout	非双曲线时差
nonideal elastic body	非理想弹性体
non-ideal gas	非理想气体
nonideal solution	非理想溶液
nonidentical	非恒等；不恒等的
non-idiogenous gas	非原生气
non-ignitibility	不可燃性
nonimaging	非成像的
non-impact screen	非冲击式筛
non-impact vibrator	无冲击振动器
non-implementation of contract	合同的不履行
non-incendive light	无焰光源
nonincremental	非递增的
nonindigenous oil	非原地生成油
non-inductive	无感的；非电感应性的
non-inductive induced polarization method	不接地激发极化法
non-inductive load	无感负载
non-inductive resistance	无感电阻
non-inductive resistor	无感电阻器
non-inductive winding	无感绕组；无感绕法
non-industrial land	非工业用地
non-industrial oil run	非工业性油流
noninflammability	不易燃性
noninflammable	不可燃的
non-inflammable hydraulic fluid	难燃液压液
non-inserted pump	非插入式泵
non-insulated overlap	非绝缘锚段关节
non-integrable component	非集成元件
non-integral quantity	非整数量
noninteracting	不人机交互的；不互相影响的
noninteractive	不相关的；非交互的
noninterconnected pore space	不连通孔隙空间
noninterconnected void	不连通空隙
noninterference	不干涉；不相互干扰
non-interfering	无干扰
non-interlocked switch	非联锁道岔
non-interlocking area	非联锁区
non-interlocking switch	非联锁道岔
non-intrusive flowmeter	不介入式流量计
non-invaded bed	未侵入层
noninvaded formation	未受侵入层
non-ionic	非离子的
non-ionic compound	非离子化合物
non-ionic derivate	非离子衍生物
non-ionic detergent	非离子型洗涤剂
non-ionic emulsifier	非离子乳化剂
non-ionic hydrocarbon surfactant	非离子烃类活性剂
non-ionic reaction	非离子反应
non-ionic surfactant	非离子表面活性剂
non-ionizing radiation	非电离辐射
non-irritating pollutant	非刺激性污染物
non-isentropic flow	非等熵流
nonisoelastic	非等弹性的
non-isometric	非等距的
non-isometric line	非等距线
non-isothermal	非等温的
non-isothermal compositional flow simulation	非等温组分流动模拟
non-isothermal filtering	非等温渗流
nonisothermal injection	非等温注入
non-isothermal jet	非等温射流
non-isothermal multiphase flow	非等温多相流
non-isothermal process	非等温过程
nonisotropic	各向异性的；非各向同性的
nonisotropic elasticity	各向异性的弹性
nonisotropic materal	各向异性材料
nonisotropic reservoir	各向异性储层
nonius	游标；游尺；游标卡尺
non-laminar flow	非层流
nonleaky anisotropic artesian aquifer	非越流各向异性承

压含水层
nonleaky artesian aquifer 非越流承压含水层
nonleaky isotropic artesian aquifer 非越流各向同性承压含水层
non-level geologic effect 非水平地质效应
nonlinear 非直线的；非线性的
nonlinear acoustical effect 非线性声效应
nonlinear acoustics 非线性声学
non-linear algebraic equation 非线性代数方程
nonlinear amplifier 非线性放大器
nonlinear analysis 非线性分析
nonlinear attenuation 非线性衰减
non-linear beam element 非线性杆件
nonlinear behaviour 非线性性状，非线性性质
nonlinear boundary condition 非线性边界条件
nonlinear boundary value 非线性边界值
nonlinear buckling 非线性屈曲
nonlinear constitutive equation 非线性本构方程
nonlinear constitutive model 非线性本构模型
non-linear constrain 非线性约束
nonlinear constrained optimization 非线性约束优化
nonlinear constraint 非线性约束
nonlinear continuum 非线性连续介质
non-linear contrast enhancement 非线性反差增强
nonlinear control system 非线性控制系统
nonlinear control theory 非线性控制理论
nonlinear correction 非线性修正
nonlinear correction term 非线性修正项
nonlinear correlation 非线性相关
nonlinear Coulomb device 非线性库仑装置
nonlinear coupled differential equation of motion 非线性耦联微分运动方程
nonlinear coupling 非线性耦联
nonlinear criterion of failure 非线性破坏准则
nonlinear damping 非线性阻尼
nonlinear deconvolution 非线性反褶积
nonlinear deformation 非线性变形
nonlinear dependence 非线性相关
nonlinear design 非线性设计
nonlinear differential equation 非线性微分方程
nonlinear diffusion 非线性扩散
nonlinear displacement 非线性位移
nonlinear distortion 非线性畸变；非线性失真
nonlinear distributed parameter 非线性分布参数
nonlinear dynamic analysis 非线性动力分析
nonlinear dynamic soil property 土壤非线性动力特性
nonlinear dynamic system 非线性动力系统
nonlinear dynamics 非线性动力学
nonlinear earthquake response 非线性地震反应；非线性地震响应
nonlinear edge enhancement method 非线性边缘增强法
nonlinear effect 非线性效应
nonlinear eigenvalue 非线性特征值
nonlinear elastic behavior 非线性弹性状
nonlinear elastic material 非线性弹性材料
nonlinear elastic model 非线弹性模型
nonlinear elasticity 非线弹性
nonlinear elasticity theory 非线弹性理论
nonlinear elastodynamics 非线弹性动力学
nonlinear elastoplastic analysis 非线弹塑性分析

nonlinear envelope 非线性包线；非线性包面
nonlinear equation 非线性方程
nonlinear equation of motion 非线性运动方程
nonlinear estimation 非线性估计
nonlinear evolution 非线性演化
nonlinear factor 非直线系数
nonlinear failure 非线性破坏
nonlinear filtering 非线性滤波
nonlinear filtering flow 非线性渗流
nonlinear filtering of gravity and magnetic anomaly 重力电磁异常非线性滤波
nonlinear finite difference equation 非线性有限差分方程
nonlinear finite element 非线性有限元
nonlinear finite element analysis 非线性有限元分析
nonlinear finite element technique 非线性有限元技术
nonlinear flow 非线性流
non-linear fluid 非线性流体
nonlinear fold region 非线状褶皱区
nonlinear force vector 非线性力向量
nonlinear forced resonant response 非线性强迫共振反应；非线性强迫谐振反应
nonlinear function 非线性函数
nonlinear geochemical data analysis 非线性地球化学数据分析
nonlinear geology 非线性地质学
nonlinear geomechanics 非线性地质力学
nonlinear geostatistics 非线性地质统计学
nonlinear governing equation 非线性控制方程
nonlinear harmonic wave 非线性谐波
nonlinear height effect 非线性高程效应
nonlinear hysteretic system 非线性滞回体系
nonlinear hysteretic vibration 非线性滞回振动
nonlinear influence coefficient 非线性影响系数
nonlinear instability 非线性不稳定性
nonlinear interpolation 非线性插值
nonlinear inversion 非线性反演
nonlinear iteration 非线性迭代
nonlinear iterative matrix solution method 非线性迭代矩阵解法
nonlinear large deflection theory of stability 非线性大挠度稳定理论
nonlinear least square method 非线性最小二乘法
nonlinear load 非线性载荷
nonlinear magnitude frequency law 非线性震级频度规律
nonlinear mapping 非线性映射
nonlinear mapping method 非线性映射方法
nonlinear mapping plot 非线性映射图
nonlinear mapping technique 非线性映射方法
nonlinear material 非线性材料
nonlinear material problem 材料非线性问题
nonlinear mathematical process 非线性数学过程；非线性数学方法；非线性数学程序
nonlinear mechanics 非线性力学
nonlinear mechanics of plate and shell 板壳非线性力学
nonlinear medium 非线性介质
nonlinear mode 非线性模态；非线性振型
nonlinear model 非线性模型
nonlinear nature 非线性特性
nonlinear operation 非线性运算
nonlinear operator 非线性算子；非线性算符

nonlinear optimization 非线性优化
nonlinear optimum design 非线性最优设计
nonlinear oscillation 非线性振动
nonlinear passive damping 非线性无源阻尼；非线性被动阻尼
non-linear polymer 非线性聚合物
nonlinear potentiometer 非线性电位计；非线性分压计
nonlinear problem 非线性问题
nonlinear problem convergence 非线性问题收敛
nonlinear program method 非线性规划法
nonlinear programming 非线性规划
nonlinear programming approach 非线性规划方法
nonlinear programming method 非线性规划法
nonlinear random vibration 非线性随机振动
nonlinear recording 非线性记录
nonlinear rectification 非线性检波；非线性校正
nonlinear recurrence 非线性递归关系
nonlinear regression 非线性回归
nonlinear regression analysis 非线性回归分析
nonlinear regression technique 非线性回归法
nonlinear relationship 非线性关系
non-linear resistance 非线性电阻
non-linear resistivity profile 非线性电阻率剖面
nonlinear resistor 非线性电阻器
nonlinear resonance 非线性共振
nonlinear response 非线性反应；非线性响应
nonlinear response analysis 非线性反应分析；非线性响应分析
non-linear response history 非线性响应历史
nonlinear response spectrum 非线性反应谱；非线性响应频谱
nonlinear restoration 非线性复原
nonlinear restoring force characteristics 非线性回复力特性
non-linear scale 非线性标度；非线性标尺
nonlinear science 非线性科学
nonlinear seepage 非线性渗透；非线性渗流
nonlinear seepage law 非线性渗透定律
nonlinear seismic behaviour 非线性地震特性
non-linear sensor response 非线性传感器的响应
non-linear shift register 非线性移位寄存器
nonlinear simultaneous equations 非线性联立方程组；非线性联立方程
nonlinear smoothing 非线性光滑；非线性平滑化
nonlinear smoothing filter 非线性平滑滤波器
nonlinear soil behavior 土的非线性性质
nonlinear soil structure interaction 非线性土-结构相互作用
nonlinear spring 非线性弹簧
nonlinear stiffness 非线性刚度
nonlinear stiffness matrix 非线性刚度矩阵
nonlinear stochastic system 非线性随机系统
nonlinear strain tensor 非线性应变张量
nonlinear stress 非线性应力
nonlinear stress strain behaviour 非线性应力-应变特性
nonlinear stress strain relationship 非线性应力-应变关系
nonlinear stretch 非线性扩展
nonlinear structural analysis 非线性结构分析
nonlinear structure 非线性结构
nonlinear superposition principle 非线性叠加原理
nonlinear surface effect 非线性表面效应
nonlinear sweep 非线性扫描

nonlinear system 非线性系统
nonlinear theory 非线性理论
nonlinear torsional spring 非线性扭转弹簧；非线性减震弹簧
nonlinear transformation 非线性变换
nonlinear transient analysis 非线性暂态分析；非线性瞬态分析
nonlinear transverse vibration 非线性横向振动
nonlinear trend 非线性趋势
nonlinear tsunami wave 非线性海啸波；非线性海啸浪
nonlinear vibration 非线性振动
nonlinear viscoelastic analysis 非线性黏弹性分析
nonlinear viscoelastic model 非线性黏弹性模型
nonlinear wave 非线性波
nonlinearities change 非线性变化
nonlinearity 非线性
nonlinearity least square 非线性最小二乘
nonlinearity of failure envelope 破坏包络线的非线性
nonlinearity test 非线性检验
nonlinearly polarized wave 非线性偏振波
non-liquefaction 非液化
nonliquefied basic material 非液化基本材料
nonliquid water 非液态水
nonliving 非生物的；无生命的
non-living resources 非生物资源
non-load bearing 非承重
non-load bearing solid 非承载固体
non-load bearing wall 非承重墙
non-load probability 空载概率
non-load-bearing wall 非承压墙
non-loaded 无载的；空载的
non-loaded shear strength 无荷载抗剪强度
non-loading shear test 无荷载抗剪强度
non-local 非局部的；非区域的
nonlocal model 非局部模型
non-localized 非局限的；非定域的
non-magmatic water 非岩浆水
nonmagnetic 非磁性的
nonmagnetic alloy 非磁性合金；无磁合金
non-magnetic alloy 非磁性合金；无磁合金
nonmagnetic compass 非磁性罗盘
non-magnetic drill collar 无磁钻铤；无磁钻环
nonmagnetic material 非磁性材料；非磁体
non-magnetic ore 非磁性矿石
nonmagnetic rod 无磁钻杆
non-magnetic solid 非磁性固体
nonmagnetic stabilizer 非磁性稳定器
non-magnetic steel 无磁性钢；非磁性钢
non-magnetic sub 无磁性接头
nonmagnetic taconite 非磁性铁燧石
nonmagnetic trailer 非磁性拖车
non-magnetic vessel 非磁性船
non-manmade source 非人为污染源
nonmarine 非海成的；陆相的
nonmarine conglomerate 非海相砾岩
nonmarine deposit 陆相沉积
nonmarine environment 陆相环境
nonmarine facies 陆相
nonmarine organic matter 非海相有机物
nonmarine organism 非海相生物

non-marine reservoir rock 非海相储集岩
nonmarine sandstone 陆相砂岩
non-marine source rock 非海相烃源岩；陆相烃源岩
nonmaskable interrupt 非屏蔽中断
non-matched data 非匹配数据
nonmechanical 非机械破碎的；非碎屑的
non-melt 非熔化的
non-membership function 非隶属函数
non-membership grade 非隶属度
non-meridional intensity 非子午线方向强度；非子午线方向亮度
nonmetal 非金属
nonmetallic 非金属的；非金属物质
nonmetallic air channel 非金属风道
nonmetallic ceramic 非金属陶瓷
non-metallic deposit 非金属矿床
nonmetallic element 非金属元素
non-metallic impurity 非金属杂质
nonmetallic inclusion 非金属夹杂物
non-metallic inorganic toxicant 非金属无机毒物
nonmetallic luster 非金属光泽
nonmetallic mineral 非金属矿物
nonmetallic mineral and rock 非金属矿物和岩石
nonmetallic mineral deposit 非金属矿床
nonmetallic phase 非金属相
nonmetallic pipe 非金属管
non-metallic tank 非金属罐
nonmetalliferous 非金属的；不含金属的
non-metalliferous ore 非金属矿石
non-metalliferous ore deposit 非金属矿床
nonmetamorphic 非变质的
non-metric 非米制的；非量测的
non-metric camera 非量测摄影机
nonmineral anomaly 非矿异常
non-minimum phase 非最小相位
non-mining damage 非开采性破坏
nonmining loss 非开采损失；非开采损耗
non-mining personnel 非采矿人员；非井下工作人员
nonmiscibility 非混溶性
nonmobile phase 非活动期
non-mobile water 不流动水
non-monumented bench mark 临时水准点；无标石水准点
nonmotorized vehicle lane 非机动车道
non-movable oil 不可动油
non-natural 非天然的
non-navigable 不能通航的
non-negative definite matrix 非负定矩阵
non-negative operator 非负算子
non-negativity constraint 非负性约束
non-negotiable 不可转让的；不可谈判的
non-new town project 新市镇以外地区工程计划
non-Newtonian crude 非牛顿原油
non-Newtonian equation 非牛顿方程
non-Newtonian flow 非牛顿流动
non-Newtonian flow characteristic 非牛顿流动特性
non-Newtonian fluid 非牛顿流体
non-Newtonian interfacial flow 非牛顿界面流动
non-Newtonian liquid 非牛顿液体
non-Newtonian multiphase flow 非牛顿多相流
non-Newtonian power-law fluid 非牛顿幂律流体

non-Newtonian viscosity behavior 非牛顿黏度特性
non-nil solubility 非零溶解度
non-nitrogen resin 不含氮树脂
non-nitro-glycerine powder 非硝化甘油炸药
nonnormal 非正态；不正交
non-normal incidence modeling 非垂直入射模拟
non-normal random variable 非正态变量；非正规随机变量
nonnormality 非正态性
non-normalized 非正规化的；非标准化的
non-nuisance technique 无公害工艺
non-null 非空
non-numerical 非数值的
non-numerical calculation 非数值计算
non-numerical data processing 非数值数据处理
non-observance 不遵从；不按惯例
non-occupation 空置；无人住用
nonoccurrence 不发生
non-official 非正式的；非法定的
non-offset bit 牙轮轴线不偏移的钻头
nonoil 非石油的
non-oleaginous 非油质的
non-operative 不工作的；不动作的；失效的
nonoperator party 非作业方
non-ore anomaly 非矿异常
non-organic soil 无机土；无机泥土
nonorientable 不能定向的
nonorienting tubing hanger 非定向油管悬挂器
non-original 非原始的
non-orogenic zone 非造山带
non-orthogonal analysis 非正交分析
non-orthogonal eigenfunction 非正交特征函数
non-orthogonal frame 非正交框架
non-orthogonal system 非正交系统
nonorthogonality 非正交性
non-orthotropic material 非正交各向异性材料
non-outcropping fracture 隐伏地裂缝
nonoverflow 非溢流
non-overflow dam 非溢流坝；封闭坝
non-overlap 非重叠时相
nonoverlapping 不相重叠的；非覆盖的
non-oxidizability 不可氧化性
non-oxidizable 不可氧化的
nonoxidizing heating 无氧化加热
non-packet terminal 非分组终端
nonpaired terrace 不对称阶地
non-paraffinic oil 不含蜡石油
nonparallel slot 两壁不平行割缝
non-parallel train diagram 非平行运行图
nonparametric classification 非参数分类
nonparametric detector 非参数检测器
non-parametric estimation 非参数估计
nonparametric modeling 非参数建模
nonparametric procedure 非参数法
nonparametric statistical method 非参数统计法
nonparametric test 非参数检定；非参数检验
non-particulate 非颗粒性的
non-particulate grout 非颗粒性灌浆
nonpay 非生产层
non-pay interval 非生产层段

nonpay zone	非生产层	non-polar polymer	非极性聚合物
nonpaying	无开采价值的；无利的	nonpolar resin	非极性树脂
non-paying mine	亏损的矿井	nonpolar solvent	非极性溶剂
non-pedestal truck	无导框式转向架	nonpolarity	非极性
non-penetrating	非穿透	nonpolarized	非极化的
nonpenetrating artesian well	非完整自流井	nonpolarizing electrode	不极化电极
non-penetrating model	不渗透模型	non-polishing	不易磨光
nonpenetrating well	不完整井；非完整井	non-polluted industrial wastewater	无污染工业废水
nonpenetrative	非透入性的	non-polluting	不污染的
non-percussive piling	非撞击式打桩工程	nonpolluting energy source	无污染能源
nonperennial	季节性的	non-polluting lubricant	无杂质润滑剂
nonperennial lake	季节湖	non-polluting substance	非污染物质
nonperennial variation	非周期性变化	non-pollution fuel	无污染燃料
non-perforated interval	未射孔井段	non-pollution technology	无污染工艺
non-performance	未履约；违约	nonporosity	无孔性
non-periodic excitation	非周期性激励	nonporous	无孔隙的
non-periodic function	非周期函数	non-porous carbonate	非孔隙性碳酸岩
nonperiodic signal	非周期信号	non-porous pressure meter	无孔压力计
nonperiodic variation	非周期性变化	non-positive	非正的
non-periodic wave	非周期波	nonpotable water	非饮用水
non-permanent deformation	非永久变形	non-power	不做功的
nonpermeable	不透水的；不渗透的；防渗漏的	nonpower reactor	非动力反应堆
non-permeable soil subgrade	非渗水土路基	non-preformed	松散的
nonpermeable stringer	不渗透夹层	non-pressure culvert	无压力式涵洞
non-perpendicularity	不垂直度	non-pressure drainage	无压排水
nonphosphatic sequence	非磷灰质岩层	non-pressure flow	无压流
nonphotographic imagery	非摄影成像	non-pressure tank	非压力罐
nonphotographic remote sensing	非摄影遥感	non-pressure tunnel	无压隧洞
nonphotographic sensor	非摄影传感器	non-pressure welding	不加压焊接；非压力焊
nonphotographic spectrum	非摄影光谱	nonpressurized	不加压的
nonphysical primary color	假原色	non-prestressed anchor	非预应力锚杆
nonpiercement	不刺穿	non-prestressed reinforcement	非预应力钢筋
nonpiercement dome	非刺穿穹隆	nonprimary reflection	非一次反射
non-piercement type	非刺穿型	non-prismatic	变截面的；非棱柱体的；非定型的
non-pillar roadway protection	无煤柱护巷	non-prismatic beam	变截面梁；非柱状梁
non-planar fracture propagation	裂缝非平面扩展	non-processed natural gas	未处理的天然气
nonplastic	无塑性	nonproducing	非生产的
non-plastic silt	无塑性粉土	nonproducing reserve	未投产储量
nonplastic soil	非塑性土壤；无塑性土	non-producing reservoir	未动用储量
non-plugging	非堵塞的	nonproducing well	非生产井；无生产能力的井
non-plugging constituent	不堵塞地层的组分	non-production platform	非生产平台
nonplugging damage	非堵塞性损害	nonproductive	非生产性的；对生产无直接关系的
non-plugging shaped charge	无堵塞聚能射孔弹	nonproductive construction	非生产性建设
non-plunging fold	非倾伏褶皱	non-productive fixed asset	非生产固定资产
non-point pollutant	非点源污染物	nonproductive formation	非生产层
non-point pollution	非点源污染	non-productive interval	非生产层段
non-point pollution source	非点污染源；面污染源	non-productive operation	辅助操作
nonpoint source	非点污染源；非点源	non-productive personnel	非生产性人员
nonpoint source pollution	非点污染源	nonproductive petroleum reservoir	无开采价值油藏
non-poisonous	无毒的	non-productive time	不生产时间；停工；窝工
non-Poissonian model	非泊松模型	non-productive wage	非生产工资
nonpolar	非极性的	non-productive well	不生产井
non-polar adsorption	非极性吸附	nonprofit organization	非营利组织；非营利性机构
non-polar atom	非极原子	non-programmed decision	非程序化决策
non-polar bond	非极性键	non-propagating crack	不扩展裂纹；停止开裂
nonpolar compound	非极性化合物	non-propelling	非自动推进式的
non-polar fluid	非极性流体	non-proportional damping	非比例阻尼
non-polar link	非极性键	nonproportionality	非线性
non-polar liquid	非极性液体	non-protective	无防护的
nonpolar molecule	非极性分子	non-proton flare	非质子耀斑

non-puddled soil	非黏闭土；非紧实土壤
nonpumping	未抽水的；非抽水的
nonpumping water level	抽水前水位；静水位
nonpumping well	未抽水的井
non-radial oscillation	非径向振荡
non-radioactive electron capture detector	无放射源电子俘获检测器
nonradiogenic lead	非放射性成因铅
nonrandom distribution	非随机分布
nonrandom variable	非随机变量
nonrandomized strategy	非随机化策略
nonreactive	不起反应的
nonreactivity	惰性；无反应性
non-realtime	非实时
non-realtime data	非实时资料；非实时数据
non-realtime processing	非实时处理
nonreciprocal	单向的
nonreciprocal observation	单向观测
nonreciprocity	非互易性
non-recommended	非推荐的
nonrecoverable	不可采出的；不可回收的；不可恢复的
nonrecoverable compaction	不可恢复压缩；不可恢复压实
nonrecoverable deformation	不可逆变形
nonrecoverable oil	非可采油；不可采油
non-recoverable reserve	非可采储量
non-rectilinear border	非直线边界
non-recurrent congestion	非重现性交通壅塞；偶发性交通拥挤
nonrecurrent point	非回归点
non-recurring maintenance	非日常维护
non-recurring repair	非日常修理
non-recyclable waste	不可循环再用的废物；不可回收垃圾
non-reducing	非还原的
nonredundant closed-loop path	非冗余闭合环路
non-reef bioherm	非礁生物丘
non-reef carbonate reservoir	非礁碳酸盐岩储集层；非礁碳酸盐岩油藏
non-reef limestone reservoir	非礁灰岩储集层；非礁灰岩油藏
nonreflecting boundary condition	无反射边界条件
nonrefractory alloy	非耐热合金
non-refuelling	不加油
non-refuelling duration	不加燃料时间
non-registered	无记录的
non-regular	非正则的
non-regular building	非正规建筑物
nonregulated discharge	非调节流量；未调节的流量
nonreinforced	未加固的；无钢筋的
nonreinforced concrete	无钢筋的混凝土
non-reinforced section	无钢筋部分
non-relay transmission distance	无中断传输距离
non-relevant indication	假象
non-removable	不可去除的
non-renewable energy	非再生能源
non-renewable energy resource	不可再生能源
non-renewable land lease	不可续期土地契约
non-renewable lease	不可续期租契
nonrenewable resource	不可再生资源；非再生资源
non-renewal of franchise clause	专营权不可续期条款
nonrepeatability	不可重复性
nonrepeated sampling	不重复抽样
nonrepeated signal	非重复信号
nonrepresentative sample	非代表性样品；不具代表性样品
nonreservoir	非储集层
non-reservoir sediment	非储集层沉积物
nonresistant strata	不稳定地层
non-resistant to earthquake	未经抗震设计的
non-resonant buffeting	非共振冲击
nonresponsible accident	非责任事故
non-restricted algorithm	非限制性算法
nonreturn damper	止回阀
non-return finger	止回装置；单向装置
non-return inlet valve	止回进给阀
non-return protecting unit	防滑机构；止回保护单元；逆止保护装置
nonreturn to zero	不归零，非回零
non-return to zero method	不归零法
non-return to zero pulse	不归零脉冲
non-return to zero recording	不归零记录
non-return to zero system	不归零制
non-return valve	单向阀；止回阀
non-reusable prop	不能反复使用的支柱；一次性支柱
nonreversibility	不可逆性
nonreversible	不可逆的
nonreversible deformation	不可逆变形
non-reversible process	不可逆过程
non-reversible reaction	不可逆反应
nonreversible thermodynamic process	不可逆的热力学过程
nonrigid	非刚性的
non-rigid behavior	非刚性性状
nonrigid plastic	软质塑料
non-rigorous adjustment	近似平差
non-rock	非岩质
non-rotating rope	不旋转钢丝绳
non-rotating rubber sleeve stabilizer	非旋转式胶皮套筒稳定器
non-rotating stabilizer	非旋转稳定器
non-rotating tail rope	不旋转钢丝尾绳
nonrotational deformation	非转动变形
nonrotational hydraulic jar	非旋转式水力振击器
non-rubbing surface	非摩擦面
non-rust steel	不锈钢
non-rusting	不锈的；防锈的
nonsaline	淡的
nonsaline sodic soil	非盐碱性土
non-saponifiable extract	非皂化萃取物
nonsaponifying	不可皂化的
nonsaturable	非饱和的；不饱和的
nonsaturated	非饱和的；不饱和的
non-saturated reservoir	未饱和油藏；不饱和油藏
non-saturated zone	非饱和带
nonscale domain	无标度域
non-scattering atmosphere	无散射大气
nonscheduled	不定期的
non-scouring velocity	止冲流速；无冲刷流速
non-sealed bearing	非密封轴承

nonsealing fault	未充填断层；未封闭断层
non-seismic area	非震区；非地震带
non-seismic design	非抗震设计
non-seismic fault motion	非地震断层运动
non-seismic modeling system	非地震模型系统
nonseismic region	无震区；非地震区
non-seismic vibration	非地震振动
nonseismic zone	无震带
non-seismological	非地震学的
non-selective	非选择性的
nonselective absorptivity	非选择性吸收
non-selective chemical plugging	非选择性化学堵水
nonselective scattering	非选择性散射
non-self adjoint operator	非自伴随算子
nonself-clearing bit	非自洁式牙轮钻头
non-self-reset push-key switch	非自复式按键开关
non-self-weight collapse loess	非自重湿陷性黄土
nonself-weight collapsible loess	非自重湿陷性黄土
non-self-weight collapsing	非自重湿陷性
non-sequence	不连续；间断
nonsequent	不连续的
nonsequent fold	限制褶皱
nonsequent folding	限制褶皱；不连续褶皱
nonsequential bed	无序层；间断层
nonsequential operation	非顺序操作
non-serviceable	不能使用的
nonsettling	不沉降的；不沉陷的
non-shaly formation	无泥质地层
non-shelf-derived	非来源于陆架的
non-shrink	防缩
nonshrink grout	不收缩水泥灌浆；不收缩水泥薄浆
non-shrink mortar	抗收缩砂浆
non-shrinkage grout	不收缩薄浆
non-shrinking concrete	不收缩混凝土
nonshrinking soil	非收缩性土
non-siliceous dust	无硅尘末
non-silicosis mine	无矽肺危险的矿
nonsilting velocity	不淤积流速；不淤积流速
nonsimultaneous	非同时的；非同步的
non-simultaneous observation loop	异步观测环；非同步观测环
nonsine-wave	非正弦型波
nonsingular	非奇异的；非退化的；满秩
nonsingular matrix	非奇异矩阵
nonsingular stress	非奇异应力
non-sinusoidal	非正弦
non-sinusoidal wave	非正弦波
non-skeletal cement reef	非骨骼胶结物生物礁
non-skeletal limestone	非骨骼灰岩
non-skeletal microbialite framework	非骨骼的微生物岩骨架
non-skeleton	非骨架相
nonskid	不滑的；防滑的；防滑装置；防滑器
nonskid chain	防滑链
non-skid device	防滑装置；防滑器；防滑法
nonskid surface treatment	防滑处理
non-slaking	不水解的
non-slaking clay	非湿化性黏土；不水解黏土
nonslip	无滑动的；防滑；不滑的
non-slip condition	无滑移条件
non-slip drive	无滑动传动装置；非滑动传动
non-slip granolithic	防滑细石混凝土
non-slip surface	防滑地面；防滑表面
non-slipping	无滑动的；不滑移
non-slipping block	防滑块
non-sludge oil	不产生酸渣的原油；非淤渣油，非酸渣抽出油
non-slumping soil	非湿陷性土
non-soapbase grease	非皂基润滑酯
non-softening	不变软的
non-soil	无黏性土
non-soil volume	土的空隙体积
non-soil-retentive softener	抗污柔软剂
nonsolid drilling fluid	无固相钻井液
non-solid space	未填入固体颗粒的空隙
non-solidified	不牢固的
nonsoluble	不溶解；不溶解物
nonsorted	未分选的
nonsorted polygon	未分选多边形土
nonsorted stripe	无分选石条
non-sparking floor	无火花地面；防爆地面
non-special	非特殊的
nonspecification	非规范
non-spectral colour	谱外色
nonspreading ridge	非扩张海岭
nonspreading rift	非扩张裂谷
non-square integrable	非平方可积
non-stability	非稳定性；不稳定性
non-stabilized	未被稳定的
nonstabilized condition	不稳定条件
nonstabilized rate	不稳定产量
nonstabilized well test	不稳定试井
nonstabilized zone	不稳定带
non-stable	非稳定的
non-staggered row	整齐的割缝排
nonstagnant basin	非滞水海盆
non-staining cement	白色水泥；无杂质水泥
non-standardized piping component	非标准管件
nonstationarity	非平稳性；非恒定性
nonstationary	非固定的；不稳定的；非平稳的
non-stationary bilinear system	非平稳双线性系统
non-stationary envelope function	非平稳包络函数
non-stationary excitation	非平稳激励
nonstationary flow	非稳定流；不稳定流
non-stationary hydroelasticity	非定常流体弹性力学
nonstationary iterative method	不定常迭代法
non-stationary process	非平稳过程；非稳定过程
non-stationary random excitation	非平稳随机激励
non-stationary random function	非平稳随机函数
non-stationary random process	非平稳随机过程
non-stationary response	非平稳反应；非稳态响应
non-stationary spectrum	非平稳谱
non-stationary stochastic model	非平稳随机模型
non-stationary stochastic signal	不定常随机信号；非平稳随机信号
non-stationary trend	非平稳趋势
non-stationary two dimensional flow	不稳定平面流
non-stationary vibration	非平稳振动
non-stationary wave	非平稳波
nonsteady	不稳定的；不定常的

nonsteady flow	不稳定流；非稳定流
nonsteady groundwater flow	非稳定地下水流
nonsteady radial flow	非稳定径向流
nonsteady state	非稳定态；不稳定状态
non-steady state simulation analogue	非稳定流的模拟比拟法
nonstick	不黏的
non-stop character	直达性；不间断的特征
non-stop dual operation	不间断的工作；不间断作业
non-stop dumping	行进卸载；不停卸载
non-stop operation	不间断的工作
nonstop-concreting	混凝土连续浇筑
non-stranded rope	无股钢丝绳
non-stratified	不成层的
non-stratified deposit	非层状矿床
non-stratified rock	非成层岩
non-stress metre	无应力计
nonstress strain meter	无应力应变计
non-stretch	无伸缩
non-stretch yarn	非弹力丝
non-stromatolitic	非叠层岩的
non-structural crack	非结构性裂缝；非构造裂缝
non-structural damage	非结构性损伤
non-structural design	非结构设计
non-structural element	非结构构件
non-structural hydrodynamic trap	非构造水动力圈闭
non-structural subsidence	非构造沉降
nonstructural trap	非构造圈闭
non-structural works	非结构工程
non-structured	非构造的
nonsupervised training computer method	非监督训练计算机法；空间积群法
non-supporting diaphragm wall	自立式地下连续墙
non-supporting retaining structure	自立式挡土结构
non-surfaced	非铺路面的；未铺面的
nonsurfactant	非表面活性剂
nonsweepable	不可波及的
non-swelling	非膨胀性的；不膨胀的
non-symbiotic	非共生的
non-symmetric	非对称的
non-symmetric loading	非对称荷载
non-symmetrical	非对称的
nonsymmetrical aquifer	非对称性含水层
nonsymmetrical flow	非对称水流
non-symmetrical loading	非对称荷载
non-synchronous	非同步的；异步的；不同步的
nonsynchronous controller	异步控制器
non-synchronous ground motion	非同步地震动；非同步地表运动
nonsystematic distortion	非系统畸变；不规则畸变
non-systematic error	非系统误差
non-systematic fracture	非系统破裂
nonsystematic joint	非系统节理；不规则节理
nontarget zone	非目的层
nontechnical	非技术性的
non-tectonic	非构造
non-tectonic factor	非构造因素
non-tectonic fault	非构造断层
non-tectonic ground crack	非构造地裂缝
non-tectonic ground rupture	非构造性地裂缝
non-tectonic influence	非地质构造影响
nontectonic joint	非构造节理
nontectonic process	非构造作用
nontectonic slumping	非构造滑塌
nontectonite	非构造岩
non-temperature compensated meter	无温度补偿流量计
non-tension type bolt	非张拉锚杆
non-terminating alarm	长鸣
non-terminating continued fraction	无尽连分数
non-terminating decimal	无穷小数；无尽小数
non-terminating fraction	无尽小数；无尽分数
nonterrain-related factor	非地形相关因数
non-territorial strait	非领海海峡
non-thermal groundwater	非热地下水；无热能地下水
non-thermal radiation	非热辐射
non-thermal spring	非温泉
nonthixotropic	非触变性的
non-threaded fastener	非螺纹紧固件
non-tilt seismograph	非倾斜地震仪
non-tilting concrete mixer	非翻转式混凝土搅拌机
nontopographic photogrammetry	非地形摄影测量
non-toxic	无毒的
non-toxic hydraulic fluid	无毒压裂液；无毒液压液
non-toxic polymeric thinner	无毒高分子稀释剂
non-tractor drill	非拖拉式钻机
non-traditional mineral fertilizer	非传统矿产肥料
nontransferable	不可转让
non-transmission	非传导性
non-transmission of an internal explosion	隔爆性
nontrivial solution	非常规解
nontronite	绿脱石；囊脱石；绿高岭石
nonturbidite	非浊流层
non-turbulent	非湍流的；非扰动的
nonturbulent flow	非紊流；非湍流
nonuniform	不一致的；不均匀的
nonuniform attenuation	非均匀性衰减
non-uniform beam	变截面梁
non-uniform capillary	不均匀毛细管
non-uniform cross-section beam	变截面梁
nonuniform deformation	非均匀变形
nonuniform distribution	不均匀分布；非均匀分布
nonuniform envelope	非均匀包络
non-uniform face	不均匀面
nonuniform field	不均匀场
non-uniform flow	不均匀流
non-uniform foundation	不均匀地基
non-uniform friction	非均匀摩擦
non-uniform function	非单值函数
nonuniform grading	不均匀级配
non-uniform grid	非均匀网格
nonuniform impulse response	非均匀脉冲响应
nonuniform magnetization	非均匀磁化
non-uniform medium	非均匀介质
non-uniform motion	变速运动；不均匀运动
non-uniform placement	不均匀充填
nonuniform pressure	不均压；不等压；非均布压力
nonuniform quantization	非均匀量化
non-uniform rock pressure	围岩偏压；不均匀岩石压力
non-uniform scale division	不均匀刻度；非均匀标尺分划

nonuniform settlement	不均匀沉降；差异沉降
non-uniform settling	不均匀沉降
nonuniform soil	不均匀土壤
non-uniform spacing	非均匀间隔
nonuniform stratification	非均质层理
non-uniform stress	非均布应力
nonuniform stress field	不均匀应力场；非均匀应力场
non-uniform surface load	非均匀表面荷载
non-uniform velocity	变速；不等速
non-uniform weighting	非均匀加权
nonuniform-grained soil	不均粒土
nonuniformity	非均质性；不均匀性
nonuniformity coefficient	不均匀系数
non-uniformly convergent	非均匀收敛的
nonuniformly weighted patterns	非均匀加权组合
non-uniplanar bending	异面弯曲
non-uniplanar straight line	异面垂直的
non-unique mapping	非唯一映射
non-uplift pile	受压桩
nonupset tubing	外平式油管
non-urban area	市区以外地区
non-urban source	非城市源
non-valid	无效的；非零的
non-vanishing	不为零
nonvariant	不变的；恒定的
nonvegetated area	非植被区
nonventilated	不通风的
non-vertical wellbore	斜井井筒
nonviolent fault creep	非剧烈断层蠕变
nonviscous flow	非黏性流；非黏滞流
non-viscous fluid	无黏性液
non-viscous lubricating distillate	非黏性润滑油馏出物
non-volatile	非挥发性的；不挥发的
nonvolatile element	非挥发性组分
nonvolatile matter	不挥发物
nonvolatile memory	非易失存储器
Non-Volatile Random Access Memory	非易失随机读写存储器
non-volcanic anticline	非火山背斜
non-volcanic continental margins	非火山大陆边缘
non-volcanic earthquake	非火山地震
non-volcanic eruption	非火山喷溢；非火山喷发
nonvolcanic geothermal region	非火山地热区
nonvolcanic mid-ocean ridge	非火山大洋中脊
nonvolcanic outer arc	非火山外弧
nonvolcanic passive continental margin	非火山型被动大陆边缘；非火山被动陆缘
nonvolcanic ridge	非火山脊岭
non-vortex	无涡流的
non-wall-building	不造壁的
nonwaste technology	无废料生产工艺
non-water-soluble polyurethane	非水溶性聚氨酯
non-waxy	不含蜡的；无蜡的
non-waxy crude	不含蜡原油；无蜡原油
nonweathering	未风化的
non-weighted code	非加权码
nonwettable	不可湿润的
nonwetted	未润湿的
nonwetting fluid	非润湿性流体
non-wetting meniscus	非润湿弯月面
nonwetting phase	非润湿相
non-wetting phase front	非润湿相前缘
non-wetting resident fluid	非润湿滞留液
nonwetting-phase recovery efficiency	非润湿相采收率
nonwhite noise	非白噪声
non-Wilson cycle	非威尔逊旋回
non-wireline logging method	无电缆测井方法
non-work	非工作的
non-working slope	非工作帮
non-working slope face	非工作帮坡面
non-working wall	非工作帮
nonwoven geotextile	无纺土工织物
non-woven geotextile	无纺土工布
nonwoven heat-bonded geotextile	无纺热塑土工织物
nonyielding	不屈服的
nonyielding arch	刚性拱；不可缩性拱
nonyielding prop	刚性支架
non-yielding retaining wall	不变形挡土墙；刚性挡土墙；不移动挡土墙
nonzero	非零的
nonzero amplitude	非零振幅
non-zero contact angle	非零接触角
nonzero digit	非零位
nonzero element	非零元素
non-zero plasma temperature	非零等离子体温度
non-zero spin quantum number	非零自旋量子数
non-zero temperature	非零温度
nonzero value	非零值
nonzero-lag	非零延迟
nonzero-offset	非零炮检距
non-zinc system	无锌体系
nook	工作面边角；煤柱的一角
no-operation	无操作指令
no-operation instruction	空指令；无作业指令；自操作指令；空操作指令
no-overtopping condition	未溢水条件；未越波条件
no-radius	零半径
Noranda method	诺兰达分段深炮眼开采法；诺兰达法
norbergite	块硅镁石
nordite	硅钠锶镧石
nordmarkite	英碱正长岩；锰十字石
no-return	无回波
noricite	绿闪石片岩
norite	苏长岩
norite ore	苏长岩矿石
norite plagioclasite	苏长斜长岩
norm	范数；规范；标准；均值
norm error	模误差
norm of material consumption	材料消耗定额
norm of output	产量定额
normal	正常的；标准的；正态的；垂直的；正交的；法线的
normal abundance	正常丰度
normal acceleration	法向加速度
normal acceleration of gravity	标准重力加速度
normal air	常态空气
normal alkane	正烷烃
normal analysis	正序分析
normal angle	法线角
normal annual runoff	正常年径流

normal anomaly　正异常
normal anticlinorium　正常复背斜
normal approximation　正态逼近；正态近似
normal assumption　正态性假设
normal atmosphere　标准大气压；常压
normal atmospheric pressure　标准大气压；常压
normal atmospheric temperature　常温
normal atom　常态原子
normal axis　垂直轴线；法线轴线
normal barometer　标准气压表
normal beam　垂直光束
normal binding force　法向约束力
normal butane　正丁烷
normal butyl alcohol　正丁醇
normal carbon chain　正碳链
normal case photography　正直摄影
normal chain　正链
normal charge　标准药包；正常装药；标准装药
normal circulation　正循环；正常循环
normal close valve　常关阀；常闭阀
normal closed contact　常闭触点
normal coldest month period　累年最冷月
normal color film　标准彩色胶片
normal color photograph　标准彩色相片
normal combustion　完全燃烧；正常燃烧
normal compaction　正常压实作用
normal component　法向分量；垂直分量
normal component of strain　法向应变分量
normal component of stress　法向应力分量
normal compound　正构化合物
normal compound cone　正常型混合火山锥
normal compressive stress　法向压应力
normal concentration　标准浓度；当量浓度；规定浓度
normal concrete　普通混凝土
normal condition　正常状态；正常条件
normal consolidated soil　正常固结土
normal consolidation　正常固结
normal consolidation line　正常固结线
normal consolidation state　正常固结状态
normal contact　定位接点
normal contour　标准等高线
normal coordinate　正则坐标；法向坐标；正交坐标
normal correction　正常校正
normal correlation　正态相关
normal correlation function　正态相关函数
normal counterforce　法向反力
normal cross section　正剖面；横截面
normal cubic meter　标准立方米
normal curvature　标准曲率
normal curve　正态曲线；正常曲线
normal cycle　正常旋回
normal cylindrical projection　正圆柱投影
normal decline fraction　正常递减率
normal deflection of the vertical　标准垂线偏差
normal deformation of joint　节理的法向变形
normal density　正态密度
normal depth　正常深度
normal derivative　法向导数
normal device　电位电极系
normal dimension　标准尺寸

normal dip　正常倾斜；正倾斜
normal dip separation　倾向正断距
normal dip slip faulting stress regime　正走滑型应力机制
normal dip-slip fault　倾滑正断层
normal direction　法向；法线方向
normal direction cosine　法向余弦
normal direction cosine for plane　平面法向余弦
normal discharge　正常流量
normal dispersion　正常频散；正常波散；正常色散
normal displacement　法向位移；倾向移距；正常位移
normal distribution　正态分布；对称分布
normal distribution curve　正态分布曲线
normal domestic sewage　一般家庭污水；正常浓度家庭污水
normal drag　正牵引
normal earthquake　正常深度地震；浅震；浅源地震
normal effect　正常效应
normal electric field　正常电场
normal electrode　标准电极
normal electromagnetic field　正常电磁场
normal elevation　正常高程；正常水位
normal epoch　正向期；正常期
normal equation　标准方程；正规方程
normal erosion　常态侵蚀；自然侵蚀；正常侵蚀
normal erosion cycle　正常侵蚀旋回
normal fan-shaped fold　正常扇形褶皱
normal fault　正断层
normal fault scarp　正断层崖
normal faulting　正断层活动；正断层作用
normal faulting stress regime　正断层型应力机制
normal field　正常场；正常磁场
normal five-point pattern　正五点井网
normal flood plain terrace　堆积阶地；正常漫滩阶地
normal flow　正常水流；正常径流；正常流量
normal flow year　平水年
normal fluid　标准液体
normal fluid pressure　标准流体压力
normal fluid static pressure　正常流体静压力
normal flux　法向流
normal fold　正褶皱；正褶曲；对称褶皱
normal force　法向力；正交力
normal force base of a slice　条块底面的法向力
normal form　标准形式；正则；范式；规格化形式
normal formation pressure　正常地层压力
normal frequency curve　正态频率曲线
normal frequency distribution　正态频率分布
normal function　正态函数
normal gauge　标准轨距
normal geomagnetic polarity epoch　正磁极期
normal gnomonic projection　正轴球心投影
normal gradient　正常梯度
normal gradient of geomagnetic field　地磁场正常梯度
normal grading　正递变
normal grain growth　正常晶粒生长
normal granite　正常花岗岩；石英二长岩
normal graph paper　标准坐标纸
normal gravel concentration　常规砾石含量
normal gravitation potential　正常引力位
normal gravitational potential　正常引力位
normal gravity　正常重力；标准重力

normal gravity correction　正常重力修正；正常场修正；纬度修正
normal gravity fault　正断层
normal gravity field　正常重力场
normal gravity formula　正常重力公式
normal gravity line　正常重力线
normal gravity potential　正常重力位
normal head　正常水头
normal height　正常高程
normal heptane　正庚烷
normal hexane　正己烷
normal high-water level　正常高水位
normal homologue　正同系物
normal horse power　额定马力
normal hottest month period　累年最热月
normal hour　标准工时
normal hydrocarbon　直链烃
normal hydrostatic pressure　正常静水压力
normal hysteresis loop　正常磁滞回线
normal incidence　法线入射；垂直入射
normal incidence absorption coefficient　法向入射吸声系数
normal incidence point　垂直入射点
normal inspection　正常检查；常规检查；一般检验
normal integral　正规积分
normal isopleth　平均等值线
normal law of error　正态误差定律
normal level ellipsoid　正常水准椭球；水准椭球
normal lighting　正常照明
normal line　法线；正态线
normal load　法向荷载；垂直荷载
normal loading　正常加载；法向加载
normal loading condition　正常荷载条件
normal log　电位电极系测井
normal logarithmic distribution　对数正态分布
normal loss　正常损耗
normal magnetic field　正常磁场
normal magnetization curve　正常磁化曲线
normal magnification　正常放大倍数
normal maintenance　日常维修
normal make-up　常规补水
normal maximum level　正常高水位
normal metamorphism　区域变质作用；正常变质作用
normal metamorphism of coal　煤正常变质
normal mode　简正振型；标准振型；正规模态
normal mode analysis　标准振型分析
normal mode approach　标准振型法
normal mode equation　标准振型方程
normal mode filter　正常型滤波器
normal mode initialization　正规模态初值化
normal mode of vibration　标准振型
normal mode theory　简正波理论
normal module　法向模数
normal moisture capacity　正常湿度
normal move out remover　动校正器；正常时差校正器
normal movement　正常移动，正常位移
normal moveout　正常时差
normal moveout correction　动校正；正常时差校正
normal moveout remover　动校正器
normal moveout velocity　正常时差速度
normal negative　标准底片

normal oblique slip fault　斜滑正断层
normal octane　正辛烷
normal oil level　正常油位
normal open valve　常开阀
normal operating condition　正常操作条件
normal operating expense　正常运转费
normal operating period　正常运转时间；正常工作时期间
normal operating state　正常操作状态
normal operation　正常运行
normal orbit　正常轨道
normal order　正常层序
normal output　正常产量；标准作业量
normal panning　井壁法向衬片
normal paraffin　正链烷烃；正构石蜡烃
normal paraffin hydrocarbon　正链烷烃
normal paraffin isomerization　正构烷烃异构比
normal paraffin maturity index　正烷烃成熟指数
normal pentane　正戊烷
normal pH range　正常酸碱值幅度
normal photography　正直摄影；标准摄影术
normal pitch　标准距；法向节距；法线螺距
normal plane　法平面；法向面
normal point load　法向集中荷载
normal polarity　正向极性
normal pool level　正常蓄水位；正常水位
normal pore canal　垂直毛细管
normal pore pressure　正常孔隙压力
normal Portland cement　普通波特兰水泥；普通硅酸盐水泥
normal position　正常顺序；原位；正常层位
normal potential　正常位势
normal precipitation　正常降水量
normal pressure　正常压力；法向压力
normal pressure and temperature　标准压力和温度
normal pressure gradient　正常压力梯度
normal probability curve　正态概率曲线；正态分布曲线
normal probability distribution　正态概率分布
normal probability function　正态概率函数
normal probability paper　正态概率纸
normal probe　直探头
normal problem　正问题
normal procedure　常规方法；正常手续
normal process　正态过程
normal profile　正常剖面；法向剖面
normal projection　正轴投影；正投影；垂直投影
normal random　正态随机数
normal random process　正态随机过程
normal random variable　正态随机变量
normal range　正常范围
normal rated power　额定功率
normal rated thrust　额定推力；标准推力
normal rating　正常定额；正常功率额定值；正常推力额定值
normal ratio　正常场比值
normal ray　正常射线；法向射线
normal reflection　正反射；垂直反射
normal resistivity device　电位电极系
normal resistivity tool　电位电极系
normal response　正常响应

normal response mode 正常响应模式
normal revolution 正常转数
normal ripple 正常波痕
normal risk 正常风险
normal rock pressure 正常岩石压力；覆盖岩层压力
normal runoff 标准径流；多年平均径流
normal salinity range 正常盐度值；正常盐度幅度；正常盐度范围
normal sampling inspection 正常抽样检查；正规抽样检验；正态样本检查
normal sand 标准砂
normal sea water 标准海水
normal section 法截面；正剖面；正截面
normal segregation 正偏析；标准偏析
normal sequence 标准剖面；标准层序
normal seven-spot 正七点
normal shear fracture 正剪切断裂
normal shear structure system 直扭构造体系
normal shift 垂直移距；法线移距；正变位；法向位移
normal shock wave 正激波
normal shrinkage 常态收缩；正常损耗，正常缺损
normal size 正常尺寸
normal slip fault 正断层；正滑断层
normal slope 正常坡度
normal slurry 普通水泥浆
normal soil 正常土壤
normal solution 规度溶液；当量溶液；正解
normal sonde 电位电极系
normal speed 正常速度；标准速度
normal standard cost 正常标准成本
normal state 标准状态；正常状态
normal stiffness 法向刚度
normal stiffness of discontinuity 不连续面法向刚度
normal stochastic process 正态随机过程
normal storage water level 正常蓄水位线
normal strain 法向应变；正应变
normal strain component 法向应变分量
normal stratigraphic sequence 正常层序
normal stress 正应力；垂直应力；法向应力
normal stress component 法向应力分量
normal strike slip fault 正-平移断层；走滑断层
normal substitution 正常取代
normal succession 正常层序
normal succession of strata 正常层序
normal superposition 正常层序
normal support 标准支承
normal surface 法面；法向曲面
normal swelling 正常膨胀
normal switch indication 道岔定位表示
normal tail water level 正常尾水位
normal temperature 常温
normal temperature and pressure 常温常压
normal temperature zone 常温层
normal test 常规试验
normal thermometer 标准温度计
normal throw 垂直断距；正落差
normal time 标准时间；正常时间
normal to a plane 平面的法线
normal to the stratification 层理面法向；垂直于层理面
normal transformation 正交变换
normal travel time 正常走时
normal traveltime curve 正常旅行时曲线；正常走时曲线
normal twin 正交双晶；垂直双晶
normal uranium 标准铀
normal value 正常值；标准值
normal value of bearing capacity 承载力标准值
normal varimax criterion 正规方差极大准则
normal vector 法向量；法向向量；法向矢量
normal velocity 法向速度
normal velocity of particle 颗粒法向速度
normal vibration 简正振动
normal viscosity 标准黏度；正常黏度
normal vitrinite 正常镜质体
normal volume 正常体积
normal water flow 稳定涌水
normal water level 正常水位；常水位
normal water yield 正常涌水量
normal wave base 正常浪底
normal well pattern 标准井网；标准布井
normal wet combustion 常规湿式地层燃烧法
normal working condition 正常工作条件
normal worn out 正常磨损
normal year 平水年
normal zero 标准零点
normal zoning 顺向分带
normal-angle aerial camera 常角航摄仪
normal-aspect map projection 正轴投影
normal-bedded 正常成层的
normal-circulation rotary drilling 正循环回转钻进；正循环旋转钻进
normal-focus earthquake 正常震源地震
normal-incidence reflection 垂直入射的反射
normal-incidence traces 垂直入射道集
normalisation 归一化
normalised arch thickness 归一化拱厚度
normalised deflection 归一化挠度
normalised room span 归一化矿房跨度
normalised saturation 归一化饱和度
normalised span 归一化跨度
normality assumption 正态假设
normality condition 正交条件
normality law 正交定律；正态分布律
normality principle 正交原理
normalization 归一化；标准化；规格化；规范化
normalization criterion 归一化标准；归一化准则
normalization factor 标准化因子；归一化因子
normalization method 标准化方法；归一化方法
normalization of data base 数据库规范化
normalize 标准化；归一化；规格化；规范化
normalized 标准化的；归一化的；规格化的；规范化的
normalized amplitude 归一化振幅
normalized apparent resistivity 归一化视电阻率
normalized autocorrelation 归一化自相关
normalized autocorrelation function 标准化自相关函数
normalized condition 归一化条件
normalized coordinate 归一化坐标
normalized correlation coefficient 归一化相关系数
normalized covariance 标准协方差
normalized crosscorrelation 归一化互相关
normalized curve 标准化曲线

northern Atlantic oscillation 北大西洋涛动
northern branch jet stream 北支急流
northern hemisphere 北半球
northern hemisphere reference line 北半球参考线
northern latitude 北纬
northern light 北极光
northern Pacific oscillation 北太平洋涛动
northern margin of South China Sea 中国南海北缘
northern polar cap 北极盖
northern Ryukyu trench 琉球海沟北部
northernmost 极北的
northern-type coal with southern-lithofacies 北型南相煤田
northfieldite 石英伟晶岩
north-finding instrument 寻北器
northing 北向; 北向坐标差; 北偏
north-seeking gyro tool 寻北陀螺仪
north-seeking instrument 正北指示器
north-seeking pole 指北极
north-south asymmetry 南北不对称
north-south structural zone 南北向构造带
northupite 菱碳钠镁石
northward 向北方; 向北的
northwest corner method 西北角法
northwest Pacific seamount 西北太平洋海山
northwester 西北风
northwestward 向西北
Norton tube well 诺顿管井
Norwegian clay 挪威黏土
Norwegian Geotechnical Institute 挪威岩土工程研究所
Norwegian Sea deep water 挪威海深层水
no-sand cushion drainage system 无砂垫层排水系统
nose 鼻状构造; 鞍形褶; 山嘴; 凸尖
nose advance 鞍形推进
nose cap 鼻罩; 锥形帽
nose down 机俯
nose drum 毛辊滚筒; 咽滚筒
nose filter 头锥式滤器
nose jack 机头起重器; 机头千斤顶
nose key 鞍头楔
nose mudflow 鼻状泥流
nose of cone 孔锥前端
nose of fold 褶皱鼻
nose pile 定位桩; 鞍桩
nose plier 嘴形钳
nose pollution 噪声污染
nose pressure 头锥压力
nose radius 顶部半径
nose rail 心轨
nose structure 鼻形结构; 鼻状构造
nose wheel 前轮
nosean 黝方石
nosean basanite 黝方石玄武岩
noseanite 黝方岩
noseanolite 黝方岩
nose-like structure 鼻状构造
noselite 黝方石
noselitite 含黝方石粗面岩
nose-out 鼻状露头
nose-type fold 鼻型褶皱; 非斜淹没式褶皱
nosh 易碎的; 松软的; 粉状的

no-shrink condition 不收缩状态
nosing structure 鼻状构造
no-slip drive 不打滑传动
no-slip flow 不滑流动
no-slump concrete 不坍落度混凝土
no-stop dynamometer 不停机功力计
nosykombite 黑云二长岩
not applicable 不适用的; 不适用; 不可
not available 不能得到的; 不能用的; 不可
not driven end 非驱动端
not due to mining 不因开采的; 不因采矿; 而
not immediately supported roof area 未能立即支护的顶板面积
not inspected and repaired 检修不良
not well maintained 维修不良; 保养差的
notable 值得注意的; 显著的; 著名的
notandum 记录事项
notar 公证人
notarial 公证的
notarial act 公证手续
notarial certificate 公证证书
notarial deed 公证证书
notarial document 公证文件
notarial fee 公证费
notarization 公证
notarize 以公证人资格证实; 公证
notary 公证人
notary organ 公证机关
notary public 公证人
notation 标志; 符号; 注释; 记号; 记法
not-cap-sensitive explosive 非雷管引爆炸药
notch 凹槽; 切口; 凹痕; 隘口; 山口; 缺口
notch base 鞍口底部
notch bend test 切口弯曲试验
notch brittleness 切口脆性; 切痕脆性
notch chamfer 刻槽
notch dam 切口溢流坝
notch drop 切口跌水
notch ductility 切口延性; 切口延展性
notch effect 切痕效应; 切口效应; 刻槽效应
notch fatigue factor 切口疲劳系数
notch filler 脱浆装置; 刻痕装置
notch impact strength 切口冲击强度; 缺口冲击强度
notch impact toughness 切口冲击韧性; 冲击韧度
notch indicator 缺隙指示器
notch nog 开槽木桩
notch plate V形开口黑漆间板
notch preparation 切削加工; 切削预备
notch root 切口根部; 缺口底部
notch sensitive 切口敏感的
notch sensitivity 切口敏感度; 缺口敏感度
notch sensitivity test 切口敏感度试验
notch shock test 切口冲击试验
notch specimen 切口试样
notch strength 切口强度
notch stress 切口应力
notch toughness 切口韧性; 切口韧度
notch wedge impact 缺口楔形冲击试验

normalized distribution	正规化分布
normalized drawdown pressure gradient	标准化水压力梯度
normalized echo intensity	归一化回波强度
normalized equation	正规方程；归一化方程
normalized factor	归一化因子
normalized force	标准化作用力；归一化作用力
normalized format	归一化格式
normalized frequency	归一化频率
normalized histogram	归一化直方图
normalized impedance	归一化阻抗
normalized matrix	正规化矩阵
normalized mean square error	归范均方误差
normalized mode	标准化振型；正规模正
normalized number	规格化数
normalized offset	归一化偏移；标准偏差
normalized ordinate	标准化纵坐标
normalized penetration rate	标准化钻速
normalized percent frequency effect	归一化百分频率效应
normalized random number	规格化随机数
normalized rate of penetration	标准化钻速
normalized resistance	归格电阻
normalized response spectrum	标准反应谱；标准响应谱
normalized saturation	标准饱和
normalized steel	正火钢；淬火钢
normalized total gradient method	归一化总梯度法
normalized total gravity gradient	归一化重力总梯度
normalized unit	正规化单位；标准化单位
normalized value	正规化数值；归一化值
normalized variate	正规化变量
normalized velocity	归一化速度
normalized wavelet	归一化子波
normalizer	正规化子
normalizing	正火化；正常化
normalizing condition	正规条件
normalizing factor	归一化因素；标准化因素
normalizing mode shape	归一化振型
normalizing parameter	标准化参数
normalizing quantity	归一化的量
normalizing treatment	正火处理
normally closed	常闭的
normally closed contact	常闭接点，常闭触点
normally consolidated	正常压紧的；正常压实的
normally consolidated soil	正常固结土
normally graded	正常分级
normally loaded soil	正常固结土
normally open contact	常开接点
normally open push button	常开按钮
normally opened contact	常开接点
normally pressured aquifer system	正常压力承压水系统
normally-compacted	正常压实的；正常压实的
normally-pressured neighbor	正常压力邻圈
normal-moveout correction	正常时差校正
normal-moveout spectrum	正常时差谱
normal-polarity interval	正常极性间隔
normal-pressure surface	正压面
normal-section line	法截线
normal-separation fault	倾滑正断层
normal-slip fault	正滑断层；正移断层
normal-stage punch	正规冲孔
normal-strike slip fault	正-平移断层；正平移动层
normal-to-bed depth	垂直层面深度
normative	标准的；规范的
normative basalt tetrahedron	标准玄武岩四面体
normative classification	标准分类
normative decision theory	规范决策论
normative forecasting technique	规范预测法
normative mineral	标准矿物
no-rope hoisting equipment	无绳提升设备
Norse miner	塔赫蒂婆米矿
norsethite	菱钡镁石
North America plate	北美板块
North American plate	北美板块
North American Shale Composite	北美页岩综合样体
North American Society for Trenchless Technology	北美洲非开挖技术学会
North American tektite	北美撒散陨石片
north Anatolian fault	北安纳托利亚断层
north arrow	指北针
North Asian Craton	北亚细亚地台
North Atlantic	北大西洋
North Atlantic deep water	北大西洋深层水
North Atlantic massif	北大西洋海底山
North Atlantic oscillation	北大西洋涛动
North Atlantic tracks	北大西洋航线
North Atlantic Treaty Organization	北大西洋公约组织
North Atlantic warm current	北大西洋暖流
north celestial pole	天北极
North China land	华北陆地
North China oldland	华北古陆
North China platform	华北地台
North Dakota cone test	北达科他州圆锥贯入试验
North East New Territories Landfill	新界东北堆填区
north elevation	北立面图
north equator warm current	北赤道暖流
north frigid zone	北寒带
north geographic pole	地理北极
north latitude	北纬
north light	北极光
north magnetic pole	磁北极
North New Hebrides Trench	北新赫布里底海沟；托里斯沟
North Pole	北极
north reference pulse	指北参考脉冲
North Sea	北海
north slope of South China Sea	中国南海北坡
north star	北极星
north temperate zone	北温带
north Tethyan orogenic zone	北特提斯造山带
north tropic	北回归线
Northeast Asian orogenic zone	东北亚造山带
northeast Atlantic	大西洋东北部
Northeast China low	东北低压
northeast pacific ridge	东北太平洋海岭
northeast trending fault	北东走向断层
northeaster	东北风
Northeast-north China tectonic stress region	东北-华北构造应力区
northern	北的；北方的
northern Adriatic Sea	亚得里亚海西北部
northern Alpine foreland	北阿尔卑斯前陆区

notch width 切口宽度；切槽宽度
notched 齿状的；锯齿形的；有刻痕的；有槽口的；有凹口的
notched bar impact test 凹口冲击试验；切口冲击试验
notched bar toughness 切口韧性
notched beam 开槽梁
notched brick 舌槽砖；开槽砖
notched safety fuse 有切痕的安全导火线
notched sill effect 齿坎效应
notched weir 开口堰
notcher 开槽器
notching 台阶开掘法；开缺口，开槽；做凹口法
notching blasting 切槽爆破
notching curve 阶梯曲线
notching joint 凹槽节
note 票据；借据；通知；便条；记录
note amplifier 音频放大器
note filter 声频滤波器
note frequency 音频；低频
note of hand 借据；期票
note on demand 即期票据
note on discount 贴现票据
note payable 应付票据
note receivable 应收票据
note to bearer 不记名票据
note to order 记名支票
notedly 显著地
notekeeper 记录员
notekeeping 记录
noteless 不引人注意的
notes for tenderer 投标须知
notes on drawing 图注
notes payable 应付票据
notes receivable 应收票据
noteworthiness 值得注意；显著
noteworthy 显著的
not-go side 不通过端；上端
notice 通知；布告；注意；简介
notice board 布告牌
notice by fax 传真通知
notice in writing 书面通知
notice of abandonment 委付通知；委弃通知
notice of acceptance 承兑通知；验收通知单
notice of alternation in rent by agreement 协议更改租金通知书
notice of assessment for property tax 物业税评税通知书
notice of authorization 授权公告
notice of award 签订合同通知书；中标通知
notice of cancellation 撤销通知书
notice of cancellation of vesting 撤销转归公告
notice of claim 索赔通知书
notice of decision 决定通知书
notice of default 违约通知
notice of defect 缺陷通知
notice of evacuation 令迁出通知书
notice of eviction 收回房舍通知书
notice of export 出口申请
notice of import 进口申请
notice of importation 进口通知
notice of increase in rent 加租通知书

notice of intention to amend valuation list 拟修正估价册记录通知书
notice of loss 损失通知
notice of objection 反对通知书
notice of payment 通知偿还
notice of protest 拒绝承付通知书
notice of removal of projections or objection 饬令拆除突出建造物或障碍物通知书
notice of resumption 收地通告
notice of tax payment 缴税通知书
notice of termination 终止租赁通知书
notice of termination of tenancy by landlord 业主终止租约通知书
notice of test 检验通知
notice of the alteration in rent 更改租金通知书
notice of transfer 转让通告
notice of unsafe building 危险房屋公告
notice of vacancy 物业空置通知书
notice of valuation 估价通知书
notice period 通知期限
notice plate 标记牌
notice to commence 开工通知
notice to mariner 航海通告
notice to navigator 航行通告
notice to proceed 施工通知
notice to quit 迁出通知书
noticeable 显著的
notifiable 应报告的；须申报的
notifiable work 应呈报工程
notification 布告；通知；通知书
notification number 公告号
notification of approval 批准通知
notification of arrival 到货通知
notification of award 签订合同通知书
notification of shipment 装船通知
notification of the contract 签订合同通知书
notification of transboundary movement of waste 废物越境移运的公告
notify 通知；宣告
notifying bank 通知银行
noting 注释法；计算法
notion 概念；见解；想法
notional building cost 估计建筑成本
notional price 假设价格；理论价格
notional sound position 象征性声源位置
not-saturated 不饱和的
not-so-minor detail 不可忽视的细节
Nottingham longwall method 诺丁汉长壁开采法
nought 无；零
noughts complement 补码；零补数；基补数
noumeite 硅镁镍矿
nourishment condition 补给条件
nourishment source 补给水源
nova 新星
novacekite 镁砷铀云母；水砷镁铀矿
novaculite 致密石英岩；微晶石英质燧石岩
novakite 砷铜银矿
novalac 酚醛树脂
novalac epoxy 酚醛环氧树脂
novel drilling method 新钻井方法

novel procedure 新颖工艺规程
novelty 新奇性；新颖
novenary 九进制的
no-voltage protection 欠压保护；失压保护；无压保护
nowadays 现今
no-wall-stick 防黏附卡钻
nowcast 临近预报
now-quiescent area 目前地震平静区
noxious 有害的；有毒的
noxious colouring matter 有害色素
noxious emission 有害排放物
noxious fume 有毒炮烟
noxious gas 有害气体；有毒气体
noxious industry 有害工业
noxious liquid substance 有害液体
noxious odour 有害气味
noxious pollutant 有害污染物
noxious substance 有害物质；毒素
noxious waste pollution 固体废弃物污染
noxiousness 有毒
nozzle bushing 喷射管；喷嘴
nozzle cap 喷口盖；出水口帽盖
nozzle casing 喷嘴罩
nozzle coefficient 管嘴系数
nozzle control 喷嘴控制
nozzle drier 热风喷嘴烘燥机
nozzle effect 喷嘴效应
nozzle efficiency 水眼效率
nozzle expansion area ratio 喷嘴扩散面积比
nozzle filter 喷嘴过滤器；喷头滤网
nozzle flapper valve 喷嘴挡板阀
nozzle needle valve 喷油嘴针阀
nozzle needle valve body 喷油嘴针阀体
nozzle orifice 喷嘴型小孔；喷嘴孔口；喷口
nozzle orifice coefficient 喷嘴流量系数
nozzle placement 喷嘴布置
nozzle pressure 喷嘴压力
nozzle ring 喷嘴环
nozzle size 喷嘴尺寸
nozzle tip 喷嘴尖
nozzle tube 喷管
nozzle valve 喷嘴阀
nozzle velocity 喷嘴速度
nozzle with iso-variable 等变速喷嘴
nozzle-axis 喷嘴轴线
nozzle-end 喷嘴端部
nozzle-fluid velocity 流体喷射速度
nozzleman 水枪工；射水机工
nozzle-plate 喷嘴板
nozzle-type water wheel 喷嘴式水轮
n-paraffin hydrocarbon 正烷烃
n-path filter N通道滤波器
NSF 国家科学基金
NSFC 国家自然科学基金
n-th differential n阶微分
n-th extension information source n次扩展信源
n-th harmonic n次谐波
n-th order n阶
n-trace seismograph n道地震仪
N-truss N形构架；N形桁架

n-tuple n元组
n-tuple integral n重积分
n-tupling n倍
N-type material N型材料
N-type semiconductor N型半导体
nub 结块
nubbin 基岩残丘；山脉残脊
nubble 小块；小丘岛
nubbly 小块状的；瘤状的
nucle 核
nuclear 核的；原子核的；核心的
nuclear activation analysis 中子活化分析
nuclear age determination 核年代测定法；放射性年代测定法
nuclear area 陆核；核区
nuclear astrophysics 核天体物理学
nuclear back scatter 核子背散射技术
nuclear basin 陆核盆地
nuclear belt weighing 胶带输送机核称量法
nuclear binding energy 核结合能
nuclear biogeochemical survey 生物核地球化学测量
nuclear biological survey 生物核测量
nuclear blast 核爆炸
nuclear blasting 核爆炸
nuclear blasting and dredging method 核爆破挖掘法
nuclear bomb 核弹
nuclear capture 核俘获
nuclear cement log 核水泥测井
nuclear charge 核电荷
nuclear chemistry 核化学
nuclear chimney for underground gasification 地下核爆炸形成的垂直筒形碎裂带
nuclear column 砥柱
nuclear corrosion 核腐蚀
nuclear counter 核计数器
nuclear data file 核数据文件
nuclear dating methods 核测年方法
nuclear decay 核衰变
nuclear densimeter 核密度计
nuclear density meter 核子密度计
nuclear densometer 同位素密度探测仪；核子密度仪
nuclear densometer method 核密度计法
nuclear detection 核探测
nuclear detection array 核爆炸检测台阵
nuclear detection element 核探测元件
nuclear detonation 核爆炸
nuclear device 核装置
nuclear disintegration 核衰变；核分裂
nuclear energy 核能
nuclear energy level 核能级
nuclear engineering 核工程
nuclear equation 核方程
nuclear exploration 核勘查
nuclear explosion 核爆炸
nuclear explosion seismology 核爆炸地震学
nuclear explosives 核炸药
nuclear facility 核设施
nuclear fission 核裂变
nuclear flolog 核流量测井
nuclear force 核力

nuclear fracturing 核爆炸震裂
nuclear fuel 核燃料
nuclear fuel waste site 核燃废料场
nuclear fusion 核聚度；核聚变；核子融合
nuclear gamma ray resonance 核伽马射线共振
nuclear gaseous survey 气体核测量
nuclear gauge 核子测定器
nuclear geochemical standard sample 核地球化学标样
nuclear geochemistry 核地球化学
nuclear geology 核地质学
nuclear geophysical exploration 核地球物理勘查
nuclear geophysics 核地球物理
nuclear hazard 核伤害
nuclear installation 核装置
nuclear instrument calibration 核仪器标定；核仪器校准
nuclear interaction 核相互作用
nuclear land 地盾
nuclear lifetime 核寿命
nuclear log 核测井；放射性测井
nuclear logging 核测井；放射性测井
nuclear magnetic alignment 核磁排列
nuclear magnetic double resonance 核磁双共振
nuclear magnetic log 核磁测井
nuclear magnetic logging 核磁测井
nuclear magnetic moment 核磁矩
nuclear magnetic relaxation 核磁弛豫
nuclear magnetic resonance 核磁共振
nuclear magnetic resonance absorption 核磁共振吸收
nuclear magnetic resonance imaging 核磁共振成像
nuclear magnetic resonance imaging logging 核磁共振成像测井
nuclear magnetic resonance log 核磁共振测井
nuclear magnetic resonance method 核磁共振法
nuclear magnetic resonance spectrometer 核磁共振波谱仪
nuclear magnetic resonance spectrometry 核磁共振波谱法
nuclear magnetic-resonance detector 核磁共振检波器
nuclear magnetisation 核激磁
nuclear magnetism log 核磁测井
nuclear magnetometer 核子磁力仪；核地磁仪
nuclear material 核材料
nuclear measurement 核子放射量测
nuclear metallurgy 核燃料冶金；核冶金学
nuclear microprobe 核子微探针
nuclear mining 核采矿
nuclear model 核模型
nuclear orientor 核定位器
nuclear paramagnetic resonance 核顺磁共振
nuclear paramagnetism 核顺磁性
nuclear physics 核物理学
nuclear plant 核工厂；核电厂；核电站
nuclear pollution 核子污染；核污染
nuclear power 核电
nuclear power generating station 核电站
nuclear power package 原子能发电装置
nuclear power plant 核电厂；原子能发电站；原子能发电厂
nuclear power plant equipment 核电厂设备
nuclear power plant reactor building 核电厂反应堆建筑物
nuclear power source 核动力
nuclear power station 核电站
nuclear powered ship 核动力船
nuclear precession magnetometer 核磁旋进磁力仪
nuclear pressure vessel 核压力容器
nuclear prospecting 核法勘探
nuclear pulse amplifier 核子脉冲放大器
nuclear radiation 核辐射
nuclear reaction 核反应
nuclear reaction analysis 核反应分析
nuclear reactor 核反应堆
nuclear reactor core 核反应堆芯
nuclear reactor design 核反应堆设计
nuclear reactor facility 核反应堆设备
nuclear reactor safety 核反应堆安全
nuclear reactor site 核反应堆场地
nuclear reactor structure 核反应堆结构
nuclear relaxation time 核弛豫时间
nuclear resonance log 核共振测井
nuclear resonance magnetometer 核子共振磁力仪
nuclear safety criterion 核安全准则
nuclear scale 核标度
nuclear scattering 核散射
nuclear science 核子学
nuclear sediment density meter 核子沉积物比重计
nuclear separation 核分离
nuclear series 放射系
nuclear soil moisture meter 同位素含水量仪；核子土壤湿度计
nuclear soil survey 土壤核测量
nuclear specific gravity indicator 核比重指示器；放射性比重指示器
nuclear spin 核自旋
nuclear stability 核稳定性
nuclear statistical equilibrium theory 核统计平衡理论
nuclear steam supply system 核蒸汽供应系统
nuclear steelmaking 核炼钢
nuclear stimulation 核爆炸增产措施
nuclear structure 核结构
nuclear target 核靶
nuclear transition 核跃迁
nuclear waste 核废料
nuclear waste disposal 核废料处理
nuclear waste repository 核废料库；核废料处置库
nuclear waste storage facility 核废料贮存设备
nuclear-aided gasification 核反应堆余热气化
nuclear-binding energy 核结合能
nuclear-explosive mining 核爆采矿
nuclear-fracturing 核爆炸压裂
nuclear-precession magnetometer 核子旋进磁力仪；质子旋进磁力仪
nuclear-spin density 核自旋密度
nucleating 核化
nucleating agent 晶核试剂；成核剂
nucleating centre 成核中心
nucleating effect 致核效应
nucleation 成核作用；核化；晶核作用
nucleation and evolution 成核与演化
nucleation energy 成核能
nucleation of folds 褶皱成核作用
nucleation rate 成核速度

nuclein 核素
nucleo 核
nucleogenesis 核起源；成核作用
nucleometer 核子计
nucleon 核子
nucleon number 核子数
nucleonic 核子的
nucleonic density control 核子密度控制法；放射性密度控制法
nucleonic density gauge 核密度计
nucleonic equipment 核设备
nucleonic guidance equipment 核子制导设备
nucleonic level indicator 核子水平指示器
nucleonic probe 核探测器
nucleonic sensing device 核子传感器
nucleonic sensor 核子传感器
nucleonic steering 核子调向
nucleonicks 原子核物理学；核子学
nucleonics 核子学
nucleophile 亲核物质
nucleophilic 亲核的；亲质子的
nucleophilic promotor 亲核性促进剂
nucleophilic reactivity 亲核反应性
nucleophilicity scale 亲核标度
nucleoplasm 核质；核浆
nucleoprotein 核蛋白
nucleor 核子心
nucleostratigraphy 核地层学；核子地层学
nucleosynthesis 核合成；核聚合
nucleotide 核苷酸
nucleus 核；核心；原子核；晶核；地核
nucleus flaw 核伤
nucleus magnetic resonance 核磁共振
nucleus number 晶核数
nucleus of continent 陆核；大陆核
nucleus of crystal 晶核
nucleus of crystallization 晶核
nucleus of the Earth 地核
nuclide 核素
nuclide abundance 核素丰度
nuclide chart 同位素表
nuclide chronology 核素年代学
nuclide cosmic abundance 核素宇宙丰度
nuclide solar system abundance 核素太阳系丰度
nuevite 铌钇矿
nuffieldite 硫铋铜铅矿
Nugget 天然金块；矿块
nugget effect 金块效应
nugget of gold 天然金块
nuggeting 寻找天然金块
nuisance 有害物；污害
nuisance analysis 公害分析
nuisance effect 公害效应
nuisance parameter 多余参数；多余参量
nuisance vibration 有害振动
nuke 核武器；核动力发电站
null 零；零的；无效的；无价值的
null and void 无效的
null array 零数组
null balance 零点平衡
null bias 零偏压
null character 零字符
null circle 点圆；零圆
null correlation 零相关
null cycle 零周期
null detector 消尽指示器；零值指示器；零值检波器
null direction 零向
null displacement 零位移
null drift 零点位移；零点漂移
null ephemeris table 零值星历表
null event 零概率事件
null indicator 零位指示器；零值指示器
null information 无信息
null instrument 平衡点测定器
null line 中立线；零线
null manifold 零流形
null matrix 零矩阵
null method 零示法
null offset 零点偏移
null point 零点
null position 零位；零点位置
null reading 零位读数
null reading instrument 零位读数仪表
null reading type magnetometer 零点读数式磁力仪
null set 空集
null setting 调零装置；调零
null shift 零点漂移；零漂
null state 零状态
null string 空行
null test 零位试验
null valence 零价
null vector 零向量
null-adjusting spring 零位调节弹簧
nullah 峡谷；沟壑；干涸的河床；间歇河床；防洪渠；明渠
nullah bridge 峡谷桥；明渠大桥；防洪渠桥
nullah bund road 防洪渠旁道路
nullah decking 加建明渠上盖
nullification 无效
nullification of exploration right 探矿权注销
nullify 使无效；废弃；取消
nulliplanomural 无平壁的
nullity 零度；无效
null-reading type magnetometer 零点读数式磁力仪
number address code 数地址码
number axis 数轴
number base 数基
number bus 数值汇流条；数码总线
number concentration 计数浓度
number converter 计数制变换器
number field 数域
number generator 数码发生器
number information 查号台查号
number language 数字语言
number normalization 数值标准化
number of blows 击数；锤击数
number of conflict 冲突点数目
number of copies 拷贝数
number of crimp 卷曲数
number of cycle 循环次数

number of degrees of freedom 自由度数
number of delay 延发段数
number of delay period 延发次数
number of divided segment 管片衬砌环分块数目
number of earthquake 地震次数
number of free radicals 自由基数
number of holes 钻眼数
number of loading cycles 加载循环数
number of phase encoding steps 相位编码步骤数
number of piles 桩数；群桩
number of points 点数
number of roller passes 碾压遍数
number of samples 样本数
number of sampling 抽样数量
number of segments 管片衬砌环分块数目
number of shot per foot 每英尺射孔数
number of sites 位数
number of space coordinates 空间配位数
number of standby stoping block 生产备用矿块数
number of steps 步数
number of strokes 冲程数
number of table break points 表中断点数
number of temporal positions 时序位置数
number of trains 列车数量
number of transfer node 换乘节点数
number of turns 匝数
number of twists 捻数
number of zero crossing 零跨数；零交次数
number percent 百分数；百分比
number plane 实数平面；数平面
number plate 号码牌
number range 数值范围
number representation system 数制
number scale 记数法
number sequence 数序
number storage 数字存储器
number system 数系
number theoretic transform 数论变换
number theory 数论
number up 列举
number-average molecular weight 数均分子量
numbered 数值数；已被编号的
numbering 位次编排；计数；编号
numbering of points 道岔编号
numbering of tracks 股道编号
numbering system 编号系统
numerable 可数的
numeral 数码；数字
numeral data acquisition 数字数据采集
numeral order 编号次序
numeral roll 数字卷筒
numeral sign 数符
numeral system 数系
numeration 命数法；计数法
numerator 分子；计数器；信号机
numeric 数字的
numeric character 数字符号
numeric coding 数字编码
numeric display 数字显示
numeric literal 数值文字

numeric projection 球心投影；心射图法；心射切面投影
numeric protection for overhead contact lines 数字式架空牵引网保护装置
numeric punch 数字穿孔
numerical 数字的；数值的
numerical analogue 数值模拟
numerical analysis 数值分析
numerical aperture 数值孔径
numerical approach 数字方法
numerical approximation 数值近似；数值逼近
numerical basin modelling 数值盆地模拟；数字盆地建模
numerical bound 数值边界
numerical boundary condition 数值边界条件
numerical calculation 数值计算
numerical calculus 数值计算；数值微积分
numerical code 数字码；数码
numerical coding 数字编码
numerical coefficient 数字系数
numerical computation 数值计算
numerical computing language 数值计算语言
numerical concentration 计数浓度
numerical constant 数值常数
numerical control 数字控制；数控；数值控制
numerical control device 数控装置
numerical control lathe 数控车床
numerical control machining 数控切削加工
numerical control part programming 数控加工零件编程
numerical control system 数控系统
numerical control tool 数控刀具
numerical control tool path 数控刀具轨迹
numerical control-servomotor 数控伺服电机
numerical damping 数值阻尼
numerical data 数字资料；数据
numerical decrement 衰减率
numerical determinant 数字行列式
numerical differential 数值微分法
numerical differentiation 数值微分
numerical diffusion 数值扩散
numerical dispersion 数值弥散
numerical dispersion model 数值散布模拟系统
numerical dissipation 数值耗散
numerical dual coding 数字的双重编码
numerical equation 数字方程；数值方程
numerical error 数值误差
numerical error constraint 数值误差约束
numerical estimate 数值估计
numerical evaluation 数值计算
numerical example 数字示例
numerical experiment 数值实验
numerical expression 数值表达式
numerical figure 数位
numerical filter 数值滤波；数字滤波
numerical filtering 数值滤波；数字滤波
numerical flux 数值通量
numerical forecasting 数值预报
numerical frequency 数字频率
numerical geomechanics 数值地质力学
numerical grid generation 数值网格生成

numerical information	数字信息
numerical instability	数值不稳定性
numerical integration	数值积分;积分计算;数值积分法
numerical integration procedure	数值积分程序
numerical intercept	截距系数;标轴系数
numerical interpolation method	数值插值法
numerical invariant	不变数
numerical inversion	数值反演
numerical key	数字键
numerical limit analysis for geological materials	岩土数值极限分析法
numerical manifold method	数值流形元法;数值流形法
numerical method	数值方法;数值计算法
numerical model	数值模型;数值模拟
numerical model of groundwater	地下水数值模型
numerical modeling	数值建模;数值模拟
numerical optimization	数值优化
numerical oriented remote sensing technique	数字遥感技术
numerical oscillation	数值振荡
numerical prediction	数值预报
numerical procedure	计算方案
numerical pumping test analysis	抽水试验数字分析
numerical range	数值范围
numerical real-out indicator	数字读数指示器
numerical relation	数量关系
numerical representation	数字表示;数值表示;数量表征
numerical reservoir simulation	油藏数值模拟
numerical search	数值搜索
numerical series	数值级数
numerical simulation	数值模拟;数值仿真
numerical simulator	数值求解器;数值模拟软件
numerical solution	数值解;近似解
numerical solution of motion equation	运动方程数值解
numerical solution of partial differential equation	微分方程数值解
numerical specimen	数值试样
numerical speed limit	数值速率限制
numerical stability	数值稳定性
numerical statement	统计
numerical strength	数值强度
numerical study	数值研究
numerical symbol	数字符号
numerical system	数系;数控系统
numerical tape	数字磁带
numerical taxonomy	数字分类
numerical technique	数值技术;数值法
numerical time-stepping method	数值时间步进法
numerical treatment	数值处理
numerical tube pressure	数值管道压强;数值管压力
numerical value	数值
numerical value method for slope stability analysis	边坡稳定分析数值法
numerical verification of method	方法的数值验证
numerical viscosity	数值黏性
numerical weather prediction	数值天气预报
numerical-analytic technique	数值-分析方法
numerical-graphic method	数值图解法
numerically integrated finite element	数值综合有限元
numerically oriented system	数字遥感系统
numerically-controlled machine	数控机床
numerically-controlled machine tool	数控机床;数字控制机床
numerical-perturbation method	数值摄动法
numeric-alphabetic	字母数字的;字符
numeric-field data	数字域数据
numeroscope	示数器;数字记录器
numerous	大批量的;许多的
numerous multivariate statistical method	多数多变量统计法
N-unit	中子剂量单位
Nunivak event	努尼瓦克正向事件;努尼瓦克亚磁极期
Nunivak polarity subchron	努尼瓦克亚时
nussierite	钙磷铅矿
nut	螺帽;螺母
nut bolt	带帽螺栓
nut coal	核级煤
nut key	螺帽扳手
nut lock	制动螺帽;固定螺帽
nut lock washer	锁紧垫圈
nut locking	螺母锁紧
nut runner	风动拧螺帽器
nut screen	核级煤筛
nut seat	螺母支承面
nut switch	螺帽开关
nut tap	螺帽丝锥
nut wrench	螺母扳手
nutcoal	核级煤
nutlike structure	核状结构
nut-lock	螺帽固定器
nut-lock washer	螺母锁紧垫圈
nut-locking device	固定装置;锁紧螺母装置
nutritious groundwater	地下肥水
nutritive groundwater	地下肥水
nutritive salt	营养盐
nutritive water	肥水;富营养水
nut-runner	上螺母器
nutsch filter	吸滤器
nutschfilter	吸滤器
nuttalite	中柱石
nutty	多硬核的
nutty structure	核状结构
nuvistor	超小型抗震管
N-value	标准贯入试验锤击数;N值
n-year flood	n年一遇洪水
nylon bearing	尼龙轴承
nylon bush	尼龙衬套
nylon cable	尼龙电缆
nylon cable carrier chain	尼龙电缆夹链
nylon coating	尼龙涂层
nylon cord	尼龙绳
nylon fabric	尼龙织物
nylon ferrule	尼龙套圈
nylon fiber	尼龙纤维
nylon filter	尼龙过滤器
nylon filter antiscour system	尼龙滤层防冲刷系统
nylon gasket	尼龙填料
nylon insulator	尼龙绝缘块
nylon plastics	尼龙塑料

nylon resin 尼龙树脂
nylon rope 尼龙绳
nylon salt 尼龙盐
nylon screen 尼龙筛网
nylon sling 尼龙吊带
nylon stopper 尼龙堵头
nylon wire 尼龙丝
nylon-epoxy adhesives 尼龙环氧黏合剂
Nyquist criterion 尼奎斯特稳定性准则
Nyquist diagram 尼奎斯特图
Nyquist folding frequency 尼奎斯特折叠频率
Nyquist frequency 尼奎斯特频率；折叠频率
Nyquist noise 尼奎斯特噪声
Nyquist rate 尼奎斯特速度
Nyquist sampling rate 奈奎斯特取样率
Nyquist theorem 尼奎斯特取样定理
Nyquist wave number 尼奎斯特波数

O

O ring　O形环
O ring recess　O形圈槽
oasis　绿洲；水草田
obducted sheet　仰冲岩席
obduction　逆冲；仰冲；上冲
obduction orogen belt　仰冲造山带
obduction plate　逆冲板块；仰冲板块
obduction sheet　仰冲岩席
obduction slab　仰冲板片；仰冲板块；逆冲板块
obduction zone　仰冲带；逆冲带
obdurability　坚硬；强韧性；坚固性
obelisk　火山柱；方尖石塔；方尖碑
object contrast　景物反差；工件对比度
object criteria　客观准则；客体判别标准
object database　对象数据库
object detection　目标探测；目标检测；对象检测
object extraction　对象提取
object function　目标函数
object glass　物镜
object height　地物高度
object identification　物体识别；目标识别；对象识别
object language　目标语言
object lens　物镜
object line　轮廓线；外形线
object machine　专用计算机；目标机；对象链接与嵌入
object modelling　实体建模
object module　目标模块；目的程序；目的组件程序
Object Oriented Analysis　面向对象的分析
Object Oriented Analysis and Design　面向对象的分析与设计
Object Oriented Design　面向对象的设计
object plate　载片
object positioning　地物定位
object profile　目标剖面
object program　目的程序；结果程序
object program library　目标程序库
object retrieval system　目标检索系统
object routine　目标程序
object selection process of control　选组
object slide　载玻片；物镜筒
object space coordinate system　物空间坐标系
object spectrum characteristic　地物波谱特性
object staff　水准标尺；准尺
object time　目标时间；目标程序执行时间
object-class hierarchy　分层结构图
objection　反对；异议；反对的理由；妨碍；缺陷；缺点
objectionable constituent　有害成分
objectionable impurity　有害物质
objective　物镜；目的，目标；客观的
objective analysis　客观分析；目标分析
objective angle of image field　像场角
objective angular field　物镜视场
objective aperture　物镜孔径
objective cap　物镜套
objective failure probability　客观破坏概率
objective forecast　客观预报
objective function　目标函数
objective grating　物镜光栅
objective interval　目的层段
objective lens　物镜镜头；物镜
objective measurement　客观测量法；客观量度
objective pattern　物体形态特征
objective prior probability distribution　客观先验概率分布
objective probability　客观概率
objective program　目的程序
objective sand　目的砂岩层
objective table　载物台
objectivity　客观，客观性，客观规律
object-line　轮廓线；外围线；地平线；等高线
object-oriented database　面向对象数据库
object-oriented programming　面向对象编程；面向对象的程序设计
object-staff　准尺
oblate　扁的；扁圆形的
oblate ellipsoid　扁椭球体
oblate spheroid　扁球
oblateness　扁率；扁圆形
oblateness of the earth　地球扁率
obligation well　承诺井
obligatory well　义务井
oblique　斜的；倾斜的
oblique aerial photograph　倾斜航摄照片；偏斜航空照片
oblique aerial photography　倾斜航空摄影
oblique aerial print　斜摄航空相片
oblique air photograph　倾斜航测照片
oblique angle　斜角
oblique arch　斜拱；轴线不垂直于其面的拱
oblique aspect　斜轴；斜方位
oblique axis　斜轴
oblique bearing pile　斜向受力桩；斜向承重桩
oblique bedding　交错层理；斜层理
oblique bending　斜弯曲
oblique borehole　斜钻孔
oblique chart　斜轴投影地图
oblique cleat　斜割理；斜劈理
oblique cleavage　斜劈理
oblique collision　斜向碰撞
oblique compression　斜压缩
oblique configuration　倾斜结构
oblique conformal projection　斜轴正形投影；斜轴保角映射
oblique convergent　斜向聚合
oblique coordinates　斜交坐标；斜角坐标
oblique crossing　斜交
oblique cut　斜眼掏槽
oblique cutting　斜刃切削；斜插

oblique cutting edge	斜割刃
oblique dimetric projection	斜二测投影
oblique displacement	斜位移
oblique distance	斜距
oblique distance projection	斜距投影
oblique drawing	斜视图
oblique eccentric loading	斜偏心荷载
oblique error	倾斜误差
oblique expansion wave	斜向扩散波
oblique exposure	斜接近；斜曝光
oblique exposure length	斜接近段长度
oblique extension	斜向拉伸
oblique face	伪倾斜工作面；伪倾斜工祖；斜工作面；斜交工作面
oblique factor	斜交因子
oblique factor solution	斜交因子解
oblique fault	斜断层；斜交断层
oblique flexural-slip fold	斜挠曲滑动褶皱
oblique flow	斜向水流；斜流
oblique fold	斜褶皱；斜卧褶皱
oblique fracture	斜断口
oblique front	对角工作线；斜工作面
oblique girdle	斜环带
oblique heading	斜的联络巷道；斜向钻进
oblique helicoid	斜螺旋面
oblique hole	斜钻孔；斜炮眼
oblique impact	斜向碰撞；斜向冲击
oblique incidence	倾斜入射
oblique interference figure	斜交干涉图
oblique joint	斜节理；斜接头
oblique lamination	斜交纹理；横纹理
oblique line	斜线
oblique load	倾斜载荷
oblique longwall mining	伪斜长壁采煤法
oblique magnetization	斜磁法；斜磁化，斜向磁化
oblique mining	伪倾斜开采；对角式回采
oblique notching	开斜槽
oblique observation	倾斜观测
oblique orbit	斜轨道
oblique orientation of spread	斜向排列；地震仪倾斜排列
oblique photograph	倾斜航摄相片
oblique photography	倾斜摄影
oblique plane	斜面
oblique plotting instrument	倾斜测图仪
oblique profile	斜剖面
oblique progradation configuration	斜交前积结构
oblique projection	斜轴投影；倾斜投影
oblique projection drawing	斜轴投影图
oblique ramp	斜断坡；斜坡；斜道
oblique road	斜巷
oblique rotation factor solution	斜交旋转因子解
oblique section	斜截面；斜切薄片
oblique seismic reflection configuration	斜交地震反射结构
oblique shear fold	斜剪切褶皱
oblique shift fault	斜滑断层
oblique shock	斜激波；斜震
oblique shock front	斜交激震波面
oblique shock wave	斜交冲击波；斜交激震波
oblique slicing	斜切分层采煤法
oblique slip	斜向滑动
oblique slip fault	斜向滑动断层；斜滑断层；斜移断层
oblique stoping	对角式回采；斜向回采
oblique stratification	斜层理
oblique subduction	斜向俯冲
oblique superposition	斜跨叠加
oblique system	单斜晶系
oblique thick bedding rock slope	斜倾厚层岩质边坡
oblique traces	斜截面法
oblique wave	斜波；斜向波
oblique weir	斜堰
oblique well seismic profile	斜井垂直地震剖面
oblique-angle projection	倾斜投影
obliquely layered medium	斜层介质
obliquely-stepped face	倾斜阶梯式工祖；倾斜阶梯式工作面
obliqueness	倾斜；斜度
oblique-projection method	斜投影法
oblique-slip	斜向滑动
obliquity	倾斜；斜向；斜交，倾度；斜度；倾角；斜角
obliquity angle	倾斜角
obliquity change	倾斜度变化
obliteration of cleavage	劈理的消失
obliteration of lake	湖泊淤塞
oblong	长方形；长椭圆形
obnoxious odour	难受的气味
oborite	鄂博矿
obscurity	模糊；不明确
obsequence	逆向
obsequent	逆向的；反向的；逆向河；反向河
obsequent fault line	逆向断层线
obsequent fault-line scarp	逆向断线崖
obsequent flow	逆向流动
obsequent rift block	逆陷断块
obsequent river	逆向河
obsequent tilt block mountain	逆掀断块山
obsequent tilt-block valley	逆掀断块谷
obsequent valley	逆向谷
obsequent-fault-line scarp	逆向断线崖
obsequent-subsequent stream	反向次成河
observability	可观测性；能观测性
observability condition	观测条件
observable	可观察的；可观测的；可探测到的
observable stratigraphic unit	可见地层单位
observable subspace	可观测子空间
observation	观察，观测；探测；检查
observation accuracy	观测精度
observation base	观测基线
observation borehole	观测钻孔
observation criteria	观测标准
observation data	观测数据
observation desk	试验台
observation equation	观测方程
observation error	测量误差；观测误差
observation field	观测场；野外观测
observation grid	观测网格
observation hole	观察孔；观测孔
observation in groups	分组观测
observation instrument	观测仪器

observation line 观测线；观测剖面线
observation management error 测量误差
observation matrix 观测矩阵
observation measurement 观测
observation monument 观测标志
observation net 观测网
observation network 观测网
observation node 观测结点
observation of density current 密度流观测
observation of earth tide 固体潮观测
observation of evaporation from soil 土壤蒸发观测
observation of evaporation from water surface 水面蒸发观测
observation of ground stress 地应力观测
observation of ground vibration effect from blasting in mine 矿山爆破地震效应观测
observation of landslide regime 滑坡动态观测
observation of slope stability 边坡稳定性观测
observation of surrounding rock deformation 围岩变形观测
observation of water temperature 水温观测
observation perforation 观测孔眼
observation pillar 观测墩
observation platform 观察台；观测平台
observation point 测点；观测点；观察点
observation port 观察口，窥孔；观察窗；观测所
observation record 观测记录
observation report 观察报告；监测记录
observation result 观测结果
observation route 观测线；观测路线
observation satellite 观测卫星
observation session 观测时段
observation set 观测集；测回
observation site 观测场
observation station 测站；观测站
observation station of ground movement 地表移动观测站
observation station of surface movement 地表移动观测站
observation system 观测系统；监测系统
observation target 测量觇标；测量观标
observation to the earth 对地观测
observation tower 观测塔；测量觇标
observation unit 观测单元
observation value 观测值
observation well 观测井；观测孔；验潮井
observation window 观测窗；观察窗；瞭望窗
observational borehole 观测孔
observational data 观测数据；观测资料
observational error 观察误差；观测误差
observational frequency 观测次数；观测频率
observational grid 观测网格
observational hole 观测孔
observational method 观测方法；观察法
observational network 观测网
observational seismology 观测地震学
observational shaft 观测井
observational synthesis 综合观测结果
observational tunneling method 观测隧道施工法
observational well 观测井；观测孔
observation-well response 观测井压力反应
observatory 天文台；气象台；观测所

observatory remote 遥测台
observe 观察；观测；注意到
observed amplitude 观测振幅
observed anomaly 观测异常
observed array 观测数组
observed azimuth 观测方位角
observed correlation 观测相关
observed cumulative production 实测累积产量
observed current 实测流
observed data 观测数据；实测资料
observed data array 观测数据数组
observed depth 实测水深；观测深度
observed dip 实测倾斜
observed direction 实测方向；观测方向；实测流向
observed discharge 实测流量
observed fall 实测落差
observed flood 实测洪水
observed frequency 观测频数；观测平率
observed geologic boundary 实测地质界线
observed gravity 观测重力
observed heat value 观测热值
observed level 实测水位
observed performance 观测动态
observed pressure 实测压力
observed profile 实测剖面
observed quantity 观测量
observed ray 观测方向
observed reading 观测读数；观测值
observed reserves 实测储量
observed result 实测结果
observed stage 实测水位，观测水位
observed temperature 实测温度
observed tidal variation 观测潮汐变化
observed value 观测值；实测值
observed water level 实测水位，观测水位
observed wavelet 观测子波
observed wellhead data 实测井口数据
observer 观测员；观测器
observing point 观测点
observing program 观测程序
observing ship station 观测船站
observing station 观测站
obsidian 黑曜石；黑曜岩
obsidian hydration dating 黑曜岩年代测定法
obsolescence 废弃；报废；作废
obsolescent 将要废弃的
obsolete 已废的；已不用的；陈旧的
obstacle 障碍；障碍物
obstacle scour marking 障碍冲刷痕
obstruct 阻止；拦挡；妨碍；堵塞
obstructed stream 阻塞河
obstruction 阻碍；闭塞；遮断；障碍物
obstruction detection system 障碍探测系统
obstruction lake 壅塞湖
obstruction to an extension 扩张的障碍
obstruction-congestion 阻碍拥挤；阻碍拥塞
obstructive ridge 保安矿柱
obtaining core 取得岩心
obturate 封严；密闭
obtuse 钝的；钝角的

obtuse angle　钝角
obtuse peak abutment　矿柱的钝角应力峰值区
occasional occurrence　非经常发生；偶然出现的
occasional storm　洼地蓄水
occlude　封闭；阻塞；滞留
occluded gas　滞留气；吸藏气体；吸留气体
occluded methane　包藏沼气；附吸沼气
occluded methane content　附吸瓦斯容量
occluded oil　滞留油；吸藏油
occluded water　吸着水；吸附水
occlusion　包藏；包藏气体作用；吸留，吸着；吸气体作用
occlusion mineral　隐蔽矿物
occlusion water　封存水
occult mineral　隐蔽矿物；不能分辨的矿物
occur　发生；出现
occurrence　产状；存象；埋藏；发生；出现；观象；事故
occurrence condition　产出条件
occurrence condition of coal seam　煤层赋存条件
occurrence condition of ground water　地下水赋存条件
occurrence frequency　发生频次
occurrence in beds　呈层状；呈层状出现
occurrence in veins　呈脉状
occurrence mechanism　发生机制
occurrence number　出现次数
occurrence of coal seam　煤层赋存情况；煤层产状
occurrence of magmatic rock　岩浆岩产状
occurrence probability　产状概率
occurrence rate　发生率
occurrence reason　发生原因
occurrence time　发生年代；发生时间
ocean　海洋；大洋
ocean acidification　海洋酸化
ocean acoustics　海洋声学
ocean basalt　大洋玄武岩
ocean basin　海洋盆地；洋盆；大洋盆地
ocean basin floor　洋盆区
ocean bed　洋底；海底
ocean bottom　洋底
ocean bottom current　洋底流
ocean bottom earthquake　海底地震
ocean bottom hydrophone instrument　海底水听仪器
ocean bottom microseism　海底微震
ocean bottom profiler　海底地层剖面仪
ocean bottom relief　海底起伏
ocean bottom scanning sonar　海底扫描声呐
ocean bottom seismograph　海底地震仪；海洋地震仪
ocean bottom seismometer　海底检波器；海底地震计，海底地震仪
ocean bottom station　海底观测站
ocean bottom topographic meter　海底地貌仪
ocean canyon　海底峡谷；洋底峡谷
ocean carbon sink　海洋碳汇
ocean chart　海洋图
ocean chemistry　海洋化学
ocean circulation　海洋环流
ocean climate　海洋性气候
ocean coal　海底煤层
ocean crust　大洋地壳

ocean crustal reservoir　洋壳储层
ocean current　海流；洋流
ocean data acquisition system　海洋资料收集系统
ocean deep　海渊
ocean delimitation survey　海洋划界测量
ocean depot　海洋油库
ocean depth　海洋深度
ocean depth curve map　海洋等深线图
ocean disposal　远洋废物处理；投海处理
ocean dredging　海底挖泥
ocean drilling　海洋钻井
Ocean Drilling Program　大洋钻探计划
ocean dumping　海洋倾倒
ocean dynamics　海洋动力学
ocean eddy　海洋涡流
ocean engineer　海洋工程师
ocean engineering　海洋工程；海洋工程学
ocean engineering construction industry　海洋工程建筑业
ocean engineering hybrid model　海洋工程复合模型
ocean engineering physical model　海洋工程物理模型
ocean engineering research vessel　海洋工程调查船
ocean environment　海洋环境
ocean exploitation　海洋开发
ocean exploration　海洋勘探；海洋探测
ocean floor　洋底；海底
ocean floor fracture zone　洋底断裂带
ocean floor geomorphology　海底地貌学
ocean floor producer　海底生产井
ocean floor spreading　海底扩张；洋底扩张
ocean front　沿海地带；海洋锋，洋锋
ocean geotechnique　海洋岩土工程；海洋土工学
ocean gravimeter　海洋重力仪
ocean hill　深海丘
Ocean Information Technology Project　海洋信息技术计划
ocean island　海中岛；洋中岛；大洋岛
ocean island basalt　洋岛玄武岩
ocean isothermal plot　海洋等温线图
ocean magnetic survey　海洋磁测
ocean mapping　绘制海洋图
ocean meteorology　海洋气象学
ocean mining　大洋采矿；深海采矿
ocean observation apparatus　海洋观测装置
ocean observation satellite　海洋观测卫星
ocean observation technology　海洋观测技术
ocean observation tower　海洋观测塔
ocean observatory　海洋观测
ocean outfall　入海河口
ocean physics　海洋物理学
ocean piping　海洋管道
ocean platform　海洋平台；海上平台
ocean pollution　海洋污染
ocean port　海港
ocean remote sensing　海洋遥感
ocean remote sensing camera　海洋遥感照相机
ocean remote sensing observation　海洋遥感观测
ocean resources　海洋资源
Ocean Resources Conservation Association　海洋资源保护协会
ocean ridge　大洋中脊；海脊，海岭
ocean ridge seismic belt　洋脊地震带

ocean ridge tholeiite 大洋脊拉斑玄武岩
ocean science 海洋科学
ocean sequestration 深海隔离
ocean shipping 海运
ocean shoreline 海岸线
ocean soil 海洋土
ocean sounding chart 大洋水深图
ocean state 海洋状态
ocean station 海洋观测站
ocean storage 深海封存；海底封存
ocean structure 海洋构筑物
ocean surface current 表层洋流
Ocean Surveillance Satellite 海洋监视卫星
ocean survey technology 海洋调查技术
ocean swell 海洋涌浪
ocean technology 海洋技术
ocean thermodynamics 海洋热力学
ocean tidal model 海潮模型
ocean tide 海潮；潮汐
ocean trench 海沟
ocean trough 海槽
ocean warming 海洋水温升高
ocean water 海水
ocean wave 海洋波动；海浪；大洋波浪
ocean weather ship 海洋气象观测船
ocean weather station 海洋气象站
ocean-atmosphere evolution 海洋大气演化
ocean-atmosphere exchange 海洋-大气交换
ocean-atmosphere interaction 海洋大气相互作用
ocean-atmosphere interface 海洋大气界面
oceanaut 潜水员；海中作业人员
ocean-bearing plate 大洋板块
ocean-bottom multiple 海底多次波
ocean-bottom seismograph 海底地震仪
ocean-bottom seismometer 海底地震仪
ocean-continent transition 海陆过渡
ocean-crust type basin 洋壳型盆地
oceaneering 海洋工程
ocean-estuary coupling 海河口耦合
ocean-floor metamorphism 洋底变质作用
ocean-floor spreading 洋底扩张
ocean-floor spreading hypothesis 海底扩张假说
oceanfront 滨海地带；海滨
ocean-geomagnetic survey 海洋地磁测量
ocean-going 远洋的
ocean-going vessel 远洋轮船
Oceania 大洋洲
Oceanian 大洋洲的
oceanic 海洋的
oceanic absorption 海洋吸收
oceanic abyss 海渊
Oceanic Anoxic Event 大洋缺氧事件
oceanic bank 大洋浅滩
oceanic basement 洋壳基底
oceanic basin 大洋盆地；洋盆
oceanic bottom 洋底
oceanic circulation 海洋环流
oceanic climate 海洋性气候
oceanic condition 海况
oceanic craton 洋台；大洋稳定地块

oceanic crust 大洋型地壳；海洋地壳；洋壳
oceanic current 洋流
oceanic delta 海洋三角洲
oceanic deposit 海洋沉积；大洋沉积
oceanic desert 大洋荒漠
oceanic earthquake 海洋地震
oceanic formation 洋盆相地层建造
oceanic fracture zone 洋底断裂带
oceanic gas hydrate 海洋天然气水合物
oceanic general circulation model 海洋环流模式
oceanic geotechnical investigation 海洋工程地质调查；海洋岩土工程勘察
oceanic heat flow 海洋热流；海底热流
oceanic hot-spot shield volcano 海洋热点盾状火山
oceanic igneous basement 海洋火成岩基底
oceanic island 海洋岛屿；大洋岛
oceanic island arc 大洋岛弧；海洋岛弧
oceanic island evolution 洋岛演化
oceanic island tholeiite 海洋岛屿拉斑玄武岩
oceanic island volcano 洋岛火山
oceanic layer 大洋层
oceanic leveling 平均海平面高度测量
oceanic lithosphere 大洋岩石圈；海洋岩石圈；洋壳
oceanic load 海洋负荷
oceanic load tide 海洋负荷潮
oceanic lysocline 海洋溶跃面；海洋溶跃层
oceanic magma series 大洋岩浆系
oceanic magnetometer 海洋磁力仪
oceanic mantle 洋幔
oceanic margin 大洋边缘
Oceanic Margin Drilling Program 大洋边缘钻探计划
oceanic massive sulfide 海底块状硫化物
oceanic meteorology 海洋气象学
oceanic microseism 海洋微震
oceanic mining 海洋采矿；海洋矿业
oceanic moderate 海洋性温和气候
oceanic outfall 海洋排污渠口；深海排污渠口
oceanic period 原始海洋纪；洪荒纪
oceanic plagiogranite 大洋斜长花岗岩
oceanic plate 大洋板块；海洋板块
oceanic plate stratigraphy 大洋板块地层
oceanic plate subduction 大洋板块俯冲作用
oceanic plateau 大洋高原；大洋海台
oceanic prediction 海况预报；海况预测
oceanic province 洋区
oceanic red beds 大洋红层
oceanic reef 大洋礁
oceanic ridge 大洋中脊；海岭
oceanic ridge basalt 洋脊玄武岩
oceanic ridge hydrothermal system 洋脊热液系统
oceanic ridge tholeiite 洋脊拉斑玄武岩
oceanic rift 大洋裂谷
oceanic rift system 大洋裂谷系
oceanic rise 洋隆；海隆；大洋中隆
Oceanic Scientific Drilling 海洋科学钻探
oceanic sediment 海洋沉积物
oceanic sequestration 海洋封存
oceanic slab 海洋板块；大洋板块
oceanic stratigraphy 大洋地层学；海洋地层学
oceanic stratosphere 大洋平流层

oceanic structure 海洋构造
oceanic terrane 海洋岩层
oceanic tholeiite 大洋拉斑玄武岩；深海拉斑玄武岩
oceanic tide 海潮
oceanic transform 海洋转换
oceanic trench 洋沟；海沟；深海槽
oceanic turbulence 大洋湍流；海洋湍流
oceanic upwelling 大洋上升流
oceanic volcano 海底火山；洋底火山；大洋型火山
oceanic wave 海浪
oceanic-atmospheric evolution on earth 地球上的海洋大气演变
oceanic-floor fracture zone 洋底破裂带
oceanic-island basalt 洋岛玄武岩
oceanic-oceanic convergence 海洋-海洋型聚敛作用
oceanics 海洋学；海洋工程学
oceanite 大洋岩；富橄暗玄岩；苦橄玄武岩
oceanity 海洋性；海洋度
oceanization 海洋化；大洋化；基性岩化
oceanizational hypothesis 大洋化假说
oceanogenic 海洋成因的
oceanogenic sedimentation 海洋沉积
oceanographer 海洋学家
oceanographic 海洋学的
oceanographic buoy 海洋浮标
oceanographic chemistry 海洋化学
oceanographic current meter 海洋流速仪
oceanographic data 海洋水文资料
oceanographic depth recorder 海洋水深记录仪
oceanographic engineering 海洋工程
oceanographic environmental survey 海洋环境调查
oceanographic hydrological data 海洋水文数据
oceanographic hydrological forecast 海洋水文预报
oceanographic information centre 海洋情报中心
oceanographic instrumentation 海洋仪器
oceanographic leveling 海洋水准测量
oceanographic model 海洋学模式
oceanographic physics 海洋物理学
oceanographic radar 海洋雷达
oceanographic remote sensing 海洋遥感
oceanographic research 海洋调查
oceanographic satellite 海洋卫星
oceanographic survey 海洋调查
oceanographic vessel 海洋调查船
oceanographic vessel for multipurpose research 海洋综合调查船
oceanographical 海洋学的
oceanographical chart 海洋图表
oceanographical dredger 洋底采样器
oceanographical engineering 海洋工程
oceanographical equator 海洋学赤道
oceanographical leveling 海洋水准测量
oceanographical remote sensing 海洋遥感
oceanographical survey 大洋调查
oceanographical vessel 海洋调查船
oceanography 海洋学
oceanologist 海洋学家
oceanology 海洋学；海洋资源研究；成因海洋学，海洋开发技术
oceanophysics 海洋物理学

oceanosphere 海洋圈
oceansat 海洋卫星
oceanside 海边
oceansphere 海洋圈
oceanward 向海的
ocellar 眼斑状的
ocellar structure 眼球构造
ocellar texture 眼斑结构；眼球结构
ocellus 单眼
ocher 赭石
ocherous 赭石的；赭土的
ocherous deposit 铁质沉积物
ochreous iron ore 赭石
ochrept 淡色始成土；寒温带始成土
ochroite 硅铈石
ochrolite 鲜黄石
octagon 八边形；八角形
octagon kelly 八角形方钻杆
octahedral 八面的；八面体的
octahedral borax 八面体硼砂
octahedral iron ore 磁铁矿
octahedral layer 八面体层
octahedral normal stress 八面体法向应力；八面体正应力
octahedral plane 八面体平面
octahedral shaft 八角钻杆
octahedral shear stress 八面体剪应力
octahedral shear stress theory 八面体剪应力理论
octahedral shearing stress 八面体剪应力
octahedral sheet 八面体片
octahedral strain 八面体应变
octahedral stress 八面体应力
octahedral stress theory 八面体应力理论
octahedral structure 八面体结构
octahedral void 八面体空隙
octahedrite 八面石；八面体陨铁；奥克铁陨石类
octahedron 八面体
octal 八进制；八进制的；八进位
octal digit 八进制数
octal notation 八进制计数法
octal number 八进制数
octal number system 八进制
octal numeral 八进制数的
octal system 八进制
octave 倍频程
octave band 倍频程；倍频带
octave band analysis 倍频程分析
octave filter 倍频程滤波器
octobolite 辉石
octonary 倍频；八进制的
octonary number system 八进制
octophyllite 八叶云母
ocular 目镜
ocular estimate 目测，目估
ocular lens 接目镜
ocular measurement 目测
ocular micrometer 目镜测微计
odd 临时的；奇数的；剩余的，多余的；奇特的
odd bit behavior 钻头工况异常
odd dip angle 异常的地层倾角

odd function	奇函数
odd harmonic function	奇次谐波函数
odd knobbing	厚煤层工作区近旁的落煤
odd location	奇数存储单元
odd multiple	奇倍数
odd number	奇数
odd numbered trace	奇数道
odd parity check	奇校验
odd peak	奇脉冲峰值
odd pinnate	奇数羽状的
odd pulse	奇脉冲
odd response	奇脉冲响应值
odd symmetry	奇对称
Odderade interglacial period	奥代拉德间冰期
odd-even	奇-偶
odd-even check	奇偶校验
odd-even check code	奇偶校验码
odd-even effect	奇偶效应
odd-even element	奇偶元素
odd-integral number	奇整数
oddity	异态；奇特
oddment	碎屑
odd-parity check	奇数奇偶校验
odds and ends	残余
odinite	钛云母；拉辉煌斑岩
Odintsovo interglacial period	奥丁特索沃间冰期
odometer	里程计；测距器；轮转计
odometry	测距法
odontolite	齿绿松石
odorous	有臭气的；有气味的
odorousness	气味浓度
odour	气味；臭味
odour pollution	臭气污染
odourless	无气味的；无臭的
oecology	生态学
oecostratigraphy	生态地层学
oedometer	固结仪；压缩仪
oedometer curve	压缩曲线
oedometer test	固结试验；压缩试验
oedometric modulus	侧限压缩模量
oedometric test	压缩试验；固结试验
oedotriaxial test	固结仪三轴仪联合试验法
Oehman's apparatus	欧曼测斜仪；欧曼氏仪
Oehman's survey instrument	欧曼氏测量仪；欧曼型钻孔测量仪
oellacherite	钡白云母
Oersted	奥斯特
oerstedmeter	磁场强度计
off analysis	成分不合格
off and on	断断续续地；不时地；间隙
off balance	失去平衡
off bottom	不接触底部的；在底部以上的
off centre	偏心；偏离中心
off coast	离岸；外海
off colour gem	次级宝石；有色痕的宝石
off gas	废气
off gate	采空区巷道
off land	离岸；在海洋
off line	离线；脱机
off line operation	离线操作；脱机操作
off peak	非最大的；非峰值的
off shore	离岸
off shore boring	海底钻进；海底钻探
off shore rig	海中钻塔
off soundings	深海区；非锤测区
off specification	超出规范的
off specification requirement	超出规范的要求
off structure	错亡
off structure well	在构造无油区钻的井；构造外的井
off technical requirement	超出技术要求
off the solid	不可靠的；失稳
off the structure	构造外的
off-angle	斜的
off-angle drilling	钻斜孔；钻斜孔法；斜钻
off-axis	轴外的；离轴的
off-axis aberration	轴外像差；离轴像差
off-axis error	离轴误差
off-axis holography	离轴全息术；离轴全息摄影
off-axis volcanism	离轴火山作用
off-balance	不平衡的；失去平衡的；倾倒的
off-balance bit	偏心钻头
off-bottom cementing	提离井底注水泥
off-bottom distance	离底距离
off-bottom rotation	离井底转动
off-bottom setting tool	不坐底坐封工具
off-bottom spacing	离孔底距离；离井底的距离
off-center	不对称的；偏的；离心的
off-center loading	偏心荷载；偏心加载
off-centre	不对称的；偏的；离心的
off-centre loading	偏心载荷；偏心加载
off-channel	不在河槽中
off-color	变色的
off-course	偏离航向的；偏离路线的
off-design	非计算的；非设计的
off-design conditions	非计算条件；非设计条件；计算偏差的条件
off-design performance	非设计性能
off-diagonal	非对角线的
off-diagonal element	非主对角线元素
off-dimension	尺寸不合格
off-duty	下班；不值班的；未运行的；备用的
off-effect	中止效应
offending sand	干扰砂层
offensive odour	厌恶性气味；难闻气味；恶臭
offer well	排水井
off-gas	废气；放出的气
off-gas cleaning equipment	废气净化设备
off-gas line	废气管
off-gas pump	排气泵
off-gauge	非标准的；不合量规的
off-grade	低级的；品位低的；等外的；不合格的
off-highway haulage	矿山运输
office analysis	室内分析；室内整理
office automation	办公室自动化
office computation	室内计算
Office Management Information System	办公信息系统
office master drawing	正式原图
office operation	室内作业
office study	室内研究；室内分析
office treatment	室内处理

office work	室内工作;内业	offset correction	偏移距校正
official	公务员,职员;官方的,正式的;公务的;业务的	offset deposit	分枝矿床
		offset direction	偏移方向
official acceptance	正式验收	offset distance	偏移距
official contract	正式合同	offset drainage	不均衡开采
official speed trial	正式试航	offset drilling	偏置钻眼;补充钻眼;边界钻进
official standard	正式标准	offset frequency	偏离频率
official test	正式试验	offset geophone	偏离检波器
offing	海面;洋面;深水海面	offset injection rate	补偿油层亏空的注入量
off-land	离岸	offset intersection	错位交叉
offlap	退覆;分错距;平移断层	offset landform	断错地貌
offlap sequence	退覆层序	offset line	补充测线
offlap slope facies	退覆斜坡相	offset log	补测的测井曲线
off-leveling	水平失调	offset method	支距法
off-limit	界限外	offset of bed	岩层分岔
off-line	脱机的;脱离主机单独工作的	offset of displacement	位移距离
off-line computation	脱机计算	offset parameter	补偿参数
off-line computer	脱机计算机	offset ridge	断错山脊;断错海岭
off-line computer processing	计算机脱机处理	offset section	炮检距剖面
off-line control	离线控制	offset seismic profile	偏移地震剖面
off-line data	偏离数据	offset space	炮检距空间;炮检距域
off-line dual operation	离线操作;脱机工作	offset staff	偏距尺
off-line equipment	脱机设备	offset stake	外移桩
offline function	脱机功能	offset stream	断移河
off-line operation	脱机作业;间接操作	offset structure	错动构造;断裂构造
off-line printing	脱机打印	offset tangent modulus	补偿切线模量;偏移切线模量
off-line processing	脱机处理	offset the pressure in a well	抵消井内的压力
off-line software	脱机软件	offset transition	时差递变
off-line unit	脱机装置	offset unidirectional spread	偏置单向排列法
off-line working	脱机工作	off-set value	偏移值
off-load	卸载;卸荷	offset vertical seismic profiling	非零井源距垂直地震剖面
off-load regulation	卸荷调节	offset well	排水井;补偿井;边界孔
off-loading	偏心荷载	offset windage	偏距游隙
offloading capacity	卸载能力	offset yield strength	补偿屈服强度
off-loading station	卸煤台	offset-dependent amplitude	随炮检距变化的振幅
off-location	离位	offset-dependent reflectivity	随炮检距变化的反射系数
off-lying	向海的;离岸的;遥远的	offset-dependent waveform	随炮检距变化的波形
off-lying sea	远海	offsetting	偏移;偏心距;支距测法
off-normal	不正常的;偏离;离位	offsetting dip	偏移倾斜
off-pattern producer	井网外生产井	offsetting injector	相邻的注入井
off-pattern well	井网外的井	offsetting well	补偿井
off-rating	不正常条件;非标准条件;不正常状态;非标准状态	offshoot	岩枝;分枝;支流;支脉
		offshore	滨海地带;海上;近海地区;离岸的;离岸地区
off-reef	礁外的;离礁的	offshore acquisition	海上数据采集
off-reef facies	礁外相	offshore activity	海上作业
off-resonance frequency	非共振频率	offshore anchorage	离岸下锚
offretite	菱钾铝矿	offshore area	近海区;近海水域;近岸海域;滩外地带
offroad prospecting	路边勘探	offshore bank	近海浅滩
offscum	废渣	offshore bar	近海沙州;滨外沙州;岸外坝;离岸坝
off-scum	废渣	offshore bar sand	滨外沙坝
off-sea	离海的;外海的	offshore basin	近海盆地
offset	偏置;偏移;支距;水平断错;支巷;支脉	offshore beach	滨外滩;离岸隐滩
offset a well	略移井位再钻井	offshore beach terrace	滨外滩阶地
offset acoustic path	偏移声波路径法	offshore boring	近海钻探;海底钻探
offset angle	偏斜角	offshore breakwater	岛式防波堤;离岸防波堤
offset arrangement	非零偏移排列	offshore cluster wells	海洋丛式井
offset bend	平移弯管	offshore coastal strip	滨外带
offset carrier	偏离载波	offshore completion	海上完井
offset cone bit	移轴牙轮钻头	offshore construction	海上施工;滨海建筑物
offset core	补取的岩心		

offshore coring drilling 近海浅钻钻探
offshore current 离岸流；沿岸流；滩外流
offshore deposit 滨外沉积；近海沉积
offshore directional well 海上定向井
offshore distance 离岸距离
offshore dock 岸架式浮坞
offshore dredging 滨外疏浚
offshore drilling 海上钻探；近海钻探；离岸钻探
offshore drilling rig 海上钻井平台；海上钻探装置；海上钻探机
offshore drilling riser 海上钻井装置；海上钻井隔水管
offshore drilling technology 浅海钻探技术
offshore earthquake 近海地震
offshore engineering 离岸工程；近海工程
off-shore engineering 近海工程
offshore environment 近海环境
offshore exploration 海上勘探
offshore exploration for oil and gas 海上油气勘探
offshore facies 近海相；滨外相
offshore facility 海上设施
offshore faulting 近海断层
offshore floating drilling 海上浮式钻井
offshore floor 近海底部
offshore flow 滨外流
offshore gas 海洋天然气
offshore gas field 海上气田
offshore gas production 海上采气
offshore geological storage 海洋地质储存
offshore geology 离岸地质
offshore geophysical and geological survey 近海物探和地质调查
offshore geophysical prospecting 海洋地球物理勘探
offshore geophysical survey 海上地球物理测量
offshore geotechnics 近海岩土工程；离岸岩土工程
offshore gravel packing operation 海洋油井砾石充填作业
offshore gravity meter 海洋重力仪
offshore hazard 近海灾害
offshore hydrocarbon reservoir evaluation 海洋油气藏评价
offshore hydrocarbon resources 海洋油气资源
offshore installation 海上装置；海上设施
offshore jack-up drilling vessel 近海自升式钻井船
offshore location 海上定位
offshore logging unit 海洋测井仪
offshore mineral resources 海岸矿产资源；近海矿产资源
offshore mining 水下开采；海底开采
offshore mooring-island structure 滨外系泊岛式建筑物
offshore morphotectonics 离岸构造地貌学；海洋构造地貌特征
offshore oil 海洋石油；海上石油
offshore oil and gas basin 海洋油气盆地
offshore oil and gas reserves 海洋油气储量
offshore oil contamination 近海石油污染
offshore oil delivery 海底油管输送；海上原油输送
offshore oil drilling 近海石油钻探；海上石油钻探
offshore oil exploitation 海上石油开采
offshore oil field development 海上油田开发
offshore oil industry 海洋石油工业
offshore oil production 海上采油

offshore oil production platform 海上采油平台
offshore oil spillage 近海溢油
offshore oil storage platform 海上储油平台
offshore oil transportation platform 海上输油平台
offshore oil wells 近海油井
offshore oil-gas bearing basin 海洋油气盆地
offshore oil-gas field 海上油气田
offshore oil-gas field development 海上油气田开发
offshore oil-gas horizontal well 海上油气水平井
offshore oil-gas pool 海上油气藏
offshore oil-gas recovery 海洋油气采收率
offshore oil-gas resource evaluation 海洋油气资源评价
offshore oil-gas ultimate recovery 海洋油气最终采收率
offshore oolite bar 滨外鲕粒沙坝
offshore operation 海上作业
offshore petroleum geophysical prospecting 海洋石油地球物理勘探
offshore petroleum resources 海洋石油资源
offshore pile 近海桩；离岸桩
offshore pipeline 海底管道；海底管线
offshore pipeline construction 近海管道建筑
off-shore placer 滨外砂矿
offshore platform 近海平台
offshore possible hydrocarbon reserves 海洋油气预测储量
offshore production facility 海上油田生产设施
offshore production system 海上生产系统
offshore production system with onshore terminal 半海半陆式海上生产系统
offshore prospecting 近海勘探
offshore proved hydrocarbon reserves 海洋油气探明储量
offshore purchases 国外采购
offshore rig 海上钻井设备；海中钻塔
offshore sand and gravel 离岸砂砾石
offshore sea-wall 近海防波堤；离岸防波堤
offshore sediment 近海沉积物；滨外沉积
offshore seismic 近海地震；海上人工地震
offshore seismic exploration 近海地震勘探
offshore set-up 海上设备
offshore shelf 近海陆架
offshore shoal 滨外沙洲
off-shore shooting 海洋地震勘探；近海地震勘探
offshore single-source and single-streamer seismic acquisition 单源单缆海上地震采集
offshore site 近海场地
offshore slide 离岸滑坡
offshore slope 滨外坡；水下岸坡；近海底坡
offshore soil mechanics 近海土力学
offshore soil mechanics and rock mass mechanics 海洋土力学与岩体力学
offshore storage unit 海上贮油装置
offshore strip 滨外带；近海带
off-shore stripping 海上剥离
offshore structure 海上建筑物；海上平台；近海结构物
offshore survey 海上勘探；近海测量
offshore terminal 近岸设施；近岸码头；浮码头
offshore terrace 滨外阶地；近海阶地
offshore trap 海上圈闭
offshore trough 滨外海槽；近岸海槽
offshore wave 离岸波
offshore wave climate 离岸波况

offshore well	海上井；海上油井
offshore well drilling platform	近海钻井平台
offshore wildcat well	海上预探井
offshore wind	陆风；向海风；离岸风
offshore work platform	海上工作平台
offshore zone	滨外带；岸外带；近海带
offshore-onshore line	海底至海滨管线
off-side	后面，反面；在非正式班；在非规定的位置
off-site work	场外工程
off-size	非规定尺寸；尺寸不合格
offspeed	转速偏离
offspur	山脉的支脉
off-standard	不合标准的
off-standard size	非标准尺寸
off-structure well	构造外的井
off-surge	外波浪
offtake	钻孔衍开的钻杆长度；抽取；开采；夺去，耗去，移去；排水沟；排水渠
offtake drift	排水平硐
off-take gallery	回风平巷
offtake pattern	开采井网
offtake pipe	排水管
off-take point	取出点
off-take potential	产能
off-take rate	产量
offtake strategy	开采策略
offtake structure	分水结构物
off-take target	开采目的层
off-take well	生产井
off-target drilling	未达靶位钻井
off-time	关井时间；非规定时间；停止时间
off-tract injection system	开采区外的注入系统
offtrend structure	垂直于走向的构造
offward	离岸的
OFS	光纤传感器
ogee	S形的；双弯形的；S形曲线；双弯曲线
ogive curve	拱形曲线；累计频率曲线
OGR	矿床地质评论
Ohio sampler	压入式取土器；俄亥俄取样器
ohm gauge	电阻计
ohm unit	阻抗元件；电抗元件；测量用标准电阻
oikocryst	主晶
oil	油；石油；润滑油；脱模油，铺路油；油分，涂油，润滑
oil accumulation	油聚集；油形成；油藏
oil advancing contact angle	油前进接触角
oil and gas	油气
oil and gas bearing basin	含油气盆地
oil and gas emplacement	油气成藏
oil and gas geological exploration	油气地质勘探
oil and gas prospect	含油气远景
oil and geothermal water exploitation	油和地热水开发；油热联产
oil and water separation	油水分离
oil area	含油面积
oil asphalt	石油沥青；油沥青
oil band	集油带；油带
oil band saturation	油带饱和度
oil basin	油田；含油盆地；贮油池；贮油器
oil bearing	含油的
oil bearing basin	含油盆地
oil bearing bed	含油地层
oil bearing formation	含油地层
oil bearing horizon	含油地层
oil bearing reservoir	储油层
oil bearing rock	油母岩；含油岩
oil bearing stratum	含油地层
oil bearing structure	储油构造
oil bed	含油层
oil belt	油带
oil car	油槽车
oil carrier	油船；油槽车
oil cellar	地下油库；油盒
oil coal	沥青煤
oil column	油柱
oil column thickness	油藏含油高度
oil company	石油公司
oil component	石油组分
oil composition	石油组分；油的组成
oil compressibility	原油压缩系数
oil concession	石油特许权
oil connectivity	油连续性
oil conservation	石油资源保护
oil consuming country	石油消费国
oil contamination	油污染
oil content	含油量
oil corrosion	油侵蚀
oil corrosion test	油腐蚀试验
oil country tubular goods	石油管材
oil cut	油侵
oil cutting	油侵
oil dehydrating	石油脱水
oil dehydrator	石油脱水器
oil delivery	石油输送
oil density	油的密度
oil deposit	石油矿床；油田，油藏
oil depot	油库
oil derrick	井架
oil detection	探油
oil development	石油开发
oil displacement agent	驱油剂
oil displacement direction	驱油方向
oil displacement efficiency	驱油效率
oil displacement pressure	驱油压力
oil displacement process	驱油过程
oil distribution	原油分布
oil dock	油港
oil domain	油域
oil drag	在边界层的润滑油黏度
oil drain	泄油
oil drain assembly	排油装置
oil drain hole	放油孔
oil drain pump	泄油泵
oil drain valve	卸油阀
oil drainage	油的排泄
oil drilling	石油钻井
oil emulsion drilling fluid	混油钻井液
oil emulsion mud	混油泥浆
oil engineer	石油工程师
oil entrapment	石油捕集作用

oil entry 出油层位	oil isoperm 油藏等渗透率图
oil entry profile 出油剖面	oil isopotential surface 油等势面
oil equipment 石油设备	oil jack 油压千斤顶
oil equipotential surface 油等势面	oil jar 油压震击器
oil equivalent 油当量	oil jetty 石油码头
oil exploitation geology 石油开发地质	oil keeper 油承；储油槽；储油器
oil exploitation strata 石油开发层系	oil layer 含油地层
oil exploration 石油勘探	oil lead 油道
oil extraction 原油开采；石油抽提	oil leak 漏油；油泄漏
oil field 油田	oil leakage 漏油；原油泄漏
oil field automation 油田自动化	oil leg 油柱
oil field development 油田开发	oil lenses 透镜状油藏
oil field equipment 油田设备	oil line 输油管线；油管
oil field production capacity 油田产能	oil loading port 装油港口
oil field water 油田水	oil log 油井记录
oil finder 石油勘探工作者	oil map 油田图
oil finding 找油；找油的	oil mass 油量
oil flecked 油斑	oil measure 油层；含油层组
oil flecked mud 油污泥浆	oil measurement 石油计量
oil floor 生油底界	oil meter 石油计量器；油量计
oil flow 油流量	oil migration 石油运移
oil flow meter 油流量计	oil mining method 石油开采方法
oil formation volume factor 原油地层体积系数	oil movability 原油可动性
oil forms of marine facies 海相生油	oil mud 油基泥浆
oil forms of terrigenous facies 陆相生油	oil offtake 产油量
oil fuel 石油；燃料油	oil off-take target 产油目的层
oil gallery 油沟	oil operated transmission 油压传动
oil gas 石油气；石油裂解气	oil origin 石油成因
oil gas contact 油气接触面	oil originally in reservoir 油层原始储油量
oil gas interface 油气界面	oil patch 含油岩屑；油斑；油渍；油田；石油产区；油层中残存的油
oil gas separation 油气分离	oil permeability 原油渗透率
oil gasification 石油气化	oil petroleum 石油沥青
oil gauge 油量计；油位表；油液比重计	oil phase 油相
oil generation 石油生成	oil phase compressibility 油相压缩系数
oil generation zone 石油生成带	oil pipeline 油管；输油管线
oil genesis 石油成因	oil pipeline system 油管系统
oil geochemistry 石油地球化学	oil piping 输油管网
oil geology 石油地质	oil piping layout 油管系统的布置
oil glut 石油过剩	oil polluted water 石油污染水域
oil gradient 油柱压力梯度	oil pollution 油污染；石油污染
oil gradient line 石油梯度线	oil pool 油田；油藏
oil horizon 含油层；井中油柱高度	oil pool outline 油藏边缘
oil hydrometer 石油比重测定仪	oil possibility 含油可能性
oil hydrosol 油水溶胶	oil potential 潜在石油储量
oil hysteresis 油滞后现象	oil potentiometric surface 油层等势面
oil immersed 浸油的	oil power 石油动力；石油动力机
oil immersion 油浸	oil press 油压机；液压机
oil immersion lens 油浸镜头	oil pressure 油压
oil immersion method 油浸法	oil pressure alarm 油压报警
oil in excess of quota 超产油	oil pressure check valve 油压止回阀
oil in place 油层中现存油量	oil pressure controller 油压控制器
oil in reserve 储存油；尚未能利用的石油储量	oil pressure damper 油压减震器
oil in sight 在望储量	oil pressure engine room 油压动力室
oil in situ 原地石油储量	oil pressure gauge 油压表；油压计
oil in storage 管道中的油	oil pressure pump 油压泵
oil incidence factor 生油率	oil pressure system 油压系统
oil industry 石油工业	oil processing plant 原油处理站
oil industry wastewater 石油工业废水	oil producer 石油生产商；产油井
oil initially in place 原始石油地质储量	oil producing area 产油区
oil insulation 油绝缘；隔油	

oil producing country	产油国
oil producing interval	产油层段
oil producing potentiality	产油潜能
oil producing region	采油区
oil producing reservoir	产油层
oil producing zone	产油层
oil product	石油产品
oil production	采油；石油开采；采油量
oil production curve	采油曲线
oil production history	采油史
oil production level	产油水平
oil production pattern	采油井网
oil production platform	采油平台
oil production rate	采油速率
oil production response	产油量反应
oil productive capacity	产油能力
oil proof	耐油的
oil property	石油性质
oil prospecting	石油勘探
oil province	含油区
oil puddle	油团
oil pump	油泵
oil pumping	泵油；抽油
oil puncher	石油经纪商
oil purification	净化
oil purification plant	净油站
oil rate performance	产油量动态
oil rate response	产油量见效
oil reclaimer	集油器
oil reclamation	废油再生
oil recoverable	可采油量
oil recovery	采油；油的回收；油再生
oil recovery efficiency	采油效率
oil recovery factor	原油采收率；原油采收系数
oil recovery mechanism	采油机理
oil recovery performance	采油动态
oil recovery process	采油过程
oil recovery projection	采油设计
oil recovery rate	采油速度
oil recovery technique	采油技术
oil refinery	炼油厂
oil refinery effluent	炼油废水
oil refinery waste	炼油厂废物
oil region	油区
oil relative permeability	油相对渗透率
oil relief valve	安全阀；减压阀；放油阀
oil remaining	剩余油
oil reserves	石油储量；石油储备
oil reservoir	油藏；储油层
oil reservoir engineering	油藏开采工程
oil reservoir rock	贮油岩层；储油岩层；集油层
oil reservoir volume factor	原油地层体积系数
oil residue	残渣油；油渣
oil residue recuperation	残油回收
oil resources	石油资源
oil return passage	油回流管道
oil rig	石油钻塔；油井设备；抽油装置
oil right	采油权
oil rock	含油岩；含油岩石；含油岩层
oil royalty	石油矿区使用费；石油开采税

oil run	在一段时期产出的油量；流出流量
oil sales line	商品原油管道
oil salvage	废油再生
oil sand	油层；油砂；含油砂层
oil sand pack	油砂层模型
oil sand performance	油层动态
oil sand processing plant	油砂加工厂
oil sand tailing	油砂残渣
oil saturation	油饱和率；含油饱和度
oil saturation distribution	含油饱和度分布
oil saturation profile	含油饱和度剖面
oil seal	油封
oil search	石油普查
oil seep	油渗漏；油苗
oil seepage	漏油；渗油；油苗
oil segment	残余油滴
oil self-sufficiency	石油自给
oil separation	油水分离；油分离
oil separator	油气分离器
oil separator well	油水分离器井
oil series	油层组
oil servicing	供油
oil servicing facilities	供油设备
oil shale	油页岩；油母页岩
oil shale fuel	页岩油
oil shale of humic origin shale	腐植成因的油页岩
oil shale processing	油页岩加工
oil shale refining	油页岩炼油
oil shale underground method	油母页岩地下蒸馏法
oil shale waste	油页岩渣
oil sheet	薄油层
oil shock absorber	油压缓冲器；油压减震器
oil shortage	石油不足
oil show	油显示
oil shrinkage	石油紧缩
oil shrinkage factor	原油紧缩率
oil slick	油斑；油膜
oil sludge	油泥；油垢
oil sludge formation	油垢形成
oil slug	油段塞
oil slurry	油基砂浆
oil soaked sand	油浸砂
oil soluble additive	油溶性添加剂
oil soluble fluid loss additive	油溶性降滤失剂
oil soluble surfactant	油溶性表面活性剂
oil source	油源
oil source rock	生油岩
oil spill	石油泄漏；油溢污染；浮油；漏出的油；海上飘油
oil spill disaster	溢油灾害
oil spill physical treatment	溢油物理处理
oil spill removal	水面除油；溢油去除
oil spill treatment	溢油治理
oil spillage	漏油；燃油溢出
oil spring	油泉
oil stained rock	浸油岩层
oil station	油站
oil stock	库存油
oil stone	油砥石；油石
oil storage	石油储存；储油库

oil storage installation	贮油装置
oil storage ship	储油船
oil storage tank	贮油箱；贮油罐
oil storage tank sludge	油库污泥
oil store	油库
oil storing	石油储存
oil stream	油流
oil string	油层套管
oil structure	储油构造；含油构造
oil suction pump	吸油泵
oil tank	油罐；油箱；油槽车
oil tank farm	燃油贮存库；油管区；油库
oil tankage	油罐容量；储油容量
oil tar	焦油
oil temperature	油温
oil temperature protection	油温保护
oil terminal	油库；石油港；石油码头
oil testing	试油作业
oil thermometer	油温表
oil thief	取油样器；油样采集管
oil threshold	生油门限
oil trace	油痕
oil transferring	石油输送
oil transmission line	原油输送管线
oil trap	石油圈闭；隔油池；集油器
oil trapping curve	原油捕获曲线
oil truck	油罐汽车
oil trucking	汽车运油
oil type gas	油型气
oil value	油量
oil viscosity	油的黏度
oil volatility	油的挥发性
oil volume factor	原油体积系数
oil waste water treatment	含油废水处理
oil well	油井
oil well casing	油井套管
oil well cement	油井水泥；堵塞水泥
oil well cementing	油田固井
oil well completion	油井完井
oil well derrick	井架
oil well drilling	石油钻井
oil well drilling bit	石油钻头
oil well drilling equipment	石油钻井设备
oil well drilling fluid	石油钻井液
oil well flooding	油井驱
oil well fracturing	油井压裂
oil well hydraulic fracturing	油井水力压裂
oil well logging	石油测井
oil well potential test	油井稳定试井
oil well production	油井产量；油井生产
oil well productivity	油井产能；井管滤网
oil well pump	抽油泵
oil well screen	油井滤管
oil well structure	油井井身结构
oil well testing	油井测试
oil well tubing	油管
oil wet	油湿的；亲油
oil wet capillary	亲油毛细管
oil wet core	油湿岩心
oil wet drainage path	亲油泄油通道
oil wet formation	油湿地层
oil wet grain	亲油颗粒
oil wet sand	油湿砂岩
oil wet surface	油湿表面
oil wettability	油湿性
oil whiting test	渗透探伤
oil window	油窗
oil withdrawal	采油
oil yield	油产量
oil zone	油带
oil-and-gas-bearing basin	含油气盆地
oil-aqueous interface	油水界面
oil-aqueous solution interface	油-水溶液界面
oil-base core	用油基泥浆钻取的岩心
oil-base fluid	油基液
oil-base mud	油基泥浆；石油基钻泥
oil-base water shutoff agent	油基堵水剂
oil-based foam	油基泡沫
oil-based mud	油基泥浆
oil-bearing	含油的
oil-bearing area	含油面积
oil-bearing block	含油块体
oil-bearing formation	含油地层
oil-bearing reservoir	储油层
oil-bearing rock	含油岩；油母岩
oil-bearing sand	油层；含油砂岩
oil-bearing series	含油层系
oil-bearing shale	含油页岩
oil-bearing strata	含油层
oil-bearing structure	储油构造
oil-brine interface	油-盐水界面
oil-chuck	液压卡盘；油压卡盘
oil-coal	沥青煤
oil-coal association	油-煤组合
oil-column height	含油柱高度
oil-containing formation	含油层
oil-content	含油量
oil-continuous completion fluid	油为连续相的完井液
oil-continuous phase	连续油相
oil-country pipe	油矿作业用的管子
oil-cut	含油量
oil-cut curve	含油百分数曲线
oil-cut mud	含油泥浆；油侵泥浆
oil-depth gauge	油面测定器；油高测定器
oil-dispersible clay	可分散在油中的黏土
oil-displacement mechanism	驱油机理
oildom	油区
oilfield brine	油田卤水；油田盐水
oil-field development	油田开发
oilfield development plan	油田开发方案
oilfield drilling	油田钻井
oil-field hydrogeological investigation	油田水文地质勘查
oilfield life	油田寿命
oilfield unit	油矿单位；油田单元
oil-field water	油田水
oilfield water technology	油田水处理工艺
oilfielder	从事钻井、采油和铺管的工人
oil-filled	充油的
oil-filled bushing	油浸套管
oil-forming	生油的

oil-forming shale 生油页岩
oil-free 不含油的
oil-free fluid 无油液体
oil-fuel depot 石油燃料库
oil-gage 油量计
oil-gas bearing structure 含油气构造
oil-gas engineering 油气工程
oil-gas field 油气田
oil-gas field development 油气田开发
oil-gas mobility ratio 油气流度比
oil-gas reservoir structure 储油气构造
oil-gas separator 油气分离器
oil-gas well 油气井
oil-gas-water separation 油气水分离
oilgear 油压传动
oil-generating window 生油窗
oil-generative assemblage 生油层系
oil-impregnation 石油浸染
oiling 润滑；上油；加油
oil-intermediate water 油层层间水
oil-leak detector 漏油检验器
oil-lower water 油层下部水
oil-measuring pump 油计量泵
oil-oil correlation 原油对比
oil-operated 液压传动的；油压传动的
oil-overflow valve 油溢流阀
oil-phase saturation 油相饱和度
oil-pressure shut-off switch 油压开关
oil-pressure unit 油压装置
oil-producing capacity 产油能力
oil-producing formation 产油层
oil-producing pay 产油层
oil-producing rate curve 采油量曲线
oil-producing sandstone 产油砂岩
oil-producing unit 采油单元；采油区
oil-production decline curve 产油量递减曲线
oil-production efficiency 采油效率
oil-prone source rock 易生油的烃源岩
oil-related gas 热成因气
oil-rich 富含石油
oil-rock correlation 油岩对比
oil-saturated reservoir rock 储油岩
oil-saturated section 含油剖面
oil-seal 油封
oil-shale kerogen 油母页岩；油母质
oil-soaked 油浸的
oil-solid interface 油-固体界面
oil-soluble 油溶的；可溶于油的
oil-soluble acid 油溶性酸
oil-soluble wall building solid 油溶性造壁固体颗粒
oil-solvent preservative 油溶性防腐剂
oil-source rock correlation 油源岩对比
oil-spring 石油井；油泉
oil-stained rock 油渍岩石
oilstone 油石
oil-to-oil booster 液压压力放大器；液压增压器
oil-transferring 石油输送；油传送
oil-treated 浸油的；油处理的
oil-treating cost 原油处理成本
oil-upper water 油层上部水

oil-water boundary 油-水边界
oil-water contact 油-水接触面
oil-water contact angle 油-水接触角
oil-water displacement efficiency 水驱油效率
oil-water displacement front 水驱油前缘
oil-water emulsion 油-水乳状液
oil-water interaction energy 油水互作用能
oil-water interface 油与水之间的交界面；油-水界面
oil-water interfacial film 油水界面膜
oil-water interfacial tension 油-水界面张力
oil-water isosaturation surface 油-水等饱和度面
oil-water level 油水界面
oil-water mixture 油水混合液
oil-water ratio 油水比
oil-water relative permeability ratio 油水相对渗透率比
oil-water surface 油水界面；油-水接触面
oil-water table 油水界面
oil-water transition zone 油-水过渡带
oil-water viscosity ratio 油水黏度比
oil-water-rock interaction 油-水-岩相互作用
oil-water-sand interface 油-水-砂界面
oil-well derrick 油井井架
oil-well packing 油井环空封隔
oil-well performance data 油井动态资料
oil-well production information 油井开采资料
oil-well shooting 油井射孔作业
oil-well wastewater 油井废水
oil-wet behavior 亲油特性
oil-wet displacement 亲油驱替
oil-wet gravel 油湿砾石
oil-wet rock 亲油岩石
oil-wet side 亲油侧
oil-wetness 亲油度
oil-wetted surface 亲油表面
oil-wetting agent 油润湿剂
oil-wetting surfactant 油湿活性剂
oily discharge 带油废水排放
oily layer 油层
oily luster 油脂光泽
oily pollutant 油质污染物
oily sewage 含油污水
oily waste 油质废物；含油废物
oily waste water 含油废水
oily water treatment 含油污水处理
oil-yielding stratum 出油层
ojosa 蜂窝构造
okaite 蓝方黄长岩
okawaite 霓辉松脂斑岩
okenite 水硅钙石
Okinawa trough 冲绳海槽
olafite 钠长石
old alluvium 老冲积层；古冲积层；冲积土
old bench flow 古阶地岩流
old channel 古河道；旧河床；废河道
old cliff 古海蚀崖
old coast 古海岸
old course 古河道
old crater 古火山口
old drift 古冰碛；古冲积层
old fault 老断层

old forms	古地形
old groundwater	古地下水
old inland marine basin	古海盆
old karst	古岩溶；古喀斯特
old lake	萎缩湖
old land	古陆
old landmass	古陆块
old landslide	老滑坡
old loess	老黄土
old oceanic trough	古洋盆遗迹
old red	老红层
old red earth	古红色土；老红壤
old red sandstone	古老红色砂岩；老红砂岩
old river bed	古河道
old sea basin	古海盆
old sediment	古代沉积
old shield	老地盾
old shoreline	古滨线；古海岸线
old sliding area	古老滑坡区
old slope	旧斜坡；老斜坡
old soil	古土壤
old stone age	旧石器时代
old territory	采空区域；旧区
old Tertiary	早第三纪；古近纪
old topography	古地形
old total depth	原油井深度
old trough	古海槽
old upland	古高地
old valley	老谷地；古峡谷
old valley filling soil	老谷地堆积土壤
old waste	采空区；老塘
old waste chamber	老矿硐
old water	封存水；古水
old well	旧井
old well plugged back	老井封堵
old well worked over	老井大修
old workings	老工作区；老塘；老巷道
old-age stream	老年期河流
older peat	老泥炭；高度分解泥炭
Oldham-Gutenburg discontinuity	奥德汉-古登堡不连续面
oldhamite	陨硫钙石
oldlake	老年湖；萎缩湖
Olduvai event	奥杜瓦地磁事件
old-well deeper drilling	老油井加深钻井
oleaginous	含油的；产油的；油质的
oleaginousness	含油量
olefiant	生油的
olefin	烯烃
olefin copolymer	烯烃共聚物
olefine hydrocarbon	烯烃
oleic permeability	油相渗透率
oleic phase	油相
oleometer	油比重计；油量计；验油计
oleophilic	亲油的
oleophobic	疏油的；憎油的
oleophobic finisher	疏油整理剂
oleophobic property	疏油性
olgite	磷锶钠石
oligiste iron	赤铁矿；红铁矿
Oligocene	渐新统；渐新世
Oligocene epoch	渐新世
Oligocene ocean	渐新世大洋
Oligocene paleogeography	渐新世的古地理
Oligocene series	渐新统
oligochronometer	微时计
oligoclase	奥长石；更长石
oligoclase andesite	奥安山岩；更长安山岩
oligoclase basalt	长石富钠玄武岩
oligoclase granite	奥长花岗岩
oligoclasite	奥长岩；奥长石
oligocrystalline	半晶质
oligodynamic	微动力的；微动作用的
oligodynamics	微动力学；微动作用
oligoelement	少量元素
oligomict	单成分的
oligomictic	单成分的；单岩碎屑的
oligomictic sediment	单岩碎屑沉积物；陆海沉积
Oligo-Miocene	渐-中新世
oligonite	菱锰铁矿；锰菱铁矿
oligopelic deposit	少泥湖底沉积
oligophyre	更长斑岩；奥长斑岩
oligophyric	少斑的；少斑晶的
oligophyric texture	少斑结构
oligorganic layer	矿质层
oligosaprobic waters	低污染水域
oligosaprobic zone	低污染带
oligosite	奥长岩
oligotaxic ocean	贫属种型大洋；少种型大洋
oligothermal	狭温性；狭低温性
oligotrophic	贫瘠的
oligotrophic bog	贫营养沼泽
oligotrophic lake	缺营养湖
oligotrophic peat	低滋育泥炭；缺营养湖泥炭
oligotrophication	贫营养化
olistoglyph	层间滑痕；重力滑动痕
olistolite	滑来岩块；滑积岩
olistolith	滑塌岩块；滑来岩块
olistostrome	重力滑动沉积；滑塌层；滑来层
olistostromic mélange	泥砾混杂岩
olivenite	橄榄铜矿
olivine	橄榄石
olivine andesine basalt	橄榄中长玄武岩
olivine basalt	橄榄玄武岩
olivine clinopyroxenite	橄榄单斜辉石岩
olivine gabbro	橄榄辉长岩
olivine gabbronorite	橄榄辉长苏长岩
olivine hornblende pyroxenite	橄榄角闪辉石岩
olivine hornblendite	橄榄角闪石岩
olivine leucitite	橄榄白榴岩
olivine melilitite	橄榄黄长岩
olivine melilitolite	橄榄黄长石岩
olivine nephelinite	橄榄霞石岩
olivine norite	橄榄苏长岩
olivine orthopyroxenite	橄榄斜方辉石岩
olivine pacificite	橄榄石太平洋岩
olivine pyroxene	橄榄辉石
olivine pyroxene hornblendite	橄榄辉石角闪石岩
olivine pyroxenite	橄榄辉岩
olivine rock	纯橄榄岩；橄榄岩

olivine tholeiite	橄榄拉斑玄武岩
olivine uncompahgrite	橄榄辉石黄长石岩
olivine websterite	橄榄二辉岩
olivine-clinopyroxenite	橄榄单斜辉石岩
olivine-diabase	橄榄辉绿岩
olivine-dolerite	橄榄粗粒玄武岩；橄榄灰绿岩
olivine-gabbro	橄榄辉长岩
olivine-leucitite	橄榄白榴岩
olivine-norite	橄榄苏长岩
olivine-pyroxenolite	橄榄辉石岩
olivine-rock	纯橄榄岩
olivinfree basalt	无橄玄武岩
olivinite	辉闪苦橄岩；苦闪橄榄岩
ollenite	绿帘金红角闪片岩
olletas	河床凹地
olshanskyite	羟硼钙石
ombrogenous bog	雨积沼泽；高位沼泽
ombrogenous peat	可变泥炭；雨积泥炭；高位泥炭
ombrology	测雨学
ombrometer	雨量计；雨量器
ombrotiphic	雨水沼泽的
Omega chart	奥米加海图
Omega navigation system	奥米加导航系统
Omega system	奥米加系统
omen	预兆；先兆
omission	缺失；省略；遗漏
omission of beds	岩层缺失
omission of earthquake predication	地震漏报
omission surface	沉积小间断面
omnidirectional	全向的；非定向的
omnidirectional characteristic	各向相同特性；无定向特性
omnidirectional coverage	全向覆盖
omnidirectional geophone pattern	全方位检波器组合
omnidirectional hydrophone	全向水平检波器
omnidirectional seismic trigger	无定向地震触发器；无定向地震报警器；无定向地震诱发力
omnilateral pressure	全侧向压力
omphacite	绿辉石
omphacite jade	绿辉石玉
omphacitite	绿辉岩
omphalopter	扁豆状层
on a dry weight basis	按干重的；干重基的
on a field basis	根据现场资料的
on bord	与主解理成直角推进的采煤工作面；采煤工作面垂直主节理推进
on bottom	接触钻孔底部的，在钻孔底部的
on center	中心间距
on depth	达到深度
on face	工作面平行于煤层主节理掘进的
on half bord	与主解理成45°角推进的采煤工作面；与主节理成45°角的煤
on job lab	工地实验室
on level	在水平上的
on line computer	联机计算机
on line data	联机数据
on line operation	联机操作
on load	负重的；有负荷的
on one trip	一次下井
on period	闭合周期
on place	原地；就地；在地层内
on plane	垂直于主节理面；在平面上
on shore	滨岸的；向岸的；在岸上；国内的
on site	原地；就地；在现场位置；在位置；在工地
on site concrete	现浇混凝土
on site work	现场工作
on slip	用卡瓦卡住钻柱
on soundings	浅海区；近岸水域；在近海
on strike	沿走向
on the cross	在主次节理面之间的
on the ground	在地上；当场
on the pressure side	压力侧
on the site computer	现场用计算机
on the spot	在现场
on the strike	沿走向
on the structure	在构造上
onboard computer	船用计算机
on-bottom stability	在海床上的稳定性
on-bottom stability design of submarine pipeline	海底管道稳定设计
once establishment control network	一次布网
once-through	单程的
once-through cracking	单程裂化
once-through method of dust sampling	连续式尘末取样法
once-through operation	一次通过操作
oncoid	似核形石
oncolite	藻灰结核；核形石
oncolith	球状叠层石类
on-deck storage	平台储存
on-dip	沿倾斜向下；沿倾斜下行的
ondograph	高频示波器；电容式波形记录器
ondometer	测波仪；波长计；波形测量器
ondometry	波形测量
ondoscope	辉光管振荡指示器
ondulation	波动
one body tilt method	一体倾斜法
one channel seismograph	单道地震仪
one cut	一次采全高
one degree-of-freedom	单自由度
one dimensional consolidation	一维固结
one dimensional flow	一维流
one dimensional lattice	一维晶格
one dimensional settlement	一维沉降
one dimensional strain	一维应变
one dimensional stress wave propagation theory	一维应力波传播理论
one direction excavation	单口掘进；单向掘进
one hundred year wave	百年一遇的海浪
one lane	单车道
one man drill	单人凿岩机
one pass	一次通过；一次走刀
one pipe system	单管系统
one receiver logging tool	单接收式测井仪
one shaft orientation	单井定向
one shaped concrete base	单锥混凝土基底
one shot multivibrator	单稳态多谐振荡器
one side earth pressure	偏压
one stage crushing	单段破碎；一段破碎
one step technique	一步法技术
one story building	单层建筑物；单层房屋

one trip	起下一次管柱
one trip perforating gravel pack system	射孔与砾石充填一次起下作业装置
one trip run	起下一次管柱作业；一次管柱作业行程
one trip technique	一步法技术
one way channel	单向通道
one way service pipe system	单向供水
one way time	单程时间
one way traffic	单向行车
one-address	单地址
one-address code	单地址码
one-address computer	单地址计算机
one-address instruction	单地址指令
one-armed excentralizer	单臂偏心器
one-atmosphere	一个大气压；常压
one-atmosphere wellhead cellar	常压井口室
one-atmosphere work station	常压工作站
one-channel seismograph	单道地震仪
one-chip computer	单片计算机
one-circle reflection goniometer	单圈反射测角仪
one-core-per-bit memory	每位-磁心存储器
one-day cycle	一天一个循环
one-dimensional	一维的；线性的；单向的
one-dimensional analysis	一维分析法；一元分析法
one-dimensional consolidation	单向固结；一维固结
one-dimensional correction	一维校正
one-dimensional element	一维单元
one-dimensional flow	一维流动；一维渗流
one-dimensional Fourier transform	一维傅里叶变换
one-dimensional Fourier transform of an image	影像-维傅里叶变换
one-dimensional heave	单向隆起
one-dimensional inverse theory	一维反演理论
one-dimensional linear array	一维线性排列
one-dimensional linear problem	一维线性问题
one-dimensional medium	一维介质
one-dimensional method	单向法
one-dimensional model	一维模式；一维模型
one-dimensional nonlinear differential equation	一维非线性微分方程
one-dimensional problem	一维问题
one-dimensional quasi-nonlinear wave propagation theory	一维准非线性波传播理论
one-dimensional reverberation	一维交混回响
one-dimensional scalar wave equation	一维标量波动方程
one-dimensional seismic trace	一维地震道
one-dimensional seismogram	一维地震记录
one-dimensional shear propagation	一维剪切传播
one-dimensional simulation	一维模拟
one-dimensional soil-moisture movement	一维土壤水运动；土中水单向运动
one-dimensional state of stress	单向应力状态
one-dimensional swelling	单向膨胀
one-dimensional water flow	一维水流
one-dimensional wave	一维波
one-directional excitation	单向激振
one-directional loading	单向加载；单向荷载
one-directional ventilation system	单向通风系统
onegite	针铁矿
one-in-100-year flood frequency	百年一遇洪水频率
one-lane	单车道
one-lane travel	单车道行驶；单向行驶
one-layer	单层；单层的；一层的
one-level address	一级地址
one-level sampling mill	单层取样车间
one-limbed flexure	单翼挠曲
one-man drill	单人钻机；单人凿岩机
one-man drilling machine	单人钻机
one-man operated unit	单人操作的设备
one-man paver	单人铺路机
one-panel stope	全高单一回采工作面；单盘区回采
one-parameter	单参数的
one-parameter criterion	单参数准则
one-parameter symmetric model	单参数对称模型
one-pass	一次通过的
one-pass dual operation	单程操作；非循环过程
one-pass image	单轨云图
one-pass machine	单程操作机
one-pass solid removal efficiency	一次过滤固体颗粒清除率
one-percent increments	百分之一增量
one-phase	单相的
one-phase design	一阶段设计
one-piece	整个的；整块的
one-piece construction	整体结构
one-piece set	单一构件支护
one-pip area	单脉冲区
one-point iteration	不动点迭代
one-point method	单点法
one-point observation	单点观测
one-point separation theory	一点分离理论
one-row blasting	单行炮眼爆破
one's complement	二进制反码
one-shot method	单液法；一步法
one-side loading	单侧装载；侧面加载
one-side spectral density function	单侧谱密度函数
one-side water pressure	单侧水压力
one-sided	单侧的
one-sided filter	单边滤波器
one-sided function	单侧函数
one-sided least-squares inverse filter	单边最小平方反滤波器
one-stage	单级
one-stage amplifier	单级放大器
one-stage crushing	单段破碎；一段破碎
one-stage grinding	单段研磨；单段磨矿
one-step cross beam	单步梁
one-step design	一阶段设计
one-step installation	一次安装
one-step method	一步法
one-step statics algorithm	一步静校正算法
one-to-one	一对一
one-to-one assembler	直接汇编程序
one-to-one correspondence	一一对应
one-to-one slope	1∶1的坡度；45°坡
one-track	单线的；单轨的
one-trip crossover packing system	一步法转换充填砾石工具
one-trip gravel pack technique	一步法砾石充填技术
one-trip gravel packing	一步法砾石充填
one-trip method	一步法

one-trip packing system 一步法充填装置
one-trip tool 一步法工具
one-valued 单值的
one-valued function 单值函数
one-variable function 单变量函数
one-way 单向的
one-way attenuation 单程衰减
one-way downgrade 单向下坡
one-way drainage 单向排水
one-way haul distance 单向运距
one-way mining 单翼开采
one-way propagation 单程传播
one-way ram 单作用千斤顶；单作用油缸
one-way ramp 单向匝道
one-way reinforcement 单向钢筋
one-way slip 单向卡瓦
one-way slope 单面坡
one-way time 单程时间
one-way trip 单行程
one-way vent 单向流出
one-way ventilation 边界式通风
one-way wave equation 单向波动方程
one-web-back system 退后一刀支护法；即时前移支架
one-winged stope 单翼回采工作面
onflow 湍流；洪流
on-gauge 标准的；合格的
on-going 正在进行的
on-going excavation 正在进行的开挖
ongoing pilot 在用的试验工厂
on-going project 正在实施的项目
ongonite 翁岗岩
onion skin weathering 葱皮状风化
onion weathering 球状风化；洋葱状风化
onion-skin structure 葱皮状构造
onion-skin weathering 葱皮状风化；洋葱状风化；球状风化
on-job 在工地的
on-job lab 工地实验室
on-job laboratory 工地实验室
on-job trial 工地试验
onkilonite 橄辉霞玄岩
onkolite 核形石
onland 陆上的
on-land drilling 陆地钻井
on-land geologic feature 陆地地质构造
onlap 上覆层；超覆；重叠
onlap fill 上超充填；上覆填土
onlap sequence 超覆层序
onlap slope facies 上超斜坡相
on-line 联机的；联用的；主机控制的，在线
on-line aerophotogrammetric triangulation 联机空中三角测量
on-line analog output 联机模拟量输出；在线模拟量输出
online analytical processing 在线分析处理
on-line analyzer 流程监测仪；在线监测仪
on-line computer 联机计算机
on-line computer control 在线计算机控制
on-line control 联机控制
on-line data base 联机数据库
on-line data handling system 直接数据处理系统
on-line data reduction 联机数据简化；在线数据处理
on-line debugging 在线调试；联机调试
on-line decision system 联机决策系统
on-line diagnostics 联机诊断；在线诊断
on-line digital output 联机数字输出
on-line digitizer 联机的数字化装置
on-line display 联机显示
on-line drilling program 联机钻井程序
on-line dual operation 联机操作；联机工作；在线操作
online error-trapping 联机错误检查
on-line expert system 在线专家系统
on-line input 联机输入
on-line maintenance 不停机检修
on-line memory 联机存储器
online monitoring 联机监视；在线监测
on-line operation 联机操作，联机工作；在线操作
online optical cable monitoring system 光纤在线监测系统
online output 联机输出
on-line plotter 联机绘图器；联机绘图仪
on-line processing 联机处理
on-line proving 在线标定
on-line real time processing 在线实时处理
on-line real-time dual operation 在线实时操作
on-line real-time operation 联机实时操作
on-line real-time processing 联机实时处理
on-line real-time system 联机实时系统
on-line retrieval 联机检索
online storage 联机存储器
on-line system 联机系统；在线系统
on-line terminal mode 联机终端模式
on-line test 联机试验
on-line test routine 在线检验程序
on-line timing plan selection 动态查表
on-line unit 联机装置
on-load 装载，加载，负载
on-load operation 负载操作
on-load regulation 载荷调节
on-load running test 负载运行测试
on-load test 带负荷试验
on-location 现场
on-off bottom pressure-drop 钻-悬循环压力差；钻-悬循环压力降
on-off connector 上卸扣连接器
onofrite 硒汞矿
on-peak 最大的；最高的；峰值的
onset 开始；发动；波至
onset time 初动时间；发震时间
onshore 滨岸的；岸上的；陆上的
onshore base 陆上基地
onshore current 向岸流；沿岸沙洲
on-shore drilling 陆地钻井
onshore gale 向岸大风，海滨大风
onshore oil field 海滨油田；陆上油田
onshore oil terminal 陆岸石油集输终端
onshore pile 岸上桩
onshore pipeline 海岸管线
onshore storage 海滨储存；陆上封存
on-shore wind 向岸风；离海风
on-site 原位；原地；现场

on-site computer	现场计算机
on-site concrete	现浇混凝土
on-site data gathering	现场数据采集
on-site data monitoring	现场数据监测
on-site data processing	现场数据处理
on-site engineer	现场工程师
on-site experience	现场经验
on-site identification of soil and rock	土石场现场鉴别法
on-site inspection	现场检验；现场检查；实地调查；现场调查
on-site interpretation	现场解释
on-site investigation	现场调查
on-site maintenance	现场维修
on-site manufacture	就地制造；当场制造
on-site monitoring	现场监测
on-site proving	现场标定
on-site rheology test	现场流变性测试
on-site sampling	现场取样
on-site supervision	现场监控；现场监管
on-site survey	现场勘测；现场调研
on-site test	实地测试；现场测试
on-site training	现场培训
on-site turbidity test	现场浊度试验
on-site work	现场工作
on-size gravel	符合尺寸要求的砾石
on-specification gravel	符合规格的砾石
onspeed	达到给定速度
on-spot strip mining	就地露天开采
on-steam instrument	流程监控器；连续监控仪
on-stream pressure	操作压力
on-structure well	构造井
ontariolite	中柱石；针柱石；钙钠柱石
on-the-job assembly	现场装配
on-the-job condition	现场操作条件
on-the-job research	现场研究
on-the-road mixer	路上拌和机
on-the-run	在运转中；在操作中；在进行中
on-the-site analysis	现场分析
on-the-solid	不掏槽爆破；炮眼利用率低的爆破；炮窝
on-the-spot	现场的
on-the-spot determination	现场测定
ontogeny of minerals	矿物发生学
on-track full section undercutting cleaner	大型全断面清筛机
on-trend well	位于油层主要渗透率方向上的井；位于预计开发井网上的井
onward movement	前向运动
on-way resistance	沿程阻力
onychite	雪花石膏；花纹石笋
onyx	缟状石灰华；缟玛瑙；缟状燧石
onyx marble	条纹状大理岩
onyx mortar	玛瑙臼
oocastic chert	空鲕状燧石
ooid	鲕粒；鲕石；鱼卵石
ooidal	鲕石的；鲕粒的
ooide	鲕石
oolicastic chert	空鲕状燧石
oolimold	空鲕粒
oolite	鲕状岩；鲕石；鲕粒；鱼卵石
oolite bar	鲕状砂坝
oolite ironstone	鱼卵状铁矿
oolite limestone	鲕状石灰岩
oolite-bearing	含鲕粒的
oolith	鲕石；鱼卵石；鲕粒
oolitic	鲕状的；鲕粒岩的
oolitic beach	鲕粒滩
oolitic biosparite	鲕状生物亮晶灰岩
oolitic chert	鲕状燧石
oolitic delta	鲕粒三角洲
oolitic dolomite	鲕状白云岩
oolitic formation	鲕状层
oolitic hematite	鲕状赤铁矿
oolitic iron ore deposit	鲕状铁矿床
oolitic limestone	鲕粒灰岩
oolitic pelmicrite	鲕状团粒微晶灰岩
oolitic porosity	鲕状孔隙性
oolitic shoal	鲕滩；鲕粒浅滩
oolitic shoal reservoir	鲕滩储层
oolitic structure	鲕状构造
oolitic texture	鲕状结构
ooliticity	鲕状性质
oolitoid	似鲕状
oomicrite	鲕状微晶灰岩；鲕状泥晶灰岩
oomicrudite	鲕粒泥晶砾屑灰岩
oomicsparite	鲕粒微晶亮晶灰岩
oomold	鲕状穴
oomoldic	鲕穴状的
oomoldic carbonate	鲕模状碳酸盐岩
oomoldic chert	鲕模状燧石
oomoldic porosity	鲕穴状孔隙
oopellet	鲕球粒
oospararenite	鲕粒亮晶砂屑灰岩
oosparite	鲕状亮晶灰岩
oosparmicrite	鲕粒亮晶泥晶灰岩
oosparrudite	鲕粒亮晶砾屑灰岩
ooze	软泥；渗出；慢慢流出
ooze film	软泥薄层
ooze mud	淤泥；软泥
ooze out	渗出
ooze turbidity	软泥浊流层
ooze water	软泥水
oozing	渗漏
oozing volcano	静火山
oozuanolite	蛋粪石
oozy	淤泥质的；软泥的；渗出的
opacity	不透明性；不透明度；浑浊度
opacity coefficient	不透明度；不透明系数
opaco	背阳坡
opal	蛋白石，猫眼石；欧泊石
opal agate	玛瑙蛋白石；蛋白玛瑙
opal glass	乳白玻璃
opal sinter	蛋白石质硅华
opaline	蛋白石的；乳白色的
opalization	蛋白石化
opalize	使呈乳白状
opalus	蛋白石
opaque	不透明体；不透明的
opaque matrix	不透明基质
opaque mineral	不透明矿物
opdalite	苏云石英闪长岩

opdar 光学定向和测距
open a hole 清理井；打井，开井
open a mine 矿山开发
open a well 打井；开井
open air 露天；户外
open air equipment 户外设备
open air model 露天模型
open air plant 露天车间；露天装置
open air station 露天遗址
open anomaly 出露异常
open arch 明拱
open area 露天场地
open area factor 开孔面积系数
open area ratio 开孔面积比
open bay 开阔海湾
open blasting 露天爆破；明探爆破
open caisson 沉井；开口沉箱
open caisson foundation 开口沉箱基础；沉井基础
open caisson method 沉井法
open caisson sinking by suction dredge 吸泥下沉沉井
open caisson wall 沉井壁
open caisson with a tailored cutting edge 高低刃脚沉井
open caisson with cross walls 格墙沉井
open caisson with two shells of wire-mesh cement 双壁钢丝网水泥沉井
open canal 明渠
open car park 露天停车场
open carrying structure 开式承载结构
open cast 露天矿
open cast brown coal mining 露天褐煤开采
open cast mining 露天开采
open casting 露天开采
open centrifugal pump 开式离心泵
open channel 明渠；明槽；开放水道
open channel diversion 明渠导流
open channel flow 明渠流；明槽流
open channel spillway 开敞式溢洪道
open channel supply system 露天水道输水系统
open circuit water supply 开路供水
open circulating 开式循环
open circulating water 开式循环水
open coal 露天煤矿
open coal mine area 打开煤炭区
open coal seam in shaft 开放煤层轴
open coast 开放海岸；开阔海岸
open cog 空心垛式支架
open column 空心柱
open combination drive 开式混合驱动
open condition of surface 可透水的表面条件；松散的表面状态
open conduit 明渠；畅通通道；明流管道
open crack 张开的裂缝
open crib 空心木垛；未填石木垛
open cribbing 间隔井框支护
open cube 立方体展开图
open cube display 展开立体图
open cut 露天矿；明沟，明堑；明挖；露天开挖；揭露剖面
open cut foundation 明挖基础
open cut method 明挖法；露天开采法
open cut mine 露天矿

open cut mining 露天开采；露天采矿
open cut ore mine 露天开采矿山
open cut runnel 明洞
open cut tunnel 明挖隧道；明洞
open cut working 露天采矿
open cutting 露天开采
open cutting wheel 开放式切削刀盘
open data link interface 开放数据链路接口
open data service 开放式数据服务 ODS
open database connectivity 开放式数据库互连
open discontinuity 大裂缝；张开或张大的裂缝
open distribution reservoir 明渠排水
open district 未建成区；房屋稀少地区；空旷地区
open ditch 明沟；明渠
open ditch drainage 明渠排水；明沟排水
open drain 明渠；排水明沟；露天排水渠
open dredging process 明挖法
open drive sampler 薄壁压入式取土器；敞开击入式取样器
open dug foundation 明挖基础；明挖地基
open end 矿柱的回采端
open end method 短柱开采法
open ended branch 开启式分支
open ended tubing method 开口油管作业法
open ending 矿柱外端的开采
open entry 开孔率；滤水管孔隙率
open entry of filter 过滤器孔隙率
open estuary 不冻河口弯；开敞河口弯
open excavation 露天挖掘工程；露天开采；开挖工程；明挖；明堑；明坑
open excavation foundation 明挖基础
open face 无支护的工作面
open face blasting 临空爆破；多面临空爆破
open face shield 敞开式盾构
open fault 张开断层；开口断层
open fissure 开口裂缝；开敞裂隙；张裂缝
open flame 明焰；无遮盖火焰；明火
open floor 明桥面
open floor duct 明沟
open flow 敞喷
open flow line 敞喷管线
open flow potential 敞喷产量；无阻流量
open flow pressure 油井敞喷时的油藏压力
open flow test 敞喷测试
open flume 敞开式水槽；明槽引水室
open flume setting 明槽式装置
open fold 敞开褶皱；开阔褶皱
open folded zone 宽缓褶皱带
open fracture 开口裂隙；张断裂；张开裂缝
open frame reef 中空的骨架生物礁
open full-face TBM 敞开式全断面隧道掘进机
open furrow 明沟
open gash 张口裂缝
open Geographic Information System 开放式地理信息系统
open geothermal reservoir 无盖层地热储集层
open GIS 开放式地理信息系统
open GIS Service 开放式地理信息服务模式
open goaf 未充填的采空区
open grain 松颗粒；大孔隙性

open grain structure	大孔结构；多孔式结构；粗粒结构；粗晶组织
open ground	裸露地层；揭露土层；剥土；开放空间
open harbour	无掩护的港；开敞港口
open height	液压支架最大高度
open hillslope landslide	山坡滑坡；边坡塌滑；山坡塌滑
open hole	裸井；裸眼；井口无套管的钻孔
open hole area	裸眼段渗流面积
open hole arrangement	裸眼井结构
open hole bridge plug	裸眼桥塞
open hole by-pass valve	裸眼旁通阀
open hole caliper	孔径仪；井径仪
open hole completion	裸眼完井
open hole displacement	裸眼驱替
open hole fishing	裸眼打捞
open hole packer	裸眼封隔器
open hole perforator	裸眼井射孔器
open hole portion	裸眼部分
open hole productivity	裸眼井产能
open hole section	裸眼井段
open hole single packer test string	裸眼单封隔器测试管柱
open hole sloughing	裸眼井段坍塌
open hole straddle packer test string	裸眼跨式双封隔器测试管柱
open hole survey	裸眼测量
open hole test	裸眼测试
open hole tester	裸眼井地层测试器
open hole well	裸眼井；无套管井
open hook	开口吊钩
open hydraulic piezometer	开敞式孔隙水压计；开敞式水压计
open impeller pump	开式叶轮泵
open installation	露天装置；露天设备
open interval	裸眼井段；开区间
open jet	线应力
open joint	开口节理；张开节理；裂开节理；裂缝接合；明缝
open jointed rock	张开裂隙岩石
open lake	有出口的湖；开口湖
open landscape layout	大空间布局
open light mine	无瓦斯矿
open loop	开路；开口回路；开环
open loop control	开环控制
open loop control system	开环控制系统
open loop numerical control system	开环数控系统
open loop system	开环系统
open loop unstable	开环非稳定性
open marginal geosyncline	陆缘地槽，地向斜
open marine facies	开阔海相
open marine platform facies	开阔海台地相
open marsh facies	开阔沼泽相
open method	开放式方法
open mining	露天开采；露天采矿
open mining of coal	煤炭露天开采
open nullah	明渠；大明渠；露天大沟渠
open ocean	公海
open off	开始进行长壁开采；开始开拓新工作区；开始掘进支巷
open oil bearing rock	不封隔的含油层；透油岩层；渗油岩层
open ore	露天矿
open overhand stope	空场上向梯段回采工作面
open overhand stoping	无支护向上台阶法
open packing	疏松堆积
open perforation	畅通的射孔孔眼
open physicochemical system	开放的物理化学体系
open pier	开口防波堤；开敞式防波堤
open piled jetty	透空式桩基突堤
open pipe piezometer	开口测压计；竖管测压计
open pipe string	开口的管柱
open pit	露天矿；露天采场；露天探坑；取土坑；露天开采
open pit and underground mining	露天和地下开采
open pit backfill	露天矿坑回填
open pit bench	露天矿台阶
open pit bottom	露天矿的底盘
open pit boundary	露天开采边界
open pit deepening	露天采场延深
open pit excavation	露天矿采掘；露天基坑开挖
open pit exploitation	露天开采
open pit footwall	露天采场底盘
open pit limit	露天矿开采极限边界
open pit method	露天开采法
open pit mine	露天矿
open pit mining	露天采矿；露天开采
open pit mining slope	露天采矿边坡
open pit oil mining	露天采油
open pit optimization	采场优化
open pit quarry	露天矿场
open pit slope	露天矿边坡
open pit slope enlarging	露天采场扩帮
open pit transportation	露天矿运输
open pit ventilation	露天矿通风
open pit working	露天采矿场；露天采矿作业
open pitting	露天开采
open plain	开阔平原
open planning	自由式平面布置；开敞布置
open platform	开阔台地
open platform facies	开阔台地相
open pore	连通孔隙
open porosity	开孔孔隙度
open porous texture	开孔结构
open power fluid system	开式动力液系统
open pressure	消耗已久的气井压力；开启压力
open ratio	筛缝面积比
open reservoir	明坑；露天水库；集水坑；敞开油储；不封闭油层
open road drain	道路排水明沟
open roadstead	开敞锚地
open rock	多孔隙岩石；多孔岩石
open roof	无支护顶板；平台；开敞式屋顶
open sand	多孔隙砂层；多孔性砂土；多孔砂；松砂
open sea	开阔海洋；公海；外海
open sea canal	通海运河
open sea pollution	公海污染
open sea tidal flat	开放的海滩涂
open seafloor mud disposal area	海上淤泥卸置区
open set	开放集，开集
open shaft	通达地面的竖井

open shelf 开放大陆架
open shield 开放式盾构；开口盾构
open shortest path first 开放最短路径优先
open sight alidade 开放式照准仪
open site 开启位置；开放平台；开放场地测试
open slot 开槽；开沟
open sluice way 明渠排水道
open sound 开阔海湾
open space 开放孔隙；连通孔隙；粒间孔隙；开放空间；空地；休憩用地
open space filling 孔隙充填
open span 空跨；敞跨
open spandrel arch bridge 空腹拱桥
open spread 开距排列；敞开排列法
open station 露天驻留地
open stone mixture 未灌浆的碎石混合物
open stope 空场回采工作面
open stope and pillar method 空场房柱法
open stope method 空场采矿法
open stope mining 空场采掘；空场回采法
open stope with random pillar 留贫矿为矿柱的空场法
open stoping 无支护采矿
open stoping method 空场采矿法
open storage 露天仓库；露天储存；露天堆场
open storage area 露天货仓；露天贮物用地
open structure 敞开构造；未闭合构造
open stulled stope 横撑支柱空场回采工作面
open subprogram 开型子程序
open subroutine 开型子程序
open surface method 明挖法
open system 开放系统；无盖层系统
open system environment 开放系统环境
open system interconnection 开放系统互连
open system interconnection reference model 开放系统互连参考模型
open system interlink 开放系统互连
open tender 一般竞争投标；公开投标
open terrace 平台；露天平台；露天台地
open terrain 开阔地区；开阔地形
open test pit 探井；探坑；试坑；样洞
open texture 多孔结构；疏松结构
open timbering 间隔支架法
open to atmosphere 通大气的
open to surrounding 与大气连通的
open to the sky 在露天的
open top chamber 开顶实验槽
open trench 明沟；明渠
open tube analysis 开管分析
open tunnel 明洞
open type geothermal reservoir 开放式地热储
open type triaxial test 开式三轴试验
open underhand stope 空场下向梯段回采工作面；下向空场采场
open underhand stoping 无支护向下台阶法
open up 投产；开发；揭露
open valley 开阔谷
open water 地表水体；无冰水面；开阔水面
open water drilling 海上钻井
open water facies 开阔水体相
open well 大口径井；开敞竖井；普通井

open well calculation method 大井计算法
open well foundation 井筒基础
open wire 明线
open wiring 明线
open work 露天矿；露天开采；户外作业
open work gravel 沉井；架空砾石层
open working 露天开采
open working area 露天工地
open yard 天井；露天货场
open-air 野外的；露天的；开启的
open-air plant 露天车间；露天装置
open-air seasoning 露天干燥；室外干燥
open-air storage 露天储存
open-air weathering test 室外耐气候牢度试验
open-bottom capsule 开底式潜水舱
open-bottom diving bell 底部敞口式潜水钟
open-bottom type 开底式
open-bottomed well 敞底井；套管底开式井
open-burning coal 长焰煤
open-caisson method 开口沉箱法；沉井法
opencast 露天开采；露天矿
opencast blasting 露天矿爆破
opencast coal 露天煤矿
opencast coal mining 露天采煤
opencast explosive 露天矿用炸药
opencast mine 露天矿
opencast mine road 露天矿山道路
opencast mining 露天采矿；露天开采
opencast mining cable 露天矿用电缆
opencast mining equipment 露天采矿设备
opencast mining machine 露天采矿机械
opencast mining machinery 露天采矿机械
opencast mining method 露天开采法
opencast mining plan 露天矿图
opencast site 露天采场
opencast survey 露天矿测量
open-casting 露天开采
open-chain hydrocarbon 开链烃
open-channel constriction 明渠收缩段
open-channel development 明渠开发
open-channel drainage 明渠排水；明沟排水
open-channel flow 明渠水流
open-channel hydraulics 明渠水力学
open-channel hydrodynamics 明渠流体动力学
open-circuit comminution 开路破碎循环
open-circuit comminution dual operation 开路破碎操作
open-circuit comminution operation 开路破碎操作
open-circuit crushing 开路破碎
open-circuit grinding 开路研磨；开路磨矿
open-circuit impedance 开路阻抗
open-circuit test 开路试验；空载试验
open-circuit water supply 开路供水
open-coast marsh 开阔岸沼泽
open-cup flash-and-fire test 开口闪点与着火点试验
open-cup test 开杯试验
open-cut 露天开采的矿山；明挖
open-cut coal mine 露天煤矿
open-cut crossing 开沟穿越
open-cut drain 明渠排水
open-cut drainage 明挖排水；路堑排水；明沟排水

open-cut excavation	明挖工程	opening for drainage	排水巷道
open-cut foundation	明挖基础	opening gap	缺口
open-cut method	露天开采法；明挖法	opening inventory	开始存量
opencut mining	露天采矿	opening location	巷道位置
open-cut mining	露天开采	opening mode	张开型
opencut mining method	露天开采法	opening of crack	裂缝开张
opencut quarry	露天采场	opening of mine	矿体开发；矿山开拓
open-cut tailrace	明挖尾水渠	opening of solution	溶洞；溶蚀洞隙
open-cut tunnel	露天坑道；明挖隧道；明洞	opening of tender	开标
open-cut volume	明挖方量	opening of the channel	孔道口
open-cutting	露天开采；露天采矿	opening pressure	开启压力
open-cut-tunnel portal	明洞门	opening ratio	孔径率，开口率；井中压力与打开防喷器心子的操作液压比
open-cycle gas turbine	开式循环燃气轮机		
open-ditch drainage	明沟排水	opening reading	开始时读数
opened oil and gas gathering process	开式油气收集工艺	opening shot	掏槽炮眼；开槽炮眼；开口炮眼
opened system	开放系统；开式系统	opening size	筛孔尺寸；开口尺寸
opened void	开启空隙	opening slot	开缝；切缝；开割槽
opened void rate	开启空隙率	opening stability	洞室稳定性
opened-up	已开拓的	opening strength	巷道强度；巷道稳定性
open-end caisson foundation	沉井基础	opening time	开启时间；施工时间
open-end hole	通孔	opening to surface	通地面的巷道
open-end lift	采空区侧不留残柱的煤柱回收	opening up	开发矿山；开矿
open-end pile	开放式桩	opening up by adit	平硐开拓
open-end pillar work	开端式短柱开采法	opening up by blind shaft	暗井开拓
open-end pillaring	开端式煤柱回采；开端式矿柱回采	opening up by inclined shaft	斜井开拓
open-end pipe pile	敞口管桩；开口管桩	opening up by tunnel	平硐开拓
open-end steel pile	开口钢桩	opening up by vertical shaft	竖井开拓
open-end system	敞露进路开采法	opening up individually	单独开拓
open-end well	管状过滤衬砌井；开底井	opening-out	开拓，切开
open-ended extraction	开端式开采	openings network	巷道网
open-ended grouting	敞口灌浆	opening-to-pillar width ratio	房-柱宽度比
open-ended line drive	开口直线驱	opening-up	开发或开矿；打井
open-ended tubing squeeze method	敞口油管挤注法	opening-up by adit	用平硐开拓
open-ending method	端面分块回收煤柱	opening-up by combined method	综合开采法；综合开拓法
open-end-pipe pile	开口管桩	opening-up by inclined shaft	用斜井开拓
opener	辅助炮眼；开启工具；开启者	opening-up by vertical shaft	立井开拓；竖井开拓
open-flow capacity	无阻流量	opening-up individually	单独开拓
open-flow potential test	无阻流量试井	opening-up lower level by blind shaft	用暗井开拓下层
open-front derrick	前敞口井架	open-joint fissure	张开节理缝；开口裂缝
open-front tower	轻便井架	open-jointed rock	张开裂隙岩石
open-gash fracture	开裂缝；张口裂隙	open-line production	无油嘴采油
open-grained	粗粒的；非级配的	open-lode	露出的矿脉
open-hole drilling	无套管钻探；敞孔钻探	open-loop control system	开环控制系统
open-hole gravel pack	裸眼井砾石充填	open-loop experiment	开环实验
open-hole inflatable packer	膨胀式裸眼封隔器	open-loop surface water system	开环式地表水循环系统
openhole injector	裸眼注水井	openness	开合度
open-hole log	裸眼井测井	open-off	开始长壁开采；开始新工作区
openhole productive zone	裸眼生产层段	open-off cut	开切眼；切割眼
open-hole shooting	裸眼井段射孔	open-packed structure	疏松堆积结构
opening	孔；口；缝；隙；矿山巷道；通煤层巷道；开度；矿井入口；开拓	open-pit	露天矿；露天采场；露天采掘的
		open-pit bench	露天矿台阶，采场台阶
opening angle	开度角；张角	open-pit boundary	露天矿边界
opening aperture	开口口径	open-pit coal mine	开坑煤矿
opening area	开孔面积	open-pit edge	露天矿边界线
opening area ratio	缝隙面积比	open-pit field	露天采场
opening bit	矫形钻头	open-pit floor	露天矿的底盘
opening delay time	开启延滞时间	open-pit floor edge	露天矿底部界限
opening driving	开始掘进；通路掘进；开拓掘进	open-pit limit	露天矿开采极限
opening Earth system	地球系统开放性	open-pit limit optimization	露天开采边界优化
opening enlargement of bridge and culvert	桥涵扩孔		

open-pit method	露天开采法
open-pit mine	露天矿
open-pit mining	露天开采；露天采矿
open-pit side	露天矿边坡
open-pit sill pillar	露天矿底柱
open-pit slope	露天矿边坡
open-pit survey	露天矿测量
open-pit top edge	露天矿地表边界线
open-pit working	露天开采工作；露天开采场
open-pressure blasting	露天压风喷砂清理法
open-source	开放源代码的
open-source program	开放源代码的程序
open-space deposit	岩石间沉积物；孔隙间沉积物；裂缝中沉积物
open-space filling	有空隙的充填物；孔隙充填
open-space structure	散开构造
open-spaced structure	连通空间构造
open-spandrel arch	空腹拱
open-spandrel arch bridge	空腹拱桥
open-stope benching	采场下向梯段开采
open-stope method	空场采矿法；敞开工作面采矿法
open-stope mining	空场采矿
open-stope system	空场采矿法
open-style winning	开式落煤顺序
open-system test	开式试验法
open-tension fracture	开口张裂隙
open-timbered stope	疏设支柱回采工作面；疏置支柱采场
open-top culvert	透顶涵洞
open-tubular capillary column	开口毛细管柱
open-type	非封闭式的；敞开式的
open-type fold	开口褶皱
open-type wharf	顺岸栈桥式码头；开敞式码头
open-type winning sequence	开式落煤顺序
open-up by combination method	综合开拓法
open-up through vertical shaft and a permanent crow	立井主要石门开拓法
open-wharf mooring island	系泊岛式码头
openwork	露天开采；露天采掘；露天矿；露天开挖
openwork gravel	架空砾石
operability	可操作性
operand	操作数；运算对象；运算域
operand specifier	操作数说明符
operate	操纵，开动；运行，驾驶；管理，经营
operate miss	操作误差
operate on	影响
operate storage	操作存储器
operate time	作用时间
operated	开动的；起动的；已运行的
operating	有效的；工作的；运行的；实施的；操作的
operating accident	操作事故
operating accident rate	运行事故率
operating advantage	操作上的优点
operating basis earthquake	运行基准地震；基准地震；容许运行地震；可运行的地震
operating bottom hole pressure	井底工作压力
operating bug	生产故障；技术故障
operating charge	操作费用
operating chart	作业图
operating condition	操作条件；作业条件；运行工况；运行条件
operating control	操纵装置
operating control point	工作基点
operating cycle figure	循环作业图表
operating data	操作数据；工作记录
operating decision	业务决策；作业决策
operating delay	操作延迟
operating depth	工作水深
operating derrick	在用井架
operating device	操作装置；控制设备
operating distance	工作距离；运营里程；作用距离
operating efficiency	操作效率；施工效率
operating engineer	施工工程师；操控工程师；绞车司机
operating environment	工作环境；作业环境
operating error	操作误差
operating failure	操作中的故障
operating floor	操作平台；工作面；运行层
operating fluid	工作液
operating fluid level	工作液面
operating frequency	运转频率
operating frequency range	工作频率范围
operating function	运算函数；操作功能
operating handle	操纵手柄；操纵杆
operating head	工作压头；运行水头
operating height	操作高度；工作高度
operating hydraulic line pressure	液压管工作压力
operating instruction	操作说明；使用说明书；操作须知；操作规程
operating instruction manual	使用说明
operating length of railway	铁路运营长度
operating level earthquake	运行级地震；容许运行地震
operating license	运行许可证；营业执照
operating life	使用期限；使用寿命
operating limitation	操作的限制
operating link assembly	操作联杆组件
operating maintenance	运行维护；日常维护；操作维修
operating management	经营管理
operating manual	操作规程；操作手册；使用指示书
operating mechanism	操作机理
operating mine	生产矿
operating mode	工况；操作模式；工作模式
operating parameter	作业参数
operating performance	操作性能；工作特性；经营业绩
operating personnel	操作人员
operating platform	工作台；操作平台
operating point	工况点；工作点
operating position	工作位置；操作位置
operating practice	操作实践
operating pressure	工作压力；操作压力；控制压力
operating principle	操作原理
operating procedure	操作程序
operating provision	运行规定；操作规定
operating radius	作用半径
operating railway	运营铁路
operating range	作用距离；有效距离；作用半径；工作范围
operating ratio	有效工作系数；运行率
operating record	操作记录
operating regime	操作规程；工况
operating region	工作区；工作范围

operating reliability 操作可靠性；运行可靠性
operating repair 日常维护检修
operating requirement 操作要求；使用要求
operating risk 作业风险；经营风险
operating room 作业空间；作业范围
operating rope 起重索
operating rule curve 水库操作规程曲线
operating rules 操作规程
operating rules for water-quality analysis of highway engineering 公路工程水质分析操作规程
operating schedule 工作时间表
operating sequence 操作顺序；工序
operating severity 操作苛刻度
operating shaft arm 操作轴臂
operating speed 运行速率；营运速率
operating stress 工作应力
operating switch 操作开关
operating system 应用程序，实用程序；辅助程序；操作系统；业务系统，经营系统
operating technique 施工技术
operating temperature 工作温度；操作温度
operating tensile stress 工作拉伸应力
operating test 运行试验
operating tool string 井下施工管柱
operating trouble 操作故障；运行故障
operating valve 操纵阀；动作阀
operating variable 操作变量
operating wave 工作波
operating wavelength 工作波长
operation 操作，作业，工序，操纵，运行；开采，回采；经营；控制，启动；运算；手术
operation against regulations 违章作业
operation agreement 运行协定
operation analysis 运算分析
operation and maintenance contract 运营维修合同
operation and maintenance manual 操作和维修手册
operation area 作业区；作业地带；营运范围
operation based earthquake 运行安全地震；运转基准地震
operation based ground motion 运行安全地震动
operation beside ditch 沟边组装
operation bill of quantities 作业工程量表
operation board 操作仪表板
operation code 指令码；操作码
operation condition 作业条件
operation decoder 操作译码器
operation desk 操作台
operation device 控制设备
operation efficiency rate 工作效率
operation error 操作误差；操作错误
operation guidance 操作指导
operation guide 操作说明书；操作指导
operation in ditch 沟底组装
operation in tandem 顺序操作
operation inspection log 操作检查记录
operation instruction 作业规程；操作指示；操作指南
operation item 操作项目；工序项目
operation length of railway 铁路运营长度；运营里程
operation license 营业执照
operation life 运行寿命
operation limit 使用限度
operation maintenance center 操作维护中心
operation management 作业管理；下级主管
operation manual 操作手册
operation method 操作方法
operation mine rehabilitation 生产矿山改造
operation mode 运行方式；操作方式
operation number 操作号码，；运算数
operation panel 控制仪表板
operation period 运营期
operation personnel 生产人员；工作人员
Operation Planning Management System 计划调度管理系统
operation platform 操作平台
operation pressure 工作压力；运行压力
operation pressure diagram 工作压力图；运行水压图
operation procedure 操作程序
operation process chart 作业流程图
operation radius 工作半径
operation regulation 运行调节
operation room 操作室
operation rule 操作规程
operation scale 操作规模；运转规模；经营规模
operation schedule 作业程序表
operation season 作业季节
operation sequence 操作次序；操作步骤
operation sheet 操作卡；运行流程
operation specification 操作规程
operation state 运行工况；操作状态
operation step 操作步骤
operation study 作业研究
operation supervisor 作业管理人员
operation symbol 运算符号
operation system 操作系统
operation test 运用试验；运行试验
operation through tubing 过油管作业
operation valve 控制阀
operation ventilation 运营通风
operational 操作的；运行的；运算的
operational amplifier 运算放大器
operational area 作业地区；操作地区；运行区
operational attribute 作业属性
operational base 工作基地；现用基地
operational calculus 运算微积分
operational center 操作中心
operational control 操作控制；经营控制
operational data 运算数据；工作数据
operational delay 运行延滞；运行延误
operational development 改进性研制
operational environment 运行环境
operational environmental satellite 同步环境卫星
operational experience 操作经验；运行经验
operational facies 工作地层相
operational factor 作用因素；运转因数；运算因数；运行参数
operational forecast 业务预报；作业预报
operational guideline 施工技术措施
operational log 运行记录
operational logbook 操作日志
operational longwall face observation 作业长壁工作面矿

压观测
operational maintenance 日常维修
operational manual 操作说明书
operational method 运算方法
operational mode 操作模式；运作模式
operational order 运算指令；操作指令
operational outfit 运行设备
operational parameter 开采参数；运行参数
operational performance 使用特性；工作性能
operational plan 操作计划书
operational planning 业务规划；作业计划
operational problem 运算问题；操作问题
operational procedure 操作程序
operational project 执行中的项目
operational record 运行记录
operational regulation 操作规程
operational reliability 操作可靠性；使用可靠性
operational safety ground motion 运行安全地震动
operational scheme 运行流程
operational sequence 工序；操作工序
operational specification 操作规范
operational speed of a computer 计算机运算速度
operational state 运行状态
operational symbol 运算符
operational taxonomic unit 运算分类单位
operational test 工作试验；操作试验；运转试验
operational test set 工作状态测试设备
operational unit 工作地层单位；实用地层单位
operational variable 作业的变化因素；操作变量
operative 工作的，运转的；有效力的；技术工人
operative condition 可正常工作状态；有效运转条件
operative constraint 有效约束
operative construction organization design 实施性施工组织设计
operative force 作用力
operative technique 操作技术
operator calculus 算子微积
ophicalcite 蛇纹大理岩；蛇纹方解石；蛇纹白云石
ophigranitone 蛇纹辉长岩
ophiolite 蛇绿岩；蛇纹大理石
ophiolite complex 蛇绿岩杂岩体
ophiolite formation 蛇绿岩建造
ophiolite suite 蛇绿岩套
ophiolite-radiolarite nappe 蛇绿岩-放射虫岩推覆体
ophiolitic assemblage 蛇绿岩组合
ophiolitic melange 蛇绿混杂岩
ophiolitic suite 蛇绿岩套；蛇绿岩系
ophiolitiferous nappe 蛇绿岩质推覆体
ophite 纤闪辉绿岩；纤闪石化辉绿岩；蛇纹石；蛇纹岩
ophitic 辉绿岩结构的
ophitic interstitial texture 辉绿填隙结构
ophitic texture 辉绿结构；纤闪辉绿岩结构
ophitone 辉绿岩
ophthalmite 眼球混合岩
ophyolitic complex 蛇绿岩复合体
opisometer 曲线计；距离测定器
opoka 蛋白岩
opportunity map 机会图
opposed current 逆流；反流
opposing movement 相对运动
opposing reaction 对抗反应
opposing torque 反向力矩
opposite 相反的；对面的；对应的；对立物
opposite angle 对角
opposite arc 反向弧
opposite change 180°的转动；逆转
opposite direction 反方向
opposite faces 对立面
opposite fold limb 反向褶皱翼
opposite force 反向力
opposite hand of face 工作面的回程采煤方向
opposite hand view 镜像；对称视图
opposite pressure 反向压力；反压力
opposite resultant force component 方向相反的合力分量
opposite sense 相反方向
opposite sequence 逆程序
opposite side 对边
opposite state 反状态
opposite tide 逆潮流
opposite to 与…相反
opposite vertex 对顶
oppositely directed 反向的
opposition 反接；反相，移相；对置，对抗
optalic metamorphism 熔热变质作用
optic angle 光轴角；视角
optic axial angle 光轴角
optic axis 光轴
optic binormal 光轴副法线
optic biradial 副光轴
optic elastic axis 光弹性轴
optic elasticity axes 光学弹性轴
optic fiber 光导纤维
optic sensor 光学传感器
optic stimulated luminescence dating 光释光测年法
optic symmetric axis 光学对称轴
optic twinning 光性双晶
optical 光学的；光的；视觉的
optical absorption spectroscopy 光学吸收光谱学
optical access network 光纤接入网
optical activity 旋光性
optical adapter 光适配器
optical air mass 大气光学质量；光学空气质量
optical alignment 光学校整；光学校直
optical analog computer 光学模拟计算机
optical analogy 光学模拟
optical angle 光轴角
optical anisotropy 光学各向异性
optical anisotropy index 光学非均质指数
optical anomaly 光性异常
optical antipode 旋光对映体
optical attenuator 光衰减器
optical automatic ranging 光学自动测距仪
optical axial angle 光轴角
optical axial plane 光轴面
optical axis 光轴
optical axis of camera 摄影主光轴
optical balance 光学天平
optical cable 光缆
optical calcite 光学方解石
optical calculating machine 光计算机

optical camouflage 光学伪装；视觉伪装
optical centering 光学定中，光学对中；光学垂准
optical centering device 光学定中器
optical centre 光心；像主点
optical channel 光通道
optical character 光学特性
optical comparator 光学比测器
optical compensator 光学补偿器
optical computer 光学计算机
optical concentration 光学浓度
optical condition 光学条件
optical constant 光学常数
optical correlation 光学相关
optical correlator 光学相关器；光相干器
optical coupler 光电转换器
optical coupling 光耦合
optical crystallography 光性结晶学；结晶光学；晶体光学
optical data highway 光数据总线；资讯高速公路
optical data processing 光学数据处理
optical densitometer 光测密度计
optical density 光学密度；光密度
optical depth 光学厚度
optical detecting and ranging 光学定向测距
optical detector 光探测器
optical detector technology 光探测技术
optical device 光学装置
optical digital section 光数字段
optical directional coupler 光学定向耦合器
optical disc memory 光盘存储器
optical disc storage 光盘存储器
optical disk servo control system 光盘伺服控制系统
optical dispersion 色散
optical distance 光程；光学距离
optical distance measurement equipment 光学测距仪
optical distortion 光学畸变
optical element 光学组件
optical emission spectroscopy 发射光谱学
optical emission spectrum 发射光谱
optical encoder 光学编码器
optical energy 光能
optical enhancement 光学增强
optical fiber 光纤；光导纤维
optical fiber amplifier 光纤放大器
optical fiber cable 光缆
optical fiber cable for mine 矿用电缆
optical fiber cable joint closure 光缆接头
optical fiber cluster 光导纤维束
optical fiber communication 光纤通信
optical fiber communication net work 光纤通信网络
optical fiber communication system 光纤通信系统
optical fiber cutter 光纤切断器
optical fiber digital line system 光纤数字线路系统
optical fiber distribution frame 光纤分配器
optical fiber fusion splicing machine 光纤融接机
optical fiber line terminal equipment 光端机；光纤线终端设备
optical fiber sensor 光纤传感器
optical fiber splice loss 光纤接续损耗
optical fiber stripper 光纤剥除机

optical fiber telemetry 光纤遥控
optical fibre 光导纤维；光学纤维；光纤
optical fibre cable 光缆
optical fibre communication 光纤通信
optical filter 半导体滤光片；滤光器
optical filtering 光学滤波
optical fixed attenuator 光纤固定衰减器
optical flat 光学平面；真实平面
optical gain 光学增益
optical gas determinator 光学瓦斯检定器
optical glass 光学玻璃
optical goniometer 光学测角计；光学测向器
optical graphical rectification 光学图解纠正
optical head 光度头
optical height finder 光学测高仪
optical hologram 光全息图
optical holography 光学全息
optical illusion 光幻视；视错觉
optical image enhancement 光学影像增强
optical image processing 光学图像处理
optical imaging instrument 光学成像仪
optical imaging system 光学成像系统
optical index 光指数
optical instrument 光学仪器
optical instrument positioning 光学定位
optical integrator 光学积分器
optical interface 光接口
optical interference 光波干涉
optical interference effect 光学干涉效应
optical interferometer 光学干涉仪；光波干涉仪
optical isolator 光隔离器
optical isomer 旋光异构体
optical isomerism 旋光异构
optical isotropy 光学各向同性
optical laminate 透光微层理
optical lane 光路
optical laser 激光；激光器
optical laser strainmeter 激光器应变计；激光应变仪
optical level 光学水准仪
optical line protection switching equipment 光线路保护切换设备
optical line terminal 光线路终端
optical line terminal equipment 光线路终端设备
optical link 光缆连接
optical log 光记录测井曲线
optical magnification 光学放大；光放大
optical margin 光边
optical mark 光学测标
optical mark reader 光学标记阅读器
optical maser 激光；激光器
optical mass 光学质量
optical means 光学装置
optical measurement 光学测量法
optical measurement of distance 光学测距法
optical mechanical rectification 光学机械纠正
optical mechanical scanner 光学机械扫描仪
optical memory 光存储器；光学存储器
optical method 光学法；光学定向法
optical micrometer 光学测微器
optical microscope 光学显微镜

optical microscopy 光学显微术
optical mineralogy 光性矿物学
optical model 光学模型
optical monitoring device 光学监测设备
optical mosaic 光学镶嵌
optical oceanography 海洋光学
optical orientation 光性方位；光学取向
optical orthophotoscope 光学正射投影纠正仪
optical parallax 光学视差
optical path 光程；光路
optical path difference 光程差
optical pattern 光图样；光学图像
optical pattern recognition 光学模式识别；光学图形识别
optical phenomenon 光学现象
optical photography 光学摄影
optical plane 光轴面
optical plumbing 光学垂准；光学投点；光学定向
optical plummet 光学对中器；光学定中仪
optical plummet centering 光学对中；光学垂准
optical power difference 光功率差
optical power level 光能级
optical power meter 光功率计
optical preamplifier 光前置放大器
optical processing 光学处理
optical projection 光学投影
optical property 光学特性；光学性质
optical pump 光泵
optical pump magnetometer 光泵磁力仪
optical pumping magnetometer 光泵磁力仪
optical pumping reaction 光泵作用
optical pyrometer 光学高温计；光测高温计
optical radar 光雷达
optical range 光学距离
optical range finder 光学测距仪；光学对距镜
optical reader 光输入机
optical receiver 光接收模块；光接收器
optical receiving component 光接收器件；光接收组件
optical reconstruction 光学再现
optical recorder 光记录仪
optical recording media 光记录媒体
optical recording system 光学记录系统
optical rectification 光学纠正，光学校正；光整流
optical regenerator section 光再生段
optical register 光学记录
optical registration 光学记录；光性记录
optical relief model 光学立体模型
optical resistor 光敏电阻
optical rotary power 旋光性
optical rotation 旋光度；旋光性
optical rotatory dispersion 旋光色散
optical scanner 光扫描仪；光扫描器
optical scanning 光学扫描
optical scanning device 光学扫描装置
optical scanning system 光学扫描系统
optical section 光切面；光学切面
optical sensor 光学遥感器；光传感器
optical shaft plumbing 竖井光学垂准
optical sight 光学瞄准；光学瞄准器
optical sign 光性符号
optical source 光源
optical spectrographic method 光谱法；光谱分析法
optical spectroscopy 分光法；光谱学
optical spectrum 光谱
optical spectrum feature 光谱特征
optical speed 光学透光率；光速
optical splitter 分光器
optical square 光学直角器；直角转光器
optical stack 光学叠加
optical stereoscopic model 光学立体模型
optical storage 光学存储
optical strain 光性变形
optical stress meter 光学应力计
optical switch 感光开关；光开关
optical system 光学系统
optical telemetry 光学测距法；光遥测术
optical testing 光学检验；显微镜检验
optical theodolite 光学经纬仪
optical thickness 光学厚度
optical throughput 光通量
optical time domain reflectometer 光时域反射仪
optical tracking system 光学跟踪系统
optical transfer function 光学传递函数
optical transform 光学变换
optical transit 光学经纬仪
optical transition energy 光学跃迁能
optical transmission mode 光传输模式
optical transmission system 光传输系统
optical wavelength 光波长
optical-electrical dust determinator 光电测尘仪；光电粉尘测定仪
optical-electronic scanner 光电扫描仪；光学扫描仪
optically active compound 具旋光性化合物；光学活性化合物
optically active gas 旋光性气体
optically anisotropic 光学异向性的
optically biaxial 双光轴的
optically denser medium 光密媒质
optically isotropic 光性均质的
optically negative 负光性的
optically positive 正光性的
optically pumped magnetometer 光泵磁力仪
optically recording accelerograph 光学记录加速度仪
optically uniaxial 单光轴的
optical-mechanical imaging scanner 光学机械成像扫描仪
optical-mechanical projection 光学机械投影
optical-mechanical rectification 光学机械纠正
optical-mechanical scanning 光学机械扫描
optical-photographic analog recorder 光学-照相模拟记录器
optical-pumping magnetometer 光泵磁力仪
optical-reading theodolite 光学读数经纬仪
opticator 光学部分；光学扭簧测微仪
opticity 旋光性；光偏振性
optic-pumping magnetometer 光泵磁力仪
optics 光学；光学器件
optics laboratory 光学试验室
optiform 光成形
optimal 最优的；最佳的；最适宜的
optimal adaptive control 最佳自适应控制

optimal adjustment 最优调整
optimal algorithm 最优算法；优化算法
optimal allocation 最优配置
optimal amount of exploration 最优勘探量
optimal approximation 最佳逼近
optimal aspect ratio 最佳长宽比
optimal band combination 最佳波段组合
optimal case 最优方案
optimal classification 最佳分类
optimal combination 优化组合
optimal control 最优控制
optimal control theory 最优控制理论
optimal controlled water level 最佳控制水位
optimal criterion 优化准则
optimal decision 最优方案；最优决策
optimal decision function 最优决策函数
optimal decision rule 最优决策规则
optimal defense intensity 最优设防烈度
optimal design 最优设计
optimal design of control network 控制网优化设计
optimal direction of horizontal drilling 最优水平井钻进方向
optimal distribution 最佳分布；最优分布
optimal estimator 最佳估计量；最优估计函数
optimal feasible solution 最优可行解
optimal field development program 最佳油田开发方案
optimal filter 最优滤波器
optimal filtering 最优滤波
optimal fit of function to solution 解的最佳拟合函数
optimal focal depth 最佳震源深度
optimal fuzzy controller 最优模糊控制器
optimal fuzzy system 最优模糊系统
optimal gradation 最佳级配
optimal limit design 极限优化设计
optimal linear predictor 最优线性预报值
optimal measure 最优测度
optimal mesh 最优网格
optimal mining yield 最佳开采量
optimal operation 最优操作
optimal order 最优阶
optimal outline 最优边界
optimal output 最佳产量
optimal path 最佳路径
optimal path selection 最佳路径选择
optimal perforation 优化射孔；最优射孔
optimal programming 最优规划
optimal reinforcement site 最优加固位置
optimal reservoir management 最佳油藏管理
optimal robustness 最佳稳健性；最优鲁棒性
optimal rolling depth 最佳压实深度；最佳轧制深度；最佳滚压量
optimal route 最佳路径
optimal salinity 最佳盐度；最适盐度；最佳含盐量
optimal sample point 最佳抽样点；最佳取样点
optimal sampling 最优采样
optimal samplingpoint 最优样本点
optimal segmentation 最优分割法
optimal seismic deconvolution 最佳地震反褶积
optimal size of business 最优经营规模
optimal solution 最优解

optimal spacing 最适间距
optimal static correction 最佳静校正
optimal stochastic control 最佳随机控制
optimal structure design 最优结构设计
optimal superconvergent sampling 最优超收敛采样
optimal trajectory 最优轨迹
optimal value 最优值
optimal velocity filter 最佳速度滤波器
optimal water distribution scheme 最佳配水方案
optimal weight 最优权重
optimal well discharge 井的最佳出水量
optimal yield 最佳开采量
optimal-adaptive 最优适应的
optimality 最优性；最佳化
optimality condition 最优条件
optimality criterion 最优性判别准则；最佳判据
optimality criterion method of energy distribution 能量分布的最优准则方法
optimality principle 最优性原理
optimeter 光电比色计；光度计；视力计；光学比较仪器
optiminimeter 光电测微器
optimistic estimate 乐观估计
optimization 最佳化；最优化；最佳选择
optimization algorithm 最优化算法
optimization approach 最优化方法
optimization control theory 最优控制理论
optimization criteria 最优化判据
optimization criterion 最优化准则
optimization design 优化设计
optimization drilling 最优化钻井
optimization for design 优化设计
optimization inversion 最优化反演
optimization method 优选法；最优化方法
optimization model 最佳模型；最优模型；优化模型
optimization of groundwater monitoring network 地下水监测网优化
optimization of parameter 最佳参数选择
optimization of perforation parameter 射孔参数的优化
optimization principle 优化原理
optimization procedure 最优化方法
optimization program 最优化程序
optimization technique 最优化技术
optimization test 优选试验
optimization theory 最佳化理论
optimization-based computer aided design system 基于优化的计算机辅助设计系统
optimized directional drilling 优化定向钻井
optimized directional well 优化定向井
optimized drilling 最优化钻井
optimized drilling program 最优化钻井程序
optimized drilling technology 优化钻井工艺
optimized migration 最佳偏移
optimized scheme of groundwater resources management 地下水资源优化管理方案
optimized working zone 最优工作区
optimizer 优化程序；最优控制器
optimizing 最佳化；最佳的；优选的
optimizing criterion 优化标准
optimizing fracturing treatment design 优化压裂处理

设计
optimizing planning 最佳规划
optimum 最好的条件；最适宜的；最好的；最佳效果；优选的
optimum allocation 最优分配；最佳配置
optimum allocation of resources 资源最佳分配
optimum amplitude response 最佳幅度响应
optimum approximation 最佳逼近
optimum assembly 最优组合
optimum band 最佳波段
optimum bit program 最优钻头程序
optimum bit weight 最佳钻压
optimum blast round 最佳炮眼组布置
optimum blast routed 最佳炮眼组布置
optimum break 最佳破碎粒度；精选粒度
optimum capacity 最佳容量，最佳生产额；最好能力，最佳能力
optimum character 最佳特性
optimum circle 最佳拟合圆
optimum combination 最优组合
optimum completion design 最优完井设计
optimum condition 最佳状态；最佳条件；最佳工况
optimum conformance 最佳波及系数
optimum control 最优控制；优化控制
optimum coupling 最佳耦合
optimum coverage 最佳覆盖
optimum criterion 最佳准则
optimum curve 涡动传导率
optimum cut-off grade 最佳边际品位
optimum damping 最佳阻尼
optimum deconvolution operator 最佳反褶积算子
optimum density 最优密度
optimum depth 最佳覆盖厚度
optimum design 最佳设计；最优设计
optimum development 最优开发
optimum dimension 最佳尺寸
optimum discharge 最佳流量；最佳排水量
optimum distribution 最佳分布
optimum drawdown 最优降深
optimum drawn ore grade 最佳出矿品位
optimum drill depth 合理钻进深度
optimum drilling efficiency 最佳钻进效率
optimum drilling program 最佳钻井程序
optimum drilling rate 最优钻速；最佳钻速
optimum drilling technique 最优钻井技术
optimum dynamic characteristic 最优动力特性；最优动态特征曲线
optimum efficiency 优化效率；最优效率
optimum estimate 最佳估计
optimum estimation value 最佳估算值
optimum feasible program 最佳可行方案
optimum field offtake 最佳油田产量；最佳油田开采方式
optimum filter 最佳滤波器
optimum filtering 最佳滤波
optimum fine aggregate percentage 最佳含砂率
optimum formula 最佳公式
optimum frequency 最佳频率
optimum gradation 最佳级配
optimum humidity 最适湿度
optimum hydraulics 最佳水力因素
optimum injection recovery pattern 最佳注采井网
optimum interpolation method 最优插值法
optimum least-squares filter 最佳最小平方滤波器
optimum level 最优值；最优程度
optimum linear prediction 最佳线性预测
optimum load factor 最优负荷系数；最佳载荷因子
optimum matching 最佳匹配
optimum mix 最佳配比
optimum moisture 最优含水率；最佳湿度
optimum moisture content 最优含水量；最佳含水量
optimum number 最佳数
optimum operation 最佳操作
optimum operator 最佳算子
optimum parameter 优化参数
optimum performance 最佳状态；最佳性能
optimum planning 最优规划
optimum prediction 最佳预测
optimum procedure 最优化程序
optimum processing parameter 最佳处理参数
optimum program 最佳程序
optimum programming 最佳程序设计
optimum range of concentration 最佳浓度范围
optimum rate of production 最优开采速度
optimum reinforcement 合理锚固；最佳配筋
optimum reinforcement layer number 最优加筋层数
optimum reservoir operating pressure 最佳油层开采压力
optimum resolution 最佳分辨率
optimum resolution bandwidth 最佳分辨带宽
optimum rigidity distribution 最优刚度分布
optimum sample 最优样本
optimum sample size 最佳样品量
optimum sampling density 最佳采样密度
optimum sampling interval 最佳采样间距
optimum section 微量过滤器
optimum seeking method 优选法
optimum seismic design 最优抗震设计
optimum shape 最优形状
optimum slope 最佳边坡
optimum solution 最优解
optimum specification 最优规范
optimum spectral band 最佳光谱带；最佳频谱带；最佳波谱段
optimum speed 临界速度；最佳速度；最佳航速
optimum stacking 最佳叠加
optimum station spacing 最佳站间距
optimum strength distribution 最优强度分布
optimum structure 最优结构
optimum temperature 最适温度
optimum temperature difference 最适宜温差
optimum temperature gradation 最佳温度梯度
optimum temperature of solubilization 最佳加溶温度
optimum temperature sequence 最优温度序列
optimum theory 最优化理论
optimum time of burst 最佳爆炸时刻
optimum use 最佳用途
optimum value 最佳值
optimum velocity 最佳速度
optimum water 最佳水量
optimum water content 最佳含水量；最优含水量

optimum water management plan	最优水管理方案
optimum weight design	最小重量设计
optimum weight on bit	最佳钻压
optimum weight stacking	最佳加权叠加
optimum well spacing	最佳井距
optimum wide band horizontal stack	最佳宽带水平叠加
optimum wide band vertical stack	最佳宽频带垂直叠加
optimum working efficiency	最佳工作效率
optimum working frequency	最佳工作频率
optimum yield	最佳收得率；最佳产量
option	选择；选择程序；选择方案；选项
option comparison	方案对比
option contract for crude oil	原油交易选择权合同；原油交易期权合同
option exercise price	选择权行使价格
optional	任意的；随意的，任选的；附加的
optional equipment	非标准装置；专用装置；附加设备
optional parameter	可选参数
optional sampling	可选采样
optional test	选定试验
optional turning	随意转向
orange jadeite	红翡
orange-peel excavator	多瓣式抓斗挖掘机
orange-peel shape dam	双曲拱坝
orangite	橙黄石
oraviczite	杂镁铝硅石
orb	天体；星球；球
orb texture	球粒状结构
orbed	球状的；圆的
orbicular	球状的；圆的
orbicular granite	球状花岗岩
orbicular rock	球状岩
orbicular structure	球状构造；球形结构
orbicule	球状体
orbiculite	球状岩
orbital movement	轨道移动
orbital parameter	轨道参数
orbital path	轨道，轨迹
orbit-disorder	有序-无序
orbite	角闪玢岩
orbitrol	液压伺服控制
orcelite	褐砷镍矿；六方砷镍矿
ordanchite	含橄蓝方碱玄岩
order code	指令码；操作码
order distribution	序态分布
order number	原子序数；订单号
order of a matrix	矩阵的阶
order of accuracy	精度等级；精确度
order of convergence	收敛阶
order of crystallization	结晶顺序；结晶次序
order of deposition	沉积顺序
order of error	误差阶
order of extraction	开采顺序
order of fringes	等色线的条纹级数
order of integration	积分阶
order of magnitude	数量级；震级大小
order of matrix	阵阶
order of mobility	活动性序列
order of precedence	优先次序
order of stability	稳定性顺序；稳定阶
order parameter	序参数
order register	指令寄存器
order set	指令系统；指令组
order-disorder	有序-无序
order-disorder polymorph	有序-无序型同质多形体
order-disorder polymorphism	有序-无序型同质多象
order-disorder structure	有序-无序结构
order-disorder transformation	有序-无序转换
order-disorder transition	有序-无序转变
ordered arrangement	有序排列
ordered lattice	有序晶格
ordered phase	有序相
ordered retrieval	顺序检索
ordered sample	有序标本；有序样品
ordered search	有序搜索
ordered set	有序集合
ordered solid solution	有序固熔体
ordered structure	有序结构
ordering	有序化；调整；排序；定序
ordering energy	有序能；成序能
ordering of grid	网格排序
orderliness	有秩序
orders of magnitude	数量级
order-writing	写出指令
ordinal	顺序的；依次的；序数
ordinal number	序数
ordinal relation	顺序关系
ordinal scale	有序尺度
ordinal scale data	有序尺度数据
ordinal scaling	顺序量表；等级分类
ordinal selection	序数选择法
ordinal structure	有序结构
ordinance datum	法定基准面
ordinance load	规定荷载
ordinary accident	一般事故
ordinary binary	普通二进制
ordinary brass	普通黄铜
ordinary carbon steel	普通碳钢
ordinary chernozem	普通黑土
ordinary chert	普通燧石
ordinary clay	普通黏土
ordinary concrete	普通混凝土
ordinary construction	一般建筑；一般构造；普通结构
ordinary deposited soil	一般堆积土
ordinary difference differential equation	常差分微分方程
ordinary difference equation	常差分方程
ordinary differential	常微分
ordinary differential equation	常微分方程；普通差方程
ordinary fault	正断层
ordinary fault rift	正断裂谷；正断层地堑
ordinary grade	普通等级
ordinary levelling	普通水准测量；普通水平测量
ordinary maceral	原生显微组分
ordinary pressure	常压
ordinary rock drill	普通凿岩机
ordinary subgrade	一般路基
ordinary subroutine	常用子程序
ordinary temperature	寻常温度；常温
ordinary theodolite	普通经纬仪

ordinary tide	正常潮位
ordinary triaxial test	常规三轴试验
ordinary water level	常水位
ordinary wave	寻常波，正常波
ordinary wear and tear	自然磨损；自然损耗
ordinate	纵坐标；纵距；弹道高度
ordinate axis	纵轴
ordinate dimension	垂直度量
ordinate scale	坐标刻度；坐标范围
ordination	排列；分类；规格
ordnance bench mark	法定水准点；水准基点
ordnance datum	国家基准点；水准基准；基本标高
ordnance survey	地形测量；军用地形测量
ordonnnance	布局；安排
Ordos platform	鄂尔多斯台地
ordosite	鄂尔多斯岩，暗霓正长岩；河套岩
Ordovician	奥陶纪；奥陶系
Ordovician limestone	奥陶纪石灰岩
Ordovician period	奥陶纪
Ordovician system	奥陶系
Ordovician water	奥陶系水
ore	矿石；矿物；矿砂
ore agglomeration property	矿石结块性
ore annealing	矿石退火；铁矿粉中退火
ore anomaly	矿异常
ore area	矿区；矿田；矿场面积
ore assay	矿石分析
ore bearing	含矿的
ore bearing structure	含矿构造
ore bearing vein	含矿脉
ore bed	矿床；矿层
ore belt	矿石带；含矿带
ore beneficiation	选矿；矿石加工
ore bin	矿石仓；储碴仓
ore blend	矿石融合
ore blending	配矿；混矿
ore block	矿块；矿段；矿石区段
ore block method	块段法
ore block out	划分矿体
ore blocked out	开采储量
ore bloom	原矿块
ore body	矿体
ore body limit	矿体界限
ore body occurrence	矿体赋存
ore boulder	矿石漂砾；需二次破碎的大块矿石
ore boundary	矿体边界
ore bowl	矿仓
ore box	矿仓
ore breaking	崩矿；破碎矿石
ore breaking plant	碎矿车间
ore brine	成矿卤水
ore bringer	运矿岩
ore briquette	团矿
ore bucket	矿斗
ore bunch	矿囊；矿袋
ore buttress	矿壁
ore car	矿石车
ore carrier	矿石船；运矿岩
ore caving	矿石崩落
ore channel	矿槽
ore charge	矿石试样剂量；矿石料
ore chimney	矿筒；矿柱；管状矿体
ore chute	溜矿井；放矿溜口；矿石泻槽
ore chute line	溜矿槽排列
ore classification	矿石分级
ore cluster	矿簇；矿群
ore coal briquette	矿煤团块
ore coke ratio	矿焦比
ore column	矿柱
ore complex	矿石复合体
ore concentrate	精矿砂；精矿
ore concentration	选矿
ore concentration plant	选矿厂
ore conduit structure	导矿构造
ore control	控矿作用；成矿控制
ore control structure	控矿构造
ore core	矿心
ore course	富矿体；矿脉走向
ore crusher	矿石破碎机；碎矿机
ore current	矿流
ore cushion	矿石垫层
ore dampness	矿石湿度
ore delineation	矿体圈定
ore deposit	矿床；矿层
ore deposit association	矿床组合
ore deposit due to magmatic segregation	岩浆分结矿床
ore deposit series	矿床系列
ore deposits	矿床；矿藏
ore developed	已完全开拓的矿体
ore developing	矿体开拓；矿床开拓
ore dilution	矿石贫化
ore dilution control	防止矿石贫化
ore distribution	矿体分布
ore distributor	矿石分配器
ore district	矿区
ore drawing	放矿；落矿；矿石回采
ore dressing	选矿
ore dressing plant	选矿厂
ore drift	沿脉
ore drilling	矿床钻孔
ore estimate	矿石评价
ore evaluation	矿石评价；矿物评价
ore expectant	预测矿藏量
ore exploration	矿床勘探；探矿
ore extraction	矿石开采
ore extraction area	采矿场
ore extraction rate	采矿回收率
ore fabric	矿石组构
ore fault	矿体断层
ore feed	给矿；给料
ore feeder	给料机
ore field	矿田
ore field structure	矿田构造
ore fines	粉矿；矿石颗粒
ore floatation promoter	浮选促进器
ore flow during draw	放矿期矿石流动
ore fluid	矿液；成矿流体
ore fold	含矿褶皱；矿体褶皱
ore formation	矿层；矿石建造
ore forming element	成矿元素

ore genesis	矿床成因
ore geochemical anomaly	成矿地球化学异常
Ore Geology Reviews	矿床地质评论
ore grade	矿石品位；矿石品级
ore guide	导矿作用；导矿的；矿化标志；找矿导引
ore handling system	矿石搬运系统
ore hardness	矿硬；矿石硬度
ore haulage	矿石运输
ore hoist	矿石提升机
ore hole	有矿钻孔
ore horizon	矿层；平矿层
ore hunting evidence	找矿征兆
ore hydrothermal system	成矿热液系统
ore impoverishment	矿化贫石
ore in paddock	堆仓矿石；堆场矿石
ore in pillar	矿角
ore in place	未采的矿体；原地的矿石
ore in sight	可见或已知储量的矿体
ore in stock	现存矿石；贮存矿石
ore indicator	找矿标志
ore industry sewage	采矿工业污水
ore injection	矿脉贯入
ore intersection	矿体穿通点
ore leaching	矿石浸出
ore lift	矿石分层崩落高度
ore locus	矿区
ore lode	矿脉
ore loss	矿石损失
ore losses rate	矿石损失率
ore magma	矿浆；含矿岩浆
ore mass	矿石质量；矿体
ore microscope	矿石显微镜；矿相显微镜
ore microscopy	矿相学；矿石显微术
ore mill	捣矿机；选矿厂
ore mineral	有用矿物；矿石矿物；金属矿物
ore mineral body	金属矿体
ore mining	采矿
ore mining industry	金属采矿工业
ore nest	矿巢
ore occurrence	矿藏；矿点；矿产地；矿石赋存
ore of sedimentation	沉积矿石；砂矿
ore outline	矿体外形
ore packing	矿石驮运；矿石黏结
ore pass	放矿溜道；放矿溜井
ore pass raise	放矿上山
ore passage way	放矿溜道
ore petrology	矿石岩石学；矿石学
ore pillar	矿柱
ore pipe	矿筒；管状脉；管状矿体
ore piping	矿石管
ore plot	精矿贮存场
ore pocket	矿石仓；矿袋；矿囊
ore pod	扁豆状矿体
ore process by remobilization	活化转移成矿作用
ore processing plant	选矿厂
ore prospecting	找矿；矿床勘探；探矿
ore province	成矿区
ore pulp	矿浆
ore recovery	矿石回收
ore recovery rate	矿石回采率
ore reduction	矿石破碎；矿石还原
ore removal	运出矿石；出矿
ore reserve	矿石埋藏量；矿产储量
ore reserve calculation	矿石储量计算
ore rib	矿柱；矿壁
ore roasting kiln	焙矿炉
ore rock	含矿岩石
ore roll	矿卷；卷状矿体；放矿溜道
ore run	矿体；矿层走向；矿层延展长度
ore scraper	扒矿机
ore self-ignition	矿石自燃
ore shaft	矿石提升井；出矿井
ore shell	保护矿壳
ore shoot	富矿柱；富矿体；柱状富矿体
ore showing	矿苗
ore sintering	矿石烧结
ore sizing	矿石筛分
ore skip	提矿箕斗
ore slime	矿泥
ore smelting	矿石熔炼
ore solution	成矿溶液；矿液
ore sort	矿石工业品级
ore sorting	选矿；拣块选矿法
ore spot	矿点
ore stamp	捣矿机
ore station	采矿站
ore stock	矿储藏量
ore stock yard	堆矿场
ore stockwork	网状脉
ore stone	矿石
ore storage	贮存矿石；矿石堆场
ore stream	矿流
ore structure	矿石构造；矿体构造
ore system	矿石系统；成矿系统
ore testing	矿石试验；矿石化验
ore texture	矿石结构
ore top	矿体顶部
ore train	运矿列车
ore transshipment plant	矿石转运装置
ore treatment	矿石处理
ore treatment plant	选矿厂
ore trend	矿体走向
ore type	矿石类型
ore vein	矿脉
ore wash	洗矿
ore washer	洗矿机
ore washery	洗矿场
ore washing	洗矿
ore zone	矿带
ore-and-rock-forming fluid	成岩成矿流体
ore-bearing	含矿的
ore-bearing hydrothermal solution	含矿热水溶液；含矿热液
ore-bearing vein	矿脉
ore-bearing zone	含矿带
orebed	矿层；矿床
orebody	矿体
orebody contour isotropy of mine product	矿体等高线图
orebody contour map	矿体等高线图
orebody isopach isotropy of mine product	矿体等厚线图

orebody thickness	矿石厚度
ore-breaking floor	崩矿层
ore-breaking plant	碎矿车间
ore-column height	矿柱高度
ore-concentration plant	选矿厂
ore-drawing order	放矿规程
ore-dressing engineer	选矿工程师
ore-dressing plant	选矿厂
ore-dressing treatment	选矿
orefield	矿区；矿产地
ore-finding	找矿
ore-forming element	成矿元素
ore-forming fluid	成矿流体
ore-forming geodynamic setting	成矿的地球动力背景
ore-forming process in groundwater	地下水成矿作用
oregonite	砷铁镍矿
ore-handling capacity	处理矿石能力
ore-hunting evidence	找矿标志
ore-limit	可采矿石金属含量的最低极限
ore-loading crane	矿车装卸吊车
ore-mass settling	矿石体下降
orendite	金云白榴斑岩
oreography	山态学；山岳形态学
orepass	放矿溜井；溜矿道
ore-pass raise	放矿天井
ore-pass system	放矿道系统；溜眼系统
ore-process	成矿作用；矿化作用
ore-reserve analysis	矿石储量分析
ore-reserve estimate	矿石储量估计
ore-reserve isotropy of mine product	矿石储量图
ore-reserve map	矿石储量图
ore-reserve-base isotropy of mine product	矿石储量工作草图
ore-roasting kiln	焙矿炉
oreshoot	富矿体
ore-solution	含矿溶液
ore-structure	含矿构造
ore-treatment plant	选矿厂
ore-waste	矿石中的杂质；矸石
ore-zone	矿带
Orford process	奥福德法；奥福德顶底分层炼镍法
organic	有机的；机构
organic accumulation	有机堆积
organic analysis	有机分析
organic anion	有机阴离子
organic bactericide	有机杀菌剂
organic bank	生物滩；有机滩
organic base	有机碱；有机盐基
organic build-up	生物建造
organic carbon	有机碳
organic cation	有机阳离子
organic chemical rock	生物化学岩
organic chemistry	有机化学
organic clay	有机黏土
organic coast	生物海岸
organic component	有机组分
organic compound	有机化合物
organic constituent	有机组分；有机成分
organic contaminant	有机污染物
organic contamination	有机污染
organic content	有机质含量
organic content of soil	土壤有机物含量
organic cycle	有机循环
organic degradation	有机降解；发霉
organic deposit	有机沉积；生物沉积；有机沉积矿床；原生岩石
organic diagenesis	有机成岩作用
organic facies	有机相
organic framework	生物格架；生物骨架
organic framework reef	生物骨架礁
organic geochemistry	有机地球化学
organic geochemistry method	有机地球化学法
organic glass	有机玻璃
organic group	有机基
organic halogen compound	有机卤化合物
organic heat induration	有机热固结
organic hieroglyph	生物遗迹；生物遗痕
organic impurity	有机杂质
organic inclusion	有机包裹体
organic inhibitor	有机缓蚀剂；有机抑制剂
organic ion	有机离子
organic ion complex	有机离子复合体
organic lattice	生物骨架；生物格架
organic limestone	有机质灰岩；生物石灰岩
organic liquid	有机液体
organic marking	有机质痕迹
organic mass	有机物质
organic mass spectrometer	有机质谱仪
organic mass spectrometry	有机质谱分析
organic material	有机物质；有机物；有机材料
organic matrix	有机基体
organic matter	有机质；有机物
organic matter basis	有机基
organic matter content	有机物质含量
organic matter degradation	有机物降解
organic matter distribution	有机物质分布
organic matter geochemistry	有机质地球化学
organic maturation	有机质成熟作用
organic metamorphism	有机变质作用
organic micro-constituent	有机显微组分
organic mineral	有机矿物
organic mineral soil	有机矿质土
organic mineralization	有机矿化
organic molecule	有机分子
organic mound	有机丘；有机物岩岗
organic mud	有机泥浆
organic nitrogen	有机氮
organic origin	有机起源；有机成因
organic origin theory	有机成因说
organic particle	有机颗粒
organic pedological feature	有机土壤物质
organic peroxide	有机过氧化物
organic petrography	有机岩石学
organic phase	有机相
organic pollutant	有机污染物
organic pollutant analysis	有机污染物分析
organic pollution	有机物污染；有机污染
organic pollution load	有机污染物量
organic pollution of the soil	土壤有机污染
organic polymer	有机聚合物

organic promoter 有机助催化剂；有机促进剂；有机助聚剂
organic reaction 有机反应
organic reef 生物礁；生物礁堆积
organic reef facies 生物礁相
organic remain 有机残余物；生物遗骸
organic resin 有机树脂
organic rock 有机岩；有机质岩；生物岩
organic secretion 有机分泌作用
organic sediment 有机沉积物；有机沉淀物
organic sedimentary rock 有机沉积岩
organic sewage 有机污水
organic silt 泥炭黏土；有机质淤泥
organic slime 腐泥炭；腐泥煤；腐泥岩
organic sludge 有机污泥
organic soil 有机土
organic soil material 有机土质
organic solvent 有机溶剂
organic source material 有机生油物质
organic sphere 生物界；生物圈
organic spring 有机质泉水
organic substance 有机物质
organic synthesis 有机合成
organic theory 有机说；有机理论
organic waste 有机废物
organic weathering 生物风化
organic-inorganic inhibitor 有机-无机缓蚀剂
organic-inorganic interaction 有机-无机相互作用
organic-lean 贫有机质
organic-rich 富含有机质的
organic-rich continental sandstone 富有机质的陆相砂岩
organic-rich shale 富含有机物页岩
organics 有机物
organigram 构造示意图
organism 有机体；有机物；生物
organization for operation 施工组织
organization of construction 建设单位；施工组织
Organization of Petroleum Exporting Countries 石油输出国组织
organization of work 生产组织；工作组织；工程组织
organo 有机；有机金属的
organo facies modelling 有机相建模
organobentonite 有机膨润土
organochemical 有机化学的
organoclay complex 有机黏土络合物
organogenic 有机生成的；有机成因的
organogenous sediment 有机沉积物；有机沉淀物
organolite 有机岩；生物岩；离子交换树脂
organo-mineral aggregates 连生体
organophilic 亲有机质的；亲有机性的
organophilic bentonite 亲有机质膨润土
organophilic clay 有机土
organosedimentary 生物沉积的；有机沉积的
organosilicon 有机硅
orictocoenosis 生物化石群落；化石群
orient 定向；朝向；方位；东方
orientability 可定向性
orientable 可定向的
oriental alabaster 条带大理岩；纹状大理岩
oriental cat's eye 金绿猫眼石
oriental jasper 鸡血石
orientate 定向；给…定位；向东；使适应
orientated 朝向；面向；定向的；确定方向；使适应
orientated section 定向切片
orientating coring 定向取心
orientating tool 定向工具
orientation 产状；方向；方位；定向；定位；倾向性
orientation analysis 定向分析；方位分析
orientation angle 定向角；方位角
orientation by geometrical method 几何定向法
orientation by magnetic method 磁性定向法
orientation by optical method 光学定向法
orientation by shaft plumbing 竖井垂线定向法；竖井铅垂测量
orientation chart 方位图
orientation connection survey 定向连接测量
orientation contrast 方位衬度；方位反差
orientation data 定向数据
orientation density distribution 方位密度分布
orientation diagram 定位图；方位图
orientation distribution function 取向分布函数；方位分布函数
orientation effect 取向效应；定向效应；顺向效应
orientation effect of magnetometer 磁力仪定向效应；磁力仪顺向效应
orientation error 定向误差
orientation factor 取向因子；取向因素
orientation index 取向指数；晶向指数
orientation latchdown sub 定向闭锁接头
orientation line 标定线
orientation maxima 方位极密
orientation measurement 方向度量；方位测量；定向测量
orientation measuring system 方位测量系统
orientation mode 定向式；定向方法
orientation of core 岩芯定向
orientation of crustal movement 地壳运动定向性；地壳运动方向
orientation of crystal 晶体定位；晶体方位
orientation of discontinuity 不连续性的定向；不连续面的方位
orientation of failure plane 破坏面的定向
orientation of joint 节理方向
orientation of plane table 平板仪定向
orientation of reference ellipsoid 参考椭球体定位；基准椭圆定位
orientation of stria 擦痕方向
orientation plane 定位面
orientation plate 定位板；定点板；定中板
orientation point 定向点
orientation population 总体方位
orientation projection 定向投影
orientation release stress 解取向应力；定向解除应力；定向应力释放
orientation response 定向反应；定向响应
orientation sub 定向接头
orientation survey 试验测量；试点测量；定向调查
orientation texture 取向结构；定向结构
orientation uniformity 取向均一性；顺向均一性
orientation variant 方位变化
orientation well 定向井

orientational coring	定向取心
orientational instrument	定向仪
orientational order	取向有序性
orientator	定向器；定向仪
oriented	面向…的；定向的；定位的
oriented adsorption	取向吸附；定向吸附；定向表面吸附
oriented bit	定向钻头
oriented blasting	定向爆破
oriented branch	定向分支
oriented core	定向岩心
oriented core barrel	定向岩芯管；定向取心筒
oriented crystallization	定向结晶
oriented diamond	定向金刚石
oriented drilling device	定向钻井装置
oriented film	定向膜
oriented geophone	定向检波器
oriented graph	定向图
oriented growth	取向生长；定向生长
oriented hole	定向井眼
oriented magnetometer	定方位磁力计
oriented manifold	定向流形
oriented medium	有向介质
oriented micro-fabric	定向微组构
oriented needle	磁针
oriented network	取向网络
oriented nucleation	取向成核；定向成核
oriented perforating	定向射孔
oriented region	取向区
oriented rock sample	定向岩样
oriented sample	定向标本；定向样品；定向取样
oriented section	定向切片；定向截面
oriented specimen	定向标本；定向试样
oriented structure	定向构造
oriented thin section	定向薄片
orienteer	定向装置；标定装置；调整装置
orienting	定向的
orienting core barrel	定向取心筒
orienting device	定向装置；定向器
orienting line	定向线；方位线；方向基线
orienting lock mandrel	定向锁紧心轴
orienting lug	导向块
orienting pipe	定向管柱
orienting sleeve	导向套
orienting sub	定向接头
orienting tool	定向工具
orienting well survey	定向测井
orientite	锰柱石
orifice	小孔；注油孔；喷嘴；洞口，孔口
orifice diameter of grouting hole	注浆孔的孔径
orifice differential	孔口压差
orifice discharge	锐孔流量
orifice effect	孔眼效应
orifice fill collar	浮箍
orifice fill shoe	浮鞋
orifice flange	孔板法兰
orifice flow	孔口流
orifice flow constant	孔板流量系数
orifice flow measurement	孔板流量计量
orifice flow meter	锐孔流量计
orifice flowmeter factor	孔板流量计系数
orifice for injection	注入孔道
orifice meter	孔板流量计；细孔流速仪
orifice meter coefficient	孔板流量计系数
orifice of ejection	火山口
orifice temperature	泉口温度
orifice tip	孔板喷嘴端
orifice viscometer	孔口黏度计
orifice well tester	孔板式天然气流量计
orifice-closed grouting method	孔口封闭灌浆法
orifice-metering coefficient	孔口流量系数
orifice-type meter	孔板式流量计
orifice-type weir	孔型堰；孔板堰
origin	成因；起源；来源；原点
origin cohesion	原始黏聚力；初始黏聚力
origin coordinate	起始坐标
origin distortion	度盘零位误差
origin location	裂源位置
origin of a datum	基准点
origin of coordinate	坐标原点
origin of earth crust	地壳的起源
origin of force	着力点；力源
origin of ground water	地下水源
origin of hydrosphere	水圈的起源
origin of longitude	经度起算点
origin of oil	石油成因
origin of petroleum	石油生成
origin of the Earth	地球起源
origin of vector	矢量原点
origin stress	初始应力
origin time	发震时刻；发生期；开始时间
origin time of earthquake	发震时间
origin type of sediments	沉积物成因类型
original	原文，原作；原来的，开始的；最初的，固有的
original accelerogram	原始加速度记录
original and secondary interstices	原生和次生空隙
original ash	原生灰分
original asphalt	天然沥青
original attitude	原生产状
original bed	原河床
original bedding	原生层理
original bench mark	水准原点；水准基点
original blasting	首次爆破；原矿体爆破
original bubble-point pressure	原始泡点压力
original bulk	初始量
original cleavage	原生劈理
original cohesion	原始内聚力
original consequent lake	原始顺向湖
original cracks	原始裂纹；原裂缝；原生裂纹
original crust of the earth	原始地壳
original data	原始资料；原始数据
original data base	原始资料数据库
original deep thermal water	深层原生热水
original design	原设计
original dip	原生倾斜；原始倾斜
original displacement	原始位移
original drawing	原图
original drilling report	原始钻进报表；原始钻井报表
original edgewater front	原始边水前缘
original effective permeability	原始有效渗透率
original estimate	原始估计

original fabric	原生组构
original field recording	野外原始记录
original fissure water	原生裂隙水
original flow structure	原生流动构造
original foliation	原生叶理
original formation volume factor	原始原油地层体积系数
original forms	原始地形
original free-gas cap	原生游离气顶
original gas in place	原始天然气地质储量
original gas-oil ratio	原始气油比
original gas-water contact	原始气-水接触面
original ground	原地面
original ground level	原地标高；天然地面标高
original horizontal stratification	原生水平层理
original horizontality	原始水平状态
original image identification	原始影像识别
original interstice	原生间隙；原生空隙；原始孔隙
original isotopic composition	原始同位素组成
original joint	原生节理
original loess	原生黄土
original magma	原岩浆；母岩浆
original mineral	原生矿物
original oil in place	原始原油地质储量
original oil volume factor	原始石油体积系数
original oil-water interface	原始油水界面
original oil-water level	原始油水界面
original order	正常产状；原层序；原始顺序
original ore	原生矿
original outcrop	原生露头
original overburden pressure	原始上覆压力
original permeability	原始渗透率
original piezometer level	初始测压水位
original piezometric level	初始测压水位
original piezometric surface	初始水压面
original plan	原来图则；原始图则
original planet	原始行星
original plot	实测原图
original porosity	原始孔隙度；原生孔隙率
original position	原始位置；起始位置
original pressure	原始压力，初始压力
original program	原始程序
original raypath	原始射线路径
original reconnaissance	初次勘测
original record	原始记录
original recorded value	原始记录值
original recoverable oil	原始可采油量
original reservoir fluid	原始油层流体；原始储层流体
original reservoir pressure	原始油层压力；原始储层压力
original reservoir wettability	油层原始润湿性；储层原始润湿性
original river	原有河道
original river bed	原河床
original rock	基岩；原岩；原生岩石；母岩
original rock pressure	原始岩层压力；原始地层压力
original rock temperature	原始岩地层温度；原始地层温度
original salinization	原生盐渍化；原生盐化作用
original sample	原样；原样品
original scale	原比例尺；原始比例尺
original scheme	原始方案
original section	原始剖面
original sedimentary facies	原始沉积相
original silt	渠道及水库中淤积物
original soil	原土；原始土
original source material	生油母质；原始油源物质
original spacing	原始井距；原始间距
original state	初始状态
original stiffness	初始刚度
original stope	最初回采工作面
original stream	顺向河
original stress	原始应力；原应力
original structure	原生构造
original structure surface	原始结构面
original subgrade	原始路基；原有路基
original surface	原地表面；起始面
original tally	原始标记
original temperature	原始温度
original texture	原生结构
original thickness	原始厚度
original time	发震时刻；初始时间
original topography	原始地形
original trace	清绘原图；全要素原图
original water content	原先含水量；初始含水量
original water level	原水位；初见水位
original water line	原始水线
original wave	原波形；基波
original-fissure water	原生裂隙水
originality	原始；独创性；原创
originate	引起，发起；起源；开头，开始；发生；创办，创设
originated input	原始输入
originating	发源；起源的
origination	出现，开始产生；起源
originative	有创造性的
origofacies	原始沉积相
O-ring	O形环；圆环；胶皮垫圈
O-ring seal	密封圈；密封胶圈
Ormsby filter	欧姆斯比滤波器
Ormsby type filter	欧姆斯比型滤波器
ornamental stone	彩石；装饰石；观赏石
ornitholite	鸟化石
ornoite	奥闪闪长岩
oroclinal folding	弯移褶皱
oroclinal warping	斜向扭曲
orocline	弯曲造山带；造山带弯曲
oroclinotath	曲张造山带
orocratic	地壳运动期的；剧烈变动期的
orocratic period	地壳运动期
oroflexural	山曲
orogen	造山带
orogen collapse	造山带塌陷
orogen in continental margin	陆缘造山带
orogen in plate margin	板缘造山带
orogene	造山带
orogenesis	造山运动；造山作用
orogenetic	造山的；造山带的
orogenetic belt	造山带
orogenetic cycle	造山旋回
orogenetic movement	造山运动
orogenic	造山的；造山带的

orogenic activity	造山活动
orogenic andesite	造山带安山岩
orogenic batholith	造山成因的岩基
orogenic belt	造山带
orogenic crumpling	造山挤压作用
orogenic cycle	造山旋回
orogenic episode	造山幕
orogenic epoch	造山期
orogenic erosion process	造山侵蚀作用
orogenic event	造山运动
orogenic facies	造山相
orogenic geothermal belt	造山地热带
orogenic hydrothermal deposit	造山热液矿床
orogenic landslip	造山滑坡
orogenic metamorphism	造山变质作用；褶皱期变质作用；区域变质作用
orogenic movement	造山运动
orogenic phase	造山期；造山幕
orogenic process	造山过程；造山运动
orogenic sedimentation	造山沉积
orogenic sequence	山地土系列
orogenic soil	山地土壤
orogenic suture	造山带缝合
orogenic unconformity	造山不整合
orogenic wedge	造山楔
orogenic welt	造山接缝
orogenic zone	造山带
orogeny	造山运动；造山作用
orogeosyncline	造山地槽
orographic	山岳；山地的；山形的；山势的；地壳运动的
orographic basin	山间盆地
orographic character	山势特性；地形特征
orographic condition	地形条件
orographic depression	地形低压
orographic divide	山岳分水界
orographic drag	地形拖曳
orographic effect	地形影响
orographic fault	山形断层
orographic precipitation	地形性降雨；地形降水
orographic rain	地形雨；山岳雨
orographic storm	山岳暴雨
orographic thunderstorm	地形雷暴
orographic uplift	地形抬升；山形隆起
orographic wave	地形波
orographical	山地形的；山岳的
orographical effect	地形影响
orographically-enhanced precipitation	山岳地形增强降水
orography	山文学；山志学；山岳形态学
orohydrographic model	山水地形模型；高山水文地理模型；山水立体地图
orohydrography	高山水文地理学；山地水文学；地形水文学
orology	山理学；山岳成因学
orometer	高山气压计；山岳气压表
orophase	造山幕
Orosirian	造山纪
orotectonic line	山岳构造线
orotrite	云闪霞斜岩
Orsat apparatus	奥氏分析仪
ort	横巷；煤门
orterde	硬磐；硬土层
orthent	正常新成土
orthid	典型旱成土
orthite	褐帘石
orthoalaskite	正白岗岩
orthoalbitophyre	正钠长斑岩
orthoalkaligneiss	碱性正片麻岩
ortho-amphibolite	正角闪岩
orthoandesite	斜辉安山岩
orthoarenite	正砂屑岩
orthoaxis	正轴；正交轴
orthobasalt	方辉玄武岩
orthobase	正长辉绿岩
orthobituminous coal	正烟煤；肥煤；中阶烟煤
orthobrannerite	斜方钛铀矿
orthocenter	垂心
orthochamosite	正鲕绿泥石
orthochem	正常沉积物；原生化学沉积
orthochemical	正化学的
orthochemical rock	正化岩；正化学岩
orthochlorite	正绿泥石
orthochronology	正年代学；标准地质年代学；常规地层年代学
orthoclase	正长石；钾长石
orthoclase porphyry	正长斑岩
orthoclase-gabbro	正长辉长岩
orthoclasite	微正长岩；细粒正长岩
orthoclasization	正长石化
orthoclastic	正节理；正交解理；沿互成直角面劈开的
orthoclastization	正长石化
orthocline	正倾型；直倾型
orthocoking coal	正焦煤
orthoconglomerate	正砾岩；原生砾岩
orthocumulate	正堆积岩
orthod	灰土
ortho-derivative	邻位衍生物
orthodiadochite	磷铁华
orthodiagonal	正轴
orthodiagonal axis	正轴
orthodolomite	正白云石；原始白云岩
orthodome	正轴坡面
orthodox face	普通工作面
orthodox face of coal	普通采煤工作面
orthodox-anthracite	中阶无烟煤
orthodox-lignite	中阶褐煤
orthodromic projection	大圆投影
orthoeluvial weathered crust	正残积风化壳
orthoeluvium	正残积层；典型淋溶层
orthoevolution	直向进化
orthofeldspar	正长石
orthofelsite	正长霏细岩
orthogabbro	正辉长岩
orthogenesis	直向演化
orthogeosynclinal system	正地槽系
orthogeosyncline	正地槽
orthogneiss	正片麻岩；火成片麻岩
orthogon	矩形；长方形
orthogonal	正交的；直角的；相互垂直的
orthogonal anisotropy	正交各向异性

orthogonal array	正交阵列；正交设计表
orthogonal axes	正交轴
orthogonal axis	正交轴
orthogonal basis	正交基
orthogonal circle	正交圆
orthogonal code	正交码
orthogonal component	正交分量
orthogonal condition	正交性条件
orthogonal coordinate system	正交坐标系
orthogonal curvilinear	正交曲线
orthogonal curvilinear coordinate	正交曲线坐标
orthogonal direction	正交方向
orthogonal drift	正交漂移
orthogonal experiment	正交试验
orthogonal experimental design method	正交实验设计方法
orthogonal fracture	正交裂缝
orthogonal fracture plane	正交断裂面
orthogonal fracturing	正交断裂
orthogonal function	正交函数
orthogonal grid	正交网格
orthogonal intersection	正交
orthogonal joint	横节理
orthogonal joint set	正交节理组
orthogonal linear transformation	正交线性变换
orthogonal lineation	横线理
orthogonal lines	正交直线组
orthogonal map projection	正交投影
orthogonal matrix	正交矩阵
orthogonal mode	正交振型；正交模式
orthogonal network	正交网络系统
orthogonal operation	正交运算
orthogonal planar frame	正交平面框架
orthogonal plane	垂直面；正交平面
orthogonal polarization	正交极化
orthogonal polynomial	正交多项式
orthogonal projection	正交投影；正投影
orthogonal reference axis	正交参考轴；正交基准轴
orthogonal regression line	正交回归线
orthogonal section	垂直剖面
orthogonal series	正交级数
orthogonal set of principal axis	正交主轴组
orthogonal signal	正交信号
orthogonal signal set	正交信号集
orthogonal sounding	正交测深
orthogonal system	正交系
orthogonal test	正交试验
orthogonal thickness	垂直层面厚度
orthogonal trajectory	正交轨线
orthogonal transform	正交变换
orthogonal transformation	正交变换
orthogonal vector	垂直向量；正交向量
orthogonality	正交性；正交状态；直交性；相互垂直
orthogonality condition	正交条件
orthogonality criterion	正交判别准则
orthogonality eigenvector	正交特征向量
orthogonality of mode	振型的正交性
orthogonality property	正交性
orthogonality property of the mode function	振型函数的正交性
orthogonality relation	正交关系
orthogonality theorem	正交定理
orthogonalization	正交化
orthogonalizing process	正交化步骤
orthogranite	正花岗岩
orthograph	正视图；正投影图
orthographic plan	平面图
orthographic projection	正射投影
orthographical	正交的；直角的
orthographical projection	正射投影；平行投影
orthographical view	正射观察
orthography	正交射影；正射投影
orthohexagonal	正六方形的
orthohydrous	正常氢含量的
orthohydrous vitrinite	正氢镜质体
ortho-isomer	邻位异构物
orthokinetic	同向的；同向移动的
orthokinetic coagulation	同向凝聚
ortholignitous coal	褐煤；正褐煤
ortholimestone	正石灰石；原生石灰岩
ortholite	云母正长岩
orthomagma	正岩浆；原岩浆
orthomagmatic	原岩浆的；正岩浆的
orthomagmatic deposit	正岩浆矿床
orthomagmatic ore deposit	正岩浆矿床
orthomagmatic stage	原岩浆期；正岩浆期
orthomarble	正大理岩
orthomatrix	正杂基；正基质
orthometabasite	正变质基性岩
orthometamorphic rock	正变质岩
orthometamorphism	正变质作用
orthometamorphite	正变质岩
orthometric correction	正高改正；竖高校正
orthometric elevation	正高
orthometric height	正视高度；正高
orthomicrite	原生微晶灰岩；正泥晶灰岩
orthomicrosparite	原生微亮晶灰岩；正泥晶亮晶灰岩
orthomigmatite	正混合岩
orthomin	正交极小化
orthomin acceleration	正交极小加速度
orthomin algorithm	正交极小化算法
orthomorphic	正形的；等角的；正形投影的
orthomorphic map projection	正形地图投影
orthomorphism	等角性；正形性
orthonormal	正规化的；标准正交的
orthonormal basis	标准正交基；正交基
orthonormal function	标准正交函数
orthonormal system	正交系
orthonormal tunnel portal	正洞门；正隧道入口
orthonormal vector	正交单位向量
orthonormality	标准化；正规化；标准正交性
orthophonite	霞石正长岩
orthophosphate	正磷酸盐
orthophosphoric acid	正磷酸
orthophoto	正射像片；正射影像
orthophoto isotropy of mine product	正射影像地图
orthophoto map	正射影像地图
orthophoto quad	正射像片标准图幅
orthophoto quadrangle	正射投影像片图幅
orthophoto stereo-mate	正射影像立体配对片；立体正射

像片
orthophoto technique 正射影像技术
orthophotograph 正射像片
orthophotography 正射摄影；正射投影技术
orthophotomap 正射影像图
orthophotomapping 制作正射影像地图
orthophotoquad 正射影像图
orthophotoscope 正射投影仪
orthophotoscopy 正射投影纠正
orthophyre 正长斑岩
orthophyric 正斑状的
orthopinacoid 正轴轴面
orthoplatform 正地台
orthopole 正交极；垂极
ortho-position 邻位；正位
orthoprism 正轴柱
orthoprojector 正射投影仪
orthopyroxene 斜方辉石类；正辉石
orthopyroxene gabbro 斜方辉石辉长岩
orthopyroxenite 斜方辉岩
orthoquartzite 正石英岩；沉积石英岩
orthorhombic 斜方的；斜方晶系的；正交的
orthorhombic crystal system 正交晶系；斜方晶系
orthorhombic pyroxene 斜方辉石；正辉石
orthorhombic symmetry fabric 斜方对称岩组
orthorhombic system 斜方晶系
orthorhyolite 正长流纹岩
orthorock 正变质岩；火成变质岩
orthoschist 火成片岩；正片岩
orthoscope 正射投影仪
orthoscopic 无畸变的
orthose 正长石
orthoselection 直向选择；自然选择
orthosequence 正层序
orthoserpentine 火成蛇纹岩
orthoshonkinite 钾质等色岩
orthosilicate 正硅酸盐；原硅酸盐
ortho-silicic acid 原硅酸
orthosite 粗粒正长岩
orthostereoscopy 正立体效应；正射投影立体观测；正立体
orthostratigraphy 标准地层学；正交地层学；正地层学
orthostyle 列柱式
orthosyenite 钾质正长岩
orthotarantulite 正长英石岩；正长白岗岩
orthotectic 岩浆的；正熔的
orthotectogenesis 正构造运动
orthotectonic Caledonides 标准构造型加里东褶皱带；标准构造型加里东造山带
orthotectonic orogen 正构造型造山带
orthotectonic region 正构造区
orthotectonics 正构造；正大地构造
orthotrachyte 正长粗面岩
orthotropic 正交各向异性的
orthotropic elasticity 正交各向异性弹性体
orthotropic material 正交各向异性材料
orthotropic media 正交各向异性介质
orthotropic particle 正交异性颗粒；正交各向异性颗粒
orthotropic plate 正交异性板
orthotropy 正交各向异性

orthovideo 正射影像
orthox 正常氧化土
ortlerite 云闪辉长玢岩；闪斑玢岩
ortstein 褐铁矿；灰质壳
ortstien soil 铁磐土；铁盘土
orvietite 响白碱玄岩；响白灰玄岩
oryctocoenose 化石群；生物化石群
oryctocoenosis 生物化石群
oryctogeology 化石地质学
oryctognosy 地物学
oryctology 化石学
osannite 钠闪石
osar delta 蛇丘三角洲
osciducer 中频偏移传感器；振荡传感器
oscillate 振动；振荡
oscillating 振荡的
oscillating axle 摆动轴
oscillating ball meter 往复球式流量计
oscillating bar 摇动筛条
oscillating beam 摆动臂
oscillating bit 振荡钻头
oscillating crystal 振荡晶体
oscillating crystal method 回摆晶体法
oscillating current 振荡电流
oscillating curve 振动曲线；振荡曲线
oscillating cutterhead 摇摆式切削刀盘
oscillating detector 振荡检测器
oscillating dryer 立式振动离心脱水机
oscillating feeder 振动给料器
oscillating flow 波动流量；脉动流
oscillating function 振荡函数
oscillating motion 摆动运动；振荡运动；振动
oscillating piston liquid meter 往复式活塞液体流量计
oscillating platform 摆动平台
oscillating plough 振动式刨煤机
oscillating point 振荡点
oscillating sampler 摆动式取样器
oscillating screen 振动筛
oscillating screening centrifuge 立式振动离心脱水机
oscillating separator 振动分选机；振动式分离器
oscillating sequence 振荡级数
oscillating table 振动台；摇床
oscillating trough 震动槽；槽式给矿机
oscillating-arm-type boring machine 摆臂式岩石掘进机
oscillating-column pump 振动泵
oscillating-disk viscometer 摆动盘式黏度计
oscillation 振荡；摆动；振动
oscillation amplitude 振幅；振荡幅度
oscillation constant 振荡常数
oscillation cross ripple mark 振荡干涉波痕
oscillation damping 消振；振动阻尼；振荡阻尼
oscillation effect 振荡效应
oscillation envelope 振荡包线
oscillation force 振荡力
oscillation frequency 振荡频率
oscillation hypothesis 振荡说
oscillation impedance 振动阻抗
oscillation instability 振荡不稳定性
oscillation loop 振荡波腹
oscillation measurement 振动测量

oscillation of ground water level	水震波
oscillation period	振荡周期
oscillation pickup	拾振器；振动传感器
oscillation ripple	振荡波；对称波痕；摆动波痕
oscillation ripple mark	摆动波痕；振荡波痕
oscillation space	振动空间
oscillation theory	颤动说；振动理论
oscillation volume	振动体积；振颤容积
oscillator	振荡器；振动子；发生器；加速器
oscillator measurement	振动测量
oscillatory	振荡的
oscillatory motion	振动；摆动
oscillatory movement	振荡运动
oscillatory pressure	脉动压力
oscillatory region	振荡区域
oscillatory rhythm	波动韵律
oscillatory ripples	对称波痕；摆动波痕
oscillatory series	振荡级数
oscillatory twinning	反复双晶
oscillatory viscometer	振动式黏度计
oscillatory wave	摆动波
oscillatory wave function	振荡波函数
oscillogram	示波图；波形图
oscillogram recording	波形记录
oscillogram trace reader	示波图读出器
oscillograph	示波器；示波仪；录波器；波形图
oscillograph record	示波器记录
oscillograph trace	示波器迹线
oscillograph vibrator	示波器振子
oscillographic commutator	示波管
oscillographic method	示波法
oscillography	示波术；示波法
oscillometer	示波仪；录波器
oscillopolarography	示波极谱
oscilloprobe	示波器测试头；示波器探头
oscilloscope	示波器；示波仪；录波器
oscilloscope pattern	示波器屏幕的图形
oscilloscope photograph	示波器照相
oscilloscope trace	示波器迹线
oscillosynchroscope	同步示波器
oscillotron	示波管
osciloscope	示波器
osculating circle	密切圆
osculating ellipse	密切椭圆
osculating helix	密切螺旋线
osculating orbit	密切轨道
osculation plane	密切面
osculation point	密接点
O-seal	O形密封圈；O形密封环
Oslo-essexite	奥斯陆碱性辉长岩
Oslograben	奥斯陆地堑
Oslo-porphyry	奥斯陆斑岩，更长斑岩
osmiridium	锇铱矿；铱锇矿
osmium	锇
osmium-osmium dating	锇-锇测年法
osmium-rhenium dating	锇铼法年龄测定；锇-铼测年法
osmo regulation	渗流调节
osmolality	渗透压度
osmolar	渗透的
osmolarity	渗透性
osmometer	渗透压力计；渗压计
osmometry	渗透压测定
osmophilic	易渗的；耐高渗透压的
osmophobic	憎渗的；不易渗的
osmoregulation	渗透压调节；渗压调节；渗透调节
osmoregulatory	调节渗透的
osmoscope	渗透试验器；渗透计
osmose	渗透；渗透性；渗析
osmose movement	渗透运动
osmose potential	渗透势
osmose pressure	渗透压力
osmosis	渗透；渗透性；渗析
osmosis coefficient	渗透系数
osmosis potential	渗透潜能；渗透势
osmosis pressure	渗析压力；渗透压力；渗透压
osmosize	渗透
osmostic pressure method	渗透压法
osmotaxis	趋渗性
osmotic	渗透的
osmotic coefficient	渗透系数
osmotic concentration	渗透浓度
osmotic effect	渗透效应
osmotic energy	渗透能
osmotic equilibrium	渗透平衡
osmotic equivalent	渗透当量
osmotic exchange	渗透交换
osmotic flow	渗透水流；渗析流
osmotic force	渗析力
osmotic gradient	渗析梯度；渗透梯度
osmotic hydration	渗透水化作用
osmotic hypothesis	渗透学说
osmotic membrane	渗透隔膜
osmotic movement	渗透运动
osmotic phenomenon	渗透现象；渗析现象
osmotic potential	渗透势
osmotic pressure	渗透压力；渗透压；渗析压力
osmotic pressure method	渗透压法
osmotic pressure potential	渗透压势
osmotic pressure suction	渗透压吸力
osmotic repulsive pressure	渗透排斥力
osmotic suction	渗透吸力
osmotic suction gradient	渗析吸力梯度
osmotic swelling	渗透膨胀
osmotic tensiometer	渗透张力计
osmotic water	渗透水
osmotically bound water	渗透结合水；渗透束缚水
osmotic-pressure potential	渗透压势
osmotic-pressure suction	渗透压吸力
ossipite	粗橄长岩
osteocolla	多孔石灰华
osteolite	土磷灰岩；磷钙土；粪化石
osteolith	骨磷化石
ostia	入水孔
ostiary	河口
ostium	入水孔；流入口
ostraconite	远洋灰岩
ostraite	磁铁尖晶辉岩
Ostwald gravimeter	奥斯特瓦尔德比重计
osumilite	大隅石
otavite	菱镉矿

ottajanite 浅白碱玄岩	outbuilding 海退建造作用
ottrelite 锰硬绿泥石	outburst 突出；崩出；喷出；破裂；脉冲
ottrelite schist 硬绿泥石片岩	outburst coal seam 突出煤层
ouachitite 黑云沸煌岩	outburst danger 突出危险
oued 枯水河；旱谷；干谷	outburst forewarning 突出警报
ouenite 细苏橄辉长岩	outburst material 突出物
oulopholite 石膏花；叶片石膏	outburst mechanism 突出机理
out alignment 不对准的	outburst of gas 瓦斯突出
out array 外部数组	outburst of gas and rock 突出的气体和岩石
out break 超欠挖	outburst of water 突然涌水，突水
out cylinder 外圆筒	outburst phenomenon 突出事故；突出现象
out device 输出设备	outburst prediction 突出预测
out drill 超钻	outburst proneness 突出倾向性；突出危险程度
out fan 冲积扇	outburst risk 突出风险
out flow 流出；外流；流出量；泄出	outburst risk prediction 突出危险性预测
out flow of methane 瓦斯涌出；沼气涌出	outburst-prone mine 易发生突出的矿井
out flow of water 水的涌出	outburst-prone strata 有突出危险的岩层
out force 外力	outburst-proneness 突出危险程度
out of adjustment 未平差的；失调	outburst-relieved seam 突出释放煤层
out of balance 不平衡的；失去平衡的	outbursts of coal and gas 煤和天然气的突出
out of condition 不良	outby rib 安全矿柱
out of control 失控的；失去控制的	outby work 靠近井底作业
out of date 过时的；落后的；陈旧的	outcome 结果；成果；产量；输出
out of door 室外的；露天的	outcome parameter 最后参数；输出参数
out of flat 不平	outcrop 露头；露出地面的岩层
out of focus 焦点失调；散焦	outcrop area 露头区；出露面积
out of gauge bit 尺寸不合标准的钻头	outcrop columnar section 露头柱状剖面图
out of gauge hole 井径不合标准的井	outcrop coring 露头取心
out of level 不平坦的；失平的	outcrop line 露头线
out of line 不在一条线上	outcrop map 露头图
out of operation 不能工作的；失效的	outcrop migration 露头迁移
out of phase 不同相；异相；失相	outcrop mine 露天矿
out of phase cross correlation 异相交叉相关	outcrop mining 露头开采
out of phase difference effect 差动影响；相位差影响	outcrop of coal seam 煤层露头
out of pit dump 外排土场	outcrop of fault 断层露头
out of plane 平面外；出平面	outcrop of water 水流出逸口
out of repair 失修的	outcrop pattern of planar feature 面状形态的露头型式
out of roundness 不圆度；椭圆度	outcrop spring 露头泉
out of run 不能使用	outcrop stripping 沿露头露天开采
out of service 停运的；不能使用的	outcrop water 露头渗出水
out of shape 形状不规则的；变形的；形状遭到破坏的	outcropped area 露头区；出露面积
out of size 尺寸不合规定的	outcropping 露头；外露的；露出地表
out of specifications 超出规范的	outcropping bed 露头层
out of square 歪斜；倾斜的；不成直角	outcropping pervious formation 出露地表的含水层；含水层露头
out of work 不能工作的；失效的	outcropping rock 露头岩石
out performance 流出动态	outcropping seam 露头煤层；露头矿层
out phase 异相；不同相	outcropping water 泉水
out rigger 悬臂梁	outdiffusion 向外扩散
out washing 冲刷作用	out-dip structural plane 外倾结构面
outage table 存油体积换算表	outdoor 室外的；野外的；露天的
outage time 运行中断时间；停工时间	outdoor electric substation 露天变电站
out-and-in bond 凹凸砌合	outdoor equipment 露天设备
out-away view 剖视图	outdoor installation 室外安装
outback 内地的；内陆的	outdoor location 室外装置；露天设备
outboard 外侧	outdoor pile 野外堆积
outboard boom 外侧摇臂	outdoor power station 露天电站
outboard profile 侧面图	outdoor storage 露天储存
outbowed 矿层露头	outdoor work 露天作业
outbreak 破裂；断裂；中断；超挖	outdoors 在户外；露天；野外
outbreak-crash 穿透性破坏	

outdoor-type	野外的；露天的
out-draw	外拉伸
outdrill	超钻
outer arc	外弧；外岛弧
outer arc basin	外弧盆地
outer arc ridge	外弧脊
outer arch	外缘隆起；海沟边缘隆起；外拱状构造
outer atmosphere	外大气圈
outer bar	外沙洲
outer barrel	外岩心筒；外套管
outer beach	外滩
outer boundary	外部边界；外边界
outer breakwater	外海防波堤
outer capillary water	外毛管水
outer clearance ratio	外间隙比
outer continental shelf	外大陆架；大陆架外缘
outer core	外核；外地核
outer core barrel	外岩心筒
outer corner	钻头转角
outer cutting bit	外管钻头
outer diameter	外径
outer dimension	外形尺寸；外部尺寸
outer drainage boundary	外泄油边界
outer edge	外缘
outer edge waterflood	边外注水
outer enveloping profile	液压气动系统
outer flank	外翼
outer grout pipe	外灌浆管
outer high	外陆缘高地
outer hole	外缘炮眼
outer island arc	外岛弧
outer island arc ridge	外弧海岭
outer iteration	外迭代
outer layer	外层
outer limit of the continental shelf	大陆架外部界限
outer limit of the territorial sea	领海外部界限；领海线
outer line of sea ice area	海冰区外缘线
outer linearization	外线性化
outer locator	外部探测器
outer mantle	地幔上部；上地幔；外地幔
outer marine subzone	外海分区
outer normal	外法线
outer orientation	外方位；外部定向
outer planet	外行星
outer pressure resistance	抗外压强度
outer product	外积；矢量积
outer radius	外径
outer sedimentary arc	外沉积弧
outer shelf	外陆架；外大陆架
outer sleeve	外套筒；外套管
outer slope	外侧坡度
outer space	外层空间；太空
outer spoil pile	外部排土场
outer stripping material	外剥离物
outer sublittoral zone	底亚滨海带；外浅海带
outer subzone	外分区
outer surface	外表面
outer surfacing waterproofing layer	外贴式防水层
outer swell	外缘隆起；海沟边缘隆起
outer thermostat	外恒温器
outer tidal delta	外潮汐三角洲
outer tubing string	外油管柱
outer vortex	外旋流；外螺旋线
outface	倾向坡
outfall	泄水口；出水口；渠口；排放口
outfall ditch	排水沟；溢水沟
outfall fan	冲积扇
outfit	设备，备用工具；配备，装配；供给，准备；成套钻机设备及工具
outflow	外流；流出量；溢出量；流量
outflow boundary	外流边界；外流端面
outflow cave	出水溶洞
outflow channel	放水渠道；溢流河床
outflow face	渗出面
outflow from reservoir	水库出水量
outflow lake	外流湖
outflow of groundwater	地下水露头
outflow of spring	泉的流量
outflow of water	涌水量
outflow pattern	面状出水点
outflow point	泉点；出流点；出口
outflow pressure	流出压力
outflow rate	涌水量；出水量
outflow spring	溢流泉
outflow surface	溢出面
outflow test	放水试验
outflow velocity	流出速度
outflow vortex	流出旋涡
outflow-end saturation	流出端饱和度
outfold	倒转褶皱
outgassing	除气；释气的；脱气作用
outgassing rate	释气率；出气率；放气速率
outgoing air	废气
outgoing carrier	输出载波
outgoing data	输出数据
outgoing ebb	落潮流
outgoing infrared radiation	射出红外辐射
outgoing long-wave radiation	向外长波辐射
outgoing manifold	发送管汇
outgoing neutron	出射中子
outgoing pipeline	发送管线
outgoing side	输出侧；流出侧
outgoing signal	输出信号
outgoing tide	退潮
outgoing wave	外向波
outgush	涌出；喷出；排出
out-hole run	上行测量
out-laying well	扩边井
outleakage	泄漏；漏出量
outlet	出口；出口管；排水孔；上风道，回风井；通地面巷道；引出线
outlet aperture	出口孔；出口孔径
outlet branch	输出支管；放水支管；放气支管
outlet cave	出水洞
outlet conduit	出水渠；出水管道；泄水道
outlet discharge	出口流量
outlet drain	去水排水渠；出口排水渠
outlet hole	出口孔；排出孔
outlet lake	开口湖
outlet loss	出口水头损失

outlet of lake	湖泊出口
outlet of thermal water	热泉
outlet opening	出口；回风口；排气孔
outlet pipe	排污管；排水管
outlet pipework	出水管
outlet pressure	出口压力；排出压力
outlet pressure of monitor	水枪出口压力
outlet pump	排水泵；排油泵
outlet regulation	河口整治
outlet shaft	上风竖井
outlet sluice	排水闸
outlet stream	泉口溪流
outlet submerged culvert	压力式涵洞
outlet temperature	出口温度
outlet tunnel	出渣隧洞；排水隧道
outlet velocity	出口速度；流出速度
outlet water	废水
outlet work	排水工程
outlet-pressure	出口压力
outlier	异常值；超限误差；外露层
outlier drawing	轮廓图；略图；草图
outlier map	轮廓图；示意图；略图
outlier of overthrust mass	飞来峰；孤残层；翻卷褶皱孤山
outlier reef	离群礁
outline	轮廓，外形；大纲，概要；回线，外线；略图；草图
outline an orebody	圈定矿体边界
outline dimension	外形尺寸
outline drawing	轮廓图；略图；草图
outline line	轮廓线
outline map	轮廓图；略图；草图
outline map for filling	填图用的略图
outline of site investigation	勘察工作纲要
outline plan	粗略规划
outline programme	工程计划概要
outline sketch	轮廓图
outlining	圈定边界；画轮廓
outlying	外围的；分离的；远离中心的
outlying area	郊区；外围地区
outlying well	外缘井
outlying zone	外围地区
out-of-gage	不合规格的
out-of-hole time	非钻进时间
out-of-limit	越限
out-of-mine trench	外部沟
out-of-phase	异相；不同相的
out-of-plane bend	面外弯曲
out-of-plane shear	面外剪切
out-of-round	不圆；失圆的
out-of-round oversized hole	大肚子井眼
out-of-service record	故障记录
out-of-size	尺寸不合格的
out-of-specification line	不符合技术要求的剖面
out-of-work	不工作的，失效的
outport	外港；输出港
outpost well	油藏边界外的井
outpour	泻出
outpouring	泻出；喷溢；流出；倾倒
outpouring facies	喷溢相
outpouring spring	涌泉；喷泉
output curve	输出曲线
output data	输出数据
output device	输出设备；输出装置
output energy	输出能量
output energy filtering	输出通量滤波
output filter	输出滤波器
output fluctuation	产量波动
output format	输出格式
output gas factor	气油比；输出气油比
output gas-oil ratio	气油比
output image	输出图像
output impedance	输出阻抗
output layer	输出层
output of raw coal	原煤产量；原煤开采量
output parameter	输出参数
output per coal face worker	每个回采工人产量
output per cycle	循环产量
output per man	每人产量
output per manshift	每人每班产量
output per man-year	每人年产量；每工年产量
output per open-pit miner	每个露天采煤工人产量
output per stope	工祖产量；每个工作面产量
output per underground worker	每个井下工人产量
output program	输出程序
output pulse	输出脉冲
output resistance	输出电阻
output signal	输出信号
output sleeve	输出套
output spectrum	输出能谱
output speed	输出速度
output statistics	产量统计
output torque	输出转矩
output tube	输出管
output unit	输出装置；输出部件；输出设备
output variable	输出变量
output wave shape	输出波形
output waveform	输出波形
output wavelet	输出子波
output well	生产井
output work	输出功
output-energy filter	输出能量滤波器
output-input ratio	输出-输入比；产出-投入比率
outreach	伸展长度；超过；胜过
outrush	急剧冲出；急速泄出
outshore	海上；向海一边；远离海滨
outside airway	旁风巷；边缘风巷
outside balance reserves	表外储量
outside bank slope	河岸外坡
outside barrel	外取心筒
outside blast	露天爆破
outside caliper	外径规；外卡钳
outside casing load	套管外载荷
outside casing voidage	套管外空隙
outside clearance ratio	外间隙比
outside diameter	外径
outside dimension	外形尺寸
outside drawing	外形图
outside dumping	外排土
out-side fleet angle	外偏角；侧偏角

outside gravel packing	管外砾石充填
outside heading	边界平巷；边缘平巷
outside micrometer	外径千分尺
outside micrometer caliper	外径千分卡齿；外径测微计
outside monkey board	折叠式井架外侧二层台
outside network	外部管网
outside of sensor array	传感器阵列外
outside perforation packing	射孔孔眼外砾石充填
outside pond	外缘湖泊
outside pool boundary	油藏外边界
outside protective case	外保护套管
outside pump blender	泵外混料器
outside radius	外半径
outside seal	外部密封
outside sealing	封井
outside storage tank	露天储罐
outside surface	外表面
outside tunnel control survey	隧道洞外控制测量
outside well	边外井
outsideanchor pier	外锚墩
outskirt of mine	矿山境界
outslope disposal method	沿水平外排法
outslope reduction method	减缓外排坡角法
outspread	伸开；延伸
outsqueezing	压出
outstanding point	明显地物点
outstanding tectonic direction	优势构造方向；主构造方向
outstep drilling	甩开钻井
outstep well	外扩井
out-stepping well	探边井
outtake	回风道；出风井；出口
outthrust	突出；突出物；刺穿
outward longwall	前进式长壁开采
outward mining	前进式开采
outward normal	外法向
outward seepage	向外渗出
outward thrust	向外推力
outward working	向外回采；前进式回采
outward-dipping	向外倾斜
outward-spreading lithospheric plate	向外扩张的岩石圈板块
outwash	冰水沉积；冰川边碛外的沉积；冲刷；冲蚀；冲积物；坡积物
outwash apron	冰水沉积扇
outwash delta	冰水沉积三角洲
outwash drift	冰水沉积；冰水漂积
outwash drop	冰水沉积
outwash facies	冰水沉积相
outwash fan	冰水冲积扇
outwash gravel	漂水砾；冰水砾
outwash plain	冰川沉积平原
outwash sediment	冰水沉积物
outwash stream	冰水河
outwash terrace	冰水沉积阶地
outwedging	尖灭
outweigh	重于；优于
outwork	野外作业
ouvala	灰岩盆地；干喀斯特宽谷
ouvarovite	钙铬榴石
oval	卵形的；椭圆
oval groove	椭圆孔型；椭圆槽
oval hole	椭圆形的井眼
oval mesh	椭圆形筛孔
oval opening	卵圆形巷道
oval shaped hole	椭圆井眼
oval socket	打捞筒
oval stone	卵石
ovality	椭圆度
ovalization	成椭圆形
ovalize	呈椭圆形
oval-shaped	椭圆形的
oven	烘箱
oven dry	烘干状态
oven dry tensile strength	绝干拉伸强度；烘干抗拉强度
oven dry weight	烘干重
oven drying	烘干
oven loss test	炉热损失试验，加热损失试验
oven method	烘箱测含水量方法
oven weathering test	烘箱风化试验
oven-dried	烘干的
oven-dried density	烘干密度
oven-dried sample	烘干样
oven-dried soil	烘干土
oven-dry	烘干
oven-dry soil	烘干土
oven-dry weight	烘干重量
oven-drying method	烘干法
ovenstone	耐火石
over arm	横臂；横杆；悬臂；悬梁
over break	掘进截面过大；超挖；超限爆破；后冲爆破
over breakage	过度破碎
over breaking	过粉碎；过粉岩；挑顶；过多崩落
over coarse-grained soil	巨粒土
over consolidation ratio	超固结比
over cored	取钻孔岩心圈的岩心
over crushing	过粉碎
over cut	挑顶；上部掏槽
over cutter	盾构超挖工具；超挖刀
over cutting	顶槽；顶部掏槽
over cutting tool	超挖工具
over damming	堤坝淹没
over development area	超采区
over dimensioning	扩大
over disperse	密集分布
over excavation	超挖
over exploitation	超采
over fault	上冲断层
over load	过量负载；过荷；过载
over load test	超载试验；过载试验
over mountain line tunnel	越岭隧道
over pressure	过压
over pull force	超拉力
over pumping of groundwater	地下水过量抽取
over pumping of underground water	地下水过量抽取
over reduction	过量减压
over reinforced	钢筋过多的；过度加强的
over relaxation	过度松弛
over size	尺寸过大
over stress	超应力

over stressed area	超应力区
over stretch	过度伸长
over stretching	超张拉；过度伸长
over temperature	超温
over tension	超张力；电压过高
over thrust plane	逆断层面
over thrusting	逆冲断层；逆掩断层
over turned fold	倒转褶皱
over water	顶部水
over weight	过重；采空区顶板沉降
overall accuracy	总精度；整体精度
overall adjustment	整体平差
overall amplification	总放大系数；总放大率
overall analysis	全面分析；全分析
overall approach	综合法
overall arrangement	总体布局
overall attenuation	总衰减；总减少
overall average	总平均；总平均值
overall balance	综合平衡；整体平衡
overall behavior	总体矿压显现
overall bias	全面偏差；整体偏压
overall buckling	整体屈曲
overall capacity	总体容量
overall character	总特性
overall coefficient	总系数
overall coefficient of heat transfer	总传热系数
overall cohesiveness of design	结构严密的整体设计
overall compression ratio	总压缩比
overall concentration	矿井开采集中
overall construction plan	施工总平面图
overall depth	总高度；总深
overall design	总体设计；整体设计
overall development	综合开发
overall development plan	总体开发方案
overall development plan of offshore oil-gas field	海上油气田总体开发方案
overall development program	总体开发方案
overall development programme	总体开发方案
overall dimension	总尺寸；全尺寸；最大尺寸；轮廓尺寸；外形尺寸
overall drilling speed	总的钻井速度；钻进总速度
overall drilling time	总钻进时间
overall drop	总落差
overall efficiency	总效率；综合效率
overall elastic modulus	总弹性模量
overall elasticity modulus	综合弹性模量
overall elongation	总延伸；总伸长
overall equilibrium	整体平衡
overall error	总体误差
overall excavation	全面开挖
overall exploitation	整体开发
overall exploitation plan	总体开采计划
overall exploration risk	总勘探风险
overall field performance	全油田生产动态
overall flaw detection sensitivity	相对探伤灵敏度
overall flexure	整体弯曲
overall framework	总体结构；总体设计框架
overall gas-oil ratio	总气油比
overall head allowance	总水头容许值
overall heat transfer coefficient	总传热系数
overall heat transmission coefficient	总传热系数
overall height	全高；总高度
overall inclination	整体倾斜
overall injection-withdrawal ratio	总注采比
overall length	总长度；总长；全长
overall length of bridge	桥梁全长
overall load	总荷载
overall loading	满载；最大负荷
overall loss	总损失；总损耗
overall mean velocity	全断面平均流速
overall minimality	全局最小性
overall mining method	全面回采法；全面采矿法
overall output	总产量；总输出
overall penetration	总穿透深度
overall performance	总性能；总指标
overall plan	总体规划；总体设计
overall planning	总体规划
overall planning method	统筹方法；统筹法
overall pressure drop	总压降
overall pressure ratio	总压比；总压力比
overall progress	整体进度
overall recovery	总采出量
overall recovery efficiency	总采收率
overall recovery factor	总采收率
overall reliability	总可靠性
overall response	总体响应
overall rock section	全部岩层截面；全部基岩
overall safety	完全安全
overall screen surface area	总筛分面积
overall shear failure	整体剪切破坏
overall shearing	总剪切
overall site layout	施工现场总平面布置图
overall size	总尺寸；轮廓尺寸
overall solution design	总体方案设计
overall stability	整体稳定性；总稳定度
overall stage of underground mine	矿井建设总工期
overall standard deviation	总标准偏差
overall strength	总强度
overall stress	总应力
overall stress system	总应力系统
overall stripping ratio	平均剥采比
overall structure	总体结构；整体结构
overall test	总试验；总体测试；综合试验
overall thickness	全厚度；总厚度；综合厚度
overall time delay	总延迟时间
overall transfer function	总传递函数
overall transfer matrix	总转换矩阵
Overall Travel Speed	总旅行速率；总传播速率
Overall travel time	全程行驶时间；总传播时间
overall treatment	综合处理
overall unstability	整体失稳
overall velocity	总速度
overall view	全景；全图
overall volume change	总体积变化
overall water injection plant	注水总站
overall width	总宽；全厚
overall yield	总回收率；总产量
overamplification	放大过度
over-and-under method	倾斜溜井上下分层掘进法
over-and-under raising	双向天井掘进

overarch	上架拱圈；成拱形
overarching weight	跨载于充填带和矿体的顶板压力；回采工作区的顶板压力
overbalance	失去平衡；不平衡；超出平衡；过平衡
overbalance perforating	过平衡射孔
overbalance pressure	正压
overbalanced drilling	超平衡钻井
overbalanced hoisting	超静力平衡提升；动力平衡提升
overbalanced slipper	过平衡滑块；超平衡滑块
overbank	大坡度转弯；漫滩；溢岸
overbank area	漫滩面积；洪水淹没地区
overbank deposit	泛溢平原沉积；漫滩沉积；越岸沉积
overbank environment	漫滩环境
overbank flow	漫滩流
overbank process	泛滥作用
overbank splay	漫滩冲积扇
overbank turbidite	漫滩浊积岩
overbend	过度弯曲
overbend region	过弯曲区
overbreak	超挖；超爆；过限爆破；台阶后冲爆破；洞顶破裂
overbreak control	超爆破控制；超挖控制
overbreak in tunnel	隧道塌落
overbreak zone	破裂带
overbreakage	超挖；超挖度
overburden	覆盖层；上覆岩层；表土；风化层；剥离物；超载；使过载；积土压力
overburden amount	覆盖岩层量；盖层厚度；剥离量
overburden bit	表土钻头
overburden cover	覆盖层厚度
overburden cover line	剥离物覆盖线
overburden density	上覆岩层密度
overburden depth	覆盖层深度；覆盖层厚度
overburden disposal	剥离物处理；覆土处理；排土
overburden drilling	土层钻进；剥离钻眼
overburden dual operation	剥离工作
overburden dump	覆土转储
overburden gradient	上覆梯度
overburden grouting	覆盖层灌浆
overburden layer	覆盖层；上覆层；表土；覆土层；剥离层
overburden layer and weathering depth in project area	枢纽区覆盖层及风化深度
overburden load	上覆岩层负荷；表土层荷载；覆盖土荷载
overburden map	覆盖层地质图
overburden material	剥离物
overburden mining	剥离工作
overburden movement	覆岩运动
overburden non-collapsible loess	非自重湿陷性黄土
overburden operation	剥离工作
overburden pressure	覆盖地层压力；上覆岩层压力；覆盖压力；自重压力
overburden ratio	剥采比；剥离比；盖挖比
overburden recasting	剥离岩石捣堆
overburden rehandle	剥离物再次移运；剥离再倒堆
overburden removal	剥离表层；剥土；剥离工作
overburden removing	剥离表层；剥土；剥离工作
overburden rock	覆盖层；上覆岩层；盖层
overburden sampling at depth	深层土壤采样
overburden sediment	上覆沉积层
overburden soil	覆盖土
overburden strata	覆岩；上覆地层
overburden stress	上覆岩层应力
overburden stripping	覆盖层剥离
overburden stripping formation	剥离层
overburden structure	盖层结构；上覆岩层结构；表土结构
overburden thickness	剥离厚度；覆盖层厚度；表土厚度
overburden weight	上覆岩层重量
overburdening	过载；表土
overburdensome	超载的；过于沉重的；不胜负担的
overburden-to-thickness of coal ratio	剥离-煤厚比；剥离物与煤层厚度比
overcapacity of pump	泵的过载能力；泵的超负荷
overcarburizing	过度渗碳
overcastting dual operation	倒堆工作
overcastting method	倒堆法
over-clayey soil	超黏土
overcompacted	超压实的；过度压实的；超固结的
overcompacted clay	超固结黏土；超压实黏土
overcompaction	过度压实；过度夯实
overconsolidated	超固结的；过度固结的；过压密
overconsolidated clay	过度压密黏土；超固结黏土
overconsolidated soil	超固结土
overconsolidation	超固结；过度压密
overconsolidation clay	超固结黏土
overconsolidation ratio	超固结比
overconsolidation soil	超固结土
overconsolidation state	超固结状态
overconvergence	过度收敛
overcore	套钻岩芯；应力解除岩芯
overcoring	套芯钻；套钻；套钻钻进；岩芯应力解除
overcoring hole	套钻取芯钻孔
overcoring method	应力解除法；套孔法；套钻法
overcoring technique	应力解除法
overcrack	过度裂化
overcracking	过度裂化
overcritical	超临界的
overcrushing	过度破碎；过碎
overcrusting	被壳构造；盖壳构造；被壳堆积作用
overcut	井径扩大；挑顶；沿煤层顶部截槽
overcutting tool	超挖工具
overdam	堤坝淹没
overdamming	过度堰塞；过度壅水
overdamped galvanometer seismograph	大阻尼电流计地震仪
overdamped gravity meter	超阻尼型重力仪
overdeepened	过量下蚀；过量加深
overdeepened valley	过蚀谷
overdeepening	过量下蚀作用
overdeeping	过量下蚀作用
overdesign	保险设计；保守的设计；过于安全的设计
overdesigned	超过设计要求的；超标准设计
overdetermined	超定的
overdetermined equation	超定方程
overdetermined set	超定方程组
overdeveloped	过度开发的
overdevelopment	过度开发
overdimensioned	超过尺寸的

overdip slope	超斜斜坡
overdisplace	替置
overdisplacement	后置液注入
overdisplacement volume	后置液用量
overdisplacing	替置过量
overdraft	过量抽水；过度抽取；过度抽汲；透支额
overdrainage	过度排水
overdrawing	过多放矿；张拉过度
overdredging	超挖
overdrilling	钻进过多；过度钻进
over-drilling depth	超钻深度
overdriven pile	超深桩；超打桩
overestimate	估计过高；高估
overestimated	估计过高的；高估的
overestimation	估计过高；高估
overexcavate	超挖
overexcavation	超挖
overexpansion	过度膨胀；过膨胀
overexploitation	过度开采
overextraction	过度抽水；过量开采
overfall	溢水；溢流；溢出口；溢水坝；过量放矿
overfall crest	溢洪堰顶
overfall dam	溢流坝；溢流堰
overfault	逆断层；上冲断层
overfeeding	送钻过量
overfill	超填
overflow	溢流；溢出；溢流量；溢口；溢洪道
overflow arch dam	溢流拱坝
overflow capacity	溢流容量
overflow chamber	溢流室
overflow channel	泛滥河道；溢流渠道
overflow cofferdam	过水围堰
overflow critical depth	漫溢临界水深
overflow dam	溢流坝
overflow diameter	溢流口直径
overflow dike	过水堤
overflow discharge	溢流量
overflow earth dike	溢流土堤
overflow earth-rock dam	溢流土石坝
overflow edge	溢流堰
overflow gate	溢流式闸门
overflow gravity dam	溢流重力坝
overflow groove	溢流槽
overflow indicator	溢流指示器
overflow land	漫水地；河滩地
overflow launder	溢流槽
overflow levee	熔岩堤
overflow meadow	泛滥草甸；河漫滩
overflow outlet	溢流口；满溢出口
overflow port	溢流口
overflow rate	溢流速率
overflow section	溢流断面
overflow spillway	溢洪道；溢流排洪道
overflow spring	溢流泉
overflow stage	洪水期
overflow stream	溢流河
overflow velocity	溢流速度
overflow weir	溢流堰；溢流坝
overflowed land	泛滥地区
overflowing	溢流；过流
overflowing bypass	溢流岔道
overflowing capacity	溢流能力
overflowing spring	溢流泉
overflowing well	溢水井
overflush injection	后置液注入
overflush viscosity	后置液黏度
overflux	超通量
overfold	倒转褶皱；倒转褶曲
overfolding	倒转褶皱
overgassing	放气过多；过度析出气体
overgauge hole	超径井眼；正偏差孔
overgetting	超额采量
overgrinding	过度碎磨
overground	地上的；研磨过度的
overgrowth fault	增生断层
overgrowth of crystal	结晶粗大化
overhand	上向梯段式的；倒台阶式的
overhand bore	向上钻孔；上向钻孔
overhand cut-and-fill	上向分层充填
overhand double stopes	上向梯段双翼回采工作面
overhand filled stope	上向梯段充填回采工作面
overhand filled stope face	上向梯段充填回采工作面
overhand longwall	上向梯段式长壁开采法
overhand longwall method	仰斜长壁采煤法
overhand method	上向梯段开采法
overhand mining	倒台阶回采；倒台阶采煤法；仰斜开采；上向梯段式回采
overhand mining method	上向台阶采矿法
overhand pillar method	仰斜柱式采煤法
overhand pillar method with caving	仰斜柱式垮落采煤法
overhand rill stope	上向倒V形梯段留矿工作面；上向倒V形梯段回采工作面
overhand shortwall method	仰斜短壁采煤法
overhand single stope	上向单翼采场
overhand step method	上向梯段开采法
overhand step system	倒台阶采煤法
overhand stepped face	倒台阶工作面
overhand stope	上向梯段回采工作面；仰采工作面；倒台阶工作面
overhand stope face	上向梯段式工作面；仰采工作面
overhand stope method	上向梯段开采法
overhand stope system	上向梯段采矿法
overhand stope toe	上向倒梯段回采面下角
overhand stoping	上向梯段回采；仰采；倒台阶回采；顶蚀作用
overhand stoping and milling system	上下向混合采矿法
overhand stoping in inclined floors	倾斜分层上向梯段回采
overhand stoping in open stope	空场上向梯段回采
overhand stoping-and-filling	上向梯段充填回采
overhand-underhand stoping	上下向梯段联合回采
overhang	倒坡；倒悬；外伸
overhang caving mining	强制放顶采煤法
overhang load	外伸荷重
overhanging	悬挂；伸出的；前探的；突出的
overhanging bank	悬岸；陡岸
overhanging beam	悬臂梁
overhanging cable	上牵引绳
overhanging cliff	悬崖；陡壁
overhanging face	突出工作面；悬吊工作面
overhanging pile driver	伸臂式打桩机

overhanging ripple　菱形波痕
overhanging rock　悬垂岩石；悬石；危岩
overhanging rock and rockfall　危岩落石
overhanging section of ventiduct　风道悬壁
overhanging siphon spillway　虹吸溢洪道
Overhauser magnetometer　欧弗豪泽磁力仪；双重核共振磁力仪
overhead caving　放顶；顶岩崩落
overhead ditch　天沟；高墙排水沟
overhead driller　向上打眼凿岩机
overhead drilling　上向钻眼；上向凿岩
overhead drilling system　顶部驱动钻井系统
overhead irrigation　人工降雨灌溉
overhead mining　上向梯段回采
overhead stope　空场上向梯段回采工作面
overhead stoping　上向梯段回采；仰采；倒台阶回采
overhead viaduct　架空行车道；高架桥
overhead view　顶视图；俯视图
overinflation　过分膨胀
overinjection　超注
overirrigation　过量灌溉；漫灌
overlade　超载的；过负荷的
overladen　过载的；超载的
overlaid　覆盖的
overlain　上覆的；覆在上面的
overland　陆上的；陆路的；越野
overland borehole　探土孔；陆上钻孔
overland flow　地面径流；地表径流；陆上径流；漫滩流；坡面流
overland runoff　地表径流；陆上径流；坡面漫流
overlap　地层超覆；逆断层；冲断层；重叠；互搭
overlap fault　冲断层；逆断层；超覆断层；重叠层错
overlap fold　超覆褶皱
overlap sequence　重叠岩序
overlap trap　超覆圈闭
overlap-fault　超覆断层
overlapped structure　超覆构造
overlapping　超覆；重叠；互搭；搭接部分
overlapping anomaly　重叠异常
overlapping average　叠加平均
overlapping block or parallel span　重叠区段或平行跨
overlapping combination　重叠组合；联合成组
overlapping curve　重合曲线
overlapping domain　重叠域
overlapping fault　重叠层错；超覆断层
overlapping fissures　相互叠加裂隙
overlapping layers　叠压层
overlapping lenses　相互超覆的透镜体
overlapping pulses　重叠脉冲
overlapping recording　重叠记录
overlapping seismometer output　地震仪的重叠记录
overlapping spectra　重叠谱
overlapping structure　逆掩构造
overlapping volcano　叠覆火山
overlay　覆；盖；铺上；覆盖层；程序段落；程序分段；叠加；重叠；共用存储区
overlay analysis　叠置分析；叠加分析
overlay chart　叠加图
overlay interpretation　叠合解释
overlay program　覆盖程序；重叠程序

overlay segment　重叠段
overlay technique　叠合法
overlaying rock　上覆岩石；覆岩
overlaying strata　上覆岩层；表层；覆盖层
overlift crude　超采原油
overload　超载；过载；超负荷；多泥沙
overload breakage　过载断裂
overload point　过载点
overload pressure　过载压力
overload test　过载试验；超负荷试验
overloaded pillar　过载矿柱
overloaded relieve valve　过载安全阀
overloaded stream　多泥沙河流
overloading effect　过载效应
overloading experiment　超载实验
overlower　过度降低
overlying　上覆的；覆盖的；上卧
overlying bed　覆盖层；上覆层
overlying confining bed　上覆隔水层
overlying crust　硬壳；上覆地壳
overlying deposit　表层沉积；上覆沉积层；上覆矿床；上覆层
overlying fluid column　上覆液柱
overlying formation　上覆地层
overlying impervious bed　上覆不透水层
overlying measures　覆盖层系；上部层系；超覆岩层
overlying model　盖层模型
overlying roadway　覆巷道
overlying rock　上覆岩石；盖层
overlying rock strata　上覆岩层
overlying sand　上覆砂层
overlying seam　上覆煤层；上覆矿层
overlying soil　上覆土层
overlying soil cover　表层土壤
overlying strata　上覆地层；表层；覆盖土层；盖层
overlying strata control　上覆岩层控制
overlying strata movement　上覆岩层移动
overlying strata movement law　覆岩移动规律
overlying strata of coal measures　煤系盖层
overlying strata spatial structure　覆岩空间结构
overlying stratum　上覆层；盖层
overlying workings　本层上方采场
overmature gas　过成熟气
overmaximal　超最大值的
overmeasure　高估；过量；剩余
overmigrate　过量偏移
overpacking　超紧密堆积作用
overpass　超越；优于；立体交叉
overpass bridge　跨线桥；立交桥；上跨路桥
overpass grade separation　上跨铁路立体交叉
overpass highway　天桥公路
overplate　仰冲板块
overplating　顶侵作用；板上作用
overplotting　叠绘
overplus　过剩；剩余
overpressure　超压；过压；异常高压；剩余压力
overpressure detection　过压检测
overpressure limit control　超压极限控制
overpressure method　过压试验法
overpressure resistant　耐压的

overpressure zone 过压带
overpressured 过压的
overpressured aquifer 超压水层
overpressured formation 超压地层
overpressured reservoir 超高压储层
overpressured tight gas reservoir 超压致密气层
overpressured tight oil reservoir 超压致密油层
overpressurization 过量增压；超压
over-prestressed beam 超预应力梁
overprint 叠加标记；重叠，叠覆
overprinted crust 叠覆地壳
over-produced well 超过预定允许产量的井
overpumping 过量抽水；超量取水
over-reinforced beam 超配筋梁
over-reinforced section 超配筋断面
overrelaxation 超松弛法
over-relaxation iterative method 超松弛迭代法
over-relaxation mechanism in blasting 爆破的超松弛机制
overridden block 被逆掩推覆岩块
override 超覆；超越；过载
override pressure 过载压力
overriding 叠置；覆盖；推覆；仰冲；逆掩构造
overriding block 逆掩断块
overriding face 逆掩面；逆掩断层面
overriding plane 仰冲面
overriding plate 推覆板块；仰冲板块；上冲板块
overriding royalty interest 开采权益
over-river-levelling 过河水准测量
overroad 跨路的
overrolling 过度碾压
over-run tunnel 备用隧道；越位/掉头隧道
oversampling 采样过密
over-sampling 重复取样；采样过密
oversanded 多砂的；含砂过量的
oversanded mix 多砂混凝土
oversanded mixture 砂量过多的混合物
oversaturated 过饱和的
oversaturated rock 过饱和岩
oversaturation 过饱和
over-saturation delay 过饱和延滞
oversea 海外
overseas works 国外工程；海外工程
oversee 监视；监督
overseer 监工，管理人
overset 倾覆；搅乱
overshoe 套鞋；防水套罩
overshoot 越过；超出；过辐射；过多装药爆炸；过调节；过调整；涨溢；逸出
overshooting interference 越区干扰
overshot 打捞工具；取管器；打捞筒；仪表指针偏转过多
overshot assembly 钻具打捞器
overshot grab 抓钩打捞筒
overshot guide 打捞筒导向装置
overshot guide shoe 打捞筒导向引鞋
overshot seal 打捞筒密封
overshot tank filling 自顶部注入油罐内
overshot tubing seal divider 打捞筒油管密封分压器
overshot water wheel 上射式水轮
overshot with bowl 带碗形卡瓦的打捞筒
overshot with slot 开窗打捞筒

oversimplification 过分简化
oversimplify 过于简化
oversize 超大体积；超大的；尺寸过大；超径
oversize coupling 大尺寸接箍；大直径钻杆接头
oversize distribution 筛上粒度分布
oversize drill collar 加大尺寸钻铤；超大型钻铤
oversize factor 筛上产品率
oversize fraction 筛上率
oversize hole 直径过大的井；超径孔；钻具偏心转动而使井身扩大的井
oversize particle 筛上粒；粗大粒子；超径颗粒
oversize piece 超粒；过限块粒；不合格大块
oversize pillar 过大煤柱；过大矿柱
oversize powder 筛上粉末
oversize product 筛上产品；筛上物；筛余物
oversized 过大的；超粒的
oversized coupling 大尺寸接箍；大直径钻杆接头
oversized gravel 尺寸过大的砾石
oversized hole 超尺寸孔
oversized material 超径料；超径材料；超过尺寸的材料
oversized washpipe 直径过大的冲管
oversizing 选择参数的裕度；扩界
oversleeve 套袖
overspecialization 过于专门化
overspill 溢出物；过剩物
overspread 漫布；铺盖；覆盖
overstability 超稳定性；过度稳定性
overstate 高估；夸大
overstatement 夸大
oversteepened bed 逆转层，倒转层
oversteepening 削峭作用；削深作用
overstep 超覆；叠覆；跨覆；逆掩断层；逾越；越过
overstep propagate 上叠式扩展
overstep sequence 超覆式逆冲顺序
overstepped graben 跨覆地堑
overstory 上层；顶层
overstrain 超限应变；过度应变；残余应变
overstrained 过度应变的
overstrained rock 过度应变岩石；过应变岩石
overstrained rock round 隧道周围过渡应变岩石
overstrength 超强度
overstress 过应力；超限应力；超屈服力
overstress load 超应力荷载；过载
overstressed area 超应力区
over-stressed pillar 超应力矿柱；应力集中矿柱
overstressed rock mass 超应力岩体
overstressing 超负载；超限应力；过应力
overstressing of reinforcing steel 钢筋受超限应力
overstretch 过拉伸；延伸过长
overstrung 过度变形的
overstuffed 填塞很多的；装填过度的
oversubmergence 淹没；覆盖
over-superelevation 过超高
overtamp 捣固过度
overtemperature 过热；温度过高
overtemperature trip device 超温限制器
overtension 过张力
over-the-roadway extraction 跨采
over-the-top gravel packing tool 皮碗式砾石充填工具
overthrow 推翻；倾覆

overthrown fold	倾倒褶皱；倒转褶皱
overthrust	上冲；仰冲；逆掩断层；上冲断层；仰冲断层
overthrust anticline	逆掩断层背斜；掩冲背斜
overthrust belt	掩冲断层带
overthrust block	逆掩推覆体；掩冲体；掩冲断块
overthrust cleat	逆掩解理；扭力解理
overthrust fault	逆掩断层；掩冲断层；上冲断层
overthrust fold	倒转褶皱
overthrust mass	逆掩体；掩冲体
overthrust mountain	逆掩断层山
overthrust nappe	逆掩推覆体；冲断推覆体
overthrust plane	逆掩断层面
overthrust plate	掩冲盘
overthrust sheet	逆掩层；掩冲席
overthrust side	上冲面；逆掩侧
overthrust slab	逆掩板块；逆掩断层
overthrust structure	掩冲构造
overthrust tectonic	逆冲断层构造；掩冲构造
overthrusting	掩冲；掩冲断层作用
overthrusting plate	上冲板块
overthrust-mountain	逆掩断层山
overthrust-plane	掩冲断层面
overtide	倍潮
overtilted	倒转的
overtipped face	翻倒的工作面；被倾覆的工作面
overtipped strata	倒转地层；倾覆岩层
overtipping	岩层倾覆；地层侧转
overtone	谐波；过谐
overtone normal mode	谐波简正振型
overtopped cofferdam	漫顶围堰
overtopped dam	溢水坝；漫水坝
overtopping	越堤；漫顶；满溢
overtopping discharge	漫顶流量
overtopping water stage	漫顶水位
overtravel	大移动；超程，超行程；越层
over-tuned point	过调谐点
overturn	倾覆；倒转；倒转褶皱
overturn bed	倒转地层
overturn fold	倒转褶皱
overturn of thrust block	倒转逆掩断块
overturned	倒转的；倾覆的
overturned anticline	倒转背斜
overturned antiform	倒转背形
overturned bed	倒转地层；倾覆地层
overturned fold	倒转褶皱
overturned succession	倒转层序
overturned succession of strata	倒转层序
overturned syncline	倒转向斜
overturned synform	倒转向形
overturning	倾覆；倒转
overturning effect	倾覆作用
overturning failure	倾覆破坏；倾覆崩塌
overturning force	倾覆力
overturning moment	倾覆力矩；倾覆弯矩
overturning moment reduction factor	倾覆力矩折减系数
overturning of beds	地层倒转
overturning of retaining wall	挡土墙倾覆
overturning or slip resistance analysis	抗倾覆、滑移验算
overturning stability	倾覆稳定性
overvalued	高估
over-vibrated	过度振动
overvibration	振动过度
overview	综述；概述
overview schedule	宏观进度计划
overwash	洪泛土壤；越堤冲岸浪；漫顶
overwash drift	冰水沉积物；冰前冲积物
overwater drilling	水上钻井
overweight	超重的；超重；优势
overwet	过湿
overwet land	过湿地；沼泽地
overwetting	过度润湿
overyear storage	多年调节库容；多年蓄水；多年储存
over-years frozen soil foundation	多年冻土地基
ovoid grain	卵形颗粒
ovulate	卵化石
ovulite	鱼卵石；鲕状岩
oxbow	弓形河曲；牛轭形弯道；牛轭湖
oxbow lake	牛轭湖；弓形湖
ox-bow lake sediment	牛轭湖沉积物
oxbow swamp	牛轭沼泽；废河道沼泽
oxidability	可氧化性
oxidant	氧化剂
oxidant consumption	氧化剂比耗
oxidase	氧化酶
oxidate	氧化
oxidated asphalt	氧化沥青
oxidates	铁锰氧化物沉淀岩类
oxidation	氧化
oxidation cleavage	氧化分裂；氧化解理
oxidation corrosion	氧化腐蚀
oxidation rate	氧化速率
oxidation ratio	氧化率
oxidation zone	氧化带
oxidation zone of sulphide deposits	硫化物矿床氧化带
oxidation-reduction	氧化还原作用；氧化-还原
oxidation-reduction reaction	氧化还原反应
oxidation-resistant	抗氧化的
oxidative	氧化的
oxide	氧化物
oxide mineral	氧化矿物
oxide sediment	氧化沉积物
oxidized area	氧化区；氧化带
oxidized asphalt	氧化沥青
oxidized coal	氧化煤
oxidized deposit	氧化矿床
oxidized ore	氧化矿石
oxidized ore zone	氧化矿带
oxidized resin	氧化树脂
oxidized zone	氧化带
oxidized zone of sulfide deposit	硫化矿床氧化带
oxidizer	助燃剂；氧化剂
oxidizing action	氧化作用
oxidizing agent	氧化剂
oxidizing environment	氧化环境
oxisol	氧化土
oxy-apatite	氧磷灰石
oxybasiophitic texture	酸基性辉绿结构；酸辉绿结构的纹理
oxybiotite	氧黑云母
oxycarbide	碳氧化物

oxychloride cement 氯氧化水泥
oxycoal 氧化煤
oxydate 氧化沉积物
oxydation 氧化
oxydation zone 氧化带
oxygen activation logging 氧活化测井
oxygen and strontium isotope 氧和锶同位素
oxygen content of coal 煤中氧
oxygen corrosion 氧腐蚀
oxygen cutter 氧气切割机；气割炬
oxygen cutting 氧气切割；气割
oxygen demand 耗氧量；需氧量
oxygen detector 氧气测定器
oxygen dissolved 溶解氧
oxygen explosive 液氧炸药
oxygen geochemical barrier 氧地球化学障
oxygen isotope 氧同位素
oxygen isotope anomaly 氧同位素异常
oxygen isotope composition 氧同位素成分
oxygen isotope curve 氧同位素曲线
oxygen isotope method 氧同位素法
oxygen isotope paleotemperature 氧同位素古温度
oxygen isotope ratio 氧同位素比值
oxygen isotope stage 氧同位素阶段
oxygen isotope stratigraphy 氧同位素地层学
oxygen isotopes mutation 氧同位素突变
oxygen isotopic stage 氧同位素期
oxygen logging 氧测井
oxygen meter 测氧计

oxygenant 氧化剂
oxygenated 充氧的；氧化的；用氧饱和的
oxygenation level 氧化界面
oxygen-balanced explosive mixture 氧平衡炸药混合物
oxygen-consuming content 耗氧量
oxygen-consuming-quantity 耗氧量
oxygen-consumption type of corrosion 耗氧腐蚀
oxygen-deficiency 缺氧的
oxygen-free 无氧的
oxygen-free environment 还原环境
oxygen-isotope method 氧同位素法
oxygen-isotope paleotemperature scale 氧同位素古温标
oxygen-isotope ratio 氧同位素比值
oxygen-isotope thermometer 氧同位素温度计
oxygen-negative balance 负氧平衡
oxygenolysis 氧化分解
oxyhornblende 玄武角闪石；氧角闪石
oxyliquit 液氧炸药
oxymagnetite 氧磁铁矿；磁赤铁矿
oxymagnite 氧磁铁矿；磁赤铁矿
oxyophitic texture 酸性辉绿结构
oxyphyre 酸性斑岩
oxyplete 超酸性岩
oxysphere 岩石圈
oyamalite 稀土锆石
oylet 孔眼；视孔
oyster reef 牡蛎礁
oyster reef coast 牡蛎礁海岸
oyster-shell lime 贝壳石灰；贝壳灰

P

paar 壳间凹陷；地壳裂陷；洼地
pabstite 硅锡钡石；硅钛钡石
Pacific orogeny 太平洋造山带；太平洋造山运动；太平洋造山作用
Pacific suite 太平洋岩套；太平洋岩组；太平洋岩群
pacing 步测；步调；踱步
pack 充填；包装；充填材料；部件，组件
pack annealing 叠式退火；装箱退火
pack builder 石垛工；煤矿井下充填工
pack rolling 叠轧；叠板轧制
packed bed 致密层；填充层
packed column 填充柱；填充塔
packed joint 堵塞接缝；包垫接头
packed material 填充的物料；填实的物料
packed nog 充填的木垛；实木垛
packed soil 夯实土；捣实土
packed stone revetment 填石护坡；堆石护岸
packed tamping 炮泥筒；炮泥卷
packed tower 填充塔；填料塔
packed waste 充填废石；砌垛废石
packer 灌浆孔塞；夯具；捣实器；封隔器；堵塞器；填料
packer grouting 栓塞灌浆；分段灌浆
packer permeability test 压水试验；压水渗透试验
packer test 压水试验；分段栓塞压水试验
packing degree 装药密度；填实度
packing rock 充填用岩石；填石
packing strip 填密片；充填带
packing washer 压紧垫圈；密封环
packsand 易破碎的细粒砂岩；软砂岩；钙质细粒砂岩
packway 马道；土路
paddle mill 桨式洗矿机；叶片式洗矿机
paddle mixer 螺旋桨叶式搅拌输送机；桨叶式搅拌机
paddock 方井采矿法；地面贮矿；井口临时堆场；砂矿挖掘船采掘场
paddy mail 运送矿工；升降人员
pagoda 塔状结构；塔
pagoda stone 宝塔灰岩；叶蜡石；宝塔石
pagodite 叶蜡石；冻石；滑石
pahoehoe lava 波纹熔岩；绳状熔岩
pahoehoe structure 绳状构造；绳状熔岩结构
paigeite 硼铁矿；硼铁石
painite 红硅硼铝钙石；硅硼钙铝石
paint line marker 划线车；划线机
paired metamorphic belts 双变质带；对偶变质带
paired terraces 配对阶地；成双阶地
palaeides 古褶皱带；古构造带
palaeochannel 古河道；古河床
palaeokarst 古岩溶；古喀斯特
palaeostructure 古构造；古地质构造
palaeo-weathered crust 古风化壳；埋藏风化壳
palagonite 玄武玻璃；橙玄玻璃
palaite 红磷锰矿；肉色锰磷石
palasome 主岩；主矿；基体

paleoautochthon 古原地岩；古推覆基底
paleochannel 古河道；埋藏河道
paleogeomorphology 古地形学；古地貌学
paleokarst 古喀斯特；古岩溶
paleomagnetism 古磁学；古地磁
paleopole 古极；古地极
paleosection 古断面；古剖面
paleosome 先成体；基体；古成体
paleotectonics 古构造地质学；古构造
paleotopography 古地形；古地形学
paleovolcanism 古火山现象；古火山作用
palification 用桩加固地基；打桩；板桩支护
palimpsest sediment 变余沉积物；再生沉积物
palimpsest structure 变余构造；残留构造
palimpsest texture 变余结构；变余组织；残留组织
paling 打桩；做栅栏；栅板；栅栏圆柱；篱笆桩；垂直外模板；围栏
palingenesis 岩浆再生作用；重熔作用
palingenetic 再生的；新生的
palinspastic map 复原图；再造图；古地理复原图
palisade 峭壁；陡崖；断崖
palisade check dam 木栅拦沙坝；木栅谷坊
pallasite 橄榄陨铁；富铁橄榄岩；石铁陨石
palsa 泥炭丘；泥炭冰核丘
palstage 古地理时期；古地理稳定期
paludal 沼泽的；沼泽沉积物的；沼泽地生的
paludification 沼泽化；沼泽沉积作用；泥炭化作用；腐泥化作用
palus 河边低地；沼泽
palustrine 沼泽环境的；沼泽沉积的
palygorskite 坡缕石；镁铝皮石
pan 盘；淘金盘；土盘；硬土层；浅水盆地；盆地凹地
pan car 轮式刮煤机；车轮式铲运机
pan conveyor 平板输送机；盘式输送机
pan feeder 板式给矿机；盘式给料机
pan man 溜槽移挪工；输送机移挪工
pan mill 研磨盘；碾盘式碾磨机
pan mixer 盘式搅拌机；盘式拌和机
pan shifter 移溜槽工；输送机移动工
pan soil 硬土；坚土
pane 锤头；金刚石磨面；窗玻璃
panel 仪表板；配电板
panel length 节间长度，节间距离；盘区长度
panel method 板块法；面元法；网格法；盘区开采法
panel stope 全高单一回采工作面；单盘区回采
panel wall 节间墙；大板墙；墙板；隔墙
paneling 盘区开采；开切盘区；镶板，嵌板
panel-type construction 框格式结构；框格式结构；墙板式结构
panfan 麓原；泛洪积扇；联合山麓侵蚀面
Pangaea 泛大陆；泛古陆；盘古大陆
Pangea 联合古陆；泛大陆；超级大陆
panic bar 太平门闩；紧急保险螺栓；紧急把手
panning 淘盘选；淘金；隔水薄板；衬片
panplain 泛平原；联合泛滥平原

pantagraph	比例画图器；缩放仪；方形受电弓
panthalassa	泛古洋；原始大洋
pantile	波形瓦；筒瓦
panting	脉动；波动；晃动
pantography	缩放图法；雷达缩放仪
pantometer	经纬测角仪；万能测角仪
pantoscope	广角照相机；广角镜头
panzer	铠装输送机；铠装的；装甲的
paper clay	薄层黏土；浆土；纸黏土
paper coal	片状褐煤；纸煤
paper shale	纸状页岩；薄层页岩
papery	纸状；薄的
papery shale	纸状页岩；片状页岩
parabolic	抛物线的；比喻的
parabolic dune	新月形沙丘；抛物线沙丘
paraboloid	抛物面；抛物体
parachor	等张比容；等张体积
paraclase	断层；断层裂缝
paraconformity	准整合；似整合
paraconglomerate	副砾岩；砾泥岩
paradiagenetic	准成岩期的；拟成岩作用的
paraferrallitic soil	次生铁铝土；铁铝型土
paraffin hydrocarbon	链烷烃；烷属烃
paraffin removal	清蜡；除蜡
paragenetic	共生组合的；共生的
paragenetic sequence	共生次序；共生系列
paragenic	共生组合的；共生的
parageosyncline	准地槽；副地槽
paragneiss	副片麻岩；水成片麻岩
paraliageosyncline	滨海地槽；陆缘地槽
paralic	海陆交互的；近海的；海滨的
paralic deposit	近海沿积；近海矿床
paralithic contact	准石化接触；准石质接触界面
parallel arrangement of holes	平行钻孔布置；平行炮眼布置
parallel cleavage	平行解理；平行劈理
parallel cut	平行空眼掏槽；平行切割；平行掏槽；直眼掏槽
parallel cut blasting	直眼掏槽爆破；平行掏槽爆破
parallel dike	顺坝；导流堤
parallel dual operation	平行作业；并联工作；并联运行
parallel entry	平行平巷；平行巷道；并行输入
parallel fault	平移断层；平行断层
parallel groove clamp	并沟线夹；接触线固定线夹
parallel grooved drum	平行绳槽滚筒；直绳槽滚筒
parallel gutter	箱形雨水管；平行雨水檐沟
parallel hole cutting	垂直眼掏槽；平行眼掏槽
parallel intergrowth	平行连晶；平行互生
parallel migration	平行运移；顺层运移
parallel of altitude	地平纬圈；等高圈
parallel operation	平行作业；并联工作；并联运行；并行操作
parallel running	并行工作；并联运转
parallel series	混联；串并联
parallel system	平行式布置；并联系统；并行系统
parallel unconformity	平行不整合；假整合
parallel winding	并联绕组；并联绕法
parallel-jaw vice	平口虎钳；平口钳
parallochthon	准外来岩体；近外来岩体
paramagnetic material	顺磁性材料；顺磁性物质
parameter adjustment	参数平差；参数调整
parameter analysis for slope stability	边坡稳定性系数分析；边坡稳定性参数分析
parameter delimiter	参数定界符；参数分隔符
parameter of material	物料参数；材料参数
parameter variation	参数变值法；参数变化；参数差异
parameters affecting ground movement	地层采动参数；影响地层移动参数
parametric model	参数模型；参数模拟
parametric sounding	参数测深；频率测深
paramorph	同质异形；副象
paramorphism	全变质作用；同质异晶现象；同质假象
paranthracite	高变质无烟煤
paraoceanic	准大洋的；副大洋的
parapet	护墙；矮墙；护栏
parapet wall	护栏；低墙；护墙
paraschist	副片岩；水成片岩
parasequence	副层序；准层序
parasitic fold	寄生褶皱；附属褶皱
parasitic strain	次应变；附加应变
paratectoclase	副构；造裂隙；同构造裂隙
paratectonic	副构造的；同构造的
paratectonics	准构造；副构造
parathethys	类特提斯；准特提斯
paravolcanic	次火山的；准火山的；类火山的
parcel boundary	地块边界；宗地边界
parent map	基本图；原图
parent metal	母体金属；基底金属
parent metal test specimen	母体金属试样；母材试件
parent rock	母岩；原生岩
pargasite	韭角闪石；纳角闪石
parget	石膏；粗涂灰泥
parogenetic	先成的；前成的
paroxysmal phase	爆喷阶段；激烈喷发幕
paroxysmist	灾变说者；激变说者
parrot coal	响煤；劣质气煤；一种苏格兰烛煤
parsettensite	红硅锰矿；锰黑硬绿泥石
parsimonious deconvolution	吝啬反褶积；强极小反褶积
part-built structure	半建成构筑物；半建成搭建物
partial analysis	简分析；偏分析；部分分析
partial coherence	局部相干性；部分相干性
partial erosion	局部冲刷；局部糜烂；偏蚀
partial extraction	部分采掘；局部开采
partial penetration	局部渗透；局部渗入
partial protection	部分保护；局部保护
partial recovery	部分回收；部分复原；部分恢复
partial saturation	不完全饱和；部分饱和
partial stowing	部分充填；局部充填
partial subsidence	局部下沉；局部沉降
partially penetrated well	不完全井；非完整井
particle	质点；颗粒；粒子；微粒
particle analysis	颗粒分析；质点分析
particle counter	颗粒计数器；粒子计数器
particle distribution	颗粒级配；颗粒分布；粒度分布
particle dynamics	质点动力学；粒子动力学
particle flow code	颗粒流程序；颗粒流模拟软件
particle fraction	粗组；颗粒组成；颗粒分数
particle grading curve	颗粒级配曲线；颗粒尺寸分布曲线
particle in cell method	质点网格法；网格内质点法；粒子云网格法

particle initiation	颗粒起动；颗粒启动
particle mechanics	质点力学；颗粒力学
particle method	质点法；粒子法；粒子方法
particle settling	粒子沉降；颗粒沉降
particle shape	颗粒形状；粉粒形状
particle size	颗粒大小；粒度
particle size analysis	粒度分析；粒径分析；颗粒级配分析
particle size distribution plot	粒度分配曲线；粒度分配图
particle sizing	颗粒分级；颗粒筛分
particle texture	颗粒质地；颗粒结构
particular	特殊的；特别的
particular data	详细数据；具体数据
particular risk	特定危险性；特定风险
particular scale	局部比例尺；特殊比例尺
particular specification	特别规格明细表；特别规格
particularity	特殊性；特质；细目
particulars	详细资料；详细数据
particulate	成粒的；分散的；散碎的
particulate loading	微粒浓度；微粒量
particulate matter	不溶性微粒；颗粒性物质
particulate size distribution	粒径分布；粒径分布曲线
particulates	空气中微小颗粒；微粒；大气尘
parting agent	分型剂；脱模剂
parting flow rate	分流速度；分流量
parting line	分型线；模缝线
parting liquid	分选液体；重液
parting plane	分离面；裂开面
parting surface	分离面；劈裂面
parting-step lineation	斜交阶梯状线理；阶状裂开线理
partition	区分；划分；分配；间壁
partition chromatography	分配色层法；分配色谱法
partition density	分离密度；分配密度，分选密度
partition law	分配定律；分配定理
partition rock	壁岩；围岩
partition size	分级粒度；分配粒度
partition wall	隔墙；隔栅
partitioned	分隔的；分开的
partitioning	配分；分离；配比
partitioning allowance	间隔限度；允许的间隔
party wall	界墙；共用墙
parvafacies	分相；同时相；副相
pass schedule	孔型安排；孔型系统
passage of a face	工作面通道；工作面走廊
passage of chip	切屑排出槽；排屑槽
passband	带通；通频带
pass-by	旁路；旁道
passing material	通过筛孔的物料；筛下品；通过的物料
passivation potential	钝化电位；钝化电势
passive	被动的；消极的；钝态的
passive continental margin	被动大陆边缘；大西洋型大陆边缘
passive failure	被动损毁；被动破坏
passive microwave sensing	无源微波遥感；被动源微波遥感
passive ocean margin	稳定大洋边缘；被动大洋边缘
passive penetrated well	不完整井；被动渗透井
passive permafrost	古永冻土；早期永冻土；原永冻土
passive pressure	被动压力；分压；局部压力
passive remote sensing	被动式遥感；无源遥感
passive remote sensing system	被动遥感系统；无源遥感系统
passive resistance	被动抗性；被动压力
passive roof duplex	顶板反冲双重构造；被动顶板双重构造
passive satellite	被动卫星；无源卫星
passive sensor	被动传感器；无源传感器
passive source of dust	钝态尘源；惰性尘源
passive system	被动系统；无源系统
passometer	计步计；步测计
pasteurized soil	消毒土壤；灭菌土壤
patand	底梁；槛
patch	斑点；斑纹
patch of ore	矿巢；窝子矿
patch reef	斑礁；斑块礁；点礁；补丁礁；片礁
patch test	小块检验；斑片试验
patching panel	配线盘；转接盘
patent	专利，专利品，专利权；新奇的
patent defect	明显的建筑缺陷；明显的建筑瑕疵
patent rights	专利权；专利证书
patent survey	矿用地测量；申请矿地测量
patented	专利的；取得专利权的
paternoster bucket elevator	多斗式提升机；链斗式提升机
path of filtration	渗透途径；渗滤路径
pathfinder element	探途元素；导向元素
pathway	通路；路径
patina	铜绿；表面氧化保护层
patio	露台；天井
pattern	型式；模式；图像；图案
pattern draft	拔模斜度；起模斜度
pattern efficiency	井网效率；布井效率
pattern flooding	井网注水；面积注水
pattern ground	多边形状土；网格状土
pattern metal	制模金属；制模合金
pattern number	模型号；型号
pattern of spacing	井网布置；布井格式；井距方案
pattern of well	井网；布井方案；钻井布置
pattern recognition	模式识别；图形识别
patterned ground	多边形土；网纹土
patterned sedimentation	韵律沉积作用；循环沉积作用
paulopost texture	后期结构；后期纹理
pause	停顿；中止
pave	铺砌料；铺砌石；铺砌面
paved area	已铺路面的地区；已铺筑地区
pavement	路面；行人道
pavement aggregate	铺面骨材；铺面粒料
pavement stratum	巷道底部地层；底板岩层
pavement structure	路面结构；基础铺砌层
paver	铺路工；铺路机；铺路材料
pavilion	隔离式结构；带装饰的帐篷；亭子；楼阁
paving	铺路；路面；铺路用的
paving stone	铺路石；铺面石板
pavior	铺路工；铺地工；铺路机；铺路石
pawl	爪；棘爪；卡子；掣子
pawl mechanism	棘轮机构；棘爪机构
pawn	抵押；典当；抵押品
pay bed	可采煤层；可采矿层；产油层
pay dirt	含矿泥砂；可采矿石；可采金矿砂
pay lead	富矿脉；富砂矿

pay limit	可采矿石极限；有利开采极限
pay line	计价线；有效开挖线；设计开挖线
pay load	净荷载；净载重；有用负荷
pay off	获得良好结果；取得利益
pay rock	可采岩石；有开采价值的矿石
pay shoot	富矿体；有价值的矿体
pay streak	富线，富矿线，矿床富集处；生产层，产油层
pay wash	值得开采的砂矿；富砂矿
pay zone	补给区；储热层；产油带；可采带；含水层
payable	可付的；应付的；有利的；可获利的
payable ore	采矿石；有经济价值的矿石
payable period	应付期；付款期
payable seam	有经济价值的煤层；有经济价值的矿层
payback	偿付；付还
payback period	回收期；资金返本期
payload	载重量；运载量；有效负荷
payment	付款；支付；支付的款项
payment line	计费开挖线；支付线
payment on term	定期付款；按协议条款付款
payout period	投资返本年限法；投资回收期
pea gravel	小砾石；豆石
pea iron	雏形铁结核；铁子
pea shingle	豆砾石；绿豆砂
peacock coal	光亮型煤；辉煤
peak	山峰；山巅；波峰；洪峰
peak amplitude	峰值振幅；最大振幅
peak bucket load	铲斗峰值载荷；铲斗最大载荷
peak capacity	最大处理量；高峰容量；最大容量
peak clipper	削波器；峰值限幅器
peak demand	峰荷要求；峰荷量；最高要求
peak diameter	最大直径；峰值直径
peak flow	洪峰流量；最大流量
peak intensity	峰值强度；峰强度
peak load	峰值载荷；最大负载；高峰负荷
peak output	尖峰流量；最大输出
peak pore pressure ratio	峰值孔压比；最大孔压比
peak power	峰值功率；最大功率
peak power output	最大输出功率；峰值输出功率
peak rate of flow	最大流量；洪峰流量
peak runoff	峰值径流量；最大径流
peak shear resistance	峰值抗剪力；最大抗剪力
peak shear strength	峰值抗剪强度；最大抗剪强度
peak stage	峰顶水位；最高水位；波峰阶段；峰顶阶段
peak time	洪峰时段；成峰时段
peak torque	最大扭矩；峰值转矩
peak traffic flow	最高交通流量；高峰交通流量
peaked trace	井喷；喷出
peakedness	峰值；尖峰度
peaking value	最大值；峰值
peaky curve	高峰曲线；尖峰曲线；巅值曲线
peaky sea	汹涌海面；激浪；三角浪；尖峰浪
peal	钟；声响声；隆隆声；敲响
peanut	花生；小数目；小企业；渺小的
pearl	珍珠；珠状物；珍品；珍珠色；碎片；小粒的
pearl mica	珍珠云母；珠光云母
pearl spar	珍珠白云石；珍珠晶石
pearlite	珍珠岩；珠光体
pearlitic	珠粒的；珠状的
pearly	珍珠状；似珍珠的
peastone	豆石；豆石砾
peat	泥炭；腐叶土；腐殖土
peat accumulating rate	泥炭堆积速率；泥炭累积率
peat bed	泥炭层；草炭层；泥煤层
peat bog	泥炭沼泽；泥炭沼
peat machine	泥炭机；泥炭采掘机；泥炭压砖机
peat marl	泥煤；泥炭
peat moor	泥炭沼泽；泥炭田
peat mould	泥炭土；泥炭质土壤
peatery	泥炭产地；泥炭沼
peatland	泥炭地；泥炭田
peaty	泥炭的；多泥煤的；泥煤似的
peaty fibrous coal	丝状炭；纤维炭
pebble	卵石；鹅卵石
pebble beach	卵石滩；砾石海滩
pebble bed	卵石层；砾石层
pebble gravel	中砾；砾石层；卵石
pebble mill	碎石磨机；砾磨机
pebble pavement	卵石盖层；卵石地坪；卵石铺面
pebble peat	堆积在卵石下的泥炭；砾下泥炭
pebble pup	地质师助手；地质系学生；业余矿物学家
pebble stone	中砾；卵石；砾石
pebble walling	卵石壁；卵砌墙
pebbled	小砾覆盖的；含砾的；多石子的
pebbly	多卵石的；中砾的
pebbly sandstone	含中砾砂岩；砾质砂岩
pecker block	支撑；支架
pectization	凝结；胶凝，胶化
peculiarity	特性；特质
pedalfer	淋余土；铁铝土
peddler	搅炼炉；捣密机；传播者
pedestal	基座；底座；支架
pedestal fire hydrant	座墩消防栓；柱形消防栓
pedestal footing	基座底脚；柱基
pedestrian path	人行小路；人行路线
pedestrian traffic	行人交通；行人往来
pedestrian tunnel	人行隧道；行人地道
pedestrian underpass	人行地道；下穿式人行过道
pedestrian way	行人路；行人径
pedestrian zone	行人专用区；步行街
pediment	山墙；麓原；山前侵蚀平原
pediment basin	山前侵蚀盆地；麓原盆地
pedimentation	山麓夷平作用；宽谷形成过程
pedimeter	步程计；间距规；里程计
pediocratic	相对稳定的；宁静期的
pedion	单面；端面
pedogenesis	成土作用；成壤作用
pedogenic	成土的；成壤的
pedogenic process	成土作用；成壤作用
pedometer	步程计；间距规；里程计
pedosphere	土壤层；土壤圈
pedrosite	奥闪石岩；钙钠闪长岩
peel	皮；剥皮；剥去
peel strength	撕裂强度；拉伸强度
peel technique	揭片法；剥皮法
peeling	脱皮；剥皮
peen	锤头的尖头；用锤尖敲打；锤头
peening	锤击；轻敲；锤平
peer review	同行审核；同行互审
peerless	唯一的；无比的；无双的
peg	定界；木桩；测标

peg out the boundary	以栓钉标明地界；量定地界
peg stake	标桩；标橛
peganite	磷铝石；磷矾土
pegmatite	伟晶岩；结晶花岗岩
pegmatitic structure	文象构造；伟晶构造
pegmatophyre	伟斑岩；文象斑岩
pelagian	深海的；远洋动物
pelagic	远洋的；深海的
pelagic deposit	深海沉积；远海沉积
pelagic facies	深海相；远洋相
pelagic organism	漂游生物；远洋生物
pelagic zone	浮游带；远洋带
pelagite	深海结核；海底锰块
pelitic texture	黏土质结构；泥质结构
pelitic tuff	泥质凝灰岩；细凝灰岩
pellet	球团矿；颗粒
pellet diameter	钻粒直径；丸粒直径
pellet impact drilling	钻粒钻眼；钻粒钻进
pellet interference	钻粒干扰；钻粒互相碰撞
pellet limestone	球粒灰岩；团粒灰岩
pellet size	钻粒粒度；团粒粒度
pelletization	球粒化作用；团矿；造球作用
pelletized ore	粒化矿石；球团矿
pelletoid	似球粒；似团粒；球粒状的；富球粒的
pellicle	薄膜；薄皮
pellicular	薄膜的；薄皮的
pellicular moisture	薄膜水；附着水
pellucid	透明的；透彻的
pellucidity	透明度；透明性
peloid	球状粒；疗效泥
pelted of porosity	孔隙率；孔积百分率
pelter	投掷器；倾注
pen stylus	电子感应笔；笔头
pencatite	水滑大理岩；水滑结晶灰岩
penchant	倾向；倾斜
pencil jet	束状喷射；密集喷射
pencil of conics	圆锥曲线束；二次曲线束
pencil ore	赤铁矿；笔铁矿
pendence	斜面；斜度
pendent	下垂物；吊悬的；下垂的
pending	悬空的；未定的；未决的
pendular water	悬着水；环状水
pendulous	吊悬的；摇摆的
pendulum	摆；挂钩；摆动式的
pendulum bucket conveyor	可摇动斗式提升机；可摇动斗式输送机
pendulum mass	摆锤；摆体
pendulum mill	悬辊式磨碎机；摆旋式研磨机
pendulum shaft	摆轴；测锤竖井
pendulum stem	钟摆钻杆；摆杆
pendulum tup	摆动式打桩锤；摆动式打桩机
penecontemporaneous	准同期的；准同生的
penecontemporaneous quicksand deformation	准同期漂砾层形变；准同期流沙变形
peneplain	准平原；侵蚀平原
peneplanation	准平原作用；蚀原作用
penetrability	穿透性；渗透性
penetrable	可渗入的；渗透的
penetrance	穿透性；贯穿
penetrant	浸透剂；贯入剂
penetrate	贯穿；穿透；渗透
penetrating cone	触探锥；贯入锥
penetrating pile	沉桩；桩的贯入
penetrating power	贯入力；渗透性
penetrating radiation	穿透辐射；贯穿辐射
penetrating water	下渗水；入渗水
penetrating wave	透入波；潜地波
penetration	焊透；穿透
penetration characteristics	渗透性能；贯入特性
penetration coefficient	穿透系数；贯入系数
penetration cone	触探锥；贯入锥
penetration depth	贯入深度；渗透深度
penetration distance	渗透距离；贯入深度
penetration index	渗透指数；针入度指数
penetration of pile	沉桩；桩贯入度
penetration rate	穿透速度；透入率；穿透率
penetration resistance test	贯入度试验；针入度试验；贯入阻力试验
penetration resistant	抗渗性；抗渗的
penetration sounding	触探；贯入触探
penetration test	贯入试验；触探试验
penetration test equipment	贯入试验设备；触探试验设备
penetration time	贯入时间；钻进时间
penetration twin	穿插双晶；贯穿双晶
penetration wave	透过波；传轮波；潜地波
penetrative	可贯入的；可穿透的
penetrative convection	贯穿性对流；侵入性对流
penetrative depth	透射深度；穿透深度
penetrative movement	贯穿运动；渗透运动
penetrator	穿入者；压模
penetrometer	触探仪；贯入器
penetrometer test	触探试验；贯入试验
penitent ice	冰柱；冰塔
pennant	吊环；吊缆；小旗；信号旗；奖旗
penning	铺石地基；垒木垛；石块铺砌
penny stone	泥铁石；平扁石
penny-shaped crack	扁平形裂纹；币形裂纹
penstock	压力管道；闸门
penthouse roof	凿井时的保护棚顶；单坡屋顶
peperino	白榴凝灰岩；碎晶凝灰岩
peperite	混积岩；熔积岩
per sorption	吸混；渗透成为多孔固体
perambulation	踏勘；查勘；巡视
perambulator	勘察者；路程计；测程轮
percent break	间断百分数；间隙百分数
percent consolidation	压密度；固结度；固结百分数
percent isogram of single mineral	单矿物百分含量等值线图；单矿物含量分布图
percent reduction	破碎率；压缩率
percent sorption	吸水率；吸湿率
percentage elongation	延伸率；伸长率
percentage extraction	回采率；萃取率
percentage humidity	饱和度；湿度
percentage of loss	损失率；损耗率
percentage of mineral dilution	矿物贫化率；矿物贫化百分数
percentage of perforation	穿孔率；射孔率
percentage of reinforcement	钢筋百分率；含钢率；配筋率
percentage of storage	蓄水程度；蓄水率

percentage of void	孔隙百分率；空隙率
percentage site coverage	覆盖百分比；覆盖率
percentile	按百分比的；百分位数；百分之一；百分比下降点
perceptibility	沉淀度；沉淀性；临界沉淀点
perceptible	可沉淀的；可淀析的；析出的
perception	感知；知觉
perceptron	视感控器；感知器
perceptual grouping	感知分组；类别视觉感受
perched aquifer	表层含水层；上层滞水含水层
perched block	坡栖漂砾；坡栖岩块；冰桌石
perched river	悬河；地上河
perched water table	栖留地下水位；上层滞水面
percolate	渗滤；渗出液；滤出液
percolating	渗流；渗透
percolating filter	滴漏滤池；渗滤池
percolating solution	渗漏溶液；渗流溶液
percolating water	渗透水；渗滤水
percolation gauge	测渗仪；渗漏仪
percolation indicator	渗滤指示器；渗滤指示剂
percolation path	渗流途径；渗透路径
percolation pit	渗坑；渗井
percolation pond	渗水池塘；渗水池
percolation rate	渗流量；渗透速率
percolation test	渗透试验；渗滤试验
percolation time	渗滤时间；渗透时间
percolation water	渗流水；渗漏水
percolation well	渗滤井；渗水井
percolation zone	渗透带；渗漏带
percolator	渗滤器；渗滤浸出器
percussion	冲击；撞击
percussion borer	风钻；凿岩机；冲击式钻机
percussion boring	冲击钻探；冲击打眼
percussion cutting machine	冲击式切割机；冲击式截煤机
percussion drill	冲击式钻机；冲击式凿岩机；潜孔钻机
percussion fuze	着发引信；碰炸引信
percussion plate	碰撞板；撞击板；反射板
percussion powder	碰炸炸药；雷汞炸药；撞击起爆药
percussion scar	击痕；撞痕
percussion table	振动台；冲击试验台
percussion welding	冲击焊；锻接
percussive action	撞击作用；破碎作用
percussive air machine	风力凿岩机；风钻
percussive drilling	冲击式钻眼；冲击式凿岩；冲击钻进
percussive force	冲击力；撞击力
perdurability	耐久性；延续时间；持久的
peremptory	绝对的；强制的；断然的
perennial base flow	常年基流；常年底流
perennial frost	永久冻土；多年冻土
perennial impounded body	常年蓄水体；常年静水体
perennial river	常年性河流；常流河
perennial stream	常年河；常流河
perfect black body	绝对黑体；理想黑体
perfect crystal	完整晶体；理想晶体
perfect dislocation	完整位错；完全位错
perfect elastic material	完全弹性材料；理想弹性材料
perfect elasticity	完全弹性；极限弹性
perfect fluid	理想流体；完全流体
perfect soil	理想土体；完全土体
perfect solution	理想溶液；完全溶液
perfection	完备；完整；圆满；完善
perfectly elastic	完全弹性的；理想弹性的
perfectly elastic body	完全弹性体；理想弹性体
perfectly elastic media	完全弹性介质；理想弹性介质
perfectly plastic body	完全塑性体；理想塑性体
perforate	穿孔；钻孔；打孔
perforated	多孔的；穿孔的
perforated baffle plate	多孔消力板；多孔折流挡板
perforated pipe	打孔管；多孔管
perforated plate	多孔板；冲孔板
perforated stone	透水石；穿孔石
perforating	钻孔；打孔；凿孔
perforating azimuth	射孔方位；射孔方位角
perforating equipment	打眼设备；穿孔设备
perforation	冲孔；穿孔；孔，孔眼
perforation length	炮眼长度；射孔段长度
perform	完成；执行；进行；表演
performance	性能；特性；表现
performance bond	履约保证书；履约保函
performance character	运行特性；工作特性
performance chart	性能曲线；性能图
performance computation	特性计算；性能计算
performance criteria	实施准则；实行标准
performance data	工作性能资料；性能数据；运用数据
performance efficiency	运行效率；工艺效能
performance evaluation	性能评价；性能评估
performance index	性能指数；特性指数
performance number	特性数；功率值
performance parameter	性能参数；动态参数
performance prediction	性能预测；性能估计
performance rating	性能额定值；性能率；成绩评定；性能评定
performance ratio	性能系数；特性比
performance specification	规范；规格
performance table	性能表；能率表
performance test	性能试验；运转试验
performance testing	运转试验；性能试验
performer	执行者；执行器
performeter	工作监视器；自动调谐控制谐振器
perfusion	灌注；充满
pergelation	永冻作用；永冻过程
perhydrous vitrinite	高氢镜质组；高氢镜质体
periclinal structure	穹状构造；围斜构造
pericline	肖钠长石；围斜构造；盆状向斜
peridot	橄榄石；贵橄榄石
periglacial	冰缘的；冰边
periglacial loess	冰缘黄土；冷黄土
perihelion	近日点；最高点
perils of the sea	海难；海上风险
perimeter	周长；周；周边；周缘；视野计
perimeter beam	圈梁；周边梁
perimeter blasting	光面爆破；周边爆破
perimeter hole	周边炮孔；周边孔
perimeter shear	周边剪力；周边剪切
perimorph	被壳矿物；包被矿物
period of construction	工程期限；施工期限
period of defects liability	保养期；保修期；维修责任期
period of oscillation	振动周期；振荡周期
period of probation	试用期；见习期

period of rise	涨洪期；上升期	permanent way	路面；路基；轨道
period of warranty	保证期；保修期	permeability	渗透率；渗透性
periodic	周期的；定期的；间歇的	permeability seal	阻渗层；不渗透隔水层
periodic calibration error	周期标定误差；周期性刻度误度	permeability test	渗透试验；渗水试验；渗透率测试
periodic condition	周期性边界；周期性条件	permeability variation coefficient	渗透率变异系数；渗透率变化系数
periodic coverage	周期性遥测区域；周期性覆盖	permeable	透水的；可渗透的
periodic inspection	定期检修；定期检查；小修	permeable aquifer	透水层；含水层
periodic roof pressure	周期性顶板压力；顶板周期来压	permeable channel	渗透渠道；渗水通道
periodic volcano	周期性火山；间隙火山	permeable formation	透水层；透水建造；渗透岩层
periodical	周期的；定期的；间歇的；期刊	permeable horizon	透水层；含水层
periodicity	周期性；周波；频率	permeable joint	透水缝；渗透缝
peripheral	周围的；外围的	permeable level	透水层；含水层
peripheral depression	边缘洼地；环状坳陷	permeable passage	透水通道；导水通道
peripheral device	外部设备；外围装置	permeable river bed	透水河床；渗水河床
peripheral equipment	外围设备；外部设备	permeable soil	渗透性土；透水土
peripheral fault	边缘断层；环周断层	permeable wall	透水墙；渗透墙
peripheral flood	边缘注水；环状注水	permeable zone	渗水带；透水带
peripheral moraine	周碛；围碛	permeameter	磁导计；导磁率计
peripheral pressure	切线压力；侧压；侧切压力	permeameter apparatus	渗透仪；渗透性试验仪
periphery	圆周；外围	permeate	渗透；弥漫
periphery hole	周边炮眼；圈定炮眼	permeation	渗透；渗透作用
periptery	围柱式建筑；周围气流区	permeation rate	渗滤速度；渗流速度
peristyle	列柱廊；柱列	permeator	渗透膜；渗透层；连通器
peritectic structure	包晶组织；包晶结构	permissibility	容许性；防爆性
peritectic temperature	包晶温度；转熔温度	permissibility test	防爆性试验；安全性试验
peritectoid	包析；包析体	permissible	容许的；许可的
perlitic texture	珍珠结构；珍珠纹理	permissible aberration	许可偏差；容许偏差
perm alloy	磁性铁镍合金；透磁钢	permissible cartridge	安全药筒；安全药卷
permafrost	永久冻土；永冻层	permissible deflection	容许偏差；容许挠度
permafrost area	永久冻土区；多年冻土区	permissible discharge	容许排放；容许泄量；容许流量
permafrost soil	永冻土壤；永久冻土	permissible equipment	防爆设备；许用设备
permafrost table	永冻面；永冻土面	permissible level	容许标准；容许含量
permanence	永久性；持久性	permissible motor	安全电动机；隔爆式电动机
permanent	永久的；长期的；固定的	permissible speed	容许速度；容许速率
permanent anomaly	永久异常；稳定异常	permissible stress	容许应力；许用应力
permanent archiving	永久性归档；永久封存	permissible tolerance	容许公差；容许剂量
permanent bed	永久河床；常流河床	permissible variation	允许偏差；容许变化
permanent chock	永久性木垛；固定式木垛	permission	允可；许可
permanent deformation	永久变形；残余变形	permissive yield	允许出水量；容许产出量
permanent extension	永久伸长；剩余伸长	permit	许可证；执照；许可；允许
permanent fault	常发故障；永久失效	permitted explosive	安全炸药；许用炸药
permanent hoist	固定提升机；永久提升机	permitted plot ratio	核准地积比率；准许地积比率
permanent lake	常年湖；稳定湖	permitted rent	准许租金；核准租金
permanent lining	永久衬砌；永久支架	permutation	置换；排列
permanent load	永久荷载；恒载；静荷载	permutator	转换开关；交换器
permanent magnet	永久磁体；永久磁铁；永磁	permute	变更；交换；置换
permanent meadow	永久性草地；永久性草原	pernicious	有害的；有毒的；致命的
permanent mould	永久性铸型；永久铸模	perpendicular	垂直的；正交的；垂线
permanent opening	永久巷道；主要巷道	perpendicular shaft	垂直轴；垂直导井
permanent peg	永久测量标志；永久标桩	perpendicular slip	垂直滑动；铅直滑距
permanent repair	日常修理；日常检修；小修	perpendicularity	垂直；正交
permanent set	永久变形；固定装置	perpetual screw	蜗杆；螺旋杆
permanent stability	恒稳性；耐久性	perpetual snow	积雪；永久积雪
permanent station	永久测站；埋石点	perpetually frozen ground	永冻土；多年冻土
permanent strain	永久应变；永久变形；残余变形	perpetually frozen soil	多年冻土；永冻土
permanent stress	永久应力；同定应力	persistence	持久性；持久度；稳定性
permanent support	永久支护；永久支架	persistence characteristic	持续特性；余辉特性
permanent water	稳定水；常年性水源	persistent	持续的；持久的
permanent water level	稳定水位；恒定水位	persistent chemical	不易分解的化学物；持久性化学品

persistent fracture	持续性裂隙；延续裂隙
persistent oscillation	持续振荡；等幅震荡
persistent pesticide	残留性农药；持久性农药
personal plotter	个人绘图仪；小型绘图仪
personality	个性；人格；品格
perspective	透视；远景；前程；透视的；透视图
perspective plan	远景计划；远景规划
perspective plane	透视面；投影面
perspective projection	透视投影；透视射影
perspective view	透视图；远景图
perspectograph	透视制图；透视仪
perspex	有机玻璃；透明塑胶
perspex window	透明塑胶观测孔；透明塑胶窗
perthitic structure	条纹构造；条纹结构
perthitic texture	条纹结构；条纹纹理
perthosite	淡钠二长岩；淡纹长岩
perturbation	扰动；干扰；波动；摄动；摄动法
perturbation procedure	扰动法；扰动程序
pervasive	透水的；可渗透的；遍布的
perveance	导电系数；导流系数
pervibrator	内部振捣器；插入式振捣器
pervious	透水的；可渗透的
pervious aquifer	透水含水层；孔隙含水层
pervious blanket	透水垫层；透水铺盖
pervious course	透水层；渗透层
pervious rock	透水层；透水岩
pervious subsoil	透水性地基；透水底土层；透水下层土
pervious zone	渗透区；透水带
perviousness	透水性；渗透性
petalite	透锂长石；叶状锂长石
peter out	尖灭；消失；贫化；减少
petrifactology	化石学；古生物学
petrification of sand	固砂；流砂固化
petrified wood	木化石；石化木
petrocalcic horizon	石化钙积层；灰结磐层
petrochemical	石油化学产品；石油化学的
petrochemical works	石油化学工程；石油化工厂
petrochemistry	岩石化学；石油化学
petrofabric	岩石组构；岩组
petrofactor	生油率；生油因子
petrogenesis	岩石成因论；岩石发展学
petroglyph	岩画；岩石雕刻
petrographic comparison	岩石对比；岩性对比
petrographic component	煤岩成分；煤岩类型
petrographic investigation	岩相研究；岩石研究，岩类研究
petrographic microscope	岩相显微镜；偏光显微镜
petrographic microscopy	岩石显微术；岩相显微术
petrographic province	岩区；岩石成因区
petrography	岩相学；岩类学
petrolatum	凡士林；石油冻
petrolene	石油烯；软沥青
petroleum deposit	油气藏；油气矿床
petroleum emulsion breaker	石油破乳装置；石油乳胶分解剂
petroleum fermentation process	石油发酵过程；石油发酵工艺
petroleum pipeline	石油管线；输油管线
petroleum spirit	汽油；汽油酒精混合燃料
petroliferous	含石油的；石油的
petroliferous area	含油区；油田
petrolift	油泵；燃料泵
petrologen	油页岩；油母质
petrothermal	岩热的；干热的
petrous	岩石的；岩状的；坚硬的
pH modifier	pH值改良剂；pH值调节剂
pH recorder	pH值记录器；氢离子浓度记录器
pH regulator	pH值调整器；氢离子浓度调整器
phacoid	透镜体；扁豆体
phacolith	岩脊；岩鞍
phanerocryst	显斑晶；斑晶
phanerocrystalline	显晶质；显晶的
phantom crystal	幻影体；阴影晶体
phantom load	人工载荷；家乡荷载
phantom view	透视图；透明内视图
phase	相；阶段；期
phase comparison	相位比较；比相
phase contrast microscope	相对比显微镜；相反衬显微镜
phase corrector	相位校正电路；相位校正器
phase defect	相位差；相位亏损
phase distortion	相位失真；相畸变
phase front	波阵面；相前
phase inversion	倒相；反相
phase margin	相位储备；相位余量
phase of seismogram	地震波记录的相位；震相
phase shifter	移相器；相位调整器；移相电路
phase splitter	分相器；分相电路
phase transformation	相变；相位变换
phase velocity	相位速度；相速度
phasing	定相；相位调整
phasitron	调频管；调相管；调相用真空管
phasor	相量；相图
phasor diagram	相量图；相矢量图
phenhydrous	水下生成的；在敞露水体中形成的
phenocrystalline	显晶质；显晶的
phenomenological approach	唯象学方法；现象学方法
philosophy of measurement	测量方法；测量原理
phlebite	脉成岩类；混脉岩
phlegm	黏液；冷凝液；迟钝
pholerite	大岭石；地开石；鳞高岭石
phonation	发声；音信号传输
phonautograph	声波记振仪；声波振动记录
phonoscope	验声器；微音器
phosphate chalk	磷灰岩；磷灰土
phosphate coating	磷酸盐处理法；磷酸盐涂层
phosphate matrix	磷灰基质；含磷钙土的岩石
phosphate rock	磷酸盐岩；磷灰岩
phosphatic nodule	磷酸盐质结核；磷质结核
phosphatization	磷化；磷酸盐化
phosphor	磷光体；磷光质
phosphorescent	发磷光的；磷光质
phosphorite	磷灰岩；磷钙土
phosphoroscope	磷光计；磷光镜
phosphuranylite	磷钙铀矿；磷铀矿
photic zone	透光带；透光层
photo coverage	像片覆盖；有航拍资料的地区
photo distance	相片上距离；摄影距离
photo reconnaissance	摄影勘测；摄影侦察
photo sensitive coating	感光层；感光涂层
photo sensitivity	感光性；光敏度

photoactivation	用光敏化；用光催化
photocell pickoff	光电元件；光电池
photochemical activity	光化学活动；光化活性
photocoagulation	光致凝结；光焊接
photoconducting cell	光敏电阻；光电管
photoconductivity	光电导性；光电导率
photoconductor	光电导体；光敏电阻
photoelastic	光测弹性的；光弹的
photoelastic load cell	光弹加载传感器；光弹测压元件
photoelastic test	光弹试验；光弹性测试
photoelasticity	光测弹性学；光弹性学
photoelectric cell	光电管；光电池
photogeologic guide	航摄地质标志；航片地质解释标志
photogeological lineament	航摄地质线；航摄地质特征
photogeomorphology	航摄地貌学；摄影地貌学
photogrammetric sketch	摄影测量草图
photogrammetry	摄影测量学；摄影制图法
photographic map	影像地图；照相地图
photographic speed	摄影速度；感光度
photographic technique	照相法；拍照法；摄影法
photokey	航摄地区略图；航空相片判读样片
photolithograph	照相平版印刷；光刻
photomap	影像地图；航空照片图
photometric calibration	光度校准；光度定标
photomicrograph	显微摄影；显微照片
photoplasticity	光塑性；光塑力学
photostereograph	立体绘图仪；立体测图仪
photostratigraphy	摄影地层学；摄影地层法
phreatic	井的；潜水的
phreatic aquifer	潜水含水层；非承压含水层
phreatic cycle	潜水位变化周期；地下水循环
phreatic discharge	地下水流量；潜水涌出量
phreatic divide	潜水分水岭；地下水分水岭
phreatic flow	潜水；潜水流；大气起源的水流
phreatic fluctuation	潜水位波动；潜水面升降
phreatic gas	次生蒸气；再生蒸气；准火山气体
phreatic groundwater	潜水；地下水；饱水带水
phreatic layer	潜水层；地下含水层
phreatic level	潜水水位；地下水位
phreatic line	浸润线；地下水位线；土中渗水线
phreatic low	低地下水位；地下水下部含水层
phreatic rise	潜水位上升；地下水位上升
phreatic surface	潜水面；地下水面
phreatic water	地下水；潜水
phreatic water discharge	地下水排泄量；地下水流量；潜水排泄量
phreatic water level	潜水位；地下水位
phreatic zone	潜水带；饱水带
phugoid	长周期振动；长周期的
phyletic evolution	线系演化；种系演化
phyllonite	千糜岩；千枚糜棱岩
phyllosilicate	层状硅酸盐；页硅酸盐
phyllovitrinite	结构镜质体；叶状镜质体
physical damage	机械损坏；物理破坏
physical environment	物理环境；自然环境
physical examination	健康检查；体格检查
physical form	天然形态；自然形状
physical geological phenomena	自然地质现象；物理地质现象
physical measurement	实际丈量；实际量度
physical model	物理模型；实体模型
physical residue	物理风化残余物；机械风化残余物
physical settling	物理沉积；物理堆积
physical solution	物理解；物理溶液
physical structure	机械构造；物理结构
physical weathering	机械风化；物理风化
physics	物理学；物理过程；物理现象；物理成分
physiochemical	生物化学的；生理化学的
physiographic	地文的；自然地理的
physiographic cycle	地文旋回；侵蚀旋回
physiographic division	地文区；地形分区
physiography	自然地理学；地文学
phytolith	植物化石；植物岩
phytopaleontology	古植物学；植物化石学
pick hammer	风镐；尖锤
pick man	风镐手；手镐工
pick point	镐尖；风镐钎子
picked coal	手选煤；手拣煤
picket fence	尖桩篱栅；木桩栅栏
picking belt	手选带；挑选输送带
picking chute	拣矸槽；手选槽
picking conveyor	拣矸输送机；手选输送机
picking table	拣矸台；手选台
picknometer	比重计；比重瓶
pickoff	传感器；敏感元件
pickrose hoist	调车绞车；调度绞车
picrite	橄榄岩；辉石
picrochromite	镁铬铁矿；铬镁尖晶石
piedmont	山麓；山前地带
piedmont depression	山前坳陷；山前洼地
pier	墩；码头；窗间墙；扶壁
pier column	墩柱；支柱
pier dam	折流堤；防砂堤；防波堤
pier foundation	墩式基础；墩式地基
piercement dome	刺穿穹丘；底辟构造
piercement fold	刺穿褶皱；底辟褶皱
piercement salt dome	刺穿盐丘；底辟盐丘
piercer	钻孔器；锥
piercing	贯穿；掘进；钻进；锐利的
piercing drill	火力钻机；热力钻机
piercing fold	刺穿褶皱；挤入褶皱；盐丘
piercing jet	热钻喷流；火力穿孔喷流
piercing mandrel	穿孔冲头；冲子
piercing mill	穿孔机；穿轧机
piercing press	穿孔压力机；冲孔机
piezoelectric seismometer	压电地震检波器；压电式地震计
piezometer	测压计；测压管
piezometer data	测压计数据；水压计数据
piezometric head	测压管水头；孔隙水头
piezometric level	测压管水位；孔隙水位
piezometric potential	测压水头；承压水头
piezometric surface	测压面；承压面
pig breaker	铁块破碎机；生铁打碎机
piggyback	驮着，背着；搭架在另机器之上的；重叠安装的
pilage	砂布；砂纸
pile	桩；堆；堆积
pile anchor	桩锚；拉力桩
pile bent	桩排架；桩桥台

pile built in place	就地灌注桩；现场灌注桩
pile cap	桩帽；桩承台
pile capacity	单桩承载力；桩承载能力
pile cluster	桩群；群桩
pile cover	桩箍；桩帽
pile curtain	桩排；桩成帷幕
pile drain	砂井；桩成排水通道
pile driving rig	打桩设备；打桩机
pile eccentricity	桩的偏心度；桩的倾斜度
pile fender	护桩；防冲桩
pile footing	桩承底脚；桩基础
pile formula	打桩公式；桩承载公式
pile foundation	桩地基；桩基础
pile grillage	桩格排；桩群；桩基承台
pile group	桩群；群桩
pile height	桩高；桩长
pile helmet	桩盔；桩帽
pile layout	桩的布置；桩的平面布置
pile leading	桩导杆；桩架带导向杆
pile overdriven	过深桩；超打桩
pile pedestal	桩底脚；打桩台座
pile pier	桩墩；桩基桥墩；高桩码头
pile puller	拔桩机；拔桩器
pile shoe	桩靴；桩鞋
pile sinking	打桩；沉桩；板桩法凿井
pile splice	桩拼接；桩接头
pile tip	桩端；桩尖
pile top	桩顶；桩盖顶
pile trestle	排桩栈道；桩构栈道
pile work	打桩工程；打桩设施
pilework	桩式建筑物；打桩工程
pilger mill	周期式轧管机；皮尔格式轧管机
piling	打桩；打桩工程
pillage	砂布；砂纸
pillar	柱；矿柱；煤柱
pillar and breast method	房柱式井采法；房柱式采煤法
pillar burst	煤柱突然破裂；矿柱突然破裂
pillar drawing	回采矿柱；回采煤柱
pillar extraction	矿柱回采；煤柱回采
pillar extraction drilling	煤柱回采钻眼；矿柱回采凿岩
pillar face	矿柱工作面；煤柱工作面
pillar leaving	留煤柱；留矿柱；矿柱保留；煤柱保留
pillar line	矿柱线；煤柱线；大后退回采煤柱的放顶线
pillar man	石垛工；煤柱回采工；矿柱开采工
pillar method	柱式开采法；留柱护顶法
pillar mining	矿柱回采；煤柱回采
pillar pulling	回采煤柱；回采矿柱
pillar rating	矿柱评价；煤柱评价
pillar remnant	矿柱残余；煤柱残余；残柱
pillar residue	残留煤柱；残留矿柱
pillar robbing	回采矿柱；回采煤柱
pillar slabbing	矿柱刷帮；煤柱刷帮
pillar spacing	煤柱间距；矿柱间隙
pillar stub	煤柱根；矿柱根
pillar support	柱基；矿柱支护
pillar taking	回采矿柱；回采煤柱
pillar thinning	矿柱部分回采；煤柱部分开采
pillar withdrawing	矿柱回采；煤柱回采
pillar work	矿柱回采工作；煤柱回采工作
pillow	轴枕；垫座；枕状构造
pilolite	镁石棉；镁坡缕石
pilot area	试验地区；前导地
pilot bore	试验孔；导洞
pilot boring	预先钻进；超前钻孔
pilot brush	控制刷；测试或辅助刷
pilot burner	长明灯；导燃器；引燃器
pilot cable	控制电缆；引示电缆
pilot drift	导洞；掌子面
pilot drilling	导向孔钻进；超前钻眼
pilot dual operation	先导控制；先导操作
pilot hole	导向钻孔；超前钻孔
pilot pile	排障桩；导桩
pilot poppet valve	先导提动阀；导向提动阀
pilot project	试点项目；试点工程；先导性项目
pilot test	小规模试验；初步试验；先导性试验
pilot tunnel	导洞；导坑；前探平硐；导向隧道
pilotaxitic texture	交织结构；交织组织
pin connection	销钉连接；螺栓连接；引线连接
pinch out	尖灭；变薄
pinched side	挤压侧；压缩侧
pioneer hole	勘探钻孔；导孔
pioneer tunnel	导坑；导洞
pipe culvert	管道涵洞；涵管；管渠
pipe line network design	管道网设计；管道线路网设计
pipe pile	管桩；管柱桩
pipe sampling	用管子取样；管取法
pipe sleeve	套管；管箍
pipe spilling	管桩；簪桩掘进法
piping	管涌；管道系统
pirate river	袭夺河；夺流河
piston air drill	活塞式风钻；活塞式风动凿岩机
piston drill	活塞式钻机；活塞式凿岩机
pit	矿井；坑；洼地；竖井
pit bottom	井底；井底车场
pit coal	坑煤；矿产煤
pit crater	矿山陷落坑；锅形火山口
pit dump	矿山废石堆；矿山矸石堆
pit exploitation	坑槽探；掘探
pit haulage	矿山运输；矿山拖运
pit head arrangement	井口建筑物；井楼；井口设备
pit head building	井口棚；井口建筑
pit hill	井口建筑；井口地面；地面拣煤台
pit limit	煤柱可采极限；露天矿场境界
pit planning	矿山设计；采场设计；采场规划
pit rail	井下运输铁路；矿用钢轨
pit sampling	探槽取样；探坑取样
pit scale	矿山秤；地磅；有基坑的秤
pit shaft	矿井；井筒
pit test	坑探；坑检
pit tip	矿山矸子堆；矿山废石堆
pit top	井口；地面；井口建筑
pit top building	井口建筑物；组合；装配；积累；加厚
pit wall	采场边坡；露天矿帮；矿井壁
pit water	矿井水；矿坑水
pitch angle	倾角；螺距角
pitch coal	沥青煤；烟煤
pitch coke	沥青焦；沥青焦炭
pitch earth	地沥青；沥青土
pitch of arch	拱矢；拱的高；跨比
pitch stiffness	纵摇过稳性；纵向安定性

pitch stone	琢石；松脂岩
pitchblende	沥青铀矿；非晶质铀矿
pitched slope seawall	向海斜建的海堤；砌石护坡海堤
pitching	工作面上砌石；路面铺粗石块；粗石块路；倾斜的；陡的
pitching bed	陡倾岩层；立槽煤层
pitching borer	开孔钻头；开眼钎子
pitching seam	倾斜煤层；倾斜矿层
pitching vein	陡脉；急倾斜脉
pithead	竖井口；竖井口平台
pitometer	皮托压差计；流量计
pitot	空速管；皮托管
pitot commutator	皮托管；总压管
pitot meter	皮托压差计；皮氏流速计
pitot tube	皮托管；流速测定管
pit-run gravel	天然级配砾石；天然料坑砂砾石
pit-run sand	坑砂；未筛砂
pitting	坑探；金属腐蚀；点蚀；麻点；露头凿浅井开采
pivot point layout	垂直定位法；测深垂线定位
pivotal fault	枢纽断层；旋转断层
pivoted weir	活动堰；翻板堰
placanticline	盾状背斜；平背斜；长垣
placation	褶皱；皱纹
place advance	工作面掘进；工作面超前
place cleaning	清理工作面；清扫工作地
place driving	工作区掘进；采场掘进
place measurement	就地测量；实体测量
placeability	可灌筑性；和易性
placeability of concrete	混凝土的和易性；混凝土的可浇筑性
placement	浇筑；铺筑；放置；部位
placement moisture	填筑湿度；填土湿度
placer	砂矿；砂积矿床；浇筑混凝土装置
placer deposit	砂矿；砂积床
placer mining	砂矿开采；砂矿淘采
plagioclase	斜长石；斜长岩
plagioliparite	斜长疗岩；斜长流纹岩
plagiotrachyte	斜长；粗面岩
plain concrete	无筋混凝土；素混凝土
plain girder	实腹梁；板梁
plain strain	简单应变；单向应变
plain washer	普通垫圈；平垫圈
plait	褶；辫绳；编织
plait point	褶点；临界点
plan	计划；设计；方案；平面图；草图
plan area	直线面；设计面积
plan map	规划图；平面图
plan of operation	项目实施计划；作业计划
plan of site	总平面图；平面布置图；总布置图
plan of wiring	线路图；布线图
plan sketch	草图；设计图；简图；平面图
planar	平面的；平的；二维的
planar structure	板状构造；面状构造
planar water	吸附水；晶面层间水
planation	夷平作用；均夷作用
plane	轨道上山；自重滑行坡；沿煤层底板的巷道；平面，面
plane bed	平底；平坦河床；平面岩层；平坦沉积层
plane failure	平面破坏；平面滑坍
plane fault	平面断层；顺层断层
plane flow	平面流；片流
plane fracture	平面裂缝；平坦断口
plane of break	破裂面；断裂面；垮落面
plane of cleavage	劈理面；解理面；裂开面；劈裂面
plane of delineation	图像平面；投影平面
plane of discontinuity	不连续面；间断面；折断面
plane of fissility	片理面；劈理面；劈裂面
plane of flattening	延展面；压扁面
plane of flexure	挠曲面；弯曲面
plane of junction	接触面；接合面
plane of parallel texture	平行构造面；平行纹理面
plane of projection	投射面；投影面；射影面
plane of reference	参考面；基础面；标志面
plane of rupture	断裂面；破裂面
plane of saturation	潜水面；饱和带水面
plane of separation	分开面；分裂面；分离面
plane of shear	剪切面；剪断面
plane of stratification	层理面；层面
plane of weakness	软弱面；脆弱面
plane reference system	平面坐标系统；平面坐标参照系
plane section	水平剖面；水平层状地层
plane source	平面震源；面源
plane state of stress	二向应力状态；平面应力状态
plane stoker	平式加煤机；普通加煤机
plane structure	平面结构；面状构造
plane surface	平面；平曲面
plane table	平板仪；平板测绘仪；平板测量仪
planer	刨床；平煤工；刨煤机
planform	平面的；平面图；平面形状
planianticline	平背斜；平原型背斜
planimegraph	比例规；缩图器
planimeter	面积仪；测面仪；求积仪
planimetric data	地物数据；平面位置数据
planimetric map	平面图；无等高线地图
planimetric method	测面积法；平面测量法
planimetry	平面几何；测面学；测面法
planing	刨平；整平
planing machine	刨床；刨煤机
planish	抛光；磨光；碾平
planisphere	平面球体图；平面天体图；星座一览图
plank check dam	木板谷坊；木板拦沙坝
plank layer	板层；垫板
plank sheathing	木板护壁；木模板
planking	铺板；板材；地板；船壳板
planned mine	待建矿山；拟建矿山
planned performance	预期性能；计划成效
planned subsidence	控制下沉；预计下沉
planning operation sequence	计划作业顺序；计划工序
planning reference	设计标准；设计参考资料
planning statement plan	规划纲领图；规划说明图
planning survey	设计测量；规划调查
planometer	测平仪；平面规；求积仪
plasma	等离子体；深绿玉髓
plasma torch	等离子体喷枪；等离子体吹管；等离子体焰炬
plasmoid	等离子体状态；等离子粒团
plasmons	等离体子；等离激元
plaster	泥饼；灰泥；熟石膏；粉刷；药膏
plaster bead	灰泥墙角护条；灰泥护角

plaster board	灰泥板；粉饰板；石膏板
plaster of paris	烧石膏；热石膏；熟石膏
plaster rendering	灰泥荡面；灰泥抹面
plaster shooting	糊炮爆破；外部装药爆破；覆土装药爆破；裸露爆破；糊泥爆破法
plaster shot	糊炮；外部装药
plaster slab	石膏板；糊炮
plasterboard	糊墙纸板；石膏灰泥板
plastered brickwork	混水墙；抹灰砖砌体
plastering	覆土爆破；糊炮；抹灰；抹面
plastic	塑性的；可塑的；有适应力的；塑料；胶质物
plastic bronze	塑性高铅轴承青铜；塑性铅青铜
plastic casing	塑料封装；塑料套管
plastic cement	塑胶；塑性水泥
plastic clay pan soil	塑性黏土；塑性硬黏土
plastic deformation	塑性变形；柔性变形
plastic design	钢或钢筋混凝土构架的塑性铰接设计；塑性设计
plastic drain	塑料板排水；塑料排水管
plastic earth pressure	塑性地压；塑性土压力
plastic failure	塑性损毁；塑性破坏
plastic fines	塑性细颗粒；塑性细粉
plastic flow	塑性流；塑流
plastic fracture	塑性断裂；塑性断口
plastic index	塑性指数；可塑指数
plastic laminated sheet	层压塑料板；防火胶板
plastic layer index	塑性；胶质层指数
plastic limit	塑性极限；塑限
plastic map	塑料地图；立体地图
plastic material	塑性材料；塑料
plastic metal	锡基轴承合金；塑性合金
plastic nature	塑性；可塑性
plastic property	塑性；可塑性
plastic range	塑性范围；可塑范围
plastic range test	塑性范围试验；塑限试验
plastic region	塑性范围；塑性区
plastic resonance	塑性共振；塑性反应
plastic soil	塑性土；可塑土
plastic stage	塑性状态；塑性阶段
plastic strain	塑性变形；塑性应变
plastic viscosity	塑性黏度；黏塑性
plastic wave	塑性波；压缩波
plastic yield	塑性变形；塑性流动；塑性屈服；塑变值
plastic yield test	塑料变形测试；塑流试验
plastication	增模；塑化作用
plasticgraph	塑度计；塑性计
plasticity	可塑性；范性；适应性；塑性力学
plasticity agent	增塑剂；塑化剂
plasticity index	塑性指数；塑性指标
plasticization	增塑作用；塑化作用
plasticizer	增塑剂；塑化剂；柔韧剂
plastid	质体；成形粒
plastograph	塑性变形记录仪；塑度计
plastomer	塑性体；塑料
plastometer	塑度计；塑性仪
plat	井底装卸台；地段图；平面图；平台
plate	极板；铜轨；镀；板块
plate accretion	板块增大；板块增生
plate arch	平板拱；板拱
plate bearing test	承载板试验；平板载荷试验
plate boundary	板块边界；板块界限；板块结合带
plate coal	板状煤；大块煤
plate convergence	板块交汇；板块汇聚
plate destruction	板块消亡；板块破坏
plate divergence	板块分歧；板块离散
plate juncture	板块边界；板块结合带
plate layer	铺铁路工；铺轨工
plate shale	板状页岩；硬泥质页岩
plate subduction	板块消亡；板块消减；板块俯冲
plate tectonic theory	板块构造学说；板块构造理论
plate tectonics theory	板块构造说；新全球构造说
plate valve	片阀；板阀
plate vibrator	平板振动机；平板振捣器
plateau	高原；台地；海台；高地
plateau basalt	溢流玄武岩；高原玄武岩
platelayer	巷道维护工；修路工
platelet	小片状体；片晶
platelike	层状的；板状的；片状的
platen	压盘；压纸卷轴；台板；屏
plate-shaped drill	盘式钻机；盘式凿岩机
plate-tectonic cycle	板块构造旋回；板块构造循环
platform	平台；站台；月台；工作台
platform drill	台式钻机；台式凿岩机
platform hoist	吊盘绞车；平台式起重机
platform reef	台地生物礁；礁坪；平顶礁
platform subside	台陷；台地沉陷
platform uprise	台隆；台地隆升
platform vibrator	板式振动器；振动台
platiniridium	铂铱齐；铂铱矿
platometer	测面仪；求积仪
platy fabric	片状组构；板状组构
platy flow structure	板状流动构造；流面构造
plauenite	钾质富斜正长岩；钾正长岩
play out	尖灭；做完；用完
playa	沙漠中的盆地，干盐湖；河口沙地
playa lake	干盐湖；雨季湖
plazolite	水钙铝榴石；无色榴石
pleistocene glaciation	更新世冰川；更新世冰期
plenum chamber	压力通风室；送气室
plenum duct	送气通道；送气道
plenum system of ventilation	压入式通风法；吹入通风法
plenum ventilation	送气式通风；强制式通风
pleonaste	镁铁尖晶石；亚铁尖晶石
plesiostratotype	补充层型；辅助层型
plesiotype	近型；近模
plica	壳褶；皱襞
plication	褶皱；皱纹；细褶皱
plinth	底座；柱基；沙丘基底；山坡基
plinthite	赤黏土；杂赤铁土；铁铝斑纹层
pliomagmatic	多岩浆的；岩浆活动强烈的；过岩浆的；优地槽岩浆的
pliomagmatic zone	优地槽；多岩浆带；强烈岩浆活动带
plot	标绘；曲线；图表；图
plot of mine	矿井平面图；矿图
plot point	标定点；展点；标绘点
plot ratio	地积比率；建造比率
plotted point	投影点；标绘点；交会点
plotter	绘图机；图形显示器；测图仪；标图员；标绘器

plotting	点绘；点绘曲线；绘曲线
plotting board	绘图板；图形显示幕；曲线板；图形输出板
plotting device	曲线绘图仪；绘图装置
plotting machine	绘图机；绘图仪器；绘图仪
plotting method	画法；测绘方法
plotting paper	方格绘图纸；坐标纸
plotting tablet	绘图板；图形输入板
plotting unit	描绘设备；绘图机
plough body	犁身；刨煤机刨体
plough chain	刨链；刨煤链
plough share	犁板；犁头
plover	校准仪；检验仪
plow	犁；煤犁；刨煤机
plucking	冰蚀；冰川拔削
plug	塞子，塞盖，栓塞；插头，插销
plug adapter	插塞式接合器；插塞式转接器
plug dome	岩颈穹丘；颈状火山穹丘
plug drill	冲击式凿岩机；冲击式岩石钻
plug drilling	冲击式钻眼；冲击式凿岩
plug hole	二次爆破小炮眼；插头孔
plugged crib	基础井框；基本垛盘
plugger	凿岩机；填塞物
plugger drill	风动凿岩机；风镐
plugging	堵塞；填塞；封孔；止水；钻孔回填
plugging ability	密封能力；堵塞能力；密封性能
plugging agent	止水剂；堵塞剂；密封剂
plumb	铅锤，测锤；铅锤测量；竖直
plumb aligner	垂准仪；铅垂仪
plumb bob	铅锤；铅球；垂球；测锤
plumb bob cord	铅锤线；测锤线
plumb level	水准仪；水平仪；绝对水平
plumb line	铅垂线；准绳；准绳
plumb point	垂准点；铅锤测点；天底点
plumb rule	垂线尺；垂规
plumb shaft	立井；直井
plumbago	石墨；石墨状矿物
plumber	白铁工；管子工；管道工
plumbing	管道工程；测深；铅垂测量；吊线；波导设备；波导管
plumbing works	水管工程；敷设水管工程
plumbism	铅毒；铅中毒
plumboferrite	铅铁矿；磁铅铁矿
plume	羽状物；羽状水流；地幔柱；地幔羽；瑕疵
plume structure	羽状构造；羽式构造
plummer	轴台；垫块
plummer block	轴台；止推轴承
Plummet	垂球；铅锤；铅垂线；对中器
plumose fracture	羽状裂隙；羽状破裂
plunge of fold	褶曲倾伏；褶曲的倾伏角；褶皱扑伏
plunger	柱塞；按塞；潜水员；阳模
plunger packing	柱塞填密；柱塞垫圈
plunger pump	柱塞泵；往复式水泵
plus point	导线加桩点；内分点
pluton	侵入体；深成岩体
plutonic cognate ejecta	深成同源喷出物；深成岩浆喷出物
plutonic earthquake	深震；深成地震
plutonic facies	深成相；深成岩相
plutonic plug	深成岩栓；岩栓
plutonic rock	深成岩；火成岩
plutonic terrain	深成侵入体；深成岩体
plutonic water	深成水；深源水；岩浆水
plutonism	火成论；岩石火成说
pluvial	洪水的；雨成的；洪积的；多雨的；雨期的
pluvial age	洪积世；洪积时期
pluvial denudation	雨水冲蚀；洪水剥蚀
pluvial facies	洪积相；洪水相
pluvial lake	洪积湖；洪水湖
pluvial period	多雨时期；洪积纪
pluviometric	量雨的；雨量器的
pluviometry	雨量测定法；测雨法
pluvioscope	雨量计；雨量器
ply	层片；股；折叠
PMMA	亚克力；有机玻璃；聚甲基丙烯酸甲酯
pneumatic brake	风力制动；气力制动；压缩空气制动；气闸
pneumatic caisson method	压气沉井掘进法；气沉箱法
pneumatic cell without pump body	无泵体的充气式浮选槽；浅型充气式浮选槽
pneumatic charging	风力装药；吹送装药
pneumatic charging apparatus	风动装药器；风力装药器
pneumatic chisel	风凿；风錾
pneumatic chuck	气动卡盘；气动夹头
pneumatic cleaning	风力精选；风选；风力选矿；风力选煤
pneumatic compactor	气夯；风动夯
pneumatic concentrator	风力选矿机；风力选煤机
pneumatic control	气动控制；压力气体控制
pneumatic cushion	气枕；气垫
pneumatic digger	风镐；风铲
pneumatic drill	气钻；风钻
pneumatic drilling	风动凿岩；风动钻进
pneumatic ejector	压气喷射器；风动喷射器
pneumatic flotation machine	压气式浮选机；充气式浮选机
pneumatic hammer	气锤；风镐
pneumatic hoist	风动绞车；风动起重机；气动绞车
pneumatic jack	气力千斤顶；压气千斤顶
pneumatic linkage	气压联动装置；风力联动装置
pneumatic maul	风锤；风动锤
pneumatic mortar	喷射用砂浆；压力喷浆
pneumatic packer	风动栓塞；充气灌浆塞
pneumatic percussion hammer	气动冲击器；风动冲击器
pneumatic perforator	气压凿孔机；压气钻孔器
pneumatic pick	松土机；风镐
pneumatic pile driver	风动打桩机；气压式压桩；压气打桩
pneumatic preparation	风选；风力选
pneumatic process	压气掘进法；压气沉箱法
pneumatic pump	压气泵；风动泵；楔形闸阀
pneumatic ram	气力夯；气动夯锤
pneumatic rammer	风动捣锤；气动夯锤
pneumatic remote control	风动遥控；风动远距离操纵
pneumatic separation	风选；风力选
pneumatic separator	风力分选机；风力分离器
pneumatic shovel	风力铲；风动铲；风铲
pneumatic table	风力淘汰盘；风力摇床
pneumatic tamper	气夯；风动打夯机
pneumatic tool	风动工具；气动工具
pneumatic transport	风力运输；气动输送

pneumatic traversing guide control	气动式横动导丝控制；气动式横动导纱控制
pneumatic tyred roller	气胎压路机；气胎碾路机
pneumatic work	风力掘进；压气工作
pneumatolytic mineral	气成矿物；气化矿物
pneumatolytic product	气化产物；气化矿物
pocket calculator	袖珍计算器；袖珍计算机
pocket penetrometer	小型贯入仪；袖珍贯入仪；携带式贯入仪
pocket shear meter	袖珍剪力仪；袖珍十字板仪
pocket spring	穴泉；轴承油槽
pocket storage	袋状贮水；洼地积水
pocket transit	袖珍经纬仪；原生土
pocket-and-fender pillaring	独头进路留柱式煤柱回采；独头进路留柱式矿柱回采
pocketclay	囊状黏土；高硅黏土
pocket-size	袖珍型的；小型的
pod form	豆荚状；透镜状
pod like	扁豆形的；透镜形的
podiform	豆荚状；透镜状
podium level	平台层；平台高度
podsol	灰化土；灰壤
podsol soil	灰壤；灰化土
podzolization	灰化作用；灰壤化作用；灰化酌
poecilitic	嵌晶状；嵌晶状的
poetsch process	盐水方式；间接热交换方式
poikilitic	嵌晶结构的；嵌晶状的
poikilitic texture	嵌晶结构；包含结构
poikiloblastic	变嵌晶状；变嵌晶状的
poikiloblastic texture	包含变晶结构；变嵌晶结构
poikilocrystallic	嵌含的；嵌晶的
point attack bit	点击式截齿；镐形齿
point bar deposit	凸岸坝沉积；边滩沉积
point bearing pile	端承桩；支承桩
point block	塔式大厦；点式住宅
point diagram	极点图；点图
point heat source	点热源；岩浆囊热源
point intersection	交会插点法；交点法
point loading	集中点装载；集中负载
point measurement	点测量；点测
point of application	作用点；施力点
point of contra flexure	反弯点；反曲点；拐点；回折点
point of contraflexure	转折点；反弯点；拐点；回折
point of curvature	拐弯点；曲线起点；曲线原点
point of detonation	爆震点；爆炸点；爆发点
point of discontinuity	曲线折点；不连续点；突变点
point of force	力的作用点；着力点；力的端点
point of force application	着力点；力作用点；荏力点
point of inflection	转折点；反弯点；反曲点；拐点；曲率变化点
point of issue	涌水点；泉口
point of no flow	无流点；始喷点
point of observation	观察点；地质点；观测点；测站
point of origin	原点；起始点；源点
point of reference	水准点；控制点；基点；参考点
point of same elevation	相同标高的点；同高程点
point of use	使用点；使用场所
point of zero flow	流量零点；断流点
point resistance	顶端阻力；桩尖阻力
point resistance force	端阻力；总端阻力
point sample	点样品；点样
point sort	选点；选排
point system	评分法；点系
point-counter analysis	点计数分析；计数器分析
pointing error	瞄准误差；指向误差
pointing of holes	钻孔布置；炮眼布置
points man	转辙工；板道工
point-to-point control	定点控制；点位控制
poise	泊；使平衡，稳定；砝码；秤锤
poisonous material	有害物质；有毒物质
poke hole	拨火孔；探视孔
poke out	伸出；拨出
polar	极，极线；极面；南极的，北极的；极性的；磁极的；极坐标的
polar coordinate positioning	极坐标定位；方位距离定位
polar curve	极坐标曲线；配极曲线
polar diagram	极坐标图；极线图
polar map	极点图；极投影地图
polar migration	极移；地极移动
polar path	磁极径迹；地极径迹；极移轨迹
polar plan meter	定极面积仪；定极求积仪
polar winding	极向缠绕法；极向卷绕
polar wobble	极移；地极晃动
polarimeter	偏振计；旋光测定计
polarimetry	旋光测定；测极化
polaris cope	旋光计；偏光镜
polariton	电磁声子；偏振子
polarity chron	极性时；极性年代
polarity reversal	极性倒转；磁极倒转；磁极反向
polarity rock-stratigraphic unit	极性岩石地层单位；磁极岩石地层单位
polarization	偏振，偏振化；极化；偏振度；极化度，极化强度；分化
polarization effect	极化效应；偏振效应
polarization ellipse	极化椭圆；偏振椭圆
polarization method	激发极化法；偏振法
polarization plane	偏振光面；极化面；偏光面；偏振面
polarized relay	极化继电器；有极继电器
polarized shear wave velocity	极化剪切波波速；偏振剪切波波速
polarizer	起偏振镜；起偏光镜；极化镜；下偏光镜；起偏器，偏振器
polarizing angle	极化角；偏角
polarographic analysis	极谱分析；极化分析
polarography	极谱法，极谱分析法；极谱学
polaroid	人造偏振片；偏振片；偏光玻璃
pole bracket	悬臂；撑架；支架
pole of face	面投影极点；面极
pole of rotation	转动极；自转极
pole of the ecliptic	黄道极；黄极
pole pitch	极距；磁极距
pole roadway	维持在留矿内的天井；运支架的通道
pole tool	钻杆冲击钻进工具；钻杆打钻工具
pole-and-chain	撤柱机；回柱机
pole-dipole array	单极-偶极排列；三极排列
pole-strength	极强；磁极强度
polianite	黝锰矿；锰矿
police	校正；警察；修正
poling board	桩板，撑板，壁板；超前支护；前探梁掘进

polished dressing	表面抛光；表面修饰
polished rod head	光杆头；抛光钻杆头
polished section	抛光片；磨光片；抛光磨片；光片
polished surface	抛光面；磨光面
polisher	抛光机；抛光工人；打磨工；高纯度水处理装置
polishing hardness	磨蚀硬度；抗磨硬度
polishing machine	磨光机；磨岩机
polishing roll	抛光辊；轧光辊
polje	灰岩盆地；岩溶盆地；喀斯特盆地
polje lake	灰岩盆地湖；溶蚀盆地湖
pollenite	碱玄质响岩；方钠霞玄岩
pollucite	铯沸石；铯榴石
polluted groundwater	污染的地下水；受污染地下水
polluted sign of water quality	水质污染征兆；水质污染
polluted surface water	污染的地表水；受污染地表水
polluted waterway	污染河流；受污染水道
polluted zone	污染带；受污染带
pollution by particulates	颗粒物污染；微粒污染
polybaric	多压的；变压的
polycarbonate sheet	聚碳酸酯纤维片；碳化纤维胶片
polychroism	多色；多色性
polychrome	彩色；多色的
polycrystalline	混染晶体；多晶的
polycrystalline material	多晶物质；多晶材料
polycycle	多旋回；多循环
polycyclic orogenesis	复相造山运动；多旋回造山运动
polydymite	辉镍矿；辉铁镍矿
polyfoam	泡沫塑料；发泡胶
polyformation	多建造；复式岩层；多次形成
polyformative	复式形成的；多期形成的
polygene	多源的；多成因的
polygenetic	多次的；复成的；多源的；复成分的
polygenetic conglomerate	复成因砾岩；复成砾岩
polygenetic enclave	多源包体；复成包体
polygenetic mineralization	复成矿化；多成因矿化
polygenetic soil	多成因土壤；多成因土
polygenetic topography	多成因地形；复成地形
polygenetic volcanic edifice	复式火山机构；复式火山机体
polygenic	多次的；多源的；复成的；复成分的
polygon	多边形；多边形土；龟裂土；多角形；封闭导线；封闭折线
polygon method	导线测量法；折线法；多边形法
polygonal ground	多边形土；龟裂地面
polygonal height traverse	三角高程导线；高程导线网
polygonal marking	多边形土；地面龟裂；龟裂标志
polygonal structure	多边形构造；龟裂构造
polygonal support	多边形支架；多角形支柱
polygonization	多边形化；多角化
polygonometry	导线测量；多边形几何学
polymer	聚合物；聚合体
polymer grid reinforcement	聚合物格网加强；土工网加固
polymer stabilization	聚合物加固；聚合物的稳定化
polymere	复矿质；多矿物的
polymerisation	聚合；聚合作用
polymerization	聚合；聚合酌
polymerizer	聚合剂；聚合器；高温焙烘机
polymictic rock	复成分碎屑岩；复矿碎屑岩
polymigmatite	多相混合岩；多混合岩
polymolecularity	多分子性；高分散性
polymorphism	多型性；同质多象；同质异象；多形现象；多晶型现象；多晶形
polymorphous	多晶型的；多形的
polymorphous series	同质多象系列；多型系列
polynia	冰沼湖；冰穴
polynigritite	细分散煤化沥青；复煤化沥青
polynomial interpolation	多项式插值；多项式差分；多项式内插法
polynuclear	多核的；多环的
polynuclear compound	多环化合物；多核化合物
polyphase	多期；多阶段；多相
polyphyly	复系；多系
polysulphide	复硫化物；多硫化合物
polysulphide sealant	多硫化物密封剂；多硫化物密封胶
polytope	多面体；可剖分空间；多胞形
polytrope	多变性；多变曲线
polytropic index	多变指数；多方指数；多层球指数
polytropy	多变现象；多变性
polyvinyl chloride pipe	聚氯乙烯水管；硬胶喉管
polyvinyl chloride wind shutter	塑胶防风板；硬胶防风板
polyxene	粗粒铂矿；自然铂
pondage correction	蓄量改正；外泄流；挖泥工
ponded stream	阻塞河；拦阻河
ponded water	拦蓄水；积水
pony	小型的；辅助的
pony truck	小车；小型转向架
ponzite	霓辉粗面岩；斜霓粗面岩类
pool area	工作面积；池塘面积
pool geyser	间歇泉池；塘式间歇泉
pool of boiling water	沸水塘；喷泉塘
poop shot	低速带测量；风化层爆炸
poor aggregate	不良级配集料；劣质骨料
poor aquifer	弱含水层；不良含水层
poor compactibility	低压实性；低成型性
poor concrete	贫混凝土；少灰混凝土；水泥少的混凝土
poor construction	施工不良；不良构造
poor lime	贫石灰；劣质石灰
poor mud	劣质泥浆；劣质钻井液；贫钻泥
poor oil	低质量油料；低质油；贫油
poor ore	贫矿石；贫矿
poor reflector	不良反瘠土；施工条件不好的土壤
poor subgrade	不良路基；松软地基；软弱地基
poor subsoil	软弱底土；软弱地基土；不良地基；劣质底土
poor value	低值；低金属矿石
poorly graded	分选不好的；分级差的；级配不良的
poorly interconnected gravel	胶结不好的砾石；胶结差的砾石
poorly permeable strata	弱透水层；弱透水岩层
pop off	压力放泄阀；突然放出；突然离去；突然爆开；爆脱
pop shooting	小炮眼爆破；二次爆破
pop shot	修整炮眼；小炮眼；二次爆破的炮眼
pop valve	突开阀；紧急阀
pop-off valve	保险阀；压力放泄阀；安全阀；泥浆泵安全阀
poppet	滑轮支座；车床头部；随转尾座；阀提动头；托架；枕木
poppet head	井架；随转尾座

popping	激发；突然鸣叫；突然跳出；二次爆破	porousness	空隙率；孔隙率；空隙度；多孔性
popping rock	岩石爆裂；无旋运动	porphyrite	玢岩；斜长斑岩；斑状的；微闪长石
popple	起光翻滚；波动；起伏	porphyrite rock	斑状岩石；斑岩
porcelain clay	瓷土；高岭土	porphyritic	斑状；斑状的
porcelain insulator	陶瓷绝缘器；陶瓷绝缘子；瓷绝缘子	porphyritic texture	斑状结构；斑状组织
porcelain jasper	瓷碧玉；白陶石	porphyroblast	斑状变晶；变斑晶
porcelainite	莫来石；瓷状岩；烧变岩	porphyroblastic texture	变斑晶结构；斑状变晶结构
porcelanite	白陶石；燧石页岩；陶瓷状变岩	porphyroclast	残碎斑晶；碎斑
porcellanite	瓷状岩；白陶土；中柱石；陶瓷状变岩	porphyroid	残斑岩；斑状变质岩，残斑变质岩；残斑状的
pore	微孔；气孔；毛孔；孔隙；空隙	porphyry copper	次生辉铜矿；斑岩铜矿；斑岩铜矿
pore air pressure	孔隙气压力；孔隙气体压力	port and harbour engineering	港湾工学；港口工程
pore compressibility	孔隙压缩性；孔隙压缩率	portabelt	轻便胶带输送机；移动式皮带运输机
pore configuration	孔隙形状；孔隙结构	portability	可携带性；轻便性；移植；可移植性
pore diameter	孔径；孔隙直径	portable air compressor	挖掘起重两用机；便携式空气压缩机；移动式空气压缩机
pore entry radius	过水孔道半径；孔隙入口半径	portable apparatus	手提式装置；轻便装置
pore filling	孔隙充填；填孔	portable barrel loader	轻便装桶机；移动式装桶机
pore geometry factor	孔隙几何系数；孔隙几何因数	portable boom conveyor	铸铁压重；移动式长臂输送机
pore pressure	孔隙压力；孔隙水压力	portable brake	轻便闸；轻便制动器
pore pressure cell	孔隙水压力计；孔隙水压力盒	portable cement pump	移动式水泥泵；堰槛
pore pressure gauge	孔隙水压力计；孔隙压力计	portable chute	轻便溜槽；移动式溜槽
pore pressure measurement	孔隙水压力测量；孔压量测	portable compactor	手提击实仪；手提夯实仪
pore pressure metre	孔隙压力计；渗压计	portable compressor	移动式压气机；移动式压缩机
pore ratio-pressure curve	孔隙率-压力曲线；孔隙比-压力关系曲线；压缩曲线	portable conveyor	移动式运输机；轻便输送机
		portable derrick crane	移动式人字起重机；游标规
pore size	孔隙尺寸；孔径；孔径大小	portable dipmeter	携带式地下水位测定仪；可携式地下水位测定仪
pore size distribution	孔隙大小分布；孔径分布		
pore space	孔隙；孔隙空间；孔腔；孔洞	portable drill	轻便钻机；移动钻；轻便钻床
pore suction	孔隙吸力；负孔压	portable drilling rig	轻便式钻机；移动式钻机
pore throat	孔喉；孔喉喉道	portable instrument	轻便仪表；便携式仪器
pore volume	孔隙空间，孔隙体积；空隙量；孔隙容积	portable loader	轻便装载机；移动式装载机
pore water	孔隙水；间隙水；软泥水	portable magnetometer	袖珍磁力仪；便携式磁力仪
pore water head	孔隙水压头；孔隙水头	portable mast	便移式井架；轻便井架
pore water tension	孔隙水张力；孔隙压力比	portable pulling machine	轻便拉出机；轻便式提管机；便移式拔管机
pore-free mass	无孔隙体；外伸叉架		
pore-solid ratio	孔隙-固体比；孔隙比；空隙比	portable pump	轻便水泵；移动式水泵
poriness	多孔性；疏松性；孔隙度；孔隙率	portable rig	移动式钻机；轻便钻机；移动式钻机支架；轻便钻探设备
porodine	非晶质岩；胶状岩		
poroelasticity	孔弹塑性；孔弹性；多孔弹性力学	portable seismograph	轻便地震仪；流动地震仪；便携式地震仪
poromechanics	多孔介质力学；孔隙力学		
pororoca	河口高潮；河口涌潮	portable shear box	轻便剪切盒；手提式剪切盒
porosimeter	孔隙率计，孔率计；岩石孔隙率仪	portable-pipe irrigation	移动式水管灌溉；总线分配器
porosimetry	孔隙度测定；孔隙度测量法；孔隙度计	portage	搬运；运输；水陆联运；运费
porosity	孔隙性；多孔性；孔隙率；气孔；砂眼	portal crane	高架起重机；龙门吊车；门式起重机；桥式启闭机
porosity factor	孔隙率；孔隙度因数		
porosity gradient	孔隙变化率；孔隙梯度	portal shed	井口棚；井口房
porosity percent	孔隙百分比；孔隙比率	portfire	点火装置；发射装置
porosity ratio	孔隙比；孔隙度；孔隙率	Portland blast furnace slag cement	矿渣硅酸盐水泥；矿渣水泥
porous	多孔的；多孔状		
porous aquifer	多孔含水层；孔隙状含水层	Portland pozzolan cement	火山灰硅酸盐水泥；火山灰水泥
porous drain	多孔排水管道；多孔排水管		
porous friction course	多孔道路面层；多孔面层	portlandite	氢氧钙石；羟钙石
porous ground	多孔隙土；多孔地层；多孔岩层	position angle	方位角；星位角
porous membrane	多孔膜；多孔滤膜	position buoy	标志浮标；定位指示浮标
porous reservoir	孔洞型储层；孔隙型储层	position data	位置数据；位置坐标
porous rock	多孔岩石；透水岩石	position head	势头；位置水头；落差
porous soil	多孔土；多孔隙土；大孔土	position sensor	位置传感器；高度传感器
porous spring	孔隙泉；多孔泉	position vector	位置矢量；位置向量；方位向量
porous stone	多孔石；透水石	positional efficiency	空气调节效率；位置效率
porous structure	多孔结构；多孔构造		

positional notation	按位记数法；位置记数法
Positional number	位置数；定位数
positioner	位置控制器；定位器
positioning	放在适当位置；定位；安放工程；配置
positioning screen plate	定位螺钉；调整螺钉
positioning system	定位系统；位置控制系统
positive acceleration	加速度；正加速度
positive action	强制作用；肯定动作
positive bias	正偏压；正偏置
positive blower	压入式扇风机；旋转鼓风机
positive carrier	正调制载波；正电载荷流子
positive confining bed	正向赋压层；隔水顶板；正封闭层；上承压层
positive displacement meter	正排量流量计；容积式流量计
positive displacement motor	正排量液压马达；正排量电动机
positive drive	强迫驱动；正传动
positive effect	明显效果；正面效应
positive element	上升构造单元；相对上升区；阳性元素
positive elongation ion	阳离子；正离子
positive elongation movement	正向运动；上升运动；海面上升
positive elongation ore	可靠的矿石储量；证实的矿石储量；肯定储量
positive feed	强迫给进；强制进料
positive intensified image	增强图像；增强正像
positive landform	正地形；隆起地形
positive mold	正压塑模；阳模
positive motion	强制运动；无滑运动
positive peak	最大正值；正峰值
positive plate	正极；阳极板；正极板
positive pole	正极；阳极
positive potential	正电位；正电势；正势；正位
positive pulse	正极性脉冲；正脉冲
positive ray	正射线；阳射线
positive reaction	阳性反应；正反应
positive reserve	动储量；探明的储量；确切储量
positive sealing	可靠密封；刚性密封
positive slope	顺坡；正斜率
positive stop	正确到位；固定挡板；固定式限位器
positive terminal	正极接线柱；阳极线接头
positive-pressure perforating	正压射孔；增压射孔
possible ore	可能的矿石；地质储量；远景储量；可能矿量
possum belly	工具箱；密封导槽；泥浆储罐
post analysis	后分析；事后分析
post and panel structure	镶板式结构；总胀升量
post drill	架式风钻；支架式凿岩机
post evaluation	后评价；后评估
post hole	埋桩坑；埋杆孔；浅钻孔；柱穴；电线杆坑；浅井
post hole digger	柱桩孔挖孔器；大型螺旋挖坑钻
post horn	顶板锚杆销；栅柱顶面上的方栓
post mineral	成矿期后的；矿化后的
post precipitation	后沉淀；沉淀后
post puncher	立柱支撑的风镐；架柱式穿孔机
post stress	后应力；残余应力
post testing	事后试验；效果检测
post-accumulative	聚集后的；堆积后的
post-analysis of survey	测后分析；调查后的分析
post-and-stall method	房柱法；房柱式开采法
post-blasting	爆破后的；放炮后的
post-buckling behavior	后期压屈特性；屈服后性状
post-Cambrian	寒武纪后的；后寒武纪的
post-construction monitoring	竣工后监测；营运期监测；完工后监测
postcure	后固化；后硬化
post-deformational stage	变形后阶段；后变形阶段
post-earthquake	地震后的；余震
postemphasis	后加重；去加重
post-event data reduction	震后数据简化；震后数据整理
post-flush production	压裂后产量；高峰后稳产值
postforming grade	后成形；热压成形
postforming sheet	后成形板材；热压成形板材
postheating	随后加热；焊后加热；加热
post-Hercynian	海西；期后的
posthole well	勘探钻孔井；浅井；浅坑；柱坑；杆坑
posthumous fold	复活褶皱；继承性褶皱
posthumous movement	复活运动；继承运动
posthumous structure	复活构造；继承性构造
posting	回收支柱；回采煤柱；安设支柱
posting hole	联络小巷；联络小石门
postkinematic	造山运动后的；构造期后的
post-mortem routine	算后检查程序；事后分析例程
postmortem transport	死亡搬运；生物死后的搬运
post-mounted drill	柱架式钻机；柱架式凿岩机
post-orogenic basin	造山期后盆地；造山运动后的盆地
postorogenic metamorphism	褶皱后变质作用；造山期后变质作用
post-placement settling	充填后；沉降
post-Pliocene	晚上新世；后上新世
postprocessor	后处理机；后处理程序
post-rift	断裂后的；后裂谷的；裂谷期后的
post-seismic deformation	震后形变；震后变形
post-stressed concrete	后张法混凝土；后张预应力混凝土
post-stressing	后加应力；后张的
post-stretching	后张拉；后张的
post-suturing	构造缝期后的；后构造缝的
post-tectonic	造山运动后的；构造期后的
post-tectonic crystallization	后构造结晶；构造期后结晶作用
post-tensioned concrete beam	后张混凝土梁；后张法预应力混凝土梁
post-tensioned concrete pile	后张法混凝土桩；后张预应力混凝土桩
post-tensioned prestressing	后张预应力；后张法预加应力
post-tensioned system	后张拉设备；后张法
post-tensioning	后张；后张拉；后加拉力
pot clay	陶土；粗陶土；白陶土
pot life	使用时限；活化寿命
potamic	河川的；江河的
potamogenic deposit	河流沉积；河口沉积；江口沉积
potamogenic rock	河成岩；河流沉积岩
potamology	河流学；河川学
potash	钾碱；碳酸钾，苛性钾；草碱，木灰；钾质的；钾云母
potash alteration	钾质蚀变；钾化作用

potash mica	钾云母；白云母
potash-bentonite	变质斑脱岩；变蒙脱石
potassa	氢氧化钾；苛性钾
potassium and magnesium salt rock	钾镁质盐岩；饱和磁化强度
potassium dating	利用钾鉴定时代；钾氩年代测定法
potassium-argon dating	钾氩年龄测定；钾氩年代测定
poteclinometer	连续井斜仪；连续测斜仪
potential area	勘探地区；可能含油地区；潜藏区
potential crimp	潜在卷曲；潜在褶皱
potential damage	潜在损害；潜在性损伤
potential deformation energy	变形位能；弯形潜势能；潜在变形能；潜在应变能
potential difference	势差；位差；电势差；电位差
potential differential	势差；位差；电势差；电位差
potential drop	势降；位降；电势降，势位降；电位降
potential electrode	测量电极；电位电极
potential evaporation	可能蒸发；潜在蒸发
potential failure surface	潜在破裂面；潜在破坏面；潜在破损面
potential fault	隐伏断层；潜在断层
potential field	位场，势场；守恒场
potential flow	潜流；势流
potential gradient	水力梯度；电位梯度；势梯度，势能梯度
potential hazard	潜在危害；潜在危险；隐患
potential head	位势水头；位能，位差；静压头
potential infiltration rate	潜在入渗强度；可能入渗强度；入侵量
potential instability	位势不稳定；污水泵
potential line	等势线；等水位线
potential power	水力蕴藏量；动力蕴藏量
potential pressure	位置高差，位差；静压头
potential problem	潜在问题；位势问题
potential production	潜产量；潜在产量
potential site	潜在坝址；可能坝址
potential supply	潜在供应量；储备量
potential test	产能测试；潜能测验
potential transformer	仪表用变压器，电压互感器；变压器
potential transpiration	可能蒸腾；可能散发
potential trouble measure	预检；预防
potential yield	可能出水量；潜在产量；潜在开采量
potentially dangerous	可能危险；潜在的危险
potentially unstable	可能不稳定；潜在的不稳定
potentiate	稳压器；加强，使更有效力
potentiometric surface	等势面；等测压面；静水压面
potholing	洞穴探险；挖坑，挖洞
potrero	岸外堆积堤；沿岸淤积埂
potstone	块滑石；粗皂石
pound stone	煤层的岩石底板；煤层的黏土底板
pounder	一磅重的物；捣具，捣锤
pounding	撞击；捣；捣碎
pour	浇注段高，一次浇注量；浇注，注，倾注
pour point	倾点；流点
pour point test	倾点试验；流点试验
pourability	流动性；可浇注性
pouring	浇注；倾倒，灌，灌注；泼
pout	坑木回收器；回柱器
powder blast developing	爆破巷道掘进；爆破开拓
powder helper	分发炸药辅助人员；送炸药工
powdered charge	装药；装药量
power	功率，电源；幂；乘方；动力
power auger	机械钻；动力钻机
power brake	机动闸；电力制动器
power cable	动力电缆；电源电缆
power channel	发电明渠；开机频道
power conduit	发电管道；动力管道
power consumption	动力消耗；功率消耗；耗电量
power cost	电力费；动力费
power cowl	动力装煤板；动力挡煤板
power crane	动力吊车；动力起重机
power delivery	动力输送；功率输出
power detector	强信号检波器；功率检波器
power distribution system	电力分配系统；配电系统
power drag scraper	动力拉索扒矿机；动力拉索刮土机
power drill	机力钻机；机力凿岩机
power drive	电力拖动；动力拖动
power driving	动力打桩；动力传动
power dual operation	机械操作；动力操作
power dump	切断电源；电力中断
power efficiency	效率；能量利用率；能量效率
power engineering	动力工程；动力学
power equipment	动力设备，电源设备；供电设备
power failure	发动机故障；供电中断，停电；电力故障
power feed	机动给进；机动进料；自动进刀
power generating station	电站；发电厂
power haulage	动力运输；动力拖运
power house	发电站；发电厂；动力车间
power interruption	电源中断；供电中断
power jacking	动力升举；机械起重
power level	功率级；幂级；电平
power line	电力线；输电线；电源网路；动力线路
power loader	装载机；长壁工作面采煤机
power loading	动力装载；动力负载
power loss	功率损耗；动力损失；能量损失
power main	电力干线；输电线
power margin	备用功率；动力裕度
power meter	瓦特表；功率表
power of test	测试能力；检验效能
power project	发电工程；动力工程
power set	动力装置；动力设备；发电机组；供电装置
power setting	动力装置；动力调整
power shovel	动力铲；机铲；电铲，掘土机；单斗挖土机
power shut-off	供电停止；停电
power source	能源；功率源；电源；动力源
power station	发电厂；发电站；动力厂
power strip	电源板；配电盘
power supply	电源；动力源；供电；动力供应
power tamper	机动夯；动力夯
power tong	动力大钳；液压大钳
power transmission	电力传输；动力传送；输电；动力传输
power valve	发射管；功率管
power wrench	动力扳钳；机械扳手；动力扳手
power-absorbing attenuator	能量吸收衰减器；吸能衰减器
powered axle	主动轴；传动轴
powered prop	机械支柱；液压支柱

powered roof support 液压支架；动力顶板支护
power-factor correction 校正功率因数；改善功率因数
powerful 有力的；强有力的
powering 功率估计；进线
power-level 电平；功率电平
power-line carrier 动力线载波电流；动力线路载波
power-off relay 故障继电器；电源切换继电器
power-operated 电力驱动的；动力驱动的；机动的；动力操作的
power-pack 电力部分；电源部分
power-plant 发电厂；动力装置
power-quantity relation 通风马力和风量的关系；风压和风量关系
power-to-volume ratio 动力-体积比；动力-重量比
pozzolan 火山灰；火山灰水泥；白榴石火灰
pozzolan cement 火山灰水泥；普索兰水泥
practical completion 实际竣工；实际完工
practical duty 实际能率；实际功率
practical formulation 实用配方；实际规划
practical partition size 实际分级粒度；分配粒度
practical porosity 实际孔隙度；有效孔隙度
practical sustained yield 实际持续开采量；实际持续出水量
practice code 实用法规；实用代码
prairie soil 草原土；普列利土；湿草原土
preaccentuator 预增强器；频率校正电路
preact 提前；超前
preadmission 预进气；提前进气
pre-aeration 预充气；预通风
pre-aeration tank 预曝气箱；预曝气池；泄水闸过流量；泄水道闸门室
preanalysis 预分析；事前分析
pre-Andean 安第斯期前的；前安第斯时期的
prearrange 预先安排；预定
preassemble 预装；预先安装；预先组装
pre-blast 爆破前的；放炮前的
precambrian 前寒武纪；前寒武系
pre-Cambrian boundary 前寒武纪边界；前寒武纪界线
pre-Cambrian era 太古代；前寒武纪
precast 预铸的，预浇注，预制的；装配式预制；预制构件
precast building method 预制混凝土方块建筑方法；预制混凝土件建筑法
precast facade 预制外墙；预制面墙
precast product 预制品；预制件
precast yard 预制场；预制件工厂；预制厂
precautionary underpinning 预防性托换；预防性基础托换
preceding settlement 前期沉降；前期决算
precession 进动；旋进；岁差
precharge 预加压力；预先充电
precious garnet 贵榴石；镁铝榴石；铁铝榴石
precipice 悬崖；峭壁
precipitated water 降水，大气降水；雨水
precipitating action 沉淀作用；沉降作用
precipitating zone 沉淀区；沉淀带
precipitation 沉淀；析出；脱溶；降雨，降水
precipitation excess 降水径流量；剩余降水
precipitation facies 沉淀相；沉积相
precipitation gauge network 雨量站网；组合梁

precipitation hardening 沉淀硬化；弥散硬化；析出硬化
precipitation tank 沉淀池；沉淀槽；沉降罐
precipitation threshold 沉淀度；临界沉淀点
precipitator 沉淀器；集尘器，聚尘器，吸尘器；沉淀剂
precision 精确，精度；精密，精密度
precision casting 精密铸件；精密铸造
precision depth recorder 精密回声测深仪；精确深度测定记录器
precision indicator of meridian 子午线精密指示器；正北指示器
precision instrument 精密仪表；精密仪器
precision measurement 精密测定；精密测量
precision tolerance 精确裕度；精度容差
preclusion 排除；遮断；阻止；预防
pre-coalified 前期煤化；预煤化；先成煤化；预炭化
preconcentration 预富集；预精选；再选
preconception 预想；先入之见
precondensator 预冷凝器；预缩合器
precondition 前提；预处理；预先安排；预调湿处理；悬浮液处理
preconditioning 预处理；预调节；预调和
preconsolidated soil 先期同结土；预固结土
preconsolidation 先期同结；预压实；预固结；前期固结；压实试验
preconsolidation pressure 先期固结压力；前期固结压力；天然固结压力
pre-consolidation settlement 先行下沉，预先下沉；预固结沉降
precontinent 大陆前缘；前大陆；大陆架
precoordination 先组式；预协调
precursor 预兆，前兆；前驱波；先驱，前驱；预先审查
precursor fault 前兆断层；初始断层
pre-cutting method 预切槽法；预拱法
predazzite 水镁结晶灰岩；水滑结晶灰岩
predeformation 前期变形；预变形；先期形变
pre-depositional porosity 沉积期前孔隙；沉积期前孔隙度
pre-development 预先开拓；预先开采
pre-diagenesis 成岩前沉积作用；预成岩作用
predict 预示；预测；预报
predictability 可预报性；可预测性；可预推性
prediction error 预报误差；预期误差
prediction method 预测方法；预计法
prediction of macroseism 强震预报；巨震预报
predictive deconvolution 滤除雷达反射波法；预测反褶积
predictor 预测器，预报器；预测者，预报员；预测值，预报因子，预测因子
prediluvian 前洪积世；洪积前的
pre-displacement 预驱替；预位移；前移
predissociation 预离解；预分离
predistorter 前置补偿器；预修正电路
predistortion 预失真；预矫正
predominance 优越；主导性；优势；显著
predominant 优势的；主要的；卓越的；占优势的
predominant current 优势流；盛行流
predominant direction 主方向；盛行方向
predominant feature 优势性能；重要性能

predominant formation	主要岩层；最厚岩层
predominant frequency	主频率；卓越频率
predominant wave	优势波；盛行波
predomination	占优势；统治；突出
pre-draft	预拉伸；预牵伸
pre-drain	预先排泄；预抽放；预先排水
pre-drawing process	预拉伸过程；预牵引过程
predrilling	打开口眼；打超前钻孔；预钻孔；超前钻进，超前钻探
pre-earthquake	震前；前震
pre-earthquake deformation	震前形变；震前变形
pre-emphasis	预加强；预加重
pre-estimate	预测；预估
pre-event memory	震前记忆；震前存储器
pre-examination	预试；预先检查
pre-existing fault	原有断层；预先存在的断层；震前原有断层
pre-exposure	预曝光；先期曝光
prefabricate	预制；预先构思；预制构件
prefabricated	预制的；预制品
prefabricated notch	预制切槽；预制裂纹
prefabricated structure	装配式结构；预制结构
prefabricated subaqueous tunnel	预制管水下隧道；预制水底隧道
prefabrication	预先制造；预加工
pre-feasibility study	预可行性研究；初步可行性研究
preferential adsorption	选择性吸附；优势吸附
preferential crystallization	优先结晶；选择性结晶
preferential flow path	优选流径；优先流槽；优势流路径
preferential orientation	最佳取向；优先定位；择优定向
preferential path	高渗透通道；优势通道
preferred	优先的；优选的；择优的
preferred orientation	择优定向；择优取向；优选方位；定向方位；最佳取向；优先取向
prefilter	预过滤器，预滤器；前置过滤器
prefiltration	初过滤；预过滤
pre-flawed	预制裂纹的；预制缺陷的
preflush	前置液；预冲
preform	预先形成；预先决定；初步加工；预先成型
preformation	预先形成；预先成型
pre-fossilization	石化前作用；前石化作用
pregeologic	地质史前的；前地质时期的；地质前的
pre-hardening	初凝；初硬化；预硬化
prehistory	史前学；史前史；史前考古学
prehnite	葡萄石；芙蓉石
pre-ignition	预点火；先期点火
preinstallation	安装前；预装；安装前试验
preliminary	初步的；初级的；预备的，预先的
preliminary adjustment	初步调整；初调
preliminary breaker	初碎机，粗碎机；预先压碎机
preliminary breaking	准备破碎；预破碎；预先压碎
preliminary computation	初步计算；概算
preliminary crusher	初碎机，粗碎机；初轧碎机
preliminary crushing	预先压碎；选前破碎；粗碎
preliminary data	预备数据；初级数据；原始数据
preliminary draft	预牵引；预拉伸
preliminary evaluation	初步评价；初步评估
preliminary exploration	初步勘探，普查；石油预探；初勘
preliminary geotechnical investigation	初步勘察；初步岩土工程勘察
preliminary grinding	预磨；初磨
preliminary grouting	预灌浆；预注浆
preliminary operation	试运转，试转；初步预先操作
preliminary phase	初步阶段；初相，初始相
preliminary shock	首震；前震
preliminary survey	初测，初步勘测；踏勘；普查；初步调查
preliminary testing pile	先期试桩；预试桩
preliminary treatment	预处理；初级处理；初步处理
preload	预先加料，预先加重；预加载；预压
preloaded soil	预压土；先期加载土
preloading	初负载；预先加重，预压
preloading compaction	预加荷载压实；预压加固
premature	未成熟的；过早的；不到期的；早熟的
premature explosion	过早爆炸；过早起爆；早爆
premature failure	过早失效；过早破坏
premature firing	过早起爆；提前点火；过早点火
premature ignition	提前点火，先期点火；先期发爆；过失放炮
premature set	先期凝固；不到位坐封
premature setting	先期凝固；不到位坐封
premature water breakthrough	过早见水；过早水窜
pre-mine stripping	初始剥离；回采前剥离
pre-mineral	矿化前的；成矿前的
pre-mineralization	预成矿作用；初期成矿作用；矿化期前
pre-mining technology	采前技术；采前技术措施
premium coal	高级煤；高价煤
premixing	预混合；顶拌
premonition	前兆；预感；预警
premonitory	前兆的；预兆的；预感的
premonitory symptom	征兆；前兆；预兆
premoulded	预铸模的；预模制的
premoulding	预压制；预制模
pre-notched	预制裂纹的；预先切槽的
preorogenic	造山运动前的；造山前的
preorogenic magmatism	造山期前岩浆活动；造山前的岩浆作用
prepacked concrete	预填骨料混凝土；压浆混凝土
preparation cost	加工费用；制备成本
preparation equipment	选煤设备；选矿设备
preparation of construction	施工设备；施工准备
prepared reserves	准备储量，预备储量；采准储量，采准矿量
prepared size	制备粒度；配制粒度
preplaced-aggregate concrete	预填骨料混凝土；灌浆混凝土
pre-plate	板块期前的；前板块期的
prepolymer	预聚合物；预聚物
preponderance	优势；优越
preponderant age	优势年龄；主要年龄
pre-pressurize	增压；密封
preprocessing	预处理；预加工；前处理的
preprocessor	预处理器，前处理器；准备器；预处理程序；前处理程序
preproduction	试制；小批量生产
pre-production cost	生产前费用；试制费用
preproduction model	试制模型；样机

prepump 预抽泵；前置泵，前级泵
pre-quake 临震的；震前的
prereduction 预先还原；预还原
pre-reinforcement 超前锚固；预加固
prescience phenomenon 预知现象
pre-selected site 预定选点；预定工作点；预定场地
preselector 预选器；高频预先滤波器
presence 存在；出现；出席；出庭
presentation 提出，提交；描述，介绍，陈述；呈现，显示，表示；外观，形式
presentation drawing 示意图；草图
preservative substance 防腐剂；防腐料
preset adjustment 预调谐；预调整
preset device 自动导航仪；程序机构；预订装置
preset mechanism 预调机构；程序机构
preset position 预调位置；预定位置
preset time 规定时间；预定时间
pre-setting 预凝，初凝；预先调节的；预先配置的
pre-shearing 预裂爆破；预剪
pre-slit method 预开沟槽法；预沟施工法
presplitting 预裂；预裂法；预裂爆破
press 压缩，压挤，冲压，压机，压力机；冲床；出版社
press filter 压滤机；压滤器
press forging 压力锻造；压锻
press quenching 压力淬火；模压淬火
pressboard 压板，压制板；绝缘用合成纤维板
pressed 加压的；压缩的
pressed peat 泥炭砖；压型的泥炭
pressing 压制；模压；冲压件；压药
pressing crack 压裂；压制裂纹
pressiometer 压力仪；旁压仪；横压仪
pressostat 恒压器；稳压器
pressure 压力；压强
pressure accumulator 蓄压器；压气罐
pressure altimeter 气压高度表；气压测高表；气压测高计
pressure altitude 气压高程；压力高度
pressure application 压力法；施压法
pressure arch 压力拱；自然平衡拱
pressure ascending zone 应力升高区；压力升高区
pressure at external boundary 外边界压力；外缘压力
pressure behavior 压力动态；压力的行为
pressure blower 压入式扇风机；压力鼓风机；空气压缩机
pressure bomb 压力计；压力室法
pressure breccia 构造角砾岩；压碎角砾岩
pressure build-up curve 压力恢复曲线；压力增加曲线
pressure burst 压裂；压碎；岩石破裂；压力岩爆，岩爆
pressure calibration 压力校准；压力标定
pressure capsule 压力计，地压计，测压计；压力传感器；压力囊，水力囊
pressure casting 压力铸造；压铸
pressure cell 压力传感器；压力盒；压敏元件；压力计，测压计
pressure chamber 气压调节室；吸气装置；压力室
pressure conduit 压力水管；有压管道
pressure container 压力容器；耐压容器
pressure control 压力控制；压力调整；地压控制；地压管理
pressure core barrel 保压岩心筒；恒压取芯管
pressure cushion 压力枕；扁平千斤顶
pressure cylinder 压力缸；增压缸
pressure damping factor 压力衰减因子；压力阻尼系数
pressure deficiency 压力衰减；压力降落
pressure depletion 压力衰减；压力降落，井压递减
pressure detector 压电检波器；压电地震检波器；压力探测器
pressure difference 压差；气压差
pressure differential meter 差压计；压差测量仪
pressure dissolution 压力融解；压溶
pressure drawdown 压力下降；压降
pressure drilling 加压钻进；压力钻进
pressure drop 降压，压力降；压力下降
pressure due to breaking wave 碎波压力；破浪电池
pressure effect 压力效应，压力影响；压力；压强雾化
pressure element 测压元件；感压元件
pressure fall 压力降；压降
pressure field 压力场；气压场
pressure flow 压力流；有压流；逆流，回流
pressure fringe 压力边；压力影
pressure gauge 压力计，压强计；压力器，增压表；压力表，测压表；压力验潮仪
pressure grade line 压头线；游丝
pressure grouting 压力灌浆；压浆
pressure history 压力历程；压力变动情况；压力变化曲线
pressure instrument 测压仪表；压力计；压力表，压力仪表
pressure integrity 压力完善性；压力密封性
pressure intensifier 增压器；压力倍增器
pressure leaching 压滤作用；加压沥滤
pressure level-off 压力趋于稳定；压力趋稳
pressure limit 压力限制；极限压力；压力范围
pressure load 压力载荷；压力荷载
pressure loss 压力损失；压力降
pressure maintenance 压力保持；保持压力
pressure manometer 差示压力计；压力计
pressure meter 压力计；压力表
pressure mold 压模；压辊
pressure of loose surrounding rock 松动围岩压力；散体地压
pressure on foundation soil 地基土承压力；基底压力
pressure parting 水压破裂煤层；受压裂隙；压裂；压力劈开
pressure peak 压力突升；压力激波；压力最大值；压力峰
pressure pile 压入桩；气压打桩；压力灌注桩
pressure plate test 压力板试验；加压板试验
pressure reducing reservoir 降压水库；压力降低的储层
pressure reduction 减压；压力减少；减压量；减震器；气压换算
pressure regulating valve 给气调整阀；调节阀
pressure regulator 调压器；压力调节器
pressure release 压力释放；压力松弛
pressure relief 压力分散；压力缩小；卸压；降压；解除压力
pressure relief valve 压力排放阀；泄压阀；减压安全阀

pressure relieve	减压；卸荷
pressure reliever	压力调节器；压力缓和装置
pressure reservoir	压力风缸；压力容器；蓄压器
pressure resistance	抗压力；压阻力，压阻；耐压性
pressure ridge	挤压脊；高压脊；冰脊
pressure ring	压力环；耐压环；压缩环
pressure sand filter	压力砂滤器；压力砂滤池
pressure shadow	压力影；压力边；压力裙
pressure solution	压力溶解作用；压溶作用；压溶
pressure spring	压力弹簧；自流泉
pressure surge	液压冲击；压力波动；压力冲花
pressure texture	压碎结构；碎裂结构；压缩结构；压力结构
pressure thief	井底压力取样器；压力取样器
pressure threshold value	压力阈值；压力门限值；压力界限值
pressure transducer	压力传感器；压强传感器
pressure transformer	压力变换器；变压器
pressure tube	测压管；耐压管；压力管
pressure tubing	压力管道；压力管道铺设
pressure unit	压力单位；增压装置；压力传感器；压力变送器
pressure variation	压力变动；气压变量
pressure vault	自然平衡拱顶；陷落平衡拱顶
pressure vessel	压力容器；高压釜；高压容器
pressured anchor	预应力锚杆；预应力锚固
pressureless groundwater	无压地下水；非承压地下水
pressuremeter	旁压仪；横压仪；压力表
pressuremeter limit pressure	旁压仪极限压力；旁压试验极限压力
pressuremeter modulus	旁压仪模量；旁压模量
pressuremeter test	旁压试验；横压试验
pressure-release jointing	压力释放节理作用；释压节理
pressure-released zone	免压带；压力释放带
pressure-sensitive adhesive tape	压敏胶黏带；压敏胶带
pressure-void ratio curve	压力孔隙比曲线；压缩曲线
pressure-volume relation	风质和风量的关系；压力和体积的关系
pressuring	加压；试压
pressurize	增压；密封
pressurizer	增压装置；保持压力装置
prestage	预级；前置级
prestratigraphy	初级地层学；基础地层学
prestress	固结前压力；预应力；预拉伸
prestressed	预加应力的；施加应力的
prestressed element	预应力元件；预应力组件；预应力构件
prestressed ground anchor	预应力锚杆；预应力地锚
prestressed reinforcement	预应力筋；预应力钢筋
prestressed rock anchor	预应力岩石锚栓；预应力岩锚
prestressing anchorage	预应力锚固；预应力台座
prestressing force	预应力；预应力作用力
prestressing tendon	预应力筋束；预应力索；预应力钢筋束
prestretching	预先拉伸；先张拉
presumable	可推测的；可假定的
presumptive bearing pressure	假设承载力；推断承载力
pre-tectonic	构造期前的；造山前的
pre-tender estimate	回标前估算；标底
pre-tension	预加力；预张力
pretensioned prestressing	先张法预应力；圆顶堰
pre-tensioning method	预应力法；先张法
pretest	试验前的；预先检验；预试；预测试
pre-tightening force	预紧力；预应力
pretreatment	预处理，预先处理；前处理
prevent	预防；防备；阻止；防止
preventer pin	保险销；安全销；止动销
prevention	防止，预防，防治；妨碍；保护；制止，阻止
prevention method of outburst	防止突出的措施；预防爆发的方法
preventive works	防护工程；预防工作
prevision	预见；预知；预测
prewhitening	预白噪声化；预白噪声化
prian	白色软黏土；软白黏土
prill	颗粒状炸药，使…变颗粒状；金属小球
primacord	传爆索；导爆索；导爆线
primary arc	原始构造弧；原生弧
primary basalt	原生玄武岩；原始玄武岩
primary blast	原体爆破；首次爆破
primary blast hole	主要炮眼；首次爆破的主要炮眼；掏槽眼
primary blasting	原岩首次爆破；原矿体首次爆破；初始爆破；岩石初爆
primary breaking	初碎；粗碎；一次爆破
primary clay	原生黏土；残余黏土
primary cleaning	主选；初选
primary coefficient of permeability	原始渗透系数；主渗透系数
primary consolidation	主固结；初始固结，初期固结，初步固结
primary contrast	原始灰度；初始对比度
primary crusher	初级碎石机；粗碎机
primary crushing	粗碎；初碎；原始压碎；初次破碎；预破碎
primary data	主要资料；原始资料；原始数据；主要数据
primary dip	原始倾斜；始倾角
primary energy	初能量；原粒子能量；一次能源；初始能量
primary environment	原生环境；第一环境
primary faulting	原生断裂作用；原断裂
primary field	原始场；一次场；初始场
primary flat joint	原生平缓节理；原生板状节理
primary fluid inclusion	原生气液包裹体；原生流态包裹体
primary formation	基底岩层；原始地形；原始结构
primary fracture	原生裂缝；原有裂缝；初始断裂
primary jaw crusher	初碎颚式破碎机；颚式初碎机
primary lineation	原生线理；原生线条
primary magma	原始岩浆；母岩浆；原生岩浆
primary migration	初次运移；初始运移
primary oil	初级油；初生石油
primary orogeny	原始造山运动；主造山运动
primary peneplain	原始准平原；始准平原
primary phase	初相；原始相
primary platform	原始地台；初生地台
primary pollutant	原生污染物；一次污染物
primary porosity	原生孔隙度；同生孔隙
primary pressure	一次压力；初始压力；原生水压

primary process 初步处理；主要流程
primary production 原始生产；主要生产；一次采油量；初级生产
primary recovery 初次回收率；主要回收率；初次开采；初次回采；一次开采
primary reflection 原始反射；主反射；一次反射波
primary risk 一级风险；头等风险
primary rock 原成岩，原成岩石；原生岩，原生岩石
primary scale 主要比例尺；基本比例尺
primary sedimentation tank 初级沉淀地；初次沉淀池
primary separation 初选；主选
primary settling tank 初沉降槽；主澄清槽
primary shear plane 主剪力面；主切力面
primary slime 原生矿泥；原生煤泥
primary soil 原生土；原始土壤
primary standard 基本标准；主要标准；一级标准
primary stoping 初段回采；初始回采
primary stratification 原生层理；原始层理
primary stress 初始应力，原始应力，一次应力；主应力
primary stress field 初始应力场；主应力场
primary structural element 原生构造要素；初次构造要素
primary structure 原生构造；原始结构；一级结构
primary support 初期支护；第一次支护，初支
primary trench 初期槽段；一期槽孔
primary waste 初级废物；基本废物
primary water 内生水；原生水；初期水；原有水
primary wave 初波，初始波；初至波；地震纵波
prime amplifier 前置放大器；前级放大器
prime coat 底漆；透层；打底
prime coking coal 最佳焦煤；主焦煤
prime contractor 主承包商；总承包商
prime cost 主要成本，生产成本，直接成本，原价；价，进货价格；购置费
prime number 质数；素数
prime rate 最优惠利率；主要利率
primer 起爆剂；起爆药包；发火器，点火剂，雷管；起动注油器
primer cap 起爆雷管；雷管；雷管帽
primer cartridge 导爆药筒；带雷管的药卷；起爆炸药管；起爆药卷
primer charge 起爆装药；点火药
primeval 原始的；古老
priming pump 起动泵；起动注水泵
primitive atmosphere 原始大气圈；上古代大气
primitive rock 原生岩；原始岩
primordial 始生纪，寒武系，寒武纪；初生的，原始的
principal charge 基本装药；主要装药
principal component 主分量；主成分
principal datum 主水平基准；主基准面；水平基准面；高程基准面
principal deformation 主形变；主要变形
principal direction 主方向；主向
principal load 主力；主荷载
principal moment of inertia 主惯性矩；惯性主矩
principal normal 主法线；主法向
principal plane 主平面；主应力面
principal point 主点；投影中心；主要点
principal river 干流；主流；主要河流

principal stream 主流；干流
principle of consistency 一贯原则；一致性原则
principle of design 设计原理；设计原则
principle of equivalence 等效原理；等值原理；等价性原理
principle of optimality 最佳性原理；最优原理；最佳则；优化原理
principle of reciprocity 互易原理；互换原理
principle of similitude 相似法则；相似原理
principle of superposition 叠加原理；层序原理
prism 棱镜；棱柱体；棱晶；三棱形，棱形
prism cutting 角柱掏槽；桶形掏槽
prism plough 棱柱式刨煤机；角柱式刨煤机
prismatic 柱状；斜方晶系；棱柱状的；棱形的
prismatic class 柱组；柱体类
prismatic cleavage 柱状解理；柱状劈理
prismatic coke 柱状焦；棱柱焦
prismatic structure 柱状构造；似柱状构造
pristine 原始时代的；原始的；古代的
probabilistic analysis 概率分析；概率性分析；概率分析定性
probabilistic risk assessment 概率风险评估；概率风险判定
probabilistic uncertainty 概率不确定性；概率不定性
probability 概率；几率；或然率，或然性；可能性
probability distribution 概率分布；几率分布
probability estimate 概率估计；概率估计值
probability of failure 破坏概率；崩塌概率；失效概率
probability of occurrence 事件概率；发生概率
probability paper 概率纸；概率坐标纸
probability plot 概率关系曲线图；概率曲线
probability scale 概率分度；概率比例尺
probable deviation 可能偏差；或然偏差；概差
probable error 或然误差；概率误差；可能误差；概差；概率可能误差
probable performance curve 可能性能曲线；预期性能曲线
probable reserve 可期储量，可能储藏量；可能储量；概略储量；近似储量；概算储量
probable value 概值；几值
probe 探针，探头，测高仪；探测器；调查；探查；钻探
probe boring 试钻；钻探
probe drilling 试钻；勘探钻进；钻探
probe hole 探测孔；探查孔
probe method 试探法；探针法
prober 探测器；探针
probing 探测，探查，测试；钻探；槽探；坑探
problem formulation 问题形成；问题描述
problematic fossil 可疑化石；未定化石
problem-oriented language 面向问题语言；面向问题的语言
procedure control 工作技术监督；工序技术检查；过程的控制
proceed signal 进行信号；允许信号
proceedings 会报；会议录；会议论文集；学报
process 过程，进程；方法，工序；手续，程序；处理，加工
process chart 工艺程序图；工艺流程图
process control instrument 流程控制仪表；程序控制

仪表
process controller 过程控制装置；工艺操作控制器
process design 流程设计；工艺设计；过程设计
process engineering 加工工艺；操作程序；操作技术；工艺过程；程序工程；工艺学
process flow diagram 工艺流程图；流程图
process instrumentation 生产过程用检测仪表；生产过程检测仪表
process integration 工艺联合；流程整合
process load 处理量；工艺负荷
process monitoring 加工过程监控；过程监控
process specification 工艺说明书；加工规格
process steam 工艺用蒸汽；工艺物料流
process unit 工艺装置；工艺设备
process water 加工物污染水；生产用水；工业用水
process-controlled system 程序控制系统；连续调整系统
processed rock 加工石料；加工过的石料
processing 处理；加工；工艺设计；工艺过程；操作；作业
processing method 加工方法；选矿方法
processing plant 加工厂；加工车间
processing technique 选矿技术；选矿工艺；加工技术；处理技术
processing unit 加工设备；处理装置；处理机
processor 加工装置；处理机；处理程序；信息处理机；加工程序
processor camera 缩微照相机；缩微摄影机
prochlorite 铁斜绿泥石；蟹绿泥石
Proctor compaction test 普罗克特击实试验；普氏击实试验；普氏压实试验；击实试验
Proctor needle 普氏贯入针；普罗克托尔密度计
Proctor needle moisture test 普罗克特针测含水量试验；普氏针测含水试验
procurement 采办；采购；购置
prodelta 前三角洲；三角洲前沉积
produced coal 采出的煤；产煤
produced ore 采出的矿石；采出矿石
producer 发生器；发生炉；生产者；生产井；采矿者
producing aquifer 开采含水层；生产含水层
producing area 生产区；开采区；集水面积
producing characteristic 生产特性；开采动态
producing cost 生产费；成本
producing depth 生产层深度；层深度
producing formation 含矿层；产油层；生产层
producing gas-oil ratio 采出气油比；目前气油比
producing horizon 生产水平；产油层；生产层
producing interval 生产层段；生产井段
producing method 生产方法；开采方法
product separation 矿石分采；矿石选采
productimeter 起点距测量仪；求积仪
production 生产；制造；开采；开采量
production acceptance testing 产品接收检验；出厂试验
production aquifer 生产层；开采层
production capacity 生产能力，生产率；开采能力
production control 生产管理；生产控制；操作控制；产量控制
production cost 制造费；生产费；生产成本；开采费用
production method 生产方法；制造方法；开采方法
production mine survey 生产矿测量；生产矿井测量
production of land 土地辟增；土地开发

production outage 生产中断；生产暂时停工
production output 产量；流量；出力
production per point 每截齿产量；每钻头产量
production quota 生产定额；生产指标；生产限额
production rate 生产率；生产能力；产量
production schedule 生产进度表；生产计划；进度计划
production unit 产区；生产组
production well 抽水井场；生产井场；开采井场
production-well yield 生产井出水量；开采量
productive aquifer 富水含水层；有产能含水层
productive capacity 生产井出水量；生产开采量；油气产能
productive head 有效压头；发电水头
productive interval 生产层段；生产井段
productive reservoir 生产层；开采层
productive series 出水岩层；生产层群；产油气层系；生产层系
productive strata 生产层；产油层；油母岩层
profile 侧面；纵断面；外观，外形，轮廓
profile angle 齿形角；齿廓角
profile board 剖面样板；规；轨距规
profile coefficient 剖面系数；断面系数
profile gradient component 剖面梯度分量；剖面坡度分量
profile hole 异形孔；周边孔
profile in elevation 有高程注记的地形剖面图；纵断面图；高程纵断面；注有标高的纵断面图
profile in plan 带等高线的平面图；平剖面
profile levelling 剖面水准测量；水准测量
profile map 剖面图；纵断面图
profile method 断面法；二维方法
profile of equilibrium 平衡曲线；均衡剖面；平衡剖面
profile plotter 剖面绘图仪；剖面绘图设备
profile projection 侧投影；侧面投影
profile section 纵截面，纵剖面；断面图，剖面图；纵断面测量；测线段
profile shooting 地震测线勘探法；地震剖面勘探；测剖面爆破
profile spacing 剖面间距；断面间距
profile survey 剖面测量，纵断面测量；线路测量，线路调查
profiled bar 异型钢材；异型材；型钢
profileometer 验平器，路面测平器；表面粗糙度测定仪，表面光度计；断面仪
profiler 剖面测量系统；剖面测量仪；断面仪；廓线仪
profiling 剖面观测，剖面测量；断面测定；剖面法，绘地震剖面图；轮廓法；靠模
profiling device 靠模装置；靠模板
profiling of boundary 边界轮廓确定；绘制边界轮廓
profilograph 表面光度计；轮廓曲线仪；断面测定仪；纵断面测绘器，断面绘图器
profilometer 表面光度仪；轮廓曲线仪；粗糙度测定计；路面平整度测定仪
profluent stream 常流河；丰水河流
profound fault 深断裂；深断层
proglacial 冰状的；冰河时代的；冰期前的
proglacial lake 冰堰湖；冰川前缘湖泊
proglacial stream 冰前河；冰前河流
prognostic 前兆；预兆的；预报的；预测的
prognostic chart 预测图；淤泥滩

prognostic map	预测图；预报地图
prognostication	预测法；预示；预测；先兆；前兆
prognosticator	预测者；预言者
progradation	进积作用；推进；延伸
prograde	进积；正转
prograding sequence	岸进层序；进积层序
program source code	程序源码；程序源代码
programme	程序；方案；计划
programmer	程序设计员，程序员；程序器，编程器；程序设计器；程序机构；程序装置
programming	编程序；程序设计；规划
programming language	程序设计语言；程序编制语言；编程语言
programming system	程序设计系统；程序系统；编程系统
progress	进步；进程，进展；进度；进尺速度
progress chart	进度表；进度曲线图；工作进度表；施工进度表
progress report	进度报告；进度报告书
progress schedule	进度计划表；进度时间表；进度计划
progression	级数；前进；进展；数列
progression agent	增效剂；改进剂
progressive	前进的；发展的；递增的；分级的；累进的
progressive aging	渐进时效；分级时效
progressive approximation	渐进近似法；渐进求近法
progressive deformation	递增变形；累积变形；连续变形；递增变形；前进变形；渐进变形
progressive development	渐进发展；滚动开发
progressive error	累进误差；行差
progressive evolution	前进性演化；渐进演化
progressive failure	渐进破坏；渐进性破坏；累进性破坏；前进破裂
progressive fault	生长断层；渐进断层
progressive formation	渐变过程；渐变作用
progressive metamorphism	前进变质作用；渐进变质作用；递增变质作用
progressive settlement	渐进性沉降，累积性沉降；性沉降；累计性沉降
progressive shear	渐进剪切；递增剪切
progressive slide	渐进性滑动；持续滑动；递增滑动；逐渐滑动
progressive solidification	逐渐固结；渐进固结作用；逐步固化
progressive strain	渐进应变；递增应变
project appraisal	工程项目评价；项目评估；项目评审；工程评价；计划审查
project manager	工程项目经理；工程负责人；工程经理；项目经理
project site plan	项目场地图；工程地盘平面图
projected profile	投影剖面；映射剖面图
projectile	抛射体；射弹；抛射的；发射的
projectile sampler	冲击式取样器；抛射式取样机
projecting	设计；突出的；投射的；投影
projecting line	投射线；投影线
projecting plane	投影面；投影面
projection	投射；投影；突出，突出物，伸出物
projection instrument	显微映像器；映射仪
projection lamp	投射灯；聚光灯；幻灯
projection line	投影线；投射线
projective geometry	射影几何；投影几何
projective transformation	投影变换；射影变换
projector	发射器，投射器，投影仪，幻灯机，幻灯；探照灯，投光灯；设计者；水下声波发射器
projector sketchmaster	投影转绘仪；单投影器
projecture	突出部分；投影
prolific zone	富矿带；高产带
prolonged agitation	持续搅动；延时搅拌
prolonged annealing	长时间退火；延期退火
prolusion	序言；绪论
proluvial	洪积的；洪积物；洪积扇；沉积物
prominent feature	突出特征，主要特征；显著特征；显著要素
promontory	海角；悬崖；峭壁；岬，岬角
promoter	促进剂，助催化剂；助聚剂，活化剂
promotor	助催化剂；赞助者；激发器
prone	易于…的，有…的倾向；陡的，倾斜的
prone area	易发震区；易受袭击区
prong	叉，尖头物；河汊，汉流；山嘴，石嘴
proof load	容许负载，容许负荷；验证荷载，检验荷载；弹性极限荷载；标准荷载
proof muck	探针；试验棒；塞尺
proof sample	验证的样品；样品
proof stress	弹性极限应力；试验压力；屈服点
proof test	校验；验证试验；验证性试验；加压试验
prop	支撑；支柱；承托支架；支撑柱
prop counter	支柱计数员；支柱验收员；立柱计数器
prop density	支柱密度；支护密度
prop housing	柱窝；立柱底座；支柱外罩
prop support	支柱；支柱支护
prop taking	回收坑木；回收支柱；回柱
prop up	支起；支持
prop wall	密集支柱；排柱
prop with an integral short bar	T型立柱；用完整的短杆支撑
prop withdrawer	回柱机；撤柱机
prop wood	坑木；柱木
prop yield	支柱屈服；支柱让压
propagate	传播；传导；扩散
propagating wave	行波；进行波；传播波
propagating wave front	行波波前；波前
propagation	传播；传导；扩展
propagation of explosion wave	爆炸波传播；爆炸冲击波传播
propagation pressure	裂缝扩展压力；屈曲扩展压力
propagation-through-air test	空气中传爆试验；诱导爆破试验
propel cylinder	推进油缸；推进千斤顶
propellant	推进剂；推进的
propellant effect	推进作用；发射作用
propellent action	推进作用；抛射作用
propeller	螺旋桨；推进器
propeller pump	螺旋泵，螺旋桨泵；轮叶泵，轴流泵，旋桨式水泵
propeller-blade current metre	旋桨流速仪；压缩
propeller-type impeller	旋桨式叶轮；推进器式叶轮
propelling machine	牵引机；推进机车
propelling motor	行走式电动机；移动式电动机
proper phase	固有相位；特征相位
proper value	固有值；本征值；特征值
property survey	地界；测绘图；地产测量

prop-free front distance 无立柱空间距离；空顶距；无支柱距
prophylactic repair 定期检修；预防检修
propjet 涡轮螺桨飞机；涡轮螺旋喷气发动机
proportion 比例，比率；配合，配合比
proportion of extraction 采出比例；回采率；回采程度
proportional cost 比例费用；比例成本
proportional limit 比例限度；比限；比例界限
proportional range 比例范围；线性范围
proportional sampling 比例抽样；按比例取样
proportionality 比值；比例；比例性；均衡性；定比性
proportionate effect 相称的效果；比例效应
proportioner 比例调节器；输送量调节装置；定量给料器；配料斗
proportioning 按比例配合；配料；配合比；配比；比例法；配量
proportioning device 比例缩放仪；配料设备
proportioning pump 分配泵；配合泵
proppant 支撑剂；支撑物
propped fracture area 有效裂缝面积；被支撑的破裂面积
propping of floor 底板加固；巷底支护
propping of fracture 破裂支撑；裂缝的支撑；裂隙的加固
propping of roof 顶板支撑；顶板支护
propping works 支撑工程；承托工程
propulsion 推进；推力；动力装置；驱动；发动机
propulsion plant 推进动力舱；机器间
propulsive force 推进力，推力；驱动力
propulsor 喷气发动机；火箭发动机；推进器；推进物
propylite 粒状安山岩；青磐岩
propylitic alteration 青磐岩蚀变；青磐岩化；绿磐岩；粒状安山岩
prospect 找矿，矿点；勘察，勘探；展望，前途，预期，期望；远景构造，远景区
prospect area 探区；勘探远景区；远景区；勘探区
prospect drilling 普查钻进；普查勘探钻进；勘探钻井；钻探
prospect hole 探井，探孔，探眼；勘探平巷；勘探孔；勘探钻孔；试坑
prospect pit 探坑；探井；勘探井；探竖井；勘探试坑
prospect shaft 勘探竖井；探井
prospected reserve 预期储量；推测储量；勘探储量
prospecting 找矿；初步勘探；勘探；探矿；探查
prospecting bore 勘探钻机；勘探钻孔
prospecting borehole 勘探井；勘探钻孔
prospecting boring 勘探钻进；勘探钻孔
prospecting by boring 钻探；无因次过程线
prospecting drift 勘探坑道；探寻巷道；勘探平巷
prospecting drill 勘探钻机；钻机
prospecting entry 勘探平巷；勘探巷道
prospecting method 找矿方法；勘探方法；勘查方法
prospecting model 勘探模式；找矿模型；找矿模式
prospecting pit 探井；勘探试坑
prospecting shaft 探井；试钻井；勘探竖井
prospecting target 找矿靶区；找矿地区；勘探对象
prospecting trench 探槽；探沟；勘探沟
prospecting work 探矿工作；勘探工作
prospection 勘探；探矿
prospective ore 预期储量；推测储量；远景矿石储量

prospector 勘探者；找矿人；探矿者；勘探人员
prospector's pan 淘金盘；淘砂盘
protected band 防护带；安全带
protecting apron 防护板；保护墙；掩护支架
protecting casing 护壁套管；外套管
protecting cover 护盖；护罩
protecting jacket 保护罩；保护套
protecting magnet 防护磁体；保护磁体
protecting means 保护装置；保险装置
protecting piece 护板；护件
protecting pillar 保安煤柱；保安矿柱
protecting wall 挡墙；防护墙；护墙
protection course 防护层；保护层
protection device 保护设备；防护设备；防护装置
protection embankment 防护围墙；防护堤；防洪堤
protection ratio 保护率；保护率；保护比
protection shield 导坑护板；导坑保护装置
protection system 安全设备；安全设施；防护系统
protective casing 保护套；技术套管；保护套管；保护罩
protective clothing 防护服；防护衣；防护衣物
protective coal seam 解放层；保护煤层
protective coat 保护层；保护面
protective coating 保护层；保护面；防护涂层；保护涂层；保护盖层；保护涂料
protective cover 覆盖保护层；保护层；保护罩；防护罩
protective covering 防护罩；护层；护面
protective device 保护装置；防护装置
protective filter 反滤层；防护过滤器；防护滤层
protective layer 保护层；防护层
protective ozone shield 臭氧防护屏；保护性臭氧屏蔽
protective pillar 保安煤柱；保安矿柱
protective pipe 护管；保护管
protective reactance 保护电抗；保护电抗器
protective screen 保护屏蔽；防护罩
protective seam 防护层；保护层
protective shield 防护板；掩护支架；护盾；掩护支架
protective structure 防护建筑；防护建筑物；防护结构
protective switch 保护开关；保险开关；防护道岔
protective zone 保护带；防护区；保护范围
protector 防护器，保安器，保护器；保护者；防腐剂；保护层；保护装置，防护装置，防护罩；导坑护板
protector cap 防护罩；防护帽
proterozoic 元古代；元古界；原生代；元古宙；元古宇
proterozoic eon 元古代；元古宙
proto ranker 薄层腐殖质土；原始粗骨土
protocontinent 最初陆地；原大陆；原始大陆
protodolomite 原白云岩；原白云石
Protodyakonov coefficient 普氏系数；岩石硬度系数
Protodyakonov coefficient of rock strength 普氏岩石强度指数；普氏系数
Protodyakonov number 普罗托季亚科诺夫数；普氏系数；岩石强度系数
Protodyakonov scale of hardness 普氏岩石硬度系数；普氏岩石硬度标
protogenic 原生的；火成的
protohistoric age 原始时代；史前期
proto-kerogen 原油母质；原干酪根
protomagmatic deposit 分凝岩浆矿床；早期岩浆矿床
protomatrix 原杂基；原基质

protomylonite 初糜岩；原生糜棱岩；原始糜棱岩
proton 质子；氕核
proton magnetometer 质子磁力仪；质子地磁仪；质子磁强计
proton processional magnetometer 质子进动磁强计；质子进动地磁仪
protostratigraphy 基础地层学；原始地层学
prototectonics 原始构造；原始大地构造学
prototype 原型，原片，原版，典型，标准，原型法；试验模型阶段；样机，机品
prototype bit 样品钻头；实验钻头
prototype drill 样品钻机；实验钻机；原型钻头
prototype engine 原型发动机；实验发动机
prototype gradation 天然级配；原级配
prototype machine 原型机；样机
prototype model 原型模型；足尺模型
protozoic 原生界；原生代
protracted 伸出的；延长的；持久的
protrusion 突出；伸出
protrusive 固态侵入的；冷侵入的
proustite 淡红银矿；硫砷银矿
prove 证明，表示；试验，验算，检验；钻探，勘探，探明
proved area 已探明地区；已勘探地区
proved ore 证实的矿石；可靠的矿石
proven reserves 证实储量；勘定储量；可靠储量；探明储量
proven territory 证实的油区；勘探证实地区；探明的油区；探明的矿区
provincial series 区系；大区统；地区统
proving 校对；试验；钻探；验证
proving hole 勘探钻孔；勘探坑道；前探钻孔；前探小巷；探抗
provitrain 结构镜煤；类镜煤
prowersite 正云煌岩；橄辉云煌岩
proximal 近源的；邻近的，近端，近似的，接近的，最近的
proximal end 基端；根部
proximate 最近的；紧邻的；近似的；贴近的
proximate analysis 近似分析；工业分析；组分分析
proximity 接近；邻近；附近；近似；周围
proximity survey 透过波法；接近度测量
proximity switch 接近开关；迫近开关
psammite 砂质岩；砂屑岩
psammyte 砂质岩；砂屑岩
psepholite 砂砾碎屑岩；粗屑岩
psephyte 砾质岩；砾岩
pseudo section 拟剖面；视剖面
pseudoclastic 假碎屑状；假碎屑状的
pseudocleavage 假劈理；假解理
pseudo-code 伪码；伪代码
pseudo-cohesion 假黏聚力；准黏聚力
pseudocritical pressure 假临界压力；准临界压力
pseudo-fuel 假燃料；高灰分燃料
pseudo-geyser 类喷泉；类间歇泉
pseudogradational bedding 倒粒序层理；假递变层理
pseudo-gravitational field 假重力场；伪重力场
pseudo-magnetic field 假磁力场；伪磁力场
pseudomatrix 假杂基；假基质
pseudo-mining damage 假性采矿损坏；非采矿损坏；非开采性破坏
pseudomorph 假象；假晶
pseudomorphous 假象的；假晶的
pseudomountain 分割高原山；侵蚀山；假石山
pseudophase 假相；伪相
pseudoplastic 假塑性体；假塑性的；伪塑性
pseudo-potential 伪势；假势
pseudo-reduced compressibility 假对比压缩性；拟对比压缩系数
pseudoscopic view 反立体图；反立体视图
pseudo-section 视剖面；准剖面；假剖面
pseudo-slickenside 假擦痕；假擦痕面
pseudo-spar 假亮晶；重结晶方解石
pseudo-stationary 假平稳；伪稳态的
pseudo-steady state seepage 拟稳态渗流；准稳定流动；半稳定流动
pseudostromatism 假层理构造；交错层理
pseudotachylyte 假玻璃熔岩；假玄武玻璃
pseudo-velocity log 虚速度测井；拟速度测井
pseudo-viscosity 假黏度；非牛顿黏度
pseudovitrinite 假镜质组；假镜质体
psychrometer 干湿球湿度计；相对湿度计；湿度表；干湿计；干湿表
psychrometric chart 湿度表；湿度计算图；温湿图；湿度图；焓湿图
psychrometric difference 干湿球温差；湿度差；干湿球差
psychrometric table 湿度表；干湿球湿度表
ptygmatical 肠状的；蠕曲的
puckering 皱纹；小褶皱
pudding 揉捏黏土；黏土质矿石洗矿
puddle 揉捏的黏土，胶泥；小塘；水坑，水洼；稠黏土浆
puddle clay 胶合黏土；捣实黏土；胶黏土
puddle cofferdam 黏土围堰；胶土围堰；夯土围堰
puddle core 黏土心墙；夯土心墙
puddle iron 搅炼锻铁；锻铁；熟铁
puddle soil 黏土；淤泥土
puddle wall 夯实土墙；干打垒墙
puddled backfill 捣实回填；夯实回填
puddled clay 搅捣的黏土；黏土制品
puddled condition 分散状态；粉末状态
puddled ditch 实底沟；填石沟
puddled soil 压实土；捣实土；黏封土；淤泥土
puddler 捣实器；捣拌机
puddling 捣实黏土；以黏土润滑；搅炼，捣成泥浆；捣成浆状；黏土夯实
puddly soil 淤泥土；泥质土；淤泥质土
puffer 提升绞车；曳引绞车；小绞车
pug mill 捏土机；搅土机；片叶式洗矿机；搅拌机；拌泥机；黏土拌合机
pugged clay 揉捏的黏土；经搅拌的黏土
puglianite 白榴辉长岩；云白辉长岩
pulaskite 斑霞正长岩；斑霞正长细晶岩
pull apart 拉张；拉裂
pull down 压低；拆除
pull drift 连接两矿脉或矿体的石门；装载平巷
pull it green 钻头未磨损或磨损不大就起出；螺纹拧得过紧
pull length 一次钻程的深度；工作面一次放炮的进度；

起钻井深；一次取心长度
pull of gravity 重力；地心引力
pull out 提出；起钻，提钻
pull strength 扯离强度；拉拔强度
pull test 拉伸试验；抗拔力试验；牵引试验
pull-apart 拉张，拉裂；拉开
pull-apart basin 拉裂盆地；拉张盆地
pull-back and quick-release 牵拉然后快速释放；自由释放
puller 拉出器；回柱机；拔出器
pulley 滑轮；滑车；皮带轮；托滚；滚筒
pulley assembly for tensioning 补偿滑轮装置；滑轮组补偿装置
pulley block 滑轮组；滑车组；滑车
pulley drive 滑轮传动；皮带轮传动
pulley hoist 滑车；滑轮；皮带轮
pulling 拉，拖，拔；拖引，拉力；回采；放出；回收；提升
pulling back 回采煤柱；回采矿柱
pulling cable 拉绳；牵引绳；头绳
pulling equipment 牵引设备；提升设备
pulling force 牵引力；拉力
pulling jack 拉力千斤顶；拉重器；回柱机
pulling machine 提取机；提升机；起拔机；起油管装置
pulling pillar 回采煤柱；回采矿柱
pulling power 牵引力；拉力
pulling resistance 抗拉；抗拔力
pulling scraper 拖拉扒矿机；拖拉刮土机
pulling yoke 起管卡；打捞钩；提升架
pulling-down 挖掘；下拉
pull-jack 拉重器；回柱机
pull-lift 葫芦；轻便提升设备
pull-off pole 双撑杆；锚桩
pull-out 拉出；拔出
pull-out capacity 拔拉承载力；抗拔力；固着力
pullout force 拔力；拉拔力
pullout resistance 抗拉出阻力；拔拉阻力
pull-out test 抗拔试验；拉拔试验；拔拉测试
pull-out torque 临界过载力矩；失步力矩；拔拉转矩
pullshovel 反铲；反向铲
pull-type 牵引式；拖曳式
pull-up 视隆起；上拉现象
pulp 矿浆；泥浆
pulp aeration 煤浆充气；矿浆充气
pulp climate 矿浆的状态；调和矿浆
pulp conditioning 煤浆准备；矿浆准备
pulp density 矿浆浓度；矿浆密度
pulper 搅拌机；打浆机
pulpy peat 浆状泥炭；沉积泥炭
pulsating spring 脉动泉；多潮泉
pulsating wave 脉冲波；脉动波
pulsation theory 脉动理论；脉动学说
pulse amplitude 脉冲振幅；脉冲幅度；脉冲幅值
pulse of the Earth 地球脉动；大地脉动
pulse polarography 脉冲电极谱；脉冲极谱
pulse width 脉冲宽度；脉冲持续时间；脉动宽度
pulsion 上升水冲动；搏跳
pulsion stroke 脉冲冲程；上向冲程
pulsometer 气压抽水机；蒸汽双缸泵
pulv 粉末；磨碎的

pulverator 锤碎机；粉碎机；制粉机
pulverescent 含尘的；粉状的；粉块；粉状
pulveriser 喷雾机；粉碎机
pulverization 粉磨；粉碎；研磨
pulverize 磨成粉状；粉磨；磨碎；研磨
pulverized coal 粉磨煤；粉煤；煤粉
pulverized fuel 粉状燃料；煤粉燃料
pulverized fuel ash 粉煤灰；飞灰；煤灰；研磨煤灰
pulverizing jet 喷烧器；烧嘴；粉碎射流
pulverulence 粉末状态；易碎性
pulverulent 易碎的；粉状的
pulverulent brown-coal 粉末状褐煤；尘状褐煤
pulverulent soil 粉质土；粉质土壤
pulverulite 细尘熔结凝灰岩；尘熔结凝灰岩
pumice 浮岩；浮石；泡沫岩；火山灰
pumice cone 浮石锥；浮石火山锥
pumice stone 浮石；轻石；浮岩；泡沫岩
pumice volcano 爆裂火山；爆发火山
pumiceous 泡沫状；浮石状的
pumiceous structure 泡沫状构造；浮石构造
pumicite 浮岩层；浮岩沉积；火山灰
pump 抽水；抽气，泵，水泵；抽水机；泵送
pump ability 可泵汲性；抽送能力；泵汲量，汲出量
pump capacity 泵出水量；泵排量，泵量；水泵出水率；抽水机出水率
pump compartment 泵隔间；管子隔间；泵房
pump control setting 泵变量；泵调节
pump delivery 泵送水量；泵排出量；泵供水量；泵输送；泵排量
pump discharge 抽水量；泵流量；水泵出水量；水泵排水量；泵排量；泵出口
pump displacement 泵排量；泵排水量
pump drain 水泵排水沟；抽水；排水
pump drainage 泵排；抽排；抽排；机泵排水
pump dredger 抽泥机；吸泥机
pump duty 泵流量；泵排量；泵量
pump efficiency 水泵效率；泵效
pump manifold 水泵集水管；泵的管汇
pump out 泵出；抽出
pump output 抽水量；水泵出水率；泵排量
pump pit 抽水机坑；水泵坑
pump pressure gauge 泵压力计；泵压表
pump priming 泵充水；泵的起动注水；灌泵
pump rod 拉杆；泵杆
pump set 泵水装置；泵；泵机组
pump shaft 泵轴；排水井；泵转轴
pump snoring bowl 水泵吸水管滤头；水泵吸水龙头
pump stage 泵段；排水分段；泵级
pump station 水泵站；抽水站；泵房；泵站
pump strainer 泵的滤器；泵的过滤器；泵口拦污网
pump suction head 泵的吸入高度；泵的吸水头
pump suction pit 泥浆沉淀池；泥浆泵汲池；泵上水池
pump sump 抽水坑；抽水集水坑；水泵集水井；水坑；废水收集池
pump tree 水泵管节；扬水管柱
pump unit 抽水机组；水泵机组；泵装置
pump water conduit 泵水排出管；泵水排出沟
pump way 排水间；排水管间
pump well 抽油井；抽水井；抽水孔；泵井
pumpability 泵送性；可泵性；泵送能力；可用泵输送

pumpage	抽汲；抽水；抽水量；泵的抽送量；泵送能力；抽送能力	punctuated equilibrium	点断平衡；间断平衡
pumpcrete	泵送混凝土；泵浇混凝土	puncture resistance	耐破坏性；抗穿刺性能；抗冲穿强度；冲穿阻力
pumped well	抽水井；机井	puncture strength	刺破强度；冲穿强度；抗刺强度；击穿强度
pumper	抽油井；泵车；司泵；司泵工，司泵员	puncture test	钻孔试验；打眼试验；击穿试验；破坏性试验；刺破试验
pumping	抽水；泵送，泵抽；抽吸；汲水现象	pungernite	地蜡；富油母质油页岩
pumping capacity	抽水量；抽水能力；泵排量；抽水容量；排水能力	punky	半硬化的；半固结的；有弹性的
pumping compartment	泵房；泵隔间	punner	打夯机；夯锤；夯具；夯
pumping cone	抽水漏斗；降落漏斗	pure coal	净煤；纯煤
pumping effect	抽水效应；泵抽效应	pure shear	纯剪切；纯剪；纯切变；纯切力；纯剪力
pumping equipment	抽水设备；排水设备；泵汲设备；泵送设备	pure slip	纯滑移；纯滑动
pumping gear	抽油机的减速装置；排水装置	purge	清除；排污
pumping head	泵压头；水头；抽水扬程	purification	提纯；净化；精炼；精制
pumping installation	抽油装置；泵送装置	purifier	净水装置；净水剂；净化器；净水剂；提纯器
pumping irrigation	抽水灌溉；机械提水灌溉	purity	纯度；纯净
pumping jack	多井联动抽油简易抽油架；深井泵的胶带传动装置	purlieu	邻近地带；范围；界限
pumping level	抽水水位；泵吸高程；抽水高程	purple ore	紫色矿石；焙烧黄铁矿
pumping lift	抽水扬程；提水高度	purpurite	磷锰石；紫磷铁锰矿
pumping line	扬水管道；水泵出水管；中间冷却器	push button control	按钮控制；按钮操纵
pumping load	泵负荷；抽水量	push contact	按钮开关触点；按钮开关
pumping pattern	抽水井布置；抽水型式	push conveyor	链板输送机；刮板输送机
pumping plan	泵送计划；矿井水情图	push loading	推力负荷；推压装载
pumping plant	抽水站；泵房；抽水厂；泵汲装置；泵装置	push rod assembly	推杆总成；推杆装配组合
pumping plant capacity	抽水站容量；抽水站抽水能力	push water	冲水；推动水
pumping power	深井泵的动力装置；泵水功率；抽水动力装置	push-button drilling	按钮钻眼；按钮钻进
pumping pressure	泵送压力；泵压	push-button mining	按钮开采；自动化开采；遥控开采
pumping rate	抽水量；抽水速度；抽水率；泵汲速度；泵浆率	push-button mining method	全自动化采煤法；遥控采煤法
pumping shaft	抽水竖井；扬水竖井；排水井	push-down	下拉现象；下推；视坳陷，拖陷
pumping test	抽水试验，扬水试验；排水试验	pusher jack	推力千斤顶；推移装置
pumping unit	抽水机；抽水设备；泵设备；泵装置	pusher leg	自动推进式气腿；气顶
pumping water level	抽水水位；排水后最低地下水位	pusher tractor	推式拖拉机；推进式拖拉机；后推机
pumping well	抽水井；抽水孔；泵抽井；抽出井；抽液孔；抽油井	push-feed drill	推进式钻机；推进式凿岩机；自动推进式钻机
pumping-depression cone	抽水降落漏斗；抽水降落锥面	push-feed drilling	自动推进式钻眼；自动推进式凿岩
pumping-in test	泵入试验；压水试验；注水试验	pushloading scraper	推式铲运机；推式装载铲运机
pumpingunit	排水设备；泵	pushover	倾覆；推覆；静力弹塑性
pumps in parallel	并联泵；并联泵组	push-pull	推挽的，推挽式的；推挽；推动的；推拉双作用的
pumps in series	串联泵组；串联泵	push-pull support system	液压支架推拉系统；推拉式支承系统
punch	冲孔，冲床，冲切，冲击，冲压机，穿孔器，打眼，穿孔，打出的孔眼，切口	push-pull wave	纵波；P波
punch card machine	卡片穿孔机；穿孔卡片机	put into dual operation	使实施；使生效；投入生产；使开始工作或运转
punch core	冲压取芯；井壁取芯	put into operation	开动；开始使用；开始经营；开通，投产；使实施，使生效，投入生产；使开始工作或运转
punch load	冲击荷载；冲剪荷载		
punch mining	山坡露天开采转地下平硐开采；移动式穿硐开采	put into service	开动；使工作；使运转；启用；运用；投入运行
punched plate	冲孔板；冲头板；冲孔筛板	put out of service	停止使用；不运用
puncher	穿孔机，凿孔机，穿孔器；冲孔机，冲孔器；冲床；冲压机	put right	修理；改正
punching	穿孔；冲孔；掏槽	put through	通过；接通
punching failure	冲剪破坏；刺入破坏	put to test	试验；检验；投入试验
punching shear	冲切剪力；冲剪破坏；冲剪；冲压剪切；冲剪力	putrefaction	腐化；腐烂；腐泥化；腐烂物
		putrid mud	腐泥；腐殖泥
punching shear failure	冲剪破坏；地基刺入剪切破坏	putrid slime	腐泥层；腐泥沉积
		putter	运煤工；推车工；小型拖运绞车
punctuality	准时性；正确；规矩	putting	运输；拖运

putting-down	放下；记下
putty	油灰；油灰胶泥
puy	残留火山锥；死火山锥
P-wave	P波；压缩波；纵波
pycnite	柱状黄晶；圆柱黄晶
pycnocline	密度跃层；密度梯度
pycnometer	比重瓶；比重管；密度瓶；比色计；比重计
pyknometer	比重瓶；比重计
pyralspite	铝榴石；铝榴石类
pyramidal	棱锥形的；角锥形；角锥状；锥状
pyramidal system	正方晶系；四方晶系
pyrargyrite	硫锑银矿；深红银矿；浓红银矿；矿硫锑银矿
pyrobitumous shale	焦性沥青页岩；沥青质页岩
pyrochroite	片水锰矿；羟锰矿
pyroclast	火山碎屑；火成碎屑；火山碎屑物
pyroclastic	火山碎屑的；火成碎屑物；火山碎屑堆积岩
pyroclastic cone	集块火山锥；火山碎屑锥
pyroclastic flow	火山碎屑流；火成碎屑岩流
pyroclastic rock	火山碎屑岩；火成碎屑岩；火成岩
pyroclastics	火山碎屑物；火山碎屑岩；火成碎屑沉积；火成碎屑物
pyroclasts	火山碎屑物；碎屑物
pyrofusinite	焦性丝质体；焦性丝
pyrogenic rock	火成岩；岩浆岩
pyrography	热解色谱；裂解色谱
pyrolysis	热解；裂解；高温分解；加热分解
pyromagma	高温岩浆；浅源岩浆
pyromeride	球粒流纹岩；球泡霏细岩
pyrometallurgy	火法冶金；高温冶金；火法冶炼
pyrometasomatic deposit	高温热液交代矿床；高热交代矿床
pyrometasomatism	高温交代作用；高热交代
pyrometer	压力计；压强计；压缩计；高温温度计
pyrometric cone	测温锥；高温锥
pyrometry	高温测定法；高温测量术；高温测定
pyromorphite	磷氯铅矿；火成结晶
pyrophanite	红钛锰矿；锰矿
pyroschist	沥青页岩；可燃油页岩；焦页岩；含油页岩
pyroscope	辐射热度计；测高温器
pyroshale	油页岩；可燃页石
pyrosphere	火界；岩浆圈；火圈
pyrotic	腐蚀剂；腐蚀的
pyroxene andesite	辉石安山岩；辉安山岩
pyroxene granulite	辉石麻粒岩；辉石麻粒岩相
pyroxene hornblendite	辉石角闪石岩；辉石角闪岩
pyroxene peridotite	辉橄岩；辉石橄榄岩
pyroxmangite	三斜锰辉石；锰三斜辉石

Q

qanat 坎儿井,地下水道,暗渠
Q-factor Q因数,品质因数
Qiafang Limestone 卡房石灰岩
Qin Terra-cotta Army Museum 秦俑博物馆
Qing 磬
Q-joint Q节理,横节理
Q-mode analysis Q型分析
Q-mode cluster analysis Q型聚类分析
Q-mode factor analysis Q型因子分析
Q-mode of eigenvalue analysis Q型特征值分析,Q型本征值分析
Q-mode space Q形空间
Q-mode statistical method Q型统计法
Q-mode vector diagram Q形向量图
Q-switch Q开关,光量开关
Q-system Q-系列
Q-type curve Q型曲线
quad 梯形图幅,标准图幅,象限;四分仪;夸特;扇体齿轮;回芯组线
quadrangle 图幅,分幅;四边形,四角形;四合院
quadrangle map 标准地形图图幅,分幅图
quadrangle method 方格法
quadrangle name 图幅名称
quadrangular 四角形的,四边形的;四棱柱
quadrant 象限;象限仪,四分仪;扇形齿轮,换向器;扇形体,扇形齿板
quadrant angle 象限角
quadrant antenna 正方形天线
quadrant bracket 扇形托架
quadrant depression 俯角
quadrant elevation 仰角
quadrant formula 平方根公式
quadrant iron 方铁,方钢,扇形钢材
quadrant pit 矩形井筒
quadrant revetment 锥体护坡
quadrant tooth 扇形轮齿
quadrant valve 扇形阀
quadrantal 象限的
quadrantal corrector 象限自差校正器
quadrantal davit 俯仰式吊艇柱
quadrantal diagram 象限图
quadrantal distribution 象限分布
quadrantal error 象限方向角误差
quadraplex cable 四芯电缆
quadrate 正方形;长方形;四等分;方块物
quadratic 二次的,平方的;二次方程式
quadratic algorithm 二阶算法
quadratic component 平方分量,平方项;二次谐波
quadratic convergence 二次收敛
quadratic criterion 二次准则
quadratic damping 平方阻尼
quadratic discriminant 二次方程判别式
quadratic element 二次单元,二次元

quadratic equation 二次方程
quadratic error 二次误差,平方误差
quadratic expansion 二次扩展
quadratic expression 二次表达式
quadratic fit 二次拟合
quadratic form 二次型
quadratic frequency-magnitude relation 频率-震级二次关系式
quadratic function 二次函数
quadratic interpolation method 二次插值法
quadratic mean 均方值;二次平均值
quadratic mean deviation 均方差
quadratic mean error 均方差
quadratic mode 二次模式
quadratic polynomial 二次多项式
quadratic regression analysis 二次回归分析
quadratic resistance 二次阻力
quadratic root 平方根
quadratic spline 二次样条
quadratic sum 平方和
quadratic surface 二次曲面
quadratic tetrahedron 二次四面体
quadratic transformation 二次变换
quadratic trend surface 二次趋势面
quadratic triangle 二次三角形
quadratic triangular element 二次三角形单元
quadratic variation 二次变分
quadrature 正交;求面积,求积分;图幅
quadrature analysis 正交分析
quadrature axes 正交轴
quadrature component 正交分量,异相分量,虚分量
quadrature filtering 正交滤波
quadrature formula 求积公式
quadrature network 正交网络;无功电路
quadrature point 正交点
quadrature spectral density 正交谱密度
quadrature spectrum 正交谱,求积谱
quadrature tide 小潮,最低潮,弦潮
quadrature trace 正交道
quadrature weight 正交权
quadrel 方形;方块石,方砖,方瓦
quadric 二次的;二次曲面的
quadric stress 曲面应力
quadric surface 二次曲面
quadrilateral 四边的;四边形;四边形地下作业区
quadrilateral element 四边形元,四边形单元
quadriplanar coordinate 四面坐标
quadripole 四极;四端电路,四端网络
quadrivalent 四价的
quadruple 四倍的,四重的;四倍仪;四叶螺旋桨
quadruple aerial survey camera 四镜头航空测量摄影机
quadruple boom 四钻机钻车托臂
quadruple camera 四镜头摄影机
quadruple-action hand pump 四作用手摇泵

quag 沼泽地，泥沼
quagmire 沼泽，沼泽地，软泥地，泥沼
quake 震，地震
quake centre 震中
quake sheet 地震崩滑层
quake-prone 经常发生地震的
quake-proof structure 抗震结构
quake-resistant 抗震的，防震的
quake-resistant structure 抗震结构
quake-resistant wall 抗震墙
quaking bog 跳动沼，颤沼，浮沼
quaking concrete 高流态混凝土，塑性混凝土
qualification 资格；合格条件，限制条件
qualification certificate 资格证书
qualification examination 资格审查
qualification rate 合格率
qualification record 鉴定记录
qualification test 鉴定试验，鉴定测验，验收试验；资格考试
qualified medium 合格介质；合格悬浮液
qualified person 合格人士；高素质人员
qualified personnel 高素质人员，高素质团队
qualified project 鉴定合格工程
qualified rate 鉴定合格率
qualified test 合格性检验
qualify 限定，把…称作…；使合格；证明…合格
qualifying examination 资格考试
qualitative 质的，性质的，定性的
qualitative analysis 定性分析
qualitative assessment 定性评定
qualitative attribute 定性属性
qualitative carbon steel 优质碳钢
qualitative change 质变
qualitative classification 定性分类
qualitative comparison 定性比较
qualitative curve 定性曲线
qualitative data 定性资料
qualitative description 定性描述
qualitative detection 定性探测
qualitative determination 定性检验
qualitative differential thermal analyzer 定性差热分析仪
qualitative Drakely formula 特雷克利定性公式
qualitative effectiveness 定性效果，定性有效性
qualitative estimate 定性评估
qualitative evaluation 定性评价
qualitative examination 定量考核
qualitative filter paper 定性滤纸
qualitative forecast 定性预报
qualitative image interpretation 定性图像解释
qualitative index 定性指标
qualitative interpretation 定性解释，定性解译
qualitative investigation 定性研究
qualitative method 定性方法
qualitative model 定性模型
qualitative observation 定性观察
qualitative perception 质感
qualitative reaction 定性反应
qualitative response data 定性响应数据
qualitative result 定性结果
qualitative scheme 定性方案

qualitative simulation 定性模拟，定性仿真
qualitative spectral analysis 定性光谱分析
qualitative spectral scan 定性光谱扫描
qualitative spectroscopic analysis 定性光谱分析
qualitative steel 优质钢
qualitative stratigraphic correlation 定性地层对比
qualitative test 定性试验，定性测试
qualitative variable 定性变量
quality 质，质量，品质，性质，特性
quality analysis 质量分析
quality assessment 质量评估，质量评价
quality assurance acceptance standard 质量保证验收标准
quality assurance manual 质保手册；质量保证手册
quality assurance program 质量保证大纲
quality assurance provision 质量保证条例
quality audit 质量审核
quality certificate 质量证书
quality certification 质量认证
quality check 质量检查
quality classification of soil sample 土样质量分类
quality coefficient 质量系数
quality concrete 优质混凝土
quality control 质量控制，质量管理
quality control chart 质量控制图
quality control inspector 质量控制检查员
quality control of earth-rock fill 土石填筑质量控制
quality control of life 全程质量控制
quality control specification 质量管理规范
quality control system 质量控制系统
quality cost 质量成本
quality cost optimization model 质量成本优化模型
quality criteria 质量标准
quality defect 质量缺陷
quality determination 质量测定
quality discrepancy record 质量不合规定记录
quality effectiveness 质量效果，质量有效性
quality error 质量误差
quality evaluation 质量评价
quality examination 质量检测
quality factor 质量因素，品质因素
quality governing 质量控制
quality grade 质量等级
quality index 质量指标，质量指数
quality index of tunnel surrounding rock 隧道围岩质量指标；巷道围岩质量指标
quality information 质量信息
quality inspection 质量检测
quality inspection of bolted connection 螺栓连接质量检测
quality inspection of masonry 砌体质量检测
quality inspection of riveted connection 铆钉连接质量检测
quality material 优质材料
quality mind 质量意识
quality monitoring 质量监测
quality monitoring station 质量监测站
quality monitoring system 质量监控系统
quality objective 质量目标，质量标准
quality of aerophotography 航摄质量
quality of air environment 空气环境质量
quality of concrete 混凝土质量

quality of design 设计质量
quality of fuel 燃料质量
quality of groundwater 地下水质量；地下水质
quality of irrigation water 灌溉水质量
quality of life 生活质量
quality of lighting 照明质量
quality of material 材料质量
quality of pump lifted water 回扬水质量
quality of river water 河水质量
quality of water 水质
quality of work 工作质量
quality reduction 质量下降
quality related cost 质量相关成本
quality requirement 质量要求
quality review 质量审查
quality standard 质量标准
quality standardization management information system 质量标准化管理信息系统
quality supervision 质量监督
quality test 质量检验
quality tracking system 质量追踪系统
quantification 量化，定量化
quantifying risk 量化风险
quantile 分位数，分位点数
quantitative 数量的，定量的
quantitative airborne infrared mapping 定量机载红外制图
quantitative analysis 定量分析
quantitative assay 定量测定
quantitative assessment 定量评估，定量评价，量化评估
quantitative change 量变
quantitative classification 定量分类
quantitative comparison 定量比较
quantitative criterion 定量标准
quantitative data 定量数据，定量资料
quantitative design 定量设计
quantitative determination 定量测定
quantitative Drakely formula 特雷克利定量公式
quantitative effect 定量效应
quantitative effectiveness 定量效果，定量有效性
quantitative environmental science 定量环境科学
quantitative estimation 定量估算
quantitative evaluation 定量评价
quantitative examination 定量检测
quantitative filter paper 定量滤纸
quantitative forecast 定量预报
quantitative geochemistry 定量地球化学
quantitative geology 定量地质学
quantitative geomorphology 定量地貌学
quantitative hydrology 定量水文学
quantitative index 定量指标
quantitative information 定量信息
quantitative interpretation 定量解译，定量解释
quantitative investigation 定量研究
quantitative lithologic data 定量岩性数据
quantitative management 定量管理；计量管理
quantitative management school 计量管理学派
quantitative measurement 定量测量
quantitative method 定量法
quantitative mineralogical analysis 矿物定量分析
quantitative modelling technique 定量模拟技术，定量建模技术
quantitative morphology 定量形态学；定量地貌学
quantitative observation 定量观测
quantitative paper chromatography 定量纸色谱法
quantitative perception 数量感
quantitative polarizing microscope 定量偏光显微镜
quantitative quality flowsheet 定量质量流程；定量质量流程图，定量质量流程表
quantitative reaction 定量反应
quantitative relation 定量关系，数量关系
quantitative remote sensing 定量遥感
quantitative reserve assessment 定量储量评定
quantitative risk assessment 定量风险评估
quantitative spectrographic analysis 定量光谱分析
quantitative spectrography 定量光谱学
quantitative test 定量测试
quantitative three dimensional information 三维定量信息
quantity 量，数量
quantity of excavation 开挖量
quantity of flow 流量
quantity of heat 热量
quantity of information 信息量
quantity of injected water 注入水量
quantity of percolation 渗水量
quantity of precipitation 降水量
quantity of seepage 渗漏量，渗透量
quantity of water 水量
quantity of water recharge 回灌量
quantity of work 工作量，工程量，作业量
quantity production 成批生产，大量生产
quantity sheet 数量表；工程量表，土方表
quantity underrun 数量差额；工程量差额
quantity-quality flowsheet 产量质量流程；产量质量流程图，产量质量流程表
quantivalence 原子价，化合价
quantization 量化；量子化
quantization error 量化误差
quantized computer 数字转换计算机
quantized signal 量化信号
quantizing 量化；整量化，取整
quantometer 光量计
quantum 量子；定量，定额；总量，总计
quantum chemistry 量子化学
quantum field theory 量子场论
Quantum geochemistry 量子地球化学
quantum mechanic modelling 量子力学建模
quantum mechanics 量子力学
quantum optics 量子光学
quantum theory 量子论
quaquaversal dome 穹隆
quaquaversal fold 穹形褶皱
quaquaversal structure 穹状构造，穹窿构造
quark 夸克，层子
quark model 夸克模型，层子模型
quarrel 菱形瓦，菱形玻璃
quarrier 采石工
quarry 采石场，露天矿场；方形玻璃；铺地砖；采石，挖掘
quarry bank 露天矿梯段

quarry blasting	采石爆破
quarry bucket	采石铲斗
quarry car	采石斗车
quarry drill	采石钻机
quarry dynamite	采石硝甘炸药
quarry engineering	采石工程
quarry equipment	采石设备
quarry face	采石工作面,采石开拓面
quarry floor	采石场底板
quarry in open cut	露天采石
quarry material	毛石料,粗石料
quarry powder	采石炸药
quarry rock	粗石,毛石
quarry rock breakwater	粗石防波堤
quarry rubbish	石碴
quarry run rock	毛石料
quarry run rockfill	毛石填筑
quarry run stone	毛石料
quarry sap	采石坑
quarry site	采石点
quarry stone masonry	粗石圬工;粗石砌体,粗石建筑
quarry water	矿场水,石坑水,石窝水
quarrying	采石;露天开采;采石工程
quarrying machine	采石机械
quarrying method	采石方法
quart	夸脱
quartation	四分法
quarter bond	一顺一丁砌砖法
quarter ditch	支沟;集水沟
quarter line	象限线
quarter octagon drill rod	圆角方断面钻杆
quarter octagon steel	圆角方钢
quarter point	四分点
quarter post	四角标桩
quarter sample	四分法缩分样;四等分取样,四分法选样
quartering	四分法取样
quartering shovel	四分取样铲
quarter-point deflection	四分点挠度
quarter-point loading	四分点荷载
quarter-section	四分之一图幅
quarter-turn ball valve	直角回转球阀
quarter-turn belt	直角回转皮带
quarter-wave filter	四分之一波长滤波器
quartet configuration	一注三采配置
quartic curve	四次曲线
quartic element	四次元
quartic polynomial	四次多项式
quartic surface	四次曲面
quartile	四分位数
quartile deviation	四分位差
quartile sorting coefficient	四分位分选系数
quartile value	四分位数值
quartimax method	四次幂极大法
quartimin method	四次幂极小法
quartz	石英
quartz clock	石英钟
quartz conglomerate	石英砾石
quartz crystal	石英晶体,水晶
quartz crystal pressure transducer	石英晶体压力传感器
quartz dust	石英尘末
quartz fiber gravimeter	石英丝重力仪
quartz fiber horizontal magnetometer	石英丝水平磁力仪
quartz filter	石英滤波器
quartz geothermometer	石英地质温度计
quartz glass	石英玻璃
quartz gravel	石英砾石
quartz lamp	石英灯
quartz lens	石英透镜
quartz lung	硅肺病
quartz magnetometer	石英磁力仪
quartz monzonite	石英二长岩
quartz oscillator	石英振荡器
quartz phyllite	石英千枚岩
quartz porphyry	石英斑岩
quartz reef	石英脉
quartz rhyolite	石英流纹岩
quartz sand	石英砂
quartz sandstone	石英砂岩
quartz saturation geothermometer	石英饱和地热温标
quartz schist	石英片岩
quartz siltite	石英粉砂岩
quartz sinter	硅华
quartz slate	石英板岩
quartz spectrograph	石英摄谱仪
quartz spring gravimeter	石英弹簧重力仪
quartz syenite	石英正长岩
quartz thermometer	石英温度计,石英温标
quartz trachyte	石英粗面岩
quartz vein	石英脉
quartz-feldspathic rock	长英质岩石
quartzite	石英岩,硅岩
quartzitic soil	石英质土
quartz-keratophyre	石英角斑岩
quartz-mica gneiss	石英云母片麻岩
quartz-mica schist	石英云母片岩
quartz-pebble conglomerate	石英砾岩
quasi-analytical method	准解析法
quasi-anisotropic media	准各向异性介质
quasi-brittle materials	准脆性材料
quasi-cleavage fracture	准解理断裂
quasi-conductor	准导体
quasi-continental crust	准大陆壳
quasi-continuous medium	准连续介质
quasi-coordinates	准坐标
quasi-criterion	准判据
quasi-crystal	准晶体
quasi-cylindrical	似柱形的
quasielastic	准弹性的
quasi-elastic scattering	准弹性散射
quasi-equilibrium	准平衡
quasi-equilibrium state	准平衡态
quasi-fluid	准流体
quasi-friction	准摩擦
quasi-geodetic datum	准大地测量基准面
quasi-geoid	准大地水准面
quasi-gradiometer	准梯度仪
quasi-gravity	准重力
quasi-group	拟群
quasi-harmonic equation	准谐方程

quasi-homogeneous	准均质的	Quaternary geological map	第四纪地质图
quasi-instantaneous	准瞬时的，准同时的	Quaternary geology	第四纪地质学
quasi-isostatic displacement	准均衡位移	Quaternary glacial period	第四纪大冰期
quasi-isotropy	准各向同性	Quaternary glaciation	第四纪冰期；第四纪冰川作用
quasi-laminar flow	拟层状流	Quaternary glacier	第四纪冰川
quasi-linear	拟线性的，准线性的	Quaternary period	第四纪
quasilinear equation	拟线性方程	quaternary research	第四纪研究
quasilinearization	拟线性化	Quaternary sediment	第四纪沉积物
quasi-longitudinal wave	准纵波	quay	岸壁，堤岸；码头
quasi-natural terrain	准天然地形	quay berth	码头泊位
quasi-Newton method	拟牛顿法	quay breastwork	码头护墙
quasi-normal distribution	拟正态分布	quay crane	码头起重机
quasi-oceanic crust	类洋壳	quay pier	岸壁式突码头；突堤码头，防波堤
quasi-one-dimensional flow	准一维流	quay shed	码头仓库
quasi-optimal solution	准优解	quay wall	岸壁，岸墙
quasi-ordered system	拟序系统	quay wall foundation	岸壁基础
quasi-orthogonal polynomial	准正交多项式	quay wall on pile foundation	桩基岸壁
quasi-orthogonal projection	准正射投影	quay wall structure	岸墙结构
quasi-overconsolidation	准超固结	quebrada	地震裂缝；粗糙地形区；峡谷，小溪；
quasi-periodic	准周期的	Queen Anne style	安妮女王式
quasi-permanent combination	准永久组合	queen post	双柱架
quasipermanent deformation	准永久变形	queen post truss	双柱架式桁架
quasi-planar fracture	准平面破裂	quench alloy steel	淬硬合金钢
quasi-plastic flow	准塑性流动	quench hardening	淬火硬化
quasi-polynomials	拟多项式	quenching	淬火
quasi-preconsolidation pressure	准前期固结压力	quenching crack	淬火裂纹，淬致裂痕
quasi-saturated soil	准饱和土	quenching reaction	猝火反应
quasi-section	似剖面，拟剖面，视剖面	quenching strain	淬火应变
quasi-shear-wave	准横波	quenching stress	淬火应力
quasi-simple wave	拟简波	quenching temperature	淬火温度
quasi-single phase flow	准单相流动	Quer-wave	奎尔波
quasi-solid	准固态的	Querwellen wave	奎威林波，横波，Q-波
quasi-sorted	半分选的	quest	探索，寻找；找矿
quasi-stability	准稳定性	questionnaire method	问卷法
quasi-stable	准稳定的	queuing model	排队模型
quasi-stable adjustment	拟稳平差	queuing theory	排队论
quasi-stable state	准稳态	quick analysis	快速分析
quasi-static	准静态的	quick blow drilling	速冲钻进
quasi-static analysis	准静态分析	quick bog	浮沼
quasi-static cone penetration	准静力触探贯入度	quick burning fuse	速燃引信
quasi-static displacement	拟静态驱替	quick cement	速凝水泥
quasi-static method	准静力法	quick clay	过敏性黏土；流黏土
quasi-stationarity	准稳定性	quick consolidation test	快速固结试验
quasi-stationary	准静态的	quick direct shear test	快速直剪试验
quasi-stationary state	准静态	quick freezing	速冻
quasi-steady	准稳定的	quick ground	松软地基
quasi-steady flow	准定常流，准稳定流	quick match	速燃引信头；速燃引信，速燃导火索
quasi-steady state	准稳态	quick sand	流砂
quasi-stratigraphic unit	准地层单位	quick sand bed	流砂层
quasi-transverse wave	类横波	quick sand foundation	流沙地基
quasi-variable	准变数	quick shear cohesion	快剪黏聚力
quasi-viscosity	准黏滞度	quick shear strength	快剪强度
quasi-viscous creep	准黏性蠕变	quick shear test	快剪试验
quasi-viscous flow	准黏性流动，半黏性流动	quick soil	浮土
quasi-viscous fluid	半黏性流体	quick soil classification	简易土壤分类
Quaternary	第四纪；第四系	quick water	急流水
Quaternary alluvium	第四纪冲积层	quick-acting explosive	速爆炸药
Quaternary climate	第四纪气候	quick-acting fuse	速燃引信
Quaternary deposit	第四纪沉积	quick-acting valve	速动阀
Quaternary geochronology	第四纪地质年代学	quick-break switch	快断开关

quick-closing valve	速闭阀
quick-drying paint	快干漆，立干漆
quick-growing plantation	速生人工林
quick-hardening cement	快硬水泥，早强水泥
quick-jellying property	速凝性
quicklime	生石灰，氧化钙
quicklime pile	石灰桩
quickly detachable	可快速拆卸的
quick-mounting	迅速安装的，迅速装配的
quick-release grip	速开卡栓
quick-release joint	速卸接头，速脱接头
quick-release pick	速卸截齿
quick-response	快速响应
quick-setting additive	快凝添加剂
quick-setting adhesive plastic cement	速凝塑性黏合水泥
quick-setting agent	快凝剂，速凝剂
quick-setting asphalt	快凝沥青
quick-setting cement	快凝水泥
quick-setting grouting	快速注浆
quick-setting level	速调水平仪
quick-setting material	速凝材料
quick-setting mixture	速凝混合料
quick-setting mortar	速凝灰浆
quick-settling particle	速沉颗粒
quid	扩孔器
quiescence spectrum	宁静光谱
quiescent consolidation	自重固结
quiescent current	静态电流
quiescent layer	静止层
quiescent level	静液面
quiescent load	静负荷，恒载
quiescent settling	静水沉淀
quiescent spring	死泉，古泉
quiescent stage	静止阶段
quiescent tank	静水沉淀池
quiescent volcano	休眠火山
quiescent water	静水
quiescently loaded	静荷载的，恒载的
quiet battery	无噪声电池
quiet magnetic zone	地磁平静区
quiet motor	低噪声电动机
quiet period	宁静期
quiet reach	静水河段
quiet solar wind	宁静太阳风
quiet volcano	静火山
quiet working method	安静作业方法
quieter	消声装置
quill	套管轴；缓燃导火线
quill bit	勺形钻；石工錾子
quilt	衬垫；垫料
quinary hole	五序孔
quincuncial pile	梅花桩
quincunx arrangement	梅花形布置
quinquangular	五角的，五边形的
quintal	公担；英担
quintic	五次的
quintic curve	五次曲线
quintuple space	五维空间，五度空间
quirk	深槽，沟
quoin	隅石；拱楔石
quota	定额，限额，分配额
quota cost	定额成本
quota of budget	预算定额
quota of capital construction	基本建设定额
quota system	定额分配制度
quotation	行情，估价单，报价单；引语，引文；引用，引证
quotient	商；份额
quotient-difference algorithm	商差算法
Q-value	Q值
Q-wave	Q波

R

rabbler 刮管器；搅拌器
rabbling 搅拌
rabbling hoe 搅耙机；搅拌机
race 急流；水道；竞赛
raceway 矿山导水沟；水道；电缆管
rack rail 齿轨
rack railway 齿轨铁路
rack-mounted 装在机架上的
rack-railroad 齿轨铁路
radar 雷达
radar altimeter 雷达测高仪；雷达高度计
radar antenna 雷达天线
radar band 雷达波带；雷达波段
radar beacon 雷达信标
radar beam 雷达波束
radar clutter 雷达的地面反射干扰
radar coverage 雷达覆盖范围
radar depression angle 雷达俯角
radar detection 雷达探测
radar echo 雷达回波
radar illumination swathe 雷达照射条带
radar imagery 雷达成像
radar imaging 雷达成像
radar indicated face 雷达显示表面
radar interaction 雷达干扰
radar interferometry 雷达干涉测量
radar jamming 雷达干扰
radar map 雷达图；雷达导航图
radar measurement 雷达测量
radar meteorological observation 雷达气象观测
radar meter speed gun 雷达测速仪
radar microwave technique 雷达微波技术
radar mosaic 雷达综合图
radar navigation 雷达导航
radar navigator 雷达导航设备
radar network composite 雷达组网
radar observation 雷达观测；雷达监视
radar overlay 雷达覆盖区
radar performance figure 雷达性能指标
radar photograph 雷达航摄照片
radar platform 雷达台
radar plotter 雷达量测仪
radar positioning 雷达定位
radar precipitation echo 雷达降水回波
radar presentation 雷达显示
radar ramark 雷达指向标
radar range 雷达测距
radar range finder 雷达测距计
radar ranging 雷达测距
radar reflector 雷达反射器
radar remote sensing 雷达遥感
radar resolution 雷达分辨率
radar return analysis 雷达回波分析
radar scanning 雷达扫描
radar set 雷达装置
radar signal 雷达信号
radar speedometer 雷达测速器
radar surveying 雷达测量
radar trace 雷达反射波
radar traverse 雷达测迹线
radar type speed control 雷达速度控制
radar-directed 雷达定向的；雷达操纵的
radargrammetry 雷达测量
radar-rock unit 雷达岩石单位
radarscope 雷达示波器
radarscope photography 雷达摄影学
radiac 放射性检测仪；辐射计
radial 径向的；放射状的；辐射的
radial acceleration 径向加速度
radial action 径向作用
radial adjustment 径向调整
radial amplitude 径向振幅
radial arm 旋臂
radial attenuation 径向衰减
radial ball bearing 径向球轴承
radial bar 径向钢筋
radial basin 辐形流域
radial basis function 径向基函数
radial bearing 径向轴承
radial beat eccentricity 径向跳动量
radial bracing 径向支撑
radial center 辐射中心
radial character 径向特性
radial characteristic 径向特性
radial check gate 弧形节制闸门
radial clearance 环向间隙；径向净空
radial closure 径向闭合
radial collector well 放射状集水井
radial columnar joint 辐射柱状节理
radial component 径向分量
radial compression 径向压缩
radial compressive stress 径向压应力
radial compressor 离心式压缩机
radial conductive heat transfer 径向热导传热
radial consolidation 径向固结
radial contraction 径向收缩
radial coordinate 径向坐标
radial crack 放射形裂缝；径向裂缝
radial crack in plate 辐板径向裂纹
radial crushing strength 中心破碎强度；径向抗压强度
radial deformation 径向变形
radial diffusion 径向扩散
radial digging force 径向挖掘力
radial direction 径向
radial displacement 径向位移
radial distance 径向距离
radial distortion 径向畸变
radial distortion of lens 透镜径向畸变

radial distribution	径向分布
radial distribution function	径向分布函数
radial drainage	辐射状水系，辐射状流域；径向排液
radial drainage pattern	辐射状水系
radial drainage system	辐射形排水系统
radial drilling machine	摇臂钻床
radial drilling pattern	辐射形炮眼排列方式；辐射形钻孔排列型式
radial end-loading chute	径向端装式溜槽
radial expansion	径向膨胀
radial fault	辐射断层；放射状断层
radial flat jack technique	径向扁千斤顶法
radial flow	径向渗流，径向流动；辐射流
radial flow pump	径流泵
radial flow settling basin	辐射流沉淀池
radial force	辐射力；径向力
radial fracture	径向裂缝
radial fracture propagation	径向裂纹扩展
radial gate	弧形闸门
radial gradient	径向梯度
radial groove	径向沟槽
radial height	径向高度
radial heterogeneous medium	径向非均匀介质
radial highway	辐射式公路
radial impulse load	径向冲击荷载
radial inertia	径向惯性
radial inflow turbine	径向内流式水轮机
radial intersection method	辐射交会法
radial inward flow	辐射内向流
radial line	放射线；辐射线
radial line plot	辐射线图
radial load	径向荷载
radial motion	径向运动
radial movement	垂直运动；放射式运动
radial normal stress	径向正应力
radial orientation	放射状排列
radial oscillation	径向振荡
radial outward flow	径向外流；辐射外流
radial pattern	放射型；辐射型
radial permeability	径向渗透性
radial permeability test	径向渗透试验；径向透气性试验；径向渗透率测试
radial permeability test index	径向渗透性试验指数
radial point interpolation method	径向基点插值法
radial position	辐向位置；径向位置
radial positioning grid	辐射线格网
radial pressure	径向压力
radial principal stress	径向主应力
radial processing	放射状处理
radial ray	径向射线
radial reactor	径向反应器
radial refraction	径向折射
radial resolving power	径向分辨能力
radial response	径向响应
radial rift	放射状断裂
radial run out	径向跳动
radial scan	径向扫描
radial scheme	辐射式系统布置
radial seal	径向密封
radial section	径向截面
radial seepage	径向渗流，径向流动；辐射流
radial sensitivity	辐向灵敏度
radial shear	径向剪切
radial shear wave	径向剪切波
radial shrinkage	径向收缩
radial stacker	转动式堆积机；旋转式堆积机
radial steady-state flow equation	径向稳定流动方程
radial strain	径向应变
radial strength	径向强度
radial stress	径向应力；辐向应力
radial structure	辐射状构造
radial strut in tunnel support	扇形支撑
radial support	径向支撑
radial support bearing	径向支承轴承
radial survey	径向观测
radial symmetry	径向对称
radial tension	径向张力
radial thickener	辐射式浓缩机
radial thrust	径向推力
radial top slicing	下行辐状分层采矿法
radial trace	径向记录道
radial transformation	辐射变换
radial triangulation	辐射三角测量
radial triangulator	辐射三角仪
radial vector	径向矢量
radial velocity	径向速度
radial velocity field	径向速度场
radial ventilation	对角式通风；辐射式通风
radial vibration	径向振动
radial well	辐射井
radian	弧度
radian frequency	角频率；弧频，弧度频率
radian frequency domain	角频率域
radian in horizontal direction	水平方向弧度
radian measure	弧度
radiance	光亮度；辐射率；辐射性能
radiance contour map	辐射外形图；发光度外形图
radiance resolution	辐射率分辨率
radiant drying	辐射法干燥
radiant emittance	辐射率；辐射度
radiant energy	辐射能
radiant field	辐射场
radiant flux	辐射通量
radiant heat	辐射热
radiant heating	辐射供暖；辐射采暖
radiant intensity	辐射强度
radiant matter	辐射物
radiant panel	辐射板
radiant power	辐射功率；辐射能力；辐射通量
radiant quantity	辐射量
radiant rays	辐射线
radiant section	辐射段
radiant structure	放射构造
radiant surface	辐射面
radiant temperature	辐射温度
radiant temperature sensitivity	辐射热感温灵敏度
radiated aggregate	放射状集合体
radiated area	辐射区
radiated field system	辐射场系统
radiated interference	辐射干扰

radiated light 辐射光
radiated logging 放射性测井
radiated microseismic energy 微震释放能量
radiated power 辐射功率
radiated road system 放射形道路系统
radiated structure 放射状构造
radiated wave 辐射波
radiating 辐射；放射
radiating aerial 辐射天线
radiating capacity 辐射能力
radiating flange 散热片
radiating heat 辐射热
radiating matter 放射物质
radiating medium 放射性介质
radiating near-field region 辐射近场区
radiating principal 辐射主体
radiating surface 辐射面
radiating well 辐射井
radiation 辐射；放射；辐射线；放射线
radiation ablation 辐射消融
radiation absorption 辐射吸收
radiation balance 辐射平衡
radiation belt 辐射带
radiation boundary 辐射边界
radiation boundary condition 径向边界条件
radiation contamination 放射性污染
radiation control 放射性控制；辐射防治
radiation damage 辐射线损伤
radiation defect 辐照缺陷；放射性缺损
radiation degradation 辐射降解
radiation density 辐射密度
radiation detection instrument 辐射探测仪
radiation detector 辐射探测器
radiation dosage 辐射剂量
radiation dose 辐射剂量
radiation dosemeter 辐射剂量计
radiation effect 辐射效应
radiation efficiency 辐射效率
radiation energy 辐射能
radiation equilibrium 辐射平衡；放射平衡
radiation exchange 辐射交换；辐射热交换
radiation hazard 放射危害
radiation intensity 辐射强度
radiation layer 辐射层
radiation leakage rate 辐射泄漏率
radiation level 辐射水平；辐射强度；辐射量
radiation logging 放射性测井
radiation loss 辐射损失；辐射损耗
radiation method 辐射法
radiation pattern 辐射模式；辐射特性图
radiation pollution 放射性污染
radiation polymerization 放射聚合
radiation pressure 辐射压力
radiation pressure perturbation 辐射压扰动
radiation protection 辐射防护
radiation pyrometer 辐射高温表；辐射高温计
radiation resistance 辐射电阻
radiation resistant finish 防辐射整理
radiation resolution 辐射分辨率
radiation section 辐射段

radiation sensitizer 辐射敏化剂
radiation sensor 辐射探测器；辐射传感器
radiation shielding 辐射屏蔽
radiation source 辐射源
radiation source safety 放射源安全
radiation standard 辐射标准
radiation station 辐射观测站
radiation strength 辐射强度
radiation survey 辐射测量
radiation survey meter 辐射测量仪
radiation temperature 辐射温度
radiation test site 辐射实验场地
radiation thermal drying 辐射热干燥法
radiation thermometer 辐射温度计
radiation transfer 辐射传输
radiation transfer equation 辐射传输方程
radiation unit 辐射单位
radiation value 辐射值
radiation wall 防辐射墙
radiation waste 放射性废物
radiational damping 辐射阻尼
radiational tide 辐射潮
radiative effect 辐射效应
radiative process 辐射过程
radiative transfer 辐射传输
radiative transfer code 辐射传输规律
radical equation 无理方程
radical expression 根式
radicand age dating 放射性年龄测定
radicand constant 放射性常数
radicand dating 放射性年龄测定
radicand decay law 放射性衰变律
radicand isotope 放射性同位素
radication 开方
radio altimeter 无线电测高计
radio autocontrol 无线电自动控制
radio carbon 碳同位素；放射性碳
radio compass 无线电罗盘；射电罗盘
radio control 无线电控制；无线电操纵
radio control relay 无线电遥控继电器
radio controlled 无线电控制的
radio detection 无线电探测；无线电警戒
radio direction finder 无线电测向器
radio distance-measurement 无线电测距
radio distance-measuring set 无线电测距仪
radio disturbance 无线电干扰
radio emission 无线电发射；射电辐射
radio engineering 无线电工程
radio environment 无线电环境
radio examination X射线透视法
radio frequency 射频
radio frequency power amplifier 射频功率放大器
radio frequency welding 射频焊接
radio geologic time scale 放射性地质年表
radio goniometer 无线电测向计；无线电定向器；无线电测角器
radio horizon 射电水平线；射电作用距离
radio hydrogeological survey 放射性水文地质测量
radio interferometer 射电干涉仪
radio interferometry 无线电干涉测量

radio levelling	无线电高程测量
radio link	无线电通信线路
radio link control	无线链路控制
radio link management	无线链路管理
radio navigation	无线电导航
radio navigation and positioning	无线电导航定位
radio network	无线通信网
radio noise	无线电噪声
radio physics	无线电物理学
radio position fixing	无线电定位
radio positioning	无线电定位
radio range fix	无线电导航定位
radio ranging	无线电测距
radio receiver	无线电接收机
radio relay station	无线电中继站
radio scanner	无线电扫描仪
radio set	无线电收音机；无线电装置
radio signal	无线电信号
radio source	射电源
radio spectrum	射频频谱
radio station	无线电台
radio survey	无线电勘探；无线电测量
radio telecontrol	无线电遥控
radio telecontrol for locomotive	机车无线电遥控
radio telecontrol for railway	铁路无线电遥控
radio telemetering	无线电遥测；无线电遥测的
radio telerecording seismograph	无线电遥测地震仪
radio tracer	放射性示踪物质
radio wastes	放射性废物
radio-acoustic ranging	电声测距法
radioactivation analysis	辐射激化分析；放射活化分析
radioactive	放射性的
radioactive age determination	放射性年龄确定
radioactive anomaly	放射性异常
radioactive assay	放射性分析
radioactive background correction	辐射本底校正
radioactive chemical element	放射性化学元素
radioactive contaminant	放射性污染物
radioactive contamination	放射性污染
radioactive debris	放射性碎屑；放射性碎片
radioactive decay	放射性衰变
radioactive decay rate	放射性衰变率
radioactive decay series	放射性衰变系列
radioactive decontamination	消除放射性
radioactive density meter	放射性密度计
radioactive deposit	放射性沉淀物
radioactive detector	放射性检测器
radioactive disintegration	放射性衰变
radioactive dust	放射性尘埃；放射性灰尘
radioactive earth	放射性土
radioactive effect	放射性效应
radioactive effluent	放射性废水；放射性流出物
radioactive element	放射性元素
radioactive element decay	放射元素衰变
radioactive exploration	放射性勘探
radioactive fallout	放射性沉降
radioactive gamma survey	放射性伽马测量
radioactive geophysical prospecting	放射性物探
radioactive groundwater	放射性地下水
radioactive half-life	放射性半衰期
radioactive halo	放射性晕
radioactive heat	放射性热
radioactive heat source	放射性热源
radioactive hydrogeochemistry	放射性水文地球化学
radioactive hydrogeology	放射性水文地质学
radioactive indicator	放射性指示剂
radioactive isotope	放射性同位素
radioactive leak	放射性泄漏
radioactive logging	放射性测井
radioactive marking	放射性同位素标记；示踪剂标记
radioactive material	辐射性物质；放射性物质
radioactive measurement	放射性测量
radioactive metal	放射性金属
radioactive method for groundwater search	放射性找水法
radioactive mineral	放射性矿物
radioactive nuclide	放射性核素
radioactive nuclide survey	放射性核素测量
radioactive ore	放射性矿石
radioactive ore detector	放射性矿石探测器
radioactive pollution	放射性污染
radioactive prospecting	放射性勘探
radioactive pulp density meter	放射性矿浆密度计
radioactive radiation	放射性辐射
radioactive rare metal	放射性稀有金属
radioactive ray	放射性射线
radioactive residue	放射性残渣
radioactive rock	放射性岩石
radioactive sensing device	放射传感器
radioactive series	放射系列
radioactive solid waste	放射性固体废物
radioactive source	放射源
radioactive substance	放射性物质
radioactive survey	放射性测量
radioactive tracer	放射性示踪剂
radioactive tracer log	放射性示踪剂测井
radioactive tracer logging	放射性示踪测井
radioactive tracing	放射性示踪
radioactive transformation	放射性转化；放射性蜕变
radioactive waste	放射性废弃物
radioactive waste disposal	放射性废物处置
radioactive waste storage	放射性废物贮存
radioactive wastewater	放射性废水
radioactive water	放射性水
radioactive well logging	放射性测井
radioactivity	放射能力；放射性；放射现象
radioactivity analysis	活化分析
radioactivity anomaly	放射性异常
radioactivity background	放射性本底
radioactivity dating	放射性年龄测定
radioactivity decay	放射性衰变
radioactivity determination	放射法测定
radioactivity equilibrium	放射性平衡
radioactivity hazards	放射性危险
radioactivity hydrogeology	放射性水文地质学
radioactivity indicator	放射性强度指示器
radioactivity level	放射性程度
radioactivity log	放射性测井
radioactivity logging	放射性测井
radioactivity logging method	放射测井法

radioactivity pollution 放射性污染
radioactivity prospecting 放射性勘探
radioactivity standard 放射性标准
radioactivity well logging 放射性测井
radioalarm equipment 无线电报警器
radiobeacon 斜度测量
radiocarbon 放射性碳
radiocarbon age 放射性碳年龄
radiocarbon chronology 放射性碳年代学
radiocarbon dating 放射性碳测定年龄
radiocarbon stratigraphy 放射性碳地层学
radiocarbon tracer 放射性碳示踪物
radiochemical analysis 放射化学分析
radiochemical method 放射化学法
radiochemistry 放射化学
radiochromatogram 放射性色谱图
radiogenic dating 放射性年龄测定
radiogenic heat 放射生成热；辐射热；放射热
radiogenic isotope 放射性同位素
radiogeochemistry 放射性地球化学
radiogeology 放射地质学；放射性地质学
radiogram 无线电报；射线照片
radiographic examination 透射检查
radiographic inspection 放射照相检验；放射线探伤
radiographic source 射线源
radiographic technique 放射线照相技术
radiographic testing 射线检验
radiography analysis 放射线照相分析
radioindicator 放射性指示剂
radioisotope 放射性同位素
radioisotope logging 同位素测井
radioisotope ratio 放射性同位素比值
radioisotope survey 放射性同位素测量
radioisotope tracer 放射性同位素示踪剂
radioisotope unit 同位素室
radioisotopic tracer 放射性同位素指示剂
radioisotopic tracing 放射性同位素示踪
radioistope 放射性同位素
radiolitic texture 放射扇状结构
radiologic hazard 放射危害
radiologic physics 放射物理学
radiologic safety control 放射性安全管理
radiological contamination 放射性污染
radiological data 辐射数据
radiological hazard 辐射危害
radiological imaging 辐射显像
radiological monitoring 辐射监控
radiological survey 辐射检查；X-射线检查
radiometer 辐计计；放射计；射线探测仪
radiometer gauge 辐射式真空计；辐射测压计；射线探测仪
radiometric age dating 放射性年龄测定
radiometric correction 辐射改正
radiometric dating 放射性年龄测定
radiometric method 放射性测量法
radiometric prospecting method 辐射勘探法
radiometric survey 放射性勘探；放射性测量
radionuclide migration 放射性核素转移
radiotracer analysis 放射示踪分析
radiotracer method 放射性示踪剂法

radius 半径
radius at bend 曲率半径
radius bar 半径杆
radius curvature 曲率半径
radius deformation 回弹变形
radius of action 作用半径；活动半径
radius of air-water interphase 空气-水界面的曲率半径
radius of circular arc 圆弧半径
radius of contractile skin 收缩膜的曲率半径
radius of convergence 收敛半径
radius of corner 圆角半径
radius of curvature 曲率半径
radius of curvature in a normal section 法截弧曲率半径
radius of curvature method 曲率半径法
radius of curve 曲线半径
radius of cut at grade 挖掘净半径
radius of disruptive action 破裂作用半径
radius of drainage 排油半径
radius of fractured zone 破裂区半径
radius of grouting spread 浆液扩散半径
radius of gyration 惯性半径；回转半径
radius of inertia 惯性半径
radius of influence 影响半径
radius of influence of well 井的影响半径
radius of interpolation circle 内插圆半径
radius of investigation 探测半径
radius of lead curve 导曲线半径
radius of plastic zone 塑性区半径
radius of protection 保护半径
radius of receptibility 可感半径；震感半径
radius of relative stiffness 相对刚度半径；刚比半径
radius of seismic danger due to blasting 爆破地震危险半径
radius ratio 半径比
radius vector 向量径；矢径
radius-depth ratio 半径-深度比
radius-vector 矢径
radix number 基数；底数
radix point 小数点
radon method 氡测法
radon survey 氡量测量
raft bridge 浮桥
raft chute 筏道；浮运水槽
raft foundation 筏基；筏形基础；浮筏基础
raft foundation under walls 墙下筏形基础
raft stiffening 筏基加固
raft way 筏道
rafter set 基坑支架
rag agenesis 断裂作用
rag bolt 棘螺栓；地脚螺栓
rag paving 石板铺面
rag rubble 粗面块石
rag rubble walling 毛石墙砌体
rag stone 粗结构硬岩石；含化石的砂质粗石灰岩；铺路石板
rag work 石板砌合，砌石板
ragged hole 孔壁不平整的钻孔
ragged terrain 崎岖地形
ragstone 硬质岩
rail corrosion 钢轨锈蚀

rail corrugation 轨头波纹磨损
rail cracks 钢轨裂纹
rail creep indication posts 钢轨位移观测桩
rail crew 铁路维修队
rail defect and failure 钢轨伤损
rail drilling machine 钢轨钻孔机
rail drilling tool 钢轨钻孔器
rail dump 铁道运输排土场
rail elevation 轨头标高
rail end breakage 轨头垂直劈裂
rail facility 铁路
rail fatigue 轨道疲劳
rail flat car 轨道平车
rail gauge 轨距
rail grinding machine 磨轨机
rail haulage 铁路运输
rail head 轨头
rail impedance 钢轨阻抗
rail insulation 钢轨绝缘
rail joint bond 钢轨连接器
rail laying 铺轨
rail laying machine with cantilever 悬臂式铺轨机
rail leakage resistance 钢轨泄漏电阻
rail level 轨面水平；轨头标高
rail link 铁路联机
rail movement joint 路轨活动接合件
rail potential to ground 钢轨对地电位
rail press 压轨机
rail profile gauge 钢轨磨损检查仪
rail profile measuring car 钢轨磨损检查车
rail section 钢轨剖面
rail shifting machine 移道机
rail splice 铁轨接合；鱼尾板接合
rail square 准轨尺或准轨距；轨道角尺
rail stress 钢轨应力
rail tensor 钢轨拉伸器
rail track 轨道
rail track installation area 轨道安装区
rail traffic 铁道交通
rail transit network planning 轨道交通线网规划
rail transport 轨道运输
rail travel 轨道运行
rail treadle 轨道踏板
rail welding machine 焊轨机
rail welding seam 钢轨焊缝
rail/wheel contact stress 轮/轨接触应力
rail-hugging stability 轨道稳定性
railroad bed 铁路路基
railroad branch 铁道支线
railroad break down due to flood 水害断道
railroad bridge 铁路桥梁
railroad building 铁路房屋；铁路建筑
railroad construction 铁路建筑
railroad crossing 铁路交叉点
railroad embankment 铁路路堤
railroad engineering 铁路工程学
railroad engineering survey 铁路工程测量
railroad excavation 铁路挖方
railroad gauge 轨距
railroad grade crossing 道口
railroad line 铁路线
railroad system 铁路系统
railroad tunnel 火车隧道；铁路隧道
railway aerial photogrammetry 铁路航空摄影测量；铁路航测
railway aerial surveying 铁路航空勘测
railway bed 铁路路基
railway boundary 铁路边界
railway bridge 铁路桥梁
railway bridge engineering 铁路桥梁工程
railway classification 铁路等级
railway connecting line 铁路连接线；铁道连接线
railway construction 铁路建筑；铁道工程
railway construction clearance 铁路建筑限界
railway construction gauge 铁路建筑限界
railway culvert 铁路涵洞
railway curve 铁路曲线；铁道曲线
railway data exchange system 铁路数据交换系统
railway date-processing system 铁路运输数据处理系统
railway embankment 铁路路堤
railway engineering 铁道工程
railway freight traffic organization 铁路货运组织
railway junction terminal 铁路枢纽
railway line 铁路线路；铁道线路
railway line engineering 线路工程
railway location 铁路选线；铁道选线
railway network 铁路网
railway operation 铁路运营
railway operation information system 铁路运营信息系统
railway passenger traffic 铁路旅客运输
railway pile driver 有轨打桩机
railway protection area 铁路保护区
railway radio communication 铁路无线电通信
railway reconnaissance 铁路勘测；铁道勘测
railway reserve 铁路专用范围
railway science 铁道科学
railway service train 路用列车
railway sign 铁路标志；铁道标志
railway special line 铁路专用线
railway station 火车站
railway subgrade 铁路路基；路基；铁道路基
railway survey 铁路测量；铁道测量
railway technology 铁路技术
railway territory 铁路用地
railway track 铁路；轨道
railway track material 铁路用材料
railway traffic organization 铁路运输组织
railway traffic profit 铁路运输利润
railway train load norm 列车重量标准
railway transfer platform 铁路转运站
railway transport 铁路运输
railway transportation 铁路运输
railway tunnel 铁路隧道
railway vehicle 铁道车辆
railway yard 铁路编组站；调车场
rain area 雨区；雨水排泄区
rain capacity 降雨量
rain channel 水蚀沟；雨水沟
rain drop effect 雨点效应
rain drop splash erosion 雨滴溅蚀

rain duration 降雨历时
rain erosion 雨蚀作用
rain free period 无雨期；干季
rain gauge 雨量计
rain graph 雨量图
rain height 降雨量
rain impression 雨痕
rain imprint 雨痕
rain intensity 降雨强度
rain precipitation 雨量；降雨
rain print 雨痕
rain proof 防雨的
rain recorder 雨量记录器
rain sampler 雨水取样器；降水收集器
rain shadow 未着雨地带；雨影
rain shelter 避雨亭
rain simulator 人工降雨设备
rain tight 防雨的
rain wash 雨水冲刷；雨蚀
raindrop erosion 溅击侵蚀；雨滴侵蚀
raindrop impression 雨滴痕
raindrop imprint 雨痕
raindrop size distribution 雨滴谱
rainer 人工降雨器
rainfall 降雨；雨量；降雨量
rainfall acidification 降雨酸化
rainfall acidity 雨水酸度
rainfall action 降雨作用
rainfall amount 降雨量
rainfall area 降雨区；降雨面积
rainfall characteristics 降雨特性
rainfall chart 雨量图
rainfall curve 降雨量曲线
rainfall data 降雨资料
rainfall density 降雨强度
rainfall depth 降雨量
rainfall depth-duration curve 雨量历时关系曲线
rainfall discharge 降雨流量
rainfall distribution 降雨分布
rainfall distribution coefficient 降雨分布系数
rainfall duration 降雨历时
rainfall efficiency 降雨效率
rainfall erosion 降雨侵蚀，雨蚀作用
rainfall erosion index 降雨冲蚀指数
rainfall flood 降雨洪水；雨洪
rainfall frequency 降雨频率
rainfall index 降雨指数
rainfall infiltration 降雨入渗；降雨入渗量
rainfall infiltration coefficient 降水入渗系数
rainfall intensity 降雨强度；雨量强度
rainfall intensity curve 降雨强度曲线
rainfall intensity frequency 降雨强度频率
rainfall intensity recorder 雨强计
rainfall intensity recurrence interval 降雨强度重现期
rainfall intensity-duration curve 降雨强度历时曲线
rainfall interception 降雨截留
rainfall loss 降雨损失
rainfall map 降水图
rainfall mass curve 降雨累积曲线
rainfall penetration 降雨渗入深度

rainfall percentage 降雨百分率
rainfall probability 降雨几率；降雨概率
rainfall province 雨区
rainfall rate 降雨量；降雨率
rainfall recharge 大气降水补给
rainfall record 雨量记录
rainfall recurrence interval 降雨重现期
rainfall regime 雨量型
rainfall runoff 降雨径流；雨水径流
rainfall season 雨季
rainfall simulator 降雨模拟器；模雨计；模拟降雨装置
rainfall station 雨量站
rainfall storm 暴雨
rainfall threshold 降雨阈值
rainfall variation 降雨变化
rainfall volume 降雨量
rainfall year 降雨水文年
rainfall-depth contour 降雨等深线
rainfall-evaporation balance 降雨蒸发平衡
rainfall-induced landslide 降雨型滑坡
rainfall-runoff correlation 降雨径流相互关系
rainless region 无雨区；干旱区
rainmaker 人工降雨设备
rain-producing system 致雨系统
rainproof 防雨的
rainstorm 暴雨
rainstorm centre 暴雨中心
rainstorm pattern 暴雨类型
rainstorm runoff 暴雨径流
rain-tight 防雨的
rainwater 雨水；软水
rainwater pipe 雨水管
rainy climate 多雨气候
rainy day 降雨日
rainy season 雨季，降雨期
rainy season construction 雨季施工
rainy tropics 多雨热带
rainy year 多雨年
raise 提升；起拱
raise advance 上山掘进
raise and winze assay 上山、下山采样分析
raise boring guidance 凿反井指向器
raise chute 溜井；漏斗天井
raise drilling 上山钻孔；上升式钻孔
raise drilling system 反井钻进系统
raise driving 上山掘进；天井掘进
raise lift 溜井高度；溜井的提升高度
raise opening 天井井口；上山开拓
raise stope 上向工作面；逆倾斜工作面
raised arch 陡拱顶
raised bench 上升阶地
raised bog 高地沼泽
raised mining 上向回采；上向掘进
raised platform 升高平台
raised shaft 反井
raising by cage lift 天井吊罐法掘进
raising of water level 水位升高
raising ventilation 天井掘进通风
raising water 溢出水；溢出水带
raising-through 向上凿穿

rake	倾斜；斜脉	ramp signal	斜坡信号
rake conveyor	刮板输送机；耙式输送机	ramp slope	斜坡坡度
rake dozer	耙式推土机	ramp step wave	斜阶跃波
raked floor	地面升起	ramp valley	对冲断层谷
raked pile	倾斜桩	ramp velocity medium	渐变速度介质
raking drain	排水斜管	ramp velocity profile	渐变速度剖面；渐变速度分布图
raking equipment	粗筛机	rampactor	跳击夯
raking pile	斜桩	rampant arch	跛拱；高低脚拱
raking shore	斜撑	randanite	硅藻土；腐殖质硅藻土
ram	撞击，填塞；千斤顶	randing	浅井勘探
ram anchor pin	千斤顶锚固销	random	随机的；无规则的；任意的
ram attachment bracket	安装千斤顶托架	random access	随机存取
ram cover	千斤顶罩；千斤顶盖	random access memory（RAM）	随机存取存贮器
ram engine	打桩机	random algorithm	随机算法
ram guide	桩锤导架	random amplitude	随机振幅
ram home	尽量捣固；夯击到底	random amplitude error	随机振幅误差
ram impact machine	捣锤冲击机；打夯机	random analysis	随机分析
ram locating bracket	千斤顶定位托架	random arrangement	随机排列
ram machine	打桩机	random array	随机排列
ram packer	冲压夯	random ashlar	乱砌料石
ram pile driver	打桩机	random autoregressive model	随机自回归模型
ram plate	千斤顶板	random bulk	料堆
ram pressure	速度压力	random character	随机特性
ram pull	千斤顶拉力；千斤顶拉回	random choice method	随机选择法
ram pump	桩塞泵	random component	随机分量
ram push	千斤顶推力；千斤顶推进	random coring	混凝土钻探抽样测试
ram sounding method	锤击触探方法	random coupling	随机耦合
ram steam pile driver	汽锤打桩机	random crack	不规则裂缝；随机裂缝
ram striker valve	千斤顶碰撞阀；千斤顶行程限制阀	random depositional texture	无序沉积结构
ram tester	撞击试验机；冲击试验机	random deviation	不规则井斜
ramble roof	随采随落的顶板	random digit	随机数字
rambling stone	松散岩石	random direction	随机方向
rammed clay	夯实黏土；捣密黏土	random displacement	随机位移
rammed earth	夯实土	random distribution	随机分布；不规则分布
rammed earth building	土建筑	random drilling	任意钻井
rammed earth construction	夯土结构	random dynamic loading	随机动荷载
rammed earth wall	夯土墙	random earthquake excitation	随机地震激振
rammed-soil foundation	夯实土基；夯土基础	random earthquake motion	随机地震动
rammed-soil pile	夯实土桩	random earthquake response	随机地震反应
rammer	夯具；夯实机；夯锤	random element	随机成分
rammer compactor	夯土机；夯实机	random element method	随机有限元法
ramming	铸砂捣实；打夯；捣实；锤击	random error	偶然误差；随机误差
ramming bar	夯实杆；捣固杆	random event	随机事件；随机现象
ramming by heavy hammer	重锤夯实	random experiment	随机试验
ramming compound	捣打混合物	random failure	随机失效
ramming depth	打夯深度	random fatigue	随机疲劳
ramming in layers	分层夯实	random fault	随机故障
ramming machine	捣固机；打夯机	random fault analysis	随机故障分析
ramming mix	捣打料	random field	随机场
ramming speed	抛砂速度；打夯速度；捣实速度	random field model	随机场模型
ramp	陡冲断层；对冲断层；斜坡道	random file	随机文件
ramp anticline	冲断层面背斜	random flow	不规则流动
ramp control	陡度调整；斜度调整	random fluctuating nature	随机波动特性
ramp function	陡度函数；倾斜度函数	random fluctuation	无规则波动；随机起伏
ramp mining	斜坡道开采	random forcing function	随机扰动函数
ramp pan	连接斜槽；过渡槽	random fractal	随机分形
ramp plate	倾斜铲煤板；倾斜溜板	random function	随机函数
ramp region	冲断层陡倾区	random geometry	任意几何状态
ramp rises along thrust	顺冲断层的对冲隆起	random impact	随机碰撞
ramp road	斜面路	random inhomogeneity	随机非均匀性

random inspection	抽查
random interference	随机干涉效应
random loading	随机荷载
random mantissas	随机尾数
random medium	随机介质
random memory access	随机存取器
random mixed-layer structure	无序混层结构
random model	随机模式；随机模拟
random motion	无规则运动，随机运动
random number	随机数
random occurrence	随机事件
random ordered sample	随机有序样本
random orientation	随机取向
random packing	乱堆充填
random parameter	随机参数
random path	随机路线
random pattern	随机模式
random pattern of well spacing	任意布置井位
random perforation orientation	炮眼不规则排列；随机射孔方位
random permutation	随机排列
random phenomenon	随机现象
random pillar	随意布置的矿柱
random process	随机过程
random processing	随机处理
random projection	任意投影
random quantity	随机量
random reading error	随机读数误差
random response	随机反应
random riprap	堆石工程；抛石工程；乱石堆
random rockfill	乱石堆筑
random rubble armored structure	不规则堆石护面结构
random rubble fill	乱石堆填；房渣堆填
random rubble wall	不规则砌石墙；乱石砌石墙
random sample	随机样本
random sampler	随机取样；随机采样
random sampling	随机抽样
random sampling distribution	随机抽样分布
random scanning	随机扫描
random search method	随机搜索方法
random search optimization algorithm	随机搜索优化算法
random sequence	随机顺序；随机时序；随机序列
random series	随机序列
random spacing	不规则间距；不规则排列
random stone work	乱砌石工
random stratified reservoir	不规则层状油层
random structural system	随机结构体系
random structure	随机结构；不规则构造；杂乱构造
random test	抽样试验；随机试验
random timbering	不规则支护
random time spacing	随机时间间隔
random traverse	任意连测导线
random trial method	随机试验法
random turbulence	不规则紊流
random uncertainty	随机不确定度
random variable	随机变量
random variable function	随机变量函数
random variate	随机变量
random vector	随机向量
random vibration	随机振动
random vibration analysis	随机振动分析
random vibration theory	随机振动理论
random walk	随机游动
randomization	随机化
randomization method	随机化方法
randomized	随机法；随机的
randomly layered model	随机呈层模型
random-media model	随机介质模型
randomness	随机性
randomness number	随机数
randomness of discontinuity spacings	不连续面间距的随机性
randomness process	随机过程
randomness test	随机性检验；随机性检测
random-perturbation optimization	随机扰动最优化
range accuracy	测距精确度
range adjustment	距离调整；范围调整；量程调整
range ambiguity	距离模糊
range constraint	限程约束
range deviation	射程偏差
range direction	距离方向
range extension	范围扩大；量程的扩展
range extraction	区域提取
range finder	测距仪
range finding apparatus	测距仪
range index	变幅指数
range indicator	距离显示器
range measurement	距离测定
range measurement system	测距系统
range meter	测距仪
range of adjustment	调节范围
range of an instrument	仪表量程
range of applicability	可用性区域
range of application	适用范围
range of control	控制范围
range of dispersion	扩散范围；分散范围
range of error	误差范围
range of fluctuation	变动范围
range of influence	影响范围
range of lost strata	地层缺失范围
range of pressure	压力范围
range of profitability	有利区域
range of regulation	调节范围；调整范围
range of scatter	扩散范围
range of stability	稳定范围
range of stage	水位变幅
range of stress	应力范围
range of thrust	推力范围
range of variation	变化范围
range of visibility	能见度
range of wave length	波段
range pole	测距标杆；标杆；视距尺
range positioning system	测距定位系统
range processing	距离数据处理
range rate	测距变化率
range resolution	距离分辨率
range restriction	范围限定
range rod	视距尺；标杆；标尺
range scale	距离刻度
range sensing	距离探测

range sweep	距离扫描
range switch	波段开关；量程选择开关；距离转换开关
range system	测距系统
range tie	测距点
range transformation	极差变换
range unit	测距装置
range zone	延限带；延伸带；范围带
rangeability	量程
rangefinder	测距仪
range-only radar	测距雷达
ranger resolution	距离分辨率
range-range determination	双距定位法
range-range positioning	双距离定位；双圆定位
ranging jack	调高千斤顶
ranging line	测距线
ranging pole	标尺测杆
ranging rod	视距尺；标杆；标尺
ranging theodolite	测距经纬仪
rank	排，列；分类；级别；煤级，煤阶
rank correlation	秩相关；等级相关
rank correlation analysis	秩相关分析
rank defect adjustment	秩亏平差
rank defect network adjustment	秩亏网平差
rank of a determinant	行列式的秩
rank of calcification	钙化程度
rank of matrix	矩阵的秩
rank parameter	品级参数
rank stage	煤级阶段；煤化阶段
ranked data	分级数据
Rankine analysis	朗肯分析法
Rankine state	朗肯状态
Rankine states of stresses	朗肯应力状态
Rankine's earth pressure theory	朗肯土压力理论
Rankine's formula	朗肯公式
ranking system	先后次序排列系统；序列方法；分级系统
rapid ascent	陡坡
rapid assessment	快速评估
rapid blow drilling	快速冲击钻进
rapid cement	快硬水泥
rapid charge	快速充电
rapid condensation	快速凝结
rapid consolidation test	快速固结试验
rapid construction	快速施工
rapid curing	快速养护
rapid deterioration	迅速变质
rapid developer	快速显像剂
rapid drawdown	水位骤降
rapid drivage	快速掘进
rapid drying	快速干燥
rapid earth flow	快速泥土流
rapid earthquake information report	地震速报
rapid erosion	快速侵蚀
rapid excavation	快速采掘；快速掘进
rapid excavation and mining	高速射弹掘进法
rapid field assessment	快速现场检定
rapid flow	急流；滩流
rapid flowage	急剧流动
rapid fluctuation of water level	水面急剧波动
rapid hardening	速凝；快速硬化
rapid hardening cement	快硬水泥；早强水泥
rapid hardening cement grouted bolt	快硬水泥锚杆
rapid hardening cement	快凝水泥
rapid impact rammer	快速冲击夯实
rapid indexing	快速分度
rapid inspection	快速检查
rapid loading	快速加载
rapid memory	快速存储器
rapid permeability	快速渗透
rapid recharge	快速补给
rapid reconnaissance	快速勘测
rapid river	湍急河流
rapid setting admixture	速凝剂
rapid setting cement	速凝水泥
rapid shaft-sinking method	快速凿井法
rapid soil classification	土的简易分类法
rapid solidification	快速凝固
rapid static surveying	静态快速测量
rapid stream	湍急河流
rapid survey	快速测量
rapid test	快速化验；快速分析
rapid test for soil	土壤速测法
rapid transit railway	高速铁路
rapid transit system	快速运输系统
rapid transit tunnel	地下铁；地下高速铁道
rapid tunnel driving	隧洞快速掘进
rapid-blow drilling	高频冲击钻
rapidly varied flow	急变流
rapidly varied unsteady flow	急变不稳定流
rapid-positioning	快速定位
rapids regulation	急滩整治
rapid-sand filter strainer	快速沙滤器
rapid-setting	快凝的；早凝的
rapid-setting cement	快凝水泥
rapid-setting grout	速凝浆液
rapid-varying	迅速变化的
rapping device	振动装置；震击装置
rare earth element (REE)	稀土元素
rare earth treatment	稀土元素处理
rare gas	稀有气体；惰性气体
rarefaction wave	膨胀波；稀疏波；余波
raster data	栅格数据
raster display	光栅显示
raster display device	光栅显示装置
raster element	光栅元
raster format	光栅格式
raster graphics	光栅图形
raster pattern	光栅模式
raster plotter	光栅绘图仪
raster plotting	栅格绘图
raster scan	光栅扫描
raster scan device	光栅扫描设备
raster scanner	光栅扫描器
raster scanning	光栅扫描
ratchet	棘爪；棘轮机构
ratchet brace	扳钻，摇钻
ratchet drill	棘轮凿岩机
ratchet jack	棘轮千斤顶
ratchet lever jack	棘轮杠杆起重器
rate	率；等级；速度；定额

rate and sequence of draw	回采速度和顺序
rate coefficient	速率系数
rate constant	速率常数
rate controlling step	速率控制阶段；速率控制步骤
rate decline curve	产量递减曲线
rate determining step	速率决定阶段
rate equation	速率方程
rate form	速率形式
rate independent damping	速率无关阻尼
rate independent theory	速率无关理论
rate indicator	速度计
rate limitation	速度限制
rate meter	速度计
rate method	速率法
rate network	比率网络
rate of advance	钻进速度；掘进速度
rate of change	变化率
rate of circulation	循环速度
rate of climate change	气候变化率；气候变化速度
rate of coastal erosion	海岸侵蚀速度
rate of collapse	倒塌率
rate of compression	压缩率；压缩比
rate of consolidation	固结速率
rate of convergence	收敛率
rate of conveyance loss	输水损失率
rate of corrosion	腐蚀速率；腐蚀速度
rate of crack propagation	裂缝扩展速度
rate of creep	蠕变速率
rate of crushing	压碎速率
rate of crust uplifting	地壳隆起速率
rate of curve	曲线斜率；曲率
rate of cutting	切削速率
rate of damping	阻尼系数；减幅系数
rate of decay	衰减速度，衰落速度，衰变率；腐烂速度，破碎程度；风化程度；声压衰减速度
rate of deceleration	减速率
rate of decline	递减速度
rate of delivery	输送率；给料速度；排液速度
rate of deposition	沉积速度；沉积速率
rate of descent	下降率；下降速度
rate of detonation	爆速；爆轰速度
rate of diagenetic transformation	成岩改造率
rate of diffusion	扩散速度；扩散速率
rate of direction change	方位变化率
rate of discharge	放电率；流量，排量；卸载率；排放速率
rate of displacement	驱替速度；置换率
rate of drying	脱水率；干燥率
rate of earthquake occurrence	地震发生率
rate of easement curvature	缓和曲线半截变更率
rate of elastic wave propagation	弹性波传播速度
rate of elongation	延伸率；伸长率
rate of emission	排放速度
rate of equipment utilization	设备利用率
rate of erosion	侵蚀率；冲刷率
rate of excavation	开采速率
rate of exchange	兑换率；交换率
rate of expansion	膨胀率；膨胀速率
rate of face advance	工作面推进速度
rate of fall	降落速度；沉降速度
rate of fill settlement	充填材料沉缩率
rate of filtration	滤速
rate of flow	渗流速率；流量
rate of free expansion	自由膨胀率
rate of gasification	气化率
rate of gradient	百分比坡度
rate of gross profit	毛利率
rate of groundwater discharge	地下水径流速率
rate of groundwater flow	地下水流动速率
rate of groundwater level decline	地下水水位下降速率
rate of grout absorption	进浆速率；泥浆吸收速率；吸浆量
rate of grout acceptance	吸浆量；泥浆吸收速率
rate of growth	增长率；生长速度
rate of handling charge	装卸费率
rate of heat exchange	热交换率
rate of heat flow	热流量；热流速率
rate of infiltration	入渗量；入渗率
rate of inflow	流入速度；流入率；进水流量
rate of loading	加荷速率；加荷速度
rate of methane emission	瓦斯泄出速率
rate of motion	运动率
rate of movement	移动速率
rate of mud content	含泥率
rate of natural increase	自然增长率
rate of net drilling	纯钻进速度
rate of occurrence	重复率
rate of operation	开工率
rate of outflow	流出速度；流出率
rate of penetration	钻进速率；贯入速率；渗漏速率
rate of penetration recorder	机械钻速记录仪
rate of percolation	渗透速率
rate of piercing	钻进速率
rate of plant coverage	植物覆盖率
rate of pressure decline	压力递减率
rate of pressure rise	压力升高比
rate of production	生产率；产量
rate of profit	利润率
rate of progress in depth	开采深度；推进深度
rate of propagation	传播速度
rate of pumping	泵送速度；抽油速度
rate of reaction	反应速率
rate of real loading capacity	实载率
rate of rebound	回弹率
rate of recovery	采出速度，回采率；产油率；回收率
rate of recovery of capital	资本回收率
rate of redemption	偿还率
rate of reduction	破碎比
rate of reinjection	回注速率
rate of return method	收益率评估法
rate of return on fixed assets	固定资产回报率
rate of return-on-investment	投资回收率
rate of revolution	转速
rate of rise	增长速度；曲线上升斜率
rate of roundtrip	行程钻速；回次钻速
rate of runoff	径流率；径流模数
rate of sampling	取样比率
rate of secondary consolidation	次固结速率
rate of sediment yield	产沙率
rate of sedimentation	沉淀速度；沉降率；沉积速率

rate of seepage 渗出率；渗出量
rate of setting 沉降速率；凝固速度
rate of shear 剪切速率
rate of shear strain 剪应变速率
rate of shearing 剪切速率
rate of sinking 掘进速度；下沉速度
rate of slip 滑动速率
rate of specific surface area 比表面积率
rate of speed 速率
rate of speed fluctuation 转速波动率
rate of spoiled 废品率
rate of strain 应变速率
rate of stress application 施加应力的速率
rate of stripping 剥离比
rate of subsidence 沉陷速度
rate of surface runoff 地表径流率；地面径流率
rate of swelling 膨胀速率
rate of swelling of soil 土的膨胀率
rate of taxation 税率
rate of temperature change 温度改变的速率；变温率；温度变化率
rate of temperature increase 增温率
rate of throughput 流过率
rate of tidal current 潮流速率
rate of transition curve 过渡曲线曲率
rate of travel 运行速度
rate of travel of flood peak 洪峰移动速度
rate of vegetation 植物覆盖率
rate of volume flow 体积流量
rate of water injection 注水速率
rate of water level rise 水位上升速率
rate of water loss 失水率
rate of water make-up 补水率
rate prediction 产量预测
rate process theory 速率过程理论
rated capacity 计算容量；额定容量；额定功率；校准能力
rated condition 规定条件；额定条件；计算条件
rated consumption 额定耗量
rated depth 额定深度
rated flow 额定流量
rated flows model 标称流量模型
rated frequency 额定频率
rated head 额定扬程，额定水头
rated hole diameter 额定钻孔直径
rated horsepower 额定马力；额定功率
rated input 额定输入量
rated load 额定荷载；额定装载量
rated payload 额定载重量
rated power 额定功率；额定动力
rated pressure 额定压力
rated pump pressure 额定泵压
rated radio frequency output power 额定射频输出功率
rated revolution 额定转速
rated rope pull 额定钢丝绳拉力；计算钢丝绳拉力
rated specific speed 额定比速
rated speed 额定速度
rated thrust 额定推力；标准推力；计算推力
rated tractive force 额定牵引力；计算牵引力
rated value 额定值

rated waste treatment capacity 额定垃圾处理量
rated yield load 额定屈服阻力；额定工作阻力
rate-maintenance capability 稳产能力
ratemeter 速率计
rate-of-failure curve 破坏率曲线；失效率曲线
rate-of-flow indicator 流速指示器
rate-of-frontal-advance equation 前缘推进速率方程
rate-of-penetration log 钻速记录
rate-of-return comparison 回收率比较法
rate-time relationship 产量-时间关系；速度-时间关系
rating curve 流量曲线；率定曲线
rating discharge curve 率定流量曲线
rating form for estimate 概算定额
rating of merit 性能评价
rating of seismic intensity 地震烈度评定
rating table 率定表
rating value 额定值；计算值
ratio 比值；比例；比率
ratio analysis 比率分析
ratio analysis method 比值分析法
ratio and phase meter 振幅和相位差计
ratio by volume 体积比
ratio combination 比值组合
ratio composite image 比值合成图像
ratio contour map 比值等值线图
ratio control 比率控制
ratio curve 比值曲线
ratio data 比值数据
ratio enhanced image 比值增强图像
ratio enhancement 比值增强；比例增强
ratio error 比例误差
ratio estimation 比值估计
ratio image 比值影像；比值图像
ratio information 比值信息
ratio inspection 比率检验
ratio map 岩石厚度比率图
ratio method 比值法
ratio of abundance 丰度比
ratio of attenuation 衰减率
ratio of closing error 闭合比
ratio of collapsed buildings 房屋倒塌率
ratio of compression 压实度；压缩比
ratio of concentration 选矿比
ratio of curvature 曲率化
ratio of damping 阻尼系数
ratio of depreciation 折旧率
ratio of division 分割比
ratio of drawing 牵伸比
ratio of ejected to injected mercury 退汞压汞比
ratio of elastic modulus 弹性模量比
ratio of elongation 伸长率
ratio of enrichment 富集比
ratio of expansion 膨胀率
ratio of floor loading 地基系数
ratio of floor loading to settlement 地基系数；地基沉陷系数
ratio of frost heaving 冻胀率
ratio of gravity to capillary force 重力-毛细管力比
ratio of great pore 大孔隙比
ratio of ions in groundwater 地下水元素比值

ratio of length to diameter 长径比
ratio of length-diameter 长径比
ratio of light weight to loading capacity 自重系数
ratio of mesh 网格比
ratio of mixture 混合比
ratio of peak strength to residual strength 峰残强度比
ratio of preparation work 采准工作量
ratio of principal stress 主应力比
ratio of reduction 破碎比
ratio of reinforcement 配筋率；含筋率
ratio of reserves to production 储采比
ratio of rigidity 刚度比
ratio of rise to span 高跨比
ratio of secant modulus 割线模量比
ratio of shear span to effective depth of section 剪跨与截面有效高度的比值
ratio of side slope 边坡坡率
ratio of similitude 相似比
ratio of size reduction 粒度缩小比；破碎比
ratio of slenderness 长细比
ratio of slip 滑动系数
ratio of slope 坡度比；坡率；边坡斜率
ratio of soil pressure to settlement 地基沉陷系数；地基系数
ratio of strength to stress 强度应力比
ratio of strength to weight 强度重量比
ratio of stress to strength 应力-强度比
ratio of transformation 变换比
ratio of void volume to total volume of body 孔隙比；孔隙度
ratio of wall thickness to size 壁厚与管径比
ratio of wear 磨损率
ratio of weight 重量比
ratio processing 比值处理
ratio recovery 采出率
ratio scale 比例标尺
ratio scaling 比例量表
ratio table 比率表
ratio test 比率检验
ratio transformation 比值变换
ratiograph 辐射比例图解图；辐射诺谟图
ratioing technique 比值法
ratiometer 比率表
ratiometric technique 比值测量技术
rational 有理的；合理的
rational analysis of hydrological data 水文资料合理性分析
rational approximation 有理逼近；有理近似
rational design 合理设计
rational development 合理开发
rational exploitation 合理开采
rational expression 有理式
rational formula 有理式；有理化公式；示构式；示性式
rational function approximation 有理函数逼近
rational horizon 真地平；真水平线
rational interpolation 有理插值
rational mechanics 理性力学，有理力学
rational method 推理法
rational mining of groundwater 地下水合理开采
rational number 有理数
rational polynominal 有理多项式
rational polynominal filter 有理多项式滤波器
rational proposal 合理化建议
rational runoff formula 推理径流公式
rational speed 合理速度
rational traffic 合理运输
rational use 合理利用
rational utilization of groundwater 地下水合理利用
rational well depth 合理井深
rational well spacing 合理井距
rationale 基本原理；理论基础
rationale for design 设计原理
rationality 可靠性；可行性；合理性；适应性
rationalization 有理化
rationalization proposal 合理化建议
rationing 定量分配；配量
raveling failure 松散破坏；塌落式破坏
raveling ground 松散地层
ravelling of pavement 路面松散
ravelly ground 破碎地层；松散地层
ravine 沟壑；峡谷；山涧
ravine stream 溪流
ravined 沟谷切割的；布满沟壑的
ravinement 冲沟侵蚀
raw coal 原煤；入厂原煤
raw coal blending 原煤配煤
raw coal blending bin 原煤混煤仓
raw coal bunker 原煤煤仓
raw coal dump 原煤堆放场
raw coal feed 入选原料
raw coal handling 原煤处理；原煤手选
raw coal parameter 原煤参数
raw coal screening 原煤筛分；预先筛分
raw coal sizing 原煤分级
raw data 原始数据
raw data acquisition 原始数据采集
raw data calibration 原始数据校验
raw data file 原始数据文件
raw effluent 未经处理的废水
raw glaze 生釉；瓷土
raw image 原始图像；未经处理的图像
raw information 原始信息
raw intensity data 原始强度数据
raw log data 原始测井数据
raw material 原料
raw mix 生混合料
raw natural pozzolan 原生天然火山灰
raw sewage 原污水；未处理污水
raw shale 生页岩；硬页岩
raw sludge 未净化污泥；原污泥
raw smalls 末原煤；入选末煤
raw soil 生土；原始土；未处理土
raw soil layer 生土层
raw stuff 原料
raw subgrade 原土路基；天然路基
raw water 生水；未经净化水
raw water storage basin 未经净化水贮水池
raw water tank 清水池
raw water transfer 原水输送系统

ray diagram 射线图
ray emergence angle 射线出射角
ray equation 射线方程
ray inversion method 射线求逆法
ray launch angle 射线辐射角
ray method 射线法
ray parameter 射线参数
ray path 射线路径；射线途程
ray plane 射线面
ray proofing 防辐射
ray structure 射线结构
ray trace 射线迹径
ray tracing 光线追迹；射线示踪
ray-equation extrapolation 射线方程外推
ray-equation migration 射线方程偏移
Rayleigh coefficient 瑞利系数
Rayleigh cross-section 瑞利散射截面
Rayleigh damping 瑞利阻尼
Rayleigh distribution 瑞利分布
Rayleigh distribution of peaks 瑞利峰值分布
Rayleigh energy method 瑞利能量法
Rayleigh frequency function 瑞利频率分布函数
Rayleigh ground roll wave 瑞利地滚波
Rayleigh group velocity 瑞利波群速度
Rayleigh limit 瑞利极限
Rayleigh number 瑞利数
Rayleigh probability distribution 瑞利概率分布
Rayleigh scattering 瑞利散射；放射性散射
Rayleigh scattering technique 瑞利散射技术
Rayleigh theorem 瑞利定理
Rayleigh wave 瑞利波；面波；R 波
Rayleigh wave detection method 瑞利波探测法
Rayleigh wave dispersion curve 瑞利面波频散曲线
Rayleigh wave method 表面波法；瑞利波法
Rayleigh-Ritz method 瑞利-李兹法
Rayleigh's criterion 瑞利标准
Rayleigh's dissipation function 瑞利耗散函数
Rayleigh's method 瑞利法
Raymond cast-in-place concrete pile 雷蒙德现场灌注混凝土桩
Raymond pile 雷蒙德桩
ray-path chart 射径图
ray-path curvature 射线曲率
ray-path distortion 射线路径畸变
ray-path trajectory 射线路径轨迹
ray-proof 防辐射的
ray-tracing forward modeling 射线追踪正演模拟
ray-tracing migration 射线追踪偏移
ray-tracing procedure 射线追踪程序；射线追踪过程
ray-tracing scheme 射线追踪方法
R-curve R 曲线；阻力曲线；裂纹增长阻力曲线
reach travel curve 河段汇流曲线
reactant differential thermal analysis 反应差热分析
reacting fluid 反应流体
reacting force 反作用力
reacting thrust 反推力
reacting value 反应值
reacting weight 反应量
reaction 反力；反作用；反应
reaction capacity 反应能力
reaction coefficient 回授系数；反应系数
reaction condition 反应条件
reaction control 反馈控制
reaction couple 反作用力偶
reaction field 反应场
reaction force 反作用力
reaction formula 反应式
reaction frame 反力框架
reaction isotherm 反应等温线
reaction kinetics 反应动力学
reaction line of support 支护反作用线
reaction locus 反力轨迹
reaction mass 反作用体
reaction mechanism 反应机理
reaction mixture 反应混合物
reaction of bearing 支承反力
reaction of earthquake 地震反应
reaction of support 支座反力；支点反力
reaction of the support 支点反力
reaction order 反应级，反应级数；反应次序
reaction pad 反力衬垫
reaction pair 反应对
reaction path 反应路径
reaction period 反应期
reaction pile 反力基桩
reaction principle 反应原理
reaction probability 反应概率
reaction rate 反应速度；反应速率；化学反应速率
reaction rate coefficient 反应速率系数
reaction rate constant 反应速率常数
reaction relation 反应关系
reaction rim 反应边
reaction rim texture 反应边结构
reaction ring 反力环
reaction series 反应系列
reaction succession 反应序列；反应顺序
reaction system 反应系统；反应体系
reaction time 反应时间；反应周期
reaction turbine 反动式涡轮机；反动式汽轮机；反唤水轮机
reaction velocity 反应速度
reaction velocity constant 反应速度常数
reaction vessel 反应器
reaction wall 反应墙
reaction water turbine 反击式水轮
reaction water wheel 反动式水轮
reaction zone 反应带
reactivated landslide 复活性滑坡
reactivated old slide 重新活动的老滑坡
reactivated slide 复活滑坡
reactivated well 重新工作井
reactivation 复活作用；重激活
reactivation cycle 活化周期
reactivation of ancient landslide 老滑坡复活
reactivation of old folds 老褶皱活化
reactivation process 活化过程
reactivation slides 复活滑动
reactivation surface 沉积再作用面；复活面；再生表面
reactivation surface structure 活化表面结构；再生表面结构；再作用面构造

reactive	活性的；反应的	ready mix truck	预拌混凝土运送车
reactive concrete aggregate	活性混凝土集料；活性混凝土骨料	ready-made	准备好的；预制的
		ready-mix concrete	已搅拌混凝土
reactive fluid flow	反应性流体的流动	ready-mix plant	预拌混凝土机；预拌混凝土厂
reactive force	反作用力	ready-mixed	掺和好的；搅拌好的
reactive load	无功负载；电抗负载	ready-mixed concrete	预拌混凝土
reactive monitoring	反应监测	ready-mixed distribution facility	预搅拌分配设备
reactive power	无效功率；无功功率	reagent	试剂，化学试剂；反应物
reactive reagent	反应性试剂	real angle	真角；实际角度
reactive solute transport	反应溶质运移	real aperture	有效孔径；实际孔径
reactive system	反应系统	real aperture radar	真实孔径雷达
reactive transport	反应运移	real axis	实轴
reactive transport modeling	反应运移模拟	real brightness	真亮度
reactor debris	反应堆的裂变产物	real constant	实常数
reactor pressure vessel	反应堆压力壳	real cost	实际成本
reactor reliability	反应堆可靠性	real earthquake	真实地震
reactor structure	反应堆结构	real fault displacement	实际断距
react-to-known-hazard principle	对已知灾害采取跟进行动的原则	real fold	真褶皱
		real function	实函数
read interpreter	读入解释程序	real image	实像
read number	读数	real load	实际负载；有效载荷；真实载荷
read only memory (ROM)	只读存储器	real loading capacity	实际载重量
read only storage	只读存储器	real matrix	实矩阵
read rate	读速度	real medium	实际介质
readability	可读性	real memory	实在存储器
readily accessible	易达到的	real multiply	实数乘
readily air-slaking	易风化的	real number	实数
readily available water	速效水	real number axis	实数轴
readily combustible	随时可燃烧	real number field	实数域
readily removable cover	易于移走的封盖	real power	实际功率
read-in	读入	real precession	归算进动
readiness model	备用模型	real rate	真实速度
reading	读数；指示数	real resistance	实电阻
reading accuracy	读数精度	real solution	实际溶液；真实溶液
reading accuracy of sounder	测深仪读数精度	real specific gravity	真密度
reading apparatus	读数装置	real storage	实在存储器
reading device	读数装置；阅读器	real structure	实际结构
reading error	读数误差	real substract	实数减
reading glass	读数镜	real support	后支撑
reading interval	读数间隔	real thermodynamic effect	实热动力学效应
reading line	读数线	real time	实时
reading magnifier	读数放大器	real time controlling	实时控制
reading plotter	读数装置；记录装置	real time correlation	实时相关
reading precision	读数精度	real time correlator	实时相关器
reading stability	读数稳定性	real time data	实时数据
reading station	观测站；读数装置；读数位置，读数台	real time data processing	实时数据处理
reading system of the gravimeter	重力仪的读数系统	real time executive routine	实时执行程序
readjust	重新调整；微调；校准	real time interpretation	实时判读；实时解译
readjustment	重新调整	real time modelling	实时模拟
readjustment of zero	重新调零	real time process	实时处理
readout box	读数显示箱	real time recording	实时记录
readout capacity	读出能力	real time simulation	实时仿真，实时模拟
readout device	读数设备；读出装置	real time system	即时制度；实时制；实时系统
Readout pump	读数泵	real time system identification	实时系统识别
read-write channel	读写通道	real variable function	实变函数
read-write check	读写校验	real velocity of seepage flow	渗流的实际流速
read-write command	读写指令	real well	实井
read-write cycle	读写周期	realign	矫直，修直；重新排列；重新对齐
read-write memory	读写存储器	realignment	重新定线；改变线向；重新勘定航道界线
read-write operation	读写操作	realistic accuracy	实际精度

realistic conceptual model 真实概念模型
realistic model 仿真模型
realistic simulation 实际模拟
realistic well condition 实际井况
reality criterion 逼真度准则
realm 区域；范围；领域
real-time analysis 实时分析
real-time analytical capability 实时分析能力
real-time analyzer 实时分析器
real-time comparison 实时对比
real-time compression 实时压缩
real-time computation 实时计算
real-time control system 实时控制系统
real-time correction 实时校正
real-time data reduction 实时数据处理
real-time data transport 数据实时传输
real-time differential global positioning system 实时差分全球定位系统
real-time display 实时显示
real-time drilling 实时钻井
real-time drilling decision 实时钻井决策
real-time estimate 实时预测
real-time follow-up 实时跟踪
real-time information 实时信息
real-time information system 实时信息系统
real-time input 实时输入
real-time input-output 实时输入输出
real-time interpretation 实时解释
real-time kinematic survey 实时动态测量
real-time measurement 实时测量
real-time monitoring 实时监控
real-time operating system 实时操作系统
real-time operation 实时操作；实时运算
real-time photogrammetry 实时摄影测量
real-time positioning 实时定位
real-time prediction mechanism 实时预测机制
realtime processing 实时处理
real-time signal processing 实时信号处理
real-time simulation 实时模拟
real-time status 实时状态
real-time steering 实时控制；实时操作
real-time telemetry 实时遥测
realty industry 房地产业
reamed well 扩孔；扩大钻孔
reamer bit 扩孔钻头
reamer plate 挡板；护板；保持板
reamer shell 扩孔器
reaming angle 刷大角
reaming bit 扩孔钻头；铰孔锥
reaming boring machine 扩孔式全断面掘进机
reaming face 扩孔刃面
reaming of hole 扩孔
reaming pilot 扩孔器导向部分
reaming ring 扩孔器
reaming shell 扩孔筒；扩孔器
reaming surface 扩孔面；扩大面
reamplify 再放大
reanalysis 预分析
reapportion 重新分配
reappraisal 重新估价

rear 后端；后方；背部
rear abutment pressure 后壁支承压力
rear axle protection 后轴护板
rear axle shaft 后轮轴；后桥轴
rear beam 后部梁
rear bearing 后轴承
rear bogie 后轮轴转向架
rear canopy 后顶梁
rear chamber 后殿；后室
rear connection 背面接线
rear conveyor 尾端输送机
rear conveyor motor 尾端输送机的电动机
rear cubicle 尾房
rear elevation 后视图；背立图
rear expansion sheet 后膨胀板
rear shield 盾构后部；后护盾
rear support 后下支承；后支承
rear view 背视图；后视图
rear-drive axle 后驱动轴
rear-dump body 后卸式车厢
rear-end 后端
rearrangement 重新排列
reason for variation 变更原因
reason machine 推理机
reasonable 合理的；适当的
reasonable development 合理开发
reasonable error 合理误差
reasonable gradient 恰当的梯度；适当的坡度，合理坡度
reasonable precautions 合理的预防措施
reasonable price 公平价格
reasonable stripping ratio 合理剥采比
reasonable use of water 合理用水
reasonably beneficial use 有合理实益用途
reasoning 推理；推论
reassemble 重新装配；再集合；再组合
reassessment 重新评定
reassorted loess 再造黄土
re-bar 钢筋；加固锚杆
rebate rabbet 切口
rebound 回弹；回跳
rebound apparatus 回弹仪
rebound curve 回弹曲线
rebound deformation 回弹变形
rebound ground 回弹地基
rebound hammer 混凝土回弹仪；回弹锤
rebound hammer test 回弹测试
rebound hardness 回跳硬度
rebound hardness test 回弹硬度试验
rebound index 回弹指数
rebound material 回弹料
rebound method 回弹法
rebound modulus 回弹模量
rebound number 回弹数
rebound of foundation 地基回弹
rebound of pile 桩的回弹
rebound of shotcrete 喷射混凝土回弹
rebound pendulum machine 回跳打桩机
rebound procedure deflection 回弹过程挠曲
rebound resilience 回弹性

rebound resiliometer	回弹性测定仪
rebound strain	回弹应变
rebound stress	回弹应力
rebound tester	回弹仪
rebound value	回弹值
rebounding device	回弹仪
rebuild	重新装修；重建
recalculating water pump	循环水泵
recalculation	重新计算
recalibrate	重新校准
recalibrator	重新校准器
recapitulation	重演
recast	另算；重做；重铸；改做
receding contact angle	后退接触角
receding difference	后退差分
receding glacier	后退冰山
receding hemicycle	退却半周期
receding interface	后退界面
receding line	退缩线
receding shock wave	后退冲击波
receding triple-phase line	后退三相线
receding water table	退落地下水位
receding wave	后退波
receivable	可收到的
receiver amplifier	接收机放大器
receiver bandwidth	接收机带宽
receiver dynamic range	接收机动态范围
receiver selection	接收器选择
receiver sensitivity	接收机灵敏度
receiver sensor	接收传感器；接收探头
receiver static correction	接收点静校正
receiver statics	接收点静校正
receiving area	承受面积；接受区，接收面积；集水面积
receiving arrangement	接收装置
receiving bin	接受仓；受煤仓；受矿仓
receiving bowl	集料箱，收集箱
receiving box	集料箱，收集箱
receiving bunker	受煤仓
receiving centre	接收中心
receiving cone	受料漏斗
receiving hopper	集料仓；接受斗仓；受料槽
receiving pool	集水池
receiving probe	接收探头
receiving region	容泄区
receiving report	收料报告
receiving station	接收站
receiving stream	容泄水流；排污河流
receiving surface	接收面
receiving system	接收系统
receiving waters	承受水体；承受水域
recementation	再胶接作用
recementing	重新胶结，再胶结；补注水泥
recent alluvium	现代冲积层
recent bed	近代沉积层
recent crust horizontal movement	现代地壳水平运动
recent crustal movement	现代地壳运动
recent earth surface movement	现代地表运动
recent epoch	近代；冰期后最新期
recent exogenetic movement	现代外生运动
recent fill	新填土
recent fluvial sediment	现代河流沉积物；新近河流泥沙
recent geological period	现代地质时期
recent landslide	现代滑坡
recent non-tectonic movement	现代非构造运动
recent orogenesis	现代造山运动
recent period	现代；近代
recent sediment	新近沉积物；现代沉积物
recent seismicity	近代地震活动
recent soil	新生土；原始土；现代土壤
recent tectonic movement	近代构造运动
recent tectonic stress field	现代构造应力场
recent tectonics	近代构造
recent transcurrent buckling	近代横向地壳弯曲
recent vertical crust movement	现代地壳垂直运动
recent vertical movement	现代垂直运动
recent volcano	近期火山
recentered projection	断裂投影
recently deposited soil	新近堆积土
receptance function	响应函数
receptance matrix	响应矩阵
reception basin	受水盆地；汇水盆地；储水池；进水池
reception basin of river	河流补给区
reception capacity	接收能力
reception facility	接收设施
recession constant	衰退常数；消退系数；退水常数
recession curve	退水曲线；递减曲线；下降曲线
recession curve of flow	水流退落曲线
recession discharge	消退排放量；枯竭排放量
recession equation	退水方程
recession flow	退缩水流，后退水流；衰减流
recession hydrograph	退水过程线
recession of beach	岸线后退；海滩后退
recession of level	水面下降；水位下降
recession of sea level	海面下降
recession of valley sides	谷岸后退
recession segment	后退部分；退水段
recession time	地面水消退时间
recession velocity	退水流速
recessional moraine	后退冰碛
recessional stage	后退阶段
recessional stream	后退河流
recessive	隐性的；逆行的
rechannelling	再处理；再搬运
recharge	补给；回灌；充电；装填
recharge area	补给区；受水区；汇水区
recharge basin	补给盆地；补给池
recharge boundary	补给边界
recharge by rainfall penetration	雨渗灌注；雨水渗透补给
recharge by seepage of stream	河水入渗补给
recharge capacity	补给能力；补给量
recharge cone	回灌锥
recharge cycle	回灌周期
recharge discharge regime	地下水补排动态
recharge efficiency	回灌效率
recharge intensity of groundwater	地下水补给强度
recharge line	线型补给井群；补给线
recharge method	回灌法
recharge modulus	地下水补给模数

recharge modulus of aquifer	含水层补给模数
recharge of basin	盆地补给
recharge of groundwater	地下水的回灌；地下水的补给
recharge pit	回填井坑；回灌井坑
recharge pressure	回灌压力
recharge quantity of single well	单井回灌量
recharge rate	补给率；回灌率
recharge system	补注系统
recharge test	回灌试验
recharge water	补给水；回灌水
recharge water source	回灌水源
recharge well	补给井；回灌井
rechargeable	可再充电的；可再装料的
recharge-discharge cycle	供-泄水循环
recharging system	回灌系统
recharging well	注水井
recheck	再次检查；重新检验
recipient	容纳器；容器；接收者
reciprocal	倒数；往复的；相互的
reciprocal action	交互作用
reciprocal agreement	互惠协定
reciprocal basis	对偶基
reciprocal bearing	反方位角；对向方位角
reciprocal contract	互惠合同
reciprocal differences	反商差分
reciprocal direction	往复方向
reciprocal distribution	对应分布
reciprocal figure	可易图形
reciprocal formation volume factor	地层体积系数的倒数
reciprocal function	互反函数
reciprocal geometric factor	逆几何因子
reciprocal gradient	反地温梯度；反地热增温率
reciprocal leveling	双程水准测量
reciprocal load	往复荷载
reciprocal motion	往复运动
reciprocal observation	对向观测
reciprocal permeability	渗透率倒数
reciprocal principle	互换原理
reciprocal proportion	倒数比；反比
reciprocal ratio	反比
reciprocal reaction	可逆反应；往复反应
reciprocal salt pair	平衡盐对
reciprocal shotpoint	互换炮点
reciprocal sight	对向照准
reciprocal square distance method	距离平方倒数法
reciprocal square root	平方根倒数
reciprocal stress	对向应力；可逆应力
reciprocal theory	互易定理；互换定理；可逆定理
reciprocal time	互换时间
reciprocal transducer	可逆换能器
reciprocal transformation	相互转化
reciprocal treatment	互惠待遇
reciprocal vertical angle	对向观测竖直角
reciprocate theory	互等定理
reciprocating	往复的
reciprocating action	往复动作
reciprocating activity	往返活动
reciprocating compressor	往复式压气机；活塞式压气机
reciprocating crusher	往复式破碎机
reciprocating drill	往复式钻机；锤击式钻机
reciprocating machine	往复式机器
reciprocating mining	往复式开采
reciprocating motion	往复运动
reciprocating motor	往复式电动机
reciprocating piston pump	往复式活塞泵
reciprocating plate	往复运动板；往复式给料机
reciprocating pressure-type crusher	往复压挤式破碎机
reciprocating pump	往复泵
reciprocating saw	往复锯；横切长锯
reciprocating scratcher	往复式刮子
reciprocating screen	往复式泥浆筛；往复筛；摇动筛；振动筛
reciprocating seal	往复运动密封件
reciprocating shaft	往复轴
reciprocating sieve	往复筛；振动筛
reciprocating time	纯钻井时间
reciprocating type refrigeration compressor	往复式制冷压缩机
reciprocating vibrator	往复式振动器；往复式振动筛
reciprocating winch	往复式绞车
reciprocity	交互作用
reciprocity calibration	互易校准
reciprocity principle	倒易原理；互换原理
reciprocity theory	互等定理
recirculate	再循环
recirculated cooling water	循环冷却水
recirculated gas	回注气
recirculated medium	循环介质
recirculated water	回注水
recirculating	再循环的
recirculating air	循环风
recirculating load	循环负载
recirculating pump	再循环泵
recirculating ratio	循环比；回注比
recirculating water	循环水
recirculating zone	回流区
recirculation damper	再循环风闸
recirculation degassing	循环真空脱气法
recirculation of air	空气再循环；风流连续循环
recirculation pump	循环泵
recirculation water	回流水
reckon	计算；核算；推算；统计
reckoning	计算；估计；判断
reclaim	回收；再生；重新使用；恢复；复原；开垦
reclaim conveyor	转载输送机
reclaimed area	复垦区域
reclaimed asphalt mixture	再生沥青混合料
reclaimed bituminous pavement	再生沥青路面
reclaimed fibre	回用纤维
reclaimed ground	填筑场地
reclaimed land	复田地，开垦地；填筑地；填海土地
reclaimed produced water	回收的采出水
reclaimed quarry	可回收碎石；可再使用碎石
reclaimed soil layer	填土层
reclaimed spoil	回用的废渣；再用的矸石
reclaimed spoil area	复垦的剥离物区域
reclaimed timber	回收的坑木
reclaimed water	回收水
reclaimed water system	循环水系统
reclaiming	回收；再生；再利用；复田

reclaiming machine	贮场装载输送机
reclaiming system	复垦法；装卸系统
reclaiming tank	回水池
reclamation	围垦；开垦；改良；填筑
reclamation area	填海区
reclamation dual operation	复田工作；废矿重采工作
reclamation hydrogeology	土壤改良水文地质学
reclamation in water area	水域围垦
reclamation level	填筑高程
reclamation of land	土地填筑
reclamation of mine water	矿井水资源化
reclamation of saline-alkali soil	盐碱土改良；盐碱地治理
reclamation of saltern	盐碱土改良；盐田改良
reclamation operation	复田工作；废矿重采工作
reclamation policy	复田政策
reclamation program	复田规划
reclamation project	围垦工程；复垦项目
reclamation revetment	填筑护岸
reclamation scheme	填海计划
reclamation survey	复垦测量
reclamation work	复田工作
reclassification	再分级
recleaner flotation	再精选
recleaner screen set	再选筛组
recleaning material	再选物料
reclined fold	斜卧褶皱
recognition capability	识别能力
recognition criteria	识别标准
recognition device	识别系统
recognition element	识别单元；识别要素
recognition feature	识别特征
recognition logic	识别逻辑
recognition range	识别范围
recognition system	识别系统
recognizable	可识别的
recognizable anomaly	可识别的异常
recognized overload	认可过载
recognized standard	认可标准
recoil fault	反冲断层
recoil force	反作用力；反冲力；反应力
recoil spring	反冲弹簧；后座弹簧
recoil stress	后退力；后坐力
recoil stroke	回程；回转冲程
recoiling motion	反冲运动；后座运动
recombination coefficient	复合系数
recombination dual radiation	复合辐射
recombination rate	复合率
recombination velocity	复合速度
recombine	复合
recombined sample	混合试样
recombining	合流
recommended route	推荐航线
recompacted fill	再压实填土；再夯实填土
recompaction of fill	填土的再夯实
recomposed rock	再生岩；再造岩
recompression curve	再压缩曲线
recompression index	再压缩指数
recompression modulus	再压缩模量
recompression ratio	再压缩比
recompression zone	再压缩带
reconnaissance	普查；踏勘；初测
reconnaissance and design stage	勘探设计阶段
reconnaissance bedrock geologic map	路线基岩地质图
reconnaissance borehole	探井
reconnaissance diagram	勘测草图
reconnaissance drill	地质普查用钻机；轻型普查钻机
reconnaissance electromagnetic system	电磁勘探法
reconnaissance evaluation	踏勘评价
reconnaissance exploration	可行性研究勘查；普查踏勘
reconnaissance for control point section	控制网选点
reconnaissance geochemical survey	普查性地球化学测量
reconnaissance geologic map	踏勘地质图
reconnaissance geological survey	地质普查
reconnaissance hydrogeological mapping	普查水文地质填图
reconnaissance information	普查资料
reconnaissance map	踏勘图；勘测图
reconnaissance mapping	路线填图；踏勘填图
reconnaissance phase	踏勘阶段
reconnaissance photography	勘测摄影；侦察摄影
reconnaissance report	调查报告
reconnaissance shooting	勘察爆破；普查炸测
reconnaissance sketch	勘测草图
reconnaissance soil map	土壤概图
reconnaissance strip	勘测航线
reconnaissance survey	草测；踏勘；勘测；选线
reconnaissance traverse	勘测导线
reconsequent drainage	再顺水系；复向水系
reconsolidation	再固结；再压实
reconstituted feed	重组给料；计算入料
reconstituted specimen	再制试件
reconstructed chart	复制图
reconstructed granite	再生花岗岩
reconstructed railway	改建铁路
reconstructed sample	重建样值
reconstruction	重建；改建；改造
reconstruction planning	重建计划
reconstruction project	改建项目
reconstruction works	重建工程
record	记录；数据；资料
record blocking	记录编块
record book	记录本
record cell	记录单元
record chart	记录曲线图
record clerk	记录员；统计员
record conversion	记录转换
record conversion apparatus	记录转换设备
record date	登记日期
record diagram	记录曲线；记录图
record file address	记录文件地址
record flood	创记录的洪水
record format	记录格式
record gap	记录间隙
record heading	记录图头
record in paper	纸上记录
record keeping	记账
record key	记录键
record layout	纪录布置
record length	记录长度

record length-dependent error	与记录长度有关的误差	records management	记录管理
record locking	记录锁定	records of deposit	沉积记录
record mark	记录标记	recover ratio	岩芯获得率；回收率
record mode	记录方式	recoverability	可恢复性；可回收性
record of deposit	矿床地质资料；矿床编录	recoverable coal	可回收的煤；可采煤
record of discharge	排放记录	recoverable compaction	可恢复压实
record of flood	最高洪水位	recoverable error	可恢复错误
record of measurement	测量记录	recoverable pillar	可回收的煤柱；可回收的矿柱
record of title deed	业权契据档案	recoverable property	可采矿地
record operator	记录员	recoverable reserves	可采储量；工业储量
record peak flood	记录最大洪水；历史最大洪水	recoverable resources	可采资源量
record photograph	记录照片	recoverable value	回收价值
record section	记录剖面	recovered strength	再生强度
record sheet	记录表	recovering	回收；复原
record storage	记录存储器	recovering head	恢复水头
record surface	记录面	recovering water level	恢复水位
recorded flood	实测洪水；记录洪水	recovery coefficient	恢复系数；回收系数；回采率
recorded hydrograph	实测水文过程线	recovery cost	收回成本
recorded information	记录信息	recovery creep	回复蠕变；复原蠕变；恢复蠕变
recorded value	实测值；观测值；记录值	recovery curve	回收率曲线；回置曲线；开采曲线
recorder and counter device	记录和计算设备	recovery cycle	回升周期
recorder apparatus	记录装置	recovery device	回收装置
recorder strip	记录带	recovery efficiency	回收效率；数量效率
recorder unit	记录装置	recovery equipment	回收用设备
recorder well	观测井；自记仪器测井	recovery factor	采收率
recording apparatus	记录器；记录仪	recovery force	恢复力
recording attachment	记录附件	recovery fraction	回采率
recording chart	自动记录图表	recovery loss	开采损失
recording density	记录密度	recovery mechanism	开采机理
recording desk	记录台	recovery method	复原法，恢复法；回放法；回收法
recording device	记录装置	recovery of coal mining	煤炭回采率
recording draft gauge	通风记录仪	recovery of core	取岩心；岩心采取率
recording drum	记录圆筒	recovery of deformation	变形恢复
recording duration	记录持续时间	recovery of elasticity	弹性恢复
recording echo sounder	自记回声测深仪	recovery of land	收回土地
recording equipment	记录设备	recovery operation	回收作业
recording flowmeter	自记流量计	recovery peg	测站基点指示桩；参考标桩
recording gas analyser	自计气体分析器	recovery percent	采收率；回收率
recording gauge	自记仪；自动记录测定计	recovery period	复原期间
recording gear	记录仪	recovery phase	恢复相
recording geometry	观测系统	recovery process	开采过程
recording hygrometer	自记湿度计；湿度记录器	recovery profile	开采剖面
recording instrument	记录仪	recovery rate	回收率；采收率；恢复速率；开采速度
recording location	记录位置	recovery ratio	回收率；采出率
recording medium	记录介质	recovery tank	回水箱
recording meter	自记计数器；记录仪	recovery time	复原时间
recording needle	记录针	recovery treatments	恢复处理
recording paper of sounder	测深仪记录纸	recovery value	回收率值；回采率值
recording period	记录时间；记录周期	recovery well	生产井；开采井
recording potentiometer	记录式电位计	rectangle coordinate	直角坐标
recording pressure gauge	记录式压力计	rectangle cross section	矩形横截面
recording rain gauge	自计雨量器；自计雨量计	rectangle lattice	矩形网络
recording range	记录量程；记录范围	rectangle slot	矩形槽
recording speed	记录速度	rectangular	长方形的；矩形的
recording station	记录站；观测站	rectangular arrangement	矩阵
recording strain gauge	自记应变仪	rectangular array	矩阵
recording thermometer	自记温度计	rectangular bar	矩形顶梁
recording truck	地震测勘汽车；操测车；仪器车	rectangular basin	矩形流域；矩形海湾；矩形盆地
recording unit	记录装置	rectangular beam	矩形截面梁
recording water-gauge	自记水位计	rectangular Cartesian coordinate	笛卡儿坐标；直角坐标

rectangular channel	矩形水槽	rectilinear motion	直线运动
rectangular chart	正交投影地图	rectilinear oscillation	线性摆动
rectangular column	矩形截面柱	rectilinear path	直线路程
rectangular contour	方格等值线	rectilinear propagation	直线传播
rectangular coordinate	直角坐标	rectilinear scanning	直线扫描
rectangular coordinate system	直角坐标系统	rectilinear shoreline	直岸线
rectangular crib groin	矩形木笼式丁坝	rectilinear stress diagram	直线应力图
rectangular cross section	矩形截面；矩形横断面	rectilinearity	直线性
rectangular cross section beam	矩形截面梁	recumbent	倾伏的；横卧的
rectangular crossing	直角交叉	recumbent anticline	伏卧背斜
rectangular distribution	矩形分布；均匀分布	recumbent fold	倾伏褶曲；平卧褶皱；伏卧褶皱
rectangular drainage	矩形排水系统	recumbent foreset	伏卧前积层
rectangular drainage pattern	直角水系型式	recumbent infold	伏卧褶皱
rectangular drainage region	矩形泄油区；矩形排水区	recumbent isocline	伏卧等斜褶皱
rectangular element	矩形单元	recumbent overfold	倒转伏褶皱
rectangular equation	直角坐标方程	recumbent overturned fold	几乎水平的倒转褶皱；伏卧褶皱
rectangular figure	矩形		
rectangular fold	长方形褶皱	recuperability	恢复能力；反馈能力
rectangular footing	矩形基础	recurrence	重现；递归；循环
rectangular foundation	矩形基础；矩形地基	recurrence algorithm	递推算法
rectangular grid	直角坐标网；矩形网格；矩形格栅	recurrence curve	递归曲线
rectangular hollow section	矩形空心截面	recurrence equation	递推方程
rectangular joint structure	直角节理构造	recurrence formula	递归公式
rectangular map projection	正交地图投影	recurrence frequency	重现频率
rectangular map subdivision	矩形分幅	recurrence horizon	再现土层；急变层；骤变层
rectangular matrix	矩形矩阵；长方阵	recurrence interval	重现期；复发间隔
rectangular mesh	矩形网格	recurrence of tectonic movement	构造运动新生性
rectangular numerical integration	矩形数值积分	recurrence period	重现周期
rectangular plane coordinate system	平面直角坐标系	recurrence rate	重现率
rectangular plate	矩形板	recurrence relation	递推关系
rectangular projection	正交投影	recurrent	重现的；递归的；循环的
rectangular section	矩形截面	recurrent association	重现组合
rectangular section tunnel	矩形隧道	recurrent community	重现群落
rectangular shaft	长方形井筒；矩形竖井	recurrent deposition	重复沉积；叠次沉积
rectangular shape tunnel	箱形隧道	recurrent fault	生长断层
rectangular shape wave	矩形波	recurrent folding	重复褶皱作用
rectangular shaped shield	矩形盾构	recurrent frequency	重复频率
rectangular shield	矩形盾构	recurrent input	经常性投入
rectangular side ditch	矩形侧沟	recurrent interval	脉冲间隔
rectangular slab	矩形板	recurrent orogenesis	重复造山作用
rectangular space coordinate	空间直角坐标	recurrent period	重现周期；循环周期
rectangular stone slab	阶条石	recurrent state	循环状态
rectangular support	矩形支架	recurring series	循环级数
rectangular triangle	直角三角形	recurring wave	循环波
rectangular tunnel	矩形隧道	recursion	递归；递推；循环
rectangular wave	矩形波	recursion filter	递归滤波器
rectangular winze	矩形矿井	recursion formula	递推公式
rectification	修整；纠正；整流	recursion function	递归函数
rectification coast	夷平海岸；平直海岸	recursion subroutine	递归子程序
rectification of a channel	渠道取直；裁弯取直	recursive	递归的；循环的
rectification of harbor approach	港湾进口水道改直	recursive analysis	递归分析
rectified coast	浪成平直海岸；夷平海岸	recursive database	递归数据库
rectified coefficient	修正系数	recursive deconvolution	递归反褶积
rectilinear	直线的	recursive filtering	递推滤波；递归滤波
rectilinear asymptote	渐近直线	recursive function	递归函数
rectilinear coordinate	直角坐标	recursive procedure	递归过程
rectilinear cross-stratification	直线交错层理	recursive subroutine	递归子程序
rectilinear drainage	线形排水	recurvate	反弯的；反曲的
rectilinear figure	直线图形	recycle	再循环；回收；重复利用
rectilinear flow	直线流	recycle back	反向循环

recycle cracking operation	循环裂化操作
recycle gas	再循环气体；循环气；回注气
recycle gas compressor	循环气压缩机
recycle ratio	循环比
recycle slurry	循环泥浆；循环悬浮体
recycle system	循环系统
recycle-water	回收水；回用水
recycling	再循环；重复利用
red bauxite	铁铝土
red bed	红色岩层；红层
red bed facies	红层岩相
red bed landslide	红层滑坡
red bed phase	红层相；红层阶段
red brick	红砖
red brittleness	热脆性
red brown Mediterranean soil	地中海褐色土
red calcareous soil	石灰性红土
red capping	铁帽
red cement	红水泥
red clay	红黏土
red desert soil	红漠土
red earth	红壤；红土
red ochre grout	红土浆
red ooze	红软泥
red rock	红层；红色沉积岩
red short steel	热脆钢
red shortness	热脆性
red soil	红壤；红土
red tide	赤潮；红潮
red water bloom	赤潮；红水花
red water effect	红水效应；锈水效应
redeposited loess	次生黄土；再积黄土
redeposited structure	再沉积构造
redeposition	再沉积作用
redesign	再设计；再绘制；重新构思；重建
redesigning	重新设计
redetermination	再测定；重新认定；重新决定
redevelopment	再发展；改造
redevelopment estate	列入重建计划的屋
redevelopment of landfill sites	垃圾填埋地的重新整修
redevelopment order	重建令
redevelopment plan	重建计划
redevelopment potential	重建潜力
rediscover	再发现
redispersion	再分散
redisplay system	再显示系统
redistribution	重分配；再分布
redistribution of earth pressure	土压力重新分布
redistribution of internal force	内力重分布
redistribution of moment	弯矩再分配
redistribution of stress	应力重分布
redox	氧化还原作用
redox buffer	氧化还原缓冲剂
redox catalyst	氧化还原催化剂
redox chemistry	氧化还原化学
redox condition	氧化还原条件
redox equilibrium	氧化还原平衡
redox logging	氧化还原测井
redox potential	氧化还原电势；氧化还原电位
redox process	氧化还原过程
redox reaction	氧化还原反应
redox state	氧化还原状态
redrill	重新钻孔；再钻
redrill bit	扩孔钻头；修井钻头
redriving	复打
redriving test	重钻试验；复打试验
reduce	减少；降低；稀释
reduce stability	降低稳定性
reduced angle of draw	折算塌陷角；折算极限角
reduced complexity modelling	降低复杂性建模
reduced compressibility	折算压缩系数
reduced correlation matrix	约化相关矩阵
reduced density	对比密度；折算密度
reduced depth	折算深度
reduced elastic modulus	折减弹性模量
reduced equation	简化方程
reduced factor	折减系数
reduced flow	收缩水流
reduced fraction	简化分数；约简分数
reduced grade	换算坡道
reduced head	折算水头
reduced height	折算高度
reduced height of soil	土骨架净高
reduced integration	降阶积分
reduced level	折减标高；降低标高
reduced level of water	降落的水位
reduced load	折算荷载
reduced mass	缩减质量
reduced mass matrix	约化质量矩阵
reduced modulus	换算模数
reduced mud	黑泥；还原泥
reduced natural frequency	折减固有频率
reduced parameter	对比参数；换算变量；简化参数；折算参数
reduced pressure	折算压力；换算压力；减压
reduced pressure coefficient	换算压力系数
reduced ratio	折合比
reduced scale	缩尺；按比例缩小；比例尺
reduced scale prototype model test	缩尺原型试验
reduced stiffness matrix	缩减刚度矩阵
reduced stone	碎石
reduced temperature	对比温度；折算温度
reduced time	对比时间
reduced turbidity	比浊度
reduced value	折算值
reduced vertical profile	缩减的垂直剖面
reduced viscosity	对比黏度；折合黏度
reduced volume	对比体积；折合体积
reduced zone	还原带
reducer	还原剂，还原器；减压阀；渐缩管；减速器；扼流圈；节流器
reducing action	还原作用
reducing environment	还原环境
reducing gear	减速齿轮；减速器
reducing joint	异径接头，缩颈接头；缩接套；异颈连接
reducing medium	还原介质
reducing piece	缩小接管；缩小接头
reducing pipe	渐缩管
reducing power	还原能力

reducing union	渐缩管接头；减小管接头
reducing unit	减速装置；减速器；还原设备；减压器
reducing value	还原值
reductant	还原剂
reductibility	还原性
reduction activator	还原活化剂
reduction agent	还原剂
reduction ascending	向上折算
reduction at source	在源头减少污染；源头污染物减排
reduction coefficient	换算系数；修正系数；折减系数；减少系数
reduction coefficient of external water pressure	外水压力折减系数
reduction crusher	碎减式破碎机；次轧碎石机
reduction density	换算密度
reduction descending	向下折算
reduction division	减数分裂
reduction factor	衰减系数；减缩因数
reduction factor for pile in group	群桩折减系数
reduction factor of pile group	群桩折减系数
reduction factor of rock modulus	岩石模量折减系数
reduction factor on number of waves	波数折减系数
reduction for distance	距离换算
reduction for principal-distance	主距改正
reduction formula	换算公式
reduction gear ratio	减速比
reduction in area	面积缩减；面积收缩；压缩率
reduction in sizes	破碎
reduction index	还原指数
reduction of area	断面缩减率；断面收缩；面积缩小
reduction of coal sample	煤样缩制；煤样缩分
reduction of cross-section	横截面收缩；径向收缩
reduction of cross-sectional area	断面缩减
reduction of data	数据处理；信息简缩变换
reduction of diameter	直径减小
reduction of gradient	坡度折减
reduction of gravity	重力归算
reduction of heat	减热
reduction of load	卸载；卸压
reduction of operation	运算换算
reduction of porosity	孔隙度降低
reduction of roadway section	巷道断面缩小量
reduction of rock	岩石破碎；岩石崩解
reduction of sample	样品缩减；样品缩分
reduction potential	还原势；还原电位
reduction process	还原过程
reduction rate	破碎率
reduction ratio	缩小比例；递减率
reduction ratio of roadway section	巷道断面缩小率
reduction stage	破碎段；碎磨段
reduction to centre	归心计算
reduction to standard condition	归算至标准条件
reduction ton	折算吨
reduction works	选矿厂；矿物加工厂
reduction zone	还原带
reductive agent	还原剂；脱氧剂
reductive dissolution	还原溶解
reductive factor	折减系数；折算系数
reductive water	还原水
redundancy	多余；剩余

redundancy analysis	冗余分析
redundant	丰余的；冗余的
redundant constraint	多余的约束条件
redundant data	多余数据
redundant drainage facility	多余的排水设施
redundant frame	超静定框架
redundant information	冗余信息
redundant observation	多余观测
redundant reaction	超静定反力
redundant structure	静不定结构；超静定结构；赘余结构
redundant support	多余支点
reduplication	重复；加倍
reed-frequency meter	簧片振动频率计
reed-sedge peat	苇菅泥炭
reef	礁石；暗礁；矿脉；矿层；露头岸；外堤
reef bin	矿石仓
reef drift	脉内平巷；脉内巷道
reef drive	在围岩中开掘平巷
reef platform karst	礁坪岩溶
reef process	成礁过程
reef rock breccia	礁岩角砾岩
reef rock classification	礁石分类体系
reef rubble	礁瓦砾
reef section	礁剖面
reef sedimentation process	礁石沉积过程
reef slope	礁坡
reef system	礁石系统；矿脉系
reef-beach system	礁滩系统
reel packer	卷筒式灌浆塞
reel winder	扁绳提升机；绞轮提升机
reeling hoisting	绕轮式提升
reentrant	凹角；凹角的
re-entrant angle	内斜角
reentrant intake	凹入式进水口
reentrant library	重入库段
reference area	基准面；起始面；参考面
reference block	标准试块
reference circle	参照圆
reference climatological station	基准气候站
reference clock	参考时钟
reference coal sample	标准煤样
reference condition	标准条件；参比条件
reference configuration	参考结构
reference data	参考资料；参考数据
reference datum	参考基准面
reference density	参考密度
reference design	参考设计
reference dimension	参考尺寸
reference drawing	参考图
reference dynamics	参照动力学
reference edge	基准边
reference effect	参照效应
reference element	参考元素
reference ellipsoid	参考椭球体
reference ellipsoid surface	参考椭球面
reference equivalent	参考当量
reference error ratio	参考差错率
reference feature	参考要素
reference field	基准面

reference flowmeter 基准流量计
reference frame 参考构架；参考系统
reference frequency 基准频率
reference fringe 参照边缘；参考边缘
reference gauge 检验量规；标准量规；校对规；校对计
reference grid 参考格网
reference horizon 标准层；对比层
reference image 标准图像
reference input 参考输入
reference instrument 标准仪器
reference intensity 参比强度
reference lamina 标志层；参照层
reference level 基准面；基准高程
reference line 基准线
reference line of turnout 道岔基线
reference locality 参考地区
reference manual 参考手册
reference mark 参考标记；基准标记；参照符号；起点读数
reference master gauge 校对控制规
reference material 标准物质；参照样品；参考资料
reference model 参考模型
reference monument 参考标石
reference number 参考数
reference orientation 基准定向
reference pattern area 参考井网面积
reference peak 参考峰
reference phase 参考相位；基准相位
reference plane 基准面；参考平面
reference point 参考点；基准点
reference position 原始位置；参考位置
reference pressure 基准压力；参考压力
reference radiation 基准辐射；参考辐射
reference radius 分度圆半径
reference record 程序编制信息记录；参考记录
reference sample 参考样品
reference section 参考剖面
reference signal 参考信号
reference size 参照粒度
reference soil system 参考土体系
reference solution 参比溶液
reference source 基准源；参考源，对照源
reference spheroid 参考椭球体
reference stake 参考标桩；水准基点
reference standard 参考标准；参照标准
reference station 参考站，参考台；基准站
reference strain 参考应变
reference surface 参考面；基准面
reference system 参照系；标准体系
reference temperature 标准温度；起始温度；基准温度
reference time 参考时间；基准时间
reference time line 参考时间线
reference transit station 参考测点
reference trend line 基准趋势线
reference value 参考值
refilling 回填；还土
refilling and grouting 回填灌浆
refilling of trench 沟槽还土；堑壕回填
refinement 修饰；精制
reflected plane wave 反射平面波

reflected ray 反射线
reflected refraction 反射折射波
reflected seismic information 反射地震资料
reflected tension wave 反射张力波
reflected wave 反射波
reflection amplitude 反射振幅
reflection angle 反射角
reflection coefficient 反射系数
reflection configuration 反射结构；反射形态
reflection factor 反射系数
reflection interval 反射间隔时间
reflection law 反射定律
reflection light 反射光
reflection matrix 反射矩阵
reflection method 反射法
reflection path 反射路径
reflection prospecting 反射法勘探
reflection seismic exploration 反射法地震勘探
reflection seismic survey 反射波地震勘探
reflection seismics 反射地震学
reflection seismograph 反射地震仪
reflection seismology 反射地震学
reflection strength 反射强度
reflection survey 反射波法地震勘探
reflection wave 反射波
reflective condition 反射情况
reflective index 反射率
reflexed fold 倒转褶皱
reflow 回流；退潮
refluence 倒流；逆流
reflux 逆流；回流
reflux rate 回流速率
reflux ratio 回流比
refolded fold 重褶褶皱
refracted wave 折射波
refraction 折射
refraction constant 折射常数
refraction correlation method 折射波对比法
refraction index 折射率
refraction law 折射定律
refraction method 折射法
refraction path 折射路线
refraction profiling 折射法勘探
refraction prospecting 折射勘探法
refraction seismograph 折射地震仪
refraction seismology 折射地震学
refraction shooting 折射法勘探
refraction survey 折射法勘探
refraction wave 折射波
refractionation 褶曲
refractive 折射的
refractive displacement 折射位移
refractive exponent 折射率
refractive index 折射系数
refractivity 折射性；折射率差
refrigerating capacity 冷冻能力；制冷能力
refrigerating chamber 冷冻室；冷藏室
refrigerating engineering 冷冻工程；制冷工程
refrigerating machine 冷冻机
refrigerating medium 冷却介质；冷却剂；制冷剂

refrigerating system	制冷系统
refrigerating unit	制冷设备；冷冻设备；冷冻单位
refrigeration	制冷；冷冻；制冷学
refrigeration cycle	制冷循环
refrigeration effect	制冷效应
refuge chamber	井下躲避所；避难硐室
refuge floor	庇护层；隔火层
refuge harbor	避风港
refuge hole	避车洞
refuge island	安全岛
refuge passageway	避难通道
refuge pocket	躲避洞；防护洞
refuge shelter	防护洞；隐蔽所；避难处
refusal criterion	拒浆标准
refusal flow rate	拒浆流率
refusal of pile	抗沉
refusal point	桩的止点
refusal pressure	拒浆压力
refuse disposal	垃圾处置；废物处置
refuse dump	废石堆；垃圾堆；废物堆
refuse landfill	垃圾填埋场
refuse loading ramp	垃圾装卸坡
refuse ore	尾矿；废矿石
refuse pile	矸石堆；废石堆
refuse recovery	矸石回收率
refuse storage area	垃圾站
refuse storage chamber	垃圾房
refuse transfer station	垃圾转运站
refuse treatment	垃圾处理
refuse trough	垃圾槽
refuse utilization	尾煤利用
refuse yard	尾矿场；矸石场
refuse-disposal system	废渣处理法；尾矿处理法；废石处理法；尾矿处理系统
regarding of slope	重整斜坡；斜坡分级
regenerated roof	再生顶板
regime	状态；方式；规范
regime of flow	流态
regime of runoff	径流状况
regime stream	稳定河流
regime theory	河流动态理论
regime type	动态类型
region	地区；区域；范围
region of activation	活化区
region of alimentation	补给区
region of anomalous seismic intensity	地震烈度异常区
region of high stress	高应力区
region of influence	影响范围；影响区
region of intake	补给区
region of integration	积分区域
region of interest	研究区；评价区
region of no-relief	无起伏地区；平原
region of outflow	排泄区
region of permanent frost	永冻土带；永冻带
regional	区域的；地区的；局部的
regional alluvial coast	区域冲积海岸
regional analysis	区域性分析
regional analysis of hydrologic data	水文资料区域分析
regional anomaly	区域异常
regional appraisal	大面积普查
regional aquifer	区域含水层
regional arching	区域背斜隆起
regional atlas	区域地图集
regional background	区域背景
regional complex	区域复合体
regional cone of depression	区域下降漏斗；区域降落漏斗
regional correlation	区域对比
regional covering stratum	区域性盖层
regional crustal stability	区域地壳稳定性
regional descent of groundwater level	区域性地下水位下降
regional dip	区域倾斜
regional earthcrust stability	区域地壳稳定性
regional earthquake	区域地震
regional earthquake-generating stress	引起地震的区域应力
regional engineering geological map	区域工程地质图
regional engineering geology	区域工程地质
regional environment	区域环境
regional environment assessment	区域环境评价
regional environmental impact assessment	区域环境影响评价
regional estimation	区域评估；区域评价
regional evaluation	区域评价
regional exploration	区域勘察
regional extent	区域范围
regional fabric	区域组构
regional flood	区域性洪水
regional flood forecast	区域洪水预报
regional flow system	区域水流系
regional forecast	区域预报
regional geochemical anomaly	区域性地球化学异常
regional geochemical background	区域地球化学背景
regional geochemical differentiation	区域地球化学分异
regional geochemical prospecting	区域地球化学勘探
regional geochemical survey	区域地球化学测量；区域化探
regional geochemistry	区域地球化学
regional geologic unit	区域地质单元
regional geological environment	区域地质环境
regional geological mapping	区域地质填图
regional geological reconnaissance	区域地质普查；区域地质调查
regional geological reserves	区域地质储量
regional geological surveying	区域地质调查；区域地质测量
regional geology	区域地质学
regional geomorphology	区域地貌学
regional geo-stress distribution	区域地应力分布
regional geotectology	区域大地构造学
regional gravity anomaly	区域重力异常
regional ground subsidence	区域性地面沉降
regional hazard map	区域灾害图
regional hydrochemistry	区域水文化学
regional hydrodynamic pattern	区域水动力型式
regional hydrogeological investigation	区域水文地质研究
regional hydrogeological survey	区域水文地质普查
regional hydrogeology	区域水文地质学
regional hydrology	区域水文学

regional key	区域性标志
regional leveling survey	区域性水准测量
regional metamorphic rock	区域变质岩
regional metamorphism	区域变质
regional planning	区域规划
regional planning investigation stage	区域规划勘察阶段
regional precipitation line	地区降水线；地区雨线
regional prediction	区域预测
regional profile	区域性剖面
regional prospecting	区域勘探
regional reconnaissance	区域普查
regional regularity	区域性规律
regional remote sensing	区域遥感
regional seismic network	区域地震台网
regional seismicity	地震活动区域性；区域地震活动
regional seismological network	区域地震台网
regional seismology	区域地震学
regional settlement	区域性沉降
regional sewerage system	区域排水系统
regional stability	区域稳定性
regional stability to tectogenesis	区域构造稳定性
regional stratigraphic scale	区域地层表
regional stratigraphy	区域地层学
regional stress	区域应力
regional stress field	区域应力场
regional stress regime	区域应力状态
regional stress tensor	区域应力张量
regional strike	区域走向
regional structural trend	区域构造走向
regional structure	区域构造
regional subsidence	区域沉降
regional survey	区域测量；区域性调查
regional synthesis of hydrological data	区域性水文资料整编
regional tectonic analysis	区域构造分析
regional tectonic movement	区域性构造运动
regional tectonic stress field	区域构造应力场
regional tectonics	区域地质构造
regional trend	区域背景；区域走向
regional unconformity	区域不整合
regional uplift	区域性隆起
regional variation	区域性变化
regional water balance	区域水平衡
regional water budget	区域水量平衡
regional water supply	区域给水
regionally metamorphosed rock	区域变质岩
registration district	注册区域
registration mast	定位柱
regolith	表土，地表土石层；浮土；风化层
regradation	退化；后退；海退；回归
regrade	重整坡度；重新分级，再分级；重新分类
regrading of slope	斜坡修整
regression	退化；海退；溯源侵蚀；回归
regression analysis	回归分析；相关分析
regression analysis method	回归分析法；因子分析法
regression coefficient	回归系数
regression correction	回归校正
regression curve	回归曲线
regression design	回归设计
regression equation	回归方程
regression erosion	海退侵蚀
regression error	回归误差
regression estimation	回归估计
regression experimental design	回归试验设计
regression function	回归函数
regression line	平衡曲线；回归线；回归曲线
regression matrix	回归矩阵
regression model	回归模型
regression parameter	回归参数
regression phase	海退相
regression plane	回归平面
regression process	回归过程
regression reasoning	回归推理
regression testing	回归测试
regression variable	回归变量
regression-type analysis	回归分析
regression-type statistical analysis	回归型统计分析
regressive	退化的；后退的；海退的
regressive consequent bedding rockslide	后退式顺层滑坡
regressive cycle	海退旋回
regressive erosion	逆向侵蚀；逆向冲刷
regressive metamorphism	退化变质
regressive offlap	海退退覆
regressive overlap	海退超覆
regressive sediment	海退沉积物
regressive sequence	海退层序
regressive series	海退岩系
regrinding return	返砂
regrinding return ratio	返砂比
regripping	换支撑
reground cuttings	重复研磨的岩屑
regrout	二次压浆
regrowth	再生长
regular	有规律的；定期的；固定的
regular arc	正弧
regular arm	令拱；厢拱
regular band model	规则带模式
regular barrel	标准桶
regular bed	稳定层；规则层
regular bedding	规则层理
regular caving zone	规则垮落带；规则冒落带
regular change	规律性变化
regular check	定期检查；定期考核
regular coal seam	稳定煤层
regular coast	规则海岸；平直岸
regular coursed rubble	正规砌筑毛石
regular dip	规则倾斜
regular diurnal variation	规则日变化
regular feed	有规则地给料；正常给料
regular forecast	定期预报
regular function	正则函数；正规函数；解析函数
regular grade	普通品类；正规品级
regular grid	规则网格
regular inspection	定期检查
regular interval	有规律间隔
regular longwall	正规长壁开采法
regular longwall along strike	走向正规长壁开采法
regular maintenance	例行维修
regular maintenance of bridge and tunnel	桥隧经常保养
regular maintenance of buildings and structures	房屋维修

regular matrix 规则矩阵
regular mesh 规则网格
regular mixed-layer structure 有序混层结构
regular monitoring of special measure 特殊设施的定期监测
regular movement 有规则移动；规律性的活动
regular mud 普通泥浆
regular net 常规网络
regular noise 规则干扰
regular observation 定时观测
regular pillar 系统排列的矿柱；匀称布置的矿柱
regular pit dewatering 经常性排水
regular Portland cement 普通波特兰水泥
regular production 正常生产
regular reflection 单向反射；规则反射
regular service condition 正常使用条件
regular size 标准尺寸
regular solution 正规溶液
regular spread 规则排列
regular stopping method 普通回采法；正规回采法
regular supply 正常供应
regular system 等轴晶系
regular use 经常利用；经常用途
regular visit 定期巡检
regular wave 规则波
regular well pattern 规则井网
regular working cycle 正规循环
regularity 规律性；规则性；整齐度
regularity attenuation 正规衰减
regularity of coal seam 煤层稳定性
regularity of distribution 分布规律
regularity property 正则性质
regularization 调整；规则化
regularization method 正则化方法
regularization of area 修正界限；修正面积
regularization of employment 职务的调整
regularization of lot boundary 修正地段界线
regularized polygon 正多边形
regulated flow 调节流量；调节径流
regulated groundwater storage 地下水调节储量
regulated set cement 调凝水泥
regulated split 调节风流分支
regulated stream 受调节水流
regulated stream flow 调节流量
regulated water stage 调节后水位
regulated-set cement 控凝水泥
regulating agent 调整剂
regulating amplifier 调节放大器
regulating apparatus 调节设备
regulating chamber 调节室
regulating commutator 稳压管；稳流管；调整管
regulating course 路面整平层
regulating dam 分流坝；分水坝
regulating device 调整装置
regulating equipment 调节设备
regulating gate 调节闸门
regulating handle 调节阀，调节手柄
regulating mechanism 调节机构；调节机制；调控机制
regulating pondage 调节容量
regulating pull rod 调节拉杆

regulating reservoir 调节水库
regulating resistance 调节电阻；电位器
regulating resistor 调节电阻器
regulating rod 控制棒；调节棒
regulating spring 调节弹簧
regulating storage 调节库容
regulating structure 调治构造物；导流建筑物
regulating unit 调节装置；调整装置
regulating valve 调节阀；控制阀
regulating variable 调节量
regulation 调节；调整；控制校准
regulation capacity 调节能力
regulation capacity of aquifer 含水层调节能力
regulation cock 调节栓
regulation curve 调整曲线
regulation degree 调准度
regulation drop-out 停止调节
regulation factor 调整系数
regulation for technical operation 技术操作规程
regulation of affairs 事务管理
regulation of level 水位调节
regulation of meandering river section 压水冷却
regulation of output 输出调整；产量调整
regulation of river 河道整治
regulation of river confluence section 河渠交汇段整治
regulation of sediment-laden stream 多沙河流整治
regulation of tidal water 潮汐控制
regulation of wandering river section 尾水管性能
regulation of water level 水位调节
regulation of wire 导线调整
regulation pace 调整步测
regulation period 调节期
regulation reserves 调节储量
regulation reservoir 调节蓄水池
regulation screen plate 调节过滤板；调节筛板
regulation shaft 调节轴；调节手柄
regulator 调整剂；调节器；调节阀；调节风门
regulator box 调压箱
regulator canal 调节渠
regulator for special purpose 专用调节器；专用校准器；专用稳压器
regulator generator 电机调整器
regulator mixture 调节混合物
regulator of level 液位调节器
regulator opening 调节孔；调节风窗
regulator valve 调整阀；气门
regulator valve rod 调整阀杆
regulatory sign 禁令标志
rehabilitation 改造；恢复；修理；复垦
rehabilitation cast 修复费用
rehabilitation of the environment 恢复环境；环境修复
rehabilitation redevelopment 改建
rehabilitation scheme 修理方案；改建方案
rehabilitation work 美化工程
rehabilitative measure 善后措施
rehandling 再处理；再整顿
rehealing 再愈合
rehealing rockfill 加固堆石体
rehealing timbering 加强支护
reheat crack 再热裂缝

reheat cycle	再热循环		
reheat turbine	中间再热式汽轮机		
reheater	中间加热器		
reheating	再加热		
reheating furnace	再加热炉；重热炉		
reheating linear change	重烧线变化；残余线变化		
rehousing programme	安置计划		
rehydration	再水化；再水合		
reidentification	重新识别；重新鉴定		
reimbursable	可偿还的；可补偿的；可报销的		
reimbursable expense	可偿还的费用；可补偿的费用		
reimburse	偿还；补偿		
reimbursed at cost	按价补偿		
reimbursement	补偿；赔偿；报销		
reimbursing bank	偿付银行		
reinforce	加强；加固；加固材料；加筋，配置钢筋		
reinforce the hole wall	加固孔壁		
reinforced	加强的；配钢筋的		
reinforced brick	配筋砖；钢筋砖		
reinforced brick construction	配筋砖结构		
reinforced brick masonry	配筋砖砌体		
reinforced brick-work wall	配筋砖墙		
reinforced concrete	钢筋混凝土		
reinforced concrete arch	钢筋混凝土拱		
reinforced concrete arch culvert	钢筋混凝土拱涵		
reinforced concrete beam	钢筋混凝土梁		
reinforced concrete beam column joint	钢筋混凝土梁柱节点		
reinforced concrete bed	钢筋混凝土地基		
reinforced concrete bolt	钢砂浆锚杆		
reinforced concrete bridge	钢筋混凝土桥		
reinforced concrete building	钢筋混凝土建筑物		
reinforced concrete bunker	钢筋混凝土贮仓		
reinforced concrete buttressed dam	钢筋混凝土支墩坝		
reinforced concrete caisson	钢筋混凝土沉箱		
reinforced concrete cap	钢筋混凝土顶梁		
reinforced concrete circular water storage tank	钢筋混凝土圆形蓄水池		
reinforced concrete coating	钢筋混凝土加层		
reinforced concrete column	钢筋混凝土柱		
reinforced concrete construction	钢筋混凝土建筑		
reinforced concrete cover	钢筋混凝土封盖		
reinforced concrete dam	钢筋混凝土坝		
reinforced concrete deep beam	混凝土深梁		
reinforced concrete dock	钢筋混凝土船坞		
reinforced concrete draught tube	钢筋混凝土尾水管		
reinforced concrete drivage	钢筋混凝土支架巷道掘进		
reinforced concrete facing rockfill dam	钢筋混凝土面板堆石坝		
reinforced concrete flume	钢筋混凝土渡槽		
reinforced concrete foundation	钢筋混凝土基础		
reinforced concrete frame	钢筋混凝土框架		
reinforced concrete framing	钢筋混凝土构架		
reinforced concrete gate	钢筋混凝土闸门		
reinforced concrete lined tunnel	钢筋混凝土衬砌隧洞		
reinforced concrete lining	钢筋混凝土衬砌		
reinforced concrete masonry construction	钢筋混凝土砌块结构		
reinforced concrete member	钢筋混凝土构件		
reinforced concrete moment resisting frame	钢筋混凝土抗弯框架		
reinforced concrete open caisson	钢筋混凝土开口沉箱		
reinforced concrete pavement	钢筋混凝土路面		
reinforced concrete penstock	钢筋混凝土压力水管		
reinforced concrete pier	钢筋混凝土闸墩		
reinforced concrete pile	钢筋混凝土桩		
reinforced concrete plinth	钢筋混凝土基座		
reinforced concrete pneumatic caisson	钢筋混凝土气压沉箱		
reinforced concrete pressure pipe	钢筋混凝土压力水管		
reinforced concrete prop	钢筋混凝土支柱		
reinforced concrete radial gate	钢筋混凝土弧形闸门		
reinforced concrete retaining wall	钢筋混凝土挡土墙		
reinforced concrete roof	钢筋混凝土屋顶		
reinforced concrete segment	钢筋混凝土砌块		
reinforced concrete shear wall	钢筋混凝土剪力墙		
reinforced concrete sheet pile	钢筋混凝土板桩		
reinforced concrete skeleton frame	钢筋混凝土骨架		
reinforced concrete slab	钢筋混凝土板		
reinforced concrete sleeper	钢筋混凝土轨枕		
reinforced concrete spiral casing	钢筋混凝土蜗壳		
reinforced concrete storage	钢筋混凝土储罐		
reinforced concrete strip foundation under columns	柱下钢筋混凝土条形基础		
reinforced concrete structure	钢筋混凝土结构		
reinforced concrete support	钢筋混凝土支护		
reinforced concrete wall	钢筋混凝土墙		
reinforced concrete wall-frame interaction	钢筋混凝土墙-框架相互作用		
reinforced concrete walling	筑钢筋混凝土壁		
reinforced earth	加筋土		
reinforced earth abutment	加筋土桥台		
reinforced earth dam	加筋土坝		
reinforced earth embankment	加筋土路堤		
reinforced earth retaining wall	加筋土挡墙		
reinforced earth wall	加筋土墙		
reinforced embankment	加筋路堤		
reinforced fiber composite glass	玻璃钢；玻璃纤维增强复合材料		
reinforced fill	加筋土；加筋填土		
reinforced fill proprietary product	加筋土的专用材料		
reinforced fill structure	加筋土结构		
reinforced fill wall	加筋挡土墙		
reinforced framing	钢筋混凝土结构；钢筋混凝土构架		
reinforced geogrid	土工格栅加筋		
reinforced gunite lining	加强的喷射灌浆衬		
reinforced hollow unit masonry	加筋空心砖石砌体		
reinforced insulation layer	加强绝缘层		
reinforced layer	加固层		
reinforced layer for pressure	耐压加强层		
reinforced lime	加筋石灰		
reinforced liner	加固支架衬砌		
reinforced masonry	配筋砌体		
reinforced masonry structure	配筋砌体结构		
reinforced masonry wall	加筋砌体墙		
reinforced plastic	增强塑料		
reinforced plastic mortar	加筋塑料灰浆		
reinforced prim cord	加固导爆线		
reinforced profile concrete parapet	钢筋混凝土纵向护栏		
reinforced prop	加固的支柱；加筋的支柱		

reinforced retaining wall	加筋土挡土墙
reinforced rib	加强筋
reinforced rock arch	加固拱；组合拱
reinforced shaft	加固的井筒
reinforced shotcrete	加筋喷射混凝土
reinforced soil retaining wall	加筋土挡土墙
reinforced soil wall	加筋土挡墙
reinforced sprayed concrete support	配筋喷射混凝土支护
reinforced strength	加固强度
reinforced stress	增强应力
reinforced structure	加固结构
reinforced stull	加强的横撑支架
reinforced support	加强支架
reinforced thermoplastic sheet	增强热塑性片材
reinforced tile lintel	加筋空心砖过梁
reinforced tyre	加固外胎
reinforced wall	配筋墙；加固墙
reinforced weld	加固焊缝
reinforcement	加强；加固；钢筋；配筋
reinforcement arrangement	钢筋排列
reinforcement cage	钢筋笼；钢筋骨架
reinforcement concrete of support legs	护基混凝土
reinforcement corrosion	钢筋腐蚀
reinforcement displacement	钢筋位移
reinforcement drawing	配筋图
reinforcement effect	补强效果
reinforcement fabric	钢筋网
reinforcement factor	钢筋比；加固系数
reinforcement for shearing	抗剪加固
reinforcement limitation	配筋限度
reinforcement mat	钢筋网
reinforcement measure	加固措施
reinforcement metal	钢筋
reinforcement meter	钢筋计
reinforcement method	加筋法
reinforcement metre	钢筋测力计
reinforcement of massive rock	岩体加固
reinforcement of ore-pass	溜井加固
reinforcement of rock and soil	岩土加固
reinforcement of rock mass	岩体加固
reinforcement of soft foundation	软基加筋
reinforcement of surrounding rock of stope	采场围岩加固
reinforcement plate	加固板；增强板
reinforcement prefabrication	钢筋预加工
reinforcement ratio	含钢率；配筋率
reinforcement ratio per unit volume	体积配筋率
reinforcement steel	钢筋
reinforcement steel area	钢筋截面积
reinforcement summary	钢筋明细表
reinforcing	加固；加强
reinforcing agent	增强剂
reinforcing bar	钢筋；加固锚杆
reinforcing bar spacer	钢筋隔块
reinforcing bar stress transducer	钢筋应力传感器
reinforcing bars	加固钢筋；锚杆
reinforcing cage	钢筋笼
reinforcing capsule	加强帽
reinforcing dam	加固堤
reinforcing element	锚固构件
reinforcing fabric	钢筋网
reinforcing filler	增强填充料
reinforcing girder	加强梁
reinforcing leg	加固井架大腿
reinforcing material	增强材料
reinforcing member	加固件
reinforcing pad	加强圈
reinforcing plate	加强板
reinforcing post	加强柱；辅助柱
reinforcing prop	加强柱
reinforcing rib	加强肋
reinforcing ring	加固环；加固圈
reinforcing rod	加固杆
reinforcing set	加强棚子
reinforcing stay	加固支撑
reinforcing steel	钢筋
reinforcing steel area	钢筋截面积
reinforcing steel cage	钢筋笼
reinforcing steel workshop	钢筋加工厂
reinforcing strut	加强支撑
reinforcing system	钢筋系
reinforcing wire	钢丝
reinjection	回灌；回注
reinjection percentage	回注百分数
reinjection rate	回注速率
reinjection well	回注井；回灌井
reinstate and make good	修葺使之恢复原状
reinstatement	回复原状；恢复原貌
reinstatement of site	修补回填
reinstatement of supply	恢复供应
reinstatement works	修复工程
reiteration method	复测法
reiteration theodolite	复测经纬仪
reiterative direct shear test	反复直剪试验
reject disposal problem	尾矿处理问题；废渣处理问题
reject elevator	废渣提升机
reject fraction	损失率；损失量
reject gate	排渣闸门
rejected ore	废弃矿石
rejected part	不合格部件
rejected work	拒绝验收的工作
rejection	废弃；排除；损失；耗损
rejection of error	误差消除
rejection of heat	散热
rejection of load	弃负荷
rejection rate	废品率；报废率；衰减率；变薄率
rejection ratio	抑制比
rejuvenate	复活；使复原；使更新
rejuvenated erosion	再生侵蚀
rejuvenated fault	复活断层
rejuvenated fault scarp	更生断层崖；复活断层崖
rejuvenated landform	地形侵蚀回春
rejuvenated ore deposit	再生矿床
rejuvenated stream	更生河
rejuvenated water	再生水
rejuvenation	回春作用；更生作用
rejuvenation of fault	断层复活
related coefficient	相关系数
related function	相关函数
related phenomenon	相关现象
related variable	有关变量

relating cost 相关成本
relation 关系；比率
relation curve 相关曲线；关系曲线
relation of identity 恒等关系
relation of inclusion 包含关系
relation of nonlinear stress-strain 非线性应力应变关系
relation of principal stress and structural feature 主应力与结构特性关系
relation of strain to stress on a borehole wall 钻孔壁应力-应变关系
relation schema 关系模式
relational data base 相关数据库
relational data structure 关系数据结构
relational database 关系数据库
relational expression 相关式
relational graph 关系图
relational matching 关系匹配
relational model 关系模型
relational operator 关系操作符；关系算子；关系运算符
relational ordering 检索；排序
relational schema 关系模式
relational structure 关系结构
relationship 关系；联系
relationship between water and sediment 水沙关系
relationship graph 关系图
relationship in contaction 接触关系
relative 相对的；相关的；有关系的
relative abundance 相对丰度
relative abundance of methane 相对瓦斯涌出量
relative acceleration 相对加速度
relative accuracy 相对精度
relative activity 相对活度
relative age 相对时代；相对年龄
relative age of groundwater 地下水相对年龄
relative age of soil 土壤相对年龄
relative air density 相对空气密度
relative air speed 相对气流速度
relative altitude 相对高程
relative amplitude 相对振幅
relative amplitude preserve 相对振幅保持
relative aperture 相对孔径
relative apparent resistivity 相对视电阻率
relative area 相对面积
relative asymmetry 相对不对称
relative attenuation 相对衰减
relative bending 相对弯曲
relative bulk strength 相对体积强度
relative capacity 相对容量
relative change of sea level 海面相对变化
relative chronology 相对年代；相对时序
relative closing error 相对闭合差
relative code 相对代码
relative coefficient 相对系数
relative coefficient of collapsibility 相对湿陷系数
relative coefficient of permeability 相对渗透系数
relative coefficient of thaw settling 相对融陷系数
relative cohesion 相对黏聚力
relative compaction 相对密实度；相对压实度；压实系数
relative compressibility 相对压缩量；相对压缩率

relative concentration 相对浓度
relative confidence interval 相对置信区间
relative consistency 稠度指数；相对稠度
relative content 相对含量
relative contrast 相对对比度
relative control 相对控制
relative conversion ratio 相对转换比
relative coordinate 相对坐标
relative coordinate system 相对坐标系
relative correction factor 相对校正因子
relative cost 相对价值
relative current 相对流
relative damping 相对阻尼
relative dampness 相对湿度
relative dating 相对年龄测定
relative deflection 相对挠曲；相对挠度
relative deflection of the vertical 相对垂线偏差
relative deformation 相对变形
relative degree of humidity 相对湿度
relative density 相对密度
relative density of soil 土的相对密度
relative density test 相对密度试验
relative departure 相对偏离
relative depth of compressive area 相对受压区高度
relative detuning 相对失调
relative deviation 相对偏差
relative dielectric constant 相对介电常数
relative discharge 相对漏电
relative displacement 相对位移
relative displacement index 相对驱替指数
relative distance 相对距离
relative drilling rate 相对钻速
relative efficiency 相对效率
relative elevation 相对高差
relative embedment 相对埋深
relative equilibrium 相对平衡
relative error 相对误差
relative error ellipse 相对误差椭圆
relative error of closure 相对闭合差
relative evaporation 相对蒸发量
relative expansion 相对膨胀
relative exposure 相对照度
relative extreme 相对极值
relative fall 相对落差
relative fall of sea level 海面相对下降
relative floatability 相对可浮性
relative flow 相对流动
relative flow rate 相对流量；流量比
relative flying height 相对航高
relative formation strength 地层相对强度
relative frequency 相对频率
relative friction coefficient 相对摩擦系数
relative gamma contour map 相对伽马等值线图
relative gas emission rate 相对瓦斯涌出量
relative gravity difference 相对重力差
relative gravity value 相对重力值
relative grindability 相对可磨性
relative hardness 相对硬度
relative height 相对高度
relative humidity 相对湿度

relative humidity of atmosphere	大气相对湿度
relative humidity of the air	大气相对湿度
relative ice content	相对含冰率
relative impedance trace	相对阻抗曲线
relative inclinometer	相对倾斜计
relative increment	相对增量
relative inductivity	相对介电常数
relative intensity	相对强度
relative load factor	相对承重因素
relative magnitude	相对值
relative mass density	相对密度
relative mean square error	相对中误差
relative mean square error of side length	边长相对中误差
relative measuring method	相对测量法
relative merit	相对价值；相对优点
relative method	比较法
relative mobility	相对移动性；相对活度；相对迁移率
relative mobility ratio	相对流度比；相对活度比；相对迁移率比
relative moisture	相对湿度
relative moisture content	相对含水量；相对湿度
relative moisture content of soil	泥土相对含水量
relative moment	转矩；转动力矩
relative movement	相对运动
relative orientation	相对定位；相对定向；相对方位
relative out flow of gas	相对瓦斯涌出量
relative out flow of methane	相对瓦斯涌出量
relative parallax	相对视差
relative performance	相对性能
relative period elongation	相对周期延长
relative permeability	相对渗透率
relative plasticity index	稠度指数；相对塑性指数
relative porosity	相对孔隙度
relative position	相对位置
relative positioning	相对定位
relative productivity	相对产能
relative refractive index	相对折射率
relative relief	相对起伏；地势起伏量；相对高差
relative response	相对响应
relative rigidity	相对刚度
relative risk	相对风险
relative roughness	相对粗糙率
relative salinity	相对矿化度；相对盐度
relative scale	相对比例尺；相对规模
relative sea level	相对海平面
relative selectivity	相对选择性
relative settlement	相对沉降量
relative shift	相对移动
relative soil moisture	土壤相对湿度
relative spacing	相对间距
relative spacing ratio	岩体主节理与顶板可悬露跨度之比；相对间距比
relative speed	相对速度
relative stability	相对安定度；相对安定性；相对稳定性
relative standard deviation	相对标准偏差
relative starting pressure gradient	相对启动压力梯度
relative stiffness	相对刚度
relative strength	相对强度
relative stress measurement	相对应力测量
relative submergence	相对沉没度
relative sunshine duration	日照百分率
relative surface	相对表面积
relative surface activity	相对表面活度
relative surface area	比表面
relative temperature	相对温度
relative thickness	相对厚度
relative tilt	相对倾斜
relative time scale	相对年代表
relative velocity	相对速度
relative viscosity	相对黏滞度
relative void ratio	相对空隙比
relative volatility	相对挥发度
relative water content	相对含水量；液性指数
relative water depth	相对水深
relative weight strength	相对重量强度
relative wettability	相对润湿性
relative wetting preference	相对润湿优先
relative-spacing ratio	相对间距比
relative-stiffness method	相对刚度法
relativism	相对论
relativistic correction	相对论改正
relativistic effect	相对论效应
relativity	相关；比较性；相对性；相对论
relaxation	松弛；弛张；弛豫；缓和
relaxation constant	松弛常数；弛张常数
relaxation curve	松弛曲线
relaxation damping	松弛阻尼
relaxation effect	张弛效应
relaxation factor	松弛因数；松弛因子
relaxation frequency	弛豫频率
relaxation function	松弛函数
relaxation loss	张弛损失
relaxation method	松弛法；逐次近似法
relaxation method of stress	应力解除法；压力卸载法
relaxation modulus	松弛模量
relaxation of prestressed tendon	预应筋松弛
relaxation of residual stress	消除残余应力
relaxation of rock	岩石应力松弛
relaxation of rock mass	岩体松弛
relaxation of stress	应力松弛
relaxation oscillation	张弛振荡
relaxation parameter	松弛参数
relaxation pressure	松弛压力
relaxation process	弛豫过程
relaxation resistance	松弛强度
relaxation shrinkage	松弛回缩
relaxation solution	松弛解
relaxation technique	松弛技术
relaxation test	松弛试验
relaxation time	弛豫时间；松弛时间
relaxation zone	松动圈；松动带
relaxed rock	卸载岩石；消除应力的岩石
relaxed rock mass	消除应力的岩体；卸载的岩体
relaxed shear modulus	张弛剪切模量
relaxed zone	弛豫带
relay	中继；传送；继电器
relay point	中继站；转播站
relay sensitivity	继电器灵敏度
relay set	继电器组；继电器装置

relay station	中继站	reliability assessment	可靠性估计
relay storage	中间储存	reliability coefficient	可靠性系数
relay tank	中间储罐	reliability demonstration	可靠性验证
relay tank farm	中继油库	reliability design	可靠性设计
release	析出；释放；放出；解除	reliability engineering	可靠性工程
release adiabat	解压绝热线	reliability estimation	可靠性估计
release agent	脱模剂	reliability evaluation	可靠度评估
release curve	释放曲线；松放曲线	reliability function	可靠性函数
release device	放落装置；释放装置	reliability growth	可靠性增长
release factor	返还系数	reliability index	可靠性指数
release force	解锁力	reliability model	可靠性模型
release fracture	释放裂隙；解压裂隙	reliability of earthquake resistance of structure	结构抗震可靠性
release gear	释放装置	reliability of forecast	预报可靠性
release hook	放松钩；释放钩	reliability of structure	结构的可靠性
release joint	释压节理；解压裂隙	reliability of test	试验的可靠性
release of heat	放热；释放热量	reliability optimization	可靠性最优化
release of in-situ stress	原位应力释放	reliability optimization design	可靠性最优设计
release of internal pressure	内压释放	reliability prediction	可靠性预计
release of pressure	减压；卸压；压力减降	reliability screening	可靠性筛选
release of stress	应力释放	reliability service	可靠运行
release pin washer	卸压销垫	reliability standard	可靠性标准
release pipe	放泄管	reliability step	可靠性等级
release plug	释放塞	reliability test	可靠性试验
release point	溢出点；释放点	reliability theory	可靠性理论
release portion	缓解部	reliability trade-off	可靠性综合标准；可靠性权衡
release pressure	卸压	reliability trials	可靠性试验
release propagation rate	缓解波速	reliability-based criteria	根据可靠性制定的准则
release rate	释放率；释放速度	reliability-based optimum design	基于可靠度的优化设计
release spring	卸压弹簧；回位弹簧；放松弹簧	reliable	可靠的；牢固的；确实的
release time	释放时间	reliable indicator	可靠指标
release value	释放值；落下值	reliable source	可靠来源
release yield valve	卸载屈服阀；卸压屈服阀	reliance	信任；信心；依靠
released grain	解离的颗粒；分离的颗粒	relic alteration	残余蚀变
released hole	泄水孔	relic area	残遗分布区
released joint	释放节理；解压节理	relic back-arc basin	残留弧后盆地
release-push	释放按钮；释放开关	relic bar	残留沙坝
releaser	释放装置；排气装置	relic bedding	残留层理
releasing bend	放开弯曲	relic fabric	残余组构
releasing force	解锁力	relic mountain	残山
releasing lever	松开拉杆	relic sediment	残留沉积物
releasing means	释放装置	relic soil	残遗土壤
relending	再混合；重混合	relic structure	残余构造
relevance	关联；相关	relic texture	残余结构
relevance ratio	相关比	relic water	残留水；封存水
relevancy	相关性；关联	relict	残留；残余物；残留体；残留生物
relevant analysis	相关分析	relict alteration	残余蚀变
relevant data	有关数据	relict basin	残留盆地
relevant department	有关部门	relict beach ridge	遗滩脊
relevant experience	有关经验	relict bedding	残留层理
relevant feature	相关要素	relict dike	残留岩墙；残余岩脉
relevant parameter	相应参数	relict dune	残遗沙丘
relevant pressure	相应压力	relict fabric	残余组构
relevant regulation	有关规定	relict joint	残留节理
relevant testimony	有关证据	relict lake sediment	残留湖积物
relevant works	有关工程	relict mineral	残余矿物
relevel	复测水准	relict paleosol	残余古土壤
relevelling	重复水准测量	relict pseudodike	残留假岩墙
reliability	可靠；可靠率；确实性	relict relief	残余地形
reliability allocation	可靠性分配	relict scar	残余断崖；残余断壁
reliability analysis	可靠度分析		

relict sediment	残留堆积物	relocatable routine	浮动程序
relict structure	残留结构	relocation	迁徙；迁移
relict texture	残留结构	relocation of route	路线重定；路线复测
reliction	海退；水退地面	relocation payment	搬迁费
relief	地形起伏,地势；减轻；解除；减压,减重	relocation project	搬迁工程
relief angle	后让角；后角；间隙角	relocation register	浮动寄器
relief block	地形模型	relocation settlement cost	搬迁安置费用
relief contour	地形等高线	relog	再次测井
relief data	地貌资料	reluctance	磁阻
relief displacement	投影差；地形偏移	reluctance regulator	磁阻调节器
relief distortion	投影变形	reluctivity	磁阻率；磁阻系数
relief ditch	放水沟；泄水闸	remagnetization	再磁化
relief drawing	地形原图	remagnetization arc	重磁化弧；再磁化弧
relief effect	立体效应	remain	剩下；留下
relief element	地貌要素	remain arc	残留弧
relief energy	起伏量	remain stationary	保持固定
relief expression	地形显示；地形表示法	remainder error	剩余误差
relief feature	地貌要素	remainder function	余项函数
relief fitting	溢流塞	remainder mountain	残山；环蚀山；蚀余山
relief form	地形形态	remainder term	余项
relief height	地面起伏高度	remainder theory	余式定理
relief hole	减载浅炮眼；释负炮眼；卸载炮眼	remaining	余度
relief intensity	起伏幅度	remaining beds	混合层；剩余层
relief inversion	地形转换	remaining benefit	剩余效益
relief map	地形图；地面起伏图	remaining liquid	残余熔料
relief mechanism	减载装置；减荷装置；救济机制	remaining reserve	剩余储量
relief model	地形模型；立体地图	remaining resource	保留地下的资源
relief of pressure	压力下降	remaining rock	残岩
relief of stress	应力解除	remaining slag	剩余熔渣
relief outlet	放水口	remaining stress	残余应力
relief plan	地形图	remaining ultimate recovery	剩余总开采量
relief platform	卸荷台	remaining value	残值；余值
relief post	补充柱；辅助柱	remaining wall	剩余厚
relief pressure	释放压力	remaining works	余下工程
relief ratio	起伏率；流域宽高比	remapping	重测图,重新勘测；重新映射；地图重测
relief representation	地形表示	remarkable earthquake	显著地震
relief roadway	卸载巷道	remeasurement	重新测量
relief sewer	溢流污水渠	remeasurement contract	按量付款工程合约
relief shading	地貌晕渲	remedial	修补用的
relief sprue	冒渣口；缓冲冒口	remedial acid job	补救性酸化作业
relief valve	安全阀；减压阀；溢流阀	remedial action	补救措施
relief volume	起伏量	remedial cementing	补注水泥
relief well	减压井；减压排水井	remedial gravel pack	补救性砾石充填
relieve	减压；卸载；解除；降低；分离；释放；替换	remedial grouting	补救灌浆；补强灌浆
relieve from	解除	remedial maintenance	补救维修
relieved zone	免压带	remedial measure	补救措施
relieving	减缓；降低；解除	remedial treatment	补强处理
relieving arch	减压拱	remedial works	整治工程；补救工程；修葺工程
relieving beam	减压梁	remedy	补救；修补
relieving cut	辅助炮眼	remedy allowance	公差
relieving device	放空孔；救援井	remedying defects	修补缺陷
relieving layer	减压层	remelting process	重熔法
relieving seam	减压层	remelting separation	重熔分离
relieving shothole	清理瞎炮眼	remeshing	重新划分网格
relieving timber	新换支架	remigration	再运移；再迁移
relieving wall	辅助墙	remining	再次开采；重新开采
reload	重新加载；重新装载；再装填	remittance against document	凭单付款
reloader	换装机；转载机	remittance by draft	汇票汇款
reloading operation	再次加载操作；再装油作业	remittance settlement	汇款结算
reloading point	再装点；转运点	remittee	收款人

remitting bank	汇款银行；托收银行
remix water	再混合用水
remixed concrete	二次搅拌混凝土
remixer chamber	二次混合器
remixing	重拌和
remnant	残余, 剩余；残余物；侵蚀残余；残山；残余矿体
remnant abutment	残留矿柱的应力集中区
remnant magnetism	剩余磁性
remnant magnetization	剩余磁化强度
remnant pillar	残留煤柱
remnant strain	残余应变
remnant stress	残余应力
remnant works	修补工程
remnants of glaciation	冰川遗迹
remobilization	再活化
remodeling	改型
remote	遥远的；远距离的；远程的
remote aerial reconnaissance	远程航空勘察, 遥感航测
remote area	遥远地区；偏远地区
remote bolter	遥控锚杆安装机
remote calibration	遥控校准；远距标定
remote control	遥控；远程控制
remote control assembly	主河道
remote control board	遥控盘；遥控台
remote control equipment	遥控设备
remote control for a section	区段遥控
remote control hydroelectric station	遥控水力发电站
remote control ignition	遥控点火, 电力放炮
remote control index	遥控指数
remote control lever	遥控杆
remote control module	远动控制模块
remote control node	远动控制节点
remote control support system	液压支架遥控系统
remote control system	遥控系统
remote control technology	远动控制技术
remote control unit	远动控制单元
remote controlled pump regulation	泵的遥控变量；泵的遥控变量机构
remote controlled seafloor drilling	遥控海底钻探
remote cut-off	遥控开关
remote data indicator	数据遥示器
remote data processing	远距离数据处理
remote data telemetry	数据遥测
remote data unit	遥控数据单元
remote debugging	远程调试
remote delivery end	远伸端
remote detector	遥控检波器
remote display	遥控显示
remote drive	远距离驱动
remote effect	间接影响；远隔效应
remote equipment	遥控装置
remote equivalent	直线运动探测等值
remote error sensing	远程误差检测
remote haulage	外牵引；远距离牵引
remote job entry	远程作业输入
remote location	偏僻地区
remote measurement	遥测
remote measurement device	遥测距仪；遥测仪表
remote measuring	遥测
remote metering	遥测
remote methane monitor	甲烷遥测仪
remote monitor	远距离监视器
remote monitoring system	遥控监测系统
remote probe	远程探测
remote program entry	远程程序输入
remote push-button control	远距离按钮操纵；按钮遥控
remote reading	遥控读数
remote reading gage	遥控读数仪
remote reading gravimeter	遥控读数重力仪
remote real-time transmit	远程实时传输
remote record	遥测记录
remote recording seismograph	遥测记录地震仪
remote reentry capability	远距再入能力
remote reference magnetotelluric method	远参考场大地电磁法
remote region	边远地区
remote regulation	遥调
remote resonance	远谐振；远共振
remote sealing system	远距离封闭火区系统
remote sensing	遥感；遥测
remote sensing application	遥感应用
remote sensing cartography	遥感制图
remote sensing data	遥感资料
remote sensing data acquisition	遥感数据获取
remote sensing device	遥感装置
remote sensing equipment	遥控仪器
remote sensing estimation	遥感估算
remote sensing geology	遥感地质；遥感地质学
remote sensing image	遥感图像
remote sensing image processing	遥感图像处理
remote sensing information	遥感信息
remote sensing information system	遥感信息系统
remote sensing interpretation	遥感解译
remote sensing mapping	遥感制图
remote sensing material	遥感资料
remote sensing method	遥感法
remote sensing of oil spills	油溢遥感
remote sensing of resources	资源遥感
remote sensing platform	遥感平台
remote sensing prospecting	遥感勘测
remote sensing rock mechanics	遥感岩石力学
remote sensing software	遥感软件
remote sensing sounding	遥感测深
remote sensing survey	遥感勘测
remote sensing technique	遥感技术
remote sensing technology	遥感技术
remote sensing test	遥感试验
remote sensor	遥感器；遥测传感器
remote signal	遥测信号；遥控信号；远程信号
remote signal digitization	遥控信号数字化
remote station	遥控台；远距台
remote stress	远程应力
remote survey	遥感测量
remote switch	遥控开关；远程开关
remote terminal	远程终端
remote thermal scanning	遥控热扫描
remote thermometer	遥测温度计
remote valve	遥控阀
remote video source	远距离影像信号源

remote water level indicator	遥感水平仪；遥感水位表
remotely controlled mining	遥控开采
remotely controlled section	遥控区段
remotely operated	遥控的；远程控制的
remotely operated longwall face	遥控长壁工作面
remotely-controlled dual operation	遥控操作
remotely-controlled operation	遥控操作
remote-position control	遥控定位；位置遥控
remoldability	重塑性；成型性
remoulded clay	重塑黏土
remoulded landslide debris	重塑的滑坡泥石
remoulded loess	重塑黄土
remoulded property	重塑土特性
remoulded sample	重塑土样；重塑样本
remoulded soil	重塑土壤；扰动土壤
remoulded soil sample	重塑土样
remoulded strength	重塑强度
remoulded undrained shear strength	重塑土不排水抗剪强度
remoulding	重塑；扰动
remoulding apparatus	重塑仪
remoulding degree	重塑度
remoulding effect	重塑效应
remoulding gain	重塑增强
remoulding index	重塑指数
remoulding loss	重塑减弱
remoulding sensitivity	重塑土灵敏性；扰动灵敏度
removable agitator	移动式搅拌器
removable anchor	可拆式锚杆
removable bottom	活动底板
removable cementer	可钻碎的注水泥器
removable drill bit	活钻头；可换式钻头
removable equipment	可卸设备
removable hatch cover	活顶盖
removable section	可拆卸部分；可移动部分
removable stopping	移动式隔挡
removal	除去；移开；迁移
removal and decoration allowance	搬迁及装修津贴
removal charge	搬迁费用
removal cost	拆迁成本
removal efficiency	去除效率
removal expense	搬运费
removal of building	建筑物迁移
removal of core	取出岩心
removal of cuttings	清除岩屑
removal of demolition waste	清理废墟
removal of fault	故障的排除
removal of ground	移土工程
removal of heat	排热
removal of load	卸载
removal of overburden	剥离
removal of timbering	拆除木支撑
removal process	消除过程；消除步骤
removal project	迁建项目
removal rate	消除速率
remove the relief	削平地形
removing	清除；迁移
rending effect	爆破作用；破裂效应
rendzina	黑色石灰土；腐殖质碳酸盐土
renewable natural resources	可再生自然资源
renewable resource	可再生资源
renewal	更新；恢复；更换；复原；修理；备件
renewal and replacement fund	固定资产更新资金
renewal and upgrading of fixed assets	固定资产更新改造
renewal cost	恢复费；更新费
renewal of contract	合同续订；合同展期
renewal part	换用零件；更换的零件
renewal process	更新过程
renewal rate	更新率
renewed attention	加倍重视
renewed consequent stream	复活顺向河
renewed fault	复活断层
renewed faulting	复活断裂活动
renewed folding	重新褶皱作用
renewed landslide	滑坡复活
rent affordability	负担租金能力
rented land	租地
reopened vein	重展矿脉
reorganization	改组；改建
repack	改装；再填；复载
repair cost	修理费
repair equipment	检修设备
repair expense	修理费
repair job	修理工作
repair link	备修链节
repair material	修理用材料
repair paint	修补用漆
repair parts	备件；备用零件；备用部件
repair procedure	检修作业程序
repair scheme	修葺计划；修理方案
repair works	修葺工程；修复工程
repairability	可修复性
repairing shift	维修班
reparability	可选性；可准备性
repayment	偿还；报答；补偿
repayment ability	偿还能力
repayment guarantee	偿还保函
repayment of advance	预付款的偿还
repayment of loan	偿还贷款
repayment on credit	信贷偿还
repayment period	偿还期
repeat coverage	重复覆盖
repeat determination	再鉴定；再测定
repeat observation	重复观测
repeatability	重复性
repeatability error	重复性误差
repeatability precision	重复精度
repeatability segment	重复段
repeatable accuracy	重复精度
repeatable recording	可重复记录
repeated addition	叠加；重复相加
repeated averaging	累次平均
repeated bend test	弯曲疲劳试验
repeated bending fatigue	反复弯曲疲劳
repeated bending stress test	反复弯曲应力试验；弯曲应力疲劳试验
repeated blading	反复平整土方
repeated check	重复检查
repeated check measurement	检查重复测量

repeated checking	重复检查
repeated combination	重新组合
repeated cycles of freezing and thawing	重复的冻融循环
repeated deformation	重复变形
repeated direct shear test	反复直剪强度试验
repeated evaluation of function	函数的重复评价
repeated fault	重叠断层
repeated folding	重复褶皱
repeated formation tester	重复式地层测试器
repeated frequency vibration	重复频率振动
repeated impact test	反复冲击试验；冲击疲劳试验
repeated integral	重积分
repeated load	重复负载；反复多次负载
repeated mapping sheet	重测图幅
repeated measurement	多次测量
repeated measurement model	重复测量模型
repeated pollution	流动污染源
repeated precipitation process	反复沉淀法
repeated reflection	重复反射
repeated root	重根
repeated sampling	重复取样
repeated selection	多次选择
repeated shear test	重复剪切试验
repeated stress	反复应力；重复应力
repeated stress fatigue	疲劳断裂
repeated stress test	反复荷载试验
repeated tension and compression test	重复拉压试验
repeated test	重复试验
repeated use	重复使用；多次利用
repeated-load test	重复荷载试验
repeater theodolite	复测经纬仪
repeating boundary	重复边界
repeating decimal	循环小数
repeating earthquake	重复地震
repeating optical transit	复测光学经纬仪
repeating pattern	重复井网
repeating resistance	耦合电阻
repeating signal	复示信号
repeating theodolite	复测经纬仪
repellent	防水布；防护剂；相斥的；防水的
repelling groin	挡水丁坝
repelling program	互斥方案
repercussion	反击；反响；回弹；反射
repetition	重复；重现；复测
repetition code	重复码
repetition coverage	重复航摄面积
repetition frequency	重复频率
repetition interval	重复周期；重复的间隔；脉冲周期
repetition measurement	复测；复量
repetition method	复测法
repetition of beds	地层重复
repetition of outcrop	露头重复
repetition period	重复周期
repetition prevention	防止重复
repetition process for fracturing	重复压裂过程
repetition rate	重复率
repetition survey of existing railway	线路复测
repetition theodolite	重测经纬仪
repetition time	重复时间
repetition work	成批加工
repetitional	重复的
repetitional load	反复荷载
repetitive and reversed stress	反复反向应力
repetitive dive capability	重复潜水能力
repetitive error	重复误差
repetitive faulting	重复断层作用
repetitive form	重复使用的模板
repetitive load	反复荷载
repetitive shock	重复冲击
repetitive strain	反复应变
repetitive stress	重复应力；反复应力
repetitive survey	施工复测
repetitive symmetry	重复对称
repetitive work	重复工作
replace	交代；置换；复位；复原
replaceability of cation	离子置换能力
replaceable cutter element	可更换的截割头部件
replaceable cutting bit	可换式截齿
replaceable underflow orifice	可拆换的底流口
replaced foundation	置换基础
replaced ground base	置换地基
replacement	替代；置换；交代作用
replacement along cleavage	顺解理置换
replacement bit	重嵌钻头
replacement breccia	交代角砾岩
replacement bucket	备用铲斗
replacement chemical	代替化学品；代用化学品
replacement cost	重置成本；更新成本
replacement cost addition	重置成本加额
replacement deposit	交代矿床
replacement dike	交代岩脉
replacement effect	置换效用
replacement housing	重置房屋
replacement material	代用材料
replacement method	换土法
replacement method of depreciation	重置成本折旧法
replacement of foundation	基础置换
replacement of soil	换土
replacement ore	交代矿
replacement orebody	交代矿体
replacement pile	排土桩；挤密桩；置换桩
replacement pressure	排替压力
replacement ratio	置换率
replacement reaction	复分解反应
replacement texture	交代结构
replacement time	置换时间；取代时间
replacement value	更新价值；重置价值
replacement velocity	置换速度
replacement well	备用井
replacing concretion	置换结核
replacing of timbering support	置换支撑
replacing overlap	交代覆蔽
replacing power	置换能力
replanning	重新规划
replat	谷肩；山肩
replenishment of groundwater	地下水补给
replica method	复型法
replicate	平行测定
replicate analysis	重复分析
replicate determination	平行测定

replicate sample	复制样品
replicate test	重复试验
replication	重复；重现；复制；复制品
replicative symbol	象形符号
report item	报表项目
report of site investigation	场址勘察报告
report on bid evaluation	评标报告
report on tender	投标审查报告
reporting	汇报；报告
reporting of data	资料的报告
reporting station	水情站
repose	安静；静止；休息；安息
repose angle	休止角；安息角
repose period	静止期
repose slope	挖沟
reposition	重新设置；重新布置；重新定位
repositioning	重新定位
representative	代表性的；代表
representative average	典型平均值
representative basin	典型流域
representative condition	典型条件
representative core	代表性岩心
representative data	典型资料
representative elementary volume	表征单元体积；代表性单元体积
representative error	代表性误差
representative method of sampling	代表性抽样法
representative observation	代表性观测
representative property	典型性质
representative sample	代表性试样
representative sampling horizon	代表性采样层
representative scale	惯用比例尺
representative section	代表性剖面
representative species	样本；标本
representative specimen	代表性试样
representative station	代表性测站
representative value	代表值
representative value of action	作用代表值
representative value of load	荷载代表值
representative volume element	代表性体积单元；代表性体元
representativeness	代表性
repress ring	加压的
repressive measures	抑制性措施
repressuring	压力回升
repressuring method	恢复压力法；再加压法
reprocessing	再加工；再处理
reprocessing cycle	再生循环
reproduce	复现；再现；重发；复制
reproducibility	可复现性；复验性
reproducibility error	重复性误差
reproducibility index	复现指数
reproducible record	可重现式记录
reproducible result	可重复的结果
reproducing unit	复制装置；再生设备
reproduction	复制；重现；再生
reproduction cost	再生产成本；重置成本
reproduction of mortar	灰泥复制
reproduction operation	复制操作
reproduction operator	复制算子
reproduction plan	复制图
reproduction probability	复制概率
reproduction quality	保真度
reproduction speed	再现速度
reproductions analysis	再现分析
reprogramming	重编程序；改编程序
reprographic	摄影制图
reprovision	重建；重配
reprovision site	重配地盘
reprovisioning assessment	重配地盘评估
reprovisioning project	重建计划
reprovisioning site	重配地盘
repudiate the contract	否认合同有效
repulsion	推斥；拒斥；斥力
repulsive force	排斥力；斥力
request for allocation	要求分配
required element size	要求的单元大小
required embankment	必要填方
required resolution	要求的分辨率
required strength	要求强度
required support line	所需支撑线
required support line calculations	所需支护线计算
required value	要求值
requirement	需要；需要量；规格，技术条件
requirement analysis	需求分析
requirement for fortification against earthquake	抗震设防要求
requirement specification	需求规格说明
requirements for support	支护要求
requisite	必要的；所需要的；必需品；要件
reregulating reservoir	再调节水库；水库反调节
resampling	重新取样；重复取样
resampling error	重复采样误差
resaturation	再饱和
rescreening	重新过筛；再筛分
rescue regulation	救护规程
rescue service	救护工作
rescue station	救护站
rescue team	救护队
rescue vehicle	救护车；工程救险车
rescue work	救护工作
research	研究；调查；探讨；勘查
research and development	研究
research effort	研究计划
research establishment	研究机构
research facility	研究设施
research finding	研究成果；研究发现
research institute	研究院；研究所
research project	研究计划；研究项目
research proposal	研究项目申请书
research room	研究室
research ship	调查船；考察船；研究船
research station	观测站
research system	研究系统
research trial	研究试验
researcher	研究工作者；研究员
reseat pressure	阀座修整压力
resected point	后方交会点
resected station	后方交会点
resection	交叉；后方交会法

resedimentation 再沉积
resedimented clast 再沉积碎屑
resedimented family 再沉积系列
resedimented rock 再沉积岩；浊流沉积岩
resemblance 相似
resequent drainage 再顺水系
resequent fault-line scarp 再顺断层线崖
resequent ridge 再顺脊
resequent rift-block mountain 再顺断块山
resequent rift-block valley 再顺断块谷
resequent river 再顺河
resequent stream 再顺河
resequent valley 再顺谷
reservation 保留；保存
reservation area 保留地
reservation of site 预留地盘
reserve 储量；埋藏量
reserve abundance 储量丰度
reserve account 准备金账户
reserve alkalinity 储备碱度
reserve area 专用范围；预留土地
reserve base 储量基础；基础储量
reserve calculating boundary 储量计算边界线
reserve calculation 储量计算
reserve cancelling 储量注销
reserve capacity 备用能力；备用容量
reserve checking and approving 储量审批
reserve control 储量控制
reserve distribution 储量分布
reserve energy capacity 能量储备能力
reserve energy reduction factor 能量储备降低系数
reserve equipment 备用设备
reserve error 储量误差
reserve estimate 储量估算
reserve estimation 气藏估计
reserve factor 储量备用系数；备用率
reserve factor of mine reserve 储量备用系数
reserve figure 储量数字
reserve for repair 维修费准备
reserve forecasting 储量预测
reserve fund 储备基金；准备金
reserve fund of project 工程预备费
reserve mud tank 备用泥浆罐
reserve not on balance sheet 表外储量
reserve of building material 建材储量
reserve pit 泥浆储备池
reserve place 后备区；保留区；备用工作面
reserve price 内定最低价；保留价；标金底价
reserve project 后备工程
reserve pump 备用水泵
reserve ratio 准备金比率
reserve replacement cost 储量接替成本；储量替换成本
reserve replacement ratio 储量接替率；储量替换率
reserve scale 储量规模
reserve stock 储备库存
reserve strength 残余强度；备用强度
reserve tank 储备罐；储水箱
reserve to output ratio method 储采比法
reserve tractive force 储备牵引力
reserve upgrading 储量升级

reserve usable in future 暂不能利用储量
reserve volume 储量
reserve way 安全出口
reserve zone 专用范围；专用区；预留地
reserved area 保留区
reserved bucket 备用铲斗
reserved pump 备用泵；储备泵
reserved second line 预留第二线
reserved settlement 预留沉落量
reserved share 保留股票
reserved stock 保留股票
reserved strength 保留强度
reservoir 油气储集层；油藏；储存器；储罐；水库；蓄水池；热储
reservoir action 调蓄作用
reservoir analysis 油层分析
reservoir anisotropism 储层非均质性；储层各向异性
reservoir area 储层面积
reservoir area survey 库区调查
reservoir area-capacity curve 水库面积库容关系曲线
reservoir axis 水库轴线
reservoir bank 水库库岸
reservoir bank caving 水库塌岸
reservoir bank collapse 水库塌岸
reservoir bank failure 水库坍岸
reservoir bank reformation 库岸再造
reservoir bank ruin 水库坍岸
reservoir bank stability 库岸稳定性
reservoir basin 蓄水盆地；蓄水区
reservoir bed 蓄水层；蓄油层；蓄气层
reservoir bed set correlation 储层层组对比
reservoir behavior 油层动态
reservoir body 储集体
reservoir boundary 油层边缘
reservoir brine 油层盐水
reservoir capacity 热储容量；储层容量；储罐容量；储水容量
reservoir capillary pressure 储层毛细管压力
reservoir characterisation 油藏描述
reservoir characteristic 储层特征
reservoir characterization 储库表征；储层刻画
reservoir choke 地层堵塞
reservoir cleaning 水库清理
reservoir communication 储层连通性
reservoir compaction 储层压实
reservoir compartmentalization 储层划分；储层分区
reservoir condition 储层条件
reservoir configuration 油藏结构；储层构型
reservoir conformance 油层波及系数
reservoir connectivity 储层连通性
reservoir continuity 储层连通性
reservoir coverage 油层波及系数
reservoir covering stratum 油气藏盖层
reservoir delineation 储集层划分
reservoir depleting curve 水库放水曲线
reservoir description 油藏描述
reservoir design 水库设计
reservoir design capacity 水库设计容量
reservoir design flood 水库设计洪水
reservoir draught 水库实际供水量

reservoir drawdown	水库水位泄降
reservoir drive energy	油层驱动能量
reservoir drive mechanism	油藏驱动机理
reservoir dynamics	油藏动力学
reservoir earthquake	水库地震
reservoir elevation	水库水位
reservoir empty	空库
reservoir emptying	水库放空
reservoir energy	储层能量
reservoir engineering	油藏工程
reservoir evaluation	储层评价
reservoir evaporation	水库蒸发
reservoir experiment station	水库实验站
reservoir filling	水库蓄水
reservoir flow unit	储层流动单元
reservoir fluid	储层流体
reservoir fluid gravity	储层流体比重
reservoir for irrigation	灌溉用水库
reservoir for overyear storage	年际调蓄水库
reservoir fracturing	地层压裂；油层水力压裂
reservoir geochemistry	油藏地球化学
reservoir geology	储层地质学
reservoir geomechanics	油藏地质力学；储层地质力学；储库地质力学
reservoir geometry	油气藏几何形状
reservoir geophysics	油藏地球物理学
reservoir group system	水库群系统
reservoir heterogeneity	储层非均质性
reservoir hydrocarbons	储层油气
reservoir hydrology	水库水文学
reservoir immersion	水库浸没
reservoir impairment	储层损害
reservoir impounding induced seismicity	水库诱发地震
reservoir impoundment	水库蓄水
reservoir induced earthquake	水库诱发地震
reservoir inflow	水库入流量
reservoir inflow hydrograph	水库入流过程线
reservoir inlet	水库入口
reservoir interval	储层层段
reservoir inundated area	水库淹没区
reservoir inundation line survey	水库淹没界线测量
reservoir investigation	水库调查
reservoir lake	蓄水湖
reservoir layering model	油藏分层模型
reservoir leakage	水库渗漏
reservoir length	水库长度
reservoir level	水库水位
reservoir life	水库使用年限
reservoir limit	油藏边界
reservoir lining	水池衬砌
reservoir lithology	储层岩性
reservoir live storage	水库有效容量；工作库容；有效库容
reservoir load	水库荷载
reservoir loading	水库负载
reservoir loss	水库损失
reservoir management	油藏管理
reservoir microstructure	储层微构造
reservoir model	油藏模型
reservoir modelling	储层建模
reservoir monitoring	储层监测
reservoir nature	储层性质
reservoir net pay thickness limit	储层有效厚度界限
reservoir of mud	泥浆池
reservoir operation	水库运营
reservoir operation chart	水库运用图表
reservoir operation curve	水库调度曲线
reservoir operation guide curve	水库运用指示线
reservoir performance	储层行为
reservoir performance assessment	油藏动态评价
reservoir peripheral area	油藏边缘区
Reservoir permeability	储层渗透率
reservoir physical parameters	储层物性参数
reservoir pollution	水库污染
reservoir pore space	储层孔隙空间
reservoir porosity	储层孔隙度
reservoir pressure	储层压力；含水层压力；水库压力
reservoir pressure depletion	储层压力衰减
reservoir pressure gradient	储层压力梯度
reservoir pressure maintenance	储层压力保持
reservoir production history	储层开采史
reservoir productivity	储层产能
reservoir property	储层性质
reservoir quality	储层质量
reservoir refill	水库重蓄满
reservoir region	库区
reservoir regulation	水库调节作用
reservoir release	水库放水
reservoir retention capacity	水库滞洪能力
reservoir rim	水库周边
reservoir rock	储集岩；储层岩石
reservoir rock opening	储集岩孔径
reservoir rock potentiality	储集岩远景
reservoir sand	储油砂层；油砂体；油层砂
reservoir sandstone	储层砂岩
reservoir seal	储层封闭层
reservoir sedimentation	储层沉积；水库淤积
reservoir sedimentology	储层沉积学
reservoir seepage	水库渗漏
reservoir seismic effect	水库地震效应
reservoir seismic stratigraphy	储层地震地层学
reservoir seismicity	储层地震
reservoir sensitivity	储层敏感性
reservoir shape factor	油藏形状因子
reservoir shoreline	水库岸线
reservoir shortening	储层变薄
reservoir shrinkage	油层收缩
reservoir silting	水库淤积；水库沉泥
reservoir simulation	热储模拟；油藏模拟
reservoir site	库址
reservoir space	储集空间
reservoir stage	库水位
reservoir storage	库容；水库容量
reservoir storage capacity	水库蓄水量
reservoir storage survey	库容测量
reservoir storage volume	水库蓄水量
reservoir stress	储层应力
reservoir stripping	水库表土剥离
reservoir structure	储层构造
reservoir survey	水库测量；储层调查

reservoir system	水库系统；储层系统
reservoir temperature	储层温度
reservoir thickness	储层厚度
reservoir tightness	储层致密度
reservoir trap	储层圈闭
reservoir trap efficiency	水库搁沙率，水库沉沙效率；储层圈闭效率
reservoir volume	水库总容积；总库容
reservoir volume ratio	地层体积系数
reservoir volume unswept	油层未波及体积
reservoir water salinity	地层水矿化度
reservoir works	水库工程
reservoir yield	水库的产水量
reservoir zonation	储层划分
reservoir-induced earthquake	水库诱发地震
reset condition	复位条件
reset function	复位功能
reset switch	复位开关
reset time	复位时间
resettability	可重调性
resetting	重放；复位；重安装；重镶嵌
resetting age	重置年龄
residence quarter	住宅区；居住区
residence time	停留时间；阻滞时间
resident engineer	驻地工程师；常驻工程师
resident fluid	地层流体
resident geologist	区段地质师；驻矿地质工作者
resident site staff	驻工地人员
residential accommodation	住宅设施
residential area	住宅区；居住区
residential building	住宅建筑物
residential density	居住密度
residential development	住宅用地发展
residential district	居住区
residential floor	居住层
residential floor area	居住建筑面积
residential land	住宅用地
residential property	住宅物业
residential purpose	住宅用途
residential quarter	住宅区
residential standard	住宅标准；居住标准
residential zone	住宅分区；住宅地带；住宅用地区域
residual	残余的；剩余的
residual amplitude	剩余振幅
residual amplitude section	剩余振幅剖面
residual analysis	残差分析
residual angle of internal friction	残余内摩擦角
residual anomaly	剩余异常
residual anticline	残余背斜
residual attenuation	剩余衰减
residual based error estimator	基于残值的误差估计
residual basin	残留盆地
residual carrying capacity	剩余承载力
residual clay	残积黏土；残积土
residual cohesion	残余凝聚力
residual cohesion intercept	残余黏聚力
residual concentration	残余浓度
residual correction	剩余校正
residual deformation	残留形变；残余变形
residual density	剩余密度；剩余质量
residual deposit	残积物；残余沉积
residual diluvial expansive soil	残坡积膨胀土
residual displacement	残余位移
residual distance error	距离残差
residual disturbance	剩余扰动
residual dome	残余穹丘
residual drawdown	剩余降深；参与水位下降
residual effect	剩余效应
residual elasticity	弹性后效
residual element	残余元素
residual environmental impact	剩余环境影响
residual equation	剩余方程
residual error	残留误差
residual error estimator	残值误差估计
residual error rate	差错漏检率
residual estimator	残差估计量
residual excitation	剩余激励
residual factor	残余系数
residual field	剩余场
residual fine dust	残留细尘
residual flow	剩余流量
residual fluid content	残余含液量
residual flux density	剩磁通量密度
residual gas	残余气；残存瓦斯
residual gas saturation	残余气饱和度
residual geosyncline	残余地槽
residual gliding force	剩余下滑力
residual gravity	剩余重力值
residual gravity anomaly	剩余重力异常
residual gravity map	剩余重力图
residual hardness	剩余硬度
residual head	剩余水头
residual heat	余热
residual index	偏差指数
residual lake	残留湖
residual land value	土地余值
residual landform	残余地形
residual landscape	残余景观
residual latosol	砖红壤残积古土
residual level	残留量
residual life	残效期；剩余寿命
residual liquid	结晶后的残余岩浆；残留液体，残液
residual load	残余荷载
residual load array	残余荷载数组
residual loss	剩余损耗
residual magma	残余岩浆
residual magnetism	残磁；剩磁
residual mass	剩余质量
residual mass inversion	剩余质量反演
residual material	残积物
residual method	余值法
residual migration	剩余偏移
residual mineral	残余矿物
residual mobility period	余动期
residual moisture	剩余水分
residual moisture content	剩余含水量
residual molten process	残余溶蚀作用
residual monomer	剩余单体
residual mountain	残余山
residual moveout	剩余时差

residual normal moveout	剩余正常时差
residual of seismic travel time	地震走时残差
residual of seismic wave travel time	地震波走时残差
residual ore	残余矿石
residual oscillation	残余摆动
residual overburden	残积层
residual peak	剩余最大值
residual phase error	剩余相位误差
residual pillar	残留煤柱；残留矿柱
residual placer	残积砂矿
residual pore pressure	剩余孔隙压力
residual pore water pressure	残余孔隙水压力
residual porosity	残留孔隙度；残余孔隙
residual pressure	剩余压力
residual product	残余产物；残油；副产品
residual rainfall	残余降雨
residual resistance	剩余阻力
residual resistance factor	残余阻力系数
residual salt feature	残余盐丘构造
residual saturation	残余饱和度
residual sediment	残余沉积物
residual settlement	残余沉降
residual shear strength	残余抗剪强度
residual shrinkage	剩余缩率
residual side force	剩余侧向力
residual sliding force	剩余下滑力
residual sludge	残余污泥；污泥残渣
residual soil	残积土
residual stability	残余稳定性
residual stability period	余定期
residual standard deviation	剩余标准差；剩余标准离差
residual static correction	剩余静校正
residual strain	残余应变
residual strain energy	残余应变能
residual strain field	残余应变场
residual strength	残余强度
residual strength criterion	残余强度准则
residual strength of rock	岩石残余强度
residual stress	残余应力
residual stress distribution	残余应力分布
residual stress field	残余应力场
residual stress in rock	岩石残余应力
residual stress relieving	消除残余应力；残余应力解除
residual structure	残余构造
residual subsidence	残余沉陷；残余下沉
residual subsidence developing	剩余沉陷作用
residual sum	余量和
residual swelling	残余膨胀
residual tack	残余黏性
residual tensile strength	残余抗拉强度
residual texture	残余结构
residual thrust method	剩余推力法
residual time curve	剩余时间曲线
residual time-shift anomaly	剩余时移异常
residual trust method	剩余推力法
residual valence	剩余价
residual valley	残谷
residual value	残值
residual variance	剩余方差
residual vertical parallax	残余上下视差
residual vibration	残余振动
residual viscosity	残留黏度
residual void	剩余空隙
residual volatile matter	残余挥发物
residual water	残留水
residual water saturation	残余水饱和度
residue	剩余，余余；残渣，残留物
residue gas	残气，干气
residue of boundary	边界余量
residue of equation	方程余量
residuum	风化壳；残余物，残积层
residuum surface	风化表层
resilience	回弹；弹回性；回能；弹能
resilience coefficient	回弹系数
resilience energy	回弹能量
resilience index	回弹指数
resilience modulus	回能模量
resilience test	弹性试验
resilience under stress	应力下的回弹力
resilient	有弹性的；有回弹力的
resilient bushing	弹性衬套
resilient coupling	弹性联轴节；弹性连接器
resilient deformation	弹性变形
resilient gear	弹性齿轮
resilient gear drive	弹性齿轮驱动
resilient material	弹性材料
resilient modulus	回弹模量
resilient seal	弹性密封
resilient sealing system	弹性密封系统
resiliometer	回弹仪
resillage	网状裂纹
resin	树脂；环氧树脂
resin adhesive	树脂胶黏剂
resin anchor	树脂锚杆
resin anchored bolt	树脂锚杆
resin based mix	树脂基浆液
resin bolt	树脂锚杆
resin bolting	灌注树脂固定锚杆法
resin cast	树脂铸模
resin cement	树脂水泥
resin compound	树脂化合物
resin dispersion	树脂弥散
resin divertor	树脂转向剂
resin emulsion paint	树脂涂料；树脂漆
resin grout	树脂浆液；树脂灌浆
resin insulator	树脂绝缘器
resin liptobiolith	树脂残殖煤
resin mortar	树脂胶浆
resin peat	树脂泥炭
resin rodlet	树脂棒
resin roof bolting	树脂锚杆
resin slurry	树脂砂浆
resin solution	树脂溶液
resinite	树脂体，树脂沥青
resinous coal	树脂质煤；低温干馏用煤
resinous lignite	树脂褐煤
resinous liptobiolith	树脂残殖煤
resinous luster	树脂光泽
resinous shale	油页岩
resist	抵抗；抗蚀剂

resistance	阻力；抗力
resistance angle	抗滑角
resistance bond	抗黏性
resistance bowl	电阻箱
resistance capacitance	电阻电容
resistance capacitance filter	阻容滤波器
resistance capacity coupled amplifier	阻容耦合放大器
resistance capacity filter	阻容滤波器
resistance closure meter	电阻闭合计
resistance coefficient	阻力系数；电阻系数
resistance constant	电阻常数
resistance coupled amplifier	电阻耦合放大器
resistance coupling	电阻耦合
resistance curve	R 曲线；阻力曲线；裂纹增长阻力曲线
resistance drop	电阻电压降
resistance dynamometer	测力仪；动力仪；功率计；测功仪
resistance effect	拦挡效应
resistance effect of trees	树木拦挡效应
resistance element	电阻元件
resistance factor	阻力系数
resistance force	阻力
resistance function	阻力函数
resistance head	摩擦压头；摩擦水头损失
resistance in parallel	并联电阻
resistance in series	串联电阻
resistance measurement	电阻测定法
resistance measurement	测阻
resistance mineral	稳定矿物
resistance modulus	抗力模量
resistance of flow	流动阻力
resistance of friction	摩擦阻力
resistance of rock	岩石抗力性
resistance of semi-confining layer	弱透水层阻力
resistance of soil	土抗力；土阻力
resistance rock mass	抗力体
resistance strain gauge	电阻应变计
resistance strain instrument	电阻应变仪
resistance survey	电阻测量
resistance thermometer	电阻温度计
resistance to acid	抗酸性
resistance to aging	抗老化性
resistance to alternate freezing and thawing	抗冻融性
resistance to alternate mechanical stress	耐交变应力性
resistance to alternating load	抗交变载荷
resistance to bending	抗弯强度
resistance to bending fatigue	抗弯曲疲劳能力
resistance to blasting	抗爆破阻力
resistance to bond	抗黏着力
resistance to breakage	抗断裂性
resistance to breaking	抗裂性
resistance to broom	桩顶抗裂能力
resistance to capsizing	抗倾覆能力
resistance to chemical deterioration	抗化学变质性
resistance to compressive strain	压缩应变阻力
resistance to corrosion	耐腐蚀性
resistance to crushing	抗破碎性
resistance to deformation	抗形变性
resistance to erosion	抗侵蚀性能
resistance to failure	抗破坏能力
resistance to fire	耐火性
resistance to fire and heat	防火隔热
resistance to flow	流动阻力
resistance to fracture	抗破裂
resistance to freezing and thawing	抗冻融性
resistance to heat	耐热性
resistance to impact	抗冲击强度
resistance to lateral bending	抗侧弯能力
resistance to oxidation	抗氧化能力
resistance to penetration	抗浸透性；耐刺穿性
resistance to repeated impact	抗重复冲击能力
resistance to rupture	抗破裂性
resistance to shock	抗震性
resistance to sliding	抗滑阻力
resistance to solvent action	抗溶解作用
resistance to splitting	抗劈力
resistance to water	耐水性；抗水能力
resistance to wear	耐磨性
resistance to weathering	耐风化强度
resistance to withdrawal	抗拔力
resistance to yield	抗屈服
resistance wire strain gauge	电阻丝应变计
resistant	耐久的；有抵抗性的
resistant mineral	耐蚀矿物；稳定矿物
resistant rock	稳固岩石；耐蚀岩石
resistant structure	抗力结构
resistant to bending	抗弯曲
resistant to corrosion	抗蚀的
resistant to pollution	耐污染的；抗污染的
resistant zone	抗拔区
resistate	抗蚀残积；残余物；耐蚀物，稳定物
resistibility	抵抗性
resisting force	抗力；阻力
resisting moment	抵抗弯矩；抵抗力矩
resisting sliding force	抗滑力
resisting strength	阻力
resistive load	电阻负载
resistivity	电阻率；电阻系数，比阻；抵抗力，抵抗性
resistivity anisotropy	电阻率各向异性
resistivity anisotropy coefficient	电阻率不均匀系数
resistivity anomaly	电阻率异常
resistivity coefficient	阻力系数
resistivity contrast	电阻率对比
resistivity cross section	电阻率截面
resistivity curve	阻力曲线
resistivity dipmeter	电阻率倾角仪
resistivity exploration	电阻法
resistivity gradient	电阻率梯度
resistivity index	电阻率指数
resistivity log	电法测井
resistivity logging	电阻率测井
resistivity logs	电阻率法测井记录
resistivity measurements	电阻率测量
resistivity meter	电阻率计
resistivity method	电阻率法
resistivity of rock	岩石电阻率
resistivity of soil	大地电阻率
resistivity profile	电阻率分布
resistivity profiling	电阻率剖面法
resistivity profiling method	电阻率剖面法

resistivity prospecting 电法勘探
resistivity reflection coefficient 电阻率反射系数
resistivity sensor 电阻率传感器
resistivity sounding 电阻率测深
resistivity sounding configuration 电阻率测深布极形式
resistivity survey 电阻测量；电阻勘探法
resistivity test point 电阻率测点
resistivity tomography 电阻率层析成像
resistivity tool 电阻率测井仪
resistor caliper 电阻式井径仪
resistor network 电阻网络
resoiling 还土；复田工程
resolidification 再凝固
resolidification temperature 再固化温度
resoluble 可溶解的；可分解的；可解决的
resolution 分辨能力；清晰度；分解；分开；决心；决议；解答；解决；消解；归结
resolution acuity 视觉分辨敏锐度
resolution capability 分辨能力，分解能力，析象能力
resolution cell 分辨单元
resolution characteristic 分辨能力
resolution deposit 再溶沉积
resolution element 分辨单元
resolution error 分辨误差
resolution factor 分辨率
resolution in bearing 方向分辨率
resolution limit 可分辨极限
resolution matrix 分辨率矩阵
resolution of anomaly 异常的分辨
resolution of force 力的分解
resolution of ground 地面分解力
resolution of stress 应力分解
resolution of vector 矢量分析
resolution of velocity 速度分解
resolution pattern 分辨能力测视图
resolution power 分辨率；分辨力
resolution principle 归结原理；消解原理
resolution ratio 分辨率
resolution requirement 分辨率要求
resolution shear stress 分解切应力
resolution target 分辨目标
resolution width 分辨宽度
resolutive clause 解除条款
resolvability 可溶解性；可解析性；可解决性
resolvable 可分解的
resolve 溶解，解析；分解，分解；解答；决定，决议
resolving ability 鉴别能力；分辨力
resolving function 分辨函数
resolving limit 分辨极限
resolving power 分辨力
resolving power of image 影像分辨率
resolving time 分解时间
resonance 共振
resonance absorption 共鸣吸收；共振吸收
resonance amplifier 谐振放大器
resonance conveyor 共振输送机
resonance cross-section 共振截面
resonance curve 共振曲线
resonance damping 共振阻尼
resonance domain 共振域
resonance effect 共振现象
resonance energy 共振能
resonance energy level 共振能级
resonance fatigue characteristic 共振疲劳特性
resonance frequency 响应频谱
resonance integral 谐振积分
resonance ionization mass spectrum 共振电离质谱
resonance level 谐振级
resonance method 谐振法；共振法
resonance modulus 共振模量
resonance peak 共振峰；谐振峰
resonance phenomenon 共振现象
resonance potential 共振电位
resonance test 共振法试验
resonant 谐振的；共鸣的；共振的
resonant column method 共振柱法
resonant column test 共振柱试验
resonant column triaxial test apparatus 共振柱三轴仪
resonant frequency 共振频率
resonant interaction 共振干扰
resonant load 谐振负载
resonant method 共振法
resonant resistance 谐振电阻
resonant scattering 共振散射
resonator 谐振器；共振器
resorption 再吸收；再溶解
resource allocation 资源分配
resource assessment 资源评价
resource assessment methodology 资源评定方法
resource base 资源基础
resource conservation 保护资源
resource development 资源开发
resource exploration 资源调查
resource information 资源信息
resource inventory 资源清单
resource management 资源管理
resource manager 资源管理程序
resource material 资源物料；资源材料
resource participative 资源管理
resource protection 资源保护
resource recovery 资源回收
respective 各自的；各个的
respectively 各自地；分别地
responding well 反应井
response 响应；反应
response amplitude 响应幅度
response analysis 响应分析
response bandwidth 响应带宽
response character 响应特性
response characteristic 响应特性；动态特性
response coefficient 响应系数
response control 反应控制
response correlation matrix 反应相关矩阵
response curve 响应曲线
response excursion 响应偏移
response factor 响应因子
response function 反应函数
response history 反应时程
response history analysis 反应时程分析
response level 响应量级

response matrix	响应矩阵
response modification factor	反应修正因数
response period	反应期
response plan	应急计划；反应计划
response quantity	反应量
response ratio	反应比
response space	反应空间
response spectrum method	反应谱法
response surface	响应面
response surface method	修平法，校平法
response time	反应时间
response to irrigation	灌溉效应
responsibility	责任；义务；可靠性
responsibility range	职责范围
responsibility system	责任制
responsible accident	责任事故
responsive	响应性；响应度
responsiveness	响应性；响应度
rest	静止；休息；支柱；剩余部分
rest level	静水位
rest period	回复期；休眠期
rest position	静止位置
rest temperature	再启动温度
rest time	静止时间
rest water level	静止水位
restart	再起动；重新起动
resting frequency	载频
restitution coefficient	回弹系数；弹力恢复系数
restitution instrument	纠正仪
restoration material	修复材料
restoration measure	修复措施
restoration plan	美化计划
restoration process	修复过程
restoration programme	修复计划
restoration project	恢复项目
restoration scheme	美化计划
restoration strategy	修复策略
restoration technique	复原技术
restoration test	修复试验
restoring force	恢复力
restoring force character	恢复力特征
restoring force characteristics	恢复力特性
restoring moment	回复力矩
restoring spring	复原弹簧
restrain	约束
restrain condition	约束条件
restrain condition of pile top	桩顶约束条件
restrain end	约束端点
restrained beam	固端梁；约束梁
restrained contraction	约束收缩；受限收缩
restrained end	固定端
restrained pile	约束桩
restrained shrinkage	约束收缩
restrained support	约束支撑
restrained torsion	约束扭转
restrained torsional constant	约束扭转常数
restraining barrier	阻挡层
restraining bend	受阻弯曲
restraining condition	约束条件
restraining device	约束装置
restraining force	约束力
restraining liner	预充填筛管
restraining mass	阻挡物
restraining moment	约束力矩
restraining order	禁止令
restraining pressure	约束压力
restraint	抑制；约束；减速
restraint capability	拴固能力
restraint coefficient	约束系数
restraint column	传力柱
restraint deformation	约束变形
restraint stress	约束应力
restricted basin	封隔盆地
restricted condition of use	使用的限制条件
restricted flow	限制性流动
restricted flow range	节流范围
restricted hours	受限制时间；管制时间
restricted instant tender	局限性现场投标
restricted method	限制法
restricted movement	限制运动
restricted output	限制产量
restricted platform	局限台地
restricted resources	限采资源
restricted restricting bushing	节流衬套
restricted restricting device	节流装置
restricted restriction	限制
restricted road	限制车辆使用的道路；限制使用的道路
restricted sea basin	局限海盆
restricted slot	受堵塞割缝
restricted subtidal zone	局限潮下带
restricted tendering	局限性投标
restricted zone	限制区
restriction	限制；限定；约束
restriction arc	限制弧
restriction device	节流装置
restriction of export	出口限制
restriction of import	进口限制
restriction of production	生产限制
restrictive	限制性的
restrictive approach	限制性途径
restrictive clause	限制条款
restrictive element	制约因素
restrictive motion	限制运动；非自由运动
restrictor valve	限流阀
restriping overburden	预先剥离；表土剥离
resulatant angle	合力角
resulatant curve	合量曲线；合力曲线
resulatant velocity	合成速度
result function	目标函数；结果函数
result integration	结果综合；结果集成
result of chemical test	化学试验结果
result of hydrogeological investigation	水文地质勘查成果
resultant	合成的；组合；合力
resultant accuracy	总精度
resultant amplitude	合成振幅
resultant bending moment	合弯矩
resultant compressive stress	合成压应力
resultant couple	合成力偶
resultant curve	综合曲线；总曲线
resultant displacement	合位移；合成位移

resultant error 总误差
resultant expansion stress 综合膨胀应力
resultant exponential curve 生成物指数曲线
resultant field 总场
resultant force 合力
resultant gradient 合成坡度
resultant ground displacement 合成地面位移
resultant intensity 总强度；合力强度
resultant load 总载荷
resultant moment 合力矩
resultant motion 合成运动
resultant of forces 力的合成；合力
resultant of shear stress 剪应力的合
resultant pitch 总节距
resultant plot 成果图像
resultant pressure 最终压力；总压力
resultant seismic coefficient 合成地震系数
resultant strain 总应变
resultant stress 合应力
resultant vector 合矢量
resultant velocity 合速度
resultant wall angle 边帮终止角
resulting display 结果显示
resulting fluid 井口流体
resulting force 合力
resulting force curve 合力曲线
resulting image 成果图像
resulting impedance 总阻抗
resulting intensity 总强度
resulting load 合成负荷
resulting pack 最终充填层
resulting pressure 最终压力
resulting side force 侧向合力
resulting solution 合成溶液
resulting stress 合成应力
resulting velocity 合速度
resumption 再取回；恢复；再继续；摘要
resumption notice 收地通告；收楼通告
resumption of land 收回土地
resumption of leased land 收回已批租土地
resumption payment 收回土地付款
resumption plan 收地图则
resumption procedure 收地程序
resurgence 再现；再生；复活
resurgent tectonics 复活构造
resurgent water 再生水
resurrect 复活；剥露；再现
resurrected fault scarp 剥露断层崖
resurrected karst 复活喀斯特
resurrected mountain 剥蚀山
resurvey 重测
resurvey of existing railway 线路复测
resuscitation 苏醒；复活；回生
resuspended sediment 悬浮泥沙
reswell 再次膨胀
resynthesis 再合成
retain 保留；保持；卡住；制动
retain a pillar 保留煤柱；保留矿柱
retained percentage 保留百分率；筛余百分率
retained permeability 保留的渗透率

retained pillar 保留煤柱
retained profit 待分配利润；留成利润
retained silt 截留泥分
retained strength 残留强度
retained water 存留水
retainer 夹具；夹圈，护圈；止动器，定位器
retainer ring 挡圈
retainer valve seat 护圈阀座；抵圈阀座
retaining balance 挡护板
retaining barrier 挡土护栏
retaining basin 集水盆地
retaining block 支块
retaining concrete plug 锚固洞
retaining dam 拦水坝；挡水坝
retaining device 固定装置；锁定装置
retaining plate 定位板；固定板；制动板；阻滞板
retaining ring 扣套；扣环；定位环
retaining spring 扣紧弹簧；止动弹簧
retaining structure 挡土结构
retaining structure of embankment 路堤支挡结构物
retaining structure of subgrade 路基支挡结构
retaining structure system 支挡结构体系
retaining structure with anchor 锚定式挡土结构
retaining valve 保持阀
retaining wall 挡墙；挡土墙；挡土支撑
retaining wall failure 挡土墙崩塌
retaining wall of embankment 路堤挡土墙
retaining wall of shoulder 路肩挡土墙
retaining wall of subgrade 路基挡土墙
retaining wall with anchors 锚杆挡墙支护
retaining wall with stepped back 踏步式挡土墙
retaining wall with surcharge 上有土坡的挡土墙
retaining walls shear failure 挡土墙剪切破坏
retaining works 蓄水工程；挡土工程
retainment and protection 支护
retapping 复打
retard 减速；延迟；阻滞
retard meter 制动力测量仪
retardance coefficient 滞流系数；阻滞系数
retardation 减速；制动；阻滞；延迟
retardation coefficient 扼流系数
retardation device 减速装置
retardation factor 延迟因子
retardation function 延迟函数
retardation limit 减速度限制
retardation path 减速行程
retardation rate 减速速率
retardation speed 减速速度
retardation test 减速试验
retardation time 滞后时间
retarded caving 延迟坍塌
retarded cement 缓凝水泥
retarded control 推迟控制
retarded creep 延迟蠕变
retarded elasticity 弹性后效
retarded subsidence 减速沉降
retarded time 延迟时间
retarded velocity 减速度
retarding basin 滞洪区；滞洪水库
retarding effect 滞洪效果；减速效应

retarding field	减速场；迟滞场
retarding force	减速力；制动力
retarding period	减速期
retarding water reducer	缓凝型减水剂
retempered concrete	二次拌和混凝土；重拌混凝土
retender	再次投标；重新投标
retention ability	持水力
retention activity	保留活性
retention capacity of soil	土壤保水能力
retention constant	保留常数
retention criterion	截留颗粒准则
retention curve	持水曲线
retention dam	挡水坝
retention device	阻挡装置
retention effect	滞流效应；滞洪效应
retention effect of lake	湖泊滞洪作用
retention efficiency	去除效率
retention elevation	壅水高程
retention factor	保留系数
retention force	滞留力
retention index	保留指数
retention level	保持水位
retention of water	水分保持
retention parameter	保留参数
retention period	保留期
retention rate	滞留率
retention reservoir	洪水调节池
retention temperature	保留温度
retention time	保留时间
retention time of recharge water	补给水停滞时间
retention volume	保留容量
retentive	有保持力的；保持湿度的
retentive soil	持水土壤
retentiveness	持水性
reticular density	面网密度
reticular drainage pattern	网状水系
reticulate drainage	网状水系
reticulate pattern	网纹
reticulated structure	网状结构
reticulated texture	网状结构
reticulated vein	网状脉
reticulation system	收集系统
retractable cantilever	伸缩式悬臂梁
retractable core barrel	活动式岩心管
retractable core bit	可缩取出的取心钻头
retractable core bit drilling	不提钻换钻头钻探
retractable plug sampler	活塞式取土器
retractable rock bit	可提换式钻头
retractable stress	收缩应力
retraction device	提引装置；牵引装置；缩进装置
retraction stress	收缩应力
retreat room mining system	后退房式开采系统
retreating cut	后退式回采
retreating face	后退式工作面
retreating glacier	后退冰川
retreating intermittent longwall	间隔留柱后退式长壁开采法
retreating longwall	后退式长壁开采法
retreating mining	后退式开采
retreating pillar	后退式回采中的煤柱；后退式回采中的矿柱
retreating shrinkage	后退式留矿开采法
retreating work	后退式开采
retrievable	可检索的
retrieval	恢复；检索
retroactive effect	反作用
retroarc basin	反拱盆地；弧后盆地
retroarc foreland basin	弧后前陆盆地
retroarc setting	弧后背景
retrocorrelation	逆相关
retrofit	加固；改建；修复
retrograde	后退的；退化的；逆行的
retrograde change	退化变化
retrograde condensation	逆凝析现象
retrograde diagenesis	退化成岩作用
retrograde metamorphism	退化变质作用
retrograde rotation	逆向旋转
retrograde vernier	逆游标；反读游标
retrograde wave	退行波；后退波；逆行波，反向波
retrograding sequence	岸退层序；退积层序
retrogressive	后退的；倒退的；退化的
retrogressive accumulation	向源堆积；溯源堆积
retrogressive erosion	向源侵蚀；溯源侵蚀
retrogressive failure	牵引式破坏
retrogressive flow slide	逆行流滑动
retrogressive landslide	牵引式滑坡
retrogressive metamorphism	退化变质作用；退变质作用
retrogressive slide	牵引滑坡；牵引式滑坡
retrogressive sliding	牵引式滑坡
retrogressive slope	反坡；逆向演进坡
retrogressive succession	逆行演替
retrogressive type	牵引型
retrogressive wave	退行波
retrospective retrieval	回溯检索
return air	循环空气；回气；回风
return air course	回风巷道
return air duct	回风管道
return air duct system	回风管道系统
return air inlet	回风进口；回气进口
return airflow	回风
return airway	工作面回风巷；回风顺槽；上顺槽
return bend	回转弯头
return bend header	回弯头箱
return compartment	回风隔间
return current	回风；回流
return device	回转装置；回返装置
return gear	回行装置；回程装置
return heading	回风巷道
return line	回流管；回浆管
return on investment	投资回报；投资收益
return passage	回路
return period	重现期
return period of earthquake	地震重复周期
return pressure	回流压力
return pulley	导轮；滑轮
return quantity	回风量
return rate	偿还率
return rock bolt	回头张拉锚杆
return rock reinforcement	回头加固
return seepage	回归渗流

return shaft	出风竖井
return shock	反冲
return slant	回风斜井
return sludge flow	轴承
return spring	回动弹簧
return strand	回空段；回行段
return stroke	回程；后退冲程
return tonnage	返砂吨量
return water temperature	回水温度
return wave	回波；反射波
return-air raise	回风天井
returned fold	倒转褶皱
returning echo	地震回波
returning flow weather	回流天气
returning fluid	返出液
returning well	恢复井
returns on investment	投资利润
reuse method of mining	先采脉壁的采矿法
reuse stoping	选采矿体；选别回采
reuse stress	合应力
reusing method	削壁充填采矿法
REV	表征单元体积；代表性单元体积
revaluation	重新估价；再估价
revenue of mine	矿山销售收入
reversal	反向；逆转；改变符号；极性变换
reversal condition	可逆条件
reversal development	反演；反转现象
reversal dip	反向倾斜
reversal event	逆转事件
reversal faulting	逆断层活动
reversal field	倒转场
reversal load	变向载荷；反转载荷
reversal magnetization	反磁化
reversal of curvature	反曲率
reversal of dip	倾角反向
reversal of geomagnetic field	地磁场倒转
reversal of load	交变荷载
reversal of poles	极性反转；极性变换
reversal of river	河流倒淌
reversal of stress	应力交变
reversal points method	逆转点法
reversal polarity event	地磁极性倒转事件
reversal positive	反转正片
reversal process	反转成像法
reversal stress	应力逆转
reversal temperature	反转温度
reversal test	倒转检查
reversal ventilation	反向通风
reverse bearing	反象限角；反方向角
reverse bend	反向弯矩
reverse bend test	反向弯矩试验
reverse bending test	反复弯曲试验；抗弯疲劳试验
reverse bias	反偏压
reverse branch	回转波
reverse circulate	反循环泥浆
reverse circulation drilling	反循环钻进
reverse circulation percussion rig	反循环冲击钻机
reverse circulation pump	反循环泵
reverse circulation rotary drill	反循环回转钻
reverse control	反向勘探；反向控制
reverse control hodograph	相遇时距曲线
reverse curvature	内曲
reverse dip slip faulting stress regime	逆走滑型应力机制
reverse direction	相反方向
reverse direction flow	反向流程
reverse displacement	反向驱替
reverse drag	反拖曳
reverse fault	逆断层
reverse faulting stress regime	挤压型
reverse flow	逆向流动
reverse flow process	逆流法
reverse flow valve	回流阀
reverse gear	回动装置
reverse graded bedding	反序粒层理；反粒级层理
reverse grading	逆递变；反向递变；反向粒序
reverse layout	逆向布置
reverse leading	反向移动
reverse leveling	逆高程测量
reverse listric fault	铲状逆断层
reverse locking	反位锁闭
reverse loop	折返环线
reverse migration	反向偏移
reverse motion	回动
reverse movement	逆转运动；反向运动
reverse osmosis	反渗透
reverse photogrammetry	逆反摄影测量
reverse polarography	反向极谱法
reverse process	可逆过程
reverse pumping	逆向泵送
reverse rotary	逆转；反转；反循环回转
reverse rotary boring	反向旋转钻井
reverse rotary drill rig	反循环回转钻机
reverse running	倒转；逆行
reverse sequence	逆层序
reverse shaft	反循环排泥管
reverse slide block	反向滑块
reverse slip fault	逆断层
reverse slope	脊坡
reverse symmetry	反对称
reverse thrust	逆冲断层；反推力
reverse torsion test	反复扭转试验
reverse trace	反向道
reverse travel path	反向传播路径
reverse turn	反转
reverse wave	逆行波；反向波
reverse weathering	反风化作用
reverse work	反向工作法
reversed	倒转的；反向的；逆的
reversed anticline	倒转背斜
reversed arrangement	反向排列
reversed bending	反向弯曲
reversed consequent stream	逆顺向河
reversed cross-bed	逆流交错层
reversed current	逆流
reversed curve	反向曲线
reversed dip	逆倾斜
reversed erosion fault scarp	逆向侵蚀断崖
reversed fault	逆断层；逆掩断层
reversed field	倒转场；反向场；反转场
reversed flow	回流；逆流

reversed fold	逆褶皱
reversed fold-fault	逆褶断层
reversed image	倒像
reversed load	反向荷载；反复荷载
reversed migration	逆偏移剖面
reversed order	逆序；倒转层序
reversed pendulum	倒垂线
reversed phase	倒相
reversed phase partition	反相分配
reversed polarity	反向极性
reversed profile	反向剖面；相遇时距曲线
reversed projection	逆投影
reversed reaction	逆反应
reversed relief	逆地形
reversed river	逆流河
reversed run	返测
reversed stratigraphic sequence	倒转层序
reversed stream	逆流河
reversed stress	反向应力
reversed time	反向时间
reversed zoning	逆向分带；反向环带
reverse-rotary method	反循环法
reverse-separation fault	逆裂断层
reverse-slip fault	逆滑断层
reverse-time extrapolation	逆时间外推
reverse-time migration	逆时偏移
reverse-time recursive filter	逆时递归滤波器
reverse-wetting mud	反转润湿泥浆
reverse-wound geophone	反绕地震检波器
reversibility	可逆性
reversible	可逆的；倒转的；反向的
reversible change	可逆变化
reversible chemical reaction	可逆化学反应
reversible expansion	可逆膨胀
reversible hammer mill	可逆转锤式破碎机
reversible hydraulic machine	可逆式水力机械
reversible impact type hammer crusher	可逆反击式锤碎机
reversible level	回转式水准仪
reversible reaction	可逆反应
reversible transformation	可逆相变；可逆变换
reversible ventilation	可逆式通风
reversing current	逆流
reversing device	反风设备
reversing dune	逆行沙丘；反向沙丘
reversing dune cross bedding	反沙丘交错层理
reversing facility	反风装置
reversing lever	回动杆
reversing of airflow	反风
reversing operation	反循环操作
reversing position	反循环位置
reversing pressure	反循环压力
reversing screen plate	回动螺旋
reversing sequence	反循环程序
reversing shear box test	往复直剪试验
reversing thermometer	翻转温度计；深海温度计
reversing type	可逆式
reversing valve	反向阀
reversion	颠倒；转换；倒转；回行；复原，恢复；退回
revetment dike	护堤
revetment of dike	堤岸护坡
revetment wall	护墙；护岸墙
revibration	重复振动
revised design	已修改的设计
revised drawing	修正图
revised earthquake intensity scale	修正地震烈度表
revised edition	修订版
revised general estimate	修正总概算
revised programmer	修正的计划
revised rent	经调整的租金
revision	修订；修改；修正
revision of topographic map	地形图更新
revision program	修正程序
revision survey	修正勘测
revision test	重复试验
revival	恢复；复活
revive	复兴；复活；还原
revived	复活的；再生的
revived fault	复活断层
revived fault scarp	复活断层崖
revived folding	重复褶皱作用
revived river	复活河
revived stream	复活河；回春河
revived structure	复活构造
revivification	复活；再生；恢复
revivification of catalyst	催化剂复活
reviving landslide	复合型滑坡
revocation of an approved plan	核准图撤销
revolution counter	往复滑动
revolution indicator	转速指示器；转速计
revolution of the Earth	地球公转
revolution speed	转速
revolution surface	回转面
revolution switch	旋转开关
revolutionary	旋转的
revolutionary shell	旋转壳
revolving bed plate	回转底盘；转盘
revolving distributor	旋转分配器
revolving drum screen	转筒筛；滚筒筛；回转筛
revolving frame	回转架
revolving fund	备用金
revolving member	回转构件
revolving mining	旋转式推进
revolving screen	旋转筛
revolving shaft	回转轴
revolving shovel	回转挖掘机
revolving stone screen	筛石转筒筛
revolving switch	旋转开关
revolving valve	回转阀
Reynolds analogy	雷诺数相似
Reynolds criterion	雷诺准则
Reynolds critical velocity	雷诺临界流速
Reynolds equation	雷诺方程
Reynolds lubrication equation	雷诺润滑方程式
Reynolds model law	雷诺模型律
Reynolds number	雷诺数
Reynolds number in fluid flow	渗流雷诺数
Reynolds resistance formula	雷诺阻力公式
Reynolds similarity law	雷诺相似定律
Reynolds stress	雷诺应力

Reynolds transport theory	雷诺输运定理
rezoning	重新分区
RFPA	岩石破裂过程分析；射频功率放大器
RFPA software	岩石破裂过程分析软件
RGG	俄罗斯地质与地球物理学报
rhegma	破裂；裂损
rheid	流变体；固流体；软流体
rheid fold	流变褶
rheid folding	固流褶皱作用
rheidity	流变度；软流度
rheologic model	流变模型
rheologic theory	流变理论
rheological	流变性
rheological analysis	流变分析
rheological behaviour	流变性；流变行为
rheological behaviour of rock	岩石流变性
rheological body	流变体
rheological characteristic	流变特征
rheological diagram	流变图
rheological equilibrium	流变平衡
rheological intrusion	流变性侵入体
rheological layer	流变层
rheological model	流变模型
rheological model of consolidation	固结的流变模型
rheological parameter	流变参数
rheological property	流变性质
rheological stratification	软流分层
rheological strength of rock	岩石流变强度
rheological test curve	流变试验曲线
rheological theory	流变理论
rheological voluminosity	流变容积度
rheologist	流变学家
rheologram	流变断面图
rheology	流变学；液流学
rheology and friction of fault zone	断层带流变与摩擦
rheology design	流变设计
rheology impact	流变性影响
rheomorphic	流变的；深流的；塑流的
rheomorphic breccia	流变角砾岩
rheomorphic effect	深流作用
rheomorphic fold	流变褶皱；固流褶皱
rheomorphic intrusion	流变侵入
rheopexy	触变性；震凝性；抗流变性
rheostatic brake	电阻制动
rheostatic control	变阻调压
rheostatic loss	变阻器损失；电阻损失
rheostatic starter	起动用变阻器
rheostriction	捏缩效应；夹紧效应
rheotaxis	液相处延性；趋液性；向流性
rheotome	周期断流器
rheotron	原子感应加速器
rheotrope	电流转换开关
rheotropic brittleness	低温脆性；流动脆性
rhodesite	纤硅碱钙石；罗德斯石；罗针沸石
rhodium	铑；自然铑；铑钌矿；立方铑钼矿
rhodolite	铁镁铝榴石；红榴石
rhomb	菱形；斜方形
rhombic	菱形的；斜方形的；正交的
rhombic dodecahedron	菱形十二面体
rhombic pyramid	斜方锥
rhombic system	斜方晶系；正交晶系
rhombic-shaped grid	菱形网格
rhombochasm	菱形断陷；平行裂开谷
rhombohedron	菱面体
rhomboid ripple mark	菱形波痕
rhomb-porphyry	菱长斑岩；碱性正长斑岩
rhombus	菱形；斜形；斜方形
rhyolite	流纹岩
rhythm	韵律；周期性
rhythm of sedimentation	沉积周期性
rhythmic	韵律的
rhythmic banding	韵律性条带
rhythmic change	韵律性变化
rhythmic crystallization	韵律结晶作用
rhythmic graded bedding	韵律性递变层理
rhythmic layering	韵律层
rhythmic movement	节奏性移动
rhythmic sedimentation	韵律性沉积
rhythmic succession	韵律层序
rhythmic unit	韵律层；韵律单位
rhythmicity	韵律性
rhythmite	韵律层；带状纹泥层
rhythmless	无节奏的
rhythmo-stratigraphy	韵律地层学
rhythmo-structure	叠层结构
rib	肋条；棱；拱肋；拱棱；加强肋；弯梁
rib bolt	帮锚杆
rib dilatation	巷帮膨胀
rib drill	煤壁钻机
rib fall	片帮；巷道壁倒塌
rib fill	带状充填
rib member	肋杆
rib mesh	肋条钢丝网
rib of column	肋柱
rib pillar	长方形煤柱
rib pillar extraction	煤柱回收
rib roadway	边为矿柱的巷道
rib sample	间柱样品
rib side pack	巷道侧的石垛墙；巷道侧的充填带
rib spalling	片帮；巷帮剥落
rib stiffener	加劲肋
rib strip	肋条
riband	条纹；条带
riband stone	条纹砂岩
rib-and-pillar	房柱式采煤法
rib-and-pillar method	房柱式开采法
ribbed	起棱的；有肋条的；含骨煤的
ribbed coal seam	含夹矸煤层；夹页岩煤层
ribbed liner	带肋衬板；带肋衬里
ribbed lining	肋形衬里
ribbing	煤带；扩大平巷
ribbon banding	带状层理
ribbon brake	带式制动器
ribbon cementation	带状胶结；薄膜胶结
ribbon clay	带状黏土
ribbon conveyor	皮带运输机
ribbon course	带状瓦层
ribbon diagram	带状图解；连续剖面图
ribbon foundation	条形基础
ribbon injection	带状贯入

ribbon rock	带状岩；条纹岩
ribbon strip	肋梁
ribbon structure	条带状构造；带状构造
ribbon tape	卷尺；带尺
ribbony bed	带状层
rich clay	重黏土；致密黏土
rich concrete	富混凝土
rich lime	肥石灰
rich ore	富矿石
Richter magnitude	里氏震级
Richter magnitude scale	里氏震级表
Ricker wavelet	雷克子波
ricochet plate	反射板；反跳板
rid up	清理井下矸石
ride meter	检路仪；量震仪
ride off	脱落；岔开；岔到
rider	层间岩石，矿体中夹石；厚煤层上方的薄煤层；含铁脉石；手动筛；游码；跟车工
ridge	脊；海岭；分水岭；冰脊
ridge beam	栋梁；脊瓦
ridge crest	山脊
ridge crossing line	越岭线
ridge fault	脊状断层
ridge field	垄作区田
ridge line	山脊线；分水岭
ridge projection	集水圈檐；防水棚
ridge push force	洋脊推力
ridge subduction	洋脊俯冲
ridge system	山脊系统
ridge tillage	垄作
ridge top	山顶
ridge type spreading center	脊状扩张中心
ridge-crest	脊波峰
ridged fault	地垒断层
ridged profile	起伏断面
ridge-like fold	梳状褶皱；脊状褶皱
ridge-pole	栋梁；屋脊
ridge-push model	洋脊推动模型
ridge-ridge transform fault	脊-脊转换断层
ridge-top trench	脊顶沟；山顶槽
ridge-transform fault	海岭-海沟转换断层
ridge-trench complex	海岭-海沟组合
ridge-trench transform fault	脊-沟转换断层
ridge-type earthquake	洋脊型地震
riding index	平稳性指标
Riedel shear	里德尔剪切
riegel	谷中岩坝；冰谷岩坎
Riemann solver	黎曼解算子
Riemann zeta function	黎曼ζ函数
riffle bowl	分样器
riffle box	分样器
riffle sampler	格槽式缩样器
riffler	沉砂槽
riffling bench	分样台；缩样台
rift	断裂，断层线；长峡谷；河流浅石滩
rift and sag model	裂谷-沉陷模式
rift axis	裂谷轴
rift basin	裂谷盆地；断裂盆地
rift belt	裂谷带
rift bulge	裂谷隆起
rift cushion	裂谷垫层
rift direction	断裂方向
rift element	裂谷单元
rift fault	裂谷断层
rift fault pattern	裂谷断层模式
rift faulting	断陷作用
rift geosyncline	裂谷地槽
rift lake	裂陷湖
rift opening	断陷缝；裂谷开口
rift structure	断裂构造
rift system	断裂系；断陷系
rift trough	裂谷
rift valley	地堑；断裂谷；深海槽；槽形断层；堑沟；沟道，水渠
rift valley basin	裂谷盆地
rift valley lake	裂谷湖；地堑湖
rift zone	裂谷带；断裂带
rift-block valley	裂谷断块谷
rift-bounding fault	裂谷边界断层
rift-drift transition	裂谷-漂移过渡
rifted	断裂的
rifted arch basin	拱顶断陷盆地
rifted continental margin	裂谷型大陆边缘
rifting	断裂作用；裂谷作用；断裂系；裂谷系
riftogeneous structure	裂谷构造；裂谷形成构造
riftogenesis	裂谷运动
riftogenic fluid	裂谷型流体
rift-valley lake	裂谷湖
rig	钻机；装备
rig boom	钻车的托臂
rig down	拆卸钻机
rig for model test	模型试验台；模型试验装置
rig operating cost	钻机操作费用
rig out	拆除；拆卸
rig release	钻机拆卸
rig repair	钻机维修
rig replaceable stabilizer	井场可修复的稳定器
rig spotting	钻机定位
rig substructure	钻机底座
rig time	钻机运转时间
rig tool	井口工具
rig up	安装；装设
rigging datum line	装配基准线
rigging down	钻机拆卸
rigging equipment	装配设备
rigging handle	操纵手柄
rigging line	吊索
right bank	右岸
right bridge	正交桥
right elevation	右视图
right fault	右行断层；右旋断层
right hand turnout	右开道岔
right handed	右旋的；右行的；右向的；右面的
right hyperbola	正双曲线
right justify	向右对齐；右边对齐
right lane	右边道；右通路
right lateral motion	右旋运动
right lateral slip	右旋滑动
right lateral strike slip fault	右旋走向滑断层
right lateral transcurrent fault	右旋平移断层

right line	直线
right of eminent domain	土地征用权
right of exoneration	免责权；免除权
right of ingress	进入权
right of patent	专利权
right of possession	占有权
right of priority	优先权
right of succession	继承权
right plane triangle	正平面三角形
right section	正剖面
right shear fault	右旋剪切断层
right shift	向右移位
right slip	右滑；顺时针滑动
right to claim	索赔权
right to occupation	占用权；占用土地权
right to use	使用权
right view	右视图
right way up	正常层序
right-angle array	直角排列
right-angle bend	直角弯管
right-angle drive	直角传动
right-angle intersection	正交叉
right-angle projection	直角投影
right-angle tee	直角三通
right-angled stream	横河
right-dipping fault	右倾断层
right-hand crusher	右式破碎机
right-hand derivative	右导数
right-hand rule	右手定则
right-hand screw rule	右手螺旋定则
right-hand switch	右侧道岔
right-hand turnout	右侧道岔
right-handed	右向的；右旋的；右手的
right-handed coordinate system	右手坐标系
right-handed form	右形
right-handed orthogonal system	右手正交系
right-handed rotation	右旋
right-handed screen plate	右旋螺钉；右转螺旋
righting arm	稳性力臂；回复力臂
righting moment	扶正力矩
right-lateral	右行的；右旋的
right-lateral arc-arc transform fault	右旋弧-弧转换断层
right-lateral deformation	右旋形变
right-lateral displacement	右旋位移
right-lateral fault	右旋断层
right-lateral movement	右旋运动
right-lateral offset	右旋断错
right-lateral ridge-arc transform fault	右旋脊-弧转换断层
right-lateral shear	右旋剪切
right-lateral shear and tensional belt	右旋剪张带
right-lateral shear fault	右旋剪切断层
right-lateral transcurrent fault	右行横推断层
right-lateral wrench fault	右旋扭动断层
right-lined	直线的
right-normal slip	右旋正倾滑
right-normal-slip fault	右旋正滑断层
right-oblique	右倾斜的
right-of-way restoration	管带地貌复原
right-reverse-slip fault	右旋逆滑断层
rights of property	产权；财产权益
right-skewed distribution	右偏斜分布
rigid	坚硬的；刚性的
rigid adherence	牢固黏附
rigid anchor	刚性锚固
rigid arch	刚性拱
rigid bar	刚性顶梁
rigid bar extensometer	刚性杆应变计
rigid base	刚性地基
rigid base support	整体刚性底座支架
rigid basement	刚性基底
rigid block	刚性地块
rigid block analysis	刚性地块分析；刚性块体分析
rigid block failure	刚性地块破坏；刚性块体破坏
rigid body	刚体；刚性体
rigid body displacement	刚体位移
rigid body kinematics	刚体运动学
rigid body kinetics	刚体动力学
rigid body mode	刚体模式
rigid body motion	刚体运动
rigid body motion approach	刚体位移法
rigid body rotation	刚体旋转
rigid body spring method	刚体弹簧方法
rigid body translation	刚性体位错；刚体变换
rigid bond	刚性联结
rigid boundary	刚体边界
rigid boundary condition	刚体边界条件
rigid computer model	刚性计算机模型
rigid conduit	刚性管道
rigid connection	刚性连接
rigid construction	刚性结构
rigid core wall	刚性心墙
rigid coupling	刚性连接
rigid crust	刚性地壳
rigid culvert	刚性涵洞
rigid demand	刚性需求
rigid diaphragm	刚性连续墙
rigid double tube	双动双层岩心管
rigid dynamics	刚体动力学
rigid element	刚性元素；刚性单元
rigid equilibrium method for slope stability analysis	边坡稳定分析刚体平衡法
rigid extensible bar	可伸刚性梁
rigid fastening	刚性扣件
rigid fill	刚性垫层
rigid fixing	刚性固定；固接
rigid flat truck	平板车
rigid fluid	刚性流体
rigid foam	硬质泡沫
rigid foamed plastics	硬质泡沫塑料
rigid foams	硬质泡沫塑料
rigid footing foundation	刚性底脚地基；刚性台阶式基础
rigid foundation	刚性基础；刚性地基
rigid frame	刚性构架，刚架
rigid frame bridge	刚架桥；刚构桥
rigid frame structure	刚架结构
rigid frame system	刚性架系统
rigid frame work	刚构架
rigid guide	刚性罐道
rigid joint	刚性缝；刚性接缝

rigid load	刚性荷载	rigidity of a structure	结构刚度
rigid material	刚性材料	rigidity of section	截面刚度
rigid matrix	坚实骨架	rigidity of track panel	轨道框架刚度
rigid mixed	刚性混合式	rigidity ratio	刚度比
rigid model	刚性模型	rigidity spectrum	刚度谱
rigid motion	刚体运动	rigidize	硬化
rigid mounting	刚性安装	rigidly connected	刚性连接的
rigid offset constraint	刚性偏移约束	rigidly positioned wheelset	刚性定位轮对
rigid pavement	刚性路面；混凝土路面	rigidly supported system	刚性支承系统
rigid penstock	刚性压力水管	rigid-perfectly plastic material	理想刚塑性材料
rigid pier	刚性墩	rigid-plastic body	刚塑性体
rigid piled	刚性桩基式	rigid-plastic model	刚塑性模型
rigid pipe	刚性管	rigid-type deformer	钻孔壁面变形测定器
rigid plastic	硬性塑胶	rigorous	严格的；精确的
rigid plastic material	刚塑性材料	rigorous adjustment	严密平差
rigid plastic resistance	刚塑性阻力	rigorous analysis	精确分析
rigid plastic structure	刚塑性结构	rigorous equation	精确方程式
rigid plastic system	刚塑性系统	rigorous examination	精确检验
rigid plate	刚性板块	rig-site utilization	现场应用
rigid plate sealing	刚性板块封闭	rig-up	配置；装配
rigid plunger block	刚性俯冲地块	rig-up time	安装时间；钻机安装时间
rigid prop	刚性支柱	rill cut mining	上向倒V形梯段回采；伪倾斜倒台阶回采
rigid proppant	刚性支撑剂	rill cut-and-fill method	上向倒V形梯段充填采矿法
rigid protection measure	刚性保护措施	rill cutting	倾斜分层回采
rigid protection mesh	刚性防护网	rill erosion	细沟侵蚀；带状侵蚀；小沟冲蚀
rigid reinforcement	刚性钢筋	rill floor	倾斜底
rigid retaining wall	刚性挡土墙	rill stope	上向倒V形梯段留矿工作面；上向倒V形梯段回采工作面
rigid road base	刚性路基		
rigid rock	坚固岩石；硬岩，硬质岩	rill stope-and-pillar system	倒V形上向梯段留柱采矿法
rigid rotor	刚性转子；刚接式旋翼；整体转筒	rill washing	细沟冲刷
rigid seal foam system	刚性密封泡沫法	rim angle	边界角；接触角
rigid set	刚性支架	rim cement	环边胶结物
rigid shaft coupling	刚性联轴节	rim cementation	环边胶结
rigid sheath	硬被筒	rim deposit	边缘沉淀；边缘矿床
rigid side frame	刚性侧架	rim elevation	露天矿边缘标高
rigid slab	刚性板	rim hole	周边孔
rigid steel arches	刚性拱形支架的钢构件	rim rock	砂矿露边底岩
rigid stratum	刚性层	rim seal	边缘密封
rigid structural plane	硬性结构面	rim space	边缘空间
rigid structural precaution	刚性结构措施	rim stripping	边缘剥离
rigid structure	刚性结构	rim syncline	边缘向斜
rigid structure measure	刚性结构措施	rim value	边缘值
rigid support	刚性支架	rimmed texture	镶边结构
rigid supporting device	刚性支护装置	rimose	龟裂的；裂缝的
rigid surface	刚性路面；刚性面层	ring and radial road system	环形辐射式道路系统
rigid temperature control	严格的温度控制	ring balance	环形水准仪；环形水平尺
rigid test	严格试验；刚性实验	ring beam	圈梁；环梁
rigid valley and arch dam	刚性河谷与拱坝	ring course	拱圈层
rigid wall	刚性墙	ring crater	环形火口
rigid wheel base	刚性轴距；固定轴距	ring cut method	环形开挖法
rigid-elastic analysis scheme	刚弹性方案	ring dike	环状岩墙
rigid-extensible prop	刚性可伸缩支柱	ring drilling	岩石钻进；环向钻孔法
rigid-frame shotcrete support	钢架喷射混凝土支护	ring fissure	环状裂缝
rigidimeter	刚度计	ring foundation	环形基础
rigidity	刚性；刚度	ring fracture	环状破裂
rigidity agent	硬化剂	ring fracture zone	环状破裂带
rigidity coefficient	刚性系数	ring gradient	炮孔排面倾角
rigidity factor	刚度因子	ring hole	扇形孔
rigidity index	刚度指数	ring levee	环形堤
rigidity modulus	刚性模量	ring line	环形管线

ring of tunnel lining	隧道衬砌环
ring pivot	环形支枢
ring puckering motion	环形皱缩运动
ring reef	环礁
ring road	环路
ring sampler	环刀
ring scalar	环形脉冲计数器
ring seal	环形密封
ring segment	环形分布的锁紧块；环形片；环形段
ring shake	环裂
ring shaped structure	环状构造
ring shear	环剪
ring shear apparatus	环状剪切仪
ring shear test	环形剪切试验；环剪试验
ring shift	循环移位
ring size	环粒度
ring spanner	环形扳手
ring spring	环簧；环形弹簧
ring step bearing	环形止推轴承
ring stress	环形应力
ring structure	环状构造
ring support	环形支架；井圈支架
ring switch	环形开关
ring syncline	环状向斜；环状坳陷
ring test	井圈压力试验
ring texture	环状结构
ring water meter	圆盘水量计；盘式流量计
ring whirl	环形涡旋
ring winding	环形绕组
ring-bolt	锚环；环端螺栓
ring-cut blasting	环形掏槽爆破
Ringelmann number	林格尔曼数
ringing	振鸣；振铃；冲击激励产生的减幅振荡；瞬变；混响；振荡效应
ringing system	振鸣系统
ring-like ore body	环状矿体
ring-like structure	环状构造
ring-shaped	环状的
ring-shear test	环剪力试验
ring-small	小于环径的
ring-type form	环形模板
ringwoodite	尖晶橄榄石
rinneite	钾铁盐
rinse water	洗涤水；冲洗水
Rio Tinto Iron Ore	力拓铁矿公司
rip	裂缝
rip bit	可卸式钻头
rip blasting	落底爆破；卧底爆破
rip current	裂流；离岸流；岸边回流；急流
rip current channel	离岸流水道
rip head	离岸流头
rip rooter	犁土机
rip saw	纵切锯
rip tide	离岸流
riparian land	滨湖土地；河岸边土地
riparian right	河岸权，滨岸权
riparian state	沿岸国家
riparian water loss	河面蒸发损失
riparian work	治水工程
rip-detector	撕裂探测器
ripe	成熟；成熟的
ripe lubricant	钢丝绳润滑剂
ripe peat	成熟泥炭
ripe sludge	熟污泥
ripe snow	软雪
ripening	成熟度
ripening of snow	雪熟
ripper	平巷掘进机；松土机；挑顶工；粗齿锯
ripper bar	采掘头；撬棍
ripper breakage	用破土机松碎
ripper-bulldozer	松土推土机
ripper-scarified	松土翻土机
ripping	煤巷挑顶；煤巷卧底；松土；开裂
ripping dirt	挑顶、卧底和刷帮矸石；采掘清出的废石
ripping edge	挑顶边界；卧底边界
ripping face	挑顶卧底工作面
ripping face support	挑顶面支护
ripping guard	采掘保护罩
ripping lip	挑顶顶板边缘；掘凿边缘；采石平巷工作面
ripping machine	割裂机；松土机；巷道掘进机；松碎式采煤机
rippings	清出的矸石；清出的废石
ripple amplitude	波纹振幅
ripple bedding	波痕层面
ripple drift	波痕迁移；交错层理
ripple dynamics	纹波动态
ripple effect	爆破的重叠作用
ripple height	波高
ripple hierarchy	纹波层次
ripple index	波痕指数
ripple lamination	波状纹理；砂纹层理
ripple mark	波痕砂纹；波纹标志；斑点；波痕
ripple mark-like structure	波痕状构造
ripple quantity	脉动量；波纹量
ripple ratio	波纹系数
ripple symmetry index	波痕对称指数
ripple tide	涨潮
ripple train	波痕列
ripple voltage	波纹电压
rippled surface	焊波表面；波皱面
riprap breakwater	抛石防波堤
riprap protection	抛石护坡；乱石护坡
riprap protection of slope	抛石护坡
riprap revetment	抛石护岸
riprap stone	乱堆石；抛积石
riprap stone revetment	堆石护岸
riprap work	抛石工程
rise and decline of water level	水位升降
rise entry	上山；天井
rise face	向上工作面；仰斜工作面
rise heading	上山
rise heading drivage	上山掘进
rise of arch	拱矢
rise of ground water level	地下水位上升高度
rise of watertable	地下水位上升高度
rise pipe	竖管
rise pit	冒水坑；隆起穴
rise rate	陡度；增长速度
rise side	上帮
rise working	上向开采；逆倾斜掘进

riser pipe	吸水管	risk investment	风险投资
rise-span ratio	矢跨比；高跨比	risk label	危险标记；危险标志
rising	上升；向上掘进；上向凿井	risk level	风险水平
rising borehole	向上钻孔	risk management	风险管理
rising coast	上升岸	risk measure	风险量度；风险标准
rising current	上升流；上向流	risk minimization	风险极小化
rising edge	上升沿	risk mitigation measure	缓解危险的措施
rising head test	上升水头试验	risk model	风险模型
rising method	上向凿井法；上向掘进法	risk monitoring	风险监测
rising period	上升时期；增长时期	risk of accidents	意外风险
rising resistance support	增阻式支架	risk of landslip	滑坡风险；山泥倾泻风险
rising shaft	竖井	risk of repairable damage	可修复破坏的风险
rising spring	涌泉	risk of total condemnation	完全无法使用的风险
rising stage	上升水位；上涨水位	risk perception	风险感知
rising tendency	增长趋势	risk premium	风险酬金
rising tide	涨潮	risk prevention	风险预防
rising velocity	上升速度	risk probability analysis	风险概率分析
rising water	上升泉	risk profile	风险剖面
rising water stage	上升水位；不位上升期	risk prognosis	风险预测
rising wave	上升波	risk prophecy	风险估计
rising whirl	上升旋涡	risk region	危险区
rising-head permeability test	升水头渗透试验	risk sharing	风险分担
risk	风险；危险	risk structure	风险结构
risk acceptance criteria	风险接受准则	risk surveillance	风险监测
risk acceptance criterion	风险接受准则	risk theory	风险论
risk adjusted discount rate	风险调整折现率	risk transfer	风险转移
risk administration	风险管理	risk value	风险值
risk allocation	风险分担	risk warning	风险预警
risk allowance	风险准备金	rither	岩层内压性裂隙；小断层
risk analysis	风险分析	Ritz method	里茨解法
risk analysis model	风险分析模型	Ritz vector	里茨向量
risk apportionment	风险分摊	Ritz-Galerkin method	里茨-加辽金法
risk area	风险大的地区	river action plan	改善河道计划
risk assessment	风险评估；风险评价	river alluvium	河流冲积物
risk assessment council	危险评估委员会	river axis	河流轴线
risk assessment of failure	破坏风险评估	river bank	河岸
risk asset ratio	风险资产比率	river bar placer	阶地砂矿
risk avoiding	风险规避	river barrage	拦河建筑物
risk bearing	风险负担	river basin	流域；河流盆地
risk capital	风险资本	river basin area	流域面积
risk categorization system	风险分类系统	river basin balance	流域平衡
risk category	风险类别	river basin development	流域开发
risk control	风险控制	river basin model	流域模型
risk decision-making	风险决策	river basin plan	流域规划
risk delineation	风险界定	river basin planning	流域规划
risk determination	风险测定	river bed	河床
risk disposal	风险处理	river bed armoring	河道粗化；河床粗化
risk distribution	风险分布	river bed change	河床演变
risk early warning	风险预警	river bed deformation	河床变形
risk effect	风险效应	river bed deposit	河床沉积物
risk estimation	风险估计	river bed evolution	河床演变
risk evaluation model	风险评价模型	river bed experimental station	河床实验站
risk factor	风险因素	river bed level	河床高程
risk function	风险函数	river bed paving	河床铺砌
risk grade	风险等级	river bed pier	河墩；河底桥墩
risk guidelines	风险指引	river bed profile	河床纵断面
risk handling	风险处理	river bend	河曲；河弯
risk hedging	风险承担	river bottom	河底
risk index	风险指数	river branch	河汊；河流支流
risk index system	风险指标体系	river branching	河道分叉

river capture	河流截夺
river cascade project	河道梯级工程
river center	河心
river channel	河道
river channel aggradation	河道淤积；河床淤积
river channel cross-section	河槽断面
river channel degradation	河道冲刷
river channel erosion	河道侵蚀
river channel pattern	河床类型
river channel sediment	河槽沉积物
river channel straightening	河道裁直
river channelization	河道渠化
river chart	江河图
river classification	河流分类
river cliff	河流陡岸；淘蚀河岸
river closure	截流
river confluence	河流汇合点
river construction	治河工程
river control	河流整治
river course	河道
river course survey	河道调查
river craft	河船
river cross section	河流断面
river crossing	渡口；横渡河流
river crossing regulation	河渠交汇段整治
river cutoff	河道裁弯取直
river delta marsh	河口沼泽
river density	河流密度
river deposit	河流沉积
river deposition	河流堆积作用
river development	河流开发
river discharge	河流流量；河量
river dissolved load	河流溶解物质含量；河流溶解量；河流溶解搬运质
river diversion	河流改道，河流分道
river diversion arrangement	导流布置
river diversion run-out flow	导流
river diversion tunnel	导流隧洞
river drift	河流冲积物
river dynamics	河流动力学
river ending	河流消失
river engineering	河流工程学
river erosion	河流侵蚀
river estuary	河口湾
river etching	河流侵蚀
river facies	河流相
river fall	河流落差
river fan	河流冲积扇
river flat	河流冲积坪；冲积平原
river flow	河流径流；河道流量
river freeze-up	河道封冻
river geomorphology	河流地貌
river gorge	河谷
river gradient	河道坡降
river gravel	河砾
river head	河源；河流源头
river hydraulics	河道水力学
river hydrology	河流水文学
river ice	河冰
river ice-drifting	河流流冰；河流淌凌；河流浮冰
river improvement	河道改善
river improvement survey	河道整治测量
river intake	河上取水口
river inversion	河流倒流
river island	河间岛；河心滩
river junction	河流汇流点
river landscape district	江河风景区
river length	河流长度
river level	河流水位
river load	河流荷重；河流泥沙
river marsh soil	河滨沼泽土
river meandering	河道的蜿曲
river mechanics	河流动力学
river model	河床模型
river morphology	河流形态学
river mouth	河口
river mouth bar	河口沙坝；拦门沙
river network	河道网
river outwash	河流冲积
river parting	河道分汊
river pattern	河流类型
river persistence	河流持续性
river pier	河墩
river piracy	河流袭夺；河流截夺
river plain	河成平原；冲积平原
river planning	河流规划；河道规划；治河规划
river pollution	河道污染
river profile	河道纵断面；河流水面线
river profile of equilibrium	河流均衡剖面
river property	河流特性
river purification	河川净化；河流净化
river realignment	河道整治；河道改道
river rectification	河道改直
river regime	河流情况
river regulation	河道整治
river runoff	河流径流量
river sand	河沙
river sediment	河流沉积物；河流淤积；河流泥沙
river shoal	河滩
river side	河岸
river side face	临水面
river side slope	临水岸坡
river slope	河道坡降
river source	河源
river span	河跨
river stage	河水位
river stage fluctuation	河水涨落
river station	河流水文站
river straightening	河道裁弯取直
river stretch	河段
river surface temperature	河面水温
river survey	江河测量；河道测量
river swamp	河流沼泽；河成沼地
river system	河系；水系
river system pattern	河系型式
river system survey	河系调查
river temperature	河水温度
river terminal	内河转运油库
river terrace	河流阶地
river tortuosity	河流的弯曲率

river trade limit	内河航限
river trade terminal	内河码头
river traffic	河道通航
river training	改善河道；治理河道；疏浚河道
river training works	河流治理工程；河道疏浚工程
river transport	河流运输
river transportation	河流搬运作用
river valley	河谷
river wall	河道导流墙；河堤
river weir	拦河坝
river width	河宽
river winding	河道的曲折
river-bank erosion	河岸冲蚀
river-bank stability	河岸稳定
river-bar placer	河成沙坝砂矿；阶地砂矿
riverbed accumulation	河床淤塞
river-bed morphometry	河床形态测量
river-bed placer	河床砂矿
river-bed scour	河床冲刷
river-bed scouring	河床冲刷
river-bed variation	河床变动
river-bed working	河床砂矿开采
river-borne material	河流挟带物
river-channel reservoir	河道储集层
river-connected lake	连河湖
river-crossing facility	跨河设施
river-crossing levelling	跨河水准测量
river-deposition coast	河流沉积海岸
river-dominated delta	河控三角洲
river-fed	河流补给的
river-fracturing treatment	清水压裂
riverhead	河道源头；河源
riverhead inversion	河流倒流
riverine	河流的；河成的；河岸的
riverine lake	河源湖
riverlet	小河；溪流
riverside bluff	河边陡岸；河边悬崖
riverside embankment	河岸堤
riverside park	河滨公园
riverside slope	外坡
riverside soil	河滨土壤
riverside source field	傍河水源地
riversideite	单硅钙石
riverwall	河堤
riverwash	河床冲积物；沿河荒地；河川侵蚀地
riveted steel girder	铆接钢梁
riveted steel structure	铆接钢结构
riveted truss	铆接桁架
riveting hammer	铆锤
riveting intrusion	贯入体
riveting machine	铆钉机
rivet-lap joint	铆接
rivulet	小河；溪流
RMRE	岩石力学与岩石工程学报
road administration	路政管理
road alignment	道理线形
road appearance	路容
road approach	桥梁引路
road area per citizen	人均道路面积
road area ratio	道路面积率
road asphalt	道路沥青
road axis	道路轴线
road base	路床
road bed	路基
road bend	路弯
road bitumen	路用沥青
road bridge	道路桥；公路桥
road capacity	道路流通量；交通流通量
road closure	道路封闭
road condition	路况
road condition survey	路况调查
road construction machine	筑路机械，筑路机
road construction works	道路建造工程
road conveyor	巷道输送机
road crossing	道路交叉点
road crown	路拱
road crust	道路面层
road cum rail bridge	道路铁路两用大桥
road cut	路堑
road cutting slope	公路边坡
road deck	道路面层
road deck surface	道路面层
road decking panel	路面覆盖
road design	道路设计
road drag	压路机
road drainage	道路排水
road drainage system	道路排水系统
road dust	巷道尘末；道路尘末
road embankment	路堤
road embankment sliding	路堤滑移
road engineering	道路工程
road engineering geology	道路工程地质
road engineering survey	公路工程测量
road feasibility study	可行性研究
road filling facility	油罐车装油设备
road finisher	路面整修机
road foundation	道路基础
road freight rate	公路运输费率
road frost boiling	道路翻浆
road frost damage	道路冻害
road frost heaving	道路冻胀
road geometric factor	道路几何因素
road geometry	路形
road grade	道路坡度
road grader	平路机
road head	巷道端头；迎头
road heading machine	平巷掘进机
road heading machinery	掘进机械
road improvement	改善工程
road intersection	道路交叉口
road junction	平巷交叉口
road level	路面水平
road lining	巷道衬砌；巷道支架
road log	野外地质路线记录
road logger	道路记录器
road making	巷道掘进；道路修筑
road map	路线图
road mark	路标
road marking	道路标记
road marking paint	路标漆

road mat 路面面层
road material 铺路碎石
road metal 筑路石碴
road miller 碾路机
road mixer 筑路拌料机
road mixing method 路拌法
road model 线路模型
road net 道路网
road net planning 线网规划
road net total length 线网总长
road network 道路网
road network planning 道路网规划
road network scale 线网规模
road network structure 线网结构
road nonlinear coefficient 线路非直线系数
road opening works 道路开掘工程
road pack 平巷石垛墙
road pavement 道路路面
road pavement bed 路基
road performance 道路修筑
road project 方案图
road realignment 更改道路定线
road reconstruction works 道路重建工程
road rerouting 道路改线
road ripper 煤巷掘进机；割土机
road roller 碾路机
road scraper 平路机；刮土机
road sign 路标
road speed 行车速度；运行速度
road spreader 筑路撒料机
road stone 筑路石；铺路石
road surface 路面
road surface thickness 路面厚度
road surfacing 道路铺面
road survey 道路测量
road test 线路试验；线路实验
road traffic sign 道路交通标志
road transport 公路运输
road trough 路槽
road tunnel 行车隧道
road way 路幅
road way shaftbottom 直线型井底车场
road width 路面阔度
road work 筑路工程
roadbase 路基
roadbed section 路基断面
roadcut 路堑
roaded catchment 道路型集水区
roadhead 巷道工作面
roadheader 综掘机；巷道掘进机
roadheader machine 巷道掘进机
roadheader shield 巷道掘进盾构
roadside 路旁；巷道侧
roadside pack wall 巷道侧的废石垛墙；巷道侧的充填带
roadside slope 路旁斜坡
roadside support 巷旁支护
roadside survey 路边调查
roadstone 筑路石
roadway 巷道；路面；道路
roadway closure 巷道闭合

roadway construction 巷道掘进；平巷掘进
roadway crossing 巷道交叉点；平巷交叉口
roadway deformation 巷道变形
roadway deterioration model 道路损坏模式
roadway drainage 路面排水系统
roadway drivage 巷道掘进
roadway excavation 巷道掘进；巷道的空间体积；巷道掘进工作量
roadway frame 巷道支架
roadway group 巷道组
roadway head 巷道尽头；巷道掘进工作面
roadway in slicing seam 分层平巷
roadway intersections 巷道交岔点
roadway junction 巷道交叉点
roadway layout 巷道布置图
roadway lighting 巷道照明
roadway maintenance 路面养护
roadway pressure arch 巷道压力拱
roadway reduction ratio 巷道断面缩小率
roadway reinforcement 巷道加固
roadway ripping machine 平巷掘进机；巷道掘进机
roadway side 平巷侧壁
roadway support 巷道支架；平巷支架
roadway surrounding control 巷道围岩控制
roadway system 巷道系统；道路系统
roadway tunneling 巷道掘进
roadway within soft rock 软岩巷道
roasting 焙烧；焙烧选矿；焙砂
roasting bed 焙烧床层；焙烧料层
roasting furnace 焙烧炉
roasting kiln 焙烧窑
roasting sample 烘样
roasting test 焙烧试验
robbing 回采矿柱
robbing a mine 掠夺式开采；选择性开采
robbing line 矿柱回采线；煤柱回采线
robbing on the retreat 后退式回采；大后退式回采全部煤柱
robbing pillar 回采煤柱；回采矿柱
robbing work 回采工作
robot 自动装置；遥控装置；机器人；遥控的
robot control system 自动控制系统
robot device 自动装置
robot engineering 机器人工程学
robot geology 遥感地质学；自动化地质学
robust 坚固耐用的；结实的；稳健的
robust estimation 稳健估计；抗差估计；鲁棒估计
robust least squares method 抗差最小二乘法；鲁棒最小二乘法
robust regression 稳定回归；稳健回归；鲁棒回归
robust regression analysis 稳健回归分析；鲁棒回归分析
robust smooth 稳健平滑
robustness 稳定性；坚固性；鲁棒性
rock abrasion platform 浪蚀岩石阶地
rock abrasiveness 岩石研磨性
rock abrasivity 岩石研磨性
rock absorption 岩石吸水率
rock abutment 岩拱脚
rock activation 岩石强化
rock adit 岩石平硐

rock affected by working 掘进作业扰动的岩石
rock analysis 岩石分析
rock anchor 岩石锚杆
rock anchoring 岩石锚固
rock and coal pillar 岩石和煤柱
rock and gas outburst 岩石和瓦斯突出
rock and mineral analysis 岩矿分析
rock and mineral identification 岩矿鉴定
rock and mineral magnetism 岩石和矿物的磁性
rock and rock mass structure 岩石与岩体结构
rock and soil 岩土
rock and soil engineering 岩土工程学
rock and soil mechanical parameter 岩土力学参数
rock and soil mechanics 岩土力学
rock asphalt 石沥青
rock assemblage 岩石组合
rock association 岩石组合
rock auger 岩石麻花钻；凿岩螺旋钻
rock avalanche 岩崩；岩石崩滑
rock ballast 石碴；石道碴
rock base 岩石基底；基岩
rock base outcropping 基岩露头
rock basin 岩盆
rock beam 石梁
rock bed 基岩；底岩；岩层
rock bedding 岩石层理
rock behaviour 岩石特性
rock bench 岩棚；岩台；海蚀台地
rock bind 岩石胶结；岩石固结
rock bit 凿岩钎头；凿岩钻头
rock blasting 岩石爆破
rock block 块岩
rock block filler 岩块填料
rock block toppling 岩块式倾倒
rock blowing 岩炮；岩石突出
rock bolt 锚杆；岩石锚杆；岩栓
rock bolt drawing dynamometer 锚杆拉拔计
rock bolt dynamometer 锚杆测力计
rock bolt setter 锚杆安装机
rock bolt support 锚杆支护
rock bolter 锚杆冲击机
rock bolting 锚杆支护；岩体锚固；岩体锚固
rock bolting drill 岩石锚杆钻机
rock bolting rig 锚杆安装机械手
rock bolting theory 岩石锚杆理论
rock bonding 岩石固结
rock bore machine 凿岩机
rock boring device 凿岩装备
rock boring machine 凿岩机
rock breakage 岩石破碎
rock breakage by explosive 爆炸破岩
rock breakage with static expansive agent 静态膨胀剂破碎岩石
rock breakdown 岩石破裂
rock breaker 岩石破碎机
rock breaking character 岩石破碎特性
rock breaking tool 碎石工具
rock breaking vessel 碎石船
rock brecciation 岩石角砾化作用
rock bridge 岩桥

rock bridge weakening 岩桥弱化
rock brushing 刷岩；挑顶
rock bucket 铲岩斗
rock bulk 岩体；整体岩石
rock bulk compressibility 岩石体积压缩性
rock bulking factor 岩石体胀系数；岩石体积扩张系数；岩石湿胀率
rock bump 岩石突出；冲击矿压
rock burst 岩爆
rock burst critical depth 岩爆临界深度
rock burst endangered seam 具有突出危险的煤层
rock burst energy index 岩爆能量指数
rock burst potential 岩爆势；岩爆潜力
rock burst process 岩爆过程
rock burst region 岩爆区域
rock burst shock 岩石突出震动
rock burst tendency 岩爆倾向性
rock cap 覆盖岩层；盖层
rock category 围岩分类
rock cave 石窟；岩洞
rock cavern 石窟；岩洞
rock cavity 岩洞；岩石硐室
rock cavity stability 岩石空穴稳定性；岩洞稳定性
rock channel 基岩河道
rock character 岩性；岩石特性
rock characteristic 岩石特性
rock classification 岩石分类
rock clay 黏土岩
rock cleavage 岩石劈理
rock cliff 石崖；悬崖
rock cliff erosion 崖体风化
rock coast 基岩海岸；岩石海岸
rock cohesion 岩石黏聚力；岩石内聚力；岩石黏结力
rock competency rating 岩层稳定性等级的划分
rock complex 杂岩
rock composition 岩石成分
rock compression 岩石压缩
rock concrete interface 岩石混凝土接触界面
rock condition 岩石条件
rock constituent 岩石组分；岩石成分
rock contraction 岩石收缩
rock control 岩石控制
rock core 岩心
rock core bit 取芯钻头
rock core drilling 岩芯钻进
rock core recovery 岩芯回收率
rock core sample 岩芯样品
rock coring 岩芯钻进
rock cover 岩石覆盖层
rock creep 岩石蠕动；岩石蠕变
rock crusher 碎石机；岩石压碎机
rock cut 岩石掘进
rock cut slope 削石坡；岩削坡
rock cutting 岩石井凿
rock cutting blasting 路堑石方爆破
rock cutting by high pressure water jet 高压水射流破碎岩石
rock cutting theory 岩石切削理论
rock dam abutment 坝肩
rock damage 岩体损伤

rock damage and healing	岩石损伤及愈合
rock debris	岩堆；岩屑
rock debris dam	石碴坝
rock decay	岩石风化，岩石腐蚀
rock deformation	岩石变形
rock deformation and stability measurement	岩石变形和稳定性测量
rock deformation parameter	岩石变形参数
rock deformation pressure	变形压力
rock deformation process	岩石变形过程
rock degradation	岩石风化
rock density	岩石密度
rock dent	石窝
rock deposit	岩石沉积
rock depression	岩层下沉
rock description	岩石的描述
rock desert	石漠
rock dike	岩墙
rock dilatation	岩石膨胀
rock discontinuity	岩石不连续
rock discontinuity structural plane	岩体结构面
rock discontinuous cellular automaton	岩体连续-非连续细胞自动机
rock discrimination	岩石类型鉴别
rock disintegration mechanics	岩石破碎力学
rock disposal	废石处理
rock disturbance	岩石扰动
rock dolomite	白云岩
rock dowel	石钉；岩石销钉
rock drift	岩石平巷
rock drift machine	岩巷掘进机
rock drifting	岩漂流
rock drill	钻机；凿岩机
rock drill bit	凿岩钻头；岩芯钻头；牙轮钻头
rock drill design	凿岩机结构
rock drill machine	凿岩机；钻机
rock drill mounting	钻架
rock drillability	岩石可钻性
rock drilling	岩石钻探
rock drilling crawler	履带式凿岩钻车
rock drilling hardness classification	岩石可钻性硬度分类
rock drilling jumbo	凿岩台车；钻孔台车
rock drilling tool	凿岩工具
rock drivage	岩巷；岩石平巷；岩石平硐
rock dump	废石堆
rock dumping yard	排矸场
rock dust	岩粉
rock dust barrier	岩粉棚
rock dynamics	岩石动力学
rock elasticity	岩石弹性
rock element	岩石要素
rock embankment	堆石堤；堆石坝
rock embedding pile	嵌岩桩作业
rock engineering	岩石工程
rock engineering system	岩石工程系统
rock excavation	岩石开挖；开采岩石
rock excavator	岩石挖掘机；挖岩机
rock explosion	岩爆
rock explosive	岩石炸药
rock exposure	岩石露头
rock fabric	岩石组构
rock facies	岩相
rock factor	岩石的抗爆阻力系数
rock failure	岩石崩落；岩石破坏
rock failure process analysis	岩石破裂过程分析
rock fall	岩石崩落；落石；滚石
rock fall deposit	岩石崩落沉积；落石沉积
rock fall earthquake	岩崩地震
rock fall fence	防石栏
rock fall prevention work	岩崩防护工程
rock family	岩石族
rock fan	基岩扇；石质扇形地
rock fill	岩石填方
rock fill material	堆石料
rock fill raise	充填天井
rock filled gabion	石笼
rock filling	堆石
rock fissure	岩石裂隙
rock flour	石粉；岩粉
rock flow	岩流
rock flowage	岩流；石流
rock fold	岩层褶皱
rock foreshore	岩滩
rock formation	岩系；岩层；成岩作用
rock formation test	岩层试验；地层测试
rock forming	造岩；造岩的
rock forming mineral	造岩矿物
rock foundation	岩石地基；岩基
rock foundation grouting	岩基灌浆
rock foundation with low dip	缓倾角基岩
rock fracture	岩石破裂；岩石裂损；岩石裂隙
rock fracture mechanics	岩石断裂力学
rock fracture zone	心轴
rock fragment	岩屑；岩石碎块
rock fragmentation	岩石破碎
rock fragmentation zone	岩石破碎区
rock frame	岩层构造；岩石骨架
rock frame stress	颗粒应力；岩石骨架应力
rock frost resistance	岩石抗冻性
rock gangway	石巷；岩巷
rock gas	天然气；天然煤气
rock glacier	石冰川；石河；石流
rock gradient temperature	地温梯度；岩石梯度温差
rock grinding	岩石磨碎
rock grout	岩石灌浆
rock grouting	岩石灌浆
rock gypsum	石膏
rock hammer drill	凿岩机；风钻
rock hardness	岩石硬度
rock hardness ratio	岩石硬度比
rock hardness reducer	岩石硬度减低剂
rock head	岩石顶层；基岩；基底
rock heading	岩石平巷
rock high slope	岩质高边坡
rock hoisting	岩石提升
rock hole	岩石钻孔；岩石炮眼；石洞；放矿岩石溜道
rock hole system	岩层溜眼开采法
rock house	岩洞
rock hydraulics	岩石水力学
rock identification	岩石鉴定，岩石鉴别

rock in place 原地岩石；本地岩石
rock instability 岩石不稳定性
rock intercalation 岩石夹层
rock joint 岩石节理
rock joint shear strength test 岩石节理剪力试验
rock jointing 岩石颗粒的结合
rock laboratory 岩石实验室
rock landslide 岩滑
rock layer interface 岩层界面；岩层接触面
rock layers contact relation method 地层接触关系法
rock leather 石棉
rock ledge 岩架；岩礁；岩面突出部
rock lifting 开凿岩石
rock lithology 岩性
rock load 岩石荷载，岩石压力
rock loader 装岩机
rock loading conveyor 装石输送机
rock loading machine 装岩机
rock loccolith 岩盖
rock machine 装岩机；抓岩机
rock magnetic property 岩石磁学性能
rock magnetism 岩石磁学
rock mantle 岩石表层，岩屑层；岩石地幔
rock mass 岩体
rock mass avalanche 岩体崩落
rock mass basic quality 岩体基本质量
rock mass classification 岩体分类
rock mass classification criteria by acoustic test 声波测试岩体分类标准表
rock mass classification scheme 岩体分类法；岩体分类表
rock mass damage assessment 岩体损伤评价
rock mass deformation 岩体变形
rock mass dynamics 岩石动力学
rock mass factor 岩石系数
rock mass failure 岩体失稳
rock mass lining support 岩体衬砌支护
rock mass mechanics 岩体力学
rock mass of side slope 边坡岩体
rock mass performance 岩体性状
rock mass property 岩体特性
rock mass quality 岩体质量，岩体质量
rock mass quality designation 岩体质量指标
rock mass rating 岩体分级；岩体评级；岩体分类
rock mass response to stope 岩体对开采响应
rock mass sliding 岩体滑塌
rock mass stability 岩体稳定性
rock mass strength 岩体强度
rock mass stress 岩体应力
rock mass stress test 岩体应力测试
rock mass structure 岩体结构
rock mass structure mechanics 岩体结构力学
rock mass weakening 岩体弱化
rock massif 岩体地块；岩体断块；岩体山丘
rock material 岩石材料
rock material slope 岩石材料坡
rock matrix compressibility 岩体压缩性；岩石基质压缩系数
rock meal 岩粉
rock mechanical test 岩石力学试验

rock mechanics 岩石力学；岩体力学
Rock Mechanics and Rock Engineering 岩石力学与岩石工程学报
rock mechanics for mine 矿山岩石力学
rock mechanics in coal mining 煤矿岩石力学
rock mechanics instrumentation 岩石机械性质测试设备
rock mechanics laboratory 岩石力学实验室
rock mechanics parameter 岩石力学参数
rock melting drill 热熔融钻
rock melting point 岩石熔点
rock member 岩段
rock miner 岩巷掘进工
rock mineral 造岩矿物
rock mineralogy 岩石的矿物成分
rock motion 基岩运动
rock mound 岩石墩；堆石
rock movement 岩层移动；岩石移动
rock of low modulus ratio 低模量比岩石
rock outburst 岩石突出；岩爆
rock outcrop 外露岩石；岩石露头
rock overlying 覆盖岩石；顶板岩石
rock pack 岩石充填带
rock parting 夹矸层
rock peeling 岩土剥落灾害
rock permeability 岩石渗透性
rock phosphorite 磷灰岩
rock physics 岩石物理学
rock physics model 岩石物理模型
rock picking 拣矸；拣选废石
rock pillar 岩柱；柱状岩礁
rock plant 采石场；采石场设备
rock plasticity 岩石塑性
rock plug 岩塞
rock plug blasting 岩塞爆破
rock pore 岩石孔隙
rock pore pressure 岩石的孔隙压力
rock porosity 岩石的孔隙度
rock powder 岩粉；石粉
rock pressure 岩石压力；山岩压力
rock pressure in hill 山体压力
rock pressure in horizontal roadway 平巷地压
rock pressure in inclined shaft 斜井地压
rock pressure in mine 在矿山的岩石压力
rock pressure in vertical shaft 立井地压
rock pressure indication 岩压迹象
rock pressure measurement technique 岩石压力测试技术
rock pressure study 岩石压力研究
rock pressure theory 岩石压力理论
rock prestressing 岩石预应力
rock prestressing theory 岩石预应力假说
rock prism 岩石棱柱体
rock product 岩石制品
rock product industry 岩石产品工业；采石工业
rock propelling 岩石抛出
rock properties test 岩石性能测试
rock property 岩石性质
rock pulverizer 岩石粉磨机
Rock Quality Designation (RQD) 岩石质量指标
rock quality index 岩石质量指数
rock quality rating 岩石质量评价

rock quarry	采石场
rock quarrying	采石场
rock rating	岩石分级
rock reduction	岩石崩解
rock refrigeration	岩石冻结
rock refuse	废石；矸石
rock reinforcement	岩石锚固
rock relaxation	岩体松弛
rock resistance	岩石抗力
rock rheology	岩石流变学
rock ripper	岩石掘进机；凿岩机；挑顶机
rock roof bolting	岩顶锚杆支撑
rock rot	岩石风化
rock rubble	断层角砾岩
rock salt	石盐；岩盐
rock salt mining	岩盐开采
rock sample	岩样
rock sample analysis	岩石试样分析
rock saw	岩石锯
rock schistosity	岩石片理
rock section	岩层截面；岩石薄片
rock series	岩系
rock shaft	充填料井；下放矸石井筒
rock shear strength	抗剪断强度
rock shear test	岩石剪力试验
rock shelter	崖洞；石洞
rock shield	防岩石损伤的护板
rock shovel	铲斗式清岩机
rock signature	岩性特征
rock site	岩石场地
rock slaking test	岩石崩解性试验；岩石湿化实验
rock slice	岩石薄片
rock slide	塌方；岩坍
rock sliding	岩体滑动
rock slip	岩体滑移；岩石塌方
rock slope	岩质边坡
rock slope engineering	岩质边坡工程
rock slope in hydropower engineering region	水电岩质边坡
rock slope stability	岩质边坡稳定性
rock socket	嵌岩孔
rock soil	岩土类土
rock solid compressibility	岩石固体压缩率
rock solidarity coefficient	岩石坚固性系数
rock solution	岩溶
rock soundness	岩石坚固性
rock specific heat	岩石比热
rock specimen	岩石试样
rock stability	岩石稳定性
rock stabilization	岩体加固
rock strata	岩层
rock strata control	岩层控制
rock strata mechanics	岩层力学
rock stratification	岩层层理
rock stratigraphy	岩石地层学
rock stratum	岩层
rock stratum movement	岩层移动
rock stream	石流
rock strength	岩石强度
rock strength character	岩石强度特性
rock strength test	岩石强度测试
rock stress	岩石应力
rock stress behavior around mine opening	巷道围岩应力显现
rock stress measurement	岩石应力测量
rock stress state	岩石应力状态
rock structure	岩石结构
rock structure analysis	岩石构造分析
rock study	岩石研究
rock subgrade	岩质地基
rock subsidence	岩石沉陷；岩层陷落；岩层移动
rock subsoil	岩质底土
rock substance	岩石物质
rock suite	岩套；岩群
rock support	岩石支护
rock support and reinforcement	岩石支护与加固
rock surface	岩面
rock system	岩石系统
rock temperature	岩石温度
rock temperature gradient	岩温率；岩石增温梯度
rock terrace	岩石阶地
rock test	岩石测试；岩石试验
rock texture	岩石组构
rock thrust	岩石压力；地压；矿山压力
rock toughness	岩石韧性；岩石韧度
rock tunnel	岩石隧道
rock tunneling	岩层平巷掘进
rock type	岩石类型
rock uplift	岩石隆起
rock wall	围岩；石墙；石垛带
rock waste	岩屑
rock weathering	岩石风化；岩石风化作用
rock weathering index	岩石风化系数；岩石风化指标
rockbolt reinforcement design	锚杆支护设计
rockbolt system	锚杆支护系统
rockbolt testing	锚杆试验
rockbolting	岩石锚固
rockbolting machine	锚杆机
rockbolt-rock mass interaction	锚杆与岩体相互作用
rock-boring machine	凿岩机；钻机
rock-bottom	岩底
rock-breaking efficiency	岩石破碎效率
rock-breaking mechanics	岩石破碎力学
rockburst	岩爆
rockburst control	岩爆控制
rockburst control measure	岩爆控制措施
rockburst damage	岩爆损坏
rockburst evolution process	岩爆演化过程
rockburst forecast	岩爆预测
rockburst hazard	岩爆灾害
rockburst monitor	岩石突出监测仪；岩爆监测仪
rockburst prediction	岩爆预报
rockburst prevention	岩爆预防
rockburst probability	岩爆概率
rockburst source	岩爆源
rock-burst valve	冲击地压阀
rock-cavity	岩洞
rock-crushing plant	碎石厂；碎石装置
rock-disposal dump	废石堆
rock-disposal site	废石场，尾矿场
rockery engineering	假山工程

rock-eval pyrolysis 岩石快速热解分析仪
rock-face 岩石立面
rock-faced dam 堆石护面坝
rock-facies analysis 岩相分析
rockfall avalanche 岩崩滑塌
rockfall deposit 岩崩沉积；岩崩堆积
rockfall hazard 坠石的危险；滚石灾害
rockfall prevention wall 岩崩防护墙
rockfill 堆石
rockfill breakwater 堆石防波堤
rockfill dam 堆石坝
rockfill diversion weir 堆石引水堰；堆石分流堰
rockfill drain 填石排水沟
rockfill riprap 堆石护坡
rockfill slope 填石坡；堆石坡
rock-flowage zone 岩石流动带；岩石蠕变带
rock-fluid system 岩石-流体系统
rock-forming component 造岩组分
rock-forming process 成岩作用；成岩过程
rock-fracture zone 岩石破裂带
rockhole 岩层钻孔；岩层联络小巷
rock-making mineral 造岩矿物
rock-mound breakwater 抛石防波堤
rock-self stability time 围岩自稳时间
rockshaft 送充填料的井筒
rockslide 岩体滑坡
rockslide avalanche 岩滑崩塌
rockslide dam 岩崩坝
rock-socketed 嵌岩
rock-soil dynamics 岩土动力学
rock-soil foundation 岩土地基
rock-soil mechanics 岩土力学
rock-stratigraphic unit 岩石地层单位
rock-stratum pollution 岩石地层污染
rock-stress measurement 岩石应力测量
rock-support interaction 岩石-支护相互作用
rock-support pressure 岩石-支护压力
rock-support pressure reaction line 岩石支护压力响应线
rock-tunnel 岩石隧洞
rock-tunneling machine 岩石隧道掘进机
Rockwell apparatus 洛氏硬度计
Rockwell figure 洛克威尔硬度值
Rockwell hardness 洛氏硬度
Rockwell hardness number 洛氏硬度指数
rocky desert 岩漠，石漠
rocky desertification 石漠化
rocky ground 岩石地面
rocky matrix 脉岩；脉石
rocky soil 石质土
rod boring 钻杆钻进
rod constant 尺常数；标尺零点差
rod correction 标尺改正
rod coupling 钻杆接头；钻杆接箍
rod extensometer 杆式伸长计；杆式应变计
rod graduation 标尺分划；标尺刻度
rod holder 钻杆夹持器
rod level 标尺水准器
rod puller 拔钻杆机
rod reading 标尺读数
rod sounding 钎探；杆探测
rod stabilizer 钻杆导向和稳定装置；抽油杆稳定器
rod thief 长杆取样器
rod vibration 钻杆振动
rod wrench 钻杆扳手；管式扳手
rodded end 钢筋加固端
rodding eye 通管孔；通渠孔；清理孔
rodding structure 杆状构造
roentgenology X射线学
Roga index 罗加黏力指数
roil 湍流河段
roll coal crusher 辊式碎煤机
roll crusher 辊式破碎机
roll damping 摇动阻尼；侧倾阻尼
roll deflection 轧辊挠度
roll displacement 摇动转角位移
roll force 轧制力
roll formed shape 冷弯型钢
roll forming 辊轧成形；成形轧制
roll forming mill 辊轧成形机
roll grinder 轧辊磨床
roll grinding 轧辊研磨
roll grinding machine 辊式碾碎机；辊子磨床
roll housing 轧机机架
roll ore body 卷式矿体；新月形矿体
roll product 辊碎机产品
roll profile 轧辊辊型
roll protection 辊筒防护
rolled 轧制的；滚动的
rolled fill dam 碾压土石坝
rolled hardcore 经碾压碎石底层
rolled iron 轧钢；轧制钢
rolled sheet iron 轧制铁皮；轧制钢板
rolled sheet metal 轧制薄板
rolled steel 型钢；轧制钢
rolled steel beam 型钢梁；轧制钢梁
rolled steel channel 轧制钢槽
rolled steel joist 辗钢工字梁
rolled stock 轧制材
rolled tube 轧制管；滚压管
rolled-earth fill 碾压填土
roller bearing 滚子轴承
roller bit 牙轮钻头
roller briquetting machine 对辊式成型机；辊式压块机
roller carriage 滚子支架；托滚架
roller chain 滚轮链
roller compacted concrete 碾压混凝土
roller compaction 碾压；碾压法
roller cone 齿轮钻头；锥形滚柱
roller conveyor 滚柱式输送机
roller cutter 滚刀掘进机球齿钻头
roller cutter for pipe 滑轮割管器
roller press 滚压机
roller shaker conveyor 滚柱式振动输送机
roller shutter 卷闸
roller side bearing 辊式侧支座
roller support 滚柱架；托辊；辊座；滚筒架
roller tractor 轮式拖拉机
roller train 滚轴式输送机
roller trough-type conveyor 滚柱式槽形输送机
roller turner 滚花轮

roller type cutter	滚柱型钻头
roller vibrator	滚轴式振动器
roller-cone bit	牙轮钻头
roller-curve	导滚曲轨
roller-type guide	滚筒导板
roller-type rotary pump	滚轴式回转泵
roller-way face	滚水坝面
rolling acceleration	摇动加速度
rolling anticline	滚动背斜；平缓背斜
rolling area	丘陵山区；起伏地区
rolling avalanche	滚动崩落
rolling axis	轧制线；滚动轴线
rolling bearing	滚动轴承；辊道
rolling compaction	碾压压实
rolling compaction test	碾压试验
rolling contact	滚动接触
rolling country	丘陵地带；准平原
rolling crack	轧制裂缝
rolling cutter bit	牙轮钻头
rolling cutter number	滚刀数
rolling cycle	轧制周期；轧制循环
rolling cylinder gate	滚动式圆筒闸门
rolling dam	滚动式活动坝
rolling damping	横摇阻尼
rolling dimensional tolerance	轧制尺寸公差
rolling direction	轧制方向；滚动方向
rolling direction of hump	驼峰溜车方向
rolling disk planimeter	转盘求积仪
rolling equipment	轧制设备
rolling force	滚动力
rolling frequency	摇动频率
rolling friction	滚动摩擦；轧制摩擦
rolling gate	滚筒闸门
rolling grade	起伏坡度
rolling ground	起伏地面
rolling hill	波状丘陵地
rolling ladder	转梯
rolling land	起伏地
rolling load	滚动荷载；车轮荷载；滚动载重
rolling lock gate	滚动式船闸闸门
rolling motion	滚动
rolling movement area	溜放部分
rolling plain	波状平原
rolling plan	连续计划
rolling plant	轧制车间
rolling pontoon	滚动浮筒
rolling power	轧制能力；轧制功率
rolling pressure	轧制压力
rolling region	起伏地区
rolling resistance	滚动阻力
rolling rig	滚动试验台
rolling ring type crusher	圆环式破碎机
rolling rock block	滚石
rolling section of hump	驼峰溜放部分
rolling shaft	滚轴
rolling strata	波痕交错纹理；波状层理
rolling support	滚动支承
rolling surface	丘陵地表
rolling temperature	轧制温度
rolling temperature region	轧制温度区
rolling terrain	微丘区；丘陵地带
rolling test	碾压试验
rolling texture	轧制织构
rolling topography	丘陵地形；起伏地形
rolling transport	滚动搬运
rolling trial	碾压试验
rollover	翻转；倾翻
rollover anticline	滚动背斜；反转背斜
rollover structure	滚动构造；反转构造
roll-up structure	旋卷构造
Roman cement	天然水泥；罗马水泥
Rontgen ray	伦琴射线，X射线
roof	顶壁；层顶；上盘
roof and ground control	顶板管理与地压控制
roof arch	顶拱
roof bar	顶梁
roof bar tip	顶梁前端
roof bed	顶板层
roof behavior	顶板矿压显现
roof bit	顶板钻头
roof block tilting	顶板岩块侧倾
roof board	顶棚板
roof boarding	屋面板
roof bolt	顶板锚杆
roof bolt plate	锚杆垫板
roof bolter	顶板锚杆安装机；锚杆钻机
roof bolting	顶板锚杆支护
roof bolting equipment	顶板锚杆安装设备
roof bolting method	洞顶锚栓法
roof bolting with bar and wire mesh	锚梁网支护
roof bracing system	屋面支撑系统
roof break	顶板断裂；顶板裂隙；顶板垮落
roof break line	放顶线
roof breaking-off line	顶板垮落线；顶板截断线
roof burst	顶板突然崩落
roof caved overriding the break line	顶板超越放顶线垮落
roof caving	顶板垮落；顶板冒落；顶板陷落
roof caving angle	顶板垮落角
roof cavity	顶板冒落空穴
roof classification	顶板分类
roof clay	隔水黏土；不透水黏土
roof close timbering	密集顶板支护
roof coal	留顶板煤
roof coating	屋面防水涂层
roof collapse	顶板崩落；冒顶
roof concave subsidence	顶板凹形下沉
roof condition	顶板状况
roof control	工作面顶板控制；顶板管理
roof control drill	顶板锚杆孔钻机
roof control line	顶板管理线；顶板控制线
roof control survey	顶板观测；顶板控制测量
roof convergence	顶板下沉；顶板收敛变形
roof convex subsidence	顶板凸形下沉
roof cover	上盖
roof crack	顶板裂缝
roof cross beam	顶板横梁
roof cut	截顶槽
roof deflection	顶板沉降；顶板挠曲
roof deformation	顶板变形
roof descent	顶板沉降

roof disaster	顶板灾害
roof exposure	顶板露出
roof failure	顶板失稳
roof fall	冒顶；顶板塌落
roof fall cavity	顶板冒落空穴
roof fall sensitivity	顶板冒落洞穴率；冒顶率；顶板冒落灵敏度
roof falling zone	顶板垮落区
roof fissure	顶板裂隙
roof flaking	顶岩剥落；端面冒顶
roof flaking ratio	顶板破碎度
roof flexibility	顶板挠性
roof forepoling	顶板超前支护
roof foundering	顶板陷落
roof holding	顶板支护
roof hole	顶板炮孔；上向炮孔
roof inspector	顶板检查员
roof instability	顶板不稳定
roof irregularity	顶板不规则性
roof jack	顶板螺旋千斤顶
roof jumbolter	顶板锚孔机
roof limb	顶翼
roof live load	顶板活载
roof load	顶板压力；顶板负载
roof monitoring	顶板监测
roof movement	顶板移动；顶板下沉
roof of an excavation in stratified rock	层状岩石巷道顶板
roof of coal seam	煤层顶板
roof of drift	平巷顶板
roof of tunnel	顶板
roof overhang	悬顶
roof pendent	顶棚悬挂体
roof penetration	顶板穿透
roof periodic caving	顶板周期冒落
roof pin	顶板锚杆
roof pitch	屋面坡度
roof plate	顶板
roof pressure	顶板压力；工作面顶板压力
roof pressure arch	顶板自然平衡拱；顶板压力拱；冒落平衡拱
roof pressure gauge	顶板压力计
roof pressure roof weight	顶板压力
roof propping	用支柱支撑顶板；支顶
roof rebound	顶板回弹
roof ripping	挑顶
roof rock	顶盖岩；盖层
roof rupture	顶板断裂；顶板破坏
roof safety	顶板安全
roof sag	顶板垂落；顶板下沉
roof shield	半盾构；顶部盾构
roof slope	顶板坡度
roof spalling	顶板剥落
roof stability	顶板稳定性
roof station	顶板观测站
roof station direction annexed traverse	顶板方向附加测量导线
roof statistical observation	顶板统计观测法
roof step	顶板阶梯；顶板台阶
roof stone	直接顶板岩石
roof strain indicator	顶板应力仪；顶板应变指示器
roof strata	顶板岩层
roof strata activity	顶板岩层活动
roof strata behaviour	顶板岩层行为
roof stratum	顶板岩层
roof strength	顶板承压能力；顶板强度
roof stress	顶板应力
roof stringer	纵向护顶背板
roof structure	屋顶结构
roof structure detection	顶板结构检测
roof support	顶板支架；顶板支护
roof support machinery	顶板支护机械
roof support pattern	顶板支护布置方式
roof support valve	液压支架用阀
roof surface	屋顶表面
roof survey plug	顶板上的测桩
roof suspension	顶板悬挂装置
roof thrust	顶冲断层；顶板逆冲断层
roof timbering	顶板支护
roof truss	顶板锚杆桁架；顶板桁架
roof truss support	桁架式支架
roof water	顶板水
roof water prevention	顶板防治水
roof weakening	顶板弱化
roof weight	顶板压力
roof weighting	顶板来压
roof weir	钻石凿井机
roof-bolt drill	顶板锚杆钻机
roof-bolt head	顶板锚头
roof-bolt hole	顶板锚杆钻孔
roof-bolt mat	锚杆共用长垫板
roof-bolt steel header	锚杆的护顶带孔钢托梁
roof-bolt steel strap	锚杆护顶钢带条
roof-bolted roadway	锚杆支护的巷道
roof-bolting machine	锚杆安装机
roof-bolting stopper	顶板锚杆钻孔机
roof-control	顶板管理；顶板控制
roof-control analysis	顶板管理分析
roof-control by caving method	用陷落法管理顶板
roof-control distance	控顶距
roof-engaging structure	接顶框架
roof-fall accident	冒顶事故
roofing	屋面；屋面材料；同巷道顶卡住
roofing board	屋面板
roofing hole	顶板孔；顶部炮眼
roofing slate	盖板岩
roof-rock slip	顶板岩石滑移
roof-testing device	顶板试验装置
roof-to-floor extensometer	顶底板伸长仪
rooftop structure	天台构筑物；天台搭建物
roof-up	急倾斜小天井
roofway	屋顶通道
room and pillar	矿房；房柱式采煤
room and pillar caving	房柱式崩落开采
room boss	煤房工作面工长；回采工作面安全工长
room capacity	矿房容积
room constant	矿房常数
room conveyor	矿房输送机；煤房输送机
room cross-cut	煤房联络小巷
room depth	矿房深度；煤房深度
room drainage method	矿房抽放方法；矿房排水法

room driving 矿房掘进；煤房掘进；硐室掘进
room entry 矿房平巷；煤房平巷
room face 矿房工作面；煤房工作面
room mining 房式开采
room on the strike 沿走向的煤房
room stability 硐室稳定性；煤房稳定性
room system with caving 放顶房柱法
room temperature 室温
room thermostat 室温控制器；恒温控制器
room timbering 矿房支护；煤房支护
room unit 房间单元
room ventilation 室内通风
room with filling 房柱式充填采煤法
room without extraction of pillar 留柱式房柱采煤法
room without pillar extraction 留柱式房柱采煤法
room work 煤房开采工作；矿房开采工作
room-and-pillar method 房柱式开采方法
room-and-rance 窄长煤柱式房柱采煤法
room-and-rance method 窄长煤柱式房柱采煤法
room-and-rance system 狭长房柱式开采系统
room-and-stoop 房柱式采煤法
root clay 根土岩；底黏土
root hole 根孔
root mean square 均方根
root mean square criterion 均方根准则
root mean square detector 均方根值检波器
root mean square deviation 均方根偏差
root mean square error 中误差
root mean square pressure 均方根压力
root of mountain 山根
root opening 根部间隙
root wedging 根劈作用
root zone 根部
root-mean-square 均方根
root-mean-square error 均方根误差
root-mean-square intensity 均方根烈度
root-mean-square misfit 均方根拟合差
root-mean-square value 均方根值
root-mean-square velocity 均方根速度
roots of quadratic algorithm 二次算法的根
root-soil composite 根土复合体
root-sum-square three-dimensional method 三维平方和开方法
rope bolt 钢索锚杆
rope boring 钢绳冲击钻进
rope boring machine 钢丝绳钻机
rope breaking test 断绳试验
rope cager 钢丝绳装罐机
rope capacity 缠绳总长度；容绳量；钢丝绳的抗断强度
rope capping 提升钢丝绳的接头
rope changing 钢丝绳的更换
rope clamp 索钳；索夹
rope control 绳索控制；绳索操纵
rope conveyor 缆索输送机
rope core 绳芯
rope coupling 钢丝绳接头；绳索连接
rope creep 钢丝绳潜伸；钢丝绳的蠕动
rope creep compensator 松绳补偿器
rope deflection 钢丝绳垂曲
rope deflection pulley 钢丝绳偏导轮

rope design 钢丝绳结构
rope dressing 钢丝绳涂油；钢丝绳保护层
rope drill 绳拉钻；钢丝绳式钻机
rope drive 绳索传动；钢索传动
rope extension 绳索接长
rope factor 提升钢丝绳系数
rope fastening 钢丝绳扎结
rope friction hoist 摩擦提升机
rope gear 钢索传动
rope haulage 钢丝绳运输；绳索运输
rope haulage roller 钢丝绳运输滚柱
rope hoisting 钢丝绳提升
rope hook 绳钩
rope incline 钢丝绳运输斜井
rope installation 安装钢丝绳
rope life 钢丝绳寿命；钢丝绳使用期限
rope lubricant 绳索润滑剂
rope lug 绳栓
rope packing 封密绳
rope pennant 吊缆
rope pull 钢丝绳拉力
rope pulley 绳轮
rope roll 绞车滚筒；绕绳筒
rope round pulley 绕绳滑轮
rope safety factor 钢丝绳安全系数
rope shackle 网丝绳钩环
rope shaker conveyor 钢丝绳振动式输送机
rope sheave 绳轮
rope sheave excavating bucket 绳轮挖斗
rope shovel 钢绳提升式挖掘机
rope side frame conveyor 钢丝绳吊挂输送机
rope slippage 绳滑
rope slipper 钢丝绳罐道导环；稳绳导环
rope spear 钢丝绳打捞钩
rope spooling 钢丝绳绕卷；缆索绕卷
rope stretch 绳索拉紧；绳索伸长
rope stretching station 紧绳站
rope swivel 缆索转环
rope tension 绳张力
rope test 钢丝绳试验
rope to pulley ratio 绳径轮径比
rope transfer 钢丝绳移送机
rope transmission 柔索传动；绳索传动；钢丝绳传动
rope transport 钢丝绳运输；缆索运输
rope tread 触绳衬面
rope treading surface 钢丝绳接触面
rope tunnel 钢丝绳运输平硐
rope winch 钢丝绳运输绞车
rope wire 钢丝绳；合股线
rope-bolt 钢丝绳锚杆
rope-clip 索夹；钢丝绳夹；抱绳器
ropy structure 绳状构造
Roscoe model 罗斯科强度模型
Roscoe surface 罗斯科面
rose curve 玫瑰线
rose diagram 玫瑰花图；玫瑰图
rose diagram of joints 节理玫瑰图
rosette 玫瑰花式；丛生；簇生
rosette displacement gauge 应变花位移计
rosette gauge 应变片丛

rosette joint diagram	节理玫瑰图
Rossi-Forel scale	罗西-福勒烈度表
rotary action	旋转作用
rotary agitator	旋转式搅拌机
rotary air drill	旋转式压气钻机；旋转式风钻
rotary air pump	旋转式空气泵
rotary air valve	跳汰机回转风阀
rotary bit	旋转式钻头
rotary boring	旋转钻进
rotary bucket excavator	斗轮式挖掘机
rotary concentrator	旋转式选矿机；旋转式选煤机
rotary core bit	旋转式岩心钻头
rotary core drill	旋转式岩芯钻机
rotary core drilling	旋转式岩芯钻进
rotary coring bit	旋转式取芯钻头
rotary coring drilling	旋转式取芯钻进
rotary cowl	旋转式挡煤板
rotary crusher	旋转压碎机；旋转式破碎机
rotary displacement compressor	旋转排量压缩机
rotary drill	旋转式钻机
rotary drilling	旋转式钻进
rotary drilling machine	旋转式钻机
rotary drilling rig	旋转钻机
rotary equipment	旋转钻井设备
rotary fault	旋转断层
rotary horsepower	旋转马力
rotary inertia	转动惯量
rotary interchange	环形立体交叉
rotary jet piercing	旋转喷焰火力钻进法
rotary kiln	旋转炉；旋转窑
rotary kiln incinerator	旋转窑式焚化炉
rotary mixer	转动搅拌机
rotary moment	转矩
rotary moment of inertia	转动惯矩
rotary motion	转动
rotary movement	转动；旋转运动
rotary mud	旋转钻进用钻泥
rotary percussion boring	旋转冲击钻探
rotary percussion drilling	旋转冲击式钻进；旋转冲击式钻眼
rotary percussive drill	旋转冲击式凿岩机；液压马达冲击式钻机
rotary piercing drill	旋转式火钻
rotary plate filler	旋转板填料机
rotary plate motion	板块旋转运动
rotary processing equipment	旋转式加工设备
rotary pump	旋转泵；回转泵
rotary sampler	旋转式取土器
rotary screen	转筒筛；滚筒筛
rotary screen press	旋转隔网压滤机
rotary screw compressor	旋转螺旋压缩机
rotary side-wall coring	旋转式孔壁取芯
rotary sidewall coring tool	旋转或旋进式井壁取芯器
rotary slide valve gear	旋转滑阀装置
rotary supply outlet	旋转送风口
rotary support frame	转盘座架
rotary swager	旋转锤锻机
rotary system of drilling	旋转钻眼法；旋转钻进法
rotary table	转台；转盘
rotary torque	旋转扭矩
rotary tripper	旋转振动翻车机
rotary vacuum pump	旋转真空泵
rotary valve	旋转阀
rotary vane water meter	翼轮式水表
rotary vibratory drilling	旋转振动钻井
rotary viscosimeter	旋转黏度计
rotary water meter	旋转式水表
rotatable transformer	可旋转变压器
rotate	旋转
rotate about a longitudinal axis	绕纵轴转动
rotate frame	转动架
rotated factor	旋转因子
rotated garnet	旋转石榴子石
rotating anode	旋转阳极
rotating antenna	旋转天线
rotating arc	旋转电弧
rotating arc welding	旋转电弧焊
rotating bending test	旋转弯曲试验
rotating casing drilling	旋转套管钻进
rotating casing drilling method	旋转套管钻进法
rotating couple	旋转力偶
rotating coupling	旋转连接器；旋转车钩
rotating crane	旋转式起重机
rotating device	旋转装置
rotating field	旋转磁场
rotating filter	旋转式过滤机
rotating gear	旋转机构
rotating gravity force couple	旋转重力偶
rotating joint	旋转接头；旋转连接
rotating journal bearing	转动轴颈轴承
rotating magnet	旋转的磁铁
rotating paddle	旋转叶片
rotating piston	旋转式活塞
rotating plane wave	旋转平面波
rotating screen	转筒筛；滚筒筛
rotating shaft	旋转轴；转轴
rotating switch	转动开关
rotating table	转盘
rotating tensioning device	转动式张力装置
rotating union	旋转联轴器
rotating valve	回转阀
rotating vector	旋转向量
rotation and translation	旋转和平移
rotation angle	旋转角
rotation axis	旋转轴
rotation centre	旋转中心；转动中心
rotation fault	旋转断层
rotation gear	旋转机构
rotation inertia	转动惯量
rotation irrigation	轮灌
rotation matrix	旋转矩阵
rotation moment	转矩
rotation momentum	自转角动量
rotation of the earth	地球自转
rotation parameters	旋转参数
rotation pole	自转极
rotation radius	旋转半径
rotation speed	旋转速度
rotation strain	旋转应变
rotation stress	旋转应力

rotation surface	旋回面
rotation tectonite	构造岩
rotation tensor	旋转张量
rotation torque	转矩
rotation viscometer	旋转黏度计
rotation wave	旋转波；横波
rotational angular velocity of the earth	地球自转角速度
rotational axis	转动轴
rotational component	转动分量
rotational deformation	旋转变形
rotational discontinuity	转动间断
rotational effect	转动效应
rotational ellipsoid	旋转椭球
rotational energy	转动能
rotational failure of soil slope	土坡的旋转破坏
rotational fault	旋转断层
rotational flow	涡流
rotational fold	扭转褶皱
rotational frequency	转动频率
rotational grain	旋转晶粒
rotational inertia	转动惯量
rotational landslide	旋转滑坡
rotational movement	旋转运动；断错运动
rotational quantum number	转动量子数
rotational scar	旋转断壁；滑塌断壁
rotational shear	旋转剪切
rotational shear failure	圆弧形剪切破坏
rotational shear tectonic system	旋扭构造体系
rotational slide	旋转滑动；石笼墙
rotational speed	旋转速度
rotational spheroid	旋转椭球
rotational stiffness	转动刚度
rotational strain	旋转应变
rotational stress	旋转应力
rotational structure	旋转构造
rotational symmetry	轴对称；旋转对称
rotational torque meter	转动转矩计
rotational type of motion	旋转型运动
rotational velocity	自转速度；转动速度
rotational viscometer	旋转黏度计
rotational viscosimeter	旋转黏度计
rotational wave	旋转波
rotation-inversion axis	倒转轴
rotative action	旋转作用；转动作用
rotative flow	旋转水流
rotator	旋转器；转子；旋转反射炉；旋翼
rotatory fault	旋转断层
rotatory movement	旋转运动
rotatory reflection axis	旋转反映轴
rotatory reflection inversion axis	旋转反伸轴
rotometer	旋转流量计
rotor displacement angle	转子位移角
rotor erection pedestal	转子
rotor float	转子浮筒
rotor hammer boring	回转冲击锤钻进
rotten	分解的；破坏的；风化的；腐烂的
rotten rock	风化岩石
rotten slime	腐泥岩
rotten stone	朽石；蚀余硅质灰岩；含硅软土
rotting	风化；分解
rough adjustment	粗调整；粗调准
rough bottom	不平的巷道底
rough coal	原煤；入厂原煤
rough country	山丘地带；难通行地区；不平地区
rough dentation	犬牙交错
rough determination	粗略测定
rough drawing	草图；略图；示意图
rough estimate	概算；粗估
rough finish	初级修整；低级光洁度
rough grading	粗修整土方
rough grain	粗糙颗粒；粗晶；粗纹理
rough grinding	粗研磨；粗磨
rough ground	崎岖地面
rough measurement	粗测
rough out	大致绘制；打草图；拟订概略计划
rough plan	初步计划；设计草图
rough positioning	初定；粗调
rough purification	初步净化
rough road	不平整道路
rough rule	粗略近似法；粗率定则
rough section	粗制截面；毛断面
rough set	粗糙集
rough sizing	粗筛分
rough sketch	草图；略图
rough stock	未加工材料；原材料
rough stone	粗石；毛石
rough surface	粗面；粗糙表面
rough survey	概测；草测
rough test	粗测定；初步试验
rough use	粗用；繁重条件下使用
rough walled fracture	粗糙壁裂缝
rough walling	毛石砌壁；毛石圬工
rough water effect	波浪效应；糙水效应
rough weight	毛重
rough-cast finish	粗注面
rougher	粗选机；初选机；粗选槽
rougher flotation	粗选
rougher tailing	粗选尾矿
rougher-scavenger flowsheet	粗选-扫选流程图
rough-faced form board	粗面模板
rough-finished	粗面的；草率制成的
rough-finished stone	粗琢石；粗制石
rough-ground	粗磨的
roughing	粗选；粗加工；初步加工
roughing machine	粗选机
roughing mill group	粗轧机组
roughing out	粗分离
roughing pass	粗轧孔型
roughing plant	粗选车间
roughing pocket	粗选室；粗选槽
roughing pump	低真空泵
roughness	粗糙度
roughness coefficient	粗糙系数；粗糙率
roughness index	粗糙指数
roughness length	粗糙长度
roughness of relief	地形起伏
roughness of the surface	表面粗糙度
roughness parameter	糙度参数
roughness ratio	粗糙度比
roughness Reynolds number	糙率雷诺数

roughness-relation to dilatancy 粗糙度与扩容性的关系
round about line 迂回线
round about line of hump 驼峰迂回线
round excavation 掘进循环
round figure 整数
round gravel 圆砾
round length 炮眼深度；循环进尺
round of holes 炮眼组；钻孔组
round rope 圆钢丝绳
round sampler 循环取样
round sand 圆粒砂
round shaft 圆形立井
round throw 炮眼组爆破
round timber 圆木
round timber chock 圆木垛
round trip 钻进操作循环；来回路程
roundabout 回旋处
roundabout open caisson method 环形沉井法
roundabout stoping 大量回采；混合回采
rounded fold 圆形褶皱
rounded grain 磨圆颗粒
rounded pebble 磨圆卵石
round-hole screen 圆孔筛
rounding 使成圆形；倒圆
roundness 磨圆度，浑圆度
roundness grade 圆度级
roundness index 圆度指数
roundness of particles 颗粒圆度
roundness ratio 圆度比
roundness tolerance 圆度公差
roundoff 舍入；凑整
round-off error 舍入误差
round-off number 取整数目
roundstone 圆砾岩
round-the-clock 昼夜不停的
round-timber bulkhead 圆木隔墙
round-timber set 圆木棚子
round-trip speed 往返行程速度
roundtrip time 来回时间
route 路线；路程；道路
route analysis 路径分析
route diversion 变更径路
route integration 路线整合
route intersection type 线路交叉形式
route leveling 线路水准测量
route location 定线
route location on paper 纸上定线
route locking 进路锁闭
route locking indication 进路锁闭表示
route map 路线图
route of road 道路路线
route parallel weaving 线路平行交织
route plan 线路规划
route reconnaissance 路线踏勘
route release 进路解锁
route selection 路线选择；选线
route setting 排列进路
route sheet 进路表
route signaling 选路制信号
route skew intersection 线路斜交
route survey 路线测量；路线调查
route survey for power transmission 输电线路测量
route survey of surface mine 露天矿线路测量
routine 常规的；例行的
routine adjustment 例行调整；定期调整
routine analysis 常规分析
routine calculation 普通计算；程序计算；典型计算
routine check 程序校验；例行检查；常规检查
routine chemical analysis 常规化学分析
routine consolidation test 常规固结试验
routine control 常规控制
routine design 常规设计
routine examination 例行顺序检查；常规检查
routine experiment 例行实验；例行试验
routine fashion 常规制法
routine inspection 日常检查；例行检查；常规检查
routine library 程序库
routine maintenance 例行维修，例行保养
routine maintenance inspection 例行维修检查
routine maintenance works 例行维修工程
routine measurement 常规测量
routine monitoring 常规监测
routine observation 常规观测
routine of work 工作制度
routine operating pressure 正常操作压力
routine practice 操作规程
routine safety 日常安全
routine sampler 常规取样器；常规取样机
routine sampling 常规取样
routine soil test 常规土工试验
routine test 例行试验；例行检验；定型试验；标准试验
routine work 例行工作；常规操作
routing 运输路线；制订路线；布线；路径选择
routing algorithm 布线算法
routing and scheduling analysis 路径及时程分析
routing method 定线方法
routing metric value 路由度量值
routing node 路由节点
row initiation 排孔起爆
row matrix 行矩阵
row of bars row 排顶梁
row of hole 成排炮孔
row of linked bar 排铰接顶梁
row of pile 排桩；板桩
row of prop 排柱
row of support 排柱
row pitch 行距；排距
row shooting 按排引爆，排间引爆
row spacing 排距
row sum method 行数总和法
row vector 行向量
row well pattern 排状井网
rubbed concrete 磨面混凝土；水磨石
rubbed finish 磨平修整
rubber 橡胶；橡皮；磨光砖；打磨工；垫托
rubber arch fender 拱形橡胶护舷
rubber band 橡皮圈；橡皮带
rubber banding 橡皮板变换
rubber belt 橡皮带；橡胶带

rubber belt conveyor	橡胶带输送机
rubber block	橡胶块
rubber buffer	橡胶减震器；橡胶缓冲器
rubber bush	橡皮垫圈；橡皮衬套；橡皮碗
rubber cell fender	鼓形橡胶护舷
rubber collar drill steel	橡胶肩钢钎
rubber collar grommet	橡皮垫环
rubber conveyor belt	输送机橡胶带
rubber dam	橡胶坝
rubber draft gear	橡胶缓冲器
rubber extension modulus	橡皮拉伸模量
rubber friction draft gear	橡胶摩擦式缓冲器
rubber grommet	橡胶孔环；橡胶索环
rubber hose	橡皮管；橡胶管
rubber insulation	橡胶绝缘
rubber pipe	橡胶管
rubber piping	橡胶活塞
rubber plug	橡皮塞
rubber railway	长胶带输送机线
rubber scraper	橡胶刮板；刮橡胶刀
rubber seal	橡胶密封圈；橡胶垫
rubber sealing strip	橡胶密封条；橡胶止水带
rubber sealing washer	橡胶密垫圈
rubber seat	橡皮座
rubber sheath	橡皮套
rubber sheeting	橡胶板
rubber shield	橡胶护板
rubber sleeve	橡皮管；软管
rubber spring	橡胶弹簧
rubber spring shackle	橡皮弹簧钩环
rubber-tired compactor	胶轮压实机
rubber-tired drill	胶轮式钻机
rubber-tired equipment	装胶轮的设备
rubber-tired excavator	轮胎式挖掘机
rubber-tired haulage	橡胶轮矿车运输
rubber-tired scraper	胶轮式铲运机
rubber-tired tractor	橡胶轮拖拉机
rubber-tired transport	橡胶轮胎运输；胶皮轮运输
rubber-tyre crane	轮胎式起重机
rubbing	摩擦
rubbing area	摩擦面积
rubbing characteristics	摩擦特性
rubbing contact	摩擦接触
rubbing fastness	耐摩擦程度
rubbing rail	滑移轨
rubbing surface	摩擦面
rubbing wear	摩擦损耗
rubbish	废石；碎屑；废物；垃圾
rubbish disposal	垃圾处理
rubbish dump	碎石堆
rubble	角砾；碎石；毛石；碎冰块
rubble aggregate	毛石集料
rubble breccia	破碎角砾岩；构造角砾岩
rubble concrete	块石混凝土
rubble dam	堆石坝；碎石坝
rubble deposition works	沉积毛石工程
rubble disposal	毛石处置
rubble drain	盲沟；块石排水沟
rubble fill	碎石填土；毛石填筑；碎石充填
rubble flow	泥石流；碎石流
rubble foundation	抛石基床
rubble mound	毛石基；毛石堆
rubble mound seawall	毛石海堤；抛石海堤
rubble retaining wall	毛石挡土墙
rubble slope	毛石堆；毛石护坡；碎石坡
rubble soling	乱石基底
rubble stone	乱石；粗石；块石
rubble stone footing	片石基础
rubble stone retaining wall	毛石挡土墙
rubble structure	毛石结构；堆石结构
rubble wall	毛石墙；堆石墙
rubble with binding material	浆砌毛石
rubble work	毛石工程
rubble-mound breakwater	抛石防波堤
rubble-mound foundation	抛石基床
rubble-mound seawall	抛石防潮堤
rubble-mound structure	抛石建筑物
rubblerock	角砾岩
rubblework	毛石圬工；毛石建筑
ruby manometer	红宝石压力计
rudaceous	砾状的；砾屑的；砾质的
rudaceous rock	粗碎屑岩
rudaceous texture	砾状结构
rude foliation	粗糙叶理
rude processing	粗处理
rudimentary	基本的
rudite	砾屑岩
ruffle	褶边；皱纹
ruffled groove cast	皱状沟铸型
rugged country	起伏地区；丘陵地；崎岖地
rugged shore	崎岖海岸
rugged terrain	起伏形；崎岖地带
rugged topography	崎岖地形
ruggedization	使粗糙；加固；增加耐磨可靠性
ruggedized construction	加固结构；耐震结构
ruggedness	强度；坚固性
ruggedness number	粗度数
rugosity	粗糙度
rugosity coefficient	粗糙系数
ruinous earthquake	破坏性地震
ruinous shock	毁灭性地震
rule for maximum yield	最高产率原则
rule of combination	组合规则
rule of inference	推理规则
rule of strain-hardening	应变硬化；应变强化
rule of thumb	经验法则；凭经验估计
rules and regulations	规章制度
ruling engine	等距线仪；划线机
ruling grade	极限坡度；最大坡度
ruling gradient	指导性坡度；基本坡度
ruling point	起点；原点；始点
run a line	拉线；定线
run a survey	测量
run ability	流动性
run back	返测
run book	运行资料；操作手册，操作说明书
run by gravity	重力流
run casing	下套管
run coal	软煤；粒状煤；原煤
run down	冲刷；流失

run error	行差	running water sediment	流水沉积物
run gravel	冲积砾石	running-ground timbering	易塌地层支护
run in depth	地下径流	running-in test	空转试验
run location	运行位置；操作位置	running-out	流走
run mode	执行方式；运行方式	running-up test	起动试验
run of a shift	工作班时间	run-of-crusher stone	拣出的废石
run of bank gravel	采石坑砾石	runoff	径流；径流量；溢出；煤柱倒塌
run of gold	富金矿脉；富金矿带	runoff amount	径流量
run of hill	崖堆	runoff analysis	径流分析
run of mine coal	原煤	runoff area	汇水区；汇水面积
run of mine ore	原矿	runoff channel	径流河床
run of quarry ore	露天开采原矿	runoff coefficient	径流系数
run of steel	钻杆长度；炮眼深度	runoff computing formula	径流计算公式
run of the bank filling	岸土流失	runoff cross-section	过水断面
run of waste drilling	不取芯钻进；清水洗孔钻进	runoff curve	径流曲线
run off	溢出，流出；放出，倾出，印出	runoff cycle	径流循环
run pipe	压力管；流通管	runoff depth	径流深度
run speed	运行速度	runoff detention	径流滞留量；径流储水量
run test	运行测试	runoff distribution	径流分配
run time	运行时间	runoff distribution curve	径流分布曲线
run up	涌上	run-off elevation	顺坡
run up elevation	上涌高程	runoff experimental station	径流实验站
run-around ramp	回返坑线	runoff factor	径流因素
run-around track	环行线；迂回线	runoff flow	径流
runaround way	绕道	runoff formula	径流公式
runaway by gravity	重力流	runoff frequency curve	径流频率曲线
runaway of bit	跑钻	runoff generating precipitation	产流降水
run-coal bin	原煤仓	runoff generating rain	产流降雨
runner casing	迂缓弯曲渠道	runoff hydrograph	径流过程线
running	运转；行程；行使；流动的	runoff in depth	地下径流
running ability	运转能力	runoff intensity	径流强度
running adhesion factor	运行黏着系数	runoff loss	径流损失
running average	追随平均；滑动平均	runoff map	径流图
running axis	工作轴	runoff mass curve	径流累积曲线
running bar	纵向顶梁	runoff model	径流模型
running charge	生产费；工作费	runoff modulus	径流模数
running cost	经常费；使用费；机械的运转费	runoff percentage	径流百分数
running efficiency	使用效率；运转效率	run-off pit	溢流池
running expense	经常费；管理费	runoff plot	径流试验小区
running ground	松软地基；流动地层；松散岩层	run-off point	流出点；溢出点；脱出点
running hour	工作小时；生产小时	runoff process	径流过程
running inspection	运行检查；行修	runoff producing precipitation	产流降水
running level	变动的水平面；运行等级	runoff producing rain	产流降雨
running light	无载运转的；空转的	runoff quantity	径流流量
running maintenance	日常维修；日常保养	runoff rate	径流强度
running mean	移动平均；连续平均	runoff regime	径流情况
running ore	放落矿石；流动矿石	runoff source	径流来源
running period	试运转时期；试车时期；运转时期	runoff variability	径流变率
running program	操作程序	runoff volume	径流总量
running quality	流动性	runoff water	径流水；流水；活水
running repair	临时修理；小修	run-out time	停工时间
running sand	松砂；流砂	runs analysis	流程分析
running soil	流土；流动地层	run-through	贯通；掘穿
running speed	行驶速度	run-through operation	连续运转
running surface	波状表面	run-time error	运行错误
running temperature	运行温度；运转温度	run-time period	开动时期；转动期；绞车的提升时期
running test	试车；试探性试验	rupture	破裂；断裂；裂缝；裂开
running time	试运转时间；试车时间；运转时间	rupture angle	破坏角
running tunnel	区间隧道	rupture area	破裂范围；破裂面积
running water	活水；自来水；长流水	rupture circle	破裂圆

rupture deformation	破裂变形
rupture diagram	破裂图
rupture elongation	破坏延伸率；断裂伸长
rupture envelope	破裂包络线
rupture front	破裂前沿
rupture grouting	破裂灌浆
rupture length	破裂长度
rupture life	持久强度；破断寿命
rupture limit	断裂极限；破坏极限
rupture line	破裂线
rupture metamorphism	压碎变质
rupture model	破裂模型
rupture modulus	折断模量；破裂模量
rupture moment	破坏力矩
rupture motion	破裂运动
rupture of ligament	韧带断裂
rupture of shaft lining	井壁破裂；井筒衬砌破裂
rupture plane	破裂面
rupture plane of slope	土滑动面
rupture point	破裂点
rupture process	破裂过程
rupture propagation	破裂传播
rupture rate	破裂速率
rupture resistance	抗破裂性；抗断强度
rupture speed	破裂速度
rupture stage	破损阶段
rupture strain	断裂应变
rupture strength	抗裂强度；极限强度
rupture stress	断裂应力；破裂应力
rupture surface	破裂面
rupture thickness	破裂厚度
rupture velocity	断裂传播速度
rupture zone	破裂带；破碎带
rupturing capacity	抗裂能力
rural architecture	农村建筑
rural area	乡郊地区
rural building lot	乡郊建屋地段；乡郊屋地
rural construction	农村建筑
rural counting program	郊区交通量调查计划
rural district	郊区
rural domestic waste	农村生活废水
rural dwelling	农村住房
rural estate	乡郊屋；郊区屋
rural extension area	乡郊扩展区
rural feeder road	郊区支路
rural highway	郊区公路
rural holding	农村土地；乡郊土地
rural housing	乡村式公屋；郊区公屋
rural housing estate	乡郊屋；郊区屋
rural improvement area	乡郊改善区
rural land	农村地；乡郊土地
rural lot	农村地段
rural planning survey	乡村规划测量
rural protection area	乡郊保护区
rural road	郊区道路
rural water supply	农村供水
rush	冲，突进，涌出；突然塌落；突然冒顶；找矿热
Russian Geology and Geophysics	俄罗斯地质与地球物理学报
rust-proofing compound	抗锈剂
rust-resisting	抗锈；抗锈的
rust-resisting material	防锈材料
rust-resisting property	抗锈性能
rut	压痕；凹槽；沟

S

S criterion　S 判据；应变能密度因子准则
S wave　S 波
Saalian glacial stage　萨埃尔冰期
Saalic orogeny　萨埃尔造山作用
sabot　桩靴；垫板；镗杆
sabugalite　铝钙铀云母；铝铀云母
sabulous　砂质的；多砂的
sabulous clay　亚砂土，砂土质土壤，砂质黏土
saccharoidal dolomite　糖粒状白云岩
saccharoidal texture　砂糖状结构
sack borer　小型钻机
sacked cement　袋装水泥
sacked concrete　袋装混凝土
sacked concrete revetment　袋装混凝土护岸
sacked gravel packing sand　袋装砾石充填砂
sack-like structure　囊状构造
sacrificial concrete　废弃混凝土
saddle　背斜，凹谷；鞍状构造，鞍状矿脉；鞍部；滑板
saddle axis　背斜轴，鞍轴
saddle back　垭口；鞍形山；鞍背；顶板起伏；底板起伏
saddle back reef　鞍状矿脉
saddle bend　鞍顶，褶皱鞍形顶部；背斜
saddle dam　副坝，副堤
saddle dolomite　鞍白云岩
saddle fold　鞍状褶皱
saddle function　鞍式函数
saddle jack　座式千斤顶
saddle pebble　马鞍石
saddle point　鞍状点
saddle vein　鞍状矿脉
saddle-shaped shell　鞍形壳
safari solution　萨费里解
safe allowable load　安全容许荷载；安全容许负载；安全荷载；容许荷载
safe allowable stress　安全容许应力
safe bearing capacity　容许承载力；安全承载能力
safe bearing load　安全负载；安全承载
safe channel capacity　河槽安全泄量；河槽安全容量
safe coefficient　安全系数
safe depth　安全深度
safe drilling　安全钻井
safe formation　安全地层
safe gap　安全空隙
safe leg load　立柱安全荷载
safe life　安全寿命，安全使用期限
safe limit　安全限度，安全极限
safe load　容许荷载，安全荷载，安全载重
safe load capacity　安全负荷量，容许负荷
safe margin　安全储备
safe measure　安全措施
safe pressure　安全压力
safe range　安全范围
safe stress　安全应力，允许应力

safe working condition　安全工作条件；安全操作状态
safe working load　容许荷载，安全工作荷载，安全负载
safe working stress　安全工作应力，容许应力
safe yield　安全开采量，安全产水量，可靠开采水量；安全供水量
safe yield of groundwater　地下水允许开采量
safe yield of stream　河的安全给水量
safeguard　防护；防护设备，安全装置
safety accessory　安全设备，安全装置
safety action　安全措施；屏蔽效应
safety against overturning　抗倾覆安全度；抗倾覆安全
safety against shear failure　抗剪切破坏安全
safety against sliding　抗滑安全；抗滑安全度
safety alarm device　安全警报装置
safety analysis　安全性分析；安全分析
safety analysis and evaluation　安全分析与评价
safety and failure probability　安全系数和失效概率
safety appliance　安全设备，安全用具，保险装置
safety arch　安全拱，分载拱，辅助拱
safety assessment in post-earthquake field　地震现场安全评估
safety assessment model　安全评估模型
safety assessment system model　安全评估系统模型
safety audit　安全核查；安全审计
safety class　安全等级
safety coefficient　安全系数
safety curtain　防护幕
safety devices of crane　起重机安全装置
safety distance　安全距离
safety drift　安全巷道；人行道
safety equipment　防护装置，安全装置
safety estimating model　安全评估模型
safety explosion　安全爆破
Safety factor　安全系数；保险系数；安全因子
safety factor against shear failure　剪切破坏安全系数
safety factor for bearing capacity　承载力安全系数
safety factor of slope　边坡安全系数
safety factor of stability　稳定安全系数
safety feature　安全特性；安全设备
safety fence　安全护栏，防护栅，防撞栏
safety glass　安全玻璃
safety grade　安全等级
safety guard　护栏，保险板
safety hole　安全洞，躲避洞
safety impedance　安全阻抗
Safety index system　安全指标体系
safety induction training　安全引导培训
safety initiative　安全倡议
safety inspection　安全检查
safety interval　安全间距；安全区间
safety level　安全等级，安全程度；安全液面；安全水位
safety limit　安全极限，安全限制
Safety line　安全线
safety load　安全负荷；容许负载

safety load carrying capacity 安全承载能力
safety load factor 安全荷载因子
safety margin 安全系数，可靠程度；安全储备；安全限度
safety measure 安全措施
safety method 安全方法，安全技术；保护措施
safety monitoring 安全监测
safety monitoring distributed analysis platform 安全监测分布式分析平台
safety operation 安全操作
safety pathway 安全通道，安全通路
safety performance 安全性能
safety pillar 保安煤柱，保安矿柱
safety pillar for avoiding water rush 防水煤岩柱
safety pillar loss 安全煤柱损失
safety pillar under water-body 防水煤岩柱
safety policy 安全政策
safety practice 安全措施
safety precaution 安全预防措施，保安措施
safety precaution measure 安全预防措施
safety promotion programme 安全推广计划
safety protection system 安全保护系统
safety reserve 安全储备
safety spacing 安全间距
safety test 安全试验
safety threshold 安全阈值；安全限值
safety valve 安全阀，保险阀
safety walkway 安全通道，安全通路
safety water head 安全水头
safety yield 可靠产量，安全产量，可靠出水量
safety zone 安全区域，安全地带
safflorite 斜方砷钴矿
sag 松弛，下垂；坳陷，凹陷
sag and swell 凹凸构造，凹凸地形；波状起伏构造，波状起伏地形
sag arch 张力拱
sag bend region 垂弯区
sag bolt 测顶板垂度锚杆
sag correction 下垂校正，垂度校正
sag crossing 下垂穿越
sag curve 挠度曲线；垂度曲线，弛垂弯曲
sag dome 张力穹
sag gauge 垂度计
sag grading 降坡，下坡，下坡
sag meter 顶板下垂测定器，顶板下沉测定计
sag of protecting coating 保护层脱落
sag of roof 顶板陷落
sag pipe 下垂管
sag pond 断陷湖，断陷塘，沉陷洼地，地堑湖；堰塞湖；裂谷湖
sag profile 弛垂断面；洼地断面
sag rate 下沉速度；下垂率
sag ratio 垂跨比；垂度和跨度比
sag structure 下陷构造
sag subduction 沉陷俯冲消亡作用
sag-and-swell topography 凹凸地形
sagenite 网金红石
Sager 烧箱，烧盆，铸件退火箱；耐火黏土
sagging 下垂，中垂，船身下垂；垂度；挠度；凹陷，地壳沉降部

sagging building 下陷建筑物
sagging condition 中垂状态
sagging curve 下垂曲线
sagging moment 中垂力矩
sagging point 下沉压弯点，下垂凹下点；软化点
sagging soil 融沉土
sagging zone 弯曲下沉带，弯曲带
sag-swell topography 波状起伏地形
sagvandite 菱镁古铜岩
Sahara 撒哈拉沙漠；荒野
sahlinite 黄砷氯铅矿
sahlite 次透辉石
sail angle 最大偏角，最后角
sailaba 泛滥平原，洪泛平原
Saint Venant compatibility condition 圣维南相容条件
Saint Venant's principle 圣维南原理
Saint-Venant body 圣维南体；等效力体
sajong soil 砂姜土
sakalavite 玻基安山岩
salable rock 易片落的岩石
salamander's hair 石棉
salammonite 卤砂
salar 干盐湖；盐坪，盐坪沉积物
salband 近围岩岩脉，矿脉壁
salcrete 煮盐的沉渣
saleite 镁磷铀云母
salesite 碘铜矿
sales-quality gas 合格天然气
salic horizon 积盐层
salic mineral 硅铝质矿物
salicylate 水杨酸盐
salicylic acid 水杨酸
salience 凸起
salient 扇形地背斜轴；凸出部；显著的，突出的
salient angle 凸角
salient relief 显著起伏地形
saliferous 盐渍化的，含盐的
saliferous clay 盐渍土
saliferous rock leaching 含盐岩淋滤
saliferous soil 盐渍土
salification 积盐，盐渍化，变咸；成盐作用
salimeter 盐度计，含盐量测定计，盐液比重计
salina 盐田，盐场，海水蒸发槽；盐碱滩，盐湖，盐沼；煮盐锅
salina solonchak 盐沼地
salinastone 盐类岩
saline 盐水；盐碱滩，盐湖，盐泉，盐皮；盐化的，含盐的，咸的；盐渍
saline alkali soil 盐碱土
saline alkaline soil 盐碱土
saline alkaline water 盐碱水
saline aquifer 咸水层
saline beneficiation 盐的精选
saline bog 盐沼泽，咸水沼泽
saline concentration 含盐度
saline connate water 含盐共生水
saline contamination 盐水污染；盐水入侵
saline current 盐水流
saline deposit 盐类沉积，盐矿床
saline dispersion halo 盐分散晕

saline dome 盐穹，盐丘
saline encroachment 盐水入侵，咸水入侵，海水入侵
saline facies 盐相
saline flotation 岩盐浮选，盐浮选
saline formation 含盐层，含盐地层；含盐建造
saline groundwater 地下咸水
saline intrusion 盐水入侵，咸水入侵
saline karst 盐岩溶
saline lake 盐湖
saline lake deposit 盐湖矿床
saline land 盐地，盐碱地；盐土
saline lick 盐类沉积；盐渍地
Saline meadow 盐化草原
saline mineral 盐类矿物
saline mud 盐水泥浆
saline pollution 盐水污染
saline rock 盐岩
saline sediment 盐沉积
saline soil 盐渍土
saline solution 盐溶液
saline solution mud fluid 盐水泥浆
saline spring 盐泉，咸水泉
saline system 咸水系统
saline water 海水，盐水
saline water intrusion 盐水渗入
salineness 含盐度，含盐性
saliniferous 含盐的，盐土的
saliniferous rock 含盐岩石
salinimeter 盐度仪，盐液密度计
salinity 盐浓度，含盐量，盐度，矿化度，咸度；渍度
salinity correction factor 盐度校正系数
salinity gradient 含盐量梯度，盐度梯度
salinity indicator ratio 矿化度指示比
salinity log 含盐度测井
salinity logging 盐度测井
salinity ore 钾石盐矿石
salinity stratification 盐度分层，盐水成层
salinity tolerance 耐盐度
salinity try 盐浓度测量法
salinization 盐渍化，盐碱化，盐化，盐渍；盐化酎
salinization damage 盐碱害
salinize 盐化
salinized chestnut soil 盐化栗钙土
salinized land 盐碱地
salinized soil 盐渍土
salinometer 盐度计，含盐量测定仪，盐液比重计，盐液密度计
salinous 咸的，盐的
salite 次透辉石
salitrite 榍石霓辉岩
salivation process 硅化凿井法
salmare 石盐
salmonsite 黄磷锰铁矿
salometer 盐液比重计
salometry 盐量测定
salse 泥火山
salsima 硅铝镁层
salt annealing 盐浴退火
salt anticline 盐背斜

salt balance 盐分平衡
salt basin 含盐盆地
salt bath 盐槽；盐浴
salt bed 盐层
salt bridge 盐桥
salt brine 盐溶液，盐水，卤水，盐卤
salt cake 硫酸氢钠，芒硝
salt cavern 盐穴
salt cement 含盐水泥
salt cementation 盐类胶结
salt collapse structure 盐陷构造，盐塌构造
salt concentration 含盐浓度
salt contamination 盐侵
salt content 含盐量；盐度
salt core 盐芯；盐核
salt core structure 盐核构造
salt corrosion 盐侵蚀
salt crust 盐壳，盐皮
salt crystal 盐晶
salt damage 盐害
salt deposit 盐类沉积；盐矿
salt desert 盐漠，盐质漠境
salt diapir 盐底辟；盐刺穿
salt diapirism 盐底辟作用
salt dome 盐丘，盐穹
salt dome oil pool 盐丘油藏
salt dome reservoir 盐丘油藏
salt dome storage 盐丘地下储库
salt dome trap 盐丘圈闭
salt efflorescence 盐霜
salt encrustation 盐壳
salt exclusion 脱盐
salt face 盐丘面
salt feature 盐丘特征
salt flank 盐岩侧翼
salt flat 盐滩
salt flotation 岩盐浮选
salt flowage 盐流
salt gauge 盐浮计，盐水比重计
salt glacier 盐川
salt graben 碱土
salt ground water 含盐地下水，地下盐水
salt karst 盐岩溶
salt lagoon facies 咸化泻湖相
salt lake 盐湖
salt lick 盐沼地
salt lime 石膏，硫酸钙
salt marsh 咸沼，盐沼，盐碱滩
salt mine 岩盐矿，盐矿
salt mud 盐水泥浆
salt pan 浅咸水湖，盐池
salt pelite 含盐泥质岩
salt pillar 盐柱
salt pit 采盐矿；岩盐矿
salt playa 干盐湖
salt pond 盐池
salt precipitation 盐沉淀，盐析
salt profile logging 含盐剖面测井
salt rampart 盐障
salt range 盐岭

salt removal	脱盐
salt residual	剩余盐丘，岩丘残渣
salt resistance	抗盐性
salt resisting	耐盐的
salt ridge	岩盐岭
salt roast	盐皮，盐壳
salt rock	岩盐，石盐
salt roller	岩盐隆起
salt secretion	盐分泌
salt soil	盐渍土
salt solution	盐溶液
salt spray	喷盐
salt spring	盐泉，盐井
salt structure	盐体构造
salt swamp	盐沼
salt table	盐台
salt tectonic	盐构造
salt tolerance	耐盐性，抗盐限度
salt tolerant	耐盐的
salt uplift	岩盐隆起
salt wall	岩盐墙
salt waste	盐质沙漠
salt water	矿化水，盐水
salt water bearing reservoir	盐水储层
salt water contamination	盐水入侵
salt water encroachment	盐水入侵，海水侵入
salt water immersion test	盐水浸渍试验
salt water interface	盐水界面
salt water intrusion	盐水侵入，海水入侵
salt water lake	咸水湖，盐湖
salt weathering	盐风化
salt well	盐井，盐水井
salt-affected soil	盐渍土壤
salt-and-water balance	水盐平衡
saltation	跃移，跳跃搬运，突变；不连续变异
saltation load	跳跃搬运物
salt-bearing	含盐的
salt-bearing rock	含盐类岩石
salt-bearing section	含盐剖面
salt-brine cement	含盐水泥
salt-cored structure	盐核构造
salt-cored trap	盐核圈闭
salt-crust method	盐壳防尘法
salt-crusted soil	盐壳土
salt-crystal cast	盐印痕
salt-crystal imprint	盐印痕
salt-dome boundary	盐丘界面
salt-dome coast	盐丘海岸
salt-dome well	盐丘井
saltern	碱土；盐田，盐场
salt-filtering	盐渗
salt-generated paleostructure	盐成古构造
saltier	硝石，钠硝
salt-impregnated carbon	盐碳
saltiness	含盐性，咸度
salting	盐碱滩，海滩盐泽，岩滩岸；盐碱化
salting of soil	土壤盐渍化，土壤盐碱化
salting-out	盐析
salt-lake basin	盐湖盆地
salt-marsh plain	盐沼平原
salt-motivated uplift	盐成隆起
saltness	盐度
saltpeter	硝石
salt-poor cement slurry	低含盐水泥浆
salt-rich cement slurry	高含盐水泥浆
salt-roof rock	盐盖层
salt-velocity method	盐液流速法
saltwater barrier	咸水屏障
salt-water clay	含盐黏土
saltwater disposal well	盐水处置井
saltwater injection	注盐水
saltwater invasion	盐水入侵，咸水入侵，海水入侵
saltwater mud	盐水泥浆
saltwater wedge	盐水楔
salt-water well	产盐水的油井
salt-wedge effect	盐楔效应
salty soil	盐渍土，含盐土壤
salvage archaeology	古物抢救工程
same facies difference time	同相异期
same-sense curve	同向曲线
samoite	多水硅铝石
sample	样品，样本，标本，试件，试块；采样，取样，抽样
sample accreditation	试样鉴定
sample analyzer	试样分析器
sample application	点样
sample autocorrelation function	样本自相关函数
sample average	样本均值，样本平均
sample axis	样本轴
sample bag	试样袋，岩屑样品袋
sample boring	取样钻孔；取样钻探
sample bowl	试样箱，煤样箱
sample box	试样箱，煤样箱，样品箱
sample capacity	样本容量
sample catcher	样品收集器，岩粉取样器；取样工
sample census	抽样普查
sample characteristic function	样本特征函数
sample collecting	采样，取样
sample collection	取样
sample concentration	试样浓缩
sample condensation	试样冷凝
sample container	样品容器，样品储存器
sample contamination	样品污染
sample correlation coefficient	样本相关系数
sample covariance	试样协变性；样本协方差
sample crusher	试样破碎机
sample cut	试样采取
sample cutter	试样截取器，岩心切断器，岩芯提断器
sample decomposition	样品分解
sample density	样品密度
sample depth	取样深度
sample design	样本设计
sample dispersion	样本离差；试验分散性
sample distribution	样品分布
sample divider	分土器，样品缩分器，样品分割器
sample divider pass	试样缩分
sample division	样品缩分
sample drawing	抽样
sample drilling	取样钻进；取样钻眼
sample dryer	试样干燥机

sample drying	试样的干燥
sample evaporation	试样气化,试样蒸发
sample evaporation chamber	试样气化室
sample extruder	推样器
sample frequency	取样频率
sample function	样本函数
sample gasification	试样气化
sample hand reducing	人工缩样,试样手工缩分
sample holder	试样夹,标本架法
sample hole	样洞
sample horizon	取样层位
sample hose	样品管
sample identification	砂样鉴定
sample increment	试样小样,子样
sample indexing	试样编号
sample injection port	试样注射口
sample injector	试样注射器
sample inquiry	抽样调查
sample inspection	抽样检查
sample interval	取样间隔,采样间隔,样品间距
sample introduction	进样
sample layout	采样布局
sample length	取样长度,样品长度
sample likelihood function	样本似然率函数
sample log	样品记录;取样剖面,采样剖面
sample mark	取样深度标
sample mean	样品平均值,取样平均
sample mean and variance	样本平均值与方差
sample mean value	样本均值
sample measurement	样品测量
sample median	样本中位值,样本中位数;
sample moment	样本矩
sample mud	岩屑录井泥浆
sample of undisturbed soil	原状土试样
sample ordination method	样本定位法
sample outcrop	样品露头
sample pan	砂样盘
sample period	取样间隔,采样周期
sample piece	样品,试样,样件
sample pit	取样探井
sample point	样本点,取样点,抽样点
sample polynomial	取样多项式
sample population	抽样总体
sample preparation	样品制备,样品加工
sample prescription bottle	砂样瓶
sample pretreatment	样品预处理
sample probe	试样探针
sample procedure	取样程序
sample processing	样品加工,试样处理
sample program	采样程序
sample quartering	四分法取样;四分缩样法
sample range	样本范围
sample rate	采样率;取样速度
sample receiver	试样容器
sample record	样本记录
sample recovery	试样回收
sample reducing	试样缩分
sample reduction flowsheet	试样缩制流程
sample regression coefficient	样本回归系数
sample retention	样品保存
sample riffle	二分缩样器
sample room	样品室
sample sack	试样袋;样品袋
sample scoop	取样勺
sample section	采样区;取样截面
sample set	样本集
sample sieve analysis	砂样筛析
sample site	取样处
sample size	样本大小,样本尺寸;样本数量
sample space	样本空间
sample spectral function	样本谱函数
sample splitter	分样器,劈样器
sample splitting	试样缩分,样品缩分
sample splitting device	试样缩分器
sample spout	取样槽
sample standard deviation	样本标准差
sample standard error	样本标准误差
Sample strength	样品强度
sample survey	抽样调查,抽样测量
sample taker	取样器
sample technique	抽样技术
sample test	抽样试验
sample thief	取样器
sample time	采样时间
sample trap	样品收集器
sample trough	泥浆临时取样盒
sample tube	采样管,取土器
sample valve	取样阀
sample variance	样本方差,样本离差
sample wave form	抽样波形曲线
sample window	采样窗口
sample work piece	样品工件
sample-assay data	矿样分析数据
sampled analog data	抽样模拟数据
sampled data	采样数据;取样数据
sampled signal	采样信号
sampled value	样本值
sampled well	取样井
sampled width	取样厚度
sample-data control	采样数据控制
sample-data system	采样数据系统
sampleite	氯磷钠铜矿
sampler	取土器,采样器,取样器;试样;采样工
sampler barrel	取样筒
sampler drill	取样钻机
sample-taking gun	取样器
sample-to-background activity ratio	样品-本底活性比
sampling	取样,采样,抽样
sampling action	取样
sampling aperture	取样孔径
sampling apparatus	取样器
sampling appliance	取样用具
Sampling area	取样区
sampling barrel	取样筒
sampling bias	取样偏差
sampling bottle	取样瓶
sampling bucket	取样勺斗
sampling by crosscutting	横切取样
sampling by filtration	过滤采样
sampling by impaction	撞击采样

sampling by stripping	剥层取样
sampling channel	取样槽
sampling cock	取样旋塞
sampling condition	取样条件
sampling configuration	采样模式
sampling control	取样控制
sampling density	采样密度
sampling depth	取样深度
sampling design	取样规格；抽样设计，取样设计
sampling device	取样装置，取样器，采样设备
sampling distribution	抽样分布；取样分布
sampling disturbance	取样扰动
sampling equipment	采样装备，取样设备
sampling error	抽样误差，采样误差
sampling flowsheet	取样流程；取样流程图
sampling frequency	采样频率，取样频率
sampling function	采样函数
sampling grid	采样网格
sampling gun	取样枪
sampling head	采样头
sampling hole	取样钻孔
sampling horizon	采样层位
sampling information	抽样信息
sampling inspection	抽样检查，采样检查；抽样试验
sampling instant	采样时刻
sampling interval	测量间隔；采样间隔，取样间距
sampling investigation	抽样调查
sampling jar	取样瓶
sampling kit	采样箱
sampling length	取样长度
sampling line	取样管线
sampling loop	采样环路
sampling machine	取样机，采样机
sampling material	试样
sampling method	取样法，采样法，抽样法
sampling normal distribution	抽样正态分布
sampling observation	取样检查，抽样检验
sampling of suspended load	悬移质取样
sampling of waste water	废水取样
sampling oscilloscope	采样示波器
sampling pair	样本对
sampling pattern	采样模式
sampling period	抽样周期
sampling pipe	取样管
sampling plan	抽样方案；取样方案
sampling point	取样点，采样点
sampling probe	采样探头，取样探头
sampling procedure	采样步骤，取样程序，取样方法
sampling program	取样程序
sampling pulse	抽样脉冲
sampling pulse generator	采样脉冲发生器
sampling rate	采样率，取样率，抽样率
sampling ratio	采样比
sampling receptacle	样品容器
sampling size	样本大小
sampling skew	取样时移
sampling spoon	取样环刀，取样勺
sampling spot	取样点
sampling station	采样站
sampling survey	样本调查
sampling system	取样系统
sampling technique	取样技术，采样技术
sampling test	样品试验，抽样试验
sampling theorem	采样定理
sampling tube	取样筒，取样管，样品管
sampling variability	抽样变化性
sampling variance	采样方差，抽样方差
sampling well	取样井
sampling with replacement	有放回抽样
sampling without replacement	无放回抽样
sampling-data system	采样数据系统
samsonite	硫锑锰银矿
San Andreas fault	圣安德烈斯断层
San Cristobal Trench	圣克里斯托瓦尔海沟；圣克立托巴海沟；南所罗门海沟
sanbornite	硅钡石
sand	砂土，砂，砂粒，粗砂，砂地，砂层；撒砂
sand abrasion	砂蚀
sand accretion	砂涨地，淤砂地；淤砂
sand accumulation	砂粒累积
sand aerator	松砂机
sand aggregate ratio	砂石比
sand anchor	砂锚
sand and gravel	砂砾石，砂砾
sand and gravel concrete	砂砾石混凝土
sand and gravel overlay	砂砾覆盖层
sand and gravel trap	砂砾沉淀池，沉砂池
sand and gravel wash	砂砾洗涤层
sand apron	砂裙
sand aquifer	含水砂层
sand arch	砂桥
sand asphalt	沥青砂，地沥青砂
sand asphalt cushion	沥青砂垫层
sand asphalt facing	沥青砂罩面，砂沥青护面
sand avalanche	砂崩
sand bag	砂袋，砂包
sand bag coffer dam	砂袋围堰
sand bag pack	砂袋充填带；砂袋充填
sand bag revetment	砂袋护坡
sand bag wall	砂袋护墙；袋装砂井
sand bag walling	砂袋筑墙
sand bailer	抽砂筒，捞砂筒
sand ballast	砂道碴，砂质道碴，砂质石碴
sand bank	沙坝，砂坝，沙滩，沙洲，砂洲
sand bar	沙洲，砂洲，沙滩；白砂
sand barite	玫瑰花状重晶石
sand barrier	沙坝，沙堤；河口沙洲，拦门沙
sand baseline	砂层基线
sand basin	沉沙池，沉砂槽
sand bath	砂浴；砂盘
sand beach	沙滩，砂质海滩
sand bed	砂垫层，砂床
sand bedding	铺砂层，砂垫层
sand bedding course	砂垫层
sand bin	支撑剂容器
sand binder	固砂植物；砂黏结剂
sand bitumen	沥青砂
sand bitumen core	沥青砂土心墙
sand blanket	砂垫层
sand blast	喷砂，砂冲；砂糊炮；喷砂器

sand blast erosion	喷砂冲蚀
sand blast machine	喷砂机
sand blast pretreatment	喷砂预处理
Sand blast test	喷砂试验
sand blasted panel	喷砂样板
sand blasting	砂冲,喷砂,喷砂处理;喷砂冲毛,喷沙打磨法
sand blow	喷砂
sand blower	喷砂器,喷砂机
sand body	砂层,砂体
sand body morphology	砂体形态
sand boil	喷砂泉;喷水冒砂,砂沸
sand boiling	砂沸,喷砂
sand boils	砂沸腾
sand bowl	砂箱
sand box	砂箱,砂槽,脱水除砂槽
sand box model	砂槽模型
sand bulking	砂体体积膨胀
sand burning	黏砂;铸砂烧结
sand calcite	砂方解石
sand capacity	油层产油能力
sand casting cement	砂型水泥
sand catcher	水中集砂器
sand cay	沙洲
sand cement	掺砂水泥
sand cement ratio	砂灰比
sand chimney	砂质垂直排水
sand clay	砂质黏土,亚黏土,壤土;砂泥混合物
sand clay liquidation	砂土液化
sand clay loam	砂质壤土
sand clean out	冲砂
sand coarse aggregate ratio	砂石率,砂石比
sand coast	砂质海岸
sand coking	地层砂焦化
sand column	砂桩,砂柱;砂井;人造岩心
sand compaction pile	挤密砂桩
sand compaction pile method	挤密砂桩加固法
sand composition	沙组成;配砂
sand cone method	灌砂法;砂锥法
sand consolidation	固砂;取样
sand consolidation anchorage	砂固结锚固
sand consolidation technique	固砂技术
sand contamination	砂侵
sand content	含砂率,含砂量
sand content apparatus	含砂量测定仪
sand content set	含砂量测定仪
sand control	型砂控制,铸砂控制;含沙量控制,治沙
sand control gravel	防砂砾石
sand control pack	防砂充填
sand core	砂芯
sand count	砂岩层总有效厚度;砂岩层总层数
sand course	沙面层
sand crushing resistance	砂子抗破碎强度
sand cushion	砂垫层
sand cushion stabilization method	砂垫层加固法
sand cut	冲砂
sand cutter	移动式砂处理机,碎砂机
sand cutting	松砂,拌砂;砂蚀;砂屑
sand deformation	砂变形
sand desert	砂质荒漠,沙漠
sand detection device	测砂器
sand detector	测砂器
sand devil	沙暴
sand dome	圆砂丘,砂穹;砂钟
sand drain	排水砂井;砂井排水
sand drain method	砂井排水法
sand drain vacuum method	砂井真空排水法
sand drift	沙丘;流沙,风沙吹移
sand drum	捞砂滚筒
sand dump valve	排砂阀
sand dune	沙丘
sand dune effect	沙丘效应
sand dune fixation	沙丘固定
sand dyke	砂脉
sand ejector	喷砂器
sand entry	砂子侵入
sand entry area	砂侵区
sand equivalent	含砂当量,沙的百分比含量
sand equivalent shaker	砂当量振动器,砂当量振动装置
sand erosion	砂蚀
sand erosion control	防止砂粒冲蚀
sand erosion damage	砂粒冲蚀破坏
sand erosion time	砂蚀时间
sand erosional effect	砂粒冲蚀效应
sand erosional plug	砂粒冲蚀探测塞
sand eruption	喷砂
sand escape	泄沙闸,排沙道
sand exclusion	防砂
sand exploration	砂石勘探
sand extraction	砂石开采
sand face pressure	井底流压
sand facies	砂相
sand fall out	砂子沉出
sand falling out	砂粒沉降
sand fill	充填砂,砂填料,水砂充填;砂堵
sand filled drainage well	砂井,填砂排水井
sand filling	砂充填,填砂,装砂
sand filter	砂滤器,矿砂脱水机;砂滤池;砂滤层
sand filter bed	砂过滤层
sand filter layer	砂过滤层
sand filtration	砂滤法;砂滤
sand fineness	砂子平均粒度
sand fixation	固砂;固砂措施
sand fixation by plantation	植物固砂
sand flag	层状细粒砂岩,层状砂岩
sand flat	沙滩,沙坪
sand flood	砂暴
sand flotation	砂浮选,砂浮法
sand flotation process	尾砂浮选法;强斯砂浮法
sand flow	出砂;砂流
sand flow ability	型砂流动性
sand flow method	压砂法
sand foundation	砂基础,砂土地基
sand fraction	砂粒粒组,砂粒级配
sand fracturing	加砂压裂
sand furrow	砂沟
sand fusion	黏砂
sand grading	砂的粒度组成,砂粒级配,砂粒分级
sand grain	砂粒
sand grain cluster	砂粒团块

sand grain contact point	砂粒间的接触点	sand pack	填砂模型
sand grain content	砂粒含量	sand pack filter	砂粒充填过滤器
sand grain count	砂粒计数	sand pack height	填砂高度
sand grain distribution	砂子粒度分布	sand packed fracture	填砂裂缝
sand grain roundness	砂粒圆度	sand pad	砂垫层
sand grain sphericity	砂粒球度	sand paper	砂纸
sand grain uniformity	砂粒均匀度	sand particle	砂粒
sand gravel	砂砾	sand particle size range	砂粒粒度范围
sand grit	粗砂岩	sand parting	砂隔层
sand grout	薄水泥浆，水泥砂浆	sand patch test	铺砂法试验，铺砂试验砂桩
sand grouting	砂层内灌浆；砂浆	sand permeability	砂层渗透性，砂层渗透率
sand hazard	砂害	sand pile	砂桩
sand heap analogy	砂堆比拟，堆砂模拟	sand pile stabilization method	砂桩加固法
sand hill	砂丘，砂岗	sand piling barge	砂桩船
sand hole	砂眼，砂孔	sand pipe	沙管
sand impinging	砂粒撞击	sand pit	沙池，砂坑，采砂场，采砂矿
sand inclusion	夹砂；砂眼	sand plain	沙原
sand inclusion mark	夹砂痕	sand plate	板夯
sand infiltration bed	砂滤层	sand plug	砂堵
sand inflow	地层砂流入	sand plugging	砂堵；砂塞
sand influx	地层砂流入	sand pocket	砂袋；砂矿囊
sand interceptor	拦砂设施	sand polygon	多边形砂
sand intrusion	砂侵	sand porosity	砂层孔隙度
sand invasion	砂侵	sand prediction	出砂预测
sand island	砂岛	sand pressure	油层压力；含油砂层压力
sand island method	砂岛法	sand pressure control	砂压控制，地层压力控制
sand jet	喷砂器，喷砂嘴	sand prevention	防砂
sand jet method	水力喷砂法	sand probe	含砂探测头
sand laden	含砂的	sand problem reservoir	出砂的产层
sand laden mud	含砂泥浆	sand producing formation	出砂地层
sand laminate	砂质薄夹层	sand producing region	出砂区域
sand layer	砂层	sand producing tendency	出砂倾向
sand lens	砂岩透镜体；透镜状砂体	sand production	砂生产；油井出砂
sand lens reservoir	砂岩透镜体储层	sand production modeling	出砂模型
sand levee	沙垅，纵向沙丘	sand production monitoring	出砂监测
sand lime brick	砂灰砖	sand production rate	出砂量
sand lime mortar	石灰砂浆	sand promoting condition	促使出砂的条件
sand lime stone	砂灰砌块	sand proportion	加砂比
sand liquefaction	砂土液化	sand protection facility	防砂设施
sand loam	肥土，壤土，砂壤土；砂质黏土，亚砂土，轻亚黏土	sand pulling	抽砂
sand lock	挡砂浆闸	sand pulp	砂浆
sand loess	砂质黄土	sand pump	抽砂泵，扬砂泵，泥泵
sand lump	砂团，砂块	sand pump sampler	泵式砂样采取器
sand map	砂层图	sand pumping	抽砂
sand mat	砂垫层；砂面	sand quality	砂粒质量
sand mat of subgrade	排水砂垫层	sand rammer	型砂捣固机
sand material	砂粒材料	sand rate	砂率
sand matrix density	砂岩骨架密度	sand ratio	砂率
sand measurement	含砂量测量	sand recovery	砂回收率
sand measuring container	量砂箱	sand reef	沙洲
sand member	砂层	sand reel	泥泵绳轮；捞砂滚筒
sand migration	砂粒运移	sand relative density test	砂的相对密实度试验
sand mobile bed	砂粒流动的河床	sand replacement method	换砂法，灌砂法
sand model	砂模型	sand replacement test	填沙法，灌砂法
sand mold	砂型	sand return	返砂
sand mortar	砂浆	sand ribbon	沙垄
sand mould	砂型	sand ridge	沙脊
sand mound	砂堆，沙堆	sand ripple	沙波纹，砂波痕
sand movement	地层砂运移；沙丘移动	sand rock	松散砂岩，砂岩

sand sampling	取砂样
sand screen	砂筛；砂滤层
sand screening	滤砂
sand screw feeding system	螺旋供砂系统
sand seal	砂封
sand sealing	砂密封
sand section	砂层段
sand separator	砂分离器
sand settling	砂粒沉降，沉砂
sand shadow	沙影
sand shaker	砂筛
sand sheet	席状砂
sand shoal	沙浅滩
sand sieve grading	砂筛分级
sand sifter	型砂筛子，砂筛，筛砂器
sand sifting	筛砂
sand silt	粉砂
sand size designation	砂粒尺寸选定
sand skeleton	砂土骨架
sand soil	砂质土，砂土
sand solubility	砂粒溶解度
sand sphericity	砂粒球度
sand spit	沙嘴；沙洲
sand spraying	撒砂
sand stability	砂岩稳定性
sand stone	砂岩
sand stop	防止砂子下落装置
sand stowing	水砂充填，砂充填
sand stratum	砂层
sand streak	簿砂岩夹层，砂质条带
sand sucker	泥砂泵，吸泥机，吸砂船
sand surface	砂面
sand suspension	砂粒悬浮
sand sweeping	回砂
sand sweeping equipment	回砂机
sand swell	沙隆
sand tank	充填用砂箱
sand tank model	砂槽模型，渗流槽
sand tempering	型砂湿度调节
sand tester	带筛孔油管的井下取样器
sand texture	沙丘，流沙堆
sand thickness	产层厚度；砂层厚度
sand transport	输砂
sand trap	沉砂池，截砂池；砂槽
sand trial	落砂试验
sand undulation	起伏沙丘
sand up	砂堵
sand vent	喷砂口
sand volcano	砂火山
sand volume	砂岩体积，砂子体积
sand washing	冲砂
sand wave	沙波，沙浪
sand well	砂井，砂岩层产油井
sand wick	袋装砂井
sand without epoxy	净砂
sand-bag dam	沙袋坝，沙袋隔墙
sandbag damming	砂袋筑堤，沙袋筑坝；沙袋隔墙
sandbergerite	锌铜矿
sand-blast finish	喷砂表面处理；喷砂后的表面
sand-blasting action	喷砂作用
sand-blasting method	喷砂处理法
sandcay	小砂岛
sand-cement coating	水泥砂浆保护层
sand-cement grout	加砂水泥浆
sand-cement mixture	水泥砂混合物
sand-cement mortar	水泥-砂灰浆
sand-cement slurry	水泥砂浆
sand-clay matrix	砂-黏土混合物，砂土混合料
sand-drift control	防砂
sanded	撒上砂的；混砂的，砂磨的
sanded cement grout	水泥砂浆
sanded up	砂堵
Sande-Tukey algorithm	Sande-Tukey算法
sandface convolution plot	井底褶积图
sandface flow rate	井底流量
sandface rate	井底流量
sandface rate convolution	井底流量褶积
sandfall	滑落面；落砂
sand-filled drainage	填砂排水井；砂井排水
sand-filling	装砂的；砂充填的；砂填充物
sandfilling chamber	注砂井
sandfilling shaft	注砂井，充填砂井
sand-flood	大流动砂体
sand-glass structure	沙钟构造
sand-grain crushing	砂粒压碎
sand-grain-size distribution	砂粒粒度分布
sand-gravel cushion	砂石垫层，砂砾垫层
sand-gravel mixture	砂-砾混合物
sand-gravel pile	砂石桩
sandiness	沙质
sanding	撒砂，出砂；透长石，玻璃长石；用沙打磨
sanding formation	出砂地层
sanding in	砂子堆集
sanding prediction	出砂预测
sanding tendency	出砂倾向
sanding up	淤砂，形成砂堵
sandish	砂的
sand-jetting method	喷砂法
sandkey	小沙岛，小沙洲
sand-laden fluid	含砂液
sand-mud transition	砂泥过渡
sand-out pressure	脱砂压力
sand-pack column	填砂柱
sand-pack flood	填砂模型水驱试验
sand-pack profile	充填砂层剖面
sand-packed model	填砂模型
sand-packed tube displacement	砂粒填充管驱替试验
sand-prone depositional facies	偏砂沉积相
sand-prone seismic facies	偏砂地震相
sand-raking capacity	耙砂容量，耙砂能力
sands	含油砂岩地层；沙滩；砂
sandshale	砂页岩
sand-shale boundary	砂页岩界面
sand-shale ratio	砂页岩比
sand-shale sequence	砂页岩层序
sand-shoal system	浅滩砂系统
sand-size grain	砂级颗粒
sand-size particle	砂级颗粒
sandstone acidizing	砂岩酸化
sandstone aquifer	砂岩含水层

sandstone band	砂岩夹层	sandy marl	砂质泥灰岩
sandstone cave	砂岩穴，崖洞，石洞	sandy mobile bed	砂质流动河床
sandstone composition	砂岩组成	sandy muck	砂质腐泥土，砂质淤泥
sandstone cylinder	砂岩柱	sandy mud	砂质泥
sandstone diagenesis	砂岩成岩作用	sandy sapropel	砂质腐泥
sandstone dike	砂岩墙	sandy seabed	砂质海床
sandstone formation	砂岩地层	sandy shale	砂质页岩
sandstone grit	粗粒砂岩	sandy silt	砂质粉土，砂质粉砂，亚砂土
sandstone in block	块状砂岩	sandy siltstone	砂质粉砂岩
sandstone petrography	砂岩岩石学	sandy slate	砂质板岩
sandstone pipe	砂岩管	sandy soil	砂类土，砂土
sand-stone ratio	砂石比	sandy soil foundation	砂土地基
sandstone reservoir	砂岩储层	sandy soil with breccia	砂质土夹角砾
sandstone seal	砂岩封闭	sandy turbidite	砂质浊积岩
sandstone soil	砂岩土	sandy-conglomeratic flysch	砂质砾岩复理石
sandstone target	砂岩射孔试验靶	sanidine	透长石，玻璃长石
sandstone-gas outburst	砂岩与瓦斯突出	sanidinite	透长岩
sandstorm	沙暴，沙尘暴	sanitary landfill	垃圾堆填，垃圾堆填区，垃圾掩埋场
sand-thickness map	砂岩等厚图，砂体等厚图	sanmartinite	钨锌矿
sand-transport efficiency	携砂效率	sannaite	霞闪正煌岩
sand-trap gate	截砂栅	santorine	杂伊利石
sand-water slurry	砂-水浆液	santorinite	钠长苏安岩；紫苏安山岩
sand-well drainage	砂井排水法	sap flow	液流
sandwich	夹层结构，层状结构	saponite	皂石；蒙脱石
sandwich belt	夹层带	sapphire	蓝宝石色
sandwich board	夹层板	sapphirine	假蓝宝石
sandwich concrete	夹层混凝土	sapping	基蚀，掏蚀，剥蚀，拔蚀；挖掘；下切
sandwich construction	夹层结构	saprokonite	细粒灰岩
sandwich glass	夹层玻璃，层压玻璃	saprolite	风化岩石，风化土；腐泥土，残积物，残余土；钠沸石；腐泥褐煤
sandwich moulding	夹心注塑	saprolite placer	腐泥砂积矿床；腐泥土砂矿
sandwich structure	夹层结构，夹层构造，叠层构造	saprolith	风化土，残余土
sandwich wall	夹层墙	saprolitic	腐泥质的
sandwich waterproofing of lining	衬砌夹层防水	saprolitic soil	腐泥土
sandwiched coal	夹层煤	sapropel clay	腐泥黏土
sandwiched-in	夹在两层之间的	sapropel-calc	腐泥钙藻岩
sandwiching	夹在两层之间的	sapropelic	腐泥的
sandy	砂质的；含砂的；砂屑的；多砂的；沙地的	sapropelic groundmass	腐泥基质
sandy aggregate	砂质集料	sapropelic kerogen oil shale	腐泥油母质油页岩
sandy aquifer	砂质含水层	sapropelic material	腐泥质
sandy beach	砂质海滩；沙滩	sapropelic oil shale	腐泥质油页岩
sandy bed	沙质河床	sapropelic rock	腐泥岩
sandy breccia	砂质角砾岩	sapropelite	腐泥岩
sandy chert	砂质燧石	sapropelitic group	腐泥化组
sandy clay	砂质黏土，瘦黏土	saprovitrite	腐泥微镜煤
sandy clay loam	砂质壤土，砂质黏壤土	sarcolite	肉色柱石；钠菱沸石
sandy coast	砂质海岸	Sardic orogeny	萨迪造山作用，萨迪造山运动；
sandy conglomerate	砂质砾岩	sardonyx	缠丝玛瑙
sandy debris flow	沙土泥石流	sarkinite	红砷锰矿
sandy desert	沙漠	sarmientite	砷铁矾
sandy formation	砂质地层	sarnaite	霓辉钙霞正长岩
sandy gravel	含砂砾石，砂砾石	sarsen	羊背石；砂岩漂砾，砂岩残块
sandy gravel layer	砂砾石垫层；砂砾层	sarsen stone	羊背石；砂岩漂砾，砂岩残块
sandy ground	沙地；砂地基；砂质土	saspachite	钾沸石
Sandy hillock	小沙丘	sassolite	天然硼砂，天然硼酸
Sandy land	沙地，砂田	satellite altimetry	卫星测高法
sandy limestone	砂质灰岩	satellite altitude	卫星高度
sandy loam	轻亚土，轻亚黏土，亚砂土；砂壤土，壤土	satellite anchor	附属锚具
sandy loam soil	砂壤土	satellite detection system	卫星探测系统
sandy loess	砂质黄土	satellite dike	附庸岩脉
sandy low-sinuosity river	砂质低弯度河流		

satellite drilling	辅助钻眼
satellite fault	附属断层
satellite fix	卫星定位
satellite frequency error	卫星频率误差
satellite gas field	周围气田，附属气田
satellite geodesy	卫星大地测量学
satellite gradiometry	卫星重力梯度测量
satellite gravimetry	卫星重力学
satellite gravity measurement	卫星重力测量
satellite gravity survey	卫星重力测量；空间重力测量
satellite ground station	卫星地面站
satellite hole	辅助炮眼；辅助钻孔
satellite image	卫星图像
satellite imagery	卫星图像，卫星影像
satellite infrared data	卫星红外资料
satellite injection	附带注入
satellite laser ranger	卫星激光测距仪
satellite laser ranging	卫星激光测距
satellite line	伴线；卫线
satellite mapping	卫星测图，卫星测绘
satellite navigation	卫星导航，卫星定位
satellite orientation	卫星定向
satellite photograph	卫星相片
satellite positioning	卫星定位
satellite positioning system	卫星定位系统
satellite remote sensing	卫星遥感；卫星遥感测量
satellite remote sensing system	卫星遥感系统
satellite reservoir	附近油藏；周围油藏
satellite sounding	卫星探测
satellite spot	卫星斑点
satellite station	卫星站；辅助站
satellite tank farm	辅助油库
satellite telemetry	卫星遥测技术
satellite triangulation method	卫星三角测量法
satellite-based	星载的
satellite-borne altimeter	星载测高计
satellite-borne sensor	星载遥感器
satin spar	纤维石
saturability	饱和度；饱和能力
saturable	可饱和的，饱和的
saturable system	可饱和系统
saturant	饱和的；饱和剂，浸渍剂
saturate	饱和的；饱和剂，浸渍剂；使饱和，浸透
saturate solution	饱和溶液
saturated	饱和的
saturated adiabatic lapse rate	饱和绝热递减率
saturated air	饱和空气
saturated aquifer	饱和含水层
saturated belt	饱和带；饱水带
saturated brine	饱和盐水
saturated bulk weight	饱水容重
saturated clay	饱和黏土
saturated clay layer	饱和黏土层
saturated coefficient	饱和系数；饱和度
saturated compound	饱和化合物
saturated core sample	饱和取芯样本
saturated coverage	饱和覆盖
saturated density	饱水密度，饱和密度
saturated flow	饱和流
saturated gas	含水天然气
saturated ground	饱和土
saturated humidity	饱和湿度
saturated hydraulic conductivity	饱和导水率
saturated in brine	盐水浸泡
saturated liquid	饱和溶液
saturated liquid line	饱和液体线
saturated moisture content	饱和含水率
saturated permeability	饱和透水性，饱和渗透率
saturated pool	饱和油藏
saturated porous medium	饱水孔隙介质；饱水多孔介质
saturated rock	饱和岩石
saturated sample	饱水样，饱和样品，饱和试样，饱和样本
saturated sand	饱和砂
saturated slope	饱水边坡
saturated soil	饱和土，饱水土壤
saturated soil layer	饱和土层
saturated solution	饱和溶液
saturated state	饱和状态
saturated steam	饱和蒸汽
saturated surface	饱和面
saturated unit weight	饱和重度，饱和容重
saturated vapour line	饱和蒸汽线
saturated water	饱和水
saturated water absorptivity	饱和吸水率
saturated water capacity	饱和水容度
saturated zone	饱和带，饱水带，饱和区
saturating capacity	饱和量
saturating point	饱和点
saturating swelling stress	饱和湿胀应力
saturating temperature	饱和温度
saturation	饱和状态；饱和度；浸润
saturation analysis	饱和分析法
saturation capacity	饱和能力；饱和含水量，饱和容量
saturation capillary head	饱和毛细管水头
saturation character	饱和特性
saturation characteristic	饱和特性
saturation charge	饱和量
saturation coefficient	饱和系数
saturation concentration	饱和浓度
saturation condition	饱和状态
saturation curve	饱和曲线
saturation deficiency	饱和差
saturation deficit	饱和差；饱和亏缺
saturation degree	饱和度
saturation distribution	饱和状态分布
saturation effect	饱和效应
saturation energy	饱和能量
saturation equation	饱和度方程
saturation exponent	饱和度指数
saturation extract	饱和浸提液；饱和浸出液
saturation factor	饱和系数；饱和因数
saturation field	饱和场
saturation force	饱和力
saturation front	饱和度前缘
saturation gradient	饱和梯度
saturation humidity ratio	饱和含湿量
saturation index	饱和指数
saturation isotherm	饱和等温线
saturation level	饱和度

saturation line	浸润线；饱和曲线
saturation magnetization	饱和磁化作用
saturation method	饱和法
saturation moisture content	饱和含水率；饱和含水量
saturation percentage	饱和百分率，饱和度
saturation plane	饱和面；自然地下水面
saturation point	饱和点
saturation pressure	饱和压力
saturation procedure	饱和方法
saturation profile	纵剖面饱和度分布图
saturation prospecting	饱和勘探
saturation ratio	饱和比
saturation regime	饱和状态
saturation region	饱和区
saturation specific humidity	饱和比湿度
saturation steam temperature	饱和蒸汽温度
saturation swelling stress	饱和湿胀应力
saturation test	饱和程度试验
saturation testing	饱和测试
saturation time	饱和时间
saturation type gravimeter	感应式重力仪，饱和式重差计
saturation value	饱和值；饱和度
saturation vapour pressure	饱和水汽压；饱和蒸汽压
saturation zone	饱和区；饱和层；饱和带
saturation-growth-rate equation	饱和增长速率方程
saturator	饱和器；饱和塔
saucer lake	浅盆湖，浸没湖
sauconite	锌蒙脱石
sault	瀑布；急流
sausage dam	铅丝石笼坝
sausage dynamite method	定向爆炸法
sausage structure	香肠构造
saussurit	钠帘石，糟化石
saussuritization	钠帘石化作用；糟化作用
saussuritize	钠帘石化
Savic orogeny	萨夫造山运动
sawback	锯齿形山脊
saw-cut	陡峻峡谷；切谷
sawtooth wave	锯齿波
sawtooth waveform	锯齿波形
sawtoothed	锯齿状
saw-toothed curve	锯齿形曲线
saw-toothed profile	锯齿形剖面
saxonite	方辉橄榄岩
scabbard fold	剑鞘褶皱
scabbling	凿花；粗琢；凿毛
scabby rock	板状岩石
scabland	崎岖地
scacchite	钙镁橄榄石
scagliola	仿云石；人造大理石
scalar	标量；无向量的
scalar equation	标量方程
scalar extension	标量扩张
scalar field	标量场
scalar function	标量函数
scalar impedance	标量阻抗
scalar intensity	标量强度
scalar magnetometer	标量磁强计，标量磁力仪；
scalar magnetotelluric method	标量大地电磁法
scalar potential	标量势；标量位
scalar pressure	标量压力
scalar property	无向量性
scalar quantity	标量，无向量，非向量；数量；
scalar seismic moment	标量地震矩
scalar wave	标量波
scalar wave equation	标量波动方程
scale accuracy	比例尺精度
scale adjustment	比例尺缩放
scale distortion	比例尺失真，比例变形
scale division	刻度，分度，分标度
scale drawing	缩尺图
scale effect	尺度效应，尺寸效应；放缩效应
scale equation	标度方程
scale error of Gauss projection	高斯投影长度变形
scale factor	比例换算系数，比例系数，比例因子，相似系数；标度因子；缩尺率
scale height	标高
scale interval	刻度的格值
scale length	标尺长度
scale line	标度线
scale load	标准负荷
scale map	比例图
scale mark	分度，刻度线，分刻度，刻度标记
scale model	成比例模型，相似模型；缩尺模型
scale model law	相似律；比例模型法
scale modeling	比例建模；比例模型
scale of earthquake intensity	地震烈度表
scale of geological reconnaissance	地质调查比例尺
scale of hardness	硬度表，硬度计
scale of height	高度比例尺，垂直比例尺
scale of intensity	烈度表，强度表
scale of seismic intensity	地震烈度表；地震强度分级
scale of seismic magnitude	震级，地震强度
scale of tectonic phenomena	构造现象规模
scale of topographic map	地形图比例尺
scale off	剥落，脱落，片帮
scale parameter	尺度参数
scale spacing	刻度间距，标度间距
scale test	比例模型试验
scale threshold	比例界限；临界
scale transformation	比例尺变换
scale up	扩展；按比例增大，递增
scale value	标度值，刻度值，格值
scale value of the gravimeter	重力仪格值；重力仪常数
scale-change theorem	标度变换定理
scaled	比例的；有刻度的；结垢的；鳞状的
scaled discriminant vector	标度判别向量
scaled factor	尺度因子，标度因数
scaled model	按比例缩小的模型
scaled model test	缩尺模型试验
scaled physical model	相似物理模型
scaled porous media model	相似多孔介质模型
scaled-division	分度
scaled-down-test	按比例缩小的模拟试验
scale-down	按比例缩小，按比例减小，递减；规模缩小
scale-down test	模拟试验；缩尺试验
scale-model investigation	比例模型研究
scaler	定标器计数器；比例系数；计数器
scale-up factor	放大时的比例因数
scaling	尺度，标度；相似性；剥落；缩放；定标，定

比例；按比例测量，按比例尺描绘
scaling chip 剥落碎屑
scaling coefficient 相似系数
scaling down 缩小比例，递减，按比例缩小；挑顶刷帮
scaling factor 换算系数；比例因子，定标因数；计数递减率，降低计算速度
scaling function 定标函数
scaling law 比例法则；尺度定律，缩尺定律，相似律
scaling law of earthquake spectra 地震强度频谱比例定律
scaling method 比例法
scaling of loose rock 整理松石
scaling of model 模型缩放
scaling procedure 标度法
scaling projection 比例投影
scaling ratio 换算系数；标度系数
scaling up 按比例增大；增大比例
scaling-off 片落，剥落；定尺寸
scall 疏松岩层；易碎岩石
scallop 海扇，弄成扇形；槽形结构，卵形洼地；粗糙度；刷帮；易碎岩石，松散地层，疏松岩层
scalloped 扇形的
scalloping 轴向变化
scalped anticline 蚀脊背斜
scalped structure 秃顶构造
scalping-out of fines 筛出细粒
scaly 鳞片状，片状；鳞状的
scaly structure 鳞片状结构
scan line 勘探线，测线
scan line survey 测线测绘
scan resolution 扫描分辨率
scanner 扫描仪，扫描器，扫描盘，扫描设备；多点测量仪；析像器
scanning 扫描；扫照；搜索
scanning angle 扫描角
scanning area 扫描面积
scanning beam 扫描射束
scanning curve 扫描曲线
scanning electron microscope 扫描电子显微镜
scanning electron microscope image analysis 扫描电镜图像分析
scanning imagery 扫描成像
scanning imaging radar 扫描成像雷达
scanning spectrometer 扫描分光计
scanning speed 扫描速度
scapolite 方柱石
scapolitization 方柱石化作用
scar 断崖，滑波边崖，陡岩坡，突岩；河曲痕，痕迹，伤疤；山崩凹屲；岩面；炉渣
scarifying 刨毛
scarp 陡坡，悬崖；马头丘；倾斜面；垂直开切
scarp face 崖面
scarp height 崖高
scarp retreat 崖坡后退
scarp slope 崖坡
scarp wall 陡坡墙
scarped plain 梯级平原
scarped ridge 单斜脊；单面山
scarplet 滑坡；小悬崖
scatter 散射，散布，扩散，分散；浸射；离散性
scatter diagram 散布图，分布图，散点图，离散图

scatter plot 散布图
scatter propagation 散射传播
scattered amplitude 散射振幅
scattered boulder 散布漂石
scattered flooding 点状注水
scattered stone 散点石
scattered vector wave 散射向量波
scattered waterflood 点状注水
scattered wave 散射波
scattered wave field 散射波场
scattered X-rays 散射X射线
scattered γ-ray logging 散射γ测井
scattergram 散点图，离散图，中心点图
scattering 钻眼偏斜；散射；散布，扩散，分散
scattering boundary 散射边界
scattering by crust inhomogeneity 地壳不均匀性散射
scattering by irregular interface 不规则界面散射
scattering characteristic 散射特性曲线
scattering coefficient 散射系数
scattering cross section 散射截面
scattering effect 散射效应
scattering law 散射定律
scattering layer 散射层
scattering stone 飞石
scattering surface 散射面，面散射
scattering topography 散射层析成像
scattering-in 内部散射
scawtite 碳硅钙石
SCB 半圆盘弯曲试样
scenario earthquake 设定地震，情景地震
scene 景物，景象；图幅；现场；出事地点
scenograph 透视图
schairerite 硫酸齿钠石
schalstein 辉绿凝灰岩
schank 褶皱翼
schefferite 锰钙辉石
scheitel 地震波峰
schematic 示意的，图解的，概略的，示意图
schematic analysis 图标分析法
schematic arrangement of anchor 锚杆布置示意
schematic design 方案设计
schematic design document 初步设计文件，方案设计文件
schematic diagram 示意图，概略图，采矿过程框图，简图；原理图
schematic drawing 简图，示意图；初步设计图
schematic figure 略图；原理图
schematic geological map 地质草图，地质略图，地质示意图
schematic layout 草图
schematic map 草图；略图
schematic model 示意模型
schematic plan 平面示意图
schematic section 示意剖面
schematic view 示意图
schernikite 红纤云母
schillerfels 顽火橄榄岩
schist 片麻岩，片岩，页岩，板岩
schist arenite 片岩屑，砂屑岩
schistic 片状的

schistoclastic structure	片岩状碎裂构造
schistoid	似片岩的，似片岩状
schistose	片理；片岩的，层状的，片状的，页状的；片岩结构的
schistose clay	片岩状黏土，板状黏土
schistose fabric	片状组构
schistose granular	片粒状
schistose grit	片状粗砂岩
schistose rock	片状岩石
schistose structure	片状构造，片状结构；片理；层状结构
schistose texture	片状结构
schistosity	劈理，片理；片岩性
schistosity cleavage	片理
schistosity plane	片理面
schistous sand	片状砂岩
schizolite	斜锰针钠钙石；锰针钠钙石
schizomorphic	压碎变形的
schlieren	条纹照相，暗线照相，纹影法；矿条，流层
Schmidt diagram	施密特井斜投影平面图
Schmidt field balance	施密特磁秤
Schmidt hammer	施密特锤；混凝土测试枪
Schmidt net	施密特网
Schmidt plot	施密特图
Schmidt plot polar	施密特极坐标图
Schmidt projection	施密特投影
Schmidt rebound test hammer	施密特回弹试验锤
schneebergite	铁锑钙石
schohartite	重晶石
schorenbergite	白霓霞斑岩；黝白霓霞斑岩
schorl	黑电气石
schorlamit	钛榴石
schorlite	黑电气石
schorlomite	钛榴石
schott	浅盐湖；回线
schriesheimite	辉闪橄榄岩
Schrodinger equation	薛定谔方程
schroeckingerite	硫铀钠钙石
schrotterite	蛋白铝英石
schuilingite	铜铅霰石
Schumann plot	休曼曲线
schuppen structure	叠瓦构造
sciatic	片岩的
science of marine engineering rock and soil	海洋工程岩土学
scissors fault	旋转断层
scissure	裂开，片裂，裂隙
sclerometer	硬度计
sclerometric hardness	刻划硬度，刮痕硬度；硬度计测定的硬度
scleropelite	硬化黏土岩
scleroscope hardness	回跳硬度
scleroscope hardness test	肖氏硬度试验
scleroscopic hardness	回跳硬度
sclerosis	硬化
sclerotic	硬化的
sclerotization	硬化作用；骨化作用
sclerotized	硬化的
scolecite	钙沸石
scooped lake	侵蚀湖
scope	范围，规模；显示器，示波器
scope of a mooring line	锚索的长深比
score	擦痕，刻痕，冰川擦痕；分数，得分
scoria cone	火山渣锥
scoria flow	火山渣流
scoriaceous	火山渣状的，渣状的
scoriaceous texture	渣状结构
scoria-tuff	火山渣凝灰岩
scoring	擦蚀作用；刻痕，划痕，冰川擦痕，划线；凹槽
scoring filamentation	断痕
scorodite	臭葱石
scorzalite	铁天蓝石
Scotch stone	苏格兰石；蛇纹石
scour	淘蚀，冲刷，冲蚀；冲砂水流；冰川侵蚀
scour action	冲刷作用，冲蚀作用
scour apron	防冲护坦
scour basin	冲刷盆地
scour below dam	坝下淘刷
scour channel	冲蚀槽；冲刷水道
scour culvert	冲沙涵洞
scour depression	冲蚀沟，冲蚀凹
scour depth	淘蚀深度，冲刷深度
scour finger	指状冲刷痕
scour gallery	冲砂槽，冲砂廊道
scour hole	冲蚀穴，冲刷坑
scour lag	冲刷滞积物
scour lineation	冲刷线条；冲刷线状脊
scour mark	冲痕
scour pad	防冲衬垫
scour pit	冲刷坑
scour pool	冲刷潭
scour prevention wall	挡水墙
scour protection	淘蚀保护
scour shoaling zone	冲刷浅水区
scour sluiceway	冲沙泄水道
scour trough	冲刷槽
scour velocity	冲刷速度
scour-and-fill	冲淤作用
scour-and-fill feature	冲淤构造
scour-and-fill structure	冲淤构造
scouring	冲刷作用，淘蚀作用，冲蚀，洗刷
scouring action	冲刷作用；冲刷，侵蚀
scouring force	冲刷力
scouring of foundation	基础冲刷
scouring prevention	防止冲刷
scout	初步勘探者；搜索；侦察，粗查
scout drilling	试钻
scout hole	普查孔；勘探孔
scouting	侦察；初步勘探；粗查；踏勘
scovillite	磷铈钇矿
scow drilling	平底船钻探技术
scrap mica	云母碎片
scraping hardness test	划痕测硬法
scraping pebble	擦痕卵石
scratch hardness	刻划硬度，划痕硬度，刮痕硬度
scratch hardness number	刮痕硬度值
scratch hardness tester	刮痕硬度试验仪，划痕硬度计
scratch mark	刮痕
scratch method	刮痕硬度试验法
scratch test	划痕试验；刻痕硬度试验
scratched boulder	擦痕漂砾

scratched pebble	擦痕卵石	sea bed mining	海底采矿
scratching	擦痕，划痕，刮痕	sea bed movement	海底移动
scree	岩屑堆，山石堆,；山麓碎石堆，山麓碎石；崩落	sea boring	海洋钻探
scree breccia	碎屑角砾岩	sea bottom radioactivity	海底放射性
scree cone	岩屑堆	sea bottom survey	海底勘测
scree debris	角砾；棱角碎屑	sea bridge	跨海桥梁；海上桥梁
screeding	砂浆底层	sea canal	通海运河
screen box	筛框；筛箱；筛选器；振动筛	sea cave	海蚀洞；海蚀穴
screen chamber	拦污栅室	sea channel	海沟，海槽
screen grid	帘栅极	sea chasm	海蚀洞；海蚀峡谷
screen guide	挡污栅导板	sea cliff	海崖，海蚀崖；陡崖；巉岩
screen map	屏幕地图	sea climate	海洋气候
screen out	筛分；筛除	sea coal	海水冲刷出的煤；海运煤；软煤；细末煤
screen oversize	筛上存留物；筛上产品	sea condition	海况
screen perforation	筛孔	sea damage	海损
screen pipe	花管；滤管，衬管，滤水管，过滤器；穿孔套管	sea defense	海岸防护
		sea dike	海堤
screen shaker	振筛机	sea drilling platform	海底钻进台
screen size	滤网孔大小；筛号；筛分粒度	sea duty	在海上钻机作业
screen sizing	筛分分级；筛分	sea dyke	海堤
screen stoppage	过滤器堵塞	sea echo	海面回波
screen surface	筛面	sea embankment	防潮堤，海堤
screen test	筛析；筛分试验	sea environment	海洋环境
screen well point	滤网式井点	sea erosion	海蚀
screenage	屏蔽；过滤	sea facies	海相
screened pipe	过滤管；衬管	sea fan	海底扇
screened sand	过筛砂	sea floor deposit	海底矿床
screened well	有过滤器的井，有过滤网的井，过滤井	sea floor exploration	海底勘探
screening analysis	筛分分析	sea floor geomorphology	海底地貌
screening curve	筛分曲线	sea floor mining	海床开采
screening inspection	筛分检查	sea floor relief	海底地形
screening rate	筛选率，筛分率	sea floor spreading	海底扩张
screening ratio	筛分比	sea floor storage facility	海底储油设施
screenings	筛余物，筛屑，滤余物	sea floor trench	海沟
screw anchor	螺钉锚固；带螺纹锚桩	sea front	海岸
screw anchor pile	螺旋锚固桩	sea gravimeter	海洋重力仪，海重力仪
screw auger	螺旋钻；麻花钻	sea height	海水高度
screw conveyer	螺旋输送机	sea horizon	海平线
screw jack	螺旋千斤顶；螺旋起重器	sea ice	海冰
screw late test	螺旋载荷试验	sea inlet	海港
screw pile	螺旋桩	sea knoll	海丘
screw plate load test	螺旋板载荷试验	sea level change	海平面变化
screw plate loading test	螺旋板载荷试验	sea level disturbance	海平面扰动
screw plate test	螺旋板载荷试验，载荷板试验	sea level elevation	海拔高，海拔高度，海拔高程
screw stairs	盘梯	sea level height	海拔高度；潮高
screw thread pipe	金属波纹管	sea level oscillations	海平面振荡
scribing	刻线，划线，刻绘，刻图	sea level rise disaster	海平面上升灾害
scrobiculate	蜂巢状	sea line	海岸线；海底管线；水下管道
scroll bar	涡形沙坝，螺旋形沙坝	sea map	海图
scroll meander	内侧堆积曲流	sea meadow	咸水沼泽
sculpture	刻蚀；雕刻，雕塑	sea mile	海里
SDEE	土壤动力学与地震工程学报	sea moat	海底环状洼地
sea air corrosion	海洋大气侵蚀	sea mud	海泥
sea anchor	垂锚，海锚	sea notch	海蚀龛；海蚀壁龛；海蚀洞
sea atmosphere corrosion	海洋大气侵蚀	sea outfall	河道出口入海处；河口
sea bank	海岸；防波堤	sea peat	海成泥炭
sea beach placer	海滨砂矿	sea pendulum	海洋重力摆仪
sea bed coring	海床取芯	sea plateau	海底高原
sea bed echo	海底回声	sea range	海底山岭
sea bed gravimeter	海底重力仪	sea retreating	海退

sea return	海面反射信号
sea ridge	海岭
sea risk	海险
sea sands	砂质海滩
sea shock	海震
sea shore	海岸，海滨
sea slide	海底滑坡
sea slope	向海陆坡
sea spray depression	浪蚀洼地
sea stack	海蚀柱；海栈
sea state	波浪等级
sea surface	海面
sea surface ghost	海面虚反射
sea surface temperature	海面温度
sea surface topography	海面地形
sea term	航海用语
sea terminal	海上浮码头
sea terrace	海成阶地
sea trail	海上试航
sea transport	海洋运输
sea triangulation	海上三角测量
sea valley	海底谷
sea valve	通海阀
sea wall	防波墙；海堤；防波堤
sea water	海水
sea water booster pump	海水增压泵
sea water corrosion test	海水腐蚀试验
sea water encroachment	淘水浸蚀地
sea water intrusion	海水浸入
sea water screen	海水隔滤网
sea wave	海浪
sea wax	软沥青
seabeach	海岸，海滩
seabeach placer	海滨砂矿，阶地砂矿，高地砂矿
seabeach scenic spot	海滨风景区
seabed	海底；海床
seabed anode	海床阳极
seabed characterisation	海底表征
seabed classification	海底分类
seabed drilling gear	海底钻井井口装置
seabed hazard	海底灾害
seabed mapping	海底绘图
seabed mining	海底采矿
seabed morphology	海底形态
seabed pipeline	海底管道
seabed preparation	海床处理
seabed reinstatement	海床修复
seabed scouring	海床冲刷
sea-bed sonar survey system	海底声呐探测装置
sea-bed topography	海底地貌
seabed transponder	海底应答器
seabed tunnel	海底隧道
seaboard	海滨；海岸
sea-borne	海运的
seaborne bottom gravimeter	海底重力仪
seaborne magnetic survey	海上磁测
sea-bottom contour	等深线
sea-bottom deposit	海底沉积，海底矿床
sea-bottom gravimeter	海底重力仪
sea-bottom multiple	海底多次波
sea-bottom plain	海底平原
sea-bottom profile	海底剖面
sea-bottom topography	海底地形
seacoast	海岸
sea-cut	浪蚀
seadrome	海上机场
seafloor anomaly	海底异常
seafloor bearing capacity	海底承压能力
seafloor channel	海底通道
seafloor density	海底密度
seafloor foundation	海底基础
seafloor geology	海底地质
sea-floor height	海底高地
seafloor hydrothermal deposit	海底热液矿床
seafloor imaging system	海底成像系统
sea-floor instability	海底不稳定性
sea-floor magnetic anomaly	海底磁异常
sea-floor magnetic survey	海底磁测
seafloor massive sulfide	海底块状硫化物
seafloor material	海底土壤物质
seafloor morphology	海底形态
sea-floor mudflow	海底泥流
seafloor observatory	海底观测
seafloor phenomena	海底现象
sea-floor sampler	海底采样；海底取样器
seafloor samples	海底样品
sea-floor section	海底地形剖面图
seafloor sediments	海底沉积物
seafloor seeps and mud volcano	海底渗漏和泥火山
sea-floor slope	海底斜坡
seafloor slope correction	海底倾斜校正
seafloor soil	海底土
seafloor spreading zone	海底扩张区
sea-floor texture	海底纹理结构
seafloor topography	海底地形
seafloor undulations	海底起伏
sea-floor valley	海底谷
seafloor video observation	海底录像观测
sea-going dredge	航海挖掘船
sea-going grab dredge	深海抓斗式挖掘船
seagoing pipeline	海底管道
sea-going tug	海上拖轮
sea-harbour	海港
seahigh	海底高地
sea-ice	海冰
sea-ice field	浮冰区
sea-ice penetrometer	插入式海冰厚度测定计
seal	盖层；密封
seal ability	密封能力
seal anchor	封锚
seal assembly	密封组件；密封装置
seal at the top of wall	墙顶封口
seal boot	封闭套
seal bore	密封孔
seal bore extension	密封筒加长短节
seal cage	密封笼
seal capacity	密封能力
Seal characterization	密封特性
seal coat	封层
seal construction rope	带芯股钢索

seal curtain	油罐浮顶密封圈装置
seal disc	密封盘
seal element	密封元件
seal extension	密封加长短节
seal flange	密封法兰
seal friction	密封摩擦
seal hanger	密封悬架
seal integrity	密封完整性
seal joint	密封接头
seal leakage	密封泄漏
seal load	密封载荷
seal nipple	密封接头
seal of hole	封孔
seal off	封堵；密封，封闭
seal packer	充填工；砌矸石墙工；毛石工
seal pad	密封板
seal pipe	主气管线中的水封阀
seal protector ring	密封防护垫环
seal receptacle	密封座
seal retainer	密封护圈
seal ring	密封环，密封圈
seal ring shank	有橡胶密封垫圈的钎尾
seal rock	密封岩
seal stack	一组密封
seal steam	汽封蒸汽
seal sub	密封接头
seal tightness test	密封性试验
seal up	密封
seal valve	密封阀
seal washer	密封垫圈
seal weld	密封焊缝
seal welding	密封焊接
sealab experiment	海洋实验室试验
sealable tank	密封罐
sea-land effect	海陆效应
sealant	止水材料；密封剂；封填料
sealed	密封的，封闭的；焊接的，焊封的
sealed against	封住
sealed area	封闭区
sealed bearing	密封轴承；油封轴承
sealed bearing rock bit	密封轴承牙轮钻头
sealed chamber	密封筒
sealed coring	密闭取芯
sealed cover	密封盖
sealed end	焊接端；封端
sealed fault	闭合断层
sealed fracture	闭合裂缝
sealed in	焊死的；封闭的
sealed joint	止水缝
sealed off	封闭的；封锁的；密封的；封离的，烫开
sealed oil filling adapter	密封注油器接头
sealed package	密封包件；密封包装
sealed porosity	封闭气孔
sealed reservoir	封闭储集层
sealed tube	密封管
sealed-level elevation	海拔高度
sealer	密封器；密封垫；保护层；检验员
sea-level change	海平面变化
sea-level contour	海水面等高线
sea-level cyclicity	海平面周期性
sea-level datum	基准海平面；海水面基准；高程基准面
sea-level elevation	海拔高度；海拔高
sea-level height	海平面高
sea-level lowstand	海平面低位
sea-level pressure	海平面气压
sea-level rise	海平面上升
sealing	密封，封堵，封闭，加封；裂隙填塞，裂隙填塞作用；盖章；焊接
sealing ability	密封能力
sealing adaptor	密封接头
sealing capacity	密封能力，封闭能力
sealing characteristics	封堵性能；密封性能
sealing coat	保护层；隔离层
sealing compound	密封剂；密封填料；绝缘胶
sealing core drill	密闭取芯
sealing element	密封元件；封闭胶芯
sealing fault	封闭断层
sealing field zone	封闭油田区
sealing film	密封薄膜
sealing fluid	密封液
sealing gasket	密封垫圈
sealing gland	密封盖，密封压盖
sealing installation	止水设施；密封设备
sealing leakage	堵漏
sealing lip	密封唇；带唇边的密封件
sealing liquid	密封液
sealing material	封闭用材料；密封用材料
sealing mechanism	密封机制
sealing of a gallery	巷道密封
sealing of cable	电缆灌胶密封
sealing of hole	封孔；钻孔封闭
sealing oil system	封油系统
sealing pad	密封盘
sealing plug	密封插头
sealing polymer	密封用聚合物
sealing process	封闭作用
sealing property	密封性能
sealing ring	密封圈；密封环；垫圈
sealing rubber strip	密封胶条
sealing run	封底焊道
sealing screen plate	密封螺钉
sealing signal	密封信号
sealing socket	密封座
sealing strip	封密条
sealing surface	密封面
sealing tape	密封带
sealing trap	密封圈闭
sealing up	止水；密闭
sealing washer	密封垫圈
sealing wax	封蜡；火漆
sealing weld	气密焊缝；封焊
sealing works	止水工程
sealing-grout	止水泥浆
sealing-in	焊接；熔接；密封
sealing-off	密封；折焊；脱焊
sealing-on	焊上
sealing-wax flow	封蜡流
sealing-wax structure	封蜡构造
seam	煤层，层间薄夹层，夹层，层，矿层，地层；叶理，节理；缝，裂缝；接合处，接合面

seam analysis	煤层分析；煤层断面分析
seam annealing	焊缝处退火
seam blast	缝隙装药爆破；缝隙爆破
seam blasting	缝隙装药爆破；缝隙爆破
seam contour	煤层等高线
seam course	煤层走向，矿层走向
seam cut	缝式换槽；龟裂换槽
seam density of coal reserve	绝对含煤率
seam distribution	煤层分布；矿层分布
seam dustiness	煤层含尘量；矿层生尘量
seam edge	矿层边缘
seam floor contour	煤层底板等高线
seam fold	矿层褶皱；煤层褶皱
seam gas content	煤层的瓦斯含量
seam height	层厚
seam inclination	煤层倾斜；矿层倾斜
seam interval	层间距离；煤层间的距离
seam outcrop	煤层露头
seam pitch	煤层倾斜；矿层倾斜
seam pressure	煤层压力，矿层压力
seam profile	煤层剖面
seam roof	煤层的顶板；顶盖岩
seam roof cleaning	清理煤层顶板
seam strike	煤层走向
seam structure	煤层构造
seam thickness	煤层厚度；层厚
seam top	煤层的顶板
seam worked to provide relief	采掘煤层卸压
seamarsh	滨海沼泽
seaming	用弯边法使两个板料连接；接缝
seamless casing	无缝套管
seamless steel pipe	无缝钢管
seamless track	无缝线路
seamless tube	无缝管
seamless tube rolling mill	无缝管轧机
seamount	海山；海峰
seamount chain	海山列；海山链
seamount group	海山群
seamount range	海底山脉
seamount subduction	俯冲海山
seam-out	无效爆破
seam-road blasting	煤巷爆破
seamy	有裂缝的
seamy rock	多裂隙岩，裂缝岩石；层状岩石，层状岩
seaport	海港；港口；海口
sea-port bulk station	海港油库
seaquake	海底地震；海震
search coil	探测线圈
search cycle	检索周期；搜索循环
search direction vector	搜索方向向量
search for coal	找煤；初步普查
search for minerals	找矿；探矿
search for oil	勘察石油
search interval	探索区间；搜寻区间
search key	检索关键字
search lamp	探照灯
search light	探照灯
search unit	探头；探测装置
searcher	搜索器
searching area	搜索区
searching unit	探头
Searle viscometer	塞尔黏度计
searlesite	硅硼钠石；水硅硼钠石
seashore saline soil	滨海盐土
season cracking	季节性开裂；季节性破裂；老化开裂
season labor	季节劳动；季节工
seasonal correction of mean sea level	平均海面季节校正；平均海面归算
seasonal creep	季节性蠕动
seasonal drainage	季节性流水
seasonal dynamics	季节性动态
seasonal effect	季节效应
seasonal fluctuation	季节性升降
seasonal frozen soil	季节冻土
seasonal grades	季节级别
seasonal growth	季节生长
seasonal lake	季节湖
seasonal load	季节性荷载
seasonal load line	季节载重线
seasonal migration	季节性迁移
seasonal periodicity	季节周期性
seasonal river	间隙河流；季节性河流
seasonal storage	季节性库容
seasonal stream	季节性河流；季节河
seasonal succession	季节性演替
seasonal variation factor	季节性变化系数
seasonal water	季节性水
seasonal wind	季风
seasonally freezing stratum	季节冻土层
seasonally frozen ground	季节冻土
seasonally frozen soil	季节性冻土，季节冻土
seasoning	风干；老化；时效
seastrand	海岸
sea-surface platform	海面工作平台
seat	基座；所在地；矿山底板；座；底座
seat angle	座角钢
seat beam	座梁
seat board	座板
seat clay	底黏土；不透水黏土；耐火黏土
seat connection of beam to column	梁柱座接法
seat earth	底板，煤层底板；底层；层底黏土；煤层底部黏土；耐火黏土；根土岩
seat frame	座架
seat of settlement	沉陷源地；沉陷影响范围
seat pad	座垫
seat plinth	座位底座
seat pocket	坐槽
seat post rod	座杆
seat rail	坐凳栏杆
seat reamer	阀座修整铰刀
seat retainer	座圈；阀座护圈
seat rim	座缘
seat ring	阀座圈
seat rock	基岩；底岩；根土岩
seat screen plate	定位螺钉，安装螺钉
seat shear pin	球座剪切销
seat stay	座撑
seat stone	根土岩
seat valve	座阀
seat wrench	阀座松紧扳手

seating	座，基座；支承；支座
seating arrangement	定位装置；复位装置
seating cup	密封皮碗
seating cup body	密封皮碗套
seating nipple	座节
seating position	坐落位置
seating shoe	底座
seawall	防波堤；海塘；海堤；防潮堤
seawall opening	海堤通孔
seawall trench	海堤壕沟
seaward	向海的
seaward slope	向海坡
seawater bronze	耐海水腐蚀青铜
seawater chemical corrosion	海水化学腐蚀
seawater corrosion	海水腐蚀
seawater encroachment	海水入侵；海水侵蚀
seawater hydraulic	海水液压
seawater intrusion	海水入侵
seawater pump	海水泵
sea-water sampler	海水取样器
sea-wave dynamics	海波动力学
seaway force	浪力
seaweed theory	海藻成油说
seaworn	海蚀的
sebastianite	黑云钙长岩
sebkha	盐沼
sebkha deposit	盐沼沉积
sebkha sediment	盐沼沉积
sebkha soil	盐沼土
secant	正割
secant bored pile	搭接钻孔桩
secant circle	正割圆
secant conic chart	割圆锥投影地图
secant conic projection	割圆锥投影；双标准纬线圆锥投影
secant curve	正割曲线
secant damping	割线阻尼
secant modulus	割线模量
secant moment	剪切模数；受剪模量
secant pile wall	搭接桩墙
secant projection	正割射影
secant shear modulus	割线剪切模量
secant stiffness	割线刚度
secant yield stress	正割屈服应力
secant Young's modulus	割线弹性模量，割线杨氏模数
seclusion	闭塞；隔离
second advance	台阶面次后推进
second arrival	伴随波；二次波；续至波
second class ore	次品矿石；需选矿石
second cycle basin	二次循环盆地
second event	续至波
second harmonic	二次谐波
second mining	第二次采，二次回采；二采；回采矿柱或煤柱
second moment of area	截面惯性矩
second phase	次相
second preliminary tremor	次波
second stage cofferdam	二期围堰
second stage concrete	二期混凝土
second strength theory	第二强度理论；最大正应变理论
second terrace	二级阶地
secondaries	次波；二次波
secondary access road	支路
secondary aftershock	第二期余震
secondary backfill	二次回填
secondary barrier	辅助岩粉棚；辅助阻隔物或墙
secondary beam	次梁
secondary blasting	二次爆破
secondary breakage	二次破碎；二次破岩
secondary breccia	次生角砾岩
secondary building unit	二级构造单元
secondary cementing	第二次注水泥
secondary clay	次生黏土；冲积黏土
secondary cleaning	再次精选
secondary cleat	次生解理；次生割理
secondary cleavage	次生劈理
secondary coal recovery	煤炭二次回收；报废时期回采
secondary coast	次生海岸
secondary consolidation	次固结；二次固结
secondary consolidation settlement	次固结沉降
secondary conveyor	支巷输送机
secondary cracking	二次裂化
secondary creep	第二阶段蠕变，二期蠕变，二级蠕变；稳态蠕变
secondary crystal	二次结晶；次生晶体
secondary crystalline fabric	次生结晶组构
secondary dam	副坝
secondary deflection	次挠曲
secondary deformation	二次变形；次要变形
secondary depletion	二次采油
secondary deposit	次生矿床
secondary dike	副堤
secondary disaster	次生灾害
secondary drilling	二次爆破钻眼
secondary earthquake	次震
secondary effect	次生效应；次生灾害
secondary energy	二次能源
secondary enrichment	次生富集；二次富集
secondary environment	次生环境
secondary equipment	辅助设备
secondary era	中生代
secondary exploitation	二次开采；二次采油
secondary explosion	二次爆炸
secondary fault	次生断层，小断层
secondary field	二次场
secondary fissure	次生裂缝
secondary fold	小褶曲，次生褶皱
secondary fossil	次生化石
secondary fracture	次生破裂；次生裂缝
secondary fragmentation	二次破裂，二次破碎
secondary free-gas cap	次生游离气顶
Secondary fumarole	次生喷气孔
secondary gas	精洗煤气
secondary gas cap	次生气顶
secondary gas recovery	二次采气
secondary gas-oil pool	次生油气藏
secondary grout hole	二次灌浆孔
secondary grouting	二次压注，二次灌浆
secondary hardness	二次硬度
secondary hazard	次生危险，次生灾害

secondary hematite 二次赤铁矿
secondary hole blasting 二次爆破
secondary hydrocyclone 再选水力旋流器
secondary hydrothermal deposit 次生热液矿床
secondary identification of liquefaction 液化的再判别
secondary inclusion 二次加热时形成的夹杂物；次生包裹体
secondary interstice 次生间隙；次生裂隙；次生孔隙
secondary ion 次级离子
secondary ion mass spectroscopy 次级离子质谱分析法
secondary joint 次生节理；小节理
secondary kaolin 次生高岭土
secondary kick 次生井涌
secondary laterite 次生砖红壤
secondary leached porosity 次生淋滤孔隙度
secondary leakage 二次泄漏
secondary levee 副堤
secondary limestone 次生石灰岩
secondary line 支线
secondary lining 二次衬砌；永久衬砌；二层支护
secondary loading 附加荷载
secondary loess 次生黄土
secondary magnetic field 次生磁场
secondary migration 二次运移
secondary mineral 次生矿物
secondary noise 次生噪声
secondary oil generation theory 二次生油说
secondary oil recovery 二次采油
secondary oil response rate 二次开采见效油产量
secondary oil shale 次生油页岩
secondary oil-gas pool 次生油气藏
secondary operating standard 二级操作标准
secondary option 次级选项，第二选项
secondary order fault 次级断层
secondary ore 次生矿
secondary origin 次生成因
secondary orogeny 次生造山运动
secondary pores 次生孔隙
secondary porosity 次生孔隙度；次生孔隙；次生孔隙率
secondary precipitate 二次沉淀
secondary pressure pulse 二次冲击压力
secondary primer 副起爆药包
secondary principal stress 次主应力，第二主应力
secondary production 二次开采
secondary pulse 二次冲击
secondary radiation 次级辐射；二次辐射
secondary radiation effect 二次辐射效应
secondary recovery 二次开采；二次回收率
secondary reduction 二次破碎
secondary reflection 次级反射；续至反射；二次反射波
secondary refraction survey 续至波折射法
secondary reserve 二次可采储量
secondary resistance 附加阻力
secondary resonance 次共振
secondary rock 次生岩
secondary rock pressure 二次岩石压力
secondary roof 第二顶板
secondary roof seal 浮顶二次密封
secondary saline soil 次生盐土
secondary schistosity 次生片理
secondary settlement 次沉降；次固结沉降
secondary shaft 暗井
secondary shear plane 次剪力面
secondary shooting 二次爆破
secondary silica cementation 次生二氧化硅胶结
secondary silica outgrowth 次生硅质增生
secondary sliding body 次生滑体
secondary slime 次生矿泥；次生煤泥
secondary slip plane 次生滑移面
secondary sludge 二次污泥
secondary soil 次生土壤
secondary state of stress 次应力状态
secondary stratification 次生层理
secondary stratigraphic trap 次生地层圈闭
secondary stress 次级应力；二次应力；次应力
secondary structure 次级构造；次生构造；派生构造；次生结构面
secondary subsidence 二次沉陷
secondary succession 次生演替
secondary support 二次支护
secondary surface rupture 次生地表破裂
secondary tectonic overlap 次生构造超覆
secondary tectonics 次生构造
secondary tectonite 次生构造岩
secondary volumetric sweep efficiency 二次采油体积波及系数
secondary waterflood reserve 注水二次采油储量
secondary wave S波；续至波；剪切波；次至波，次波，地震横波
secondary zone 次生矿物带
secondary-recovery dual operation 二次回采工作
second-order bench mark 二等水准点
second-order fault zone 二级断层带
second-order fold 次级褶皱
second-order thrust 二级冲断层
second-order trilateration 二等三边测量
second-phase fold 第二幕褶皱
second-row holes 二排炮眼组
seconds Saybolt Furol 赛氏重油黏度秒
secret dovetailing 暗楔接合
secret fountain 隐头喷泉
secretion 分泌；分泌物；空隙充填构造，空隙充填；隐藏
sectile 可切的；可剖开的
section 剖面，断面，截面，切面；部分，节，段，组，部门，部件，零件；断面图；薄片；区域，采区，地区，区间，区段；切开，分区
section airway 区段风巷；分区风巷；采区风巷
section analysis 截面分析
section area of reinforcement 钢筋横断面
section belt 采区输送机；采区胶带输送机
section boundary 采区边界
section conveyor 采区输送机
section deformation 截面变形
section diagram 横断面图
section gauge 剖面规；截面测量
section gauge logging 井径测井
section house 工区房屋
section line 剖面线；分区线；截线
section map 断面图；剖面图

section method 截面法
section mill 截面铣鞋
section mill cutter 套管磨铣工具
section model 剖面模型
section modulus 截面模量；截面模数，剖面模数
section moment 截面矩
section multiple 层间多次反射
section of a seam 矿层断面
section of earth-work 土方截面
section of heading 导坑断面
section of pass 孔型轮廓；孔型断面
section of reservoir 地层的单元；油层剖面
section of sufficient grade 紧坡地段
section of unsufficient grade 非紧坡地段
section plan 水平投影；平面图
section plane 剖面
section plastic modulus 塑性截面矩
section plotter 剖面仪
section point 电分段装置
section road 区段巷道
section steel 型钢；型材
section steel concrete structure 型钢混凝土结构
section survey 断面测量
section tailings sampler 尾矿分段取样机
section well 吸水井
section wire 异形钢丝
section wire method 断面索法
sectional area 截面积，横截面积；断面积
sectional blasting method 分段爆破法
sectional construction 预制构件拼装结构，预制构件结构
sectional core-barrel 多节岩芯管
sectional dimension 截面尺寸
sectional drawing 剖面图；截面图；断面图
sectional drill rod 可卸钻杆
sectional elevation 剖面图，剖视图；立剖面；截视立面图
sectional installation 分段安装
sectional line 剖切线
sectional material 型材
sectional ploughing method 分段刨煤法
sectional retaining wall 分段挡土墙
sectional ring rubbers 多段橡皮环
sectional sizing variables 截面尺寸变量
sectional steel drilling 接杆钻眼；接杆凿岩；接杆钻进
sectional tamping rod 接长炮棍
sectional track 可拆式轨道，移动式轨道
sectional tunnel 区间隧道
sectional turbodrill 多级涡轮钻具
sectional type 分段式
sectional view 剖视图；断面图；截视图
sectional view of bars 钢筋断面图
sectionally pressing water test 分段压水试验
sectionally smooth 分段光滑
sectional-type conveyor 分节式输送机
section-chord technique 截面弦法
sectioned drilling mast 分块式钻井井架
sectioning 电分段装置；电分段
sectionlization 分段
sector 扇形，扇形面；区段；部门；扇区；两脚规

sector boundary 扇形边界
sector coordinate 扇形坐标
sector coordinate system 扇形坐标系统
sector dam 扇形闸堰；扇形坝
sector display 扇形扫描
sector distortion 扇形失真
sector distribution 扇形分布
sector fault 扇形断层
sector gate 弧形闸门；扇开闸门
sector gear 扇形齿轮
sector method 扇区法
sector planning 行业规划
sector rack 扇形齿板
sector scanning 扇区扫描
sector shaft 扇形齿轮轴
sector structure 扇形结构
sector theory 扇形理论
sector velocity function 扇形速度函数
sector weir 弧形堰
sectorial geometric property 扇形几何性质
sectorial harmonics 扇球函数
sectorial moment of inertia 扇形惯性矩
sectorial normal stress 扇形正应力
sectorial pipe-coupling method 分段连接管道法
sectorial shear stress 扇形剪应力
sectorial statical moment of area 扇形静面矩
sectorial stress distribution 扇形应力分布
sectrometer 真空管滴定计
secular change 长期变化
secular compaction 缓慢压实作用
secular crustal deformation 地壳缓慢变形
secular decay 长期风化；长期腐烂；长期衰变
secular drift 缓慢漂移
secular effect 长期效应
secular equilibrium 长期平衡
secular gravity measurement 长期重力测量
secular gravity variation 长期性重力变化
secular instability 长期不稳定性
secular iron ore 镜铁矿
secular motion 长期缓慢运动
secular movement 长期缓慢运动；地壳缓慢升降运动
secular rise 长期上升
secular sinking 长期下沉
secular stability 长期稳定度；长期稳定性
secular structure 民用结构
secular subsidence 缓慢沉陷；长期缓慢下沉，长期缓慢沉降
secular trend 长期趋势
secular upheaval 缓慢隆起；长期缓慢上升
secular value 长年值；长期值
secular variation 长期变化；世纪变化；时效变化；长期缓慢变化
secure 使安全；保证；安全的，可靠的，牢固的；保护；紧固
securing band 安全带
securing dowel 紧固定位销
securing hook 紧固
securing nut 锁紧螺母；扣紧螺母；紧固螺帽
securing pin 固定销
securing plate 紧固板

securing ring	紧固环
securing rod	紧固连杆
securing screw	紧固螺钉
securing strap	紧固索带
security factor	安全系数
security fence	防栏，安全挡
security fencing	保安围栏
security gate	保安闸；防盗闸
security level	安全水平
security of data	数据的安全性
security risk	安全风险
security system	安全防范系统，安全系统
sedarenite	沉积砂屑岩
sedentary	残积的；残积
sedentary deposit	原地沉积
sedentary product	风化产物；残积物
sedentary sediment	原地沉积
sedentary soil	原地土壤，原地土；定积；原生土；残积土
sedge land	苔草丛
sedge moor	苔青沼泽
sediment	沉积层，冲积物，沉淀物，淤积物；沉积，沉淀；泥沙；油脚；残渣
sediment accumulation	沉积物堆积；泥沙淤积
sediment acoustic classification	沉积物声学分类
sediment analysis	沉积物分析；泥沙分析；沉淀分析；沉淀法粒度分析
sediment and species richness	沉积物和物种丰富度
sediment availability	沉积物的可用性
sediment bar	边缘沙洲
sediment barrier	防沙堤；拦沙堤
sediment box	沉淀箱
sediment budget	泥沙预算
sediment burial	泥沙掩埋
sediment by extraction	抽提法沉淀试验
sediment bypassing	泥沙绕过
sediment chemistry	沉积物化学
sediment column	沉积柱
sediment compaction	沉积物压实作用
sediment composition	沉积物组成
sediment concentration	含沙量
sediment consistency	沉积物的一致性
sediment content	沉积物含量；泥沙含量
sediment density	泥沙密度
sediment deposition	泥沙淤积；泥沙沉降
sediment discharge	输沙量；排放沉积物；含沙率；输沙率
sediment dispersal	泥沙扩散
sediment dynamics	沉积动力学；泥沙动力学
sediment ejection	排沙；放淤
sediment ejector	截沙设施；排沙设施；截沙槽
sediment entrainment	泥沙启动；挟带泥沙
sediment exchange	沉积物交换
sediment fluidization	泥沙流态化
sediment flushing efficiency	冲沙效率
sediment flux	泥沙通量
sediment fore-arc	沙弧前
sediment formation	沉积物的形成；沉积地层
sediment gate	防淤闸
sediment geochemistry	沉积物地球化学
sediment grain size	沉积物粒度
sediment gravity flow	沉积物重力流
sediment incrustation	沉积水垢；积垢
sediment instability	沉积物不稳定性
sediment load	输沙量；河流泥沙
sediment loading	泥沙负荷
sediment loss	沉积物流失；泥沙流失
sediment meter	沉淀计
sediment mobilization	沉积物移动
sediment movement	泥沙运动
sediment pathway	泥沙途径
sediment percentage	含泥量
sediment pile	沉积物堆积
sediment plume	悬浮物卷流
sediment process	沉积过程
sediment provenance	沉积物物源
sediment quality	沉积物质量
sediment record	沉积记录
sediment release	底泥释放
sediment sampler	沉积物采样器；底泥采样器
sediment sampling	沉积物取样
sediment slide	沉积物滑动
sediment source	沉积物源；沙源
sediment sphericity	泥沙球形度；沉积物颗粒圆度
sediment stability	沉积物稳定性
sediment stratification	泥沙层理
sediment supply	沉积物供给
sediment thickness	沉积厚度
sediment toxicity	沉积物毒性
sediment tracers	沉积物示踪
sediment transport	泥沙输移；沉积物运输，沉积物运移，沉积物搬移；输沙；泥沙输移情况；泥沙流移
sediment transport pathways	输沙途径
sediment trap	沉积区；沉积物收集器；沉积捕集器
sediment tube	取粉管
sediment tunnel	排砂洞
sediment yield	产沙
sedimental	沉积物的
sedimentary	沉积；沉积岩类；沉淀的
sedimentary basin	沉积盆地
sedimentary blanket	沉积盖层
sedimentary body	沉积体
sedimentary breccia	沉积角砾岩
sedimentary brine	沉积盐水；沉积卤水
sedimentary characteristic	沉积特征
sedimentary clast	沉积碎屑
sedimentary clay	沉积黏土
sedimentary column	沉积柱
sedimentary compaction	沉积压实作用
sedimentary compensation	沉积补偿
sedimentary complement	沉积杂层
sedimentary complex	沉积杂层
sedimentary context	沉积环境
sedimentary cover	沉积盖层
sedimentary cycle	沉积旋回
sedimentary density current	沉积密度流
sedimentary deposit	成层沉积；沉积矿床；沉积层
sedimentary derivation	沉积来源
sedimentary differentiation	沉积分异作用
sedimentary discontinuity	沉积不连面；沉积结构面

sedimentary dyke	沉积堤坝
sedimentary dynamics	沉积动力学
sedimentary ecology	沉积生态学
sedimentary environment	沉积环境
sedimentary fabric	沉积组构
sedimentary facies	沉积相
sedimentary fault	同生断层；生长断层
sedimentary feature	沉积特征
sedimentary flux	沉积通量；颗粒通量
sedimentary formation	沉积岩层；沉积建造；沉积地层
sedimentary framework	沉积地层构造
sedimentary gap	沉积缺失
sedimentary geochemical facies	沉积地球化学相
sedimentary geochemistry	沉积地球化学
sedimentary geology	沉积地质学
Sedimentary Geology	沉积地质学报
sedimentary gneiss	沉积片麻岩
sedimentary history	沉积史
sedimentary infilling	沉积充注
sedimentary interstice	沉积岩孔隙；沉积生成孔隙
sedimentary lag	沉积滞后
sedimentary layer	沉积层
sedimentary loading	沉积载荷，沉积负荷
sedimentary mantle	沉积盖层，沉积覆盖层
sedimentary microstructure	沉积微结构
sedimentary mineralogy	沉积矿物学
sedimentary model	沉积模式
sedimentary nappe	沉积推覆体
sedimentary organic matter	沉积有机质
sedimentary organism	沉积生物
sedimentary origin	沉积成因
sedimentary outer arc	沉积外弧
sedimentary overlap	沉积超覆
sedimentary palaeoenvironment	沉积古环境
sedimentary pathway	沉积途径
sedimentary patterns	沉积模式
sedimentary peat	沉积泥炭
sedimentary petrography	沉积岩相学；沉积岩石学，沉积岩学，沉积岩类学
sedimentary petrology	沉积岩石学
sedimentary pile	沉积堆
sedimentary porosity	沉积孔隙度
sedimentary process	沉积过程
sedimentary province	沉积区
sedimentary rate	沉积速率
sedimentary record	沉积记录
sedimentary region	沉积区
sedimentary relict	沉积体残余
sedimentary ridge	沉积脊
sedimentary ripple	沉积波纹
sedimentary rock	沉积岩；水成岩
sedimentary section	沉积剖面
sedimentary sequence	沉积顺序；沉积层序；沉积序列
sedimentary series	沉积层系
sedimentary soft rock	沉积软岩
sedimentary soil	沉积土
sedimentary source	沉积源
sedimentary strata	沉积岩层
sedimentary stratum	沉积岩层
sedimentary structure	沉积构造；沉积结构
sedimentary succession	沉积层序
sedimentary supply	沉积物补给
sedimentary system	沉积体系；沉积系
sedimentary tectonics	沉积构造；沉积大地构造
sedimentary terrain	沉积地形
sedimentary texture	沉积结构
sedimentary topographic relief	沉积地形
sedimentary trap	沉积圈闭
sedimentary veneer	沉积盖层
sedimentary volcanism	沉积火山作用
sedimentary water	沉积水；埋藏水；残存水
sedimentary wedge	沉积楔
sedimentary-exhalative system	喷流沉积系统
sedimentary-metamorphic deposit	沉积变质矿床
sedimentation	沉积作用；沉淀，沉淀作用；沉降；淤积；沉积法
sedimentation analysis	沉积分析；沉降分析；沉积分析法
sedimentation balance	沉积分析天平；沉积平衡
sedimentation basin	沉积盆地；沉淀池；沉降池
sedimentation centrifuge	沉降式离心机
sedimentation coefficient	沉积系数
sedimentation concentration	沉积物浓度
sedimentation constant	沉积常数
sedimentation curve	沉降曲线；沉积曲线
sedimentation cylinder	沉降筒
sedimentation diameter	沉积直径
sedimentation disturbance	沉积干扰
sedimentation engineering	泥沙工程
sedimentation equilibrium	沉积平衡
sedimentation gradient	沉积梯度
sedimentation in reservoir	水库淤积
sedimentation load	沉积载荷
sedimentation method	沉降法
sedimentation model	沉积模式
sedimentation of particulates	粒子的沉降
sedimentation patterns	沉积模式
sedimentation potential	沉降电势差；沉积电势；沉积电位
sedimentation potential method	沉积电位测定法
sedimentation rate	沉积速率；沉降率，沉积率；沉淀速度，沉降速度
sedimentation rate test	沉积速率试验
sedimentation sample	沉淀式采样器
sedimentation sizing	沉积分级；沉积分粒
sedimentation tank	沉降罐
sedimentation test	沉降分析
sedimentation type	沉积类型
sedimentation unit	沉积单元
sedimentation velocity	沉积速度
sedimentation volume	沉积体积
sedimentation-equivalent particle	沉积等效颗粒
sedimentation-filled	沉积物充填的
sedimentationist	沉积岩学者
sedimentation-size analysis	沉积量值分析
sedimentator	沉淀器
sediment-carrying	挟沙的
sediment-induced deformation	沉积物诱发变形
sediment-laden pipeline	充满沉积物的管道
sedimentogeneous rock	沉积岩

sedimentogenesis 沉积成因；沉积物形成作用
sedimentogenic 沉积成因的
sedimentography 沉积岩相学
sedimentological 沉积学的
sedimentological characteristic 沉积特征
sedimentologist 沉积学者
sedimentology 沉积学；泥沙学
sedimentometer 沉积测定仪；沉积计
sedimentometry 沉积测量法
sedimentophile 亲沉积的
sedimentous 沉积的
sediment-size frequency distribution 沉积物粒度频率分布
Seebeck effect 塞贝克热电效应；塞贝克效应
seed crystal 晶种；籽晶
seeded slope 植草护坡
seeded strip 绿化分隔带
seedholder 籽晶夹持器
seeding 播种；引晶技术
seeding lawn 播种草坪
seeding polymerization 接种聚合
seeding within framework 骨架内植草
seep 渗透，渗漏，渗出；小泉
seep carbonate 冷泉碳酸盐岩
seep hole 渗水孔；冒水孔
seep oil 渗至地面的油
seep proof screen 止水墙
seep water 渗出水
seep zone 渗流区
seepage 渗流，渗出，渗水，渗漏；渗透；渗出物，渗出量；油苗
seepage along paleocourse 古河道渗漏
seepage analysis 渗漏分析
seepage area 渗漏区
seepage area of well 井的渗流面积
seepage around abutment 绕坝渗漏
seepage boil 渗流管涌
seepage calculation 渗透计算
seepage coefficient 渗透系数
seepage collar 减渗环
seepage continuity equation 渗流连续方程
seepage control 渗流控制
seepage deformation 渗透变形
seepage discharge 渗流量；渗透流量
seepage equation of motion 渗流运动方程
seepage erosion 渗流侵蚀
seepage exit 渗流出口；渗流逸出点
seepage face 浸润线；渗流面
seepage failure 渗透破坏
seepage field 渗流场
seepage flow 渗流；渗透水流；渗流量
seepage flow mechanics 渗流力学
seepage flow net 渗流网
seepage force 渗透力；渗流压力
seepage gauge 渗流计
seepage gradient 渗透梯度；渗透坡降
seepage gutter 渗水沟
seepage index 渗流指数
seepage interception 渗流截断
seepage isolation dike 截渗堤
seepage law 渗透定律
seepage line 浸润线；渗流线；渗透线；渗漏线
seepage liquefaction 渗漏液化
seepage loss 渗漏损失；渗出损失；渗透损失
seepage measurement 渗流测定
seepage meter 渗流计
seepage monitoring 渗流监测
seepage of storm flow 暴雨渗流
seepage oil 油苗
seepage path 渗透路径，渗透途径，渗流路径，渗漏路径
seepage pit 渗水坑
seepage point 渗水点；渗水口
seepage pressure 渗透压力
seepage prevention 防渗
seepage profile 渗透剖面图
seepage proof curtain 防渗帷幕
seepage quantity 渗流量
seepage rate 渗透率
seepage reflux 渗透回流
seepage resistance 渗流阻力
seepage speed 渗透速度
seepage spring 渗出泉
seepage state 渗流形态
seepage stress 渗透应力
seepage structure 渗水建筑物
seepage surface 渗流面；渗出面
seepage tests of field permeability 现场渗水试验
seepage tracking 渗流路径
seepage up-dip 沿倾斜向上的渗出
seepage uplift 渗透扬压力
seepage velocity 渗流速度，渗透速度
seepage water 渗流水；渗透水
seepage well 渗水井；渗井
seepage zone 渗出带
seepage-flow model 渗流模型
seepage-off 渗出
seep-in grouting 渗入性灌浆
seeping 渗入作用
seeping dam 滤水坝
seeping discharge 渗流量
seepology 油苗学
seepy 透水的
seepy material 透水物料，透水材料，渗透物料
seesaw motion 往复运动
Seger cone 西格测温锥；标准锥；塞氏测热熔锥
seggar 煤层底部黏土；耐火黏土
seggar clay 耐火黏土；火泥
segment joint 管片接头
segment model 分节模型
segment of a circle 弓形
segment of curve 曲线段
segment of leveling 水准测段
segment reef 节状礁
segmental arch culvert 弓形拱涵
segmental averaging 分段平均
segmental baffle 缺圆折流板
segmental girder 弓形梁
segmental lining 大弧板支护；预制块支护
segmental method 分段法
segmental orifice plate 缺圆孔板

segmental steel lining	钢瓦拼合件井壁	seismic arrival-time	地震波到时
segmental steel plate	弓形钢板	seismic attenuation	地震波衰减
segmental truss	弓形桁架	seismic attribute	地震属性
segmentation	分段；分割	seismic band	地震频带
segmented bond tool	分区水泥胶结测井仪	seismic basic intensity	地震基本烈度
segmented copolymer	嵌段共聚物	seismic basic standard	地震基础标准
segmented spring ring	扇形弹簧圈	seismic bearing capacity of structure	结构抗震承载能力
segment-tooth roll	节段齿辊	seismic bedrock motion	基岩地震动
segregability	离析性	seismic behavior factor	抗震性态系数
segregate	分结；分凝	seismic belt	地震带，地震条带
segregated ash content	游离灰分含量，游离灰量	seismic blasting	地震勘探爆破
segregated completion	分层完井	seismic boat	地震船
segregated oil reservoir	重力分离油藏	seismic body wave	地震体波
segregated vein	分凝脉	seismic borehole tomography	井中地震层析成像
segregation	分离，分隔，分凝；离析，离析现象；离散；偏析	seismic boundary reflection	地震的界面反射
		seismic bracing	抗震支撑
segregation band	偏析带	seismic bright spot technology	地震亮点技术
segregation banding	分结条带；分凝条带	seismic cable	地震电缆
segregation boundary layer	偏析境界层	seismic camera	地震照相记录仪
segregation bunker	隔离仓	seismic capacity	抗震能力
segregation drive	重力驱动	seismic catalogue	地震目录
segregation drive index	重力驱指数	seismic category of engineering structure	工程结构抗震类别
segregation foliation	分凝叶理	seismic center	震源中心，震中
segregation of concrete	混凝土离析	seismic channel	地震波信道；地震道
segregation of mortar	砂浆分层度	seismic character	地震特性
segregation of waste	废液分流	seismic charge	震源药包
seif	纵向沙丘	seismic cluster	地震群
seis	测震仪；地震检波器，检波器；地震仪	seismic coda	地震尾波
seis spacing	检波器间距	seismic code	抗震规范
seiscrop	地震波等时图；三维地震水平切片图	seismic coefficient	地震系数
seisline	检波线	seismic compilation	地震剖面编绘器
seislith section	地震岩性剖面	seismic compiler	地震编译器
seislog	地震测井	seismic condition	地震条件；地震状况
seislog section	合成地震测井剖面	seismic constant	地震常数
seislog trace	合成地震测井道	seismic constraint	地震约束
seisloop	地震环线	seismic country	地震区
seism	地震	seismic coupling	地震耦合
seismal	地震的	seismic creep	地震蠕变
seismic	地震	seismic crosshole method	井间地震方法
seismic absorption band	地震吸收带	seismic cross-section	地震剖面
seismic acceleration	地震加速度；震动加速度	seismic cycle	地震轮回，地震周期
seismic acceleration spectral density	地震加速度谱密度	seismic damage anomaly	震害异常
seismic acoustic impedance log	地震波阻抗测井曲线	seismic damage assessment	震害评估
seismic acquisition configuration	地震采集系统	seismic damage control	震害控制
seismic action	地震作用	seismic damage from blasting	大爆破地震破坏
seismic activity	地震活动；地震活动性	seismic damage index	震害指数
seismic aftershock	余震	seismic damage prediction	震害预测
seismic amplifier	地震放大器	seismic data	地震资料；地震数据
seismic amplitude	地震振幅	seismic datum	地震基准面
seismic analog amplifier	地震模拟放大器	seismic decoupling	地震屏蔽；地震解耦
seismic analysis	震动分析；地震分析	seismic deformation	地震变形
seismic and acoustic method	地震声测法	seismic design	抗震设计
seismic and volcanic hazard	地震和火山灾害	seismic design coefficient	抗震设计系数
seismic anisotropy	地震各向异性；地震各相异性	seismic design decision analysis	抗震设计决策分析
seismic anomaly	地震异常	seismic design intensity	地震设计烈度
seismic apparatus	地震仪器；地震仪	seismic design regionalization map	地震区划图
seismic area	地震带；震区，地震区	seismic destructiveness	地震破坏性；地震破坏
seismic arrangement	地震排列	seismic detection	地震探测
seismic array	地震台阵；地震阵列	seismic detection network	地震监测网
seismic arrival	地震波至		

seismic detection of nuclear explosion	核爆炸的地震检测
seismic detector	地震仪；地震检波器
seismic digital amplifier	地震数字放大器
seismic digital processing system	地震数字处理系统
seismic direction	地震方向
seismic directivity	地震方向性
seismic disaster planning	抗震防灾规划
seismic discontinuity	地震不连续面；地震界面
seismic discontinuous surface	地震不连续面
seismic discrimination	地震鉴别
seismic dislocation	地震位错；断层位错
seismic disturbance	地震扰动
seismic disturbance pulse	地震扰动脉冲
seismic disturbance range of amplitude	地震扰动的振幅范围
seismic disturbance range of frequency	地震扰动的频率范围
seismic downlap	地震下超；地震下覆
seismic drill	地震勘探钻机
seismic drilling rig	地震钻机
seismic drop hammer	地震落锤
seismic earth pressure	地震动土压力
seismic effect	地震效应
seismic effect field	地震效应场
seismic effect of explosion	爆炸的地震效应
seismic efficiency	地震效率
seismic energy	地震能量
seismic energy absorption	地震能量吸收
seismic energy dissipation	地震能量耗散
seismic energy flux	地震能量通量
seismic energy release	地震能量释放
seismic energy scattering	地震能量散射
seismic energy source	地震能源
seismic energy transmission	地震能量传递
seismic energy wavefront	地震能量波前
seismic engineering	地震工程
seismic environment	地震环境
seismic equipment	地震检测设备
seismic event	地震事件
seismic event counter	地震事件计数器
seismic event identification	地震事件识别
seismic excitation	地震激发，地震作用
seismic experiment	地震试验
seismic exploration	地震勘探；地震探矿
seismic explosion	地震勘探爆破
seismic exposure	地震威胁
seismic facies	地震相
seismic facies analysis	地震相分析；震测相分析
seismic facies map	地震相图；震测相图
seismic facies unit	地震相单元；震测相单元
seismic failure	地震破坏
seismic fatigue intensity	地震疲劳强度
seismic fault	地震断层；发震断层
seismic fault plane solution	地震断层面机制解
seismic field data	地震野外数据
seismic filtering	地震滤波
seismic foci location	震源位置
seismic focus	震源
seismic force	地震力
seismic force resisting system	抗地震力体系
seismic fortification measure	抗震设防措施
seismic forward model	地震正演模型
seismic fracture zone	地震断裂带
seismic frequency	地震频度
seismic gap	地震空区，地震空白区
seismic gelatin	地震胶质炸药
seismic general standard	地震通用标准
seismic geography	地震地理学
seismic geologic condition	地震地质条件
seismic geologic engineering	地震地质工程
seismic geology	地震地质学
seismic geophone	地震检波器
seismic geophysical method	地震地球物理方法，地震物探法
seismic geophysical survey	地震物探法
seismic ghost reflection	地震虚反射
seismic grid	地震测网
seismic ground crack	地震裂缝
seismic ground motion	地震动
seismic ground motion parameter	地震动参数
seismic ground motion parameter zonation map	地震动参数区划图
seismic hardening	抗震加固
seismic hazard	地震灾害；震灾；地震危险性
seismic head wave	地震首波
seismic history	地震史
seismic holographic exploration	全息地震勘探
seismic holography	地震全息；全息地震
seismic horizon	地震层位
seismic impact	地震冲击
seismic impedance	地震阻抗
seismic impedance log	地震阻抗测井
seismic impulse holography	地震脉冲全息
seismic inactivity	地震不活动性
seismic index	地震波指数
seismic influence coefficient	地震影响系数
seismic information	地震情报；地震信息
seismic instrument	地震仪，地震仪器
seismic intensity	地震烈度；地震强度
seismic intensity expectancy map	预期地震烈度图
seismic intensity field	地震烈度场
seismic intensity increment	地震烈度增量
seismic intensity microregionalization	地震烈度小区域划分
seismic intensity regionalization	地震烈度区划
seismic intensity scale	地震烈度表
seismic intensity zonation map	地震烈度区划图
seismic intensity zoning	地震烈度区划
seismic interface	地震界面
seismic interference noise	地震干涉噪声
seismic interpretation	地震解释；地震说明
seismic interpretation aided by computer	计算机辅助地震解释；彩色地震显示系统
seismic inversion	地震反演
seismic investigation	地震研究；地震调查
seismic isolation	隔震
seismic isolation design	隔震设计
seismic isolation retrofitting	隔震加固
seismic joint	抗震缝
seismic landslide and collapse	地震震滑

seismic lateral resolution	地震横向分辨率
seismic level	地震能级
seismic line	地震测线
seismic liquefaction	地震液化
seismic lithologic modeling	地震岩性模拟
seismic lithology	地震岩性学
seismic load	地震荷载；地震载重
seismic load analysis	地震荷载分析
seismic location	地震定位
seismic log	地震测井
seismic longitudinal wave	地震纵波
seismic macro-survey	地震宏观调查
seismic magnitude	地震震级
seismic map	地震图；地震等值线图
seismic marker	地震标准层
seismic marker bed	地震标准层；震测标准层
seismic marker horizon	地震标准层；震测标准层
seismic mass	地震量
seismic measurement	地震测量
seismic method	地震勘探法；地震方法
seismic metrology	地震计量
seismic microzonation	地震小区划
seismic microzoning	地震小区划
seismic model	地震模型
seismic module	地震模块
seismic moment	地震矩
seismic moment tensor	地震力矩张量；地震矩张量
seismic monitoring system	地震监测系统
seismic motion	地震运动
seismic motion compensation	地震运动补偿
seismic network	地震台网；地震网络
seismic noise	地震噪声
seismic numerical model	地震数值模型
seismic observation	地震观测
seismic onlap	地震上超
seismic operating system	地震操作系统
seismic origin	震源；地震成因
seismic oscillation	地震振动；地震动
seismic parameter	地震参数
seismic pattern	地震模式；地震型式
seismic pattern recognition	地震图形识别；地震模式识别
seismic peak acceleration	地震动峰值加速度
seismic peak ground acceleration	地震动峰值加速度
seismic peak value	地震动峰值
seismic pendulum	地震摆
seismic penetration	地震穿透力
seismic performance	地震性能
seismic performance category	抗震性能类别
seismic period	地震周期
seismic phase	震相
seismic phase analysis	震相分析
seismic phenomena	地震现象
seismic phenomenon	地震现象
seismic physical modeling	地震物理模拟
seismic pick-up	地震检波器；地震仪；地震车
seismic picture	地震图
seismic play-back apparatus	地震记录回放装置；地震回放仪
seismic predominant period	地震卓越周期
seismic preprocessing system	地震预处理系统
seismic probability estimation	地震概率估计
seismic processing technique	地震处理技术
seismic product standard	地震产品标准
seismic profile	地震剖面
seismic profiler	地震剖面测量仪；地震剖面
seismic profiling	地震波剖面测定法
seismic program	地震勘探程序
seismic propagation path	地震波传播路径
seismic property	抗震性能
seismic prospecting	地震勘探，地震波勘探，地震探测；地震探矿
seismic protection	防震
seismic pulse	地震脉冲
seismic qualification	地震鉴定
seismic quiescence	地震平静期
seismic ray direction anomaly	地震射线方向异常
seismic ray paths	地震射线路径
seismic ray theory	地震射线理论
seismic rays	地震波射线
seismic recharge	震时互易原理
seismic reciprocity	地震互换性
seismic reconnaissance	地震普查
seismic record	地震记录
seismic recording instrument	地震仪
seismic recording streamer	地震记录等浮电缆
seismic recording system	地震记录系统
seismic recurrence interval	地震复发间隔
seismic reduction factor	地震折减系数
seismic reference datum	地震参考基准面
seismic reflection amplifier	地震反射放大器
seismic reflection crew	反射地震队
seismic reflection measurement	地震反射测量
seismic reflection method	地震波反射法，地震反射波法；地震反射法
seismic reflection profile	地震反射剖面
seismic reflection tomography	地震反射层析
seismic refraction	地震波折射；地震折射
seismic refraction tomography	地震折射层析成像
seismic regime	震源机制
seismic region	地震区；地震范围
seismic regionalization	地震区域划分
seismic rehabilitation	震后修复，震后复原
seismic report	震测报告
seismic requirement	抗震要求
seismic research observatory	地震研究观测台
seismic reservoir study	储层地震勘探研究
seismic resistance capacity	抗震能力
seismic resolution	地震分辨率
seismic response	地震反应；地震响应
seismic response analysis	地震反应分析
seismic response coefficient	地震反应系数
seismic response control	地震反应控制
seismic response spectra	地震反应谱
seismic restraint of equipment	设备抗震
seismic rigidity of ground	土体抗震刚度
seismic risk	地震风险；地震危险性
seismic rupture	地震破裂
seismic safety	地震安全
seismic safety evaluation	地震安全性评价

seismic safety factor of rock slope 岩质边坡的地震安全系数
seismic safety net 地震安全网
seismic safety of building 建筑的抗震安全性
seismic scale 地震烈度表
seismic scaling law 地震定比定律
seismic scattering 地震波散射
seismic scattering tomography 地震散射层析成像
seismic sea wave 海啸，地震海啸；地震海浪
seismic sea wave warning 海啸警报
seismic seawave 海震；地震海啸
seismic secondary disaster 地震次生灾害
seismic section 地震剖面
seismic sedimentology 地震沉积学
seismic sensor 地震传感器
seismic sensor cluster 地震传感器组
seismic sequence 地震序列
seismic sequence analysis 地震层序分析
seismic settlement 震陷
seismic shear 地震剪切力；地震剪力
seismic ship 地震船
seismic shock wave 地震冲击波
seismic signal 地震信号
seismic signal coherence 地震信号相干
seismic similarity attribute 地震相似属性
seismic site intensity 地震场地烈度
seismic site survey 地震现场调查
seismic sliding 地震滑移
seismic slowness 地震慢度
seismic smear 地震模糊
seismic sounding 地震测深
seismic source dynamics 震源动力学
seismic source function 地震震源函数，震源函数
seismic source geometry 震源几何学
seismic source kinematics 震源运动学
seismic source location 震源定位
seismic source mechanism 震源机制
seismic source model 震源模型
seismic source parameter 震源参数
seismic sources 震源
seismic spectrum 地震波谱
seismic spread 地震勘探排列
seismic stability 抗震稳定性
seismic standard system 地震标准体系
seismic standardization 地震标准化
seismic station 地震台；地震站
seismic status 地震状态
seismic stimulus 地震激发
seismic strain energy 地震应变能
seismic stratigraphic analysis 地震地层学分析
seismic stratigraphic interpretation 地震地层解释
seismic stratigraphic model 地震地层模型
seismic stratigraphy 地震地层学
seismic streamer 地震勘探等浮电缆
seismic streamer cable 地震勘探等浮电缆
seismic streamer tracking system 地震等浮电缆跟踪系统
seismic strength 地震强度
seismic strengthening 抗震加固
seismic stress drop 地震应力降
seismic stripping 地震剥离法

seismic structural map 地震构造图
seismic structural wall 抗震墙
seismic structure control 结构抗震控制
seismic structure zone 地震构造区
seismic subsidence 震陷
seismic surface wave 地震面波
seismic surge 海啸
seismic surveillance 地震监测
seismic survey 地震调查；地震勘探，震波勘测；地震探测；地震查勘；地震测量
seismic surveying 地震测量
seismic synthetic profile 合成地震剖面
seismic synthetic record 合成地震记录
seismic system 地震系统
seismic system model 地震系统模型
seismic technique 地震方法；地震技术
seismic tectonics 地震构造学
seismic telemetry 地震遥测
seismic telemetry buoy 地震遥测浮标
seismic test 地震波试验；地震试验
seismic threat 地震威胁
seismic thumper 液压震动发生器
seismic tomography 地震层析成像，地震层析成像法；层析地震成像
seismic trace 地震记录线；地震道
seismic transient pulse 地震瞬时脉冲
seismic transmission prospecting 地震透射勘探
seismic transversal wave 地震横波
seismic travel time 地震传播时间；地震波走时；地震波旅行时间
seismic tremor 地震震动
seismic trigger 地震触发器
seismic upgrading 抗震加固
seismic velocity 地震波速度，地震波速
seismic velocity ratio 地震波速比
seismic velocity structure 地震波速度结构
seismic vertical 震中；震源至震中的连线
seismic vertical resolution 地震垂向分辨率
seismic water wave 地震水波
seismic wave 地震波；海啸
seismic wave attenuation 地震波的衰减
seismic wave energy 地震波能量
seismic wave field 地震波场
seismic wave generator 地震波发生器
seismic wave path 地震波路径
seismic wave propagation 地震波传播
seismic wave reflection 反射地震波
seismic wave theory 地震波理论
seismic wave travel curve 地震波走时曲线
seismic wave velocity 地震波速度，地震波速
seismic waveform 地震波形
seismic waveguide 地震波导
seismic wavelet 地震子波
seismic waveshape 地震波形
seismic weathering 地震风化层
seismic while drilling 以井底钻头声能为震源的地震方法；随钻地震勘探
seismic wipeout 地震剖面上无反射的垂直区域
seismic world map 世界地震图
seismic zonation 地震区划

seismic zone due to blasting 爆破震动区
seismic zone factor 地震区系数
seismic zoning 地震区域划分，地震区划
seismic zoning map 地震区划图
seismic-acoustic device 地震测听器
seismically active area 地震活动区
seismically active belt 地震活动带
seismically active period 地震活跃期
seismically active structure 地震活动构造
seismically induced ground failure 地震引起的地面失效
seismically quiet period 地震平静期
seismically quiet site 地震平静场地
seismic-electric effect 震电效应
seismicity 地震活动，地震活动性；震级；地震
seismicity anomaly 地震活动性异常
seismicity coefficient 地震活动性系数
seismicity degree 震度
seismicity distribution 地震活动分布
seismicity gap 地震活动空区；地震空区；地震活动间歇
seismicity induced by impounding of water 蓄水诱发地震活动性
seismicity map 地震活动分布图；地震活动图
seismicity parameter 地震活动性参数
seismicity pattern 地震活动性图像
seismicity peak 地震活动高峰
seismicity rate 地震活动率
seismicity triggering 地震活动触发
seismic-like event 似地震事件
seismic-method of prospecting 地震勘探法
seismics 地震学；地震探测法；地震勘探法
seismic-tectonic zone 地震构造带
seismic-velocity discontinuity 地震速度不连续性
seismic-wave attenuation 地震波衰减
seismic-wave dispersion 地震波频散
seismism 地震现象
seismo-acoustic reflection survey 震声反射测量
seismoacoustics 地震声学
seismo-active fault 地震活动断层
seismochronograph 地震计时器
seismo-electric effect 震电效应
seismogenesis 地震成因
seismogenic fault 发震断层；发震断裂
seismogenic faulting 地震断层作用
seismogenic province 地震构造区；孕震区
seismogenic structure 发震构造
seismogenic zone 孕震区
seismogeological map 地震地质图
seismogeology 地震地质学
seismogragh 地震仪
seismogram 地震图；震波图，地震波图；地震波曲线；地震波记录
seismogram record 地震波记录；地震波曲线
seismogram section 震波图剖面
seismogram synthesis 地震记录合成
seismogram trace 地震图记录线
seismograph 地震仪；测震仪；地震检波器
seismograph explosive 地震探矿用炸药
seismograph instrument 地震仪器
seismograph monitoring network 地震仪监测台网

seismograph network 地震仪台网
seismograph parameter 地震仪参数
seismograph station 地震台站；测震台
seismograph tape recorder 地震磁带记录设备
seismograph trace 地震记录线
seismographer 地震学者
seismographic disturbance 地震干扰
seismographic prospecting 地震勘探
seismographic record 地震图；地震记录；震波图
seismographic station 地震台站
seismographs 地震仪
seismography 测震学；地震学；地震记录；地震仪器学
seismo-impedance 地震阻抗
seismologic 地震学的
seismologic atlas 地震图集
seismologic laboratory 地震实验室
seismologic observation 地震观测
seismologic pulse 地震脉冲
seismologic record 地震记录
seismologic report 地震报告
seismologic surveillance 地震监测
seismologic zone 地震区域
seismological 地震学的
seismological array 地震台阵
seismological background 地震背景
seismological center 地震中心
seismological consideration 地震会商
seismological data 地震数据
seismological engineering 地震工程
seismological evasion 地震疏散
seismological field survey 地震现场调查
seismological instrument 地震仪器
seismological interpretation 地震解释
seismological network 地震台网
seismological notes 地震简报
seismological observation 地震观测
seismological observation data 地震观测数据
seismological observatory 地震观测台
seismological parameter 地震参数
seismological prediction 地震预报
seismological report 地震报告
seismological station 地震台站
seismological surveillance 地震监测
seismological table 走时表；地震表
seismological zoning 地震区划
seismologist 地震学家；地震工作者
seismology 地震学；地震
seismology model 地震模型
seismo-magnetic data 地震磁力记录
seismo-magnetic effect 地震磁力效应
seismometer 地震计；地震仪；测震仪；地震检波器
seismometer amplifier system 地震仪放大系统
seismometer array 地震检波器组合
seismometer cable 地震检波器电缆
seismometer cable assembly 大线组
seismometer coil 地震检波器线圈
seismometer coupling 地震检波器耦合
seismometer free period 地震检波器自由振荡周期
seismometer galvanometer 地震检波器检流计
seismometer group recorder 地震检波器组记录系统

seismometer pattern layout	检波器组合方式
seismometer resonant frequency	地震检波器谐振频率
seismometer testing apparatus	地震检波器测试仪器
seismometric observation	地震观测
seismometrical station	地震台站
seismometry	测震学；用地震仪记录或研究地震现象；地震波探矿
seismo-phantom map	地震假想层构造图
seismophone	地震检波器
seismophysics	地震物理学
seismopickup	地震车；地震仪
seismoreceiver	地震检波器
seismos	地震；震动
seismoscope	验震器；地震示波仪；地震计
seismoscope data	地震计数据
seismoscope record	地震计记录
seismosociology	地震社会学
seismostation	地震台
seismostatistical data	地震统计资料
seismostatistical method	地震统计方法
seismotectonic	地震构造的
seismotectonic effect	地震大地构造效应
seismotectonic line	地震构造线
seismotectonic map	地震构造图，地震大地构造图；震测构造图
seismotectonic province	地震构造区；发震构造区
seismotectonic regime	地震构造状态；发震构造机制
seismo-tectonic regime	地震构造状态
seismotectonic zoning	地震构造分区
seismotectonics	地震构造学，地震大地构造学；地震构造
seismo-tectono-magmatic belt	地震构造岩浆带
seismotelegram	地震电报
seisms	地震
seisphone	检波器
seisquare	方网地震采集系统
seistrach	地震曳引法
seisviewer	井下声波电视；地震观测仪
selagite	黑云粗面岩
selbergite	白霓霞岩
selected filling	精选的填料
selected granular material	特选砂石料
selected rockfill	精选堆石
selected zone	选择层
selecting sampling	重要抽样
selection criteria	选择标准
selection of project site	厂地选择
selection of route	道路选线
selection of underground mining methods	地下采矿方法选择
selection operation	选择操作；选择算子
selection or urban land	城市用地选择
selection transportation	选择搬运
selective	选择的，选择性的
selective abrasion	选择磨蚀
selective absorption	选择性吸收；选择吸收作用
selective action	选择作用
selective adsorption	选择吸附作用；选择性吸附
selective agglomeration	选择性结团
selective blockage	选择性封堵
selective breaking	选择性破碎
selective breaking-out	选择崩落
selective chopper radiometer	选择斩波辐射计
selective completion tool	选择性完井工具
selective crushing	选择性破碎
selective crystallization	选择性结晶；优先结晶
selective dual completion	选择性双层完井
selective erosion	选择性侵蚀
selective excavation	选择性挖掘，选择性开挖
selective extraction	选择性抽提；选择提取
selective firing	选择性射孔；依次射孔
selective flooding	选择性注水
selective formation tester	选择性地层测试器
selective fracturing	选择性压裂
selective gamma log	选择性伽马测井
selective gas shut-off	选择性封堵气层
selective grinding	优先磨磨；选择性研磨；选择性磨矿
selective injection	选择性注入
selective interference	窄带干扰
selective leaching	选择沥滤；选择浸滤
selective loading	选择装载；按类装载
selective metamorphism	选择变质
selective mining	选择性开采，选采
selective perforating	选择性射孔
selective permeability	选择透性
selective permeation	选择性渗透
selective plugging	选择性封堵
selective position tool	选择性定位工具
selective precipitation	选择性沉淀；优先沉淀
selective production scheduling	选择性开采程序
selective protection	选择性保护
selective radiation	选择辐射
selective reaction	选择反应
selective reagent	选择性试剂；选择试剂
selective reduced quadrature	选择性缩减正交
selective reduction	选择性还原
selective reflection	选择反射
selective replacement	选择交代；选择置换；分别交代
selective repressuring	选择性恢复压力
selective retrieval	选择检索
selective road header	部分断面掘进机
selective sampling	选择抽样；重要抽样
selective sand control	选择性防砂
selective scattering	选择散射
selective sequence calculator	选择程序计算机
selective shaped charge	选发聚能弹
selective shut-off	选择性封堵
selective solvent	选择性溶剂
selective sorbent	选择性吸着剂
selective sorption	选择性吸附
selective sorting	选择分选
selective stacking	选择叠加
selective stimulation	选择性增产措施
selective top-down	自顶向下选择
selective tracing routine	选择追踪程序
selective trapping	选择性圈闭作用
selective water shut-off	选择性堵水
selective weathering	选择风化
selective wetting	选择性润湿
selective zonal injection	选择性分层注水
selective zone	选择层
selective zone anchor	选层锚；分层锚固

selective zone production	分层开采	self-compacting concrete	自压实混凝土
selective γ-γ logging	选择 γ-γ 测井	self-compensating pressure system	压力自动补偿系统
selective-cutting machine	选择式截煤机	self-conjugate vector space	自共轭向量空间
selectively blocking off	选择性封堵	self-consolidation	自重固结
selectively fired gun	选发式射孔器	self-contained drill	独立式钻机；自给式钻机
selectively flocculate	选择性絮凝	self-contained drilling platform	自足式海洋钻井平台
selectivity	选择性	self-contained platform	自足式平台
selenodesy	月面测量学；月球学	self-contained wireline	一车装试井设备
selenotectonics	月球构造学	self-dumping cage	自翻式升降车
self cleaning drilling	自动除粉钻眼	self-dumping car	自动翻卸车
self clearing car	自卸式矿车	self-dumping truck	自卸式卡车
self consistent	自相容的	self-elevating drilling platform	自升式钻井平台
self discharging car	自卸式矿车	self-elevating platform	自升式平台
self dumping cage	自翻式升降车；自翻罐笼	self-elevating substructure	自升式底座
self dumping car	自动翻卸车	self-elevation unit	自升式平台
self elevating drilling platform	桩脚式钻探平台	self-emptying borer	自动出屑钻
self energizing	自身供给能量的；自激的；自馈的；带自备电源；自加力效应	self-equilibrating	自平衡
		self-erection rig	自立的钻机
self feeder	自动给矿机	self-feeder	自动给矿机
self healing soil	自愈性土料	self-feeding portable conveyor	自动给料轻便输送机
self heating	自热	self-floating jacket	自浮式导管架
self ignition	自燃	self-flowing grout	自流灌浆
self inductance	自感	self-flowing well	自流井
self inflammability	自燃发火性	self-fluxed sinter	自熔性烧结
self loading conveyor	自动装载输送机	self-fluxing ore	自熔矿
self locking frictional prop	自销式摩擦支柱	self-focusing	自动调焦
self organizing	自组；自编	self-fusible ore	自熔矿
self potential	自然电位	self-generating extinguisher	自生式灭火器
self propelled scraper	自动刮土机	self-generating mud-acid	自生土酸
self propelled vibro-plate	自行式振动板	self-geosyncline	自地槽
self repairing material	自修复材料	self-grown volume deformation	自生体积变形
self rescuer	自救器	self-guided	自动导向的
self sharpening bit	自磨式钻头	self-hardening slurry cut-off wall	自凝灰浆防渗墙
self tipping car	自动翻卸车	self-hauling drill	自行式钻机
self weight	自重	self-jetting wellpoint	自射式井点
self weight collapse loess	自重湿陷性黄土	self-leveling laser	自动调平激光器
self weight non-collapse loess	非自重湿陷性黄土	self-leveling rod	简单水准尺
self weight stress	自重应力	self-loading conveyor	自动装载输送机
self-absorption	自吸收	self-loading head	自装头；鸭嘴输送机头
self-actuated sampler	自动取样器	self-loading scraper	自装式铲运机
self-adaptive mesh	自适应网格	self-locating geophone	自定位检波器
self-adjusting leveling instrument	自动安平水准仪	self-locking connection	自锁接头
self-advancing support	自动超前支护；自移式支架	self-locking frictional prop	自锁式摩擦支柱
self-aligning level	自动安平水准仪	self-locking gear	自动闭锁装置
self-balancing	自动平衡的	self-locking nut	自锁螺母
self-bias	自偏压	self-locking pin	自锁销
self-biased off	自动偏压截止	self-lubricating bearing	自润滑轴承；含油轴承
self-boring pressuremeter	自钻式旁压仪	self-lubrication	自动润滑
self-boring type torsion shear apparatus	自钻式扭剪仪	self-made quicksand	自成悬浮体
self-calibration procedure	自校程序	self-magnetization	自磁化
self-cement for stressing	自胶结应力水泥	self-matching deconvolution	自匹配反褶积
self-cementing fill	自行胶结充填	self-matching filter	自匹配滤波器
self-cleaning bit	自除钻粉钻头；自清式钻头；自洗式钻头	self-measuring device	自动计量装置
		self-mobile	自动的
self-cleaning connectors	自清洗接头	self-modulation	自调制
self-cleaning drilling	自动除粉钻进；自动除粉钻眼	self-oiling bearing	自润滑轴承
self-cleaning filter	自清洗式过滤器	self-operated control	自行控制
self-cleaning nozzle	自清洗喷嘴	self-operated controller	自动操作控制器；自控式调节器
self-cleaning strainer	自洁式过滤器	self-optimizing	最佳自动的；自动优化
self-clearing bit	自洁式牙轮钻头	self-orientation	自动定向

self-orienting geophone	自动定向检波器
self-oscillation	自激振荡
self-potential logging	自然电位测井
self-potential prospecting	自然电位勘探
self-potential survey	自然电位测量
self-powered scraper	自行式铲运机
self-prestressed cement	自应力水泥
self-priming pump	自吸起动泵；自吸泵
self-propelled blade grader	自动推板式平路机
self-propelled blader	自动推土机
self-propelled conveyor	自推进输送机
self-propelled drill mounting	自动推进式钻机车
self-propelled drill vessel	自航式钻井装置
self-propelled driller	自进式钻机；自进式凿岩机
self-propelled grader	自行式平土机
self-propelled plough	自行式刨煤机
self-propelled reversible tripper car	自动可翻式卸料车
self-propelled rig	自行式钻机
self-propelled scraper	自动刮土机；自行式刮管器；自行式铲运机
self-puring of aquifer	含水层自净作用
self-reading level rod	自读式水准尺
self-recorder	自动记录器；自记器
self-releasing	自动卸开的
self-relieving	自动泄压的
self-sealing coupling	自封联管节
self-sharpening bit	自锐钻头；自磨式钻头
self-sharpening tooth	自锐齿
self-similar propagation	自相似扩展
self-similarity	自相似性；自相似
self-similarity dimension	自相似维数
self-similarity transformation	自相似变换
self-stabilization time of tunnel	坑道自稳时间
self-stiffness	固有刚度
self-stressing concrete	自应力混凝土
self-stressing method	自应力法
self-supported opening	不需支架的巷道；工作面无支架回采法
self-supporting	自稳的；自支撑的
self-supporting capacity of surrounding rock	围岩自承能力
self-supporting opening method	自撑式开采法；无支架开采法
self-supporting rock	自稳的岩石
self-supporting static pressed pile	自承式静压桩
self-supporting wall	自承墙
self-travelling wheeled jumbo	自行轮胎式钻车
self-wallowing soil	天然波状土
selfward-dipping fault	自倾断层
self-weight	自重
self-weight collapsibility	自重湿陷性
self-weight stress	自重应力
selvage	断层泥；边缘带，岩体边缘带；脉壁泥
SEM	扫描电镜
semi horizon mining	分段回采
semi-anthracite coal	半无烟煤
semiarid	半干旱的
semi-automatic hydraulic mining	半自动化水采法
semiautomatic well-test unit	半自动油井测试装置
semibasement	半地下室
semicircular arch	半圆拱
semi-circular bending specimen	半圆盘弯曲试样
semi-column	半圆柱
semiconcealed coalfield	半隐伏式煤田；半隐伏煤田
semi-confined aquifer	半承压水层
semi-consolidated	半固结的
semi-consolidated sand	胶结性弱的砂岩
semi-consolidated soil	半固结土
semi-continuous haulage	半连续运输
semi-continuous mill	半连续式轧机
semi-continuous mining technology	半连续开采工艺
semi-continuous surface mining	露天矿半连续开采
semi-continuous-duty equipment	半连续工作的设备
semicraton	准稳定地块
semideep-deep lacustrine facies	中深-深湖相
semidesert	半荒漠；半沙漠
semi-discontinuous rock	半连续性岩石
semi-dry concrete	半干混凝土
semi-dry friction	半干摩擦
semidry gas	半干煤气
semi-dry mining	半干式防尘开采
semidull coal	半暗煤
semi-elliptical surface crack	半椭圆形表面裂纹
semi-empirical	半经验的
semi-enclosed basin	半封闭盆地
semifixed sand dune	半固定沙丘
semifossil	半化石
semifusite	微半丝煤
semi-gelatinous explosive	半胶质炸药
semi-girder	悬臂梁
semi-graphic panel	半图式面板
semi-gravity earth-retaining wall	半重力式挡土墙
semi-gravity wall	半重力式墙
semi-hard	半坚硬的
semi-hard packing	半硬填料
semi-hard rock	半坚硬岩石；中硬岩
semi-horizon mining	次水平开采；近水平开采
semi-infinite confined aquifer	半无限承压含水层
semi-infinite earth	半无限地层
semi-infinite elastic solid	半无限弹性固体
semi-infinite half-space	半无限半空间
semi-infinite plane problem	半无限平面问题
semi-infinite porous medium	半无限多孔介质
semi-infinite reservoir	半无限水库；半无限储层
semi-infinite stratified medium	半无限层状介质
semi-liquid flushing	半液态冲洗；半液体冲洗
semilog data plot	半对数坐标曲线图
semilongitudinal	偏斜走向的
semilongitudinal fault	斜断层；偏斜走向断层
semilongwall face	短壁工作面
semi-mechanized shield	半机械化盾构
semi-mobile crusher	半可移式破碎机
semi-mobile operation	半移动式作业方式
semi-open stoping method	上向分层半敞工作面采矿法；上向分层柱式采矿法
semi-pelagic facies	半深海相
semipelitic	半黏土质
semipermeable rock	半透水岩体
semipetrified	半岩化；半石质
semi-rigid clay cement diaphragm wall	自凝式黏土水泥

防渗墙
semi-rigid type base 半刚性基层；半刚性底座
semi-rock soil 半岩石土
semiround nose bit 半圆形冠部钻头
semi-saturation 半饱和
semischist 半片岩；准片岩
semi-shield 半盾构
semi-silica brick 半硅砖
semi-splint coal 中亮条带煤；半暗硬煤
semi-steep 缓倾斜的
semi-submersible drilling platform 半潜式钻探平台；半潜式钻井平台
semi-submersible drilling unit 半潜式钻井装置
semi-submersible production facility 半潜式开采设施
semi-submersible rig 半潜式钻井平台；半潜式钻机
semi-theoretical approach 半理论方法
semitight well 资料不多的井
semitrailer mounted rig 半拖车装钻机
semitransverse 偏斜倾向的
semitransverse fault 偏斜走向断层；斜断层
semi-water gas 半水煤气
semiwildcat 半探井
semseyite 单斜铅锑矿
senaite 铅钛铁矿
senarmontite 方锑矿
sendust 铝硅铁粉；铁硅铝磁合金
senile form 老年地形；终期地形
senile river 老年河
senile soil 老年土；衰竭土
senile topography 老年地形
senility 老年期；侵蚀终期
senior geologist 高级地质师；高级地质工作者
senior staff geologist 高级地质师
senior surveyor 高级测量师
sensator 压力转换器
sense displacement 位移方向
sense finding 测向
sense of movement 运动方向
sense of orientation 定向
sense of rotation 旋转方向
sense of tectonic transport 构造迁移方向
sense shock 有感地震
sensibility analysis 敏感性分析；灵敏度分析
sensibility index 灵敏度指数
sensibility of shield 盾构灵敏度
sensible horizon 视平地圈
sensible shock 有感地震；可感地震
sensing capillary 传感毛细管
sensing point 探测点
sensing probe 传感探头
sensitive analysis 灵敏度分析
sensitive aquifer 灵敏含水层
sensitive clay 对结构破坏敏感的黏土；敏感性黏土；水敏性黏土
sensitive formation 敏感性地层
sensitive system of the gravimeter 重力仪的灵敏系统
sensitivity coefficient 敏感系数
sensitivity of gravimeter 重力仪灵敏度
sensitivity of magnetometer 磁力仪灵敏度
sensitivity of soil 土的灵敏度
sensitivity ratio 灵敏度比
sensitizing agent 敏化剂
sensitizing explosive 敏化炸药
sensor information 遥感信息
separate completion 分层完井
separate energy channel 分立能通道
separate event method 分离事件法
separate excitation sample 分离样品
separate gas cap 游离气顶
separate high speed colloidal mixer 分体高速胶体浆液搅拌机
separate hoisting 分类提升
separate injection well 分层注水井
separate minilayer fracturing 细分层压裂
separate sewer system 分流排污系统
separate subzone development 分小层开发
separate transport and hoisting 分级运提
separate unit 分选设备
separate ventilation 单独通风
separate zone production 分层开采
separate zone water injection 分层注水
separated accumulator 隔离式蓄能器
separated fault 开裂断层，张开断层
separated flow 独立水流
separated layer logging data 分层测井数据
separated layer water-flooding 分层注水
separate-interval pumping test 分层抽水试验
separate-layer hydrofracturing 分层水力压裂
separate-layer water shutoff 分层堵水
separate-zone water plugging 分区堵水
separating 分选；分隔作用
separating cone 圆锥选矿机；圆锥选煤机
separation coal 精选煤
separation plane of bed 岩层界面
separation wall 隔墙
sepiolite 海泡石
septarian structure 龟甲构造
sequence analysis 序列分析
sequence blasting 串联爆破；顺序爆破
sequence boundary 层序界面
sequence chron 层序年龄；层序时间；序列时
Sequence chronostratigraphy 序列年代地层
sequence etching 依次浸蚀；顺序浸蚀
sequence of crystallization 结晶顺序
sequence of cyclic stress 循环应力序列
sequence of filling 填筑顺序，充填顺序
sequence of folds 褶皱序列
sequence of grouting 灌浆顺序
sequence of injection 注浆顺序
sequence of mining 开采程序
sequence of pours 浇筑顺序
sequence of sedimentation 沉积层序
sequence of strata 地层次序；层序
sequence of zoning 分带序列
sequence stratigraphy 层序地层学
sequential deposition 连续沉积
sequential fault 次序层错
sequential form 后成地形
sequential sampling 顺序取样
sequential support method 顺序支护法

sequential test	序贯检定
sequential time delay	顺序时间延迟
sequestration	吸集；储存；封存；隔离
serendibite	蓝硅硼钙石
serial construction	连续施工
serial gravity concentration	连续重力选矿法
serial sampling	系列取样；连续取样
serial section	连续剖面
serial-type pore	连续式孔隙
sericite	绢云母
sericite schist	绢云片岩
sericitic alteration	绢云母蚀变
sericitization	绢云母化
series car incline	串车斜井
series index	岩系指数
series of strata	层系；岩系
series shotfiring	串联爆破
series softening	多级软化
series-parallel hole connection	炮眼串并联
series-reactor-start method	串联电感器起动法
serokinematic granite	晚构造期花岗岩
serorogenic granite	晚造山期花岗岩
serozem soil	灰钙土
serpentine	蛇纹石，蛇纹岩
serpentine boulder	蛇纹石
serpentine kame	蛇形丘
serpentine reservoir rock	蛇纹岩储集岩
serpentine seamount	蛇纹石海山
serpentine waterways	蛇形水道
serpentine-chlorite	蛇纹石绿泥石
serpentinite	蛇纹岩
serpentinization	蛇纹石化；蛇纹岩化
serpentinized mantle	蛇纹石化地幔
serpophite	胶蛇纹石
serrate ridge	齿形脊
serrate suture	锯状缝
serrated boundary	锯齿状边界
serrated picture	锯齿状图像
serrated topography	锯齿状地形
service access	维修设施用通道
service basin	储水池
service bench	修理台
service bridge	专用桥
service compartment	井筒辅助隔间；井筒服务隔间
service conduit	设施管道
service drift	辅助平巷
service efficiency	使用效率
service engineering	维修工程；维护工程学；运行工程
service entry	辅助平巷；临时巷道
service equipment	修理工具
service experience	运转经验；工作经验
service facility	服务设施；配套设施
service factor	机器基础工作因数；运转系数
service frequency	服务频率
service gallery	辅助坑道
service hoist	辅助提升机；副井提升机
service horse power	工作马力；运行马力
service inclined shaft	副斜井
service instruction	操作说明；操作须知；操作规程；使用说明书
service lane	通道巷；通道；后巷
service length	使用年限；使用时期；服务年限
service life	使用期限，使用年限，使用寿命，服务期限
service lifetime	使用寿命，使用年限
service limit	使用寿命
service load	实用负荷；实用负载；工作负载；有效负载
service machine	修井机
service manual	维修说明书，维护手册；维护和使用细则；维修程序表；服务项目说明书
service openings	辅助巷道；辅助硐室
service performance	服务性能；操作性能
service pipe	供水管道；给水管；供水管线
service port	工作油孔
service position	工作位置
service power	厂用电力
service pressure	工作压力；用时压力
service program	服务程序
service raise	人行天井；辅助天井
service rake	坡路；斜路
service regulation	服务工作规程；维护、保养或操作规程；业务规程
service rig	修井设备
service road	运料平巷；便道；辅助道路
service roadway	辅助巷道
service room	水电表房；工作房；机房
service routine	服务程序
service selector panel	测井服务项目选择器面板
service shaft	辅助竖井；交通井
service spillway	正常溢洪道
service stress	使用应力
service table	测井项目表
service test	操作试验
service tool	施工工具
service truck	维修车，供应车
service water	工业水
service wear	工作磨损
service well	辅助井
serviceability limit state	耐久性极限状态；使用功能极限状态；正常使用极限状态
serviceability stage	正常使用阶段
serviceable life	使用寿命
serviced land	具备公用设施的土地；已敷设公用设施的土地
serviceman	技术人员；服务人员
servicer	燃料加注车
services catalog	测井项目目录
servo accelerometer	伺服型加速度计
servo analog computer	伺服模拟计算机
servo analyser	伺服随动系统分析器
servo assisted braking	伺服辅助制动
servo brake	伺服制动；伺服制动器；伺服闸
servo controlled testing machine	伺服控制试验机
servo cylinder regulator	伺服油缸式变量机构；伺服油缸式调节器
servo drive	伺服传动；伺服驱动装置
servo dynamics	伺服动力学
servo gear	伺服装置
servo jack	伺服千斤顶；从动千斤顶；补助千斤顶
servo loop	伺服系统；伺服回路

servo platform	伺服平台
servo relief valve	伺服安全阀
servo system	伺服系统
servo-controlled sensor	伺服控制传感器
servo-hydraulic actuator	伺服液压激振器
servo-manometer	伺服压力计
servo-pump	伺服泵
sesquioxide clay	铁铝性黏土
sesseralite	刚闪辉长岩
sessile dislocation	不滑动位错；不动位错
sessile drop method	液滴法；卧滴法；悬滴法
seston	悬浮有机质
set	集合；设置，调整，调节；台，套，组；装置
set accelerator	促凝剂；速凝剂
set bit	嵌镶钻头
set bolt	固定螺栓
set cap	横梁；顶梁；棚梁
set cement	凝固水泥
set coal	风化煤
set collar	棚梁；顶梁；固定轴环，固定环
set copper	凹铜；饱和铜
set diameter	嵌镶钻头的内径或外径
set dimension	规定尺寸；固定尺寸
set forward force	前冲力
set girt	木梁横撑
set grout	灌浆结石
set hammer	击平锤
set interval	棚子间距
set mining method	规定的采矿法
set of arch	拱架组数
set of faults	断层组
set of joints	节理系
set of pile	桩的贯入度
set out observation point	设定观测点
set out well	扩边井
set penetration	贯入度
set point control	定值控制
set ram	手夯锤
set retarding admixture	缓凝剂
set strength	凝固强度
set time	固化时间；凝固时间
set up the tool joint	配装钻杆接头
set with hard alloy	镶嵌硬合金的
set-retarding admixture	缓凝剂
set-stretching	拉伸定形
settable orifice	可调节流孔
setting accelerator	速凝剂；促凝剂
setting accuracy	调整精度；安平精度
setting basin	沉降盆地；沉淀池；沉沙池
setting block	镶嵌夹具
setting cap	垫帽
setting chamber	沉降室
setting claw	钩形紧固器
setting collar	坐放箍
setting condition	凝固状况
setting depth	坐封深度
setting depth of pump	泵挂深度
setting device	落刀装置；设定装置；调节用具
setting expansion	凝结膨胀
setting force	坐封力
setting gauge	金刚石钻头镶嵌规
setting gun	注液枪
setting hammer	击平锤
setting jack	安装千斤顶；安装
setting load	预压负荷；初撑力；设定负载，额定载荷
setting load density	加载密度设计，设置加载密度；初撑力强度，初撑强度
setting mechanism	坐封机构
setting of cement	胶结物凝固
setting of concrete	混凝土凝结
setting of ground	地基沉降；地面下沉
setting of mortar	灰浆凝结
setting of plane-table	平板仪安装
setting out of axis	轴线放样
setting out of foundation	基础放样
setting out plan	定位图，放样图
setting out rod	定线桩
setting out survey	放样测量
setting pipe	沉淀管
setting point	坐定点；凝固点
setting pressure	设定压力；标定压力；初撑压力
setting procedure	坐封程序
setting process	凝固过程
setting rate	凝结速度
setting ratio	充填沉缩率
setting rig	调整、装配试验台
setting ring	金刚石钻头的嵌镶圈
setting shrinkage	凝结收缩；凝固收缩
setting sleeve	坐封套
setting stake	定线桩
setting strength	凝固强度
setting tank	沉淀罐
setting time	沉淀时间，凝固时间，凝结时间；下沉时间
setting up agent	凝固剂
setting up procedure	启动步骤
setting value	设定值；给定值
setting velocity	沉降速度
setting wedge	固紧楔
setting winch	安装绞车；沉淀
setting-out center line of roadway	巷道中线标定
setting-out of cross line through shaft	立井十字中线标定
setting-out of junction	交叉点放样
setting-out of main axis	主轴线测设
setting-out of reservoir flooded line	水库淹没线测设
setting-out of roadway gradient	巷道坡度线标定
setting-out of shaft center	立井中心标定
setting-up procedure	调整顺序；装配过程
settle	沉降，下沉，沉陷；沉积，沉淀；安排；解决；支付
settle contour	沉降等值线
settle tank	沉淀槽
settleability	可沉性
settleable solids	可沉淀固体颗粒
settled dust	沉落尘末；落尘
settled gravel	沉降的砾石
settled ground	沉实地层；停止沉陷地表
settled material	沉淀物料；沉积物料
settled production	规定生产；制定生产；井初期自喷产量下降后的平均产量

settled proppant bank	砂堤
settled proppant bed	支撑剂沉淀层
settled sludge	沉积污泥
settled solution	澄清溶液
settlement	沉降，沉淀，沉积，下沉；沉降量，沉积物；地陷，沉陷
settlement action	沉降作用
settlement after completion	工后沉降
settlement angle	崩落角
settlement basin	沉降盆地
settlement calculation	沉陷计算
settlement calculation by specification	规范中的沉降计算法
settlement calculation depth	沉降计算深度
settlement casing	沉降盒
settlement cell	沉降盒
settlement change	下沉变化
settlement computation	沉降计算
settlement control	沉降控制
settlement correction factor	沉降计算经验系数；沉降修正系数，沉降校正系数
settlement crack	沉降裂缝；陷落裂缝
settlement crater	沉陷坑
settlement criterion	沉降准则
settlement curve	沉降曲线
settlement device	沉陷测定装置
settlement difference	沉降差
settlement due to consolidation	固结沉降
settlement effects chart	相邻影响曲线图
settlement environment	聚落环境
settlement factor	沉降系数；沉降比选
settlement failure	沉陷破坏
settlement forecast	沉降预测
settlement gauge	沉降计；沉降测量仪，沉降仪
settlement isoline	沉降等值线
settlement joint	沉降缝
settlement measurement	沉降测量
settlement monitoring for slope surface	坡面沉降监测
settlement observation	沉降观测
settlement of arch crown	拱顶下沉
settlement of compacted soil	土体压后沉降量
settlement of embankment	路堤沉陷
settlement of fill	填土沉陷
settlement of foundation	地基沉降
settlement of pile group	群桩沉降量
settlement of single pile	地基沉降量；单桩沉降量
settlement of subsoil	地基沉降量
settlement plate	沉降板
settlement rate	沉降速率
settlement ratio	沉降比
settlement ratio of stability	稳定性沉降比
settlement record	沉降记录
settlement separate	沉降分离
settlement space	沉降空间
settlement stress	沉降应力；沉陷应力
settlement velocity	沉降速度
settlement-load curve	沉降-荷载曲线
settler	沉降罐；沉淀器
settling	沉陷；沉淀；沉降；下沉
settling area	沉降面积；沉积面积
settling basin	沉淀池；沉降池
settling bowl	沉淀箱；沉淀槽
settling box	沉淀桶
settling centrifuge	沉降离心机
settling chamber	沉淀室，沉淀箱；沉降室
settling classification	沉积分级
settling condition	沉降条件
settling cone	圆锥沉淀槽
settling curve	沉淀曲线；沉降曲线
settling ditch	澄清沟；沉淀沟
settling equipment	沉淀装置
settling hopper	沉淀仓
settling lag	沉降滞后
settling lagoon	沉淀湖
settling of support	支座下沉
settling out	沉析；沉降分离
settling pipe	沉降测管
settling pit	矿水沉淀坑；沉淀槽；沉砂池
settling plant	沉淀池；沉淀装置
settling pond	沉淀池；沉积池；澄清池
settling process	沉降过程
settling promoter	凝结促进剂；沉降促进剂；沉淀催化剂
settling quality	沉降性；下沉性
settling rate	沉降速度，下沉速度；沉淀速率，沉降速率
settling ratio	沉降系数；沉降比，下沉比
settling region	沉淀区
settling speed	沉降速度
settling system	沉砂系统
settling tank	沉降槽；沉淀池
settling time	沉降时间
settling velocity	沉降速度，下沉速度；污泥沉降比
settling vessel	沉淀器
settling volume	沉降体积
setup	安装，装配；定置仪器
set-up diagram	装置图
setup entry	开切眼
setup error	安置误差
set-up expense	设备安装费
set-up procedure	准备过程；安装过程；调定程序
setup room	切眼
set-up time	扫描时间
severe contamination	严重污染
severe crooked hole	严重弯曲井眼
severe drought	大旱
severe stress	危险应力
severe test	在苛刻条件下的试验
severely destructive earthquake	严重破坏性地震
severe-service piping	恶劣条件下工作的管线
severity rate	严重事故率
sewage	污物；污水，阴沟水；下水道，下水道系统；污水渠
sewage discharge	污水排放
sewage disposal	污水处理；污水排放
sewage disposal system	污水处理系统
sewage drain	排污管
sewage drainage standard	污水排放标准
sewage ejector	污水射流泵；污水压气喷射器
sewage gas	污水分解的气体；沼气

sewage line	污水管道
sewage loading	污水负荷量
sewage outlet	污水出口
sewage pipe	污水管
sewage plant	污水站
sewage pump	排污泵；污水泵
sewage pump house	污水泵房
sewage purification	污水净化
sewage sludge	污水污泥，污泥
sewage sump	污水坑；集污槽
sewage system	污水系统
sewage tank	污水柜；污水池
sewage treatment	污水处理
sewage treatment plant	污水处理站；污水处理厂
sewage treatment system	污水处理系统
sewage treatment tank	污水处理池
sewage treatment works	污水处理厂
sewage tunnel protection area	污水隧道保护区
sewage tunnel works	污水隧道工程
sewage water	污水
sewage works	下水道工程，污水工程
sewage-farm	污水处理场
sewer	污水管，下水道，污水渠，暗沟，排水管
sewer distribution line	污水线网
sewer gas	下水道气体
sewer invert	排水管之弯曲段
sewer line	排污管线；污水管道
sewer pipe cleaner	下水管线清扫机
sewer system	排水系统
sewerage	污水工程；排水工程；下水道
sewerage outfall	污水排放口，污水渠口
sewerage reticulation system	网状污水渠系统
sewerage system	污水系统
sewerage treatment	污水处理
sewerage works	污水渠工程
sextant	六分仪
sextant chart	六分仪后方交会解算图
shale oil	页岩油
shallow-focus earthquake	浅源地震
shear band	剪切带
shear bond	抗剪固结力
shear deformation	剪切变形
shear dilatancy effect	剪胀效应；剪切扩容效应
shear expended volume	剪切扩容
shear modulus	剪切模量
Shenhua Group Corporation	神华集团公司
short rod specimen	短棒试样；短圆棒试样
SHPB	分离式霍普金森压杆；霍普金森压杆
SI unit	公制单位；国际单位；国际单位制单位
SI unit system	标准国际单位制
sial	硅铝带，硅铝层
sialic	硅铝质的；硅铝层的
sialic crust	硅铝壳
sialic layer	硅铝层
sialic underplating	硅铝壳的板底作用；硅铝壳的底侵作用
siallite	硅铝土；硅铝质岩；黏土矿物；黏土类
siallitic laterite	硅铝砖红壤
siallitic soil	硅铝土
siallitization	黏土矿物化；黏土化
sialma	硅铝镁带；硅铝镁层
sialsima	硅铝镁层；硅铝镁带
sialsphere	硅铝层；硅铝圈
sialsphere fault	硅铝层断裂
siberite	紫电气石；紫碧玺
Sicilian Stage	西西里阶
side	旁，边，面，侧，帮，翼；直煤壁；采空区直壁；脉壁；开采区；岸
side abutment	侧面支承压力带
side abutment pressure	侧支承压力；边墩压力
side adhesion	侧向黏聚力
side adit	支平硐
side adjustment	横调；边平差
side anchor	边锚
side and front slope	边仰坡
side bank	旁堆积；边滩；岸坡
side bar	侧板；侧杆；侧砂坝
side beam	侧梁
side bearing	旁承
side bearing loading	旁承承载
side benching	横向开采；侧向阶梯式开采
side bend specimen	侧弯试件；侧弯试样
side bend test	侧弯试验；侧面弯曲试验
side bolting	边帮锚杆支护
side borehole	周边钻孔；周边眼
side borrow	边沟取土坑
side bumper	边挡
side camber	侧边弯曲
side compartment	井筒旁隔间
side compression	侧压力；侧压
side condition	附带条件；限制条件
side constraint	界限约束；边际约束
side coring	井壁取芯；井壁取样
side cover	侧盖；边盖
side cut	帮槽；侧面掏槽；侧面开采
side cutting	山边开挖；侧向开挖；堤旁借土；侧向切削；侧面采掘；侧面掏槽
side cutting displacement	侧向切削偏移量
side discharge	侧部排料；侧面卸货；侧向卸料
side discharge bit	侧泄钻头
side ditch	边沟；侧沟；路边明沟；排水沟
side drain	侧沟；边沟；路边排水沟；侧向排水；试样壁排水条
side drainage	路边排水
side drift	平硐；平巷；横贯全矿的平巷；分支巷道；边部掏槽；侧壁导坑；侧壁导流坑
side drift method	侧导坑法；侧导流法；侧壁导坑先墙后拱法
side driller	水平孔钻机
side effect	副作用；副效应；边缘效应
side elevation	侧视图；侧面；端视
side erosion	横向冲刷；侧向侵蚀
side excavation	侧向开挖
side face	侧面
side falling accident	巷道片帮事故
side force	侧向力；侧力
side friction	侧面摩擦；侧向摩擦力
side friction factor	侧向摩擦系数
side friction resistance of pile	桩侧摩阻力
side gutter	侧沟；边沟

side heading	侧导坑；侧导坑超前掘进
side heading method	侧壁导坑法
side hill	山坡
side hill canal	山边渠道
side hill cut	边坡开挖
side hill quarry	山坡采石场
side hole	边缘炮眼；边缘钻孔；侧孔
side intake	旁侧进水口
side intersection	侧方交会；侧方交会法
side lacing	边帮背板
side lagging	边帮背板
side lap	旁向重叠
side line	边线；侧线；抢泥船侧向稳定缆
side load	边负载；边载重；横向载荷；侧向载荷
side loading	侧面加载；侧加料
side looking	侧视
side looking airborne radar	机载侧视雷达
side looking radar	侧视雷达
side looking radar image	侧视雷达图像
side of an anticline	背斜的一翼
side of fault	断层盘
side of foundation pit	基坑侧壁
side opening	侧孔
side pack	废石垛墙
side pipe	旁管
side port	侧孔
side pressure	侧压力；侧压；旁压力；偏压
side producer	边部生产井
side pull	侧拉力
side rake angle	侧倾角
side roadway	侧面巷道
side sampler	井壁取样器
side scan sonar	侧扫声呐；水平扫描声呐
side scan sonar survey	旁测声呐测量
side scanner	侧向扫描仪
side scanning sonar	旁侧扫描声呐
side seal	侧水封；侧密封
side seepage	侧渗透；侧渗流
side shoal	边滩
side shot	旁测点；侧向爆炸
side shothole	侧面炮眼；帮眼；周边炮眼
side skid	侧向滑移
side skidding	侧向滑动
side slide block	侧向滑块
side slip	侧滑
side slope	边坡
side slope coefficient	边坡斜率
side slope grade	边坡坡度
side slope of cut	路堑边坡
side slope of cutting	路堑边坡
side slope of embankment	路堤边坡
side slope of trackbed	道床边坡
side slope shape	边坡坡形
side slope stability	边坡稳定
side slopes of canal	渠道边坡
side sonar	旁侧声呐
side span	边跨；旁跨；边跨径；侧孔
side spillway	旁侧溢洪道
side spillway dam	旁侧溢流坝
side stoping	水平推进上向梯段回采
side stream	支流
side support	侧支撑；侧撑
side swipe	侧击波
side tension	侧向张力；侧拉力
side tester	侧壁试验器；侧壁测试仪
side thrust	侧推力；侧向推力；侧压
side tie	岩石平巷，围岩平巷；脉外平巷
side tracking	侧钻；避开；转入旁轨
side tracking point	侧钻点
side trench	边沟
side view	侧视图
side wall	边墙；侧墙；帮部；井壁；侧壁
side wall concrete	边墙混凝土
side wall coring	井壁取样
side wall coring gun	射入式取芯器；井壁取芯枪
side wall cutting sampler	切割式取芯器
side wall first lining method	先墙后拱法
side wall sampling	井壁取样
side water hole	侧面水孔
side wave	侧面波
side well	侧钻井眼
side-bed correction	围岩校正
side-cutting rate	侧向切削速度
side-delta lake	支流三角洲湖
side-draw cut	上向侧面掏槽
side-drive drilling system	侧向驱动钻井系统
side-effect	边界效应
side-entry	侧面进入，侧面入口；侧面的
side-fill cut	半堑半堤
side-flow weir	侧流堰
side-friction coefficient	侧向摩擦系数
side-friction force	侧向摩擦力
side-hill	山坡
side-hill cut	山坡露天矿
side-hill fill	半堑堑；山边填土
sidehill quarry	山坡采石场
sidehill seepage	山坡渗流；山边渗流
side-hill work	山边土石方工程
side-hole bit	侧孔式钻头
sideline	横线；旁线；边界线；附带工作；副业
sideling	斜的
sidelooking aerial radar	航空侧视雷达
side-looking airborne radar	机载侧向雷达
side-looking image radar	侧视制图雷达
side-looking radar	侧视雷达
side-looking radar imagery	旁视雷达影像
side-looking sonar	旁侧声呐
side-pilot method	侧壁导坑；侧壁导坑法
siderazote	氮铁矿
siderazotite	氮铁矿
siderite	菱铁矿；陨铁；蓝石英
siderite concentrate	菱铁精矿
sideritization	菱铁矿化
sideroferrite	自然铁
siderolite	石铁陨石；铁陨石；陨铁石
siderophile affinity	亲铁性
siderophile element	亲铁元素
siderophyllite	铁叶云母
sideroplesite	镁菱铁矿
siderotil	铁矾

sides of fracture 裂缝壁
sides of hole 孔壁
sidescan 侧向扫描设备
side-scan sonar 侧扫声呐；侧向扫描声呐；旁侧扫描声呐
side-scanning sonar 侧向扫描声呐
side-scattered waves 侧面散射波
sideslip 横滑；侧滑
sidesway 侧移；侧倾
sidesway moment 侧移弯矩
sidesway stiffness 侧移刚度
side-swept 擦撞
sideswipe 侧向反射
sideswipe reflection 侧向岩面多次反射
side-tipping 侧倾的
sidetrack drilling 侧钻
sidetrack horizontal well 侧钻水平井
sidetracked 偏向钻孔
side-tracked hole 侧钻井眼
side-tracking 侧钻；偏向钻进
side-tracking bit 侧钻钻头
sidetracking drilling 侧钻井
sidetracking hole 侧钻井眼
sidetracking pocket 侧钻套
sidetracking whipstock 侧钻造斜器
sidewall 侧帮，侧壁；井壁
sidewall core 井壁岩芯
sidewall core cutter 切割式井壁取芯器
sidewall corer 井壁取芯器
sidewall coring 井壁取芯
sidewall coring gun 井壁取芯枪；井壁取芯器
sidewall coring tool 井壁取芯器
sidewall cutting sampler 井壁切割取芯器
sidewall density log 井壁密度测井
sidewall drill 侧壁钻机；侧壁钻机
sidewall driller 水平钻机；侧壁钻机
sidewall neutron detector 井壁中子探测器
sidewall neutron log 井壁中子测井
sidewall neutron logging 井壁中子测井；贴井壁的极板
sidewall pad 井壁压紧器；侧壁
sidewall rock 侧帮；井壁岩石
sidewall sample 井壁岩样
sidewall sample taker 井壁取芯器
sidewall sampler 井壁取芯器
sidewall sampling 井壁取样；井壁取芯
sidewall sampling apparatus 侧壁取样器
sidewall scaling 片帮；井壁剥落
sidewall slabbing 片帮
sidewall spalling 片帮
sidewall stress in stope 采场边墙应力
sideward motion of earthquake 地震侧向运动；地震水平动
sideway force coefficient 侧向力系数
sideway movement 侧向运动
sideway skid resistance 侧向抗滑力
sideway stiffness 侧斜刚度
sideways force 横向力
sidewise bending 旁弯；侧弯
sidewise buckling 侧向屈曲；侧向挠曲；侧向翘曲
sidewise cut 侧面掏槽

sidewise restraint 侧向限制
siding-over 矿柱中的短巷道
Sieberg seismic intensity scale 西伯格地震烈度表
siegenite 硫镍钴矿
sienite 正长岩
sierozem 灰钙土；灰色沙漠土
sierra 锯齿状山脊；呈齿状起伏的山脉
sieve analysis 筛分析；筛分法
sieve aperture size 筛孔尺寸
sieve classification 筛分分级；筛析
sieve correction 筛分析校正
sieve curve 筛分曲线
sieve diameter 筛孔直径
sieve fraction 筛分粒级
sieve machine 筛分机
sieve mapping 筛网制图
sieve mesh 标准网目；筛孔
sieve method 筛分法
sieve net 筛网
sieve residue log 岩屑录井；筛屑
sieve scale 筛比
sieve size 筛分粒度
sieve sizing 筛分分级
sieve test 筛分试验
sieve testing 筛分试验；筛分分析
sieving analysis 筛分析；筛样机
sieving curve 筛分曲线
sieving machine 筛选机
sievite 玻质安山岩系
SIF 应力强度因子
SIF handbook 应力强度因子手册
SIF manual 应力强度因子手册
sifema 硅铁镁层
sight alidade 封闭式照准仪
sight angle 视线角
sight axis 视轴
sight bracket 视准尺架
sight cell 观察器
sight compass 矿用罗盘
sight distance 视距
sight gauge 观测计
sight glass 观察玻片；窥镜
sight hole 检查孔；探视孔；窥视孔
sight inspection 目测；照准线；外观检查
sight line 瞄准中心挂线；瞄准线；视线
sight plane 视准面；视点
sight point 瞄准点
sight pump 排量可见泵
sight rail 视准轨；标尺
sight rod 视距标尺
sight sample 目视试样
sight size 透光孔口
sight triangle 视距三角形
sight vane of fore side 前视准板
sight window 观察孔
sight-hole 观察孔；照准
sighting 观察；照准；测视
sighting bar 瞄准杆；觇板
sighting board 测视板；觇板
sighting centring 照准点归心

sighting compass	觇板罗盘
sighting condition	通视条件
sighting cylinder	照准圆筒
sighting distance	视距
sighting error	照准误差；瞄准线
sighting gear	照准器
sighting line	视线；瞄准线
sighting line method	瞄直法
sighting mark error	照准误差
sighting pendant	瞄准锤
sighting point	照准点
sighting straight method	瞄直法
sighting target	照准靶
sight-line	侧线，视线；视线
sight-reading chart	模拟系统图
sigma log	中子俘获截面测井
sigmoidal en echelon tension gash	S形斜列张裂缝
sigmoidal fold	S形褶皱；反曲褶皱；肠曲褶皱
sigmoidal grain-size distribution	S形粒度分布
sigmoidal joint	S形节理
sigmoid-progradation seismic facies	S形前积地震相
sign	标志；标记；符号；迹象；签字；签约
sign bit	符号位
sign bit seismograph	符号位地震仪；标志牌
signal amplification	信号放大
signal amplifier	信号放大器
signal amplitude	信号振幅
signal analyzer	信号分析器
signal aspect and indication	信号显示
signal attenuation	信号衰减
signal averager	信号平均器
signal carrier	载波
signal channel	信号传输通道
signal compression	信号压缩
signal conditioner	信号调整器
signal conditioning package	信息处理装置
signal correction	信号校正
signal delay	信号延迟
signal detection	信号检测
signal device	信号设备
signal distortion	信号失真
signal duration	信号持续时间
signal encoding	信号编码
signal enhancement	信号增强
signal extraction	信号提取
signal fault	信号故障
signal flag	行波
signal frequency	信号频率
signal frequency shift	信号频率偏移
signal generation	信号发生
signal generator	信号发生器
signal identification	信号识别
signal imitation	信号模拟
signal indicator	信号指示器
signal intensity	信号强度
signal interpretation	信号译码；信号解释
signal inverter	信号变换器
signal lamp	信号灯
signal noise	信号噪声
signal noise ratio	信噪比
signal phase	信号相位
signal processing	信号处理
signal processing seismograph	信号处理地震仪
signal processor	信号处理机
signal quality	信号品质
signal quality detection	信号质量检测
signal quantization	信号量化
signal ratio	信号比
signal recognition	信号识别
signal recorder	信号记录器
signal sampling	信号采样技术；信号取样
signal selector	信号选择器
signal shaping	信号整形技术
signal simulator	信号模拟器
signal smoothing	信号平滑
signal source	信号源
signal synthesis	信号合成
signal to distortion ratio	信号失真比
signal to interference ratio	信号干扰比
signal to noise ratio	信噪比
signal transformation	信号转换
signal transmission	信号传递
signal wave	信号波
signal waveform	信号波形
signal-noise ratio	信噪比
signal-to-noise	信噪比
signal-to-noise ratio	信噪比
signature analysis	符号差分析；特征分析
signature deconvolution	特征反褶积
signature identification	特征识别
signature log	特征测井
signature processing	震源信号处理
signature record	子波记录
signature turbulence	特征紊流
signature waveform	特征波形
significance analysis	显著性分析
significance function	显著性函数
significance level	显著水平
significance of correlation coefficient	相关系数显著性
significance of difference	差数显著性
significance of impact	影响程度
significance of regression coefficient	回归系数显著性
significance test	显著性检验
significant damage	严重破坏
significant depth	有效深度
significant detrital buildups	有效碎屑堆积
significant distance	有效距离
significant earthquake	重要地震
significant height	有效高度
significant hydraulic parameter	特征水力参数
significant level	特性层
significant long-drift motion	大幅度长周期漂移运动
significant surface	有效面积
significant wave height	有效波高
significant wave length	有效波长
significant wave method	有效波法
significant wave period	有效波周期
signs of distress	损坏迹象，破坏迹象
silcrete	硅结砾岩；硅质壳层
silent earthquake	寂静地震；无声地震

silent piling	静力压桩
Silesia method	西里西亚采煤法
Silesia system	西里西亚厚煤层开采法
Silesian	西里西亚阶
Silesian Stage	西里西亚阶
silex	硅石；石英；燧石
silex glass	石英玻璃；火成石英岩
silexite	硅石岩；英石岩
silfbergite	锰磁铁矿
silica	硅石；二氧化硅
silica alumina ratio	硅铝比
silica budget	氧化硅收支
silica cement	硅土水泥砂浆
silica conglomerate	硅质砾岩
silica diagenesis	二氧化硅成岩作用
silica dust	矽尘；石英粉尘
silica earth	硅质土
silica fines	硅质细粒土
silica flour	石英粉
silica fume	硅粉
silica fume concrete	硅粉混凝土
silica fume modified concrete	硅粉改性混凝土
silica loam	硅质壤土
silica rock	硅质岩
silica sand	硅质砂；石英砂
silica sheet	硅氧片层
silica soil	硅土
silica stone	硅质岩石
silica tetrahedral sheet	片状硅四面体
silica tetrahedron	硅氧四面体
silica-lime cement	石英砂石灰水泥
silicalite	硅质岩
silicarenite	石英砂岩；石英砂屑岩；硅质砂岩
silicareous ooze	硅质软泥
silica-replaced fossil	硅交代化石
silicastone	硅质岩；硅质沉积岩
silicastone deposit	硅石矿床
silicate	硅酸盐
silicate clay	硅化土
silicate melt	硅酸盐熔体
silicate mineral	硅酸盐矿物
silicate rock	硅酸盐岩石；硅酸盐岩
silicates	硅酸盐类
silicate's Earth	硅酸盐地球
silication	硅化作用；硅酸盐化作用
silicatization	硅化作用；硅化固结
silicatization process	硅化过程
siliceous	硅质的；含硅的
siliceous aggregate	硅质骨料
siliceous cement	硅质胶结物
siliceous clay	硅质黏粒
siliceous clayey ooze	硅质黏土软泥
siliceous composition	硅质成分
siliceous concretion	硅质结核
siliceous crust type	硅质壳型
siliceous earth	硅土；硅藻土
siliceous lava	硅质熔岩
siliceous limestone	硅质灰岩；硅质石灰岩
siliceous marl	硅质泥灰岩
siliceous ooze	硅质软泥
siliceous phyllite	硅质千枚岩
siliceous proppant	硅质支撑剂
siliceous reservoir rock	硅质储集岩层
siliceous rock	硅质岩石
siliceous sandstone	硅砂岩
siliceous schist	硅质片岩
siliceous sediment	硅质沉积；硅质沉积物
siliceous shale	硅质页岩
siliceous shale rock	硅质页岩岩石
siliceous soil	硅质土壤
siliceous stratum	硅质岩层
silicic	硅质的；富硅质的
silicic acid	硅酸
silicic rock	酸性岩；酸性火山岩
silicic volcanic rock	硅质火山岩
siliciclastic	硅质碎屑的
siliciclastic deposit	硅质碎屑沉积
siliciclastic sediment	硅质碎屑的沉积物
siliciclastic-carbonate	硅质碎屑碳酸盐
silicide	硅化物
silicification	硅化；硅化加固
silicification grouting	硅化灌浆
silicified	硅化的
silicified fossil	硅化化石
silicified fracture zone	硅化断裂带
silicified mudstones	硅化泥岩
silicified rock	硅化岩
silicified tectonite	硅化构造岩
silicilith	沉积石英岩
silicilutite	泥屑石英岩
silicilyte	沉积石英岩
silicious	硅质的
silicirudite	砾屑石英岩
silicite	硅质岩
silicium	硅
silicium-carbide	碳化硅
silicobiolith	硅质生物岩
silicocarbonatite	硅质碳酸岩
silicolite	硅质岩
silicon	硅
silicon carbide	金刚砂；碳化硅
silicon chip	硅片
silicon dioxide	二氧化硅
silicon earth	硅土
silicon pellet	硅片
silicon wafer	硅片
silicon-calcium ratio log	硅-钙比测井
siliconizing	硅化；渗硅
silico-rudite	硅砾岩
silicotic mine	有矽肺危险的矿
silcrete	硅质层
silky	丝的；丝状断裂面；丝绢状
silky fracture	丝状断口
silky luster	绢丝光泽
sill	岩床；海底山脊；基石；底梁
sill anchor	地脚锚栓
sill basin	海槛盆地
sill depth	海槛深度
sill drift	最低水平巷道
sill floor	方框支架的最下层；基底水平

sill for jump control 控制水跃的底槛
sill level 底平；基底水平
sill mining 拉底回采
sill of ore 矿床
sill of river 河床
sill pillar 底柱；底矿柱；底煤柱
sill stope 岩床采场；底柱回采工作面
sill structure of stope 采场底部结构
sill timbering 底板支护
sillenite 软铋矿
silliciophite 杂蛋白蛇纹石
sillimanite 矽线石；硅线石
silling-out 刷大放矿漏斗
sillite 辉绿玢岩
sill-like 席状
sills 岩床
silt 粉土；粉砂；淤泥；煤粉
silt agglomeration 粉粒聚积
silt arrester 淤地坝
silt arresting by explosion 爆破排淤
silt basin 拦沙池；淤积池
silt bench 泥沙台地
silt carrying capacity 泥沙挟带量；挟砂能力
silt clay 粉质黏土
silt coal 煤泥
silt coast 淤泥质海岸；粉砂淤泥质海岸；粉砂海岸
silt concentration 泥沙含量；泥沙浓度
silt content 泥沙含量；粉粒含量；含泥量
silt contourite 粉砂质等深积岩
silt control 泥沙控制
silt covered 淤泥覆盖的
silt deposit 泥沙沉积
silt discharge 输沙量
silt discharge capacity 排沙能力
silt discharge rating 输沙量率定
silt drain 淤泥泄出口
silt ejector 泥沙喷射泵
silt factor 泥沙系数
silt flow 泥沙流
silt fraction 粉粒粒组；粉土粒组
silt grain 粉粒
silt granularity 泥沙粒度
silt gravel 粉砾
silt index 粉砂指数
silt laden river 含泥沙河流
silt layer 粉沙层
silt loam 粉沙壤土；粉质壤土
silt lodging 泥沙沉积
silt material 泥沙物质
silt movement 泥沙运动
silt particle 粉砂颗粒；泥沙颗粒
silt plug 泥沙淤塞
silt pressure 泥沙压力
silt remover 除泥器
silt rock 粉砂岩
silt sampler 泥沙取样器；取砂器
silt seam 淤泥层
silt separator 除泥分离器
silt size 粉土粒径；粉砂粒径
silt sluicing 冲积层流槽冲洗，淤泥流槽冲洗

silt slurry 淤土浆
silt soil 粉砂土；粉砂质土壤；粉粒土
silt stone 粉砂岩
silt storage capacity 拦沙能力
silt stratification 淤土层；粉砂层理
silt suspension 泥沙悬浮物
silt test 含泥量测定
silt texture 粉砂结构
silt transport 泥沙搬运
silt transportation 泥沙搬运
silt up 淤塞；淤积
siltage 粉砂体
siltation 淤积；淤塞
siltation in estuary 河口淤塞
siltation of reservoir 水库淤积
siltation rate 淤积率
silt-covered 淤泥覆盖的；淤塞的
silted 淤积的；淤塞的
silted land 淤积地
silted-river lake 河淤塞湖
silting 淤积；淤泥沉积；淤塞；沉沙
silting deposit 淤积层
silting in watercourse 河道淤积
silting method 淤填法，淤积法
silting of backwater area 回水区淤积
silting up 淤塞；充填
siltite 粉砂岩
siltized intercalation 粉砂质泥化夹层；泥化夹层
silt-laden 含泥沙的；富泥沙的
silt-laden flow 含泥水流
silt-laden river 含沙河流
siltpelite 粉砂泥岩；泥质岩
siltrock 粉砂岩
silt-seam 粉土薄夹层
silt-sized particle 粉砂级颗粒
siltstone 粉砂岩
siltstone reservoir 粉砂岩储集层
silt-up channel 幼年河
silty 粉土质的；粉砂质的；淤泥的
silty breccia 粉砂质角砾
silty clay 粉状黏土；粉质黏土；粉砂质黏土；亚黏土
silty clay loam 粉质壤土；粉质黏壤土；粉质黏壤土
silty fine sand 粉幼砂；粉质细砂
silty gravel 粉质土砾
silty lime 粉砂石灰
silty lutite 粉砂质泥屑岩
silty material 粉土；粉质沙
silty mudstone 粉砂质泥岩
silty sand 粉质砂土；粉砂；淤泥砂
silty sandstone 粉砂质砂岩
silty shale 粉砂质页岩
silty soil 粉质土；粉性土
silundum 硅碳刚石
Silurian 志留纪；志留系
Silurian Hot Shale 志留系热页岩
Silurian period 志留纪
Silurian sequence 志留纪序列
Silurian system 志留系
silver 银
silver glance 辉银矿

silver lead ore 银方铅矿
silver mineral 银矿类
silver ore 银矿
silver sand 银白色细砂
silveriness 银白；银色光泽
sima 硅镁层；硅镁圈；硅镁带
sima sphere 硅镁带；硅镁层
simasphere fault 硅镁层断裂
simatic 硅镁质的；硅镁层的
simatic crust 硅镁壳
simatic rock 硅镁质岩石
simblosite 蜂窝状结核
similar coefficient 相似系数
similar fault 相似断层
similar fold 相似褶曲；相似褶皱
similar folding 相似褶皱作用
similar fossil valley 相似古山谷
similar law 相似律
similar material 相似材料
similar material model 相似材料模型
similar material simulation 相似材料模拟
similar material test 相似材料模拟试验
similar matrice 相似矩阵
similar physical model experiment 相似物理模型试验
similar physical simulation 类似的物理模拟
similar simulation test 类似的模拟试验
similar solution 相似性解
similarity 相似性
similarity coefficient 相似系数
similarity criterion 相似准则；相似性判据
similarity index 相似性指数
similarity level 相似水平
similarity parameter 相似参数
similarity principle 相似性原理
similarity rule 相似性法则
similarity search 相似性搜索
similarity solution 相似解
similarity theory 相似性理论；相似性定理
similarity transformation 相似变换
similar-type fold 相似型褶皱
similar-type structure 相似形构造
similicoronilithus 似花冠石
similitude 相似；相似模拟
similitude method 相似法
simple analysis 简分析
simple arithmetic mean 简单算术平均数
simple average method 简单平均法
simple beam 简支梁
simple bending 纯弯曲；纯挠曲
simple boundary 简支边界
simple compression 单纯压缩
Simple correlation 简单相关；简单立方点阵
simple cross-bedding 简单交错层理；沉积交错层理
simple crosscorrelation 互相关
simple curve 单曲线
simple dimensional method 单量纲法
simple fault 单断层；简单断层
simple fault scarp 单断崖
simple fixed-point iteration 简单的不动点迭代
simple flow 简单流

simple glide 简单滑移
simple hand excavation 人力开挖
simple harmonic motion 简谐运动
simple harmonic progressive wave 简谐前进波
simple harmonic vibration 简谐振动
simple harmonic wave 简谐波
simple hydrogeological observation in well 钻孔简易水文地质观测
simple implicit method 简单隐式方法
simple lattice 简单晶格；简单点阵
simple leveling 简易水准测量
simple lineament 简单线性构造
simple linear regression 简单线性回归
simple loading 简单加载
simple movement rate curve 地层移动速度曲线
simple multiple 全程多次波
simple multiple reflection 简单多次反射
simple ore 单矿
simple pumping test 简易抽水试验
simple retaining 简易挡土墙
simple sample 单一样品
simple shear 纯剪；单剪
simple shear apparatus 单剪仪
simple shear test 单剪试验
simple shear-strain 单剪应变
simple sill 单岩床
simple sliding friction 纯滑动摩擦
simple slip 纯滑移
simple sluice 简单洗矿槽
simple stock 单岩株
simple strain 单纯应变；单应变
simple stress 简单应力状态；单应力
simple structure 静定结构；简单构造
simple supported beam 简支梁
simple supported girder bridge 简支梁桥
simple supported system 简支体系
simple vein 单矿脉
simple volcano 单火山
simple water testing 简易压水试验
simple wave 简单波
simple wedge failure 单楔体破坏
simple-supported beam 简支梁
simple-type wave generator 简单式波浪发生器
simplex 单的，单体
simplex algorithm 单纯型算法
simplex jack 单式千斤顶柱
simplex wave 单波
simplicity 简单；简化
simplification 简化；单一化
simplification principle 简化原理
simplified 简化的
simplified approach 简化法
simplified calculation 简化计算
simplified facies-model 简化相模式
simplified instrument 简易仪器
simplified method 简化方法
simplified method of analysis 简化分析法
simplified method of slice 简化条分法
simplified model 简化模型
simplified network model 简化网络模型

simplified solution 简化解
simplify 简化
simplifying assumption 简化用的假设
simply supported beam 简支梁
simply supported end 简支端
Simpson formula 辛普公式
Simpson one third rule 辛普森三分之一法则
Simpson's composite formula 辛普森复合公式
Simpson's formula 辛普森公式
Simpson's rule 辛普森法则
SIMS dating 离子探针定年
simulate 模拟；模型化；制作模型；模型试验
simulate test 模拟试验
simulate thermal equilibrium 仿真热平衡
simulated condition 模拟条件；相似条件
simulated data 模拟数据
simulated domain 模拟区域
simulated earthquake 模拟地震
simulated earthquake load 合成地震荷载
simulated fault motion 合成断层运动
simulated field condition 模拟现场条件
simulated flow 模拟水流
simulated formation 模拟地层
simulated function test 模拟功能试验
simulated ground motion 模拟地震动
simulated leakage 模拟泄漏
simulated line 仿真线
simulated motion 模拟运动
simulated normal color image 模拟真彩色影像；模拟天然色图像
simulated program 仿真程序
simulated response spectrum 合成反应谱
simulated section 模拟剖面
simulated seismic wave 合成地震波
simulated seismogram 模拟地震记录
simulated series 模拟序列
simulated service test 模拟使用试验
simulated sonic log 模拟声波测井
simulated strong motion record 合成强震记录
simulated test 模拟试验
simulating experiment 模拟实验
simulating material modeling of rock engineering 岩体工程相似材料模拟
simulating signal 模拟信号
simulating test 模拟试验
simulation 模拟；仿真
simulation analogue 模拟比拟法
simulation analysis 仿真分析；模拟分析；合成分析
simulation condition 模拟条件
simulation equation 模拟方程
simulation equipment 模拟装置
simulation experiment 仿真实验
simulation experiment system 仿真实验系统
simulation job 模拟作业
simulation lab 仿真实验室，模拟实验室
simulation language 仿真语言
simulation layer 模拟层
simulation manipulation 模拟处理
simulation method 模拟法
simulation model 仿真模型；模拟模型

simulation of groundwater regimes 地下水动态的模拟
simulation of mineral exploration 矿床勘探模拟
simulation of ore drawing 模拟放矿
simulation of stage excavation 模拟阶段开挖
simulation package 模拟程序包
simulation program 仿真程序；模拟程序
simulation rule 模拟准则
simulation scale 模拟比例尺
simulation system 模拟系统
simulation technique 模拟技术
simulation training 模拟训练
simulative 模拟的
simulative generator 模拟振荡器，模拟发生器
simulative network 模拟网络
simulator 模拟器；模拟计算机；模拟电路；模拟设备；模拟程序
simulator program 模拟程序
simultaneous adjustment 联合平差
simultaneous blasting 同时爆破；同步爆破
simultaneous bullet gun perforator 联动子弹射孔器
simultaneous computer 同时操作计算机
simultaneous cross-folding 同时交叉褶皱作用
simultaneous development 同时开发
simultaneous differential equation 联立微分方程组
simultaneous draw texturing machine 同时拉伸变形机
simultaneous drifting 一次成巷
simultaneous drilling 双筒钻井
simultaneous drilling and mucking 钻眼；装岩平行作业
simultaneous earthquake 同时地震
simultaneous equations 联立方程式
simultaneous extraction of several seams 多煤层同步开采
simultaneous faulting 同期断裂作用
simultaneous fill 同时充填；立即充填
simultaneous firing 同时爆破
simultaneous grouting 同时回填注浆；盾构同步压浆
simultaneous injection 同时注入
simultaneous injection well 多层注入井
simultaneous iteration 同时迭代；联立迭代
simultaneous iterative reconstructive technique 同时迭代重建技术
simultaneous level line 双线水准测量；双转点水准测量路线
simultaneous linear algebraic equations 联立线性代数方程组
simultaneous linear equations 联立线性方程组
simultaneous logging 同时测井；同步录井
simultaneous measurement of groundwater level 地下水位统测
simultaneous mining 同时挖掘
simultaneous mode 同步模式观测法
simultaneous observation 同步观测
simultaneous processing 同时处理
simultaneous pumping 同时泵送
simultaneous shafting 一次成井
simultaneous shaft-sinking 一次成井
simultaneous shot 同时爆破；同步激发
simultaneous shotfiring 同时爆破
simultaneous test 对比试验
simultaneous triggering 同时触发
simultaneous walling and sinking 井筒掘砌平行作业

simultaneous work	平行作业	single grained structure	单粒结构
simultaneous-operation computer	同时操作计算机	single heading	单工作面掘进
sine	正弦	single hole	单孔
sine curve	正弦曲线	single hole method	单孔法
sine curve-shaped structure	正弦曲线式构造	single hole open caisson	单孔沉井
sine function	正弦函数	single hole recorder	单孔记录仪
sine integral	正弦积分	single hole recording system	单孔记录系统
sine series	正弦级数	single hydraulic prop	单体液压支柱
sine transform	正弦变换	single hydraulic support	单体液压支柱
sine wave	正弦波	single job	一次操作
sine weighting	正弦加权	single layer	单层
sine-shaped	正弦形的	single layer folding	单层褶皱作用
sine-wave carrier	正弦载波	single line grout curtain	单排灌浆帷幕
sine-wave generator	正弦波发生器	single load	集中荷载；单一载荷
sine-wave oscillation	正弦振荡	single non-central load	偏心集中荷载
sine-wave oscillator	正弦波振荡器；作响	single normal distribution	一元正态分布
single acting centrifugal pump	单动式离心泵	single notching	单面刻槽
single acting hydraulic capsule	单作用液压千斤顶	single observation	单次观测
single acting pump	单动泵	single opening	单一巷道
single action centrifugal pump	单动式离心泵	single pile	单桩
single amplitude	单振幅	single pipe perforation completion	单管射孔完井
single anticline	单背斜；单岛弧	single plane of weakness theory	单个软弱面理论
single aquifer	单层含水层	single point	单点
single axle compressive creep test	单轴压缩蠕变试验	single point extensometer	单点位移计
single bend test	单向弯曲试验	single point load	集中负载；单点负载
single blasting	单眼爆破	single point measurement	单点量测
single borehole	单孔	single point positioning	单点定位
single breast	单工作面；单工祖	single precision	单精度
single chisel bit	一字形钎头；单刃钻头	single purpose computer	专用计算机
single commutator	单式岩芯管	single random variable	简单随机变量
single completion	单层完井	single receiver velocity log	单接收器速度测井
single component	单组分	single reflecting horizon	单反射层
single conductor	单个的传感器	single seam operation	单煤层作业
single core barrel	单管岩芯筒	single section turbodrill	单节式涡轮钻具
single core dynamic method	单岩芯动力法	single seismometer trace	单检波器地震道
single core tube	单管岩芯取样器	single selective completion	分层开采完井
single crystal	单晶体；单晶	single shaft	单轴
single crystal diffractometer	单晶衍射计	single shaft extension	单轴伸长
single crystal growing	单结晶生长	single shear	单剪
single crystal particle	单晶颗粒	single shield TBM	单护盾隧道掘进机
single crystalline materials	单晶体材料	single shot camera	单点测斜照相机
single curvature arch dam	单曲拱坝	single shot directional surveying instrument	单点井斜方位测量仪
single degree of freedom	单自由度	single shot firing	单独炮眼放炮
single degree of freedom system	单自由度体系	single shot grouting	单液灌浆
single dimensional	一维的	single shot tool	单点测量井斜仪
single dipping layer	单倾斜层	single side heading method	单侧导坑法
single direction thrusted pier	单向推力墩	single slice caving	单层崩落采矿法
single drainage	单向排水	single span	单跨
single drive	单个驱动；单独传动	single stall	单房采煤法
single drum tilt rig	单滚筒倾斜钻机	single stall method	单进路柱式采矿法
single dual operation	单一作业	single stope	单面工作面；单独工作面
single effect	单向作用	single stratum	单层
single end	单端	single syncline	单向斜
single fault	单断层	single track	单轨
single flow	单向流动	single track drift	单轨巷道
single fluid grouting	单液注浆	single track railway	单线铁路
single fluid process	单液法	single track tunnel	单线隧道
single force couple	单力偶	single tube core barrel	单层岩芯管
single grain	单粒	single unit face	一个单位的长壁工祖
single grain structure	单粒结构		

single way time	单程时间
single well	单井
single well backflow tracer test	单井示踪剂回流试验
single well injection	单井注水；单井注入
single well oil-production system	单井采油系统
single well pumping	单孔抽水
single well pumping test	单孔抽水试验
single zone compaction	单层充填
single zone completion	单层完井
single zone embankment	均质土堤
single zone gravel pack	单层砾石充填
single zone injection well	单层注入井
single/multiphase flow	单相/多相流
single-arch dam	单拱坝
single-bench open-pit mine	单台阶露天矿
single-boom drill rig	单臂钻车
single-boom jumbo	单臂钻车；单钻机钻车
single-boundary reflection	单界面反射
single-cell porosimeter	单室孔隙度计
single-channel	单信道的
single-channel predictive deconvolution	单道预测反褶积
single-channel seismograph	单道地震仪；单刃钻头
single-chisel bit	一字钻头；单刃钻头；一字钎头
single-circuit	单回路的
single-circuit hydraulic transmission	单循环液力传动
single-coil geophone	单线圈检波器
single-crystalline array	单晶排列
single-curvature arch dam	单曲率拱坝
single-curvature surface	单曲率曲面
single-cycle cratonic basin	单旋回克拉通盆地
single-degree-of-freedom system	单自由度体系
single-end break	单端破裂
single-end wall packer	单头井壁封隔器
single-entry method	单平巷法
single-entry system	单平巷系统
single-grained	单粒的
single-grained arrangement	单粒排列
single-grained structure	单粒结构；散粒结构
single-hand drilling	单人凿岩
single-interval test	单层试井
single-jack bar	单柱式钻架
single-layer	单层的
single-layer fold	单层褶皱
single-layer structure	单层结构
single-line clamshell	单绳式蛤式抓岩机
single-line rock cutting	单线石质路堑
single-line soil cutting	单线土质路堑
single-lined fault	单断层
single-model stereoscopic plotter	单模型立体测图仪
single-packer test	单封隔器地层试验
single-perforation penetration test	单孔射孔穿透试验
single-phase fluid flow	单相渗流
single-phase reservoir	单相储层
single-phase seepage	单相渗流
single-phase thrust	单幕冲断层
single-point test	一点法试井
single-point-entry perforating technique	单点入口射孔技术
single-road stall method	窄房式采煤法；窄场子采煤法
single-row spacing	单排布置
single-run leveling	单程水准测量
singles rig	单根钻机
single-screw column drill mounting	单螺旋柱钻机架
singles-drilling system	单根钻机系统
single-seam dual operation	单一煤层作业
single-shaft development system	单井筒开拓法
single-shaft system	单井筒开拓法
single-shear test	单面剪切试验
single-shot gather	单炮点道集
single-shot inclinometer	单点井中测斜仪；单点摄影测斜仪
single-shot spread	单点爆炸排列法
single-shot survey	单点测斜
single-shot tool	单点井斜仪
single-stage centrifugal booster pump	单级离心式增压泵
single-stage centrifugal pump	单级离心泵
single-stage compression	单级压缩
single-stage crushing	一段破碎
single-stage grouting	一次灌浆
single-stage packing	一步法砾石充填
single-stage pump	单级泵
single-stage pumping	单阶段泵水
single-stage shaft	单水平提升立井
single-stall method	房柱式开采法
single-stall system	无联络巷房柱式开采法
single-step	单步
single-step algorithm	单步算法
single-step procedure	单步过程
single-step recurrence relation	单步迭代关系
single-string injector	单管注水井
single-string selective zone production	单管分层开采
single-stroke deep-well pump	单冲程深井泵
single-suction centrifugal pump	单吸式离心泵
single-target drilling	单目的层钻进
single-track drift	单轨平巷
single-track heading	单线平巷
single-tube core barrel	单岩芯管
single-unit face	一个单位的长壁工作面
single-use bit	一次性钻头
single-use trench	单沟
single-value decomposition	奇异值分解
single-valued function	单值函数
single-variable optimization	单变量优化
single-volute pump	单螺旋泵
single-wall cofferdam	单壁围堰
single-well	单井
single-well formation evaluation	单井地层评价
single-well radial model	单井径向流模型
single-well reservoir	单井油藏
single-well test	单井试井
single-well test analysis method	单井试井分析方法
single-well transient testing	单井不稳定试井
single-zone embankment	均质土堤
singular	单一的；单数的；非正常的；奇异的
singular boundary condition	奇异边界条件
singular eigenfunction	奇异特征函数
singular element	奇异单元
singular element by mapping	奇异单元映射
singular element matrix	奇异单元矩阵
singular element mode	奇异单元模式

singular equation	奇异方程	sinker bar	加重杆；重钻杆；冲击钻杆；撞杆
singular function	奇函数；奇异函数	sinker bar guide	冲击式钻杆导向器
singular integral	奇异积分	sinker drill	凿井用凿岩机；下向凿岩机；冲钻；凿井冲击钻机
singular k matrix	奇异 k 矩阵	sinker leg	立井凿岩机气腿
singular matrix	奇异矩阵；退化矩阵	sinkhole	落水洞；渗坑；灰岩坑；岩溶坑
singular mode	奇异模式	sinkhole deformation	塌陷
singular phase	单相	sinkhole formation	塌陷层
singular point	奇点；奇异点	sinkhole lake	溶坑湖；溶陷湖；落水洞湖
singular solution	奇异解	sinking	下沉；沉落；沉没；沉陷；下挖；凿井；建井；下降流
singular value	奇异值	sinking advance	凿井速度
singular value decomposition	奇异值分解	sinking and drifting	井巷施工
singularities	奇异点	sinking and drifting face	井巷工作面
singularity	奇点；奇异性；奇异点；特殊性	sinking and driving engineering	井巷工程
singularity gradient	奇异性梯度	sinking basin	下沉盆地；下沉洼地；塌陷盆地；陷落区
singularity theory method	奇异性理论方法	sinking by benching	阶段工作面凿井法
sinhalite	硼铝镁石	sinking by boring method	钻井法
Sinian	震旦纪；震旦系	sinking by jetting	射水下沉法
Sinian Epoch	震旦世	sinking by raising	自下向上凿井
Sinian massif	震旦地块	sinking caisson	凿井沉箱；沉箱
Sinian Period	震旦纪	sinking coast	下沉海岸；海侵海岸
Sinian system	震旦系	sinking compartment	延深间，接井间
sinidine	透长石	sinking creek	落水溪；隐没河
sinistral	左侧的；左旋的；左行的	sinking cycle	凿井工作循环
sinistral block	左旋断块；左行位移	sinking cylinder foundation	沉柱基础
sinistral displacement	左旋位移；左行走滑	sinking down	加深
sinistral fault	左旋断层	sinking drum	沉井
sinistral faulting	左旋断层作用	sinking drum method	沉井法
sinistral fold	左旋褶皱	sinking dual operation	井筒掘进；凿井工程
sinistral motion	左旋运动；左行运动	sinking engineer	凿井工程师
sinistral rotation	左旋	sinking equipment	凿井设备
sinistral sense of displacement	左旋位移；左行位移	sinking gang	凿井组
sinistral shear fault	左旋剪切断层	sinking head frame	凿井井架
sinistral shear strain profile	左旋剪切应变剖面，左行剪切应变剖面；左行位移	sinking headgear	凿井井架
sinistral shift	左旋变位；左行位移	sinking jumbo	凿井钻台；凿井钻车
sinistral slip	左旋滑动；左行滑动	sinking lithosphere plate	下沉岩石圈板块
sinistral strike slip	左旋走向滑动	sinking machine	凿井机；井筒开凿机
sinistral strike slip fault	左旋走向滑动断层；左行走向滑动断层	sinking method	凿井法
sinistral transcurrent movement	左旋横推运动	sinking of bore hole	钻孔；凿井
sinistral vergence	左旋倾没	sinking of filling	填土沉陷
sinistral-slip fault	左行滑动断层	sinking of open caisson on filled up island	筑岛沉井
sink	落水洞；渗坑；凿井；下沉；沉降；洗涤盆；水槽；漏水池	sinking operation	井筒掘进
sink a hole	钻孔；钻眼；打钻	sinking pile by water jet	水冲沉桩
sink a shaft	凿井	sinking piling by water jet	水冲沉桩
sink a well	钻一口井	sinking pipe	沉管
sink flow	汇流	sinking pit	开凿中的井筒
sink hole	落水洞；渗坑；渗孔；地面塌陷坑；陷穴；陷落柱；无碳柱	sinking plant	凿井设备
sink holes in carbonate rock	碳酸盐岩体渗坑	sinking platform	凿井吊盘
sink lake	岩溶湖；溶陷湖；喀斯特湖；渗坑湖；落水洞湖	sinking rate	凿井速度
		sinking resistance	下沉阻力
sink line	线汇；汇线	sinking river	伏河；暗河
sink point	点汇；汇点	sinking speed	沉降速度
sink pump	潜水泵	sinking stream	隐没河；伏流
sink well	掘井；沉井	sinking technique	凿井法；凿井技术
sinkage	下沉；凹穴；坑	sinking tower	凿井井架
sinker	轻型风钻；手持式凿岩机；轻型凿岩机；冲钻；凿井工；沉锤	sinking velocity	沉降速度
		sinking vertical shaft by convention	立井普通掘进
		sinking water	下渗水
		sinking well	沉井

sinking with piling	板桩法凿井	siphonage	虹吸；虹吸能力
sinking with pilot shaft	超前小井凿井法	siphonage test	虹吸试验
sinking work	沉井作业	siphonic water collection	虹吸集水
sinking-type tunnel	沉埋式隧道	siphoning	虹吸；虹吸作用
sinkman	凿井工	siphoning effect	虹吸效应
sink-source image method	汇源反映法	siphon-operated lock	虹吸充水式水闸
Sino-Korean massif	中朝地块	sismondine	硬绿泥石
Sino-Korean platform	中朝地台	sismondinite	镁硬绿泥片岩
SINOPEC	中国石化集团；中国石化	sismondite	硬绿泥石
Sinopec Group	中国石油化工集团公司；中国石油化工集团	sitaparite	方锰铁矿
		site	现场；地盘，工地，场地；事发地点；位置；晶位
sinopis	铁铝英土	site amplification	场地放大作用
sinople	铁水铝英石；铁石英	site amplification characteristic	场地放大特性
Sinotype fold	震旦式褶皱	site analysis	建设场地分析
sinter	硅华；泉华；熔渣；矿渣；炉渣；烧结物；烧结	site appraisal	工程地址鉴定；工程场地评价；坝址评价
sinter breccia	硅华角砾岩	site arrangement	现场布置；工地布置
sinter cavity	泉华洞穴	site boundary	工地界线，工地范围；场地边界
sinter cone	泉华丘	site characterisation study	场地特征研究
sinter deposit	泉华沉积	site characterization	场址特性判定；场地特征；场址特性评价；场址评价
sintered bit	烧结的钻头		
sintered carbide	硬质合金	site classification	场地类别；场地分类
sintered carbide-tipped pick	镶焊硬合金尖截齿	site cleaning	场地清理；清理现场
sintered tungsten carbide compact bit	镶碳化钨硬合金齿钻头	site clearance	现场清理
		site coefficient	场地系数
sintered tungsten-carbide insert	烧结碳化钨硬质合金镶齿	site complexity	场址复杂程度；场地复杂程度
		site condition	场地情况；场地条件；场址条件
sintering coal	炼焦煤；焦煤	site confirmation	场址确定
sinterite	泉华沉积	site control	现场控制
sinuation	波状；弯曲	site coverage	建筑区；上盖面积；建筑密度
sinuosity	弯曲；弯痕；曲折；曲折度	site data	现场资料
sinuosity index	弯度指数	site defect	晶位缺陷
sinuous	弯曲的；曲折的；波状的	site density	测点密度
sinuous channel	曲折河道	site design earthquake	厂址设计地震
sinuous course	蜿蜒的河道	site designation memorandum	选址意见书
sinuous flow	乱流；曲折水流	site diary	工地工程日志；值班日志
sinuousness	弯曲；曲折；错综复杂；弯曲度	site disposal	现场处理；野外处置
sinus	海湾；凹地；穴	site distortion	位置畸变；晶位畸变
sinusoid	正弦曲线；正弦波；正弦形的	site distribution	晶位分布
sinusoidal curve	正弦曲线	site drainage	场地排水
sinusoidal distribution	正弦分布	site drawing	区段图；地区图
sinusoidal fold	正弦曲线状褶皱	site effect	场地效应；现场效果
sinusoidal function	正弦函数	site engineer	现场工程师
sinusoidal motion	正弦运动	site equipment	工地设备
sinusoidal projection	正弦投影	site expense	现场管理费
sinusoidal structure	正弦状构造	site exploration	现场勘探；场地勘察
sinusoidal vibration	正弦振动	site facility	工地设施
sinusoidal wave	正弦波	site factor	场地因数
siphon	虹吸；虹吸管；U形山道	site feasibility study	场地可行性研究
siphon action	虹吸作用	site filtering characteristic	场地滤波特性
siphon gauge	虹吸表，虹吸压力计	site geologic condition	现场地质条件
siphon method	虹吸法	site geology	现场地质；场地地质
siphon mine drainage	矿井虹吸排水	site height	测站高程；场地高程
siphon pipe	虹吸管	site influence	场地影响
siphon pit	虹吸坑	site inspection	实地视察；现场勘察；实地检查
siphon pump	虹吸泵；虹吸水泵	site intensity	场地烈度
siphon sluice	虹吸式泄水闸	site investigation	场地勘察；场址勘察；实地勘测；现场调研；现场勘察；现场调查
siphon spillway	虹吸式溢洪道；虹吸溢水道		
siphon string	虹吸管柱	site investigation works	地盘勘测工程
siphon suction	虹吸抽水	site laboratory	工地试验室
siphon suction height	虹吸高度	site load	施工荷载
siphon well	虹吸井		

site location 定位
site management 现场管理
site measurement 现场量度；工量测度；现场测试，实测
site model 现场模型
site monitoring 场地监测，现场监测
site observation 现场观察
site of construction 建筑工地；施工场地
site of medium complexity 中等复杂场地
site of special archaeological interest 具特殊考古价值的地点
site of special scientific interest 具特殊科学价值的地点
site of work 工地
site operation 现场作业
site organization 现场施工组织
site parameter 场地参数
site pile 就地浇灌桩
site plan 场地平面图；现场平面图；总平面图
site planning 总平面设计
site preparation work 现场预备工程
site property 场地特性
site reconnaissance 踏勘；场地踏勘；场地查勘
site record 现场记录
site remanence 原地剩磁
site response 现场响应；场地反应
site response analysis 场地反应分析
site security 现场安全
site selection 选址；场址选择；基地选择
site selection investigation 选址查勘
site settlement 土地沉降
site slope of excavation 挖方边坡
site soil 场地土
site stability 场址稳定性；场地稳定性
site strength 实测现场强度
site stripping 表土剥除
site structure interaction coefficient 场地结构相互作用系数
site surface water drainage system 场地排水系统；工地排水系统
site survey 现场踏勘；场地调查；场地测量；现场勘测；坝址勘测
site survey plan 地盘测量图
site test 现场试验
site trial 实地试验；工地试验
site work 现场施工；现场作业；现场工作
site-acceleration-magnitude procedure 场地-加速度-震级方法
site-dependent response spectrum 场地相关反应谱
site-specific 针对场地的
site-specific response spectrum 场地相关反应谱
siting 选址；定位；定线；定点
situ 就地；在原处；地点；位置
situated 位于…的
six arm dipmeter 六臂地层倾角仪
six boom jumbo 六臂伞形钻架
six point bottom hole reamer 六点式井底扩眼器
six-bladed grab 六瓣式抓岩机
six-entry method 六巷法
sixfold 六次覆盖
six-fold coverage 六次覆盖

sixfold twin 六连晶
six-hole rock drilling machine 六孔钻岩机
six-point bit 六刃钎头；六刃钻头
six-post type headframe 六柱型井架
six-wheel truck 三轴载重卡车；三轴转向架
sizable man-made slope 较大的人造斜坡；大型的人造斜坡
size 粒度；粒级；大小；尺寸；筛分
size analysis 粒径分析；粒度分析；筛分分析；筛分试验
size analysis by sedimentation 沉降法粒度分析
size category fraction 粒级
size character 粒度特性；筛分特性
size characteristics 粒径特性；尺寸特性
size classification 按粒度分选；粒度分类
size composition 大小组成；粗径组成；粒度组成
size control 粒度控制；尺寸控制；尺寸控制器
size cut 筛分粒度
size data 粒度数据；粒径数据
size degradation 粒度减小；块粒碎裂；磨细，粉碎
size deviation 尺寸偏差
size distribution 粒度分布；粒度组成
size distribution curve 粒度分布曲线
size distribution of rock particle 岩石颗粒大小的分布
size division 粒度级别；粒级
size effect 尺寸效应
size effect of rock mass 岩体尺寸效应
size effect of rock strength 岩石强度尺寸效应
size fraction 粒度级别；粒级
size fraction analysis 粒度分析
size fraction experiment 粒度试验
size frequency 粒度百分比数
size frequency curve 颗粒大小分布曲线；粒径曲线
size grade scale 粒径标准
size graded layer 粒级递变层
size grading 粒度分级；粒径级配；粒度组成
size group 尺寸组
size of grain 颗粒大小；粒径；粒度
size of mesh 筛孔尺寸
size of particle 粒径
size of pipe and tubing 管道标称尺寸；管子公称尺寸
size of the hole 井眼尺寸
size parameter 粒度参数
size range 粒度范围，粒度上下限；大小范围，大小限度，尺寸限度
size ratio 粒径比
size reduction 破碎
size reduction dual operation 破碎矿物作业
size reduction ratio 粒度减小的比率
size scale 粒度比；分级比
size separation 分粒；粒析；筛分；分级
size separation process 粒度分级
size sorting 粒度分选
size specification 粒度规格；尺寸规格
size test 筛分试验；筛别试验；筛分分析
size testing 筛分试验；筛别试验；筛分分析
size-consist 粒度组成
size-consist determination 粒度组成测定
sized ore 分级矿石，过筛矿石
size-distribution curve 粒度特性曲线；筛分曲线

size-fraction analysis	粒度分析
size-frequency analysis	粒径频率分析
size-frequency curve	粒径频率曲线
size-frequency distribution	粒度频数分布
size-grade distribution	级配；粒级分配；粒度分布
size-strength classification	尺寸强度双因素岩石分类
size-strength system	尺寸-强度分类系统
size-weight ratio	尺寸-重量比
sizing	筛分；分级；测定尺寸；测定大小
sizing analysis	粒径分析；粒度分析
sizing assay test	粒度分析试验
sizing curve	粒度曲线；筛分曲线
sizing device	筛分设备；分级设备
sizing efficiency	筛分效率；颗粒分级效率
sizing test	筛分试验；筛别试验；筛析
sizing-sorting-assay test	矿物分级、分类和分析试验
S-joint	S节理
skarn	矽卡岩；砂卡岩
skarn mineral	矽卡岩矿物
skarn type deposit	矽卡岩矿床
skarnization	矽卡岩化
skeletal	骨骼的；骨架的；轮廓的；骸晶的
skeletal calcarenite	骸晶砂屑石灰岩
skeletal density	骨架密度
skeletal detritus	骨质碎屑
skeletal diagram	三相图
skeletal facies	骸骨相
skeletal limestone	骨粒灰岩；骸晶灰岩
skeletal sand	骨屑砂
skeletal soil	粗骨土；石质土
skeletal structure	格架状结构，骨架状结构
skeleton construction	骨架构造
skeleton curve	干线；滞变干线；骨架曲线
skeleton diagram	概略图，总图；方框图，方块图
skeleton facies	骨架相
skeleton frame	骨架；框架
skeleton grain	骨粒，颗粒
skeleton layout	草图；初步布置
skeleton line	骨架线
skeleton map	略图
skeleton material	骨骼物质
skeleton model	骨架模型
skeleton reinforced concrete girder	型钢混凝土梁；劲性骨架混凝土梁
skeleton shrinkage	防散矿堆采矿法
skeleton soil	粗骨土；砾质土
skeleton structure	骨架结构
skeleton texture	骸晶构造
skeleton type bucket	骨架式铲斗
skeleton view	透视图
skeletonization	绘制草图
skeletonization process	骨架化过程
Skempton's ultimate bearing capacity formula	斯肯普顿极限承载力公式
skene arch	弧拱
skerry	砂夹层；残岛
sketch	素描；草图；简图，略图；素描；概要
sketch contour	草绘等高线
sketch design	设计草图
sketch drawing	绘制草图；示意图；略图；草图
sketch for the colluvial soils	崩积土图样
sketch geology	素描地质学
sketch isotropy -of mine product	草图；略图
sketch map	示意图；略图，草图
sketch plan	草图，简图；初步计划；草绘平面
sketch projector	投影转绘仪
sketch reference	参考草图
sketch survey	草测
sketched contours	手勾等高线；等高线草图
sketched geological profile	素描地质剖面图
sketching board	图板
sketchy diagram	草图；略图
skew angle	倾角；斜拱角；斜角
skew arch	斜交拱；斜拱
skew brick	斜砖；拱脚砖
skew bridge	斜交桥；斜桥
skew coordinates	斜坐标
skew correction	偏斜校正
skew curve	斜曲线；不对称曲线
skew distribution	非正态分布；偏斜分布
skew hole	斜孔
skew intersection	斜交叉
skew joint	斜接口
skew matrix	斜对称矩阵；扭矩阵
skew plate	斜板
skew supports constraint	斜支撑约束
skew symmetric matrix	反对称矩阵
skew symmetric tensor	斜对称张量
skew tunnel portal	斜洞门
skewback	拱座；起拱石
skewed distribution	非正态分布
skewed four-spot injection pattern	歪四点注水井网
skewed normal distribution	偏正态分布
skewed pattern	不规则井网
skewed slab	泄洪道
skewed slot	斜槽；斜沟
skewed storage	错位存储器
skewing	偏移；时滞；相位差；倾斜；弯曲
skewness	失真，畸变；偏斜；分布不均匀；偏斜度；非对称性
skew-symmetric matrix	圆筒式旋涡
skid force	滑动力
skid pile-driver	滑动打桩机
skid plate	撬板；滑板
skid prevention	防滑
skid resistance	防滑；抗滑度
skid resistance test	防滑试验
skid rig	滑橇式钻机
skid road	滑行路
skidproof surface	防滑面
skid-resistance	抗滑性
skid-resistant surface	防滑面层；防滑表面
skid-resisting capacity	抗滑能力
ski-jump energy dissipation	挑流消能
ski-jump spillway	挑流式溢洪道
skill level	熟练程度
skilled labour	熟练工人；技工
skilled personnel	技术熟练人员
skilled work	技术性工作
skilled worker	熟练工人

skimming wear	滑动磨损
skin bearing coefficient	侧摩擦承载系数
skin current	表面流
skin damage	表皮堵塞
skin depth	皮厚；表皮效应深度；穿透深度
skin drying	表面干燥法
skin effect	趋表效应；集肤效应；表皮效应
skin equation	表皮效应方程
skin factor	表皮因子
skin friction	表面摩擦；表面摩擦力
skin friction coefficient	表面摩擦系数
skin friction force	表面摩擦力
skin friction loss	表面摩擦损失
skin friction of driven pile	打入桩摩阻力
skin friction of pile	桩的表面摩擦力
skin friction pile	摩擦桩
skin grouting	表皮喷浆
skin hardness	表面硬度
skin of shield	盾构外壳
skin resistance	表面摩阻力
skin stress	表层应力
skin temperature	表层温度
skin to skin support	紧密支架
skin wall	附加表墙；表面墙；贴面
skin zone	表皮效应地带
skin-friction drag	表面摩擦阻力
skin-frictional resistance	表面摩擦阻力
skin-to-skin	紧靠着的
skip bucket	翻斗，箕斗
skip capacity	箕斗容量
skip continuous thumper method	跳跃式连续锤击法
skip crane	吊斗起重机
skip door	箕斗门
skip dump	箕斗翻卸器
skip grading curve	不连续级配曲线；断续级配曲线
skip guide	箕斗罐道
skip hoist system	箕斗提升系统
skip hoisting	箕斗提升
skip hoisting plant	箕斗提升装置
skip installation	箕斗提升装置
skip loading arrangement	箕斗装载设备
skip winder	箕斗提升机
skirt resolution	边缘分辨力
skirted foundation	裙桩基础
skirt-supported pressure vessel	裙式支承压力容器
slab and block slide	整体平移式滑坡
slab and girder	板梁结构
slab blasting	片层爆破；刷帮爆破
slab box beam	板箱梁
slab bridge	板桥
slab buckling	平板屈曲
slab culvert	平板式涵管；盖板涵；盖板暗沟
slab dam	平板坝
slab effect	层板作用
slab entry	加宽的巷道
slab face	加宽的工作面
slab footing	平板底脚
slab foundation	板式基础
slab hole	板块崩落炮眼
slab joint	面板接缝
slab jointing	板状节理
slab lagging	板皮背板
slab layer	石板层；平板层
slab matrix block	层状基质岩块
slab off	片帮，片落，层状剥落
slab shed gallery	盖板式棚洞
slab shed tunnel	盖板式棚洞
slab slides	板状滑动
slab staggering	错位
slab stone	石板
slab structure	板块构造，板片构造
slab track	板式轨道
slab vibrator	板振捣器
slab wax	板状石蜡；食品用蜡
slabbed construction	大板结构
slabbed core	岩芯切片
slabbing	片落；剥落；岩板；切块；衬板
slabbing action	劈裂作用
slabbing cut	分片爆破
slabbing method of pillar recovery	刷帮法回收煤柱
slabbing off-cut	衬板下脚料；切余板
slabby	板状的；层状的；稠黏的
slabby coal	层状煤
slabby ore	板状矿石
slabby rock	板状岩石
slab-column frame	板柱框架
slab-column system	板柱结构
slab-like rock mass	板状岩体
slabstone	易成板片岩石；石板；片石；石板岩
slab-surface fault	板状表面断层
slack action	间隙效应
slack adjuster	松紧调整器；松紧螺钉；间隙调整器
slack lime	熟石灰
slack loop	膨胀圈
slack tide	滞潮；平潮；缓流；憩流；憩潮
slack time	延缓时间；闲弛时间；对差；闲置时间
slack variable	松弛变量
slack water	憩流；平潮；滞水；死水
slack water time	平潮期
slacking character	风化特性；崩解特性
slacking index	风化指数；破碎指数
slackness	松弛
slacktip	崩塌；崩解；崩坍
slag	矿渣
slag aggregate	矿渣集料；熔渣骨料
slag cement	炉渣水泥；矿渣水泥
slag concrete	炉渣混凝土
slag expansive cement	矿渣膨胀水泥
slag fill	矿渣填料
slag gypsum cement	石膏矿渣水泥
slag Portland cement	矿渣波特兰水泥
slag sand	熔渣砂
slag stone	熔渣石；矿渣石
slag sulphate cement	石膏矿渣水泥；硫酸盐矿渣水泥
slaggy agglomerate	多孔集块角砾岩
slag-sulfate cement	硫酸盐矿渣水泥
slag-tap type boiler furnace	出渣式锅炉
slake	湿化；水解，崩解
slake durability test	抗崩解持久性试验
slake durability test results of soil material	土料湿化试验

成果统计表
slaked lime 熟石灰；消石灰
slake-durability 耐崩解性；崩解耐久性
slake-durability index of rock 岩石的耐崩解性指标
slaking 崩解性；水解；熟化
slaking characteristic 崩解特性
slaking characteristic of rock 岩石的崩解性
slaking clay 消解黏土；湿化性黏土；水解黏土
slaking of soil 土的崩解性
slaking property 水解性
slaking property of soil 土的崩解性
slaking resistance 崩解阻力
slaking test 崩解试验；湿化试验；水化试验；熟化时间
slaking value 水化度；水化值
slaking water content 消解含水量；湿化含水量
slalom line profile 弯曲测线剖面
slant angle bearing 倾斜方位角
slant chute 斜溜槽
slant crack 斜裂
slant distance 倾斜距离
slant hole 斜井；斜孔；斜直井眼
slant range 斜距
slant shear test 斜剪试验
slant stack 倾斜叠加
slant stack migration 倾斜叠加偏移
slant visibility 倾斜可见距离
slant well 斜井；定向井；倾斜井
slanted strut 斜撑
slanting 倾斜的
slanting cut 斜切
slanting face 伪倾斜工作面；斜工作面；斜交工作面
slanting longwall 伪倾斜长壁开采法
slanting support 斜撑；撑杆
slanting toe 倾斜壁座
slate 板岩，黏板岩；石板，面砖石板；石片；高灰煤；煤中页岩，页岩，泥层页岩；废渣
slate band 板岩夹层
slate clay 板黏土；泥板岩；页岩黏土；黏板岩
slate hanging 墙面铺石板
slate intercalation 板岩夹层
slate pit 页岩露天矿坑
slaty 板状
slaty cleavage 板劈理；片理，板岩劈理；片状黏土；黏板岩
slaty fracture 板状断口
slaty limestone 板状石灰岩
slaty soil 板岩土壤；层状土
slaty structure 板状构造
sledge hammer 消波结构；手用大锤
sledged stone 锤碎岩石
sleeper beam 枕梁
sleeper bolt 轨枕螺栓
sleeper pitch 轨枕间距
sleeper positioning 轨枕定位
sleeper production equipment 轨枕生产设备
sleeper screw 轨枕螺钉
sleeper tamping 枕木捣固
sleeper wall 地垄墙
sleeve bolt 胀壳式锚杆

sleeve coupling 套筒联轴节
sleeve grout 阀套浆液
sleeve grouting 套管灌浆
sleeve joint 套筒接头；套接
sleeve piece 壳筒；嵌环
sleeve pipe 轴套；联轴节；套管；预留管
sleeve pipe grouting 套管法灌浆
sleeve sample chamber 滑套取样器
sleeve tube 套阀管
sleeve valve 套阀；套筒阀
sleeve with clevis and ring 套管铰环
sleeved injection commutator 带橡皮管的注浆管
sleeved pile 套管桩
slender 细长的；微少的
slender beam 引水隧洞
slender body 细长体
slender column 细长柱
slender flexible structure 高柔构筑物结构
slender pointed hydrofoil 组合结构
slender proportion 细长比
slender ratio 长细比
slenderness 细长度；细长
slenderness effect 细长效应
slenderness ratio 预报改正；细长比；长径比；长度直径比
slew rate 转换速度
slewability 可旋转性；可摇摆性；可振荡性
slewable 可回转的；可旋转的
slewable boom 旋转式悬臂
slewable conveyor 旋转式输送机
slewing crane 旋臂起重机；回转式吊机
slewing discharge conveyor 回转卸载运输机
slewing gear ring 转盘
slewing motor 回转机构电动机
slewing radium 回转半径
slewing rate 转换速度
slewing roller 转辊；转盘托辊
slewing speed 回转速度
slice 岩片；切片，薄片，层片，尾片；分层开采；土岩切片；电流层
slice and package 部分与组合
slice base 条带底部
slice correlation 切片对比法
slice cross-section 切片断面
slice cross-section ratio 切片断面比
slice drift 分层平巷；分段平巷
slice height 分层厚度；堆高
slice map 分层图；切片图
slice method 条分法
slice mining 分层开采
slice roadway 分层平巷
slice stoping 分层采矿法
sliced feldspar 页片长石
sliced gateway 分层巷道
slice-dividing method 条分法
slicing 分层开采；切片，薄片；分片作用；分裂作用；下行分层陷落开采法
slicing method 分层开采法
slicing system 分层开采法
slicing-and-caving 分层崩落开采法

slicing-and-filling	分层充填开采法
slick	光滑的；平滑的；光滑层面；擦痕面
slick line	溜灰管
slick pipeline	光滑管道
slick top	光滑顶板；平顺顶板
slicken fibre	擦痕纤维
slicken line	擦线
slickened contact	平坦接触面
slickenside	擦痕面；断层擦痕面；镜岩
slickenside grooving	擦痕槽
slickenside line	擦痕线
slickenside lineation	擦痕线理
slickenside striation	擦痕
slickensided surface	擦痕面
slickensiding	擦痕面
slick-line placer	滑管混凝土浇筑装置
slickness	光滑性
slickolite	直立断续擦痕
slick-water injection test	减水阻注入试验
slidable	可滑动的
slide	滑动；滑移
slide angle	滑动角
slide area	滑动地区；滑坡区
slide axis	滑坡主轴
slide balancer	滑块平衡装置
slide bar	滑杆，导杆
slide bed	滑床，滑坡床
slide block	滑块；滑动断块
slide breccia	滑动角砾岩；滑坡角砾岩
slide calliper	游标卡尺
slide cast	滑动铸型
slide cliff	滑坡后壁
slide coefficient	滑动系数
slide conglomerate	滑动砾岩
slide contact	滑动接触
slide control	滑动控制
slide crack	滑坡裂缝
slide deformation	滑坡变形
slide deposit	滑坡沉积
slide drill	滑道式钻机
slide drilling	滑动钻井
slide extension rail	活轨；滑伸的轨道
slide face	滑动面
slide failure	滑动破坏
slide fault	滑动断裂；滑动断层
slide fit	滑动配合
slide floor	滑动地板
slide form	滑动模板
slide fracture	滑动裂隙
slide gauge	滑动卡尺；游标卡尺
slide gear	滑动齿轮
slide load	滑坡荷载
slide loader	滑斗装载机；耙斗装载机
slide mark	滑动痕；滑痕
slide mass	滑坡体
slide path	坍滑路径
slide plane	滑动面
slide plate	滑床板；滑板
slide prediction	滑坡预测
slide prevention	防止滑移
slide rail	滑轨；防滑围栏
slide rock	坡积岩；崖屑堆
slide rule	计算尺
slide sediment	滑坡沉积；滑坡堆积物
slide sediments facies	滑坡堆积相
slide slope	滑坍边坡
slide structure	滑动构造
slide surface	滑动面；滑脱面；滑面
slide test	抗滑试验
slide theocrat	滑动变阻器
slide thrust force	滑坡推力
slide tongue	滑坡舌
slide track	滑轨
slide treatment	滑坡处理
slide triangle	滑动楔体
slide type expansion joint	套筒伸缩器
slide valve	滑阀
slide valve buckle	滑阀套
slide valve gear	滑阀装置
slide wave	滑行波
slide wedge	滑动楔体
slide wedge method	滑动楔体法
slide weight	游码
slide wire	滑动导线
slide zone	滑动带
slide-and-guide	滑动导向
slide-bar system	滑杆系统
slider	滑块
slider crank mechanism	滑块曲柄机构
slide-resistant pile	抗滑桩
slide-retaining wall	抗滑挡土墙
slide-roll ratio	滑滚系数
slide-type calculator	计算尺式计算装置
slideway	导轨；滑块；滑槽；滑路
sliding	滑动，滑落；下滑，滑移；平移；重力滑动
sliding advance of the support	擦边移架，带压移架
sliding angle	摩擦角；滑动角
sliding axis	滑动轴线
sliding bank	滑岸
sliding bearing	滑动轴承
sliding bed	滑坡床
sliding block	滑动断块；滑块
sliding body	滑动体
sliding boundary	滑坡边界
sliding caliper	游标卡尺
sliding circle	滑动圆弧
sliding circle analysis	圆弧滑动分析
sliding coefficient	滑动系数
sliding collar	滑动轴环
sliding contact	滑动接触
sliding contact connection	滑动接触连接
sliding crown bar	滑动式拱顶纵梁
sliding crown-block	可平移的天车
sliding damper	滑动挡板
sliding deposit	滑坡沉积
sliding displacement mechanism	滑移机理
sliding down of face parting	工作面片帮
sliding earth mass	地滑体；滑动土体
sliding erosion	滑塌侵蚀
sliding face	滑面

sliding factor	滑动因数；滑动系数
sliding failure	滑动破坏；滑移图
sliding figure	滑移图
sliding fill	移动式充填
sliding fit	滑动配合
sliding floor	滑动底盘
sliding force	滑动力
sliding force monitoring	滑动力监测
sliding form	滑动模板
sliding formwork	滑动模板
sliding fracture	滑动裂隙
sliding friction	滑动摩擦
sliding gate	滑动闸
sliding in cut	路堑滑坡；挖方滑坡；坍方
sliding index	游标
sliding integration	滑动求和
sliding isolation	滑移隔震
sliding joint	滑动接头；滑动缝；滑动关节
sliding layer	滑动层
sliding line	滑动线
sliding liquefaction	滑动液化
sliding locus	滑动轨迹
sliding mass	滑坡体；滑动土体
sliding mode	滑开型
sliding motion	滑动；滑移
sliding movement	滑移；滑动
sliding neighborhood	滑动邻域
sliding of embankment	路堤滑移
sliding of foundation	基础滑动
sliding of roadbed	路基滑移
sliding oscillation	滑移振动
sliding parallel-plate viscometer	滑动平行板式黏度计
sliding plane	滑动面；坍滑面
sliding pole	滑杆
sliding reaction	滑动反应；连续反应
sliding resistance	抗滑力；抗滑阻力，滑动阻力
sliding resistor	滑动变阻器
sliding rod	滑杆
sliding rule	计算尺
sliding scale	计算尺；滑尺；物价计酬法；滑准法
sliding seal	滑动式密封装置
sliding seat	滑动座
sliding shear failure	滑动剪切破坏
sliding sheet	滑席
sliding shoe	滑动块；滑瓦
sliding slope	滑动边坡
sliding socket joint	滑动套筒接合
sliding stability	滑动稳定性
sliding stage	滑动阶段
sliding strain	滑动应变
sliding structure	滑移结构；滑动构造
sliding support	可缩支撑
sliding surface	滑面；滑动面
sliding surface algorithm	滑动曲面算法
sliding table	滑台
sliding theory	滑动理论
sliding thrust sheet	滑动冲断岩席
sliding tripod	伸缩三脚架
sliding type building	可滑动式建筑物
sliding under load	载荷作用下滑动
sliding valve	滑阀
sliding vane meter	滑片式流量计
sliding vector	滑动向量
sliding way	滑槽；导轨
sliding wedge	滑楔
sliding wedge roof bolt	滑楔式锚杆；楔缝锚杆
sliding zone	滑动带
sliding-buckling of slope	斜坡滑移-弯曲
sliding-compressive tensile fracturing of slope	斜坡滑移-压致拉裂
sliding-susceptible	易滑坡的
sliding-wedge bolt	可滑动的楔形锚杆
sliding-wedge method	滑楔法
slight damage	轻微破坏
slight pitch	缓倾斜；微倾斜
slight rain	最大落潮流
slight shock	微震
slight slack	细微间距
slightly acidic water	弱酸性水
slightly active	低放射性的
slightly compact	微坚实
slightly compressible fluid	弱可压缩流体
slightly dipping	缓倾斜的
slightly gassy	含微量瓦斯的
slightly gassy mine	低瓦斯矿
slightly metallized	弱矿化
slightly permeable	弱透水的；弱渗透的
slightly sandy shale	细砂质页岩
slightly soluble salt test	难溶盐试验
slightly viscoelastic fluid	微黏弹性流体
slightly water-wet	弱亲水
slightly weathered	弱风化的；微风化的
slightly weathered zone	微风化带
slim hole drilling	小口径钻探
slim hole joint	小眼接头
slime	泥渣，泥浆，沉渣，废渣；岩粉淤泥；矿泥，煤泥，淤泥，黏土
slime water	泥浆水；矿浆水；煤泥水
slime-peat soil	淤泥泥炭土
slimes analysis	煤泥试验；煤泥分析；矿泥试验
slimes dam	泥浆池；矿浆池
slime-swamp soil	淤泥沼泽土
slim-hole well	小口径钻井
sling psychomotor	摇转湿度计
slip	滑动
slip acceleration	滑动加速度
slip along the fault	沿断层滑动
slip amplitude	滑移幅度
slip angle	边坡角；坡变角；滑动角
slip anomaly	坍落异常；滑移异常
slip band	滑移带；平移带；滑动带
slip bed	滑坡基底；滑床
slip bedding	滑动层理
slip block	滑动岩体，滑动块体；滑塌体
slip boundary	滑动边界
slip casting	粉浆浇注；粉浆浇注法
slip circle	滑弧
slip circle analysis	滑动圆弧分析
slip cleavage	滑劈理；微劈理
slip cliff	滑坡壁；滑坡后壁

slip coefficient	抗滑移系数
slip cover	滑顶
slip crack	滑移裂纹
slip deformation	滑动变形
slip direction	滑移方向，滑动方向
slip direction fitting method	滑动方向拟合法
slip dock	桩身
slip dust	煤层裂缝的尘末
slip erosion	滑坡侵蚀
slip face	背风面；背流面；滑落面
slip factor	滑动比
slip fault	滑动断层
slip flow	滑流
slip fold	滑褶皱
slip form	滑模；活动板模；滑动模板
slip form liners	滑模浇注补壁
slip foundation	滑坡基座
slip fracture	滑动破裂；剪切破裂
slip function	滑移函数
slip indicator	滑坡指示器
slip joint	滑节理，剪切节理；套筒连接；伸缩连接；滑动接合
slip line	滑移线；滑动线；滑落线
slip line field	滑动线场
slip mark	滑痕
slip mass	滑体，滑坡体
slip mechanism	滑动机理
slip metre	重量百分率
slip motion	滑动
slip multiple	顺差复接
slip of bar	钢筋滑动
slip of fault	断距；滑距
slip phenomena	滑流现象；滑脱现象
slip plane	滑面；滑移面，滑动面
slip potential	滑动潜力
slip process	滑移过程；滑移操作
slip rate	滑动率；滑行速率
slip rate of fault	断层滑动速率
slip ratio	滑脱比；转差率；滑动系数
slip resistance	滑动阻力
slip ring	集电环；汇电环；滑环
slip scar	滑痕；滑动痕迹
slip scratch	滑痕
slip sheet	滑动层；滑动岩片
slip sheet structure	层滑构造；滑片构造
slip slope	背流面；滑坡
slip soil	易滑土
slip spider	卡盘
slip structure	滑塌构造；滑陷构造；崩滑构造；滑移构造
slip surface	滑面；滑动面；流滑面
slip surface of failure	破坏滑动面
slip switch	滑动道岔；十字形道岔；交叉道岔
slip system	滑移系；滑动系统
slip tectonite	滑构造岩
slip tendency	滑动趋势
slip throw	断层滑距，滑距；纵断距
slip tongue	滑坡舌
slip vector	断层滑动矢量；滑动向量
slip vector of fault	断层滑动矢量
slip vector residual	剩余滑动矢量
slip velocity	下滑速度；沉降速度；滑脱速度
slip volume	气体滑脱体积
slip weakening model	滑动弱化模型
slip zone	滑动带
slip-free drive	自植被土
slip-joint casing pipe	滑动接头套管
slip-off slope	冲积坡；游荡坡，游荡滩
slippage	滑动；滑动量；滑移量；滑程
slippage coefficient	滑脱系数
slippage cracks	滑动裂缝
slippage effect	滑脱效应
slippage factors	滑脱因子
slippage phenomenon	滑脱现象
slippage prevention device	防滑装置
slippage velocity	滑脱速度
slipped area of the fault	断层滑动面积
slipped mass	滑动岩体；滑动块体
slipper mould lining	滑模套壁
slipper pad pump	滑靴泵；滑履泵
slipper piping	滑块活塞
slipper ring	滑环
slippery formation	打滑岩层
slippery seepage of gas	气体滑渗
slipping	滑动；移动；平移；滑移
slipping anchor	滑动锚固
slipping bank	滑塌岸
slipping bar composite support	滑移顶梁支架
slipping fracture	滑动破裂
slipping joint	滑动节理
slipping of abutment footing	桥台基础滑移
slipping of embankment	路堤溜塌；路堤坍滑
slipping of foundation pit of pier and abutment	墩台基坑滑塌
slipping of roadbed slope	路基边坡滑坍
slipping plane	滑动面
slipping surface	滑动面
slipping zone	滑移区
slip-plane medium	滑面介质
slips deformation	斜坡变形
slip-stick effect	黏滑效应
slipstream	滑流；向后气流
slipway	滑道
slit	裂口；裂缝
slit bolt	楔缝式锚杆
slit commutator	有缝管
slit erosion	自动称重仪
slit reinforced concrete shear wall	开缝钢筋混凝土剪力墙
slit trough simulation	隙缝槽模拟
slit up	切开；割缝
slit viscometer	缝隙式黏度计
slit wedge type rock bolt	楔缝式锚杆
slit-wedge rock bolt	楔缝式锚杆
slop cutting	削坡
slop exposure	坡地朝向；坡地方位
slope	斜坡，边坡；斜率；倾斜角；坡度
slope analysis	坡度分析；斜度分析
slope and height factor	斜高比
slope and stope rock	边坡和采场围岩
slope angle	坡度角，坡角；倾斜角

slope aspect	坡度，坡向
slope back	斜坡后部
slope base	坡脚
slope basin	斜坡盆地；汇水盆地；汇水流域
slope benching	坡级；坡台
slope board	坡板
slope break	坡度转折
slope catalogue	斜坡记录册
slope category map	坡度等级图
slope changes	坡度变化
slope character	边坡质地
slope circle	坡面圆
slope class	坡级
slope classification	边坡分类
slope closed	坡面封闭
slope coefficient	斜率；边坡系数
slope compacting machine	边坡夯实机
slope compaction	边坡压实
slope configuration	边坡构形
slope construction	边坡施工
slope control	电流升降调节；陡度调整；斜度调整；边坡控制
slope conveyor	倾斜输送机
slope correction	斜长换算；倾斜校正
slope creeping	斜坡蠕动
slope crest	边坡边缘；坡顶
slope current	倾斜流；信号系统
slope curvature	坡曲率
slope curve	斜率曲线
slope cutting	边坡开挖；切坡，刷坡，削坡
slope debris	边坡滑落物
slope debris flow	山坡型泥石流
slope deflection method	倾角挠度法
slope deformation	边坡变形
slope deformation monitoring	边坡变形监测
slope deposit	坡积物
slope depressurization	边坡减压
slope design	边坡设计；斜坡设计
slope destabilized	陆坡失稳
slope destabilizing	边坡失稳
slope dimension	边坡规模
slope displacement	边坡位移
slope distance	斜距
slope distance of level	阶段斜长
slope drain	山坡排水沟
slope ecological protection	边坡生态防护
slope engineering	边坡工程，斜坡工程
slope environment	边坡环境
slope erosion	边坡冲蚀；坡面冲刷；坡地侵蚀
slope erosion control	坡面冲刷治理
slope erosion treatment	边坡冲刷防治
slope excavation	边坡开挖
slope face	斜井掘进工作面；边坡工作面；坡面
slope failure	边坡失稳；斜坡失稳；边坡破坏；边坡坍塌
slope failure mechanism	边坡破坏机理；滑坡机理
slope fan	斜坡扇
slope flaking	边坡剥落
slope flattening	坡度削平
slope foot	坡脚
slope formation works	斜坡平整工程
slope geometry	边坡几何
slope grade	坡度
slope gradient	边坡倾斜度；坡度
slope grading	边坡坡度减缓；坡面平整
slope immersion	边坡浸水
slope indicator	测斜仪；倾斜指示器
slope information system	斜坡资讯系统
slope instability	边坡不稳定性
slope intercept form	斜截式
slope inversion analysis	边坡反分析
slope land development	山坡地开发与水土保持
slope length	坡长
slope level	测斜仪；测坡度计
slope leveling	斜度测量
slope limit analysis method	斜坡极限分析法
slope line	坡度线；示坡线
slope loess	坡地黄土
slope maintenance	斜坡维护
slope maintenance audit	斜坡维护审核
slope manway	人行坡道；人行上山
slope map	坡度图；斜坡图
slope materials	坡积物
slope meter	坡度计；斜度仪
slope method	坡度法
slope mine	斜井矿山
slope monitoring	边坡监测
slope mouth	斜井口
slope movement	边坡移动
slope near river	临河边坡
slope networking	编篱护坡工程
slope number	斜率
slope of a curve	曲线斜率
slope of deflection	挠曲坡度
slope of easy flow direction	顺向坡
slope of end wall of portal	洞门端墙面坡
slope of formation	地层斜坡
slope of ground	地面斜坡
slope of repose	休止坡度；休止角
slope of river	河床坡降
slope of river bank	河岸坡度
slope of river bed	河床坡降；河底比降；底坡
slope of roof	屋顶坡度
slope of syncline	向斜翼；向斜坡
slope of the stowed material	充填材料斜坡
slope of unity	正切等于1的坡度；45°斜坡
slope of water surface	水面坡度
slope parameter	坡度
slope pattern	坡型
slope paving	斜面铺砌
slope paving machine	网络
slope peg	坡桩
slope phase	陆坡相
slope pillar	倾斜矿柱；斜井保安煤柱，斜井矿柱
slope pipeline	有坡度管线
slope planting	护坡绿化
slope post	坡桩
slope preventive works	斜坡巩固工程
slope profile	边坡外形；边坡轮廓
slope protection	护坡；坡面防护

slope protection by concrete	混凝土护坡
slope protection by dry stone pitching	干砌石护坡
slope protection by gridding	网格护坡
slope protection by masonry mortar stone	浆砌石护坡
slope protection of subgrade	路基坡面防护
slope protection with dry rubble	干砌片石护坡
slope protection with framework	骨架护坡
slope protection with mortar rubble	浆砌片石护坡
slope protection with slurry masonry	浆砌石护坡
slope protection works	护坡工程
slope ramp	斜坡道；引道；泄水隧洞
slope ratio	斜率比；坡比，坡度
slope ratio method	斜率法
slope regarding	斜坡削缓
slope registration	斜坡登记
slope reinforcement	边帮加固；边坡加固
slope reliability	边坡可靠度
slope remedial works	斜坡修葺工程；修补斜坡工程
slope resistance	坡道阻力；坡度阻力
slope retaining	边坡支护
slope revetment	边坡护面
slope safety curve	边坡安全曲线
slope safety factor	边帮安全系数
slope safety hotline	斜坡安全热线
slope scale	边坡规模
slope scale ratio	坡降比尺
slope sedimentation	边坡沉降
slope sinking	斜井掘进
slope slide	滑坡
slope soil	坡面土壤
slope stability	边坡稳定性；斜坡稳定性；斜坡稳定加固
slope stability analysis	边坡稳定分析；斜坡稳定性分析
slope stability factor	边坡稳定系数
slope stability monitoring	边帮稳定性监测
slope stability works	巩固斜坡工程；边坡加固工程
slope stabilization	边坡加固；滑坡防治
slope stabilization by solid bed	固床稳坡
slope stabilization works	斜坡稳定工程；斜坡巩固工程
slope stake	边桩；斜拉桩
slope stake location	边坡桩测设
slope station	坡地测站
slope steepness	边坡陡度；边坡坡度急降带
slope structure	坡体结构
slope subgrade	斜坡路基；坡地型地基
slope supported by vegetation	植物护坡
slope surface	坡面
slope surface erosion	坡面冲蚀
slope surface protection works	斜坡护面工程
slope surveillance	边坡监测
slope tamping	边坡夯实
slope taping	斜距丈量
slope theodolite	坡面经纬仪
slope thrust method	滑坡推力法
slope to zero	坡度平缓到零
slope toe	边坡坡脚；坡底
slope toe level	坡脚标高
slope toe wall	坡脚墙
slope top	边坡顶线，坡顶；斜井口
slope tunnel	斜坡隧道
slope upgrading work	斜坡加固工程
slope value	边坡值
slope wall	斜井井壁；边墙；护坡墙
slope wall angle	边坡角
slope wash	坡积物，坡积土；斜坡侵蚀；坡面冲蚀
slope wash deluvium	坡积物
slope wash facies	坡积相
slope waters	陆坡水域；陆坡水区
slope wind	坡风
slope works	斜坡工程
slope-and-strain-free basin of subsidence	缓坡无应变沉陷盆地
slope-basin facies	斜坡盆地相
slope-change trend	倾斜变化趋势
slope-covered terrace	坡下阶地
sloped footing	斜坡式基脚
sloped foundation	斜坡基础
sloped pile	斜桩
slope-deflection equation	边坡挠度方程
slope-discharge curve	水面坡降流量曲线
slope-distance rule	斜距定则
slope-over-wall structure	坡面消蚀构造
sloping apron	扬程
sloping aquifer	倾斜含水层
sloping baffle	倾斜挡板
sloping base	倾斜基底
sloping bed	倾斜地层；斜层
sloping breakwater	斜坡式防坡堤
sloping core	斜心墙
sloping core earth dam	斜墙挡土坝
sloping core earth-rock dam	黏土斜墙土石坝
sloping difference table	斜坡式差分表
sloping face	斜面
sloping faced breakwater	斜坡式防波堤
sloping fault	倾斜断层
sloping field conflux	坡地汇流
sloping foundation	倾斜基础
sloping ground	倾斜地面
sloping land	坡地
sloping plate	斜盘
sloping platform	斜平台
sloping quay	斜坡式码头
sloping reflector	倾斜反射面
sloping roof	斜面屋顶；倾斜顶板
sloping seam	倾斜煤层
sloping side	斜边
sloping step	斜阶梯
sloping water table	倾斜潜水面；倾斜地下水面
sloping-prone coal face	容易片帮的采煤工作面
sloping-wall-type breakwater	中热水泥
sloshing	晃动
sloshing effect	晃动效应
sloshing frequency	晃动频率
slot	缝
slot and mass method blasting	切槽和大体积爆破法
slot width	缝宽；槽宽
slot-and-wedge bolt	楔缝式锚杆
slot-flow equation	槽流方程
slotted bucket	齿槽式挑流鼻坎
slotted gravity dam	宽缝重力坝

slotted pipe 条缝滤水管；长孔衬管；割缝衬管
slotted roof bolt 开缝顶板锚杆
slotting 掏槽
slough 泥沼；泥坑；崩坍；片帮
sloughing bank 坍岸
sloughing bank of reservoir 水库坍岸
sloughing in earth cut 路堑边坡滑坍；路堑坍方
sloughing of a cutting 开挖坍方
sloughing of embankment 堤岸坍塌
sloughing of sand 砂土崩落
sloughing shale 膨胀性页岩；坍塌页岩
slow ascending flow 缓慢上升的水流
slow banking method 缓慢堆堤法；慢速筑堤法
slow cement 慢凝水泥
slow descending flow 缓慢下降的水流
slow direct shear test 慢剪直剪试验
slow formation 倾斜地层
slow freezing 缓冻；缓凝
slow motion 缓慢运动
slow release 缓释放；缓放出；缓解离
slow sand filter 慢沙滤池
slow setting cement 缓凝水泥
slow shear 慢剪
slow shear test 慢剪试验
slow slippage 缓慢滑动
slow surface wave 慢表面波
slow test 泄水剪力试验
slowing water 流动水
slow-motion screw 下陷率
slow-moving depression 慢移低气压
slow-moving landslide 缓动式滑坡
slow-moving traffic 慢行交通
slowness method 慢度法
slowness-porosity relationship 慢度-孔隙度关系
slow-speed motor 低速发动机
slow-speed plough 低速刨煤机
slow-speed powder 慢速炸药；缓作用炸药
slow-speed roll 低速辊碎机
sludge 矿泥；污泥；泥浆；淤渣；最优分配
sludge bank 淤泥沉积；淤泥滩
sludge circulation 泥浆循环
sludge collecting 矿泥收集
sludge concentration 污泥浓缩
sludge conditioning 淤渣处理
sludge content 污泥含量
sludge density 矿泥浓度；矿泥密度
sludge density index 矿泥浓度指数
sludge deposit 淤渣沉积
sludge discharge pipe 排泥管
sludge drain tube 排泥管
sludge pit 泥浆池；水仓；泥浆沉淀槽；钻泥坑
sludge pump 砂泵；泥浆泵；排泥泵
sludge sample 岩粉样品；污泥样品
sludge sampling 取淤泥样
sludge treatment 泥浆处理；淤塞处理
sludge utilization 污泥利用
sludge volume index 污泥体积指数
sludgy soil 淤泥质粉砂土
sluffing 岩石剥落
slug flow 迟滞流

slugged relay 延时继电器
sluggish flow 缓慢流动；黏滞流
sluggish river 向岸流
sluggish stream 线流；旋转；外冲
sluice 流水槽；流矿槽；掏洗；水槽搬运；水门，水闸；水力冲采
sluice collar 席冲断层
sluice dam 闸坝
sluice door 阻力
sluice entrance 岩台
sluice gate 水闸；冲沙闸门；闸门
sluice gate chamber 砖托梁
sluice operating chamber 有刺铁丝网
sluice operating gallery 有限差分方程
sluice pier 闸墩
sluice pipe 冲洗管；排灌管
sluiced fill 水力填方
sluiceway 冲沙道；泄水道
sluicing 水力冲刷开采；水冲开挖法
sluicing capacity 泄水闸流量；冲刷量；洼地
sluicing chamber 闸室；冲沙室
sluicing-siltation dam 冲填淤积坝；水坠坝
sluicing-siltation method 水坠法
slum 黑滑硬黏土；页岩状煤层；采矿废物
slump 滑移；沉陷；坍落度；湿陷，塌陷；衰落
slump basin 滑塌盆地；滑移盆地
slump bedding 滑动层理
slump belt 崩落地带
slump block 坍塌块体
slump breccia 滑塌角砾岩
slump coherent 黏结滑塌
slump cone 坍落度试验锥；稠度试验锥；坍落度筒
slump consistency test 坍落稠度试验
slump constant 坍落度常数，坍落度
slump deposit 滑坡沉积物；崩移沉积物
slump failure 沉陷；塌陷；错落
slump fan 滑塌扇
slump fault 滑动断层；重力断层
slump fold 滑动褶皱
slump in 井壁坍塌
slump incoherent 松散滑塌
slump limitation 坍落度限值
slump loss 坍落度损失
slump meter 坍落度计
slump seismic facies 滑塌地震相
slump structure 滑动构造；塌滑构造，滑陷构造；水底滑坡构造
slump test 坍落度试验；坍落度法
slumped mass 塌落体；错落体
slumping 滑动；崩塌；坍落；陷落；错落
slumping loess 湿陷性黄土
slumping of fill 填土陷落
slumping settlement 湿陷性沉降；湿陷量
slumping slide 塌落式滑坡，错动式滑坡；圆弧形滑坡
slumping sliding 推动式滑坡
slumping soil 湿陷性土
slump-mass 崩滑体
slurry 泥浆；矿浆；黏土悬浮液；泥状炸药
slurry bored pile 泥浆护壁钻孔灌注桩
slurry coat method 护壁泥浆法

slurry component	砂浆组分	small diameter multiple completion	小井眼多层完井
slurry composition	砂浆成分	small discontinuity	裂隙
slurry concentration	泥浆浓度	small displacement pile	小位移桩
slurry concretion	矿泥固结	small format aerial photography	小像幅航空摄影
slurry consistency	砂浆稠度	small load	小荷载
slurry cut-off	泥浆防渗墙	small pore	小孔隙
slurry density	砂浆密度	small sample	小样本
slurry discharging pipe	泥浆排出管	small scale	小比例尺；小规模的，小尺寸的
slurry drilled pile	泥浆钻孔桩	small scale experiment	小比例试验；小尺寸试验
slurry drilling method	泥浆固壁钻孔法	small scale yielding	小范围屈服
slurry equipment	拌浆设备	small shock	小震动；小冲击
slurry explosive	塑胶炸药；浆状炸药	small strain	小应变
slurry feeding pipe	泥浆输送管	small strain amplitude	小应变幅
slurry flocculation	矿泥絮凝；泥浆絮凝	small water storage project	小型蓄水工程
slurry for preventing collapse of borehole	泥浆护壁	small watershed	小流域
slurry for stability	稳定液	small-bore deep-hole method	小径深孔开采法
slurry grouting	泥浆灌浆	smallest horizontal stress	最小水平应力
slurry handling	泥浆处理	smallest limit	下极限
slurry hardening time	砂浆固化时间	small-felt-area earthquake	小区域地震
slurry injection	泥浆灌注	small-scale chart	小比例尺图
slurry method	泥浆护孔法；泥浆套法	small-scale discontinuity	裂隙
slurry mix density	砂浆混合物密度	small-scale map	小比例尺图
slurry of ceramic	陶瓷泥浆	small-scale temporary project	小型临时工程
slurry pool	泥浆池	small-scale test	小型试验
slurry pressure	泥浆压力	small-scale vibration table	小型振动台
slurry pump	泥浆泵；输浆泵	small-size mechanical rammer	小型夯土机
slurry purification system	泥浆净化设备	smeared shear band model	钝剪切带模型
slurry reclaiming equipment	泥浆回收装置	smearing	玷污
slurry sampler	泥浆取样筒	smearing effect	润滑作用
slurry sedimentation	泥浆沉淀	smectite	蒙脱石；绿胶埃洛石；蒙皂石
slurry separator	泥水分离	SMFE	土力学与地基工程学报
slurry shield	泥水加压盾构	smoke outlet	排烟口
slurry slump	淤泥崩滑	smoke pipe	烟囱
slurry TBM	泥水加压隧道掘进机	smoke prevention and dust control	消烟除尘
slurry technique	泥浆技术	smooth blasting	光面爆破
slurry transport	浆料输送	smooth curve	平滑曲线；平顺曲线
slurry treatment plant	泥水处理设备	smooth faced rubble armored structure	光滑护面堆石结构
slurry trench cutoff	泥浆槽黏土截水墙；泥浆截水墙	smooth joint model	光滑节理模型
slurry trench method	泥浆槽法；槽壁法	smooth laminar flow	平滑层流
slurry trench wall	泥浆防渗墙；地下连续墙	smooth lining	平面井壁；平面衬砌；平滑衬里
slurry viscosity	泥浆黏度	smooth motion	均匀运动；平稳运动
slurry wall	地下连续墙；泥浆槽防渗墙	smooth performance	平稳特性
slurry-fall fill dam	水冲淤坝	smooth profile	平滑断面
slurry-reinforced overburden	泥浆增强的覆盖层	smooth relief	平缓地形
slurry-trench	泥沟	smooth sagging zone	缓慢下沉带
slurry-trench method	泥浆槽法；槽壁法	smooth surface blasting	光面爆破
slurry-wall	泥浆墙	smooth wheel roller	平碾
slush	淤泥	smoothed exponential mean	平滑指数平均数
slush-grout	泥浆灌浆	smoothness	平滑；光滑度
slushing	水力冲填	snap grid	抽点格网
slushing system	水砂充填法；水力充填法	snap-to-grid	格网取整
small amplitude	小振幅；微幅	snow avalanche	雪崩
small amplitude low gradient magnetic anomaly	微幅低缓磁异常	snow classification	徐变率
		snow clearing equipment	清雪设备
small cross section	小断面	snow climate	雪原气候
small damage	轻微破坏	snow course	测雪取样路线
small deformation theory	小变形理论	snow erosion	雪蚀
small diameter bored pile	小直径钻孔桩	snow flood	融雪洪水
small diameter hole	小口径钻孔	snow gauge	量雪计
small diameter hole drilling	小口径钻进	snow hazard	雪害

snow hedge	挡雪屏障
snow load	雪荷载
snow mantle	雪盖；积雪
snow plough	扫雪机
snow protection facility	防雪设施
snow sampler	取雪管
snow slide	雪崩
snow survey	雪深及密度测量
snow sweeper	扫雪器
snow water flood	融雪洪水
snowball structure	雪球构造；旋球构造
snowbank digging	雪堤掘进
snow-drift control	防雪
snow-limit	雪线；雪区
snowline fluctuation	雪线升降
snowmelt	融雪
snowmelt area	融雪面积
snowmelt coefficient	融雪系数
snowmelt erosion	融雪侵蚀
snowmelt flood	融雪洪水
snowmelt hydrograph	融雪径流过程线
snowmelt release	融雪水流
snowmelt runoff	融雪径流
snowpack	絮凝结构
snowstorm	暴风雪；雪暴
soak process	湿化过程
soak solution	浸泡液
soak time	浸泡时间
soakage pit	渗水井
soakaway	渗滤坑；渗水坑
soaked soil	潜育土
soaker	浸渍槽；浸渍剂；均热炉；裂化反应室
soaking retaining wall	浸水挡土墙
soaking sample	浸水试样
soaking subgrade	浸水路基
soaking surface	浸润面
soap test	皂液试验
soapstone	滑石；皂石；块滑石
soapy clay	油性黏土；肥黏土
sobralite	铁锰辉石
social acceptable earthquake mortality	社会可接受地震死亡率
social effect of earthquake	地震社会影响
socket foundation	杯口状基础
socketed pile	嵌岩桩
socketed pipe	套接管
socle	基脚
socle beam	悬臂梁
sod revetment	草皮护坡
sod strip	带状草皮
soda feldspar	钠长石
soda lake	碱湖
soda minette	钠云煌岩
soda mint	碳酸氢钠
soda niter	钠硝石
soda rhyolite	钠疗岩
soda solonetz	苏打碱土
sodaclase	钠长石
soda-lime	碱石灰
soda-lime glass	钠钙玻璃
sodalite	方钠石；钠沸石
sodalitite	方钠石岩
sodalitophyre	方钠斑岩
sodalumite	钠茂
sodar	声雷达；声波大气探测
soda-saline soil	苏打盐土
soda-solonchak	苏打盐土
soddite	硅铀矿
soddy podzolic soil	草生灰化土
sodic metasomatism	钠质交代酌
sodic soil	碱土
sodium absorption ratio	钠吸附比
sodium bentonite	钠膨润土
sodium bentonite mud	钠膨润土泥浆
sodium clay	钠质黏粒；钠质黏土
sodium dichromate	重络酸钠
sodium soil	钠质土
sodium uranospinite	钠砷钙铀矿
soffit	拱腹；涵管内顶面
soffit formwork	底模板
soffit lining	拱底衬砌
soffit of girder	梁腹
soffit scaffolding	拱腹架
soft aggregate	软骨料
soft alluvium loam	软冲积亚砂土
soft base effect	软弱基座效应
soft bottom	软底
soft clay	软黏土
soft clay ground	软土地基
soft coal	软煤；烟煤；劣质烟煤
soft coal seam	软煤层
soft cohesive soil	软黏性土
soft deposit	软土层
soft first story	柔首层
soft floor	软底板
soft formation	软岩层；软地层
soft formation bit	软岩用钻头
soft formation cutter	软岩层切削尺
soft foundation	软土地基
soft ground	软弱地基；软土地面；不稳定地层
soft ground excavation	软基开挖
soft ground floor	柔性底层
soft layer	软层；软弱层
soft lignite	软褐煤
soft mineral	软矿石
soft organic clay	淤泥，软有机黏土
soft packing	柔软填料
soft particle	土粒
soft particle content	软颗粒含量
soft pitch	软沥青
soft rock	软岩
soft rock composite roof	软岩复合顶板
soft rock drift	软岩巷道
soft rock mine	软岩煤矿
soft rock roadway	软岩巷道
soft rock strata	软弱夹层
soft roof	软顶板
soft shale	软页岩
soft shear	软性剪切的
soft shut-in	软关井

soft sludge	软污泥
soft soil	软土
soft soil characteristics	软土特性
soft soil foundation	软土地基
soft soil layer	软弱土层
soft soil section	软土地段
soft surrounding rock	软弱围岩
softened rock	软化岩石
softening coefficient of rock	岩石软化系数
softening effect	软化作用
softening index	软化系数
softening of surrounding rock	围岩软化
softening point	软化点
softening point in hot air	干热软化点
softening point test	软化点试验
softening point tester	软化点仪
softening property of rock	岩石软化性
softening range	软化极限；软化范围
softening temperature	软化温度
soggendalite	多辉粗玄岩
soil	土壤；土层
soil absorption capacity	土的吸附能力
soil aeration	土壤通气性
soil age	土壤年龄
soil aggregate	土粒集合
soil air	土壤空气
soil amplification effect	土壤放大效应
soil analysis	土壤分析
soil anchor	土层锚杆
soil and rock mechanics	岩土力学
soil and rock testing	土及岩石试验
soil and water conservation	水土保持
soil and water loss	水土流失
soil anomaly	土壤异常
soil arch effect	土拱效应
soil assessment	土壤评价
soil auger	土钻；土壤螺旋钻
soil bank	土壤库；土堤
soil bearing capacity	土壤承载能力
soil bearing value	土承载力值
soil behaviour	土壤性能
soil binder	土的胶结物
soil bitumen	沥青土
soil blister	土胀
soil body	土体
soil bolt	土层锚杆
soil boundary	土壤界线
soil building	造土
soil capillarity	土壤毛细管作用
soil carbon	土壤碳
soil cartography	土壤制图学
soil cast	土壤样模
soil category	土壤类型
soil cement	掺土水泥
soil cement processing	水泥土加固法
soil characteristic	土壤特性
soil chemistry	土壤化学
soil class	土纲
soil classification	土壤分类
soil classification triangle	土分类三角图
soil clod	土块
soil cofferdam	土围堰
soil cohesive strength	土凝聚强度
soil collapse	土的塌陷
soil colloid	土胶体
soil color	土色
soil column	土桩
soil column stabilization method	土桩加固法
soil compactibility	土壤密实度
soil compaction	土壤压缩性
soil compaction by lime-soil pile	灰土挤密桩加固法
soil compaction by stone pile	挤密碎石桩加固法
soil compaction index	土壤紧实度指标
soil compaction pile	土桩
soil compaction pile method	挤密土桩法
soil composition	土的成分
soil concrete	黏土混凝土
soil condition	土壤条件；土壤环境
soil conditioner	土壤结构改良剂
soil conductivity measurement	土壤电导率测量
soil conservation	水土保持；土壤保持
soil conservation district	土壤保持区
soil consistence	土壤结持度
soil consistence constant	土壤结持常数
soil consolidation	土固结
soil constituent	土壤成分；土壤组成部分
soil constitution	土壤结构；土壤成分
soil contamination	土壤污染
soil core	土壤柱；土样；土心；钻探土样
soil corrosion	土壤腐蚀
soil cover	土壤覆盖层
soil crack	土壤裂隙
soil creep	土体蠕变；土滑；土爬
soil cut slope	削土坡
soil cutting	土质路堑
soil cycle	土壤循环
soil damping	土的阻尼
soil deformation parameter	土的变形参数
soil degradation	土壤退化
soil densification	土压密
soil density	土壤密度
soil denudation	土体侵蚀作用
soil depletion	土壤耗竭
soil deposit	土壤沉积层
soil deterioration	土体破坏
soil drainage	土体排水
soil dressing	土壤改良
soil drill	土壤钻杆；土钻
soil drought	土壤干旱
soil dynamic property	土动力特性
soil dynamics	土动力学
Soil Dynamics and Earthquake Engineering	土壤动力学与地震工程学报
Soil ecology	土壤生态学
soil element	土单元
soil engineer	土工工程师
soil engineering	土工工程
soil engineering fabric	土工织物；土工构造，土工建筑物
soil erosion	土壤侵蚀

soil erosion index 土壤侵蚀指标；土壤侵蚀指数
soil erosion map 土壤侵蚀图
soil evaporation 土壤蒸发
soil excavation 土方开挖
soil exploration 土质勘探；土质调查
soil fabric 土的组构
soil fines 土的细粒部分
soil flow 流土
soil formation 土体形成
soil freezing prevention 土壤防冻
soil genesis 成土过程
soil geography 土壤地理学
soil geology 土壤地质学
soil gradation factor 土壤级配系数
soil grain 土粒
soil grain distribution test 土壤粒径分布试验
soil grain size accumulation curve 累计土壤粒度曲线
soil granulation 土壤团粒作用；土团粒
soil hardness scale 土壤硬度计
soil heterogeneity 土壤差异性；土壤不均一性
soil horizon 土层
soil humification 土壤腐殖化
soil improvement 地基处理，岩土加固；土质改良
soil in situ 原位土
soil investigation 土质勘察；土质调查
soil investigation works 探土工程
soil lab 土工实验室
soil lathe 修样器
soil layer 土层
soil layer response calculation 土层反应计算
soil legend 土的图例
soil lines of equal pore pressure 土内等孔隙水压线
soil liquefaction 土体液化，土壤液化
soil lump 土块
soil mass 土体
soil mass stability 土体的稳定
soil matrix 土壤基质，土基
soil mechanics 土力学
soil microbiology 土壤微生物学
soil microorganism 土壤微生物
soil mineral 土壤矿物
soil mineralogy 土矿物学；土矿物成分
soil moisture 土壤含水量；土壤湿度
soil moisture belt 土壤水带
soil moisture content 土壤含水量；土壤含水率
soil moisture deficiency 土壤饱和差
soil moisture deficit 土壤的水分不足，土壤缺水
soil moisture film 土的薄膜水
soil moisture meter 土壤湿度计
soil moisture stress 土的水分应力；土的含水应力
soil moisture suction 土壤水分吸力
soil moisture tension 土壤水分张力
soil nail 土钉
soil nailing 土钉法；装设土钉
soil nailing retaining wall 土钉墙
soil nailing wall 土钉墙
soil nutrient 土壤养分
soil organic carbon 土壤有机碳
soil organic matter 土壤有机质
soil packer 压土器

soil paraform 多种土壤组合
soil parameter 土工参数
soil particle 土粒
soil paste 土浆
soil pattern 土样
soil penetration test 土壤贯入试验
soil permeability 土壤渗透性
soil phase 土相
soil physical and chemical property 土壤的物理化学性质
soil physicochemistry 土壤物理化学
soil physics 土壤物理学
soil pile 土桩
soil pile system 土-桩体系
soil pipe 表层套管
soil piping 土的管涌
soil plasma 土壤细粒物质
soil pore 土壤孔隙
soil porosity 土壤孔隙度
soil pressure 土压力
soil pressure cell 土壤测压计
soil profile 土壤剖面
soil property 土壤性质
soil pulverizer 碎土机
soil pusher 推土器
soil quality 土壤质量
soil reaction 地基反力
soil regime 土壤状况
soil reinforcement 土的加筋法，土的加固
soil relief 土方开挖
soil replacement 换土
soil resistance 土壤阻力
soil resistivity 土壤电阻率
soil retardant 防污整理剂
soil road 土路
soil sack cofferdam 土袋围堰
soil salination 土壤盐碱化
soil salinity 土壤盐渍度
soil salinization 土壤盐渍化
soil sample 土壤样品；土样
soil sample analysis sieve 土样分析筛
soil sample barrel 取土器
soil sampler 取土器
soil sampling tube 取土管
soil saving dike 拦土堤
soil section 土壤剖面
soil section sheet 土壤剖面图
soil separate 土壤粒组；土壤颗粒分组
soil separator 土选分机
soil series 土系
soil settlement 土沉陷；地基沉陷
soil shifter 推土机
soil shredder 碎土机
soil skeleton 土骨架
soil slide 土质滑坡
soil slope 土质边坡
soil sloping core 土质斜芯墙
soil sounding device 土触探装置
soil sphericity factor 土壤球度系数
soil stabilization 土体稳定；地基加固
soil stabilization method by sheet pile 板桩加固法

soil stabilization works 土壤加固工程；加固工程
soil stabilizer 土加固剂
soil strain 土应变
soil strata 土层
soil stratification 土的层理；土壤层序
soil stratum 土层
soil strengthening 土的加固
soil stress 土应力
soil stress measurement 土层应力测量
soil stressed zone 地基受力层；地基持力层
soil structure 土的结构
soil subgrade 土质地基
soil suction 土的负孔隙压力；土壤毛管吸力
soil suction potential 土吸力势
soil supporting layer 地基持力层
soil surface 土壤表层
soil survey 土质调查
soil suspension 土的悬浮液
soil swamping 沼泽化
soil swelling 土体膨胀
soil system classification 土壤系统分类
soil tank 装土箱
soil tank modeling 土槽模拟法
soil technology 土工技术
soil temperature 土壤温度
soil test 土工试验
soil test boring 土质试验钻孔
soil testing 土壤试验
soil texture classification 土壤质地分类
soil texture map 土壤质地图
soil theology 土体流变学
soil thermal conductivity 土壤导热系数
soil thermometer 土壤温度表
soil tone 土的色调
soil transfer coefficient 土传递系数
soil treatment 土处理
soil triaxial testing 土的三轴试验
soil type 土类
soil ulminite 腐殖质
soil utilization map 土壤利用图
soil variability 土的变异性
soil water 土壤水
soil water movement 土中水分流动
soil water retention curve 土壤持水曲线
soil water tension 土的水压力
soil water zone 土体饱和带
soil wedge 土楔
Soil Mechanics and Foundation Engineering 土力学与地基工程学报
Soils and Foundations 土和地基学报
solar azimuth 太阳方位角
solder 钎料；焊料；焊锡
solder bead 焊点
solder connection 焊接
solder glass method 熔融玻璃胶结法
solder strip 焊带
soldered joint 焊接头
soldering 锡焊
soldering coal 炼焦煤
soldering iron 焊铁

solderless 无焊料的
solderless joint 无焊连接；扭接；机械连接
soldier arch 立砌平拱
soldier beam 立柱
soldier course 立砌砖层
soldier pile 支护桩
soldier pile with lateral lagging 立桩横板挡土墙
sole 底部
sole fault 基底断层；底部断层
sole flue 底烟道
sole mark 底痕
sole marking 底模
sole of thrust 底基冲断面
sole piece 钢筋填块；垫块；垫木
sole plane 底部冲断面
sole plate 支撑底板，支架底板
sole structure 底面构造
sole thrust 基底逆冲断层，冲断层基底活动面
sole timber 垫木
sole weight 自重
soled boulder 有擦痕巨砾
solfataric clay 温泉余土
solid 固体；固体的
solid angle 立体角
solid arch 实体拱
solid asphalt 石腊青
solid axle 实心轴
solid bearing 固定轴承
solid bed 坚硬层
solid bedrock 坚硬基岩
solid bitumen 固体沥青
solid bituminite 固体沥青质
solid blockage 固体堵塞
solid body 固体
solid brick 实心砖
solid bridging material 固体堵漏材料
solid carbon 固体碳
solid carbon dioxide 固体二氧化碳
solid carbonate deposit 固体碳酸盐沉积
solid carrying capacity 携带固体颗粒能力
solid cement 固结水泥
solid clay unit 实心黏土砌块
solid coal 实体煤
solid colloid 固体胶体
solid column 实体柱
solid combustible mineral 固体可燃矿产
solid concentration 固相含量；固体浓度
solid connection 紧密连接
solid content 固体含量
solid control 固相控制
solid copper wire 硬铜线
solid core recovery 岩芯采取率
solid coupling 刚性联轴节
solid cribbing 密置井框支架；密集井框支护
solid cryogen 固态致冷剂
solid dam 实体坝；重力坝
solid damping 固体阻尼
solid deposit 坚实沉积物；岩石沉积
solid detrital material 固体碎屑物质
solid diffusion 固体扩散

solid displacement	固态位移
solid drawn tube	无缝管
solid floor	实肋板
solid formation	胶结地层
solid frame	实肋骨
solid friction loss	固体摩擦损失
solid fuel	固体燃料
solid fuel gasification	固体燃料气化
solid gravel column	密实的砾石充填柱
solid gravel seal	密实的砾石充填层间密封
solid gravity dam	重力坝
solid gravity wall	重力式挡土墙
solid ground	稳固地层
solid hardness	实体硬度
solid joint	刚性接合；固定接合
solid liquid	稠液
solid load	固相含量
solid masonry infill wall	实心砖石填充墙
solid mass	固体
solid material	固体材料
solid matrix	骨架
solid matter	固体物质
solid measure	体积
solid medium	固体介质
solid mineral fuel	固体矿物燃料
solid model	实体模型
solid modeling	实体建模；实体模型化
solid of revolution	回转体；实体旋转
solid ore	未采矿石；原矿体，整体矿石
solid packing	密实充填；全部充填
solid paraffin	固体石蜡
solid particle	固体颗粒
solid phase	固相
solid phase content	固相含量
solid pier	实体桥墩
solid pier foundation	实体支墩基础
solid pillar	实心支柱
solid pipeline	输送固体的管线
solid plug	实心堵头
solid plugging	固相堵塞
solid raft	实体筏基
solid removal	固体颗粒脱除；固相清除
solid removal equipment	固体颗粒消除设备
solid residue	固体残渣
solid retention	防砂
solid roof	稳固底板
solid skeleton	固体骨架
solid snap thread gage	整体螺纹卡规
solid stowing	密实充填
solid stress	固体应力
solid surface	固体表面
solid tube	整体取土管
solid upper member	固体上顶梁；可靠的上顶梁
solid wall	实体墙
solid walled structure	承重墙结构
solid web girder	实腹梁
solid weighting material	固体加重材料
solid wire	实线；单股线
solid wire electrode	实心焊条
solid wireline	实心钢丝
solidification	固化，固结
solidification cracking	硬化裂纹
solidification grouting	硅化灌浆
solidified carbon dioxide	固化二氧化碳
solidifying	硬化
solidifying and stabilizing effect	固化与加固效果
solidifying oil sludge	固化油泥
solidity	坚固性；固态
solidity ratio	硬度比
solidly packed tunnel	充填密实的炮眼孔道
solidness	硬度；硬性
solids precipitation	颗粒沉淀
solidus	固相曲线
solifluction	融冻泥流
solodic soil	脱碱化土壤
solodization	脱碱化
solonetz	碱土
soluble silicate	水玻璃
solute transfer model	地下水溶质运移模型
solution	溶蚀
solution breccia	溶塌角砾岩
solution cave	溶蚀洞穴
solution channel	溶蚀槽
solution channel spring	溶洞泉
solution creep	溶解蠕变
solution ditch	溶沟
solution erosion	溶蚀
solution groove	溶沟，溶蚀沟
solution lake	溶蚀湖
solution mining	溶液采矿
solution opening	溶洞
solution platform	溶蚀台地
solution polymerization	溶液聚合
solution porosity	溶解孔隙度
solution pressure	溶解压；溶解压强
solution sink	落水洞；溶坑
solution sinkhole	溶蚀落水洞
solution slot	溶槽
solution transfer	溶移
solutional phase	溶蚀相
solvsbergite	细碱辉正长岩
sonar	声呐
sonar beacon	声呐信标
sonar caliper	声呐井径仪
sonar dome	声呐导流罩
sonar equipment	声呐装置
sonar image	声呐图像
sonar indicator	声呐指示器
sonar navigation	声呐导航
sonar operating range	声呐作用距离
sonar operational performance	声呐工作性能
sonar reference intensity	声呐参考强度
sonar reflector	声呐反射器
sonar signal processing	声呐信号处理
sonar surveillance system	声呐监测系统
sonar sweeping	声呐扫海
sonar target	声呐目标
sonar target classification	声呐目标分类
sonar test set	声呐测试仪
sonar transducer	声呐换能器

sonar unit	声呐装置
sonde	探头
sonde axis	下井仪轴线
sonde body	探头体；线圈系
sonde error	线圈系误差
sonde velocity correction	探测器速度校正
sonic altimeter	声波测高计
sonic analyzer	声波探伤仪
sonic apparent formation factor	声波视地层因素
sonic approach	声波法
sonic boom effect	声震音响效应
sonic cleaning	超声波清理
sonic delay line	声延迟线
sonic depth finder	声波测深仪
sonic detector	声波探测仪
sonic digitizer	声控数字化器
sonic drill	振动钻孔机；声波钻头
sonic echo depth finder	回声测深仪
sonic echo sounder	回声测深仪
sonic fluid-level sounder	回声测液面仪
sonic formation attenuation factor	声波地层衰减系数
sonic frequency response correction	声波频率响应校正
sonic gauge	音响传感器；声波应变计
sonic inspection	声波检验
sonic interface detector	回声界面探测器
sonic leak detection	声法测漏
sonic leak pinpointing unit	声波检漏装置
sonic location system	声波定位系统
sonic locator	声定位仪
sonic log	声波测井；声波测录
sonic logging	声波测井
sonic logging cartridge	声波测井下井仪电子线路部分
sonic logging panel	声波测井面板
sonic logging sonde	声波测井探头
sonic nozzle flow prover	声速喷嘴流量检定装置
sonic pile driver	声波打桩机；振动打桩机
sonic porosity	声波孔隙度
sonic positioning method	声波定位法
sonic probe	声波探头
sonic profile	声波剖面
sonic prospecting	声波勘探；声波法探测
sonic pulse technique	声波脉冲技术
sonic ranging	声波测距
sonic receiver	声波接收机
sonic sand detection	声波探砂
sonic sand detector	声波探砂器
sonic scanner	声波扫描仪
sonic seismogram	声波地震曲线
sonic sensor	声波传感器
sonic shear stability	声速剪切稳定性
sonic sounding	回声测深；声波测深
sonic sounding gear	回声测深仪
sonic test	声波测试
sonic tide gauge	声波验潮仪
sonic tool	声波测井仪
sonic transducer	声传感器
sonic velocity	声速
sonic volumetric scan	声波体积扫描
sonic wave train	声波波列
sonication	声裂法
sonicator	近距离声波定位器
sonigage	超声波金属厚度测量仪
sonigauge	超声波测厚仪
soniscope	脉冲式超声波探伤仪；声测仪
sonobuoy	声呐浮标，声呐浮筒
sonobuoy array	浮标台阵
sonochemistry	超声波化学
sonogram	频时图
sonograph	地震波谱分析仪；声谱仪
sonometer	振动式频率计
sonoprobe	声波探测器
sordawalite	玄武玻璃
Sorel cement	索瑞尔水泥；镁石水泥
sorption of carbon-dioxide	二氧化碳吸附
sorption of methane	甲烷吸附
sorted bedding	分选层理
sorted landslide debris	经分选的滑坡泥石
sorted polygon	形成龟裂纹
sorting belt	拣矸区，手选带
sorting coefficient	分选系数
sorting coefficient of pores	孔隙分选系数
sotch	灰岩洼地
soughing bank	塌岸；容易变得危险
sound bedrock	坚硬岩基
sound borer	土钻
sound cement	安定性水泥
sound foundation	坚实地基
sound rock	完整岩石
sound unaltered rock	坚固完整岩石；坚固新鲜岩石
sounding	探测
sounding borer	试探器；触探钻
sounding datum	深度基准面
sounding equipment	测深仪
sounding lead	测深锤
sounding level	测时水位
sounding line	测深线
sounding machine	测深仪
sounding method	触探法
sounding pipe	测深管
sounding well	探井
sounding wire	测深线
soundness classification of rocks in-situ	原地岩石弹性模数比分类法
soundness of rock	岩石的坚硬性
soundness test	坚固性试验
soupy mortar	稠灰浆
sour crude	含硫原油
sour crude oil	含硫原油
sour earth	酸性土
sour formation	含硫层
sour gas	含硫气
sour gas line pipe	含硫气体管线管
sour gas well	含硫气井
sour natural gas	酸性天然气
sour product	含硫油品
sour water stripping	含硫污水汽提
Source acceleration	震源加速度
source array	震源组合
source bed	生油层；源层
source data	源数据

source depth	震源深度
source dimension	震源大小
Source displacement	震源位移
source distance	震源距
source earthquake	源地震
source effect	震源效应
source field	震源场
source geology	震源地质
source ghost	震源虚反射
source kinematics	震源运动学
source level	震源强度
source location	震源位置
source mechanism	震源机制
source mechanism array	震源机制台阵
source model	震源模型
source moment	震源矩
source of collapse	崩塌物源
source of secondary disaster of earthquake	地震次生灾害源
source of seismic energy	地震能量来源
source parameter	震源参数
source pattern	震源组合
source radius	震源半径
source region	震源区
source residual static	震源剩余静校正
source rock	烃源岩
source rock evaluation	油源岩评价
source rock of petroleum	生油岩
source sag	生油凹陷
source sampling	源采样
source sequence	生油层系
source sink term	源汇项
source spectrum	震源谱
source strength	震源强度
source tensor	震源张量
Source velocity	震源速度
sourceless seismic exploration	无源地震勘探
sources of tectonic stress	构造力源
sources sag	生油凹陷
source-sink matching model	源汇匹配模型
sourdine	弱音器；静噪器
South America plate	南美板块
South China tectonic stress region	华南构造应力区
south geographical pole	地理南极
Southern chernozem	南方黑钙土
souzalite	基性磷铝铁石
sovite	黑云碳酸岩
sowback	底板隆起；马脊岭，等斜山脊
spacer bar	隔离杆；间隔条
spacer fluid	隔离液
spacer lug	间隔凸块
spacer remote sensing	航天遥感
spacer tubing	间隔油管
spacing boss	定位突
spacing interval	井距
spacing lug	间隔凸块
spacing of bars	钢筋间距
spacing of blast hole	炮眼间距
spacing of discontinuities	不连续面间距；结构面间距
spacing of reinforcement	钢筋间距
spacing of stirrup legs	箍筋肢距
spacing of tanks	油罐间隔
spacing of wells	井的布置
spacing pattern	井网型式
spacing pulse	间隔脉冲
spacing ratio	井距比
spaciousness	宽广
spad	矿山井下测量用钉
spadaite	红硅镁石
spader	挖土机
spading	铲掘
spading machine	铲土机
spall drain	碎石排水沟
spalling	剥落；劈裂
spalling concrete	混凝土剥落
spalling floor	手锤碎矿场
spalling hammer	碎矿锤；碎石锤
spalling hazard	剥落病害
spalling of rock	落石，掉块
spalling off	剥落，片帮
span	跨度；间距
span by span method	移动支架逐跨施工法
span clearance	跨度净空
span dip	跨距弧曲度
span length	巷道宽度
span of an opening	硐室跨度
span of arch	拱跨
span of beams	梁跨度
span of jaw	钳口开度
span potentiometer	量程电位器
span rope	绷绳；牵引钢丝绳
spandite	锰钙铁榴石
spandrel	拱肩；上下层窗空间
spandrel arch	腹拱；拱肩拱
spandrel beam	外墙托架；系梁；圈梁
spandrel column	拱肩柱
spandrel structure	拱上结构
spandrel wall	拱上侧墙；窗肚墙
spanline	悬跨管线
spanner wrench	扳手
spanning	跨越
spanning of river	穿越河流
spar	晶石
spar platform	立柱浮筒式平台；筒状平台
sparagmite	破片砂岩
spare anchor	备用锚
spare bucket	备用铲斗
spare equipment	备用设备
spare face	备用回采工作面
spare module	备用组件
spare parts inventory	备件库存
sparge	喷射
sparge water	冲洗水
sparger	喷洒器；分布器；配电器；起泡装置
sparite	亮晶；亮晶方解石；亮晶灰岩
spark array	电火花组合
spark arrester	阻火器
spark arrester netting	火星网
spark chamber	火花室
spark drill	电火花钻井

spark erosion 电火花腐蚀
spark micrometer 火花放电显微计
spark over 绝缘击穿
sparker survey 电火花震源的地震勘探；浅层气勘探
sparry allochemical rock 亮晶异化灰岩
sparry calcite 亮晶方解石
sparry coal 裂缝含方解石脉的煤
sparry intraclastic calcarenite 亮晶内碎屑砂屑灰岩
sparry limestone 颗粒大理石
sparse coefficient array 稀疏系数矩阵
spartaite 含锰灰岩
spartalite 红锌矿
spathic iron 菱铁矿
spatial analysis 空间分析
spatial attribute 空间特征；空间属性
spatial clustering 空间集群法
spatial data infrastructure 空间数据基础设施
spatial data processing 空间数据处理
spatial data structure 空间数据结构
spatial data transfer 空间数据转换
spatial discretization 空间离散化
spatial distribution 空间分布
spatial distribution of hypocenter 震源的空间分布
spatial distribution of seismic load 地震荷载的空间分布
spatial domain 空间域
spatial feature 空间特征
spatial geologic pattern 空间地质图像
spatial information system 空间信息系统
spatial model 空间模型
spatial modeling 空间建模；空间模拟
spatial pattern recognition 立体图形识别
spatial phase spectrum 空间相位谱
spatial polar coordinate 球面坐标
spatial resolution 空间分辨率；空间分辨力
spatial sampling 空间采样
spatial scale 空间尺度
spatial variability 空间变异性
spatter cone 溅落熔岩锥
spatter loss 飞溅损失
spatter shield 防溅挡板
special adapter 专用接头
special casing 专用套管
special cement 特种水泥
special character of deposit 矿床的开采和机械特性
special core analysis 专项岩芯分析
special depth 特殊水深
special drivage 特殊掘进法
special engineering geology 专门工程地质学
special engineering structure 特种工程结构
special first-order triangulation 专用一等三角测量
special foundation 特殊基础
special geophysical processor 专用地球物理处理机
special hydrogeological map 专门性水文地质图
special inspection 专检
special marine engineering geologic map 专用海洋工程地质图
special marine engineering geology 专门海洋工程地质学
special rotating cutting machine 专用旋转切削机
special shaft sinking method 特殊向下凿井法
special shaped column 异型柱
special slide rule 专用计算尺
special soil 特殊土
special treatment for redevelopment 土地整治复田的补充处理
special use map 专用地图
specific absorption 吸收率；比吸水量
specific acoustic impedance 比声阻抗
specific area 比面积
specific capacity 比容量；比功率；功率系数；比充电率
specific density 比重；相对密度
specific density of solid particle 土粒相对密度
specific digging resistance 单位挖掘阻力
specific discharge of rainfall 单位面积降雨流量
specific disintegration 比破碎体积；单位功所破碎的岩石体积
specific drawdown 单位降深
specific gravity curve 比重曲线
specific gravity hydrometer 液体比重计
specific gravity of rock 岩石容重
specific gravity of soil 土的比重
specific gravity of soil particle 土粒相对密度
specific gravity test 比重试验
specific grout absorption 单位吸浆量；比吸浆量
specific load 单位载荷；比载荷
specific moisture content 单位含水率
specific penetration resistance 比贯入阻力
specific permeability 比渗透率
specific power 比功率
specific pressure 比压；单位压力
specific ratio 比率
specific rock removal 比破岩量
specific storage of aquifer 含水层单位储存量
specific storativity 单位贮水系数；单位释水系数
specific strength 单位强度
specific surface 比表面
specific surface area measurement 比表面积测试
specific surface area of particle 颗粒比表面积
specific volume 比容
specific volume power 单位体积功率
specific volumetric fracture work 单位体积破碎功
specific water absorption 单位吸水量
specific water capacity 单位容水度
specific water yield 给水度
specific weight 比重
specific yield 出水率；单位产水量；单位给水量；给水度
specification error 特征误差
specified concrete 特种混凝土
specified criterion 技术规范
specified data 修正数据；确定数据
specified gravel particle 规定的砾石颗粒
specified gravel size range 规定的砾石尺寸范围
specified load 额定荷载；设计荷载
specified minimum yield strength 额定最小屈服强度
specified mix 指定配比
specified penetration 指定贯入度
specimen chamber 样室
specimen coordinate system 样品坐标系
specimen copy 样本

specimen geometry 试样几何形状
specimen mould 试件模
specimen of ore 矿石标本
specimen ore 矿样
specimen preparation 试样制备；样品制备
specimen test 样品试验
specimen thickness 试样厚度
speckstone 滑石
spectacles type tunnelling method 眼镜式开挖法
spectra of seismic waves 地震波谱
spectral log clay evaluation 用能谱测井资料评价黏土
spectral photography 光谱摄影学
spectral resolution 光谱分辨率
spectral stratigraphy 光谱地层学
spectral technique 频谱技术
spectroscopic analysis 光谱分析
spectrum character of ground feature 地物波谱特征
spectrum of composite seismic signal 合成地震信号谱
spectrum resolving power 光谱分辨力
specular alabaster 镜雪花石膏
specular coal 镜煤
specular hematite 镜铁矿
specular iron ore 镜铁矿
specularite 镜铁矿
speed cement 快速水泥
speed concreting 快速浇筑混凝土
speed fluid drive 液压变速器
speed hump 缓冲路拱
speed log 速度计
speed reducer 减速路拱
speedlight 闪光管
speedmuller 快速混砂机
speedomax 电子自动电势计
speedy drivage 快速掘进
speedy floatation 快速浮选
spelean deposit 洞穴沉积
speleogenesis 洞穴形成过程
speleothem 洞穴堆积物；钟乳石
spelter 粗锌
spencerite 单斜磷锌矿
spencite 硅硼钙钇矿
spending beach 消波滩
spending rate 消耗速度
spent acid 废酸；余酸
spent fuel storage facility 核废料储存设施
spent gas 废气
spent liquor 废液
spent radioactive source 废放射源
spent reagent 废试剂
spergenite 鲕粒生物碎屑灰岩；鲕粒灰岩
sperone 黑榴白榴岩
sperrylite 砷铂矿
spessartine 锰铝榴石；闪斜煌斑岩
sphaerite 菱磷铝石
sphaerocobaltite 菱钴矿
sphaerosiderite 球菱铁矿
sphagnum bog 水藓沼泽
sphagnum peat 水藓泥炭
sphalerite 闪锌矿
sphene 楣石

sphenitite 钛辉楣石岩
sphenoconformity 楔状整合
sphenolith 岩楔
sphenolithus 楔形石
sphere gap 球间隙
sphere ore 环状矿石
sphere packing 球粒人造多孔介质模型
sphere stress tensor 球应力张量
sphere tee 通球形三通
spheric function 球函数
spheric grain 球形颗粒
spheric stress 球应力
spheric structure 球状构造
spherical buoy 球形浮标
spherical cap tripod 球面盘三脚架
spherical cavity 球形空腔；球形空穴
spherical component of stress 球形应力分量
spherical coordinate 球面坐标；球坐标
spherical crystal 球状晶体
spherical dam 球面防水墙
spherical distance 球面距离
spherical disturbance 球面扰动
spherical projection 球面投影
spherical radial flow 球面径向渗流
spherical radial seepage 球面径向渗流
spherical splitting experiment 球形劈裂试验
sphericity test 球形检验
spherite 球形重矿物；沉积球粒
spheroclast 圆碎屑
spherocobaltite 菱钴矿
spherocrystal 球晶
spheroid cavity 球状洞穴
spheroid tank 扁球形油罐
spheroidal geodesic 椭球面测地线
spheroidizing annealing 球化退火
spherolite 球粒
spherolitic 球粒状的
spherophyre 球粒斑岩
spherosiderite 球菱铁矿
spiauterite 纤维锌矿
spiculite 锤雏晶
spider beams 辐式横梁
spider guard 十字护板
spider line 十字瞄准线
spiegel 镜铁
spigot discharge 排砂
spiling 插板桩
spilite 细碧岩
spilitic texture 细碧结构；细碧组织
spilitization 细碧岩化
spill plate 防溢板，挡煤板，挡板
spill weir 溢流堰
spilling breaker 崩顶碎波
spilling plane 溢出面
spilling water 溢水；溢流
spillover 溢出；溢出量；信息漏失；泄漏放电；附带结果
spillway 溢洪道；泄洪道
spillway dam 溢流坝
spillway gate 溢洪闸；溢水闸

spillway structure 溢流结构
spillway tunnel 泄洪隧洞
spilosite 绿点板岩
spindle breaker 回转式破碎机
spindle type drill 立轴式钻机
spine road 知平巷
spinel 尖晶石
spiral arrangement of holes 螺旋形炮眼布置
spiral bit 麻花钻头
spiral blade stabilizer 螺旋翼片稳定器
spiral classifier 螺旋分级机
spiral coal cleaner 螺旋选煤器
spiral collar 带螺旋形沟槽的钻铤
spiral concentrator 螺旋选矿机；螺旋分选机
spiral densifier 螺旋浓缩机；螺旋稠化机
spiral deviation 钻孔螺转偏斜
spiral flow mixer 涡流式搅拌机
spiral grooved drum 螺旋绳槽滚筒
spiral hole follower 螺旋状钻孔随动器
spiral reinforcement 螺纹钢筋；螺旋钢筋
spiral sensor 测扭仪
spiral separator 螺旋选矿机
spiral separator chute 螺旋选矿机的溜槽
spiral stabilizer 螺旋式稳定器
spiral stairway 盘梯
spiral staple chute 暗井螺旋溜槽
spiral steel 螺旋钢筋
spiral stirrup 螺旋箍筋
spirit gauge 酒精比重计
spirit level 气泡水准仪
spirit level axis 水准器轴
spirit level wind 水准器偏差
spirit leveling 水准测高
spirit leveling instrument 气泡水准测量仪
spiroffite 碲锌锰矿
splay fault 八字形断层系
splay sandstone 决口扇砂岩
splay structure 分支构造
splayed abutment 八字形桥台
splayed arch 八字墙拱
splayed footing 八字形基脚
splayed jamb 八字形侧墙
splent coal 暗硬煤
splice bar 拼接钢筋；鱼尾板
splice welding 绑条焊接
spliced pile 拼接桩
splicing detonating cord 铰接导爆索
spline joint 填实缝，方栓接缝
splint 暗硬煤
splintered fault scarp 参差断层崖
splintery fracture 片状断口
split barrel 对开取土器；可拆式岩芯管
split blasting 劈裂爆破；碎裂爆破
split coal 夹层煤
split fault 开口断层
split Hopkinson pressure bar 分离式霍普金森压杆；霍普金森压杆
split kelly bushing 拼合方钻杆补芯
split level rig 双高度钻机
split process 劈裂作用；劈裂过程

split seam 分岔煤层
split spoon sampler 对开式取样器
split test 劈裂试验；巴西试验
split tube sampler 对开管式取样器
split wedge roof bolt 叉楔顶板锚杆
splitting and peeling off 劈裂剥落
splitting board 风障
splitting chisel 开尾凿；分叉钎头
splitting failure 劈裂破坏
splitting force 破碎力
splitting of heavy emulsion 重乳状液的破裂
splitting parameter 分裂参数
splitting resistance 抗裂性
splitting strength 劈裂强度
splitting tensile strength 劈裂抗拉强度
splitting tensile test 劈裂抗拉试验
split-wedge roof bolt 缝楔顶板锚杆
spodic horizon 灰化层；灰壤淀积层
spodiophyllite 叶石
spodiosite 氟磷钙石
spodite 长石玻屑火山灰
spodumene 锂辉石
spoil 弃土
spoil area 废料场
spoil bank 弃土堆
spoil bank conveyor 废石堆输送机
spoil chute 废石溜槽
spoil coal 脏煤
spoil disposal 废土弃置；排泥
spoil disposal track 排土线
spoil dump slip 排土场滑坡
spoil ground 弃土场
spoil heap 废石堆
spoil heap edge 矸石堆边缘
spoil heap incline 废石堆斜面
spoil heap rail track 废石堆轨道
spoil heap settlement 废石堆沉降
spoil pile 矸石堆，废石堆
spoil reclaiming 矸石堆复田；排土场复田；尾矿再选
spoil site 弃碴场
spoil stability 排土场稳定性
spoil terrace 排土台阶
spoil toe 弃碴坡脚
spoiling 出碴
sponge chert 海绵骨针燧石
sponge coke 海绵状石油焦
sponge core barrel 海绵套岩芯筒
sponge coring 海绵取芯
sponge filter 海绵滤器
sponge oil 贫油
sponge rubber 海绵橡皮
sponginess 海绵状
spongodiatomite 骨针硅藻岩
spongolite 海绵岩
spongy filter cake 海绵状泥饼
spongy layer 海绵状层
spongy parenchyma 海绵组织
spongy rock 多孔岩石
spongy soil 松软土
spongy structure 海绵状结构

spontaneous caving	自然垮落；自然崩落
spontaneous combustion	自燃
spontaneous combustion coal	自燃性煤
spontaneous combustion of heap	矸石山自燃
spontaneous combustion period	自燃发火期
spontaneous decomposition	自然分解；自发分解
spontaneous earthquake	天然地震
spontaneous expansion	自行膨胀
spontaneous explosion	自然爆炸；自爆
spontaneous fault rupture	自发断层破裂
spontaneous fire prevention	自燃预防
spontaneous fission	自然分裂
spontaneous gas emission	瓦斯突出
spontaneous heating	自然升温；自热
spontaneous ignition	自燃，自然发火
spontaneous magnetization	自发磁化
spontaneous mine fire	矿内自燃火灾
spontaneous potential logging	自然电位测井
spoon	匙状物；挖土机；吊斗
spoon bit	勺形钻头
spoon penetration test	土勺灌入试验
sporadic geomagnetic storm	偶发地磁暴
spore coal	孢子煤
sporite	微孢子煤
spot analysis	点滴分析
spot check	现场抽查
spot correlation	跳点对比
spot elevation	高程点；高程注记
spot frequency	标定频率
spot height	高程点；标高点
spot level	水准点
spot mapping	现场绘图
spot sample	个别试样
spot test	就地试验；抽查试验；斑点试验
spotted clay structure	斑状黏土结构；斑点黏土结构
spotted oil territory	斑点状分布的含油地区
spotted ore	斑点矿
spotted rock	斑点状岩
spotted slate	斑点板岩
spotted structure	斑点状构造
spotter	测位仪；搜索雷达；去污机；定心钻
spotting fluid	解卡液
spotting holes	布置炮眼
spotty deposit	断续矿床；扁豆状矿床
spotty oil saturation	斑状油饱和度
spout	排水口；喷口；龙卷风
spout of stock	喷浆
spouted bed	喷射床
spouter	喷油井；气井
spouting	喷射
spouting velocity	喷射出口速度
sprawling dike	河网上的分布堤
spray application of concrete	混凝土喷射
spray concrete	喷射混凝土
spray deposition	喷射沉积
sprayed boring tool	棱锥形钻头
sprayed coating	喷涂敷层
sprayed concrete	喷浆；喷射混凝土
sprayed drill	棱锥形钻头
sprayed tank	淋水式油罐
sprayed thermal insulant	喷涂绝热材料
spraying nozzle	喷嘴
spraying painting	喷漆
spraying plant	喷雾装置
spraying unit	洒水装置
spread angle	扩散角
spread angle of pressure in subsoil	地基压力扩散角
spread blasting	辅助爆破
spread bottom foundation	扩底基础
spread correction	排列校正；正常时差校正
spread footing	扩展基础
spread geometry	排列形状
spread length	地震仪间距
spread time	卸料时间
spreader	撒料器；扩张器
spreader bar	横撑杆；扩杆
spreader beam	担梁
spreader bowl	给矿箱
spreader plate	分流板
spreader plough	堆积刨煤机
spreader strip	分流防冲带
spreadhead	预处理液；前置液
spreading area	分水面积
spreading basin	扇形流域
spreading belt	扩张带
spreading coefficient	扩展系数；流散系数
spreading edge	扩张边缘
spreading footing foundation	扩展的独立基础
spreading force	蔓延力
spreading in layers	层铺法
spreading loss	发散损耗
spreading of groundwater	地下水扩展
spreading pressure	扩张压力
spreading ridge	扩张脊；扩张洋脊
spreading rift	扩张裂谷
spreading velocity of recharge water	补给水扩散速度
spring abutment	弹簧支座
spring bumper	弹簧缓冲器；弹簧避震垫
spring catcher	弹簧捕捉器
spring clamp	弹簧夹；弹簧压板；弹簧卡子
spring cone penetrometer	弹簧锥贯入度计
spring damper	弹簧式减振器
spring dart	打捞工具
spring dashpot system	弹簧黏性阻尼器体系
spring guide	弹簧导槽；弹簧导向装置；弹簧导承
spring igniter	弹簧点火器
spring leaf	弹簧片；板簧
spring line	起拱线
spring link	弹簧连杆
spring of arch	拱脚
spring pole drilling	手动冲击钻进
spring scale	弹簧秤
spring seismometer	弹簧式地震仪
spring shock absorber	弹簧减振器
spring steel	弹簧钢
spring support	弹簧支架
spring suspension	弹簧悬挂
spring tension	弹簧张力
spring tide	大潮
spring type gravimeter	弹簧式重力仪

spring washer	弹簧垫圈	squeezability	可压缩性
springiness	弹性	squeeze	顶板下沉，底板隆起；挤，压
springing	弹性	squeeze cementing	压力灌浆；通过有孔套管的油井压力灌浆
springing blast	扩孔底爆破	squeeze grouting	挤压灌浆
springing block	拱基；拱脚；拱底石	squeeze indenter	压头
springing centre	起拱中心	squeeze injection	挤压注入
springing course	起拱层	squeeze out	挤出，挤压出
springing of intrados	拱腹起拱线	squeeze out of plastic ground	塑性挤出
springline	起拱线	squeeze pressure	挤压力
springset	簧片组	squeeze test	挤压试验
sprue puller	浇道推出杆	squeeze up	熔岩挤出部
spruing	打浇口	squeezed anticline	挤压背斜
sprung blasthole	扩孔底炮眼	squeezed fluid	挤压液
sprung hole	药壶炮眼	squeezed gun	挤压式射孔器
spud bar	罐底积污清除铲	squeezer	压榨者；压榨机；弯板机；颚式破碎机
spud base	海底基板	squeezing	挤压
spud bit	开眼钻头	squeezing ground	挤压性地盘；塑性流动土
spud can	定位桩罐	squeezing off	挤注水泥
spud leg	桩腿	squeezing out of soft soil	软土的挤出
spud mud	开钻泥浆	squeezing tool	挤注工具
spud pile	固定桩	squegging	间歇振荡器的振荡模式
spud rope	泥铲吊索	squib blasting	导火线爆破
spud spear	钢丝绳打捞钩	squib shooting	用小炸药包爆炸
spud well	桩腿通孔	squib shot	小型硝化甘油炸弹
spudder	铲凿	squibbing	药线爆破；掏壶，扩孔底；油井爆炸
spudding	开钻	squinch	内角拱
spudding action	钻具上下活动	squirt	喷；喷射器；喷出的液体；喷湿
spudding beam	平衡杆	squirt can	弹性注油器
spudding bit	开眼钻头	squirt flow	喷射流
spudding bore	由亚土层到坚硬岩层的钻孔	squitter	断续振荡器
spudding well	开钻井	SR	短棒试样；短圆棒试样
spun pipe	离心浇铸管	SRV	储层改造体积；储层压裂体积
spur dike	丁坝	SRV fracturing	体积压裂
spur leveling line	水准支线	SSY	小范围屈服
spur line	铁路支线，矿区铁路专用线	St. Venant model	圣维南模型
spur pile	斜桩	St. Venant's principle	圣维南原理
spur protection	护岸丁坝	stab hole	不装药炮孔；直眼掏槽
spurrite	灰硅钙石	stabilise	使稳定
spurt	脉冲	stabiliser	稳定器；稳定剂；钻杆护箍；平衡器；稳压器
spurt distance	瞬时滤失距离	stabilising agent	稳定剂
spurt loss	初滤失	stability	稳定，稳定性，稳定度；耐久性；牢固性；强度；刚度；自由面
spurt loss coefficient	初滤失系数	stability against collapse	抗倒塌稳定性
spurt value	瞬时值；初损值	stability against heave	抗隆起稳定性
spurt volume	初损量	stability against overturning	抗倾稳定；抗倾覆稳定性
square cut	楔形掏槽	stability against sliding	抗滑稳定；抗滑稳定性
square girt	方框支架横撑	stability against tilting	抗倾稳定性；抗倾覆稳定性
square groove weld	直坡口焊缝	stability analysis	稳定分析
square joint	平头接合	stability analysis of diaphragm trench	槽壁稳定性验算
square pattern	方形炮眼排列法	stability analysis of limit equilibrium	极限平衡稳定性分析法
square setting	方框支架开采法	stability analysis of rock slope	岩坡稳定分析
square tie	方箍	stability analysis of slope	土坡稳定分析
square timber support	井框支架；方木支架	stability and interaction factor	稳定性及相互作用因素
square up holes	整边炮孔	stability assessment	稳定性评估；稳定性评价
squaring	井框	stability at end-of-construction	竣工期稳定性
squaring shear	剪边机	stability augmentation system	稳定性加强系统
squash	压碎	stability calculation	稳定计算；预防措施
squash plot	压缩剖面		
squat wall	矮抗剪墙		
squeegee pig	橡皮滚子清管器		
squeegee pump	挤压泵		

stability check 稳定性检查
stability checking computation 稳定验算
stability classification 稳定性分类
stability coefficient 稳定性系数；稳定系数
stability condition 稳定条件
stability constant 稳定常数；应力矩阵
stability criteria 稳定准则；稳定性标准；稳定性
stability criterion 稳定性准则；稳定度判据；稳定标准
stability curve 稳定性曲线；稳线
stability domain 稳定区
stability equation 稳定方程；稳定性方程
stability evaluation 稳定性评价
stability factor 稳定性因素；稳定系数；稳定性因素；稳定性因子
stability factor against overturning 倾覆稳定系数
stability factor of slope 边坡稳定安全系数；斜坡稳定系数
stability field 稳定范围；稳定场
stability formula 稳定性计算公式
stability investigation 稳定性勘察
stability limit 稳定极限；稳定性限制；自由边界
stability matrix 稳定矩阵
stability moment 稳定力矩；恢复力矩
stability number 稳定数值；稳定数
stability of a slope 边坡稳定性
stability of a system 系统稳定性
stability of cement grout 水泥浆稳定性
stability of dam foundation 坝基稳定性
stability of earth crust 地壳稳定性
stability of explosives 炸药安定性
stability of float caissons 浮式沉井稳定性
stability of foundation rock 岩基稳定性
stability of foundation soil 基础稳定性
stability of general algorithm 一般算法的稳定性
stability of magnets 磁稳定性
stability of mixed approximation 混合近似的稳定性
stability of rating curve 率定曲线稳定性
stability of remnant magnetization 剩磁稳定性
stability of reservoir slope 水库库岸斜坡稳定性；水库岸坡稳定性
stability of retaining wall 挡土墙稳定性
stability of roadway 巷道稳定性
stability of rock mass 岩体稳定性
stability of slope 岸坡稳定性；边坡稳定性
stability of soil slope 土坡稳定性
stability of stage-discharge relation 水位流最关系稳定性
stability of subsoil 地基稳定性
stability of surge tank 调压井的稳定性
stability of surrounding rock 围岩稳定性
stability of suspension 悬浮液稳定性
stability of truss 桁架稳定性
stability parameter 稳定性参数，稳定特性
stability range 稳定范围
stability region 稳定区
stability study 稳定性研究
stability test 稳定特性试验；稳定度试验
stability theory 稳定性理论
stabilization 固化；加固；稳定，稳定化，稳定作用
stabilization by compaction 压实稳定
stabilization by consolidation 固结稳定

stabilization by densification 挤密加固
stabilization by sand drains and pre-loading 砂井排水预压加固
stabilization by vibroflotation 振冲加固
stabilization method by mixture 拌合加固法
stabilization moment 稳定力矩
stabilization of borehole 钻孔的稳定性
stabilization of deep subgrade 深层地基加固
stabilization of landside 滑坡加固
stabilization of superficial subgrade 浅层地基加固法
stabilization pond 稳定池；酸化池
stabilization soil mixing machine 稳定土拌合机
stabilization with cement 水泥稳定法
stabilization works 加固工程；修葺工程；巩固工程；稳固工程
stabilized back-fill 稳定回填土
stabilized base 稳定基层
stabilized course 稳定层
stabilized grade 稳定坡度
stabilized grain structure 稳定粒状结构
stabilized platform 稳定平台；稳定地台
stabilized sod 稳定土
stabilized soil 经稳定处理的土；稳定土
stabilized soil base course 稳定土基层
stabilizer 稳定剂；稳定器；稳定计；消振器；加固剂；稳压器，扶正器，平衡器，稳定塔
stabilizer commutator 稳压管
stabilizer printed circuit board 稳定器
stabilizing ability 安定性；稳定性
stabilizing adjustment 稳定调整
stabilizing agent 稳定剂
stabilizing annealing 稳定性退火
stabilizing berm 稳定护道；反压道
stabilizing fluid 稳定液；固壁泥浆
stabilizing grout 加固灌浆；固结灌浆；稳定浆液
stabilizing moment 稳定弯矩；稳定力矩
stabilizing process 稳定法
stabilizing resistor 稳定电阻器
stabilizing treatment 稳定处理
stabilometer 稳定仪；土壤稳定计；稳定性测量仪；稳定仪
stable area of the earth's crust 地壳稳定区
stable bank 稳定河岸
stable bed-rock 稳定岩层
stable block 稳定地块
stable borehole 井壁稳定
stable carbon isotope 稳定碳同位素
stable coast 稳定海岸
stable continental margin 大陆稳定边缘
stable crack growth 稳定裂纹扩展
stable crack propagation 裂纹稳定扩展
stable crustal block 稳定地块
stable detonation 稳定起爆
stable diagram 平衡图；状态图
stable equilibrium 稳定平衡
stable excavation spans 稳定开挖跨度
stable foam 稳定泡沫
stable grout 稳定性浆液
stable hole 壁龛；切口；机窝；旁洞
stable isotope 稳定性同位素

stable isotope geochemistry 稳定同位素地球化学
stable isotope standard 稳定同位素标准
stable isotope stratigraphy 稳定同位素地层学
stable oscillation 稳定摆动；稳定振荡
stable range 稳定范围
stable region 稳定区域
stable rock 稳定岩石
stable rock bed 稳定岩基
stable servo 稳定伺服系统；稳定随动系统
stable shield 稳定地盾
stable sliding 蠕动；稳定滑动
stable slope 稳定坡；稳定边坡
stable state 稳定状态；稳定态
stable structure 稳定结构
stable structured soil 稳定结构土
stable wave 稳定波
stable yield of a well point 井点稳定出水量
stable-hole machine 机窝开切机
stable-type gravimeter 稳定式重差计
stablized arch 稳定拱
stably ore 板状矿石
staccato explosions 断续爆炸
staccato injection 多次注入
stack 叠加，堆，堆叠
stack effluent 烟道气；炉气
stack filling 堆填法
stack load 堆放荷载
stack processing program 叠加处理程序
stack structure 堆垛结构
stacked plate 堆积式承载板
stacked profiles map 叠加剖面图
stacker 堆料机；堆煤工，堆积工，堆积机，转载输送机；废料堆置机；存卡片柜
stacking 堆放，堆积；叠加；堆垛；分层盘旋飞行；堆积干燥法
stacking capability 堆码能力
stacking fault 堆垛层错
stacking height 堆高
stacking system 堆垛法；堆存法
stack-yard 堆料场
staddle 支撑物；承架；牵条；支柱
stadia 视距尺，视距仪；视距法；视距测量；经纬仪；临时测站
stadia addition constant 视距加常数
stadia constant 视距常数
stadia diagram 视距图表
stadia formula 视距公式
stadia hair 视距线；视距丝
stadia hand level 手持式视距水准仪
stadia intercept 视距标尺截距
stadia intercept surveying 视距标尺截距测量
stadia leveling 视距水准测量；视距高程测量；视距测高
stadia line 视距线；视距丝
stadia measurement 视距测量
stadia method 视距法；视距测量法
stadia multiplication constant 视距乘常数
stadia plane-table survey 平板仪视距测量
stadia point 视距测量点；视距点
stadia reduction diagram 视距折算图表
stadia rod 视距标尺；视距杆；测杆
stadia slide rule 视距计算尺
stadia survey 视距测量；平板仪测量法
stadia surveying 视距测量
stadia table 视距表
stadia theodolite 视距经纬仪
stadia transit 视距经纬仪
stadia transit survey 视距经纬仪测量
stadia traverse 视距导线
stadia trigonometric leveling 视距三角高程测量
stadimeter 测距仪
staff error 标尺误差；水尺误差
staff gage 标杆；量杆；测杆；测尺
staff gauge 水位尺；测深杆
staff gauge station 水文站；测流站
stage 冰段；亚冰期
stage and packer grouting 分段堵塞灌浆
stage being grouted 在灌段
stage breaking 分段破碎；分段粉岩
stage cementing technology 分段固并技术
stage compaction 分级压密；分级压缩
stage compressor 复式压缩机；多级压缩机
stage construction 分期施工
stage crushing 分段破碎
stage cut 分段开挖法
stage development 分级开发
stage discharge record 水位流量记录
stage duration curve 水位历时曲线
stage earthquake prediction 阶段性地震预测
stage entry 阶段巷道
stage fluctuation 水位变化；水位涨落
stage fluctuation range 水位起伏幅度
stage gauge 水位计；水位标尺
stage glass 载物玻璃片；载物片
stage grinding 分段研磨；阶段磨矿
stage grouting 分级灌浆
stage hydrograph 水位线；水位图
stage indicator 水位指示器；水位计
stage level 水位
stage loader 转载机
stage micrometer 目镜测微计
stage observation 水位观测
stage of coalification 煤化阶段
stage of declining 衰老期
stage of growth 生长期
stage of hydrogeological exploration 水文地质勘探阶段
stage of maturity 壮年期；成熟期
stage of old age 老年期
stage of peatification 泥炭化阶段
stage of putrefaction 腐泥化阶段
stage of reconnaissance survey 普查阶段；踏勘阶段
stage of river 河流水位
stage of weathering 风化阶段
stage of youth 幼年期
stage planking 铺脚手架板
stage plumbing 分段垂准定向法
stage process 分段过程
stage pumping 分阶段泵水；多阶段泵水
stage record 水位记录；水位-流量关系表
stage recorder 水位记录仪；自记水位计

stage reduction	分段碎磨
stage rod	水位尺
stage temperature riprap	级温升
stage treatment	分级处理
stage working	阶梯式开采；台阶式开采
stage-capacity curve	水位库容曲线；水位容量曲线
staged	分段的
staged fracturing	分段压裂
staged limited-entry fracturing	分段限流量压裂
staged limited-entry technique	分段限流量法
stage-discharge curve	水位流量曲线
stage-discharge record	水位流量记录
stage-discharge relationship	水位-流量关系
stage-duration curve	水位-时间变化曲线；水位变化曲线
stage-loader advancing unit	转载机推进装置
stage-plunger pump	多级柱塞泵
stagewise penetration	阶梯式渗透
stagger arrangement	错列
staggered arrangement of piles	桩的梅花式排列
staggered building arrangement	交叉组合
staggered check-board spaced	棋盘式排列
staggered feed	不均匀进给
staggered freezing	长短管冻结
staggered hole	交错排列的炮眼
staggered joint	相错式接头；相互式接头
staggered junction	错位交叉
staggered magnet configuration	交错排列的磁铁布置
staggered mesh	交错网格
staggered pattern	交错排列方式
staggered piling	跳打；打梅花桩；错列打桩
staggered riveting	错列铆接
staggered row	错开排列的排
staggered structure	错列构造；棋盘式构造
staggered well	交错排列井；错列布井
staggered-mesh	交错网络
staggering	交错；参差调谐；交错的
staggering cut	参差割缝
stagger-tuned amplifier	参差调谐放大器
staging plank	脚手架板
stagmalite	滴水石；石笋
stagnancy number	停滞次数
stagnant area	静水区；滞流区；停滞区
stagnant glacier	停滞冰川；死冰川
stagnant groundwater	上层滞水；静止地下水
stagnant point	静止点；驻点
stagnant swamp facies	闭流沼泽相
stagnant temperature	临界温度
stagnant water	积水；停滞水；上层滞水
stagnation	停滞
stagnation point	临界点；静止点；驻点；停滞点；水位滞点；扬水管道
stagnation pressure	临界压力；驻点压力；滞点压力
stained stone	改色宝石
stainierite	水钴矿
staining method	染色法
staining test	染色试验；染色法
stainless bar	不锈钢筋
stainless mild steel plate	不锈软钢板
stainless steel	不锈钢
stainless steel cloth	不锈钢丝布
stains	着色剂
staircase fault	阶状断层
stair-stepping	阶梯状的
stake	木桩；测点桩；测量标杆；立杆
stake anchor	桩锚
stake line	测杆线；标桩线
stake resistance	接地电阻
staker	锚栓立柱；锚固柱；桩支撑；稳柱
stake-setting	放样；定线；立桩
staking girder	锚固梁
staking out	打桩放样；定线；立桩；标出
staking out in survey	测量放样
staking pin	标桩；标杆
stalactite	钟乳石；钟乳石状沉淀物；石钟乳
stalactite grotto	钟乳石洞
stalactite stalagmite	石柱
stalactitic	钟乳石的；钟乳状的
stalactitic aggregate	钟乳状集合体
stalagmite	石笋；熔岩笋
stalagmitic	石笋的；石笋状的
stalagmitic marble	石笋大理石
stalagmometer	滴重计
stalinite	钢化玻璃
stalk	茎；杆；叶梗饰；茎状饰；高烟囱
stall prop	锚固支柱
stall roadway	由回采区通主巷的运输巷道
stall-and-breast method	房柱法
stalled flow	失速气流
stalling range	失速范围
stalloy	硅钢；硅铁合金；薄钢片
stalogometer	滴径表面张力计
stamina	耐力；精力
stamp	印模；痕迹；标志
stamp battery	捣矿机的锤组；捣矿杵
stamp copper	捣矿机解离的精铜矿
stamp die	捣矿机砧；压模
stamp pulling	捣磨矿浆
stamp rock	要捣碎精选的矿石
stamped concrete	已夯实混凝土；已捣固混凝土
stamper	压模；捣矿机；捣击机；杵
stamper bowl	捣矿机的臼槽
stamping	冲压；捣碎，捣矿；冲压制品；捣固；捣实
stamping die	冲模；模具
stamping form	捣固模
stamping hammer	机械捣锤
stamping mill	捣矿厂；捣矿机
stanch air	窒息气体；碳酸气
stanchion	柱子；立柱；立撑；支柱
stanchion base	支柱基础
stand	台；架；钻杆组；机座；三节管组
stand clarification	静止澄清法；静止脱泥法
stand pipe	导向管；立管；井口管
standage	聚水坑；水仓容量；大水仓
standard	标准，规格，规范，标准的；合规格的；标准体，样品
standard accessory	标准零件
standard amount	标准数量；额定数量
standard atmosphere	标准大气压；标准大气
standard atmospheric condition	标准大气状况
standard atmospheric pressure	标准大气压

standard axial loading	标准轴载
standard base	标准底图
standard bit	标准钻头
standard cable drilling method	标准吊绳冲击钻眼法
standard candle	标准烛光
standard class	标准级别
standard classification	标准类别
standard clause	标准条款
standard coal	标准煤
standard coal equivalent	标准煤当量
standard compaction	标准击实
standard compaction method	标准击实法
standard compaction test	标准压实试验
standard concrete consistometer	标准工业黏度计
standard condition	标准状态；标准条件；标准工况
standard cone	标准锥；标准型
standard cone crusher	标准型圆锥破碎机
standard cost	标准成本；简算成本
standard curing	标准养护
standard curve	标准曲线
standard damped least-squares method	标准阻尼最小二乘方法
standard design	标准设计
standard detail	标准详图
standard deviation	标准差
standard dimension	标准尺寸
standard discrete system	标准离散系统
standard drill rod	标准钻杆
standard earth	标准地球模型
standard eigenvalue problem	标准特征值问题
standard electric logging	标准电测井
standard electrode log	标准电极测井
standard equation	标准方程
standard equipment	标准设备
standard error	标准误差；标准差
standard error ellipsoid	均方根差椭球
standard error of estimate	标准估计误差；标准误差估计
standard error of the mean	均值标准差
standard field logging system	标准野外测井系统
standard field-mapping technique	标准野外填图技术
standard for data exchange	数据交换标准
standard for potable water	饮用水标准
standard for surveying and mapping	测绘标准
standard form	标准型
Standard format	标准格式
standard frequency	标准频率
standard frost depth	标准冻结深度
standard frost penetration	标准冻深
standard frozen depth	标准冻深
standard gauge	标准规；标准计
standard Gauss quadrature	标准高斯求积法
standard global chronostratigraphic scale	标准世界年代地层表
standard global geochronologic scale	标准世界地质年代表
standard gravity field	标准重力场
standard horsepower	标准马力
standard ignition test	煤尘爆炸极限试验
standard instrument	标准仪器；校准用仪器
standard isobaric surface	标准等压面
standard linear model	标准线性模型
standard lines	标准线
standard lining plate	标准衬砌环
standard load	荷载标准值
standard load test	标准载荷试验
standard mean ocean water	标准平均海洋水
standard meridian	标准子午线
standard meter	标准仪表；线纹米尺
standard metering base	标准计量基准
standard metering orifice	标准流量计锐孔
standard method	标准方法
standard method of measurement	标准量测方法；标准计量方法
standard mineral	标准矿物
standard mix concrete	标准配比混凝土
standard model	标准样品
standard mould	标准混凝土砖模
standard normal distribution	标准正态分布
standard notation	标准符号表示法
standard of conduct	行为准则
standard of construction	施工标准
standard of reflectance	反射率标准
standard of reflectivity	反射率标准
standard of structural stability	结构稳定性标准
standard of surveying and mapping	测绘标准
standard operation procedures	标准操作程序
standard ordering	标准排列
standard orifice	标准孔
standard output	额定产量；标准产量；标准输出
standard panel length	标准节长度
standard parallel	标准纬线；基准纬线；零变纬线
standard parallel-shaft speed reducer	标准平行轴减速器
standard penetration	标准贯入度
standard penetration number	标准贯入击数
standard penetration resistance	标准贯入阻力；标准贯入抗力
standard penetration test	标准贯入试验；重型动力触探
standard penetration value	标准贯入值
standard penetrometer	标准贯入仪
standard performance	标准性能；标准工作强度
standard pitch	标准螺距
standard plough	标准刨煤机
standard polarity	标准极性
standard polynomial	标准多项式
standard powder	标准炸药
standard practice	标准作业法；标准操作规程；常规作法
standard pressure	标准压力
standard pressure and temperature	标准气压和温度
standard price method	标准价格法
standard procedure	标准程序；标准法
standard Proctor's compaction test	标准葡氏击实试验；轻锤击实试验
standard profile	标准剖面
standard program	标准程序
standard programme	标准程序
standard radioactive source	标准放射源
standard rate	标准费率；规定标准；规定定额
standard ratio method	标准比率法
standard recipe	标准配方

standard reference material　标准参考物质
standard rig　标准钻具
standard risk level　标准风险水平
standard routine　标准程序
standard sample　标准试样；标准样品
standard sand　标准砂
standard screen　标准筛
standard sea level　基准海平面
standard seawater　标准海水
standard section　标准剖面；标准地层断面；标准河段
standard segment　标准管片
standard seismic coefficient　标准地震系数
standard seismograph　标准地震仪
standard seismometer　标准地震计
standard series　标准系列；矿井风量标准
standard shunting sensitivity　标准分路灵敏度；标准分流感度
standard sieve　标准筛
standard sieve series　标准筛号
standard size　标准尺寸
standard soil sampler　标准取土器
standard specification　标准规格；标准规范
standard specimen for rocks　岩石标准试件
standard state　标准状态
standard station　基准台
standard steam pressure　标准蒸汽压
standard system　标准体系
standard system of levels　标准高程系统
standard temperature　标准温度
standard thermometer　标准温度计
standard time　标准时间
standard tolerance　标准公差
standard tomographic technique　标准层析成像技术
standard tool　标准工具；校准工具
standard tool joint　标准钻杆接头
standard truck vibrator　标准车装可控震源
standard type contour package　标准型等高线程序组
standard unit　标准单位
standard value of a geotechnical parameter　岩土参数标准值
standard wire gauge　标准线规
standard-height analysis　标准高度分析
standardization　标准化；规格化；标定法；校准；检定
standardization engineering　标准化工程学
standardization of data base　数据库标准化
standardization system　标准化系统
standardized design　标准设计
standardized fitness　标准适应值
standardized holes　标准化炮眼组
standardized normal distribution　标准正态分布
standardized solution　标准溶液
standardized unit　标准化单位
standardized variable　标准化变量
standardizing　规格化
standardizing area　标准化领域
standardizing number　规格化数
standardizing order　标准化指令
standards clay mineral　标准黏土矿物
standard-sized　标准尺寸的
stand-by equipment　备用设备
stand-by power source　备用电源
stand-by unit　备用装备；辅助装置
standing　停止的；不流动的；直立的；长期的；常设的；不再运转的
standing balance　静力平衡
standing barrel rod pump　固定泵筒杆式泵
standing charge　固定费用
standing fluid level　静液面
standing ground　坚实地层；稳定地层；稳定地层
standing half wave　驻半波
standing lake　停蓄湖
standing level　静水位；稳定水位
standing loss　储存损耗
standing on end　直立岩层
standing order　规程；操作规程
standing oscillation　稳定振荡
standing pipe　竖管；立管
standing platform　固定平台
standing sea level　基本水准面；基本水平面
standing storage　长期储存
standing support　常规支护；支撑式支架
standing tank　固定罐
standing valve　固定阀
standing water　静止水；滞流水；停滞水
standing water level　静水位
standing wave antenna　驻波天线
standing wave line　驻波线
standing well　备用井
standing-wave flume　钻孔柱状图
stand-mounted drill　柱架式凿岩机
stand-off collar　偏距轴环
stand-off distance　偏距；投射距离
standpipe　立管，竖管；疏水器；圆筒形水塔；压力管；测压管；测水管，水位计
standpipe piezometer　竖管式测压计
standpipe pressure　立管压力；竖管压力
standpipe-type well　立管式井
standstill　停止；停顿；静止；稳定状态
stand-up formation　直立地层
stand-up time　自承时间；自稳时间；自持时间；站立时间
stanfieldite　磷镁钙矿
stannary　锡矿；锡厂
stannate　锡酸盐；黄锡矿
stannic　锡的；四价锡
stanniferous ore　锡矿石
stannite　黄锡矿
stannolite　锡石
stannous　亚锡的；二价锡的；含锡的
stannous oxide　氧化亚锡；一氧化锡
stantienite　黑琥珀；黑树脂石
staple bolt　卡钉；夹子；箍；套环
staple cutter　纤维切断机；短纤维切断机
staple fibre　短纤维
staple fibre apparatus　纤维长度试验仪
staple pit　暗井
staple shaft　暗立井；盲立井
staple type joint　卡式接头
staple-shaft hoist　暗井提升机
star antimony　精炼锑

star atlas	星图
star catalogue	星表
star centralizer	星形中心器
star chart	星图
star cluster	星团
star configuration	涡流层
star controller	星形挡车器
star dune	星状沙丘
star gate	星形给料器；星形给矿器
star map	星图
star observation	星观测
star reamer	星形扩孔器
star ruby	星彩红宝石
star sapphire	星彩蓝宝石
star structure	星状结构
star system	银河系
star transmission network	星状监控传输网
star wheel extractor	星轮排渣器
star wheel type feeder	星形轮给料机
starch flocculant	淀粉絮凝剂
starlite	蓝锆石
starolite	星石英
star-pattern drilling bit	十字钻头
star-plot	星状图；星形-点
star-shaped bit	星形钎头；星形钻头
starter bar	预留搭接钢筋；露头钢筋
starter magneto	起动磁电机
starter pedal	启动踏板
starter pinion	起动小齿轮
starter size	开孔钻头直径；开孔尺寸；开口钎子尺寸；开口钎子直径
starting adhesion factor	起动黏着系数
starting compensator	起动补偿器
starting condition	初始条件；启动条件
starting control	起始控制点
starting crank	起动摇把
starting cut	开切巷道
starting datum mark	起测基点
starting diameter	开孔直径
starting engine	启动发动机
starting fluid	启动燃料
starting impulse	启动脉冲
starting material	原材料；入洗原料；给料；入料
starting outfit	启动设备
starting plane	起始面
starting point	起始点；出发点
starting position	开动位置；原始位置；初始位置
starting potential	起始电压
starting power	启动功率；起动动力；启动电源
starting pressure	启动压力
starting pressure gradient	启动压力梯度
starting procedure	起动程序
starting push-button	起动按钮
starting resistance	起动阻力；起动电阻
starting rod	引弧棒
starting saturation	起始饱和度
starting shaft	出发竖井
starting shot	掏槽炮眼；开槽炮眼；开口炮眼
starting switch	起动开关
starting torque	启动扭矩；启动转矩
starting value	开始值，初始值
startling moment	起动力矩
start-of-record	记录开始
start-of-run temperature	运转初期温度
start-stop	起停的；起止的，启闭的
start-stop transmission	起止传输
start-up	启动；开工；开动
start-up pressure	启动压力
start-up procedure	起动程序
start-up strainer	开工投产用过滤筒
starve a pump	泵吸入端压头太小
starved basin	非补偿盆地；未补偿盆地
starved ripple mark	不完全波痕
starved side	无补偿侧
star-wheel axle controller	星轮式控轴器
stassfurtite	纤硼石
state boundary surface	状态边界面
state classification	状态分类
State coordinate system	国家坐标系；州坐标系
state diagram	状态图
state equation	状态方程
state estimation	状态估计
state of capillary equilibrium	毛细管平衡状态
state of elastic equilibrium	弹性平衡状态
state of failure	破坏状态
state of limit equilibrium	极限平衡状态
state of loading	载荷状态
state of nature	自然状态
state of plane stress	平面应力状态
state of plastic equilibrium	塑性平衡状态
state of plastic-elastic equilibrium	塑弹性平衡状态
state of risk	风险状态
state of saturation	饱和状态
state of strain	应变状态
state of stress	应力状态
state of vector	状态向量
state parameter	状态参数
state plane coordinate system	国家平面坐标系
state plane grid	国家平面坐标格网
state seismological network	国家地震台网
state space	状态空间
state space filtering	状态空间滤波
state standard	国家标准
state transition function	状态迁移函数
state transition time	状态过渡时间
state variable	状态变量
state vector	状态向量
stated accuracy	标定精度
statement of equilibrium	平衡状态
statement of flow	流态
statement of stagnancy	停滞状态
state-of-the-art down hole tool	现代井下工具
state-of-the-art facility	现代化设备
states of equivalent	平衡状态
states of stress	应力状态
state-space method	状态空间法
state-variable model	可变状态模型
static	静力的；静态的；静学的；静电的；天电
static action	静态作用
static adsorption	静吸附

static advancing contact angle　静止前进接触角
static allocation　静态分配
static analysis　静态分析；静校正分析；静定分析法
static and dynamic pressures　静和动压力
static and dynamic reservoir modelling　静态和动态油藏建模
static attribute　静态特征
static balance　静平衡
static balancing　静态平衡；静力平衡；静平衡
static bath　静力分选槽
static behavior　静态
static bottom hole pressure　井底静压
static budget　静态预算
static calculation　静力计算
static casing pressure　关井时的套管压力
static chain　静电导链
static character　静态特性
static characteristic　静态特性；静态特性曲线
static charge　静电荷
static chip hold down pressure　压住钻屑的静压力
static compressive modulus of elasticity　静力抗压弹模
static computation　静校正计算
static condition　静力条件
static cone penetration resistance　静力锥形贯入抗力
static cone penetration test　静力触探试验；静锥贯入试验
static cone sounding　静力触探
static cone test　静力触探试验
static contact angle　静止接触角
static control　静态控制
static cool down　静置冷却
static correction　静校正
static correction method　静校正法
static cracking　静态破碎
static decomposition　静态分解
static deflection　静载挠度；静变位；静载弯曲
static deformation　静力变形
static density gradient　静密度梯度
static determinate structure　静定结构
static discharge　静电放电
static discharge head　排送静压头
static displacement　静态位移
static disturbance　静干扰
static drilling model　静态钻井模式
static earth pressure　静态土压力
static effect　静效应
static electricity　静电
static eliminator　静电干扰消除器
static engine　固定式发动机
static equilibrium　静力平衡；静定平衡
static error　静态误差
static failure matrix methodology　静态破坏矩阵法
static fatigue　静态疲劳
static field　静力场
static file　静态文件
static filtration rate　静失水率
static fixing　定点定位
static flow　层流；平静流
static fluid column pressure　静液柱压力
static fluid gradient　静液柱压力梯度
static fluid level　静液面
static fluid loss test　静态滤失试验
static force　静力
static force penetration curve　静压钻穿曲线
static formation pressure　地层静压力
static formula　静力公式
static formula of pile　桩的静力公式
static friction　静摩擦
static friction coefficient　静摩擦系数
static gas test　静态气体试验
static gradient　静力梯度
static gravimeter　静力重差计；静重力仪
static ground water level　静地下水位
static groundwater level　静地下水位
static haulage chain　固定牵引链
static head　静水头，静压头；扬程；静水位
static hoisting moment　静提升力矩
static hold-down　静压制
static impedance　静态阻抗
static implosion　静力内向挤聚
static indentation test　静凹痕试验
static indeterminacy　静不定
static indeterminate structure　静不定结构
static inscribing　静力内接
static instability　静态不稳定；静态不稳定性
static ionization method　静电离子化法
static leakage　静漏量
static level　静液面；静水位
static line　防静电接地线
static liquid level　静力水准系统；静液面
static load　静荷载；不动荷载；恒载；静重
static load deflection　静荷载挠度
static load movement　静压负荷运动
static load test　静载荷试验；静载试验
static load test of pile　单桩竖向静载荷试验
static load test-rig　静载试验装置；静载试验台；平稳的
static loading　静力加载
static loading test　静力载荷试验；静载试验
static loading test machine　静载试验机
static magnetic field　静磁场
static measurement　静态测定
static metamorphism　静力变质
static meter　无定向地磁仪；助动重力仪
static method　静力法
static mode　静态
static model　静态模型；静力模型
static modulus of elasticity　静力弹性模量；静弹性模量
static moment　静矩；静力矩
static orifice　静水头测量孔
static output power of fan　通风机静压输出功率
static penetration test　静贯入试验；静力触探试验
static pile loading　静力压桩
static pile press-extract machine　静力压拔桩机
static point resistance　静触探探头阻力；静力触探锥尖阻力
static Poisson's ratio　静泊松比
static preload　静力预压
static pressure　静压力；静水压力
static pressure energy　静压力能量
static pressure line　静水压线

static pressure transducer 静压传感器
static probing 静力触探
static probing test 静力探测试验
static properties of intact rock 原岩静态特性
static protection 静态保护
static reading 静态读数
static receding contact angle 静止后退接触角
static register 静态寄存器
static regulator 无静差调整器
static relay 摆动继电器
static relocation 静态重分配
static reserve 净储量
static reservoir pressure 油气层静压
static resistance 静态电阻；静阻力
static resistance to penetration of pile 桩贯入的静阻力
static response 静力响应
static rheological parameter 静态流变参数
static routine 静态程序
static safety factor 静力安全系数
static seal 静密封；固定密封
static sedimentation 静态沉积
static self potential 静止自然电位
static shift 静态时移；静校正时移
static skew 静态扭曲
static solution 固定解
static sounding 静力触探
static source parameter 静态震源参数
static spontaneous potential 静止自然电位
static stability 静态稳定性；静力稳定度
static state 静态
static statistics 静态统计学
static stereo photography 静态立体摄影
static stiffness 静力刚度
static storage cell 静态存储单元
static store 静态存储
static strength 静止强度
static strength test rack 静强度试验台
static stress change 静压力变
static stress drop 静应力降
static structural analysis 结构静力分析
static submergence 静沉没
static suction head 吸引静压头
static survey 静态测量
static tab 静态标记
static temperature 静态温度
static test 静力试验；静态试验；静止试验
static test load 静态试验载荷
static test of pressure 静态强度试验
static thrust 静推力
static time shift 静态时移
static torque 静力矩
static triaxial shear test 静三轴剪切试验
static type packer 固定式封隔器
static undrained test 静力不排水试验
static voltage 静电压
static water level 静水位；静压水平
static water pressure 静水压力
static water table 静水面；静水位
static wellhead pressure 井口静压
static yield stress 静态屈服应力

static Young's modulus of elasticity 静杨氏弹性模量
static-aging 静态老化
statical energy 静能
statical equilibrium 静力平衡
statical friction 静摩擦
statical head 静压头
statical lifting capacity 静止载重量
statical load 静载荷；静荷重
statical moment 静力矩
statical solution 静力学解
statically balanced 静平衡的
statically determinate 静定的
statically determinate frame 静定构架
statically determinate problem 静定问题
statically determinate structure 静定结构
statically determinate truss 静定构架
statically equivalent 静力等效
statically indeterminate 超静定的；静不定的
statically indeterminate beam 静不定梁
statically indeterminate structure 静不定结构；超静定结构
statically indeterminate system 超静定体系
statically stable 静力稳定的
statically undetermined 超静定
static-dynamic probing 静-动态探测
static-interface survey 静止界面测量
static-pressure loss 静压损失
statics 静力学；静电干扰；天电干扰
statics correction 静校正
statics estimation 静校正估算
statics of fluid 流体静力学
statics range 静校正量范围
static-type plough 静力型刨煤机
station 站，车站；位置；发电厂；测站，测点；所，台
station adjustment 测站平差
station bin 主要水平贮仓
station buoy 位置标志浮标
station centring 测站对中；测站归心
station coordinate 台站坐标
station cutting 开辟井底车场
station elevation 测站高程；台站海拔
station error 台站误差；测点误差；垂线偏差
station of triangulation 三角测量点
station peg 测站标桩
station point 测站点
station pole 测量标杆；测量准尺；标杆
station pump 排水站泵；主排水泵
station repeatability 测点重复性
station spacing 点距；站间距
station tunnel 车站隧道
station type dip log 点测式地层倾角测井
stationarity 平稳性；稳定性；固定性
stationarity index 不动指标
stationary 稳定的；固定的；静止的；不变的
stationary bed-saltation flow pattern 稳定的砂床波动流动型式
stationary block 稳定地块
stationary concrete mixer 固定式混凝土搅拌机
stationary condition 稳态条件

stationary coordinate	固定坐标
stationary creep	稳定蠕动
stationary crushing plate	固定破碎板
stationary curve	平稳曲线
stationary cylinder	固定圆筒
stationary distribution	平稳分布
stationary dog	固定爪
stationary electric concrete mixer	固定式电动混凝土搅拌机
stationary electrical equipment	固定式电气设备
stationary element	静环
stationary engine	固定式发动机
stationary excitation	稳定激励
stationary face	备用工作面；停工作业面
stationary field	恒定场
stationary filter	稳态滤波器
stationary fit	静配合
stationary flow	稳态流
stationary flow field	稳态流场
stationary function	稳态函数
stationary Gaussian process	稳态高斯过程
stationary hypothesis	平稳假设
stationary inner-tube core barrel	单动双层岩芯管
stationary installation	固定装置
stationary iteration	定常迭代
stationary layer	稳定层
stationary liquid	静态液体，固定液体
stationary liquid method	静态液体法
stationary magnet	固定磁铁
stationary mass	静止质量；稳定体；原地岩体
stationary measurement	点测
stationary medium	固定介质；稳定介质
stationary monitoring station	固定监测站
stationary mortar mixer	固定式灰浆搅拌机
stationary noise	平稳噪声
stationary phase	固定相；静止相
stationary phase approximation	稳相近似法
stationary phase method	稳相法
stationary piston drive sampler	固定活塞打入式取土器
stationary piston-type sampler	固定活塞式取样器
stationary plastic flow	柔性静流，塑性静流
stationary platform	固定平台
stationary point	驻点
stationary point on a curve	曲线平稳点
stationary pollution source	固定污染源
stationary power unit	固定式发电机组
stationary process	平稳过程
stationary pulse number	静脉冲数
stationary pump	固定泵
stationary random function	平稳随机函数
stationary random process	平稳随机过程
stationary random vibration	平稳随机振动
stationary reading	定点读数法
stationary response	稳态响应
stationary satellite	静止卫星
stationary satellite navigation system	静止卫星导航系统
stationary scanner	固定式扫描仪
stationary seal ring	静环
stationary shaft	固定轴
stationary slip	静止卡瓦
stationary snubber	静止夹持器
stationary solution	稳态解
stationary source of pollutant	污染物固定源
stationary stage	静止阶段，间歇阶段
stationary state	静止状态；稳态
stationary state probability distribution	稳态概率分布
stationary stochastic process	平稳随机过程
stationary stream	稳态流；稳定流；平稳流
stationary structure	固定式结构
stationary temperature	恒温度
stationary time series	平稳时间序列
stationary tubesheet	固定管板
stationary value	平稳值；静叶片
stationary wave	驻波；定常波
stationary wave theory	立波理论；注液站
stationary-epicentre distance	震中距
stationary-piston sampler	固定活塞式取样器
station-keeping	位置固定
station-keeping ability	定位能力
statistic	统计学的
statistic correlation method	统计相关法
statistic law	统计规律
statistic output	统计产量
statistic recognition	统计识别
statistic soil mechanics	统计土力学
statistical	统计的；统计学的
statistical accuracy	统计精度
statistical analysis	统计分析
statistical analysis of anomalies	异常统计分析
statistical arbitration	统计判优法
statistical average	统计平均
statistical calculation	统计计算
statistical calculation library	统计计算程序库
statistical chart	统计图
statistical check	统计检验
statistical computer	统计计算机
statistical computing	统计计算
statistical control	统计控制
statistical control of compaction	压实统计控制
statistical correlation	统计相关
statistical counting method	统计计数法
statistical damage theory	统计损伤理论
statistical data	统计数据；统计资料
statistical decision function	统计决策函数
statistical decision method	统计判定法
statistical decision theory	统计决策理论
statistical dependence	统计相关；统计相关性
statistical discrepancy	统计差异
statistical distribution	统计分布
statistical entropy	统计熵
statistical equilibrium	统计平衡
statistical error	统计误差
statistical estimate of error	统计误差估计值
statistical estimation	统计估值
statistical evaluation	统计学评价
statistical figure	统计数字
statistical fluctuation	统计波动
statistical fluid mechanics	统计流体学
statistical forecast	统计预测；统计预报
statistical frequency	统计频率

statistical function 统计函数
statistical geochemistry 统计地球化学
statistical geometry 统计几何学
statistical graph 统计图
statistical homogeneity 统计均一性
statistical hydrology 统计水文学
statistical hypothesis 统计假设
statistical independence 统计独立性
statistical inference 统计推断
statistical information 统计资料
Statistical learning theory 统计学习理论
statistical lithofacies 统计岩相
statistical measurement 统计计量
statistical measures 统计度量
statistical mechanics 统计力学
statistical method 比例法；统计法；统计方法
statistical model 统计模型
statistical model of data 数据的统计模型
statistical noise 统计噪声
statistical parameter 统计参数
statistical permeability model 统计渗透率模型
statistical precision 统计精度
statistical probability 统计概率
statistical property 统计特征
statistical range 统计范围
statistical reasoning 统计推理
statistical regression analysis 统计回归分析
statistical regularity 统计规律性
statistical relative frequency 统计相对频率
statistical sample 统计样本
statistical sampling 统计抽样
statistical sampling technique 统计取样技术
statistical seismology 统计地震学
statistical significance 统计显著性
statistical simulation 统计模拟
statistical soil mechanics 统计土力学
statistical stacking 统计叠加
statistical study 统计分析；数理统计研究；统计学研究
statistical technique 统计法；统计技术
statistical test 统计检验；统计试验
statistical testing method 统计检验法
statistical theory 统计理论
statistical thermodynamics 统计热力学
statistical tolerance limit 统计容许限度
statistical treatment 统计处理
statistical treatment of joint attitudes 节理产状的统计学处理
statistical uncertainty 统计误差；统计不确定性
statistical variable 统计变量
statistical variance 统计方差
statistical variation 统计上的波动
statistical velocity analysis 统计性速度分析
statistical weight 统计加权；统计权重
statistical zonation technique 层段划分统计技术
statistically biased stack 统计有偏叠加
statistics 统计学；统计法；统计资料；统计量；统计表；统计数字
statistics in stratigraphic palynology 地层孢粉统计学
statistics of round off 误差统计

statistics of search 探查统计学
statistic-thermodynamic analysis 统计热力学分析
statobiolith 原地生物礁岩
statohydral metamorphism 静压含水变质作用
statoscope 变压计；高差仪；微动气压计；高度灵敏气压计
statothermal metamorphism 静压热力变质作用
status diagram 状态图
staurolite 十字石；交沸石
stauroscope 十字镜
stavrite 黑云角闪岩
stay bolt 锚杆；锚栓
stay pile 锚桩；拉索桩
stay rod 拉线杆；拉线桩
stay rope 锚索；拉索；拉条
staybolt 锚栓
stayed structure 拉索结构
STDEV 标准差
steady 稳态的；稳定的；经常的；使稳固，使稳定；中心架，托架
steady arm 定位器
steady buff and puff method 蒸汽吞吐法
steady coal 锅炉煤
steady crack growth 定常裂纹扩展；裂纹稳定扩展
steady creep 稳定蠕变
steady creep stage 等速蠕变阶段
steady discharge 稳定流量
steady dowel 定位销；紧固销
steady drag force 稳定阻力
steady draw 稳定放矿
steady field 蒸气田；汽田；冒汽地面
steady field demagnetization 恒稳磁场退磁
steady flow 稳定流；稳定流动
steady flow energy equation 稳定流能量方程
steady flow model 周期应力
steady fraction 干度；蒸汽馏分；汽水比
steady geyser 稳态喷泉
steady gradient 连续坡度；稳定坡度
steady ground water level 稳定地下水位
steady groundwater flow 稳定地下水流
steady heat transfer 稳定传热
steady jet 喷气孔
steady load 稳定负载；稳恒荷载
steady mass 静止质量；稳定体；原地岩体
steady motion 稳恒运动；稳定运动
steady periodic oscillation 稳态周期振荡
steady pumping test 稳定流抽水试验
steady rise of pressure 压力稳定上升
steady secondary ground pressure 二次地压
steady seepage 稳定渗流；预制拉杆
steady seepage field 稳定渗流场
steady soak 蒸汽浸泡法
steady state 稳定态；稳态；稳定状态
steady state air diffusion 稳态空气扩散
steady state air flow 稳态空气流
steady state condition 稳态条件
steady state constraint 稳态约束
steady state creep 稳态蠕变
steady state distribution 稳态分布
steady state flow 稳态流；稳态流动

steady state iteration 定态迭代
steady state method 稳定状态法；稳态法
steady state model 稳恒态模型
steady state phase 稳定状态阶段
steady state resonance curve 稳态共振曲线
steady state response 稳态反应
steady state seepage 稳态渗流
steady state sinusoidal excitation 稳态正弦波激振
steady state solution 稳态解
steady state transfer 稳定状态传递
steady state vibration 稳态振动
steady state vibration isolation 稳态振动隔离
steady stress 静应力；稳恒应力
steady turbidity current 恒定异重流
steady uniform flow 稳定均匀流动
steady water level 稳定水位
steady water table 稳定水位
steady-flow pumping test 稳定流抽水试验；稳流泵抽试验
steadying bracket 铅锤摆动稳定夹
steady-slip active fault 稳滑型活断层
steady-stage algorithm 稳定阶段算法
steady-state character 稳态特性；定态特性
steady-state conditions 稳定状态；稳态条件；稳定状况
steady-state creep 稳态蠕变
steady-state creep of rock 岩石稳定蠕变
steady-state growth 稳定增长
steady-state oil relative permeability 稳定态油相对渗透率
steady-state response 稳态响应
steady-state saturation 稳定态饱和度
steady-state stream 静态河
steady-state sweep efficiency 稳态波及系数
steady-state test 稳态试验
steady-state theory 稳态理论
steady-state vibration 稳态振动
steady-state water-table 稳态水位
steam coal 蒸汽煤，动力煤；锅炉煤；烟煤和无烟煤之间的煤
steam crane 蒸汽起重机；汽力起重机
steam cycle 注蒸汽周期
steam cylinder 汽缸
steam deflector 蒸汽偏导器
steam direct drive 蒸汽直接传动
steam drying efficiency 蒸汽干燥效率
steam front stability 蒸汽前缘稳定性
steam gauge 蒸汽压力计；汽压计
steam generator 蒸汽发生器；蒸汽发电机；锅炉
steam gun 蒸汽枪
steam heater 蒸汽加热器；蒸汽配箱
steam heating coil 蒸汽加热盘管
steam hoist 蒸汽提升机；蒸汽绞车
steam huff and puff 蒸汽吞吐
steam infusion 注入蒸汽
steam injection 注蒸汽
steam injection maturing 注蒸汽接近完成
steam injection pressure 注蒸汽压力
steam injection well 注蒸汽井
steam jacket 蒸汽套；加温套
steam jet 蒸汽喷射；蒸汽喷头

steam limit curve 蒸汽界线
steam line 蒸汽管线
steam pile driver 蒸汽打桩机
steam pile hammer 打桩汽锤；汽桩锤；蒸汽打桩锤
steam piston 蒸汽活塞
steam plateau 蒸汽平稳阶段
steam power 蒸汽能
steam pressure 蒸汽压力
steam reforming process 蒸汽转化过程
steam tension 蒸汽压力
steam wetness 蒸气湿度
steam-driven excavator 蒸汽挖掘机
steaming ground 冒汽地面
steam-jet injector 蒸汽喷射器
steam-turbine generator 汽轮发电机
stearine pitch 紊流场
steatite 冻石；皂石；块滑石；滑石
steel alloy 合金钢
steel arch 拱形钢支架；钢拱支架
steel arch support 钢拱支撑
steel arched rigid support 拱形刚性金属支架
steel arched support 拱顶纵梁；钢拱支护；钢拱支撑
steel area 钢筋截面积
steel ball 钢球
steel band 黄铁矿夹层；钢带
steel band tape 钢卷尺；钢带尺
steel bar 钢筋条；钢棒
steel bar bender 钢筋弯曲机
steel bar cold draw 钢筋冷拔
steel bar cold-extruding 钢筋冷拔机
steel bar heading press 钢筋冷镦机
steel bar heading press machine 钢筋冷墩机
steel beam 钢梁
steel beam support 钢支撑；钢支架
steel belt 钢带；钢带传动带；钢丝绳芯胶带
steel bender 钢筋工；弯钢筋工具
steel bit 钢钻头
steel box pile 箱式钢板桩
steel breakage 钎子断裂
steel bridge 钢桥
steel brush 钢丝刷
steel bunker 钢造贮仓
steel bunton 钢横梁
steel cable 钢丝绳；钢缆
steel cable core 钢芯
steel caisson 钢沉箱
steel cap 钢顶梁
steel car 全钢车；运钎车
steel casing 钢套管；铸钢井壁
steel cast 铸钢
steel change 更换钎子
steel channel 槽钢
steel chock 钢制支垛；金属垛
steel collar 钎肩
steel composite construction 劲性钢筋混凝土结构
steel composite pile 复合钢桩
steel concrete 钢筋混凝土
steel construction 钢结构
steel core column 钢芯混凝土柱
steel crib 钢棚架

steel cutting shield	钢刃盾构
steel cylindrical column	钢管柱
steel extension machine	钢筋拉伸机
steel fabric	钢筋网
steel fabricating yard	钢制件工场
steel fiber concrete	钢纤混凝土
steel fiber reinforced concrete	钢纤维混凝土
steel fiber reinforced shotcrete	喷射钢纤维混凝土
steel fiber shotcrete	喷射钢纤维混凝土
steel figure	钢字码
steel form	金属模板；钢模板
steel form jumbo	钢模台车
steel formwork	钢模板；钢模
steel frame	钢框架；钢支撑，钢支架
steel framework	钢框架；钢构架
steel frontal frame	防撞钢构架
steel grit	钢砂
steel guide	钢罐道
steel H bearing pile	H型钢承重桩
steel hand tape	钢卷尺；小钢尺
steel headframe	钢井架
steel headgear	钢井架
steel holder	夹钎器
steel hoop	钢箍
steel hose	钢制软管
steel H-pile	H型钢桩
steel insert	钢埋件
steel jack	钢制千斤顶柱
steel jacket	钢制导管架
steel jacket platform	钢导管架平台
steel jacket structure	钢护套结构
steel lagging	钢背板；钢隔板
steel leg	钢支腿
steel line	钢丝
steel liner	钢衬
steel lining	钢井壁；钢支架；钢衬砌
steel loop	钢环
steel magnet	磁钢
steel measuring tape	钢卷尺；钢身拉尺
steel mesh	钢筋网；钢丝网
steel open caisson	钢沉井
steel open caisson with double-shell	双壁钢沉井
steel piling	打钢桩
steel pipe	钢管
steel pipe assembly	钢管组合
steel pipe pile	钢管桩
steel pipe post	钢管支柱
steel pipe sheeting	钢管桩挡土墙
steel plate	钢板
steel plate conveyor	钢板输送机
steel plate liner	钢板衬砌
steel platform	钢质平台
steel product	钢材
steel prop	钢柱；铁柱
steel rack	钻杆架
steel rail	钢轨
steel rail guide	钢轨罐道
steel ratio	钢筋百分率；配筋率
steel ratio of concrete	混凝土的配筋比
steel reinforced concrete	钢筋混凝土
steel reinforcement	钢筋；钢条
steel reinforcement and concrete material	钢筋及混凝土材料
steel reinforcement cage	钢筋笼
steel reinforcing segment	加强钢肋
steel rim	钢轮圈
steel ring	环形钢支架；钢拱
steel rip bit	凿岩钢钎头；凿岩钢钻头
steel rule	钢直尺
steel section	型钢
steel segment	钢制管片
steel set	钢支架
steel shaft lining	钢井壁
steel shapes	型钢
steel sheet pile	钢板桩
steel sheet pile bulkhead	钢板桩墙护岸
steel sheet pile cofferdam	钢板桩围堰
steel sheet piled quay wall	钢板桩式岸壁
steel sheet piling	钢板桩墙；打钢板桩
steel sheeting	钢板桩
steel sheet-pile cell	钢板桩圈
steel shoe	钢桩靴
steel shot abrasive	钢砂
steel shot blasting	钢粒喷净法
steel shot drilling	钢粒钻进；钢粒钻眼；钢砂钻井
steel slag pile	钢渣桩
steel solid bit	实心钢钎；实心钢钻头
steel spring shock sub	钢质弹簧减震器
steel square	钢曲尺
steel storage	地面钢罐内储油
steel straighten machine	钢筋调直机
steel strand	钢绞线
steel stranded wires	钢绞线
steel stretcher	钢筋冷拉机
steel structure	钢结构
steel support	钢支撑；钢支座；钢圈支架；金属支架
steel tape	钢卷尺
steel taper drift	钢尖冲
steel tooth	钢齿
steel tooth bit	钢齿钻头
steel tower	标；钢塔；铁塔
steel truss girder	钢桁梁
steel tube	钢管
steel tube pile	钢管桩
steel tube section guide	中空型钢罐道；管状型钢罐道
steel tubular goods	钢制管材
steel tubular prop	钢管支柱
steel wire	钢丝；钢线
steel wire cloth	钢丝筛网
steel wire reinforced belt	钢丝芯胶带
steel wire rope	钢丝绳
steel-band conveyor	钢带输送机
steel-box pile	钢箱型桩
steel-cable belt	钢丝绳胶带
steel-framed footbridge	钢框架步行桥
steel-H-pile	H型钢桩
steel-mesh lagging	金属网假顶；钢丝网衬板
steel-pipe column	钢管柱
steel-sheet-pile wall	钢板桩墙
steel-wood prop	钢木混合支柱

steelwork assembly　装备组装
steep　陡峭的；悬崖；浸渍；浸染；大锥度；耐火衬料
steep access　陡沟
steep angled conveyor　陡坡胶带输送机
steep angled reflection　陡倾斜反射
steep ascent　陡斜的提升；陡斜的上升；陡坡度
steep bank　陡岸
steep bed　陡倾岩层；急倾斜层
steep breast　急倾斜煤房；急倾斜工祖
steep coast　绝崖海岸；陡岸
steep conveyor　陡坡胶带输送机
steep curve　陡曲线，锐曲线
steep descent　陡坡
steep dip　陡倾；急倾斜
steep dipping　陡降的
steep dipping bed　陡立地层带
steep drift　急倾斜巷道
steep face　陡壁
Steep fault　陡倾断层
steep formation　急倾斜地层
steep grade　陡坡
steep gradient　大坡度；陡坡度；陡坡
steep head　溶蚀谷；陡头谷
steep incidence　陡峭入射
steep lift　陡斜提升
steep measures　倾斜层系
steep mountain　峻岭
steep normal fault　陡倾正断层
steep pitch　陡倾；急倾斜
steep ramp　陡斜路陡坡道
steep reflection　陡反射地震波
steep ridge　陡峭山脊
steep seam　急倾斜煤层；急倾斜矿层
steep slope　陡峭山坡；陡峭山坡；陡坡；陡斜率
steep terrain　陡峭地带
steep topography　峻峭地形；起伏大的地形
steep trench　陡堑沟；急倾斜沟道
steep valley　急沟；陡谷
steep vein　陡倾脉；急倾斜脉
steep-deeping　急倾斜的，陡坡的
steep-dipping　急倾斜的；陡坡的
steep-dipping bed　陡倾斜岩层
steep-grade mine　急倾斜层矿山
steeply dipping　陡倾；陡倾斜
steeply dipping formation　陡峭倾斜地层
steeply dipping raise　急倾斜上山
steeply dipping seam　大倾角煤层
steeply dipping zone　陡立地层带
steeply inclined fold　陡倾斜褶曲
steeply inclined reversed fault　陡倾逆断层
steeply inclined seam　急倾斜层；急倾斜煤层
steeply pitching coal　急倾斜煤层
steeply plunging fold　陡倾伏褶层
steeply sloping　急倾斜
steeply sloping seam　急倾斜煤层；急倾斜矿层
steeply-folded anticline　急褶皱背斜
steepness　陡度；斜度；坡度；陡峭
steepness of slope　边坡陡度
steepness ratio　陡度比率
steep-pitch method　急倾斜煤层开采法

steep-seam plough　倾斜煤层刨煤机
steep-sided cone　陡倾锥
steep-sided dome　陡边穹隆
steep-sided valley　峭壁峡谷
steering device　盾构掘进位置控制装置
steering logic　控制逻辑
steinheilite　堇青石
stele　中心柱；石碑，石柱
stell prop　锚固支柱；硬合金支柱
stellar astronomy　恒星天文学
stellar material　星球物质
stellar population　星族
stellar wind　恒星风
stellarite　土沥青煤；沥青煤；脉沥青
stellated　星状
stellated drainage　星状水系；星点状水系
stellated structure　星状构造
stellerite　淡红沸石
stem　柄，杆，堵炮泥；堵住，塞住；干，茎；钻杆；炮泥
stem deflection　钻杆弯曲
stem stream　主流
stemming bag　封眼袋；炮泥袋
stemming cartridge　封泥卷；炮泥筒
stemming deck　分隔炮泥
stemming material　炮泥；炮眼堵塞料；炮泥材料
stenode　晶体滤波的中频放大器
step adit developing　阶梯平硐开拓方式
step and-platform topography　阶梯平台状地形
step arrangement　台阶布置；梯段布置；阶段布置
step bit　阶梯式钻头
step block　阶状断块；阶状岩块
step by step　逐步
step by step procedure　逐次逼近法
step by step process　逐次逼近求解过程
step change　阶跃变化
step control　逐步控制；分级控制
step curve　阶梯曲线
step cutting　台阶式工作面掘进
step discontinuity　分段不连续
step disk　支承盘
step down　逐步下降；降低；降压
step fault　阶梯状断层；阶状断层
step fault scarp　阶状断层崖
step faulting　阶状断层作用
step fold　阶状褶皱
step forward　步进
step fracture　阶梯状断口
step function　阶梯跃函数；阶跃函数
step generator　阶跃发生器
step glacier　阶状冰川
step graben　阶状地堑
step grade　间断级配
step grinding　分段磨削
step horst　阶状地垒
step index　步长指数
step instruction　步进指令
step length　步长
step load　阶跃载荷
step motor　步进马达

step out 步测
step profile 台阶状剖面
step profile of invasion 台阶型侵入剖面
step pulley 级轮；塔轮
step quenching 分段淬火；分级淬火
step rate injectivity 阶跃注入率
step relief 阶状起伏；阶状地形
step response 瞬态特性；过滤特性；阶跃响应
step ring 支承环
step sedimentation model 阶梯沉积模式
step size 步长
step system of longwall 台阶式长壁采煤法
step taper pile 多节变截面桩
step terrace 梯状阶地
step topography 阶状地形
step transformer 升降压变压器
step transmission 有级变速传动
step type bit 阶梯形钻头
step up pressure 升压
step valve 级阀
step vein 阶状脉
step width 步长
step-and-platform topograph 阶梯平台状地形
stepback 回步
step-by-step approach to final adjustment 逐渐逼近平差法
step-by-step carry 逐位进位
step-by-step construction 逐次近似法
step-by-step control 步进控制
step-by-step design 分段设计
step-by-step formulation 逐次逼近公式
step-by-step integration 逐步积分
step-by-step method 逐步法；步进法
step-by-step operation 步进操作
step-by-step plan 分期规划
step-by-step procedure 逐步过程；逐步法
step-by-step process 逐次逼近求解过程
step-by-step program 分步方案
step-by-step test 阶段试验；逐步试验
step-by-step time integration 逐步时间积分
step-down method 逐渐减少分配法
step-down ratio 降压比
step-down substructure 阶梯式基座
step-drawdown test 分段降深试验
step-fault scarp 阶状断层崖
step-iterative method 逐步迭代法
step-length 步长
stepless 无级的；连续的
stepless control 连续控制
steplike profile 阶梯状断面
steplike series 阶状系列
steplike surface 阶状面
steplike terrace 梯状阶地
step-out time analysis 时差分析
stepover 阶跃；台跳
steppe ecosystem 干草原生态系统
stepped 阶梯式的
stepped annealing 阶段退火；分段退火
stepped appearance 阶梯形
stepped bit 阶梯钻头
stepped channel 梯级渠；
stepped crescent cliff 阶状新月形陡崖
stepped face 台阶式采煤工作面；梯段式工作面
stepped face mining 梯段式工作面开采，台阶式工作面开采
stepped face working 台阶法采掘作业
stepped factor of safety 分段变动安全系数
stepped fault 阶状断层
stepped feeding 分段进水
stepped footing 挡土墙台阶基础；阶梯形底脚
stepped foundation 阶梯形地基；阶式底座
stepped gradient 梯级降坡
stepped landform 层状地形；台阶状地形
stepped longwall 台阶式长壁开采法
stepped longwall advancing on strike 前进式走向台阶长壁开采法
stepped longwall stoping 长壁梯段回采法
stepped pool 梯级式水池
stepped quarry 分段采石场；台阶式采石场
stepped retaining wall 阶式挡土墙
stepped retreat line 梯级式后退工作面线
stepped section 阶形截面
stepped side-wall 阶式边墙
stepped stope 梯段回采工作面；台阶工作面
stepped wall footing 阶梯式墙基；台阶式墙基础
stepped-knife plough 多段刀片式刨煤机
stepped-up infill drilling 逐步加密钻井
Stepper 步进器；分档器
stepper motor 步进马达
stepper motor driven 步进电机驱动的
stepping down dual operation 降段作业
stepping method 分段开挖法
stepping stone method 踏脚石法
step-rate 呈台阶状变化的产量
step-rate analysis 台阶状流量试井分析
step-rate injectivity test 台阶状流量注入试井
step-rate testing 台阶状产量试井
step-scan 步进扫描
step-shaped 分段的
step-size control algorithm 步长控制算法
step-sizing dual operation 分段筛分作业
step-tapered-cylindrical structure 阶梯状锥形结构
step-test procedure 逐步检验方法
step-type pumping test 分段抽水试验
stepwise 逐步的；阶式的；分段的
stepwise backwards method 逐步向后法
stepwise computation 逐步计算
stepwise countercurrent extraction 分段逆流抽提
stepwise cracking 阶式破裂
stepwise cyclic loading curve 逐级循环加载曲线
stepwise discriminant analysis 逐步判别分析
stepwise fault 阶状断层
stepwise forwards method 逐步向前法
stepwise iterative process 按步迭代法
stepwise liberation 依次释出；依次分解；逐级分解
stepwise method 分段依次计算法
stepwise regression 逐步回归
stepwise regression analysis 逐步回归分析
stepwise trend analysis 逐步趋势分析
steradian 球面度；立体角

steradiancy 球面发射强度
stercorite 磷钠铵石
stereo binocular 双目实体显微镜；双目显微镜；实体显微镜
stereo camera 立体摄影机
stereo elevation 立体高程模型
stereo image 立体图像
stereo imagery 立体成像
stereo microscope 立体显微镜
stereo model 立体模型
stereo viewer 立体观察镜
stereo wave 立体波
stereoautograph 立体自动测图仪；体视绘图仪
stereobate 基座，基础；土台；建筑物地下室；台基；无柱底基；立体基线
stereoblock adjustment 立体区域网平差
stereoblock polymer 立体嵌段聚合物
stereocamera 立体摄影机
stereocartograph 立体测图仪
stereochemical formula 立体化学式
stereochemistry 立体化学
stereocomparagraph 立体坐标测图仪；立体镜
stereocomparator 立体坐标量测仪
stereocompilation 立体测图
stereocoverage 立体覆盖
stereogoniometer 体视量角化
stereogram 极射赤平投影图，赤平图；立体图；三维透视图；立体照片，立体相片
stereogram analysis 极射赤平分析
stereogrammetric method 立体测量法
stereogrammetry 立体摄影测量学
stereograph 立体测图仪；立体平面图；立体像对
stereograph map 立体图
stereograph method 赤平投影法
stereograph net 极射赤平网
stereograph polar projection 极射赤平投影
stereographic azimuthal projection 方位球面投影
stereographic chart 立体平面投影图
stereographic diagram 球极平面投影图
stereographic horizon map projection 地平球面投影
stereographic map 立体图
stereographic method 球面投影法；立体观测法
stereographic net 球面投影网；赤平投影网
stereographic polar projection 极射赤平投影
stereographic projection 极射赤平投影；立体投影；球面投影
stereographic projection technique 赤平投影技术
stereographic projection-conformal property 赤平极射投影的保角特性
stereographic projection-decomposition of a vector 矢量分解的赤平极射投影
stereographic projection-dot product 矢量点积的赤平极射投影
stereographic survey 球面投影测量
stereographic technique 立体测量技术；极射赤平技术
stereography 立体画法；立体摄影术；立体测图
stereo-image alternator 立体图像转换器
stereoinspection 体视镜检验；立体镜检查
stereointerpretoscope 立体判读仪
stereomapping 立体测图

stereometer 体积计；视差测图镜；立体量测仪；自动立体测图仪；立体异构体；体积计
stereomethod 立体测量法
stereometric 立体测量的
stereometric camera 立体摄影机
stereometric map 立体测绘地形图
stereometrical map 立体测绘地形图
stereometry 立体几何；立体几何学；体积测定法；比重测定法
stereomicroscope 实体显微镜；立体显微镜
stereomodel 立体模型
stereonet 赤平投影网；球面投影网；赤平极射投影网
stereo-orthophoto technique 立体-正射投影技术
stereopair 立体像对；立体照片对
stereophony 立体声
stereophotogrammetric survey 立体摄影测量
stereophotogrammetrical survey 立体摄影测量
stereophotogrammetry 立体摄影测量学
stereophotograph 立体摄影；立体相片
stereophotographic map 立体测绘地形图
stereophotography 立体摄影；立体摄影术
stereophototopography 立体摄影地形测量学
stereophytic 固体起源的
stereopicture 立体图像
stereoplanigraph 精密立体测仪
stereoplot 赤平极射投影；立体测图
stereoplotter 立体绘图仪；立体测图仪；立体影像绘制仪
stereoplotting 立体测图
stereopticon 投影放大器；透射投影仪
stereoradar 立体雷达
stereorandom copolymer 立构无规共聚物
stereoregular polymer 立体规整聚合物
stereosat 立体卫星；立体观测卫星
stereoscope 立体镜；体视镜；立体显微镜；双眼照相机
stereoscopic aerial photograph 立体航空相片
stereoscopic analysis 立体分析
stereoscopic effect 立体效应
stereoscopic image 立体影像
stereoscopic interpretation 立体判读
stereoscopic monitoring 立体监测
stereoscopic monitoring network 立体监测网
stereoscopic observation 立体观测
stereoscopic parallax 立体视差
stereoscopic perception 立体感
stereoscopic photograph 立体照片
Stereoscopic photography 立体摄影术
stereoscopic plotter 立体测图仪
stereoscopic projection 体视投影；立体投影
stereoscopic sensation 立体感
stereoscopic technique 立体观察技术
stereoscopic topography instrument 立体地形仪
stereoscopic vision 立体观察；立体视力；立体视觉
stereoscopical 立体的；体视的
stereoscopical map 立体地图
stereoscopical path mapping 立体路线填图
stereoscopy 立体观察；立体观测；体视法；体视学
stereoselective polymerization 立构选择聚合
stereosphere 固结圈；岩石圈；坚固圈；刚性圈

stereostatic	地静压的	stiff frame	刚架；刚性架；刚性骨架
stereotelemeter	立体测距仪；立体遥测仪	stiff gantry	刚性台架
stereotheodolite	立体经纬仪；立视经纬仪	stiff layer	刚性层；坚硬层
stereotopographic isotropy of mine product	立体测绘地形图	stiff reinforcement	刚性钢筋
		stiff slope	陡坡
stereotopographic map	立体测绘地形图；立体地形图	stiff soil	硬土；黏结性土壤
stereovector	立体矢量	stiff stratum	硬地层
stereoviewer	立体观察	stiff structure	刚性结构
steric effect	空间效应	stiff test	刚性试验
steric hindrance	位阻现象	stiff testing machine	刚性试验机
steric isomer	立体异构体	stiffen	加固；稠化；硬化；加强
steric regularity	立构规整度	stiffened	加劲的；加固的
sterile coal	碳质页岩	stiffener ring	加强圈
sterilized coal	煤层的不采部分	stiffening effect	刚度效应
sternbergite	硫银铁矿	stiffening limit	硬化极限
stevensite	硅镁石	stiffening piece	加固片；助力片
stewartite	斜磷锰矿	stiffening plate	加强板；加劲板；补强板
stibarsen	砷锑矿	stiffening rib	加强肋；加劲肋
stibiconite	黄锑华；黄锑矿	stiffening ring	加强圈
stibiocolumbite	锑铌矿	stiffening sheet	加固板
stibiopalladinite	锑钯矿	stiffening truss	加劲桁架
stibiotantalite	锑钽矿	stiff-fissured clay	坚硬裂隙黏土
stibium	锑	stiff-leg derrick	刚性柱架
stibium bloom	锑华	stiff-leg platform crane	刚性腿平台起重机
stibnite	辉锑矿	stiffness	刚度，劲度；硬度；稳定性
stichtite	碳酸镁铬矿	stiffness analysis	刚度分析
stick count	药包量	stiffness approach	刚度法
stick force	黏附力；外附力	stiffness array	刚度数组
stick powder	卷装炸药；装包炸药	stiffness calculation	刚度计算
stick shear modulus	黏接剪切模量	stiffness change	刚性变化
stick slip	黏滑；黏着滑动	stiffness coefficient	刚度系数
stick slip earthquake	黏滑地震	stiffness coupling	刚度耦合
stick slip fault	黏滑断层	stiffness curve	刚度曲线
stick slip motion	黏滑运动	stiffness factor	刚度系数
stickability	黏性；黏着力	stiffness mat foundation	刚性板式基础
stickiness	黏性；黏结；胶黏性	stiffness matrix	刚度矩阵
sticking	黏附；黏着；卡住	stiffness method	刚度法
sticking factor	黏滞系数	stiffness modulus	刚度模量；劲度
stickogram	反射系数图；滤波图	stiffness of a structure	结构刚度
stick-slip	黏着滑动；黏滑	stiffness of foundation soil	地基刚度
stick-slip active fault	黏滑型活断层	stiffness of pile	桩的刚度
stick-slip instability	黏滑不稳定性	stiffness of support	支护刚度
sticky clay	胶黏黏土	stiffness property	刚度特性
sticky coal	黏煤；黏性煤炭	stiffness ratio	刚度比
sticky formation	黏性地层	stiff-plastic	硬塑性的
sticky limit	黏着界限；黏限；黏韧度；黏着界限	stilbite	辉沸石；束沸石
sticky material	黏性物料	still basin	静水池；消力池
sticky oil	高黏度石油	still stand	停滞；静止
sticky ore	黏结矿石	still water	静水
sticky shale	黏性页岩	still water deposition	静水沉积作用
sticky slip	黏滑	still water head	静水头
stiction	静摩擦；静态阻力	still water level	静水位；静水面
stiff assembly	刚性组合	stilling basin	消能池；静水池
stiff beam	刚性梁	stilling chamber	消能室；调压井
stiff brush	刚性刷子；硬刷	stilling pond overflow	消力池溢流
stiff cement grout	稠水泥浆	stilling pool	沉淀池；沉沙池
stiff clay	低塑性黏土；硬黏土；硬泥	stilling well	静水井；调压井
stiff concrete	硬稠混凝土	stillroom	蒸馏室；储藏室
stiff drill rod	刚性钻杆	still-stand	停滞
stiff equations	刚度方程	stillstand period	静止期

stilpnomelane	黑硬绿泥石
stilt	支撑物；支柱；支架的让压部分；可缩性拱脚
stimulated reservoir volume	储层改造体积；储层压裂体积
stimulation of well	油井增产措施
sting period	启动期
stinkkalk	臭灰岩
stinkstone	臭灰岩
stipite	富含黄铁矿瘦煤；黄铁矿化煤
stir up	搅拌；摇晃
stirred bed	搅动床层
stirrer bar	搅棒
stirring arm	搅拌桨叶
stirring leaching test	搅拌浸出试验
stirring machine	搅拌机
stirring mechanism	搅动机构；搅拌机构
stirring motion	湍流；涡流；搅拌运动
stirring rate	搅拌速度
stirring screw conveyor	搅拌螺旋输送机，螺式搅拌输送机
stirring-type mixer	搅拌式混合机
stirrup feeder	摆动式给料机
stochastic	随机的
stochastic acceleration	随机加速
stochastic allocation	随机分布
stochastic approximation method	随机近似法
stochastic automation	随机自动机
stochastic characterization	随机特征
stochastic clustering model	随机震群模型
stochastic component	随机分量
stochastic congested network	随机性拥挤路网
stochastic control	随机控制
stochastic convergence	随机收敛
stochastic design earthquake	随机设计地震
stochastic differential equation	随机微分方程
stochastic distributed parameter system	随机分布参数系统
stochastic distribution	随机分布
stochastic distributive parameter systems	随机分布参数系统
stochastic disturbance	随机扰动
stochastic equation	随机方程
stochastic error	随机误差
stochastic event	随机事件
stochastic excitation	随机激励
stochastic fault model	随机断层模型
stochastic function	随机函数
stochastic game	随机对策
stochastic hydraulics	随机水力学
stochastic incident	随机事件
stochastic independence	随机独立
stochastic inversion	随机反演
stochastic language	随机语言
stochastic linear discrete system	随机线性离散系统
stochastic linear system	随机线性系统
stochastic loading	随机加载
stochastic matrix	随机矩阵
stochastic medium	随机介质
stochastic medium equation	随机介质方程
stochastic medium theory	随机介质理论
stochastic model	随机模型
stochastic modeling	随机建模
stochastic modeling for mining subsidence	岩石移动的随机模型；开采塌陷的随机模型
stochastic nature of ground motion	地面运动随机性质
stochastic network pore model	随机网络孔隙模型
stochastic non-linear system	随机非线性系统
stochastic observability	随机可观测性
stochastic occurrence	发震随机过程
stochastic porphyritic texture	随机斑状结构
stochastic prediction	随机预测
stochastic process	随机过程
stochastic programming	随机规划
stochastic property	随机性
stochastic response	随机反应
stochastic response characteristic	随机反应特性
stochastic response surface method	随机响应面法
stochastic retrieval	随机检索
stochastic sampling	随机采样；随机抽样
stochastic seismic excitation	随机地震激发
stochastic seismic response analysis	随机地震反应分析
stochastic shale	随机页岩
stochastic signal	随机信号
stochastic simulation	随机模拟
stochastic system	随机系统
stochastic value	随机值
stochastic variable	随机变量
stochastic variance	随机变化；随机方差
stochastic vector	随机向量
stochastic vibration	随机振动
stochastically independent component	随独立分量
stock	岩株；岩干；矿筒；储备；库存；资源
stock anchor	有杆锚；普通锚
stock antivibrator	虎钳；钳工台
stock assessment	资源评价
stock bin	贮料仓
stock dump	堆场；贮矿场
stock fluctuation	储量变化
stock ground	料场
stock heap	矿堆；煤堆；料堆
stock house	贮料房；仓库
stock mineralization hypothesis	岩钟成矿说
stock oil	库存油；原料油
stock pile	储料堆；排土堆；排土场
stock pile area	堆料场
stock removing	切削
stock room	材料库；贮料室；储藏室
stock size	标准尺寸；常规尺寸
stock tank	油库油罐；库存罐；原料罐
stock tank oil	地面脱气原油
stock tank vapor recovery	储罐蒸汽回收
stock withdrawal	动用库存
stock yard	存放场；堆料场；存料场
stocked ore	库存矿石
stockinette sampler	具有衬套的取土器
stockless anchor	无杆锚
stockpile	储矿堆；储料堆；资源；储备
stockpile area	堆渣场
stock-pile moisture	堆存物料水分
stockpile reclamation	恢复堆料区原貌

stock-pile sample 料堆取样
stockpiling 堆存；储备
stockpiling area 堆料区；贮存区
stockpiling dual operation 推土作业
stockpiling operation 推土作业
Stock's formula 斯托克斯公式；斯托克斯积分公式
Stockton dam 斯托克顿坝
stockwork 网状脉；网状矿脉
stockwork deposit 网状脉矿床；细脉浸染型矿床
stockwork ore body 网状矿体
stockyard 储煤场；堆料场，存料场
Stodola method 斯托多拉法
stoichiometric 按化学式计算的；化学计量的
stoichiometric balance 化学计算平衡
stoichiometric calculation 化学计算
stoichiometric chemistry 计量化学
stoichiometric composition 化学计量成分
stoichiometric equation 化学计算方程式
stoichiometric point 当量点；化学计量点
stoichiometry 化学计算；化学计量；化学计量学
stoker 给煤机；加煤机
stoker coal 加煤机用煤
stoker grade 加煤机用煤炭粒度分级
stoker prop 锚柱
Stokes diameter 斯托克斯直径
Stokes flow 斯托克斯流体
Stokes formula 斯托克斯公式
Stokes law 斯托克斯定律
Stokes theorem 斯托克斯定理
Stokes theory 斯托克斯理论
stokesite 硅钙锡石
stolzite 钨铅矿
stomatal structure 气孔构造
stomatal transpiration 气孔蒸腾；气孔散发
stone 石；石头；石材；石料；岩石；宝石
Stone Age 石器时代
stone ballast 石碴；铁路碎石道碴
stone band 矸石；夹矸石；夹石层
stone base 石基础
stone basket 石笼；结石网篮
stone bin 石碴仓；废石仓
stone bind 夹页砂岩；粉砂岩
stone block 石块；大块矸石
stone block pavement 石块路面
stone bolt 锚杆；锚栓
stone bound 界标石
stone bowl 废石仓
stone breaker 碎石机
stone breaking plant 碎石厂；碎石装置
stone breakwater 石防波堤
stone bridge 石桥
stone bud 石芽；石蕊
stone cage cofferdam 石笼围堰
stone cake 石饼
stone canal 石渠
stone cavern 石洞
stone chip 石片；石屑
stone chisel 石凿
stone circle 石环；环状列石
stone coal 石煤；块状无烟煤；硬煤

stone column 碎石桩
stone column replacement ratio 碎石桩置换率
stone concrete 碎石混凝土；石子混凝土
stone construction 石构造
stone counterfort 石扶垛；石撑壁
stone crusher 碎石机
stone crushing plant 碎石厂
stone culvert 石砌涵洞；石块暗渠
stone cutter 石工；割石机；切石机
stone cutting machine 切石机；石材切割机
stone dam 石坝
stone deposit 石料层
stone desert 石漠
stone desertization 石漠化
stone dike 砌石堤；石堤
stone drain 砌石排水沟
stone drainage 底铺碎石沟排水；填石排水
stone drainage ditch 石砌排水沟
stone dresser 石工
stone dressing 琢石
stone drift 岩巷；岩石平巷
stone drifter 岩石巷道掘进工；凿岩工；架式凿岩机
stone drifting 岩巷掘进
stone drill 岩石钻机；凿岩机；岩石钎子
stone dump 倾石场
stone dust 岩粉
stone dust dispersibility 岩尘扩散性
stone dust plan 沉积岩粉采样区段图
stone dyke 砌石堤
stone embankment 堆石堤坝；石堤
stone facing 砌石护面
stone fan 石扇
stone fields 岩海
stone fill 填石
stone filled crib 石笼；填石木垛
stone filled drain 填石排水沟；盲沟
stone filled trench 填石沟
stone filling 填石
stone flag 石板
stone flower bed 石花台
stone footing 毛石基础
stone forest 石林
stone fork 扒石耙；石碴耙
stone foundation 石基；毛石基础
stone fragment 石屑；石碴；石质碎屑
stone gall 黏土结核
stone gobbing 石料充填
stone grinding machine 石材研磨机
stone hammer 碎石锤
stone head 岩石平巷；岩巷；流砂下第一层硬岩
stone hill 石丘
stone imprint 雹痕
stone jetty 石突堤式码头；石防波堤
stone lane 采石区
stone lifting bolt 吊石栓
stone line 碎石带
stone lined 块石衬砌的
stone lining of shaft 石砌井壁
stone lintel 条石梁；石过梁
stone loader 装岩机

stone man	石工；岩石掘进工	stone-dust barrier	岩粉棚
stone mason	石匠；石工；凿石工作	stone-faced bank	砌石堤
stone masonry	砌石工程；石砌体；砌石	stone-filled crib dam	填石木框坝
stone masonry dam	砌石坝	stone-filled drain	盲沟，填石排水沟
stone masonry structure	石砌体结构	stone-filled timber crib groyne	术笼填石丁坝
stone material	山石材料；石料	stone-filled trench	填石暗沟
stone mattress	填石沉排；石料褥层	stone-filled works	堆石工程
stone mesh apron	钢丝网填石沉排	stone-flax	石棉；纤蛇纹石
stone mesh groyne	钢丝网填石丁坝	Stoneley wave	斯通利面波
stone mesh mattress	钢丝网填石沉排	stone-like coal	石煤
stone mill	碎石机；磨石机；岩石破碎机	stone-picking machine	除石机；选石机
stone mulch	石幕；石覆盖	stone-pit	采石矿；石坑
stone oil	石油	stone-pitched jetty	砌石堤
stone pavement	石块路面	stones laying	叠石
stone pavement road	石铺路	stone-sizing plant	石料筛分厂
stone paving	石块地面；砌石护面	stone-slab correction	中间层校正
stone peat	坚实泥炭；致密泥炭；石状泥炭	stonewall	石墙；石垣；石壁
stone picker	除石机；选石机	stoneware	石器；石制品
stone pile	碎石桩	stoneware drain	排水陶管
stone pillar	石柱	stoneware pipe	粗陶管；陶土管
stone pine	南欧松；瑞士松	stonework	砌石；石方工程
stone pit	采石坑；采石场	stoning	碎石护岸；河床铺石
stone pitch	石沥青	stony	多石的；石质的；含石的
stone pitched facing	砌石护面	stony clay	含碎石黏土
stone pitching	砌石护坡；砌石护面	stony desert	石漠
stone pitching surface	砌石保护面	stony drainage	石砌排水沟
stone plums	大块石	stony ground	含石土壤；石质地；多石地
stone pockets of concrete	混凝土蜂窝状气孔	stony iron	陨铁石
stone polygon	石多边形；成形多边形土	stony meteorite	石陨石，无铁陨石
stone powder	岩粉	stony soil	石质土；含石土
stone quarry	采石场	stony stream	石流
stone raft	石筏	stony tundra	石质冻原
stone revetment	干砌石护坡；石护岸	stony-iron meteorite	石铁陨石
stone riprap	乱抛石块；抛石体	stoop	煤柱；保安矿柱；安全煤柱
stone riprap bulk rockfill	抛石体充填	stoop and room	房柱式开采法
stone river	石河	stoop and room system	房柱式开采法
stone roll	石辊	stoop method	柱式开采法
stone sand	碎石砂	stoop road	柱式采场内巷道
stone shaft	废石充填井	stoop roadway	实煤中的平巷
stone slab	石板	stoop system	柱式开采法
stone slab bridge	石板桥	stoop-and-room method	房柱式开采法
stone slab correction	中间层校正	stop and deposit field	停淤场
stone step	石级；石阶	stop band	抑制频率；阻带
stone structure	石结构	stop bank	堤；堤岸
stone support	料石砌碹支护	stop board	挡板
stone surfacing	石护面	stop cock	截流旋塞；龙头；活塞
stone teeth	石牙	stop dike	阻挡堤
stone tie	加固条石；条石加固	stop end	堵头；封端；封堵模板
stone wall	石墙	stop fold	陡斜褶皱
stone ware clay	陶土；粗陶土	stop pulse	停止脉冲
stone ware pipe	粗陶管	stop water loss	堵漏
stone weir	石坝；石堰；堆石坝；堆石堰	stop-and-go mining	断续式开采
stone wheel	砂轮	stop-and-waste valve	关闭泄漏阀
stone work	石方工程；石工	stopcock	旋塞阀；活塞
stone workings	石洞；岩层巷道	stop-cock	旋塞阀
stone yellow	黄赭石	stopcocking	周期关井
stone-concrete footing	毛石混凝土基础	stope	采场；采矿场；回采工作面；采矿工作面
stonecutter	石工；石匠；切石机；琢石机	stope and pillar design	采场和矿柱设计
stone-cutting machine	切石机	stope and pillar mining	仓柱式开采
stone-dragon lava	石龙熔岩	stope area	回采区面积；矿房面积

stope assay 采场采样分析
stope assay plan 采矿场取样品味分析图
stope back 回采工作面上帮；回采工作面上部梯段
stope book 回采工作面记录簿
stope bowl 工作面溜口
stope caving method 崩落采矿法
stope cut 台阶掏槽；底槽
stope design 采场设计
stope drift 回采平巷
stope drill 伸缩式凿岩机
stope engineer 回采工程师
stope face 回采工作面
stope filling 采场充填
stope filling operations 采场充填施工
stope hammer 小型回采凿岩机
stope heel 底梯段；顶梯段
stope hole 回采区炮眼
stope layout 采场布置
stope leaching 原地碎矿石沥滤；采场沥滤
stope of stepped face 梯段工作面采场
stope ore pass 回采区放矿溜道
stope pillar 场内矿柱；矿房矿柱
stope preparation 采准；采场准备
stope pressure 采场压力
stope rejection 回采区废石
stope sidewall stress 采矿边墙应力
stope support 回采工作面支架
stope surrounding rock 采场围岩
stope survey 采场测量
stope ventilation 采场通风
stope wall 矿壁
stope-caving method 崩落采矿法
stoped out 采空的
stoped out area 采空区
stoped out workings 采空区
stoped-out area 采空区
stope-mining sequence 回采顺序
stoper 向上凿岩机；上向式凿岩机；伸缩式凿岩机
stoper hammer 凿岩机
stopes intersect 小的中间矿柱
stop-grouting plug 止浆塞
stoping 回采；采矿法；顶蚀作用
stoping and filling 充填回采
stoping area 采区
stoping cycle 回采循环
stoping drift 回采平巷
stoping drill 伸缩式凿岩机
stoping dual operation 回采操作；回采工作
stoping entry 水采回采巷道
stoping face 回采工作面；梯段回采工作面
stoping ground 可准区段；备采区段；回采工作面
stoping height 回采高度
stoping hole 扩槽炮孔；回采炮孔；回采炮眼
stoping layout 采矿设计；采矿工作图
stoping level 回采水平
stoping limit 临界品位；品位极限
stoping machine 采矿机
stoping method 回采方法；阶段采矿法
stoping reserve 备采矿量
stoping-and-filling 充填回采

stoppage cleaning 清除堵塞
stoppage of formation sand 地层砂的阻挡
stopper drill 套筒式凿岩机
stoppered bottle 带阀取样器
stopping procedure 回采顺序
stopple coupon 取样管
stop-start control 停止-启动控制
stop-start frequency 停止-起动频率
storability 可储藏性；可存储性；储量；储量系数
storable 可储存的；耐储层物品
storage 存储；贮存；封存；仓库；存储量
storage access 存储器存取
storage allocation 库容分配；存储分配
storage area 蓄水面积；储藏区；存储区
storage balance 蓄水平衡
storage basin 蓄水池；封存盆地
storage bunker 储煤仓
storage canal 贮存沟
storage capacity 封存能力；存储容量；储水能力；库容；蓄水量
storage capacity curve 蓄量曲线；库容曲线
storage capacity of drainage basin 流域盆地蓄水量
storage capacity of reservoir 水库库容；储层封存能力
storage capacity of watershed 流域的蓄水容量；盆地蓄水量
storage cell 存储单元；蓄电池
storage characters 存储特性
storage coefficient 贮水系数；储水系数；库容系数；存储系数
storage constant for aquifer 含水层蓄水常数
storage counter 累积式计数器；存储计数器
storage curve 容量曲线；库容曲线；储量曲线
storage cycle 存贮周期；蓄水周期
storage dam 蓄水坝；拦水坝；贮水堰
storage density 存储密度
storage depot 储油库
storage device 存储装置；存储设备；存储系统
storage discharge curve 槽蓄流量曲线
storage drift 储矿平巷
storage dual operation 存库操作；贮藏工作
storage element 存贮元件
storage equation 储量方程
storage evaluation 储集能力评价
storage excavation ratio 蓄水量与开挖量之比
storage facility 贮存装置；储存设备；储藏设施；器材库；贮藏室；拦蓄设施
storage factor 储存系数；库容系数；储能因数
storage format 存储格式
storage fragmentation 存储碎片
storage fund 储备资金
storage gage 贮水式雨量计
storage gas 地层天然气
storage head 蓄水水头
storage hierarchy 存储器层次结构
storage in depression 洼地积水；洼地储水
storage in reach 河段蓄水量
storage instruction 存储指令
storage irrigation 蓄流灌溉
storage jug 地下液化石油气储穴
storage layer 封存层；存储层

storage level	水库水位
storage life	储存期限
storage life of gasoline	汽油的可储存期限
storage limit register	存储界限寄存器
storage location	存储器单元；存储单元；存储位置；存储场所
storage magazine	贮藏库；仓库
storage mapping	存储映像
storage matrix	存储矩阵
storage medium	存储媒介
storage modulus	储存模量；储能模量；储存系数
storage of surplus water	剩余水量的蓄存
storage of water in aquifer	含水层储水
storage operation	存库操作；贮藏工作
storage oscilloscope	存储示波器
storage outflow curve	储水量-泄流量关系曲线
storage percentage	存储系数；存储百分比
storage piping installation	油库管线
storage pond	储蓄池；蓄水池；沉淀池
storage pond for dewatering	脱水沉淀池
storage pore	不连通的孔隙；封闭孔隙
storage porosity	封闭孔隙度
storage potential	蓄水的势能；封存潜力；封存容量
storage power	储存能力
storage program computer	程序存储式计算机；内存程序计算机
storage protection	存储器保护；存储保护
storage pump	蓄能泵；蓄水泵
storage ratio	调蓄率
storage register	存储寄存器
storage reservoir	水库；蓄水库；蓄水池；封存储层
storage routing	调洪计算；库容演算
storage site	储藏地；储藏场所；封存场址
storage space	存储空间
storage stability	储存稳定性；耐储存性
storage station	蓄能水电站
storage system	存储系统；存储器存储系统
storage tank	储池；储罐；储水箱
storage tank farm	储油库
storage terminal	转运仓库
storage test	储存试验；封存试验
storage time	存储时间；封存时间
storage under pressure	压力储存
storage unit	存储器；存储单元；记忆单元
storage volume	储藏容积；储存容量；储存量；水库容积；封存容量；封存容积
storage water	储存水
storage water structure	储水构造
storage writing speed	存储记录速度
storage yard	存贮场
storage-capacity curve	库容曲线
storage-discharge curve	库容-流量关系曲线
storage-elevation curve	库容-高程关系曲线
storage-means	存储方法
storage-production terminal	储油-采油浮码头
storage-production vessel	储油-采油浮式装置；储油-采油平台
storage-tank piping	储罐管网
storascope	存储式同步示波器
storativity	储水性；储水系数；贮水系数
storativity of aquifer	含水层贮水系数
storativity ratio	存储比
store access controller	存储访问控制器
store access cycle	存取周期
store access time	存储器的存取时间
store cycle time	存储器的周期时间
store life	贮存期限
store room	储藏室
store shed	贮存房
store through	全存储
stored energy	储存能
stored ore	堆存矿石
stored program	内存程序；存储程序
stored program computer	内存程序式计算机；存储程序式计算机
stored route	备用线路
stored routine	存入程序
stored schema	存储模式
stored water	封存水；储存的水
stored-program computer	存储程序计算机
storer	存储器；仓库保管员
storey deformation	层间变形
storey displacement	层间位移
storey stiffness	层间刚度
Storgruve iron ore	斯托格洛夫磁铁矿
storied cambium	叠生形成层
storied structure	叠生构造
storing	储存；储藏；蓄积
storing device	存储装置
storing mechanism	存储机理
storm back-flow	风暴回流
storm beach	暴风浪滩脊；风暴海滩
storm belt	暴风雨地带；风暴区
storm centre	风暴中心
storm choke	油井控制阀；油管安全阀
storm collector	雨水沟渠；暴雨排水设备
storm current	风暴潮
storm delta	浪成三角洲；越浪堆积
storm deposit	风暴沉积
storm disaster	风暴灾害
storm distribution pattern	暴雨分布类型
storm door	防风暴门
storm drain	雨水道；暴雨排水道；暴雨下水道
storm drainage	排泄雨水；暴雨排泄
storm duration	暴雨历时
storm eye	风暴眼
storm flood	暴雨洪水；暴雨洪泛
storm flow	暴雨流量；暴雨径流
storm frequency	暴雨频率
storm gate	防洪闸
storm inlet	雨水进水口；雨水井
storm intensity	暴雨强度；风暴强度
storm intensity pattern	风暴强度图；暴雨强度类型
storm investigation	暴雨调查
storm lane	风暴路径
storm lantern	汽灯
storm loss	雨量损失
storm maximization	暴雨极大化；暴雨放大
storm mechanism	暴雨机制
storm model	风暴模式

storm morphology 磁暴形态学
storm oil 防波油
storm outfall sewer 暴雨排水管
storm overflow 暴雨溢水槽；暴雨溢流排水管
storm overflow well 暴雨溢流井
storm path 风暴路径
storm pavement 防暴雨护面
storm profile 风暴示意图
storm proof 防风暴的；暴风雨防备
storm rainfall 暴雨量；风暴降雨
storm region 风暴区
storm roller 风暴卷浪
storm runoff 暴雨径流
storm runoff forecast 暴雨径流预报
storm sample 风暴样品
storm sea 风暴海况
storm seepage 暴雨渗流
storm sewage 雨水；暴雨污水
storm sewer 暴雨水沟；暴雨排水沟；暴雨水管
storm sudden commencement 磁暴急始
storm surge 风暴大潮；风暴潮；风暴大浪；风暴波涌；暴风雨涌潮
storm surge protection breakwater 高潮防波堤；风暴潮防波堤；防暴潮堤
storm survey 暴雨调查
storm tide 风暴波浪；风暴潮
storm tide disaster 风暴潮灾害
storm tide gate 挡潮闸；暴潮门
storm tide process 风暴潮地质作用
storm track 风暴路径；风暴轴
storm transposition 风暴转移
storm wall 防浪墙
storm water 暴雨水
storm water pollution 暴雨水污染
storm water runoff 洪水径流
storm wave 风暴潮；暴风浪
storm wave base 风暴浪底
storm wind 暴风
storm zone 暴雨地带
Stormer viscosimeter 斯氏黏度计
storminess 风暴度
storm-proof 防暴风雨的
storm-sewage diversion weir 雨水-污水分流渠
storm-sewer 雨水沟
storm-surge current 风暴潮流
storm-surge forecasting 风暴潮预报
storm-time proton belt 暴时质子带
storm-time variation 暴时变化
stormwater box culvert 箱形雨水暗渠
stormwater channel 排雨水渠道
stormwater collecting 暴雨水的收集
stormwater discharge 雨水排放
stormwater drain 雨水渠
stormwater drainage 雨水疏导渠道；雨水排放渠道；雨水排放
stormwater drainage system 雨水排放系统；雨水疏导系统
stormwater inlet 暴雨进水口
stormwater main drain 雨水排放主渠
stormwater outfall 雨水渠排水口

stormwater overflow 暴雨溢流
stormwater overflow chamber 雨水溢流室；雨水溢流井
stormwater runoff 雨水径流
stormwater runoff reduction system 减少雨水溢流系统
stormwater sewer 雨水渠；暴雨下水道；暴雨排水道
stormwater system 雨水排放系统
stormwater tank 暴雨水流储水池
stormy weather 暴风雨天气；风暴天气
story drift 层偏移
story drift ratio 层侧移比
story shear 层剪力
stoss slope 迎风坡；向冰川面的坡；迎冰川的坡
stoss-and-lee topograph 鼻状地形；不对称丘
stove 烘箱；温室；烘干
stoving 烘干
stow 充填；装填，装载
stowage 充填采空区；充填；贮藏，蓄积；货栈；库房
stowage material 充填料；填密材料
stowage space 装载空间；有效空间
stowage unit 充填装置；充填组
stowed goaf 回填废矿；回填采空区；已充填的采空区
stowing 充填；回填
stowing capacity 充填能力
stowing density 填充密度；回填密度
stowing dual operation 充填作业
stowing factor 充填率；充填系数
stowing interval 充填步距
stowing machine 充填机械
stowing material 充填物；充填材料
stowing method 充填法；充填开采法
stowing operation 充填过程
stowing pipe 充填管
stowing pressure 充填压力
stowing pulp 充填泥浆
stowing raise 充填溜井
stowing ratio 充填率；充采比
stowing screen 充填屏
stowing shift 充填转移；充填替换
stowing space interval 充填的间距
stowing up 回填
stowing-up method 充填开采法
straddle 支柱；井圈支柱；跨装
straddle packer 跨式双封隔器中的一个封隔器
straddle spread 中间放炮排列
straddle test 双封隔器选择性地层测试
straddle-packer test 跨式双封隔器地层测试
straddling 支撑
straight asphalt 纯沥青
straight beam 直射波；直梁
straight beam method 直射法；直梁法
straight bit 一字形钻头
straight bituminous filler 纯沥青填塞料
straight bolt 直螺栓；直锚杆
straight cement 纯水泥；无杂质水泥
straight cement mortar 纯水泥浆；纯水泥砂浆
straight chain compound 直链化合物
straight chain hydrocarbon 直链烃
straight coal 褐煤；黑色褐煤
straight coast 平直海岸
straight coastline 平直海岸线

straight cross	等径四通	straighten	矫直
straight cut	直眼掏槽；直边割缝	straightener	矫直器
straight dynamite	纯炸药	straightening	矫直
straight embedment of anchorage	钢筋直端锚固	straightening cliff	平直崖壁
straight ends and wall	房柱式采煤法	straightening device	矫直装置
straight extinction	直消光	straightening machine	矫直机
straight face	直线工作面；连续工作面	straightening of hole	井身矫直；孔身矫直
straight fault	直线断层	straightening of kinked rail	矫直钢轨
straight flotation	直接浮选	straightening of river channel	河道裁弯取直
straight flow pump	直流泵	straightening press	压直机；校直机
straight freezing	冻透	straightening vanes	整流叶片
straight girder	直梁	straightening-cutting machine	矫直切断机；矫直切割机
straight girder beam	直梁	straight-freezing	一次冻透
straight gneiss	强直片麻岩	straight-hole downhole motor	直井井下马达
straight hole	垂直孔；直井眼	straight-hole drilling	钻直井
straight injection	连续注入	straight-hole turbodrill	直井涡轮钻具
straight line approximation	直线逼近	straight-in hole	垂直于工作面的炮眼；垂直掏槽炮眼
straight line correlation	直线相关	straight-legged arch	直壁拱；直腿拱形支架
straight line depot	直线折旧法	straight-line basis	基线
straight line source	直线源	straight-line code	直接式程序
straight lined vessel	直线型船	straight-line correlation	直线相互关系
straight method of depreciation	直线折旧法	straight-line decline	直线递减
straight mineral oil	纯矿物油；纯矿油	straight-line depreciation	直线折旧
straight parallel	并联	straight-line equation	直线方程
straight pipe prover	直管式标准体积管	straight-line extrapolation	直线外推
straight pipeline	直管线	straight-line flow	直线流动
straight polymer	纯聚合物	straight-line gravimeter	直线重力仪；直线比重计
straight pull	直拉	straight-line interpolation	直线内插
straight ramp	直斜路；直坡道	straight-line method	直线递减法
straight ray inversion	直射线反演	straight-line path	直线途径
straight ray path tomographic inversion	直射线路径层析成像反演	straight-line portion	直线段
		straight-line progression	直线增进；直线上升
straight reach	直河段	straight-line regression	直线回归
straight reaming	直接扩孔	straight-line relation	线性关系
straight reciprocating motion	直线往复运动	straight-line segment	直线段
straight reinforcement bar	直钢筋条	straight-line tomographic method	直射线层析成像法
straight ring	标准环	straight-line-spectrum	直线频谱
straight river	顺直性河流	straightness accuracy	直线度
straight river deposit	平直河沉积	straightness error	直线度误差
straight rotary drilling	正循环回转钻进	straight-path approximation method	直线路程近似法
straight rule	直尺；划线尺	straight-penetration method	直接贯入法
straight run coal tar	直馏煤焦油沥青	straight-ray approximation	直射线近似
straight segment	直线段	straight-sided arch	直臂拱
straight selective flotation	直接选择性浮选；直接优先浮选	straight-tension rod	直拉钢筋
		straight-through process	直通过程
straight series	串联	straightway pump	直流泵
straight sheet pile	板桩	strain	应变；变形；拉紧；加压
straight side casting	直接侧面倒堆	strain acceleration	应变加速度
straight sided slot	直边割缝	strain accelerometer	应变加速度计
straight slip surface	平滑面；直滑动面	strain accumulation	应变积累
straight slope	直线坡；直线斜率	strain aging	应变时效
straight wall bit	直孔空心钻头	strain amplitude	应变幅
straight water	不掺添加剂的水	strain analysis	应变分析
straight waterflooding	单纯注水	strain and stress in shell	壳体的应变和应力
straight well	直井	strain antiplane	应变反平面
straight-ahead drilling	顺利续钻	strain at failure	破坏应变
straight-cement concrete	纯水泥混凝土	strain axis	应变轴
straight-chain paraffin	直链烷烃	strain band	应变带
straight-chain polymer	直链聚合物	strain bar	应变棒
straight-edge leveling	直尺水准测量法	strain break	张力裂隙；张力破裂

strain burst	应变突出；应变岩爆
strain bursting	应变突出；应变岩爆
strain by bending	弯曲应变
strain by compression	压缩应变
strain capacity	应变量；变形度；应变能力
strain cell	应变计；应变盒
strain centre	应变中心
strain clamp	紧定夹持器；拉线线夹
strain coefficient	应变系数
strain compatibility equation	应变协调方程
strain component	应变分量；分应变
strain concentration	应变集中
strain contour diagram	应变等值线图
strain control	应变控制
strain control triaxial compression apparatus	应变控制式三轴压缩仪
strain controlled test	应变控制试验
strain coordination factor	应变协调因子
strain corresponding to the maximum stress	峰值应力对应的应变
strain crack	应变裂缝；应变裂纹
strain cracking	应变裂缝；变形开裂
strain creep	应变蠕变
strain curve	应变曲线
strain cycle	应变循环
strain deformation	应变形变；应变变形
strain determination in three dimension	三维应变测量
strain deviation	应变偏差
strain deviator tensor	应变偏斜张量
strain diagram	应变图
strain discontinuity	应变不连续
strain displacement equation	位移方程式
strain displacement relation	应变-位移关系
strain distribution	应变分布
strain due to torsion	扭曲应变；扭应变
strain element	应变单元
strain ellipse	应变椭圆
strain ellipsoid	应变椭球；应变椭球体
strain energy	应变能
strain energy density	应变能密度
strain energy density factor	应变能密度因子
strain energy density function	应变能密度函数
strain energy function	应变能函数
strain energy release rate	应变能释放率
strain engineering	应变工程
strain facies	应变相
strain failure	应变破坏，破坏应变
strain fatigue	应变疲劳
strain field	应变场
strain figure	应变图
strain foil	应变片
strain forces	应变力；变形力
strain formulation	应变公式
strain fringe value	应变条纹值
strain gage	应变仪
strain gage rosette	应变片花
strain gauge	应变计；应变仪；电阻应变片
strain gauge logger	应变记录仪
strain gauge rosette	应变片花
strain gauge tensiometer	应变张力计
strain gauge-type load cell	应变仪式测力传感器
strain gradient	应变梯度
strain gradient plasticity theory	应变梯度塑性理论
strain hardening	应变硬化
strain history	应变时程；应变史
strain in rock	岩石中的应变
strain increment	应变增量
strain indicator	应变指示器；应变仪
strain influence factor	应变影响因素；应变影响因子
strain intensity	应变强度
strain invariant	应变不变量
strain kinetics	应变动力学
strain lag of added concrete	应变滞后
strain lattice	应变晶格
strain level	应变大小；应变水平
strain limit failure criteria	应变极限破坏准则
strain line	应变线
strain localization	应变局部化
strain marker	应变标志；应变标记
strain matrix	应变矩阵
strain measurement	应变测量
strain meter	应变计；应变仪
strain monitoring	应变监测
strain net method	变形网格法
strain observation in horizontal tunnel	水平巷洞应变观测
strain of flexure	挠曲应变
strain path	应变路径
strain pattern	应变图像；应变模式
strain phenomenon	应变现象
strain pressure transducer	应变式压力传感器
strain pulse	应变脉冲
strain quadric	应变二次曲面
strain rate	应变率；应变比；应变速率；变形速度
strain rate control	应变速率控制
strain rate effect	应变速率影响
strain rate history	应变率时程
strain rate sensitivity	应变率敏感性
strain ratio	应变率；应变比
strain ratio method	应变比法
strain rebound	应变回跳
strain recovery	应变恢复
strain recovery ratio	应变恢复率
strain recrystallization	应变重结晶作用
strain relaxation	应变松弛
strain release	应变释放；应变解除
strain release acceleration	应变释放加速度
strain release map	应变释放分布图
strain relief	应变释放；应变解除
strain relief method	岩石应变解除法
strain resolution	应变分辨率
strain rockburst	应变型岩爆
strain rosette	应变片花
strain seismogram	应变地震图
strain seismograph	应变地震仪
strain seismometer	应变地震仪
strain sensitivity	应变灵敏度；应变分辨率
strain shadow	应变影；波动消光
strain sheet	应变图
strain slip	应变滑移
strain slip cleavage	应变滑劈理；错动劈理

strain softening	应变软化
strain softening pillar	应变软化煤柱
strain space	应变空间
strain spectrum	应变谱
strain spherical tensor	应变球面张量
strain state	应变状态
strain state at a point	一点处的应变状态
strain step	应变阶跃
strain surface	应变曲面
strain tensor	应变张量
strain test	应变试验
strain tolerance	容许应变；极限应变
strain trajectory	应变迹线
strain transducer	应变传感器
strain transformation	应变变换
strain triangular element	应变三角形单元体
strain vector	应变向量；应变矢量
strain vector in finite element method	有限元法应变矢量
strain volumetric	应变体积的
strain wave	应变波
strain wire	应变导线；振弦；钢弦应变仪
strain-aged steel	应变时效钢
strain-aging	应变老化；应变时效；弥散硬化
strain-analysis	应变-分析
strain-control shear apparatus	应变控制剪切仪
strain-controlled direct shear apparatus	应变控制式直剪仪
strain-controlled load test	应变控制荷载试验
strain-controlled testing method	可控应变试验法
strain-displacement	应变-位移
strained	应变的；变形的
strained folding	滑动褶皱；应变滑褶皱
strained region	应变区
strain-energy method	应变能法
strainer	滤器；滤网；滤水管
strainer cartridge	滤网筒；粗滤片筒
strainer casing	过滤箱
strainer check valve	滤尘止回阀
strainer filter	粗滤器；线网滤器；滤片过滤器
strainer head	过滤器水头
strainer injection	滤网注浆
strainer screen	网式滤器；网状过滤器
strainer tube	过滤管
strainer well	过滤井
strainer-gage type displacement transducer	应变式位移传感器
strain-free zone	无应变区域
strain-gauge dynamometer	应变片式动力测力计
strain-gauge rosette	应变片花
strain-gauge test	应变仪测试
strain-gauge-type pressure transducer	应变仪式压力传感器
strain-hardened material	形变硬化材料；应变硬化材料
strain-hardening	应变硬化
strain-hardening behaviour of rock	岩石的应变硬化性
strain-hardening modulus	应变硬化模量
strain-hardening stage	应变强化阶段
straining filter	粗滤池
straining meter	应变仪；应变计
straining ring	夹紧圈；应变圈
strainless	无应变的
strainless ring	无应变环
strainmeter	应变仪；应变计
strainometer	伸长计；应变仪
strain-optic sensitivity	应变光学灵敏度
strain-recovery characteristics	应变恢复特性
strain-release	应变释放
strain-relief approach	应变释放测定法；应变解除法
strain-relief measurement	应变解除测定法；弛张测定法
strain-slip cleavage	错动劈理；应变滑动劈理
strain-slip folding	滑动褶皱；曲滑褶皱；应变滑动褶皱
strain-softening	应变软化
strain-softening behaviour of rock	岩石的应变软化性
strain-stress curve	应变-应力曲线
strain-stress diagram	应变-应力图解
strain-stress loop	应变-应力环；应变-应力循环
strain-structure slip rockburst	应变-结构面滑移型岩爆
strain-time curve	应变-时间曲线
strain-time response	应变时间响应
strain-to-failure	应变造成的损坏
strain-transducer	应变传感器
strain-wire gauge	电阻应变仪；有线应变仪
strait	海峡；峡谷；通道；狭窄的
strait of Gibraltar	直布罗陀海峡
strait of Malacca	马六甲海峡
strake	洗矿槽；轮箍；箍条；条纹
strand	海滨；潮间带；沿岸带；触礁；搁浅
strand core	钢绞线芯
strand cutting	切割预应力钢缆
strand deposit	沿滨沉积；海滨沉积物
strand dune	海岸沙丘
strand flat	沿海台地；潮间坪；海滨滩
strand line	滨线；岸边线；海岸线；搁浅线
strand plain	海滨平原；滨海平原；潮间平原
strand tapered anchorage	钢缆式凹形锚；夹片式锚具
strand tendon	钢绞线锚索
stranded	绞成股的
stranded cable	绞合电缆；股绞钢丝绳
stranded caisson	缆吊沉箱；出水沉箱
stranded conductor	绞线；绞合导线
stranded galvanized steel	镀锌钢绞线
stranded geothermal energy	滞留地热能
stranded rope	股绳；多股钢丝绳
stranded steel wire	钢绞线
strandflat	潮间坪；海滨坪；海滨滩
stranding harbor	浅水港
stranding harbour	浅水港
strand-line	海滨线
strandline pool	海岸线油藏
strand-line trend	滨线趋向；滨线走向
strandplain	海滨平原
strangeness	奇异性
strap footing	条形基础
strap iron	条钢
strap jet gun	带状聚能射孔器
strap lift	绳缆扶手
strap steel	扁钢；带钢
strapping	箍条
strapping of tank	油罐围测标定
strata	层；岩层；地层

strata alternation	层的交互变化
strata and surface movement	岩层和地表移动
strata behavior	岩层动态；岩层性质；岩压显现；矿山压力显现
strata behavior around coal face	采煤工作面周围矿压显现；采煤工作面周围岩层性质
strata behaviour	地层特性
strata bolting	岩层锚固；岩层锚杆支护
strata bridge	岩桥
strata cohesion	岩层内聚力；岩层或地层黏结强度
strata configuration	地层结构
strata control	地压控制；岩层控制；顶板控制
strata control engineer	顶板管理工程师；岩层控制工程师
strata control measurement	矿压观测
strata control observation	矿压观测
strata correlation	地层对比；标志层对比
strata deposit	层控矿床
strata destruction	地层破坏
strata dip	地层倾斜
strata displacement	地层位移；岩层移动；岩移
strata failure	岩层破坏；岩层失稳
strata gap	地层间断
strata group	地层群
strata grouting	地层灌浆；地层注浆
strata grouting reinforcement	注浆加固地层
strata inclination	岩层倾斜；地层倾斜度
strata model	地层或岩层模型；岩体模型
strata movement	岩层移动；岩移；地层移动
strata of coal measures	煤系地层
strata pinching	地层变薄；地层尖灭
strata pressure	地层压力；地压
strata prevention	岩层注水防尘
strata profile	地层侧面图；地质剖面图
strata relation	地层关系
strata section	地层剖面；地层横断面
strata sensing	判断岩层性质
strata sequence	地层次序；层序
strata stress	地层应力
strata strike	地层走向
strata subsidence theory	地层下沉理论
strata succession	地层层序
strata superposition method	地层对比法
strata surface	地层面；层面
strata tilting	地层倾斜；地层倾倒
strata time	层时
strata water	层状水；层状水体；层间水
strata-binding process	层控作用
strata-bolting	地层锚固
stratabound	层控的
strata-bound	层控的
stratacontrol	岩层控制；地压控制
strata-control analysis	矿压分析
stratafrac	地层压裂
stratal	地层的；层的
stratal architecture	地层结构
stratal configuration	地层组合；地层结构
stratal continuity	地层连续性
stratal fracture water	层间裂隙水
stratal pinch-out	地层尖灭
stratal surface	层面；岩层层面
stratal water pressure	地层水压力
stratameter	地层仪；检层器
strata-punch	气动井壁取芯器
stratascope	地层仪；岩层观察镜
strategic landfill	重点堆填区
strategic mineral	战略矿物
strategic oil inventory	石油战略储备
strategic oil reserve	石油战略储备
strategic reserve	战略储备
strategic road network	重要道路网
strategic sewage disposal scheme	策略性污水排放计划
strath	宽谷；平底谷
strath terrace	宽谷阶地；平底河谷阶地
strath valley	平底谷；平地河谷
stratic	地层的；层序的
straticulate	薄层的；分层的；成层的；带状构造
stratification	层理；分层现象；成层作用；层化作用
stratification cross-section	地层横剖面
stratification effect	分层效应
stratification foundation	层状地基
stratification index	层理指数；成层指数
stratification joint	层理面
stratification line	层理线；分层线
stratification of temperature and salinity	温度与盐度的分层状况
stratification plane	层理面；层面
stratification ratio	成层比
stratification sampling	分层取样
stratification section	地层剖面
stratification testing	分层测试
stratification-foliation	叶理地层
stratified	成层的；分层的；层状的；层理的
stratified alluvium	成层冲积层
stratified atmosphere	分层大气；成层大气
stratified clastic structure	层状碎裂结构；层状碎屑结构
stratified coal seam sample	分层煤样；分层煤矿层样品
stratified cone	层状火山锥；复式火山锥
stratified current	层流
stratified deposit	层状沉积；层状堆积；成层矿床；层状矿床；沉积矿床；层状油藏
stratified drift	成层漂碛；成层冰碛；漂碛层
stratified finger	层状指进
stratified fissure water	层状裂隙水
stratified flow	层流；分层流
stratified fluid	分层流体
stratified formation	层状地层；层状层系
stratified foundation	成层地基；层状地基
stratified geologic body	层状地质体
stratified heterogeneous reservoir	层状非均质油藏
stratified in thick beds	厚层分层
stratified in thin beds	薄层分层
stratified mass	成层岩土体
stratified material	层状材料
stratified media	成层介质；层状介质
stratified medium	成层介质；层状介质
stratified model	层状模型
stratified ocean	分层的海洋
stratified ore body	分层矿体
stratified ore deposit	成层矿床；层状矿床

stratified pumping 分层取水
stratified random sampling 分层取样；分层随机取土样
stratified relief 层状地貌
stratified reservoir 层状油气藏；层状储油层
stratified rock 层状岩石
stratified rock mass 层状岩体
stratified rock slope 层状岩质边坡
stratified roof 成层顶板
stratified sampler 分层取样；分层抽样
stratified sampling 分层取样；分层采样；分层抽样
stratified sand 成层砂；层夹砂
stratified sediment 层状沉积物
stratified sedimentation 分层沉积；成层沉积；重叠沉积
stratified slope deposit 层状坡积物
stratified soil 层状土；成层土
stratified structure 层状结构；层状构造
stratified subgrade 成层路基；成层地基
stratified system 层状体系
stratified till 层状冰碛
stratified volcano 层状火山
stratified water 成层水
stratified waterflooding 层状注水
stratified wavy flow 分层波浪流
stratiform 层状的；成层的
stratiform cloud 成层云
stratiform deposit 成层矿床；层状矿床
stratiform intrusion 层状侵入体
stratiform manganese deposit 层状锰沉积；层状锰矿床
stratiform ore deposit 层状矿床
stratiform stromatolite 叠层石；层状叠层石
stratiform water 层状水；成层水
stratify 使成层；使分层
stratigrapher 地层学家
stratigraphic 地层的；地层学的
stratigraphic analysis 地层分析
stratigraphic anomaly 地层异常
stratigraphic arrangement 地层排列
stratigraphic boundary 地层边界
stratigraphic break 地层间断；地层缺失
stratigraphic category 地层类别
stratigraphic classification 地层划分；地层分类
stratigraphic code 地层规范
stratigraphic column 地层柱状图；柱状剖面图
stratigraphic condensation 化石杂聚层位
stratigraphic contact 地层接触；地层控制
stratigraphic correlation 地层对比
stratigraphic criteria 地层标志；地层准则
stratigraphic cross-section 地层剖面；地层横剖面
stratigraphic datum 地层的基准面
stratigraphic deconvolution 地层反褶积
stratigraphic dip 地层倾角
stratigraphic distribution 地层分布
stratigraphic division 地层划分；地层区划
stratigraphic effect 地层效应
stratigraphic equipment 地层测量设备
stratigraphic equivalence 相当地层层位
stratigraphic erosional reservoir 地层剥蚀油气藏
stratigraphic evolution 地层演化
stratigraphic facies 地层相
stratigraphic fluctuation 地层变化

stratigraphic forward modelling 地层正演模拟
stratigraphic gap 地层间断；地层缺失
stratigraphic geology 地层学；地层地质学
stratigraphic geophysics 地层地球物理学
stratigraphic guide 地层标志
stratigraphic heave 地层隆起；地层水平移动；地层平错
stratigraphic hiatus 地层缺失；缺层；地层间断
stratigraphic high resolution dipmeter tool 高分辨率地层倾角测井仪
stratigraphic history 地层史
stratigraphic horizon 地层层位
stratigraphic information 地层信息；地层资料
stratigraphic interpretation 地层解释；地层说明
stratigraphic interval 地层间距
stratigraphic lithologic and palaeogeographic interpretation 地层岩相古地理分析
stratigraphic map 地层图
stratigraphic marker 地层标志
stratigraphic model 地层模式
stratigraphic modelling 地层建模
stratigraphic nomenclature 地层命名
stratigraphic oil pool 地层油藏
stratigraphic order 地层层序
stratigraphic overlap 地层超覆
stratigraphic palaeontology 地层古生物学
stratigraphic paleontology 地层古生物学
stratigraphic piercing-point 地层刺穿点
stratigraphic pitch-out 地层尖灭
stratigraphic procedure 地层程序
stratigraphic profile 地层剖面；地层纵断面
stratigraphic prognosis 地层推断
stratigraphic range 时代-地层跨度；地层延续时限
stratigraphic record 地层记录
stratigraphic reef 地层礁；层礁
stratigraphic repetition 地层重复
stratigraphic reservoir 地层油气藏
stratigraphic resolution 地层分辨率
stratigraphic rig 地质钻探钻机
stratigraphic scale 地层表；地层年代表；地层时序表
stratigraphic screened oil accumulation 地层遮挡油藏
stratigraphic section 地层剖面；地质剖面
stratigraphic seismic 地层地震的
stratigraphic separation 地层离距；地层断距；地层落差
stratigraphic sequence 层序；地层层序；地层序列
stratigraphic study 地层研究
stratigraphic subdivision 地层划分
stratigraphic succession 地层层序
stratigraphic symbol 地层标志
stratigraphic terminology 地层术语
stratigraphic test well 地层探井
stratigraphic throw 地层落差；地层断距
stratigraphic time table 地层年代表
stratigraphic timescale 地层年代表
stratigraphic time-scale 地层年代表
stratigraphic trap 圈闭层；地层圈闭
stratigraphic type oil pool 地层油藏
stratigraphic unconformity 地层不整合；假整合
stratigraphic unconformity reservoir 地层不整合油气藏

stratigraphic unit	地层单位；地层分层单位	stray pay	零散可采层
stratigraphic well	基准井；参数井	stray resistance	杂散电阻
stratigraphic window	地层窗	stray resonance	杂散谐振
stratigraphic younging	地层变新	stray sand	零散砂层；钻井中不期而遇的砂层
stratigraphic zonal sequence	地层带序列	streak	条痕；条纹；矿物痕色；矿脉；夹层
stratigraphic zonation	地层分带性	streak line	条纹线
stratigraphic zone	地层带	streak photograph	纹影摄影
stratigraphical	地层的；地层学的	streak test	刻痕试验
stratigraphical break	地层断缺；地层缺失	streaked-out ripple	纹影波痕
stratigraphical column	地层柱状图	streaking	条纹状的；条纹
stratigraphical correlation	地层对比	streakline	纹线
stratigraphical discordance	地层不整合	streaky	有条纹的；似条纹的
stratigraphical division	地层划分	streaky structure	条纹构造；条痕构造
stratigraphical drilling	地层钻探	stream	流；河流；水流；溪流
stratigraphical evidence	出土层位；地层物据	stream abstraction	河流合流作用；退流作用
stratigraphical gap	地层间断	stream action	河流作用；水流作用
stratigraphical heave	地层平错	stream adjustment	河道调整
stratigraphical hiatus	地层间断	stream anchor	雪水补给河流
stratigraphical map	地层图	stream aquifer	河源含水层
stratigraphical overlap	地层超覆	stream azimuth	河流方位
stratigraphical profile	地层纵断面；地层剖面	stream band erosion	河岸侵蚀
stratigraphical relation	地层关系	stream bank	河岸
stratigraphical section	地层剖面	stream bank erosion control	河岸防冲
stratigraphical separation	地层隔距；地层落差	stream bank stability	河岸稳定
stratigraphical sequence	地层顺序；层序	stream bank stabilization	河岸稳定；河岸加固
stratigraphical throw	地层落差	stream basin	河流盆地
stratigraphical time-scale	地层时序表；地层年代表	stream bed	河床
stratigraphical timetable	地层时序表；地层年代表	stream bed aggradation	河床淤积；河床加积
stratigraphic-sedimentologic	地层沉积学的	stream bed degradation	河床刷深
stratigraphic-sedimentologic framework	地层沉积岩石构架；沉积地层格架	stream bed erosion	河床冲蚀
		stream bed gradient	河床比降
stratigraphic-type reservoir	地层油气藏	stream bed intake	河底取水口
stratigraphy	地层组合；区域地层；地层学	stream bed protection	河床保护
stratigraphy interpretation	地层解译；地层判读	stream betrunking	河流干涸
stratinomy	层序学	stream biota	河流生物群
stratofabric	成层组构；层状组构	stream bottom soil	河底土壤
stratographic	色谱的	stream boundary	流域界；流域边界
stratographic analysis	色谱分析	stream built terrace	河成阶地；冲积阶地
stratography	色层分离	stream capacity	河流最大输沙能力；河流搬运力
stratoid structure	似层状构造	stream capture	河流袭夺；河流抢水
stratoisohypse	地层等厚线	stream centre line	水流轴线；中泓线
stratometric surveying	地层测量	stream channel	河槽；河道；河床
stratoscope	潜孔观测镜	stream channel pattern	水网类型；水系类型；河道类型
stratosphere	同温层；平流层	stream channel slope	河道边坡
stratotectonic	地层构造学的	stream closure	截流
stratotype	地层类型	stream closure hydraulics	截流水力学
stratous	成层的	stream competence	水流强度
stratovolcano	层状火山；成层火山；复合火山	stream control works	河流整治工程
stratovolcano edifice	层状火山机构；层状火山体	stream course	河道；水道
stratum	层；岩层；地层	stream course clearance	河道清理工作
stratum compactum	致密层	stream crossing	河流交叉点；小河流穿越；跨河桥
stratum contour	地层等高线	stream current	河流；水流；洋流
stratum media	地层介质	stream data transmission	数据流传输
stratum movement	覆岩层运动	stream deposit	河流沉积；河流堆积
stratum of sand	砂层	stream deposition	河流沉积
stratum pressure	地层压力	stream development	河流发育
stratum spring	接触泉；地层泉	stream digitizing	流式数字化
stratum water	层状水；地层水	stream dimension	水流尺度
straw and earth cofferdam	草土围堰	stream discharge	河流量；河流搬运量
stray parameter	随机参数；随机变量	stream diversion	河流袭夺；河流分流

stream drainage pattern	水系类型
stream dynamics	河流动力学
stream enclosure	封闭式输水道；地下输水道
stream energy	河流能量
stream erosion	河流侵蚀；河流冲刷；河流冲蚀；河道侵蚀
stream fall	河流落差；水面落差；河流坡度
stream filling	河流沉积；河流沉积物；河流填积物
stream flow	河道流量；河川径流；水流
stream flow gaging	河川流量测量；河道流量测量
stream flow gauging	河川流量测量；河道流量测量
stream flow hydrograph	河流水文图；流量过程线；河川流量过程线
stream flow hydrograph diagram method	河流水文图分解法；流量过程分割法
stream fluting	河流槽蚀
stream frequency	河流密度；河床频率；河流频率；水流密度频率
stream function	流函数；流量函数
stream function theory	流函数理论
stream function-vorticity method	流函数-涡旋法
stream gauging	河道测流量；流速测量；河道水文测量
stream gauging network	河流测站网
stream gauging operation	水文测验作用
stream gauging section	水文测验断面
stream gauging station	流量站；水文测量站
stream geochemistry	水系沉积物地球化学
stream gold	沙金；河金
stream gradation	河流均夷作用
stream gradient	河道坡度；河流坡降；河流比降
stream gravel	河流砾石；河砾
stream gravel veneer	河流砾石层
stream length ratio	河长比
stream level	河流高程；河流标高
stream line	流线；流程线；中锚索
stream line motion	流线运动
stream load	河流泥沙量；河流负荷
stream load saturation	泥沙流饱和度
stream measurement station	流量站；水文站
stream measurer	河流测量员
stream mode	流方式；批处理方式
stream mode digitizing	流式数字化
stream morphology	河流地貌学；河流形态学
stream of fast plasma	快速等离子体流
stream of stress	应力流
stream order	水系级别；河流等级；河流分级；支流级次
stream ore	砂矿
stream outlet	河口
stream pattern	河流类型；河流形式；河网图
stream piracy	河流袭夺；河道夺流
stream placer	河砂矿
stream pollution	河流污浊；河流污染
stream potential	流动势
stream pressure probe	水流压力传感器
stream processing	流水式处理；流处理
stream profile	河道纵剖面；河流纵剖面
stream reach	河区；流域
stream regime	河流状况
stream rise	河水上涨
stream robber	河流袭夺
stream robbery	河道夺流
stream routing	河道洪水演算；河川水流演算；河道流量演算
stream runoff	地表径流；河川径流；河径流
stream sediment	河流沉积物；水系沉积；河川泥沙
stream seepage	河流渗漏；河川渗漏
stream segment	河段
stream seismography	应变地震仪
stream self-purification	河流自净
stream sensitivity	应变灵敏度
stream set	流向
stream slope	河流坡降；河道坡度
stream source area	河源地区
stream splitting	水流分支；气流分支
stream stage	河水位
stream straightening	河流取直；河段整直
stream survey	水系沉积物测量
stream system	河系；水系
stream terrace	河流阶地；河成阶地
stream tin	砂锡矿
stream traction	河流推移作用；水流挟带作用
stream transportation	河流搬运
stream trenching	河流下切；河流切蚀作用
stream underflow	河底潜流
stream valley	河谷
stream velocity	河流流速；水流速度
stream velocity fluctuation	流速脉动；流速波动
stream water	河水；溪水
stream width	河宽；溪宽
stream-actuated	液压驱动的
stream-bank erosion	河岸冲蚀；河岸冲刷
stream-bank stabilization	河岸加固；稳定河岸
stream-bed degradation	河床刷深；河床加深
stream-bed erosion	河床冲刷
stream-borne material	河流挟带物
stream-built terrace	冲积阶地；河成阶地
stream-channel deposit	河道沉积物
stream-channel erosion	河道冲刷；河槽冲刷；河槽冲蚀
stream-cut terrace	切割阶地；河切阶地
stream-deposited	河流沉积
stream-discharge record	河道流量记录
streamer	海洋地震拖缆；海上电缆；等浮电缆
streamer cable system	等浮电缆系统
streamer depth indicator	等浮电缆深度指示器
streamer digitizing module	等浮电缆数字化模块
streamer hydrophone	等浮电缆检波器
streamer polygon	等浮电缆位置图
streamer positioning	等浮电缆定位
streamer reel	等浮电缆卷筒
streamer track	等浮电缆轨迹
streamer tracking	等浮电缆跟踪
stream-eroded	河流侵蚀的
streamer-towed hydrophones	等浮电缆检波器
streamflow	河水流量；河流流量；河川径流
streamflow control	流量控制；径流控制
streamflow data	流量资料；河道流量数据；河川流量资料
streamflow data compilation	汇编流量资料
streamflow forecasting	流量预报

streamflow gauging 河流流量测量
streamflow hydrograph 河水水文曲线；河流水文图
streamflow measurement station 流量站；河道测流站
streamflow record 流量记录；测流记录；河道流量记录
streamflow record extension 河道流量记录延长；河川径流资料延长
streamflow regime 水流情况；水流状态
streamflow regulation 流量调节；径流调节
streamflow routing 河道洪水演算；河川水流演算；河道流量演算
streamflow separation 河流分叉；水流分离
streamflow wave 河川径流波
stream-gauging network 流量站网；河道测站网；河流测站网
stream-gauging station 河道测量站
streamhead 河源；河头
streaming 连续流动
streaming bowl 流洗槽；洗矿槽
streaming box 洗矿槽
streaming flow 连续水流；冰川水流；稳流；缓流
streaming gradation 河流均夷作用
streaming lineation 水流线理
streaming potential 流势；水流势能；泳动电势
streaming potential method 流动电势法
streaming-potential 水流势能
stream-laid deposit 河流堆积；河成冲积层；河流沉积
streamlet 小河，小溪
streamline 流水线；流束，流线
streamline analysis 流线分析
streamline board 流线体型
streamline flow 线流；流线状流动；层流
streamline measurement station 水文站
streamline measurer 流速计
streamline molded form 流线型地形
streamline morphology 河流地貌学
stream-line motion 流线运动；无旋流
streamline order 河流等级
streamline outlet 河口
streamline stratification 流线分层
streamline topographic form 流线型地形
streamlined 流线型的
streamlined flow 直线流；层流
streamlined motion 流线型运动
streamlined pressure tank 流线型压力储罐
streamlined pump 流线型抽油泵
streamlined spillway surface 流线型溢流面
streamliner 流线型装置
streamlining 成流线型
streamplain facies 河流平原相
stream's self-purification 河流自净
stream-sediment anomaly 水系沉积物异常
streamway 河流水道；河道
streamwise flow condition 沿程水流状况
streamwise section 水流沿程剖面
street excavation 道路挖掘；掘路
street furniture 路政设施；街道设施
strength 强度
strength analysis 强度分析
strength anisotropy 强度各向异性
strength anisotropy index 强度各向异性指标
strength anomaly 强度异常
strength beam 强梁；加强梁
strength calculation 强度计算
strength ceiling 极限强度
strength characteristics 强度特性
strength class 强度分级
strength classes of concrete 混凝土强度等级
strength classes of masonry unit 块体强度等级
strength classes of mortar 砂浆强度等级
strength classes of prestressed tendon 预应力筋强度等级
strength classes of steel bar 普通钢筋强度等级
strength classes of structural steel 钢材强度等级
strength condition 强度条件
strength constraint 强度约束
strength contour graph 等强度线图
strength criterion 强度判据；强度准则
strength criterion for isotropic rock material 各向同性岩石材料强度准则
strength criterion for rock mass 岩体强度准则
strength criterion of slope safety 边坡安全性标准
strength curve surface of rock 岩石强度曲面
strength decrease 强度递减
strength deficiency 强度不足
strength degradation 强度衰减；强度退化；强度下降
strength design criterion 强度设计准则
strength envelope 强度包络线
strength factor 强度因数；强度系数
strength factor of material 材料强度系数
strength failure 强度破坏
strength grade 强度等级
strength grade of concrete 混凝土强度等级
strength grading of mortar 砂浆强度等级
strength in undisturbed state 原状强度
strength index 强度指数
strength limit 强度极限
strength loss 强度损失
strength loss ratio of rock 岩石强度损失率
strength member 强度构件；重要构件
strength model 强度模型
strength of a pillar 矿柱强度
strength of acid 酸强度
strength of anisotropic rock material in triaxial compression 三轴压缩各向异性岩石材料的强度
strength of axially-loaded member 轴向受力构件强度
strength of cement 水泥强度
strength of cement grout 水泥灌浆强度
strength of compression 抗压强度
strength of concrete 混凝土强度
strength of connection 连接强度
strength of discontinuous rock mass 不连续岩体强度
strength of ebb interval 最大落潮流间隙
strength of extension 抗拉强度
strength of flood interval 最大涨潮流间隙
strength of joint 连接强度；节理强度
strength of material 材料强度
strength of mortar 砂浆强度
strength of pillar 矿柱强度
strength of pressure 压力强度
strength of reflection 反射强度
strength of rock 岩石强度

strength of rock mass 岩体强度
strength of shearing 抗剪强度
strength of structure 结构强度
strength of testing 试验强度
strength of vortex 涡流强度
strength of wall 墙体强度
strength of water drive 水驱强度
strength parameter 强度参数
strength parameter of discontinuity 岩体弱面强度参数；不连续面强度参数
strength parameters 强度参数
strength ratio 强度比
strength recovery 强度恢复
strength reduction 强度折减
strength reduction factor 强度折减系数
strength reduction FEM 有限元强度折减法
strength reduction finite element method 有限元强度折减法
strength reduction method 强度折减法
strength reduction to account for jointing 节理作用引起的强度折减
strength regain 强度恢复
strength requirement 强度要求
strength reserve 备用强度
strength retrogression 强度递降；强度衰退
strength safety coefficient 强度储备系数
strength softening 强度软化
strength test 强度试验；强度测试
strength tester 强度测试机；强度测试试验机
strength theory 强度理论
strength theory of rock 岩石强度理论
strength theory of shape strain energy 形状应变能强度理论
strength to density ratio 强度-密度比
strength under peripheral pressure 三向压缩强度；有侧限强度；围压下的强度
strength under shock 冲击强度
strength under sustained load 长期荷载作用下的强度
strength value 强度值
strength-aging relationship 强度-年限关系
strength-deformation characteristics 强度-变形特性
strength-density ratio 强度-密度比
strengthen 加强；加固；强化；补强
strengthen layer 补强层
strengthened member 加强构件
strengthened section for tunnel 隧道加强长度
strengthened stage 强化阶段
strengthening 加强；加固；强化；补强
strengthening action 加固作用；补强作用
strengthening building 建筑物加固
strengthening by struts 支顶加固
strengthening existing structure 现有建筑物加固
strengthening grout 补强灌浆
strengthening grouting 补强灌浆
strengthening layer 补强层
strengthening measure 加固措施
strengthening measure for earthquake resistance 抗震加固措施
strengthening method 加固方法
strengthening method with prestressed brace bar 预应力撑杆加固法
strengthening of a dam 大坝加固
strengthening of steel bridge 钢梁加固
strengthening of structure 结构加固；结构补强
strengthening of structure under loading 带负荷加固法
strengthening procedure 加固方法
strengthening ring 加强环；锁紧圈
strengthening soil 加固土
strengthening stability 强化稳定性
strengthening works 加固工程
strength-producing material 加强材料
strength-stress ratio 强度-应力比
strength-to-weight ratio 强度-重量比
strength-weight ratio 强度-重量比
stress 应力
stress accommodation 张紧夹具
stress accumulation 应力积累，应力积聚
stress accumulation and release 应力积累与释放
stress adjustment 应力调整
stress along bolt 锚杆应力
stress alternation 应力变换；应力交替
stress amplitude 应力幅值；应力幅
stress analysis 应力分析
stress analysis using seismology method 地震学应力分析法
stress and displacement 应力和位移
stress and displacement distributions around circular excavation 圆形巷道周围的应力和位移分布
stress and permeability 应力和渗透率
stress and strain response 应力及应变响应
stress anisotropy 应力各向异性
stress approximation 应力近似
stress arch 应力拱
stress around borehole 井口周围应力；钻孔周向应力
stress axial cross 应力轴交叉
stress axis 应力轴
stress balance 应力图
stress beyond yield strength 超过屈服强度的应力
stress block 应力块，应力区
stress boundary condition 应力边界条件
stress bridge 应力桥
stress buildup 应力积累；应力积聚
stress change 应力变化
stress change indicator 应力变化指示器
stress chart 应力图
stress circle 应力圆
stress compensation 应力补偿
stress compensator 应力补偿器；应力代偿装置
stress component 应力分量
stress concentration 应力集中
stress concentration diffusion 应力集中的扩散
stress concentration factor 应力集中系数；应力集中因子
stress concentration in rock mass 岩体应力集中
stress concentration ratio 应力集中比；应力集中度
stress concentrator 应力集中器
stress condition 应力状态；应力条件
stress cone angle 应力锥角
stress constraint 应力约束
stress contour 应力等值线；应力云图

stress contrast	应力差
stress control	应力控制
stress control method	应力控制方法
stress control triaxial compression apparatus	应力控制式三轴压缩仪
stress controlled load test	应力控制载荷试验
stress controlled test	应力控制试验
stress corrosion	应力腐蚀
stress corrosion crack	应力腐蚀裂纹；应力腐蚀裂缝
stress corrosion cracking	应力腐蚀开裂
stress couple	应力偶
stress coupling	应力耦合
stress crack	应力裂缝；应力开裂；应力龟裂
stress cracking	应力开裂
stress curved surface	应力曲面
stress cycle	应力循环
stress cycling	应力循环
stress dependency of physical property	物理性质的应力依赖性
stress dependent	应力依赖的；取决于应力的
stress depletion	应力衰减
stress detector	应力计；应力测定器
stress development	应力发展
stress deviation	应力偏量
stress deviator	应力偏量；偏应力
stress deviatoric tensor	应力偏斜张量
stress diagram	应力图；应力分布图
stress difference	应力差
stress dilatancy theory	应力剪胀理论；应力膨胀理论
stress direction	应力方向
stress discontinuity	应力间断
stress dislocation	应力位错
stress dispersion	应力扩散
stress distribution	应力分布
stress distribution characteristics	应力分布特征
stress distribution pattern	应力分布图
stress divergence	应力散度；应力发散
stress driving effect	应力驱动作用；应力驱动效应
stress drop	应力降
stress due to friction	摩擦应力
stress due to hammer driving	锤击应力
stress due to temperature difference	温差应力；温度应力
stress ellipse	应力椭圆
stress ellipsoid	应力椭球
stress endurance limit	应力持久极限
stress engineering	应力工程师；应力分析工程师
stress envelope	应力包络线
stress equilibrium	应力平衡
stress erosion	应力腐蚀
stress evaluation	应力评估
stress evanesce	应力消散
stress evolution	应力演变
stress extrapolation	应力外插；应力外推
stress fatigue	应力疲劳
stress field	应力场
stress field expansion	应力场扩展
stress field of in-situ rock	原岩应力场
stress field type	应力场类型
stress flow	应力流
stress fluctuation	应力波动；应力脉动
stress for temporary loading	短期荷载应力
stress free method	应力释放法
stress free rail temperature	零应力轨温
stress freezing effect	应力冻结效应
stress freezing method	应力冻结法
stress fringe	应力边缘
stress function	应力函数
stress function formulation	应力函数公式
stress function of bending	弯矩应力函数；弯曲应力函数
stress function of torsion	扭转应力函数
stress gage	应力计
stress gauge	应力计
stress glut	应力过量
stress gradient	应力梯度
stress grid diagram	应力网络图；应力轨迹图
stress history	应力历史
stress in a pillar	矿柱应力
stress in fault plane	断层面应力
stress in ground	地基应力
stress in original rock	原岩应力
stress in prestressing tendon at transfer or anchorage	传力锚固应力
stress in rock mass	岩体应力
stress in soil mass	地基应力；土体中的应力
stress in surrounding rock	围岩应力
stress in the earth's crust	地应力
stress in the surrounding rock	围岩应力
stress in three dimensions	三维应力
stress in virgin rock mass	原岩应力
stress increment	应力增量
stress induced	应力诱发
stress induced corrosion	应力诱发腐蚀
stress induced during block caving	自然崩落引起的应力
stress induced fracture	应力引起的断裂
stress influence function	应力影响函数
stress intensification factor	应力增大系数；应力强化因数
stress intensity	应力强度
stress intensity factor	应力强度因子
stress invariant	应力不变量
stress inversion	应力反转；应力反演
stress joint	应力节
stress level	应力水平
stress liberation	放散温度应力
stress limit	应力极限；极限应力
stress limitation	应力极限
stress line	应力线
stress loading	应力载荷
stress locus	应力轨迹
stress loss correction	应力恢复
stress matrix	应力矩阵
stress measurement	应力测量
stress measurement in rock mass	岩体应力测量；岩体应力测试法
stress meter	应力计；应力仪
stress meter for reinforcement	钢筋应力计
stress method	应力法
stress metre	应力计；应力仪
stress mineral	应力矿物
stress modeling	应力模拟

stress modified method	应力修正法
stress modulus	受应力影响的模量；应力模量
stress monitoring	应力监测
stress notation	应力符号；应力表示法
stress of rupture	断裂应力
stress of soil moisture	土壤水分应力
stress of surrounding rock	围岩应力
stress on the wall of a borehole	钻孔壁应力
stress path	应力路径
stress path method	应力路径法
stress path model	应力路径模型
stress path test	应力路径试验
stress pattern	应力图；应力分布图
stress peak	应力峰值；应力最大值
stress plane	应力平面
stress point	应力点
stress prediction	应力预测
stress pulse	应力波动；应力脉冲
stress raiser	形成应力集中的因素；应力集中器
stress range	应力范围；应力变化范围
stress ratio	应力比
stress ratio between horizontal and vertical	水平与垂直应力比
stress ratio method	应力比法
stress ratio of liquefaction	液化应力比
stress recorder	应力记录仪
stress recovery method	应力恢复法
stress redistribution	应力重分布
stress redistribution in rock mass	岩体应力重分布
stress reduction factor	应力折减系数
stress regime	应力结构；应力状态
stress regulation	应力调整
stress relax meter	应力松弛仪
stress relaxation	应力松弛；应力弛豫
stress relaxation degree	应力释放程度
stress relaxation method	应力解除法
stress relaxation under constant load	恒载下的应力松弛
stress release	应力释放；卸荷
stress release crack	卸荷裂隙
stress release joint	卸荷节理
stress release rate	应力释放率
stress releasing borehole	应力解除钻孔
stress relief	应力消除；应力解除
stress relief and stress measurement	应力释放和应力测量
stress relief coalbed methane	应力消除煤层气
stress relief core	应力解除岩芯
stress relief in rock mass	岩体应力解除
stress relief method	应力解除法
stress relief technique	应力解除法
stress relieving	应力解除；消除应力
stress relieving method	应力解除
stress repetition	应力反复
stress response	应激反应；压力反应
stress restoration	应力恢复
stress restoration method	应力恢复法
stress reversal	应力反向
stress reversal tolerance	应力逆转容限
stress ring	巷道横断面地压应力线；应力环
stress riser	应力梯级
stress rotation matrix	应力旋转矩阵
stress route	应力路径
stress rupture	应力破坏
stress sensitivity	应力敏感性
stress share ratio	应力分担比
stress sheet	应力图；应力表
stress shell	应力壳
stress sign convention	应力符号的规定；应力符号法则
stress singularity	应力奇异性
stress slacking	应力松弛
stress smoothing	应力光顺；应力平滑化
stress space	应力空间
stress spectrum	应力谱
stress spherical tensor	应力球面张量
stress state	应力状态
stress state approach	应力状态方法
stress state at a point	一点应力状态
stress state variable	应力状态变量
stress stiffening	应力刚化；应力硬化
stress stiffness matrix	应力刚度矩阵
stress strain curve	应力-应变曲线
stress strain diagram	应力应变图
stress strain relation	应力应变关系
stress strength factor	应力强度因子
stress superposition	应力叠加
stress surface	应力面；受力面
stress system	应力系统
stress tensor	应力张量
stress tensor gauge	应力张量计
stress tensor transformation	应力张量变换
stress test	应力测试
stress theory of shear strain energy	剪应变能强度理论
stress threshold	应力门槛；应力阈值
stress to rupture	断裂应力
stress to strain ratio	应力-应变比
stress tolerance	抗逆性；抗压能力
stress traction relation	应力-牵引力关系
stress trajectory	应力轨迹；应力迹线
stress transfer	应力转移；应力传递
stress transfer technology	应力转移技术
stress transfer theory	应力传递理论
stress transformation	应力变换
stress transformation formula	应力变换公式
stress transmission	应力传递
stress triggering	应力触发
stress triggering and stress shadow	应力触发与应力影区
stress under impact	冲击应力
stress uniformity	应力均匀
stress value	应力值
stress variation	应力变化
stress vector	应力向量；应力矢量
stress wave	应力波
stress wave attenuation	应力波衰减
stress wave dispersion	应力波弥散
stress wave in rock mass	岩体中的应力波
stress wave in soils	土中应力波
stress wave propagation	应力波传播
stress wave response	应力波响应
stress wave theory	应力波理论
stress within a finite element	有限单元应力
stress wrinkle	应力皱纹

stress zone	应力带
stress-assisted localized corrosion	应力辅助局部腐蚀
stresscoat	应力涂料
stress-concentrated area	应力增高区；应力集中区
stress-concentrated zone	应力增高区；应力集中区
stress-concentration factor	应力集中系数
stress-constrained conditions	应力约束条件
stress-control shear apparatus	应力控制剪切仪
stress-controlled	应力控制的
stress-controlled direct shear apparatus	应力控制式直剪仪
stress-controlled load test	压力控制加载试验
Stress-controlled sorption	应力控制式吸附作用
stress-controlled test	应力控制测试
stress-controlled testing method	可控应力试验法；应力控制测试法
stress-damage coupling	应力损伤耦合
stress-deformation characteristic	应力-变形特征
stress-deformation diagram	应力-变形图
stress-dependent permeability	应力依赖的渗透率；随应力变化的渗透率
stresse and displacement induced by line load	线荷载引起的应力与位移
stresse and displacement induced by linear load	线性荷载引起的应力和位移
stresse and displacement induced by point load	由点荷载引起的应力和位移
stressed collar	剪力环
stressed shell construction	张壳构造
stressed skin action	应力表层作用；应力蒙皮作用
stressed skin construction	表层受力式结构；预应力薄壳结构
stress-equilibrium method	应力平衡法
stresses combination	应力组合
stresses in lining	支护体中应力
stresses in soil-rock mass	岩土体中的应力
stresses initial	原始应力
stresses residual	应力残值；残余应力
stress-free	无压力；无应力
stress-free completion	无应力完井
stress-free method	应力释放法
stress-free strain meter	无应力应变计；纯应变计
stress-freezing effect	应力冻结效应
stressful region	受到压力区域
stress-induced crack	应力致裂
stress-induced instability	应力引起的失稳
stress-induced outburst	应力引发突出
stress-induced wellbore breakout	应力诱发的钻孔剥落
stressing device	张拉设备；施加预应力装置
stressing method	先张拉法；施加预应力法
stressing process	张拉过程
stressing state	应力状态
stress-laid deposit	河流沉积
stressless	无应力的
stress-measuring device	应力测量装置
stress-meter monitoring	应力计监测
stress-number curve	应力-循环次数曲线
stress-number diagram	疲劳试验曲线
stressometer	应力计
stress-optic constant	光弹性系数；光学力学常数
stress-orientation line	应力方向线
stress-path	应力途径
stress-path settlement analysis	应力路径沉降分析
stress-penetration curve	应力-贯入度曲线
stress-producing force	引起应力的外力
stress-raising	应力提高
stress-ratio coefficient of original rock	原岩应力比值系数
stress-ratio design	应力比设计
stress-ratio method	应力比法
stress-reduced protective zone	卸压保护带
stress-relaxed area	应力降低区
stress-release channel	应力解除槽
stress-released zone	应力降低区
stress-relief	应力释放；应力解除
stress-relief annealing	消除内应力退火
stress-relief blast	卸荷岩爆；卸荷冲击波；去应力爆破；卸压爆破
stress-relief by blasting	爆炸应力解除
stress-relief heat treatment	消除应力的热处理
stress-relieved ground	无压地层；应力解除的地层
stress-rupture test	应力断裂试验
stress-rupture testing	应力-破裂试验
stress-rupture testing of tube	管子的应力-破裂试验
stress-strain	应力-应变
stress-strain behavior	应力-应变性状
stress-strain characteristics	应力-应变特性
stress-strain contour	应力-应变等值线
stress-strain curve	应力-应变曲线
stress-strain curve in tension	受拉应力-应变曲线
stress-strain diagram	应力-应变图
stress-strain equations	应力应变方程
stress-strain field	应力-应变场
stress-strain graph	应力-应变图
stress-strain loop	应力-应变回线
stress-strain matrix	应力应变矩阵
stress-strain modulus	应力-应变模量
stress-strain ratio	应力-应变比
stress-strain relation	应力-应变关系
stress-strain relationship	应力-应变关系
stress-strain relationship of mortar	砂浆的应力-应变关系
stress-strain relationship of soil under cyclic loading	循环荷载下土的应力-应变关系
stress-strain state	应力应变状态
stress-strain tensor	应力-应变张量
stress-strain-time behaviour	应力-应变-时间性状
stress-strain-time function	应力-应变-时间函数
stress-strain-time relation	应力-应变-时间关系
stretch	拉伸；拉紧；伸长；延展；范围；限度
stretch chart	拉伸图表
stretch coefficient	伸长系数
stretch elongation	拉伸变形；伸长
stretch fabric	拉伸组构
stretch fault	引伸断层；伸展断层；拖曳断层
stretch forming	张拉成形法
stretch modulus	拉伸模量
stretch of river	河段
stretch proportion	延伸率
stretch resistance	抗拉强度
stretch section	拉伸段
stretch slide	拉伸滑动

stretch tensor	拉伸张量
stretch thrust	引伸冲断层；拖曳冲断层；拖引逆断层
stretchability	拉伸性；延性
stretch-breaking technology	牵切技术
stretched outer arc	拉伸外弧
stretched pebble	拉长卵石；伸长卵石
stretched wire	引张线
stretched zone	张拉区
stretched-out field	伸展磁场
stretched-out view	展开图
stretcher bar	伸张杆；伸缩式经纬仪支杆
stretcher bond	错缝接合；顺砖式砌合
stretcher jack	深井泵拉杆的拉紧器
stretcher strain	拉伸变形；拉伸应变
stretching	拉伸；伸展；延伸；伸长
stretching bolt	拉紧螺栓
stretching device	伸张装置；伸张器；拉伸装置
stretching direction	伸展方向
stretching fault	引张断层；拖曳断层
stretching force	张力；拉力
stretching in batches	成组张拉
stretching lineation	拉伸线理
stretching process	张拉过程
stretching screw	扩张螺丝；拉紧螺丝
stretching strain	拉伸应变；张拉应变
stretching stress	拉伸应力；张拉应力
stretching surface	扩张面
stretching vibration	伸缩振动；弹性振动
stretching wire	张拉钢筋
stretch-out view	展开图
stretch-twister	拉伸加捻机
strewing sand	铺砂；撒砂
strewn field	熔融石场；玻陨石散落区；玻陨石场
stria	条痕；条纹；擦痕
striate	线状的；有条纹的；有细槽的
striate boulder	擦痕漂砾
striate pavement	冰川擦痕面
striate pebble	擦痕卵石
striated	条纹状的；条痕的；擦痕的
striated coal	有擦纹的煤；有条纹的煤
striated pattern	条纹图形；线性光栅
striated pavement	冰川擦痕面
striated pebble	擦痕卵石
striated rock surface	擦痕面；擦痕岩面
striated structure	线理状结构；条纹构造
striation	条痕；擦痕；条纹；脊线
striation cast	条纹模；擦痕铸型
striation lineation	擦痕线理；条痕线理
striatus	带状；条纹状
striction	收缩；拉紧；变窄；限制；约束
striction strain	颈缩应变
strictly concave function	严格凹函数
strictly convex function	严格凸函数
strictly decrease monotonically	严格单调递减
strictly increase monotonically	严格单调递增
strictly stationary process	强平稳过程；严格平稳过程
strictly upper triangular matrix	严格上三角矩阵
stride	步测
stride scale	步测比例尺
strided distance	跨距；跨度
striding level	跨水准管
strigovite	柱绿泥石
strike	走向；撞击；冲击
strike a bed	遇到煤层；遇到矿层
strike a lead	发现矿脉；找到富矿
strike breakthrough	沿走向的联络横巷
strike cleat	走向割理
strike closure	走向闭合
strike diagram of joint	节理玫瑰图
strike dip	走向滑距
strike direction	走向
strike drift	走向平巷
strike entry	沿走向的平巷
strike face	沿走向推进的工作面
strike fault	走向断层
strike fold	纵褶皱；走向褶皱
strike joining	走向节理
strike joint	走向节理
strike length	走向长度
strike limit	走向可采极限
strike line	走向线
strike line of discontinuities	不连续面走向线
strike of alignment	线路走向；沿走向排列
strike of bed	地层走向
strike of rock strata	岩层走向
strike oil	找到石油；探井见油
strike orientation	走向定向
strike pillar	走向矿柱
strike pitch	走向倾角
strike plate	防冲击板
strike ridge	走向脊；走向山脊
strike road	走向平巷
strike separation	走向断距
strike separation fault	横移断层；走向分离断层；侧向断层
strike shift	走向移位；走向变位
strike slip	走向滑动，走滑；走向滑距；走向位移
strike slip fault	走向滑动断层；平移断层；横冲断层
strike slip faulting	走滑断层作用
strike slip focal mechanism	走向滑动震源机制
strike slip motion	走滑运动
strike slip normal fault	平移正断层
strike slip reverse fault	平移逆断层
strike stream	走向河；次成河
strike through	击穿
strike valley	走向谷；纵谷
strike waste pack	沿走向的废石垛墙
strike-and-dip symbol map	走向和倾角符号图
strike-dip survey	产状测量
strike-line	走向线
strike-overlap	走向超覆
striker	撞击弹；撞针；撞击杆
striker bar	撞击杆
strike-ridge	走向脊
strike-shift fault	平移断层；走向移距断层
strike-slip	走向滑距的；走滑
strike-slip component	走向滑动分量
strike-slip displacement	走向滑动位移
strike-slip earthquake	走向滑动地震
strike-slip fault	平移断层；走滑断层

strike-slip microfault 走向滑动微断层
strike-slip movement 平移运动；走向滑动
strike-slip plate 走向滑移板块
strike-slip pull apart basin 走滑拉分盆地
strike-slip source mechanism 走向滑动震源机制
strike-slip thrust 走向滑动逆冲断层
striking angle 转角；冲击角
striking contrast 显著对比
striking distance 起弧距离
striking end of shank 钎子的冲击端
striking energy 冲击功；冲击能
striking lever 操纵杆；变速杆
striking mottled pattern 明显斑点图形
striking off 勾缝；嵌缝
striking off lines 勾划线
striking order 点火次序；引爆次序
string 细矿脉；钻具组；钻杆柱；线；绳；弦；检波器串
string control 钻柱控制
string description 钻柱说明
string dynamometer 弦式测压计
string failure 钻杆折断
string level 线测水平；悬挂式水准器
string line 定线线；标线
string lining of curve 绳正法整正曲线
string manipulation 字符串处理；行处理
string measurement 绳测
string of casing 套管柱
string of deposits 条带状油气藏
string of rods 抽油杆柱
string of tank cars 油槽列车
string of tools 钻柱
string of tubing 油管柱
string operation 串操作
string pendulum 线摆
string processing language 串处理语言
string reamer 钻柱扩眼器
string rod 水泵连杆
string shot 孔内爆破炸药包；井下爆破器
string shot assembly 解卡爆震装置
string shot backoff 爆震倒扣
string stabilizer 钻柱稳定器
string suspending weight 悬重
string wire 钢弦
string-directed hole 绳量定位炮眼
stringer 纵梁；细脉；细矿脉；沉积薄层；薄岩层
stringer bracing 纵梁联结系
stringer conveyor 纵梁式胶带机
stringer lead 小矿脉
stringer lode 网状细矿脉；细复成脉
stringer vein 细脉
stringing coupling 接管管箍
stringing deal 井圈吊板
string-wire concrete 钢弦混凝土
strip 狭条；带状地带；剥露；剥离；露天开采；航线
strip a well 从井内同时起抽油杆和油管
strip abutment 带状矿柱的应力集中区
strip adjustment 航线网平差；航线平差
strip adjustment with bundle method 光束法航线网平差
strip adjustment with independent model method 独立模型法航线网平差
strip aerial triangulation 航带法空中三角测量
strip aerotriangulation adjustment 航带法空中三角测量平差
strip borer 水平钻孔机；露天矿水平炮眼钻机
strip camera 条幅摄影机；航线式相机
strip carrier capsule gun 钢带托架座舱式射孔器
strip chart 长图；条形记录纸
strip chart apparatus 记录带打印机
strip chart recorder 条图记录仪
strip coal mine 露天煤矿
strip coal pillar 条带煤柱
strip coating 可剥离的覆盖层
strip coordinate 带坐标；航线坐标
strip deformation 航线弯曲度
strip footing 带状基础；条形基础
strip foundation 条形基础；条形地基
strip foundation under wall 墙下条形基础
strip inclined drift 分带斜巷
strip index 航线相片索引图
strip indication light 光带式表示
strip irrigation 带状灌溉
strip layering 条状层
strip line 带状线
strip load 条形荷载；带状荷载
strip log 柱状剖面图；柱状图；条带录井图
strip longwall 条带长壁采煤法
strip main inclined drift 分带集中斜巷
strip map 航线地图
strip matrix 带形矩阵
strip method 充填带砌筑法；剥离法
strip method of adjustment 航线平差法
strip mine 露天矿
strip mine explosive 露天矿用炸药
strip mine survey 露天矿测量
strip miner 露天矿矿工
strip mining 露天开采；露天剥离
strip mining method 条带开采法；剥离法；刀柱式开采
strip mosaic 航线相片镶嵌图
strip off 剥离；剥采
strip pack 带状充填；矸石带；废石充填带
strip packing 带状充填
strip photography 航线摄影术
strip pillar 条带煤柱
strip pillar mining 剥离柱式开采；条带开采
strip pit 露天矿
strip prediction 航线预报
strip radial plot 航线辐射三角测量
strip radial triangulation 航线辐射三角测量
strip recorder 带式记录仪
strip region 带形区域
strip roadway 分带斜巷
strip stowing 带状充填
strip strain gauge 条形应变片
strip tensile test 条带拉伸试验；狭条拉伸试验
strip the drill pipe 提出钻杆
strip thrust 挤压冲断层；脱顶逆断层
strip topographic map 带状地形图
strip triangulation 航带法三角测量
strip width 航带宽

strip-area 带形地区
strip-beam recorder 条带记录器
strip-borer drill 露天矿水平眼钻机
strip-chart recorder 带形图表记录器；纸带记录器
strip-chart seismogram 带状地震图
stripe 条带；条纹
stripe process 粗砂槽洗法
stripe-chart recorder 图线记录器
striped 条带状的
striped bedding plane 条带状层面
striped ground 条带状土；带状土
striped magnetic anomalies 磁异常条带
striped structural surface 条带状构造面
stripmine 露天矿
strip-mine explosive 露天矿用炸药
stripmine production 露天开采
strip-mined lands 条带开采破坏的土地
strip-mining 露天开采
strippable coal 可露天开采的煤
strippable deposit 可露天开采的矿床
strippant 洗涤剂
stripped 被剥离的；剥去的；卸下的；拆开的；拉断的
stripped bedding plane 剥露层面；条带状层面
stripped fossil plain 剥露古平原
stripped joint 深槽缝
stripped oil 脱去汽油的石油
stripped peneplain 剥露准平原
stripped structural surface 剥露的构造面
stripped surface 清基面；表层剥除的地面
stripped well 低产井；贫油井
stripper 刨土机；剥离挖掘机；拆卸器
stripper development well 低产开发井
stripper field 低产油田
stripper head 封井头
stripper pool 低产油藏
stripper shovel 剥离挖掘机；倒堆铲
stripper stage 枯竭阶段
stripper stopper 刮泥器止动块
stripper tube 封井管
stripper type preventer 环状橡胶心子防喷器
stripper well 低产井
stripping 清理场地；拆除；剥离；露天开采
stripping line 清基线
stripping a mine 剥离；露天开采；只采富矿
stripping and mining engineering profile 采剥工程断面图；采剥工程剖面图
stripping assembly 拆卸装置
stripping balance 剥采比均衡
stripping bank 剥离台阶；剥离阶段
stripping belt 剥离胶带输送机
stripping bench 剥离台阶；倒堆作业台阶；剥离阶段
stripping depth 剥离深度
stripping down 剥离；休整工作面
stripping dual operation 剥离工作
stripping equipment 剥离设备；露天开采设备
stripping excavator 剥离挖掘机
stripping face 剥离工祖；剥离工作面
stripping filtering 剥离滤波
stripping front 剥离工祖；剥离工作面
stripping in and out of the hole 用防喷器在压力下强行起下钻

stripping job 同时起出抽油杆和油管作业
stripping limit 剥离极限
stripping line 开段沟；清基线
stripping machine 剥离机；露天矿采掘机；带楔形落矿装置的联合采矿机
stripping method 剥离法；露天开采法；地层剥除法
stripping mining method 露天开采法
stripping of reservoir 清库工作
stripping off 剥离；剥采；剥开
stripping operation 剥离工作；拆模作业
stripping ratio 剥采比；采剥比
stripping ratio of pit limit 露天开采境界剥采比
stripping shovel 剥离挖掘机；倒堆铲；表土层剥离机铲
stripping system 露天开采法
stripping to crude ore ratio 原矿采剥比
stripping topsoil 剥除表土
stripping-in operation 强行起下钻作业
stripping-to-ore ratio 采矿剥离系数；剥采比
strobe interrupt 选通中断
strobe pulse 选通脉冲
strobe release time 选通释放时间
strobing pulse 选通脉冲
strobophonometer 爆震测声计
stroboscope 闪频观测仪
stroboscopic effect 闪光效应；频闪效应
stroke 行程；冲程；打；敲击；桩棰落高；振幅
stroke adjuster 冲程调整器
stroke coefficient 冲程系数
stroke diagram 行程图；冲程图
stroke limiter 冲程限制器
stroke loss 冲程损失
stroke mark 冲程标记
stroke of bit 钻头冲击；钻头冲程；钻头行程
stroke of piston 活塞冲程
strokemeter 泵冲次计
strokes per fill 每次灌泥浆冲数
strokes per minute 每分钟冲数
strokes to fill 灌泥浆冲数
stroke-time diagram 冲程-时间图；行程-时间图
stroma 基质
stromatactis 平底晶洞构造
stromatite 叠层混合岩；层状混合岩
stromatoid 叠层石类
stromatolite 叠层石
stromatolith 叠层；叠层石；叠层混合岩；层石藻
stromatolithic 叠层状的；叠层的
stromatolithic facies 叠层石相
stromatolithic limestone 叠层灰岩
stromatolithic structure 叠层构造
stromatolitic 叠层岩的
stromatolitic framework 叠层石骨架
stromatolitic limestone 叠层石灰岩
stromatolitic structure 叠层构造
stromatology 叠层学；地层学
stromatoporoid limestone 层孔虫灰岩
strombolian type eruption 斯特朗博利型火山喷发
stromeyerite 硫铜银矿
stromnite 碳酸钡锶矿
stromoconolith 成层锥状体

strong acid	强酸
strong aftershock	强余震
strong alkaline solution	强碱溶液
strong aqua	浓氨水
strong axis	强轴
strong back	厚基座
strong base	强碱
strong bonding	胶结良好
strong breeze	强风
strong brine	浓盐水
strong caking coal	强黏结煤
strong caustic	强碱的
strong cement grout	稠水泥浆
strong clay	强黏土;重黏土;油性黏土
strong coal	硬煤;强煤
strong compacting	压的非常紧密;致密压实
strong concrete	高强混凝土
strong convergence	强收敛
strong coupling	强耦合
strong coupling theory	强耦合理论
strong discontinuity	强间断;强不连续性
strong dispersion	强分散;强弥散
strong earthquake	强地震;强震
strong earthquake motion	强地震运动
strong earthquake record	强震记录
strong eddy	强旋涡
strong expansive fusion caking	强膨胀熔融黏结
strong extremum	强极值
strong fold	强烈褶皱
strong fracture	大裂缝;强烈破裂
strong frontal zone	强锋带
strong geomagnetic activity	强地磁活动
strong girder-weak brace	强梁-弱撑
strong ground motion	强地面运动;强地动
strong ground pressure	强大的地面压力
strong jump	强水跃
strong lining	坚固衬砌;坚固井壁
strong magnetic die	强磁力打捞器
strong magnetic material	强磁性材料
strong metal-support interaction	金属-载体强相互作用
strong motion	强震;强震地面运动
strong motion accelerogram	强震加速度图
strong motion accelerograph	强震加速度仪
strong motion accelerometer	强震加速度计
strong motion array in Taiwan number 1	台湾强地动一号台阵
strong motion displacement	强震位移
strong motion earthquake	强烈地震
strong motion earthquake instrumentation	强震观测;强地震测量仪器
strong motion instrument	强震仪器;强震仪
strong motion network	强震网络
strong motion observation	强震观测
strong motion observation array	强震观测台阵
strong motion observation network	强震观测台网
strong motion record	强震记录
strong motion response spectrum	强震反应谱
strong motion seismogram	强地震图
strong motion seismograph	强震仪
strong motion seismology	强地动地震学;强震学
strong motion seismometer	强震计
strong motion station network	强震动台网
strong motion telemetric network	强震遥测台网
strong motion wave	强震波
strong nonlinearity	强非线性
strong oil-wet	强亲油
strong oxidizer	强氧化剂
strong pulse	强脉冲
strong reflection	强反射
strong refraction	强折射
strong response signal	强响应信号
strong reverse wetting effect	强润湿反转效应
strong rock	硬质岩石;坚硬岩石
strong rod	刚性钻杆
strong roof	坚固顶板;坚硬顶板;稳固顶板
strong sand-dust flood	强沙尘暴;强尘暴
strong sand-dust storm	强沙尘暴;强尘暴
strong schistosity	完善片理
strong seismic motion station network	强震台网
strong sewage	浓污水
strong shock	强震;强激波
strong solution	浓溶液;强解
strong strata	坚硬岩层;坚硬地层
strong stratum	坚硬岩层;坚硬地层
strong water drive	强水驱
strong well	高压井
strong-acid number	强酸中和值
strongaxis	强轴
strong-base number	强碱中和值
strong-column	强柱
strong-fissured stratum	裂隙发育地层
strongly acid solution	强酸溶液
strongly alkaline soil	强碱性土
strongly bound water	强结合水
strongly coking coal	强黏性煤
strongly implicit	强隐式
strongly magnetic	强磁性的
strongly magnetic mineral	强磁性矿物
strongly oil-wet	强亲油的
strongly saline soil	重盐土
strongly shaken area	强震区
strongly soluble salt test	易溶盐试验
strongly stationary	强平稳
strongly stationary random process	强平稳随机过程
strongly water-wetting system	强亲水系统
strongly wetted media	强润湿介质
strongly wetting liquid	强润湿液
strongly-fissured stratum	裂隙发育地层
strongly-soluble salt test	易溶盐试验
strongly-weathered zone	强风化带
strong-motion accelerograph	强震加速度仪
strong-motion earthquake	强烈地震;强震
strong-motion information retrieval system	强震信息检索系统
strong-motion instrument array	强震仪台阵
strong-motion measurement	强震观测
strong-motion seismograph	强震仪;强动地震仪
strong-motion seismology	强地动地震学
strong-motion sensor	强震传感器
strong-soft-strong structure model	强-软-强结构模型;三

明治结构模型
strong-tide estuary 强潮河口
strong-tide river mouth 强潮河口
strong-weathered zone 强风化带
strontia 氧化锶；氢氧化锶
strontianite 菱锶矿
strontium 锶
strontium carbonate 碳酸锶矿
strontium isotope 锶同位素
strontium isotope dating 锶同位素年龄测定
strontium mineral 锶矿类
strontium sulfate 硫酸锶
strontium sulfate scale 硫酸锶垢
strontium-isotope stratigraphy 锶同位素地层学
strop inclined drift 分带斜巷
strophotron 超高频振荡器；多次反射振荡器
struck joint 外斜缝；刮平缝
struck pipe 被卡钻杆套管
structural 构造的；结构的
structural abnormal pressure 构造异常压力；构造型异常高压
structural accommodation 构造调节
structural adequacy assessment 结构足够性评估
structural alignment 构造排列线；结构校准
structural alteration 结构上的改动；结构变化
structural analysis 构造分析；结构分析
structural analysis program 结构分析程序
structural anisotropy 构造非均质性；结构各向异性
structural anomaly 构造异常
structural antiseismic test 结构抗震测试
structural appraisal 结构评估；结构检定；结构勘测评估
structural appraisal methodology 结构评估法
structural approach limit of tunnel 隧道建筑限界
structural architecture 房屋建筑；体系结构
structural arrangement 结构布置
structural attitude 构造特征；构造产状
structural bar 构造筋
structural base moment 结构基底弯矩
structural base shear 结构基底剪力
structural basin 构造盆地
structural behaviour 结构性能；结构性状；构造动态
structural belt 构造带
structural bench 构造阶地
structural block 结构体；构造断块
structural body 结构体
structural bond 结构黏结；结构键
structural borne noise 经结构传送的噪声；结构噪声
structural break 结构断裂
structural bridge 构造桥
structural bulge 构造凸起
structural capacity 结构容量；结构能力；构造形成能力
structural casing 结构套管
structural characteristic 结构特征；结构特性
structural chemistry 结构化学
structural classification 构造分类
structural clay 结构性软土
structural closure 构造闭合度
structural coefficient 结构系数
structural cohesion 结构黏聚力；结构黏聚力

structural column 结构柱
structural component 构件
structural concordance 构造整合
structural condition survey 结构状况勘测
structural conditions monitoring system 楼宇结构监察系统
structural configuration 结构性配置；构造轮廓
structural conservation 结构防腐；结构保护
structural constant 结构常数；构造常数
structural continuity 构造连续性；结构连续性
structural contour 构造等高线
structural contour map 构造等高线图
structural control 构造控制；结构控制
structural correction 构造校正
structural cracking 结构裂缝；内部裂缝
structural creeping 构造蠕变
structural cross section 构造横剖面
structural cross-section 构造横剖面
structural crystallography 构造结晶学
structural cycle 构造旋回
structural damage 结构破坏；结构损伤
structural damage assessment 结构损伤评估
structural data 结构数据；构造资料
structural decoupling 构造分离作用；构造去耦作用
structural deficiency 构造缺陷；构造缺失；结构性缺陷
structural deformation 结构变形；构造变形
structural depression 构造洼地；构造坳陷
structural depth 构造深度
structural design 结构设计；土建设计
structural design and construction specification for ceramsite concrete 陶粒混凝土结构设计及施工暂行规程
structural design requirement 结构设计要求
structural detail 结构元件；结构详图；构造细节；构造细部
structural developing 结构改进
structural diagram 构造图；结构图
structural differentiation 构造分异；结构差异
structural dip 构造倾斜
structural discontinuity 结构不连续；构造间断
structural discordance 构造不整合
structural disharmony 构造不谐调
structural dislocation 构造错位
structural disorder 构造无序性；结构无序
structural displacement 结构位移
structural disturbance 结构扰动
structural divergence 构造离散度
structural division 构造划分
structural domain 构造域
structural dome 构造穹隆；构造圆丘
structural drawing 结构图；构造图
structural drawing of construction joint waterproofing 嵌缝防水构造图
structural drawing of lining waterproofing 衬砌防水构造图
structural drawing of portal waterproofing 洞门防水构造图
structural drawings 结构图；建筑图
structural dynamics 结构动力学
structural earthquake 构造地震
structural effect 构造效应；结构效应

structural effect of rock mass 岩体结构效应
structural element 构造单元；结构成分；构件；结构元件；结构要素
structural elevation 结构标高
structural encrustation 构造薄膜
structural engineer 结构工程师
structural engineering 建筑工程；结构工程
Structural Engineers Registration Committee 结构工程师注册事务委员会
Structural Engineers Registration Committee Panel 结构工程师注册事务委员团
structural environment 构造环境；构造条件
structural equation model 结构方程模型
structural erosion 构造侵蚀
structural evolution 构造演化
structural examination 结构测试；结构分析
structural excavation 结构开挖
structural fabric 构造组构；构造岩组
structural factor 构造因素；结构因素；结构因子
structural failure 结构失效
structural fall 结构斜度
structural fatigue failure 结构疲劳损坏
structural faults 结构缺陷；结构层错；构造断陷
structural feature 构造特征；构造形迹；构造形态；结构特征
structural features of a rock mass 岩体结构特征
structural fill 人工填筑垫层
structural fissure water 构造裂隙水
structural flexibility 结构柔性
structural flow 结构流
structural form 构造形式；构造形态；构造地形；结构形态
structural form surface map 构造形面图
structural formula 结构式；化学结构式；分子式
structural fortification 结构加固工程
structural frame 结构构架；结构框架
structural framework 构造格局；构造格架；结构骨架
structural gas pool 构造气藏
structural generation 构造世代；构造次序
structural geochemical anomaly 构造地球化学异常
structural geological analysis 构造地质分析
structural geological surveying of photo 相片构造地质测量
structural geology 构造地质学；构造地质
structural geometry 构造几何形态；构造几何学
structural geomorphology 构造地貌学
structural grain 构造条理；构造方向
structural group 结构族；构造群，构造组
structural growth 结构生长
structural health monitoring 结构健康监测
structural high 构造高地；构造高区；构造隆起
structural history 构造发育史
structural homogeneity 构造均匀性；构造均质性
structural horst 构造地垒
structural idealization 结构理想化
structural identification 结构识别；机构鉴定
structural image 构造图像；结构图像
structural imperfection 结构不完整性
structural improvement works 结构改善工程
structural index 结构指数；结构指标

structural indication 构造指示层；构造标志层
structural inference 结构推断；构造推断
structural influence factor 结构影响系数
structural information 构造信息；构造资料；结构信息
structural inheritance 构造继承性
structural instability 结构不稳定性
structural integrity 结构完整性；结构完整程度
structural interface 结构界面；构造界面
structural interpretation 构造解译；构造判释
structural interpretation of photo 相片地质构造判读
structural intersection 构造交切；构造交叉
structural inversion 构造反转
structural investigation 结构勘察
structural iron 结构型铁
structural isomeride 结构同分异构体
structural joint 结构缝；构造节理；构造缝
structural joint gap 构造节理间隙
structural kinematics 构造运动学
structural landform 构造地形；构造地貌
structural landscape 构造景观；构造地形
structural layer 构造层；结构面层
structural layout 结构布置
structural layout plan 结构布置图
structural length 全长
structural lenticular body 构造透镜体
structural level 构造面；构造层；构造部位
structural line 构造线
structural lineament 区域构造线
structural lineation 构造线条；区域构造线
structural load 结构荷载；结构负载；结构重量
structural load-carrying capacity 结构承载能力
structural localization 构造圈定
structural location 构造位置
structural low 构造洼地；构造低地；构造低点
structural map 构造图；构造等高线图；结构图
structural mapping 构造填图；勾画结构
structural mass 结构体
structural mast 桁架结构桅式井架
structural material 结构材料
structural mechanics 结构力学
structural member 结构构件；结构杆件
structural member strengthening 构件加固
structural model 结构模型；构造型式
structural model recognition 结构模式识别
structural model test 结构模型试验
structural nose 构造鼻
structural offset 构造偏移
structural oil pool 构造油藏
structural optical mineralogy 结构光性矿物学
structural optimization 结构优化；结构最优化
structural overburden 构造覆盖层
structural parameter 结构参数
structural part 结构部件
structural pattern 构造模式；构造型式；构造类型；结构形式
structural pattern of rock mass 岩体结构类型
structural permeability 结构渗透性；结构渗透率
structural petrology 构造岩石学
structural pillar 构造柱
structural pipe 结构管

structural plain	构造平原
structural plane	构造面；结构面
structural plane of rock mass	岩体结构面
structural planning	结构规划
structural plateau	构造高原
structural platform	构造台地
structural pool	构造油气藏；构造热储
structural porosity	结构孔隙度
structural position	构造层位；结构位置
structural pressed coal	构造挤压煤
structural profile	结构剖面；构造剖面
structural property	构造性质；构造特性；结构特性
structural property of soil	土壤结构特性
structural province	构造区；地质构造区
structural qualities redundancy	结构的质量储备
structural rearrangement	结构重构
structural reconnaissance	构造勘测
structural regime	构造形态；构造状况；构造状态
structural relationship	结构关系
structural reliability	结构可靠性
structural relief	构造起伏；构造高差；构造地形
structural repair	结构补强
structural response	结构响应
structural response coefficient	结构响应系数
structural risk minimization principle	结构风险最小准则
structural rock bench	构造阶地；构造岩石阶地
structural rock event	构造岩石事件
structural rock unit	构造岩石单位
structural saddle	构造沉降部分；构造鞍部
structural safety	结构安全性
structural safety evaluation	结构安全度评定
structural salient	构造凸起
structural section	构造断面；构造剖面；结构断面
structural segmentation	构造分割作用
structural sequence	构造顺序；构造序列
structural setting	构造背景；构造环境
structural set-up	构造配置
structural shale	结构泥岩；结构页岩
structural shape	结构形状
structural sheet	结构层
structural silicified zone	构造硅化带
structural simulation	结构模拟；构造模拟
structural situation	构造位置；构造环境
structural skeleton	结构骨架
structural sketch	构造纲要图
structural soil	有结构土壤
structural span	结构跨度
structural spring	构造泉
structural stability	结构稳定性
structural stability analysis	结构稳定性分析
structural stability condition	结构稳定性条件
structural stage	构造层；构造运动阶段
structural static mechanics	结构静力学
structural steel	钢结构
structural steel erector	结构钢架工
structural steel member	结构钢构件
structural steelwork	结构钢筋工程
structural stiffness matrix	结构刚度矩阵
structural stillstand	构造宁静期
structural stochastic response	结构随机反应
structural strain	结构应变
structural stratigraphic event	构造地层事件
structural stratigraphic section	构造地层剖面
structural stratigraphic trap	构造-地层圈闭
structural stratigraphic unit	构造地层单位
structural strength	结构强度
structural strength of soil	土的结构强度
structural strengthening	结构加固
structural style	构造型式；构造类型
structural sub-area	构造亚区
structural subsidence	构造沉降
structural succession	构造顺序；构造序列
structural surface	构造面
structural survey	构造测量；结构勘测
structural system	构造体系；结构系统
structural system composed of bar	杆系结构
structural system composed of plate	板系结构
structural system identification	结构系统识别
structural system strengthening	结构体系加固
structural terrace	构造阶地
structural test	结构试验；结构测试
structural theory	结构理论；构造理论
structural transition	构造转变；构造过渡
structural trap	构造圈闭
structural trapping	构造圈闭
structural trend	构造方向；构造走向
structural truncation	构造削截；构造截断；构造截切
structural type	构造型式；构造类型
structural type of subgrade bed	路基基床结构型式
structural types of rock mass	岩体结构类型
structural uncertainty	结构不确定性
structural unconformity	构造不整合；角度不整合
structural unit	构造单元；结构构件
structural use	结构用途
structural valley	构造谷
structural vibration	结构振动
structural viscosity	结构黏度
structural vitrain	结构镜煤
structural wall	承重墙；结构墙
structural water	结构水；构造水；化合水
structural weakness plane	结构弱面
structural works	结构工程
structural zonation	构造分带
structural zone	构造带
structural-formation analysis	构造-建造分析
structural-genetic connection	构造成因联系
structural-stratigraphic trap	构造-地层圈闭
structural-tectonic analysis	结构构造分析
structural-wall tile	承重墙用空心砖
structure	构造；结构；建筑物；组织
structure amplitude	结构振幅
structure analysis	构造分析；结构分析
structure analysis equation	结构分析方程
structure area	建筑面积
structure basin	构造盆地
structure block	结构体；结构块体
structure body	结构体
structure bond	结构联结；结构连接；构造联结
structure capacity	结构能力
structure chart	结构图

structure clay	结构黏土
structure closure	构造闭合度
structure coefficient	结构系数
structure collapse	结构崩溃
structure conformity	构造整合
structure contour	构造等高线
structure contour interval	构造等高距
structure contour map	构造等高线图
structure damage	结构破坏
structure deformation	结构变形；构造变形
structure design	结构设计
structure deterioration	结构破坏
structure differentiation	构造差异作用
structure drilling	构造钻探；构造钻井
structure ductility	结构延性；结构延展性
structure earthquake	构造地震
structure effect of rock mass	岩体结构效应
structure element	结构体；结构单元
structure elucidation	构造解释；机构鉴定；结构解析
structure excavation	建筑物基坑挖掘
structure expansion solution	结构膨胀溶液
structure face	结构面
structure facing	构造面向；构造朝向
structure failure	结构损坏；结构破坏
structure for cooling water	冷却水构筑物
structure form of anchor cable	锚索结构形式
structure foundation	结构基础
structure geology	构造地质学
structure geometry	构造几何形状
structure grain	构造纹理
structure grumeleuse	构造凝块；凝块构造
structure in space	空间桁架；空间结构
structure index	结构指数
structure indicator	标准层；构造标志
structure information	构造信息；构造资料
structure isomer	结构同分异构体
structure isomeride	结构同分异构体
structure joint	构造节理
structure line	构造线
structure map	构造图
structure material	结构材料
structure mechanics model	结构力学模型
structure model	结构模型；构造模式
structure nature mudflow	结构性泥石流
structure of bar system	杆系结构
structure of basement	基底构造；地下室结构
structure of coal	煤构造
structure of energy consumption	能源消费结构
structure of oil field	油田构造
structure of ore	矿石构造
structure of rock	岩石构造；岩石结构
structure of rock mass	岩体结构
structure of sedimentary cover	沉积盖层构造
structure of soil	土的结构
structure of subgrade bed	基床结构
structure of trace fossil	生物遗迹构造
structure outline map	构造纲要图
structure parameter	结构参数
structure pattern	构造模式；结构型式
structure pattern of rock mass	岩体结构类型
structure plane	构造面；结构面
structure plot	结构标绘图
structure polar	构造极地
structure profile	结构剖面；构造剖面
structure programming	结构程序设计
structure realm	构造区域
structure repair	结构修复
structure rigidity-strengthening measures	结构刚度强化措施
structure rock stratigraphic unit	构造岩石地层单位
structure safety	结构安全度
structure section	结构断面；构造剖面
structure sequence	构造顺序；构造序列
structure softening	结构弱化
structure stability	结构稳定性
structure stage	构造层
structure strength	结构强度
structure surface	岩体结构面
structure surface continuity	结构面连续性
structure system	构造体系
structure test hole	地质构造探测孔
structure trap	构造圈闭
structure trunk	建筑物本体；堤体
structure type	结构类型；构造类型
structure unconformity	构造不整合
structure under construction	在建结构
structure unit	构件；结构部件；构造单位；构造单元；构造单位
structure vulnerability classification	结构易损性分类
structure vulnerability index	结构易损性指数
structure wall	结构墙
structure yielding measures	柔性结构措施
structure-contour map	构造等高线图
structured design	结构化设计
structured material	结构材料
structured packing	规整填料
structured programming	结构化程序设计
structured water	结合水
structure-equipment system	结构-设备体系
structure-fissure water	构造裂隙水
structure-fluid interaction	结构-流体相互作用
structure-fluid-foundation interaction	结构-流体-基础相互作用
structure-foundation interaction	结构-基础相互作用
structure-foundation-soil interaction analysis	上部结构-基础-地基共同作用分析
structureless soil	无结构土壤；无结构性土
structureless vitrain	无结构镜煤
structure's life	结构物使用期限；结构物寿命
structures of all kinds	各种结构
structure-sensitive conductivity	构造敏感传导率
structure-soil interaction	结构-土相互作用
structure-soil system	结构-土体系
structure-structure interaction	结构间相互作用
structuring scheme	结构化方案
structurology	结构学；构造学
strut	支柱；支撑；抗压构件；压杆
strut beam	支梁
strut bracing	支撑
strut excavation	支撑开挖

strut joint	水平撑接头
strut load	支撑荷载
strut pole	撑杆
strut rail	横撑；横挡
strut support	柱式支架
strut thrust	支撑冲断层；强剪逆断层
strut to	给加支撑；膨胀
strut-framed beam	撑架式梁
strut-framed bridge	撑架式桥
strutted excavation	支撑开挖
strutting	加固；支撑物；横撑；支撑
struverite	钛铌钽矿
struvite	鸟粪石
stub	短支柱
stub axle	短轴
stub column	突出短柱
stub cutoff wall	防渗矮墙
stub drift	支平巷；不通的平巷
stub drilling	先钻；超钻
stub entry	独头平巷
stub heading	独头平巷
stub pillar	保巷煤柱；保巷矿柱
stub traverse	支导线
stub wall	矮墙
stubachite	蛇异橄榄岩
stucco	灰墁；灰泥；粉刷
stuck casing	受卡套管
stuck fish	被卡的落物
stuck freeing spotting fluid	解卡液
stuck jumper	卡住的冲击钻杆
stuck pipe	卡钻；被卡的钻杆
stuck pipe log	卡管测井
stuck pump	卡泵
stuck rod	抽油杆被黏卡
stuck tool	被卡的钻具
stuck-in-hole point	井内的卡点数
stuck-point	卡点
stuck-point indicator tool	卡点指示仪
stud	双端螺栓；双头螺栓；销杆
stud bolt	双头螺栓；螺柱
stud cutter assembly	切削齿总成
stud driller	大班司钻
stud driver	双头螺栓扳手；柱螺栓传动轮
stud gear	变速轮
stud horse	钻井技师
stud link chain	有挡锚链
stud nut	柱螺栓螺母
stud partition	框架板间墙；隔墙架
stud pin	柱螺栓销；大头钉
stud registration	销钉定位法
stud remover	双头螺栓拧出器
stud ring	齿环
stud type chain	日字链；挡环式链
stud wall	撑柱墙；立柱墙
studded-flange connection	带柱螺栓的法兰连接
student engineer	见习技术员
studtite	铀霞石；水丝铀矿
study area	试验地区；调查地区；研究地区
study brief	研究概要
study of East Asia tectonics and resources	东亚构造与资源研究计划
study of mineral deposit	矿床学
study of neotectonics	新构造学
study of ore field structure	矿田构造学
study of the potential use of underground space	地下空间的可能用途研究
study on harbour reclamation and urban growth	海港填海及市区发展研究
study site	试验场地；研究场地
stuff	脉石中的矿石；矿产；材料；原料；填料；充填
stuff gauge	标尺
stuffed basin	填塞盆地
stuffed structure	充填构造
stuffing	填料；填塞
stuffing function	填充函数
stuffing material	填料
stull	横梁；横撑；支柱；顶梁；采掘台；采空区
stull floor	横撑上的铺板
stull piece	横撑木
stull stoping method	横撑支柱回采法
stulled open stoping	横撑支柱回采法
stulled raise	横撑支护的溜井
stull-floor method	横撑支柱底板开采法
stull-laced raise	横撑支护天井
stull-level method	横撑支柱阶段平巷留矿开采法
stull-set method	横撑支柱开采法
stump	残留煤柱；小煤柱；矿柱；树桩
stump blasting	小煤柱爆破；小矿柱爆破
stump burning	煤柱燃烧
stump dozer	除桩机
stump drawing	回采矿柱；回采煤柱
stump extraction	回采煤柱；小煤柱回采
stump leaving	留煤柱；煤柱保留
stump pillar	主运输道与煤房间的煤柱
stump pulling	回采煤柱
stump recovery	矿柱回收
stumping	回采煤柱
sturdily constructed	构造坚实的；结构坚固的
sturdiness	坚固性；稳定性
stylolite	缝合线；缝合岩面；柱状构造
stylolitic	缝合的；柱状
stylolitic cleavage	缝合线状劈理
stylolitic joint	缝合线节理；柱状结点
stylolitic structure	缝合构造；柱状构造
stylolitization	缝合作用；柱状化作用
stylotypite	柱形石；黝铜矿
Styrian orogeny	斯蒂里亚造山运动
suanite	遂安石
sub bank	分台阶；分阶段
sub drift	中间平巷
sub entry	中间平巷；顺槽；次要平巷
sub pressure gradient	低压力梯度
sub slicing	分层陷落开采
subaerial	陆上的
subaerial block	陆上断块
subaerial delta	陆上三角洲
subaerial denuation	陆相剥蚀
subaerial deposit	陆上沉积物
subaerial erosion	陆上侵蚀
subaerial fluvial system	陆上河系

subaerial landslide 陆上滑坡；地面滑坡
subaerial spring 地表泉
subaerial unconformity 陆成不整合
subage 亚纪；亚期；亚代
subagent 次要营力；次作用力；副代理人
subalkalic rock 次碱性岩
subalpine belt 亚高山带
subalpine peat 亚高山泥炭
subalpine rendzina 亚高山黑色石灰土
subaluminous rock 次铝质岩
subangular 次棱角状
subangular grain 次棱角状颗粒
subangular gravel 次棱角状砾石
subangular particle 次棱角颗粒
subangular-grained 次棱角状颗粒的
subaquatic delta 水下三角洲
subaqueous concrete 水下混凝土
subaqueous concreting 水下混凝土施工，水下灌筑混凝土
subaqueous debris flow 水下泥石流
subaqueous deposit 水底沉积
subaqueous disposal 水下处置
subaqueous foundation 水下基础
subaqueous glide 水底滑动
subaqueous moraines 水下冰碛
subaqueous mudflow deposits 水下泥流沉积
subaqueous raw soil 水底原始土壤
subaqueous rockfall 水下岩崩
subaqueous shock 水下振动
subaqueous slide 水下滑坡
subaqueous slump 水下滑塌；水下滑移；水下滑坡
subaqueous slump-mass 水下崩滑体
subaqueous spring 水下泉
subaqueous tunnel 水下隧道；沉埋隧道
subaqueous volcanism 水下火山活动
sub-aquifer 亚含水层
sub-arch 副拱
subarctic 亚寒带；近北极的
subarea 子区；分区；次区域
subarid 半干旱的
subarkose 次长石砂岩；亚长石砂岩
subarkosic 亚长石砂岩质的
sub-artesian ground water 半承压地下水
sub-artesian groundwater 半承压地下水
subartesian well 半自流井；地面下自流井；近自流
subatmospheric hydraulic model 负压水工模型；减压水工模型
subatmospheric pressure 负压
subatmospheric weathering 负压风化作用
subautochthonous batholith 半原地岩基
subautomorphic 半自形的
subaxial fracturing 亚轴向破坏
sub-ballast 底碴，底层道碴
subballast consolidating machine 道床底碴夯实机
sub-bank 分台阶
subbase 底基层，地基；基础下卧层；子基
subbase course 底垫层；底基层；基层
subbase level 局部基准面
subbasement rock 基底下岩石
subbasin 海盆；次盆地；子流域；水区

subbench 中间阶段
sub-bentonite 次膨润土
subbing 地下渗灌；潜灌
subbituminous 亚烟煤的
sub-bituminous 次烟煤
subbituminous coal 亚烟煤，次烟煤，副烟煤
sub-bottom 基底
subbottom depth recorder 海底深度记录仪
subbottom profile runs 海底地层剖面调查
subbottom profiler 海底地层剖面仪
sub-bottom profiler 浅底地层剖面仪，浅地层剖面仪，海底地层剖面仪，地层剖面仪
subbottom profiler prospecting 浅地层剖面仪勘探，海底剖面仪探测
sub-bottom profiling system 海底浅层剖面仪
sub-bottom seismic profiling 海底地震剖面图
subbottom tunnel 海底隧道
subbrackish 微咸的
subcalcic amphiboles 亚钙角闪石
subcannel coal 低变质烛煤；次烛煤；副烛煤
subcapillary interstice 次毛细孔隙，亚毛细孔隙
subcatchment 支流集水处
subclass 亚纲；子集
sub-clause 子条款；副条款
sub-compacted 欠压实的
subconchoidal fracture 不明显的贝壳状断口；次贝壳状断口
subconsequent stream 次顺向河；后成河
subcontinent 次大陆；次陆地
subcontinental crust 次大陆地壳
subcontinental denudation 陆相剥蚀
subcontract 分包合同；转包合同；分包
sub-contract 分包工程合约；分包合约；分包合同；分包
subcontractor 分包商，分包者，分包单位；转包商
subcooling condenser 低温冷却器；低温冷凝器；过冷冷凝器
sub-coupling 变径接箍
subcritical 临界点之下的；亚临界的
sub-critical anneal 亚临界退火
sub-critical area of extraction 亚临界开采区
subcritical crack growth 亚临界裂纹增长；亚临界裂纹扩展
subcritical crack propagation 亚临界裂纹增长；亚临界裂纹扩展
subcritical extraction 非充分采动；非临界开采；次临界开采
subcritical flow 次临界流；缓流，缓流流速小于临界值的流动；亚临界流动
subcritical flow installation 缓流设施
subcritical reflection 临界角前的反射
subcritical speed 亚临界速度
sub-critical temperature 亚临界温度
subcriticality 次临界度，亚临界度
subcrop 隐伏露头；地下古露头
subcrop trap 隐伏露头圈闭
subcrust 地壳内层；路面底层；地壳下地层
subcrustal 壳下的；路面底层的
subcrustal earthquake 壳下地震
subcrustal layer 壳下地层

subcrystalline 部分结晶的；亚晶态的；结晶不明显的
subcycle 次旋回；次循环；子循环
subdelta 小三角洲
subdendritic pattern 次树枝状水系
subdeposit drift 矿体下部平巷；脉下平巷
sub-diagonal 副斜杆
sub-diapir 次刺穿褶皱；次底辟
sub-diapir trap 次级挤入圈闭
subdivision 细分；分部；分支；小块土地
subdivision of reservoir 储层细分；细分层
subdivision plat 详细地区图；地籍图
subdivision survey 地籍测量；重划测量
subdouble 二分之一的
subdrain 地下排水管；次级排水沟
subdrainage 地下排水；地下排水系统
subdrift 中间平巷
sub-duct assembly 防回流装置
subducted oceanic crust 俯冲洋壳
subducted zone 俯冲带
subducting edge 俯冲边界
subducting lithosphere 消减的岩石圈
subducting plate 俯冲板块；消减板块
subduction 俯冲；消减；隐没；减法；除去；扣除
subduction belt 消减带；俯冲带
subduction complex 俯冲复合体；俯冲带杂岩
subduction component 俯冲成分
subduction erosion 俯冲侵蚀；构造侵蚀；隐没侵蚀
subduction mantle source 俯冲地幔源
subduction process 消减/俯冲作用；消减/俯冲过程
subduction type orogeny 俯冲型造山运动
subduction zone 俯冲带；潜没带；消减带
subduction-type geothermal belt 消减型地热带
subdued 平缓的
subdued mountain 平缓山
subdued relief 平坦地形，削平地形
subdued topography 平坦地形
subdued uplift 缓和的隆起
sub-economic production 不经济的开采
subeffusive dike 半喷发岩墙
subelongate 次伸长状
subepoch 亚世
subequant 次等轴状
subequant crystal 次等轴晶体
subequatorial belt 副赤道带
subera 亚代
suberain 木栓质煤
suberathem 亚界
suberinitic liptobiolite 树皮残植煤
subfabric 亚组构；次组构
subface 地层底面
subfacies 亚相
subfeldspathic 次长石质的
subfeldspathic lithic arenite 亚长石质岩屑砂屑岩
subfloor deposit 海床下面的沉积层
subfluvial 水下的；河底的；水下产生的
subflysch 次复理石
subform 从属形式；衍生形式；亚品种
subfossil 亚化石；半化石；准化石
subfoundation 下层基础
sub-fracturing pressure 低于地层破裂压力的压力

subfrigid zone 亚寒带
subgeneric 亚属的
subgeoanticline 次地背斜
subgeostrophic wind 次地转风
subgeosyncline 次地槽
subglacial 冰川下的；冰期后的；冰川底部的
subglacial channel 冰下河道；冰下水道
subglacial drainage 冰下水系
subglacial eruption 冰下喷发
subglacial moraine 冰底碛
subglacial zone 冰内带
subglobular 次球形的，接近球形的
subgrade 路基；地基；基床
subgrade bearing capacity 地基承载力
subgrade bed damage 基床病害
subgrade bed of cutting 路堑基床
subgrade bed of embankment 路堤基床
subgrade bed softening 基床软化
subgrade bed soil 基床土
subgrade coefficient method 基床系数法
subgrade compacted soil 路基压实土
subgrade cooling 地基冷却
subgrade cutting 向下挖掘
subgrade damage 路基病害
subgrade design of railway 铁路路基设计
subgrade digging 下向挖掘
subgrade drainage 路基排水
subgrade drilling 超钻；钻孔加深；钻透
subgrade elevation 路肩高程
subgrade engineering 路基工程
subgrade erosion 路基冲刷，路基侵蚀
subgrade in slide section 滑坡地段路基
subgrade in snow damage area 雪害地区路基
subgrade in soft soil section 软土地区路基
subgrade in windy and dusty area 风沙地区路基
subgrade of karst area 岩溶路基
subgrade of railway 铁路路基
subgrade of special condition 特殊条件路基
subgrade of special rock and soil area 特殊土路基
subgrade protection 路基防护
subgrade reaction 地基反力；路基反力；基床反力
subgrade reaction theory 基床反力理论
subgrade restraint 路基约束；基床约束作用
subgrade retaining 路基支挡
subgrade settlement 地基沉降；路基沉降
subgrade stabilization 路基稳定
subgrade standard 路基标准
subgrade strengthening 路基加固
subgrade subsidence 路基下沉，路基沉降
subgrade surface 路基面
subgrade works 路基工程
subgradient wind 次梯度风
subgrain 亚晶粒；亚颗粒
sub-grain boundary 亚颗粒边界
subgraphic structure 亚文象构造
subgraphite 次石墨；亚石墨；高煤化无烟煤
subgreywacke 亚杂砂岩
subground condition 地下条件
subhedral 半自形
subhedral crystal 半自形晶

subhedron 半自形晶
Subhercynian movement 新海西运动
Subhercynian orogeny 新海西造山运动
subhorizon 亚层；心土层
subhorizontal isocline 近水平的等斜褶皱
subhorizontal lineation 近水平线理
subhumid climate 半湿润气候
subhumid region 半湿润区
subhumid soil 半湿土
sub-humid tropic 湿副热带；湿亚热带
subhumid zone 半湿润带
subhydrous coal 低氢煤
subidiomorphic 半自形的
subindividual 晶片
subinterval 子区间；小音程
subintrusive dyke 下层侵入岩墙
subirrigation 地下灌溉
subisocline fold 准等斜褶皱
subjacent bed 下卧层；下伏层
subjacent intrusion 深部侵入体
subjacent karst 地下岩溶；地下喀斯特
subjacent seam 下卧层
subject index 主题索引
subject reservoir 目的层
subject to contract 以合同为准
subject to the law 依照法律；受法律约束
subjection to bending 受折弯
subjection to pressure 受压
subjection to tension 受拉
subjoint 次级节理
sublacustrine fan 湖底扇
sublateral 暗沟分支；分支管
sublateral canal 分支渠
sublayer 分层；次层；子层；下卧层；回采分层；底层
sublayer lamina 次层流
sublet 转包；分包
sublevation 海底侵蚀
sublevel 分段；小阶段；亚阶段；次能级
sublevel back stoping 亚阶段上向梯段回采
sublevel benching 分段阶梯式开采
sublevel blast hole caving 深炮眼分段；崩落采矿法
sublevel caving 分段崩落采矿法
sublevel caving hydraulic support 放顶煤支架
sublevel caving method 分段崩落开采法
sublevel caving mining 分段崩落采矿；放顶煤开采
sublevel centralized entry 区段集中平巷
sublevel crosscut 中间水平石门
sublevel design of layout 分段崩落法；采场结构设计
sub-level drift 分段平巷；区段平巷；中间平巷
sublevel drive 分段掘进
sublevel entry 分段平巷
sublevel interval 分段间距；小阶段间距；顺槽间距
sublevel long hole benching 分层台阶深孔开采
sublevel method 分段开采法
sublevel mining 分段采矿；分段开采；分段回采
sublevel mother gate 小阶段集中巷
sublevel open stope method 分段无充填工作面采矿法；分段空场采矿法
sublevel open stoping 分段空场采矿法
sublevel roadway 分阶段平巷

sublevel stoping 分段回采法；分段采矿法
sublevel system 分段开采法
sublevel undercut caving method 分段下部掏槽崩落开采
sublimate deposit 升华矿床
sublimation temperature 升华温度
sublitharenite 亚岩屑砂岩
sublithwacke 亚岩屑瓦克岩
sublittoral 次大陆架的；潮下的；近岸的；亚沿岸生长的
sublittoral region 亚潮带
sublittoral zone 潮下带；浅海地带；亚海岸带；亚潮间带
submaceral 亚显微亚组分
submain 次干管；辅助干线
submarginal 接近边缘的；近边缘的；限界以下的
submarginal mineral body 边界品位以下的矿体
submarginal producer 亚边际性生产井
submarginal resource 接近边际的资源
submarginal stream 近冰缘河
submarine 水下的
submarine abrasion 海底海蚀
submarine alluvial fan 海底冲积扇
submarine apron 海底扇
submarine arc volcano 海底火山弧
submarine bank 海底滩
submarine bar 水下坝；海底沙洲；海底沙坝
submarine basalt 海底玄武岩
submarine bench 深海/海底平台；深海平坦面
submarine blasting 水下爆破
submarine blasting gelatine 水下用硝甘炸药
submarine bottom scanner 海底扫描器
submarine bulge 海底凸起
submarine canyon 海底峡谷；海底谷地；水下峡谷
submarine canyon evolution 海底峡谷演化
submarine canyon morphology 海底峡谷形态
submarine cementation 海底胶结作用
submarine channels 海底通道
submarine chimney 海底烟囱
submarine cliff 海底悬崖
submarine coalfield 海底煤田
submarine composite salt glacier 海底复合盐冰川
submarine cone 海底锥
submarine construction survey 海底施工测量
submarine continental margin 海底大陆边缘
submarine control network 海底控制网
submarine core drill 海底岩芯钻
submarine coring drilling rig 浅钻；海底岩芯钻机
submarine creep 海底蠕变
submarine delta 海底三角洲；海底沙洲
submarine denudation 海底剥蚀
submarine deposit 海底矿床
submarine depression 海底洼地；海底深陷区
submarine detector 海底探测器
submarine dispenser 海底扩散器
submarine drilling 水下钻孔；海底钻孔
submarine dune 海底沙丘
submarine earthquake 海底地震；海下地震
submarine electric detonator 水下电雷管
submarine environment 海底环境
submarine equilibrium profile 海底平衡剖面；海底平衡

断面
submarine erosion 海底侵蚀
submarine exploration 海底勘探
submarine fan 海底扇；海底陆源沉积扇；海底扇形地
submarine fault 海底断层
submarine flank 海底侧面
submarine fracture zone 海底断裂带
submarine fumarole 洋底喷气孔
submarine gas field 海上气田
submarine gas hydrate 海底天然气水合物
submarine geology 海洋地质学；海底地质学；海底地质
submarine geomorphologic chart 海底地貌图
submarine geomorphological map 海底地貌图
submarine geomorphology 海底地貌学；海底地貌
submarine geomorphy 海底地貌
submarine geophysics 海底地球物理
submarine gravity survey 水下重力测量
submarine groundwater 海底地下水
submarine gully 海底沟壑
submarine hot spring 海底热泉；洋底热泉
submarine humus 海底腐殖质
submarine hydrothermal deposit 海底热液矿床
submarine hydrothermal solution 海底热液
submarine lahar 海底泥流
submarine landform map 海底地势/地形图
submarine landforms 海底地貌
submarine landslide 海底滑坡
submarine landslide deposits 海底滑坡沉积
submarine lateral reflection 海底侧面反射
submarine lava flow 海底火山熔岩流
submarine laying 海底铺设
submarine line crossing 水下管线穿越
submarine magnetic field 海底磁场
submarine magnetometer 海底磁力仪
submarine magnetotelluric field 海底大地电磁场
submarine mast system 水下杆柱系统
submarine mine 海底矿；海底采矿
submarine mining 海底开采；海底采矿
submarine morphology 海底地形学
submarine mounds 海底土墩/土丘
submarine oil field 海底油田；海上油田
submarine oil formation 海底含油地层
submarine oscillator 水下振动器
submarine outfall 海底排水口；海底渠口；排放口
submarine photography 海底摄影
submarine photometer 水下光度计
submarine pipeline 水底管道；水下管道；海底管道
submarine plain 海底侵蚀平原
submarine plateau 海底高原；海底高地；海台
submarine platform 海底侵蚀台地；海底台地
submarine pressure hull 水下承压外壳；潜艇承压壳
submarine range 海底山脉
submarine relief 海底地貌/地形；海底起伏
submarine resources 海底资源
submarine ridge 海脊；海岭
submarine rift 海底裂谷/断裂
submarine saline deposits 海底盐矿
submarine salt glacier 海底盐冰川
submarine sampling 海底取样
submarine sand wave 水下沙波

submarine sandy soil 海底砂质土壤
submarine scarp 海底崖，陡崖
submarine sediment 海底沉积物
submarine sedimentology 海底沉积学
submarine seismic sources 海底地震源
submarine seismograph 海底地震仪
submarine sewage pipeline 海底污水管道
submarine shallow-buried gas 海底浅层天然气
submarine shock 海底地震
submarine sill 海槛；海底岩床
submarine situation chart 海底地势图
submarine slide 海底滑坡；海底滑动
submarine slope failure 海底斜坡失稳
submarine slope of coast 水下岸坡
submarine slump 海底滑坡；海底崩移
submarine slumping 海底崩坍
submarine snowline 海底雪线
submarine structural chart 海底地质构造图，海底结构图
submarine swell 海底隆起
submarine tectonic geology 海底构造地质学
submarine tectonics 海底构造学
submarine terrace 海底阶地，海底台地；水下阶地；陆架阶地
submarine tholeiitic basalt 海底拉斑玄武岩
submarine topographic map 海底地形图
submarine trench 海沟
submarine tunnel 海底隧道；水底隧道
submarine tunnel survey 海底隧道测量
submarine valley 海底山谷，海底谷
submarine volcanism 海底火山活动，海底火山作用
submarine volcano 海底火山
submarine volcano morphology 海底火山形态
submarine weathering 海底风化作用；海解作用
submarine-canyon fill 海底峡谷充填
submarine-canyon sediment 海底峡谷沉积
submaritime 近海岸的
submeander 河道小曲流；次生河曲
submerge 浸没，淹没，沉没
submerged arc welding 埋弧焊
submerged bank 水下堤
submerged bar 暗沙洲
submerged beach 下沉海滩，沉没海滩
submerged breakwater 潜堤；水下防波堤；潜水坝
submerged buoy 沉没式浮筒；潜标
submerged coast 溺岸；沉没岸；下沉海岸
submerged combustion 浸没燃烧，沉没燃烧
submerged contour 水底等高线
submerged crib 潜没滤水栅；潜式进水口
submerged current 潜流
submerged dam 潜坝
submerged delta 水下三角洲
submerged delta front 水下三角洲前缘
submerged density 浸没密度；浮密度
submerged digging 水下挖掘
submerged dike 潜堤，潜堤
submerged discharge valve 淹没式泄水阀
submerged displacement 水下排出量，排水量
submerged efflux 潜没出流，淹没射流
submerged filter 淹没式滤器

submerged floating tunnel 水下悬浮隧道
submerged flow 潜流
submerged forest 沉埋森林
submerged inlet 潜没式取水口
submerged isothermal current 等温潜流
submerged jet 淹没射流；潜喷射流；潜浸喷射
submerged jetty 潜堤
submerged lubrication 浸入式润滑
submerged marine delta 沉没海相三角洲
submerged non-free jet 淹没非自由射流
submerged outfall 海底排污渠
submerged peneplain 水下准平原；水下侵蚀平原
submerged pier 暗墩
submerged pipeline 海底管线
submerged platform 水下工作平台
submerged production system 水下采油系统
submerged pump 潜水泵
submerged reef 沉没礁；暗礁
submerged sharp-crested weir 尖顶潜堰
submerged shore 溺岸
submerged sluice 潜水闸
submerged soil 浸没土
submerged speed 水下航速
submerged storage tank 沉没式储罐
submerged structure 水下构筑物
submerged tank 海底油罐；水下油库；水下罐
submerged tunnel 沉埋隧道；水下隧道
submerged under water zone 淹没带
submerged unit weight 浮容重；浸没容重
submerged valley 溺谷；潜谷；沉没谷
submerged wellhead jacket 沉没式井口导管架
submerged wellhead platform system 沉没式井口平台系统
submerged zone 长期浸水部分；浸水带，沉水带
submergence depth 浸入深度；沉没度
submergence of ground 地面下沉
submergible deep well pump 深井潜水泵
submergible roller dam 淹没式定轮闸门坝
submersed filter 浸液滤油器
submersible 沉入的，沉没的
submersible barge 坐底式钻井船
submersible bridge 漫水桥，浸水桥
submersible decompression chamber 水下减压舱
submersible deep well pump 深井潜水泵
submersible diving chamber 潜水舱
submersible drilling platform 坐底式钻井平台
submersible drilling rig 坐底式钻井平台
submersible electric pump 电动潜油泵
submersible monitor 水底监测仪器
submersible non-clog centrifugal pump 潜水防塞离心泵
submersible pipe alignment rig 水下对管机
submersible platform 潜水平台
submersible pump 潜水泵；水下泵
submersible service 水下应用
submersible sewage pump 可沉浸的水泵；潜污泵
submersible Sump 地下泵
submersible turbine pump 潜油透平泵
submersible work chamber 潜水作业舱
submersion 沉没，淹没，浸没，下沉
submetallic luster 半金属光泽

submicroearthquake 亚微震
submicron 亚微细粒
submicroscopic 亚微观的
submicroscopic crystal 亚显微晶体
submicroscopic damage track 次显微的损坏径迹
submicroscopic size 亚微观大小
submicroscopic structure 亚显微结构
submineering 超小型工程
subminiature geophone 超小型检波器
subminiaturization 超小型化
submission of bid 投标
submission to arbitration 提请仲裁
submountain region 山前地区
subnival belt 高寒带
subnormal formation pressure 低于正常的地层压力
subnormal pressure 低异常压力
suboceanic basaltic rock 洋底玄武岩质岩石
suboceanic earthquake 海底地震
suboceanic layer 洋底层
subophitic structure 亚辉绿构造
subophitic texture 次辉绿结构
subordinate displacement 二次驱替
subordinate fold 从属褶皱
subordinate geochemical landscape 从属的地球化学景观
subordinate phase of production 次要开采阶段
subore halo 矿下晕；近矿晕
suboutcrop 隐蔽露头，隐伏露头，埋藏露头
subparallel anticlinal ridge 近平行背斜脊
subparallel crustal ridge 近平行地壳隆起
subparallel fault 近平行断层
sub-parallel orientation 近似平行方向
subparallel seismic reflection configuration 亚平行地震反射结构
subperiod 亚纪
subpermafrost water 冻土底水；永久冻土下位水
subphase 亚幕
subphyllarenite 亚叶砂屑岩
subplate 次板块
subplate boundary 板块下边界
subpolar zone 亚寒带；副极带
subpressure 低异常压力
sub-rail foundation 轨下基础
subrecent eruption 新近喷发
subregion 次区域；亚区
subregional unconformity 小区域不整合
subrosion 地下侵蚀
subrounded gravel 次圆砾石
subrounded particle 次浑圆土粒
subsaline 微咸的
subsalt imaging 盐下成像
subsalt structure 基性盐构造
subsalt well 含盐油井
subsample 二次取样
subsaturation 亚饱和；半饱和
subsea 海底的，在海底
subsea apron 海底扇
subsea atmospheric system 水下常压系统
subsea bedrock 滨海基岩
subsea bedrock mining 海底基岩开采
subsea blowout prevented system 海底防喷器系统

subsea blowout-preventer stack	水下防喷器装置
subsea choke	水下阻流器
subsea completion system	水下完井系统
subsea cone	海底锥
subsea drilling equipment	水下钻具
subsea elevation	海底隆起
subsea enclosure	海底封罩
subsea multiwell template	水下多井底盘
subsea pipeline	海底管道
subsea production	水下采油
subsea production station	海底采油站
subsea production system	水下采油系统
subsea production tree	水下采油树
subsea reeled tubing system	水下挠管修井系统
subsea riser	立管
subsea storehouse	海底仓库
subsea telemanipulation	海底遥控操作
subsea tunnel	海底隧道
subsea TV	水下电视
subsea valve arrangement	海底阀组
subsea well completion	海底完井
subsea wellhead system	水下井口系统
subsea work enclosure	水下工作舱
sub-seabed	小海床
subsection	细目
subsequence	准层序
subsequent deposit	后成矿床
subsequent fill	随后充填
subsequent fold	后成褶皱
subsequent ridge	后成山脊
subsequent river	后成河
subsequent settlement	附加沉降量
subsequent thrusting	后成冲断层作用
subsequent valley	后成谷；次成谷
subseries	亚统；子群列；次分类
subshell	支壳层
subshot	重复放炮；分炮
subside	下沉；沉降；沉陷
subsided coast	沉降海岸
subsided water area	塌陷水域
subsidence	下沉；下陷；沉降；塌陷
subsidence area	沉陷区，沉降区，塌陷区，陷落区；下降面积
subsidence basin	地表移动盆地；地表沉陷槽；地表下沉盆地
subsidence break	地层沉裂
subsidence caving	沉陷崩落
subsidence coefficient	沉降系数
subsidence contours	地面沉陷等值线
subsidence control	塌陷控制，沉陷控制岩层移动控制
subsidence curve	沉降曲线
subsidence damage	沉降破坏，沉陷损坏
subsidence difference	沉陷差值
subsidence draw	地面塌陷超出地下采空区的延伸部分
subsidence earthquake	沉陷地震
subsidence extraction	非充公采动
subsidence factor	沉陷系数；下沉系数
subsidence factor of dump	排土场下沉系数
subsidence factor of multi-mining	重复采动下沉系数
subsidence fracture	沉降龟裂
subsidence inversion	下沉逆温
subsidence level network	控制地面沉降的水准网
subsidence mechanism	沉陷机制；沉降机制
subsidence monitoring	沉降监测
subsidence of ground	地面沉降
subsidence of pier-abutment foundation	墩台地基沉陷
subsidence of reservoir bottom	库底沉陷
subsidence of roadbed	路基沉陷
subsidence of top	顶板沉陷
subsidence pattern	沉降模式
subsidence percentage	下沉率
subsidence prediction	沉降预测
subsidence prevention	塌陷预防
subsidence profile	地表沉陷剖面；沉陷纵断面
subsidence rate	沉降速度，沉降速率
subsidence slope	沉陷坡度
subsidence strain curve	沉陷应变曲线
subsidence trough	地表移动盆地；下沉凹槽；塌陷坑
subsidence trough reclamation	塌陷区复垦
subsidence zone	沉陷带，沉降带
subsider	沉降槽
subsidiary	辅助的；副的
subsidiary anticline	次级背斜
subsidiary closure	次级闭合构造
subsidiary developing	辅助开拓工程
subsidiary development drivage	采准掘进；准备巷道
subsidiary dome	附属穹窿
subsidiary earthquake effect	次生地震效应
subsidiary fault	分支断层；副断层；附生断层
subsidiary fold	次级褶皱
subsidiary level	辅助水平
subsidiary massif	次级地块
subsidiary shaft	副井
subsidiary shot hole	辅助炮眼
subsiding basin	沉降盆地
subsiding belt	沉降带
subsiding by matching weights	配重下沉法
subsiding coal accumulating basin	坳陷型聚煤盆地
subsiding coast	沉降海岸
subsiding delta	淹没三角洲
subsiding geosyncline	沉降地槽
subsiding water	退水
subsieve	微粒的；亚筛；微粒
subsiliceous rock	基性岩
subsilicic rock	基性岩
subsoil	地基土；下层土；底土
subsoil drain	地下排水；地下排水管道
subsoil drainage	地下排水
subsoil exploration	次层土探测
subsoil formation	土壤下伏岩层
subsoil investigation	地基勘探；下层土勘探
subsoil karst	地下岩溶
subsoil permissible load	天然地基允许荷载
subsoil profile	地基土剖面图；下层土剖面
subsoil storage	底土存储
sub-soil water drain	地下水排水渠
subsoil waterproofing	地下防水
subsolid	半固体
subsolidus	亚固相线
subspecies	亚种

substage 分期；亚期
sub-standard concrete 未符标准的混凝土
sub-standard slope 不合标准斜坡
substantial completion 实质性竣工；基本完工
substantial repair work 大修
substantiation of claim 索赔证明
substation 变电所；变电站
substep 小台阶，分台阶
substitute column approach 替代柱法
substitute energy 替代能源
substitution 置换，替代，替换，代用
substitution excavation 置换开挖
substoping and caving 分阶段崩落采矿
substratal lineation 层底线状印痕
substrate 基质；基底；底层；地层；水道底层物质
substratification 分层
substratum 下卧层；下伏地层；底土层；基底
substructure 下部结构,；基础结构；井架底座；底层结构
substructure works 基础工程
subsurface 地下的，表面下的；底面
subsurface anomaly 地下异常
subsurface barrier 地下遮挡；地下不渗透层
subsurface bedrock 地表下基岩
subsurface benchmark 井下基准
subsurface biosphere 地下生物圈
subsurface blowout 地下井喷
sub-surface building works 地下建筑工程
subsurface circulation 次表层环流；海面下环流
subsurface condition 地表下的条件；地质情况
subsurface construction 地下建筑
subsurface contour line 地下等深线
subsurface contour map 地下等深线图
subsurface correlation 地下地层对比
subsurface coverage 地下覆盖段
subsurface cross section 地下横剖面
subsurface current 潜流
subsurface deposit 地下油藏
subsurface disposal 地下排放；地下处置
subsurface disposal of refuse 废料地下处理
subsurface divide 地下分水界
subsurface drainage 地下排水；地下径流；地下水系
subsurface drainage gallery 地下排水巷道
subsurface dynamograph 井下测力计
subsurface erosion 地下潜蚀；地下侵蚀
subsurface excavation 地下开挖；地下采掘
subsurface excavation method 暗挖法
subsurface exploration 地基勘探；地下勘探；浅层勘探
subsurface fault 地下断层
subsurface filtration 地下渗透
subsurface flood performance 地下注入动态
subsurface flow 地下水流
subsurface flow control 地下流量控制
subsurface fluid 地下流体
subsurface fracture 地下破裂
subsurface geological map 地下地质图
subsurface geology 地下地质学
subsurface geology condition 地下地质情况
sub-surface ground 地表下土壤
subsurface heat flow 地下热流
subsurface hydraulics 地下水力学
subsurface hydrology 地下水文学
subsurface information 井下资料
subsurface interface 地下界面
subsurface level 次表层水位
subsurface map 地下地貌图
subsurface mapping 地下填图
subsurface motion 地下运动
subsurface nuclear explosion 地下核爆炸
subsurface occurrence 地下产状
subsurface opening 地下洞室
subsurface pipeline 地下管线
subsurface pressure 地下压力
subsurface reservoir 地下储层
subsurface runoff 地下径流；潜水径流；壤中径流；次表层径流
subsurface sample 地下取样；钻孔中的样品
subsurface sampler 井下取样器
subsurface seepage flow 地下渗流
subsurface sewage disposal 地下污水处理
subsurface soil 地下土；亚表土
subsurface source 地下震源
subsurface storage 地下储存，地下封存
subsurface structural map 地下构造图
subsurface structure 地下构造
subsurface temperature 井下温度
subsurface transient saturated flow 非稳定饱和地下水流
subsurface unconformity 地下不整合
subsurface water 地下水
subsurface water balance 地下水均衡；地下水剩余量
subsurface water basin 地下水盆地
subsurface water regime 地下水动态
subsurface water resource 地下水资源
subsurface water storage 地下水储存量
subtended angle 对角；对弦角；对弧角；视差角
subtense bar 横基尺；视距尺；横距尺
subtense method distance measurement 视差法测距
subtense method with horizontal staff 横基尺视差法
subtense method with vertical staff 竖基尺视差法
subterrane 下伏基岩
subterranean cavern 地下洞穴
subterranean crossing 地下穿越
subterranean divide 地下分水线
subterranean drainage 地下水系
subterranean drainage basin 地下排泄盆地
subterranean flow 地下径流
subterranean flow modulus method 地下径流模数法
subterranean formation 地层
subterranean gas storage 地下储气库，地下天然气储集库
subterranean gasification method of mining 地下气化采矿法
subterranean ice 地下冰
subterranean karst 地下岩溶
subterranean leak 地下渗漏
subterranean outcrop 隐伏露头
subterranean river 地下河；暗河
subterranean room 地下室
subterranean stream 地下河
subterranean thermal water 地下热水

subterranean tunnel	地下隧道	successive extension	顺序外延
subterranean water	地下水；潜水	successive failure	渐进性破坏
subterranean water parting	地下分水界	successive fold episodes	连续褶皱幕
subterranean works	地下工程	successive imbricate thrust	连续叠瓦式冲断层
subterraneous outcrop	隐伏露头	successive induction	逐次归纳法
subterraneous quarry	地下采石场	successive injection	连续注入
subterrestrial	地球内的；地下的	successive integration method	逐步积分法
subtidal	潮下带	successive iteration	逐次迭代
subtidal community	潮下群落	successive launching method	顶推法
subtidal deposit	潮滩沉积	successive layers	连续层次
subtidal facies	潮下相	successive line overrelaxation	逐排超松弛
subtidal high-energy zone	潮下高能带	successive orthogonalization	逐次正交化
subtidal line	低潮线	successive overrelaxation	逐次超松弛
subtle banding	微夹层	successive oxidation	顺序氧化
subtle pool	隐蔽油气藏	successive reduction	逐次简化
subtle reservoir	隐蔽油气藏	successive row overrelaxation	逐行超松弛
subtle trap	隐蔽圈闭	successive sampling method	逐次取样法
subtractive color	减色	successive sedimentation	连续沉积
subtranslucent	微透明的	successive slides	逐级滑动
subtrusion	下层侵入	successive slip	渐进性滑动
subtype	亚类，亚型	successive steps	逐步
subunconformity	隐伏不整合	successive substitution	逐步取代法
subunconformity reservoir	隐伏不整合储集层	successive subtraction method	递减法
subunit	亚组；副族；亚单位；子群	successive sweep method	逐步淘汰法
subvariety	亚变种；从属品种	successive term	逐项
subvertical	陡倾的；近垂直的	successive trials	逐次试探
subvertical shaft	暗竖井	successor	继承人；继任者；后续事件；紧后工序
subvolcanic	浅成的	succinite	琥珀；钙铝榴石；琥珀色
subvolcanic deposit	地下火山矿床	sucker	吸管；吸入活塞；吸入器
subvolcanic rock	次火山岩	sucker pole	抽油杆
subvolcanism	地下火山作用	sucker rod	抽油杆
subvolcano	潜火山	sucker rod float	抽油杆浮子
subwater pipeline	水下管道	sucker rod guide	抽油杆导向器
subwatering	地下给水	sucker rod hanger	抽油杆悬挂器
subway	地下铁道，地铁；地下管道；地道	sucker rod hook	抽油杆吊钩
subway construction	地下道施工	sucker rod jack	抽油杆千斤顶
subway survey	地下铁道测量	sucker rod jar	抽油杆震击器
subway tunnel	地铁隧道	sucker rod joint	抽油杆接箍
subweathered zone	亚风化层	sucker rod load	抽油杆负荷
subweathering velocity	风化层底面速度	sucker rod pump	有杆泵
subzone	亚带；分区	sucker rod socket	抽油杆打捞筒
succeeding cycle	后期旋回	sucker rod spear	抽油杆打捞矛
successful bid	中标	sucker rod string	抽油杆柱
successful bidder	中标者	sucker rod substitute	抽油杆接头
successful stack	成功叠加	sucker rod tightening torque	抽油杆旋紧扭矩
succession	演替；连续；继承	sucker rod wrench	抽油杆钳
succession appearance zone	顺序出现带	sucker-rod coupling	抽油杆接箍
succession disappearance zone	顺序消失带	sucker-rod pumping	有杆泵抽油
succession of beds	层序，地层层序	sucking action	抽吸作用；吸入作用
succession of formation	建造层序	sucrosic texture	糖粒状结构
succession of strata	层序；地层层序	suction	抽汲；吸取；吸力；吸入
succession tectonics	继承性构造	suction anchor	吸力锚
successive adjustment	序贯平差	suction blast	逆爆破
successive approximation	逐次渐近法，逐步逼近法；迭代反演法	suction capacity	自吸能力
		suction draft	抽风；吸气通风；吸气通风；吸入送风
successive block overrelaxation	逐块超松弛	suction effect	吸入作用；虹吸作用；吸力效应
successive difference	递差；逐次差分	suction fan	抽风机；抽出式扇风机
successive differentiation	逐次微分	suction filter	吸滤器
successive displacement method	逐次替换法	suction force	吸力；负压力
successive elimination	逐次消示	suction gauge	吸力计；吸入压力表

suction gradient	抽吸力梯度
suction head	扬程；吸入水头；负压水头
suction lift	吸引高度；吸引升力
suction plate	负压力板
suction plate apparatus	负压力板仪
suction pore pressure	吸湿孔隙水压力
suction pressure	负压力；吸力
suction pressure membrane	吸湿压力薄膜
suction pump	吸入泵
suction specific speed	抽吸比速度
suction water barrel	抽水筒
suction well	抽汲井
sudburite	倍长苏玄岩
sudden change	突变
sudden collapse	突然塌陷；突然崩落
sudden commencement	急始
sudden commencement magnetic storm	急始磁暴
sudden displacement	突然位移
sudden drawdown	骤降；水位突降
sudden emission of gas	瓦斯突然涌出
sudden fall	突然崩落；突然坍塌
sudden flooding	突然涌水
sudden frequency deviation	频率急偏
sudden gas outburst	瓦斯突出
sudden impulse	急脉冲
sudden increase of ionization	电离突增
sudden ionospheric disturbance	突发电离层骚扰
sudden load	突加荷载
sudden occurrence	突发事件
sudden phase anomaly	突发相位异常
sudden pressure drop	突然压降
sudden release of energy	能量突然释放
sudden rupture	突然破裂
sudden settlement	突然沉陷
sudden transition	突变段
suddenly generated hazard	突发性灾害
sue	起诉
suevite	陨磺砾岩；冲积凝灰角砾岩
sufficiency of contract price	合同价格的充分性
sufficiency of tender	投标书的完备性
sufficient and necessary condition	充要条件
sufficient clay	黏泥
sufficient condition	充分条件
sufficient condition for convergence	收敛的充分条件
sufficient estimate	充分估计量
sufficient statistic	充分统计量
suffosion	潜蚀作用；管流现象
suffosion funnel	潜蚀漏斗
sugary dolomite	糖粒状白云岩
suit	控告；诉讼；适合
suitability	适应性
suitability number	适用系数
suitable formation	适合的地层
sulcate	具沟的
sulcus	中槽，沟，裂缝
suldenite	闪安岩
sulfate	硫酸盐
sulfate aerosol	硫酸盐气溶胶；硫酸盐悬浮微粒
sulfate reduction	硫酸盐还原
sulfatic cancrinite	硫钙霞石
sulfation	硫化作用；硫酸化
sulfatizing roasting	硫酸化焙烧
sulfidal	胶状硫
sulfidation	硫化作用
sulfidation corrosion	硫蚀
sulfide	硫化物
sulfide ore	硫化矿石
sulfide oxidation	硫化物氧化
sulfite	亚硫酸盐
sulfofication	硫化作用
sulfonate	磺化；磺酸盐
sulfonated lignite	磺化褐煤
sulfonating agent	磺化剂
sulfur capture	固硫
sulfur capture agent	固硫剂
sulfur compounds	含硫化合物
sulfur corrosion	硫腐蚀
sulfur determination	硫测定
sulfur isotope	硫同位素
sulfur removal	脱硫
sulfur sequestration	硫封存
sulfuric acid	硫酸
sulfurized organic matter	硫化有机物质
sulfurous acid	亚硫酸
sullage	污物；渣滓；淤泥；沉积的淤泥
sullage waste	沟渠废物
sullage water	淤泥水；污水
sulphate	硫酸盐
sulphate content	硫质含量，含硫量；硫酸盐含量
sulphate deposition	硫酸盐沉降
sulphate ion	硫酸离子
sulphate resisting Portland cement	耐硫酸盐水泥
sulphate-resistant concrete	耐硫酸盐混凝土
sulphating	硫酸盐化；硫酸酯化
sulphide	硫化物
sulphide concentrate	硫化矿精矿；硫精矿砂
sulphide inclusion	硫化物包体
sulphide ore	硫化矿
sulphide-vein deposit	硫化脉状矿床
sulphidization	硫化作用
sulpho-aluminous cement	硫酸铝水泥
sulphoborite	硫硼镁石；硼镁矾
sulphur	硫
sulphur coal	高硫煤
sulphur dioxide	二氧化硫
sulphur hydride	氢硫化合物
sulphur spring	含硫泉
sulphur-containing impurity	含硫杂质
sulphuric acid	硫酸
sum insured	投保金额
sum of squares	平方和
sum total	总计
sumacoite	安山碱玄岩；磁铁碱玄岩
sum-and-difference system	和差系统
Sumatera-Java island arc	苏门答腊-爪哇岛弧
summary	概要；摘要；总结
summary of earning	盈余总汇表
summary of operation	经营概况
summary of tender	投标报价汇总表
summation curve	累积曲线；累计曲线

summation of subdivided layer settlement 分层总和法
summation percentage 累计百分率
summed seismogram 叠加地震记录
summing method in layer 分层总和法
summit crater 山顶火山口
summit eruption 山顶喷发
summit level 峰顶面
summit plane 峰顶平面
sump 集水坑，污水坑，水坑，池；机油箱；挖深
sump cleaner 水仓清理工
sump cut 侧面楔形掏槽
sump gallery 排水平台，集水平台；排水平巷
sump hole 集水孔，集水窝；超前凿孔，超前掘孔；掏槽炮眼
sump packer 沉砂封隔器
sump pit 集水坑；排液槽
sump pump 潜水泵；深井泵；集水井泵
sump pump line 循环水泵管路
sump pumping 集水坑抽水
sump shaft 排水井
sump tank 污物沉淀池
sumper 掏槽炮眼
sun angle 太阳仰角；日照角
sun azimuth 太阳方位角
sun compass 太阳罗盘
sun crack 干裂
sun elevation 太阳仰角；太阳高度角
sun proton monitor 太阳质子监视仪
sun sensor 太阳方位传感器
sun-flower pattern 葵花状布井
sungulite 蠕状蛇纹石
sunk 沉没的；水底的；地中的；凹下去的
sunk caisson well foundation 沉井基础
sunk cost 沉没成本；旁置成本
sunk foundation 埋设基础
sunk land 沉陷地块
sunk shaft 沉井
sunk shaft foundation 沉井基础
sunk well foundation 沉井基础
sunken block 地堑
sunken caldera 陷落火山；陷落火山口；陷落火山石
sunken reef 暗礁
sunken tube 沉埋管段；沉管
sunken tube method 沉管法
sunken well 钻好的井
sunken-ship breakwater 沉船防波堤
sunny slope 阳坡；向阳面
sunrise and sunset transition 日出与日落过渡期
sunshine deformation survey 日照变形测量
sunstone 猫睛石；日长石；太阳石
super conducting magnetometer 超导磁力仪
super conductive gravimeter 超导重力仪
super elevation 超高
super entropy 超熵
super face 高产工作面
super profit 超额利润
super pure 超纯净的
super pure coal 超纯净煤
super stratum 上覆层
superactivity 超活性

superannuated building 老朽建筑
superanthracite 高煤化无烟煤
superbituminous coal 半烟煤
supercapillary 超毛细管；超毛细现象
supercapillary interstice 超毛细孔隙
supercapillary percolation 超毛细渗透作用；超毛细管渗透
supercapillary porosity 超毛细管孔隙率
supercapillary seepage 超毛细管渗流
supercapital 副柱头
supercavitating flow 超空化流；超空蚀水流
supercavitating hydrofoil 超空化水翼
supercharge 过载；过量装药；增压
supercharge load 超载荷重
supercharge storage 超高库容
supercharger 增压器
super-class gas tunnel 超级瓦斯隧道
supercompressibility 超压缩性
superconductivity 超导电性
superconductor 超导体
supercontinent 超大陆
supercooled soil 过冷土
supercooling 过冷
supercritical 超临界的
supercritical adsorption 超临界吸附
supercritical carbon dioxide 超临界二氧化碳
supercritical CO_2 超临界二氧化碳
supercritical extraction 超临界开采；超临界萃取；超临界抽提
supercritical flow 超临界流
supercritical fluid 超临界流体
supercritical speed 超临界速度
supercritical state 超临界状态
supercrop trap 隐覆露头圈闭
superdeep drillhole 超深钻孔
superdeep drilling 超深度钻井
super-elasticity 超弹性
superelevated curve 超高曲线
superelevation 超高
superelevation runoff 超高缓和段
superface 顶面
superficial 表面的；地面的；浅薄的
superficial area 表面积
superficial coat 表层
superficial compaction 表层压实
superficial constituent 表生组分
superficial crust 表层地壳
superficial degradation 表层剥蚀
superficial deposit 表层沉积；表土沉积物
superficial deterioration 表面剥蚀
superficial expansion 表面膨胀
superficial fault 盖层断裂
superficial fire 地面火灾
superficial fluid velocity 表面流体速度
superficial fold 表层褶皱
superficial landslide 浅层滑坡
superficial layer 表层
superficial Rockwell hardness tester 轻压力洛氏硬度试验机
superficial sliding 表层滑动

superficial stratum	表土层
superficial superflood	特大洪水
superficial tectonics	表层构造
superficial tension	表面张力
superficial water	地表水
superfine	超细的；超细粉末
superfluid	超流体；超流动的
superfluous water	过剩水分
superfractionation	超精馏
supergene	表生的，浅成生的；浅成矿床
supergene deposit	浅成矿床
supergene enrichment	表生富集，浅生富集
supergene migration	表生迁移
supergene structure	表生构造
supergenesis	表生成因；表生作用
supergiant	超大的
superglacial moraine	表碛石，表碛
superglacial till	表碛
supergradient	超梯度
supergradient wind	超梯度风
supergravity	超重力，超引力
superhard	超硬度的
superheat	过热
superheavy	超重的
superheavy dynamic sounding	超重型动力触探
superhigh frequency	超高频
super-huge landslide	特大型滑坡
superhypabyssal rock	超浅成岩
superimpose	重叠
superimposed anomaly	叠加异常
superimposed arch	层叠拱
superimposed basin	重叠盆地，叠合盆地
superimposed bed	上覆层，覆盖层
superimposed dead load	叠加净荷载；超静载
superimposed deformation	叠加变形
superimposed drainage	重叠排水系统；叠置水系；重叠水系
superimposed flow	叠流
superimposed fold	横褶皱；叠加褶皱；交错褶皱
superimposed load	附加荷载；超载
superimposed metamorphism	叠加变质作用
superimposed penetrative structure	叠加穿插构造
superimposed profile	重叠剖面图
superimposed reservoir	叠置储集层
superimposed ripple-mark	叠加波痕，叠置波痕
superimposed seam	重叠煤层
superimposed soil	上覆土
superimposed strain	叠加应变
superimposed stress	附加应力；叠加应力
superimposed subsequent stream	叠置后生河
superimposed tectonic basin	叠加构造盆地
superimposed tectonic deformation	叠架构造变形
superimposed valley	叠置谷
superincumbent	上覆的
superincumbent bed	上覆层；覆盖层
superindividual	超单晶；超个体的
superior limit	上限，最大限度
superiority	优势；优越性
superjacent	上覆的
superjacent bed	覆盖层；顶板；悬帮
superjacent stagnant water	上层滞水
superlattice	超点阵
superload	超载；附加荷载
super-long source	超长震源
supermarine	海面的
supermature	发育好的；过成熟的
supermicroanalysis	超微量分析
super-microscopic paleontology	超微古生物学
superminiature	超小型的
supernatant	表层漂浮物；上层清液
supernormal pressure	超常压力
superperformance	超级性能，良好性能
superpermafrost water	冻结层上水；永久冻土上的水
superplasticity	超塑性
superplasticizer	超塑化剂，超增塑剂，强塑剂；高效减水剂
superposed	上叠的，叠加的，叠覆的
superposed basin	上叠盆地，叠置盆地
superposed beam	叠合梁
superposed drainage	重叠排水系统；叠置水系；重叠水系
superposed field	叠加场
superposed fluid	叠加流体
superposed fold	叠加褶皱；重叠褶皱
superposed river	原生河
superposed tectonics	上叠构造
superposed-epoch analysis	时间叠加分析
superposition	叠加
superposition of stress system	应力系统叠加
superposition principle	叠加原理
superposition specific heat	叠加比热
superposition theorem	叠加原理
superpower	超功率
superpressured gas reservoir	超压气层
supersaturated	过饱和的
supersaturated liquid	过饱和溶液
supersaturated solid solution	过饱和固溶体
supersaturated solution	过饱和溶液
super-saturation	过饱和
superscript	上标；标在上方的
supersede	替代，取代；比…优先
supersequence	上覆层序，叠覆层序，重复层序；超序列
supersession	代替，取代
supersonic	超声波的
supersonic crack detector	超声波探伤仪
supersonic echo sounder	超声波测深仪
supersonic generator	超声波发生器
supersonic inspection	超声波检查；超声波探伤法
supersonic reflectoscope	超声波反射探伤仪
supersonic sounding	超声波测深法
supersonic test	超声波检验
supersonic thickness meter	超声波测厚仪
supersonic wave drill	超声波钻井
supersound	超声波
superstage	超阶
superstratum	覆盖层，上覆层
superstructure	上部结构；上部构造
supersulphated cement	富硫酸盐水泥；高硫酸盐水泥
supertension	超高压；过应力
superterranean	地表上的

superthermal	超热的
superturbulent flow	超紊流；超湍流
supervise	管理，监督，监管，督导
supervised classification	监督分类法
supervising engineer	监理工程师
supervision charge	监督费
supervision of construction	工程监理
supervision of project	工程监理
supervision plan	监工计划书；监理规划
supervision scheme	监管计划；监理方案
supervisory board	监督委员会，监事会
supervisory control	监控
supervoltage	超高压
super-wide-angle projection	超广角投影
supplement angle	补角
supplemental contract	补充合同
supplemental control point	补充控制点
supplemental hoisting	辅助提升
supplemental recovery efficiency	附加采收率
supplementary bar	辅助钢筋
supplementary blasting	二次爆破，补充爆破
supplementary budget	追加预算；补充预算
supplementary contour	半距等高线；补充等高线
supplementary exploration	补充勘探
supplementary instrument	辅助仪器
supplementary means	辅助手段
supplementary pressure	附加压力
supplementary provisions	附则
supplementary tax	附加税
supplementary unit	辅助单位
supplementary water injection	补充注水
supply air fan	供气风机
supply air temperature difference	送风温差
supply channel	引水渠；供水渠
supply failure	供应中断
supply gate	运料巷道
supply line	供给线；进给线
supply provenance	供给区
supply raise	运料天井
supply track	运料轨道
supply tunnel	供水隧洞
supply-cage compartment	井筒运料罐笼隔间
support	支护；支撑；支架
support advance sequence	支撑推进工序
support arch	拱形支架
support backfilling	架后充填
support bar	支承杆，支撑杆；架立筋
support capacity	支护能力，支撑能力
support characteristics	支架特性
support column	门腿；支撑管，支撑柱
support constraint	支承约束
support density	支架密度
support design	支护设计，支撑设计，支座设计
support displacement	支座位移
support element	支架构件；支架部件
support frame	框式支架；支撑架；支架
support lag	支护滞后
support load bearing capacity	支架负荷能力；支架承载能力
support mechanisms	支护机理
support moment	支承力矩，支座力矩，支点力矩
support optimization design	支护优化设计
support parameter optimization	支护参数优化
support pattern	支护方式；支护型式
support pillar	支承矿柱；支撑柱
support plate	底板
support pressure	支护压力；支承压力；支点压力
support principle	支护原则
support reaction	支承反力；支点反力
support regulations	支护规程
support resistance	支架阻力；支护阻力
support ring	环形支架；支撑环
support safety factor	支护安全系数
support setting	支架安设
support settlement	支座沉降；支点沉陷
support shield	盾构支护
support stiffness	支护刚度
support system	支护体系；支护系统
support technology	支护技术
support to stabilize wall	固壁支撑
support unit	支护系统；支护设备
support withdrawing	支架回收
supporter	支架
supporting and blocking	支挡
supporting course	持力层
supporting force	支承力；支承反力
supporting frame	支承构架；承重构架
supporting layer	持力层；受压层
supporting performance	支护性能
supporting platform	底座；支承平台
supporting protecting engineering	支护工程
supporting rigidity	支护刚度
supporting ring	垫环
supporting surface	支承面
supposed fault	推测断层
supposition	想象；推测；假定
suppressant	抑制剂；灭火剂
suppressed contraction	不完全收缩
suppressed layer	被压制层
supraconductivity	超导电性
supracrustal	上地壳的；地表的
supracrustal erosion	上地壳侵蚀
supracrustal formation	上地壳建造
suprafan	上叠扇；扇谷下端隆起带
supralittoral zone	潮上带
supraore halo	矿上晕；前缘晕；远矿晕
suprapermafrost layer	永冻层上土层
suprapermafrost water	永冻层上水
supratenuous fold	同沉积褶皱；薄顶褶皱
supratidal	潮上的
supratidal deposit	潮上沉积
supratidal flat	潮上坪
supra-unconformity	叠覆不整合
supremum	上确界；最小上界
surcharge	过载，超载；超高水深；洪水超高
surcharge preloading	超载预压
surety	保证人，担保人；保证；抵押物
surety bond	保证书；担保书；担保债券
suretyship	担保人身份
surf zone	破浪带；碎波带

surface	表面；表面积
surface abrasion	表面磨蚀
surface action	表面作用
surface active	表面活性的
surface active agent	表面活性剂
surface active substance	表面活性物质
surface activity	表面活性；表面活度
surface adsorption	表面吸附
surface air temperature	地面气温
surface anomaly	地表异常
surface area of mineral	矿物的表面积
surface area of particle	颗粒表面积
surface avalanche	表层雪崩
surface bedrock	地表基岩；露头基岩
surface behaviour	表面作用；表面性质
surface blasting	露天爆破；表面爆破
surface blemish	表面缺陷
surface boundary condition	地表边界条件
surface break	地面断裂，地面陷落
surface breakage	地表断裂
surface breakthrough	贯通地面的巷道
surface brightness	表面亮度
surface casing	表层套管；孔口套管，第一节套管
surface cauldron subsidence	地表环状沉陷
surface chaining	井口罐座；井口罐托
surface channel	地表排水沟；排水明渠
surface charge	采矿地面费用；矿山地面费用
surface check	地面裂缝；表面裂缝
surface coal mine	露天煤矿
surface coal mining	露天开采
surface collapse	地面塌陷
surface compaction	表面压实；表层压密
surface concentration	表面浓度；表面冷凝
surface condensation	表面凝结
surface condenser	表面冷凝机
surface configuration	地表形态
surface construction	地面建筑
surface contact	面接触，表面接触
surface correction	地表校正
surface course	面层
surface cover	护面
surface crack	表面裂纹
surface crater	露天采矿漏斗
surface creep	表层蠕动
surface curvature	地表曲率；表面曲率；表面弯曲
surface damage	地面破坏
surface decarbonization	表面脱碳
surface defect	表面缺陷
surface deformation	表面变形；地表变形
surface density	面密度；地表密度
surface deposit	地表沉积；浅藏矿床
surface depression	地面凹陷，地表下陷
surface detention	地表滞留；地面滞流；积涝
surface diffusion	表面扩散
surface displacement vector	表层位移矢量
surface doming	地表隆起
surface drain	明沟
surface drainage	地表排水
surface drainage channel	地表排水渠，排水明渠
surface drainage system	地表排水系统；地表水系
surface drainage water	地面排水
surface eddy	表面旋涡
surface elastic wave	弹性面波
surface elevation	地面高程
surface energy	表面能
surface erosion	地表冲刷，表面侵蚀；水土流失
surface evaporation	地表蒸发
surface evenness	路面平整度
surface exposure	地面裸露
surface fault	地表断层
surface fault trace	表面断层迹线
surface faulting	表面断层作用
surface feature	表面特征
surface finish	表面抛光，表面光洁度
surface fitting	曲面拟合；曲面拟合法
surface float	水面浮标
surface flooding	地面泛流
surface flooding irrigation	地表漫灌
surface flow	地表径流
surface focus	表面震源
surface folding	表层褶皱；表面褶皱
surface footing	明置基础
surface for fitting	内插曲面
surface force	表面力，面力
surface foreman	地面作业班长
surface free-energy	表面自由能
surface friction	表面摩擦
surface frost heave	路面冻胀
surface geology	地表地质学
surface geothermal manifestation	地表地热显示
surface gloss	表面光泽度
surface gradient	表面坡度
surface gravel pack	地面倾倒式砾石充填
surface gravity anomaly	表面重力异常
surface grouting	接触灌浆
surface hardening	表面硬化处理；表面硬化
surface hardness	表面硬度
surface heat flow	地表热流
surface hodograph	时距曲面
surface horizon	地表土层
surface ice	表层冰
surface impedance	表面阻抗
surface imperfection	表面缺陷
surface infiltration	地面入渗
surface injection pressure	地面注入压力
surface installation	地表构筑物；地面装置；地面设施
surface integral	面积分，球面积分
surface interference	地表异常
surface irregularity	地面起伏
surface layer	表层；地表
surface layout	地面布置
surface leakage	地面漏风；矿井外部漏风；表面漏泄
surface level	表层；地表水准
surface lift	地面隆起
surface load	表面荷载
surface magnitude	面波震级
surface manifestation	地表显示
surface mapping	地表填图
surface mark	地面标志
surface measuring equipment	表面粗糙度测量仪

surface mine	露天矿	surface roughness	表面粗糙度
surface mine capacity	露天矿规模	surface runoff	地表径流；地表水流量
surface mine design	露天矿设计	surface rupture	地表破坏，表面破裂
surface mine development	露天矿开拓	surface S wave	横波型面波
surface mine life	露天矿服务年限	surface sample	地表样品
surface mine survey	露天矿测量	surface sampler	表层采样器
surface mine ventilation	露天矿通风	surface scattering	表面散射
surface mine water	露天矿坑水	surface sealing	表面密封
surface miner	露天矿矿工；露天采矿机	surface sediments	表层沉积物
surface mining	露天开采；露天采矿学	surface seepage	地表渗出
surface modification	表面改性	surface seepage of gas	气体表面渗流
surface moisture	表面水分	surface seismic method	地面地震法
surface moraine	表碛石	surface settlement	地表沉陷
surface morphology	表面形态	surface shear viscosity	表面剪切黏度
surface normal	曲面法线	surface sliding	表层滑移
surface nullah	明渠	surface slope	地表倾斜；地面坡度，表面坡度
surface observation	地表观测	surface soil	表土
surface occurrence	地表显示；地面产状；地表露头	surface soil stabilization	浅层土加固
surface of contact	接触面	surface sound pressure level	表面声压级
surface of deposition	沉积面	surface source	面源；地面震源
surface of discontinuity	不连续面；间断面；突变面	surface specific adsorption	比表面吸附
surface of equipressure	等压面	surface spreading	表面扩张
surface of reference ellipsoid	参考椭球面	surface spring	泉水；地面泉
surface of revolution	旋转曲面	surface stabilizer	土面坚固剂
surface of seepage	渗出面	surface storage	地面蓄水；地表蓄水
surface of shear	剪切面	surface storm flow	地面暴雨径流
surface of sliding	滑动面	surface strain indicator	表面应变仪
surface of stability	稳定面	surface stratum	表层
surface of the rupture	破裂面	surface stripping	条状表土剥露；剥槽；地表剥离
surface of unconformity	不整合面	surface structure	表层构造
surface of underground water	地下水面	surface structure contour map	表层构造图
surface of weakness	软弱面	surface subsidence	地面沉降
surface oscillation	表面振荡	surface survey	地面测量
surface outcropping	地表露头	surface temperature anomaly	地表温度异常
surface permeameter	路面透水度测定仪	surface tension	表面张力
surface phenomena	地表现象	surface tension of moisture film	水薄膜的表面张力
surface pitting	表面点蚀；表面凹坑	surface texture	表面纹理；表面结构
surface plan	地面平面图	surface thrust	侵蚀冲断层
surface polarization	表面极化	surface tilt	表面倾斜
surface pore	表面孔隙	surface topography	表面形貌；表面形貌学
surface pressure	表面压力；地面气压；井口压力	surface trace	地面形迹；地表显示
surface production system	地面生产系统	surface traction	表面引力
surface profile	表面轮廓	surface treatment	表面处理
surface property	表面性质	surface undulation	地面起伏；地表起伏；水面起伏
surface protection	护面，表面防护	surface uplift	地面抬升
surface radiation	表面辐射	surface vertical well	表面垂直井
surface reading	地面读数	surface vibration	地表振动
surface receiver	地面接收器；水面接收器	surface viscometer	表面黏度计
surface reconstruction	表面重构	surface viscosity	表面黏性；表面黏度
surface recorder	地面记录仪	surface visibility	地面能见度
surface reference datum	地面参考基准面	surface wash	片蚀
surface reflection	表面反射	surface wash off	地表冲刷
surface relief	地面起伏，地表起伏；地形，地势	surface water	地表水，地面水
surface relief method	表面应力解除法	surface water channel	地面水渠道；排水明渠
surface resistivity	表面电阻率	surface water drain	地面排水渠
surface retention	地表滞水	surface water leakage	地表水漏失
surface reverberation	表面交混回响	surface water recharge	地表水补给
surface rheology	表面流变学	surface water supply	地面给水；地面水补给
surface rock	地表岩石	surface watershed	地面分水岭
surface rolling measure	地表碾压措施	surface wave	面波，表面波

surface wave effect	面波效应
surface wave magnitude	面波震级
surface weathering	表层风化
surface wetting characteristic	表面润湿特性
surface wiring	明线
surface working	露天作业；露天开采场
surface-creep transport	推移搬运
surface-curvature apparatus	路面曲率半径测定仪
surface-guided wave	表面波；导波
surfaceness	表面粗糙度
surface-tension equilibrium	表面张力平衡
surface-to-surface contact	面-面接触
surface-to-volume ratio	比表面积
surface-truth field study	地面实况野外研究
surface-water diversion	地面水引流
surface-water resource	地表水资源
surfacing	铺筑路面；表面修整；平面切削
surfacing cut	表面切削
surfacing design	路面设计
surfacing foundation	路基
surfacing laying	路面铺筑
surfacing restoration	路面修复
surfactant	表面活化剂
surfactant activity	表面活性剂效能
surfactant additive	表面活性添加剂
surfactant mud	活性剂泥浆
surfactivity	表面活性
surficial	地表的；地面的
surficial belt	地表带
surficial creep	表层蠕动
surficial deposit	表层沉积
surficial geology	地表地质学
surficial sediment	表层沉积物
surficial soil	表土
surficial water	地表水
surf-shaken beach	强浪蚀滩
surge bunker	缓冲仓；调节仓
surge capacity	缓冲容量，贮存能力
surge damper	减震器；缓冲器
surge diverter	浪涌分流器；巨波转换器；避雷针；冲击转向器
surge effect	涌浪效应
surge generator	脉冲发生器；冲击发生器
surge impedance	波阻抗
surge peak head	峰值负荷
surge pile	储料堆
surge pipe	平压竖管
surge pressure	涌浪压力；峰值压力；冲击压力
surge reliever	水击泄压阀；减压装置
surge time	缓冲时间
surge wave	涌波，涌浪
surge-type	急流型
surgical filtering	切除滤波
surging breaker	激散破波
surging force	脉冲力
surging water	潮水
surging well	间歇自喷井
surmisable	可推测的
surmounted arch	超半圆拱
surplus driven force	剩余下滑力

surplus sliding force	剩余下滑力
surplus sludge	过剩污泥
surplus valve	溢流阀
surrounding bed	围岩层
surrounding rock	围岩
surrounding rock classification	围岩分类
surrounding rock control	围岩控制
surrounding rock deformation	围岩变形
surrounding rock failure	围岩破坏
surrounding rock loosing zone	围岩松动区
surrounding rock stability	围岩稳定性
surrounding rock stress	围岩应力
surroundings	环境
sursassite	红帘石；锰帘石
surtax	附加税
surveillance	监视
surveillance network	观测台网
surveillance system	观测系统
survey	测量；勘测；调查
survey adjustment	测量平差
survey beacon	测量标志
survey by aerial photograph	航空摄影测量
survey control	测量控制
survey country	测量地区，地质调查地区
survey grid	控制网格；测量格网
survey gyroscope	陀螺经纬仪
survey in mining panel	采区测量
survey instrument	测量仪器
survey launch	测量艇
survey line	测线
survey mark	测量标志；测量点
survey marker	觇标；方位标
survey monument	测量标桩，测量标石
survey party	测量队，勘测队，勘探队
survey peg	测量标桩
survey photography	地形测图摄影
survey plan	测量图；调查方案
survey point	测点
survey record plan	测量记录图
survey report	检验报告
survey sheet	测量图
survey signal	测量觇标，测量标志
survey stake	测量木桩
survey station	测点；测站
survey steering tool	测斜仪
surveyed	已测的
surveyed area	测区
surveying	测绘；勘察；测量
surveying accuracy	勘探精度
surveying altimeter	测高计
surveying and mapping	测绘学；测绘科学与技术
surveying arrow	测针；测钎
surveying camera	测量摄影机；测量相机
surveying collar	测量用钻铤
surveying compass	测量罗盘
surveying control network	测量控制网
surveying for geological mapping	地质图测绘
surveying panel	平板仪
surveying reference point	测量参考点；测量基准点
surveying rod	测杆；测量标尺；标杆

survey-net adjuster 测网调整器
surveyor's level 水准仪
surveyor's pole 测量标杆
surveyor's rod 标尺
surveyor's table 平板仪
surveyor's transit 经纬仪
survival probability 幸存概率
susannite 硫碳酸铅矿
susceptibility 灵敏度；磁化率
susceptibility to frost action 易冻结性
suspend talks 中止谈判
suspended 悬浮的
suspended beam 挂梁
suspended cable structure 悬索结构
suspended current 浊流
suspended deck structure 承台结构；支撑面板结构
suspended frame weir 活动式悬架堰
suspended load 悬浮泥沙；悬浮载荷
suspended material 悬浮物质
suspended mixture 悬浮混合物
suspended particle 悬浮颗粒
suspended particulate 悬浮粒子
suspended railway 高架铁路
suspended sediment 悬浮沉积物，悬浮沉淀物；悬浮泥沙
suspended solid 悬浮固体
suspended solid discharge standard 悬浮固体排放标准
suspended solid material 悬浮固体
suspended spring 悬挂泉；季节泉
suspended state 悬浮态
suspended structure 悬挂结构
suspended support 悬吊支承
suspended system 悬吊体系
suspended truss 悬式桁架；悬式构架
suspended valley 悬谷
suspended velocity 悬浮速度
suspended water 包气带水；上层滞水
suspended-bed load ratio 悬浮物与推移质比值
suspender 吊杆，悬吊带
suspending agent 悬浮剂
suspending coal dust test 悬浮煤尘试验
suspending weight 悬重
suspensate 悬移质，悬浮质
suspense account 暂记账户；待清理账户
suspension 悬浮液，悬浮，悬移
suspension box cofferdam 吊箱围堰；浮箱式围堰
suspension bridge 悬索桥
suspension cable 吊索；悬索
suspension current 浊流；悬浮流；悬移质流
suspension girder 悬梁
suspension grout 悬液浆；悬液灌浆
suspension of contract 中止合同
suspension of cuttings 岩屑悬浮
suspension of disbursement 暂停拨款
suspension of payment 暂停付款
suspension property 悬浮性能
suspension pulley 悬吊滑轮
suspension transport 悬浮搬运；悬移
suspensoid 悬浮质；悬胶体
sussexite 白硼镁锰矿；硼锰矿

sustainability 可持续性
sustainable development 可持续发展
sustainable energy 可持续能源
sustained acceleration 持续加速度
sustained load 恒载；持续负荷；长期负载
sustained loading procedure 持续加荷方法
sustained rock strength 岩石长期强度
sustained stress 持续应力
sustained yield 持续涌水量；持续开采量
sustainer 支点
sustaining wall 扶墙
suttle weight 净重
sutural texture 缝合结构
suture contact 缝合接触
suture line 缝合线
suture line of fault block 断块缝合线
suture orogen belt 接合造山带
suture zone 缝合带
svabite 砷灰石
svanbergite 硫磷铝锶矿
Svecofennian 瑞典芬兰系
Svecofennide belt 瑞典-芬兰造山带
swag 顶板下沉；挠度；垂度
swaging process 型锻方法；冲压方法；锻造方法
swale 洼地；沼泽地
swale fill deposit 滩槽沉积
swallet 落水洞；灰岩渗水坑，石缝，矿井涌水；伏流；落渗河
swallet hole 落水洞；陷阱；斗淋
swallet river 暗河；地下河
swallet stream 伏流；暗流
swallow 落水洞；溶岩洞
swallow hole 石灰坑；灰岩坑；落水洞；陷落坑
swallow hole drainage pattern 落水洞排水形式
swamp 沼泽；泥沼；渍水沼泽
swamp community 沼泽群落
swamp deposit 沼泽沉积物
swamp ditch 沼泽排水沟
swamp drainage 渍水沼泽排水
swamp facies 沼泽相
swamp formation 沼泽化；沼泽形成〔作用〕
swamp gas 沼气
swamp lake 沼泽湖
swamp muck 沼地泥炭；腐殖土
swamp seismic exploration 沼泽地震勘探
swamp soil 渍水沼泽土；沼泽土
swamp stream 沼泽河
swamp vegetation 沼泽植被
swampiness 沼泽化
swamping 沼泽化
swamping lagoon facies 沼泽化泻湖相
swamping of soil 土壤沼泽化
swampland 沼泽地
swampy 多沼泽的
swampy basin 沼泽盆地
swampy ground 沼泽地；低湿地
swampy soil 沼泽土
swarm activity 震群活动
swarm earthquake 群发地震，震群，群震
swarm of earthquake 地震群

swarm of shock 地震群
swarm sequence 震群序列
swarm-type 震群型
swash 冲流，冲溅；冲流水道；浅滩；冲刷滩
swash bar 冲流沙坝
swash cross-bedding 冲洗交错层理；海滩加积层理
swash height 波浪爬高
swash mark 波痕；冲痕
swash width 扫描行距
swash zone 冲刷带；破浪带
S-wave anisotropy S波各向异性；横波各向异性
S-wave splitting analysis S波分裂分析法；横波分裂分析法
sway angle 摇摆角度
sway brace 竖向支撑
sway bracing 斜向支撑
sweating action 浸润作用
Swedish circle method 瑞典圆弧法，瑞典圆弧滑动法
Swedish cone penetration test 瑞典触探试验
Swedish fall-cone method 瑞典落锥法
Swedish method 瑞典法；条分法
Swedish mining compass 瑞典式矿山罗盘
Swedish slip circle method 瑞典滑弧法
Swedish sounding 瑞典式触探
Swedish timber set 瑞典式框式支架
sweep away 冲走
sweep frequency radar 扫描频率雷达
sweeping survey 扫测
sweep-out 清除；清扫
sweet soil 非酸性土
sweet water 饮用水
sweetening of soil 土壤中和
swell 涌浪；膨胀；隆起
swell factor 膨胀系数；松胀系数；泡胀系数；岩石碎胀系数
swell index 膨胀指数；湿胀指数
swell potential 膨胀势
swell pressure 膨胀压力
swell process 膨胀过程
swell ratio 膨胀比
swell soil 膨胀土壤
swell value 膨胀值
swellability 可膨胀性
swellable clay 膨胀性黏土
swellable formation 可膨胀地层
swell-and-swale topography 波状碛原地形
swelled dowel anchor 胀销式锚杆
swelled dowel system 胀壳式锚栓系统
swelled ground 膨胀土
swelled mudstone 膨胀泥岩
swelling 涌浪；膨胀；隆起
swelling agent 膨胀剂
swelling amount 膨胀量
swelling capacity 膨胀能力；膨胀量
swelling capacity of coal 煤膨润度
swelling capacity test 膨胀量试验
swelling chlorite 膨胀绿泥石
swelling clay 膨胀性黏土
swelling coal 膨胀煤
swelling coefficient 膨胀系数
swelling curve 膨胀曲线
swelling dome 胀裂丘
swelling earth pressure 膨胀性土压力
swelling fissure 膨胀裂缝
swelling force 膨胀力
swelling force test 膨胀力试验
swelling ground 冻胀土；膨胀性地层
swelling index 回弹指数；膨胀指数
swelling limit 胀限；膨胀限度
swelling moisture content 膨胀含水率
swelling number 膨胀数
swelling of floor 地鼓
swelling of shale 页岩膨胀
swelling out of surrounding rock 膨胀内鼓
swelling percentage 膨胀百分数
swelling potential 膨胀势
swelling power 膨胀力
swelling pressure 膨胀压力
swelling property 膨胀性；湿胀性
swelling rate 膨胀速度
swelling rate test 膨胀率试验
swelling ratio 膨胀率
swelling rock 鼓胀性岩石
swelling shale 膨胀性页岩
swelling soil 膨胀土
swelling stage 膨胀阶段
swelling strain 膨胀应变
swelling strain index 膨胀应变指数；膨胀应变指标
swelling stress 膨胀应力
swelling study 膨胀性研究
swelling test 膨胀试验
swelling value 膨胀值
swelling-shrinkage characteristics 胀缩特征
swelling-shrinkage of coal 煤的膨胀-收缩，煤的胀缩特征
swelling-shrinkage test result 胀缩性试验统计表
swell-shrink characteristics 胀缩特性
swell-shrinking behaviour 胀缩性状
swell-shrinking soil 膨缩土
swept reservoir volume 储层波及体积
swing arm type TBM 摇臂式掘进机
swing crushing 振动破碎
swing excavator 回转式挖掘机
swing hammer crusher 摆锤式破碎机；旋锤式破碎机
swing sledge mill 摆动锤碎机
swing speed 转动速度
swinging feeder 摇动式给矿机；摆动给料器
swinging gear 回转机构
swinging screen 摇动筛；摆动筛
switch-back curve 回头曲线
swivel joint 转环接合
swiveling drill 旋转式钻机
swiveling drill machine 旋转式钻机
swiveling pile driver 旋转打桩机
swollen soil 隆胀土；冻胀土
syderolite 陶土
syenite 正长岩
symbol 符号
symbol designation 符号标志
symbol for geological map 地质图的图例

symbol for rocks and soil 岩石和土的符号
symbol library 符号库
symbol map 符号图；字符图；花样图
symmetric 相称性的，均衡的
symmetric anomaly 对称异常
symmetric anticline 对称背斜
symmetric array 对称组合
symmetric barycentre 对称重心
symmetric bedding 对称层理
symmetric building 对称建筑
symmetric Cartesian stress tensor 对称笛卡尔应力张量
symmetric closed system 对称封闭系统
symmetric coordinate 对称坐标
symmetric coupling form 对称耦合形式
symmetric eigenproblem 对称特征问题
symmetric figure 对称形状；对称图形
symmetric form 对称型
symmetric free motion method 对称式自由运动法
symmetric fuzzy relation 对称模糊关系
symmetric half-controlled bridge rectifier 对称半控桥式整流器
symmetric idempotent matrix 对称幂等矩阵
symmetric line 对称线
symmetric offset spread 对称排列
symmetric operator 对称算子
symmetric parameter 对称参数
symmetric profile method 对称剖面法
symmetric section 对称剖面
symmetric tensor 对称张量
symmetrical 对称的
symmetrical amplifier 对称放大器
symmetrical antenna 对称天线
symmetrical anticline 对称背斜
symmetrical axis 对称轴
symmetrical balance 对称平衡
symmetrical banded structure 对称带状构造
symmetrical banding 对称条带
symmetrical bedding 对称层理
symmetrical body 对称体
symmetrical curve 对称曲线
symmetrical determinant 对称行列式
symmetrical difference scheme 对称差分格式
symmetrical distribution 对称分布
symmetrical double turnout 复式对称道岔
symmetrical equations 对称方程组
symmetrical fast Fourier transformation 对称快速傅里叶变换
symmetrical fold 对称褶皱
symmetrical limiting 对称限幅
symmetrical load 对称荷载
symmetrical loading 对称加载
symmetrical matrix 对称矩阵
symmetrical mode 对称振型
symmetrical profiling 对称剖面法
symmetrical ripple mark 对称波痕
symmetrical scattering 对称散射
symmetrical slope 对称边坡
symmetrical state 对称状态
symmetrical stress system 对称应力系统
symmetrical structure 对称结构

symmetrical syncline 对称向斜
symmetrical vein 对称矿脉
symmetrical vibration 对称振动
symmetry argument 对称分析
symmetry axis 对称轴
symmetry axis of rotation 旋转对称轴
symmetry boundary condition 对称边界条件
symmetry direction 对称轴方向
symmetry element 对称要素；并组单元
symmetry plane 对称面
symmetry principle 对称原理
symmetry type 对称型
sympathetic detonation 感应起爆
sympathetic fault 同步断层
sympathetic vibration 共振
symphitic deposit 同生矿床
symposium 学术讨论会；论文集；论丛
symptomatic 征兆的；症状的
symptomatic mineral 特征矿物
synaeresis 脱水收缩作用
synaeresis crack 收缩裂隙
synchronal 同步的；同时的
synchrone 同时性
synchronic 同时的
synchronism 同步性；同步；同时性
synchronization 同步
synchronized construction 同步建设
synchronized control 同步控制
synchronized operation 同步运转
synchronous 同时的；同步的
synchronous deposit 同时沉积；同期沉积
synchronous detection 同步检测
synchronous Earth-observing satellite 同步地球观测卫星
synchronous error 同步误差
synchronous pile 成孔灌注同步桩
synchronous processing 同步处理
synchronous satellite navigation system 同步卫星导航系统
synchronous speed 同步速度
synchronous transmission 同步传输
synchronous vibration 同步振动
synchrony 同时性
synclinal axis 向斜轴
synclinal basin 向斜盆地
synclinal bend 向斜槽
synclinal fault 向斜断层
synclinal flexure 向斜挠曲
synclinal fold 向斜褶曲
synclinal limb 向斜翼
synclinal mountain 向斜山
synclinal oil pool 向斜油藏
synclinal ridge 向斜山脊
synclinal river 向斜河流
synclinal spring 向斜泉
synclinal stratum 向斜层
synclinal stream 向斜河流
synclinal structure 向斜构造
synclinal trap 向斜圈闭
synclinal valley 向斜谷
synclinal zone 向斜带

syncline 向斜；向斜构造
syncline diagenesis 同生成岩作用
syncline slope 向斜翼
syn-compressional subsidence 同步压缩沉降
syndeformational 同变形的；变形同期的
syndeposition 同期沉积作用；同生沉积作用
syndepositional 同生沉积的；沉积同期的；与沉积同时的
syndepositional deformation 同沉积变形
syndepositional fold 同沉积褶皱
syndepositional slump structure 同生滑移构造
syndepositional structure 同沉积构造
syndiagenesis 同生成岩作用；早期成岩作用
syneresis crack 脱水收缩裂隙
synergetics 协同学
synergistic effect 协同作用，协同效应
synergistic evaluation 协作评价
syngenetic structure 同生构造
synisotropization 共同均质化
synoptic diagram 示意图；概要图
synorogenesis 同造山期运动
synorogenic metamorphism 褶皱期变质作用
synorogenic period 同造山期
synsedimentary fault 同沉积断层
synsedimentary structure 同沉积构造
synsedimentary tectonics 同沉积构造
synsomatic texture 原生结构
synthem 构造层
synthesis 综合
synthesis database 综合数据库
synthesis method 综合法
synthesis wave 合成波
synthesizer 综合器；合成器；综合程序
synthetic accelerogram 合成加速度图；人造加速度图
synthetic aperture 合成孔径
synthetic aperture radar 合成孔径雷达
synthetic aperture side looking radar 合成孔径侧视雷达
synthetic aperture system 合成孔径系统
synthetic fault 次级同向断层，同级断层
synthetic hydrocarbon 合成烃；合成碳氢化合物
synthetic hydrograph 综合水文过程线
synthetic hydrology 综合水文学
synthetic index 标准指标；综合指标
synthetic interferometer radar 合成孔径干涉雷达
synthetic map 组合图；综合图
synthetic membrane liners 合成薄膜衬层
synthetic model of geodynamics 地球动力学的综合模型
synthetic natural gas 合成天然气
synthetic plan of striping and mining 采剥工程综合平面图
synthetic radar 合成雷达
synthetic resin 合成树脂
synthetic rock motion 合成基岩运动
synthetic seismic section 合成地震剖面
Synthetic seismogram 合成地震记录；合成地震图
synthetic soil 人工合成土；人造土
synthetic standard 人工标准
synthetic strong motion record 合成强震记录
synthetic study 综合研究
synthetic time history 合成时间历程

synthetic unit hydrograph 综合水文曲线
synthetic wave form 合成波形
synthetical strength 综合强度；合成强度
syntony 谐振；调谐；共振
syphonage 虹吸作用
system analysis 系统分析
system anchor bolt 系统锚杆
system approach 综合方法；系统方法；总体途径
system architecture 系统结构
system chart 系统流程图
system construction 系统构造；系统构建；制度建设
system damping 系统阻尼
system design 系统设计；总体设计
system dynamics 系统动力学
system emulation 系统仿真；系统模拟
system engineering 系统工程；系统工程学；总体工程学
system error 系统误差
system evaluation 系统评价
system for analysis of soil-structure interaction 土-结构相互作用分析系统
system for digital mapping 数字测图系统
system for monitoring 监测系统
system function 系统功能
system identification 系统识别；系统鉴别
system impedance 系统阻抗
system management 系统管理
system manager 系统管理员；系统管理器
system model 系统模型
system of anchor bar 系统锚杆
system of conjugated fracture 共轭破裂系
system of coordinate 坐标系
system of crack 裂隙系统
system of drilling 钻井方法
system of equation 方程组
system of fault 断层系
system of fissure 裂隙系统
system of fold 褶皱系
system of force 力系
system of joint 节理组
system of linear equation 线性方程组
system of map representation 图示系
system of mass-spring 质量弹簧系统
system of mining 开采系统；开采方法，开采方式
system of ordinary differential equation 常微分方程组
system of rock fissure 岩石裂隙系统
system operator 系统操作员；系统运算符
system optimization 系统优化
system parameter 系统参数
system performance 系统特性；系统性能
system performance criteria 系列性能判据
system planning 系统规划
system programming 系统程序设计
system region 系统区
system reliability 系统可靠性
system response 系统反应
system science 系统科学
system simulation 系统模拟，系统仿真
system software 系统软件
system space 系统空间
system stability 系统稳定性

system state 系统状态
system stiffness 系统刚度
system support 系统支持
system test 系统测试；系统试验
system testing 系统测试
system theory 系统论
system time 系统时间
system transfer function 系统传递函数；系统传输函数
system triangular 三角形系统
system under test 试验中的系统
system variable 系统变量

system with nonlinear damping 具有非线性阻尼的系统
system with one degree of freedom 单自由度系统
system with quadratic nonlinearity 二次非线性系统
systematic analysis of groundwater resource 地下水资源系统分析
systematic error 系统误差
systematic generalization 系统归纳
systematic orientation 系统定向
systematic risk 系统风险
systematical analysis of water resources 水资源系统分析
systematization 系统化；规则化

T

tabbgite 硬沥青
tabby 平纹
tabbyite 韧沥青
tabergite 纯镁叶绿泥石；钠里蛭石
tabetisol 层间不冻层，冰原夹土；不冻地，不冻土
table feeder 盘式送料机
table iceberg 平顶冰山
table land 高原；台地
table mountain 平顶山；桌状山
table of correction 校正表
table of hydrometric data 水文测验成果表
table reef 平顶礁，桌礁，台礁
table rock 平顶岩
table shore 低平海岸
table tailings 摇床尾矿；淘汰盘尾矿
table vibrator 振动台
table water aquifer 潜水含水层
tabled scarf 叠嵌接合
tableknoll 海底平顶山；平顶山丘
tableland 高原；台地
tablemount 平顶海山；桌状海山
tabling 嵌合；列表、造册
Tabo Formation 踏波组
tabular 表格式的；平板状的
tabular cross stratification 板状交错层理
tabular cross-bedding 平坦斜层理；板状交错层理
tabular deposit 板状矿床
tabular dike 板状岩脉
tabular fault 平台断层
tabular formation 板状地层
tabular joint 板状节理
tabular mass 层状体
tabular ore body 板状矿体
tabular spar 硅灰石
tabular spring 溶洞泉
tabular structure 板状构造
tabulate 制表；使成平板状；平板状的
tabulated data 列表数据；制表资料
tabulation 制表；列表；表格
tabulation analysis 列表分析
tabulation method 列表法
tabulator 制表键
tacheometer 快速测距仪；准距计；视距仪；流速计
tacheometer alidade 视距照准仪
tacheometer staff 视距标尺
tacheometer traverse 视距导线
tacheometric rule 视距计算尺
tacheometric surveying 视距测量；视距测距
tacheometric table 视距计算表
tacheometry 视距测量法；视距测量
tacho-generator 测速发电机；测速传感器
tachograph 速度记录器；自动回转速度计
tachometer 流速计

tachometer generator 测速发电机；转速计传感器
tachometric 快速测距的
tachoscope 转速表
tachyaphalite 硅钍锆石
tachydrite 溢晶石
tachygenesis 加速发生；简捷发生
tachylite 玄武玻璃；玄武黑曜岩
tachylite basalt 玻璃玄武岩
tachymeter 速测仪；视距仪；准距仪
tachymetric method 视距法
tachymetric surveying 视距测量；测速
tachymetric triangle 视距测量三角形
tack coat 黏结层
tack rivet 平头铆钉
tack weld 点固焊；定位焊；临时点焊；间断焊
tackifier 胶黏剂；黏结剂
tackiness 黏着性；胶黏性
tackle 滑车；滑轮
tacky 胶黏的
taconite 铁燧岩；角岩
tactite 接触岩；接触碳酸岩；接触变质岩
Tade'ertake Group 塔得尔塔克群
tadpole plot 蝌蚪图；箭头图
taele 冻土
tafelberg 平顶山
tafelkop 平顶孤丘
Tafeng Formation 塔峰组
taffoni 蜂窝洞
tafoni 侵蚀孔；风化穴
tafrogenic 地裂的
tafrogeny 地裂运动
tag 标签；辅助信息
tag water 示踪水
Tage'er Formation 塔哥儿组
Tagg method 泰格法
tagged 有标记的；加标签的；示踪的
tagged atom 示踪原子
tagged element 示踪元素
tagger 薄铁皮；加标签者；标记器
tagging method 标记法；跟踪法
tagilite 假孔雀石；纤磷铜矿
tag-line method 牵线法
Tagou conglomerate 塔沟砾岩
tahitite 蓝方粗安岩
Taibei Formation 台北组
Taidong Earthquake 台东地震
tail 尾水
tail assay 尾料分析
tail bay 尾水池
tail bay wall 船闸后庭墙
tail beam 端梁；尾梁
tail bearing 尾轴承
tail block 末端滑轮；滑轮活轴；床尾
tail cement slurry 末端水泥浆

tail clearance 尾部间隙
tail conveyor 尾端输送机
tail end 末端，尾端
tail erosion 底部侵蚀
tail gas analyzer 废气分析仪
tail gate 上部通风平巷；采区辅助平巷；尾顺槽；上顺槽
tail house 尾矿场
tail load 尾部负载
tail mill 尾矿处理厂
tail of shield 盾构尾部
tail of water 积水边缘
tail pulley 尾轮，尾滑轮；尾部滚筒
tail rope 尾绳，尾索；平衡绳
tail rope balancing 尾绳平衡
tail rope haulage 尾绳运输
tail rope hoisting 尾绳提升
tail seal 盾尾密封
tail shield 尾盾
tail stinger system 尾刺系统
tail structure 尾部结构
tail trimmer 墙端短托梁；短横梁
tail void 盾构净空；盾尾空隙
tail water 下游段；尾水；废水
tail water level 尾水位
tail wind 顺风
tailentry 回风平巷；上顺槽
tailgate 下闸门；尾巷；回风巷；下游闸门
tailing 砖石嵌入墙中部分；屑；尾矿
Tailing Breccia 太岭角砾岩
tailing cake 尾矿滤饼
tailing classifier 尾矿分级机
tailing dam 尾矿坝
tailing dump 尾矿堆
tailing impoundments 尾矿库
tailing out 尖灭
tailing pile 尾矿堆
tailing pond 尾矿池
tailing pump 尾矿泵
tailing race 尾矿排出沟
tailing reservoir 尾矿池
tailing sample 尾矿试样
tailing sand 尾矿砂
tailing site 尾煤场；尾矿场
tailings 尾矿，尾渣，尾浆
tailings disposal 尾矿处理
tailings disposal area 尾矿处理区
tailings disposal plant 尾矿处理装置
tailings fill dam 尾矿坝
tailings grade 尾矿品位
tailings loss 尾矿中的损失
tailings pond 尾矿池；尾煤沉淀池
tailings sampler 尾矿取样
tailo 阳坡
tailor-made unit 定做装置
tailpiece 墙端短梁；尾管；附属物
tailrace channel 尾水渠
tailrace divider wall 尾水隔墙
tailrace outlet 尾水出口
tailrace tunnel 尾水洞

taimyrite 英钠粗面岩；黝方透长岩
tainter gate 弧形闸门
take a core 提取岩心
take a range 测距
take down 拆卸；记录结果；摄影
take down the top 破坏顶板
take level 进行水准测量
take off the gangue 清理矸石
take over 接收；接管
take over for use 征用
take sample 取样
take sights 抄平；瞄准
take the elevation 测量标高
take up 吸收
take up load 承受负载
takedown 拆卸；可拆卸的
take-in device 夹具
taken bottom 爆破的底板
take-off angle 离源角；出射角
take-off tower 出线塔
take-out channel 退水渠
take-up 拉紧；拉紧装置
take-up by gravity 重力拉紧；重锤拉紧
taking back 后退式开采
taking bearing 定方位
taking down 向下开采；崩落
taking of sample 采样，取样
taking top 挑顶
taking up 向上开采
taking-over 移交；接管
takovite 水铝镍石
Taku Glacial Stage 大姑冰期
takyr soil 龟裂土
talat 干冲沟
talc 滑石；云母
talc schist 滑石片岩
talc slate 滑石板岩
talc talcum 滑石
talc-chlorite schist 滑石绿泥片岩
talcite 滑块石；块白云母
talcose 滑石的；含滑石的
talcose minerals 滑石矿类
talcous 滑石的
Tali glaciation 大理冰期
talik 融区；层间不冻层
tall buildings 高耸建筑物
tall slender structure 高柔结构
talmessite 砷酸镁钙石
taltalite 电气石；绿碧玺
talus 山麓堆积；岩屑
talus apron 坡积裙
talus breccia 塌坡角砾岩；倒石锥角砾岩
talus cone 岩屑堆；岩锥
talus creep 岩屑下滑；岩屑蠕滑
talus deposit 岩屑堆积；岩锥堆积土
talus facies 塌积相
talus fan 坡积扇，坡积裙
talus material 山麓堆积物；坡积物
talus slide 堆积体滑坡
talus spring 堆积坡泉

talus wall	堆石护岸		分流水轮机
taluvium	岩屑崩塌物	tangential point load	切向点荷载
talweg	河道中泓线；河流谷底线	tangential Poisson's ratio	切线泊松比
tamanite	斜磷钙铁矿；三斜磷钙铁矿	tangential pressure stress	切向压应力
Tamar Basin Reclamation	添马填海区	tangential section	切向断面
tamaraite	辉闪霞煌岩	tangential shear force	切力
tamarite	云母铜矿；绿钠闪石	tangential shrinkage	切向收缩
tamarugite	三斜钠明矾	tangential slip	切向滑距
tamp	夯实；填塞	tangential stiffness	剪切模量
tamped concrete	捣实混凝土；夯实混凝土	tangential strain	切向应变
tamped finish	夯实整修面	tangential stress	切向应力
tamper	夯；夯具；捣棒	tangential stress concentration	切向应力集中
tamping	捣固；填塞炮泥；炮泥	tangential thrust	切向推力
tamping area	夯击面积；夯击范围	tangential velocity	切线速度，切向速度
tamping energy	夯击能	tangential wave	S 波，横波
tamping in layers	分层夯实	tangential Young's modulus	切向杨氏模量
tamping iron	铁夯；铁撞柱	Tangfang sandstone	塘房砂岩
tamping machine	打夯机	tangiwaite	透蛇纹石
tamping plate	夯板；捣板	tangue	浅湾贝壳沉积；浅湾钙质泥
tamping roller	夯击式碾；夯实压路机	tank foundation	罐基础
tamping speed	打夯速度	tank loading test	坑槽载荷试验
tanatarite	斜铝石	tank model	槽式模型
tangaite	铬磷铝石	tankage	池容量，槽容量
tangeite	钒钙铜矿	tanking	防水层，地窖防水层
tangenite	多钛钙铀矿	tankite	红钙长石
tangent	切线，正切	tantalite	钽铁矿
tangent angle	切线角，正切角	tantalum tracer	钽示踪剂
tangent bend	双弯曲	tanzanite	坦桑黝帘石
tangent between curves	夹直线	tap	水龙头；阀门；塞子
tangent circle	相切圆	tap cock	水管栓
tangent conic map projection	切圆锥地图投影	tap drill	螺纹底孔钻
tangent curve	正切曲线	tap hole	塞孔
tangent damping	切线阻尼	tape gauge	传送式水尺；活动水尺
tangent distance	切线距离	tape measure	卷尺
tangent force	切向力	tape person	地质采集员；调查人员的助手
tangent friction force	切线摩擦力	taper	锥形物；锥形化
tangent modulus	切线模量	taper angle	锥角；尖角
tangent modulus of deformation	切线变形模量	taper foot roller	羊足碾
tangent modulus of elasticity	切线弹性模量	taper out	尖灭
tangent offset	切线垂距；切线支距	tapered	斜的
tangent pile	支护桩	tapered arrangement	锥形排列
tangent pore pressure parameter	切线孔隙压力系数	tapered beam	楔形变截面梁
tangent stiffness	切线刚度	tapered ledge reamer	锥形扩孔钻头
tangential	切线方向的；水平的	tapered opening	锥形开口
tangential acceleration	切向加速度	tapered pile	锥形桩
tangential component	切向分量	tapered wedge	锥形楔
tangential compression	切向压缩	taper-foot roller	锥形足碾
tangential coordinate	切线坐标	tapering	削成锥形；削尖
tangential cross-bedding	切向交错层理	tapering gutter	斜槽沟；端宽沟
tangential deformation	切向变形	taphocoenose	埋藏群落
tangential direction	切向	taphofacies	埋藏相
tangential dislocation	水平变动	taphoglyph	生物埋藏印痕
tangential displacement	切向位移	taphrogen	地裂带
tangential distortion	切向畸变	taphrogenesis	地裂作用；断裂作用
tangential dynamic stress	切向动应力	taphrogeny	分裂运动；张陷断裂运动
tangential force	切向力	taphrogeosyncline	地堑式地槽；断裂地槽
tangential frost-heave force	冻胀切力	taping	卷尺测量
tangential line load	切向线荷载	tapoon	轻便水坝
tangential longitudinal strain	切面纵应变	tapping hammer	叩顶锤
tangential partial turbine	切向非整周进水式轮机；切向	tapping machine	顶板取样机；老塘穿孔机

tapping method 敲击法
tapping of groundwater 潜水集水井
tapping old workings 排放老采区积水或瓦斯
tapping point 地下水出水点
tapping sleeve 分流套管
tapping well 集水井
tar coal 沥青褐煤，焦油煤
tar concrete 柏油混凝土；焦油混凝土
tar macadamed road 沥青碎石路
taramite 绿铁闪石
taranakite 磷钾铝石
tarantulite 白岗英石岩
tarapacaite 黄铬钾石
taraspite 白云石
tarbuttite 三斜磷锌矿
target 目标
target acquisition 目标搜索
target advance 计划进度
target analysis 目标分析
target aquifer 目标层；回灌层；目标含水层
target area 目标区，靶区
target depth 设计深度
target location 目标定位
target output 产量指标
target point 靶点
target production 预期产量，设计能力
target stratum 目的层
Tarim Basin 塔里木盆地
Tarim massif 塔里木地块
Tarim oldland 塔里木古陆
tarn 山间小湖；冰斗湖
tarnowitzite 铅霰石
tarra firma 稳固地基
taseometer 应力计
task equipment 专用设备
task master 工头；监工
Tasman Basin 塔斯曼海盆
Tasman Sea 塔斯曼海
Tasmanian Seaway 塔斯马尼亚海道
tasmanite 辉沸岩
taspinite 杂块花岗岩
tatarian 阶
Tatarian 鞑靼的；鞑靼人的
tau-gamma mapping τ-γ 变换
tau-P mapping τ-P 变换；倾斜叠加
taurite 霓流纹岩
tautirite 霞石粗安岩
tautness 拉紧
tautness meter 拉力计；紧度计
tawmawite 铬绿帘石
taxa 分类单元
taxite 斑杂岩
taxitic ophitic texture 斑杂辉绿结构
taxitic structure 斑杂构造
taxoite 蛇纹石
taxon 生物分类单元
Taylor method 泰勒法
Taylor series 泰勒级数
Tayloran 泰勒阶
taylorite 铵钾矾；硫铵钾石

Taylor's stability chart 泰勒稳定性图
Taylor's stability number 泰勒稳定数
T-dike 丁字坝
tear 撕裂
tear fault 横推断层
tear resistance 抗扯性；抗扯强度
tear-and-wear allowance 磨损容差
tearing away of bank 河岸滑坡
tearing down 拆卸
tearing force 撕裂力
tearing mode 撕开型
tearing modulus 撕裂模量
tearing strain 张应变
tearing strength 抗扯强度
technical application 工程应用
technical audit 工程审核
technical code 技术规范
technical cohesion limit 真正断裂应力；真正断裂强度
technical data 技术资料
technical economic index 技术经济指标
technical error 技术误差
technical evaluation 技术评价
technical feasibility 技术可行性
technical information 技术资料；技术情报
technical inspection report 技术检验报告
technical jargon 技术俚语；专业俗语
technical journal 科技刊物；技术杂志
technical know-how 技术诀窍；专门技术
technical mission 技术代表团
technical norms 技术标准；技术规范
technical order 技术规程
technical organization 技术机构
technical paper 技术文献
technical parameter 技术参数
technical performance 技术性能
technical petrology 工艺岩石学
technical progress 技术进展；技术发展
technical proposal 工程建议书
technical reference 工程证明书
technical regulation 技术规程
Technical regulation for dynamic sounding 动力触探技术规定
Technical regulation for engineering geological boring of electrical power engineering 电力工程地质钻探技术规定
Technical regulation for existing railway survey 既有铁路测量技术规则
technical regulations 技术规程
technical report 技术报告
technical research 技术研究
technical responsibility 技术责任
technical review and approval 技术审查与批准
technical service 技术服务
technical specification 技术规范；技术说明
Technical specification for building pile foundations 建筑桩基技术规程
Technical specification for electrical cone penetration test 静力触探技术规程
Technical specification for geotechnical centrifuge model tests 土工离心模型试验规程
Technical specification for jet grouting 高压旋喷注浆技术

规程
Technical specification for low strain dynamic testing of piles 基桩低应变动力检测技术规程
Technical specification for retaining and protection of building foundation excavations 建筑基坑支护技术规程
Technical specification for soil nailing in foundation excavations 基坑土钉支护技术规程
Technical specification for underwater foundations treatment by means of blasting method 爆炸法处理水下地基和基础技术规程
Technical specification for waterproof of shield-driven tunnel 盾构法隧道防水技术规程
technical staffs 技术人员
technical standard 技术标准
Technical standard for boring of geotechnical investigation of buildings 建筑工程地质钻探技术标准
Technical standard for highway engineering 公路工程技术标准
Technical Standard for sampling of undisturbed soils 原状土取样技术标准
technical standard of road 道路技术标准
Technical standard of static sounding 静力触探技术规则
technical standards 技术标准
technical supervision 技术监理，技术监督
technical tolerance 技术公差；技术容限
technical viscometer 工程黏度计
technicals 技术细则
technico-construction investigation stage 技术施工勘察阶段
technique data 技术资料
technique of preparing 选矿技术
techno-economic verification 技术经济论证
technogenic 人工造成的地质、地貌景观
technological breakthrough 技术上突破
technological chain 工艺过程
technological face 工艺基准面
technology 工艺；技术
technology assessment 技术评定
technology of ultradeep drilling 超深井钻抽技术
technology transfer 技术转让；技术转移
technosphere 技术圈
tectocline 深坳带；大地构造槽
tectofacies 构造相
tectogene 深坳槽；海沟，海渊
tectogene hypothesis 构造槽说；构造根假说
tectogenesis 构造运动；造山运动
tectogenetic 构造成因的；构造运动的
tectogenetic movement 构造运动；造山运动
tecto-lithostratigraphic unit 构造地层单位
tectomorphic 熔蚀变形的
tectonic 构造的
tectonic accretion 构造增生；大陆增生
tectonic active area 构造活动区域
tectonic activity 构造活动性
tectonic analysis 构造解析
tectonic arkose 构造长石砂岩
tectonic assemblage diagram 构造组合图解
tectonic association 构造组合
tectonic axis 构造轴
tectonic basin 构造盆地
tectonic belt 构造带

tectonic breccia 构造角砾岩
tectonic characteristics 构造特征
tectonic clastic 构造碎裂
tectonic coal 构造煤
tectonic coast 构造海岸
tectonic community 构造群落
tectonic complex 构造杂岩
tectonic contact 构造接触
tectonic creep 构造蠕动
tectonic crustal phenomena 地壳构造现象
tectonic culmination 构造运动顶峰
tectonic cycle 构造旋回；造山旋回
tectonic deformation 构造变形
tectonic deformation facies 构造变形相
tectonic deformation field 构造变形场
tectonic deformation models 构造变形模式
tectonic deformation sequence 构造变形序列
tectonic denudation 构造剥蚀
tectonic depression 构造洼地
tectonic differential movement 构造差异运动
tectonic discontinuity 构造不连续面
tectonic disturbance 构造扰动
tectonic division 构造划分
tectonic domain 构造域
tectonic dynamics 构造动力学
tectonic earthquake 构造地震
tectonic element 构造要素；构造单元
tectonic emplacement 构造侵位
tectonic energy 构造能量
tectonic episode 构造幕
tectonic event 构造事件
tectonic evolution 构造演变
tectonic fabric 构造组构
tectonic facies 构造相
tectonic feature 构造特征
tectonic fissure 构造裂隙
tectonic flow 构造迁移
tectonic fold 构造褶皱
tectonic force 构造力
tectonic fracture 构造缝
tectonic framework 构造格架
tectonic gap 滞后断层
tectonic generation 构造世代；构造幕
tectonic geology 构造地质
tectonic geomorphology 构造地貌学
tectonic grain refinement 构造细粒化；构造细晶化
tectonic ground crack 构造性地裂缝
tectonic history 构造历史
tectonic homogenization 构造均化
tectonic inclusion 构造包体
tectonic inheritance 构造继承
tectonic intensity 构造运动强度
tectonic inversion 构造反转
tectonic joint 构造节理
tectonic joint system 构造节理系
tectonic landform 构造地形；构造地貌
tectonic level 构造层位
tectonic line 构造线
tectonic magmatic belt 构造岩浆岩带
tectonic magnitude 构造等级

tectonic map	大地构造图
tectonic massif	构造岩块
tectonic melange	构造混杂岩
tectonic mesh line	构造网线
tectonic meshworm	构造网眼
tectonic metamorphism	构造变质
tectonic migmatite	构造混合岩
tectonic models	构造模式
tectonic movement	构造运动
tectonic oil gas basin	构造油气盆地
tectonic order	构造级别；构造序次
tectonic origin	构造成因
tectonic outlier	构造外露层
tectonic pattern	构造模式
tectonic phase	构造相
tectonic plain	构造平原
tectonic plate	地壳板块；地层
tectonic position	构造部位
tectonic pressure	构造压力
tectonic process	构造过程
tectonic profile	构造剖面
tectonic quiet region	构造稳定区
tectonic relief	构造起伏
tectonic rock	构造岩
tectonic rotation	构造旋转
tectonic scale	构造规模
tectonic sequence	构造序列
tectonic setting	构造背景
tectonic sinking	构造沉降
tectonic strain	构造应变
tectonic stress	构造应力
tectonic stress field	构造应力场
tectonic structure	大地构造
tectonic style	构造类型
tectonic stylolites	构造缝合线
tectonic subsidence	构造沉陷
tectonic superposition	构造叠加
tectonic synthesis	构造综合分析
tectonic system	构造系统
tectonic termination	构造中断
tectonic terrace	构造阶地
tectonic transport	构造迁移
tectonic triangle zone	构造三角带
tectonic unconformity	构造不整合
tectonic unit	构造单元
tectonic uplift	构造隆起
tectonic wedging	构造楔入
tectonic window	构造窗
tectonic zone	区域构造带
tectonically active region	构造活动区
tectonically deformation	构造变形
tectonically denudation	构造剥蚀
tectonically differential movement	构造差异运动
tectonically discordance	构造不整合
tectonically earthquake	构造地震
tectonically stable margin	构造稳定界限
tectonic-pressure release	构造压力释放
tectonics	大地构造学；地质构造，区域地质构造
tectonism	区域构造作用
tectonite	构造岩
tectonite fabric	构造岩组构
tectonization	造山运动
tectonoblastic	构造变晶的
tectonoclastic rock	构造碎裂岩
tectonodynamics	构造动力学
tectono-eustasy	构造海面升降
tectono-eustatism	地壳变动海平面升降；构造海平面升降
tectono-karstic depression	构造岩溶洼地
tectono-lithologic map	构造-岩性图
tectonomagmatic belt	构造岩浆带
tectonophysics	地壳构造物理学
tectonosphere	地壳结构层
tectono-stratigraphic unit	构造地层单位
tectonostratigraphy	构造地层学
tectorium	疏松层
tectosequent	反映构造的；构造表象的
tectosilicate	架状硅酸盐
tectosome	构造体
tectosphere	构造圈；构造层
tectostratigraphic	构造地层的
tectotope	构造境；构造地带
tee girder	丁字形大梁
tee iron	丁字形铁件
tee joint	三通接头
tee-beam and slab construction	肋形板结构
teeth bar	齿条
tektite	玻璃陨石；似黑曜岩；熔融石
tektonite	构造岩
telain	结构镜煤；胞煤
telalginite	圆形藻质体；结构藻类体
telechemic mineral	最早结晶矿物
teleclinograph	遥测钻孔偏斜仪
telecontrol	遥控
telegauge	遥测计，测远仪
telemagmatic	远岩浆的
telemagmatic deposit	远岩浆矿床
telemagmatic metamorphism	区域岩浆热变质
telemagnatic hydrothermal deposit	远岩浆热液矿床
telemagnatic thermal metamorphism	远岩浆热变质作用
telemeter	遥测仪，测距仪
telemetered seismic net	遥测地震台网
telemetering seismograph	遥测地震仪
telemetering system	遥测装置，遥测系统
telemetric	远距离测量的
telemetry data	遥测数据
telemetry network	遥测台网
telescope	望远镜
telescope bubble	望远镜水准器
telescope direct	盘左；正镜
telescope grouting method	套管式灌浆法
telescope reverse	盘右；倒镜
telescope structure	嵌入构造
telescoped deposit	套叠矿床
telescoped intrusive body	套叠式侵入体
telescopic	套筒式的；望远镜的；可伸缩的
telescopic boom	伸缩式悬臂
telescopic bucket reamer	伸缩式扩孔机
telescopic caisson	套筒式沉箱
telescopic chimney	套筒式烟囱
telescopic chute	伸缩式溜槽；套筒倾卸槽

telescopic drilling post	伸缩式钻机柱	temperature cycling test	温度循环试验；冷热交变试验
telescopic feed	伸缩式推进器	temperature deformation	温度变形
telescopic joint	套管接头；伸缩式接头	temperature difference	温差
telescopic loader	伸缩式装载机	temperature drop load	温降荷载
telescopic range finder	钻孔定向器	temperature effect	温度效应
telescopic ripper	伸缩式掘进机	temperature expansion	温度膨胀
telescopic shaft	伸缩轴	temperature field	温度场
telescopic shield	伸缩护盾	temperature indicator	温度计
telescopic shuttle car	伸缩车身式梭车	temperature inversion	逆温
telescoping	可伸缩的；套叠作用	temperature joint	温度缝；伸缩缝
telescoping tube settlement gauge	分层沉降管	temperature load	温度荷载
teleseism	远震	temperature logging	井温测井
teleseismic data	远震资料	temperature map	地温图
teleseismic distance	远震距离	temperature measurement	温度测量
teleseismic magnitude	远震震级	temperature of fusion	熔点
teleseismic network	远震台网	temperature of solidification	凝固点
teleseismic record	远震记录	temperature or flux contour	温度或流量云图
teleseismic signal	远震信号	temperature pressure diagram	温度压力图
teleseismic station	远震台站	temperature profile	温度剖面
teleseismic wave	远震地震波	temperature scale	温度标
teleseismology	遥测地震学	temperature sensor	温度传感器
telethermal deposit	远成热液矿床	temperature strain	温度应变
televiewer	井下电视	temperature stress	温度应力
television borehole camera	钻孔电视摄影机	temperature survey	地温测井
Telford base	块石基层；大块石底层	temperature well logging	温度测井
telgsten	绿泥石	temperature-sensitive paint	热敏油漆，热敏涂料
telinite	结构镜质体；结构凝胶体	tempering	混合
telltale rod	应变杆；信号杆	tempest	风暴
tellurian	地球的	template	模板
telluric current method	大地电流法	template beam	垫梁
telluric current prospecting	大地电流法勘探	template frame	放样架
telluric electric field	大地电流场	temporal sequence	时间层序
telluric electromagnetic profile	大地电磁剖面	temporary	临时的，暂时的
telluric electromagnetic sounding	大地电磁测深	temporary anchor	临时性锚杆
telluric magnetic force	地磁	temporary ballast	临时压载；临时压载物
telluric water	大气水	temporary bank protection	临时性护岸
telluroid	近似地形面；似地球面	temporary base level	临时基准面
telmatic peat	浅沼泥炭	temporary base level of erosion	临时侵蚀基准面
telmatology	湿地学，沼泽学	temporary casing	临时井壁
telocollinite	均质镜质体	temporary cofferdam	临时围堰
telogenesis	表成作用；后期形成作用	temporary dam	临时挡水坝
telogenetic	表生的	temporary excavation	临时开挖
telogenetic porosity	后期形成孔隙性	temporary export	临时出口
telogenetic stage	晚期表成阶段	temporary flash board	临时挡水板
temblor	地震	temporary hardness	暂时硬度
temoin	挖方土柱	temporary lining	临时衬砌，临时支护
temper	韧度	temporary loading	临时荷载
temperate	适度的	temporary loading condition	临时荷载条件
temperate climate	温带气候	temporary measure	临时措施
temperate karst	温带岩溶	temporary project	临时工程
temperate zone	温带	temporary prop	临时支柱
temperature action	温度作用	temporary reinstatement	临时修复工程
temperature anomaly	温度异常	temporary rock support	临时岩体支护
temperature coefficient	温度系数	temporary saturated zone	暂时饱和区
temperature conductivity	导温率	temporary seismograph station	临时地震台站
temperature contour	等温线	temporary staging	临时工作架；脚手架
temperature contrast	温差	temporary strain	瞬时应变
temperature correction	温度校正	temporary stream	间歇河
temperature crack	温度裂缝	temporary strength	短期强度
temperature cracking	温度开裂	temporary strengthening	临时加固

temporary stress	短时应力
temporary structure	临时结构
temporary stull	临时横撑
temporary support	临时支护
temporary torrent	季节性河流
temporary trench support	临时槽支撑；临时坑壁支撑
temporary wall	封闭墙；临时墙
temporary water	季节性地下水
temporary waterbearing layer	上层滞水
temporary works	临时工程
tenacious	强韧的；黏的；有附着力的
tenacious clay	黏土
tenacity	黏滞性，韧性
tendency	倾向
tendency forecasting	趋势预测
tendency of stage variation	水位变化趋势
tendency toward sliding	滑动趋势
tender	招标
tender accepted	中标
tender documents	招标文件
tender dossiers	招标材料；招标文件
tender in date	收标日期；截标日期
tender invitation	招标
tender notice	招标公告
tender procedure	招标程序
tender report	标书评估报告
tender roof	不稳顶板；软弱顶板
tender shaft	风井；通风井
tender submission	投标申请书；投标建议书
tender vertical shaft	立斜相连的井
tendon	钢筋束；锚索
tendon control	锚索控制；钢筋束控制
tennantite	砷黝铜矿；砷铜矿
Tennesseean	田纳西统
tenor	品位；金属含量
tenor of ore	矿石品位
tenorite	黑铜矿
tense	拉紧的
tensibility	伸长性，伸张性
tensile	抗拉的；张性的
tensile adhesion property	拉伸黏结性能
tensile area	受拉区
tensile breccia	张性角砾岩
tensile capacity	受拉承载力
tensile crack	张性破裂；张裂隙
tensile curve	拉伸曲线
tensile damage	拉伸损伤
tensile deformation	拉伸变形
tensile depth	拉裂深度
tensile dislocation	张位错
tensile ductility	拉伸韧性
tensile elasticity	拉伸弹性
tensile elongation	拉长；延长率
tensile end anchorage	张拉端锚具
tensile failure	拉伸破裂；拉断
tensile fatigue test	拉伸疲劳试验
tensile figure	抗张值；拉伸图
tensile fixation	张拉固定
tensile force	拉力
tensile fracture	张裂隙；拉伸断裂
tensile fracture collapsing	张裂塌落
tensile impact	张力冲击
tensile impact test	抗拉冲击试验
tensile load	拉伸载荷
tensile modulus	拉伸模量
tensile modulus of elasticity	抗拉弹性模量
tensile normal stresses	法向拉应力
tensile pile	抗拔桩
tensile region	受拉区
tensile reinforcement	抗拉钢筋
tensile rigidity	拉伸刚度
tensile splitting strength	劈裂抗拉强度
tensile strain	拉应变，张应变
tensile strain pulse	张力脉冲
tensile strength	抗拉强度
tensile strength limit	极限抗拉强度
tensile strength of rock	岩石抗拉强度
tensile strength test	抗张强度试验
tensile stress	拉应力；张应力
tensile structural plane	张性结构面
tensile structure	张拉结构
tensile surface	拉裂面
tensile test	拉伸疲劳试验
tensile testing machine	拉力试验机
tensile trajectory	受拉轨迹线
tensile wave	张力波
tensile yield point	拉力屈服点
tensile yield strength	抗拉屈服强度
tensility	可拉伸性
tensiometer	拉力计
tension	拉力，张力
tension area	受拉面积；抗张面积
tension axis	张力轴
tension band	张紧带；箍圈
tension bolt	张力锚杆
tension brace	拉撑
tension by flexure	弯曲受拉
tension coefficient	拉伸系数
tension crack	拉裂缝
tension crack zone	张裂带
tension cut-off	张拉截断
tension deep fault	张性深断裂
tension deep fracture	张性深断裂
tension dynamometer	拉力测功力计
tension fault	张性断裂
tension fissure	张裂隙，张裂缝
tension fold	张褶皱
tension fracture	张性破裂
tension gash	张裂缝
tension increment	张力增量
tension intensity	张力强度
tension joint	张节理
tension member	受拉杆件，受拉构件
tension meter	张力计
tension piles	拉力桩
tension reinforcement	受拉钢筋
tension ring	拉力环；拉力测力计
tension strain	拉伸应变，张应变
tension strength	抗拉强度
tension stress	张应力

tension test	抗拉试验
tension wave front	张力波前沿，拉力波前沿
tension zone	张力带
tensional fault	张断层；张性断裂
tensional fracture	张性断裂
tensional graben	张性地堑
tensional rift valley	张性裂谷
tensional rigidity	抗拉刚度
tensional stress	拉应力
tensional tectonics	张性构造
tensometer	伸长计；张力计
tensor	张量
tensor analysis	张量分析
tensor calculus	张量计算
tensor field	张量场
tensor of stress	应力张量
tensor transformation	张量变换
tensorial set	张量集
tenso-shear structure plane	张扭性结构面
Tentaculites	竹节石属
tentaculitids	竹节石类
tentative data	初始数据；试验数据
tentative method	暂行办法
tentative provision	暂行条例
tentative scheme	示意图
tentative standard	暂行标准
tentative valuation	初步估计
tenting	拱起；隆起
tepee	锥状构造；圆锥丘
tepee structure	圆锥形构造；帐篷构造
tephra	火山碎屑；火山灰
tephra chronology	火山灰年代学
tephrite	碱玄岩
tephroite	锰橄榄石
tephros	火山灰
Teratan Stage	特雷坦阶
tergal	背面的
terlinguaite	黄氯汞矿
term	术语
term of validity	有效期间；有效
terminal deposit	终端沉积
terminal face	端面
terminal load	末端荷载
terminal manhole	终端沙井；尾井
terminal moraine	终碛；尾碛
terminal pressure	边界压力
terminal speed	临界速度
terminal water saturation	最终含水饱和度
termination criterion	终止条件
termination line	停采线
termination of grouting	灌浆终结
terminology	术语
ternary	三相的
ternary diagram	三相图
ternary sediment	三元沉积物；三相沉积
ternary system	三元系
ternary systems	三元体系
terra	陆地
terra alba	硬石膏；白土
terra cariosa	硅藻土
terra firma	大地；稳固地基
terra nera	黑土
terra rossa	红色石灰土；红土
terra verte	绿土
terrace	阶地
terrace aggradation	阶地填积
terrace bench	阶地
terrace channel	地埂槽沟
terrace correlation	阶地对比
terrace cover	阶地沉积
terrace cusp	阶地陡坎
terrace deformation	阶地变形
terrace deposit	阶地沉积
terrace edge	阶地边沿
terrace fault	阶梯断层
terrace formation	阶地地层
terrace grade	地埂坡度
terrace gravel	阶地砾石；阶地砂矿
terrace height	阶地高度
terrace interval	阶地间距
terrace mining	多台阶开采
terrace pattern	阶地类型
terrace ridge	台地脊线
terrace scarp	阶地崖
terrace sediment	阶地沉积物
terrace spring	阶地泉
terrace wall	梯状挡土墙，台阶式挡墙
terrace width	阶地宽度
terraced platform	梯状平台
terraced slope	梯状斜坡
terracette	土滑小阶坎
terrace-type open-pit	台阶式露天矿
terracing	阶地形成
terra-cotta clay	硬陶土；赤陶土
terrain	地形，地势，地带；地域；地体
terrain analysis	地形分析
terrain classification map	地形分类图
terrain clearance	离地高度
terrain coefficient	地面利用系数
terrain condition	地形条件
terrain configuration	地表外形，地形特征，地貌结构
terrain correction	地形校正
terrain details	地形要素
terrain element	地面要素
terrain estimation	地面估测
terrain evaluation	场址评价
terrain evaluation mapping	地形评估勘测
terrain factor	地形因素；地形因子
terrain feature	地形特征
terrain interpretation	土质鉴定
terrain line	地形线
terrain model	地形模型
terrain movement	地层移动
terrain photo interpretation	地面摄影像片解译
terrain profile photography	地形剖面摄影
terrain reflectance	地面反射率；地物反射率
terrain sampling	地形采样
terrain surface	地表
terrain surveying	地形测量
terrane	地块；岩层

terra-probe	水下振砂器
terrastatic pressure	地静压力
terratolite	密高岭土
terrazzo	水磨石
terrene	地球表面
terres rares	稀土元素
terrestrial	地球的；陆地的
terrestrial age	陆地年龄
terrestrial bed	陆成层
terrestrial camera	地面摄影机
terrestrial coordinates	大地坐标
terrestrial deposit	陆相沉积
terrestrial ecology	陆地生态学
terrestrial ecosystem	陆地生态系统
terrestrial effect	地球效应；地面弯曲效应
terrestrial ellipsoid	地球椭球体
terrestrial environment	陆地环境
terrestrial equator	地球赤道
terrestrial erosion	地表侵蚀
terrestrial event	地球事件
terrestrial facies	陆相
terrestrial formation	陆相地层
terrestrial globe	地球；地球仪
terrestrial gravitation	地球重力，地球引力
terrestrial heat	地热
terrestrial heat flow	大地热流
terrestrial heat sources	地球热源
terrestrial hydrology	陆地水文学
terrestrial impact structures	地面撞击构造
terrestrial magnetic field	地磁场
terrestrial magnetic pole	地磁极
terrestrial magnetism	地磁学；地磁
terrestrial matter	地球物质
terrestrial meridian	地面子午圈；地球经线
terrestrial mining	陆地采矿
terrestrial movement	大陆运动
terrestrial parallel	地面纬圈；地球纬度圈
terrestrial peat	陆相泥煤；陆相泥炭
terrestrial photogrammetry	地面摄影测量学；地面摄影测量
terrestrial photograph	地面摄影照片
terrestrial pole	地理极
terrestrial radiation	地球辐射
terrestrial sediment	陆相沉积物
terrestrial space	近地空间
terrestrial spectrograph	地面摄谱仪
terrestrial stereoplotter	地面立体测图仪
terrestrial stress	地应力
terrestrial triangulation	地面三角测量
terrestrial-limnal facies	陆生湖泊相；陆生湖沼相
terrigenous	陆源的
terrigenous clast	陆源碎屑
terrigenous clastic rock	陆源碎屑岩
terrigenous component	陆源组分
terrigenous deposit	陆源沉积
terrigenous lake	陆源湖
terrigenous sediment	陆源沉积；陆源沉积物
terrigenous turbidite	陆源浊积岩
territorial	区域性的
territorial planning	国土规划
territorial sea	领海
territoriality	大陆性
territory	版图
Tertiary	第三纪
Tertiary clastic	第三纪碎屑岩
tertiary creep	第三级蠕变；三重蠕变
Tertiary deposit	第三纪沉积
Tertiary period	第三纪
Tertiary system	第三系
Tertiary volcanism	第三纪火山活动
Terzaghi bearing capacity theory	太沙基承载力理论
Terzaghi consolidation theory	太沙基固结理论
Terzaghi theory	太沙基理论
Terzaghi-Rendulic diffusion equation	太沙基-伦杜列克扩散方程
Terzaghi's bearing capacity formula	太沙基承载力公式
Terzaghi's effective stress equation	太沙基有效应力公式
Terzaghi's model	太沙基模型
Terzaghi's solution	太沙基解
teschenite	沸绿岩
tesselation	镶嵌状格局；镶嵌作用
test adit	试坑
test bench	试验台
test blasting	试验爆破
test bore	试验孔
test boring	试验性取芯；试钻
test boring record	试井记录
test cabinet	测试室；化验室
test code	试验规则，试验规程
test condition	试验条件
test constant	试验常数
test core	试验岩心
test cube	试件；立方体试块
test data	试验数据
test desk	测试台，试验台
test drilling	试钻
test driving	试验打桩
test duration	试验持续时间
test embankment	试验填方；试验堤；碾压试验
test excavation	试验开挖
test expression	表达式
test fill	试验填方
test flow chart	测试流程图
test flume	试验槽
test for convergence	收敛检验
test for losses of prestress	预应力损失试验
test groin	试验丁坝
test grouting	灌浆试验
test hole	探井；试验孔
test hole drilling	试验孔钻探
test interval	测试间隔时间
test level	校核水准
test load	试验荷载
test material preparation	试件制备
test method	试验方法
Test methods of soils	土工试验规程
test mine	试验矿井
test model	试验模型
test of aging	老化试验
test of normality	正态性检验

test paper	试纸
test period	试验周期
test piece	试样
test piece model	试件模型；试样模型
test pile	试桩
test piling	试打桩
test pit	试坑
test pressure	试验压力
test procedure	试验程序；试验方法
test program	试验程序；试验计划
test pumping	试抽水
test report	试验报告
test result	试验成果
test revision	校核试验
test rig	试验台；实验设备
test sample	试样
test sample set	测试样本集；检样本集
test shaft	试验井
test simulation	模拟测试
test site	试验场地
test specification	试验规范
test specimen	试件，试样
test stand	试验台
test station	试验站
test trench	探槽
test tube	试管
test unit	试验装置
test value	测试值
test well	探井
test yield	试验出水量
tester	探针
tester for randomness	随机性检验
tester hole	试验孔，探井
tester hole pattern	勘探孔布置
tester house	试验站
test-hole pattern	勘探钻孔排列方式
testing accuracy	检验精确度
testing counter	试验台
testing error	试验误差
testing for strength	强度试验
testing gallery	试验巷道
testing ground	试验场地
testing laboratory	试验室
testing load	试验荷载
testing machine	试验机
testing method	试验方法
testing observation station	试验观测站
testing of algorithm	算法测试
testing of soil	土质检验
testing pressure	试验压力
testing procedure	试验程序；试验方法
testing regulation	试验规程
testing scheme	测试方案
testing sequence	试验程序，试验步骤
testing station	试验站
testing technique	试验技术
testing-sieve shaker	试验筛振动器，振动试验筛
test-plant work	大型试验研究
test-run	试运行
Tethyan fault system	特提斯断裂体系
Tethyan ocean	特提斯洋
Tethyan orogenic region	特提斯造山区
Tethyan realm	特提斯生物区
Tethyan rifting	特提斯裂谷
Tethyan tectonic domain	特提斯构造域
Tethyan tectonics	特提斯构造
Tethys	特提斯海
Tethys Himalayan metallogenic province	特提斯-喜马拉雅成矿域
Tethysian metallogenic domain	特提斯成矿域
tetradymite	辉碲铋矿
tetragonal	四方的；正方的
tetragonal axis	正方轴
tetragonal prism	四方柱；正方柱
tetrahedral	四面体的
tetrahedral sheet	四面体层
tetrahedral void	四面体空隙
tetrahedral wedge	四面体楔块
tetrahedrite	黝铜矿
tetrahedron	四面体
tetrahedron block revetment	四面块护岸
textinite	木质结构体
Textulariina	串珠虫亚目
textural analysis	结构分析
textural anisotropy	结构非均质性；异向性
textural anomaly	结构异常
textural classification of soil	土的质地分类
textural difference	结构差异
textural element	结构要素
textural maturity	结构成熟度
textural porosity	结构孔隙度
texture	构造，结构
texture plane	结构面
thalassic	深海的
thalassocratic movement	造海运动
thalassocratic period	海洋期
thalassocratic sea level	海洋扩张期海面
thalassocraton	海洋稳定地块；海洋克拉通
thalassogenesis	造海作用；海洋化
thalassogenic movement	海底运动；造海运动
thalassogenic sedimentation	海底沉积
thalenite	红钇矿
thallite	绿帘石
thalweg	河道中泓线；河流谷底线；海谷底线
thaumasite	硅灰石膏
thaw	融化
thaw bowl	融化盘
thaw bulb	融坑
thaw coefficient	融化系数
thaw collapse	热融滑塌；融沉
thaw collapsibility	融化塌陷；融陷性
thaw compression	冻融固结
thaw compression factor	融化压缩因数
thaw compression test	融化压缩试验
thaw consolidation	冻融固结
thaw depth	融化深度
thaw lake	融陷湖；融沉湖
thaw period	融化期；解冻期
thaw point	融点
thaw settlement	融沉陷；融陷性

thaw slumping	热融滑塌	thermal convection	热对流
thaw subsidence factor	融沉系数	thermal crack	热裂纹
thaw unstable permafrost	融化不稳定的永冻土	thermal creep	热蠕变
thaw-collapse soil	融陷土	thermal cubic expansion coefficient	体积膨胀系数
thawed layer	融化层	thermal deformation	热变形
thawed soil	融土	thermal degradation	热降解
thawed zone	融冻带	thermal differential	温差
thawing	融化	thermal diffusion	热扩散
thawing index	融化指数	thermal diffusion coefficient	热扩散系数
thawing layer	融冻层	thermal diffusivity	导温系数；散热率；散热系数
thawing season	融季	thermal dilatation	热膨胀
thawing water	雪水；融解水	thermal discharge	热排放；排热；散热
thawing water irrigation	融水灌溉	thermal dissipation	热耗散
thematic atlas	专题地图集	thermal distortion	热变形
thematic cartography	专题地图学	thermal drift	热漂移
thematic data	专题数据	thermal drilling	热力钻进
thematic map	主题图	thermal drilling process	热力钻眼法
theme	主题	thermal dryer	热力干燥机
theodolite	经纬仪	thermal drying	热力干燥；热能干燥；加热干燥
theodolite with stadia lines	视距经纬仪	thermal dynamics	热动力学
theorem	定理	thermal effect	热效应
theorem of errors	误差理论	thermal excavation	热力开挖
theorem of induced cleavage	采动裂隙理论；诱生裂隙理论	thermal expansion	热膨胀
theoretical model	理论模型	thermal expansion coefficient	热膨胀系数
theoretical porosity	理论孔隙度	thermal expansivity	热膨胀系数
theory of blasting	爆破理论	thermal geotechnics	热岩土工程学
theory of consolidation	固结理论	thermal gradient	热梯度
theory of continental drift	大陆漂移说，大陆漂移论	thermal groundwater	地下热水
theory of elastic rebound	弹性回跳理论	thermal inertia	热惰性；热惯性
theory of elastic thin shell	弹性薄壳理论	thermal infrared	热红外
theory of elasto-plastic deformation	弹塑性变形理论	thermal instability	热不稳定性
theory of fault block tectonics	断块构造说	thermal insulating course	隔温层
theory of information	信息论	thermal insulating layer	隔热层
theory of locking-up	锁固理论	thermal insulation	隔热性
theory of material strength	材料强度理论	thermal insulator	绝热体
theory of maximum strain energy	最大变形能理论	thermal island	热岛
theory of plasticity	塑性理论	thermal load	热负荷；热应力
theory of plates tectonics	板块构造学说	thermal logging	温度测井
theory of polycycle	多旋回说	thermal metamorphic rock	热变质岩
theory of rupture	破裂理论	thermal metamorphism	热力变质作用
theory of seepage	渗流理论	thermal pollution	热污染
theory of strength	强度理论	thermal reduction	热还原
theory of thin shell	薄壳理论	thermal resistivity	热阻率
theory of yield surface of soil	土的屈服面理论	thermal resistor	热敏电阻；热变电阻
theralite	霞斜岩；企猎岩	thermal resolution	温度分辨率
thermal abrasion	热蚀	thermal spring	温泉
thermal alteration	热力蚀变作用	thermal stabilization	热加固
thermal anomaly	热异常	thermal strain	热应变
thermal belt	地热带	thermal stress	热应力
thermal boring	热力钻孔	thermal stress analysis	热应力分析
thermal capacity	热容量	thermal structure	热力构造
thermal coefficient of expansion	热膨胀系数	thermal transmission	热传导
thermal conductance rate	热导率	thermal turbulence	热扰动
thermal conduction	热传导	thermal weathering	热力风化
thermal conduction capability	导热能力	thermic boring	热力钻孔
thermal conduction coefficient	导热系数	thermisopleth	变温等值线
thermal conductivity	热导率；导热系数	thermo abrasion coast	热力侵蚀海岸
thermal conductivity sensors	热传导传感器	thermo-capillary movement	温差毛细管水运动
thermal constant	热常数	thermochemistry	热化学
thermal contraction	热收缩	thermochronology	热年代学

thermocouple psychrometers	热电偶湿度计
thermocurrent	热电流
thermodynamic analogy	热力模拟
thermodynamic circulation	热力环流
thermodynamic equilibrium	热力平衡
thermodynamic potential	热力电位
thermodynamic property	热动力学性质
thermoelectric couple	热电偶
thermoelectric effect	热电效应
thermoelectric power generation	热力发电
thermo-geotechnology	热岩土工程学
thermogram	差热曲线
thermoisopleth	等温线
thermokarst	冰融喀斯特，热岩溶
thermokarst topography	热喀斯特地形
thermologging	温度测井
thermoluminescence	热发光
thermomagnetic preparation	热磁选煤法
thermomechanical behavior	热力学行为
thermometamorphism	热力变质；热变质作用
thermometer	温度计
thermometry	测温法；温度测量
thermonuclear	热核的
thermonuclear explosion	热核爆炸
thermoosmosis	热渗
thermophone	传声温度计；热致发声器
thermoplastic	热塑塑料；热塑的
thermosiphon	热虹吸管
thermosounder	热探测仪
thick alluvium	厚冲积层
thick bed method	厚垫法；厚铺砌法
thick deposit	厚矿床，厚矿层
thick grout	浓浆；稠浆
thick overburden	巨厚盖层
thick sanded grout	稠砂浆
thick skinned structure	厚皮构造
thick wall sampler	厚壁取土器
thick-cylinder theory	厚壁圆筒理论
thickening	变厚；浓缩
thickening of formation	地层增厚
thickening of mud	泥浆浓化
thickness	厚度
thickness contour	等厚线
thickness map	等厚图
thickness measurement	厚度测量
thickness of coal seam	煤层厚度
thickness of confined aquifer	承压含水层厚度
thickness of covering soil	覆土的厚度
thickness of overburden	覆盖层厚度
thickness of spraying	喷射厚度
thickness of subgrade bed	基床厚度
thickness ratio	厚度比
thick-walled	厚壁的
thief	取样
thief formation	非常透水的岩层；漏水地层
thief zone	漏失带
thin	薄的
thin arch dam	薄拱坝
thin bed	薄层
thin core bit	薄壁取芯钻头
thin core dam	薄心墙坝
thin cut-off wall	薄截水墙
thin deposit	薄矿层；薄矿床
thin layer	薄层
thin out	尖灭
thin out of coal seam	煤层尖灭
thin out trap	尖灭圈闭
thin overburden	薄覆盖层
thin seam	薄煤层
thin seam support	薄煤层支架
thin section	薄片；薄切片
thin section analysis	薄片分析
thin skinned tectonics	薄皮构造
thin stratum	薄层
thin wall sampler	薄壁取土器
thin wall tube	薄壁取土管
thin-bedded alternation	薄层交互层
thin-cylinder formula	薄壁圆筒公式
thinly interlayered bedding	薄的互层层理
thinly laminated	薄层状
thinning out	尖灭
thinning out of strata	地层尖灭
thin-section of a rock	岩石薄片
thin-shell structure	薄壳结构
thin-wall construction	薄壁结构
thiorsauite	钙长石
thirl	穿孔；钻孔；开掘联络巷
thirsty soil	干燥土
thixotropic	触变性
thixotropic effect	触变效应
thixotropic hardening	触变硬化
thixotropic injection	触变灌浆
thixotropy	触变性
tholeiite	拉斑玄武岩
tholeiitic series	拉斑系列
tholeiitic texture	拉斑玄武结构
thomsonite	杆沸石，镁沸石
thoreaulite	钽锡矿
thorianite	方钍石
thortveitite	钪钇石
thoruraninite	钍铀矿
thread interval	螺距
thread twisting method	搓条法
threading	穿入拱；穿入套管
three centre arch	三心拱
three dimensional	三维的；立体的
three dimensional city model	城市三维建模
three dimensional consolidation	三维固结
three dimensional entity	三维实体
three dimensional fault	三维断层
three dimensional map	立体图
three dimensional seismic method	三维地震勘探法
three dimensional stress	三维应力
three dimensional visualization	三维可视化
three dimensional wedge failure	三维楔体滑落
Three Gorges Project	三峡工程
three hinged arch	三铰拱
three phase diagram	三相图
three phases of soil	土壤三相
three-dimension display	三维显示

three-dimension fracturing	三维压裂
three-dimension map	三维地图；立体地图
three-dimensional consolidation	三向固结
three-dimensional loading	三维加荷
three-dimensional model	立体模型，三维模型
three-dimensional nonlinear analysis	三维非线性分析
three-dimensional stress analysis	三维应力分析
three-dimensional terrain simulation	三维地景仿真
three-directional layerwise summation method	三向分层总和法
three-edge bearing test	三边支承试验；三向载荷试验
three-fold	三倍的，三重的
three-phase	三相的
three-point method	三点法
three-view drawing	三视图
three-way elbow	三向弯头
three-way flat slab	三向无梁楼板
three-way reinforcement	三向配筋
threshold	开始点，临界值；限度；阈
threshold analysis	临界分析
threshold effects	阈值效应
threshold magnitude	临界震级
threshold of moisture	临界湿度
threshold pressure	界限压力
threshold saturation	临界持水度；饱和阈值；将饱和
threshold state	临界状态
threshold value	临界值；阈值
throat	喉道；巷道口；入口
throat ring	节流环；填料衬环
throttle	风门；节流阀
throttling resistance	节流阻抗
through bridge	下承式桥
through crack	贯穿裂缝
through cut	贯穿开挖；路堑
through fault	下落断层
through span	穿过式桥跨
through stone	穿墙石
through structural plane	贯通结构面
through valley	贯通谷
throughout capacity	总生产能力
throw	垂直断距
throw blasting	抛掷爆破
throw fault	下落断层；正断层
throwaway bit	一次性使用钻头
throwing stones to retain embankment	抛石护岸
thrown	错动的；降落的
thrown fault	下落断层
thrown side wall	断层下降盘
thrown slip	下落滑动
thrust	冲断层
thrust at springer	拱端推力，拱脚推力
thrust block	推覆体
thrust borer	凿进机，隧道凿岩机；冲击式凿岩机；冲击式钎子钻机
thrust boring	顶管法；冲击钻孔
thrust capacity	支承能力
thrust fault	逆冲断层
thrust force	推力
thrust imbrication	逆掩叠瓦作用
thrust load	推力负载
thrust metre	推力计
thrust moment	推力力矩
thrust nappe	逆冲推覆体，冲掩推覆体
thrust of arch	拱的推力
thrust of earth	土压力
thrust plane	冲断面；逆冲断层面
thrust plate	推覆体
thrust roof fall	推垮型冒顶
thrust scarp	逆冲断崖
thrust sheet	逆掩盘；上冲盘；推覆体
thrust slip	冲断滑动
thrust slip fault	冲断层
thrust test	推力试验
thrust wedge	楔状冲断体；推断楔形体
thrust-fault plane	逆掩面
thrusting action	推压作用
thrusting force	推力
thrust-strike slip fault	逆-平移断层
thump	强冲击
thuresite	钠闪正长岩
thuringite	鳞绿泥石
thwack	捣实
tidal action	潮汐作用
tidal amplitude	潮幅；潮汐振幅
tidal atlas	潮汐图
tidal backwater	潮水顶托回水
tidal ball	报潮球
tidal bank	潮滩
tidal barrage	挡潮闸
tidal basin	潮船坞；潮港池
tidal bedding	潮汐层理
tidal bench mark	潮汐水准点
tidal bore	潮津波；潮水上涨；潮水涌浪
tidal coast	潮汐海岸
tidal constant	潮汐常数
tidal correction	潮汐校正
tidal creek	潮沟
tidal current	潮流
tidal current scour	潮流冲蚀
tidal datum	潮准线；潮汐基准面
tidal delta	潮汐三角洲
tidal deposit	潮积物
tidal depositional cycle	潮汐沉积旋回
tidal divide	潮道分水岭；分潮岭
tidal double ebb	双低潮；双重低潮
tidal drift	潮汐漂移
tidal effect	潮汐效应；潮汐影响
tidal embankment	挡潮堤
tidal energy	潮汐能
tidal envelope	潮汐包线
tidal epoch	潮汐迟角；潮滞
tidal factor	潮汐因子
tidal flat	潮滩；潮成平地；潮汐滩地
tidal flat deposit	潮滩沉积
tidal force	起潮力
tidal influx	进潮量
tidal lag	潮汐滞后；潮滞
tidal lamination	潮汐分层
tidal land	潮间地
tidal load	潮压

tidal lock	潮汐船闸；挡潮船闸
tidal lockup	潮汐托顶；潮壅
tidal observation	验潮；潮汐观测
tidal quay	潮汐岸壁式码头；高潮码头
tidal range	潮差；潮位变幅
tidal regime	潮汐状况；潮汐特性
tidal ridge	潮流脊
tidal scouring channel	潮流冲沟
tidal stand	平潮；停潮
tidal strain	潮汐应变
tidal stress	潮汐应力
tidalite	潮积岩；潮积物
tide	潮汐
tide amplitude	潮汐振幅
tide disaster	潮灾
tide gauge	测潮杆，测潮器，潮位计，测潮仪
tide prime	初潮
tide priming	潮时提前
tide range	潮差；潮位变幅
tide rip	激潮
tide-generating force	引潮力，起潮力，涨潮力
tide-generating potential	引潮位
tie back retaining structure	锚杆支挡结构
tie back wall	锚定挡墙
tie beam	系梁；水平拉杆
tie bolt	锚固螺栓
tie distance	枕木间距，轨枕间距
tie plug	枕木塞
tie point	连接点
tie rod	拉杆
tie rod anchored retaining wall	锚杆式挡土墙
tieback	锚杆；锚固
tie-back cable	锚索；后拉索
tied island	陆连岛
tied retaining wall	锚定挡墙
tied sheet pile wall	锚定板桩墙
tie-into	嵌入
tiering	堆叠
tierra templada	高山温带区
tiger eye	虎睛宝石；虎眼石
tight	致密的
tight blasting	挤压爆破
tight fissure	紧闭裂缝，致密裂隙
tight fold	紧闭褶皱
tight formation	低渗透性地层；致密坚硬的岩层
tight fracture	闭合裂隙
tight gas	致密气
tight gas reservoirs	致密气储层
tight joint	紧密结合
tight pack	密实充填
tight rock	致密岩
tight roof	坚固顶板
tight sand	致密沙层
tight sandstone	致密砂岩
tight sandstone gas	致密砂岩气
tight zone	致密层
tight-confining bed	强隔水层
tighten	拉紧；固定
tightly bound water	强结合水
tilaite	透橄岩

tile subdrain	地下瓦管
tile underdrain	地下排水瓦管
tilestone	石板；扁石
till	冰碛物
till angle	倾斜角
till plain	冰碛平原
tilleyite	碳硅钙石
tillite	冰碛岩
tilloid	冰积石；类冰碛物；似冰碛岩
tilly	冰碛的
tilt	倾斜
tilt adjustment	纠偏扶正，倾斜纠正
tilt angle	倾角
tilt block	掀斜断块；翘起断块
tilt block basin	掀断盆地
tilt block tectonics	掀斜断块构造
tilt displacement	倾斜位移
tilt effect	倾斜效应
tilt gauge	倾斜仪
tilt meter	测斜仪
tilt monitoring device	倾斜监测装置
tilt observation	倾斜观测
tilt reference	倾斜基准
tilt stability	倾覆稳定性
tilt-angle method	倾角法
tilted aquifer	倾斜含水层
tilted block	倾斜地块
tilted fault block	掀斜断块；抬斜断块
tilted mountain	掀斜山
tilted strata	倾斜地层
tilted structure	掀斜构造
tilted terrace	倾斜阶地
tilted trees	马刀树
tilted up	倒转的
tilting	掀斜，倾斜
tilting load	倾覆荷载
tilting axis	倾斜轴
tilting failure	倾倒破坏
tilting load	倾斜荷载
tilting moment	倾覆力矩
tilting motion	翻转运动；倾倒
tilting of strata	地层倒转；地层倾斜；地层掀起
tilting platform	倾卸平台
tilting stiffness	抗斜刚度
tilting survey	倾斜测量
timazite	闪安绿磐岩
timber	坑木；支架
timber caisson	木沉箱
timber cofferdam	木围堰
timber dam	木坝
timber damage	支架破坏
timber extraction	回收坑木；支柱回收
timber fender	防撞护木；护木
timber groin	木丁坝
timber groyne	木丁坝
timber guide crib	导水木笼
timber pile	木桩
timber recovery	回收坑木
timber seal	木质水封
timber set	框架支护

timber sheet pile	木板桩
timber structure	木结构
timber support	木支撑，木支护
timber withdrawing	支架回收；坑木回收
timbered drift	木支护的平巷
timbered shaft	木支架井筒；木支护竖井
timbering	木支架，木支撑
timberless	无圆木支护的
timber-sheet piling	木板桩；打木板桩
timber-trussed roof	以木架支撑的屋顶
time accuracy	时间准确度
time analysis	时间分析
time anomaly	时间异常
time classification of strata	地层时代分类
time constant	时间常数
time cross-section	时间剖面图
time curve	时间曲线；时距曲线
time delay	时间延迟
time delay compensation	时滞补偿
time delayed rockburst	时滞型岩爆
time dependency	时间依存性；时间相关性
time depth conversion	时深转换
time depth curve	时间深度曲线
time derivative	时间导数
time difference	时差
time distance curve	走时曲线；时距曲线
time edge effect	时间边缘效应
time effect	时间效应
time epoch	时刻
time exceeded	超时
time expand	延时
time factor	时间因子；时间因素
time for completion	竣工时间
time interval	时段；时间间隔
time intervals	间歇
time lag	时间滞后
time lapse seismology	时间推移地震法
time lead	时间超前
time limitation	时限
time of consolidation	固结时间
time of earthquake at epicenter	震中震时
time of filling	注水时间
time of final setting	终凝时间
time of hardening	硬化时间
time of initial setting	初凝时间
time of loading	装载时间
time of observation	观测时间
time of occurrence	发生时间
time of one round	循环时间
time of oscillation	振动周期
time of slaking	崩解时间；熟化时间
time of travel	走时
time piece	时间间隔
time profile	时间剖面
time pulse	时间脉冲
time range	时间间隔
time schedule control	时序控制
time section	时间剖面
time sequence	时间序列
time sequence model	时间序列模型
time sequential	时序的
time series	时间序列
time series analysis	时间序列分析
time splitting method	时间分步法
time step	时间步长
time stepping	时步
time stepping scheme	时步法
time stratigraphy	年代地层学
time to depth conversion	时深转换
time to failure	破坏时间
time zone	时区
time-consolidation curve	时间-固结曲线
time-creep curve	时间-蠕动曲线
time-delayed	时间落后的，时滞的，时间延迟的
time-dependent	随时间变化的
time-dependent effect	时间效应；依时效应
time-dependent subsidence	随时间沉陷；依时沉陷
time-depth chart	垂直时距曲线图
time-displacement curve	时间-位移曲线
time-distance curve	时间-距离曲线，时距曲线
time-distance graph	时距曲线图
time-drawdown curve	时间-降深曲线；时间-沉降曲线
time-integration sampling	积时法取样
time-intensity graph	时间-强度关系图
time-interval method	时间间隔法
time-lag action	延时作用，时滞作用
timeous	及时的；适时的
time-path curve	时距曲线
timeproof	耐久的
time-related factors	时间因素
time-reversibility	时间可逆性
time-settlement curve	时间-沉降曲线
time-stratigraphic facies	时代地层相
time-stratigraphic unit	时间地层单位
timestratigraphy	年代地层学
time-subsidence curve	时间-沉陷曲线
timetable	时间表
time-transgressive	不同时代的
time-transitional	时代过渡的
time-travel curve	时程曲线
time-variable filtering	时变滤波
time-variant seepage	不稳定渗流
time-yield	蠕变
timing sampling	定时取样
tin minerals	锡矿类
tin ore	锡矿石
tincal	粗硼砂；原硼砂
tinguaite	细霞霓岩；丁古岩
tinlode	锡矿脉
tiny	微小的，细小的
tiny crack	微裂缝
tiny void	细微孔隙
tinzenite	锰蓝宝石
tip	倾斜，翻倒，倾覆
tip effect	梢端效应；尖部效应
tip of pile	桩端
tip point	倾卸点；卸载点
tip resistance	桩端阻力
tip-back	向后倾斜；后翻
tip-line fold	尖顶褶皱

tipping	翻转；翻卸；倾斜；倾卸的
tip-tilting	掀斜作用
titanaugite	钛辉石
title block	工程图明细表；标题栏
titrant	滴定剂；滴定标准液
titration	滴定
titration analysis	滴定分析
tiwa	地洼
tjaele	冻结层；冻土
toadstone	蟾蜍岩；玄武斑岩
to-and-fro method	往返法
todorokite	钡镁锰矿
toe	坡脚；墙脚
toe block	护脚块体；坡脚砌块
toe circle	坡趾圆
toe cut	底槽
toe drain	背水面坡脚排水；坝趾排水体
toe drainage	堤趾排水；坡趾排水
toe excavation	坡趾挖方；基趾开挖；坡脚开挖
toe failure	坡脚破坏，坡趾破坏
toe filter	坝趾滤层
toe of a slope	斜坡脚；坡脚
toe of dam	坝趾
toe of fill	填土坡脚；路堤坡脚
toe of slope	坡脚
toe of wall	墙脚
toe pressed by backfill	回填压脚
toe protection	坡脚保护
toe slope	冲刷坡；山麓坡
toe stress	坝趾应力
toe wall	坝趾齿墙；坡脚墙
toe weight	坡脚压重；堤脚压重
toellite	英云闪玢岩
tolerable concentration	可容许浓度
tolerable limit	容许限度
tolerable settlement	容许沉降量
tolerance	耐受力；容许误差
tolerance analysis	公差分析，容差分析
tolerance limit	容许极限
tollite	英长云闪玢岩
tombolo	连岛沙洲；陆连岛
tomite	藻煤
tomography	层析成像
ton	吨
tonal anomaly	色调异常
tonalite	英云闪长岩
tone range	色阶；灰阶；色调范围；明暗亮度范围
tone signature	色调特征
Tonge-Karmadec Trench	汤加-克马德克海沟
Tongrian Stage	通格里阶
tongue	舌状体
tongue and groove	企口缝；凹凸缝
tongue of landslide	滑坡舌部
tongue tear test	舌形扯破试验
tonnage	吨位；吨数
tonnage noting	载重计算法
tonstein	黏土岩
tool extractor	钻具打捞器
tool grab	钻具打捞器
tool mark	压刻痕
tool stand	工具台；工具架
tooth	软土刮刀
top	山顶；顶板
top and bottom heading method	上下导坑法
top and twin-side bottom heading method	品字形导坑法
top beam	顶梁
top bed	顶部岩层
top bench	上部阶段
top benching	上梯段式开采
top capacity	极限能力
top circle	外圆；顶圆
top coal	顶煤
top coal caving	放顶煤开采
top concordance	顶整合
top conglomerate	顶砾岩
top contraction	顶部收缩
top convergence	顶板沉陷
top cover	顶板岩层；顶盖
top crown	拱顶；顶冠
top drift	顶部掏槽
top failure	顶板破坏；落顶，冒顶
top heading	上导洞；顶部导坑
top heading and bench method	顶导坑及平台法，坑台法
top heading excavation	上导洞开挖
top heading method	上导坑法
top impermeable layer	隔水顶板
top layer	顶层
top longitudinal bracing	上弦纵向水平支撑
top of cutting	堑顶
top of slope	边坡顶
top parallel entry	上部平行平巷
top performance	最高性能；最高生产率
top picking	挑顶
top plate	顶板
top pressure	顶部压力；最大压力
top road	上部平巷
top rock	顶板岩层
top sand	粗粒砂
top seal	盖层
top seam	顶部煤层
top shooting	顶板爆破
top slicing	分层崩落采矿法；上部剥落
top slicing and cover caving	顶部分层崩落采矿法；下行分层陷落开采法
top slope	顶帮
top soil	表层土
top soil erosion	表土侵蚀
top squeeze	顶板下沉
top storey	顶层
top stratum	地表岩层
top stripping	表层剥离
top subsidence	顶板沉陷
top surface	顶面
top testing	问顶；敲顶
top thickness	顶厚
top view	顶视图，俯视图
top wall	上盘；上升盘
top water-level	最高蓄水位
topaz	黄晶；黄玉
topazite	黄玉岩

topazization	黄玉化
topazolite	黄榴石
top-coal caving	放顶煤
top-discordance	顶部不整合
top-draw cut	上部掏槽，顶槽
topo beacon	地形测量标志
topographic adjustment	地形校正
topographic adolescence	地形青年期；地形成长期
topographic analysis	地形分析
topographic condition	地形条件；地形状况
topographic contour	地形等高线
topographic control point	地形控制点
topographic control survey	地形控制测量
topographic correction	地形校正
topographic crest	山脊；分水岭
topographic depression	凹陷
topographic divide	分水岭
topographic drawing	地形图
topographic expression	地形表示法
topographic feature	地形要素
topographic feature survey	地貌调查
topographic inequality	地形起伏
topographic infancy	地形幼年期
topographic interpretation	地形解译
topographic map	地形图
topographic map representation	地形图表示法
topographic map revision	地形图修测
topographic map subdivision	地形图分幅
topographic maturity	壮年地形
topographic old age	老年地形
topographic profile	地形剖面
topographic reconnaissance	地形踏勘，地形普查
topographic relief	地形起伏
topographic relief map	地势图
topographic survey	地形测量
topographic symbols	地形图图例
topographic trough	地形槽
topographic unconformity	地形不整合
topographical crest	山脊线
topographical cycle	地形旋回
topographical map	地形图
topographical maturity	地形壮年期
topographical old age	地形老年期
topographical profile	地形剖面
topographical survey	地形测量
topographical unconformity	地形不整合
topographical youth	地形幼年期
topography	地貌，地形；地形学
topography interpretation	地形判读
topography note	地形测量记录
topological	拓扑的
topometry	地形测量
toposequence	地形系列
topple	倾覆；坍塌
toppling	倾倒，溃曲
toppling failure	倾倒破坏，倾覆破坏
toppling mass	倾倒体
toppling pressure	最高压力
top-quality	最优质的
top-ranking	最高级的
topsailite	中辉煌岩
topset	顶积层
topset deposit	顶积沉积
topsoil excavation	表土开挖
topsoil stripping	表土剥离
torpedo gravel	尖砾石；粗粒砂
torpedo sand	粗粒砂
torque	转矩；扭转力矩
torque failure	扭转破坏
torque force	扭力
torque moment	扭矩
torrent	山洪；湍流
torrent tract	上游段
torrential	洪流的；急流的
torrential burst	山洪暴发
torrential cross bedding	急流交错层理
torrential current	急流
torrential flood	山洪；湍流
torrential rain	暴雨
torrential sediment	洪积物
torrential stream	急流
torrential wash	急流冲刷
torreyite	锰镁锌矾
torsel	承梁木
torsiogram	扭矩图；扭振图
torsion	扭力；扭转
torsion angle	扭转角
torsion fault	扭断层
torsion force	扭力
torsion joint	扭节理
torsion modulus	扭曲模量；扭转模量
torsion moment	扭转力矩
torsion resistance	抗扭；抗扭强度
torsion rigidity	扭转刚度
torsion ring shear apparatus	环形试件扭剪仪
torsion shear	扭转剪切
torsion stiffness	抗扭刚度
torsion strength	抗扭强度
torsion stress	扭转应力；抗扭应力
torsional	扭转的
torsional angle	扭角
torsional bending	扭转弯曲
torsional buckling	扭转屈曲
torsional capacity	抗扭承载力
torsional cyclic load triaxial apparatus	扭转周期荷载三轴仪
torsional deformation	扭曲变形
torsional displacement	扭转位移
torsional endurance limit	抗扭抗疲劳限度
torsional failure	扭转破坏
torsional fatigue limit	扭转疲劳极限
torsional force	扭力
torsional oscillation	扭转摆动
torsional reinforcement	抗扭钢筋
torsional restraint	扭转约束
torsional rigidity	扭转刚度；抗扭刚度
torsional shear strength	扭转剪切强度
torsional shear test	扭转剪切试验
torsional strain	扭应变
torsional strength	抗扭强度
torsional stress	扭应力

torsional vibrations	扭转振动
torsionmeter shear test	扭转剪切试验
torso mountain	残山
tortuosity	弯曲度
tortuous flow	紊流，乱流
torvane	小型十字板仪
toryhillite	多霞钠长岩
total air quantity	总风量
total analysis	全分析；全项分析
total annual precipitation	年降水总量
total chemical analysis of water	水化学全分析
total cohesion	总黏聚力
total cohesion intercept	总黏聚力截距
total cohesion method	总黏聚力法
total collapse	总塌陷量
total content	总含量
total core recovery	总岩芯采取率
total deformation	总变形
total depth	总深度
total displacement	总位移
total evaporation	总蒸发量
total extraction	全部采出
total flow	总流量
total flux	总流量
total groundwater resource	地下水资源总量
total groundwater yield	地下水总出水量
total grout take	总灌浆量
total hardness	总硬度
total hardness of groundwater	地下水的总硬度
total head	总水头；全水头
total head gradient	总水头梯度
total heave	总隆起量
total hole depth	全孔深
total intensity	总强度
total intensity of magnetic anomaly	总磁异常强度
total load	总负载
total mineralization	总矿化度
total mineralization of groundwater	地下水总矿化度
total moisture content	总含水量
total normal stress	法向总应力
total organic content	有机质含量
total oxygen demand (TOD)	总需氧量
total penetration of pile	打桩总深度
total permissible error	总允许误差
total porosity	总孔隙度
total potential energy	总势能
total precipitation	总降水量
total pressure	总压力
total quality control (TQC)	全面质量管理
total reduction ratio	总破碎比
total reflection	全反射
total reservoir storage	水库总容量
total run	总行程
total runoff	总径流量
total settlement	总沉降量
total shaft resistance	总侧面阻力
total shearing force	总剪切力
total stiffness matrix	总刚度矩阵
total strain	总应变
total stream load	总输沙量
total strength envelope	总强度包线
total stress	总应力
total stress analysis	总应力分析
total stress approach	总应力法
total stress circle	总应力圆
total stress failure envelope	总应力破坏包线
total stress path	总应力路径
total stress strength index	总应力强度指标
total stress strength parameter	总应力强度参数
total suction	总吸力
total theory of plasticity	塑性全量理论
total throw	总落差
total up to	合计
total volume	总容积
total volume of rain	总雨量
total water cement ratio	总水灰比
total water requirement	总需水量
total water saturation	总含水饱和度
total weight	总重量
totality theory	全量理论
totalized load	总负荷
totally lost circulation	全部漏失
touchstone	试金石
tough	坚韧的；坚固的
toughness	韧度；韧性
toughness factor	韧性因数
toughness test	韧性试验
tourmaline	电气石
tourmalite	电英岩
tow	拖拉，曳，牵引
tow wall	矮墙
towed grader	牵引式平土机
tower excavator	塔式挖土机；塔式挖掘机
tower karst	塔状岩溶，塔式喀斯特
tower structure	塔式结构
tower support	塔架支座
tower-type headframe	塔式井架；井塔
towing	拖曳
towing force	拖曳力，拉力，牵引力
towing plate	牵引板
town planning	城市规划
town planning layout	城市规划详细蓝图
town-lot drilling	城区钻探
toxic mine gas	矿山毒性瓦斯
trace	痕迹；迹线
trace amount	微量
trace atom	示踪原子
trace element	示踪元素；微量元素
trace fossil	遗迹化石
trace indicator	示踪指示剂
trace inversion	地震踪反演
trace mode	追踪方式
trace of fault	断层迹线
trace of matrix	矩阵的迹
tracer	示踪剂，追踪物
tracer amount	示踪量
tracer determination	示踪测定
tracer isotope	示踪同位素；同位素指示剂
tracer log	示踪测井
tracer mineral	示踪矿物

tracer technique	示踪技术
trace-slip fault	迹线滑移断层
trachyandesite	粗安岩
trachybasalt	粗面玄武岩，粗玄岩
Trachyceras	粗菊石属
trachydacite	粗面英安岩
trachydiscontinuity	凹凸不整合
trachyliparite	粗面流纹岩
Trachysphaeridium	粗面球形藻属
trachyte	粗面岩
trachyte andesite	粗面安山岩
trachyte porphyry	粗面斑岩
trachytic	粗面状的
trachytic texture	粗面结构
trachytoid texture	似粗面状结构
tracing	描图，示踪
tracing experiment	示踪试验
tracing instrument	描图仪器
tracing joint	追踪节理
tracing paper	描图纸
tracing sands	示踪沙
track	示踪径迹
track bearing capacity	轨道承载力
track bed	道床；轨道路基
track buckling	轨道鼓出
track disorder	轨道变形
track displacement	轨道位移
track dynamics	轨道动力学
track elevation	轨道标高
track entry	有轨平巷
track failure	轨道失效
track mechanics	轨道力学
track plan	线路平面图
track profile	线路纵断面；轨道剖面
track resistance	轨道阻力
track section	线路区段
track shimming	垫冻害垫板
track stability	轨道稳定性
track strength	轨道强度
track stresses	轨道应力
trackless mine	无轨矿井
trackless pit	无轨露天矿
trackless transport	无轨运输
trackway	行迹
tract	地区，地带，地域
traction	牵引；牵引力
traction force	牵引力
traction load	牵引荷载
traction phenomena	牵引现象
traction resistance	牵引阻力
traction test	牵引试验
tractive	牵引的
tractive capacity	牵引能力
tractive characteristic	牵引特性
tractive coefficient	牵引系数
tractive effort	牵引力；牵引作用
tractive force	牵引力
trade effluent	工业污水，工业废水
trade refuse	工业废物
trade sewage	工业污水
trade waste sewage	工业废水
traditional geotechnical engineering	传统岩土工程
traditional strip foundation	常规条形基础
traffic engineering	交通工程
traffic pan	车压硬土层
traffic tunnel	运输隧道
trail	痕迹，轨迹
trail of fault	断层迹
trail pumping	试验抽水
trailing load	牵引载荷
trailing vortex	后缘涡流
train overturning	列车颠覆
train resistance	列车运行阻力
training dike	导流坝；导水堤
training for sediment	导沙
training jetty	导堤；导水突堤
training levee	导流堤
training mole	导流堤
training wall	导流堤；导水墙
trainroad	临时铁路
trajectories of principal stresses	主应力轨迹
trajectory	轨迹
trajectory of boulder fall	堕石落径
trajectory of strain	应变轨迹
tramming drift	运输平巷
Tran-Eurasian seismic belt	横贯欧亚地震带
tranquil regime	平稳状态
transaction	学报；会刊
transcendental	超越的
transcendental curve	超越曲线
transcendental equation	超越方程
transcendental function	超越函数
transcontinental	横贯大陆的
transcrystalline	横穿晶的
transcrystallization	横晶集；穿晶
transcurrent	横过的，横贯的，横向延伸的
transcurrent fault	平移断层
transcurrent motion	横推运动
transcurrent thrust	横冲断层
transducer	换能器
transect	横切，横断；横断面
transect cleavage	横切劈理
transfer	泥沙运移
transfer bond	传递黏结力
transfer coefficient	传递系数
transfer fault	转换断层
transfer function	转换函数；传递函数
transfer strength	传递强度
transference	迁移；搬运；输送
transference of load	荷载传递
transform	转换，变换，转变
transform analysis	变换分析
transform boundary	转换边界
transform fault	转换断层
transform mechanism of landslide-debris flow	滑坡泥石流转化机制
transform plate boundary	转换型板块边界
transform strike slip boundary	转换边缘
transformable	可变换的；可变形的
transformation coefficient	变换系数

transformation factor	变换因子，转换因子
transformation function	转换函数，变换函数
transformation load	变换荷载
transformation matrix	变换矩阵，转换矩阵
transformation of magnetization direction	磁化方向换算
transformation of strain	应变变换
transformation of stress	应力变换
transformation of stress space	应力空间转换
transformation period	半衰期
transformation point	临界点
transformation ratio	转化率
transformation twin	转变双晶，转变孪晶
transformational faulting	转换断层作用
transformed section	换算截面
transformer	变压器
transgression	海进，海侵
transgression facies	海侵相；海进相
transgression of sea water	海水入侵
transgression phase	海侵相；海进相
transgression sea	侵陆海
transgressive injection	不整合侵入
transgressive intrusion	贯层侵入，不整合侵入
transgressive overlap	海侵超覆
transgressive sequence	海进层序
transgressive series	海侵岩系
transgressive surface	海侵面
transgressive unconformity	海侵不整合
transient	瞬时的；瞬态的
transient action	瞬时作用
transient analysis	瞬态分析
transient cavity pressure	洞室瞬态压力
transient change	瞬时变化
transient computation	瞬态计算
transient creep	暂态蠕变
transient deformation	瞬时变形
transient displacement	瞬时位移
transient distortion	瞬态失真
transient electromagnetic field	瞬变电磁场
transient electromagnetic method	瞬变电磁勘探法；脉冲电磁法
transient electromagnetic sounding	瞬变电磁测深
transient field method	瞬变场法
transient flow	瞬变流，瞬态流
transient load	瞬变荷载
transient motion	瞬时运动，瞬变运动
transient performance	瞬态特性
transient pumping test	瞬时抽水试验
transient response	瞬态响应
transient seepage flow	不稳定渗流
transient strain	瞬态应变
transient test	瞬态试验
transient unloading	瞬态卸荷
transient variation	瞬时变化
transient vertical loading	瞬时竖向荷载
transient vibration	瞬态振动
transient waveform analysis	瞬态波形分析
transit	经纬仪
transit compass	经纬仪罗盘
transit level	经纬水准仪
transit station	经纬仪测站
transit traverse	经纬仪导线
transition beds	过渡层
transition belt	过渡带
transition bond	过渡联结
transition curve	缓和曲线；过渡曲线
transition element	过渡元素
transition flow regime	过渡流态
transition grading zone	缓和坡段
transition layer	过渡层
transition phase	过渡相
transition point section	过渡段
transition structure	过渡结构
transition type	过渡型
transitional belt	过渡带
transitional fault	滑动断层
transitional form	过渡类型
transitional series	过渡岩系
transitional slide	顺坡滑动
transitional type	过渡类型
transitional zone	过渡带；过渡区
translation axis	平移轴
translation control	平移控制
translation coordinates	平移坐标
translation direction	平移方向
translation fault	平移断层
translation of rigid body	刚体平移
translation plane	平移面
translational	平移的；直移的
translational failure	平移破坏
translational landslide	平推式滑坡
translational scar	平移断壁
translational slides	平移性滑动
translatory motion	平移运动
translithospheric fracture	超岩石圈断裂
translocation	移位
translucence	半透明
transmissibility	导水性
transmissibility coefficient	导水系数
transmission	输电；输送
transmission capacity	渗透能力
transmission characteristic	传递特性
transmission coefficient	透射系数
transmission dynamometer	传动测力计
transmission electron microscope (TEM)	透射电子显微镜，透射电镜
transmission of load	荷载传递
transmission of pressure	压力传递
transmissivity	导水率；导水系数
transmissivity contour map	导水系数等值线图
transmissivity of aquifer	含水层输水能力
transmitter	传感器；变送器；传送器
transmountain diversion	穿山引水
transmutation	转变；演变
transoceanic	越洋的，横渡大洋的
transparence	透明度；透明性；透明
transparent	透明的
transparent groundmass	透明基质
transport capacity	搬运能力
transport concentration	输沙率
transport of sediment	泥沙搬运

transportable	可搬运的
transportation	搬运；搬运作用；运输
transportation by glacier	冰川搬运
transportation by leaps and bound	跳跃式输移
transportation deposit	搬运沉积
transportation mechanisms of sediment	沉积物的搬运机制
transportation medium	搬运介质
transportation mineralization	转移成矿作用
transportation of sea and lake	海湖搬运作用
transportation of sediments	沉积物搬运
transported	被搬运的；搬运过的
transported deposit	搬运堆积物；搬运沉积
transported soil	运积土
transporting action of running water	流水搬运作用
transporting agent	搬运营力
transporting capacity	搬运能力
transpose matrix	转置矩阵
transposition cleavage	变换劈理
transposition foliation	置换剥理；置换面理
transpositional structure	转位构造
transpression	扭压作用；转换挤压作用
transtensional basin	扭张盆地
transtensional fault	转换拉张断层
transudation	渗滤作用
transversal	横向的；横断的
transversal beam	横梁
transversal cross-section	横断面
transversal erosion	横向侵蚀
transversal fault	横推断层；横冲断层
transversal jointing	横节理；交错节理
transversal section	横断面
transversal shrinkage	横向收缩
transversal strain	横向应变
transversal surface	横截面
transversal thrust	横冲断层
transversal valley	横谷
transversal wave	横波；剪切波
transversal wave frequency	横波频率
transverse	横向的
transverse anisotropy	横观各向异性
transverse arch	横向拱
transverse axis	横轴
transverse bar	横向沙坝；横向沙洲
transverse bending	横向弯曲
transverse bending strength	横向抗弯强度
transverse bending test	横向弯曲试验
transverse cleavage	交错劈理；横劈理
transverse compression wave	横向压缩波
transverse crack	横向裂缝
transverse creep	横向蠕动
transverse cutting plane	横向切割面
transverse deformation	侧向变形，横向变形
transverse dislocation	横向位错
transverse displacement	横向位移
transverse distribution	横向分布
transverse elasticity	横向弹性
transverse elasticity test	横向弹性试验
transverse expansion joint	横向伸缩缝
transverse extension	横向扩张
transverse extensometer	横向应变计
transverse fault	横推断层
transverse fissure	横裂缝
transverse fold	横褶皱
transverse force	横向力，剪力
transverse gallery	横洞；横向平巷
transverse isotropy	横观各向同性
transverse joint	横节理
transverse load	横向荷载
transverse motion	横向运动
transverse permeability	横向渗透性
transverse pitch	横向坡度
transverse prestress	横向预应力
transverse profile	横断面
transverse projection	横轴投影；横向投影
transverse reinforcement	横向钢筋；横向加固
transverse section	横截面；横断面
transverse shear	横向剪切
transverse shear modulus	横向剪切模量
transverse shearing stress	横向剪应力
transverse slide	横向滑移
transverse strength	横向强度；弯曲强度
transverse stress	横向应力
transverse thrust	横向断层；横向推力
transverse to the strike	垂直于走向
transverse valley	横谷
transverse wedge	横楔
transversely isotropic	横观各向同性
trap	陷阱
trap dike	暗色岩脉
trap down	下落断层
trap dyke	暗色岩脉
trap efficiency	拦沙效率；截淤效率
trap rock	暗色岩
trap up	逆断层
trapeziform	梯形
trapezium	梯形
trapezoid	梯形的
trapezoid cross section	梯形截面
trapezoid projection	梯形投影
trapezoidal foundation	梯形基础
trapezoidal groin	梯形丁坝
trapezoidal groyne	梯形丁坝
trapezoid-shaped load	梯形荷载
trapp	暗色岩
trappide	暗色岩
trapping of ground water	地下水截流
trash	碎屑
trash baffle	拦污墙
trash chute	排污斜槽
trash rack	护板；挡泥板；拦污栅
trash sluice	排污闸
trass	浮石凝灰岩；火山土
trass cement	火山灰水泥；粗面凝灰岩水泥
trass concrete	火山灰混凝土
travel of underground water	地下水的运动
travel stone	漂石
travel time	走时
travel time curve	走时曲线
travel time difference	走时差
travelable	可移动的

traveled soil	运积土	tremor epicenter	震源
traveling load	移动荷载	tremor source parameter	震源参数
traveling wave	行波	trench	沟，沟渠，沟槽；海沟，海槽
travelling disturbance	运动性扰动	trench backfill	挖槽回填
travelling dune	移动沙丘	trench cut method	挖沟法
travelling roadway	运输巷道	trench digger	挖掘机
traverse	导线；横穿	trench excavation	开挖沟槽，挖沟
traverse adjustment	导线平差	trench excavator	挖沟机，挖坑机
traverse angle	导线折角	trench fault	沟状断层
traverse closure	导线闭合	trench method	沉埋法
traverse distance	导线距离；导线长	trench refilling	堑壕回填
traverse error of closure	导线闭合差	trench sample	刻槽样品
traverse geological map	路线地质图	trench sampler	探槽取样；槽沟取样
traverse layout	导线布设	trench sediment	海沟沉积
traverse leg	导线边	trench structure	海沟结构
traverse line	测线；导线；方向线	trench test	槽探
traverse map	导线图	trench timbering	挡土支护，挡土支撑
traverse measurement	导线量测	trench wall	槽壁，沟壁
traverse method	穿越法	trench-arc-basin system	沟弧盆系
traverse network	导线控制网	trenching	槽探
traverse point	导线点	trenching method	掘坑覆盖法
traverse station	导线站	trenching shot	开沟爆破
traverse survey	导线测量	trenching-slope basin	海沟-大陆坡盆地
traverse table	小平板	trenchless technology	暗挖技术
traversellite	透辉石	trend	走向；倾向
travertine	钙华，石灰华，凝灰石	trend analysis	趋势分析
travertine sinter	钙华；泉华	trend chart	趋势图
travertine terrace	石灰华阶地，石灰华台地	trend line	构造线
treacherous roof	不可靠的顶板	trend map	趋势图
treated	精选的	trend of a line	线的走向
treated soil	加固土；改良土	trend surface	趋势面
treatment by sand-bath	喷砂处理	trend surface analysis	趋势面分析
treatment of dam foundation	坝基处理	trepanning	钻削；打眼；穿孔
treatment of fault-fracture zone	断层带处理	Tresca criterion	特雷斯卡准则
treatment of negative skin friction	桩负摩阻处理	Tresca failure criteria	特雷斯卡破坏准则
treatment of rock	矸石装载	Tresca yield condition	特雷斯卡屈服条件
treatment of wastewater	废水处理；污水处理	trestle pile	支架桩
treatment of water	水的处理	trevorite	镍磁铁矿，铁镍矿
treatment process	处理过程；加工过程；精选过程	trial anchor	试锚
treatment scheme	处理方案；精选方案	trial blast	试验爆破
treatment with short pile	短桩处理	trial bore hole	试探钻孔
treble	三倍的；三重的	trial boring	试钻
tree diagram	树形图	trial drilling	试验钻探，试钻
treelike drainage pattern	树状水系	trial driving	试验打桩，试桩
trellis	格构；棚架	trial excavation	试挖
trellis arch	格构拱；钢花拱；格棚拱	trial hole	试孔
trellis diagram	格状水系；格子图	trial load	试验荷载
trellis drainage	格状水系	trial pile	试验桩
trellis drainage pattern	格状水系	trial pit	试坑
trellis girder	格构梁	trial pumping	抽水试验
trellised drainage	格状排水系统	trial shaft	勘探井
tremie	漏斗管；混凝土导管	trial shots	试验爆破
tremie concrete	水下浇注混凝土；导管灌筑混凝土	trial speed	试验速度
tremie concrete seal	水下混凝土封底	trial trench	探槽；探沟
tremie concrete wall	导管灌注混凝土墙	trial wedge method	滑楔试算法
tremie method	导管法	trial-and-error	反复试验；逐次逼近法；试算法
tremie placing	导管浇筑	triangle	三角形
tremie seal	水下封底	triangle closing error	三角形闭合差
tremolite	透闪石	triangle load distribution	三角形荷载分布
tremor	震动；沟槽	triangular	三角的；三角形的

triangular arch	三角拱
triangular cut	三角形掏槽
triangular dam profile	三角形坝体断面
triangular distribution	三角形分布
triangular distribution load	三角形荷载
triangular elastic wedge	三角形弹性楔
triangular element	三角形单元
triangular facet of fault	断层三角面
triangular method	三角形法
triangularly distributed strip load	三角形式长条形荷载
triangulated irregular network (TIN)	不规则三角网
triangulation	三角测量
triangulation survey	三角测量
Trias	三叠纪；三叠系
Triassic	三叠系；三叠纪
Triassic period	三叠纪
Triassic rifting	三叠纪裂谷
Triassic subduction complex	三叠纪俯冲复合体
Triassic system	三叠系
Triassic unconformity	三叠纪不整合
triaxial	三轴的
triaxial apparatus	三轴试验仪，三轴试验机
triaxial cell	三轴压力室
triaxial compression	三轴压缩
triaxial compression test	三轴压缩试验
triaxial compressive strength	三轴抗压强度
triaxial consolidation test	三轴固结试验
triaxial dissipation test	三轴消散试验
triaxial equipment	三轴仪
triaxial extension test	三轴拉伸试验
triaxial loading	三向加载
triaxial photoelastic measurement	三轴光弹量测法
triaxial quick test	三轴快剪试验
triaxial seismograph	三轴地震仪
triaxial seismometer	三轴地震计
triaxial shear equipment	三轴剪力仪
triaxial shear test	三轴剪切试验
triaxial shrinkage test	三轴收缩试验
triaxial state of stress	三向应力状态
triaxial strain	三向应变
triaxial strain cell	三轴应变盒
triaxial strength	三轴强度
triaxial strength factor	三轴强度因子
triaxial stress	三轴应力
triaxial stress field	三轴应力场
triaxial stress state	三轴应力状态
triaxial stress system	三轴应力系统
triaxial tension test	三轴拉伸试验
triaxial unloading test	三轴卸载试验
tribar	三柱块体
tributary	支流
tributary ditch	进水沟；支沟
tributary fault	分支断层
trickle drain	滴流排水管
tricone bit	三锥形钻头；三牙轮钻头
tridiagonal matrix	三对角矩阵
tridimensional	三维的，立体的
trigger	触发
trigger action	触发作用
trigger mechanism	触发机制
trigger model of earthquake occurrence	地震发生的触发模型
trigger outburst	诱导突出
trigger slide	触发滑坡
triggered earthquake	触发地震
triggering area of debris flow	泥石流源区
trigonal	三角的；三角形的；三方晶系的
trigonal prism	三方柱
trigonal pyramid	三方锥
trigonal pyramidal class	三方单锥晶类
trigonal system	三方晶系
trigonometric heighting	三角高程测量
trigonometric leveling	三角高程测量
trilateration	三边测量
trilinear diagrams	三角图解
trilobite	三叶虫
trim	吃水差
trimming of shoulder	路肩修整
trimming of slope	斜坡修整
trimming of specimen	试件修整
trimming the working face	修整工作面
trimorph	同质三象；三形体
trip casing spear	套管打捞矛
triphase	三相，三相的
triphase soil	三相土
triple	三倍的；三重的；三元的
triple entry	三平巷
triple tube core barrel	三重岩心管
triplite	氟磷铁锰矿
tripod	三脚架
tripod drill	三角架式钻机；三角架式凿岩机
tripoli earth	硅藻土
tripuhyite	锑铁矿
troctolite	橄长岩
trona	天然碱
trondhjemite	奥长花岗岩
Tropic of Cancer	北回归线
Tropic of Capricorn	南回归线
tropic zone	热带
tropical island effect	热岛效应
tropical soil	热带土壤
tropical weathering	热带风化
trouble area	难开采区，不稳定岩层区
troublesome soil	难处理土
troublesome zone	扰动带
trough	槽，深海槽；谷
trough core	向斜中心
trough cross bedding	槽状交错层理
trough earthquake	海沟地震
trough fault	地堑
trough line of fold	褶皱槽线
trough of a syncline	向斜槽部
trough plane of fold	褶皱槽面
trough subsidence	下沉盆地
trough valley	槽谷
trough weathering	槽状风化
truck mounted drill	车装钻机
truck mounted rig	汽车钻
truck-mounted drill rig	车载钻机
truck-mounted percussion drill rig	车载冲击钻机
true angle of internal friction	真内摩擦角

true azimuth	真方位角
true bearing	真方位角；真方位
true bearing capacity	实际承载力
true cohesion	真黏聚力
true dip angle	真倾角
true drawdown	实际降深
true effective porosity	真有效孔隙度
true error	真误差
true groundwater velocity	地下水实际流速
true hardness	真硬度
true lode	裂缝矿脉；真矿脉
true slip	真断距，真滑距
true strain	真应变
true stress	真应力
true triaxial apparatus	真三轴仪
true triaxial shear test	真三轴剪切试验
true triaxial test	真三轴试验
true triaxial unloading	真三轴卸载
truncated anticline	削峰背斜
truncated normal distribution	截尾正态分布
truncated soil	剥蚀土壤
truncated soil profile	土壤剥蚀剖面
truncation	削截；削蚀；侵蚀
truss	桁架；构架
truss arch	桁架拱
truss bridge	桁架桥
truss spacing	桁架间距
truss span	桁架跨度
tsunami	海啸
tsunami earthquake	海啸地震
tsunami energy	海啸能量
tsunami erosion	海啸侵蚀
tsunamigenic earthquake	海啸地震
Tsushima Strait	对马海峡
tube caisson foundation	管柱基础
tube cross section area	隧洞横截面积
tube pile	管桩
tube pressure	隧道气压
tube railroad	地下铁道
tube railway	地下铁道
tube roughness	隧洞粗糙度
tube sample	取土管试样
tube sampler	管式取样器
tube sinking method	沉管法
tube structure	筒式结构
tube support	管支撑
tube wall	管壁
tubing catcher	捞管器
tubing spear	套管矛，套管打捞矛
tubular	管的，管状的
tubular brick	筒砖；空心砖
tubular column foundation	管柱基础
tubular construction	筒体结构
tubular drill	管钻
tubular foundation	管状基础
tubular pile	管桩
tubular railway	地下铁路
tubular spring	管状泉
tubular well	管井
tuff	火山灰；凝灰岩
tuff agglomerate-lava	凝灰集块熔岩
tuff ball	凝灰岩球
tuff breccia	凝灰角砾岩
tuff cone	凝灰岩锥，凝灰锥
tuffaceous	凝灰质的
tuffaceous breccia	凝灰角砾岩
tuffaceous clastic rock	凝灰质碎屑岩
tuffaceous conglomerate	凝灰质砾岩
tuffaceous phyllite	凝灰千枚岩
tuffaceous rhyolite	凝灰质流纹岩
tuffaceous sandstone	凝灰质砂岩
tuffaceous schist	凝灰片岩
tuffaceous shale	凝灰质页岩
tuffaceous siltstone	凝灰质粉砂岩
tuffaceous texture	凝灰结构
tufflava	凝灰熔岩
tundra	冻原；冻土地带
tungspat	重晶石
tungsten carbide	硬质合金；碳化钨
tungstenite	辉钨矿
tungstite	钨华
tunnel	平硐；隧道，隧洞；隧洞
tunnel alignment	隧洞定线
tunnel arch	隧道拱圈
tunnel arch top settlement	隧道拱顶下沉
tunnel blasting	隧洞爆破
tunnel blasting method	矿山法
tunnel borer	隧道掘进机
tunnel boring machine method	掘进机法
tunnel boring machine（TBM）	隧道掘进机
tunnel boring works	隧道挖掘工程
tunnel construction	隧道施工
tunnel drainage	隧洞排水
tunnel drill	隧洞凿岩机
tunnel drivage	平硐掘进；隧道掘进
tunnel driving	平巷掘进；隧道掘进
tunnel effect	隧道效应
tunnel engineering	隧道工程
tunnel entrance	隧道口
tunnel excavation	隧道开挖
tunnel face	隧道工作面
tunnel floor heave	隧道底鼓
tunnel ground subsidence	隧道地表沉陷
tunnel grouting	隧洞灌浆
tunnel heading	隧道掘进；隧洞导坑
tunnel headroom	隧道净空
tunnel horizontal deformation	隧道水平变形
tunnel invert	隧道掘进用钻车
tunnel jumbo	隧道掘进用钻车
tunnel leak	隧道漏水
tunnel lining	隧道衬砌
tunnel maintenance	隧道养护
tunnel mechanics	隧道力学
tunnel mining	平硐开采
tunnel muck	隧洞弃渣；隧道崩落的岩石
tunnel perimeter deflection	隧道周边位移
tunnel piercing	隧道掘进
tunnel portal	平硐口；隧道口
tunnel reconstruction	隧道改建
tunnel roof	隧道顶板

tunnel roof settlement 隧道拱顶下沉
tunnel seismic prediction (TSP) 隧道地震勘探法超前预报
tunnel seismic tomography 隧道地震CT法
tunnel shaft 竖井
tunnel shield 隧道盾构
tunnel side wall 隧道边墙
tunnel spoil 隧道弃渣
tunnel structure 隧道结构
tunnel subgrade 隧道路基
tunnel support 隧洞支护
tunnel surrounding rock 隧道围岩
tunnel survey 隧道测量
tunnel under pressure 有压隧洞
tunnel vault 隧洞顶拱
tunnel ventilation 隧道通风
tunnel without cover 明洞
tunnel work 隧道工程
tunneling 开凿隧道
tunneling index 岩巷掘进指数
tunneling quality index 隧道围岩质量指标
tunneling shield 隧道掩护支架；隧道盾构
tunneling shield machine 隧道盾构机
tunnels lined with concrete 混凝土支护的隧道
turbidite 浊流沉积岩，浊积岩
turbidite facies 浊积相
turbidite sequence 浊积岩层序
turbidity 浊度
turbidity flow 异重流，浊流
turbo drill 涡轮钻机
turbulence 湍流，紊流
turbulence condition 紊流状态
turbulence intensity 紊流强度，湍流强度
turbulent current 紊流，湍流
turbulent diffusion 湍流扩散
turbulent diffusion coefficient 紊流扩散系数
turbulent diffusion process 紊流扩散过程
turbulent flow 湍流，紊流
turbulent resistance 紊流阻力，湍流阻力
turbulent stress 紊流应力
turf 草皮；泥炭
turfing 铺草皮
turgite 水赤铁矿
turjaite 云霞黄长岩
turjite 方解沸石榴云岩
Turkey-Aegean Sea plate 土耳其-爱琴海板块
turmaline 碧玺，电气石
turn angle 回转测角，转角
turn table 转台，回转台
turned-vertical shaft 直井-斜井复合井
turning 转弯
turning angle 回转角
turning axle 转轴
turning moment 转矩，动力矩
turning piece 单拱架
turning point 转折点
turning radius 转变半径
turnover job 大修
Turonian 土仑阶
turquoise 绿松石
Turrilites 塔菊石

twin axis 双晶轴，共轭轴
twin boundary 双晶边界，李晶界
twin crystal 双晶；李晶
twin drift 双平巷
twin entry 双平行平巷布置
twin entry mining 双平巷开采
twin jetties 双边突堤；双导流堤
twin side heading method 双侧导坑法
twist 扭曲，扭转
twist angle 扭转角
twist auger 螺旋钻
twist bit 麻花钻；蛇形钻
twist drill 螺旋钻；麻花钻
twist strain 扭应变
twist strength 抗扭强度
twist test 扭转试验
twisted strata 扭曲地层
twisting couple 扭转力偶
twisting deformation 扭转变形
twisting force 扭力
twisting moment 扭转力矩
twisting motion 扭转运动，扭动
twisting resistance 扭转阻力；抗扭力
twisting rigidity 抗扭刚度
twisting strength 抗扭强度
twisting stress 扭应力
twisting test 扭力试验，扭转试验
twist-off 扭断
two dimension approximation 二维近似
two dimensional analysis 二维分析
two dimensional consolidation 二维固结
two dimensional finite strain 二维有限应变
two layer mineral 双层结构矿物
two mica granite 二云母花岗岩
two-cone bit 双锥牙轮钻头
two-deck jumbo 双层钻车
two-dimension flow 二向流
two-dimensional 二维的；二向的
two-dimensional analysis 二维分析
two-dimensional consolidation 二维固结
two-dimensional flow 二元水流
two-dimensional space 二维空间
two-dimensional state of stress 二维应力状态
two-dimensional stress 二维应力
two-fluid process 双液法
two-hinged arch 双铰拱
two-layer mineral 双层矿物
two-phase flow 两相流动
two-point bit 双刃钻头
two-row method 双排炮眼爆破法
two-shot grouting 双液法
two-way 双向的
two-way drainage 双向排水
two-way mining 两翼开采
two-wedge slip 双楔体滑动
type curve 标准曲线；理论曲线
type distribution 类型分布
type formation 标准层
type fossil 标准化石
type locality 标准地点

type of cementation 胶结类型
type of cracking 破裂类型
type of cross-section 断面类型
type of dam 坝型
type of earthquake 地震类型
type of flow 流态
type of groundwater regime 地下水动态类型
type of rock mass 岩体类型
type region 典型地区
type sample 标准样品
type section 典型剖面
type specimen 标准样品
type symbol 典型符号
type test 典型试验
types of creep 蠕变类型
types of engineering geological mapping 测绘种类
types of investigation working 勘察工作种类
types of soil 土的类型
types of survey 勘察类型
types of undergrounds 地下结构物类型
typhonic rock 深源岩
typhoon 台风
typhoon guard 防风板
typhoon surge 台风涌浪
typical curve method 标准曲线法
typical design 标准设计
typical detail 标准详图
typical flat 典型单位；代表性单位
typical layout 典型布置
typical mineral 标型矿物；典型矿物
typical road section 典型道路截面图
typical sample 典型试样，典型样本
typical section 标准剖面；典型剖面
typomorphic mineral 标型矿物
tyrolite 铜泡石
tyuyamunite 钒钙铀矿

U

uadi 干谷
U-anchor U形锚栓
ubac 阴坡，背阳坡，山阴
U-bar 槽钢
U-bearing mineral 含铀矿物
U-bend U形河湾；U形弯
U-bit U形钻头
U-bolt U形螺栓
U-capacitor U型电容器
U-channel U形水道
U-clamp U形夹具
udalf 湿淋溶土
udent 湿新成土
udert 湿变性土
udoll 湿软土
udox 湿氧化土
udult 湿老成土
U-form tube U形管
U-free 无铀的
U-gully U形沟
uintahite 硬沥青
U-iron 槽铁，槽钢
U-joint 万向节
U-leather ring U形皮圈
uliginous 沼泽的，淤泥的
ullage gage 储罐空高检尺
Ullrich separator 乌尔里希型分离器
ulmification 泥炭化，腐殖化
ulrichite 晶质铀矿，方铀矿，沥青铀矿，碱长霞霓岩
ultimate analysis 全分析，最终分析；元素分析
ultimate balance theory 极限平衡理论
ultimate base level 最低基准面；终极侵蚀基准面
ultimate bearing capacity 极限承载能力
ultimate bearing capacity of subsoil 地基土极限承载力
ultimate bearing pressure 极限承载压力
ultimate bearing resistance 极限承载阻力
ultimate bearing stress 极限承载应力
ultimate bending moment 极限弯曲力矩
ultimate bending strength 极限抗弯强度
ultimate bond strength 极限黏结强度
ultimate breaking load 极限破坏荷载
ultimate breaking strength 极限抗断强度
ultimate capacity 最终产量；极限能力；极限容量
ultimate carrying capacity 极限承载能力
ultimate carrying capacity of single pile 单桩极限承载力
ultimate composition 基本成分，基本组成
ultimate compressive strain 极限压应变
ultimate compressive strength 极限抗压强度
ultimate condition 极限状态
ultimate deformation 极限形变
ultimate design 极限设计
ultimate destabilization 极限失稳
ultimate earthquake 极限地震
ultimate factor of safety 极限安全系数，最大安全系数
ultimate fatigue strength 极限疲劳强度
ultimate flexural capacity 极限弯曲能力
ultimate friction angle 极限摩擦角
ultimate installed capacity 最终装机容量
ultimate lateral resistance of single pile 单桩极限侧阻力
ultimate load 极限荷载，最大荷载，临界载荷
ultimate load design 极限荷载设计
ultimate load method 极限荷载法
ultimate load of pile 桩的极限荷载
ultimate load of subsoil 地基土极限荷载
ultimate long-term strength 极限长期强度
ultimate output 最终产量，总产量，极限产量
ultimate penetration 极限贯入度
ultimate pile load 极限桩荷载
ultimate pit slope 基坑最终边帮，露天采场最终边帮
ultimate plain 终极平原
ultimate pressure 极限压力
ultimate principle 基本原理
ultimate production 总产量，总开采量
ultimate pullout capacity 极限抗拔力
ultimate recoverable reserves 最终可采储量
ultimate recovery 最终采收率，最终开采量
ultimate resistance 极限阻力；极限电阻
ultimate resolution 最大分辨率
ultimate resources 总资源量
ultimate seismic-resistant capacity 极限抗震能力
ultimate settlement 终极沉降
ultimate shaft resistance 极限侧阻力
ultimate shear strength 极限抗剪强度
ultimate shear stress 极限剪应力，最大剪应力
ultimate shearing resistance 极限抗剪力
ultimate shearing strength 极限抗剪强度
ultimate shrinkage 极限收缩
ultimate single pile bearing capacity 单桩极限承载力
ultimate sliding resistance 极限抗滑力
ultimate spacing pattern 最终井距网格，最大井距网格
ultimate stable temperature 最终稳定温度
ultimate state 极限状态
ultimate strain 极限应变
ultimate strength 极限强度
ultimate strength design 极限强度设计
ultimate strength of rupture 极限破裂强度
ultimate strength theory 极限强度理论
ultimate strength value 强度终值
ultimate stress 极限应力
ultimate stress circle 极限应力圆
ultimate stress state 极限应力状态
ultimate subsidence 极限沉陷，最深沉陷
ultimate tensile strain 极限拉应变
ultimate tensile strength 极限抗拉强度，极限抗张强度
ultimate tensile stress 极限拉伸应力
ultimate tip resistance 极限端阻力，极限桩端阻力
ultimate value 极限值
ultimate yield 最终产量，极限产量；最终出水量

ultisol	老成土
ultra deep well	超深井
ultra-abyssal	超深海的
ultra-acid rock	超酸性岩
ultracentrifugation	超速离心分离
ultracentrifuge	超速离心机
ultra-clay	超微黏粒
ultra-clean air system	超净空气系统
ultra-close distance thin coal seam	超近距薄煤层
ultradeep drilling	超深钻探
ultradeep exploration	超深勘探
ultradeep geothermal development	超深地热能开发
ultradeep geothermal drilling	超深地热钻探
ultradeep prospecting well	超深预探井
ultradeep shaft	超深立井
ultrafiltration	超过滤；超滤法
ultrafiltration membrane	超滤膜
ultrafine cement	超细水泥
ultrafine cement grout	超细水泥浆
ultrafine dust	超细粉尘
ultrafine fly ash	超细粉煤灰
ultrafine grain	超细颗粒
ultrafine particle	超微颗粒
ultrafine powder	超细粉末
ultra-flotation	超细浮选
ultra-heavy duty rig	超重型钻机
ultrahigh frequency	超高频
ultrahigh frequency percussion	超高频冲击
ultrahigh pressure metamorphism	超高压变质作用
ultrahigh purity	超高纯度
ultrahigh speed railway	超高速铁路
ultrahigh strength steel	超高强度钢
ultrahigh vacuum	超高真空
ultrahigh voltage transformer oil	超高压变压器油
ultrajet	高效能射孔器
ultralarge crude carrier	超大型油轮
ultralight alloy	超轻合金
ultralight element	超轻元素
ultralight weight fill	特轻质填方
ultralong period seismograph	超长周期地震仪
ultralong spaced electric log	超长距电测井
ultralong wave	超长波
ultralow density polyethylene	超低密度聚乙烯
ultralow frequency	超低频
ultralow permeability	超低渗透率
ultralow velocity zone	超低波速带
ultralow viscosity oil	超低黏度油
ultramafic deposit	超基性矿床
ultramafic rock	超镁铁质岩
ultramicro-earthquake	超微地震
ultramicro-element	超微量元素
ultramicroscope	超高倍显微镜
ultramodern	超新型的，超现代的
ultramylonite	超糜棱岩
ultramylonite texture	超糜棱结构
ultrapaleontology	超微古生物学
ultrapure metal	超纯金属
ultrapure water	超纯水
ultraray	宇宙线
ultrared	红外的；红外线
ultrared ray	红外线
ultrared spectrometry	红外光谱测量法
ultraselective cracking	超选择性裂解
ultrasensitive clay	超敏感黏土
ultrasensitive pressure gauge	高灵敏压力计
ultrasensitive seismometer	极灵敏地震计
ultrashort baseline positioning	超短基线定位
ultrashort distance seams	超短距离缝隙；超短距离煤层
ultrashort pulse	超短脉冲
ultrashort wave	超短波
ultraslim hole rig	超小井眼钻机
ultrasoft clay	超软黏土
ultrasoft X-ray	超软 X 射线
ultrasonator	超声振荡器
ultrasonic atomizer	超声雾化器
ultrasonic bias	超声偏移
ultrasonic casing inspection tool	超声套管检测仪
ultrasonic cement analyzer	超声水泥分析器
ultrasonic chemical process	超声化学工艺
ultrasonic cleaning	超声清洗；超声消磁
ultrasonic cleaning agent	超声清洗剂
ultrasonic crack detection	超声裂缝检测
ultrasonic current metre	超声流速仪
ultrasonic defect detector	超声探伤仪
ultrasonic degradation	超声降解
ultrasonic delay line	超声延迟线
ultrasonic depolymerization	超声解聚合作用
ultrasonic depth finder	超声测深仪
ultrasonic detector	超声探测器
ultrasonic drawing	超声拉拔
ultrasonic drill	超声钻机
ultrasonic drilling	超声钻探
ultrasonic echo method	超声回声法
ultrasonic effect	超声效应
ultrasonic electrostatic sprayer	超声静电喷涂机
ultrasonic emission	超声发射
ultrasonic examination	超声检验
ultrasonic extraction	超声萃取
ultrasonic extractor	超声拔桩机；超声萃取器；超声抽气机
ultrasonic fine screener	超声细分析筛
ultrasonic flaw detector	超声探伤仪
ultrasonic flowmeter	超声波流量计
ultrasonic frequency	超声波频率
ultrasonic gauge	超声波计
ultrasonic generator	超声波发生器
ultrasonic hologram	超声全息像片
ultrasonic image logging	超声成像测井
ultrasonic leak detection	超声检漏
ultrasonic pulse	超声波脉冲
ultrasonic pulse probe	超声脉冲探头
ultrasonic pulse test	超声波脉冲测试
ultrasonic radar	超声雷达
ultrasonic ranging	超声波测距
ultrasonic rock drill	超声波钻岩机
ultrasonic seismic scattering tomography	超声地震散射层析成像法
ultrasonic sensor	超声波传感器
ultrasonic test	超声波检验
ultrasonic thickness indicator	超声波厚度指示器

ultrasonic transmitter 超声波发射器
ultrasonic velocity test 超声波速度测试
ultrasonic vibrating screen 超声振动筛
ultrasonic vibration drilling 超声振动钻探
ultrasonic vibrator 超声波振动器
ultrasonic wave 超声波
ultrasonic welding inspection 超声波焊缝检查
ultrasonics 超声波学
ultrasonography 超声波探测
ultrasonomicroscope 超声显微镜
ultrasound 超声
ultrasound tomography 超声层析成像法
ultrastability 超稳定性
ultrastrength material 超强度材料
ultrathin film 超薄薄膜
ultrathin whitetopping 超薄混凝土加铺
ultratrace element 超痕量元量
ultraviolet 紫外的；紫外线的；紫外线辐射
ultraviolet A radiation（UVA） 紫外线 A 波段辐射，长波紫外线
ultraviolet absorber 紫外线吸收剂
ultraviolet absorber fixative 紫外线吸收剂固定剂
ultraviolet absorption 紫外吸收
ultra-violet absorption spectrometry 紫外线吸收光谱法
ultraviolet absorption spectrophotometry 紫外吸收分光光度学
ultraviolet absorption spectroscopy 紫外吸收光谱学
ultraviolet astronomy 紫外线天文学
ultraviolet B radiation（UVB） 紫外线 B 波段辐射，短波紫外线
ultraviolet band 紫外区
ultraviolet C radiation（UVC） 紫外线 C 波段辐射，短波紫外线
ultraviolet flux 紫外线通量
ultraviolet generator 紫外线发生器
ultraviolet ion source 紫外离子源
ultraviolet lamp 紫外光灯
ultraviolet light 紫外光
ultraviolet microscope 紫外光显微镜
ultraviolet ozone cleaning 紫外线臭氧清洗
ultraviolet photoelectron spectroscopy 紫外光电子能谱学
ultraviolet protecting agent 紫外线防护剂
ultraviolet radiation stability 紫外线辐射稳定性
ultraviolet ray 紫外线
ultraviolet remote sensing 紫外遥感
ultraviolet resistance 抗紫外线性
ultraviolet screening agent 紫外线屏蔽剂
ultraviolet spectral analysis 紫外光谱分析
ultraviolet spectrograph 紫外光谱仪
ultraviolet spectrometer 紫外线分光仪
ultraviolet spectrophotometer 紫外分光光度计
ultraviolet spectroscopy 紫外线光谱学
ultraviolet spectrum 紫外光谱
ultraviolet video system 紫外视频系统
umber 棕土；铁锰质土
umbilical line 操纵缆绳；集成管束
umbrella arch 套拱，拱伞形
umbrella effect 伞形效应
umbrella pre-reinforcement method 伞形预加固法
umbrept 暗始成土

umbrult 暗色老成土
unactivated state 未激活状态，未启动状态
unadulterated sample 纯净样品
unaka 残丘；残丘群
unaltered rock 未蚀变岩石
unanchored sheet piling 无锚板桩
unary 一元系，一组分体系
unary diagram 一元相图
unascertained clustering method 未确知聚类法
unattended remote-control power station 无人值守遥控电站
unattended substation 无人值守变电所
unavailable moisture 无效水分
unavailable soil water 无效土壤水
unavailable water 无效水，不可能利用的水
unbalance 失衡，不平衡
unbalance factor 不平衡系数
unbalance vibrating screen 不平衡振动筛
unbalanced moment 不平衡力矩
unbalanced pump 不平衡泵
unbalanced system 不平衡系统
unbalanced thrust transmission method 不平衡推力传递法
unbalanced vane pump 不平衡式叶片泵
unbalancedness 不平衡性
unbalanced-throw screen 惯性筛，不平衡冲程筛
unban air pollution concentration 城市空气污染浓度
unbedded 不成层的
unbiased conditions 无偏条件
unbiased error 无偏误差
unbiased estimation 无偏估计
unbiased polynomial coefficient 无偏多项式系数
unbiased statistics 无偏统计量
unbiased test 无偏检验
unbiased variance 无偏方差，均方差
unblocking field 解阻场
unblocking temperature 解阻温度
unboarded derrick 无塔布钻塔，无遮挡井架
unbonded concrete overlay 非黏结混凝土加铺
unbonded member 无黏结构件
unbonded prestressed bar 无黏接预应力钢筋
unbonded prestressing 无黏结预加应力
unbonded strand 无黏结钢绞线
unbonded tendon 无黏结筋
unbound water 非结合水
unbounded domain 无界域
unbounded fracture 无边界破裂
unbounded function 无界函数
unbounded optimization problem 无界优化问题
unbounded problem 无界问题
unbounded solution 无界解
unbounded stream 无界水流
unbraced excavation 无支撑开挖
unbraced length of column 无支撑柱长
unbranched stream 无分叉水流，无支流河，独流河
unbroken curve 连续曲线
unbroken layer 连续层
unbroken rock 未破碎岩石
unbroken wave 未碎波
unburnt brick 欠火砖，未烧透砖
uncased bore hole 无套管钻孔
uncased concrete pile 无套管混凝土灌注桩

uncased drill hole 无套管钻孔
uncased piling 无套管灌桩
uncased well 无套管井，未加固井
uncemented deposit 未胶结沉积
uncemented fault 未胶结断层
uncertain factor 不确定因素
uncertain linguistic variable 不确定语言变量
uncertain reasoning 不确定性推理
uncertain system 不确定性系统
uncertainty 不确定性
uncertainty analysis 不确定性分析
uncertainty principle 测不准原理，不确定性原理
uncertainty range 不确定范围
uncertainty relation 不确定关系
uncertainty risk 不确定风险
unchanging bench mark 固定水准基点
uncharged hole 未装药的炮眼
uncharged particle 不带电粒子
uncharged structure 未充填结构；无负荷结构
unchoking 清除堵塞
unclassified excavation 不分类开挖，混挖
uncoated electrode 裸焊条，无药焊条
uncohesive 不黏结的，无内聚力的
uncombined water 非化合水
uncomformity 不整合
uncompacted rockfill 未压实堆石，未压实填石
uncompacted sand 未压实砂
uncompacted shale 欠压实页岩
uncompensated load 无补偿载荷，不平衡载荷
uncompensated pump 无补偿泵，不平衡泵
uncomplete explosion 不完全爆破
unconcentrated flow 漫流，散流，片状流
unconditional convergence 无条件收敛
unconditional probability 无条件概率
unconditional stability 无条件稳定性
unconditionally stable 绝对稳定的
unconfined blasting 无侧限爆破，外部爆破
unconfined compression 无侧限压缩
unconfined compression apparatus 无侧限压缩仪
unconfined compression strength 无侧限抗压强度
unconfined compression strength of remolded soil 重塑土的无侧限抗压强度
unconfined compression test 无侧限压缩试验
unconfined compressive strength 无侧限抗压强度
unconfined compressive strength test 无侧限抗压强度试验
unconfined cylindrical sample 无侧限圆柱体试样
unconfined flow 非承压流
unconfined ground water 非承压地下水
unconfined interlayer water 无压层间水
unconfined shear strength 无侧限剪切强度
unconfined shot 糊炮，外部爆破
unconfined specimen 无侧限试样
unconfined undrained test 无侧限不排水试验
unconfined uniaxial loading 无侧限单轴加载
unconfined water 非承压水，潜水
unconfined water body 非承压水体
unconfined well 非承压井
unconformability 不整合
unconformable contact 不整合接触
unconformable strata 不整合地层

unconformable stratification 不整合层理
unconformity 不整合；不一致
unconformity bounded stratigraphic unit 不整合围限地层单位
unconformity by erosion 侵蚀不整合
unconformity interface 不整合界面
unconformity spring 不整合泉
unconformity surface 不整合面
unconformity trap 不整合圈闭
unconsolidated cover 未固结盖层，松散盖层
unconsolidated deposit 未固结沉积，松散沉积
unconsolidated overburden 疏松覆盖层
unconsolidated porous medium 松散多孔介质
unconsolidated rock 未固结岩石
unconsolidated sediment 未固结沉积
unconsolidated soil material 未固结土料
unconsolidated stratum 未固结地层
unconsolidated surface layer 未固结表层
unconsolidated-drained test 不固结排水试验
unconsolidated-undrained shear test 不固结不排水剪切试验
unconsolidated-undrained test 不固结不排水试验
unconsolidated-undrained triaxial compression strength 不固结不排水三轴抗压强度
unconsolidated-undrained triaxial shear test 不固结不排水三轴剪切试验
unconsolidated-undrained triaxial test 不固结不排水三轴试验
uncontaminated air 未污染空气
uncontaminated mud 未污染泥
uncontaminated soil 未污染土
uncontaminated water 未污染水
uncontaminated zone 未污染带
uncontracted weir 不收缩堰
uncontrolled spillway 无控溢洪道
uncontrolled storage 无控水库
uncontrolled weir 无控制堰
unconventional gas reservoir 非常规气藏
unconventional reservoir 非常规油气藏
unconventional resources 非常规资源
uncorrelated function 不相关函数
uncorrelated random force 不相关随机力
uncorrelated variable 不相关变量
uncoupled differential equation 非耦合微分方程
uncoupled equation of motion 非耦合运动方程
uncoupled formulation 非耦合配方
uncoupled mode 非耦合模式
uncoupled structure 非耦合结构
uncoupled wave 非耦合波
uncovered blasting 裸露爆破
uncovered canal 明渠
uncovered digging method 明挖法
uncovered map 基岩图
uncovered well 裸井，无套管井
uncovering of foundation 基础揭露
uncropped location 荒地
unctuous clay 油性黏土
uncultivated land 未垦地，荒地
uncurrent 非现行的，非现时的，非时兴的，非流行的
unda 浪波作用环境，浪蚀环境，浪蚀带，浪蚀底
unda environment 浪蚀环境

undaform 浪蚀地形，波浪形地面
undamped forced vibration 无阻尼受迫振动
undamped free vibration 无阻尼自由振动
undamped natural frequency 无阻尼自然频率，无阻尼固有频率
undamped wave 无阻尼波，无衰减波，等幅波
undathem 浪成岩，浪成地层，浪蚀底岩层
undation theory 波动说
undeformed 未变形的，无形变的
undepleted aquifer 未枯竭的含水层
underbalanced drilling 低压钻探
underbeam girder 梁下纵桁
underbracing 下支撑
underbreak in tunnel excavation 隧道欠挖
underbridge 桥底部；桥下
undercapacity factor 欠载系数
underclay 底黏土
undercomposed explosive 未分解炸药
underconsolidated clay 欠固结黏土
underconsolidated soil 欠固结土
underconsolidated soil deposit 欠固结土层
underconsolidation state 欠固结状态
undercurrent 暗流，潜流；底流，下层流；低电流
undercut 下切，淘底，底切，拉底；掏槽；凹口
undercut anchor 暗掘锚固
undercut stope 底部掏槽回采工作面，掏底工作面
undercut-and-fill stoping 下部掏槽充填回采法
undercut-caving system 下部掏槽崩落法
undercutting 下切，底切，拉底，底部掏槽
undercutting chamber 拉底硐室
undercutting level 拉底水平；拉底平巷
undercutting method 暗挖法，硐挖法，底切法
underdamping 欠阻尼，阻尼不足；弱衰减
underdeterminant 子行列式
underdetermined equation 欠定方程
underdeveloped area 欠发达地区
underdrain 暗渠，暗沟，地下渠道
underdrain pipe 地下排水管
underdrainage 暗沟排水，地下排水
underearth 底土，地面下层土；地球内部
underestimation 低估
underexpansion 不完全膨胀
underfeed 补给不足；下部进料
underfill method 下填法
underflow 地下流，潜流，底流，下溢
underflow conduit 潜流水道
underflow density current 水下异重流
underflow gate 底流闸门
underflow spring 水下泉
underflow turbidity current 水下混浊流
underflow zone 潜流带
underflowing irrigation 地下水流灌溉
underframe mounted traction motor 有底架的牵引电动机
underframe with center sill 有中梁的底架
underframe without center sill 无中梁的底架
undergauge bit 缩径钻头
underground access 地下通道
underground air raid shelter 地下防空洞
underground alignment 地下放线，坑道定线
underground batterycharging room 井下充电室

underground blast hole drilling 井下炮眼钻进
underground blasting 地下爆破，井下爆破
underground blowout 地下井喷
underground boring 井下钻进，坑道钻进
underground building 地下建筑
underground cable 地下电缆
underground cave detection 地下洞穴探测
underground cavern 地下洞穴；地下洞室
underground cavity 地下洞室
underground central substation room 井下主变电室
underground channel 地下渠道
underground cistern 地下蓄水池，地下液槽，地下水窖
underground clay mining 地下采土
underground CO_2 storage 二氧化碳地下储存
underground coal gasification 煤炭地下气化
underground coal gasification station 煤炭地下气化站
underground complex 地下综合建筑
underground conduit 地下管道
underground construction 地下建筑，地下工程
underground control room 井下调度室
underground crusher 井下破碎机，井下碎矿机
underground dam 地下坝
underground diaphragm wall 地下连续墙，地下防渗墙
underground drain 地下排水；地下排水管
underground drainage 地下排水；地下排水设施；地下水系
underground electromagnetic wave perradiator 坑道电磁波透视仪
underground energy storage 地下储能
underground engineering 地下工程
underground excavation 地下开挖，地下工程开挖
underground excavation method 暗挖法，硐挖法
underground exploitation 地下开采
underground exploration 地下勘探
underground explosion 地下爆破
underground fire risk 地下火灾风险
underground flooding 地下泛水；地下注水
underground fluid 地下流体
underground freight transport 地下物流运输
underground garage 地下车库
underground gas storage 地下储气
underground gasification 地下气化
underground gasifier 地下煤气气化炉
underground geologic map 地下地质图，坑道地质图
underground geological radar 矿井地质雷达
underground geophysical prospecting 地下地球物理勘探，地下物探
underground grouting works 地下灌浆工程
underground heat 地下热
underground heated line 地下加热管线
underground horizontal control survey 井下平面控制测量
underground ice 地下冰
underground industry 地下产业
underground installation 地下设施
underground karst landform 地下岩溶景观
underground lake 地下湖，暗湖
underground leakage 地下渗流，地下渗漏
underground leveling 坑道水准测量
underground lifeline system 地下生命线系统
underground liquefied petroleum gas storage 地下液化石油气

储存
underground locomotive repair room 井下机车修理间
underground loss 地下漏水,地下渗失,地下损耗
underground main substation 井下主变电所,井下中央变电所
underground methane monitoring system 井下甲烷监测系统
underground mine 矿井
underground mine construction 矿井建设
underground mine design 矿井设计
underground mine development 矿井开发
underground mine fire 矿井火灾
underground mine gas emission 矿井瓦斯涌出
underground mine life 矿井寿命
underground mine pillar 地下矿柱
underground mining 地下开采,地下采矿,坑采
underground nuclear explosion 地下核爆炸
underground nuclear test 地下核试验
underground opening 地下洞室;地下开挖
underground parking garage 地下停车场
underground pass 地下通道;地下通过
underground passage 地下通道
underground passageway 地下通道
underground percolation 地下渗漏,地下渗流
underground pipe line 地下管线
underground pipe network 地下管网
underground pollution 地下污染
underground power station 地下电站
underground powerhouse 地下电厂
underground pressure 地压,地下压力,岩层压力
underground quarry 地下采石场
underground railway 地下铁道,地铁
underground research laboratory 地下研究实验室
underground residue 地下残留物
underground resources 地下资源
underground river 地下河,暗河
underground river system 地下河网,暗河水系
underground runoff 地下径流
underground rupture plane 地下破裂面
underground shelter 地下防空洞,地下掩体
underground space 地下空间
underground store for explosives 地下炸药库
underground stream 地下河,暗河,伏流
underground structure 地下结构;地下建筑物
underground substation 地下变电站
underground tank 地下水池;地下储罐
underground temperature 地下温度
underground thermal pollution 矿井热污染
underground thermal water 地下热水
underground tow vehicle 井下牵引车
underground trackless haulage 井下无轨运输
underground tractor 井下拖拉机,地下牵引车
underground traffic 地下交通
underground traverse survey 地下导线测量
underground tunnel 地下隧道
underground utilities 地下公用设施
underground water 地下水,潜水
underground water basin 地下水域
underground water course 地下水道
underground water divide 地下水分水线
underground water level 地下水位,地下水面,潜水面

underground water parting 地下分水线
underground water resources 地下水资源
underground water stock 地下水储存
underground water surface 地下水面
underground workings 地下井巷
underhand double stope 下向双翼回采工作面
underhand stoping 下向梯段回采
underice tunnel 冰下隧道
underlay 下伏,位于下部;下伏层,底层;向下延伸矿体
underlayer 底层,下层;子层,副层,次层,亚层;通向下部开采水平的竖井
underlaying bed 下伏层
underlier 底板,下帮,下盘,下伏层
underloaded stream 轻载河流,少砂石河流
underlying bed 下伏层;底床,河床
underlying bedrock 下伏基岩
underlying geologic formation 下伏地质建造
underlying layer 下伏层
underlying mass 下伏岩体
underlying rock 下伏岩石,底岩,基岩
underlying soft soil 下伏软土
underlying soil 下伏土
underlying stratum 下伏地层
underlying surface 下伏面
underlying topography 下伏地形
undermass 下伏岩体
undermining 淘蚀,基蚀,底部冲刷;潜挖,拉底,下采
undermining blast 潜挖爆破,坑道爆破
undermining of supporting soil 持力层土的掏挖
undermixing 混合不均,搅拌不足
underpass approach 地下通道引道
underpinned pier 托换墩
underpinned pile 托换桩,托底桩
underpinning 托换,基础托换
underpinning by vibroflotation 振冲法托换
underpinning method 托换法,基础托换法
underpinning technique 托换技术
underplate 底板,垫板
underplating 垫托;底侵
underpour type gate 下泄式闸门
underprestressed beam 低预应力梁
underprop 支撑物,支柱
underrail foundation 轨下基础
underreamed bored pile 扩底钻孔桩
underreamed foundation 扩底基础
underreamed pier 扩底墩
underreamed pile 扩底桩
underreaming 井下扩眼,孔内扩孔
underreaming bit 扩孔钻头,扩底钻头
underreinforced beam 低配筋梁
undersanded 含砂量不足的
undersaturated permafrost 不饱和永冻土
undersaturated soil 不饱和土
undersaturation 不饱和,不饱和
undersea device 海底设备,水下设备
undersea laboratory 海底实验室
undersea military base 海底军事基地
undersea mining 海底开采,海底采矿
undersea pipeline 海底管道

undersea power station	海底电站
undersea sonar mapping	海底声呐测图
undersea tunnel	海底隧道
undersea vehicle	潜水器
undersealing	底封
underseam	下伏岩层；底部煤层
underseepage	地下渗流
undershot head gate	下泄水头闸门
undershot water wheel	底射式水轮
undershrub	矮灌木
underside	下侧；下面，底面
undersize aggregate	尺寸不足的骨料
undersluice	底泄水道
underslung conveyor	悬架式输送机
undersoil	底土，下层土
understratum	下层
understructure	下层结构
undersurface filling	液面下灌注
undersurface loading	液面下灌装
underthrust	俯冲；俯冲断层
underthrust plate	俯冲板块
underthrust zone	俯冲带
undertow	回流；底流；离岸流
undervoltage protection	低电压保护
undervoltage relay	低压继电器
underwall	下盘；底帮
underwashing	冲刷，冲洗
underwater	水下的，水底的；地下水，边缘水；下游
underwater acoustic positioning	水声定位
underwater acoustic telemeter	水声遥测器
underwater acoustics	水声学
underwater apron	水下沉积裙
underwater archaeology	水下考古学
underwater bank	水下浅滩
underwater blasting	水下爆破
underwater bulk weight	水下容重
underwater bulldozer	水下推土机
underwater color TV camera	水下彩色电视摄影机
underwater communication system	海底通信系统
underwater compass	水下罗盘
underwater completion	水下完井
underwater concreting	水下混凝土浇筑
underwater connection	水下联接
underwater construction	水下施工
underwater contour	等深线，水深线
underwater coring	水下取芯
underwater crane	水下升降机
underwater crawler	水下曳引车
underwater digging	水下挖掘
underwater disturbance	水下扰动
underwater drilling	水下钻探
underwater driving	水下打桩
underwater earthquake	水下地震
underwater engineering	水下工程
underwater excavation	水下开挖
underwater excavator	水下掘削机
underwater exploration	水下勘探
underwater explosion	水下爆破
underwater fill	水下填土
underwater foundation	水下基础
underwater gradient	水下坡降
underwater gravimeter	水下重力仪
underwater hammer	水下打桩机
underwater ice	水下冰
underwater illumination	水下照明
underwater landslide	水下滑坡
underwater laser radar	水下激光雷达
underwater laser surveying system	水下激光测量系统
underwater lighting	水下照明
underwater noise	水下噪声
underwater nuclear power station	海底核电站
underwater observation	水下观察
underwater operation	水下作业
underwater outfall	水下排放口
underwater paint	水下漆，水下涂料
underwater photogrammetry	水下摄影测量
underwater pile driver	水下打桩机
underwater pipeline	水下管线
underwater platform	水下平台
underwater positioning	水下定位
underwater power plant	水下发电厂
underwater pump	潜水泵
underwater rig	潜水钻机
underwater robot	水下机器人
underwater rock drill	水下钻岩机
underwater sampler	水下取样器
underwater sound gear	水声设备
underwater spring	水下泉
underwater storage tank	水下储罐
underwater structure	水下结构；水下建筑物
underwater survey	水下测量
underwater technology	水下工艺学
underwater telemeter	水下遥测器
underwater telemetry	水下遥测
underwater television	水下电视，海底电视
underwater testing	水下试验
underwater topographic survey	水下地形测量
underwater topography	水下地形
underwater tracking system	水下跟踪系统
underwater tunnel	水底隧道
underwater vehicle	潜水器
underwater wave guide	水下波导
underwater welding	水下焊接
underwater well-head control system	水下井口控制系统
underwater works	水下工程
underwork	支承结构，根基
underworkings	地下巷道
undesirable component	有害组分，不良组分
undesired signal	干扰信号
undetermined coefficient	未定系数
undetermined fault	性质不明的断层
undetermined multiplier	待定因子
undetermined parameters	待定参数
undeveloped area	不发达地区，未开发地区
undifferentiated alluvium	均匀冲积物
undifferentiated igneous rock	未分异火成岩
undigested sludge	未分解的污泥，生污泥
undirectional flow	不定向流
undiscovered possible reserves	未发现的可能储量
undiscovered possible reserves resource	未发现的潜在储量

资源
undiscovered resources 待发现资源
undissociated molecule 未离解的分子
undissolved residue 不溶残余
undistorted mode 无畸变模式
undistorted model 无畸变模型
undisturbed clay 未扰动黏土，原状黏土
undisturbed explosive 未分解炸药，未变质炸药
undisturbed flow 稳流，无扰动流
undisturbed formation 未扰动地层建造，原状岩层
undisturbed ground 未扰动土地，原状土地
undisturbed jet 无扰动射流
undisturbed rock 原状岩
undisturbed sample 原状样品，未搅动样品
undisturbed sand 原状砂
undisturbed soil sample 无扰动土样，原状土样
undisturbed soil shrinkage test 原状土收缩试验
undisturbed structure 未扰动结构
undisturbed velocity 无扰动速度
undisturbed water 无扰动水
undrained behavior 不排水性状
undrained creep 不排水蠕变
undrained lake 不排水湖
undrained load condition 不排水荷载条件
undrained loading 不排水加载
undrained pressure 不排水压力
undrained quick shear test 不排水快剪试验
undrained settlement 不排水沉降，瞬时沉降
undrained shear 不排水剪切
undrained shear strength 不排水抗剪强度
undrained state 不排水状态
undrained strength parameter 不排水强度参数
undrained test 不排水试验
undrained triaxial compression test 不排水三轴压缩试验
undrained triaxial test 不排水三轴试验
undulate fold 波状褶皱
undulated anticline 波状背斜
undulated extinction 波状消光
undulated peneplain 波状准平原
undulated seam 波状层；波状煤层
undulated structure 波纹构造
undulating ground 波状起伏地；丘陵地
undulating topography 丘陵地形，波状地形
undulation 波动；起伏
undulation hypothesis 波动说
undulation period 波动周期
undulatory layer 波状起伏层
undulatory motion 波状运动
undulatory theory 波动理论
unelasticity 非弹性
unenriched ore 未富集矿
unequal angle steel 不等边角钢
unequal diameter nozzles 不等径喷嘴
unequal precision measurement 不等精度测量
unequal pressure 不等压
unequal settlement 不均匀沉降，差异沉降
unequally spaced data 不等间距数据
unequally spaced data point 非等间距数据点
unequigranular 不等粒状的
uneven bulging 不均匀胀形

uneven distribution 不均匀分布
uneven pressure 不均匀压力
uneven settlement 不均匀沉陷
uneven surface 粗糙面
uneven wear 不均匀磨损
unevenly compressible foundation 不均匀压缩地基
unevenness 不平整度；不均匀性
unexpected flooding 意外涌水
unexploded charge 瞎炮
unexploded explosive 残余炸药
unexplored region 未勘探地区
unexplosive dynamite disposing technique 盲炮处理技术
unfailing spring 不涸泉
unfamiliar feature 未知要素
unfavorable condition 不利条件
unfavorable geological condition 不良地质条件
unfilled caisson foundation 无充填沉井基础
unfilled porosity 无充填孔隙度
unfired hole 未爆炮孔
unfixed sand 流沙
unfolding drawing 展开图
unforeseeable event 不可预见事件
unformed 无模板的；未成形的
unfossiliferous 不含化石的
unfree water 死水
unfrozen lake 不冻湖
unfrozen layer 不冻层
unfrozen soil 不冻土
unfrozen water content 不冻水含量
unfrozen zone 不冻带
ungated flow 敞开泄流
ungauged river 未施测河流
ungauged stream 未施测河流
ungauged watershed 未施测流域
ungraded aggregate 无级配骨料
ungraded material 无级配材料
ungrouted hole 未灌浆孔
unhardened concrete 未硬化混凝土
unhomogeneous rock 非均质岩石
unhydrated ion 未水化离子
unhydrated plaster 干灰膏，未水化灰泥
uniaxial anisotropy 单轴各向异性
uniaxial compaction modulus 单轴压实模量
uniaxial compression 单轴压缩
uniaxial compression force 单轴压缩力
uniaxial compression strain 单轴压缩应变
uniaxial compression strength 单轴压缩强度
uniaxial compression test 单轴压缩试验
uniaxial compressive loading test 单轴压缩载荷试验
uniaxial compressive strength 单轴抗压强度
uniaxial compressive strength test 单轴抗压强度试验
uniaxial compressive stress 单轴压应力
uniaxial confined compression test 单轴侧限压缩试验
uniaxial cyclic loading 单轴循环加载
uniaxial deformation 单轴变形
uniaxial dynamic analysis 单轴动力分析
uniaxial eccentricity 单轴偏心
uniaxial loading 单轴加载
uniaxial orientation 单轴取向
uniaxial pressure 单轴压力

uniaxial static creep test	单轴静载蠕变试验
uniaxial strain	单轴应变
uniaxial strain test	单轴应变试验
uniaxial stress	单轴应力
uniaxial tensile strength	单轴抗拉强度
uniaxial tensile strength test	单轴抗拉强度试验
uniaxial tensile stress	单轴拉应力
uniaxial tensile test	单轴拉伸试验
uniaxial tension	单轴拉伸
uniaxial tension strength test	单轴抗拉强度试验
uniaxial tension test	单轴拉伸试验
unicircuit	集成电路
uniclinal fold	单斜褶皱
uniclinal structure	单斜构造
unicomponent system	一元系，单组分系统
unidentified flying object (UFO)	不明飞行物
unidimensional	一维的，一度空间的，线性的
unidimensional consolidation	单向固结，一维固结
unidimensional displacement	一维驱替
unidirectional flow	单向水流
unidirectional flux	单向通量
unidirectional loading	单向加载
unidirectional motion	单向运动
unidirectional pressure	单向压力
unidirectional pressure temporary plugging additive	单向压力暂堵剂
unidirectional shaking	单向振动
unidirectional spread	单向排列，单边排列
unidirectivity	单向性
uni-element map	单元素图
uniextensional	单向延伸的
unifacial	单面的
unified classification system	统一分类系统
unified dilatation	均匀膨胀
unified distribution	均匀分布
unified field theory	统一场论
unified soil classification system	统一土壤分类法
uniflow	单向流；顺流
uniform accelerated motion	等加速运动，匀加速度
uniform approximation	一致逼近
uniform asymptotic formula	均匀渐近公式
uniform beam	等截面梁
uniform bed	均匀层
uniform building code	统一建筑规范
uniform channel	等截面渠道
uniform compaction	均匀压实
uniform compression	均匀压缩
uniform contact	均匀接触
uniform convergence	均匀收敛，一致收敛
uniform corrosion	均匀腐蚀
uniform cross-section beam	等截面梁
uniform cross-section jacket	等截面导管架
uniform decompression	均匀减压
uniform delay	均匀延滞
uniform deposit	均匀沉积
uniform dilatancy	均匀扩容
uniform distributed load	均布荷载
uniform electric field	均匀电场
uniform face	均匀面
uniform field	均匀场
uniform flow	均匀流
uniform fluid	均匀流体
uniform geothermal gradient	不均匀地热梯度
uniform gradation	均匀级配
uniform grade	均匀坡度
uniform grade channel	均坡河槽
uniform grid	正方形网格，均匀网格
uniform intergranular porosity	均匀粒间孔隙度
uniform internal diameter	等内径，均匀内径
uniform load	均布载荷
uniform medium	均匀介质
uniform mix	均匀混合；均匀混合料
uniform mutation	均匀变异
uniform peripheral pressure	均布周边压力
uniform pressure	均匀压力
uniform rainfall	均匀降雨
uniform random number	均匀随机数
uniform recharge	均匀补给
uniform rectilinear flow	均匀直线流
uniform resistance	均匀阻力
uniform rotation	匀速转动
uniform sandstone	均质砂岩
uniform settlement	均匀沉降
uniform size aggregate	均匀粒径骨料
uniform slope	均匀斜坡；等坡度
uniform soil	均质土
uniform strain	均匀应变
uniform stream	均匀流
uniform strength	均匀强度，等强度
uniform stress	均匀应力，均布应力
uniform stress element	均匀应力元，常应力元
uniform stress field	均匀应力场
uniform surcharge	均匀超载
uniform system	均一体系
uniform variable motion	均匀变速运动
uniform velocity	匀速，等速度
uniform vorticity fluid	均匀涡旋流体
uniform wettability	均匀润湿性
uniform-grained soil	均粒土
uniformitarianism	均变论，均变说
uniformity coefficient	均匀系数
uniformity modulus	均质模数
uniformization	单值化，均匀化，一致化
uniformly bounded variation	均匀有界变差
uniformly regulated discharge	均匀调节的流量
uniformly variable motion	匀变速运动
uniformly varying load	均变荷载
uniformly-distributed circular load	均布圆形荷载
uniformly-distributed load	均布荷载
uniformly-distributed strip load	均布条形荷载
uniformly-graded aggregate	均匀级配骨料
uniformly-graded soil	等粒级土
uniformly-yielding foundation	均匀沉降基础
unilateral	单方面的，一方的；单边的，单侧的；片面的
unilateral constraint	单侧约束
unilateral continuity	单侧连续性
unilateral impedance	单向阻抗
unilateral moving dislocation	单向移动位错
unilateral stretching	单向拉伸

unimodal distribution 单峰分布
unimodal frequency curve 单峰频率曲线
unimodal function 单峰函数
unimodal fuzzy set 单峰模糊集
unimolecular 单分子的
uninsulated over-lap 不绝缘搭接
uninsulated transition mast 不绝缘转换柱
uninterrupted concreting 连续混凝土浇筑
uninterrupted power supply 不中断电源
uninterrupted pumping 连续泵送
uninterruptible power system 不间断供电系统
union coupling sleeve 联管接箍套管
union joint 联管接头
union nut 联管螺母
union purchase system 双吊杆操作系统；双杆联吊起重系统
union wrench 联管节扳手
unionmelt welding 合熔式自动电焊
uniphaser 单相交流发电机
unipivot 单支枢，单枢轴
uniplanar flow 单面流
unipod 独脚架
unipolar armature 单极电枢
unipolar dynamo 单极发电机
unipolar machine 单极电机
unipolar magnetic region 单极磁区
unipump 摩托泵
unique solution 唯一解
uniqueness 唯一性
unit 单位；单元；组；组合
unit absorption 单位吸水量
unit amplitude 单位振幅
unit area 单位面积
unit area loading 单位面积加载
unit area resistance 单位面积阻力
unit auxiliary transformer 机组辅助变压器
unit capacity 单位容量；机组容量
unit circle 单位圆
unit clamp 专用夹具
unit complex forcing function 单位复抗力函数
unit consumption of energy 单位能耗
unit consumption of water 单位用水量
unit contact pressure 单位接触压力
unit conversion 单位换算
unit conversion factor 单位转换系数
unit cost 单价；单位成本
unit deformation 单位变形
unit delay operator 单位延迟算子
unit discharge 单位流量
unit dislocation 单位位错
unit displacement 单位位移
unit drawdown 单位降深
unit dry weight 干重度，干容重
unit elongation 单位伸长
unit error 单位误差
unit force 单位力
unit ground pressure 单位地压
unit hydraulic gradient 单位水力梯度
unit hydrograph 单位水文曲线
unit impulse function 单位脉冲函数
unit impulse response 单位脉冲响应
unit length 单位长度
unit load 单位载荷
unit mass 单位质量
unit master control 机组主控制
unit matrix 单位矩阵
unit normal vector 单位法向矢量
unit of advance 进尺单位
unit of area 面积单位
unit of consistency 稠度单位
unit of error 误差单元
unit of measurement 计量单位
unit of stress 应力单位
unit of structure 构件；结构单元
unit of time 时间单位
unit of water measurement 水量度单位
unit power plant 成套动力装置
unit pressure 单位压力
unit project 单位工程
unit pulse 单位脉冲
unit recovery factor 单元采收率
unit resistance 单位阻力
unit shaft resistance 单位侧面阻力
unit skin friction 单位表面摩擦力
unit specific gravity 单位比重
unit speed 单位速度
unit strain 单位应变
unit stratotype 单元层型
unit strength 单位强度
unit stress 单位应力
unit system 单位制
unit take 单位吸浆量
unit tamping energy 单位夯击能量
unit temperature gradient 单位温度梯度
unit time 单位时间
unit trench section 单元槽段
unit value 单位值
unit vector 单位矢量，单位向量
unit virtual force 单位虚力
unit volume 单位体积，单位容积
unit volume change 单位容积变化
unit volume expansion 单位体积膨胀
unit water content 单位含水量
unit weight 容重，单位重量；单位权
unit weight test 容重试验
United Nations Atomic Energy Commission（UNAEC） 联合国原子能委员会
United Nations Development Programme（UNDP） 联合国开发计划署
United Nations Educational, Scientific and Cultural Organization（UNESCO） 联合国教育、科学及文化组织，联合国教科文组织
United Nations Environment Programme（UNEP） 联合国环境总署，联合国环境规划
United Nations Organization（UNO） 联合国组织
United Nations Technical Assistance Administration（UNTAA） 联合国技术援助局
United Nations（UN） 联合国
united polders and merged polders 联圩并圩
unitized solid-block tree 整体式采油树

unity 一，单一，唯一；一致，统一；单位，元素；同质，均一
unity gain amplifier 单位增益放大器
unity ratio equation 统一比值方程
univalent ion 单价离子，一价离子
univariance 单变量
univariant assemblage 单变组合
univariant curve 单变曲线
univariate 单变量
univariate analysis 一元分析，单变量分析
universal 宇宙的；通用的；万能的；普遍的
universal abundance 宇宙丰度
universal ball joint 球形万向节
universal beam 通用钢梁
universal beam mill 通用钢梁轧机
universal bender 通用弯管机
universal bridge 通用电桥
universal buoyage 通用浮标
universal chain wrench 万能链钳
universal character set 通用字符集
universal classification 通用分类
universal coiler 通用卷取机
universal comparator 通用比长仪
universal compass 万能罗盘
universal compensator 万能补偿器
universal constant 通用常数
universal coupling 通用联轴节
universal coupling shaft 万向联结轴
universal crane 万能吊车，万能起重机
universal cutting plier 万能剪钳
universal decimal classification 国际十进位分类法
universal digital computer 通用数字计算机
universal distinct element code 通用离散元程序
universal drafting machine 通用绘图机
universal driving shaft 通用驱动轴
universal excavator 通用挖掘机，万能挖掘机
universal gas constant 普通气体常数
universal gravitation 万有引力
universal gravitational constant 万有引力常数
universal grinder 万能磨床
universal indicator 通用指示剂
universal indicator element 通用指示元素
universal indicator plant 通用指示植物
universal input module 通用输入模件
universal instrument 通用仪表
universal joint 万向节，万向接头
universal joint cross 万向节十字头
universal joint spider 万向节十字头
universal language 通用语言
universal lathe 万能车床；万能机床
universal level 通用水准器
universal material testing machine 万能试验机
universal meter 万用表
universal method mapping 全能法测图
universal miller 万能铣床
universal mining machine 通用采煤机
universal motor 通用电动机
universal mounting 通用托架
universal piling rig 通用打桩机
universal polar stereographic projection 通用极射赤平投影
universal program 通用程序
universal rail gage 万能道尺
universal ring type crusher 通用环式破碎机
universal rolling 万能轧制
universal scale 通用比例尺
universal seismograph 通用地震仪
universal shaft 万向轴
universal soil loss equation 通用土壤流失方程
universal space rectangular coordinate system 通用空间直角坐标系
universal spindle coupling 万向轴连接器
universal stage 费氏台，旋转台
universal stand 通用机架
universal standard 通用标准
universal stratigraphic classification 统一地层划分
universal stratigraphic scale 统一地层表
universal strength tester 通用强度试验机
universal switch 通用开关
universal table 万能工作台
universal testing machine 万能试验机
universal theodolite 通用经纬仪
universal time（UT） 世界时；国际标准时间
universal trailer 万能拖车
universal transverse Mercator projection 通用横轴墨卡托投影
universal turbulence constant 通用湍流常数
universe 宇宙；万物；世界；全人类
universe abundance 宇宙丰度
univibrator 单稳多谐振荡器
univoltage 单电位
uniwafer 单晶片
unjacked test 无护套测试
unknown number 未知数
unknown quantity 未知量
unknown reaction force 未知反力
unknown term 未知项
unlabelled basic statement 无标号基本语句
unlabelled block 无标号信息块
unleaded petrol 无铅汽油
unlimited flow deformation 无限流动变形
unlimited flow strain 无限流动应变
unlined canal 无衬砌渠道
unlined shaft 无砌壁立井
unlined tunnel 无衬砌隧道，毛洞
unloaded hole 未装药炮眼
unloader 卸载机，卸料机
unloading 卸载
unloading auger 螺旋卸载机
unloading condition 卸载条件
unloading conveyor 卸载运输机
unloading curve 卸载曲线，卸荷曲线
unloading fractured rock mass 卸荷破裂岩体；卸荷裂隙岩体
unloading joint 卸载节理
unloading modulus 卸载模量
unloading plough 犁式卸载机
unloading point 卸载点
unloading pressure by blasting 爆破卸压
unloading process 卸载过程

unloading pump	卸油泵
unloading relaxation	卸载松弛
unloading test	卸载试验
unloading tipper hopper	卸载翻斗车
unloading valve	卸载阀
unloading wave	卸载波
unloading well	自流井，自喷井，卸压井
unloading zone of slope	卸载斜坡区
unmagnetized plasma	非磁化等离子体
unmanned power station	无人电站
unmanned satellite	不载人卫星
unmanned spacecraft	无人空间飞行器
unmanned station	无人台站，无人泵站
unmanned underwater vehicle	无人潜水器
unmanned vehicle	无人驾驶运载工具；无人驾驶车辆
unmapped fault	未填出断层
unmapping area	未填图区
unmatched terraces	不匹配阶地
unmeasurable	不可测量的；无法计量的；不可度量的
unmeasured	未测定的
unmechanized hump yard equipment	非机械化编组站驼峰设备
unmetamorphosed	未变质的
unmigrated stack section	无偏移叠加剖面
unminable coal seam	不可采煤层；非采煤层
unmixing of solid solution	固溶体分离
unmixing system	不混合系统
unmixing texture	不混合结构
unmounted drill	手持式风钻，手持式凿岩机
unnotched specimen	无切口试样
unpaired terraces	不成对阶地
unparallel bedding	不平行层理
unpermeability	不透水性
unplasticised polyvinyl chloride	低塑性聚氯乙烯，非塑化聚氯乙烯
unploughed land	未耕土地
unpolarizing	去极化作用，去偏振作用
unpolished	未磨光的，粗糙的
unpolluted water	未污染水
unpolluted zone	未污染带
unpredictability	不可预测性
unpredictable element	不可预见因素
unpressurised diving system	常压潜水系统
unprocessed data	未处理的数据
unproductive interval	非生产层段
unprospected area	未勘探区
unprotected river bed	无护岸河床
unprotected tool joint	无保护钻杆接头
unpublished data	未发表数据，未发表资料
unrammed concrete	未捣实混凝土
unrated gauging station	未分级水位站
unreactive aggregate	惰性骨料
unrecoverable oil	不可采石油
unrecoverable reserves	不可采储量
unrecovered oil	残留石油
unreduced coefficient	不可约系数
unreduced matrix	不可约矩阵
unrefinable crude oil	不适于炼制的原油
unregulated bed	未整治河床
unregulated flood flow	未调节的洪水流量
unregulated river reach	未整治的河段
unregulated stream	未整治的溪流
unreinforced brick building	无筋砖砌建筑
unreinforced masonry	无筋砌体
unreinforced section	无筋断面
unreliability	不可靠性
unrelieved stress	残余应力
unrestricted flow	无约束流动，自由流动
unrestricted plastic flow	无限制塑性流动
unrestricted random sampling	无限制随机抽样
unrolling	展开图
unroofing	去顶，揭顶
unsafe building	不安全建筑
unsafe condition	不安全条件
unsafe depth foundation protection	不安全深度地基保护，浅地基保护
unsafe foundation depth	不安全基础深度
unsaturated acid	不饱和酸
unsaturated bond	不饱和键
unsaturated coefficient of permeability	不饱和渗透系数
unsaturated flow	不饱和流
unsaturated hydraulic conductivity	不饱和导水率，不饱和渗透系数
unsaturated hydrocarbon	不饱和烃
unsaturated intact loess	不饱和原状黄土
unsaturated liquid-solid coupling	不饱和液固耦合
unsaturated polyester	不饱和聚酯
unsaturated pore space	不饱和孔隙空间
unsaturated pore volume	不饱和孔隙体积
unsaturated rock	不饱和岩
unsaturated sample	不饱和样
unsaturated soil	非饱和土
unsaturated soil mechanics	非饱和土力学
unsaturated solution	非饱和溶液
unsaturated steady flow	不饱和稳定流
unsaturated zone	不饱和带，充气带
unsaturation	不饱和，未饱和
unsaturation test	不饱和试验
unsawn timber	粗材
unscreened gravel	未过筛砾石
unscrewing	拧松
unsealing	开封，未密封
unseasoned timber	未干燥木材
unset concrete	未凝固混凝土
unsettled weather	不稳定天气
unshared composite beam	非共享组合梁
unshielded arc welding	无屏蔽电弧焊
unsized ore	未分级矿石
unslaked lime	生石灰
unslaked lime pile	生石灰桩
unsmooth	不光滑的
unsoaked	未浸湿的
unsodded	未铺草皮的
unsoldering	脱焊
unsolvable	不可解决的；不可溶解的
unsorted	未分选的
unsound aggregate	不良粒料
unsound aquifer	不良含水层
unsound cement	不良水泥，变质水泥
unstable association	不稳定组合

unstable channel	不稳定河槽
unstable combustion	不稳定燃烧
unstable component	不稳定组分
unstable compound	不稳定化合物
unstable condition	不稳定条件
unstable creep	不稳定蠕动
unstable equilibrium	不稳定平衡
unstable failure	不稳定破坏
unstable floor	不稳定底板
unstable flow	不稳定流
unstable grout	不稳定浆液
unstable heavy mineral	不稳定重矿物
unstable hydrocarbon	不稳定烃
unstable isotope	不稳定同位素
unstable mineral	不稳定矿物
unstable protobitumen	不稳定原沥青
unstable region	不稳定区域
unstable relict mineral	不稳残余矿物
unstable rock	浮石；危岩
unstable rock mass	危岩体
unstable slope	不稳定边坡
unstable soil	不稳定土
unstable solution	不稳定解
unstable state	不稳定状态
unstable state flow	不稳定状态流
unstable stratification	不稳定层理
unstable structure	不稳定结构
unstable wave	不稳定波
unstainable	无瑕疵的，不可玷污的
unsteadiness	不稳定性
unsteady boundary layer	不稳定边界层
unsteady drag	不稳定阻力
unsteady flow	不稳定流，非稳态流，非定常流
unsteady flow pattern	不稳定流型
unsteady flow pumping test	不稳定流抽水试验
unsteady groundwater hydraulics	不稳定地下水水力学
unsteady heat conduction	非稳态热传导
unsteady motion	不稳定运动，非稳态运动
unsteady non-uniform flow	不稳定非均匀流，非稳态不均匀流
unsteady pumping test	不稳定态抽水试验
unsteady seepage	不稳定渗透，非稳态渗流
unsteady state	不稳定状态，非稳态
unsteady wave	不稳定波
unsteady-state displacement	非稳态驱替
unsteady-state flow	非稳态流
unsteady-state heat transfer	非稳态传热
unsteady-state radial flow	非稳态辐射流
unsteady-state seepage	非稳态渗流
unsteamed concrete	未蒸养混凝土
unstemmed explosive	无炮泥炸药
unstemmed shot	无炮泥爆破
unstrained member	无应变件
unstrained pile head	无变形桩头
unstratified deposit	非成层矿床
unstratified drift	非层状堆积物，漂砾泥
unstratified ore body	非层状矿体
unstratified rock	非层状岩，不成层岩
unstratified soil	非层状土
unstratified structure	非层状构造，不成层构造
unstructural creeping	非构造蠕变
unsubmerged weir	不淹没堰
unsuitable fill	不适当填土
unsuitable material	不适用物料
unsulfured	不含硫的；未硫化的
unsupported back	无支护拱冠
unsupported back span	无支护拱冠跨度，空顶距
unsupported distance	悬空距离
unsupported excavations	无支撑开挖
unsupported height	悬空高度，无支撑高度，自由高度
unsupported length	无支撑长度
unsupported width	无支护宽度，无支护跨度，悬空宽度
unsurfaced road	土路，未敷路面的道路
unswept region	未波及区
unsymmetric beam	不对称梁
unsymmetric stiffness matrix	非对称刚度矩阵
unsymmetrical balance	不对称平衡
unsymmetrical circuit	不对称回路；不对称电路
unsymmetrical crown	不对称路拱
unsymmetrical disposal	不对称处置
unsymmetrical footing	不对称基础
unsymmetrical load	不对称荷载
unsymmetrical loading tunnel	偏压隧道
unsymmetrical matrix	不对称矩阵
unsymmetrical pressure	不对称压力，偏压
unsymmetrically loading lining	偏压衬砌
unsymmetry	不对称性
untamped backfill	未夯实回填土
untapered pile	无尖桩，不削尖桩，直桩
untested	未测试的
unthreaded pipe	无螺纹管
untimbered rill stope	无支撑倒V形梯段回采工作面
untimbered rill stoping	无支撑倒V形梯段回采
untimbered tunnel	无支撑隧道
untreated effluent	未处理废水
untreated foundation	未处理地基
untreated sewage	未处理污水
untreated water	未处理水，原样水
untrimmed sample	未加工的样品，原样
untubed well	未下油管的井
untwisting torgue	退捻力矩
untwist-retwist method	退捻加捻法
unused land	荒置地
unusual casing	特殊套管
unusual loss	非常损失
unusual soil	特殊土
U-nut	U形螺母
unvegetated	无植被的
unventilated	不通风的
unverifiable	不可验证的
unvulcanized rubber	未硫化橡胶
unwatering	排水，疏干
unwatering borehole	排水钻孔
unwatering coefficient	疏干系数
unwatering conduit	排水管道
unwatering gallery	排水廊道
unwatering header	排水集管
unwatering pump	排水泵
unwatering sump	排水坑
unwatering system	排水系统

unweathered mineral soil	未风化矿质土壤
unweathered rock	未风化岩石
unweighted average	未加权平均数
unworkable concrete	和易性差的混凝土
unworkable deposit	不可开采的矿床
unworked country	未开采区
unyielding	不屈服的，非让压的
unyielding foundation	无沉陷地基
unyielding prop	非让压支柱，刚性支柱
unyielding support	非让压支架，刚性支架
up and down hill run	上行和下行
up-and-down motion	上下运动
U-Pb dating	铀-铅测年
U-Pb method	铀铅法测年
up-boom dredge	上挖式挖掘船
upbuilding	建立，筑起；加积作用，叠积作用
upcast	上风井，出风井，上风口；逆断层；上投，上升盘
upcast shaft	出风井
update	使现代化，革新；修改，校正；最新资料
updigging	上挖
updip pinchout	上倾尖灭
up-dip portion	上山部分
updirection	上行方向
updriving hammer	后继打桩锤
upflow	上升流，顶流
upgrade	提高品位，精选；改良，使升级
upgrading	提高品位，精选；加积作用，叠积作用
upgrading ratio	选矿率
upgrading stream	加积河
upgrowth	成长，发育，发展
upheaval	隆起，鼓起
upheaval of road pavement	路面隆起
uphill	上坡的，上山的
uphill line	上山管线
uphill opening	上山巷道
uphill side	上山侧
uphill slope	上山坡
uphill sloping profile	上坡倾斜剖面
uphill water diversion	引水上山
uphill way	上山路
uphole	上向钻孔，仰孔；沿井身向上
uphole amplifier	井口放大器
uphole drilling	上向钻进，仰孔钻进
uphole geophone	井口检波器
uphole method	井口法
uphole seis	井口检波器
uphole shooting	井炮测井，微地震测井
uphole stack	垂直叠加；井口到达时校正后叠加
uphole stacking time	井口时间
uphole survey	井炮测井
uphole time	井口时间
uphole time correction	井口到达时间校正
uphole velocity	上返速度
uphole work	仰孔作业，上向炮眼作业
upholstery	室内装饰
U-pipe	U形管
up-jarring	向上震击
upkeep	保养；维修费；维持
upkeep cost	维修保养费
upkeep obligation	维修契约；保养维修义务
upkeep undertaking	维修任务
upland	高地，山地
upland bog	高地泥沼
upland catchment	高地流域
upland coast	高地海岸
upland earthquake	山地地震；高地地震
upland erosion	高地侵蚀
upland field	旱田
upland moor	高地沼泽；高位泥沼
upland peat	高地泥炭
upland plain	高平原
upland river catchment	山地流域；高地流域
upland rural area	高地乡郊地区
upland soil	高原土
upland swamp	高地沼泽
upland water	上游来水，地表水
uplap	叠超
upleakage	向上越流
upleap	断层上升侧
uplift	隆起；浮托力；扬压力
uplift beam	隆起梁
uplift blasting	上升爆破
uplift block	上升地块
uplift coast	上升海岸
uplift force	上举力；浮力
uplift history	抬升历史
uplift load test of pile	桩的抗拔试验
uplift monitoring device	上浮力监测仪器
uplift of granitoid bodies	隆起花岗岩体
uplift pile	抗拔柱
uplift pressure	扬压力；浮托力
uplift reduction coefficient	扬压力折减系数
uplift resistance	抗拔力
uplift speed	提升速率
uplift test	抗拔试验
uplift velocity	提升速度
uplifted area	隆起区
uplifted block	上升地块
uplifted horst	上升地垒
uplifted peneplain	上升准平原
uplifted plain	上升平原
uplifted side	上升侧；上升盘
uplifted structure	隆起构造
uplifted wall	上升盘
uplimb thrust fault	沿翼上冲断层
upline	入站线；上行线
up-link transmitter	上行发送器
upload	加载；负载；上载
up-logging	向上提时测井
upmost	最上的；最高的；最主要的
upnormal pressure gradient	高压压力梯度
up-over bench stoping	仰挖台阶
upover borehole	上向钻孔，仰孔
upper	上面；上头；上部；上限；上面的；较高的
upper air	高空，高层空气
upper air chart	高空图
upper air circulation	高空环流
upper air current	高空气流
upper air data	高空资料

upper air front　高空锋
upper air observation　高空观测
upper air synoptic station　高空气象站
upper anchor head　上锚头
upper and lower limit criterion　上下限准则
upper and lower with the coal mining　上部和下部与煤炭开采
upper anticyclone　高空反气旋
upper apex of fold　褶皱顶部
upper approach channel　上游引航道
upper Aptian　上阿普第期
upper asthenosphere　重软流圈上部
upper atmosphere　高层大气
upper atmosphere dynamics　高层大气动力学
upper atmosphere pollution　高层大气污染
Upper Atmosphere Research Satellite（UARS）　高层大气研究卫星
upper atmospheric physics　高层大气物理学
upper bainite　上贝茵体；上贝氏体
upper bank　上游坝体
upper base　上底
upper bed　上层
upper bench　上台阶，上梯段
upper bend　背斜；背斜顶部；上托力计；向上拱
upper bend section　上弯段
upper bounce　上界
upper bound　上界；上限
upper bound constraint　上界约束
upper bound theorem　上限定理
upper branch　上子午圈
upper break　上部缺失，上部间断
upper Cambrian　上寒武统；上层寒武纪
upper cantilever　拉杆
upper Carboniferous　上石炭统；晚石炭世
upper Carboniferous coalfields　上石炭纪煤田
upper case　大写字体
upper catchment area　上段集水区
upper cave culture　山顶洞文化
upper cave culture cave man　山顶洞人
upper chamber　上气室
upper chord　上弦
upper chromosphere　色球高层
upper coal　上部煤层
upper cofferdam　上游围堰；上游水池
upper cold front　高空冷锋
upper column　上立柱
upper confidence limit　置信上限
upper confined bed　隔水顶板；上承压层
upper confined bed crust　上地壳
upper confining bed　上覆隔水层；隔水顶板
upper control limit　控制上限
upper course　上游；上游段；上层
upper Cretaceous　上白垩统
upper critical cooling rate　上部临界冷却速率
upper crust　上地壳；地壳上部
upper culmination　上游闸门；上中天
upper curved section　上弯段
upper cut-off frequency　上限截止频率
upper cyma recta　上裟
upper dead center　上死点

upper deck of bridge　上层桥面
upper density limit　密度上限
upper deviation　上偏差；上差
upper Devonian　上泥盆统；晚泥盆世
upper drain plug　上排泄封塞
upper drift　上风向
upper drive mechanism　上部驱动装置
upper edge　上游心墙
upper elastic limit　弹性上限
upper end　上端
upper end of reach　河段上端
upper entry　上部平巷；上平巷
upper envelope　上包络
upper expansion chamber surge tank　上扩式调压井
upper extreme　上限
upper fan　上部扇
upper fillet and fascia　上枋
upper flange of girder　梁上翼
upper floor　上层楼面
upper floor shop　楼上铺位
upper floor unit　高层单位上层单位
upper flow regime　上部水流动态；高流态
upper formation　上部地层
upper frequency limit　频率上限
upper front　高空锋
upper gangway　上部平巷；上平巷
upper ground floor　地下高层
upper gudgeon　顶枢轴；上耳轴
upper guidevane ring　上导叶环
upper half-plane　上半面
upper hemisphere　上半球
upper hemisphere projection　上半球赤平极射投影；上半球赤平投影
upper horizon soil　上层土
upper horizontal section　上平洞；上平段
upper hybrid frequency　高混杂频率
upper income band　上层入息组别
upper intertidal facies　上潮间带相
upper Jurassic　上侏罗统
upper Jurassic period　上侏罗纪
upper kelly cock　方钻杆上旋塞
upper layer　上层
upper ledge　上框
upper level　上部水平；上水平；上水平，上部平巷
upper level loading　上装
upper level streamline　高空流线
upper limb　上缘；上翼
upper limit　上限尺寸；上限
upper limit for coal mining　煤炭开采上限
upper limit in code　规范上限
upper limit magnitude　震级上限
upper limit of plasticity　塑性上限
upper littoral zone　沿岸带上部
upper lock　上游船闸
upper lock chamber　上游闸室
upper longitudinal strut　上纵撑
upper low　高空低压
upper mantle　上地幔
Upper Mantle Committee（UMC）　上地幔委员会
upper mantle project　上地幔计划

upper mantle tectonics	上地幔构造
upper mantle xenolith	上地幔捕房体
upper mesosphere	高中层
upper Miocene	晚中新世
upper miter wall	上游人字墙
upper Oligocene	晚渐新世
upper Ordovician	上奥陶统；晚奥陶世
upper pack	上部充填层
upper Paleolithic	旧石器时代晚期
upper Paleozoic	上古生界
upper Paleozoic Era	上古生代
upper Paleozoic Erathem	上古生界
upper pickup	游车上升最高点
upper pile	上部桩
upper plastic limit	塑性上限；上塑限
upper plate	顶板
upper platform	上层平台
upper Pleistocene	晚更新世
upper Pliocene-Pleistocene	上新世
upper plug	上塞
upper pond	上游铺盖
upper pool	船闸前池；上水池；上游河湾
upper prop	上柱
upper range	上限
upper reach	上游，上游河段
upper reaches	上游
upper river	上游
upper roof	老顶；主顶板
upper soil layer	土壤表层
upper storey	上层
upper surface	顶面
upper triangular matrix	上三角矩阵
upper tropical troposphere	热带高对流层
upper wall	上盘；顶壁
upper water	上层水
upper wing wall	上游翼墙
upper working seam to provide relief	上解放层
upper workings	上部巷道，上部采区；上部采场
upper Yangtze oldland	上扬子古陆
upper yield limit	屈服上限
upper yield point	屈服点上限；塑性上限；蠕变上限
upper zone of dual	两层完井的上层
upper-air prognostic chart	高空预报图
upper-case	大写字母；大写体；大写的；用大写字母排印
upper-frame	顶架；顶框
upper-frequency limit	频率上限
upper-ignition honey combed briquette	上点火蜂窝煤
upper-ignition honeycomb briquette	上点燃蜂窝煤
upper-level jet stream	高空急流
upper-lower limit of slope protection	护坡的上下限
upper-mantle structure	上地幔结构
uppermost	最上的；最高的；最主要的；最上，最高
uppermost crust	最上部地壳
uppermost layer	表层，最上层，顶层
uppermost storey	最高楼层
upper-phase microemulsion	高含油微乳液
upper-shoreface	上临滨
upraise	上升，升起
upraise drift	上坡平巷；天井巷道
upraise shaft	由下向上开凿的立井
upraising	上向掘进
up-ramp	上行坡道
uprated	大功率的
upright	垂直，竖立；垂直的，直立的；笔直，柱
upright anticline	直立背斜
upright breakwater	直立式防波堤
upright converter	竖式吹炉；立吹炉
upright course	立砌层
upright drill	钻床
upright fan structure	直立扇状构造
upright fold	直立褶皱；垂直褶皱
upright position	垂直位置；垂直状态
upright projection	垂直投影；垂直剖面图；侧视图
upright resistance	垂直阻力
upright stanchion	立杆；支柱
upright tank	立式罐
upright tubular boiler	立式管式锅炉
upright wall	直立墙
uprighting force	扶正力
uprighting moment	扶正力矩
uprise	涌高
uprise of salt	盐分上升
uprisnig	上向掘进
uprush	上冲；冲岸浪；垂直急流；强上升流
uprush limit	上爬界限
UPS distribution panel	不间断电源配电盘
ups-and-downs	高低起伏
upscaling	升尺度；网格粗化；升级
upscaling extent	升级程度
upscattering	导致能量增加的散射
upset and threaded male end	加厚的阳螺纹端
upset bolt	膨径螺栓
upset casing	加厚套管
upset condition	不正常条件
up-set condition	不正常状态
upset diameter	镦锻直径
upset drill pipe	端部加厚的钻杆
upset end	加厚端
upset end casing	端部镦粗套管
upset end joint	加厚接头
upset forging	顶锻；镦锻
upset kelly	端部加厚的方钻杆
upset premium	标金底价
upset price	拍卖底价
upset rent	租金底价
upset test	顶锻试验
upset tube	镦粗管
upset tubing	端部加厚油管
upset underfill	加厚端未充满
upset welding	电阻对接焊
upset wrinkles	加厚端褶皱
upset-end casing	端部加厚的套管
upsetter	镦锻机
upsetting couple	倾覆力偶
upsetting machine	顶锻机；镦粗机
upsetting moment	倾覆力矩
upsetting point	倾覆点
upsetting press	镦头机；镦锻压力机
upside	上面；上部；上盘；上升盘

upslope	上坡，逆坡；往上，向上坡
upslope fog	上坡雾
upslope ripple	上坡波痕
upstage	顶级
upstairs	楼上；在楼上；在高空；楼上的
upstand beam	直立梁
upstanding block	隆起断块
upstock	造斜器
upstrain shear release	上提管柱剪切丢手
upstream	向上游的；逆流而上的；上游领域；上游；向上游的；逆流的
upstream apron	上游堤岸
upstream batter	迎水坡；上游坡
upstream bay	上游河汊
upstream benefits	上游面板
upstream berm	迎水面马道
upstream blanket	上游铺盖；泄洪排沙系统
upstream bottom current	向源底层流
upstream catchment	上游进水池
upstream cofferdam	上游围堰
upstream core	上游填土体
upstream deposit	河流上源沉积
upstream elevation	上游高程
upstream end	上游鼻端
upstream facing	上游流域
upstream fill	上游坝脚
upstream floor	上游铺砌
upstream gain	上游效益
upstream gate	上游围堰
upstream heel	上游坝踵
upstream inflow	上游护坦
upstream integration	上游一体化
upstream line	吸入管道
upstream maximum water level	上游最高水位
upstream method	上游法
upstream minimum water level	上游最低水位
upstream normal water depth	上游正常水深
upstream plant	上游电站
upstream pressure	上游压力
upstream pumping unit	上游泵送设备
upstream radius of crest	坝顶上游面曲率半径
upstream reach	上游河段
upstream river	上游端
upstream scheme	迎风格式；迎风差格式
upstream section	上游段；上流段
upstream sector	上游部门
upstream shell	上游坝壳
upstream shen	上涌现象
upstream shock	上流激波
upstream side	上游侧
upstream slope	上游坡；迎水面坡度；上游坡面；上游坡度
upstream slope protection	上游护坡
upstream toe	上行船
upstream toe of dam	上游坝趾
upstream total head	上游总水头
upstream water	上游水；上层带水；上游防冲铺砌
upstream water level	上游水位
upstream water line	上游水位线
upstream water pressure	上游水压力
upstream well	布置在地下水流上游的井
upstreambound fish	上溯鱼
upstreanung ion	上流离子
upstroke	上划；活塞上行
upstroke speed	上提速度；提升速度
upstroke twisting	上行式捻线
upstructure	沿构造向上
upstructure well	构造顶部井
upsupported height	无支撑高度
upsurge	高潮；增长；上涌浪
upsweep	向上扫描；升频扫描
upswell	胀起；隆起
upswelling	上升，上涌；涌出，溢出
uptake	咽喉；烟箱；垂直向上的管道；吸收，摄取
up-the-hole testing	自下而上试井
upthnist	上推；上冲；仰冲断层
upthown fault	逆冲断层
upthrow	上投；上冲上升垂直位移；隆起；上升盘
upthrust bearing	上推力轴承
upthrust fault scarp	上冲断层崖
up-to-date map	最新地图
upturned bed	倒转层
up-underreaming	向上扩眼
upward annular velocity of drilling mud	钻井泥浆向上回流速度
upward borehole	向上钻孔，仰孔
upward buckling	向上弯曲
upward compatibility	向上兼容
upward component	向上分量
upward continuation	向上延拓
upward conveyor	上向输送机
upward cumulation	向上累积
upward digging	向上挖掘
upward flow	上升流，向上流
upward grouting	上行式注浆
upward hole	向上钻孔
upward hydraulic pressure	浮托力，扬压力，向上水压力
upward hydrostatic pressure	向上静水压力
upward irrigation	底水灌溉
upward landscape	仰视景观
upward mining	上行式开采
upward movement	向上移动
upward percolation	向上渗滤
upward pressure	向上压力
upward shaft deepening method	上向井筒延深法
upward steepening fault	向上变陡的断层
upward velocity	向上速度，上升速度
upward view	仰视图
upward welding	上行焊，向上焊接
upward-current classifier	上升水流分级机
upward-fining	向上变细的
upwarp	上拱，上翘，隆起
upwarped district	翘曲区，隆起区
upwarping	向上挠曲
upwelling	上涌，上升；涌出，溢出
upwelling current	上升流
upwelling hot flow zone	上涌热流带
upwelling water	上涌泉水
upwind	逆风

up-wind finite element	迎风有限元
up-wind infinite element	迎风无限元
up-wind side	上风面
uraconite	土硫铀矿，水铀钒
uralite	水泥石棉板；纤闪石
uramphite	铀铵磷石
urania	氧化铀
uranic	铀的，含铀的，六价铀的
uranic oxide	三氧化铀
uraniferous sediment	含铀沉积
uraninite	晶质铀矿，沥青铀矿
uranite	铀云母，云母铀矿
uranium	铀
uranium barren	贫铀的
uranium bearing shale	含铀页岩
uranium black	铀黑
uranium carbide	碳化铀
uranium-complexes	铀络合物
uranium compound	铀化合物
uranium content	含铀量
uranium deposit	铀矿床
uranium dioxide	二氧化铀
uranium disequilibrium dating	铀系不平衡测年
uranium family	铀族
uranium fuel fabrication facility	铀燃料加工设施
uranium group	铀组
uranium hexafluoride	六氟化铀
uranium hexafluoride production facility	六氟化铀生产设施
uranium isotope age	铀同位素年龄
uranium lead age	铀铅年龄
uranium lead age method	铀铅年龄测定法
uranium lead dating	铀铅测年
uranium mine	铀矿
uranium mineral	铀矿物
uranium ochre	铀华
uranium ore	铀矿石
uranium ore assay	铀矿石分析
uranium ore leaching	铀矿浸出
uranium oxidized zone	铀氧化带
uranium poisoning	铀中毒
uranium radium series	铀镭系
uranium reactor	铀反应堆
uranium reduction zone	铀还原带
uranium resources	铀资源
uranium rich	富铀的
uranium separation	铀的分离
uranium series	铀系
uranium silicalite	含铀硅岩
uranium source	铀源
uranium tetrafluoride	四氟化铀
uranium trioxide	三氧化铀
uranium xenon method	铀氙法
uranium yellow	铀黄
uranium-actinium series	锕铀系
uranium-bearing	含铀的
uranium-helium dating	铀-氦测年
uranium-potassium ratio	铀钾比
uranium-series dating	铀系定年
uranium-thorium-lead dating	铀-钍-铅测年
uranmicrolite	铀细晶石
uranocircite	钡铀云母
uranolite	陨石
uranology	天文学
uranometry	天体测量
uranophane	硅钙铀矿
uranopilite	铀矾，铀钙矾
uranospathite	水磷铀矿
uranospherite	纤铀铋矿
uranospinite	砷钙铀矿
uranotantalite	钽铀矿
uranothallite	铀钙石，铀灰石
uranothorianite	铀方钍矿
uranothorite	铀钍矿
uranotile	硅钙铀矿
uranous oxide	二氧化铀
urao	天然碱
urban aesthetics	城市美学
urban air pollution	城市空气污染
urban air pollution concentration	城市空气污染浓度
urban air pollution source	城市空气污染源
urban and rural planning	城乡规划
urban architecture	城市建筑
urban area	市区
urban conservation	城市保护
urban construction	城市建设
urban data base	城市数据库
urban demography	城市人口学
urban disaster prevention	城市防灾
urban ecological planning	城市生态规划
urban ecological system	城市生态系统
urban ecology	城市生态学
urban economics	城市经济学
urban ecosystem	城市生态系统
urban electricity supply	城市供电
urban electricity system	城市电力系统
urban engineering geological investigation	城市工程地质勘察
urban engineering geological mapping	城市工程地质填图
urban engineering geology	城市工程地质
urban environment protection	城市环境保护
urban environmental decay	城市环境恶化
urban evaluation	城市评价
urban expansion	城区扩展
urban expressway	城市快速道路
urban facilities	城市设施
urban flood control	城市防洪
urban garden	城市花园
urban gas supply system	城市燃气供应系统
urban geographical information system	城市地理信息系统
urban geography	城市地理学
urban geological hazard	城市地质灾害
urban geology	城市地质，城市地质学
urban green cover ratio	城市绿化覆盖率
urban green space	城市绿地
urban ground subsidence	城市地面沉降
urban health	城镇卫生
urban heat island	城市热岛
urban heat island effect	城市热岛效应

urban heat supply	城市供热
urban heating system	城市供热系统
urban highway	城市道路，市区公路
urban hydrology	城市水文学
urban information system	城镇信息系统
urban land use	城市土地利用
urban landscape	城市景观
urban layout	城市布局
urban living-space	城市居住空间
urban mapping	城市制图
urban mass transit	城市轨道交通
urban morphology	城市形态
urban noise	城市噪声
urban park system	城市公园系统
urban pattern	城市格局
urban physical environment	城市物理环境
urban planning	城市规划
urban planning information system	城市规划信息系统
urban planning theory	城市规划理论
urban population	城市人口
urban public transportation system	城市公共交通系统
urban radiation	城市辐射力
urban railway	城区铁路
urban reconstruction	城市改造
urban refuse	城市垃圾；城市废弃物
urban remote sensing	城市遥感
urban renewal	旧城更新
urban road network	城市道路网
urban road system	城市道路系统
urban runoff	城市径流
urban sanitation	城市环境卫生
urban science	城市科学
urban sewage	城市污水，城市废水
urban sewage and drainage system	城市污水与排水系统
urban sociology	城市社会学
urban space	城市空间
urban sprawl	城市扩张
urban structure	城市结构
urban survey	城市测量
urban thoroughfare	城市干道；城市大道
urban tissue	城市组织
urban topographic survey	城市地形测量
urban traffic	城市交通
urban traffic management	都市交通管理
urban traffic noise	城市交通噪声
urban utilities	城市公用事业
urban waste	城市废物
urban water management	城市水管理
urban water supply	城市供水
urban water system	城市供水系统
urban weather	城市天气
urban wind	城市风
urbanization	城市化
urbanology	城市学
urea-formaldehyde grout	脲醛树脂浆
urgent measures for earthquake emergency	地震应急措施
usable capacity	可用库容；有效产能；可用能力
usable component	有用组分
usable flow	有效流量
usable hole	有效钻孔，合格钻孔
usable land	可用土地
usable life	有效期
usable production well	有效生产井
usage factor	利用系数，利用率
used equipment	旧设备
used heat	废热；余热
used oil	废油
useful area	有效面积，使用面积
useful associated component	有益伴生组分
useful capacity	有效容量；有效容积
useful component	有用组分，有益组分
useful constituent	有用组分，有益组分
useful cross section	有效截面
useful current	有效电流
useful effect	有效效应
useful efficiency	有效效率
useful height	有效高度
useful horsepower	有效马力
useful information	有用信息
useful life	使用期限，使用寿命
useful load	有效荷载，工作荷载
useful mineral	有用矿物
useful output	有效产量
useful power	有效功率
useful thermal efficiency	有效热效率
useful volume	有效容积
useful water capacity	有用水容量
useful work	有用功
useless reserves	表外储量
user	用户，使用者
user interface	用户界面，用户接口
user key	用户键
user library	用户库
user manual	用户手册
user partition	用户分区
user program	用户程序
U-series dating	铀系测年
U-shaped abutment	U形桥台
U-shaped magnet	马蹄形磁铁
U-shaped valley	U形谷；平底谷
ustalf	干淋溶土
ustent	干新成土
ustert	干变性土
ustisol	老成土
ustoll	干软土
ustox	干氧化土
ustult	干老成土
usual practice	惯例
usufruct outlay of mineral exploration right	探矿权使用费
U-Th-Pb dating	铀钍铅定年
utility	公用事业，公用设施；有用，实用
utility building	公用设施建筑
utility factor	利用率
utility program	应用程序，实用程序
utility routine	辅助程序，应用程序
utility subprogram	实用子程序
utility theory	效用理论
utility value	效用值；实用价值；利用价值
utility winch	杂用绞车，辅助绞车
utilizable flow	可用流量

utilization 利用
utilization coefficient 利用系数
utilization efficiency 利用效率
utilization factor 利用率，利用系数
utilization of debris 废品利用
utilization of tidal power 潮汐力利用
utilization of underground space 地下空间利用
utilization of waste material 废料利用
utilization rate 利用率
utilization rate of raw materials 原材料利用率
utilization ratio 利用率
utilization ratio of machine handling 机械装卸利用率
utilized river section 已开发河段
U-trap 存水湾；虹吸管
U-tube U形管

U-tube manometer U形管压力计
U-tube principle U形管原理
UU-test 不固结不排水试验，UU试验
UV degradation 紫外光降解
UV radiation 紫外光辐射
UV spectroscope 紫外线分光器
UV stabilizer 紫外光稳定器
uval 波状丘陵地
uvala 干宽谷，岩溶谷地，喀斯特谷地，灰岩谷地；溶蚀洼地；灰岩盆
uvanite 钒铀矿
UV-detector 紫外光检测器
U-vortex 马蹄形涡流
UV-scanner 紫外线扫描器
Uzawa algorithm 乌拉算法

V

V cut　V形掏槽，楔形掏槽
V sheave　三角皮带轮
V thread　V形螺纹
V weld　V形焊
vacancy　空；空白；空处，空地；空位，空穴
vacant　空的，空缺的，空位的
vacant line　虚线
vacant lot　空地
vacant position　空位
vacant run　空载运行
vacant site　空置地盘
vacant tenement　空置住宅，空置物业
vacuate　抽真空；抽稀
vacuity　真空，真空度；空隙，空间
vacuole　空穴；空泡；气泡；液泡
vacuometer　真空计
vacuous　真空的
vacuseal　真空密封
vacuseal grease　真空密封树脂；真空密封脂
vacuseal wax　真空密封蜡
vacuum　真空；真空的
vacuum air pump　真空泵
vacuum annealing　真空退火
vacuum arc furnace　真空电弧炉
vacuum arc melting　真空电弧熔炼
vacuum arc remelting　真空电弧重熔
vacuum blasting　真空喷砂法
vacuum brake　真空制动闸；真空制动器
vacuum breaker　真空断路器；真空开关；真空破坏器
vacuum breaker valve　真空开关阀；真空断路阀
vacuum breaking system　真空破除系统
vacuum capillary viscometer　真空毛细管黏度计
vacuum carbonitriding　真空碳氮共渗
vacuum case　真空箱，真空室
vacuum casting　真空铸造，真空浇铸
vacuum centrifugal pump　离心式真空泵
vacuum chamber　真空室，真空箱
vacuum cleaner　真空吸尘器；真空清洁器
vacuum cleaning line　真空除尘管线
vacuum cleaning plant　真空除尘设备
vacuum collector　真空集尘器
vacuum consolidation　真空固结法
vacuum contactor　真空接触器
vacuum control check valve　真空控制止回阀；真空止回阀
vacuum corer　真空取样器；真空岩芯提取器
vacuum deaerator　真空脱气机
vacuum degasification　真空脱气
vacuum degasser　真空脱气机
vacuum degassing　真空脱气；真空除气
vacuum deposition　真空沉淀
vacuum desiccator　真空干燥器
vacuum dewatering　真空脱水
vacuum diffusion pump　真空扩散泵
vacuum distillation　真空蒸馏
vacuum drainage　真空排水
vacuum dryer　真空干燥器
vacuum drying　真空干燥
vacuum drying of pipeline　管线真空干燥
vacuum drying of timber　木材真空干燥
vacuum drying oven　真空干燥炉
vacuum dynamic consolidation　真空动力固结法
vacuum etching　真空腐蚀
vacuum evaporation　真空蒸发；真空蒸镀；真空涂膜
vacuum evaporator　真空蒸发器；真空镀膜台；真空涂膜器
vacuum exhaust　真空排气
vacuum extraction　真空吸出，真空提取
vacuum extraction equipment　抽真空设备
vacuum fan pump　扇叶形真空泵
vacuum filter　真空过滤器
vacuum filtration　真空过滤
vacuum flasher　真空闪蒸器
vacuum flashing　真空闪蒸
vacuum floatation　真空浮选
vacuum fluorescent lamp　真空荧光灯
vacuum form　真空模板
vacuum forming　真空成形
vacuum fusion　真空熔化
vacuum gauge　真空计，真空表
vacuum grouting technology　真空压浆工艺
vacuum heat treatment　真空热处理
vacuum heating　真空加热；真空采暖
vacuum helium leak detector　真空氦探漏仪
vacuum hydrolysis　真空水解
vacuum impregnation　真空浸渍
vacuum impregnation method　真空浸渍法
vacuum induction melting　真空感应熔化，真空感应熔炼
vacuum injection　真空灌注
vacuum lightning protector　真空避雷器
vacuum line　真空管线
vacuum loader　真空装载机
vacuum manometer　真空压力计
vacuum melting　真空熔炼
vacuum metallurgy　真空冶金
vacuum meter　真空计
vacuum method of preloading　真空预压法
vacuum monitor　真空监视器
vacuum oil recovery　真空采油
vacuum outgas　真空脱气
vacuum packing　真空包装
vacuum pipe　真空管道
vacuum plating　真空电镀
vacuum polymerization　真空聚合
vacuum preloading　真空预压
vacuum pressure impregnation　真空压力浸渍
vacuum process　真空法；真空处理
vacuum pump　真空泵

vacuum recharge	真空回灌	valency charge	价电荷
vacuum refining	真空精炼	valency state	价态
vacuum refrigeration	真空制冷	valid	有效的，正当的，有根据的
vacuum relief valve	真空减压阀	valid data	有效数据
vacuum return line system	真空回水系统	valid figure	有效数字
vacuum rolling mill	真空轧机	valid slip surface	有效滑动面
vacuum sampling tube	真空取样管	valid time of forecast	预报时效
vacuum saturating	真空饱水；真空饱和	validation period	有效期
vacuum seal	真空密封	validity	确实性，有效性
vacuum sintering	真空烧结	validity check	有效性检验
vacuum slag cleaning	真空除渣	validity limit	有效限度；有效范围
vacuum space	真空空间	validity period	有效期
vacuum spraying	真空喷涂	valley	谷；谷地，河谷，山谷；流域
vacuum system	真空系统	valley and ridge tectonics	谷岭构造
vacuum tempering	真空回火	valley basin	河谷盆地
vacuum test machine	真空试验机	valley bottom	谷底
vacuum thermocouple	真空热电偶	valley bulging	河谷横扩
vacuum tightness	真空密封度	valley cross section	河谷断面
vacuum treatment	真空处理	valley dam	谷坝
vacuum triaxial test	真空三轴试验	valley fen	河谷沼泽
vacuum tube	真空管	valley fill	河谷沉填，河谷堆积
vacuum ultraviolet spectrometer	真空紫外光谱仪	valley flat	河漫滩，谷地小平原
vacuum under nappe	水舌下的真空	valley floor	谷底
vacuum valve	真空阀	valley floor basement	谷底基岩面
vacuum vaporiser	真空蒸发器	valley floor plain	谷底平原
vacuum viscometer	真空黏度计	valley floor step	谷底台阶
vacuum welding	真空焊接	valley glacier	山谷冰川
vacuum well	真空井	valley gradient	河谷坡度，河谷比降
vacuum well point	真空井点	valley gravel	河谷砾石
vacuum well system	真空井系统	valley groundwater reservoir	峡谷地下水库
vacuum-irradiated	真空辐照的	valley head	谷源，谷头
vacuum-treated concrete	真空处理混凝土	valley head region	谷源区
vacuum-tube detector	真空管侦检器	valley impoundment	河谷蓄水
vacuum-tube frequence multiplier	电子管倍频器	valley in valley	迭谷，谷中谷
vacuum-type tensiometer	真空式张力计	valley incision	河谷下切
vadose	渗流的；包气带的	valley line	最深谷底线，谷中心线
vadose canyon	渗流峡谷，渗流凹地	valley loess	谷黄土
vadose cave system	渗流洞系	valley meander	河谷曲流，深切曲流
vadose cement	渗流胶结物	valley migration	河谷迁移
vadose region	渗流区	valley moor	河谷沼泽
vadose seepage	包气带渗流	valley of corrugation	瓦楞槽
vadose solution	渗溶	valley of denudation	剥蚀谷
vadose spring	渗流泉，上层滞水泉	valley of elevation	隆起谷，背斜谷
vadose trickle	包气带细流	valley of subsidence	沉降谷，向斜谷
vadose water	渗流水，包气带水，上层滞水，循环水	valley pattern	河谷型式
vadose water discharge	渗流水排出；渗流水排出量	valley placer	河谷砂矿
vadose zone	渗流带，包气带，上层滞水带	valley plain	谷底，泛滥平原
vagrant benthos	游移底栖生物	valley profile	河谷纵剖面
vagueness	模糊性；模糊；隐晦	valley project	流域规划；流域开发工程
valchovite	褐煤树脂，聚合醇树脂	valley slope	河谷边坡
vale	溪谷，槽，沟	valley soil	谷地土壤
valence	价，化合价，原子价	valley spring	河谷泉，侵蚀泉
valence bond	价键	valley storage	河谷蓄水，谷槽蓄水
valence bond structure	价键结构	valley terrace	河谷阶地
valence charge	价电荷	valley tract	谷区
valence effect	价效应	valley train	谷碛；谷缘冰水沉积；谷边碛
valence electron	价电子	valley wall	谷坡，谷壁
valence force	原子价力	valley wind	谷风
valence link	价键	valley-in-valley structure	迭谷构造
valency	价；化合价；原子价	valley-type debris flow	河谷型泥石流

valley-type glacier 山谷型冰川
valuable mineral 有用矿物
valuable ore 有用矿石，有价值的矿石
valuation 估价，评价
valuation at cost 按成本计价
value 值，量；价值；意义；估价，评价
value engineering 价值工程
value of calculation 计算值
value of elasticity 弹性量
value of expectation 期望值
value of hydraulic jump 水跃值
value of orientation 定向值
valve 阀，阀门，活门；壳；电子管
valve adjuster 阀调节器
valve assembly 阀门总成；活门组件；气门组件
valve ball 球阀
valve block 阀块，阀体，阀组
valve body 阀身，阀体
valve bonnet seal 阀盖密封
valve bushing 阀套，阀衬
valve cap 阀盖
valve cover 阀罩
valve cup grab 阀座打捞器
valve guide 阀导套，气门导管
valve header 阀头，气门头
valve housing 阀箱，气门室；阀体；阀罩
valve key 阀门键；阀门扳手；气门锁片
valve lever 阀杆
valve lifter 起阀器；气门挺杆；气门提升器
valve opening pressure 阀开启压力
valve operating gear 阀动齿轮，气门分配齿轮；阀操纵装置
valve orifice 阀孔
valve piston 阀门活塞
valve pit 阀门井；阀坑；阀槽
valve positioner 阀位限位器
valve push rod 气门推杆，阀推杆，活门顶杆
valve rotator 阀门转子
valve seat 阀座；气门座
valve shaft 阀轴
valve spring 阀弹簧；气门弹簧
valve stem 阀杆
valve stroke 阀冲程
valve system 阀门系统
valve tap 阀打捞锥
valve tappet 阀挺杆
valve tee 带阀三通
valve tower 进水塔，分水塔，取水阀塔
valve tree 阀门树，井口装置
valve tube 阀管；电子管
valve washer 阀门垫圈；阀座垫圈；滑阀垫板
valve well 阀阱
valve wide open 阀门大开
valve wide opin 全开阀
valve with ball seat 球座阀
valve with inclined stem 斜杆阀
valveless 无阀门的；无瓣的；无叶片的
valveless filter 无阀滤池
valvelet 小阀
valving 装设阀门；分瓣

van 洗矿铲；淘矿，洗矿；风扇；有篷卡车，行李车；冰斗
Van der Waals bonding force 范德华键力
Van der Waals equation 范德华方程
Van der Waals force 范德华力
Van Mater sampler 冯马特尔型直路取样机
vanadine 钒土
vanadium 钒
vandenbrandeite 绿铀矿；水铀铜矿
vandendriesscheite 橙黄铀矿
vane 风向标；视准器；叶轮，叶片
vane apparatus 十字板剪力仪
vane borer 涡轮钻机，透平钻机；钻孔验土器
vane compressor 滑片式压缩机；叶片式压缩机
vane motor 叶轮液压马达，叶片电动机
vane penetrometer 十字板贯入仪
vane pump 叶片泵
vane shear 十字板剪切
vane shear apparatus 十字板剪力仪
vane shear strength 十字板抗剪强度
vane shear test 十字板剪切试验
vane shear test apparatus 十字板剪力试验仪
vane strength 十字板抗剪强度
vane water meter 叶片式水量计，叶片式水表
vane wheel 叶轮
vane-type motor 叶片式发动机，风动发动机
vanishing 尖灭，消没，消失，消散
vanishing angle 消没角
vanishing line 投影线，消失线
vanishing point 灭点，合点
vanishing point perspective of building 灭点法建筑透视图
vanishing tide 缺潮，失潮
vanishing-point condition 合点条件
vanner 淘矿机；淘选带
vanner concentration 带式摇床精选
vanning 淘选，淘矿
vanning jig 振筛跳汰机
vanstoning 凸缘接合，范斯通接合
vapor balancer 蒸气平衡阀
vapor barrier 隔气层，阻凝层
vapor blasting 蒸气喷射；湿法喷砂
vapor bubble 蒸气泡
vapor compression 蒸气压缩
vapor compression evaporator 蒸气压缩蒸发器
vapor compression machine 蒸气压缩机
vapor compression refrigerator 蒸气压缩制冷机
vapor concentration 水汽浓度
vapor condensation 水汽凝聚
vapor condensation zone 蒸气冷凝带
vapor condenser 蒸气冷凝器
vapor cooler 蒸气冷却器
vapor curve 蒸气曲线
vapor density 水汽密度
vapor deposition 气相淀积；蒸镀
vapor diffusion 水汽扩散
vapor diffusivity 气体扩散性；气体扩散系数
vapor dominant reservoir 蒸气型热储
vapor expansion 蒸气膨胀
vapor flow 水汽流

vapor fractionation 蒸气分馏
vapor gun 蒸气枪
vapor phase 蒸气相
vapor pressure 蒸气压，水汽压
vapor pressure curve 蒸气压曲线
vapor pressure difference 蒸气压差，水汽压差
vapor pressure gradient 蒸气压梯度，水汽压梯度
vapor pump 蒸气泵
vapor seal 气封
vapor state 气态
vapor transmission 蒸气输送
vapor zone 蒸气带
vaporability 汽化性，挥发性
vaporable 可气化的
vaporation 蒸发
vapor-dominated geothermal reservoir 蒸汽型地热储
vaporimeter 气压计
vaporizability 可汽化性
vaporization 汽化作用，蒸发
vaporization coefficient 蒸发系数
vaporization heat 汽化热
vaporization loss 蒸发损失，汽化损失
vaporization point 蒸发点，蒸发温度
vaporization rate 汽化率
vaporization temperature 蒸发温度
vaporization-condensation cycle 气化冷凝循环
vaporized chemical 汽化剂
vaporizer 蒸发器，汽化器
vaporizing 蒸发，汽化
vaporizing tape 蒸发带
vapor-liquid chromatography 气-液色谱法
vapor-liquid equilibrium 气-液平衡
vapor-oil-water separator 气-油-水分离器
vaporous 蒸汽的，气态的
vaporous water 气态水
vapor-pressure test 蒸气压试验
vaporproof closure 防蒸发罩
vapor-recovery system 蒸气回收系统
vaportight 不透气的，气密的
vapour 蒸汽，气
variability 可变性，变异性，变化率
variability index 变化指数
variability law 变异定律
variability of spring water 泉水变化率
variability of stream flow 流量变化率
variable 变量，变数，变元；变化的，可变的，不定的；各种各样的
variable accelerated motion 变加速运动
variable acceleration 变加速
variable area flowmeter 变截面流量计
variable binding 可变连结；变量结合
variable bit weight 可变钻压
variable capacitor 可变电容器
variable cell 可变单元
variable channel 不稳定水道
variable coefficient 可变系数
variable cohesive-strength theory 可变内聚强度理论
variable concentrated load 可变集中荷载
variable condenser 可变电容器
variable coupling 可变耦合
variable cross section 可变横截面
variable damper 可变阻尼器
variable density 变密度
variable density log 变密度测井
variable difference 变差
variable dimension 可变维
variable domain 可变域
variable Dunlap method 邓禄普可变系数法
variable element 变量元素；可变元素
variable end-point 可变端点
variable error 变量误差；可变误差
variable factor 可变因子
variable flow 可变流，变速流
variable flow hydraulic pump 变流量液压泵
variable focal length camera 变焦相机
variable force 可变力
variable frequency 可变频率
variable frequency control 变频控制
variable frequency drive 变频驱动
variable frequency induced polarization 变频激发极化；变频诱发极化
variable frequency sinusoidal excitation 变频率正弦波激振
variable frequency vibrator 变频振动器；变频振捣器
variable gain 可变增益
variable gear 变速齿轮
variable gradient 可变梯度
variable head 变水头
variable head permeability test 变水头渗透试验
variable head permeameter 变水头渗透仪
variable height cutting 变高度掏槽
variable identifier 变量标识符
variable impedance transformer 可调阻抗变压器
variable inductor 可变电感器
variable intensity log 变强度测井
variable linear system 可变线性系统
variable load 可变荷载
variable mesh size 可变网距，可变步长
variable metric method 变尺度法
variable module model 变模量模型
variable modulus of elasticity 可变弹性模量
variable motion 变速运动
variable multivibrator 可调多谐振荡器
variable name 变量名
variable normalization 变量正规化
variable pair 变量对
variable parameters model 变参数模型
variable phase 可变相位
variable point 变量点；可变点
variable pressure capillary viscometer 变压毛细管黏度计
variable pressure transducer 变压式传感器
variable quantity 改变量；可变量
variable resistance 可变电阻；可变阻力；变阻
variable resistor 变阻器
variable section 可变截面，变截面
variable set 变量集
variable slope 变坡度，变坡降
variable speed 变速，可变速度
variable speed case 变速箱
variable speed drive 变速驱动

variable speed fluid drive 变速液压传动
variable speed scanning 变速扫描
variable spring 变流量泉，多态泉
variable spring hanger 可变弹簧吊架
variable standardization 变量标准化
variable state 可变状态
variable step size 可变步长
variable stiffness 可变刚度
variable stiffness member 可变刚度杆件
variable storage 可变储存量
variable stroke pump 变冲程泵
variable structure control system 可变结构控制系统
variable systematic error 可变系统误差
variable temperature drop method 不等温降法
variable time increment 可变时间增量
variable transformer 可调变压器
variable vector 变向量
variable velocity 变速，可变速度
variable voltage 可变电压
variable word length computer 可变字长计算机
variable-aperture flowmeter 变孔径流量计
variable-aperture seismic array 变孔径地震台阵
variable-capacity pump 可调流量泵
variable-flow pump 变流量泵
variable-grade channel 变坡渠
variable-mass dynamics 变质量动力学
variable-number-node element 节点数可变元
variable-pitch turbine 变螺距轮机
variable-radius arch dam 变半径拱坝
variable-slope pulse modulation 变斜率脉冲调制
variable-speed generator 变速发电机
variable-speed motor 变速马达
variable-speed pump 变速泵
variable-speed transmission 变速传动
variable-speed turbine 变速轮机
variable-thickness arch 变厚度拱
variable-voltage generator 调压发电机
variance 变化，变动；方差，离差，离散；偏差，差异，变异
variance analysis 方差分析，离差分析
variance contribution 方差贡献
variance law 方差律
variance matrix 方差矩阵
variance of unit weight 单位权方差；方差因子
variance ratio 方差比
variance ratio transformation 方差比变换
variance test 方差检验
variance-covariance matrix 方差-协方差矩阵
variance-covariance propagation law 方差-协方差传播律
variant 变体，变种；变数，变量；不同的
variate 变数，变量，变元
variate difference analysis 变差分析
variate difference method 变量差分法
variation 变迁，变异，变动；变分，变差，变量；变种
variation coefficient 变差系数；变异系数
variation curve 变化曲线
variation diagram 变异图解
variation equation 变分方程
variation in cross section 沿断面变化
variation in depth 随深度变化
variation in discharge 流量变化
variation in thickness 厚度变化
variation in water level 水位变化
variation iteration 变分迭代法
variation method 变分法
variation of azimuth 方位变化
variation of precipitation 降水量变化
variation of water quality 水质变化
variation of works 工程变更
variation order 变更命令
variation principle 变分原理
variational approach 变分法
variational calculus 变分学；变分法
variational difference scheme 变分差分格式
variational finite element method 变分有限元法
variational form 变分形式
variational formulation 变分列式
variational function 变分函数
variational method 变分法
variational principle 变分原理
variator 变速器；聚束栅
varicolored clay 杂色黏土
varicoloured claystone 杂色黏土岩
varied flow function 变流函数
variegated sandstone 杂色砂岩
variety 种类；变种；多样性
varigradation 差异均夷
varigrained 不同粒度的
varimax method 最大方差法
varimin method 最小方差法
varindor 变感器
variocoupler 可变耦合器；可变结合器
variode 变容二极管
variogram 方差图，变差图，变异图方差曲线
variolite 球粒玄武岩
variolitization 球粒化
variolosser 可控损耗器
variomatic 可变自动程序的
variometer 磁变仪；变感器
varioplex 可变多路传输器，变路转换器
varioscale projection 变比例投影
various 不同的；各种各样的
variplotter 自动曲线绘制器；自动制图仪
Variscian cycle 瓦力西旋回
Variscian movement 瓦力西运动
Variscian orogeny 瓦力西造山运动
vari-size grained 不同粒度的
vari-speed drive 无级变速传动装置
varistor 变阻器；压敏电阻
varistor modulator 变阻器调制器；压敏电阻调制器
varistructured array 可变结构阵列
varisweep technique 变频带扫描技术
varnish 清漆，罩光漆，绝缘漆；釉
varnish base 漆底
varnish paint 清漆；透冕
varnished cloth 漆布
varnished wire 漆包线
varve 纹泥，季候泥
varve chronology 纹泥年代学

varve sediment 纹泥沉积物，季节性沉积物
varve slate 纹泥板岩
varve-clay 季候泥
varved clay 层状黏土，纹泥，季候泥
varved deposit 纹泥沉积
varved slate 纹层板岩
varying acceleration 变加速度
varying capacity 变容量
varying cross section 变截面
varying head 变水头
varying load 变负荷，不定荷载
varying mass distribution 变质量分布
varying rigidity 变刚度
varying stress 变应力
vase 粉砂
vaseline 凡士林；石油冻，矿脂
vast scale 大比例尺
vault 拱穹，拱顶，筒拱；地下室
vault and invert formwork 顶拱及底拱模板
vault head 圆拱顶盖
vault roof 拱顶
vault settlement 拱顶下沉
vaulted dam 拱坝
vaulted mud crack 拱起泥裂
vaulted passage 甬道
vaulted roof 拱状屋顶
vaulted shell 穹壳
V-belt drive 三角皮带传动
V-belt pulley 三角皮带轮
V-body car V形矿车
V-coal 煤尘
V-conveyor V形滚柱输送机
V-cylinder V形汽缸
V-door 井架人字形大门
V-drill 尖钻
Vean diagram 维恩图
vectograph 矢量图，向量图
vectopluviometer 风向雨量计；雨向计
vector 矢量；向量
vector addition 矢量加法
vector adjustment 矢量平差
vector algebra 矢量代数
vector algorithm 矢量算法
vector analogue computer 矢量模拟计算机
vector analysis 矢量分析；向量分析
vector bundle 矢量丛
vector cellular automaton 矢量细胞自动机
vector chart 矢量图
vector component 矢量分量
vector computer 矢量计算机
vector correlation 矢量相关
vector data 矢量数据
vector data structure 矢量数据结构
vector description 矢量描述
vector diagram 矢量图表
vector element 矢元
vector equation 矢量方程
vector expression 向量表示式
vector field 矢量场
vector function 矢量函数
vector gauge 矢量测雨器
vector generator 矢量发生器
vector geographic information system 矢量地理信息系统
vector graphics 矢量图；矢量绘图
vector graphics language 矢量图形语言
vector gravimetry 矢量重力测量
vector image 矢量图像
vector impedance 矢量阻抗
vector locus 矢量轨迹
vector magnetic meter 矢量磁力仪
vector magnitude 矢量大小，矢量幅度
vector map 矢量图
vector Markov process 矢量马尔可夫过程
vector matrix 矢量矩阵
vector mean 矢量平均值
vector method 矢量法
vector multiplication 矢量乘法
vector norms 矢量范数
vector notation 矢量记法
vector of acceleration 加速度矢量
vector of displacement 位移矢量
vector of generalized coordinates 广义坐标矢量
vector of unit length 单位长度矢量
vector of velocity 速度矢量
vector operation 矢量运算
vector plot 矢量图
vector plotting 矢量绘图，矢量标绘
vector polygon 向量多边形
vector potential 矢量势，矢量位
vector product 矢量积，叉积
vector projection 矢量投影
vector resolution 矢量分解
vector space 矢量空间
vector strength 矢量强度
vector stress 矢量应力
vector stress function 矢量应力函数
vector structure 矢量结构
vector subtraction 向量减法
vector sum 矢量和
vector superposition method 矢量叠加法
vector topology 矢量拓扑学
vector variable 矢量变量
vector wave equation 矢量波动方程
vectorial adjustment 矢量平差
vectorial angle 矢量角
vectorial difference 矢量差
vectorial operator 向量算子
vectorial property 矢量性
vectorization 矢量化
vee crack V形裂缝，扩展裂缝
vee cut V形掏槽，楔形掏槽
vee drain V形排水沟
vee thread V形螺纹
vee trough V形槽
vegetable 植物；植物性的
vegetable fiber rope 植物纤维绳
vegetable garbage 植物性垃圾
vegetable gum 植物胶
vegetable lubricant 植物性润滑剂；植物性润滑油
vegetable mould 腐殖土；沼泽泥

vegetable organic matter	植物有机质
vegetable origin theory	植物成因说
vegetable remains	植物残余
vegetable soil	种植土，熟化土
vegetal cover	植被
vegetated area	植被区
vegetation	植物；植被
vegetation anomaly	植被异常
vegetation boundary	植被界线
vegetation cover	植被
vegetation damage	植被损伤
vegetation geochemical anomaly	植物地球化学异常
vegetation index	植被指数
vegetation loss	植被损失
vegetation map	植被图
vegetation mapping	植被制图
vegetation on slope	植被护坡
vegetation planting	绿化
vegetation protection measures	植被防护措施
vegetation reproduction	植物繁殖
vegetation restoration	植被恢复
vegetation species	植被种类
vegetation succession	植被演替
vegetation survey	植被调查
vegetation type	植被类型
vegetation zone	植被带
vegetational type	植物类型
vegetation-covered area	植被覆盖区
vegetative chain	营养链
vegetative cover	植被
vegetative propagation	营养繁殖，无性繁殖
vehicle	车辆，机动车，汽车；运载器，飞行器；火箭，导弹
vehicle access road	车辆通道
vehicle capacity	车辆载重量
vehicle design standard	车辆设计标准
vehicle earth station	车载地球站
vehicle emission standard	车辆废气排放标准
vehicle empty weight	车辆空载重量，车辆自重
vehicle energy consumption	车辆能源消耗
vehicle equivalence	车辆当量
vehicle fuel standard	汽车燃料标准
vehicle identification number	车辆辨识号码
vehicle inspection	车辆检验
vehicle insurance	车辆保险
vehicle kilometrage	车辆行驶里程
vehicle length	车长
vehicle management	车辆管理
vehicle model	车辆模型，汽车模型
vehicle navigation system	车辆导航系统
vehicle noise	车辆噪声
vehicle odometer	车辆里程表
vehicle parapet	车辆护栏
vehicle stream	车流
vehicle tax	车辆税
vehicle thrust	车辆推力
vehicle vibration	车辆振动
vehicle weigh station	秤车站
vehicle without tax	免税车辆
vehicle-mounted laboratory	车载实验室
vehicular access	车路，车辆进出通道
vehicular border link	车辆过境通道
vehicular bridge	公路桥；行车桥
vehicular corridor	行车走廊
vehicular gap	车间距
vehicular load	车辆荷载
vehicular pollution	机动车污染
vehicular tunnel	行车隧道；公路隧道
veil	膜；罩；幕
vein	矿脉，岩脉；窄水道
vein breccia	脉状角砾岩
vein deposit	脉状矿床
vein dyke	脉状岩墙
vein filling	矿脉充填
vein fissure	脉状裂隙
vein flow	脉状水流
vein gold	脉金
vein groundwater	脉状地下水
vein intersection	矿脉交切
vein material	脉质
vein matter	脉质
vein mesh	脉网
vein mineral	脉石矿物
vein of spotty character	品位分布不规则的矿脉
vein ore	矿脉，脉状矿体
vein quartz	脉石英
vein rock	脉岩
vein stone	脉石
vein structure	脉状构造
vein system	脉系
vein texture	脉状结构
vein wall	脉壁
vein water	脉状水
veined	脉状的
veined marble	纹理大理石
veined structure	脉状构造
veined-fissure water	脉状裂隙水
veinlet	细脉；小静脉
veinlet dissemination deposit	细脉浸染型矿床
vein-like mass	脉状岩体
veinule	细脉
velocimeter	速度计
velocity	速度
velocity ambiguity	速度模糊
velocity analysis	速度分析
velocity anisotropy	速度各向异性
velocity anomaly	速度异常
velocity azimuth display	速度方位显示
velocity azimuth range display	速度方位距离显示
velocity change	速度变化
velocity contrast	速度对比，速度差
velocity correction	速度校正
velocity curve	速度曲线
velocity defect law	速度亏损律
velocity derived porosity	声速测井确定的孔隙度
velocity diagram	速度图解，流速图解
velocity discontinuity	速度不连续
velocity dispersion	速度弥散；速度色散
velocity distribution	速度分布，流速分布
velocity factor	速度因子，流速系数

velocity feedback	速度反馈	velocity spectrum	速度谱
velocity field	速度场	velocity structure	速度结构
velocity filter	速度滤波器	velocity sweeping	速度扫描
velocity filtering	速度滤波	velocity time diagram	速度时间图
velocity flowmeter	速度流量计	velocity time history	速度时程
velocity focusing	速度聚焦	velocity to prevent erosion	防冲蚀流速
velocity function	速度函数	velocity to prevent plant growth	防长草流速
velocity gauge	速度计	velocity triangle	速度三角形
velocity gradient	速度梯度	velocity variation	速度变化
velocity head	流速水头，速度落差	velocity vector	流速矢量
velocity head difference	流速水头差	velocity wavelet	速度子波
velocity head probe	速头探测器	velocity-area gauging station	流速面积测量站
velocity histogram	速度直方图	velocity-area method	流速面积法
velocity index	速度指数	velocity-azimuth-depth assembly	流速流向水深仪
velocity interface	速度界面	velocity-controlled safety valve	控速安全阀
velocity interpolation formula	速度插值公式	velocity-depth gradient	速度深度梯度
velocity inversion	速度反演	velocity-head coefficient	流速水头系数
velocity inversion procedure	速度反演方法	velocity-head correction factor	流速水头校正因子
velocity jump	速度跳跃	velocity-sensing receiver	速度传感接收器
velocity layer	速度层	velometer	测速仪
velocity limit	速度极限	veneer	薄层；表层；风化层，沙漠漆；饰面，镶面，贴面
velocity log	速度测井		
velocity measurement	速度测量	veneer board	贴面板；胶合板
velocity meter	速度计	veneer of soil	表层土
Velocity model	速度模型	veneered brick	陶面砖，釉面砖
velocity modulation	速度调制	veneered wall	镶面墙
velocity of discharge	排放速度；输水速度	veneered with brick	用砖镶面的
velocity of escape	逃逸速度	veneering	胶合；镶饰
velocity of flow	水流速度，流速	veneering press	胶合压力机
velocity of groundwater flow	地下水流速	Vening Meinesz isostasy	韦宁迈内兹均衡
velocity of longitudinal wave	纵波速度	Vening-Meinesz formula	韦宁迈内兹公式
velocity of particle motion	质点运动速度	venite	脉状混合岩
velocity of penetration	穿透速度	venom	毒液；毒物
velocity of permeability	渗透速度	vent	排放；排放口，通气口，通风平巷；火山口；泉口
velocity of primary wave	初至波速度，头波速度		
velocity of propagation	传播速度	vent adit	风道，通风平硐
velocity of pulsation	脉动速度	vent agglomerate	火山口集块岩
velocity of rotation	旋转速度	vent hole	通风孔
velocity of secondary wave	次波速度	vent line	通风管线
velocity of seismic wave	地震波速度	vent opening	通风口
velocity of shear wave	剪切波速度	vent pipe	通风管
velocity of transversal wave	横波速度	vent shaft	通风井
velocity of wave	波速	vent tunnel	通风隧道
velocity of wave propagation	波传播速度	ventage	孔隙；排气孔
velocity pickup	速度拾取器	vented explosion	泄爆
velocity potential	速势，速度势	vented front-head machine	带放气孔的湿式凿岩机
velocity potential function	速势函数	ventifact	风棱石，风磨石
velocity pressure	速压，速度压	ventilate	通风，换气
velocity profile	速度剖面，流速剖面	ventilated	通风的
velocity profile curve	速度廓线，流速廓线	ventilated box car	通风棚车
velocity pulsation	速度脉动，流速脉动	ventilated case	通风柜
velocity ratio	速度比	ventilated corridor	通风走廊
velocity resonance	速度共振	ventilated lobby	通风门厅
velocity resonance frequency	速度共振频率	ventilated motor	风冷式电动机
velocity response envelope spectrum	速度反应包络谱	ventilating	通风
velocity response spectrum	速度响应谱	ventilating air	新鲜空气
velocity rod	测速杆	ventilating device	通风装置
velocity scanning	速度扫描	ventilating dryer	通风干燥机
velocity seismograph	速度地震记录仪；地震速度计	ventilating duct	通风管道
velocity shear boundary	速度切变界限	ventilating entry	通风平巷

ventilating facilities	通风设备
ventilating fan	通风扇
ventilating machine	通风机
ventilating machinery	通风机械
ventilating opening	通风口
ventilating pipe	通风管
ventilating raise	通风天井
ventilating roadway	风巷
ventilating shaft	通气井
ventilation	通气
ventilation by air passage	风道式通风
ventilation by duct	巷道式通风
ventilation by pipe	风管式通风
ventilation capacity	通风能力
ventilation coefficient	通风系数
ventilation connection	通风联络巷
ventilation curtain	通风帘幕；风障
ventilation district	通风区
ventilation efficiency	通风效率
ventilation flow	通风流量
ventilation model	通风模型
ventilation modulus	通气模量
ventilation monitoring	通风监测
ventilation network	通风网络；通风系统
ventilation pressure	通风压力
ventilation resistance	通风阻力
venting	放喷；洗井；通风
venting valve	通风阀；泄流阀
ventless delay electric cap	无烟延发电雷管
venture	风险
venture business	风险企业
venture capital	风险资本；风险投资
venture evaluation	风险评估
venture investment	风险投资
Venturi blower	文丘里风机
Venturi deduster	文丘里除尘器
Venturi effect	文丘里效应
Venturi flume	文丘里试槽
Venturi metre	文丘里流量计
venturi nozzle	文丘里喷嘴
venturi scrubber	文丘里洗涤器
Venturi tube	文丘里管
verandah	走廊；游廊；阳台
verbal contract	口头约定
verde salt	无水芒硝
verfication	校准，检定；验证，核实
verge	边缘；路肩；山墙突瓦；檐口瓦
verge angle	边缘角
vergence	聚散度；朝向，指向，面向
vergence belt	倒转带
vergence direction	倒转方向；倾斜方向
verifiability	可证实性
verification	检定，验证；确认
verify	核实；证实
vermicular	蠕虫状的，弯曲的；蠕动的
vermicular structure	蠕虫状构造
vermiculite	蛭石；金精石
vermiculite concrete	蛭石混凝土
vermilion	辰砂；橘红色石榴子石
vermin	害虫；有害动物
verminproof	防虫的，防菌的
vernacular construction	乡土建筑，地方建筑
vernacular housing	地方民居
vernier	游标，微调装置
vernier adjustment knob	微调旋钮
vernier bevel protractor	活动游标量角器
vernier caliper	游标卡尺
vernier circle	游标盘
vernier compass	游标罗盘仪
vernier depth gauge	游标深度尺
vernier dial	游标刻度盘
vernier division	游标分度
vernier drive	微变传动
vernier graduation	游标刻度
vernier micrometer	游标千分尺；游标测微计
vernier microscope	游标显微镜
vernier protractor	游标量角器；游标分度器
vernier reading	游标读数
vernier scale	游标尺
vernier theodolite	游标经纬仪
vernier zero	游标零点
versant	山坡，山侧
versatile	万能的，通用的；多方面的；易变的
versatile data structure	通用数据结构
versatile digital computer	通用数字计算机
versatile multichannel magnetic tape cassette	通用多道盒式磁带机
versatile pile frame	多用途桩架
versatile user interface	多用途用户接口
vertebrate paleontology	古脊椎动物学
vertex	顶，顶点；天顶
vertex angle	顶角
vertex distance	顶点距离；顶点距
vertical	垂直的；直立的；垂直线；纵向
vertical acceleration	垂直加速度
vertical accuracy	垂直精度
vertical additional force	纵向附加力
vertical aerial photography	垂直航空摄影
vertical alignment	垂直定线；垂直定中校准
vertical allowable load capacity of single pile	单桩竖向抗压极限承载力
vertical anchor	垂直锚固，竖锚
vertical and batter pile group	垂直及倾斜的桩群
vertical angle	垂直角，仰角，高度角
vertical angulation	垂直三角测高法，三角高程测量
vertical anticline	直立背斜
vertical average degree of consolidation	竖向平均固结度
vertical axis	直立轴，纵轴，竖轴
vertical axis current metre	竖轴流速仪
vertical barrier	垂直隔层
vertical bearing test	垂直载荷试验
vertical blasthole	垂直爆破孔
vertical blind ditch	竖向盲沟
vertical blind shaft	暗井，盲竖井
vertical boom	立式悬臂，竖吊杆
vertical boom trenching machine	立臂挖沟机
vertical booster pump	立式增压泵
vertical boring tool	立式钻具
vertical bottle depth integrating sampler	竖瓶积深式采样器

vertical bracing	垂直支撑；竖向支持
vertical calibration	垂直校准
vertical cave	竖井；竖洞
vertical centrifugal pump	立式离心泵
vertical circle	垂直度盘
vertical clamp	垂向固定夹；垂直制动螺旋
vertical claw rock loader	直立爪式装岩机
vertical clearance	竖向净空，垂直净空；垂直间隙
vertical climatic zone	垂直气候带
vertical closure	垂直封闭；立堵截流
vertical coaxial coils system	垂直同轴线圈系统
vertical collimator	垂准器，垂直准直仪
vertical compaction	垂向压实
vertical component	垂向分量，垂直分力
vertical component seismograph	垂直分量地震记录仪
vertical compressive stress	垂直压应力
vertical compressor	立式压缩机
vertical concrete spiral	直立混凝土蜗壳
vertical coning	竖向锥进
vertical control	高程控制；垂直控制；垂向控制
vertical control datum	高程控制基准
vertical control network	高程控制网
vertical control point	高程控制点
vertical control survey	高程控制测量
vertical convective mixing	竖向对流混合
vertical coordinate	纵坐标，垂直坐标
vertical coplanar coil system	垂直共面线圈系统
vertical coverage	垂直视域；垂直覆盖范围
vertical crater retreat mining method	垂直漏斗状后退式采矿法
vertical cross section	垂直剖面，垂直断面
vertical crustal deformation measurement	垂直地壳形变测量
vertical curb	立式路缘石
vertical curve	竖曲线
vertical cut	竖直挖方，垂直开挖；垂直掏槽
vertical cut-and-fill	垂直分条充填开采法
vertical datum	高程基准面
vertical deflection	垂直偏转，垂直偏差
vertical deformation	垂直形变；竖向变形
vertical density profile	垂直密度剖面
vertical depth	垂直深度
vertical depth sounding	垂直测深
vertical deviation	垂直偏差
vertical diffusion	垂直扩散
vertical direction	垂直方向
vertical discharge	垂直排放
vertical discharge rate	垂直排放率
vertical displacement	垂直位移；垂向驱替
vertical distance	垂直距离
vertical distributed load	垂直分布荷载
vertical distribution	垂直分布
vertical down pipe	直立式落水管
vertical drain	垂直排水
vertical drainage	垂直排水
vertical drainage well	垂直排水井
vertical drilling	垂直钻进
vertical dynamic load	垂直动载荷
vertical dynamic stiffness	垂直动刚度
vertical eddy	竖直旋涡
vertical electrical sounding	电测深，垂直电探
vertical energy flux	垂直能量通量
vertical engine	立式发动机
vertical epipolar line	垂核线
vertical epipolar plane	垂核面
vertical erosion	垂直侵蚀，向下侵蚀
vertical error	垂直误差，高程误差
vertical exaggeration	垂直放大
vertical exaggeration ratio	垂直放大比
vertical excavation	垂直开挖，竖直开挖
vertical expansion rate	垂直膨胀率；纵向扩张率
vertical exploration well	垂直探井
vertical exploratory opening	垂向探掘，勘探竖井
vertical extent	垂直深度，垂直距离
vertical extruder	立式挤出机，立式挤压机
vertical face stoping	垂直向上梯段回采，倒台阶回采
vertical fault	垂直断层，直立断层
vertical fault scarp	垂直断层崖
vertical filtering	垂直滤波
vertical fissure	垂直裂缝，垂直裂隙
vertical flood	垂向注水
vertical flow	垂向渗流
vertical flux	垂直通量
vertical fold	直立褶皱；垂直褶皱
vertical force	垂直力
vertical force of foundation top	基础顶面垂向力
vertical formwork	垂直模板
vertical fracture	垂直破裂；纵裂
vertical geological section	垂直地质剖面
vertical geophone	垂直检波器
vertical gradient	垂直梯度
vertical gradient of gravity	垂直重力梯度
vertical gravity anomaly	垂直重力异常
vertical greening	垂直绿化
vertical ground heat exchanger	垂直地热交换器；垂直地埋管换热器
vertical hair	十字丝竖丝
vertical height	垂直高度
vertical heterogeneity	垂向非均质性
vertical hole	垂直钻孔
vertical hydrogeochemical zoning	垂直水文地球化学分带
vertical imposed load	垂直外加荷载
vertical inclinometer	垂直测斜仪
vertical instability	垂直不稳定性
vertical instrument	竖直仪
vertical intensity	垂直强度
vertical interval	垂直间距；垂向间距；等高线间距
vertical joint	竖缝；垂直节理
vertical justification	垂直对齐；纵向对齐
vertical ladder	竖梯；直爬梯
vertical leakage	垂向渗漏；垂向渗漏量
vertical lift bridge	垂直升降桥
vertical line	垂直线，铅垂线
vertical load	垂直荷载
vertical loading test	垂直加载试验
vertical loading test of pile	单桩竖向加载试验
vertical magnetic intensity	垂直磁强度
vertical mass curve	垂向累积曲线；垂直质量曲线
vertical migration	垂直运移
vertical migration parameter	垂直运移参数

vertical monitoring control network	垂直监测网；高程监测网
vertical motion	垂直运动
vertical movement	垂直运动
vertical multi-stage pump	立式多级泵，多级立泵
vertical offset	垂直断错；垂直偏移
vertical opening	垂直巷道，竖井，垂直开拓
vertical pair	竖直象对
vertical parallax	垂直视差，上下视差
vertical partition wall	直立隔断墙
vertical pass	直立溜井；立轧孔型
vertical permeability	垂向渗透
vertical photography	垂直摄影
vertical pile	竖直桩
vertical pressure	垂直压力
vertical pressure gradient	垂直气压梯度
vertical profile	垂直剖面
vertical profiling	测深；垂直剖面测量
vertical projection	垂直投影
vertical recharge	垂直补给
vertical recharge intensity	垂直补给强度
vertical reference line	垂直基准线；垂直参考线
vertical refraction coefficient	垂直折光系数
vertical refraction error	垂直折光误差
vertical relevance tree	垂直关联树技术
vertical resistance	垂直阻力
vertical rib	垂直加强筋，纵肋
vertical sand drain	垂直排水砂井
vertical saturation gradient	垂直饱和梯度
vertical scale	垂直比例尺
vertical seawall	直立式海堤
vertical second derivative of gravity	重力垂向二阶导数
vertical section	垂直剖面
vertical seismic profile	垂直地震剖面
vertical separation	垂直分离
vertical sequence	垂直层序
vertical shaft type mixer	直立筒形搅拌机
vertical shear beam model	垂直剪切梁模型
vertical shear method	垂直剪切法
vertical shear vector	垂直剪切矢量
vertical shear wave	垂直剪切波
vertical shearing stress	垂直剪切应力
vertical sheeting	直立板桩墙；直立板撑
vertical shift	垂直移距
vertical shore	垂直支撑；直立海岸
vertical shrinkage	垂直收缩率
vertical slice and fill	垂直分层充填开采法
vertical sounding	测深；垂直探测，立面探测
vertical stability	垂直稳定性
vertical strain	垂直应变
vertical stress	垂直应力
vertical surrounding rock pressure	垂直围岩压力
vertical tectonic joint	垂直构造节理
vertical temperature gradient	垂直温度梯度
vertical temperature profile radiometer	垂直温度剖面辐射仪
vertical thermal conductivity	垂直热导率
vertical thermal gradient	垂直热力梯度
vertical thickness	垂直厚度
vertical throw	垂向落差；竖向落差
vertical transportation	垂直运输
vertical tube settlement gauge	垂直管式沉降计
vertical turbine pump	立式叶轮泵
vertical turret lathe	立式转塔车床
vertical ultimate carrying capacity of single pile	单桩垂向抗压极限承载力
vertical ultimate load capacity of pile group	群桩垂向极限承载力
vertical ultimate uplift resistance of single pile	单桩垂向抗拔极限承载力
vertical upward welding	向上立焊
vertical V-cut	垂直V形掏槽，垂直楔形掏槽
vertical velocity gradient	垂直速度梯度
vertical vibration	垂直振动
vertical view	俯视图
vertical wave	垂直波
vertical well	竖井
vertical zonation	垂直分带
vertical zoning	垂直分带
vertical-beam ditcher	竖杆挖沟机
vertical-circle left	盘左
vertical-circle right	盘右
vertical-dipping bed	垂直岩层
vertical-face stope	垂直工作面采场
verticality	垂直度，垂直
vertical-lateral expansion ratio	纵横膨胀比
vertical-lateral shrinkage ratio	纵横收缩比
vertically distorted model	垂直畸变模型
vertically high cut slope	垂直高切坡
vertically oriented fracture	垂直定向破裂
vertically perforated brick	垂直多孔砖
vertical-sections mining	垂直分条采矿
very angular	多棱角的；极锐利的
very coarse columnar aggregate	极粗柱状团聚体
very fine columnar aggregate	极细柱状团聚体
very fine crumb aggregate	极细团粒状团聚体
very fine granular aggregate	极细粒状团聚体
very fine particle	极细颗粒
very fine prismatic aggregate	极细棱柱状团聚体
very fine sand	极细砂
very finely crystalline	极细晶粒的
very gassy mine	超级瓦斯矿
very hard water	极硬水
very high frequency	甚高频，特高频，超高频
very high frequency radar	甚高频雷达
very high liquid limit soil	极高液限土
very high plasticity soil	极高塑性土
very high radiation area	极高辐射区
very high resolution radiometer	极高分辨率辐射仪
very high risk	极高风险
very high salinity water	极高盐度水；极高矿化水
very high temperature reactor	极高温反应堆
very long baseline interferometry	极长基线干涉测量
very long baseline radio interferometer	极长基线射电干涉测量仪
very low density polyethylene	极低密度聚乙烯
very low frequency band radiated field system	甚低频带辐射场系统
very poorly drained	排水极差的
very rapid permeability	极速渗透

very shallow earthquake 极浅地震
very shallow water wave 极浅水波浪
very short range weather forecast 甚短期天气预报
very short wave 超短波，甚短波
very slow permeability 极慢渗透率
very soft ground 极软弱地基
very soft silty clay 极软粉质黏土
very strong acid 极强酸
very thin platy aggregate 极薄片状团聚体
very-low-frequency aeroelectromagnetic survey 甚低频航空电磁测量
very-low-frequency positioning system 甚低频定位系统
vesicle 气泡；气孔；水泡；囊泡
vesicle water 气孔水
Vesic's ultimate bearing capacity formula 魏锡克极限承载力公式
vesicular 多孔的，多泡的，蜂窝状的
vesicular filling 气孔填充
vesicular lava 多孔状熔岩
vesicular rock 多孔岩石
vesicular structure 多孔构造
vesicular texture 多孔结构
vessel 容器；船舱
vestigial fossil 遗迹化石，足迹化石
V-G meter 黏度-比重计
V-gear V形齿轮
V-gully V形沟；V形侵蚀沟
viaduct pier 高架桥桥墩
viameter 路程计；测震仪；路面平整度测量仪
viberation deterioration 振动劣化
vibracorer 振动取芯器
vibrafeeder 振动给料机
vibrameter 振动计
vibra-pack 振动充填
vibrated concrete 振实混凝土
vibrated soil 振实土
vibrating 振动，振捣
vibrating ball mill 振动球磨机
vibrating base plate compactor 振动基板压实机
vibrating board 平板振捣器；振动板
vibrating compaction 振动压实
vibrating compactor 振动压实机
vibrating concrete float 混凝土振平器
vibrating crushing test 振动粉碎试验
vibrating extractor 振动拔桩机
vibrating feeder 振动供料机
vibrating grizzly 振动栅筛
vibrating hammer compaction test 振动锤实试验
vibrating impacting boring machine 振动冲击钻机
vibrating load 振动荷载
vibrating pile driver 振动打桩机
vibrating pile hammer 振动打桩锤
vibrating plate compactor 振动板压实机
vibrating plate tamper 振动板夯实机
vibrating stressmeter 振动应力计
vibrating string extensometer 振弦伸长计；振弦引伸计
vibrating string strain meter 振弦应变计
vibrating string-type gravimeter 振弦式重力仪
vibrating strip piezometer 振条渗压计
vibrating table 振动台

vibrating tamper 振动夯
vibrating wire load cell 振弦载荷计
vibrating wire piezometer 振弦孔压计
vibrating wire pressure cell 振弦压力盒
vibrating wire sensor 振弦传感器
vibrating wire strain gauge 振弦应变计
vibration 振动
vibration absorber 减振器
vibration acceleration 振动加速度
vibration acceleration level 振动加速度水平
vibration analysis 震动分析，振动分析
vibration attenuation 振动衰减
vibration axis 振动轴
vibration blasting 振动爆破
vibration compactor 振动碾压机
vibration convergence 振荡收敛
vibration curve 振动曲线
vibration damper 减振器
vibration damping 振动阻尼
vibration detector 振动检测器
vibration direction 振动方向
vibration displacement 振动位移
vibration drilling 振动钻进
vibration dynamics 振动动力学
vibration effect of explosion 爆破振动效应
vibration exciter 激振器
vibration frequency 振动频率
vibration grinding 振动研磨
vibration isolation 隔振
vibration isolation material 隔振材料
vibration isolation of foundation 基础隔震
vibration isolator 隔振器
vibration load 振动荷载
vibration machine 振动机
vibration measurement 振动测量
vibration mechanics 振动力学
vibration metre 测振计
vibration mill 振动磨矿机
vibration mode 振动模式
vibration monitor 振动监测器
vibration monitoring 振动监测
vibration mount 隔振座
vibration of blasting 爆破振动
vibration period 振动周期
vibration pick-up 拾振器
vibration pile driver 振动打桩机
vibration pollution 振动污染，振动公害
vibration proof 防震的，抗震的
vibration proof design 防振设计
vibration proof foundation 防震基础
vibration proof material 防振材料
vibration reducing measures 减振措施
vibration reduction 减振
vibration replacement stone column 振冲碎石桩
vibration roller 振动碾
vibration screen 振动筛
vibration screening 振动筛分
vibration sensor 振动传感器
vibration signal 振动信号
vibration simulator 振动模拟器

vibration spectrum 振动谱
vibration strength 振动强度
vibration stress 振动应力
vibration test 振动试验
vibration tolerance 振动容许值
vibration triaxial apparatus 振动三轴仪
vibration velocity 振动速度
vibration wave 振动波
vibration-absorbing base 减振基础；吸振基础
vibrational bucket 振动铲斗
vibrational coupling effect 振动耦合效应
vibrational energy 振动能
vibrational frequency 振动频率
vibrational mode 振动模式
vibration-control device 控制控制装置
vibrator 振动器，振捣器
vibrator pile hammer 振动桩锤
vibrator sunk pile 振沉桩
vibratory bored pile 振动成孔灌注桩
vibratory centrifuge 振动离心机
vibratory compacting roller 振动压实碾
vibratory compaction 振动压实
vibratory crusher 振动粉碎机
vibratory densification 振动密实
vibratory drilling 振动钻进
vibratory driver 振动驱动器；振动打桩机
vibratory impulse 振动脉冲
vibratory liquefaction 振动液化
vibro drilling 振动钻进
vibro-boring 振动钻进
vibrocap 电雷管
vibro-classifier 振动分级机
vibro-compaction 振动压实
vibro-concrete pile 振动混凝土桩
vibro-core cutter 振动岩芯切割器
vibrocorer 振动取样器
vibrodriver 振动驱动器；振动打桩机
vibroflot 振冲器，振浮压实机
vibrograph 示振仪，振动显示器
vibro-hammer 振动夯
vibroll 振动压路机
vibrometer 振动计
vibronite 维勃罗奈特炸药
vibropile 振动灌注桩
vibro-pile driver 振动打桩机
vibroplate 振动板，平板振捣器
vibroplatform 振动台
vibroprobe 振动探头
vibro-pulverization 振动粉碎，振动粉化
vibrorecord 振动记录
vibroreplacement 振动置换
vibroreplacement stone column 振冲碎石桩
vibro-roller 振动碾
vibrosieve 振动筛
vibrosinking pile 振动沉桩
vibrospade 振动铲
vibrostand 振动台
Vicat penetrometer 维卡特透度计
vice 老虎钳，台钳
vice bench 钳工台

vicious circle 恶性循环；循环论证
Vickers hardness 维氏硬度
Vickers hardness test 维氏硬度试验
video 视频，录像
video camera 摄像机，摄像头
video conference 视频会议
video display 视频显示
video monitoring fixation 视频监控装置
video output 视频输出
video processing 视频处理
video recording 录像
video signal 视频信号
video tape recorder 磁带录像机
video target 录像靶
video terminal 视频终端
video transmission network 视频传输网络
video transmitter 视频发射机
videocorder 录像机，视频信号编码器
videophone 可视电话，视频电话
videoplayer 播放机
videotape 录像带
vierendeel girder 空腹大梁，带上下桁条的梁
vierendeel truss 空腹桁架，连框桁架
view 景观；视图；视野；观点
view angle 视角
viewfinder 取景器，瞄准器；检像器
viewing angle 观察角，视角
viewing angle transformation 视角变换
viewing aperture 取景孔
viewing distance 观察距离
viewing field 视野；视场
viewing port 观察孔；观察窗
vigorite 维戈赖特炸药
villa 别墅；乡间庄园
village flood protection works 乡村防洪工程
village water supply 农村供水
village-type development 乡村式发展
vine-and-melon type irrigation system 藤瓜式灌溉系统
vinometer 酒精比重计
vinyl resin paint 乙烯基树脂漆
vinyl-asbestos tile 石棉聚氯乙烯瓦
vinylon 维尼纶
violent earthquake 大地震
violent outburst 猛烈突出
violent shock 强烈震动
violently weathered zone 剧烈风化带
violet quartz 紫石英
virga 雨幡
virgin 自然的，原始的，未开发的，未经变动的
virgin compression 原始压缩
virgin compression curve 原始压缩曲线
virgin compression ratio 原始压缩比
virgin conditions 原始条件
virgin flow 自然水流
virgin forest 原生林；原始森林
virgin ground 处女地；原始地
virgin isotropic consolidation line 原始各向固结曲线
virgin land 处女地，生荒地
virgin pressure 初压，原始压力
virgin rock 原岩

virgin rock mass	原岩体
Virgin rock stress	原岩应力
virgin rock temperature	原岩温度
virgin soil	生荒地土壤
virgin state	原始状态
virgin state of stress	原始应力状态
virgin stress	原始应力
virgin sulfur	自然硫
virgin wood land	原始林地
virginal cohesion	原始黏聚力
virginal overburden pressure	原始上覆压力
virtual address	虚拟地址
virtual asymptotic line	假渐近线；虚渐近线
virtual beam method	虚梁法
virtual data	虚拟数据
virtual deformation	虚变形
virtual displacement	虚位移
virtual force	虚力
virtual friction angle	虚摩擦角
virtual geomagnetic pole	虚地磁极
virtual gravity	假重力
virtual image	虚像
virtual load	虚荷载
virtual map	虚拟地图
virtual mass	虚质量
virtual medium	虚拟介质
virtual memory	虚拟存储
virtual moment	虚力矩
virtual network	虚拟网络
virtual preconsolidation pressure	视预固结压力
virtual reality	虚拟现实
virtual reality modelling language	虚拟现实模拟语言
virtual reality technology	虚拟现实技术
virtual space	虚拟空间
virtual storage	虚拟存储
virtual strain	虚应变
virtual stress	虚应力
virtual value	有效值；虚拟价值
virtual velocity	虚速度
virtual viscosity	有效黏度
virtual voltage	有效电压
virtual work	虚功
virucide	杀病毒剂
virulence	毒性
virulent	剧毒的；致命的；恶性的
virus	病毒
viscidity	黏性；黏度
viscoelastic	黏弹性的
viscoelastic analysis	黏弹性分析
viscoelastic attenuation	黏弹性衰减
viscoelastic behavior	黏弹性状
viscoelastic body	黏弹性体
viscoelastic damper	黏弹性阻尼器
viscoelastic deformation	黏弹性变形
viscoelastic deformational structure	黏弹性变形结构
viscoelastic foundation	黏弹性地基
viscoelastic mass	黏弹性体
viscoelastic material	黏弹性材料
viscoelastic model	黏弹性模型
viscoelastic modulus	黏弹性模量
viscoelastic property	黏弹性质
viscoelastic response analysis	黏弹性响应分析
viscoelastic transient initialization	黏弹性瞬态初始化
viscoelasticity	黏弹性
viscoelastoplastic	黏弹塑性的
viscoelasto-plastic material	黏弹塑性材料
viscoelasto-plastic soil	黏弹塑性土
viscogel	黏性凝胶
viscometer	黏度计
viscometer calibration	黏度计标定
viscoplastic failure	黏塑性破坏
viscoplastic flow	黏塑性流
viscoplastic fluid	黏塑性流体
viscoplastic material	黏塑性材料
visco-plastic medium	黏塑性介质
viscoplastic structure	黏塑性结构
viscoplasticity	黏塑性
visco-plasto-elastic mass	黏塑弹性体
viscosimeter	黏度计
viscosimetry	黏度测定法
viscosity	黏度，黏性，黏滞性
viscosity coefficient	黏滞系数；黏度系数
viscosity constant	黏滞常数
viscosity effect	黏滞效应
viscosity factor	黏滞因子
viscosity flow	黏性流
viscosity force	黏滞力
viscosity index	黏度指数
viscosity ratio	黏度比
viscosity resistance	黏滞阻力；黏性阻力
viscosity test	黏度试验
viscoslip fault	黏滑断层
viscous	黏性的，黏滞的，黏稠的
viscous behavior	黏滞性状；黏滞行为
viscous body	黏性体
viscous boundary	黏性边界
viscous cement grout	黏性水泥浆
viscous coupling	黏性耦合
viscous coupling coefficient	黏性耦合系数
viscous creep	黏性蠕变
viscous crude	稠油
viscous damper	黏滞阻尼器
viscous damping	黏滞阻尼
viscous debris flow	黏性泥石流
viscous deformation	黏滞变形
viscous displacement	黏滞位移
viscous dissipation	黏滞耗散
viscous drag	黏滞拖曳，黏滞阻力
viscous drag force	黏滞拖曳力
viscous effect	黏滞效应
viscous elastic	黏弹性的
viscous elastic mechanics	黏弹性力学
viscous factor	黏滞因子
viscous flow	黏性流
viscous flow equation	黏滞流方程
viscous flow model	黏滞流模型
viscous flow modulus	黏性流动模量
viscous fluid	黏滞流体
viscous fluid model	黏滞流体模型
viscous fluid simulation	黏滞流模拟

viscous force	黏滞力
viscous fracture	黏性破坏
viscous friction	黏滞摩擦；黏性摩擦
viscous gel	黏胶
viscous glide	黏性滑移
viscous hysteresis	黏性滞后现象；黏迟滞
viscous instability	黏滞不稳定性
viscous layer	黏滞层
viscous liquid	黏性液体
viscous loss	黏滞损失
viscous material	黏滞材料
viscous motion	黏滞运动
viscous pressure	黏滞压力
viscous resistance	黏滞阻力
viscous shear	黏性剪切
viscous similarity	黏滞相似性
viscous state	黏性状态
viscous strain	黏滞应变
viscous stress	黏滞应力
viscous trailing vortex	黏滞后缘涡流
viscous turbulent flow	黏滞紊流
viscous water	黏着水
viscous-damping coefficient	黏滞阻尼系数
Visec's bearing capacity formula	魏锡克承载力公式
visibility	能见度，可见度
visibility analysis	可视性分析
visibility distance	视距
visibility limit	能见度极限
visibility range	能见范围
visible	可见的，明显的
visible absorption spectrum	可见光吸收谱
visible crack	可见裂缝
visible flaw	可见缺陷
visible horizon	视地平线
visible light	可见光
visible reserves	视储量
visible spectral remote sensing	可见光遥感
visible spectrum	可见光谱
visible thermal infrared radiometer	可见光热红外辐射计
visible waveband	可见光波段
visible wavelength	可见光波长
vision	视线；视觉，视力；观察；远见
vision angle	视角
vision colorimetry	目视比色法
vision effect	视觉效应
vision estimation	目估
vision field	视野，视域
vision inspection	肉眼检测
vision measurement	目测
vision observation	肉眼观察
vision plane	视平面
vision resolution	目视分辨率
visual	可见的，直观的
visual alignment	目视调准，视觉准直
visual angle	视角
visual assessment	目测评估
visual field	视野
visual identification	肉眼鉴定
visual illusion	视错觉
visual inspection	肉眼检测
visual interpretation	目视判读
visual line	视线
visual measurement	目测
visual observation	目视观测
visual perception	视觉
visual pollution	视觉污染
visual range	视觉范围
visual ray	视线
visual sensitivity	视觉敏感性
visual sharpness	清晰度
visual signal	视觉信号
visual simulation	视觉仿真
visual soil classification	土的肉眼分类
visual technique	可视化技术
visualization	可视化，形象化
visual-manual procedure	目视手感法
vitasphere	生命圈
vitiation	损坏；失效；污染
vitreous	玻璃质的；透明的
vitreous texture	玻璃状结构
vitreous tile	釉瓷瓦，玻璃瓦
vitreous ware	玻璃器皿
vitric	玻璃的，玻璃质的，玻璃状的
vitric fragment	玻屑
vitric tuff	玻璃质凝灰岩
vitrification	玻璃化
vitrified brick	玻璃砖，陶砖
vitrified pipe	陶管
vitrified tile pipe	陶土瓦管
vitriol	硫酸盐；矾类；刻薄
vitriolic acid	硫酸
vitriolic clay	含矾黏土
vitriolization	硫酸处理；溶于硫酸
vitrobasalt	玻璃玄武岩；玻基玄武岩
vitroclastic	玻屑状的
V-notch weir	V形量水堰，三角堰
voe	小海湾
voice	声音
voice frequency	音频
voice-activated control unit	声敏控制装置
void	空的；空位，孔隙，空隙；空白；无效的
void cement ratio	孔灰比
void content	空隙度，空隙含量
void determination	孔隙测定
void index	孔隙指数
void measurement apparatus	孔隙测定仪
void plugging	孔隙堵封，空隙堵漏
void rate	孔隙率
void ratio	孔隙比
void ratio in densest state	紧密状态孔隙比
void ratio in loosest state	松散状态孔隙比
void ratio of soil	土的孔隙比
void ratio test	孔隙比测试
void ratio-pressure curve	孔隙比压力曲线
void ratio-strength curve	孔隙比强度曲线
void ratio-time curve	孔隙比时间曲线
void space	空隙，间隙
void volume	孔隙体积，空隙体积
void water	空隙水
void water pressure	空隙水压力

voidage 空隙率；空隙度
voided material 多孔材料
voided slab 空心板；空心桥板
void-filling aggregate 填隙骨料
void-free 无孔隙的，无空隙的
void-strength curve 孔隙-强度曲线
Voigt model 沃伊特模型
Voigt solid 佛格特体
Voigt wave 佛格特波
Voigt wave equation 佛格特波动方程
Voigt-Kelvin damping 伏格特-开尔文阻尼
Voigt-Kelvin model 伏格特-开尔文模型
volatile 挥发性的，易挥发的，不稳定的；挥发分，挥发物
volatile component 挥发性组分
volatile content 挥发分含量
volatile element 挥发性元素
volatile grade 挥发等级
volatile matter 挥发性物质
volatile metal 挥发性金属
volatile oil 挥发油
volatile oil simulator 挥发油模拟模型
volatile pollutant 挥发性污染物
volatile residue 挥发性残渣
volatile solid 挥发性固体
volatile solvent 挥发性溶剂
volatile storage 易失存储器
volatile sulfur 挥发硫
volatile suspended solid 挥发性悬浮固体
volatile test 挥发性试验
volatility 挥发性，挥发度
volatility index 挥发度指数；波动度指数
volatility range 挥发范围
volatilization 挥发
volatilization loss 挥发损失
volatilization roasting 挥发焙烧
volatilizer 蒸发器
volcanic 火山的；喷出的
volcanic accumulation 火山堆积
volcanic action 火山作用，火山活动
volcanic activity 火山活动
volcanic agglomerate 火山集块岩，火山烧结物
volcanic apparatus 火山机构
volcanic arc 火山弧
volcanic ash 火山灰
volcanic ash deposit 火山灰层，火山灰堆积
volcanic barrier lake 火山阻塞湖
volcanic basin 火山盆地
volcanic belt 火山带
volcanic block 火山岩块
volcanic bomb 火山弹
volcanic breccia 火山角砾岩
volcanic butte 火山孤峰，火山残丘
volcanic cinder 火山渣
volcanic cinder brick 火山渣轻骨料砌块
volcanic clastic mineral 火山碎屑矿物
volcanic clay 火山黏土
volcanic cluster 火山群
volcanic coarse-grained soil 火山粗粒土
volcanic cohesive soil 火山黏性土

volcanic conduit 火山通道；火山孔道
volcanic cone 火山锥
volcanic conglomerate 火山砾岩
volcanic crater 火山口
volcanic damage 火山灾害
volcanic debris 火山碎屑
volcanic deposit 火山沉积；火山矿床
volcanic depression 火山洼地
volcanic detritus soil 火山岩屑土
volcanic disaster 火山灾害
volcanic dome 火山丘，火山穹隆
volcanic earth 火山灰；火山泥
volcanic earthquake 火山地震
volcanic edifice 火山堆积；火山机构
volcanic eruption 火山喷发
volcanic facies 火山相
volcanic foam 浮岩
volcanic fragment 火山碎屑
volcanic front 火山前缘
volcanic gas 火山气体
volcanic geology 火山地质学
volcanic geothermal system 火山地热系统
volcanic glass 火山玻璃
volcanic graben 火山地堑
volcanic gravel 火山砾
volcanic group 火山群
volcanic hazard 火山灾害
volcanic landform 火山地貌，火山地形
volcanic manifestation 火山显示
volcanic massif 火山地块
volcanic mud 火山泥
volcanic mudflow 火山泥流
volcanic neck 火山颈
volcanic neck reservoir 火山颈油气藏
volcanic rock 火山岩
volcanic sand 火山砂
volcanic saprolite 火山风化土
volcanic scoria 火山熔渣
volcanic sediment 火山沉积物
volcanic sedimentary ore deposit 火山沉积矿床
volcanic tuff 凝灰岩；火山凝灰岩
volcanic type uranium deposit 火山岩型铀矿
volcanic vent 火山口
volcanic water 火山原生水，岩浆水
volcanic zone 火山带
volcanics 次火山岩；火山岩；火山物质
volcanic-sedimentary rock 火山沉积岩
volcanism 火山活动，火山作用，火山现象
volcanist 火山学家
volcanite 火山岩
volcano 火山
volcano-clastic rock 火山碎屑岩
volcano-clastic sediment 火山碎屑沉积
volcanogenic deposit 火山成因矿床
volcanogenic soil 火山土
volcano-geothermal region 火山地热区
volcano-karst 火山溶洞，火山喀斯特
volcanology 火山学
volcano-mudrock flow deposit 火山-泥石流沉积
volcano-sedimentary deposit 火山沉积矿床

volcano-stratigraphy 火山地层学
volcano-tectonic collapse 火山构造塌陷
volcano-tectonic depression 火山构造凹地
volcano-tectonic graben 火山构造地堑
volcano-tectonic horst 火山构造地垒
vole drill 竖式旋转冲击钻机
volt 伏，伏特
volt cell 伏打电池
volt gradient 电压梯度
voltage 电压；伏特数
voltage adjuster 电压调节器
voltage amplification coefficient 电压放大系数
voltage amplifier 电压放大器
voltage balancer 均压器
voltage breakdown 电压击穿
voltage change switch 电压转换开关
voltage changer 电压转换装置；变压器
voltage compensation 电压补偿
voltage conditioner 稳压器
voltage distribution 电压分布
voltage divider 分压器；分压电路
voltage drop 电压降
voltage dropping resistor 降压电阻器
voltage gradient 电压梯度
voltage impulse 电压脉冲
voltage limitator 电压限制器
voltage loop 电压回路
voltage multiplier 倍压器
voltage protector 电压保护装置
voltage ratio 电压比
voltage regulation 电压调节
voltage regulation mode 调压模式
voltage regulator 电压调节器
voltage relay 电压继电器
voltage resolution 电压分辨率
voltage resonance 电压谐振；电压共振
voltage sensitivity 电压敏感度
voltage sensor 电压传感器
voltage sharing resistor 电压均分电阻器，分压电阻器
voltage stabilizer 稳压器
voltage table 电压表
voltage transformer 变压器；电压互感器
voltage tripler 电压三倍器
voltage-sensitive 电压敏感的
voltage-stabilized source 稳压电源
voltaic cell 伏达电池，一次电池
volt-ampere 伏安，伏特安培
volt-ampere-hour 伏安小时
Volterra integral equation 伏尔泰拉积分方程
voltmeter 伏特计，电压表
volt-milliamperemeter 伏特毫安计
volume 体积，容积，容量；音量；流量，强度；柱，栏；卷，册
volume analysis 体积分析
volume and flow controller 容量-流量控制器
volume batching 按体积配料
volume capacity 容量
volume change index 体积变化指数
volume change modulus 体积变化模量
volume coefficient 容积系数

volume compressibility 体积压缩率；体积压缩性
volume compression coefficient 体积压缩系数
volume compression test 体积压缩试验
volume concentration 体积浓度
volume conductance 体积电导性
volume contraction 体积收缩
volume control 容量控制
volume control device 容量控制装置
volume conversion factor 体积换算系数
volume coordinates 体积坐标
volume correction factor 体积校正系数
volume curve 容积曲线
volume decrease 体积减小
volume density 体密度
volume diffusion 体积扩散
volume distribution 体积分布
volume efficiency 体积效率
volume elasticity 体积弹性，容积弹性
volume element 体积元
volume energy 体积能量
volume expansion 体积膨胀
volume expansion coefficient 体膨胀系数
volume extraction ratio 矿床总体积和采空区总体积之比
volume factor 体积系数，容积系数
volume flow rate 体积流量
volume forecasting 流量预测
volume index 体积指数
volume integral 体积积分
volume law 体积法则
volume limit 容积极限
volume meter 容量计
volume method 体积法
volume mix 体积配合法
volume modulus 体积模数
volume of earthwork 土方工程量
volume of entrainment 夹带量
volume of excavation 挖方量
volume of explosion gas 爆炸瓦斯体积
volume of flood 洪水量；注水量
volume of flow 流量
volume of production 生产量
volume of reservoir 库容
volume of runoff 径流量
volume of stock 储备量
volume of void 空隙体积，孔隙体积
volume of water input 注水量
volume percent 容积百分数，体积百分数
volume percentage 容积百分率，体积百分率
volume polarization 体积极化
volume production 批量生产
volume ratio 体积比，容积比
volume recovery 体积回收率
volume reduction 减容，体积缩减
volume reduction factor 体积减缩系数，减容因子
volume shrinkage 体积收缩，容积收缩
volume shrinking rate 体积收缩率，容积收缩率
volume thermal expansion coefficient 体积热膨胀系数
volume unit 容积单位，体积单位
volume withdrawl 体积采出量，容积采出量
volume-diffusion creep 体积扩散蠕变

volume-pressure coefficient	体积-压力系数
volume-temperature coefficient	体积-温度系数
volumetric analysis	容量分析，容量法
volumetric apparatus	容量器
volumetric average pressure	体积平均压力
volumetric balance equation	体积平衡方程
volumetric batching	按体积配料
volumetric coefficient	体积系数，容积系数
volumetric coefficient of solubility	体积可溶性系数
volumetric composition	容积组成
volumetric compression test	体积压缩试验
volumetric contraction	体积收缩
volumetric correction factor	容量校准因子
volumetric creep	体积蠕变
volumetric cylinder	量筒
volumetric deformation	体积变形
volumetric deformation coefficient	体积变形系数
volumetric deformation modulus	体积变形模量
volumetric expansion	体积膨胀
volumetric flask	量瓶
volumetric flow rate	容积流量
volumetric flowmeter	体积流量仪
volumetric flux	容积通量
volumetric heat capacity	体积热容量
volumetric initial strain	体积初始应变
volumetric joint count	体积节理模数
volumetric modulus	体积模量
volumetric modulus of elasticity	体积弹性模量
volumetric moisture content	体积含水量
volumetric molar concentration	体积摩尔浓度
volumetric parameter	体积参数
volumetric porosity	体孔隙率
volumetric pressure plate extractor	体积压力板提取器
volumetric proportion	容积比，体积比
volumetric quantity	容积量
volumetric ratio	体积比
volumetric reserves	体积储量
volumetric reservoir	封闭储层，圈闭油层
volumetric response	体积响应
volumetric scan well logging	体积扫描测井
volumetric shrinkage	体积收缩；体积收缩量
volumetric shrinkage limit	体积收缩极限
volumetric solution	滴定液
volumetric specific heat	体积比热
volumetric standard	容量标准
volumetric strain	体应变，体积应变
volumetric strain energy	体积应变能
volumetric stress	体应力，体积应力
volumetric swell	体积膨胀
volumetric water content	体积含水量
volumetric weight	容重
volumetrically weighted average	体积加权平均值
volumetry	容量测定，容量分析
voluminal	体积的，容积的
voluminal compressibility	体积压缩性
voluminal volatile matter	体积挥发分
voluminous	容积大的，体积大的；大量的；多卷的，长篇的
volute	涡管；涡壳；螺旋形
volute centrifugal pump	蜗壳式离心泵
von Karman constant	冯卡曼常数
von Karman vortex trail	冯卡曼涡流尾流
von Mises criterion	冯米塞斯准则
von Mises yield condition	冯米塞斯屈服条件
von Neuman architecture	冯诺依曼结构
von Neumann condition	冯诺伊曼条件
von Schmidt wave	冯施密特波，首波
vortex	涡流；旋涡，旋卷
vortex band	涡流带；涡流层
vortex breakdown	涡旋破碎
vortex cast	旋涡铸型
vortex cavitation	涡流空蚀；涡空化
vortex density	涡流密度
vortex dryer	旋涡干燥机
vortex ejector	涡流喷射器
vortex equation	涡流方程
vortex field	涡流场
vortex filament	涡流流线，涡丝
vortex finder	旋涡溢流管；涡流探测器
vortex fitting	涡拟合
vortex flow	涡流
vortex flowmeter	涡流流量计
vortex generator	涡流发生器
vortex interference	涡流干扰
vortex layer	涡流层
vortex line	涡流线，涡线
vortex motion	涡动
vortex pair	涡对，双旋涡
vortex period	旋涡周期
vortex pump	涡流泵
vortex ring	涡环
vortex sheet	涡片
vortex strength	涡强度
vortex structure	旋卷构造；涡结构
vortex suppressing basin	消涡池
vortex surface	涡面
vortex trail	涡旋尾流
vortex tube	涡旋管
vortex velocity anemometer	涡旋速率风速计
vortex-free discharge	无旋涡排水
vortex-free motion	无旋涡运动
vortex-siphon spillway	旋涡虹吸式溢洪道
vortical surface	旋回面
vorticity	涡旋状态；涡度，涡量，涡旋强度
vorticity equation	涡度方程；涡量方程
vorticity field	涡旋场
vorticity initiation	旋涡初起
vorticity meter	涡量计
vorticity transfer	旋涡转移；涡度传送
vorticity transport theory	涡输送理论；涡输运理论
voussoir	楔形拱石
voussoir arching action	楔拱作用
voussoir beam	楔拱梁
voussoir brick	楔形砖
V-rope	V形钢索
V-rope drive	V形钢索传动
V-section jackscrew	V形截面起重螺杆
V-shape cut	V形掏槽，楔形掏槽
V-shaped	V形的
V-shaped inclined strut	V形倾斜撑杆

V-shaped pier　V形墩
V-shaped valley　V形谷
V-type eight cylinder engine　V型八缸发动机
vug　岩穴，溶洞；孔洞；晶洞
vug porosity　溶洞孔隙度
vuggy　多孔的；晶洞的
vuggy formation　多孔地层建造，多孔地层
vuggy rock　多孔岩石
vugular　空穴的
vugular pore space　空穴孔隙空间
vugular-solution porosity　空穴溶蚀孔隙度
vulcanization　硫化；硬化

vulcanization accelerator　硫化促进剂；硫化加速剂
vulcanizator　硫化剂
vulcanized rubber　硫化橡胶
vulcanizing agent　硫化剂
vulcanology　火山学
vulnerability　易损性，脆弱性
vulnerability analysis　易损性分析
vulnerability assessment　易损性评价
vulnerable　易损的；易受攻击的
vulnerable slope　易生危险的斜坡
V-valley　V形谷
V-weld　V形焊接

W

wacke 玄武质砂岩，瓦克砂岩；玄武土，玄土，玄武岩风化物
wadi 旱谷，干谷
wadi type depression 干谷型凹地，干谷洼地，干谷低地
wading pool 浅水池
waferer 切片机
waffle raft 格栅筏板
waffle slab 格栅肋板，井字形密肋楼板
wage by the piece 计件工资
wagon 矿车，货车
wagon balance 轨道衡，地磅
wagon drill 移动式钻机，汽车钻机，车装钻机，凿岩台车
wagon jack 起重车
wagon mounted drill 凿岩台车，钻车
waist bottling 缩颈
waisting pile 缩颈桩
waiting on cement 水泥候凝时间
waiting time theory 排队论
wake 尾流，伴流
wake flow theory 尾流理论
wake resistance 尾流阻力
wake strength 尾流强度
wake volute 尾涡
wale 横撑，腰梁，箍条；条状隆起部；凸起的条纹
wale piece 横撑板
waling 围令，支撑；支腰梁，横挡，横向护木；挖槽横撑
walk path 人行道，步道
walker 迈步式挖掘机
walker dragline 迈步式索斗铲
walker river 不定床河流
walking crane 步行式吊车
walking dredge 迈步式挖掘机，移动式挖土机
walking prop 移动式支柱，迈步式支架
walking scoop dredge 移动式铲斗挖泥船
walking side 人行侧道
walking sprinkler 移动式喷水机
walking support 自进式支架，迈步式支架
walk-over survey 踏勘
wall 墙，壁；巷道壁，帮；盘，侧
wall adhesion 墙附着力，墙黏着力
wall area index 墙面积系数
wall back 墙背
wall bearing construction 承重墙结构
wall bed 墙基
wall body 墙体
wall bolting 洞壁锚栓加固，巷壁锚杆支护
wall bracket crane 井场悬臂起重机
wall brick 墙砖
wall bushing 穿墙套管
wall compression stress 墙压应力
wall construction 墙体结构
wall crane 壁装式起重机

wall downthrow 断层下落盘
wall drain 墙排水孔
wall effect 墙效应，边壁效应，洞壁效应
wall flow 管壁流；壁面流动
wall footing 墙脚
wall foundation 墙基
wall frame interaction 墙-框架相互作用
wall frame structure 墙-框架结构
wall length 工作面长度
wall map 挂图
wall pipe 穿墙管
wall rock 围岩
wall rock alteration 围岩蚀变
wall sampler 井壁取样器
wall sill 窗台
wall stabilization with clear water 清水护壁
wall static pressure 壁面静压
wall strength 井壁强度
wall temperature 壁温
wall temperature sensor 壁温传感器
wall temperature transmitter 壁温传送器
wall thickness 壁厚
wall tie 壁锚
wall toe 墙脚
Wallace agitator 瓦莱斯搅拌机
wallboard 墙板，墙面板，壁板
walled frame 壁式框架
walling 筑墙
walling scaffold 砌壁吊盘，砌壁脚手架
walling stage 砌井壁吊盘
wallpaper 壁纸，墙纸
wall-protecting slurry 护壁泥浆
wandering 漂移，迁移；偏斜
wandering channel 游荡河槽
wandering dune 流动沙丘，迁移沙丘
wandering electron 游离电子
wandering mark 浮动测标
wandering river 游移河流
wandering water 渗流水
waning slope 凹坡
waning stage 衰退期
war mineral 战略矿物
ware 制品，成品，商品；仪器
warehouse 仓库，货栈
warehouse hoist 备件库吊车
warehousing 库存
warehousing and storage activity 仓储作业
warm air heating system 热风采暖系统
warm area 暖区；微热区
warm climate 温暖气候
warm current 暖流
warm front 暖锋
warm ground 微热地面，温热地面
warm groundwater 地下热水

warm humid climate　温湿气候
warm lake　温泉湖，温水湖
warm loess　暖黄土
warm mine　高温矿井
warm plasma　暖等离子体
warm spring　温泉
warm start　温态启动
warm swamp　热水沼泽
warm water　温水
warm water port　不冻港
warm working　温加工
warmed soil　加温土
warming　变暖；加热
warm-temperate region　温暖地区
warm-temperate zone　暖温地带
warm-ventilation　暖通风
warm-wet climate　湿热气候
warner　报警器；警告者，预告者
warning　警告，警报，预报
warning based on deformation　基于变形的预警
warning based on micro-seismicity　基于微震的预警
warning bell　警铃
warning board　警告板
warning device　警报装置
warning lamp　警报灯
warning level　预警水平
warning message　警告信息
warning of air pollution　空气污染警报
warning of danger　危险警报
warning pipe　溢水管，溢流管
warning seismonet　地震警报网
warning sign　警告标志
warning signal　警报信号
warning system　预警系统，警报系统
warning water level　警戒水位
warp　淤积；扭曲，挠曲，弯曲
warp land　放淤地
warp soil　淤积土
warped basin　翘曲盆地
warped clay　淤积黏土，淤泥
warped element　扭曲单元
warped fault　翘曲断层
warped surface　翘曲面
warping function　翘曲函数
warping irrigation　淤灌
warping rigidity　翘曲刚度
warping stress　翘曲应力，扭曲应力
warping works　放淤工程
warrenite　石蜡石油
Warren-Root model　沃伦-鲁特模型
wart hog　钻具修整器
wash　冲洗，洗矿；冲刷，侵蚀；冲积物，冲积砂矿
wash boring　冲洗钻探，清水钻进，水冲式钻探，回水钻探
wash drilling　冲洗钻探，清水钻进，水冲式钻探，回水钻探
wash load　冲刷搬运的泥沙
wash off soil　侵蚀土
wash pan　淘金盘
wash plain　淤积平原，冲积平原
wash point penetrometer　水冲式贯入仪
wash sample　冲洗样
wash slope　重力坡脚，崖脚缓坡，剥蚀基坡
wash thickener　洗矿浓缩机
wash trommel　洗矿滚筒，洗矿圆筒筛
washability　可洗性，可选性
washability curve　可选性曲线
washability data　可选性数据
washability test　可选性试验
wash-and-drive method　边冲出钻屑边打入套管钻进法
washbox　跳汰机
washed coal　精煤，洗煤
washed off soil　侵蚀土壤，淋洗土壤
washed ore　洗矿，精选矿
washed-sieve analysis　水冲筛分析
washer　洗矿机，洗煤机；洗涤器，洗衣机
washery refuse　洗渣，选矿渣
washing agent　洗涤剂
washing elevator　洗矿塔
washing machine　洗矿机械
washing screen　洗矿筛
washland　河漫滩，洪泛地
washout of flood　洪水冲刷
washover　冲溢沉积，越浪堆积
wastage　废物，废料，废水，弃土，垃圾；消耗，损失，磨损
waste bank　弃土堆，矸石堆，废石场
waste battery　废旧电池
waste bin　废渣仓，废石仓，矸石仓
waste blasting　放顶爆破；挑顶取料
waste collection vehicle　废物收集车，垃圾车
waste collection vessel　废物收集船
waste containment　废料污染
waste conversion technique　废物转化技术
waste discharge　废水排放
waste disposal facility　废料处理设施
waste disposal with bulldozer　推土机排土
waste disposal with over-burden spreader　排土机排土
waste disposal with plough　排土犁排土
waste disposal with power shovel　单斗挖掘机排土
waste drainage　采空区瓦斯排放，老峒通风
waste fluid　废液
waste gas　废气
waste gas analysis　废气分析
waste gas emission standard　废气排放标准
waste gas purification　废气净化
waste gas treatment　废气处理
waste gas utilization　废气利用
waste geotechnics　废料岩土工学
waste grout　废浆
waste heap reclamation　矸石山复垦
waste heat　废热
waste heat boiler　废热锅炉
waste heat recovery　废热回收，废热利用
waste heat refrigeration　余热制冷
waste heat treatment　废热处理
waste incinerator　垃圾焚烧炉，废物焚化炉
waste isolation pilot plant　废物隔离中间试验厂
waste land　荒地，废地
waste material　废料

waste minimization	废物最少化
waste product	废品；废物；人体排泄物
waste reclamation	废物回收利用
waste recycling	废物回收
waste residue	废渣
waste rock	废石，矸石
waste rock pile	废石堆；矸石堆
waste shaft	矸石井
waste soil stabilization	废土加固
waste solid	固体废物
waste steam	废蒸气
waste storage farm	废物储存场
waste treatment	废物处理
waste utilization	废物利用
waste vapor	废蒸汽
waste vegetable oil	废植物油
waste water	废水，污水
waste weir	废水堰
waste well	废水井，废水渗井
waste-fill mining	废石充填开采
waste-lifting stoping	挖底回采法
wastepaper	废纸
waster	废品
waste-rock pile	矸石堆，废石堆
wastewater disinfection	废水消毒
wastewater disposal	污水处理，废水处置
wastewater effluent	废水流
wastewater engineering	废水处理工程
wastewater from living	生活废水
wastewater from nitrogenous fertilizer plant	氮肥厂废水
wastewater liquid residual	废水液态残余物
wastewater plant	废水处理厂
wastewater purification	废水净化
wastewater reinjection	污水回注
wastewater renovation	废水净化回收，废水重复使用
wastewater sewer	废水下水道，废水管道
wastewater system	污水系统
wastewater treatment	废水处理，污水处理
wasteway	废水道
wasteyard	废物场，垃圾场
watch-dog	监视器
water abrasion	水蚀，水力冲蚀
water absorbability	吸水性
water abundance	富水性
water acidification	水体酸化
water adsorption	吸水；吸水率
water affinity	亲水性
water analysis	水质分析
water atomizer	喷水器，喷雾器
water backfill	水回填，水封
water balance	水均衡
water barrier	防水矿柱，防水煤柱；水棚
water basin protection	流域保护
water blast	突水
water blocking	堵水
water board	水闸板
water body	水体
water body pollution	水体污染
water booster pump	升压水泵
water borehole logging	水井录井；水井钻孔测井
water bottom	水体底部；水垫，垫底水
water box	水箱；运水车，洒水车
water breakthrough	水窜，冒水
water burst	突水
water bursting coefficient	突水系数
water catchment area	集水区
water cement ratio	水灰比
water cement slurry	水泥浆
water circulation	水循环
water column	水柱
water concentration	水选
water conducted zone	导水带
water conservation	水保持；水利
water conservation project	水利工程
water constructional works	水工建筑工程
water consumption	水量消耗，耗水量，用水量
water contamination	水污染
water contamination detection	水污染检测
water content	含水量，含水率
water content index	含水量指数
water content ratio	含水比
water content test	含水率试验
water conveyance	输水
water conveyance structure	输水建筑物
water cooled wall	水冷壁
water cooling	水冷却
water corrosion	水腐蚀
water course	水路，水道，河道；水流；水系
water creep	管涌，渗水
water crop	出水量，产水量
water crossing	跨水
water curtain	水幕
water curtain system	水幕系统
water cushion	水垫
water cut	冲沟，冲刷坑；含水率；水侵
water cut-off	截水；截水墙
water cycle	水循环
water delivery	给水，供水；排水量
water dependent well	水控井
water development	水开发
water diffusion	水分扩散；水分子扩散
water discharge	排水，泄流；排水量，泄流量
water discharge piping	管道排水，管道泄流
water discharge temperature	排水温度
water displacement	水置换，水驱
water displacement method	水置换法，水驱法
water displacement recovery	水驱采收率
water disposal	水处理
water disposal well	水处理井
water distribution	配水
water distribution system	配水系统；配水网
water diversion	引水，调水；引水调水工程
water divide	分水岭
water door	防水闸门，防水门
water drain	排水道，下水道
water drain pipe	排水管
water drain valve	放水阀
water drainage	排水
water drainage works	排水工程
water draining	排水

water drilling	水洗式钻进，水洗式凿岩
water drip	洞穴滴水；滴水槽，滴水线；集水井
water drive index	水驱指数
water drive pool	水驱油藏
water drive recovery efficiency	水驱采收率
water drive reservoir	水驱油层
water drop	跌水，水位降落；水滴
water drum	下汽包，水包
water equivalent	水当量
water erosion	水冲蚀，水蚀
water escape	泄水道
water evaluation	水质评价
water exclusion	水层封闭
water exhaust	水耗；水耗量
water exit interval	出水层段
water extraction	抽水，取水，汲水；水萃取
water fall	瀑布，跌水
water fall erosion	瀑布侵蚀，跌水侵蚀
water film theory	水膜理论
water filter	滤水器
water fingering	指状水侵
water fissure	含水裂隙
water flood field	注水油田
water flood recovery	注水采油；注水采油量
water flooded interval	水淹层段
water flooding	注水法；水驱法
water flush boring	水冲钻探
water flush drill	湿式凿岩机；水冲钻机
water for domestic use	生活用水
water for firefighting	消防用水
water fountain	饮水喷头，喷泉，喷水池
water fracturing	水致压裂
water front	岸线，岸边线；沿岸，水岸，水边；海旁用地，海旁区；
water front construction	沿岸建筑
water furnishing ability	供水能力
water furrow	明沟，排水沟
water garden	水景花园
water gas	水煤气
water gauge	水位标尺，量水表
water gel explosive	水胶炸药
water geochemistry	水地球化学
water geology	水地质学
water glass	水玻璃，硅酸钠
water grade	水质等级；水力坡度，引水坡度
water gradient	水柱压力梯度；水分梯度
water grinding stone	水磨石
water gun	水枪
water gushing	涌水
water hammer	水锤
water hammer effect	水锤效应
water hardening	水淬硬化
water hardening steel	水淬硬化钢
water hardness	水质硬度
water hazard	水害
water hazard control	水害防治
water head	水头，水位差，落差
water head loss	水头损失
water holding capacity	保水能力，持水能力
water holding rate	保水率
water horizon	含水层位，蓄水层位；含水层，蓄水层
water hysteresis	水滞后现象
water inflow	涌水量
water infusion	注水
water infusion blasting	水封爆破，注水爆破
water injection	注水
water injection test hole	注水试验孔
water inrush danger	突水危险
water insoluble matter	水不溶物
water intake	进水，取水；取水口
water intake tunnel	引水隧洞
water intensity	水强度
water invasion	水侵，突然浸水
water ionization limit	水电离限
water jet	喷水；喷水器；水柱
water jet method of pile driving	射水沉桩法
water jet tunneling machine	射水隧道掘进机
water jetting	射水
water jetting at high pressure	高压射水
water joint	防水接头
water law	水法
water leakage	漏水；漏水量
water leakage zone	渗漏带
water level	水位；水准面，水平面；水平尺，水准测量仪；
water level weathering	水位风化作用
water leveling	水位测量
water lift	扬水工具
water line	水位线，吃水线；水管线路
water load	水荷载，水载重
water loading test	水压法试验
water locating device	水位测定装置
water lock	水闸，水封，水锁，水堰
water loss	失水，失水量，水量损失
water loss and soil erosion	水土流失
water loss test	漏水试验
water lowering	水位下降
water meter	水量计，水表
water microbiology	水域微生物学
water molecule	水分子
water of constitution	组成水
water of crystallization	结晶水
water of dehydration	脱水水
water of infiltration	渗入水，渗透水
water oil mixture	水油混合物
water opening	出水口
water outlet	出水口；出水量
water overflow settlement gauge	溢水式沉降计
water parting	水道岔口
water passage	水道；水眼
water penetration	漏水，渗水，透水
water permeability	渗水性，渗透性
water permeability test	透水性试验
water phase	水相，水态
water pipe	水管
water piping	铺设水管
water pit	水坑
water plane	水面；潜水面，地下水面，地下水位；吃水面，水线面
water planning	水利规划

water plant 水厂，供水车间；水泵房
water plugging cement 堵水水泥
water pollution 水污染，水质污染
water pollution analysis 水污染分析
water pollution control 水污染防治
water pollution index 水污染指数
water pollution monitoring 水污染监测
water pollution source 水污染源
water pool 水塘
water potential 水势能，水潜能
water potential gradient 水势梯度
water power 水力，水能；水力发电
water power dam 水力发电坝
water power engineering 水电工程
water power plant 水力发电厂
water power station 水力发电站
water power survey 水力调查
water pressure 水压力
water pressure gauge 水压计
water pressure technique 水压法
water pressure test 水压试验，压水试验
water pressure test hole 压水试验孔
water pretreatment 水预处理
water probe 水探头
water production 出水量，产水量
water productivity 出水能力，产水能力
water project 水工程，水利工程，给水工程
water proof polymer fabrics 防水聚合织物
water prospecting 探水
water pulverization 喷水，喷雾，水雾化
water pulverizer 喷水器，喷雾器
water pump 水泵，抽水机
water pumping test 抽水试验
water pumping test hole 抽水试验孔
water purification 水净化
water purifier 净水器
water purifying tank 净水罐，净水池
water quality 水质
water quality analysis 水质分析
water quality and hydraulic mathematical model 水质及水力数学模型
water quality and pollution load information system 水质及污染量信息系统
water quality criterion 水质标准
water quality index 水质指数
water quality management 水质管理
water quality model 水质模型
water quality monitor 水质监测器
water quality monitoring 水质监测
water quality observation 水质观测
water quality parameter 水质参数
water quality protection 水质保护
water quality regulation 水质调节
water quality specification 水质规范
water quality standard 水质标准
water quality stratification 水质分层
water quality surveillance 水质监测
water quenching 水淬火
water resource and hydropower 水利水电
water resource balance method 水资源均衡法

water resource evaluation 水资源评价
water resource investigation 水资源调查
water resource management 水资源管理
water resource management database 水资源管理数据库
water resource management information system 水资源管理信息系统
water resource policy 水资源政策
water resource protection 水资源保护
water resource remote sensing 水资源遥感
water resource system planning 水资源系统规划
water resources 水资源，水利资源，水力资源
water retention curve 保水曲线
water retentivity 保水度，保水性
water reuse 水的再利用；废水利用
water right 用水权，水权
water riser pipe 给水立管
water runoff 径流，活水；水流量，径流水量
water salination 水盐化
water sample 水样
water sampler 取水样器
water seal 水封
water seepage 渗水，漏水
water sensitivity 水敏性
water shooting 水下爆炸
water shortage 水量不足，水短缺
water shutoff 堵水，封水
water shutoff test 堵水试验
water sink 落水洞，渗水坑
water slaked lime 水熟石灰，熟石灰，消石灰
water slot 流水槽
water slug 水栓
water soluble precondensate 水溶性预凝液
water soluble-oil emulsion 水溶性油乳胶，水溶性乳化油
water solution 水溶液
water source 水源
water source protection 水源保护
water space 水域
water spouting 喷水，射水
water spray system 喷水系统，喷雾防尘系统
water spraying 喷水，喷雾
water spraying and sand emitting 喷水冒砂
water spreading 漫流
water stemming bag 水炮泥袋
water stop 止水；止水栓，止水塞，止水条
water stop joint 止水接头
water stop plate for lining 衬砌止水板
water stop tie 止水带
water stopper 止水条
water stopping 止水
water storage 蓄水；蓄水量
water storage works 蓄水工程
water substance 水体
water supply 供水，给水；给水工程
water supply and sewage works 给水排水工程，上下水道工程
water supply engineering 给水工程
water supply hydrogeology 供水水文地质学
water supply line 供水管线，上水道
water supply network 供给水网
water supply pipe 供水管，水管

water supply potential	供水潜力	water-based explosive	水基炸药
water supply scheme	供水方案	water-based mud	水基泥浆
water supply source	供水水源	water-based oil emulsion mud	水基油乳泥浆
water supply survey	供水勘测，供水调查	water-based paint	水基涂料
water supply tunnel	供水隧洞	water-based rotary drilling fluid	旋转钻进用水基泥浆
water supply well	供水井	water-bearing bed	含水层
water surface	水面；水位	water-bearing capacity	含水量
water surface evaporation	水面蒸发	water-bearing explosive	水基炸药
water surface slope	水面坡度，水面比降	water-bearing formation	含水建造，含水岩组，含水层
water survey	水测量，水调查	water-bearing fracture zone	含水破裂带
water swelling	湿胀，水胀	water-bearing ground	含水层
water system survey	水系调查	water-bearing horizon	含水层；含水层位
water table	水位，水位面；地下水位，潜水面	water-bearing interval	含水层段
water tagging	水追踪剂标记	water-bearing layer	含水层
water tank modeling	水槽模拟	water-bearing level	含水层位
water tap	水龙头	water-bearing medium	含水介质
water temperature	水温	water-bearing rock	含水岩石
water test	压水试验，水压试验；水质试验；渗水测试	water-bearing sand	含水砂
water to earth ratio	蓄水量与挖土方量之比	water-bearing soil	含水土壤
water to oil area	水油过渡区	water-bearing stone	含水石
water tower	水塔	water-bearing stratum	含水地层，含水层；蓄水层
water tracer	水示踪剂	water-bearing structure	含水构造
water trading	水权交易	water-bearing subsoil	含水底土
water transfer pump	输水泵	water-bearing zone	含水带
water transmission beyond basins	跨流域调水	water-blasting face	水力爆破工作面
water transmitting ability	输水能力	water-borehole logging	水钻孔测井
water trap	集水坑	waterborne endemic disease	水致地方病
water treatment	水处理，水净化，净水处理，软水处理	waterborne material	水成物质
water treatment feed water pump	水处理供水泵	water-borne noise	水生噪声
water trough	水槽	waterborne sediment	水成沉积物
water tube boiler	水管锅炉	water-bound	与水连接的，四周环水的
water tunnel	引水隧道，水工隧道	water-bound macadam	水结碎石路面
water turbine	水力涡轮，水轮机，水力透平	water-budget calculation	水平衡计算
water turnover time	供水周转时间	water-carriage system	输水系统
water type	水类型，水型	water-carrying pipe	输水管
water vapor	水汽，水蒸气	water-carrying services	输水设施
water vapor pressure	水汽压	water-carrying tunnel	输水隧洞，引水隧道，排水平硐
water vein	裂隙地下水，水脉	water-CO_2-coal system	水-二氧化碳-煤系统
water void ratio	水隙比	water-collecting area	集水面积
water wash pump	清洗泵	water-collecting gallery	集水廊道
water washing	水洗作用	water-collecting header	集水联箱；集水总管
water well	水井	water-collecting structure	集水构筑物
water wet core	水湿岩芯	water-collecting tunnel	集水隧道
water wheel	水轮，水车	water-column anomaly	水柱异常
water wing	挡水翼墙	water-column manometer	水柱压力计
water year	水文年；水年	water-cone breakthrough	水锥突破
water yield	出水量，开采水量	water-confined regime	承压水动态
water yield coefficient	出水系数	water-coning rate	水锥进速率
water yield formation	产水层，出水层，出水地层建造	water-conservation measure in agriculture	农田水利措施
water-absorbing capacity	吸水能力，吸水量	water-content gradient	含水率梯度
water-adit	排水平硐	water-content profile	含水量剖面
water-air boundary	水气界线	water-controlled field	水控油田
water-balance area	水均衡区	water-controlled reservoir	水控油层
water-balance element	水均衡要素	water-cooled bearing	水冷轴承
water-balance equation	水均衡方程	water-cooled condenser	水冷冷凝器
water-balance method	水均衡法	water-cooled diesel engine	水冷柴油机
water-balance simulation	水均衡模拟	water-cooled electric motor	水冷电动机
water-bar	阻水栏栅	water-cooled engine	水冷发动机
water-based carrier fluid	水基携砂液	water-cooling spraying system	水冷却喷雾系统
water-based drilling fluid	水基钻井液	water-cooling tower	水冷却塔，喷淋冷却塔

watercourse management	水道管理，河道管理
watercourse monitoring	水道监控
watercourse training	水道疏浚
water-covered area	水覆盖区
water-deficient	缺水的
water-depth line	水深线
water-detecting in advance	超前探水
water-discharge pipe	排水管
water-distillation unit	水蒸馏装置
water-drive field	水驱油田
watered gas reservoir	水淹气藏
watered gas well	水淹气井
watered oil	含水石油
watered-out area	水淹区
water-fast	耐水的
water-fed rock drilling	湿式凿岩，水冲岩芯钻探
water-filled aggregate	充水填料
water-filled porosity	含水孔隙度
water-film deduster	水膜除尘器
water-float upward erection method	水浮正装法
waterflood	洪水；注水，水驱
waterflood bulkhead	防洪护岸
waterflood candidate	水驱开发油层
waterflood channel	水驱孔道
water-flood decline curve	注水产量递减曲线
waterflood development	注水开发，水驱开发
waterflood efficiency	水驱效率
waterflood enhanced recovery	注水提高的采收率
waterflood front	水驱前缘
waterflood mobility ratio	水驱流度比
waterflood packaged plant	组装注水站
waterflood path	注水路线
waterflood permeability	水驱渗透率
waterflood pump	注水泵
waterflood recoverable reserve	注水可采储量
waterflood response	注水反应
waterflood stage	注水开发阶段
waterflood susceptibility test	注水敏感性试验
waterflood sweep pattern	注水波及型式
water-floodability	可注水性，可水驱性
waterflooded area	水淹区
waterflooded reservoir	水淹油层
water-flooded zone	水淹带
waterflooding extraction	注水提取，水驱提取
waterflooding program	注水开发方案
waterflood-transmission line	注水管线
waterflow	水流
water-free gas rate	无水产气量
water-free oil	无水原油
water-free recovery factor	无水采收率
water-free soil	干土，不含水土壤
water-free well	无水井，干井；无水油井
waterfront industrial area	沿海工业区，沿河工业区
waterfront structure	沿岸结构，驳岸结构，水边建筑物
water-gas condenser	水煤气冷凝器
water-gas generator	水煤气发生器
water-gas interface	水气界面
water-gas pitch	水煤气沥青
water-gel blasting agent	水胶炸药
water-geochemical survey	水文地球化学调查
water-glass grout	水玻璃注浆材料
water-glass paint	水玻璃漆
water-glycol type hydraulic fluid	水-乙二醇型液压液
water-granulated slag	水淬粒状矿渣
water-hammer arrester	水锤制动器
water-holding material	含水材料
water-holding pore	持水孔隙
water-holding power	持水力
water-holding structure	挡水建筑物
waterhole	水坑，水池
water-hydrocarbon interface	水-烃界面
water-immiscible	与水不混溶的
water-impregnated stratum	充水地层，保水地层
water-impulse separator	水脉冲分选机
water-indicating plant	示水植物
watering period	灌水周期
watering place	泉口
watering system	洒水系统
water-injection capability	注水能力
water-injection drill	注水式钻机，注水式凿岩机
water-injection efficiency	注水效率
water-injection intake level	注水吸水层位
water-injection mode	注水模式
water-injection test	注水试验
water-in-oil emulsion mud	油包水乳化泥浆
water-inrush accident	涌水事故
water-inrush aquifer	突水含水层
water-inrush prediction	突水预测
water-inrush vulnerability	易突水性
water-intake interval	取水层段
water-jacket	水套，水衣
water-jet drill	射水钻机
water-jet driving	射水沉桩
water-jet ejector	射水抽气器；喷水器
water-jet pump	射水泵
water-laid deposit	水成沉积
water-land boundary	水陆边界
water-layer reverberation	水层混响
water-layer scattered noise	水层散射噪声
water-leach	水浸出，沥水
water-level alarm	水位警报器
water-level control	水位控制
water-level float	水位浮标
water-level fluctuation	水位波动
water-level gauge	水位计
water-level mark	水位痕
water-level profile	水位剖面
water-level recovery	水位恢复
water-level regime	水位动态
water-level rise	水位上升
water-level scale	水位尺
water-level survey	水准测量
water-level transfer	水位转换
water-level trend	水位趋势
water-level-decline contour map	水位降等值线图
water-lifting pipe	升水管，扬水管
waterlogged depression	积水洼地
waterlogged karst	积水岩溶
waterlogged land	水浸地
waterlogged region	涝渍地域；洪涝灾害区

waterlogged soil 渍水土壤
water-loving plant 亲水植物
Waterman method 华特曼法
Waterman ring analysis 华特曼环分析
water-mark 水位标志，吃水标志
water-measuring device 量水设备
watermelon mill 椭圆形磨铣工具
water-oil contact 油水接触面
water-oil displacement 水驱替油
water-oil front 油-水前缘
water-oil interface 油水界面
water-oil mobility ratio 水油流度比
water-oil ratio 水油比
water-phase corrosion 水相腐蚀
water-physical property 水理性质
waterplane area 水线面面积
water-plasticity ratio 流性指数，液化指数
water-pollution-prone agricultural chemicals 水质污染性农药
water-pressure tunnel 有水压隧道
water-pressure type leak detector 水压式渗透仪
water-producing horizon 出水层位；出水层
water-producing zone 出水带
waterproof 防水；防水的，不透水的
waterproof agent 防水剂
waterproof barrier 防水层
waterproof belt 耐水皮带
waterproof board 防水板
waterproof casing 隔水套管，隔水罩
waterproof cement 防水水泥
waterproof charge 防水药包
waterproof cloth 防水布
water-proof coating 防水涂层
waterproof concrete 防水混凝土
waterproof dam 防水坝
waterproof electric blasting cap 防水电雷管
waterproof fuse 防水导火线
waterproof grease 防水脂
waterproof layer 防水层；隔水层
waterproof lining 防水衬砌
water-proof paint 防水漆，防水涂料
waterproof sheet 防水板
waterproof socket 防水插座
waterproof stratum 不透水层，隔水层，防水层
waterproofed explosive 防水炸药
waterproofing admixture 防水外加剂
waterproofing gum mortar 防水胶浆
waterproofing material 防水材料
waterproofing measure 防水措施
waterproofing membrane 防水薄膜
waterproofing mortar 防水砂浆
waterproofing paste 防水油膏
waterproofing system 防水系统
waterproofing with asphalt 沥青防水
waterproofing works 防水工程
water-purification plant 水净化厂
water-purification unit 水净化设备
water-quality prediction 水质预测，水质预报
water-quality stabilization technology 水质稳定技术
water-raising capacity 扬水高程，扬水能力

water-raising engine 扬水机
water-reducing accelerator 减水加强剂
water-reducing admixture 减水掺加剂
water-reducing agent 减水剂
water-reducing and set-retarding admixture 减水缓凝剂
water-repellent agent 防水剂
water-resistant concrete 防水混凝土
water-resisting layer 隔水层
water-resisting property 抗水性，隔水性
water-resisting rock 隔水岩石
water-retaining wall 挡水墙
water-rock interaction 水岩作用；水岩相互作用
water-rock reaction 水岩反应
waters 水域，水体
water-saturated bed 水饱和层
water-saturated curve 水饱和曲线
water-saturated formation 饱水地层建造，饱水层
water-saturated layer 饱水层
water-saturated rock 饱水岩石
water-saturated sand 饱水砂
water-saturated soil 饱水土
water-saturated stratum 饱水地层
water-saturated test 饱水试验
water-saturation coefficient 饱水系数
water-sealed bearing 防水轴承；密封轴承
water-sealing curtain 防水帷幕
water-sensitive formation 水敏岩层，水敏地层建造
water-sensitive sand 水敏砂
water-sensitive shale 水敏性页岩
water-sensitive soil 水敏性土，湿陷性土
watershed 流域，集水区；分水界，分水岭
watershed area 流域面积
watershed boundary line 流域界线
watershed divide 分水岭
watershed ecosystem 流域生态系统
watershed elevation 分水岭高程
watershed hydrology 流域水文学
watershed lag 流域滞时，流域集水滞后时间
watershed leakage 流域渗漏，流域越流
watershed line 分水线
watershed management 流域管理
watershed protection and flood prevention project 流域保护及防洪工程
watershed restoration 流域恢复
watershed slope 流域坡降
watershed tunnel 越岭隧道
watershed variable 流域变数
water-side land 堤外土地
water-soaked 水漫的，水淹的，水浸的
water-soaking process 水浸法
water-softening by heating 加热水软化法
water-solid ratio 水固比
water-soluble acrylamide 水溶性丙烯酰胺
water-soluble cellulose derivative 水溶性纤维素衍生物
water-soluble gas 水溶性气体
water-soluble material 水溶性物质
water-soluble organic matter 水溶有机物
water-soluble phenoplast 水溶性酚醛塑料
water-soluble polymer 水溶性聚合物
water-soluble preservative 水溶性防腐剂

water-soluble salt 水溶盐
water-soluble tracer 水溶性示踪剂
water-sorption ability 吸水性，吸水能力
watersource control 水源控制
waterspout 排水口，槽口；水龙卷，海龙卷
water-stable 耐水的，水稳定的
water-stage recorder 自记水位计
waterstead 河床
water-stilling device 水消能设备，水静止设备
waterstop blade 止水片
water-storage capacity 蓄水能力
water-storage dam 蓄水坝
water-storage facility 蓄水设施
water-storage reservoir 蓄水库，水库
water-storage reservoir shore 水库边岸
water-storage tank 蓄水池，贮水池
water-storing capacity 储水量
water-supply district 供水区
water-supply facility 给水设施
water-surface profile 水面线
water-surface reflection coefficient 水面反射系数
water-swelling clay 遇水膨胀黏土
water-table aquifer 潜水含水层，潜水层
water-table contour 等水位线
water-table decline 水位下降
water-table depression curve 水位下降曲线
water-table divide 分水岭
water-table isobath 水位等深线，地下水位等深线
water-table isohypse 水位等高线，等水位线，地下水位等高线
water-table outcrop 地下水露头
water-table profile 水位剖面，地下水位剖面
water-table slope 水面坡度，水面比降，水面斜率，地下水面坡度，地下水面比降，地下水面斜率
water-table spring 潜水泉，地下水面出露泉
water-table stream 潜水流
water-table well 潜水井
water-tamped explosion 水坑爆炸
water-thermometer 水温表
watertight 不漏水的，不透水的，隔水的，防水的，止水的
watertight concrete 抗渗混凝土
watertight core 防渗心墙
watertight curtain 止水帷幕
watertight diaphragm 隔水膜，防水膜，防渗膜
watertight facing 防渗面层，防渗斜墙
watertight facing arch 防渗护面拱
watertight integrity 水密封完全性
watertight joint 防水接缝
watertight layer 防水层，不透水层，隔水层
watertight lining 防水衬砌
watertight rubber membrane 防水橡皮膜
watertight screen 防渗帷幕，防渗墙；阻渗板桩
watertight shutter 挡水板
watertight stratum 不透水地层，隔水地层
watertight structure 不透水结构
watertight wall 不透水墙，防水墙
watertightness 防水性，不透水性
watertightness test 水密性试验
water-timed sampler 水定时式取样机

water-to-cement ratio 水灰比
water-transmission line 输水管线
water-wastewater system 上下水系统
water-wasting pressure regulator 弃水式压力调节器
water-wave erosion 水波冲蚀
waterway 水路，水道，航道；水渠，水沟
waterway engineering 水道工程
waterway tunnel 输水隧道
water-wet formation 亲水湿地层，亲水地层建造
water-wet reservoir 水湿油层
water-wet sand 水湿砂
water-wetted 水湿的，亲水的
water-wetting 亲水，水湿润
water-wheel generator 水轮发电机
waterworks 水厂，自来水厂，水处理厂；水务工程，水务设施，水工建筑物
waterworks item 水务工程项目
waterworks sludge 水厂污泥
waterworn 流水冲刷的，水蚀的
waterworn sand 水蚀砂
watery 多水分的，湿的
watery phase 水相
water-yielding opening 出水口
water-yielding potential 出水潜力
water-yielding stratum 涌水层
water-yielding zone 出水带
watt 瓦，瓦特
wattage 瓦特数
wattage rating 额定功率消耗
watt-hour 瓦时，瓦特-小时
watt-hour consumption 瓦时数，瓦时消耗量
watt-hour meter 瓦时计，电度表
wattmeter 瓦特计，功率计
waughammer drill 架式凿岩机
waughoist 架式绞车；塔式起重机
wave 波；波浪；起伏，波浪形
wave beam angle 波束角
wave crest 波峰
wave cutting 浪蚀，波蚀
wave cyclone 波动性气旋
wave decay 波浪衰减
wave depth 波底，浪蚀基面，浪蚀水深
wave detector 检波器
wave differential equation 波动微分方程
wave diffraction 波的衍射；波的绕射
wave direction 波向
wave direction meter 波向仪
wave dispersion 波频散，波弥散
wave dissipating concrete block 消波混凝土块体
wave divergence 波扩散
wave dominated delta 浪控三角洲
wave drag 波阻，波浪阻力
wave drift 波浪漂移
wave duration 波时，波浪持续时间
wave dynamics 波动力学
wave energy 波浪能
wave energy coefficient 波能系数
wave envelope 波包络
wave equation 波动方程
wave equation analysis 波动方程分析

wave equation migration	波动方程偏移
wave erosion	浪蚀
wave erosion base level	浪蚀基面
wave erosion coast	浪蚀海岸
wave erosion model	浪蚀模型
wave face	波面
wave field	波浪场，波场
wave filter	滤波器
wave flow	波浪流
wave force	波浪力
wave forecasting	波浪预报
wave form distortion	波形畸变
wave frequency	波频
wave front	波前；波阵面
wave function	波函数
wave height	波高
wave height coefficient	波高系数
wave height forecast	波高预报
wave height meter	波高仪
wave hindcasting	波浪后报，波浪追算
wave hollow	波谷
wave intensity	波强度
wave interference	波干扰
wave invasion	波浪侵袭
wave laminar flow	波状层流
wave length filtering	波长滤波
wave liner	波纹衬板
wave lining	波状衬里
wave loop	波腹
wave making machine	造波机
wave mark	波痕
wave mechanics	波动力学
wave noise	波噪声
wave normal	波法线
wave number	波数
wave observation radar	测波雷达
wave of compression	压缩波
wave of condensation	压密波
wave of condensation and rarefaction	疏密波
wave of dilatation	膨胀波
wave of distortion	畸变波
wave of explosion	爆炸波
wave of finite amplitude	有限振幅波
wave of infinitely small amplitude	无限小振幅波
wave of oscillation	振荡波
wave of stress	应力波
wave of translation	转移波
wave parapet	防浪墙
wave prediction	波浪预报
wave pressure	波压
wave propagation law	波传播规律
wave propagation rule	波传播规律
wave rear	波尾
wave reflection	波反射
wave refraction	波折射
wave ridge	波峰
wave ripple	波痕
wave run-up	波浪爬高，波浪上冲
wave scale	浪级
wave solution	波动解
wave spectrum	波谱
wave speed	波速
wave staff	测波标杆
wave steepness	波陡
wave stress	波应力
wave surface	波面
wave tail	波尾
wave theoretical tomographic method	波动理论层析成像法
wave theory	波动说
wave trough	波谷
wave uplift pressure	波浪浮托力
wave vector	波矢量
wave velocity	波速
wave velocity method	波速法
wave wall	防浪墙
wave wash levee	防波堤
wave width	波宽
wave zone	波浪区
wave-absorbing structure	消浪建筑
wave-beaten beach	浪蚀海滩
wave-breaker block	消浪块体
wave-built cross bedding	浪成交错层理
wave-built platform	浪成平台，波积台地
wave-built terrace	浪成阶地
wave-cut beach	波蚀滩地
wave-cut chasm	浪蚀峡谷，浪蚀裂缝
wave-cut cliff	浪蚀崖
wave-cut notch	浪蚀峡谷，浪蚀槽
wave-cut plain	浪蚀平原
wave-cut scarp	浪蚀崖
waveform	波形
waveform analysis	波形分析
waveform generator	波形发生器
waveform log	波形测井
waveform measurement	波形测量
waveform modeling	波形模拟
waveform monitor	波形监视器
waveform parameter	波形参数
waveform recording	波形记录
waveform test	波形试验
wavefront angle	波前角
wavefront arrival	波前初至
wavefront chart	波前量板
wavefront method	波前法
waveguide	波导；电磁波导
wave-height record	波高记录
wave-height spectrum	波高谱
wave-induced oscillation	波激振动
wave-induced scour	波浪冲刷
wavelength	波长
wavelength constant	波长常数
wavelength modulation	波长调制
wavelength spectrometer	波长分光计
wavelength-dispersive X-ray analysis	波长色散 X-射线分析
wavelet	小波，子波
wavelet analysis	小波分析
wavelet approach	小波算法
wavelet bandwidth	子波带宽

wavelet denoising	小波去噪
wavelet filter	小波滤波器
wavelet polarity	子波极性
wavelet processing	子波处理
wavelet resolution	小波分辨率
wavelet theory	小波理论
wavelet transform	小波变换
wave-like advance	波状前进
wave-like lamination	波状纹理
wave-like uplift	波状隆起
wave-measuring buoy	测波浮标
wavemeter	波长计
wavenumber filter	波数滤波器
wavenumber-frequency spectrum	波数-频率谱
wave-particle duality	波粒二象性
wave-particle interaction	波粒相互作用
wave-path equation	波径方程
wave-reflection method	波反射法
wave-ripple bedding	波痕层理
wave-ripple cross-bedding	波痕交错层理
waveshape	波形
wave-vector filtering	波向量滤波
waviness	起伏度；波浪状
wavy arch	波形拱
wavy bedding	波状层理
wavy boundary	波状边界
wavy extinction	波状消光
wavy structural plane	波状结构面
wavy surface	波状面
wavy texture	波状结构
wax	蜡，石蜡，蜡状物；打蜡，上蜡
wax containing asphalt	含蜡沥青，含蜡石油沥青
wax content	含蜡量
wax cracking	石蜡裂化
wax deoiling	蜡脱油
wax filter	蜡过滤器
wax pattern	蜡模
wax press	蜡压滤机
wax primer	涂蜡起爆药包
wax removal	除蜡
wax seal	蜡封
wax sealing method	蜡封法
wax separator	蜡分离器
wax shale	蜡页岩，油页岩
wax stone	粗蜡
wax-bearing crude	含蜡原油
wax-containing crude	含蜡原油
waxen	蜡制的，上蜡的
wax-free crude	无蜡原油
waxing	打蜡，上蜡
wax-resinite	蜡树脂体，蜡质体
wax-sealed sample	蜡封样品，蜡封土样
waxy coal	蜡煤
waxy crude	含蜡原油
waxy distillate	含蜡馏分
waxy hydrocarbon	含蜡烃
waxy luster	蜡状光泽
waxy oil	蜡质油
way	道路，途径；方法，方式
way right	通行权，路权

way shaft	小暗井，人行天井
way up	地层由老至新的方向
wayboard	薄夹层
wayside	路旁
wayside transmitter	路旁传送器
weak acid	弱酸
weak acid gel	弱酸凝胶
weak activating spring	弱弹性弹簧
weak anomaly	弱异常
weak aquifer influence	弱水驱
weak beam	弱梁
weak bed	弱岩层
weak bind	松软黏土夹层
weak bonding	胶结疏松
weak boundary condition	弱边界条件
weak caustic solution	弱苛性碱溶液
weak coal	脆煤
weak coherence	弱黏结性
weak column	强梁，弱柱
weak concrete	低强混凝土，低标号混凝土
weak convergence	弱收敛
weak coupling theory	弱耦合理论
weak current	弱电流
weak damping	弱阻尼，小阻尼
weak definition	弱定义
weak earthquake	弱震
weak echo	弱回波
weak eddy	弱旋涡
weak electrolyte	弱电解质
weak extremum	弱极值
weak form	弱形式
weak form heat conduction	弱形式热传导
weak formation	软弱地层建造，软弱层
weak foundation	软弱地基
weak framework	弱骨架
weak gas drive	弱气驱
weak ground	软弱地基
weak hydrochloric acid	淡盐酸
weak intercalated layer	软弱夹层
weak jump	弱跃变
weak limit	弱限
weak low pressure	弱低压
weak mineralized groundwater	弱矿化地下水
weak patch test	弱小片实验，弱斑贴试验
weak plane	脆弱面，软弱结构面
weak point	脆弱点
weak reflector	弱反射层
weak response	弱响应
weak rock	软弱岩石，软岩
weak salty soil	弱盐渍土
weak seismicity region	弱震区
weak shock	弱震
weak singularity	弱奇异性
weak soil	软土
weak solution	弱解；稀溶液
weak stratum	松软岩层
weak structural plane	软弱结构面
weak structure	脆弱结构
weak subgrade	软弱地基，软弱路基
weak subsoil	软弱地基土，软弱下卧层

weak surrounding rock 软弱围岩
weak weathering 弱风化
weak zone 软弱带
weak-acid-soluble 弱酸能溶的
weak-cementing matrix 弱胶结性基质
weaken 变弱；使…稀薄
weakening 弱化
weakening model 弱化模型
weak-line approximation 弱谱线近似
weakly alkaline soil 弱碱性土
weakly bound water 弱结合水
weakly leached brown soil 轻度淋溶棕壤
weakly magnet 弱磁体
weakly magnetic material 弱磁性物质
weakly magnetic mineral 弱磁性矿物
weakly nonlinear system 弱非线性系统
weakly penetrating radiation 弱贯穿辐射
weakly podzolic soil 弱灰化土
weakly stationary process 弱稳定过程
weakly stationary random process 弱稳定随机过程
weakly weathered layer 弱风化层
weakly weathered rock 弱风化岩石
weakly weathered zone 弱风化带
weakly-permeable boundary 弱透水边界
weakness 软弱性，软弱，脆弱，弱点
wear 磨损，磨耗
wear allowance 容许磨耗，磨损留量
wear coefficient 磨损系数
wear flat 磨损面，磨平处
wear hardness 抗磨硬度
wear limit 磨损极限
wear pad 耐磨垫片
wear print 磨损痕迹
wear rate 磨损率
wear resistant cast iron 耐磨铸铁
wear resistant rail 耐磨轨
wear resistant steel 耐磨钢
wear test 磨损试验
wearability 磨损性，耐磨性
wearable sensor 耐用传感器
wearing capacity 磨损量
wearing coat 耐磨层
wearing coefficient 磨损系数
wearing course 磨耗层
wearing depth 磨损深度
wearing part 磨损零件
wearing strength 抗磨强度
wearing surface 磨损面，磨平处
wearproof 耐磨的，抗磨的
wear-resistant material 抗磨材料
wear-resistant metal 耐磨金属
wear-resisting strength 抗磨强度
weather 天气，气象；风化
weather anchor 防风锚
weather chart 天气图
weather condition 气候条件
weather door 调节风门
weather echo 天气回波
weather effect 天气影响
weather element 天气要素

weather forecast 天气预报，气象预测
weather grade of forest 森林火险天气等级
weather map 气象图
weather modification 人工影响天气
weather monitoring 天气监测
weather prediction＋ 天气预报
weather radar 天气雷达
weather satellite 气象卫星
weather stain 风化染色
weather station 气象台
weather symbol 天气符号
weather window 气候窗
weatherability 耐风化性
weathercock 风标
weathered bedrock 风化基岩
weathered clay 风化黏土
weathered crust 风化壳
weathered escarpment 风化崖
weathered fissure water 风化裂隙水
weathered granite 风化花岗石
weathered horizon 风化层
weathered intensity 风化强度
weathered lava 风化熔岩
weathered layer 风化层
weathered mantle 风化壳
weathered material 风化物
weathered matrix 风化母岩
weathered mudrock 风化泥岩
weathered outcrop 风化露头
weathered rock 风化岩石
weathered rock-soil 风化岩土料
weathered schist 风化片岩
weathered state 风化状态
weathered stratum 风化地层
weathered zone 风化带
weather-geohazard forecast 气象地质灾害预报
weatherglass 晴雨计；气压计；温度计
weathering 风化
weathering agency 风化营力
weathering agent 风化营力
weathering classification 风化分类
weathering condition 风化条件
weathering crust 风化壳
weathering crust deposit 风化壳矿床
weathering crust reservoir 风化壳储集层
weathering degree 风化程度
weathering deposit 风化矿床
weathering depth 风化层厚度；低速带厚度
weathering disintegration 风化崩解
weathering grade 风化度
weathering index 风化指数
weathering intensity 风化强度
weathering joint 风化节理，风化裂缝
weathering landform 风化地貌
weathering map 风化层厚度图；低速带厚度图
weathering potential 风化潜力
weathering process 风化过程
weathering product 风化产物
weathering profile 风化剖面
weathering rate 风化速率

weathering residual deposit 风化残积型矿床
weathering residue 风化残余物
weathering zone 风化带
weather-proof 全天候的，防风雨的
weather-protected 不受天气影响的，防风雨的
weaving 交织；编织；构成，设计
weaving point 交织点
weaving section 交织路段
web 网络，互联网，蛛网；腹板，梁腹；轨腰，轨腹；
web beam 强横梁
web member 腹杆
web of things 物联网
web plate 腹板
web reinforcement 腹筋；抗剪钢筋
web splice 梁腹拼接
web stiffener 梁腹加劲板
web texture 蛛网结构
web-based information 基于网络的信息
webbed bit 蹼式钻头
webbing 条带织物
web-conveyor digger 输送链式挖掘机
weber 韦伯
weber number 韦伯数
websterite 矾石；二辉岩
web-stiffened structure 腹板加劲结构
wedge 楔，楔状体；楔入，挤入
wedge absorber 吸声尖劈；楔形吸声器
wedge anchorage 楔形锚具，楔块锚
wedge aquifer 楔形含水层
wedge block 楔块
wedge brick 楔形砖
wedge cut 楔形掏槽
wedge edge 尖灭
wedge edge trap 楔边圈闭
wedge effect 楔效应
wedge failure 楔形破坏
wedge failure-kinematic test 岩楔破坏的运动学试验
wedge failure-stability analysis 岩楔破坏稳定性分析
wedge flow 楔形流
wedge gate 楔形闸门
wedge joint 楔形结合，楔形接头
wedge method 楔形体法
wedge out 尖灭，变薄
wedge pile 楔形桩
wedge plate 楔板
wedge prop 楔锁式支柱
wedge reaming bit 扩孔钻头
wedge shaped graben 楔状地堑
wedge shaped horst 楔状地垒
wedge shaped sheet pile 楔形板桩
wedge stability analysis 楔形体稳定分析
wedge structure 楔形构造，楔形结构
wedge tenon 楔榫
wedge test piece 楔形试样
wedge theory 楔体理论
wedge timber-sheet-pile 楔形木板桩
wedge-clamp-type prop 楔钳固定式支柱
wedge-nut expansion shell bolt 楔螺母胀壳锚杆
wedge-shaped aquifer 楔形含水层
wedge-shaped assembly 楔形组件

wedge-shaped curb 楔形路缘石
wedge-shaped fracture 楔形破裂
wedge-tube bolt 楔管式锚杆
wedge-type gate valve 楔形闸阀
wedge-wire screen 楔形丝筛网
wedging 楔入，嵌入；楔固
wedging effect 楔入效应
wedging out 尖灭
wedging reamer 锥进扩孔器
weekly storage plant 周调节电站
weekly storage reservoir 周调节水库
Weeks manometer 韦克斯流体压力计
weep drain 渗水管，排水沟
weep trough 渗水槽
weepage 渗水，渗漏
weep-hole 泄水孔
weeping 滴水，泌水
weeping formation 渗水地层建造，渗水层
weeping rock 渗水岩石
weeping spring 细泉，渗出泉
weeping zone 渗水带
Wegener hypothesis 魏格纳大陆漂移说
Weibull distribution 韦布尔分布
Weibull probability density function 韦布尔概率密度函数
weigh bridge 地秤，地磅，桥秤
weigh-batching 按重量配料
weighing 称重，加权；权衡
weighing bottle 称量瓶
weighing burette 滴定管，量管
weighing factor 加权系数
weighing sensor 称重传感器
weighing-in-water method 水中称重法
weight 重，重量；砝码；权重
weight arithmetic mean 加权算术平均
weight batching 按重量配料
weight coefficient 加权系数；重量系数
weight density 重量密度
weight dropping technique 落锤法
weight function 权函数；加权函数
weight hopper 称量斗
weight house 磅房
weight indicator 重量指示器，指重表；钻压表
weight loss 失重
weight loss curve 失重曲线
weight matrix 权矩阵
weight number 权数
weight on bit 钻压，钻头压力
weight percent 重量百分数
weight potential 重力势
weight ratio 重量比
weight reciprocal 权倒数
weight-carrying capacity 承重能力
weighted approximation 加权近似值
weighted array 加权组合
weighted average 加权平均值
weighted average error 加权平均误差
weighted average grade 加权平均品位
weighted average porosity 加权平均孔隙度
weighted average value 加权平均值

weighted coefficient	加权系数
weighted drilling mud	加重钻泥
weighted error	加权误差
weighted least square	加权最小二乘法
weighted least square fit	加权最小二乘拟合
weighted mean value	加权平均值
weighted mud	加重泥浆
weighted one-rank local-region method	加权一阶局域法
weighted ordinate method	加权纵坐标法
weighted pair-group average linkage method	加权对组均连法
weighted pair-group average method	加权对组法
weighted residual	加权残值
weighted residual finite element	加权残值有限单元
weighted residual method	加权残量法
weighted residual p-step algorithm	加权残值 p 步算法
weighted stack	加权叠加
weighted strain energy	加权应变能
weighted sum	加权和
weighted velocity head	加权速度头
weighted-moving average	加权移动平均
weighting	称量，衡量；加权；加重
weighting agent	增重剂
weighting factor	加权因子
weighting function	加权函数
weighting material	加重剂
weighting parameter	加权参数
weir	堰，小溢流坝；堰口，水口
weir crest	堰顶
weir dam	溢流坝
weir flow	堰流；溢流量
weir formula	堰流公式
weir gauge	堰顶水位计
weir head	堰顶水头
weir lip	堰唇，堰前缘
weir meter	溢流表，溢流计
weir notch	堰孔
weir pier	堰墩
weir plate	堰板
weir section	堰段
weir sill	堰槛
weir-type fish pass	堰式鱼道
weir-type spillway	堰式溢洪道
weld	焊接，熔接
weld assembly	焊接组件
weld bead	焊道，焊珠
weld buildup	堆焊
weld crack	焊接裂纹
weld crater	焊口
weld length	焊接长度
weld pitch	焊缝距
weld ripple	焊缝波纹
weld strength	焊缝强度
weld thickness	焊缝厚度
weldability	可焊性
weldability test	焊接性试验
weldable strain gage	焊接式应变计
welded agglomerate	熔结集块岩
welded drill rod	焊接钻杆
welded framework	焊接骨架
welded lapillistone	熔结角砾岩
welded rigidity	焊接刚度
welded rotor	焊接转子
welded seam	焊缝
welded splice	熔接连接
welded steel beam	焊接钢梁
welded steel girder	焊接钢梁
welded steel pipe	焊接钢管
welded steel structure	焊接钢结构
welder	电焊机；电焊工
welding arc	焊弧
welding bar	电焊条
welding cable	电焊电缆
welding current	焊接电流
welding defect	焊接缺陷
welding electrode	焊接电极；电焊条
welding flux	焊剂
welding ground	焊接地线
welding gun	焊枪
welding helmet	焊工面罩
welding machine	焊接机
welding metallurgy	焊接冶金学
welding powder	焊粉
welding pressure	焊接压力
welding residual stress	焊接残余应力
welding robot	焊接机器人
welding rod	焊条
welding seam	焊缝
welding shield	头盔，焊罩
welding shrinkage	焊接收缩量
welding slag	焊渣
welding stress	焊接应力
welding transformer	电焊变压器
welding wire	焊丝
weldless steel tube	无缝钢管
weldment	焊件
weldmesh screen	防盗网
weldwood	胶板
well	井，水井，钻井，油井，钻井
well alignment	一字型排列井，成排井，井排
well array	井组，井排；布井
well auger	凿井螺旋钻
well blowout	井喷
well bore imaging tool	井筒成像工具
well borer	钻机；钻工
well boring	钻井作业
well casing	钻井套管，下钻井套管
well casing anchor	套管锚
well cementation	固井
well cementing operation	固井作业
well completing test	完井测试
well completion	完井
well completion interval	完井层段
well completion technology	成井工艺
well construction technology	建井技术
well control algorithm	井控算法
well control fluid	压井液
well control information	井控信息
well core	岩心
well deflection	井斜；井眼偏斜

well deformation 井变形
well depth 井深
well diameter 钻孔直径
well drill 钻机，打井机
well effluent 井流出物，油气井涌出物
well eruption 井喷
well field 井场，井区，井田，井群区
well field equipment 井场装置
well flooding 井泄洪
well flow 井流
well fluid logging 井液测井，泥浆测井
well flushing 洗井
well foundation 井筒基础；沉井基础
well fracturing 油气井水力压裂
well function 井函数
well geochemistry 钻井地球化学
well graded soil 级配良好土
well grain-size distribution 良好粒径级配
well group 井群，井组
well head 井口
well head assembly 井口装置
well head control valve 井口控制阀
well head elevation 井口高程，井口标高
well head flowing pressure 井口流压
well head gas 井口气
well head pressure 井口压力
well head separator 井口气水分离器
well head temperature 井口温度
well hydraulics 井孔水力学
well hydrograph 井孔水文曲线，井孔水文图
well hydrostatic pressure 井筒静水压力
well in operation 工作井
well infiltration area 井内入渗区
well intake 井内进水区
well intake pressure 井口注入压力
well interference 井干扰
well irrigation 井灌
well killing 压井
well killing fluid 压井液
well leakage 井漏
well log 钻孔柱状图，钻井剖面，钻井记录，测井曲线
well log analysis 测井曲线分析
well logger 测井仪
well logging 测井
well logging during drilling 随钻测井
well logging interpretation 测井解释
well logging on digital magnetic tape 数字磁带测井
well logging truck apparatus 车载测井设备
well mouth 井口
well of complete penetration 完全贯穿井
well of partial penetration 非完全贯穿井
well off 停井，关井
well out of control 失控井
well pattern 井型
well perforation 钻井射孔
well perforator 钻井射孔器
well performance testing 钻井生产测试
well pit 钻井基坑，井坑
well point 过滤器，有孔管，井点
well point dewatering 井点去水

well point drain 井点排水
well point pumping system 井点抽水系统
well pressure 井压
well pulling machine 钻井起管机
well pump 井泵；井用泵
well pumping test 钻井抽水试验
well radioactivity logging 放射性测井
well response test 钻井响应试验
well salt 井盐
well sample 钻井样品
well sampling 钻井取样
well screen 钻井滤网
well section 钻井断面
well seepage area 井内渗漏区
well shooting 地震测井；井下爆破；油井射孔
well sinking 沉井
well site 井场，井位
well site arrangement 井场布置
well slough 井塌
well spacing 井距
well spacing chart 井网图
well straightening 井身矫直
well structure 井身结构
well television 井下电视
well temperature 井温
well temperature log 井温曲线
well test 试井
well thermometer 井温计
well top 井口
well trajectory 井轨迹
well tube 井孔套管
well washing 洗井
well water 井水
well water level 井水水位
well water pump 井水泵
well workover 井孔维修
well zone 井区
well-bedded 成层良好的，层理显明的
wellbore 井身，井筒，钻井
wellbore cement 井筒胶结剂
wellbore collapse pressure 井筒坍塌压力
wellbore dynamics 井筒动力学
wellbore failure 井筒故障
wellbore hydraulics 井孔水力学
wellbore instability 井筒不稳定性
wellbore model 井筒模型
wellbore shrinkage 井筒收缩
wellbore stability 井筒稳定性
wellbore stress 井筒应力
well-cemented 胶结良好的
well-cemented rock 胶结良好的岩石
well-data system 钻井数据系统
well-drained soil 排水良好的土
well-flow theory 井流理论
well-graded aggregate 级配良好的骨料
wellhead pressure 井口压力
wellhole 井孔
wellhole blasting 井孔爆破
welling 涌水的，自流的
welling out 涌出，流出

well-log data	测井数据	wet-subsidence for classification	分级湿陷量
well-loss constant	井损常数	wettability	润湿性
well-point method	井点法，集井挖土法	wettability hysteresis	润湿滞后
well-recharge area	井补给区	wettability index	润湿指数
well-resistivity log	钻井电阻率测井	wettable	可湿润的
well-rounded	磨圆度良好的	wetted area	受潮面积
well-sorted	分选良好的	wetted cross section	湿润断面
well-stimulation technology	油气井激发工艺	wetted perimeter	湿润周界
well-type	井型	wetting agent	湿润剂
well-water quality	井水水质	wetting and drying cycle	干湿循环
well-watered	充分灌水的，多水的	wetting and drying test	干湿试验
Wemco drum-type separator	韦姆克筒式分选机	wetting angle	润湿角
Wemco lab flotation machine	韦姆克实验室用浮选机	wetting band approach	浸湿带法
Wenner array	温纳排列，温纳阵	wetting coefficient	潮湿系数
Wenstrom separator	温斯特罗姆型磁选机	wetting contact angle	润湿接触角
Wentworth grade scale	温特沃兹粒度表	wetting front	湿润线，湿润锋面
Wentzel-Kramers-Brillouin-Jeffreys method WKBJ法		wetting heat	湿润热
Wernerian theory	魏尔纳学说	wetting period	湿润周期
Westergaard equation	韦斯特格德方程	wetting phase	非润湿相
Westergaard theory	韦斯特格德理论	wetting power	湿润力
Westphalia fast plough	威斯法利亚型高速刨煤机	wetting water	润湿水
wet	湿的，潮湿的；湿气，水分	wetting zone	湿润带
wet air compressor	湿空气压缩机	wet-type roof bolting machine	湿式锚杆安装机
wet analysis	湿法分析	wet-way analysis	湿法分析
wet assay	湿法分析	wharf	码头
wet automatic sprinkler system	湿式自动喷水灭火系统	wharf wall	码头岸墙
wet avalanche	湿崩坍	wheel	轮，车轮
wet blasting	湿法爆破	wheel axle	轮轴
wet boring	湿法钻进	wheel band	轮箍
wet bulb thermometer	湿球温度计	wheel brake	轮闸
wet climate	潮湿气候	wheel crane	轮式起重机
wet compaction	湿压实	wheel ditcher	轮式挖沟机
wet concentration	湿选，水选	wheel dozer	轮式推土机
wet concrete mix	混凝土湿拌	wheel dredge	轮式挖掘船
wet corrosion	浸蚀	wheel excavator	轮式挖掘机
wet crushing	湿法破碎	wheel hook	制动钩
wet cutting	湿挖	wheel load	车轮荷载
wet density	湿密度	wheel loader	轮式装载机
wet deposition	湿沉降	wheel loading shovel	轮式装载铲
wet drawing	湿法拉拔	wheel mounted belt conveyor	轮架式胶带输送机
wet drilling	湿式钻进	wheel plough	轮式刨煤机
wet dust extraction	湿法除尘	wheel scraper	轮式铲运机，轮式刮煤机
wet dust removal	湿式除尘	wheel track vehicle	轮轨车辆
wet excavation	湿挖，水下开挖	wheel tracking test	车辙试验
wet extraction	湿法提取	wheel tractor	轮式牵引车，轮式拖拉机
wet well	湿井	wheel tractor scraper	轮式牵引铲运机
wet-and-dry screening	干湿联合筛分	wheel trench excavator	轮式挖沟机
wet-bulb isotherm	湿球等温线	wheel-and-axle assembly	轮轴组合
wet-chemical analysis	湿化学分析	wheelbarrow	手推车；独轮车
wet-dry cycle	干湿循环	wheeled diesel crane	轮式柴油起重机
wetland	湿地	wheeled excavator	轮式挖土机
wetland ecosystem	湿地生态系统	wheeled tractor	轮式牵引车，轮式拖拉机
wetland restoration	湿地恢复	wheel-mounted dragline	轮架式挖土机
wetness	湿度，潮湿	wheel-mounted mucker	轮架式装岩机
wetness factor	湿度系数	wheel-mounted power shovel	轮架式动力铲
wetness index	湿度指数	wheel-mounted scraper	轮架式铲运机
wet-proof	防湿的	wheel-rail dynamics	轮轨动力学
wet-pulp distributor	矿浆分配器	wheel-rail noise	轮轨噪声
wet-strength	湿强度	wheel-tractor mounted bulldozer	轮式牵引推土机
wet-subsidence due to overburden	自重湿陷量	wheel-type trenching machine	轮胎式挖沟机

whetstone	油石，磨石，砥石；暗色岩	wicking	芯吸；毛细作用
whipstock	造斜器	wide angle camera	广角相机
whirl	旋转，回转；旋涡，涡流；旋风	wide angle common depth point	广角共深度点
whirl jet drill rig	旋喷钻机	wide angle lens	广角镜头
whirl structure	旋涡状构造	wide angle scanning	广角扫描
whirler crane	回转式起重机	wide area communication	广域通信
whirlpool	旋涡	wide area network	广域网
whirlwind	旋风	wide band digital seismograph	宽频带数字地震仪
whistle	笛，汽笛；笛声，哨声，啸声；鸣笛	wide band spectra	宽带谱
whistle board	鸣笛标	wide base tyre	宽基轮胎
whistle buoy	鸣笛浮标	wide flange steel section	宽翼工字钢
whistle-pipe sampler	笛管取样机	wide line profile	宽线剖面
whistling signal	啸声信号	wide range orifice meter	宽量程孔板流量计
white alkali soil	白碱土，盐土	wide strip foundation	宽条形基础
white alloy	白合金，巴氏合金	wide strip tensile test	宽条拉伸试验
white asphalt	白沥青	wide-angle aerial camera	广角航摄仪
white cast iron	白口铸铁	wide-angle photography	广角摄影
white cement	白水泥	wide-angle reflection	广角反射，大角度反射
white clay	高岭土	wide-angle stack	广角叠加
white cold	白冷	wide-angle telescope	广角望远镜
white heat	白热	wide-aperture seismic reflection	广角地震反射
white light image processing	白光影像处理	wide-band acoustic logging	宽频带声波测井
white lime	熟石灰	wideband data transmission system	宽带数据传输系统
white manifold paper	复印纸	wide-band earthquake loading	宽频带地震荷载
white marble	汉白玉；白色大理石	wide-band feedback seismometer	宽频带反馈地震计
white mica	白云母	wide-flange beam	宽缘梁
white noise	白噪声	wide-flange girder	宽缘大梁；宽缘桁架
white noise coefficient	白噪系数	wide-meshed screen	大孔筛
white oil	凝析油	wide-mouth bottle	广口瓶
white peat	硅藻土	widened channel	拓宽河段
white Portland cement	白硅酸盐水泥	widened section of viaduct	高架桥加宽段
white powdery filling	白色粉末状填充物	widened value of subgrade	路基加宽值
white radiation	白色辐射	wide-room mining	宽房式开采
white smoker	海底白烟筒	wide-span beam	大跨度梁
white vitriol	皓矾，白矾	Widia bit	威地亚型硬质合金钻头
white wax	白蜡	width	宽度，跨度；宽大，宽阔
whitener	增白剂	width index	宽度指数
whiteness	白度，洁白度	width of ballast bed	道床宽度
whitening agent	增白剂	width of fault trace	断层迹宽度
white-noise signal	白噪声讯号	width of foundation base	基础底面宽度
whiteprint	白印，晒图	width of joint	缝宽
whiteruss	液体石蜡	width of joint opening	节理张开宽度
whitewash	白灰浆，白涂料；粉饰，刷白	width of outcrop	露头宽度
whitish soil	白土，漂白土	width of pavement	路面宽度
whizzer	离心分离机，离心机	width of subgrade	路基宽度
whole core analysis	全岩心分析	Wiechert seismograph	维歇特地震仪
whole current method	全流法	Wiener filter	维纳滤波器
whole lead age method	全铅测年法	Wiener inverse operator	维纳反滤波算子
whole number	整数	Wiener optimal design criterion	维纳最优设计准则
whole rock age	全岩年龄	Wiener-Hopf optimum filter	维纳-霍普夫最佳滤波器
whole rock isochron age	全岩等时线年龄	Wiener's algorithm	维纳算法
whole-body dose	全身剂量	Wien's displacement law	维恩位移定律
whole-rock dating	全岩年龄测定	WiFi	无线互联技术
whole-rock isochron	全岩等时线	WiFi localization	无线互联定位
wholesale business	批发业，批发业务	Wigner effect	魏格纳效应
wholesale price	批发价	Wikipedia	维基百科
wholly filled	全部充填的	wild animal protection area	野生动物保护区
wholly saturated	完全饱和的	wild cat well	野猫井，初探井
wick drain	排水板，排水芯板；袋装砂井	wild flooding irrigation	自由淹灌，漫灌
wicker	柳条，枝条；搭接头	wild land	荒地，野地，未开垦的土地

wild mine	未登记的矿山，掠夺性开采的矿山
wild outburst	爆喷
wild plant	野生植物
wild population	野生种群
wild river	自然河流；急流河
wild rubber	天然橡胶
wild well	失控井，猛喷井
wild wood	自然林
wild-cat area	投机性探采区
wild-cat drilling	冒险性钻探
wildlife conservation	野生生物保护
willow groin	柳梢丁坝
willow-box groyne	柳梢捆丁坝
Wilson algorithm	威尔逊算法
winch	卷扬机，绞车
wind abrasion	风蚀，风蚀作用
wind action	风力作用
wind arrow	风矢
wind backwater	风壅水
wind beam	抗风梁
wind belt	风带；防风林
wind bent	防风排架
wind blocking coefficient	风阻系数
wind borne sediment	风成沉积
wind box	风箱
wind brace	防风斜撑
wind channel	风道
wind cone	风向锥，圆锥风标
wind corrosion	风蚀
wind current ripple	风成波纹
wind damage	风害，风灾
wind deposition	风积作用
wind deposition coast	风积海岸
wind desert	风成沙漠
wind desertization	风沙化
wind desiccated soil	风干土
wind direction	风向
wind disturbance	风扰动
wind divide	风分界线
wind drying soil	风干土
wind electric power generation	风力发电
wind energy	风能
wind energy density	风能密度
wind engineering	防风工程
wind erosion	风蚀
wind factor	风力因数
wind field	风场
wind force	风力
wind force scale	风级
wind frame	防风桁架
wind friction	风摩擦
wind frictional drag	风摩擦拖力
wind gap	风隙，风口
wind gauge	风速器
wind guard	风挡
wind indicator	风向指示器
wind induced oscillation	风生震荡
wind laid deposit	风积层
wind load	风荷载
wind load capacity	风载能力
wind niche	风蚀穴
wind power	风力
wind pressure	风压
wind resistance	风阻
wind resistance coefficient	风阻力系数
wind resistant design	防风设计
wind resisting truss	抗风桁架
wind ripple mark	风成波痕
wind rose map	风玫瑰图
wind scale	风级
wind separation	风力分选
wind shear	风切变；乱流
wind shield	风挡
wind shield glass	风挡玻璃
wind signal pole	风讯信号杆
wind site assessment	风场评价
wind sleeve	风向袋
wind soil	风积土
wind speed	风速
wind speed alarm	风速警报
wind speed scale	风速表，风速盘
wind spread sterility	风沙瘠化
wind spun vortex	风动涡旋
wind storm	风暴
wind stress	风应力
wind stress factor	风压系数
wind structure	风区结构，风流结构；防风建筑
wind truss	防风桁架
wind tunnel	风道，风洞
wind tunnel test	风洞试验
wind vane	风向标
wind vector	风矢量
wind velocity	风速
wind velocity gradient	风速梯度
wind velocity rose	风速玫瑰图
wind wall	挡风墙
wind wave	风浪
wind worn pebble	风蚀砾石
wind zone	风带
windage	空气阻力，风阻；风致偏差；游隙
windage effect	风阻效应
windage loss	通风损失
wind-aggregated sand	风积砂
windblown accumulation	风成堆积
wind-blown deposit	风成沉积，风积层
windblown sand	流砂，风积砂，风成砂
wind-blown silt	风积粉土
wind-blown soil	风积土
wind-borne sand deposit	风成砂沉积
wind-borne soil	风成土
wind-break wall	挡风墙
wind-cut stone	风棱石，风磨石
wind-deposited soil	风积土
wind-drift sand	流砂
wind-driven circulation	风成环流
wind-driven water pump	风动水泵
wind-eroded basin	风蚀盆地
wind-erosion lake	风蚀湖
wind-excited vibration	风激振动
wind-faceted stone	风棱石

wind-finding radar	测风雷达
wind-formed bar	风成沙洲
wind-generated gravity wave	风生重力波
wind-generated sea wave	风成海浪
wind-grooved boulder	风蚀巨砾
wind-induced random wave	风生随机波
wind-induced turbulence	风生湍流
winding engine	卷扬机；提升机
winding pulley	绕线滑轮
winding roadway	弯曲巷道
winding-up works	结尾工程
wind-laid soil	风积土
windlass	卷扬机，绞盘，小绞车，辘轳
windless	无风的
wind-polished rock	风磨岩
wind-protection plantation	防风林
windrow	长条形土料堆
wind-scoured hollow	风蚀凹地
windshield wiper	刮雨器
windtight	不透风的
wind-to-jet difference	风-射差
windward	向风的，迎风的；向风面，迎风面
windward drift	迎风漂流
windward side	迎风面
windward slope	迎风坡
wind-wave flume	风洞造波水槽
wind-worn stone	风棱石
wing	翼；叶片
wing abutment	翼墩，翼座
wing angle	刃角
wing crack	翼型裂纹
wing dam	翼坝
wing form	翼形
wing levee	翼堤
wing pile	翼桩
wing pump	轮叶泵
wing rail	翼轨
wing screen	翼式拦污栅
wing screw	翼形螺钉
wing tear test	叶片撕裂试验，翼撕裂试验
wing tensile crack	翼型拉伸裂纹
wing thickness	刃厚
wing tip vortex	翼梢涡流
wing trench	侧翼截水槽
wing tripper	卸料侧板
wing type method	翼形撕裂法
wing wall	翼墙
wing wall tunnel portal	翼墙式隧道口
wing-tip system	翼梢系统
Winkler foundation	温克勒地基
Winkler foundation model	温克勒地基模型
Winkler's assumption	温克勒假定
winning	回采，开采；超前平巷；备采煤区
winning area	回采区
winning assembly	采矿机组，采煤机组
winning district	采区
winning-block	采垛
winnowing	风选，簸选，漂选，吹蚀
winter concreting	冬季混凝土浇筑
winter construction	冬季施工
winterization	安装防寒装置
winze	下山，暗井
winzing	开凿下山，开凿暗井
wiper	擦拭巾，擦具；刮水器
WIPP	废物隔离中间试验厂
wire	丝，金属线，电线，导线
wire basket	铁丝笼
wire bolster	铁石笼
wire bridge	悬索桥
wire brush	钢丝刷
wire cable	钢丝绳，钢缆
wire cableway	索道，缆道
wire chase	线槽
wire communication	有线通讯
wire cutter	钢丝切割器，电线切割器
wire depth	电线埋深
wire drawing	拉丝
wire drawing machine	拉丝机
wire entanglement	铁丝网
wire fence	铁丝栅栏
wire flattening	线材压扁
wire fuse	保险丝
wire gabion dam	铅丝网石龙坝
wire gauze	铁丝网
wire mesh	金属丝网，钢丝网
wire mesh installation and shotcrete protection	挂网喷护
wire mesh installation and shotcrete rockbolt	挂网喷锚
wire mesh-reinforced shotcrete	钢丝网加固喷射混凝土
wire nail	圆铁钉
wire net	钢丝网
wire number	导线编号，线材编号
wire pliers	剪丝钳，克丝钳
wire pointer	钢丝压尖机
wire roll	钢筋盘条
wire rope	钢缆，钢索，钢丝绳
wire rope detector	钢丝绳探伤器
wire rope pulley	钢丝绳轮
wire rope reel	钢线绳卷筒
wire saw	钢丝锯
wire screen	金属丝筛
wire sieve	方孔筛，金属丝筛
wire sounding	钢丝绳测深
wire strain gauge	金属丝应变计
wire stripper	剥线钳
wire suspension bridge	悬索桥
wire symbol	线材号
wire tapping	线抽头
wire tensile and torsion tester	金属丝张扭试验器
wire tramway	架空索道，架空电车道
wire video recorder	钢丝录音机
wired remote control	有线遥控
wire-electrode cutting	线切割
wireless fidelity	无线保真
wireless mobile internet	无线移动互联网
wireline core barrel	钢缆岩心管
wireline core drilling	钢缆岩心钻进
wireline coring	钢缆取芯
wireline drill string	钢缆钻柱
wire-mesh hurdle	铁丝网栏
wiremesh-shotcrete-rock bolt support	钢筋网喷射混凝土

岩石锚杆支护
wire-rope way 钢缆索道
wire-type bore hole extensometer 弦式钻孔应变仪
wire-weight gauge 悬锤水尺
wiring box 接线盒
wiring diagram 接线图
wiring layout 布线图，接线图
wiring list 接线表
wiring scheme 接线图，布线图
wiring tree diagram 布线树状图
withdrawal diameter 后撤直径
withdrawal length 拔出长度
withdrawal load 拔出荷载
withdrawal of casing 套管回收
withdrawal of pile 拔桩
withdrawal resistance 拔出阻力
withdrawal speed 拔出速度
withdrawal test of pile 拔桩试验
withstand voltage test 耐压试验
witness 目击者；见证人；证据
wobble crank 摆动曲柄
wobble pump 手摇泵
Wodfson's trapping theory 乌尔夫逊俘获说
Wohlwill process 沃尔威尔法
wolfram 钨；钨矿
wolframite 黑钨矿
Wolf's definition 乌尔夫定义
wollastonite 硅灰石；钙硅石；硅酸钙岩矿
wolsendorfite 红铅铀矿
wood borer 木钻
wood brick 木砖，木块
wood chip 木屑
wood chip board 木屑板
wood chisel 木凿
wood coal 木煤，褐煤
wood column 木柱
wood construction 木结构；木建筑
wood derrick 木制井架
wood dust 木粉尘
wood file 木锉
wood floor 木地板
wood glue powder 木胶粉
wood hammer handle 木锤柄
wood industry 木材工业
wood joint 木接头，木榫头
wood meal 木粉
wood mosaic 木镶块，拼花木地板块
wood nail 木钉，木螺钉
wood nog 木榫，木栓
wood peat 本质泥炭
wood pile 木桩
wood pile cluster 木桩群
wood plug 木塞
wood products 木产品
wood ruler 木尺
wood screw 木螺丝
wood soil 林地土
wood step-taper pile 变截面木桩，变直径木桩
wood tar 木焦油
wood turpentine 松节油

wood waste concrete 木屑混凝土
Wood-Anderson seismograph 伍德-安德森地震仪
woodblock floor 木砖地面，木砖地坪
wood-cased well 木套管井
wood-charcoal 木炭
wooden beam 木梁
wooden bracket 木制牛腿
wooden cage cofferdam 木笼围堰
wooden chamber 木椁墓
wooden chock 木楔
wooden concrete form 木制混凝土模板
wooden dam 防水木墙，木坝
wooden dowel 木销钉，木榫钉
wooden fame 木框架
wooden former 木制模板
wooden framed building 木构架建筑物
wooden funeral chamber 木椁室
wooden headframe 木井架
wooden marking gauge 木墨斗
wooden maul 木锤
wooden peg 木栓，木桩
wooden pile 木桩
wooden pin 木钉
wooden pipe 木管
wooden plug 木塞
wooden pole 木杆，木柱
wooden prop 木支柱
wooden rammer 木夯
wooden sheet pile 木板桩
wooden sheet pile cofferdam 木板桩围堰
wooden sleeper adzing machine 木枕削平机
wooden stage 木吊盘
wooden support 木支架
wooden wedge 木楔
woodland 林地
woodland ecosystem 林地生态系统
woodland soil 林地土
wood-preserving oil 木材防腐油
woodstone 硅化木
woodwork construction 细木工程；木结构
word code 文字代码
word code processing 文字处理
word length 字长
Worden gravimeter 渥尔登重力仪
work accident 工伤事故，操作事故
work bag 工具袋
work bench 工作台
work capacity 工作能力；工作量
work capacity evaluation 工作能力评估
work contracting 工程承包
work defect 工程缺陷
work face 工作面
work factor 工作系数；功系数
work function 功函数
work hardened steel 加工硬化钢
work light 工作照明，工作灯
work organization plan 施工组织计划
work overtime 加班
work package 工作包
work plan 工作计划

work progress control 工程进度管理
work report 工作报告
work sheet 工单
work site 工地；地盘
work softening 加工软化
work standardization 工作标准化
work station 工作站
work stone 料石
work suspension 停工
workable bed 可采层
workable concrete 和易性好的混凝土
workable grade 可采品位
workable reserves 可采储量
workable soil 可耕土壤
workable surface mine reserves 露天矿可采储量
workable thickness 可采厚度
workable underground mine reserves 矿井可采储量
workable deposit 可开采矿床
workblank 毛坯
workbook 工作手册
work-done factor 做功系数
worked out area 采空区
working anchor 工作锚
working anchor clamp slice 工作锚夹片
working anchoring plate 工作锚板
working barrel pump 工作筒式泵，工作深井泵
working bench and warehouse 工作台和备件库
working bond stress between anchor and rock 锚索与岩石间黏接工作应力
working capacity 工作量；劳动能力，工作能力
working chamber of caisson 沉箱工作室
working chamber of pneumatic caisson 气压沉箱作业室
working current 工作电流
working data 工作数据
working day 工作日
working design 施工设计
working district boundary 采区边界；工作区边界
working efficiency 工作效率
working environment 工作环境
working expenditure 工作费用；经营费用
working face 工作面，掌子面
working fund 流动资金；周转资金；营运资金
working grounding 工作接地
working hypothesis 工作假说
working instruction 操作说明，操作须知，使用说明书
working language 工作语言
working level of piling rig 打桩机工作面
working lever 操纵杆
working life 有效期，使用期，使用寿命
working load 工作荷载
working load design 工作荷载设计
working load limit 工作负荷极限
working map 施工图
working medium 工作介质
working method 生产方法
working model 工作模型
working pile 工作桩
working pit 生产井
working place 工作场地，施工地点
working plan 工作计划，施工图

working platform 操作平台，工作平台
working pressure 作用压力
working pressure head 工作压力水头
working principle 工作原理
working radius 工作半径
working reference plane 工作基准面
working routine 工作程序
working rule 操作规则
working season 工作季节
working section 工作剖面；采区，采段
working section recovery 采区回采率
working shaft 施工竖井，生产立井
working signal 施工标志信号
working slope face 工作帮面
working specification 操作规程
working standard 工作标准
working substance 工质，工作介质，工作物
working system 工作制度
working temperature 工作温度
working voltage 工作电压
workings 坑道；山地工程
workings subject to dynamic pressure 动压巷道
workings subject to static pressure 静压巷道
workmanship 手艺；工艺；工艺质量
workover 大修；修井
workpiece 工件
workroom 工作室
works acceptance 工程验收
works category 工程类别
workshop 专题讨论会；工作组；车间，工场，工厂
world atlas 世界地图集
World Bank 世界银行
World Count 国际法庭
world cultural heritage 世界文化遗产
world geodetic system 全球大地测量系统
world geopark 世界地质公园
world heritage 世界遗产
world natural heritage 世界自然遗产
world standard 世界标准
world time 世界时
World Trade Organization 世界贸易组织
world water balance 全球水平衡
World Wetland Day 世界湿地日
World Wide Web 全球网络，万维网
worldwide potential gradient 全球电位梯度
world-wide standard seismographic network 世界标准地震台网
worm auger 麻花钻，螺纹钻
worm bearing adjuster 蜗杆轴承调整器
worm conveyor 螺旋输送机
worm drive 蜗杆传动
worm feeder 螺旋给料机
worm gear 蜗轮，蜗轮传动装置
worm gear feed 蜗轮给料输送机
worm gear hob 蜗轮滚刀
worm gear hobbing 蜗轮滚削
worm gear speed reducer 蜗轮减速器
worm geared hoist 蜗轮卷扬机
worm mesh 蜗杆啮合
worm motor 蜗轮马达

worm pipe 蜗形管
worm press 螺旋压力机
worm rack 蜗杆齿条
worm shaft oil seal 蜗杆油封
worm thread 蜗杆螺纹
worm type oil pump 螺杆油泵
worm wheel 蜗轮
wormkalk 竹叶状灰岩
worms 蠕虫动物
worm's-eye map 仰视露头图，上覆层揭露图
worm's-eye view 仰视图
worm-wheel drive 蜗轮传动
worn bit 磨钝的钻头
worn-out parts 损坏件
worst case design 最坏情况设计
worst credible groundwater condition 最不可置信地下水条件
worst error 最大误差
worst-case noise pattern 最坏情况噪声模式
wortex-excited oscillation 涡流激振
wound rotor 线绕转子
wound-rotor motor 绕线转子电机
woven geotextile 土工织物
woven glass tape 玻璃纤维编织带
woven metal wire cloth 金属丝编织布
woven wire mesh 金属丝编织网
wrapped underdrain system 包缠式地下排水系统
wrapping machine 绕带机

wrapping test 卷缠试验
wrecking bar 撬棍
wrecking car 抢险车
wrecking crew 抢险队
wrench 扳手，扳钳
wrench opening 扳手开度
wrench set 成套扳手
wrench socket 扳手套筒
writable control storage 可写控制存储器
write head 写入磁头
write store 写入存储器
writer 写入程序；作者，编写者
writing galvanometer 自记灵敏电流计
wrought iron 锻铁，熟铁
wrought timber 刨光木料
wrought-iron pipe 熟铁管
W-settling tank W形沉淀槽
W-shaped steel belt W型钢带
W-shaped valley W形谷
WTO 世界贸易组织
Wulff grid 乌尔夫网格
Wulff net 乌尔夫网，乌氏网
Wulff projection 乌尔夫投影
Wulff stereographic net 乌尔夫极射赤平投影
wye branch Y形支管
wye junction Y形连接短管
wye level Y形水准仪，回转水准仪

X

X conversion　X转换
X format　X格式
X format code　X格式编码
X radiation　X光辐射，X辐射，伦琴辐射
x-alloy　铜铝轴承合金
xalsonte　粗粒砂；砾石堆积
xanthan gel　黄原凝胶，生物凝胶
xanthan gum　黄原胶，生物胶
xanthan viscosifier　黄原胶增黏剂
xanthate　黄酸盐，黄原酸酯，黄原酸盐，黄药
xanthic acid　黄原酸
xanthocoite　黄银矿
xanthogenate　黄盐酸，黄原酸盐
xanthogenic acid　黄原酸
xanthosiderite　黄针铁矿，针铁矿
X-axis　X-轴，横坐标轴
X-axis amplifier　X轴信号放大器，水平信号放大器
X-band　X波段
X-band coherent radar　X波段相干雷达
X-band image　X波段图像
X-band radar　X波段雷达
X-bearing　X形支座
X-bit　X形钻头
X-bound part　X约束部分
X-brace　交叉支撑，十字支撑
X-braced frictional energy dissipator　X形支撑摩擦耗能器
X-bracing　交叉支撑，剪力撑；交叉联接
X-bridge　X型电桥
X-chisel　X形钻头
X-chisel bit　X-型冲击钻头
X-component　X分量
X-conn　十字接头
X-coordinate　X坐标，横坐标
X-crack　X形裂缝
X-cross member　X形横梁
X-cut　X切割；X形掏槽
X-datum line　X基准线，穿孔卡片基准线
X-direction　X轴方向
X-direction resolution　X轴方向分辨率
xenbucin　联苯丁酸
xenene　联苯
xenobiology　外层空间生物学，宇宙生物学
xenoblast　他形变晶
xenoblastic texture　他形变晶结构
xenoclastolava　捕虏碎屑熔岩
xenocryst　重硅线石，重矽线石
xenogenous　体外性的，异体的
xenolith　捕虏体，捕虏岩；重夕线石
xenolithic enclave　捕虏岩包体
xenology　氙年代学，氙测年法
xenomorphic crystal　他形晶
xenomorphic granular structure　他形粒状构造
xenomorphic granular texture　他形粒状结构
xenomorphic texture　他形结构

xenomorphism　他形，异形现象
xenon arc weatherometer　氙弧老化试验机
xenon build-up after shutdown　停堆后氙积累
xenon concentration　氙浓度
xenon dating　氙测年
xenon isotope method　氙同位素法
xenon lamp　氙灯
xenon oscillation　氙振荡
xenon poisoning　氙中毒
xenon predictor　氙预测计
xenon reactivity　氙中毒反应性
xenon reactivity override　氙反应性补偿
xenon-iodine dating　氙碘法年龄测定
xenon-xenon dating　氙-氙法测年
xenothermal deposit　浅成高温热液矿床
xeralf　干热淋溶土
xerarch　在干旱地发展的；旱生演替
xerasium　旱涝演替
xerert　季节性干旱变性土
xeric　干旱的，旱生的
xerochase　干裂
xerochore　无水沙漠区
xerocline　旱坡，干坡
xerocole　喜旱的；喜旱植物；旱生植物
xerocopy　静电复印本，干印本
xerodrymium　旱生森林群落
xerographic printer　静电复印机
xerography　静电印刷术，干印术
xeroll　干热软土
xeromorph　旱生型；旱生植物
xeromorphic　旱性结构的，旱生形态的
xeromorphic vegetation　旱生植被
xeromorphism　旱生形态，旱性形态
xerophile　旱生植物；喜旱的，适旱的
xerophilous　喜旱的，适旱的
xerophilous plant　喜旱植物
xerophobous　避旱的
xerophytia　旱地群落，旱生植物群落
xerophytic vegetation　旱生植被
xeropoium　干草原，荒芜群落
xeroprinting　静电印刷术，干印术
xerorendsina　旱黑石灰土，干草原土
xerosol　干旱土
xerotherm　干热植物
xerothermic index　干热指数
xerothermic period　干热期；干温期
xerothermic plant　干热植物
xerox　静电复印，干印；静电复印机
xerult　干热老成土
X-frame　X形构架，交叉形构架
X-frame brace　方框支架的对角撑木
X-gear　变位齿轮
X-gripper　X-型支撑
Xi'an Geodetic Coordinate system 1980　1980年西安大地

坐标系
Xinjiang Kucha Grottoes Research Institute　新疆龟兹石窟研究所
Xinjiang Uygur Autonomous Region　新疆维吾尔自治区
X-intercept　X 轴上的截距
XIR extreme infrared　超红外
Xixia Mausoleum　西夏陵
X-mas tree　采油树，采气树；井口装置
X-member　X 形梁，交叉形梁
X-motion　沿 X 轴方向的运动
X-offset　X 轴方向偏移
X-over　转换接头，配合接头
X-parallax　X 视差，左右视差，横视差
X-particle　X 粒子
X-radiation　X 射线辐射
X-radiography　X 射线照相术
X-ray　X 射线，伦琴射线，X 光
X-ray absorption　X 射线吸收
X-ray absorption analysis　X 射线吸收分析
X-ray absorption spectrum　X 射线吸收谱
X-ray activation analysis　X 射线活化分析
X-ray analysis　X 射线分析，X 光分析
X-ray astronomy　X 射线天文学
X-ray background　X 射线本底，X 射线背景
X-ray beam　X 射线束
X-ray camera　X 射线照相机，X 射线摄像机
X-ray computerized tomography　X 射线层析成像法
X-ray core　X 射线岩心
X-ray counting　X 射线计数
X-ray crack detector　X 射线探伤仪；X 射线裂痕探测仪，X 射线裂缝探测仪
X-ray crystallographic analysis　X 射线结晶学分析
X-ray crystallography　X 射线结晶学
X-ray defect detector　X 射线探伤仪
X-ray defectoscopy　X 射线探伤法
X-ray detector　X 射线探测仪
X-ray detectoscope　X 射线探伤仪
X-ray diagnosis　X 射线诊断
X-ray diagram　X 射线图
X-ray diffraction　X 射线衍射
X-ray diffraction analysis　X 射线衍射分析
X-ray diffraction data　X 射线衍射数据
X-ray diffraction method　X 射线衍射法
X-ray diffraction pattern　X 射线衍射图谱
X-ray diffraction spectrum　X 射线衍射谱
X-ray diffractogram　X 射线衍射图
X-ray diffractometer　X 射线衍射仪
X-ray dispersion　X 射线色散
X-ray effect　X 射线效应
X-ray electron spectroscopy　X 射线电子光谱学
X-ray emission　X 射线发射
X-ray emission gauge　X 射线发射计
X-ray emission microanalysis　X 射线发射显微分析
X-ray emission spectroscopy　X 射线发射光谱学
X-ray energy dispersive analyzer　X 射线能谱分析仪
X-ray energy spectroscopy　X 射线能谱学
X-ray examination　X 射线检查，X 光鉴定
X-ray film　X 射线照片，X 光片
X-ray filter　X 射线滤波器
X-ray flaw detection　X 射线探伤
X-ray fluid analysis tool　X 射线流体分析仪
X-ray fluorescence analysis　X 射线荧光分析
X-ray fluorescence instrument　X 射线荧光仪
x-ray fluorescence spectrometer　X 射线荧光光谱仪
X-ray fluorescence spectrometry　X 射线荧光谱法
X-ray fluorescence spectroscopy　X 射线荧光光谱学
X-ray fluorescence technique　X 射线荧光技术
X-ray fluorimeter　X 射线荧光仪
X-ray fluoroscopy　X 射线荧光透视，X 射线荧光检查
X-ray flux　X 射线通量
X-ray generator　X 射线发生器
X-ray goniometer　X 射线定向仪
X-ray identification　X 射线鉴定
X-ray image　X 射线图像
X-ray injury　X 射线伤害
X-ray inspecting　X 射线检验
X-ray inspection　X 射线检验
X-ray intensity meter　X 射线强度计
X-ray investigation　X 射线调查
X-ray machine　X 光机，X 射线机
X-ray metallography　X 射线金相学
X-ray micro-analysis　X 射线显微分析
X-ray micro-analysis system　X 射线显微分析系统
X-ray microradiography　X 射线显微放射照相法
X-ray microscope　X 射线显微镜
X-ray microscopy　X 射线显微术
X-ray monochromator　X 射线单色仪
X-ray optics　X 射线光学
X-ray pattern　X 射线图谱
X-ray phase analysis　X 射线物相分析
X-ray photoelectron spectrometry　X 射线光电子能谱法
X-ray photogram　X 射线照片
X-ray photogrammetry　X 射线摄影测量
X-ray photograph　X 射线照片
X-ray photography　X 射线照相
X-ray photometer　X 射线光度计
X-ray photometry　X 射线光度法
X-ray photon spectroscopy　X 射线光子光谱学
X-ray picture　X 射线图片
X-ray powder diffraction　X 射线粉末衍射
X-ray powder method　X 射线粉末法
X-ray powder pattern　X 射线粉末图谱
X-ray powdered crystal diagram method　X 射线粉晶图解法
X-ray powdered crystal diffraction method　X 射线粉末晶体衍射法
X-ray protection　X 射线防护
X-ray spectrochemical analysis　X 射线光谱化学分析
X-ray spectrometer　X 射线分光计，X 射线谱仪
X-ray spectroscopic analysis　X 射线波谱分析，X 射线看谱分析
X-ray spectrum analysis　X 线光谱分析
X-ray tomography　X 射线层析成像法
X-ray tube　X 射线管
X-rayed　X 射线检查的
X-raying　X 射线照射
XRD X-ray diffraction　X 射线衍射
XRF X-ray fluorescence　X 射线荧光分析
X-ring　X 形油封，双唇油封
X-rotation　绕 X 轴旋转

XRT X-ray technique　X射线技术
X-scale　横向比例尺，X方向比例尺
X-section　横截面，交叉截面
X-shaped metallic energy dissipator　X型金属耗能器
X-spread　十字排列，交叉排列
X-threaded　错扣的
X-type structure　X型构造
X-typed anchorage　X-型锚具
Xumishan Grottoes　须弥山石窟
X-unit　X射线波单位
X-wave　X波，横波，轴向波，异常波
X-Y caliper logging　X-Y轴井径测井
X-Y mount antenna　X-Y座架天线
X-Y plotter　X-Y绘图机
X-Y reader　X-Y坐标阅读器
X-Y recorder　X-Y记录器，纵横坐标记录器
X-Y scope　X-Y示波器
XY switch　XY开关
X-Y synchroscope　X-Y同步示波器
XY/t recorder　XY/t记录仪
xylain　木质结构镜煤，闪镜煤，木煤
xylan　木聚糖，多木糖
xylanthrax　木煤，木炭
xylem　木质部
xylene　二甲苯
xylene chloride　氯代二甲苯
xylene isomerization catalyst　二甲苯异构化催化剂
xylenol　二甲苯酚，二甲酚
xylidin　二甲基苯胺
xylinite　木煤体，木煤
xylinoid　木质的
xylinoid group　木质组

xylite　木质煤，木质褐煤；铁石棉
xylith　镜煤质褐煤
xylitol　木糖醇
xylochlore　鱼眼石
xylocryptite　木晶蜡
xylofusinite　木质丝煤，木质丝炭
xylogen　木质纤维素
xyloid　木质的，类木质的，似木的
xyloid coal　木煤，木质煤，木质褐煤
xyloid coal lignite　木质褐煤
xylokryptit　板晶蜡
xylol　混合二甲苯
xylolite　木花板；菱苦土木屑板
xylometer　木材比重计
xylon　木质，木纤维
xylonite　赛璐珞
xylopal　木蛋白石，木化石
xylophagous　食木的，蚀木的，蛀木的
xylophilous　适木的，喜木的
xylophyta　木本植物
xylose　木糖
xylotelinite　木质结构镜质体，木质结构凝胶体
xylotile　铁石棉，海泡石棉
xylovitrain　木质镜煤，无结构镜煤
xylovitrite　镜煤
xylovitrofusinite　木质镜丝煤
xylulose　木酮糖
xylyl　甲苄基，二甲苯基
xylylene　苯二甲基
Xynetics plotter　赛乃绘图仪
XYZ chromaticity diagram　XYZ色度图
XYZ color coordinate　XYZ彩色坐标

Y

Yager's operator 亚格算子
Yale brass 耶鲁锡黄铜
Yale bronze 耶鲁低锡青铜
Yale lock 耶鲁弹簧锁
Yale University 耶鲁大学
Yalu river 鸭绿江
Yalu Tsangpo river 雅鲁藏布江
Yangtze paraplatform 扬子准地台
Yangtze River 长江
Yanshan movement 燕山运动
yard 英尺，码；工场，厂区，现场；露天堆场，仓库；天井
yard crane 场地起重机，移动吊车
yard hydrant 工业消防栓
yard pipe 一码长管子
yard plan 地面图；地面建筑平面图；货场计划
yard station 土方站
yard store 堆场
yardage 土方；土方码数，土方数
yardage distribution 土方码数分配，土方分配
yardang 雅丹地貌；风蚀土脊；白龙堆
yardstick 尺度；标准；码尺
Yates correction 耶茨校正
yaw 偏航，侧滑；偏航角，侧滑角
yaw angle 船头摆角，偏航角
yaw guy 系塔索
yaw instability 偏航不稳定性
yaw rudder 方向舵
yawer 偏航控制器
yawing 偏航，摇摆；左右偏差
yawing axis 偏航轴
yawing mode of vibration 摇摆振动模式
yawing moment 偏航力矩
yawing oscillation 偏航振荡
yawmeter 偏航计
yawn 裂口，裂开，间隙，缝隙
Y-axis Y轴，纵轴，纵坐标轴
Y-axis modulation image Y轴调制图像
Yazoo drainage pattern 伴支水系，亚祖式水系
Yazoo stream 支流干流平行河流，亚祖式河流
Yazoo tributary 伴支流，亚祖式支流
Y-bend 分叉弯头，三通，Y管
Y-branch Y形支管；Y形接线
Y-channel Y信道
Y-connection 星形连接，Y形接法
Y-coordinate Y坐标，纵坐标
Y-curve 叉形曲线，y形曲线
Y-diagonal network Y-角联网路
Y-direction Y轴方向
Y-direction resolution Y方向分辨率
year 年，年份
year book 年鉴，年报，年刊
year ring 年轮
year under review 审查年度，报告年度

year with low water 枯水年，少水年
year-climate 年气候
year-end adjustment 年终调整
year-end audit 年终审计
year-end bonus 年终奖
year-end integrate plan of opencut 露天矿年末综合平面图
year-end report 年终报表
yearlong stream 常年河流
yearly administration cost 年度管理费
yearly amortization cost 年度折旧费
yearly budget 年度预算
yearly consumption 年度耗量
yearly decline factor 年度递减率
yearly depreciation 年度折旧
yearly displacement 年位移
yearly maintenance 年度维修
yearly maximum 年最大值
yearly minimum 年最小值
yearly off-take rate 年开采速度
yearly output 年产量
yearly plant 一年生植物
yearly production 年产量
yearly production forecast 年产量预测
yearly rainfall 年降雨量
yearly renewable term 每年更新期
yearly runoff 年径流量
yearly sea level 年平均海平面
yearly storage 年蓄水量；年储备量
yearly survey 年度检验，年检
yearly temperature difference 年温差
yearly variation 逐年变化，年变化
yearly variation of runoff 径流年变化
yearly water level 年平均水位
year-round 整年的，全年的
year-round exploratory drilling 全年勘探钻井
year-round surface 全天通车路面
year-to-year variation 逐年变化，年变化
yeast 酵母，发酵粉；泡沫
yellow arsenic 雌黄
yellow brass 黄铜
yellow chrome 钼铅矿
yellow cinnamon soil 黄褐土
yellow clay 黄黏土
yellow copper ore 黄铜矿
yellow deal 黄松板
yellow dirt 金
yellow earth 黄土，黄色黏土，黄壤，赭黄土
yellow forest soil 黄色森林土
yellow ground 黄地
yellow laterite 黄色砖红壤
yellow lead primer 黄底漆
yellow metal 黄色金属，黄铜，黄金
yellow mud 黄泥

yellow ore 黄矿
yellow podzolic soil 灰化黄壤，黄色灰化土
yellow pyrite 黄铜矿
Yellow River 黄河
yellow sand 黄沙
Yellow Sea 黄海
Yellow Sea Datum 黄海高程基准面
Yellow Sea mean sea level 黄海平均海面
yellow soil 黄色土壤，黄壤
yellow wax 黄石蜡
yellow-brown forest soil 黄棕色森林土，黄棕壤
yellowcake 黄饼
yellowish brown lateritic soil 黄棕色砖红壤
yellowness index 黄度指数
yermosol 漠境土
yes-no classification 是-否分类
yes-no data 是-否数据
yes-no radiation monitor 有-无辐射监测器
yield 产生；产量，流量；收获率，回收率；屈服，让压，弯曲；沉陷
yield acceleration in cohesionless soil 无黏性土屈服加速度
yield condition 屈服条件
yield criterion 屈服准则，屈服判据
yield curve 产量曲线
yield deformation 屈服变形
yield failure 屈服破坏
yield function 屈服函数
yield limit 屈服极限
yield line 屈服线；塑性变形线
yield load 屈服荷载
yield load density 屈服荷载密度
yield locus 屈服轨迹；破坏包络线
yield loss 产量损失
yield mechanism 屈服机制
yield modulus 屈服模量
yield moment 屈服矩，屈服力矩，屈服弯矩
yield of bore well 井孔出水量
yield of catchment 流域流量
yield of drainage basin 泄水区流量
yield of foundation 基础沉陷
yield of ground water 地下水出水量
yield of radiation 辐射量
yield of single well 单井出水量
yield of spring 泉流量，泉涌水量
yield of water 给水度，出水量
yield of well point 井点出水量
yield optimization 产量最佳化
yield output 产出量
yield point 屈服点
yield point elongation 屈服点伸长
yield point strain 屈服点应变
yield point stress 屈服点应力
yield point value of displacement 屈服点位移值
yield power 生产力
yield pressure 屈服压力
yield range 屈服范围
yield rate 收益率
yield ratio 屈服比
yield ratio response spectrum 屈服比响应谱
yield ring support 环形让压支架，环形可伸缩支架
yield soil 软土，流动土
yield spreading 屈服扩散
yield stage 屈服阶段
yield stiffness ratio 屈服刚度比
yield strain 屈服应变
yield strength 屈服强度，抗屈强度，软化强度
yield strength ratio 屈服强度比
yield stress 屈服应力
yield stress model 屈服应力模型
yield surface 屈服面
yield test 屈服试验
yield timbering 可伸缩支架，让压支架
yield time diagram 塑性变形时间图
yield value 屈服值，塑变值；产值
yield zone 屈服区，变形区
yieldability 可屈服性；沉陷性
yieldable amount 可屈服量
yieldable arch 让压拱
yieldable arch assembly 让压拱形支架
yieldable bolt 可伸缩锚杆
yieldable ring set 让压圆形支架
yieldable support 可伸缩性架，让压支架
yield-ash curve 产量-灰分曲线
yielding 屈服；产生，形成；沉陷；屈服的，让压的，可弯曲的，易变形的
yielding arch 可伸缩拱，让压拱
yielding capacity 生产量；生产能力
yielding clamp plate 让压夹板
yielding curvature 屈服曲率
yielding displacement 屈服位移
yielding flow 塑性流，塑流
yielding ground 松软土地，易沉陷土地
yielding hydraulic jack 可伸缩液压千斤顶
yielding leg 可伸缩支腿
yielding limit 屈服极限
yielding lining 可伸缩衬砌，让压衬砌
yielding material 可伸缩材料，让压材料
yielding mechanism 屈服机制
yielding moment 屈服矩
yielding of foundation 基础沉陷
yielding of reinforcing 钢筋屈服
yielding of support 支座下沉
yielding pattern 屈服型式
yielding pillar 让压矿柱
yielding plate 让压板
yielding point 屈服点，屈服值
yielding pressure 屈服压力
yielding prop 可伸缩支柱，让压支柱，恒阻支柱
yielding property 屈服性，让压性
yielding ring support 让压环形支架
yielding roadway arch 让压巷道拱
yielding rubber 减振橡皮
yielding soil 流动土，软土
yielding speed 屈服速度
yielding steel prop 让压钢柱
yielding strength 屈服强度
yielding structure 屈服结构
yielding support 让压支架
yield-part of support 支架让压部分

Y-intercept　Y 截距
Y-intersection　Y 形交叉
Y-joint　叉形接头，Y 形接头
Y-junction　Y 形接头
Y-level　Y 型水准仪，回转水准仪，活镜水准仪
Y-line　Y 轴线，纵轴线
Y-motion　沿 y 轴方向运动
YMP　尤卡山项目
yoke　套箍，轭，座，支架；偏转线圈；偏转系统；磁头组
yoke arm　轭臂
yoke connection　轭连接
yoke method　极间法
yoke plate　轭板
yoke support　轭架
yoked basin　共轭盆地；配合盆地
yoked lake　共轭湖
yolk coal　软煤，沥青煤
young alluvium　新冲积层
young clay　新黏土
young coast　幼年海岸
young deposit　新沉积
young fault scarp　年青断层崖
young fold belt　年轻褶皱带，新褶皱带
young ice　初期冰
young industrial country　新兴工业国家
young karst　幼年岩溶，幼年喀斯特
young lake　幼年湖
young loess　新黄土
young marine soil　幼年海成土
young mountain　幼年山脉
young peat　幼年泥炭
young plain　幼年平原
young red earth　幼红壤
young river　幼年河
young topography　幼年地形
young valley　幼年期谷
young volcano　幼年期火山
young yellow earth　幼黄壤
younger oil　初期石油
younger sand body　幼年砂岩体
Young's dynamic modulus　弹性动力模量，杨氏动力模量
Young's equation　杨氏方程
Young's modulus　杨氏模量，弹性模量
youthful mountain range　年青期山脉，新山脉
youthful relief　年青期地形起伏
youthful topography　幼年期地形
yo-yo　上下移动
yo-yo driller　顿钻司钻
yo-yo drilling　钢丝绳冲击钻进，顿钻钻进
yo-yo effect　振动效应
yo-yo fashion　上下移动方式
yo-yo technique　电缆收放技术
yo-yo winch　正反转绞车
Y-pipe　Y 形管，三通管，叉形管，斜角支管
Y-plane　Y 形面
Y-shaped　Y 形的，叉形的
Y-shaped fracture system　Y 型破裂体系
Y-shaped valley　Y 形谷
Y-strainer　Y 型管道过滤器；Y 型拉紧器
Y-support　Y 型支座
Y-track　分叉线
ytterbium　镱
yttrialite　钍钇矿，硅钍钇矿，硅酸钍钇矿
yttrium　钇
yttrogummite　钇铅铀矿
Y-tube　Y 型管，裤衩管
Y-type anchorage　Y 形锚具
Y-type cylinder filter　Y 型过滤筒
Y-type pipe　Y 形管，叉管，三叉管
Yuan　元，人民币单位
Yucca Mountain Project　尤卡山项目
Yulin Grottoes　榆林石窟
Yungang Grottoes　云冈石窟
Y-valley　Y 形谷
Y-Y connection　Y-Y 形接线，双星形接法，Y-Y 接法

Z

zahn viscosimeter 锥板黏度计
zahn viscosity 锥板黏度
zala 硼砂
Z-angle iron Z形角铁
zarnec 雄黄，雌黄，硫砷矿
z-axis Z轴，垂直轴
Z-bar Z形钢
Z-beam Z形梁
Z-beam torsion balance Z形扭秤
Z-bit Z形钻头
Z-coordinate Z坐标
Z-crank Z形曲柄
Z-crossplot Z值交会图
Z-direction Z轴方向，沿Z坐标
zebra crossing 斑马纹人行横道，斑马线
zebra dolomite 条带白云岩
zebra layering 条带状层理
zebra limestone 条带灰岩
zebra line 斑马线
zebra time 格林尼治平均时，世界时
zebraic 条带状的
zed Z形钢；Z形铁
zee-beam Z形梁
Zeeman effect 塞曼效应
Zeeman modulated experiment 塞曼调制实验
Zeeman splitting 塞曼分裂
zee-transform Z变换，离散变换
zee-type piling Z形板桩
Zeigler catalyst 齐格勒催化剂
Zeiss universal theodolite 蔡司万能经纬仪
zellerite 碳钙铀矿，菱钙铀矿
Zengmu Shoal 曾母暗沙
zenith 天顶；顶点，极点
zenith angle 天顶角
zenith blind area 天顶盲区
zenith camera 天顶摄影仪
zenith column 大气气柱
zenith distance 天顶距
zenith instrument 天顶仪
zenith point 天顶点
zenith telescope 天顶望远镜
zenithal chart 方位投影海图
zenithal leveling instrument 天顶水准测量仪
zenithal projection 天顶投影，方位投影
zenolite 矽线石；硅线石
Zeo-Dur 硅酸盐阳离子交换剂
zeolite 沸石；分子筛
zeolite catalyst 沸石催化剂
zeolite softener 沸石软水剂
zeolite softening 沸石软化，离子交换软化
zeolitization 沸石化
zerk 加油嘴，注油嘴
zero 零
zero access storage 立即存取存储器

zero adjustment 对准零点，归零
zero air void 零气隙
zero air void density 零气隙密度
zero air void ratio 零气隙比，饱和气隙比
zero air void unit weight 零气隙容重，饱和孔隙容重
zero alignment 零位校正，对准零位
zero allowance 无容差，无留量
zero altitude 零海拔，零高程
zero amplitude 零振幅
zero axial stress cross-section 零轴向应力截面
zero axial stress point 零轴向应力点
zero balance 零点平衡；余额为零
zero Bessel function 零阶贝塞尔函数
zero capillary-pressure plane 零毛细管压力面，自由水面
zero carrier 零振幅载波
zero charge 零电荷；哑药包
zero check 零点核对
zero circuit 零电路
zero clearance 零间隙，无间隙
zero clearance rolled joint 零间距滚轧接合
zero clearing 清零
zero complement 零补码
zero compression 零压缩
zero condition 零条件，零状态
zero contact angle 零接触角
zero contour 零等高线，起点等高线
zero correction 零位校正
zero correlation 零相关
zero creep 零蠕变
zero crossing 零交叉
zero damping 零阻尼
zero date 起算日期
zero datum 零基准面
zero decrement 零衰减
zero defect management 零缺陷管理
zero deflection 零偏斜
zero degree 零度；零次
zero depth 零深度，地表
zero difference 零差异，无差异
zero dilatancy 零膨胀性
zero dimension 零维，无量纲，无因次
zero dimensional 零维的，无因次的
zero discharge 零排放；零出料
zero discharge rig 零排放钻机
zero distortion 零畸变
zero divisor 零因子
zero drawdown 零降深
zero drift 零点飘移，零点偏移
zero droop governor 无差调速器
zero elevation 零高程
zero elimination 消零
zero error 零误差
zero fill and cut 零填挖，不填不挖
zero flow 零流量

zero flow velocity	零流速
zero fluidity	零流动性
zero flux plane	零通量面
zero flux plane method	零通量面法
zero free water	零游离水，零析水
zero free water slurry	零游离水泥浆
zero frequency	零频率
zero gain relay	零增益继电器
zero gas saturation	零气饱和度
zero graduation	零刻度
zero heel stress	坝踵零应力
zero impedance filter	零阻抗滤波器
zero indicator	零点指示器
zero inflow curve	零进水曲线
zero initial data	零初始数据
zero isochron	零等时线
zero isopach	零等厚线
zero isotherm	零度等温线
zero lateral strain test	零侧向应变试验
zero level	零水位；零标高；零级
zero line	零位线，基准线，零线
zero load	零载，空载，无载
zero load method	零载法
zero load test	空载试验，无载试验
zero lot line	地界零线
zero mark	零点标志，零位记号；零位刻度，零点标准，零度
zero mass	零质量
zero matrix	零矩阵
zero mean	零均值
zero meridian	本初子午线，零子午线
zero method	零点法，零位法
zero moisture index	零湿度指数
zero momentum	零动量
zero of elevation	高程零点
zero of gauge	水尺零点
zero oil production	零产油量
zero overbalance	零值超平衡
zero oxygen balance	零氧平衡
zero padding	补零
zero period acceleration	零周期加速度
zero phase	零相位
zero point	零点，度值；坐标起始点，坐标原点
zero porosity	零孔隙度
zero position	零位，起始位置
zero power factor	零功率因子
zero power reactor	零动力反应堆
zero power test	零功率试验
zero probability of failure	零概率破坏
zero rate of penetration	零贯入度
zero reading	起点读数，零点读数
zero reading method	零点读数法
zero reference	零基准
zero release	零排放
zero sequence	零序
zero setting	调零，调至零点，对准零位
zero shift	零点漂移，零点迁移
zero signal input	零信号输入
zero slip point	零滑动点
zero slope	零斜率
zero slump concrete	零坍落度混凝土
zero standard	零标准
zero stiffness	零刚度
zero strain state	零应变状态
zero strain velocity	零应变速度
zero surface	零位面，零面
zero temperature	零温度
zero thrust pitch	无推力螺距
zero time	零时间，零时，时间零点；起始瞬间
zero to cut-off	零截止
zero valence	零价
zero vector	零矢量，零向量
zero vertical pressure drop	零垂向压力降
zero water level	水位零点，零水位
zero water pollution	零水质污染
zero water saturation line	零水饱和度线
zero zone	零区；零时区
zero-address	零地址
zero-axis component	零轴分量
zero-base budget	零基预算，零基点预算
zero-bias	零偏压
zero-capillary pressure	零毛细管压力
zero-dimensional model	零维模型
zero-energy coast	零能海岸
zero-energy element mode	零能单元模式
zero-energy mode	零能模式
zero-frequency vibration	零频振动
zero-gravity	零重力，失重
zero-growth	零增长
zero-head range	零水头区
zero-hour mixing	零间歇拌合
zeroing	调零，调至零点，对准零位
zeroize	补零，填零，充零，零位调整
zero-lag	零延迟
zero-lag correlation	零延迟相关，零滞后相关
zero-lag filter	零滞后滤波器
zero-length spring gravimeter	零长弹簧重力仪
zero-magnetic state	零磁状态
zero-number detonator	即发雷管
zero-order approximation	零阶近似
zero-order geodetic network	零级大地控制网
zero-order Hankel transform	零阶汉凯尔变换
zero-order traversing	零级导线测量
zero-output position	零输出位置
zero-phase filter	零相位滤波器
zero-phase filtering	零相位滤波
zero-phase-shift amplifier	零相移放大器
zero-point adjustment	对零，调零位，零点调整
zero-point drift	零点漂移
zero-sequence current	零序电流
zero-temperature level	零温度层
zeroth	第零的
zeroth order	零次；零阶
zeroth period	第零周期
zero-valent compound	零价化合物
zero-velocity	零速度
zeta effect	ζ 效应
zeta function	ζ 函数
zeta meter	ζ 仪
zeta-type structure	ζ 字形构造

zeunerite	铜砷铀云母，翠砷铜铀矿
Zhangheng seismoscope	张衡地动仪
zianite	蓝晶石
Ziegler picker	齐格勒型拣矸机
Ziegler-Natta catalyst	齐格勒-纳塔催化剂
zietrisikite	高温地蜡
zigger	渗漏，滴出，渗出
zigzag	曲折的，锯齿状的，Z字形的，之字形的
zigzag arch	曲折拱
zigzag bridge	曲桥
zigzag connection	锯齿形连接，Z形连接
zigzag crack	锯齿裂缝，不规则裂缝
zigzag cross bedding	锯齿状交错层理
zigzag drainage	曲折排水系统
zigzag fault	锯齿状断层，之字形断层
zigzag fold	锯齿状褶皱，曲折褶皱
zigzag girder	三角孔梁，曲折梁
zigzag irrigation	迂回灌溉
zigzag line	曲折线
zigzag raise	之字形天井，之字形上山
zigzag ramp	之字形坡道，曲折坡道
zigzag riveting	交错铆接
zigzag roadway	曲折巷道
zigzag rule	折尺，曲尺
zigzag sill	曲折围槛，曲折底坎
zigzag superheater	蛇形管过热器
zigzag trajectory	Z形轨迹，Z形轨道，Z形弹道；往返重复测线
zigzag watershed	锯齿状分水岭
zigzag wave	锯齿形波
zigzag weld	交错焊
zigzag-type pit bottom	折返型井底车场
zigzag-type shear wall	交错型剪力墙
zinc	锌，自然锌
zinc alloy	锌合金
zinc amalgam	锌汞合金，锌汞齐
zinc anode	锌阳极
zinc base alloy	锌基合金
zinc blend	闪锌矿
zinc chromate primer	锌铬黄底漆，铬底丹
zinc coating	镀锌；镀锌层
zinc coating wastes	镀锌废料
zinc concentrate	锌精矿，精炼锌
zinc deposit	锌矿床
zinc desilverization	加锌提银法
zinc distillation	锌蒸馏
zinc electrolyte	锌电解液
zinc flower	水锌矿，锌华
zinc green	锌绿
zinc ore deposit	锌矿床
zinc oxide	氧化锌，锌白
zinc oxide-zirconium oxide catalyst	氧化锌-氧化锆催化剂
zinc plate	锌板，镀锌板
zinc plating	镀锌
zinc powder	锌粉
zinc protector	防蚀锌板
zinc rich paint	富锌油漆
zinc solder	锌焊料
zinc storage battery	锌蓄电池
zinc white	锌白，氧化锌
zinc yellow	锌黄，锌铬黄
zinced iron	镀锌铁
zincosite	锌矾
zinc-oxide arrester	氧化锌避雷器
zinc-silicate glass	硅酸锌玻璃
zinnober	辰砂，朱砂，一硫化汞
zip-fastener	拉链；闪光环
zippeite	水铀矾，铀华
zipper	拉链
zipper conveyor	拉链式输送机
zircaloy	锆合金，锆锡合金
zircon	锆石，锆英石
zircon cement	耐火水泥
zircon dating	锆石测年
zircon fission track	锆石裂变径迹
zircon fission-track dating	锆石裂变径迹测年
zircon geochronology	锆石地质年代学
zircon U-Pb dating	锆石铀-铅测年
zirconia	氧化锆
zirconite	锆石，锆英石
zirconium	锆
zirconium alloy	锆合金
zirconium base alloy	锆基合金
zirconium bronze	锆青铜
zirconium ferrosilicon	锆硅铁
zirconium mineral	锆类矿物
zirconium ore	锆矿
zirconium oxide	氧化锆
Z-iron	Z形铁，Z形钢
z-motion	沿Z轴方向的运动
Zn isotope	锌同位素
Z-number	原子序数
zodiacal dust	宇宙尘
Zoelly turbine	多级压力式汽轮机；佐利汽轮机
Zoeppritz's equation	佐伊普里兹方程
zoic age	生物时代
zoisite	黝帘石
Zollner suspension	措尔纳悬挂法
zonal	带状的，分带的，分区的
zonal alteration	带状蚀变
zonal anisotropy	带状各向异性
zonal circulation	纬向环流
zonal combustion	区域燃烧，带状燃烧
zonal curve	带状曲线；晶带曲线
zonal embankment	分区填筑；带状堤坝
zonal extinction	带状消光
zonal flow	带状流
zonal fossil	分带化石
zonal fracture water	带状裂隙水
zonal growth	带状生长，环带生长
zonal guide fossil	分带标准化石
zonal injection	分层注水
zonal isolation	分区隔离，层位封隔
zonal mineral	分带矿物
zonal power	分区功率
zonal profile	分带剖面；综合剖面
zonal sampling	分带取样，区域抽样
zonal soil	区域土，区域性土，地带性土壤
zonal testing	分层测试
zonal texture	带状组织，带状结构

zonal theory	分带成矿说，矿床带状分布学说
zonal wind	纬向风
zonal withdrawal	分层开采
zonality	地带性，分带性，地区性
zonality coefficient	分带系数
zonality index	分带指数
zonality sequence	分带序列
zonary structure	带状构造，环带结构
zonation	分带，分区，区划
zonation criterion	区划准则
zonation map	区划图
zonation test	分层测试
zonational	分带的，分区的，成带的
zone	带，地带，地区
zone calibration factor	区域校准因子
zone coefficient	区系数，带系数
zone control	区控制，带控制
zone fossil	带化石，指带化石，分带化石
zone label	层段标记
zone meridian	时区子午线
zone method	区域法，段法
zone of ablation	消融带，消融区
zone of accumulation	堆积带，淀积带
zone of aeration	包气带，充气带
zone of aerodynamic shadow	空气动力阴影区
zone of affected overburden	地表采动影响带
zone of anchorage	锚固区
zone of attraction	收敛域
zone of bank storage	河岸调蓄带
zone of capillarity	毛细管作用带
zone of capillary saturation	毛细管水饱和带
zone of capillary water	毛细管水带
zone of cave	崩落区
zone of cementation	胶结带
zone of compression	压缩带
zone of constant temperature	恒温带，等温带
zone of contact	接触带
zone of continuous porosity	连续孔隙带
zone of convergence	交汇带
zone of crush	破碎带
zone of deformation	变形带
zone of deformation around pile	桩周变形区
zone of differential settlement	不均匀沉陷区
zone of diffuse scattering	漫散射带
zone of diffusion	扩散带
zone of discharge	排放带，泄水带
zone of discontinuous soil moisture	不连续土壤水分带
zone of eluviation	淋滤带
zone of enrichment	富集带
zone of erosion	侵蚀带
zone of excavation influence	开挖影响区
zone of expansion	膨胀带
zone of fault	断层带
zone of flow	流动带，塑流带
zone of fold	褶皱带
zone of fracture	破碎带，破裂带
zone of freezing	冰冻带
zone of full subsidence	完全陷落区
zone of gradual subsidence	缓慢下沉带
zone of heat liberation	热释放带
zone of heating	加热层；加热区
zone of illuviation	淋积带
zone of incomplete stowing density	不完全回填密度带
zone of intense fracturing	强断裂带
zone of intensive water circulation	强水循环带
zone of intermittent saturation	间歇饱和带
zone of invariable temperature	恒温带，常温带
zone of investigation	勘察区，调查区
zone of leaching	淋滤带，溶滤区
zone of least resistance	最小阻力区
zone of loosening	松动带
zone of loss	漏失带，漏水带
zone of mild pollution	轻污染带
zone of mining	采矿区
zone of mobility	活动带
zone of negative pressure	负压区
zone of normal distribution	正态分布带
zone of occasional distribution	偶然分布带
zone of oxidation	氧化带
zone of percolation	渗滤带
zone of plastic flow	塑性流动区
zone of plasticity	塑性区
zone of positive pressure	正压区
zone of potential bursting	潜在岩爆带
zone of primary deposit	原生沉积带
zone of primary ore	原生矿带
zone of rainfall	降雨带
zone of recompression	再压缩带
zone of reduction	还原带
zone of regional gravity gradient	区域重力梯度带
zone of regular subsidence	均匀沉陷带，均匀下沉带
zone of reverse flow	逆流区，倒流区
zone of rock flowage	岩石塑流带；岩流带
zone of rock fracture	岩石破裂带
zone of saturation	饱和带，饱水带
zone of seasonal change	季节性变化带
zone of secondary enrichment	次生富集带
zone of silence	静默区，静默带
zone of slip	滑移区
zone of soil water	土壤水带
zone of subduction	俯冲带，消减带；下沉区，沉降带
zone of suspended water	悬浮水带
zone of swelling	膨胀区
zone of tectonic disturbance	构造扰动带，构造变动带
zone of trade wind	信风带
zone of transition	过渡带
zone of uplift	上升地带，隆起带
zone of upwelling	上涌带
zone of vadose water	渗流水带
zone of variable phreatic level	潜水位变化带
zone of vegetation	植被带
zone of water-conducting fissures	导水裂隙带
zone of weakness	软弱带
zone of weathered coal seam	风化煤层带
zone of weathering	风化带
zone phenomenon	带状现象
zone position indicator	区位显示器
zone pumping	局部扬水，分带抽水
zone sampler	区层取样器
zone segregation	区域偏析，熔区偏析；层间隔离

zone settling	区域沉降，分区沉降，成层沉淀
zone time	地方时，区时，带时
zone ventilation system	分区通风系统
zone-area method	分区面积法
zone-breaking species	越带种
zone-by-zone analysis	逐层分析
zone-by-zone production testing	逐层生产测试
zoned	分区的，分带的
zoned air	分区送风
zoned array	分区阵
zoned crystal	带状晶体
zoned dam	分区坝
zoned deposit	带状矿床
zoned earth and rockfill embankment	分区填筑土石堤
zoned earth dam	分区式土坝，非均质土坝
zoned embankment dam	分区填筑坝
zoned parameter	分带参数
zoning	分带，分区，区划
zoning coefficient	分区系数，区域系数
zoning district	区划区
zoning index	分带指数
zoning map	分区图，分带图，区划图
zoning of anomaly	异常分带
zoning of ore deposits	矿床分带
zoning of primary halo	原生晕分带
zoning order	分带等级
zoning plan	区划图
zoning scheme	区划方案
zonite	燧石，花碧玉
zonular space	小区间隔，小带间隙
zonule	小带；小区域，小城区
zoobiocenose	动物群落
zoogenic rock	动物岩
zoogenous rock	动物岩
zoogeographical region	动物地理区
zoolith	动物岩，动物化石
zoom	缩放；图形变比；变频距；变焦距；图像电子放大，缩放
zoom in	放大；移向
zoom lens	变焦镜头，变焦透镜
zoom macroscope	变焦显微镜
zoom out	缩小；移离
zoom stereoscope	变焦立体镜
zoomar	可变焦距透镜系统，镜头
zoomer	可变焦距透镜系统，镜头
zoonichnia	动物迹类
zoophyte	植物形动物，植虫
zoosphere	动物圈
Zoppritz-Turner travel time table	佐普利兹-特纳走时表
Z-pile	Z形桩
Z-section	Z形剖面，Z形截面
Z-shaped fracture system	Z形断裂体系
Z-steel	Z字钢
Z-transform	Z变换
z-truss	Z形桁架
Z-type groove	Z形坡口，Z形沟槽
Z-type pile	Z形桩
Zublin bit	苏柏林钻头，独牙轮钻头
Zublin differential bit	苏柏林差异钻头，独牙轮差异钻头
Zublin drill guide	苏柏林钻具导向器，独牙轮钻具导向器
Z-variometer	垂直磁力变感器，垂直磁变记录仪
zwitterion	两性离子
zwitterionic compound	两性离子化合物
zyglo	荧光探伤器；荧光透视法
zymolysis	发酵，酶解

附录1 岩石力学领域著名学报（期刊）索引

No.	Title	SJR	Country
1	Earthquake Engineering and Structural Dynamics	2.921	United Kingdom
2	Geotechnique	2.837	United Kingdom
3	Journal of Geotechnical and Geoenvironmental Engineering-ASCE	2.344	United States
4	ShiyouKantan Yu Kaifa/Petroleum Exploration and Development	2.236	China
5	International Journal of Rock Mechanics and Minings Sciences	2.134	United Kingdom
6	Canadian Geotechnical Journal	2.093	Canada
7	Computers and Geotechnics	2.033	United Kingdom
8	Tunnelling and Underground Space Technology	2.023	United Kingdom
9	International Journal of Physical Modelling in Geotechnics	1.964	United Kingdom
10	Rock Mechanics and Rock Engineering	1.939	Austria
11	ActaGeotechnica	1.818	Germany
12	Engineering Geology	1.81	Netherlands
13	International Journal for Numerical and Analytical Methods in Geomechanics	1.676	United Kingdom
14	Soils and Foundations	1.623	Japan
15	Landslides	1.612	Germany
16	Soil Dynamics and Earthquake Engineering	1.516	United Kingdom
17	Earthquake Spectra	1.46	United States
18	Bulletin of Earthquake Engineering	1.395	Netherlands
19	Georisk	1.299	United Kingdom
20	Geotechnical Testing Journal	1.289	United States
21	Journal of Earthquake Engineering	1.272	United Kingdom
22	Geothermics	1.272	United Kingdom
23	Journal of GeoEngineering	1.205	Taiwan
24	IEEE Geoscience and Remote Sensing Letters	1.203	United States
25	Minerals Engineering	1.164	United Kingdom
26	Yanshilixue Yu GongchengXuebao/Chinese Journal of Rock Mechanics and Engineering	1.073	China
27	International Journal of Mining Science and Technology	1.057	Netherlands
28	Cold Regions, Science and Technology	1.001	Netherlands
29	SPE Journal	1	United States
30	Journal of Rock Mechanics and Geotechnical Engineering	0.992	China
31	Petroleum Exploration and Development	0.972	Netherlands
32	Minerals and Metallurgical Processing	0.954	United States
33	Journal of Petroleum Science and Engineering	0.891	Netherlands
34	International Journal of Mineral Processing	0.852	Netherlands

续表

No.	Title	SJR	Country
35	YantuLixue/Rock and Soil Mechanics	0.843	China
36	ZhongguoKuangyeDaxueXuebao/Journal of China University of Mining & Technology	0.817	China
37	YantuGongchengXuebao/Chinese Journal of Geotechnical Engineering	0.781	China
38	MeitanXuebao/Journal of the China Coal Society	0.748	China
39	TianranqiGongye/Natural Gas Industry	0.737	China
40	Proceedings of the Institution of Civil Engineers: Ground Improvement	0.732	United Kingdom
41	Structure and Infrastructure Engineering	0.699	United Kingdom
42	Proceedings of the Institution of Civil Engineers: Geotechnical Engineering	0.697	United Kingdom
43	Oil Shale	0.675	Estonia
44	Mineral Processing and Extractive Metallurgy Review	0.632	United Kingdom
45	Xinan Shiyou Xueyuan Xuebao/Journal of Southwestern Petroleum Institute	0.605	China
46	International Journal of Disaster Risk Reduction	0.599	United Kingdom
47	Zhongguo Shiyou Daxue Xuebao(Ziran Kexue Ban)/Journal of China University of Petroleum(Edition of Natural Science)	0.598	China
48	Minerals	0.58	Switzerland
49	Geomechanics and Geoengineering	0.558	United Kingdom
50	Bulletin of Engineering Geology and the Environment	0.525	Germany
51	International Journal of Coal Preparation and Utilization	0.522	United Kingdom
52	Geotechnical and Geological Engineering	0.51	Netherlands
53	Geothermal Energy	0.499	United States
54	Quarterly Journal of Engineering Geology and Hydrogeology	0.493	United Kingdom
55	International Journal of Geotechnical Engineering	0.485	United Kingdom
56	Earthquake Engineering and Engineering Vibration	0.475	China
57	Geotechnik	0.442	Germany
58	Environmental and Engineering Geoscience	0.43	United States
59	Shiyou DiqiuWuli Kantan/Oil Geophysical Prospecting	0.419	China
60	Shock and Vibration	0.406	Netherlands
61	Marine Georesources and Geotechnology	0.403	United Kingdom
62	Petroleum Science	0.401	China
63	Journal of Earthquake and Tsunami	0.4	Singapore
64	Acta Geodynamica et Geomaterialia	0.391	Czech Republic
65	Journal of Environmental and Engineering Geophysics	0.386	United States
66	Mine Water and the Environment	0.382	Germany
67	International Journal of Mining, Reclamation and Environment	0.379	United Kingdom
68	Jilin DaxueXuebao (DiqiuKexue Ban)/Journal of Jilin University (Earth Science Edition)	0.375	China
69	Geomechanics and Engineering	0.368	South Korea
70	Eurasian Mining	0.363	Russian Federation
71	Bulletin of the New Zealand Society for Earthquake Engineering	0.362	New Zealand

No.	Title	SJR	Country
72	Earthquake Science	0.36	China
73	PeriodicaPolytechnica:Civil Engineering	0.351	Hungary
74	Indian Geotechnical Journal	0.347	India
75	Acta Montanistica Slovaca	0.344	Slovakia
76	IngegneriaSismica	0.331	Italy
77	Bulletin of Mineralogy Petrology and Geochemistry	0.32	China
78	Journal of Mining Science	0.317	United States
79	Australian Geomechanics Journal	0.317	Australia
80	Madencilik	0.316	Turkey
81	Archives of Mining Sciences	0.31	Poland
82	Geomechanik und Tunnelbau	0.303	Germany
83	Journal of the Southern African Institute of Mining and Metallurgy	0.3	South Africa
84	Journal of Petroleum Exploration and Production Technology	0.298	Germany
85	Polish Geological Institute Special Papers	0.297	Poland
86	Petroleum Science and Technology	0.282	United States
87	Transportation Geotechnics	0.281	United Kingdom
88	Zhongnan Daxue Xuebao(ZiranKexue Ban)/Journal of Central South University (Science and Technology)	0.267	China
89	World Information on Earthquake Engineering	0.266	China
90	Journal of Cold Regions Engineering-ASCE	0.255	United States
91	Journal of Hazardous, Toxic, and Radioactive Waste	0.252	United States
92	International Journal of GEOMATE	0.243	Japan
93	Transactions of the Institutions of Mining and Metallurgy, Section B: Applied Earth Science	0.242	United Kingdom
94	Geotechnical Engineering	0.24	Thailand
95	Italian Journal of Engineering Geology and Environment	0.238	Italy
96	Geotechnical Fabrics Report	0.235	United States
97	Acta Geotechnica Slovenica	0.235	Slovenia
98	Exploration and Mining Geology	0.223	Canada
99	Oilfield Review	0.214	United Kingdom
100	Electronic Journal of Geotechnical Engineering	0.205	United States
101	Iranian Journal of Science and Technology-Transactions of Civil Engineering	0.204	Iran
102	World Journal of Engineering	0.19	United Kingdom
103	International Journal of Mining and Mineral Engineering	0.189	Switzerland
104	Journal of Coal Science and Engineering	0.188	China
105	Earthquake	0.179	China
106	Soil Mechanics and Foundation Engineering	0.165	United States
107	International Journal of Geotechnical Earthquake Engineering	0.164	United States
108	RivistaItaliana di Geotecnica	0.16	Italy

No.	Title	SJR	Country
109	Bulletin of the Mineral Research and Exploration	0.159	Turkey
110	Mining Engineering	0.152	United States
111	Geotechnical News	0.13	Canada
112	Scientific Review Engineering and Environmental Sciences	0.128	Poland
113	Journal of Mines, Metals and Fuels	0.122	India
114	Soils and Rocks	0.118	Brazil
115	AusIMM Bulletin	0.113	Australia
116	Structural Engineering/Earthquake Engineering	0.112	Japan
117	World of Mining-Surface and Underground	0.111	Germany
118	Engineering and Mining Journal	0.109	United States
119	Dams and Reservoirs	0.106	United Kingdom
120	Civil Engineering	0.104	United States
121	Geodrilling International	0.102	United Kingdom
122	International Water Power and Dam Construction	0.102	United Kingdom
123	Mining Report	0.102	United Kingdom
124	Canadian Mining Journal	0.101	Canada
125	Mining Magazine	0.1	United Kingdom
126	Australian Mining	0.1	United States
127	Bulletin of the International Institute of Seismology and Earthquake Engineering	0.1	Japan
128	Coal International	0.1	United Kingdom

根据 SJR 2015 年度数据（2016.6.31 发布）整理

SJR，是 Scimago Journal Rank 的缩写，是基于 Scopus 数据库的期刊评价新指数。

附录 2　ISRM 建议方法索引

1 SITE CHARACTERIZATION
　　SM for Quantitative Description of Discontinuities in Rock Masses-1978 ［EUR8］
　　SM for Geophysical Logging of Boreholes-1981 ［EUR 4］
　　Part 1-Technical Introduction
　　Part 2-SM for Single-Point Resistance and Conventional Resistivity Logs
　　Part 3-SM for the Spontaneous Potential Log
　　Part 4-SM for the Induction Log
　　Part 5-SM for the Gamma-Ray Log
　　Part 6-SM for the Neutron Log
　　Part 7-SM for the Gamma-Gamma Density Log
　　Part 8-SM for the Acoustic or Sonic Log
　　Part 9-SM for the Caliper Log
　　Part 10-SM for the Temperature Log
　　Part 11- References

2 LABORATORY TESTING
　　SM for Petrographic Description of Rocks-1978 ［EUR 4］
　　SM for Determining Water Content, Porosity, Density, Absorption and Related Properties and Swelling and Slake-Durability Index Properties-1977 ［EUR 4］
　　Part 1-SM for Determining Water Content, Porosity, Density, Absorption and Related Properties
　　　　SM for Determination of the Water Content of a Rock Sample
　　　　SM for Porosity/Density Determination Using Saturation and Caliper Techniques
　　　　SM for Porosity/Density Determination Using Saturation and Buoyancy Techniques
　　　　SM for Porosity/Density Determination Using Mercury Displacement and Grain Specific Gravity Techniques
　　　　SM for Porosity/Density Determination Using Mercury Displacement and Boyle's Law Techniques
　　　　SM for Void Index Determination Using the Quick Absorption Technique
　　Part 2-SM for Determining Swelling and Slake-Durability Index Properties
　　　　SM for Determination of the Swelling Pressure Index Under Conditions of Zero Volume Change
　　　　SM for Determination of the Swelling Strain Index for a Radially Confined Specimen With Axial Surcharge
　　　　SM for Determination of the Swelling Strain Developed in an Unconfined Rock Specimen
　　　　SM for Determination of the Slake-Durability Index
　　SM for Determining Hardness and Abrasiveness of Rocks-1978 ［EUR 4］
　　Part 1-Introduction and Review
　　Part 2-SM for Determining the Resistance to Abrasion of Aggregate by Use of the Los Angeles Machine
　　Part 3-SM for Determination of the Schmidt Rebound Hardness
　　SM for Determining the Shore Hardness Value for Rock-2006 ［EUR 4］
　　SM for Determining Sound Velocity-1978 ［EUR 4］
　　SM for Determining Point Load Strength-1985 ［EUR 4］
　　SM for Determining the Indentation Hardness Index of Rock Materials-1998 ［EUR 4］
　　SM for Determining Block Punch Strength Index (BPI) -2001 ［EUR 4］
　　SM for Determining the Uniaxial Compressive Strength and Deformability of Rock Materials-1979 ［EUR 4］
　　Part 1-SM for Determination of the Uniaxial Compressive Strength of Rock Materials
　　Part 2-SM for Determining Deformability of Rock Materials in Uniaxial Compression
　　SM for Determining the Strength of Rock Materials in Triaxial Compression-1978 ［EUR4］
　　SM for Determining Shear Strength-1974 ［EUR 4］
　　Part 1-SM for In Situ Determination of Direct Shear Strength
　　Part 2-SM for Laboratory Determination of Direct Shear Strength
　　Part 3-SM for In Situ Determination of Shear Strength Using a Torsional Shear Test
　　SM for Determining Tensile Strength of Rock Materials-1978 ［EUR 4］
　　Part 1-SM for Determining Direct Tensile Strength
　　Part 2-SM for Determining Indirect Tensile Strength by the Brazil Test
　　SM for Laboratory Testing of Argillaceous Swelling Rock-1989 ［EUR 4］
　　Part 1-SM for Sampling, Storage and Preparation of Test Specimens
　　Part 2-SM for Determining Maximum Axial Swelling Stress
　　Part 3-SM for Determining Axial and Radial Free Swelling Strain

Part 4-SM for Determining Axial Swelling Stress as a Function of Axial Swelling Strain
SM for Laboratory Testing of Swelling Rocks-1999 [EUR 4]
Part 1-SM for Sampling, Storage and Preparation of Test Specimens
Part 2-SM for Determining Axial Swelling Stress
Part 3-SM for Determining Axial and Radial Free Swelling Strain
Part 4-SM for Determining Axial Swelling Stress as a Function of Axial Swelling Strain
SM for the Complete Stress-Strain Curve for Intact Rock in Uniaxial Compression-1999 [EUR 4]
SM for Determining the Fracture Toughness of Rock-1988 [EUR 4]
Part 1-SM for Determining Fracture Toughness Using Chevron Bend Specimens
Part 2-SM for Determining Fracture Toughness Using Short Rod Specimens
  SM for Determining Mode I Fracture Toughness Using Cracked Chevron Notched Brazilian Disc (CCNBD) Specimens-1995 [EUR 4]

3 FIE
3.1 DEFORMABILITY TESTS
SM for Determining In Situ Deformability of Rock-1979 [EUR 4]
Part 1-SM for Deformability Determination Using a Plate Test (Superficial Loading)
Part 2-SM for Field Deformability Determination Using a Plate Test Down a Borehole
Part 3-SM for Measuring Rock Mass Deformability Using a Radial Jacking Test
SM for Deformability Determination Using a Large Flat Jack Technique 1986 [EUR 4]
SM for Deformability Determination Using a Flexible Dilatometer-1987 [EUR 4]
SM for Deformability Using a Flexible Dilatometer with Volume Change Measurements
SM for Deformability Using a Flexible Dilatometer with Radial Displacement Measurements
SM for Deformability Determination Using a Stiff Dilatometer-1996 [EUR 4]

3.2 IN SITU TRESS MEASUREMENTS
SM for Rock Stress Determination-1987 [EUR 4]
SM for Rock Stress Determination Using a Flatjack Technique
SM for Rock Stress Determination Using the Hydraulic Fracturing Technique
SM for Rock Stress Determination Using a USBM-Type Drillhole Deformation Gauge
SM for Rock Stress Determination Using a CSIR- or CSIRO-Type Cell with 9 or 12 Strain Gauges
SM for In Situ Stress Measurement Using the Compact Conical-Ended Borehole Overcoring (CCBO) Technique-1999 [EUR 4]
SM for Rock Stress Estimation-2003
Part 1: Strategy for Rock Stress Estimation [EUR 4]
Part 2: Overcoming Methods [EUR 4]
Part 3: Hydraulic Fracturing (HF) and/or Hydraulic Testing of Pre-existing Fractures (HTPF) [EUR 4]
Part 4: Quality Control of Rock Stress Estimation [EUR 4]

3.3 GEOPHYSICAL TESTING
SM for Seismic Testing Within and Between Boreholes-1988 [EUR 4]
Part 1-Technical Introduction
Part 2-SM for Seismic Testing Within a Borehole
Part 3-SM for Seismic Testing Between Boreholes
Part 4-SM for Seismic Tomography
SM for Land Geophysics in Rock Engineering-2004 [EUR 4]
Seismic Refraction
Shallow Seismic Reflexion
Electrical
Electromagnetic
Ground Penetration Radar
Gravity
Radiometric
SM for Borehole Geophysics in Rock Engineering-2006 [EUR 8]
Velocity Measurements Along a Borehole
Electric and Electromagnetic Logging
Nuclear Logging
Vertical Seismic Profiling
Seismic Tomography
Resistivity Tomography
Seismic Ahead of a Tunnel Face

3.4 OTHER TESTS
SM for Rapid Field Identification of Swelling and Slaking Rocks-1994 [EUR 4]

SM for Large Scale Sampling and Triaxial Testing of Jointed Rock-1989 [EUR 4]

3.5 BOLTING AND ANCHORING TESTS
SM for Rockbolt Testing-1974 [EUR 4]
Part 1-SM for Determining the Strength of a Rockbolt Anchor (Pull Test)
Part 2-SM of Determining Rockbolt Tension Using a Torque Wrench
Part 3-SM for Monitoring Rockbolt Tension Using Load Cells
SM for Rock Anchorage Testing-1985 [EUR 4]

4 MONITORING
SM for Monitoring Rock Movements Using Borehole Extensometers-1978 [EUR 4]
SM for Monitoring Rock Movements Using Inclinometers and Tiltmeters-1977 [EUR 4]
Part 1-SM for Monitoring Rock Movements Using a Probe Inclinometer
Part 2-SM for Monitoring Rock Movements Using Fixed-in-Place Inclinometers
Part 3-SM for Monitoring Rock Movements Using Tiltmeters
SM for Pressure Monitoring Using Hydraulic Cells-1980 [EUR 4]
SM for Surface Monitoring of Movements Across Discontinuities-1984 [EUR 4]
SM for Monitoring Movements Across Discontinuities Using Glass Plates
SM for Monitoring Movements Across Discontinuities Using Pins and a Tape
SM for Monitoring Movements Across Discontinuities Using a Portable Mechanical Gauge
SM for Monitoring Movements Across Discontinuities Using a Remote Reading Electrical Jointmeter
SM for Blast Vibration Monitoring-1992 [EUR 4]

DISCLAIMER: Most of the Suggested Methods documents were originally published as articles in the International Journal of Rock Mechanics and Mining Sciences. Accordingly, the ISRM cannot supply the original documents (which are in the Journal issues). All the items above are supplied as photocopies.

The Orange Book
The Orange Book" The ISRM Suggested Methods for Rock Characterization, Testing and Monitoring: 2007-2014" can be purchased directly from the editor.

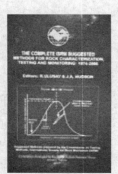

The Blue Book
The Blue Book:" The Complete ISRM Suggested Methods for Rock Characterization, Testing and Monitoring: 1974-2006", Edited by R. Ulusay and J. A. Hudson.

附录3 历届国际岩石力学大会索引

1966 September	1st ISRM Congress	Lisbon, PORTUGAL
1970 September	2nd ISRM Congress	Belgrade, YUGOSLAVIA
1974 September	3rd ISRM Congress	Denver, USA
1979 September	4th ISRM Congress	Montreux, SWITZERLAND
1983 April	5th ISRM Congress	Melbourne, AUSTRALIA
1987 September	6th ISRM Congress	Montreal, CANADA
1991 September	7th ISRM Congress	Aachen, GERMANY
1995 September	8th ISRM Congress	Tokyo, JAPAN
1999 August	9th ISRM Congress	Paris, FRANCE
2003 September	10th ISRM Congress	Sandton, SOUTH AFRICA
2007 July	11th ISRM Congress	Lisbon, PORTUGAL
2011 October	12th ISRM Congress	Beijing, CHINA
2015 May	13th ISRM Congress	Montreal, CANADA
2019 September	14th ISRM Congress	Foz do Iguaçu, BRAZIL

附录4 国际岩石力学学会简介

The International Society for Rock Mechanics was founded in Salzburg in 1962 as a result of the enlargement of the "Salzburger Kreis". Its foundation is mainly owed to Prof. Leopold Müller who acted as President of the Society till September 1966. The ISRM is a non-profit scientific association supported by the fees of the members and grants that do not impair its free action. In 2016 the Society has over 8,000 members and 60 National Groups.

The field of Rock Mechanics is taken to include all studies relative to the physical and mechanical behaviour of rocks and rock masses and the applications of this knowledge for the better understanding of geological processes and in the fields of Engineering (ISRM Statutes).

The main objectives and purposes of the Society are:
- to encourage international collaboration and exchange of ideas and information between Rock Mechanics practitioners;
- to encourage teaching, research, and advancement of knowledge in Rock Mechanics;
- to promote high standards of professional practice among rock engineers so that civil, mining and petroleum engineering works might be safer, more economic and less disruptive to the environment.

The main activities carried out by the Society in order to achieve its objectives are:
- to hold International Congresses at intervals of four years;
- to sponsor International and Regional Symposia, organised by the National Groups the Society;
- to publish a News Journal to provide information about technology related to Rock Mechanics and up-to-date news on activities being carried out in the Rock Mechanics community;
- to operate Commissions for studying scientific and technical matters of concern to the Society;
- to award the Rocha medal for an outstanding doctoral thesis, every year, and the Müller award in recognition of distinguished contributions to the profession of Rock Mechanics and Rock Engineering, once every four years;
- to cooperate with other international scientific associations.

The Society is ruled by a Council, consisting of representatives of the National Groups, the Board and the Past Presidents. The current President is Dr Eda F. de Quadros, from Brazil.

The ISRM Secretariat has been headquartered in Lisbon, Portugal, at the Laboratório Nacional de Engenharia Civil-LNEC since 1966, date of the first ISRM Congress, when Prof. Manuel Rocha was elected as President of the Society.

附录5　国际岩石力学学会历届主席团成员名单

1962—1966	President	Prof. Leopold Müller, AUSTRIA
	Members	Prof. G. Bilkenroth, GERMANY
		Dr Laurits Bjerrum, NORWAY
		Prof. Eberhard Clar, AUSTRIA
		Prof. Don U. Deere, USA
		Prof. Charles Fairhurst, USA
		Prof. Ernst Gottstein, GERMANY
		Dr T. Hagerman, SWEDEN
		Prof. Helmut Hoffmann, GERMANY
		Dr R. Kvapil, SWEDEN
		Dr Harald Lauffer, AUSTRIA
		Prof. Hans Leussink, GERMANY
		Dr H. J. Martini, GERMANY
		Dr G. Oberti, ITALY
		Dr Mario Pancini, ITALY
		Prof. Edward L. J. Potts, UK
		Dr Johann Scheiblauer, GERMANY
		Prof. G. TerStepanian, ARMENIA
		Dr A. Watznauer, GERMANY
1966—1970	President	Prof. Manuel Rocha, PORTUGAL
	Vice Presidents	Africa-Dr G. Denkhaus, SOUTH AFRICA
		Asia-Dr M. Yoshida, JAPAN
		Australasia-Mr L. Endersbee, AUSTRALIA
		Europe-Dr Armin von Moos, SWITZERLAND
		North America-Prof. W. Judd, USA
		South America-Prof. Costa Nunes, BRAZIL
1970—1974	President	Dr L. Obert, USA
	Vice Presidents	Africa-Mr A. Chaoui, MOROCCO
		Asia-Prof. B. Aisenstein, ISRAEL
		Australasia-Prof. J. C. Jaeger, AUSTRALIA
		Europe-Prof. B. Kujundzic, YUGOSLAVIA
		North America-Dr D. F. Coates, CANADA
		South America-Prof. V. de Mello, BRAZIL
1974—1979	President	Prof. P. Habib, FRANCE
	Vice Presidents	Africa-Dr Z. T. Bieniawski, SOUTH AFRICA
		Asia-Prof. Y. Hiramatsu, JAPAN
		Australasia-Dr A. J. Hargraves, AUSTRALIA
		Europe-Prof. E. Fumagalli, ITALY
		North America-Dr J. Handin, USA
		South America-Dr M. Kanji, BRAZIL

1979—1983	President	Prof. W. Wittke, GERMANY
	Vice Presidents	Africa-Mr A. Chaoui, MOROCCO
		Asia-Dr M. Yoshida, JAPAN
		Australasia-Mr W. E. Bamford, AUSTRALIA
		Europe-Prof. S. Uriel Romero, SPAIN
		North America-Dr T. C. Atchison, USA
		South America-Prof. O. Moretto, ARGENTINA
1983—1987	President	Prof. E. T. Brown, UK
	Vice Presidents	Africa-Dr H. Wagner, SOUTH AFRICA
		Asia-Prof. Tan Tjong, Kie CHINA
		Australasia-Mr W. E. Bamford, AUSTRALIA
		Europe-Dr S. A. G. Bjurström, SWEDEN
		North America-Mr A. A. Bello, MEXICO
		South America-Prof. F. H. Tinoco, VENEZUELA
1987—1991	President	Dr John A. Franklin, CANADA
	Vice Presidents	Africa-Dr Oskar K. H. Steffen, SOUTH AFRICA
		Asia-Prof. T. Ramamurthy, INDIA
		Australasia-Dr Ian W. Johnston, AUSTRALIA
		Europe-Dr Marc Panet, FRANCE
		North America-Dr James Coulson, USA
		South America-Dr Carlos Dinis da Gama, BRAZIL
		At Large-Prof. Shun-suke Sakurai, JAPAN
		At Large-Academician Eugeny J. Shemyakin, RUSSIA
1991—1995	President	Prof. Charles Fairhurst, USA
	Vice Presidents	Africa-Dr T. R. (Dick) Stacey, SOUTH AFRICA
		Asia-Prof. Koichi Sassa, JAPAN
		Australasia-Prof. Michael John Pender, NEW ZEALAND
		Europe-Prof. Ove Stephansson, SWEDEN
		North America-Dr P. K. Kaiser, CANADA
		South America-Dr Oscar A. Vardé, ARGENTINA
		At Large-Dr R. Widmann, AUSTRIA
1995—1999	President	Prof. S. Sakurai, JAPAN
	Vice Presidents	Vice Presidents
		Africa-Dr Nielen van der Merwe, SOUTH AFRICA
		Asia-Dr Ou Chin-Der, CHINA
		Australasia-Mr Garry R. Mostyn, AUSTRALIA
		Europe-Prof. Giovanni Barla, ITALY
		North America-Prof. Herbert H. Einstein, USA
		South America-Prof. Michel L. van Sint Jan F., CHILE
		At Large-Prof. John A. Hudson, UK
		At large-Prof. Sun Jun, CHINA

续表

1999—2003	President	Prof. Marc Panet, FRANCE
	Vice Presidents	Africa-Dr Güner Gürtunca, SOUTH AFRICA
		Asia-Prof. Chung-In Lee, KOREA
		Australasia-Prof. Chris M. Haberfield, AUSTRALIA
		Europe-Prof. Pekka Särkkä, FINLAND
		North America-Mr Alfredo Sanchez Gomez, MEXICO
		South America-Prof. Eurípedes Vargas, BRAZIL
		At Large-Prof. Wulf Schubert, Austria
		At Large-Dr Satoshi Hibino, JAPAN
2003—2007	President	Dr Nielen van der Merwe, SOUTH AFRICA
	Vice Presidents	Africa-Dr M. J. Pretorius, SOUTH AFRICA
		Asia-Prof. Jian Zhao, SINGAPORE
		Australasia-Dr John St George, NEW ZEALAND
		Europe-Dr. Claus Erichsen, GERMANY
		North America-Dr François Heuzé, USA
		South America-Dr Eda Freitas de Quadros, BRAZIL
		At Large-Prof. Qian Qihu, CHINA
		At Large-Prof. Luís Ribeiro e Sousa, PORTUGAL
2007—2011	President	Prof. John Hudson, UNITED KINGDOM
	Vice Presidents	Africa-Dr François Malan, SOUTH AFRICA
		Asia-Prof. Abdolhadi Ghazvinian, IRAN
		Australasia-Dr Antkony Meyers, AUSTRALIA
		Europe-Dr. Nuno Grossmann, PORTUGAL
		North America-Prof. Derek Martin, CANADA
		South America-Prof. Álvaro Gonzalez-Garcia, COLOMBIA
		At Large-Dr. Claus Erichsen, GERMANY
		At Large-Prof. Xia-Ting Feng, CHINA
2011—2015	President	Prof. Xia-ting Feng
	Vice Presidents	Africa-Mr. Jacques Lucas,
		Asia-Dr. Yingxin Zhou, SINGAPORE
		Australasia-Dr. David Beck, AUSTRILIA
		Europe-Prof. Frederic Pellet, FRANCE
		North America-Dr. John Tinucci, USA
		South America-Dr. Antonio Samaniego, PERU
		At Large-Prof. Yuzo Ohnishi, JAPAN
		At Large-Dr. Manoj Verman, INDIA
		At Large-Prof. Ivan Vrklijan, HUNGERY

2015—2019	President	Dr Eda Freitas de Quadros, BRAZIL
	Vice Presidents	Africa- Mr William Joughin, SOUTH AFRICA
		Asia-Prof. Seokwon Jeon, KOREA
		Australasia-Mr Stuart Read, NEW ZEALAND
		Europe-Prof. Charlie Chunlin Li, SWEDEN
		North America-Dr Doug Stead, CANADA
		South America-Prof. Sergio A. B. da Fontoura, BRAZIL
		At Large-Prof. Norikazu Shimizu, JAPAN
		At Large-Prof. Manchao He, CHINA
		At Large-Dr Petr Konicek, CZECH REPUBLIC
		Secretary General-DrLuis Lamas, PORTUGAL

附录 6 ISRM 专业委员会和联合技术委员会

The ISRM Board has appointed several Commissions for the period 2015-2019 in order that they may study some scientific and technical matters of topical interest to the Society. Other proposals are currently being analysed.

For proposing the creation of a Commission, the proposed Commission President shall fill up an application form and send it to the ISRM President, with copy to the Secretariat. To download the application form click here.

Please note that all ISRM Commissions only run for the ISRM Presidential period during which they have been approved, i. e. for the 4-year period between the ISRM Congresses. However, some ISRM Commissions can be extended into a new 4-year period. This form should be completed by the Commission President (proposed or existing) for the purpose of starting or continuing an ISRM Commission.

It is important to note that Commissions can only be approved if it is clear that the Commission is well organised and that there will be a significant product-which is both useful to the ISRM and produced on time. Note that this means that the Commission's report should be prepared in the year before the 2019 International Congress, at Foz do Iguaçu, Brazil. It is expected that Commission meetings will be held in association with the ISRM International Symposia: i. e., the EUROCK 2016 to be held in Capadoccia, Turkey, in August 2016, the AfriRock to be held in Cape Town, in February 2017, at the venue for 2018 which is not yet established, and of course the 2019 ISRM Congress in Brazil.

In accordance with the ISRM By-law No. 3 " Rules to be followed by Commissions and Joint Commissions", ISRM Members wishing to participate in the work of any of the Commissions shall contact the respective President.

The current list of ISRM Commissions is the following:
- Commission on Application of Geophysics to Rock Engineering
- Commission on Crustal Stress and Earthquake
- Commission on Design Methodology
- Commission on Discontinuous Deformation Analysis-DDA
- Commission on Education
- Commission on Evolution of Eurocode 7
- Commission on Grouting
- Commission on Hard Rock Excavation
- Commission on Petroleum Geomechanics
- Commission on Preservation of Ancient Sites
- Commission on Radioactive Waste Disposal
- Commission on Rock Dynamics
- Commission on Soft Rocks
- Commission on Subsea Tunnels
- Commission on Testing Methods
- Commission on Underground Nuclear Power Plants
- Commission on Underground Research Laboratory (URL) Networking

Three Joint Technical Committees operate under the umbrella of the Federation of International Geo-engineering Societies-FedIGS, which join the effort of the Sister Societies IAEG, IGS, ISRM and ISSMGE.

The current list of JTCs is the following:
- JTC 1-Joint Technical Committee on Natural Slopes and Landslides
- JTC 2-Representation of Geo-engineering Data in Electronic Form
- JTC 3-Education and Training

The ISRM and the International Tunnelling and Underground Space Association-ITA celebrated a Memorandum of Understanding, whereby ISRM may appoint members to selected ITA Working Groups and ITA may appoint members to selected ISRM Commissions.

附录7　国际岩石力学学会 Muller 奖及历届获奖者

The Müller Award

The ISRM Board decided at its Pau meeting in September 1989, to institute an award to honour the memory of Prof. Leopold Müller, the founder and first President of the Society. Prof. Müller, as the enthusiastic founder of the Society, took the initiative to aggregate to the Salzburger Group scientists from all over the world interested in the new-born branch of science, rock mechanics, with the purpose to bring together and to give unity not only to the scattered knowledge obtained by groups working more or less isolated on problems posed by rock masses, but even to knowledge contributed by those pursuing other aims but with interest in the field.

The award is made once every four years in recognition of distinguished contributions to the profession of rock mechanics and rock engineering and consists of a special Müller Lecture to be delivered at the ISRM International Congresses, a work of art typical of the culture of the country hosting the Congress and a silver medallion with a portrait of Leopold Müller.

历届"缪勒奖"获奖者

年份	获奖者	国家
1991	Evert Hoek	CANADA
1995	Neville Cook	USA
1999	Herbert Einstein	USA
2003	Charles Fairhurst	USA
2007	Edwin T. Brown	AUSTRALIA
2011	Nick Barton	UNITED KINGDOM
2015	John A. Hudson	UNITED KINGDOM

附录8　ISRM Rocha 奖历届获奖者

The Rocha Medal

The ISRM Board decided at its Tokyo meeting in September 1981, to institute an annual prize with a view to honour the memory of Past-President Prof. Manuel Rocha. Profiting from the basis established by Prof. Müller, Prof. Rocha organized the first ISRM Congress and he was the leader that was responsible for transforming the international collaboration carried out in an amateurish way into a real international scientific association, having for the purpose settled the fundamental lines that have guided and supported the ISRM activity along the years.

The Rocha Medal is intended to stimulate young researchers in the field of rock mechanics. The prize, a bronze medal and a cash prize, have been annually awarded since 1982 for an outstanding doctoral thesis selected by a Committee appointed for the purpose.

Starting with the 2010 award, one or two runner up certificates (ProximeAccessit certificates) may also be awarded.

Rocha Medal (approved by the Council in Tehran IRAN on 23 November 2008)

Rocha 奖历届获奖者

年份	获奖者	国家	论文题目
1982	A. P. Cunha	PORTUGAL	Mathematical Modelling of Rock Tunnels
1983	S. Bandis	GREECE	Experimental Studies of Scale Effects on Shear Strength and Deformation of Rock Joints
1984	B. Amadei	FRANCE	The Influence of Rock Anisotropy on Measurement of Stresses in Situ
1985	P. M. Dight	AUSTRALIA	Improvements to the Stability of Rock Walls in Open Pit Mines
1986	W. Purrer	AUSTRIA	Calculation Model for the Behaviour of a Deep-Lying Seam Roadway in a Solid (but cut by Bedding Planes) Surrounding Rock Mass, taking into Consideration the Failure Mechanisms of the Soft Layer Determined In-Situ on Models
1987	D. Elsworth	UNITED KINGDOM	Laminar and Turbulent Flow in Rock Fissures and Fissure Networks
1988	S. Gentier	FRANCE	Morphology and Hydromechanical Behaviour of a Natural Fracture in a Granite, under Normal Stress-Experimental and Theoretical Study
1989	B. Fröhlich	GERMANY	Anisotropic Swelling Behaviour of Diagenetically Consolidated Claystones
1990	R. K. Brummer	SOUTH AFRICA	Fracturing and Deformation at the Edges of Tabular Gold Mining Excavations and the Development of a Numerical Model describing such Phenomena
1991	T. H. Kleine	AUSTRALIA	A Mathematical Model of the Rock Breakage by Blasting
1992	A. Ghosh	INDIA	Fractal and Numerical Models of Explosive Rock Fragmentation
1993	O. Reyes W.	PHILIPPINES	Experimental Study and Analytical Modelling of Compressive Fracture in Brittle Materials
1994	S. Akutagawa	JAPAN	A Back Analysis Program System for Geomechanics Application
1995	C. Derek Martin	CANADA	The Strength of Massive Lac du Bonnet Granite around Underground Openings
1996	M. P. Board	USA	Numerical Examination of Mining-Induced Seismicity
1997	M. Brudy	GERMANY	Determination of In-Situ Stress Magnitude and Orientation of 9 km Depth at the KTB Site

Year	Name	Country	Title
1998	F. MacGregor	AUSTRALIA	The Rippability of Rock
1999	A. Daehnke	SOUTH AFRICA	Stress Wave and Fracture Propagation in Rock
2000	P. Cosenza	FRANCE	Coupled Effects between Mechanical Behaviour and Mass Transfer Phenomena in Rock Salt
2001	D. F. Malan	SOUTH AFRICA	An Investigation into the Identification and Modelling of Time-Dependent Behaviour of Deep Level Excavations in Hard Rock
2002	M. S. Diederichs	CANADA	Instability of Hard Rockmasses: the Role of Tensile Damage and Relaxation
2003	L. M. Andersen	SOUTH AFRICA	A Relative Moment Tensor Inversion Technique applied to Seismicity Induced by Mining
2004	G. Grasselli	ITALY	Shear Strength of Rock Joints based on the Quantified Surface Description
2005	M. Hildyard	UNITED KINGDOM	Wave Interaction with Underground Openings in Fractured Rock
2006	D. Ask	SWEDEN	New Developments of the Integrated Stress Determination Method and Application to the ÄSPÖ Hard Rock Laboratory, Sweden
2007	H. Yasuhara	JAPAN	Thermo-Hydro-Mechano-Chemical Couplings that Define the Evolution of Permeability in Rock Fractures
2008	Z. Z. Liang	CHINA	Three Dimensional Numerical Modelling of Rock Failure Process
2009	Li Gang	CHINA	Experimental and Numerical Study for Stress Measurement by Jack Fracturing and Estimation of Stress Distribution in Rock Mass
2010	J. Christer Andersson	SWEDEN	Rock Mass Response to Coupled Mechanical Thermal Loading. Äspö Pillar Stability Experiment, Sweden
2011	D. Park	KOREA	Reduction of Blast-Induced Vibration in Tunnelling Using Barrier Holes and Air-deck
2012	M. T. Zandarin	ARGENTINA	Thermo-Hydro-Mechanical Analysis of Joints-A Theoretical and Experimental Study
2013	M. Pierce	CANADA	A model for gravity flow of fragmented rock in block caving mines
2014	M. S. A. Perera	AUSTRALIA	Investigation of the Effect of Carbon Dioxide Sequestration on Coal Seams: A Coupled Hydro-Mechanical behaviour
2015	AndreaLisjak Bradley	ITALY	Investigating the influence of mechanical anisotropy on the fracturing behaviour of brittle clay shales with application to deep geological repositories
2016	Chia Weng Boon	MALASYA	Distinct Element Modelling of Jointed Rock Masses: Algorithms and Their Verification

Rocha Award Runner Up Certificates

Year	Name	Country	Title
2010	Yoon Jeongseok	KOREA	Hydro-Mechanical Coupling of Shear-Induced Rock Fracturing by Bonded Particle Modeling
2010	Abbas Taheri	IRAN	Properties of Rock Masses by In-situ and Laboratory Testing Methods
2011	Bo Li	CHINA	Coupled Shear-flow Properties of Rock Fractures
2012	B. P Watson	SOUTH AFRICA	Rock Behaviour of the Bushveld Merensky Reef and the Design of Crush Pillars
2012	J. Taron	USA	Geophysical and Geochemical Analyses of Flow and Deformation in Fractured Rock
2013	L. He	CHINA	Three dimensional numerical manifold method and rock engineering applications
2013	A. Perino	ITALY	Wave propagation through discontinuous media in rock engineering
2014	R. Resende	PORTUGAL	An Investigation of Stress Wave Propagation Through Rock Joints and Rock Masses
2014	S. Dehkhoda	IRAN	Experimental and Numerical Study of Rock breakage by Pulsed Water Jets
2015	Ahmadreza Hedayat	IRAN	Mechanical and Geophysical Characterization of Damage in Rocks
2016	Qianbing Zhang	CHINA	Mechanical Behaviour of Rock Materials under Dynamic Loading
2016	Silvia Duca	ITALY	Design of an experimental procedure and set up for the detection of ice segregation phenomena in rock by acoustic emissions

附录 9　国际岩石力学学会 "会士"
(ISRM Fellow)

ISRM Fellow

The ISRM Council decided at its New Delhi meeting in October 2010, to create the status of Fellow, as the highest and most senior grade of membership of the ISRM. It is conferred on individuals, affiliated with the ISRM, who have achieved outstanding accomplishment in the field of rock mechanics and/or rock engineering and who have contributed to the professional community through the ISRM.

An ISRM Fellowship is a lifetime position. It is intended that the title of ISRM Fellow will carry a clear recognition of his or her achievements. Thus, the induction of ISRM Fellows creates a group of experts that can provide strong support and advice to the ISRM, and can be called upon as appropriate for ISRM activities.

The appointment of ISRM Fellows is made by the ISRM Board in accordance with the Guideline for the selection of ISRM Fellows, approved by the Board in May 2011.

国际岩石力学学会 "会士" 名单

ISRM Fellows inducted in 2011 in Beijing, China

The first group of Fellows was inducted in Beijing, during the 12th International Congress on Rock Mechanics.

Ted Brown
Charles Fairhurst
John A. Franklin
Pierre Habib
Marc Panet
Shunsuke Sakurai
Nielen van der Merwe
Walter Wittke

ISRM Fellows inducted in 2013 in Wroklaw, Poland

A second group of fellows was inducted in Wroklaw, Poland, during the Eurock 2013:

Giovanni Barla
Herbert Einstein
John Hudson
Milton Kanji
Chung-In Lee
Qian Qihu
Dick Stacey
Ove Stephansson

ISRM Fellows inducted in 2015 in Montréal, Canada

A third group of fellows was inducted in Montréal, Canada, during the 13th ISRM Congress:

Nick Barton
José Delgado Rodrigues
Nuno Feodor Grossmann
Jun Sun
Peter Kaiser
Temura Ramamurthy
Koichi Sassa
Wulf Schubert
ReşatUlusay
Jian Zhao

附录 10　中国岩石力学与工程学会简介

中国岩石力学与工程学会（CSRME）是全国岩石力学与岩土工程科技工作者的学术性群众团体，1985 年依法经民政部注册登记。涉及的专业领域包括水利水电、地质矿业、铁道交通、国防工程、灾害控制、环境保护等，是国内具有影响力的跨行业、跨部门、跨学科的社团组织。截至 2015 年，学会设 6 个工作委员会、10 个专业委员会和 11 个分会，以及 20 个地方学会。

学会成立 30 年来不断壮大和发展，国内注册个人会员接近 1 万人，团体会员 40 个；国际会员超过 1300 人，国际团体会员 20 个。个人会员和团体会员数在国际岩石力学学会中持续多年保持第一。学会为广大会员提供了坚实的学术交流与发展的平台，注重学术活动的顶层设计，形成系列化、多方位的学术交流格局。对国内"精品学术会议"实施品牌建设，如全国岩石动力学学术大会、全国青年岩石力学与工程学术大会等均取得了广泛影响。学会积极实施"人才托举工程"，落实青年人才的发现和培养计划。在中国科协领导下，学会将团结广大会员，改革发展、奋发向上，在服务科技创新的征程上继续攀登新的高峰。

分支机构（27 个）：
中国岩石力学与工程学会评议工作委员会
中国岩石力学与工程学会教育工作委员会
中国岩石力学与工程学会编辑工作委员会
中国岩石力学与工程学会咨询工作委员会
中国岩石力学与工程学会青年工作委员会
中国岩石力学与工程学会国际交流工作委员会
中国岩石力学与工程学会高温高压岩石力学专业委员会
中国岩石力学与工程学会岩体物理数学模拟专业委员会
中国岩石力学与工程学会古遗址保护与加固工程专业委员会
中国岩石力学与工程学会岩石力学测试专业委员会
中国岩石力学与工程学会岩石破碎工程专业委员会
中国岩石力学与工程学会工程实例专业委员会
中国岩石力学与工程学会深层岩石力学专业委员会
中国岩石力学与工程学会废物地下处置专业委员会
中国岩石力学与工程学会动力学专业委员会
中国岩石力学与工程学会地壳应力与地震专业委员会
中国岩石力学与工程学会地面岩石工程专业委员会
中国岩石力学与工程学会软岩工程与深部灾害控制分会
中国岩石力学与工程学会锚固与注浆分会
中国岩石力学与工程学会隧道掘进机工程应用分会
中国岩石力学与工程学会地下空间分会
中国岩石力学与工程学会地下工程分会
中国岩石力学与工程学会工程安全与防护分会
中国岩石力学与工程学会环境岩土工程分会
中国岩石力学与工程学会岩土工程信息技术与应用分会
中国岩石力学与工程学会大同分会
中国岩石力学与工程学会东北分会

已加入国际组织（3 个）：
国际岩石力学学会
国际地下空间研究中心
国际地下物流研究中心

公开出版刊物（3 种）：
《岩石力学与工程学会》
《Journal of Rock Mechanics and Geotechnical Engineering》
《地下空间与工程学报》

附录11　中国岩石力学与工程学会历届理事长、副理事长、秘书长名单

第一届：
　　理事长：陈宗基
　　副理事长：牛锡倬、王武林
　　秘书长：梅剑云、牛锡倬

第二届：
　　理事长：陈宗基
　　副理事长：牛锡倬、王武林、李克、哈秋舲、潘家铮
　　秘书长：傅冰骏

第三届：
　　理事长：孙钧
　　副理事长：王思敬、白世伟、宋永津、刘天泉、刘宝琛、刘同友、哈秋舲、徐文耀、张清
　　秘书长：傅冰骏

第四届：
　　理事长：王思敬
　　副理事长：白世伟、刘宝琛、刘天泉、刘同友、钱七虎、王福柱、杨林德、张超然、赵经彻
　　秘书长：杨志法

第五届：
　　理事长：钱七虎
　　副理事长：白世伟、冯夏庭、顾金才、黄鼎成、刘同友、唐春安、谢和平、杨林德、张超然
　　秘书长：杨志法

第六届：
　　理事长：钱七虎
　　副理事长：冯夏庭、顾金才、唐春安、谢和平、陈祖煜、伍法权、王金华、何满潮、李宁、李永盛、袁亮、韩凤险、樊启祥、蔡美峰
　　秘书长：伍法权

第七届：
　　理事长：冯夏庭、钱七虎
　　副理事长：樊启祥、蔡美峰、龚晓南、郭熙灵、何满潮、李宁、李术才、潘一山、宋胜武、王金华、王明洋、杨强、袁亮、郑炳旭、朱合华
　　秘书长：刘大安

主要参考文献

[1] 刘新建．英汉矿业词汇（第二次增订本）．北京：煤炭工业出版社，1997.
[2] 魏中明，杨劲松，凌明亮．汉英水利水电技术词典．北京：中国水利水电出版社，1993.
[3] 《英汉岩土工程词典》编委会编．英汉岩土工程词典．北京：中国建筑工业出版社，2011.
[4] 中国地质调查局，地质出版社．简明英汉地质词典．北京：地质出版社，2008.
[5] 谢礼立．英汉汉英地震工程学词汇．北京：地震出版社，1983.
[6] 马中骥，李勇，马林．工程制图英汉汉英两用词典．北京：中国质检出版社，中国标准出版社，2012.
[7] 《英汉地质词典》编辑组．英汉地质词典．北京：地质出版社，1993.
[8] 《英汉石油大辞典》编委会．英汉石油大辞典．北京：石油工业出版社，2001.
[9] 张世良，张炯，姚彦之．英汉遥感地质学词汇．北京：中国科学技术出版社，1992.
[10] 潘荣成，房立政．英汉汉英矿业工程技术词汇手册．上海：上海外语教育出版社，2012.
[11] 赵明华，郑刚，刘汉龙．简明英汉汉英岩土工程词汇手册．北京：中国建筑工业出版社，2008.
[12] 刘东．简明建设工程管理英汉汉英词汇手册．中国建材工业出版社，2007.
[13] 戴光荣．英汉汉英环境科学与工程词汇手册．上海：上海外语教育出版社，2011.
[14] 庄起敏，张芳芳，武小莉．英汉汉英水利水电词汇手册．上海：上海外语教育出版社，2012.
[15] 孙东云．英汉汉英地球科学词汇手册．上海：上海外语教育出版社，2012.
[16] 严勇，王颖．英汉汉英测绘学词汇手册．上海：上海外语教育出版社，2009.
[17] 张守信．中国地层名称．北京：科学出版社，2011.
[18] 李健．英汉汉英地质学词汇手册．上海：上海外语教育出版社，2010.
[19] 张志良，张岚．英汉汉英地基处理与基础工程常用词汇．北京：中国水利水电出版社，2007.
[20] 中国科学院南京土壤研究所．英汉土壤学词汇．北京：科学出版社，1975.
[21] 陈伟，顾士才，郭飞，宫辉．英汉汉英海洋与水文词汇手册．上海：上海外语教育出版社，2012.
[22] 康昱．汉英考古分类词汇．哈尔滨：黑龙江科学技术出版社，1992.
[23] 吴春荣，张锡濂．英汉地质学缩写词汇．北京：科学出版社，1978.
[24] A. Streckeisen. 火成岩分类及术语辞典．北京：地质出版社，1991.
[25] 郭鹏，陈芳．英汉汉英土木建筑词汇手册．上海：上海外语教育出版社，2009.
[26] 薛雁，岑莉，田卉．英汉汉英能源科学技术词汇手册．上海：上海外语教育出版社，2012.
[27] 张守信．英汉现代地层学词典．北京：科学出版社，1983.
[28] 张敏波，顾定兰，陈养桃．英汉汉英工程与技术词汇手册．上海：上海外语教育出版社，2009.
[29] 徐申生．英汉汉英市政工程词汇手册．上海：上海外语教育出版社，2010.
[30] 倪慧，杨霞．英汉汉英信息与系统科学词汇手册．上海：上海外语教育出版社，2012.
[31] 傅冰骏，张镜剑．隧道掘进机（TBM）英汉简明专业词汇．南水北调西线工程深埋、长大隧道关键技术及掘进机应用国际研讨会论文集，2005.
[32] 郑雨天，傅冰骏，卢世宗等．岩石力学试验建议方法（上集）．北京：煤炭工业出版社，1982.

跋

《英汉岩石力学与工程大辞典》包含 14 万余词条，涵盖：水利、水电、核电、天然气、石油、煤层气、页岩气、采矿、冶金、铁路、交通、国防、地球科学、城乡建设、环境保护、古遗址保护、CO_2 地质封存、岩爆冲击地压、人工智能、工程管理等多个领域。这是从 81 万词条源中精选出来的。词条源包括各专业委员会专题组提交的涵盖 18 个专业领域的 5 万词条（详见"部分词条分类及责任编委"表）；35 种纸质辞典和 15 种电子辞典的 44.7 万词条；网络搜索下载的 26.3 万电子词条；以及通过文献计量采集的 5 万词条。

大辞典编纂工作历时三年多，经历了总体规划、词条收集、词条录入、词条初选、词条精选、反复核查修改、二次外部审查、数次复审及终审和定稿审查等阶段，由国内外 80 余位专家参与编写、审查，此外还有众多的由专家带领的小组人员参加了选词工作。

大辞典编纂极其艰难，需要逐词逐字地审查修改，推陈出新。既要收纳新词，又要修改翻译不当的错词、剔除不合时宜的旧词。既要选词精炼，又要覆盖全面，力图能成为岩石力学各相关学科和工程领域都能使用的工具书。这需要有高度的责任感、耐心和业务水平。为此，编审组通过不懈的努力，落实责任到人，通力合作，取得了预期成果，我们真诚希望广大使用者提出宝贵修改意见，让大辞典可以不断改进，以更好地满足大家的要求。

作为这项重要工程后期具体工作的主要组织者，首先，要诚挚感谢所有参与者，包括已列入和未列入编审者名单的全体人员，所付出的大量辛勤劳动！感谢学会理事会、各二级机构和秘书处全体同事的大力支持。特别感谢傅冰骏先生，在其耄耋之年，仍不辞辛劳，精心负责组织了大辞典的编纂工作，连续做了三年细致的审查和修订，并为大辞典的集体审查提供了样板。编审组的三个年轻人，白冰博士和崔振东博士在本职科研工作很忙的情况下，自始至终参加了大辞典的编纂工作，并辅助傅先生进行了全面编审，牛晶蕊博士快速接手并主要完成了出版前的大量编纂和定稿，没有他们的努力和付出，大辞典也不可能呈现在读者面前。

刘大安

中国岩石力学与工程学会秘书长

2016 年夏于北京

部分词条分类及责任编委

序号	词条分类	责任编委
1	工程地质、水文地质、地质力学、构造地质、地球物理、地球化学	刘大安、尚彦军、伍法权、底青云、崔振东、张勇慧
2	数值分析方法、极限分析法、多场耦合	杨强、唐春安、晏飞、郑宏、周创兵
3	人工智能、风险评估、设计方法	冯夏庭、潘鹏志
4	岩石力学基本性质与测试技术、物理模拟	邬爱清、李仲奎、李术才、张强勇
5	地下工程与地下空间（包括 TBM、钻爆法施工技术、地质超前预报海底隧道）	李术才、朱维申、陈志龙
6	地面岩石工程	盛谦、李宁、周创兵
7	岩石动力学	黄理兴、李海波、王明洋
8	锚固与注浆	程良奎、杨晓东、李术才、邬爱清、张娜、傅冰骏、姚海波
9	软岩工程与支护、煤矿开采、地下冻结工程、煤层气开发	何满潮、康红普、杨晓杰、张群、方志明、姜鹏飞、李贵红
10	石油天然气开采、石油天然气地下储层、页岩气开发、天然气水合物开发	陈勉、杨春和、李晓、贾宇星、韦昌富
11	环境岩土工程、CO_2地质封存	刘汉龙、王媛、李小春、李琦
12	金属矿山、露天开采、地壳应力与地震工程	蔡美峰、谢富仁、李小春、石磊、李宏
13	放射性及其他废弃物地下处置	王驹、刘月妙
14	古遗址保护	王旭东、李最雄、郭青林
15	岩爆、冲击地压、矿震	冯夏庭
16	海洋工程	廖红建
17	地热开发岩石力学与工程、EGS	白冰、魏宁
18	岩石力学与岩石工程管理	傅冰骏、刘大安

Subject categories and corresponding Contributors

No.	Subject categories	Corresponding contributors
1	Engineering geology, Hydrogeology, Structural Geology, Geomechanics, Geophysics, Geochemistry	Liu Da'an, Sang Yanjun, Wu Faquan, Di Qingyun, Cui Zhendong, Zhang Yonghui
2	Numerical analysis, Analysis of limit equilibrium, Multi-coupling analysis	Yang Qiang, Tang Chun'an, Yan Fei, Zheng Hong, Zhou Changbing
3	Artificial intelligence analysis, Risk Analysis, Design methodology	Xiating Feng, Pan Pengzhi)
4	Mechanical properties of rocks Testing Methods, Physical modeling	Wu Aiqing, Li Zhongkui, Li Shucai, Zhang Qiangyong
5	Underground works and Underground Space, including TBM, D&B, geological prediction, submarine tunnel	Li Shucai, Zhu Weishen, Chen Zhilong
6	Surface rock engineering	Shengqian, Li Ning, Zhou Chuangbing
7	Dynamics of rocks	Huang Lixing, Li Haibo, Wang Mingyang
8	Anchoring and grouting	Cheng Liangkui, Yang Xiaodong, Li Shucai, Wu Aiqing, Zhang Na, Fu Bingjun, Yao Haibo
9	Weak rock engineering and supporting method, coal mining, underground freezing engineering, coal bed methane development	He Manchao, Kang Hongpu, Yang Xiaojie, Zhang qun, Fang Zhiming, Jiang Pengfei, Li Guihong
10	Petro-natural gas exploitation and underground storage shale gas, and natural gas hydrate development	Chen Mian, Yang Chunhe, Li Xiao, Ben Yuxing, Wei Changfu
11	Environmental geotechnics, CO_2 under-ground capture and storage	Liu Hanlong. Wang Yuan, Li Xiaochun, Li Qi
12	Metallic mining engineering, Open pit mining, rock stress and seismic engineering	Cai Meifeng, Xie Furen, Li Xiaochun, Shi Lei, Li Hong
13	Underground disposal of radioactive and other waste	Wang Ju, Liu Yuemiao
14	Preservation of ancient sites	Wang Xudong, Li Zuixiong, Guo Qinglin
15	Rock burst, shock bump, mine seismicity	Xiating Feng
16	Oceanographic engineering	Liao Hongjian
17	Geothermal development related to rock mechanics and Engineering, Enhanced geothermal system	Bai Bing, Wei Ning
18	Business administration related to rock mechanics and rock engineering	Fu Bingjun, Liu Da'an